DERIVATIVES AND INTEGRALS

D0162981

Basic Differentiation Rules

1. $\frac{d}{dx}[cu] = cu'$
2. $\frac{d}{dx}[u \pm v] = u' \pm v'$
3. $\frac{d}{dx}[uv] = uv' + vu'$
4. $\frac{d}{dx}\left[\frac{u}{v}\right] = \frac{vu' - uv'}{v^2}$
5. $\frac{d}{dx}[c] = 0$
6. $\frac{d}{dx}[u^n] = nu^{n-1}u'$
7. $\frac{d}{dx}[x] = 1$
8. $\frac{d}{dx}[|u|] = \frac{u}{|u|}(u'), \quad u \neq 0$
9. $\frac{d}{dx}[\ln u] = \frac{u'}{u}$
10. $\frac{d}{dx}[e^u] = e^u u'$
11. $\frac{d}{dx}[\log_a u] = \frac{u'}{(\ln a)u}$
12. $\frac{d}{dx}[a^u] = (\ln a)a^u u'$
13. $\frac{d}{dx}[\sin u] = (\cos u)u'$
14. $\frac{d}{dx}[\cos u] = -(\sin u)u'$
15. $\frac{d}{dx}[\tan u] = (\sec^2 u)u'$
16. $\frac{d}{dx}[\cot u] = -(\csc^2 u)u'$
17. $\frac{d}{dx}[\sec u] = (\sec u \tan u)u'$
18. $\frac{d}{dx}[\csc u] = -(\csc u \cot u)u'$
19. $\frac{d}{dx}[\arcsin u] = \frac{u'}{\sqrt{1 - u^2}}$
20. $\frac{d}{dx}[\arccos u] = \frac{-u'}{\sqrt{1 - u^2}}$
21. $\frac{d}{dx}[\arctan u] = \frac{u'}{1 + u^2}$
22. $\frac{d}{dx}[\operatorname{arccot} u] = \frac{-u'}{1 + u^2}$
23. $\frac{d}{dx}[\operatorname{arcsec} u] = \frac{u'}{|u|\sqrt{u^2 - 1}}$
24. $\frac{d}{dx}[\operatorname{arccsc} u] = \frac{-u'}{|u|\sqrt{u^2 - 1}}$
25. $\frac{d}{dx}[\sinh u] = (\cosh u)u'$
26. $\frac{d}{dx}[\cosh u] = (\sinh u)u'$
27. $\frac{d}{dx}[\tanh u] = (\operatorname{sech}^2 u)u'$
28. $\frac{d}{dx}[\coth u] = -(\operatorname{csch}^2 u)u'$
29. $\frac{d}{dx}[\operatorname{sech} u] = -(\operatorname{sech} u \tanh u)u'$
30. $\frac{d}{dx}[\operatorname{csch} u] = -(\operatorname{csch} u \coth u)u'$
31. $\frac{d}{dx}[\sinh^{-1} u] = \frac{u'}{\sqrt{u^2 + 1}}$
32. $\frac{d}{dx}[\cosh^{-1} u] = \frac{u'}{\sqrt{u^2 - 1}}$
33. $\frac{d}{dx}[\tanh^{-1} u] = \frac{u'}{1 - u^2}$
34. $\frac{d}{dx}[\coth^{-1} u] = \frac{u'}{1 - u^2}$
35. $\frac{d}{dx}[\operatorname{sech}^{-1} u] = \frac{-u'}{u\sqrt{1 - u^2}}$
36. $\frac{d}{dx}[\operatorname{csch}^{-1} u] = \frac{-u'}{|u|\sqrt{1 + u^2}}$

Basic Integration Formulas

1. $\int kf(u)\,du = k\int f(u)\,du$
2. $\int [f(u) \pm g(u)]\,du = \int f(u)\,du \pm \int g(u)\,du$
3. $\int du = u + C$
4. $\int u^n\,du = \frac{u^{n+1}}{n+1} + C, \quad n \neq -1$
5. $\int \frac{du}{u} = \ln|u| + C$
6. $\int e^u\,du = e^u + C$
7. $\int a^u\,du = \left(\frac{1}{\ln a}\right)a^u + C$
8. $\int \sin u\,du = -\cos u + C$
9. $\int \cos u\,du = \sin u + C$
10. $\int \tan u\,du = -\ln|\cos u| + C$
11. $\int \cot u\,du = \ln|\sin u| + C$
12. $\int \sec u\,du = \ln|\sec u + \tan u| + C$
13. $\int \csc u\,du = -\ln|\csc u + \cot u| + C$
14. $\int \sec^2 u\,du = \tan u + C$
15. $\int \csc^2 u\,du = -\cot u + C$
16. $\int \sec u \tan u\,du = \sec u + C$
17. $\int \csc u \cot u\,du = -\csc u + C$
18. $\int \frac{du}{\sqrt{a^2 - u^2}} = \arcsin \frac{u}{a} + C$
19. $\int \frac{du}{a^2 + u^2} = \frac{1}{a}\arctan \frac{u}{a} + C$
20. $\int \frac{du}{u\sqrt{u^2 - a^2}} = \frac{1}{a}\operatorname{arcsec} \frac{|u|}{a} + C$

TRIGONOMETRY

Definition of the Six Trigonometric Functions

Right triangle definitions, where $0 < \theta < \pi/2$.

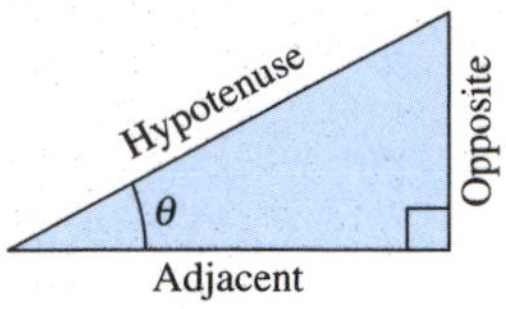

$$\sin\theta = \frac{\text{opp}}{\text{hyp}} \qquad \csc\theta = \frac{\text{hyp}}{\text{opp}}$$

$$\cos\theta = \frac{\text{adj}}{\text{hyp}} \qquad \sec\theta = \frac{\text{hyp}}{\text{adj}}$$

$$\tan\theta = \frac{\text{opp}}{\text{adj}} \qquad \cot\theta = \frac{\text{adj}}{\text{opp}}$$

Circular function definitions, where θ is any angle.

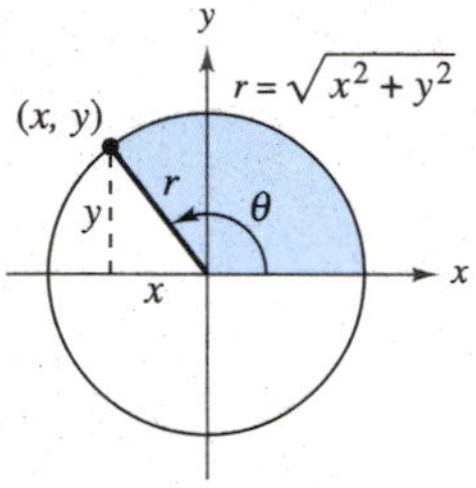

$$\sin\theta = \frac{y}{r} \qquad \csc\theta = \frac{r}{y}$$

$$\cos\theta = \frac{x}{r} \qquad \sec\theta = \frac{r}{x}$$

$$\tan\theta = \frac{y}{x} \qquad \cot\theta = \frac{x}{y}$$

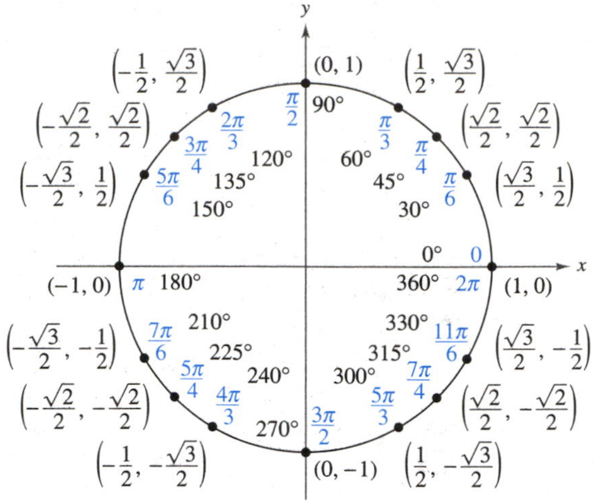

Reciprocal Identities

$$\sin x = \frac{1}{\csc x} \qquad \sec x = \frac{1}{\cos x} \qquad \tan x = \frac{1}{\cot x}$$

$$\csc x = \frac{1}{\sin x} \qquad \cos x = \frac{1}{\sec x} \qquad \cot x = \frac{1}{\tan x}$$

Quotient Identities

$$\tan x = \frac{\sin x}{\cos x} \qquad \cot x = \frac{\cos x}{\sin x}$$

Pythagorean Identities

$$\sin^2 x + \cos^2 x = 1$$

$$1 + \tan^2 x = \sec^2 x \qquad 1 + \cot^2 x = \csc^2 x$$

Cofunction Identities

$$\sin\left(\frac{\pi}{2} - x\right) = \cos x \qquad \cos\left(\frac{\pi}{2} - x\right) = \sin x$$

$$\csc\left(\frac{\pi}{2} - x\right) = \sec x \qquad \tan\left(\frac{\pi}{2} - x\right) = \cot x$$

$$\sec\left(\frac{\pi}{2} - x\right) = \csc x \qquad \cot\left(\frac{\pi}{2} - x\right) = \tan x$$

Even/Odd Identities

$$\sin(-x) = -\sin x \qquad \cos(-x) = \cos x$$

$$\csc(-x) = -\csc x \qquad \tan(-x) = -\tan x$$

$$\sec(-x) = \sec x \qquad \cot(-x) = -\cot x$$

Sum and Difference Formulas

$$\sin(u \pm v) = \sin u \cos v \pm \cos u \sin v$$

$$\cos(u \pm v) = \cos u \cos v \mp \sin u \sin v$$

$$\tan(u \pm v) = \frac{\tan u \pm \tan v}{1 \mp \tan u \tan v}$$

Double-Angle Formulas

$$\sin 2u = 2 \sin u \cos u$$

$$\cos 2u = \cos^2 u - \sin^2 u = 2\cos^2 u - 1 = 1 - 2\sin^2 u$$

$$\tan 2u = \frac{2 \tan u}{1 - \tan^2 u}$$

Power-Reducing Formulas

$$\sin^2 u = \frac{1 - \cos 2u}{2}$$

$$\cos^2 u = \frac{1 + \cos 2u}{2}$$

$$\tan^2 u = \frac{1 - \cos 2u}{1 + \cos 2u}$$

Sum-to-Product Formulas

$$\sin u + \sin v = 2 \sin\left(\frac{u + v}{2}\right) \cos\left(\frac{u - v}{2}\right)$$

$$\sin u - \sin v = 2 \cos\left(\frac{u + v}{2}\right) \sin\left(\frac{u - v}{2}\right)$$

$$\cos u + \cos v = 2 \cos\left(\frac{u + v}{2}\right) \cos\left(\frac{u - v}{2}\right)$$

$$\cos u - \cos v = -2 \sin\left(\frac{u + v}{2}\right) \sin\left(\frac{u - v}{2}\right)$$

Product-to-Sum Formulas

$$\sin u \sin v = \frac{1}{2}[\cos(u - v) - \cos(u + v)]$$

$$\cos u \cos v = \frac{1}{2}[\cos(u - v) + \cos(u + v)]$$

$$\sin u \cos v = \frac{1}{2}[\sin(u + v) + \sin(u - v)]$$

$$\cos u \sin v = \frac{1}{2}[\sin(u + v) - \sin(u - v)]$$

AP* Edition

Calculus of a Single Variable

10e

AP*Edition

Calculus of a Single Variable

10e

Ron Larson
The Pennsylvania State University
The Behrend College

Bruce Edwards
University of Florida

BROOKS/COLE
CENGAGE Learning™

Australia • Brazil • Japan • Korea • Mexico • Singapore • Spain • United Kingdom • United States

BROOKS/COLE
CENGAGE Learning™

Calculus of a Single Variable
Tenth Edition, AP* Edition
Ron Larson

Publisher: Liz Covello
Senior Development Editor: Carolyn Lewis
Assistant Editor: Liza Neustaetter
Editorial Assistant: Stephanie Kreuz
Associate Media Editor: Guanglei Zhang
VP, Director of Advanced and Elective Products Program: Alison Zetterquist
Editoral Coordinator, Advanced and Elective Products Program: Jean Woy
Senior Content Project Manager: Jessica Rasile
Art Director: Linda May
Rights Acquisition Specialist: Shalice Shah-Caldwell
Manufacturing Planner: Doug Bertke
Text/Cover Designer: Larson Texts, Inc.
Compositor: Larson Texts, Inc.
Cover Image: Larson Texts, Inc.

Library of Congress Control Number: 2012950714

ISBN-13: 978-1-285-06033-0
ISBN-10: 1-285-06033-4

Brooks/Cole
20 Channel Center Street
Boston, MA 02210
USA

Cengage Learning is a leading provider of customized learning solutions with office locations around the globe, including Singapore, the United Kingdom, Australia, Mexico, Brazil, and Japan. Locate your local office at: **international.cengage.com/region**

Cengage Learning products are represented in Canada by Nelson Education, Ltd.

Cengage Learning products are represented in high schools by Holt McDougal, a division of Houghton Mifflin Harcourt.

For your course and learning solutions, visit **www.cengage.com.**

To find online supplements and other instructional support, please visit **www.cengagebrain.com.**

Instructors: Please visit **login.cengage.com** and log in to access instructor-specific resources.

Printed in the United States of America
2 3 4 5 6 7 16 15 14

Contents

Appendices

*Available at the text-specific website *www.cengagebrain.com*

Preface

Welcome to *Calculus*, Tenth Edition, AP* Edition. We are proud to present this new edition to you. As with all editions, we have been able to incorporate many useful comments from you, our user. For this edition, we have introduced some new features and revised others. You will still find what you expect – a pedagogically sound, mathematically precise, and comprehensive textbook.

We are pleased and excited to offer you something brand new with this edition – a companion website at LarsonCalculus.com. This site offers many resources that will help you as you study calculus. All of these resources are just a click away.

Our goal for every edition of this textbook is to provide you with the tools you need to master calculus. We hope that you find the changes in this edition, together with **LarsonCalculus.com,** will accomplish just that.

In each exercise set, be sure to notice the reference to **CalcChat.com.** At this free site, you can download a step-by-step solution to any odd-numbered exercise. Also, you can talk to a tutor, free of charge, during the hours posted at the site. Over the years, thousands of students have visited the site for help. We use all of this information to help guide each revision of the exercises and solutions.

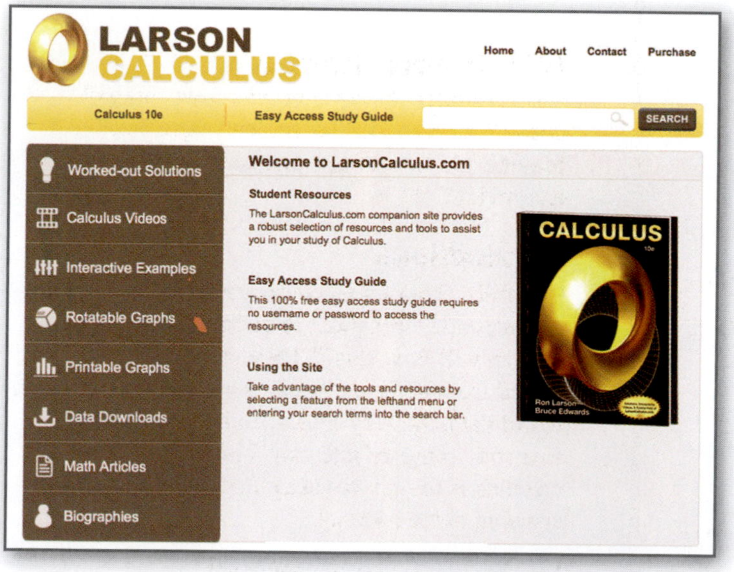

New To This Edition

NEW LarsonCalculus.com

This companion website offers multiple tools and resources to supplement your learning. Access to these features is free. Watch videos explaining concepts or proofs from the book, explore examples, view three-dimensional graphs, download articles from math journals and much more.

NEW AP* Exam Tips

Throughout the book, these tips offer important information about material that commonly appears on the AP* exam.

NEW Interactive Examples

Examples throughout the book are accompanied by Interactive Examples at LarsonCalculus.com. These interactive examples use Wolfram's free CDF Player and allow you to explore calculus by manipulating functions or graphs, and observing the results.

NEW Proof Videos

Watch videos of co-author Bruce Edwards as he explains the proofs of theorems in *Calculus*, Tenth Edition at LarsonCalculus.com.

**AP and Advanced Placement Program are registered trademarks of the College Entrance Examination Board, which was not involved in the production of, and does not endorse, this product.*

NEW How Do You See It?

The How Do You See It? feature in each section presents a real-life problem that you will solve by visual inspection using the concepts learned in the lesson. This exercise is excellent for classroom discussion or test preparation.

118. **HOW DO YOU SEE IT?** The figure shows the graphs of the position, velocity, and acceleration functions of a particle.

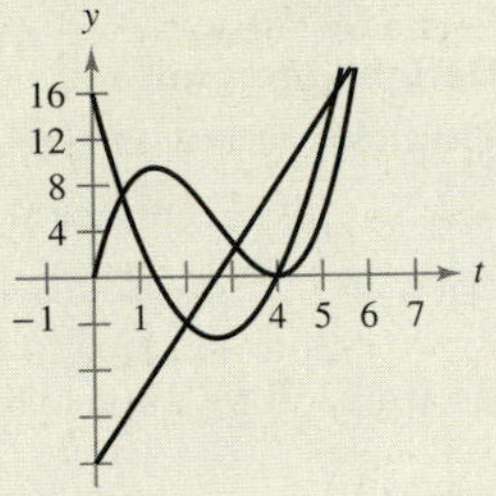

(a) Copy the graphs of the functions shown. Identify each graph. Explain your reasoning. To print an enlarged copy of the graph, go to *MathGraphs.com*.

(b) On your sketch, identify when the particle speeds up and when it slows down. Explain your reasoning.

REVISED Remark

These hints and tips reinforce or expand upon concepts, help you learn how to study mathematics, caution you about common errors, address special cases, or show alternative or additional steps to a solution of an example.

REVISED Exercise Sets

The exercise sets have been carefully and extensively examined to ensure they are rigorous and relevant and include all topics our users have suggested. The exercises have been reorganized and titled so you can better see the connections between examples and exercises. Multi-step, real-life exercises reinforce problem-solving skills and mastery of concepts by giving students the opportunity to apply the concepts in real-life situations.

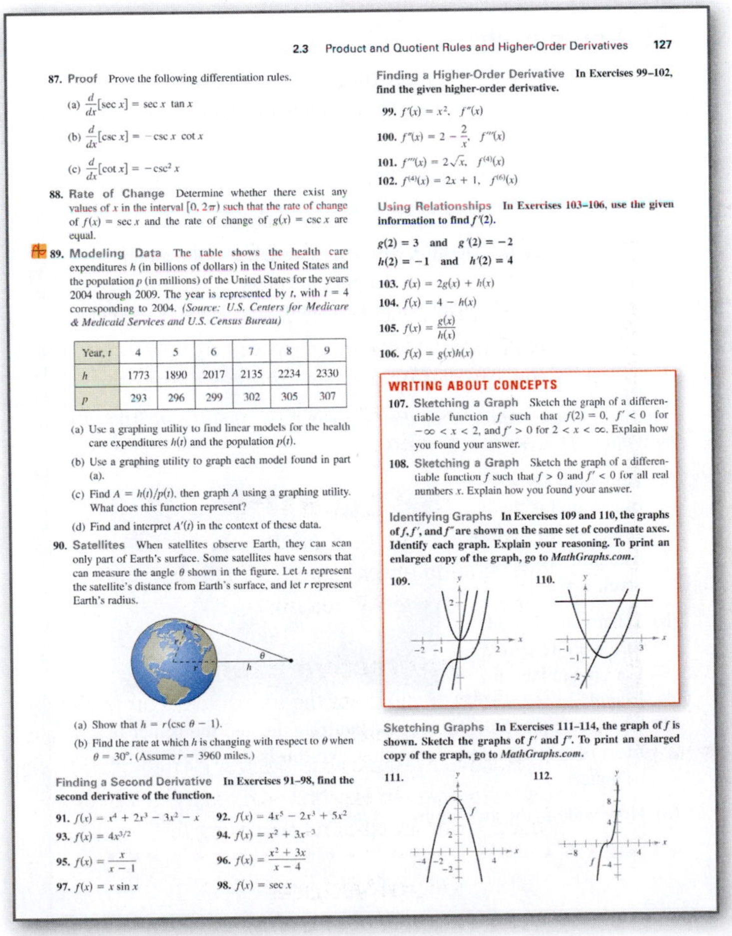

2.3 Product and Quotient Rules and Higher-Order Derivatives 127

87. **Proof** Prove the following differentiation rules.

(a) $\frac{d}{dx}[\sec x] = \sec x \tan x$

(b) $\frac{d}{dx}[\csc x] = -\csc x \cot x$

(c) $\frac{d}{dx}[\cot x] = -\csc^2 x$

88. **Rate of Change** Determine whether there exist any values of x in the interval $[0, 2\pi)$ such that the rate of change of $f(x) = \sec x$ and the rate of change of $g(x) = \csc x$ are equal.

89. **Modeling Data** The table shows the health care expenditures h (in billions of dollars) in the United States and the population p (in millions) of the United States for the years 2004 through 2009. The year is represented by t, with $t = 4$ corresponding to 2004. *(Source: U.S. Centers for Medicare & Medicaid Services and U.S. Census Bureau)*

Year, t	4	5	6	7	8	9
h	1773	1890	2017	2135	2234	2330
p	293	296	299	302	305	307

(a) Use a graphing utility to find linear models for the health care expenditures $h(t)$ and the population $p(t)$.

(b) Use a graphing utility to graph each model found in part (a).

(c) Find $A = h(t)/p(t)$, then graph A using a graphing utility. What does this function represent?

(d) Find and interpret $A'(t)$ in the context of these data.

90. **Satellites** When satellites observe Earth, they can scan only part of Earth's surface. Some satellites have sensors that can measure the angle θ shown in the figure. Let h represent the satellite's distance from Earth's surface, and let r represent Earth's radius.

(a) Show that $h = r(\csc \theta - 1)$.

(b) Find the rate at which h is changing with respect to θ when $\theta = 30°$. (Assume $r = 3960$ miles.)

Finding a Second Derivative In Exercises 91–98, find the second derivative of the function.

91. $f(x) = x^4 + 2x^3 - 3x^2 - x$
92. $f(x) = 4x^5 - 2x^3 + 5x^2$
93. $f(x) = 4x^{3/2}$
94. $f(x) = x^2 + 3x^{-3}$
95. $f(x) = \frac{x}{x-1}$
96. $f(x) = \frac{x^2 + 3x}{x - 4}$
97. $f(x) = x \sin x$
98. $f(x) = \sec x$

Finding a Higher-Order Derivative In Exercises 99–102, find the given higher-order derivative.

99. $f'(x) = x^2, \quad f''(x)$
100. $f''(x) = 2 - \frac{2}{x}, \quad f'''(x)$
101. $f'''(x) = 2\sqrt{x}, \quad f^{(4)}(x)$
102. $f^{(4)}(x) = 2x + 1, \quad f^{(6)}(x)$

Using Relationships In Exercises 103–106, use the given information to find $f'(2)$.

$g(2) = 3$ and $g'(2) = -2$
$h(2) = -1$ and $h'(2) = 4$

103. $f(x) = 2g(x) + h(x)$
104. $f(x) = 4 - h(x)$
105. $f(x) = \frac{g(x)}{h(x)}$
106. $f(x) = g(x)h(x)$

WRITING ABOUT CONCEPTS

107. **Sketching a Graph** Sketch the graph of a differentiable function f such that $f(2) = 0$, $f' < 0$ for $-\infty < x < 2$, and $f' > 0$ for $2 < x < \infty$. Explain how you found your answer.

108. **Sketching a Graph** Sketch the graph of a differentiable function f such that $f > 0$ and $f' < 0$ for all real numbers x. Explain how you found your answer.

Identifying Graphs In Exercises 109 and 110, the graphs of f, f', and f'' are shown on the same set of coordinate axes. Identify each graph. Explain your reasoning. To print an enlarged copy of the graph, go to *MathGraphs.com*.

109.
110.

Sketching Graphs In Exercises 111–114, the graph of f is shown. Sketch the graphs of f' and f''. To print an enlarged copy of the graph, go to *MathGraphs.com*.

111.
112.

Table of Content Changes

Appendix A (Proofs of Selected Theorems) now appears in video format at LarsonCalculus.com. The proofs also appear in text form at CengageBrain.com.

Trusted Features

AP* Review Questions

Modeled after the types of questions you will encounter on the AP* exam, these questions provide additional opportunities for practice and review.

Applications

Carefully chosen applied exercises and examples are included throughout to address the question, "When will I use this?" These applications are pulled from diverse sources, such as current events, world data, industry trends, and more, and relate to a wide range of interests. Understanding where calculus is (or can be) used promotes fuller understanding of the material.

Writing about Concepts

Writing exercises at the end of each section are designed to test your understanding of basic concepts in each section, encouraging you to verbalize and write answers and promote technical communication skills that will be invaluable in your future careers.

Theorems

Theorems provide the conceptual framework for calculus. Theorems are clearly stated and separated from the rest of the text by boxes for quick visual reference. Key proofs often follow the theorem and can be found at LarsonCalculus.com.

Definition of Definite Integral

If f is defined on the closed interval $[a, b]$ and the limit of Riemann sums over partitions Δ

$$\lim_{\|\Delta\| \to 0} \sum_{i=1}^{n} f(c_i)\,\Delta x_i$$

exists (as described above), then f is said to be **integrable** on $[a, b]$ and the limit is denoted by

$$\lim_{\|\Delta\| \to 0} \sum_{i=1}^{n} f(c_i)\,\Delta x_i = \int_a^b f(x)\,dx.$$

The limit is called the **definite integral** of f from a to b. The number a is the **lower limit** of integration, and the number b is the **upper limit** of integration.

Definitions

As with theorems, definitions are clearly stated using precise, formal wording and are separated from the text by boxes for quick visual reference.

Explorations

Explorations provide unique challenges to study concepts that have not yet been formally covered in the text. They allow you to learn by discovery and introduce topics related to ones presently being studied. Exploring topics in this way encourages you to think outside the box.

Historical Notes and Biographies

Historical Notes provide you with background information on the foundations of calculus. The Biographies introduce you to the people who created and contributed to calculus.

Technology

Throughout the book, technology boxes show you how to use technology to solve problems and explore concepts of calculus. These tips also point out some pitfalls of using technology.

Section Projects

Projects appear in selected sections and encourage you to explore applications related to the topics you are studying. They provide an interesting and engaging way for you and other students to work and investigate ideas collaboratively.

Putnam Exam Challenges

Putnam Exam questions appear in selected sections. These actual Putnam Exam questions will challenge you and push the limits of your understanding of calculus.

SECTION PROJECT

St. Louis Arch

The Gateway Arch in St. Louis, Missouri, was constructed using the hyperbolic cosine function. The equation used for construction was

$$y = 693.8597 - 68.7672 \cosh 0.0100333x,$$
$$-299.2239 \le x \le 299.2239$$

where x and y are measured in feet. Cross sections of the arch are equilateral triangles, and (x, y) traces the path of the centers of mass of the cross-sectional triangles. For each value of x, the area of the cross-sectional triangle is

$$A = 125.1406 \cosh 0.0100333x.$$

(Source: Owner's Manual for the Gateway Arch, Saint Louis, MO, by William Thayer)

(a) How high above the ground is the center of the highest triangle? (At ground level, $y = 0$.)

(b) What is the height of the arch? (*Hint:* For an equilateral triangle, $A = \sqrt{3}c^2$, where c is one-half the base of the triangle, and the center of mass of the triangle is located at two-thirds the height of the triangle.)

(c) How wide is the arch at ground level?

Additional Resources

Student Resources

- **Student Solutions Manual for Calculus of a Single Variable**
  (Chapters P–10 of *Calculus*): ISBN 1-285-08571-X

  Student Solutions Manual for Multivariable Calculus
  (Chapters 11–16 of *Calculus*): ISBN 1-285-08575-2

  These manuals contain worked-out solutions for all odd-numbered exercises. Ask your sales representative about available discounts.

- **Fast Track to a 5 ISBN 1-285-06326-0—**This test preparation manual provides valuable test-taking strategies, review, and full-length practice exams. Keyed to *Calculus*, it helps students efficiently and effectively prepare for the AP exam. Ask your sales representative about available discounts. *The Fast Track to a 5* may be purchased with the text or separately.

CourseMate

CourseMate is a perfect study tool for bringing concepts to life with interactive learning, study, and exam preparation tools that support the printed textbook. CourseMate includes: an interactive eBook, videos, quizzes, flashcards, and more! Ask your sales representative about available discounts.

Teacher Resources

ENHANCED WebAssign *www.webassign.net*

Exclusively from Cengage Learning, Enhanced WebAssign offers an extensive online program for *Calculus*, 10e to encourage the practice that is so critical for concept mastery. The meticulously crafted pedagogy and exercises in our proven texts become even more effective in Enhanced WebAssign, supplemented by multimedia tutorial support and immediate feedback as students complete their assignments. Ask your sales representative about available discounts.

- **Complete Solutions Manual for Calculus of a Single Variable, Volume 1**
  (Chapters P–6 of *Calculus*): ISBN 1-285-08576-0

  Complete Solutions Manual for Calculus of a Single Variable, Volume 2
  (Chapters 7–10 of *Calculus*): ISBN 1-285-08577-9

  Complete Solutions Manual for Multivariable Calculus
  (Chapters 11–16 of *Calculus*): ISBN 1-285-08580-9

  The *Complete Solutions Manuals* contain worked-out solutions to all exercises in the text.

- **Solution Builder** (www.cengage.com/solutionbuilder)— This online teacher database offers complete worked-out solutions to all exercises in the text, allowing you to create customized, secure solutions printouts (in PDF format) matched exactly to the problems you assign in class.

- **PowerLecture** (ISBN 1-285-09445-X)—This comprehensive DVD includes resources such as an electronic version of the Teacher's Resource Guide, complete pre-built PowerPoint® lectures, all art from the text in both jpeg and PowerPoint formats, ExamView® algorithmic computerized testing software, JoinIn™ content for audience response systems (clickers), and a link to Solution Builder.
- **ExamView Computerized Testing—** Create, deliver, and customize tests in print and online formats with ExamView®, an easy-to-use assessment and tutorial software. ExamView for *Calculus*, 10e contains hundreds of algorithmic multiple-choice and short answer test items. ExamView® is available on the PowerLecture DVD.
- **Teacher's Resource Guide** (ISBN 1-285-09445-X)—This robust manual contains an abundance of resources keyed to the textbook by chapter and section, including chapter summaries and teaching strategies. An electronic version of the Teacher's Resource Guide is available on the PowerLecture DVD.

CourseMate

CourseMate is a perfect study tool for students, and requires no set up from you. CourseMate brings course concepts to life with interactive learning, study, and exam preparation tools that support the printed textbook. CourseMate for *Calculus*, 10e includes: an interactive eBook, videos, quizzes, flashcards, and more! For teachers, CourseMate includes Engagement Tracker, a first-of-its kind tool that monitors student engagement. Ask your sales representative about available discounts.

Acknowledgements

We would like to thank the many people who have helped us at various stages of *Calculus* over the last 39 years. Their encouragement, criticisms, and suggestions have been invaluable.

Reviewers of the Tenth Edition

Denis Bell, *University of Northern Florida*; Abraham Biggs, *Broward Community College*; Jesse Blosser, *Eastern Mennonite School;* Mark Brittenham, *University of Nebraska*; Mingxiang Chen, *North Carolina A & T State University*; Marcia Kleinz, *Atlantic Cape Community College*; Maxine Lifshitz, *Friends Academy*; Bill Meisel, *Florida State College at Jacksonville*; Martha Nega, *Georgia Perimeter College*; Laura Ritter, *Southern Polytechnic State University*; Chia-Lin Wu, *Richard Stockton College of New Jersey*

Reviewers of Previous Editions

Stan Adamski, *Owens Community College*; Alexander Arhangelskii, *Ohio University;* Seth G. Armstrong, *Southern Utah University;* Jim Ball, *Indiana State University;* Marcelle Bessman, *Jacksonville University;* Linda A. Bolte, *Eastern Washington University;* James Braselton, *Georgia Southern University;* Harvey Braverman, *Middlesex County College;* Tim Chappell, *Penn Valley Community College;* Oiyin Pauline Chow, *Harrisburg Area Community College;* Julie M. Clark, *Hollins University;* P.S. Crooke, *Vanderbilt University;* Jim Dotzler, *Nassau Community College;* Murray Eisenberg, *University of Massachusetts at Amherst;* Donna Flint, *South Dakota State University;* Michael Frantz, *University of La Verne;* Sudhir Goel, *Valdosta State University;* Arek Goetz, *San Francisco State University;* Donna J. Gorton, *Butler County Community College;* John Gosselin, *University of Georgia;* Shahryar Heydari, *Piedmont College;* Guy Hogan, *Norfolk State University;* Ashok Kumar, *Valdosta State University;* Kevin J. Leith, *Albuquerque Community College;* Douglas B. Meade, *University of South Carolina;* Teri Murphy, *University of Oklahoma;* Darren Narayan, *Rochester Institute of Technology;* Susan A. Natale, *The Ursuline School, NY;* Terence H. Perciante, *Wheaton College;* James Pommersheim, *Reed College;* Leland E. Rogers, *Pepperdine University;* Paul Seeburger, *Monroe Community College;* Edith A. Silver, *Mercer County Community College;* Howard Speier, *Chandler-Gilbert Community College;* Desmond Stephens, *Florida A&M University;* Jianzhong Su, *University of Texas at Arlington;* Patrick Ward, *Illinois Central College;* Diane Zych, *Erie Community College*

Many thanks to Robert Hostetler, The Behrend College, The Pennsylvania State University, and David Heyd, The Behrend College, The Pennsylvania State University, for their significant contributions to previous editions of this text.

We would also like to thank the staff at Larson Texts, Inc., who assisted in preparing the manuscript, rendering the art package, typesetting, and proofreading the pages and supplements.

On a personal level, we are grateful to our wives, Deanna Gilbert Larson and Consuelo Edwards, for their love, patience, and support. Also, a special note of thanks goes out to R. Scott O'Neil.

If you have suggestions for improving this text, please feel free to write to us. Over the years we have received many useful comments from both instructors and students, and we value these very much.

Ron Larson

Bruce Edwards

Your Course. Your Way.

Calculus Textbook Options

The traditional calculus course is available in a variety of textbook configurations to address the different ways instructors teach—and students take—their classes.

TOPICS COVERED	APPROACH			
	Late Transcendental Functions	**Early Transcendental Functions**	**Accelerated coverage**	**Integrated coverage**
3-semester	Calculus 10e AP* Edition 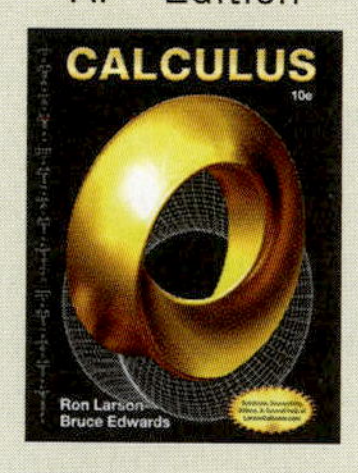	Calculus Early Transcendental Functions 5e 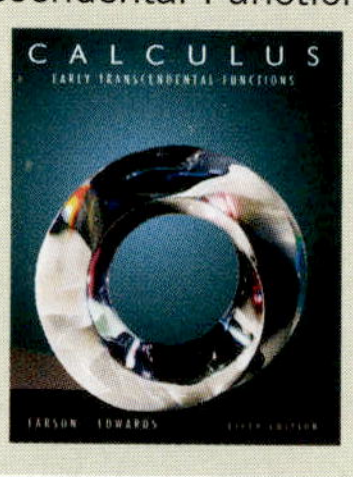	Essential Calculus 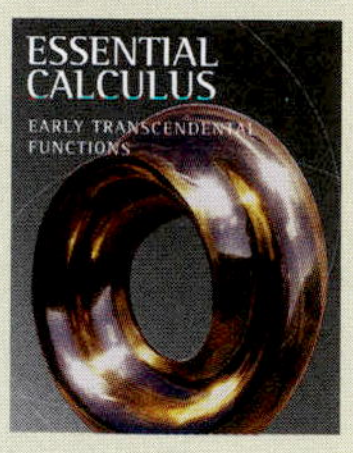	
Single Variable Only	Calculus 10e Single Variable AP* Edition 	Calculus: Early Transcendental Functions 5e Single Variable 		Calculus I with Precalculus 3e 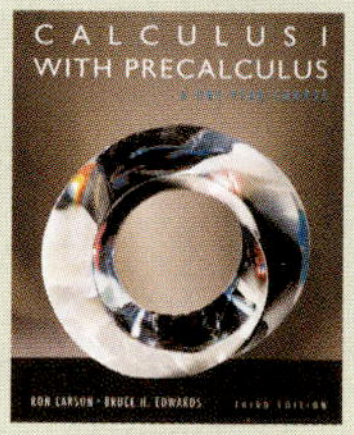
Multivariable	Calculus 10e Multivariable 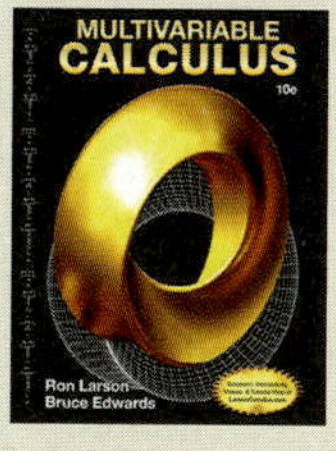	Calculus 10e Multivariable 		

Correlation of AP* Course Topics to Calculus, 10e

AB Topics

Functions, Graphs, and Limits

(A) Analysis of Graphs	
(i) Interplay between geometric and analytic information.	1.2–1.5, 3.5
(ii) Use of calculus to predict and explain the observed local and global behavior of a function.	3.1 - 3.6

(B) Limits of Functions (including one-sided limits)	
(i) An intuitive understanding of the limiting process.	1.2–1.4
(ii) Calculating limits using algebra.	1.3–1.5, 3.5
(iii) Estimating limits from graphs or tables of data.	1.2, 1.5

(C) Asymptotic and Unbounded Behavior	
(i) Understanding asymptotes in terms of graphical behavior.	1.5, 3.5
(ii) Describing asymptotic behavior in terms of limits involving infinity.	1.5, 3.5
(iii) Comparing relative magnitudes of functions and their rates of change (for example, contrasting exponential growth, polynomial growth, and logarithmic growth).	3.5, 5.1, 5.5, 8.7

(D) Continuity as a Property of Functions	
(i) An intuitive understanding of continuity. (The function values can be made as close as desired by taking sufficiently close values of the domain.)	1.4
(ii) Understanding continuity in terms of limits.	1.4, 1.5
(iii) Geometric understanding of graphs of continuous functions (Intermediate Value Theorem and Extreme Value Theorem).	1.4, 3.1

Derivatives

(A) Concept of the Derivative	
(i) Derivative presented graphically, numerically, and analytically.	2.1
(ii) Derivative interpreted as an instantaneous rate of change.	2.1, 2.2, 2.6, 6.2
(iii) Derivative defined as the limit of the difference quotient.	2.1
(iv) Relationship between differentiability and continuity.	2.1

(B) Derivative at a Point	
(i) Slope of a curve at a point. Examples are emphasized, including points at which there are vertical tangents and points at which there are no tangents.	2.1, 2.2, 6.1
(ii) Tangent line to a curve at a point and local linear approximation.	2.1–2.5, 6.1
(iii) Instantaneous rate of change as the limit of average rate of change.	2.2
(iv) Approximate rate of change from graphs and tables of values.	2.2–2.4

(C) Derivative as a Function	
(i) Corresponding characteristics of graphs of f and f'.	2.1, 3.1–3.6
(ii) Relationship between the increasing and decreasing behavior of f and the sign of f'.	3.3
(iii) The Mean Value Theorem and its geometric interpretation.	3.2
(iv) Equations involving derivatives. Verbal descriptions are translated into equations involving derivatives and vice versa.	2.6, 3.7, 6.2

(D) Second Derivatives	
(i) Corresponding characteristics of graphs of f, f', and f''.	3.4
(ii) Relationship between the concavity of f and the sign of f''.	3.4
(iii) Points of inflection as places where concavity changes.	3.4

(E) Applications of Derivatives	
(i) Analysis of curves, including the notions of monotonicity and concavity.	3.1–3.6
(ii) Optimization, both absolute (global) and relative (local) extrema.	3.1–3.6
(iii) Modeling rates of change, including related rates problems.	2.6
(iv) Use of implicit differentiation to find the derivative of an inverse function.	5.3
(v) Interpretation of the derivative as a rate of change in varied applied contexts, including velocity, speed, and acceleration.	2.3, 3.2
(vi) Geometric interpretation of differential equations via slope fields and the relationship between slope fields and solution curves for differential equations.	4.1, 5.1, 6.1
(F) Computation of Derivatives	
(i) Knowledge of derivatives of basic functions, including power, exponential, logarithmic, trigonometric, and inverse trigonometric functions.	2.2, 2.3, 5.1, 5.3–5.6
(ii) Derivative rules for sums, products, and quotients of functions.	2.2, 2.3
(iii) Chain rule and implicit differentiation.	2.4, 2.5

Integrals

(A) Interpretations and Properties of Definite Integrals	
(i) Definite integral as a limit of Riemann sums.	4.3
(ii) Definite integral of the rate of change of a quantity over an interval interpreted as the change of the quantity over the interval: $\int_a^b f'(x)\,dx = f(b) - f(a)$	4.4
(iii) Basic properties of definite integrals (examples include additivity and linearity).	4.3
(B) Applications of Integrals	
(i) Use of integrals to model physical, biological, or economic situations.	4.4, 5.5, 6.2, 6.3, 7.4, 7.5
(ii) Use of the method of setting up an approximating Riemann sum and representing its limit as a definite integral.	4.3, 7.1, 7.2
(iii) Finding the area of a region, the volume of a solid with known cross sections, the average value of a function, the distance traveled by a particle along a line, and accumulated change from a rate of change.	4.2–4.6, 7.1, 7.2
(C) Fundamental Theorem of Calculus	
(i) Use of the Fundamental Theorem to evaluate definite integrals.	4.4
(ii) Use of the Fundamental Theorem to represent a particular antiderivative, and the analytical and graphical analysis of functions so defined.	4.5
(D) Techniques of Antidifferentiation	
(i) Antiderivatives following directly from derivatives of basic functions.	4.1, 4.4, 5.2, 5.4
(ii) Antiderivatives by substitution of variables (including change of limits for definite integrals).	4.5
(E) Applications of Antidifferentiation	
(i) Finding specific antiderivatives using initial conditions, including applications to motion along a line.	4.1, 4.4, 6.1, 6.2
(ii) Solving separable differential equations and using them in modeling (including the study of the equation $y' = ky$ and exponential growth).	6.1–6.3
(F) Numerical Approximations to Definite Integrals	
(i) Use of Riemann sums (using left, right, and midpoint evaluation points) and trapezoidal sums to approximate definite integrals of functions represented algebraically, graphically, and by tables of values.	4.2, 4.3, 4.6

BC Only Topics

The AP BC exam includes all AB topics in addition to the following topics. Please note that the Chapter 12 vector topics are also covered in Appendix G: AP BC Vector Topics (Web).

Functions, Graphs, and Limits

(A) Parametric, Polar, and Vector Functions	
(i) The analysis of planar curves includes those given in parametric form, polar form, and vector form.	10.1–10.4, Appendix G

Derivatives

(A) Applications of Derivatives	
(i) Analysis of planar curves given in parametric form, polar form, and vector form, including velocity and acceleration.	10.2–10.5, Appendix G
(ii) Numerical solution of differential equations using Euler's method.	6.1
(iii) L'Hospital's Rule, including its use in determining limits and convergence of improper integrals and series.	8.7, 8.8

(B) Computation of Derivatives	
(i) Derivatives of parametric, polar, and vector functions.	10.3, 10.4, Appendix G

Integrals

(A) Applications of Integrals	
(iii) Finding accumulated change (including a region bounded by polar curves) and the length of a curve (including a curve given in parametric form).	7.4, 10.3, 10.5

(B) Techniques of Antidifferentiation	
(i) Antiderivatives by substitution of variables (including change of limits for definite integrals), parts, and simple fractions (nonrepeating linear factors only).	4.5, 8.1–8.6
(ii) Improper integrals (as limits of definite integrals).	8.8

(C) Applications of Antidifferentiation	
(i) Solving logistic differential equations and using them in modeling.	6.3

Polynomial Approximations and Series

(A) Concept of Series	
(i) Definition of a series as a sequence of partial sums.	9.2
(ii) Definition of convergence in terms of the limit of the sequence of partial sums.	9.2
(iii) Use of technology to explore convergence and divergence.	9.2

(B) Series of Constants	
(i) Motivating examples, including decimal expansion.	9.2
(ii) Geometric series with applications.	9.2
(iii) The harmonic series.	9.3
(iv) Alternating series with error bound.	9.5
(v) Terms of series as areas of rectangles and their relationship to improper integrals, including the integral test and its use in testing the convergence of p-series.	9.3
(vi) The ratio test for convergence and divergence.	9.6
(vii) Comparing series to test for convergence or divergence.	9.4

(C) Taylor Series	
(i) Taylor polynomial approximation with graphical demonstration of convergence (for example, viewing graphs of various Taylor polynomials of the sine function approximating the sine curve).	9.7
(ii) Maclaurin series and the general Taylor series centered at $x = a$.	9.7–9.10
(iii) Maclaurin series for the functions e^x, $\sin x$, $\cos x$, and $\frac{1}{1-x}$.	9.7–9.10
(iv) Formal manipulation of Taylor series and shortcuts to computing Taylor series, including substitution, differentiation, antidifferentiation, and the formation of new series from known series.	9.8–9.10
(v) Functins defined by power series.	9.8–9.10
(vi) Radius and interval of convergence of power series.	9.8
(vii) Lagrange error bound for Taylor polynomials.	9.7

Preparing for the AP* Calculus Examination

By: Sharon Cade, *Mathematics Consultant*; Rhea Caldwell, *Providence Day School, Charlotte, North Carolina*; and Jeff Lucia, *Providence Day School, Charlotte, North Carolina*

Advanced Placement can be exhilarating. Whether you are taking an AP course at your school or you are working on the AP curriculum independently, the stage is set for a great intellectual experience.

But sometime in the spring, when the examination begins to loom on a very real horizon, Advanced Placement can seem intimidating. It is a normal feeling to be nervous about the test; you are in good company.

The best way to deal with an AP examination is to master it, not let it master you. You should think of this examination as a way to show off how much calculus you know. Attitude *does* help. But, no matter what type of math student you are, there is still a lot you can do to prepare for the exam. Focused review and practice time will help you master the examination so that you can walk in with confidence and earn a great score.

What's in the *Calculus* Textbook that will help you prepare

As you work through the textbook, there are some things that you should do to get the most out of the text.

- Brush up on your precalculus skills by reviewing Appendix C (Web). Success in calculus comes not only from understanding the calculus but being able to do the mechanics, so the skills covered in review materials are key.
- Visit the LarsonCalculus.com website. The site contains a variety of tools and resources that will help you master critical calculus concepts.
- As you read the sections,
    - take time reading the examples,
    - do the conceptual exercises so that you can check your understanding of the material, and
    - notice that the author begins each set of exercises with the basic concepts and skills and proceeds to more challenging problems and applications; be sure to try some of each level.
- After finishing a chapter,
    - work through the Review Exercises and the P.S. Problem Solving Exercises and
    - complete the AP* Review Questions to practice your test-taking skills.

In calculus, reading the text is an essential part of the learning curve and will, in the end, save you time in understanding and mastering the material. There is more to the study of calculus than just being able to do some mathematics; you must understand the concepts and how they fit together. You will also learn to broaden your thinking as well as think logically if you allow yourself to try to see mathematics in a new way, not just as a set of algorithms.

**AP and Advanced Placement Program are registered trademarks of the College Entrance Examination Board, which was not involved in the production of, and does not endorse, this product.*

How to get the most out of your Calculus class

- Know your advanced algebra skills:
  - Linear equations
  - Quadratics (factoring)
  - Functions (transformations, piecewise, odd, even, domain, range)
  - Polynomials (zeros, end behavior)
  - Exponential and logarithmic curves
  - Rational and radical equations
  - Direct, inverse relations
  - Conics (for BC)
- Know your trigonometry:
  - Unit circle (0, 30, 45, 60, and 90 degrees and equivalent radian measures)
  - Symmetry around the unit circle
  - Basic identities
- Know basic sequences and series and when to apply which sequence or series formula.
- Have some knowledge of vectors (BC only).
- Know your calculator:
  - Plot the graph of a function in an appropriate window.
  - Find the zeros of a function (solving the equation numerically).
  - Know how to evaluate a function at a specific value of x: on the graph, 2nd calc, value; on the home screen, $y1$(value).
  - Graph your function and analyze it as a comparison to the algebra you do.

Remember that you have to analyze what the calculator gives you. For example, know how to tell when the graph has a "hole:" graph $y = \frac{x-2}{x^2-4}$ and look for the value of y at $x = 2$.

Setting up a review schedule

The AP Calculus courses are concerned with developing a student's understanding of the concepts while providing experiences with its methods and applications. Both the AB and BC courses require a depth of understanding. If you have been doing your homework steadily and keeping up with the course work, you are in good shape. Organize your notes, homework, and handouts from class by topic. For example, keep a set of notes on precalculus topics (no longer tested on the AP exam but essential to your success with calculus), limits, derivative rules and applications, integral rules and applications, series and sequences techniques and methods, and major theorems. If you can summarize the main information in a few pages, by topic, you will find reviewing much easier. Refer to these materials as you begin to prepare for the exam. Use your textbook to get more detail as needed.

AP Information before the examination

In February

- Make sure that you are **registered** to take the test. Some schools take care of the paperwork and handle the fees for their AP students, but check with your teacher or the AP coordinator to make sure that you are registered. This is especially important if you have a documented disability and need test accommodations. If you are studying AP independently, or if your school does not have an AP coordinator, call AP Services at the College Board at (609) 771-7300 or (888) 225-5427 (toll-free in the United States and Canada). You can also email *apexams@info.collegeboard.org* for the name of the local AP coordinator, who will help you through the registration process. It is suggested that you call AP Services or contact your AP coordinator by March 1.

- Check on the eligibility of your **calculator.** Go online to *http://www.collegeboard.com/student/testing/ap/calculus_ab/calc.html?calcab#list* early enough so that if you need a different calculator, you will have time to get one and become familiar with it.

By Mid-March

- Begin your review process; set a schedule for yourself that you can follow.

Night before

- Put all of your **materials** in one place.
- Relax and get a good night's **rest** (this alone could improve your score because you will be able to think more clearly throughout the test).

Things to have on test day

- **Approved graphing calculator** with fresh batteries (you may have a second calculator as a backup, but it must also be a graphing calculator). The calculator must not have a typewriter-style (QWERTY) keyboard, nor can it be a non-graphing scientific calculator. Calculator memories are not cleared for the exam.

AP* Exam Tip

Be sure your calculator is set in radian mode (pi radians, or approximately three radians, is half the circle; 3 degrees is an angle just barely above the x-axis).

- #2 pencils (at least 2) with good erasers.
- A watch (to monitor your pace, but turn off the alarm if it has one).
- A bottle of water and a snack (fruit or power bar).
- Social Security number (if you choose to include it on the forms).
- The College Board school code.
- Photo identification and the admissions ticket.
- Comfortable clothes and a sweatshirt or sweater in case the room is cold.

Schools may have your admissions ticket at the testing site; a photo identification may not be needed at your own school, but check with your AP coordinator prior to test day.

On the day of the examination, it is wise to eat a good breakfast. Studies show that students who eat a hot breakfast before testing get higher scores. Breakfast can give you the energy you need to power through the test and more. You will spend some time waiting while everyone is seated in the right room for the right test. That's before the test has even begun. With a short break between Section I and Section II, the AP Calculus exam can last almost four hours.

Now go get a 5!

Taking the AP* Calculus Examination

To do well on the AP Calculus examination, you should be able to:

- Understand and work with the connections between the graphical, numerical, analytical, and verbal representations of functions.
- Use derivatives to solve a variety of problems and understand the meaning of a derivative in terms of rate of change and local linearity.
- Use integrals to solve a variety of problems and understand the meaning of a definite integral in terms of the limit of Riemann sums as well as the net accumulation of change.
- A student also needs to:
    - Understand both parts of the Fundamental Theorem of Calculus.
    - Communicate mathematics in written sentences.
    - Appropriately model a physical situation.
    - Use technology correctly and efficiently.
    - Determine whether solutions are reasonable and understand them in terms of units of measurement, size, and so on.

It is important to realize that a student who is in AP Calculus is expected to have studied all of the prerequisite material. A student should have a mastery of functions and their properties and an understanding of algebra, graphs, and the language of domain, range, symmetry, periodicity, and so on. The student should also understand trigonometry and have a mastery of the basic values in the unit circle and the basic trigonometric identities.

Exam Format

The AP Calculus examination currently consists of two major sections, each of which has two parts. All sections test proficiency on a variety of topics.

Multiple Choice: Section I has two sets of multiple-choice questions. Part A has 28 questions with an allotted time of 55 minutes and does not allow the use of a calculator. Part B has 17 questions with an allotted time of 50 minutes and does contain some questions for which a graphing calculator is needed. The multiple choice section score is based on the number of questions answered correctly; no points will be deducted for incorrect answers and no points are awarded for unanswered questions.

Free Response: Section II has six free-response questions, and it is broken into two parts. Part A consists of two problems, in which some parts of the problems may require a graphing calculator. You will be allowed 30 minutes for Part A. Part B has four problems, in which a calculator is not permitted. You will be allowed 60 minutes for Part B. Although you may continue working on Part A problems during this 60-minute session, you may no longer use a calculator. Thus, when working on Part A, you must be sure to answer the questions requiring a calculator during that first 30-minute period.

The grade for the examination is equally weighted between the multiple-choice and free-response sections. You can possibly earn a 5 on the exam even if you miss an entire free-response question. Students taking the BC exam will also receive an AB subscore grade.

The free-response questions and solutions are published annually after the AP Reading is completed and can be found at *apcentral.com*.

**AP and Advanced Placement Program are registered trademarks of the College Entrance Examination Board, which was not involved in the production of, and does not endorse, this product.*

General AP Test-Taking Strategies

Strategize the test question. Begin somewhere. Ask "What do I need?" and then "How do I get there?" Start with a clear definition; for example, for question 6 about continuity on the 2003 AB test, students needed to have a clear definition of continuity to answer the question fully.

- Know what the required tools on your calculator are and know how to access and use them:
    - Plot the graph of a function within a viewing window.
    - Find the zeros of a function (numerically solve equations).
    - Numerically calculate the derivative of a function.
    - Numerically calculate the value of a definite integral.
- Know the relationships between f, f', and f''.
- Know your differentiation and integration rules.
- Underline key components of the questions.
- Treat units carefully.
- Set the calculator to at least THREE decimal places and properly use the store key for intermediate steps (if you round too soon, your final answer will not be correct to the requisite three decimal places). Although you only see three decimal places on the calculator, it actually retains the complete value in its memory; if you just write the three numbers down and then use them, you are rounding too soon. If you choose to write your answer with more than three decimal places, remember only the first three places are read as your answer—you may truncate or round.

Strategies for the Multiple-Choice Section

To Look or not to look at the answers: Often you are better off trying to work through the problem before you look at the answer choices. However, there are times when seeing the answer choices may help you determine a method of solution (see 1998 multiple-choice test question 17 where only in the answer do you see that you have to compare $f(1)$ to $f'(1)$ to $f''(1)$).

Read the question carefully: Pressured for time, many students make the mistake of reading the questions too quickly or merely skimming them. By reading a question carefully, you may already have some idea about the correct answer. Careful reading is especially important in EXCEPT questions. After you solve the problem and have a solution, reread the question to be sure your answer you solved for actually answers the question. For example, you may have solved for where the maximum occurred (the x-value), but the question actually asks for the maximum value of f (the y-value), and thus you need one more step to complete the problem.

Eliminate any answer you know is wrong: You can write on the multiple-choice questions in the test book. As you read through the responses, draw a line through any answer you know is wrong. Do as much scratch work as is necessary in the exam book, but be sure to mark your solution choice on the answer sheet in the corresponding oval.

Read all of the possible answers, then choose the most accurate response: AP examinations are written to test your precise knowledge of a subject. Some of the responses may be partially correct, but there will only be one response that is completely true.

Be careful of absolute responses. These answers often include the words *always* or *never*. They could be correct, but you should try to think of counterexamples to disprove them.

Skip tough questions: Skip them in the first go-through, but be sure that you mark them in the margin so you can come back to them later if you possibly can. Make sure you skip those questions on your answer sheet, too.

There is no penalty for guessing: Thus, at the end, try to narrow down your choices by eliminating answers you think are incorrect and make an educated guess.

Additional Thoughts

- The exact numerical answer may not be among the choices given. You will have to choose the solution that best approximates the exact numerical value.
- The domain of a function f is assumed to be the set of all real numbers x, where $f(x)$ is a real number, unless specified otherwise.
- f^{-1} or the prefix arc- indicates the inverse of a trigonometric function ($\cos^{-1} x = \arccos x$).

Types of Multiple-Choice Questions

All kinds of topics will be covered in the multiple-choice section; your skills and vocabulary will be tested as well as your ability to do multi-step problem solving. Phrases like *average value*, the *definition of continuity*, *extremum (relative and absolute)*, *the definition of a derivative in its two forms*, *differential equations*, *graphical interpretations*, and *slope fields* are just a sampling of terms that identify the kinds of problems you will see. Read through this text and do the practice problems to familiarize yourself with the way the questions are framed.

Multiple-choice questions will be formatted in two basic ways. You will find classic questions where there are just five answer choices. This is the most common type of problem; it requires you to read the question and select the correct answer. Strategies for solving this type of problem include

- reading the question carefully,
- eliminating known wrong answers,
- solving the problem and then interpreting your solution to fit the question, and
- on occasion, testing each solution choice to see which one is correct.

There will also be problems that could be called "list" and "group," where you may be asked "Which of the following is true about g?" You will be given choices such as I, II, and III, and the multiple-choice answers might appear as:

(a) None
(b) I only
(c) II only
(d) I and II only
(e) I, II, and III

This kind of problem requires a clear understanding of a concept or definition. To approach this kind of question,

- eliminate known wrong answers,
- recall necessary theorems or definitions to help you interpret the question, and
- reread the problem to check your solution's accuracy.

Strategies for the Free-Response Section

ALL work needs to be shown IN the test booklet, not on the question sheets. The readers must read only solutions in the test booklet itself. Be sure to write the solutions on the correct pages (answer for question 1 on the designated page). It is easy to write a solution on the wrong page. If this happens, make a note of it on the correct page and the reader will read with you; do not take the time to erase and redo the problem.

- Scan all of the questions in the section you are working in. First solve the problems that you think you can do easily. You can mark and come back to the harder problems later. During Part A (the calculator section with only 2 problems), be sure to tackle any parts of questions that requires the use of the calculator first, because after the 30 minutes, although you can continue to work on the problems, you will not be allowed to use your calculator.
- Show all of your work. Partial credit will be awarded for problems if the correct work is shown, even if the answer is not present or is incorrect. Although not required, it can be helpful to the reader if you circle your final answer.
- Cross out incorrect answers with an "X" rather than spending time erasing. Crossed-out or erased work will not be graded. However, don't cross out or erase work unless you have replaced it. Let the reader see what you tried; it may be worth some points.
- Be clear, neat, and organized in your work. If a reader cannot clearly understand your work, you may not receive full credit.
- Some free-response questions have several parts, such as a, b, c, and d. Attempt to solve each part. Even if your answer to part "a" is incorrect, you still may be awarded points for the remaining parts of the question if the work is correct for those parts. Remember, the answers may not depend on an earlier response and that is why it is important to try each part. If you work with your incorrect answer (as long as it is reasonably derived) from a previous part, the reader will read with you in a later part, checking your numerical answer on the basis of your incorrect input.
- Units are important in your answer. Keeping track of your units throughout calculations and performing unit cancellations, where possible, will help guide you to your answer. Points will be deducted for missing or incorrect units in the answer if the question asked that units be given.
- Don't just write equations or numbers in hopes of finding the correct answer. Extraneous or incorrect information could lead to a lower score. Don't make up work that is trivial, but do try the problem and the reader will read with you.
- You do not need to answer the questions in order, but be sure the answer is entered in the correct section.
- When you use a table or a graph from one section (part a) in another part of the problem (part c, for example), be sure to refer to it in some way—state your use of it or draw an arrow back to it. If you inadvertently put a response in the wrong part of the problem, note it clearly to the reader.
- Show all your work.
    - Clearly label any functions [if the problem uses $g(x)$, don't call it $f(x)$].
    - Label your sign charts accurately, for example, f' or f'' for the derivative tests. However, these by themselves do not count as a justification. No credit is given for sign charts, but they can help develop your analysis and subsequent answer to the question.
    - Label all graphs with appropriate notation including numeric intervals (by 1s or 10s, for example) and the names for the x- and y-axes (like distance and time).
    - Label all tables or other objects that you use to show your work.
    - Show standard mathematical (noncalculator) notation. For example, you must show the integral as $\int_1^3 (x + 2)\, dx$, not as fnInt $(x + 2, x, 1, 3)$.

Remember: You are not required to simplify your answer; you can save both time and the opportunity to make an error by leaving an answer in an unsimplified form. For example, $y - 2.3 = -6(x + 5.4)$ is an appropriate equation of a tangent line; there is no need to simplify it to slope-intercept form.

- Decimals require an accuracy of three decimal places in the solution. Thus be sure to understand how to carry (store in your calculator) the intermediate steps of a problem until you round to three decimal places at the end of the problem. If you do multiple calculations and each calculation is rounded to three decimal places prior to the next calculation, your final solution will not have the required accuracy. The third digit in the final solution can be either rounded or truncated.

Scoring for Free-Response Questions

The free-response sections are graded on a scale of 0–9 with a dash (–) given for no math on the page. The chief reader is ultimately responsible not only for working through the solution and alternate solutions for each problem, but also for assigning points on a 9-point scale to each problem. This varies from problem to problem, based on how many parts are in the problem as well as the complexity of the problem.

For example, in a problem that asks for units, units are generally assigned 1 point for the whole problem. In other words, if you do units correctly in part a, but incorrectly in part c, you would not be awarded the 1 point for units.

If a problem requires an explanation, or reasoning, it generally earns 1–2 points.

In a typical area and volume problem, the integral is often worth 1 point and the answer is worth an additional 1 point. Sometimes the limits of integration are also worth 1 point. Thus, it is important for you to at least get started on the problem because often some points are earned on the setup, even if the solution is not there or is incorrect.

A "bald" answer is one that has no supporting work or documentation, like "yes" or just a number. A bald answer is seldom awarded a point.

To learn calculus and best prepare for the examination, read the text, take risks, ask questions, and look for the connections between the algebraic, numeric, and graphical approaches to similar problems. As much as possible, graph every problem to enhance your understanding of the concepts involved.

P Preparation for Calculus

Hours of Daylight
(Example 3, p. 33)

Automobile Aerodynamics *(Exercise 96, p. 30)*

Conveyor Design *(Exercise 23, p. 16)*

Cell Phone Subscribers
(Exercise 68, p. 9)

Modeling Carbon Dioxide Concentration *(Example 6, p. 7)*

P.1 Graphs and Models

- Sketch the graph of an equation.
- Find the intercepts of a graph.
- Test a graph for symmetry with respect to an axis and the origin.
- Find the points of intersection of two graphs.
- Interpret mathematical models for real-life data.

RENÉ DESCARTES (1596–1650)

Descartes made many contributions to philosophy, science, and mathematics. The idea of representing points in the plane by pairs of real numbers and representing curves in the plane by equations was described by Descartes in his book *La Géométrie*, published in 1637.

See LarsonCalculus.com to read more of this biography.

The Graph of an Equation

In 1637, the French mathematician René Descartes revolutionized the study of mathematics by combining its two major fields—algebra and geometry. With Descartes's coordinate plane, geometric concepts could be formulated analytically and algebraic concepts could be viewed graphically. The power of this approach was such that within a century of its introduction, much of calculus had been developed.

The same approach can be followed in your study of calculus. That is, by viewing calculus from multiple perspectives—*graphically*, *analytically*, and *numerically*—you will increase your understanding of core concepts.

Consider the equation $3x + y = 7$. The point $(2, 1)$ is a **solution point** of the equation because the equation is satisfied (is true) when 2 is substituted for x and 1 is substituted for y. This equation has many other solutions, such as $(1, 4)$ and $(0, 7)$. To find other solutions systematically, solve the original equation for y.

$$y = 7 - 3x \qquad \text{Analytic approach}$$

Then construct a **table of values** by substituting several values of x.

x	0	1	2	3	4
y	7	4	1	-2	-5

Numerical approach

From the table, you can see that $(0, 7)$, $(1, 4)$, $(2, 1)$, $(3, -2)$, and $(4, -5)$ are solutions of the original equation $3x + y = 7$. Like many equations, this equation has an infinite number of solutions. The set of all solution points is the **graph** of the equation, as shown in Figure P.1. Note that the sketch shown in Figure P.1 is referred to as the graph of $3x + y = 7$, even though it really represents only a *portion* of the graph. The entire graph would extend beyond the page.

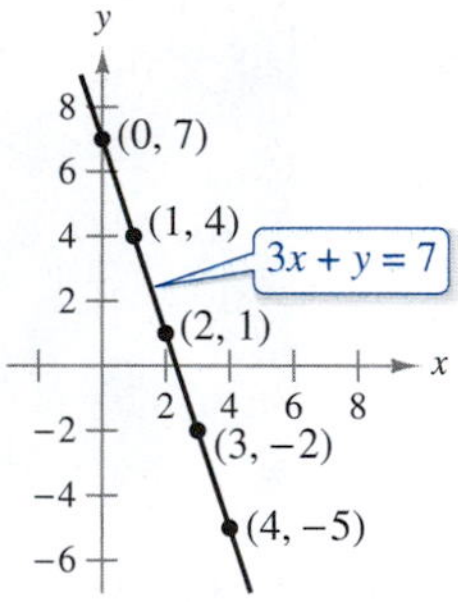

Graphical approach: $3x + y = 7$
Figure P.1

In this course, you will study many sketching techniques. The simplest is point plotting—that is, you plot points until the basic shape of the graph seems apparent.

EXAMPLE 1 Sketching a Graph by Point Plotting

To sketch the graph of $y = x^2 - 2$, first construct a table of values. Next, plot the points shown in the table. Then connect the points with a smooth curve, as shown in Figure P.2. This graph is a **parabola.** It is one of the conics you will study in Chapter 10.

x	-2	-1	0	1	2	3
y	2	-1	-2	-1	2	7

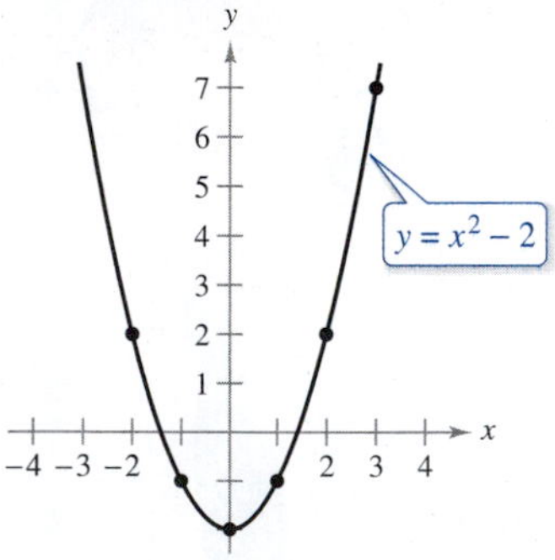

The parabola $y = x^2 - 2$
Figure P.2

*AP and Advanced Placement Program are registered trademarks of the College Entrance Examination Board, which was not involved in the production of, and does not endorse, this product.

One disadvantage of point plotting is that to get a good idea about the shape of a graph, you may need to plot many points. With only a few points, you could badly misrepresent the graph. For instance, to sketch the graph of

$$y = \frac{1}{30}x(39 - 10x^2 + x^4)$$

you plot five points:

$$(-3, -3), \quad (-1, -1), \quad (0, 0), \quad (1, 1), \quad \text{and} \quad (3, 3)$$

as shown in Figure P.3(a). From these five points, you might conclude that the graph is a line. This, however, is not correct. By plotting several more points, you can see that the graph is more complicated, as shown in Figure P.3(b).

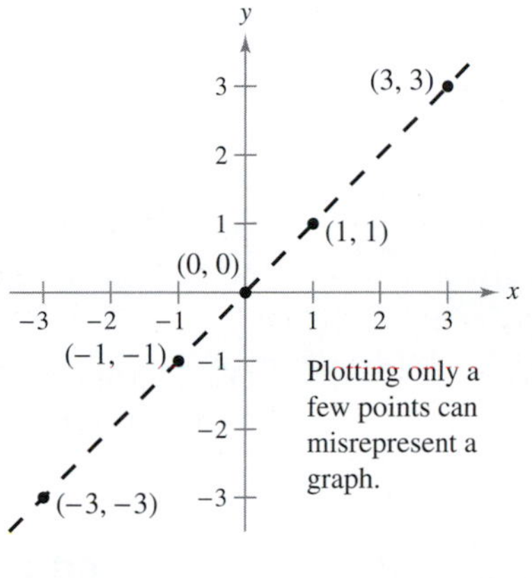

(a)

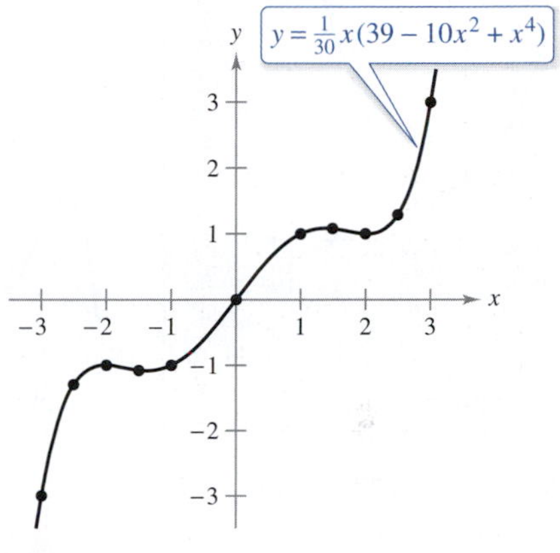

(b)

Figure P.3

Exploration

Comparing Graphical and Analytic Approaches Use a graphing utility to graph each equation. In each case, find a viewing window that shows the important characteristics of the graph.

a. $y = x^3 - 3x^2 + 2x + 5$

b. $y = x^3 - 3x^2 + 2x + 25$

c. $y = -x^3 - 3x^2 + 20x + 5$

d. $y = 3x^3 - 40x^2 + 50x - 45$

e. $y = -(x + 12)^3$

f. $y = (x - 2)(x - 4)(x - 6)$

A purely graphical approach to this problem would involve a simple "guess, check, and revise" strategy. What types of things do you think an analytic approach might involve? For instance, does the graph have symmetry? Does the graph have turns? If so, where are they? As you proceed through Chapters 1, 2, and 3 of this text, you will study many new analytic tools that will help you analyze graphs of equations such as these.

TECHNOLOGY Graphing an equation has been made easier by technology. Even with technology, however, it is possible to misrepresent a graph badly. For instance, each of the graphing utility* screens in Figure P.4 shows a portion of the graph of

$$y = x^3 - x^2 - 25.$$

From the screen on the left, you might assume that the graph is a line. From the screen on the right, however, you can see that the graph is not a line. So, whether you are sketching a graph by hand or using a graphing utility, you must realize that different "viewing windows" can produce very different views of a graph. In choosing a viewing window, your goal is to show a view of the graph that fits well in the context of the problem.

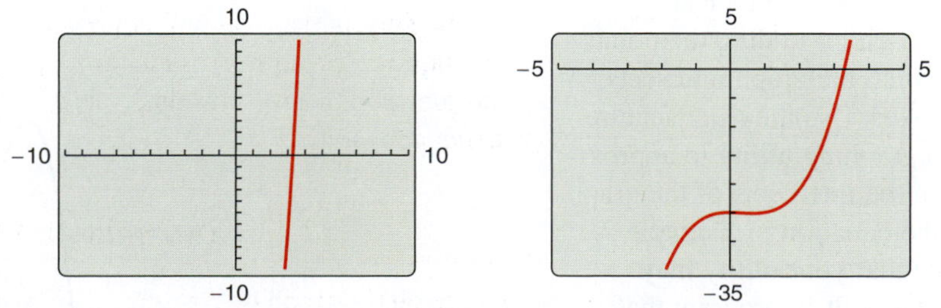

Graphing utility screens of $y = x^3 - x^2 - 25$

Figure P.4

*In this text, the term *graphing utility* means either a graphing calculator, such as the *TI-Nspire*, or computer graphing software, such as *Maple* or *Mathematica*.

Intercepts of a Graph

REMARK Some texts denote the x-intercept as the x-coordinate of the point $(a, 0)$ rather than the point itself. Unless it is necessary to make a distinction, when the term *intercept* is used in this text, it will mean either the point or the coordinate.

Two types of solution points that are especially useful in graphing an equation are those having zero as their x- or y-coordinate. Such points are called **intercepts** because they are the points at which the graph intersects the x- or y-axis. The point $(a, 0)$ is an ***x*-intercept** of the graph of an equation when it is a solution point of the equation. To find the x-intercepts of a graph, let y be zero and solve the equation for x. The point $(0, b)$ is a ***y*-intercept** of the graph of an equation when it is a solution point of the equation. To find the y-intercepts of a graph, let x be zero and solve the equation for y.

It is possible for a graph to have no intercepts, or it might have several. For instance, consider the four graphs shown in Figure P.5.

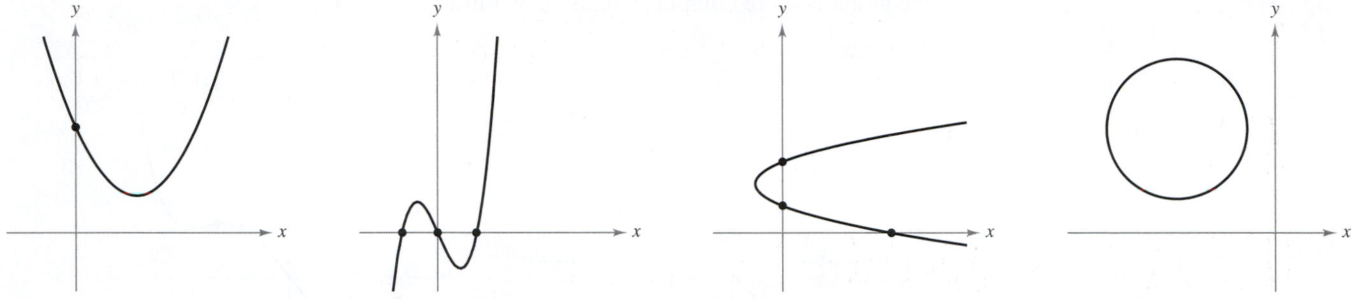

No x-intercepts, One y-intercept | Three x-intercepts, One y-intercept | One x-intercept, Two y-intercepts | No intercepts

Figure P.5

EXAMPLE 2 Finding *x*- and *y*-Intercepts

Find the x- and y-intercepts of the graph of $y = x^3 - 4x$.

Solution To find the x-intercepts, let y be zero and solve for x.

$$x^3 - 4x = 0 \quad \text{Let } y \text{ be zero.}$$
$$x(x - 2)(x + 2) = 0 \quad \text{Factor.}$$
$$x = 0, 2, \text{ or } -2 \quad \text{Solve for } x.$$

Because this equation has three solutions, you can conclude that the graph has three x-intercepts:

$$(0, 0), \quad (2, 0), \quad \text{and} \quad (-2, 0). \qquad x\text{-intercepts}$$

To find the y-intercepts, let x be zero. Doing this produces $y = 0$. So, the y-intercept is

$$(0, 0). \qquad y\text{-intercept}$$

(See Figure P.6.)

TECHNOLOGY Example 2 uses an analytic approach to finding intercepts. When an analytic approach is not possible, you can use a graphical approach by finding the points at which the graph intersects the axes. Use the *trace* feature of a graphing utility to approximate the intercepts of the graph of the equation in Example 2. Note that your utility may have a built-in program that can find the x-intercepts of a graph. (Your utility may call this the *root* or *zero* feature.) If so, use the program to find the x-intercepts of the graph of the equation in Example 2.

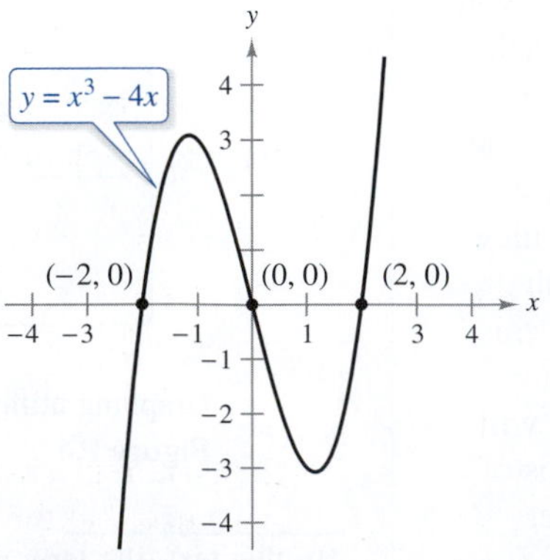

Intercepts of a graph
Figure P.6

Symmetry of a Graph

Knowing the symmetry of a graph before attempting to sketch it is useful because you need only half as many points to sketch the graph. The three types of symmetry listed below can be used to help sketch the graphs of equations (see Figure P.7).

1. A graph is **symmetric with respect to the y-axis** if, whenever (x, y) is a point on the graph, then $(-x, y)$ is also a point on the graph. This means that the portion of the graph to the left of the y-axis is a mirror image of the portion to the right of the y-axis.
2. A graph is **symmetric with respect to the x-axis** if, whenever (x, y) is a point on the graph, then $(x, -y)$ is also a point on the graph. This means that the portion of the graph below the x-axis is a mirror image of the portion above the x-axis.
3. A graph is **symmetric with respect to the origin** if, whenever (x, y) is a point on the graph, then $(-x, -y)$ is also a point on the graph. This means that the graph is unchanged by a rotation of 180° about the origin.

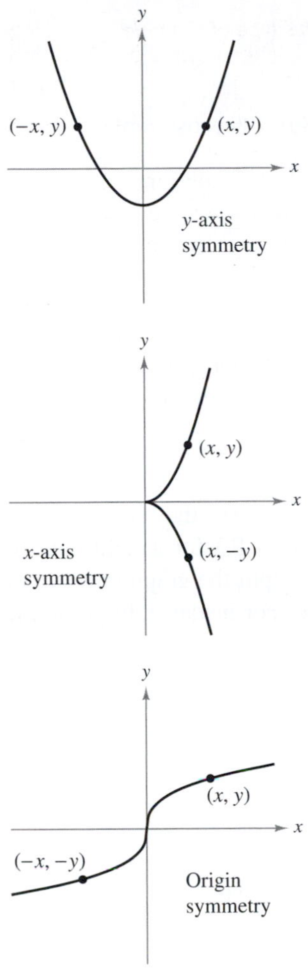

Figure P.7

Tests for Symmetry

1. The graph of an equation in x and y is symmetric with respect to the y-axis when replacing x by $-x$ yields an equivalent equation.
2. The graph of an equation in x and y is symmetric with respect to the x-axis when replacing y by $-y$ yields an equivalent equation.
3. The graph of an equation in x and y is symmetric with respect to the origin when replacing x by $-x$ and y by $-y$ yields an equivalent equation.

The graph of a polynomial has symmetry with respect to the y-axis when each term has an even exponent (or is a constant). For instance, the graph of

$$y = 2x^4 - x^2 + 2$$

has symmetry with respect to the y-axis. Similarly, the graph of a polynomial has symmetry with respect to the origin when each term has an odd exponent, as illustrated in Example 3.

EXAMPLE 3 Testing for Symmetry

Test the graph of $y = 2x^3 - x$ for symmetry with respect to (a) the y-axis and (b) the origin.

Solution

a. $y = 2x^3 - x$ — Write original equation.

$y = 2(-x)^3 - (-x)$ — Replace x by $-x$.

$y = -2x^3 + x$ — Simplify. It is not an equivalent equation.

Because replacing x by $-x$ does *not* yield an equivalent equation, you can conclude that the graph of $y = 2x^3 - x$ is *not* symmetric with respect to the y-axis.

b. $y = 2x^3 - x$ — Write original equation.

$-y = 2(-x)^3 - (-x)$ — Replace x by $-x$ and y by $-y$.

$-y = -2x^3 + x$ — Simplify.

$y = 2x^3 - x$ — Equivalent equation

Because replacing x by $-x$ and y by $-y$ yields an equivalent equation, you can conclude that the graph of $y = 2x^3 - x$ is symmetric with respect to the origin, as shown in Figure P.8.

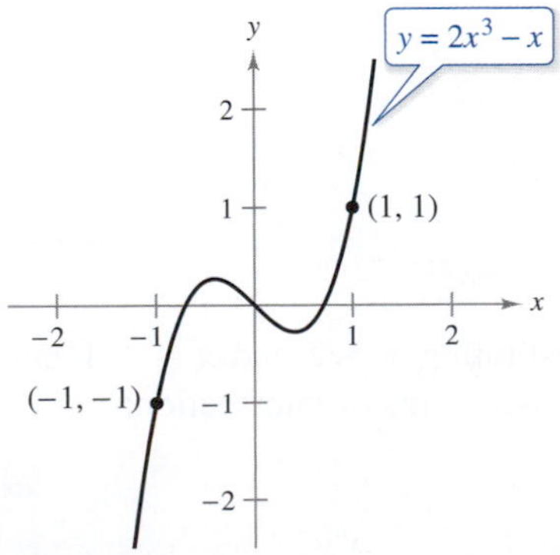

Origin symmetry
Figure P.8

EXAMPLE 4 Using Intercepts and Symmetry to Sketch a Graph

See LarsonCalculus.com for an interactive version of this type of example.

Sketch the graph of $x - y^2 = 1$.

Solution The graph is symmetric with respect to the x-axis because replacing y by $-y$ yields an equivalent equation.

$$x - y^2 = 1 \quad \text{Write original equation.}$$
$$x - (-y)^2 = 1 \quad \text{Replace } y \text{ by } -y.$$
$$x - y^2 = 1 \quad \text{Equivalent equation}$$

This means that the portion of the graph below the x-axis is a mirror image of the portion above the x-axis. To sketch the graph, first plot the x-intercept and the points above the x-axis. Then reflect in the x-axis to obtain the entire graph, as shown in Figure P.9.

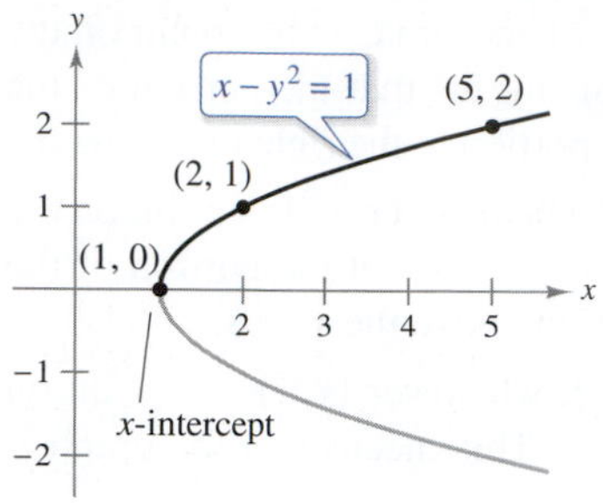

Figure P.9

TECHNOLOGY Graphing utilities are designed so that they most easily graph equations in which y is a function of x (see Section P.3 for a definition of **function**). To graph other types of equations, you need to split the graph into two or more parts *or* you need to use a different graphing mode. For instance, to graph the equation in Example 4, you can split it into two parts.

$$y_1 = \sqrt{x - 1} \quad \text{Top portion of graph}$$
$$y_2 = -\sqrt{x - 1} \quad \text{Bottom portion of graph}$$

Points of Intersection

A **point of intersection** of the graphs of two equations is a point that satisfies both equations. You can find the point(s) of intersection of two graphs by solving their equations simultaneously.

EXAMPLE 5 Finding Points of Intersection

Find all points of intersection of the graphs of

$$x^2 - y = 3 \quad \text{and} \quad x - y = 1.$$

Solution Begin by sketching the graphs of both equations in the *same* rectangular coordinate system, as shown in Figure P.10. From the figure, it appears that the graphs have two points of intersection. You can find these two points as follows.

$$y = x^2 - 3 \quad \text{Solve first equation for } y.$$
$$y = x - 1 \quad \text{Solve second equation for } y.$$
$$x^2 - 3 = x - 1 \quad \text{Equate } y\text{-values.}$$
$$x^2 - x - 2 = 0 \quad \text{Write in general form.}$$
$$(x - 2)(x + 1) = 0 \quad \text{Factor.}$$
$$x = 2 \text{ or } -1 \quad \text{Solve for } x.$$

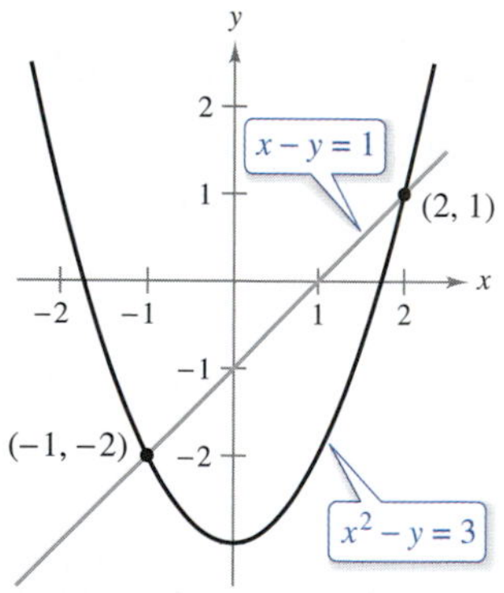

Two points of intersection
Figure P.10

The corresponding values of y are obtained by substituting $x = 2$ and $x = -1$ into either of the original equations. Doing this produces two points of intersection:

$$(2, 1) \quad \text{and} \quad (-1, -2). \quad \text{Points of intersection}$$

You can check the points of intersection in Example 5 by substituting into *both* of the original equations or by using the *intersect* feature of a graphing utility.

Mathematical Models

Real-life applications of mathematics often use equations as **mathematical models.** In developing a mathematical model to represent actual data, you should strive for two (often conflicting) goals: accuracy and simplicity. That is, you want the model to be simple enough to be workable, yet accurate enough to produce meaningful results. Section P.4 explores these goals more completely.

The Mauna Loa Observatory in Hawaii has been measuring the increasing concentration of carbon dioxide in Earth's atmosphere since 1958.

EXAMPLE 6 Comparing Two Mathematical Models

The Mauna Loa Observatory in Hawaii records the carbon dioxide concentration y (in parts per million) in Earth's atmosphere. The January readings for various years are shown in Figure P.11. In the July 1990 issue of *Scientific American*, these data were used to predict the carbon dioxide level in Earth's atmosphere in the year 2035, using the quadratic model

$$y = 0.018t^2 + 0.70t + 316.2$$ Quadratic model for 1960–1990 data

where $t = 0$ represents 1960, as shown in Figure P.11(a). The data shown in Figure P.11(b) represent the years 1980 through 2010 and can be modeled by

$$y = 1.68t + 303.5$$ Linear model for 1980–2010 data

where $t = 0$ represents 1960. What was the prediction given in the *Scientific American* article in 1990? Given the new data for 1990 through 2010, does this prediction for the year 2035 seem accurate?

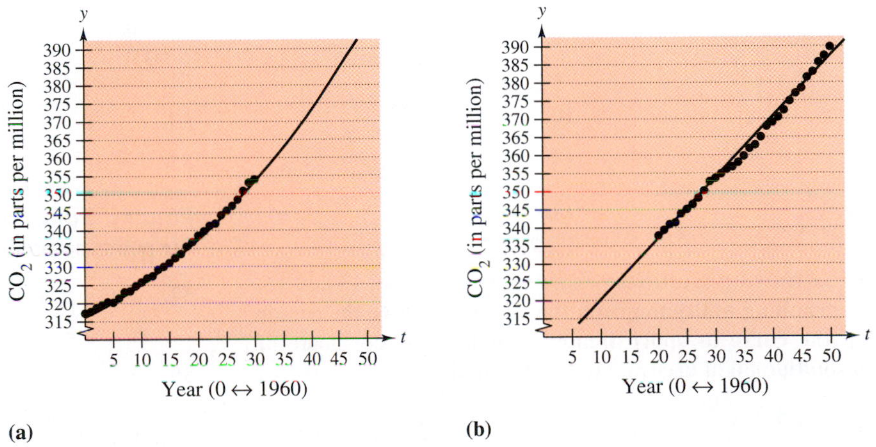

Figure P.11

Solution To answer the first question, substitute $t = 75$ (for 2035) into the quadratic model.

$$y = 0.018(75)^2 + 0.70(75) + 316.2 = 469.95$$ Quadratic model

So, the prediction in the *Scientific American* article was that the carbon dioxide concentration in Earth's atmosphere would reach about 470 parts per million in the year 2035. Using the linear model for the 1980–2010 data, the prediction for the year 2035 is

$$y = 1.68(75) + 303.5 = 429.5.$$ Linear model

So, based on the linear model for 1980–2010, it appears that the 1990 prediction was too high.

The models in Example 6 were developed using a procedure called *least squares regression* (see Section 13.9). The quadratic and linear models have correlations given by $r^2 \approx 0.997$ and $r^2 \approx 0.994$, respectively. The closer r^2 is to 1, the "better" the model.

P.1 Exercises

See CalcChat.com for tutorial help and worked-out solutions to odd-numbered exercises.

Matching In Exercises 1–4, match the equation with its graph. [The graphs are labeled (a), (b), (c), and (d).]

(a)

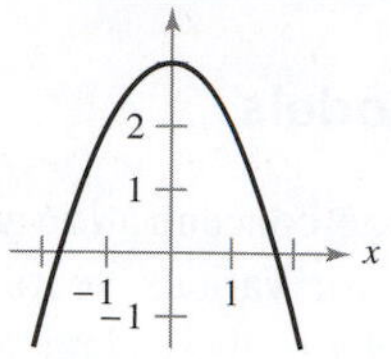

(b)

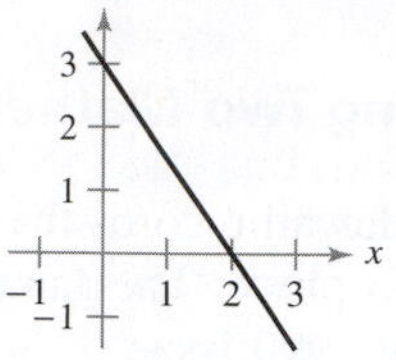

(c)

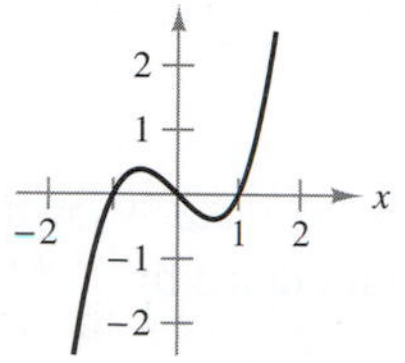

(d)

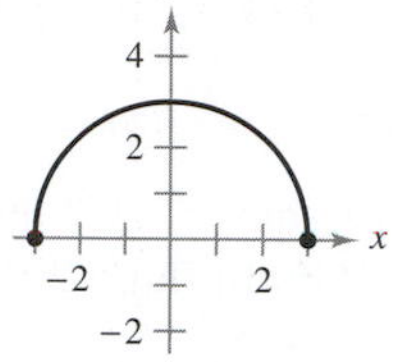

1. $y = -\frac{3}{2}x + 3$
2. $y = \sqrt{9 - x^2}$
3. $y = 3 - x^2$
4. $y = x^3 - x$

Sketching a Graph by Point Plotting In Exercises 5–14, sketch the graph of the equation by point plotting.

5. $y = \frac{1}{2}x + 2$
6. $y = 5 - 2x$
7. $y = 4 - x^2$
8. $y = (x - 3)^2$
9. $y = |x + 2|$
10. $y = |x| - 1$
11. $y = \sqrt{x} - 6$
12. $y = \sqrt{x + 2}$
13. $y = \dfrac{3}{x}$
14. $y = \dfrac{1}{x + 2}$

Approximating Solution Points In Exercises 15 and 16, use a graphing utility to graph the equation. Move the cursor along the curve to approximate the unknown coordinate of each solution point accurate to two decimal places.

15. $y = \sqrt{5 - x}$
    (a) $(2, y)$
    (b) $(x, 3)$
16. $y = x^5 - 5x$
    (a) $(-0.5, y)$
    (b) $(x, -4)$

Finding Intercepts In Exercises 17–26, find any intercepts.

17. $y = 2x - 5$
18. $y = 4x^2 + 3$
19. $y = x^2 + x - 2$
20. $y^2 = x^3 - 4x$
21. $y = x\sqrt{16 - x^2}$
22. $y = (x - 1)\sqrt{x^2 + 1}$
23. $y = \dfrac{2 - \sqrt{x}}{5x + 1}$
24. $y = \dfrac{x^2 + 3x}{(3x + 1)^2}$
25. $x^2y - x^2 + 4y = 0$
26. $y = 2x - \sqrt{x^2 + 1}$

Testing for Symmetry In Exercises 27–38, test for symmetry with respect to each axis and to the origin.

27. $y = x^2 - 6$
28. $y = x^2 - x$
29. $y^2 = x^3 - 8x$
30. $y = x^3 + x$
31. $xy = 4$
32. $xy^2 = -10$
33. $y = 4 - \sqrt{x + 3}$
34. $xy - \sqrt{4 - x^2} = 0$
35. $y = \dfrac{x}{x^2 + 1}$
36. $y = \dfrac{x^2}{x^2 + 1}$
37. $y = |x^3 + x|$
38. $|y| - x = 3$

Using Intercepts and Symmetry to Sketch a Graph In Exercises 39–56, find any intercepts and test for symmetry. Then sketch the graph of the equation.

39. $y = 2 - 3x$
40. $y = \frac{2}{3}x + 1$
41. $y = 9 - x^2$
42. $y = 2x^2 + x$
43. $y = x^3 + 2$
44. $y = x^3 - 4x$
45. $y = x\sqrt{x + 5}$
46. $y = \sqrt{25 - x^2}$
47. $x = y^3$
48. $x = y^2 - 4$
49. $y = \dfrac{8}{x}$
50. $y = \dfrac{10}{x^2 + 1}$
51. $y = 6 - |x|$
52. $y = |6 - x|$
53. $y^2 - x = 9$
54. $x^2 + 4y^2 = 4$
55. $x + 3y^2 = 6$
56. $3x - 4y^2 = 8$

Finding Points of Intersection In Exercises 57–62, find the points of intersection of the graphs of the equations.

57. $x + y = 8$
    $4x - y = 7$
58. $3x - 2y = -4$
    $4x + 2y = -10$
59. $x^2 + y = 6$
    $x + y = 4$
60. $x = 3 - y^2$
    $y = x - 1$
61. $x^2 + y^2 = 5$
    $x - y = 1$
62. $x^2 + y^2 = 25$
    $-3x + y = 15$

Finding Points of Intersection In Exercises 63–66, use a graphing utility to find the points of intersection of the graphs. Check your results analytically.

63. $y = x^3 - 2x^2 + x - 1$
    $y = -x^2 + 3x - 1$
64. $y = x^4 - 2x^2 + 1$
    $y = 1 - x^2$
65. $y = \sqrt{x + 6}$
    $y = \sqrt{-x^2 - 4x}$
66. $y = -|2x - 3| + 6$
    $y = 6 - x$

The symbol indicates an exercise in which you are instructed to use graphing technology or a symbolic computer algebra system. The solutions of other exercises may also be facilitated by the use of appropriate technology.

67. Modeling Data The table shows the Gross Domestic Product, or GDP (in trillions of dollars), for selected years. *(Source: U.S. Bureau of Economic Analysis)*

Year	1980	1985	1990	1995
GDP	2.8	4.2	5.8	7.4

Year	2000	2005	2010
GDP	10.0	12.6	14.5

(a) Use the regression capabilities of a graphing utility to find a mathematical model of the form $y = at^2 + bt + c$ for the data. In the model, y represents the GDP (in trillions of dollars) and t represents the year, with $t = 0$ corresponding to 1980.

(b) Use a graphing utility to plot the data and graph the model. Compare the data with the model.

(c) Use the model to predict the GDP in the year 2020.

68. Modeling Data

The table shows the numbers of cellular phone subscribers (in millions) in the United States for selected years. *(Source: CTIA-The Wireless)*

Year	1995	1998	2001	2004	2007	2010
Number	34	69	128	182	255	303

(a) Use the regression capabilities of a graphing utility to find a mathematical model of the form $y = at^2 + bt + c$ for the data. In the model, y represents the number of subscribers (in millions) and t represents the year, with $t = 5$ corresponding to 1995.

(b) Use a graphing utility to plot the data and graph the model. Compare the data with the model.

(c) Use the model to predict the number of cellular phone subscribers in the United States in the year 2020.

69. Break-Even Point Find the sales necessary to break even $(R = C)$ when the cost C of producing x units is $C = 2.04x + 5600$ and the revenue R from selling x units is $R = 3.29x$.

70. Copper Wire The resistance y in ohms of 1000 feet of solid copper wire at 77°F can be approximated by the model

$$y = \frac{10{,}770}{x^2} - 0.37, \quad 5 \le x \le 100$$

where x is the diameter of the wire in mils (0.001 in.). Use a graphing utility to graph the model. By about what factor is the resistance changed when the diameter of the wire is doubled?

71. Using Solution Points For what values of k does the graph of $y = kx^3$ pass through the point?

(a) $(1, 4)$ (b) $(-2, 1)$ (c) $(0, 0)$ (d) $(-1, -1)$

72. Using Solution Points For what values of k does the graph of $y^2 = 4kx$ pass through the point?

(a) $(1, 1)$ (b) $(2, 4)$ (c) $(0, 0)$ (d) $(3, 3)$

WRITING ABOUT CONCEPTS

Writing Equations In Exercises 73 and 74, write an equation whose graph has the indicated property. (There may be more than one correct answer.)

73. The graph has intercepts at $x = -4$, $x = 3$, and $x = 8$.

74. The graph has intercepts at $x = -\frac{3}{2}$, $x = 4$, and $x = \frac{5}{2}$.

75. Proof

(a) Prove that if a graph is symmetric with respect to the x-axis and to the y-axis, then it is symmetric with respect to the origin. Give an example to show that the converse is not true.

(b) Prove that if a graph is symmetric with respect to one axis and to the origin, then it is symmetric with respect to the other axis.

76. HOW DO YOU SEE IT? Use the graphs of the two equations to answer the questions below.

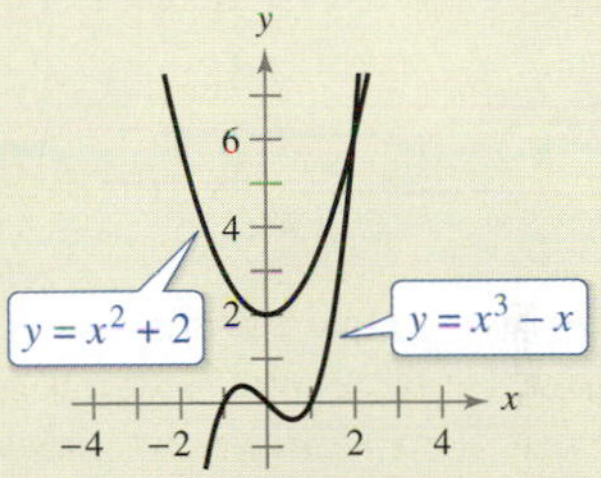

(a) What are the intercepts for each equation?

(b) Determine the symmetry for each equation.

(c) Determine the point of intersection of the two equations.

True or False? In Exercises 77–80, determine whether the statement is true or false. If it is false, explain why or give an example that shows it is false.

77. If $(-4, -5)$ is a point on a graph that is symmetric with respect to the x-axis, then $(4, -5)$ is also a point on the graph.

78. If $(-4, -5)$ is a point on a graph that is symmetric with respect to the y-axis, then $(4, -5)$ is also a point on the graph.

79. If $b^2 - 4ac > 0$ and $a \neq 0$, then the graph of $y = ax^2 + bx + c$ has two x-intercepts.

80. If $b^2 - 4ac = 0$ and $a \neq 0$, then the graph of $y = ax^2 + bx + c$ has only one x-intercept.

Andy Dean Photography/Shutterstock.com

P.2 Linear Models and Rates of Change

- **Find the slope of a line passing through two points.**
- **Write the equation of a line with a given point and slope.**
- **Interpret slope as a ratio or as a rate in a real-life application.**
- **Sketch the graph of a linear equation in slope-intercept form.**
- **Write equations of lines that are parallel or perpendicular to a given line.**

The Slope of a Line

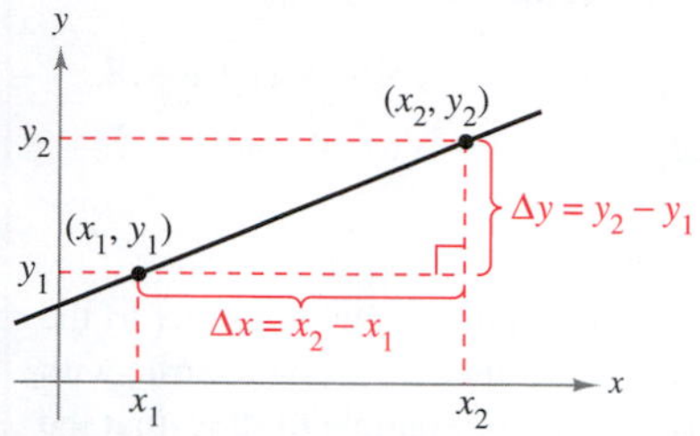

$\Delta y = y_2 - y_1 =$ change in y
$\Delta x = x_2 - x_1 =$ change in x
Figure P.12

The **slope** of a nonvertical line is a measure of the number of units the line rises (or falls) vertically for each unit of horizontal change from left to right. Consider the two points (x_1, y_1) and (x_2, y_2) on the line in Figure P.12. As you move from left to right along this line, a vertical change of

$$\Delta y = y_2 - y_1 \qquad \text{Change in } y$$

units corresponds to a horizontal change of

$$\Delta x = x_2 - x_1 \qquad \text{Change in } x$$

units. (Δ is the Greek uppercase letter *delta*, and the symbols Δy and Δx are read "delta y" and "delta x.")

Definition of the Slope of a Line

The **slope** m of the nonvertical line passing through (x_1, y_1) and (x_2, y_2) is

$$m = \frac{\Delta y}{\Delta x} = \frac{y_2 - y_1}{x_2 - x_1}, \quad x_1 \neq x_2.$$

Slope is not defined for vertical lines.

When using the formula for slope, note that

$$\frac{y_2 - y_1}{x_2 - x_1} = \frac{-(y_1 - y_2)}{-(x_1 - x_2)} = \frac{y_1 - y_2}{x_1 - x_2}.$$

So, it does not matter in which order you subtract *as long as* you are consistent and both "subtracted coordinates" come from the same point.

Figure P.13 shows four lines: one has a positive slope, one has a slope of zero, one has a negative slope, and one has an "undefined" slope. In general, the greater the absolute value of the slope of a line, the steeper the line. For instance, in Figure P.13, the line with a slope of -5 is steeper than the line with a slope of $\frac{1}{5}$.

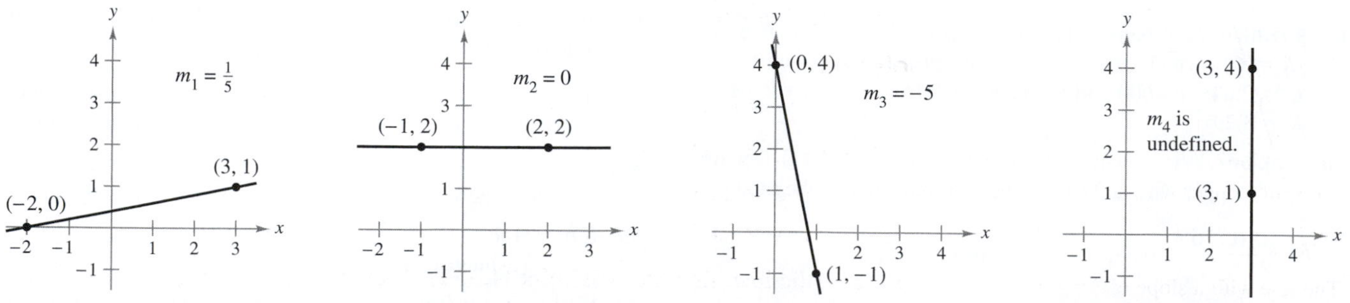

If m is positive, then the line rises from left to right.

If m is zero, then the line is horizontal.

If m is negative, then the line falls from left to right.

If m is undefined, then the line is vertical.

Figure P.13

Equations of Lines

Any two points on a nonvertical line can be used to calculate its slope. This can be verified from the similar triangles shown in Figure P.14. (Recall that the ratios of corresponding sides of similar triangles are equal.)

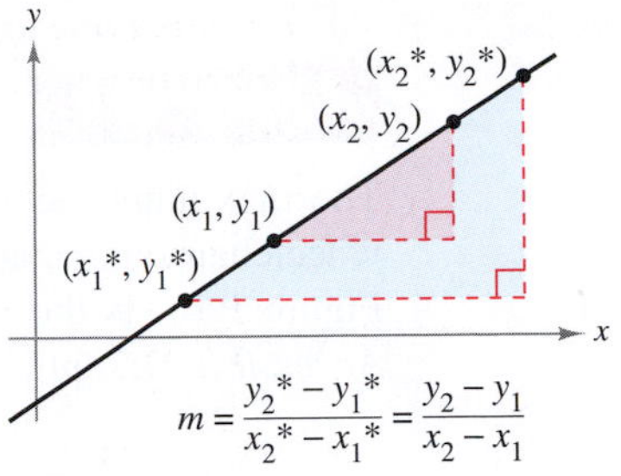

Any two points on a nonvertical line can be used to determine its slope.
Figure P.14

Exploration

Investigating Equations of Lines Use a graphing utility to graph each of the linear equations. Which point is common to all seven lines? Which value in the equation determines the slope of each line?

a. $y - 4 = -2(x + 1)$
b. $y - 4 = -1(x + 1)$
c. $y - 4 = -\frac{1}{2}(x + 1)$
d. $y - 4 = 0(x + 1)$
e. $y - 4 = \frac{1}{2}(x + 1)$
f. $y - 4 = 1(x + 1)$
g. $y - 4 = 2(x + 1)$

Use your results to write an equation of a line passing through $(-1, 4)$ with a slope of m.

If (x_1, y_1) is a point on a nonvertical line that has a slope of m and (x, y) is *any other* point on the line, then

$$\frac{y - y_1}{x - x_1} = m.$$

This equation in the variables x and y can be rewritten in the form

$$y - y_1 = m(x - x_1)$$

which is the **point-slope form** of the equation of a line.

Point-Slope Form of the Equation of a Line

The **point-slope form** of the equation of the line that passes through the point (x_1, y_1) and has a slope of m is

$$y - y_1 = m(x - x_1).$$

REMARK Remember that only nonvertical lines have a slope. Consequently, vertical lines cannot be written in point-slope form. For instance, the equation of the vertical line passing through the point $(1, -2)$ is $x = 1$.

EXAMPLE 1 Finding an Equation of a Line

Find an equation of the line that has a slope of 3 and passes through the point $(1, -2)$. Then sketch the line.

Solution

$y - y_1 = m(x - x_1)$	Point-slope form
$y - (-2) = 3(x - 1)$	Substitute -2 for y_1, 1 for x_1, and 3 for m.
$y + 2 = 3x - 3$	Simplify.
$y = 3x - 5$	Solve for y.

To sketch the line, first plot the point $(1, -2)$. Then, because the slope is $m = 3$, you can locate a second point on the line by moving one unit to the right and three units upward, as shown in Figure P.15.

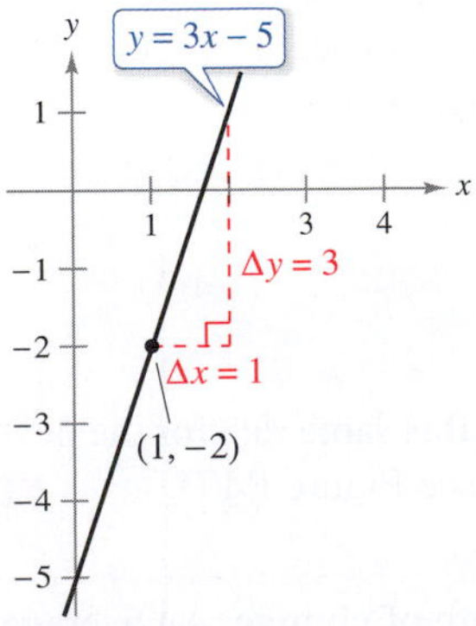

The line with a slope of 3 passing through the point $(1, -2)$
Figure P.15

Ratios and Rates of Change

The slope of a line can be interpreted as either a *ratio* or a *rate*. If the x- and y-axes have the same unit of measure, then the slope has no units and is a **ratio.** If the x- and y-axes have different units of measure, then the slope is a rate or **rate of change.** In your study of calculus, you will encounter applications involving both interpretations of slope.

EXAMPLE 2 Using Slope as a Ratio

The maximum recommended slope of a wheelchair ramp is $\frac{1}{12}$. A business installs a wheelchair ramp that rises to a height of 22 inches over a length of 24 feet, as shown in Figure P.16. Is the ramp steeper than recommended? *(Source: ADA Standards for Accessible Design)*

Figure P.16

Solution The length of the ramp is 24 feet or $12(24) = 288$ inches. The slope of the ramp is the ratio of its height (the rise) to its length (the run).

$$\begin{aligned}\text{Slope of ramp} &= \frac{\text{rise}}{\text{run}}\\ &= \frac{22 \text{ in.}}{288 \text{ in.}}\\ &\approx 0.076\end{aligned}$$

Because the slope of the ramp is less than $\frac{1}{12} \approx 0.083$, the ramp is not steeper than recommended. Note that the slope is a ratio and has no units.

EXAMPLE 3 Using Slope as a Rate of Change

The population of Colorado was about 4,302,000 in 2000 and about 5,029,000 in 2010. Find the average rate of change of the population over this 10-year period. What will the population of Colorado be in 2020? *(Source: U.S. Census Bureau)*

Solution Over this 10-year period, the average rate of change of the population of Colorado was

$$\begin{aligned}\text{Rate of change} &= \frac{\text{change in population}}{\text{change in years}}\\ &= \frac{5{,}029{,}000 - 4{,}302{,}000}{2010 - 2000}\\ &= 72{,}700 \text{ people per year.}\end{aligned}$$

Assuming that Colorado's population continues to increase at this same rate for the next 10 years, it will have a 2020 population of about 5,756,000 (see Figure P.17). ■

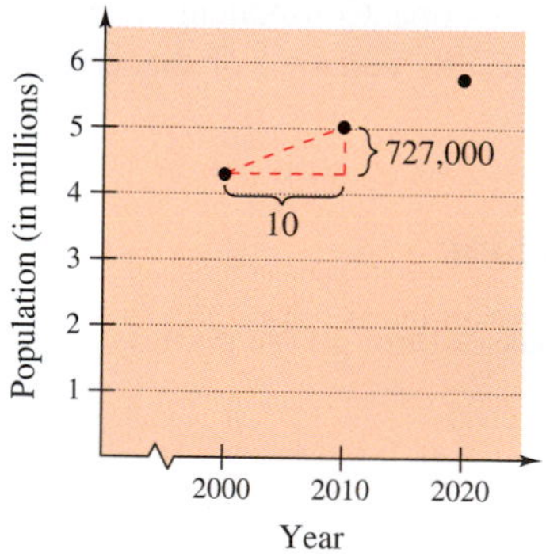

Population of Colorado
Figure P.17

The rate of change found in Example 3 is an **average rate of change.** An average rate of change is always calculated over an interval. In this case, the interval is [2000, 2010]. In Chapter 2, you will study another type of rate of change called an *instantaneous rate of change.*

Graphing Linear Models

Many problems in coordinate geometry can be classified into two basic categories.

1. Given a graph (or parts of it), find its equation.
2. Given an equation, sketch its graph.

For lines, problems in the first category can be solved by using the point-slope form. The point-slope form, however, is not especially useful for solving problems in the second category. The form that is better suited to sketching the graph of a line is the **slope-intercept** form of the equation of a line.

The Slope-Intercept Form of the Equation of a Line

The graph of the linear equation

$$y = mx + b \qquad \text{Slope-intercept form}$$

is a line whose slope is m and whose y-intercept is $(0, b)$.

EXAMPLE 4 Sketching Lines in the Plane

Sketch the graph of each equation.

a. $y = 2x + 1$

b. $y = 2$

c. $3y + x - 6 = 0$

Solution

a. Because $b = 1$, the y-intercept is $(0, 1)$. Because the slope is $m = 2$, you know that the line rises two units for each unit it moves to the right, as shown in Figure P.18(a).

b. By writing the equation $y = 2$ in slope-intercept form

$$y = (0)x + 2$$

you can see that the slope is $m = 0$ and the y-intercept is $(0, 2)$. Because the slope is zero, you know that the line is horizontal, as shown in Figure P.18(b).

c. Begin by writing the equation in slope-intercept form.

$$3y + x - 6 = 0 \qquad \text{Write original equation.}$$

$$3y = -x + 6 \qquad \text{Isolate } y\text{-term on the left.}$$

$$y = -\tfrac{1}{3}x + 2 \qquad \text{Slope-intercept form}$$

In this form, you can see that the y-intercept is $(0, 2)$ and the slope is $m = -\frac{1}{3}$. This means that the line falls one unit for every three units it moves to the right, as shown in Figure P.18(c).

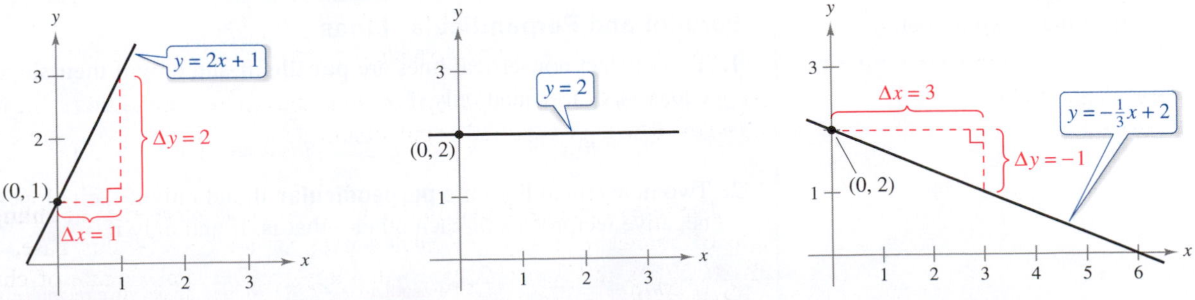

(a) $m = 2$; line rises (b) $m = 0$; line is horizontal (c) $m = -\frac{1}{3}$; line falls

Figure P.18

Because the slope of a vertical line is not defined, its equation cannot be written in slope-intercept form. However, the equation of any line can be written in the **general form**

$$Ax + By + C = 0$$ General form of the equation of a line

where A and B are not *both* zero. For instance, the vertical line

$$x = a$$ Vertical line

can be represented by the general form

$$x - a = 0.$$ General form

SUMMARY OF EQUATIONS OF LINES

1. General form: $Ax + By + C = 0$
2. Vertical line: $x = a$
3. Horizontal line: $y = b$
4. Slope-intercept form: $y = mx + b$
5. Point-slope form: $y - y_1 = m(x - x_1)$

Parallel and Perpendicular Lines

The slope of a line is a convenient tool for determining whether two lines are parallel or perpendicular, as shown in Figure P.19. Specifically, nonvertical lines with the same slope are parallel, and nonvertical lines whose slopes are negative reciprocals are perpendicular.

REMARK In mathematics, the phrase "if and only if" is a way of stating two implications in one statement. For instance, the first statement at the right could be rewritten as the following two implications.

a. If two distinct nonvertical lines are parallel, then their slopes are equal.

b. If two distinct nonvertical lines have equal slopes, then they are parallel.

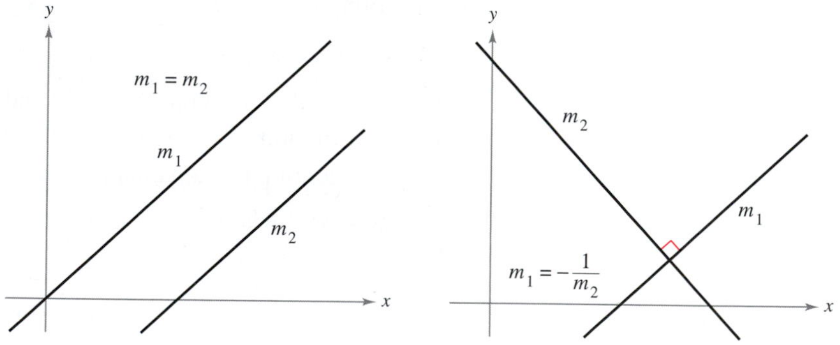

Parallel lines Perpendicular lines
Figure P.19

Parallel and Perpendicular Lines

1. Two distinct nonvertical lines are **parallel** if and only if their slopes are equal—that is, if and only if

$$m_1 = m_2.$$ Parallel ⟺ Slopes are equal.

2. Two nonvertical lines are **perpendicular** if and only if their slopes are negative reciprocals of each other—that is, if and only if

$$m_1 = -\frac{1}{m_2}.$$ Perpendicular ⟺ Slopes are negative reciprocals.

EXAMPLE 5 Finding Parallel and Perpendicular Lines

See LarsonCalculus.com for an interactive version of this type of example.

Find the general forms of the equations of the lines that pass through the point $(2, -1)$ and are (a) parallel to and (b) perpendicular to the line $2x - 3y = 5$.

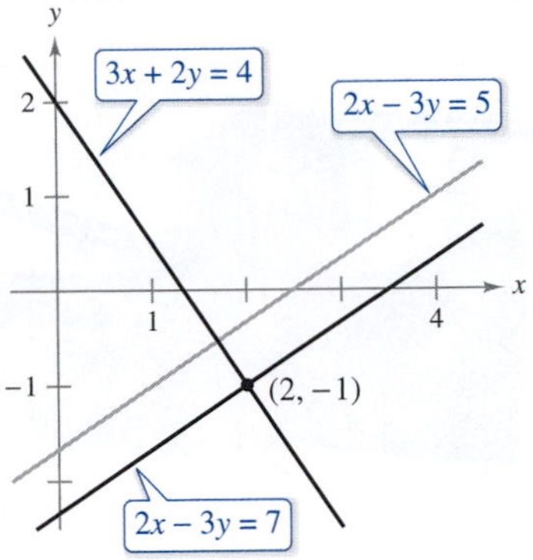

Lines parallel and perpendicular to $2x - 3y = 5$

Figure P.20

Solution Begin by writing the linear equation $2x - 3y = 5$ in slope-intercept form.

$$2x - 3y = 5 \qquad \text{Write original equation.}$$
$$y = \frac{2}{3}x - \frac{5}{3} \qquad \text{Slope-intercept form}$$

So, the given line has a slope of $m = \frac{2}{3}$. (See Figure P.20.)

a. The line through $(2, -1)$ that is parallel to the given line also has a slope of $\frac{2}{3}$.

$$y - y_1 = m(x - x_1) \qquad \text{Point-slope form}$$
$$y - (-1) = \frac{2}{3}(x - 2) \qquad \text{Substitute.}$$
$$3(y + 1) = 2(x - 2) \qquad \text{Simplify.}$$
$$3y + 3 = 2x - 4 \qquad \text{Distributive Property}$$
$$2x - 3y - 7 = 0 \qquad \text{General form}$$

Note the similarity to the equation of the given line, $2x - 3y = 5$.

b. Using the negative reciprocal of the slope of the given line, you can determine that the slope of a line perpendicular to the given line is $-\frac{3}{2}$.

$$y - y_1 = m(x - x_1) \qquad \text{Point-slope form}$$
$$y - (-1) = -\frac{3}{2}(x - 2) \qquad \text{Substitute.}$$
$$2(y + 1) = -3(x - 2) \qquad \text{Simplify.}$$
$$2y + 2 = -3x + 6 \qquad \text{Distributive Property}$$
$$3x + 2y - 4 = 0 \qquad \text{General form}$$

TECHNOLOGY PITFALL The slope of a line will appear distorted if you use different tick-mark spacing on the x- and y-axes. For instance, the graphing utility screens in Figures P.21(a) and P.21(b) both show the lines

$$y = 2x \quad \text{and} \quad y = -\frac{1}{2}x + 3.$$

Because these lines have slopes that are negative reciprocals, they must be perpendicular. In Figure P.21(a), however, the lines don't appear to be perpendicular because the tick-mark spacing on the x-axis is not the same as that on the y-axis. In Figure P.21(b), the lines appear perpendicular because the tick-mark spacing on the x-axis is the same as on the y-axis. This type of viewing window is said to have a *square setting*.

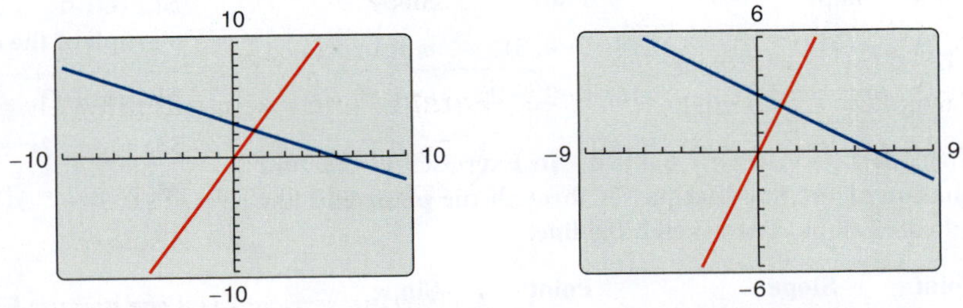

(a) Tick-mark spacing on the x-axis is not the same as tick-mark spacing on the y-axis.

(b) Tick-mark spacing on the x-axis is the same as tick-mark spacing on the y-axis.

Figure P.21

P.2 Exercises

See CalcChat.com for tutorial help and worked-out solutions to odd-numbered exercises.

Estimating Slope **In Exercises 1–4, estimate the slope of the line from its graph. To print an enlarged copy of the graph, go to *MathGraphs.com*.**

1.
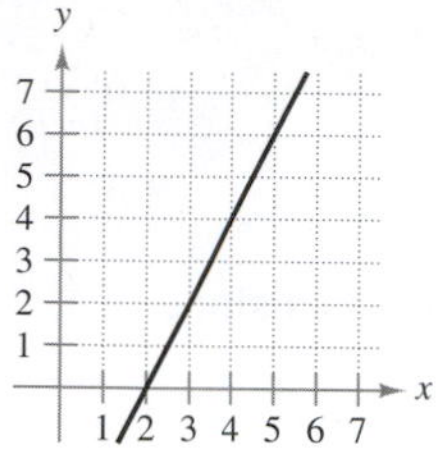

2.
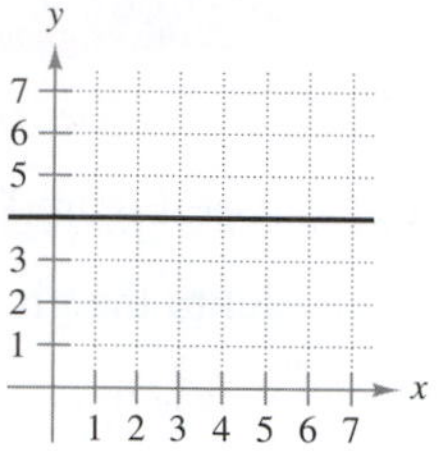

3.
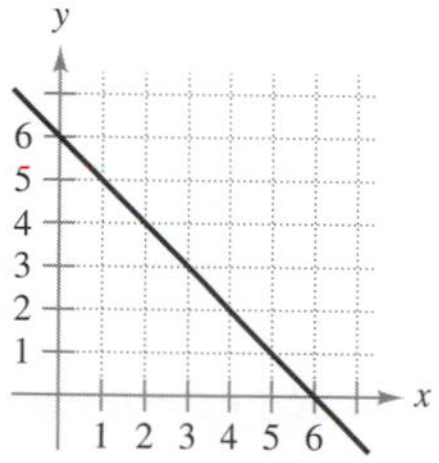

4.
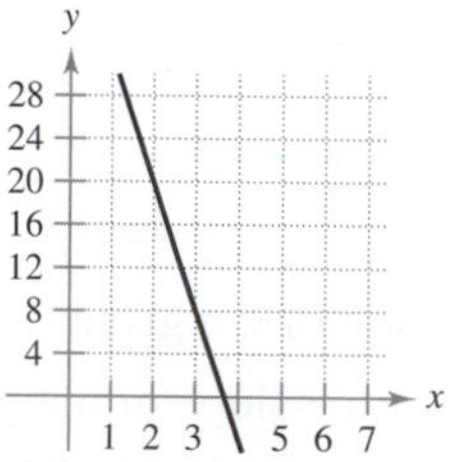

Finding the Slope of a Line **In Exercises 5–10, plot the pair of points and find the slope of the line passing through them.**

5. $(3, -4), (5, 2)$ **6.** $(1, 1), (-2, 7)$

7. $(4, 6), (4, 1)$ **8.** $(3, -5), (5, -5)$

9. $\left(-\frac{1}{2}, \frac{2}{3}\right), \left(-\frac{3}{4}, \frac{1}{6}\right)$ **10.** $\left(\frac{7}{8}, \frac{3}{4}\right), \left(\frac{5}{4}, -\frac{1}{4}\right)$

Sketching Lines **In Exercises 11 and 12, sketch the lines through the point with the indicated slopes. Make the sketches on the same set of coordinate axes.**

	Point	Slopes			
11.	$(3, 4)$	(a) 1	(b) -2	(c) $-\frac{3}{2}$	(d) Undefined
12.	$(-2, 5)$	(a) 3	(b) -3	(c) $\frac{1}{3}$	(d) 0

Finding Points on a Line **In Exercises 13–16, use the point on the line and the slope of the line to find three additional points that the line passes through. (There is more than one correct answer.)**

	Point	Slope		Point	Slope
13.	$(6, 2)$	$m = 0$	**14.**	$(-4, 3)$	m is undefined.
15.	$(1, 7)$	$m = -3$	**16.**	$(-2, -2)$	$m = 2$

Finding an Equation of a Line **In Exercises 17–22, find an equation of the line that passes through the point and has the indicated slope. Then sketch the line.**

	Point	Slope		Point	Slope
17.	$(0, 3)$	$m = \frac{3}{4}$	**18.**	$(-5, -2)$	m is undefined.
19.	$(0, 0)$	$m = \frac{2}{3}$	**20.**	$(0, 4)$	$m = 0$
21.	$(3, -2)$	$m = 3$	**22.**	$(-2, 4)$	$m = -\frac{3}{5}$

xtrekx/Shutterstock.com

23. Conveyor Design

A moving conveyor is built to rise 1 meter for each 3 meters of horizontal change.

(a) Find the slope of the conveyor.

(b) Suppose the conveyor runs between two floors in a factory. Find the length of the conveyor when the vertical distance between floors is 10 feet.

24. Modeling Data The table shows the populations y (in millions) of the United States for 2004 through 2009. The variable t represents the time in years, with $t = 4$ corresponding to 2004. *(Source: U.S. Census Bureau)*

t	4	5	6	7	8	9
y	293.0	295.8	298.6	301.6	304.4	307.0

(a) Plot the data by hand and connect adjacent points with a line segment.

(b) Use the slope of each line segment to determine the year when the population increased least rapidly.

(c) Find the average rate of change of the population of the United States from 2004 through 2009.

(d) Use the average rate of change of the population to predict the population of the United States in 2020.

Finding the Slope and *y*-Intercept **In Exercises 25–30, find the slope and the *y*-intercept (if possible) of the line.**

25. $y = 4x - 3$ **26.** $-x + y = 1$

27. $x + 5y = 20$ **28.** $6x - 5y = 15$

29. $x = 4$ **30.** $y = -1$

Sketching a Line in the Plane **In Exercises 31–38, sketch a graph of the equation.**

31. $y = -3$ **32.** $x = 4$

33. $y = -2x + 1$ **34.** $y = \frac{1}{3}x - 1$

35. $y - 2 = \frac{3}{2}(x - 1)$ **36.** $y - 1 = 3(x + 4)$

37. $2x - y - 3 = 0$ **38.** $x + 2y + 6 = 0$

Finding an Equation of a Line **In Exercises 39–46, find an equation of the line that passes through the points. Then sketch the line.**

39. $(0, 0), (4, 8)$ **40.** $(-2, -2), (1, 7)$

41. $(2, 8), (5, 0)$ **42.** $(-3, 6), (1, 2)$
43. $(6, 3), (6, 8)$ **44.** $(1, -2), (3, -2)$
45. $\left(\frac{1}{2}, \frac{7}{2}\right), \left(0, \frac{3}{4}\right)$ **46.** $\left(\frac{7}{8}, \frac{3}{4}\right), \left(\frac{5}{4}, -\frac{1}{4}\right)$

47. Find an equation of the vertical line with x-intercept at 3.

48. Show that the line with intercepts $(a, 0)$ and $(0, b)$ has the following equation.

$$\frac{x}{a} + \frac{y}{b} = 1, \quad a \neq 0, b \neq 0$$

Writing an Equation in General Form **In Exercises 49–54, use the result of Exercise 48 to write an equation of the line in general form.**

49. x-intercept: $(2, 0)$
y-intercept: $(0, 3)$

50. x-intercept: $\left(-\frac{2}{3}, 0\right)$
y-intercept: $(0, -2)$

51. Point on line: $(1, 2)$
x-intercept: $(a, 0)$
y-intercept: $(0, a)$
$(a \neq 0)$

52. Point on line: $(-3, 4)$
x-intercept: $(a, 0)$
y-intercept: $(0, a)$
$(a \neq 0)$

53. Point on line: $(9, -2)$
x-intercept: $(2a, 0)$
y-intercept: $(0, a)$
$(a \neq 0)$

54. Point on line: $\left(-\frac{2}{3}, -2\right)$
x-intercept: $(a, 0)$
y-intercept: $(0, -a)$
$(a \neq 0)$

Finding Parallel and Perpendicular Lines **In Exercises 55–62, write the general forms of the equations of the lines through the point (a) parallel to the given line and (b) perpendicular to the given line.**

	Point	Line		Point	Line
55.	$(-7, -2)$	$x = 1$	**56.**	$(-1, 0)$	$y = -3$
57.	$(2, 5)$	$x - y = -2$	**58.**	$(-3, 2)$	$x + y = 7$
59.	$(2, 1)$	$4x - 2y = 3$	**60.**	$\left(\frac{5}{6}, -\frac{1}{2}\right)$	$7x + 4y = 8$
61.	$\left(\frac{3}{4}, \frac{7}{8}\right)$	$5x - 3y = 0$	**62.**	$(4, -5)$	$3x + 4y = 7$

Rate of Change **In Exercises 63–66, you are given the dollar value of a product in 2012 *and* the rate at which the value of the product is expected to change during the next 5 years. Write a linear equation that gives the dollar value V of the product in terms of the year t. (Let $t = 0$ represent 2010.)**

	2012 Value	Rate
63.	\$1850	\$250 increase per year
64.	\$156	\$4.50 increase per year
65.	\$17,200	\$1600 decrease per year
66.	\$245,000	\$5600 decrease per year

Collinear Points **In Exercises 67 and 68, determine whether the points are collinear. (Three points are *collinear* if they lie on the same line.)**

67. $(-2, 1), (-1, 0), (2, -2)$

68. $(0, 4), (7, -6), (-5, 11)$

WRITING ABOUT CONCEPTS

Finding Points of Intersection **In Exercises 69–71, find the coordinates of the point of intersection of the given segments. Explain your reasoning.**

69.

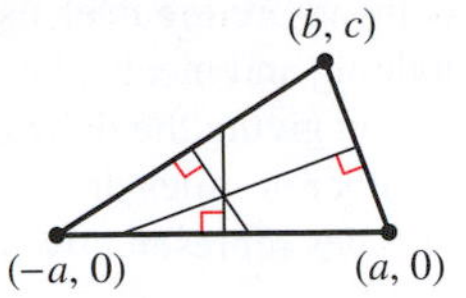

Perpendicular bisectors

70.

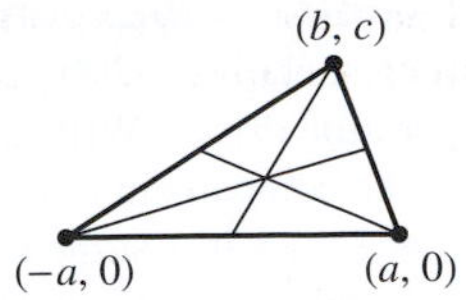

Medians

71.

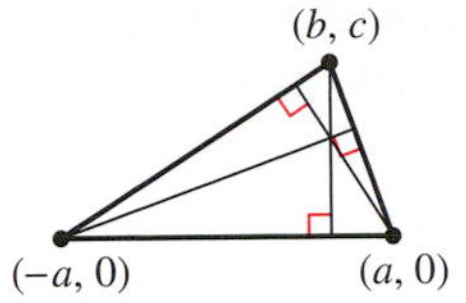

Altitudes

72. Show that the points of intersection in Exercises 69, 70, and 71 are collinear.

73. Analyzing a Line A line is represented by the equation $ax + by = 4$.

(a) When is the line parallel to the x-axis?

(b) When is the line parallel to the y-axis?

(c) Give values for a and b such that the line has a slope of $\frac{5}{8}$.

(d) Give values for a and b such that the line is perpendicular to $y = \frac{2}{5}x + 3$.

(e) Give values for a and b such that the line coincides with the graph of $5x + 6y = 8$.

74. HOW DO YOU SEE IT? Use the graphs of the equations to answer the questions below.

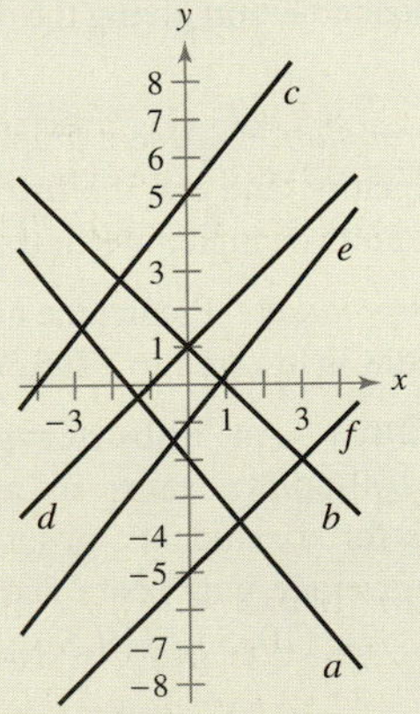

(a) Which lines have a positive slope?

(b) Which lines have a negative slope?

(c) Which lines appear parallel?

(d) Which lines appear perpendicular?

75. Temperature Conversion Find a linear equation that expresses the relationship between the temperature in degrees Celsius C and degrees Fahrenheit F. Use the fact that water freezes at 0°C (32°F) and boils at 100°C (212°F). Use the equation to convert 72°F to degrees Celsius.

76. Reimbursed Expenses A company reimburses its sales representatives \$200 per day for lodging and meals plus \$0.51 per mile driven. Write a linear equation giving the daily cost C to the company in terms of x, the number of miles driven. How much does it cost the company if a sales representative drives 137 miles on a given day?

77. Choosing a Job As a salesperson, you receive a monthly salary of \$2000, plus a commission of 7% of sales. You are offered a new job at \$2300 per month, plus a commission of 5% of sales.

(a) Write linear equations for your monthly wage W in terms of your monthly sales s for your current job and your job offer.

(b) Use a graphing utility to graph each equation and find the point of intersection. What does it signify?

(c) You think you can sell \$20,000 worth of a product per month. Should you change jobs? Explain.

78. Straight-Line Depreciation A small business purchases a piece of equipment for \$875. After 5 years, the equipment will be outdated, having no value.

(a) Write a linear equation giving the value y of the equipment in terms of the time x (in years), $0 \le x \le 5$.

(b) Find the value of the equipment when $x = 2$.

(c) Estimate (to two-decimal-place accuracy) the time when the value of the equipment is \$200.

79. Apartment Rental A real estate office manages an apartment complex with 50 units. When the rent is \$780 per month, all 50 units are occupied. However, when the rent is \$825, the average number of occupied units drops to 47. Assume that the relationship between the monthly rent p and the demand x is linear. (*Note:* The term *demand* refers to the number of occupied units.)

(a) Write a linear equation giving the demand x in terms of the rent p.

(b) *Linear extrapolation* Use a graphing utility to graph the demand equation and use the *trace* feature to predict the number of units occupied when the rent is raised to \$855.

(c) *Linear interpolation* Predict the number of units occupied when the rent is lowered to \$795. Verify graphically.

80. Modeling Data An instructor gives regular 20-point quizzes and 100-point exams in a mathematics course. Average scores for six students, given as ordered pairs (x, y), where x is the average quiz score and y is the average exam score, are (18, 87), (10, 55), (19, 96), (16, 79), (13, 76), and (15, 82).

(a) Use the regression capabilities of a graphing utility to find the least squares regression line for the data.

(b) Use a graphing utility to plot the points and graph the regression line in the same viewing window.

(c) Use the regression line to predict the average exam score for a student with an average quiz score of 17.

(d) Interpret the meaning of the slope of the regression line.

(e) The instructor adds 4 points to the average exam score of everyone in the class. Describe the changes in the positions of the plotted points and the change in the equation of the line.

81. Tangent Line Find an equation of the line tangent to the circle $x^2 + y^2 = 169$ at the point (5, 12).

82. Tangent Line Find an equation of the line tangent to the circle $(x - 1)^2 + (y - 1)^2 = 25$ at the point $(4, -3)$.

Distance **In Exercises 83–86, find the distance between the point and line, or between the lines, using the formula for the distance between the point (x_1, y_1) and the line $Ax + By + C = 0$.**

$$\text{Distance} = \frac{|Ax_1 + By_1 + C|}{\sqrt{A^2 + B^2}}$$

83. Point: $(-2, 1)$
Line: $x - y - 2 = 0$

84. Point: $(2, 3)$
Line: $4x + 3y = 10$

85. Line: $x + y = 1$
Line: $x + y = 5$

86. Line: $3x - 4y = 1$
Line: $3x - 4y = 10$

87. Distance Show that the distance between the point (x_1, y_1) and the line $Ax + By + C = 0$ is

$$\text{Distance} = \frac{|Ax_1 + By_1 + C|}{\sqrt{A^2 + B^2}}.$$

88. Distance Write the distance d between the point (3, 1) and the line $y = mx + 4$ in terms of m. Use a graphing utility to graph the equation. When is the distance 0? Explain the result geometrically.

89. Proof Prove that the diagonals of a rhombus intersect at right angles. (A rhombus is a quadrilateral with sides of equal lengths.)

90. Proof Prove that the figure formed by connecting consecutive midpoints of the sides of any quadrilateral is a parallelogram.

91. Proof Prove that if the points (x_1, y_1) and (x_2, y_2) lie on the same line as (x_1^*, y_1^*) and (x_2^*, y_2^*), then

$$\frac{y_2^* - y_1^*}{x_2^* - x_1^*} = \frac{y_2 - y_1}{x_2 - x_1}.$$

Assume $x_1 \neq x_2$ and $x_1^* \neq x_2^*$.

92. Proof Prove that if the slopes of two nonvertical lines are negative reciprocals of each other, then the lines are perpendicular.

True or False? **In Exercises 93–96, determine whether the statement is true or false. If it is false, explain why or give an example that shows it is false.**

93. The lines represented by $ax + by = c_1$ and $bx - ay = c_2$ are perpendicular. Assume $a \neq 0$ and $b \neq 0$.

94. It is possible for two lines with positive slopes to be perpendicular to each other.

95. If a line contains points in both the first and third quadrants, then its slope must be positive.

96. The equation of any line can be written in general form.

P.3 Functions and Their Graphs

- **Use function notation to represent and evaluate a function.**
- **Find the domain and range of a function.**
- **Sketch the graph of a function.**
- **Identify different types of transformations of functions.**
- **Classify functions and recognize combinations of functions.**

Functions and Function Notation

A **relation** between two sets X and Y is a set of ordered pairs, each of the form (x, y), where x is a member of X and y is a member of Y. A **function** from X to Y is a relation between X and Y that has the property that any two ordered pairs with the same x-value also have the same y-value. The variable x is the **independent variable,** and the variable y is the **dependent variable.**

Many real-life situations can be modeled by functions. For instance, the area A of a circle is a function of the circle's radius r.

$$A = \pi r^2 \qquad \text{A is a function of r.}$$

In this case, r is the independent variable and A is the dependent variable.

Definition of a Real-Valued Function of a Real Variable

Let X and Y be sets of real numbers. A **real-valued function f of a real variable x** from X to Y is a correspondence that assigns to each number x in X exactly one number y in Y.

The **domain** of f is the set X. The number y is the **image** of x under f and is denoted by $f(x)$, which is called the **value of f at x.** The **range** of f is a subset of Y and consists of all images of numbers in X (see Figure P.22).

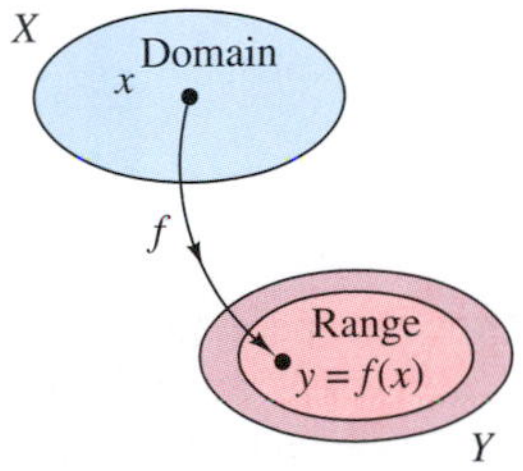

A real-valued function f of a real variable
Figure P.22

Functions can be specified in a variety of ways. In this text, however, you will concentrate primarily on functions that are given by equations involving the dependent and independent variables. For instance, the equation

$$x^2 + 2y = 1 \qquad \text{Equation in implicit form}$$

defines y, the dependent variable, as a function of x, the independent variable. To **evaluate** this function (that is, to find the y-value that corresponds to a given x-value), it is convenient to isolate y on the left side of the equation.

$$y = \tfrac{1}{2}(1 - x^2) \qquad \text{Equation in explicit form}$$

Using f as the name of the function, you can write this equation as

$$f(x) = \tfrac{1}{2}(1 - x^2). \qquad \text{Function notation}$$

The original equation

$$x^2 + 2y = 1$$

implicitly defines y as a function of x. When you solve the equation for y, you are writing the equation in **explicit** form.

Function notation has the advantage of clearly identifying the dependent variable as $f(x)$ while at the same time telling you that x is the independent variable and that the function itself is "f." The symbol $f(x)$ is read "f of x." Function notation allows you to be less wordy. Instead of asking "What is the value of y that corresponds to $x = 3$?" you can ask "What is $f(3)$?"

FUNCTION NOTATION

The word *function* was first used by Gottfried Wilhelm Leibniz in 1694 as a term to denote any quantity connected with a curve, such as the coordinates of a point on a curve or the slope of a curve. Forty years later, Leonhard Euler used the word "function" to describe any expression made up of a variable and some constants. He introduced the notation $y = f(x)$.

In an equation that defines a function of x, the role of the variable x is simply that of a placeholder. For instance, the function

$$f(x) = 2x^2 - 4x + 1$$

can be described by the form

$$f(\blacksquare) = 2(\blacksquare)^2 - 4(\blacksquare) + 1$$

where rectangles are used instead of x. To evaluate $f(-2)$, replace each rectangle with -2.

$$\begin{aligned} f(-2) &= 2(-2)^2 - 4(-2) + 1 && \text{Substitute } -2 \text{ for } x. \\ &= 2(4) + 8 + 1 && \text{Simplify.} \\ &= 17 && \text{Simplify.} \end{aligned}$$

Although f is often used as a convenient function name and x as the independent variable, you can use other symbols. For instance, these three equations all define the same function.

$$\begin{aligned} f(x) &= x^2 - 4x + 7 && \text{Function name is } f, \text{ independent variable is } x. \\ f(t) &= t^2 - 4t + 7 && \text{Function name is } f, \text{ independent variable is } t. \\ g(s) &= s^2 - 4s + 7 && \text{Function name is } g, \text{ independent variable is } s. \end{aligned}$$

EXAMPLE 1 Evaluating a Function

For the function f defined by $f(x) = x^2 + 7$, evaluate each expression.

a. $f(3a)$ **b.** $f(b - 1)$ **c.** $\dfrac{f(x + \Delta x) - f(x)}{\Delta x}$

Solution

a.
$$\begin{aligned} f(3a) &= (3a)^2 + 7 && \text{Substitute } 3a \text{ for } x. \\ &= 9a^2 + 7 && \text{Simplify.} \end{aligned}$$

b.
$$\begin{aligned} f(b - 1) &= (b - 1)^2 + 7 && \text{Substitute } b - 1 \text{ for } x. \\ &= b^2 - 2b + 1 + 7 && \text{Expand binomial.} \\ &= b^2 - 2b + 8 && \text{Simplify.} \end{aligned}$$

c.
$$\begin{aligned} \frac{f(x + \Delta x) - f(x)}{\Delta x} &= \frac{[(x + \Delta x)^2 + 7] - (x^2 + 7)}{\Delta x} \\ &= \frac{x^2 + 2x\Delta x + (\Delta x)^2 + 7 - x^2 - 7}{\Delta x} \\ &= \frac{2x\Delta x + (\Delta x)^2}{\Delta x} \\ &= \frac{\cancel{\Delta x}(2x + \Delta x)}{\cancel{\Delta x}} \\ &= 2x + \Delta x, \quad \Delta x \neq 0 \end{aligned}$$

REMARK The expression in Example 1(c) is called a *difference quotient* and has a special significance in calculus. You will learn more about this in Chapter 2.

In calculus, it is important to specify the domain of a function or expression clearly. For instance, in Example 1(c), the two expressions

$$\frac{f(x + \Delta x) - f(x)}{\Delta x} \quad \text{and} \quad 2x + \Delta x, \quad \Delta x \neq 0$$

are equivalent because $\Delta x = 0$ is excluded from the domain of each expression. Without a stated domain restriction, the two expressions would not be equivalent.

The Domain and Range of a Function

The domain of a function can be described explicitly, or it may be described *implicitly* by an equation used to define the function. The implied domain is the set of all real numbers for which the equation is defined, whereas an explicitly defined domain is one that is given along with the function. For example, the function

$$f(x) = \frac{1}{x^2 - 4}, \quad 4 \le x \le 5$$

has an explicitly defined domain given by $\{x: 4 \le x \le 5\}$. On the other hand, the function

$$g(x) = \frac{1}{x^2 - 4}$$

has an implied domain that is the set $\{x: x \ne \pm 2\}$.

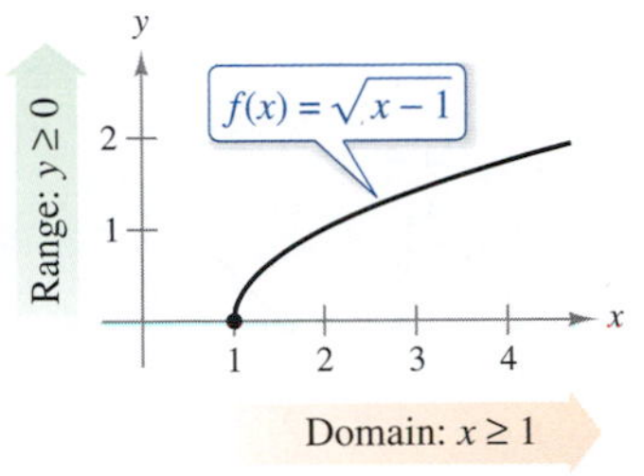

(a) The domain of f is $[1, \infty)$, and the range is $[0, \infty)$.

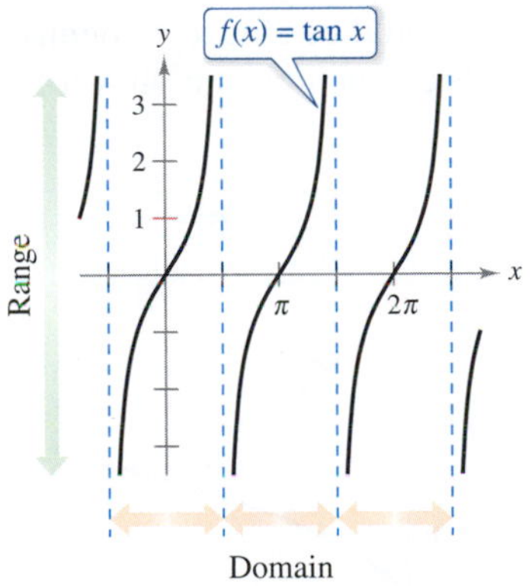

(b) The domain of f is all x-values such that $x \ne \frac{\pi}{2} + n\pi$, and the range is $(-\infty, \infty)$.

Figure P.23

EXAMPLE 2 Finding the Domain and Range of a Function

a. The domain of the function

$$f(x) = \sqrt{x - 1}$$

is the set of all x-values for which $x - 1 \ge 0$, which is the interval $[1, \infty)$. To find the range, observe that $f(x) = \sqrt{x - 1}$ is never negative. So, the range is the interval $[0, \infty)$, as shown in Figure P.23(a).

b. The domain of the tangent function

$$f(x) = \tan x$$

is the set of all x-values such that

$$x \ne \frac{\pi}{2} + n\pi, \quad n \text{ is an integer.} \qquad \text{Domain of tangent function}$$

The range of this function is the set of all real numbers, as shown in Figure P.23(b). For a review of the characteristics of this and other trigonometric functions, see Appendix C.

EXAMPLE 3 A Function Defined by More than One Equation

For the piecewise-defined function

$$f(x) = \begin{cases} 1 - x, & x < 1 \\ \sqrt{x - 1}, & x \ge 1 \end{cases}$$

f is defined for $x < 1$ and $x \ge 1$. So, the domain is the set of all real numbers. On the portion of the domain for which $x \ge 1$, the function behaves as in Example 2(a). For $x < 1$, the values of $1 - x$ are positive. So, the range of the function is the interval $[0, \infty)$. (See Figure P.24.)

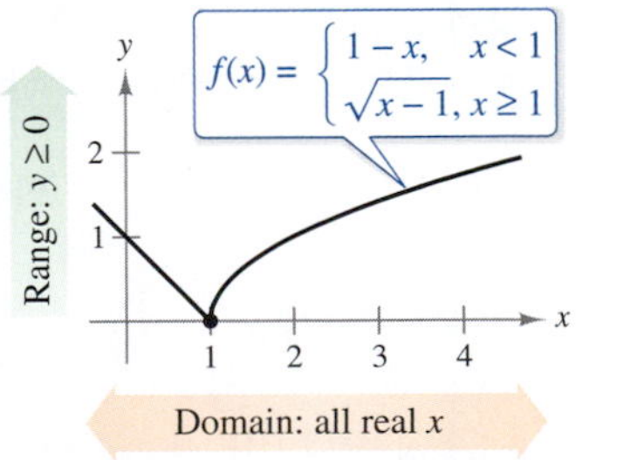

The domain of f is $(-\infty, \infty)$, and the range is $[0, \infty)$.

Figure P.24

A function from X to Y is **one-to-one** when to each y-value in the range there corresponds exactly one x-value in the domain. For instance, the function in Example 2(a) is one-to-one, whereas the functions in Examples 2(b) and 3 are not one-to-one. A function from X to Y is **onto** when its range consists of all of Y.

The Graph of a Function

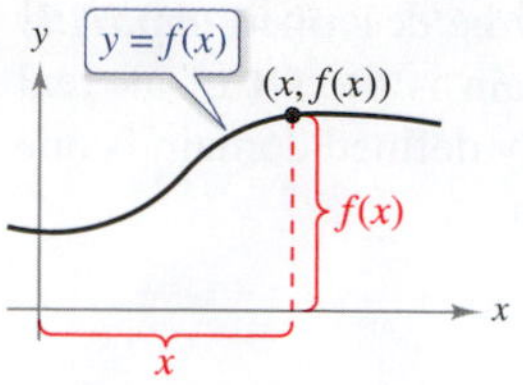

The graph of a function
Figure P.25

The graph of the function $y = f(x)$ consists of all points $(x, f(x))$, where x is in the domain of f. In Figure P.25, note that

$$x = \text{the directed distance from the } y\text{-axis}$$

and

$$f(x) = \text{the directed distance from the } x\text{-axis.}$$

A vertical line can intersect the graph of a function of x at most *once*. This observation provides a convenient visual test, called the **Vertical Line Test,** for functions of x. That is, a graph in the coordinate plane is the graph of a function of x if and only if no vertical line intersects the graph at more than one point. For example, in Figure P.26(a), you can see that the graph does not define y as a function of x because a vertical line intersects the graph twice, whereas in Figures P.26(b) and (c), the graphs do define y as a function of x.

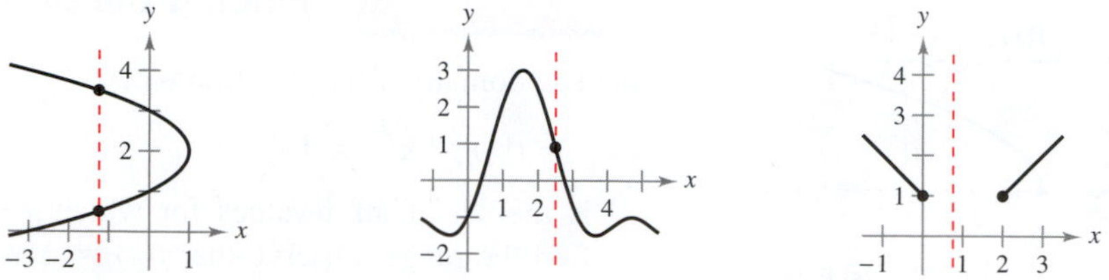

(a) Not a function of x **(b)** A function of x **(c)** A function of x
Figure P.26

Figure P.27 shows the graphs of eight basic functions. You should be able to recognize these graphs. (Graphs of the other four basic trigonometric functions are shown in Appendix C.)

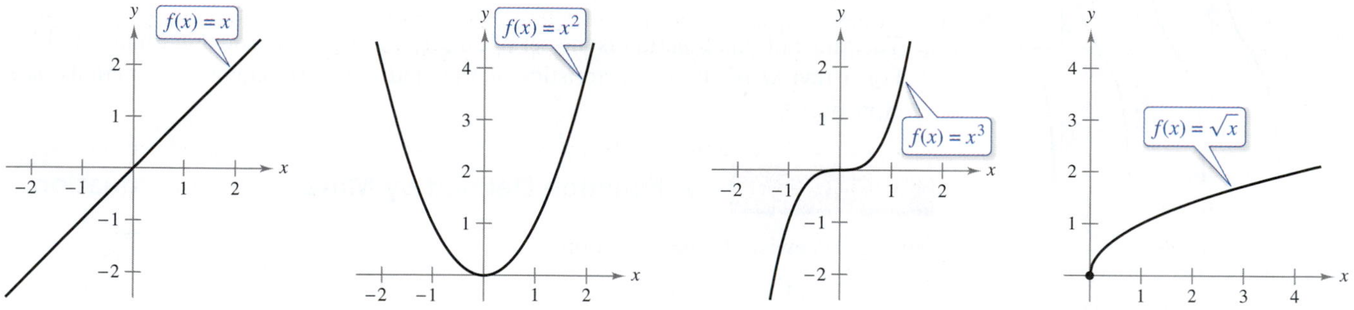

Identity function Squaring function Cubing function Square root function

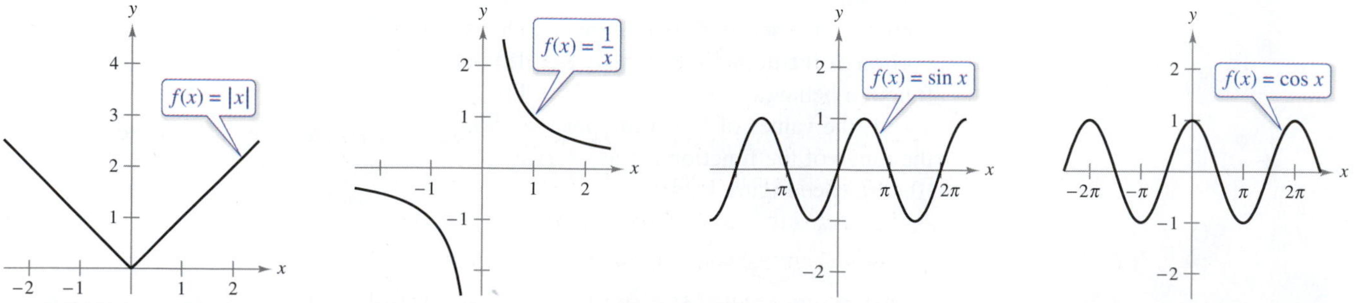

Absolute value function Rational function Sine function Cosine function

The graphs of eight basic functions
Figure P.27

Transformations of Functions

Some families of graphs have the same basic shape. For example, compare the graph of $y = x^2$ with the graphs of the four other quadratic functions shown in Figure P.28.

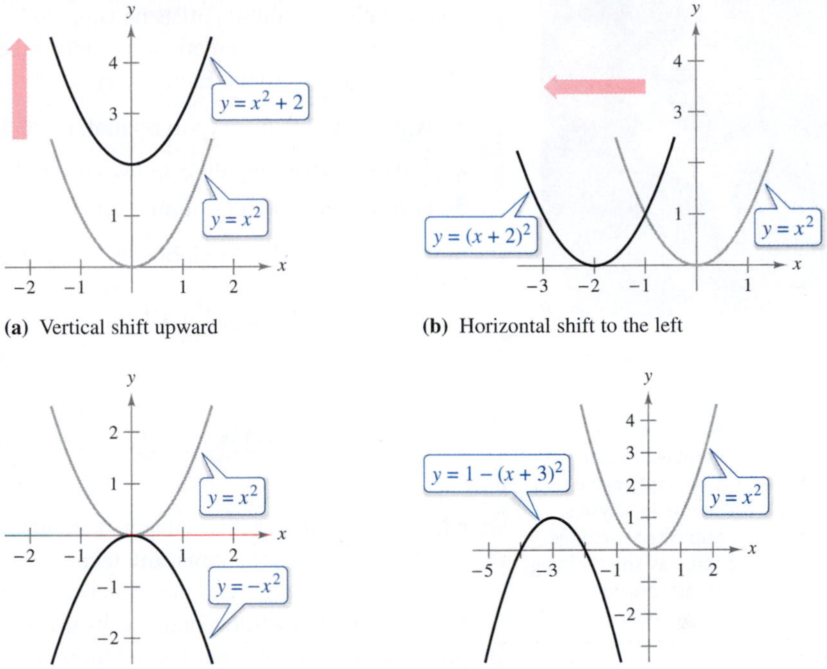

(a) Vertical shift upward

(b) Horizontal shift to the left

(c) Reflection

(d) Shift left, reflect, and shift upward

Figure P.28

Each of the graphs in Figure P.28 is a **transformation** of the graph of $y = x^2$. The three basic types of transformations illustrated by these graphs are vertical shifts, horizontal shifts, and reflections. Function notation lends itself well to describing transformations of graphs in the plane. For instance, using

$f(x) = x^2$ Original function

as the original function, the transformations shown in Figure P.28 can be represented by these equations.

a. $y = f(x) + 2$ Vertical shift up two units

b. $y = f(x + 2)$ Horizontal shift to the left two units

c. $y = -f(x)$ Reflection about the x-axis

d. $y = -f(x + 3) + 1$ Shift left three units, reflect about the x-axis, and shift up one unit

Basic Types of Transformations $(c > 0)$	
Original graph:	$y = f(x)$
Horizontal shift c units to the **right:**	$y = f(x - c)$
Horizontal shift c units to the **left:**	$y = f(x + c)$
Vertical shift c units **downward:**	$y = f(x) - c$
Vertical shift c units **upward:**	$y = f(x) + c$
Reflection (about the x-axis):	$y = -f(x)$
Reflection (about the y-axis):	$y = f(-x)$
Reflection (about the origin):	$y = -f(-x)$

Classifications and Combinations of Functions

The modern notion of a function is derived from the efforts of many seventeenth- and eighteenth-century mathematicians. Of particular note was Leonhard Euler, who introduced the function notation $y = f(x)$. By the end of the eighteenth century, mathematicians and scientists had concluded that many real-world phenomena could be represented by mathematical models taken from a collection of functions called **elementary functions.** Elementary functions fall into three categories.

1. Algebraic functions (polynomial, radical, rational)
2. Trigonometric functions (sine, cosine, tangent, and so on)
3. Exponential and logarithmic functions

You can review the trigonometric functions in Appendix C. The other nonalgebraic functions, such as the inverse trigonometric functions and the exponential and logarithmic functions, are introduced in Chapter 5.

The most common type of algebraic function is a **polynomial function**

$$f(x) = a_n x^n + a_{n-1}x^{n-1} + \cdot \cdot \cdot + a_2x^2 + a_1x + a_0$$

where n is a nonnegative integer. The numbers a_i are **coefficients,** with a_n the **leading coefficient** and a_0 the **constant term** of the polynomial function. If $a_n \neq 0$, then n is the **degree** of the polynomial function. The zero polynomial $f(x) = 0$ is not assigned a degree. It is common practice to use subscript notation for coefficients of general polynomial functions, but for polynomial functions of low degree, these simpler forms are often used. (Note that $a \neq 0$.)

Zeroth degree: $f(x) = a$ Constant function

First degree: $f(x) = ax + b$ Linear function

Second degree: $f(x) = ax^2 + bx + c$ Quadratic function

Third degree: $f(x) = ax^3 + bx^2 + cx + d$ Cubic function

Although the graph of a nonconstant polynomial function can have several turns, eventually the graph will rise or fall without bound as x moves to the right or left. Whether the graph of

$$f(x) = a_n x^n + a_{n-1}x^{n-1} + \cdot \cdot \cdot + a_2x^2 + a_1x + a_0$$

eventually rises or falls can be determined by the function's degree (odd or even) and by the leading coefficient a_n, as indicated in Figure P.29. Note that the dashed portions of the graphs indicate that the **Leading Coefficient Test** determines *only* the right and left behavior of the graph.

LEONHARD EULER (1707–1783)

In addition to making major contributions to almost every branch of mathematics, Euler was one of the first to apply calculus to real-life problems in physics. His extensive published writings include such topics as shipbuilding, acoustics, optics, astronomy, mechanics, and magnetism.

See LarsonCalculus.com to read more of this biography.

FOR FURTHER INFORMATION For more on the history of the concept of a function, see the article "Evolution of the Function Concept: A Brief Survey" by Israel Kleiner in *The College Mathematics Journal*. To view this article, go to *MathArticles.com*.

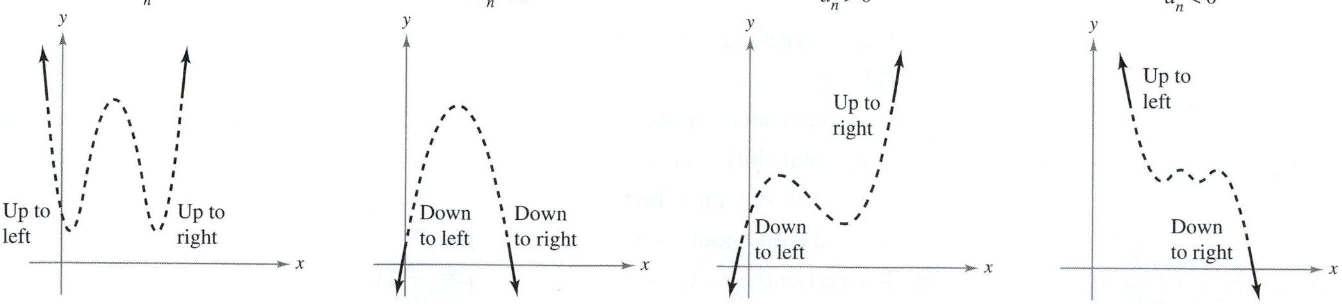

Graphs of polynomial functions of even degree

Graphs of polynomial functions of odd degree

The Leading Coefficient Test for polynomial functions

Figure P.29

North Wind Picture Archives/Alamy

Just as a rational number can be written as the quotient of two integers, a **rational function** can be written as the quotient of two polynomials. Specifically, a function f is rational when it has the form

$$f(x) = \frac{p(x)}{q(x)}, \quad q(x) \neq 0$$

where $p(x)$ and $q(x)$ are polynomials.

Polynomial functions and rational functions are examples of **algebraic functions.** An algebraic function of x is one that can be expressed as a finite number of sums, differences, multiples, quotients, and radicals involving x^n. For example,

$$f(x) = \sqrt{x + 1}$$

is algebraic. Functions that are not algebraic are **transcendental.** For instance, the trigonometric functions are transcendental.

Two functions can be combined in various ways to create new functions. For example, given $f(x) = 2x - 3$ and $g(x) = x^2 + 1$, you can form the functions shown.

$$(f + g)(x) = f(x) + g(x) = (2x - 3) + (x^2 + 1) \quad \text{Sum}$$
$$(f - g)(x) = f(x) - g(x) = (2x - 3) - (x^2 + 1) \quad \text{Difference}$$
$$(fg)(x) = f(x)g(x) = (2x - 3)(x^2 + 1) \quad \text{Product}$$
$$(f/g)(x) = \frac{f(x)}{g(x)} = \frac{2x - 3}{x^2 + 1} \quad \text{Quotient}$$

You can combine two functions in yet another way, called **composition.** The resulting function is called a **composite function.**

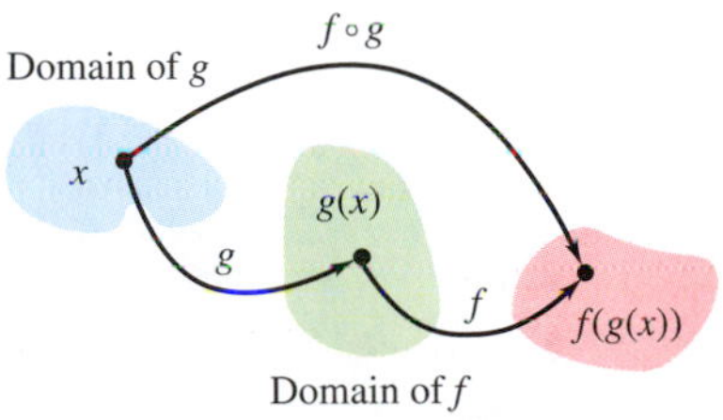

The domain of the composite function $f \circ g$
Figure P.30

Definition of Composite Function

Let f and g be functions. The function $(f \circ g)(x) = f(g(x))$ is the **composite** of f with g. The domain of $f \circ g$ is the set of all x in the domain of g such that $g(x)$ is in the domain of f (see Figure P.30).

The composite of f with g is generally not the same as the composite of g with f. This is shown in the next example.

EXAMPLE 4 Finding Composite Functions

See LarsonCalculus.com for an interactive version of this type of example.

For $f(x) = 2x - 3$ and $g(x) = \cos x$, find each composite function.

a. $f \circ g$ **b.** $g \circ f$

Solution

a. $(f \circ g)(x) = f(g(x))$ — Definition of $f \circ g$

$= f(\cos x)$ — Substitute $\cos x$ for $g(x)$.

$= 2(\cos x) - 3$ — Definition of $f(x)$

$= 2 \cos x - 3$ — Simplify.

b. $(g \circ f)(x) = g(f(x))$ — Definition of $g \circ f$

$= g(2x - 3)$ — Substitute $2x - 3$ for $f(x)$.

$= \cos(2x - 3)$ — Definition of $g(x)$

Note that $(f \circ g)(x) \neq (g \circ f)(x)$. ■

Exploration

Use a graphing utility to graph each function. Determine whether the function is *even, odd,* or *neither.*

$f(x) = x^2 - x^4$

$g(x) = 2x^3 + 1$

$h(x) = x^5 - 2x^3 + x$

$j(x) = 2 - x^6 - x^8$

$k(x) = x^5 - 2x^4 + x - 2$

$p(x) = x^9 + 3x^5 - x^3 + x$

Describe a way to identify a function as odd or even by inspecting the equation.

In Section P.1, an x-intercept of a graph was defined to be a point $(a, 0)$ at which the graph crosses the x-axis. If the graph represents a function f, then the number a is a **zero** of f. In other words, *the zeros of a function f are the solutions of the equation* $f(x) = 0$. For example, the function

$$f(x) = x - 4$$

has a zero at $x = 4$ because $f(4) = 0$.

In Section P.1, you also studied different types of symmetry. In the terminology of functions, a function is **even** when its graph is symmetric with respect to the y-axis, and is **odd** when its graph is symmetric with respect to the origin. The symmetry tests in Section P.1 yield the following test for even and odd functions.

Test for Even and Odd Functions

The function $y = f(x)$ is **even** when

$$f(-x) = f(x).$$

The function $y = f(x)$ is **odd** when

$$f(-x) = -f(x).$$

EXAMPLE 5 Even and Odd Functions and Zeros of Functions

Determine whether each function is even, odd, or neither. Then find the zeros of the function.

a. $f(x) = x^3 - x$ **b.** $g(x) = 1 + \cos x$

Solution

a. This function is odd because

$$f(-x) = (-x)^3 - (-x) = -x^3 + x = -(x^3 - x) = -f(x).$$

The zeros of f are

$$x^3 - x = 0 \quad \text{Let } f(x) = 0.$$
$$x(x^2 - 1) = 0 \quad \text{Factor.}$$
$$x(x - 1)(x + 1) = 0 \quad \text{Factor.}$$
$$x = 0, 1, -1. \quad \text{Zeros of } f$$

See Figure P.31(a).

b. This function is even because

$$g(-x) = 1 + \cos(-x) = 1 + \cos x = g(x). \quad \cos(-x) = \cos(x)$$

The zeros of g are

$$1 + \cos x = 0 \quad \text{Let } g(x) = 0.$$
$$\cos x = -1 \quad \text{Subtract 1 from each side.}$$
$$x = (2n + 1)\pi, \ n \text{ is an integer.} \quad \text{Zeros of } g$$

See Figure P.31(b).

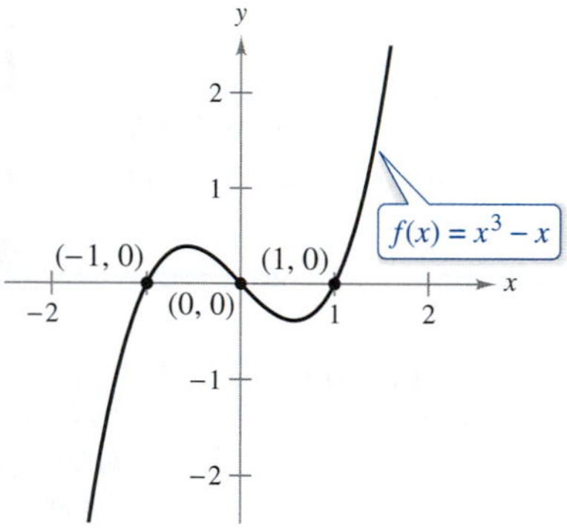

(a) Odd function

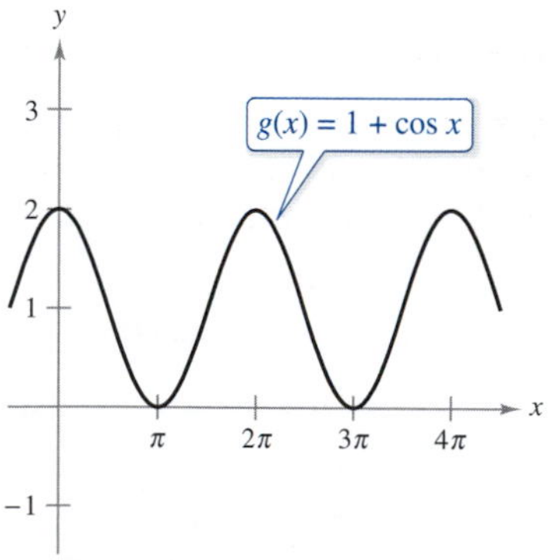

(b) Even function

Figure P.31

Each function in Example 5 is either even or odd. However, some functions, such as

$$f(x) = x^2 + x + 1$$

are neither even nor odd.

P.3 Exercises

See CalcChat.com for tutorial help and worked-out solutions to odd-numbered exercises.

Evaluating a Function In Exercises 1–10, evaluate the function at the given value(s) of the independent variable. Simplify the results.

1. $f(x) = 7x - 4$
(a) $f(0)$ (b) $f(-3)$
(c) $f(b)$ (d) $f(x-1)$

2. $f(x) = \sqrt{x+5}$
(a) $f(-4)$ (b) $f(11)$
(c) $f(4)$ (d) $f(x + \Delta x)$

3. $g(x) = 5 - x^2$
(a) $g(0)$ (b) $g(\sqrt{5})$
(c) $g(-2)$ (d) $g(t-1)$

4. $g(x) = x^2(x-4)$
(a) $g(4)$ (b) $g(\frac{3}{2})$
(c) $g(c)$ (d) $g(t+4)$

5. $f(x) = \cos 2x$
(a) $f(0)$ (b) $f\left(-\frac{\pi}{4}\right)$
(c) $f\left(\frac{\pi}{3}\right)$ (d) $f(\pi)$

6. $f(x) = \sin x$
(a) $f(\pi)$ (b) $f\left(\frac{5\pi}{4}\right)$
(c) $f\left(\frac{2\pi}{3}\right)$ (d) $f\left(-\frac{\pi}{6}\right)$

7. $f(x) = x^3$
$\dfrac{f(x+\Delta x) - f(x)}{\Delta x}$

8. $f(x) = 3x - 1$
$\dfrac{f(x) - f(1)}{x - 1}$

9. $f(x) = \dfrac{1}{\sqrt{x-1}}$
$\dfrac{f(x) - f(2)}{x - 2}$

10. $f(x) = x^3 - x$
$\dfrac{f(x) - f(1)}{x - 1}$

Finding the Domain and Range of a Function In Exercises 11–22, find the domain and range of the function.

11. $f(x) = 4x^2$
12. $g(x) = x^2 - 5$
13. $f(x) = x^3$
14. $h(x) = 4 - x^2$
15. $g(x) = \sqrt{6x}$
16. $h(x) = -\sqrt{x+3}$
17. $f(x) = \sqrt{16 - x^2}$
18. $f(x) = |x - 3|$
19. $f(t) = \sec \dfrac{\pi t}{4}$
20. $h(t) = \cot t$
21. $f(x) = \dfrac{3}{x}$
22. $f(x) = \dfrac{x-2}{x+4}$

Finding the Domain of a Function In Exercises 23–28, find the domain of the function.

23. $f(x) = \sqrt{x} + \sqrt{1-x}$
24. $f(x) = \sqrt{x^2 - 3x + 2}$
25. $g(x) = \dfrac{2}{1 - \cos x}$
26. $h(x) = \dfrac{1}{\sin x - (1/2)}$
27. $f(x) = \dfrac{1}{|x+3|}$
28. $g(x) = \dfrac{1}{|x^2 - 4|}$

Finding the Domain and Range of a Piecewise Function In Exercises 29–32, evaluate the function as indicated. Determine its domain and range.

29. $f(x) = \begin{cases} 2x+1, & x < 0 \\ 2x+2, & x \ge 0 \end{cases}$
(a) $f(-1)$ (b) $f(0)$ (c) $f(2)$ (d) $f(t^2 + 1)$

30. $f(x) = \begin{cases} x^2+2, & x \le 1 \\ 2x^2+2, & x > 1 \end{cases}$
(a) $f(-2)$ (b) $f(0)$ (c) $f(1)$ (d) $f(s^2 + 2)$

31. $f(x) = \begin{cases} |x|+1, & x < 1 \\ -x+1, & x \ge 1 \end{cases}$
(a) $f(-3)$ (b) $f(1)$ (c) $f(3)$ (d) $f(b^2 + 1)$

32. $f(x) = \begin{cases} \sqrt{x+4}, & x \le 5 \\ (x-5)^2, & x > 5 \end{cases}$
(a) $f(-3)$ (b) $f(0)$ (c) $f(5)$ (d) $f(10)$

Sketching a Graph of a Function In Exercises 33–40, sketch a graph of the function and find its domain and range. Use a graphing utility to verify your graph.

33. $f(x) = 4 - x$
34. $g(x) = \dfrac{4}{x}$
35. $h(x) = \sqrt{x-6}$
36. $f(x) = \frac{1}{4}x^3 + 3$
37. $f(x) = \sqrt{9 - x^2}$
38. $f(x) = x + \sqrt{4 - x^2}$
39. $g(t) = 3 \sin \pi t$
40. $h(\theta) = -5 \cos \dfrac{\theta}{2}$

WRITING ABOUT CONCEPTS

41. Describing a Graph The graph of the distance that a student drives in a 10-minute trip to school is shown in the figure. Give a verbal description of the characteristics of the student's drive to school.

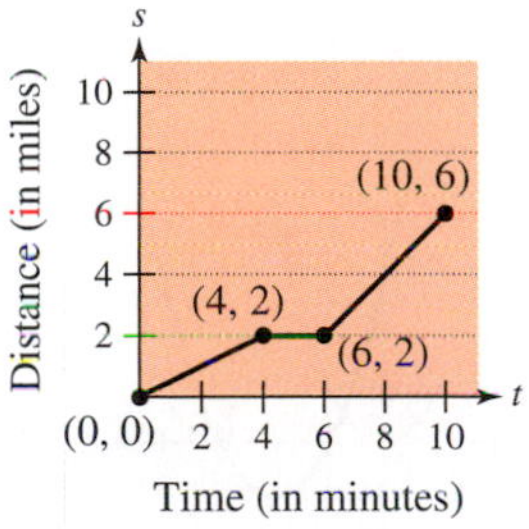

42. Sketching a Graph A student who commutes 27 miles to attend college remembers, after driving a few minutes, that a term paper that is due has been forgotten. Driving faster than usual, the student returns home, picks up the paper, and once again starts toward school. Sketch a possible graph of the student's distance from home as a function of time.

Using the Vertical Line Test In Exercises 43–46, use the Vertical Line Test to determine whether y is a function of x. To print an enlarged copy of the graph, go to *MathGraphs.com*.

43. $x - y^2 = 0$

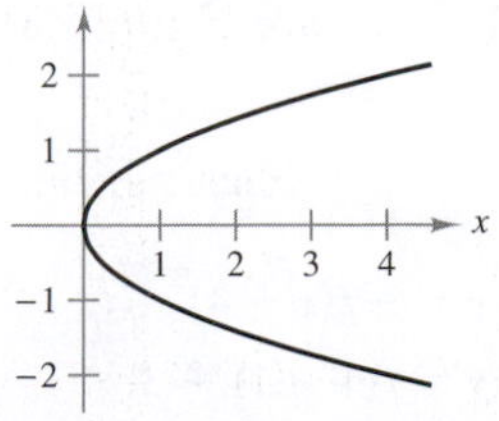

44. $\sqrt{x^2 - 4} - y = 0$

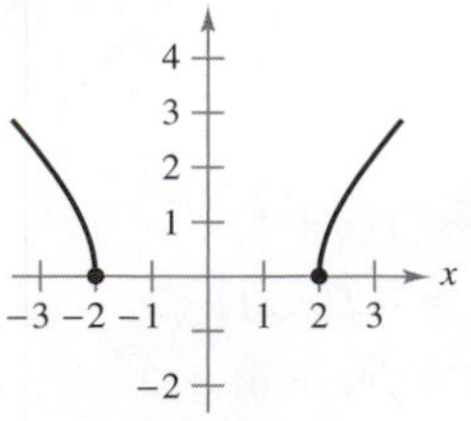

45. $y = \begin{cases} x + 1, & x \le 0 \\ -x + 2, & x > 0 \end{cases}$

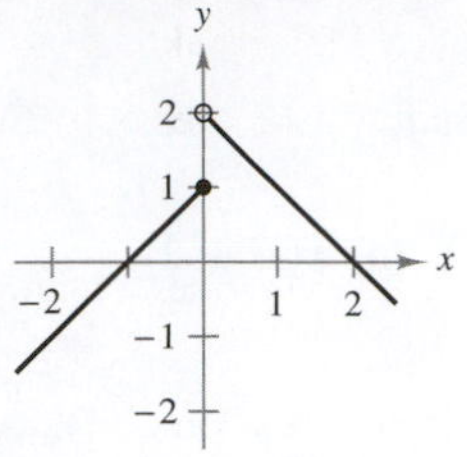

46. $x^2 + y^2 = 4$

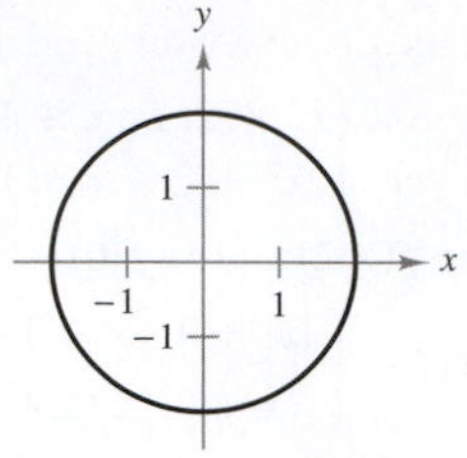

Deciding Whether an Equation Is a Function

In Exercises 47–50, determine whether y is a function of x.

47. $x^2 + y^2 = 16$

48. $x^2 + y = 16$

49. $y^2 = x^2 - 1$

50. $x^2y - x^2 + 4y = 0$

Transformation of a Function

In Exercises 51–54, the graph shows one of the eight basic functions on page 22 and a transformation of the function. Describe the transformation. Then use your description to write an equation for the transformation.

51.

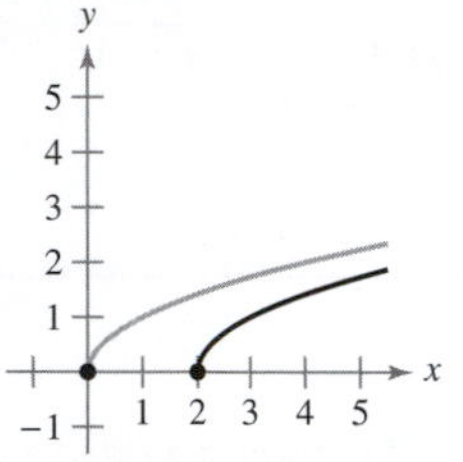

52.

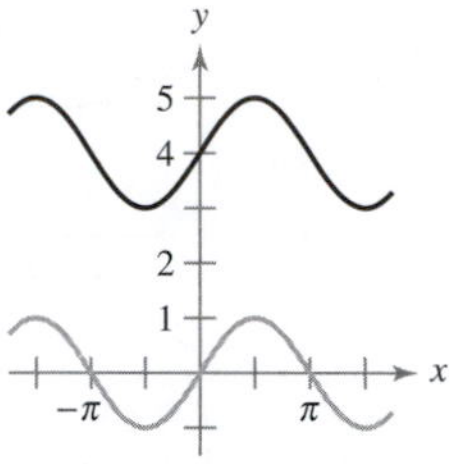

53.

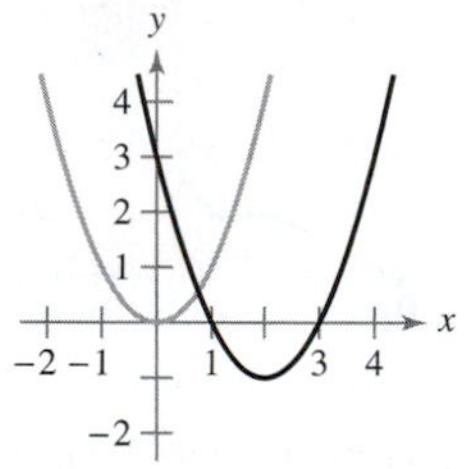

54.

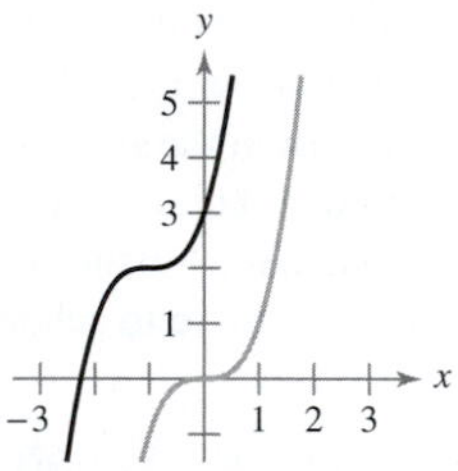

Matching

In Exercises 55–60, use the graph of $y = f(x)$ to match the function with its graph.

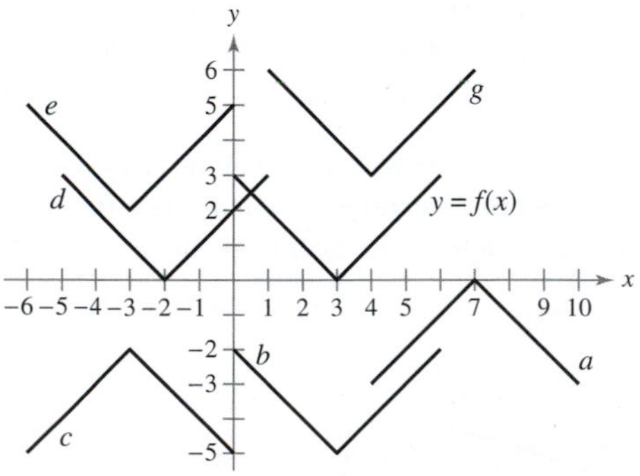

55. $y = f(x + 5)$

56. $y = f(x) - 5$

57. $y = -f(-x) - 2$

58. $y = -f(x - 4)$

59. $y = f(x + 6) + 2$

60. $y = f(x - 1) + 3$

61. Sketching Transformations Use the graph of f shown in the figure to sketch the graph of each function. To print an enlarged copy of the graph, go to *MathGraphs.com*.

(a) $f(x + 3)$ (b) $f(x - 1)$
(c) $f(x) + 2$ (d) $f(x) - 4$
(e) $3f(x)$ (f) $\frac{1}{4}f(x)$
(g) $-f(x)$ (h) $-f(-x)$

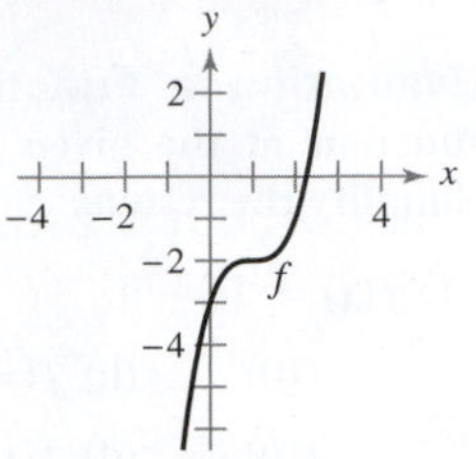

62. Sketching Transformations Use the graph of f shown in the figure to sketch the graph of each function. To print an enlarged copy of the graph, go to *MathGraphs.com*.

(a) $f(x - 4)$ (b) $f(x + 2)$
(c) $f(x) + 4$ (d) $f(x) - 1$
(e) $2f(x)$ (f) $\frac{1}{2}f(x)$
(g) $f(-x)$ (h) $-f(x)$

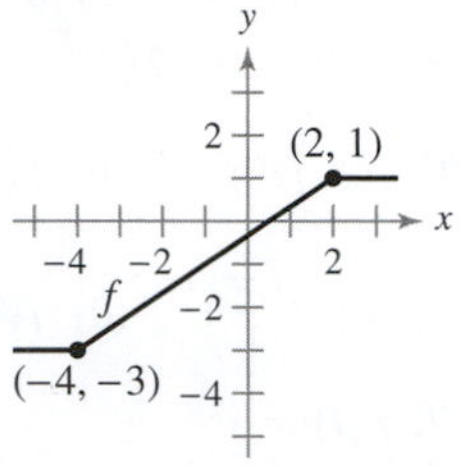

Combinations of Functions

In Exercises 63 and 64, find (a) $f(x) + g(x)$, (b) $f(x) - g(x)$, (c) $f(x) \cdot g(x)$, and (d) $f(x)/g(x)$.

63. $f(x) = 3x - 4$
$g(x) = 4$

64. $f(x) = x^2 + 5x + 4$
$g(x) = x + 1$

65. Evaluating Composite Functions Given $f(x) = \sqrt{x}$ and $g(x) = x^2 - 1$, evaluate each expression.

(a) $f(g(1))$ (b) $g(f(1))$ (c) $g(f(0))$
(d) $f(g(-4))$ (e) $f(g(x))$ (f) $g(f(x))$

66. Evaluating Composite Functions Given $f(x) = \sin x$ and $g(x) = \pi x$, evaluate each expression.

(a) $f(g(2))$ (b) $f\left(g\left(\frac{1}{2}\right)\right)$ (c) $g(f(0))$
(d) $g\left(f\left(\frac{\pi}{4}\right)\right)$ (e) $f(g(x))$ (f) $g(f(x))$

Finding Composite Functions

In Exercises 67–70, find the composite functions $f \circ g$ and $g \circ f$. Find the domain of each composite function. Are the two composite functions equal?

67. $f(x) = x^2$, $g(x) = \sqrt{x}$

68. $f(x) = x^2 - 1$, $g(x) = \cos x$

69. $f(x) = \dfrac{3}{x}$, $g(x) = x^2 - 1$

70. $f(x) = \dfrac{1}{x}$, $g(x) = \sqrt{x + 2}$

71. Evaluating Composite Functions Use the graphs of f and g to evaluate each expression. If the result is undefined, explain why.

(a) $(f \circ g)(3)$ (b) $g(f(2))$
(c) $g(f(5))$ (d) $(f \circ g)(-3)$
(e) $(g \circ f)(-1)$ (f) $f(g(-1))$

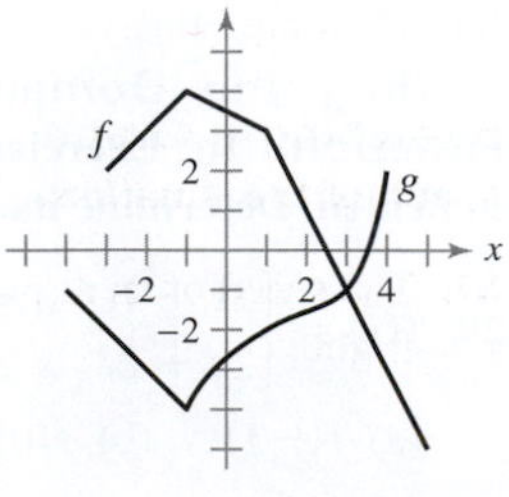

72. Ripples A pebble is dropped into a calm pond, causing ripples in the form of concentric circles. The radius (in feet) of the outer ripple is given by $r(t) = 0.6t$, where t is the time in seconds after the pebble strikes the water. The area of the circle is given by the function $A(r) = \pi r^2$. Find and interpret $(A \circ r)(t)$.

Think About It **In Exercises 73 and 74, $F(x) = f \circ g \circ h$. Identify functions for f, g, and h. (There are many correct answers.)**

73. $F(x) = \sqrt{2x - 2}$ **74.** $F(x) = -4 \sin(1 - x)$

Think About It **In Exercises 75 and 76, find the coordinates of a second point on the graph of a function f when the given point is on the graph and the function is (a) even and (b) odd.**

75. $\left(-\frac{3}{2}, 4\right)$ **76.** $(4, 9)$

77. Even and Odd Functions The graphs of f, g, and h are shown in the figure. Decide whether each function is even, odd, or neither.

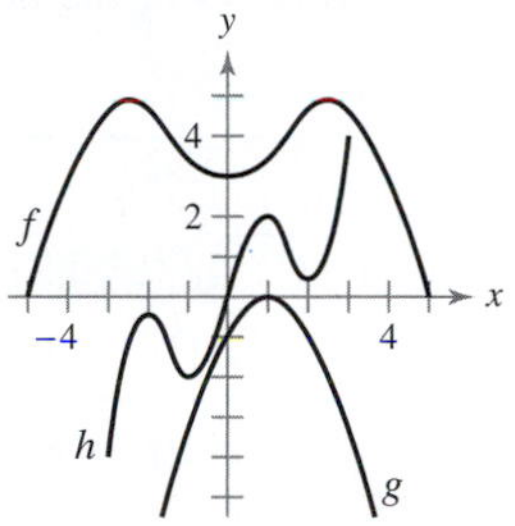

Figure for 77

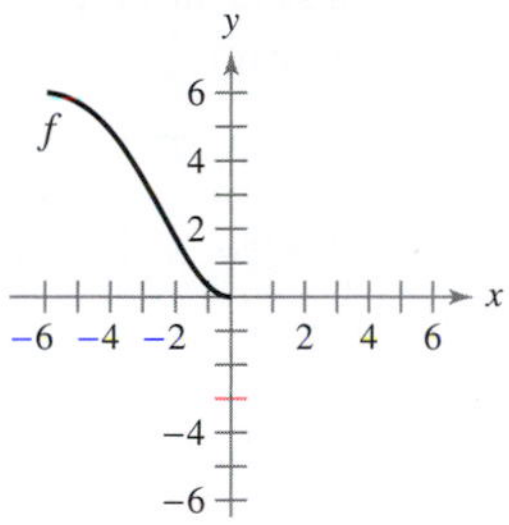

Figure for 78

78. Even and Odd Functions The domain of the function f shown in the figure is $-6 \le x \le 6$.

(a) Complete the graph of f given that f is even.

(b) Complete the graph of f given that f is odd.

Even and Odd Functions and Zeros of Functions **In Exercises 79–82, determine whether the function is even, odd, or neither. Then find the zeros of the function. Use a graphing utility to verify your result.**

79. $f(x) = x^2(4 - x^2)$ **80.** $f(x) = \sqrt[3]{x}$

81. $f(x) = x \cos x$ **82.** $f(x) = \sin^2 x$

Writing Functions **In Exercises 83–86, write an equation for a function that has the given graph.**

83. Line segment connecting $(-2, 4)$ and $(0, -6)$

84. Line segment connecting $(3, 1)$ and $(5, 8)$

85. The bottom half of the parabola $x + y^2 = 0$

86. The bottom half of the circle $x^2 + y^2 = 36$

Sketching a Graph **In Exercises 87–90, sketch a possible graph of the situation.**

87. The speed of an airplane as a function of time during a 5-hour flight

88. The height of a baseball as a function of horizontal distance during a home run

89. The amount of a certain brand of sneaker sold by a sporting goods store as a function of the price of the sneaker

90. The value of a new car as a function of time over a period of 8 years

91. Domain Find the value of c such that the domain of $f(x) = \sqrt{c - x^2}$ is $[-5, 5]$.

92. Domain Find all values of c such that the domain of

$$f(x) = \frac{x + 3}{x^2 + 3cx + 6}$$

is the set of all real numbers.

93. Graphical Reasoning An electronically controlled thermostat is programmed to lower the temperature during the night automatically (see figure). The temperature T in degrees Celsius is given in terms of t, the time in hours on a 24-hour clock.

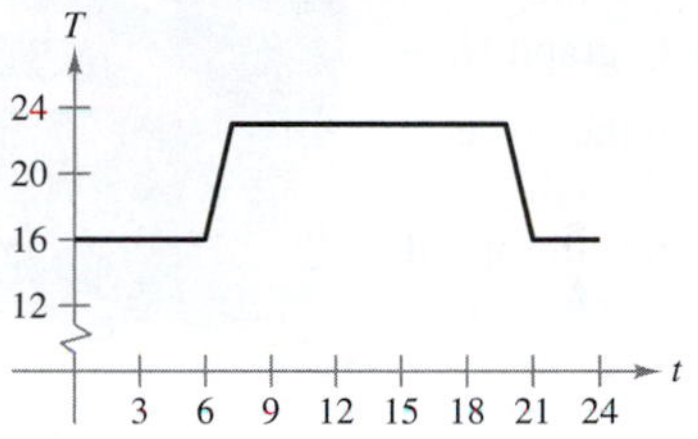

(a) Approximate $T(4)$ and $T(15)$.

(b) The thermostat is reprogrammed to produce a temperature $H(t) = T(t - 1)$. How does this change the temperature? Explain.

(c) The thermostat is reprogrammed to produce a temperature $H(t) = T(t) - 1$. How does this change the temperature? Explain.

94. **HOW DO YOU SEE IT?** Water runs into a vase of height 30 centimeters at a constant rate. The vase is full after 5 seconds. Use this information and the shape of the vase shown to answer the questions when d is the depth of the water in centimeters and t is the time in seconds (see figure).

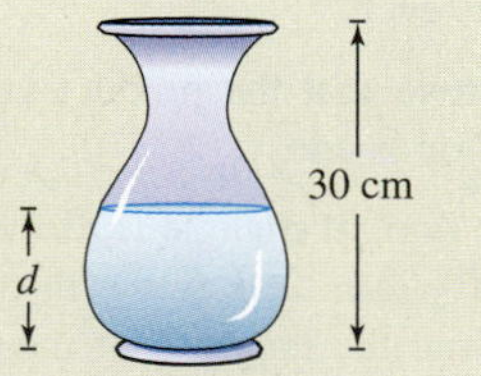

(a) Explain why d is a function of t.

(b) Determine the domain and range of the function.

(c) Sketch a possible graph of the function.

(d) Use the graph in part (c) to approximate $d(4)$. What does this represent?

95. Modeling Data The table shows the average numbers of acres per farm in the United States for selected years. *(Source: U.S. Department of Agriculture)*

Year	1960	1970	1980	1990	2000	2010
Acreage	297	374	429	460	436	418

(a) Plot the data, where A is the acreage and t is the time in years, with $t = 0$ corresponding to 1960. Sketch a freehand curve that approximates the data.

(b) Use the curve in part (a) to approximate $A(25)$.

96. Automobile Aerodynamics

The horsepower H required to overcome wind drag on a certain automobile is approximated by

$H(x) = 0.002x^2 + 0.005x - 0.029, \quad 10 \le x \le 100$

where x is the speed of the car in miles per hour.

(a) Use a graphing utility to graph H.

(b) Rewrite the power function so that x represents the speed in kilometers per hour. [Find $H(x/1.6)$.]

97. Think About It Write the function $f(x) = |x| + |x - 2|$ without using absolute value signs. (For a review of absolute value, see Appendix C.)

98. Writing Use a graphing utility to graph the polynomial functions $p_1(x) = x^3 - x + 1$ and $p_2(x) = x^3 - x$. How many zeros does each function have? Is there a cubic polynomial that has no zeros? Explain.

99. Proof Prove that the function is odd.

$$f(x) = a_{2n+1}x^{2n+1} + \cdots + a_3x^3 + a_1x$$

100. Proof Prove that the function is even.

$$f(x) = a_{2n}x^{2n} + a_{2n-2}x^{2n-2} + \cdots + a_2x^2 + a_0$$

101. Proof Prove that the product of two even (or two odd) functions is even.

102. Proof Prove that the product of an odd function and an even function is odd.

103. Length A right triangle is formed in the first quadrant by the x- and y-axes and a line through the point $(3, 2)$ (see figure). Write the length L of the hypotenuse as a function of x.

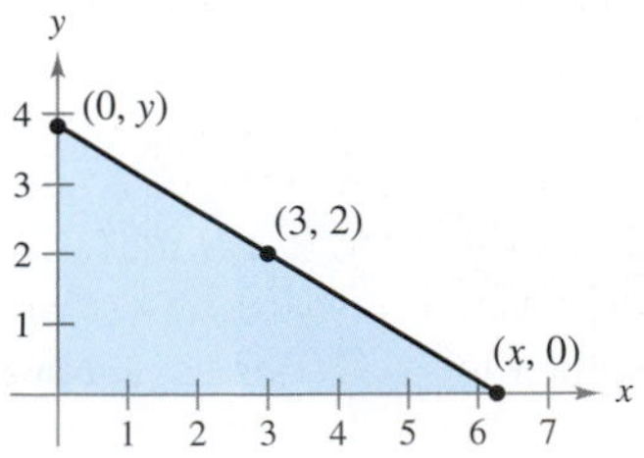

104. Volume An open box of maximum volume is to be made from a square piece of material 24 centimeters on a side by cutting equal squares from the corners and turning up the sides (see figure).

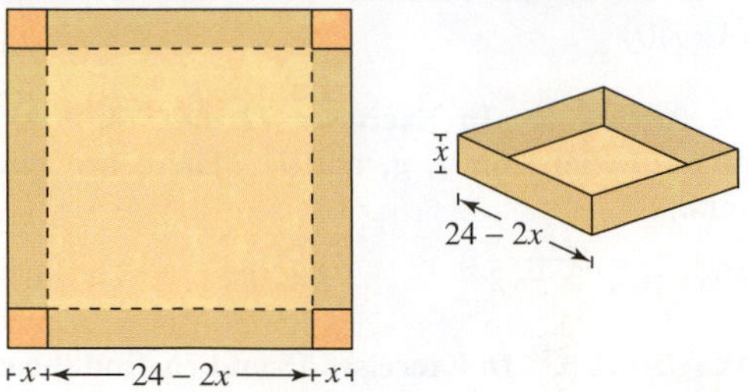

(a) Write the volume V as a function of x, the length of the corner squares. What is the domain of the function?

(b) Use a graphing utility to graph the volume function and approximate the dimensions of the box that yield a maximum volume.

(c) Use the *table* feature of a graphing utility to verify your answer in part (b). (The first two rows of the table are shown.)

Height, x	Length and Width	Volume, V
1	$24 - 2(1)$	$1[24 - 2(1)]^2 = 484$
2	$24 - 2(2)$	$2[24 - 2(2)]^2 = 800$

True or False? **In Exercises 105–110, determine whether the statement is true or false. If it is false, explain why or give an example that shows it is false.**

105. If $f(a) = f(b)$, then $a = b$.

106. A vertical line can intersect the graph of a function at most once.

107. If $f(x) = f(-x)$ for all x in the domain of f, then the graph of f is symmetric with respect to the y-axis.

108. If f is a function, then

$$f(ax) = af(x).$$

109. The graph of a function of x cannot have symmetry with respect to the x-axis.

110. If the domain of a function consists of a single number, then its range must also consist of only one number.

PUTNAM EXAM CHALLENGE

111. Let R be the region consisting of the points (x, y) of the Cartesian plane satisfying both $|x| - |y| \le 1$ and $|y| \le 1$. Sketch the region R and find its area.

112. Consider a polynomial $f(x)$ with real coefficients having the property $f(g(x)) = g(f(x))$ for every polynomial $g(x)$ with real coefficients. Determine and prove the nature of $f(x)$.

These problems were composed by the Committee on the Putnam Prize Competition.

P.4 Fitting Models to Data

- Fit a linear model to a real-life data set.
- Fit a quadratic model to a real-life data set.
- Fit a trigonometric model to a real-life data set.

Fitting a Linear Model to Data

A basic premise of science is that much of the physical world can be described mathematically and that many physical phenomena are predictable. This scientific outlook was part of the scientific revolution that took place in Europe during the late 1500s. Two early publications connected with this revolution were *On the Revolutions of the Heavenly Spheres* by the Polish astronomer Nicolaus Copernicus and *On the Fabric of the Human Body* by the Belgian anatomist Andreas Vesalius. Each of these books was published in 1543, and each broke with prior tradition by suggesting the use of a scientific method rather than unquestioned reliance on authority.

One basic technique of modern science is gathering data and then describing the data with a mathematical model. For instance, the data in Example 1 are inspired by Leonardo da Vinci's famous drawing that indicates that a person's height and arm span are equal.

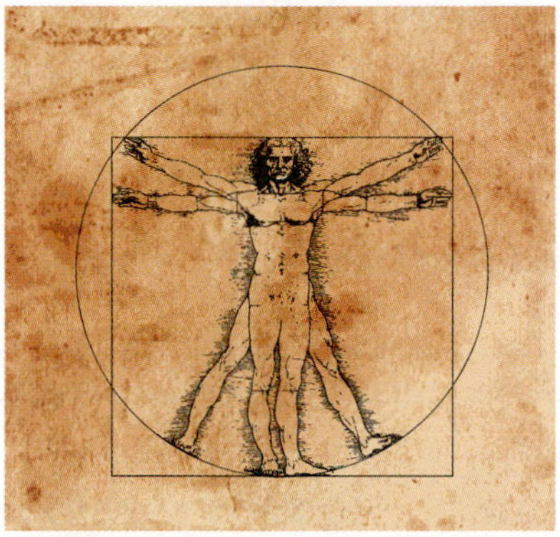

A computer graphics drawing based on the pen and ink drawing of Leonardo da Vinci's famous study of human proportions, called *Vitruvian Man*

EXAMPLE 1 **Fitting a Linear Model to Data**

See LarsonCalculus.com for an interactive version of this type of example.

A class of 28 people collected the data shown below, which represent their heights x and arm spans y (rounded to the nearest inch).

$$(60, 61), (65, 65), (68, 67), (72, 73), (61, 62), (63, 63), (70, 71),$$
$$(75, 74), (71, 72), (62, 60), (65, 65), (66, 68), (62, 62), (72, 73),$$
$$(70, 70), (69, 68), (69, 70), (60, 61), (63, 63), (64, 64), (71, 71),$$
$$(68, 67), (69, 70), (70, 72), (65, 65), (64, 63), (71, 70), (67, 67)$$

Find a linear model to represent these data.

Solution There are different ways to model these data with an equation. The simplest would be to observe that x and y are about the same and list the model as simply $y = x$. A more careful analysis would be to use a procedure from statistics called linear regression. (You will study this procedure in Section 13.9.) The least squares regression line for these data is

$$y = 1.006x - 0.23. \quad \text{Least squares regression line}$$

The graph of the model and the data are shown in Figure P.32. From this model, you can see that a person's arm span tends to be about the same as his or her height.

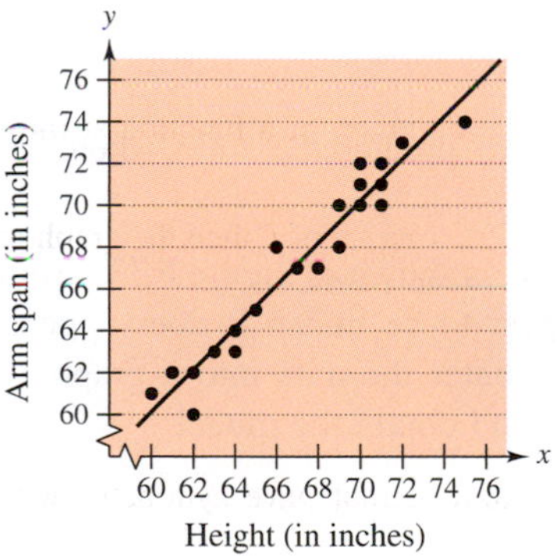

Linear model and data
Figure P.32

TECHNOLOGY Many graphing utilities have built-in least squares regression programs. Typically, you enter the data into the calculator and then run the linear regression program. The program usually displays the slope and y-intercept of the best-fitting line and the *correlation coefficient* r. The correlation coefficient gives a measure of how well the data can be modeled by a line. The closer $|r|$ is to 1, the better the data can be modeled by a line. For instance, the correlation coefficient for the model in Example 1 is $r \approx 0.97$, which indicates that the linear model is a good fit for the data. If the r-value is positive, then the variables have a positive correlation, as in Example 1. If the r-value is negative, then the variables have a negative correlation.

Fitting a Quadratic Model to Data

A function that gives the height s of a falling object in terms of the time t is called a *position function*. If air resistance is not considered, then the position of a falling object can be modeled by

$$s(t) = \frac{1}{2}gt^2 + v_0t + s_0$$

where g is the acceleration due to gravity, v_0 is the initial velocity, and s_0 is the initial height. The value of g depends on where the object is dropped. On Earth, g is approximately -32 feet per second per second, or -9.8 meters per second per second.

To discover the value of g experimentally, you could record the heights of a falling object at several increments, as shown in Example 2.

EXAMPLE 2 Fitting a Quadratic Model to Data

A basketball is dropped from a height of about $5\frac{1}{4}$ feet. The height of the basketball is recorded 23 times at intervals of about 0.02 second. The results are shown in the table.

Time	0.0	0.02	0.04	0.06	0.08	0.099996
Height	5.23594	5.20353	5.16031	5.0991	5.02707	4.95146

Time	0.119996	0.139992	0.159988	0.179988	0.199984	0.219984
Height	4.85062	4.74979	4.63096	4.50132	4.35728	4.19523

Time	0.23998	0.25993	0.27998	0.299976	0.319972	0.339961
Height	4.02958	3.84593	3.65507	3.44981	3.23375	3.01048

Time	0.359961	0.379951	0.399941	0.419941	0.439941
Height	2.76921	2.52074	2.25786	1.98058	1.63488

Find a model to fit these data. Then use the model to predict the time when the basketball will hit the ground.

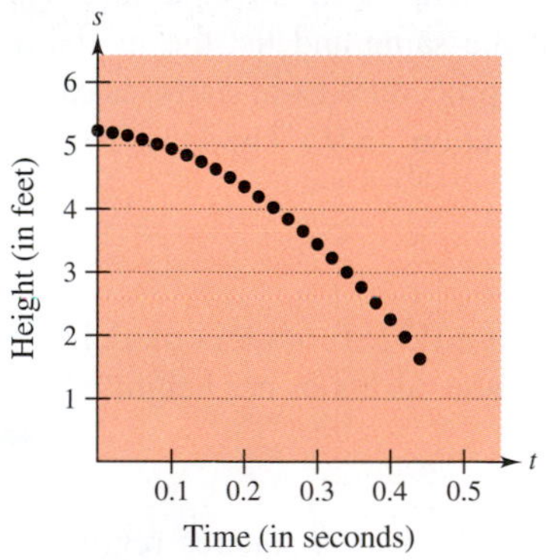

Scatter plot of data
Figure P.33

Solution Begin by sketching a scatter plot of the data, as shown in Figure P.33. From the scatter plot, you can see that the data do not appear to be linear. It does appear, however, that they might be quadratic. To check this, enter the data into a graphing utility that has a quadratic regression program. You should obtain the model

$$s = -15.45t^2 - 1.302t + 5.2340. \qquad \text{Least squares regression quadratic}$$

Using this model, you can predict the time when the basketball hits the ground by substituting 0 for s and solving the resulting equation for t.

$$0 = -15.45t^2 - 1.302t + 5.2340 \qquad \text{Let } s = 0.$$

$$t = \frac{-b \pm \sqrt{b^2 - 4ac}}{2a} \qquad \text{Quadratic Formula}$$

$$t = \frac{-(-1.302) \pm \sqrt{(-1.302)^2 - 4(-15.45)(5.2340)}}{2(-15.45)} \qquad \text{Substitute } a = -15.45,\ b = -1.302, \text{ and } c = 5.2340.$$

$$t \approx 0.54 \qquad \text{Choose positive solution.}$$

The solution is about 0.54 second. In other words, the basketball will continue to fall for about 0.1 second more before hitting the ground. (Note that the experimental value of g is $\frac{1}{2}g = -15.45$, or $g = -30.90$ feet per second per second.) ■

Fitting a Trigonometric Model to Data

What is mathematical modeling? This is one of the questions that is asked in the book *Guide to Mathematical Modelling*. Here is part of the answer.*

1. Mathematical modeling consists of applying your mathematical skills to obtain useful answers to real problems.
2. Learning to apply mathematical skills is very different from learning mathematics itself.
3. Models are used in a very wide range of applications, some of which do not appear initially to be mathematical in nature.
4. Models often allow quick and cheap evaluation of alternatives, leading to optimal solutions that are not otherwise obvious.
5. There are no precise rules in mathematical modeling and no "correct" answers.
6. Modeling can be learned only by *doing*.

The amount of daylight received by locations on Earth varies with the time of year.

EXAMPLE 3 Fitting a Trigonometric Model to Data

The number of hours of daylight on a given day on Earth depends on the latitude and the time of year. Here are the numbers of minutes of daylight at a location of 20°N latitude on the longest and shortest days of the year: June 21, 801 minutes; December 22, 655 minutes. Use these data to write a model for the amount of daylight d (in minutes) on each day of the year at a location of 20°N latitude. How could you check the accuracy of your model?

Solution Here is one way to create a model. You can hypothesize that the model is a sine function whose period is 365 days. Using the given data, you can conclude that the amplitude of the graph is $(801 - 655)/2$, or 73. So, one possible model is

$$d = 728 - 73 \sin\left(\frac{2\pi t}{365} + \frac{\pi}{2}\right).$$

REMARK For a review of trigonometric functions, see Appendix C.

In this model, t represents the number of each day of the year, with December 22 represented by $t = 0$. A graph of this model is shown in Figure P.34. To check the accuracy of this model, a weather almanac was used to find the numbers of minutes of daylight on different days of the year at the location of 20°N latitude.

Date	Value of t	Actual Daylight	Daylight Given by Model
Dec 22	0	655 min	655 min
Jan 1	10	657 min	656 min
Feb 1	41	676 min	672 min
Mar 1	69	705 min	701 min
Apr 1	100	740 min	739 min
May 1	130	772 min	773 min
Jun 1	161	796 min	796 min
Jun 21	181	801 min	801 min
Jul 1	191	799 min	800 min
Aug 1	222	782 min	785 min
Sep 1	253	752 min	754 min
Oct 1	283	718 min	716 min
Nov 1	314	685 min	681 min
Dec 1	344	661 min	660 min

You can see that the model is fairly accurate.

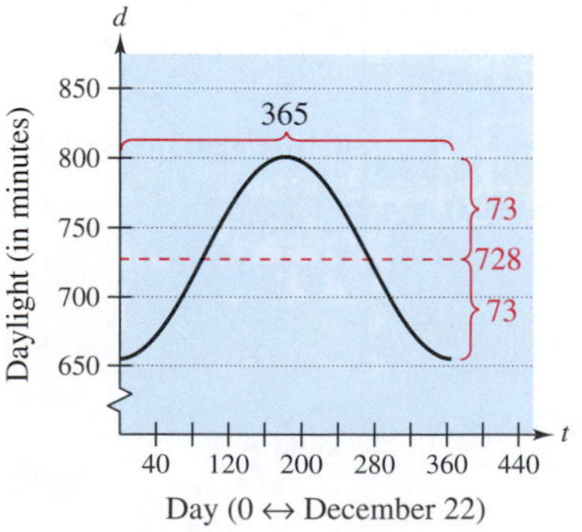

Graph of model
Figure P.34

* Text from Dilwyn Edwards and Mike Hamson, *Guide to Mathematical Modelling* (Boca Raton: CRC Press, 1990), p. 4. Used by permission of the authors.

P.4 Exercises

See CalcChat.com for tutorial help and worked-out solutions to odd-numbered exercises.

1. **Wages** Each ordered pair gives the average weekly wage x for federal government workers and the average weekly wage y for state government workers for 2001 through 2009. *(Source: U.S. Bureau of Labor Statistics)*

   (941, 727), (1001, 754), (1043, 770), (1111, 791), (1151, 812), (1198, 844), (1248, 883), (1275, 923), (1303, 937)

   (a) Plot the data. From the graph, do the data appear to be approximately linear?

   (b) Visually find a linear model for the data. Graph the model.

   (c) Use the model to approximate y when $x = 1075$.

2. **Quiz Scores** The ordered pairs represent the scores on two consecutive 15-point quizzes for a class of 15 students.

   (7, 13), (9, 7), (14, 14), (15, 15), (10, 15), (9, 7), (11, 14), (7, 14), (14, 11), (14, 15), (8, 10), (15, 9), (10, 11), (9, 10), (11, 10)

   (a) Plot the data. From the graph, does the relationship between consecutive scores appear to be approximately linear?

   (b) If the data appear to be approximately linear, find a linear model for the data. If not, give some possible explanations.

3. **Hooke's Law** Hooke's Law states that the force F required to compress or stretch a spring (within its elastic limits) is proportional to the distance d that the spring is compressed or stretched from its original length. That is, $F = kd$, where k is a measure of the stiffness of the spring and is called the *spring constant*. The table shows the elongation d in centimeters of a spring when a force of F newtons is applied.

F	20	40	60	80	100
d	1.4	2.5	4.0	5.3	6.6

   (a) Use the regression capabilities of a graphing utility to find a linear model for the data.

   (b) Use a graphing utility to plot the data and graph the model. How well does the model fit the data? Explain.

   (c) Use the model to estimate the elongation of the spring when a force of 55 newtons is applied.

4. **Falling Object** In an experiment, students measured the speed s (in meters per second) of a falling object t seconds after it was released. The results are shown in the table.

t	0	1	2	3	4
s	0	11.0	19.4	29.2	39.4

   (a) Use the regression capabilities of a graphing utility to find a linear model for the data.

   (b) Use a graphing utility to plot the data and graph the model. How well does the model fit the data? Explain.

   (c) Use the model to estimate the speed of the object after 2.5 seconds.

5. **Energy Consumption and Gross National Product** The data show the per capita energy consumptions (in millions of Btu) and the per capita gross national incomes (in thousands of U.S. dollars) for several countries in 2008. *(Source: U.S. Energy Information Administration and The World Bank)*

Argentina	(81, 7.19)	India	(17, 1.04)
Australia	(274, 40.24)	Italy	(136, 35.46)
Bangladesh	(6, 0.52)	Japan	(172, 38.13)
Brazil	(54, 7.30)	Mexico	(66, 9.99)
Canada	(422, 43.64)	Poland	(101, 11.73)
Ecuador	(35, 3.69)	Turkey	(57, 9.02)
Hungary	(110, 12.81)	Venezuela	(121, 9.23)

   (a) Use the regression capabilities of a graphing utility to find a linear model for the data. What is the correlation coefficient?

   (b) Use a graphing utility to plot the data and graph the model.

   (c) Interpret the graph in part (b). Use the graph to identify the three countries that differ most from the linear model.

   (d) Delete the data for the three countries identified in part (c). Fit a linear model to the remaining data and give the correlation coefficient.

6. **HOW DO YOU SEE IT?** Determine whether the data can be modeled by a linear function, a quadratic function, or a trigonometric function, or that there appears to be no relationship between x and y.

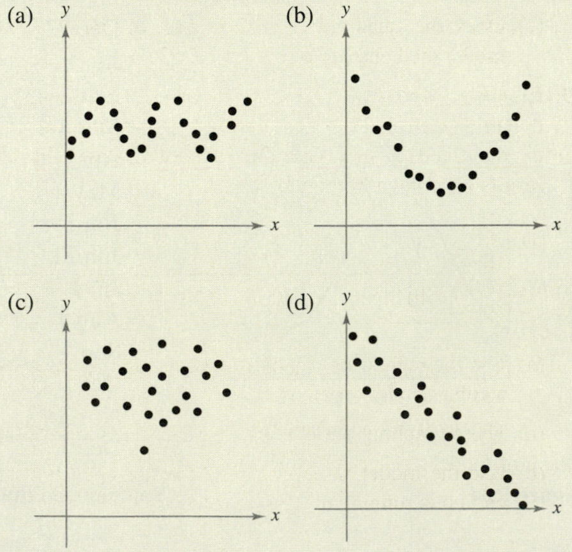

7. Beam Strength Students in a lab measured the breaking strength S (in pounds) of wood 2 inches thick, x inches high, and 12 inches long. The results are shown in the table.

x	4	6	8	10	12
S	2370	5460	10,310	16,250	23,860

(a) Use the regression capabilities of a graphing utility to find a quadratic model for the data.

(b) Use a graphing utility to plot the data and graph the model.

(c) Use the model to approximate the breaking strength when $x = 2$.

(d) How many times greater is the breaking strength for a 4-inch-high board than for a 2-inch-high board?

(e) How many times greater is the breaking strength for a 12-inch-high board than for a 6-inch-high board? When the height of a board increases by a factor, does the breaking strength increase by the same factor? Explain.

8. Car Performance The time t (in seconds) required to attain a speed of s miles per hour from a standing start for a Volkswagen Passat is shown in the table. *(Source: Car & Driver)*

s	30	40	50	60	70	80	90
t	2.7	3.8	4.9	6.3	8.0	9.9	12.2

(a) Use the regression capabilities of a graphing utility to find a quadratic model for the data.

(b) Use a graphing utility to plot the data and graph the model.

(c) Use the graph in part (b) to state why the model is not appropriate for determining the times required to attain speeds of less than 20 miles per hour.

(d) Because the test began from a standing start, add the point $(0, 0)$ to the data. Fit a quadratic model to the revised data and graph the new model.

(e) Does the quadratic model in part (d) more accurately model the behavior of the car? Explain.

9. Engine Performance A V8 car engine is coupled to a dynamometer, and the horsepower y is measured at different engine speeds x (in thousands of revolutions per minute). The results are shown in the table.

x	1	2	3	4	5	6
y	40	85	140	200	225	245

(a) Use the regression capabilities of a graphing utility to find a cubic model for the data.

(b) Use a graphing utility to plot the data and graph the model.

(c) Use the model to approximate the horsepower when the engine is running at 4500 revolutions per minute.

10. Boiling Temperature The table shows the temperatures T (in degrees Fahrenheit) at which water boils at selected pressures p (in pounds per square inch). *(Source: Standard Handbook for Mechanical Engineers)*

p	5	10	14.696 (1 atmosphere)	20
T	162.24°	193.21°	212.00°	227.96°

p	30	40	60	80	100
T	250.33°	267.25°	292.71°	312.03°	327.81°

(a) Use the regression capabilities of a graphing utility to find a cubic model for the data.

(b) Use a graphing utility to plot the data and graph the model.

(c) Use the graph to estimate the pressure required for the boiling point of water to exceed 300°F.

(d) Explain why the model would not be accurate for pressures exceeding 100 pounds per square inch.

11. Automobile Costs The data in the table show the variable costs of operating an automobile in the United States for 2000 through 2010, where t is the year, with $t = 0$ corresponding to 2000. The functions y_1, y_2, and y_3 represent the costs in cents per mile for gas, maintenance, and tires, respectively. *(Source: Bureau of Transportation Statistics)*

t	y_1	y_2	y_3
0	6.9	3.6	1.7
1	7.9	3.9	1.8
2	5.9	4.1	1.8
3	7.2	4.1	1.8
4	6.5	5.4	0.7
5	9.5	4.9	0.7
6	8.9	4.9	0.7
7	11.7	4.6	0.7
8	10.1	4.6	0.8
9	11.4	4.5	0.8
10	12.3	4.4	1.0

(a) Use the regression capabilities of a graphing utility to find cubic models for y_1 and y_3, and a quadratic model for y_2.

(b) Use a graphing utility to graph y_1, y_2, y_3, and $y_1 + y_2 + y_3$ in the same viewing window. Use the model to estimate the total variable cost per mile in 2014.

12. Health Maintenance Organizations The bar graph shows the numbers of people N (in millions) receiving care in HMOs for the years 1994 through 2008. *(Source: HealthLeaders-InterStudy)*

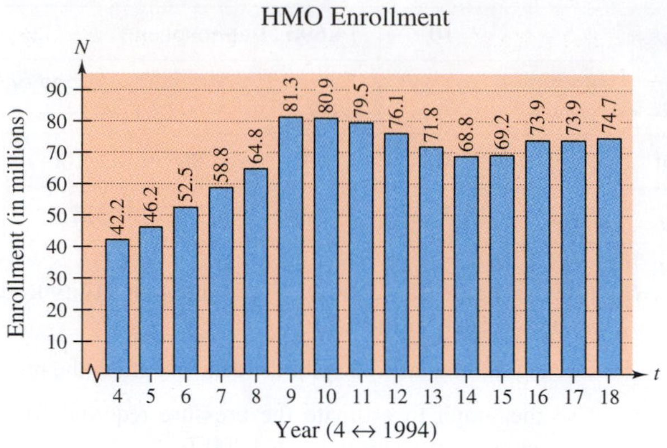

(a) Let t be the time in years, with $t = 4$ corresponding to 1994. Use the regression capabilities of a graphing utility to find linear and cubic models for the data.

(b) Use a graphing utility to plot the data and graph the linear and cubic models.

(c) Use the graphs in part (b) to determine which is the better model.

(d) Use a graphing utility to find and graph a quadratic model for the data. How well does the model fit the data? Explain.

(e) Use the linear and cubic models to estimate the number of people receiving care in HMOs in the year 2014. What do you notice?

(f) Use a graphing utility to find other models for the data. Which models do you think best represent the data? Explain.

13. Harmonic Motion The motion of an oscillating weight suspended by a spring was measured by a motion detector. The data collected and the approximate maximum (positive and negative) displacements from equilibrium are shown in the figure. The displacement y is measured in centimeters, and the time t is measured in seconds.

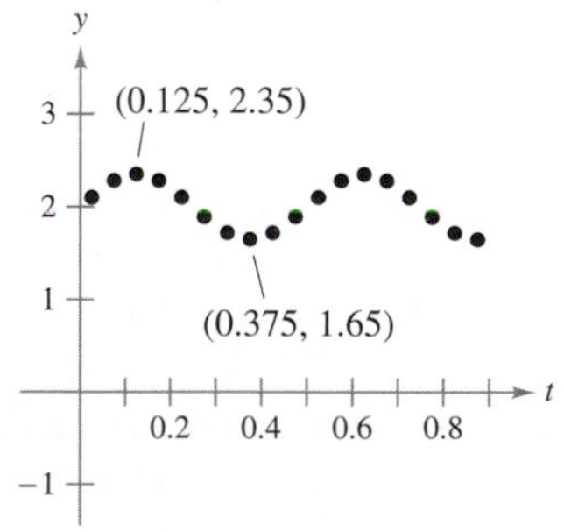

(a) Is y a function of t? Explain.

(b) Approximate the amplitude and period of the oscillations.

(c) Find a model for the data.

(d) Use a graphing utility to graph the model in part (c). Compare the result with the data in the figure.

14. Temperature The table shows the normal daily high temperatures for Miami M and Syracuse S (in degrees Fahrenheit) for month t, with $t = 1$ corresponding to January. *(Source: National Oceanic and Atmospheric Administration)*

t	1	2	3	4	5	6
M	76.5	77.7	80.7	83.8	87.2	89.5
S	31.4	33.5	43.1	55.7	68.5	77.0

t	7	8	9	10	11	12
M	90.9	90.6	89.0	85.4	81.2	77.5
S	81.7	79.6	71.4	59.8	47.4	36.3

(a) A model for Miami is

$$M(t) = 83.70 + 7.46 \sin(0.4912t - 1.95).$$

Find a model for Syracuse.

(b) Use a graphing utility to plot the data and graph the model for Miami. How well does the model fit?

(c) Use a graphing utility to plot the data and graph the model for Syracuse. How well does the model fit?

(d) Use the models to estimate the average annual temperature in each city. Which term of the model did you use? Explain.

(e) What is the period of each model? Is it what you expected? Explain.

(f) Which city has a greater variability in temperature throughout the year? Which factor of the models determines this variability? Explain.

WRITING ABOUT CONCEPTS

Modeling Data **In Exercises 15 and 16, describe a possible real-life situation for each data set. Then describe how a model could be used in the real-life setting.**

15.

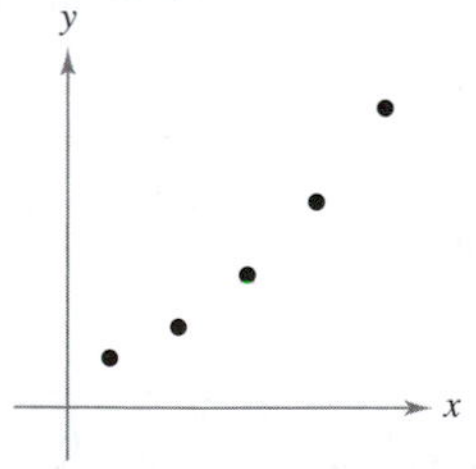

16.

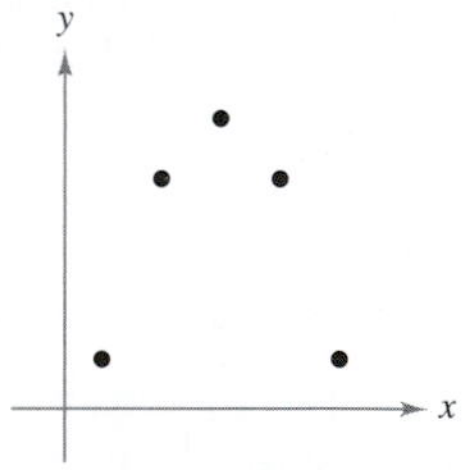

PUTNAM EXAM CHALLENGE

17. For $i = 1, 2$, let T_i be a triangle with side lengths a_i, b_i, c_i, and area A_i. Suppose that $a_1 \le a_2$, $b_1 \le b_2$, $c_1 \le c_2$, and that T_2 is an acute triangle. Does it follow that $A_1 \le A_2$?

Review Exercises

See CalcChat.com for tutorial help and worked-out solutions to odd-numbered exercises.

Finding Intercepts **In Exercises 1–4, find any intercepts.**

1. $y = 5x - 8$
2. $y = x^2 - 8x + 12$
3. $y = \dfrac{x-3}{x-4}$
4. $y = (x-3)\sqrt{x+4}$

Testing for Symmetry **In Exercises 5–8, test for symmetry with respect to each axis and to the origin.**

5. $y = x^2 + 4x$
6. $y = x^4 - x^2 + 3$
7. $y^2 = x^2 - 5$
8. $xy = -2$

Using Intercepts and Symmetry to Sketch a Graph **In Exercises 9–14, sketch the graph of the equation. Identify any intercepts and test for symmetry.**

9. $y = -\frac{1}{2}x + 3$
10. $y = -x^2 + 4$
11. $y = x^3 - 4x$
12. $y^2 = 9 - x$
13. $y = 2\sqrt{4-x}$
14. $y = |x - 4| - 4$

Finding Points of Intersection **In Exercises 15–18, find the points of intersection of the graphs of the equations.**

15. $5x + 3y = -1$
    $x - y = -5$
16. $2x + 4y = 9$
    $6x - 4y = 7$
17. $x - y = -5$
    $x^2 - y = 1$
18. $x^2 + y^2 = 1$
    $-x + y = 1$

Finding the Slope of a Line **In Exercises 19 and 20, plot the points and find the slope of the line passing through them.**

19. $\left(\frac{3}{2}, 1\right), \left(5, \frac{5}{2}\right)$
20. $(-7, 8), (-1, 8)$

Finding an Equation of a Line **In Exercises 21–24, find an equation of the line that passes through the point and has the indicated slope. Then sketch the line.**

	Point	Slope		Point	Slope
21.	$(3, -5)$	$m = \frac{7}{4}$	22.	$(-8, 1)$	m is undefined.
23.	$(-3, 0)$	$m = -\frac{2}{3}$	24.	$(5, 4)$	$m = 0$

Sketching Lines in the Plane **In Exercises 25–28, use the slope and y-intercept to sketch a graph of the equation.**

25. $y = 6$
26. $x = -3$
27. $y = 4x - 2$
28. $3x + 2y = 12$

Finding an Equation of a Line **In Exercises 29 and 30, find an equation of the line that passes through the points. Then sketch the line.**

29. $(0, 0), (8, 2)$
30. $(-5, 5), (10, -1)$

31. **Finding Equations of Lines** Find equations of the lines passing through $(-3, 5)$ and having the following characteristics.
    (a) Slope of $\frac{7}{16}$
    (b) Parallel to the line $5x - 3y = 3$
    (c) Perpendicular to the line $3x + 4y = 8$
    (d) Parallel to the y-axis

32. **Finding Equations of Lines** Find equations of the lines passing through $(2, 4)$ and having the following characteristics.
    (a) Slope of $-\frac{2}{3}$
    (b) Perpendicular to the line $x + y = 0$
    (c) Passing through the point $(6, 1)$
    (d) Parallel to the x-axis

33. **Rate of Change** The purchase price of a new machine is \$12,500, and its value will decrease by \$850 per year. Use this information to write a linear equation that gives the value V of the machine t years after it is purchased. Find its value at the end of 3 years.

34. **Break-Even Analysis** A contractor purchases a piece of equipment for \$36,500 that costs an average of \$9.25 per hour for fuel and maintenance. The equipment operator is paid \$13.50 per hour, and customers are charged \$30 per hour.
    (a) Write an equation for the cost C of operating this equipment for t hours.
    (b) Write an equation for the revenue R derived from t hours of use.
    (c) Find the break-even point for this equipment by finding the time at which $R = C$.

Evaluating a Function **In Exercises 35–38, evaluate the function at the given value(s) of the independent variable. Simplify the result.**

35. $f(x) = 5x + 4$
    (a) $f(0)$
    (b) $f(5)$
    (c) $f(-3)$
    (d) $f(t + 1)$
36. $f(x) = x^3 - 2x$
    (a) $f(-3)$
    (b) $f(2)$
    (c) $f(-1)$
    (d) $f(c - 1)$
37. $f(x) = 4x^2$
    $\dfrac{f(x + \Delta x) - f(x)}{\Delta x}$
38. $f(x) = 2x - 6$
    $\dfrac{f(x) - f(-1)}{x - 1}$

Finding the Domain and Range of a Function **In Exercises 39–42, find the domain and range of the function.**

39. $f(x) = x^2 + 3$
40. $g(x) = \sqrt{6 - x}$
41. $f(x) = -|x + 1|$
42. $h(x) = \dfrac{2}{x + 1}$

Using the Vertical Line Test **In Exercises 43–46, sketch the graph of the equation and use the Vertical Line Test to determine whether y is a function of x.**

43. $x - y^2 = 6$

44. $x^2 - y = 0$

45. $y = \dfrac{|x - 2|}{x - 2}$

46. $x = 9 - y^2$

47. Transformations of Functions Use a graphing utility to graph $f(x) = x^3 - 3x^2$. Use the graph to write a formula for the function g shown in the figure. To print an enlarged copy of the graph, go to *MathGraphs.com.*

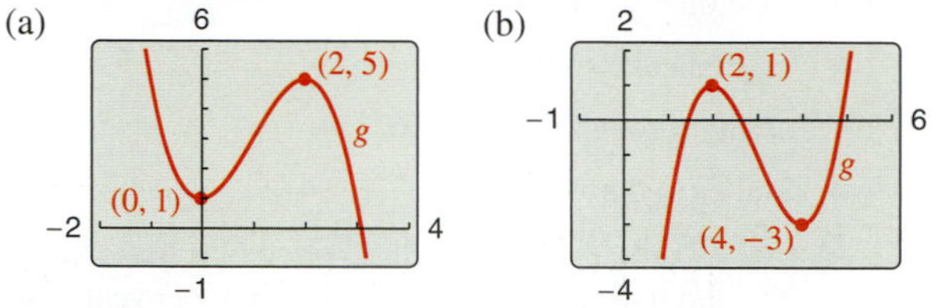

48. Conjecture

(a) Use a graphing utility to graph the functions f, g, and h in the same viewing window. Write a description of any similarities and differences you observe among the graphs.

Odd powers: $f(x) = x$, $g(x) = x^3$, $h(x) = x^5$

Even powers: $f(x) = x^2$, $g(x) = x^4$, $h(x) = x^6$

(b) Use the result in part (a) to make a conjecture about the graphs of the functions $y = x^7$ and $y = x^8$. Use a graphing utility to verify your conjecture.

49. Think About It Use the results of Exercise 48 to guess the shapes of the graphs of the functions f, g, and h. Then use a graphing utility to graph each function and compare the result with your guess.

(a) $f(x) = x^2(x - 6)^2$

(b) $g(x) = x^3(x - 6)^2$

(c) $h(x) = x^3(x - 6)^3$

50. Think About It What is the minimum degree of the polynomial function whose graph approximates the given graph? What sign must the leading coefficient have?

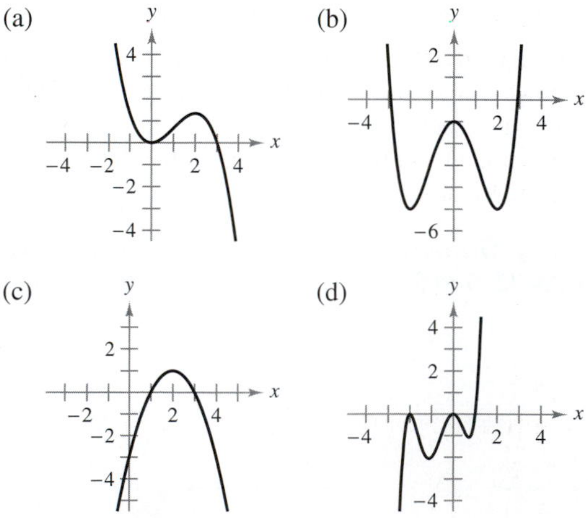

51. Stress Test A machine part was tested by bending it x centimeters 10 times per minute until the time y (in hours) of failure. The results are recorded in the table.

x	3	6	9	12	15
y	61	56	53	55	48

x	18	21	24	27	30
y	35	36	33	44	23

(a) Use the regression capabilities of a graphing utility to find a linear model for the data.

(b) Use a graphing utility to plot the data and graph the model.

(c) Use the graph to determine whether there may have been an error made in conducting one of the tests or in recording the results. If so, eliminate the erroneous point and find the model for the remaining data.

52. Median Income The data in the table show the median income y (in thousands of dollars) for males of various ages x in the United States in 2009. *(Source: U.S. Census Bureau)*

x	20	30	40	50	60	70
y	10.0	31.9	42.2	44.7	41.3	25.9

(a) Use the regression capabilities of a graphing utility to find a quadratic model for the data.

(b) Use a graphing utility to plot the data and graph the model.

(c) Use the model to approximate the median income for a male who is 26 years old.

(d) Use the model to approximate the median income for a male who is 34 years old.

53. Harmonic Motion The motion of an oscillating weight suspended by a spring was measured by a motion detector. The data collected and the approximate maximum (positive and negative) displacements from equilibrium are shown in the figure. The displacement y is measured in feet, and the time t is measured in seconds.

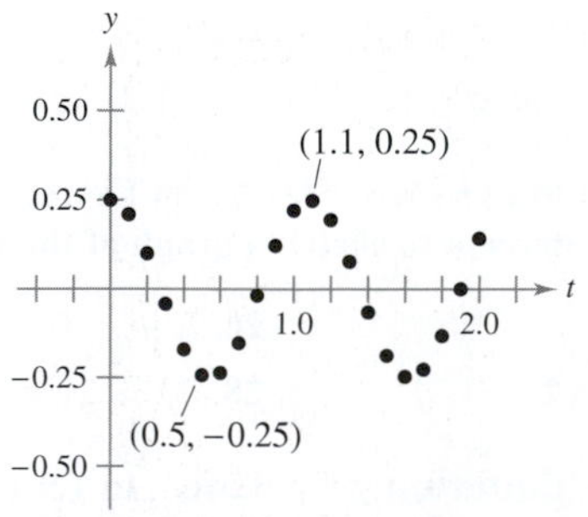

(a) Is y a function of t? Explain.

(b) Approximate the amplitude and period of the oscillations.

(c) Find a model for the data.

(d) Use a graphing utility to graph the model in part (c). Compare the result with the data in the figure.

P.S. Problem Solving

See CalcChat.com for tutorial help and worked-out solutions to odd-numbered exercises.

1. Finding Tangent Lines Consider the circle

$$x^2 + y^2 - 6x - 8y = 0,$$

as shown in the figure.

(a) Find the center and radius of the circle.

(b) Find an equation of the tangent line to the circle at the point $(0, 0)$.

(c) Find an equation of the tangent line to the circle at the point $(6, 0)$.

(d) Where do the two tangent lines intersect?

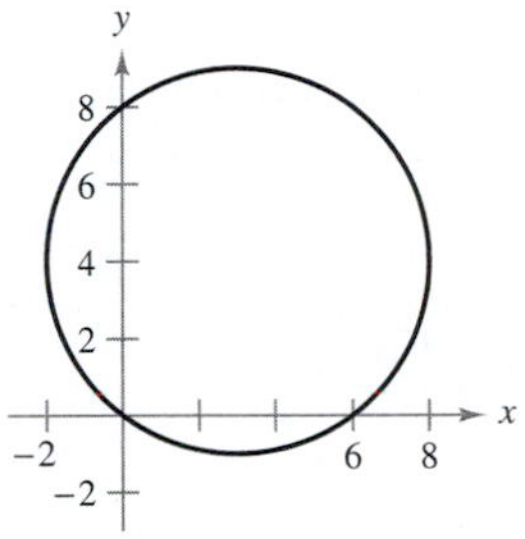

Figure for 1

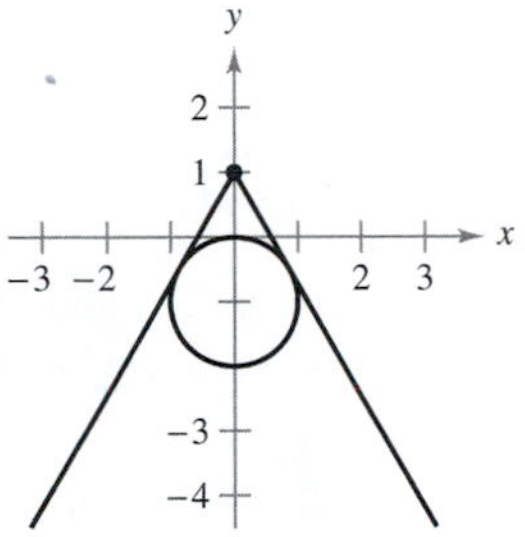

Figure for 2

2. Finding Tangent Lines There are two tangent lines from the point $(0, 1)$ to the circle $x^2 + (y + 1)^2 = 1$ (see figure). Find equations of these two lines by using the fact that each tangent line intersects the circle at *exactly* one point.

3. Heaviside Function The Heaviside function $H(x)$ is widely used in engineering applications.

$$H(x) = \begin{cases} 1, & x \geq 0 \\ 0, & x < 0 \end{cases}$$

Sketch the graph of the Heaviside function and the graphs of the following functions by hand.

(a) $H(x) - 2$ (b) $H(x - 2)$ (c) $-H(x)$

(d) $H(-x)$ (e) $\frac{1}{2}H(x)$ (f) $-H(x - 2) + 2$

OLIVER HEAVISIDE (1850–1925)

Heaviside was a British mathematician and physicist who contributed to the field of applied mathematics, especially applications of mathematics to electrical engineering. The *Heaviside function* is a classic type of "on-off" function that has applications to electricity and computer science.

4. Sketching Transformations Consider the graph of the function f shown below. Use this graph to sketch the graphs of the following functions. To print an enlarged copy of the graph, go to *MathGraphs.com*.

(a) $f(x + 1)$ (b) $f(x) + 1$

(c) $2f(x)$ (d) $f(-x)$

(e) $-f(x)$ (f) $|f(x)|$

(g) $f(|x|)$

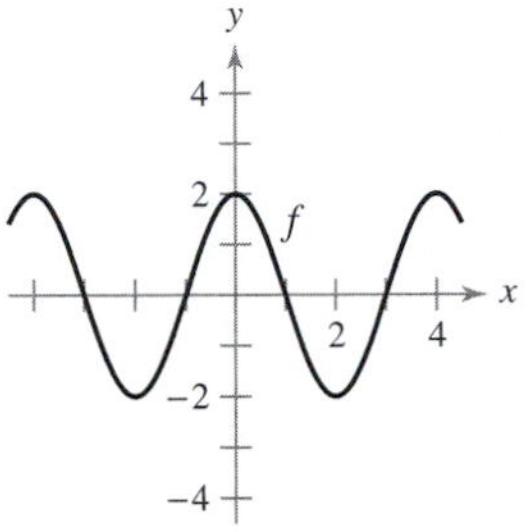

5. Maximum Area A rancher plans to fence a rectangular pasture adjacent to a river. The rancher has 100 meters of fencing, and no fencing is needed along the river (see figure).

(a) Write the area A of the pasture as a function of x, the length of the side parallel to the river. What is the domain of A?

(b) Graph the area function and estimate the dimensions that yield the maximum amount of area for the pasture.

(c) Find the dimensions that yield the maximum amount of area for the pasture by completing the square.

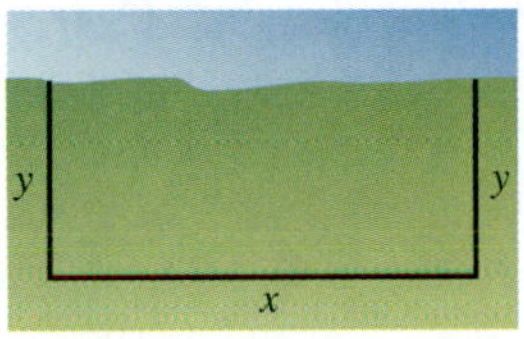

Figure for 5

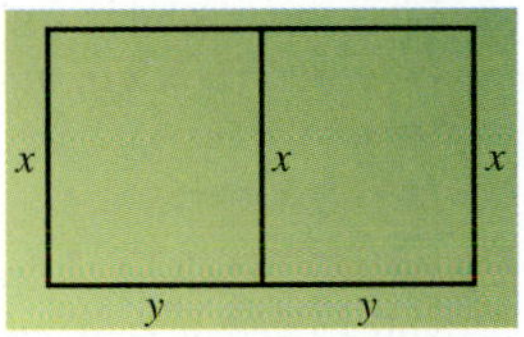

Figure for 6

6. Maximum Area A rancher has 300 feet of fencing to enclose two adjacent pastures (see figure).

(a) Write the total area A of the two pastures as a function of x. What is the domain of A?

(b) Graph the area function and estimate the dimensions that yield the maximum amount of area for the pastures.

(c) Find the dimensions that yield the maximum amount of area for the pastures by completing the square.

7. Writing a Function You are in a boat 2 miles from the nearest point on the coast. You are to go to a point Q located 3 miles down the coast and 1 mile inland (see figure). You can row at 2 miles per hour and walk at 4 miles per hour. Write the total time T of the trip as a function of x.

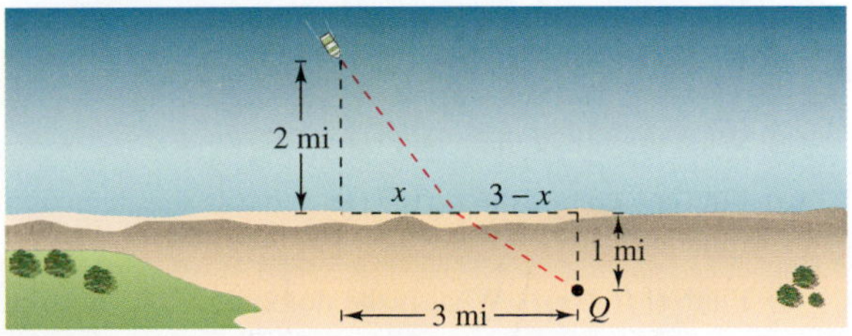

8. Average Speed You drive to the beach at a rate of 120 kilometers per hour. On the return trip, you drive at a rate of 60 kilometers per hour. What is your average speed for the entire trip? Explain your reasoning.

9. Slope of a Tangent Line One of the fundamental themes of calculus is to find the slope of the tangent line to a curve at a point. To see how this can be done, consider the point (2, 4) on the graph of $f(x) = x^2$ (see figure).

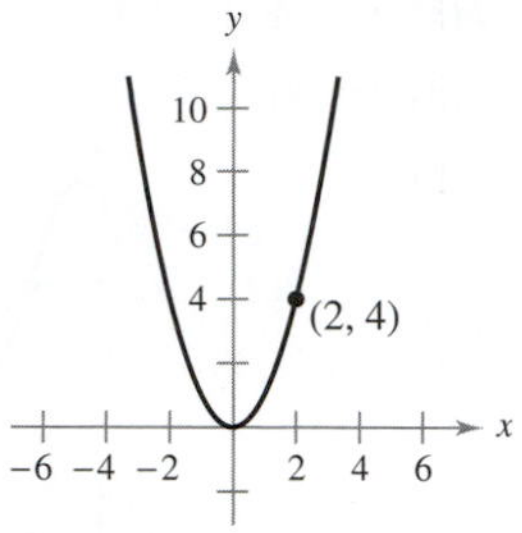

(a) Find the slope of the line joining (2, 4) and (3, 9). Is the slope of the tangent line at (2, 4) greater than or less than this number?

(b) Find the slope of the line joining (2, 4) and (1, 1). Is the slope of the tangent line at (2, 4) greater than or less than this number?

(c) Find the slope of the line joining (2, 4) and (2.1, 4.41). Is the slope of the tangent line at (2, 4) greater than or less than this number?

(d) Find the slope of the line joining (2, 4) and $(2 + h, f(2 + h))$ in terms of the nonzero number h. Verify that $h = 1, -1$, and 0.1 yield the solutions to parts (a)–(c) above.

(e) What is the slope of the tangent line at (2, 4)? Explain how you arrived at your answer.

10. Slope of a Tangent Line Sketch the graph of the function $f(x) = \sqrt{x}$ and label the point (4, 2) on the graph.

(a) Find the slope of the line joining (4, 2) and (9, 3). Is the slope of the tangent line at (4, 2) greater than or less than this number?

(b) Find the slope of the line joining (4, 2) and (1, 1). Is the slope of the tangent line at (4, 2) greater than or less than this number?

(c) Find the slope of the line joining (4, 2) and (4.41, 2.1). Is the slope of the tangent line at (4, 2) greater than or less than this number?

(d) Find the slope of the line joining (4, 2) and $(4 + h, f(4 + h))$ in terms of the nonzero number h.

(e) What is the slope of the tangent line at (4, 2)? Explain how you arrived at your answer.

11. Composite Functions Let $f(x) = \dfrac{1}{1 - x}$.

(a) What are the domain and range of f?

(b) Find the composition $f(f(x))$. What is the domain of this function?

(c) Find $f(f(f(x)))$. What is the domain of this function?

(d) Graph $f(f(f(x)))$. Is the graph a line? Why or why not?

12. Graphing an Equation Explain how you would graph the equation

$$y + |y| = x + |x|.$$

Then sketch the graph.

13. Sound Intensity A large room contains two speakers that are 3 meters apart. The sound intensity I of one speaker is twice that of the other, as shown in the figure. (To print an enlarged copy of the graph, go to *MathGraphs.com.*) Suppose the listener is free to move about the room to find those positions that receive equal amounts of sound from both speakers. Such a location satisfies two conditions: (1) the sound intensity at the listener's position is directly proportional to the sound level of a source, and (2) the sound intensity is inversely proportional to the square of the distance from the source.

(a) Find the points on the x-axis that receive equal amounts of sound from both speakers.

(b) Find and graph the equation of all locations (x, y) where one could stand and receive equal amounts of sound from both speakers.

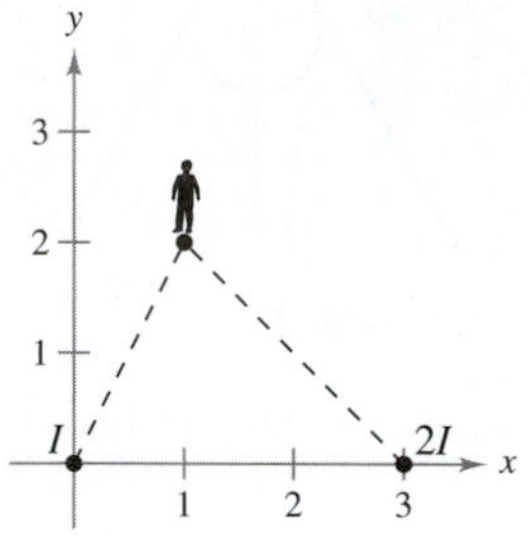

Figure for 13

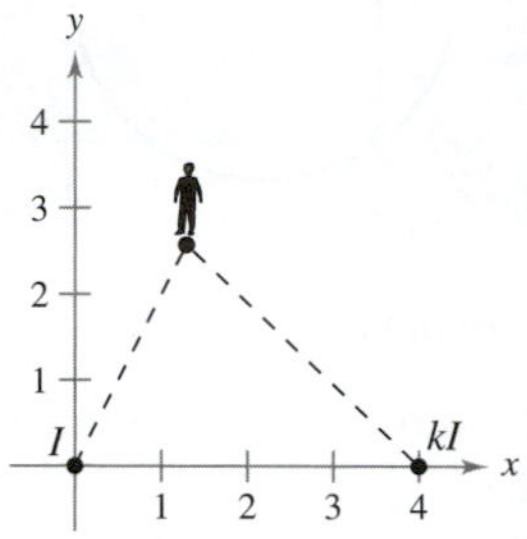

Figure for 14

14. Sound Intensity Suppose the speakers in Exercise 13 are 4 meters apart and the sound intensity of one speaker is k times that of the other, as shown in the figure. To print an enlarged copy of the graph, go to *MathGraphs.com.*

(a) Find the equation of all locations (x, y) where one could stand and receive equal amounts of sound from both speakers.

(b) Graph the equation for the case $k = 3$.

(c) Describe the set of locations of equal sound as k becomes very large.

15. Lemniscate Let d_1 and d_2 be the distances from the point (x, y) to the points $(-1, 0)$ and $(1, 0)$, respectively, as shown in the figure. Show that the equation of the graph of all points (x, y) satisfying $d_1 d_2 = 1$ is

$$(x^2 + y^2)^2 = 2(x^2 - y^2).$$

This curve is called a **lemniscate.** Graph the lemniscate and identify three points on the graph.

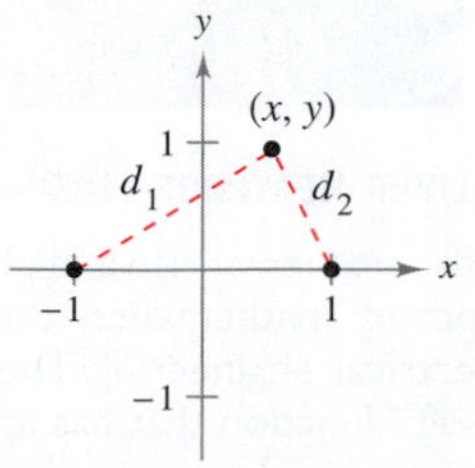

1 Limits and Their Properties

Inventory Management *(Exercise 110, p. 81)*

Average Speed *(Exercise 62, p. 89)*

Free-Falling Object *(Exercises 101 and 102, p. 69)*

Sports *(Exercise 62, p. 57)*

Bicyclist *(Exercise 3, p. 47)*

Clockwise from top left, Christian Delbert/Shutterstock.com; WendellandCarolyn/iStockphoto.com; Tony Bowler/Shutterstock.com; Ljupco Smokovski/Shutterstock.com; Kevin Fleming/Corbis

1.1 A Preview of Calculus

- **Understand what calculus is and how it compares with precalculus.**
- **Understand that the tangent line problem is basic to calculus.**
- **Understand that the area problem is also basic to calculus.**

REMARK As you progress through this course, remember that learning calculus is just one of your goals. Your most important goal is to learn how to use calculus to model and solve real-life problems. Here are a few problem-solving strategies that may help you.

- Be sure you understand the question. What is given? What are you asked to find?
- Outline a plan. There are many approaches you could use: look for a pattern, solve a simpler problem, work backwards, draw a diagram, use technology, or any of many other approaches.
- Complete your plan. Be sure to answer the question. Verbalize your answer. For example, rather than writing the answer as $x = 4.6$, it would be better to write the answer as, "The area of the region is 4.6 square meters."
- Look back at your work. Does your answer make sense? Is there a way you can check the reasonableness of your answer?

What Is Calculus?

Calculus is the mathematics of change. For instance, calculus is the mathematics of velocities, accelerations, tangent lines, slopes, areas, volumes, arc lengths, centroids, curvatures, and a variety of other concepts that have enabled scientists, engineers, and economists to model real-life situations.

Although precalculus mathematics also deals with velocities, accelerations, tangent lines, slopes, and so on, there is a fundamental difference between precalculus mathematics and calculus. Precalculus mathematics is more static, whereas calculus is more dynamic. Here are some examples.

- An object traveling at a constant velocity can be analyzed with precalculus mathematics. To analyze the velocity of an accelerating object, you need calculus.
- The slope of a line can be analyzed with precalculus mathematics. To analyze the slope of a curve, you need calculus.
- The curvature of a circle is constant and can be analyzed with precalculus mathematics. To analyze the variable curvature of a general curve, you need calculus.
- The area of a rectangle can be analyzed with precalculus mathematics. To analyze the area under a general curve, you need calculus.

Each of these situations involves the same general strategy—the reformulation of precalculus mathematics through the use of a limit process. So, one way to answer the question "What is calculus?" is to say that calculus is a "limit machine" that involves three stages. The first stage is precalculus mathematics, such as the slope of a line or the area of a rectangle. The second stage is the limit process, and the third stage is a new calculus formulation, such as a derivative or integral.

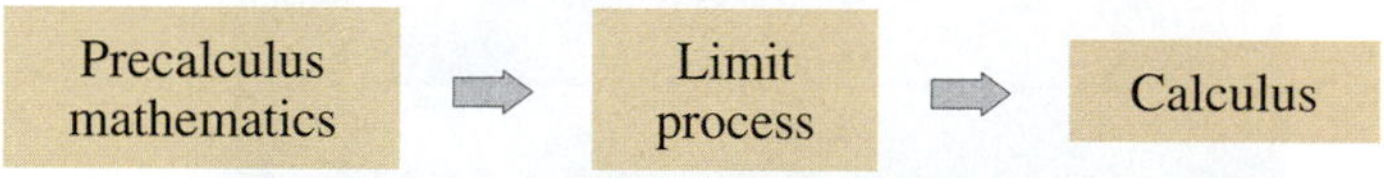

Some students try to learn calculus as if it were simply a collection of new formulas. This is unfortunate. If you reduce calculus to the memorization of differentiation and integration formulas, you will miss a great deal of understanding, self-confidence, and satisfaction.

On the next two pages are listed some familiar precalculus concepts coupled with their calculus counterparts. Throughout the text, your goal should be to learn how precalculus formulas and techniques are used as building blocks to produce the more general calculus formulas and techniques. Don't worry if you are unfamiliar with some of the "old formulas" listed on the next two pages—you will be reviewing all of them.

As you proceed through this text, come back to this discussion repeatedly. Try to keep track of where you are relative to the three stages involved in the study of calculus. For instance, note how these chapters relate to the three stages.

Chapter P: Preparation for Calculus	Precalculus
Chapter 1: Limits and Their Properties	Limit process
Chapter 2: Differentiation	Calculus

This cycle is repeated many times on a smaller scale throughout the text.

Without Calculus		With Differential Calculus	
Value of $f(x)$ when $x = c$	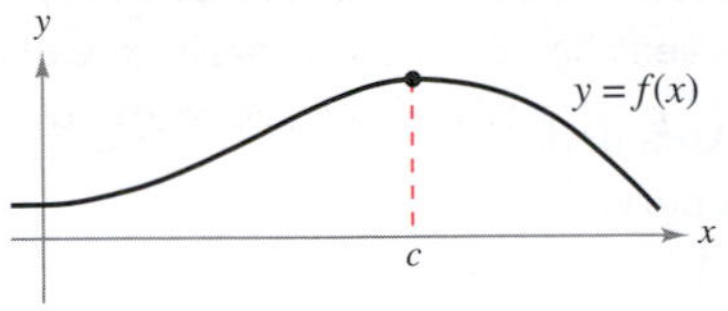	Limit of $f(x)$ as x approaches c	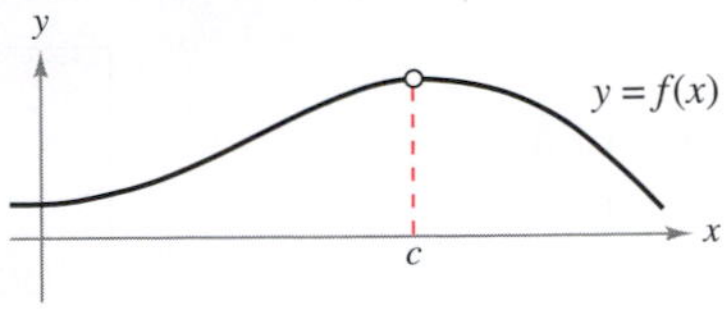
Slope of a line	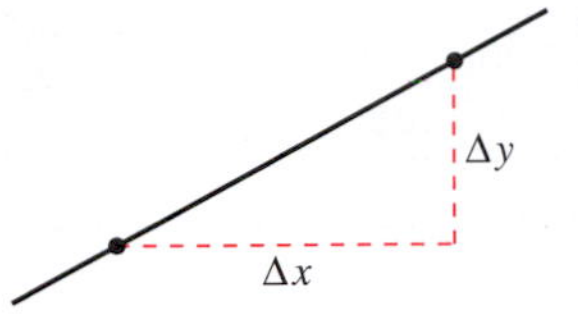	Slope of a curve	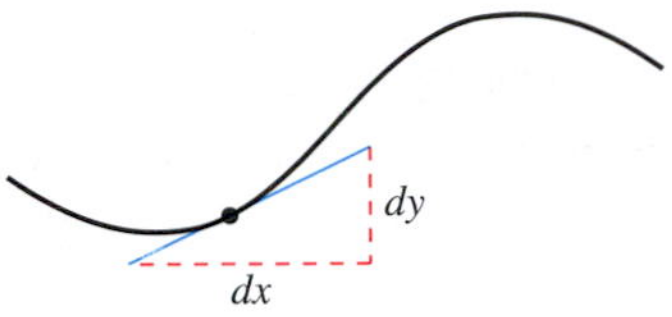
Secant line to a curve	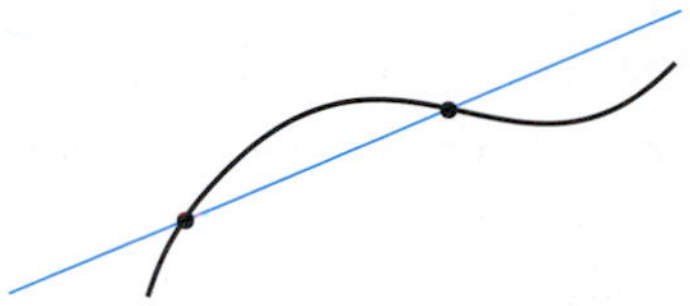	Tangent line to a curve	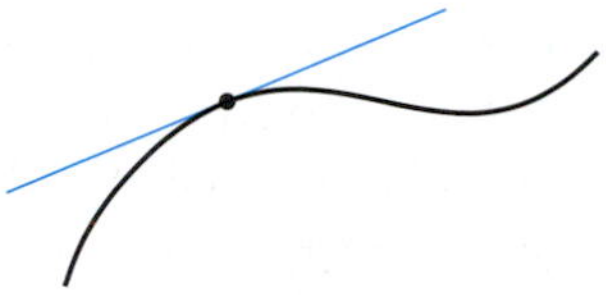
Average rate of change between $t = a$ and $t = b$		Instantaneous rate of change at $t = c$	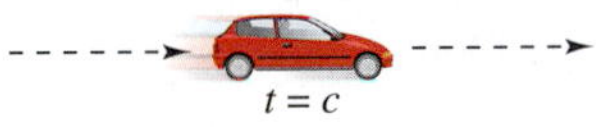
Curvature of a circle	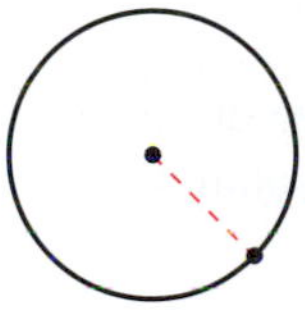	Curvature of a curve	
Height of a curve when $x = c$	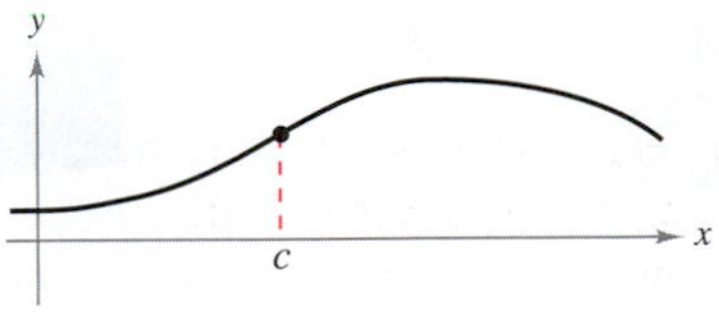	Maximum height of a curve on an interval	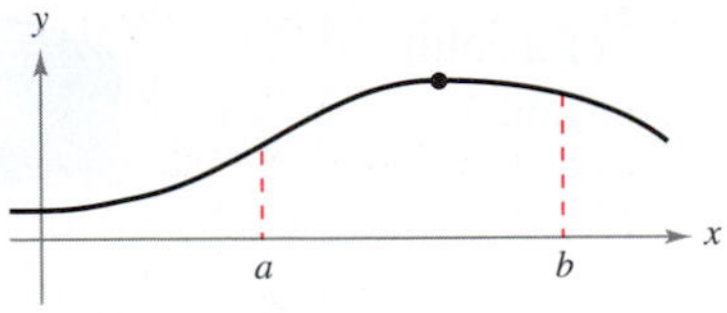
Tangent plane to a sphere		Tangent plane to a surface	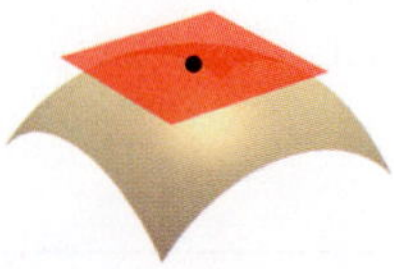
Direction of motion along a line	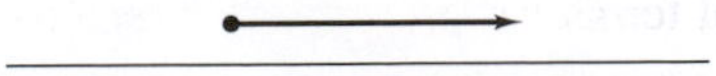	Direction of motion along a curve	

Without Calculus		**With Integral Calculus**	
Area of a rectangle		Area under a curve	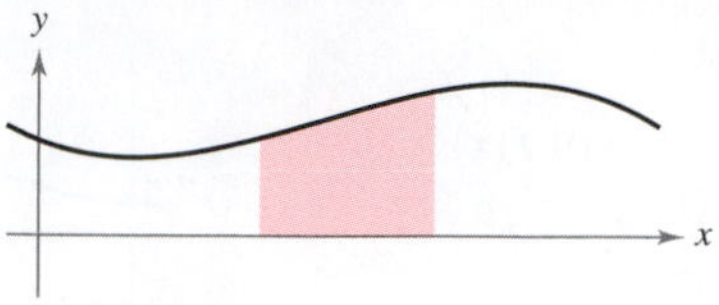
Work done by a constant force	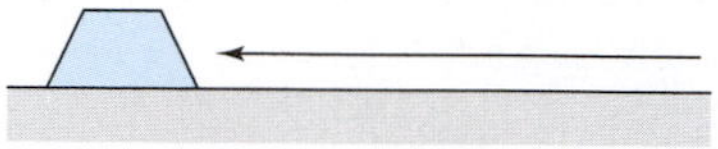	Work done by a variable force	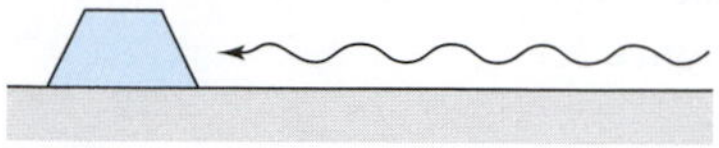
Center of a rectangle	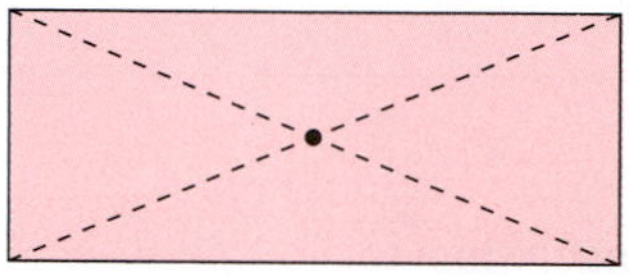	Centroid of a region	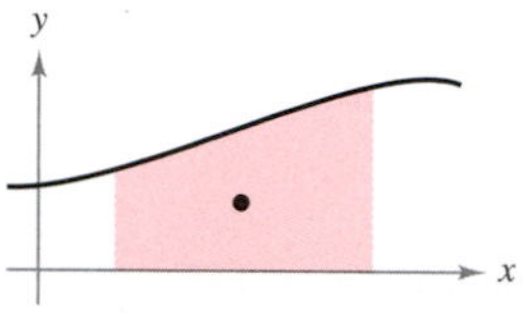
Length of a line segment	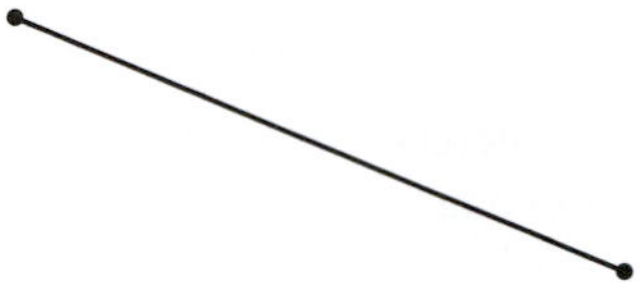	Length of an arc	
Surface area of a cylinder		Surface area of a solid of revolution	
Mass of a solid of constant density		Mass of a solid of variable density	
Volume of a rectangular solid		Volume of a region under a surface	
Sum of a finite number of terms	$a_1 + a_2 + \cdots + a_n = S$	Sum of an infinite number of terms	$a_1 + a_2 + a_3 + \cdots = S$

The Tangent Line Problem

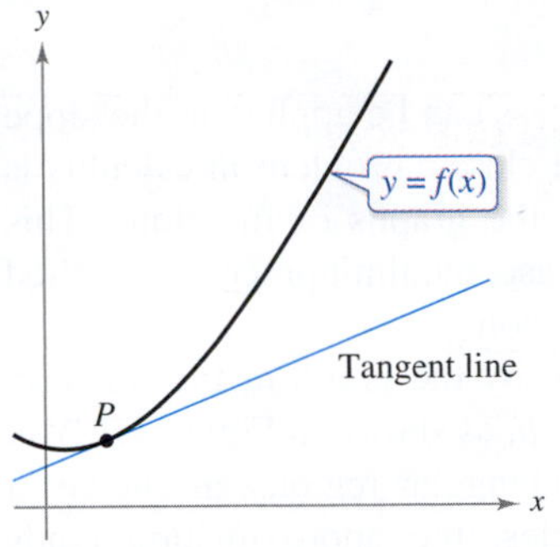

The tangent line to the graph of f at P
Figure 1.1

The notion of a limit is fundamental to the study of calculus. The following brief descriptions of two classic problems in calculus—*the tangent line problem* and *the area problem*—should give you some idea of the way limits are used in calculus.

In the tangent line problem, you are given a function f and a point P on its graph and are asked to find an equation of the tangent line to the graph at point P, as shown in Figure 1.1.

Except for cases involving a vertical tangent line, the problem of finding the **tangent line** at a point P is equivalent to finding the *slope* of the tangent line at P. You can approximate this slope by using a line through the point of tangency and a second point on the curve, as shown in Figure 1.2(a). Such a line is called a **secant line.** If $P(c, f(c))$ is the point of tangency and

$$Q(c + \Delta x, f(c + \Delta x))$$

is a second point on the graph of f, then the slope of the secant line through these two points can be found using precalculus and is

$$m_{sec} = \frac{f(c + \Delta x) - f(c)}{c + \Delta x - c} = \frac{f(c + \Delta x) - f(c)}{\Delta x}.$$

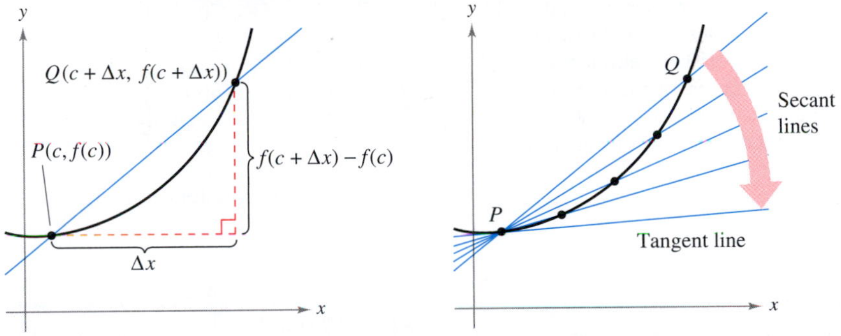

(a) The secant line through $(c, f(c))$ and $(c + \Delta x, f(c + \Delta x))$

(b) As Q approaches P, the secant lines approach the tangent line.

Figure 1.2

GRACE CHISHOLM YOUNG (1868–1944)

Grace Chisholm Young received her degree in mathematics from Girton College in Cambridge, England. Her early work was published under the name of William Young, her husband. Between 1914 and 1916, Grace Young published work on the foundations of calculus that won her the Gamble Prize from Girton College.

As point Q approaches point P, the slopes of the secant lines approach the slope of the tangent line, as shown in Figure 1.2(b). When such a "limiting position" exists, the slope of the tangent line is said to be the **limit** of the slopes of the secant lines. (Much more will be said about this important calculus concept in Chapter 2.)

Exploration

The following points lie on the graph of $f(x) = x^2$.

$$Q_1(1.5, f(1.5)),\quad Q_2(1.1, f(1.1)),\quad Q_3(1.01, f(1.01)),$$
$$Q_4(1.001, f(1.001)),\quad Q_5(1.0001, f(1.0001))$$

Each successive point gets closer to the point $P(1, 1)$. Find the slopes of the secant lines through Q_1 and P, Q_2 and P, and so on. Graph these secant lines on a graphing utility. Then use your results to estimate the slope of the tangent line to the graph of f at the point P.

The Area Problem

In the tangent line problem, you saw how the limit process can be applied to the slope of a line to find the slope of a general curve. A second classic problem in calculus is finding the area of a plane region that is bounded by the graphs of functions. This problem can also be solved with a limit process. In this case, the limit process is applied to the area of a rectangle to find the area of a general region.

As a simple example, consider the region bounded by the graph of the function $y = f(x)$, the x-axis, and the vertical lines $x = a$ and $x = b$, as shown in Figure 1.3. You can approximate the area of the region with several rectangular regions, as shown in Figure 1.4. As you increase the number of rectangles, the approximation tends to become better and better because the amount of area missed by the rectangles decreases. Your goal is to determine the limit of the sum of the areas of the rectangles as the number of rectangles increases without bound.

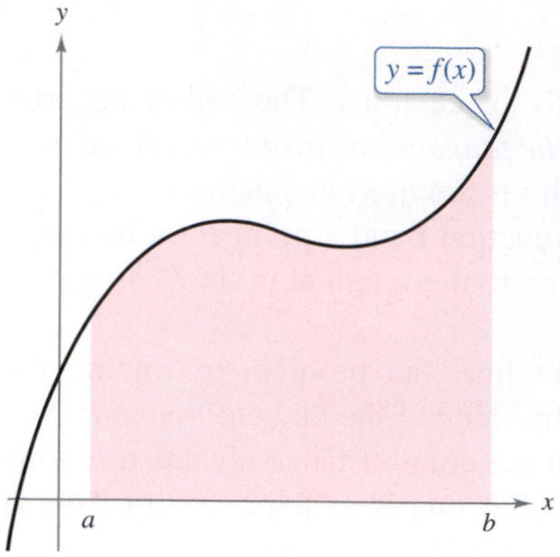

Area under a curve
Figure 1.3

HISTORICAL NOTE

In one of the most astounding events ever to occur in mathematics, it was discovered that the tangent line problem and the area problem are closely related. This discovery led to the birth of calculus. You will learn about the relationship between these two problems when you study the Fundamental Theorem of Calculus in Chapter 4.

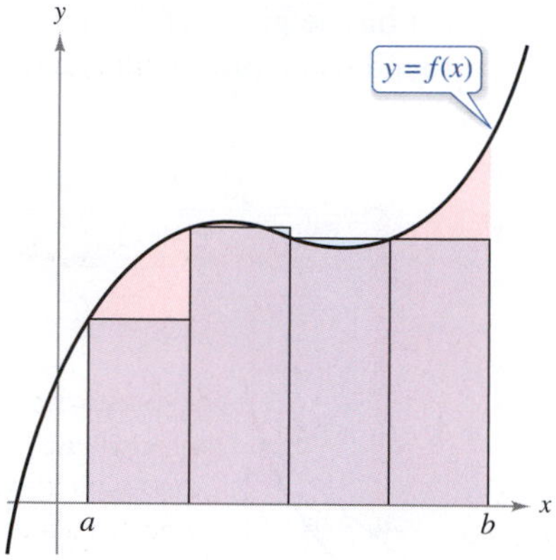

Approximation using four rectangles
Figure 1.4

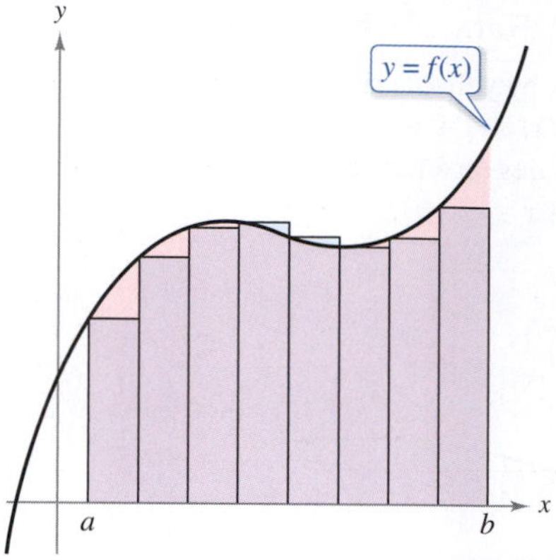

Approximation using eight rectangles

Exploration

Consider the region bounded by the graphs of

$$f(x) = x^2, \quad y = 0, \quad \text{and} \quad x = 1$$

as shown in part (a) of the figure. The area of the region can be approximated by two sets of rectangles—one set inscribed within the region and the other set circumscribed over the region, as shown in parts (b) and (c). Find the sum of the areas of each set of rectangles. Then use your results to approximate the area of the region.

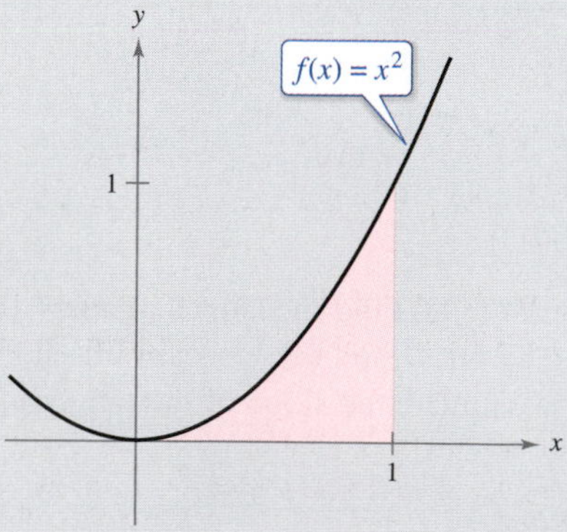

(a) Bounded region

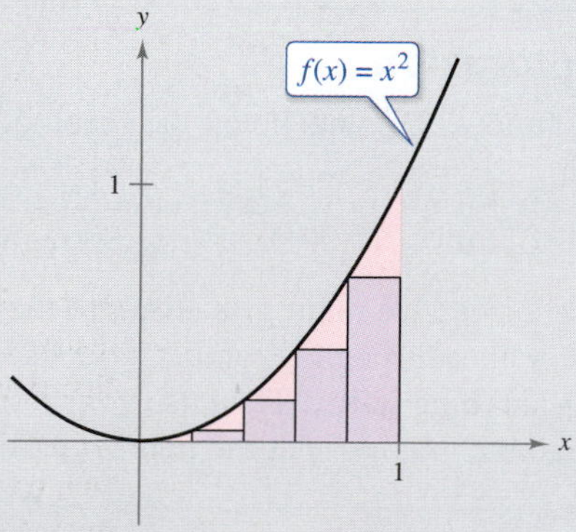

(b) Inscribed rectangles

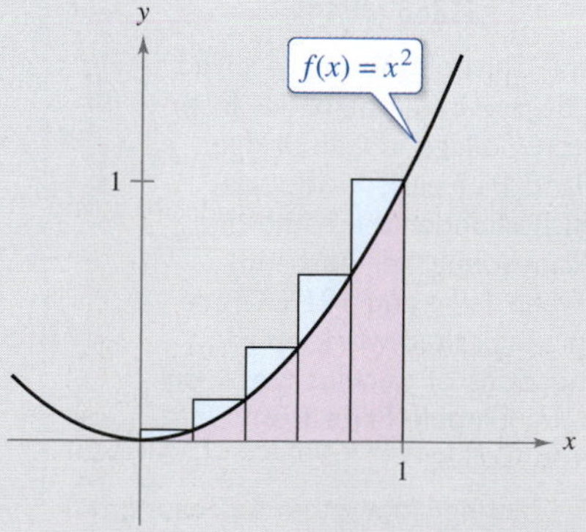

(c) Circumscribed rectangles

1.1 Exercises

See CalcChat.com for tutorial help and worked-out solutions to odd-numbered exercises.

Precalculus or Calculus **In Exercises 1–5, decide whether the problem can be solved using precalculus or whether calculus is required. If the problem can be solved using precalculus, solve it. If the problem seems to require calculus, explain your reasoning and use a graphical or numerical approach to estimate the solution.**

1. Find the distance traveled in 15 seconds by an object traveling at a constant velocity of 20 feet per second.

2. Find the distance traveled in 15 seconds by an object moving with a velocity of $v(t) = 20 + 7 \cos t$ feet per second.

3. **Rate of Change** A bicyclist is riding on a path modeled by the function $f(x) = 0.04(8x - x^2)$, where x and $f(x)$ are measured in miles (see figure). Find the rate of change of elevation at $x = 2$.

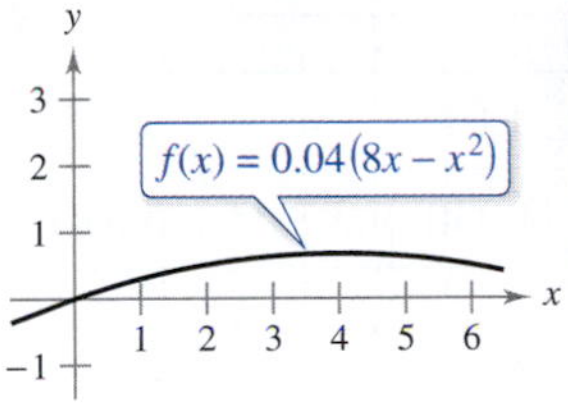

4. A bicyclist is riding on a path modeled by the function $f(x) = 0.08x$, where x and $f(x)$ are measured in miles (see figure). Find the rate of change of elevation at $x = 2$.

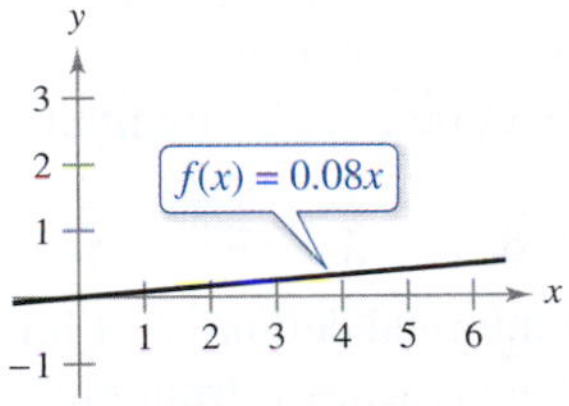

5. Find the area of the shaded region.

(a)

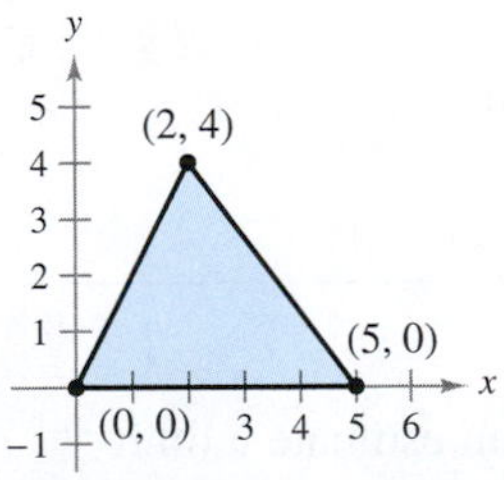

(b)

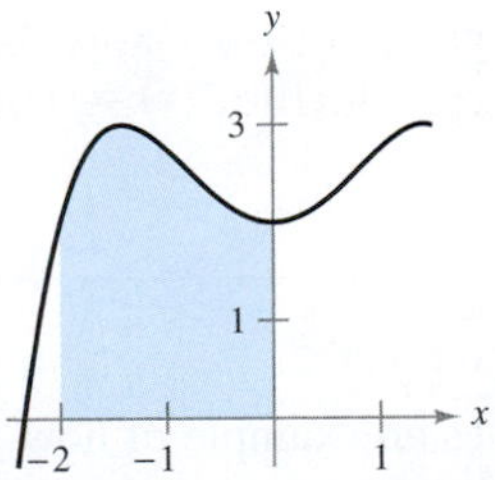

6. **Secant Lines** Consider the function

$$f(x) = \sqrt{x}$$

and the point $P(4, 2)$ on the graph of f.

(a) Graph f and the secant lines passing through $P(4, 2)$ and $Q(x, f(x))$ for x-values of 1, 3, and 5.

(b) Find the slope of each secant line.

(c) Use the results of part (b) to estimate the slope of the tangent line to the graph of f at $P(4, 2)$. Describe how to improve your approximation of the slope.

7. **Secant Lines** Consider the function $f(x) = 6x - x^2$ and the point $P(2, 8)$ on the graph of f.

(a) Graph f and the secant lines passing through $P(2, 8)$ and $Q(x, f(x))$ for x-values of 3, 2.5, and 1.5.

(b) Find the slope of each secant line.

(c) Use the results of part (b) to estimate the slope of the tangent line to the graph of f at $P(2, 8)$. Describe how to improve your approximation of the slope.

8. **HOW DO YOU SEE IT?** How would you describe the instantaneous rate of change of an automobile's position on a highway?

9. **Approximating Area** Use the rectangles in each graph to approximate the area of the region bounded by $y = 5/x$, $y = 0$, $x = 1$, and $x = 5$. Describe how you could continue this process to obtain a more accurate approximation of the area.

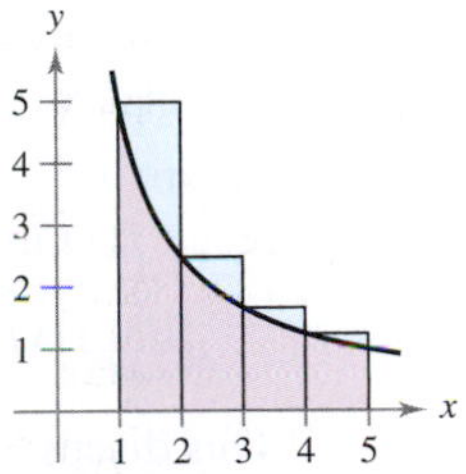

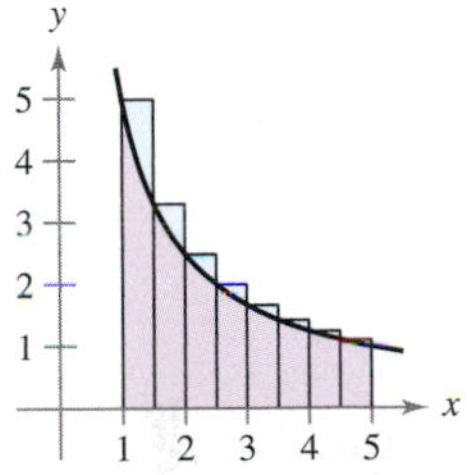

WRITING ABOUT CONCEPTS

10. **Approximating the Length of a Curve** Consider the length of the graph of $f(x) = 5/x$ from $(1, 5)$ to $(5, 1)$.

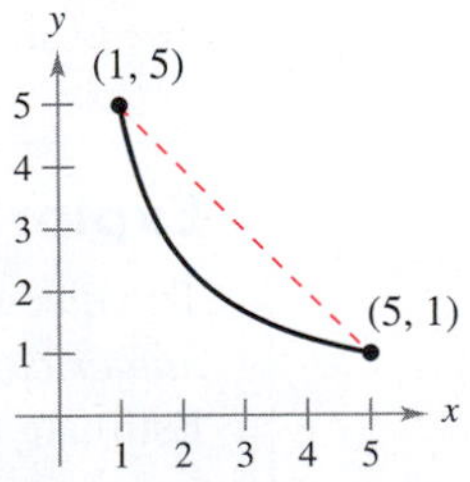

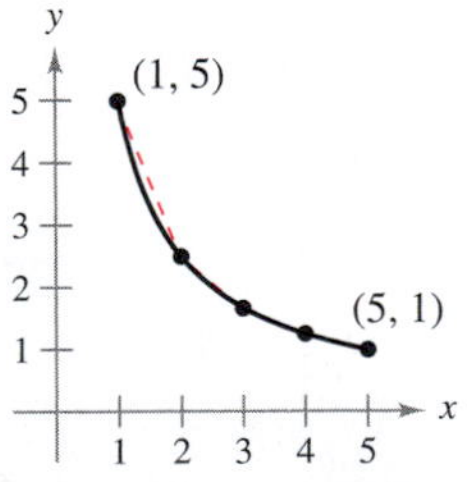

(a) Approximate the length of the curve by finding the distance between its two endpoints, as shown in the first figure.

(b) Approximate the length of the curve by finding the sum of the lengths of four line segments, as shown in the second figure.

(c) Describe how you could continue this process to obtain a more accurate approximation of the length of the curve.

1.2 Finding Limits Graphically and Numerically

- Estimate a limit using a numerical or graphical approach.
- Learn different ways that a limit can fail to exist.
- Study and use a formal definition of limit.

An Introduction to Limits

To sketch the graph of the function

$$f(x) = \frac{x^3 - 1}{x - 1}$$

for values other than $x = 1$, you can use standard curve-sketching techniques. At $x = 1$, however, it is not clear what to expect. To get an idea of the behavior of the graph of f near $x = 1$, you can use two sets of x-values—one set that approaches 1 from the left and one set that approaches 1 from the right, as shown in the table.

x approaches 1 from the left. → ← *x* approaches 1 from the right.

x	0.75	0.9	0.99	0.999	1	1.001	1.01	1.1	1.25
$f(x)$	2.313	2.710	2.970	2.997	?	3.003	3.030	3.310	3.813

$f(x)$ approaches 3. → ← $f(x)$ approaches 3.

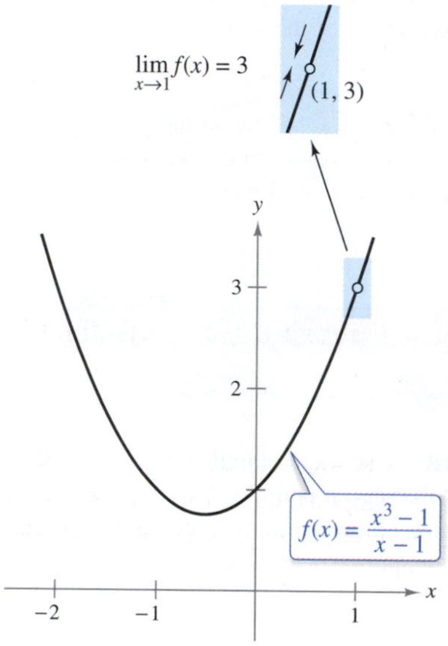

The limit of $f(x)$ as x approaches 1 is 3.
Figure 1.5

The graph of f is a parabola that has a gap at the point $(1, 3)$, as shown in Figure 1.5. Although x cannot equal 1, you can move arbitrarily close to 1, and as a result $f(x)$ moves arbitrarily close to 3. Using limit notation, you can write

$$\lim_{x\to 1} f(x) = 3.$$ This is read as "the limit of $f(x)$ as x approaches 1 is 3."

This discussion leads to an informal definition of limit. If $f(x)$ becomes arbitrarily close to a single number L as x approaches c from either side, then the **limit** of $f(x)$, as x approaches c, is L. This limit is written as

$$\lim_{x\to c} f(x) = L.$$

AP* Tips

Limits are tested primarily on the multiple choice section of the AP Exam.

Exploration

The discussion above gives an example of how you can estimate a limit *numerically* by constructing a table and *graphically* by drawing a graph. Estimate the following limit numerically by completing the table.

$$\lim_{x\to 2} \frac{x^2 - 3x + 2}{x - 2}$$

x	1.75	1.9	1.99	1.999	2	2.001	2.01	2.1	2.25
$f(x)$	?	?	?	?	?	?	?	?	?

Then use a graphing utility to estimate the limit graphically.

EXAMPLE 1 Estimating a Limit Numerically

Evaluate the function $f(x) = x/(\sqrt{x+1} - 1)$ at several x-values near 0 and use the results to estimate the limit

$$\lim_{x \to 0} \frac{x}{\sqrt{x+1} - 1}.$$

Solution The table lists the values of $f(x)$ for several x-values near 0.

x approaches 0 from the left. → ← x approaches 0 from the right.

x	-0.01	-0.001	-0.0001	0	0.0001	0.001	0.01
$f(x)$	1.99499	1.99950	1.99995	?	2.00005	2.00050	2.00499

$f(x)$ approaches 2. → ← $f(x)$ approaches 2.

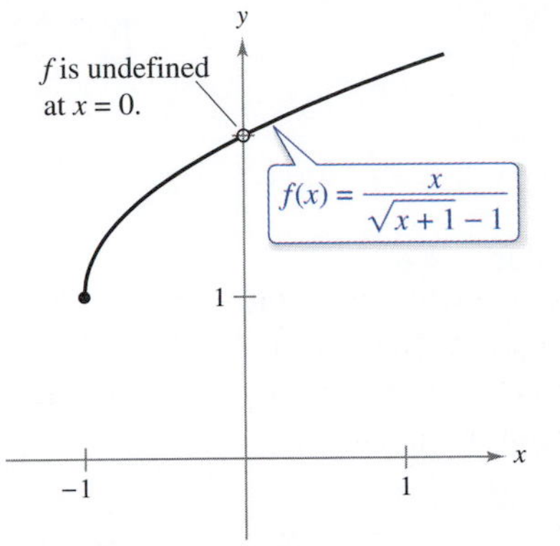

The limit of $f(x)$ as x approaches 0 is 2.
Figure 1.6

From the results shown in the table, you can estimate the limit to be 2. This limit is reinforced by the graph of f (see Figure 1.6). ■

In Example 1, note that the function is undefined at $x = 0$, and yet $f(x)$ appears to be approaching a limit as x approaches 0. This often happens, and it is important to realize that *the existence or nonexistence of $f(x)$ at $x = c$ has no bearing on the existence of the limit of $f(x)$ as x approaches c.*

EXAMPLE 2 Finding a Limit

Find the limit of $f(x)$ as x approaches 2, where

$$f(x) = \begin{cases} 1, & x \neq 2 \\ 0, & x = 2 \end{cases}.$$

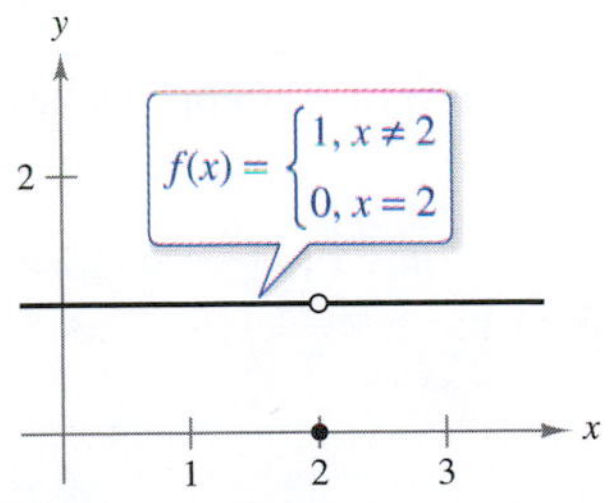

The limit of $f(x)$ as x approaches 2 is 1.
Figure 1.7

Solution Because $f(x) = 1$ for all x other than $x = 2$, you can estimate that the limit is 1, as shown in Figure 1.7. So, you can write

$$\lim_{x \to 2} f(x) = 1.$$

The fact that $f(2) = 0$ has no bearing on the existence or value of the limit as x approaches 2. For instance, as x approaches 2, the function

$$g(x) = \begin{cases} 1, & x \neq 2 \\ 2, & x = 2 \end{cases}$$

has the same limit as f. ■

So far in this section, you have been estimating limits numerically and graphically. Each of these approaches produces an estimate of the limit. In Section 1.3, you will study analytic techniques for evaluating limits. Throughout the course, try to develop a habit of using this three-pronged approach to problem solving.

1. Numerical approach — Construct a table of values.
2. Graphical approach — Draw a graph by hand or using technology.
3. Analytic approach — Use algebra or calculus.

Limits That Fail to Exist

In the next three examples, you will examine some limits that fail to exist.

EXAMPLE 3 Different Right and Left Behavior

Show that the limit $\lim_{x\to 0} \frac{|x|}{x}$ does not exist.

Solution Consider the graph of the function

$$f(x) = \frac{|x|}{x}.$$

In Figure 1.8 and from the definition of absolute value,

$$|x| = \begin{cases} x, & x \geq 0 \\ -x, & x < 0 \end{cases}$$ Definition of absolute value

you can see that

$$\frac{|x|}{x} = \begin{cases} 1, & x > 0 \\ -1, & x < 0 \end{cases}.$$

So, no matter how close x gets to 0, there will be both positive and negative x-values that yield $f(x) = 1$ or $f(x) = -1$. Specifically, if δ (the lowercase Greek letter *delta*) is a positive number, then for x-values satisfying the inequality $0 < |x| < \delta$, you can classify the values of $|x|/x$ as

$$(-\delta, 0) \quad \text{or} \quad (0, \delta).$$

Negative x-values yield $|x|/x = -1$.

Positive x-values yield $|x|/x = 1$.

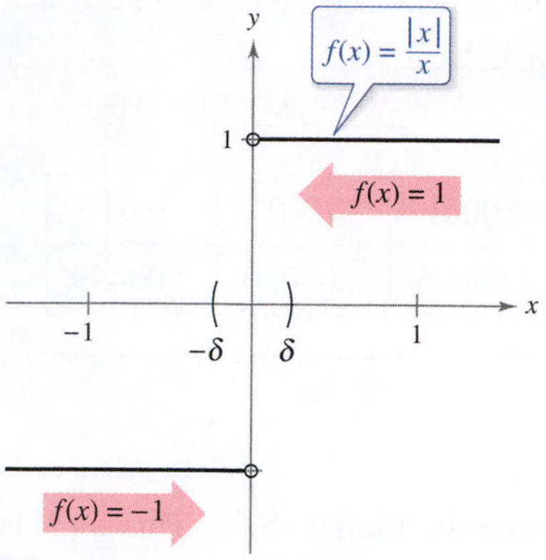

$\lim_{x\to 0} f(x)$ does not exist.
Figure 1.8

Because $|x|/x$ approaches a different number from the right side of 0 than it approaches from the left side, the limit $\lim_{x\to 0} (|x|/x)$ does not exist.

AP* Tips

The AP Exam, especially the free response section, stresses the major applications of calculus, rather than the foundational limit concept.

EXAMPLE 4 Unbounded Behavior

Discuss the existence of the limit $\lim_{x\to 0} \frac{1}{x^2}$.

Solution Consider the graph of the function

$$f(x) = \frac{1}{x^2}.$$

In Figure 1.9, you can see that as x approaches 0 from either the right or the left, $f(x)$ increases without bound. This means that by choosing x close enough to 0, you can force $f(x)$ to be as large as you want. For instance, $f(x)$ will be greater than 100 when you choose x within $\frac{1}{10}$ of 0. That is,

$$0 < |x| < \frac{1}{10} \Rightarrow f(x) = \frac{1}{x^2} > 100.$$

Similarly, you can force $f(x)$ to be greater than 1,000,000, as shown.

$$0 < |x| < \frac{1}{1000} \Rightarrow f(x) = \frac{1}{x^2} > 1{,}000{,}000$$

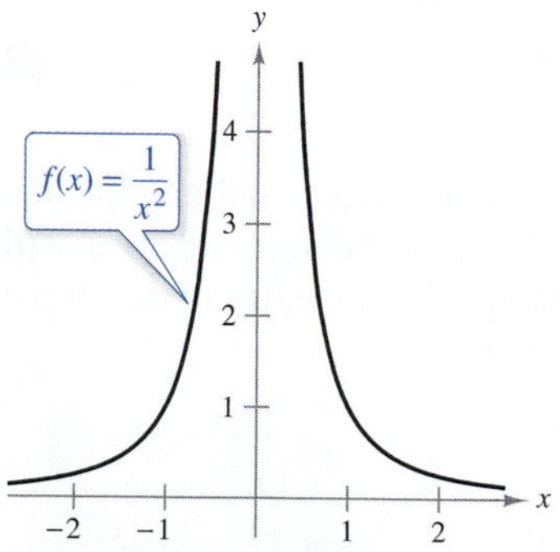

$\lim_{x\to 0} f(x)$ does not exist.
Figure 1.9

Because $f(x)$ does not become arbitrarily close to a single number L as x approaches 0, you can conclude that the limit does not exist. ■

EXAMPLE 5 Oscillating Behavior

• • • • ▷ See LarsonCalculus.com for an interactive version of this type of example.

Discuss the existence of the limit $\lim_{x \to 0} \sin \frac{1}{x}$.

Solution Let $f(x) = \sin(1/x)$. In Figure 1.10, you can see that as x approaches 0, $f(x)$ oscillates between -1 and 1. So, the limit does not exist because no matter how small you choose δ, it is possible to choose x_1 and x_2 within δ units of 0 such that $\sin(1/x_1) = 1$ and $\sin(1/x_2) = -1$, as shown in the table.

x	$\frac{2}{\pi}$	$\frac{2}{3\pi}$	$\frac{2}{5\pi}$	$\frac{2}{7\pi}$	$\frac{2}{9\pi}$	$\frac{2}{11\pi}$	$x \to 0$
$\sin \frac{1}{x}$	1	-1	1	-1	1	-1	Limit does not exist.

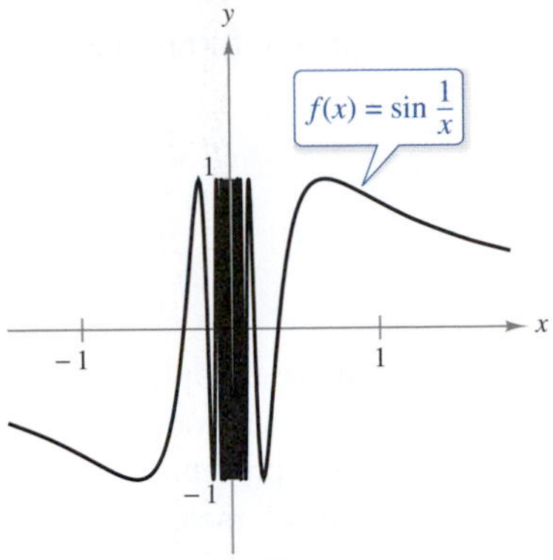

$\lim_{x \to 0} f(x)$ does not exist.

Figure 1.10

Common Types of Behavior Associated with Nonexistence of a Limit

1. $f(x)$ approaches a different number from the right side of c than it approaches from the left side.
2. $f(x)$ increases or decreases without bound as x approaches c.
3. $f(x)$ oscillates between two fixed values as x approaches c.

There are many other interesting functions that have unusual limit behavior. An often cited one is the *Dirichlet function*

$$f(x) = \begin{cases} 0, & \text{if } x \text{ is rational} \\ 1, & \text{if } x \text{ is irrational} \end{cases}.$$

Because this function has *no limit* at any real number c, it is *not continuous* at any real number c. You will study continuity more closely in Section 1.4.

PETER GUSTAV DIRICHLET (1805–1859)

In the early development of calculus, the definition of a function was much more restricted than it is today, and "functions" such as the Dirichlet function would not have been considered. The modern definition of function is attributed to the German mathematician Peter Gustav Dirichlet.

See LarsonCalculus.com to read more of this biography.

INTERFOTO/Alamy

▷ **TECHNOLOGY PITFALL** When you use a graphing utility to investigate the behavior of a function near the x-value at which you are trying to evaluate a limit, remember that you can't always trust the pictures that graphing utilities draw. When you use a graphing utility to graph the function in Example 5 over an interval containing 0, you will most likely obtain an incorrect graph such as that shown in Figure 1.11. The reason that a graphing utility can't show the correct graph is that the graph has infinitely many oscillations over any interval that contains 0.

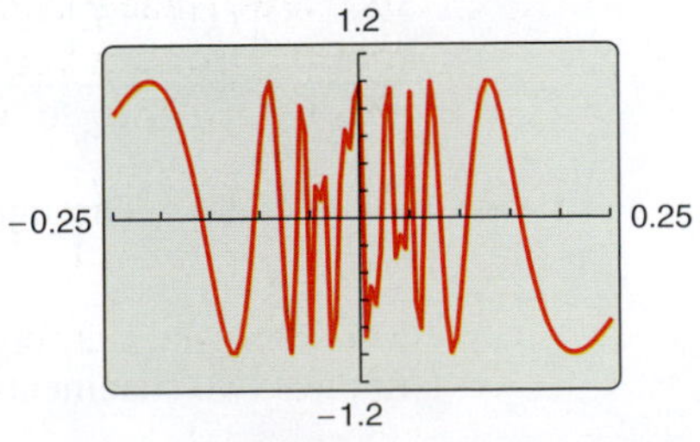

Incorrect graph of $f(x) = \sin(1/x)$

Figure 1.11

FOR FURTHER INFORMATION For more on the introduction of rigor to calculus, see "Who Gave You the Epsilon? Cauchy and the Origins of Rigorous Calculus" by Judith V. Grabiner in *The American Mathematical Monthly*. To view this article, go to *MathArticles.com*.

A Formal Definition of Limit

Consider again the informal definition of limit. If $f(x)$ becomes arbitrarily close to a single number L as x approaches c from either side, then the limit of $f(x)$ as x approaches c is L, written as

$$\lim_{x\to c} f(x) = L.$$

At first glance, this definition looks fairly technical. Even so, it is informal because exact meanings have not yet been given to the two phrases

"$f(x)$ becomes arbitrarily close to L"

and

"x approaches c."

The first person to assign mathematically rigorous meanings to these two phrases was Augustin-Louis Cauchy. His **ε-δ definition of limit** is the standard used today.

In Figure 1.12, let ε (the lowercase Greek letter *epsilon*) represent a (small) positive number. Then the phrase "$f(x)$ becomes arbitrarily close to L" means that $f(x)$ lies in the interval $(L - \varepsilon, L + \varepsilon)$. Using absolute value, you can write this as

$$|f(x) - L| < \varepsilon.$$

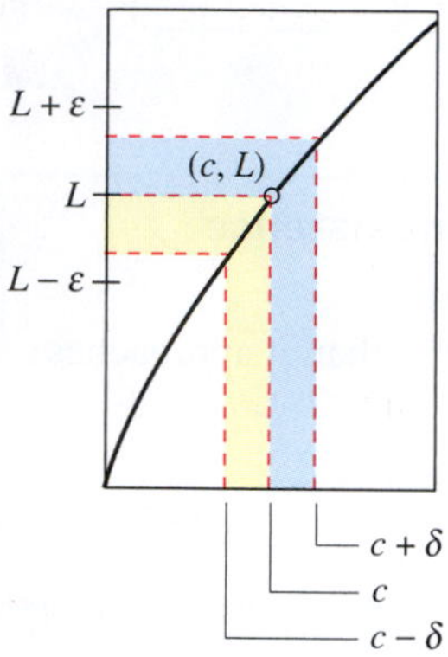

The ε-δ definition of the limit of $f(x)$ as x approaches c
Figure 1.12

Similarly, the phrase "x approaches c" means that there exists a positive number δ such that x lies in either the interval $(c - \delta, c)$ or the interval $(c, c + \delta)$. This fact can be concisely expressed by the double inequality

$$0 < |x - c| < \delta.$$

The first inequality

$$0 < |x - c| \qquad \text{The distance between } x \text{ and } c \text{ is more than 0.}$$

expresses the fact that $x \neq c$. The second inequality

$$|x - c| < \delta \qquad x \text{ is within } \delta \text{ units of } c.$$

says that x is within a distance δ of c.

Definition of Limit

Let f be a function defined on an open interval containing c (except possibly at c), and let L be a real number. The statement

$$\lim_{x\to c} f(x) = L$$

means that for each $\varepsilon > 0$ there exists a $\delta > 0$ such that if

$$0 < |x - c| < \delta$$

then

$$|f(x) - L| < \varepsilon.$$

REMARK Throughout this text, the expression

$$\lim_{x\to c} f(x) = L$$

implies two statements—the limit exists *and* the limit is L.

Some functions do not have limits as x approaches c, but those that do cannot have two different limits as x approaches c. That is, *if the limit of a function exists, then the limit is unique* (see Exercise 75).

The next three examples should help you develop a better understanding of the ε-δ definition of limit.

EXAMPLE 6 Finding a δ for a Given ε

Given the limit

$$\lim_{x\to 3}(2x-5)=1$$

find δ such that

$$|(2x-5)-1|<0.01$$

whenever

$$0<|x-3|<\delta.$$

Solution In this problem, you are working with a given value of ε—namely, $\varepsilon = 0.01$. To find an appropriate δ, try to establish a connection between the absolute values

$$|(2x-5)-1| \quad \text{and} \quad |x-3|.$$

Notice that

$$|(2x-5)-1| = |2x-6| = 2|x-3|.$$

Because the inequality $|(2x-5)-1|<0.01$ is equivalent to $2|x-3|<0.01$, you can choose

$$\delta = \tfrac{1}{2}(0.01) = 0.005.$$

REMARK In Example 6, note that 0.005 is the *largest* value of δ that will guarantee

$$|(2x-5)-1|<0.01$$

whenever

$$0<|x-3|<\delta.$$

Any *smaller* positive value of δ would also work.

This choice works because

$$0<|x-3|<0.005$$

implies that

$$|(2x-5)-1| = 2|x-3| < 2(0.005) = 0.01.$$

As you can see in Figure 1.13, for x-values within 0.005 of 3 ($x \neq 3$), the values of $f(x)$ are within 0.01 of 1.

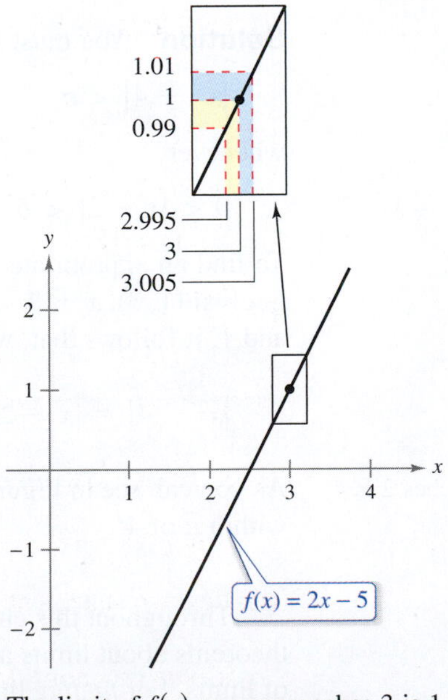

The limit of $f(x)$ as x approaches 3 is 1.
Figure 1.13

In Example 6, you found a δ-value for a *given* ε. This does not prove the existence of the limit. To do that, you must prove that you can find a δ for *any* ε, as shown in the next example.

EXAMPLE 7 Using the ε-δ Definition of Limit

Use the ε-δ definition of limit to prove that

$$\lim_{x \to 2} (3x - 2) = 4.$$

Solution You must show that for each $\varepsilon > 0$, there exists a $\delta > 0$ such that

$$|(3x - 2) - 4| < \varepsilon$$

whenever

$$0 < |x - 2| < \delta.$$

Because your choice of δ depends on ε, you need to establish a connection between the absolute values $|(3x - 2) - 4|$ and $|x - 2|$.

$$|(3x - 2) - 4| = |3x - 6| = 3|x - 2|$$

So, for a given $\varepsilon > 0$, you can choose $\delta = \varepsilon/3$. This choice works because

$$0 < |x - 2| < \delta = \frac{\varepsilon}{3}$$

implies that

$$|(3x - 2) - 4| = 3|x - 2| < 3\left(\frac{\varepsilon}{3}\right) = \varepsilon.$$

As you can see in Figure 1.14, for x-values within δ of 2 $(x \neq 2)$, the values of $f(x)$ are within ε of 4.

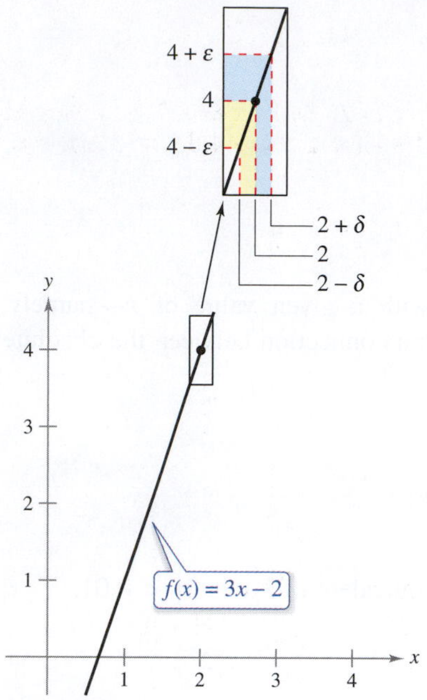

The limit of $f(x)$ as x approaches 2 is 4.
Figure 1.14

EXAMPLE 8 Using the ε-δ Definition of Limit

Use the ε-δ definition of limit to prove that

$$\lim_{x \to 2} x^2 = 4.$$

Solution You must show that for each $\varepsilon > 0$, there exists a $\delta > 0$ such that

$$|x^2 - 4| < \varepsilon$$

whenever

$$0 < |x - 2| < \delta.$$

To find an appropriate δ, begin by writing $|x^2 - 4| = |x - 2||x + 2|$. For all x in the interval $(1, 3)$, $x + 2 < 5$ and thus $|x + 2| < 5$. So, letting δ be the minimum of $\varepsilon/5$ and 1, it follows that, whenever $0 < |x - 2| < \delta$, you have

$$|x^2 - 4| = |x - 2||x + 2| < \left(\frac{\varepsilon}{5}\right)(5) = \varepsilon.$$

As you can see in Figure 1.15, for x-values within δ of 2 $(x \neq 2)$, the values of $f(x)$ are within ε of 4. ■

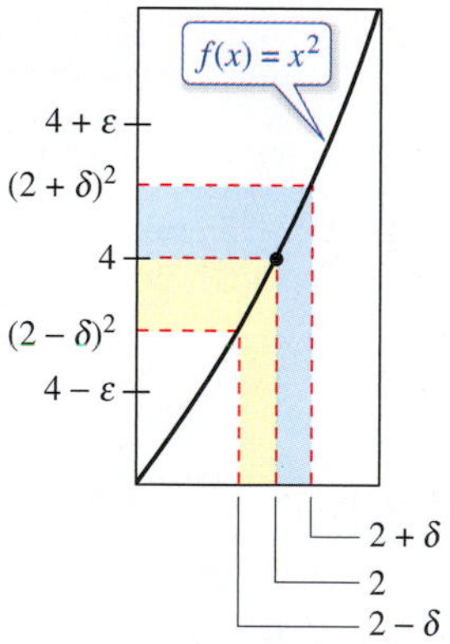

The limit of $f(x)$ as x approaches 2 is 4.
Figure 1.15

Throughout this chapter, you will use the ε-δ definition of limit primarily to prove theorems about limits and to establish the existence or nonexistence of particular types of limits. For *finding* limits, you will learn techniques that are easier to use than the ε-δ definition of limit.

1.2 Exercises

See CalcChat.com for tutorial help and worked-out solutions to odd-numbered exercises.

Estimating a Limit Numerically In Exercises 1–6, complete the table and use the result to estimate the limit. Use a graphing utility to graph the function to confirm your result.

1. $\lim_{x\to 4} \dfrac{x-4}{x^2-3x-4}$

x	3.9	3.99	3.999	4	4.001	4.01	4.1
$f(x)$				?			

2. $\lim_{x\to 3} \dfrac{x-3}{x^2-9}$

x	2.9	2.99	2.999	3	3.001	3.01	3.1
$f(x)$				?			

3. $\lim_{x\to 0} \dfrac{\sqrt{x+1}-1}{x}$

x	−0.1	−0.01	−0.001	0	0.001	0.01	0.1
$f(x)$				?			

4. $\lim_{x\to 3} \dfrac{[1/(x+1)]-(1/4)}{x-3}$

x	2.9	2.99	2.999	3	3.001	3.01	3.1
$f(x)$				?			

5. $\lim_{x\to 0} \dfrac{\sin x}{x}$

x	−0.1	−0.01	−0.001	0	0.001	0.01	0.1
$f(x)$				?			

6. $\lim_{x\to 0} \dfrac{\cos x - 1}{x}$

x	−0.1	−0.01	−0.001	0	0.001	0.01	0.1
$f(x)$				?			

Estimating a Limit Numerically In Exercises 7–14, create a table of values for the function and use the result to estimate the limit. Use a graphing utility to graph the function to confirm your result.

7. $\lim_{x\to 1} \dfrac{x-2}{x^2+x-6}$

8. $\lim_{x\to -4} \dfrac{x+4}{x^2+9x+20}$

9. $\lim_{x\to 1} \dfrac{x^4-1}{x^6-1}$

10. $\lim_{x\to -3} \dfrac{x^3+27}{x+3}$

11. $\lim_{x\to -6} \dfrac{\sqrt{10-x}-4}{x+6}$

12. $\lim_{x\to 2} \dfrac{[x/(x+1)]-(2/3)}{x-2}$

13. $\lim_{x\to 0} \dfrac{\sin 2x}{x}$

14. $\lim_{x\to 0} \dfrac{\tan x}{\tan 2x}$

Finding a Limit Graphically In Exercises 15–22, use the graph to find the limit (if it exists). If the limit does not exist, explain why.

15. $\lim_{x\to 3} (4-x)$

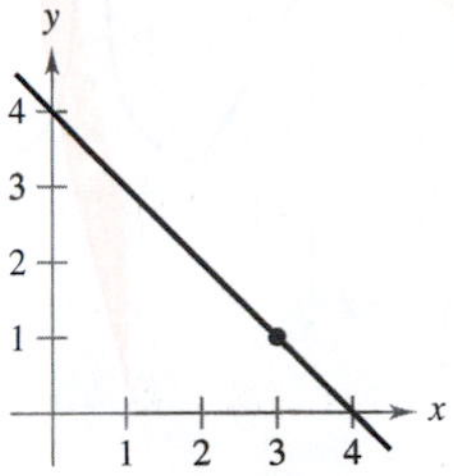

16. $\lim_{x\to 0} \sec x$

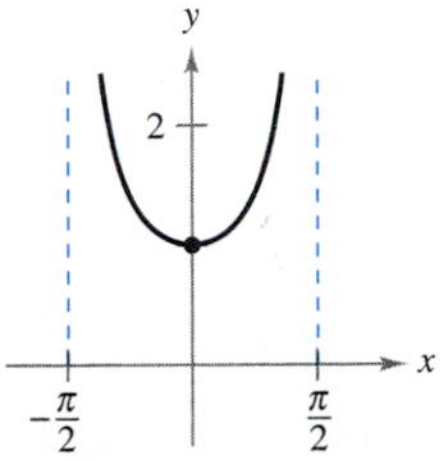

17. $\lim_{x\to 2} f(x)$

$$f(x) = \begin{cases} 4-x, & x \neq 2 \\ 0, & x = 2 \end{cases}$$

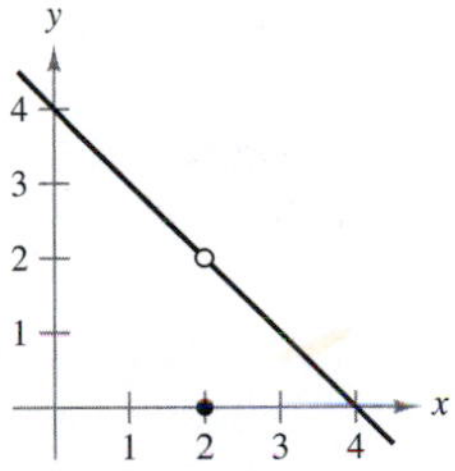

18. $\lim_{x\to 1} f(x)$

$$f(x) = \begin{cases} x^2+3, & x \neq 1 \\ 2, & x = 1 \end{cases}$$

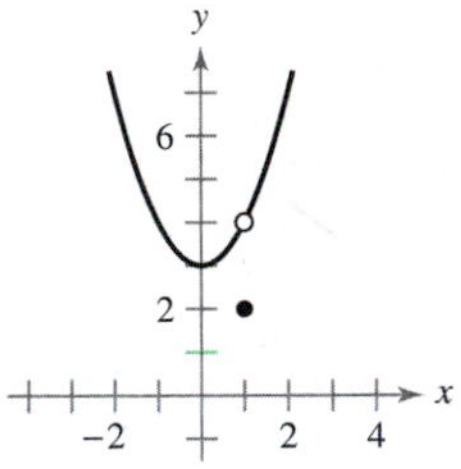

19. $\lim_{x\to 2} \dfrac{|x-2|}{x-2}$

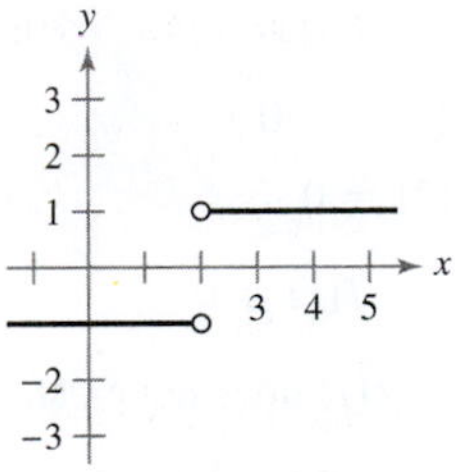

20. $\lim_{x\to 5} \dfrac{2}{x-5}$

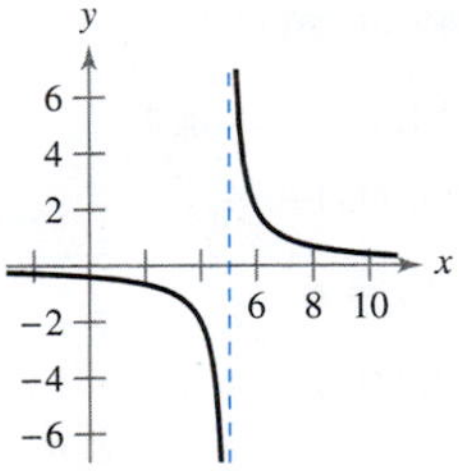

21. $\lim_{x\to 0} \cos \dfrac{1}{x}$

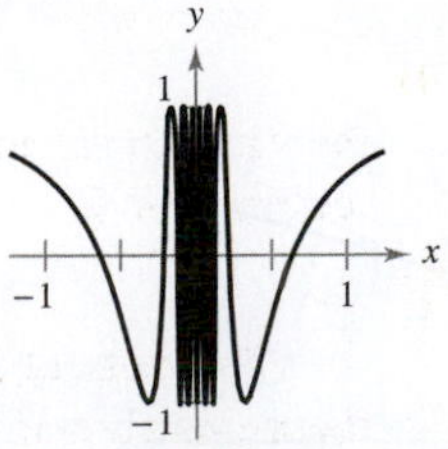

22. $\lim_{x\to \pi/2} \tan x$

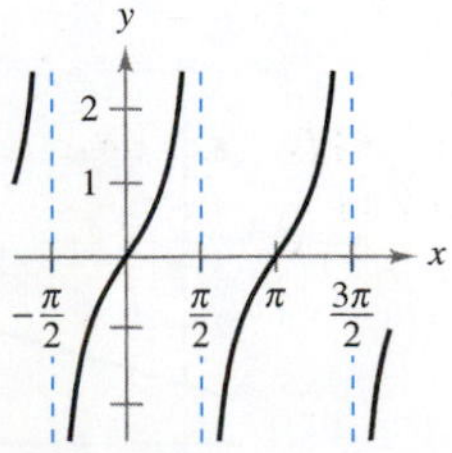

Graphical Reasoning **In Exercises 23 and 24, use the graph of the function f to decide whether the value of the given quantity exists. If it does, find it. If not, explain why.**

23. (a) $f(1)$

(b) $\lim_{x\to 1} f(x)$

(c) $f(4)$

(d) $\lim_{x\to 4} f(x)$

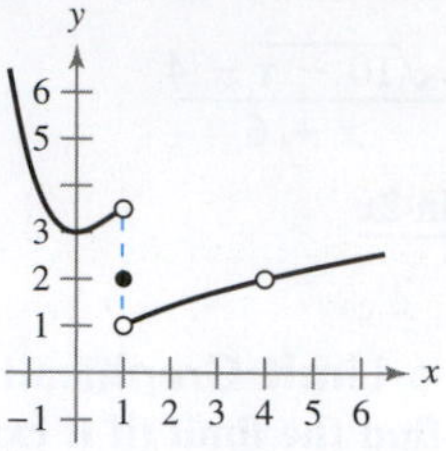

24. (a) $f(-2)$

(b) $\lim_{x\to -2} f(x)$

(c) $f(0)$

(d) $\lim_{x\to 0} f(x)$

(e) $f(2)$

(f) $\lim_{x\to 2} f(x)$

(g) $f(4)$

(h) $\lim_{x\to 4} f(x)$

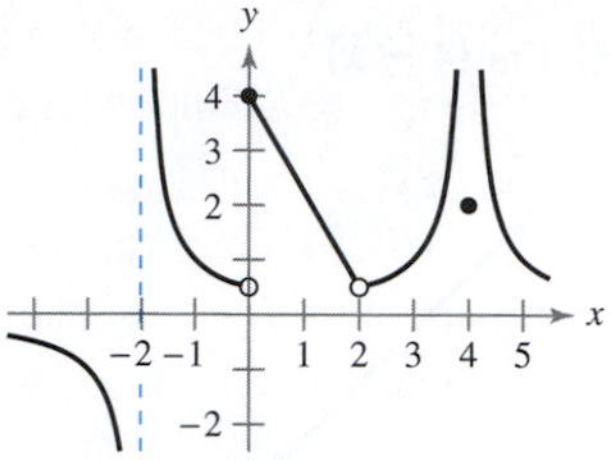

Limits of a Piecewise Function **In Exercises 25 and 26, sketch the graph of f. Then identify the values of c for which $\lim_{x\to c} f(x)$ exists.**

25. $f(x) = \begin{cases} x^2, & x \le 2 \\ 8 - 2x, & 2 < x < 4 \\ 4, & x \ge 4 \end{cases}$

26. $f(x) = \begin{cases} \sin x, & x < 0 \\ 1 - \cos x, & 0 \le x \le \pi \\ \cos x, & x > \pi \end{cases}$

Sketching a Graph **In Exercises 27 and 28, sketch a graph of a function f that satisfies the given values. (There are many correct answers.)**

27. $f(0)$ is undefined.

$\lim_{x\to 0} f(x) = 4$

$f(2) = 6$

$\lim_{x\to 2} f(x) = 3$

28. $f(-2) = 0$

$f(2) = 0$

$\lim_{x\to -2} f(x) = 0$

$\lim_{x\to 2} f(x)$ does not exist.

29. Finding a δ for a Given ε The graph of $f(x) = x + 1$ is shown in the figure. Find δ such that if $0 < |x - 2| < \delta$, then $|f(x) - 3| < 0.4$.

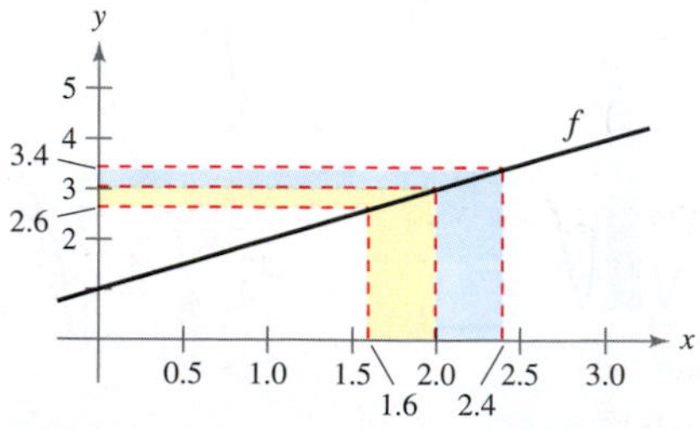

30. Finding a δ for a Given ε The graph of

$$f(x) = \frac{1}{x - 1}$$

is shown in the figure. Find δ such that if $0 < |x - 2| < \delta$, then $|f(x) - 1| < 0.01$.

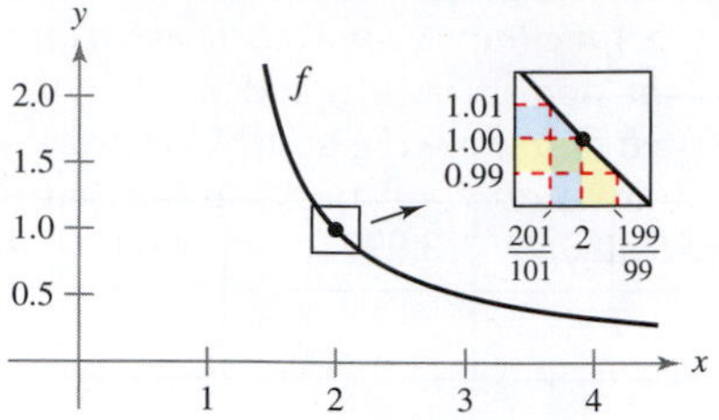

31. Finding a δ for a Given ε The graph of

$$f(x) = 2 - \frac{1}{x}$$

is shown in the figure. Find δ such that if $0 < |x - 1| < \delta$, then $|f(x) - 1| < 0.1$.

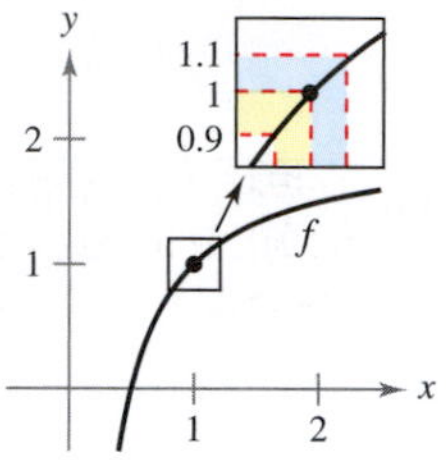

Figure for 31

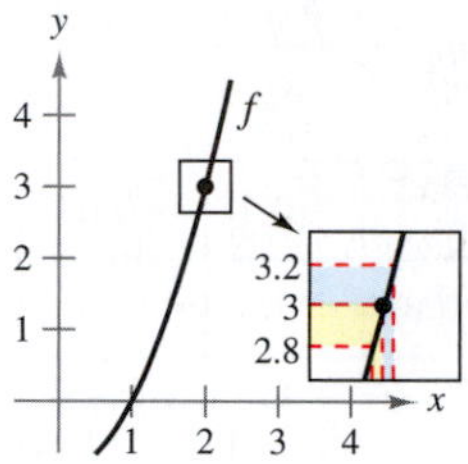

Figure for 32

32. Finding a δ for a Given ε The graph of

$$f(x) = x^2 - 1$$

is shown in the figure. Find δ such that if $0 < |x - 2| < \delta$, then $|f(x) - 3| < 0.2$.

Finding a δ for a Given ε **In Exercises 33–36, find the limit L. Then find $\delta > 0$ such that $|f(x) - L| < 0.01$ whenever $0 < |x - c| < \delta$.**

33. $\lim_{x\to 2} (3x + 2)$

34. $\lim_{x\to 6} \left(6 - \frac{x}{3}\right)$

35. $\lim_{x\to 2} (x^2 - 3)$

36. $\lim_{x\to 4} (x^2 + 6)$

Using the ε-δ Definition of Limit **In Exercises 37–48, find the limit L. Then use the ε-δ definition to prove that the limit is L.**

37. $\lim_{x\to 4} (x + 2)$

38. $\lim_{x\to -2} (4x + 5)$

39. $\lim_{x\to -4} \left(\frac{1}{2}x - 1\right)$

40. $\lim_{x\to 3} \left(\frac{3}{4}x + 1\right)$

41. $\lim_{x\to 6} 3$

42. $\lim_{x\to 2} (-1)$

43. $\lim_{x\to 0} \sqrt[3]{x}$

44. $\lim_{x\to 4} \sqrt{x}$

45. $\lim_{x\to -5} |x - 5|$

46. $\lim_{x\to 3} |x - 3|$

47. $\lim_{x\to 1} (x^2 + 1)$

48. $\lim_{x\to -4} (x^2 + 4x)$

49. Finding a Limit What is the limit of $f(x) = 4$ as x approaches π?

50. Finding a Limit What is the limit of $g(x) = x$ as x approaches π?

Writing In Exercises 51–54, use a graphing utility to graph the function and estimate the limit (if it exists). What is the domain of the function? Can you detect a possible error in determining the domain of a function solely by analyzing the graph generated by a graphing utility? Write a short paragraph about the importance of examining a function analytically as well as graphically.

51. $f(x) = \dfrac{\sqrt{x+5}-3}{x-4}$

$\lim\limits_{x \to 4} f(x)$

52. $f(x) = \dfrac{x-3}{x^2-4x+3}$

$\lim\limits_{x \to 3} f(x)$

53. $f(x) = \dfrac{x-9}{\sqrt{x}-3}$

$\lim\limits_{x \to 9} f(x)$

54. $f(x) = \dfrac{x-3}{x^2-9}$

$\lim\limits_{x \to 3} f(x)$

55. Modeling Data For a long distance phone call, a hotel charges \$9.99 for the first minute and \$0.79 for each additional minute or fraction thereof. A formula for the cost is given by

$C(t) = 9.99 - 0.79[\![-(t-1)]\!]$

where t is the time in minutes.

(*Note:* $[\![x]\!]$ = greatest integer n such that $n \le x$. For example, $[\![3.2]\!] = 3$ and $[\![-1.6]\!] = -2$.)

(a) Use a graphing utility to graph the cost function for $0 < t \le 6$.

(b) Use the graph to complete the table and observe the behavior of the function as t approaches 3.5. Use the graph and the table to find $\lim\limits_{t \to 3.5} C(t)$.

t	3	3.3	3.4	3.5	3.6	3.7	4
C				?			

(c) Use the graph to complete the table and observe the behavior of the function as t approaches 3.

t	2	2.5	2.9	3	3.1	3.5	4
C				?			

Does the limit of $C(t)$ as t approaches 3 exist? Explain.

56. Repeat Exercise 55 for

$C(t) = 5.79 - 0.99[\![-(t-1)]\!]$.

WRITING ABOUT CONCEPTS

57. Describing Notation Write a brief description of the meaning of the notation

$\lim\limits_{x \to 8} f(x) = 25$.

58. Using the Definition of Limit The definition of limit on page 52 requires that f is a function defined on an open interval containing c, except possibly at c. Why is this requirement necessary?

59. Limits That Fail to Exist Identify three types of behavior associated with the nonexistence of a limit. Illustrate each type with a graph of a function.

60. Comparing Functions and Limits

(a) If $f(2) = 4$, can you conclude anything about the limit of $f(x)$ as x approaches 2? Explain your reasoning.

(b) If the limit of $f(x)$ as x approaches 2 is 4, can you conclude anything about $f(2)$? Explain your reasoning.

61. Jewelry A jeweler resizes a ring so that its inner circumference is 6 centimeters.

(a) What is the radius of the ring?

(b) The inner circumference of the ring varies between 5.5 centimeters and 6.5 centimeters. How does the radius vary?

(c) Use the ε-δ definition of limit to describe this situation. Identify ε and δ.

62. Sports

A sporting goods manufacturer designs a golf ball having a volume of 2.48 cubic inches.

(a) What is the radius of the golf ball?

(b) The volume of the golf ball varies between 2.45 cubic inches and 2.51 cubic inches. How does the radius vary?

(c) Use the ε-δ definition of limit to describe this situation. Identify ε and δ.

63. Estimating a Limit Consider the function

$f(x) = (1+x)^{1/x}$.

Estimate

$\lim\limits_{x \to 0} (1+x)^{1/x}$

by evaluating f at x-values near 0. Sketch the graph of f.

The symbol indicates an exercise in which you are instructed to use graphing technology or a symbolic computer algebra system. The solutions of other exercises may also be facilitated by the use of appropriate technology.

64. Estimating a Limit Consider the function

$$f(x) = \frac{|x+1| - |x-1|}{x}.$$

Estimate

$$\lim_{x\to 0} \frac{|x+1| - |x-1|}{x}$$

by evaluating f at x-values near 0. Sketch the graph of f.

65. Graphical Analysis The statement

$$\lim_{x\to 2} \frac{x^2-4}{x-2} = 4$$

means that for each $\varepsilon > 0$ there corresponds a $\delta > 0$ such that if $0 < |x-2| < \delta$, then

$$\left|\frac{x^2-4}{x-2} - 4\right| < \varepsilon.$$

If $\varepsilon = 0.001$, then

$$\left|\frac{x^2-4}{x-2} - 4\right| < 0.001.$$

Use a graphing utility to graph each side of this inequality. Use the *zoom* feature to find an interval $(2-\delta, 2+\delta)$ such that the graph of the left side is below the graph of the right side of the inequality.

66. HOW DO YOU SEE IT? Use the graph of f to identify the values of c for which $\lim_{x\to c} f(x)$ exists.

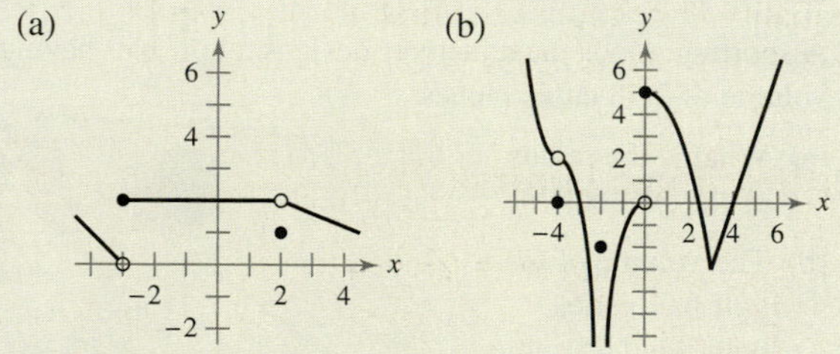

True or False? **In Exercises 67–70, determine whether the statement is true or false. If it is false, explain why or give an example that shows it is false.**

67. If f is undefined at $x = c$, then the limit of $f(x)$ as x approaches c does not exist.

68. If the limit of $f(x)$ as x approaches c is 0, then there must exist a number k such that $f(k) < 0.001$.

69. If $f(c) = L$, then $\lim_{x\to c} f(x) = L$.

70. If $\lim_{x\to c} f(x) = L$, then $f(c) = L$.

Determining a Limit **In Exercises 71 and 72, consider the function $f(x) = \sqrt{x}$.**

71. Is $\lim_{x\to 0.25} \sqrt{x} = 0.5$ a true statement? Explain.

72. Is $\lim_{x\to 0} \sqrt{x} = 0$ a true statement? Explain.

73. Evaluating Limits Use a graphing utility to evaluate

$$\lim_{x\to 0} \frac{\sin nx}{x}$$

for several values of n. What do you notice?

74. Evaluating Limits Use a graphing utility to evaluate

$$\lim_{x\to 0} \frac{\tan nx}{x}$$

for several values of n. What do you notice?

75. Proof Prove that if the limit of $f(x)$ as x approaches c exists, then the limit must be unique. [*Hint:* Let $\lim_{x\to c} f(x) = L_1$ and $\lim_{x\to c} f(x) = L_2$ and prove that $L_1 = L_2$.]

76. Proof Consider the line $f(x) = mx + b$, where $m \neq 0$. Use the ε-δ definition of limit to prove that $\lim_{x\to c} f(x) = mc + b$.

77. Proof Prove that

$$\lim_{x\to c} f(x) = L$$

is equivalent to

$$\lim_{x\to c} [f(x) - L] = 0.$$

78. Proof

(a) Given that

$$\lim_{x\to 0} (3x+1)(3x-1)x^2 + 0.01 = 0.01$$

prove that there exists an open interval (a, b) containing 0 such that $(3x+1)(3x-1)x^2 + 0.01 > 0$ for all $x \neq 0$ in (a, b).

(b) Given that $\lim_{x\to c} g(x) = L$, where $L > 0$, prove that there exists an open interval (a, b) containing c such that $g(x) > 0$ for all $x \neq c$ in (a, b).

PUTNAM EXAM CHALLENGE

79. Inscribe a rectangle of base b and height h in a circle of radius one, and inscribe an isosceles triangle in a region of the circle cut off by one base of the rectangle (with that side as the base of the triangle). For what value of h do the rectangle and triangle have the same area?

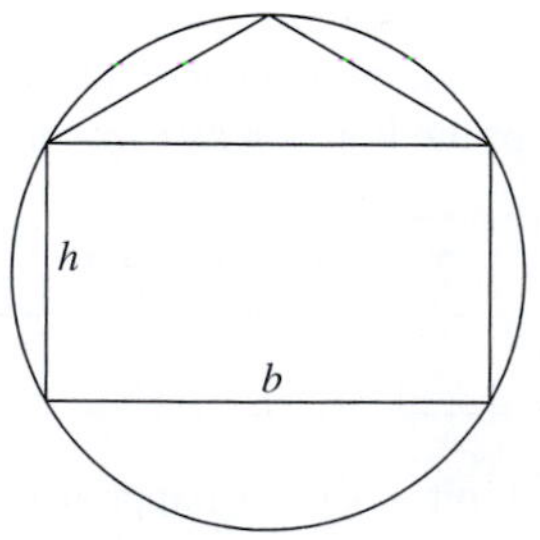

80. A right circular cone has base of radius 1 and height 3. A cube is inscribed in the cone so that one face of the cube is contained in the base of the cone. What is the side-length of the cube?

1.3 Evaluating Limits Analytically

- Evaluate a limit using properties of limits.
- Develop and use a strategy for finding limits.
- Evaluate a limit using the dividing out technique.
- Evaluate a limit using the rationalizing technique.
- Evaluate a limit using the Squeeze Theorem.

Properties of Limits

In Section 1.2, you learned that the limit of $f(x)$ as x approaches c does not depend on the value of f at $x = c$. It may happen, however, that the limit is precisely $f(c)$. In such cases, the limit can be evaluated by **direct substitution.** That is,

$$\lim_{x\to c} f(x) = f(c). \qquad \text{Substitute } c \text{ for } x.$$

Such *well-behaved* functions are **continuous at *c*.** You will examine this concept more closely in Section 1.4.

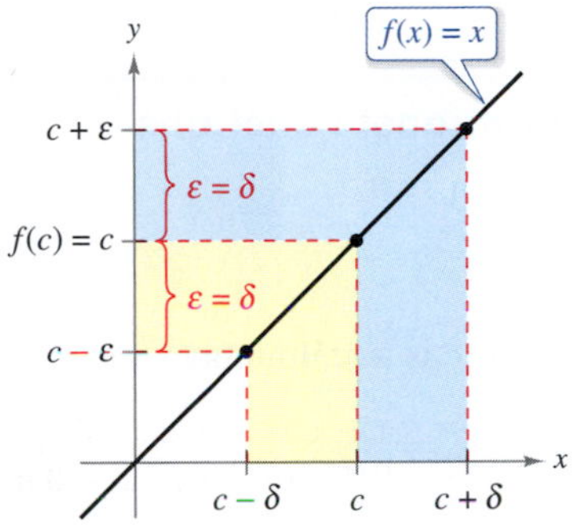

Figure 1.16

THEOREM 1.1 Some Basic Limits

Let b and c be real numbers, and let n be a positive integer.

1. $\lim_{x\to c} b = b$ 2. $\lim_{x\to c} x = c$ 3. $\lim_{x\to c} x^n = c^n$

Proof The proofs of Properties 1 and 3 of Theorem 1.1 are left as exercises (see Exercises 107 and 108). To prove Property 2, you need to show that for each $\varepsilon > 0$ there exists a $\delta > 0$ such that $|x - c| < \varepsilon$ whenever $0 < |x - c| < \delta$. To do this, choose $\delta = \varepsilon$. The second inequality then implies the first, as shown in Figure 1.16.

See LarsonCalculus.com for Bruce Edwards's video of this proof.

REMARK When encountering new notations or symbols in mathematics, be sure you know how the notations are read. For instance, the limit in Example 1(c) is read as "the limit of x^2 as x approaches 2 is 4."

EXAMPLE 1 Evaluating Basic Limits

a. $\lim_{x\to 2} 3 = 3$ **b.** $\lim_{x\to -4} x = -4$ **c.** $\lim_{x\to 2} x^2 = 2^2 = 4$

REMARK The proof of Property 1 is left as an exercise (see Exercise 109).

THEOREM 1.2 Properties of Limits

Let b and c be real numbers, let n be a positive integer, and let f and g be functions with the limits

$$\lim_{x\to c} f(x) = L \quad \text{and} \quad \lim_{x\to c} g(x) = K.$$

1. Scalar multiple: $\lim_{x\to c} [b f(x)] = bL$
2. Sum or difference: $\lim_{x\to c} [f(x) \pm g(x)] = L \pm K$
3. Product: $\lim_{x\to c} [f(x)g(x)] = LK$
4. Quotient: $\lim_{x\to c} \frac{f(x)}{g(x)} = \frac{L}{K}, \quad K \neq 0$
5. Power: $\lim_{x\to c} [f(x)]^n = L^n$

A proof of this theorem is given in Appendix A.

See LarsonCalculus.com for Bruce Edwards's video of this proof.

EXAMPLE 2 The Limit of a Polynomial

Find the limit: $\lim_{x\to 2}(4x^2 + 3)$.

Solution

$$\begin{aligned}\lim_{x\to 2}(4x^2+3) &= \lim_{x\to 2} 4x^2 + \lim_{x\to 2} 3 && \text{Property 2, Theorem 1.2}\\ &= 4\left(\lim_{x\to 2} x^2\right) + \lim_{x\to 2} 3 && \text{Property 1, Theorem 1.2}\\ &= 4(2^2) + 3 && \text{Properties 1 and 3, Theorem 1.1}\\ &= 19 && \text{Simplify.}\end{aligned}$$

In Example 2, note that the limit (as x approaches 2) of the *polynomial function* $p(x) = 4x^2 + 3$ is simply the value of p at $x = 2$.

$$\lim_{x\to 2} p(x) = p(2) = 4(2^2) + 3 = 19$$

This *direct substitution* property is valid for all polynomial and rational functions with nonzero denominators.

AP* Tips

Algebraic limit evaluation methods are not explicitly tested on the free response section of the AP Exam. They may be helpful on some multiple choice questions.

THEOREM 1.3 Limits of Polynomial and Rational Functions

If p is a polynomial function and c is a real number, then

$$\lim_{x\to c} p(x) = p(c).$$

If r is a rational function given by $r(x) = p(x)/q(x)$ and c is a real number such that $q(c) \neq 0$, then

$$\lim_{x\to c} r(x) = r(c) = \frac{p(c)}{q(c)}.$$

EXAMPLE 3 The Limit of a Rational Function

Find the limit: $\lim_{x\to 1} \dfrac{x^2 + x + 2}{x + 1}$.

Solution Because the denominator is not 0 when $x = 1$, you can apply Theorem 1.3 to obtain

$$\lim_{x\to 1} \frac{x^2 + x + 2}{x + 1} = \frac{1^2 + 1 + 2}{1 + 1} = \frac{4}{2} = 2.$$

Polynomial functions and rational functions are two of the three basic types of algebraic functions. The next theorem deals with the limit of the third type of algebraic function—one that involves a radical.

THE SQUARE ROOT SYMBOL

The first use of a symbol to denote the square root can be traced to the sixteenth century. Mathematicians first used the symbol $\surd$, which had only two strokes. This symbol was chosen because it resembled a lowercase *r*, to stand for the Latin word *radix*, meaning root.

THEOREM 1.4 The Limit of a Function Involving a Radical

Let n be a positive integer. The limit below is valid for all c when n is odd, and is valid for $c > 0$ when n is even.

$$\lim_{x\to c} \sqrt[n]{x} = \sqrt[n]{c}$$

A proof of this theorem is given in Appendix A.

See LarsonCalculus.com for Bruce Edwards's video of this proof.

The next theorem greatly expands your ability to evaluate limits because it shows how to analyze the limit of a composite function.

THEOREM 1.5 The Limit of a Composite Function

If f and g are functions such that $\lim_{x\to c} g(x) = L$ and $\lim_{x\to L} f(x) = f(L)$, then

$$\lim_{x\to c} f(g(x)) = f\left(\lim_{x\to c} g(x)\right) = f(L).$$

A proof of this theorem is given in Appendix A.

See LarsonCalculus.com for Bruce Edwards's video of this proof.

EXAMPLE 4 The Limit of a Composite Function

See LarsonCalculus.com for an interactive version of this type of example.

Find the limit.

a. $\lim_{x\to 0} \sqrt{x^2+4}$ **b.** $\lim_{x\to 3} \sqrt[3]{2x^2-10}$

Solution

a. Because

$$\lim_{x\to 0} (x^2+4) = 0^2 + 4 = 4 \quad \text{and} \quad \lim_{x\to 4} \sqrt{x} = \sqrt{4} = 2$$

you can conclude that

$$\lim_{x\to 0} \sqrt{x^2+4} = \sqrt{4} = 2.$$

b. Because

$$\lim_{x\to 3} (2x^2 - 10) = 2(3^2) - 10 = 8 \quad \text{and} \quad \lim_{x\to 8} \sqrt[3]{x} = \sqrt[3]{8} = 2$$

you can conclude that

$$\lim_{x\to 3} \sqrt[3]{2x^2-10} = \sqrt[3]{8} = 2.$$

You have seen that the limits of many algebraic functions can be evaluated by direct substitution. The six basic trigonometric functions also exhibit this desirable quality, as shown in the next theorem (presented without proof).

THEOREM 1.6 Limits of Trigonometric Functions

Let c be a real number in the domain of the given trigonometric function.

1. $\lim_{x\to c} \sin x = \sin c$ **2.** $\lim_{x\to c} \cos x = \cos c$ **3.** $\lim_{x\to c} \tan x = \tan c$

4. $\lim_{x\to c} \cot x = \cot c$ **5.** $\lim_{x\to c} \sec x = \sec c$ **6.** $\lim_{x\to c} \csc x = \csc c$

EXAMPLE 5 Limits of Trigonometric Functions

a. $\lim_{x\to 0} \tan x = \tan(0) = 0$

b. $\lim_{x\to \pi} (x \cos x) = \left(\lim_{x\to \pi} x\right)\left(\lim_{x\to \pi} \cos x\right) = \pi \cos(\pi) = -\pi$

c. $\lim_{x\to 0} \sin^2 x = \lim_{x\to 0} (\sin x)^2 = 0^2 = 0$

A Strategy for Finding Limits

On the previous three pages, you studied several types of functions whose limits can be evaluated by direct substitution. This knowledge, together with the next theorem, can be used to develop a strategy for finding limits.

> **THEOREM 1.7 Functions That Agree at All but One Point**
>
> Let c be a real number, and let $f(x) = g(x)$ for all $x \neq c$ in an open interval containing c. If the limit of $g(x)$ as x approaches c exists, then the limit of $f(x)$ also exists and
>
> $$\lim_{x\to c} f(x) = \lim_{x\to c} g(x).$$
>
> A proof of this theorem is given in Appendix A.
>
> *See LarsonCalculus.com for Bruce Edwards's video of this proof.*

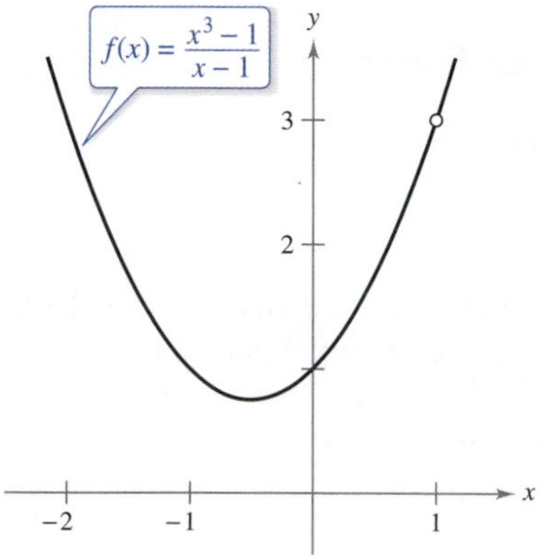

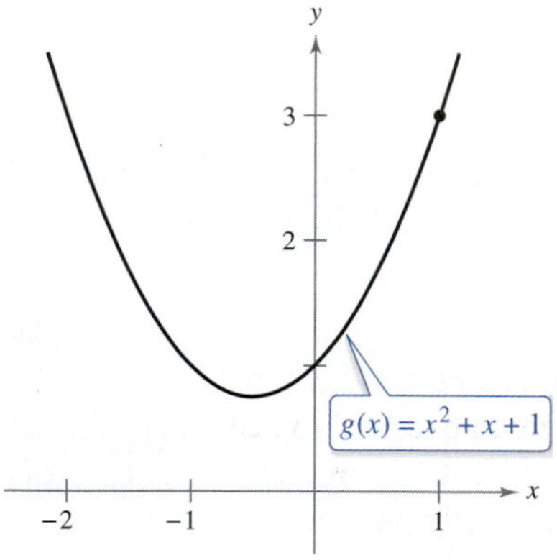

f and *g* agree at all but one point.
Figure 1.17

EXAMPLE 6 Finding the Limit of a Function

Find the limit.

$$\lim_{x\to 1} \frac{x^3 - 1}{x - 1}$$

Solution Let $f(x) = (x^3 - 1)/(x - 1)$. By factoring and dividing out like factors, you can rewrite f as

$$f(x) = \frac{\cancel{(x - 1)}(x^2 + x + 1)}{\cancel{(x - 1)}} = x^2 + x + 1 = g(x), \quad x \neq 1.$$

So, for all x-values other than $x = 1$, the functions f and g agree, as shown in Figure 1.17. Because $\lim_{x\to 1} g(x)$ exists, you can apply Theorem 1.7 to conclude that f and g have the same limit at $x = 1$.

$$\lim_{x\to 1} \frac{x^3 - 1}{x - 1} = \lim_{x\to 1} \frac{(x - 1)(x^2 + x + 1)}{x - 1} \qquad \text{Factor.}$$

$$= \lim_{x\to 1} \frac{\cancel{(x - 1)}(x^2 + x + 1)}{\cancel{x - 1}} \qquad \text{Divide out like factors.}$$

$$= \lim_{x\to 1} (x^2 + x + 1) \qquad \text{Apply Theorem 1.7.}$$

$$= 1^2 + 1 + 1 \qquad \text{Use direct substitution.}$$

$$= 3 \qquad \text{Simplify.}$$

REMARK When applying this strategy for finding a limit, remember that some functions do not have a limit (as x approaches c). For instance, the limit below does not exist.

$$\lim_{x\to 1} \frac{x^3 + 1}{x - 1}$$

> **A Strategy for Finding Limits**
>
> 1. Learn to recognize which limits can be evaluated by direct substitution. (These limits are listed in Theorems 1.1 through 1.6.)
> 2. When the limit of $f(x)$ as x approaches c *cannot* be evaluated by direct substitution, try to find a function g that agrees with f for all x other than $x = c$. [Choose g such that the limit of $g(x)$ *can* be evaluated by direct substitution.] Then apply Theorem 1.7 to conclude *analytically* that
>
> $$\lim_{x\to c} f(x) = \lim_{x\to c} g(x) = g(c).$$
>
> 3. Use a *graph* or *table* to reinforce your conclusion.

Dividing Out Technique

One procedure for finding a limit analytically is the **dividing out technique.** This technique involves dividing out common factors, as shown in Example 7.

EXAMPLE 7 Dividing Out Technique

See LarsonCalculus.com for an interactive version of this type of example.

Find the limit: $\lim_{x\to -3} \dfrac{x^2 + x - 6}{x + 3}$.

Solution Although you are taking the limit of a rational function, you *cannot* apply Theorem 1.3 because the limit of the denominator is 0.

$$\lim_{x\to -3} \frac{x^2 + x - 6}{x + 3} \quad \begin{cases} \lim_{x\to -3} (x^2 + x - 6) = 0 \\ \lim_{x\to -3} (x + 3) = 0 \end{cases}$$

Direct substitution fails.

Because the limit of the numerator is also 0, the numerator and denominator have a *common factor* of $(x + 3)$. So, for all $x \neq -3$, you can divide out this factor to obtain

$$f(x) = \frac{x^2 + x - 6}{x + 3} = \frac{\cancel{(x + 3)}(x - 2)}{\cancel{x + 3}} = x - 2 = g(x), \quad x \neq -3.$$

Using Theorem 1.7, it follows that

$$\lim_{x\to -3} \frac{x^2 + x - 6}{x + 3} = \lim_{x\to -3} (x - 2) \qquad \text{Apply Theorem 1.7.}$$

$$= -5. \qquad \text{Use direct substitution.}$$

This result is shown graphically in Figure 1.18. Note that the graph of the function f coincides with the graph of the function $g(x) = x - 2$, except that the graph of f has a gap at the point $(-3, -5)$.

REMARK In the solution to Example 7, be sure you see the usefulness of the Factor Theorem of Algebra. This theorem states that if c is a zero of a polynomial function, then $(x - c)$ is a factor of the polynomial. So, when you apply direct substitution to a rational function and obtain

$$r(c) = \frac{p(c)}{q(c)} = \frac{0}{0}$$

you can conclude that $(x - c)$ must be a common factor of both $p(x)$ and $q(x)$.

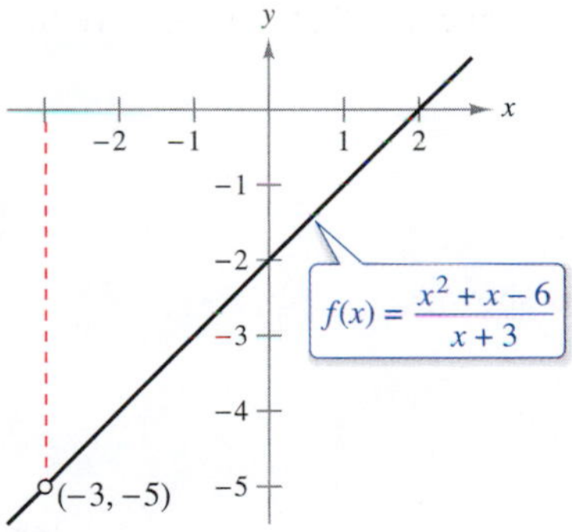

f is undefined when $x = -3$.
Figure 1.18

In Example 7, direct substitution produced the meaningless fractional form 0/0. An expression such as 0/0 is called an **indeterminate form** because you cannot (from the form alone) determine the limit. When you try to evaluate a limit and encounter this form, remember that you must rewrite the fraction so that the new denominator does not have 0 as its limit. One way to do this is to *divide out like factors*. Another way is to use the *rationalizing technique* shown on the next page.

TECHNOLOGY PITFALL A graphing utility can give misleading information about the graph of a function. For instance, try graphing the function from Example 7

$$f(x) = \frac{x^2 + x - 6}{x + 3}$$

on a standard viewing window (see Figure 1.19). On most graphing utilities, the graph appears to be defined at every real number. However, because f is undefined when $x = -3$, you know that the graph of f has a hole at $x = -3$. You can verify this on a graphing utility using the *trace* or *table* feature.

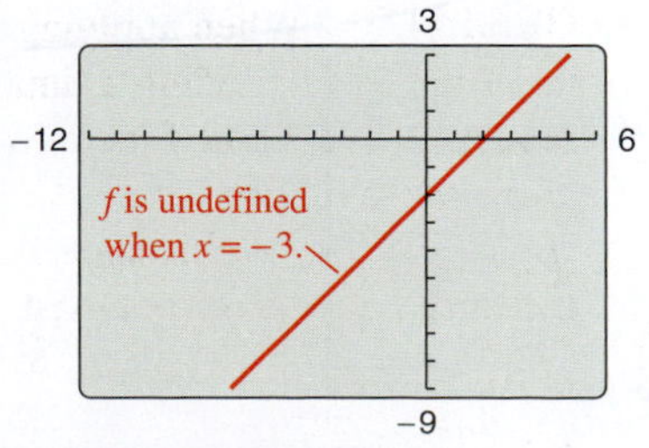

Misleading graph of *f*
Figure 1.19

Rationalizing Technique

Another way to find a limit analytically is the **rationalizing technique,** which involves rationalizing the numerator of a fractional expression. Recall that rationalizing the numerator means multiplying the numerator and denominator by the conjugate of the numerator. For instance, to rationalize the numerator of

$$\frac{\sqrt{x}+4}{x}$$

multiply the numerator and denominator by the conjugate of $\sqrt{x}+4$, which is

$$\sqrt{x}-4.$$

EXAMPLE 8 Rationalizing Technique

Find the limit: $\lim\limits_{x\to 0}\dfrac{\sqrt{x+1}-1}{x}$.

Solution By direct substitution, you obtain the indeterminate form 0/0.

$$\lim_{x\to 0}\frac{\sqrt{x+1}-1}{x} \quad \begin{cases} \lim\limits_{x\to 0}\left(\sqrt{x+1}-1\right)=0 \\ \lim\limits_{x\to 0} x = 0 \end{cases}$$

Direct substitution fails.

In this case, you can rewrite the fraction by rationalizing the numerator.

$$\begin{aligned}\frac{\sqrt{x+1}-1}{x} &= \left(\frac{\sqrt{x+1}-1}{x}\right)\left(\frac{\sqrt{x+1}+1}{\sqrt{x+1}+1}\right)\\ &= \frac{(x+1)-1}{x\left(\sqrt{x+1}+1\right)}\\ &= \frac{\cancel{x}}{\cancel{x}\left(\sqrt{x+1}+1\right)}\\ &= \frac{1}{\sqrt{x+1}+1}, \quad x \neq 0\end{aligned}$$

Now, using Theorem 1.7, you can evaluate the limit as shown.

$$\begin{aligned}\lim_{x\to 0}\frac{\sqrt{x+1}-1}{x} &= \lim_{x\to 0}\frac{1}{\sqrt{x+1}+1}\\ &= \frac{1}{1+1}\\ &= \frac{1}{2}\end{aligned}$$

A table or a graph can reinforce your conclusion that the limit is $\frac{1}{2}$. (See Figure 1.20.)

x approaches 0 from the left. → ← *x* approaches 0 from the right.

x	−0.25	−0.1	−0.01	−0.001	0	0.001	0.01	0.1	0.25
$f(x)$	0.5359	0.5132	0.5013	0.5001	?	0.4999	0.4988	0.4881	0.4721

$f(x)$ approaches 0.5. → ← $f(x)$ approaches 0.5.

REMARK The rationalizing technique for evaluating limits is based on multiplication by a convenient form of 1. In Example 8, the convenient form is

$$1 = \frac{\sqrt{x+1}+1}{\sqrt{x+1}+1}.$$

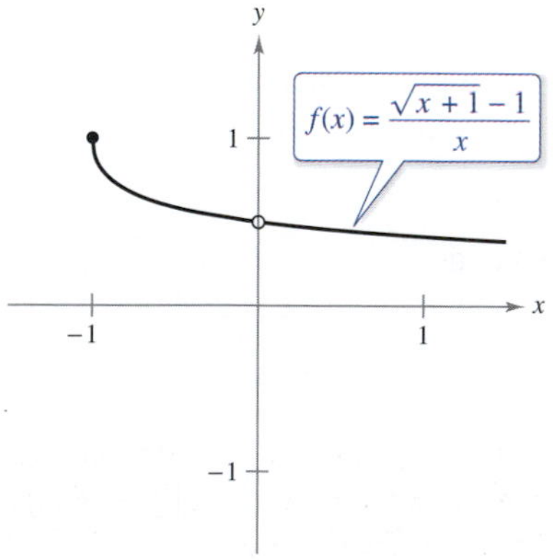

The limit of $f(x)$ as x approaches 0 is $\frac{1}{2}$.
Figure 1.20

The Squeeze Theorem

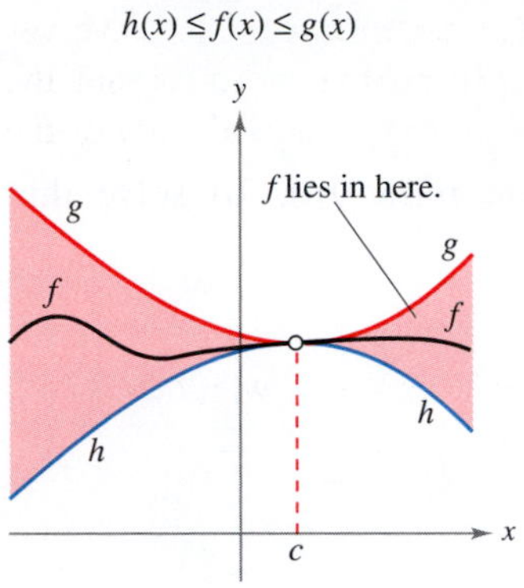

The Squeeze Theorem
Figure 1.21

The next theorem concerns the limit of a function that is squeezed between two other functions, each of which has the same limit at a given x-value, as shown in Figure 1.21.

> **THEOREM 1.8 The Squeeze Theorem**
>
> If $h(x) \le f(x) \le g(x)$ for all x in an open interval containing c, except possibly at c itself, and if
>
> $$\lim_{x\to c} h(x) = L = \lim_{x\to c} g(x)$$
>
> then $\lim_{x\to c} f(x)$ exists and is equal to L.
>
> A proof of this theorem is given in Appendix A.
>
> *See LarsonCalculus.com for Bruce Edwards's video of this proof.*

You can see the usefulness of the Squeeze Theorem (also called the Sandwich Theorem or the Pinching Theorem) in the proof of Theorem 1.9.

> **THEOREM 1.9 Two Special Trigonometric Limits**
>
> **1.** $\lim_{x\to 0} \dfrac{\sin x}{x} = 1$ **2.** $\lim_{x\to 0} \dfrac{1 - \cos x}{x} = 0$

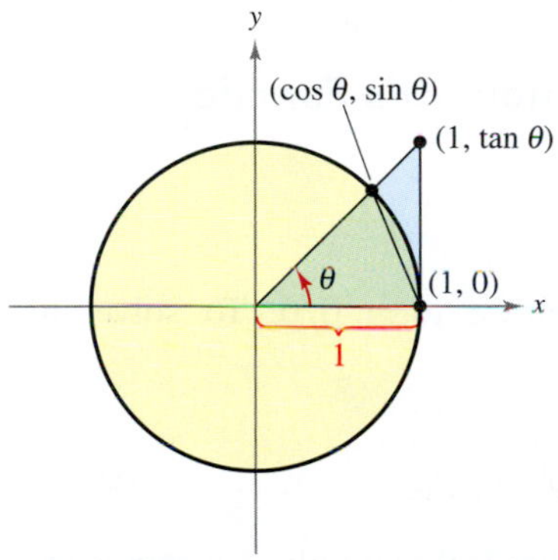

A circular sector is used to prove Theorem 1.9.
Figure 1.22

Proof The proof of the second limit is left as an exercise (see Exercise 121). To avoid the confusion of two different uses of x, the proof of the first limit is presented using the variable θ, where θ is an acute positive angle *measured in radians*. Figure 1.22 shows a circular sector that is squeezed between two triangles.

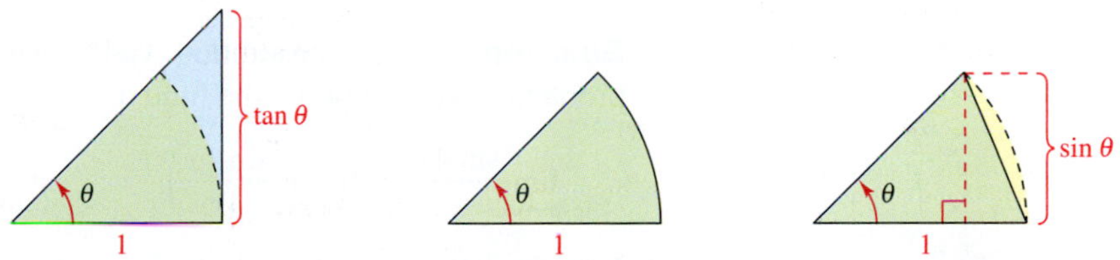

Area of triangle ≥ Area of sector ≥ Area of triangle

$$\frac{\tan\theta}{2} \ge \frac{\theta}{2} \ge \frac{\sin\theta}{2}$$

Multiplying each expression by $2/\sin\theta$ produces

$$\frac{1}{\cos\theta} \ge \frac{\theta}{\sin\theta} \ge 1$$

and taking reciprocals and reversing the inequalities yields

$$\cos\theta \le \frac{\sin\theta}{\theta} \le 1.$$

Because $\cos\theta = \cos(-\theta)$ and $(\sin\theta)/\theta = [\sin(-\theta)]/(-\theta)$, you can conclude that this inequality is valid for *all* nonzero θ in the open interval $(-\pi/2, \pi/2)$. Finally, because $\lim_{\theta\to 0} \cos\theta = 1$ and $\lim_{\theta\to 0} 1 = 1$, you can apply the Squeeze Theorem to conclude that $\lim_{\theta\to 0} (\sin\theta)/\theta = 1$. *See LarsonCalculus.com for Bruce Edwards's video of this proof.* ∎

EXAMPLE 9 A Limit Involving a Trigonometric Function

Find the limit: $\lim_{x\to 0} \dfrac{\tan x}{x}$.

Solution Direct substitution yields the indeterminate form 0/0. To solve this problem, you can write $\tan x$ as $(\sin x)/(\cos x)$ and obtain

$$\lim_{x\to 0} \frac{\tan x}{x} = \lim_{x\to 0} \left(\frac{\sin x}{x}\right)\left(\frac{1}{\cos x}\right).$$

Now, because

$$\lim_{x\to 0} \frac{\sin x}{x} = 1$$

and

$$\lim_{x\to 0} \frac{1}{\cos x} = 1$$

you can obtain

$$\begin{aligned}\lim_{x\to 0} \frac{\tan x}{x} &= \left(\lim_{x\to 0} \frac{\sin x}{x}\right)\left(\lim_{x\to 0} \frac{1}{\cos x}\right)\\ &= (1)(1)\\ &= 1.\end{aligned}$$

(See Figure 1.23.)

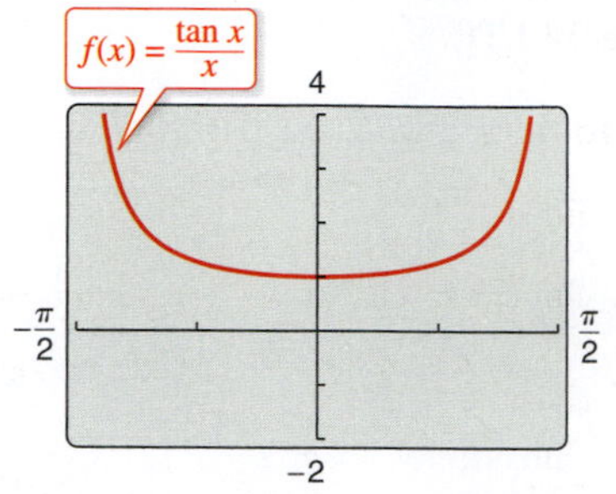

The limit of $f(x)$ as x approaches 0 is 1.
Figure 1.23

REMARK Be sure you understand the mathematical conventions regarding parentheses and trigonometric functions. For instance, in Example 10, $\sin 4x$ means $\sin(4x)$.

EXAMPLE 10 A Limit Involving a Trigonometric Function

Find the limit: $\lim_{x\to 0} \dfrac{\sin 4x}{x}$.

Solution Direct substitution yields the indeterminate form 0/0. To solve this problem, you can rewrite the limit as

$$\lim_{x\to 0} \frac{\sin 4x}{x} = 4\left(\lim_{x\to 0} \frac{\sin 4x}{4x}\right). \qquad \text{Multiply and divide by 4.}$$

Now, by letting $y = 4x$ and observing that x approaches 0 if and only if y approaches 0, you can write

$$\begin{aligned}\lim_{x\to 0} \frac{\sin 4x}{x} &= 4\left(\lim_{x\to 0} \frac{\sin 4x}{4x}\right)\\ &= 4\left(\lim_{y\to 0} \frac{\sin y}{y}\right) && \text{Let } y = 4x.\\ &= 4(1) && \text{Apply Theorem 1.9(1).}\\ &= 4.\end{aligned}$$

(See Figure 1.24.)

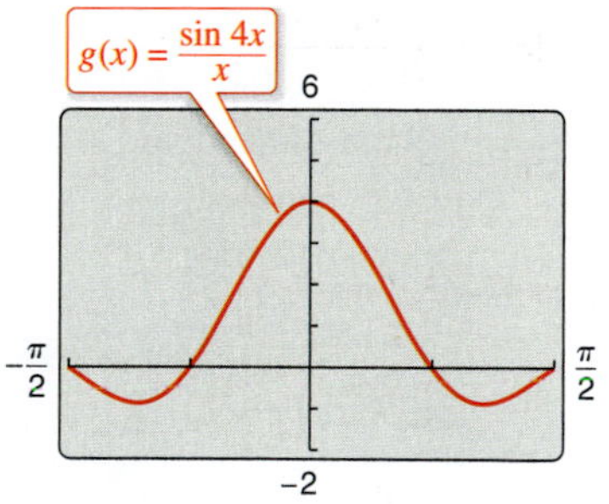

The limit of $g(x)$ as x approaches 0 is 4.
Figure 1.24

TECHNOLOGY Use a graphing utility to confirm the limits in the examples and in the exercise set. For instance, Figures 1.23 and 1.24 show the graphs of

$$f(x) = \frac{\tan x}{x} \quad \text{and} \quad g(x) = \frac{\sin 4x}{x}.$$

Note that the first graph appears to contain the point (0, 1) and the second graph appears to contain the point (0, 4), which lends support to the conclusions obtained in Examples 9 and 10.

1.3 Exercises

See CalcChat.com for tutorial help and worked-out solutions to odd-numbered exercises.

Estimating Limits In Exercises 1–4, use a graphing utility to graph the function and visually estimate the limits.

1. $h(x) = -x^2 + 4x$
   (a) $\lim_{x\to 4} h(x)$
   (b) $\lim_{x\to -1} h(x)$

2. $g(x) = \dfrac{12(\sqrt{x} - 3)}{x - 9}$
   (a) $\lim_{x\to 4} g(x)$
   (b) $\lim_{x\to 9} g(x)$

3. $f(x) = x\cos x$
   (a) $\lim_{x\to 0} f(x)$
   (b) $\lim_{x\to \pi/3} f(x)$

4. $f(t) = t|t - 4|$
   (a) $\lim_{t\to 4} f(t)$
   (b) $\lim_{t\to -1} f(t)$

Finding a Limit In Exercises 5–22, find the limit.

5. $\lim_{x\to 2} x^3$
6. $\lim_{x\to -3} x^4$
7. $\lim_{x\to 0} (2x - 1)$
8. $\lim_{x\to -4} (2x + 3)$
9. $\lim_{x\to -3} (x^2 + 3x)$
10. $\lim_{x\to 2} (-x^3 + 1)$
11. $\lim_{x\to -3} (2x^2 + 4x + 1)$
12. $\lim_{x\to 1} (2x^3 - 6x + 5)$
13. $\lim_{x\to 3} \sqrt{x + 1}$
14. $\lim_{x\to 2} \sqrt[3]{12x + 3}$
15. $\lim_{x\to -4} (x + 3)^2$
16. $\lim_{x\to 0} (3x - 2)^4$
17. $\lim_{x\to 2} \dfrac{1}{x}$
18. $\lim_{x\to -5} \dfrac{5}{x + 3}$
19. $\lim_{x\to 1} \dfrac{x}{x^2 + 4}$
20. $\lim_{x\to 1} \dfrac{3x + 5}{x + 1}$
21. $\lim_{x\to 7} \dfrac{3x}{\sqrt{x + 2}}$
22. $\lim_{x\to 3} \dfrac{\sqrt{x + 6}}{x + 2}$

Finding Limits In Exercises 23–26, find the limits.

23. $f(x) = 5 - x,\ g(x) = x^3$
   (a) $\lim_{x\to 1} f(x)$ (b) $\lim_{x\to 4} g(x)$ (c) $\lim_{x\to 1} g(f(x))$

24. $f(x) = x + 7,\ g(x) = x^2$
   (a) $\lim_{x\to -3} f(x)$ (b) $\lim_{x\to 4} g(x)$ (c) $\lim_{x\to -3} g(f(x))$

25. $f(x) = 4 - x^2,\ g(x) = \sqrt{x + 1}$
   (a) $\lim_{x\to 1} f(x)$ (b) $\lim_{x\to 3} g(x)$ (c) $\lim_{x\to 1} g(f(x))$

26. $f(x) = 2x^2 - 3x + 1,\ g(x) = \sqrt[3]{x + 6}$
   (a) $\lim_{x\to 4} f(x)$ (b) $\lim_{x\to 21} g(x)$ (c) $\lim_{x\to 4} g(f(x))$

Finding a Limit of a Trigonometric Function In Exercises 27–36, find the limit of the trigonometric function.

27. $\lim_{x\to \pi/2} \sin x$
28. $\lim_{x\to \pi} \tan x$
29. $\lim_{x\to 1} \cos \dfrac{\pi x}{3}$
30. $\lim_{x\to 2} \sin \dfrac{\pi x}{2}$
31. $\lim_{x\to 0} \sec 2x$
32. $\lim_{x\to \pi} \cos 3x$
33. $\lim_{x\to 5\pi/6} \sin x$
34. $\lim_{x\to 5\pi/3} \cos x$
35. $\lim_{x\to 3} \tan\left(\dfrac{\pi x}{4}\right)$
36. $\lim_{x\to 7} \sec\left(\dfrac{\pi x}{6}\right)$

Evaluating Limits In Exercises 37–40, use the information to evaluate the limits.

37. $\lim_{x\to c} f(x) = 3$
   $\lim_{x\to c} g(x) = 2$
   (a) $\lim_{x\to c} [5g(x)]$
   (b) $\lim_{x\to c} [f(x) + g(x)]$
   (c) $\lim_{x\to c} [f(x)g(x)]$
   (d) $\lim_{x\to c} \dfrac{f(x)}{g(x)}$

38. $\lim_{x\to c} f(x) = 2$
   $\lim_{x\to c} g(x) = \frac{3}{4}$
   (a) $\lim_{x\to c} [4f(x)]$
   (b) $\lim_{x\to c} [f(x) + g(x)]$
   (c) $\lim_{x\to c} [f(x)g(x)]$
   (d) $\lim_{x\to c} \dfrac{f(x)}{g(x)}$

39. $\lim_{x\to c} f(x) = 4$
   (a) $\lim_{x\to c} [f(x)]^3$
   (b) $\lim_{x\to c} \sqrt{f(x)}$
   (c) $\lim_{x\to c} [3f(x)]$
   (d) $\lim_{x\to c} [f(x)]^{3/2}$

40. $\lim_{x\to c} f(x) = 27$
   (a) $\lim_{x\to c} \sqrt[3]{f(x)}$
   (b) $\lim_{x\to c} \dfrac{f(x)}{18}$
   (c) $\lim_{x\to c} [f(x)]^2$
   (d) $\lim_{x\to c} [f(x)]^{2/3}$

Finding a Limit In Exercises 41–46, write a simpler function that agrees with the given function at all but one point. Then find the limit of the function. Use a graphing utility to confirm your result.

41. $\lim_{x\to 0} \dfrac{x^2 + 3x}{x}$
42. $\lim_{x\to 0} \dfrac{x^4 - 5x^2}{x^2}$
43. $\lim_{x\to -1} \dfrac{x^2 - 1}{x + 1}$
44. $\lim_{x\to -2} \dfrac{3x^2 + 5x - 2}{x + 2}$
45. $\lim_{x\to 2} \dfrac{x^3 - 8}{x - 2}$
46. $\lim_{x\to -1} \dfrac{x^3 + 1}{x + 1}$

Finding a Limit In Exercises 47–62, find the limit.

47. $\lim_{x\to 0} \dfrac{x}{x^2 - x}$
48. $\lim_{x\to 0} \dfrac{2x}{x^2 + 4x}$
49. $\lim_{x\to 4} \dfrac{x - 4}{x^2 - 16}$
50. $\lim_{x\to 5} \dfrac{5 - x}{x^2 - 25}$
51. $\lim_{x\to -3} \dfrac{x^2 + x - 6}{x^2 - 9}$
52. $\lim_{x\to 2} \dfrac{x^2 + 2x - 8}{x^2 - x - 2}$
53. $\lim_{x\to 4} \dfrac{\sqrt{x + 5} - 3}{x - 4}$
54. $\lim_{x\to 3} \dfrac{\sqrt{x + 1} - 2}{x - 3}$
55. $\lim_{x\to 0} \dfrac{\sqrt{x + 5} - \sqrt{5}}{x}$
56. $\lim_{x\to 0} \dfrac{\sqrt{2 + x} - \sqrt{2}}{x}$
57. $\lim_{x\to 0} \dfrac{[1/(3 + x)] - (1/3)}{x}$
58. $\lim_{x\to 0} \dfrac{[1/(x + 4)] - (1/4)}{x}$

59. $\lim\limits_{\Delta x \to 0} \dfrac{2(x + \Delta x) - 2x}{\Delta x}$

60. $\lim\limits_{\Delta x \to 0} \dfrac{(x + \Delta x)^2 - x^2}{\Delta x}$

61. $\lim\limits_{\Delta x \to 0} \dfrac{(x + \Delta x)^2 - 2(x + \Delta x) + 1 - (x^2 - 2x + 1)}{\Delta x}$

62. $\lim\limits_{\Delta x \to 0} \dfrac{(x + \Delta x)^3 - x^3}{\Delta x}$

Finding a Limit of a Trigonometric Function **In Exercises 63–74, find the limit of the trigonometric function.**

63. $\lim\limits_{x \to 0} \dfrac{\sin x}{5x}$

64. $\lim\limits_{x \to 0} \dfrac{3(1 - \cos x)}{x}$

65. $\lim\limits_{x \to 0} \dfrac{\sin x(1 - \cos x)}{x^2}$

66. $\lim\limits_{\theta \to 0} \dfrac{\cos \theta \tan \theta}{\theta}$

67. $\lim\limits_{x \to 0} \dfrac{\sin^2 x}{x}$

68. $\lim\limits_{x \to 0} \dfrac{\tan^2 x}{x}$

69. $\lim\limits_{h \to 0} \dfrac{(1 - \cos h)^2}{h}$

70. $\lim\limits_{\phi \to \pi} \phi \sec \phi$

71. $\lim\limits_{x \to \pi/2} \dfrac{\cos x}{\cot x}$

72. $\lim\limits_{x \to \pi/4} \dfrac{1 - \tan x}{\sin x - \cos x}$

73. $\lim\limits_{t \to 0} \dfrac{\sin 3t}{2t}$

74. $\lim\limits_{x \to 0} \dfrac{\sin 2x}{\sin 3x}$ $\left[\textit{Hint: } \text{Find } \lim\limits_{x \to 0} \left(\dfrac{2 \sin 2x}{2x}\right)\left(\dfrac{3x}{3 \sin 3x}\right).\right]$

Graphical, Numerical, and Analytic Analysis **In Exercises 75–82, use a graphing utility to graph the function and estimate the limit. Use a table to reinforce your conclusion. Then find the limit by analytic methods.**

75. $\lim\limits_{x \to 0} \dfrac{\sqrt{x + 2} - \sqrt{2}}{x}$

76. $\lim\limits_{x \to 16} \dfrac{4 - \sqrt{x}}{x - 16}$

77. $\lim\limits_{x \to 0} \dfrac{[1/(2 + x)] - (1/2)}{x}$

78. $\lim\limits_{x \to 2} \dfrac{x^5 - 32}{x - 2}$

79. $\lim\limits_{t \to 0} \dfrac{\sin 3t}{t}$

80. $\lim\limits_{x \to 0} \dfrac{\cos x - 1}{2x^2}$

81. $\lim\limits_{x \to 0} \dfrac{\sin x^2}{x}$

82. $\lim\limits_{x \to 0} \dfrac{\sin x}{\sqrt[3]{x}}$

Finding a Limit **In Exercises 83–88, find**

$$\lim_{\Delta x \to 0} \frac{f(x + \Delta x) - f(x)}{\Delta x}.$$

83. $f(x) = 3x - 2$

84. $f(x) = -6x + 3$

85. $f(x) = x^2 - 4x$

86. $f(x) = \sqrt{x}$

87. $f(x) = \dfrac{1}{x + 3}$

88. $f(x) = \dfrac{1}{x^2}$

Using the Squeeze Theorem **In Exercises 89 and 90, use the Squeeze Theorem to find $\lim\limits_{x \to c} f(x)$.**

89. $c = 0$

$4 - x^2 \le f(x) \le 4 + x^2$

90. $c = a$

$b - |x - a| \le f(x) \le b + |x - a|$

Using the Squeeze Theorem **In Exercises 91–94, use a graphing utility to graph the given function and the equations $y = |x|$ and $y = -|x|$ in the same viewing window. Using the graphs to observe the Squeeze Theorem visually, find $\lim\limits_{x \to 0} f(x)$.**

91. $f(x) = |x| \sin x$

92. $f(x) = |x| \cos x$

93. $f(x) = x \sin \dfrac{1}{x}$

94. $h(x) = x \cos \dfrac{1}{x}$

WRITING ABOUT CONCEPTS

95. Functions That Agree at All but One Point

(a) In the context of finding limits, discuss what is meant by two functions that agree at all but one point.

(b) Give an example of two functions that agree at all but one point.

96. Indeterminate Form What is meant by an indeterminate form?

97. Squeeze Theorem In your own words, explain the Squeeze Theorem.

98. HOW DO YOU SEE IT? Would you use the dividing out technique or the rationalizing technique to find the limit of the function? Explain your reasoning.

(a) $\lim\limits_{x \to -2} \dfrac{x^2 + x - 2}{x + 2}$

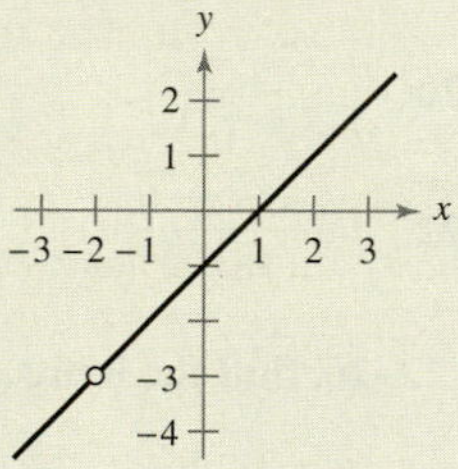

(b) $\lim\limits_{x \to 0} \dfrac{\sqrt{x + 4} - 2}{x}$

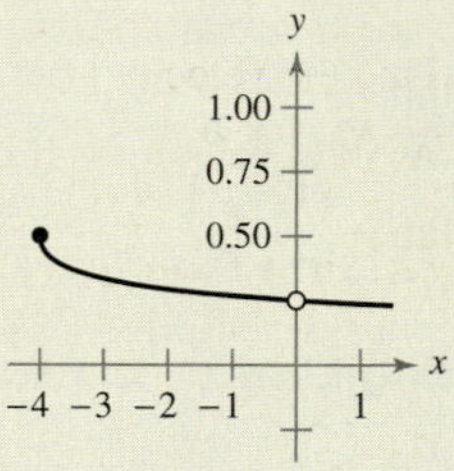

99. Writing Use a graphing utility to graph

$$f(x) = x, \quad g(x) = \sin x, \quad \text{and} \quad h(x) = \frac{\sin x}{x}$$

in the same viewing window. Compare the magnitudes of $f(x)$ and $g(x)$ when x is close to 0. Use the comparison to write a short paragraph explaining why

$$\lim_{x \to 0} h(x) = 1.$$

100. Writing Use a graphing utility to graph

$$f(x) = x, \quad g(x) = \sin^2 x, \quad \text{and} \quad h(x) = \frac{\sin^2 x}{x}$$

in the same viewing window. Compare the magnitudes of $f(x)$ and $g(x)$ when x is close to 0. Use the comparison to write a short paragraph explaining why

$$\lim_{x \to 0} h(x) = 0.$$

Free-Falling Object In Exercises 101 and 102, use the position function $s(t) = -16t^2 + 500$, which gives the height (in feet) of an object that has fallen for t seconds from a height of 500 feet. The velocity at time $t = a$ seconds is given by

$$\lim_{t \to a} \frac{s(a) - s(t)}{a - t}.$$

101. A construction worker drops a full paint can from a height of 500 feet. How fast will the paint can be falling after 2 seconds?

102. A construction worker drops a full paint can from a height of 500 feet. When will the paint can hit the ground? At what velocity will the paint can impact the ground?

Free-Falling Object In Exercises 103 and 104, use the position function $s(t) = -4.9t^2 + 200$, which gives the height (in meters) of an object that has fallen for t seconds from a height of 200 meters. The velocity at time $t = a$ seconds is given by

$$\lim_{t \to a} \frac{s(a) - s(t)}{a - t}.$$

103. Find the velocity of the object when $t = 3$.

104. At what velocity will the object impact the ground?

105. Finding Functions Find two functions f and g such that $\lim_{x \to 0} f(x)$ and $\lim_{x \to 0} g(x)$ do not exist, but

$$\lim_{x \to 0} [f(x) + g(x)]$$

does exist.

106. Proof Prove that if $\lim_{x \to c} f(x)$ exists and $\lim_{x \to c} [f(x) + g(x)]$ does not exist, then $\lim_{x \to c} g(x)$ does not exist.

107. Proof Prove Property 1 of Theorem 1.1.

108. Proof Prove Property 3 of Theorem 1.1. (You may use Property 3 of Theorem 1.2.)

109. Proof Prove Property 1 of Theorem 1.2.

110. Proof Prove that if $\lim_{x \to c} f(x) = 0$, then $\lim_{x \to c} |f(x)| = 0$.

111. Proof Prove that if $\lim_{x \to c} f(x) = 0$ and $|g(x)| \le M$ for a fixed number M and all $x \ne c$, then $\lim_{x \to c} f(x)g(x) = 0$.

112. Proof

(a) Prove that if $\lim_{x \to c} |f(x)| = 0$, then $\lim_{x \to c} f(x) = 0$.

(*Note:* This is the converse of Exercise 110.)

(b) Prove that if $\lim_{x \to c} f(x) = L$, then $\lim_{x \to c} |f(x)| = |L|$.

[*Hint:* Use the inequality $||f(x)| - |L|| \le |f(x) - L|$.]

113. Think About It Find a function f to show that the converse of Exercise 112(b) is not true. [*Hint:* Find a function f such that $\lim_{x \to c} |f(x)| = |L|$ but $\lim_{x \to c} f(x)$ does not exist.]

114. Think About It When using a graphing utility to generate a table to approximate

$$\lim_{x \to 0} \frac{\sin x}{x}$$

a student concluded that the limit was 0.01745 rather than 1. Determine the probable cause of the error.

True or False? In Exercises 115–120, determine whether the statement is true or false. If it is false, explain why or give an example that shows it is false.

115. $\lim_{x \to 0} \frac{|x|}{x} = 1$

116. $\lim_{x \to \pi} \frac{\sin x}{x} = 1$

117. If $f(x) = g(x)$ for all real numbers other than $x = 0$, and $\lim_{x \to 0} f(x) = L$, then $\lim_{x \to 0} g(x) = L$.

118. If $\lim_{x \to c} f(x) = L$, then $f(c) = L$.

119. $\lim_{x \to 2} f(x) = 3$, where $f(x) = \begin{cases} 3, & x \le 2 \\ 0, & x > 2 \end{cases}$

120. If $f(x) < g(x)$ for all $x \ne a$, then $\lim_{x \to a} f(x) < \lim_{x \to a} g(x)$.

121. Proof Prove the second part of Theorem 1.9.

$$\lim_{x \to 0} \frac{1 - \cos x}{x} = 0$$

122. Piecewise Functions Let

$$f(x) = \begin{cases} 0, & \text{if } x \text{ is rational} \\ 1, & \text{if } x \text{ is irrational} \end{cases}$$

and

$$g(x) = \begin{cases} 0, & \text{if } x \text{ is rational} \\ x, & \text{if } x \text{ is irrational} \end{cases}.$$

Find (if possible) $\lim_{x \to 0} f(x)$ and $\lim_{x \to 0} g(x)$.

123. Graphical Reasoning Consider $f(x) = \dfrac{\sec x - 1}{x^2}$.

(a) Find the domain of f.

(b) Use a graphing utility to graph f. Is the domain of f obvious from the graph? If not, explain.

(c) Use the graph of f to approximate $\lim_{x \to 0} f(x)$.

(d) Confirm your answer to part (c) analytically.

124. Approximation

(a) Find $\lim_{x \to 0} \dfrac{1 - \cos x}{x^2}$.

(b) Use your answer to part (a) to derive the approximation $\cos x \approx 1 - \frac{1}{2}x^2$ for x near 0.

(c) Use your answer to part (b) to approximate $\cos(0.1)$.

(d) Use a calculator to approximate $\cos(0.1)$ to four decimal places. Compare the result with part (c).

Kevin Fleming/Corbis

1.4 Continuity and One-Sided Limits

AP* Tips

You must know the precise definition of continuity for the multiple choice section (rarely the free response section) of the AP Exam, and be able to understand which condition of the definition is not met if a function is not continuous at a particular point.

Exploration

Informally, you might say that a function is *continuous* on an open interval when its graph can be drawn with a pencil without lifting the pencil from the paper. Use a graphing utility to graph each function on the given interval. From the graphs, which functions would you say are continuous on the interval? Do you think you can trust the results you obtained graphically? Explain your reasoning.

Function	Interval
a. $y = x^2 + 1$	$(-3, 3)$
b. $y = \dfrac{1}{x-2}$	$(-3, 3)$
c. $y = \dfrac{\sin x}{x}$	$(-\pi, \pi)$
d. $y = \dfrac{x^2-4}{x+2}$	$(-3, 3)$

FOR FURTHER INFORMATION For more information on the concept of continuity, see the article "Leibniz and the Spell of the Continuous" by Hardy Grant in *The College Mathematics Journal*. To view this article, go to *MathArticles.com*.

- **Determine continuity at a point and continuity on an open interval.**
- **Determine one-sided limits and continuity on a closed interval.**
- **Use properties of continuity.**
- **Understand and use the Intermediate Value Theorem.**

Continuity at a Point and on an Open Interval

In mathematics, the term *continuous* has much the same meaning as it has in everyday usage. Informally, to say that a function f is continuous at $x = c$ means that there is no interruption in the graph of f at c. That is, its graph is unbroken at c, and there are no holes, jumps, or gaps. Figure 1.25 identifies three values of x at which the graph of f is *not* continuous. At all other points in the interval (a, b), the graph of f is uninterrupted and **continuous.**

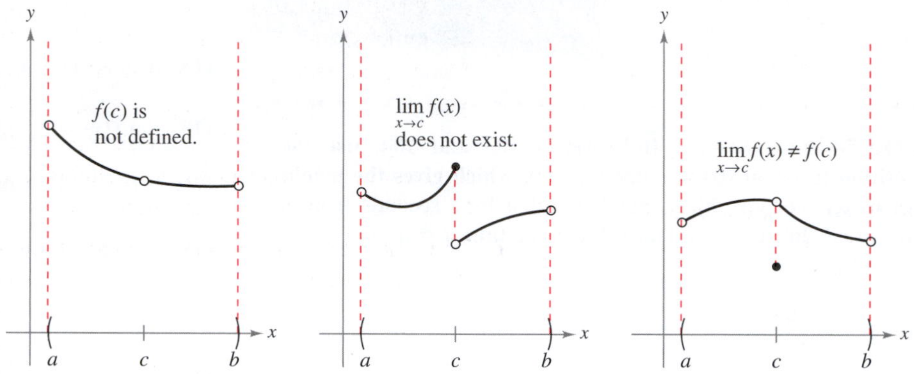

Three conditions exist for which the graph of f is not continuous at $x = c$.
Figure 1.25

In Figure 1.25, it appears that continuity at $x = c$ can be destroyed by any one of three conditions.

1. The function is not defined at $x = c$.
2. The limit of $f(x)$ does not exist at $x = c$.
3. The limit of $f(x)$ exists at $x = c$, but it is not equal to $f(c)$.

If *none* of the three conditions is true, then the function f is called **continuous at c,** as indicated in the important definition below.

Definition of Continuity

Continuity at a Point
A function f is **continuous at c** when these three conditions are met.

1. $f(c)$ is defined.
2. $\lim_{x\to c} f(x)$ exists.
3. $\lim_{x\to c} f(x) = f(c)$

Continuity on an Open Interval
A function is **continuous on an open interval (a, b)** when the function is continuous at each point in the interval. A function that is continuous on the entire real number line $(-\infty, \infty)$ is **everywhere continuous.**

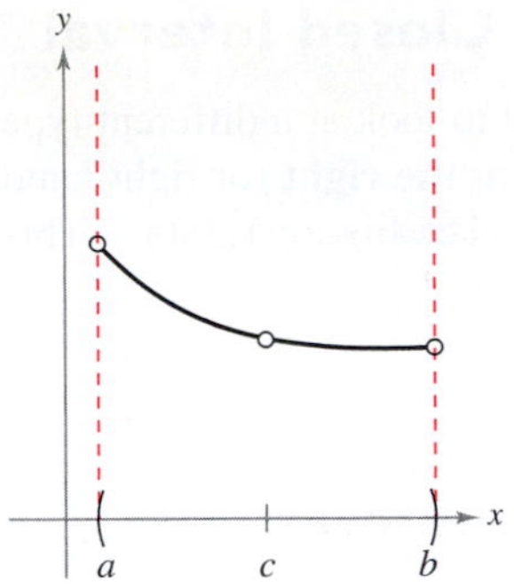

(a) Removable discontinuity

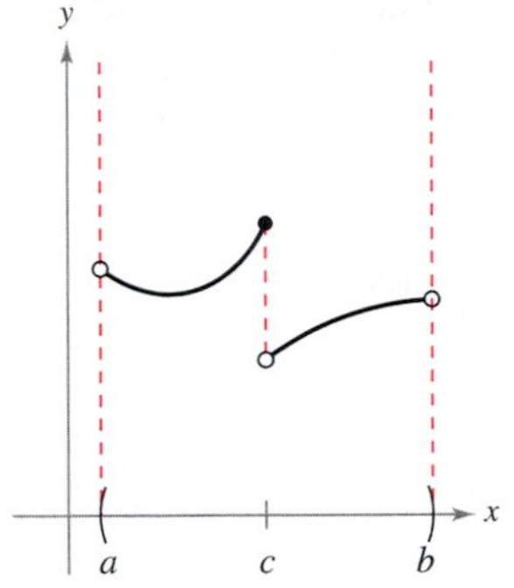

(b) Nonremovable discontinuity

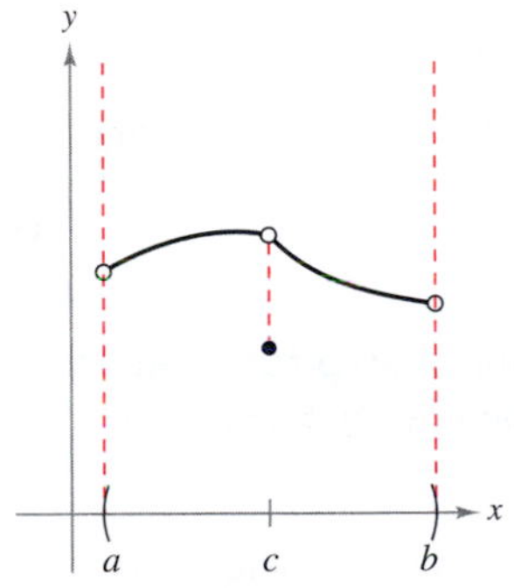

(c) Removable discontinuity

Figure 1.26

Consider an open interval I that contains a real number c. If a function f is defined on I (except possibly at c), and f is not continuous at c, then f is said to have a **discontinuity** at c. Discontinuities fall into two categories: **removable** and **nonremovable.** A discontinuity at c is called removable when f can be made continuous by appropriately defining (or redefining) $f(c)$. For instance, the functions shown in Figures 1.26(a) and (c) have removable discontinuities at c and the function shown in Figure 1.26(b) has a nonremovable discontinuity at c.

EXAMPLE 1 Continuity of a Function

Discuss the continuity of each function.

a. $f(x) = \dfrac{1}{x}$ **b.** $g(x) = \dfrac{x^2 - 1}{x - 1}$ **c.** $h(x) = \begin{cases} x + 1, & x \le 0 \\ x^2 + 1, & x > 0 \end{cases}$ **d.** $y = \sin x$

Solution

a. The domain of f is all nonzero real numbers. From Theorem 1.3, you can conclude that f is continuous at every x-value in its domain. At $x = 0$, f has a nonremovable discontinuity, as shown in Figure 1.27(a). In other words, there is no way to define $f(0)$ so as to make the function continuous at $x = 0$.

b. The domain of g is all real numbers except $x = 1$. From Theorem 1.3, you can conclude that g is continuous at every x-value in its domain. At $x = 1$, the function has a removable discontinuity, as shown in Figure 1.27(b). By defining $g(1)$ as 2, the "redefined" function is continuous for all real numbers.

c. The domain of h is all real numbers. The function h is continuous on $(-\infty, 0)$ and $(0, \infty)$, and, because

$$\lim_{x \to 0} h(x) = 1$$

h is continuous on the entire real number line, as shown in Figure 1.27(c).

d. The domain of y is all real numbers. From Theorem 1.6, you can conclude that the function is continuous on its entire domain, $(-\infty, \infty)$, as shown in Figure 1.27(d).

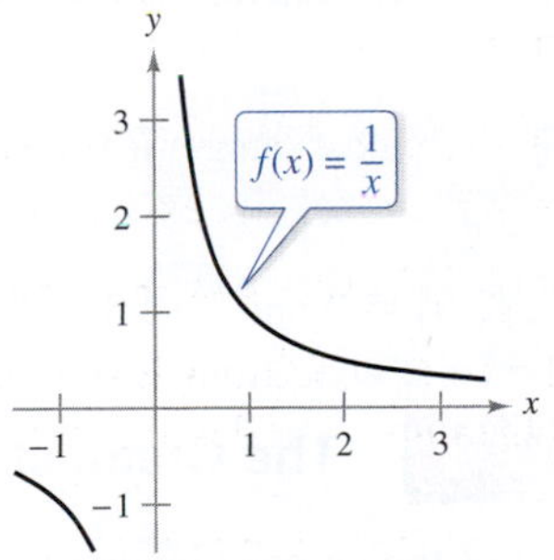

(a) Nonremovable discontinuity at $x = 0$

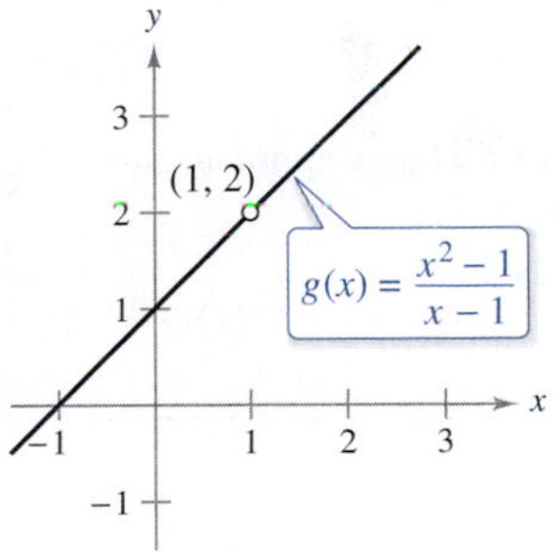

(b) Removable discontinuity at $x = 1$

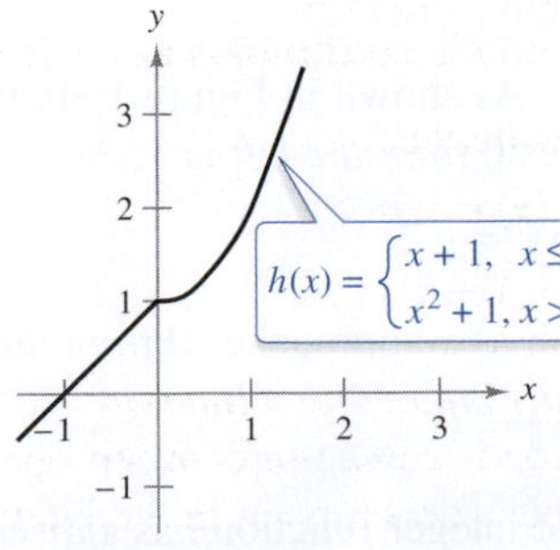

(c) Continuous on entire real number line

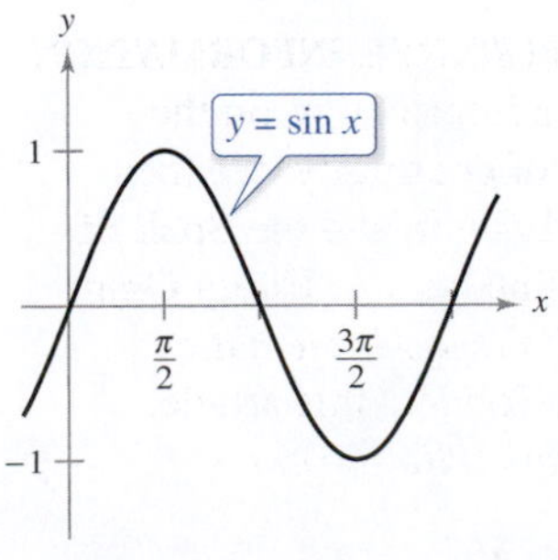

(d) Continuous on entire real number line

Figure 1.27

REMARK Some people may refer to the function in Example 1(a) as "discontinuous." We have found that this terminology can be confusing. Rather than saying that the function is discontinuous, we prefer to say that it has a discontinuity at $x = 0$.

One-Sided Limits and Continuity on a Closed Interval

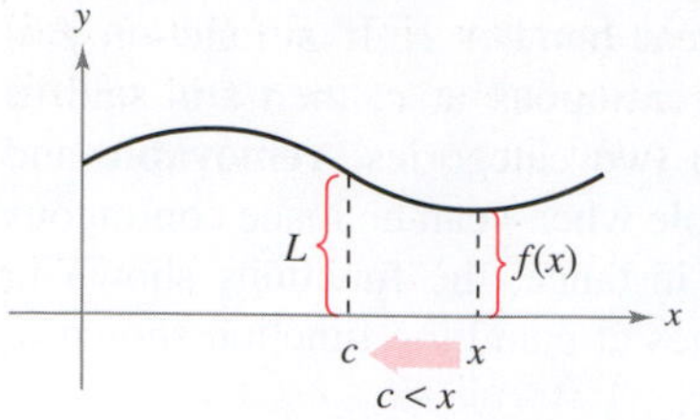

(a) Limit as x approaches c from the right.

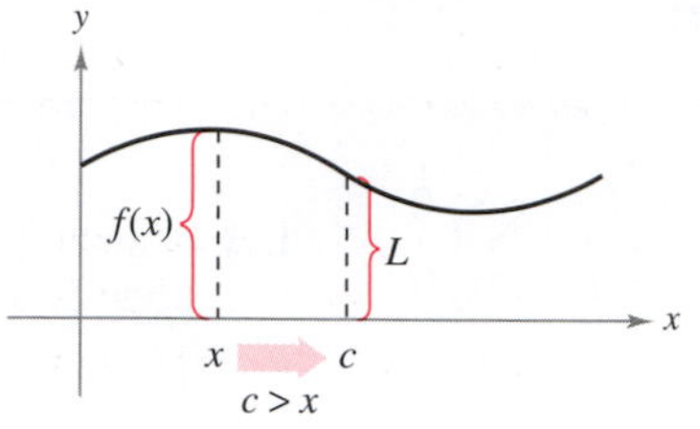

(b) Limit as x approaches c from the left.

Figure 1.28

To understand continuity on a closed interval, you first need to look at a different type of limit called a **one-sided limit.** For instance, the **limit from the right** (or right-hand limit) means that x approaches c from values greater than c [see Figure 1.28(a)]. This limit is denoted as

$$\lim_{x\to c^+} f(x) = L.$$ Limit from the right

Similarly, the **limit from the left** (or left-hand limit) means that x approaches c from values less than c [see Figure 1.28(b)]. This limit is denoted as

$$\lim_{x\to c^-} f(x) = L.$$ Limit from the left

One-sided limits are useful in taking limits of functions involving radicals. For instance, if n is an even integer, then

$$\lim_{x\to 0^+} \sqrt[n]{x} = 0.$$

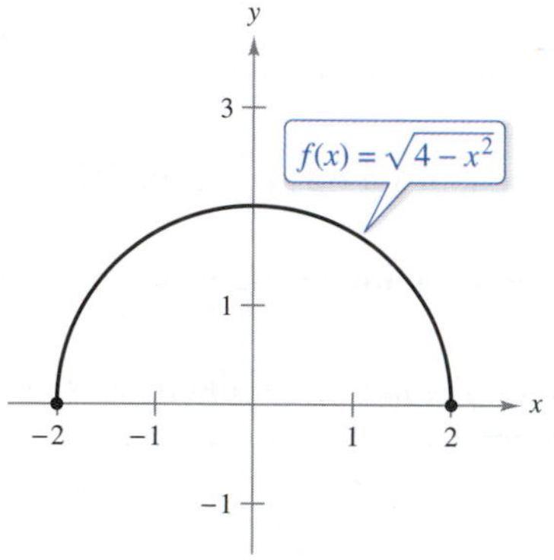

The limit of $f(x)$ as x approaches -2 from the right is 0.

Figure 1.29

EXAMPLE 2 A One-Sided Limit

Find the limit of $f(x) = \sqrt{4 - x^2}$ as x approaches -2 from the right.

Solution As shown in Figure 1.29, the limit as x approaches -2 from the right is

$$\lim_{x\to -2^+} \sqrt{4 - x^2} = 0.$$

One-sided limits can be used to investigate the behavior of **step functions.** One common type of step function is the **greatest integer function** $[\![x]\!]$, defined as

$$[\![x]\!] = \text{greatest integer } n \text{ such that } n \le x.$$ Greatest integer function

For instance, $[\![2.5]\!] = 2$ and $[\![-2.5]\!] = -3$.

EXAMPLE 3 The Greatest Integer Function

Find the limit of the greatest integer function $f(x) = [\![x]\!]$ as x approaches 0 from the left and from the right.

Solution As shown in Figure 1.30, the limit as x approaches 0 *from the left* is

$$\lim_{x\to 0^-} [\![x]\!] = -1$$

and the limit as x approaches 0 *from the right* is

$$\lim_{x\to 0^+} [\![x]\!] = 0.$$

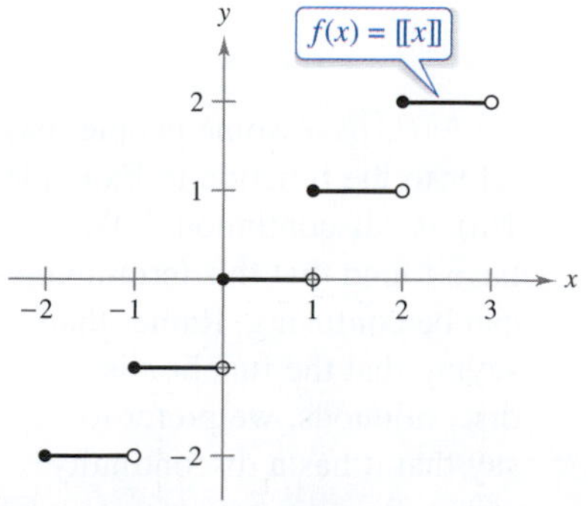

Greatest integer function

Figure 1.30

The greatest integer function has a discontinuity at zero because the left- and right-hand limits at zero are different. By similar reasoning, you can see that the greatest integer function has a discontinuity at any integer n.

When the limit from the left is not equal to the limit from the right, the (two-sided) limit *does not exist*. The next theorem makes this more explicit. The proof of this theorem follows directly from the definition of a one-sided limit.

THEOREM 1.10 The Existence of a Limit

Let f be a function, and let c and L be real numbers. The limit of $f(x)$ as x approaches c is L if and only if

$$\lim_{x\to c^-} f(x) = L \quad \text{and} \quad \lim_{x\to c^+} f(x) = L.$$

The concept of a one-sided limit allows you to extend the definition of continuity to closed intervals. Basically, a function is continuous on a closed interval when it is continuous in the interior of the interval and exhibits one-sided continuity at the endpoints. This is stated formally in the next definition.

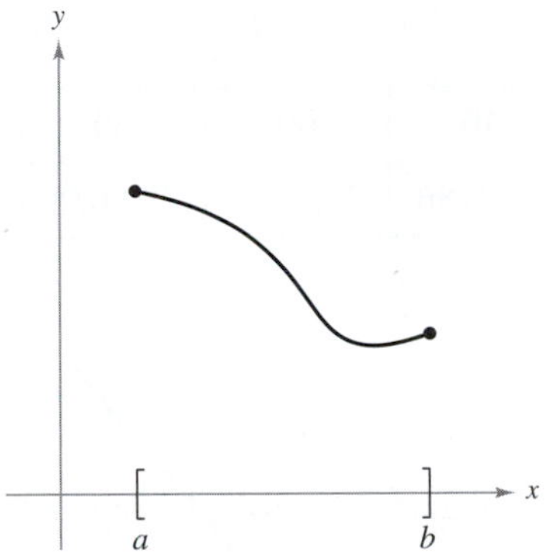

Continuous function on a closed interval
Figure 1.31

Definition of Continuity on a Closed Interval

A function f is **continuous on the closed interval $[a, b]$** when f is continuous on the open interval (a, b) and

$$\lim_{x\to a^+} f(x) = f(a)$$

and

$$\lim_{x\to b^-} f(x) = f(b).$$

The function f is **continuous from the right** at a and **continuous from the left** at b (see Figure 1.31).

Similar definitions can be made to cover continuity on intervals of the form $(a, b]$ and $[a, b)$ that are neither open nor closed, or on infinite intervals. For example,

$$f(x) = \sqrt{x}$$

is continuous on the infinite interval $[0, \infty)$, and the function

$$g(x) = \sqrt{2 - x}$$

is continuous on the infinite interval $(-\infty, 2]$.

EXAMPLE 4 Continuity on a Closed Interval

Discuss the continuity of

$$f(x) = \sqrt{1 - x^2}.$$

Solution The domain of f is the closed interval $[-1, 1]$. At all points in the open interval $(-1, 1)$, the continuity of f follows from Theorems 1.4 and 1.5. Moreover, because

$$\lim_{x\to -1^+} \sqrt{1 - x^2} = 0 = f(-1) \quad \text{Continuous from the right}$$

and

$$\lim_{x\to 1^-} \sqrt{1 - x^2} = 0 = f(1) \quad \text{Continuous from the left}$$

you can conclude that f is continuous on the closed interval $[-1, 1]$, as shown in Figure 1.32. ■

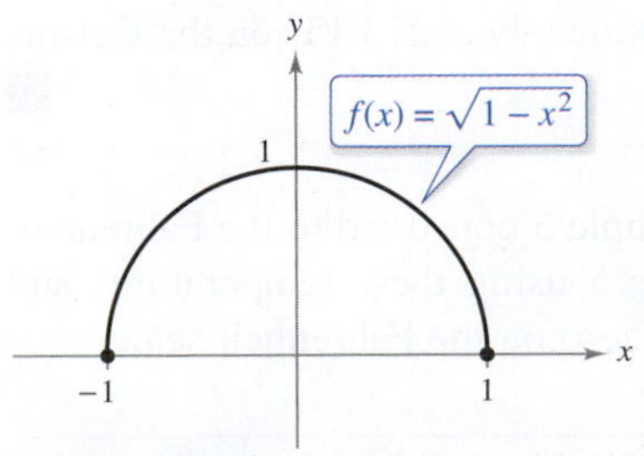

f is continuous on $[-1, 1]$.
Figure 1.32

The next example shows how a one-sided limit can be used to determine the value of absolute zero on the Kelvin scale.

REMARK Charles's Law for gases (assuming constant pressure) can be stated as

$$V = kT$$

where V is volume, k is a constant, and T is temperature.

EXAMPLE 5 Charles's Law and Absolute Zero

On the Kelvin scale, *absolute zero* is the temperature 0 K. Although temperatures very close to 0 K have been produced in laboratories, absolute zero has never been attained. In fact, evidence suggests that absolute zero *cannot* be attained. How did scientists determine that 0 K is the "lower limit" of the temperature of matter? What is absolute zero on the Celsius scale?

Solution The determination of absolute zero stems from the work of the French physicist Jacques Charles (1746–1823). Charles discovered that the volume of gas at a constant pressure increases linearly with the temperature of the gas. The table illustrates this relationship between volume and temperature. To generate the values in the table, one mole of hydrogen is held at a constant pressure of one atmosphere. The volume V is approximated and is measured in liters, and the temperature T is measured in degrees Celsius.

T	−40	−20	0	20	40	60	80
V	19.1482	20.7908	22.4334	24.0760	25.7186	27.3612	29.0038

The points represented by the table are shown in Figure 1.33. Moreover, by using the points in the table, you can determine that T and V are related by the linear equation

$$V = 0.08213T + 22.4334.$$

Solving for T, you get an equation for the temperature of the gas.

$$T = \frac{V - 22.4334}{0.08213}$$

By reasoning that the volume of the gas can approach 0 (but can never equal or go below 0), you can determine that the "least possible temperature" is

$$\lim_{V\to 0^+} T = \lim_{V\to 0^+} \frac{V - 22.4334}{0.08213}$$

$$= \frac{0 - 22.4334}{0.08213} \quad \text{Use direct substitution.}$$

$$\approx -273.15.$$

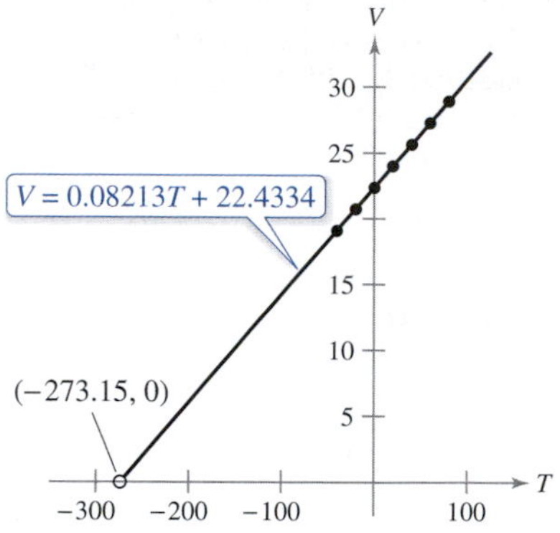

The volume of hydrogen gas depends on its temperature.
Figure 1.33

So, absolute zero on the Kelvin scale (0 K) is approximately −273.15° on the Celsius scale.

In 2003, researchers at the Massachusetts Institute of Technology used lasers and evaporation to produce a super-cold gas in which atoms overlap. This gas is called a Bose-Einstein condensate. They measured a temperature of about 450 pK (picokelvin), or approximately −273.14999999955°C. *(Source: Science magazine, September 12, 2003)*

The table below shows the temperatures in Example 5 converted to the Fahrenheit scale. Try repeating the solution shown in Example 5 using these temperatures and volumes. Use the result to find the value of absolute zero on the Fahrenheit scale.

T	−40	−4	32	68	104	140	176
V	19.1482	20.7908	22.4334	24.0760	25.7186	27.3612	29.0038

AUGUSTIN-LOUIS CAUCHY (1789–1857)

The concept of a continuous function was first introduced by Augustin-Louis Cauchy in 1821. The definition given in his text *Cours d'Analyse* stated that indefinite small changes in y were the result of indefinite small changes in x. "… $f(x)$ will be called a *continuous* function if … the numerical values of the difference $f(x + \alpha) - f(x)$ decrease indefinitely with those of α …."
See LarsonCalculus.com to read more of this biography.

Properties of Continuity

In Section 1.3, you studied several properties of limits. Each of those properties yields a corresponding property pertaining to the continuity of a function. For instance, Theorem 1.11 follows directly from Theorem 1.2.

THEOREM 1.11 Properties of Continuity

If b is a real number and f and g are continuous at $x = c$, then the functions listed below are also continuous at c.

1. Scalar multiple: bf
2. Sum or difference: $f \pm g$
3. Product: fg
4. Quotient: $\dfrac{f}{g}$, $g(c) \neq 0$

A proof of this theorem is given in Appendix A.
See LarsonCalculus.com for Bruce Edwards's video of this proof.

It is important for you to be able to recognize functions that are continuous at every point in their domains. The list below summarizes the functions you have studied so far that are continuous at every point in their domains.

1. Polynomial: $p(x) = a_n x^n + a_{n-1}x^{n-1} + \cdot \cdot \cdot + a_1 x + a_0$
2. Rational: $r(x) = \dfrac{p(x)}{q(x)}$, $q(x) \neq 0$
3. Radical: $f(x) = \sqrt[n]{x}$
4. Trigonometric: $\sin x$, $\cos x$, $\tan x$, $\cot x$, $\sec x$, $\csc x$

By combining Theorem 1.11 with this list, you can conclude that a wide variety of elementary functions are continuous at every point in their domains.

EXAMPLE 6 Applying Properties of Continuity

See LarsonCalculus.com for an interactive version of this type of example.

By Theorem 1.11, it follows that each of the functions below is continuous at every point in its domain.

$$f(x) = x + \sin x, \quad f(x) = 3\tan x, \quad f(x) = \frac{x^2 + 1}{\cos x}$$

The next theorem, which is a consequence of Theorem 1.5, allows you to determine the continuity of *composite* functions such as

$$f(x) = \sin 3x, \quad f(x) = \sqrt{x^2 + 1}, \quad \text{and} \quad f(x) = \tan\frac{1}{x}.$$

REMARK One consequence of Theorem 1.12 is that when f and g satisfy the given conditions, you can determine the limit of $f(g(x))$ as x approaches c to be

$$\lim_{x \to c} f(g(x)) = f(g(c)).$$

THEOREM 1.12 Continuity of a Composite Function

If g is continuous at c and f is continuous at $g(c)$, then the composite function given by $(f \circ g)(x) = f(g(x))$ is continuous at c.

Proof By the definition of continuity, $\lim_{x \to c} g(x) = g(c)$ and $\lim_{x \to g(c)} f(x) = f(g(c))$. Apply Theorem 1.5 with $L = g(c)$ to obtain $\lim_{x \to c} f(g(x)) = f\left(\lim_{x \to c} g(x)\right) = f(g(c))$. So, $(f \circ g)(x) = f(g(x))$ is continuous at c.
See LarsonCalculus.com for Bruce Edwards's video of this proof.

EXAMPLE 7 Testing for Continuity

Describe the interval(s) on which each function is continuous.

a. $f(x) = \tan x$ **b.** $g(x) = \begin{cases} \sin \frac{1}{x}, & x \neq 0 \\ 0, & x = 0 \end{cases}$ **c.** $h(x) = \begin{cases} x \sin \frac{1}{x}, & x \neq 0 \\ 0, & x = 0 \end{cases}$

Solution

a. The tangent function $f(x) = \tan x$ is undefined at

$$x = \frac{\pi}{2} + n\pi, \quad n \text{ is an integer.}$$

At all other points, f is continuous. So, $f(x) = \tan x$ is continuous on the open intervals

$$\ldots, \left(-\frac{3\pi}{2}, -\frac{\pi}{2}\right), \left(-\frac{\pi}{2}, \frac{\pi}{2}\right), \left(\frac{\pi}{2}, \frac{3\pi}{2}\right), \ldots$$

as shown in Figure 1.34(a).

b. Because $y = 1/x$ is continuous except at $x = 0$ and the sine function is continuous for all real values of x, it follows from Theorem 1.12 that

$$y = \sin \frac{1}{x}$$

is continuous at all real values except $x = 0$. At $x = 0$, the limit of $g(x)$ does not exist (see Example 5, Section 1.2). So, g is continuous on the intervals $(-\infty, 0)$ and $(0, \infty)$, as shown in Figure 1.34(b).

c. This function is similar to the function in part (b) except that the oscillations are damped by the factor x. Using the Squeeze Theorem, you obtain

$$-|x| \leq x \sin \frac{1}{x} \leq |x|, \quad x \neq 0$$

and you can conclude that

$$\lim_{x \to 0} h(x) = 0.$$

So, h is continuous on the entire real number line, as shown in Figure 1.34(c).

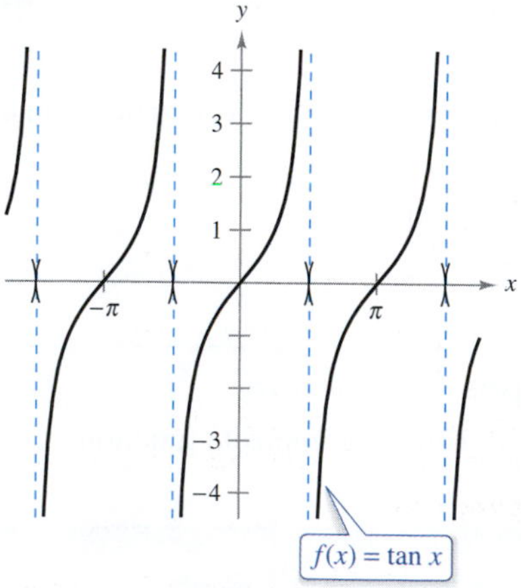

(a) f is continuous on each open interval in its domain.

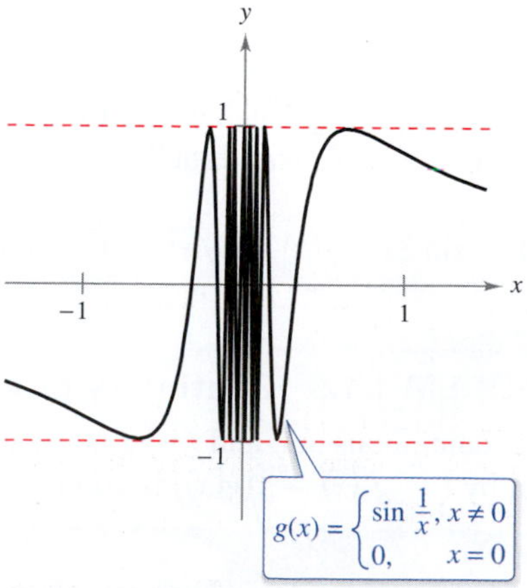

(b) g is continuous on $(-\infty, 0)$ and $(0, \infty)$.

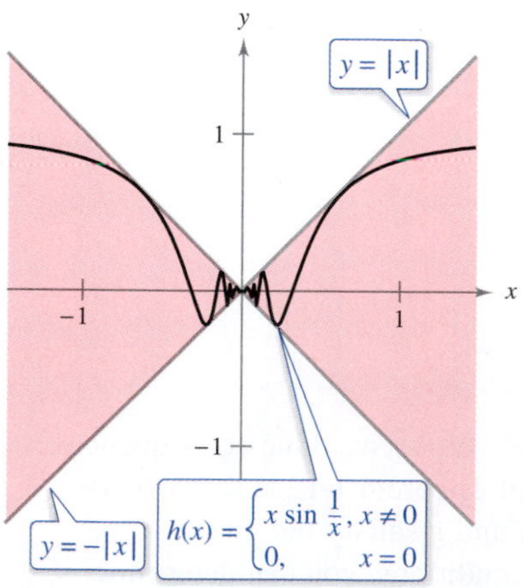

(c) h is continuous on the entire real number line.

Figure 1.34

AP* Tips
You must understand the Intermediate Value Theorem and be able to apply it for both the multiple choice and free response sections of the AP Exam.

The Intermediate Value Theorem

Theorem 1.13 is an important theorem concerning the behavior of functions that are continuous on a closed interval.

> **THEOREM 1.13 Intermediate Value Theorem**
>
> If f is continuous on the closed interval $[a, b]$, $f(a) \neq f(b)$, and k is any number between $f(a)$ and $f(b)$, then there is at least one number c in $[a, b]$ such that
>
> $$f(c) = k.$$

REMARK The Intermediate Value Theorem tells you that at least one number c exists, but it does not provide a method for finding c. Such theorems are called **existence theorems.** By referring to a text on advanced calculus, you will find that a proof of this theorem is based on a property of real numbers called *completeness*. The Intermediate Value Theorem states that for a continuous function f, if x takes on all values between a and b, then $f(x)$ must take on all values between $f(a)$ and $f(b)$.

AP* Tips
For the free response section of the AP Exam, you should be able to apply the Intermediate Value Theorem, whether the function is presented as an equation or by a table.

As an example of the application of the Intermediate Value Theorem, consider a person's height. A girl is 5 feet tall on her thirteenth birthday and 5 feet 7 inches tall on her fourteenth birthday. Then, for any height h between 5 feet and 5 feet 7 inches, there must have been a time t when her height was exactly h. This seems reasonable because human growth is continuous and a person's height does not abruptly change from one value to another.

The Intermediate Value Theorem guarantees the existence of *at least one* number c in the closed interval $[a, b]$. There may, of course, be more than one number c such that

$$f(c) = k$$

as shown in Figure 1.35. A function that is not continuous does not necessarily exhibit the intermediate value property. For example, the graph of the function shown in Figure 1.36 jumps over the horizontal line

$$y = k$$

and for this function there is no value of c in $[a, b]$ such that $f(c) = k$.

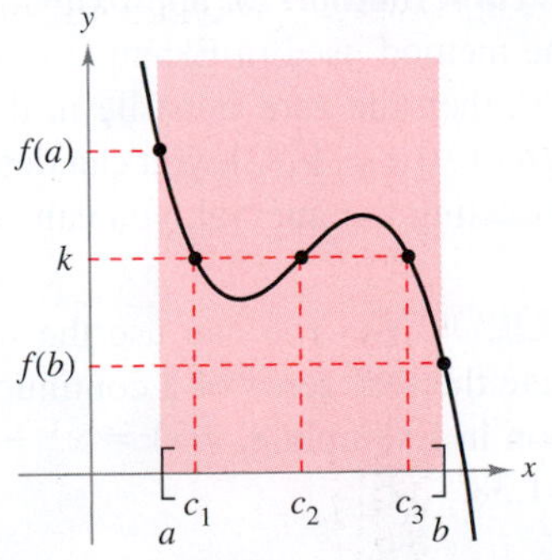

f is continuous on $[a, b]$.
[There exist three c's such that $f(c) = k$.]
Figure 1.35

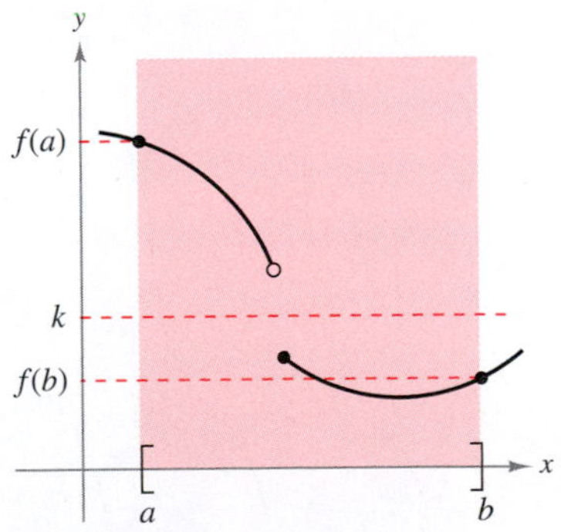

f is not continuous on $[a, b]$.
[There are no c's such that $f(c) = k$.]
Figure 1.36

The Intermediate Value Theorem often can be used to locate the zeros of a function that is continuous on a closed interval. Specifically, if f is continuous on $[a, b]$ and $f(a)$ and $f(b)$ differ in sign, then the Intermediate Value Theorem guarantees the existence of at least one zero of f in the closed interval $[a, b]$.

EXAMPLE 8 **An Application of the Intermediate Value Theorem**

Use the Intermediate Value Theorem to show that the polynomial function

$$f(x) = x^3 + 2x - 1$$

has a zero in the interval $[0, 1]$.

Solution Note that f is continuous on the closed interval $[0, 1]$. Because

$$f(0) = 0^3 + 2(0) - 1 = -1 \quad \text{and} \quad f(1) = 1^3 + 2(1) - 1 = 2$$

it follows that $f(0) < 0$ and $f(1) > 0$. You can therefore apply the Intermediate Value Theorem to conclude that there must be some c in $[0, 1]$ such that

$$f(c) = 0 \qquad f \text{ has a zero in the closed interval } [0, 1].$$

as shown in Figure 1.37.

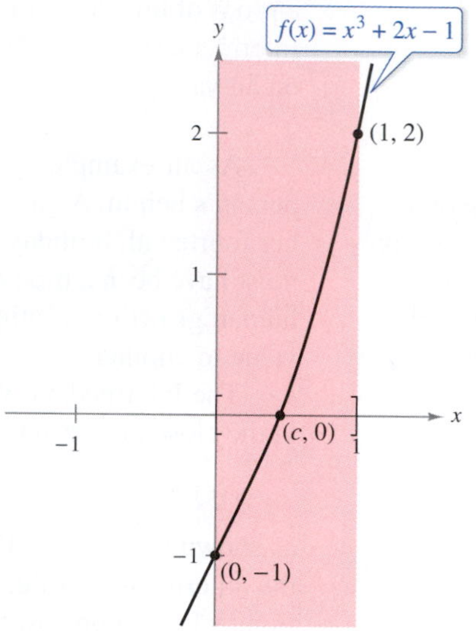

f is continuous on $[0, 1]$ with $f(0) < 0$ and $f(1) > 0$.
Figure 1.37

The **bisection method** for approximating the real zeros of a continuous function is similar to the method used in Example 8. If you know that a zero exists in the closed interval $[a, b]$, then the zero must lie in the interval $[a, (a + b)/2]$ or $[(a + b)/2, b]$. From the sign of $f([a + b]/2)$, you can determine which interval contains the zero. By repeatedly bisecting the interval, you can "close in" on the zero of the function.

TECHNOLOGY You can use the *root* or *zero* feature of a graphing utility to approximate the real zeros of a continuous function. Using this feature, the zero of the function in Example 8, $f(x) = x^3 + 2x - 1$, is approximately 0.453, as shown in Figure 1.38.

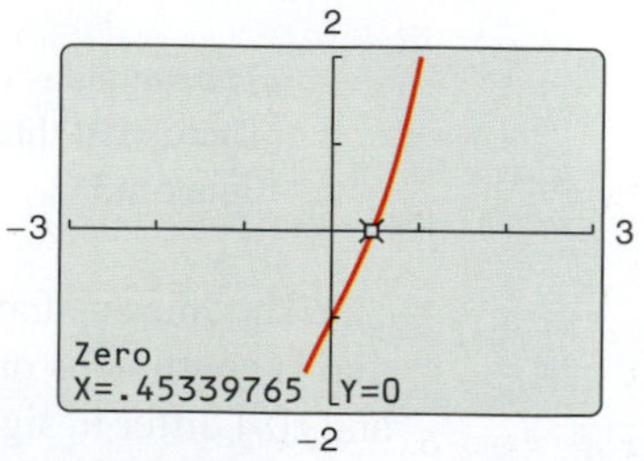

Zero of $f(x) = x^3 + 2x - 1$
Figure 1.38

1.4 Exercises

See CalcChat.com for tutorial help and worked-out solutions to odd-numbered exercises.

Limits and Continuity In Exercises 1–6, use the graph to determine the limit, and discuss the continuity of the function.

(a) $\lim\limits_{x\to c^+} f(x)$ (b) $\lim\limits_{x\to c^-} f(x)$ (c) $\lim\limits_{x\to c} f(x)$

1.

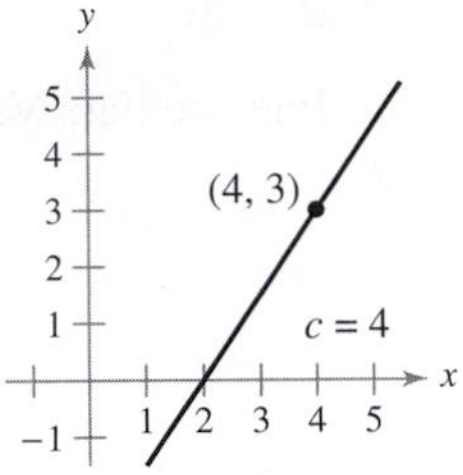

2.

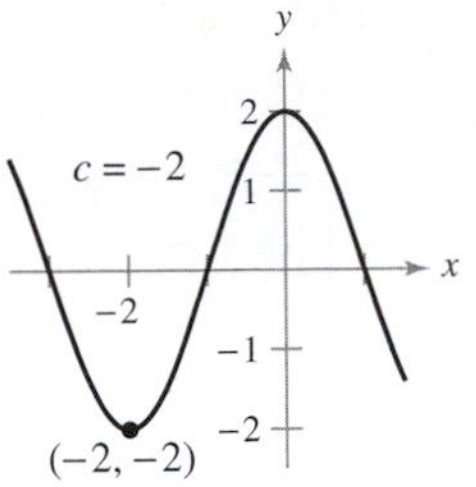

3.

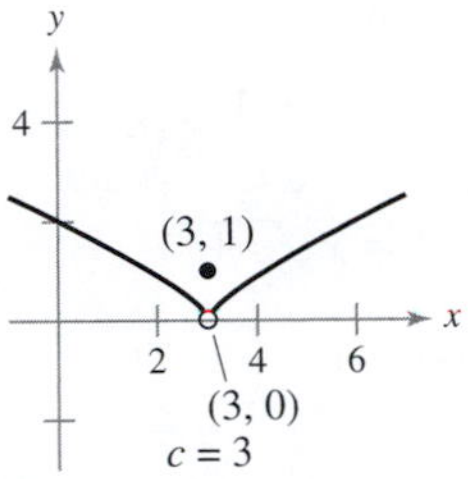

4.

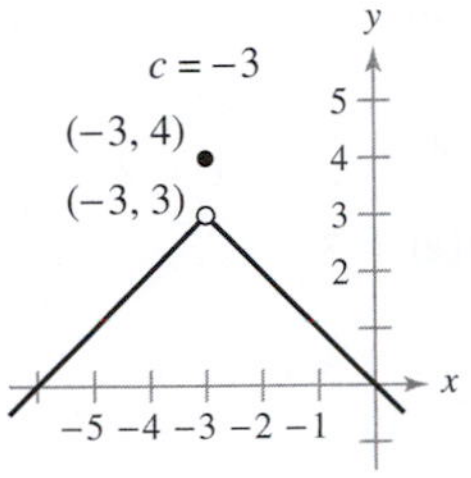

5.

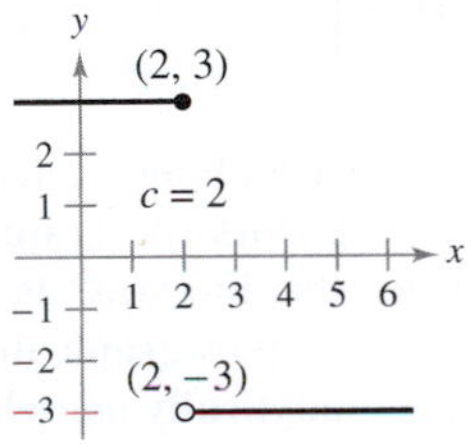

6.

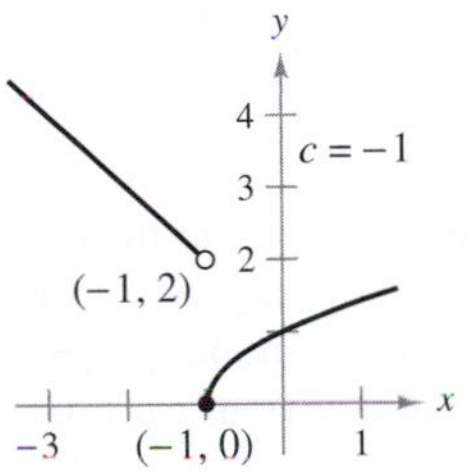

Finding a Limit In Exercises 7–26, find the limit (if it exists). If it does not exist, explain why.

7. $\lim\limits_{x\to 8^+} \dfrac{1}{x+8}$

8. $\lim\limits_{x\to 2^-} \dfrac{2}{x+2}$

9. $\lim\limits_{x\to 5^+} \dfrac{x-5}{x^2-25}$

10. $\lim\limits_{x\to 4^+} \dfrac{4-x}{x^2-16}$

11. $\lim\limits_{x\to -3^-} \dfrac{x}{\sqrt{x^2-9}}$

12. $\lim\limits_{x\to 4^-} \dfrac{\sqrt{x}-2}{x-4}$

13. $\lim\limits_{x\to 0^-} \dfrac{|x|}{x}$

14. $\lim\limits_{x\to 10^+} \dfrac{|x-10|}{x-10}$

15. $\lim\limits_{\Delta x\to 0^-} \dfrac{\dfrac{1}{x+\Delta x} - \dfrac{1}{x}}{\Delta x}$

16. $\lim\limits_{\Delta x\to 0^+} \dfrac{(x+\Delta x)^2 + x + \Delta x - (x^2+x)}{\Delta x}$

17. $\lim\limits_{x\to 3^-} f(x)$, where $f(x) = \begin{cases} \dfrac{x+2}{2}, & x \le 3 \\ \dfrac{12-2x}{3}, & x > 3 \end{cases}$

18. $\lim\limits_{x\to 3} f(x)$, where $f(x) = \begin{cases} x^2-4x+6, & x < 3 \\ -x^2+4x-2, & x \ge 3 \end{cases}$

19. $\lim\limits_{x\to 1} f(x)$, where $f(x) = \begin{cases} x^3+1, & x < 1 \\ x+1, & x \ge 1 \end{cases}$

20. $\lim\limits_{x\to 1^+} f(x)$, where $f(x) = \begin{cases} x, & x \le 1 \\ 1-x, & x > 1 \end{cases}$

21. $\lim\limits_{x\to \pi} \cot x$

22. $\lim\limits_{x\to \pi/2} \sec x$

23. $\lim\limits_{x\to 4^-} (5[\![x]\!] - 7)$

24. $\lim\limits_{x\to 2^+} (2x - [\![x]\!])$

25. $\lim\limits_{x\to 3} (2 - [\![-x]\!])$

26. $\lim\limits_{x\to 1} \left(1 - \left[\!\!\left[-\dfrac{x}{2}\right]\!\!\right]\right)$

Continuity of a Function In Exercises 27–30, discuss the continuity of each function.

27. $f(x) = \dfrac{1}{x^2-4}$

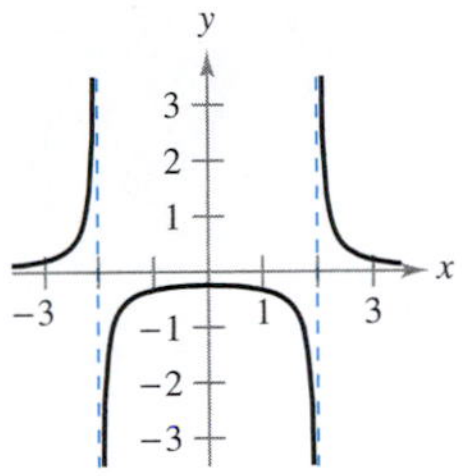

28. $f(x) = \dfrac{x^2-1}{x+1}$

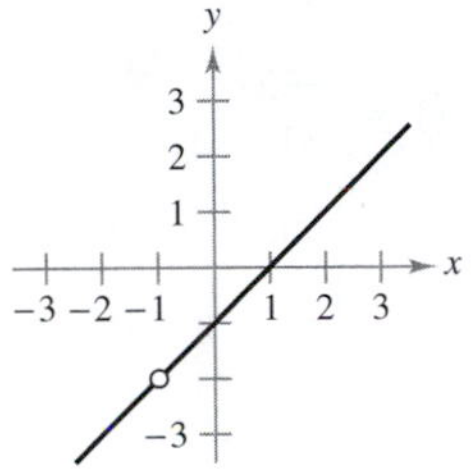

29. $f(x) = \frac{1}{2}[\![x]\!] + x$

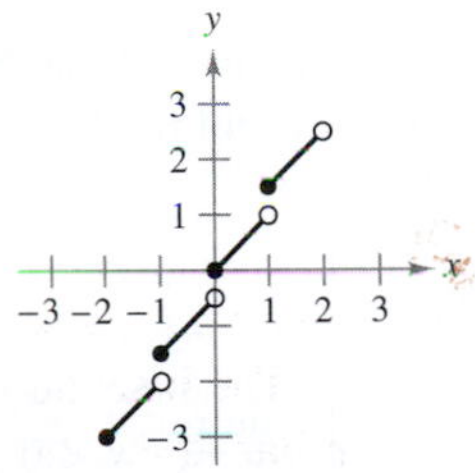

30. $f(x) = \begin{cases} x, & x < 1 \\ 2, & x = 1 \\ 2x-1, & x > 1 \end{cases}$

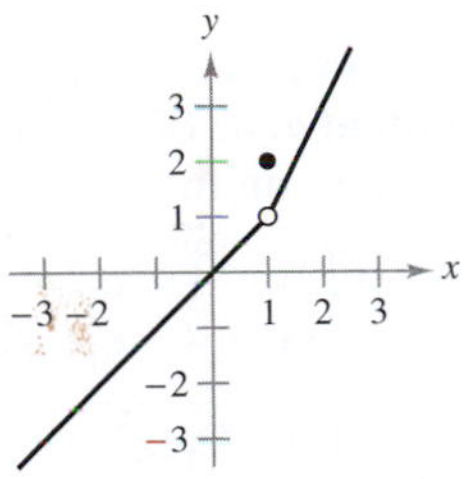

Continuity on a Closed Interval In Exercises 31–34, discuss the continuity of the function on the closed interval.

Function	Interval
31. $g(x) = \sqrt{49-x^2}$	$[-7, 7]$
32. $f(t) = 3 - \sqrt{9-t^2}$	$[-3, 3]$
33. $f(x) = \begin{cases} 3-x, & x \le 0 \\ 3+\frac{1}{2}x, & x > 0 \end{cases}$	$[-1, 4]$
34. $g(x) = \dfrac{1}{x^2-4}$	$[-1, 2]$

Removable and Nonremovable Discontinuities In Exercises 35–60, find the x-values (if any) at which f is not continuous. Which of the discontinuities are removable?

35. $f(x) = \dfrac{6}{x}$

36. $f(x) = \dfrac{4}{x-6}$

37. $f(x) = x^2 - 9$

38. $f(x) = x^2 - 4x + 4$

39. $f(x) = \dfrac{1}{4 - x^2}$

40. $f(x) = \dfrac{1}{x^2 + 1}$

41. $f(x) = 3x - \cos x$

42. $f(x) = \cos \dfrac{\pi x}{2}$

43. $f(x) = \dfrac{x}{x^2 - x}$

44. $f(x) = \dfrac{x}{x^2 - 4}$

45. $f(x) = \dfrac{x}{x^2 + 1}$

46. $f(x) = \dfrac{x - 5}{x^2 - 25}$

47. $f(x) = \dfrac{x + 2}{x^2 - 3x - 10}$

48. $f(x) = \dfrac{x + 2}{x^2 - x - 6}$

49. $f(x) = \dfrac{|x + 7|}{x + 7}$

50. $f(x) = \dfrac{|x - 5|}{x - 5}$

51. $f(x) = \begin{cases} x, & x \le 1 \\ x^2, & x > 1 \end{cases}$

52. $f(x) = \begin{cases} -2x + 3, & x < 1 \\ x^2, & x \ge 1 \end{cases}$

53. $f(x) = \begin{cases} \frac{1}{2}x + 1, & x \le 2 \\ 3 - x, & x > 2 \end{cases}$

54. $f(x) = \begin{cases} -2x, & x \le 2 \\ x^2 - 4x + 1, & x > 2 \end{cases}$

55. $f(x) = \begin{cases} \tan \dfrac{\pi x}{4}, & |x| < 1 \\ x, & |x| \ge 1 \end{cases}$

56. $f(x) = \begin{cases} \csc \dfrac{\pi x}{6}, & |x - 3| \le 2 \\ 2, & |x - 3| > 2 \end{cases}$

57. $f(x) = \csc 2x$

58. $f(x) = \tan \dfrac{\pi x}{2}$

59. $f(x) = [\![x - 8]\!]$

60. $f(x) = 5 - [\![x]\!]$

Making a Function Continuous **In Exercises 61–66, find the constant a, or the constants a and b, such that the function is continuous on the entire real number line.**

61. $f(x) = \begin{cases} 3x^2, & x \ge 1 \\ ax - 4, & x < 1 \end{cases}$

62. $f(x) = \begin{cases} 3x^3, & x \le 1 \\ ax + 5, & x > 1 \end{cases}$

63. $f(x) = \begin{cases} x^3, & x \le 2 \\ ax^2, & x > 2 \end{cases}$

64. $g(x) = \begin{cases} \dfrac{4 \sin x}{x}, & x < 0 \\ a - 2x, & x \ge 0 \end{cases}$

65. $f(x) = \begin{cases} 2, & x \le -1 \\ ax + b, & -1 < x < 3 \\ -2, & x \ge 3 \end{cases}$

66. $g(x) = \begin{cases} \dfrac{x^2 - a^2}{x - a}, & x \ne a \\ 8, & x = a \end{cases}$

Continuity of a Composite Function **In Exercises 67–72, discuss the continuity of the composite function $h(x) = f(g(x))$.**

67. $f(x) = x^2$
$g(x) = x - 1$

68. $f(x) = 5x + 1$
$g(x) = x^3$

69. $f(x) = \dfrac{1}{x - 6}$
$g(x) = x^2 + 5$

70. $f(x) = \dfrac{1}{\sqrt{x}}$
$g(x) = x - 1$

71. $f(x) = \tan x$
$g(x) = \dfrac{x}{2}$

72. $f(x) = \sin x$
$g(x) = x^2$

Finding Discontinuities **In Exercises 73–76, use a graphing utility to graph the function. Use the graph to determine any x-values at which the function is not continuous.**

73. $f(x) = [\![x]\!] - x$

74. $h(x) = \dfrac{1}{x^2 + 2x - 15}$

75. $g(x) = \begin{cases} x^2 - 3x, & x > 4 \\ 2x - 5, & x \le 4 \end{cases}$

76. $f(x) = \begin{cases} \dfrac{\cos x - 1}{x}, & x < 0 \\ 5x, & x \ge 0 \end{cases}$

Testing for Continuity **In Exercises 77–84, describe the interval(s) on which the function is continuous.**

77. $f(x) = \dfrac{x}{x^2 + x + 2}$

78. $f(x) = \dfrac{x + 1}{\sqrt{x}}$

79. $f(x) = 3 - \sqrt{x}$

80. $f(x) = x\sqrt{x + 3}$

81. $f(x) = \sec \dfrac{\pi x}{4}$

82. $f(x) = \cos \dfrac{1}{x}$

83. $f(x) = \begin{cases} \dfrac{x^2 - 1}{x - 1}, & x \ne 1 \\ 2, & x = 1 \end{cases}$

84. $f(x) = \begin{cases} 2x - 4, & x \ne 3 \\ 1, & x = 3 \end{cases}$

Writing **In Exercises 85 and 86, use a graphing utility to graph the function on the interval $[-4, 4]$. Does the graph of the function appear to be continuous on this interval? Is the function continuous on $[-4, 4]$? Write a short paragraph about the importance of examining a function analytically as well as graphically.**

85. $f(x) = \dfrac{\sin x}{x}$

86. $f(x) = \dfrac{x^3 - 8}{x - 2}$

Writing **In Exercises 87–90, explain why the function has a zero in the given interval.**

Function	Interval
87. $f(x) = \frac{1}{12}x^4 - x^3 + 4$	$[1, 2]$
88. $f(x) = x^3 + 5x - 3$	$[0, 1]$
89. $f(x) = x^2 - 2 - \cos x$	$[0, \pi]$
90. $f(x) = -\dfrac{5}{x} + \tan\left(\dfrac{\pi x}{10}\right)$	$[1, 4]$

Using the Intermediate Value Theorem **In Exercises 91–94, use the Intermediate Value Theorem and a graphing utility to approximate the zero of the function in the interval $[0, 1]$. Repeatedly "zoom in" on the graph of the function to approximate the zero accurate to two decimal places. Use the *zero* or *root* feature of the graphing utility to approximate the zero accurate to four decimal places.**

91. $f(x) = x^3 + x - 1$

92. $f(x) = x^4 - x^2 + 3x - 1$

63. $f(x) = \begin{cases} x^3, & x \le 2 \\ ax^2, & x > 2 \end{cases}$

$x = 2$ $\lim_{x \to 2^+} ax^2 = \lim_{x \to 2^-} x^3$

$\frac{ax^2}{x^2} = \frac{x^3}{x^2}$ $a = \frac{x^3}{x^2}$ $a = 2$

64. $g(x) = \begin{cases} \frac{4\sin x}{x}, & x < 0 \\ a - 2x, & x \ge 0 \end{cases}$

$x = 0$

$\lim_{x \to 0^-} \frac{4\sin x}{x}$ $\lim_{x \to 0^+} a - 2x$

$\frac{4\sin x}{x} = a - 2x$

$4\sin x = ax - 2x^2 = 0$

65. $f(x) = \begin{cases} 2 & x \le -1 \\ ax+b & -1 < x < 3 \\ -2 & x \ge 3 \end{cases}$

$x = 3$

$\lim_{x \to 3^-} ax+b$ $\lim_{x \to 3^+} -2$

$ax + b = -2$

$3a + b = -2$

$a = -\frac{2}{3} - \frac{1}{3}b$

$x = -1$

$\lim_{x \to -1^-} 2$ $\lim_{x \to -1^+} ax+b$

$2 = ax + b$

$2 = 3a + b$

$a = -\frac{1}{3}b + \frac{2}{3}$

66. $g(x) = \begin{cases} \frac{x^2 - a^2}{x - a}, & x \ne a \\ 8, & x = a \end{cases}$

$x = a$ $\frac{(x-a)(x+a)}{x-a} = 8$ $a + a = 8$

$a = 4$

Andrew Love

36. $F(x) = \frac{4}{x-6}$ $x = 6$ removable

$\lim_{x \to 6^-} F(x) = -\infty$ $\lim_{x \to 6^+} F(x) = \infty$

38. $F(x) = x^2 - 4x + 4$ $F(x) = (x-2)^2$ $x = 2$

$\lim_{x \to 2^-} F(x) = \infty$ $\lim_{x \to 2^+} F(x) = \infty$ removable

40. $F(x) = \frac{1}{x^2+1}$ $x \in \mathbb{R}$

44. $f(x) = \frac{x}{x^2-4}$ $x \neq \pm 2$

removable discontinuities at $x = \pm 2$

46. $F(x) = \frac{x+2}{x^2-x-6}$ $x = 3$, removable discontinui[ty] at $x = 2$

52. $F(x) = \begin{cases} -2x+3, & x < 1 \\ x^2, & x \geq 1 \end{cases}$ There are no values which f is not continous

$x = 1$ $\lim_{x \to 1} (x^2)^- = 1$

54. $F(x) = \begin{cases} -2x, & x \leq 2 \\ x^2 - 4x + 1, & x > 2 \end{cases}$ $F(2) = -4$

$\lim_{x \to 2^+} f(x) = -3$ $\lim_{x \to 2^-} f(x) = -$

59. $F(x) = [x-8]$ discontinious at each integer all non-remaovable

93. $g(t) = 2\cos t - 3t$

94. $h(\theta) = \tan\theta + 3\theta - 4$

Using the Intermediate Value Theorem **In Exercises 95–98, verify that the Intermediate Value Theorem applies to the indicated interval and find the value of c guaranteed by the theorem.**

95. $f(x) = x^2 + x - 1,\quad [0, 5],\quad f(c) = 11$

96. $f(x) = x^2 - 6x + 8,\quad [0, 3],\quad f(c) = 0$

97. $f(x) = x^3 - x^2 + x - 2,\quad [0, 3],\quad f(c) = 4$

98. $f(x) = \dfrac{x^2 + x}{x - 1},\quad \left[\dfrac{5}{2}, 4\right],\quad f(c) = 6$

WRITING ABOUT CONCEPTS

99. Using the Definition of Continuity State how continuity is destroyed at $x = c$ for each of the following graphs.

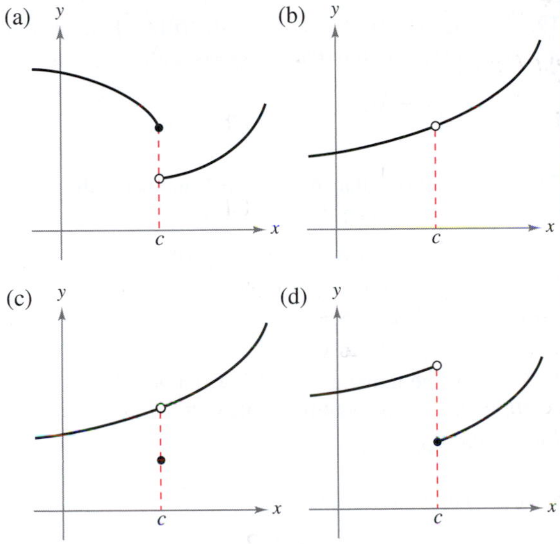

100. Sketching a Graph Sketch the graph of any function f such that

$$\lim_{x\to 3^+} f(x) = 1 \quad \text{and} \quad \lim_{x\to 3^-} f(x) = 0.$$

Is the function continuous at $x = 3$? Explain.

101. Continuity of Combinations of Functions If the functions f and g are continuous for all real x, is $f + g$ always continuous for all real x? Is f/g always continuous for all real x? If either is not continuous, give an example to verify your conclusion.

102. Removable and Nonremovable Discontinuities Describe the difference between a discontinuity that is removable and one that is nonremovable. In your explanation, give examples of the following descriptions.

(a) A function with a nonremovable discontinuity at $x = 4$

(b) A function with a removable discontinuity at $x = -4$

(c) A function that has both of the characteristics described in parts (a) and (b)

True or False? **In Exercises 103–106, determine whether the statement is true or false. If it is false, explain why or give an example that shows it is false.**

103. If $\lim_{x\to c} f(x) = L$ and $f(c) = L$, then f is continuous at c.

104. If $f(x) = g(x)$ for $x \neq c$ and $f(c) \neq g(c)$, then either f or g is not continuous at c.

105. A rational function can have infinitely many x-values at which it is not continuous.

106. The function

$$f(x) = \frac{|x - 1|}{x - 1}$$

is continuous on $(-\infty, \infty)$.

107. Think About It Describe how the functions

$$f(x) = 3 + [\![x]\!] \quad \text{and} \quad g(x) = 3 - [\![-x]\!]$$

differ.

108. HOW DO YOU SEE IT? Every day you dissolve 28 ounces of chlorine in a swimming pool. The graph shows the amount of chlorine $f(t)$ in the pool after t days. Estimate and interpret $\lim_{t\to 4^-} f(t)$ and $\lim_{t\to 4^+} f(t)$.

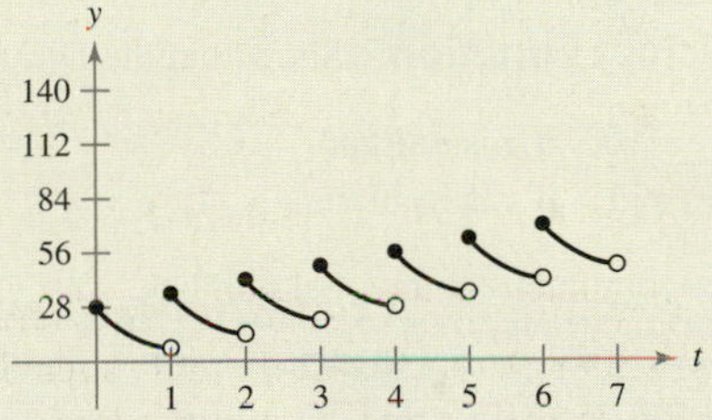

109. Telephone Charges A long distance phone service charges \$0.40 for the first 10 minutes and \$0.05 for each additional minute or fraction thereof. Use the greatest integer function to write the cost C of a call in terms of time t (in minutes). Sketch the graph of this function and discuss its continuity.

110. Inventory Management

The number of units in inventory in a small company is given by

$$N(t) = 25\left(2\left[\!\!\left[\frac{t + 2}{2}\right]\!\!\right] - t\right)$$

where t is the time in months. Sketch the graph of this function and discuss its continuity. How often must this company replenish its inventory?

Christian Delbert/Shutterstock.com

111. Déjà Vu At 8:00 A.M. on Saturday, a man begins running up the side of a mountain to his weekend campsite (see figure). On Sunday morning at 8:00 A.M., he runs back down the mountain. It takes him 20 minutes to run up, but only 10 minutes to run down. At some point on the way down, he realizes that he passed the same place at exactly the same time on Saturday. Prove that he is correct. [*Hint:* Let $s(t)$ and $r(t)$ be the position functions for the runs up and down, and apply the Intermediate Value Theorem to the function $f(t) = s(t) - r(t)$.]

Saturday 8:00 A.M. Sunday 8:00 A.M.

112. Volume Use the Intermediate Value Theorem to show that for all spheres with radii in the interval $[5, 8]$, there is one with a volume of 1500 cubic centimeters.

113. Proof Prove that if f is continuous and has no zeros on $[a, b]$, then either

$$f(x) > 0 \text{ for all } x \text{ in } [a, b] \quad \text{or} \quad f(x) < 0 \text{ for all } x \text{ in } [a, b].$$

114. Dirichlet Function Show that the Dirichlet function

$$f(x) = \begin{cases} 0, & \text{if } x \text{ is rational} \\ 1, & \text{if } x \text{ is irrational} \end{cases}$$

is not continuous at any real number.

115. Continuity of a Function Show that the function

$$f(x) = \begin{cases} 0, & \text{if } x \text{ is rational} \\ kx, & \text{if } x \text{ is irrational} \end{cases}$$

is continuous only at $x = 0$. (Assume that k is any nonzero real number.)

116. Signum Function The **signum function** is defined by

$$\operatorname{sgn}(x) = \begin{cases} -1, & x < 0 \\ 0, & x = 0. \\ 1, & x > 0 \end{cases}$$

Sketch a graph of $\operatorname{sgn}(x)$ and find the following (if possible).

(a) $\lim_{x \to 0^-} \operatorname{sgn}(x)$ (b) $\lim_{x \to 0^+} \operatorname{sgn}(x)$ (c) $\lim_{x \to 0} \operatorname{sgn}(x)$

117. Modeling Data The table lists the speeds S (in feet per second) of a falling object at various times t (in seconds).

t	0	5	10	15	20	25	30
S	0	48.2	53.5	55.2	55.9	56.2	56.3

(a) Create a line graph of the data.

(b) Does there appear to be a limiting speed of the object? If there is a limiting speed, identify a possible cause.

118. Creating Models A swimmer crosses a pool of width b by swimming in a straight line from $(0, 0)$ to $(2b, b)$. (See figure.)

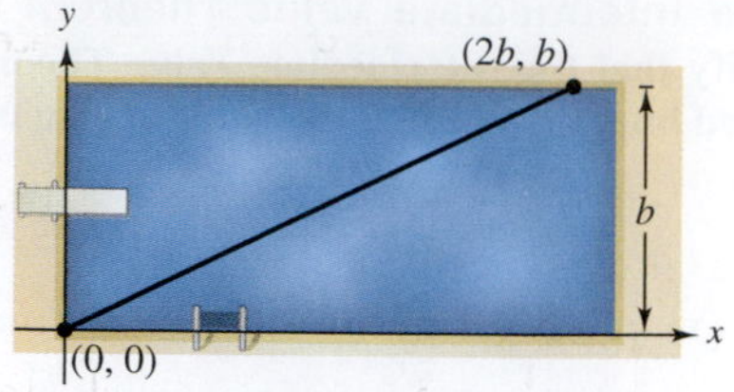

(a) Let f be a function defined as the y-coordinate of the point on the long side of the pool that is nearest the swimmer at any given time during the swimmer's crossing of the pool. Determine the function f and sketch its graph. Is f continuous? Explain.

(b) Let g be the minimum distance between the swimmer and the long sides of the pool. Determine the function g and sketch its graph. Is g continuous? Explain.

119. Making a Function Continuous Find all values of c such that f is continuous on $(-\infty, \infty)$.

$$f(x) = \begin{cases} 1 - x^2, & x \le c \\ x, & x > c \end{cases}$$

120. Proof Prove that for any real number y there exists x in $(-\pi/2, \pi/2)$ such that $\tan x = y$.

121. Making a Function Continuous Let

$$f(x) = \frac{\sqrt{x + c^2} - c}{x}, \quad c > 0.$$

What is the domain of f? How can you define f at $x = 0$ in order for f to be continuous there?

122. Proof Prove that if

$$\lim_{\Delta x \to 0} f(c + \Delta x) = f(c)$$

then f is continuous at c.

123. Continuity of a Function Discuss the continuity of the function $h(x) = x[\![x]\!]$.

124. Proof

(a) Let $f_1(x)$ and $f_2(x)$ be continuous on the closed interval $[a, b]$. If $f_1(a) < f_2(a)$ and $f_1(b) > f_2(b)$, prove that there exists c between a and b such that $f_1(c) = f_2(c)$.

(b) Show that there exists c in $[0, \pi/2]$ such that $\cos x = x$. Use a graphing utility to approximate c to three decimal places.

PUTNAM EXAM CHALLENGE

125. Prove or disprove: If x and y are real numbers with $y \ge 0$ and $y(y + 1) \le (x + 1)^2$, then $y(y - 1) \le x^2$.

126. Determine all polynomials $P(x)$ such that

$$P(x^2 + 1) = (P(x))^2 + 1 \text{ and } P(0) = 0.$$

1.5 Infinite Limits

- **Determine infinite limits from the left and from the right.**
- **Find and sketch the vertical asymptotes of the graph of a function.**

Infinite Limits

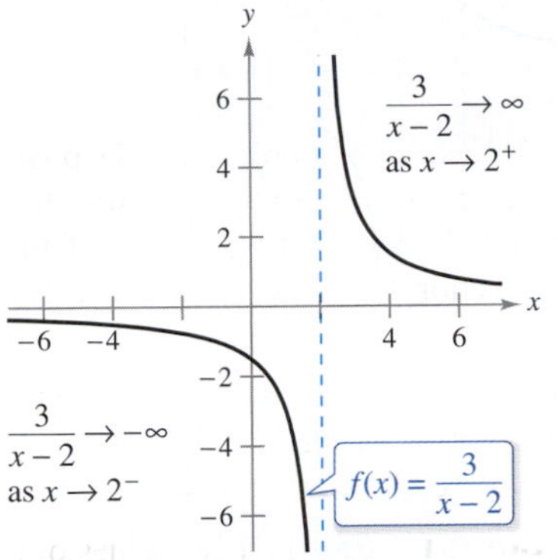

$f(x)$ increases and decreases without bound as x approaches 2.

Figure 1.39

Consider the function $f(x) = 3/(x-2)$. From Figure 1.39 and the table, you can see that $f(x)$ *decreases without bound* as x approaches 2 from the left, and $f(x)$ *increases without bound* as x approaches 2 from the right.

x approaches 2 from the left. → ← x approaches 2 from the right.

x	1.5	1.9	1.99	1.999	2	2.001	2.01	2.1	2.5
$f(x)$	-6	-30	-300	-3000	?	3000	300	30	6

$f(x)$ decreases without bound. → ← $f(x)$ increases without bound.

This behavior is denoted as

$$\lim_{x \to 2^-} \frac{3}{x-2} = -\infty \qquad f(x) \text{ decreases without bound as } x \text{ approaches 2 from the left.}$$

and

$$\lim_{x \to 2^+} \frac{3}{x-2} = \infty. \qquad f(x) \text{ increases without bound as } x \text{ approaches 2 from the right.}$$

The symbols ∞ and $-\infty$ refer to positive infinity and negative infinity, respectively. These symbols do not represent real numbers. They are convenient symbols used to describe unbounded conditions more concisely. A limit in which $f(x)$ increases or decreases without bound as x approaches c is called an **infinite limit.**

Definition of Infinite Limits

Let f be a function that is defined at every real number in some open interval containing c (except possibly at c itself). The statement

$$\lim_{x \to c} f(x) = \infty$$

means that for each $M > 0$ there exists a $\delta > 0$ such that $f(x) > M$ whenever $0 < |x - c| < \delta$ (see Figure 1.40). Similarly, the statement

$$\lim_{x \to c} f(x) = -\infty$$

means that for each $N < 0$ there exists a $\delta > 0$ such that $f(x) < N$ whenever

$$0 < |x - c| < \delta.$$

To define the **infinite limit from the left,** replace $0 < |x - c| < \delta$ by $c - \delta < x < c$. To define the **infinite limit from the right,** replace $0 < |x - c| < \delta$ by $c < x < c + \delta$.

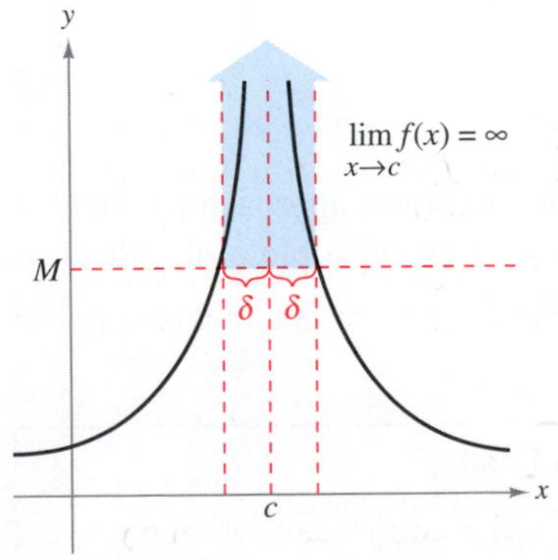

Infinite limits

Figure 1.40

Be sure you see that the equal sign in the statement $\lim f(x) = \infty$ does not mean that the limit exists! On the contrary, it tells you how the limit **fails to exist** by denoting the unbounded behavior of $f(x)$ as x approaches c.

> **Exploration**
>
> Use a graphing utility to graph each function. For each function, analytically find the single real number c that is not in the domain. Then graphically find the limit (if it exists) of $f(x)$ as x approaches c from the left and from the right.
>
> **a.** $f(x) = \dfrac{3}{x - 4}$
>
> **b.** $f(x) = \dfrac{1}{2 - x}$
>
> **c.** $f(x) = \dfrac{2}{(x - 3)^2}$
>
> **d.** $f(x) = \dfrac{-3}{(x + 2)^2}$

EXAMPLE 1 Determining Infinite Limits from a Graph

Determine the limit of each function shown in Figure 1.41 as x approaches 1 from the left and from the right.

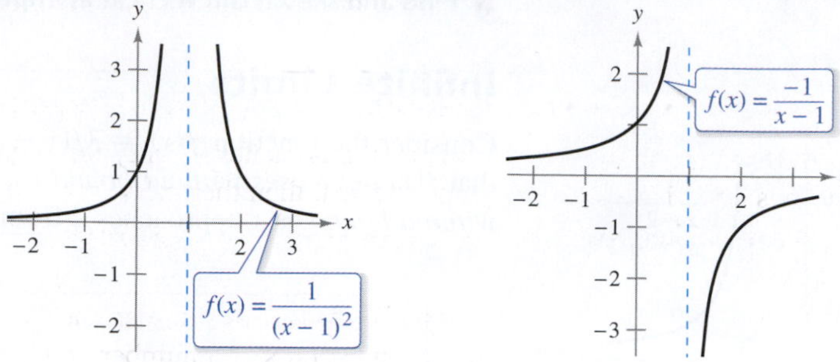

(a) **(b)**

Each graph has an asymptote at $x = 1$.

Figure 1.41

Solution

a. When x approaches 1 from the left or the right, $(x - 1)^2$ is a small positive number. Thus, the quotient $1/(x - 1)^2$ is a large positive number, and $f(x)$ approaches infinity from each side of $x = 1$. So, you can conclude that

$$\lim_{x \to 1} \frac{1}{(x - 1)^2} = \infty. \qquad \text{Limit from each side is infinity.}$$

Figure 1.41(a) confirms this analysis.

b. When x approaches 1 from the left, $x - 1$ is a small negative number. Thus, the quotient $-1/(x - 1)$ is a large positive number, and $f(x)$ approaches infinity from the left of $x = 1$. So, you can conclude that

$$\lim_{x \to 1^-} \frac{-1}{x - 1} = \infty. \qquad \text{Limit from the left side is infinity.}$$

When x approaches 1 from the right, $x - 1$ is a small positive number. Thus, the quotient $-1/(x - 1)$ is a large negative number, and $f(x)$ approaches negative infinity from the right of $x = 1$. So, you can conclude that

$$\lim_{x \to 1^+} \frac{-1}{x - 1} = -\infty. \qquad \text{Limit from the right side is negative infinity.}$$

Figure 1.41(b) confirms this analysis.

TECHNOLOGY Remember that you can use a numerical approach to analyze a limit. For instance, you can use a graphing utility to create a table of values to analyze the limit in Example 1(a), as shown in Figure 1.42.

Enter x-values using *ask* mode.

X	Y1
.9	100
.99	10000
.999	1E6
1	ERROR
1.001	1E6
1.01	10000
1.1	100

X=1

As x approaches 1 from the left, $f(x)$ increases without bound.

As x approaches 1 from the right, $f(x)$ increases without bound.

Figure 1.42

Use a graphing utility to make a table of values to analyze the limit in Example 1(b).

Vertical Asymptotes

If it were possible to extend the graphs in Figure 1.41 toward positive and negative infinity, you would see that each graph becomes arbitrarily close to the vertical line $x = 1$. This line is a **vertical asymptote** of the graph of f. (You will study other types of asymptotes in Sections 3.5 and 3.6.)

REMARK If the graph of a function f has a vertical asymptote at $x = c$, then f is *not continuous* at c.

Definition of Vertical Asymptote

If $f(x)$ approaches infinity (or negative infinity) as x approaches c from the right or the left, then the line $x = c$ is a **vertical asymptote** of the graph of f.

In Example 1, note that each of the functions is a *quotient* and that the vertical asymptote occurs at a number at which the denominator is 0 (and the numerator is not 0). The next theorem generalizes this observation.

THEOREM 1.14 Vertical Asymptotes

Let f and g be continuous on an open interval containing c. If $f(c) \neq 0$, $g(c) = 0$, and there exists an open interval containing c such that $g(x) \neq 0$ for all $x \neq c$ in the interval, then the graph of the function

$$h(x) = \frac{f(x)}{g(x)}$$

has a vertical asymptote at $x = c$.

A proof of this theorem is given in Appendix A.

See LarsonCalculus.com for Bruce Edwards's video of this proof.

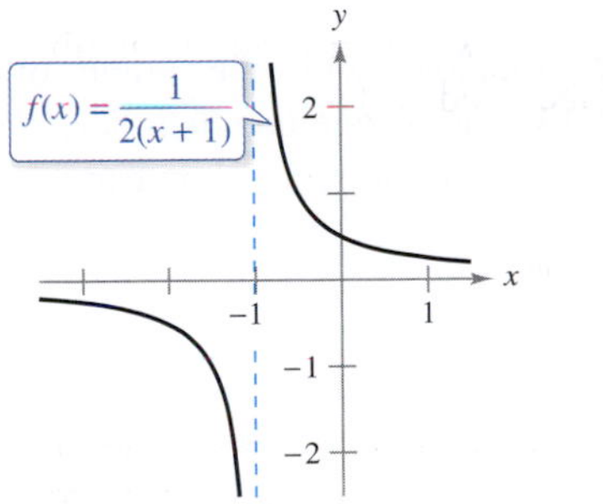

(a)

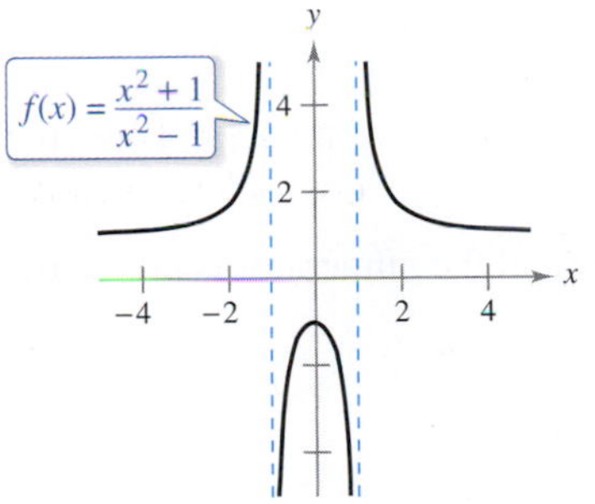

(b)

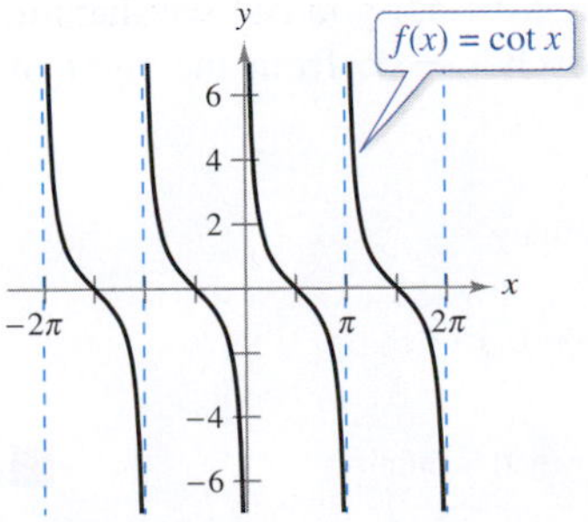

(c)

Functions with vertical asymptotes

Figure 1.43

EXAMPLE 2 Finding Vertical Asymptotes

See LarsonCalculus.com for an interactive version of this type of example.

a. When $x = -1$, the denominator of

$$f(x) = \frac{1}{2(x+1)}$$

is 0 and the numerator is not 0. So, by Theorem 1.14, you can conclude that $x = -1$ is a vertical asymptote, as shown in Figure 1.43(a).

b. By factoring the denominator as

$$f(x) = \frac{x^2+1}{x^2-1} = \frac{x^2+1}{(x-1)(x+1)}$$

you can see that the denominator is 0 at $x = -1$ and $x = 1$. Also, because the numerator is not 0 at these two points, you can apply Theorem 1.14 to conclude that the graph of f has two vertical asymptotes, as shown in Figure 1.43(b).

c. By writing the cotangent function in the form

$$f(x) = \cot x = \frac{\cos x}{\sin x}$$

you can apply Theorem 1.14 to conclude that vertical asymptotes occur at all values of x such that $\sin x = 0$ and $\cos x \neq 0$, as shown in Figure 1.43(c). So, the graph of this function has infinitely many vertical asymptotes. These asymptotes occur at $x = n\pi$, where n is an integer.

Theorem 1.14 requires that the value of the numerator at $x = c$ be nonzero. When both the numerator and the denominator are 0 at $x = c$, you obtain the *indeterminate form* $0/0$, and you cannot determine the limit behavior at $x = c$ without further investigation, as illustrated in Example 3.

EXAMPLE 3 A Rational Function with Common Factors

Determine all vertical asymptotes of the graph of

$$f(x) = \frac{x^2 + 2x - 8}{x^2 - 4}.$$

Solution Begin by simplifying the expression, as shown.

$$\begin{aligned} f(x) &= \frac{x^2 + 2x - 8}{x^2 - 4} \\ &= \frac{(x + 4)\cancel{(x - 2)}}{(x + 2)\cancel{(x - 2)}} \\ &= \frac{x + 4}{x + 2}, \quad x \neq 2 \end{aligned}$$

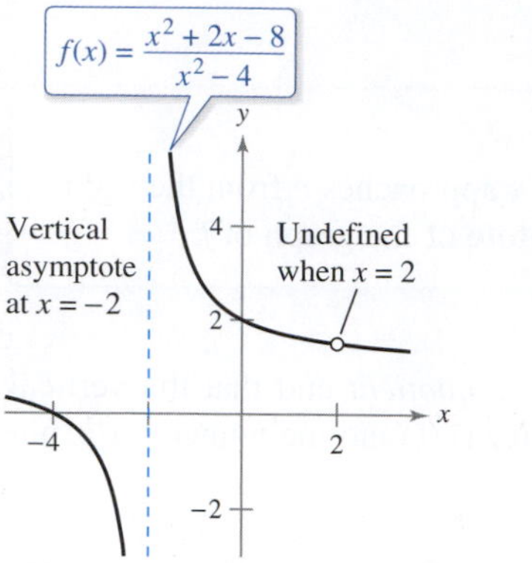

$f(x)$ increases and decreases without bound as x approaches -2.
Figure 1.44

At all x-values other than $x = 2$, the graph of f coincides with the graph of $g(x) = (x + 4)/(x + 2)$. So, you can apply Theorem 1.14 to g to conclude that there is a vertical asymptote at $x = -2$, as shown in Figure 1.44. From the graph, you can see that

$$\lim_{x \to -2^-} \frac{x^2 + 2x - 8}{x^2 - 4} = -\infty \quad \text{and} \quad \lim_{x \to -2^+} \frac{x^2 + 2x - 8}{x^2 - 4} = \infty.$$

Note that $x = 2$ is *not* a vertical asymptote.

EXAMPLE 4 Determining Infinite Limits

Find each limit.

$$\lim_{x \to 1^-} \frac{x^2 - 3x}{x - 1} \quad \text{and} \quad \lim_{x \to 1^+} \frac{x^2 - 3x}{x - 1}$$

Solution Because the denominator is 0 when $x = 1$ (and the numerator is not zero), you know that the graph of

$$f(x) = \frac{x^2 - 3x}{x - 1}$$

has a vertical asymptote at $x = 1$. This means that each of the given limits is either ∞ or $-\infty$. You can determine the result by analyzing f at values of x close to 1, or by using a graphing utility. From the graph of f shown in Figure 1.45, you can see that the graph approaches ∞ from the left of $x = 1$ and approaches $-\infty$ from the right of $x = 1$. So, you can conclude that

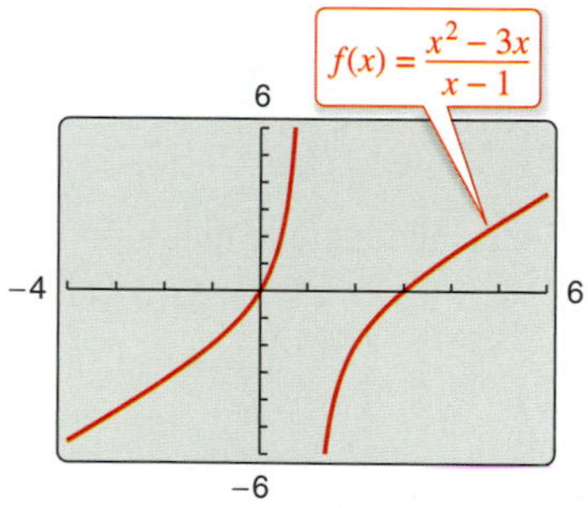

f has a vertical asymptote at $x = 1$.
Figure 1.45

$$\lim_{x \to 1^-} \frac{x^2 - 3x}{x - 1} = \infty \qquad \text{The limit from the left is infinity.}$$

and

$$\lim_{x \to 1^+} \frac{x^2 - 3x}{x - 1} = -\infty. \qquad \text{The limit from the right is negative infinity.}$$

TECHNOLOGY PITFALL When using a graphing utility, be careful to interpret correctly the graph of a function with a vertical asymptote—some graphing utilities have difficulty drawing this type of graph.

THEOREM 1.15 Properties of Infinite Limits

Let c and L be real numbers, and let f and g be functions such that

$$\lim_{x \to c} f(x) = \infty \quad \text{and} \quad \lim_{x \to c} g(x) = L.$$

1. Sum or difference: $\lim_{x \to c} [f(x) \pm g(x)] = \infty$
2. Product: $\lim_{x \to c} [f(x)g(x)] = \infty, \quad L > 0$

   $\lim_{x \to c} [f(x)g(x)] = -\infty, \quad L < 0$
3. Quotient: $\lim_{x \to c} \dfrac{g(x)}{f(x)} = 0$

Similar properties hold for one-sided limits and for functions for which the limit of $f(x)$ as x approaches c is $-\infty$ [see Example 5(d)].

Proof Here is a proof of the sum property. (The proofs of the remaining properties are left as an exercise [see Exercise 70].) To show that the limit of $f(x) + g(x)$ is infinite, choose $M > 0$. You then need to find $\delta > 0$ such that $[f(x) + g(x)] > M$ whenever $0 < |x - c| < \delta$. For simplicity's sake, you can assume L is positive. Let $M_1 = M + 1$. Because the limit of $f(x)$ is infinite, there exists δ_1 such that $f(x) > M_1$ whenever $0 < |x - c| < \delta_1$. Also, because the limit of $g(x)$ is L, there exists δ_2 such that $|g(x) - L| < 1$ whenever $0 < |x - c| < \delta_2$. By letting δ be the smaller of δ_1 and δ_2, you can conclude that $0 < |x - c| < \delta$ implies $f(x) > M + 1$ and $|g(x) - L| < 1$. The second of these two inequalities implies that $g(x) > L - 1$, and, adding this to the first inequality, you can write

$$f(x) + g(x) > (M + 1) + (L - 1) = M + L > M.$$

So, you can conclude that

$$\lim_{x \to c} [f(x) + g(x)] = \infty.$$

See LarsonCalculus.com for Bruce Edwards's video of this proof.

EXAMPLE 5 Determining Limits

a. Because $\lim_{x \to 0} 1 = 1$ and $\lim_{x \to 0} \dfrac{1}{x^2} = \infty$, you can write

$$\lim_{x \to 0} \left(1 + \frac{1}{x^2}\right) = \infty. \qquad \text{Property 1, Theorem 1.15}$$

b. Because $\lim_{x \to 1^-} (x^2 + 1) = 2$ and $\lim_{x \to 1^-} (\cot \pi x) = -\infty$, you can write

$$\lim_{x \to 1^-} \frac{x^2 + 1}{\cot \pi x} = 0. \qquad \text{Property 3, Theorem 1.15}$$

c. Because $\lim_{x \to 0^+} 3 = 3$ and $\lim_{x \to 0^+} \cot x = \infty$, you can write

$$\lim_{x \to 0^+} 3 \cot x = \infty. \qquad \text{Property 2, Theorem 1.15}$$

d. Because $\lim_{x \to 0^-} x^2 = 0$ and $\lim_{x \to 0^-} \dfrac{1}{x} = -\infty$, you can write

$$\lim_{x \to 0^-} \left(x^2 + \frac{1}{x}\right) = -\infty. \qquad \text{Property 1, Theorem 1.15}$$

REMARK Note that the solution to Example 5(d) uses Property 1 from Theorem 1.15 for which the limit of $f(x)$ as x approaches c is $-\infty$.

1.5 Exercises

See CalcChat.com for tutorial help and worked-out solutions to odd-numbered exercises.

Determining Infinite Limits from a Graph **In Exercises 1–4, determine whether $f(x)$ approaches ∞ or $-\infty$ as x approaches -2 from the left and from the right.**

1. $f(x) = 2\left|\dfrac{x}{x^2-4}\right|$

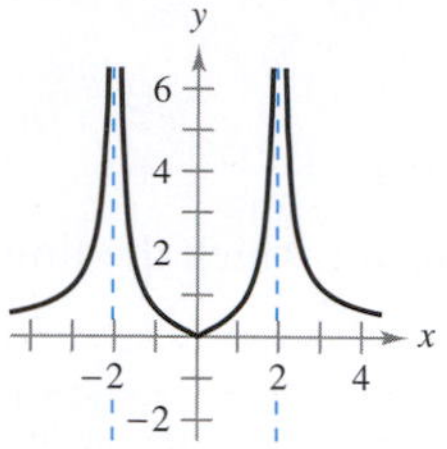

2. $f(x) = \dfrac{1}{x+2}$

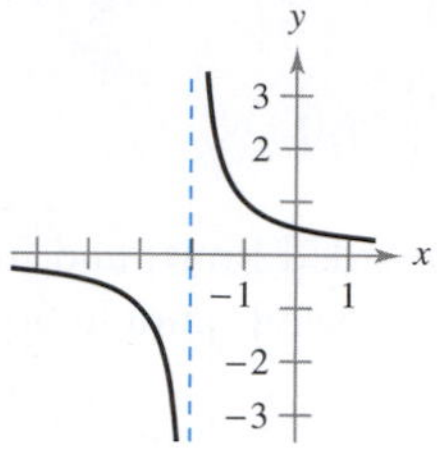

3. $f(x) = \tan \dfrac{\pi x}{4}$

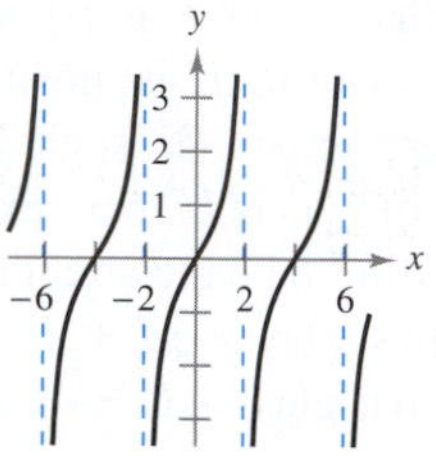

4. $f(x) = \sec \dfrac{\pi x}{4}$

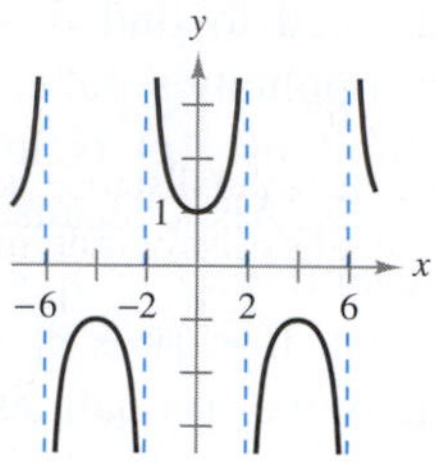

Determining Infinite Limits **In Exercises 5–8, determine whether $f(x)$ approaches ∞ or $-\infty$ as x approaches 4 from the left and from the right.**

5. $f(x) = \dfrac{1}{x-4}$

6. $f(x) = \dfrac{-1}{x-4}$

7. $f(x) = \dfrac{1}{(x-4)^2}$

8. $f(x) = \dfrac{-1}{(x-4)^2}$

Numerical and Graphical Analysis **In Exercises 9–12, determine whether $f(x)$ approaches ∞ or $-\infty$ as x approaches -3 from the left and from the right by completing the table. Use a graphing utility to graph the function to confirm your answer.**

x	-3.5	-3.1	-3.01	-3.001	-3
$f(x)$					?

x	-2.999	-2.99	-2.9	-2.5
$f(x)$				

9. $f(x) = \dfrac{1}{x^2-9}$

10. $f(x) = \dfrac{x}{x^2-9}$

11. $f(x) = \dfrac{x^2}{x^2-9}$

12. $f(x) = \cot \dfrac{\pi x}{3}$

Finding Vertical Asymptotes **In Exercises 13–28, find the vertical asymptotes (if any) of the graph of the function.**

13. $f(x) = \dfrac{1}{x^2}$

14. $f(x) = \dfrac{2}{(x-3)^3}$

15. $f(x) = \dfrac{x^2}{x^2-4}$

16. $f(x) = \dfrac{3x}{x^2+9}$

17. $g(t) = \dfrac{t-1}{t^2+1}$

18. $h(s) = \dfrac{3s+4}{s^2-16}$

19. $f(x) = \dfrac{3}{x^2+x-2}$

20. $g(x) = \dfrac{x^3-8}{x-2}$

21. $f(x) = \dfrac{4x^2+4x-24}{x^4-2x^3-9x^2+18x}$

22. $h(x) = \dfrac{x^2-9}{x^3+3x^2-x-3}$

23. $f(x) = \dfrac{x^2-2x-15}{x^3-5x^2+x-5}$

24. $h(t) = \dfrac{t^2-2t}{t^4-16}$

25. $f(x) = \csc \pi x$

26. $f(x) = \tan \pi x$

27. $s(t) = \dfrac{t}{\sin t}$

28. $g(\theta) = \dfrac{\tan \theta}{\theta}$

Vertical Asymptote or Removable Discontinuity **In Exercises 29–32, determine whether the graph of the function has a vertical asymptote or a removable discontinuity at $x = -1$. Graph the function using a graphing utility to confirm your answer.**

29. $f(x) = \dfrac{x^2-1}{x+1}$

30. $f(x) = \dfrac{x^2-2x-8}{x+1}$

31. $f(x) = \dfrac{x^2+1}{x+1}$

32. $f(x) = \dfrac{\sin(x+1)}{x+1}$

Finding a One-Sided Limit **In Exercises 33–48, find the one-sided limit (if it exists).**

33. $\displaystyle\lim_{x\to -1^+} \frac{1}{x+1}$

34. $\displaystyle\lim_{x\to 1^-} \frac{-1}{(x-1)^2}$

35. $\displaystyle\lim_{x\to 2^+} \frac{x}{x-2}$

36. $\displaystyle\lim_{x\to 2^-} \frac{x^2}{x^2+4}$

37. $\displaystyle\lim_{x\to -3^-} \frac{x+3}{x^2+x-6}$

38. $\displaystyle\lim_{x\to (-1/2)^+} \frac{6x^2+x-1}{4x^2-4x-3}$

39. $\displaystyle\lim_{x\to 0^-} \left(1+\frac{1}{x}\right)$

40. $\displaystyle\lim_{x\to 0^+} \left(6-\frac{1}{x^3}\right)$

41. $\displaystyle\lim_{x\to -4^-} \left(x^2+\frac{2}{x+4}\right)$

42. $\displaystyle\lim_{x\to 3^+} \left(\frac{x}{3}+\cot\frac{\pi x}{2}\right)$

43. $\displaystyle\lim_{x\to 0^+} \frac{2}{\sin x}$

44. $\displaystyle\lim_{x\to (\pi/2)^+} \frac{-2}{\cos x}$

45. $\displaystyle\lim_{x\to \pi^+} \frac{\sqrt{x}}{\csc x}$

46. $\displaystyle\lim_{x\to 0^-} \frac{x+2}{\cot x}$

47. $\lim_{x \to (1/2)^-} x \sec \pi x$

48. $\lim_{x \to (1/2)^+} x^2 \tan \pi x$

One-Sided Limit **In Exercises 49–52, use a graphing utility to graph the function and determine the one-sided limit.**

49. $f(x) = \dfrac{x^2 + x + 1}{x^3 - 1}$

$\lim_{x \to 1^+} f(x)$

50. $f(x) = \dfrac{x^3 - 1}{x^2 + x + 1}$

$\lim_{x \to 1^-} f(x)$

51. $f(x) = \dfrac{1}{x^2 - 25}$

$\lim_{x \to 5^-} f(x)$

52. $f(x) = \sec \dfrac{\pi x}{8}$

$\lim_{x \to 4^+} f(x)$

WRITING ABOUT CONCEPTS

53. Infinite Limit In your own words, describe the meaning of an infinite limit. Is ∞ a real number?

54. Asymptote In your own words, describe what is meant by an asymptote of a graph.

55. Writing a Rational Function Write a rational function with vertical asymptotes at $x = 6$ and $x = -2$, and with a zero at $x = 3$.

56. Rational Function Does the graph of every rational function have a vertical asymptote? Explain.

57. Sketching a Graph Use the graph of the function f (see figure) to sketch the graph of $g(x) = 1/f(x)$ on the interval $[-2, 3]$. To print an enlarged copy of the graph, go to *MathGraphs.com.*

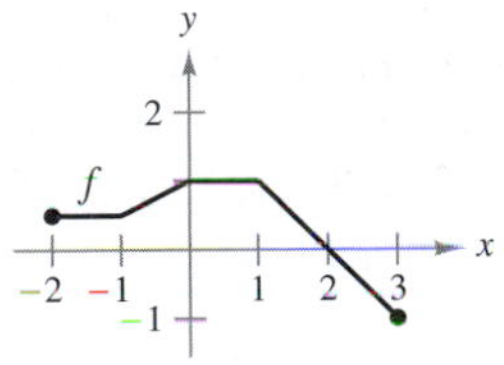

58. Relativity According to the theory of relativity, the mass m of a particle depends on its velocity v. That is,

$$m = \frac{m_0}{\sqrt{1 - (v^2/c^2)}}$$

where m_0 is the mass when the particle is at rest and c is the speed of light. Find the limit of the mass as v approaches c from the left.

59. Numerical and Graphical Analysis Use a graphing utility to complete the table for each function and graph each function to estimate the limit. What is the value of the limit when the power of x in the denominator is greater than 3?

x	1	0.5	0.2	0.1	0.01	0.001	0.0001
$f(x)$							

(a) $\lim_{x \to 0^+} \dfrac{x - \sin x}{x}$

(b) $\lim_{x \to 0^+} \dfrac{x - \sin x}{x^2}$

(c) $\lim_{x \to 0^+} \dfrac{x - \sin x}{x^3}$

(d) $\lim_{x \to 0^+} \dfrac{x - \sin x}{x^4}$

60. HOW DO YOU SEE IT? For a quantity of gas at a constant temperature, the pressure P is inversely proportional to the volume V. What is the limit of P as V approaches 0 from the right? Explain what this means in the context of the problem.

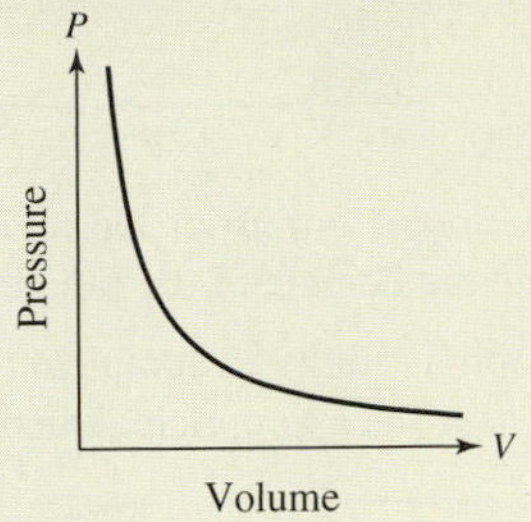

61. Rate of Change A 25-foot ladder is leaning against a house (see figure). If the base of the ladder is pulled away from the house at a rate of 2 feet per second, then the top will move down the wall at a rate of

$$r = \frac{2x}{\sqrt{625 - x^2}} \text{ ft/sec}$$

where x is the distance between the base of the ladder and the house, and r is the rate in feet per second.

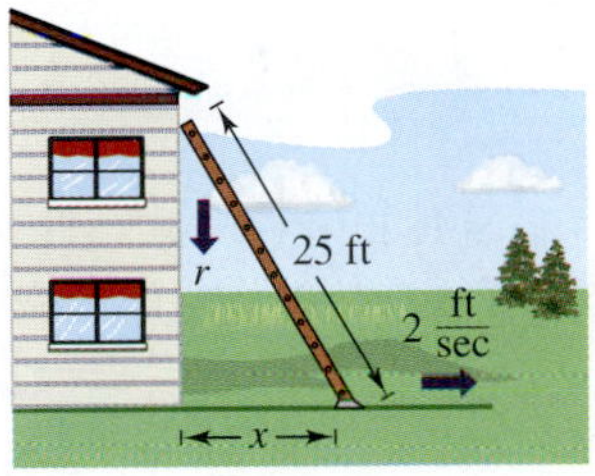

(a) Find the rate r when x is 7 feet.

(b) Find the rate r when x is 15 feet.

(c) Find the limit of r as x approaches 25 from the left.

62. Average Speed

On a trip of d miles to another city, a truck driver's average speed was x miles per hour. On the return trip, the average speed was y miles per hour. The average speed for the round trip was 50 miles per hour.

(a) Verify that

$$y = \frac{25x}{x - 25}.$$

What is the domain?

(b) Complete the table.

x	30	40	50	60
y				

Are the values of y different than you expected? Explain.

(c) Find the limit of y as x approaches 25 from the right and interpret its meaning.

WendellandCarolyn/iStockphoto.com

63. Numerical and Graphical Analysis Consider the shaded region outside the sector of a circle of radius 10 meters and inside a right triangle (see figure).

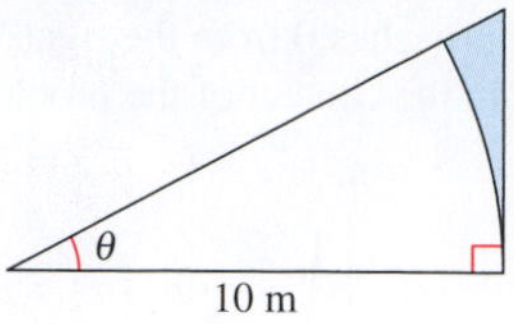

(a) Write the area $A = f(\theta)$ of the region as a function of θ. Determine the domain of the function.

(b) Use a graphing utility to complete the table and graph the function over the appropriate domain.

θ	0.3	0.6	0.9	1.2	1.5
$f(\theta)$					

(c) Find the limit of A as θ approaches $\pi/2$ from the left.

64. Numerical and Graphical Reasoning A crossed belt connects a 20-centimeter pulley (10-cm radius) on an electric motor with a 40-centimeter pulley (20-cm radius) on a saw arbor (see figure). The electric motor runs at 1700 revolutions per minute.

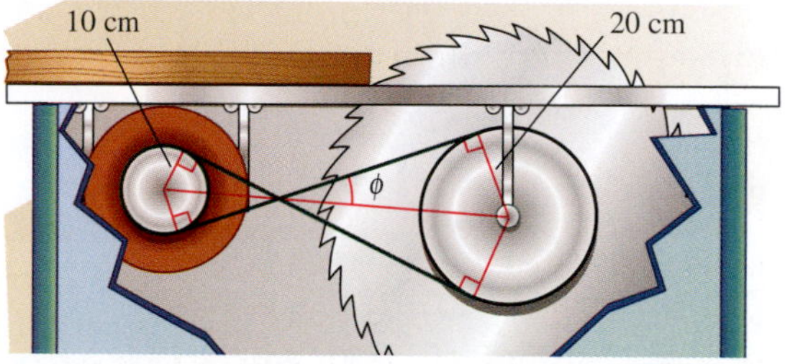

(a) Determine the number of revolutions per minute of the saw.

(b) How does crossing the belt affect the saw in relation to the motor?

(c) Let L be the total length of the belt. Write L as a function of ϕ, where ϕ is measured in radians. What is the domain of the function? (*Hint:* Add the lengths of the straight sections of the belt and the length of the belt around each pulley.)

(d) Use a graphing utility to complete the table.

ϕ	0.3	0.6	0.9	1.2	1.5
L					

(e) Use a graphing utility to graph the function over the appropriate domain.

(f) Find $\lim_{\phi \to (\pi/2)^-} L$. Use a geometric argument as the basis of a second method of finding this limit.

(g) Find $\lim_{\phi \to 0^+} L$.

True or False? **In Exercises 65–68, determine whether the statement is true or false. If it is false, explain why or give an example that shows it is false.**

65. The graph of a rational function has at least one vertical asymptote.

66. The graphs of polynomial functions have no vertical asymptotes.

67. The graphs of trigonometric functions have no vertical asymptotes.

68. If f has a vertical asymptote at $x = 0$, then f is undefined at $x = 0$.

69. Finding Functions Find functions f and g such that $\lim_{x \to c} f(x) = \infty$ and $\lim_{x \to c} g(x) = \infty$, but $\lim_{x \to c} [f(x) - g(x)] \neq 0$.

70. Proof Prove the difference, product, and quotient properties in Theorem 1.15.

71. Proof Prove that if $\lim_{x \to c} f(x) = \infty$, then $\lim_{x \to c} \frac{1}{f(x)} = 0$.

72. Proof Prove that if

$$\lim_{x \to c} \frac{1}{f(x)} = 0$$

then $\lim_{x \to c} f(x)$ does not exist.

Infinite Limits **In Exercises 73 and 74, use the ε-δ definition of infinite limits to prove the statement.**

73. $\lim_{x \to 3^+} \frac{1}{x - 3} = \infty$

74. $\lim_{x \to 5^-} \frac{1}{x - 5} = -\infty$

SECTION PROJECT

Graphs and Limits of Trigonometric Functions

Recall from Theorem 1.9 that the limit of $f(x) = (\sin x)/x$ as x approaches 0 is 1.

(a) Use a graphing utility to graph the function f on the interval $-\pi \le x \le \pi$. Explain how the graph helps confirm this theorem.

(b) Explain how you could use a table of values to confirm the value of this limit numerically.

(c) Graph $g(x) = \sin x$ by hand. Sketch a tangent line at the point $(0, 0)$ and visually estimate the slope of this tangent line.

(d) Let $(x, \sin x)$ be a point on the graph of g near $(0, 0)$, and write a formula for the slope of the secant line joining $(x, \sin x)$ and $(0, 0)$. Evaluate this formula at $x = 0.1$ and $x = 0.01$. Then find the exact slope of the tangent line to g at the point $(0, 0)$.

(e) Sketch the graph of the cosine function $h(x) = \cos x$. What is the slope of the tangent line at the point $(0, 1)$? Use limits to find this slope analytically.

(f) Find the slope of the tangent line to $k(x) = \tan x$ at $(0, 0)$.

Review Exercises

See CalcChat.com for tutorial help and worked-out solutions to odd-numbered exercises.

Precalculus or Calculus **In Exercises 1 and 2, determine whether the problem can be solved using precalculus or whether calculus is required. If the problem can be solved using precalculus, solve it. If the problem seems to require calculus, explain your reasoning and use a graphical or numerical approach to estimate the solution.**

1. Find the distance between the points $(1, 1)$ and $(3, 9)$ along the curve $y = x^2$.
2. Find the distance between the points $(1, 1)$ and $(3, 9)$ along the line $y = 4x - 3$.

Estimating a Limit Numerically **In Exercises 3 and 4, complete the table and use the result to estimate the limit. Use a graphing utility to graph the function to confirm your result.**

3. $\lim\limits_{x\to 3} \dfrac{x-3}{x^2-7x+12}$

x	2.9	2.99	2.999	3	3.001	3.01	3.1
$f(x)$				?			

4. $\lim\limits_{x\to 0} \dfrac{\sqrt{x+4}-2}{x}$

x	−0.1	−0.01	−0.001	0	0.001	0.01	0.1
$f(x)$				?			

Finding a Limit Graphically **In Exercises 5 and 6, use the graph to find the limit (if it exists). If the limit does not exist, explain why.**

5. $h(x) = \dfrac{4x - x^2}{x}$

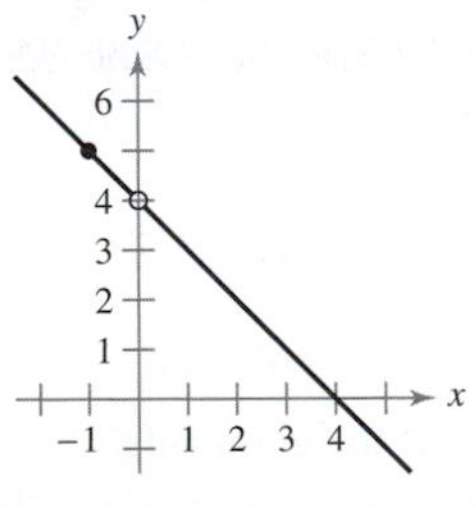

(a) $\lim\limits_{x\to 0} h(x)$ (b) $\lim\limits_{x\to -1} h(x)$

6. $g(x) = \dfrac{-2x}{x-3}$

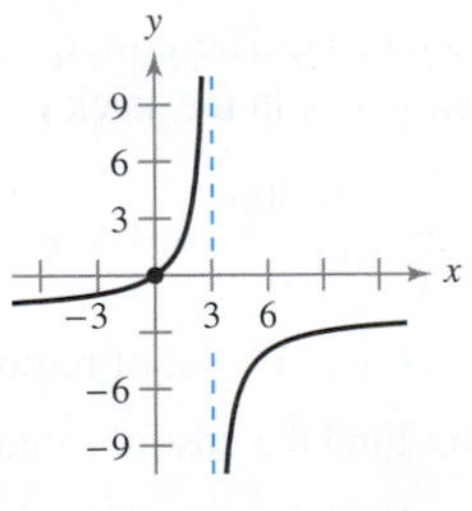

(a) $\lim\limits_{x\to 3} g(x)$ (b) $\lim\limits_{x\to 0} g(x)$

Using the ε-δ Definition of a Limit **In Exercises 7–10, find the limit L. Then use the ε-δ definition to prove that the limit is L.**

7. $\lim\limits_{x\to 1} (x + 4)$
8. $\lim\limits_{x\to 9} \sqrt{x}$
9. $\lim\limits_{x\to 2} (1 - x^2)$
10. $\lim\limits_{x\to 5} 9$

Finding a Limit **In Exercises 11–28, find the limit.**

11. $\lim\limits_{x\to -6} x^2$
12. $\lim\limits_{x\to 0} (5x - 3)$
13. $\lim\limits_{t\to 4} \sqrt{t+2}$
14. $\lim\limits_{x\to -5} \sqrt[3]{x-3}$
15. $\lim\limits_{x\to 6} (x-2)^2$
16. $\lim\limits_{x\to 7} (x-4)^3$
17. $\lim\limits_{x\to 4} \dfrac{4}{x-1}$
18. $\lim\limits_{x\to 2} \dfrac{x}{x^2+1}$
19. $\lim\limits_{x\to -2} \dfrac{t+2}{t^2-4}$
20. $\lim\limits_{x\to 4} \dfrac{t^2-16}{t-4}$
21. $\lim\limits_{x\to 4} \dfrac{\sqrt{x-3}-1}{x-4}$
22. $\lim\limits_{x\to 0} \dfrac{\sqrt{4+x}-2}{x}$
23. $\lim\limits_{x\to 0} \dfrac{[1/(x+1)]-1}{x}$
24. $\lim\limits_{s\to 0} \dfrac{\left(1/\sqrt{1+s}\right)-1}{s}$
25. $\lim\limits_{x\to 0} \dfrac{1-\cos x}{\sin x}$
26. $\lim\limits_{x\to \pi/4} \dfrac{4x}{\tan x}$
27. $\lim\limits_{\Delta x\to 0} \dfrac{\sin[(\pi/6)+\Delta x]-(1/2)}{\Delta x}$

    [*Hint:* $\sin(\theta + \phi) = \sin\theta\cos\phi + \cos\theta\sin\phi$]
28. $\lim\limits_{\Delta x\to 0} \dfrac{\cos(\pi+\Delta x)+1}{\Delta x}$

    [*Hint:* $\cos(\theta + \phi) = \cos\theta\cos\phi - \sin\theta\sin\phi$]

Evaluating a Limit **In Exercises 29–32, evaluate the limit given $\lim\limits_{x\to c} f(x) = -6$ and $\lim\limits_{x\to c} g(x) = \frac{1}{2}$.**

29. $\lim\limits_{x\to c} [f(x)g(x)]$
30. $\lim\limits_{x\to c} \dfrac{f(x)}{g(x)}$
31. $\lim\limits_{x\to c} [f(x) + 2g(x)]$
32. $\lim\limits_{x\to c} [f(x)]^2$

Graphical, Numerical, and Analytic Analysis **In Exercises 33–36, use a graphing utility to graph the function and estimate the limit. Use a table to reinforce your conclusion. Then find the limit by analytic methods.**

33. $\lim\limits_{x\to 0} \dfrac{\sqrt{2x+9}-3}{x}$
34. $\lim\limits_{x\to 0} \dfrac{[1/(x+4)]-(1/4)}{x}$
35. $\lim\limits_{x\to -5} \dfrac{x^3+125}{x+5}$
36. $\lim\limits_{x\to 0} \dfrac{\cos x - 1}{x}$

Free-Falling Object **In Exercises 37 and 38, use the position function $s(t) = -4.9t^2 + 250$, which gives the height (in meters) of an object that has fallen for t seconds from a height of 250 meters. The velocity at time $t = a$ seconds is given by**

$$\lim_{t\to a} \frac{s(a)-s(t)}{a-t}.$$

37. Find the velocity of the object when $t = 4$.
38. At what velocity will the object impact the ground?

Finding a Limit **In Exercises 39–48, find the limit (if it exists). If it does not exist, explain why.**

39. $\lim\limits_{x\to 3^+} \dfrac{1}{x+3}$
40. $\lim\limits_{x\to 6^-} \dfrac{x-6}{x^2-36}$
41. $\lim\limits_{x\to 4^-} \dfrac{\sqrt{x}-2}{x-4}$
42. $\lim\limits_{x\to 3^-} \dfrac{|x-3|}{x-3}$

43. $\lim_{x\to 2} f(x)$, where $f(x) = \begin{cases} (x-2)^2, & x \le 2 \\ 2-x, & x > 2 \end{cases}$

44. $\lim_{x\to 1^+} g(x)$, where $g(x) = \begin{cases} \sqrt{1-x}, & x \le 1 \\ x+1, & x > 1 \end{cases}$

45. $\lim_{t\to 1} h(t)$, where $h(t) = \begin{cases} t^3+1, & t < 1 \\ \frac{1}{2}(t+1), & t \ge 1 \end{cases}$

46. $\lim_{s\to -2} f(s)$, where $f(s) = \begin{cases} -s^2-4s-2, & s \le -2 \\ s^2+4s+6, & s > -2 \end{cases}$

47. $\lim_{x\to 2^-} (2[\![x]\!]+1)$ **48.** $\lim_{x\to 4} [\![x-1]\!]$

Removable and Nonremovable Discontinuities **In Exercises 49–54, find the x-values (if any) at which f is not continuous. Which of the discontinuities are removable?**

49. $f(x) = x^2 - 4$ **50.** $f(x) = x^2 - x + 20$

51. $f(x) = \dfrac{4}{x-5}$ **52.** $f(x) = \dfrac{1}{x^2-9}$

53. $f(x) = \dfrac{x}{x^3-x}$ **54.** $f(x) = \dfrac{x+3}{x^2-3x-18}$

55. Making a Function Continuous Determine the value of c such that the function is continuous on the entire real number line.

$$f(x) = \begin{cases} x+3, & x \le 2 \\ cx+6, & x > 2 \end{cases}$$

56. Making a Function Continuous Determine the values of b and c such that the function is continuous on the entire real number line.

$$f(x) = \begin{cases} x+1, & 1 < x < 3 \\ x^2+bx+c, & |x-2| \ge 1 \end{cases}$$

Testing for Continuity **In Exercises 57–62, describe the intervals on which the function is continuous.**

57. $f(x) = -3x^2 + 7$

58. $f(x) = \dfrac{4x^2+7x-2}{x+2}$

59. $f(x) = \sqrt{x-4}$

60. $f(x) = [\![x+3]\!]$

61. $f(x) = \begin{cases} \dfrac{3x^2-x-2}{x-1}, & x \ne 1 \\ 0, & x = 1 \end{cases}$

62. $f(x) = \begin{cases} 5-x, & x \le 2 \\ 2x-3, & x > 2 \end{cases}$

63. Using the Intermediate Value Theorem Use the Intermediate Value Theorem to show that $f(x) = 2x^3 - 3$ has a zero in the interval $[1, 2]$.

64. Delivery Charges The cost of sending an overnight package from New York to Atlanta is \$12.80 for the first pound and \$2.50 for each additional pound or fraction thereof. Use the greatest integer function to create a model for the cost C of overnight delivery of a package weighing x pounds. Sketch the graph of this function and discuss its continuity.

65. Finding Limits Let

$$f(x) = \frac{x^2-4}{|x-2|}.$$

Find each limit (if it exists).

(a) $\lim_{x\to 2^-} f(x)$ (b) $\lim_{x\to 2^+} f(x)$ (c) $\lim_{x\to 2} f(x)$

66. Finding Limits Let $f(x) = \sqrt{x(x-1)}$.

(a) Find the domain of f.

(b) Find $\lim_{x\to 0^-} f(x)$.

(c) Find $\lim_{x\to 1^+} f(x)$.

Finding Vertical Asymptotes **In Exercises 67–72, find the vertical asymptotes (if any) of the graph of the function.**

67. $f(x) = \dfrac{3}{x}$ **68.** $f(x) = \dfrac{5}{(x-2)^4}$

69. $f(x) = \dfrac{x^3}{x^2-9}$ **70.** $h(x) = \dfrac{6x}{36-x^2}$

71. $g(x) = \dfrac{2x+1}{x^2-64}$ **72.** $f(x) = \csc \pi x$

Finding a One-Sided Limit **In Exercises 73–82, find the one-sided limit (if it exists).**

73. $\lim_{x\to 1^-} \dfrac{x^2+2x+1}{x-1}$ **74.** $\lim_{x\to (1/2)^+} \dfrac{x}{2x-1}$

75. $\lim_{x\to -1^+} \dfrac{x+1}{x^3+1}$ **76.** $\lim_{x\to -1^-} \dfrac{x+1}{x^4-1}$

77. $\lim_{x\to 0^+} \left(x - \dfrac{1}{x^3}\right)$ **78.** $\lim_{x\to 2^-} \dfrac{1}{\sqrt[3]{x^2-4}}$

79. $\lim_{x\to 0^+} \dfrac{\sin 4x}{5x}$ **80.** $\lim_{x\to 0^+} \dfrac{\sec x}{x}$

81. $\lim_{x\to 0^+} \dfrac{\csc 2x}{x}$ **82.** $\lim_{x\to 0^-} \dfrac{\cos^2 x}{x}$

83. Environment A utility company burns coal to generate electricity. The cost C in dollars of removing $p\%$ of the air pollutants in the stack emissions is

$$C = \frac{80{,}000p}{100-p}, \quad 0 \le p < 100.$$

(a) Find the cost of removing 15% of the pollutants.

(b) Find the cost of removing 50% of the pollutants.

(c) Find the cost of removing 90% of the pollutants.

(d) Find the limit of C as p approaches 100 from the left and interpret its meaning.

84. Limits and Continuity The function f is defined as shown.

$$f(x) = \frac{\tan 2x}{x}, \quad x \ne 0$$

(a) Find $\lim_{x\to 0} \dfrac{\tan 2x}{x}$ (if it exists).

(b) Can the function f be defined at $x = 0$ such that it is continuous at $x = 0$?

P.S. Problem Solving

See CalcChat.com for tutorial help and worked-out solutions to odd-numbered exercises.

1. **Perimeter** Let $P(x, y)$ be a point on the parabola $y = x^2$ in the first quadrant. Consider the triangle $\triangle PAO$ formed by P, $A(0, 1)$, and the origin $O(0, 0)$, and the triangle $\triangle PBO$ formed by P, $B(1, 0)$, and the origin.

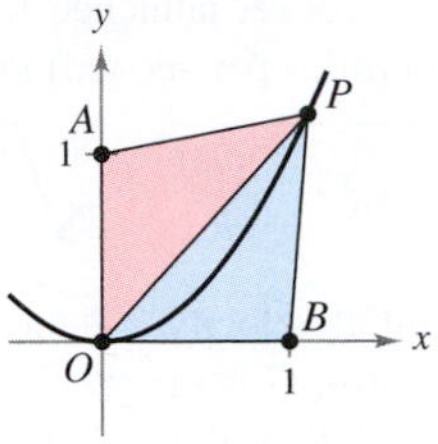

(a) Write the perimeter of each triangle in terms of x.

(b) Let $r(x)$ be the ratio of the perimeters of the two triangles,

$$r(x) = \frac{\text{Perimeter } \triangle PAO}{\text{Perimeter } \triangle PBO}.$$

Complete the table. Calculate $\lim_{x \to 0^+} r(x)$.

x	4	2	1	0.1	0.01
Perimeter $\triangle PAO$					
Perimeter $\triangle PBO$					
$r(x)$					

2. **Area** Let $P(x, y)$ be a point on the parabola $y = x^2$ in the first quadrant. Consider the triangle $\triangle PAO$ formed by P, $A(0, 1)$, and the origin $O(0, 0)$, and the triangle $\triangle PBO$ formed by P, $B(1, 0)$, and the origin.

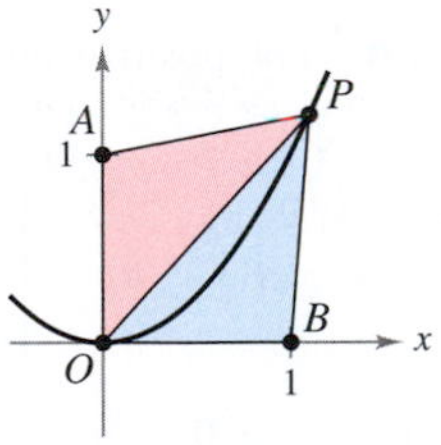

(a) Write the area of each triangle in terms of x.

(b) Let $a(x)$ be the ratio of the areas of the two triangles,

$$a(x) = \frac{\text{Area } \triangle PBO}{\text{Area } \triangle PAO}.$$

Complete the table. Calculate $\lim_{x \to 0^+} a(x)$.

x	4	2	1	0.1	0.01
Area $\triangle PAO$					
Area $\triangle PBO$					
$a(x)$					

3. **Area of a Circle**

(a) Find the area of a regular hexagon inscribed in a circle of radius 1. How close is this area to that of the circle?

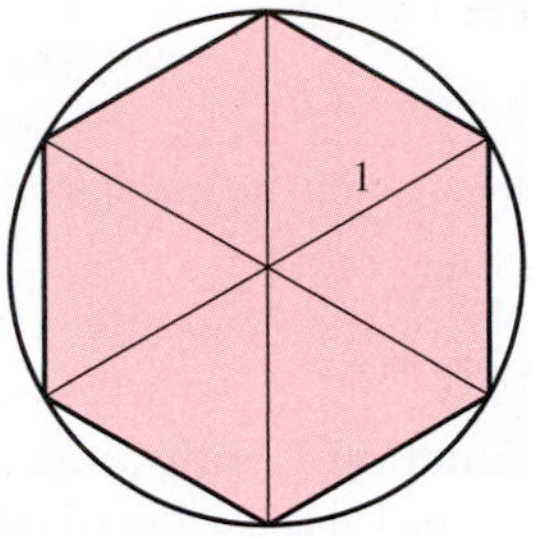

(b) Find the area A_n of an n-sided regular polygon inscribed in a circle of radius 1. Write your answer as a function of n.

(c) Complete the table. What number does A_n approach as n gets larger and larger?

n	6	12	24	48	96
A_n					

4. **Tangent Line** Let $P(3, 4)$ be a point on the circle $x^2 + y^2 = 25$.

(a) What is the slope of the line joining P and $O(0, 0)$?

(b) Find an equation of the tangent line to the circle at P.

(c) Let $Q(x, y)$ be another point on the circle in the first quadrant. Find the slope m_x of the line joining P and Q in terms of x.

(d) Calculate $\lim_{x \to 3} m_x$. How does this number relate to your answer in part (b)?

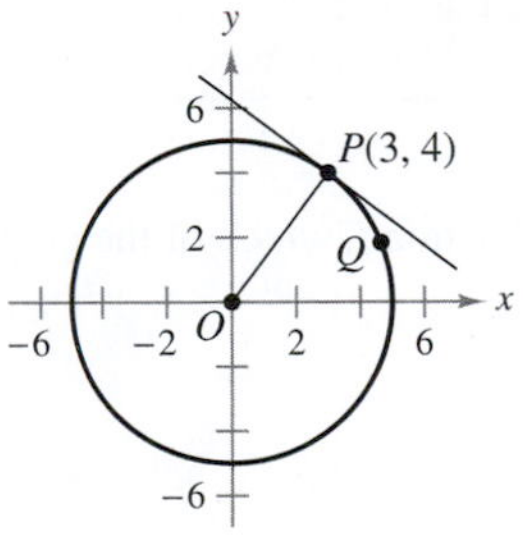

Figure for 4

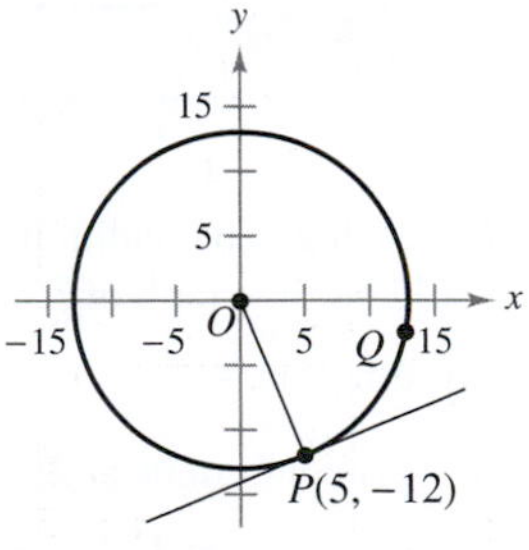

Figure for 5

5. **Tangent Line** Let $P(5, -12)$ be a point on the circle $x^2 + y^2 = 169$.

(a) What is the slope of the line joining P and $O(0, 0)$?

(b) Find an equation of the tangent line to the circle at P.

(c) Let $Q(x, y)$ be another point on the circle in the fourth quadrant. Find the slope m_x of the line joining P and Q in terms of x.

(d) Calculate $\lim_{x \to 5} m_x$. How does this number relate to your answer in part (b)?

6. Finding Values Find the values of the constants a and b such that

$$\lim_{x\to 0} \frac{\sqrt{a+bx} - \sqrt{3}}{x} = \sqrt{3}.$$

7. Finding Limits Consider the function

$$f(x) = \frac{\sqrt{3 + x^{1/3}} - 2}{x - 1}.$$

(a) Find the domain of f.

(b) Use a graphing utility to graph the function.

(c) Calculate $\lim_{x\to -27^+} f(x)$.

(d) Calculate $\lim_{x\to 1} f(x)$.

8. Making a Function Continuous Determine all values of the constant a such that the following function is continuous for all real numbers.

$$f(x) = \begin{cases} \dfrac{ax}{\tan x}, & x \ge 0 \\ a^2 - 2, & x < 0 \end{cases}$$

9. Choosing Graphs Consider the graphs of the four functions g_1, g_2, g_3, and g_4.

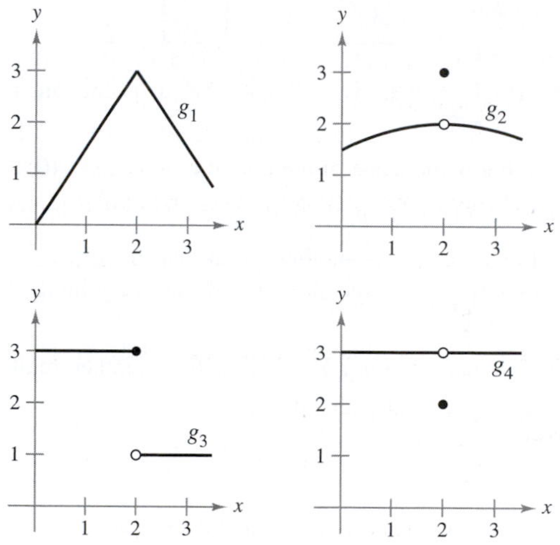

For each given condition of the function f, which of the graphs could be the graph of f?

(a) $\lim_{x\to 2} f(x) = 3$

(b) f is continuous at 2.

(c) $\lim_{x\to 2^-} f(x) = 3$

10. Limits and Continuity Sketch the graph of the function

$$f(x) = \left[\!\left[\frac{1}{x} \right]\!\right].$$

(a) Evaluate $f(\frac{1}{4})$, $f(3)$, and $f(1)$.

(b) Evaluate the limits $\lim_{x\to 1^-} f(x)$, $\lim_{x\to 1^+} f(x)$, $\lim_{x\to 0^-} f(x)$, and $\lim_{x\to 0^+} f(x)$.

(c) Discuss the continuity of the function.

11. Limits and Continuity Sketch the graph of the function $f(x) = [\![x]\!] + [\![-x]\!]$.

(a) Evaluate $f(1)$, $f(0)$, $f(\frac{1}{2})$, and $f(-2.7)$.

(b) Evaluate the limits $\lim_{x\to 1^-} f(x)$, $\lim_{x\to 1^+} f(x)$, and $\lim_{x\to 1/2} f(x)$.

(c) Discuss the continuity of the function.

12. Escape Velocity To escape Earth's gravitational field, a rocket must be launched with an initial velocity called the **escape velocity.** A rocket launched from the surface of Earth has velocity v (in miles per second) given by

$$v = \sqrt{\frac{2GM}{r} + v_0^2 - \frac{2GM}{R}} \approx \sqrt{\frac{192{,}000}{r} + v_0^2 - 48}$$

where v_0 is the initial velocity, r is the distance from the rocket to the center of Earth, G is the gravitational constant, M is the mass of Earth, and R is the radius of Earth (approximately 4000 miles).

(a) Find the value of v_0 for which you obtain an infinite limit for r as r approaches zero. This value of v_0 is the escape velocity for Earth.

(b) A rocket launched from the surface of the moon has velocity v (in miles per second) given by

$$v = \sqrt{\frac{1920}{r} + v_0^2 - 2.17}.$$

Find the escape velocity for the moon.

(c) A rocket launched from the surface of a planet has velocity v (in miles per second) given by

$$v = \sqrt{\frac{10{,}600}{r} + v_0^2 - 6.99}.$$

Find the escape velocity for this planet. Is the mass of this planet larger or smaller than that of Earth? (Assume that the mean density of this planet is the same as that of Earth.)

13. Pulse Function For positive numbers $a < b$, the **pulse function** is defined as

$$P_{a,b}(x) = H(x-a) - H(x-b) = \begin{cases} 0, & x < a \\ 1, & a \le x < b \\ 0, & x \ge b \end{cases}$$

where $H(x) = \begin{cases} 1, & x \ge 0 \\ 0, & x < 0 \end{cases}$ is the Heaviside function.

(a) Sketch the graph of the pulse function.

(b) Find the following limits:

(i) $\lim_{x\to a^+} P_{a,b}(x)$ (ii) $\lim_{x\to a^-} P_{a,b}(x)$

(iii) $\lim_{x\to b^+} P_{a,b}(x)$ (iv) $\lim_{x\to b^-} P_{a,b}(x)$

(c) Discuss the continuity of the pulse function.

(d) Why is $U(x) = \dfrac{1}{b-a} P_{a,b}(x)$ called the **unit pulse function?**

14. Proof Let a be a nonzero constant. Prove that if $\lim_{x\to 0} f(x) = L$, then $\lim_{x\to 0} f(ax) = L$. Show by means of an example that a must be nonzero.

AP* Review Questions for Chapter 1

1. (no calculator)

The function f is defined as follows: $f(x) = \dfrac{x^2 + 5x + 6}{2x^2 + 7x + 3}$.

(a) State the value(s) of x for which f is not continuous.

(b) Evaluate $\lim_{x \to -3} f(x)$. Show the work that leads to your answer.

(c) State the equation(s) for the vertical asymptote(s) for the graph of $y = f(x)$.

(d) State the equation(s) for the horizontal asymptote(s) for the graph of $y = f(x)$. Show the work that leads to your answer.

2.

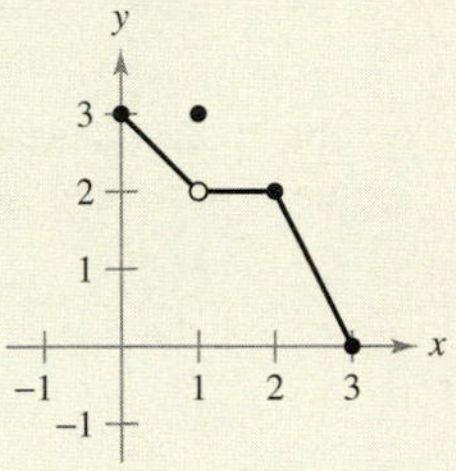

Graph of f

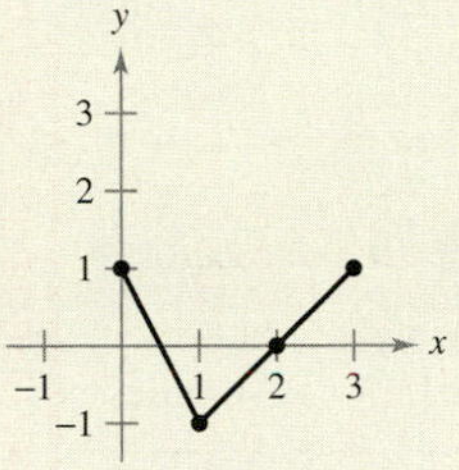

Graph of g

The graphs of functions f and g are shown above. Evaluate each limit using the graphs provided. Show the computations that lead to your answer.

(a) $\lim_{x \to 1} (f(x) + 4)$

(b) $\lim_{x \to 3} \dfrac{5}{g(x)}$

(c) $\lim_{x \to 2} (f(x) \cdot g(x))$

(d) $\lim_{x \to 3^-} \dfrac{f(x)}{g(x) - 1}$ (Assume that f and g are linear on the interval $[2, 3]$.)

3. A hot cup of tea is placed on a counter and left to cool. The temperature of the tea, in degrees Fahrenheit (correct to the nearest degree), x minutes after the cup is placed on the counter is modeled by a continuous function $T(x)$ for $0 \le x < 10$. Values of $T(x)$ at various times x are shown in the table below.

x	0	3	4	6	8	9
$T(x)$	180°	174°	172°	168°	164°	162°

(a) Evaluate: $\lim_{x \to 4} T(x)$. Justify your answer.

(b) Using the data in the table, find the average rate of change in the temperature of the tea for $3 \le x \le 8$. Include units in your final answer.

(c) Identify, using the times listed in the table, the shortest interval during which there must exist a time x for which the temperature of the tea is 166.5°. Justify your answer.

(d) Use the data in the table to find the best estimate of the slope of the line tangent to the graph of T at $x = 8$.

4. (no calculator)

Find the value of each of these limits, or else explain why the limit does not exist. Show the computations which lead to your answers.

(a) $\lim_{x \to \infty} \dfrac{3x^4 - 6x^3 + x^2 - x - 1}{2x^3 - 9x^4 - 5}$

(b) $\lim_{x \to 0} \dfrac{\sqrt{x + 5} - \sqrt{5}}{5x}$

(c) $\lim_{x \to 0} \dfrac{2 - 2\cos^2 x}{x \sin x}$

(d) $\lim_{x \to -2} \dfrac{\frac{5}{x} + \frac{5}{2}}{x + 2}$

5. (no calculator)

The position function $s(t) = -4.9t^2 + 396.9$ gives the height (in meters) of an object that has fallen from a height of 396.9 meters after t seconds.

(a) Explain why there must exist a time t, $1 < t < 2$, at which the height of the object must be 382 meters above the ground.

(b) Find the time at which the object hits the ground.

(c) Find the average rate of change in s over the interval $t \in [8, 9]$. Include units of measure. Explain why this is a good estimate of the velocity at which the object hits the ground. How can this estimate be improved?

(d) Evaluate $\lim_{t \to 3} \dfrac{s(t) - s(3)}{t - 3}$. Show the work that leads to your answer. Include units.

6. Let a and b represent real numbers. Define

$$f(x) = \begin{cases} ax^2 + x - b & \text{if } x \le 2 \\ ax + b & \text{if } 2 < x < 5. \\ 2ax - 7 & \text{if } x \ge 5 \end{cases}$$

(a) Find the values of a and b such that f is continuous everywhere.

(b) Evaluate $\lim_{x \to 3} f(x)$.

(c) Let $g(x) = \dfrac{f(x)}{x - 1}$.

Evaluate $\lim_{x \to 1} g(x)$.

7.

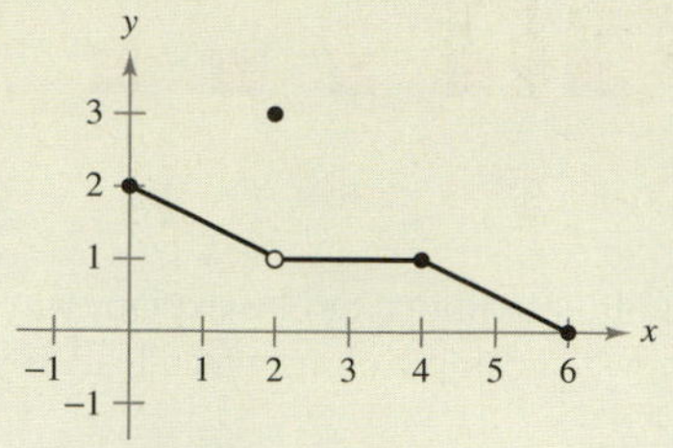

The graph of function g is shown above. Which of the following is true?

I. $\lim_{x \to 2} g(x) = 1$

II. $\lim_{x \to 2} g(x) = g(2)$

III. g is continuous at $x = 3$.

(A) I only
(B) I and II only
(C) I and III only
(D) III only
(E) I, II and III

8.

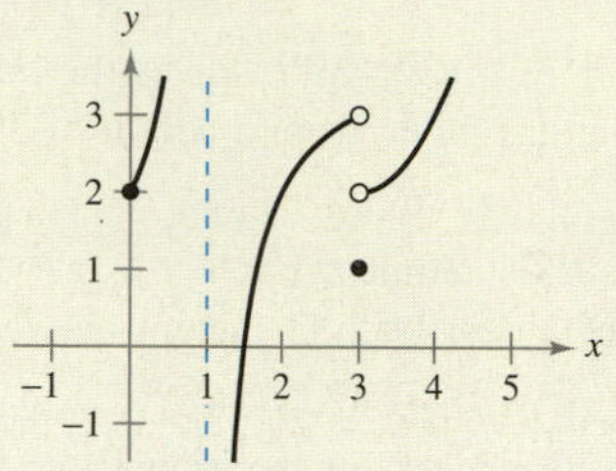

The graph of the function f is shown above. The line $x = 1$ is a vertical asymptote. Which of the following statements about f is true?

(A) $\lim_{x \to 1} = \infty$

(B) $\lim_{x \to 3} = 1$

(C) $\lim_{x \to 3^-} f(x) < \lim_{x \to 3^+} f(x)$

(D) $\lim_{x \to 4} f(x)$ does not exist

(E) $\lim_{x \to 0^+} f(x) = \lim_{x \to 3^+} f(x)$

9. Define $f(x) = \begin{cases} \dfrac{x^2 + 4x - 32}{x^2 - 2x - 8} & \text{if } x \neq -2, 4 \\ 8 & \text{if } x = 4 \end{cases}$

Which of the following statements about f are true?

I. f is not continuous at $x = 4$.

II. $\lim_{x \to \infty} f(x) = 4$

III. $x = 4$ is a vertical asymptote of the graph of $y = f(x)$.

(A) None
(B) I only
(C) I and II only
(D) I and III only
(E) I, II and III

10.

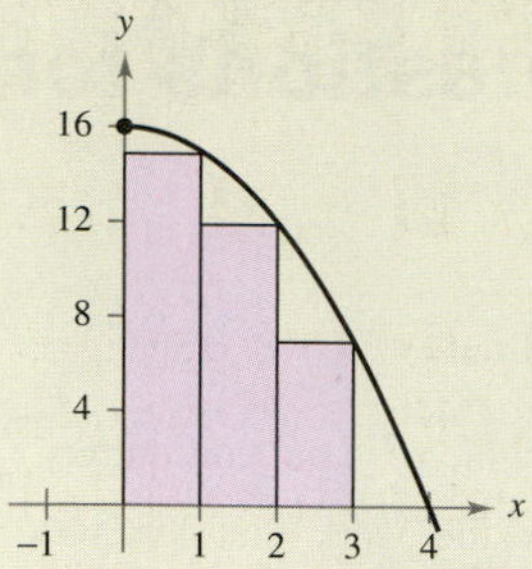

The figure above shows three rectangles each with a vertex on the graph of $y = 16 - x^2$. Find the sum of the areas of these rectangles

(A) 42 sq. units
(B) 40 sq. units
(C) 34 sq. units
(D) 33 sq. units
(E) 29 sq. units

2 Differentiation

Rate of Change (Example 2, p. 149)

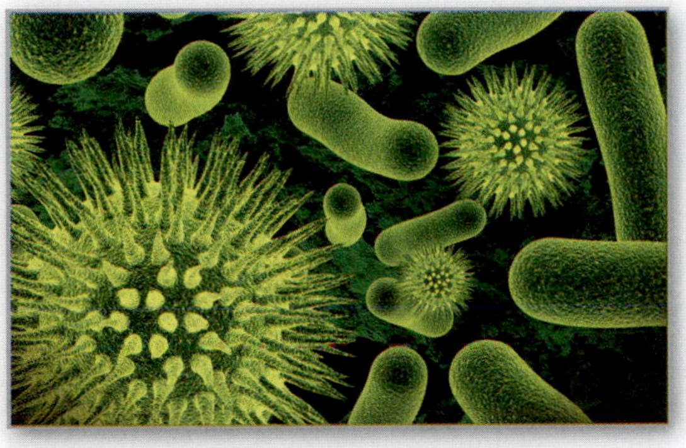

Bacteria (Exercise 111, p. 139)

Acceleration Due to Gravity (Example 10, p. 124)

Velocity of a Falling Object (Example 9, p. 112)

Stopping Distance (Exercise 107, p.117)

2.1 The Derivative and the Tangent Line Problem

- **Find the slope of the tangent line to a curve at a point.**
- **Use the limit definition to find the derivative of a function.**
- **Understand the relationship between differentiability and continuity.**

The Tangent Line Problem

Calculus grew out of four major problems that European mathematicians were working on during the seventeenth century.

1. The tangent line problem (Section 1.1 and this section)
2. The velocity and acceleration problem (Sections 2.2 and 2.3)
3. The minimum and maximum problem (Section 3.1)
4. The area problem (Sections 1.1 and 4.2)

Each problem involves the notion of a limit, and calculus can be introduced with any of the four problems.

A brief introduction to the tangent line problem is given in Section 1.1. Although partial solutions to this problem were given by Pierre de Fermat (1601–1665), René Descartes (1596–1650), Christian Huygens (1629–1695), and Isaac Barrow (1630–1677), credit for the first general solution is usually given to Isaac Newton (1642–1727) and Gottfried Leibniz (1646–1716). Newton's work on this problem stemmed from his interest in optics and light refraction.

ISAAC NEWTON (1642–1727)

In addition to his work in calculus, Newton made revolutionary contributions to physics, including the Law of Universal Gravitation and his three laws of motion.
See LarsonCalculus.com to read more of this biography.

What does it mean to say that a line is tangent to a curve at a point? For a circle, the tangent line at a point P is the line that is perpendicular to the radial line at point P, as shown in Figure 2.1.

For a general curve, however, the problem is more difficult. For instance, how would you define the tangent lines shown in Figure 2.2? You might say that a line is tangent to a curve at a point P when it touches, but does not cross, the curve at point P. This definition would work for the first curve shown in Figure 2.2, but not for the second. *Or* you might say that a line is tangent to a curve when the line touches or intersects the curve at exactly one point. This definition would work for a circle, but not for more general curves, as the third curve in Figure 2.2 shows.

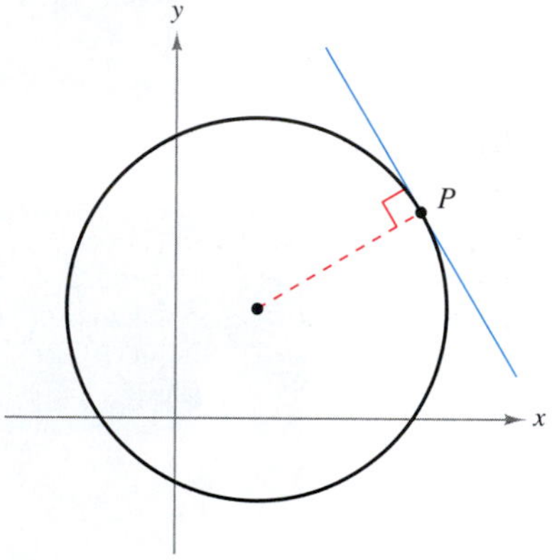

Tangent line to a circle
Figure 2.1

Exploration

Use a graphing utility to graph $f(x) = 2x^3 - 4x^2 + 3x - 5$. On the same screen, graph $y = x - 5$, $y = 2x - 5$, and $y = 3x - 5$. Which of these lines, if any, appears to be tangent to the graph of f at the point $(0, -5)$? Explain your reasoning.

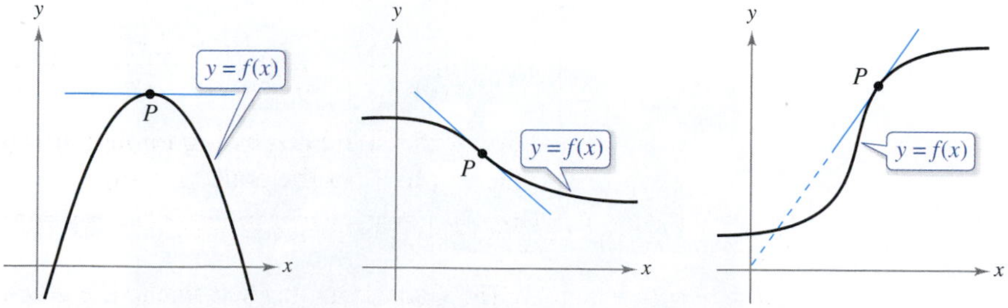

Tangent line to a curve at a point
Figure 2.2

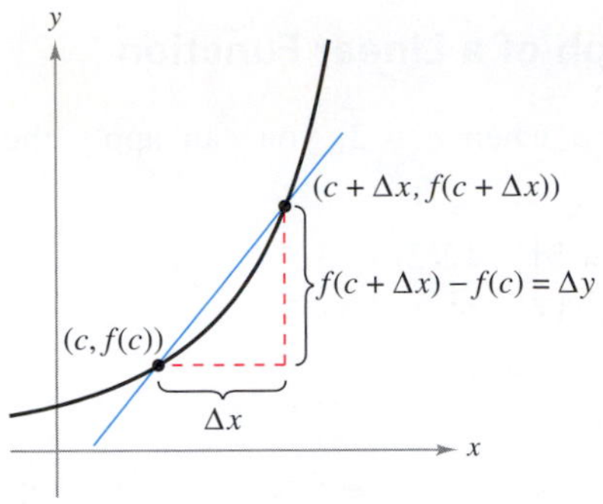

The secant line through $(c, f(c))$ and $(c + \Delta x, f(c + \Delta x))$
Figure 2.3

Essentially, the problem of finding the tangent line at a point P boils down to the problem of finding the *slope* of the tangent line at point P. You can approximate this slope using a **secant line*** through the point of tangency and a second point on the curve, as shown in Figure 2.3. If $(c, f(c))$ is the point of tangency and

$$(c + \Delta x, f(c + \Delta x))$$

is a second point on the graph of f, then the slope of the secant line through the two points is given by substitution into the slope formula

$$m = \frac{y_2 - y_1}{x_2 - x_1}$$

$$m_{\text{sec}} = \frac{f(c + \Delta x) - f(c)}{(c + \Delta x) - c} \qquad \frac{\text{Change in } y}{\text{Change in } x}$$

$$m_{\text{sec}} = \frac{f(c + \Delta x) - f(c)}{\Delta x}. \qquad \text{Slope of secant line}$$

The right-hand side of this equation is a **difference quotient.** The denominator Δx is the **change in x,** and the numerator

$$\Delta y = f(c + \Delta x) - f(c)$$

is the **change in y.**

The beauty of this procedure is that you can obtain more and more accurate approximations of the slope of the tangent line by choosing points closer and closer to the point of tangency, as shown in Figure 2.4.

> **THE TANGENT LINE PROBLEM**
>
> In 1637, mathematician René Descartes stated this about the tangent line problem:
>
> "And I dare say that this is not only the most useful and general problem in geometry that I know, but even that I ever desire to know."

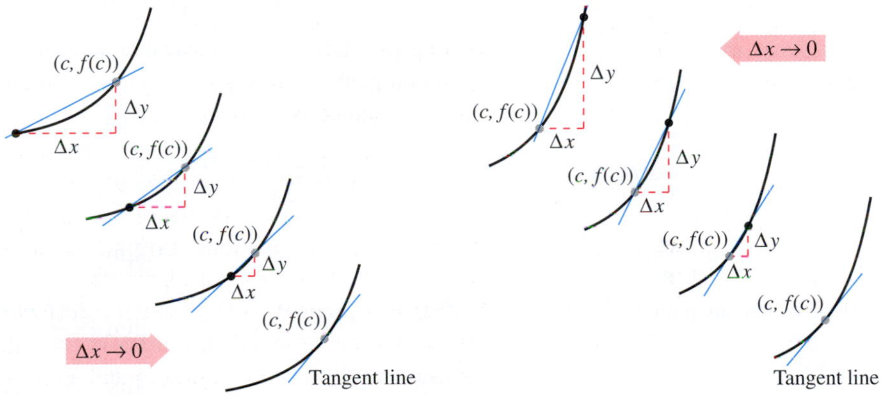

Tangent line approximations
Figure 2.4

> **Definition of Tangent Line with Slope m**
>
> If f is defined on an open interval containing c, and if the limit
>
> $$\lim_{\Delta x \to 0} \frac{\Delta y}{\Delta x} = \lim_{\Delta x \to 0} \frac{f(c + \Delta x) - f(c)}{\Delta x} = m$$
>
> exists, then the line passing through $(c, f(c))$ with slope m is the **tangent line** to the graph of f at the point $(c, f(c))$.

The slope of the tangent line to the graph of f at the point $(c, f(c))$ is also called the **slope of the graph of f at $x = c$.**

* This use of the word *secant* comes from the Latin *secare*, meaning to cut, and is not a reference to the trigonometric function of the same name.

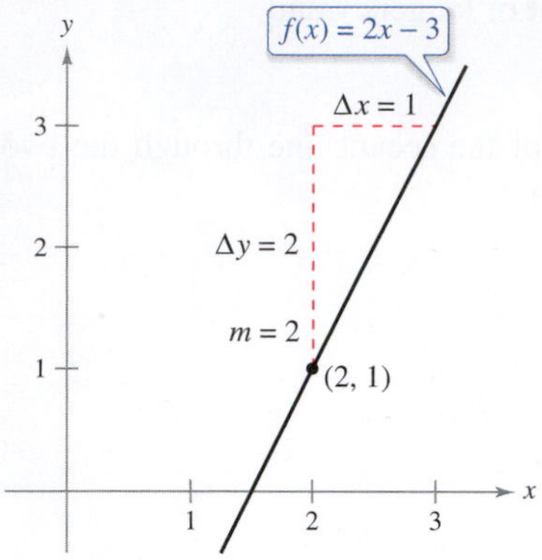

The slope of f at $(2, 1)$ is $m = 2$.
Figure 2.5

EXAMPLE 1 The Slope of the Graph of a Linear Function

To find the slope of the graph of $f(x) = 2x - 3$ when $c = 2$, you can apply the definition of the slope of a tangent line, as shown.

$$\begin{aligned}\lim_{\Delta x \to 0} \frac{f(2 + \Delta x) - f(2)}{\Delta x} &= \lim_{\Delta x \to 0} \frac{[2(2 + \Delta x) - 3] - [2(2) - 3]}{\Delta x} \\ &= \lim_{\Delta x \to 0} \frac{4 + 2\Delta x - 3 - 4 + 3}{\Delta x} \\ &= \lim_{\Delta x \to 0} \frac{2\Delta x}{\Delta x} \\ &= \lim_{\Delta x \to 0} 2 \\ &= 2\end{aligned}$$

The slope of f at $(c, f(c)) = (2, 1)$ is $m = 2$, as shown in Figure 2.5. Notice that the limit definition of the slope of f agrees with the definition of the slope of a line as discussed in Section P.2. ■

The graph of a linear function has the same slope at any point. This is not true of nonlinear functions, as shown in the next example.

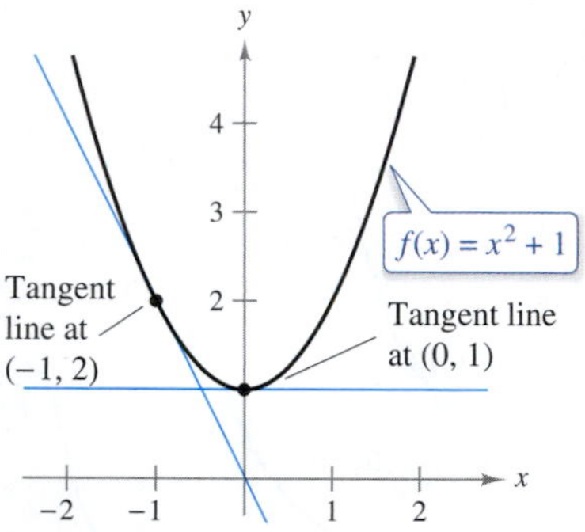

The slope of f at any point $(c, f(c))$ is $m = 2c$.
Figure 2.6

EXAMPLE 2 Tangent Lines to the Graph of a Nonlinear Function

Find the slopes of the tangent lines to the graph of $f(x) = x^2 + 1$ at the points $(0, 1)$ and $(-1, 2)$, as shown in Figure 2.6.

Solution Let $(c, f(c))$ represent an arbitrary point on the graph of f. Then the slope of the tangent line at $(c, f(c))$ can be found as shown below. [Note in the limit process that c is held constant (as Δx approaches 0).]

$$\begin{aligned}\lim_{\Delta x \to 0} \frac{f(c + \Delta x) - f(c)}{\Delta x} &= \lim_{\Delta x \to 0} \frac{[(c + \Delta x)^2 + 1] - (c^2 + 1)}{\Delta x} \\ &= \lim_{\Delta x \to 0} \frac{c^2 + 2c(\Delta x) + (\Delta x)^2 + 1 - c^2 - 1}{\Delta x} \\ &= \lim_{\Delta x \to 0} \frac{2c(\Delta x) + (\Delta x)^2}{\Delta x} \\ &= \lim_{\Delta x \to 0} (2c + \Delta x) \\ &= 2c\end{aligned}$$

So, the slope at *any* point $(c, f(c))$ on the graph of f is $m = 2c$. At the point $(0, 1)$, the slope is $m = 2(0) = 0$, and at $(-1, 2)$, the slope is $m = 2(-1) = -2$. ■

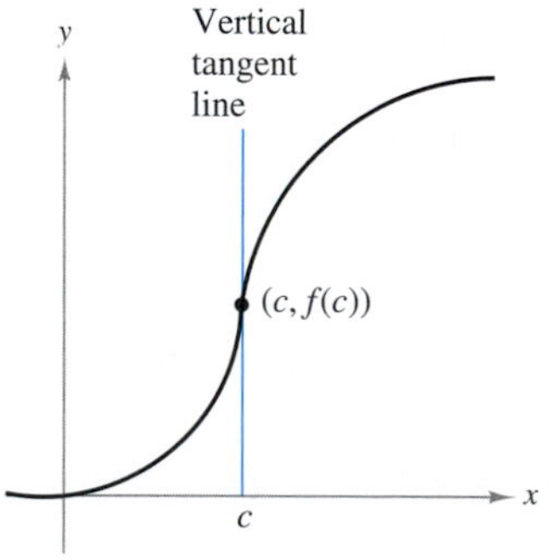

The graph of f has a vertical tangent line at $(c, f(c))$.
Figure 2.7

The definition of a tangent line to a curve does not cover the possibility of a vertical tangent line. For vertical tangent lines, you can use the following definition. If f is continuous at c and

$$\lim_{\Delta x \to 0} \frac{f(c + \Delta x) - f(c)}{\Delta x} = \infty \quad \text{or} \quad \lim_{\Delta x \to 0} \frac{f(c + \Delta x) - f(c)}{\Delta x} = -\infty$$

then the vertical line $x = c$ passing through $(c, f(c))$ is a **vertical tangent line** to the graph of f. For example, the function shown in Figure 2.7 has a vertical tangent line at $(c, f(c))$. When the domain of f is the closed interval $[a, b]$, you can extend the definition of a vertical tangent line to include the endpoints by considering continuity and limits from the right (for $x = a$) and from the left (for $x = b$).

AP* Tips
The definition of derivative is tested primarily on the multiple choice section of the AP Exam. You should master both the standard and alternative definitions and their geometric interpretations.

The Derivative of a Function

You have now arrived at a crucial point in the study of calculus. The limit used to define the slope of a tangent line is also used to define one of the two fundamental operations of calculus—**differentiation.**

Definition of the Derivative of a Function

The **derivative** of f at x is

$$f'(x) = \lim_{\Delta x \to 0} \frac{f(x + \Delta x) - f(x)}{\Delta x}$$

provided the limit exists. For all x for which this limit exists, f' is a function of x.

REMARK The notation $f'(x)$ is read as "f prime of x."

Be sure you see that the derivative of a function of x is also a function of x. This "new" function gives the slope of the tangent line to the graph of f at the point $(x, f(x))$, provided that the graph has a tangent line at this point. The derivative can also be used to determine the **instantaneous rate of change** (or simply the **rate of change**) of one variable with respect to another.

The process of finding the derivative of a function is called **differentiation.** A function is **differentiable** at x when its derivative exists at x and is **differentiable on an open interval (a, b)** when it is differentiable at every point in the interval.

In addition to $f'(x)$, other notations are used to denote the derivative of $y = f(x)$. The most common are

$$f'(x),\quad \frac{dy}{dx},\quad y',\quad \frac{d}{dx}[f(x)],\quad D_x[y].$$

Notation for derivatives

FOR FURTHER INFORMATION For more information on the crediting of mathematical discoveries to the first "discoverers," see the article "Mathematical Firsts—Who Done It?" by Richard H. Williams and Roy D. Mazzagatti in *Mathematics Teacher*. To view this article, go to *MathArticles.com*.

The notation dy/dx is read as "the derivative of y *with respect to* x" or simply "dy, dx." Using limit notation, you can write

$$\frac{dy}{dx} = \lim_{\Delta x \to 0} \frac{\Delta y}{\Delta x} = \lim_{\Delta x \to 0} \frac{f(x + \Delta x) - f(x)}{\Delta x} = f'(x).$$

EXAMPLE 3 Finding the Derivative by the Limit Process

See LarsonCalculus.com for an interactive version of this type of example.

To find the derivative of $f(x) = x^3 + 2x$, use the definition of the derivative as shown.

$$\begin{aligned} f'(x) &= \lim_{\Delta x \to 0} \frac{f(x + \Delta x) - f(x)}{\Delta x} && \text{Definition of derivative} \\ &= \lim_{\Delta x \to 0} \frac{(x + \Delta x)^3 + 2(x + \Delta x) - (x^3 + 2x)}{\Delta x} \\ &= \lim_{\Delta x \to 0} \frac{x^3 + 3x^2\Delta x + 3x(\Delta x)^2 + (\Delta x)^3 + 2x + 2\Delta x - x^3 - 2x}{\Delta x} \\ &= \lim_{\Delta x \to 0} \frac{3x^2\Delta x + 3x(\Delta x)^2 + (\Delta x)^3 + 2\Delta x}{\Delta x} \\ &= \lim_{\Delta x \to 0} \frac{\cancel{\Delta x}[3x^2 + 3x\Delta x + (\Delta x)^2 + 2]}{\cancel{\Delta x}} \\ &= \lim_{\Delta x \to 0} [3x^2 + 3x\Delta x + (\Delta x)^2 + 2] \\ &= 3x^2 + 2 \end{aligned}$$

REMARK When using the definition to find a derivative of a function, the key is to rewrite the difference quotient so that Δx does not occur as a factor of the denominator.

REMARK Remember that the derivative of a function f is itself a function, which can be used to find the slope of the tangent line at the point $(x, f(x))$ on the graph of f.

EXAMPLE 4 Using the Derivative to Find the Slope at a Point

Find $f'(x)$ for $f(x) = \sqrt{x}$. Then find the slopes of the graph of f at the points $(1, 1)$ and $(4, 2)$. Discuss the behavior of f at $(0, 0)$.

Solution Use the procedure for rationalizing numerators, as discussed in Section 1.3.

$$f'(x) = \lim_{\Delta x \to 0} \frac{f(x + \Delta x) - f(x)}{\Delta x} \qquad \text{Definition of derivative}$$

$$= \lim_{\Delta x \to 0} \frac{\sqrt{x + \Delta x} - \sqrt{x}}{\Delta x}$$

$$= \lim_{\Delta x \to 0} \left(\frac{\sqrt{x + \Delta x} - \sqrt{x}}{\Delta x}\right)\left(\frac{\sqrt{x + \Delta x} + \sqrt{x}}{\sqrt{x + \Delta x} + \sqrt{x}}\right)$$

$$= \lim_{\Delta x \to 0} \frac{(x + \Delta x) - x}{\Delta x\left(\sqrt{x + \Delta x} + \sqrt{x}\right)}$$

$$= \lim_{\Delta x \to 0} \frac{\Delta x}{\Delta x\left(\sqrt{x + \Delta x} + \sqrt{x}\right)}$$

$$= \lim_{\Delta x \to 0} \frac{1}{\sqrt{x + \Delta x} + \sqrt{x}}$$

$$= \frac{1}{2\sqrt{x}}, \quad x > 0$$

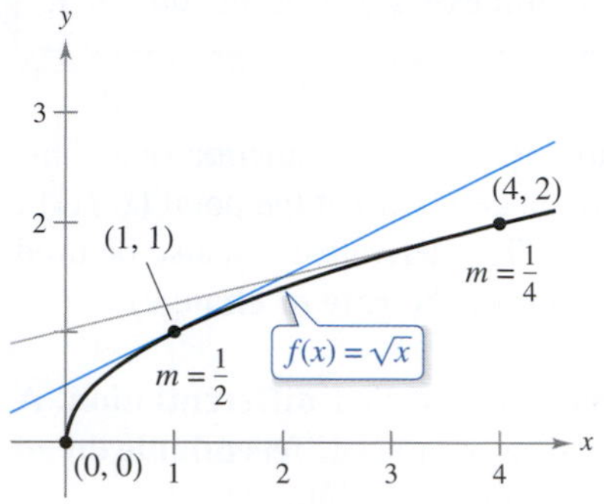

The slope of f at $(x, f(x))$, $x > 0$, is $m = 1/(2\sqrt{x})$.
Figure 2.8

At the point $(1, 1)$, the slope is $f'(1) = \frac{1}{2}$. At the point $(4, 2)$, the slope is $f'(4) = \frac{1}{4}$. See Figure 2.8. At the point $(0, 0)$, the slope is undefined. Moreover, the graph of f has a vertical tangent line at $(0, 0)$.

REMARK In many applications, it is convenient to use a variable other than x as the independent variable, as shown in Example 5.

EXAMPLE 5 Finding the Derivative of a Function

See LarsonCalculus.com for an interactive version of this type of example.

Find the derivative with respect to t for the function $y = 2/t$.

Solution Considering $y = f(t)$, you obtain

$$\frac{dy}{dt} = \lim_{\Delta t \to 0} \frac{f(t + \Delta t) - f(t)}{\Delta t} \qquad \text{Definition of derivative}$$

$$= \lim_{\Delta t \to 0} \frac{\dfrac{2}{t + \Delta t} - \dfrac{2}{t}}{\Delta t} \qquad f(t + \Delta t) = \frac{2}{t + \Delta t} \text{ and } f(t) = \frac{2}{t}$$

$$= \lim_{\Delta t \to 0} \frac{\dfrac{2t - 2(t + \Delta t)}{t(t + \Delta t)}}{\Delta t} \qquad \text{Combine fractions in numerator.}$$

$$= \lim_{\Delta t \to 0} \frac{-2\Delta t}{\Delta t(t)(t + \Delta t)} \qquad \text{Divide out common factor of } \Delta t.$$

$$= \lim_{\Delta t \to 0} \frac{-2}{t(t + \Delta t)} \qquad \text{Simplify.}$$

$$= -\frac{2}{t^2}. \qquad \text{Evaluate limit as } \Delta t \to 0.$$

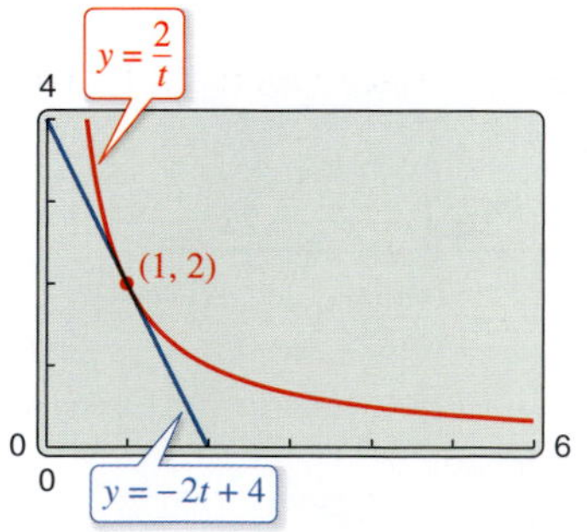

At the point $(1, 2)$, the line $y = -2t + 4$ is tangent to the graph of $y = 2/t$.
Figure 2.9

TECHNOLOGY A graphing utility can be used to reinforce the result given in Example 5. For instance, using the formula $dy/dt = -2/t^2$, you know that the slope of the graph of $y = 2/t$ at the point $(1, 2)$ is $m = -2$. Using the point-slope form, you can find that the equation of the tangent line to the graph at $(1, 2)$ is

$$y - 2 = -2(t - 1) \quad \text{or} \quad y = -2t + 4$$

as shown in Figure 2.9.

Differentiability and Continuity

The alternative limit form of the derivative shown below is useful in investigating the relationship between differentiability and continuity. The derivative of f at c is

$$f'(c) = \lim_{x \to c} \frac{f(x) - f(c)}{x - c}$$ Alternative form of derivative

provided this limit exists (see Figure 2.10).

REMARK A proof of the equivalence of the alternative form of the derivative is given in Appendix A.

See LarsonCalculus.com for Bruce Edwards's video of this proof.

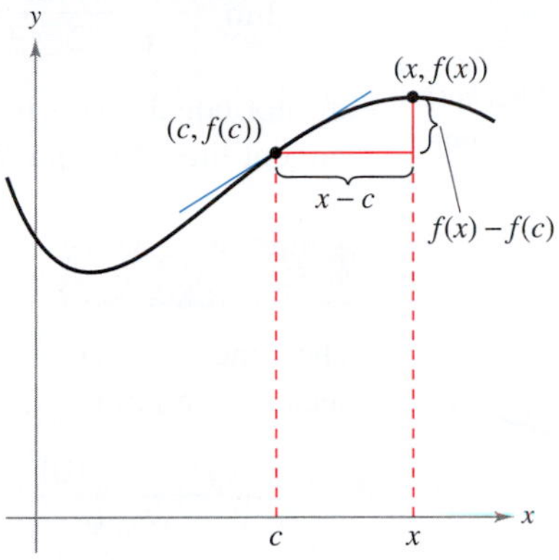

As x approaches c, the secant line approaches the tangent line.
Figure 2.10

Note that the existence of the limit in this alternative form requires that the one-sided limits

$$\lim_{x \to c^-} \frac{f(x) - f(c)}{x - c}$$

and

$$\lim_{x \to c^+} \frac{f(x) - f(c)}{x - c}$$

exist and are equal. These one-sided limits are called the **derivatives from the left and from the right,** respectively. It follows that f is **differentiable on the closed interval $[a, b]$** when it is differentiable on (a, b) and when the derivative from the right at a and the derivative from the left at b both exist.

When a function is not continuous at $x = c$, it is also not differentiable at $x = c$. For instance, the greatest integer function

$$f(x) = [\![x]\!]$$

is not continuous at $x = 0$, and so it is not differentiable at $x = 0$ (see Figure 2.11). You can verify this by observing that

$$\lim_{x \to 0^-} \frac{f(x) - f(0)}{x - 0} = \lim_{x \to 0^-} \frac{[\![x]\!] - 0}{x} = \infty$$ Derivative from the left

and

$$\lim_{x \to 0^+} \frac{f(x) - f(0)}{x - 0} = \lim_{x \to 0^+} \frac{[\![x]\!] - 0}{x} = 0.$$ Derivative from the right

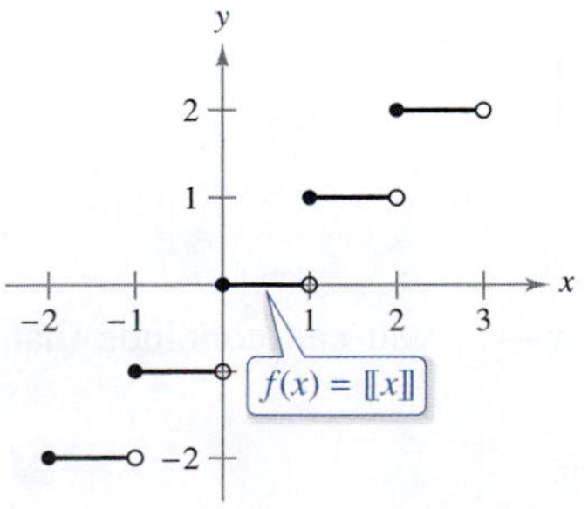

The greatest integer function is not differentiable at $x = 0$ because it is not continuous at $x = 0$.
Figure 2.11

Although it is true that differentiability implies continuity (as shown in Theorem 2.1 on the next page), the converse is not true. That is, it is possible for a function to be continuous at $x = c$ and *not* differentiable at $x = c$. Examples 6 and 7 illustrate this possibility.

EXAMPLE 6 A Graph with a Sharp Turn

See LarsonCalculus.com for an interactive version of this type of example.

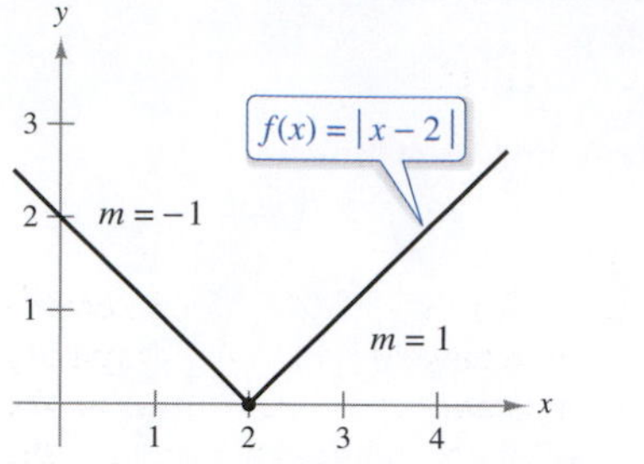

f is not differentiable at $x = 2$ because the derivatives from the left and from the right are not equal.
Figure 2.12

The function $f(x) = |x - 2|$, shown in Figure 2.12, is continuous at $x = 2$. The one-sided limits, however,

$$\lim_{x \to 2^-} \frac{f(x) - f(2)}{x - 2} = \lim_{x \to 2^-} \frac{|x - 2| - 0}{x - 2} = -1 \qquad \text{Derivative from the left}$$

and

$$\lim_{x \to 2^+} \frac{f(x) - f(2)}{x - 2} = \lim_{x \to 2^+} \frac{|x - 2| - 0}{x - 2} = 1 \qquad \text{Derivative from the right}$$

are not equal. So, f is not differentiable at $x = 2$ and the graph of f does not have a tangent line at the point $(2, 0)$.

EXAMPLE 7 A Graph with a Vertical Tangent Line

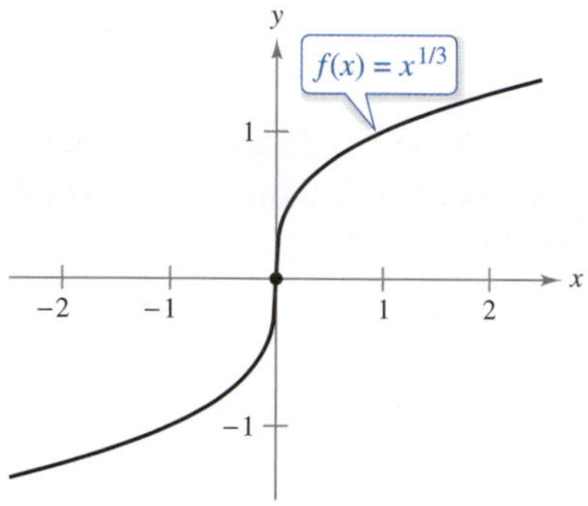

f is not differentiable at $x = 0$ because f has a vertical tangent line at $x = 0$.
Figure 2.13

The function $f(x) = x^{1/3}$ is continuous at $x = 0$, as shown in Figure 2.13. However, because the limit

$$\lim_{x \to 0} \frac{f(x) - f(0)}{x - 0} = \lim_{x \to 0} \frac{x^{1/3} - 0}{x} = \lim_{x \to 0} \frac{1}{x^{2/3}} = \infty$$

is infinite, you can conclude that the tangent line is vertical at $x = 0$. So, f is not differentiable at $x = 0$.

From Examples 6 and 7, you can see that a function is not differentiable at a point at which its graph has a sharp turn *or* a vertical tangent line.

THEOREM 2.1 Differentiability Implies Continuity

If f is differentiable at $x = c$, then f is continuous at $x = c$.

Proof You can prove that f is continuous at $x = c$ by showing that $f(x)$ approaches $f(c)$ as $x \to c$. To do this, use the differentiability of f at $x = c$ and consider the following limit.

$$\begin{aligned}
\lim_{x \to c} [f(x) - f(c)] &= \lim_{x \to c} \left[(x - c)\left(\frac{f(x) - f(c)}{x - c}\right)\right] \\
&= \left[\lim_{x \to c} (x - c)\right]\left[\lim_{x \to c} \frac{f(x) - f(c)}{x - c}\right] \\
&= (0)[f'(c)] \\
&= 0
\end{aligned}$$

Because the difference $f(x) - f(c)$ approaches zero as $x \to c$, you can conclude that $\lim_{x \to c} f(x) = f(c)$. So, f is continuous at $x = c$.

See LarsonCalculus.com for Bruce Edwards's video of this proof.

TECHNOLOGY Some graphing utilities, such as *Maple*, *Mathematica*, and the *TI-nspire*, perform symbolic differentiation. Others perform *numerical differentiation* by finding values of derivatives using the formula

$$f'(x) \approx \frac{f(x + \Delta x) - f(x - \Delta x)}{2\Delta x}$$

where Δx is a small number such as 0.001. Can you see any problems with this definition? For instance, using this definition, what is the value of the derivative of $f(x) = |x|$ when $x = 0$?

The relationship between continuity and differentiability is summarized below.

1. If a function is differentiable at $x = c$, then it is continuous at $x = c$. So, differentiability implies continuity.
2. It is possible for a function to be continuous at $x = c$ and not be differentiable at $x = c$. So, continuity does not imply differentiability (see Example 6).

2.1 Exercises

See CalcChat.com for tutorial help and worked-out solutions to odd-numbered exercises.

Estimating Slope In Exercises 1 and 2, estimate the slope of the graph at the points (x_1, y_1) and (x_2, y_2).

1.

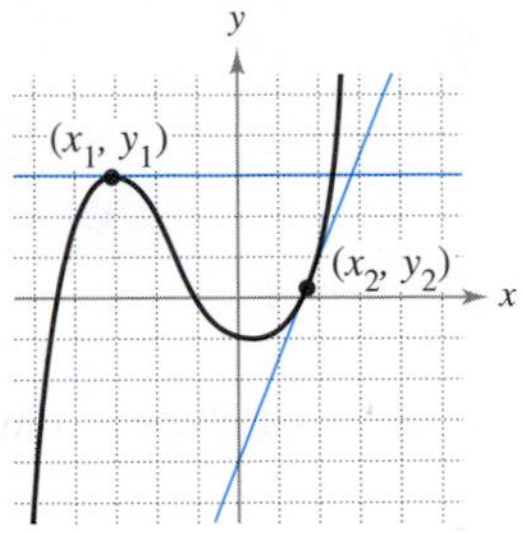

2.

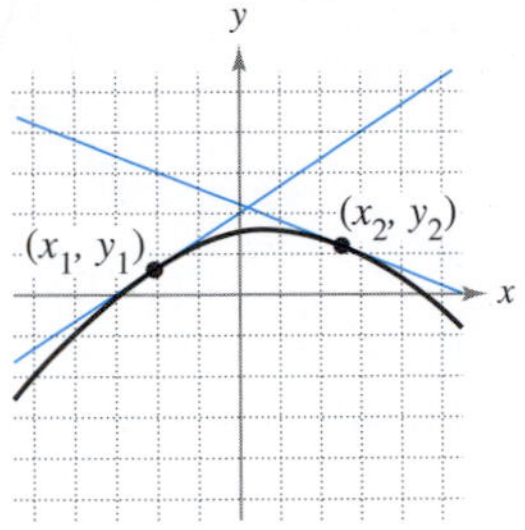

Slopes of Secant Lines In Exercises 3 and 4, use the graph shown in the figure. To print an enlarged copy of the graph, go to *MathGraphs.com*.

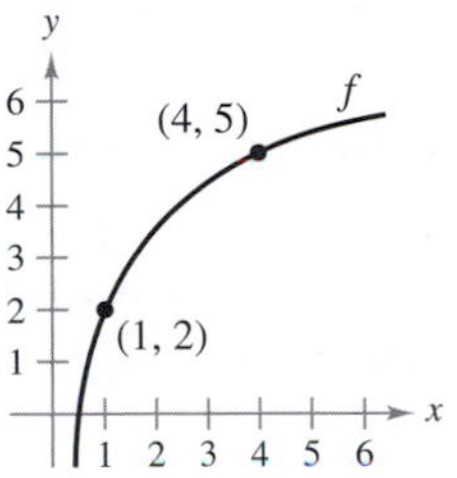

3. Identify or sketch each of the quantities on the figure.

(a) $f(1)$ and $f(4)$ (b) $f(4) - f(1)$

(c) $y = \dfrac{f(4) - f(1)}{4 - 1}(x - 1) + f(1)$

4. Insert the proper inequality symbol (< or >) between the given quantities.

(a) $\dfrac{f(4) - f(1)}{4 - 1}$ ▒ $\dfrac{f(4) - f(3)}{4 - 3}$

(b) $\dfrac{f(4) - f(1)}{4 - 1}$ ▒ $f'(1)$

Finding the Slope of a Tangent Line In Exercises 5–10, find the slope of the tangent line to the graph of the function at the given point.

5. $f(x) = 3 - 5x$, $(-1, 8)$ **6.** $g(x) = \frac{3}{2}x + 1$, $(-2, -2)$

7. $g(x) = x^2 - 9$, $(2, -5)$ **8.** $f(x) = 5 - x^2$, $(3, -4)$

9. $f(t) = 3t - t^2$, $(0, 0)$ **10.** $h(t) = t^2 + 4t$, $(1, 5)$

Finding the Derivative by the Limit Process In Exercises 11–24, find the derivative of the function by the limit process.

11. $f(x) = 7$ **12.** $g(x) = -3$

13. $f(x) = -10x$ **14.** $f(x) = 7x - 3$

15. $h(s) = 3 + \frac{2}{3}s$ **16.** $f(x) = 5 - \frac{2}{3}x$

17. $f(x) = x^2 + x - 3$ **18.** $f(x) = x^2 - 5$

19. $f(x) = x^3 - 12x$ **20.** $f(x) = x^3 + x^2$

21. $f(x) = \dfrac{1}{x - 1}$ **22.** $f(x) = \dfrac{1}{x^2}$

23. $f(x) = \sqrt{x + 4}$ **24.** $f(x) = \dfrac{4}{\sqrt{x}}$

Finding an Equation of a Tangent Line In Exercises 25–32, (a) find an equation of the tangent line to the graph of f at the given point, (b) use a graphing utility to graph the function and its tangent line at the point, and (c) use the *derivative* feature of a graphing utility to confirm your results.

25. $f(x) = x^2 + 3$, $(-1, 4)$ **26.** $f(x) = x^2 + 2x - 1$, $(1, 2)$

27. $f(x) = x^3$, $(2, 8)$ **28.** $f(x) = x^3 + 1$, $(-1, 0)$

29. $f(x) = \sqrt{x}$, $(1, 1)$ **30.** $f(x) = \sqrt{x - 1}$, $(5, 2)$

31. $f(x) = x + \dfrac{4}{x}$, $(-4, -5)$ **32.** $f(x) = \dfrac{6}{x + 2}$, $(0, 3)$

Finding an Equation of a Tangent Line In Exercises 33–38, find an equation of the line that is tangent to the graph of f and parallel to the given line.

	Function	Line
33.	$f(x) = x^2$	$2x - y + 1 = 0$
34.	$f(x) = 2x^2$	$4x + y + 3 = 0$
35.	$f(x) = x^3$	$3x - y + 1 = 0$
36.	$f(x) = x^3 + 2$	$3x - y - 4 = 0$
37.	$f(x) = \dfrac{1}{\sqrt{x}}$	$x + 2y - 6 = 0$
38.	$f(x) = \dfrac{-1}{\sqrt{x - 1}}$	$x + 2y + 7 = 0$

WRITING ABOUT CONCEPTS

Sketching a Derivative In Exercises 39–44, sketch the graph of f'. Explain how you found your answer.

39.

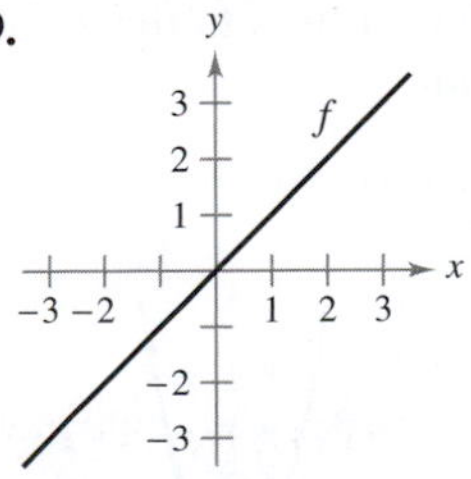

40.

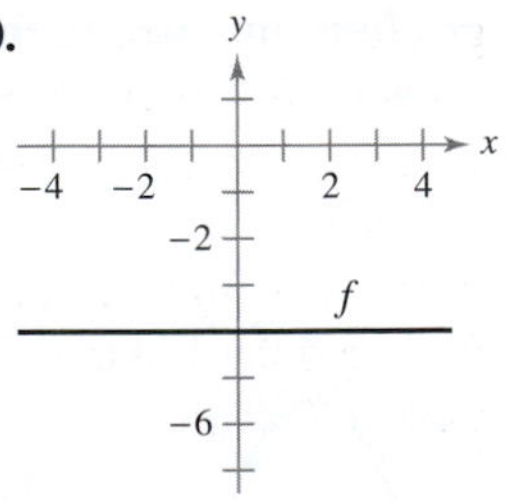

41.

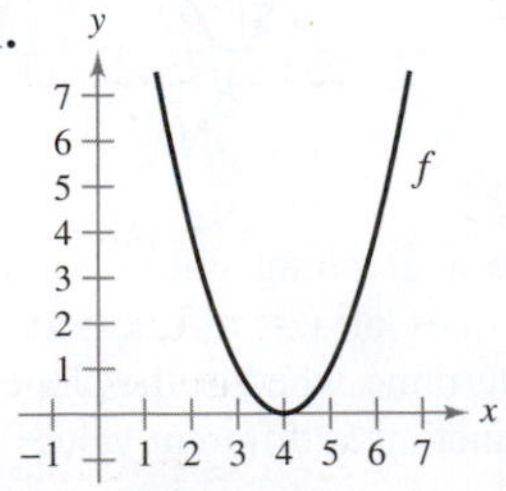

42.

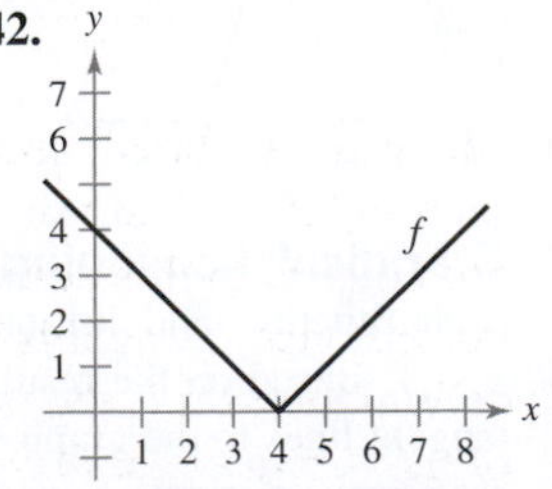

WRITING ABOUT CONCEPTS (continued)

43.

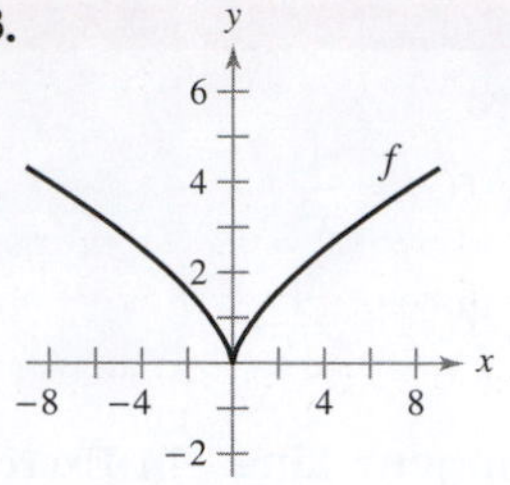

44.

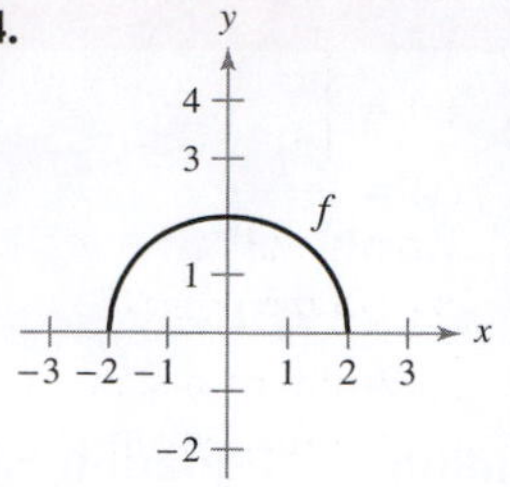

45. Sketching a Graph Sketch a graph of a function whose derivative is always negative. Explain how you found the answer.

46. Sketching a Graph Sketch a graph of a function whose derivative is always positive. Explain how you found the answer.

47. Using a Tangent Line The tangent line to the graph of $y = g(x)$ at the point $(4, 5)$ passes through the point $(7, 0)$. Find $g(4)$ and $g'(4)$.

48. Using a Tangent Line The tangent line to the graph of $y = h(x)$ at the point $(-1, 4)$ passes through the point $(3, 6)$. Find $h(-1)$ and $h'(-1)$.

Working Backwards **In Exercises 49–52, the limit represents $f'(c)$ for a function f and a number c. Find f and c.**

49. $\lim\limits_{\Delta x \to 0} \dfrac{[5 - 3(1 + \Delta x)] - 2}{\Delta x}$ **50.** $\lim\limits_{\Delta x \to 0} \dfrac{(-2 + \Delta x)^3 + 8}{\Delta x}$

51. $\lim\limits_{x \to 6} \dfrac{-x^2 + 36}{x - 6}$ **52.** $\lim\limits_{x \to 9} \dfrac{2\sqrt{x} - 6}{x - 9}$

Writing a Function Using Derivatives **In Exercises 53 and 54, identify a function f that has the given characteristics. Then sketch the function.**

53. $f(0) = 2;\ f'(x) = -3$ for $-\infty < x < \infty$

54. $f(0) = 4;\ f'(0) = 0;\ f'(x) < 0$ for $x < 0;\ f'(x) > 0$ for $x > 0$

Finding an Equation of a Tangent Line **In Exercises 55 and 56, find equations of the two tangent lines to the graph of f that pass through the indicated point.**

55. $f(x) = 4x - x^2$ **56.** $f(x) = x^2$

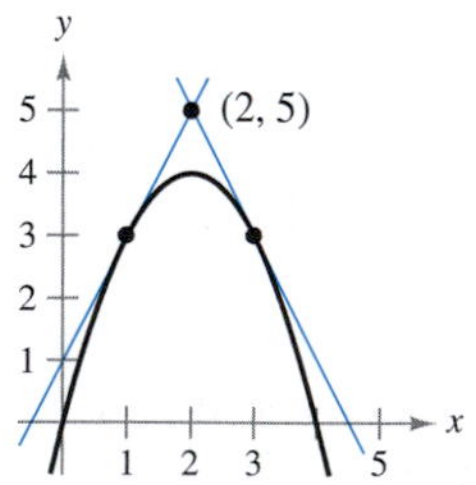

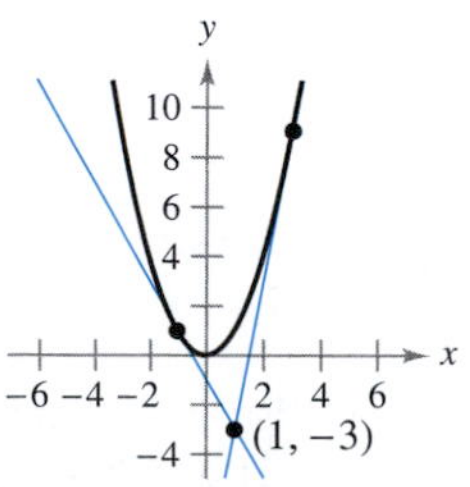

57. Graphical Reasoning Use a graphing utility to graph each function and its tangent lines at $x = -1$, $x = 0$, and $x = 1$. Based on the results, determine whether the slopes of tangent lines to the graph of a function at different values of x are always distinct.

(a) $f(x) = x^2$ (b) $g(x) = x^3$

58. HOW DO YOU SEE IT? The figure shows the graph of g'.

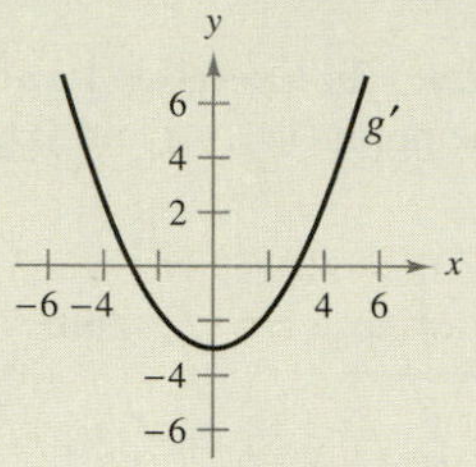

(a) $g'(0) = \square$ (b) $g'(3) = \square$

(c) What can you conclude about the graph of g knowing that $g'(1) = -\frac{8}{3}$?

(d) What can you conclude about the graph of g knowing that $g'(-4) = \frac{7}{3}$?

(e) Is $g(6) - g(4)$ positive or negative? Explain.

(f) Is it possible to find $g(2)$ from the graph? Explain.

59. Graphical Reasoning Consider the function $f(x) = \frac{1}{2}x^2$.

(a) Use a graphing utility to graph the function and estimate the values of $f'(0)$, $f'\left(\frac{1}{2}\right)$, $f'(1)$, and $f'(2)$.

(b) Use your results from part (a) to determine the values of $f'\left(-\frac{1}{2}\right)$, $f'(-1)$, and $f'(-2)$.

(c) Sketch a possible graph of f'.

(d) Use the definition of derivative to find $f'(x)$.

60. Graphical Reasoning Consider the function $f(x) = \frac{1}{3}x^3$.

(a) Use a graphing utility to graph the function and estimate the values of $f'(0)$, $f'\left(\frac{1}{2}\right)$, $f'(1)$, $f'(2)$, and $f'(3)$.

(b) Use your results from part (a) to determine the values of $f'\left(-\frac{1}{2}\right)$, $f'(-1)$, $f'(-2)$, and $f'(-3)$.

(c) Sketch a possible graph of f'.

(d) Use the definition of derivative to find $f'(x)$.

Graphical Reasoning **In Exercises 61 and 62, use a graphing utility to graph the functions f and g in the same viewing window, where**

$$g(x) = \frac{f(x + 0.01) - f(x)}{0.01}.$$

Label the graphs and describe the relationship between them.

61. $f(x) = 2x - x^2$ **62.** $f(x) = 3\sqrt{x}$

Approximating a Derivative **In Exercises 63 and 64, evaluate $f(2)$ and $f(2.1)$ and use the results to approximate $f'(2)$.**

63. $f(x) = x(4 - x)$ **64.** $f(x) = \frac{1}{4}x^3$

Using the Alternative Form of the Derivative **In Exercises 65–74, use the alternative form of the derivative to find the derivative at $x = c$ (if it exists).**

65. $f(x) = x^2 - 5,\ \ c = 3$ **66.** $g(x) = x^2 - x,\ \ c = 1$

67. $f(x) = x^3 + 2x^2 + 1,\ \ c = -2$

68. $f(x) = x^3 + 6x, \quad c = 2$

69. $g(x) = \sqrt{|x|}, \quad c = 0$

70. $f(x) = 3/x, \quad c = 4$

71. $f(x) = (x - 6)^{2/3}, \quad c = 6$

72. $g(x) = (x + 3)^{1/3}, \quad c = -3$

73. $h(x) = |x + 7|, \quad c = -7$

74. $f(x) = |x - 6|, \quad c = 6$

Determining Differentiability **In Exercises 75–80, describe the x-values at which f is differentiable.**

75. $f(x) = \dfrac{2}{x - 3}$

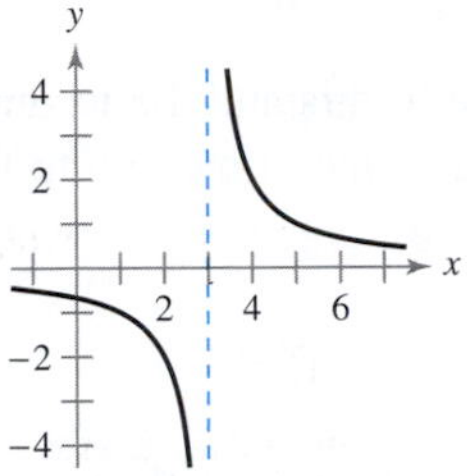

76. $f(x) = |x^2 - 9|$

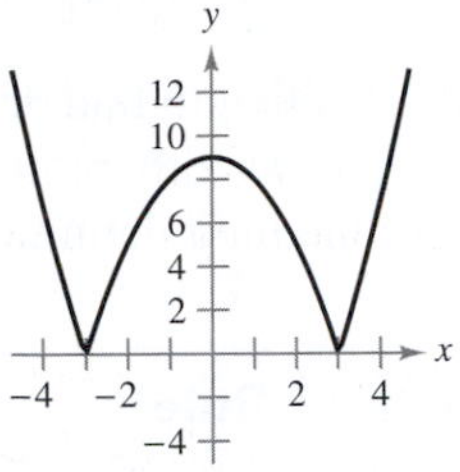

77. $f(x) = (x + 4)^{2/3}$

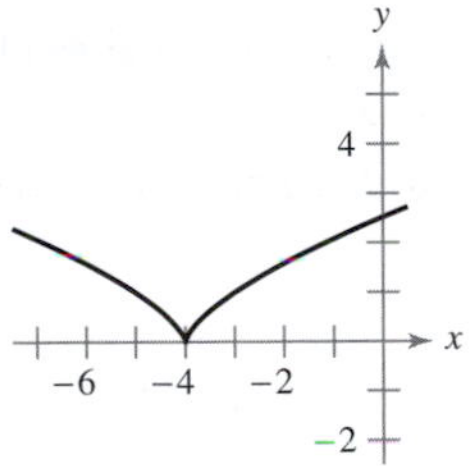

78. $f(x) = \dfrac{x^2}{x^2 - 4}$

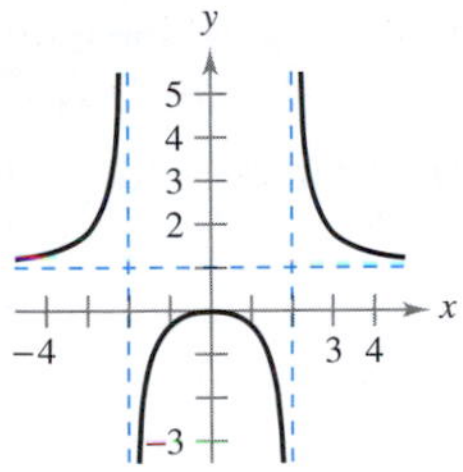

79. $f(x) = \sqrt{x - 1}$

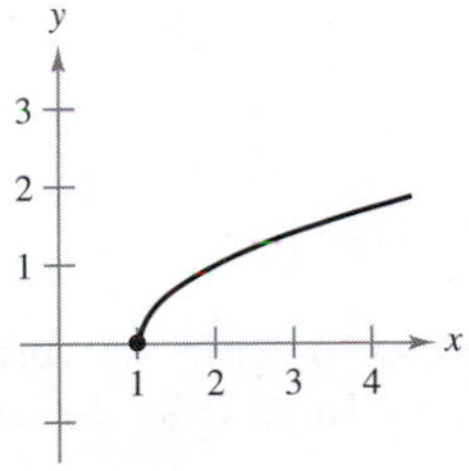

80. $f(x) = \begin{cases} x^2 - 4, & x \le 0 \\ 4 - x^2, & x > 0 \end{cases}$

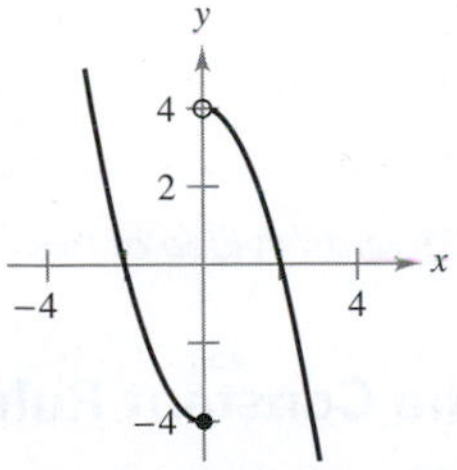

Graphical Reasoning **In Exercises 81–84, use a graphing utility to graph the function and find the x-values at which f is differentiable.**

81. $f(x) = |x - 5|$

82. $f(x) = \dfrac{4x}{x - 3}$

83. $f(x) = x^{2/5}$

84. $f(x) = \begin{cases} x^3 - 3x^2 + 3x, & x \le 1 \\ x^2 - 2x, & x > 1 \end{cases}$

Determining Differentiability **In Exercises 85–88, find the derivatives from the left and from the right at $x = 1$ (if they exist). Is the function differentiable at $x = 1$?**

85. $f(x) = |x - 1|$

86. $f(x) = \sqrt{1 - x^2}$

87. $f(x) = \begin{cases} (x - 1)^3, & x \le 1 \\ (x - 1)^2, & x > 1 \end{cases}$

88. $f(x) = \begin{cases} x, & x \le 1 \\ x^2, & x > 1 \end{cases}$

Determining Differentiability **In Exercises 89 and 90, determine whether the function is differentiable at $x = 2$.**

89. $f(x) = \begin{cases} x^2 + 1, & x \le 2 \\ 4x - 3, & x > 2 \end{cases}$

90. $f(x) = \begin{cases} \frac{1}{2}x + 1, & x < 2 \\ \sqrt{2x}, & x \ge 2 \end{cases}$

91. Graphical Reasoning A line with slope m passes through the point $(0, 4)$ and has the equation $y = mx + 4$.

(a) Write the distance d between the line and the point $(3, 1)$ as a function of m.

(b) Use a graphing utility to graph the function d in part (a). Based on the graph, is the function differentiable at every value of m? If not, where is it not differentiable?

92. Conjecture Consider the functions $f(x) = x^2$ and $g(x) = x^3$.

(a) Graph f and f' on the same set of axes.

(b) Graph g and g' on the same set of axes.

(c) Identify a pattern between f and g and their respective derivatives. Use the pattern to make a conjecture about $h'(x)$ if $h(x) = x^n$, where n is an integer and $n \ge 2$.

(d) Find $f'(x)$ if $f(x) = x^4$. Compare the result with the conjecture in part (c). Is this a proof of your conjecture? Explain.

True or False? **In Exercises 93–96, determine whether the statement is true or false. If it is false, explain why or give an example that shows it is false.**

93. The slope of the tangent line to the differentiable function f at the point $(2, f(2))$ is

$$\frac{f(2 + \Delta x) - f(2)}{\Delta x}.$$

94. If a function is continuous at a point, then it is differentiable at that point.

95. If a function has derivatives from both the right and the left at a point, then it is differentiable at that point.

96. If a function is differentiable at a point, then it is continuous at that point.

97. Differentiability and Continuity Let

$$f(x) = \begin{cases} x \sin \dfrac{1}{x}, & x \ne 0 \\ 0, & x = 0 \end{cases}$$

and

$$g(x) = \begin{cases} x^2 \sin \dfrac{1}{x}, & x \ne 0 \\ 0, & x = 0 \end{cases}.$$

Show that f is continuous, but not differentiable, at $x = 0$. Show that g is differentiable at 0, and find $g'(0)$.

98. Writing Use a graphing utility to graph the two functions $f(x) = x^2 + 1$ and $g(x) = |x| + 1$ in the same viewing window. Use the *zoom* and *trace* features to analyze the graphs near the point $(0, 1)$. What do you observe? Which function is differentiable at this point? Write a short paragraph describing the geometric significance of differentiability at a point.

2.2 Basic Differentiation Rules and Rates of Change

- Find the derivative of a function using the Constant Rule.
- Find the derivative of a function using the Power Rule.
- Find the derivative of a function using the Constant Multiple Rule.
- Find the derivative of a function using the Sum and Difference Rules.
- Find the derivatives of the sine function and of the cosine function.
- Use derivatives to find rates of change.

The Constant Rule

In Section 2.1, you used the limit definition to find derivatives. In this and the next two sections, you will be introduced to several "differentiation rules" that allow you to find derivatives without the *direct* use of the limit definition.

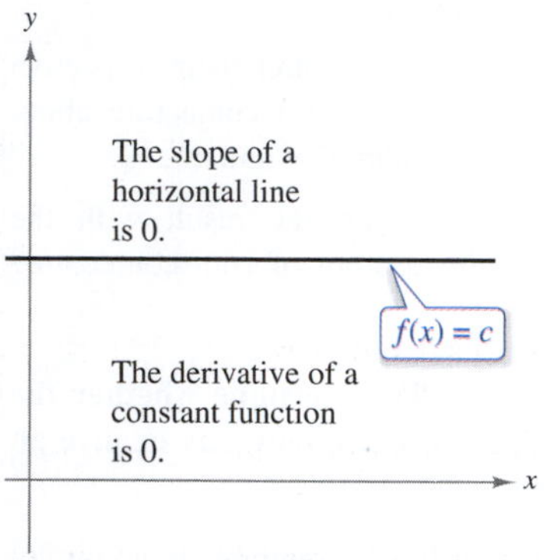

Notice that the Constant Rule is equivalent to saying that the slope of a horizontal line is 0. This demonstrates the relationship between slope and derivative.
Figure 2.14

THEOREM 2.2 The Constant Rule

The derivative of a constant function is 0. That is, if c is a real number, then

$$\frac{d}{dx}[c] = 0. \qquad \text{(See Figure 2.14.)}$$

Proof Let $f(x) = c$. Then, by the limit definition of the derivative,

$$\begin{aligned}\frac{d}{dx}[c] &= f'(x)\\ &= \lim_{\Delta x \to 0} \frac{f(x + \Delta x) - f(x)}{\Delta x}\\ &= \lim_{\Delta x \to 0} \frac{c - c}{\Delta x}\\ &= \lim_{\Delta x \to 0} 0\\ &= 0.\end{aligned}$$

See LarsonCalculus.com for Bruce Edwards's video of this proof.

EXAMPLE 1 Using the Constant Rule

	Function	Derivative
a.	$y = 7$	$dy/dx = 0$
b.	$f(x) = 0$	$f'(x) = 0$
c.	$s(t) = -3$	$s'(t) = 0$
d.	$y = k\pi^2$, k is constant	$y' = 0$

Exploration

Writing a Conjecture Use the definition of the derivative given in Section 2.1 to find the derivative of each function. What patterns do you see? Use your results to write a conjecture about the derivative of $f(x) = x^n$.

a. $f(x) = x^1$ **b.** $f(x) = x^2$ **c.** $f(x) = x^3$
d. $f(x) = x^4$ **e.** $f(x) = x^{1/2}$ **f.** $f(x) = x^{-1}$

The Power Rule

Before proving the next rule, it is important to review the procedure for expanding a binomial.

$$(x + \Delta x)^2 = x^2 + 2x\Delta x + (\Delta x)^2$$
$$(x + \Delta x)^3 = x^3 + 3x^2\Delta x + 3x(\Delta x)^2 + (\Delta x)^3$$
$$(x + \Delta x)^4 = x^4 + 4x^3\Delta x + 6x^2(\Delta x)^2 + 4x(\Delta x)^3 + (\Delta x)^4$$
$$(x + \Delta x)^5 = x^5 + 5x^4\Delta x + 10x^3(\Delta x)^2 + 10x^2(\Delta x)^3 + 5x(\Delta x)^4 + (\Delta x)^5$$

The general binomial expansion for a positive integer n is

$$(x + \Delta x)^n = x^n + nx^{n-1}(\Delta x) + \underbrace{\frac{n(n-1)x^{n-2}}{2}(\Delta x)^2 + \cdots + (\Delta x)^n}_{(\Delta x)^2 \text{ is a factor of these terms.}}.$$

This binomial expansion is used in proving a special case of the Power Rule.

THEOREM 2.3 The Power Rule

If n is a rational number, then the function $f(x) = x^n$ is differentiable and

$$\frac{d}{dx}[x^n] = nx^{n-1}.$$

For f to be differentiable at $x = 0$, n must be a number such that x^{n-1} is defined on an interval containing 0.

REMARK From Example 7 in Section 2.1, you know that the function $f(x) = x^{1/3}$ is defined at $x = 0$, but is not differentiable at $x = 0$. This is because $x^{-2/3}$ is not defined on an interval containing 0.

Proof If n is a positive integer greater than 1, then the binomial expansion produces

$$\begin{aligned}
\frac{d}{dx}[x^n] &= \lim_{\Delta x \to 0} \frac{(x + \Delta x)^n - x^n}{\Delta x} \\
&= \lim_{\Delta x \to 0} \frac{x^n + nx^{n-1}(\Delta x) + \frac{n(n-1)x^{n-2}}{2}(\Delta x)^2 + \cdots + (\Delta x)^n - x^n}{\Delta x} \\
&= \lim_{\Delta x \to 0} \left[nx^{n-1} + \frac{n(n-1)x^{n-2}}{2}(\Delta x) + \cdots + (\Delta x)^{n-1}\right] \\
&= nx^{n-1} + 0 + \cdots + 0 \\
&= nx^{n-1}.
\end{aligned}$$

This proves the case for which n is a positive integer greater than 1. It is left to you to prove the case for $n = 1$. Example 7 in Section 2.3 proves the case for which n is a negative integer. In Exercise 71 in Section 2.5, you are asked to prove the case for which n is rational. (In Section 5.5, the Power Rule will be extended to cover irrational values of n.)

See LarsonCalculus.com for Bruce Edwards's video of this proof.

When using the Power Rule, the case for which $n = 1$ is best thought of as a separate differentiation rule. That is,

$$\frac{d}{dx}[x] = 1. \qquad \text{Power Rule when } n = 1$$

This rule is consistent with the fact that the slope of the line $y = x$ is 1, as shown in Figure 2.15.

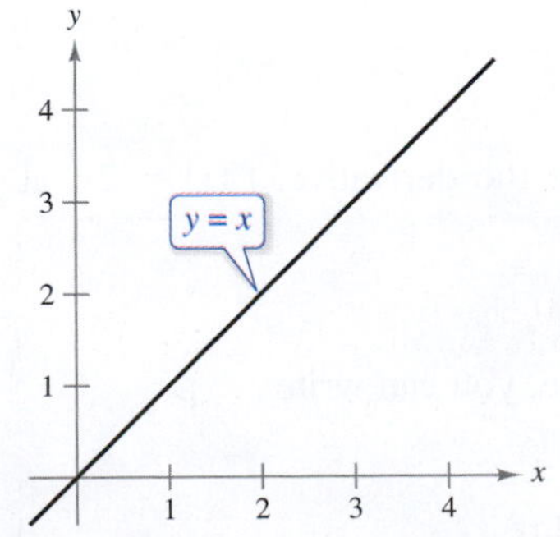

The slope of the line $y = x$ is 1.
Figure 2.15

EXAMPLE 2 Using the Power Rule

Function	Derivative
a. $f(x) = x^3$	$f'(x) = 3x^2$
b. $g(x) = \sqrt[3]{x}$	$g'(x) = \dfrac{d}{dx}[x^{1/3}] = \dfrac{1}{3}x^{-2/3} = \dfrac{1}{3x^{2/3}}$
c. $y = \dfrac{1}{x^2}$	$\dfrac{dy}{dx} = \dfrac{d}{dx}[x^{-2}] = (-2)x^{-3} = -\dfrac{2}{x^3}$

In Example 2(c), note that *before* differentiating, $1/x^2$ was rewritten as x^{-2}. Rewriting is the first step in *many* differentiation problems.

Given: $y = \dfrac{1}{x^2}$ ⇨ Rewrite: $y = x^{-2}$ ⇨ Differentiate: $\dfrac{dy}{dx} = (-2)x^{-3}$ ⇨ Simplify: $\dfrac{dy}{dx} = -\dfrac{2}{x^3}$

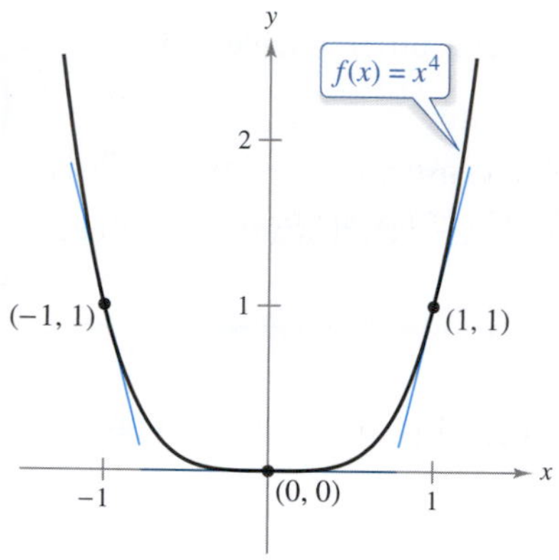

Note that the slope of the graph is negative at the point $(-1, 1)$, the slope is zero at the point $(0, 0)$, and the slope is positive at the point $(1, 1)$.
Figure 2.16

EXAMPLE 3 Finding the Slope of a Graph

See LarsonCalculus.com for an interactive version of this type of example.

Find the slope of the graph of

$$f(x) = x^4$$

for each value of x.

a. $x = -1$ **b.** $x = 0$ **c.** $x = 1$

Solution The slope of a graph at a point is the value of the derivative at that point. The derivative of f is $f'(x) = 4x^3$.

a. When $x = -1$, the slope is $f'(-1) = 4(-1)^3 = -4$. Slope is negative.

b. When $x = 0$, the slope is $f'(0) = 4(0)^3 = 0$. Slope is zero.

c. When $x = 1$, the slope is $f'(1) = 4(1)^3 = 4$. Slope is positive.

See Figure 2.16.

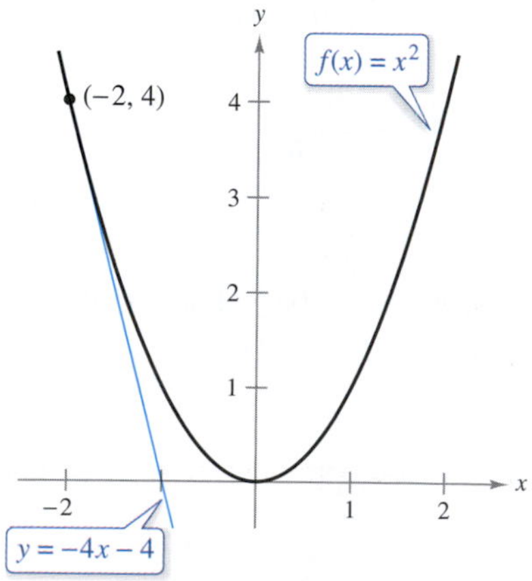

The line $y = -4x - 4$ is tangent to the graph of $f(x) = x^2$ at the point $(-2, 4)$.
Figure 2.17

EXAMPLE 4 Finding an Equation of a Tangent Line

See LarsonCalculus.com for an interactive version of this type of example.

Find an equation of the tangent line to the graph of $f(x) = x^2$ when $x = -2$.

Solution To find the *point* on the graph of f, evaluate the original function at $x = -2$.

$$(-2, f(-2)) = (-2, 4) \qquad \text{Point on graph}$$

To find the *slope* of the graph when $x = -2$, evaluate the derivative, $f'(x) = 2x$, at $x = -2$.

$$m = f'(-2) = -4 \qquad \text{Slope of graph at } (-2, 4)$$

Now, using the point-slope form of the equation of a line, you can write

$$y - y_1 = m(x - x_1) \qquad \text{Point-slope form}$$
$$y - 4 = -4[x - (-2)] \qquad \text{Substitute for } y_1, m, \text{ and } x_1.$$
$$y = -4x - 4. \qquad \text{Simplify.}$$

See Figure 2.17.

The Constant Multiple Rule

> **THEOREM 2.4 The Constant Multiple Rule**
>
> If f is a differentiable function and c is a real number, then cf is also differentiable and $\frac{d}{dx}[cf(x)] = cf'(x)$.

Proof

$$\begin{aligned}\frac{d}{dx}[cf(x)] &= \lim_{\Delta x \to 0} \frac{cf(x + \Delta x) - cf(x)}{\Delta x} && \text{Definition of derivative}\\ &= \lim_{\Delta x \to 0} c\left[\frac{f(x + \Delta x) - f(x)}{\Delta x}\right]\\ &= c\left[\lim_{\Delta x \to 0} \frac{f(x + \Delta x) - f(x)}{\Delta x}\right] && \text{Apply Theorem 1.2.}\\ &= cf'(x)\end{aligned}$$

See LarsonCalculus.com for Bruce Edwards's video of this proof. ■

Informally, the Constant Multiple Rule states that constants can be factored out of the differentiation process, even when the constants appear in the denominator.

$$\frac{d}{dx}[cf(x)] = c\frac{d}{dx}[f(x)] = cf'(x)$$

$$\frac{d}{dx}\left[\frac{f(x)}{c}\right] = \frac{d}{dx}\left[\left(\frac{1}{c}\right)f(x)\right] = \left(\frac{1}{c}\right)\frac{d}{dx}[f(x)] = \left(\frac{1}{c}\right)f'(x)$$

EXAMPLE 5 Using the Constant Multiple Rule

Function	**Derivative**
a. $y = 5x^3$	$\frac{dy}{dx} = \frac{d}{dx}[5x^3] = 5\frac{d}{dx}[x^3] = 5(3)x^2 = 15x^2$
b. $y = \frac{2}{x}$	$\frac{dy}{dx} = \frac{d}{dx}[2x^{-1}] = 2\frac{d}{dx}[x^{-1}] = 2(-1)x^{-2} = -\frac{2}{x^2}$
c. $f(t) = \frac{4t^2}{5}$	$f'(t) = \frac{d}{dt}\left[\frac{4}{5}t^2\right] = \frac{4}{5}\frac{d}{dt}[t^2] = \frac{4}{5}(2t) = \frac{8}{5}t$
d. $y = 2\sqrt{x}$	$\frac{dy}{dx} = \frac{d}{dx}[2x^{1/2}] = 2\left(\frac{1}{2}x^{-1/2}\right) = x^{-1/2} = \frac{1}{\sqrt{x}}$
e. $y = \frac{1}{2\sqrt[3]{x^2}}$	$\frac{dy}{dx} = \frac{d}{dx}\left[\frac{1}{2}x^{-2/3}\right] = \frac{1}{2}\left(-\frac{2}{3}\right)x^{-5/3} = -\frac{1}{3x^{5/3}}$
f. $y = -\frac{3x}{2}$	$y' = \frac{d}{dx}\left[-\frac{3}{2}x\right] = -\frac{3}{2}(1) = -\frac{3}{2}$ ■

REMARK Before differentiating functions involving radicals, rewrite the function with rational exponents.

The Constant Multiple Rule and the Power Rule can be combined into one rule. The combination rule is

$$\frac{d}{dx}[cx^n] = cnx^{n-1}.$$

EXAMPLE 6 **Using Parentheses When Differentiating**

	Original Function	Rewrite	Differentiate	Simplify
a.	$y = \dfrac{5}{2x^3}$	$y = \dfrac{5}{2}(x^{-3})$	$y' = \dfrac{5}{2}(-3x^{-4})$	$y' = -\dfrac{15}{2x^4}$
b.	$y = \dfrac{5}{(2x)^3}$	$y = \dfrac{5}{8}(x^{-3})$	$y' = \dfrac{5}{8}(-3x^{-4})$	$y' = -\dfrac{15}{8x^4}$
c.	$y = \dfrac{7}{3x^{-2}}$	$y = \dfrac{7}{3}(x^2)$	$y' = \dfrac{7}{3}(2x)$	$y' = \dfrac{14x}{3}$
d.	$y = \dfrac{7}{(3x)^{-2}}$	$y = 63(x^2)$	$y' = 63(2x)$	$y' = 126x$

The Sum and Difference Rules

THEOREM 2.5 The Sum and Difference Rules

The sum (or difference) of two differentiable functions f and g is itself differentiable. Moreover, the derivative of $f + g$ (or $f - g$) is the sum (or difference) of the derivatives of f and g.

$$\frac{d}{dx}[f(x) + g(x)] = f'(x) + g'(x) \qquad \text{Sum Rule}$$

$$\frac{d}{dx}[f(x) - g(x)] = f'(x) - g'(x) \qquad \text{Difference Rule}$$

Proof A proof of the Sum Rule follows from Theorem 1.2. (The Difference Rule can be proved in a similar way.)

$$\begin{aligned}
\frac{d}{dx}[f(x) + g(x)] &= \lim_{\Delta x \to 0} \frac{[f(x + \Delta x) + g(x + \Delta x)] - [f(x) + g(x)]}{\Delta x} \\
&= \lim_{\Delta x \to 0} \frac{f(x + \Delta x) + g(x + \Delta x) - f(x) - g(x)}{\Delta x} \\
&= \lim_{\Delta x \to 0} \left[\frac{f(x + \Delta x) - f(x)}{\Delta x} + \frac{g(x + \Delta x) - g(x)}{\Delta x}\right] \\
&= \lim_{\Delta x \to 0} \frac{f(x + \Delta x) - f(x)}{\Delta x} + \lim_{\Delta x \to 0} \frac{g(x + \Delta x) - g(x)}{\Delta x} \\
&= f'(x) + g'(x)
\end{aligned}$$

See LarsonCalculus.com for Bruce Edwards's video of this proof.

The Sum and Difference Rules can be extended to any finite number of functions. For instance, if $F(x) = f(x) + g(x) - h(x)$, then $F'(x) = f'(x) + g'(x) - h'(x)$.

REMARK In Example 7(c), note that before differentiating,

$$\frac{3x^2 - x + 1}{x}$$

was rewritten as

$$3x - 1 + \frac{1}{x}.$$

EXAMPLE 7 **Using the Sum and Difference Rules**

	Function	Derivative
a.	$f(x) = x^3 - 4x + 5$	$f'(x) = 3x^2 - 4$
b.	$g(x) = -\dfrac{x^4}{2} + 3x^3 - 2x$	$g'(x) = -2x^3 + 9x^2 - 2$
c.	$y = \dfrac{3x^2 - x + 1}{x} = 3x - 1 + \dfrac{1}{x}$	$y' = 3 - \dfrac{1}{x^2} = \dfrac{3x^2 - 1}{x^2}$

Derivatives of the Sine and Cosine Functions

In Section 1.3, you studied the limits

$$\lim_{\Delta x \to 0} \frac{\sin \Delta x}{\Delta x} = 1 \quad \text{and} \quad \lim_{\Delta x \to 0} \frac{1 - \cos \Delta x}{\Delta x} = 0.$$

These two limits can be used to prove differentiation rules for the sine and cosine functions. (The derivatives of the other four trigonometric functions are discussed in Section 2.3.)

FOR FURTHER INFORMATION For the outline of a geometric proof of the derivatives of the sine and cosine functions, see the article "The Spider's Spacewalk Derivation of sin′ and cos′" by Tim Hesterberg in *The College Mathematics Journal*. To view this article, go to *MathArticles.com*.

THEOREM 2.6 Derivatives of Sine and Cosine Functions

$$\frac{d}{dx}[\sin x] = \cos x \qquad \frac{d}{dx}[\cos x] = -\sin x$$

Proof Here is a proof of the first rule. (The proof of the second rule is left as an exercise [see Exercise 118].)

$$\begin{aligned}
\frac{d}{dx}[\sin x] &= \lim_{\Delta x \to 0} \frac{\sin(x + \Delta x) - \sin x}{\Delta x} && \text{Definition of derivative}\\
&= \lim_{\Delta x \to 0} \frac{\sin x \cos \Delta x + \cos x \sin \Delta x - \sin x}{\Delta x}\\
&= \lim_{\Delta x \to 0} \frac{\cos x \sin \Delta x - (\sin x)(1 - \cos \Delta x)}{\Delta x}\\
&= \lim_{\Delta x \to 0} \left[(\cos x)\left(\frac{\sin \Delta x}{\Delta x}\right) - (\sin x)\left(\frac{1 - \cos \Delta x}{\Delta x}\right)\right]\\
&= \cos x \left(\lim_{\Delta x \to 0} \frac{\sin \Delta x}{\Delta x}\right) - \sin x \left(\lim_{\Delta x \to 0} \frac{1 - \cos \Delta x}{\Delta x}\right)\\
&= (\cos x)(1) - (\sin x)(0)\\
&= \cos x
\end{aligned}$$

This differentiation rule is shown graphically in Figure 2.18. Note that for each x, the *slope* of the sine curve is equal to the value of the cosine.

See LarsonCalculus.com for Bruce Edwards's video of this proof.

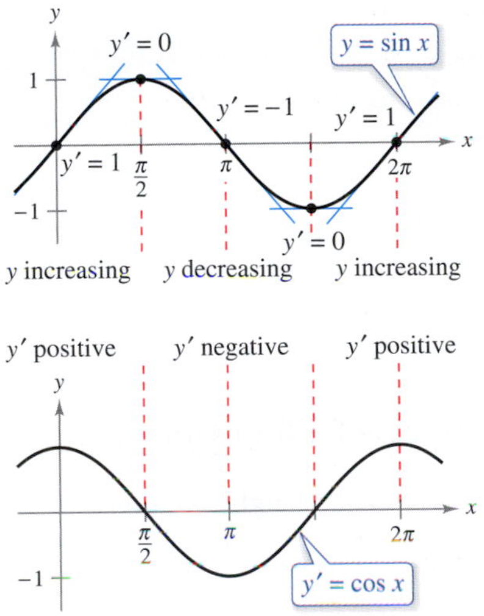

The derivative of the sine function is the cosine function.
Figure 2.18

EXAMPLE 8 Derivatives Involving Sines and Cosines

See LarsonCalculus.com for an interactive version of this type of example.

Function	**Derivative**
a. $y = 2 \sin x$	$y' = 2 \cos x$
b. $y = \dfrac{\sin x}{2} = \dfrac{1}{2} \sin x$	$y' = \dfrac{1}{2} \cos x = \dfrac{\cos x}{2}$
c. $y = x + \cos x$	$y' = 1 - \sin x$
d. $\cos x - \dfrac{\pi}{3} \sin x$	$-\sin x - \dfrac{\pi}{3} \cos x$

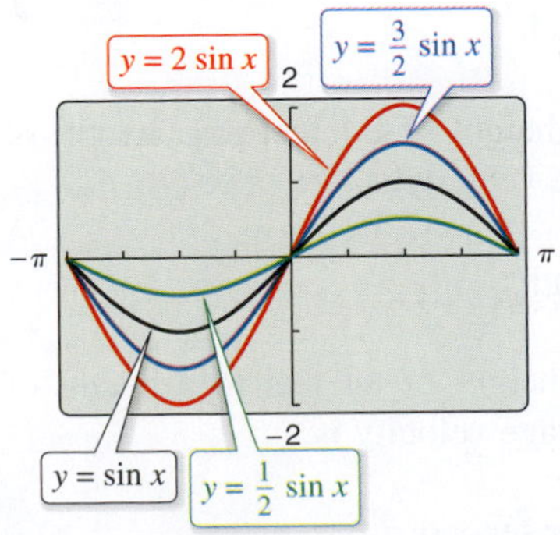

$\frac{d}{dx}[a \sin x] = a \cos x$
Figure 2.19

TECHNOLOGY A graphing utility can provide insight into the interpretation of a derivative. For instance, Figure 2.19 shows the graphs of

$$y = a \sin x$$

for $a = \frac{1}{2}, 1, \frac{3}{2}$, and 2. Estimate the slope of each graph at the point (0, 0). Then verify your estimates analytically by evaluating the derivative of each function when $x = 0$.

AP* Tips

The AP Exam requires that you have equal facility with using a function's equation, table of values, or graph in finding the average velocity or average rate of change.

Rates of Change

You have seen how the derivative is used to determine slope. The derivative can also be used to determine the rate of change of one variable with respect to another. Applications involving rates of change, sometimes referred to as instantaneous rates of change, occur in a wide variety of fields. A few examples are population growth rates, production rates, water flow rates, velocity, and acceleration.

A common use for rate of change is to describe the motion of an object moving in a straight line. In such problems, it is customary to use either a horizontal or a vertical line with a designated origin to represent the line of motion. On such lines, movement to the right (or upward) is considered to be in the positive direction, and movement to the left (or downward) is considered to be in the negative direction.

The function s that gives the position (relative to the origin) of an object as a function of time t is called a **position function.** If, over a period of time Δt, the object changes its position by the amount

$$\Delta s = s(t + \Delta t) - s(t)$$

then, by the familiar formula

$$\text{Rate} = \frac{\text{distance}}{\text{time}}$$

the **average velocity** is

$$\frac{\text{Change in distance}}{\text{Change in time}} = \frac{\Delta s}{\Delta t}. \qquad \text{Average velocity}$$

EXAMPLE 9 **Finding Average Velocity of a Falling Object**

Time-lapse photograph of a free-falling billiard ball

A billiard ball is dropped from a height of 100 feet. The ball's height s at time t is the position function

$$s = -16t^2 + 100 \qquad \text{Position function}$$

where s is measured in feet and t is measured in seconds. Find the average velocity over each of the following time intervals.

a. $[1, 2]$ **b.** $[1, 1.5]$ **c.** $[1, 1.1]$

Solution

a. For the interval $[1, 2]$, the object falls from a height of $s(1) = -16(1)^2 + 100 = 84$ feet to a height of $s(2) = -16(2)^2 + 100 = 36$ feet. The average velocity is

$$\frac{\Delta s}{\Delta t} = \frac{36 - 84}{2 - 1} = \frac{-48}{1} = -48 \text{ feet per second.}$$

b. For the interval $[1, 1.5]$, the object falls from a height of 84 feet to a height of $s(1.5) = -16(1.5)^2 + 100 = 64$ feet. The average velocity is

$$\frac{\Delta s}{\Delta t} = \frac{64 - 84}{1.5 - 1} = \frac{-20}{0.5} = -40 \text{ feet per second.}$$

c. For the interval $[1, 1.1]$, the object falls from a height of 84 feet to a height of $s(1.1) = -16(1.1)^2 + 100 = 80.64$ feet. The average velocity is

$$\frac{\Delta s}{\Delta t} = \frac{80.64 - 84}{1.1 - 1} = \frac{-3.36}{0.1} = -33.6 \text{ feet per second.}$$

Note that the average velocities are *negative*, indicating that the object is moving downward. ■

Richard Megna/Fundamental Photographs

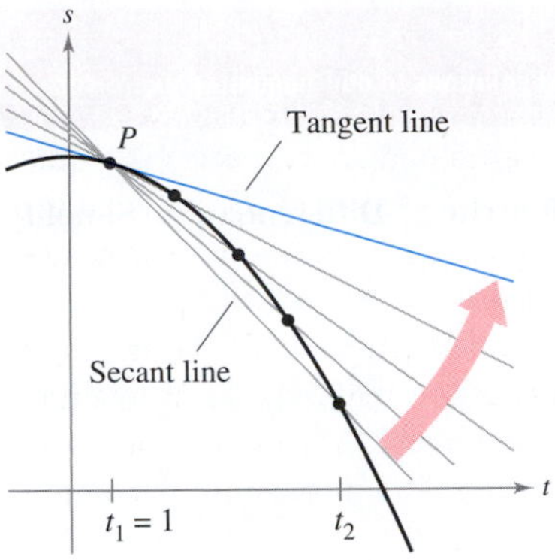

The average velocity between t_1 and t_2 is the slope of the secant line, and the instantaneous velocity at t_1 is the slope of the tangent line.
Figure 2.20

Suppose that in Example 9, you wanted to find the *instantaneous* velocity (or simply the velocity) of the object when $t = 1$. Just as you can approximate the slope of the tangent line by calculating the slope of the secant line, you can approximate the velocity at $t = 1$ by calculating the average velocity over a small interval $[1, 1 + \Delta t]$ (see Figure 2.20). By taking the limit as Δt approaches zero, you obtain the velocity when $t = 1$. Try doing this—you will find that the velocity when $t = 1$ is -32 feet per second.

In general, if $s = s(t)$ is the position function for an object moving along a straight line, then the **velocity** of the object at time t is

$$v(t) = \lim_{\Delta t \to 0} \frac{s(t + \Delta t) - s(t)}{\Delta t} = s'(t).$$ Velocity function

In other words, the velocity function is the derivative of the position function. Velocity can be negative, zero, or positive. The **speed** of an object is the absolute value of its velocity. Speed cannot be negative.

The position of a free-falling object (neglecting air resistance) under the influence of gravity can be represented by the equation

$$s(t) = \frac{1}{2}gt^2 + v_0t + s_0$$ Position function

where s_0 is the initial height of the object, v_0 is the initial velocity of the object, and g is the acceleration due to gravity. On Earth, the value of g is approximately -32 feet per second per second or -9.8 meters per second per second.

EXAMPLE 10 Using the Derivative to Find Velocity

At time $t = 0$, a diver jumps from a platform diving board that is 32 feet above the water (see Figure 2.21). Because the initial velocity of the diver is 16 feet per second, the position of the diver is

$$s(t) = -16t^2 + 16t + 32$$ Position function

where s is measured in feet and t is measured in seconds.

a. When does the diver hit the water?

b. What is the diver's velocity at impact?

Solution

a. To find the time t when the diver hits the water, let $s = 0$ and solve for t.

$$-16t^2 + 16t + 32 = 0$$ Set position function equal to 0.

$$-16(t + 1)(t - 2) = 0$$ Factor.

$$t = -1 \text{ or } 2$$ Solve for t.

Because $t \geq 0$, choose the positive value to conclude that the diver hits the water at $t = 2$ seconds.

b. The velocity at time t is given by the derivative

$$s'(t) = -32t + 16.$$ Velocity function

So, the velocity at time $t = 2$ is

$$s'(2) = -32(2) + 16 = -48 \text{ feet per second.}$$

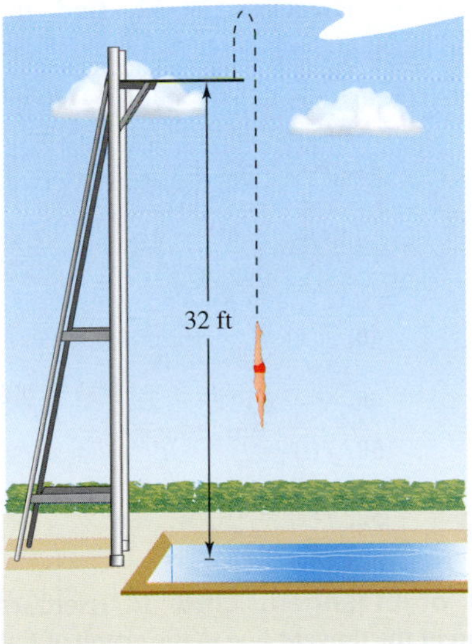

Velocity is positive when an object is rising, and is negative when an object is falling. Notice that the diver moves upward for the first half-second because the velocity is positive for $0 < t < \frac{1}{2}$. When the velocity is 0, the diver has reached the maximum height of the dive.
Figure 2.21

2.2 Exercises

See CalcChat.com for tutorial help and worked-out solutions to odd-numbered exercises.

Estimating Slope **In Exercises 1 and 2, use the graph to estimate the slope of the tangent line to $y = x^n$ at the point (1, 1). Verify your answer analytically. To print an enlarged copy of the graph, go to** ***MathGraphs.com.***

1. (a) $y = x^{1/2}$

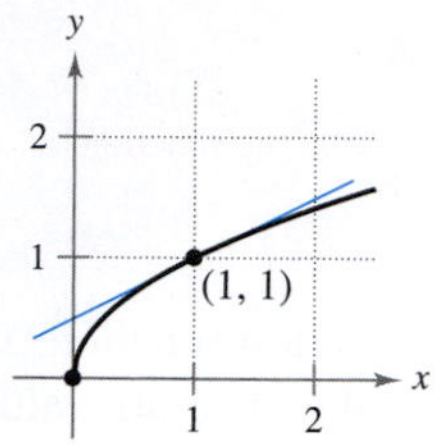

(b) $y = x^3$

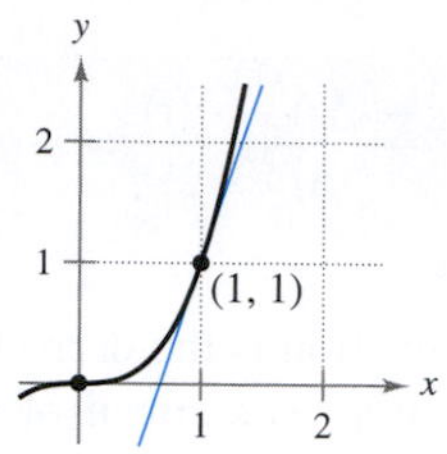

2. (a) $y = x^{-1/2}$

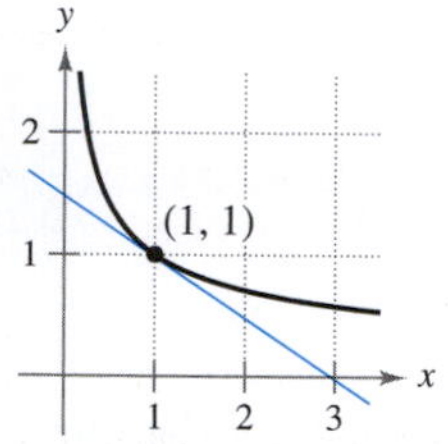

(b) $y = x^{-1}$

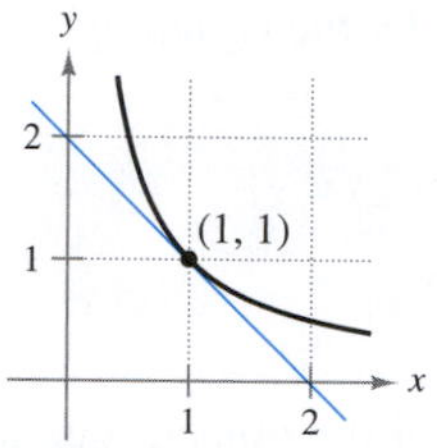

Finding a Derivative **In Exercises 3–24, use the rules of differentiation to find the derivative of the function.**

3. $y = 12$

4. $f(x) = -9$

5. $y = x^7$

6. $y = x^{12}$

7. $y = \dfrac{1}{x^5}$

8. $y = \dfrac{3}{x^7}$

9. $f(x) = \sqrt[5]{x}$

10. $g(x) = \sqrt[4]{x}$

11. $f(x) = x + 11$

12. $g(x) = 6x + 3$

13. $f(t) = -2t^2 + 3t - 6$

14. $y = t^2 - 3t + 1$

15. $g(x) = x^2 + 4x^3$

16. $y = 4x - 3x^3$

17. $s(t) = t^3 + 5t^2 - 3t + 8$

18. $y = 2x^3 + 6x^2 - 1$

19. $y = \dfrac{\pi}{2}\sin\theta - \cos\theta$

20. $g(t) = \pi\cos t$

21. $y = x^2 - \frac{1}{2}\cos x$

22. $y = 7 + \sin x$

23. $y = \dfrac{1}{x} - 3\sin x$

24. $y = \dfrac{5}{(2x)^3} + 2\cos x$

Rewriting a Function Before Differentiating **In Exercises 25–30, complete the table to find the derivative of the function.**

Original Function	Rewrite	Differentiate	Simplify
25. $y = \dfrac{5}{2x^2}$			
26. $y = \dfrac{3}{2x^4}$			
27. $y = \dfrac{6}{(5x)^3}$			
28. $y = \dfrac{\pi}{(3x)^2}$			
29. $y = \dfrac{\sqrt{x}}{x}$			
30. $y = \dfrac{4}{x^{-3}}$			

Finding the Slope of a Graph **In Exercises 31–38, find the slope of the graph of the function at the given point. Use the** ***derivative*** **feature of a graphing utility to confirm your results.**

Function	Point
31. $f(x) = \dfrac{8}{x^2}$	$(2, 2)$
32. $f(t) = 2 - \dfrac{4}{t}$	$(4, 1)$
33. $f(x) = -\frac{1}{2} + \frac{7}{5}x^3$	$\left(0, -\frac{1}{2}\right)$
34. $y = 2x^4 - 3$	$(1, -1)$
35. $y = (4x + 1)^2$	$(0, 1)$
36. $f(x) = 2(x - 4)^2$	$(2, 8)$
37. $f(\theta) = 4\sin\theta - \theta$	$(0, 0)$
38. $g(t) = -2\cos t + 5$	$(\pi, 7)$

Finding a Derivative **In Exercises 39–52, find the derivative of the function.**

39. $f(x) = x^2 + 5 - 3x^{-2}$

40. $f(x) = x^3 - 2x + 3x^{-3}$

41. $g(t) = t^2 - \dfrac{4}{t^3}$

42. $f(x) = 8x + \dfrac{3}{x^2}$

43. $f(x) = \dfrac{4x^3 + 3x^2}{x}$

44. $f(x) = \dfrac{2x^4 - x}{x^3}$

45. $f(x) = \dfrac{x^3 - 3x^2 + 4}{x^2}$

46. $h(x) = \dfrac{4x^3 + 2x + 5}{x}$

47. $y = x(x^2 + 1)$

48. $y = x^2(2x^2 - 3x)$

49. $f(x) = \sqrt{x} - 6\sqrt[3]{x}$

50. $f(t) = t^{2/3} - t^{1/3} + 4$

51. $f(x) = 6\sqrt{x} + 5\cos x$

52. $f(x) = \dfrac{2}{\sqrt[3]{x}} + 3\cos x$

Finding an Equation of a Tangent Line **In Exercises 53–56, (a) find an equation of the tangent line to the graph of** ***f*** **at the given point, (b) use a graphing utility to graph the function and its tangent line at the point, and (c) use the** ***derivative*** **feature of a graphing utility to confirm your results.**

Function	Point
53. $y = x^4 - 3x^2 + 2$	$(1, 0)$
54. $y = x^3 - 3x$	$(2, 2)$
55. $f(x) = \dfrac{2}{\sqrt[4]{x^3}}$	$(1, 2)$
56. $y = (x - 2)(x^2 + 3x)$	$(1, -4)$

Horizontal Tangent Line **In Exercises 57–62, determine the point(s) (if any) at which the graph of the function has a horizontal tangent line.**

57. $y = x^4 - 2x^2 + 3$

58. $y = x^3 + x$

59. $y = \frac{1}{x^2}$

60. $y = x^2 + 9$

61. $y = x + \sin x, \quad 0 \le x < 2\pi$

62. $y = \sqrt{3}x + 2\cos x, \quad 0 \le x < 2\pi$

Finding a Value **In Exercises 63–68, find k such that the line is tangent to the graph of the function.**

Function	Line
63. $f(x) = k - x^2$	$y = -6x + 1$
64. $f(x) = kx^2$	$y = -2x + 3$
65. $f(x) = \frac{k}{x}$	$y = -\frac{3}{4}x + 3$
66. $f(x) = k\sqrt{x}$	$y = x + 4$
67. $f(x) = kx^3$	$y = x + 1$
68. $f(x) = kx^4$	$y = 4x - 1$

69. Sketching a Graph Sketch the graph of a function f such that $f' > 0$ for all x and the rate of change of the function is decreasing.

70. HOW DO YOU SEE IT? Use the graph of f to answer each question. To print an enlarged copy of the graph, go to *MathGraphs.com*.

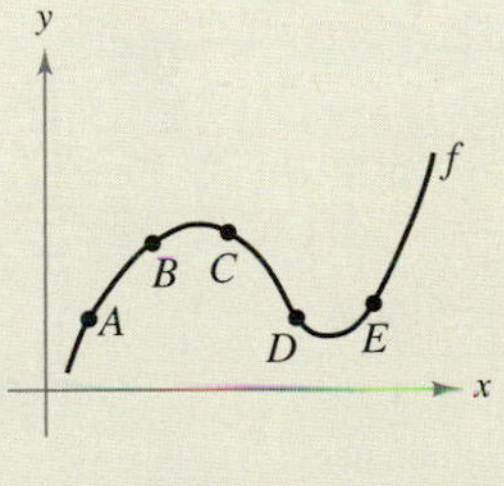

(a) Between which two consecutive points is the average rate of change of the function greatest?

(b) Is the average rate of change of the function between A and B greater than or less than the instantaneous rate of change at B?

(c) Sketch a tangent line to the graph between C and D such that the slope of the tangent line is the same as the average rate of change of the function between C and D.

WRITING ABOUT CONCEPTS

Exploring a Relationship **In Exercises 71–74, the relationship between f and g is given. Explain the relationship between f' and g'.**

71. $g(x) = f(x) + 6$

72. $g(x) = 2f(x)$

73. $g(x) = -5f(x)$

74. $g(x) = 3f(x) - 1$

WRITING ABOUT CONCEPTS (continued)

A Function and Its Derivative **In Exercises 75 and 76, the graphs of a function f and its derivative f' are shown in the same set of coordinate axes. Label the graphs as f or f' and write a short paragraph stating the criteria you used in making your selection. To print an enlarged copy of the graph, go to *MathGraphs.com*.**

75.

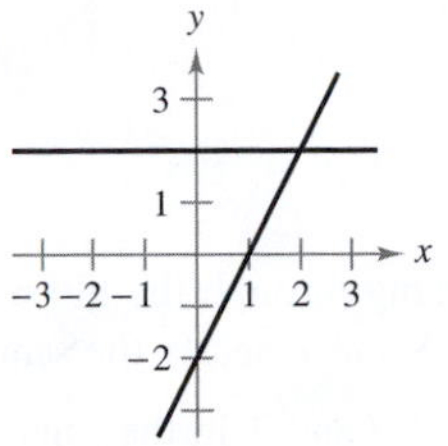

76.

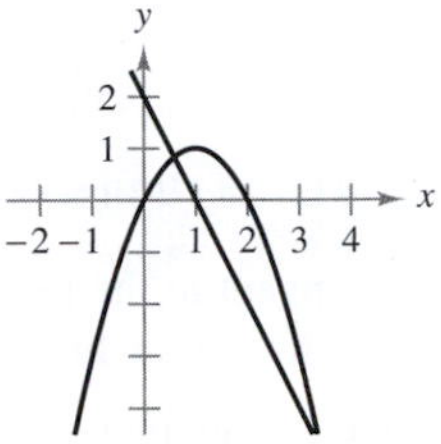

77. Finding Equations of Tangent Lines Sketch the graphs of $y = x^2$ and $y = -x^2 + 6x - 5$, and sketch the two lines that are tangent to both graphs. Find equations of these lines.

78. Tangent Lines Show that the graphs of the two equations

$$y = x \quad \text{and} \quad y = \frac{1}{x}$$

have tangent lines that are perpendicular to each other at their point of intersection.

79. Tangent Line Show that the graph of the function

$$f(x) = 3x + \sin x + 2$$

does not have a horizontal tangent line.

80. Tangent Line Show that the graph of the function

$$f(x) = x^5 + 3x^3 + 5x$$

does not have a tangent line with a slope of 3.

Finding an Equation of a Tangent Line **In Exercises 81 and 82, find an equation of the tangent line to the graph of the function f through the point (x_0, y_0) not on the graph. To find the point of tangency (x, y) on the graph of f, solve the equation**

$$f'(x) = \frac{y_0 - y}{x_0 - x}.$$

81. $f(x) = \sqrt{x}$

$(x_0, y_0) = (-4, 0)$

82. $f(x) = \frac{2}{x}$

$(x_0, y_0) = (5, 0)$

83. Linear Approximation Use a graphing utility with a square window setting to zoom in on the graph of

$$f(x) = 4 - \tfrac{1}{2}x^2$$

to approximate $f'(1)$. Use the derivative to find $f'(1)$.

84. Linear Approximation Use a graphing utility with a square window setting to zoom in on the graph of

$$f(x) = 4\sqrt{x} + 1$$

to approximate $f'(4)$. Use the derivative to find $f'(4)$.

85. Linear Approximation Consider the function $f(x) = x^{3/2}$ with the solution point (4, 8).

(a) Use a graphing utility to graph f. Use the *zoom* feature to obtain successive magnifications of the graph in the neighborhood of the point (4, 8). After zooming in a few times, the graph should appear nearly linear. Use the *trace* feature to determine the coordinates of a point near (4, 8). Find an equation of the secant line $S(x)$ through the two points.

(b) Find the equation of the line

$$T(x) = f'(4)(x - 4) + f(4)$$

tangent to the graph of f passing through the given point. Why are the linear functions S and T nearly the same?

(c) Use a graphing utility to graph f and T in the same set of coordinate axes. Note that T is a good approximation of f when x is close to 4. What happens to the accuracy of the approximation as you move farther away from the point of tangency?

(d) Demonstrate the conclusion in part (c) by completing the table.

Δx	-3	-2	-1	-0.5	-0.1	0
$f(4 + \Delta x)$						
$T(4 + \Delta x)$						

Δx	0.1	0.5	1	2	3
$f(4 + \Delta x)$					
$T(4 + \Delta x)$					

86. Linear Approximation Repeat Exercise 85 for the function $f(x) = x^3$, where $T(x)$ is the line tangent to the graph at the point (1, 1). Explain why the accuracy of the linear approximation decreases more rapidly than in Exercise 85.

True or False? **In Exercises 87–92, determine whether the statement is true or false. If it is false, explain why or give an example that shows it is false.**

87. If $f'(x) = g'(x)$, then $f(x) = g(x)$.

88. If $f(x) = g(x) + c$, then $f'(x) = g'(x)$.

89. If $y = \pi^2$, then $dy/dx = 2\pi$.

90. If $y = x/\pi$, then $dy/dx = 1/\pi$.

91. If $g(x) = 3f(x)$, then $g'(x) = 3f'(x)$.

92. If $f(x) = \dfrac{1}{x^n}$, then $f'(x) = \dfrac{1}{nx^{n-1}}$.

Finding Rates of Change **In Exercises 93–96, find the average rate of change of the function over the given interval. Compare this average rate of change with the instantaneous rates of change at the endpoints of the interval.**

93. $f(t) = 4t + 5$, $[1, 2]$

94. $f(t) = t^2 - 7$, $[3, 3.1]$

95. $f(x) = \dfrac{-1}{x}$, $[1, 2]$

96. $f(x) = \sin x$, $\left[0, \dfrac{\pi}{6}\right]$

Vertical Motion **In Exercises 97 and 98, use the position function $s(t) = -16t^2 + v_0 t + s_0$ for free-falling objects.**

97. A silver dollar is dropped from the top of a building that is 1362 feet tall.

(a) Determine the position and velocity functions for the coin.

(b) Determine the average velocity on the interval [1, 2].

(c) Find the instantaneous velocities when $t = 1$ and $t = 2$.

(d) Find the time required for the coin to reach ground level.

(e) Find the velocity of the coin at impact.

98. A ball is thrown straight down from the top of a 220-foot building with an initial velocity of -22 feet per second. What is its velocity after 3 seconds? What is its velocity after falling 108 feet?

Vertical Motion **In Exercises 99 and 100, use the position function $s(t) = -4.9t^2 + v_0 t + s_0$ for free-falling objects.**

99. A projectile is shot upward from the surface of Earth with an initial velocity of 120 meters per second. What is its velocity after 5 seconds? After 10 seconds?

100. To estimate the height of a building, a stone is dropped from the top of the building into a pool of water at ground level. The splash is seen 5.6 seconds after the stone is dropped. What is the height of the building?

Think About It **In Exercises 101 and 102, the graph of a position function is shown. It represents the distance in miles that a person drives during a 10-minute trip to work. Make a sketch of the corresponding velocity function.**

101.

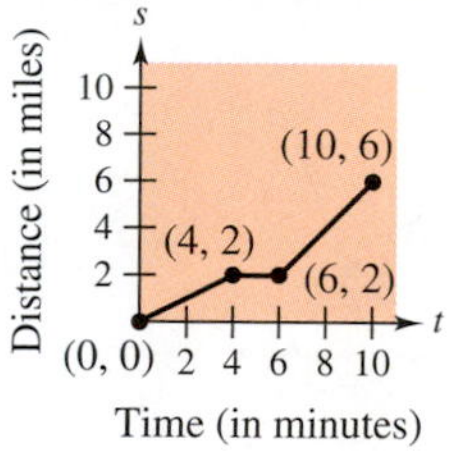

102.

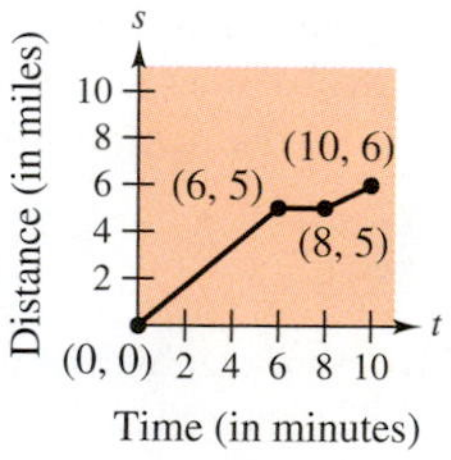

Think About It **In Exercises 103 and 104, the graph of a velocity function is shown. It represents the velocity in miles per hour during a 10-minute trip to work. Make a sketch of the corresponding position function.**

103.

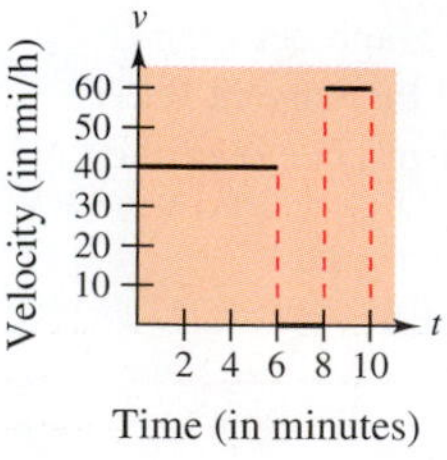

104.

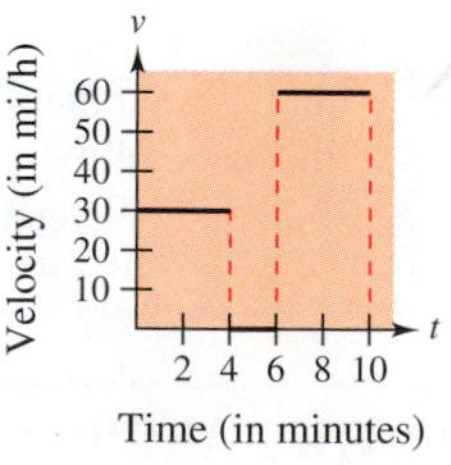

105. Volume The volume of a cube with sides of length s is given by $V = s^3$. Find the rate of change of the volume with respect to s when $s = 6$ centimeters.

106. Area The area of a square with sides of length s is given by $A = s^2$. Find the rate of change of the area with respect to s when $s = 6$ meters.

107. Modeling Data The stopping distance of an automobile, on dry, level pavement, traveling at a speed v (in kilometers per hour) is the distance R (in meters) the car travels during the reaction time of the driver plus the distance B (in meters) the car travels after the brakes are applied (see figure). The table shows the results of an experiment.

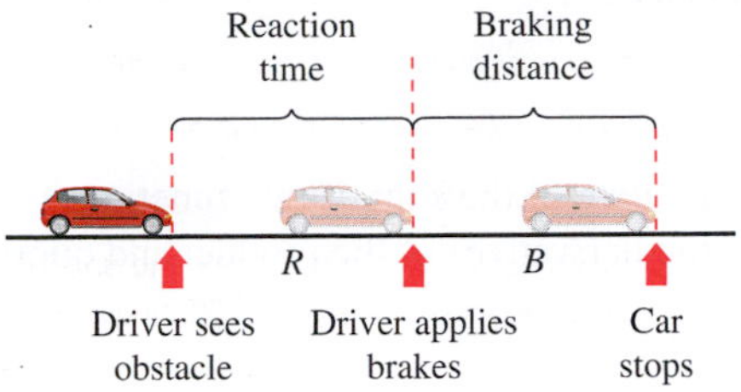

Speed, v	20	40	60	80	100
Reaction Time Distance, R	8.3	16.7	25.0	33.3	41.7
Braking Time Distance, B	2.3	9.0	20.2	35.8	55.9

(a) Use the regression capabilities of a graphing utility to find a linear model for reaction time distance R.

(b) Use the regression capabilities of a graphing utility to find a quadratic model for braking time distance B.

(c) Determine the polynomial giving the total stopping distance T.

(d) Use a graphing utility to graph the functions R, B, and T in the same viewing window.

(e) Find the derivative of T and the rates of change of the total stopping distance for $v = 40$, $v = 80$, and $v = 100$.

(f) Use the results of this exercise to draw conclusions about the total stopping distance as speed increases.

108. Fuel Cost A car is driven 15,000 miles a year and gets x miles per gallon. Assume that the average fuel cost is \$3.48 per gallon. Find the annual cost of fuel C as a function of x and use this function to complete the table.

x	10	15	20	25	30	35	40
C							
dC/dx							

Who would benefit more from a one-mile-per-gallon increase in fuel efficiency—the driver of a car that gets 15 miles per gallon, or the driver of a car that gets 35 miles per gallon? Explain.

109. Velocity Verify that the average velocity over the time interval $[t_0 - \Delta t, t_0 + \Delta t]$ is the same as the instantaneous velocity at $t = t_0$ for the position function

$$s(t) = -\frac{1}{2}at^2 + c.$$

110. Inventory Management The annual inventory cost C for a manufacturer is

$$C = \frac{1{,}008{,}000}{Q} + 6.3Q$$

where Q is the order size when the inventory is replenished. Find the change in annual cost when Q is increased from 350 to 351, and compare this with the instantaneous rate of change when $Q = 350$.

111. Finding an Equation of a Parabola Find an equation of the parabola $y = ax^2 + bx + c$ that passes through $(0, 1)$ and is tangent to the line $y = x - 1$ at $(1, 0)$.

112. Proof Let (a, b) be an arbitrary point on the graph of $y = 1/x$, $x > 0$. Prove that the area of the triangle formed by the tangent line through (a, b) and the coordinate axes is 2.

113. Finding Equation(s) of Tangent Line(s) Find the equation(s) of the tangent line(s) to the graph of the curve $y = x^3 - 9x$ through the point $(1, -9)$ not on the graph.

114. Finding Equation(s) of Tangent Line(s) Find the equation(s) of the tangent line(s) to the graph of the parabola $y = x^2$ through the given point not on the graph.

(a) $(0, a)$ (b) $(a, 0)$

Are there any restrictions on the constant a?

Making a Function Differentiable In Exercises 115 and 116, find a and b such that f is differentiable everywhere.

115. $f(x) = \begin{cases} ax^3, & x \le 2 \\ x^2 + b, & x > 2 \end{cases}$

116. $f(x) = \begin{cases} \cos x, & x < 0 \\ ax + b, & x \ge 0 \end{cases}$

117. Determining Differentiability Where are the functions $f_1(x) = |\sin x|$ and $f_2(x) = \sin |x|$ differentiable?

118. Proof Prove that $\frac{d}{dx}[\cos x] = -\sin x$.

FOR FURTHER INFORMATION For a geometric interpretation of the derivatives of trigonometric functions, see the article "Sines and Cosines of the Times" by Victor J. Katz in *Math Horizons*. To view this article, go to *MathArticles.com*.

PUTNAM EXAM CHALLENGE

119. Find all differentiable functions $f : \mathbb{R} \rightarrow \mathbb{R}$ such that

$$f'(x) = \frac{f(x + n) - f(x)}{n}$$

for all real numbers x and all positive integers n.

This problem was composed by the Committee on the Putnam Prize Competition.

2.3 Product and Quotient Rules and Higher-Order Derivatives

- **Find the derivative of a function using the Product Rule.**
- **Find the derivative of a function using the Quotient Rule.**
- **Find the derivative of a trigonometric function.**
- **Find a higher-order derivative of a function.**

The Product Rule

In Section 2.2, you learned that the derivative of the sum of two functions is simply the sum of their derivatives. The rules for the derivatives of the product and quotient of two functions are not as simple.

REMARK A version of the Product Rule that some people prefer is

$$\frac{d}{dx}[f(x)g(x)] = f'(x)g(x) + f(x)g'(x).$$

The advantage of this form is that it generalizes easily to products of three or more factors.

THEOREM 2.7 The Product Rule

The product of two differentiable functions f and g is itself differentiable. Moreover, the derivative of fg is the first function times the derivative of the second, plus the second function times the derivative of the first.

$$\frac{d}{dx}[f(x)g(x)] = f(x)g'(x) + g(x)f'(x)$$

Proof Some mathematical proofs, such as the proof of the Sum Rule, are straightforward. Others involve clever steps that may appear unmotivated to a reader. This proof involves such a step—subtracting and adding the same quantity—which is shown in color.

$$\begin{aligned}
\frac{d}{dx}[f(x)g(x)] &= \lim_{\Delta x\to 0}\frac{f(x+\Delta x)g(x+\Delta x) - f(x)g(x)}{\Delta x}\\
&= \lim_{\Delta x\to 0}\frac{f(x+\Delta x)g(x+\Delta x) - f(x+\Delta x)g(x) + f(x+\Delta x)g(x) - f(x)g(x)}{\Delta x}\\
&= \lim_{\Delta x\to 0}\left[f(x+\Delta x)\frac{g(x+\Delta x)-g(x)}{\Delta x} + g(x)\frac{f(x+\Delta x)-f(x)}{\Delta x}\right]\\
&= \lim_{\Delta x\to 0}\left[f(x+\Delta x)\frac{g(x+\Delta x)-g(x)}{\Delta x}\right] + \lim_{\Delta x\to 0}\left[g(x)\frac{f(x+\Delta x)-f(x)}{\Delta x}\right]\\
&= \lim_{\Delta x\to 0} f(x+\Delta x)\cdot\lim_{\Delta x\to 0}\frac{g(x+\Delta x)-g(x)}{\Delta x} + \lim_{\Delta x\to 0} g(x)\cdot\lim_{\Delta x\to 0}\frac{f(x+\Delta x)-f(x)}{\Delta x}\\
&= f(x)g'(x) + g(x)f'(x)
\end{aligned}$$

Note that $\lim_{\Delta x\to 0} f(x+\Delta x) = f(x)$ because f is given to be differentiable and therefore is continuous.

See LarsonCalculus.com for Bruce Edwards's video of this proof. ■

REMARK The proof of the Product Rule for products of more than two factors is left as an exercise (see Exercise 137).

The Product Rule can be extended to cover products involving more than two factors. For example, if f, g, and h are differentiable functions of x, then

$$\frac{d}{dx}[f(x)g(x)h(x)] = f'(x)g(x)h(x) + f(x)g'(x)h(x) + f(x)g(x)h'(x).$$

So, the derivative of $y = x^2 \sin x \cos x$ is

$$\begin{aligned}
\frac{dy}{dx} &= 2x\sin x\cos x + x^2\cos x\cos x + x^2\sin x(-\sin x)\\
&= 2x\sin x\cos x + x^2(\cos^2 x - \sin^2 x).
\end{aligned}$$

THE PRODUCT RULE

When Leibniz originally wrote a formula for the Product Rule, he was motivated by the expression

$$(x + dx)(y + dy) - xy$$

from which he subtracted $dx\,dy$ (as being negligible) and obtained the differential form $x\,dy + y\,dx$. This derivation resulted in the traditional form of the Product Rule. *(Source: The History of Mathematics by David M. Burton)*

The derivative of a product of two functions is not (in general) given by the product of the derivatives of the two functions. To see this, try comparing the product of the derivatives of

$$f(x) = 3x - 2x^2$$

and

$$g(x) = 5 + 4x$$

with the derivative in Example 1.

EXAMPLE 1 Using the Product Rule

Find the derivative of $h(x) = (3x - 2x^2)(5 + 4x)$.

Solution

$$\begin{aligned} h'(x) &= \underbrace{(3x - 2x^2)}_{\text{First}} \underbrace{\frac{d}{dx}[5 + 4x]}_{\text{Derivative of second}} + \underbrace{(5 + 4x)}_{\text{Second}} \underbrace{\frac{d}{dx}[3x - 2x^2]}_{\text{Derivative of first}} \qquad \text{Apply Product Rule.} \\ &= (3x - 2x^2)(4) + (5 + 4x)(3 - 4x) \\ &= (12x - 8x^2) + (15 - 8x - 16x^2) \\ &= -24x^2 + 4x + 15 \end{aligned}$$

In Example 1, you have the option of finding the derivative with or without the Product Rule. To find the derivative without the Product Rule, you can write

$$\begin{aligned} D_x[(3x - 2x^2)(5 + 4x)] &= D_x[-8x^3 + 2x^2 + 15x] \\ &= -24x^2 + 4x + 15. \end{aligned}$$

In the next example, you must use the Product Rule.

EXAMPLE 2 Using the Product Rule

Find the derivative of $y = 3x^2 \sin x$.

Solution

$$\begin{aligned} \frac{d}{dx}[3x^2 \sin x] &= 3x^2 \frac{d}{dx}[\sin x] + \sin x \frac{d}{dx}[3x^2] \qquad \text{Apply Product Rule.} \\ &= 3x^2 \cos x + (\sin x)(6x) \\ &= 3x^2 \cos x + 6x \sin x \\ &= 3x(x \cos x + 2 \sin x) \end{aligned}$$

REMARK In Example 3, notice that you use the Product Rule when both factors of the product are variable, and you use the Constant Multiple Rule when one of the factors is a constant.

EXAMPLE 3 Using the Product Rule

Find the derivative of $y = 2x \cos x - 2 \sin x$.

Solution

$$\begin{aligned} \frac{dy}{dx} &= \underbrace{(2x)\left(\frac{d}{dx}[\cos x]\right) + (\cos x)\left(\frac{d}{dx}[2x]\right)}_{\text{Product Rule}} - \underbrace{2\frac{d}{dx}[\sin x]}_{\text{Constant Multiple Rule}} \\ &= (2x)(-\sin x) + (\cos x)(2) - 2(\cos x) \\ &= -2x \sin x \end{aligned}$$

The Quotient Rule

THEOREM 2.8 The Quotient Rule

The quotient f/g of two differentiable functions f and g is itself differentiable at all values of x for which $g(x) \neq 0$. Moreover, the derivative of f/g is given by the denominator times the derivative of the numerator minus the numerator times the derivative of the denominator, all divided by the square of the denominator.

$$\frac{d}{dx}\left[\frac{f(x)}{g(x)}\right] = \frac{g(x)f'(x) - f(x)g'(x)}{[g(x)]^2}, \quad g(x) \neq 0$$

REMARK From the Quotient Rule, you can see that the derivative of a quotient is not (in general) the quotient of the derivatives.

AP* Tips

Extremely complex examples of the product and quotient rules do not appear on the AP Exam. The harder problems here will help you to master and remember the concept.

Proof As with the proof of Theorem 2.7, the key to this proof is subtracting and adding the same quantity.

$$\frac{d}{dx}\left[\frac{f(x)}{g(x)}\right] = \lim_{\Delta x \to 0} \frac{\dfrac{f(x+\Delta x)}{g(x+\Delta x)} - \dfrac{f(x)}{g(x)}}{\Delta x} \qquad \text{Definition of derivative}$$

$$= \lim_{\Delta x \to 0} \frac{g(x)f(x+\Delta x) - f(x)g(x+\Delta x)}{\Delta x g(x)g(x+\Delta x)}$$

$$= \lim_{\Delta x \to 0} \frac{g(x)f(x+\Delta x) - f(x)g(x) + f(x)g(x) - f(x)g(x+\Delta x)}{\Delta x g(x)g(x+\Delta x)}$$

$$= \frac{\displaystyle\lim_{\Delta x \to 0} \frac{g(x)[f(x+\Delta x) - f(x)]}{\Delta x} - \lim_{\Delta x \to 0} \frac{f(x)[g(x+\Delta x) - g(x)]}{\Delta x}}{\displaystyle\lim_{\Delta x \to 0} [g(x)g(x+\Delta x)]}$$

$$= \frac{g(x)\left[\displaystyle\lim_{\Delta x \to 0} \frac{f(x+\Delta x) - f(x)}{\Delta x}\right] - f(x)\left[\displaystyle\lim_{\Delta x \to 0} \frac{g(x+\Delta x) - g(x)}{\Delta x}\right]}{\displaystyle\lim_{\Delta x \to 0} [g(x)g(x+\Delta x)]}$$

$$= \frac{g(x)f'(x) - f(x)g'(x)}{[g(x)]^2}$$

Note that $\lim_{\Delta x \to 0} g(x+\Delta x) = g(x)$ because g is given to be differentiable and therefore is continuous.

See LarsonCalculus.com for Bruce Edwards's video of this proof.

TECHNOLOGY A graphing utility can be used to compare the graph of a function with the graph of its derivative. For instance, in Figure 2.22, the graph of the function in Example 4 appears to have two points that have horizontal tangent lines. What are the values of y' at these two points?

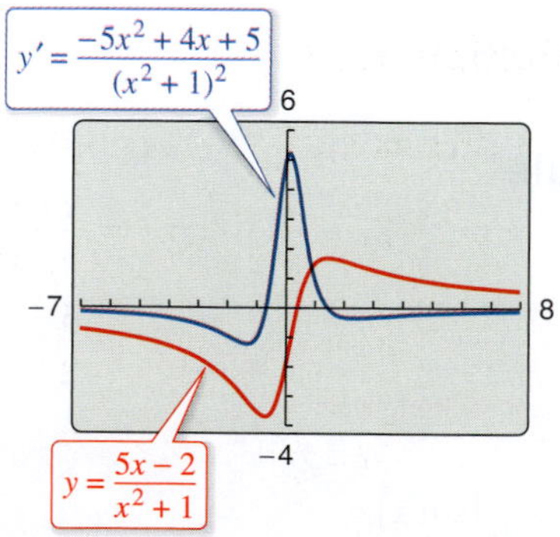

Graphical comparison of a function and its derivative

Figure 2.22

EXAMPLE 4 Using the Quotient Rule

Find the derivative of $y = \dfrac{5x - 2}{x^2 + 1}$.

Solution

$$\frac{d}{dx}\left[\frac{5x-2}{x^2+1}\right] = \frac{(x^2+1)\dfrac{d}{dx}[5x-2] - (5x-2)\dfrac{d}{dx}[x^2+1]}{(x^2+1)^2} \qquad \text{Apply Quotient Rule.}$$

$$= \frac{(x^2+1)(5) - (5x-2)(2x)}{(x^2+1)^2}$$

$$= \frac{(5x^2+5) - (10x^2-4x)}{(x^2+1)^2}$$

$$= \frac{-5x^2+4x+5}{(x^2+1)^2}$$

Note the use of parentheses in Example 4. A liberal use of parentheses is recommended for *all* types of differentiation problems. For instance, with the Quotient Rule, it is a good idea to enclose all factors and derivatives in parentheses, and to pay special attention to the subtraction required in the numerator.

When differentiation rules were introduced in the preceding section, the need for rewriting *before* differentiating was emphasized. The next example illustrates this point with the Quotient Rule.

EXAMPLE 5 Rewriting Before Differentiating

Find an equation of the tangent line to the graph of $f(x) = \dfrac{3 - (1/x)}{x + 5}$ at $(-1, 1)$.

Solution Begin by rewriting the function.

$$f(x) = \frac{3 - (1/x)}{x + 5} \qquad \text{Write original function.}$$

$$= \frac{x\left(3 - \frac{1}{x}\right)}{x(x + 5)} \qquad \text{Multiply numerator and denominator by } x.$$

$$= \frac{3x - 1}{x^2 + 5x} \qquad \text{Rewrite.}$$

Next, apply the Quotient Rule.

$$f'(x) = \frac{(x^2 + 5x)(3) - (3x - 1)(2x + 5)}{(x^2 + 5x)^2} \qquad \text{Quotient Rule}$$

$$= \frac{(3x^2 + 15x) - (6x^2 + 13x - 5)}{(x^2 + 5x)^2}$$

$$= \frac{-3x^2 + 2x + 5}{(x^2 + 5x)^2} \qquad \text{Simplify.}$$

To find the slope at $(-1, 1)$, evaluate $f'(-1)$.

$$f'(-1) = 0 \qquad \text{Slope of graph at } (-1, 1)$$

Then, using the point-slope form of the equation of a line, you can determine that the equation of the tangent line at $(-1, 1)$ is $y = 1$. See Figure 2.23. ■

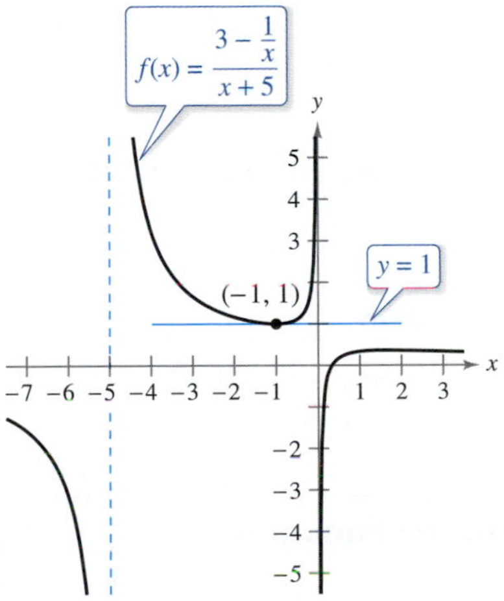

The line $y = 1$ is tangent to the graph of $f(x)$ at the point $(-1, 1)$.
Figure 2.23

Not every quotient needs to be differentiated by the Quotient Rule. For instance, each quotient in the next example can be considered as the product of a constant times a function of x. In such cases, it is more convenient to use the Constant Multiple Rule.

REMARK To see the benefit of using the Constant Multiple Rule for some quotients, try using the Quotient Rule to differentiate the functions in Example 6—you should obtain the same results, but with more work.

EXAMPLE 6 Using the Constant Multiple Rule

	Original Function	Rewrite	Differentiate	Simplify
a.	$y = \dfrac{x^2 + 3x}{6}$	$y = \dfrac{1}{6}(x^2 + 3x)$	$y' = \dfrac{1}{6}(2x + 3)$	$y' = \dfrac{2x + 3}{6}$
b.	$y = \dfrac{5x^4}{8}$	$y = \dfrac{5}{8}x^4$	$y' = \dfrac{5}{8}(4x^3)$	$y' = \dfrac{5}{2}x^3$
c.	$y = \dfrac{-3(3x - 2x^2)}{7x}$	$y = -\dfrac{3}{7}(3 - 2x)$	$y' = -\dfrac{3}{7}(-2)$	$y' = \dfrac{6}{7}$
d.	$y = \dfrac{9}{5x^2}$	$y = \dfrac{9}{5}(x^{-2})$	$y' = \dfrac{9}{5}(-2x^{-3})$	$y' = -\dfrac{18}{5x^3}$

■

In Section 2.2, the Power Rule was proved only for the case in which the exponent n is a positive integer greater than 1. The next example extends the proof to include negative integer exponents.

EXAMPLE 7 **Power Rule: Negative Integer Exponents**

If n is a negative integer, then there exists a positive integer k such that $n = -k$. So, by the Quotient Rule, you can write

$$\begin{aligned}\frac{d}{dx}[x^n] &= \frac{d}{dx}\left[\frac{1}{x^k}\right]\\ &= \frac{x^k(0) - (1)(kx^{k-1})}{(x^k)^2} && \text{Quotient Rule and Power Rule}\\ &= \frac{0 - kx^{k-1}}{x^{2k}}\\ &= -kx^{-k-1}\\ &= nx^{n-1}. && n = -k\end{aligned}$$

So, the Power Rule

$$\frac{d}{dx}[x^n] = nx^{n-1} \qquad \text{Power Rule}$$

is valid for any integer. In Exercise 71 in Section 2.5, you are asked to prove the case for which n is any rational number.

Derivatives of Trigonometric Functions

Knowing the derivatives of the sine and cosine functions, you can use the Quotient Rule to find the derivatives of the four remaining trigonometric functions.

THEOREM 2.9 Derivatives of Trigonometric Functions

$$\frac{d}{dx}[\tan x] = \sec^2 x \qquad \frac{d}{dx}[\cot x] = -\csc^2 x$$

$$\frac{d}{dx}[\sec x] = \sec x \tan x \qquad \frac{d}{dx}[\csc x] = -\csc x \cot x$$

REMARK In the proof of Theorem 2.9, note the use of the trigonometric identities

$$\sin^2 x + \cos^2 x = 1$$

and

$$\sec x = \frac{1}{\cos x}.$$

These trigonometric identities and others are listed in Appendix C and on the formula cards for this text.

Proof Considering $\tan x = (\sin x)/(\cos x)$ and applying the Quotient Rule, you obtain

$$\begin{aligned}\frac{d}{dx}[\tan x] &= \frac{d}{dx}\left[\frac{\sin x}{\cos x}\right]\\ &= \frac{(\cos x)(\cos x) - (\sin x)(-\sin x)}{\cos^2 x} && \text{Apply Quotient Rule.}\\ &= \frac{\cos^2 x + \sin^2 x}{\cos^2 x}\\ &= \frac{1}{\cos^2 x}\\ &= \sec^2 x.\end{aligned}$$

See LarsonCalculus.com for Bruce Edwards's video of this proof.

The proofs of the other three parts of the theorem are left as an exercise (see Exercise 87).

EXAMPLE 8 Differentiating Trigonometric Functions

See LarsonCalculus.com for an interactive version of this type of example.

Function	**Derivative**
a. $y = x - \tan x$	$\dfrac{dy}{dx} = 1 - \sec^2 x$
b. $y = x \sec x$	$y' = x(\sec x \tan x) + (\sec x)(1)$ $= (\sec x)(1 + x \tan x)$

REMARK Because of trigonometric identities, the derivative of a trigonometric function can take many forms. This presents a challenge when you are trying to match your answers to those given in the back of the text.

EXAMPLE 9 Different Forms of a Derivative

Differentiate both forms of

$$y = \frac{1 - \cos x}{\sin x} = \csc x - \cot x.$$

Solution

First form: $y = \dfrac{1 - \cos x}{\sin x}$

$$y' = \frac{(\sin x)(\sin x) - (1 - \cos x)(\cos x)}{\sin^2 x}$$

$$= \frac{\sin^2 x - \cos x + \cos^2 x}{\sin^2 x}$$

$$= \frac{1 - \cos x}{\sin^2 x} \qquad \sin^2 x + \cos^2 x = 1$$

Second form: $y = \csc x - \cot x$

$$y' = -\csc x \cot x + \csc^2 x$$

To show that the two derivatives are equal, you can write

$$\frac{1 - \cos x}{\sin^2 x} = \frac{1}{\sin^2 x} - \frac{\cos x}{\sin^2 x}$$

$$= \frac{1}{\sin^2 x} - \left(\frac{1}{\sin x}\right)\left(\frac{\cos x}{\sin x}\right)$$

$$= \csc^2 x - \csc x \cot x.$$

The summary below shows that much of the work in obtaining a simplified form of a derivative occurs *after* differentiating. Note that two characteristics of a simplified form are the absence of negative exponents and the combining of like terms.

	$f'(x)$ After Differentiating	$f'(x)$ After Simplifying
Example 1	$(3x - 2x^2)(4) + (5 + 4x)(3 - 4x)$	$-24x^2 + 4x + 15$
Example 3	$(2x)(-\sin x) + (\cos x)(2) - 2(\cos x)$	$-2x \sin x$
Example 4	$\dfrac{(x^2 + 1)(5) - (5x - 2)(2x)}{(x^2 + 1)^2}$	$\dfrac{-5x^2 + 4x + 5}{(x^2 + 1)^2}$
Example 5	$\dfrac{(x^2 + 5x)(3) - (3x - 1)(2x + 5)}{(x^2 + 5x)^2}$	$\dfrac{-3x^2 + 2x + 5}{(x^2 + 5x)^2}$
Example 6	$\dfrac{(\sin x)(\sin x) - (1 - \cos x)(\cos x)}{\sin^2 x}$	$\dfrac{1 - \cos x}{\sin^2 x}$

Higher-Order Derivatives

Just as you can obtain a velocity function by differentiating a position function, you can obtain an **acceleration** function by differentiating a velocity function. Another way of looking at this is that you can obtain an acceleration function by differentiating a position function *twice*.

$$s(t) \qquad \text{Position function}$$
$$v(t) = s'(t) \qquad \text{Velocity function}$$
$$a(t) = v'(t) = s''(t) \qquad \text{Acceleration function}$$

The function $a(t)$ is the **second derivative** of $s(t)$ and is denoted by $s''(t)$.

The second derivative is an example of a **higher-order derivative.** You can define derivatives of any positive integer order. For instance, the **third derivative** is the derivative of the second derivative. Higher-order derivatives are denoted as shown below.

First derivative:	y',	$f'(x)$,	$\dfrac{dy}{dx}$,	$\dfrac{d}{dx}[f(x)]$,	$D_x[y]$
Second derivative:	y'',	$f''(x)$,	$\dfrac{d^2y}{dx^2}$,	$\dfrac{d^2}{dx^2}[f(x)]$,	$D_x^2[y]$
Third derivative:	y''',	$f'''(x)$,	$\dfrac{d^3y}{dx^3}$,	$\dfrac{d^3}{dx^3}[f(x)]$,	$D_x^3[y]$
Fourth derivative:	$y^{(4)}$,	$f^{(4)}(x)$,	$\dfrac{d^4y}{dx^4}$,	$\dfrac{d^4}{dx^4}[f(x)]$,	$D_x^4[y]$
$\vdots$					
nth derivative:	$y^{(n)}$,	$f^{(n)}(x)$,	$\dfrac{d^ny}{dx^n}$,	$\dfrac{d^n}{dx^n}[f(x)]$,	$D_x^n[y]$

REMARK The second derivative of a function is the derivative of the first derivative of the function.

The moon's mass is 7.349×10^{22} kilograms, and Earth's mass is 5.976×10^{24} kilograms. The moon's radius is 1737 kilometers, and Earth's radius is 6378 kilometers. Because the gravitational force on the surface of a planet is directly proportional to its mass and inversely proportional to the square of its radius, the ratio of the gravitational force on Earth to the gravitational force on the moon is

$$\frac{(5.976 \times 10^{24})/6378^2}{(7.349 \times 10^{22})/1737^2} \approx 6.0.$$

NASA

EXAMPLE 10 Finding the Acceleration Due to Gravity

Because the moon has no atmosphere, a falling object on the moon encounters no air resistance. In 1971, astronaut David Scott demonstrated that a feather and a hammer fall at the same rate on the moon. The position function for each of these falling objects is

$$s(t) = -0.81t^2 + 2$$

where $s(t)$ is the height in meters and t is the time in seconds, as shown in the figure at the right. What is the ratio of Earth's gravitational force to the moon's?

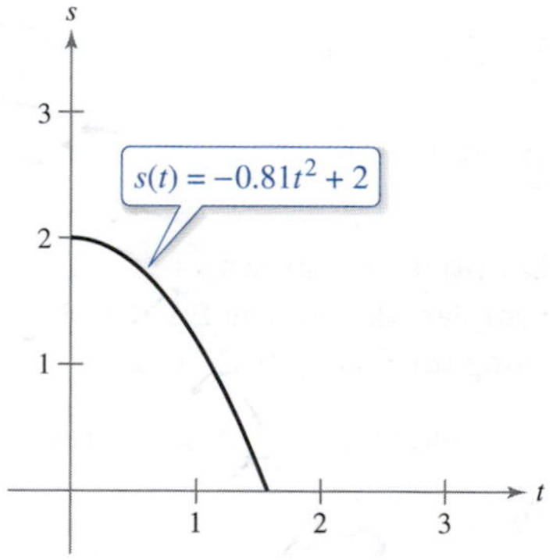

Solution To find the acceleration, differentiate the position function twice.

$$s(t) = -0.81t^2 + 2 \qquad \text{Position function}$$
$$s'(t) = -1.62t \qquad \text{Velocity function}$$
$$s''(t) = -1.62 \qquad \text{Acceleration function}$$

So, the acceleration due to gravity on the moon is -1.62 meters per second per second. Because the acceleration due to gravity on Earth is -9.8 meters per second per second, the ratio of Earth's gravitational force to the moon's is

$$\frac{\text{Earth's gravitational force}}{\text{Moon's gravitational force}} = \frac{-9.8}{-1.62} \approx 6.0.$$

2.3 Exercises

See CalcChat.com for tutorial help and worked-out solutions to odd-numbered exercises.

Using the Product Rule In Exercises 1–6, use the Product Rule to find the derivative of the function.

1. $g(x) = (x^2 + 3)(x^2 - 4x)$
2. $y = (3x - 4)(x^3 + 5)$
3. $h(t) = \sqrt{t}(1 - t^2)$
4. $g(s) = \sqrt{s}(s^2 + 8)$
5. $f(x) = x^3 \cos x$
6. $g(x) = \sqrt{x} \sin x$

Using the Quotient Rule In Exercises 7–12, use the Quotient Rule to find the derivative of the function.

7. $f(x) = \dfrac{x}{x^2 + 1}$
8. $g(t) = \dfrac{3t^2 - 1}{2t + 5}$
9. $h(x) = \dfrac{\sqrt{x}}{x^3 + 1}$
10. $f(x) = \dfrac{x^2}{2\sqrt{x} + 1}$
11. $g(x) = \dfrac{\sin x}{x^2}$
12. $f(t) = \dfrac{\cos t}{t^3}$

Finding and Evaluating a Derivative In Exercises 13–18, find $f'(x)$ and $f'(c)$.

Function	Value of c
13. $f(x) = (x^3 + 4x)(3x^2 + 2x - 5)$	$c = 0$
14. $y = (x^2 - 3x + 2)(x^3 + 1)$	$c = 2$
15. $f(x) = \dfrac{x^2 - 4}{x - 3}$	$c = 1$
16. $f(x) = \dfrac{x - 4}{x + 4}$	$c = 3$
17. $f(x) = x \cos x$	$c = \dfrac{\pi}{4}$
18. $f(x) = \dfrac{\sin x}{x}$	$c = \dfrac{\pi}{6}$

Using the Constant Multiple Rule In Exercises 19–24, complete the table to find the derivative of the function without using the Quotient Rule.

Function	Rewrite	Differentiate	Simplify
19. $y = \dfrac{x^2 + 3x}{7}$			
20. $y = \dfrac{5x^2 - 3}{4}$			
21. $y = \dfrac{6}{7x^2}$			
22. $y = \dfrac{10}{3x^3}$			
23. $y = \dfrac{4x^{3/2}}{x}$			
24. $y = \dfrac{2x}{x^{1/3}}$			

Finding a Derivative In Exercises 25–38, find the derivative of the algebraic function.

25. $f(x) = \dfrac{4 - 3x - x^2}{x^2 - 1}$
26. $f(x) = \dfrac{x^2 + 5x + 6}{x^2 - 4}$
27. $f(x) = x\left(1 - \dfrac{4}{x + 3}\right)$
28. $f(x) = x^4\left(1 - \dfrac{2}{x + 1}\right)$
29. $f(x) = \dfrac{3x - 1}{\sqrt{x}}$
30. $f(x) = \sqrt[3]{x}(\sqrt{x} + 3)$
31. $h(s) = (s^3 - 2)^2$
32. $h(x) = (x^2 + 3)^3$
33. $f(x) = \dfrac{2 - \dfrac{1}{x}}{x - 3}$
34. $g(x) = x^2\left(\dfrac{2}{x} - \dfrac{1}{x + 1}\right)$
35. $f(x) = (2x^3 + 5x)(x - 3)(x + 2)$
36. $f(x) = (x^3 - x)(x^2 + 2)(x^2 + x - 1)$
37. $f(x) = \dfrac{x^2 + c^2}{x^2 - c^2}$, c is a constant
38. $f(x) = \dfrac{c^2 - x^2}{c^2 + x^2}$, c is a constant

Finding a Derivative of a Trigonometric Function In Exercises 39–54, find the derivative of the trigonometric function.

39. $f(t) = t^2 \sin t$
40. $f(\theta) = (\theta + 1) \cos \theta$
41. $f(t) = \dfrac{\cos t}{t}$
42. $f(x) = \dfrac{\sin x}{x^3}$
43. $f(x) = -x + \tan x$
44. $y = x + \cot x$
45. $g(t) = \sqrt[4]{t} + 6 \csc t$
46. $h(x) = \dfrac{1}{x} - 12 \sec x$
47. $y = \dfrac{3(1 - \sin x)}{2 \cos x}$
48. $y = \dfrac{\sec x}{x}$
49. $y = -\csc x - \sin x$
50. $y = x \sin x + \cos x$
51. $f(x) = x^2 \tan x$
52. $f(x) = \sin x \cos x$
53. $y = 2x \sin x + x^2 \cos x$
54. $h(\theta) = 5\theta \sec \theta + \theta \tan \theta$

Finding a Derivative Using Technology In Exercises 55–58, use a computer algebra system to find the derivative of the function.

55. $g(x) = \left(\dfrac{x + 1}{x + 2}\right)(2x - 5)$
56. $f(x) = \left(\dfrac{x^2 - x - 3}{x^2 + 1}\right)(x^2 + x + 1)$
57. $g(\theta) = \dfrac{\theta}{1 - \sin \theta}$
58. $f(\theta) = \dfrac{\sin \theta}{1 - \cos \theta}$

Evaluating a Derivative **In Exercises 59–62, evaluate the derivative of the function at the given point. Use a graphing utility to verify your result.**

Function	Point
59. $y = \dfrac{1 + \csc x}{1 - \csc x}$	$\left(\dfrac{\pi}{6}, -3\right)$
60. $f(x) = \tan x \cot x$	$(1, 1)$
61. $h(t) = \dfrac{\sec t}{t}$	$\left(\pi, -\dfrac{1}{\pi}\right)$
62. $f(x) = \sin x(\sin x + \cos x)$	$\left(\dfrac{\pi}{4}, 1\right)$

Finding an Equation of a Tangent Line **In Exercises 63–68, (a) find an equation of the tangent line to the graph of f at the given point, (b) use a graphing utility to graph the function and its tangent line at the point, and (c) use the *derivative* feature of a graphing utility to confirm your results.**

63. $f(x) = (x^3 + 4x - 1)(x - 2)$, $(1, -4)$

64. $f(x) = (x - 2)(x^2 + 4)$, $(1, -5)$

65. $f(x) = \dfrac{x}{x + 4}$, $(-5, 5)$

66. $f(x) = \dfrac{x + 3}{x - 3}$, $(4, 7)$

67. $f(x) = \tan x$, $\left(\dfrac{\pi}{4}, 1\right)$

68. $f(x) = \sec x$, $\left(\dfrac{\pi}{3}, 2\right)$

Famous Curves **In Exercises 69–72, find an equation of the tangent line to the graph at the given point. (The graphs in Exercises 69 and 70 are called *Witches of Agnesi*. The graphs in Exercises 71 and 72 are called *serpentines*.)**

69.

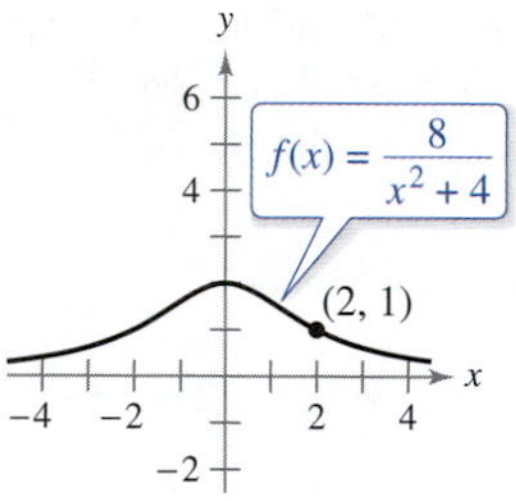

70.

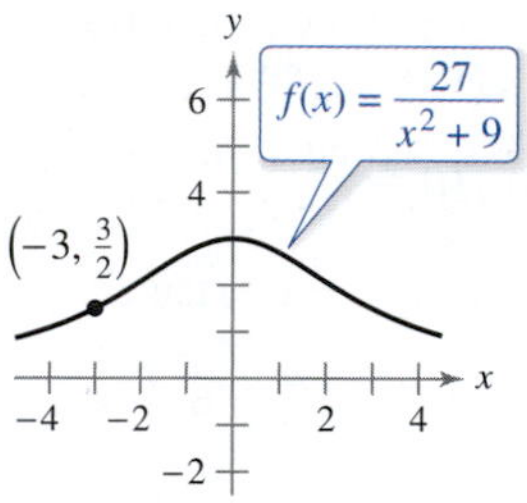

71.

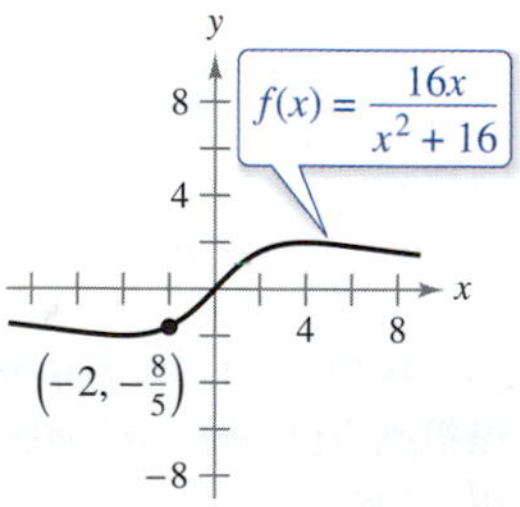

72.

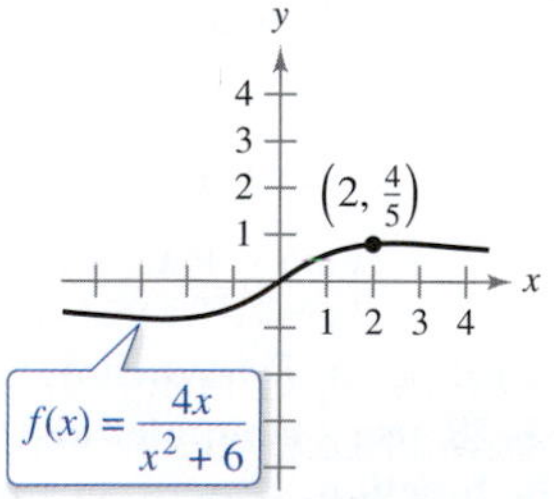

Horizontal Tangent Line **In Exercises 73–76, determine the point(s) at which the graph of the function has a horizontal tangent line.**

73. $f(x) = \dfrac{2x - 1}{x^2}$

74. $f(x) = \dfrac{x^2}{x^2 + 1}$

75. $f(x) = \dfrac{x^2}{x - 1}$

76. $f(x) = \dfrac{x - 4}{x^2 - 7}$

77. Tangent Lines Find equations of the tangent lines to the graph of $f(x) = (x + 1)/(x - 1)$ that are parallel to the line $2y + x = 6$. Then graph the function and the tangent lines.

78. Tangent Lines Find equations of the tangent lines to the graph of $f(x) = x/(x - 1)$ that pass through the point $(-1, 5)$. Then graph the function and the tangent lines.

Exploring a Relationship **In Exercises 79 and 80, verify that $f'(x) = g'(x)$, and explain the relationship between f and g.**

79. $f(x) = \dfrac{3x}{x + 2}$, $g(x) = \dfrac{5x + 4}{x + 2}$

80. $f(x) = \dfrac{\sin x - 3x}{x}$, $g(x) = \dfrac{\sin x + 2x}{x}$

Evaluating Derivatives **In Exercises 81 and 82, use the graphs of f and g. Let $p(x) = f(x)g(x)$ and $q(x) = f(x)/g(x)$.**

81. (a) Find $p'(1)$.
(b) Find $q'(4)$.

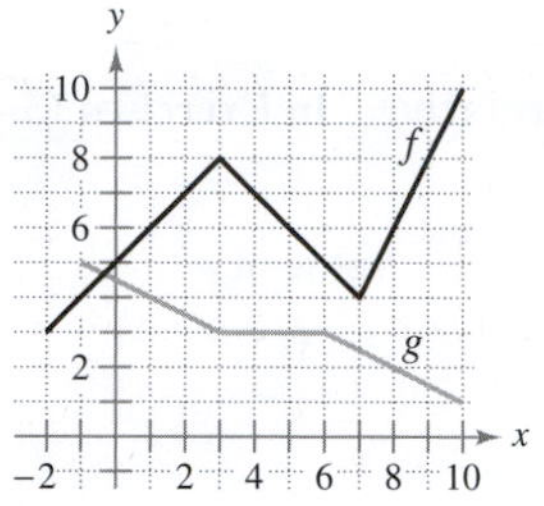

82. (a) Find $p'(4)$.
(b) Find $q'(7)$.

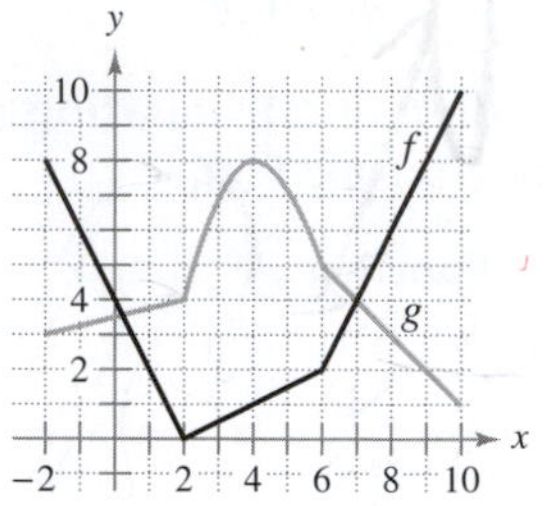

83. Area The length of a rectangle is given by $6t + 5$ and its height is $\sqrt{t}$, where t is time in seconds and the dimensions are in centimeters. Find the rate of change of the area with respect to time.

84. Volume The radius of a right circular cylinder is given by $\sqrt{t + 2}$ and its height is $\frac{1}{2}\sqrt{t}$, where t is time in seconds and the dimensions are in inches. Find the rate of change of the volume with respect to time.

85. Inventory Replenishment The ordering and transportation cost C for the components used in manufacturing a product is

$$C = 100\left(\frac{200}{x^2} + \frac{x}{x + 30}\right), \quad x \geq 1$$

where C is measured in thousands of dollars and x is the order size in hundreds. Find the rate of change of C with respect to x when (a) $x = 10$, (b) $x = 15$, and (c) $x = 20$. What do these rates of change imply about increasing order size?

86. Population Growth A population of 500 bacteria is introduced into a culture and grows in number according to the equation

$$P(t) = 500\left(1 + \frac{4t}{50 + t^2}\right)$$

where t is measured in hours. Find the rate at which the population is growing when $t = 2$.

87. Proof Prove the following differentiation rules.

(a) $\dfrac{d}{dx}[\sec x] = \sec x \tan x$

(b) $\dfrac{d}{dx}[\csc x] = -\csc x \cot x$

(c) $\dfrac{d}{dx}[\cot x] = -\csc^2 x$

88. Rate of Change Determine whether there exist any values of x in the interval $[0, 2\pi)$ such that the rate of change of $f(x) = \sec x$ and the rate of change of $g(x) = \csc x$ are equal.

89. Modeling Data The table shows the health care expenditures h (in billions of dollars) in the United States and the population p (in millions) of the United States for the years 2004 through 2009. The year is represented by t, with $t = 4$ corresponding to 2004. *(Source: U.S. Centers for Medicare & Medicaid Services and U.S. Census Bureau)*

Year, t	4	5	6	7	8	9
h	1773	1890	2017	2135	2234	2330
p	293	296	299	302	305	307

(a) Use a graphing utility to find linear models for the health care expenditures $h(t)$ and the population $p(t)$.

(b) Use a graphing utility to graph each model found in part (a).

(c) Find $A = h(t)/p(t)$, then graph A using a graphing utility. What does this function represent?

(d) Find and interpret $A'(t)$ in the context of these data.

90. Satellites When satellites observe Earth, they can scan only part of Earth's surface. Some satellites have sensors that can measure the angle θ shown in the figure. Let h represent the satellite's distance from Earth's surface, and let r represent Earth's radius.

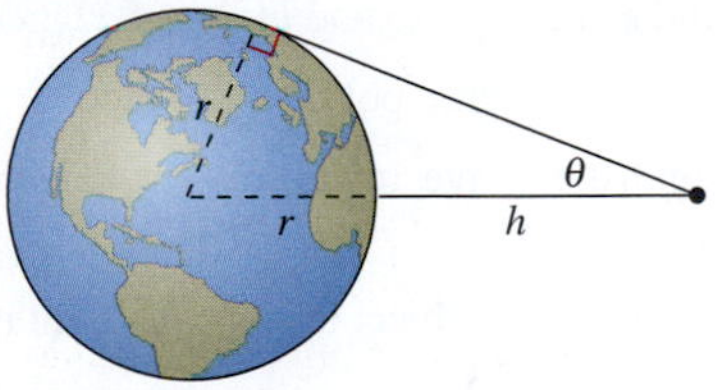

(a) Show that $h = r(\csc \theta - 1)$.

(b) Find the rate at which h is changing with respect to θ when $\theta = 30°$. (Assume $r = 3960$ miles.)

Finding a Second Derivative **In Exercises 91–98, find the second derivative of the function.**

91. $f(x) = x^4 + 2x^3 - 3x^2 - x$

92. $f(x) = 4x^5 - 2x^3 + 5x^2$

93. $f(x) = 4x^{3/2}$

94. $f(x) = x^2 + 3x^{-3}$

95. $f(x) = \dfrac{x}{x - 1}$

96. $f(x) = \dfrac{x^2 + 3x}{x - 4}$

97. $f(x) = x \sin x$

98. $f(x) = \sec x$

Finding a Higher-Order Derivative **In Exercises 99–102, find the given higher-order derivative.**

99. $f'(x) = x^2, \quad f''(x)$

100. $f''(x) = 2 - \dfrac{2}{x}, \quad f'''(x)$

101. $f'''(x) = 2\sqrt{x}, \quad f^{(4)}(x)$

102. $f^{(4)}(x) = 2x + 1, \quad f^{(6)}(x)$

Using Relationships **In Exercises 103–106, use the given information to find $f'(2)$.**

$g(2) = 3$ and $g'(2) = -2$

$h(2) = -1$ and $h'(2) = 4$

103. $f(x) = 2g(x) + h(x)$

104. $f(x) = 4 - h(x)$

105. $f(x) = \dfrac{g(x)}{h(x)}$

106. $f(x) = g(x)h(x)$

WRITING ABOUT CONCEPTS

107. Sketching a Graph Sketch the graph of a differentiable function f such that $f(2) = 0$, $f' < 0$ for $-\infty < x < 2$, and $f' > 0$ for $2 < x < \infty$. Explain how you found your answer.

108. Sketching a Graph Sketch the graph of a differentiable function f such that $f > 0$ and $f' < 0$ for all real numbers x. Explain how you found your answer.

Identifying Graphs **In Exercises 109 and 110, the graphs of f, f', and f'' are shown on the same set of coordinate axes. Identify each graph. Explain your reasoning. To print an enlarged copy of the graph, go to *MathGraphs.com*.**

109.

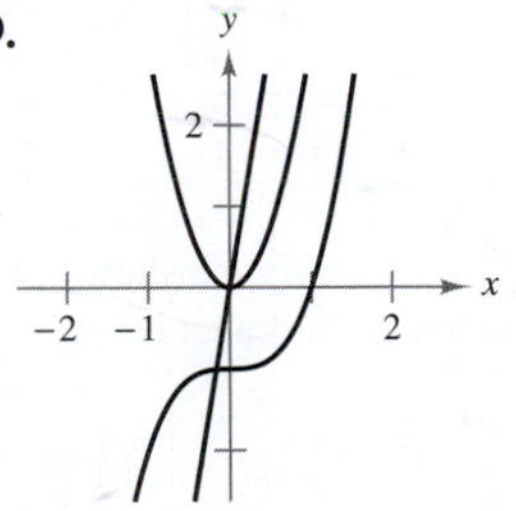

110.

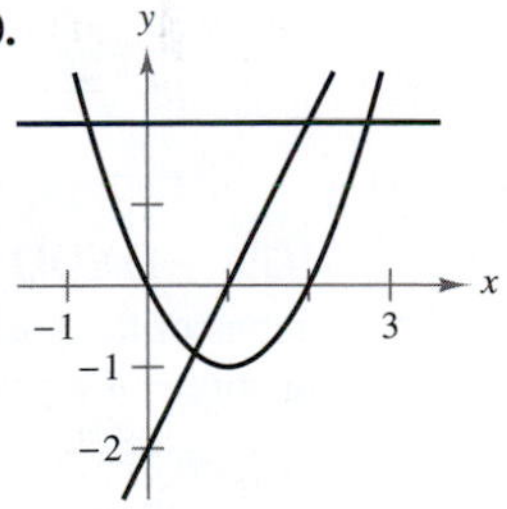

Sketching Graphs **In Exercises 111–114, the graph of f is shown. Sketch the graphs of f' and f''. To print an enlarged copy of the graph, go to *MathGraphs.com*.**

111.

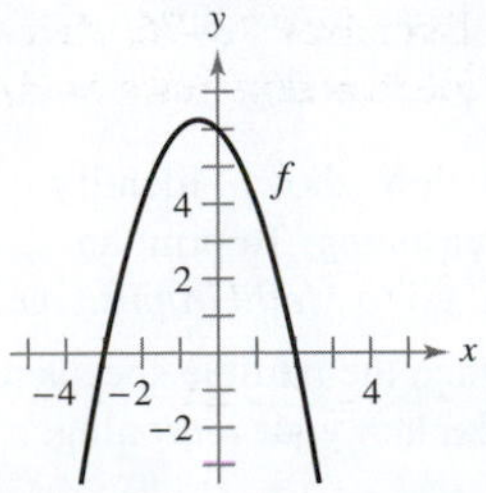

112.

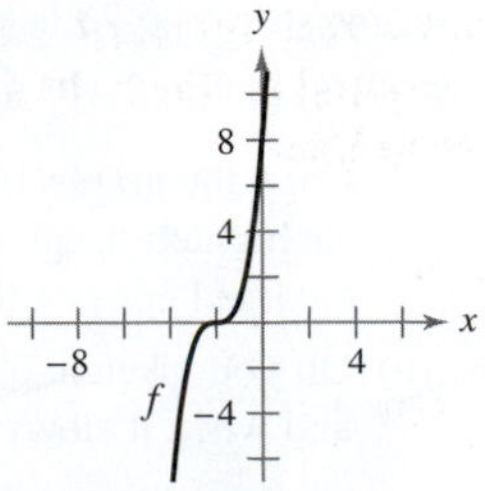

113.

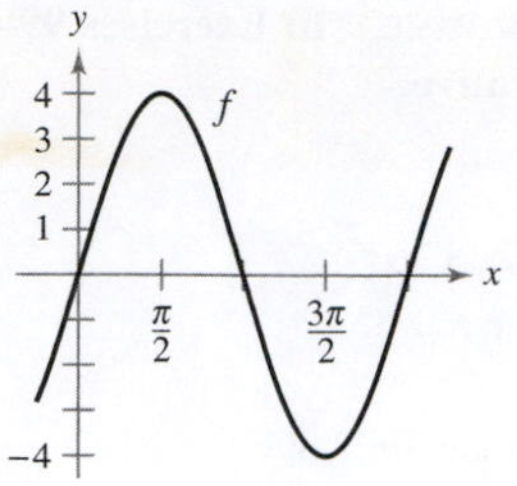

114.

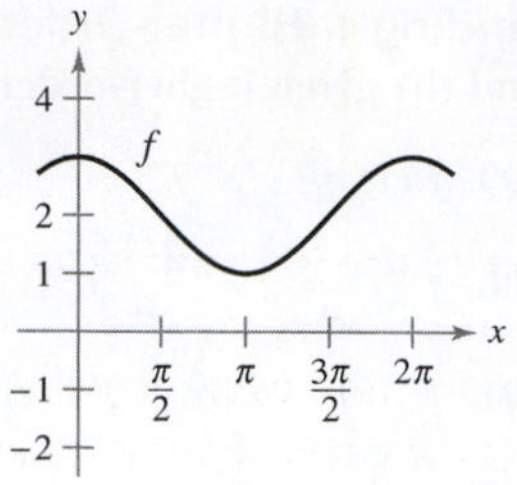

115. Acceleration The velocity of an object in meters per second is

$$v(t) = 36 - t^2$$

for $0 \le t \le 6$. Find the velocity and acceleration of the object when $t = 3$. What can be said about the speed of the object when the velocity and acceleration have opposite signs?

116. Acceleration The velocity of an automobile starting from rest is

$$v(t) = \frac{100t}{2t + 15}$$

where v is measured in feet per second. Find the acceleration at (a) 5 seconds, (b) 10 seconds, and (c) 20 seconds.

117. Stopping Distance A car is traveling at a rate of 66 feet per second (45 miles per hour) when the brakes are applied. The position function for the car is $s(t) = -8.25t^2 + 66t$, where s is measured in feet and t is measured in seconds. Use this function to complete the table, and find the average velocity during each time interval.

t	0	1	2	3	4
$s(t)$					
$v(t)$					
$a(t)$					

118. HOW DO YOU SEE IT? The figure shows the graphs of the position, velocity, and acceleration functions of a particle.

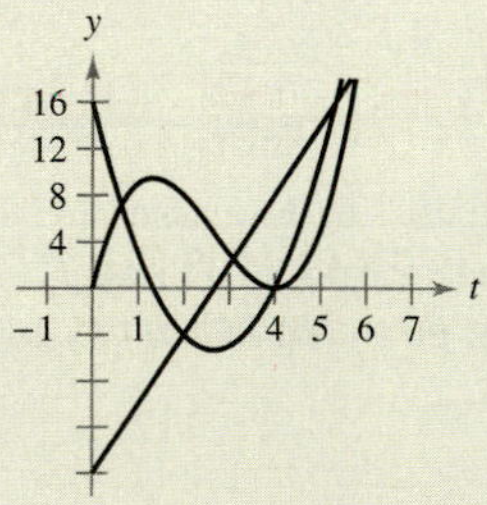

(a) Copy the graphs of the functions shown. Identify each graph. Explain your reasoning. To print an enlarged copy of the graph, go to *MathGraphs.com*.

(b) On your sketch, identify when the particle speeds up and when it slows down. Explain your reasoning.

Finding a Pattern In Exercises 119 and 120, develop a general rule for $f^{(n)}(x)$ given $f(x)$.

119. $f(x) = x^n$ **120.** $f(x) = \dfrac{1}{x}$

121. Finding a Pattern Consider the function $f(x) = g(x)h(x)$.

(a) Use the Product Rule to generate rules for finding $f''(x)$, $f'''(x)$, and $f^{(4)}(x)$.

(b) Use the results of part (a) to write a general rule for $f^{(n)}(x)$.

122. Finding a Pattern Develop a general rule for $[xf(x)]^{(n)}$, where f is a differentiable function of x.

Finding a Pattern In Exercises 123 and 124, find the derivatives of the function f for $n = 1, 2, 3,$ and 4. Use the results to write a general rule for $f'(x)$ in terms of n.

123. $f(x) = x^n \sin x$ **124.** $f(x) = \dfrac{\cos x}{x^n}$

Differential Equations In Exercises 125–128, verify that the function satisfies the differential equation.

Function	Differential Equation
125. $y = \dfrac{1}{x}, x > 0$	$x^3 y'' + 2x^2 y' = 0$
126. $y = 2x^3 - 6x + 10$	$-y''' - xy'' - 2y' = -24x^2$
127. $y = 2 \sin x + 3$	$y'' + y = 3$
128. $y = 3 \cos x + \sin x$	$y'' + y = 0$

True or False? In Exercises 129–134, determine whether the statement is true or false. If it is false, explain why or give an example that shows it is false.

129. If $y = f(x)g(x)$, then $\dfrac{dy}{dx} = f'(x)g'(x)$.

130. If $y = (x + 1)(x + 2)(x + 3)(x + 4)$, then $\dfrac{d^5y}{dx^5} = 0$.

131. If $f'(c)$ and $g'(c)$ are zero and $h(x) = f(x)g(x)$, then $h'(c) = 0$.

132. If $f(x)$ is an nth-degree polynomial, then $f^{(n+1)}(x) = 0$.

133. The second derivative represents the rate of change of the first derivative.

134. If the velocity of an object is constant, then its acceleration is zero.

135. Absolute Value Find the derivative of $f(x) = x|x|$. Does $f''(0)$ exist? (*Hint:* Rewrite the function as a piecewise function and then differentiate each part.)

136. Think About It Let f and g be functions whose first and second derivatives exist on an interval I. Which of the following formulas is (are) true?

(a) $fg'' - f''g = (fg' - f'g)'$ (b) $fg'' + f''g = (fg)''$

137. Proof Use the Product Rule twice to prove that if f, g, and h are differentiable functions of x, then

$$\frac{d}{dx}[f(x)g(x)h(x)] = f'(x)g(x)h(x) + f(x)g'(x)h(x) + f(x)g(x)h'(x).$$

2.4 The Chain Rule

- **Find the derivative of a composite function using the Chain Rule.**
- **Find the derivative of a function using the General Power Rule.**
- **Simplify the derivative of a function using algebra.**
- **Find the derivative of a trigonometric function using the Chain Rule.**

The Chain Rule

This text has yet to discuss one of the most powerful differentiation rules—the **Chain Rule.** This rule deals with composite functions and adds a surprising versatility to the rules discussed in the two previous sections. For example, compare the functions shown below. Those on the left can be differentiated without the Chain Rule, and those on the right are best differentiated with the Chain Rule.

Without the Chain Rule	**With the Chain Rule**
$y = x^2 + 1$	$y = \sqrt{x^2 + 1}$
$y = \sin x$	$y = \sin 6x$
$y = 3x + 2$	$y = (3x + 2)^5$
$y = x + \tan x$	$y = x + \tan x^2$

Basically, the Chain Rule states that if y changes dy/du times as fast as u, and u changes du/dx times as fast as x, then y changes $(dy/du)(du/dx)$ times as fast as x.

EXAMPLE 1 **The Derivative of a Composite Function**

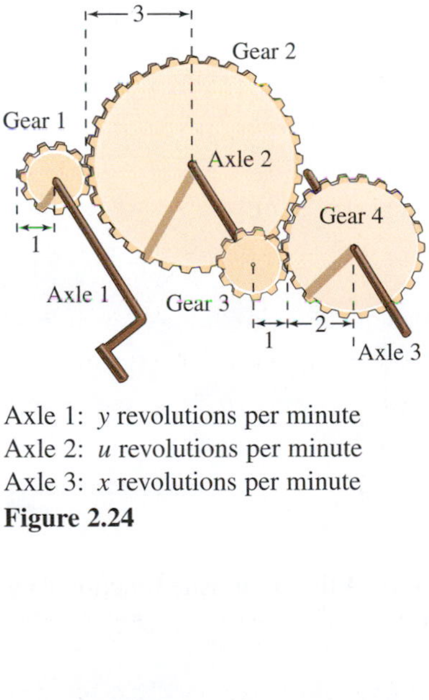

Axle 1: y revolutions per minute
Axle 2: u revolutions per minute
Axle 3: x revolutions per minute
Figure 2.24

A set of gears is constructed, as shown in Figure 2.24, such that the second and third gears are on the same axle. As the first axle revolves, it drives the second axle, which in turn drives the third axle. Let y, u, and x represent the numbers of revolutions per minute of the first, second, and third axles, respectively. Find dy/du, du/dx, and dy/dx, and show that

$$\frac{dy}{dx} = \frac{dy}{du} \cdot \frac{du}{dx}.$$

Solution Because the circumference of the second gear is three times that of the first, the first axle must make three revolutions to turn the second axle once. Similarly, the second axle must make two revolutions to turn the third axle once, and you can write

$$\frac{dy}{du} = 3 \quad \text{and} \quad \frac{du}{dx} = 2.$$

Combining these two results, you know that the first axle must make six revolutions to turn the third axle once. So, you can write

$$\begin{aligned}\frac{dy}{dx} &= \boxed{\begin{array}{l}\text{Rate of change of first axle}\\ \text{with respect to second axle}\end{array}} \cdot \boxed{\begin{array}{l}\text{Rate of change of second axle}\\ \text{with respect to third axle}\end{array}}\\ &= \frac{dy}{du} \cdot \frac{du}{dx}\\ &= 3 \cdot 2\\ &= 6\\ &= \boxed{\begin{array}{l}\text{Rate of change of first axle}\\ \text{with respect to third axle}\end{array}}.\end{aligned}$$

In other words, the rate of change of y with respect to x is the product of the rate of change of y with respect to u and the rate of change of u with respect to x. ■

> **Exploration**
>
> ***Using the Chain Rule*** Each of the following functions can be differentiated using rules that you studied in Sections 2.2 and 2.3. For each function, find the derivative using those rules. Then find the derivative using the Chain Rule. Compare your results. Which method is simpler?
>
> **a.** $\dfrac{2}{3x + 1}$
>
> **b.** $(x + 2)^3$
>
> **c.** $\sin 2x$

Example 1 illustrates a simple case of the Chain Rule. The general rule is stated in the next theorem.

> **THEOREM 2.10 The Chain Rule**
>
> If $y = f(u)$ is a differentiable function of u and $u = g(x)$ is a differentiable function of x, then $y = f(g(x))$ is a differentiable function of x and
>
> $$\frac{dy}{dx} = \frac{dy}{du} \cdot \frac{du}{dx}$$
>
> or, equivalently,
>
> $$\frac{d}{dx}[f(g(x))] = f'(g(x))g'(x).$$

Proof Let $h(x) = f(g(x))$. Then, using the alternative form of the derivative, you need to show that, for $x = c$,

$$h'(c) = f'(g(c))g'(c).$$

An important consideration in this proof is the behavior of g as x approaches c. A problem occurs when there are values of x, other than c, such that

$$g(x) = g(c).$$

Appendix A shows how to use the differentiability of f and g to overcome this problem. For now, assume that $g(x) \neq g(c)$ for values of x other than c. In the proofs of the Product Rule and the Quotient Rule, the same quantity was added and subtracted to obtain the desired form. This proof uses a similar technique—multiplying and dividing by the same (nonzero) quantity. Note that because g is differentiable, it is also continuous, and it follows that $g(x)$ approaches $g(c)$ as x approaches c.

REMARK The alternative limit form of the derivative was given at the end of Section 2.1.

$$h'(c) = \lim_{x \to c} \frac{f(g(x)) - f(g(c))}{x - c} \qquad \text{Alternative form of derivative}$$

$$= \lim_{x \to c} \left[\frac{f(g(x)) - f(g(c))}{x - c} \cdot \frac{g(x) - g(c)}{g(x) - g(c)} \right], \quad g(x) \neq g(c)$$

$$= \lim_{x \to c} \left[\frac{f(g(x)) - f(g(c))}{g(x) - g(c)} \cdot \frac{g(x) - g(c)}{x - c} \right]$$

$$= \left[\lim_{x \to c} \frac{f(g(x)) - f(g(c))}{g(x) - g(c)} \right] \left[\lim_{x \to c} \frac{g(x) - g(c)}{x - c} \right]$$

$$= f'(g(c))g'(c)$$

See LarsonCalculus.com for Bruce Edwards's video of this proof.

AP* Tips

Very complex chain rule examples do not appear on the AP Exam. However, forgetting the chain rule is a frequent cause of loss of credit on a problem. The difficult problems here will help reinforce the concept.

When applying the Chain Rule, it is helpful to think of the composite function $f \circ g$ as having two parts—an inner part and an outer part.

Outer function

$$y = f(g(x)) = f(u)$$

Inner function

The derivative of $y = f(u)$ is the derivative of the outer function (at the inner function u) *times* the derivative of the inner function.

$$y' = f'(u) \cdot u'$$

EXAMPLE 2 **Decomposition of a Composite Function**

$y = f(g(x))$	$u = g(x)$	$y = f(u)$
a. $y = \dfrac{1}{x+1}$	$u = x + 1$	$y = \dfrac{1}{u}$
b. $y = \sin 2x$	$u = 2x$	$y = \sin u$
c. $y = \sqrt{3x^2 - x + 1}$	$u = 3x^2 - x + 1$	$y = \sqrt{u}$
d. $y = \tan^2 x$	$u = \tan x$	$y = u^2$

EXAMPLE 3 **Using the Chain Rule**

Find dy/dx for

$$y = (x^2 + 1)^3.$$

Solution For this function, you can consider the inside function to be $u = x^2 + 1$ and the outer function to be $y = u^3$. By the Chain Rule, you obtain

$$\frac{dy}{dx} = \underbrace{3(x^2 + 1)^2}_{\frac{dy}{du}}\underbrace{(2x)}_{\frac{du}{dx}} = 6x(x^2 + 1)^2.$$

REMARK You could also solve the problem in Example 3 without using the Chain Rule by observing that

$$y = x^6 + 3x^4 + 3x^2 + 1$$

and

$$y' = 6x^5 + 12x^3 + 6x.$$

Verify that this is the same as the derivative in Example 3. Which method would you use to find

$$\frac{d}{dx}(x^2 + 1)^{50}?$$

The General Power Rule

The function in Example 3 is an example of one of the most common types of composite functions, $y = [u(x)]^n$. The rule for differentiating such functions is called the **General Power Rule,** and it is a special case of the Chain Rule.

THEOREM 2.11 The General Power Rule

If $y = [u(x)]^n$, where u is a differentiable function of x and n is a rational number, then

$$\frac{dy}{dx} = n[u(x)]^{n-1}\frac{du}{dx}$$

or, equivalently,

$$\frac{d}{dx}[u^n] = nu^{n-1}u'.$$

Proof Because $y = [u(x)]^n = u^n$, you apply the Chain Rule to obtain

$$\frac{dy}{dx} = \left(\frac{dy}{du}\right)\left(\frac{du}{dx}\right)$$
$$= \frac{d}{du}[u^n]\frac{du}{dx}.$$

By the (Simple) Power Rule in Section 2.2, you have $D_u[u^n] = nu^{n-1}$, and it follows that

$$\frac{dy}{dx} = nu^{n-1}\frac{du}{dx}.$$

See LarsonCalculus.com for Bruce Edwards's video of this proof.

EXAMPLE 4 Applying the General Power Rule

Find the derivative of $f(x) = (3x - 2x^2)^3$.

Solution Let $u = 3x - 2x^2$. Then

$$f(x) = (3x - 2x^2)^3 = u^3$$

and, by the General Power Rule, the derivative is

$$f'(x) = \overset{n}{3}\overbrace{(3x - 2x^2)^2}^{u^{n-1}}\overbrace{\frac{d}{dx}[3x - 2x^2]}^{u'}$$ Apply General Power Rule.

$$= 3(3x - 2x^2)^2(3 - 4x).$$ Differentiate $3x - 2x^2$.

EXAMPLE 5 Differentiating Functions Involving Radicals

Find all points on the graph of

$$f(x) = \sqrt[3]{(x^2 - 1)^2}$$

for which $f'(x) = 0$ and those for which $f'(x)$ does not exist.

Solution Begin by rewriting the function as

$$f(x) = (x^2 - 1)^{2/3}.$$

Then, applying the General Power Rule (with $u = x^2 - 1$) produces

$$f'(x) = \overset{n}{\frac{2}{3}}\overbrace{(x^2 - 1)^{-1/3}}^{u^{n-1}}\overbrace{(2x)}^{u'}$$ Apply General Power Rule.

$$= \frac{4x}{3\sqrt[3]{x^2 - 1}}.$$ Write in radical form.

So, $f'(x) = 0$ when $x = 0$, and $f'(x)$ does not exist when $x = \pm 1$, as shown in Figure 2.25.

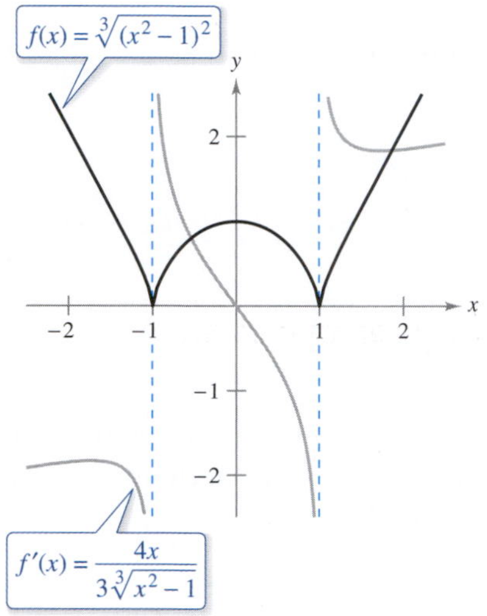

The derivative of f is 0 at $x = 0$ and is undefined at $x = \pm 1$.
Figure 2.25

EXAMPLE 6 Differentiating Quotients: Constant Numerators

Differentiate the function

$$g(t) = \frac{-7}{(2t - 3)^2}.$$

Solution Begin by rewriting the function as

$$g(t) = -7(2t - 3)^{-2}.$$

Then, applying the General Power Rule (with $u = 2t - 3$) produces

$$g'(t) = \underbrace{(-7)}_{\text{Constant Multiple Rule}}\overbrace{(-2)}^{n}\overbrace{(2t - 3)^{-3}}^{u^{n-1}}\overbrace{(2)}^{u'}$$ Apply General Power Rule.

$$= 28(2t - 3)^{-3}$$ Simplify.

$$= \frac{28}{(2t - 3)^3}.$$ Write with positive exponent.

REMARK Try differentiating the function in Example 6 using the Quotient Rule. You should obtain the same result, but using the Quotient Rule is less efficient than using the General Power Rule.

Simplifying Derivatives

The next three examples demonstrate techniques for simplifying the "raw derivatives" of functions involving products, quotients, and composites.

EXAMPLE 7 Simplifying by Factoring Out the Least Powers

Find the derivative of $f(x) = x^2\sqrt{1 - x^2}$.

Solution

$$\begin{aligned}
f(x) &= x^2\sqrt{1 - x^2} && \text{Write original function.}\\
&= x^2(1 - x^2)^{1/2} && \text{Rewrite.}\\
f'(x) &= x^2\frac{d}{dx}[(1 - x^2)^{1/2}] + (1 - x^2)^{1/2}\frac{d}{dx}[x^2] && \text{Product Rule}\\
&= x^2\left[\frac{1}{2}(1 - x^2)^{-1/2}(-2x)\right] + (1 - x^2)^{1/2}(2x) && \text{General Power Rule}\\
&= -x^3(1 - x^2)^{-1/2} + 2x(1 - x^2)^{1/2} && \text{Simplify.}\\
&= x(1 - x^2)^{-1/2}[-x^2(1) + 2(1 - x^2)] && \text{Factor.}\\
&= \frac{x(2 - 3x^2)}{\sqrt{1 - x^2}} && \text{Simplify.}
\end{aligned}$$

EXAMPLE 8 Simplifying the Derivative of a Quotient

$$\begin{aligned}
f(x) &= \frac{x}{\sqrt[3]{x^2 + 4}} && \text{Original function}\\
&= \frac{x}{(x^2 + 4)^{1/3}} && \text{Rewrite.}\\
f'(x) &= \frac{(x^2 + 4)^{1/3}(1) - x(1/3)(x^2 + 4)^{-2/3}(2x)}{(x^2 + 4)^{2/3}} && \text{Quotient Rule}\\
&= \frac{1}{3}(x^2 + 4)^{-2/3}\left[\frac{3(x^2 + 4) - (2x^2)(1)}{(x^2 + 4)^{2/3}}\right] && \text{Factor.}\\
&= \frac{x^2 + 12}{3(x^2 + 4)^{4/3}} && \text{Simplify.}
\end{aligned}$$

▷ **TECHNOLOGY** Symbolic differentiation utilities are capable of differentiating very complicated functions. Often, however, the result is given in unsimplified form. If you have access to such a utility, use it to find the derivatives of the functions given in Examples 7, 8, and 9. Then compare the results with those given in these examples.

EXAMPLE 9 Simplifying the Derivative of a Power

▷ *See LarsonCalculus.com for an interactive version of this type of example.*

$$y = \left(\frac{3x - 1}{x^2 + 3}\right)^2 \qquad \text{Original function}$$

$$\begin{aligned}
y' &= \underbrace{2}_{n}\underbrace{\left(\frac{3x - 1}{x^2 + 3}\right)}_{u^{n-1}}\underbrace{\frac{d}{dx}\left[\frac{3x - 1}{x^2 + 3}\right]}_{u'} && \text{General Power Rule}\\
&= \left[\frac{2(3x - 1)}{x^2 + 3}\right]\left[\frac{(x^2 + 3)(3) - (3x - 1)(2x)}{(x^2 + 3)^2}\right] && \text{Quotient Rule}\\
&= \frac{2(3x - 1)(3x^2 + 9 - 6x^2 + 2x)}{(x^2 + 3)^3} && \text{Multiply.}\\
&= \frac{2(3x - 1)(-3x^2 + 2x + 9)}{(x^2 + 3)^3} && \text{Simplify.}
\end{aligned}$$

Trigonometric Functions and the Chain Rule

The "Chain Rule versions" of the derivatives of the six trigonometric functions are shown below.

$$\frac{d}{dx}[\sin u] = (\cos u)u' \qquad \frac{d}{dx}[\cos u] = -(\sin u)u'$$

$$\frac{d}{dx}[\tan u] = (\sec^2 u)u' \qquad \frac{d}{dx}[\cot u] = -(\csc^2 u)u'$$

$$\frac{d}{dx}[\sec u] = (\sec u \tan u)u' \qquad \frac{d}{dx}[\csc u] = -(\csc u \cot u)u'$$

EXAMPLE 10 The Chain Rule and Trigonometric Functions

a. $y = \sin \overbrace{2x}^{u}$ $\qquad y' = \overbrace{\cos 2x}^{\cos u} \overbrace{\frac{d}{dx}[2x]}^{u'} = (\cos 2x)(2) = 2 \cos 2x$

b. $y = \cos\overbrace{(x - 1)}^{u}$ $\qquad y' = \overbrace{-\sin(x-1)}^{-(\sin u)} \overbrace{\frac{d}{dx}[x-1]}^{u'} = -\sin(x - 1)$

c. $y = \tan \overbrace{3x}^{u}$ $\qquad y' = \overbrace{\sec^2 3x}^{(\sec^2 u)} \overbrace{\frac{d}{dx}[3x]}^{u'} = (\sec^2 3x)(3) = 3 \sec^2(3x)$ ■

Be sure you understand the mathematical conventions regarding parentheses and trigonometric functions. For instance, in Example 10(a), $\sin 2x$ is written to mean $\sin(2x)$.

EXAMPLE 11 Parentheses and Trigonometric Functions

a. $y = \cos 3x^2 = \cos(3x^2)$ $\qquad y' = (-\sin 3x^2)(6x) = -6x \sin 3x^2$

b. $y = (\cos 3)x^2$ $\qquad y' = (\cos 3)(2x) = 2x \cos 3$

c. $y = \cos(3x)^2 = \cos(9x^2)$ $\qquad y' = (-\sin 9x^2)(18x) = -18x \sin 9x^2$

d. $y = \cos^2 x = (\cos x)^2$ $\qquad y' = 2(\cos x)(-\sin x) = -2 \cos x \sin x$

e. $y = \sqrt{\cos x} = (\cos x)^{1/2}$ $\qquad y' = \frac{1}{2}(\cos x)^{-1/2}(-\sin x) = -\frac{\sin x}{2\sqrt{\cos x}}$ ■

To find the derivative of a function of the form $k(x) = f(g(h(x)))$, you need to apply the Chain Rule twice, as shown in Example 12.

EXAMPLE 12 Repeated Application of the Chain Rule

$$
\begin{aligned}
f(t) &= \sin^3 4t && \text{Original function}\\
&= (\sin 4t)^3 && \text{Rewrite.}\\
f'(t) &= 3(\sin 4t)^2 \frac{d}{dt}[\sin 4t] && \text{Apply Chain Rule once.}\\
&= 3(\sin 4t)^2(\cos 4t)\frac{d}{dt}[4t] && \text{Apply Chain Rule a second time.}\\
&= 3(\sin 4t)^2(\cos 4t)(4) &&\\
&= 12 \sin^2 4t \cos 4t && \text{Simplify.}
\end{aligned}
$$

■

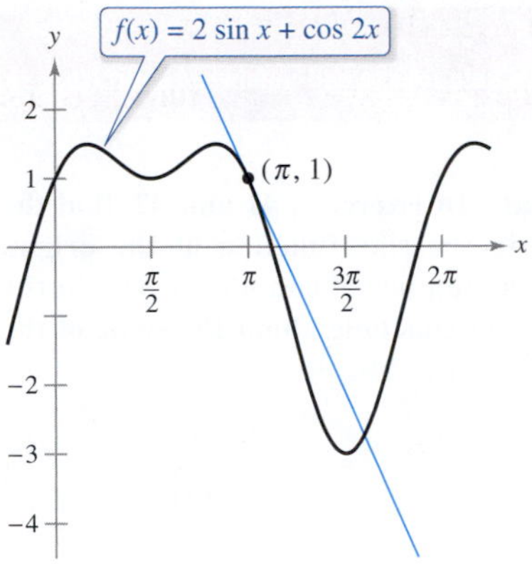

Figure 2.26

EXAMPLE 13 Tangent Line of a Trigonometric Function

Find an equation of the tangent line to the graph of $f(x) = 2\sin x + \cos 2x$ at the point $(\pi, 1)$, as shown in Figure 2.26. Then determine all values of x in the interval $(0, 2\pi)$ at which the graph of f has a horizontal tangent.

Solution Begin by finding $f'(x)$.

$$f(x) = 2\sin x + \cos 2x \quad \text{Write original function.}$$
$$f'(x) = 2\cos x + (-\sin 2x)(2) \quad \text{Apply Chain Rule to } \cos 2x.$$
$$= 2\cos x - 2\sin 2x \quad \text{Simplify.}$$

To find the equation of the tangent line at $(\pi, 1)$, evaluate $f'(\pi)$.

$$f'(\pi) = 2\cos \pi - 2\sin 2\pi \quad \text{Substitute.}$$
$$= -2 \quad \text{Slope of graph at } (\pi, 1)$$

Now, using the point-slope form of the equation of a line, you can write

$$y - y_1 = m(x - x_1) \quad \text{Point-slope form}$$
$$y - 1 = -2(x - \pi) \quad \text{Substitute for } y_1, m, \text{ and } x_1.$$
$$y = 1 - 2x + 2\pi. \quad \text{Equation of tangent line at } (\pi, 1)$$

You can then determine that $f'(x) = 0$ when $x = \frac{\pi}{6}, \frac{\pi}{2}, \frac{5\pi}{6}$, and $\frac{3\pi}{2}$. So, f has horizontal tangents at $x = \frac{\pi}{6}, \frac{\pi}{2}, \frac{5\pi}{6}$, and $\frac{3\pi}{2}$.

This section concludes with a summary of the differentiation rules studied so far. To become skilled at differentiation, you should memorize each rule in words, not symbols. As an aid to memorization, note that the cofunctions (cosine, cotangent, and cosecant) require a negative sign as part of their derivatives.

SUMMARY OF DIFFERENTIATION RULES

General Differentiation Rules Let f, g, and u be differentiable functions of x.

Constant Multiple Rule: $\frac{d}{dx}[cf] = cf'$

Sum or Difference Rule: $\frac{d}{dx}[f \pm g] = f' \pm g'$

Product Rule: $\frac{d}{dx}[fg] = fg' + gf'$

Quotient Rule: $\frac{d}{dx}\left[\frac{f}{g}\right] = \frac{gf' - fg'}{g^2}$

Derivatives of Algebraic Functions

Constant Rule: $\frac{d}{dx}[c] = 0$

(Simple) Power Rule: $\frac{d}{dx}[x^n] = nx^{n-1}, \quad \frac{d}{dx}[x] = 1$

Derivatives of Trigonometric Functions

$\frac{d}{dx}[\sin x] = \cos x \qquad \frac{d}{dx}[\tan x] = \sec^2 x \qquad \frac{d}{dx}[\sec x] = \sec x \tan x$

$\frac{d}{dx}[\cos x] = -\sin x \qquad \frac{d}{dx}[\cot x] = -\csc^2 x \qquad \frac{d}{dx}[\csc x] = -\csc x \cot x$

Chain Rule

Chain Rule: $\frac{d}{dx}[f(u)] = f'(u)\,u'$

General Power Rule: $\frac{d}{dx}[u^n] = nu^{n-1}\,u'$

2.4 Exercises

See CalcChat.com for tutorial help and worked-out solutions to odd-numbered exercises.

Decomposition of a Composite Function In Exercises 1–6, complete the table.

$y = f(g(x))$	$u = g(x)$	$y = f(u)$
1. $y = (5x - 8)^4$		
2. $y = \dfrac{1}{\sqrt{x + 1}}$		
3. $y = \sqrt{x^3 - 7}$		
4. $y = 3\tan(\pi x^2)$		
5. $y = \csc^3 x$		
6. $y = \sin \dfrac{5x}{2}$		

Finding a Derivative In Exercises 7–34, find the derivative of the function.

7. $y = (4x - 1)^3$

8. $y = 5(2 - x^3)^4$

9. $g(x) = 3(4 - 9x)^4$

10. $f(t) = (9t + 2)^{2/3}$

11. $f(t) = \sqrt{5 - t}$

12. $g(x) = \sqrt{4 - 3x^2}$

13. $y = \sqrt[3]{6x^2 + 1}$

14. $f(x) = \sqrt{x^2 - 4x + 2}$

15. $y = 2\sqrt[4]{9 - x^2}$

16. $f(x) = \sqrt[3]{12x - 5}$

17. $y = \dfrac{1}{x - 2}$

18. $s(t) = \dfrac{1}{4 - 5t - t^2}$

19. $f(t) = \left(\dfrac{1}{t - 3}\right)^2$

20. $y = -\dfrac{3}{(t - 2)^4}$

21. $y = \dfrac{1}{\sqrt{3x + 5}}$

22. $g(t) = \dfrac{1}{\sqrt{t^2 - 2}}$

23. $f(x) = x^2(x - 2)^4$

24. $f(x) = x(2x - 5)^3$

25. $y = x\sqrt{1 - x^2}$

26. $y = \frac{1}{2}x^2\sqrt{16 - x^2}$

27. $y = \dfrac{x}{\sqrt{x^2 + 1}}$

28. $y = \dfrac{x}{\sqrt{x^4 + 4}}$

29. $g(x) = \left(\dfrac{x + 5}{x^2 + 2}\right)^2$

30. $h(t) = \left(\dfrac{t^2}{t^3 + 2}\right)^2$

31. $f(v) = \left(\dfrac{1 - 2v}{1 + v}\right)^3$

32. $g(x) = \left(\dfrac{3x^2 - 2}{2x + 3}\right)^3$

33. $f(x) = ((x^2 + 3)^5 + x)^2$

34. $g(x) = (2 + (x^2 + 1)^4)^3$

Finding a Derivative Using Technology In Exercises 35–40, use a computer algebra system to find the derivative of the function. Then use the utility to graph the function and its derivative on the same set of coordinate axes. Describe the behavior of the function that corresponds to any zeros of the graph of the derivative.

35. $y = \dfrac{\sqrt{x} + 1}{x^2 + 1}$

36. $y = \sqrt{\dfrac{2x}{x + 1}}$

37. $y = \sqrt{\dfrac{x + 1}{x}}$

38. $g(x) = \sqrt{x - 1} + \sqrt{x + 1}$

39. $y = \dfrac{\cos \pi x + 1}{x}$

40. $y = x^2 \tan \dfrac{1}{x}$

Slope of a Tangent Line In Exercises 41 and 42, find the slope of the tangent line to the sine function at the origin. Compare this value with the number of complete cycles in the interval $[0, 2\pi]$. What can you conclude about the slope of the sine function $\sin ax$ at the origin?

41. (a)

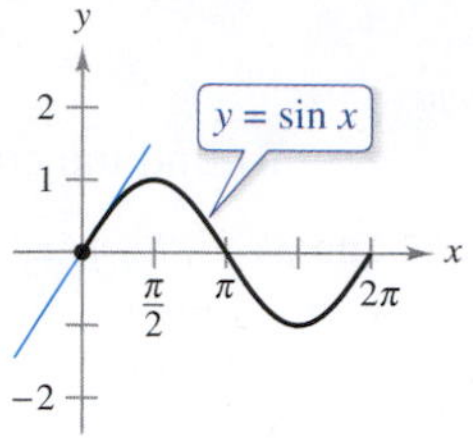

(b)

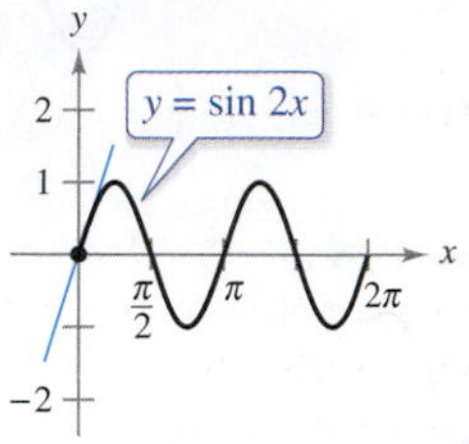

42. (a)

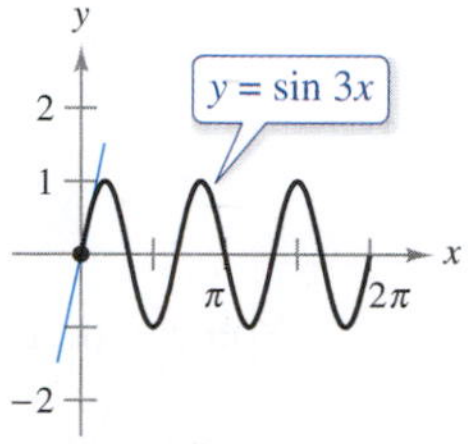

(b)

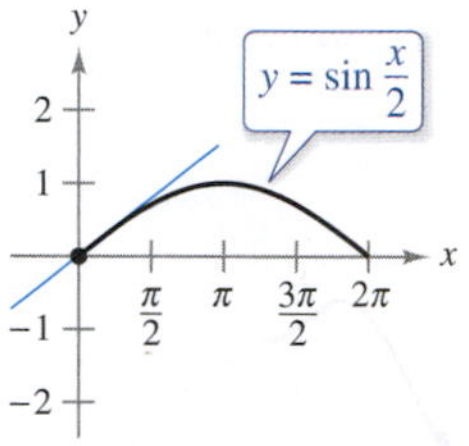

Finding a Derivative In Exercises 43–64, find the derivative of the function.

43. $y = \cos 4x$

44. $y = \sin \pi x$

45. $g(x) = 5 \tan 3x$

46. $h(x) = \sec x^2$

47. $y = \sin(\pi x)^2$

48. $y = \cos(1 - 2x)^2$

49. $h(x) = \sin 2x \cos 2x$

50. $g(\theta) = \sec\left(\frac{1}{2}\theta\right)\tan\left(\frac{1}{2}\theta\right)$

51. $f(x) = \dfrac{\cot x}{\sin x}$

52. $g(v) = \dfrac{\cos v}{\csc v}$

53. $y = 4\sec^2 x$

54. $g(t) = 5\cos^2 \pi t$

55. $f(\theta) = \tan^2 5\theta$

56. $g(\theta) = \cos^2 8\theta$

57. $f(\theta) = \frac{1}{4}\sin^2 2\theta$

58. $h(t) = 2\cot^2(\pi t + 2)$

59. $f(t) = 3\sec^2(\pi t - 1)$

60. $y = 3x - 5\cos(\pi x)^2$

61. $y = \sqrt{x} + \frac{1}{4}\sin(2x)^2$

62. $y = \sin \sqrt[3]{x} + \sqrt[3]{\sin x}$

63. $y = \sin(\tan 2x)$

64. $y = \cos\sqrt{\sin(\tan \pi x)}$

Evaluating a Derivative In Exercises 65–72, find and evaluate the derivative of the function at the given point. Use a graphing utility to verify your result.

65. $y = \sqrt{x^2 + 8x}$, $(1, 3)$

66. $y = \sqrt[5]{3x^3 + 4x}$, $(2, 2)$

67. $f(x) = \dfrac{5}{x^3 - 2}$, $\left(-2, -\dfrac{1}{2}\right)$

68. $f(x) = \dfrac{1}{(x^2 - 3x)^2}$, $\left(4, \dfrac{1}{16}\right)$

69. $f(t) = \dfrac{3t + 2}{t - 1}$, $(0, -2)$

70. $f(x) = \dfrac{x + 4}{2x - 5}$, $(9, 1)$

71. $y = 26 - \sec^3 4x$, $(0, 25)$

72. $y = \dfrac{1}{x} + \sqrt{\cos x}$, $\left(\dfrac{\pi}{2}, \dfrac{2}{\pi}\right)$

Finding an Equation of a Tangent Line **In Exercises 73–80, (a) find an equation of the tangent line to the graph of f at the given point, (b) use a graphing utility to graph the function and its tangent line at the point, and (c) use the *derivative* feature of the graphing utility to confirm your results.**

73. $f(x) = \sqrt{2x^2 - 7}$, $(4, 5)$

74. $f(x) = \frac{1}{3}x\sqrt{x^2 + 5}$, $(2, 2)$

75. $y = (4x^3 + 3)^2$, $(-1, 1)$

76. $f(x) = (9 - x^2)^{2/3}$, $(1, 4)$

77. $f(x) = \sin 2x$, $(\pi, 0)$

78. $y = \cos 3x$, $\left(\frac{\pi}{4}, -\frac{\sqrt{2}}{2}\right)$

79. $f(x) = \tan^2 x$, $\left(\frac{\pi}{4}, 1\right)$

80. $y = 2\tan^3 x$, $\left(\frac{\pi}{4}, 2\right)$

Famous Curves **In Exercises 81 and 82, find an equation of the tangent line to the graph at the given point. Then use a graphing utility to graph the function and its tangent line in the same viewing window.**

81. Top half of circle

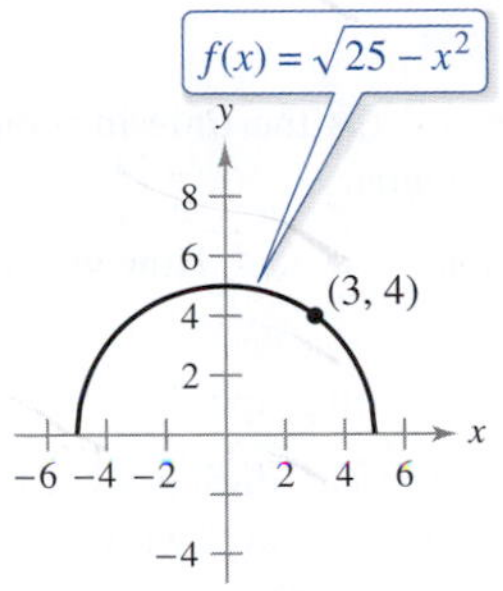

82. Bullet-nose curve

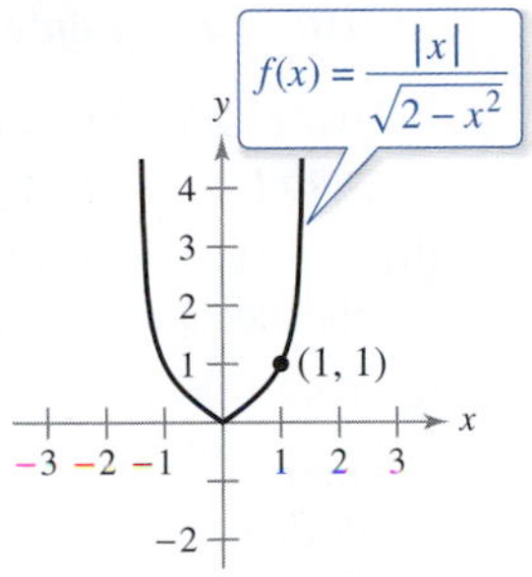

83. Horizontal Tangent Line Determine the point(s) in the interval $(0, 2\pi)$ at which the graph of

$$f(x) = 2\cos x + \sin 2x$$

has a horizontal tangent.

84. Horizontal Tangent Line Determine the point(s) at which the graph of

$$f(x) = \frac{x}{\sqrt{2x - 1}}$$

has a horizontal tangent.

Finding a Second Derivative **In Exercises 85–90, find the second derivative of the function.**

85. $f(x) = 5(2 - 7x)^4$

86. $f(x) = 6(x^3 + 4)^3$

87. $f(x) = \dfrac{1}{x - 6}$

88. $f(x) = \dfrac{8}{(x - 2)^2}$

89. $f(x) = \sin x^2$

90. $f(x) = \sec^2 \pi x$

Evaluating a Second Derivative **In Exercises 91–94, evaluate the second derivative of the function at the given point. Use a computer algebra system to verify your result.**

91. $h(x) = \frac{1}{9}(3x + 1)^3$, $\left(1, \frac{64}{9}\right)$

92. $f(x) = \dfrac{1}{\sqrt{x + 4}}$, $\left(0, \frac{1}{2}\right)$

93. $f(x) = \cos x^2$, $(0, 1)$

94. $g(t) = \tan 2t$, $\left(\frac{\pi}{6}, \sqrt{3}\right)$

WRITING ABOUT CONCEPTS

Identifying Graphs **In Exercises 95–98, the graphs of a function f and its derivative f' are shown. Label the graphs as f or f' and write a short paragraph stating the criteria you used in making your selection. To print an enlarged copy of the graph, go to *MathGraphs.com*.**

95.

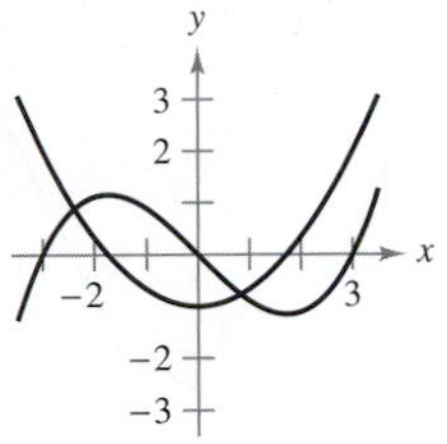

96.

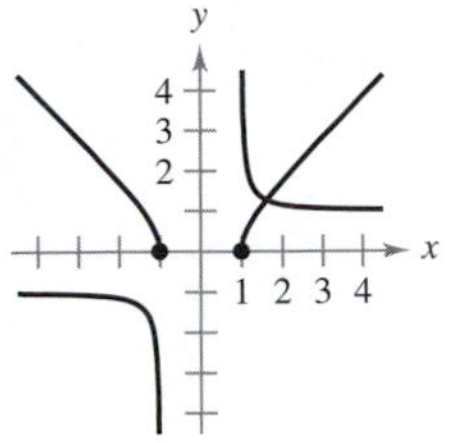

97.

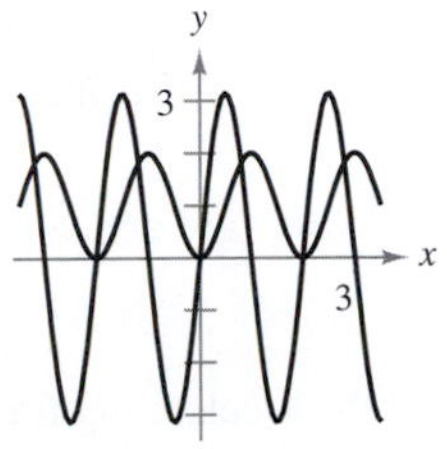

98.

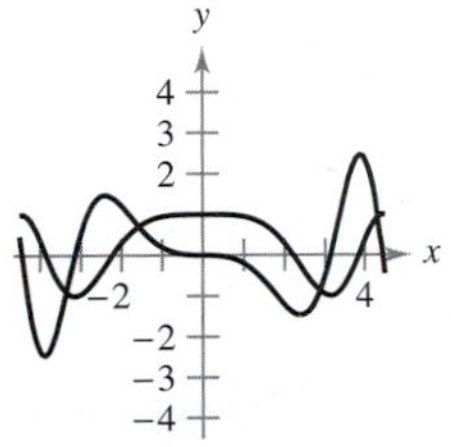

Describing a Relationship **In Exercises 99 and 100, the relationship between f and g is given. Explain the relationship between f' and g'.**

99. $g(x) = f(3x)$

100. $g(x) = f(x^2)$

101. Think About It The table shows some values of the derivative of an unknown function f. Complete the table by finding the derivative of each transformation of f, if possible.

(a) $g(x) = f(x) - 2$

(b) $h(x) = 2f(x)$

(c) $r(x) = f(-3x)$

(d) $s(x) = f(x + 2)$

x	-2	-1	0	1	2	3
$f'(x)$	4	$\frac{2}{3}$	$-\frac{1}{3}$	-1	-2	-4
$g'(x)$						
$h'(x)$						
$r'(x)$						
$s'(x)$						

102. Using Relationships Given that $g(5) = -3$, $g'(5) = 6$, $h(5) = 3$, and $h'(5) = -2$, find $f'(5)$ for each of the following, if possible. If it is not possible, state what additional information is required.

(a) $f(x) = g(x)h(x)$

(b) $f(x) = g(h(x))$

(c) $f(x) = \dfrac{g(x)}{h(x)}$

(d) $f(x) = [g(x)]^3$

Finding Derivatives **In Exercises 103 and 104, the graphs of f and g are shown. Let $h(x) = f(g(x))$ and $s(x) = g(f(x))$. Find each derivative, if it exists. If the derivative does not exist, explain why.**

103. (a) Find $h'(1)$.

(b) Find $s'(5)$.

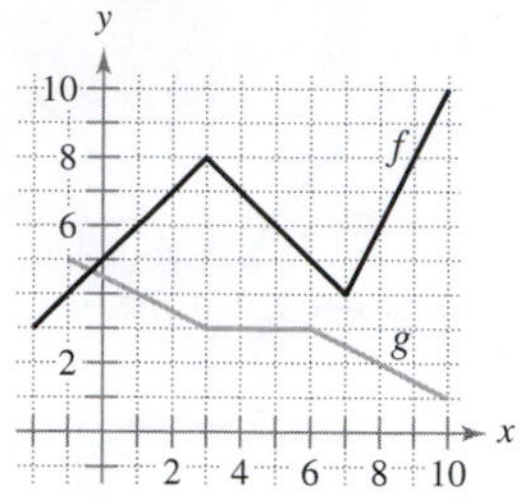

104. (a) Find $h'(3)$.

(b) Find $s'(9)$.

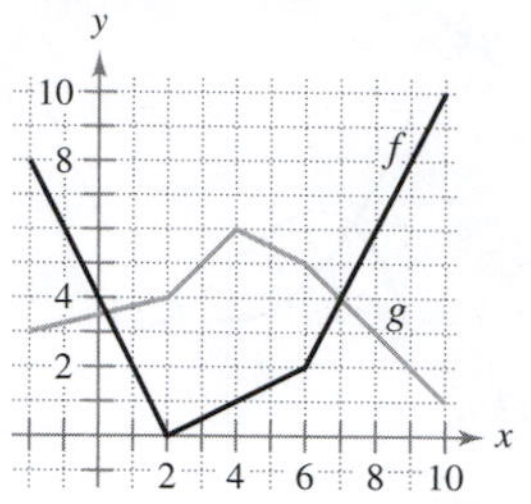

105. Doppler Effect The frequency F of a fire truck siren heard by a stationary observer is

$$F = \frac{132{,}400}{331 \pm v}$$

where $\pm v$ represents the velocity of the accelerating fire truck in meters per second (see figure). Find the rate of change of F with respect to v when

(a) the fire truck is approaching at a velocity of 30 meters per second (use $-v$).

(b) the fire truck is moving away at a velocity of 30 meters per second (use $+v$).

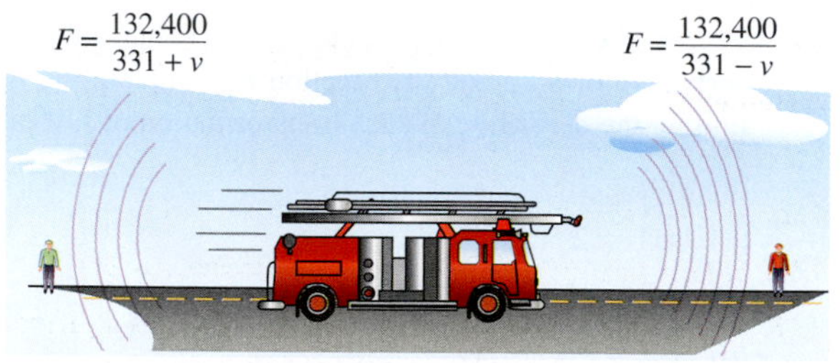

106. Harmonic Motion The displacement from equilibrium of an object in harmonic motion on the end of a spring is

$$y = \frac{1}{3}\cos 12t - \frac{1}{4}\sin 12t$$

where y is measured in feet and t is the time in seconds. Determine the position and velocity of the object when $t = \pi/8$.

107. Pendulum A 15-centimeter pendulum moves according to the equation $\theta = 0.2 \cos 8t$, where θ is the angular displacement from the vertical in radians and t is the time in seconds. Determine the maximum angular displacement and the rate of change of θ when $t = 3$ seconds.

108. Wave Motion A buoy oscillates in simple harmonic motion $y = A\cos \omega t$ as waves move past it. The buoy moves a total of 3.5 feet (vertically) from its low point to its high point. It returns to its high point every 10 seconds.

(a) Write an equation describing the motion of the buoy if it is at its high point at $t = 0$.

(b) Determine the velocity of the buoy as a function of t.

109. Modeling Data The normal daily maximum temperatures T (in degrees Fahrenheit) for Chicago, Illinois, are shown in the table. *(Source: National Oceanic and Atmospheric Administration)*

Month	Jan	Feb	Mar	Apr
Temperature	29.6	34.7	46.1	58.0

Month	May	Jun	Jul	Aug
Temperature	69.9	79.2	83.5	81.2

Month	Sep	Oct	Nov	Dec
Temperature	73.9	62.1	47.1	34.4

(a) Use a graphing utility to plot the data and find a model for the data of the form

$$T(t) = a + b\sin(ct - d)$$

where T is the temperature and t is the time in months, with $t = 1$ corresponding to January.

(b) Use a graphing utility to graph the model. How well does the model fit the data?

(c) Find T' and use a graphing utility to graph the derivative.

(d) Based on the graph of the derivative, during what times does the temperature change most rapidly? Most slowly? Do your answers agree with your observations of the temperature changes? Explain.

110. HOW DO YOU SEE IT? The cost C (in dollars) of producing x units of a product is $C = 60x + 1350$. For one week, management determined that the number of units produced x at the end of t hours can be modeled by $x = -1.6t^3 + 19t^2 - 0.5t - 1$. The graph shows the cost C in terms of the time t.

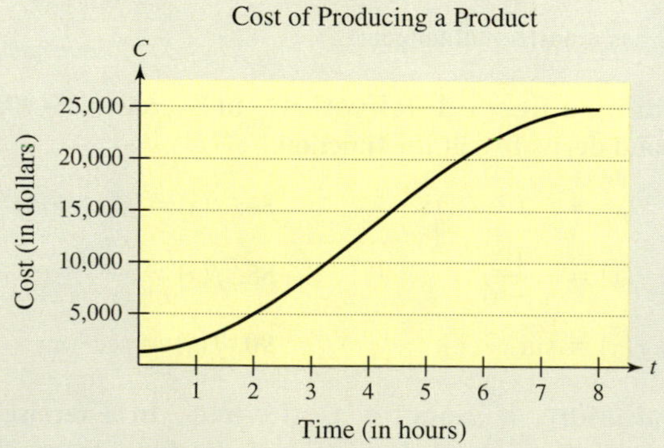

(a) Using the graph, which is greater, the rate of change of the cost after 1 hour or the rate of change of the cost after 4 hours?

(b) Explain why the cost function is not increasing at a constant rate during the eight-hour shift.

111. Biology The number N of bacteria in a culture after t days is modeled by

$$N = 400\left[1 - \frac{3}{(t^2 + 2)^2}\right].$$

Find the rate of change of N with respect to t when (a) $t = 0$, (b) $t = 1$, (c) $t = 2$, (d) $t = 3$, and (e) $t = 4$. (f) What can you conclude?

112. Depreciation The value V of a machine t years after it is purchased is inversely proportional to the square root of $t + 1$. The initial value of the machine is \$10,000.

(a) Write V as a function of t.

(b) Find the rate of depreciation when $t = 1$.

(c) Find the rate of depreciation when $t = 3$.

113. Finding a Pattern Consider the function $f(x) = \sin \beta x$, where β is a constant.

(a) Find the first-, second-, third-, and fourth-order derivatives of the function.

(b) Verify that the function and its second derivative satisfy the equation $f''(x) + \beta^2 f(x) = 0$.

(c) Use the results of part (a) to write general rules for the even- and odd-order derivatives $f^{(2k)}(x)$ and $f^{(2k-1)}(x)$. [*Hint:* $(-1)^k$ is positive if k is even and negative if k is odd.]

114. Conjecture Let f be a differentiable function of period p.

(a) Is the function f' periodic? Verify your answer.

(b) Consider the function $g(x) = f(2x)$. Is the function $g'(x)$ periodic? Verify your answer.

115. Think About It Let $r(x) = f(g(x))$ and $s(x) = g(f(x))$, where f and g are shown in the figure. Find (a) $r'(1)$ and (b) $s'(4)$.

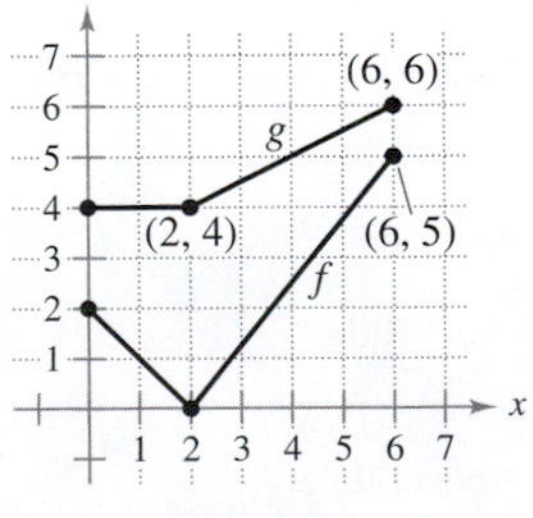

116. Using Trigonometric Functions

(a) Find the derivative of the function $g(x) = \sin^2 x + \cos^2 x$ in two ways.

(b) For $f(x) = \sec^2 x$ and $g(x) = \tan^2 x$, show that

$$f'(x) = g'(x).$$

117. Even and Odd Functions

(a) Show that the derivative of an odd function is even. That is, if $f(-x) = -f(x)$, then $f'(-x) = f'(x)$.

(b) Show that the derivative of an even function is odd. That is, if $f(-x) = f(x)$, then $f'(-x) = -f'(x)$.

118. Proof Let u be a differentiable function of x. Use the fact that $|u| = \sqrt{u^2}$ to prove that

$$\frac{d}{dx}[|u|] = u'\frac{u}{|u|}, \quad u \neq 0.$$

Using Absolute Value **In Exercises 119–122, use the result of Exercise 118 to find the derivative of the function.**

119. $g(x) = |3x - 5|$

120. $f(x) = |x^2 - 9|$

121. $h(x) = |x| \cos x$

122. $f(x) = |\sin x|$

Linear and Quadratic Approximations **The linear and quadratic approximations of a function f at $x = a$ are**

$$P_1(x) = f'(a)(x - a) + f(a) \quad \text{and}$$

$$P_2(x) = \tfrac{1}{2}f''(a)(x - a)^2 + f'(a)(x - a) + f(a).$$

In Exercises 123 and 124, (a) find the specified linear and quadratic approximations of f, (b) use a graphing utility to graph f and the approximations, (c) determine whether P_1 or P_2 is the better approximation, and (d) state how the accuracy changes as you move farther from $x = a$.

123. $f(x) = \tan x; \quad a = \dfrac{\pi}{4}$

124. $f(x) = \sec x; \quad a = \dfrac{\pi}{6}$

True or False? **In Exercises 125–128, determine whether the statement is true or false. If it is false, explain why or give an example that shows it is false.**

125. If $y = (1 - x)^{1/2}$, then $y' = \frac{1}{2}(1 - x)^{-1/2}$.

126. If $f(x) = \sin^2(2x)$, then $f'(x) = 2(\sin 2x)(\cos 2x)$.

127. If y is a differentiable function of u, and u is a differentiable function of x, then y is a differentiable function of x.

128. If y is a differentiable function of u, u is a differentiable function of v, and v is a differentiable function of x, then

$$\frac{dy}{dx} = \frac{dy}{du}\frac{du}{dv}\frac{dv}{dx}.$$

PUTNAM EXAM CHALLENGE

129. Let $f(x) = a_1 \sin x + a_2 \sin 2x + \cdots + a_n \sin nx$, where $a_1, a_2, \ldots, a_n$ are real numbers and where n is a positive integer. Given that $|f(x)| \leq |\sin x|$ for all real x, prove that $|a_1 + 2a_2 + \cdots + na_n| \leq 1$.

130. Let k be a fixed positive integer. The nth derivative of $\dfrac{1}{x^k - 1}$ has the form $\dfrac{P_n(x)}{(x^k - 1)^{n+1}}$ where $P_n(x)$ is a polynomial. Find $P_n(1)$.

These problems were composed by the Committee on the Putnam Prize Competition.

Tischenko Irina/Shutterstock.com

2.5 Implicit Differentiation

- Distinguish between functions written in implicit form and explicit form.
- Use implicit differentiation to find the derivative of a function.

Implicit and Explicit Functions

Up to this point in the text, most functions have been expressed in **explicit form.** For example, in the equation $y = 3x^2 - 5$, the variable y is explicitly written as a function of x. Some functions, however, are only implied by an equation. For instance, the function $y = 1/x$ is defined **implicitly** by the equation

$$xy = 1. \qquad \text{Implicit form}$$

To find dy/dx for this equation, you can write y explicitly as a function of x and then differentiate.

Implicit Form	Explicit Form	Derivative
$xy = 1$	$y = \frac{1}{x} = x^{-1}$	$\frac{dy}{dx} = -x^{-2} = -\frac{1}{x^2}$

This strategy works whenever you can solve for the function explicitly. You cannot, however, use this procedure when you are unable to solve for y as a function of x. For instance, how would you find dy/dx for the equation

$$x^2 - 2y^3 + 4y = 2?$$

For this equation, it is difficult to express y as a function of x explicitly. To find dy/dx, you can use **implicit differentiation.**

To understand how to find dy/dx implicitly, you must realize that the differentiation is taking place *with respect to* x. This means that when you differentiate terms involving x alone, you can differentiate as usual. However, when you differentiate terms involving y, you must apply the Chain Rule, because you are assuming that y is defined implicitly as a differentiable function of x.

EXAMPLE 1 Differentiating with Respect to x

a. $\frac{d}{dx}[x^3] = 3x^2$ — Variables agree: use Simple Power Rule.

Variables agree

b. $\frac{d}{dx}[\overbrace{y^3}^{u^n}] = \overbrace{3y^2}^{nu^{n-1}}\overbrace{\frac{dy}{dx}}^{u'}$ — Variables disagree: use Chain Rule.

Variables disagree

c. $\frac{d}{dx}[x + 3y] = 1 + 3\frac{dy}{dx}$ — Chain Rule: $\frac{d}{dx}[3y] = 3y'$

d. $\frac{d}{dx}[xy^2] = x\frac{d}{dx}[y^2] + y^2\frac{d}{dx}[x]$ — Product Rule

$= x\left(2y\frac{dy}{dx}\right) + y^2(1)$ — Chain Rule

$= 2xy\frac{dy}{dx} + y^2$ — Simplify.

Implicit Differentiation

GUIDELINES FOR IMPLICIT DIFFERENTIATION

1. Differentiate both sides of the equation *with respect to x*.
2. Collect all terms involving dy/dx on the left side of the equation and move all other terms to the right side of the equation.
3. Factor dy/dx out of the left side of the equation.
4. Solve for dy/dx.

In Example 2, note that implicit differentiation can produce an expression for dy/dx that contains both x and y.

EXAMPLE 2 Implicit Differentiation

Find dy/dx given that $y^3 + y^2 - 5y - x^2 = -4$.

Solution

1. Differentiate both sides of the equation with respect to x.

$$\frac{d}{dx}[y^3 + y^2 - 5y - x^2] = \frac{d}{dx}[-4]$$

$$\frac{d}{dx}[y^3] + \frac{d}{dx}[y^2] - \frac{d}{dx}[5y] - \frac{d}{dx}[x^2] = \frac{d}{dx}[-4]$$

$$3y^2\frac{dy}{dx} + 2y\frac{dy}{dx} - 5\frac{dy}{dx} - 2x = 0$$

2. Collect the dy/dx terms on the left side of the equation and move all other terms to the right side of the equation.

$$3y^2\frac{dy}{dx} + 2y\frac{dy}{dx} - 5\frac{dy}{dx} = 2x$$

3. Factor dy/dx out of the left side of the equation.

$$\frac{dy}{dx}(3y^2 + 2y - 5) = 2x$$

4. Solve for dy/dx by dividing by $(3y^2 + 2y - 5)$.

$$\frac{dy}{dx} = \frac{2x}{3y^2 + 2y - 5}$$

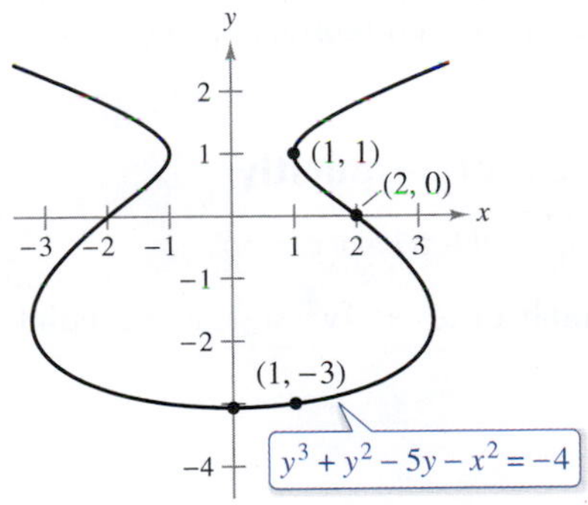

Point on Graph	Slope of Graph
(2, 0)	$-\frac{4}{5}$
(1, −3)	$\frac{1}{8}$
$x = 0$	0
(1, 1)	Undefined

The implicit equation

$$y^3 + y^2 - 5y - x^2 = -4$$

has the derivative

$$\frac{dy}{dx} = \frac{2x}{3y^2 + 2y - 5}.$$

Figure 2.27

To see how you can use an *implicit derivative*, consider the graph shown in Figure 2.27. From the graph, you can see that y is not a function of x. Even so, the derivative found in Example 2 gives a formula for the slope of the tangent line at a point on this graph. The slopes at several points on the graph are shown below the graph.

TECHNOLOGY With most graphing utilities, it is easy to graph an equation that explicitly represents y as a function of x. Graphing other equations, however, can require some ingenuity. For instance, to graph the equation given in Example 2, use a graphing utility, set in *parametric* mode, to graph the parametric representations $x = \sqrt{t^3 + t^2 - 5t + 4}$, $y = t$, and $x = -\sqrt{t^3 + t^2 - 5t + 4}$, $y = t$, for $-5 \le t \le 5$. How does the result compare with the graph shown in Figure 2.27?

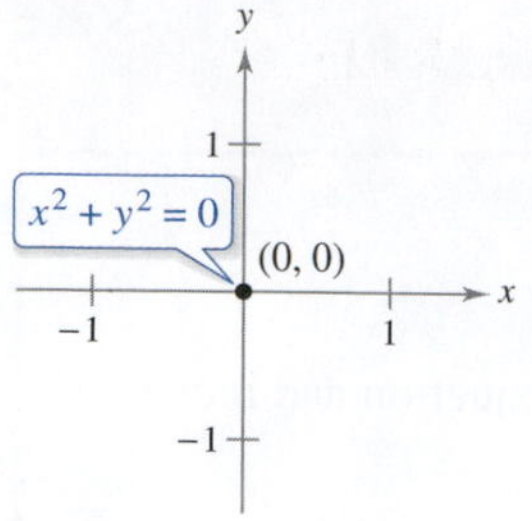

(a)

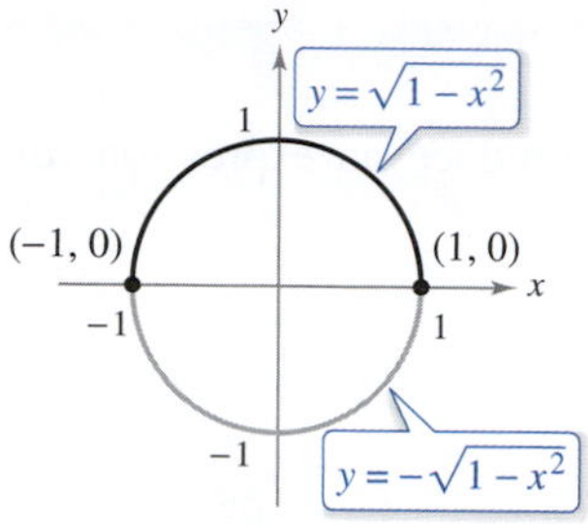

(b)

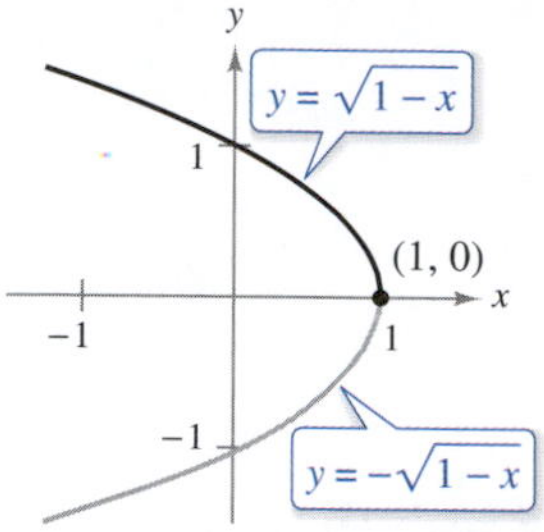

(c)

Some graph segments can be represented by differentiable functions.

Figure 2.28

It is meaningless to solve for dy/dx in an equation that has no solution points. (For example, $x^2 + y^2 = -4$ has no solution points.) If, however, a segment of a graph can be represented by a differentiable function, then dy/dx will have meaning as the slope at each point on the segment. Recall that a function is not differentiable at (a) points with vertical tangents and (b) points at which the function is not continuous.

EXAMPLE 3 Graphs and Differentiable Functions

If possible, represent y as a differentiable function of x.

a. $x^2 + y^2 = 0$ **b.** $x^2 + y^2 = 1$ **c.** $x + y^2 = 1$

Solution

a. The graph of this equation is a single point. So, it does not define y as a differentiable function of x. See Figure 2.28(a).

b. The graph of this equation is the unit circle centered at $(0, 0)$. The upper semicircle is given by the differentiable function

$$y = \sqrt{1 - x^2}, \quad -1 < x < 1$$

and the lower semicircle is given by the differentiable function

$$y = -\sqrt{1 - x^2}, \quad -1 < x < 1.$$

At the points $(-1, 0)$ and $(1, 0)$, the slope of the graph is undefined. See Figure 2.28(b).

c. The upper half of this parabola is given by the differentiable function

$$y = \sqrt{1 - x}, \quad x < 1$$

and the lower half of this parabola is given by the differentiable function

$$y = -\sqrt{1 - x}, \quad x < 1.$$

At the point $(1, 0)$, the slope of the graph is undefined. See Figure 2.28(c).

EXAMPLE 4 Finding the Slope of a Graph Implicitly

See LarsonCalculus.com for an interactive version of this type of example.

Determine the slope of the tangent line to the graph of $x^2 + 4y^2 = 4$ at the point $(\sqrt{2}, -1/\sqrt{2})$. See Figure 2.29.

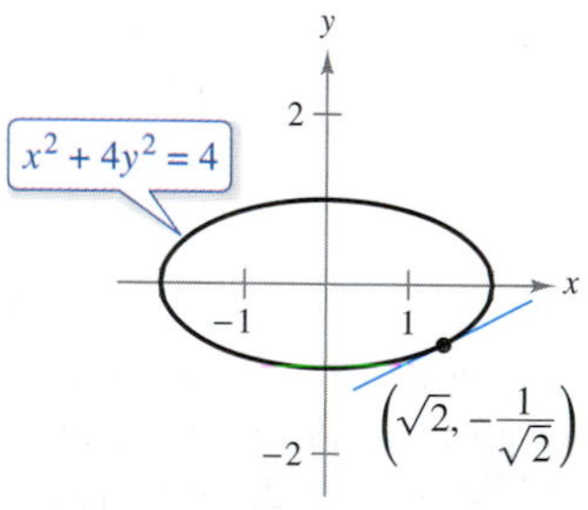

Figure 2.29

Solution

$$x^2 + 4y^2 = 4 \qquad \text{Write original equation.}$$

$$2x + 8y\frac{dy}{dx} = 0 \qquad \text{Differentiate with respect to } x.$$

$$\frac{dy}{dx} = \frac{-2x}{8y} \qquad \text{Solve for } \frac{dy}{dx}.$$

$$= \frac{-x}{4y} \qquad \text{Simplify.}$$

So, at $(\sqrt{2}, -1/\sqrt{2})$, the slope is

$$\frac{dy}{dx} = \frac{-\sqrt{2}}{-4/\sqrt{2}} = \frac{1}{2}. \qquad \text{Evaluate } \frac{dy}{dx} \text{ when } x = \sqrt{2} \text{ and } y = -\frac{1}{\sqrt{2}}.$$

REMARK To see the benefit of implicit differentiation, try doing Example 4 using the explicit function $y = -\frac{1}{2}\sqrt{4 - x^2}$.

EXAMPLE 5 **Finding the Slope of a Graph Implicitly**

Determine the slope of the graph of

$$3(x^2 + y^2)^2 = 100xy$$

at the point $(3, 1)$.

Solution

$$\frac{d}{dx}[3(x^2 + y^2)^2] = \frac{d}{dx}[100xy]$$

$$3(2)(x^2 + y^2)\left(2x + 2y\frac{dy}{dx}\right) = 100\left[x\frac{dy}{dx} + y(1)\right]$$

$$12y(x^2 + y^2)\frac{dy}{dx} - 100x\frac{dy}{dx} = 100y - 12x(x^2 + y^2)$$

$$[12y(x^2 + y^2) - 100x]\frac{dy}{dx} = 100y - 12x(x^2 + y^2)$$

$$\frac{dy}{dx} = \frac{100y - 12x(x^2 + y^2)}{-100x + 12y(x^2 + y^2)}$$

$$= \frac{25y - 3x(x^2 + y^2)}{-25x + 3y(x^2 + y^2)}$$

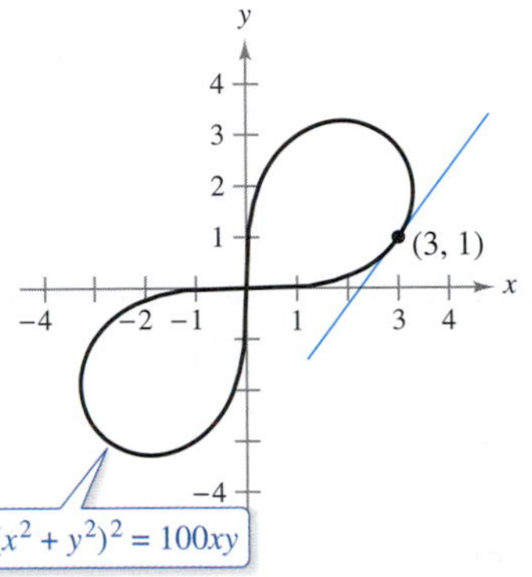

Lemniscate
Figure 2.30

At the point $(3, 1)$, the slope of the graph is

$$\frac{dy}{dx} = \frac{25(1) - 3(3)(3^2 + 1^2)}{-25(3) + 3(1)(3^2 + 1^2)} = \frac{25 - 90}{-75 + 30} = \frac{-65}{-45} = \frac{13}{9}$$

as shown in Figure 2.30. This graph is called a **lemniscate.**

EXAMPLE 6 **Determining a Differentiable Function**

Find dy/dx implicitly for the equation $\sin y = x$. Then find the largest interval of the form $-a < y < a$ on which y is a differentiable function of x (see Figure 2.31).

Solution

$$\frac{d}{dx}[\sin y] = \frac{d}{dx}[x]$$

$$\cos y\frac{dy}{dx} = 1$$

$$\frac{dy}{dx} = \frac{1}{\cos y}$$

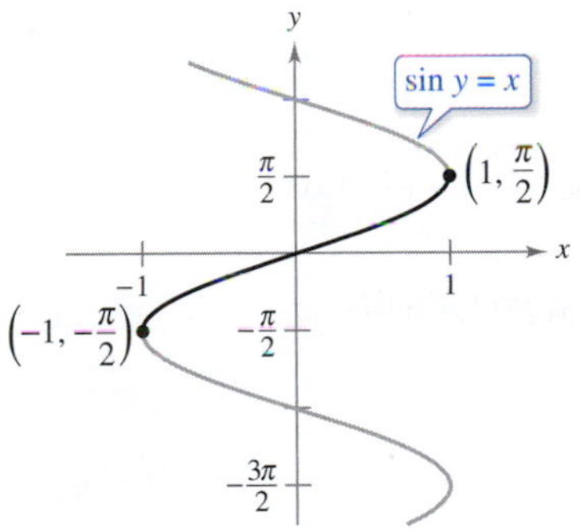

The derivative is $\dfrac{dy}{dx} = \dfrac{1}{\sqrt{1 - x^2}}$.

Figure 2.31

The largest interval about the origin for which y is a differentiable function of x is $-\pi/2 < y < \pi/2$. To see this, note that $\cos y$ is positive for all y in this interval and is 0 at the endpoints. When you restrict y to the interval $-\pi/2 < y < \pi/2$, you should be able to write dy/dx explicitly as a function of x. To do this, you can use

$$\cos y = \sqrt{1 - \sin^2 y}$$

$$= \sqrt{1 - x^2}, \quad -\frac{\pi}{2} < y < \frac{\pi}{2}$$

and conclude that

$$\frac{dy}{dx} = \frac{1}{\sqrt{1 - x^2}}.$$

You will study this example further when inverse trigonometric functions are defined in Section 5.6. ■

ISAAC BARROW (1630–1677)

The graph in Figure 2.32 is called the **kappa curve** because it resembles the Greek letter kappa, κ. The general solution for the tangent line to this curve was discovered by the English mathematician Isaac Barrow. Newton was Barrow's student, and they corresponded frequently regarding their work in the early development of calculus.

See LarsonCalculus.com to read more of this biography.

With implicit differentiation, the form of the derivative often can be simplified (as in Example 6) by an appropriate use of the *original* equation. A similar technique can be used to find and simplify higher-order derivatives obtained implicitly.

EXAMPLE 7 Finding the Second Derivative Implicitly

Given $x^2 + y^2 = 25$, find $\dfrac{d^2y}{dx^2}$.

Solution Differentiating each term with respect to x produces

$$2x + 2y\frac{dy}{dx} = 0$$

$$2y\frac{dy}{dx} = -2x$$

$$\frac{dy}{dx} = \frac{-2x}{2y}$$

$$= -\frac{x}{y}.$$

Differentiating a second time with respect to x yields

$$\frac{d^2y}{dx^2} = -\frac{(y)(1) - (x)(dy/dx)}{y^2} \qquad \text{Quotient Rule}$$

$$= -\frac{y - (x)(-x/y)}{y^2} \qquad \text{Substitute } -\frac{x}{y} \text{ for } \frac{dy}{dx}.$$

$$= -\frac{y^2 + x^2}{y^3} \qquad \text{Simplify.}$$

$$= -\frac{25}{y^3}. \qquad \text{Substitute 25 for } x^2 + y^2.$$

EXAMPLE 8 Finding a Tangent Line to a Graph

Find the tangent line to the graph of $x^2(x^2 + y^2) = y^2$ at the point $\left(\sqrt{2}/2, \sqrt{2}/2\right)$, as shown in Figure 2.32.

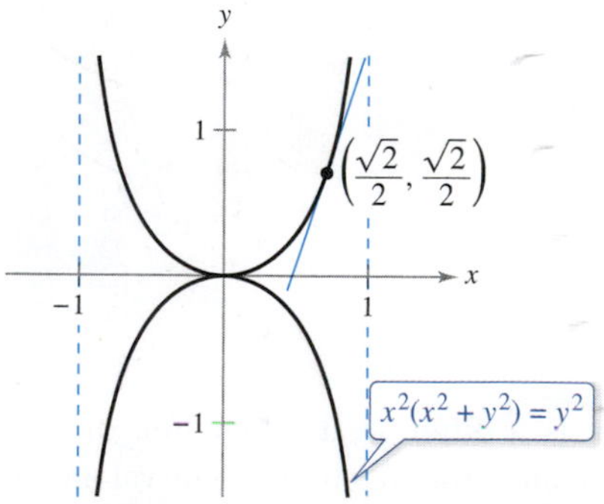

The kappa curve
Figure 2.32

Solution By rewriting and differentiating implicitly, you obtain

$$x^4 + x^2y^2 - y^2 = 0$$

$$4x^3 + x^2\left(2y\frac{dy}{dx}\right) + 2xy^2 - 2y\frac{dy}{dx} = 0$$

$$2y(x^2 - 1)\frac{dy}{dx} = -2x(2x^2 + y^2)$$

$$\frac{dy}{dx} = \frac{x(2x^2 + y^2)}{y(1 - x^2)}.$$

At the point $\left(\sqrt{2}/2, \sqrt{2}/2\right)$, the slope is

$$\frac{dy}{dx} = \frac{\left(\sqrt{2}/2\right)[2(1/2) + (1/2)]}{\left(\sqrt{2}/2\right)[1 - (1/2)]} = \frac{3/2}{1/2} = 3$$

and the equation of the tangent line at this point is

$$y - \frac{\sqrt{2}}{2} = 3\left(x - \frac{\sqrt{2}}{2}\right)$$

$$y = 3x - \sqrt{2}.$$

The Granger Collection, New York

2.5 Exercises

See CalcChat.com for tutorial help and worked-out solutions to odd-numbered exercises.

Finding a Derivative In Exercises 1–16, find dy/dx by implicit differentiation.

1. $x^2 + y^2 = 9$
2. $x^2 - y^2 = 25$
3. $x^{1/2} + y^{1/2} = 16$
4. $2x^3 + 3y^3 = 64$
5. $x^3 - xy + y^2 = 7$
6. $x^2y + y^2x = -2$
7. $x^3y^3 - y = x$
8. $\sqrt{xy} = x^2y + 1$
9. $x^3 - 3x^2y + 2xy^2 = 12$
10. $4 \cos x \sin y = 1$
11. $\sin x + 2 \cos 2y = 1$
12. $(\sin \pi x + \cos \pi y)^2 = 2$
13. $\sin x = x(1 + \tan y)$
14. $\cot y = x - y$
15. $y = \sin xy$
16. $x = \sec \dfrac{1}{y}$

Finding Derivatives Implicitly and Explicitly In Exercises 17–20, (a) find two explicit functions by solving the equation for y in terms of x, (b) sketch the graph of the equation and label the parts given by the corresponding explicit functions, (c) differentiate the explicit functions, and (d) find dy/dx implicitly and show that the result is equivalent to that of part (c).

17. $x^2 + y^2 = 64$
18. $25x^2 + 36y^2 = 300$
19. $16y^2 - x^2 = 16$
20. $x^2 + y^2 - 4x + 6y + 9 = 0$

Finding and Evaluating a Derivative In Exercises 21–28, find dy/dx by implicit differentiation and evaluate the derivative at the given point.

21. $xy = 6$, $(-6, -1)$
22. $y^3 - x^2 = 4$, $(2, 2)$
23. $y^2 = \dfrac{x^2 - 49}{x^2 + 49}$, $(7, 0)$
24. $x^{2/3} + y^{2/3} = 5$, $(8, 1)$
25. $(x + y)^3 = x^3 + y^3$, $(-1, 1)$
26. $x^3 + y^3 = 6xy - 1$, $(2, 3)$
27. $\tan(x + y) = x$, $(0, 0)$
28. $x \cos y = 1$, $\left(2, \dfrac{\pi}{3}\right)$

Famous Curves In Exercises 29–32, find the slope of the tangent line to the graph at the given point.

29. Witch of Agnesi:

$(x^2 + 4)y = 8$

Point: $(2, 1)$

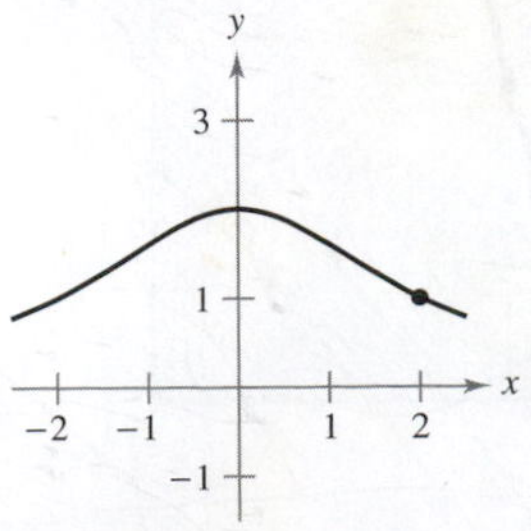

30. Cissoid:

$(4 - x)y^2 = x^3$

Point: $(2, 2)$

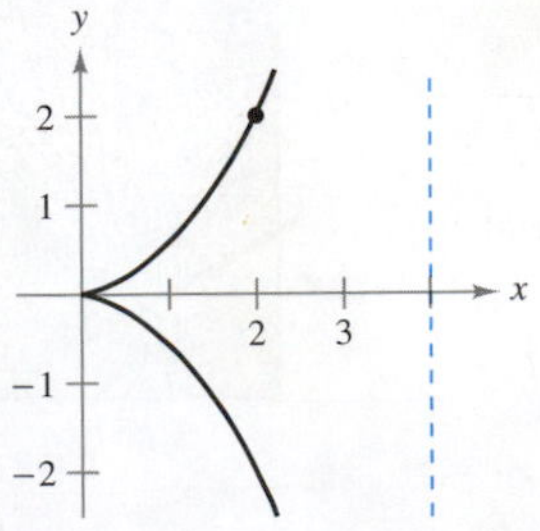

31. Bifolium:

$(x^2 + y^2)^2 = 4x^2y$

Point: $(1, 1)$

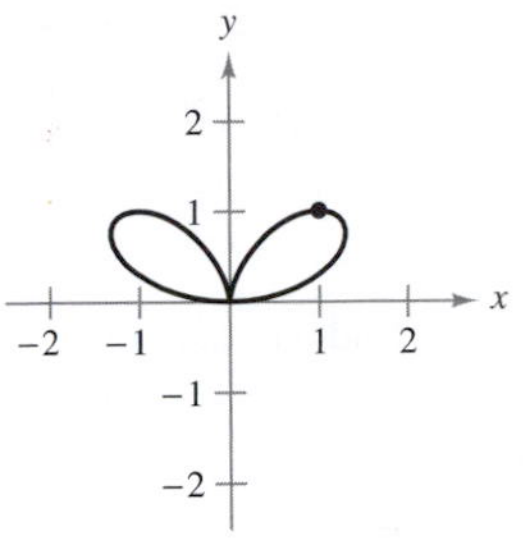

32. Folium of Descartes:

$x^3 + y^3 - 6xy = 0$

Point: $\left(\frac{4}{3}, \frac{8}{3}\right)$

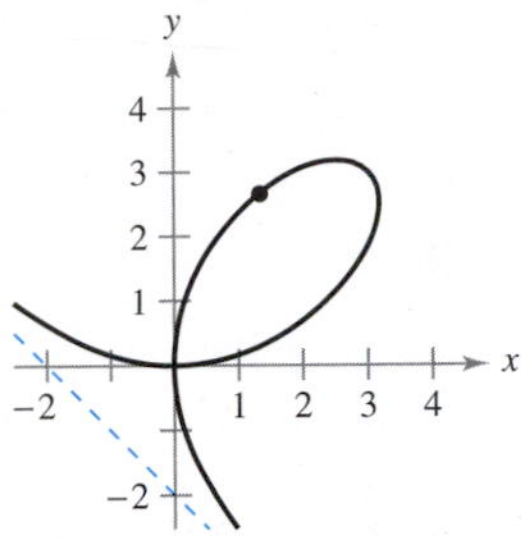

Famous Curves In Exercises 33–40, find an equation of the tangent line to the graph at the given point. To print an enlarged copy of the graph, go to *MathGraphs.com*.

33. Parabola

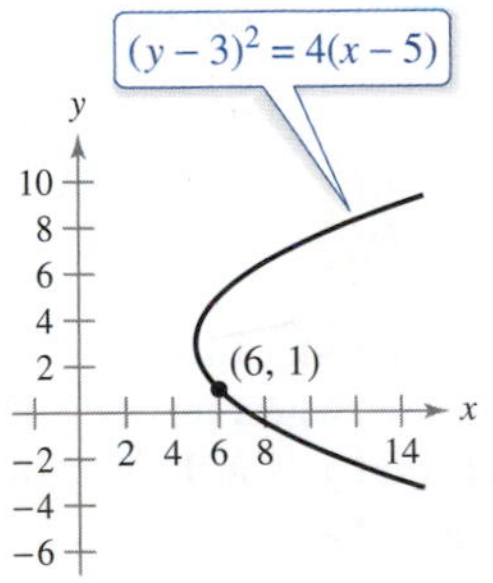

34. Circle

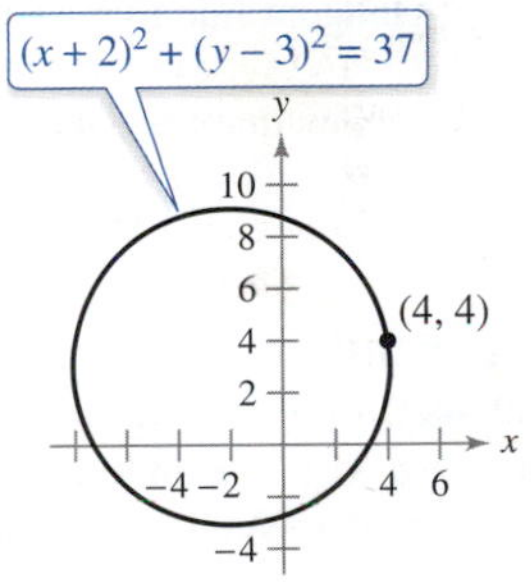

35. Rotated hyperbola

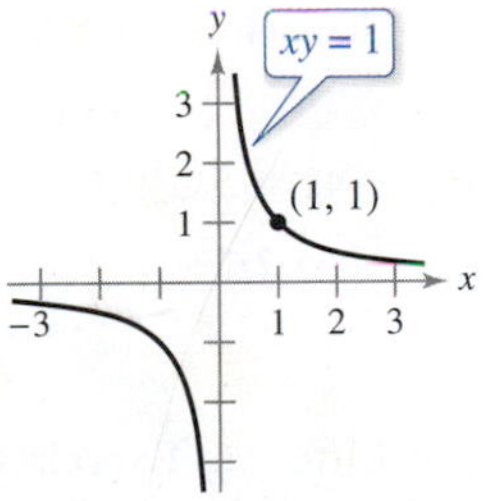

36. Rotated ellipse

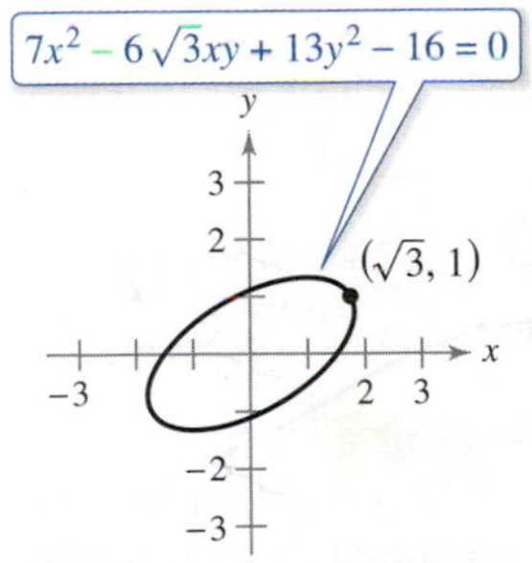

37. Cruciform

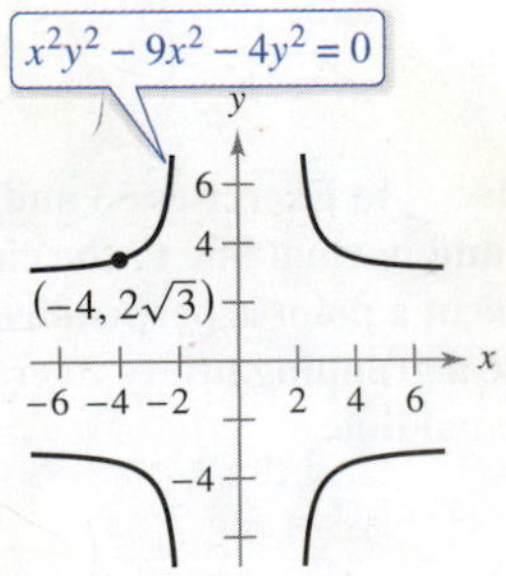

38. Astroid

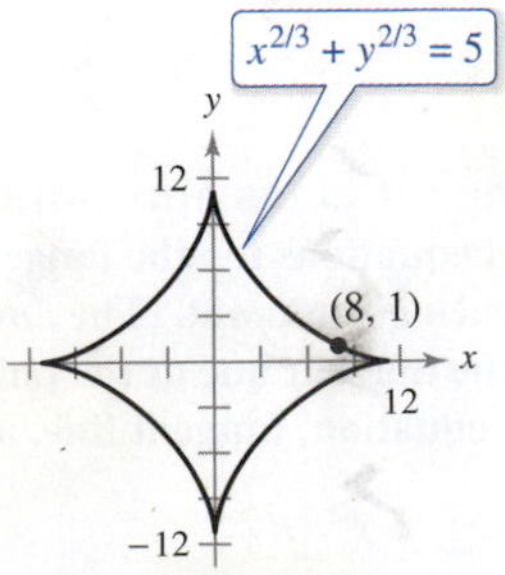

39. Lemniscate

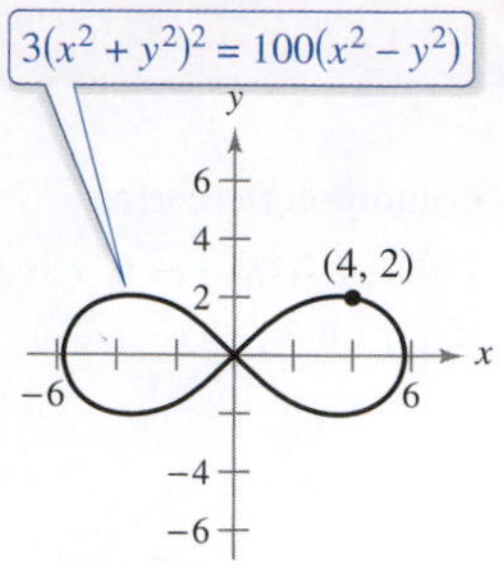

40. Kappa curve

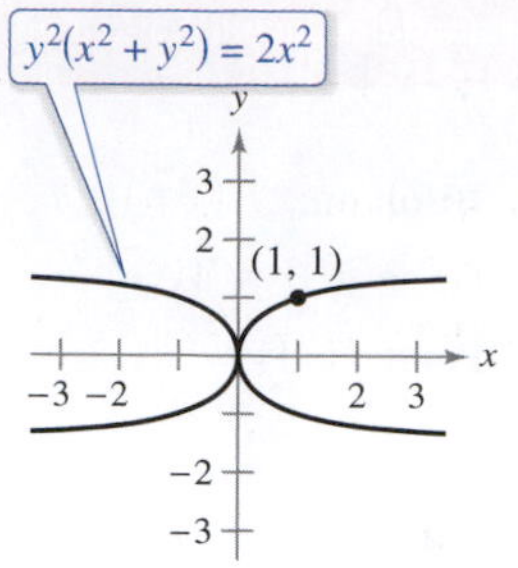

41. Ellipse

(a) Use implicit differentiation to find an equation of the tangent line to the ellipse $\dfrac{x^2}{2} + \dfrac{y^2}{8} = 1$ at $(1, 2)$.

(b) Show that the equation of the tangent line to the ellipse $\dfrac{x^2}{a^2} + \dfrac{y^2}{b^2} = 1$ at (x_0, y_0) is $\dfrac{x_0 x}{a^2} + \dfrac{y_0 y}{b^2} = 1$.

42. Hyperbola

(a) Use implicit differentiation to find an equation of the tangent line to the hyperbola $\dfrac{x^2}{6} - \dfrac{y^2}{8} = 1$ at $(3, -2)$.

(b) Show that the equation of the tangent line to the hyperbola $\dfrac{x^2}{a^2} - \dfrac{y^2}{b^2} = 1$ at (x_0, y_0) is $\dfrac{x_0 x}{a^2} - \dfrac{y_0 y}{b^2} = 1$.

Determining a Differentiable Function **In Exercises 43 and 44, find dy/dx implicitly and find the largest interval of the form $-a < y < a$ or $0 < y < a$ such that y is a differentiable function of x. Write dy/dx as a function of x.**

43. $\tan y = x$

44. $\cos y = x$

Finding a Second Derivative **In Exercises 45–50, find d^2y/dx^2 implicitly in terms of x and y.**

45. $x^2 + y^2 = 4$

46. $x^2y - 4x = 5$

47. $x^2 - y^2 = 36$

48. $xy - 1 = 2x + y^2$

49. $y^2 = x^3$

50. $y^3 = 4x$

Finding an Equation of a Tangent Line **In Exercises 51 and 52, use a graphing utility to graph the equation. Find an equation of the tangent line to the graph at the given point and graph the tangent line in the same viewing window.**

51. $\sqrt{x} + \sqrt{y} = 5, \quad (9, 4)$

52. $y^2 = \dfrac{x - 1}{x^2 + 1}, \quad \left(2, \dfrac{\sqrt{5}}{5}\right)$

Tangent Lines and Normal Lines **In Exercises 53 and 54, find equations for the tangent line and normal line to the circle at each given point. (The *normal line* at a point is perpendicular to the tangent line at the point.) Use a graphing utility to graph the equation, tangent line, and normal line.**

53. $x^2 + y^2 = 25$
$(4, 3), (-3, 4)$

54. $x^2 + y^2 = 36$
$(6, 0), (5, \sqrt{11})$

55. Normal Lines Show that the normal line at any point on the circle $x^2 + y^2 = r^2$ passes through the origin.

56. Circles Two circles of radius 4 are tangent to the graph of $y^2 = 4x$ at the point $(1, 2)$. Find equations of these two circles.

Vertical and Horizontal Tangent Lines **In Exercises 57 and 58, find the points at which the graph of the equation has a vertical or horizontal tangent line.**

57. $25x^2 + 16y^2 + 200x - 160y + 400 = 0$

58. $4x^2 + y^2 - 8x + 4y + 4 = 0$

Orthogonal Trajectories **In Exercises 59–62, use a graphing utility to sketch the intersecting graphs of the equations and show that they are orthogonal. [Two graphs are *orthogonal* if at their point(s) of intersection, their tangent lines are perpendicular to each other.]**

59. $2x^2 + y^2 = 6$
$y^2 = 4x$

60. $y^2 = x^3$
$2x^2 + 3y^2 = 5$

61. $x + y = 0$
$x = \sin y$

62. $x^3 = 3(y - 1)$
$x(3y - 29) = 3$

Orthogonal Trajectories **In Exercises 63 and 64, verify that the two families of curves are orthogonal, where C and K are real numbers. Use a graphing utility to graph the two families for two values of C and two values of K.**

63. $xy = C, \quad x^2 - y^2 = K$

64. $x^2 + y^2 = C^2, \quad y = Kx$

WRITING ABOUT CONCEPTS

65. Explicit and Implicit Functions Describe the difference between the explicit form of a function and an implicit equation. Give an example of each.

66. Implicit Differentiation In your own words, state the guidelines for implicit differentiation.

67. Orthogonal Trajectories The figure below shows the topographic map carried by a group of hikers. The hikers are in a wooded area on top of the hill shown on the map, and they decide to follow the path of steepest descent (orthogonal trajectories to the contours on the map). Draw their routes if they start from point A and if they start from point B. Their goal is to reach the road along the top of the map. Which starting point should they use? To print an enlarged copy of the map, go to *MathGraphs.com*.

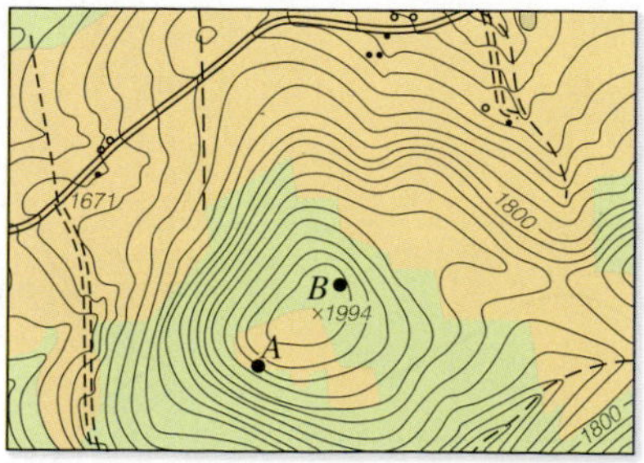

68. HOW DO YOU SEE IT? Use the graph to answer the questions.

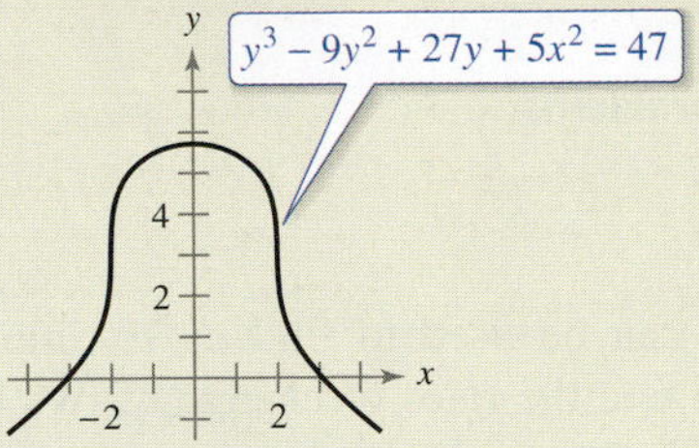

(a) Which is greater, the slope of the tangent line at $x = -3$ or the slope of the tangent line at $x = -1$?

(b) Estimate the point(s) where the graph has a vertical tangent line.

(c) Estimate the point(s) where the graph has a horizontal tangent line.

69. Finding Equations of Tangent Lines Consider the equation $x^4 = 4(4x^2 - y^2)$.

(a) Use a graphing utility to graph the equation.

(b) Find and graph the four tangent lines to the curve for $y = 3$.

(c) Find the exact coordinates of the point of intersection of the two tangent lines in the first quadrant.

70. Tangent Lines and Intercepts Let L be any tangent line to the curve

$$\sqrt{x} + \sqrt{y} = \sqrt{c}.$$

Show that the sum of the x- and y-intercepts of L is c.

71. Proof Prove (Theorem 2.3) that

$$\frac{d}{dx}[x^n] = nx^{n-1}$$

for the case in which n is a rational number. (*Hint:* Write $y = x^{p/q}$ in the form $y^q = x^p$ and differentiate implicitly. Assume that p and q are integers, where $q > 0$.)

72. Slope Find all points on the circle $x^2 + y^2 = 100$ where the slope is $\frac{3}{4}$.

73. Tangent Lines Find equations of both tangent lines to the graph of the ellipse $\frac{x^2}{4} + \frac{y^2}{9} = 1$ that pass through the point $(4, 0)$ not on the graph.

74. Normals to a Parabola The graph shows the normal lines from the point $(2, 0)$ to the graph of the parabola $x = y^2$. How many normal lines are there from the point $(x_0, 0)$ to the graph of the parabola if (a) $x_0 = \frac{1}{4}$, (b) $x_0 = \frac{1}{2}$, and (c) $x_0 = 1$? For what value of x_0 are two of the normal lines perpendicular to each other?

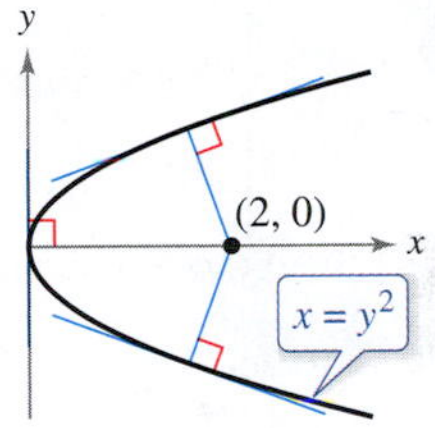

75. Normal Lines (a) Find an equation of the normal line to the ellipse $\frac{x^2}{32} + \frac{y^2}{8} = 1$ at the point $(4, 2)$. (b) Use a graphing utility to graph the ellipse and the normal line. (c) At what other point does the normal line intersect the ellipse?

SECTION PROJECT

Optical Illusions

In each graph below, an optical illusion is created by having lines intersect a family of curves. In each case, the lines appear to be curved. Find the value of *dy/dx* for the given values of *x* and *y*.

(a) Circles: $x^2 + y^2 = C^2$

$x = 3, y = 4, C = 5$

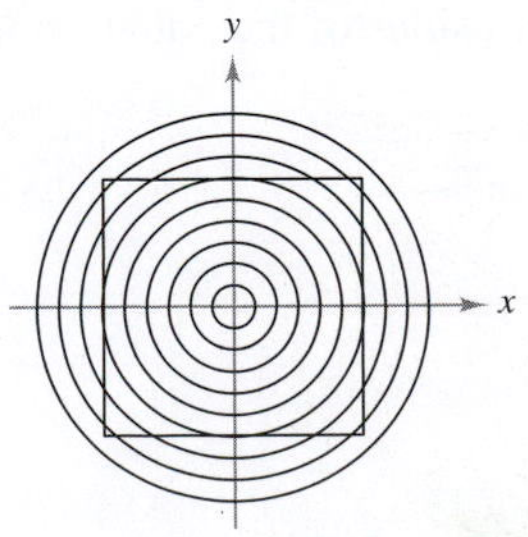

(b) Hyperbolas: $xy = C$

$x = 1, y = 4, C = 4$

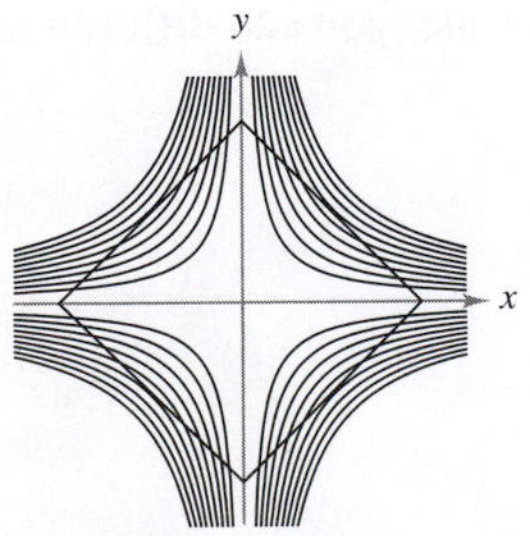

(c) Lines: $ax = by$

$x = \sqrt{3}, y = 3,$

$a = \sqrt{3}, b = 1$

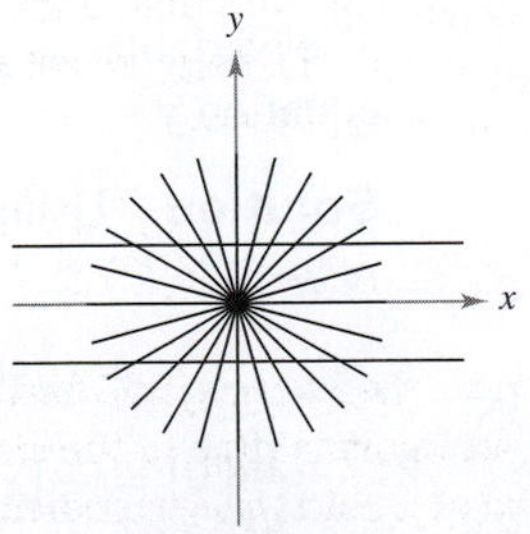

(d) Cosine curves: $y = C \cos x$

$x = \frac{\pi}{3}, y = \frac{1}{3}, C = \frac{2}{3}$

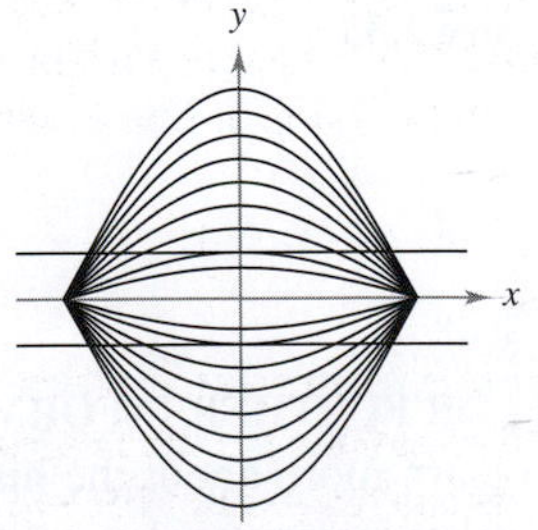

FOR FURTHER INFORMATION For more information on the mathematics of optical illusions, see the article "Descriptive Models for Perception of Optical Illusions" by David A. Smith in *The UMAP Journal*.

2.6 Related Rates

- Find a related rate.
- Use related rates to solve real-life problems.

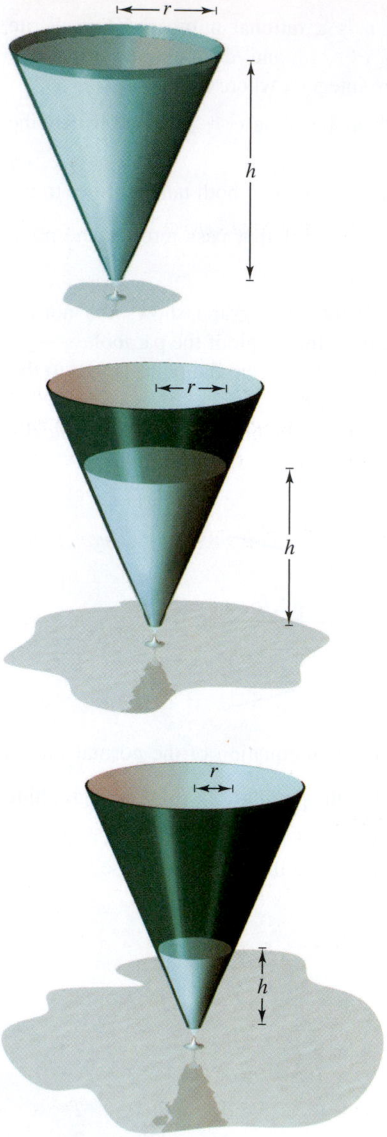

Volume is related to radius and height.
Figure 2.33

Finding Related Rates

You have seen how the Chain Rule can be used to find dy/dx implicitly. Another important use of the Chain Rule is to find the rates of change of two or more related variables that are changing with respect to *time*.

For example, when water is drained out of a conical tank (see Figure 2.33), the volume V, the radius r, and the height h of the water level are all functions of time t. Knowing that these variables are related by the equation

$$V = \frac{\pi}{3} r^2 h \qquad \text{Original equation}$$

you can differentiate implicitly with respect to t to obtain the **related-rate** equation

$$\frac{d}{dt}[V] = \frac{d}{dt}\left[\frac{\pi}{3} r^2 h\right]$$

$$\frac{dV}{dt} = \frac{\pi}{3}\left[r^2 \frac{dh}{dt} + h\left(2r\frac{dr}{dt}\right)\right] \qquad \text{Differentiate with respect to } t.$$

$$= \frac{\pi}{3}\left(r^2 \frac{dh}{dt} + 2rh\frac{dr}{dt}\right).$$

From this equation, you can see that the rate of change of V is related to the rates of change of both h and r.

> **Exploration**
>
> ***Finding a Related Rate*** In the conical tank shown in Figure 2.33, the height of the water level is changing at a rate of -0.2 foot per minute and the radius is changing at a rate of -0.1 foot per minute. What is the rate of change in the volume when the radius is $r = 1$ foot and the height is $h = 2$ feet? Does the rate of change in the volume depend on the values of r and h? Explain.

EXAMPLE 1 Two Rates That Are Related

The variables x and y are both differentiable functions of t and are related by the equation $y = x^2 + 3$. Find dy/dt when $x = 1$, given that $dx/dt = 2$ when $x = 1$.

Solution Using the Chain Rule, you can differentiate both sides of the equation *with respect to t*.

$$y = x^2 + 3 \qquad \text{Write original equation.}$$

$$\frac{d}{dt}[y] = \frac{d}{dt}[x^2 + 3] \qquad \text{Differentiate with respect to } t.$$

$$\frac{dy}{dt} = 2x\frac{dx}{dt} \qquad \text{Chain Rule}$$

When $x = 1$ and $dx/dt = 2$, you have

$$\frac{dy}{dt} = 2(1)(2) = 4.$$

■ **FOR FURTHER INFORMATION** To learn more about the history of related-rate problems, see the article "The Lengthening Shadow: The Story of Related Rates" by Bill Austin, Don Barry, and David Berman in *Mathematics Magazine*. To view this article, go to *MathArticles.com*.

Problem Solving with Related Rates

In Example 1, you were *given* an equation that related the variables x and y and were asked to find the rate of change of y when $x = 1$.

Equation: $y = x^2 + 3$

Given rate: $\frac{dx}{dt} = 2$ when $x = 1$

Find: $\frac{dy}{dt}$ when $x = 1$

In each of the remaining examples in this section, you must *create* a mathematical model from a verbal description.

EXAMPLE 2 Ripples in a Pond

Total area increases as the outer radius increases.
Figure 2.34

A pebble is dropped into a calm pond, causing ripples in the form of concentric circles, as shown in Figure 2.34. The radius r of the outer ripple is increasing at a constant rate of 1 foot per second. When the radius is 4 feet, at what rate is the total area A of the disturbed water changing?

Solution The variables r and A are related by $A = \pi r^2$. The rate of change of the radius r is $dr/dt = 1$.

Equation: $A = \pi r^2$

Given rate: $\frac{dr}{dt} = 1$

Find: $\frac{dA}{dt}$ when $r = 4$

With this information, you can proceed as in Example 1.

$$\frac{d}{dt}[A] = \frac{d}{dt}[\pi r^2] \quad \text{Differentiate with respect to } t.$$

$$\frac{dA}{dt} = 2\pi r \frac{dr}{dt} \quad \text{Chain Rule}$$

$$= 2\pi(4)(1) \quad \text{Substitute 4 for } r \text{ and 1 for } \frac{dr}{dt}.$$

$$= 8\pi \text{ square feet per second} \quad \text{Simplify.}$$

When the radius is 4 feet, the area is changing at a rate of 8π square feet per second.

REMARK When using these guidelines, be sure you perform Step 3 before Step 4. Substituting the known values of the variables before differentiating will produce an inappropriate derivative.

GUIDELINES FOR SOLVING RELATED-RATE PROBLEMS

1. Identify all *given* quantities and quantities *to be determined.* Make a sketch and label the quantities.
2. Write an equation involving the variables whose rates of change either are given or are to be determined.
3. Using the Chain Rule, implicitly differentiate both sides of the equation *with respect to time t.*
4. *After* completing Step 3, substitute into the resulting equation all known values for the variables and their rates of change. Then solve for the required rate of change.

The table below lists examples of mathematical models involving rates of change. For instance, the rate of change in the first example is the velocity of a car.

Verbal Statement	Mathematical Model
The velocity of a car after traveling for 1 hour is 50 miles per hour.	$x =$ distance traveled $\frac{dx}{dt} = 50$ mi/h when $t = 1$
Water is being pumped into a swimming pool at a rate of 10 cubic meters per hour.	$V =$ volume of water in pool $\frac{dV}{dt} = 10\ \text{m}^3/\text{h}$
A gear is revolving at a rate of 25 revolutions per minute (1 revolution $= 2\pi$ rad).	$\theta =$ angle of revolution $\frac{d\theta}{dt} = 25(2\pi)$ rad/min
A population of bacteria is increasing at a rate of 2000 per hour.	$x =$ number in population $\frac{dx}{dt} = 2000$ bacteria per hour

EXAMPLE 3 An Inflating Balloon

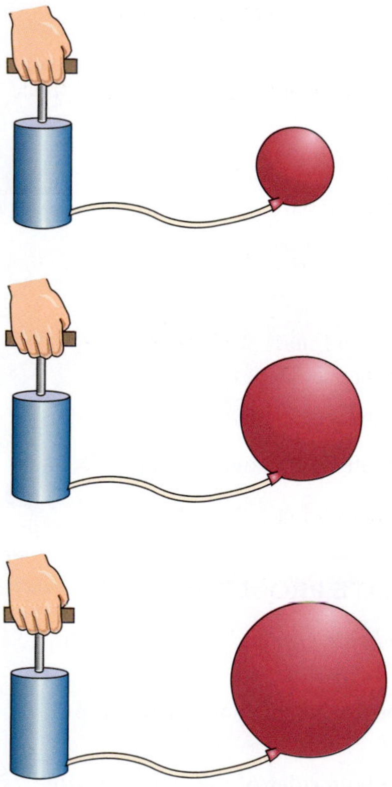

Inflating a balloon
Figure 2.35

Air is being pumped into a spherical balloon (see Figure 2.35) at a rate of 4.5 cubic feet per minute. Find the rate of change of the radius when the radius is 2 feet.

Solution Let V be the volume of the balloon, and let r be its radius. Because the volume is increasing at a rate of 4.5 cubic feet per minute, you know that at time t the rate of change of the volume is $dV/dt = \frac{9}{2}$. So, the problem can be stated as shown.

Given rate: $\frac{dV}{dt} = \frac{9}{2}$ (constant rate)

Find: $\frac{dr}{dt}$ when $r = 2$

To find the rate of change of the radius, you must find an equation that relates the radius r to the volume V.

Equation: $V = \frac{4}{3}\pi r^3$ Volume of a sphere

Differentiating both sides of the equation with respect to t produces

$$\frac{dV}{dt} = 4\pi r^2 \frac{dr}{dt} \quad \text{Differentiate with respect to } t.$$

$$\frac{dr}{dt} = \frac{1}{4\pi r^2}\left(\frac{dV}{dt}\right). \quad \text{Solve for } \frac{dr}{dt}.$$

Finally, when $r = 2$, the rate of change of the radius is

$$\frac{dr}{dt} = \frac{1}{4\pi(2)^2}\left(\frac{9}{2}\right) \approx 0.09 \text{ foot per minute.}$$

In Example 3, note that the volume is increasing at a *constant* rate, but the radius is increasing at a *variable* rate. Just because two rates are related does not mean that they are proportional. In this particular case, the radius is growing more and more slowly as t increases. Do you see why?

EXAMPLE 4 The Speed of an Airplane Tracked by Radar

See LarsonCalculus.com for an interactive version of this type of example.

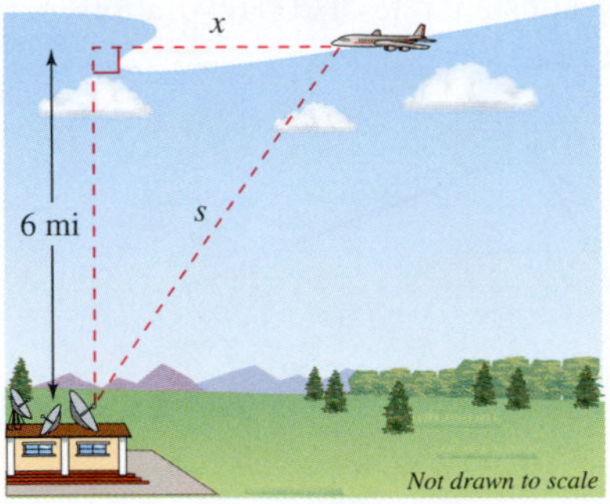

An airplane is flying at an altitude of 6 miles, s miles from the station.
Figure 2.36

An airplane is flying on a flight path that will take it directly over a radar tracking station, as shown in Figure 2.36. The distance s is decreasing at a rate of 400 miles per hour when $s = 10$ miles. What is the speed of the plane?

Solution Let x be the horizontal distance from the station, as shown in Figure 2.36. Notice that when $s = 10$, $x = \sqrt{10^2 - 36} = 8$.

Given rate: $ds/dt = -400$ when $s = 10$

Find: dx/dt when $s = 10$ and $x = 8$

You can find the velocity of the plane as shown.

Equation:

$$x^2 + 6^2 = s^2 \qquad \text{Pythagorean Theorem}$$

$$2x\frac{dx}{dt} = 2s\frac{ds}{dt} \qquad \text{Differentiate with respect to } t.$$

$$\frac{dx}{dt} = \frac{s}{x}\left(\frac{ds}{dt}\right) \qquad \text{Solve for } \frac{dx}{dt}.$$

$$= \frac{10}{8}(-400) \qquad \text{Substitute for } s, x, \text{ and } \frac{ds}{dt}.$$

$$= -500 \text{ miles per hour} \qquad \text{Simplify.}$$

Because the velocity is -500 miles per hour, the *speed* is 500 miles per hour. ■

REMARK The velocity in Example 4 is negative because x represents a distance that is decreasing.

EXAMPLE 5 A Changing Angle of Elevation

Find the rate of change in the angle of elevation of the camera shown in Figure 2.37 at 10 seconds after lift-off.

Solution Let θ be the angle of elevation, as shown in Figure 2.37. When $t = 10$, the height s of the rocket is $s = 50t^2 = 50(10)^2 = 5000$ feet.

Given rate: $ds/dt = 100t$ = velocity of rocket

Find: $d\theta/dt$ when $t = 10$ and $s = 5000$

Using Figure 2.37, you can relate s and θ by the equation $\tan\theta = s/2000$.

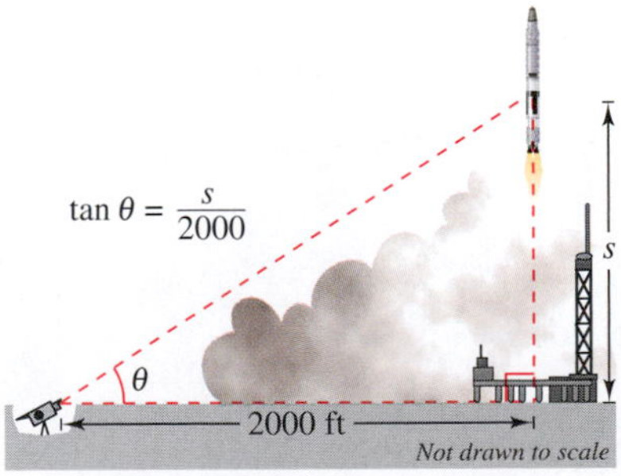

A television camera at ground level is filming the lift-off of a rocket that is rising vertically according to the position equation $s = 50t^2$, where s is measured in feet and t is measured in seconds. The camera is 2000 feet from the launch pad.
Figure 2.37

Equation:

$$\tan\theta = \frac{s}{2000} \qquad \text{See Figure 2.37.}$$

$$(\sec^2\theta)\frac{d\theta}{dt} = \frac{1}{2000}\left(\frac{ds}{dt}\right) \qquad \text{Differentiate with respect to } t.$$

$$\frac{d\theta}{dt} = \cos^2\theta\,\frac{100t}{2000} \qquad \text{Substitute } 100t \text{ for } \frac{ds}{dt}.$$

$$= \left(\frac{2000}{\sqrt{s^2 + 2000^2}}\right)^2 \frac{100t}{2000} \qquad \cos\theta = \frac{2000}{\sqrt{s^2 + 2000^2}}$$

When $t = 10$ and $s = 5000$, you have

$$\frac{d\theta}{dt} = \frac{2000(100)(10)}{5000^2 + 2000^2} = \frac{2}{29} \text{ radian per second.}$$

So, when $t = 10$, θ is changing at a rate of $\frac{2}{29}$ radian per second. ■

EXAMPLE 6 The Velocity of a Piston

In the engine shown in Figure 2.38, a 7-inch connecting rod is fastened to a crank of radius 3 inches. The crankshaft rotates counterclockwise at a constant rate of 200 revolutions per minute. Find the velocity of the piston when $\theta = \pi/3$.

AP* Tips

Related rate problems, since they represent a powerful application of implicit derivatives, make frequent appearances on the free response section of the AP Exam.

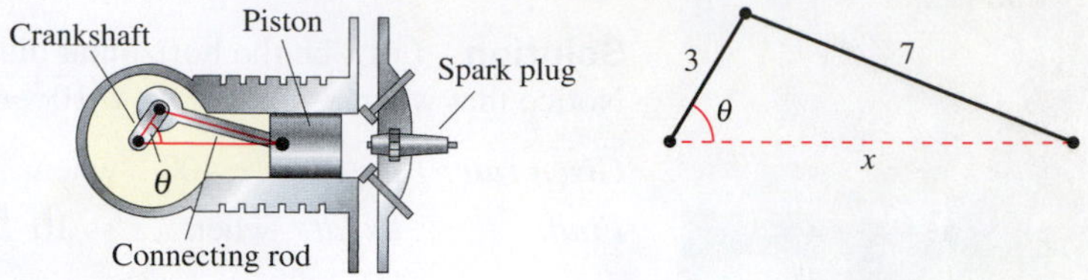

The velocity of a piston is related to the angle of the crankshaft.
Figure 2.38

Solution Label the distances as shown in Figure 2.38. Because a complete revolution corresponds to 2π radians, it follows that $d\theta/dt = 200(2\pi) = 400\pi$ radians per minute.

Given rate: $\dfrac{d\theta}{dt} = 400\pi$ (constant rate)

Find: $\dfrac{dx}{dt}$ when $\theta = \dfrac{\pi}{3}$

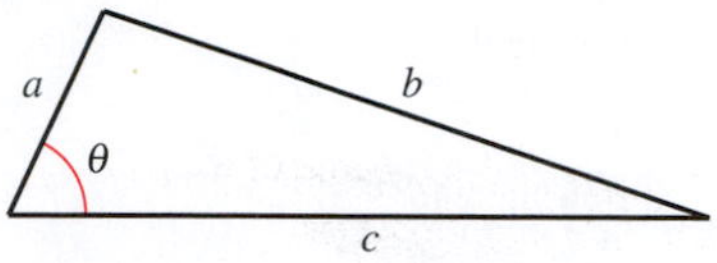

Law of Cosines:
$b^2 = a^2 + c^2 - 2ac\cos\theta$
Figure 2.39

You can use the Law of Cosines (see Figure 2.39) to find an equation that relates x and θ.

Equation:

$$7^2 = 3^2 + x^2 - 2(3)(x)\cos\theta$$

$$0 = 2x\frac{dx}{dt} - 6\left(-x\sin\theta\frac{d\theta}{dt} + \cos\theta\frac{dx}{dt}\right)$$

$$(6\cos\theta - 2x)\frac{dx}{dt} = 6x\sin\theta\frac{d\theta}{dt}$$

$$\frac{dx}{dt} = \frac{6x\sin\theta}{6\cos\theta - 2x}\left(\frac{d\theta}{dt}\right)$$

When $\theta = \pi/3$, you can solve for x as shown.

$$7^2 = 3^2 + x^2 - 2(3)(x)\cos\frac{\pi}{3}$$

$$49 = 9 + x^2 - 6x\left(\frac{1}{2}\right)$$

$$0 = x^2 - 3x - 40$$

$$0 = (x - 8)(x + 5)$$

$$x = 8$$ Choose positive solution.

So, when $x = 8$ and $\theta = \pi/3$, the velocity of the piston is

$$\frac{dx}{dt} = \frac{6(8)(\sqrt{3}/2)}{6(1/2) - 16}(400\pi)$$

$$= \frac{9600\pi\sqrt{3}}{-13}$$

$$\approx -4018 \text{ inches per minute.}$$

REMARK The velocity in Example 6 is negative because x represents a distance that is decreasing.

2.6 Exercises

See CalcChat.com for tutorial help and worked-out solutions to odd-numbered exercises.

Using Related Rates **In Exercises 1–4, assume that x and y are both differentiable functions of t and find the required values of dy/dt and dx/dt.**

Equation	Find	Given
1. $y = \sqrt{x}$	(a) $\frac{dy}{dt}$ when $x = 4$	$\frac{dx}{dt} = 3$
	(b) $\frac{dx}{dt}$ when $x = 25$	$\frac{dy}{dt} = 2$
2. $y = 3x^2 - 5x$	(a) $\frac{dy}{dt}$ when $x = 3$	$\frac{dx}{dt} = 2$
	(b) $\frac{dx}{dt}$ when $x = 2$	$\frac{dy}{dt} = 4$
3. $xy = 4$	(a) $\frac{dy}{dt}$ when $x = 8$	$\frac{dx}{dt} = 10$
	(b) $\frac{dx}{dt}$ when $x = 1$	$\frac{dy}{dt} = -6$
4. $x^2 + y^2 = 25$	(a) $\frac{dy}{dt}$ when $x = 3, y = 4$	$\frac{dx}{dt} = 8$
	(b) $\frac{dx}{dt}$ when $x = 4, y = 3$	$\frac{dy}{dt} = -2$

Moving Point **In Exercises 5–8, a point is moving along the graph of the given function at the rate dx/dt. Find dy/dt for the given values of x.**

5. $y = 2x^2 + 1$; $\frac{dx}{dt} = 2$ centimeters per second

   (a) $x = -1$ (b) $x = 0$ (c) $x = 1$

6. $y = \frac{1}{1 + x^2}$; $\frac{dx}{dt} = 6$ inches per second

   (a) $x = -2$ (b) $x = 0$ (c) $x = 2$

7. $y = \tan x$; $\frac{dx}{dt} = 3$ feet per second

   (a) $x = -\frac{\pi}{3}$ (b) $x = -\frac{\pi}{4}$ (c) $x = 0$

8. $y = \cos x$; $\frac{dx}{dt} = 4$ centimeters per second

   (a) $x = \frac{\pi}{6}$ (b) $x = \frac{\pi}{4}$ (c) $x = \frac{\pi}{3}$

WRITING ABOUT CONCEPTS

9. **Related Rates** Consider the linear function

   $$y = ax + b.$$

   If x changes at a constant rate, does y change at a constant rate? If so, does it change at the same rate as x? Explain.

10. **Related Rates** In your own words, state the guidelines for solving related-rate problems.

11. **Area** The radius r of a circle is increasing at a rate of 4 centimeters per minute. Find the rates of change of the area when (a) $r = 8$ centimeters and (b) $r = 32$ centimeters.

12. **Area** The included angle of the two sides of constant equal length s of an isosceles triangle is θ.

    (a) Show that the area of the triangle is given by $A = \frac{1}{2}s^2 \sin \theta$.

    (b) The angle θ is increasing at the rate of $\frac{1}{2}$ radian per minute. Find the rates of change of the area when $\theta = \pi/6$ and $\theta = \pi/3$.

    (c) Explain why the rate of change of the area of the triangle is not constant even though $d\theta/dt$ is constant.

13. **Volume** The radius r of a sphere is increasing at a rate of 3 inches per minute.

    (a) Find the rates of change of the volume when $r = 9$ inches and $r = 36$ inches.

    (b) Explain why the rate of change of the volume of the sphere is not constant even though dr/dt is constant.

14. **Volume** A spherical balloon is inflated with gas at the rate of 800 cubic centimeters per minute. How fast is the radius of the balloon increasing at the instant the radius is (a) 30 centimeters and (b) 60 centimeters?

15. **Volume** All edges of a cube are expanding at a rate of 6 centimeters per second. How fast is the volume changing when each edge is (a) 2 centimeters and (b) 10 centimeters?

16. **Surface Area** All edges of a cube are expanding at a rate of 6 centimeters per second. How fast is the surface area changing when each edge is (a) 2 centimeters and (b) 10 centimeters?

17. **Volume** At a sand and gravel plant, sand is falling off a conveyor and onto a conical pile at a rate of 10 cubic feet per minute. The diameter of the base of the cone is approximately three times the altitude. At what rate is the height of the pile changing when the pile is 15 feet high? (*Hint:* The formula for the volume of a cone is $V = \frac{1}{3}\pi r^2 h$.)

18. **Depth** A conical tank (with vertex down) is 10 feet across the top and 12 feet deep. Water is flowing into the tank at a rate of 10 cubic feet per minute. Find the rate of change of the depth of the water when the water is 8 feet deep.

19. **Depth** A swimming pool is 12 meters long, 6 meters wide, 1 meter deep at the shallow end, and 3 meters deep at the deep end (see figure). Water is being pumped into the pool at $\frac{1}{4}$ cubic meter per minute, and there is 1 meter of water at the deep end.

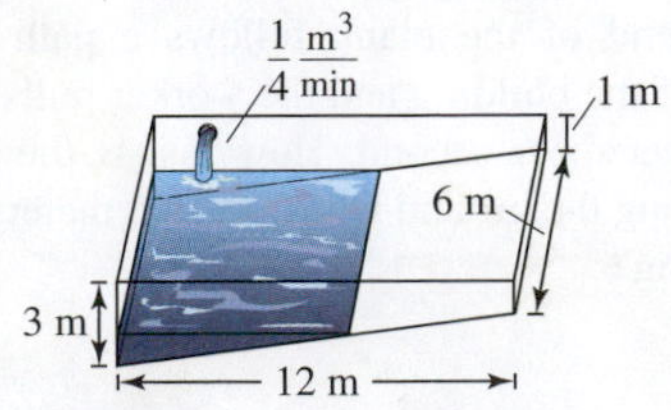

    (a) What percent of the pool is filled?

    (b) At what rate is the water level rising?

20. Depth A trough is 12 feet long and 3 feet across the top (see figure). Its ends are isosceles triangles with altitudes of 3 feet.

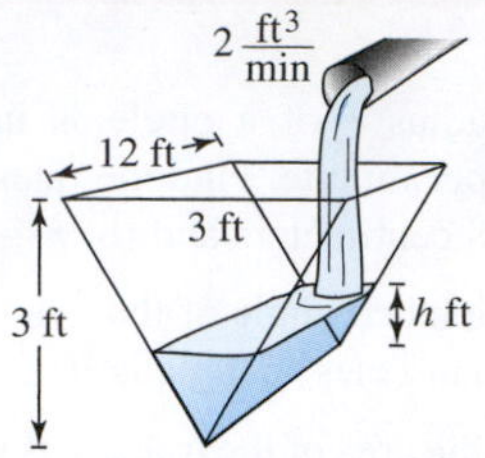

(a) Water is being pumped into the trough at 2 cubic feet per minute. How fast is the water level rising when the depth h is 1 foot?

(b) The water is rising at a rate of $\frac{3}{8}$ inch per minute when $h = 2$. Determine the rate at which water is being pumped into the trough.

21. Moving Ladder A ladder 25 feet long is leaning against the wall of a house (see figure). The base of the ladder is pulled away from the wall at a rate of 2 feet per second.

(a) How fast is the top of the ladder moving down the wall when its base is 7 feet, 15 feet, and 24 feet from the wall?

(b) Consider the triangle formed by the side of the house, the ladder, and the ground. Find the rate at which the area of the triangle is changing when the base of the ladder is 7 feet from the wall.

(c) Find the rate at which the angle between the ladder and the wall of the house is changing when the base of the ladder is 7 feet from the wall.

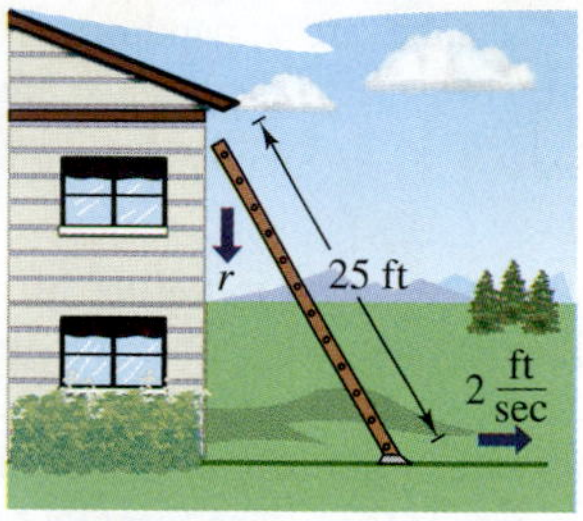

Figure for 21

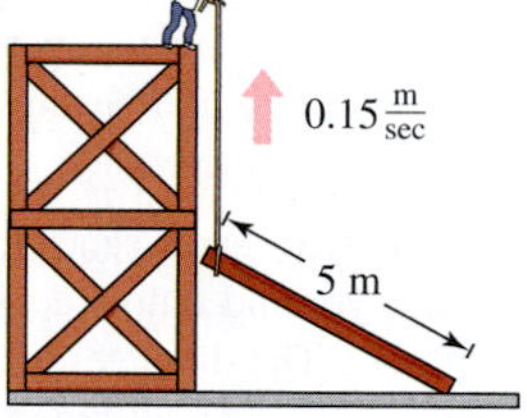

Figure for 22

FOR FURTHER INFORMATION For more information on the mathematics of moving ladders, see the article "The Falling Ladder Paradox" by Paul Scholten and Andrew Simoson in *The College Mathematics Journal*. To view this article, go to *MathArticles.com*.

22. Construction A construction worker pulls a five-meter plank up the side of a building under construction by means of a rope tied to one end of the plank (see figure). Assume the opposite end of the plank follows a path perpendicular to the wall of the building and the worker pulls the rope at a rate of 0.15 meter per second. How fast is the end of the plank sliding along the ground when it is 2.5 meters from the wall of the building?

23. Construction A winch at the top of a 12-meter building pulls a pipe of the same length to a vertical position, as shown in the figure. The winch pulls in rope at a rate of -0.2 meter per second. Find the rate of vertical change and the rate of horizontal change at the end of the pipe when $y = 6$.

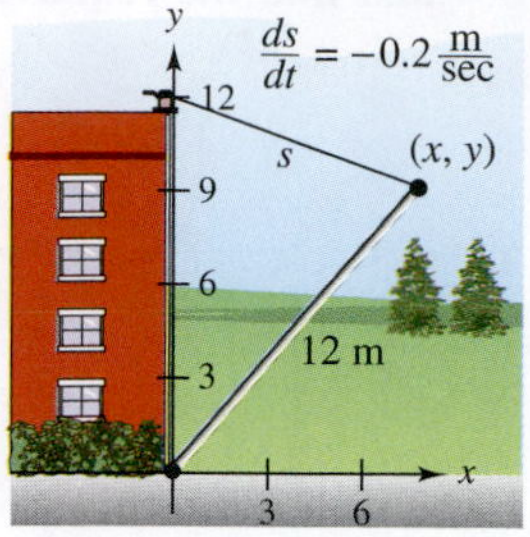

Figure for 23

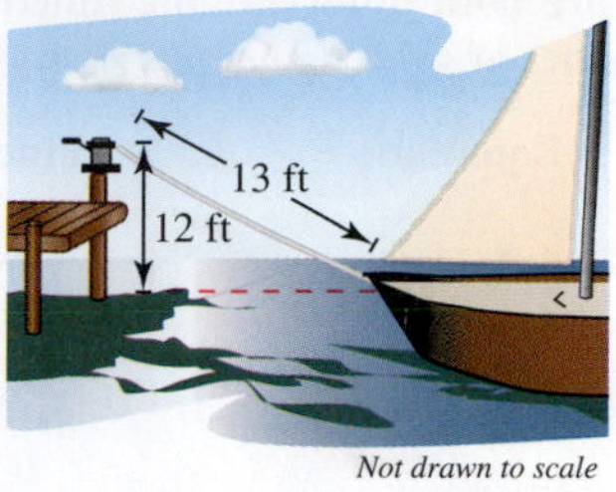

Figure for 24

24. Boating A boat is pulled into a dock by means of a winch 12 feet above the deck of the boat (see figure).

(a) The winch pulls in rope at a rate of 4 feet per second. Determine the speed of the boat when there is 13 feet of rope out. What happens to the speed of the boat as it gets closer to the dock?

(b) Suppose the boat is moving at a constant rate of 4 feet per second. Determine the speed at which the winch pulls in rope when there is a total of 13 feet of rope out. What happens to the speed at which the winch pulls in rope as the boat gets closer to the dock?

25. Air Traffic Control An air traffic controller spots two planes at the same altitude converging on a point as they fly at right angles to each other (see figure). One plane is 225 miles from the point moving at 450 miles per hour. The other plane is 300 miles from the point moving at 600 miles per hour.

(a) At what rate is the distance between the planes decreasing?

(b) How much time does the air traffic controller have to get one of the planes on a different flight path?

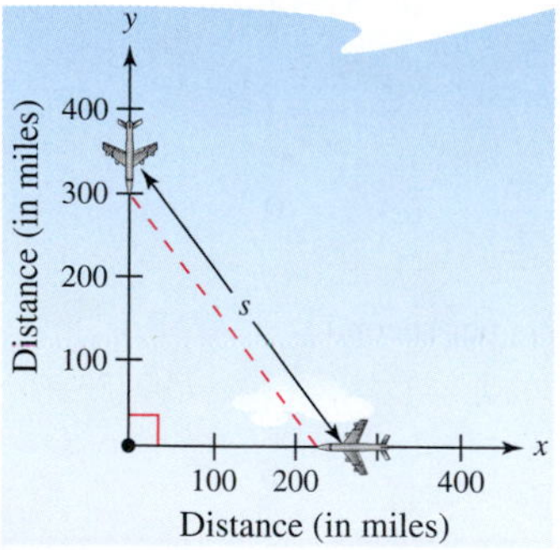

Figure for 25

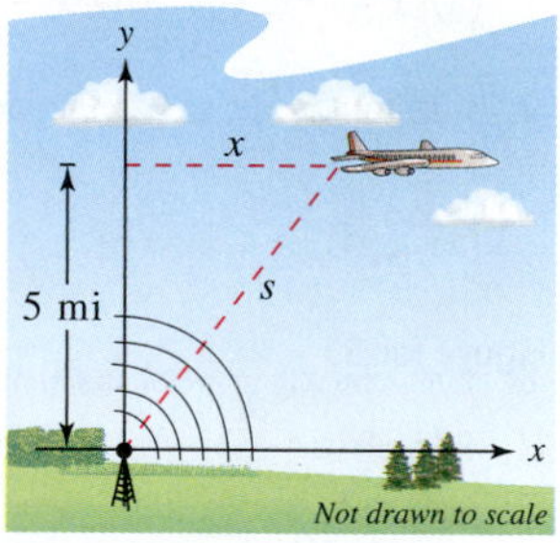

Figure for 26

26. Air Traffic Control An airplane is flying at an altitude of 5 miles and passes directly over a radar antenna (see figure). When the plane is 10 miles away ($s = 10$), the radar detects that the distance s is changing at a rate of 240 miles per hour. What is the speed of the plane?

27. Sports A baseball diamond has the shape of a square with sides 90 feet long (see figure). A player running from second base to third base at a speed of 25 feet per second is 20 feet from third base. At what rate is the player's distance s from home plate changing?

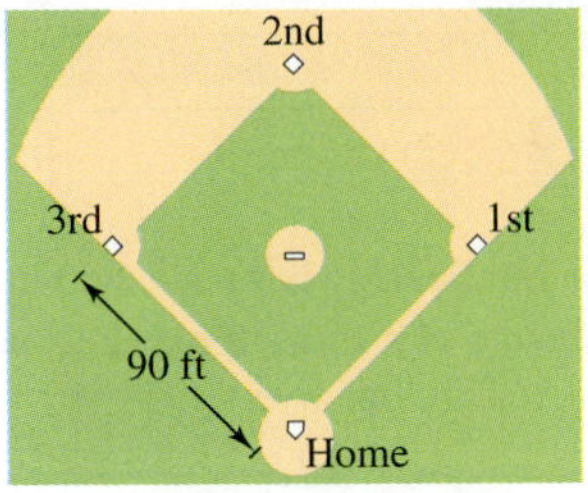

Figure for 27 and 28

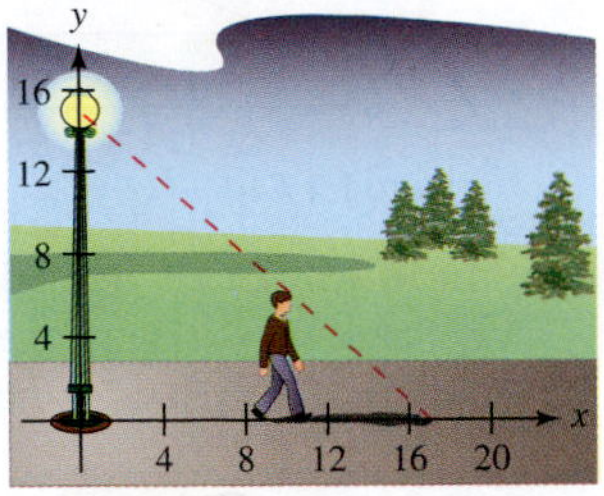

Figure for 29

28. Sports For the baseball diamond in Exercise 27, suppose the player is running from first base to second base at a speed of 25 feet per second. Find the rate at which the distance from home plate is changing when the player is 20 feet from second base.

29. Shadow Length A man 6 feet tall walks at a rate of 5 feet per second away from a light that is 15 feet above the ground (see figure).

(a) When he is 10 feet from the base of the light, at what rate is the tip of his shadow moving?

(b) When he is 10 feet from the base of the light, at what rate is the length of his shadow changing?

30. Shadow Length Repeat Exercise 29 for a man 6 feet tall walking at a rate of 5 feet per second *toward* a light that is 20 feet above the ground (see figure).

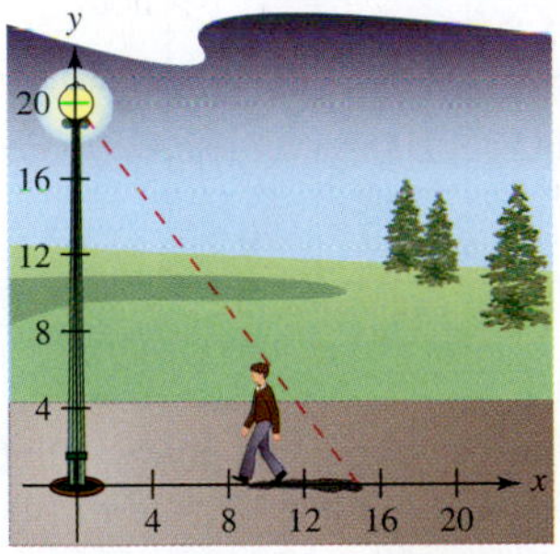

Figure for 30

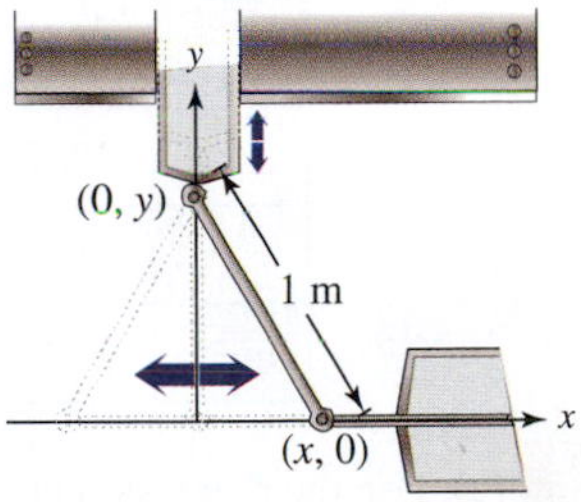

Figure for 31

31. Machine Design The endpoints of a movable rod of length 1 meter have coordinates $(x, 0)$ and $(0, y)$ (see figure). The position of the end on the x-axis is

$$x(t) = \frac{1}{2}\sin\frac{\pi t}{6}$$

where t is the time in seconds.

(a) Find the time of one complete cycle of the rod.

(b) What is the lowest point reached by the end of the rod on the y-axis?

(c) Find the speed of the y-axis endpoint when the x-axis endpoint is $\left(\frac{1}{4}, 0\right)$.

32. Machine Design Repeat Exercise 31 for a position function of $x(t) = \frac{3}{5}\sin \pi t$. Use the point $\left(\frac{3}{10}, 0\right)$ for part (c).

33. Evaporation As a spherical raindrop falls, it reaches a layer of dry air and begins to evaporate at a rate that is proportional to its surface area $(S = 4\pi r^2)$. Show that the radius of the raindrop decreases at a constant rate.

34. HOW DO YOU SEE IT? Using the graph of f, (a) determine whether dy/dt is positive or negative given that dx/dt is negative, and (b) determine whether dx/dt is positive or negative given that dy/dt is positive.

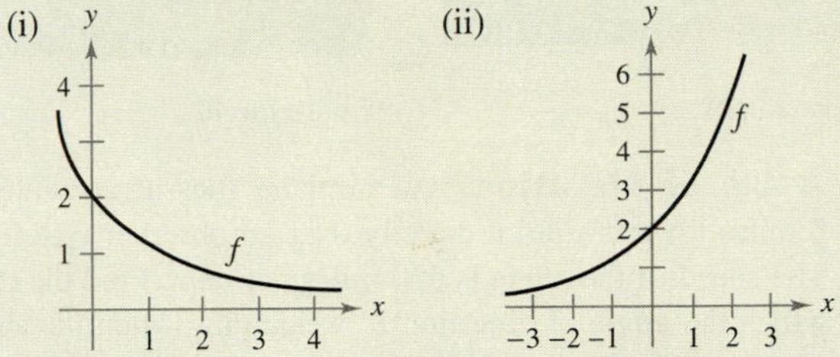

35. Electricity The combined electrical resistance R of two resistors R_1 and R_2, connected in parallel, is given by

$$\frac{1}{R} = \frac{1}{R_1} + \frac{1}{R_2}$$

where R, R_1, and R_2 are measured in ohms. R_1 and R_2 are increasing at rates of 1 and 1.5 ohms per second, respectively. At what rate is R changing when $R_1 = 50$ ohms and $R_2 = 75$ ohms?

36. Adiabatic Expansion When a certain polyatomic gas undergoes adiabatic expansion, its pressure p and volume V satisfy the equation $pV^{1.3} = k$, where k is a constant. Find the relationship between the related rates dp/dt and dV/dt.

37. Roadway Design Cars on a certain roadway travel on a circular arc of radius r. In order not to rely on friction alone to overcome the centrifugal force, the road is banked at an angle of magnitude θ from the horizontal (see figure). The banking angle must satisfy the equation $rg \tan\theta = v^2$, where v is the velocity of the cars and $g = 32$ feet per second per second is the acceleration due to gravity. Find the relationship between the related rates dv/dt and $d\theta/dt$.

38. Angle of Elevation A balloon rises at a rate of 4 meters per second from a point on the ground 50 meters from an observer. Find the rate of change of the angle of elevation of the balloon from the observer when the balloon is 50 meters above the ground.

39. Angle of Elevation A fish is reeled in at a rate of 1 foot per second from a point 10 feet above the water (see figure). At what rate is the angle θ between the line and the water changing when there is a total of 25 feet of line from the end of the rod to the water?

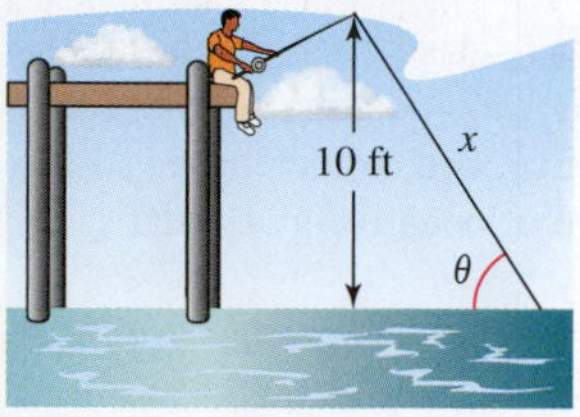

Figure for 39

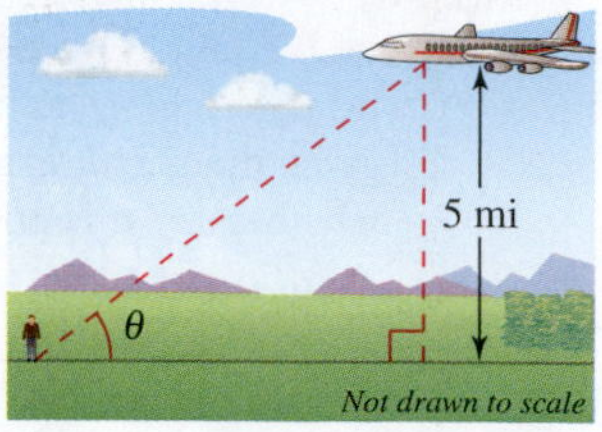

Figure for 40

40. Angle of Elevation An airplane flies at an altitude of 5 miles toward a point directly over an observer (see figure). The speed of the plane is 600 miles per hour. Find the rates at which the angle of elevation θ is changing when the angle is (a) $\theta = 30°$, (b) $\theta = 60°$, and (c) $\theta = 75°$.

41. Linear vs. Angular Speed A patrol car is parked 50 feet from a long warehouse (see figure). The revolving light on top of the car turns at a rate of 30 revolutions per minute. How fast is the light beam moving along the wall when the beam makes angles of (a) $\theta = 30°$, (b) $\theta = 60°$, and (c) $\theta = 70°$ with the perpendicular line from the light to the wall?

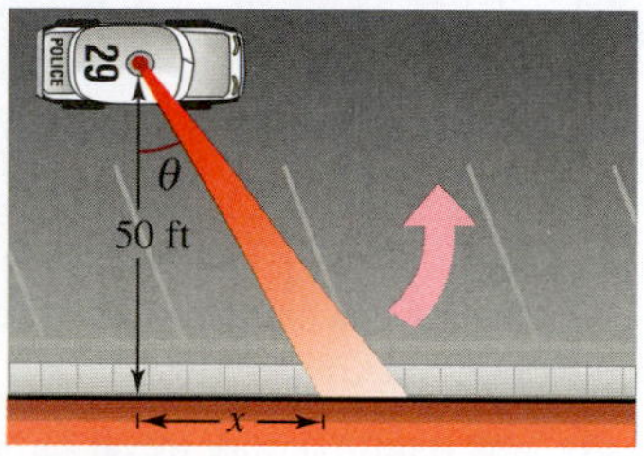

Figure for 41

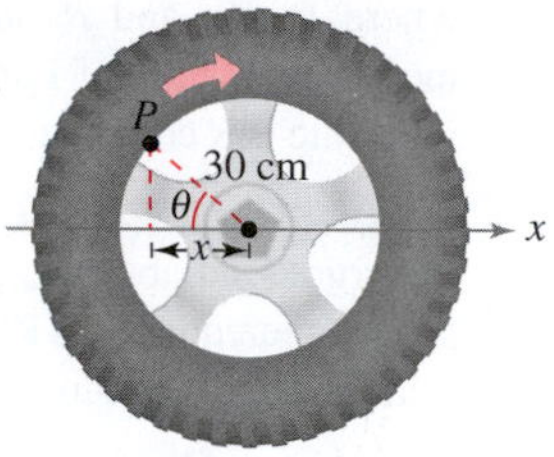

Figure for 42

42. Linear vs. Angular Speed A wheel of radius 30 centimeters revolves at a rate of 10 revolutions per second. A dot is painted at a point P on the rim of the wheel (see figure).

(a) Find dx/dt as a function of θ.

(b) Use a graphing utility to graph the function in part (a).

(c) When is the absolute value of the rate of change of x greatest? When is it least?

(d) Find dx/dt when $\theta = 30°$ and $\theta = 60°$.

43. Flight Control An airplane is flying in still air with an airspeed of 275 miles per hour. The plane is climbing at an angle of 18°. Find the rate at which it is gaining altitude.

44. Security Camera A security camera is centered 50 feet above a 100-foot hallway (see figure). It is easiest to design the camera with a constant angular rate of rotation, but this results in recording the images of the surveillance area at a variable rate. So, it is desirable to design a system with a variable rate of rotation and a constant rate of movement of the scanning beam along the hallway. Find a model for the variable rate of rotation when $|dx/dt| = 2$ feet per second.

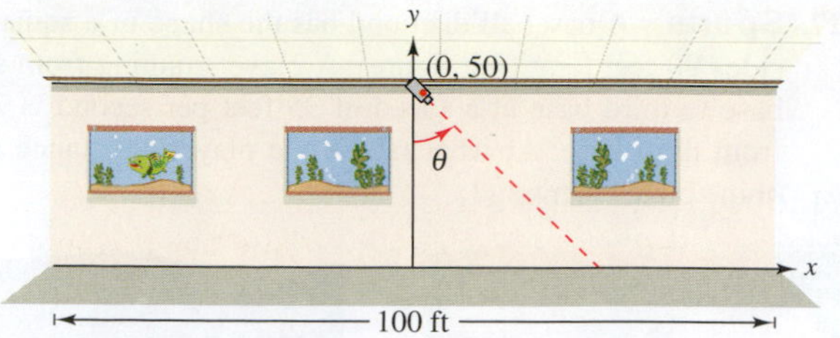

Figure for 44

45. Think About It Describe the relationship between the rate of change of y and the rate of change of x in each expression. Assume all variables and derivatives are positive.

(a) $\dfrac{dy}{dt} = 3\dfrac{dx}{dt}$ (b) $\dfrac{dy}{dt} = x(L - x)\dfrac{dx}{dt}, \quad 0 \le x \le L$

Acceleration **In Exercises 46 and 47, find the acceleration of the specified object. (*Hint:* Recall that if a variable is changing at a constant rate, its acceleration is zero.)**

46. Find the acceleration of the top of the ladder described in Exercise 21 when the base of the ladder is 7 feet from the wall.

47. Find the acceleration of the boat in Exercise 24(a) when there is a total of 13 feet of rope out.

48. Modeling Data The table shows the numbers (in millions) of single women (never married) s and married women m in the civilian work force in the United States for the years 2003 through 2010. *(Source: U.S. Bureau of Labor Statistics)*

Year	2003	2004	2005	2006
s	18.4	18.6	19.2	19.5
m	36.0	35.8	35.9	36.3

Year	2007	2008	2009	2010
s	19.7	20.2	20.2	20.6
m	36.9	37.2	37.3	36.7

(a) Use the regression capabilities of a graphing utility to find a model of the form $m(s) = as^3 + bs^2 + cs + d$ for the data, where t is the time in years, with $t = 3$ corresponding to 2003.

(b) Find dm/dt. Then use the model to estimate dm/dt for $t = 7$ when it is predicted that the number of single women in the work force will increase at the rate of 0.75 million per year.

49. Moving Shadow A ball is dropped from a height of 20 meters, 12 meters away from the top of a 20-meter lamppost (see figure). The ball's shadow, caused by the light at the top of the lamppost, is moving along the level ground. How fast is the shadow moving 1 second after the ball is released? *(Submitted by Dennis Gittinger, St. Philips College, San Antonio, TX)*

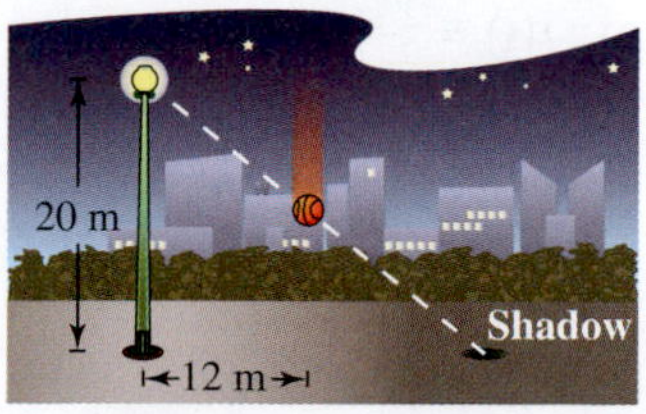

Review Exercises

See CalcChat.com for tutorial help and worked-out solutions to odd-numbered exercises.

Finding the Derivative by the Limit Process In Exercises 1–4, find the derivative of the function by the limit process.

1. $f(x) = 12$
2. $f(x) = 5x - 4$
3. $f(x) = x^2 - 4x + 5$
4. $f(x) = \dfrac{6}{x}$

Using the Alternative Form of the Derivative In Exercises 5 and 6, use the alternative form of the derivative to find the derivative at $x = c$ (if it exists).

5. $g(x) = 2x^2 - 3x, \quad c = 2$
6. $f(x) = \dfrac{1}{x + 4}, \quad c = 3$

Determining Differentiability In Exercises 7 and 8, describe the x-values at which f is differentiable.

7. $f(x) = (x - 3)^{2/5}$

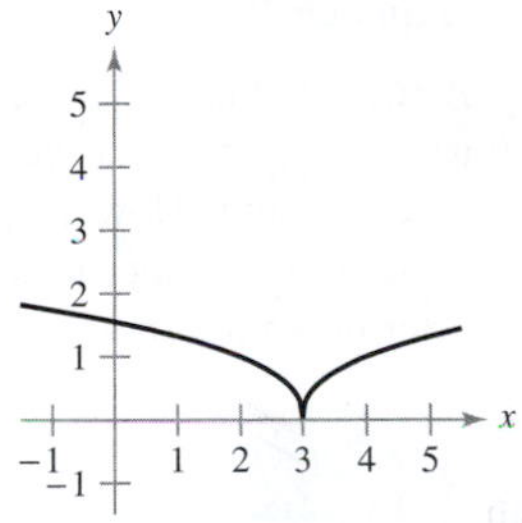

8. $f(x) = \dfrac{3x}{x + 1}$

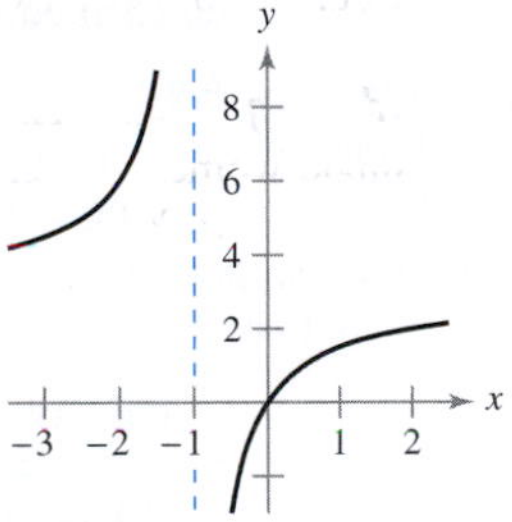

Finding a Derivative In Exercises 9–20, use the rules of differentiation to find the derivative of the function.

9. $y = 25$
10. $f(t) = 4t^4$
11. $f(x) = x^3 - 11x^2$
12. $g(s) = 3s^5 - 2s^4$
13. $h(x) = 6\sqrt{x} + 3\sqrt[3]{x}$
14. $f(x) = x^{1/2} - x^{-1/2}$
15. $g(t) = \dfrac{2}{3t^2}$
16. $h(x) = \dfrac{8}{5x^4}$
17. $f(\theta) = 4\theta - 5 \sin \theta$
18. $g(\alpha) = 4 \cos \alpha + 6$
19. $f(\theta) = 3 \cos \theta - \dfrac{\sin \theta}{4}$
20. $g(\alpha) = \dfrac{5 \sin \alpha}{3} - 2\alpha$

Finding the Slope of a Graph In Exercises 21–24, find the slope of the graph of the functions at the given point.

21. $f(x) = \dfrac{27}{x^3}, \quad (3, 1)$
22. $f(x) = 3x^2 - 4x, \quad (1, -1)$
23. $f(x) = 2x^4 - 8, \quad (0, -8)$
24. $f(\theta) = 3 \cos \theta - 2\theta, \quad (0, 3)$

25. **Vibrating String** When a guitar string is plucked, it vibrates with a frequency of $F = 200\sqrt{T}$, where F is measured in vibrations per second and the tension T is measured in pounds. Find the rates of change of F when (a) $T = 4$ and (b) $T = 9$.

26. **Volume** The surface area of a cube with sides of length ℓ is given by $S = 6\ell^2$. Find the rates of change of the surface area with respect to ℓ when (a) $\ell = 3$ inches and (b) $\ell = 5$ inches.

Vertical Motion In Exercises 27 and 28, use the position function $s(t) = -16t^2 + v_0 t + s_0$ for free-falling objects.

27. A ball is thrown straight down from the top of a 600-foot building with an initial velocity of -30 feet per second.
    (a) Determine the position and velocity functions for the ball.
    (b) Determine the average velocity on the interval $[1, 3]$.
    (c) Find the instantaneous velocities when $t = 1$ and $t = 3$.
    (d) Find the time required for the ball to reach ground level.
    (e) Find the velocity of the ball at impact.

28. To estimate the height of a building, a weight is dropped from the top of the building into a pool at ground level. The splash is seen 9.2 seconds after the weight is dropped. What is the height (in feet) of the building?

Finding a Derivative In Exercises 29–40, use the Product Rule or the Quotient Rule to find the derivative of the function.

29. $f(x) = (5x^2 + 8)(x^2 - 4x - 6)$
30. $g(x) = (2x^3 + 5x)(3x - 4)$
31. $h(x) = \sqrt{x} \sin x$
32. $f(t) = 2t^5 \cos t$
33. $f(x) = \dfrac{x^2 + x - 1}{x^2 - 1}$
34. $f(x) = \dfrac{2x + 7}{x^2 + 4}$
35. $y = \dfrac{x^4}{\cos x}$
36. $y = \dfrac{\sin x}{x^4}$
37. $y = 3x^2 \sec x$
38. $y = 2x - x^2 \tan x$
39. $y = x \cos x - \sin x$
40. $g(x) = 3x \sin x + x^2 \cos x$

Finding an Equation of a Tangent Line In Exercises 41–44, find an equation of the tangent line to the graph of f at the given point.

41. $f(x) = (x + 2)(x^2 + 5), \quad (-1, 6)$
42. $f(x) = (x - 4)(x^2 + 6x - 1), \quad (0, 4)$
43. $f(x) = \dfrac{x + 1}{x - 1}, \quad \left(\dfrac{1}{2}, -3\right)$
44. $f(x) = \dfrac{1 + \cos x}{1 - \cos x}, \quad \left(\dfrac{\pi}{2}, 1\right)$

Finding a Second Derivative In Exercises 45–50, find the second derivative of the function.

45. $g(t) = -8t^3 - 5t + 12$
46. $h(x) = 6x^{-2} + 7x^2$
47. $f(x) = 15x^{5/2}$
48. $f(x) = 20\sqrt[5]{x}$
49. $f(\theta) = 3 \tan \theta$
50. $h(t) = 10 \cos t - 15 \sin t$

51. Acceleration The velocity of an object in meters per second is $v(t) = 20 - t^2$, $0 \le t \le 6$. Find the velocity and acceleration of the object when $t = 3$.

52. Acceleration The velocity of an automobile starting from rest is

$$v(t) = \frac{90t}{4t + 10}$$

where v is measured in feet per second. Find the acceleration at (a) 1 second, (b) 5 seconds, and (c) 10 seconds.

Finding a Derivative **In Exercises 53–64, find the derivative of the function.**

53. $y = (7x + 3)^4$

54. $y = (x^2 - 6)^3$

55. $y = \dfrac{1}{x^2 + 4}$

56. $f(x) = \dfrac{1}{(5x + 1)^2}$

57. $y = 5\cos(9x + 1)$

58. $y = 1 - \cos 2x + 2\cos^2 x$

59. $y = \dfrac{x}{2} - \dfrac{\sin 2x}{4}$

60. $y = \dfrac{\sec^7 x}{7} - \dfrac{\sec^5 x}{5}$

61. $y = x(6x + 1)^5$

62. $f(s) = (s^2 - 1)^{5/2}(s^3 + 5)$

63. $f(x) = \dfrac{3x}{\sqrt{x^2 + 1}}$

64. $h(x) = \left(\dfrac{x + 5}{x^2 + 3}\right)^2$

Evaluating a Derivative **In Exercises 65–70, find and evaluate the derivative of the function at the given point.**

65. $f(x) = \sqrt{1 - x^3}$, $(-2, 3)$

66. $f(x) = \sqrt[3]{x^2 - 1}$, $(3, 2)$

67. $f(x) = \dfrac{4}{x^2 + 1}$, $(-1, 2)$

68. $f(x) = \dfrac{3x + 1}{4x - 3}$, $(4, 1)$

69. $y = \dfrac{1}{2}\csc 2x$, $\left(\dfrac{\pi}{4}, \dfrac{1}{2}\right)$

70. $y = \csc 3x + \cot 3x$, $\left(\dfrac{\pi}{6}, 1\right)$

Finding a Second Derivative **In Exercises 71–74, find the second derivative of the function.**

71. $y = (8x + 5)^3$

72. $y = \dfrac{1}{5x + 1}$

73. $f(x) = \cot x$

74. $y = \sin^2 x$

75. Refrigeration The temperature T (in degrees Fahrenheit) of food in a freezer is

$$T = \frac{700}{t^2 + 4t + 10}$$

where t is the time in hours. Find the rate of change of T with respect to t at each of the following times.

(a) $t = 1$ (b) $t = 3$ (c) $t = 5$ (d) $t = 10$

76. Harmonic Motion The displacement from equilibrium of an object in harmonic motion on the end of a spring is

$$y = \frac{1}{4}\cos 8t - \frac{1}{4}\sin 8t$$

where y is measured in feet and t is the time in seconds. Determine the position and velocity of the object when $t = \pi/4$.

Finding a Derivative **In Exercises 77–82, find dy/dx by implicit differentiation.**

77. $x^2 + y^2 = 64$

78. $x^2 + 4xy - y^3 = 6$

79. $x^3y - xy^3 = 4$

80. $\sqrt{xy} = x - 4y$

81. $x\sin y = y\cos x$

82. $\cos(x + y) = x$

Tangent Lines and Normal Lines **In Exercises 83 and 84, find equations for the tangent line and the normal line to the graph of the equation at the given point. (The *normal line* at a point is perpendicular to the tangent line at the point.) Use a graphing utility to graph the equation, the tangent line, and the normal line.**

83. $x^2 + y^2 = 10$, $(3, 1)$

84. $x^2 - y^2 = 20$, $(6, 4)$

85. Rate of Change A point moves along the curve $y = \sqrt{x}$ in such a way that the y-value is increasing at a rate of 2 units per second. At what rate is x changing for each of the following values?

(a) $x = \frac{1}{2}$ (b) $x = 1$ (c) $x = 4$

86. Surface Area All edges of a cube are expanding at a rate of 8 centimeters per second. How fast is the surface area changing when each edge is 6.5 centimeters?

87. Linear vs. Angular Speed A rotating beacon is located 1 kilometer off a straight shoreline (see figure). The beacon rotates at a rate of 3 revolutions per minute. How fast (in kilometers per hour) does the beam of light appear to be moving to a viewer who is $\frac{1}{2}$ kilometer down the shoreline?

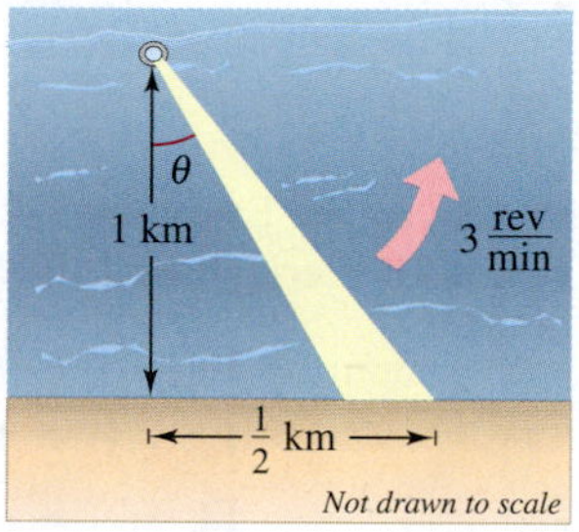

88. Moving Shadow A sandbag is dropped from a balloon at a height of 60 meters when the angle of elevation to the sun is 30° (see figure). The position of the sandbag is

$$s(t) = 60 - 4.9t^2.$$

Find the rate at which the shadow of the sandbag is traveling along the ground when the sandbag is at a height of 35 meters.

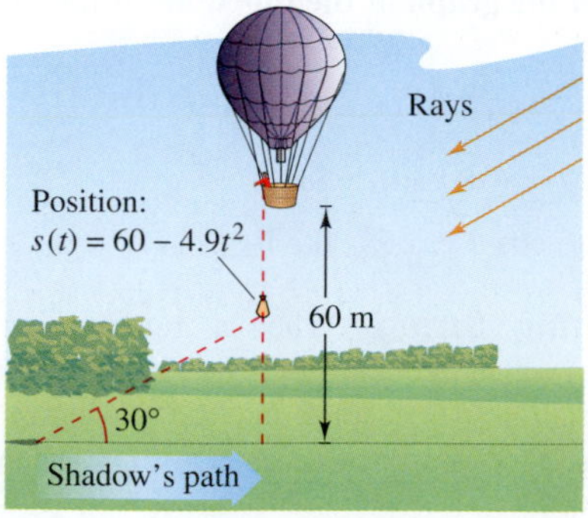

P.S. Problem Solving

See CalcChat.com for tutorial help and worked-out solutions to odd-numbered exercises.

1. Finding Equations of Circles Consider the graph of the parabola $y = x^2$.

(a) Find the radius r of the largest possible circle centered on the y-axis that is tangent to the parabola at the origin, as shown in the figure. This circle is called the **circle of curvature** (see Section 12.5). Find the equation of this circle. Use a graphing utility to graph the circle and parabola in the same viewing window to verify your answer.

(b) Find the center $(0, b)$ of the circle of radius 1 centered on the y-axis that is tangent to the parabola at two points, as shown in the figure. Find the equation of this circle. Use a graphing utility to graph the circle and parabola in the same viewing window to verify your answer.

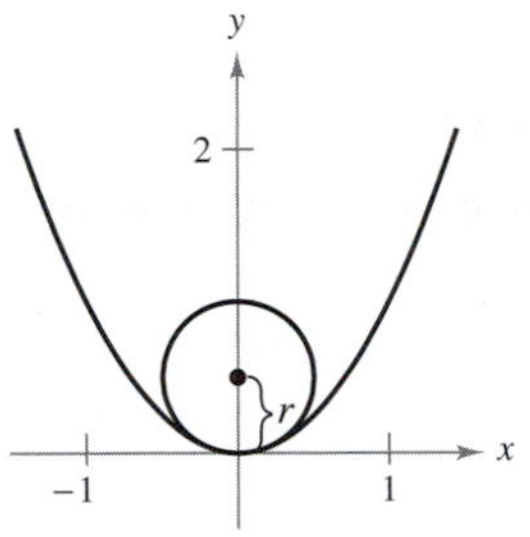

Figure for 1(a)

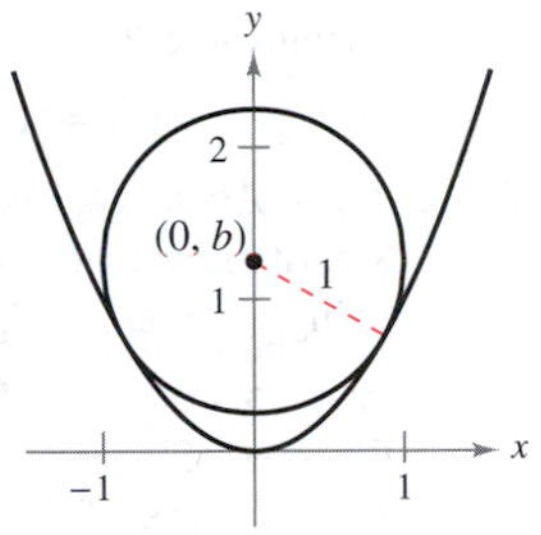

Figure for 1(b)

2. Finding Equations of Tangent Lines Graph the two parabolas

$$y = x^2 \quad \text{and} \quad y = -x^2 + 2x - 5$$

in the same coordinate plane. Find equations of the two lines that are simultaneously tangent to both parabolas.

3. Finding a Polynomial Find a third-degree polynomial $p(x)$ that is tangent to the line $y = 14x - 13$ at the point $(1, 1)$, and tangent to the line $y = -2x - 5$ at the point $(-1, -3)$.

4. Finding a Function Find a function of the form $f(x) = a + b \cos cx$ that is tangent to the line $y = 1$ at the point $(0, 1)$, and tangent to the line

$$y = x + \frac{3}{2} - \frac{\pi}{4}$$

at the point $\left(\frac{\pi}{4}, \frac{3}{2}\right)$.

5. Tangent Lines and Normal Lines

(a) Find an equation of the tangent line to the parabola $y = x^2$ at the point $(2, 4)$.

(b) Find an equation of the normal line to $y = x^2$ at the point $(2, 4)$. (The *normal line* at a point is perpendicular to the tangent line at the point.) Where does this line intersect the parabola a second time?

(c) Find equations of the tangent line and normal line to $y = x^2$ at the point $(0, 0)$.

(d) Prove that for any point $(a, b) \neq (0, 0)$ on the parabola $y = x^2$, the normal line intersects the graph a second time.

6. Finding Polynomials

(a) Find the polynomial $P_1(x) = a_0 + a_1x$ whose value and slope agree with the value and slope of $f(x) = \cos x$ at the point $x = 0$.

(b) Find the polynomial $P_2(x) = a_0 + a_1x + a_2x^2$ whose value and first two derivatives agree with the value and first two derivatives of $f(x) = \cos x$ at the point $x = 0$. This polynomial is called the second-degree Taylor polynomial of $f(x) = \cos x$ at $x = 0$.

(c) Complete the table comparing the values of $f(x) = \cos x$ and $P_2(x)$. What do you observe?

x	-1.0	-0.1	-0.001	0	0.001	0.1	1.0
$\cos x$							
$P_2(x)$							

(d) Find the third-degree Taylor polynomial of $f(x) = \sin x$ at $x = 0$.

7. Famous Curve The graph of the **eight curve**

$$x^4 = a^2(x^2 - y^2), \quad a \neq 0$$

is shown below.

(a) Explain how you could use a graphing utility to graph this curve.

(b) Use a graphing utility to graph the curve for various values of the constant a. Describe how a affects the shape of the curve.

(c) Determine the points on the curve at which the tangent line is horizontal.

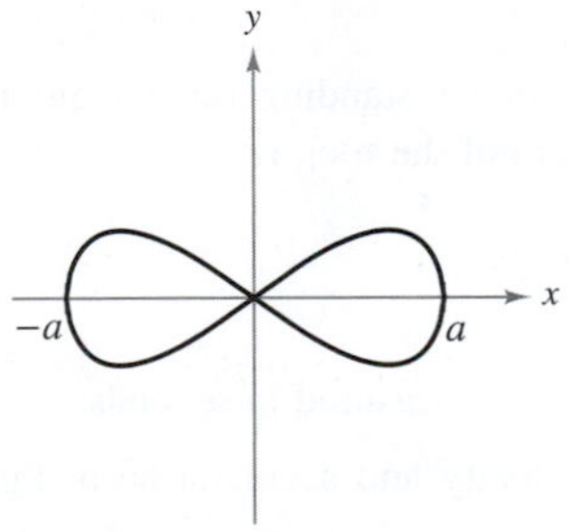

Figure for 7

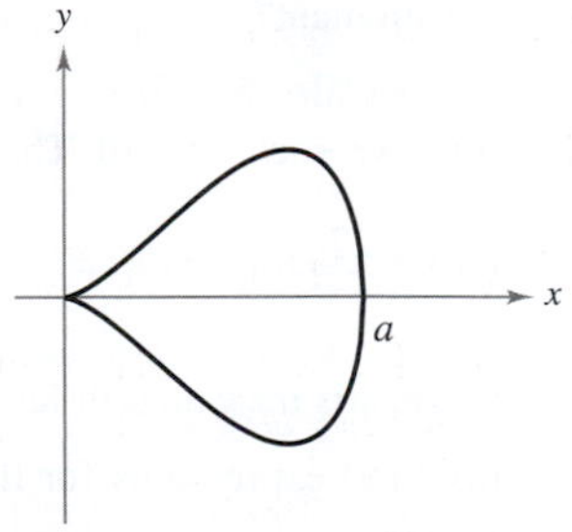

Figure for 8

8. Famous Curve The graph of the **pear-shaped quartic**

$$b^2y^2 = x^3(a - x), \quad a, b > 0$$

is shown above.

(a) Explain how you could use a graphing utility to graph this curve.

(b) Use a graphing utility to graph the curve for various values of the constants a and b. Describe how a and b affect the shape of the curve.

(c) Determine the points on the curve at which the tangent line is horizontal.

9. Shadow Length A man 6 feet tall walks at a rate of 5 feet per second toward a streetlight that is 30 feet high (see figure). The man's 3-foot-tall child follows at the same speed, but 10 feet behind the man. At times, the shadow behind the child is caused by the man, and at other times, by the child.

(a) Suppose the man is 90 feet from the streetlight. Show that the man's shadow extends beyond the child's shadow.

(b) Suppose the man is 60 feet from the streetlight. Show that the child's shadow extends beyond the man's shadow.

(c) Determine the distance d from the man to the streetlight at which the tips of the two shadows are exactly the same distance from the streetlight.

(d) Determine how fast the tip of the man's shadow is moving as a function of x, the distance between the man and the streetlight. Discuss the continuity of this shadow speed function.

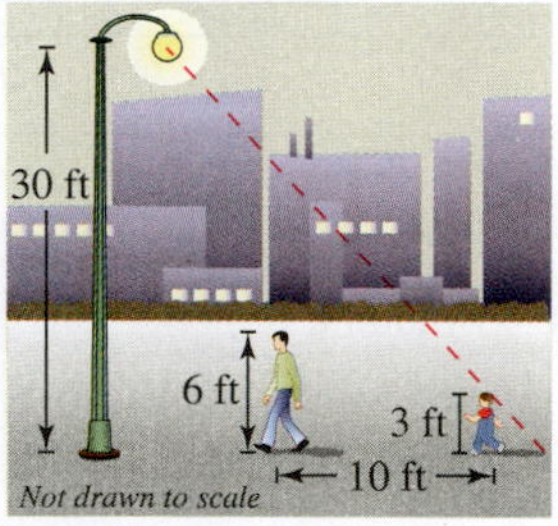

Figure for 9

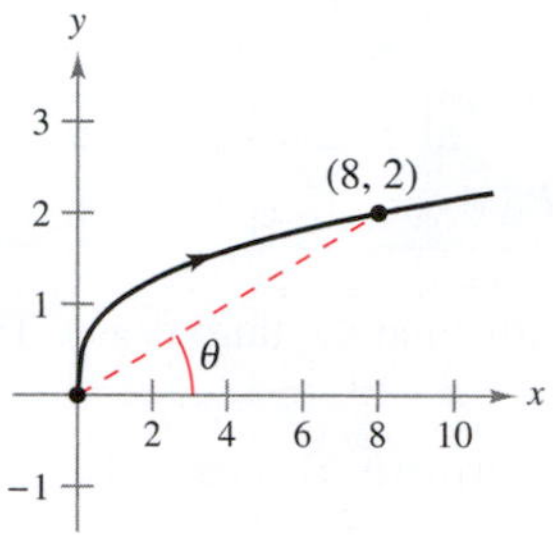

Figure for 10

10. Moving Point A particle is moving along the graph of $y = \sqrt[3]{x}$ (see figure). When $x = 8$, the y-component of the position of the particle is increasing at the rate of 1 centimeter per second.

(a) How fast is the x-component changing at this moment?

(b) How fast is the distance from the origin changing at this moment?

(c) How fast is the angle of inclination θ changing at this moment?

11. Projectile Motion An astronaut standing on the moon throws a rock upward. The height of the rock is

$$s = -\frac{27}{10}t^2 + 27t + 6$$

where s is measured in feet and t is measured in seconds.

(a) Find expressions for the velocity and acceleration of the rock.

(b) Find the time when the rock is at its highest point by finding the time when the velocity is zero. What is the height of the rock at this time?

(c) How does the acceleration of the rock compare with the acceleration due to gravity on Earth?

12. Proof Let E be a function satisfying $E(0) = E'(0) = 1$. Prove that if $E(a + b) = E(a)E(b)$ for all a and b, then E is differentiable and $E'(x) = E(x)$ for all x. Find an example of a function satisfying $E(a + b) = E(a)E(b)$.

13. Proof Let L be a differentiable function for all x. Prove that if $L(a + b) = L(a) + L(b)$ for all a and b, then $L'(x) = L'(0)$ for all x. What does the graph of L look like?

14. Radians and Degrees The fundamental limit

$$\lim_{x \to 0} \frac{\sin x}{x} = 1$$

assumes that x is measured in radians. Suppose you assume that x is measured in degrees instead of radians.

(a) Set your calculator to *degree* mode and complete the table.

z (in degrees)	0.1	0.01	0.0001
$\frac{\sin z}{z}$			

(b) Use the table to estimate

$$\lim_{z \to 0} \frac{\sin z}{z}$$

for z in degrees. What is the exact value of this limit? (*Hint:* $180° = \pi$ radians)

(c) Use the limit definition of the derivative to find

$$\frac{d}{dz} \sin z$$

for z in degrees.

(d) Define the new functions $S(z) = \sin(cz)$ and $C(z) = \cos(cz)$, where $c = \pi/180$. Find $S(90)$ and $C(180)$. Use the Chain Rule to calculate

$$\frac{d}{dz} S(z).$$

(e) Explain why calculus is made easier by using radians instead of degrees.

15. Acceleration and Jerk If a is the acceleration of an object, then the *jerk* j is defined by $j = a'(t)$.

(a) Use this definition to give a physical interpretation of j.

(b) Find j for the slowing vehicle in Exercise 117 in Section 2.3 and interpret the result.

(c) The figure shows the graphs of the position, velocity, acceleration, and jerk functions of a vehicle. Identify each graph and explain your reasoning.

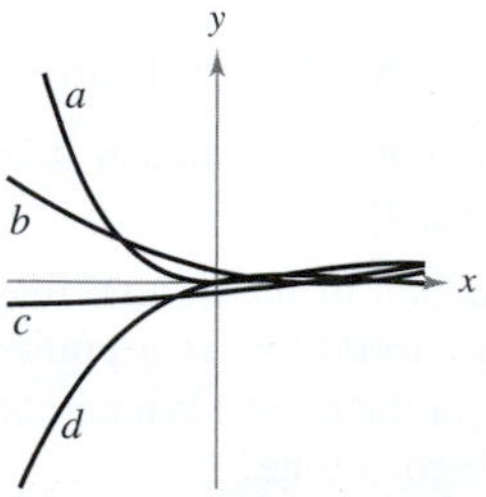

AP* Review Questions for Chapter 2

1. (no calculator)

Let $f(x) = (x^2 - 3)^4$.

(a) Write an equation of the line tangent to the graph of f at $x = 2$.

(b) Find the values of x for which the graph of f has a horizontal tangent.

(c) Find $f''(x)$.

2. (no calculator)

Let $f(x) = \sqrt{4x - 3}$ and $g(x) = \dfrac{f(x)}{x}$.

(a) What is the slope of the graph of f at $x = 3$? Show the work that leads to your answer.

(b) Write an equation of the line tangent to the graph of g at $x = 3$.

(c) What is the slope of the line normal to the graph of g at $x = 3$?

3. (no calculator)

Evaluate each limit analytically.

(*Note: Finding the answer should not involve a lengthy algebraic process.*)

(a) $\displaystyle\lim_{h\to 0} \frac{\sin(x + h) - \sin x}{h}$

(b) $\displaystyle\lim_{h\to 0} \frac{\sqrt[3]{x + h} - \sqrt[3]{x}}{h}$

(c) $\displaystyle\lim_{h\to 0} \frac{\sqrt{16 + h} - 4}{h}$

(d) $\displaystyle\lim_{h\to 0} \frac{\dfrac{1}{5 + h} - \dfrac{1}{5}}{h}$

4. Given:

x	$f(x)$	$f'(x)$	$g(x)$	$g'(x)$
2	-3	1	5	-2
5	4	7	-1	2

(a) If $h(x) = \dfrac{f(x)}{g(x)}$, find $h'(2)$.

(b) If $j(x) = f(g(x))$, find $j'(2)$.

(c) If $k(x) = \sqrt{f(x)}$, find $k'(5)$.

5. (no calculator)

Given: $f(x) = x^2$

(a) Find the slope of the normal line to the graph of f at $x = -3$.

(b) Two lines passing through the point $(3, 8)$ will be tangent to the graph of f. Find an equation for each of these lines.

6. The accompanying diagram shows the graph of the velocity in $\dfrac{\text{ft}}{\text{sec}}$ for a particle moving along the line $x = 4$.

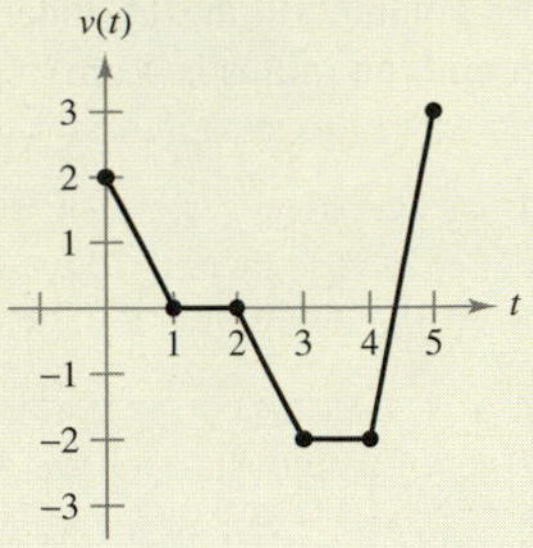

(a) During which time interval is the particle:

(i) moving upward?

(ii) moving downward?

(iii) at rest?

(b) State the acceleration of the particle at the specified times. Include units.

(i) $t = 0.75$

(ii) $t = 4.2$

7. (no calculator)

Given: $g(x) = f(x) \cdot \tan x + kx$, where k is a real number.

f is differentiable for all x; $f\left(\dfrac{\pi}{4}\right) = 4$; $f'\left(\dfrac{\pi}{4}\right) = -2$.

(a) For what values of x, if any, in the interval $0 < x < 2\pi$ will the derivative of g fail to exist? Justify your answer.

(b) If $g'\left(\dfrac{\pi}{4}\right) = 6$, find the value of k.

8. The table provided below shows the position of a particle, S, at several times, t, as the particle moves along a straight line, where t is measured in seconds and S is measured in meters.

t	2.0	2.7	3.2	3.8
$S(t)$	5.2	7.8	10.6	12.2

Which of the following best estimates the velocity of the particle at $t = 3$?

(a) $9.2\,\dfrac{\text{m}}{\text{s}}$ (b) $7.8\,\dfrac{\text{m}}{\text{s}}$ (c) $5.6\,\dfrac{\text{m}}{\text{s}}$

9. (no calculator)

If $2y^3 - 3xy + x^2 = 4$, then $\dfrac{dy}{dx} =$

(A) $-\dfrac{2x}{6y^2 - 3}$ (B) $\dfrac{2x - 3y}{3x - 6y^2}$ (C) $\dfrac{2x - 3}{6y^2}$

(D) $-\dfrac{2x}{6y^2 - 3x}$ (E) $\dfrac{3y - 2x}{6y^2}$

10. (no calculator)

The volume of a cylinder with radius r and height h is given by $V = \pi r^2 h$. The radius and height of the cylinder are increasing at constant rates. The radius is expanding at $\frac{1}{3}\frac{\text{cm}}{\text{sec}}$ and the height is increasing at $\frac{1}{2}\frac{\text{cm}}{\text{sec}}$. At what rate, in cubic *cm* per second, is the volume of the cylinder increasing when its height is 9 cm and the radius is 4 cm?

(A) 32π

(B) 6π

(C) $\frac{8\pi}{3}$

(D) $\frac{4\pi}{3}$

(E) $\frac{\pi}{18}$

3 Applications of Differentiation

Estimation of Error *(Example 3, p. 233)*

Offshore Oil Well *(Exercise 39, p. 222)*

Engine Efficiency *(Exercise 85, p. 204)*

Path of a Projectile *(Example 5, p. 182)*

Speed *(Exercise 57, p.175)*

3.1 Extrema on an Interval

- **Understand the definition of extrema of a function on an interval.**
- **Understand the definition of relative extrema of a function on an open interval.**
- **Find extrema on a closed interval.**

Extrema of a Function

In calculus, much effort is devoted to determining the behavior of a function f on an interval I. Does f have a maximum value on I? Does it have a minimum value? Where is the function increasing? Where is it decreasing? In this chapter, you will learn how derivatives can be used to answer these questions. You will also see why these questions are important in real-life applications.

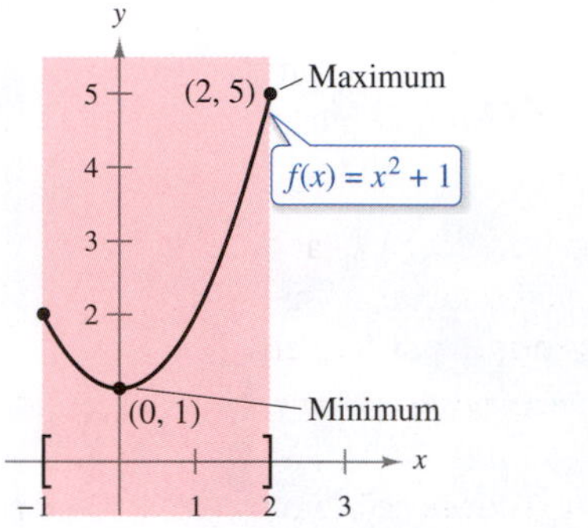

(a) f is continuous, $[-1, 2]$ is closed.

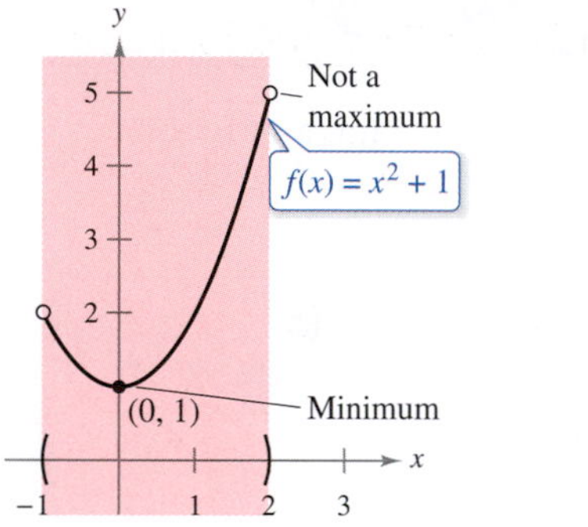

(b) f is continuous, $(-1, 2)$ is open.

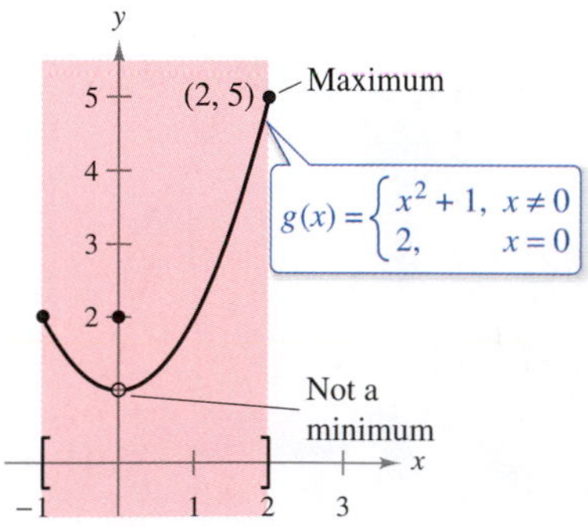

(c) g is not continuous, $[-1, 2]$ is closed.

Figure 3.1

Definition of Extrema

Let f be defined on an interval I containing c.

1. $f(c)$ is the **minimum of f on I** when $f(c) \leq f(x)$ for all x in I.
2. $f(c)$ is the **maximum of f on I** when $f(c) \geq f(x)$ for all x in I.

The minimum and maximum of a function on an interval are the **extreme values,** or **extrema** (the singular form of extrema is extremum), of the function on the interval. The minimum and maximum of a function on an interval are also called the **absolute minimum** and **absolute maximum,** or the **global minimum** and **global maximum,** on the interval. Extrema can occur at interior points or endpoints of an interval (see Figure 3.1). Extrema that occur at the endpoints are called **endpoint extrema.**

A function need not have a minimum or a maximum on an interval. For instance, in Figure 3.1(a) and (b), you can see that the function $f(x) = x^2 + 1$ has both a minimum and a maximum on the closed interval $[-1, 2]$, but does not have a maximum on the open interval $(-1, 2)$. Moreover, in Figure 3.1(c), you can see that continuity (or the lack of it) can affect the existence of an extremum on the interval. This suggests the theorem below. (Although the Extreme Value Theorem is intuitively plausible, a proof of this theorem is not within the scope of this text.)

THEOREM 3.1 The Extreme Value Theorem

If f is continuous on a closed interval $[a, b]$, then f has both a minimum and a maximum on the interval.

Exploration

Finding Minimum and Maximum Values The Extreme Value Theorem (like the Intermediate Value Theorem) is an *existence theorem* because it tells of the existence of minimum and maximum values but does not show how to find these values. Use the *minimum* and *maximum* features of a graphing utility to find the extrema of each function. In each case, do you think the x-values are exact or approximate? Explain your reasoning.

a. $f(x) = x^2 - 4x + 5$ on the closed interval $[-1, 3]$

b. $f(x) = x^3 - 2x^2 - 3x - 2$ on the closed interval $[-1, 3]$

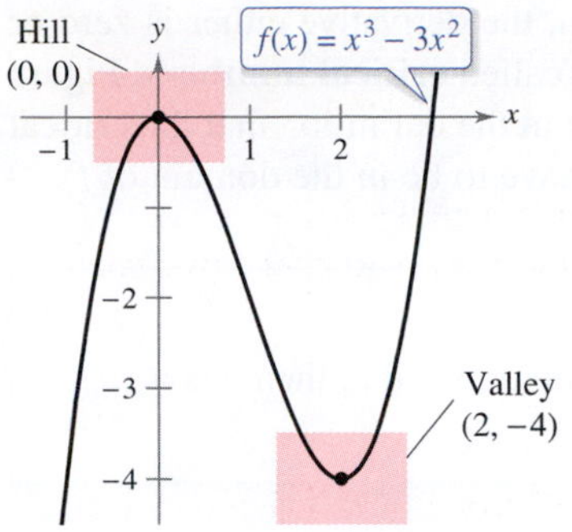

f has a relative maximum at (0, 0) and a relative minimum at (2, −4).
Figure 3.2

Relative Extrema and Critical Numbers

In Figure 3.2, the graph of $f(x) = x^3 - 3x^2$ has a **relative maximum** at the point (0, 0) and a **relative minimum** at the point (2, −4). Informally, for a continuous function, you can think of a relative maximum as occurring on a "hill" on the graph, and a relative minimum as occurring in a "valley" on the graph. Such a hill and valley can occur in two ways. When the hill (or valley) is smooth and rounded, the graph has a horizontal tangent line at the high point (or low point). When the hill (or valley) is sharp and peaked, the graph represents a function that is not differentiable at the high point (or low point).

Definition of Relative Extrema

1. If there is an open interval containing c on which $f(c)$ is a maximum, then $f(c)$ is called a **relative maximum** of f, or you can say that f has a **relative maximum at $(c, f(c))$.**
2. If there is an open interval containing c on which $f(c)$ is a minimum, then $f(c)$ is called a **relative minimum** of f, or you can say that f has a **relative minimum at $(c, f(c))$.**

The plural of relative maximum is relative maxima, and the plural of relative minimum is relative minima. Relative maximum and relative minimum are sometimes called **local maximum** and **local minimum,** respectively.

Example 1 examines the derivatives of functions at *given* relative extrema. (Much more is said about *finding* the relative extrema of a function in Section 3.3.)

EXAMPLE 1 The Value of the Derivative at Relative Extrema

Find the value of the derivative at each relative extremum shown in Figure 3.3.

Solution

a. The derivative of $f(x) = \dfrac{9(x^2 - 3)}{x^3}$ is

$$f'(x) = \frac{x^3(18x) - (9)(x^2 - 3)(3x^2)}{(x^3)^2} \qquad \text{Differentiate using Quotient Rule.}$$

$$= \frac{9(9 - x^2)}{x^4}. \qquad \text{Simplify.}$$

At the point (3, 2), the value of the derivative is $f'(3) = 0$ [see Figure 3.3(a)].

b. At $x = 0$, the derivative of $f(x) = |x|$ *does not exist* because the following one-sided limits differ [see Figure 3.3(b)].

$$\lim_{x \to 0^-} \frac{f(x) - f(0)}{x - 0} = \lim_{x \to 0^-} \frac{|x|}{x} = -1 \qquad \text{Limit from the left}$$

$$\lim_{x \to 0^+} \frac{f(x) - f(0)}{x - 0} = \lim_{x \to 0^+} \frac{|x|}{x} = 1 \qquad \text{Limit from the right}$$

c. The derivative of $f(x) = \sin x$ is

$$f'(x) = \cos x.$$

At the point $(\pi/2, 1)$, the value of the derivative is $f'(\pi/2) = \cos(\pi/2) = 0$. At the point $(3\pi/2, -1)$, the value of the derivative is $f'(3\pi/2) = \cos(3\pi/2) = 0$ [see Figure 3.3(c)].

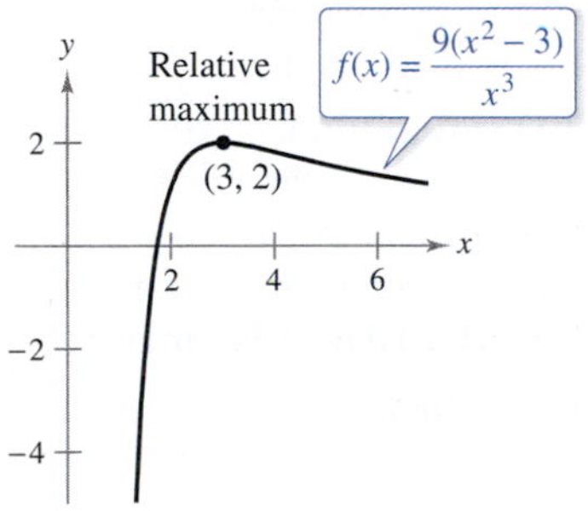

(a) $f'(3) = 0$

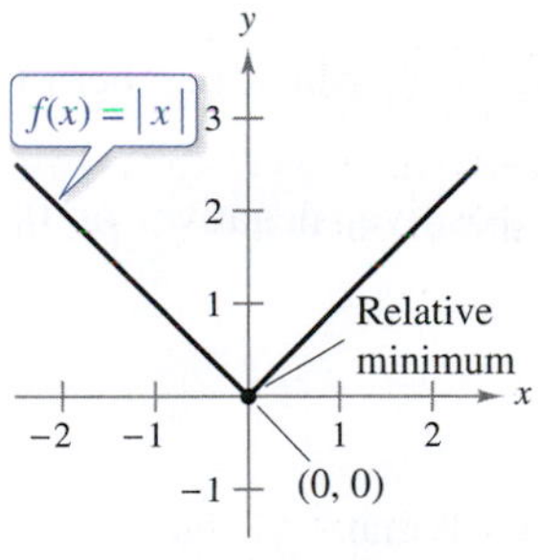

(b) $f'(0)$ does not exist.

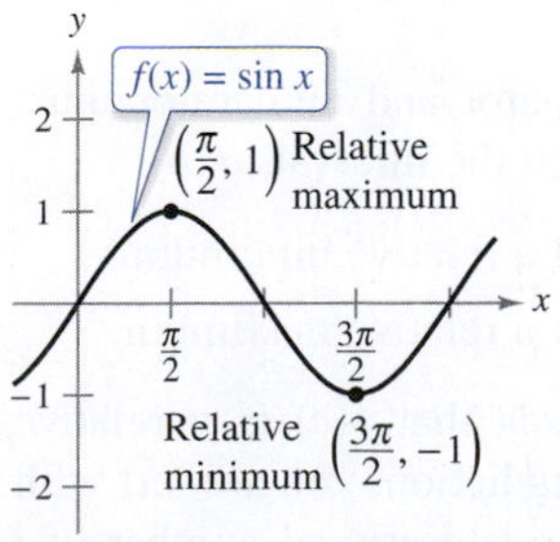

(c) $f'\left(\frac{\pi}{2}\right) = 0$; $f'\left(\frac{3\pi}{2}\right) = 0$
Figure 3.3

Note in Example 1 that at each relative extremum, the derivative either is zero or does not exist. The x-values at these special points are called **critical numbers.** Figure 3.4 illustrates the two types of critical numbers. Notice in the definition that the critical number c has to be in the domain of f, but c does not have to be in the domain of f'.

Definition of a Critical Number

Let f be defined at c. If $f'(c) = 0$ or if f is not differentiable at c, then c is a **critical number** of f.

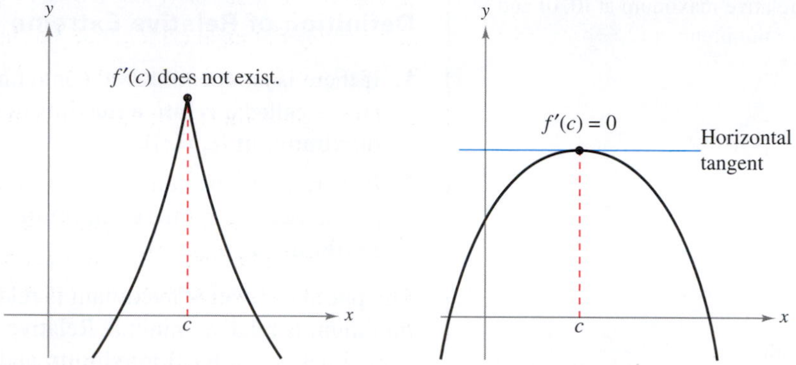

c is a critical number of f.
Figure 3.4

PIERRE DE FERMAT (1601–1665)

For Fermat, who was trained as a lawyer, mathematics was more of a hobby than a profession. Nevertheless, Fermat made many contributions to analytic geometry, number theory, calculus, and probability. In letters to friends, he wrote of many of the fundamental ideas of calculus, long before Newton or Leibniz. For instance, Theorem 3.2 is sometimes attributed to Fermat.
See LarsonCalculus.com to read more of this biography.

THEOREM 3.2 Relative Extrema Occur Only at Critical Numbers

If f has a relative minimum or relative maximum at $x = c$, then c is a critical number of f.

Proof

Case 1: If f is *not* differentiable at $x = c$, then, by definition, c is a critical number of f and the theorem is valid.

Case 2: If f is differentiable at $x = c$, then $f'(c)$ must be positive, negative, or 0. Suppose $f'(c)$ is positive. Then

$$f'(c) = \lim_{x \to c} \frac{f(x) - f(c)}{x - c} > 0$$

which implies that there exists an interval (a, b) containing c such that

$$\frac{f(x) - f(c)}{x - c} > 0, \text{ for all } x \neq c \text{ in } (a, b).$$

[See Exercise 78(b), Section 1.2.]

Because this quotient is positive, the signs of the denominator and numerator must agree. This produces the following inequalities for x-values in the interval (a, b).

Left of c: $x < c$ and $f(x) < f(c)$ ⇨ $f(c)$ is not a relative minimum.

Right of c: $x > c$ and $f(x) > f(c)$ ⇨ $f(c)$ is not a relative maximum.

So, the assumption that $f'(c) > 0$ contradicts the hypothesis that $f(c)$ is a relative extremum. Assuming that $f'(c) < 0$ produces a similar contradiction, you are left with only one possibility—namely, $f'(c) = 0$. So, by definition, c is a critical number of f and the theorem is valid.

See LarsonCalculus.com for Bruce Edwards's video of this proof. ■

The Print Collector/Alamy

Finding Extrema on a Closed Interval

Theorem 3.2 states that the relative extrema of a function can occur *only* at the critical numbers of the function. Knowing this, you can use the following guidelines to find extrema on a closed interval.

GUIDELINES FOR FINDING EXTREMA ON A CLOSED INTERVAL

To find the extrema of a continuous function f on a closed interval $[a, b]$, use these steps.

1. Find the critical numbers of f in (a, b).
2. Evaluate f at each critical number in (a, b).
3. Evaluate f at each endpoint of $[a, b]$.
4. The least of these values is the minimum. The greatest is the maximum.

The next three examples show how to apply these guidelines. Be sure you see that finding the critical numbers of the function is only part of the procedure. Evaluating the function at the critical numbers *and* the endpoints is the other part.

EXAMPLE 2 Finding Extrema on a Closed Interval

Find the extrema of

$$f(x) = 3x^4 - 4x^3$$

on the interval $[-1, 2]$.

Solution Begin by differentiating the function.

$f(x) = 3x^4 - 4x^3$ Write original function.

$f'(x) = 12x^3 - 12x^2$ Differentiate.

To find the critical numbers of f in the interval $(-1, 2)$, you must find all x-values for which $f'(x) = 0$ and all x-values for which $f'(x)$ does not exist.

$12x^3 - 12x^2 = 0$ Set $f'(x)$ equal to 0.

$12x^2(x - 1) = 0$ Factor.

$x = 0, 1$ Critical numbers

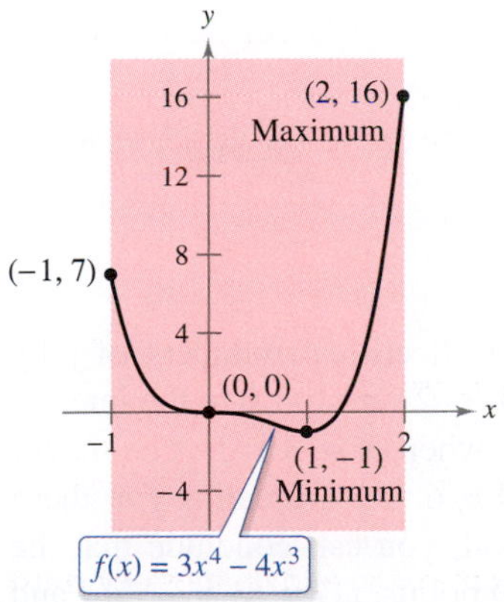

On the closed interval $[-1, 2]$, f has a minimum at $(1, -1)$ and a maximum at $(2, 16)$.
Figure 3.5

Because f' is defined for all x, you can conclude that these are the only critical numbers of f. By evaluating f at these two critical numbers and at the endpoints of $[-1, 2]$, you can determine that the maximum is $f(2) = 16$ and the minimum is $f(1) = -1$, as shown in the table. The graph of f is shown in Figure 3.5.

Left Endpoint	Critical Number	Critical Number	Right Endpoint
$f(-1) = 7$	$f(0) = 0$	$f(1) = -1$ Minimum	$f(2) = 16$ Maximum

In Figure 3.5, note that the critical number $x = 0$ does not yield a relative minimum or a relative maximum. This tells you that the converse of Theorem 3.2 is not true. In other words, *the critical numbers of a function need not produce relative extrema.*

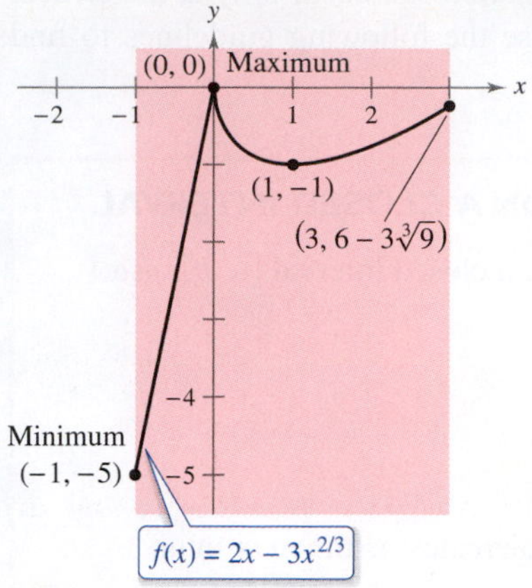

On the closed interval $[-1, 3]$, f has a minimum at $(-1, -5)$ and a maximum at $(0, 0)$.
Figure 3.6

EXAMPLE 3 Finding Extrema on a Closed Interval

Find the extrema of $f(x) = 2x - 3x^{2/3}$ on the interval $[-1, 3]$.

Solution Begin by differentiating the function.

$$f(x) = 2x - 3x^{2/3} \qquad \text{Write original function.}$$

$$f'(x) = 2 - \frac{2}{x^{1/3}} \qquad \text{Differentiate.}$$

$$= 2\left(\frac{x^{1/3} - 1}{x^{1/3}}\right) \qquad \text{Simplify.}$$

From this derivative, you can see that the function has two critical numbers in the interval $(-1, 3)$. The number 1 is a critical number because $f'(1) = 0$, and the number 0 is a critical number because $f'(0)$ does not exist. By evaluating f at these two numbers and at the endpoints of the interval, you can conclude that the minimum is $f(-1) = -5$ and the maximum is $f(0) = 0$, as shown in the table. The graph of f is shown in Figure 3.6.

Left Endpoint	Critical Number	Critical Number	Right Endpoint
$f(-1) = -5$ Minimum	$f(0) = 0$ Maximum	$f(1) = -1$	$f(3) = 6 - 3\sqrt[3]{9} \approx -0.24$

AP* Tips

If an AP question asks for an absolute extremum of a function on a closed interval, be sure to check the y-coordinates at the endpoints of the interval.

EXAMPLE 4 Finding Extrema on a Closed Interval

See LarsonCalculus.com for an interactive version of this type of example.

Find the extrema of

$$f(x) = 2 \sin x - \cos 2x$$

on the interval $[0, 2\pi]$.

Solution Begin by differentiating the function.

$$f(x) = 2 \sin x - \cos 2x \qquad \text{Write original function.}$$

$$f'(x) = 2 \cos x + 2 \sin 2x \qquad \text{Differentiate.}$$

$$= 2 \cos x + 4 \cos x \sin x \qquad \sin 2x = 2 \cos x \sin x$$

$$= 2(\cos x)(1 + 2 \sin x) \qquad \text{Factor.}$$

Because f is differentiable for all real x, you can find all critical numbers of f by finding the zeros of its derivative. Considering $2(\cos x)(1 + 2 \sin x) = 0$ in the interval $(0, 2\pi)$, the factor $\cos x$ is zero when $x = \pi/2$ and when $x = 3\pi/2$. The factor $(1 + 2 \sin x)$ is zero when $x = 7\pi/6$ and when $x = 11\pi/6$. By evaluating f at these four critical numbers and at the endpoints of the interval, you can conclude that the maximum is $f(\pi/2) = 3$ and the minimum occurs at *two* points, $f(7\pi/6) = -3/2$ and $f(11\pi/6) = -3/2$, as shown in the table. The graph is shown in Figure 3.7.

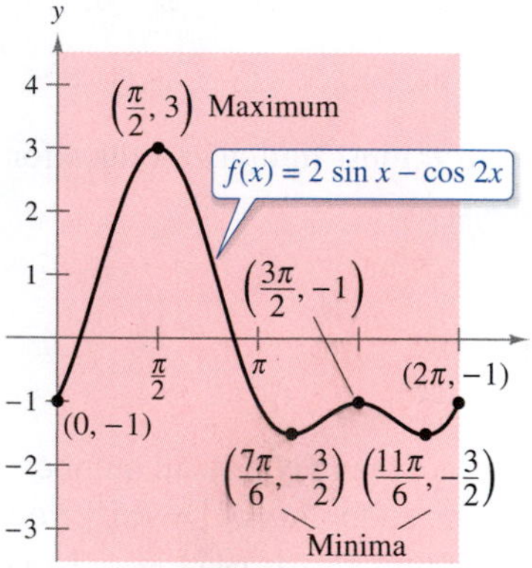

On the closed interval $[0, 2\pi]$, f has two minima at $(7\pi/6, -3/2)$ and $(11\pi/6, -3/2)$ and a maximum at $(\pi/2, 3)$.
Figure 3.7

Left Endpoint	Critical Number	Critical Number	Critical Number	Critical Number	Right Endpoint
$f(0) = -1$	$f\left(\frac{\pi}{2}\right) = 3$ Maximum	$f\left(\frac{7\pi}{6}\right) = -\frac{3}{2}$ Minimum	$f\left(\frac{3\pi}{2}\right) = -1$	$f\left(\frac{11\pi}{6}\right) = -\frac{3}{2}$ Minimum	$f(2\pi) = -1$

3.1 Exercises

See CalcChat.com for tutorial help and worked-out solutions to odd-numbered exercises.

Finding the Value of the Derivative at Relative Extrema

In Exercises 1–6, find the value of the derivative (if it exists) at each indicated extremum.

1. $f(x) = \dfrac{x^2}{x^2 + 4}$

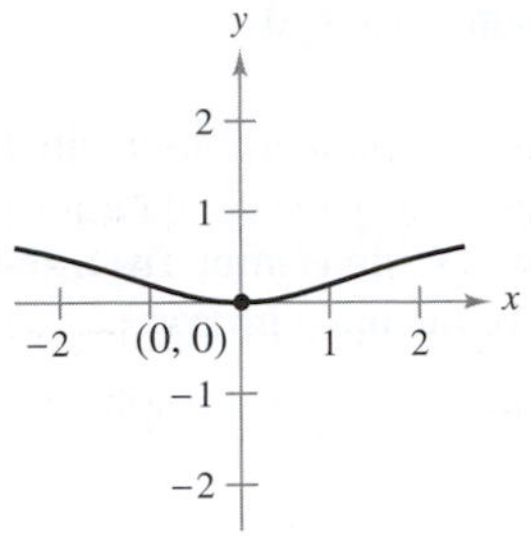

2. $f(x) = \cos \dfrac{\pi x}{2}$

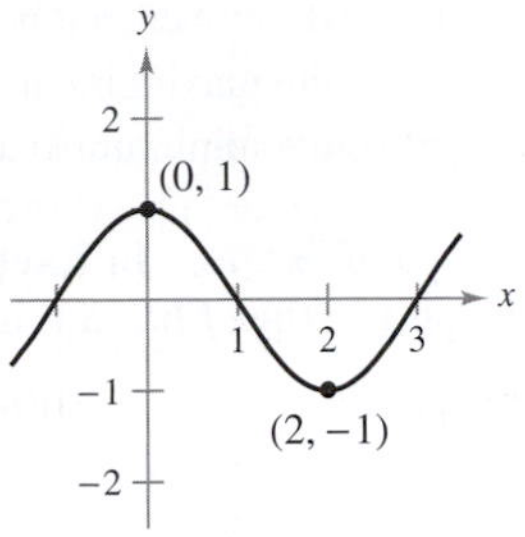

3. $g(x) = x + \dfrac{4}{x^2}$

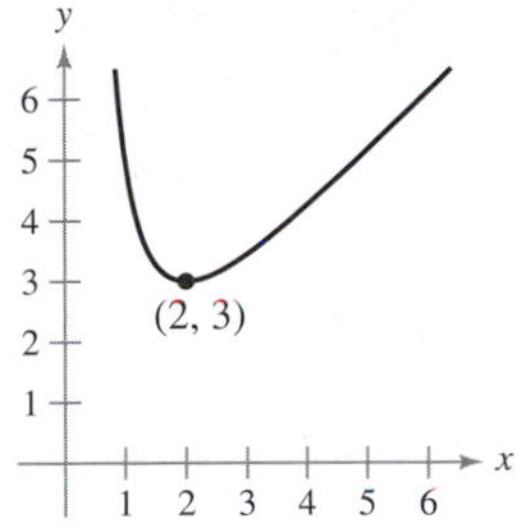

4. $f(x) = -3x\sqrt{x + 1}$

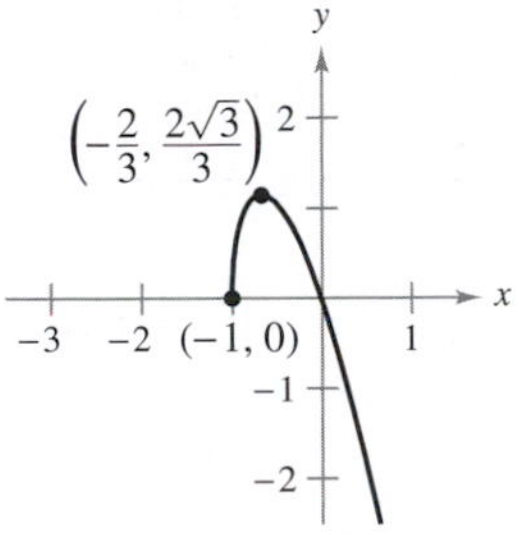

5. $f(x) = (x + 2)^{2/3}$

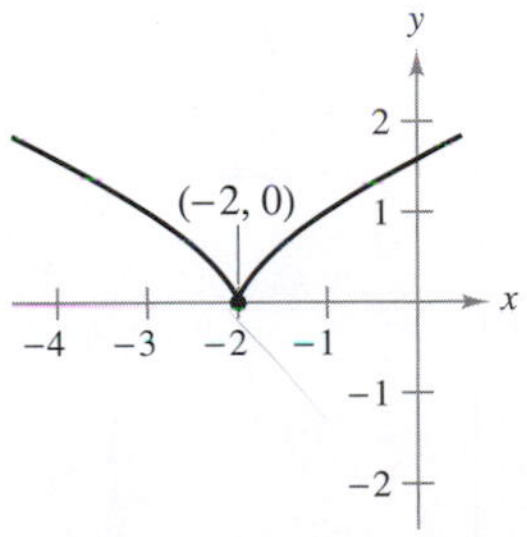

6. $f(x) = 4 - |x|$

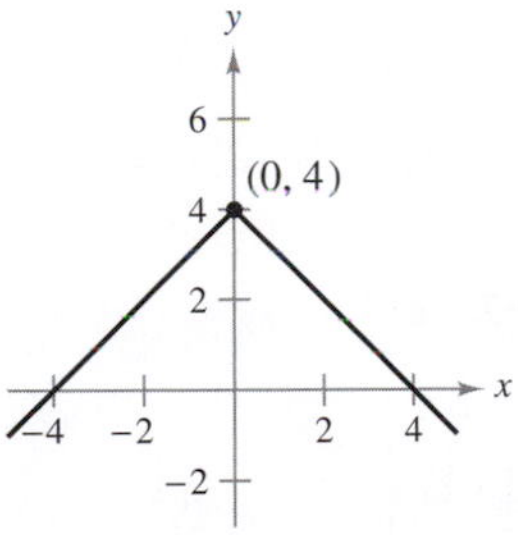

Approximating Critical Numbers

In Exercises 7–10, approximate the critical numbers of the function shown in the graph. Determine whether the function has a relative maximum, a relative minimum, an absolute maximum, an absolute minimum, or none of these at each critical number on the interval shown.

7.

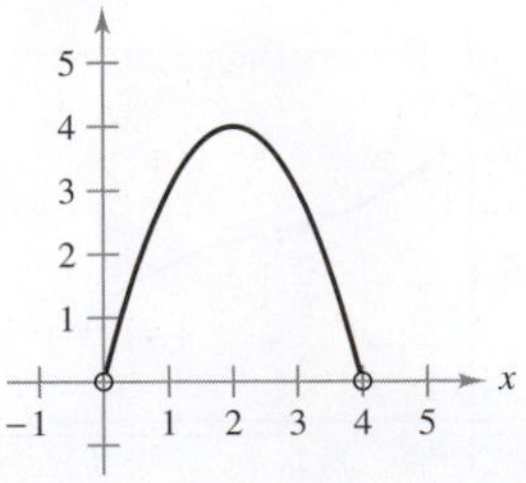

8.

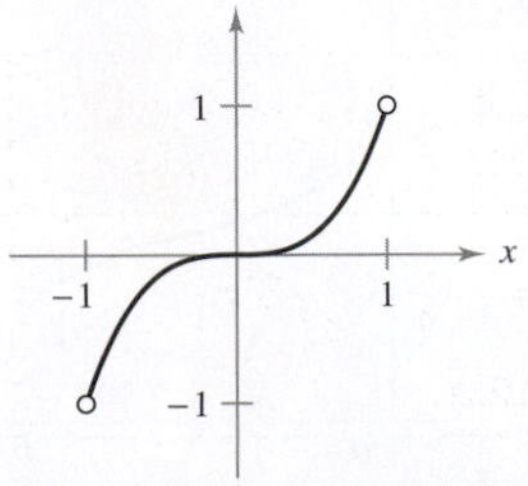

9.

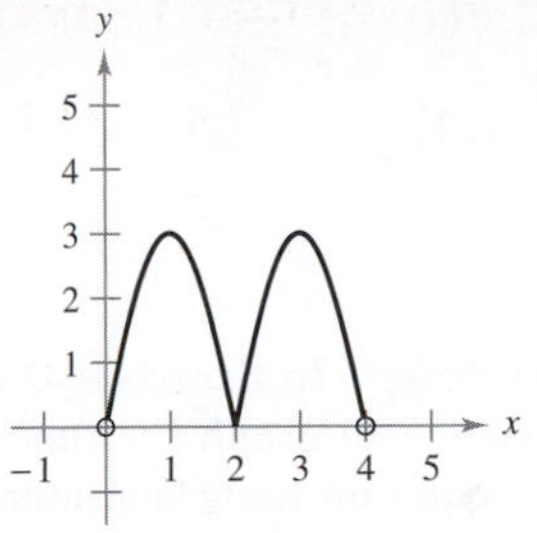

10.

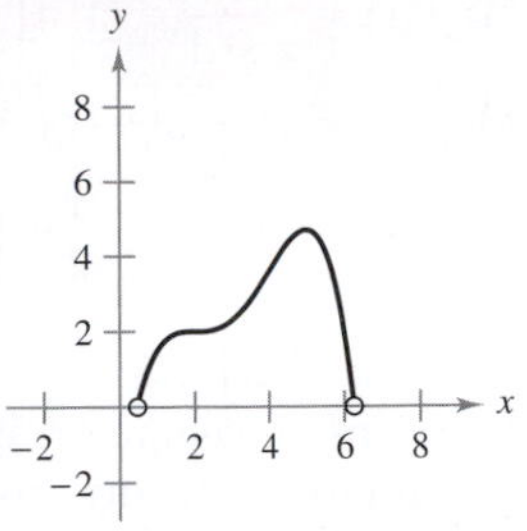

Finding Critical Numbers

In Exercises 11–16, find the critical numbers of the function.

11. $f(x) = x^3 - 3x^2$

12. $g(x) = x^4 - 8x^2$

13. $g(t) = t\sqrt{4 - t},\ t < 3$

14. $f(x) = \dfrac{4x}{x^2 + 1}$

15. $h(x) = \sin^2 x + \cos x$
$0 < x < 2\pi$

16. $f(\theta) = 2 \sec \theta + \tan \theta$
$0 < \theta < 2\pi$

Finding Extrema on a Closed Interval

In Exercises 17–36, find the absolute extrema of the function on the closed interval.

17. $f(x) = 3 - x,\ [-1, 2]$

18. $f(x) = \dfrac{3}{4}x + 2,\ [0, 4]$

19. $g(x) = 2x^2 - 8x,\ [0, 6]$

20. $h(x) = 5 - x^2,\ [-3, 1]$

21. $f(x) = x^3 - \dfrac{3}{2}x^2,\ [-1, 2]$

22. $f(x) = 2x^3 - 6x,\ [0, 3]$

23. $y = 3x^{2/3} - 2x,\ [-1, 1]$

24. $g(x) = \sqrt[3]{x},\ [-8, 8]$

25. $g(t) = \dfrac{t^2}{t^2 + 3},\ [-1, 1]$

26. $f(x) = \dfrac{2x}{x^2 + 1},\ [-2, 2]$

27. $h(s) = \dfrac{1}{s - 2},\ [0, 1]$

28. $h(t) = \dfrac{t}{t + 3},\ [-1, 6]$

29. $y = 3 - |t - 3|, [-1, 5]$

30. $g(x) = |x + 4|,\ [-7, 1]$

31. $f(x) = [\![x]\!],\ [-2, 2]$

32. $h(x) = [\![2 - x]\!],\ [-2, 2]$

33. $f(x) = \sin x,\ \left[\dfrac{5\pi}{6}, \dfrac{11\pi}{6}\right]$

34. $g(x) = \sec x,\ \left[-\dfrac{\pi}{6}, \dfrac{\pi}{3}\right]$

35. $y = 3 \cos x,\ [0, 2\pi]$

36. $y = \tan\left(\dfrac{\pi x}{8}\right),\ [0, 2]$

Finding Extrema on an Interval

In Exercises 37–40, find the absolute extrema of the function (if any exist) on each interval.

37. $f(x) = 2x - 3$
(a) $[0, 2]$ (b) $[0, 2)$
(c) $(0, 2]$ (d) $(0, 2)$

38. $f(x) = 5 - x$
(a) $[1, 4]$ (b) $[1, 4)$
(c) $(1, 4]$ (d) $(1, 4)$

39. $f(x) = x^2 - 2x$
(a) $[-1, 2]$ (b) $(1, 3]$
(c) $(0, 2)$ (d) $[1, 4)$

40. $f(x) = \sqrt{4 - x^2}$
(a) $[-2, 2]$ (b) $[-2, 0)$
(c) $(-2, 2)$ (d) $[1, 2)$

Finding Absolute Extrema In Exercises 41–44, use a graphing utility to graph the function and find the absolute extrema of the function on the given interval.

41. $f(x) = \dfrac{3}{x-1}, \quad (1, 4]$

42. $f(x) = \dfrac{2}{2-x}, \quad [0, 2)$

43. $f(x) = x^4 - 2x^3 + x + 1, \quad [-1, 3]$

44. $f(x) = \sqrt{x} + \cos\dfrac{x}{2}, \quad [0, 2\pi]$

Finding Extrema Using Technology In Exercises 45 and 46, (a) use a computer algebra system to graph the function and approximate any absolute extrema on the given interval. (b) Use the utility to find any critical numbers, and use them to find any absolute extrema not located at the endpoints. Compare the results with those in part (a).

45. $f(x) = 3.2x^5 + 5x^3 - 3.5x, \quad [0, 1]$

46. $f(x) = \dfrac{4}{3}x\sqrt{3-x}, \quad [0, 3]$

Finding Maximum Values Using Technology In Exercises 47 and 48, use a computer algebra system to find the maximum value of $|f''(x)|$ on the closed interval. (This value is used in the error estimate for the Trapezoidal Rule, as discussed in Section 4.6.)

47. $f(x) = \sqrt{1 + x^3}, \quad [0, 2]$

48. $f(x) = \dfrac{1}{x^2 + 1}, \quad \left[\dfrac{1}{2}, 3\right]$

Finding Maximum Values Using Technology In Exercises 49 and 50, use a computer algebra system to find the maximum value of $|f^{(4)}(x)|$ on the closed interval. (This value is used in the error estimate for Simpson's Rule, as discussed in Section 4.6.)

49. $f(x) = (x + 1)^{2/3}, \quad [0, 2]$

50. $f(x) = \dfrac{1}{x^2 + 1}, \quad [-1, 1]$

51. Writing Write a short paragraph explaining why a continuous function on an open interval may not have a maximum or minimum. Illustrate your explanation with a sketch of the graph of such a function.

52. HOW DO YOU SEE IT? Determine whether each labeled point is an absolute maximum or minimum, a relative maximum or minimum, or none of these.

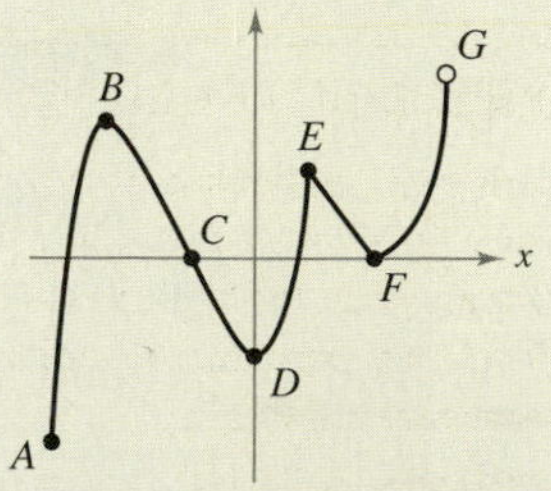

WRITING ABOUT CONCEPTS

Creating the Graph of a Function In Exercises 53 and 54, graph a function on the interval $[-2, 5]$ having the given characteristics.

53. Absolute maximum at $x = -2$
Absolute minimum at $x = 1$
Relative maximum at $x = 3$

54. Relative minimum at $x = -1$
Critical number (but no extremum) at $x = 0$
Absolute maximum at $x = 2$
Absolute minimum at $x = 5$

Using Graphs In Exercises 55–58, determine from the graph whether f has a minimum in the open interval (a, b).

55. (a)

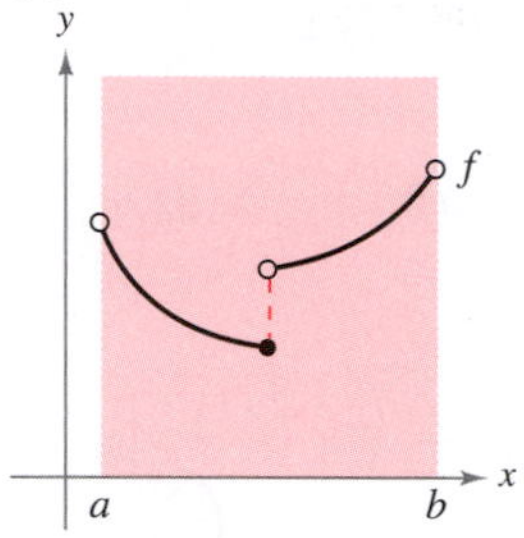

(b)

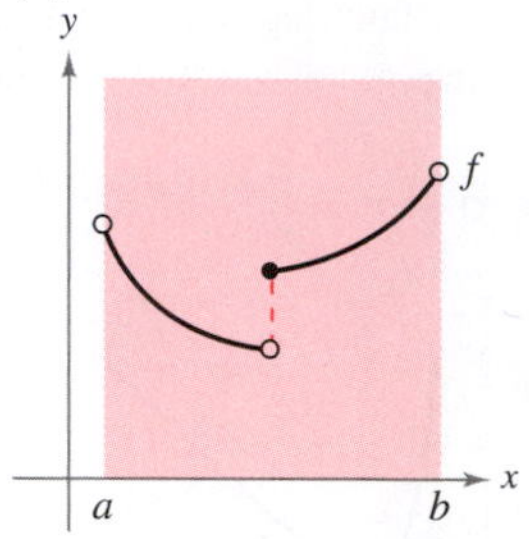

56. (a)

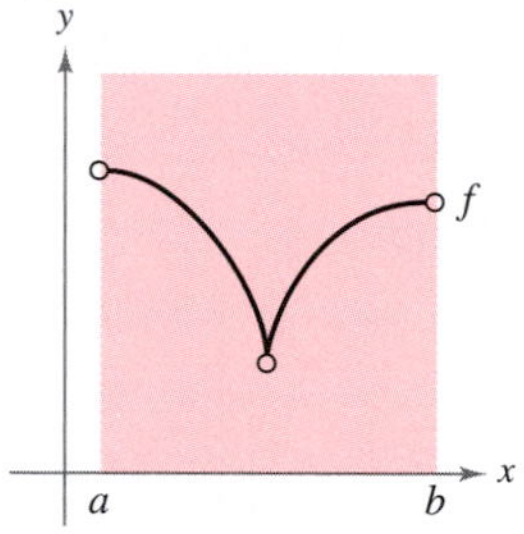

(b)

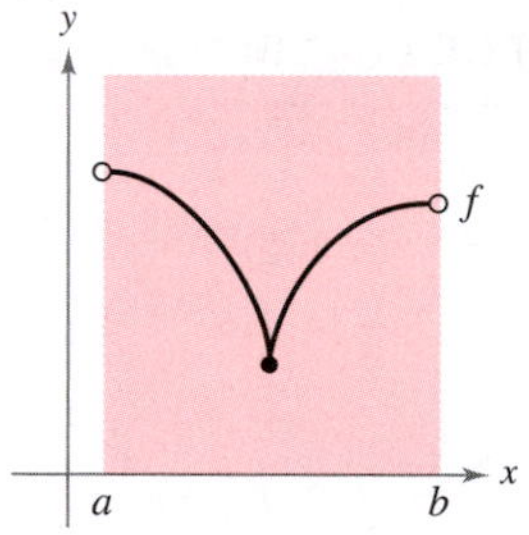

57. (a)

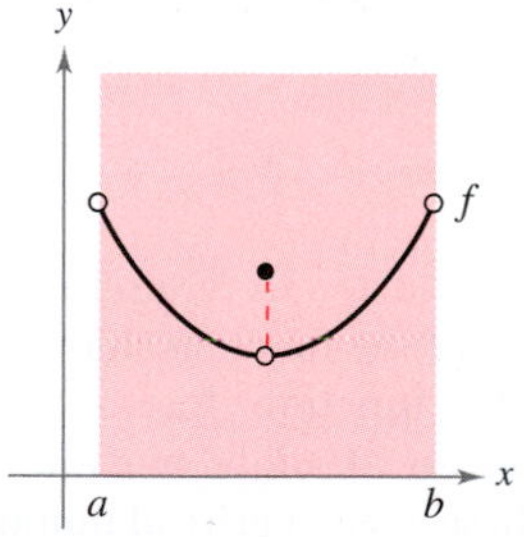

(b)

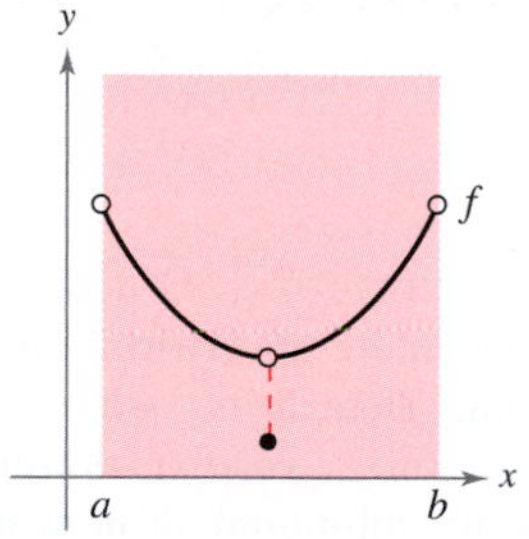

58. (a)

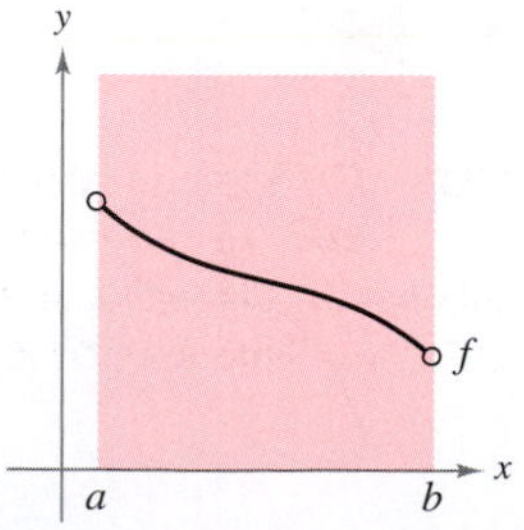

(b)

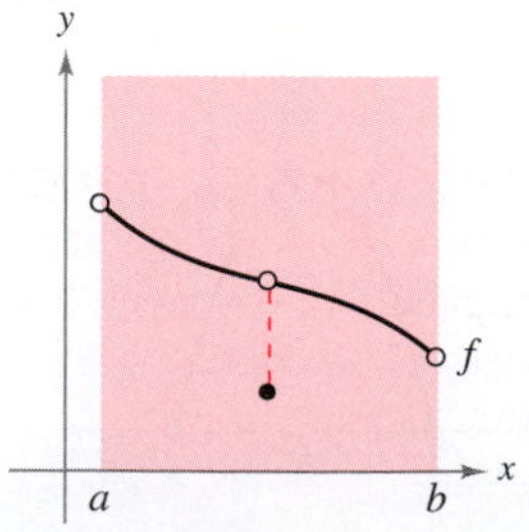

59. Power The formula for the power output P of a battery is

$$P = VI - RI^2$$

where V is the electromotive force in volts, R is the resistance in ohms, and I is the current in amperes. Find the current that corresponds to a maximum value of P in a battery for which $V = 12$ volts and $R = 0.5$ ohm. Assume that a 15-ampere fuse bounds the output in the interval $0 \le I \le 15$. Could the power output be increased by replacing the 15-ampere fuse with a 20-ampere fuse? Explain.

60. Lawn Sprinkler A lawn sprinkler is constructed in such a way that $d\theta/dt$ is constant, where θ ranges between 45° and 135° (see figure). The distance the water travels horizontally is

$$x = \frac{v^2 \sin 2\theta}{32}, \quad 45° \le \theta \le 135°$$

where v is the speed of the water. Find dx/dt and explain why this lawn sprinkler does not water evenly. What part of the lawn receives the most water?

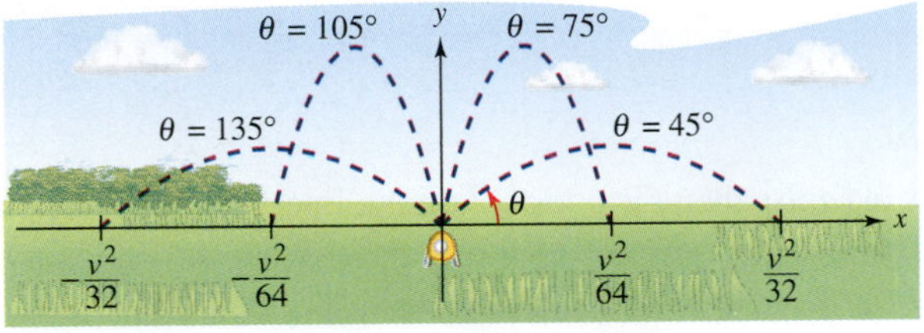

Water sprinkler: $45° \le \theta \le 135°$

FOR FURTHER INFORMATION For more information on the "calculus of lawn sprinklers," see the article "Design of an Oscillating Sprinkler" by Bart Braden in *Mathematics Magazine*. To view this article, go to *MathArticles.com*.

61. Honeycomb The surface area of a cell in a honeycomb is

$$S = 6hs + \frac{3s^2}{2}\left(\frac{\sqrt{3} - \cos\theta}{\sin\theta}\right)$$

where h and s are positive constants and θ is the angle at which the upper faces meet the altitude of the cell (see figure). Find the angle θ $(\pi/6 \le \theta \le \pi/2)$ that minimizes the surface area S.

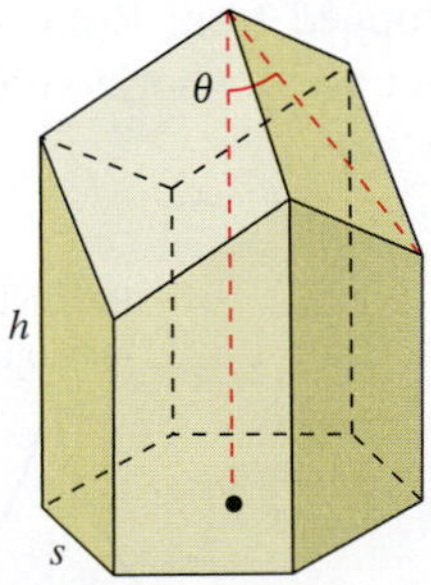

FOR FURTHER INFORMATION For more information on the geometric structure of a honeycomb cell, see the article "The Design of Honeycombs" by Anthony L. Peressini in UMAP Module 502, published by COMAP, Inc., Suite 210, 57 Bedford Street, Lexington, MA.

62. Highway Design In order to build a highway, it is necessary to fill a section of a valley where the grades (slopes) of the sides are 9% and 6% (see figure). The top of the filled region will have the shape of a parabolic arc that is tangent to the two slopes at the points A and B. The horizontal distances from A to the y-axis and from B to the y-axis are both 500 feet.

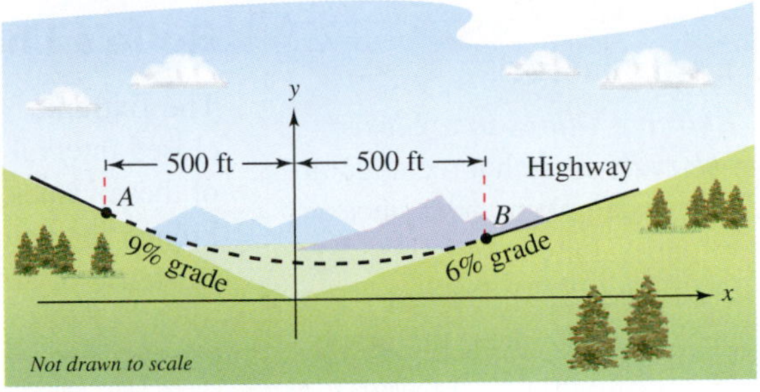

(a) Find the coordinates of A and B.

(b) Find a quadratic function $y = ax^2 + bx + c$ for $-500 \le x \le 500$ that describes the top of the filled region.

(c) Construct a table giving the depths d of the fill for $x = -500, -400, -300, -200, -100, 0, 100, 200, 300, 400$, and 500.

(d) What will be the lowest point on the completed highway? Will it be directly over the point where the two hillsides come together?

True or False? **In Exercises 63–66, determine whether the statement is true or false. If it is false, explain why or give an example that shows it is false.**

63. The maximum of a function that is continuous on a closed interval can occur at two different values in the interval.

64. If a function is continuous on a closed interval, then it must have a minimum on the interval.

65. If $x = c$ is a critical number of the function f, then it is also a critical number of the function $g(x) = f(x) + k$, where k is a constant.

66. If $x = c$ is a critical number of the function f, then it is also a critical number of the function $g(x) = f(x - k)$, where k is a constant.

67. Functions Let the function f be differentiable on an interval I containing c. If f has a maximum value at $x = c$, show that $-f$ has a minimum value at $x = c$.

68. Critical Numbers Consider the cubic function $f(x) = ax^3 + bx^2 + cx + d$, where $a \ne 0$. Show that f can have zero, one, or two critical numbers and give an example of each case.

PUTNAM EXAM CHALLENGE

69. Determine all real numbers $a > 0$ for which there exists a nonnegative continuous function $f(x)$ defined on $[0, a]$ with the property that the region $R = \{(x, y); 0 \le x \le a, 0 \le y \le f(x)\}$ has perimeter k units and area k square units for some real number k.

This problem was composed by the Committee on the Putnam Prize Competition.

3.2 Rolle's Theorem and the Mean Value Theorem

- Understand and use Rolle's Theorem.
- Understand and use the Mean Value Theorem.

Exploration

Extreme Values in a Closed Interval Sketch a rectangular coordinate plane on a piece of paper. Label the points $(1, 3)$ and $(5, 3)$. Using a pencil or pen, draw the graph of a differentiable function f that starts at $(1, 3)$ and ends at $(5, 3)$. Is there at least one point on the graph for which the derivative is zero? Would it be possible to draw the graph so that there *isn't* a point for which the derivative is zero? Explain your reasoning.

Rolle's Theorem

The Extreme Value Theorem (see Section 3.1) states that a continuous function on a closed interval $[a, b]$ must have both a minimum and a maximum on the interval. Both of these values, however, can occur at the endpoints. **Rolle's Theorem,** named after the French mathematician Michel Rolle (1652–1719), gives conditions that guarantee the existence of an extreme value in the *interior* of a closed interval.

THEOREM 3.3 Rolle's Theorem

Let f be continuous on the closed interval $[a, b]$ and differentiable on the open interval (a, b). If $f(a) = f(b)$, then there is at least one number c in (a, b) such that $f'(c) = 0$.

Proof Let $f(a) = d = f(b)$.

Case 1: If $f(x) = d$ for all x in $[a, b]$, then f is constant on the interval and, by Theorem 2.2, $f'(x) = 0$ for all x in (a, b).

Case 2: Consider $f(x) > d$ for some x in (a, b). By the Extreme Value Theorem, you know that f has a maximum at some c in the interval. Moreover, because $f(c) > d$, this maximum does not occur at either endpoint. So, f has a maximum in the *open* interval (a, b). This implies that $f(c)$ is a *relative* maximum and, by Theorem 3.2, c is a critical number of f. Finally, because f is differentiable at c, you can conclude that $f'(c) = 0$.

Case 3: When $f(x) < d$ for some x in (a, b), you can use an argument similar to that in Case 2, but involving the minimum instead of the maximum.

See LarsonCalculus.com for Bruce Edwards's video of this proof. ■

ROLLE'S THEOREM

French mathematician Michel Rolle first published the theorem that bears his name in 1691. Before this time, however, Rolle was one of the most vocal critics of calculus, stating that it gave erroneous results and was based on unsound reasoning. Later in life, Rolle came to see the usefulness of calculus.

From Rolle's Theorem, you can see that if a function f is continuous on $[a, b]$ and differentiable on (a, b), and if $f(a) = f(b)$, then there must be at least one x-value between a and b at which the graph of f has a horizontal tangent [see Figure 3.8(a)]. When the differentiability requirement is dropped from Rolle's Theorem, f will still have a critical number in (a, b), but it may not yield a horizontal tangent. Such a case is shown in Figure 3.8(b).

AP* Tips

On some AP free response questions, there may be more than one way of applying derivatives and the theorems of Chapters 1 and 3 to justify your answer.

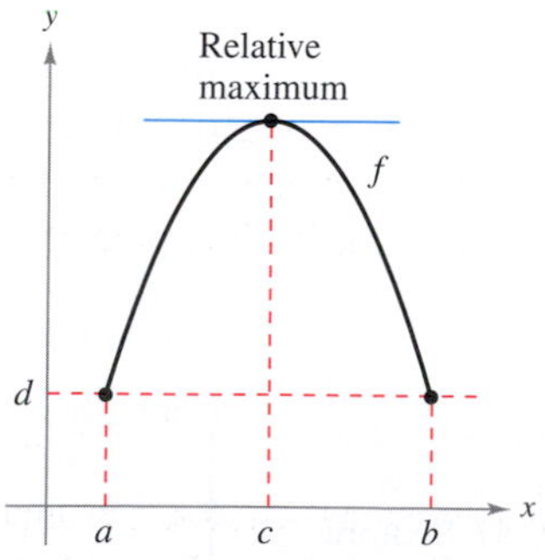

(a) f is continuous on $[a, b]$ and differentiable on (a, b).

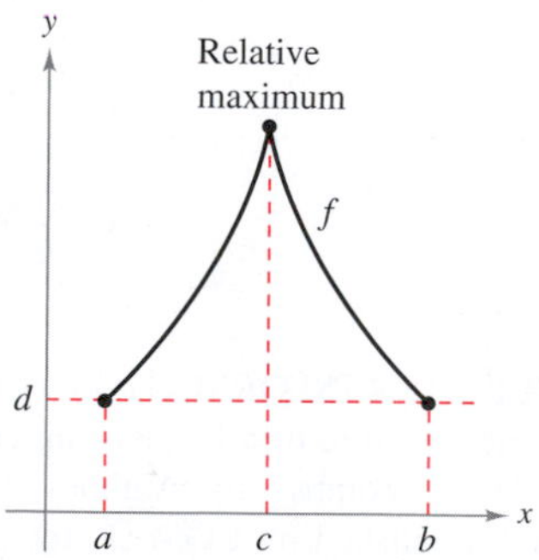

(b) f is continuous on $[a, b]$.

Figure 3.8

EXAMPLE 1 **Illustrating Rolle's Theorem**

Find the two x-intercepts of

$$f(x) = x^2 - 3x + 2$$

and show that $f'(x) = 0$ at some point between the two x-intercepts.

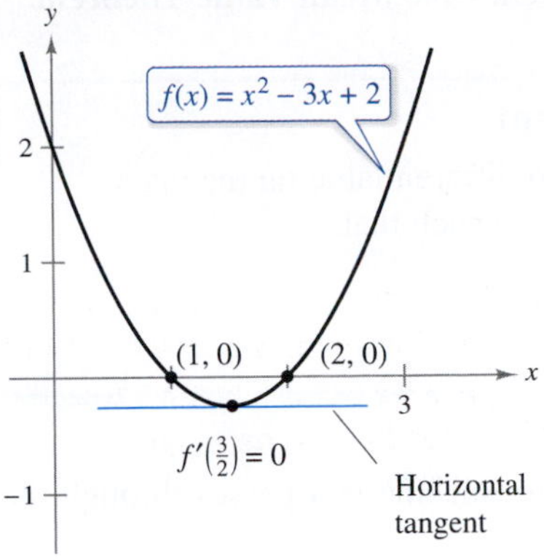

The x-value for which $f'(x) = 0$ is between the two x-intercepts.
Figure 3.9

Solution Note that f is differentiable on the entire real number line. Setting $f(x)$ equal to 0 produces

$$x^2 - 3x + 2 = 0 \quad \text{Set } f(x) \text{ equal to 0.}$$
$$(x - 1)(x - 2) = 0 \quad \text{Factor.}$$
$$x = 1, 2. \quad x\text{-values for which } f'(x) = 0$$

So, $f(1) = f(2) = 0$, and from Rolle's Theorem you know that there *exists* at least one c in the interval $(1, 2)$ such that $f'(c) = 0$. To *find* such a c, differentiate f to obtain

$$f'(x) = 2x - 3 \quad \text{Differentiate.}$$

and then determine that $f'(x) = 0$ when $x = \frac{3}{2}$. Note that this x-value lies in the open interval $(1, 2)$, as shown in Figure 3.9. ■

Rolle's Theorem states that when f satisfies the conditions of the theorem, there must be *at least* one point between a and b at which the derivative is 0. There may, of course, be more than one such point, as shown in the next example.

EXAMPLE 2 **Illustrating Rolle's Theorem**

Let $f(x) = x^4 - 2x^2$. Find all values of c in the interval $(-2, 2)$ such that $f'(c) = 0$.

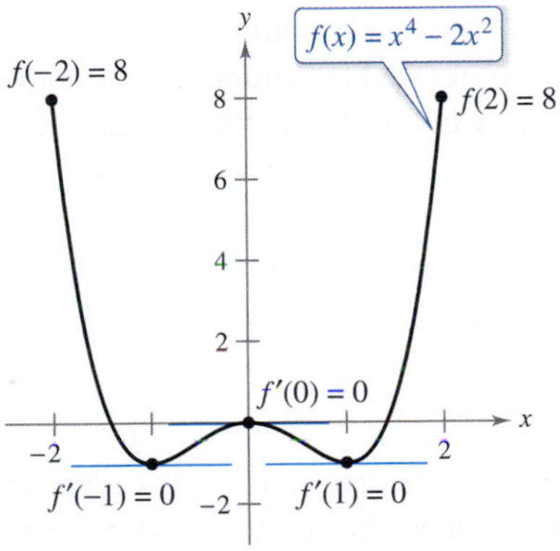

$f'(x) = 0$ for more than one x-value in the interval $(-2, 2)$.
Figure 3.10

Solution To begin, note that the function satisfies the conditions of Rolle's Theorem. That is, f is continuous on the interval $[-2, 2]$ and differentiable on the interval $(-2, 2)$. Moreover, because $f(-2) = f(2) = 8$, you can conclude that there exists at least one c in $(-2, 2)$ such that $f'(c) = 0$. Because

$$f'(x) = 4x^3 - 4x \quad \text{Differentiate.}$$

setting the derivative equal to 0 produces

$$4x^3 - 4x = 0 \quad \text{Set } f'(x) \text{ equal to 0.}$$
$$4x(x - 1)(x + 1) = 0 \quad \text{Factor.}$$
$$x = 0, 1, -1. \quad x\text{-values for which } f'(x) = 0$$

So, in the interval $(-2, 2)$, the derivative is zero at three different values of x, as shown in Figure 3.10. ■

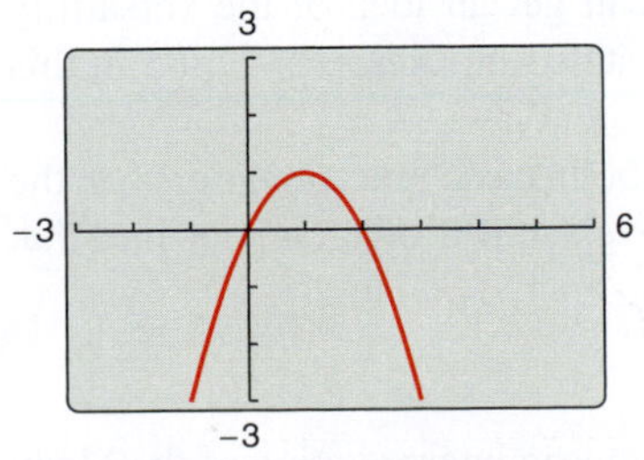

Figure 3.11

TECHNOLOGY PITFALL A graphing utility can be used to indicate whether the points on the graphs in Examples 1 and 2 are relative minima or relative maxima of the functions. When using a graphing utility, however, you should keep in mind that it can give misleading pictures of graphs. For example, use a graphing utility to graph

$$f(x) = 1 - (x - 1)^2 - \frac{1}{1000(x - 1)^{1/7} + 1}.$$

With most viewing windows, it appears that the function has a maximum of 1 when $x = 1$ (see Figure 3.11). By evaluating the function at $x = 1$, however, you can see that $f(1) = 0$. To determine the behavior of this function near $x = 1$, you need to examine the graph analytically to get the complete picture.

The Mean Value Theorem

Rolle's Theorem can be used to prove another theorem—the **Mean Value Theorem.**

REMARK The "mean" in the Mean Value Theorem refers to the mean (or average) rate of change of f on the interval $[a, b]$.

THEOREM 3.4 The Mean Value Theorem

If f is continuous on the closed interval $[a, b]$ and differentiable on the open interval (a, b), then there exists a number c in (a, b) such that

$$f'(c) = \frac{f(b) - f(a)}{b - a}.$$

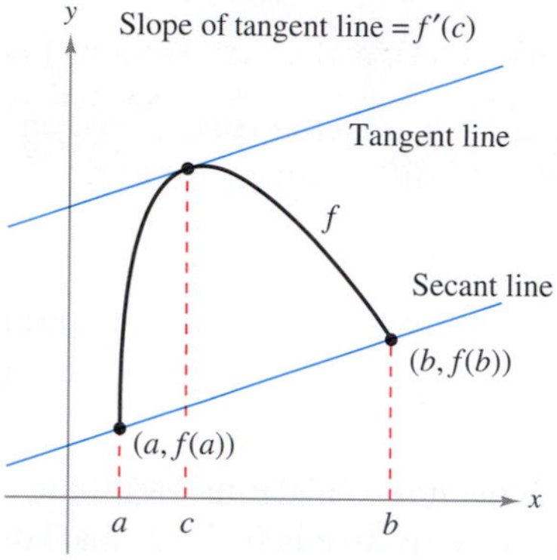

Figure 3.12

Proof Refer to Figure 3.12. The equation of the secant line that passes through the points $(a, f(a))$ and $(b, f(b))$ is

$$y = \left[\frac{f(b) - f(a)}{b - a}\right](x - a) + f(a).$$

Let $g(x)$ be the difference between $f(x)$ and y. Then

$$\begin{aligned} g(x) &= f(x) - y \\ &= f(x) - \left[\frac{f(b) - f(a)}{b - a}\right](x - a) - f(a). \end{aligned}$$

By evaluating g at a and b, you can see that

$$g(a) = 0 = g(b).$$

Because f is continuous on $[a, b]$, it follows that g is also continuous on $[a, b]$. Furthermore, because f is differentiable, g is also differentiable, and you can apply Rolle's Theorem to the function g. So, there exists a number c in (a, b) such that $g'(c) = 0$, which implies that

$$\begin{aligned} g'(c) &= 0 \\ f'(c) - \frac{f(b) - f(a)}{b - a} &= 0. \end{aligned}$$

So, there exists a number c in (a, b) such that

$$f'(c) = \frac{f(b) - f(a)}{b - a}.$$

See LarsonCalculus.com for Bruce Edwards's video of this proof.

AP* Tips

Be able to apply the Mean Value Theorem on the AP Exam. It may be referred to directly, or it may be necessary to use the theorem to justify your answer.

JOSEPH-LOUIS LAGRANGE (1736–1813)

The Mean Value Theorem was first proved by the famous mathematician Joseph-Louis Lagrange. Born in Italy, Lagrange held a position in the court of Frederick the Great in Berlin for 20 years. *See LarsonCalculus.com to read more of this biography.*

Although the Mean Value Theorem can be used directly in problem solving, it is used more often to prove other theorems. In fact, some people consider this to be the most important theorem in calculus—it is closely related to the Fundamental Theorem of Calculus discussed in Section 4.4. For now, you can get an idea of the versatility of the Mean Value Theorem by looking at the results stated in Exercises 77–85 in this section.

The Mean Value Theorem has implications for both basic interpretations of the derivative. Geometrically, the theorem guarantees the existence of a tangent line that is parallel to the secant line through the points

$$(a, f(a)) \quad \text{and} \quad (b, f(b)),$$

as shown in Figure 3.12. Example 3 illustrates this geometric interpretation of the Mean Value Theorem. In terms of rates of change, the Mean Value Theorem implies that there must be a point in the open interval (a, b) at which the instantaneous rate of change is equal to the average rate of change over the interval $[a, b]$. This is illustrated in Example 4.

EXAMPLE 3 Finding a Tangent Line

See LarsonCalculus.com for an interactive version of this type of example.

For $f(x) = 5 - (4/x)$, find all values of c in the open interval $(1, 4)$ such that

$$f'(c) = \frac{f(4) - f(1)}{4 - 1}.$$

Solution The slope of the secant line through $(1, f(1))$ and $(4, f(4))$ is

$$\frac{f(4) - f(1)}{4 - 1} = \frac{4 - 1}{4 - 1} = 1. \quad \text{Slope of secant line}$$

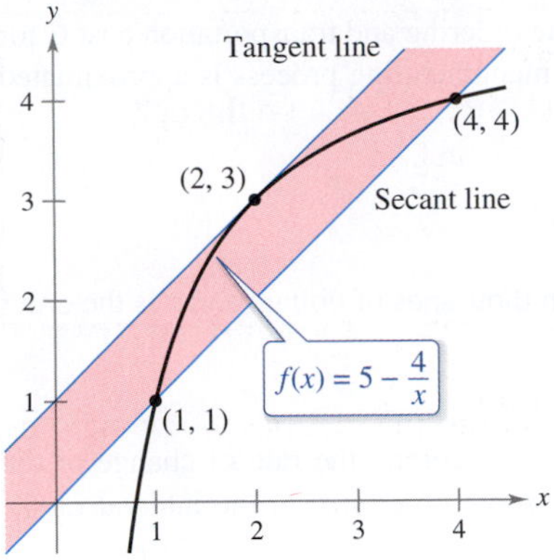

The tangent line at $(2, 3)$ is parallel to the secant line through $(1, 1)$ and $(4, 4)$.
Figure 3.13

Note that the function satisfies the conditions of the Mean Value Theorem. That is, f is continuous on the interval $[1, 4]$ and differentiable on the interval $(1, 4)$. So, there exists at least one number c in $(1, 4)$ such that $f'(c) = 1$. Solving the equation $f'(x) = 1$ yields

$$\frac{4}{x^2} = 1 \quad \text{Set } f'(x) \text{ equal to 1.}$$

which implies that

$$x = \pm 2.$$

So, in the interval $(1, 4)$, you can conclude that $c = 2$, as shown in Figure 3.13.

EXAMPLE 4 Finding an Instantaneous Rate of Change

Two stationary patrol cars equipped with radar are 5 miles apart on a highway, as shown in Figure 3.14. As a truck passes the first patrol car, its speed is clocked at 55 miles per hour. Four minutes later, when the truck passes the second patrol car, its speed is clocked at 50 miles per hour. Prove that the truck must have exceeded the speed limit (of 55 miles per hour) at some time during the 4 minutes.

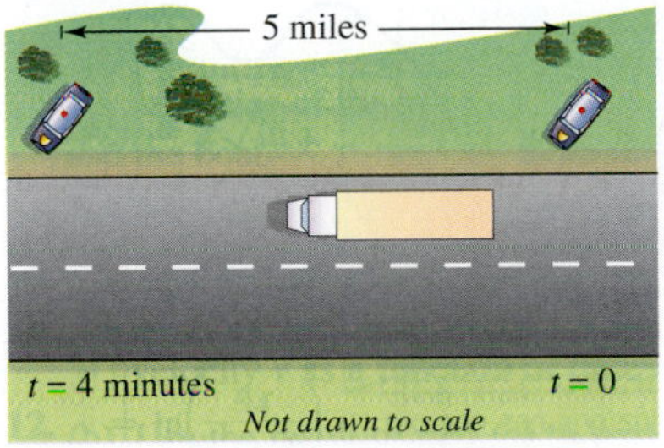

At some time t, the instantaneous velocity is equal to the average velocity over 4 minutes.
Figure 3.14

Solution Let $t = 0$ be the time (in hours) when the truck passes the first patrol car. The time when the truck passes the second patrol car is

$$t = \frac{4}{60} = \frac{1}{15} \text{ hour.}$$

By letting $s(t)$ represent the distance (in miles) traveled by the truck, you have $s(0) = 0$ and $s\left(\frac{1}{15}\right) = 5$. So, the average velocity of the truck over the five-mile stretch of highway is

$$\text{Average velocity} = \frac{s(1/15) - s(0)}{(1/15) - 0} = \frac{5}{1/15} = 75 \text{ miles per hour.}$$

Assuming that the position function is differentiable, you can apply the Mean Value Theorem to conclude that the truck must have been traveling at a rate of 75 miles per hour sometime during the 4 minutes.

A useful alternative form of the Mean Value Theorem is: If f is continuous on $[a, b]$ and differentiable on (a, b), then there exists a number c in (a, b) such that

$$f(b) = f(a) + (b - a)f'(c). \quad \text{Alternative form of Mean Value Theorem}$$

When doing the exercises for this section, keep in mind that polynomial functions, rational functions, and trigonometric functions are differentiable at all points in their domains.

3.2 Exercises

See CalcChat.com for tutorial help and worked-out solutions to odd-numbered exercises.

Writing **In Exercises 1–4, explain why Rolle's Theorem does not apply to the function even though there exist a and b such that $f(a) = f(b)$.**

1. $f(x) = \left|\frac{1}{x}\right|, \quad [-1, 1]$
2. $f(x) = \cot\frac{x}{2}, \quad [\pi, 3\pi]$
3. $f(x) = 1 - |x - 1|, \quad [0, 2]$
4. $f(x) = \sqrt{(2 - x^{2/3})^3}, \quad [-1, 1]$

Intercepts and Derivatives **In Exercises 5–8, find the two x-intercepts of the function f and show that $f'(x) = 0$ at some point between the two x-intercepts.**

5. $f(x) = x^2 - x - 2$
6. $f(x) = x^2 + 6x$
7. $f(x) = x\sqrt{x + 4}$
8. $f(x) = -3x\sqrt{x + 1}$

Using Rolle's Theorem **In Exercises 9–22, determine whether Rolle's Theorem can be applied to f on the closed interval $[a, b]$. If Rolle's Theorem can be applied, find all values of c in the open interval (a, b) such that $f'(c) = 0$. If Rolle's Theorem cannot be applied, explain why not.**

9. $f(x) = -x^2 + 3x, \quad [0, 3]$
10. $f(x) = x^2 - 8x + 5, \quad [2, 6]$
11. $f(x) = (x - 1)(x - 2)(x - 3), \quad [1, 3]$
12. $f(x) = (x - 4)(x + 2)^2, \quad [-2, 4]$
13. $f(x) = x^{2/3} - 1, \quad [-8, 8]$
14. $f(x) = 3 - |x - 3|, \quad [0, 6]$
15. $f(x) = \dfrac{x^2 - 2x - 3}{x + 2}, \quad [-1, 3]$
16. $f(x) = \dfrac{x^2 - 1}{x}, \quad [-1, 1]$
17. $f(x) = \sin x, \quad [0, 2\pi]$
18. $f(x) = \cos x, \quad [0, 2\pi]$
19. $f(x) = \sin 3x, \quad \left[0, \frac{\pi}{3}\right]$
20. $f(x) = \cos 2x, \quad [-\pi, \pi]$
21. $f(x) = \tan x, \quad [0, \pi]$
22. $f(x) = \sec x, \quad [\pi, 2\pi]$

Using Rolle's Theorem **In Exercises 23–26, use a graphing utility to graph the function on the closed interval $[a, b]$. Determine whether Rolle's Theorem can be applied to f on the interval and, if so, find all values of c in the open interval (a, b) such that $f'(c) = 0$.**

23. $f(x) = |x| - 1, \quad [-1, 1]$
24. $f(x) = x - x^{1/3}, \quad [0, 1]$
25. $f(x) = x - \tan \pi x, \quad \left[-\frac{1}{4}, \frac{1}{4}\right]$
26. $f(x) = \dfrac{x}{2} - \sin\dfrac{\pi x}{6}, \quad [-1, 0]$

27. **Vertical Motion** The height of a ball t seconds after it is thrown upward from a height of 6 feet and with an initial velocity of 48 feet per second is $f(t) = -16t^2 + 48t + 6$.
    (a) Verify that $f(1) = f(2)$.
    (b) According to Rolle's Theorem, what must the velocity be at some time in the interval $(1, 2)$? Find that time.

28. **Reorder Costs** The ordering and transportation cost C for components used in a manufacturing process is approximated by

$$C(x) = 10\left(\frac{1}{x} + \frac{x}{x + 3}\right)$$

where C is measured in thousands of dollars and x is the order size in hundreds.
    (a) Verify that $C(3) = C(6)$.
    (b) According to Rolle's Theorem, the rate of change of the cost must be 0 for some order size in the interval $(3, 6)$. Find that order size.

Mean Value Theorem **In Exercises 29 and 30, copy the graph and sketch the secant line to the graph through the points $(a, f(a))$ and $(b, f(b))$. Then sketch any tangent lines to the graph for each value of c guaranteed by the Mean Value Theorem. To print an enlarged copy of the graph, go to *MathGraphs.com*.**

29.
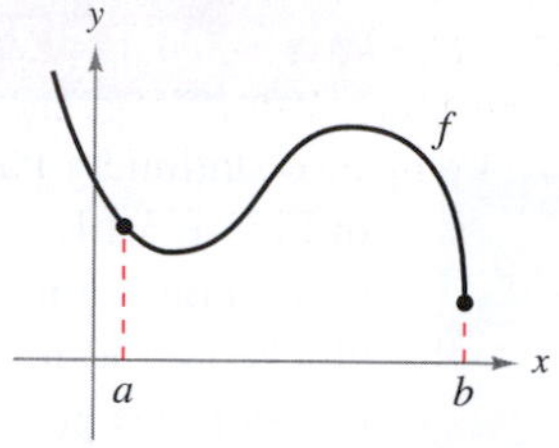

30.
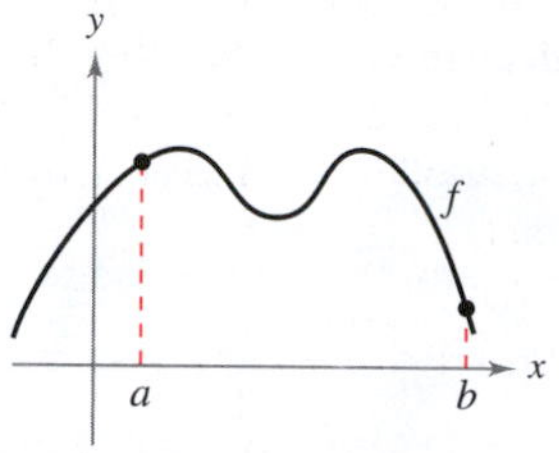

Writing **In Exercises 31–34, explain why the Mean Value Theorem does not apply to the function f on the interval $[0, 6]$.**

31.
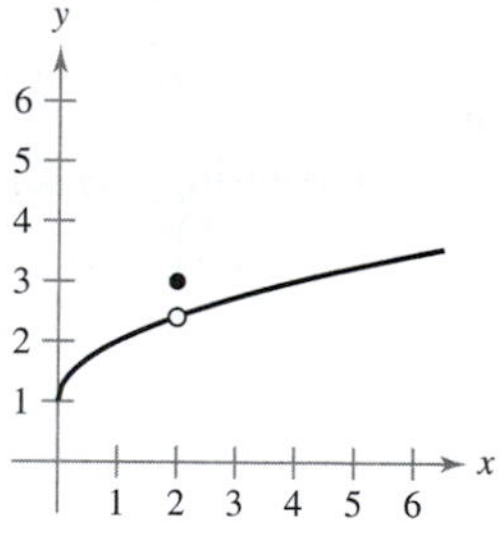

32.
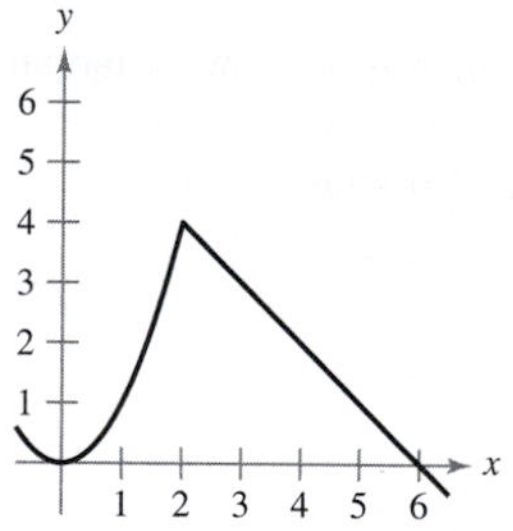

33. $f(x) = \dfrac{1}{x - 3}$
34. $f(x) = |x - 3|$

35. **Mean Value Theorem** Consider the graph of the function $f(x) = -x^2 + 5$ (see figure on next page).
    (a) Find the equation of the secant line joining the points $(-1, 4)$ and $(2, 1)$.
    (b) Use the Mean Value Theorem to determine a point c in the interval $(-1, 2)$ such that the tangent line at c is parallel to the secant line.
    (c) Find the equation of the tangent line through c.
    (d) Then use a graphing utility to graph f, the secant line, and the tangent line.

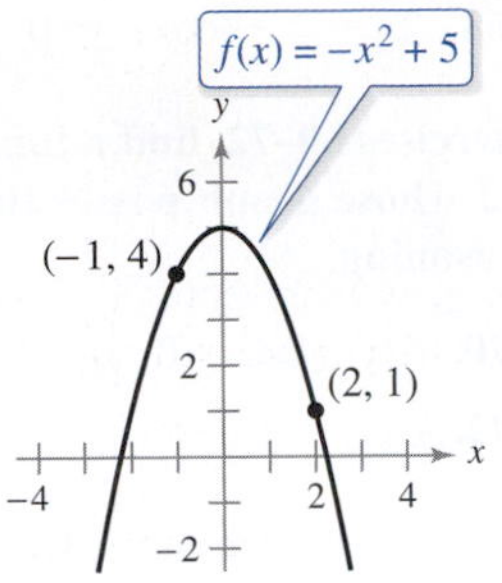

Figure for 35

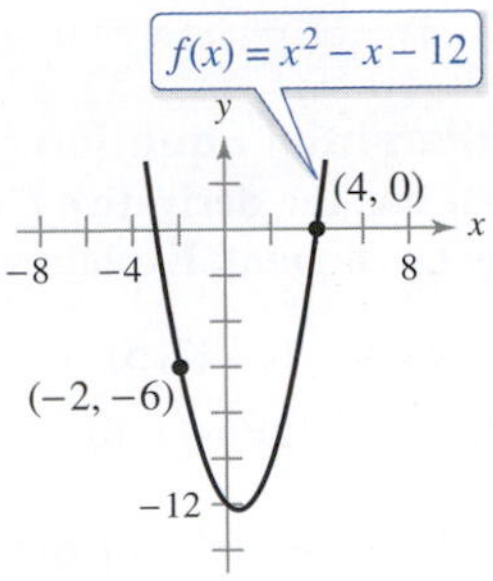

Figure for 36

36. Mean Value Theorem Consider the graph of the function $f(x) = x^2 - x - 12$ (see figure).

(a) Find the equation of the secant line joining the points $(-2, -6)$ and $(4, 0)$.

(b) Use the Mean Value Theorem to determine a point c in the interval $(-2, 4)$ such that the tangent line at c is parallel to the secant line.

(c) Find the equation of the tangent line through c.

(d) Then use a graphing utility to graph f, the secant line, and the tangent line.

Using the Mean Value Theorem **In Exercises 37–46, determine whether the Mean Value Theorem can be applied to f on the closed interval $[a, b]$. If the Mean Value Theorem can be applied, find all values of c in the open interval (a, b) such that**

$$f'(c) = \frac{f(b) - f(a)}{b - a}.$$

If the Mean Value Theorem cannot be applied, explain why not.

37. $f(x) = x^2, \quad [-2, 1]$

38. $f(x) = 2x^3, \quad [0, 6]$

39. $f(x) = x^3 + 2x, \quad [-1, 1]$

40. $f(x) = x^4 - 8x, \quad [0, 2]$

41. $f(x) = x^{2/3}, \quad [0, 1]$

42. $f(x) = \dfrac{x + 1}{x}, \quad [-1, 2]$

43. $f(x) = |2x + 1|, \quad [-1, 3]$

44. $f(x) = \sqrt{2 - x}, \quad [-7, 2]$

45. $f(x) = \sin x, \quad [0, \pi]$

46. $f(x) = \cos x + \tan x, \quad [0, \pi]$

Using the Mean Value Theorem **In Exercises 47–50, use a graphing utility to (a) graph the function f on the given interval, (b) find and graph the secant line through points on the graph of f at the endpoints of the given interval, and (c) find and graph any tangent lines to the graph of f that are parallel to the secant line.**

47. $f(x) = \dfrac{x}{x + 1}, \quad \left[-\dfrac{1}{2}, 2\right]$

48. $f(x) = x - 2\sin x, \quad [-\pi, \pi]$

49. $f(x) = \sqrt{x}, \quad [1, 9]$

50. $f(x) = x^4 - 2x^3 + x^2, \quad [0, 6]$

51. Vertical Motion The height of an object t seconds after it is dropped from a height of 300 meters is

$$s(t) = -4.9t^2 + 300.$$

(a) Find the average velocity of the object during the first 3 seconds.

(b) Use the Mean Value Theorem to verify that at some time during the first 3 seconds of fall, the instantaneous velocity equals the average velocity. Find that time.

52. Sales A company introduces a new product for which the number of units sold S is

$$S(t) = 200\left(5 - \frac{9}{2 + t}\right)$$

where t is the time in months.

(a) Find the average rate of change of $S(t)$ during the first year.

(b) During what month of the first year does $S'(t)$ equal the average rate of change?

WRITING ABOUT CONCEPTS

53. Converse of Rolle's Theorem Let f be continuous on $[a, b]$ and differentiable on (a, b). If there exists c in (a, b) such that $f'(c) = 0$, does it follow that $f(a) = f(b)$? Explain.

54. Rolle's Theorem Let f be continuous on $[a, b]$ and differentiable on (a, b). Also, suppose that $f(a) = f(b)$ and that c is a real number in the interval such that $f'(c) = 0$. Find an interval for the function g over which Rolle's Theorem can be applied, and find the corresponding critical number of g (k is a constant).

(a) $g(x) = f(x) + k$ (b) $g(x) = f(x - k)$

(c) $g(x) = f(kx)$

55. Rolle's Theorem The function

$$f(x) = \begin{cases} 0, & x = 0 \\ 1 - x, & 0 < x \le 1 \end{cases}$$

is differentiable on $(0, 1)$ and satisfies $f(0) = f(1)$. However, its derivative is never zero on $(0, 1)$. Does this contradict Rolle's Theorem? Explain.

56. Mean Value Theorem Can you find a function f such that $f(-2) = -2$, $f(2) = 6$, and $f'(x) < 1$ for all x? Why or why not?

57. Speed

A plane begins its take-off at 2:00 P.M. on a 2500-mile flight. After 5.5 hours, the plane arrives at its destination. Explain why there are at least two times during the flight when the speed of the plane is 400 miles per hour.

58. Temperature When an object is removed from a furnace and placed in an environment with a constant temperature of 90°F, its core temperature is 1500°F. Five hours later, the core temperature is 390°F. Explain why there must exist a time in the interval when the temperature is decreasing at a rate of 222°F per hour.

59. Velocity Two bicyclists begin a race at 8:00 A.M. They both finish the race 2 hours and 15 minutes later. Prove that at some time during the race, the bicyclists are traveling at the same velocity.

60. Acceleration At 9:13 A.M., a sports car is traveling 35 miles per hour. Two minutes later, the car is traveling 85 miles per hour. Prove that at some time during this two-minute interval, the car's acceleration is exactly 1500 miles per hour squared.

61. Using a Function Consider the function

$$f(x) = 3\cos^2\left(\frac{\pi x}{2}\right).$$

(a) Use a graphing utility to graph f and f'.

(b) Is f a continuous function? Is f' a continuous function?

(c) Does Rolle's Theorem apply on the interval $[-1, 1]$? Does it apply on the interval $[1, 2]$? Explain.

(d) Evaluate, if possible, $\lim_{x\to 3^-} f'(x)$ and $\lim_{x\to 3^+} f'(x)$.

62. HOW DO YOU SEE IT? The figure shows two parts of the graph of a continuous differentiable function f on $[-10, 4]$. The derivative f' is also continuous. To print an enlarged copy of the graph, go to *MathGraphs.com*.

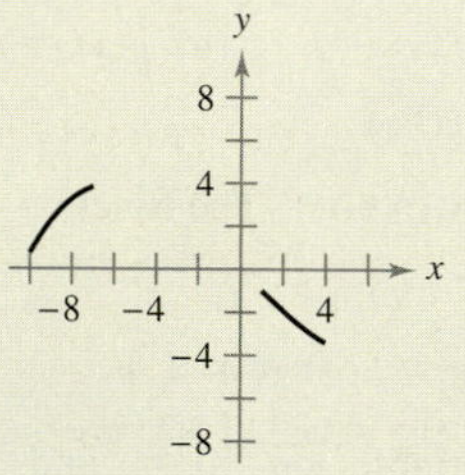

(a) Explain why f must have at least one zero in $[-10, 4]$.

(b) Explain why f' must also have at least one zero in the interval $[-10, 4]$. What are these zeros called?

(c) Make a possible sketch of the function with one zero of f' on the interval $[-10, 4]$.

Think About It **In Exercises 63 and 64, sketch the graph of an arbitrary function f that satisfies the given condition but does not satisfy the conditions of the Mean Value Theorem on the interval $[-5, 5]$.**

63. f is continuous on $[-5, 5]$.

64. f is not continuous on $[-5, 5]$.

Finding a Solution **In Exercises 65–68, use the Intermediate Value Theorem and Rolle's Theorem to prove that the equation has exactly one real solution.**

65. $x^5 + x^3 + x + 1 = 0$ **66.** $2x^5 + 7x - 1 = 0$

67. $3x + 1 - \sin x = 0$ **68.** $2x - 2 - \cos x = 0$

Differential Equation **In Exercises 69–72, find a function f that has the derivative $f'(x)$ and whose graph passes through the given point. Explain your reasoning.**

69. $f'(x) = 0$, $(2, 5)$ **70.** $f'(x) = 4$, $(0, 1)$

71. $f'(x) = 2x$, $(1, 0)$ **72.** $f'(x) = 6x - 1$, $(2, 7)$

True or False? **In Exercises 73–76, determine whether the statement is true or false. If it is false, explain why or give an example that shows it is false.**

73. The Mean Value Theorem can be applied to

$$f(x) = \frac{1}{x}$$

on the interval $[-1, 1]$.

74. If the graph of a function has three x-intercepts, then it must have at least two points at which its tangent line is horizontal.

75. If the graph of a polynomial function has three x-intercepts, then it must have at least two points at which its tangent line is horizontal.

76. If $f'(x) = 0$ for all x in the domain of f, then f is a constant function.

77. Proof Prove that if $a > 0$ and n is any positive integer, then the polynomial function $p(x) = x^{2n+1} + ax + b$ cannot have two real roots.

78. Proof Prove that if $f'(x) = 0$ for all x in an interval (a, b), then f is constant on (a, b).

79. Proof Let $p(x) = Ax^2 + Bx + C$. Prove that for any interval $[a, b]$, the value c guaranteed by the Mean Value Theorem is the midpoint of the interval.

80. Using Rolle's Theorem

(a) Let $f(x) = x^2$ and $g(x) = -x^3 + x^2 + 3x + 2$. Then $f(-1) = g(-1)$ and $f(2) = g(2)$. Show that there is at least one value c in the interval $(-1, 2)$ where the tangent line to f at $(c, f(c))$ is parallel to the tangent line to g at $(c, g(c))$. Identify c.

(b) Let f and g be differentiable functions on $[a, b]$ where $f(a) = g(a)$ and $f(b) = g(b)$. Show that there is at least one value c in the interval (a, b) where the tangent line to f at $(c, f(c))$ is parallel to the tangent line to g at $(c, g(c))$.

81. Proof Prove that if f is differentiable on $(-\infty, \infty)$ and $f'(x) < 1$ for all real numbers, then f has at most one fixed point. A fixed point of a function f is a real number c such that $f(c) = c$.

82. Fixed Point Use the result of Exercise 81 to show that $f(x) = \frac{1}{2}\cos x$ has at most one fixed point.

83. Proof Prove that $|\cos a - \cos b| \le |a - b|$ for all a and b.

84. Proof Prove that $|\sin a - \sin b| \le |a - b|$ for all a and b.

85. Using the Mean Value Theorem Let $0 < a < b$. Use the Mean Value Theorem to show that

$$\sqrt{b} - \sqrt{a} < \frac{b - a}{2\sqrt{a}}.$$

3.3 Increasing and Decreasing Functions and the First Derivative Test

- **Determine intervals on which a function is increasing or decreasing.**
- **Apply the First Derivative Test to find relative extrema of a function.**

Increasing and Decreasing Functions

In this section, you will learn how derivatives can be used to *classify* relative extrema as either relative minima or relative maxima. First, it is important to define increasing and decreasing functions.

Definitions of Increasing and Decreasing Functions

A function f is **increasing** on an interval when, for any two numbers x_1 and x_2 in the interval, $x_1 < x_2$ implies $f(x_1) < f(x_2)$.

A function f is **decreasing** on an interval when, for any two numbers x_1 and x_2 in the interval, $x_1 < x_2$ implies $f(x_1) > f(x_2)$.

A function is increasing when, *as x moves to the right*, its graph moves up, and is decreasing when its graph moves down. For example, the function in Figure 3.15 is decreasing on the interval $(-\infty, a)$, is constant on the interval (a, b), and is increasing on the interval (b, ∞). As shown in Theorem 3.5 below, a positive derivative implies that the function is increasing, a negative derivative implies that the function is decreasing, and a zero derivative on an entire interval implies that the function is constant on that interval.

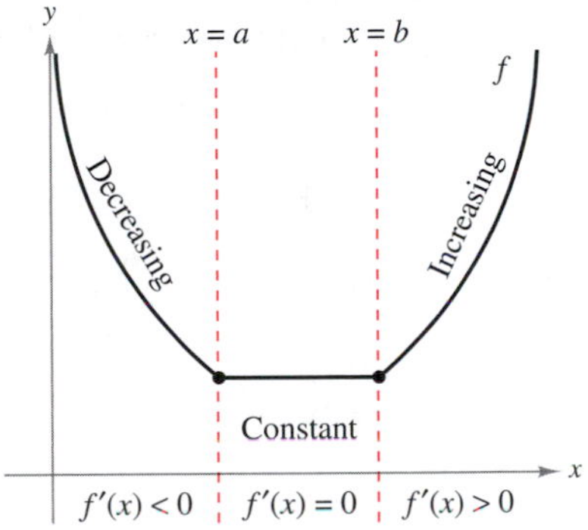

The derivative is related to the slope of a function.
Figure 3.15

THEOREM 3.5 Test for Increasing and Decreasing Functions

Let f be a function that is continuous on the closed interval $[a, b]$ and differentiable on the open interval (a, b).

1. If $f'(x) > 0$ for all x in (a, b), then f is increasing on $[a, b]$.
2. If $f'(x) < 0$ for all x in (a, b), then f is decreasing on $[a, b]$.
3. If $f'(x) = 0$ for all x in (a, b), then f is constant on $[a, b]$.

REMARK The conclusions in the first two cases of Theorem 3.5 are valid even when $f'(x) = 0$ at a finite number of x-values in (a, b).

Proof To prove the first case, assume that $f'(x) > 0$ for all x in the interval (a, b) and let $x_1 < x_2$ be any two points in the interval. By the Mean Value Theorem, you know that there exists a number c such that $x_1 < c < x_2$, and

$$f'(c) = \frac{f(x_2) - f(x_1)}{x_2 - x_1}.$$

Because $f'(c) > 0$ and $x_2 - x_1 > 0$, you know that $f(x_2) - f(x_1) > 0$, which implies that $f(x_1) < f(x_2)$. So, f is increasing on the interval. The second case has a similar proof (see Exercise 97), and the third case is a consequence of Exercise 78 in Section 3.2.

See LarsonCalculus.com for Bruce Edwards's video of this proof.

AP* Tips

Sign charts are useful organization tools, but will not be graded on the AP Exam. To receive full credit, you must justify your responses in complete sentences.

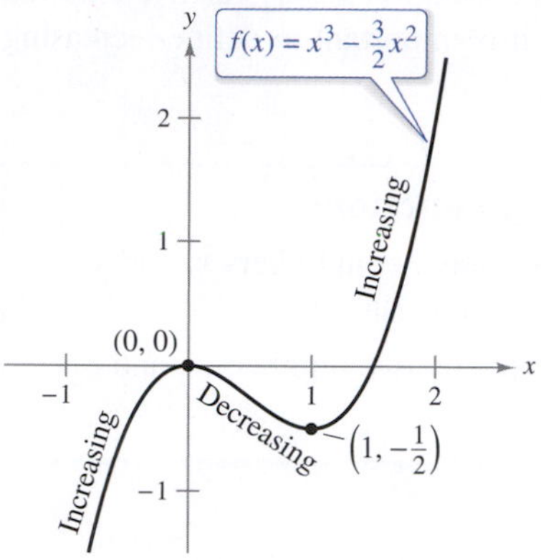

Figure 3.16

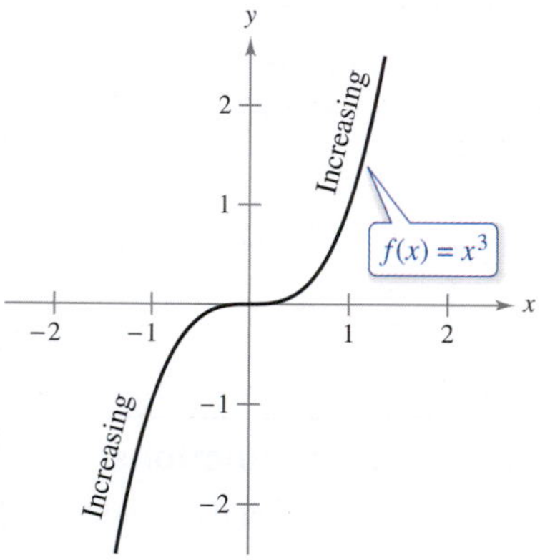

(a) Strictly monotonic function

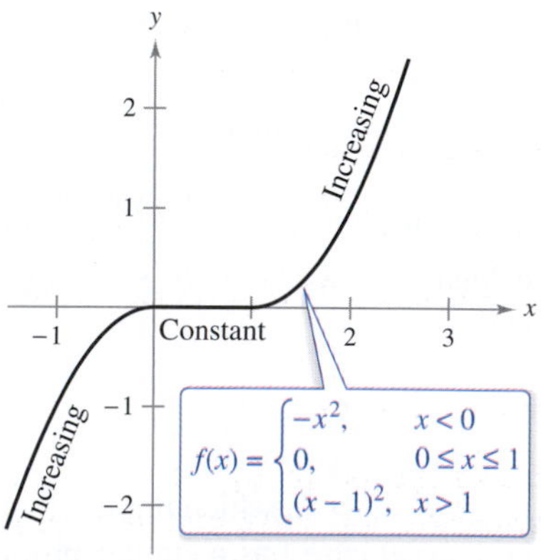

(b) Not strictly monotonic
Figure 3.17

EXAMPLE 1 Intervals on Which *f* Is Increasing or Decreasing

Find the open intervals on which $f(x) = x^3 - \frac{3}{2}x^2$ is increasing or decreasing.

Solution Note that f is differentiable on the entire real number line and the derivative of f is

$f(x) = x^3 - \frac{3}{2}x^2$ Write original function.

$f'(x) = 3x^2 - 3x.$ Differentiate.

To determine the critical numbers of f, set $f'(x)$ equal to zero.

$3x^2 - 3x = 0$ Set $f'(x)$ equal to 0.

$3(x)(x - 1) = 0$ Factor.

$x = 0, 1$ Critical numbers

Because there are no points for which f' does not exist, you can conclude that $x = 0$ and $x = 1$ are the only critical numbers. The table summarizes the testing of the three intervals determined by these two critical numbers.

Interval	$-\infty < x < 0$	$0 < x < 1$	$1 < x < \infty$
Test Value	$x = -1$	$x = \frac{1}{2}$	$x = 2$
Sign of $f'(x)$	$f'(-1) = 6 > 0$	$f'(\frac{1}{2}) = -\frac{3}{4} < 0$	$f'(2) = 6 > 0$
Conclusion	Increasing	Decreasing	Increasing

By Theorem 3.5, f is increasing on the intervals $(-\infty, 0)$ and $(1, \infty)$ and decreasing on the interval $(0, 1)$, as shown in Figure 3.16. ■

Example 1 gives you one instance of how to find intervals on which a function is increasing or decreasing. The guidelines below summarize the steps followed in that example.

GUIDELINES FOR FINDING INTERVALS ON WHICH A FUNCTION IS INCREASING OR DECREASING

Let f be continuous on the interval (a, b). To find the open intervals on which f is increasing or decreasing, use the following steps.

1. Locate the critical numbers of f in (a, b), and use these numbers to determine test intervals.
2. Determine the sign of $f'(x)$ at one test value in each of the intervals.
3. Use Theorem 3.5 to determine whether f is increasing or decreasing on each interval.

These guidelines are also valid when the interval (a, b) is replaced by an interval of the form $(-\infty, b)$, (a, ∞), or $(-\infty, \infty)$.

A function is **strictly monotonic** on an interval when it is either increasing on the entire interval or decreasing on the entire interval. For instance, the function $f(x) = x^3$ is strictly monotonic on the entire real number line because it is increasing on the entire real number line, as shown in Figure 3.17(a). The function shown in Figure 3.17(b) is not strictly monotonic on the entire real number line because it is constant on the interval $[0, 1]$.

The First Derivative Test

After you have determined the intervals on which a function is increasing or decreasing, it is not difficult to locate the relative extrema of the function. For instance, in Figure 3.18 (from Example 1), the function

$$f(x) = x^3 - \frac{3}{2}x^2$$

has a relative maximum at the point (0, 0) because f is increasing immediately to the left of $x = 0$ and decreasing immediately to the right of $x = 0$. Similarly, f has a relative minimum at the point $\left(1, -\frac{1}{2}\right)$ because f is decreasing immediately to the left of $x = 1$ and increasing immediately to the right of $x = 1$. The next theorem makes this more explicit.

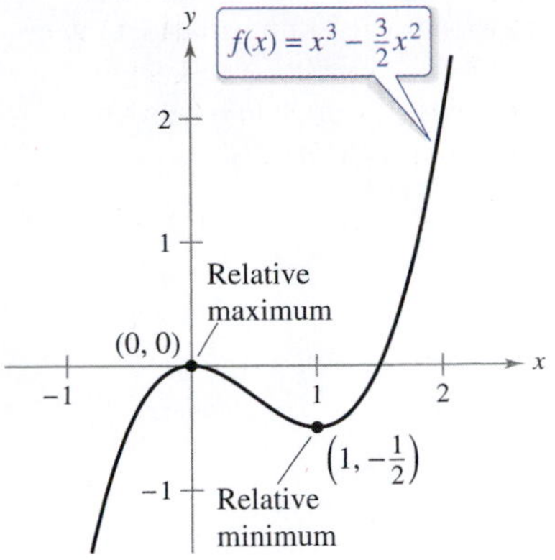

Relative extrema of f
Figure 3.18

THEOREM 3.6 The First Derivative Test

Let c be a critical number of a function f that is continuous on an open interval I containing c. If f is differentiable on the interval, except possibly at c, then $f(c)$ can be classified as follows.

1. If $f'(x)$ changes from negative to positive at c, then f has a *relative minimum* at $(c, f(c))$.
2. If $f'(x)$ changes from positive to negative at c, then f has a *relative maximum* at $(c, f(c))$.
3. If $f'(x)$ is positive on both sides of c or negative on both sides of c, then $f(c)$ is neither a relative minimum nor a relative maximum.

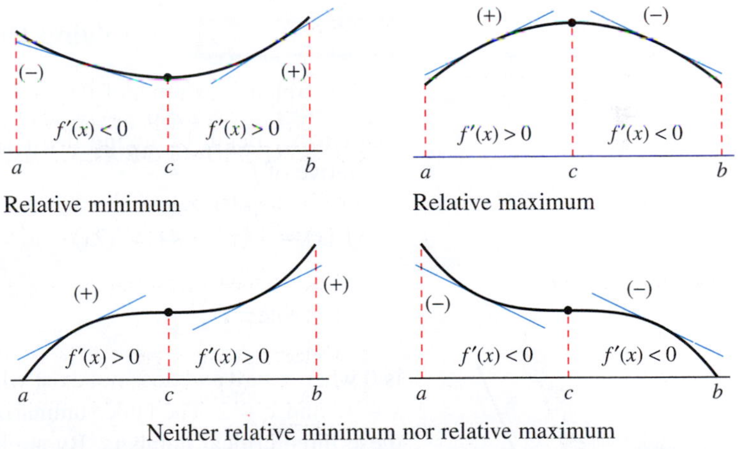

Proof Assume that $f'(x)$ changes from negative to positive at c. Then there exist a and b in I such that

$$f'(x) < 0 \text{ for all } x \text{ in } (a, c) \quad \text{and} \quad f'(x) > 0 \text{ for all } x \text{ in } (c, b).$$

By Theorem 3.5, f is decreasing on $[a, c]$ and increasing on $[c, b]$. So, $f(c)$ is a minimum of f on the open interval (a, b) and, consequently, a relative minimum of f. This proves the first case of the theorem. The second case can be proved in a similar way (see Exercise 98). ■

See LarsonCalculus.com for Bruce Edwards's video of this proof.

EXAMPLE 2 Applying the First Derivative Test

Find the relative extrema of $f(x) = \frac{1}{2}x - \sin x$ in the interval $(0, 2\pi)$.

Solution Note that f is continuous on the interval $(0, 2\pi)$. The derivative of f is $f'(x) = \frac{1}{2} - \cos x$. To determine the critical numbers of f in this interval, set $f'(x)$ equal to 0.

$$\frac{1}{2} - \cos x = 0 \qquad \text{Set } f'(x) \text{ equal to 0.}$$

$$\cos x = \frac{1}{2}$$

$$x = \frac{\pi}{3}, \frac{5\pi}{3} \qquad \text{Critical numbers}$$

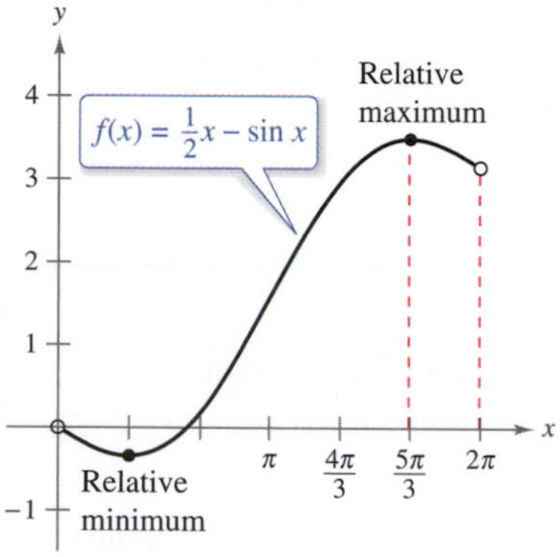

A relative minimum occurs where f changes from decreasing to increasing, and a relative maximum occurs where f changes from increasing to decreasing.
Figure 3.19

Because there are no points for which f' does not exist, you can conclude that $x = \pi/3$ and $x = 5\pi/3$ are the only critical numbers. The table summarizes the testing of the three intervals determined by these two critical numbers. By applying the First Derivative Test, you can conclude that f has a relative minimum at the point where $x = \pi/3$ and a relative maximum at the point where $x = 5\pi/3$, as shown in Figure 3.19.

Interval	$0 < x < \frac{\pi}{3}$	$\frac{\pi}{3} < x < \frac{5\pi}{3}$	$\frac{5\pi}{3} < x < 2\pi$
Test Value	$x = \frac{\pi}{4}$	$x = \pi$	$x = \frac{7\pi}{4}$
Sign of $f'(x)$	$f'\left(\frac{\pi}{4}\right) < 0$	$f'(\pi) > 0$	$f'\left(\frac{7\pi}{4}\right) < 0$
Conclusion	Decreasing	Increasing	Decreasing

EXAMPLE 3 Applying the First Derivative Test

Find the relative extrema of $f(x) = (x^2 - 4)^{2/3}$.

Solution Begin by noting that f is continuous on the entire real number line. The derivative of f

$$f'(x) = \frac{2}{3}(x^2 - 4)^{-1/3}(2x) \qquad \text{General Power Rule}$$

$$= \frac{4x}{3(x^2 - 4)^{1/3}} \qquad \text{Simplify.}$$

is 0 when $x = 0$ and does not exist when $x = \pm 2$. So, the critical numbers are $x = -2$, $x = 0$, and $x = 2$. The table summarizes the testing of the four intervals determined by these three critical numbers. By applying the First Derivative Test, you can conclude that f has a relative minimum at the point $(-2, 0)$, a relative maximum at the point $(0, \sqrt[3]{16})$, and another relative minimum at the point $(2, 0)$, as shown in Figure 3.20.

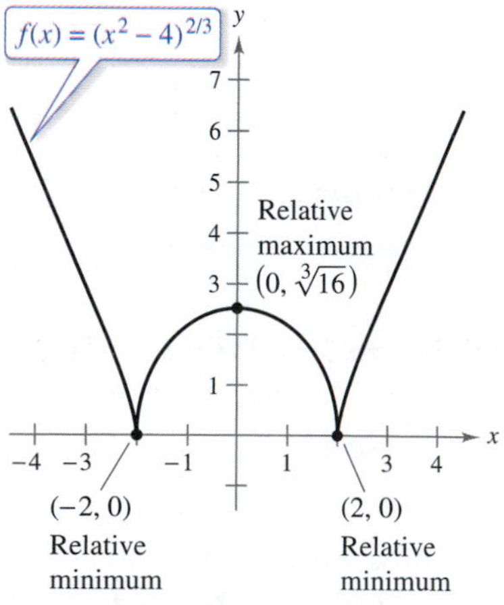

Figure 3.20

Interval	$-\infty < x < -2$	$-2 < x < 0$	$0 < x < 2$	$2 < x < \infty$
Test Value	$x = -3$	$x = -1$	$x = 1$	$x = 3$
Sign of $f'(x)$	$f'(-3) < 0$	$f'(-1) > 0$	$f'(1) < 0$	$f'(3) > 0$
Conclusion	Decreasing	Increasing	Decreasing	Increasing

Note that in Examples 1 and 2, the given functions are differentiable on the entire real number line. For such functions, the only critical numbers are those for which $f'(x) = 0$. Example 3 concerns a function that has two types of critical numbers—those for which $f'(x) = 0$ and those for which f is not differentiable.

When using the First Derivative Test, be sure to consider the domain of the function. For instance, in the next example, the function

$$f(x) = \frac{x^4 + 1}{x^2}$$

is not defined when $x = 0$. This x-value must be used with the critical numbers to determine the test intervals.

EXAMPLE 4 Applying the First Derivative Test

See LarsonCalculus.com for an interactive version of this type of example.

Find the relative extrema of $f(x) = \dfrac{x^4 + 1}{x^2}$.

Solution Note that f is not defined when $x = 0$.

$$f(x) = x^2 + x^{-2} \qquad \text{Rewrite original function.}$$

$$f'(x) = 2x - 2x^{-3} \qquad \text{Differentiate.}$$

$$= 2x - \frac{2}{x^3} \qquad \text{Rewrite with positive exponent.}$$

$$= \frac{2(x^4 - 1)}{x^3} \qquad \text{Simplify.}$$

$$= \frac{2(x^2 + 1)(x - 1)(x + 1)}{x^3} \qquad \text{Factor.}$$

So, $f'(x)$ is zero at $x = \pm 1$. Moreover, because $x = 0$ is not in the domain of f, you should use this x-value along with the critical numbers to determine the test intervals.

$$x = \pm 1 \qquad \text{Critical numbers, } f'(\pm 1) = 0$$

$$x = 0 \qquad 0 \text{ is not in the domain of } f.$$

The table summarizes the testing of the four intervals determined by these three x-values. By applying the First Derivative Test, you can conclude that f has one relative minimum at the point $(-1, 2)$ and another at the point $(1, 2)$, as shown in Figure 3.21.

Interval	$-\infty < x < -1$	$-1 < x < 0$	$0 < x < 1$	$1 < x < \infty$
Test Value	$x = -2$	$x = -\frac{1}{2}$	$x = \frac{1}{2}$	$x = 2$
Sign of $f'(x)$	$f'(-2) < 0$	$f'\left(-\frac{1}{2}\right) > 0$	$f'\left(\frac{1}{2}\right) < 0$	$f'(2) > 0$
Conclusion	Decreasing	Increasing	Decreasing	Increasing

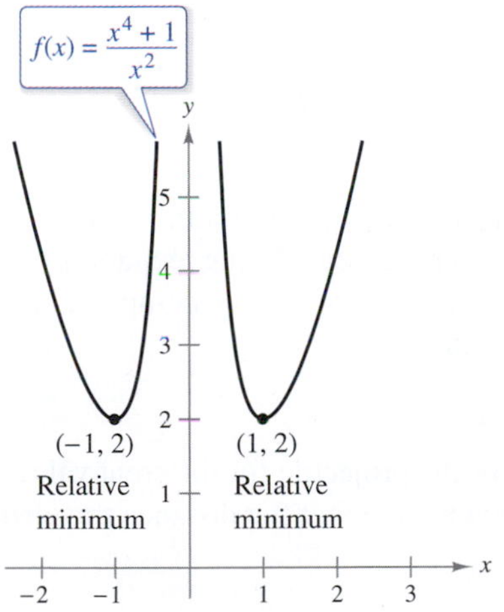

x-values that are not in the domain of f, as well as critical numbers, determine test intervals for f'.
Figure 3.21

TECHNOLOGY The most difficult step in applying the First Derivative Test is finding the values for which the derivative is equal to 0. For instance, the values of x for which the derivative of

$$f(x) = \frac{x^4 + 1}{x^2 + 1}$$

is equal to zero are $x = 0$ and $x = \pm\sqrt{\sqrt{2} - 1}$. If you have access to technology that can perform symbolic differentiation and solve equations, use it to apply the First Derivative Test to this function.

When a projectile is propelled from ground level and air resistance is neglected, the object will travel farthest with an initial angle of 45°. When, however, the projectile is propelled from a point above ground level, the angle that yields a maximum horizontal distance is not 45° (see Example 5).

AP* Tips

Questions that involve velocity or position functions are common on the AP Exam.

EXAMPLE 5 The Path of a Projectile

Neglecting air resistance, the path of a projectile that is propelled at an angle θ is

$$y = \frac{g \sec^2 \theta}{2v_0^2}x^2 + (\tan \theta)x + h, \quad 0 \le \theta \le \frac{\pi}{2}$$

where y is the height, x is the horizontal distance, g is the acceleration due to gravity, v_0 is the initial velocity, and h is the initial height. (This equation is derived in Section 12.3.) Let $g = -32$ feet per second per second, $v_0 = 24$ feet per second, and $h = 9$ feet. What value of θ will produce a maximum horizontal distance?

Solution To find the distance the projectile travels, let $y = 0$, $g = -32$, $v_0 = 24$, and $h = 9$. Then substitute these values in the given equation as shown.

$$\frac{g \sec^2 \theta}{2v_0^2}x^2 + (\tan \theta)x + h = y$$

$$\frac{-32 \sec^2 \theta}{2(24^2)}x^2 + (\tan \theta)x + 9 = 0$$

$$-\frac{\sec^2 \theta}{36}x^2 + (\tan \theta)x + 9 = 0$$

Next, solve for x using the Quadratic Formula with $a = -\sec^2 \theta/36$, $b = \tan \theta$, and $c = 9$.

$$x = \frac{-b \pm \sqrt{b^2 - 4ac}}{2a}$$

$$x = \frac{-\tan \theta \pm \sqrt{(\tan \theta)^2 - 4(-\sec^2 \theta/36)(9)}}{2(-\sec^2 \theta/36)}$$

$$x = \frac{-\tan \theta \pm \sqrt{\tan^2 \theta + \sec^2 \theta}}{-\sec^2 \theta/18}$$

$$x = 18 \cos \theta\left(\sin \theta + \sqrt{\sin^2 \theta + 1}\right), \quad x \ge 0$$

At this point, you need to find the value of θ that produces a maximum value of x. Applying the First Derivative Test by hand would be very tedious. Using technology to solve the equation $dx/d\theta = 0$, however, eliminates most of the messy computations. The result is that the maximum value of x occurs when

$$\theta \approx 0.61548 \text{ radian}, \quad \text{or} \quad 35.3°.$$

This conclusion is reinforced by sketching the path of the projectile for different values of θ, as shown in Figure 3.22. Of the three paths shown, note that the distance traveled is greatest for $\theta = 35°$.

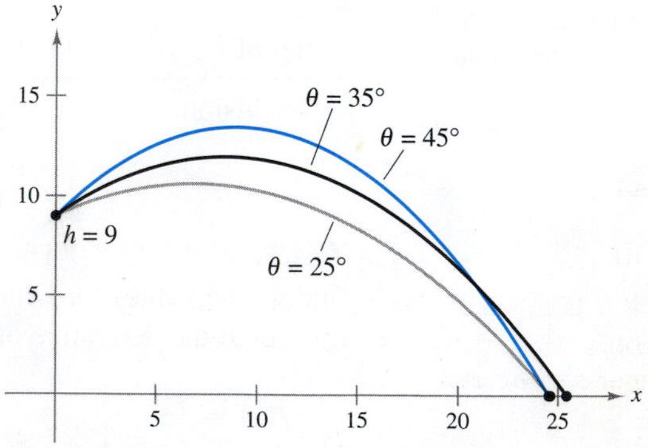

The path of a projectile with initial angle θ
Figure 3.22

3.3 Exercises

See **CalcChat.com** for tutorial help and worked-out solutions to odd-numbered exercises.

Using a Graph In Exercises 1 and 2, use the graph of f to find (a) the largest open interval on which f is increasing, and (b) the largest open interval on which f is decreasing.

1.

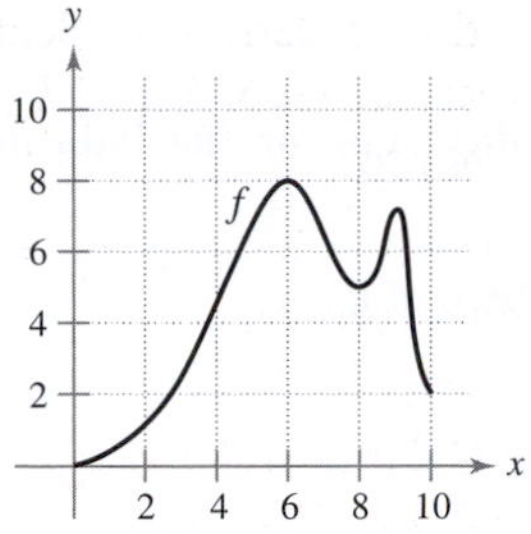

2.

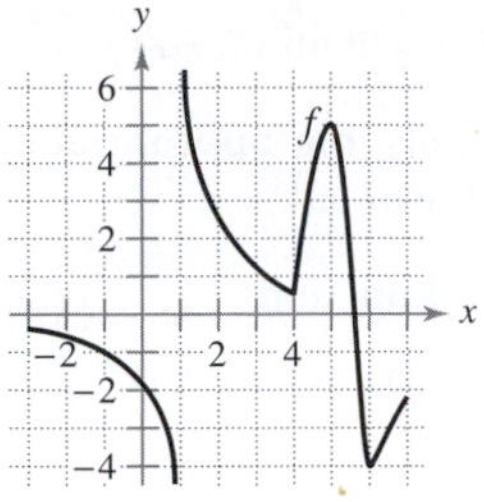

Using a Graph In Exercises 3–8, use the graph to estimate the open intervals on which the function is increasing or decreasing. Then find the open intervals analytically.

3. $f(x) = x^2 - 6x + 8$

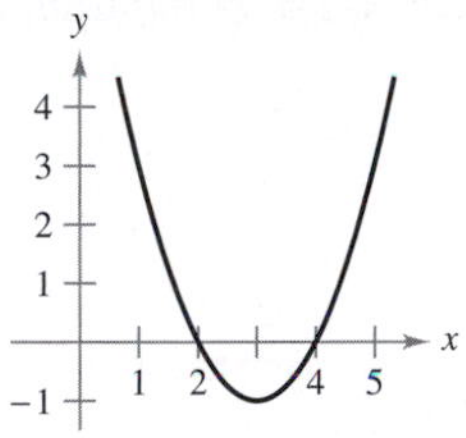

4. $y = -(x + 1)^2$

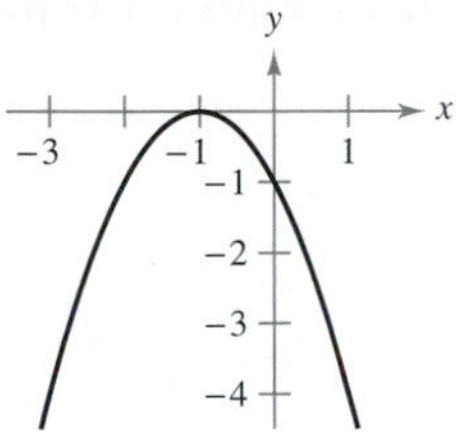

5. $y = \dfrac{x^3}{4} - 3x$

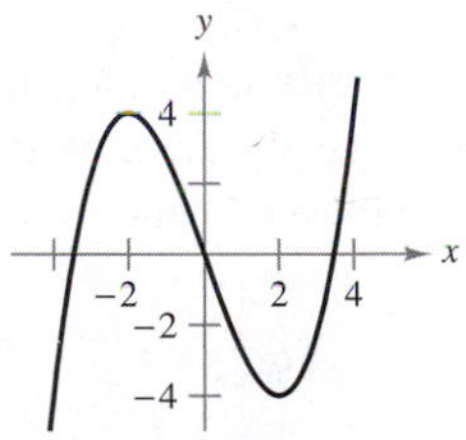

6. $f(x) = x^4 - 2x^2$

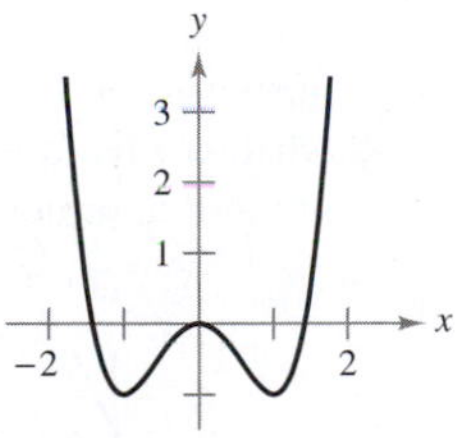

7. $f(x) = \dfrac{1}{(x + 1)^2}$

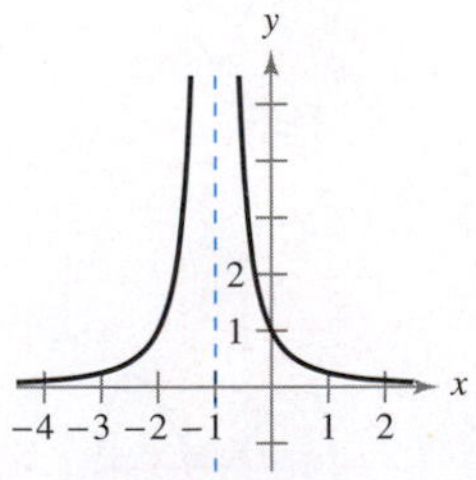

8. $y = \dfrac{x^2}{2x - 1}$

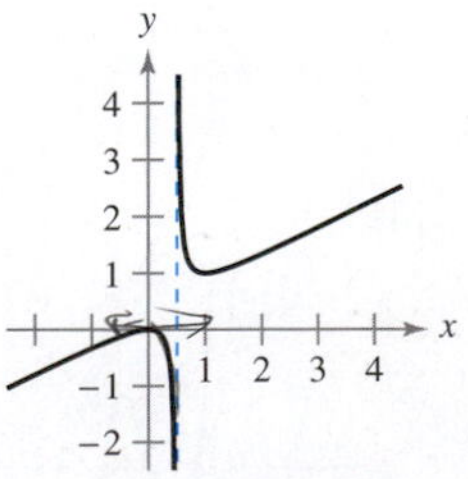

Intervals on Which f Is Increasing or Decreasing In Exercises 9–16, identify the open intervals on which the function is increasing or decreasing.

9. $g(x) = x^2 - 2x - 8$

10. $h(x) = 12x - x^3$

11. $y = x\sqrt{16 - x^2}$

12. $y = x + \dfrac{9}{x}$

13. $f(x) = \sin x - 1, \quad 0 < x < 2\pi$

14. $h(x) = \cos \dfrac{x}{2}, \quad 0 < x < 2\pi$

15. $y = x - 2\cos x, \quad 0 < x < 2\pi$

16. $f(x) = \sin^2 x + \sin x, \quad 0 < x < 2\pi$

Applying the First Derivative Test In Exercises 17–40, (a) find the critical numbers of f (if any), (b) find the open interval(s) on which the function is increasing or decreasing, (c) apply the First Derivative Test to identify all relative extrema, and (d) use a graphing utility to confirm your results.

17. $f(x) = x^2 - 4x$

18. $f(x) = x^2 + 6x + 10$

19. $f(x) = -2x^2 + 4x + 3$

20. $f(x) = -3x^2 - 4x - 2$

21. $f(x) = 2x^3 + 3x^2 - 12x$

22. $f(x) = x^3 - 6x^2 + 15$

23. $f(x) = (x - 1)^2(x + 3)$

24. $f(x) = (x + 2)^2(x - 1)$

25. $f(x) = \dfrac{x^5 - 5x}{5}$

26. $f(x) = x^4 - 32x + 4$

27. $f(x) = x^{1/3} + 1$

28. $f(x) = x^{2/3} - 4$

29. $f(x) = (x + 2)^{2/3}$

30. $f(x) = (x - 3)^{1/3}$

31. $f(x) = 5 - |x - 5|$

32. $f(x) = |x + 3| - 1$

33. $f(x) = 2x + \dfrac{1}{x}$

34. $f(x) = \dfrac{x}{x - 5}$

35. $f(x) = \dfrac{x^2}{x^2 - 9}$

36. $f(x) = \dfrac{x^2 - 2x + 1}{x + 1}$

37. $f(x) = \begin{cases} 4 - x^2, & x \le 0 \\ -2x, & x > 0 \end{cases}$

38. $f(x) = \begin{cases} 2x + 1, & x \le -1 \\ x^2 - 2, & x > -1 \end{cases}$

39. $f(x) = \begin{cases} 3x + 1, & x \le 1 \\ 5 - x^2, & x > 1 \end{cases}$

40. $f(x) = \begin{cases} -x^3 + 1, & x \le 0 \\ -x^2 + 2x, & x > 0 \end{cases}$

Applying the First Derivative Test In Exercises 41–48, consider the function on the interval $(0, 2\pi)$. For each function, (a) find the open interval(s) on which the function is increasing or decreasing, (b) apply the First Derivative Test to identify all relative extrema, and (c) use a graphing utility to confirm your results.

41. $f(x) = \dfrac{x}{2} + \cos x$

42. $f(x) = \sin x \cos x + 5$

43. $f(x) = \sin x + \cos x$

44. $f(x) = x + 2\sin x$

45. $f(x) = \cos^2(2x)$

46. $f(x) = \sin x - \sqrt{3}\cos x$

47. $f(x) = \sin^2 x + \sin x$

48. $f(x) = \dfrac{\sin x}{1 + \cos^2 x}$

Finding and Analyzing Derivatives Using Technology

In Exercises 49–54, (a) use a computer algebra system to differentiate the function, (b) sketch the graphs of f and f' on the same set of coordinate axes over the given interval, (c) find the critical numbers of f in the open interval, and (d) find the interval(s) on which f' is positive and the interval(s) on which it is negative. Compare the behavior of f and the sign of f'.

49. $f(x) = 2x\sqrt{9 - x^2}, \quad [-3, 3]$

50. $f(x) = 10\left(5 - \sqrt{x^2 - 3x + 16}\right), \quad [0, 5]$

51. $f(t) = t^2 \sin t, \quad [0, 2\pi]$

52. $f(x) = \dfrac{x}{2} + \cos\dfrac{x}{2}, \quad [0, 4\pi]$

53. $f(x) = -3 \sin\dfrac{x}{3}, \quad [0, 6\pi]$

54. $f(x) = 2 \sin 3x + 4 \cos 3x, \quad [0, \pi]$

Comparing Functions

In Exercises 55 and 56, use symmetry, extrema, and zeros to sketch the graph of f. How do the functions f and g differ?

55. $f(x) = \dfrac{x^5 - 4x^3 + 3x}{x^2 - 1}$

$g(x) = x(x^2 - 3)$

56. $f(t) = \cos^2 t - \sin^2 t$

$g(t) = 1 - 2 \sin^2 t$

Think About It

In Exercises 57–62, the graph of f is shown in the figure. Sketch a graph of the derivative of f. To print an enlarged copy of the graph, go to *MathGraphs.com*.

57.

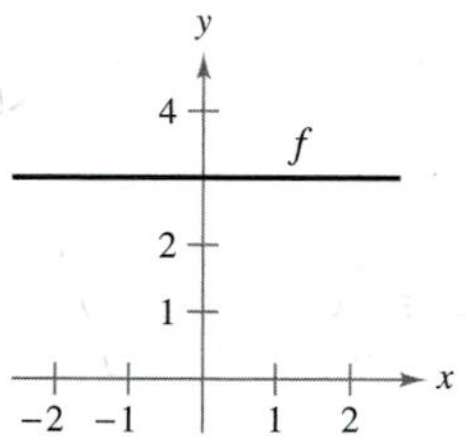

58.

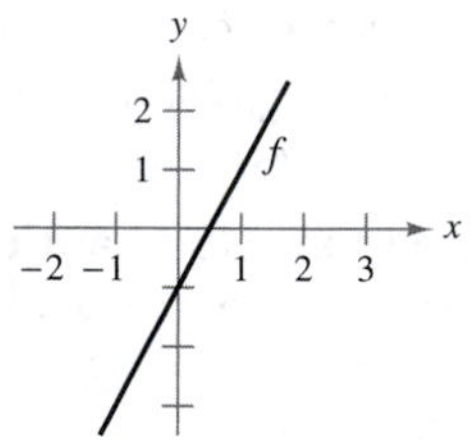

59.

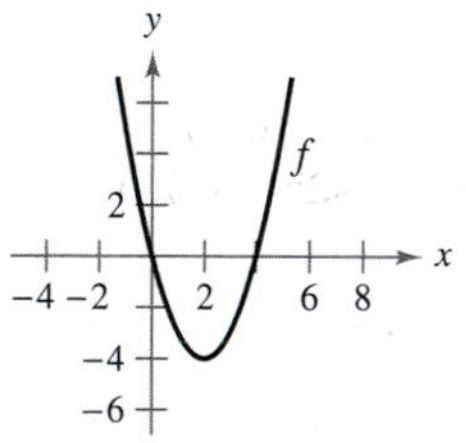

60.

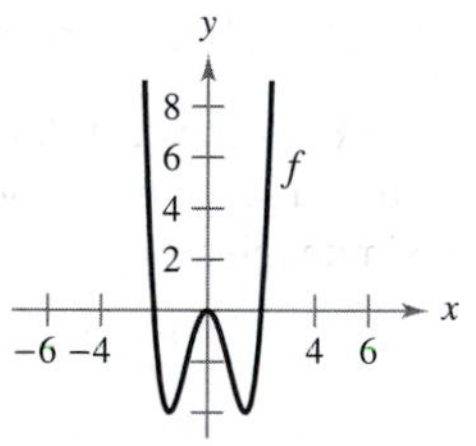

61.

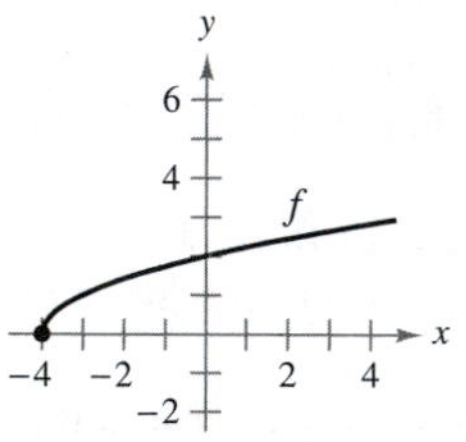

62.

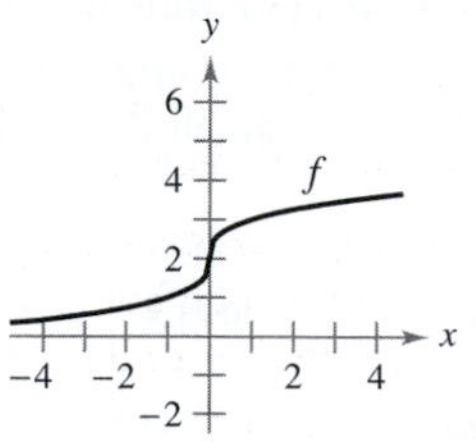

WRITING ABOUT CONCEPTS

Transformations of Functions

In Exercises 63–68, assume that f is differentiable for all x. The signs of f' are as follows.

$f'(x) > 0$ on $(-\infty, -4)$

$f'(x) < 0$ on $(-4, 6)$

$f'(x) > 0$ on $(6, \infty)$

Supply the appropriate inequality sign for the indicated value of c.

	Function	Sign of $g'(c)$
63.	$g(x) = f(x) + 5$	$g'(0) \;\square\; 0$
64.	$g(x) = 3f(x) - 3$	$g'(-5) \;\square\; 0$
65.	$g(x) = -f(x)$	$g'(-6) \;\square\; 0$
66.	$g(x) = -f(x)$	$g'(0) \;\square\; 0$
67.	$g(x) = f(x - 10)$	$g'(0) \;\square\; 0$
68.	$g(x) = f(x - 10)$	$g'(8) \;\square\; 0$

69. Sketching a Graph Sketch the graph of the arbitrary function f such that

$$f'(x)\begin{cases} > 0, & x < 4 \\ \text{undefined}, & x = 4. \\ < 0, & x > 4 \end{cases}$$

70. HOW DO YOU SEE IT? Use the graph of f' to (a) identify the critical numbers of f, (b) identify the open interval(s) on which f is increasing or decreasing, and (c) determine whether f has a relative maximum, a relative minimum, or neither at each critical number.

(i)

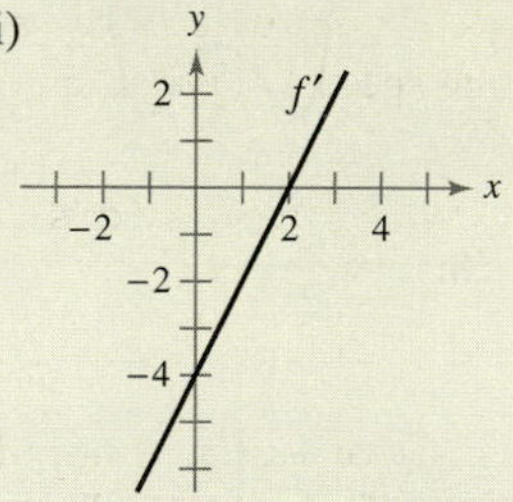

(ii)

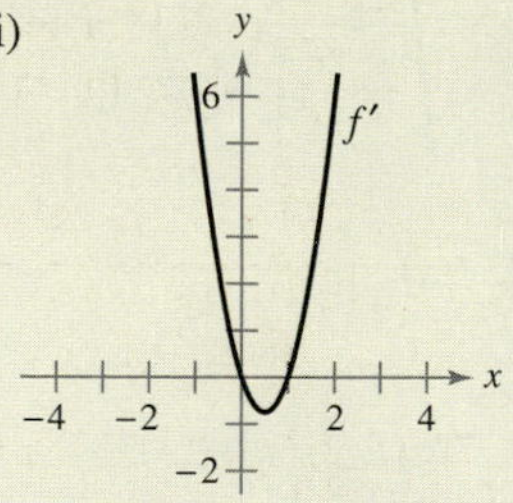

(iii)

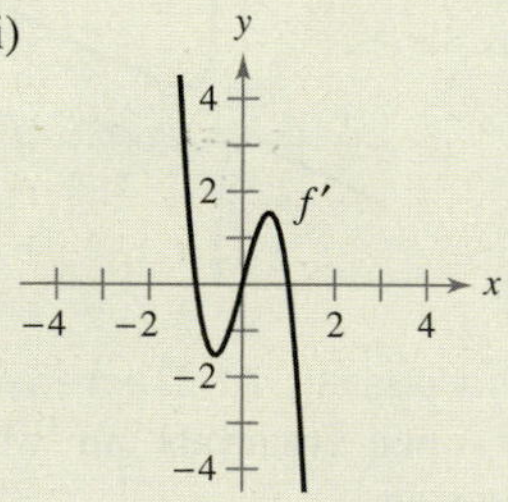

(iv)

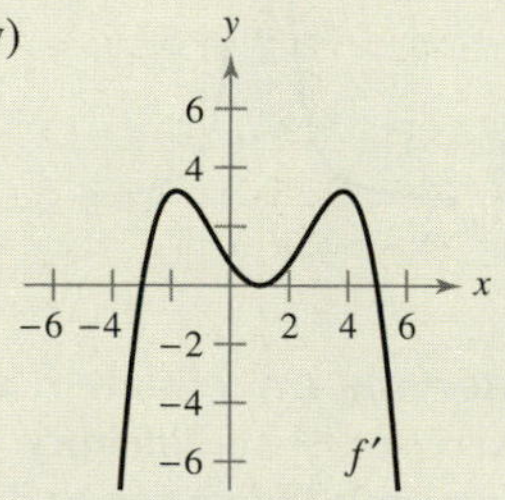

71. Analyzing a Critical Number A differentiable function f has one critical number at $x = 5$. Identify the relative extrema of f at the critical number when $f'(4) = -2.5$ and $f'(6) = 3$.

72. Analyzing a Critical Number A differentiable function f has one critical number at $x = 2$. Identify the relative extrema of f at the critical number when $f'(1) = 2$ and $f'(3) = 6$.

Think About It In Exercises 73 and 74, the function f is differentiable on the indicated interval. The table shows $f'(x)$ for selected values of x. (a) Sketch the graph of f, (b) approximate the critical numbers, and (c) identify the relative extrema.

73. f is differentiable on $[-1, 1]$.

x	-1	-0.75	-0.50	-0.25	0
$f'(x)$	-10	-3.2	-0.5	0.8	5.6

x	0.25	0.50	0.75	1
$f'(x)$	3.6	-0.2	-6.7	-20.1

74. f is differentiable on $[0, \pi]$.

x	0	$\pi/6$	$\pi/4$	$\pi/3$	$\pi/2$
$f'(x)$	3.14	-0.23	-2.45	-3.11	0.69

x	$2\pi/3$	$3\pi/4$	$5\pi/6$	π
$f'(x)$	3.00	1.37	-1.14	-2.84

75. Rolling a Ball Bearing A ball bearing is placed on an inclined plane and begins to roll. The angle of elevation of the plane is θ. The distance (in meters) the ball bearing rolls in t seconds is $s(t) = 4.9(\sin\theta)t^2$.

(a) Determine the speed of the ball bearing after t seconds.

(b) Complete the table and use it to determine the value of θ that produces the maximum speed at a particular time.

θ	0	$\pi/4$	$\pi/3$	$\pi/2$	$2\pi/3$	$3\pi/4$	π
$s'(t)$							

76. Modeling Data The end-of-year assets of the Medicare Hospital Insurance Trust Fund (in billions of dollars) for the years 1999 through 2010 are shown.

1999: 141.4; 2000: 177.5; 2001: 208.7; 2002: 234.8;
2003: 256.0; 2004: 269.3; 2005: 285.8; 2006: 305.4
2007: 326.0; 2008: 321.3; 2009: 304.2; 2010: 271.9

(Source: U.S. Centers for Medicare and Medicaid Services)

(a) Use the regression capabilities of a graphing utility to find a model of the form $M = at^4 + bt^3 + ct^2 + dt + e$ for the data. (Let $t = 9$ represent 1999.)

(b) Use a graphing utility to plot the data and graph the model.

(c) Find the maximum value of the model and compare the result with the actual data.

77. Numerical, Graphical, and Analytic Analysis The concentration C of a chemical in the bloodstream t hours after injection into muscle tissue is

$$C(t) = \frac{3t}{27 + t^3}, \quad t \geq 0.$$

(a) Complete the table and use it to approximate the time when the concentration is greatest.

t	0	0.5	1	1.5	2	2.5	3
$C(t)$							

(b) Use a graphing utility to graph the concentration function and use the graph to approximate the time when the concentration is greatest.

(c) Use calculus to determine analytically the time when the concentration is greatest.

78. Numerical, Graphical, and Analytic Analysis Consider the functions $f(x) = x$ and $g(x) = \sin x$ on the interval $(0, \pi)$.

(a) Complete the table and make a conjecture about which is the greater function on the interval $(0, \pi)$.

x	0.5	1	1.5	2	2.5	3
$f(x)$						
$g(x)$						

(b) Use a graphing utility to graph the functions and use the graphs to make a conjecture about which is the greater function on the interval $(0, \pi)$.

(c) Prove that $f(x) > g(x)$ on the interval $(0, \pi)$. [*Hint*: Show that $h'(x) > 0$, where $h = f - g$.]

79. Trachea Contraction Coughing forces the trachea (windpipe) to contract, which affects the velocity v of the air passing through the trachea. The velocity of the air during coughing is

$$v = k(R - r)r^2, \quad 0 \leq r < R$$

where k is a constant, R is the normal radius of the trachea, and r is the radius during coughing. What radius will produce the maximum air velocity?

80. Electrical Resistance The resistance R of a certain type of resistor is

$$R = \sqrt{0.001T^4 - 4T + 100}$$

where R is measured in ohms and the temperature T is measured in degrees Celsius.

(a) Use a computer algebra system to find dR/dT and the critical number of the function. Determine the minimum resistance for this type of resistor.

(b) Use a graphing utility to graph the function R and use the graph to approximate the minimum resistance for this type of resistor.

Motion Along a Line **In Exercises 81–84, the function $s(t)$ describes the motion of a particle along a line. For each function, (a) find the velocity function of the particle at any time $t \ge 0$, (b) identify the time interval(s) in which the particle is moving in a positive direction, (c) identify the time interval(s) in which the particle is moving in a negative direction, and (d) identify the time(s) at which the particle changes direction.**

81. $s(t) = 6t - t^2$

82. $s(t) = t^2 - 7t + 10$

83. $s(t) = t^3 - 5t^2 + 4t$

84. $s(t) = t^3 - 20t^2 + 128t - 280$

Motion Along a Line **In Exercises 85 and 86, the graph shows the position of a particle moving along a line. Describe how the particle's position changes with respect to time.**

85.

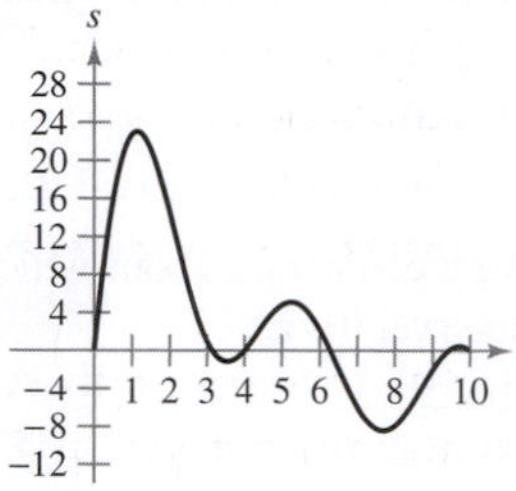

86.

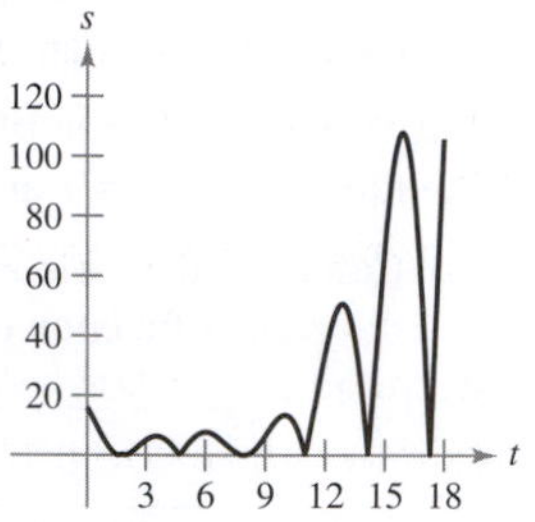

Creating Polynomial Functions **In Exercises 87–90, find a polynomial function**

$$f(x) = a_n x^n + a_{n-1}x^{n-1} + \cdots + a_2x^2 + a_1x + a_0$$

that has only the specified extrema. (a) Determine the minimum degree of the function and give the criteria you used in determining the degree. (b) Using the fact that the coordinates of the extrema are solution points of the function, and that the x-coordinates are critical numbers, determine a system of linear equations whose solution yields the coefficients of the required function. (c) Use a graphing utility to solve the system of equations and determine the function. (d) Use a graphing utility to confirm your result graphically.

87. Relative minimum: $(0, 0)$; Relative maximum: $(2, 2)$

88. Relative minimum: $(0, 0)$; Relative maximum: $(4, 1000)$

89. Relative minima: $(0, 0)$, $(4, 0)$; Relative maximum: $(2, 4)$

90. Relative minimum: $(1, 2)$; Relative maxima: $(-1, 4)$, $(3, 4)$

True or False? **In Exercises 91–96, determine whether the statement is true or false. If it is false, explain why or give an example that shows it is false.**

91. The sum of two increasing functions is increasing.

92. The product of two increasing functions is increasing.

93. Every nth-degree polynomial has $(n - 1)$ critical numbers.

94. An nth-degree polynomial has at most $(n - 1)$ critical numbers.

95. There is a relative maximum or minimum at each critical number.

96. The relative maxima of the function f are $f(1) = 4$ and $f(3) = 10$. Therefore, f has at least one minimum for some x in the interval $(1, 3)$.

97. Proof Prove the second case of Theorem 3.5.

98. Proof Prove the second case of Theorem 3.6.

99. Proof Use the definitions of increasing and decreasing functions to prove that $f(x) = x^3$ is increasing on $(-\infty, \infty)$.

100. Proof Use the definitions of increasing and decreasing functions to prove that

$$f(x) = \frac{1}{x}$$

is decreasing on $(0, \infty)$.

PUTNAM EXAM CHALLENGE

101. Find the minimum value of

$$|\sin x + \cos x + \tan x + \cot x + \sec x + \csc x|$$

for real numbers x.

This problem was composed by the Committee on the Putnam Prize Competition.

SECTION PROJECT

Rainbows

Rainbows are formed when light strikes raindrops and is reflected and refracted, as shown in the figure. (This figure shows a cross section of a spherical raindrop.) The Law of Refraction states that

$$\frac{\sin \alpha}{\sin \beta} = k$$

where $k \approx 1.33$ (for water). The angle of deflection is given by $D = \pi + 2\alpha - 4\beta$.

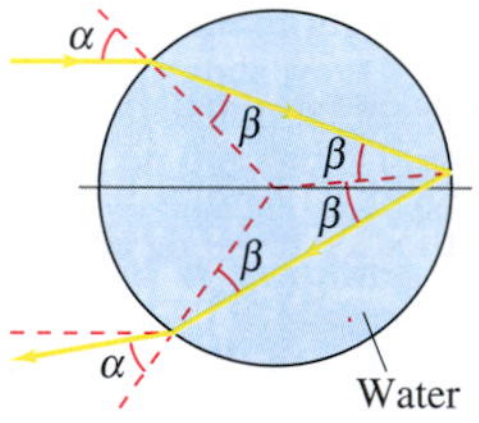

(a) Use a graphing utility to graph

$$D = \pi + 2\alpha - 4\sin^{-1}\left(\frac{\sin \alpha}{k}\right), \quad 0 \le \alpha \le \frac{\pi}{2}.$$

(b) Prove that the minimum angle of deflection occurs when

$$\cos \alpha = \sqrt{\frac{k^2 - 1}{3}}.$$

For water, what is the minimum angle of deflection $D_{\min}$? (The angle $\pi - D_{\min}$ is called the *rainbow angle*.) What value of α produces this minimum angle? (A ray of sunlight that strikes a raindrop at this angle, α, is called a *rainbow ray*.)

FOR FURTHER INFORMATION For more information about the mathematics of rainbows, see the article "Somewhere Within the Rainbow" by Steven Janke in *The UMAP Journal*.

3.4 Concavity and the Second Derivative Test

- Determine intervals on which a function is concave upward or concave downward.
- Find any points of inflection of the graph of a function.
- Apply the Second Derivative Test to find relative extrema of a function.

Concavity

You have already seen that locating the intervals in which a function f increases or decreases helps to describe its graph. In this section, you will see how locating the intervals in which f' increases or decreases can be used to determine where the graph of f is *curving upward* or *curving downward.*

> **Definition of Concavity**
>
> Let f be differentiable on an open interval I. The graph of f is **concave upward** on I when f' is increasing on the interval and **concave downward** on I when f' is decreasing on the interval.

The following graphical interpretation of concavity is useful. (See Appendix A for a proof of these results.) *See LarsonCalculus.com for Bruce Edwards's video of this proof.*

1. Let f be differentiable on an open interval I. If the graph of f is concave *upward* on I, then the graph of f lies *above* all of its tangent lines on I. [See Figure 3.23(a).]
2. Let f be differentiable on an open interval I. If the graph of f is concave *downward* on I, then the graph of f lies *below* all of its tangent lines on I. [See Figure 3.23(b).]

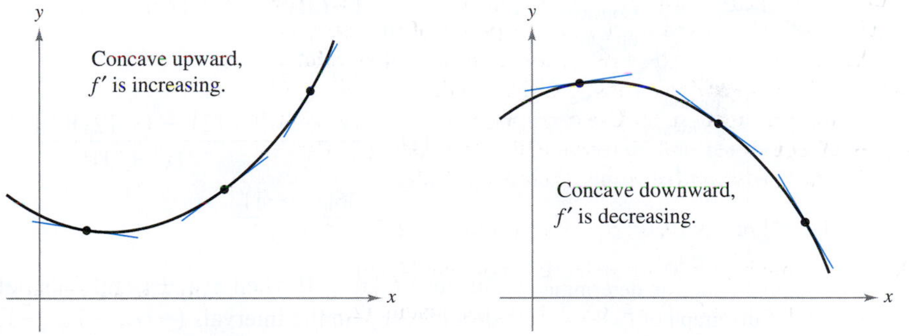

(a) The graph of f lies above its tangent lines. (b) The graph of f lies below its tangent lines.
Figure 3.23

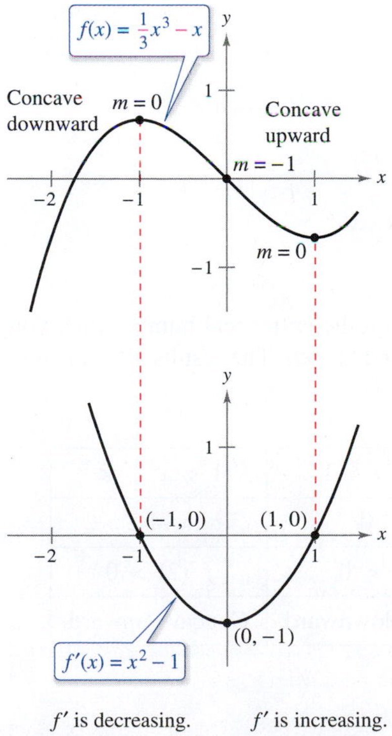

The concavity of f is related to the slope of the derivative.
Figure 3.24

To find the open intervals on which the graph of a function f is concave upward or concave downward, you need to find the intervals on which f' is increasing or decreasing. For instance, the graph of

$$f(x) = \frac{1}{3}x^3 - x$$

is concave downward on the open interval $(-\infty, 0)$ because

$$f'(x) = x^2 - 1$$

is decreasing there. (See Figure 3.24.) Similarly, the graph of f is concave upward on the interval $(0, \infty)$ because f' is increasing on $(0, \infty)$.

The next theorem shows how to use the *second* derivative of a function f to determine intervals on which the graph of f is concave upward or concave downward. A proof of this theorem follows directly from Theorem 3.5 and the definition of concavity.

REMARK A third case of Theorem 3.7 could be that if $f''(x) = 0$ for all x in I, then f is linear. Note, however, that concavity is not defined for a line. In other words, a straight line is neither concave upward nor concave downward.

THEOREM 3.7 Test for Concavity

Let f be a function whose second derivative exists on an open interval I.

1. If $f''(x) > 0$ for all x in I, then the graph of f is concave upward on I.
2. If $f''(x) < 0$ for all x in I, then the graph of f is concave downward on I.

A proof of this theorem is given in Appendix A.

See LarsonCalculus.com for Bruce Edwards's video of this proof.

To apply Theorem 3.7, locate the x-values at which $f''(x) = 0$ or f'' does not exist. Use these x-values to determine test intervals. Finally, test the sign of $f''(x)$ in each of the test intervals.

EXAMPLE 1 Determining Concavity

Determine the open intervals on which the graph of

$$f(x) = \frac{6}{x^2 + 3}$$

is concave upward or downward.

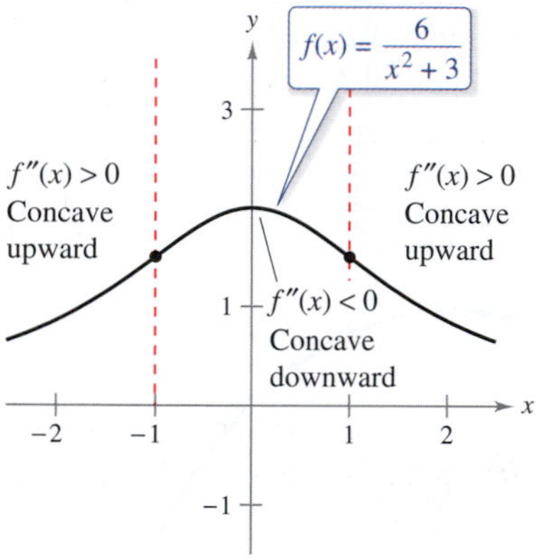

From the sign of f'', you can determine the concavity of the graph of f.

Figure 3.25

Solution Begin by observing that f is continuous on the entire real number line. Next, find the second derivative of f.

$$f(x) = 6(x^2 + 3)^{-1} \quad \text{Rewrite original function.}$$

$$f'(x) = (-6)(x^2 + 3)^{-2}(2x) \quad \text{Differentiate.}$$

$$= \frac{-12x}{(x^2 + 3)^2} \quad \text{First derivative}$$

$$f''(x) = \frac{(x^2 + 3)^2(-12) - (-12x)(2)(x^2 + 3)(2x)}{(x^2 + 3)^4} \quad \text{Differentiate.}$$

$$= \frac{36(x^2 - 1)}{(x^2 + 3)^3} \quad \text{Second derivative}$$

Because $f''(x) = 0$ when $x = \pm 1$ and f'' is defined on the entire real number line, you should test f'' in the intervals $(-\infty, -1)$, $(-1, 1)$, and $(1, \infty)$. The results are shown in the table and in Figure 3.25.

Interval	$-\infty < x < -1$	$-1 < x < 1$	$1 < x < \infty$
Test Value	$x = -2$	$x = 0$	$x = 2$
Sign of $f''(x)$	$f''(-2) > 0$	$f''(0) < 0$	$f''(2) > 0$
Conclusion	Concave upward	Concave downward	Concave upward

The function given in Example 1 is continuous on the entire real number line. When there are x-values at which the function is not continuous, these values should be used, along with the points at which $f''(x) = 0$ or $f''(x)$ does not exist, to form the test intervals.

EXAMPLE 2 **Determining Concavity**

Determine the open intervals on which the graph of

$$f(x) = \frac{x^2 + 1}{x^2 - 4}$$

is concave upward or concave downward.

Solution Differentiating twice produces the following.

$$f(x) = \frac{x^2 + 1}{x^2 - 4} \qquad \text{Write original function.}$$

$$f'(x) = \frac{(x^2 - 4)(2x) - (x^2 + 1)(2x)}{(x^2 - 4)^2} \qquad \text{Differentiate.}$$

$$= \frac{-10x}{(x^2 - 4)^2} \qquad \text{First derivative}$$

$$f''(x) = \frac{(x^2 - 4)^2(-10) - (-10x)(2)(x^2 - 4)(2x)}{(x^2 - 4)^4} \qquad \text{Differentiate.}$$

$$= \frac{10(3x^2 + 4)}{(x^2 - 4)^3} \qquad \text{Second derivative}$$

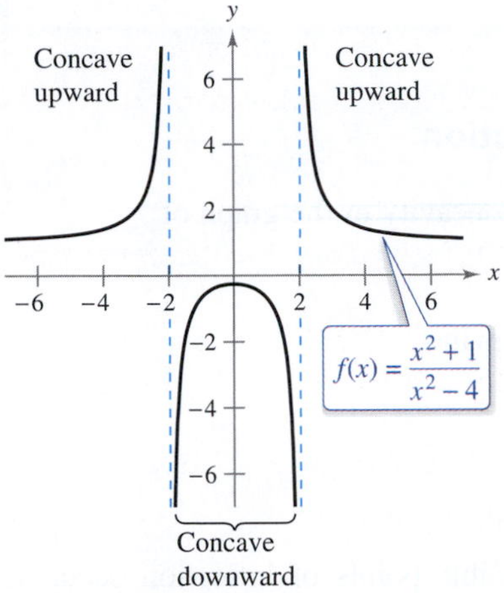

Figure 3.26

There are no points at which $f''(x) = 0$, but at $x = \pm 2$, the function f is not continuous. So, test for concavity in the intervals $(-\infty, -2)$, $(-2, 2)$, and $(2, \infty)$, as shown in the table. The graph of f is shown in Figure 3.26.

Interval	$-\infty < x < -2$	$-2 < x < 2$	$2 < x < \infty$
Test Value	$x = -3$	$x = 0$	$x = 3$
Sign of $f''(x)$	$f''(-3) > 0$	$f''(0) < 0$	$f''(3) > 0$
Conclusion	Concave upward	Concave downward	Concave upward

Points of Inflection

The graph in Figure 3.25 has two points at which the concavity changes. If the tangent line to the graph exists at such a point, then that point is a **point of inflection.** Three types of points of inflection are shown in Figure 3.27.

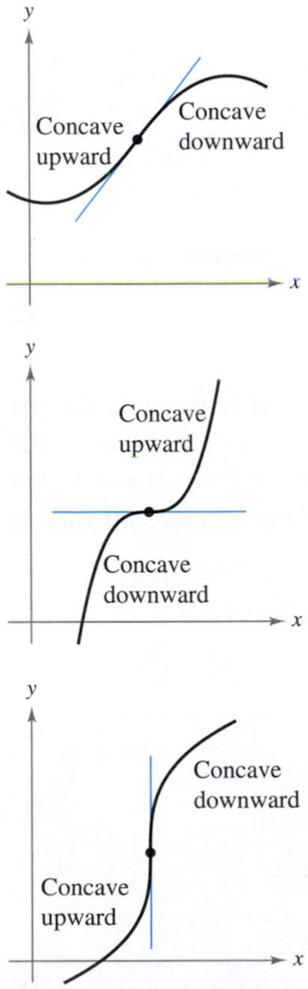

The concavity of f changes at a point of inflection. Note that the graph crosses its tangent line at a point of inflection.

Figure 3.27

Definition of Point of Inflection

Let f be a function that is continuous on an open interval, and let c be a point in the interval. If the graph of f has a tangent line at this point $(c, f(c))$, then this point is a **point of inflection** of the graph of f when the concavity of f changes from upward to downward (or downward to upward) at the point.

REMARK The definition of *point of inflection* requires that the tangent line exists at the point of inflection. Some books do not require this. For instance, we do not consider the function

$$f(x) = \begin{cases} x^3, & x < 0 \\ x^2 + 2x, & x \geq 0 \end{cases}$$

to have a point of inflection at the origin, even though the concavity of the graph changes from concave downward to concave upward.

To locate *possible* points of inflection, you can determine the values of x for which $f''(x) = 0$ or $f''(x)$ does not exist. This is similar to the procedure for locating relative extrema of f.

> **THEOREM 3.8 Points of Inflection**
>
> If $(c, f(c))$ is a point of inflection of the graph of f, then either $f''(c) = 0$ or f'' does not exist at $x = c$.

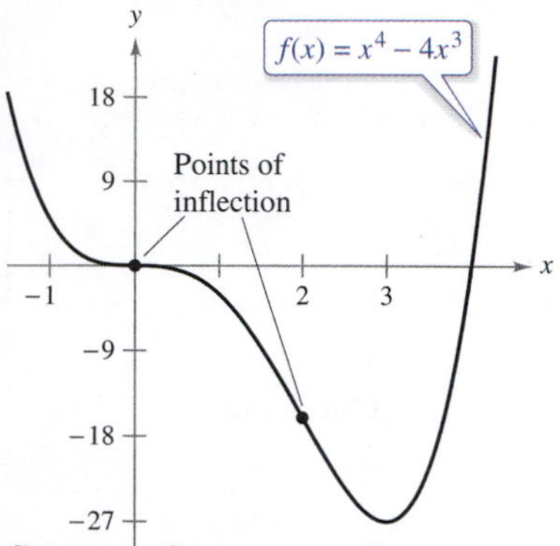

Points of inflection can occur where $f''(x) = 0$ or f'' does not exist.
Figure 3.28

EXAMPLE 3 Finding Points of Inflection

Determine the points of inflection and discuss the concavity of the graph of

$$f(x) = x^4 - 4x^3.$$

Solution Differentiating twice produces the following.

$$f(x) = x^4 - 4x^3 \qquad \text{Write original function.}$$

$$f'(x) = 4x^3 - 12x^2 \qquad \text{Find first derivative.}$$

$$f''(x) = 12x^2 - 24x = 12x(x - 2) \qquad \text{Find second derivative.}$$

Setting $f''(x) = 0$, you can determine that the possible points of inflection occur at $x = 0$ and $x = 2$. By testing the intervals determined by these x-values, you can conclude that they both yield points of inflection. A summary of this testing is shown in the table, and the graph of f is shown in Figure 3.28.

Interval	$-\infty < x < 0$	$0 < x < 2$	$2 < x < \infty$
Test Value	$x = -1$	$x = 1$	$x = 3$
Sign of $f''(x)$	$f''(-1) > 0$	$f''(1) < 0$	$f''(3) > 0$
Conclusion	Concave upward	Concave downward	Concave upward

> **Exploration**
>
> Consider a general cubic function of the form
>
> $$f(x) = ax^3 + bx^2 + cx + d.$$
>
> You know that the value of d has a bearing on the location of the graph but has no bearing on the value of the first derivative at given values of x. Graphically, this is true because changes in the value of d shift the graph up or down but do not change its basic shape. Use a graphing utility to graph several cubics with different values of c. Then give a graphical explanation of why changes in c do not affect the values of the second derivative.

The converse of Theorem 3.8 is not generally true. That is, it is possible for the second derivative to be 0 at a point that is *not* a point of inflection. For instance, the graph of $f(x) = x^4$ is shown in Figure 3.29. The second derivative is 0 when $x = 0$, but the point $(0, 0)$ is not a point of inflection because the graph of f is concave upward in both intervals $-\infty < x < 0$ and $0 < x < \infty$.

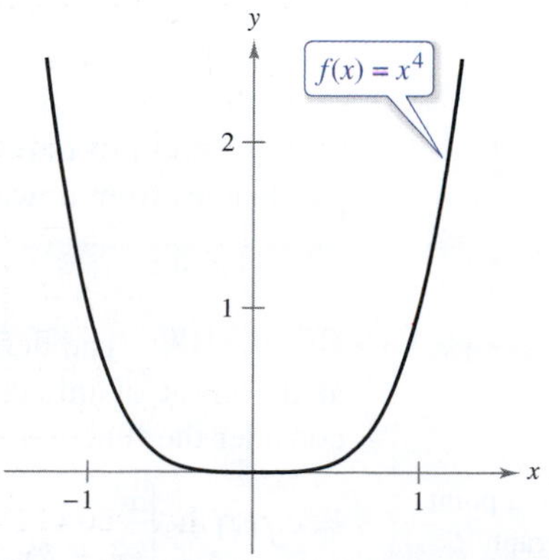

$f''(x) = 0$, but $(0, 0)$ is not a point of inflection.
Figure 3.29

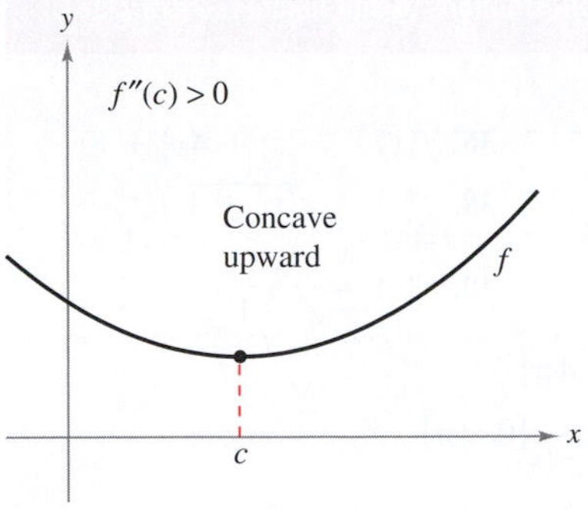

If $f'(c) = 0$ and $f''(c) > 0$, then $f(c)$ is a relative minimum.

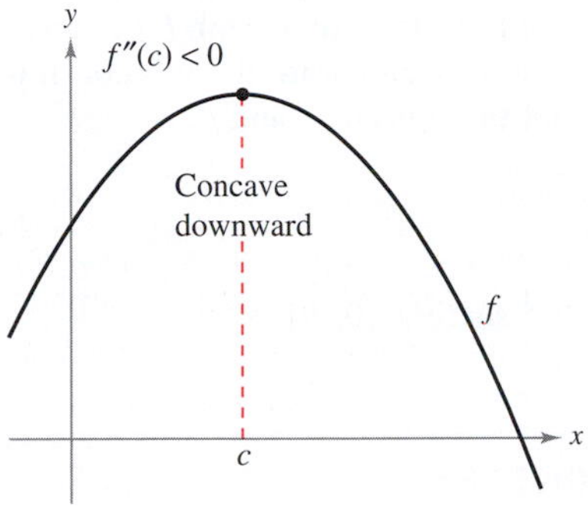

If $f'(c) = 0$ and $f''(c) < 0$, then $f(c)$ is a relative maximum.
Figure 3.30

The Second Derivative Test

In addition to testing for concavity, the second derivative can be used to perform a simple test for relative maxima and minima. The test is based on the fact that if the graph of a function f is concave upward on an open interval containing c, and $f'(c) = 0$, then $f(c)$ must be a relative minimum of f. Similarly, if the graph of a function f is concave downward on an open interval containing c, and $f'(c) = 0$, then $f(c)$ must be a relative maximum of f (see Figure 3.30).

> **THEOREM 3.9 Second Derivative Test**
>
> Let f be a function such that $f'(c) = 0$ and the second derivative of f exists on an open interval containing c.
>
> **1.** If $f''(c) > 0$, then f has a relative minimum at $(c, f(c))$.
> **2.** If $f''(c) < 0$, then f has a relative maximum at $(c, f(c))$.
>
> If $f''(c) = 0$, then the test fails. That is, f may have a relative maximum, a relative minimum, or neither. In such cases, you can use the First Derivative Test.

Proof If $f'(c) = 0$ and $f''(c) > 0$, then there exists an open interval I containing c for which

$$\frac{f'(x) - f'(c)}{x - c} = \frac{f'(x)}{x - c} > 0$$

for all $x \neq c$ in I. If $x < c$, then $x - c < 0$ and $f'(x) < 0$. Also, if $x > c$, then $x - c > 0$ and $f'(x) > 0$. So, $f'(x)$ changes from negative to positive at c, and the First Derivative Test implies that $f(c)$ is a relative minimum. A proof of the second case is left to you. *See LarsonCalculus.com for Bruce Edwards's video of this proof.*

AP* Tips

On the AP Exam, be prepared to apply the second derivative test to justify a point being a local maximum or minimum.

EXAMPLE 4 Using the Second Derivative Test

See LarsonCalculus.com for an interactive version of this type of example.

Find the relative extrema of

$$f(x) = -3x^5 + 5x^3.$$

Solution Begin by finding the first derivative of f.

$$f'(x) = -15x^4 + 15x^2 = 15x^2(1 - x^2)$$

From this derivative, you can see that $x = -1, 0,$ and 1 are the only critical numbers of f. By finding the second derivative

$$f''(x) = -60x^3 + 30x = 30x(1 - 2x^2)$$

you can apply the Second Derivative Test as shown below.

Point	$(-1, -2)$	$(0, 0)$	$(1, 2)$
Sign of $f''(x)$	$f''(-1) > 0$	$f''(0) = 0$	$f''(1) < 0$
Conclusion	Relative minimum	Test fails	Relative maximum

Because the Second Derivative Test fails at $(0, 0)$, you can use the First Derivative Test and observe that f increases to the left and right of $x = 0$. So, $(0, 0)$ is neither a relative minimum nor a relative maximum (even though the graph has a horizontal tangent line at this point). The graph of f is shown in Figure 3.31.

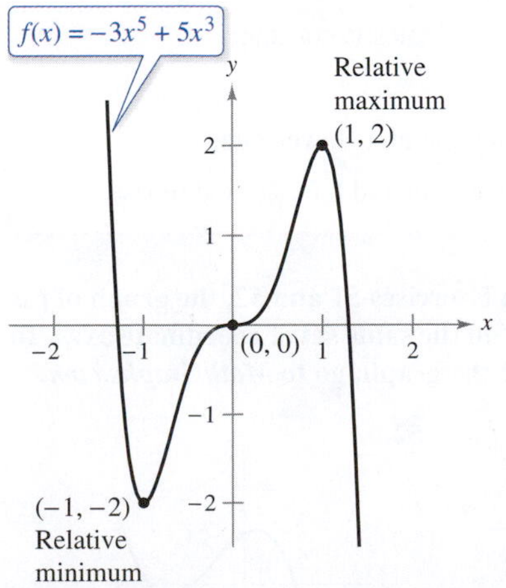

(0, 0) is neither a relative minimum nor a relative maximum.
Figure 3.31

3.4 Exercises

See CalcChat.com for tutorial help and worked-out solutions to odd-numbered exercises.

Using a Graph In Exercises 1 and 2, the graph of f is shown. State the signs of f' and f'' on the interval $(0, 2)$.

1.

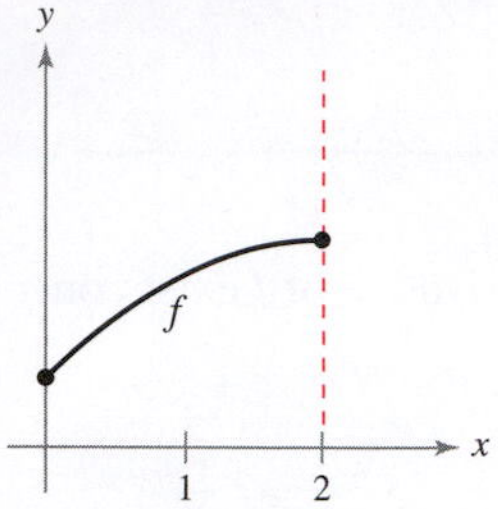

2.

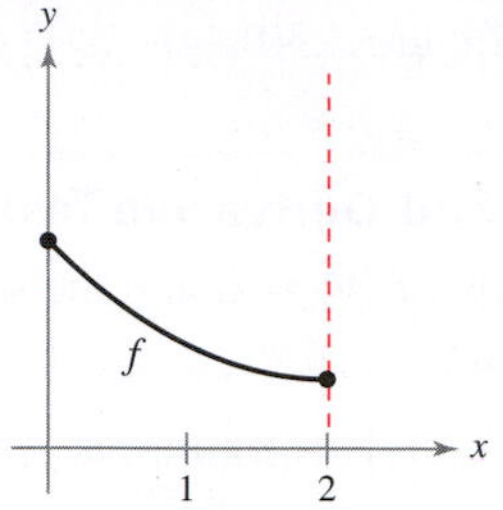

Determining Concavity In Exercises 3–14, determine the open intervals on which the graph is concave upward or concave downward.

3. $y = x^2 - x - 2$

4. $g(x) = 3x^2 - x^3$

5. $f(x) = -x^3 + 6x^2 - 9x - 1$

6. $h(x) = x^5 - 5x + 2$

7. $f(x) = \dfrac{24}{x^2 + 12}$

8. $f(x) = \dfrac{2x^2}{3x^2 + 1}$

9. $f(x) = \dfrac{x^2 + 1}{x^2 - 1}$

10. $y = \dfrac{-3x^5 + 40x^3 + 135x}{270}$

11. $g(x) = \dfrac{x^2 + 4}{4 - x^2}$

12. $h(x) = \dfrac{x^2 - 1}{2x - 1}$

13. $y = 2x - \tan x, \quad \left(-\dfrac{\pi}{2}, \dfrac{\pi}{2}\right)$

14. $y = x + \dfrac{2}{\sin x}, \quad (-\pi, \pi)$

Finding Points of Inflection In Exercises 15–30, find the points of inflection and discuss the concavity of the graph of the function.

15. $f(x) = x^3 - 6x^2 + 12x$

16. $f(x) = -x^3 + 6x^2 - 5$

17. $f(x) = \frac{1}{2}x^4 + 2x^3$

18. $f(x) = 4 - x - 3x^4$

19. $f(x) = x(x - 4)^3$

20. $f(x) = (x - 2)^3(x - 1)$

21. $f(x) = x\sqrt{x + 3}$

22. $f(x) = x\sqrt{9 - x}$

23. $f(x) = \dfrac{4}{x^2 + 1}$

24. $f(x) = \dfrac{x + 3}{\sqrt{x}}$

25. $f(x) = \sin \dfrac{x}{2}, \quad [0, 4\pi]$

26. $f(x) = 2 \csc \dfrac{3x}{2}, \quad (0, 2\pi)$

27. $f(x) = \sec\left(x - \dfrac{\pi}{2}\right), \quad (0, 4\pi)$

28. $f(x) = \sin x + \cos x, \quad [0, 2\pi]$

29. $f(x) = 2 \sin x + \sin 2x, \quad [0, 2\pi]$

30. $f(x) = x + 2 \cos x, \quad [0, 2\pi]$

Using the Second Derivative Test In Exercises 31–42, find all relative extrema. Use the Second Derivative Test where applicable.

31. $f(x) = 6x - x^2$

32. $f(x) = x^2 + 3x - 8$

33. $f(x) = x^3 - 3x^2 + 3$

34. $f(x) = -x^3 + 7x^2 - 15x$

35. $f(x) = x^4 - 4x^3 + 2$

36. $f(x) = -x^4 + 4x^3 + 8x^2$

37. $f(x) = x^{2/3} - 3$

38. $f(x) = \sqrt{x^2 + 1}$

39. $f(x) = x + \dfrac{4}{x}$

40. $f(x) = \dfrac{x}{x - 1}$

41. $f(x) = \cos x - x, \quad [0, 4\pi]$

42. $f(x) = 2 \sin x + \cos 2x, \quad [0, 2\pi]$

Finding Extrema and Points of Inflection Using Technology In Exercises 43–46, use a computer algebra system to analyze the function over the given interval. (a) Find the first and second derivatives of the function. (b) Find any relative extrema and points of inflection. (c) Graph f, f', and f'' on the same set of coordinate axes and state the relationship between the behavior of f and the signs of f' and f''.

43. $f(x) = 0.2x^2(x - 3)^3, \quad [-1, 4]$

44. $f(x) = x^2\sqrt{6 - x^2}, \quad \left[-\sqrt{6}, \sqrt{6}\right]$

45. $f(x) = \sin x - \frac{1}{3} \sin 3x + \frac{1}{5} \sin 5x, \quad [0, \pi]$

46. $f(x) = \sqrt{2x} \sin x, \quad [0, 2\pi]$

WRITING ABOUT CONCEPTS

47. Sketching a Graph Consider a function f such that f' is increasing. Sketch graphs of f for (a) $f' < 0$ and (b) $f' > 0$.

48. Sketching a Graph Consider a function f such that f' is decreasing. Sketch graphs of f for (a) $f' < 0$ and (b) $f' > 0$.

49. Sketching a Graph Sketch the graph of a function f that does *not* have a point of inflection at $(c, f(c))$ even though $f''(c) = 0$.

50. Think About It S represents weekly sales of a product. What can be said of S' and S'' for each of the following statements?

(a) The rate of change of sales is increasing.

(b) Sales are increasing at a slower rate.

(c) The rate of change of sales is constant.

(d) Sales are steady.

(e) Sales are declining, but at a slower rate.

(f) Sales have bottomed out and have started to rise.

Sketching Graphs In Exercises 51 and 52, the graph of f is shown. Graph f, f', and f'' on the same set of coordinate axes. To print an enlarged copy of the graph, go to *MathGraphs.com*.

51.

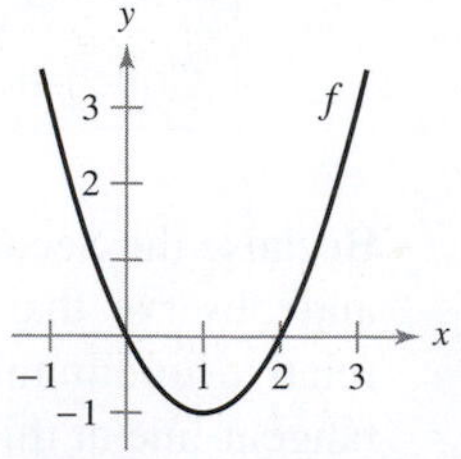

52.

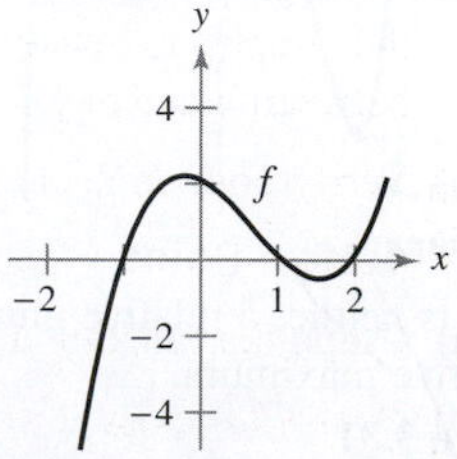

Think About It **In Exercises 53–56, sketch the graph of a function f having the given characteristics.**

53. $f(2) = f(4) = 0$

$f'(x) < 0$ for $x < 3$

$f'(3)$ does not exist.

$f'(x) > 0$ for $x > 3$

$f''(x) < 0, x \neq 3$

54. $f(0) = f(2) = 0$

$f'(x) > 0$ for $x < 1$

$f'(1) = 0$

$f'(x) < 0$ for $x > 1$

$f''(x) < 0$

55. $f(2) = f(4) = 0$

$f'(x) > 0$ for $x < 3$

$f'(3)$ does not exist.

$f'(x) < 0$ for $x > 3$

$f''(x) > 0, x \neq 3$

56. $f(0) = f(2) = 0$

$f'(x) < 0$ for $x < 1$

$f'(1) = 0$

$f'(x) > 0$ for $x > 1$

$f''(x) > 0$

57. Think About It The figure shows the graph of f''. Sketch a graph of f. (The answer is not unique.) To print an enlarged copy of the graph, go to *MathGraphs.com*.

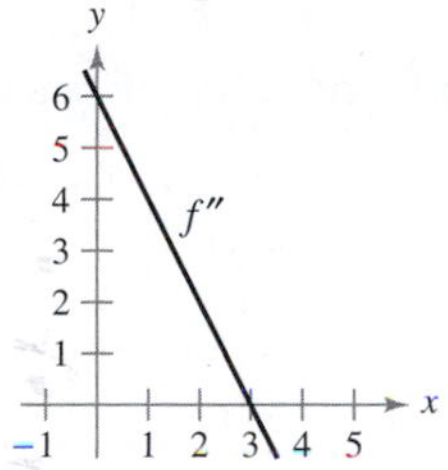

58. HOW DO YOU SEE IT? Water is running into the vase shown in the figure at a constant rate.

(a) Graph the depth d of water in the vase as a function of time.

(b) Does the function have any extrema? Explain.

(c) Interpret the inflection points of the graph of d.

59. Conjecture Consider the function

$f(x) = (x - 2)^n$.

(a) Use a graphing utility to graph f for $n = 1, 2, 3$, and 4. Use the graphs to make a conjecture about the relationship between n and any inflection points of the graph of f.

(b) Verify your conjecture in part (a).

60. Inflection Point Consider the function $f(x) = \sqrt[3]{x}$.

(a) Graph the function and identify the inflection point.

(b) Does $f''(x)$ exist at the inflection point? Explain.

Finding a Cubic Function **In Exercises 61 and 62, find a, b, c, and d such that the cubic**

$$f(x) = ax^3 + bx^2 + cx + d$$

satisfies the given conditions.

61. Relative maximum: $(3, 3)$

Relative minimum: $(5, 1)$

Inflection point: $(4, 2)$

62. Relative maximum: $(2, 4)$

Relative minimum: $(4, 2)$

Inflection point: $(3, 3)$

63. Aircraft Glide Path A small aircraft starts its descent from an altitude of 1 mile, 4 miles west of the runway (see figure).

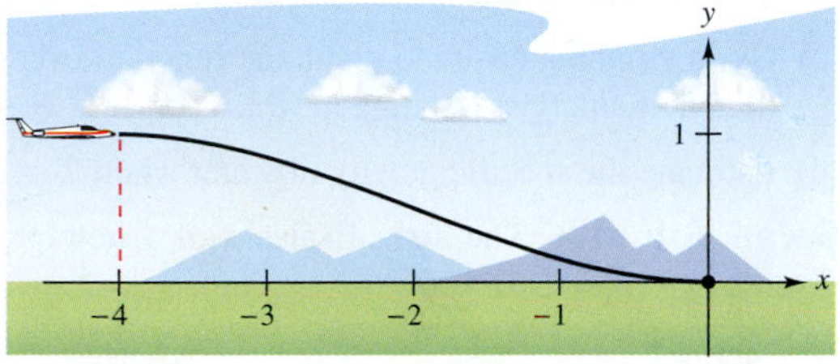

(a) Find the cubic $f(x) = ax^3 + bx^2 + cx + d$ on the interval $[-4, 0]$ that describes a smooth glide path for the landing.

(b) The function in part (a) models the glide path of the plane. When would the plane be descending at the greatest rate?

FOR FURTHER INFORMATION For more information on this type of modeling, see the article "How Not to Land at Lake Tahoe!" by Richard Barshinger in *The American Mathematical Monthly*. To view this article, go to *MathArticles.com*.

64. Highway Design A section of highway connecting two hillsides with grades of 6% and 4% is to be built between two points that are separated by a horizontal distance of 2000 feet (see figure). At the point where the two hillsides come together, there is a 50-foot difference in elevation.

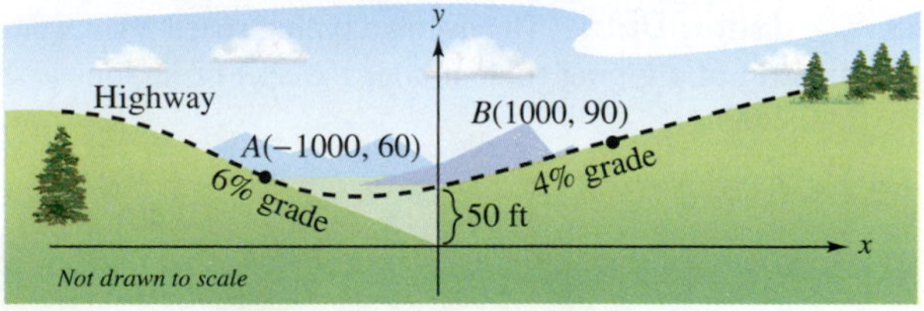

(a) Design a section of highway connecting the hillsides modeled by the function

$$f(x) = ax^3 + bx^2 + cx + d, \quad -1000 \leq x \leq 1000.$$

At points A and B, the slope of the model must match the grade of the hillside.

(b) Use a graphing utility to graph the model.

(c) Use a graphing utility to graph the derivative of the model.

(d) Determine the grade at the steepest part of the transitional section of the highway.

65. Average Cost A manufacturer has determined that the total cost C of operating a factory is

$$C = 0.5x^2 + 15x + 5000$$

where x is the number of units produced. At what level of production will the average cost per unit be minimized? (The average cost per unit is C/x.)

66. Specific Gravity A model for the specific gravity of water S is

$$S = \frac{5.755}{10^8}T^3 - \frac{8.521}{10^6}T^2 + \frac{6.540}{10^5}T + 0.99987, \; 0 < T < 25$$

where T is the water temperature in degrees Celsius.

(a) Use the second derivative to determine the concavity of S.

(b) Use a computer algebra system to find the coordinates of the maximum value of the function.

(c) Use a graphing utility to graph the function over the specified domain. (Use a setting in which $0.996 \le S \le 1.001$.)

(d) Estimate the specific gravity of water when $T = 20°$.

67. Sales Growth The annual sales S of a new product are given by

$$S = \frac{5000t^2}{8 + t^2}, \quad 0 \le t \le 3$$

where t is time in years.

(a) Complete the table. Then use it to estimate when the annual sales are increasing at the greatest rate.

t	0.5	1	1.5	2	2.5	3
S						

(b) Use a graphing utility to graph the function S. Then use the graph to estimate when the annual sales are increasing at the greatest rate.

(c) Find the exact time when the annual sales are increasing at the greatest rate.

68. Modeling Data The average typing speed S (in words per minute) of a typing student after t weeks of lessons is shown in the table.

t	5	10	15	20	25	30
S	38	56	79	90	93	94

A model for the data is

$$S = \frac{100t^2}{65 + t^2}, \quad t > 0.$$

(a) Use a graphing utility to plot the data and graph the model.

(b) Use the second derivative to determine the concavity of S. Compare the result with the graph in part (a).

(c) What is the sign of the first derivative for $t > 0$? By combining this information with the concavity of the model, what inferences can be made about the typing speed as t increases?

Linear and Quadratic Approximations **In Exercises 69–72, use a graphing utility to graph the function. Then graph the linear and quadratic approximations**

$$P_1(x) = f(a) + f'(a)(x - a)$$

and

$$P_2(x) = f(a) + f'(a)(x - a) + \tfrac{1}{2}f''(a)(x - a)^2$$

in the same viewing window. Compare the values of f, P_1, and P_2 and their first derivatives at $x = a$. How do the approximations change as you move farther away from $x = a$?

Function	Value of a
69. $f(x) = 2(\sin x + \cos x)$	$a = \frac{\pi}{4}$
70. $f(x) = 2(\sin x + \cos x)$	$a = 0$
71. $f(x) = \sqrt{1 - x}$	$a = 0$
72. $f(x) = \frac{\sqrt{x}}{x - 1}$	$a = 2$

73. Determining Concavity Use a graphing utility to graph

$$y = x \sin \frac{1}{x}.$$

Show that the graph is concave downward to the right of

$$x = \frac{1}{\pi}.$$

74. Point of Inflection and Extrema Show that the point of inflection of

$$f(x) = x(x - 6)^2$$

lies midway between the relative extrema of f.

True or False? **In Exercises 75–78, determine whether the statement is true or false. If it is false, explain why or give an example that shows it is false.**

75. The graph of every cubic polynomial has precisely one point of inflection.

76. The graph of

$$f(x) = \frac{1}{x}$$

is concave downward for $x < 0$ and concave upward for $x > 0$, and thus it has a point of inflection at $x = 0$.

77. If $f'(c) > 0$, then f is concave upward at $x = c$.

78. If $f''(2) = 0$, then the graph of f must have a point of inflection at $x = 2$.

Proof **In Exercises 79 and 80, let f and g represent differentiable functions such that $f'' \neq 0$ and $g'' \neq 0$.**

79. Show that if f and g are concave upward on the interval (a, b), then $f + g$ is also concave upward on (a, b).

80. Prove that if f and g are positive, increasing, and concave upward on the interval (a, b), then fg is also concave upward on (a, b).

3.5 Limits at Infinity

- **Determine (finite) limits at infinity.**
- **Determine the horizontal asymptotes, if any, of the graph of a function.**
- **Determine infinite limits at infinity.**

Limits at Infinity

This section discusses the "end behavior" of a function on an *infinite* interval. Consider the graph of

$$f(x) = \frac{3x^2}{x^2 + 1}$$

as shown in Figure 3.32. Graphically, you can see that the values of $f(x)$ appear to approach 3 as x increases without bound or decreases without bound. You can come to the same conclusions numerically, as shown in the table.

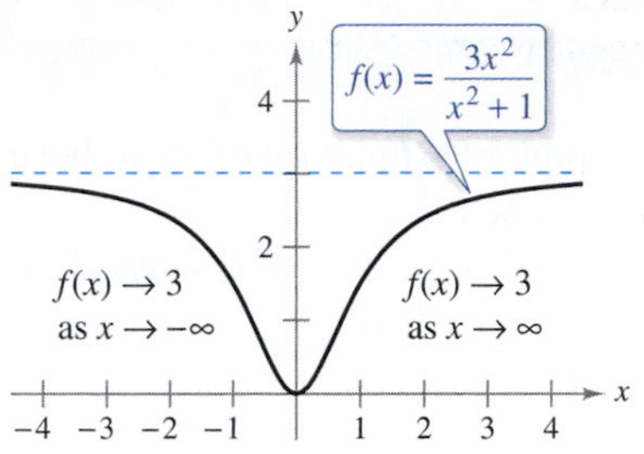

The limit of $f(x)$ as x approaches $-\infty$ or ∞ is 3.
Figure 3.32

← *x* decreases without bound. | *x* increases without bound. →

x	$-\infty \leftarrow$	-100	-10	-1	0	1	10	100	$\to \infty$
$f(x)$	$3 \leftarrow$	2.9997	2.9703	1.5	0	1.5	2.9703	2.9997	$\to 3$

← $f(x)$ approaches 3. | $f(x)$ approaches 3. →

The table suggests that the value of $f(x)$ approaches 3 as x increases without bound ($x \to \infty$). Similarly, $f(x)$ approaches 3 as x decreases without bound ($x \to -\infty$). These **limits at infinity** are denoted by

$$\lim_{x \to -\infty} f(x) = 3 \qquad \text{Limit at negative infinity}$$

and

$$\lim_{x \to \infty} f(x) = 3. \qquad \text{Limit at positive infinity}$$

REMARK The statement $\lim_{x \to -\infty} f(x) = L$ or $\lim_{x \to \infty} f(x) = L$ means that the limit exists *and* the limit is equal to L.

To say that a statement is true as x increases *without bound* means that for some (large) real number M, the statement is true for *all* x in the interval $\{x: x > M\}$. The next definition uses this concept.

Definition of Limits at Infinity

Let L be a real number.

1. The statement $\lim_{x \to \infty} f(x) = L$ means that for each $\varepsilon > 0$ there exists an $M > 0$ such that $|f(x) - L| < \varepsilon$ whenever $x > M$.
2. The statement $\lim_{x \to -\infty} f(x) = L$ means that for each $\varepsilon > 0$ there exists an $N < 0$ such that $|f(x) - L| < \varepsilon$ whenever $x < N$.

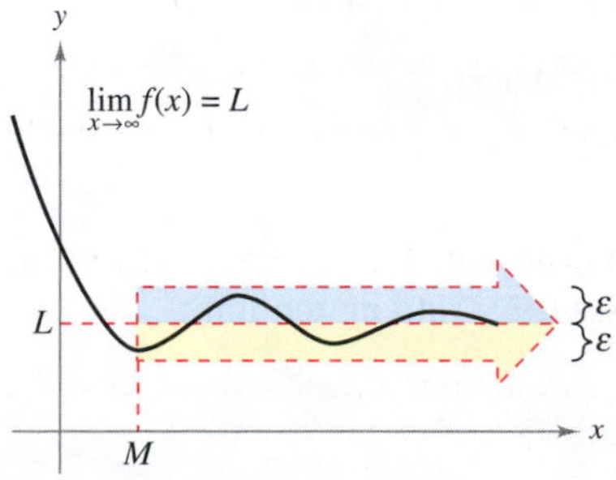

$f(x)$ is within ε units of L as $x \to \infty$.
Figure 3.33

The definition of a limit at infinity is shown in Figure 3.33. In this figure, note that for a given positive number ε, there exists a positive number M such that, for $x > M$, the graph of f will lie between the horizontal lines

$$y = L + \varepsilon \quad \text{and} \quad y = L - \varepsilon.$$

> **Exploration**
>
> Use a graphing utility to graph
>
> $$f(x) = \frac{2x^2 + 4x - 6}{3x^2 + 2x - 16}.$$
>
> Describe all the important features of the graph. Can you find a single viewing window that shows all of these features clearly? Explain your reasoning.
>
> What are the horizontal asymptotes of the graph? How far to the right do you have to move on the graph so that the graph is within 0.001 unit of its horizontal asymptote? Explain your reasoning.

Horizontal Asymptotes

In Figure 3.33, the graph of f approaches the line $y = L$ as x increases without bound. The line $y = L$ is called a **horizontal asymptote** of the graph of f.

> **Definition of a Horizontal Asymptote**
>
> The line $y = L$ is a **horizontal asymptote** of the graph of f when
>
> $$\lim_{x\to-\infty} f(x) = L \quad \text{or} \quad \lim_{x\to\infty} f(x) = L.$$

Note that from this definition, it follows that the graph of a *function* of x can have at most two horizontal asymptotes—one to the right and one to the left.

Limits at infinity have many of the same properties of limits discussed in Section 1.3. For example, if $\lim_{x\to\infty} f(x)$ and $\lim_{x\to\infty} g(x)$ both exist, then

$$\lim_{x\to\infty} [f(x) + g(x)] = \lim_{x\to\infty} f(x) + \lim_{x\to\infty} g(x)$$

and

$$\lim_{x\to\infty} [f(x)g(x)] = \left[\lim_{x\to\infty} f(x)\right]\left[\lim_{x\to\infty} g(x)\right].$$

Similar properties hold for limits at $-\infty$.

When evaluating limits at infinity, the next theorem is helpful.

AP* Tips

The AP Exam frequently uses limits at infinity as a way of describing horizontal asymptotes.

> **THEOREM 3.10 Limits at Infinity**
>
> If r is a positive rational number and c is any real number, then
>
> $$\lim_{x\to\infty} \frac{c}{x^r} = 0.$$
>
> Furthermore, if x^r is defined when $x < 0$, then
>
> $$\lim_{x\to-\infty} \frac{c}{x^r} = 0.$$
>
> A proof of this theorem is given in Appendix A.
>
> *See LarsonCalculus.com for Bruce Edwards's video of this proof.*

EXAMPLE 1 Finding a Limit at Infinity

Find the limit: $\lim_{x\to\infty} \left(5 - \frac{2}{x^2}\right)$.

Solution Using Theorem 3.10, you can write

$$\begin{aligned}\lim_{x\to\infty} \left(5 - \frac{2}{x^2}\right) &= \lim_{x\to\infty} 5 - \lim_{x\to\infty} \frac{2}{x^2} && \text{Property of limits}\\ &= 5 - 0\\ &= 5.\end{aligned}$$

So, the line $y = 5$ is a horizontal asymptote to the right. By finding the limit

$$\lim_{x\to-\infty} \left(5 - \frac{2}{x^2}\right) \qquad \text{Limit as } x \to -\infty.$$

you can see that $y = 5$ is also a horizontal asymptote to the left. The graph of the function is shown in Figure 3.34.

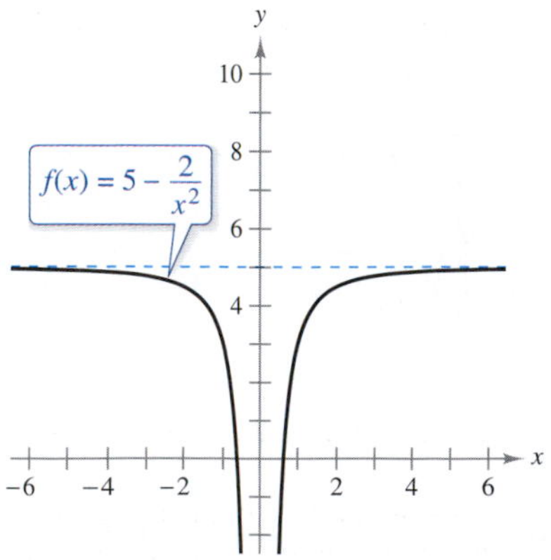

$y = 5$ is a horizontal asymptote.
Figure 3.34

EXAMPLE 2 Finding a Limit at Infinity

Find the limit: $\lim_{x\to\infty} \dfrac{2x-1}{x+1}$.

Solution Note that both the numerator and the denominator approach infinity as x approaches infinity.

$$\lim_{x\to\infty} \frac{2x-1}{x+1} \qquad \begin{cases} \lim_{x\to\infty} (2x-1) \to \infty \\ \lim_{x\to\infty} (x+1) \to \infty \end{cases}$$

This results in $\dfrac{\infty}{\infty}$, an **indeterminate form.** To resolve this problem, you can divide both the numerator and the denominator by x. After dividing, the limit may be evaluated as shown.

$$\lim_{x\to\infty} \frac{2x-1}{x+1} = \lim_{x\to\infty} \frac{\frac{2x-1}{x}}{\frac{x+1}{x}} \qquad \text{Divide numerator and denominator by } x.$$

$$= \lim_{x\to\infty} \frac{2-\frac{1}{x}}{1+\frac{1}{x}} \qquad \text{Simplify.}$$

$$= \frac{\lim_{x\to\infty} 2 - \lim_{x\to\infty} \frac{1}{x}}{\lim_{x\to\infty} 1 + \lim_{x\to\infty} \frac{1}{x}} \qquad \text{Take limits of numerator and denominator.}$$

$$= \frac{2-0}{1+0} \qquad \text{Apply Theorem 3.10.}$$

$$= 2$$

So, the line $y = 2$ is a horizontal asymptote to the right. By taking the limit as $x \to -\infty$, you can see that $y = 2$ is also a horizontal asymptote to the left. The graph of the function is shown in Figure 3.35.

REMARK When you encounter an indeterminate form such as the one in Example 2, you should divide the numerator and denominator by the highest power of x in the *denominator*.

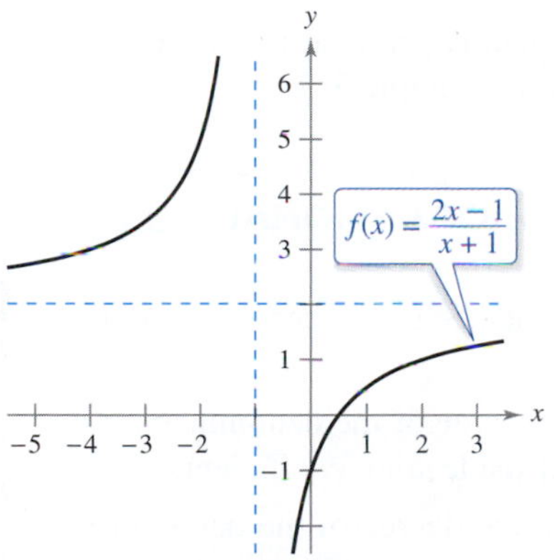

$y = 2$ is a horizontal asymptote.
Figure 3.35

TECHNOLOGY You can test the reasonableness of the limit found in Example 2 by evaluating $f(x)$ for a few large positive values of x. For instance,

$$f(100) \approx 1.9703, \quad f(1000) \approx 1.9970, \quad \text{and} \quad f(10{,}000) \approx 1.9997.$$

Another way to test the reasonableness of the limit is to use a graphing utility. For instance, in Figure 3.36, the graph of

$$f(x) = \frac{2x-1}{x+1}$$

is shown with the horizontal line $y = 2$. Note that as x increases, the graph of f moves closer and closer to its horizontal asymptote.

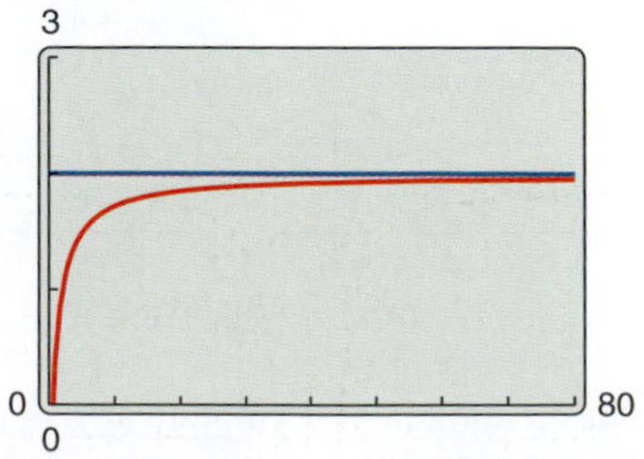

As x increases, the graph of f moves closer and closer to the line $y = 2$.
Figure 3.36

EXAMPLE 3 A Comparison of Three Rational Functions

See LarsonCalculus.com for an interactive version of this type of example.

Find each limit.

a. $\lim_{x\to\infty} \dfrac{2x + 5}{3x^2 + 1}$ **b.** $\lim_{x\to\infty} \dfrac{2x^2 + 5}{3x^2 + 1}$ **c.** $\lim_{x\to\infty} \dfrac{2x^3 + 5}{3x^2 + 1}$

Solution In each case, attempting to evaluate the limit produces the indeterminate form ∞/∞.

a. Divide both the numerator and the denominator by x^2.

$$\lim_{x\to\infty} \frac{2x + 5}{3x^2 + 1} = \lim_{x\to\infty} \frac{(2/x) + (5/x^2)}{3 + (1/x^2)} = \frac{0 + 0}{3 + 0} = \frac{0}{3} = 0$$

b. Divide both the numerator and the denominator by x^2.

$$\lim_{x\to\infty} \frac{2x^2 + 5}{3x^2 + 1} = \lim_{x\to\infty} \frac{2 + (5/x^2)}{3 + (1/x^2)} = \frac{2 + 0}{3 + 0} = \frac{2}{3}$$

c. Divide both the numerator and the denominator by x^2.

$$\lim_{x\to\infty} \frac{2x^3 + 5}{3x^2 + 1} = \lim_{x\to\infty} \frac{2x + (5/x^2)}{3 + (1/x^2)} = \frac{\infty}{3}$$

You can conclude that the limit *does not exist* because the numerator increases without bound while the denominator approaches 3.

MARIA GAETANA AGNESI (1718–1799)

Agnesi was one of a handful of women to receive credit for significant contributions to mathematics before the twentieth century. In her early twenties, she wrote the first text that included both differential and integral calculus. By age 30, she was an honorary member of the faculty at the University of Bologna.

See LarsonCalculus.com to read more of this biography.

For more information on the contributions of women to mathematics, see the article "Why Women Succeed in Mathematics" by Mona Fabricant, Sylvia Svitak, and Patricia Clark Kenschaft in *Mathematics Teacher*. To view this article, go to *MathArticles.com*.

Example 3 suggests the guidelines below for finding limits at infinity of rational functions. Use these guidelines to check the results in Example 3.

GUIDELINES FOR FINDING LIMITS AT $\pm\infty$ OF RATIONAL FUNCTIONS

1. If the degree of the numerator is *less than* the degree of the denominator, then the limit of the rational function is 0.
2. If the degree of the numerator is *equal to* the degree of the denominator, then the limit of the rational function is the ratio of the leading coefficients.
3. If the degree of the numerator is *greater than* the degree of the denominator, then the limit of the rational function does not exist.

The guidelines for finding limits at infinity of rational functions seem reasonable when you consider that for large values of x, the highest-power term of the rational function is the most "influential" in determining the limit. For instance,

$$\lim_{x\to\infty} \frac{1}{x^2 + 1}$$

is 0 because the denominator overpowers the numerator as x increases or decreases without bound, as shown in Figure 3.37.

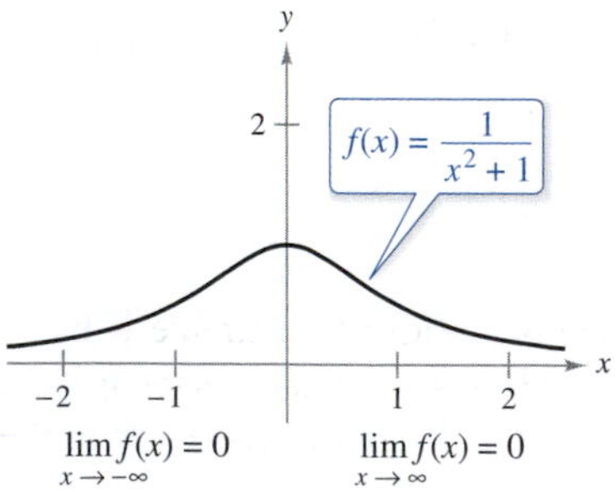

f has a horizontal asymptote at $y = 0$.
Figure 3.37

The function shown in Figure 3.37 is a special case of a type of curve studied by the Italian mathematician Maria Gaetana Agnesi. The general form of this function is

$$f(x) = \frac{8a^3}{x^2 + 4a^2} \qquad \text{Witch of Agnesi}$$

and, through a mistranslation of the Italian word *vertéré*, the curve has come to be known as the Witch of Agnesi. Agnesi's work with this curve first appeared in a comprehensive text on calculus that was published in 1748.

In Figure 3.37, you can see that the function

$$f(x) = \frac{1}{x^2 + 1}$$

approaches the same horizontal asymptote to the right and to the left. This is always true of rational functions. Functions that are not rational, however, may approach different horizontal asymptotes to the right and to the left. This is demonstrated in Example 4.

EXAMPLE 4 A Function with Two Horizontal Asymptotes

Find each limit.

a. $\lim\limits_{x\to\infty} \dfrac{3x - 2}{\sqrt{2x^2 + 1}}$ **b.** $\lim\limits_{x\to-\infty} \dfrac{3x - 2}{\sqrt{2x^2 + 1}}$

Solution

a. For $x > 0$, you can write $x = \sqrt{x^2}$. So, dividing both the numerator and the denominator by x produces

$$\frac{3x - 2}{\sqrt{2x^2 + 1}} = \frac{\dfrac{3x - 2}{x}}{\dfrac{\sqrt{2x^2 + 1}}{\sqrt{x^2}}} = \frac{3 - \dfrac{2}{x}}{\sqrt{\dfrac{2x^2 + 1}{x^2}}} = \frac{3 - \dfrac{2}{x}}{\sqrt{2 + \dfrac{1}{x^2}}}$$

and you can take the limit as follows.

$$\lim_{x\to\infty} \frac{3x - 2}{\sqrt{2x^2 + 1}} = \lim_{x\to\infty} \frac{3 - \dfrac{2}{x}}{\sqrt{2 + \dfrac{1}{x^2}}} = \frac{3 - 0}{\sqrt{2 + 0}} = \frac{3}{\sqrt{2}}$$

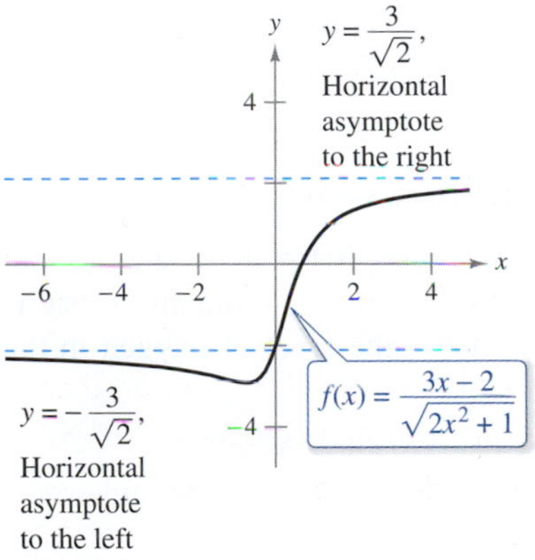

Functions that are not rational may have different right and left horizontal asymptotes.

Figure 3.38

b. For $x < 0$, you can write $x = -\sqrt{x^2}$. So, dividing both the numerator and the denominator by x produces

$$\frac{3x - 2}{\sqrt{2x^2 + 1}} = \frac{\dfrac{3x - 2}{x}}{\dfrac{\sqrt{2x^2 + 1}}{-\sqrt{x^2}}} = \frac{3 - \dfrac{2}{x}}{-\sqrt{\dfrac{2x^2 + 1}{x^2}}} = \frac{3 - \dfrac{2}{x}}{-\sqrt{2 + \dfrac{1}{x^2}}}$$

and you can take the limit as follows.

$$\lim_{x\to-\infty} \frac{3x - 2}{\sqrt{2x^2 + 1}} = \lim_{x\to-\infty} \frac{3 - \dfrac{2}{x}}{-\sqrt{2 + \dfrac{1}{x^2}}} = \frac{3 - 0}{-\sqrt{2 + 0}} = -\frac{3}{\sqrt{2}}$$

The graph of $f(x) = (3x - 2)/\sqrt{2x^2 + 1}$ is shown in Figure 3.38. ■

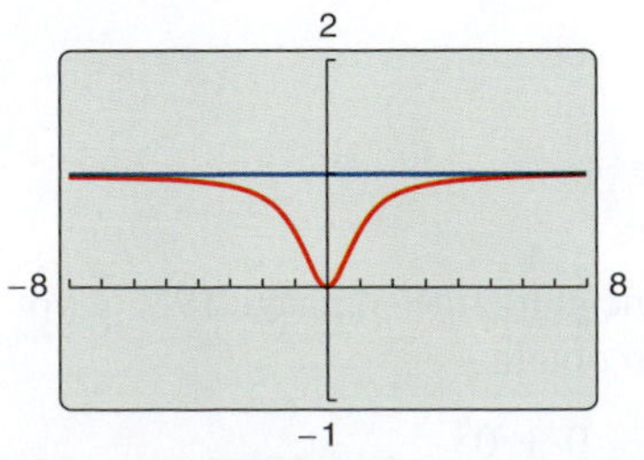

The horizontal asymptote appears to be the line $y = 1$, but it is actually the line $y = 2$.

Figure 3.39

TECHNOLOGY PITFALL If you use a graphing utility to estimate a limit, be sure that you also confirm the estimate analytically—the pictures shown by a graphing utility can be misleading. For instance, Figure 3.39 shows one view of the graph of

$$y = \frac{2x^3 + 1000x^2 + x}{x^3 + 1000x^2 + x + 1000}.$$

From this view, one could be convinced that the graph has $y = 1$ as a horizontal asymptote. An analytical approach shows that the horizontal asymptote is actually $y = 2$. Confirm this by enlarging the viewing window on the graphing utility.

In Section 1.3 (Example 9), you saw how the Squeeze Theorem can be used to evaluate limits involving trigonometric functions. This theorem is also valid for limits at infinity.

EXAMPLE 5 Limits Involving Trigonometric Functions

Find each limit.

a. $\lim_{x\to\infty} \sin x$ **b.** $\lim_{x\to\infty} \frac{\sin x}{x}$

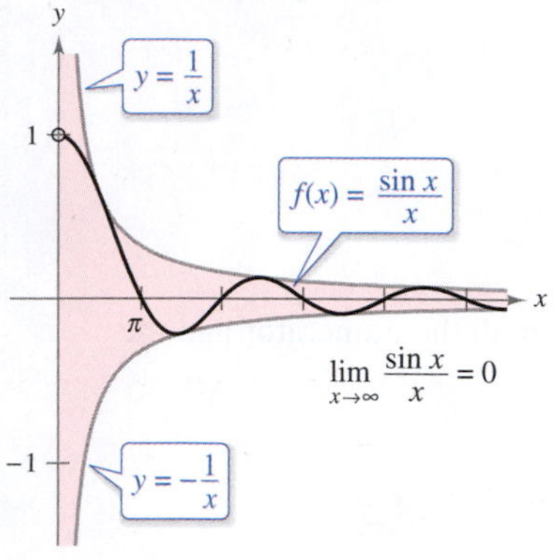

As x increases without bound, $f(x)$ approaches 0.
Figure 3.40

Solution

a. As x approaches infinity, the sine function oscillates between 1 and -1. So, this limit does not exist.

b. Because $-1 \le \sin x \le 1$, it follows that for $x > 0$,

$$-\frac{1}{x} \le \frac{\sin x}{x} \le \frac{1}{x}$$

where

$$\lim_{x\to\infty}\left(-\frac{1}{x}\right) = 0 \quad \text{and} \quad \lim_{x\to\infty}\frac{1}{x} = 0.$$

So, by the Squeeze Theorem, you can obtain

$$\lim_{x\to\infty}\frac{\sin x}{x} = 0$$

as shown in Figure 3.40.

EXAMPLE 6 Oxygen Level in a Pond

Let $f(t)$ measure the level of oxygen in a pond, where $f(t) = 1$ is the normal (unpolluted) level and the time t is measured in weeks. When $t = 0$, organic waste is dumped into the pond, and as the waste material oxidizes, the level of oxygen in the pond is

$$f(t) = \frac{t^2 - t + 1}{t^2 + 1}.$$

What percent of the normal level of oxygen exists in the pond after 1 week? After 2 weeks? After 10 weeks? What is the limit as t approaches infinity?

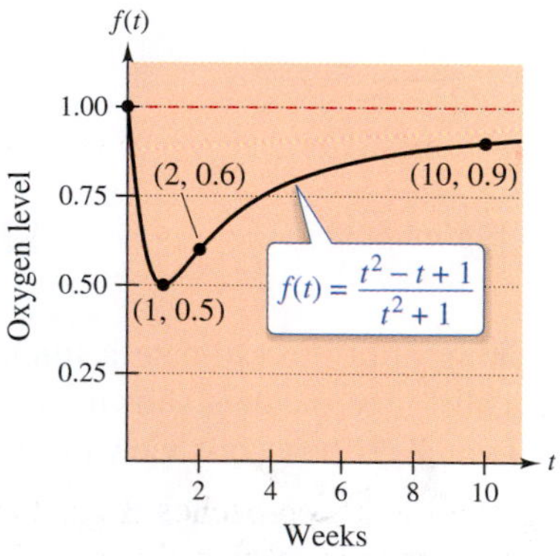

The level of oxygen in a pond approaches the normal level of 1 as t approaches ∞.
Figure 3.41

Solution When $t = 1$, 2, and 10, the levels of oxygen are as shown.

$$f(1) = \frac{1^2 - 1 + 1}{1^2 + 1} = \frac{1}{2} = 50\% \qquad \text{1 week}$$

$$f(2) = \frac{2^2 - 2 + 1}{2^2 + 1} = \frac{3}{5} = 60\% \qquad \text{2 weeks}$$

$$f(10) = \frac{10^2 - 10 + 1}{10^2 + 1} = \frac{91}{101} \approx 90.1\% \qquad \text{10 weeks}$$

To find the limit as t approaches infinity, you can use the guidelines on page 198, or you can divide the numerator and the denominator by t^2 to obtain

$$\lim_{t\to\infty}\frac{t^2 - t + 1}{t^2 + 1} = \lim_{t\to\infty}\frac{1 - (1/t) + (1/t^2)}{1 + (1/t^2)} = \frac{1 - 0 + 0}{1 + 0} = 1 = 100\%.$$

See Figure 3.41.

Infinite Limits at Infinity

Many functions do not approach a finite limit as x increases (or decreases) without bound. For instance, no polynomial function has a finite limit at infinity. The next definition is used to describe the behavior of polynomial and other functions at infinity.

REMARK Determining whether a function has an infinite limit at infinity is useful in analyzing the "end behavior" of its graph. You will see examples of this in Section 3.6 on curve sketching.

Definition of Infinite Limits at Infinity

Let f be a function defined on the interval (a, ∞).

1. The statement $\lim_{x\to\infty} f(x) = \infty$ means that for each positive number M, there is a corresponding number $N > 0$ such that $f(x) > M$ whenever $x > N$.
2. The statement $\lim_{x\to\infty} f(x) = -\infty$ means that for each negative number M, there is a corresponding number $N > 0$ such that $f(x) < M$ whenever $x > N$.

Similar definitions can be given for the statements

$$\lim_{x\to-\infty} f(x) = \infty \quad \text{and} \quad \lim_{x\to-\infty} f(x) = -\infty.$$

EXAMPLE 7 **Finding Infinite Limits at Infinity**

Find each limit.

a. $\lim_{x\to\infty} x^3$ **b.** $\lim_{x\to-\infty} x^3$

Solution

a. As x increases without bound, x^3 also increases without bound. So, you can write

$$\lim_{x\to\infty} x^3 = \infty.$$

b. As x decreases without bound, x^3 also decreases without bound. So, you can write

$$\lim_{x\to-\infty} x^3 = -\infty.$$

The graph of $f(x) = x^3$ in Figure 3.42 illustrates these two results. These results agree with the Leading Coefficient Test for polynomial functions as described in Section P.3.

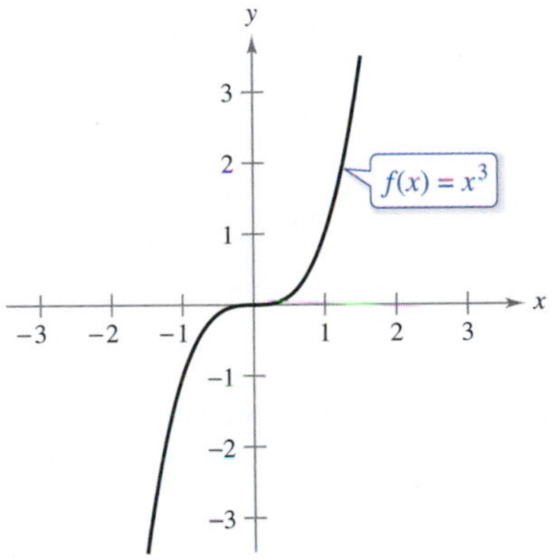

Figure 3.42

EXAMPLE 8 **Finding Infinite Limits at Infinity**

Find each limit.

a. $\lim_{x\to\infty} \dfrac{2x^2 - 4x}{x + 1}$ **b.** $\lim_{x\to-\infty} \dfrac{2x^2 - 4x}{x + 1}$

Solution One way to evaluate each of these limits is to use long division to rewrite the improper rational function as the sum of a polynomial and a rational function.

a. $$\lim_{x\to\infty} \frac{2x^2 - 4x}{x + 1} = \lim_{x\to\infty}\left(2x - 6 + \frac{6}{x + 1}\right) = \infty$$

b. $$\lim_{x\to-\infty} \frac{2x^2 - 4x}{x + 1} = \lim_{x\to-\infty}\left(2x - 6 + \frac{6}{x + 1}\right) = -\infty$$

The statements above can be interpreted as saying that as x approaches $\pm\infty$, the function $f(x) = (2x^2 - 4x)/(x + 1)$ behaves like the function $g(x) = 2x - 6$. In Section 3.6, you will see that this is graphically described by saying that the line $y = 2x - 6$ is a slant asymptote of the graph of f, as shown in Figure 3.43.

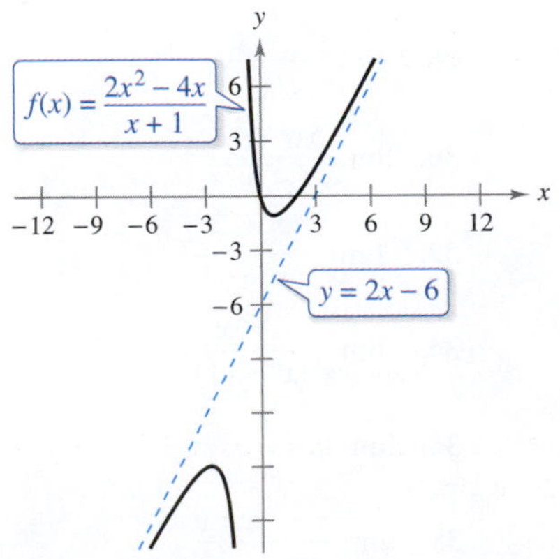

Figure 3.43

3.5 Exercises

See **CalcChat.com** for tutorial help and worked-out solutions to odd-numbered exercises.

Matching In Exercises 1–6, match the function with one of the graphs [(a), (b), (c), (d), (e), or (f)] using horizontal asymptotes as an aid.

(a)

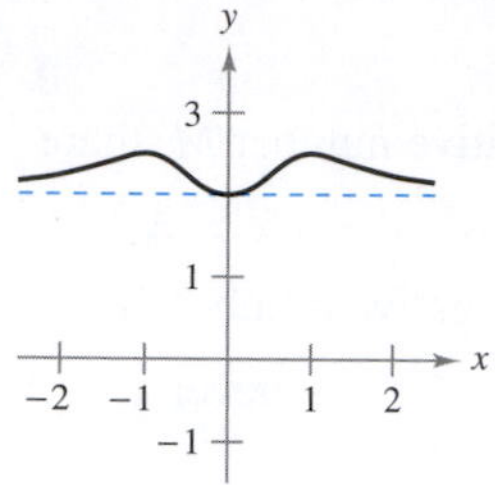

(b)

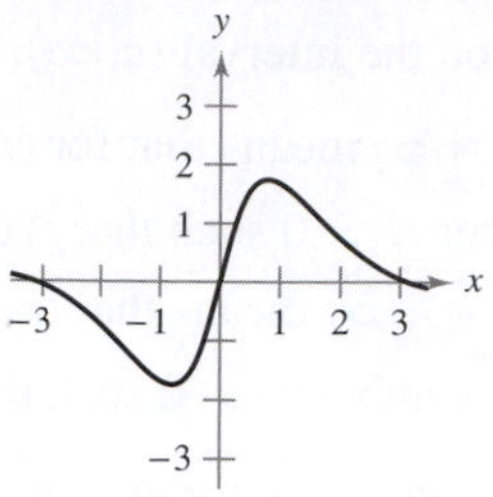

(c)

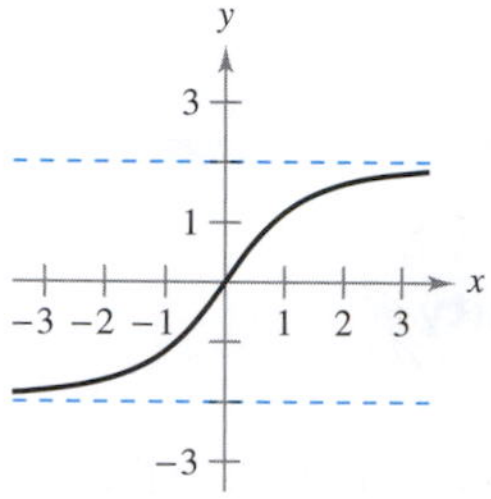

(d)

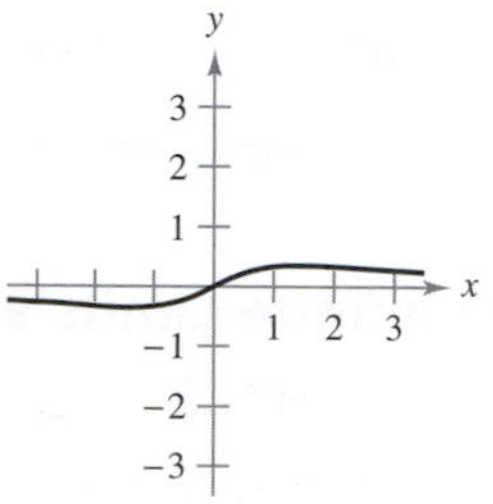

(e)

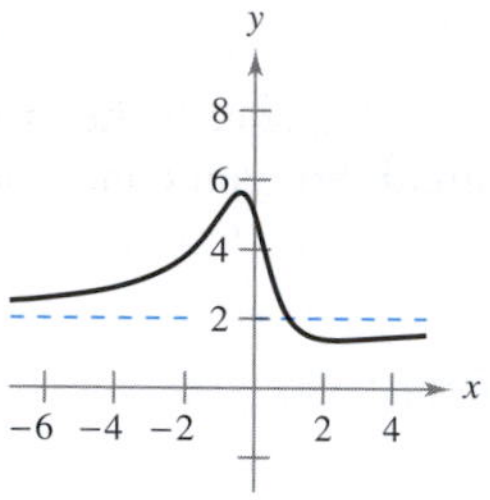

(f)

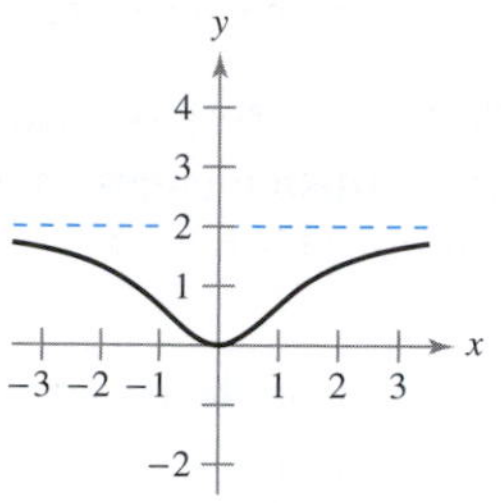

1. $f(x) = \dfrac{2x^2}{x^2+2}$

2. $f(x) = \dfrac{2x}{\sqrt{x^2+2}}$

3. $f(x) = \dfrac{x}{x^2+2}$

4. $f(x) = 2 + \dfrac{x^2}{x^4+1}$

5. $f(x) = \dfrac{4 \sin x}{x^2+1}$

6. $f(x) = \dfrac{2x^2 - 3x + 5}{x^2+1}$

Numerical and Graphical Analysis In Exercises 7–12, use a graphing utility to complete the table and estimate the limit as x approaches infinity. Then use a graphing utility to graph the function and estimate the limit graphically.

x	10^0	10^1	10^2	10^3	10^4	10^5	10^6
$f(x)$							

7. $f(x) = \dfrac{4x+3}{2x-1}$

8. $f(x) = \dfrac{2x^2}{x+1}$

9. $f(x) = \dfrac{-6x}{\sqrt{4x^2+5}}$

10. $f(x) = \dfrac{10}{\sqrt{2x^2-1}}$

11. $f(x) = 5 - \dfrac{1}{x^2+1}$

12. $f(x) = 4 + \dfrac{3}{x^2+2}$

Finding Limits at Infinity In Exercises 13 and 14, find $\lim_{x\to\infty} h(x)$, if possible.

13. $f(x) = 5x^3 - 3x^2 + 10x$
   (a) $h(x) = \dfrac{f(x)}{x^2}$
   (b) $h(x) = \dfrac{f(x)}{x^3}$
   (c) $h(x) = \dfrac{f(x)}{x^4}$

14. $f(x) = -4x^2 + 2x - 5$
   (a) $h(x) = \dfrac{f(x)}{x}$
   (b) $h(x) = \dfrac{f(x)}{x^2}$
   (c) $h(x) = \dfrac{f(x)}{x^3}$

Finding Limits at Infinity In Exercises 15–18, find each limit, if possible.

15. (a) $\lim_{x\to\infty} \dfrac{x^2+2}{x^3-1}$
   (b) $\lim_{x\to\infty} \dfrac{x^2+2}{x^2-1}$
   (c) $\lim_{x\to\infty} \dfrac{x^2+2}{x-1}$

16. (a) $\lim_{x\to\infty} \dfrac{3-2x}{3x^3-1}$
   (b) $\lim_{x\to\infty} \dfrac{3-2x}{3x-1}$
   (c) $\lim_{x\to\infty} \dfrac{3-2x^2}{3x-1}$

17. (a) $\lim_{x\to\infty} \dfrac{5-2x^{3/2}}{3x^2-4}$
   (b) $\lim_{x\to\infty} \dfrac{5-2x^{3/2}}{3x^{3/2}-4}$
   (c) $\lim_{x\to\infty} \dfrac{5-2x^{3/2}}{3x-4}$

18. (a) $\lim_{x\to\infty} \dfrac{5x^{3/2}}{4x^2+1}$
   (b) $\lim_{x\to\infty} \dfrac{5x^{3/2}}{4x^{3/2}+1}$
   (c) $\lim_{x\to\infty} \dfrac{5x^{3/2}}{4\sqrt{x}+1}$

Finding a Limit In Exercises 19–38, find the limit.

19. $\lim_{x\to\infty} \left(4 + \dfrac{3}{x}\right)$

20. $\lim_{x\to-\infty} \left(\dfrac{5}{x} - \dfrac{x}{3}\right)$

21. $\lim_{x\to\infty} \dfrac{2x-1}{3x+2}$

22. $\lim_{x\to-\infty} \dfrac{4x^2+5}{x^2+3}$

23. $\lim_{x\to\infty} \dfrac{x}{x^2-1}$

24. $\lim_{x\to\infty} \dfrac{5x^3+1}{10x^3-3x^2+7}$

25. $\lim_{x\to-\infty} \dfrac{5x^2}{x+3}$

26. $\lim_{x\to-\infty} \dfrac{x^3-4}{x^2+1}$

27. $\lim_{x\to-\infty} \dfrac{x}{\sqrt{x^2-x}}$

28. $\lim_{x\to-\infty} \dfrac{x}{\sqrt{x^2+1}}$

29. $\lim_{x\to-\infty} \dfrac{2x+1}{\sqrt{x^2-x}}$

30. $\lim_{x\to\infty} \dfrac{5x^2+2}{\sqrt{x^2+3}}$

31. $\lim_{x\to\infty} \dfrac{\sqrt{x^2-1}}{2x-1}$

32. $\lim_{x\to-\infty} \dfrac{\sqrt{x^4-1}}{x^3-1}$

33. $\lim_{x\to\infty} \dfrac{x+1}{(x^2+1)^{1/3}}$

34. $\lim_{x\to-\infty} \dfrac{2x}{(x^6-1)^{1/3}}$

35. $\lim_{x\to\infty} \dfrac{1}{2x+\sin x}$

36. $\lim_{x\to\infty} \cos\dfrac{1}{x}$

37. $\lim_{x\to\infty} \dfrac{\sin 2x}{x}$

38. $\lim_{x\to\infty} \dfrac{x-\cos x}{x}$

Horizontal Asymptotes In Exercises 39–42, use a graphing utility to graph the function and identify any horizontal asymptotes.

39. $f(x) = \frac{|x|}{x+1}$

40. $f(x) = \frac{|3x+2|}{x-2}$

41. $f(x) = \frac{3x}{\sqrt{x^2+2}}$

42. $f(x) = \frac{\sqrt{9x^2-2}}{2x+1}$

Finding a Limit In Exercises 43 and 44, find the limit. (*Hint:* Let $x = 1/t$ and find the limit as $t \to 0^+$.)

43. $\lim_{x\to\infty} x \sin \frac{1}{x}$

44. $\lim_{x\to\infty} x \tan \frac{1}{x}$

Finding a Limit In Exercises 45–48, find the limit. (*Hint:* Treat the expression as a fraction whose denominator is 1, and rationalize the numerator.) Use a graphing utility to verify your result.

45. $\lim_{x\to-\infty} \left(x + \sqrt{x^2+3}\right)$

46. $\lim_{x\to\infty} \left(x - \sqrt{x^2+x}\right)$

47. $\lim_{x\to-\infty} \left(3x + \sqrt{9x^2-x}\right)$

48. $\lim_{x\to\infty} \left(4x - \sqrt{16x^2-x}\right)$

Numerical, Graphical, and Analytic Analysis In Exercises 49–52, use a graphing utility to complete the table and estimate the limit as x approaches infinity. Then use a graphing utility to graph the function and estimate the limit. Finally, find the limit analytically and compare your results with the estimates.

x	10^0	10^1	10^2	10^3	10^4	10^5	10^6
$f(x)$							

49. $f(x) = x - \sqrt{x(x-1)}$

50. $f(x) = x^2 - x\sqrt{x(x-1)}$

51. $f(x) = x \sin \frac{1}{2x}$

52. $f(x) = \frac{x+1}{x\sqrt{x}}$

WRITING ABOUT CONCEPTS

Writing In Exercises 53 and 54, describe in your own words what the statement means.

53. $\lim_{x\to\infty} f(x) = 4$

54. $\lim_{x\to-\infty} f(x) = 2$

55. **Sketching a Graph** Sketch a graph of a differentiable function f that satisfies the following conditions and has $x = 2$ as its only critical number.

$f'(x) < 0$ for $x < 2$

$f'(x) > 0$ for $x > 2$

$\lim_{x\to-\infty} f(x) = 6$

$\lim_{x\to\infty} f(x) = 6$

56. **Points of Inflection** Is it possible to sketch a graph of a function that satisfies the conditions of Exercise 55 and has *no* points of inflection? Explain.

WRITING ABOUT CONCEPTS (continued)

57. **Using Symmetry to Find Limits** If f is a continuous function such that $\lim_{x\to\infty} f(x) = 5$, find, if possible, $\lim_{x\to-\infty} f(x)$ for each specified condition.

(a) The graph of f is symmetric with respect to the y-axis.

(b) The graph of f is symmetric with respect to the origin.

58. **A Function and Its Derivative** The graph of a function f is shown below. To print an enlarged copy of the graph, go to *MathGraphs.com.*

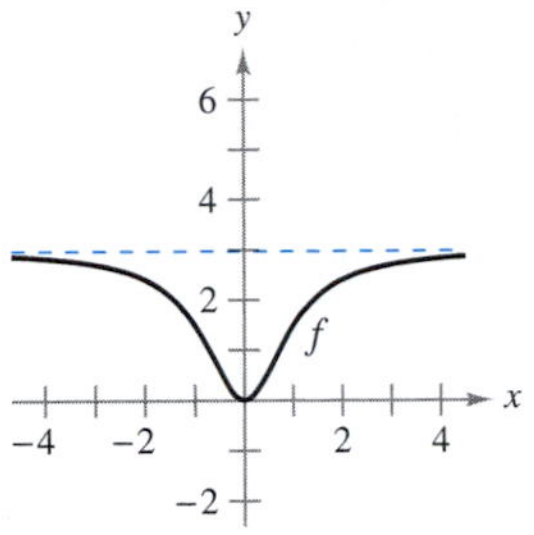

(a) Sketch f'.

(b) Use the graphs to estimate $\lim_{x\to\infty} f(x)$ and $\lim_{x\to\infty} f'(x)$.

(c) Explain the answers you gave in part (b).

Sketching a Graph In Exercises 59–74, sketch the graph of the equation using extrema, intercepts, symmetry, and asymptotes. Then use a graphing utility to verify your result.

59. $y = \frac{x}{1-x}$

60. $y = \frac{x-4}{x-3}$

61. $y = \frac{x+1}{x^2-4}$

62. $y = \frac{2x}{9-x^2}$

63. $y = \frac{x^2}{x^2+16}$

64. $y = \frac{2x^2}{x^2-4}$

65. $xy^2 = 9$

66. $x^2y = 9$

67. $y = \frac{3x}{x-1}$

68. $y = \frac{3x}{1-x^2}$

69. $y = 2 - \frac{3}{x^2}$

70. $y = 1 - \frac{1}{x}$

71. $y = 3 + \frac{2}{x}$

72. $y = \frac{4}{x^2} + 1$

73. $y = \frac{x^3}{\sqrt{x^2-4}}$

74. $y = \frac{x}{\sqrt{x^2-4}}$

Analyzing a Graph Using Technology In Exercises 75–82, use a computer algebra system to analyze the graph of the function. Label any extrema and/or asymptotes that exist.

75. $f(x) = 9 - \frac{5}{x^2}$

76. $f(x) = \frac{1}{x^2-x-2}$

77. $f(x) = \frac{x-2}{x^2-4x+3}$

78. $f(x) = \frac{x+1}{x^2+x+1}$

79. $f(x) = \dfrac{3x}{\sqrt{4x^2 + 1}}$

80. $g(x) = \dfrac{2x}{\sqrt{3x^2 + 1}}$

81. $g(x) = \sin\left(\dfrac{x}{x - 2}\right), \quad x > 3$

82. $f(x) = \dfrac{2 \sin 2x}{x}$

Comparing Functions **In Exercises 83 and 84, (a) use a graphing utility to graph f and g in the same viewing window, (b) verify algebraically that f and g represent the same function, and (c) zoom out sufficiently far so that the graph appears as a line. What equation does this line appear to have? (Note that the points at which the function is not continuous are not readily seen when you zoom out.)**

83. $f(x) = \dfrac{x^3 - 3x^2 + 2}{x(x - 3)}$

$g(x) = x + \dfrac{2}{x(x - 3)}$

84. $f(x) = -\dfrac{x^3 - 2x^2 + 2}{2x^2}$

$g(x) = -\dfrac{1}{2}x + 1 - \dfrac{1}{x^2}$

85. Engine Efficiency

The efficiency of an internal combustion engine is

$$\text{Efficiency (\%)} = 100\left[1 - \frac{1}{(v_1/v_2)^c}\right]$$

where v_1/v_2 is the ratio of the uncompressed gas to the compressed gas and c is a positive constant dependent on the engine design. Find the limit of the efficiency as the compression ratio approaches infinity.

86. Average Cost A business has a cost of $C = 0.5x + 500$ for producing x units. The average cost per unit is

$$\overline{C} = \frac{C}{x}.$$

Find the limit of $\overline{C}$ as x approaches infinity.

87. Physics Newton's First Law of Motion and Einstein's Special Theory of Relativity differ concerning a particle's behavior as its velocity approaches the speed of light c. In the graph, functions N and E represent the velocity v, with respect to time t, of a particle accelerated by a constant force as predicted by Newton and Einstein, respectively. Write limit statements that describe these two theories.

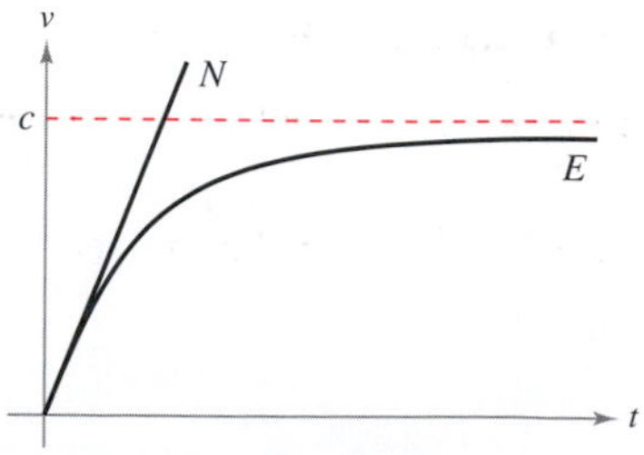

88. HOW DO YOU SEE IT? The graph shows the temperature T, in degrees Fahrenheit, of molten glass t seconds after it is removed from a kiln.

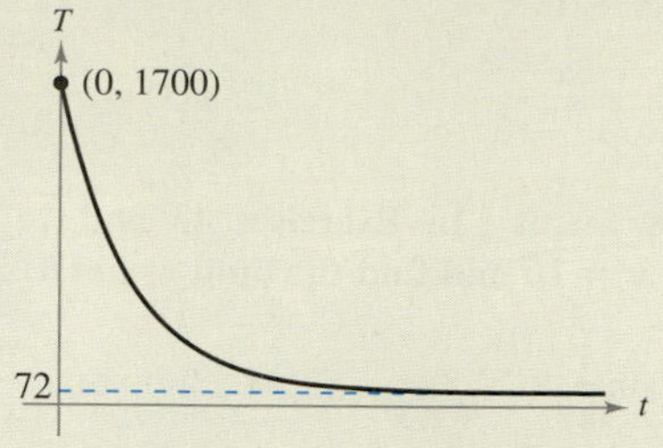

(a) Find $\lim_{t \to 0^+} T$. What does this limit represent?

(b) Find $\lim_{t \to \infty} T$. What does this limit represent?

(c) Will the temperature of the glass ever actually reach room temperature? Why?

89. Modeling Data The average typing speeds S (in words per minute) of a typing student after t weeks of lessons are shown in the table.

t	5	10	15	20	25	30
S	28	56	79	90	93	94

A model for the data is $S = \dfrac{100t^2}{65 + t^2}, \; t > 0.$

(a) Use a graphing utility to plot the data and graph the model.

(b) Does there appear to be a limiting typing speed? Explain.

90. Modeling Data A heat probe is attached to the heat exchanger of a heating system. The temperature T (in degrees Celsius) is recorded t seconds after the furnace is started. The results for the first 2 minutes are recorded in the table.

t	0	15	30	45	60
T	25.2°	36.9°	45.5°	51.4°	56.0°

t	75	90	105	120
T	59.6°	62.0°	64.0°	65.2°

(a) Use the regression capabilities of a graphing utility to find a model of the form $T_1 = at^2 + bt + c$ for the data.

(b) Use a graphing utility to graph T_1.

(c) A rational model for the data is

$$T_2 = \frac{1451 + 86t}{58 + t}.$$

Use a graphing utility to graph T_2.

(d) Find $T_1(0)$ and $T_2(0)$.

(e) Find $\lim_{t \to \infty} T_2$.

(f) Interpret the result in part (e) in the context of the problem. Is it possible to do this type of analysis using T_1? Explain.

91. Using the Definition of Limits at Infinity The graph of

$$f(x) = \frac{2x^2}{x^2 + 2}$$

is shown.

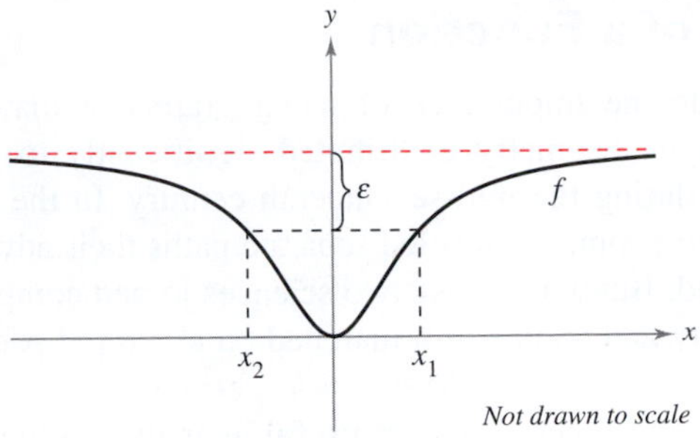

(a) Find $L = \lim_{x\to\infty} f(x)$.

(b) Determine x_1 and x_2 in terms of ε.

(c) Determine M, where $M > 0$, such that $|f(x) - L| < \varepsilon$ for $x > M$.

(d) Determine N, where $N < 0$, such that $|f(x) - L| < \varepsilon$ for $x < N$.

92. Using the Definition of Limits at Infinity The graph of

$$f(x) = \frac{6x}{\sqrt{x^2 + 2}}$$

is shown.

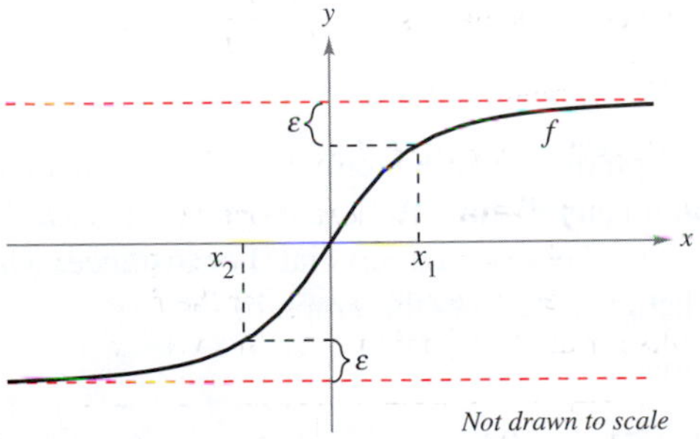

(a) Find $L = \lim_{x\to\infty} f(x)$ and $K = \lim_{x\to-\infty} f(x)$.

(b) Determine x_1 and x_2 in terms of ε.

(c) Determine M, where $M > 0$, such that $|f(x) - L| < \varepsilon$ for $x > M$.

(d) Determine N, where $N < 0$, such that $|f(x) - K| < \varepsilon$ for $x < N$.

93. Using the Definition of Limits at Infinity Consider

$$\lim_{x\to\infty} \frac{3x}{\sqrt{x^2 + 3}}.$$

(a) Use the definition of limits at infinity to find values of M that correspond to $\varepsilon = 0.5$.

(b) Use the definition of limits at infinity to find values of M that correspond to $\varepsilon = 0.1$.

94. Using the Definition of Limits at Infinity Consider

$$\lim_{x\to-\infty} \frac{3x}{\sqrt{x^2 + 3}}.$$

(a) Use the definition of limits at infinity to find values of N that correspond to $\varepsilon = 0.5$.

(b) Use the definition of limits at infinity to find values of N that correspond to $\varepsilon = 0.1$.

Proof **In Exercises 95–98, use the definition of limits at infinity to prove the limit.**

95. $\lim_{x\to\infty} \frac{1}{x^2} = 0$

96. $\lim_{x\to\infty} \frac{2}{\sqrt{x}} = 0$

97. $\lim_{x\to-\infty} \frac{1}{x^3} = 0$

98. $\lim_{x\to-\infty} \frac{1}{x - 2} = 0$

99. Distance A line with slope m passes through the point $(0, 4)$.

(a) Write the shortest distance d between the line and the point $(3, 1)$ as a function of m.

(b) Use a graphing utility to graph the equation in part (a).

(c) Find $\lim_{m\to\infty} d(m)$ and $\lim_{m\to-\infty} d(m)$. Interpret the results geometrically.

100. Distance A line with slope m passes through the point $(0, -2)$.

(a) Write the shortest distance d between the line and the point $(4, 2)$ as a function of m.

(b) Use a graphing utility to graph the equation in part (a).

(c) Find $\lim_{m\to\infty} d(m)$ and $\lim_{m\to-\infty} d(m)$. Interpret the results geometrically.

101. Proof Prove that if

$$p(x) = a_n x^n + \cdots + a_1 x + a_0$$

and

$$q(x) = b_m x^m + \cdots + b_1 x + b_0$$

where $a_n \neq 0$ and $b_m \neq 0$, then

$$\lim_{x\to\infty} \frac{p(x)}{q(x)} = \begin{cases} 0, & n < m \\ \dfrac{a_n}{b_m}, & n = m. \\ \pm\infty, & n > m \end{cases}$$

102. Proof Use the definition of infinite limits at infinity to prove that $\lim_{x\to\infty} x^3 = \infty$.

True or False? **In Exercises 103 and 104, determine whether the statement is true or false. If it is false, explain why or give an example that shows it is false.**

103. If $f'(x) > 0$ for all real numbers x, then f increases without bound.

104. If $f''(x) < 0$ for all real numbers x, then f decreases without bound.

3.6 A Summary of Curve Sketching

- Analyze and sketch the graph of a function.

Analyzing the Graph of a Function

It would be difficult to overstate the importance of using graphs in mathematics. Descartes's introduction of analytic geometry contributed significantly to the rapid advances in calculus that began during the mid-seventeenth century. In the words of Lagrange, "As long as algebra and geometry traveled separate paths their advance was slow and their applications limited. But when these two sciences joined company, they drew from each other fresh vitality and thenceforth marched on at a rapid pace toward perfection."

So far, you have studied several concepts that are useful in analyzing the graph of a function.

- x-intercepts and y-intercepts (Section P.1)
- Symmetry (Section P.1)
- Domain and range (Section P.3)
- Continuity (Section 1.4)
- Vertical asymptotes (Section 1.5)
- Differentiability (Section 2.1)
- Relative extrema (Section 3.1)
- Concavity (Section 3.4)
- Points of inflection (Section 3.4)
- Horizontal asymptotes (Section 3.5)
- Infinite limits at infinity (Section 3.5)

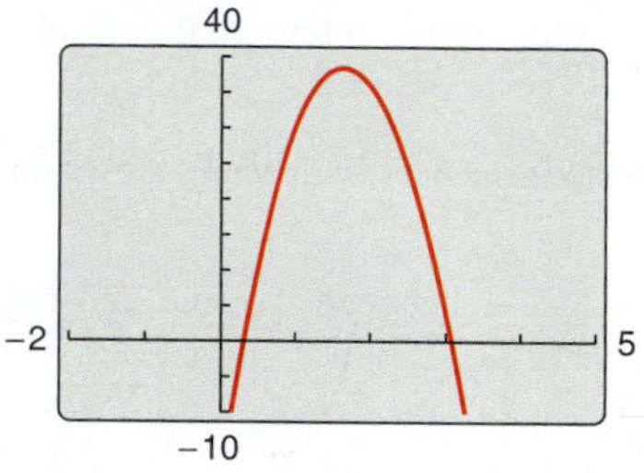

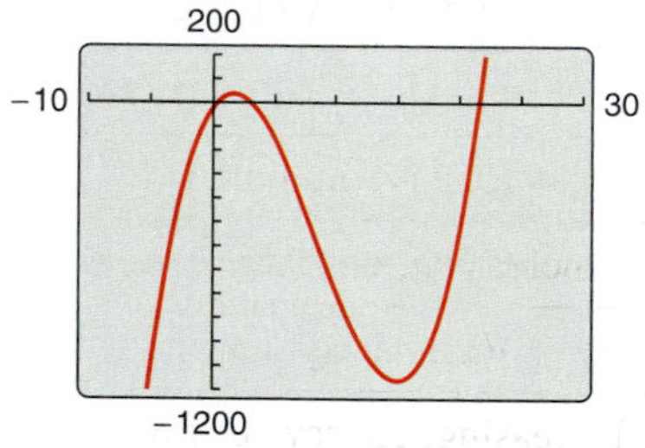

Different viewing windows for the graph of $f(x) = x^3 - 25x^2 + 74x - 20$
Figure 3.44

When you are sketching the graph of a function, either by hand or with a graphing utility, remember that normally you cannot show the *entire* graph. The decision as to which part of the graph you choose to show is often crucial. For instance, which of the viewing windows in Figure 3.44 better represents the graph of

$$f(x) = x^3 - 25x^2 + 74x - 20?$$

By seeing both views, it is clear that the second viewing window gives a more complete representation of the graph. But would a third viewing window reveal other interesting portions of the graph? To answer this, you need to use calculus to interpret the first and second derivatives. Here are some guidelines for determining a good viewing window for the graph of a function.

GUIDELINES FOR ANALYZING THE GRAPH OF A FUNCTION

1. Determine the domain and range of the function.
2. Determine the intercepts, asymptotes, and symmetry of the graph.
3. Locate the x-values for which $f'(x)$ and $f''(x)$ either are zero or do not exist. Use the results to determine relative extrema and points of inflection.

REMARK In these guidelines, note the importance of *algebra* (as well as calculus) for solving the equations

$$f(x) = 0, \quad f'(x) = 0, \quad \text{and} \quad f''(x) = 0.$$

EXAMPLE 1 Sketching the Graph of a Rational Function

Analyze and sketch the graph of

$$f(x) = \frac{2(x^2 - 9)}{x^2 - 4}.$$

Solution

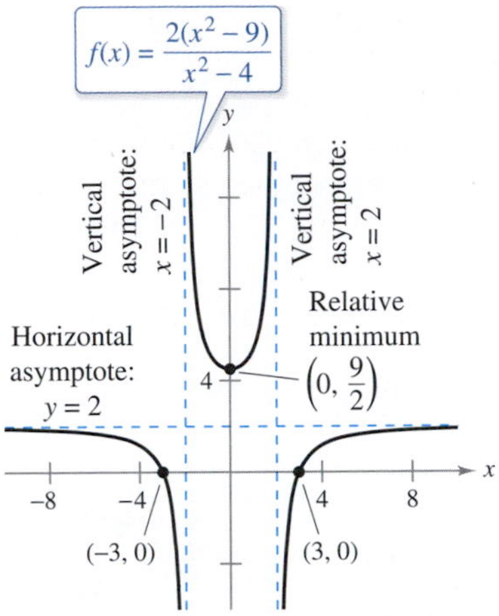

Using calculus, you can be certain that you have determined all characteristics of the graph of f.

Figure 3.45

First derivative: $f'(x) = \dfrac{20x}{(x^2 - 4)^2}$

Second derivative: $f''(x) = \dfrac{-20(3x^2 + 4)}{(x^2 - 4)^3}$

x-intercepts: $(-3, 0), (3, 0)$

y-intercept: $\left(0, \frac{9}{2}\right)$

Vertical asymptotes: $x = -2, x = 2$

Horizontal asymptote: $y = 2$

Critical number: $x = 0$

Possible points of inflection: None

Domain: All real numbers except $x = \pm 2$

Symmetry: With respect to y-axis

Test intervals: $(-\infty, -2), (-2, 0), (0, 2), (2, \infty)$

The table shows how the test intervals are used to determine several characteristics of the graph. The graph of f is shown in Figure 3.45.

	$f(x)$	$f'(x)$	$f''(x)$	Characteristic of Graph
$-\infty < x < -2$		$-$	$-$	Decreasing, concave downward
$x = -2$	Undef.	Undef.	Undef.	Vertical asymptote
$-2 < x < 0$		$-$	$+$	Decreasing, concave upward
$x = 0$	$\frac{9}{2}$	0	$+$	Relative minimum
$0 < x < 2$		$+$	$+$	Increasing, concave upward
$x = 2$	Undef.	Undef.	Undef.	Vertical asymptote
$2 < x < \infty$		$+$	$-$	Increasing, concave downward

FOR FURTHER INFORMATION For more information on the use of technology to graph rational functions, see the article "Graphs of Rational Functions for Computer Assisted Calculus" by Stan Byrd and Terry Walters in *The College Mathematics Journal*. To view this article, go to *MathArticles.com*.

Be sure you understand all of the implications of creating a table such as that shown in Example 1. By using calculus, you can be *sure* that the graph has no relative extrema or points of inflection other than those shown in Figure 3.45.

TECHNOLOGY PITFALL Without using the type of analysis outlined in Example 1, it is easy to obtain an incomplete view of a graph's basic characteristics. For instance, Figure 3.46 shows a view of the graph of

$$g(x) = \frac{2(x^2 - 9)(x - 20)}{(x^2 - 4)(x - 21)}.$$

From this view, it appears that the graph of g is about the same as the graph of f shown in Figure 3.45. The graphs of these two functions, however, differ significantly. Try enlarging the viewing window to see the differences.

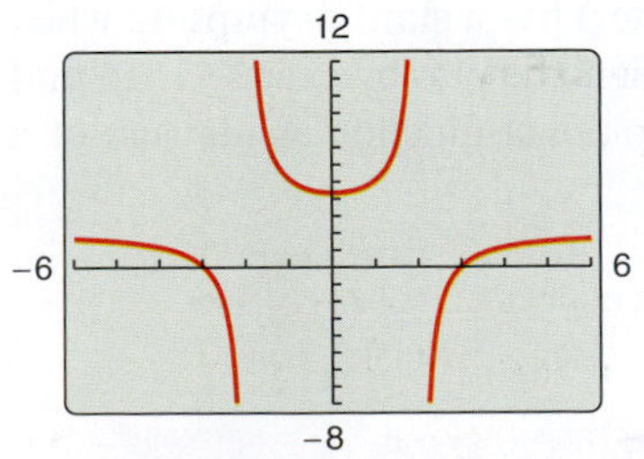

By not using calculus, you may overlook important characteristics of the graph of g.

Figure 3.46

EXAMPLE 2 Sketching the Graph of a Rational Function

Analyze and sketch the graph of $f(x) = \dfrac{x^2 - 2x + 4}{x - 2}$.

Solution

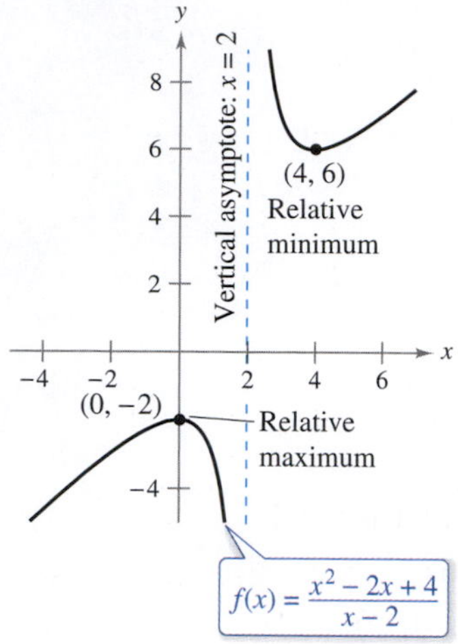

Figure 3.47

First derivative: $f'(x) = \dfrac{x(x-4)}{(x-2)^2}$

Second derivative: $f''(x) = \dfrac{8}{(x-2)^3}$

x-intercepts: None

y-intercept: $(0, -2)$

Vertical asymptote: $x = 2$

Horizontal asymptotes: None

End behavior: $\lim_{x\to-\infty} f(x) = -\infty, \ \lim_{x\to\infty} f(x) = \infty$

Critical numbers: $x = 0, x = 4$

Possible points of inflection: None

Domain: All real numbers except $x = 2$

Test intervals: $(-\infty, 0), (0, 2), (2, 4), (4, \infty)$

The analysis of the graph of f is shown in the table, and the graph is shown in Figure 3.47.

	$f(x)$	$f'(x)$	$f''(x)$	Characteristic of Graph
$-\infty < x < 0$		$+$	$-$	Increasing, concave downward
$x = 0$	-2	0	$-$	Relative maximum
$0 < x < 2$		$-$	$-$	Decreasing, concave downward
$x = 2$	Undef.	Undef.	Undef.	Vertical asymptote
$2 < x < 4$		$-$	$+$	Decreasing, concave upward
$x = 4$	6	0	$+$	Relative minimum
$4 < x < \infty$		$+$	$+$	Increasing, concave upward

Although the graph of the function in Example 2 has no horizontal asymptote, it does have a slant asymptote. The graph of a rational function (having no common factors and whose denominator is of degree 1 or greater) has a **slant asymptote** when the degree of the numerator exceeds the degree of the denominator by exactly 1. To find the slant asymptote, use long division to rewrite the rational function as the sum of a first-degree polynomial and another rational function.

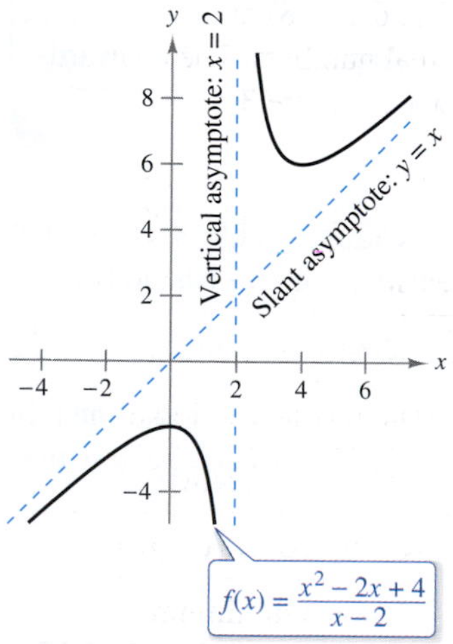

A slant asymptote
Figure 3.48

$$f(x) = \frac{x^2 - 2x + 4}{x - 2} \qquad \text{Write original equation.}$$

$$= x + \frac{4}{x - 2} \qquad \text{Rewrite using long division.}$$

In Figure 3.48, note that the graph of f approaches the slant asymptote $y = x$ as x approaches $-\infty$ or ∞.

EXAMPLE 3 **Sketching the Graph of a Radical Function**

Analyze and sketch the graph of $f(x) = \dfrac{x}{\sqrt{x^2 + 2}}$.

Solution

$$f'(x) = \frac{2}{(x^2 + 2)^{3/2}} \qquad \text{Find first derivative.}$$

$$f''(x) = -\frac{6x}{(x^2 + 2)^{5/2}} \qquad \text{Find second derivative.}$$

The graph has only one intercept, $(0, 0)$. It has no vertical asymptotes, but it has two horizontal asymptotes: $y = 1$ (to the right) and $y = -1$ (to the left). The function has no critical numbers and one possible point of inflection (at $x = 0$). The domain of the function is all real numbers, and the graph is symmetric with respect to the origin. The analysis of the graph of f is shown in the table, and the graph is shown in Figure 3.49.

	$f(x)$	$f'(x)$	$f''(x)$	Characteristic of Graph
$-\infty < x < 0$		+	+	Increasing, concave upward
$x = 0$	0	$\frac{1}{\sqrt{2}}$	0	Point of inflection
$0 < x < \infty$		+	−	Increasing, concave downward

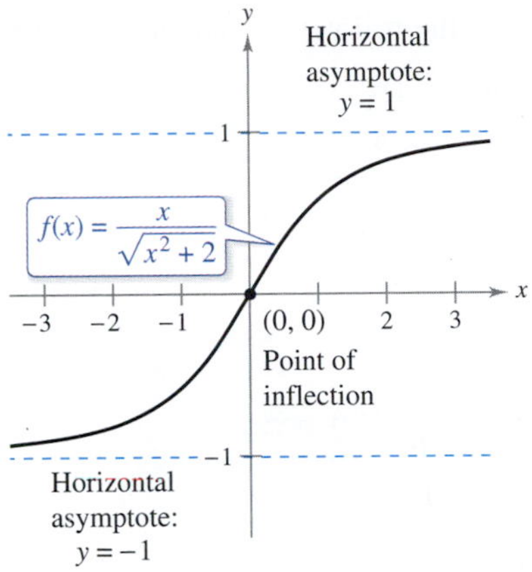

Figure 3.49

EXAMPLE 4 **Sketching the Graph of a Radical Function**

Analyze and sketch the graph of $f(x) = 2x^{5/3} - 5x^{4/3}$.

Solution

$$f'(x) = \frac{10}{3}x^{1/3}(x^{1/3} - 2) \qquad \text{Find first derivative.}$$

$$f''(x) = \frac{20(x^{1/3} - 1)}{9x^{2/3}} \qquad \text{Find second derivative.}$$

The function has two intercepts: $(0, 0)$ and $\left(\frac{125}{8}, 0\right)$. There are no horizontal or vertical asymptotes. The function has two critical numbers ($x = 0$ and $x = 8$) and two possible points of inflection ($x = 0$ and $x = 1$). The domain is all real numbers. The analysis of the graph of f is shown in the table, and the graph is shown in Figure 3.50.

	$f(x)$	$f'(x)$	$f''(x)$	Characteristic of Graph
$-\infty < x < 0$		+	−	Increasing, concave downward
$x = 0$	0	0	Undef.	Relative maximum
$0 < x < 1$		−	−	Decreasing, concave downward
$x = 1$	−3	−	0	Point of inflection
$1 < x < 8$		−	+	Decreasing, concave upward
$x = 8$	−16	0	+	Relative minimum
$8 < x < \infty$		+	+	Increasing, concave upward

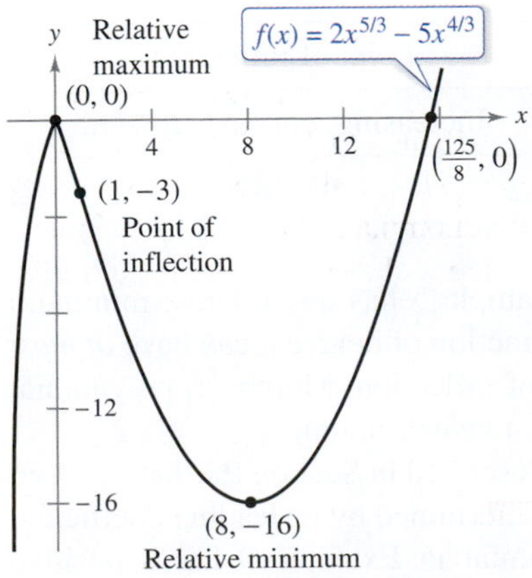

Figure 3.50

EXAMPLE 5 Sketching the Graph of a Polynomial Function

See LarsonCalculus.com for an interactive version of this type of example.

Analyze and sketch the graph of

$$f(x) = x^4 - 12x^3 + 48x^2 - 64x.$$

Solution Begin by factoring to obtain

$$\begin{aligned} f(x) &= x^4 - 12x^3 + 48x^2 - 64x \\ &= x(x-4)^3. \end{aligned}$$

Then, using the factored form of $f(x)$, you can perform the following analysis.

First derivative: $f'(x) = 4(x-1)(x-4)^2$

Second derivative: $f''(x) = 12(x-4)(x-2)$

x-intercepts: $(0, 0), (4, 0)$

y-intercept: $(0, 0)$

Vertical asymptotes: None

Horizontal asymptotes: None

End behavior: $\lim_{x\to-\infty} f(x) = \infty, \ \lim_{x\to\infty} f(x) = \infty$

Critical numbers: $x = 1, x = 4$

Possible points of inflection: $x = 2, x = 4$

Domain: All real numbers

Test intervals: $(-\infty, 1), (1, 2), (2, 4), (4, \infty)$

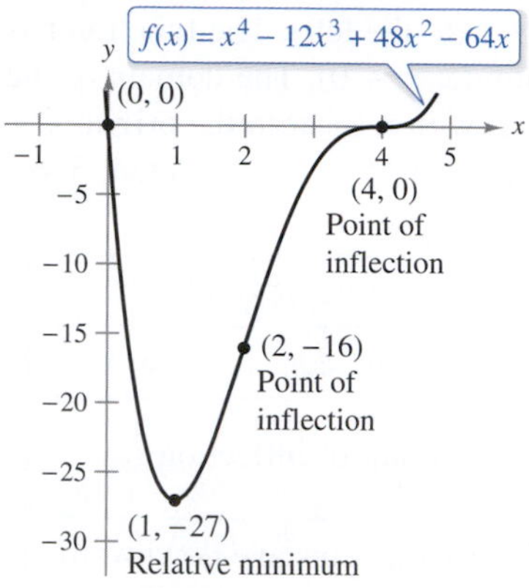

(a)

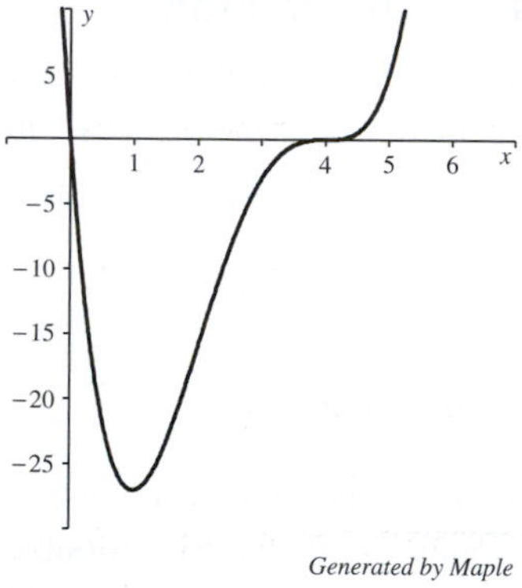

(b)

A polynomial function of even degree must have at least one relative extremum.

Figure 3.51

The analysis of the graph of f is shown in the table, and the graph is shown in Figure 3.51(a). Using a computer algebra system such as *Maple* [see Figure 3.51(b)] can help you verify your analysis.

	$f(x)$	$f'(x)$	$f''(x)$	Characteristic of Graph
$-\infty < x < 1$		$-$	$+$	Decreasing, concave upward
$x = 1$	-27	0	$+$	Relative minimum
$1 < x < 2$		$+$	$+$	Increasing, concave upward
$x = 2$	-16	$+$	0	Point of inflection
$2 < x < 4$		$+$	$-$	Increasing, concave downward
$x = 4$	0	0	0	Point of inflection
$4 < x < \infty$		$+$	$+$	Increasing, concave upward

The fourth-degree polynomial function in Example 5 has one relative minimum and no relative maxima. In general, a polynomial function of degree n can have *at most* $n - 1$ relative extrema, and *at most* $n - 2$ points of inflection. Moreover, polynomial functions of even degree must have *at least* one relative extremum.

Remember from the Leading Coefficient Test described in Section P.3 that the "end behavior" of the graph of a polynomial function is determined by its leading coefficient and its degree. For instance, because the polynomial in Example 5 has a positive leading coefficient, the graph rises to the right. Moreover, because the degree is even, the graph also rises to the left.

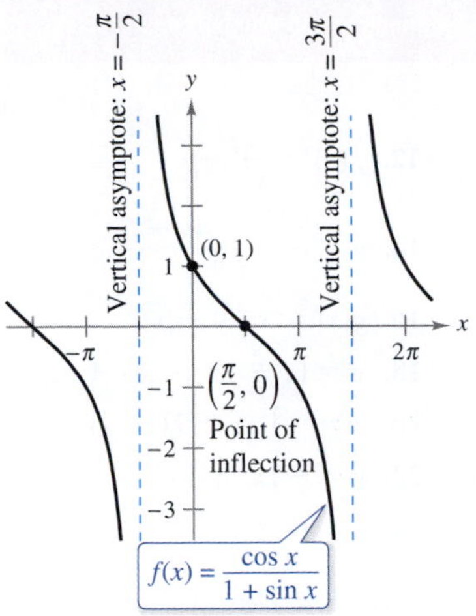

(a)

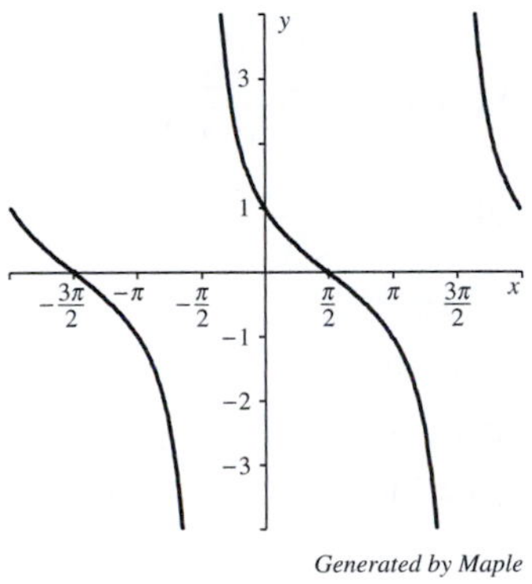

(b)

Figure 3.52

AP* Tips

Curve sketching in and of itself is not the aim of AP free response questions. However, the exercises here will help reinforce your ability to harvest information from derivatives.

EXAMPLE 6 Sketching the Graph of a Trigonometric Function

Analyze and sketch the graph of $f(x) = (\cos x)/(1 + \sin x)$.

Solution Because the function has a period of 2π, you can restrict the analysis of the graph to any interval of length 2π. For convenience, choose $(-\pi/2, 3\pi/2)$.

First derivative: $f'(x) = -\dfrac{1}{1 + \sin x}$

Second derivative: $f''(x) = \dfrac{\cos x}{(1 + \sin x)^2}$

Period: 2π

x-intercept: $\left(\dfrac{\pi}{2}, 0\right)$

y-intercept: $(0, 1)$

Vertical asymptotes: $x = -\dfrac{\pi}{2}, x = \dfrac{3\pi}{2}$ See Remark below.

Horizontal asymptotes: None

Critical numbers: None

Possible points of inflection: $x = \dfrac{\pi}{2}$

Domain: All real numbers except $x = \dfrac{3 + 4n}{2}\pi$

Test intervals: $\left(-\dfrac{\pi}{2}, \dfrac{\pi}{2}\right), \left(\dfrac{\pi}{2}, \dfrac{3\pi}{2}\right)$

The analysis of the graph of f on the interval $(-\pi/2, 3\pi/2)$ is shown in the table, and the graph is shown in Figure 3.52(a). Compare this with the graph generated by the computer algebra system *Maple* in Figure 3.52(b).

	$f(x)$	$f'(x)$	$f''(x)$	Characteristic of Graph
$x = -\dfrac{\pi}{2}$	Undef.	Undef.	Undef.	Vertical asymptote
$-\dfrac{\pi}{2} < x < \dfrac{\pi}{2}$		$-$	$+$	Decreasing, concave upward
$x = \dfrac{\pi}{2}$	0	$-\frac{1}{2}$	0	Point of inflection
$\dfrac{\pi}{2} < x < \dfrac{3\pi}{2}$		$-$	$-$	Decreasing, concave downward
$x = \dfrac{3\pi}{2}$	Undef.	Undef.	Undef.	Vertical asymptote

REMARK By substituting $-\pi/2$ or $3\pi/2$ into the function, you obtain the form $0/0$. This is called an indeterminate form, which you will study in Section 8.7. To determine that the function has vertical asymptotes at these two values, rewrite f as

$$f(x) = \frac{\cos x}{1 + \sin x} = \frac{(\cos x)(1 - \sin x)}{(1 + \sin x)(1 - \sin x)} = \frac{(\cos x)(1 - \sin x)}{\cos^2 x} = \frac{1 - \sin x}{\cos x}.$$

In this form, it is clear that the graph of f has vertical asymptotes at $x = -\pi/2$ and $3\pi/2$.

3.6 Exercises

See CalcChat.com for tutorial help and worked-out solutions to odd-numbered exercises.

Matching **In Exercises 1–4, match the graph of f in the left column with that of its derivative in the right column.**

Graph of f | **Graph of f'**

1.

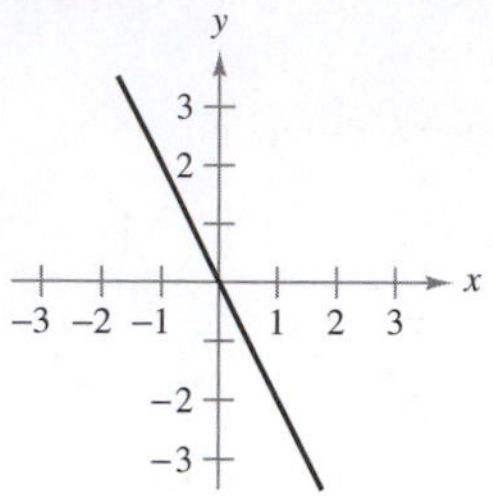

(a)

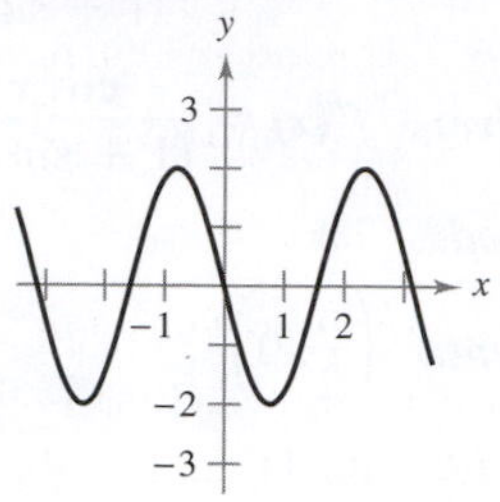

2.

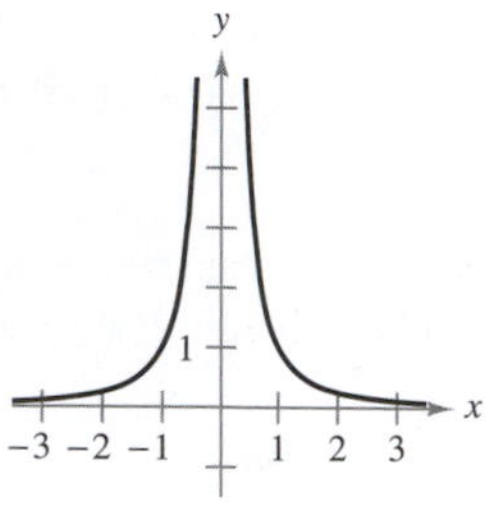

(b)

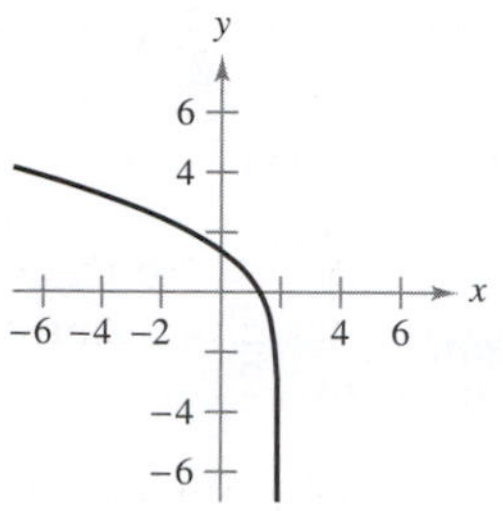

3.

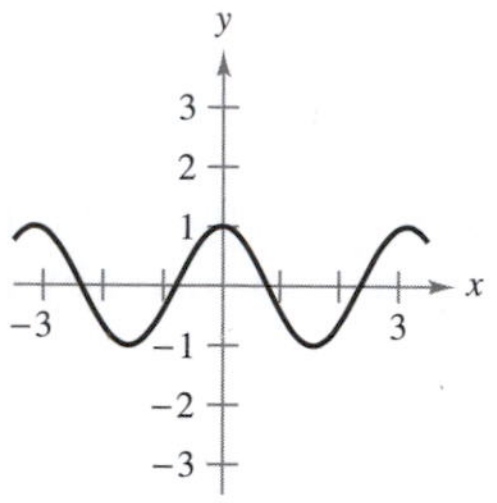

(c)

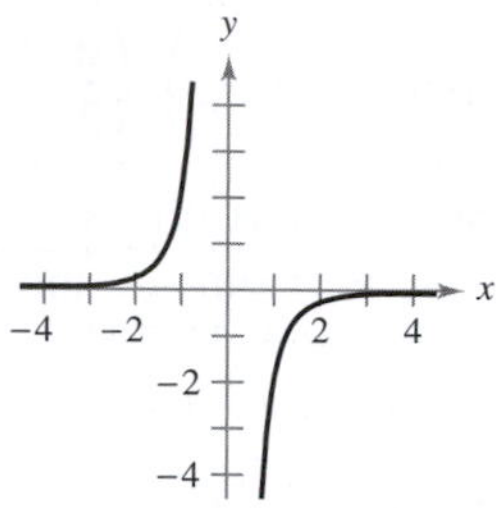

4.

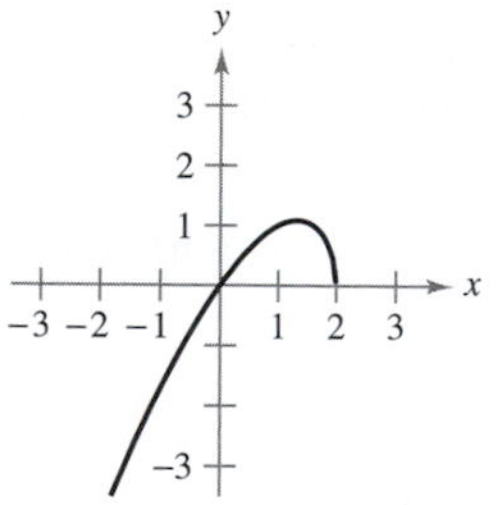

(d)

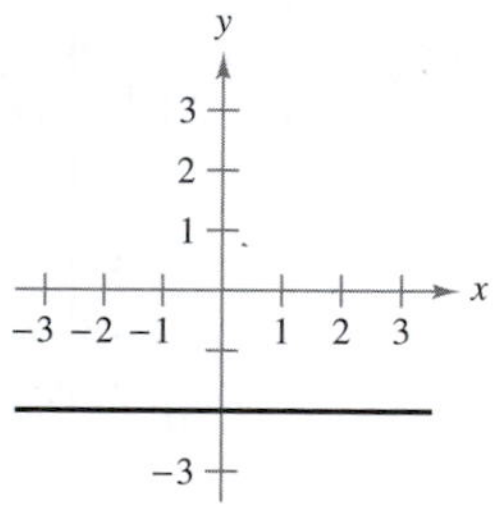

Analyzing the Graph of a Function **In Exercises 5–24, analyze and sketch a graph of the function. Label any intercepts, relative extrema, points of inflection, and asymptotes. Use a graphing utility to verify your results.**

5. $y = \dfrac{1}{x-2} - 3$

6. $y = \dfrac{x}{x^2+1}$

7. $y = \dfrac{x^2}{x^2+3}$

8. $y = \dfrac{x^2+1}{x^2-4}$

9. $y = \dfrac{3x}{x^2-1}$

10. $f(x) = \dfrac{x-3}{x}$

11. $f(x) = x + \dfrac{32}{x^2}$

12. $f(x) = \dfrac{x^3}{x^2-9}$

13. $y = \dfrac{x^2-6x+12}{x-4}$

14. $y = \dfrac{-x^2-4x-7}{x+3}$

15. $y = x\sqrt{4-x}$

16. $g(x) = x\sqrt{9-x^2}$

17. $y = 3x^{2/3} - 2x$

18. $y = (x+1)^2 - 3(x+1)^{2/3}$

19. $y = 2 - x - x^3$

20. $y = -\frac{1}{3}(x^3 - 3x + 2)$

21. $y = 3x^4 + 4x^3$

22. $y = -2x^4 + 3x^2$

23. $y = x^5 - 5x$

24. $y = (x-1)^5$

Analyzing the Graph of a Function Using Technology **In Exercises 25–34, use a computer algebra system to analyze and graph the function. Identify any relative extrema, points of inflection, and asymptotes.**

25. $f(x) = \dfrac{20x}{x^2+1} - \dfrac{1}{x}$

26. $f(x) = x + \dfrac{4}{x^2+1}$

27. $f(x) = \dfrac{-2x}{\sqrt{x^2+7}}$

28. $f(x) = \dfrac{4x}{\sqrt{x^2+15}}$

29. $f(x) = 2x - 4\sin x, \quad 0 \le x \le 2\pi$

30. $f(x) = -x + 2\cos x, \quad 0 \le x \le 2\pi$

31. $y = \cos x - \frac{1}{4}\cos 2x, \quad 0 \le x \le 2\pi$

32. $y = 2x - \tan x, \quad -\dfrac{\pi}{2} < x < \dfrac{\pi}{2}$

33. $y = 2(\csc x + \sec x), \quad 0 < x < \dfrac{\pi}{2}$

34. $g(x) = x\cot x, \quad -2\pi < x < 2\pi$

WRITING ABOUT CONCEPTS

35. **Using a Derivative** Let $f'(t) < 0$ for all t in the interval $(2, 8)$. Explain why $f(3) > f(5)$.

36. **Using a Derivative** Let $f(0) = 3$ and $2 \le f'(x) \le 4$ for all x in the interval $[-5, 5]$. Determine the greatest and least possible values of $f(2)$.

Identifying Graphs **In Exercises 37 and 38, the graphs of f, f', and f'' are shown on the same set of coordinate axes. Which is which? Explain your reasoning. To print an enlarged copy of the graph, go to *MathGraphs.com*.**

37.

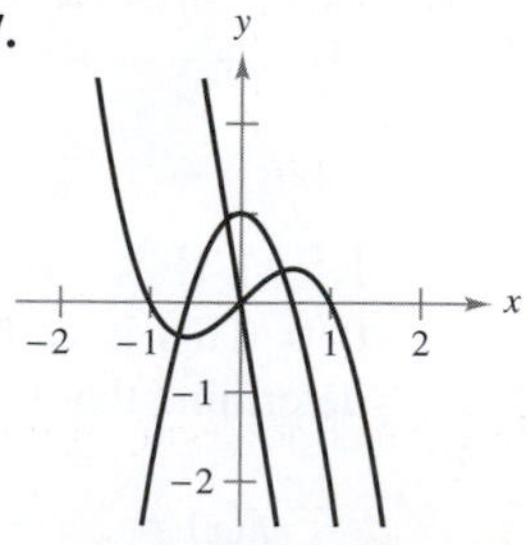

38.

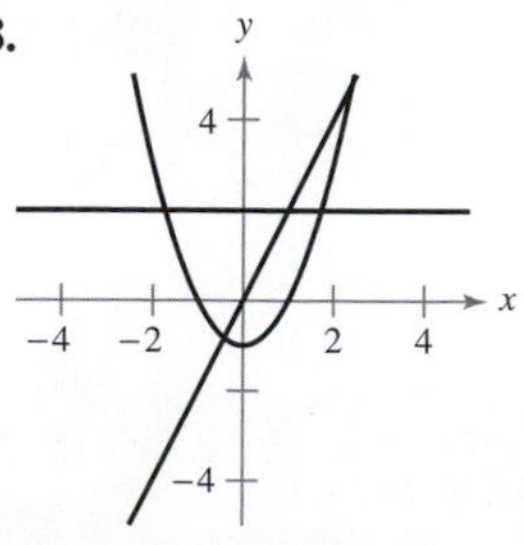

WRITING ABOUT CONCEPTS (continued)

Horizontal and Vertical Asymptotes **In Exercises 39–42, use a graphing utility to graph the function. Use the graph to determine whether it is possible for the graph of a function to cross its horizontal asymptote. Do you think it is possible for the graph of a function to cross its vertical asymptote? Why or why not?**

39. $f(x) = \dfrac{4(x-1)^2}{x^2 - 4x + 5}$

40. $g(x) = \dfrac{3x^4 - 5x + 3}{x^4 + 1}$

41. $h(x) = \dfrac{\sin 2x}{x}$

42. $f(x) = \dfrac{\cos 3x}{4x}$

Examining a Function **In Exercises 43 and 44, use a graphing utility to graph the function. Explain why there is no vertical asymptote when a superficial examination of the function may indicate that there should be one.**

43. $h(x) = \dfrac{6 - 2x}{3 - x}$

44. $g(x) = \dfrac{x^2 + x - 2}{x - 1}$

Slant Asymptote **In Exercises 45–48, use a graphing utility to graph the function and determine the slant asymptote of the graph. Zoom out repeatedly and describe how the graph on the display appears to change. Why does this occur?**

45. $f(x) = -\dfrac{x^2 - 3x - 1}{x - 2}$

46. $g(x) = \dfrac{2x^2 - 8x - 15}{x - 5}$

47. $f(x) = \dfrac{2x^3}{x^2 + 1}$

48. $h(x) = \dfrac{-x^3 + x^2 + 4}{x^2}$

Graphical Reasoning **In Exercises 49–52, use the graph of f' to sketch a graph of f and the graph of f''. To print an enlarged copy of the graph, go to *MathGraphs.com*.**

49.

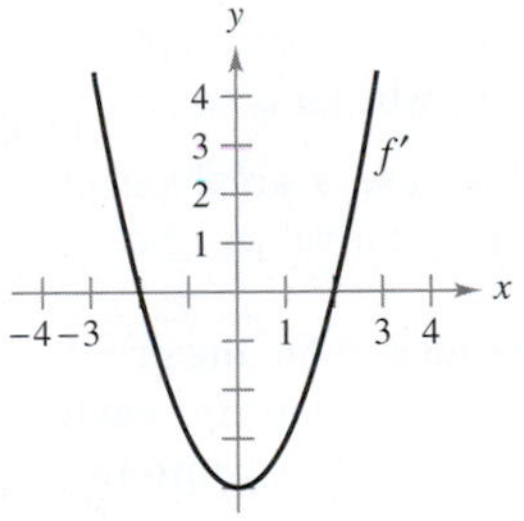

50.

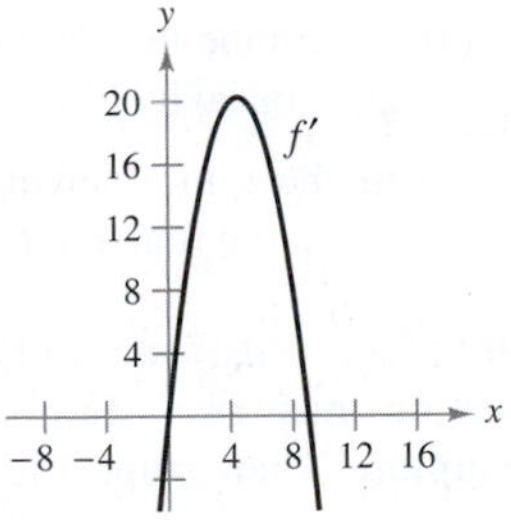

51.

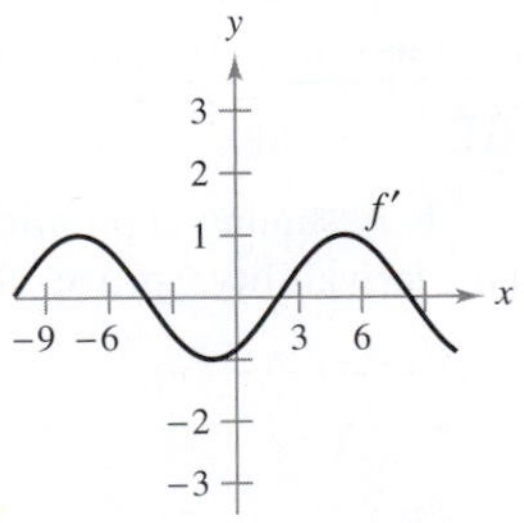

52.

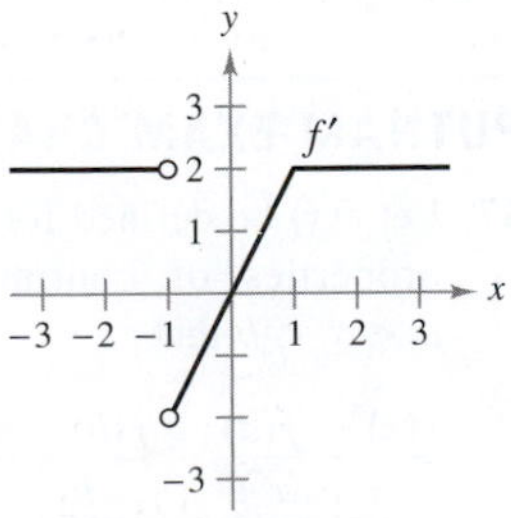

(Submitted by Bill Fox, Moberly Area Community College, Moberly, MO)

53. Graphical Reasoning Consider the function

$$f(x) = \frac{\cos^2 \pi x}{\sqrt{x^2 + 1}}, \quad 0 < x < 4.$$

(a) Use a computer algebra system to graph the function and use the graph to approximate the critical numbers visually.

(b) Use a computer algebra system to find f' and approximate the critical numbers. Are the results the same as the visual approximation in part (a)? Explain.

54. Graphical Reasoning Consider the function

$$f(x) = \tan(\sin \pi x).$$

(a) Use a graphing utility to graph the function.

(b) Identify any symmetry of the graph.

(c) Is the function periodic? If so, what is the period?

(d) Identify any extrema on $(-1, 1)$.

(e) Use a graphing utility to determine the concavity of the graph on $(0, 1)$.

Think About It **In Exercises 55–58, create a function whose graph has the given characteristics. (There is more than one correct answer.)**

55. Vertical asymptote: $x = 3$

Horizontal asymptote: $y = 0$

56. Vertical asymptote: $x = -5$

Horizontal asymptote: None

57. Vertical asymptote: $x = 3$

Slant asymptote: $y = 3x + 2$

58. Vertical asymptote: $x = 2$

Slant asymptote: $y = -x$

59. Graphical Reasoning Identify the real numbers x_0, x_1, x_2, x_3, and x_4 in the figure such that each of the following is true.

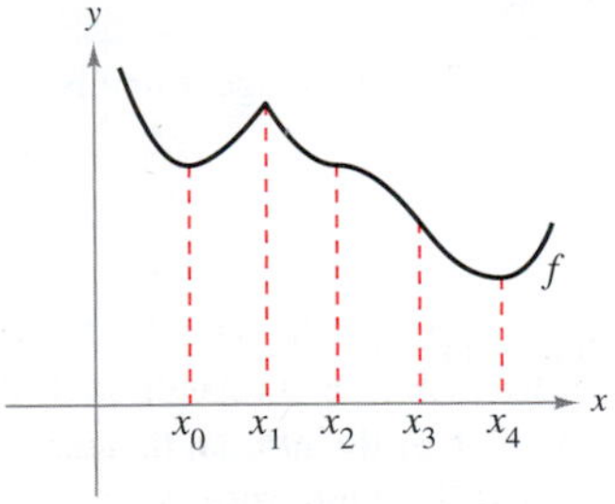

(a) $f'(x) = 0$

(b) $f''(x) = 0$

(c) $f'(x)$ does not exist.

(d) f has a relative maximum.

(e) f has a point of inflection.

60. HOW DO YOU SEE IT? The graph of f is shown in the figure.

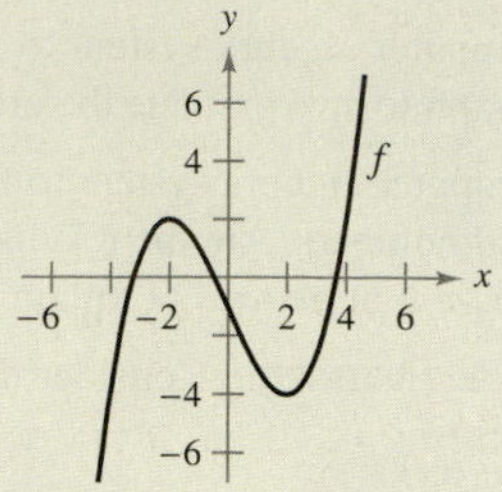

(a) For which values of x is $f'(x)$ zero? Positive? Negative? What do these values mean?

(b) For which values of x is $f''(x)$ zero? Positive? Negative? What do these values mean?

(c) On what open interval is f' an increasing function?

(d) For which value of x is $f'(x)$ minimum? For this value of x, how does the rate of change of f compare with the rates of change of f for other values of x? Explain.

61. Investigation Let $P(x_0, y_0)$ be an arbitrary point on the graph of f such that $f'(x_0) \neq 0$, as shown in the figure. Verify each statement.

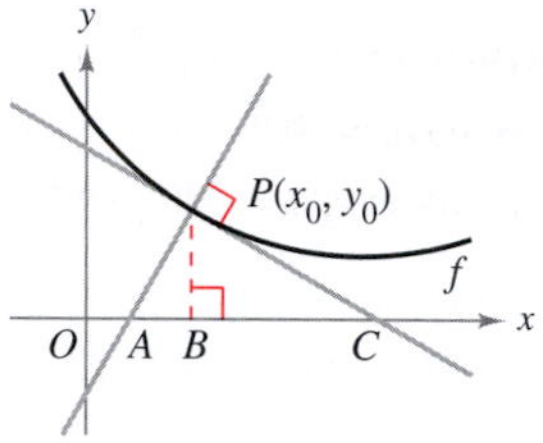

(a) The x-intercept of the tangent line is

$$\left(x_0 - \frac{f(x_0)}{f'(x_0)}, 0\right).$$

(b) The y-intercept of the tangent line is

$$(0, f(x_0) - x_0 f'(x_0)).$$

(c) The x-intercept of the normal line is

$$(x_0 + f(x_0) f'(x_0), 0).$$

(d) The y-intercept of the normal line is

$$\left(0, y_0 + \frac{x_0}{f'(x_0)}\right).$$

(e) $|BC| = \left|\dfrac{f(x_0)}{f'(x_0)}\right|$

(f) $|PC| = \left|\dfrac{f(x_0)\sqrt{1 + [f'(x_0)]^2}}{f'(x_0)}\right|$

(g) $|AB| = |f(x_0) f'(x_0)|$

(h) $|AP| = |f(x_0)|\sqrt{1 + [f'(x_0)]^2}$

62. Investigation Consider the function

$$f(x) = \frac{2x^n}{x^4 + 1}$$

for nonnegative integer values of n.

(a) Discuss the relationship between the value of n and the symmetry of the graph.

(b) For which values of n will the x-axis be the horizontal asymptote?

(c) For which value of n will $y = 2$ be the horizontal asymptote?

(d) What is the asymptote of the graph when $n = 5$?

(e) Use a graphing utility to graph f for the indicated values of n in the table. Use the graph to determine the number of extrema M and the number of inflection points N of the graph.

n	0	1	2	3	4	5
M						
N						

63. Graphical Reasoning Consider the function

$$f(x) = \frac{ax}{(x - b)^2}.$$

Determine the effect on the graph of f as a and b are changed. Consider cases where a and b are both positive or both negative, and cases where a and b have opposite signs.

64. Graphical Reasoning Consider the function

$$f(x) = \frac{1}{2}(ax)^2 - ax, \quad a \neq 0.$$

(a) Determine the changes (if any) in the intercepts, extrema, and concavity of the graph of f when a is varied.

(b) In the same viewing window, use a graphing utility to graph the function for four different values of a.

Slant Asymptotes **In Exercises 65 and 66, the graph of the function has two slant asymptotes. Identify each slant asymptote. Then graph the function and its asymptotes.**

65. $y = \sqrt{4 + 16x^2}$

66. $y = \sqrt{x^2 + 6x}$

PUTNAM EXAM CHALLENGE

67. Let $f(x)$ be defined for $a \leq x \leq b$. Assuming appropriate properties of continuity and derivability, prove for $a < x < b$ that

$$\frac{\dfrac{f(x) - f(a)}{x - a} - \dfrac{f(b) - f(a)}{b - a}}{x - b} = \frac{1}{2} f''(\varepsilon),$$

where ε is some number between a and b.

This problem was composed by the Committee on the Putnam Prize Competition.

3.7 Optimization Problems

■ **Solve applied minimum and maximum problems.**

Applied Minimum and Maximum Problems

One of the most common applications of calculus involves the determination of minimum and maximum values. Consider how frequently you hear or read terms such as greatest profit, least cost, least time, greatest voltage, optimum size, least size, greatest strength, and greatest distance. Before outlining a general problem-solving strategy for such problems, consider the next example.

EXAMPLE 1 Finding Maximum Volume

A manufacturer wants to design an open box having a square base and a surface area of 108 square inches, as shown in Figure 3.53. What dimensions will produce a box with maximum volume?

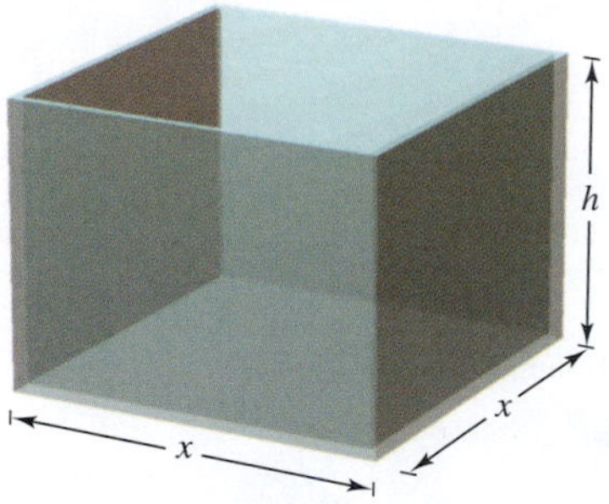

Open box with square base: $S = x^2 + 4xh = 108$

Figure 3.53

Solution Because the box has a square base, its volume is

$$V = x^2 h. \quad \text{Primary equation}$$

This equation is called the **primary equation** because it gives a formula for the quantity to be optimized. The surface area of the box is

$$S = (\text{area of base}) + (\text{area of four sides})$$
$$108 = x^2 + 4xh. \quad \text{Secondary equation}$$

Because V is to be maximized, you want to write V as a function of just one variable. To do this, you can solve the equation $x^2 + 4xh = 108$ for h in terms of x to obtain $h = (108 - x^2)/(4x)$. Substituting into the primary equation produces

$$V = x^2 h \quad \text{Function of two variables}$$
$$= x^2\left(\frac{108 - x^2}{4x}\right) \quad \text{Substitute for } h.$$
$$= 27x - \frac{x^3}{4}. \quad \text{Function of one variable}$$

Before finding which x-value will yield a maximum value of V, you should determine the *feasible domain*. That is, what values of x make sense in this problem? You know that $V \geq 0$. You also know that x must be nonnegative and that the area of the base $(A = x^2)$ is at most 108. So, the feasible domain is

$$0 \leq x \leq \sqrt{108}. \quad \text{Feasible domain}$$

To maximize V, find the critical numbers of the volume function on the interval $(0, \sqrt{108})$.

$$\frac{dV}{dx} = 27 - \frac{3x^2}{4} \quad \text{Differentiate with respect to } x.$$
$$27 - \frac{3x^2}{4} = 0 \quad \text{Set derivative equal to 0.}$$
$$3x^2 = 108 \quad \text{Simplify.}$$
$$x = \pm 6 \quad \text{Critical numbers}$$

So, the critical numbers are $x = \pm 6$. You do not need to consider $x = -6$ because it is outside the domain. Evaluating V at the critical number $x = 6$ and at the endpoints of the domain produces $V(0) = 0$, $V(6) = 108$, and $V(\sqrt{108}) = 0$. So, V is maximum when $x = 6$, and the dimensions of the box are 6 inches by 6 inches by 3 inches. ■

▷ **TECHNOLOGY** You can verify your answer in Example 1 by using a graphing utility to graph the volume function

$$V = 27x - \frac{x^3}{4}.$$

Use a viewing window in which $0 \leq x \leq \sqrt{108} \approx 10.4$ and $0 \leq y \leq 120$, and use the *maximum* or *trace* feature to determine the maximum value of V.

In Example 1, you should realize that there are infinitely many open boxes having 108 square inches of surface area. To begin solving the problem, you might ask yourself which basic shape would seem to yield a maximum volume. Should the box be tall, squat, or nearly cubical?

You might even try calculating a few volumes, as shown in Figure 3.54, to see if you can get a better feeling for what the optimum dimensions should be. Remember that you are not ready to begin solving a problem until you have clearly identified what the problem is.

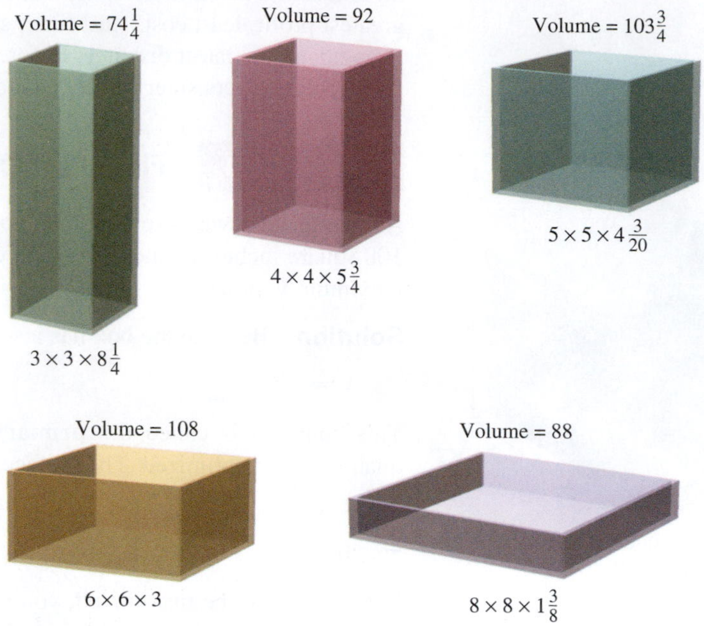

Which box has the greatest volume?
Figure 3.54

Example 1 illustrates the following guidelines for solving applied minimum and maximum problems.

GUIDELINES FOR SOLVING APPLIED MINIMUM AND MAXIMUM PROBLEMS

1. Identify all *given* quantities and all quantities *to be determined.* If possible, make a sketch.
2. Write a **primary equation** for the quantity that is to be maximized or minimized. (A review of several useful formulas from geometry is presented inside the back cover.)
3. Reduce the primary equation to one having a *single independent variable.* This may involve the use of **secondary equations** relating the independent variables of the primary equation.
4. Determine the feasible domain of the primary equation. That is, determine the values for which the stated problem makes sense.
5. Determine the desired maximum or minimum value by the calculus techniques discussed in Sections 3.1 through 3.4.

REMARK For Step 5, recall that to determine the maximum or minimum value of a continuous function f on a closed interval, you should compare the values of f at its critical numbers with the values of f at the endpoints of the interval.

EXAMPLE 2 Finding Minimum Distance

See LarsonCalculus.com for an interactive version of this type of example.

Which points on the graph of $y = 4 - x^2$ are closest to the point $(0, 2)$?

Solution Figure 3.55 shows that there are two points at a minimum distance from the point $(0, 2)$. The distance between the point $(0, 2)$ and a point (x, y) on the graph of $y = 4 - x^2$ is

$$d = \sqrt{(x - 0)^2 + (y - 2)^2}. \qquad \text{Primary equation}$$

Using the secondary equation $y = 4 - x^2$, you can rewrite the primary equation as

$$\begin{aligned} d &= \sqrt{x^2 + (4 - x^2 - 2)^2} \\ &= \sqrt{x^4 - 3x^2 + 4}. \end{aligned}$$

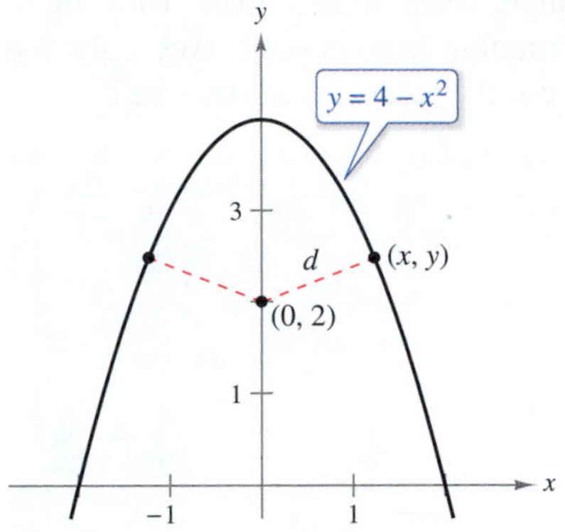

The quantity to be minimized is distance: $d = \sqrt{(x - 0)^2 + (y - 2)^2}$.
Figure 3.55

Because d is smallest when the expression inside the radical is smallest, you need only find the critical numbers of $f(x) = x^4 - 3x^2 + 4$. Note that the domain of f is the entire real number line. So, there are no endpoints of the domain to consider. Moreover, the derivative of f

$$\begin{aligned} f'(x) &= 4x^3 - 6x \\ &= 2x(2x^2 - 3) \end{aligned}$$

is zero when

$$x = 0, \sqrt{\frac{3}{2}}, -\sqrt{\frac{3}{2}}.$$

Testing these critical numbers using the First Derivative Test verifies that $x = 0$ yields a relative maximum, whereas both $x = \sqrt{3/2}$ and $x = -\sqrt{3/2}$ yield a minimum distance. So, the closest points are $\left(\sqrt{3/2}, 5/2\right)$ and $\left(-\sqrt{3/2}, 5/2\right)$.

EXAMPLE 3 Finding Minimum Area

A rectangular page is to contain 24 square inches of print. The margins at the top and bottom of the page are to be $1\frac{1}{2}$ inches, and the margins on the left and right are to be 1 inch (see Figure 3.56). What should the dimensions of the page be so that the least amount of paper is used?

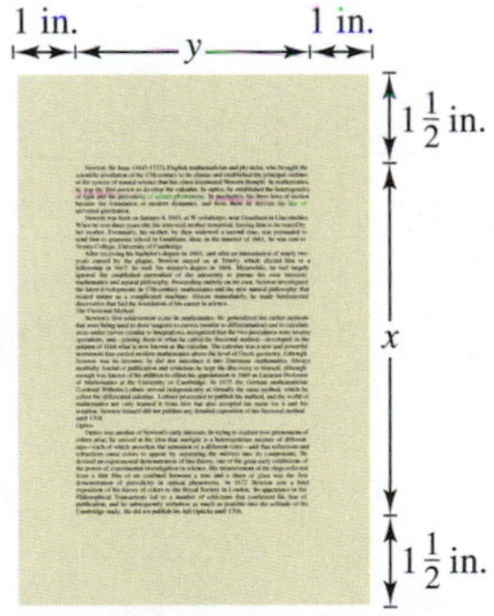

The quantity to be minimized is area: $A = (x + 3)(y + 2)$.
Figure 3.56

Solution Let A be the area to be minimized.

$$A = (x + 3)(y + 2) \qquad \text{Primary equation}$$

The printed area inside the margins is

$$24 = xy. \qquad \text{Secondary equation}$$

Solving this equation for y produces $y = 24/x$. Substitution into the primary equation produces

$$A = (x + 3)\left(\frac{24}{x} + 2\right) = 30 + 2x + \frac{72}{x}. \qquad \text{Function of one variable}$$

Because x must be positive, you are interested only in values of A for $x > 0$. To find the critical numbers, differentiate with respect to x

$$\frac{dA}{dx} = 2 - \frac{72}{x^2}$$

and note that the derivative is zero when $x^2 = 36$, or $x = \pm 6$. So, the critical numbers are $x = \pm 6$. You do not have to consider $x = -6$ because it is outside the domain. The First Derivative Test confirms that A is a minimum when $x = 6$. So, $y = \frac{24}{6} = 4$ and the dimensions of the page should be $x + 3 = 9$ inches by $y + 2 = 6$ inches.

EXAMPLE 4 Finding Minimum Length

Two posts, one 12 feet high and the other 28 feet high, stand 30 feet apart. They are to be stayed by two wires, attached to a single stake, running from ground level to the top of each post. Where should the stake be placed to use the least amount of wire?

Solution Let W be the wire length to be minimized. Using Figure 3.57, you can write

$$W = y + z. \quad \text{Primary equation}$$

In this problem, rather than solving for y in terms of z (or vice versa), you can solve for both y and z in terms of a third variable x, as shown in Figure 3.57. From the Pythagorean Theorem, you obtain

$$x^2 + 12^2 = y^2$$
$$(30 - x)^2 + 28^2 = z^2$$

which implies that

$$y = \sqrt{x^2 + 144}$$
$$z = \sqrt{x^2 - 60x + 1684}.$$

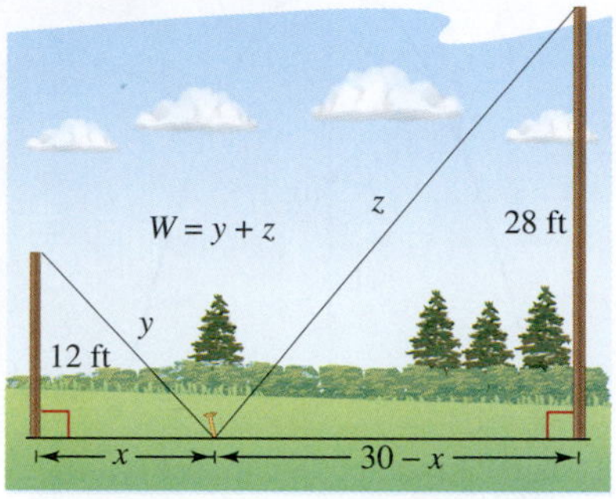

The quantity to be minimized is length. From the diagram, you can see that x varies between 0 and 30.
Figure 3.57

So, you can rewrite the primary equation as

$$\begin{aligned} W &= y + z \\ &= \sqrt{x^2 + 144} + \sqrt{x^2 - 60x + 1684}, \quad 0 \le x \le 30. \end{aligned}$$

Differentiating W with respect to x yields

$$\frac{dW}{dx} = \frac{x}{\sqrt{x^2 + 144}} + \frac{x - 30}{\sqrt{x^2 - 60x + 1684}}.$$

By letting $dW/dx = 0$, you obtain

$$\begin{aligned} \frac{x}{\sqrt{x^2 + 144}} + \frac{x - 30}{\sqrt{x^2 - 60x + 1684}} &= 0 \\ x\sqrt{x^2 - 60x + 1684} &= (30 - x)\sqrt{x^2 + 144} \\ x^2(x^2 - 60x + 1684) &= (30 - x)^2(x^2 + 144) \\ x^4 - 60x^3 + 1684x^2 &= x^4 - 60x^3 + 1044x^2 - 8640x + 129{,}600 \\ 640x^2 + 8640x - 129{,}600 &= 0 \\ 320(x - 9)(2x + 45) &= 0 \\ x &= 9, -22.5. \end{aligned}$$

Because $x = -22.5$ is not in the domain and

$$W(0) \approx 53.04, \quad W(9) = 50, \quad \text{and} \quad W(30) \approx 60.31$$

you can conclude that the wire should be staked at 9 feet from the 12-foot pole. ■

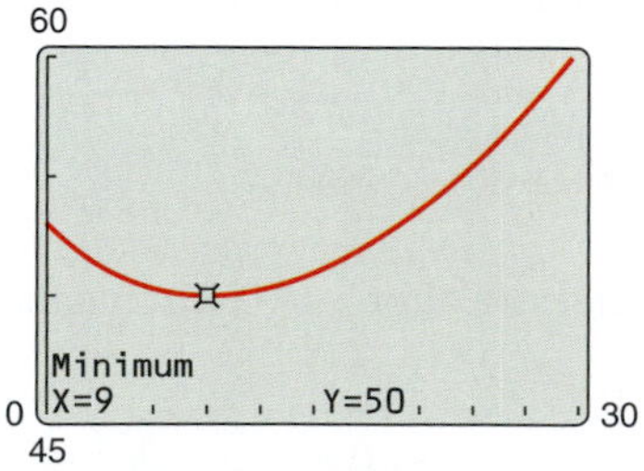

You can confirm the minimum value of W with a graphing utility.
Figure 3.58

TECHNOLOGY From Example 4, you can see that applied optimization problems can involve a lot of algebra. If you have access to a graphing utility, you can confirm that $x = 9$ yields a minimum value of W by graphing

$$W = \sqrt{x^2 + 144} + \sqrt{x^2 - 60x + 1684}$$

as shown in Figure 3.58.

In each of the first four examples, the extreme value occurred at a critical number. Although this happens often, remember that an extreme value can also occur at an endpoint of an interval, as shown in Example 5.

EXAMPLE 5 An Endpoint Maximum

Four feet of wire is to be used to form a square and a circle. How much of the wire should be used for the square and how much should be used for the circle to enclose the maximum total area?

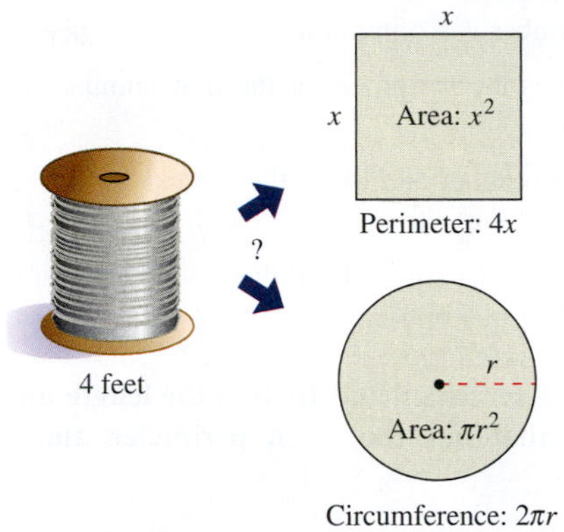

The quantity to be maximized is area: $A = x^2 + \pi r^2$.
Figure 3.59

Solution The total area (see Figure 3.59) is

$$A = (\text{area of square}) + (\text{area of circle})$$
$$A = x^2 + \pi r^2. \qquad \text{Primary equation}$$

Because the total length of wire is 4 feet, you obtain

$$4 = (\text{perimeter of square}) + (\text{circumference of circle})$$
$$4 = 4x + 2\pi r.$$

So, $r = 2(1 - x)/\pi$, and by substituting into the primary equation you have

$$A = x^2 + \pi\left[\frac{2(1-x)}{\pi}\right]^2$$
$$= x^2 + \frac{4(1-x)^2}{\pi}$$
$$= \frac{1}{\pi}[(\pi + 4)x^2 - 8x + 4].$$

The feasible domain is $0 \le x \le 1$, restricted by the square's perimeter. Because

$$\frac{dA}{dx} = \frac{2(\pi + 4)x - 8}{\pi}$$

the only critical number in $(0, 1)$ is $x = 4/(\pi + 4) \approx 0.56$. So, using

$$A(0) \approx 1.273, \quad A(0.56) \approx 0.56, \quad \text{and} \quad A(1) = 1$$

you can conclude that the maximum area occurs when $x = 0$. That is, *all* the wire is used for the circle. ■

Exploration

What would the answer be if Example 5 asked for the dimensions needed to enclose the *minimum* total area?

Before doing the section exercises, review the primary equations developed in the first five examples. As applications go, these five examples are fairly simple, and yet the resulting primary equations are quite complicated.

$$V = 27x - \frac{x^3}{4} \qquad \text{Example 1}$$
$$d = \sqrt{x^4 - 3x^2 + 4} \qquad \text{Example 2}$$
$$A = 30 + 2x + \frac{72}{x} \qquad \text{Example 3}$$
$$W = \sqrt{x^2 + 144} + \sqrt{x^2 - 60x + 1684} \qquad \text{Example 4}$$
$$A = \frac{1}{\pi}[(\pi + 4)x^2 - 8x + 4] \qquad \text{Example 5}$$

You must expect that real-life applications often involve equations that are *at least as complicated* as these five. Remember that one of the main goals of this course is to learn to use calculus to analyze equations that initially seem formidable.

3.7 Exercises

See CalcChat.com for tutorial help and worked-out solutions to odd-numbered exercises.

1. Numerical, Graphical, and Analytic Analysis Find two positive numbers whose sum is 110 and whose product is a maximum.

(a) Analytically complete six rows of a table such as the one below. (The first two rows are shown.)

First Number, x	Second Number	Product, P
10	$110 - 10$	$10(110 - 10) = 1000$
20	$110 - 20$	$20(110 - 20) = 1800$

(b) Use a graphing utility to generate additional rows of the table. Use the table to estimate the solution. (*Hint:* Use the table feature of the graphing utility.)

(c) Write the product P as a function of x.

(d) Use a graphing utility to graph the function in part (c) and estimate the solution from the graph.

(e) Use calculus to find the critical number of the function in part (c). Then find the two numbers.

2. Numerical, Graphical, and Analytic Analysis An open box of maximum volume is to be made from a square piece of material, 24 inches on a side, by cutting equal squares from the corners and turning up the sides (see figure).

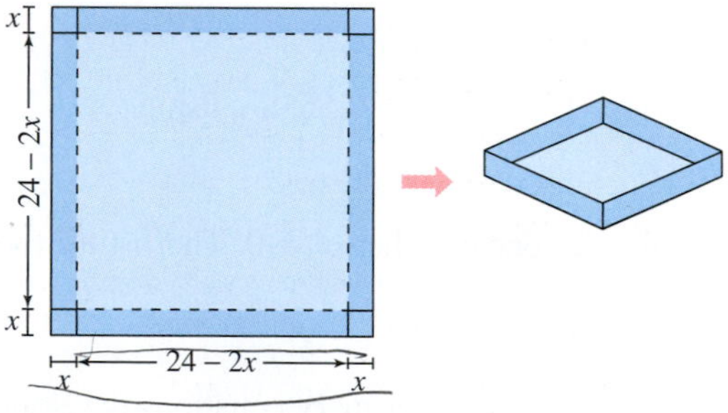

(a) Analytically complete six rows of a table such as the one below. (The first two rows are shown.) Use the table to guess the maximum volume.

Height, x	Length and Width	Volume, V
1	$24 - 2(1)$	$1[24 - 2(1)]^2 = 484$
2	$24 - 2(2)$	$2[24 - 2(2)]^2 = 800$

(b) Write the volume V as a function of x.

(c) Use calculus to find the critical number of the function in part (b) and find the maximum value.

(d) Use a graphing utility to graph the function in part (b) and verify the maximum volume from the graph.

Finding Numbers **In Exercises 3–8, find two positive numbers that satisfy the given requirements.**

3. The sum is S and the product is a maximum.

4. The product is 185 and the sum is a minimum.

5. The product is 147 and the sum of the first number plus three times the second number is a minimum.

6. The second number is the reciprocal of the first number and the sum is a minimum.

7. The sum of the first number and twice the second number is 108 and the product is a maximum.

8. The sum of the first number squared and the second number is 54 and the product is a maximum.

Maximum Area **In Exercises 9 and 10, find the length and width of a rectangle that has the given perimeter and a maximum area.**

9. Perimeter: 80 meters

10. Perimeter: P units

Minimum Perimeter **In Exercises 11 and 12, find the length and width of a rectangle that has the given area and a minimum perimeter.**

11. Area: 32 square feet

12. Area: A square centimeters

Minimum Distance **In Exercises 13–16, find the point on the graph of the function that is closest to the given point.**

13. $f(x) = x^2,\ \left(2, \frac{1}{2}\right)$

14. $f(x) = (x - 1)^2,\ (-5, 3)$

15. $f(x) = \sqrt{x},\ (4, 0)$

16. $f(x) = \sqrt{x - 8},\ (12, 0)$

17. Minimum Area A rectangular page is to contain 30 square inches of print. The margins on each side are 1 inch. Find the dimensions of the page such that the least amount of paper is used.

18. Minimum Area A rectangular page is to contain 36 square inches of print. The margins on each side are $1\frac{1}{2}$ inches. Find the dimensions of the page such that the least amount of paper is used.

19. Minimum Length A farmer plans to fence a rectangular pasture adjacent to a river (see figure). The pasture must contain 245,000 square meters in order to provide enough grass for the herd. No fencing is needed along the river. What dimensions will require the least amount of fencing?

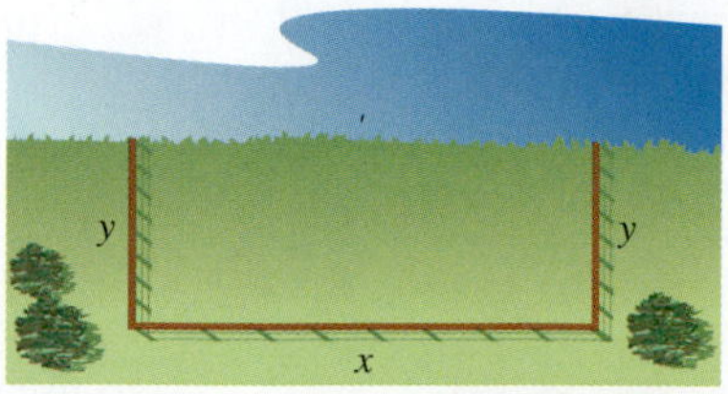

20. Maximum Volume A rectangular solid (with a square base) has a surface area of 337.5 square centimeters. Find the dimensions that will result in a solid with maximum volume.

21. Maximum Area A Norman window is constructed by adjoining a semicircle to the top of an ordinary rectangular window (see figure). Find the dimensions of a Norman window of maximum area when the total perimeter is 16 feet.

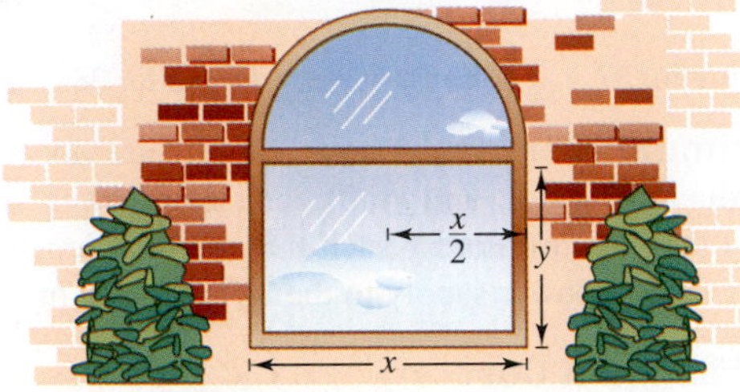

22. Maximum Area A rectangle is bounded by the x- and y-axes and the graph of $y = (6 - x)/2$ (see figure). What length and width should the rectangle have so that its area is a maximum?

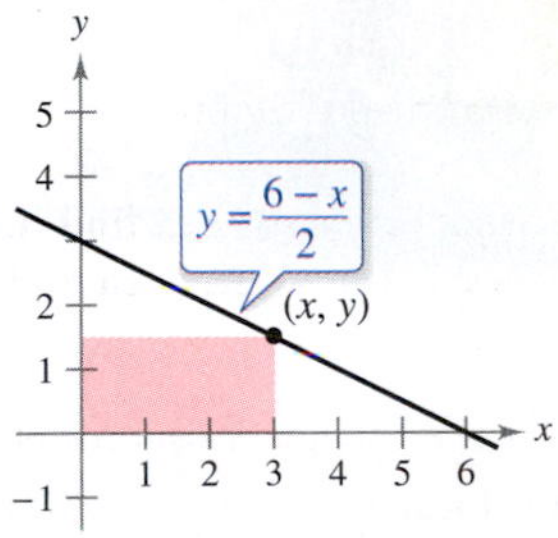

Figure for 22

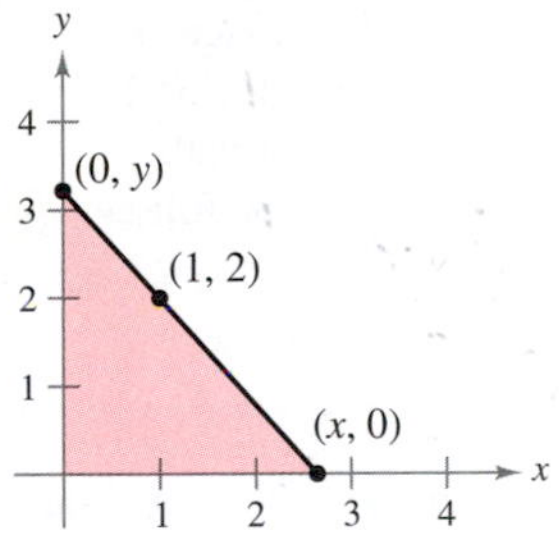

Figure for 23

23. Minimum Length and Minimum Area A right triangle is formed in the first quadrant by the x- and y-axes and a line through the point $(1, 2)$ (see figure).

(a) Write the length L of the hypotenuse as a function of x.

(b) Use a graphing utility to approximate x graphically such that the length of the hypotenuse is a minimum.

(c) Find the vertices of the triangle such that its area is a minimum.

24. Maximum Area Find the area of the largest isosceles triangle that can be inscribed in a circle of radius 6 (see figure).

(a) Solve by writing the area as a function of h.

(b) Solve by writing the area as a function of α.

(c) Identify the type of triangle of maximum area.

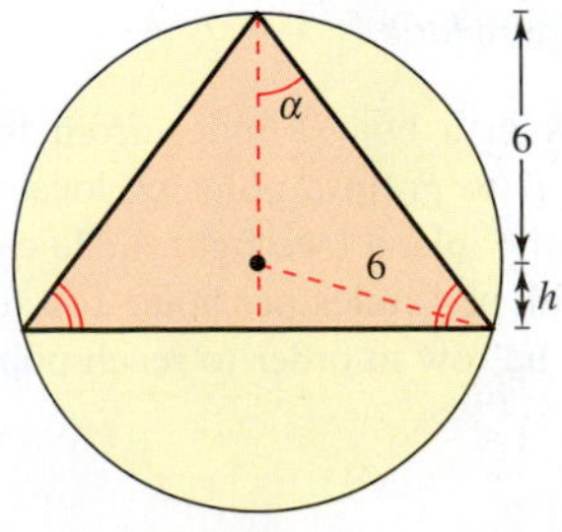

Figure for 24

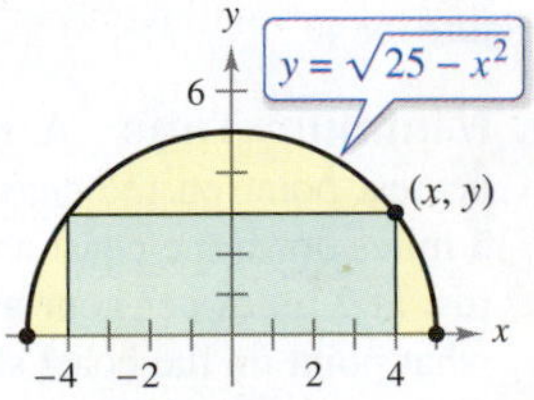

Figure for 25

25. Maximum Area A rectangle is bounded by the x-axis and the semicircle

$$y = \sqrt{25 - x^2}$$

(see figure). What length and width should the rectangle have so that its area is a maximum?

26. Maximum Area Find the dimensions of the largest rectangle that can be inscribed in a semicircle of radius r (see Exercise 25).

27. Numerical, Graphical, and Analytic Analysis An exercise room consists of a rectangle with a semicircle on each end. A 200-meter running track runs around the outside of the room.

(a) Draw a figure to represent the problem. Let x and y represent the length and width of the rectangle.

(b) Analytically complete six rows of a table such as the one below. (The first two rows are shown.) Use the table to guess the maximum area of the rectangular region.

Length, x	Width, y	Area, xy
10	$\frac{2}{\pi}(100 - 10)$	$(10)\frac{2}{\pi}(100 - 10) \approx 573$
20	$\frac{2}{\pi}(100 - 20)$	$(20)\frac{2}{\pi}(100 - 20) \approx 1019$

(c) Write the area A as a function of x.

(d) Use calculus to find the critical number of the function in part (c) and find the maximum value.

(e) Use a graphing utility to graph the function in part (c) and verify the maximum area from the graph.

28. Numerical, Graphical, and Analytic Analysis A right circular cylinder is designed to hold 22 cubic inches of a soft drink (approximately 12 fluid ounces).

(a) Analytically complete six rows of a table such as the one below. (The first two rows are shown.)

Radius, r	Height	Surface Area, S
0.2	$\frac{22}{\pi(0.2)^2}$	$2\pi(0.2)\left[0.2 + \frac{22}{\pi(0.2)^2}\right] \approx 220.3$
0.4	$\frac{22}{\pi(0.4)^2}$	$2\pi(0.4)\left[0.4 + \frac{22}{\pi(0.4)^2}\right] \approx 111.0$

(b) Use a graphing utility to generate additional rows of the table. Use the table to estimate the minimum surface area. (*Hint:* Use the *table* feature of the graphing utility.)

(c) Write the surface area S as a function of r.

(d) Use a graphing utility to graph the function in part (c) and estimate the minimum surface area from the graph.

(e) Use calculus to find the critical number of the function in part (c) and find dimensions that will yield the minimum surface area.

29. Maximum Volume A rectangular package to be sent by a postal service can have a maximum combined length and girth (perimeter of a cross section) of 108 inches (see figure). Find the dimensions of the package of maximum volume that can be sent. (Assume the cross section is square.)

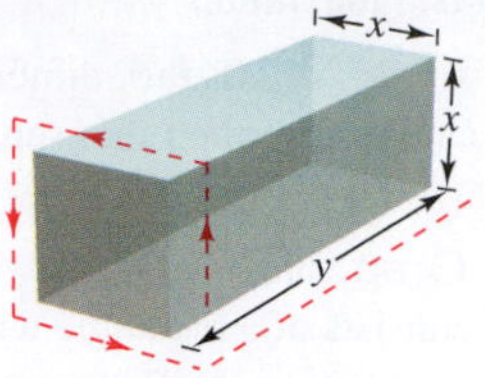

30. Maximum Volume Rework Exercise 29 for a cylindrical package. (The cross section is circular.)

WRITING ABOUT CONCEPTS

31. Surface Area and Volume A shampoo bottle is a right circular cylinder. Because the surface area of the bottle does not change when it is squeezed, is it true that the volume remains the same? Explain.

32. Area and Perimeter The perimeter of a rectangle is 20 feet. Of all possible dimensions, the maximum area is 25 square feet when its length and width are both 5 feet. Are there dimensions that yield a minimum area? Explain.

33. Minimum Surface Area A solid is formed by adjoining two hemispheres to the ends of a right circular cylinder. The total volume of the solid is 14 cubic centimeters. Find the radius of the cylinder that produces the minimum surface area.

34. Minimum Cost An industrial tank of the shape described in Exercise 33 must have a volume of 4000 cubic feet. The hemispherical ends cost twice as much per square foot of surface area as the sides. Find the dimensions that will minimize cost.

35. Minimum Area The sum of the perimeters of an equilateral triangle and a square is 10. Find the dimensions of the triangle and the square that produce a minimum total area.

36. Maximum Area Twenty feet of wire is to be used to form two figures. In each of the following cases, how much wire should be used for each figure so that the total enclosed area is maximum?

(a) Equilateral triangle and square

(b) Square and regular pentagon

(c) Regular pentagon and regular hexagon

(d) Regular hexagon and circle

What can you conclude from this pattern? {*Hint:* The area of a regular polygon with n sides of length x is $A = (n/4)[\cot(\pi/n)]x^2$.}

37. Beam Strength A wooden beam has a rectangular cross section of height h and width w (see figure). The strength S of the beam is directly proportional to the width and the square of the height. What are the dimensions of the strongest beam that can be cut from a round log of diameter 20 inches? (*Hint:* $S = kh^2w$, where k is the proportionality constant.)

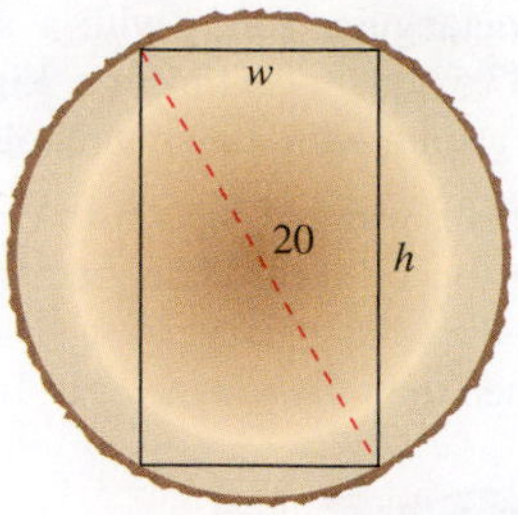

Figure for 37

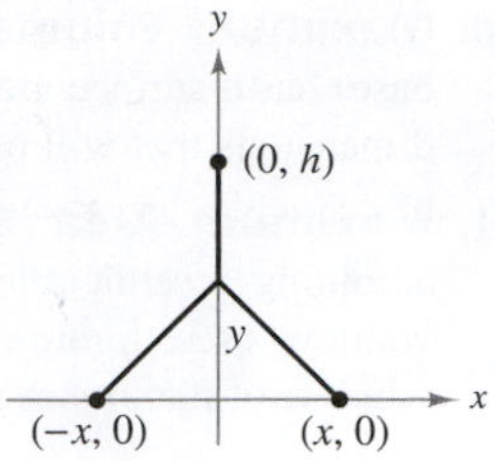

Figure for 38

38. Minimum Length Two factories are located at the coordinates $(-x, 0)$ and $(x, 0)$, and their power supply is at $(0, h)$ (see figure). Find y such that the total length of power line from the power supply to the factories is a minimum.

39. Minimum Cost

An offshore oil well is 2 kilometers off the coast. The refinery is 4 kilometers down the coast. Laying pipe in the ocean is twice as expensive as laying it on land. What path should the pipe follow in order to minimize the cost?

40. Illumination A light source is located over the center of a circular table of diameter 4 feet (see figure). Find the height h of the light source such that the illumination I at the perimeter of the table is maximum when

$$I = \frac{k \sin \alpha}{s^2}$$

where s is the slant height, α is the angle at which the light strikes the table, and k is a constant.

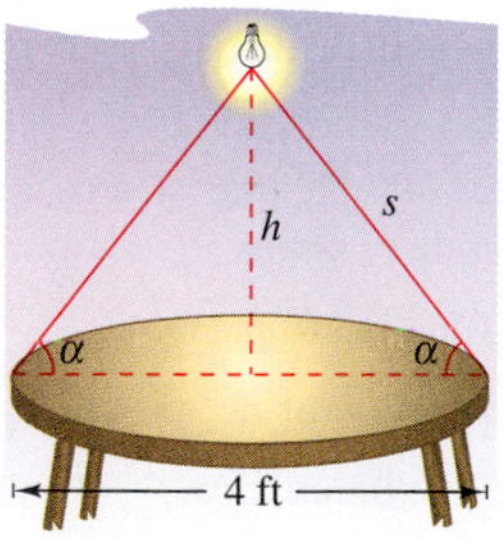

Figure for 40

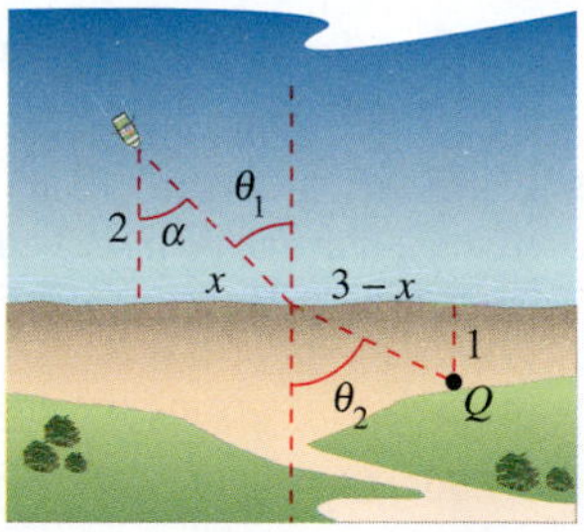

Figure for 41

41. Minimum Time A man is in a boat 2 miles from the nearest point on the coast. He is to go to a point Q, located 3 miles down the coast and 1 mile inland (see figure). He can row at 2 miles per hour and walk at 4 miles per hour. Toward what point on the coast should he row in order to reach point Q in the least time?

42. Minimum Time The conditions are the same as in Exercise 41 except that the man can row at v_1 miles per hour and walk at v_2 miles per hour. If θ_1 and θ_2 are the magnitudes of the angles, show that the man will reach point Q in the least time when

$$\frac{\sin \theta_1}{v_1} = \frac{\sin \theta_2}{v_2}.$$

43. Minimum Distance Sketch the graph of $f(x) = 2 - 2 \sin x$ on the interval $[0, \pi/2]$.

(a) Find the distance from the origin to the y-intercept and the distance from the origin to the x-intercept.

(b) Write the distance d from the origin to a point on the graph of f as a function of x. Use your graphing utility to graph d and find the minimum distance.

(c) Use calculus and the *zero* or *root* feature of a graphing utility to find the value of x that minimizes the function d on the interval $[0, \pi/2]$. What is the minimum distance?

(Submitted by Tim Chapell, Penn Valley Community College, Kansas City, MO)

44. Minimum Time When light waves traveling in a transparent medium strike the surface of a second transparent medium, they change direction. This change of direction is called *refraction* and is defined by **Snell's Law of Refraction,**

$$\frac{\sin \theta_1}{v_1} = \frac{\sin \theta_2}{v_2}$$

where θ_1 and θ_2 are the magnitudes of the angles shown in the figure and v_1 and v_2 are the velocities of light in the two media. Show that this problem is equivalent to that in Exercise 42, and that light waves traveling from P to Q follow the path of minimum time.

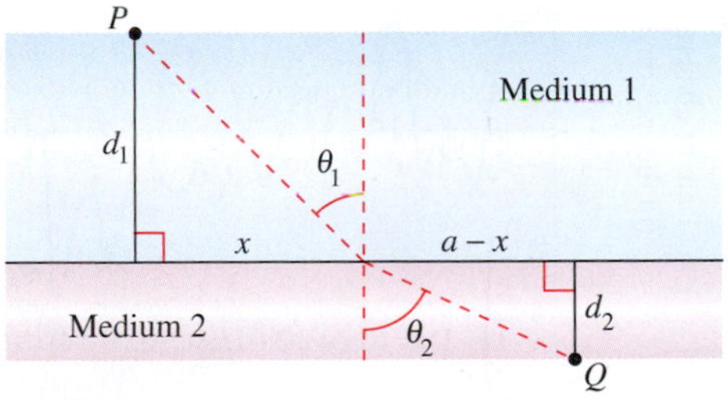

45. Maximum Volume A sector with central angle θ is cut from a circle of radius 12 inches (see figure), and the edges of the sector are brought together to form a cone. Find the magnitude of θ such that the volume of the cone is a maximum.

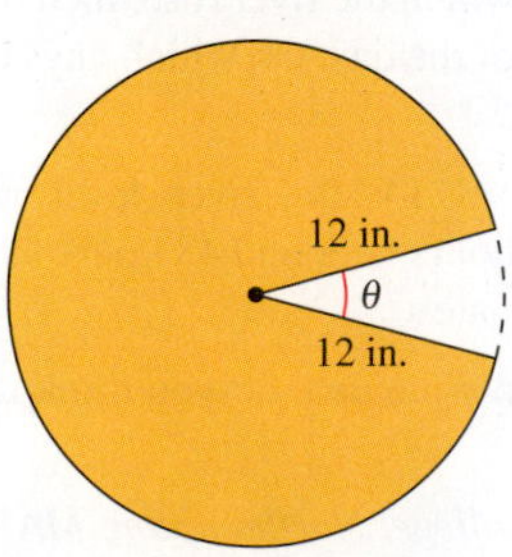

Figure for 45

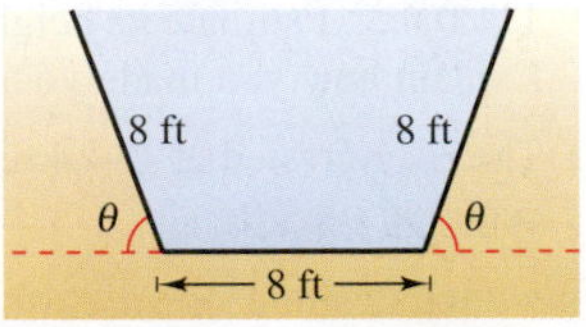

Figure for 46

46. Numerical, Graphical, and Analytic Analysis The cross sections of an irrigation canal are isosceles trapezoids of which three sides are 8 feet long (see figure). Determine the angle of elevation θ of the sides such that the area of the cross sections is a maximum by completing the following.

(a) Analytically complete six rows of a table such as the one below. (The first two rows are shown.)

Base 1	Base 2	Altitude	Area
8	$8 + 16 \cos 10°$	$8 \sin 10°$	≈ 22.1
8	$8 + 16 \cos 20°$	$8 \sin 20°$	≈ 42.5

(b) Use a graphing utility to generate additional rows of the table and estimate the maximum cross-sectional area. (*Hint:* Use the *table* feature of the graphing utility.)

(c) Write the cross-sectional area A as a function of θ.

(d) Use calculus to find the critical number of the function in part (c) and find the angle that will yield the maximum cross-sectional area.

(e) Use a graphing utility to graph the function in part (c) and verify the maximum cross-sectional area.

47. Maximum Profit Assume that the amount of money deposited in a bank is proportional to the square of the interest rate the bank pays on this money. Furthermore, the bank can reinvest this money at 12%. Find the interest rate the bank should pay to maximize profit. (Use the simple interest formula.)

48. HOW DO YOU SEE IT? The graph shows the profit P (in thousands of dollars) of a company in terms of its advertising cost x (in thousands of dollars).

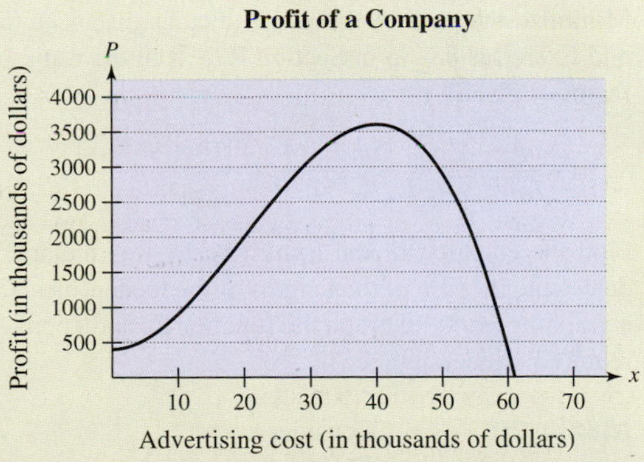

(a) Estimate the interval on which the profit is increasing.

(b) Estimate the interval on which the profit is decreasing.

(c) Estimate the amount of money the company should spend on advertising in order to yield a maximum profit.

(d) The *point of diminishing returns* is the point at which the rate of growth of the profit function begins to decline. Estimate the point of diminishing returns.

Minimum Distance **In Exercises 49–51, consider a fuel distribution center located at the origin of the rectangular coordinate system (units in miles; see figures). The center supplies three factories with coordinates (4, 1), (5, 6), and (10, 3). A trunk line will run from the distribution center along the line $y = mx$, and feeder lines will run to the three factories. The objective is to find m such that the lengths of the feeder lines are minimized.**

49. Minimize the sum of the squares of the lengths of the vertical feeder lines (see figure) given by

$$S_1 = (4m - 1)^2 + (5m - 6)^2 + (10m - 3)^2.$$

Find the equation of the trunk line by this method and then determine the sum of the lengths of the feeder lines.

50. Minimize the sum of the absolute values of the lengths of the vertical feeder lines (see figure) given by

$$S_2 = |4m - 1| + |5m - 6| + |10m - 3|.$$

Find the equation of the trunk line by this method and then determine the sum of the lengths of the feeder lines. (*Hint:* Use a graphing utility to graph the function S_2 and approximate the required critical number.)

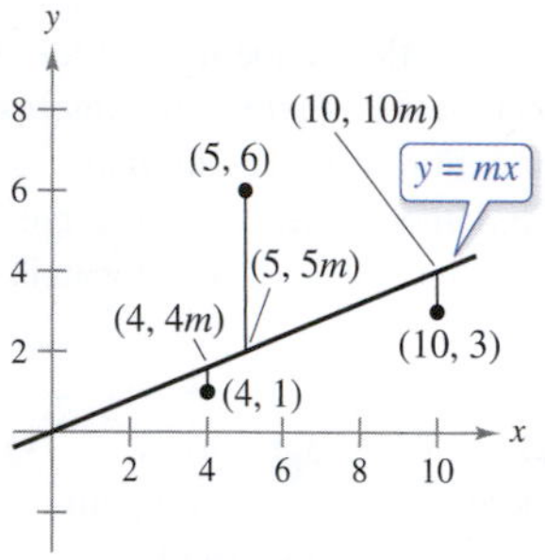

Figure for 49 and 50

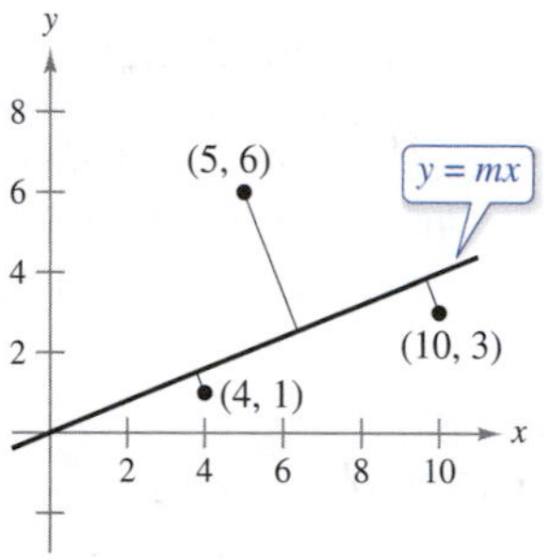

Figure for 51

51. Minimize the sum of the perpendicular distances (see figure and Exercises 83–86 in Section P.2) from the trunk line to the factories given by

$$S_3 = \frac{|4m - 1|}{\sqrt{m^2 + 1}} + \frac{|5m - 6|}{\sqrt{m^2 + 1}} + \frac{|10m - 3|}{\sqrt{m^2 + 1}}.$$

Find the equation of the trunk line by this method and then determine the sum of the lengths of the feeder lines. (*Hint:* Use a graphing utility to graph the function S_3 and approximate the required critical number.)

52. Maximum Area Consider a symmetric cross inscribed in a circle of radius r (see figure).

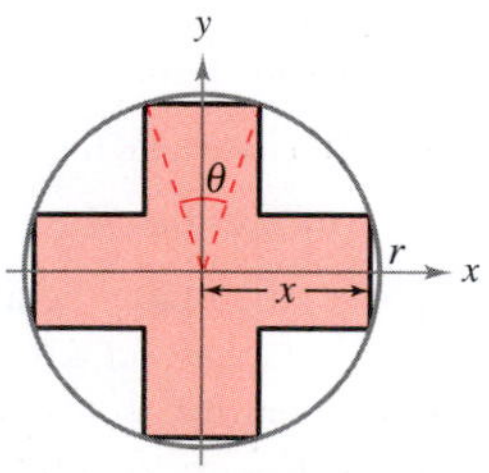

(a) Write the area A of the cross as a function of x and find the value of x that maximizes the area.

(b) Write the area A of the cross as a function of θ and find the value of θ that maximizes the area.

(c) Show that the critical numbers of parts (a) and (b) yield the same maximum area. What is that area?

PUTNAM EXAM CHALLENGE

53. Find, with explanation, the maximum value of $f(x) = x^3 - 3x$ on the set of all real numbers x satisfying $x^4 + 36 \le 13x^2$.

54. Find the minimum value of

$$\frac{(x + 1/x)^6 - (x^6 + 1/x^6) - 2}{(x + 1/x)^3 + (x^3 + 1/x^3)} \quad \text{for} \quad x > 0.$$

SECTION PROJECT

Connecticut River

Whenever the Connecticut River reaches a level of 105 feet above sea level, two Northampton, Massachusetts, flood control station operators begin a round-the-clock river watch. Every 2 hours, they check the height of the river, using a scale marked off in tenths of a foot, and record the data in a log book. In the spring of 1996, the flood watch lasted from April 4, when the river reached 105 feet and was rising at 0.2 foot per hour, until April 25, when the level subsided again to 105 feet. Between those dates, their log shows that the river rose and fell several times, at one point coming close to the 115-foot mark. If the river had reached 115 feet, the city would have closed down Mount Tom Road (Route 5, south of Northampton).

The graph below shows the rate of change of the level of the river during one portion of the flood watch. Use the graph to answer each question.

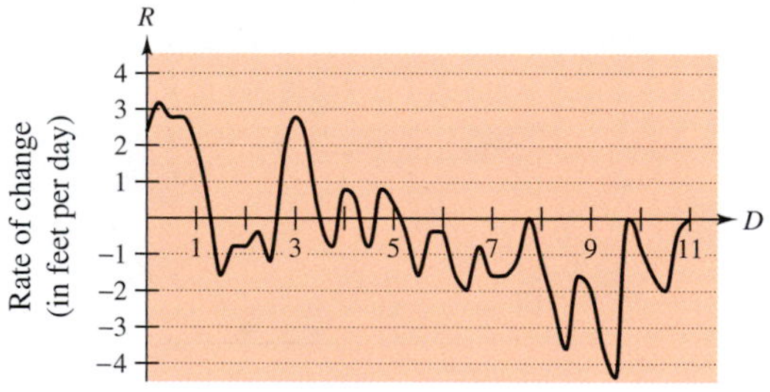

Day (0 ↔ 12:01 A.M. April 14)

(a) On what date was the river rising most rapidly? How do you know?

(b) On what date was the river falling most rapidly? How do you know?

(c) There were two dates in a row on which the river rose, then fell, then rose again during the course of the day. On which days did this occur, and how do you know?

(d) At 1 minute past midnight, April 14, the river level was 111.0 feet. Estimate its height 24 hours later and 48 hours later. Explain how you made your estimates.

(e) The river crested at 114.4 feet. On what date do you think this occurred?

(Submitted by Mary Murphy, Smith College, Northampton, MA)

3.8 Newton's Method

■ **Approximate a zero of a function using Newton's Method.**

Newton's Method

In this section, you will study a technique for approximating the real zeros of a function. The technique is called **Newton's Method,** and it uses tangent lines to approximate the graph of the function near its x-intercepts.

To see how Newton's Method works, consider a function f that is continuous on the interval $[a, b]$ and differentiable on the interval (a, b). If $f(a)$ and $f(b)$ differ in sign, then, by the Intermediate Value Theorem, f must have at least one zero in the interval (a, b). To estimate this zero, you choose

$$x = x_1 \qquad \text{First estimate}$$

as shown in Figure 3.60(a). Newton's Method is based on the assumption that the graph of f and the tangent line at $(x_1, f(x_1))$ both cross the x-axis at *about* the same point. Because you can easily calculate the x-intercept for this tangent line, you can use it as a second (and, usually, better) estimate of the zero of f. The tangent line passes through the point $(x_1, f(x_1))$ with a slope of $f'(x_1)$. In point-slope form, the equation of the tangent line is

$$y - f(x_1) = f'(x_1)(x - x_1)$$
$$y = f'(x_1)(x - x_1) + f(x_1).$$

Letting $y = 0$ and solving for x produces

$$x = x_1 - \frac{f(x_1)}{f'(x_1)}.$$

So, from the initial estimate x_1, you obtain a new estimate

$$x_2 = x_1 - \frac{f(x_1)}{f'(x_1)}. \qquad \text{Second estimate [See Figure 3.60(b).]}$$

You can improve on x_2 and calculate yet a third estimate

$$x_3 = x_2 - \frac{f(x_2)}{f'(x_2)}. \qquad \text{Third estimate}$$

Repeated application of this process is called Newton's Method.

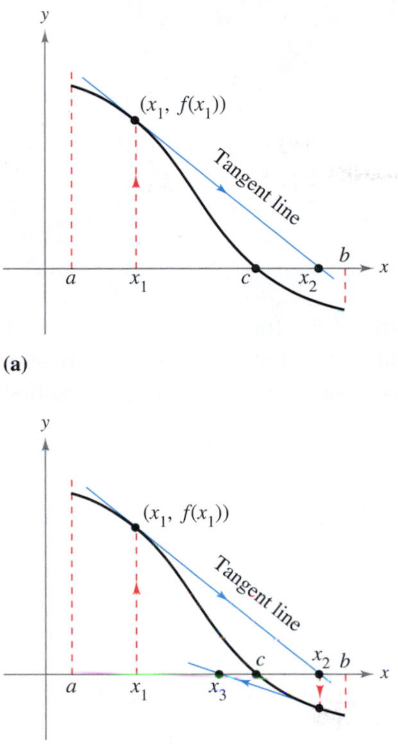

(b)
The x-intercept of the tangent line approximates the zero of f.
Figure 3.60

> **NEWTON'S METHOD**
>
> Isaac Newton first described the method for approximating the real zeros of a function in his text *Method of Fluxions*. Although the book was written in 1671, it was not published until 1736. Meanwhile, in 1690, Joseph Raphson (1648–1715) published a paper describing a method for approximating the real zeros of a function that was very similar to Newton's. For this reason, the method is often referred to as the Newton-Raphson method.

Newton's Method for Approximating the Zeros of a Function

Let $f(c) = 0$, where f is differentiable on an open interval containing c. Then, to approximate c, use these steps.

1. Make an initial estimate x_1 that is close to c. (A graph is helpful.)
2. Determine a new approximation

$$x_{n+1} = x_n - \frac{f(x_n)}{f'(x_n)}.$$

3. When $|x_n - x_{n+1}|$ is within the desired accuracy, let x_{n+1} serve as the final approximation. Otherwise, return to Step 2 and calculate a new approximation.

Each successive application of this procedure is called an **iteration.**

REMARK For many functions, just a few iterations of Newton's Method will produce approximations having very small errors, as shown in Example 1.

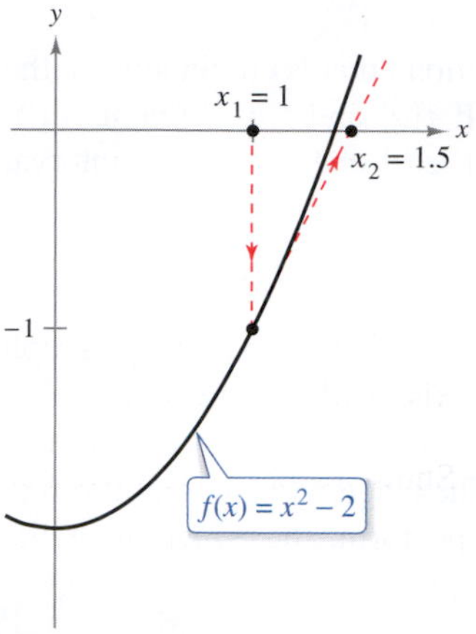

The first iteration of Newton's Method
Figure 3.61

EXAMPLE 1 Using Newton's Method

Calculate three iterations of Newton's Method to approximate a zero of $f(x) = x^2 - 2$. Use $x_1 = 1$ as the initial guess.

Solution Because $f(x) = x^2 - 2$, you have $f'(x) = 2x$, and the iterative formula is

$$x_{n+1} = x_n - \frac{f(x_n)}{f'(x_n)} = x_n - \frac{x_n^2 - 2}{2x_n}.$$

The calculations for three iterations are shown in the table.

n	x_n	$f(x_n)$	$f'(x_n)$	$\frac{f(x_n)}{f'(x_n)}$	$x_n - \frac{f(x_n)}{f'(x_n)}$
1	1.000000	−1.000000	2.000000	−0.500000	1.500000
2	1.500000	0.250000	3.000000	0.083333	1.416667
3	1.416667	0.006945	2.833334	0.002451	1.414216
4	1.414216				

Of course, in this case you know that the two zeros of the function are $\pm\sqrt{2}$. To six decimal places, $\sqrt{2} = 1.414214$. So, after only three iterations of Newton's Method, you have obtained an approximation that is within 0.000002 of an actual root. The first iteration of this process is shown in Figure 3.61.

EXAMPLE 2 Using Newton's Method

See LarsonCalculus.com for an interactive version of this type of example.

Use Newton's Method to approximate the zeros of

$$f(x) = 2x^3 + x^2 - x + 1.$$

Continue the iterations until two successive approximations differ by less than 0.0001.

Solution Begin by sketching a graph of f, as shown in Figure 3.62. From the graph, you can observe that the function has only one zero, which occurs near $x = -1.2$. Next, differentiate f and form the iterative formula

$$x_{n+1} = x_n - \frac{f(x_n)}{f'(x_n)} = x_n - \frac{2x_n^3 + x_n^2 - x_n + 1}{6x_n^2 + 2x_n - 1}.$$

The calculations are shown in the table.

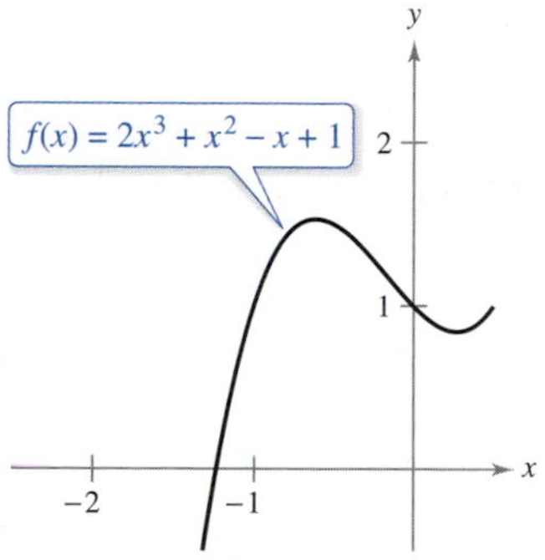

After three iterations of Newton's Method, the zero of f is approximated to the desired accuracy.
Figure 3.62

n	x_n	$f(x_n)$	$f'(x_n)$	$\frac{f(x_n)}{f'(x_n)}$	$x_n - \frac{f(x_n)}{f'(x_n)}$
1	−1.20000	0.18400	5.24000	0.03511	−1.23511
2	−1.23511	−0.00771	5.68276	−0.00136	−1.23375
3	−1.23375	0.00001	5.66533	0.00000	−1.23375
4	−1.23375				

Because two successive approximations differ by less than the required 0.0001, you can estimate the zero of f to be -1.23375.

When, as in Examples 1 and 2, the approximations approach a limit, the sequence $x_1, x_2, x_3, \ldots, x_n, \ldots$ is said to **converge.** Moreover, when the limit is c, it can be shown that c must be a zero of f.

FOR FURTHER INFORMATION For more on when Newton's Method fails, see the article "No Fooling! Newton's Method Can Be Fooled" by Peter Horton in *Mathematics Magazine*. To view this article, go to *MathArticles.com.*

Newton's Method does not always yield a convergent sequence. One way it can fail to do so is shown in Figure 3.63. Because Newton's Method involves division by $f'(x_n)$, it is clear that the method will fail when the derivative is zero for any x_n in the sequence. When you encounter this problem, you can usually overcome it by choosing a different value for x_1. Another way Newton's Method can fail is shown in the next example.

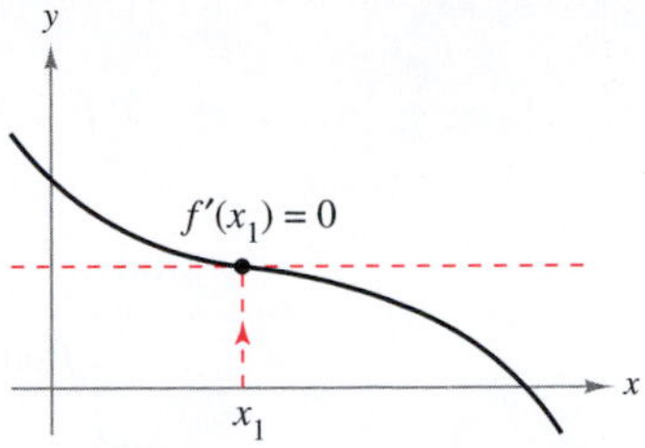

Newton's Method fails to converge when $f'(x_n) = 0$.
Figure 3.63

EXAMPLE 3 An Example in Which Newton's Method Fails

The function $f(x) = x^{1/3}$ is not differentiable at $x = 0$. Show that Newton's Method fails to converge using $x_1 = 0.1$.

Solution Because $f'(x) = \frac{1}{3}x^{-2/3}$, the iterative formula is

$$x_{n+1} = x_n - \frac{f(x_n)}{f'(x_n)} = x_n - \frac{x_n^{1/3}}{\frac{1}{3}x_n^{-2/3}} = x_n - 3x_n = -2x_n.$$

The calculations are shown in the table. This table and Figure 3.64 indicate that x_n continues to increase in magnitude as $n \to \infty$, and so the limit of the sequence does not exist.

n	x_n	$f(x_n)$	$f'(x_n)$	$\frac{f(x_n)}{f'(x_n)}$	$x_n - \frac{f(x_n)}{f'(x_n)}$
1	0.10000	0.46416	1.54720	0.30000	−0.20000
2	−0.20000	−0.58480	0.97467	−0.60000	0.40000
3	0.40000	0.73681	0.61401	1.20000	−0.80000
4	−0.80000	−0.92832	0.3680	−2.40000	1.60000

REMARK In Example 3, the initial estimate $x_1 = 0.1$ fails to produce a convergent sequence. Try showing that Newton's Method also fails for every other choice of x_1 (other than the actual zero).

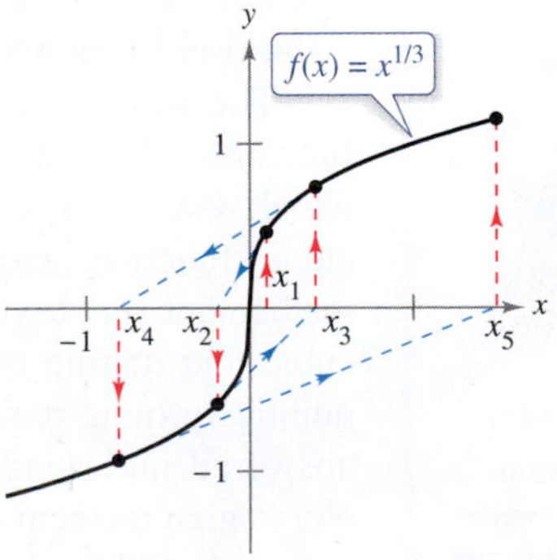

Newton's Method fails to converge for every x-value other than the actual zero of f.
Figure 3.64

It can be shown that a condition sufficient to produce convergence of Newton's Method to a zero of f is that

$$\left|\frac{f(x)f''(x)}{[f'(x)]^2}\right| < 1 \qquad \text{Condition for convergence}$$

on an open interval containing the zero. For instance, in Example 1, this test would yield

$$f(x) = x^2 - 2, \quad f'(x) = 2x, \quad f''(x) = 2,$$

and

$$\left|\frac{f(x)f''(x)}{[f'(x)]^2}\right| = \left|\frac{(x^2-2)(2)}{4x^2}\right| = \left|\frac{1}{2} - \frac{1}{x^2}\right|. \qquad \text{Example 1}$$

On the interval $(1, 3)$, this quantity is less than 1 and therefore the convergence of Newton's Method is guaranteed. On the other hand, in Example 3, you have

$$f(x) = x^{1/3}, \quad f'(x) = \frac{1}{3}x^{-2/3}, \quad f''(x) = -\frac{2}{9}x^{-5/3}$$

and

$$\left|\frac{f(x)f''(x)}{[f'(x)]^2}\right| = \left|\frac{x^{1/3}(-2/9)(x^{-5/3})}{(1/9)(x^{-4/3})}\right| = 2 \qquad \text{Example 3}$$

which is not less than 1 for any value of x, so you cannot conclude that Newton's Method will converge.

You have learned several techniques for finding the zeros of functions. The zeros of some functions, such as

$$f(x) = x^3 - 2x^2 - x + 2$$

can be found by simple algebraic techniques, such as factoring. The zeros of other functions, such as

$$f(x) = x^3 - x + 1$$

cannot be found by *elementary* algebraic methods. This particular function has only one real zero, and by using more advanced algebraic techniques, you can determine the zero to be

$$x = -\sqrt[3]{\frac{3 - \sqrt{23/3}}{6}} - \sqrt[3]{\frac{3 + \sqrt{23/3}}{6}}.$$

Because the *exact* solution is written in terms of square roots and cube roots, it is called a **solution by radicals.**

The determination of radical solutions of a polynomial equation is one of the fundamental problems of algebra. The earliest such result is the Quadratic Formula, which dates back at least to Babylonian times. The general formula for the zeros of a cubic function was developed much later. In the sixteenth century, an Italian mathematician, Jerome Cardan, published a method for finding radical solutions to cubic and quartic equations. Then, for 300 years, the problem of finding a general quintic formula remained open. Finally, in the nineteenth century, the problem was answered independently by two young mathematicians. Niels Henrik Abel, a Norwegian mathematician, and Evariste Galois, a French mathematician, proved that it is not possible to solve a *general* fifth- (or higher-) degree polynomial equation by radicals. Of course, you can solve particular fifth-degree equations, such as

$$x^5 - 1 = 0$$

but Abel and Galois were able to show that no general *radical* solution exists.

NIELS HENRIK ABEL (1802–1829)

EVARISTE GALOIS (1811–1832)

Although the lives of both Abel and Galois were brief, their work in the fields of analysis and abstract algebra was far-reaching.

See LarsonCalculus.com to read a biography about each of these mathematicians.

3.8 Exercises

See CalcChat.com for tutorial help and worked-out solutions to odd-numbered exercises.

Using Newton's Method **In Exercises 1–4, complete two iterations of Newton's Method to approximate a zero of the function using the given initial guess.**

1. $f(x) = x^2 - 5, \quad x_1 = 2.2$
2. $f(x) = x^3 - 3, \quad x_1 = 1.4$
3. $f(x) = \cos x, \quad x_1 = 1.6$
4. $f(x) = \tan x, \quad x_1 = 0.1$

Using Newton's Method **In Exercises 5–14, approximate the zero(s) of the function. Use Newton's Method and continue the process until two successive approximations differ by less than 0.001. Then find the zero(s) using a graphing utility and compare the results.**

5. $f(x) = x^3 + 4$
6. $f(x) = 2 - x^3$
7. $f(x) = x^3 + x - 1$
8. $f(x) = x^5 + x - 1$
9. $f(x) = 5\sqrt{x - 1} - 2x$
10. $f(x) = x - 2\sqrt{x + 1}$
11. $f(x) = x^3 - 3.9x^2 + 4.79x - 1.881$
12. $f(x) = x^4 + x^3 - 1$
13. $f(x) = 1 - x + \sin x$
14. $f(x) = x^3 - \cos x$

Finding Point(s) of Intersection **In Exercises 15–18, apply Newton's Method to approximate the *x*-value(s) of the indicated point(s) of intersection of the two graphs. Continue the process until two successive approximations differ by less than 0.001. [*Hint:* Let $h(x) = f(x) - g(x)$.]**

15. $f(x) = 2x + 1$

$g(x) = \sqrt{x + 4}$

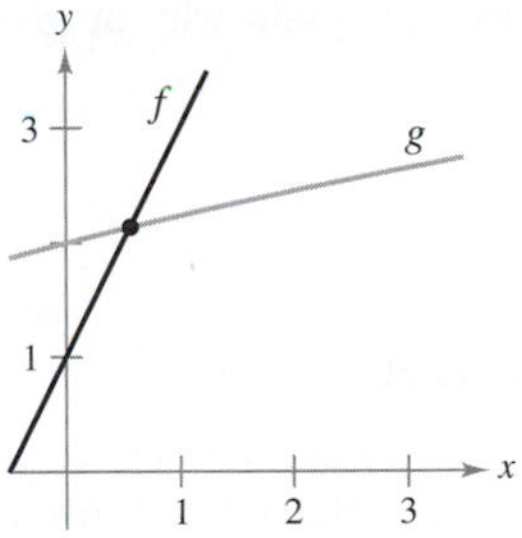

16. $f(x) = 3 - x$

$g(x) = \dfrac{1}{x^2 + 1}$

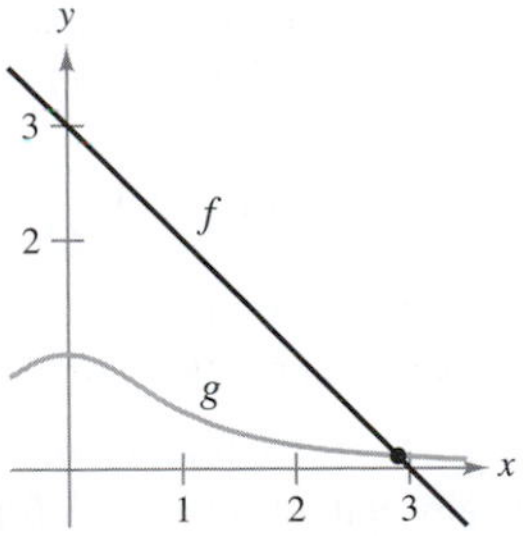

17. $f(x) = x$

$g(x) = \tan x$

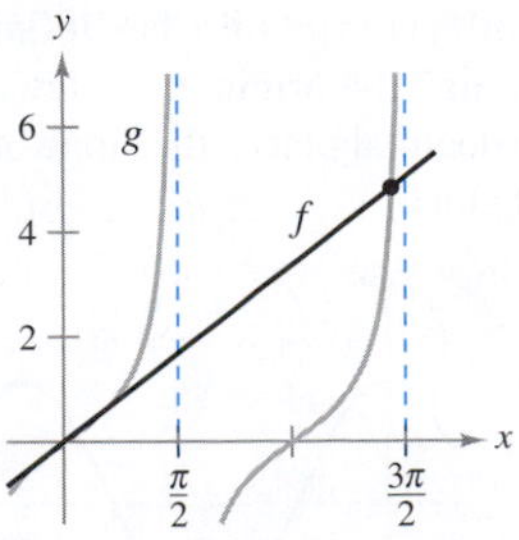

18. $f(x) = x^2$

$g(x) = \cos x$

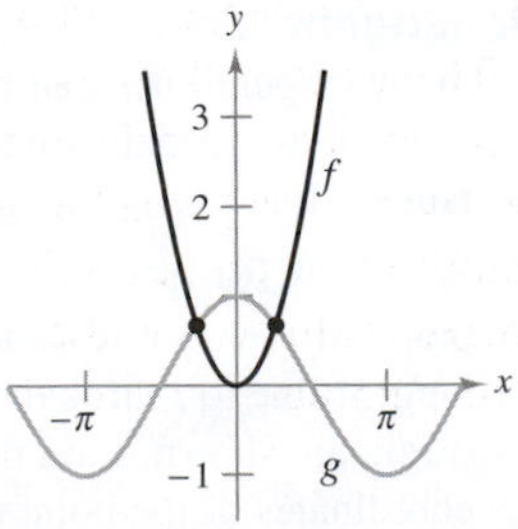

19. **Mechanic's Rule** The Mechanic's Rule for approximating $\sqrt{a}$, $a > 0$, is

$$x_{n+1} = \frac{1}{2}\left(x_n + \frac{a}{x_n}\right), \quad n = 1, 2, 3 \ldots$$

where x_1 is an approximation of $\sqrt{a}$.

(a) Use Newton's Method and the function $f(x) = x^2 - a$ to derive the Mechanic's Rule.

(b) Use the Mechanic's Rule to approximate $\sqrt{5}$ and $\sqrt{7}$ to three decimal places.

20. **Approximating Radicals**

(a) Use Newton's Method and the function $f(x) = x^n - a$ to obtain a general rule for approximating $x = \sqrt[n]{a}$.

(b) Use the general rule found in part (a) to approximate $\sqrt[4]{6}$ and $\sqrt[3]{15}$ to three decimal places.

Failure of Newton's Method **In Exercises 21 and 22, apply Newton's Method using the given initial guess, and explain why the method fails.**

21. $y = 2x^3 - 6x^2 + 6x - 1, \quad x_1 = 1$

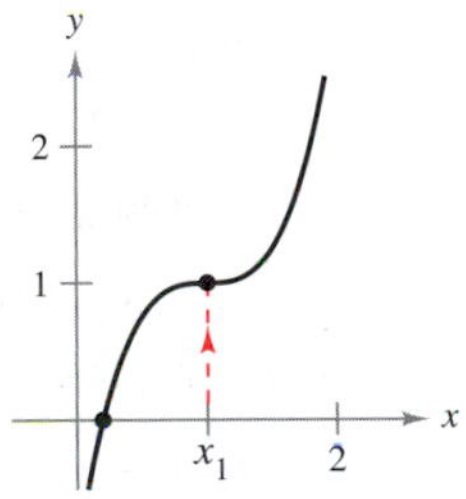

Figure for 21

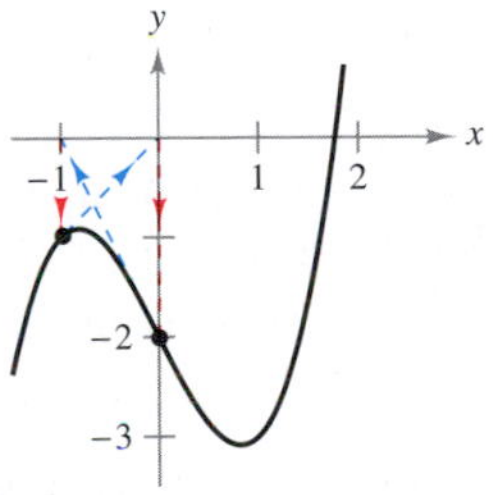

Figure for 22

22. $y = x^3 - 2x - 2, \quad x_1 = 0$

Fixed Point **In Exercises 23 and 24, approximate the fixed point of the function to two decimal places. [A *fixed point* x_0 of a function *f* is a value of *x* such that $f(x_0) = x_0$.]**

23. $f(x) = \cos x$
24. $f(x) = \cot x, \quad 0 < x < \pi$

25. **Approximating Reciprocals** Use Newton's Method to show that the equation

$$x_{n+1} = x_n(2 - ax_n)$$

can be used to approximate $1/a$ when x_1 is an initial guess of the reciprocal of a. Note that this method of approximating reciprocals uses only the operations of multiplication and subtraction. (*Hint:* Consider

$$f(x) = \frac{1}{x} - a.)$$

26. **Approximating Reciprocals** Use the result of Exercise 25 to approximate (a) $\frac{1}{3}$ and (b) $\frac{1}{11}$ to three decimal places.

WRITING ABOUT CONCEPTS

27. Using Newton's Method Consider the function $f(x) = x^3 - 3x^2 + 3$.

(a) Use a graphing utility to graph f.

(b) Use Newton's Method to approximate a zero with $x_1 = 1$ as an initial guess.

(c) Repeat part (b) using $x_1 = \frac{1}{4}$ as an initial guess and observe that the result is different.

(d) To understand why the results in parts (b) and (c) are different, sketch the tangent lines to the graph of f at the points $(1, f(1))$ and $\left(\frac{1}{4}, f\left(\frac{1}{4}\right)\right)$. Find the x-intercept of each tangent line and compare the intercepts with the first iteration of Newton's Method using the respective initial guesses.

(e) Write a short paragraph summarizing how Newton's Method works. Use the results of this exercise to describe why it is important to select the initial guess carefully.

28. Using Newton's Method Repeat the steps in Exercise 27 for the function $f(x) = \sin x$ with initial guesses of $x_1 = 1.8$ and $x_1 = 3$.

29. Newton's Method In your own words and using a sketch, describe Newton's Method for approximating the zeros of a function.

30. HOW DO YOU SEE IT? For what value(s) will Newton's Method fail to converge for the function shown in the graph? Explain your reasoning.

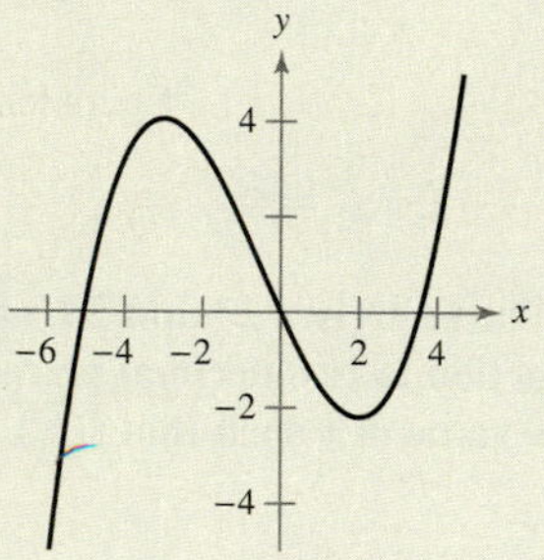

Using Newton's Method **Exercises 31–33 present problems similar to exercises from the previous sections of this chapter. In each case, use Newton's Method to approximate the solution.**

31. Minimum Distance Find the point on the graph of $f(x) = 4 - x^2$ that is closest to the point $(1, 0)$.

32. Medicine The concentration C of a chemical in the bloodstream t hours after injection into muscle tissue is given by

$$C = \frac{3t^2 + t}{50 + t^3}.$$

When is the concentration the greatest?

33. Minimum Time You are in a boat 2 miles from the nearest point on the coast (see figure). You are to go to a point Q that is 3 miles down the coast and 1 mile inland. You can row at 3 miles per hour and walk at 4 miles per hour. Toward what point on the coast should you row in order to reach Q in the least time?

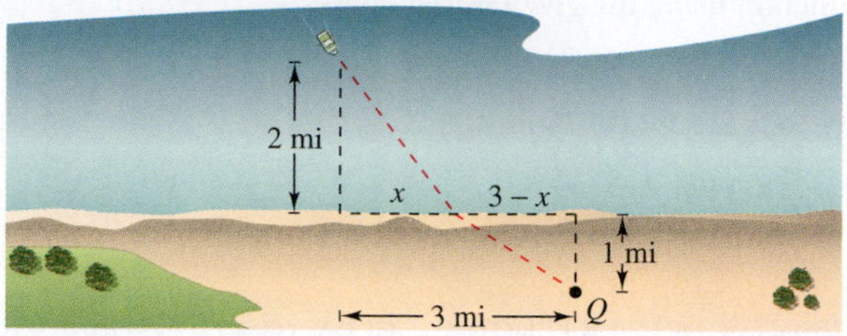

34. Crime The total number of arrests T (in thousands) for all males ages 15 to 24 in 2010 is approximated by the model

$$T = 0.2988x^4 - 22.625x^3 + 628.49x^2 - 7565.9x + 33{,}478$$

for $15 \le x \le 24$, where x is the age in years (see figure). Approximate the two ages that had total arrests of 300 thousand. *(Source: U.S. Department of Justice)*

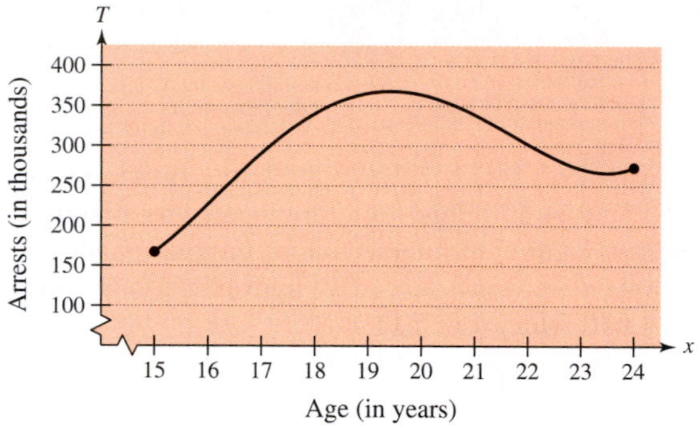

True or False? **In Exercises 35–38, determine whether the statement is true or false. If it is false, explain why or give an example that shows it is false.**

35. The zeros of $f(x) = \dfrac{p(x)}{q(x)}$ coincide with the zeros of $p(x)$.

36. If the coefficients of a polynomial function are all positive, then the polynomial has no positive zeros.

37. If $f(x)$ is a cubic polynomial such that $f'(x)$ is never zero, then any initial guess will force Newton's Method to converge to the zero of f.

38. The roots of $\sqrt{f(x)} = 0$ coincide with the roots of $f(x) = 0$.

39. Tangent Lines The graph of $f(x) = -\sin x$ has infinitely many tangent lines that pass through the origin. Use Newton's Method to approximate to three decimal places the slope of the tangent line having the greatest slope.

40. Point of Tangency The graph of $f(x) = \cos x$ and a tangent line to f through the origin are shown. Find the coordinates of the point of tangency to three decimal places.

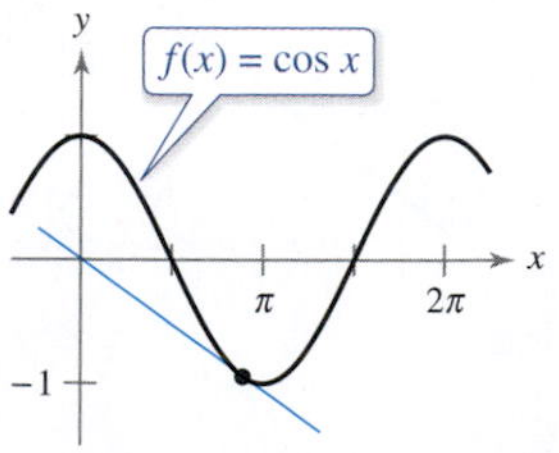

3.9 Differentials

- **Understand the concept of a tangent line approximation.**
- **Compare the value of the differential, dy, with the actual change in y, Δy.**
- **Estimate a propagated error using a differential.**
- **Find the differential of a function using differentiation formulas.**

Exploration

Tangent Line Approximation Use a graphing utility to graph $f(x) = x^2$. In the same viewing window, graph the tangent line to the graph of f at the point $(1, 1)$. Zoom in twice on the point of tangency. Does your graphing utility distinguish between the two graphs? Use the *trace* feature to compare the two graphs. As the x-values get closer to 1, what can you say about the y-values?

Tangent Line Approximations

Newton's Method (Section 3.8) is an example of the use of a tangent line to approximate the graph of a function. In this section, you will study other situations in which the graph of a function can be approximated by a straight line.

To begin, consider a function f that is differentiable at c. The equation for the tangent line at the point $(c, f(c))$ is

$$y - f(c) = f'(c)(x - c)$$

$$y = f(c) + f'(c)(x - c)$$

and is called the **tangent line approximation** (or **linear approximation**) **of f at c.** Because c is a constant, y is a linear function of x. Moreover, by restricting the values of x to those sufficiently close to c, the values of y can be used as approximations (to any desired degree of accuracy) of the values of the function f. In other words, as x approaches c, the limit of y is $f(c)$.

EXAMPLE 1 **Using a Tangent Line Approximation**

See LarsonCalculus.com for an interactive version of this type of example.

Find the tangent line approximation of $f(x) = 1 + \sin x$ at the point $(0, 1)$. Then use a table to compare the y-values of the linear function with those of $f(x)$ on an open interval containing $x = 0$.

Solution The derivative of f is

$f'(x) = \cos x.$ First derivative

So, the equation of the tangent line to the graph of f at the point $(0, 1)$ is

$y = f(0) + f'(0)(x - 0)$

$y = 1 + (1)(x - 0)$

$y = 1 + x.$ Tangent line approximation

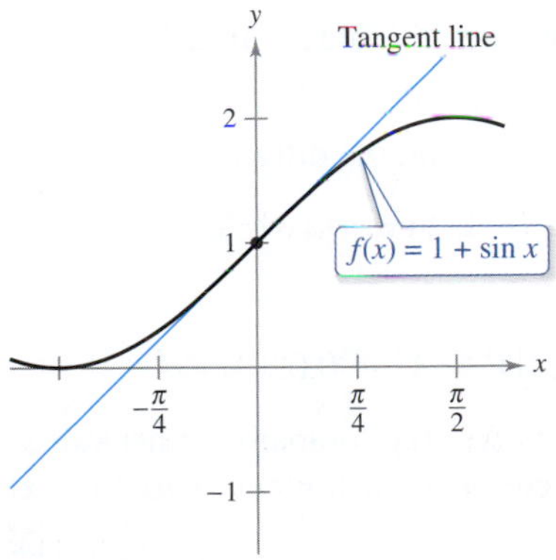

The tangent line approximation of f at the point $(0, 1)$
Figure 3.65

The table compares the values of y given by this linear approximation with the values of $f(x)$ near $x = 0$. Notice that the closer x is to 0, the better the approximation. This conclusion is reinforced by the graph shown in Figure 3.65.

x	−0.5	−0.1	−0.01	0	0.01	0.1	0.5
$f(x) = 1 + \sin x$	0.521	0.9002	0.9900002	1	1.0099998	1.0998	1.479
$y = 1 + x$	0.5	0.9	0.99	1	1.01	1.1	1.5

REMARK Be sure you see that this linear approximation of $f(x) = 1 + \sin x$ depends on the point of tangency. At a different point on the graph of f, you would obtain a different tangent line approximation.

Differentials

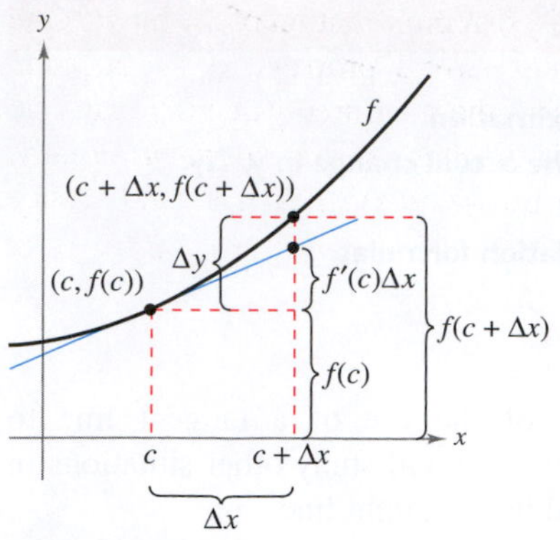

When Δx is small, $\Delta y = f(c + \Delta x) - f(c)$ is approximated by $f'(c)\Delta x$.
Figure 3.66

When the tangent line to the graph of f at the point $(c, f(c))$

$$y = f(c) + f'(c)(x - c) \quad \text{Tangent line at } (c, f(c))$$

is used as an approximation of the graph of f, the quantity $x - c$ is called the change in x, and is denoted by Δx, as shown in Figure 3.66. When Δx is small, the change in y (denoted by Δy) can be approximated as shown.

$$\Delta y = f(c + \Delta x) - f(c) \quad \text{Actual change in } y$$
$$\approx f'(c)\Delta x \quad \text{Approximate change in } y$$

For such an approximation, the quantity Δx is traditionally denoted by dx, and is called the **differential of x.** The expression $f'(x)\,dx$ is denoted by dy, and is called the **differential of y.**

Definition of Differentials

Let $y = f(x)$ represent a function that is differentiable on an open interval containing x. The **differential of x** (denoted by dx) is any nonzero real number. The **differential of y** (denoted by dy) is

$$dy = f'(x)\,dx.$$

AP* Tips

Local linear approximation, and whether such an approximation over- or under-estimates a function value, is commonly tested on the AP free response section.

In many types of applications, the differential of y can be used as an approximation of the change in y. That is,

$$\Delta y \approx dy \quad \text{or} \quad \Delta y \approx f'(x)\,dx.$$

EXAMPLE 2 Comparing Δy and dy

Let $y = x^2$. Find dy when $x = 1$ and $dx = 0.01$. Compare this value with Δy for $x = 1$ and $\Delta x = 0.01$.

Solution Because $y = f(x) = x^2$, you have $f'(x) = 2x$, and the differential dy is

$$dy = f'(x)\,dx = f'(1)(0.01) = 2(0.01) = 0.02. \quad \text{Differential of } y$$

Now, using $\Delta x = 0.01$, the change in y is

$$\Delta y = f(x + \Delta x) - f(x) = f(1.01) - f(1) = (1.01)^2 - 1^2 = 0.0201.$$

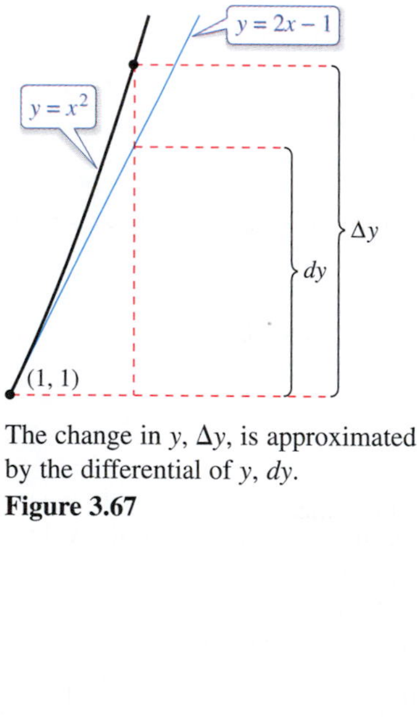

The change in y, Δy, is approximated by the differential of y, dy.
Figure 3.67

Figure 3.67 shows the geometric comparison of dy and Δy. Try comparing other values of dy and Δy. You will see that the values become closer to each other as dx (or Δx) approaches 0.

In Example 2, the tangent line to the graph of $f(x) = x^2$ at $x = 1$ is

$$y = 2x - 1. \quad \text{Tangent line to the graph of } f \text{ at } x = 1.$$

For x-values near 1, this line is close to the graph of f, as shown in Figure 3.67 and in the table.

x	0.5	0.9	0.99	1	1.01	1.1	1.5
$f(x) = x^2$	0.25	0.81	0.9801	1	1.0201	1.21	2.25
$y = 2x - 1$	0	0.8	0.98	1	1.02	1.2	2

Error Propagation

Physicists and engineers tend to make liberal use of the approximation of Δy by dy. One way this occurs in practice is in the estimation of errors propagated by physical measuring devices. For example, if you let x represent the measured value of a variable and let $x + \Delta x$ represent the exact value, then Δx is the *error in measurement*. Finally, if the measured value x is used to compute another value $f(x)$, then the difference between $f(x + \Delta x)$ and $f(x)$ is the **propagated error.**

$$\underbrace{f(x + \overbrace{\Delta x}^{\text{Measurement error}})}_{\text{Exact value}} - \underbrace{f(x)}_{\text{Measured value}} = \overbrace{\Delta y}^{\text{Propagated error}}$$

EXAMPLE 3 **Estimation of Error**

The measured radius of a ball bearing is 0.7 inch, as shown in the figure. The measurement is correct to within 0.01 inch. Estimate the propagated error in the volume V of the ball bearing.

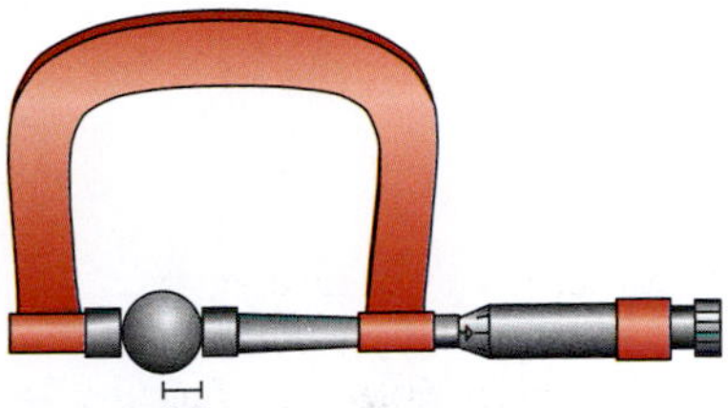

Ball bearing with measured radius that is correct to within 0.01 inch.

Solution The formula for the volume of a sphere is

$$V = \frac{4}{3}\pi r^3$$

where r is the radius of the sphere. So, you can write

$$r = 0.7 \qquad \text{Measured radius}$$

and

$$-0.01 \le \Delta r \le 0.01. \qquad \text{Possible error}$$

To approximate the propagated error in the volume, differentiate V to obtain $dV/dr = 4\pi r^2$ and write

$$\begin{aligned} \Delta V &\approx dV && \text{Approximate } \Delta V \text{ by } dV. \\ &= 4\pi r^2\,dr \\ &= 4\pi(0.7)^2(\pm 0.01) && \text{Substitute for } r \text{ and } dr. \\ &\approx \pm 0.06158 \text{ cubic inch.} \end{aligned}$$

So, the volume has a propagated error of about 0.06 cubic inch. ■

Would you say that the propagated error in Example 3 is large or small? The answer is best given in *relative* terms by comparing dV with V. The ratio

$$\begin{aligned} \frac{dV}{V} &= \frac{4\pi r^2\,dr}{\frac{4}{3}\pi r^3} && \text{Ratio of } dV \text{ to } V \\ &= \frac{3\,dr}{r} && \text{Simplify.} \\ &\approx \frac{3}{0.7}(\pm 0.01) && \text{Substitute for } dr \text{ and } r. \\ &\approx \pm 0.0429 \end{aligned}$$

is called the **relative error.** The corresponding **percent error** is approximately 4.29%.

Calculating Differentials

Each of the differentiation rules that you studied in Chapter 2 can be written in **differential form.** For example, let u and v be differentiable functions of x. By the definition of differentials, you have

$$du = u'\,dx$$

and

$$dv = v'\,dx.$$

So, you can write the differential form of the Product Rule as shown below.

$$\begin{aligned} d[uv] &= \frac{d}{dx}[uv]\,dx && \text{Differential of } uv. \\ &= [uv' + vu']\,dx && \text{Product Rule} \\ &= uv'\,dx + vu'\,dx \\ &= u\,dv + v\,du \end{aligned}$$

Differential Formulas

Let u and v be differentiable functions of x.

Constant multiple: $d[cu] = c\,du$

Sum or difference: $d[u \pm v] = du \pm dv$

Product: $d[uv] = u\,dv + v\,du$

Quotient: $d\left[\frac{u}{v}\right] = \frac{v\,du - u\,dv}{v^2}$

EXAMPLE 4 Finding Differentials

Function	Derivative	Differential
a. $y = x^2$	$\frac{dy}{dx} = 2x$	$dy = 2x\,dx$
b. $y = \sqrt{x}$	$\frac{dy}{dx} = \frac{1}{2\sqrt{x}}$	$dy = \frac{dx}{2\sqrt{x}}$
c. $y = 2\sin x$	$\frac{dy}{dx} = 2\cos x$	$dy = 2\cos x\,dx$
d. $y = x\cos x$	$\frac{dy}{dx} = -x\sin x + \cos x$	$dy = (-x\sin x + \cos x)\,dx$
e. $y = \frac{1}{x}$	$\frac{dy}{dx} = -\frac{1}{x^2}$	$dy = -\frac{dx}{x^2}$

The notation in Example 4 is called the **Leibniz notation** for derivatives and differentials, named after the German mathematician Gottfried Wilhelm Leibniz. The beauty of this notation is that it provides an easy way to remember several important calculus formulas by making it seem as though the formulas were derived from algebraic manipulations of differentials. For instance, in Leibniz notation, the *Chain Rule*

$$\frac{dy}{dx} = \frac{dy}{du}\frac{du}{dx}$$

would appear to be true because the du's divide out. Even though this reasoning is *incorrect*, the notation does help one remember the Chain Rule.

GOTTFRIED WILHELM LEIBNIZ (1646–1716)

Both Leibniz and Newton are credited with creating calculus. It was Leibniz, however, who tried to broaden calculus by developing rules and formal notation. He often spent days choosing an appropriate notation for a new concept.

See LarsonCalculus.com to read more of this biography.

EXAMPLE 5 **Finding the Differential of a Composite Function**

$$y = f(x) = \sin 3x \qquad \text{Original function}$$
$$f'(x) = 3\cos 3x \qquad \text{Apply Chain Rule.}$$
$$dy = f'(x)\,dx = 3\cos 3x\,dx \qquad \text{Differential form}$$

EXAMPLE 6 **Finding the Differential of a Composite Function**

$$y = f(x) = (x^2 + 1)^{1/2} \qquad \text{Original function}$$
$$f'(x) = \frac{1}{2}(x^2 + 1)^{-1/2}(2x) = \frac{x}{\sqrt{x^2 + 1}} \qquad \text{Apply Chain Rule.}$$
$$dy = f'(x)\,dx = \frac{x}{\sqrt{x^2 + 1}}\,dx \qquad \text{Differential form}$$

Differentials can be used to approximate function values. To do this for the function given by $y = f(x)$, use the formula

$$f(x + \Delta x) \approx f(x) + dy = f(x) + f'(x)\,dx$$

REMARK This formula is equivalent to the tangent line approximation given earlier in this section.

which is derived from the approximation

$$\Delta y = f(x + \Delta x) - f(x) \approx dy.$$

The key to using this formula is to choose a value for x that makes the calculations easier, as shown in Example 7.

EXAMPLE 7 **Approximating Function Values**

Use differentials to approximate $\sqrt{16.5}$.

Solution Using $f(x) = \sqrt{x}$, you can write

$$f(x + \Delta x) \approx f(x) + f'(x)\,dx = \sqrt{x} + \frac{1}{2\sqrt{x}}\,dx.$$

Now, choosing $x = 16$ and $dx = 0.5$, you obtain the following approximation.

$$f(x + \Delta x) = \sqrt{16.5} \approx \sqrt{16} + \frac{1}{2\sqrt{16}}(0.5) = 4 + \left(\frac{1}{8}\right)\left(\frac{1}{2}\right) = 4.0625$$

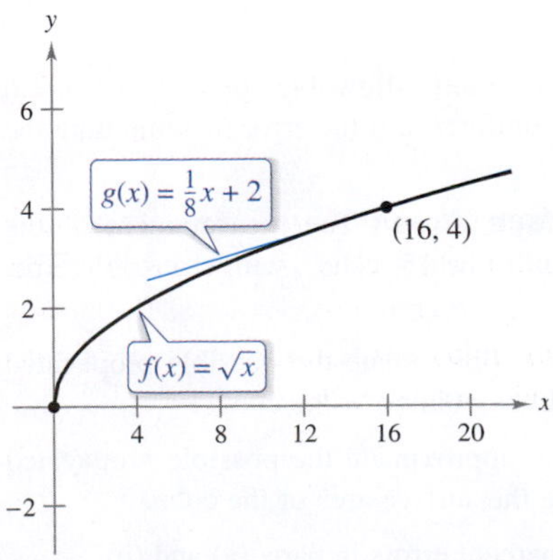

Figure 3.68

The tangent line approximation to $f(x) = \sqrt{x}$ at $x = 16$ is the line $g(x) = \frac{1}{8}x + 2$. For x-values near 16, the graphs of f and g are close together, as shown in Figure 3.68. For instance,

$$f(16.5) = \sqrt{16.5} \approx 4.0620$$

and

$$g(16.5) = \frac{1}{8}(16.5) + 2 = 4.0625.$$

In fact, if you use a graphing utility to zoom in near the point of tangency (16, 4), you will see that the two graphs appear to coincide. Notice also that as you move farther away from the point of tangency, the linear approximation becomes less accurate.

3.9 Exercises

See CalcChat.com for tutorial help and worked-out solutions to odd-numbered exercises.

Using a Tangent Line Approximation **In Exercises 1–6, find the tangent line approximation T to the graph of f at the given point. Use this linear approximation to complete the table.**

x	1.9	1.99	2	2.01	2.1
$f(x)$					
$T(x)$					

1. $f(x) = x^2, \quad (2, 4)$

2. $f(x) = \dfrac{6}{x^2}, \quad \left(2, \dfrac{3}{2}\right)$

3. $f(x) = x^5, \quad (2, 32)$

4. $f(x) = \sqrt{x}, \quad (2, \sqrt{2})$

5. $f(x) = \sin x, \quad (2, \sin 2)$

6. $f(x) = \csc x, \quad (2, \csc 2)$

Comparing Δy and dy **In Exercises 7–10, use the information to evaluate and compare Δy and dy.**

	Function	x-Value	Differential of x
7.	$y = x^3$	$x = 1$	$\Delta x = dx = 0.1$
8.	$y = 6 - 2x^2$	$x = -2$	$\Delta x = dx = 0.1$
9.	$y = x^4 + 1$	$x = -1$	$\Delta x = dx = 0.01$
10.	$y = 2 - x^4$	$x = 2$	$\Delta x = dx = 0.01$

Finding a Differential **In Exercises 11–20, find the differential dy of the given function.**

11. $y = 3x^2 - 4$

12. $y = 3x^{2/3}$

13. $y = x \tan x$

14. $y = \csc 2x$

15. $y = \dfrac{x+1}{2x-1}$

16. $y = \sqrt{x} + \dfrac{1}{\sqrt{x}}$

17. $y = \sqrt{9 - x^2}$

18. $y = x\sqrt{1 - x^2}$

19. $y = 3x - \sin^2 x$

20. $y = \dfrac{\sec^2 x}{x^2 + 1}$

Using Differentials **In Exercises 21 and 22, use differentials and the graph of f to approximate (a) $f(1.9)$ and (b) $f(2.04)$. To print an enlarged copy of the graph, go to *MathGraphs.com*.**

21.

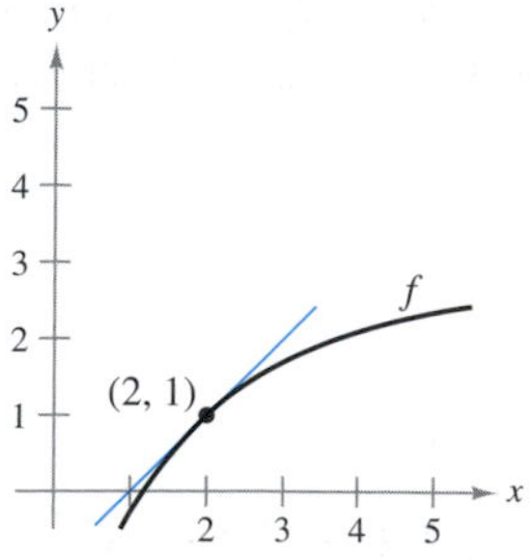

22.

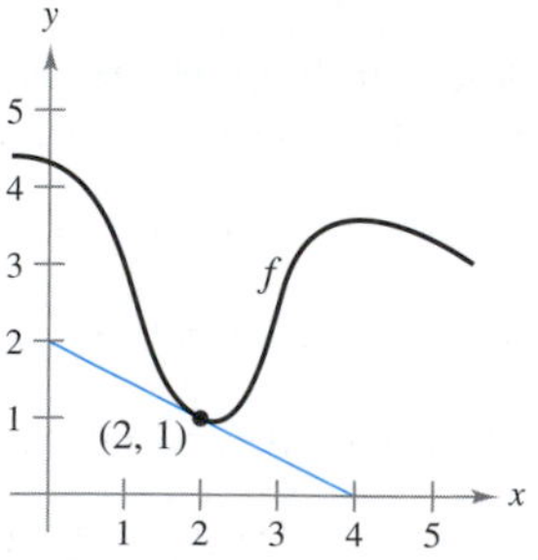

Using Differentials **In Exercises 23 and 24, use differentials and the graph of g' to approximate (a) $g(2.93)$ and (b) $g(3.1)$ given that $g(3) = 8$.**

23.

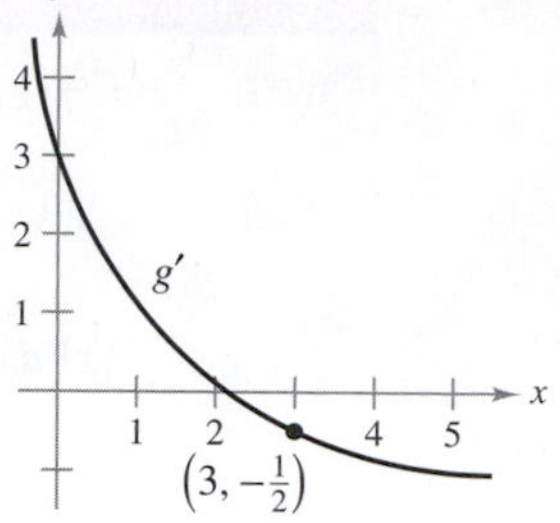

24.

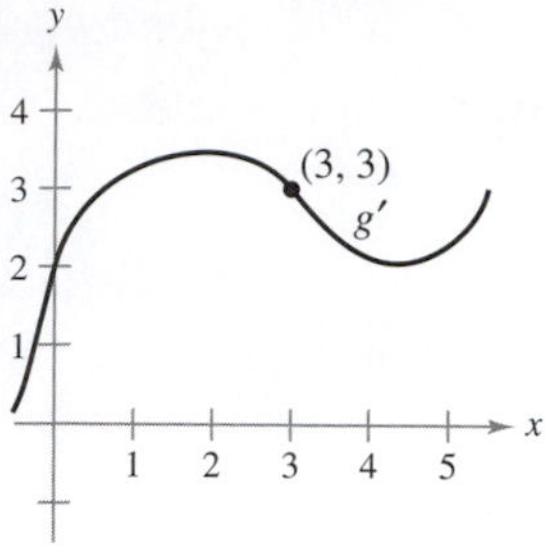

25. Area The measurement of the side of a square floor tile is 10 inches, with a possible error of $\frac{1}{32}$ inch.

(a) Use differentials to approximate the possible propagated error in computing the area of the square.

(b) Approximate the percent error in computing the area of the square.

26. Area The measurement of the radius of a circle is 16 inches, with a possible error of $\frac{1}{4}$ inch.

(a) Use differentials to approximate the possible propagated error in computing the area of the circle.

(b) Approximate the percent error in computing the area of the circle.

27. Area The measurements of the base and altitude of a triangle are found to be 36 and 50 centimeters, respectively. The possible error in each measurement is 0.25 centimeter.

(a) Use differentials to approximate the possible propagated error in computing the area of the triangle.

(b) Approximate the percent error in computing the area of the triangle.

28. Circumference The measurement of the circumference of a circle is found to be 64 centimeters, with a possible error of 0.9 centimeter.

(a) Approximate the percent error in computing the area of the circle.

(b) Estimate the maximum allowable percent error in measuring the circumference if the error in computing the area cannot exceed 3%.

29. Volume and Surface Area The measurement of the edge of a cube is found to be 15 inches, with a possible error of 0.03 inch.

(a) Use differentials to approximate the possible propagated error in computing the volume of the cube.

(b) Use differentials to approximate the possible propagated error in computing the surface area of the cube.

(c) Approximate the percent errors in parts (a) and (b).

30. Volume and Surface Area The radius of a spherical balloon is measured as 8 inches, with a possible error of 0.02 inch.

(a) Use differentials to approximate the possible propagated error in computing the volume of the sphere.

(b) Use differentials to approximate the possible propagated error in computing the surface area of the sphere.

(c) Approximate the percent errors in parts (a) and (b).

31. Stopping Distance The total stopping distance T of a vehicle is

$$T = 2.5x + 0.5x^2$$

where T is in feet and x is the speed in miles per hour. Approximate the change and percent change in total stopping distance as speed changes from $x = 25$ to $x = 26$ miles per hour.

32. HOW DO YOU SEE IT? The graph shows the profit P (in dollars) from selling x units of an item. Use the graph to determine which is greater, the change in profit when the production level changes from 400 to 401 units or the change in profit when the production level changes from 900 to 901 units. Explain your reasoning

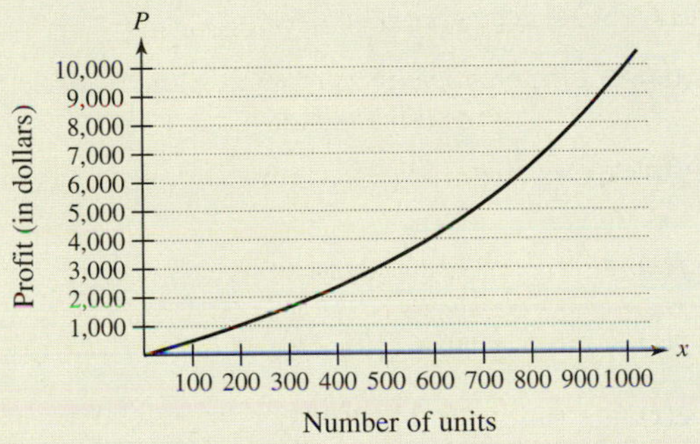

33. Pendulum The period of a pendulum is given by

$$T = 2\pi\sqrt{\frac{L}{g}}$$

where L is the length of the pendulum in feet, g is the acceleration due to gravity, and T is the time in seconds. The pendulum has been subjected to an increase in temperature such that the length has increased by $\frac{1}{2}\%$.

(a) Find the approximate percent change in the period.

(b) Using the result in part (a), find the approximate error in this pendulum clock in 1 day.

34. Ohm's Law A current of I amperes passes through a resistor of R ohms. **Ohm's Law** states that the voltage E applied to the resistor is

$$E = IR.$$

The voltage is constant. Show that the magnitude of the relative error in R caused by a change in I is equal in magnitude to the relative error in I.

35. Projectile Motion The range R of a projectile is

$$R = \frac{v_0^2}{32}(\sin 2\theta)$$

where v_0 is the initial velocity in feet per second and θ is the angle of elevation. Use differentials to approximate the change in the range when $v_0 = 2500$ feet per second and θ is changed from 10° to 11°.

36. Surveying A surveyor standing 50 feet from the base of a large tree measures the angle of elevation to the top of the tree as 71.5°. How accurately must the angle be measured if the percent error in estimating the height of the tree is to be less than 6%?

Approximating Function Values **In Exercises 37–40, use differentials to approximate the value of the expression. Compare your answer with that of a calculator.**

37. $\sqrt{99.4}$ **38.** $\sqrt[3]{26}$

39. $\sqrt[4]{624}$ **40.** $(2.99)^3$

Verifying a Tangent Line Approximation **In Exercises 41 and 42, verify the tangent line approximation of the function at the given point. Then use a graphing utility to graph the function and its approximation in the same viewing window.**

Function	Approximation	Point
41. $f(x) = \sqrt{x + 4}$	$y = 2 + \frac{x}{4}$	$(0, 2)$
42. $f(x) = \tan x$	$y = x$	$(0, 0)$

WRITING ABOUT CONCEPTS

43. Comparing Δy and dy Describe the change in accuracy of dy as an approximation for Δy when Δx is decreased.

44. Describing Terms When using differentials, what is meant by the terms *propagated error*, *relative error*, and *percent error*?

Using Differentials **In Exercises 45 and 46, give a short explanation of why the approximation is valid.**

45. $\sqrt{4.02} \approx 2 + \frac{1}{4}(0.02)$ **46.** $\tan 0.05 \approx 0 + 1(0.05)$

True or False? **In Exercises 47–50, determine whether the statement is true or false. If it is false, explain why or give an example that shows it is false.**

47. If $y = x + c$, then $dy = dx$.

48. If $y = ax + b$, then $\frac{\Delta y}{\Delta x} = \frac{dy}{dx}$.

49. If y is differentiable, then $\lim_{\Delta x \to 0} (\Delta y - dy) = 0$.

50. If $y = f(x)$, f is increasing and differentiable, and $\Delta x > 0$, then $\Delta y \geq dy$.

Review Exercises

See CalcChat.com for tutorial help and worked-out solutions to odd-numbered exercises.

Finding Extrema on a Closed Interval **In Exercises 1–8, find the absolute extrema of the function on the closed interval.**

1. $f(x) = x^2 + 5x, \quad [-4, 0]$

2. $f(x) = x^3 + 6x^2, \quad [-6, 1]$

3. $f(x) = \sqrt{x} - 2, \quad [0, 4]$

4. $h(x) = 3\sqrt{x} - x, \quad [0, 9]$

5. $f(x) = \dfrac{4x}{x^2 + 9}, \quad [-4, 4]$

6. $f(x) = \dfrac{x}{\sqrt{x^2 + 1}}, \quad [0, 2]$

7. $g(x) = 2x + 5 \cos x, \quad [0, 2\pi]$

8. $f(x) = \sin 2x, \quad [0, 2\pi]$

Using Rolle's Theorem **In Exercises 9–12, determine whether Rolle's Theorem can be applied to f on the closed interval $[a, b]$. If Rolle's Theorem can be applied, find all values of c in the open interval (a, b) such that $f'(c) = 0$. If Rolle's Theorem cannot be applied, explain why not.**

9. $f(x) = 2x^2 - 7, \quad [0, 4]$

10. $f(x) = (x - 2)(x + 3)^2, \quad [-3, 2]$

11. $f(x) = \dfrac{x^2}{1 - x^2}, \quad [-2, 2]$

12. $f(x) = \sin 2x, \quad [-\pi, \pi]$

Using the Mean Value Theorem **In Exercises 13–18, determine whether the Mean Value Theorem can be applied to f on the closed interval $[a, b]$. If the Mean Value Theorem can be applied, find all values of c in the open interval (a, b) such that**

$$f'(c) = \frac{f(b) - f(a)}{b - a}.$$

If the Mean Value Theorem cannot be applied, explain why not.

13. $f(x) = x^{2/3}, \quad [1, 8]$

14. $f(x) = \dfrac{1}{x}, \quad [1, 4]$

15. $f(x) = |5 - x|, \quad [2, 6]$

16. $f(x) = 2x - 3\sqrt{x}, \quad [-1, 1]$

17. $f(x) = x - \cos x, \quad \left[-\dfrac{\pi}{2}, \dfrac{\pi}{2}\right]$

18. $f(x) = \sqrt{x} - 2x, \quad [0, 4]$

19. **Mean Value Theorem** Can the Mean Value Theorem be applied to the function

$$f(x) = \frac{1}{x^2}$$

on the interval $[-2, 1]$? Explain.

20. **Using the Mean Value Theorem**

(a) For the function $f(x) = Ax^2 + Bx + C$, determine the value of c guaranteed by the Mean Value Theorem on the interval $[x_1, x_2]$.

(b) Demonstrate the result of part (a) for $f(x) = 2x^2 - 3x + 1$ on the interval $[0, 4]$.

Intervals on Which f Is Increasing or Decreasing **In Exercises 21–26, identify the open intervals on which the function is increasing or decreasing.**

21. $f(x) = x^2 + 3x - 12$

22. $h(x) = (x + 2)^{1/3} + 8$

23. $f(x) = (x - 1)^2(x - 3)$

24. $g(x) = (x + 1)^3$

25. $h(x) = \sqrt{x}(x - 3), \quad x > 0$

26. $f(x) = \sin x + \cos x, \quad [0, 2\pi]$

Applying the First Derivative Test **In Exercises 27–34, (a) find the critical numbers of f (if any), (b) find the open interval(s) on which the function is increasing or decreasing, (c) apply the First Derivative Test to identify all relative extrema, and (d) use a graphing utility to confirm your results.**

27. $f(x) = x^2 - 6x + 5$

28. $f(x) = 4x^3 - 5x$

29. $h(t) = \dfrac{1}{4}t^4 - 8t$

30. $g(x) = \dfrac{x^3 - 8x}{4}$

31. $f(x) = \dfrac{x + 4}{x^2}$

32. $f(x) = \dfrac{x^2 - 3x - 4}{x - 2}$

33. $f(x) = \cos x - \sin x, \quad (0, 2\pi)$

34. $g(x) = \dfrac{3}{2} \sin\left(\dfrac{\pi x}{2} - 1\right), \quad [0, 4]$

Finding Points of Inflection **In Exercises 35–40, find the points of inflection and discuss the concavity of the graph of the function.**

35. $f(x) = x^3 - 9x^2$

36. $f(x) = 6x^4 - x^2$

37. $g(x) = x\sqrt{x + 5}$

38. $f(x) = 3x - 5x^3$

39. $f(x) = x + \cos x, \quad [0, 2\pi]$

40. $f(x) = \tan \dfrac{x}{4}, \quad (0, 2\pi)$

Using the Second Derivative Test **In Exercises 41–46, find all relative extrema. Use the Second Derivative Test where applicable.**

41. $f(x) = (x + 9)^2$

42. $f(x) = 2x^3 + 11x^2 - 8x - 12$

43. $g(x) = 2x^2(1 - x^2)$

44. $h(t) = t - 4\sqrt{t + 1}$

45. $f(x) = 2x + \dfrac{18}{x}$

46. $h(x) = x - 2\cos x, \quad [0, 4\pi]$

Think About It **In Exercises 47 and 48, sketch the graph of a function f having the given characteristics.**

47. $f(0) = f(6) = 0$
$f'(3) = f'(5) = 0$
$f'(x) > 0$ for $x < 3$
$f'(x) > 0$ for $3 < x < 5$
$f'(x) < 0$ for $x > 5$
$f''(x) < 0$ for $x < 3$ or $x > 4$
$f''(x) > 0$ for $3 < x < 4$

48. $f(0) = 4, \ f(6) = 0$
$f'(x) < 0$ for $x < 2$ or $x > 4$
$f'(2)$ does not exist.
$f'(4) = 0$
$f'(x) > 0$ for $2 < x < 4$
$f''(x) < 0$ for $x \neq 2$

49. **Writing** A newspaper headline states that "The rate of growth of the national deficit is decreasing." What does this mean? What does it imply about the graph of the deficit as a function of time?

50. **Inventory Cost** The cost of inventory C depends on the ordering and storage costs according to the inventory model

$$C = \left(\frac{Q}{x}\right)s + \left(\frac{x}{2}\right)r.$$

Determine the order size that will minimize the cost, assuming that sales occur at a constant rate, Q is the number of units sold per year, r is the cost of storing one unit for one year, s is the cost of placing an order, and x is the number of units per order.

51. **Modeling Data** Outlays for national defense D (in billions of dollars) for selected years from 1970 through 2010 are shown in the table, where t is time in years, with $t = 0$ corresponding to 1970. *(Source: U.S. Office of Management and Budget)*

t	0	5	10	15	20
D	81.7	86.5	134.0	252.7	299.3

t	25	30	35	40
D	272.1	294.4	495.3	693.6

(a) Use the regression capabilities of a graphing utility to find a model of the form

$$D = at^4 + bt^3 + ct^2 + dt + e$$

for the data.

(b) Use a graphing utility to plot the data and graph the model.

(c) For the years shown in the table, when does the model indicate that the outlay for national defense was at a maximum? When was it at a minimum?

(d) For the years shown in the table, when does the model indicate that the outlay for national defense was increasing at the greatest rate?

52. **Modeling Data** The manager of a store recorded the annual sales S (in thousands of dollars) of a product over a period of 7 years, as shown in the table, where t is the time in years, with $t = 6$ corresponding to 2006.

t	6	7	8	9	10	11	12
S	5.4	6.9	11.5	15.5	19.0	22.0	23.6

(a) Use the regression capabilities of a graphing utility to find a model of the form

$$S = at^3 + bt^2 + ct + d$$

for the data.

(b) Use a graphing utility to plot the data and graph the model.

(c) Use calculus and the model to find the time t when sales were increasing at the greatest rate.

(d) Do you think the model would be accurate for predicting future sales? Explain.

Finding a Limit **In Exercises 53–62, find the limit.**

53. $\displaystyle\lim_{x\to\infty}\left(8 + \frac{1}{x}\right)$

54. $\displaystyle\lim_{x\to-\infty}\frac{1-4x}{x+1}$

55. $\displaystyle\lim_{x\to\infty}\frac{2x^2}{3x^2+5}$

56. $\displaystyle\lim_{x\to\infty}\frac{4x^3}{x^4+3}$

57. $\displaystyle\lim_{x\to-\infty}\frac{3x^2}{x+5}$

58. $\displaystyle\lim_{x\to-\infty}\frac{\sqrt{x^2+x}}{-2x}$

59. $\displaystyle\lim_{x\to\infty}\frac{5\cos x}{x}$

60. $\displaystyle\lim_{x\to\infty}\frac{x^3}{\sqrt{x^2+2}}$

61. $\displaystyle\lim_{x\to-\infty}\frac{6x}{x+\cos x}$

62. $\displaystyle\lim_{x\to-\infty}\frac{x}{2\sin x}$

Horizontal Asymptotes **In Exercises 63–66, use a graphing utility to graph the function and identify any horizontal asymptotes.**

63. $f(x) = \dfrac{3}{x} - 2$

64. $g(x) = \dfrac{5x^2}{x^2+2}$

65. $h(x) = \dfrac{2x+3}{x-4}$

66. $f(x) = \dfrac{3x}{\sqrt{x^2+2}}$

Analyzing the Graph of a Function **In Exercises 67–76, analyze and sketch a graph of the function. Label any intercepts, relative extrema, points of inflection, and asymptotes. Use a graphing utility to verify your results.**

67. $f(x) = 4x - x^2$

68. $f(x) = 4x^3 - x^4$

69. $f(x) = x\sqrt{16 - x^2}$

70. $f(x) = (x^2 - 4)^2$

71. $f(x) = x^{1/3}(x+3)^{2/3}$

72. $f(x) = (x-3)(x+2)^3$

73. $f(x) = \dfrac{5-3x}{x-2}$

74. $f(x) = \dfrac{2x}{1+x^2}$

75. $f(x) = x^3 + x + \dfrac{4}{x}$

76. $f(x) = x^2 + \dfrac{1}{x}$

77. Maximum Area A rancher has 400 feet of fencing with which to enclose two adjacent rectangular corrals (see figure). What dimensions should be used so that the enclosed area will be a maximum?

78. Maximum Area Find the dimensions of the rectangle of maximum area, with sides parallel to the coordinate axes, that can be inscribed in the ellipse given by

$$\frac{x^2}{144} + \frac{y^2}{16} = 1.$$

79. Minimum Length A right triangle in the first quadrant has the coordinate axes as sides, and the hypotenuse passes through the point (1, 8). Find the vertices of the triangle such that the length of the hypotenuse is minimum.

80. Minimum Length The wall of a building is to be braced by a beam that must pass over a parallel fence 5 feet high and 4 feet from the building. Find the length of the shortest beam that can be used.

81. Maximum Length Find the length of the longest pipe that can be carried level around a right-angle corner at the intersection of two corridors of widths 4 feet and 6 feet.

82. Maximum Length A hallway of width 6 feet meets a hallway of width 9 feet at right angles. Find the length of the longest pipe that can be carried level around this corner. [*Hint:* If L is the length of the pipe, show that

$$L = 6 \csc \theta + 9 \csc\left(\frac{\pi}{2} - \theta\right)$$

where θ is the angle between the pipe and the wall of the narrower hallway.]

83. Maximum Volume Find the volume of the largest right circular cone that can be inscribed in a sphere of radius r.

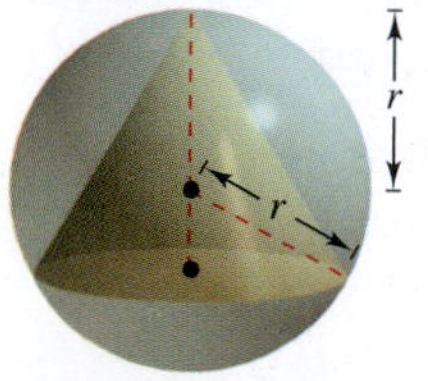

84. Maximum Volume Find the volume of the largest right circular cylinder that can be inscribed in a sphere of radius r.

Using Newton's Method **In Exercises 85–88, approximate the zero(s) of the function. Use Newton's Method and continue the process until two successive approximations differ by less than 0.001. Then find the zero(s) using a graphing utility and compare the results.**

85. $f(x) = x^3 - 3x - 1$

86. $f(x) = x^3 + 2x + 1$

87. $f(x) = x^4 + x^3 - 3x^2 + 2$

88. $f(x) = 3\sqrt{x - 1} - x$

Finding Point(s) of Intersection **In Exercises 89 and 90, apply Newton's Method to approximate the x-value(s) of the indicated point(s) of intersection of the two graphs. Continue the process until two successive approximations differ by less than 0.001. [*Hint:* Let $h(x) = f(x) - g(x)$.]**

89. $f(x) = 1 - x$
$g(x) = x^5 + 2$

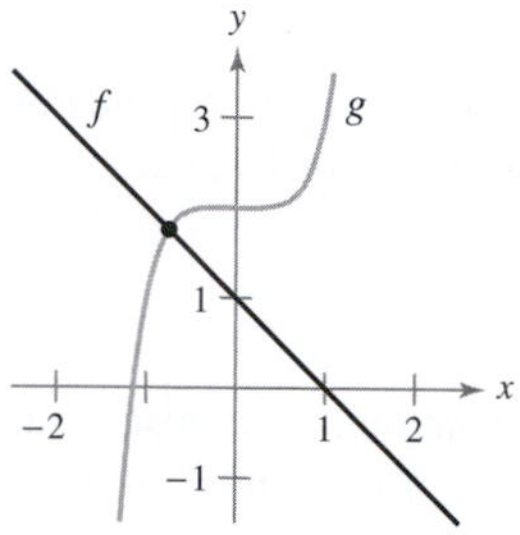

90. $f(x) = \sin x$
$g(x) = x^2 - 2x + 1$

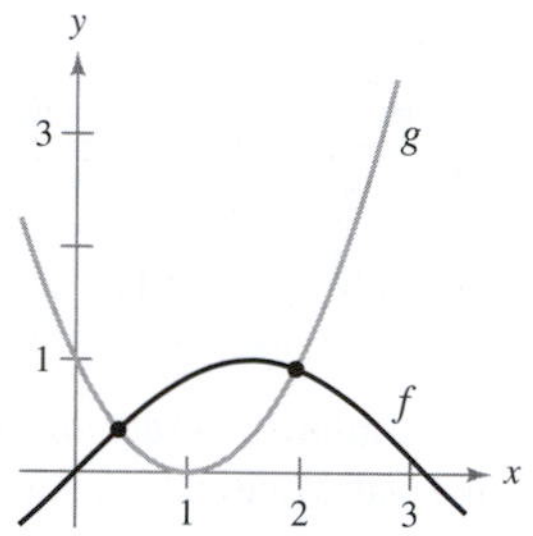

Comparing Δy and dy **In Exercises 91 and 92, use the information to evaluate and compare Δy and dy.**

	Function	x-Value	Differential of x
91.	$y = 0.5x^2$	$x = 3$	$\Delta x = dx = 0.01$
92.	$y = x^3 - 6x$	$x = 2$	$\Delta x = dx = 0.1$

Finding a Differential **In Exercises 93 and 94, find the differential dy of the given function.**

93. $y = x(1 - \cos x)$

94. $y = \sqrt{36 - x^2}$

95. Volume and Surface Area The radius of a sphere is measured as 9 centimeters, with a possible error of 0.025 centimeter.

(a) Use differentials to approximate the possible propagated error in computing the volume of the sphere.

(b) Use differentials to approximate the possible propagated error in computing the surface area of the sphere.

(c) Approximate the percent errors in parts (a) and (b).

96. Demand Function A company finds that the demand for its commodity is

$$p = 75 - \frac{1}{4}x$$

where p is the price in dollars and x is the number of units. Find and compare the values of Δp and dp as x changes from 7 to 8.

P.S. Problem Solving

See CalcChat.com for tutorial help and worked-out solutions to odd-numbered exercises.

1. Relative Extrema Graph the fourth-degree polynomial

$$p(x) = x^4 + ax^2 + 1$$

for various values of the constant a.

(a) Determine the values of a for which p has exactly one relative minimum.

(b) Determine the values of a for which p has exactly one relative maximum.

(c) Determine the values of a for which p has exactly two relative minima.

(d) Show that the graph of p cannot have exactly two relative extrema.

2. Relative Extrema

(a) Graph the fourth-degree polynomial $p(x) = ax^4 - 6x^2$ for $a = -3, -2, -1, 0, 1, 2,$ and 3. For what values of the constant a does p have a relative minimum or relative maximum?

(b) Show that p has a relative maximum for all values of the constant a.

(c) Determine analytically the values of a for which p has a relative minimum.

(d) Let $(x, y) = (x, p(x))$ be a relative extremum of p. Show that (x, y) lies on the graph of $y = -3x^2$. Verify this result graphically by graphing $y = -3x^2$ together with the seven curves from part (a).

3. Relative Minimum Let

$$f(x) = \frac{c}{x} + x^2.$$

Determine all values of the constant c such that f has a relative minimum, but no relative maximum.

4. Points of Inflection

(a) Let $f(x) = ax^2 + bx + c$, $a \neq 0$, be a quadratic polynomial. How many points of inflection does the graph of f have?

(b) Let $f(x) = ax^3 + bx^2 + cx + d$, $a \neq 0$, be a cubic polynomial. How many points of inflection does the graph of f have?

(c) Suppose the function $y = f(x)$ satisfies the equation

$$\frac{dy}{dx} = ky\left(1 - \frac{y}{L}\right)$$

where k and L are positive constants. Show that the graph of f has a point of inflection at the point where $y = L/2$. (This equation is called the **logistic differential equation.**)

5. Extended Mean Value Theorem Prove the following **Extended Mean Value Theorem.** If f and f' are continuous on the closed interval $[a, b]$, and if f'' exists in the open interval (a, b), then there exists a number c in (a, b) such that

$$f(b) = f(a) + f'(a)(b - a) + \frac{1}{2}f''(c)(b - a)^2.$$

6. Illumination The amount of illumination of a surface is proportional to the intensity of the light source, inversely proportional to the square of the distance from the light source, and proportional to $\sin\theta$, where θ is the angle at which the light strikes the surface. A rectangular room measures 10 feet by 24 feet, with a 10-foot ceiling (see figure). Determine the height at which the light should be placed to allow the corners of the floor to receive as much light as possible.

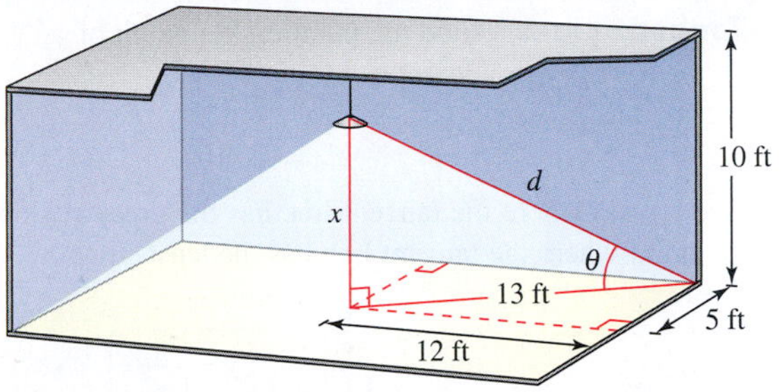

7. Minimum Distance Consider a room in the shape of a cube, 4 meters on each side. A bug at point P wants to walk to point Q at the opposite corner, as shown in the figure. Use calculus to determine the shortest path. Explain how you can solve this problem without calculus. (*Hint:* Consider the two walls as one wall.)

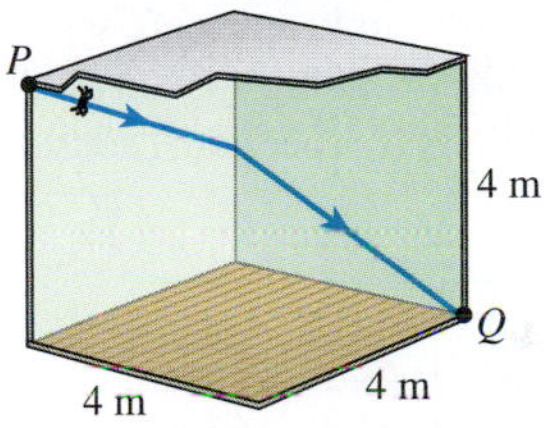

Figure for 7

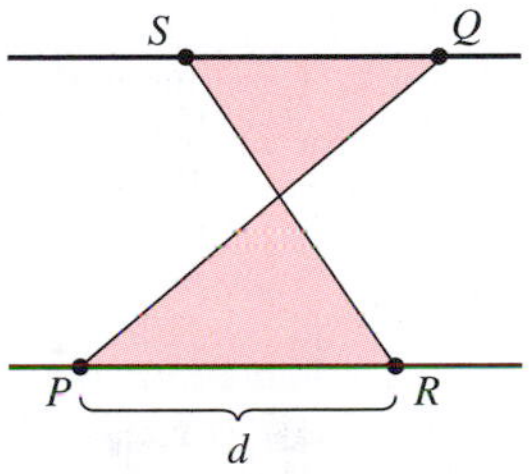

Figure for 8

8. Areas of Triangles The line joining P and Q crosses the two parallel lines, as shown in the figure. The point R is d units from P. How far from Q should the point S be positioned so that the sum of the areas of the two shaded triangles is a minimum? So that the sum is a maximum?

9. Mean Value Theorem Determine the values a, b, and c such that the function f satisfies the hypotheses of the Mean Value Theorem on the interval $[0, 3]$.

$$f(x) = \begin{cases} 1, & x = 0 \\ ax + b, & 0 < x \leq 1 \\ x^2 + 4x + c, & 1 < x \leq 3 \end{cases}$$

10. Mean Value Theorem Determine the values a, b, c, and d such that the function f satisfies the hypotheses of the Mean Value Theorem on the interval $[-1, 2]$.

$$f(x) = \begin{cases} a, & x = -1 \\ 2, & -1 < x \leq 0 \\ bx^2 + c, & 0 < x \leq 1 \\ dx + 4, & 1 < x \leq 2 \end{cases}$$

11. Proof Let f and g be functions that are continuous on $[a, b]$ and differentiable on (a, b). Prove that if $f(a) = g(a)$ and $g'(x) > f'(x)$ for all x in (a, b), then $g(b) > f(b)$.

12. Proof

(a) Prove that $\lim_{x\to\infty} x^2 = \infty$.

(b) Prove that $\lim_{x\to\infty} \left(\frac{1}{x^2}\right) = 0$.

(c) Let L be a real number. Prove that if $\lim_{x\to\infty} f(x) = L$, then

$$\lim_{y\to 0^+} f\left(\frac{1}{y}\right) = L.$$

13. Tangent Lines Find the point on the graph of

$$y = \frac{1}{1 + x^2}$$

(see figure) where the tangent line has the greatest slope, and the point where the tangent line has the least slope.

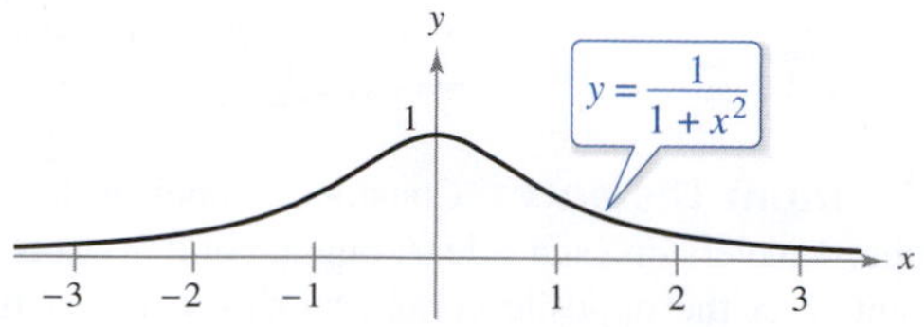

14. Stopping Distance The police department must determine the speed limit on a bridge such that the flow rate of cars is maximum per unit time. The greater the speed limit, the farther apart the cars must be in order to keep a safe stopping distance. Experimental data on the stopping distances d (in meters) for various speeds v (in kilometers per hour) are shown in the table.

v	20	40	60	80	100
d	5.1	13.7	27.2	44.2	66.4

(a) Convert the speeds v in the table to speeds s in meters per second. Use the regression capabilities of a graphing utility to find a model of the form $d(s) = as^2 + bs + c$ for the data.

(b) Consider two consecutive vehicles of average length 5.5 meters, traveling at a safe speed on the bridge. Let T be the difference between the times (in seconds) when the front bumpers of the vehicles pass a given point on the bridge. Verify that this difference in times is given by

$$T = \frac{d(s)}{s} + \frac{5.5}{s}.$$

(c) Use a graphing utility to graph the function T and estimate the speed s that minimizes the time between vehicles.

(d) Use calculus to determine the speed that minimizes T. What is the minimum value of T? Convert the required speed to kilometers per hour.

(e) Find the optimal distance between vehicles for the posted speed limit determined in part (d).

15. Darboux's Theorem Prove Darboux's Theorem: Let f be differentiable on the closed interval $[a, b]$ such that $f'(a) = y_1$ and $f'(b) = y_2$. If d lies between y_1 and y_2, then there exists c in (a, b) such that $f'(c) = d$.

16. Maximum Area The figures show a rectangle, a circle, and a semicircle inscribed in a triangle bounded by the coordinate axes and the first-quadrant portion of the line with intercepts $(3, 0)$ and $(0, 4)$. Find the dimensions of each inscribed figure such that its area is maximum. State whether calculus was helpful in finding the required dimensions. Explain your reasoning.

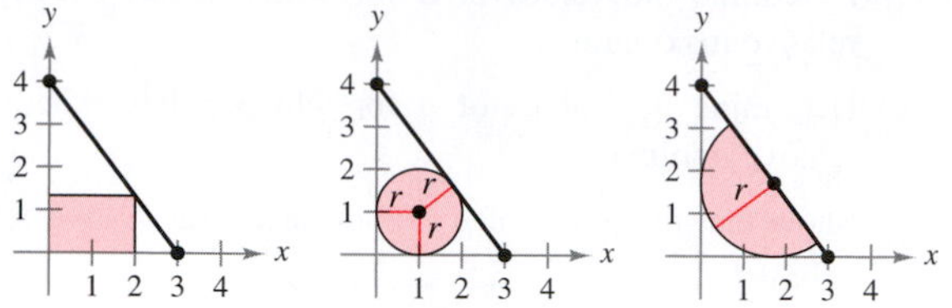

17. Point of Inflection Show that the cubic polynomial $p(x) = ax^3 + bx^2 + cx + d$ has exactly one point of inflection (x_0, y_0), where

$$x_0 = \frac{-b}{3a} \quad \text{and} \quad y_0 = \frac{2b^3}{27a^2} - \frac{bc}{3a} + d.$$

Use this formula to find the point of inflection of $p(x) = x^3 - 3x^2 + 2$.

18. Minimum Length A legal-sized sheet of paper (8.5 inches by 14 inches) is folded so that corner P touches the opposite 14-inch edge at R (see figure). (*Note:* $PQ = \sqrt{C^2 - x^2}$.)

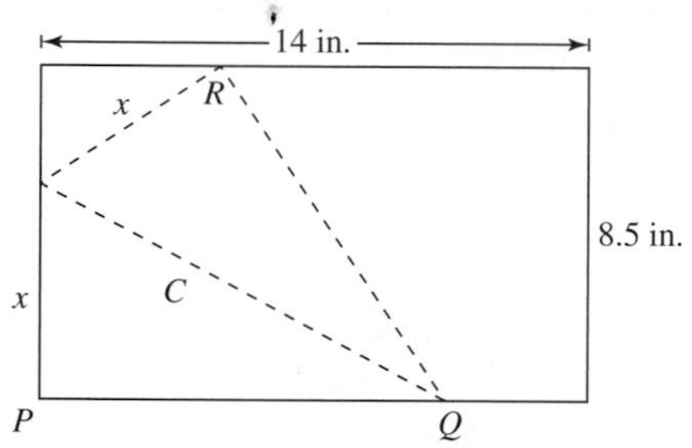

(a) Show that $C^2 = \dfrac{2x^3}{2x - 8.5}$.

(b) What is the domain of C?

(c) Determine the x-value that minimizes C.

(d) Determine the minimum length C.

19. Quadratic Approximation The polynomial

$$P(x) = c_0 + c_1(x - a) + c_2(x - a)^2$$

is the quadratic approximation of the function f at $(a, f(a))$ when $P(a) = f(a)$, $P'(a) = f'(a)$, and $P''(a) = f''(a)$.

(a) Find the quadratic approximation of

$$f(x) = \frac{x}{x + 1}$$

at $(0, 0)$.

(b) Use a graphing utility to graph $P(x)$ and $f(x)$ in the same viewing window.

AP* Review Questions for Chapter 3

1. (no calculator)

Given: $g(x) = (2x + 4)^3(x - 6)$

(a) Find the critical numbers of g.

(b) For what values of x is g increasing? Justify your answer.

(c) Identify the x-coordinate of the critical points at which g has a relative minimum. Justify your answer.

2. (no calculator)

Let $f(x) = 2x + \cos(2x)$.

(a) Find the maximum value of f for $0 \le x \le \pi$. Justify your answer.

(b) Explain how the conditions of the Mean Value Theorem are satisfied by f for $0 \le x \le \pi$. Find the value of x, $0 \le x \le \pi$, whose existence is guaranteed by the Mean Value Theorem.

3. (no calculator)

Let $f(x) = \dfrac{1 - 4x^2}{x}$.

(a) State $f'(x)$ and identify the value(s) of x for which f' does not exist.

(b) For what values of x is f decreasing? Justify your answer.

(c) For what values of x is the graph of f concave downward? Show the work that leads to your answer.

(d) Does the graph of f contain an inflection point? Justify your answer.

4.

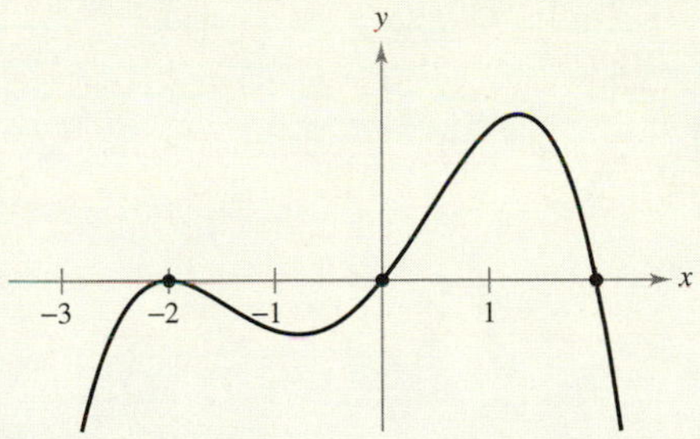

In the figure above, f', the *derivative* of function f, is shown. f is a twice differentiable function on $x \in (-\infty, \infty)$. $f''(-0.8) = 0$ and $f''(1.3) = 0$.

(a) Name the value(s) of x for which f has a relative minimum. Justify your answer.

(b) For what values of x is f increasing? Justify your answer.

(c) For what values of x is the graph of f concave downward? Justify your answer.

(d) Is $\dfrac{f(-0.5) - f(0)}{-0.5 - 0}$ positive or negative? Justify your answer.

5. The depth of the water at the end of a pier is shown in the table below and is modeled by differentiable function D for $t \ge 0$. Selected values of D are shown in the table below. D is expressed in meters, and t is the number of hours since midnight ($t = 0$).

t (hours)	0	2	5	7	8	9	12
$D(t)$ (meters)	3.0	6.7	4.9	2.3	3.1	4.9	6.7

(a) Use the data in the table to estimate the rate at which the depth of the water is changing at 3:30 A.M. and 7:40 A.M. Include units.

(b) What is the least number of times in the interval $0 < t < 12$ for which $D'(t) = 0$? Justify your answer.

(c) Use the method of linear approximation to estimate the depth of the water at 2:30 A.M. ($t = 2.5$). Show the work that leads to your answer

6. (no calculator)

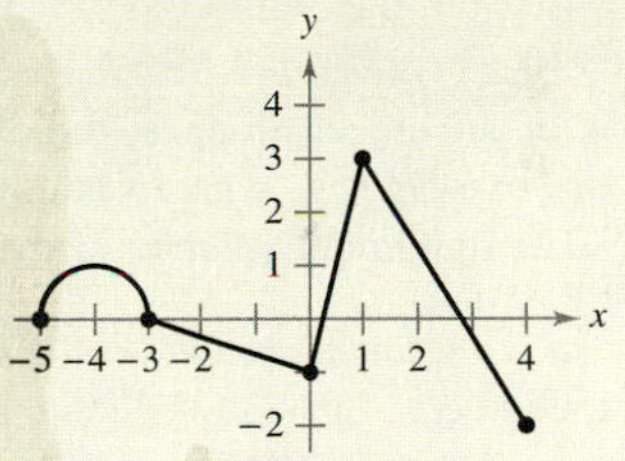

The graph of f', the derivative of f, is shown above. The function f is differentiable on the interval $-5 \le x \le 4$. $f''(-4) = 0$.

(a) Find $f'(-1)$.

(b) Find $f''(-1)$.

(c) Find the x-coordinate of each inflection point for the graph of f on the interval $-5 < x < 4$.

(d) If $g(x) = f(x) + \sin^2 x$, is g increasing or decreasing at $x = -\dfrac{\pi}{4}$? Justify your answer.

7. Given: f is continuous for $x \in (-\infty, \infty)$;

$f(2) = 4$; $\lim\limits_{x\to\infty} f(x) = 0$

	$x < 4$	$x = 4$	$x > 4$
$f'(x)$	positive	does not exist	negative
$f''(x)$	negative	does not exist	positive

(A) For what values of x is f increasing?

(B) Does f have a relative maximum at $x = 4$? Explain.

(C) If possible, name the x-coordinate of an inflection point on the graph of f. Justify your answer.

(D) Does the Mean Value Theorem apply over the interval [3, 5]? Justify your answer.

(E) Sketch a possible graph of f using the information from the table.

8.

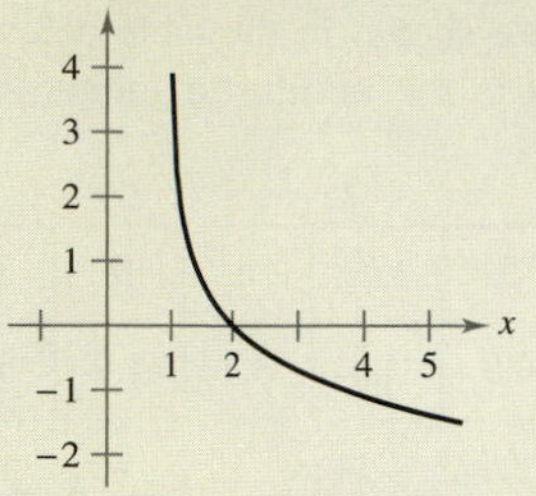

Consider the graph of $y = f(x)$ shown above. If f is a function such that f' and f'' are defined in a region around $x = 2$, then which of the following must be true?

(A) $f''(2) < f(2)$

(B) $f''(2) < f'(2)$

(C) $f(2) = f'(2)$

(D) $f''(2) > f(2)$

(E) $f(2) = f''(2)$

9. (no calculator)

The position of an object along a vertical line is given by $s(t) = -t^3 + 3t^2 + 9t + 5$, where s is measured in feet and t in seconds. The maximum *velocity* of the object in the time interval $0 \leq t \leq 4$ is

(A) $32\frac{\text{ft}}{\text{sec}}$

(B) $16\frac{\text{ft}}{\text{sec}}$

(C) $12\frac{\text{ft}}{\text{sec}}$

(D) $9\frac{\text{ft}}{\text{sec}}$

(E) $-15\frac{\text{ft}}{\text{sec}}$

10. (no calculator)

Which of the following is true for the graph of $f(x) = \dfrac{4 - x}{x - 2}$?

I. $x = 2$ is a vertical asymptote of the graph of f.

II. f is decreasing for $x \in (-\infty, \infty)$.

III. f is concave down for $x \in (-\infty, 2)$.

(A) None

(B) I and II only

(C) I and III only

(D) III only

(E) I, II and III

4 Integration

Surveying (Exercise 39, p. 311)

Electricity (Exercise 84, p. 303)

The Speed of Sound (Example 5, p. 282)

Amount of Chemical Flowing into a Tank (Example 9, p. 286)

Grand Canyon (Exercise 58, p. 252)

4.1 Antiderivatives and Indefinite Integration

- **Write the general solution of a differential equation and use indefinite integral notation for antiderivatives.**
- **Use basic integration rules to find antiderivatives.**
- **Find a particular solution of a differential equation.**

> **Exploration**
>
> ***Finding Antiderivatives***
> For each derivative, describe the original function F.
>
> **a.** $F'(x) = 2x$
>
> **b.** $F'(x) = x$
>
> **c.** $F'(x) = x^2$
>
> **d.** $F'(x) = \dfrac{1}{x^2}$
>
> **e.** $F'(x) = \dfrac{1}{x^3}$
>
> **f.** $F'(x) = \cos x$
>
> What strategy did you use to find F?

Antiderivatives

To find a function F whose derivative is $f(x) = 3x^2$, you might use your knowledge of derivatives to conclude that

$$F(x) = x^3 \quad \text{because} \quad \frac{d}{dx}[x^3] = 3x^2.$$

The function F is an *antiderivative* of f.

> **Definition of Antiderivative**
>
> A function F is an **antiderivative** of f on an interval I when $F'(x) = f(x)$ for all x in I.

Note that F is called *an* antiderivative of f, rather than *the* antiderivative of f. To see why, observe that

$$F_1(x) = x^3, \quad F_2(x) = x^3 - 5, \quad \text{and} \quad F_3(x) = x^3 + 97$$

are all antiderivatives of $f(x) = 3x^2$. In fact, for any constant C, the function $F(x) = x^3 + C$ is an antiderivative of f.

> **THEOREM 4.1 Representation of Antiderivatives**
>
> If F is an antiderivative of f on an interval I, then G is an antiderivative of f on the interval I if and only if G is of the form $G(x) = F(x) + C$, for all x in I where C is a constant.

Proof The proof of Theorem 4.1 in one direction is straightforward. That is, if $G(x) = F(x) + C$, $F'(x) = f(x)$, and C is a constant, then

$$G'(x) = \frac{d}{dx}[F(x) + C] = F'(x) + 0 = f(x).$$

To prove this theorem in the other direction, assume that G is an antiderivative of f. Define a function H such that

$$H(x) = G(x) - F(x).$$

For any two points a and b $(a < b)$ in the interval, H is continuous on $[a, b]$ and differentiable on (a, b). By the Mean Value Theorem,

$$H'(c) = \frac{H(b) - H(a)}{b - a}$$

for some c in (a, b). However, $H'(c) = 0$, so $H(a) = H(b)$. Because a and b are arbitrary points in the interval, you know that H is a constant function C. So, $G(x) - F(x) = C$ and it follows that $G(x) = F(x) + C$.

See LarsonCalculus.com for Bruce Edwards's video of this proof. ■

Using Theorem 4.1, you can represent the entire family of antiderivatives of a function by adding a constant to a *known* antiderivative. For example, knowing that

$$D_x[x^2] = 2x$$

you can represent the family of *all* antiderivatives of $f(x) = 2x$ by

$$G(x) = x^2 + C \qquad \text{Family of all antiderivatives of } f(x) = 2x$$

where C is a constant. The constant C is called the **constant of integration.** The family of functions represented by G is the **general antiderivative** of f, and $G(x) = x^2 + C$ is the **general solution** of the *differential equation*

$$G'(x) = 2x. \qquad \text{Differential equation}$$

A **differential equation** in x and y is an equation that involves x, y, and derivatives of y. For instance,

$$y' = 3x \quad \text{and} \quad y' = x^2 + 1$$

are examples of differential equations.

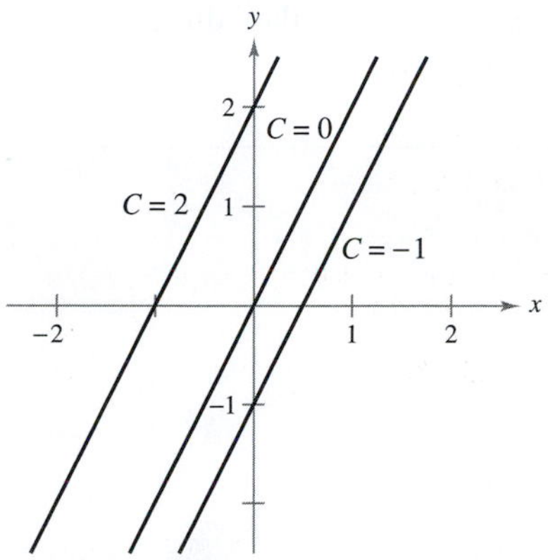

Functions of the form $y = 2x + C$
Figure 4.1

EXAMPLE 1 Solving a Differential Equation

Find the general solution of the differential equation $y' = 2$.

Solution To begin, you need to find a function whose derivative is 2. One such function is

$$y = 2x. \qquad 2x \text{ is } \textit{an} \text{ antiderivative of } 2.$$

Now, you can use Theorem 4.1 to conclude that the general solution of the differential equation is

$$y = 2x + C. \qquad \text{General solution}$$

The graphs of several functions of the form $y = 2x + C$ are shown in Figure 4.1. ■

When solving a differential equation of the form

$$\frac{dy}{dx} = f(x)$$

it is convenient to write it in the equivalent differential form

$$dy = f(x)\,dx.$$

The operation of finding all solutions of this equation is called **antidifferentiation** (or **indefinite integration**) and is denoted by an integral sign $\int$. The general solution is denoted by

Variable of integration → dx; Constant of integration → C

$$y = \int f(x)\,dx = F(x) + C.$$

Integrand → $f(x)$; An antiderivative of $f(x)$ → $F(x)$

REMARK In this text, the notation $\int f(x)\,dx = F(x) + C$ means that F is an antiderivative of f *on an interval.*

The expression $\int f(x)\,dx$ is read as the *antiderivative of f with respect to x.* So, the differential dx serves to identify x as the variable of integration. The term **indefinite integral** is a synonym for antiderivative.

Basic Integration Rules

The inverse nature of integration and differentiation can be verified by substituting $F'(x)$ for $f(x)$ in the indefinite integration definition to obtain

$$\int F'(x)\,dx = F(x) + C.$$

Integration is the "inverse" of differentiation.

Moreover, if $\int f(x)\,dx = F(x) + C$, then

$$\frac{d}{dx}\left[\int f(x)\,dx\right] = f(x).$$

Differentiation is the "inverse" of integration.

These two equations allow you to obtain integration formulas directly from differentiation formulas, as shown in the following summary.

Basic Integration Rules

Differentiation Formula	Integration Formula	
$\frac{d}{dx}[C] = 0$	$\int 0\,dx = C$	
$\frac{d}{dx}[kx] = k$	$\int k\,dx = kx + C$	
$\frac{d}{dx}[kf(x)] = kf'(x)$	$\int kf(x)\,dx = k\int f(x)\,dx$	
$\frac{d}{dx}[f(x) \pm g(x)] = f'(x) \pm g'(x)$	$\int [f(x) \pm g(x)]\,dx = \int f(x)\,dx \pm \int g(x)\,dx$	
$\frac{d}{dx}[x^n] = nx^{n-1}$	$\int x^n\,dx = \frac{x^{n+1}}{n+1} + C, \quad n \neq -1$	Power Rule
$\frac{d}{dx}[\sin x] = \cos x$	$\int \cos x\,dx = \sin x + C$	
$\frac{d}{dx}[\cos x] = -\sin x$	$\int \sin x\,dx = -\cos x + C$	
$\frac{d}{dx}[\tan x] = \sec^2 x$	$\int \sec^2 x\,dx = \tan x + C$	
$\frac{d}{dx}[\sec x] = \sec x \tan x$	$\int \sec x \tan x\,dx = \sec x + C$	
$\frac{d}{dx}[\cot x] = -\csc^2 x$	$\int \csc^2 x\,dx = -\cot x + C$	
$\frac{d}{dx}[\csc x] = -\csc x \cot x$	$\int \csc x \cot x\,dx = -\csc x + C$	

Note that the Power Rule for Integration has the restriction that $n \neq -1$. The evaluation of

$$\int \frac{1}{x}\,dx$$

must wait until the introduction of the natural logarithmic function in Chapter 5.

REMARK In Example 2, note that the general pattern of integration is similar to that of differentiation.

Original integral
↓
Rewrite
↓
Integrate
↓
Simplify

EXAMPLE 2 Describing Antiderivatives

$$\int 3x\,dx = 3\int x\,dx \qquad \text{Constant Multiple Rule}$$

$$= 3\int x^1\,dx \qquad \text{Rewrite } x \text{ as } x^1.$$

$$= 3\left(\frac{x^2}{2}\right) + C \qquad \text{Power Rule } (n = 1)$$

$$= \frac{3}{2}x^2 + C \qquad \text{Simplify.}$$

The antiderivatives of $3x$ are of the form $\frac{3}{2}x^2 + C$, where C is any constant.

When indefinite integrals are evaluated, a strict application of the basic integration rules tends to produce complicated constants of integration. For instance, in Example 2, the solution could have been written as

$$\int 3x\,dx = 3\int x\,dx = 3\left(\frac{x^2}{2} + C\right) = \frac{3}{2}x^2 + 3C.$$

Because C represents *any* constant, it is both cumbersome and unnecessary to write $3C$ as the constant of integration. So, $\frac{3}{2}x^2 + 3C$ is written in the simpler form $\frac{3}{2}x^2 + C$.

TECHNOLOGY Some software programs, such as *Maple* and *Mathematica*, are capable of performing integration symbolically. If you have access to such a symbolic integration utility, try using it to evaluate the indefinite integrals in Example 3.

EXAMPLE 3 Rewriting Before Integrating

See LarsonCalculus.com for an interactive version of this type of example.

	Original Integral	Rewrite	Integrate	Simplify
a.	$\int \frac{1}{x^3}\,dx$	$\int x^{-3}\,dx$	$\frac{x^{-2}}{-2} + C$	$-\frac{1}{2x^2} + C$
b.	$\int \sqrt{x}\,dx$	$\int x^{1/2}\,dx$	$\frac{x^{3/2}}{3/2} + C$	$\frac{2}{3}x^{3/2} + C$
c.	$\int 2\sin x\,dx$	$2\int \sin x\,dx$	$2(-\cos x) + C$	$-2\cos x + C$

REMARK The basic integration rules allow you to integrate any polynomial function.

EXAMPLE 4 Integrating Polynomial Functions

a. $$\int dx = \int 1\,dx \qquad \text{Integrand is understood to be 1.}$$

$$= x + C \qquad \text{Integrate.}$$

b. $$\int (x + 2)\,dx = \int x\,dx + \int 2\,dx$$

$$= \frac{x^2}{2} + C_1 + 2x + C_2 \qquad \text{Integrate.}$$

$$= \frac{x^2}{2} + 2x + C \qquad C = C_1 + C_2$$

The second line in the solution is usually omitted.

c. $$\int (3x^4 - 5x^2 + x)\,dx = 3\left(\frac{x^5}{5}\right) - 5\left(\frac{x^3}{3}\right) + \frac{x^2}{2} + C$$

$$= \frac{3}{5}x^5 - \frac{5}{3}x^3 + \frac{1}{2}x^2 + C$$

REMARK Before you begin the exercise set, be sure you realize that one of the most important steps in integration is *rewriting the integrand* in a form that fits one of the basic integration rules.

EXAMPLE 5 Rewriting Before Integrating

$$\int \frac{x+1}{\sqrt{x}}\,dx = \int \left(\frac{x}{\sqrt{x}} + \frac{1}{\sqrt{x}}\right) dx \qquad \text{Rewrite as two fractions.}$$

$$= \int (x^{1/2} + x^{-1/2})\,dx \qquad \text{Rewrite with fractional exponents.}$$

$$= \frac{x^{3/2}}{3/2} + \frac{x^{1/2}}{1/2} + C \qquad \text{Integrate.}$$

$$= \frac{2}{3}x^{3/2} + 2x^{1/2} + C \qquad \text{Simplify.}$$

$$= \frac{2}{3}\sqrt{x}(x+3) + C$$

When integrating quotients, do not integrate the numerator and denominator separately. This is no more valid in integration than it is in differentiation. For instance, in Example 5, be sure you understand that

$$\int \frac{x+1}{\sqrt{x}}\,dx = \frac{2}{3}\sqrt{x}(x+3) + C$$

is not the same as

$$\frac{\int (x+1)\,dx}{\int \sqrt{x}\,dx} = \frac{\frac{1}{2}x^2 + x + C_1}{\frac{2}{3}x\sqrt{x} + C_2}.$$

EXAMPLE 6 Rewriting Before Integrating

$$\int \frac{\sin x}{\cos^2 x}\,dx = \int \left(\frac{1}{\cos x}\right)\left(\frac{\sin x}{\cos x}\right) dx \qquad \text{Rewrite as a product.}$$

$$= \int \sec x \tan x\,dx \qquad \text{Rewrite using trigonometric identities.}$$

$$= \sec x + C \qquad \text{Integrate.}$$

EXAMPLE 7 Rewriting Before Integrating

	Original Integral	Rewrite	Integrate	Simplify
a.	$\int \frac{2}{\sqrt{x}}\,dx$	$2\int x^{-1/2}\,dx$	$2\left(\frac{x^{1/2}}{1/2}\right) + C$	$4x^{1/2} + C$
b.	$\int (t^2+1)^2\,dt$	$\int (t^4 + 2t^2 + 1)\,dt$	$\frac{t^5}{5} + 2\left(\frac{t^3}{3}\right) + t + C$	$\frac{1}{5}t^5 + \frac{2}{3}t^3 + t + C$
c.	$\int \frac{x^3+3}{x^2}\,dx$	$\int (x + 3x^{-2})\,dx$	$\frac{x^2}{2} + 3\left(\frac{x^{-1}}{-1}\right) + C$	$\frac{1}{2}x^2 - \frac{3}{x} + C$
d.	$\int \sqrt[3]{x}(x-4)\,dx$	$\int (x^{4/3} - 4x^{1/3})\,dx$	$\frac{x^{7/3}}{7/3} - 4\left(\frac{x^{4/3}}{4/3}\right) + C$	$\frac{3}{7}x^{7/3} - 3x^{4/3} + C$

As you do the exercises, note that you can check your answer to an antidifferentiation problem by differentiating. For instance, in Example 7(a), you can check that $4x^{1/2} + C$ is the correct antiderivative by differentiating the answer to obtain

$$D_x[4x^{1/2} + C] = 4\left(\frac{1}{2}\right)x^{-1/2} = \frac{2}{\sqrt{x}}. \qquad \text{Use differentiation to check antiderivative.}$$

Initial Conditions and Particular Solutions

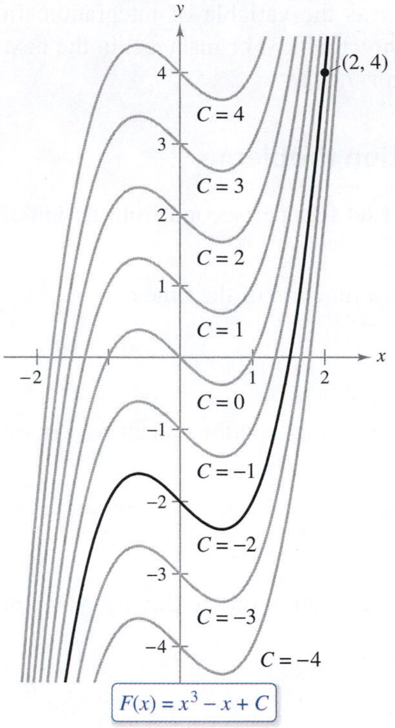

The particular solution that satisfies the initial condition $F(2) = 4$ is $F(x) = x^3 - x - 2$.
Figure 4.2

You have already seen that the equation $y = \int f(x)\,dx$ has many solutions (each differing from the others by a constant). This means that the graphs of any two antiderivatives of f are vertical translations of each other. For example, Figure 4.2 shows the graphs of several antiderivatives of the form

$$y = \int (3x^2 - 1)\,dx = x^3 - x + C \qquad \text{General solution}$$

for various integer values of C. Each of these antiderivatives is a solution of the differential equation

$$\frac{dy}{dx} = 3x^2 - 1.$$

In many applications of integration, you are given enough information to determine a **particular solution.** To do this, you need only know the value of $y = F(x)$ for one value of x. This information is called an **initial condition.** For example, in Figure 4.2, only one curve passes through the point $(2, 4)$. To find this curve, you can use the general solution

$$F(x) = x^3 - x + C \qquad \text{General solution}$$

and the initial condition

$$F(2) = 4. \qquad \text{Initial condition}$$

By using the initial condition in the general solution, you can determine that

$$F(2) = 8 - 2 + C = 4$$

which implies that $C = -2$. So, you obtain

$$F(x) = x^3 - x - 2. \qquad \text{Particular solution}$$

EXAMPLE 8 Finding a Particular Solution

Find the general solution of

$$F'(x) = \frac{1}{x^2}, \quad x > 0$$

and find the particular solution that satisfies the initial condition $F(1) = 0$.

Solution To find the general solution, integrate to obtain

$$F(x) = \int \frac{1}{x^2}\,dx \qquad F(x) = \int F'(x)\,dx$$

$$= \int x^{-2}\,dx \qquad \text{Rewrite as a power.}$$

$$= \frac{x^{-1}}{-1} + C \qquad \text{Integrate.}$$

$$= -\frac{1}{x} + C, \quad x > 0. \qquad \text{General solution}$$

Using the initial condition $F(1) = 0$, you can solve for C as follows.

$$F(1) = -\frac{1}{1} + C = 0 \implies C = 1$$

So, the particular solution, as shown in Figure 4.3, is

$$F(x) = -\frac{1}{x} + 1, \quad x > 0. \qquad \text{Particular solution}$$

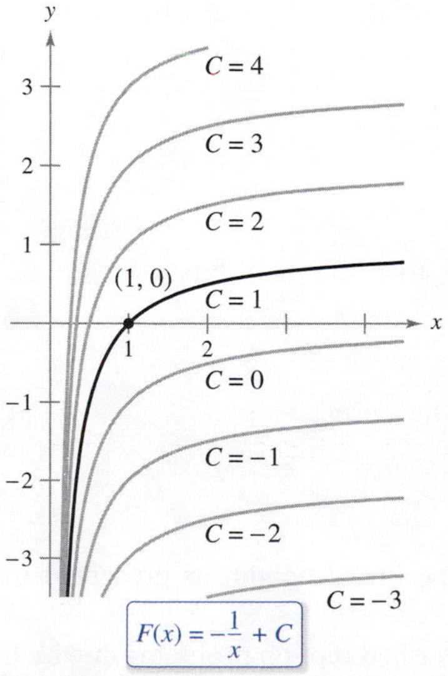

The particular solution that satisfies the initial condition $F(1) = 0$ is $F(x) = -(1/x) + 1, x > 0$.
Figure 4.3

So far in this section, you have been using x as the variable of integration. In applications, it is often convenient to use a different variable. For instance, in the next example, involving *time*, the variable of integration is t.

EXAMPLE 9 Solving a Vertical Motion Problem

A ball is thrown upward with an initial velocity of 64 feet per second from an initial height of 80 feet.

a. Find the position function giving the height s as a function of the time t.

b. When does the ball hit the ground?

Solution

a. Let $t = 0$ represent the initial time. The two given initial conditions can be written as follows.

$$s(0) = 80 \qquad \text{Initial height is 80 feet.}$$

$$s'(0) = 64 \qquad \text{Initial velocity is 64 feet per second.}$$

Using -32 feet per second per second as the acceleration due to gravity, you can write

$$s''(t) = -32$$

$$s'(t) = \int s''(t)\,dt = \int -32\,dt = -32t + C_1.$$

Using the initial velocity, you obtain $s'(0) = 64 = -32(0) + C_1$, which implies that $C_1 = 64$. Next, by integrating $s'(t)$, you obtain

$$s(t) = \int s'(t)\,dt = \int (-32t + 64)\,dt = -16t^2 + 64t + C_2.$$

Using the initial height, you obtain

$$s(0) = 80 = -16(0^2) + 64(0) + C_2$$

which implies that $C_2 = 80$. So, the position function is

$$s(t) = -16t^2 + 64t + 80. \qquad \text{See Figure 4.4.}$$

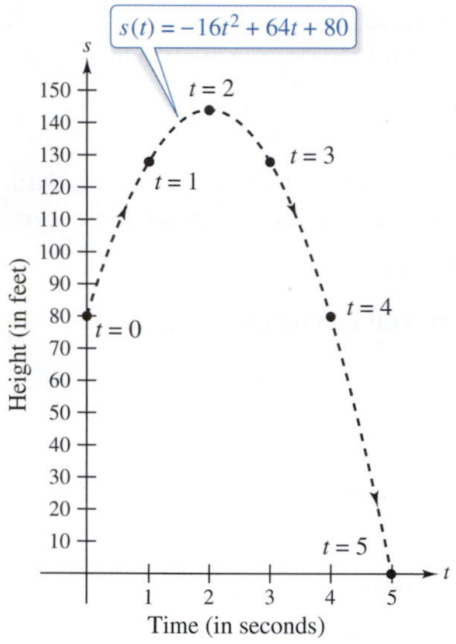

Height of a ball at time t
Figure 4.4

b. Using the position function found in part (a), you can find the time at which the ball hits the ground by solving the equation $s(t) = 0$.

$$-16t^2 + 64t + 80 = 0$$

$$-16(t + 1)(t - 5) = 0$$

$$t = -1, 5$$

Because t must be positive, you can conclude that the ball hits the ground 5 seconds after it was thrown.

In Example 9, note that the position function has the form

$$s(t) = \frac{1}{2}gt^2 + v_0t + s_0$$

where $g = -32$, v_0 is the initial velocity, and s_0 is the initial height, as presented in Section 2.2.

Example 9 shows how to use calculus to analyze vertical motion problems in which the acceleration is determined by a gravitational force. You can use a similar strategy to analyze other linear motion problems (vertical or horizontal) in which the acceleration (or deceleration) is the result of some other force, as you will see in Exercises 61–68.

4.1 Exercises

See CalcChat.com for tutorial help and worked-out solutions to odd-numbered exercises.

Integration and Differentiation In Exercises 1 and 2, verify the statement by showing that the derivative of the right side equals the integrand of the left side.

1. $\displaystyle\int\left(-\frac{6}{x^4}\right)dx = \frac{2}{x^3} + C$

2. $\displaystyle\int\left(8x^3 + \frac{1}{2x^2}\right)dx = 2x^4 - \frac{1}{2x} + C$

Solving a Differential Equation In Exercises 3–6, find the general solution of the differential equation and check the result by differentiation.

3. $\dfrac{dy}{dt} = 9t^2$

4. $\dfrac{dy}{dt} = 5$

5. $\dfrac{dy}{dx} = x^{3/2}$

6. $\dfrac{dy}{dx} = 2x^{-3}$

Rewriting Before Integrating In Exercises 7–10, complete the table to find the indefinite integral.

	Original Integral	Rewrite	Integrate	Simplify
7.	$\displaystyle\int \sqrt[3]{x}\,dx$			
8.	$\displaystyle\int \frac{1}{4x^2}\,dx$			
9.	$\displaystyle\int \frac{1}{x\sqrt{x}}\,dx$			
10.	$\displaystyle\int \frac{1}{(3x)^2}\,dx$			

Finding an Indefinite Integral In Exercises 11–32, find the indefinite integral and check the result by differentiation.

11. $\displaystyle\int (x + 7)\,dx$

12. $\displaystyle\int (13 - x)\,dx$

13. $\displaystyle\int (x^5 + 1)\,dx$

14. $\displaystyle\int (8x^3 - 9x^2 + 4)\,dx$

15. $\displaystyle\int (x^{3/2} + 2x + 1)\,dx$

16. $\displaystyle\int \left(\sqrt{x} + \frac{1}{2\sqrt{x}}\right)dx$

17. $\displaystyle\int \sqrt[3]{x^2}\,dx$

18. $\displaystyle\int \left(\sqrt[4]{x^3} + 1\right)dx$

19. $\displaystyle\int \frac{1}{x^5}\,dx$

20. $\displaystyle\int \frac{3}{x^7}\,dx$

21. $\displaystyle\int \frac{x + 6}{\sqrt{x}}\,dx$

22. $\displaystyle\int \frac{x^4 - 3x^2 + 5}{x^4}\,dx$

23. $\displaystyle\int (x + 1)(3x - 2)\,dx$

24. $\displaystyle\int (4t^2 + 3)^2\,dt$

25. $\displaystyle\int (5\cos x + 4\sin x)\,dx$

26. $\displaystyle\int (t^2 - \cos t)\,dt$

27. $\displaystyle\int (1 - \csc t\cot t)\,dt$

28. $\displaystyle\int (\theta^2 + \sec^2\theta)\,d\theta$

29. $\displaystyle\int (\sec^2\theta - \sin\theta)\,d\theta$

30. $\displaystyle\int \sec y(\tan y - \sec y)\,dy$

31. $\displaystyle\int (\tan^2 y + 1)\,dy$

32. $\displaystyle\int (4x - \csc^2 x)\,dx$

Sketching a Graph In Exercises 33 and 34, the graph of the derivative of a function is given. Sketch the graphs of *two* functions that have the given derivative. (There is more than one correct answer.) To print an enlarged copy of the graph, go to *MathGraphs.com.*

33.

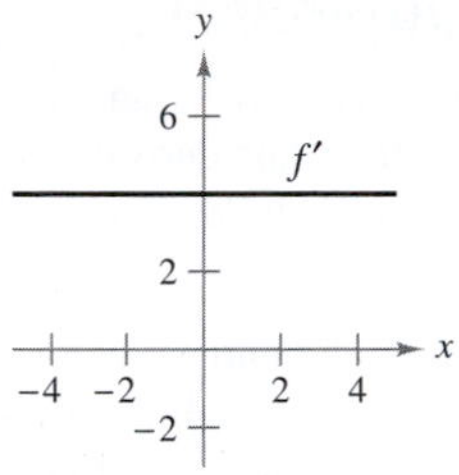

34.

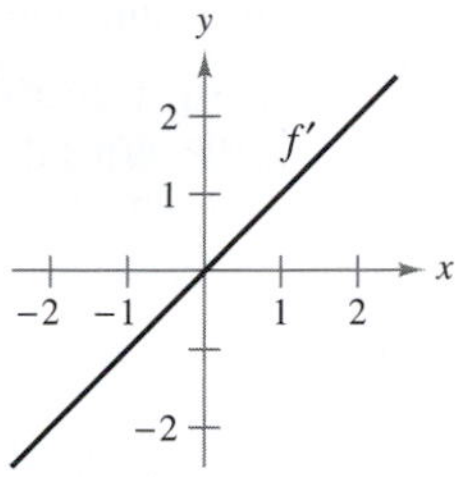

Finding a Particular Solution In Exercises 35–42, find the particular solution that satisfies the differential equation and the initial condition.

35. $f'(x) = 6x,\ f(0) = 8$

36. $g'(x) = 4x^2,\ g(-1) = 3$

37. $h'(t) = 8t^3 + 5,\ h(1) = -4$

38. $f'(s) = 10s - 12s^3,\ f(3) = 2$

39. $f''(x) = 2,\ f'(2) = 5,\ f(2) = 10$

40. $f''(x) = x^2,\ f'(0) = 8,\ f(0) = 4$

41. $f''(x) = x^{-3/2},\ f'(4) = 2,\ f(0) = 0$

42. $f''(x) = \sin x,\ f'(0) = 1,\ f(0) = 6$

Slope Field In Exercises 43 and 44, a differential equation, a point, and a slope field are given. A *slope field* (or *direction field*) consists of line segments with slopes given by the differential equation. These line segments give a visual perspective of the slopes of the solutions of the differential equation. (a) Sketch two approximate solutions of the differential equation on the slope field, one of which passes through the indicated point. (To print an enlarged copy of the graph, go to *MathGraphs.com.*) (b) Use integration to find the particular solution of the differential equation and use a graphing utility to graph the solution. Compare the result with the sketches in part (a).

43. $\dfrac{dy}{dx} = x^2 - 1,\ (-1, 3)$

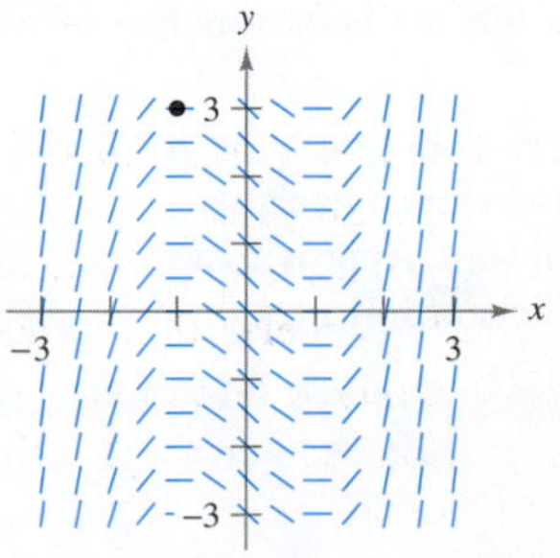

44. $\dfrac{dy}{dx} = -\dfrac{1}{x^2},\ x > 0,\ (1, 3)$

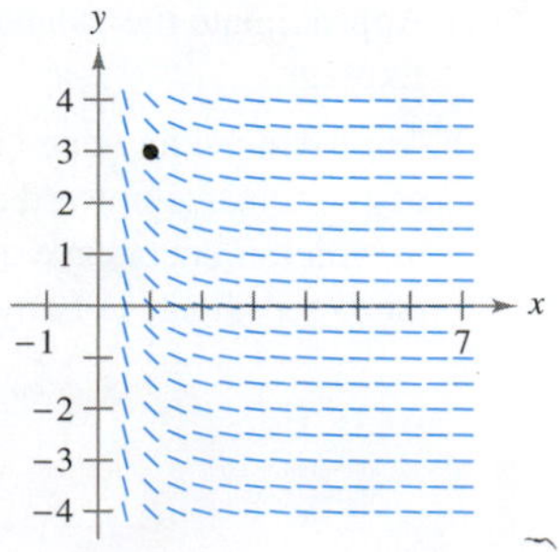

Slope Field **In Exercises 45 and 46, (a) use a graphing utility to graph a slope field for the differential equation, (b) use integration and the given point to find the particular solution of the differential equation, and (c) graph the solution and the slope field in the same viewing window.**

45. $\frac{dy}{dx} = 2x, (-2, -2)$ **46.** $\frac{dy}{dx} = 2\sqrt{x}, (4, 12)$

WRITING ABOUT CONCEPTS

47. Antiderivatives and Indefinite Integrals What is the difference, if any, between finding the antiderivative of $f(x)$ and evaluating the integral $\int f(x)\,dx$?

48. Comparing Functions Consider $f(x) = \tan^2 x$ and $g(x) = \sec^2 x$. What do you notice about the derivatives of $f(x)$ and $g(x)$? What can you conclude about the relationship between $f(x)$ and $g(x)$?

49. Sketching Graphs The graphs of f and f' each pass through the origin. Use the graph of f'' shown in the figure to sketch the graphs of f and f'. To print an enlarged copy of the graph, go to *MathGraphs.com*.

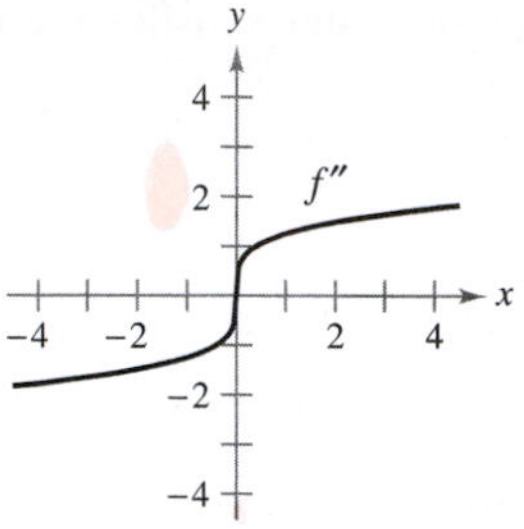

50. HOW DO YOU SEE IT? Use the graph of f' shown in the figure to answer the following.

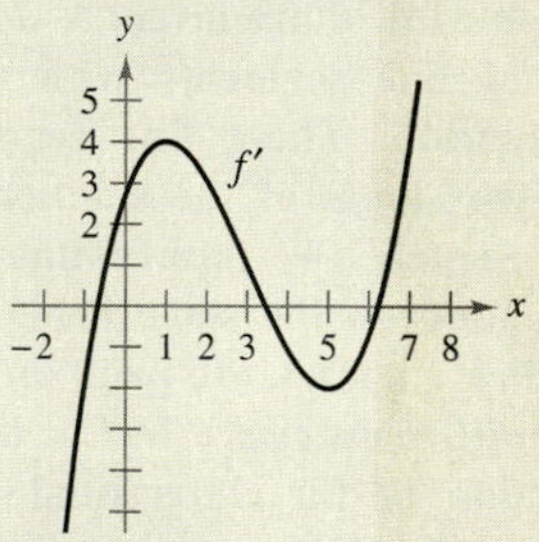

(a) Approximate the slope of f at $x = 4$. Explain.

(b) Is it possible that $f(2) = -1$? Explain.

(c) Is $f(5) - f(4) > 0$? Explain.

(d) Approximate the value of x where f is maximum. Explain.

(e) Approximate any open intervals in which the graph of f is concave upward and any open intervals in which it is concave downward. Approximate the x-coordinates of any points of inflection.

51. Tree Growth An evergreen nursery usually sells a certain type of shrub after 6 years of growth and shaping. The growth rate during those 6 years is approximated by $dh/dt = 1.5t + 5$, where t is the time in years and h is the height in centimeters. The seedlings are 12 centimeters tall when planted ($t = 0$).

(a) Find the height after t years.

(b) How tall are the shrubs when they are sold?

52. Population Growth The rate of growth dP/dt of a population of bacteria is proportional to the square root of t, where P is the population size and t is the time in days ($0 \le t \le 10$). That is,

$$\frac{dP}{dt} = k\sqrt{t}.$$

The initial size of the population is 500. After 1 day the population has grown to 600. Estimate the population after 7 days.

Vertical Motion **In Exercises 53–55, use $a(t) = -32$ feet per second per second as the acceleration due to gravity. (Neglect air resistance.)**

53. A ball is thrown vertically upward from a height of 6 feet with an initial velocity of 60 feet per second. How high will the ball go?

54. With what initial velocity must an object be thrown upward (from ground level) to reach the top of the Washington Monument (approximately 550 feet)?

55. A balloon, rising vertically with a velocity of 16 feet per second, releases a sandbag at the instant it is 64 feet above the ground.

(a) How many seconds after its release will the bag strike the ground?

(b) At what velocity will it hit the ground?

Vertical Motion **In Exercises 56–58, use $a(t) = -9.8$ meters per second per second as the acceleration due to gravity. (Neglect air resistance.)**

56. A baseball is thrown upward from a height of 2 meters with an initial velocity of 10 meters per second. Determine its maximum height.

57. With what initial velocity must an object be thrown upward (from a height of 2 meters) to reach a maximum height of 200 meters?

58. Grand Canyon

The Grand Canyon is 1800 meters deep at its deepest point. A rock is dropped from the rim above this point. Write the height of the rock as a function of the time t in seconds. How long will it take the rock to hit the canyon floor?

59. Lunar Gravity On the moon, the acceleration due to gravity is -1.6 meters per second per second. A stone is dropped from a cliff on the moon and hits the surface of the moon 20 seconds later. How far did it fall? What was its velocity at impact?

60. Escape Velocity The minimum velocity required for an object to escape Earth's gravitational pull is obtained from the solution of the equation

$$\int v\,dv = -GM\int \frac{1}{y^2}\,dy$$

where v is the velocity of the object projected from Earth, y is the distance from the center of Earth, G is the gravitational constant, and M is the mass of Earth. Show that v and y are related by the equation

$$v^2 = v_0^2 + 2GM\left(\frac{1}{y} - \frac{1}{R}\right)$$

where v_0 is the initial velocity of the object and R is the radius of Earth.

Rectilinear Motion **In Exercises 61–64, consider a particle moving along the x-axis where $x(t)$ is the position of the particle at time t, $x'(t)$ is its velocity, and $x''(t)$ is its acceleration.**

61. $x(t) = t^3 - 6t^2 + 9t - 2, \quad 0 \le t \le 5$

(a) Find the velocity and acceleration of the particle.

(b) Find the open t-intervals on which the particle is moving to the right.

(c) Find the velocity of the particle when the acceleration is 0.

62. Repeat Exercise 61 for the position function

$x(t) = (t-1)(t-3)^2, \quad 0 \le t \le 5.$

63. A particle moves along the x-axis at a velocity of $v(t) = 1/\sqrt{t}$, $t > 0$. At time $t = 1$, its position is $x = 4$. Find the acceleration and position functions for the particle.

64. A particle, initially at rest, moves along the x-axis such that its acceleration at time $t > 0$ is given by $a(t) = \cos t$. At the time $t = 0$, its position is $x = 3$.

(a) Find the velocity and position functions for the particle.

(b) Find the values of t for which the particle is at rest.

65. Acceleration The maker of an automobile advertises that it takes 13 seconds to accelerate from 25 kilometers per hour to 80 kilometers per hour. Assume the acceleration is constant.

(a) Find the acceleration in meters per second per second.

(b) Find the distance the car travels during the 13 seconds.

66. Deceleration A car traveling at 45 miles per hour is brought to a stop, at constant deceleration, 132 feet from where the brakes are applied.

(a) How far has the car moved when its speed has been reduced to 30 miles per hour?

(b) How far has the car moved when its speed has been reduced to 15 miles per hour?

(c) Draw the real number line from 0 to 132. Plot the points found in parts (a) and (b). What can you conclude?

67. Acceleration At the instant the traffic light turns green, a car that has been waiting at an intersection starts with a constant acceleration of 6 feet per second per second. At the same instant, a truck traveling with a constant velocity of 30 feet per second passes the car.

(a) How far beyond its starting point will the car pass the truck?

(b) How fast will the car be traveling when it passes the truck?

68. Acceleration Assume that a fully loaded plane starting from rest has a constant acceleration while moving down a runway. The plane requires 0.7 mile of runway and a speed of 160 miles per hour in order to lift off. What is the plane's acceleration?

True or False? **In Exercises 69–74, determine whether the statement is true or false. If it is false, explain why or give an example that shows it is false.**

69. The antiderivative of $f(x)$ is unique.

70. Each antiderivative of an nth-degree polynomial function is an $(n+1)$th-degree polynomial function.

71. If $p(x)$ is a polynomial function, then p has exactly one antiderivative whose graph contains the origin.

72. If $F(x)$ and $G(x)$ are antiderivatives of $f(x)$, then

$F(x) = G(x) + C.$

73. If $f'(x) = g(x)$, then $\int g(x)\,dx = f(x) + C$.

74. $\int f(x)g(x)\,dx = \int f(x)\,dx \int g(x)\,dx$

75. Horizontal Tangent Find a function f such that the graph of f has a horizontal tangent at $(2, 0)$ and $f''(x) = 2x$.

76. Finding a Function The graph of f' is shown. Find and sketch the graph of f given that f is continuous and $f(0) = 1$.

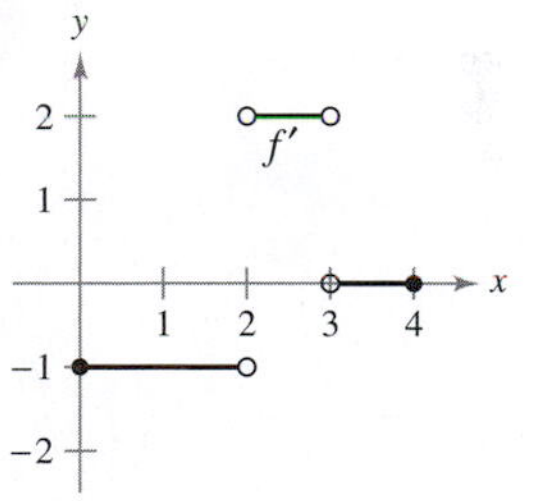

77. Proof Let $s(x)$ and $c(x)$ be two functions satisfying $s'(x) = c(x)$ and $c'(x) = -s(x)$ for all x. If $s(0) = 0$ and $c(0) = 1$, prove that $[s(x)]^2 + [c(x)]^2 = 1$.

PUTNAM EXAM CHALLENGE

78. Suppose f and g are non-constant, differentiable, real-valued functions defined on $(-\infty, \infty)$. Furthermore, suppose that for each pair of real numbers x and y,

$$f(x+y) = f(x)f(y) - g(x)g(y) \quad \text{and}$$
$$g(x+y) = f(x)g(y) + g(x)f(y).$$

If $f'(0) = 0$, prove that $(f(x))^2 + (g(x))^2 = 1$ for all x.

This problem was composed by the Committee on the Putnam Prize Competition.

4.2 Area

- Use sigma notation to write and evaluate a sum.
- Understand the concept of area.
- Approximate the area of a plane region.
- Find the area of a plane region using limits.

Sigma Notation

In the preceding section, you studied antidifferentiation. In this section, you will look further into a problem introduced in Section 1.1—that of finding the area of a region in the plane. At first glance, these two ideas may seem unrelated, but you will discover in Section 4.4 that they are closely related by an extremely important theorem called the Fundamental Theorem of Calculus.

This section begins by introducing a concise notation for sums. This notation is called **sigma notation** because it uses the uppercase Greek letter sigma, written as Σ.

Sigma Notation

The sum of n terms $a_1, a_2, a_3, \ldots, a_n$ is written as

$$\sum_{i=1}^{n} a_i = a_1 + a_2 + a_3 + \cdots + a_n$$

where i is the **index of summation,** a_i is the ***i*th term** of the sum, and the **upper and lower bounds of summation** are n and 1.

REMARK The upper and lower bounds must be constant with respect to the index of summation. However, the lower bound doesn't have to be 1. Any integer less than or equal to the upper bound is legitimate.

EXAMPLE 1 Examples of Sigma Notation

a. $\displaystyle\sum_{i=1}^{6} i = 1 + 2 + 3 + 4 + 5 + 6$

b. $\displaystyle\sum_{i=0}^{5} (i + 1) = 1 + 2 + 3 + 4 + 5 + 6$

c. $\displaystyle\sum_{j=3}^{7} j^2 = 3^2 + 4^2 + 5^2 + 6^2 + 7^2$

d. $\displaystyle\sum_{j=1}^{5} \frac{1}{\sqrt{j}} = \frac{1}{\sqrt{1}} + \frac{1}{\sqrt{2}} + \frac{1}{\sqrt{3}} + \frac{1}{\sqrt{4}} + \frac{1}{\sqrt{5}}$

e. $\displaystyle\sum_{k=1}^{n} \frac{1}{n}(k^2 + 1) = \frac{1}{n}(1^2 + 1) + \frac{1}{n}(2^2 + 1) + \cdots + \frac{1}{n}(n^2 + 1)$

f. $\displaystyle\sum_{i=1}^{n} f(x_i)\,\Delta x = f(x_1)\,\Delta x + f(x_2)\,\Delta x + \cdots + f(x_n)\,\Delta x$

From parts (a) and (b), notice that the same sum can be represented in different ways using sigma notation.

FOR FURTHER INFORMATION For a geometric interpretation of summation formulas, see the article "Looking at $\sum_{k=1}^{n} k$ and $\sum_{k=1}^{n} k^2$ Geometrically" by Eric Hegblom in *Mathematics Teacher*. To view this article, go to *MathArticles.com.*

Although any variable can be used as the index of summation, i, j, and k are often used. Notice in Example 1 that the index of summation does not appear in the terms of the expanded sum.

THE SUM OF THE FIRST 100 INTEGERS

A teacher of Carl Friedrich Gauss (1777–1855) asked him to add all the integers from 1 to 100. When Gauss returned with the correct answer after only a few moments, the teacher could only look at him in astounded silence. This is what Gauss did:

$$\begin{array}{r} 1 + 2 + 3 + \cdots + 100 \\ 100 + 99 + 98 + \cdots + 1 \\ \hline 101 + 101 + 101 + \cdots + 101 \end{array}$$

$$\frac{100 \times 101}{2} = 5050$$

This is generalized by Theorem 4.2, Property 2, where

$$\sum_{i=1}^{100} i = \frac{100(101)}{2} = 5050.$$

The properties of summation shown below can be derived using the Associative and Commutative Properties of Addition and the Distributive Property of Addition over Multiplication. (In the first property, k is a constant.)

1. $\displaystyle\sum_{i=1}^{n} ka_i = k\sum_{i=1}^{n} a_i$

2. $\displaystyle\sum_{i=1}^{n} (a_i \pm b_i) = \sum_{i=1}^{n} a_i \pm \sum_{i=1}^{n} b_i$

The next theorem lists some useful formulas for sums of powers.

THEOREM 4.2 Summation Formulas

1. $\displaystyle\sum_{i=1}^{n} c = cn$, c is a constant

2. $\displaystyle\sum_{i=1}^{n} i = \frac{n(n+1)}{2}$

3. $\displaystyle\sum_{i=1}^{n} i^2 = \frac{n(n+1)(2n+1)}{6}$

4. $\displaystyle\sum_{i=1}^{n} i^3 = \frac{n^2(n+1)^2}{4}$

A proof of this theorem is given in Appendix A.

See LarsonCalculus.com for Bruce Edwards's video of this proof.

EXAMPLE 2 Evaluating a Sum

Evaluate $\displaystyle\sum_{i=1}^{n} \frac{i+1}{n^2}$ for $n = 10$, 100, 1000, and 10,000.

Solution

$$\sum_{i=1}^{n} \frac{i+1}{n^2} = \frac{1}{n^2}\sum_{i=1}^{n}(i+1) \quad \text{Factor the constant } 1/n^2 \text{ out of sum.}$$

$$= \frac{1}{n^2}\left(\sum_{i=1}^{n} i + \sum_{i=1}^{n} 1\right) \quad \text{Write as two sums.}$$

$$= \frac{1}{n^2}\left[\frac{n(n+1)}{2} + n\right] \quad \text{Apply Theorem 4.2.}$$

$$= \frac{1}{n^2}\left[\frac{n^2+3n}{2}\right] \quad \text{Simplify.}$$

$$= \frac{n+3}{2n} \quad \text{Simplify.}$$

Now you can evaluate the sum by substituting the appropriate values of n, as shown in the table below.

n	10	100	1000	10,000
$\displaystyle\sum_{i=1}^{n} \frac{i+1}{n^2} = \frac{n+3}{2n}$	0.65000	0.51500	0.50150	0.50015

In the table, note that the sum appears to approach a limit as n increases. Although the discussion of limits at infinity in Section 3.5 applies to a variable x, where x can be any real number, many of the same results hold true for limits involving the variable n, where n is restricted to positive integer values. So, to find the limit of $(n+3)/2n$ as n approaches infinity, you can write

$$\lim_{n\to\infty} \frac{n+3}{2n} = \lim_{n\to\infty}\left(\frac{n}{2n} + \frac{3}{2n}\right) = \lim_{n\to\infty}\left(\frac{1}{2} + \frac{3}{2n}\right) = \frac{1}{2} + 0 = \frac{1}{2}.$$

Area

In Euclidean geometry, the simplest type of plane region is a rectangle. Although people often say that the *formula* for the area of a rectangle is

$$A = bh$$

it is actually more proper to say that this is the *definition* of the **area of a rectangle.**

From this definition, you can develop formulas for the areas of many other plane regions. For example, to determine the area of a triangle, you can form a rectangle whose area is twice that of the triangle, as shown in Figure 4.5. Once you know how to find the area of a triangle, you can determine the area of any polygon by subdividing the polygon into triangular regions, as shown in Figure 4.6.

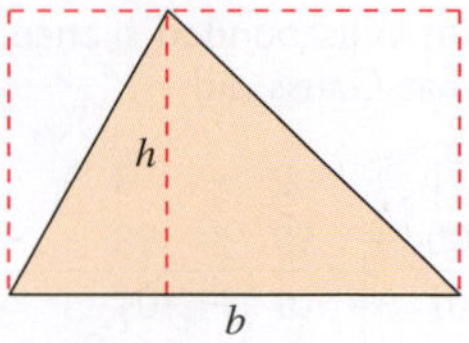

Triangle: $A = \frac{1}{2}bh$
Figure 4.5

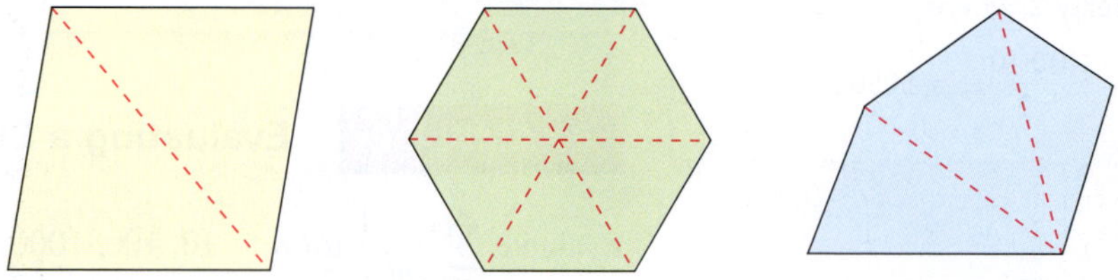

Parallelogram Hexagon Polygon
Figure 4.6

ARCHIMEDES (287–212 B.C.)

Archimedes used the method of exhaustion to derive formulas for the areas of ellipses, parabolic segments, and sectors of a spiral. He is considered to have been the greatest applied mathematician of antiquity.
See LarsonCalculus.com to read more of this biography.

Finding the areas of regions other than polygons is more difficult. The ancient Greeks were able to determine formulas for the areas of some general regions (principally those bounded by conics) by the *exhaustion* method. The clearest description of this method was given by Archimedes. Essentially, the method is a limiting process in which the area is squeezed between two polygons—one inscribed in the region and one circumscribed about the region.

For instance, in Figure 4.7, the area of a circular region is approximated by an n-sided inscribed polygon and an n-sided circumscribed polygon. For each value of n, the area of the inscribed polygon is less than the area of the circle, and the area of the circumscribed polygon is greater than the area of the circle. Moreover, as n increases, the areas of both polygons become better and better approximations of the area of the circle.

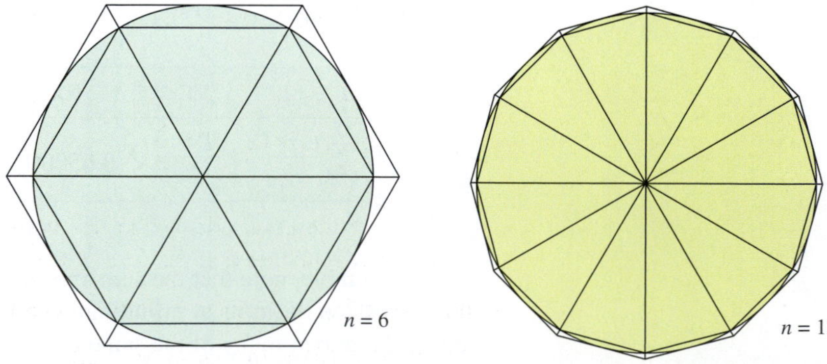

The exhaustion method for finding the area of a circular region
Figure 4.7

FOR FURTHER INFORMATION For an alternative development of the formula for the area of a circle, see the article "Proof Without Words: Area of a Disk is πR^2" by Russell Jay Hendel in *Mathematics Magazine.* To view this article, go to *MathArticles.com.*

A process that is similar to that used by Archimedes to determine the area of a plane region is used in the remaining examples in this section.

The Area of a Plane Region

Recall from Section 1.1 that the origins of calculus are connected to two classic problems: the tangent line problem and the area problem. Example 3 begins the investigation of the area problem.

EXAMPLE 3 Approximating the Area of a Plane Region

Use the five rectangles in Figure 4.8(a) and (b) to find *two* approximations of the area of the region lying between the graph of

$$f(x) = -x^2 + 5$$

and the x-axis between $x = 0$ and $x = 2$.

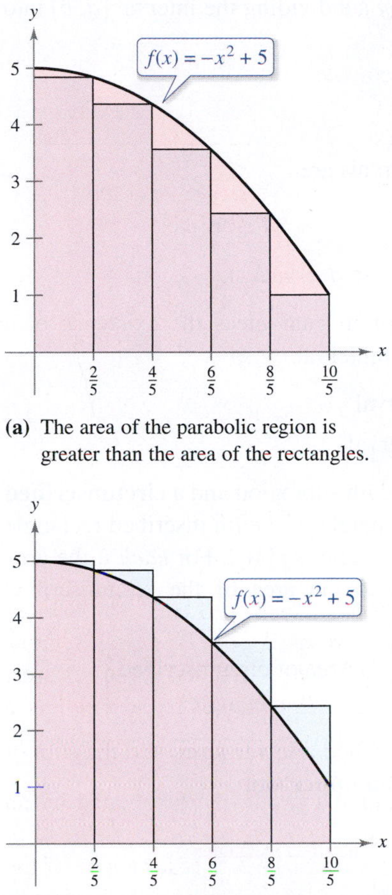

(a) The area of the parabolic region is greater than the area of the rectangles.

(b) The area of the parabolic region is less than the area of the rectangles.

Figure 4.8

Solution

a. The right endpoints of the five intervals are

$$\frac{2}{5}i \qquad \text{Right endpoints}$$

where $i = 1, 2, 3, 4, 5$. The width of each rectangle is $\frac{2}{5}$, and the height of each rectangle can be obtained by evaluating f at the right endpoint of each interval.

$$\left[0, \frac{2}{5}\right], \left[\frac{2}{5}, \frac{4}{5}\right], \left[\frac{4}{5}, \frac{6}{5}\right], \left[\frac{6}{5}, \frac{8}{5}\right], \left[\frac{8}{5}, \frac{10}{5}\right]$$

Evaluate f at the right endpoints of these intervals.

The sum of the areas of the five rectangles is

$$\sum_{i=1}^{5} \underbrace{f\left(\frac{2i}{5}\right)}_{\text{Height}} \underbrace{\left(\frac{2}{5}\right)}_{\text{Width}} = \sum_{i=1}^{5}\left[-\left(\frac{2i}{5}\right)^2 + 5\right]\left(\frac{2}{5}\right) = \frac{162}{25} = 6.48.$$

Because each of the five rectangles lies inside the parabolic region, you can conclude that the area of the parabolic region is greater than 6.48.

b. The left endpoints of the five intervals are

$$\frac{2}{5}(i - 1) \qquad \text{Left endpoints}$$

where $i = 1, 2, 3, 4, 5$. The width of each rectangle is $\frac{2}{5}$, and the height of each rectangle can be obtained by evaluating f at the left endpoint of each interval. So, the sum is

$$\sum_{i=1}^{5} \underbrace{f\left(\frac{2i - 2}{5}\right)}_{\text{Height}} \underbrace{\left(\frac{2}{5}\right)}_{\text{Width}} = \sum_{i=1}^{5}\left[-\left(\frac{2i - 2}{5}\right)^2 + 5\right]\left(\frac{2}{5}\right) = \frac{202}{25} = 8.08.$$

Because the parabolic region lies within the union of the five rectangular regions, you can conclude that the area of the parabolic region is less than 8.08.

By combining the results in parts (a) and (b), you can conclude that

$$6.48 < (\text{Area of region}) < 8.08.$$

By increasing the number of rectangles used in Example 3, you can obtain closer and closer approximations of the area of the region. For instance, using 25 rectangles of width $\frac{2}{25}$ each, you can conclude that

$$7.1712 < (\text{Area of region}) < 7.4912.$$

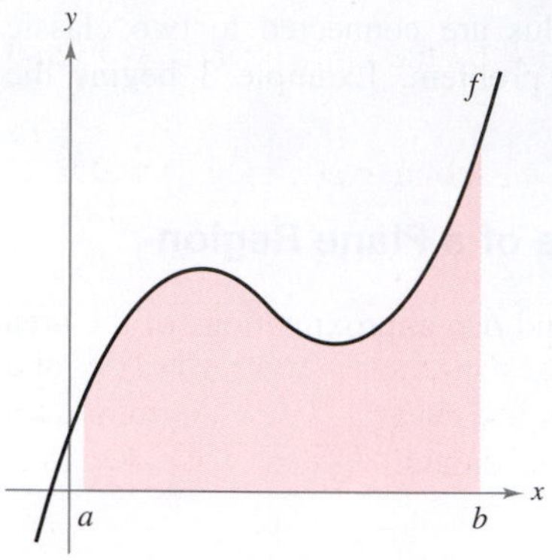

The region under a curve
Figure 4.9

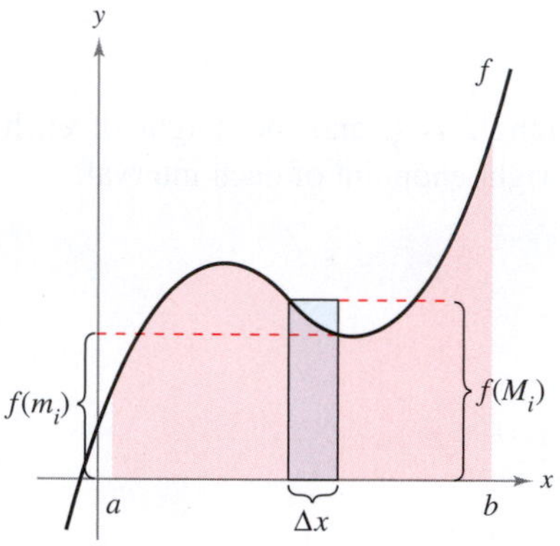

The interval $[a, b]$ is divided into n subintervals of width $\Delta x = \dfrac{b-a}{n}$.
Figure 4.10

Upper and Lower Sums

The procedure used in Example 3 can be generalized as follows. Consider a plane region bounded above by the graph of a nonnegative, continuous function

$$y = f(x)$$

as shown in Figure 4.9. The region is bounded below by the x-axis, and the left and right boundaries of the region are the vertical lines $x = a$ and $x = b$.

To approximate the area of the region, begin by subdividing the interval $[a, b]$ into n subintervals, each of width

$$\Delta x = \frac{b-a}{n}$$

as shown in Figure 4.10. The endpoints of the intervals are

$$\overbrace{a + 0(\Delta x)}^{a = x_0} < \overbrace{a + 1(\Delta x)}^{x_1} < \overbrace{a + 2(\Delta x)}^{x_2} < \cdot \cdot \cdot < \overbrace{a + n(\Delta x)}^{x_n = b}.$$

Because f is continuous, the Extreme Value Theorem guarantees the existence of a minimum and a maximum value of $f(x)$ in *each* subinterval.

$f(m_i) =$ Minimum value of $f(x)$ in ith subinterval

$f(M_i) =$ Maximum value of $f(x)$ in ith subinterval

Next, define an **inscribed rectangle** lying *inside* the ith subregion and a **circumscribed rectangle** extending *outside* the ith subregion. The height of the ith inscribed rectangle is $f(m_i)$ and the height of the ith circumscribed rectangle is $f(M_i)$. For *each i*, the area of the inscribed rectangle is less than or equal to the area of the circumscribed rectangle.

$$\begin{pmatrix}\text{Area of inscribed} \\ \text{rectangle}\end{pmatrix} = f(m_i)\,\Delta x \le f(M_i)\,\Delta x = \begin{pmatrix}\text{Area of circumscribed} \\ \text{rectangle}\end{pmatrix}$$

The sum of the areas of the inscribed rectangles is called a **lower sum,** and the sum of the areas of the circumscribed rectangles is called an **upper sum.**

$$\text{Lower sum} = s(n) = \sum_{i=1}^{n} f(m_i)\,\Delta x \qquad \text{Area of inscribed rectangles}$$

$$\text{Upper sum} = S(n) = \sum_{i=1}^{n} f(M_i)\,\Delta x \qquad \text{Area of circumscribed rectangles}$$

From Figure 4.11, you can see that the lower sum $s(n)$ is less than or equal to the upper sum $S(n)$. Moreover, the actual area of the region lies between these two sums.

$$s(n) \le (\text{Area of region}) \le S(n)$$

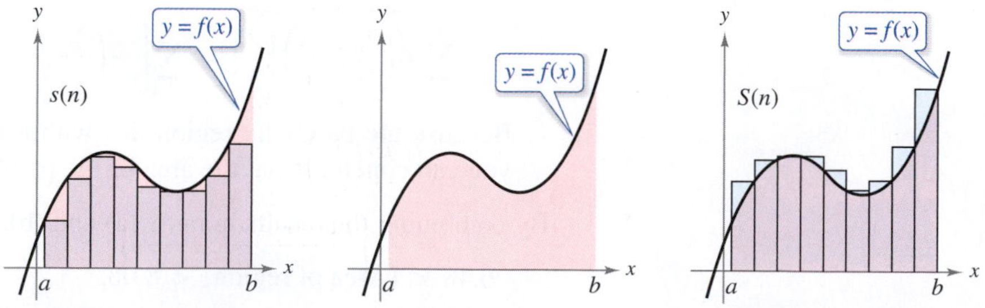

Area of inscribed rectangles is less than area of region.

Area of region

Area of circumscribed rectangles is greater than area of region.

Figure 4.11

EXAMPLE 4 Finding Upper and Lower Sums for a Region

Find the upper and lower sums for the region bounded by the graph of $f(x) = x^2$ and the x-axis between $x = 0$ and $x = 2$.

Solution To begin, partition the interval $[0, 2]$ into n subintervals, each of width

$$\Delta x = \frac{b - a}{n} = \frac{2 - 0}{n} = \frac{2}{n}.$$

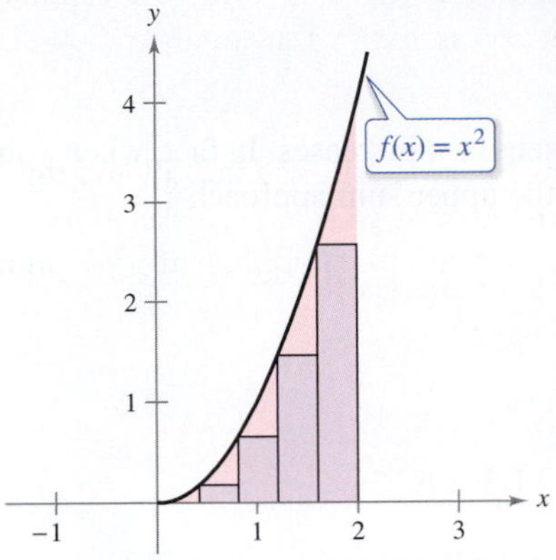

Inscribed rectangles

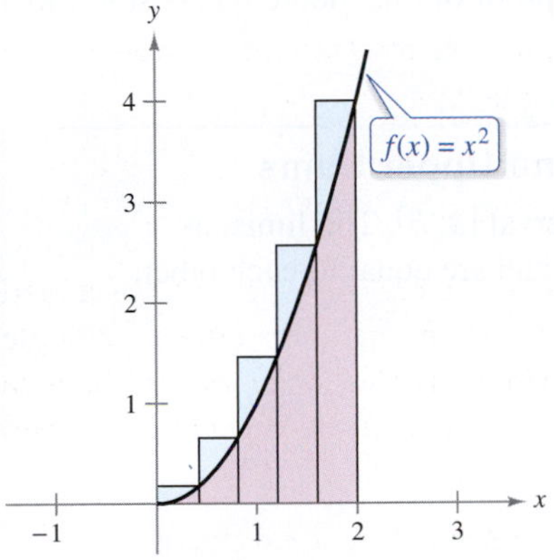

Circumscribed rectangles

Figure 4.12

Figure 4.12 shows the endpoints of the subintervals and several inscribed and circumscribed rectangles. Because f is increasing on the interval $[0, 2]$, the minimum value on each subinterval occurs at the left endpoint, and the maximum value occurs at the right endpoint.

Left Endpoints

$$m_i = 0 + (i - 1)\left(\frac{2}{n}\right) = \frac{2(i - 1)}{n}$$

Right Endpoints

$$M_i = 0 + i\left(\frac{2}{n}\right) = \frac{2i}{n}$$

Using the left endpoints, the lower sum is

$$\begin{aligned}
s(n) &= \sum_{i=1}^{n} f(m_i)\,\Delta x \\
&= \sum_{i=1}^{n} f\left[\frac{2(i-1)}{n}\right]\left(\frac{2}{n}\right) \\
&= \sum_{i=1}^{n} \left[\frac{2(i-1)}{n}\right]^2\left(\frac{2}{n}\right) \\
&= \sum_{i=1}^{n} \left(\frac{8}{n^3}\right)(i^2 - 2i + 1) \\
&= \frac{8}{n^3}\left(\sum_{i=1}^{n} i^2 - 2\sum_{i=1}^{n} i + \sum_{i=1}^{n} 1\right) \\
&= \frac{8}{n^3}\left\{\left[\frac{n(n+1)(2n+1)}{6}\right] - 2\left[\frac{n(n+1)}{2}\right] + n\right\} \\
&= \frac{4}{3n^3}(2n^3 - 3n^2 + n) \\
&= \frac{8}{3} - \frac{4}{n} + \frac{4}{3n^2}. \qquad \text{Lower sum}
\end{aligned}$$

Using the right endpoints, the upper sum is

$$\begin{aligned}
S(n) &= \sum_{i=1}^{n} f(M_i)\,\Delta x \\
&= \sum_{i=1}^{n} f\left(\frac{2i}{n}\right)\left(\frac{2}{n}\right) \\
&= \sum_{i=1}^{n} \left(\frac{2i}{n}\right)^2\left(\frac{2}{n}\right) \\
&= \sum_{i=1}^{n} \left(\frac{8}{n^3}\right)i^2 \\
&= \frac{8}{n^3}\left[\frac{n(n+1)(2n+1)}{6}\right] \\
&= \frac{4}{3n^3}(2n^3 + 3n^2 + n) \\
&= \frac{8}{3} + \frac{4}{n} + \frac{4}{3n^2}. \qquad \text{Upper sum}
\end{aligned}$$

Exploration

For the region given in Example 4, evaluate the lower sum

$$s(n) = \frac{8}{3} - \frac{4}{n} + \frac{4}{3n^2}$$

and the upper sum

$$S(n) = \frac{8}{3} + \frac{4}{n} + \frac{4}{3n^2}$$

for $n = 10$, 100, and 1000. Use your results to determine the area of the region.

Example 4 illustrates some important things about lower and upper sums. First, notice that for any value of n, the lower sum is less than (or equal to) the upper sum.

$$s(n) = \frac{8}{3} - \frac{4}{n} + \frac{4}{3n^2} < \frac{8}{3} + \frac{4}{n} + \frac{4}{3n^2} = S(n)$$

Second, the difference between these two sums lessens as n increases. In fact, when you take the limits as $n \to \infty$, both the lower sum and the upper sum approach $\frac{8}{3}$.

$$\lim_{n\to\infty} s(n) = \lim_{n\to\infty}\left(\frac{8}{3} - \frac{4}{n} + \frac{4}{3n^2}\right) = \frac{8}{3}$$ Lower sum limit

and

$$\lim_{n\to\infty} S(n) = \lim_{n\to\infty}\left(\frac{8}{3} + \frac{4}{n} + \frac{4}{3n^2}\right) = \frac{8}{3}$$ Upper sum limit

The next theorem shows that the equivalence of the limits (as $n \to \infty$) of the upper and lower sums is not mere coincidence. It is true for all functions that are continuous and nonnegative on the closed interval $[a, b]$. The proof of this theorem is best left to a course in advanced calculus.

THEOREM 4.3 Limits of the Lower and Upper Sums

Let f be continuous and nonnegative on the interval $[a, b]$. The limits as $n \to \infty$ of both the lower and upper sums exist and are equal to each other. That is,

$$\begin{aligned}\lim_{n\to\infty} s(n) &= \lim_{n\to\infty} \sum_{i=1}^{n} f(m_i)\,\Delta x \\ &= \lim_{n\to\infty} \sum_{i=1}^{n} f(M_i)\,\Delta x \\ &= \lim_{n\to\infty} S(n)\end{aligned}$$

where $\Delta x = (b - a)/n$ and $f(m_i)$ and $f(M_i)$ are the minimum and maximum values of f on the subinterval.

In Theorem 4.3, the same limit is attained for both the minimum value $f(m_i)$ and the maximum value $f(M_i)$. So, it follows from the Squeeze Theorem (Theorem 1.8) that the choice of x in the ith subinterval does not affect the limit. This means that you are free to choose an *arbitrary* x-value in the ith subinterval, as shown in the *definition of the area of a region in the plane.*

Definition of the Area of a Region in the Plane

Let f be continuous and nonnegative on the interval $[a, b]$. (See Figure 4.13.) The area of the region bounded by the graph of f, the x-axis, and the vertical lines $x = a$ and $x = b$ is

$$\text{Area} = \lim_{n\to\infty} \sum_{i=1}^{n} f(c_i)\,\Delta x$$

where $x_{i-1} \le c_i \le x_i$ and

$$\Delta x = \frac{b - a}{n}.$$

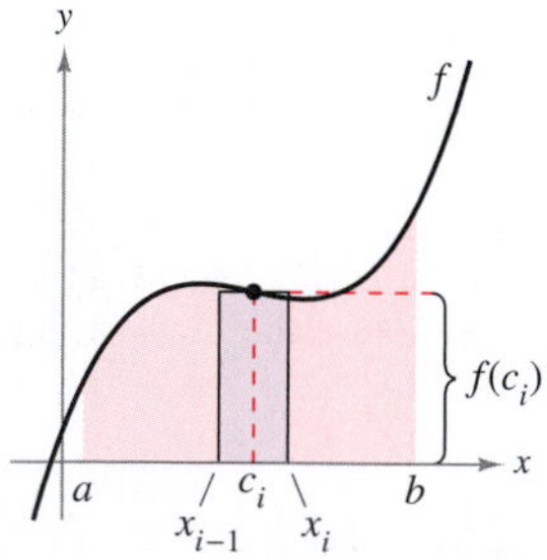

The width of the ith subinterval is $\Delta x = x_i - x_{i-1}$.
Figure 4.13

EXAMPLE 5 **Finding Area by the Limit Definition**

Find the area of the region bounded by the graph $f(x) = x^3$, the x-axis, and the vertical lines $x = 0$ and $x = 1$, as shown in Figure 4.14.

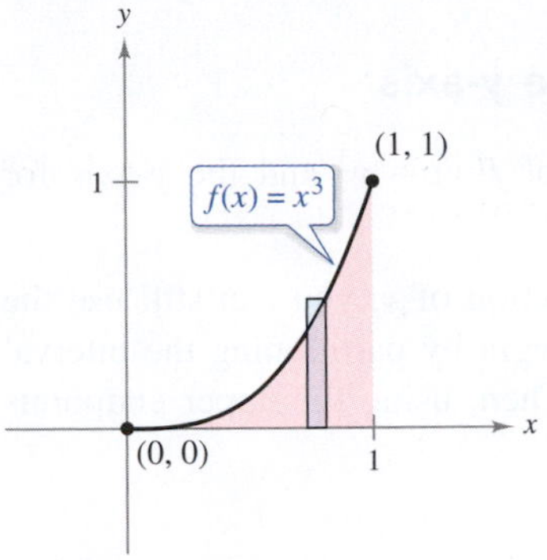

The area of the region bounded by the graph of f, the x-axis, $x = 0$, and $x = 1$ is $\frac{1}{4}$.
Figure 4.14

Solution Begin by noting that f is continuous and nonnegative on the interval $[0, 1]$. Next, partition the interval $[0, 1]$ into n subintervals, each of width $\Delta x = 1/n$. According to the definition of area, you can choose any x-value in the ith subinterval. For this example, the right endpoints $c_i = i/n$ are convenient.

$$\begin{aligned}
\text{Area} &= \lim_{n\to\infty} \sum_{i=1}^{n} f(c_i)\,\Delta x \\
&= \lim_{n\to\infty} \sum_{i=1}^{n} \left(\frac{i}{n}\right)^3 \left(\frac{1}{n}\right) && \text{Right endpoints: } c_i = \frac{i}{n} \\
&= \lim_{n\to\infty} \frac{1}{n^4} \sum_{i=1}^{n} i^3 \\
&= \lim_{n\to\infty} \frac{1}{n^4} \left[\frac{n^2(n+1)^2}{4}\right] \\
&= \lim_{n\to\infty} \left(\frac{1}{4} + \frac{1}{2n} + \frac{1}{4n^2}\right) \\
&= \frac{1}{4}
\end{aligned}$$

The area of the region is $\frac{1}{4}$.

EXAMPLE 6 **Finding Area by the Limit Definition**

See LarsonCalculus.com for an interactive version of this type of example.

Find the area of the region bounded by the graph of $f(x) = 4 - x^2$, the x-axis, and the vertical lines $x = 1$ and $x = 2$, as shown in Figure 4.15.

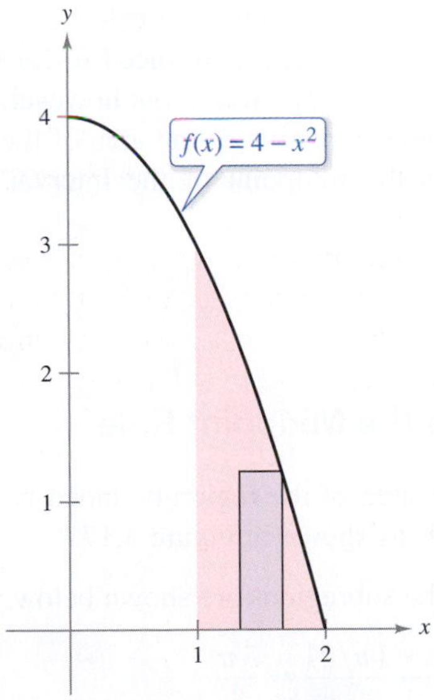

The area of the region bounded by the graph of f, the x-axis, $x = 1$, and $x = 2$ is $\frac{5}{3}$.
Figure 4.15

Solution Note that the function f is continuous and nonnegative on the interval $[1, 2]$. So, begin by partitioning the interval into n subintervals, each of width $\Delta x = 1/n$. Choosing the right endpoint

$$c_i = a + i\,\Delta x = 1 + \frac{i}{n} \qquad \text{Right endpoints}$$

of each subinterval, you obtain

$$\begin{aligned}
\text{Area} &= \lim_{n\to\infty} \sum_{i=1}^{n} f(c_i)\,\Delta x \\
&= \lim_{n\to\infty} \sum_{i=1}^{n} \left[4 - \left(1 + \frac{i}{n}\right)^2\right]\left(\frac{1}{n}\right) \\
&= \lim_{n\to\infty} \sum_{i=1}^{n} \left(3 - \frac{2i}{n} - \frac{i^2}{n^2}\right)\left(\frac{1}{n}\right) \\
&= \lim_{n\to\infty} \left(\frac{1}{n}\sum_{i=1}^{n} 3 - \frac{2}{n^2}\sum_{i=1}^{n} i - \frac{1}{n^3}\sum_{i=1}^{n} i^2\right) \\
&= \lim_{n\to\infty} \left[3 - \left(1 + \frac{1}{n}\right) - \left(\frac{1}{3} + \frac{1}{2n} + \frac{1}{6n^2}\right)\right] \\
&= 3 - 1 - \frac{1}{3} \\
&= \frac{5}{3}.
\end{aligned}$$

The area of the region is $\frac{5}{3}$.

The next example looks at a region that is bounded by the y-axis (rather than by the x-axis).

EXAMPLE 7 A Region Bounded by the y-axis

Find the area of the region bounded by the graph of $f(y) = y^2$ and the y-axis for $0 \le y \le 1$, as shown in Figure 4.16.

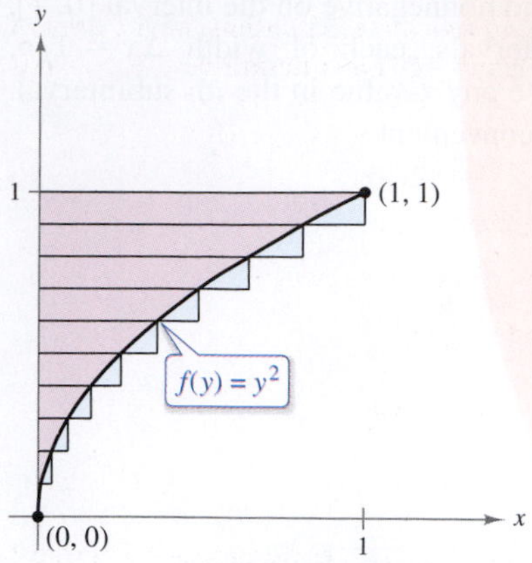

The area of the region bounded by the graph of f and the y-axis for $0 \le y \le 1$ is $\frac{1}{3}$.

Figure 4.16

Solution When f is a continuous, nonnegative function of y, you can still use the same basic procedure shown in Examples 5 and 6. Begin by partitioning the interval $[0, 1]$ into n subintervals, each of width $\Delta y = 1/n$. Then, using the upper endpoints $c_i = i/n$, you obtain

$$\begin{aligned}
\text{Area} &= \lim_{n\to\infty} \sum_{i=1}^{n} f(c_i)\,\Delta y \\
&= \lim_{n\to\infty} \sum_{i=1}^{n} \left(\frac{i}{n}\right)^2 \left(\frac{1}{n}\right) \qquad \text{Upper endpoints: } c_i = \frac{i}{n} \\
&= \lim_{n\to\infty} \frac{1}{n^3} \sum_{i=1}^{n} i^2 \\
&= \lim_{n\to\infty} \frac{1}{n^3}\left[\frac{n(n+1)(2n+1)}{6}\right] \\
&= \lim_{n\to\infty} \left(\frac{1}{3} + \frac{1}{2n} + \frac{1}{6n^2}\right) \\
&= \frac{1}{3}.
\end{aligned}$$

The area of the region is $\frac{1}{3}$.

REMARK You will learn about other approximation methods in Section 4.6. One of the methods, the Trapezoidal Rule, is similar to the Midpoint Rule.

In Examples 5, 6, and 7, c_i is chosen to be a value that is convenient for calculating the limit. Because each limit gives the exact area for *any* c_i, there is no need to find values that give good approximations when n is small. For an *approximation*, however, you should try to find a value of c_i that gives a good approximation of the area of the ith subregion. In general, a good value to choose is the midpoint of the interval, $c_i = (x_i + x_{i-1})/2$, and apply the **Midpoint Rule.**

$$\text{Area} \approx \sum_{i=1}^{n} f\left(\frac{x_i + x_{i-1}}{2}\right)\Delta x. \qquad \text{Midpoint Rule}$$

EXAMPLE 8 Approximating Area with the Midpoint Rule

Use the Midpoint Rule with $n = 4$ to approximate the area of the region bounded by the graph of $f(x) = \sin x$ and the x-axis for $0 \le x \le \pi$, as shown in Figure 4.17.

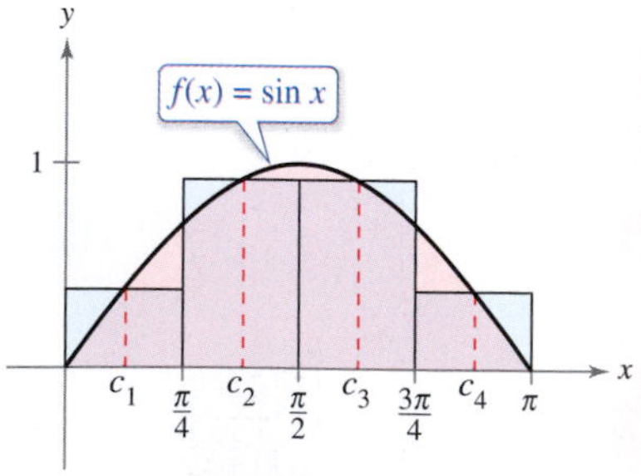

The area of the region bounded by the graph of $f(x) = \sin x$ and the x-axis for $0 \le x \le \pi$ is about 2.052.

Figure 4.17

Solution For $n = 4$, $\Delta x = \pi/4$. The midpoints of the subregions are shown below.

$$c_1 = \frac{0 + (\pi/4)}{2} = \frac{\pi}{8} \qquad c_2 = \frac{(\pi/4) + (\pi/2)}{2} = \frac{3\pi}{8}$$

$$c_3 = \frac{(\pi/2) + (3\pi/4)}{2} = \frac{5\pi}{8} \qquad c_4 = \frac{(3\pi/4) + \pi}{2} = \frac{7\pi}{8}$$

So, the area is approximated by

$$\text{Area} \approx \sum_{i=1}^{n} f(c_i)\,\Delta x = \sum_{i=1}^{4} (\sin c_i)\left(\frac{\pi}{4}\right) = \frac{\pi}{4}\left(\sin\frac{\pi}{8} + \sin\frac{3\pi}{8} + \sin\frac{5\pi}{8} + \sin\frac{7\pi}{8}\right)$$

which is about 2.052.

4.2 Exercises

See CalcChat.com for tutorial help and worked-out solutions to odd-numbered exercises.

Finding a Sum In Exercises 1–6, find the sum. Use the summation capabilities of a graphing utility to verify your result.

1. $\sum_{i=1}^{6}(3i + 2)$
2. $\sum_{k=3}^{9}(k^2 + 1)$
3. $\sum_{k=0}^{4}\frac{1}{k^2 + 1}$
4. $\sum_{j=4}^{6}\frac{3}{j}$
5. $\sum_{k=1}^{4}c$
6. $\sum_{i=1}^{4}[(i - 1)^2 + (i + 1)^3]$

Using Sigma Notation In Exercises 7–12, use sigma notation to write the sum.

7. $\frac{1}{5(1)} + \frac{1}{5(2)} + \frac{1}{5(3)} + \cdots + \frac{1}{5(11)}$
8. $\frac{9}{1+1} + \frac{9}{1+2} + \frac{9}{1+3} + \cdots + \frac{9}{1+14}$
9. $\left[7\left(\frac{1}{6}\right) + 5\right] + \left[7\left(\frac{2}{6}\right) + 5\right] + \cdots + \left[7\left(\frac{6}{6}\right) + 5\right]$
10. $\left[1 - \left(\frac{1}{4}\right)^2\right] + \left[1 - \left(\frac{2}{4}\right)^2\right] + \cdots + \left[1 - \left(\frac{4}{4}\right)^2\right]$
11. $\left[\left(\frac{2}{n}\right)^3 - \frac{2}{n}\right]\left(\frac{2}{n}\right) + \cdots + \left[\left(\frac{2n}{n}\right)^3 - \frac{2n}{n}\right]\left(\frac{2}{n}\right)$
12. $\left[2\left(1 + \frac{3}{n}\right)^2\right]\left(\frac{3}{n}\right) + \cdots + \left[2\left(1 + \frac{3n}{n}\right)^2\right]\left(\frac{3}{n}\right)$

Evaluating a Sum In Exercises 13–20, use the properties of summation and Theorem 4.2 to evaluate the sum. Use the summation capabilities of a graphing utility to verify your result.

13. $\sum_{i=1}^{12}7$
14. $\sum_{i=1}^{30}-18$
15. $\sum_{i=1}^{24}4i$
16. $\sum_{i=1}^{16}(5i - 4)$
17. $\sum_{i=1}^{20}(i - 1)^2$
18. $\sum_{i=1}^{10}(i^2 - 1)$
19. $\sum_{i=1}^{15}i(i - 1)^2$
20. $\sum_{i=1}^{25}(i^3 - 2i)$

Evaluating a Sum In Exercises 21–24, use the summation formulas to rewrite the expression without the summation notation. Use the result to find the sums for $n = 10, 100, 1000$, and $10,000$.

21. $\sum_{i=1}^{n}\frac{2i + 1}{n^2}$
22. $\sum_{j=1}^{n}\frac{7j + 4}{n^2}$
23. $\sum_{k=1}^{n}\frac{6k(k - 1)}{n^3}$
24. $\sum_{i=1}^{n}\frac{2i^3 - 3i}{n^4}$

Approximating the Area of a Plane Region In Exercises 25–30, use left and right endpoints and the given number of rectangles to find two approximations of the area of the region between the graph of the function and the x-axis over the given interval.

25. $f(x) = 2x + 5$, $[0, 2]$, 4 rectangles
26. $f(x) = 9 - x$, $[2, 4]$, 6 rectangles
27. $g(x) = 2x^2 - x - 1$, $[2, 5]$, 6 rectangles
28. $g(x) = x^2 + 1$, $[1, 3]$, 8 rectangles
29. $f(x) = \cos x$, $\left[0, \frac{\pi}{2}\right]$, 4 rectangles
30. $g(x) = \sin x$, $[0, \pi]$, 6 rectangles

Using Upper and Lower Sums In Exercises 31 and 32, bound the area of the shaded region by approximating the upper and lower sums. Use rectangles of width 1.

31.

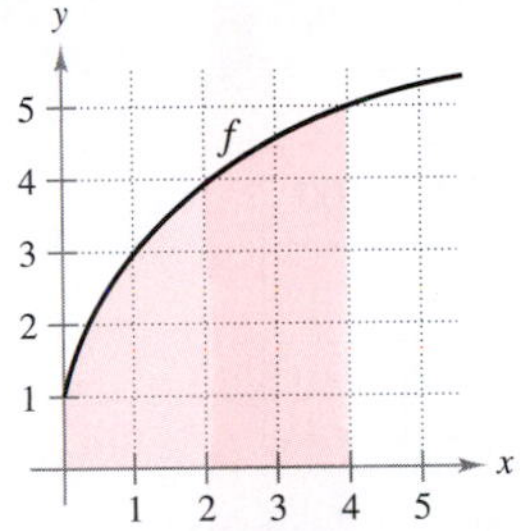

32.

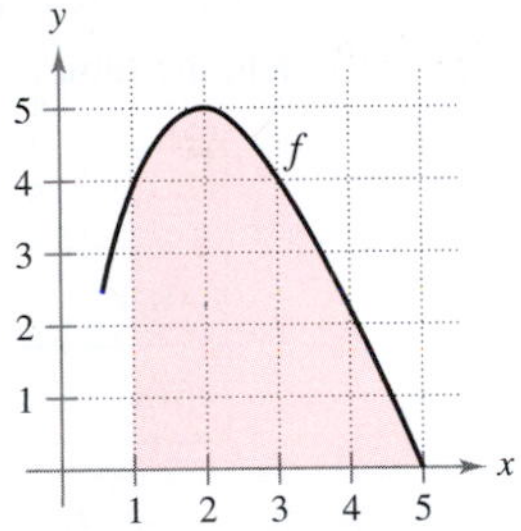

Finding Upper and Lower Sums for a Region In Exercises 33–36, use upper and lower sums to approximate the area of the region using the given number of subintervals (of equal width).

33. $y = \sqrt{x}$

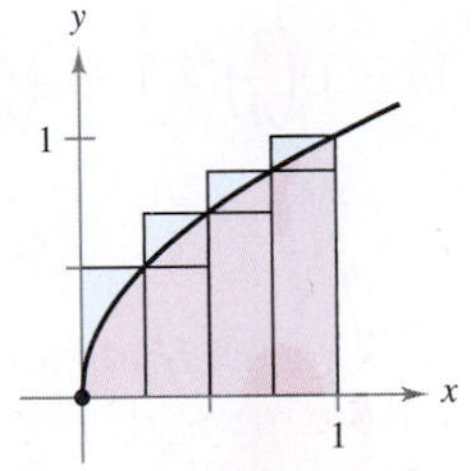

34. $y = \sqrt{x} + 2$

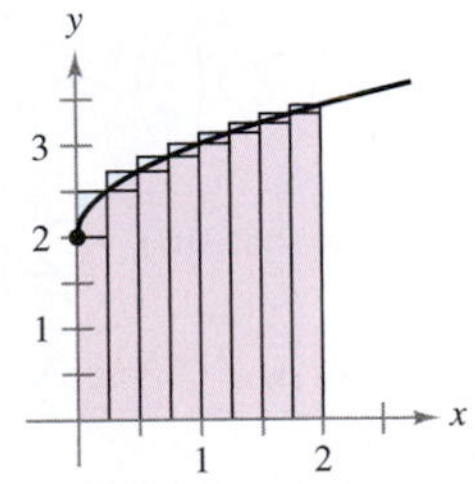

35. $y = \frac{1}{x}$

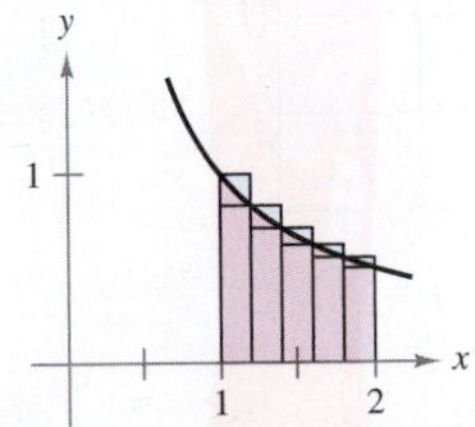

36. $y = \sqrt{1 - x^2}$

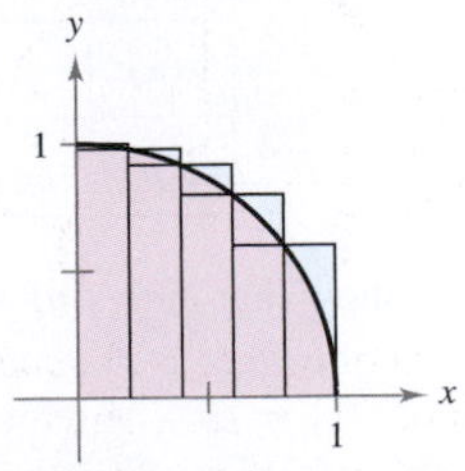

Finding a Limit **In Exercises 37–42, find a formula for the sum of n terms. Use the formula to find the limit as $n \to \infty$.**

37. $\lim_{n\to\infty} \sum_{i=1}^{n} \frac{24i}{n^2}$

38. $\lim_{n\to\infty} \sum_{i=1}^{n} \left(\frac{3i}{n}\right)\left(\frac{3}{n}\right)$

39. $\lim_{n\to\infty} \sum_{i=1}^{n} \frac{1}{n^3}(i-1)^2$

40. $\lim_{n\to\infty} \sum_{i=1}^{n} \left(1+\frac{2i}{n}\right)^2\left(\frac{2}{n}\right)$

41. $\lim_{n\to\infty} \sum_{i=1}^{n} \left(1+\frac{i}{n}\right)\left(\frac{2}{n}\right)$

42. $\lim_{n\to\infty} \sum_{i=1}^{n} \left(2+\frac{3i}{n}\right)^3\left(\frac{3}{n}\right)$

43. Numerical Reasoning Consider a triangle of area 2 bounded by the graphs of $y = x$, $y = 0$, and $x = 2$.

(a) Sketch the region.

(b) Divide the interval $[0, 2]$ into n subintervals of equal width and show that the endpoints are

$$0 < 1\left(\frac{2}{n}\right) < \cdot \cdot \cdot < (n-1)\left(\frac{2}{n}\right) < n\left(\frac{2}{n}\right).$$

(c) Show that $s(n) = \sum_{i=1}^{n} \left[(i-1)\left(\frac{2}{n}\right)\right]\left(\frac{2}{n}\right)$.

(d) Show that $S(n) = \sum_{i=1}^{n} \left[i\left(\frac{2}{n}\right)\right]\left(\frac{2}{n}\right)$.

(e) Complete the table.

n	5	10	50	100
$s(n)$				
$S(n)$				

(f) Show that $\lim_{n\to\infty} s(n) = \lim_{n\to\infty} S(n) = 2$.

44. Numerical Reasoning Consider a trapezoid of area 4 bounded by the graphs of $y = x$, $y = 0$, $x = 1$, and $x = 3$.

(a) Sketch the region.

(b) Divide the interval $[1, 3]$ into n subintervals of equal width and show that the endpoints are

$$1 < 1 + 1\left(\frac{2}{n}\right) < \cdot \cdot \cdot < 1 + (n-1)\left(\frac{2}{n}\right) < 1 + n\left(\frac{2}{n}\right).$$

(c) Show that $s(n) = \sum_{i=1}^{n} \left[1 + (i-1)\left(\frac{2}{n}\right)\right]\left(\frac{2}{n}\right)$.

(d) Show that $S(n) = \sum_{i=1}^{n} \left[1 + i\left(\frac{2}{n}\right)\right]\left(\frac{2}{n}\right)$.

(e) Complete the table.

n	5	10	50	100
$s(n)$				
$S(n)$				

(f) Show that $\lim_{n\to\infty} s(n) = \lim_{n\to\infty} S(n) = 4$.

Finding Area by the Limit Definition **In Exercises 45–54, use the limit process to find the area of the region bounded by the graph of the function and the x-axis over the given interval. Sketch the region.**

45. $y = -4x + 5$, $[0, 1]$

46. $y = 3x - 2$, $[2, 5]$

47. $y = x^2 + 2$, $[0, 1]$

48. $y = 3x^2 + 1$, $[0, 2]$

49. $y = 25 - x^2$, $[1, 4]$

50. $y = 4 - x^2$, $[-2, 2]$

51. $y = 27 - x^3$, $[1, 3]$

52. $y = 2x - x^3$, $[0, 1]$

53. $y = x^2 - x^3$, $[-1, 1]$

54. $y = 2x^3 - x^2$, $[1, 2]$

Finding Area by the Limit Definition **In Exercises 55–60, use the limit process to find the area of the region bounded by the graph of the function and the y-axis over the given y-interval. Sketch the region.**

55. $f(y) = 4y$, $0 \le y \le 2$

56. $g(y) = \frac{1}{2}y$, $2 \le y \le 4$

57. $f(y) = y^2$, $0 \le y \le 5$

58. $f(y) = 4y - y^2$, $1 \le y \le 2$

59. $g(y) = 4y^2 - y^3$, $1 \le y \le 3$

60. $h(y) = y^3 + 1$, $1 \le y \le 2$

Approximating Area with the Midpoint Rule **In Exercises 61–64, use the Midpoint Rule with $n = 4$ to approximate the area of the region bounded by the graph of the function and the x-axis over the given interval.**

61. $f(x) = x^2 + 3$, $[0, 2]$

62. $f(x) = x^2 + 4x$, $[0, 4]$

63. $f(x) = \tan x$, $\left[0, \frac{\pi}{4}\right]$

64. $f(x) = \cos x$, $\left[0, \frac{\pi}{2}\right]$

WRITING ABOUT CONCEPTS

Approximation **In Exercises 65 and 66, determine which value best approximates the area of the region between the x-axis and the graph of the function over the given interval. (Make your selection on the basis of a sketch of the region, not by performing calculations.)**

65. $f(x) = 4 - x^2$, $[0, 2]$

(a) -2 (b) 6 (c) 10 (d) 3 (e) 8

66. $f(x) = \sin \frac{\pi x}{4}$, $[0, 4]$

(a) 3 (b) 1 (c) -2 (d) 8 (e) 6

67. Upper and Lower Sums In your own words and using appropriate figures, describe the methods of upper sums and lower sums in approximating the area of a region.

68. Area of a Region in the Plane Give the definition of the area of a region in the plane.

69. Graphical Reasoning Consider the region bounded by the graphs of $f(x) = 8x/(x + 1)$, $x = 0$, $x = 4$, and $y = 0$, as shown in the figure. To print an enlarged copy of the graph, go to *MathGraphs.com.*

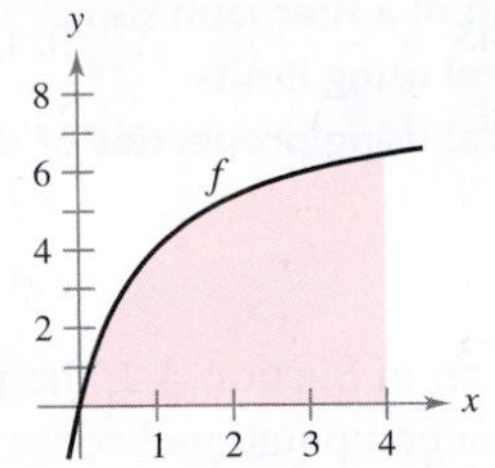

(a) Redraw the figure, and complete and shade the rectangles representing the lower sum when $n = 4$. Find this lower sum.

(b) Redraw the figure, and complete and shade the rectangles representing the upper sum when $n = 4$. Find this upper sum.

(c) Redraw the figure, and complete and shade the rectangles whose heights are determined by the functional values at the midpoint of each subinterval when $n = 4$. Find this sum using the Midpoint Rule.

(d) Verify the following formulas for approximating the area of the region using n subintervals of equal width.

Lower sum: $s(n) = \sum_{i=1}^{n} f\left[(i - 1)\frac{4}{n}\right]\left(\frac{4}{n}\right)$

Upper sum: $S(n) = \sum_{i=1}^{n} f\left[(i)\frac{4}{n}\right]\left(\frac{4}{n}\right)$

Midpoint Rule: $M(n) = \sum_{i=1}^{n} f\left[\left(i - \frac{1}{2}\right)\frac{4}{n}\right]\left(\frac{4}{n}\right)$

(e) Use a graphing utility to create a table of values of $s(n)$, $S(n)$, and $M(n)$ for $n = 4, 8, 20, 100$, and 200.

(f) Explain why $s(n)$ increases and $S(n)$ decreases for increasing values of n, as shown in the table in part (e).

70. HOW DO YOU SEE IT? The function shown in the graph below is increasing on the interval $[1, 4]$. The interval will be divided into 12 subintervals.

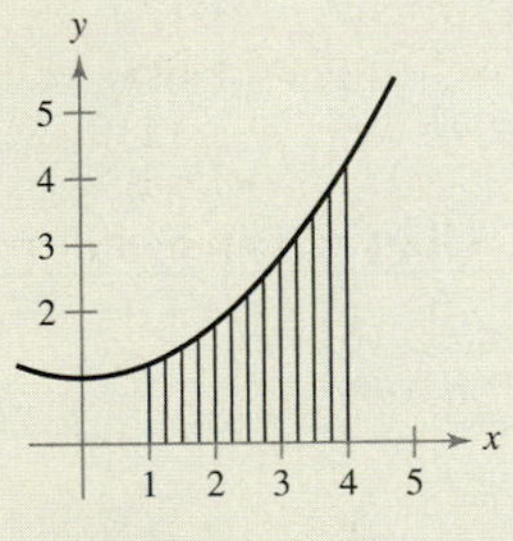

(a) What are the left endpoints of the first and last subintervals?

(b) What are the right endpoints of the first two subintervals?

(c) When using the right endpoints, do the rectangles lie above or below the graph of the function?

(d) What can you conclude about the heights of the rectangles when the function is constant on the given interval?

True or False? **In Exercises 71 and 72, determine whether the statement is true or false. If it is false, explain why or give an example that shows it is false.**

71. The sum of the first n positive integers is $n(n + 1)/2$.

72. If f is continuous and nonnegative on $[a, b]$, then the limits as $n \to \infty$ of its lower sum $s(n)$ and upper sum $S(n)$ both exist and are equal.

73. Writing Use the figure to write a short paragraph explaining why the formula $1 + 2 + \cdot \cdot \cdot + n = \frac{1}{2}n(n + 1)$ is valid for all positive integers n.

Figure for 73

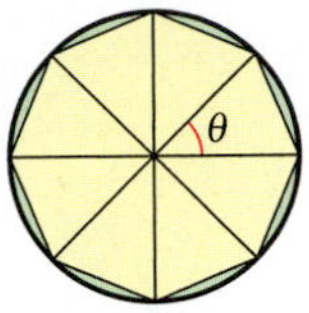

Figure for 74

74. Graphical Reasoning Consider an n-sided regular polygon inscribed in a circle of radius r. Join the vertices of the polygon to the center of the circle, forming n congruent triangles (see figure).

(a) Determine the central angle θ in terms of n.

(b) Show that the area of each triangle is $\frac{1}{2}r^2 \sin \theta$.

(c) Let A_n be the sum of the areas of the n triangles. Find $\lim_{n\to\infty} A_n$.

75. Building Blocks A child places n cubic building blocks in a row to form the base of a triangular design (see figure). Each successive row contains two fewer blocks than the preceding row. Find a formula for the number of blocks used in the design. (*Hint:* The number of building blocks in the design depends on whether n is odd or even.)

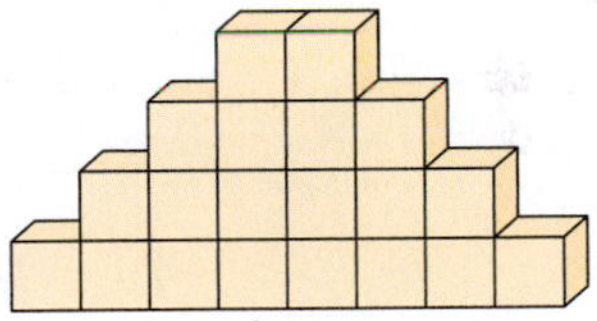

n is even.

76. Proof Prove each formula by mathematical induction. (You may need to review the method of proof by induction from a precalculus text.)

(a) $\sum_{i=1}^{n} 2i = n(n + 1)$ (b) $\sum_{i=1}^{n} i^3 = \frac{n^2(n + 1)^2}{4}$

PUTNAM EXAM CHALLENGE

77. A dart, thrown at random, hits a square target. Assuming that any two parts of the target of equal area are equally likely to be hit, find the probability that the point hit is nearer to the center than to any edge. Write your answer in the form $(a\sqrt{b} + c)/d$, where a, b, c, and d are integers.

This problem was composed by the Committee on the Putnam Prize Competition.

4.3 Riemann Sums and Definite Integrals

- **Understand the definition of a Riemann sum.**
- **Evaluate a definite integral using limits.**
- **Evaluate a definite integral using properties of definite integrals.**

Riemann Sums

In the definition of area given in Section 4.2, the partitions have subintervals of *equal width*. This was done only for computational convenience. The next example shows that it is not necessary to have subintervals of equal width.

EXAMPLE 1 A Partition with Subintervals of Unequal Widths

Consider the region bounded by the graph of

$$f(x) = \sqrt{x}$$

and the x-axis for $0 \le x \le 1$, as shown in Figure 4.18. Evaluate the limit

$$\lim_{n\to\infty} \sum_{i=1}^{n} f(c_i)\,\Delta x_i$$

where c_i is the right endpoint of the partition given by $c_i = i^2/n^2$ and Δx_i is the width of the ith interval.

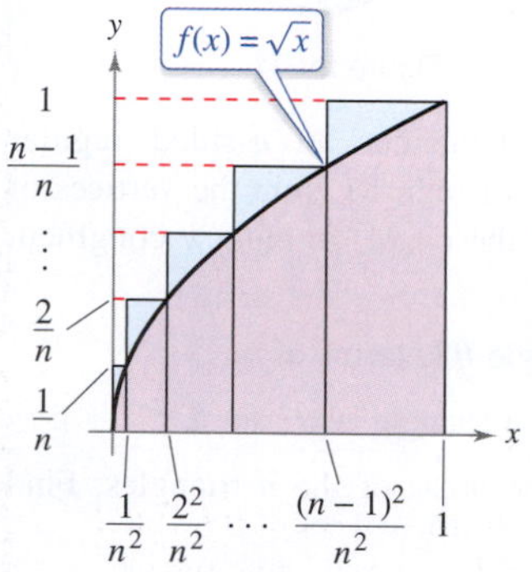

The subintervals do not have equal widths.
Figure 4.18

Solution The width of the ith interval is

$$\begin{aligned}\Delta x_i &= \frac{i^2}{n^2} - \frac{(i-1)^2}{n^2}\\ &= \frac{i^2 - i^2 + 2i - 1}{n^2}\\ &= \frac{2i-1}{n^2}.\end{aligned}$$

So, the limit is

$$\begin{aligned}\lim_{n\to\infty} \sum_{i=1}^{n} f(c_i)\,\Delta x_i &= \lim_{n\to\infty} \sum_{i=1}^{n} \sqrt{\frac{i^2}{n^2}}\left(\frac{2i-1}{n^2}\right)\\ &= \lim_{n\to\infty} \frac{1}{n^3} \sum_{i=1}^{n} (2i^2 - i)\\ &= \lim_{n\to\infty} \frac{1}{n^3}\left[2\left(\frac{n(n+1)(2n+1)}{6}\right) - \frac{n(n+1)}{2}\right]\\ &= \lim_{n\to\infty} \frac{4n^3 + 3n^2 - n}{6n^3}\\ &= \frac{2}{3}.\end{aligned}$$

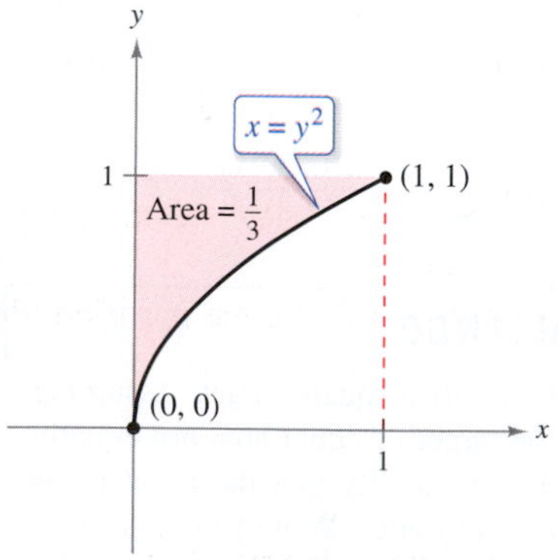

The area of the region bounded by the graph of $x = y^2$ and the y-axis for $0 \le y \le 1$ is $\frac{1}{3}$.
Figure 4.19

From Example 7 in Section 4.2, you know that the region shown in Figure 4.19 has an area of $\frac{1}{3}$. Because the square bounded by $0 \le x \le 1$ and $0 \le y \le 1$ has an area of 1, you can conclude that the area of the region shown in Figure 4.18 has an area of $\frac{2}{3}$. This agrees with the limit found in Example 1, even though that example used a partition having subintervals of unequal widths. The reason this particular partition gave the proper area is that as n increases, the *width of the largest subinterval approaches zero.* This is a key feature of the development of definite integrals.

GEORG FRIEDRICH BERNHARD RIEMANN (1826–1866)

German mathematician Riemann did his most famous work in the areas of non-Euclidean geometry, differential equations, and number theory. It was Riemann's results in physics and mathematics that formed the structure on which Einstein's General Theory of Relativity is based.

See LarsonCalculus.com to read more of this biography.

In Section 4.2, the limit of a sum was used to define the area of a region in the plane. Finding area by this means is only one of *many* applications involving the limit of a sum. A similar approach can be used to determine quantities as diverse as arc lengths, average values, centroids, volumes, work, and surface areas. The next definition is named after Georg Friedrich Bernhard Riemann. Although the definite integral had been defined and used long before Riemann's time, he generalized the concept to cover a broader category of functions.

In the definition of a Riemann sum below, note that the function f has no restrictions other than being defined on the interval $[a, b]$. (In Section 4.2, the function f was assumed to be continuous and nonnegative because you were finding the area under a curve.)

Definition of Riemann Sum

Let f be defined on the closed interval $[a, b]$, and let Δ be a partition of $[a, b]$ given by

$$a = x_0 < x_1 < x_2 < \ . \ . \ . \ < x_{n-1} < x_n = b$$

where Δx_i is the width of the ith subinterval

$$[x_{i-1}, x_i].$$ *i*th subinterval

If c_i is *any* point in the ith subinterval, then the sum

$$\sum_{i=1}^{n} f(c_i)\,\Delta x_i, \quad x_{i-1} \le c_i \le x_i$$

is called a **Riemann sum** of f for the partition Δ. (The sums in Section 4.2 are examples of Riemann sums, but there are more general Riemann sums than those covered there.)

The width of the largest subinterval of a partition Δ is the **norm** of the partition and is denoted by $\|\Delta\|$. If every subinterval is of equal width, then the partition is **regular** and the norm is denoted by

$$\|\Delta\| = \Delta x = \frac{b-a}{n}.$$ Regular partition

For a general partition, the norm is related to the number of subintervals of $[a, b]$ in the following way.

$$\frac{b-a}{\|\Delta\|} \le n$$ General partition

So, the number of subintervals in a partition approaches infinity as the norm of the partition approaches 0. That is, $\|\Delta\| \to 0$ implies that $n \to \infty$.

The converse of this statement is not true. For example, let Δ_n be the partition of the interval $[0, 1]$ given by

$$0 < \frac{1}{2^n} < \frac{1}{2^{n-1}} < \cdot \cdot \cdot < \frac{1}{8} < \frac{1}{4} < \frac{1}{2} < 1.$$

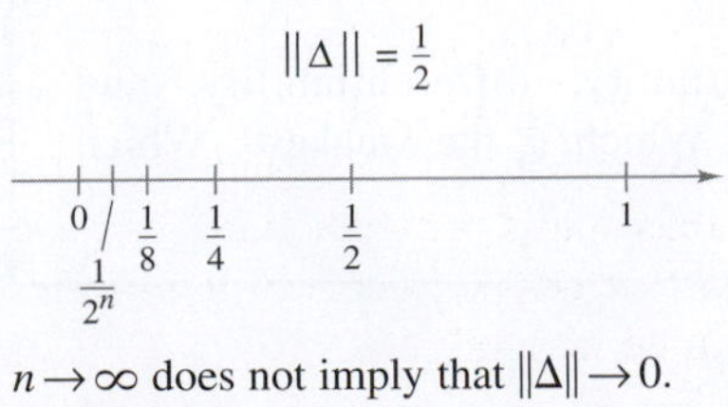

$n \to \infty$ does not imply that $\|\Delta\| \to 0$.
Figure 4.20

As shown in Figure 4.20, for any positive value of n, the norm of the partition Δ_n is $\frac{1}{2}$. So, letting n approach infinity does not force $\|\Delta\|$ to approach 0. In a regular partition, however, the statements

$$\|\Delta\| \to 0 \quad \text{and} \quad n \to \infty$$

are equivalent.

INTERFOTO/Alamy

Definite Integrals

FOR FURTHER INFORMATION For insight into the history of the definite integral, see the article "The Evolution of Integration" by A. Shenitzer and J. Stepr$\bar{a}$ns in *The American Mathematical Monthly.* To view this article, go to *MathArticles.com.*

To define the definite integral, consider the limit

$$\lim_{\|\Delta\| \to 0} \sum_{i=1}^{n} f(c_i)\,\Delta x_i = L.$$

To say that this limit exists means there exists a real number L such that for each $\varepsilon > 0$, there exists a $\delta > 0$ such that for every partition with $\|\Delta\| < \delta$, it follows that

$$\left| L - \sum_{i=1}^{n} f(c_i)\,\Delta x_i \right| < \varepsilon$$

regardless of the choice of c_i in the ith subinterval of each partition Δ.

Definition of Definite Integral

If f is defined on the closed interval $[a, b]$ and the limit of Riemann sums over partitions Δ

$$\lim_{\|\Delta\| \to 0} \sum_{i=1}^{n} f(c_i)\,\Delta x_i$$

exists (as described above), then f is said to be **integrable** on $[a, b]$ and the limit is denoted by

$$\lim_{\|\Delta\| \to 0} \sum_{i=1}^{n} f(c_i)\,\Delta x_i = \int_a^b f(x)\,dx.$$

The limit is called the **definite integral** of f from a to b. The number a is the **lower limit** of integration, and the number b is the **upper limit** of integration.

REMARK Later in this chapter, you will learn convenient methods for calculating $\int_a^b f(x)\,dx$ for continuous functions. For now, you must use the limit definition.

It is not a coincidence that the notation for definite integrals is similar to that used for indefinite integrals. You will see why in the next section when the Fundamental Theorem of Calculus is introduced. For now, it is important to see that definite integrals and indefinite integrals are different concepts. A definite integral is a *number*, whereas an indefinite integral is a *family of functions.*

Though Riemann sums were defined for functions with very few restrictions, a sufficient condition for a function f to be integrable on $[a, b]$ is that it is continuous on $[a, b]$. A proof of this theorem is beyond the scope of this text.

THEOREM 4.4 Continuity Implies Integrability

If a function f is continuous on the closed interval $[a, b]$, then f is integrable on $[a, b]$. That is, $\int_a^b f(x)\,dx$ exists.

Exploration

The Converse of Theorem 4.4 Is the converse of Theorem 4.4 true? That is, when a function is integrable, does it have to be continuous? Explain your reasoning and give examples.

Describe the relationships among continuity, differentiability, and integrability. Which is the strongest condition? Which is the weakest? Which conditions imply other conditions?

EXAMPLE 2 Evaluating a Definite Integral as a Limit

Evaluate the definite integral $\int_{-2}^{1} 2x\, dx$.

Solution The function $f(x) = 2x$ is integrable on the interval $[-2, 1]$ because it is continuous on $[-2, 1]$. Moreover, the definition of integrability implies that any partition whose norm approaches 0 can be used to determine the limit. For computational convenience, define Δ by subdividing $[-2, 1]$ into n subintervals of equal width

$$\Delta x_i = \Delta x = \frac{b-a}{n} = \frac{3}{n}.$$

Choosing c_i as the right endpoint of each subinterval produces

$$c_i = a + i(\Delta x) = -2 + \frac{3i}{n}.$$

So, the definite integral is

$$\begin{aligned}
\int_{-2}^{1} 2x\, dx &= \lim_{\|\Delta\|\to 0} \sum_{i=1}^{n} f(c_i)\, \Delta x_i \\
&= \lim_{n\to\infty} \sum_{i=1}^{n} f(c_i)\, \Delta x \\
&= \lim_{n\to\infty} \sum_{i=1}^{n} 2\left(-2 + \frac{3i}{n}\right)\left(\frac{3}{n}\right) \\
&= \lim_{n\to\infty} \frac{6}{n} \sum_{i=1}^{n} \left(-2 + \frac{3i}{n}\right) \\
&= \lim_{n\to\infty} \frac{6}{n}\left(-2\sum_{i=1}^{n} 1 + \frac{3}{n}\sum_{i=1}^{n} i\right) \\
&= \lim_{n\to\infty} \frac{6}{n}\left\{-2n + \frac{3}{n}\left[\frac{n(n+1)}{2}\right]\right\} \\
&= \lim_{n\to\infty} \left(-12 + 9 + \frac{9}{n}\right) \\
&= -3.
\end{aligned}$$

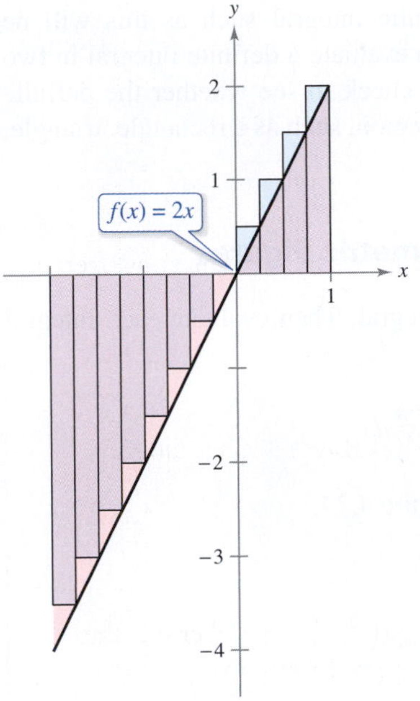

Because the definite integral is negative, it does not represent the area of the region.
Figure 4.21

Because the definite integral in Example 2 is negative, it *does not* represent the area of the region shown in Figure 4.21. Definite integrals can be positive, negative, or zero. For a definite integral to be interpreted as an area (as defined in Section 4.2), the function f must be continuous and nonnegative on $[a, b]$, as stated in the next theorem. The proof of this theorem is straightforward—you simply use the definition of area given in Section 4.2, because it is a Riemann sum.

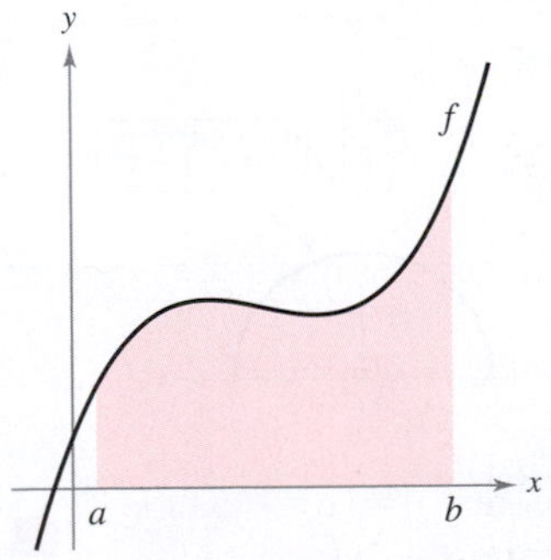

You can use a definite integral to find the area of the region bounded by the graph of f, the x-axis, $x = a$, and $x = b$.
Figure 4.22

THEOREM 4.5 The Definite Integral as the Area of a Region

If f is continuous and nonnegative on the closed interval $[a, b]$, then the area of the region bounded by the graph of f, the x-axis, and the vertical lines $x = a$ and $x = b$ is

$$\text{Area} = \int_{a}^{b} f(x)\, dx.$$

(See Figure 4.22.)

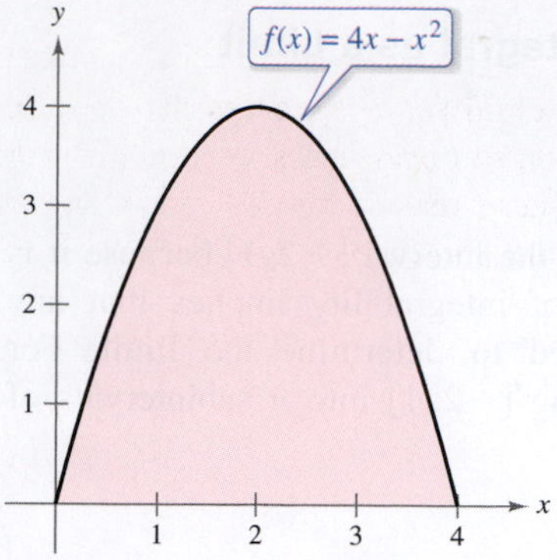

$\text{Area} = \int_0^4 (4x - x^2)\,dx$

Figure 4.23

As an example of Theorem 4.5, consider the region bounded by the graph of

$$f(x) = 4x - x^2$$

and the x-axis, as shown in Figure 4.23. Because f is continuous and nonnegative on the closed interval $[0, 4]$, the area of the region is

$$\text{Area} = \int_0^4 (4x - x^2)\,dx.$$

A straightforward technique for evaluating a definite integral such as this will be discussed in Section 4.4. For now, however, you can evaluate a definite integral in two ways—you can use the limit definition *or* you can check to see whether the definite integral represents the area of a common geometric region, such as a rectangle, triangle, or semicircle.

EXAMPLE 3 Areas of Common Geometric Figures

Sketch the region corresponding to each definite integral. Then evaluate each integral using a geometric formula.

a. $\int_1^3 4\,dx$ **b.** $\int_0^3 (x + 2)\,dx$ **c.** $\int_{-2}^2 \sqrt{4 - x^2}\,dx$

Solution A sketch of each region is shown in Figure 4.24.

a. This region is a rectangle of height 4 and width 2.

$$\int_1^3 4\,dx = (\text{Area of rectangle}) = 4(2) = 8$$

b. This region is a trapezoid with an altitude of 3 and parallel bases of lengths 2 and 5. The formula for the area of a trapezoid is $\frac{1}{2}h(b_1 + b_2)$.

$$\int_0^3 (x + 2)\,dx = (\text{Area of trapezoid}) = \frac{1}{2}(3)(2 + 5) = \frac{21}{2}$$

c. This region is a semicircle of radius 2. The formula for the area of a semicircle is $\frac{1}{2}\pi r^2$.

$$\int_{-2}^2 \sqrt{4 - x^2}\,dx = (\text{Area of semicircle}) = \frac{1}{2}\pi(2^2) = 2\pi$$

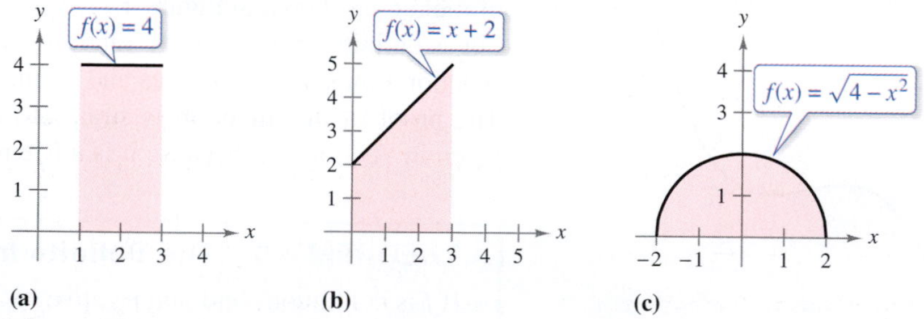

Figure 4.24

The variable of integration in a definite integral is sometimes called a *dummy variable* because it can be replaced by any other variable without changing the value of the integral. For instance, the definite integrals

$$\int_0^3 (x + 2)\,dx \quad \text{and} \quad \int_0^3 (t + 2)\,dt$$

have the same value.

Properties of Definite Integrals

The definition of the definite integral of f on the interval $[a, b]$ specifies that $a < b$. Now, however, it is convenient to extend the definition to cover cases in which $a = b$ or $a > b$. Geometrically, the next two definitions seem reasonable. For instance, it makes sense to define the area of a region of zero width and finite height to be 0.

AP* Tips

Understanding the properties of integrals will help you develop graphical interpretation skills emphasized on the AP Exam.

Definitions of Two Special Definite Integrals

1. If f is defined at $x = a$, then $\int_a^a f(x)\,dx = 0$.

2. If f is integrable on $[a, b]$, then $\int_b^a f(x)\,dx = -\int_a^b f(x)\,dx$.

EXAMPLE 4 **Evaluating Definite Integrals**

See LarsonCalculus.com for an interactive version of this type of example.

Evaluate each definite integral.

a. $\int_\pi^\pi \sin x\,dx$ **b.** $\int_3^0 (x + 2)\,dx$

Solution

a. Because the sine function is defined at $x = \pi$, and the upper and lower limits of integration are equal, you can write

$$\int_\pi^\pi \sin x\,dx = 0.$$

b. The integral $\int_3^0 (x + 2)\,dx$ is the same as that given in Example 3(b) except that the upper and lower limits are interchanged. Because the integral in Example 3(b) has a value of $\frac{21}{2}$, you can write

$$\int_3^0 (x + 2)\,dx = -\int_0^3 (x + 2)\,dx = -\frac{21}{2}.$$

In Figure 4.25, the larger region can be divided at $x = c$ into two subregions whose intersection is a line segment. Because the line segment has zero area, it follows that the area of the larger region is equal to the sum of the areas of the two smaller regions.

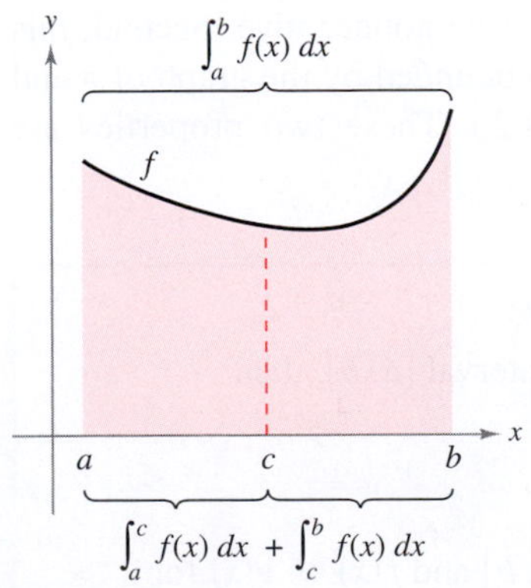

Figure 4.25

THEOREM 4.6 Additive Interval Property

If f is integrable on the three closed intervals determined by a, b, and c, then

$$\int_a^b f(x)\,dx = \int_a^c f(x)\,dx + \int_c^b f(x)\,dx.$$

EXAMPLE 5 **Using the Additive Interval Property**

$$\int_{-1}^1 |x|\,dx = \int_{-1}^0 -x\,dx + \int_0^1 x\,dx \quad \text{Theorem 4.6}$$

$$= \frac{1}{2} + \frac{1}{2} \quad \text{Area of a triangle}$$

$$= 1$$

Because the definite integral is defined as the limit of a sum, it inherits the properties of summation given at the top of page 255.

THEOREM 4.7 Properties of Definite Integrals

If f and g are integrable on $[a, b]$ and k is a constant, then the functions kf and $f \pm g$ are integrable on $[a, b]$, and

1. $\displaystyle\int_a^b kf(x)\,dx = k\int_a^b f(x)\,dx$

2. $\displaystyle\int_a^b \left[f(x) \pm g(x)\right] dx = \int_a^b f(x)\,dx \pm \int_a^b g(x)\,dx.$

REMARK Property 2 of Theorem 4.7 can be extended to cover any finite number of functions (see Example 6).

EXAMPLE 6 Evaluation of a Definite Integral

Evaluate $\displaystyle\int_1^3 (-x^2 + 4x - 3)\,dx$ using each of the following values.

$$\int_1^3 x^2\,dx = \frac{26}{3}, \qquad \int_1^3 x\,dx = 4, \qquad \int_1^3 dx = 2$$

Solution

$$\begin{aligned}\int_1^3 (-x^2 + 4x - 3)\,dx &= \int_1^3 (-x^2)\,dx + \int_1^3 4x\,dx + \int_1^3 (-3)\,dx\\ &= -\int_1^3 x^2\,dx + 4\int_1^3 x\,dx - 3\int_1^3 dx\\ &= -\left(\frac{26}{3}\right) + 4(4) - 3(2)\\ &= \frac{4}{3}\end{aligned}$$

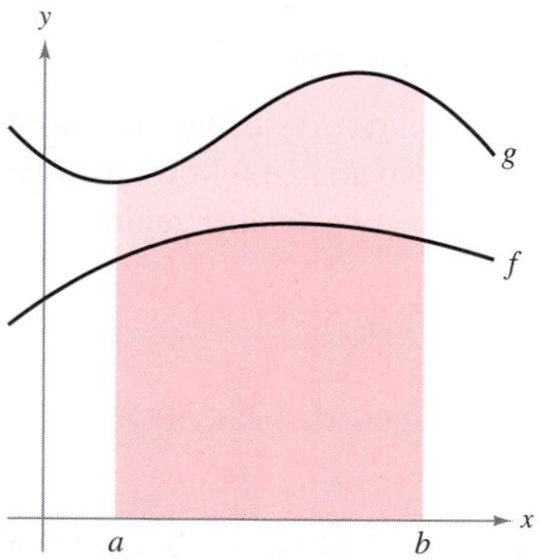

$\displaystyle\int_a^b f(x)\,dx \le \int_a^b g(x)\,dx$
Figure 4.26

If f and g are continuous on the closed interval $[a, b]$ and $0 \le f(x) \le g(x)$ for $a \le x \le b$, then the following properties are true. First, the area of the region bounded by the graph of f and the x-axis (between a and b) must be nonnegative. Second, this area must be less than or equal to the area of the region bounded by the graph of g and the x-axis (between a and b), as shown in Figure 4.26. These two properties are generalized in Theorem 4.8.

THEOREM 4.8 Preservation of Inequality

1. If f is integrable and nonnegative on the closed interval $[a, b]$, then

$$0 \le \int_a^b f(x)\,dx.$$

2. If f and g are integrable on the closed interval $[a, b]$ and $f(x) \le g(x)$ for every x in $[a, b]$, then

$$\int_a^b f(x)\,dx \le \int_a^b g(x)\,dx.$$

A proof of this theorem is given in Appendix A.

See LarsonCalculus.com for Bruce Edwards's video of this proof.

4.3 Exercises

See CalcChat.com for tutorial help and worked-out solutions to odd-numbered exercises.

Evaluating a Limit **In Exercises 1 and 2, use Example 1 as a model to evaluate the limit**

$$\lim_{n\to\infty} \sum_{i=1}^{n} f(c_i)\,\Delta x_i$$

over the region bounded by the graphs of the equations.

1. $f(x) = \sqrt{x}, \quad y = 0, \quad x = 0, \quad x = 3$

$\left(\textit{Hint: } \text{Let } c_i = \dfrac{3i^2}{n^2}.\right)$

2. $f(x) = \sqrt[3]{x}, \quad y = 0, \quad x = 0, \quad x = 1$

$\left(\textit{Hint: } \text{Let } c_i = \dfrac{i^3}{n^3}.\right)$

Evaluating a Definite Integral as a Limit **In Exercises 3–8, evaluate the definite integral by the limit definition.**

3. $\displaystyle\int_2^6 8\,dx$

4. $\displaystyle\int_{-2}^3 x\,dx$

5. $\displaystyle\int_{-1}^1 x^3\,dx$

6. $\displaystyle\int_1^4 4x^2\,dx$

7. $\displaystyle\int_1^2 (x^2 + 1)\,dx$

8. $\displaystyle\int_{-2}^1 (2x^2 + 3)\,dx$

Writing a Limit as a Definite Integral **In Exercises 9–12, write the limit as a definite integral on the interval [a, b], where c_i is any point in the ith subinterval.**

	Limit	Interval
9.	$\displaystyle\lim_{\|\Delta\|\to 0} \sum_{i=1}^{n} (3c_i + 10)\,\Delta x_i$	$[-1, 5]$
10.	$\displaystyle\lim_{\|\Delta\|\to 0} \sum_{i=1}^{n} 6c_i(4 - c_i)^2\,\Delta x_i$	$[0, 4]$
11.	$\displaystyle\lim_{\|\Delta\|\to 0} \sum_{i=1}^{n} \sqrt{c_i^2 + 4}\,\Delta x_i$	$[0, 3]$
12.	$\displaystyle\lim_{\|\Delta\|\to 0} \sum_{i=1}^{n} \left(\frac{3}{c_i^2}\right)\Delta x_i$	$[1, 3]$

Writing a Definite Integral **In Exercises 13–22, set up a definite integral that yields the area of the region. (Do not evaluate the integral.)**

13. $f(x) = 5$

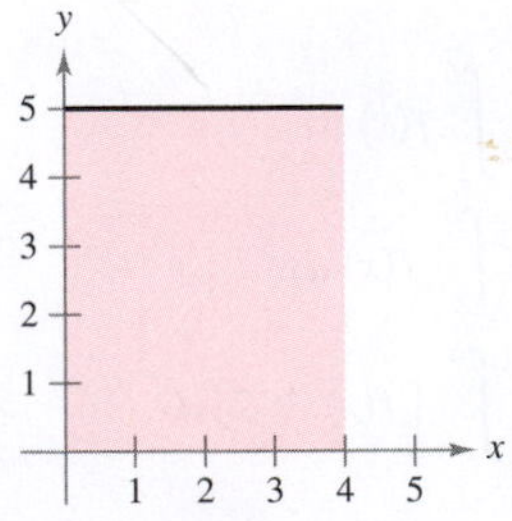

14. $f(x) = 6 - 3x$

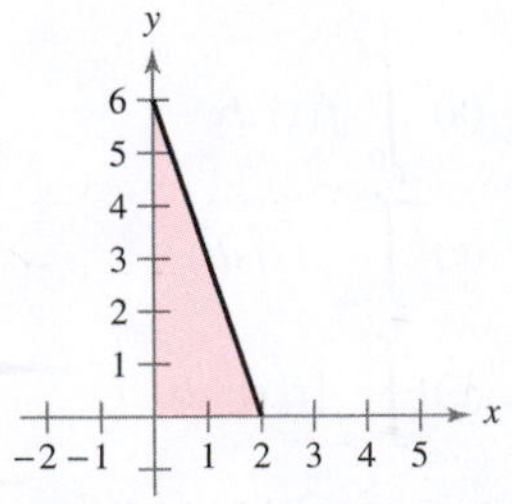

15. $f(x) = 4 - |x|$

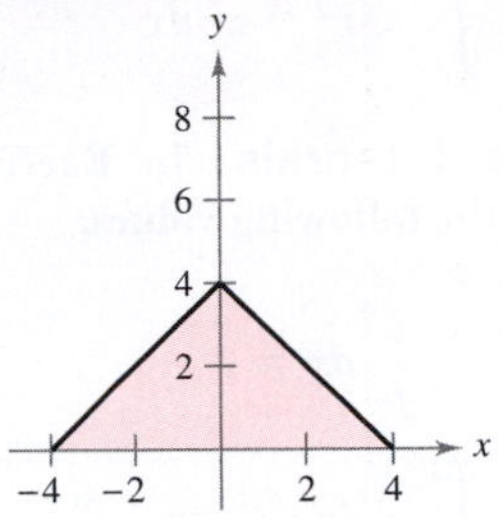

16. $f(x) = x^2$

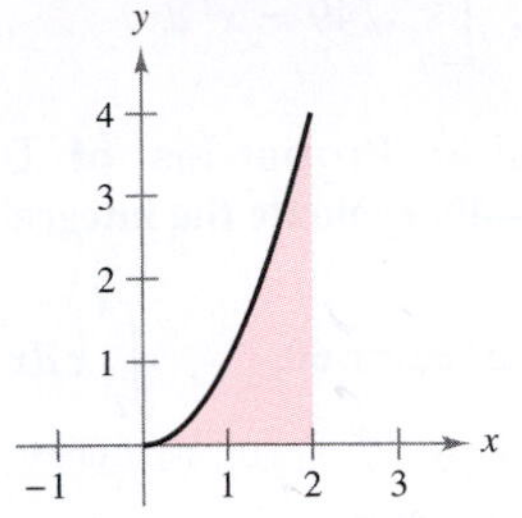

17. $f(x) = 25 - x^2$

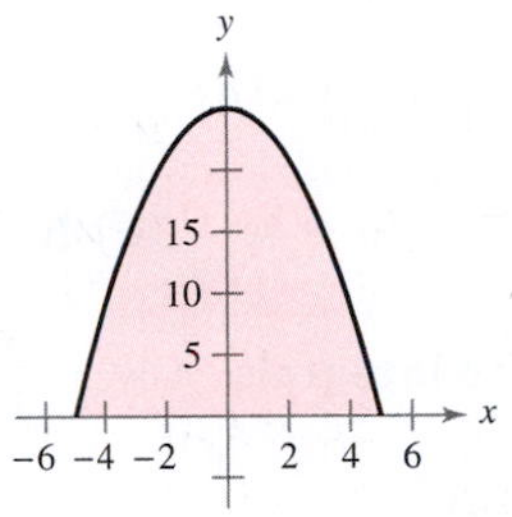

18. $f(x) = \dfrac{4}{x^2 + 2}$

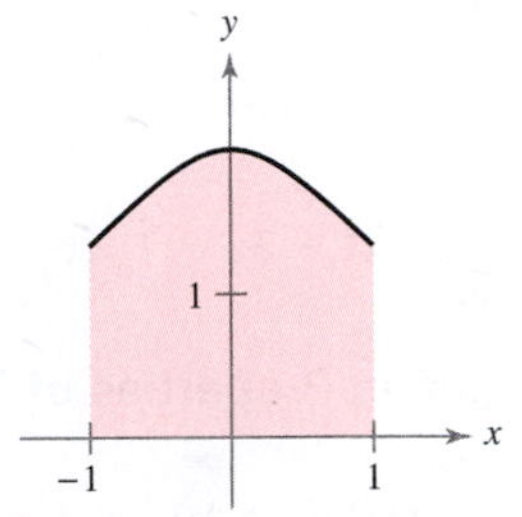

19. $f(x) = \cos x$

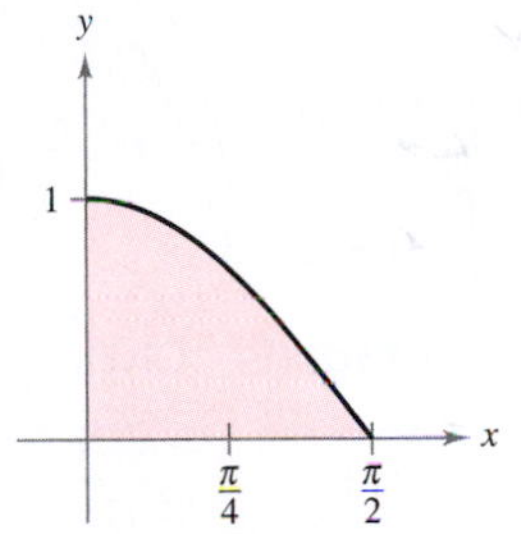

20. $f(x) = \tan x$

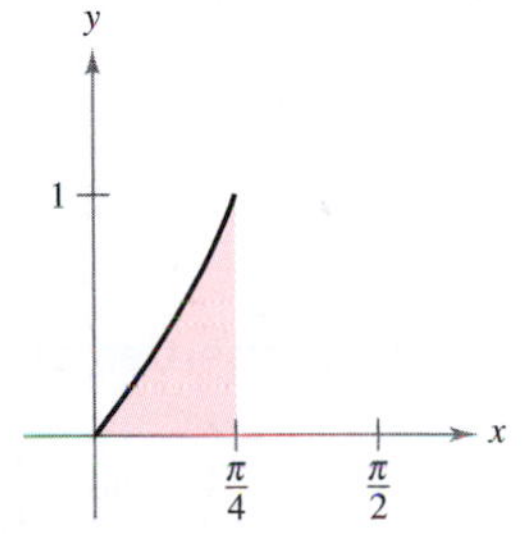

21. $g(y) = y^3$

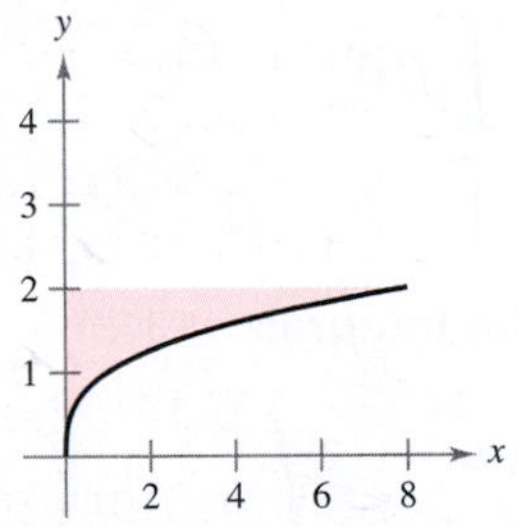

22. $f(y) = (y - 2)^2$

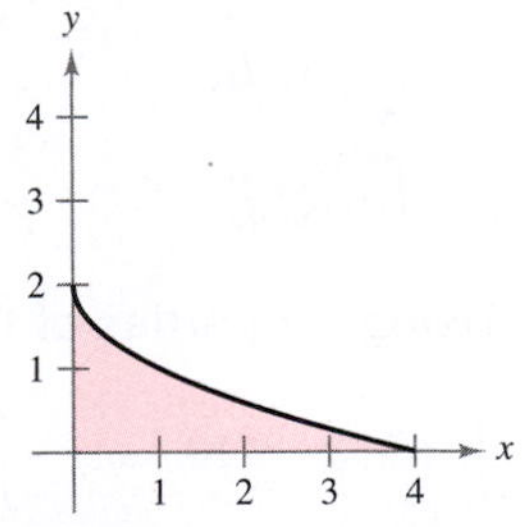

Evaluating a Definite Integral Using a Geometric Formula **In Exercises 23–32, sketch the region whose area is given by the definite integral. Then use a geometric formula to evaluate the integral ($a > 0, r > 0$).**

23. $\displaystyle\int_0^3 4\,dx$

24. $\displaystyle\int_{-4}^6 6\,dx$

25. $\displaystyle\int_0^4 x\,dx$

26. $\displaystyle\int_0^8 \frac{x}{4}\,dx$

27. $\int_0^2 (3x + 4)\, dx$ **28.** $\int_0^3 (8 - 2x)\, dx$

29. $\int_{-1}^1 (1 - |x|)\, dx$ **30.** $\int_{-a}^a (a - |x|)\, dx$

31. $\int_{-7}^7 \sqrt{49 - x^2}\, dx$ **32.** $\int_{-r}^r \sqrt{r^2 - x^2}\, dx$

Using Properties of Definite Integrals In Exercises 33–40, evaluate the integral using the following values.

$$\int_2^4 x^3\, dx = 60, \quad \int_2^4 x\, dx = 6, \quad \int_2^4 dx = 2$$

33. $\int_4^2 x\, dx$ **34.** $\int_2^2 x^3\, dx$

35. $\int_2^4 8x\, dx$ **36.** $\int_2^4 25\, dx$

37. $\int_2^4 (x - 9)\, dx$ **38.** $\int_2^4 (x^3 + 4)\, dx$

39. $\int_2^4 \left(\frac{1}{2}x^3 - 3x + 2\right) dx$ **40.** $\int_2^4 (10 + 4x - 3x^3)\, dx$

41. Using Properties of Definite Integrals Given

$$\int_0^5 f(x)\, dx = 10 \quad \text{and} \quad \int_5^7 f(x)\, dx = 3$$

evaluate

(a) $\int_0^7 f(x)\, dx.$ (b) $\int_5^0 f(x)\, dx.$

(c) $\int_5^5 f(x)\, dx.$ (d) $\int_0^5 3f(x)\, dx.$

42. Using Properties of Definite Integrals Given

$$\int_0^3 f(x)\, dx = 4 \quad \text{and} \quad \int_3^6 f(x)\, dx = -1$$

evaluate

(a) $\int_0^6 f(x)\, dx.$ (b) $\int_6^3 f(x)\, dx.$

(c) $\int_3^3 f(x)\, dx.$ (d) $\int_3^6 -5f(x)\, dx.$

43. Using Properties of Definite Integrals Given

$$\int_2^6 f(x)\, dx = 10 \quad \text{and} \quad \int_2^6 g(x)\, dx = -2$$

evaluate

(a) $\int_2^6 [f(x) + g(x)]\, dx.$ (b) $\int_2^6 [g(x) - f(x)]\, dx.$

(c) $\int_2^6 2g(x)\, dx.$ (d) $\int_2^6 3f(x)\, dx.$

44. Using Properties of Definite Integrals Given

$$\int_{-1}^1 f(x)\, dx = 0 \quad \text{and} \quad \int_0^1 f(x)\, dx = 5$$

evaluate

(a) $\int_{-1}^0 f(x)\, dx.$ (b) $\int_0^1 f(x)\, dx - \int_{-1}^0 f(x)\, dx.$

(c) $\int_{-1}^1 3f(x)\, dx.$ (d) $\int_0^1 3f(x)\, dx.$

45. Estimating a Definite Integral Use the table of values to find lower and upper estimates of

$$\int_0^{10} f(x)\, dx.$$

Assume that f is a decreasing function.

x	0	2	4	6	8	10
$f(x)$	32	24	12	−4	−20	−36

46. Estimating a Definite Integral Use the table of values to estimate

$$\int_0^6 f(x)\, dx.$$

Use three equal subintervals and the (a) left endpoints, (b) right endpoints, and (c) midpoints. When f is an increasing function, how does each estimate compare with the actual value? Explain your reasoning.

x	0	1	2	3	4	5	6
$f(x)$	−6	0	8	18	30	50	80

47. Think About It The graph of f consists of line segments and a semicircle, as shown in the figure. Evaluate each definite integral by using geometric formulas.

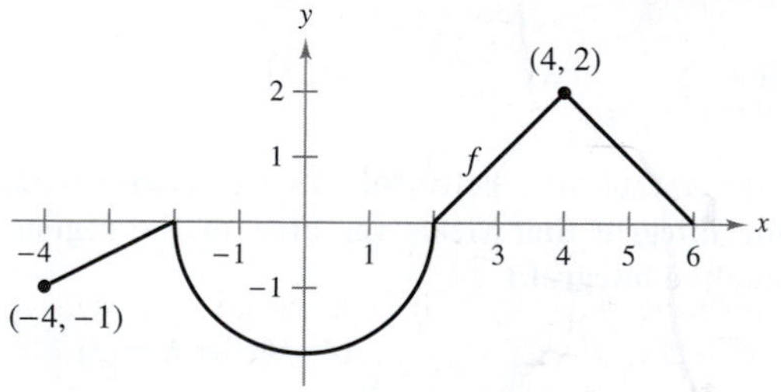

(a) $\int_0^2 f(x)\, dx$ (b) $\int_2^6 f(x)\, dx$

(c) $\int_{-4}^2 f(x)\, dx$ (d) $\int_{-4}^6 f(x)\, dx$

(e) $\int_{-4}^6 |f(x)|\, dx$ (f) $\int_{-4}^6 [f(x) + 2]\, dx$

48. Think About It The graph of f consists of line segments, as shown in the figure. Evaluate each definite integral by using geometric formulas.

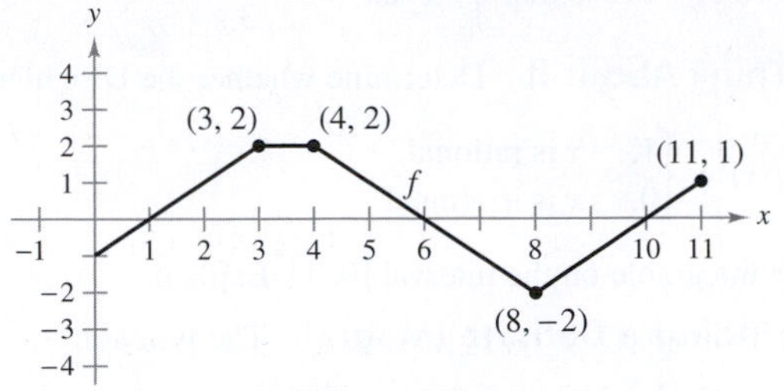

(a) $\int_0^1 -f(x)\,dx$ (b) $\int_3^4 3f(x)\,dx$

(c) $\int_0^7 f(x)\,dx$ (d) $\int_5^{11} f(x)\,dx$

(e) $\int_0^{11} f(x)\,dx$ (f) $\int_4^{10} f(x)\,dx$

49. Think About It Consider the function f that is continuous on the interval $[-5, 5]$ and for which

$$\int_0^5 f(x)\,dx = 4.$$

Evaluate each integral.

(a) $\int_0^5 [f(x) + 2]\,dx$ (b) $\int_{-2}^3 f(x + 2)\,dx$

(c) $\int_{-5}^5 f(x)\,dx$ (f is even.) (d) $\int_{-5}^5 f(x)\,dx$ (f is odd.)

50. HOW DO YOU SEE IT? Use the figure to fill in the blank with the symbol <, >, or =. Explain your reasoning.

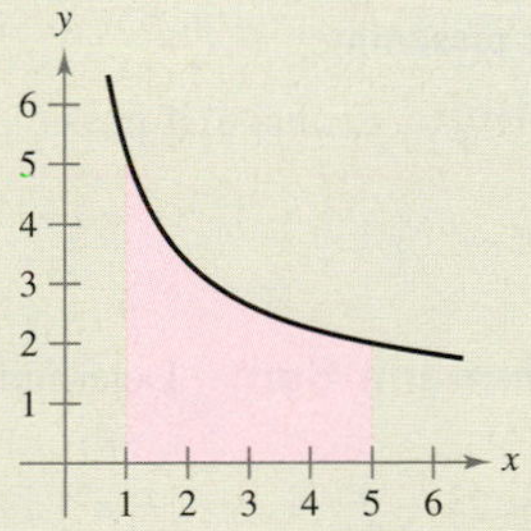

(a) The interval $[1, 5]$ is partitioned into n subintervals of equal width Δx, and x_i is the left endpoint of the ith subinterval.

$$\sum_{i=1}^{n} f(x_i)\,\Delta x \quad \_\_\_ \quad \int_1^5 f(x)\,dx$$

(b) The interval $[1, 5]$ is partitioned into n subintervals of equal width Δx, and x_i is the right endpoint of the ith subinterval.

$$\sum_{i=1}^{n} f(x_i)\,\Delta x \quad \_\_\_ \quad \int_1^5 f(x)\,dx$$

51. Think About It A function f is defined below. Use geometric formulas to find $\int_0^8 f(x)\,dx$.

$$f(x) = \begin{cases} 4, & x < 4 \\ x, & x \geq 4 \end{cases}$$

52. Think About It A function f is defined below. Use geometric formulas to find $\int_0^{12} f(x)\,dx$.

$$f(x) = \begin{cases} 6, & x > 6 \\ -\frac{1}{2}x + 9, & x \leq 6 \end{cases}$$

WRITING ABOUT CONCEPTS

Approximation **In Exercises 53–56, determine which value best approximates the definite integral. Make your selection on the basis of a sketch.**

53. $\int_0^4 \sqrt{x}\,dx$

(a) 5 (b) -3 (c) 10 (d) 2 (e) 8

54. $\int_0^{1/2} 4\cos \pi x\,dx$

(a) 4 (b) $\frac{4}{3}$ (c) 16 (d) 2π (e) -6

55. $\int_0^1 2\sin \pi x\,dx$

(a) 6 (b) $\frac{1}{2}$ (c) 4 (d) $\frac{5}{4}$

56. $\int_0^9 \left(1 + \sqrt{x}\right)dx$

(a) -3 (b) 9 (c) 27 (d) 3

57. Determining Integrability Determine whether the function

$$f(x) = \frac{1}{x - 4}$$

is integrable on the interval $[3, 5]$. Explain.

58. Finding a Function Give an example of a function that is integrable on the interval $[-1, 1]$, but not continuous on $[-1, 1]$.

Finding Values **In Exercises 59–62, find possible values of a and b that make the statement true. If possible, use a graph to support your answer. (There may be more than one correct answer.)**

59. $\int_{-2}^1 f(x)\,dx + \int_1^5 f(x)\,dx = \int_a^b f(x)\,dx$

60. $\int_{-3}^3 f(x)\,dx + \int_3^6 f(x)\,dx - \int_a^b f(x)\,dx = \int_{-1}^6 f(x)\,dx$

61. $\int_a^b \sin x\,dx < 0$

62. $\int_a^b \cos x\,dx = 0$

True or False? **In Exercises 63–68, determine whether the statement is true or false. If it is false, explain why or give an example that shows it is false.**

63. $\int_a^b [f(x) + g(x)]\, dx = \int_a^b f(x)\, dx + \int_a^b g(x)\, dx$

64. $\int_a^b f(x)g(x)\, dx = \left[\int_a^b f(x)\, dx\right]\left[\int_a^b g(x)\, dx\right]$

65. If the norm of a partition approaches zero, then the number of subintervals approaches infinity.

66. If f is increasing on $[a, b]$, then the minimum value of $f(x)$ on $[a, b]$ is $f(a)$.

67. The value of

$$\int_a^b f(x)\, dx$$

must be positive.

68. The value of

$$\int_2^2 \sin(x^2)\, dx$$

is 0.

69. Finding a Riemann Sum Find the Riemann sum for $f(x) = x^2 + 3x$ over the interval $[0, 8]$, where

$$x_0 = 0,\quad x_1 = 1,\quad x_2 = 3,\quad x_3 = 7,\quad \text{and} \quad x_4 = 8$$

and where

$$c_1 = 1,\quad c_2 = 2,\quad c_3 = 5,\quad \text{and} \quad c_4 = 8.$$

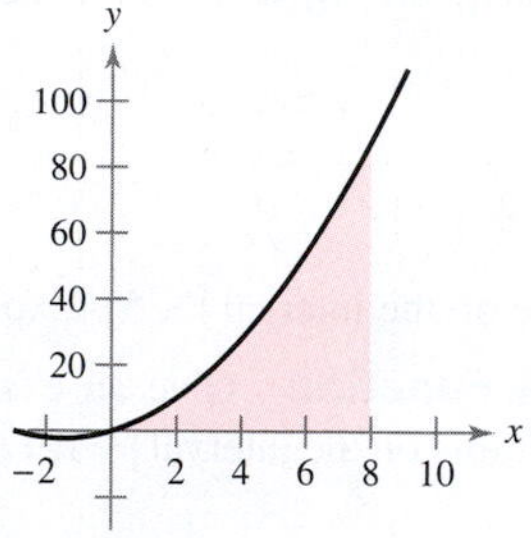

70. Finding a Riemann Sum Find the Riemann sum for $f(x) = \sin x$ over the interval $[0, 2\pi]$, where

$$x_0 = 0,\quad x_1 = \frac{\pi}{4},\quad x_2 = \frac{\pi}{3},\quad x_3 = \pi,\quad \text{and} \quad x_4 = 2\pi,$$

and where

$$c_1 = \frac{\pi}{6},\quad c_2 = \frac{\pi}{3},\quad c_3 = \frac{2\pi}{3},\quad \text{and} \quad c_4 = \frac{3\pi}{2}.$$

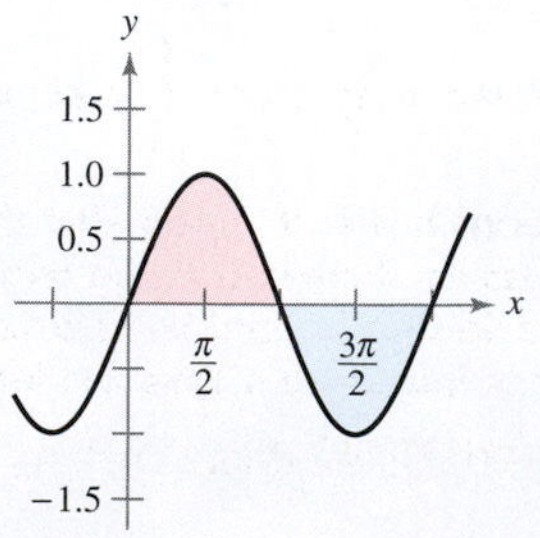

71. Proof Prove that $\int_a^b x\, dx = \dfrac{b^2 - a^2}{2}$.

72. Proof Prove that $\int_a^b x^2\, dx = \dfrac{b^3 - a^3}{3}$.

73. Think About It Determine whether the Dirichlet function

$$f(x) = \begin{cases} 1, & x \text{ is rational} \\ 0, & x \text{ is irrational} \end{cases}$$

is integrable on the interval $[0, 1]$. Explain.

74. Finding a Definite Integral The function

$$f(x) = \begin{cases} 0, & x = 0 \\ \dfrac{1}{x}, & 0 < x \le 1 \end{cases}$$

is defined on $[0, 1]$, as shown in the figure. Show that

$$\int_0^1 f(x)\, dx$$

does not exist. Why doesn't this contradict Theorem 4.4?

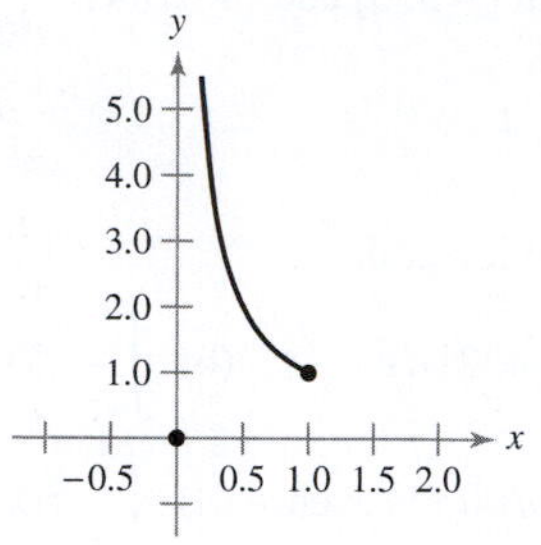

75. Finding Values Find the constants a and b that maximize the value of

$$\int_a^b (1 - x^2)\, dx.$$

Explain your reasoning.

76. Step Function Evaluate, if possible, the integral

$$\int_0^2 [\![x]\!]\, dx.$$

77. Using a Riemann Sum Determine

$$\lim_{n\to\infty} \frac{1}{n^3}[1^2 + 2^2 + 3^2 + \cdots + n^2]$$

by using an appropriate Riemann sum.

PUTNAM EXAM CHALLENGE

78. For each continuous function $f\colon [0, 1] \to \mathbb{R}$, let

$$I(f) = \int_0^1 x^2 f(x)\, dx \quad \text{and} \quad J(x) = \int_0^1 x(f(x))^2\, dx.$$

Find the maximum value of $I(f) - J(f)$ over all such functions f.

4.4 The Fundamental Theorem of Calculus

- **Evaluate a definite integral using the Fundamental Theorem of Calculus.**
- **Understand and use the Mean Value Theorem for Integrals.**
- **Find the average value of a function over a closed interval.**
- **Understand and use the Second Fundamental Theorem of Calculus.**
- **Understand and use the Net Change Theorem.**

The Fundamental Theorem of Calculus

You have now been introduced to the two major branches of calculus: differential calculus (introduced with the tangent line problem) and integral calculus (introduced with the area problem). So far, these two problems might seem unrelated—but there is a very close connection. The connection was discovered independently by Isaac Newton and Gottfried Leibniz and is stated in the **Fundamental Theorem of Calculus.**

Informally, the theorem states that differentiation and (definite) integration are inverse operations, in the same sense that division and multiplication are inverse operations. To see how Newton and Leibniz might have anticipated this relationship, consider the approximations shown in Figure 4.27. The slope of the tangent line was defined using the *quotient* $\Delta y/\Delta x$ (the slope of the secant line). Similarly, the area of a region under a curve was defined using the *product* $\Delta y \Delta x$ (the area of a rectangle). So, at least in the primitive approximation stage, the operations of differentiation and definite integration appear to have an inverse relationship in the same sense that division and multiplication are inverse operations. The Fundamental Theorem of Calculus states that the limit processes (used to define the derivative and definite integral) preserve this inverse relationship.

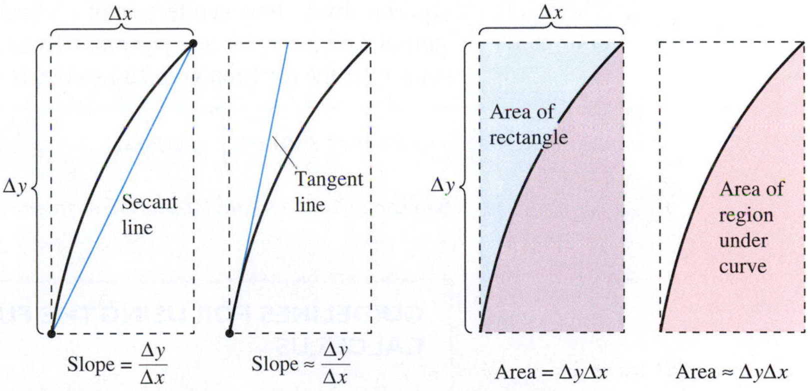

(a) Differentiation **(b)** Definite integration

Differentiation and definite integration have an "inverse" relationship.

Figure 4.27

> **ANTIDIFFERENTIATION AND DEFINITE INTEGRATION**
>
> Throughout this chapter, you have been using the integral sign to denote an antiderivative (a family of functions) and a definite integral (a number).
>
> Antidifferentiation: $\int f(x)\,dx$ Definite integration: $\int_a^b f(x)\,dx$
>
> The use of the same symbol for both operations makes it appear that they are related. In the early work with calculus, however, it was not known that the two operations were related. The symbol $\int$ was first applied to the definite integral by Leibniz and was derived from the letter S. (Leibniz calculated area as an infinite sum, thus, the letter S.)

AP* Tips

The First Fundamental Theorem of Calculus is a foundational concept that is tested throughout the AP Exam by means of analytic, verbal, numerical, and graphical approaches.

THEOREM 4.9 The Fundamental Theorem of Calculus

If a function f is continuous on the closed interval $[a, b]$ and F is an antiderivative of f on the interval $[a, b]$, then

$$\int_a^b f(x)\,dx = F(b) - F(a).$$

Proof The key to the proof is writing the difference $F(b) - F(a)$ in a convenient form. Let Δ be any partition of $[a, b]$.

$$a = x_0 < x_1 < x_2 < \cdot \cdot \cdot < x_{n-1} < x_n = b$$

By pairwise subtraction and addition of like terms, you can write

$$\begin{aligned} F(b) - F(a) &= F(x_n) - F(x_{n-1}) + F(x_{n-1}) - \cdot \cdot \cdot - F(x_1) + F(x_1) - F(x_0) \\ &= \sum_{i=1}^{n} [F(x_i) - F(x_{i-1})]. \end{aligned}$$

By the Mean Value Theorem, you know that there exists a number c_i in the ith subinterval such that

$$F'(c_i) = \frac{F(x_i) - F(x_{i-1})}{x_i - x_{i-1}}.$$

Because $F'(c_i) = f(c_i)$, you can let $\Delta x_i = x_i - x_{i-1}$ and obtain

$$F(b) - F(a) = \sum_{i=1}^{n} f(c_i)\,\Delta x_i.$$

This important equation tells you that by repeatedly applying the Mean Value Theorem, you can always find a collection of c_i's such that the *constant* $F(b) - F(a)$ is a Riemann sum of f on $[a, b]$ for any partition. Theorem 4.4 guarantees that the limit of Riemann sums over the partition with $\|\Delta\| \to 0$ exists. So, taking the limit (as $\|\Delta\| \to 0$) produces

$$F(b) - F(a) = \int_a^b f(x)\,dx.$$

See LarsonCalculus.com for Bruce Edwards's video of this proof. ■

GUIDELINES FOR USING THE FUNDAMENTAL THEOREM OF CALCULUS

1. *Provided you can find* an antiderivative of f, you now have a way to evaluate a definite integral without having to use the limit of a sum.
2. When applying the Fundamental Theorem of Calculus, the notation shown below is convenient.

$$\int_a^b f(x)\,dx = F(x)\Big]_a^b = F(b) - F(a)$$

For instance, to evaluate $\int_1^3 x^3\,dx$, you can write

$$\int_1^3 x^3\,dx = \frac{x^4}{4}\Big]_1^3 = \frac{3^4}{4} - \frac{1^4}{4} = \frac{81}{4} - \frac{1}{4} = 20.$$

3. It is not necessary to include a constant of integration C in the antiderivative.

$$\int_a^b f(x)\,dx = \Big[F(x) + C\Big]_a^b = [F(b) + C] - [F(a) + C] = F(b) - F(a)$$

EXAMPLE 1 Evaluating a Definite Integral

See LarsonCalculus.com for an interactive version of this type of example.

Evaluate each definite integral.

a. $\int_1^2 (x^2 - 3)\, dx$ **b.** $\int_1^4 3\sqrt{x}\, dx$ **c.** $\int_0^{\pi/4} \sec^2 x\, dx$

Solution

a. $\int_1^2 (x^2 - 3)\, dx = \left[\frac{x^3}{3} - 3x\right]_1^2 = \left(\frac{8}{3} - 6\right) - \left(\frac{1}{3} - 3\right) = -\frac{2}{3}$

b. $\int_1^4 3\sqrt{x}\, dx = 3\int_1^4 x^{1/2}\, dx = 3\left[\frac{x^{3/2}}{3/2}\right]_1^4 = 2(4)^{3/2} - 2(1)^{3/2} = 14$

c. $\int_0^{\pi/4} \sec^2 x\, dx = \tan x\Big]_0^{\pi/4} = 1 - 0 = 1$

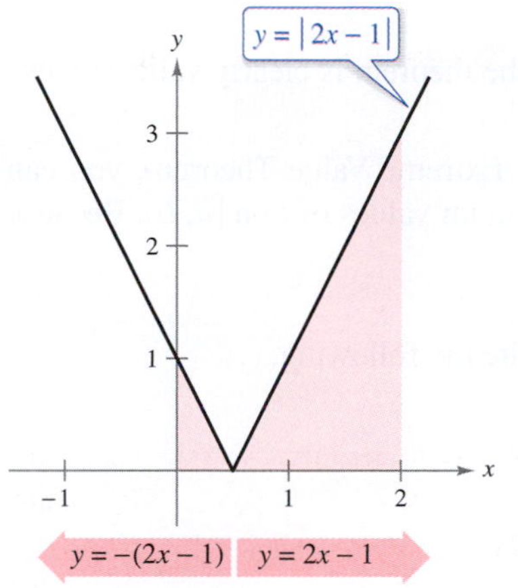

The definite integral of y on $[0, 2]$ is $\frac{5}{2}$.
Figure 4.28

EXAMPLE 2 A Definite Integral Involving Absolute Value

Evaluate $\int_0^2 |2x - 1|\, dx.$

Solution Using Figure 4.28 and the definition of absolute value, you can rewrite the integrand as shown.

$$|2x - 1| = \begin{cases} -(2x - 1), & x < \frac{1}{2} \\ 2x - 1, & x \geq \frac{1}{2} \end{cases}$$

From this, you can rewrite the integral in two parts.

$$\begin{aligned}\int_0^2 |2x - 1|\, dx &= \int_0^{1/2} -(2x - 1)\, dx + \int_{1/2}^2 (2x - 1)\, dx \\ &= \Big[-x^2 + x\Big]_0^{1/2} + \Big[x^2 - x\Big]_{1/2}^2 \\ &= \left(-\frac{1}{4} + \frac{1}{2}\right) - (0 + 0) + (4 - 2) - \left(\frac{1}{4} - \frac{1}{2}\right) \\ &= \frac{5}{2}\end{aligned}$$

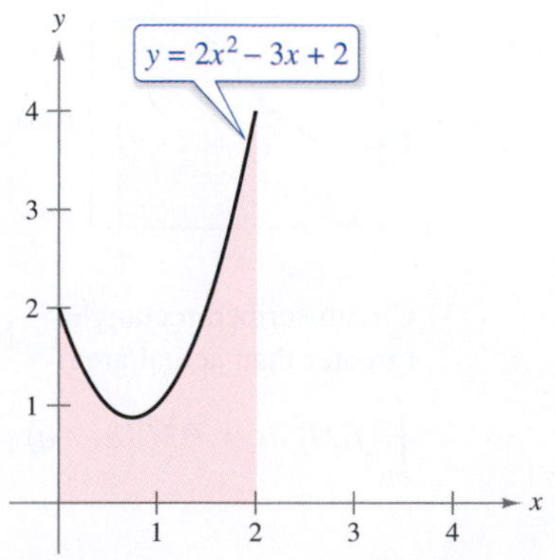

The area of the region bounded by the graph of y, the x-axis, $x = 0$, and $x = 2$ is $\frac{10}{3}$.
Figure 4.29

EXAMPLE 3 Using the Fundamental Theorem to Find Area

Find the area of the region bounded by the graph of

$$y = 2x^2 - 3x + 2$$

the x-axis, and the vertical lines $x = 0$ and $x = 2$, as shown in Figure 4.29.

Solution Note that $y > 0$ on the interval $[0, 2]$.

$$\text{Area} = \int_0^2 (2x^2 - 3x + 2)\, dx \qquad \text{Integrate between } x = 0 \text{ and } x = 2.$$

$$= \left[\frac{2x^3}{3} - \frac{3x^2}{2} + 2x\right]_0^2 \qquad \text{Find antiderivative.}$$

$$= \left(\frac{16}{3} - 6 + 4\right) - (0 - 0 + 0) \qquad \text{Apply Fundamental Theorem.}$$

$$= \frac{10}{3} \qquad \text{Simplify.}$$

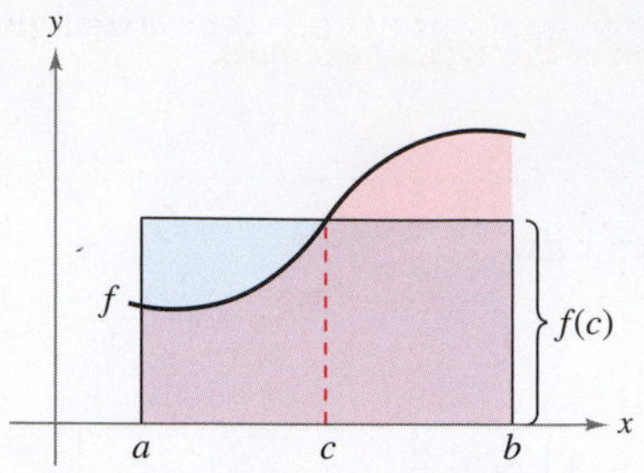

Mean value rectangle:

$$f(c)(b-a)=\int_a^b f(x)\,dx$$

Figure 4.30

The Mean Value Theorem for Integrals

In Section 4.2, you saw that the area of a region under a curve is greater than the area of an inscribed rectangle and less than the area of a circumscribed rectangle. The Mean Value Theorem for Integrals states that somewhere "between" the inscribed and circumscribed rectangles, there is a rectangle whose area is precisely equal to the area of the region under the curve, as shown in Figure 4.30.

> **THEOREM 4.10 Mean Value Theorem for Integrals**
>
> If f is continuous on the closed interval $[a, b]$, then there exists a number c in the closed interval $[a, b]$ such that
>
> $$\int_a^b f(x)\,dx = f(c)(b-a).$$

Proof

Case 1: If f is constant on the interval $[a, b]$, then the theorem is clearly valid because c can be any point in $[a, b]$.

Case 2: If f is not constant on $[a, b]$, then, by the Extreme Value Theorem, you can choose $f(m)$ and $f(M)$ to be the minimum and maximum values of f on $[a, b]$. Because

$$f(m) \le f(x) \le f(M)$$

for all x in $[a, b]$, you can apply Theorem 4.8 to write the following.

$$\int_a^b f(m)\,dx \le \int_a^b f(x)\,dx \le \int_a^b f(M)\,dx$$ See Figure 4.31.

$$f(m)(b-a) \le \int_a^b f(x)\,dx \le f(M)(b-a)$$ Apply Fundamental Theorem.

$$f(m) \le \frac{1}{b-a}\int_a^b f(x)\,dx \le f(M)$$ Divide by $b - a$.

From the third inequality, you can apply the Intermediate Value Theorem to conclude that there exists some c in $[a, b]$ such that

$$f(c) = \frac{1}{b-a}\int_a^b f(x)\,dx \quad \text{or} \quad f(c)(b-a) = \int_a^b f(x)\,dx.$$

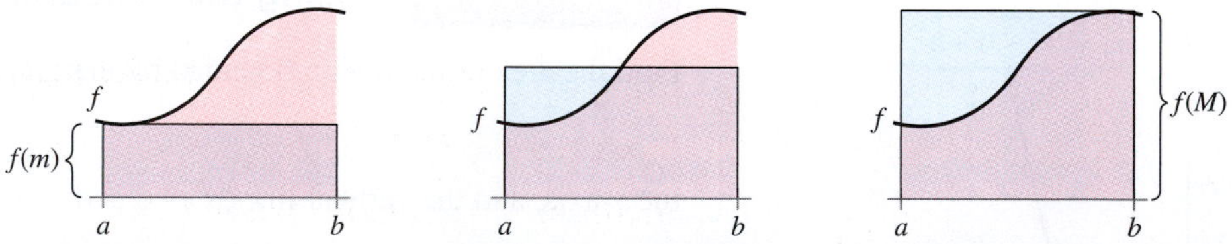

Inscribed rectangle (less than actual area)

$$\int_a^b f(m)\,dx = f(m)(b-a)$$

Mean value rectangle (equal to actual area)

$$\int_a^b f(x)\,dx$$

Circumscribed rectangle (greater than actual area)

$$\int_a^b f(M)\,dx = f(M)(b-a)$$

Figure 4.31

See LarsonCalculus.com for Bruce Edwards's video of this proof. ■

Notice that Theorem 4.10 does not specify how to determine c. It merely guarantees the existence of at least one number c in the interval.

Average Value of a Function

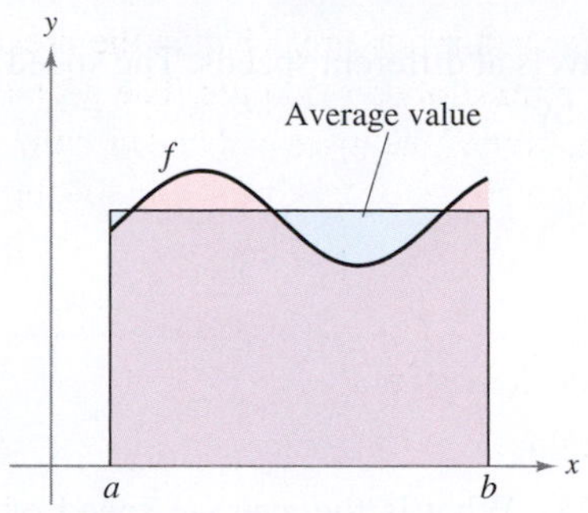

Average value $= \dfrac{1}{b-a}\displaystyle\int_a^b f(x)\,dx$

Figure 4.32

The value of $f(c)$ given in the Mean Value Theorem for Integrals is called the **average value** of f on the interval $[a, b]$.

> **Definition of the Average Value of a Function on an Interval**
>
> If f is integrable on the closed interval $[a, b]$, then the **average value** of f on the interval is
>
> $$\frac{1}{b-a}\int_a^b f(x)\,dx.$$
>
> See Figure 4.32.

To see why the average value of f is defined in this way, partition $[a, b]$ into n subintervals of equal width

$$\Delta x = \frac{b-a}{n}.$$

If c_i is any point in the ith subinterval, then the arithmetic average (or mean) of the function values at the c_i's is

$$a_n = \frac{1}{n}[f(c_1) + f(c_2) + \cdot\cdot\cdot + f(c_n)].$$

Average of $f(c_1), . . . , f(c_n)$

By multiplying and dividing by $(b - a)$, you can write the average as

$$\begin{aligned} a_n &= \frac{1}{n}\sum_{i=1}^{n} f(c_i)\left(\frac{b-a}{b-a}\right) \\ &= \frac{1}{b-a}\sum_{i=1}^{n} f(c_i)\left(\frac{b-a}{n}\right) \\ &= \frac{1}{b-a}\sum_{i=1}^{n} f(c_i)\,\Delta x. \end{aligned}$$

Finally, taking the limit as $n \to \infty$ produces the average value of f on the interval $[a, b]$, as given in the definition above. In Figure 4.32, notice that the area of the region under the graph of f is equal to the area of the rectangle whose height is the average value.

This development of the average value of a function on an interval is only one of many practical uses of definite integrals to represent summation processes. In Chapter 7, you will study other applications, such as volume, arc length, centers of mass, and work.

EXAMPLE 4 Finding the Average Value of a Function

Find the average value of $f(x) = 3x^2 - 2x$ on the interval $[1, 4]$.

Solution The average value is

$$\begin{aligned} \frac{1}{b-a}\int_a^b f(x)\,dx &= \frac{1}{4-1}\int_1^4 (3x^2 - 2x)\,dx \\ &= \frac{1}{3}\Big[x^3 - x^2\Big]_1^4 \\ &= \frac{1}{3}[64 - 16 - (1 - 1)] \\ &= \frac{48}{3} \\ &= 16. \end{aligned}$$

See Figure 4.33.

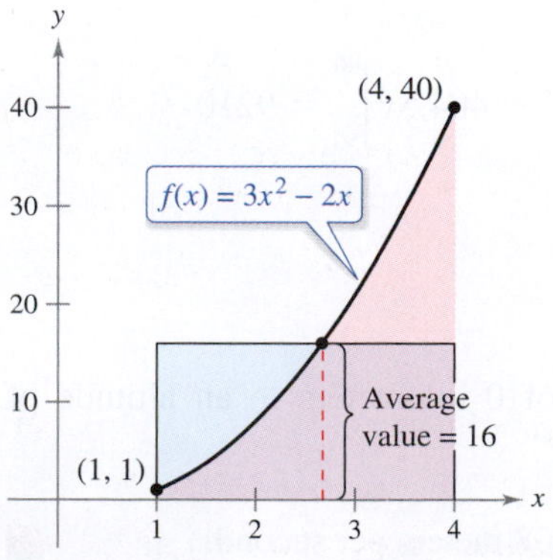

Figure 4.33

The first person to fly at a speed greater than the speed of sound was Charles Yeager. On October 14, 1947, Yeager was clocked at 295.9 meters per second at an altitude of 12.2 kilometers. If Yeager had been flying at an altitude below 11.275 kilometers, this speed would not have "broken the sound barrier." The photo shows an F/A-18F Super Hornet, a supersonic twin-engine strike fighter. A "green Hornet" using a 50/50 mixture of biofuel made from camelina oil became the first U.S. naval tactical aircraft to exceed 1 mach.

EXAMPLE 5 **The Speed of Sound**

At different altitudes in Earth's atmosphere, sound travels at different speeds. The speed of sound $s(x)$ (in meters per second) can be modeled by

$$s(x) = \begin{cases} -4x + 341, & 0 \le x < 11.5 \\ 295, & 11.5 \le x < 22 \\ \frac{3}{4}x + 278.5, & 22 \le x < 32 \\ \frac{3}{2}x + 254.5, & 32 \le x < 50 \\ -\frac{3}{2}x + 404.5, & 50 \le x \le 80 \end{cases}$$

where x is the altitude in kilometers (see Figure 4.34). What is the average speed of sound over the interval [0, 80]?

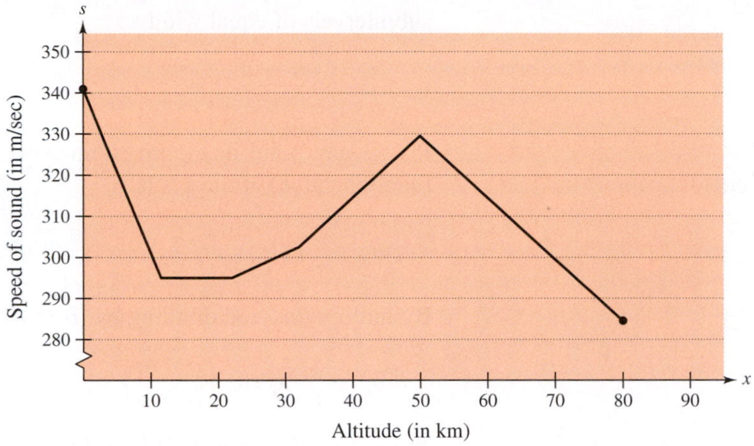

Speed of sound depends on altitude.
Figure 4.34

Solution Begin by integrating $s(x)$ over the interval [0, 80]. To do this, you can break the integral into five parts.

$$\int_0^{11.5} s(x)\,dx = \int_0^{11.5} (-4x + 341)\,dx = \Big[-2x^2 + 341x\Big]_0^{11.5} = 3657$$

$$\int_{11.5}^{22} s(x)\,dx = \int_{11.5}^{22} 295\,dx = \Big[295x\Big]_{11.5}^{22} = 3097.5$$

$$\int_{22}^{32} s(x)\,dx = \int_{22}^{32} \left(\tfrac{3}{4}x + 278.5\right)dx = \Big[\tfrac{3}{8}x^2 + 278.5x\Big]_{22}^{32} = 2987.5$$

$$\int_{32}^{50} s(x)\,dx = \int_{32}^{50} \left(\tfrac{3}{2}x + 254.5\right)dx = \Big[\tfrac{3}{4}x^2 + 254.5x\Big]_{32}^{50} = 5688$$

$$\int_{50}^{80} s(x)\,dx = \int_{50}^{80} \left(-\tfrac{3}{2}x + 404.5\right)dx = \Big[-\tfrac{3}{4}x^2 + 404.5x\Big]_{50}^{80} = 9210$$

By adding the values of the five integrals, you have

$$\int_0^{80} s(x)\,dx = 24{,}640.$$

So, the average speed of sound from an altitude of 0 kilometers to an altitude of 80 kilometers is

$$\text{Average speed} = \frac{1}{80}\int_0^{80} s(x)\,dx = \frac{24{,}640}{80} = 308 \text{ meters per second.}$$

The Second Fundamental Theorem of Calculus

Earlier you saw that the definite integral of f on the interval $[a, b]$ was defined using the constant b as the upper limit of integration and x as the variable of integration. However, a slightly different situation may arise in which the variable x is used in the upper limit of integration. To avoid the confusion of using x in two different ways, t is temporarily used as the variable of integration. (Remember that the definite integral is *not* a function of its variable of integration.)

The Definite Integral as a Number

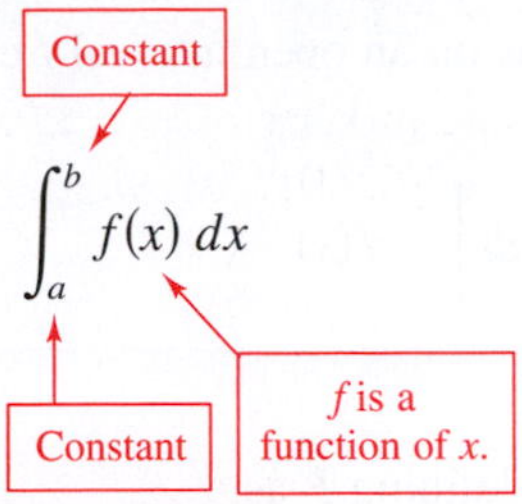

The Definite Integral as a Function of x

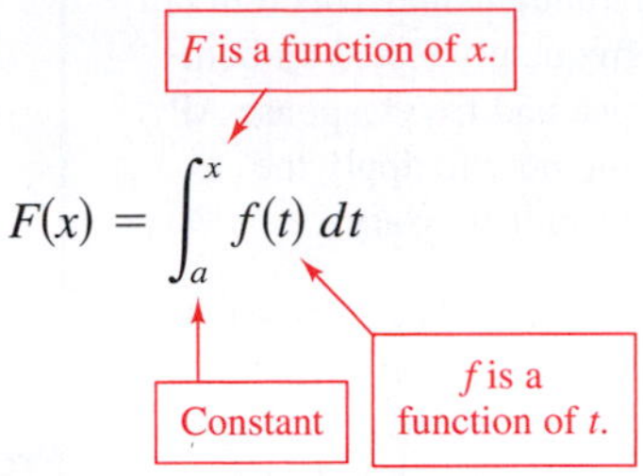

Exploration

Use a graphing utility to graph the function

$$F(x) = \int_0^x \cos t\,dt$$

for $0 \le x \le \pi$. Do you recognize this graph? Explain.

EXAMPLE 6 The Definite Integral as a Function

Evaluate the function

$$F(x) = \int_0^x \cos t\,dt$$

at $x = 0, \frac{\pi}{6}, \frac{\pi}{4}, \frac{\pi}{3}$, and $\frac{\pi}{2}$.

Solution You could evaluate five different definite integrals, one for each of the given upper limits. However, it is much simpler to fix x (as a constant) temporarily to obtain

$$\int_0^x \cos t\,dt = \sin t\Big]_0^x = \sin x - \sin 0 = \sin x.$$

Now, using $F(x) = \sin x$, you can obtain the results shown in Figure 4.35.

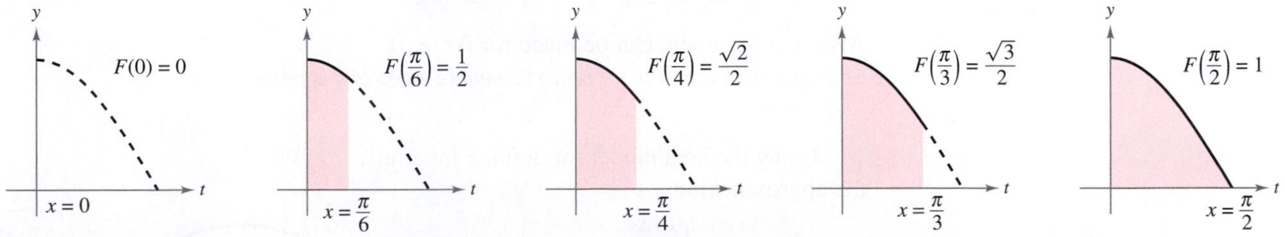

$F(x) = \int_0^x \cos t\,dt$ is the area under the curve $f(t) = \cos t$ from 0 to x.

Figure 4.35

You can think of the function $F(x)$ as *accumulating* the area under the curve $f(t) = \cos t$ from $t = 0$ to $t = x$. For $x = 0$, the area is 0 and $F(0) = 0$. For $x = \pi/2$, $F(\pi/2) = 1$ gives the accumulated area under the cosine curve on the entire interval $[0, \pi/2]$. This interpretation of an integral as an **accumulation function** is used often in applications of integration.

In Example 6, note that the derivative of F is the original integrand (with only the variable changed). That is,

$$\frac{d}{dx}[F(x)] = \frac{d}{dx}[\sin x] = \frac{d}{dx}\left[\int_0^x \cos t\, dt\right] = \cos x.$$

This result is generalized in the next theorem, called the **Second Fundamental Theorem of Calculus.**

AP* Tips

The Second Fundamental Theorem of Calculus is frequently tested on both multiple choice and free response AP sections. Remember to apply the Chain Rule when necessary.

THEOREM 4.11 The Second Fundamental Theorem of Calculus

If f is continuous on an open interval I containing a, then, for every x in the interval,

$$\frac{d}{dx}\left[\int_a^x f(t)\, dt\right] = f(x).$$

Proof Begin by defining F as

$$F(x) = \int_a^x f(t)\, dt.$$

Then, by the definition of the derivative, you can write

$$\begin{aligned} F'(x) &= \lim_{\Delta x \to 0} \frac{F(x + \Delta x) - F(x)}{\Delta x} \\ &= \lim_{\Delta x \to 0} \frac{1}{\Delta x}\left[\int_a^{x+\Delta x} f(t)\, dt - \int_a^x f(t)\, dt\right] \\ &= \lim_{\Delta x \to 0} \frac{1}{\Delta x}\left[\int_a^{x+\Delta x} f(t)\, dt + \int_x^a f(t)\, dt\right] \\ &= \lim_{\Delta x \to 0} \frac{1}{\Delta x}\left[\int_x^{x+\Delta x} f(t)\, dt\right]. \end{aligned}$$

From the Mean Value Theorem for Integrals (assuming $\Delta x > 0$), you know there exists a number c in the interval $[x, x + \Delta x]$ such that the integral in the expression above is equal to $f(c)\,\Delta x$. Moreover, because $x \le c \le x + \Delta x$, it follows that $c \to x$ as $\Delta x \to 0$. So, you obtain

$$F'(x) = \lim_{\Delta x \to 0}\left[\frac{1}{\Delta x} f(c)\,\Delta x\right] = \lim_{\Delta x \to 0} f(c) = f(x).$$

A similar argument can be made for $\Delta x < 0$.

See LarsonCalculus.com for Bruce Edwards's video of this proof.

Using the area model for definite integrals, the approximation

$$f(x)\,\Delta x \approx \int_x^{x+\Delta x} f(t)\, dt$$

can be viewed as saying that the area of the rectangle of height $f(x)$ and width Δx is approximately equal to the area of the region lying between the graph of f and the x-axis on the interval

$$[x, x + \Delta x]$$

as shown in the figure at the right.

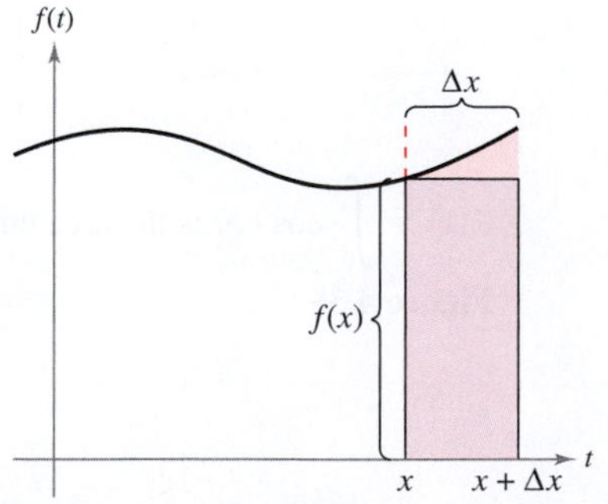

$$f(x)\,\Delta x \approx \int_x^{x+\Delta x} f(t)\, dt$$

Note that the Second Fundamental Theorem of Calculus tells you that when a function is continuous, you can be sure that it has an antiderivative. This antiderivative need not, however, be an elementary function. (Recall the discussion of elementary functions in Section P.3.)

EXAMPLE 7 **The Second Fundamental Theorem of Calculus**

Evaluate $\dfrac{d}{dx}\left[\displaystyle\int_0^x \sqrt{t^2+1}\,dt\right]$.

Solution Note that $f(t) = \sqrt{t^2+1}$ is continuous on the entire real number line. So, using the Second Fundamental Theorem of Calculus, you can write

$$\frac{d}{dx}\left[\int_0^x \sqrt{t^2+1}\,dt\right] = \sqrt{x^2+1}.$$

The differentiation shown in Example 7 is a straightforward application of the Second Fundamental Theorem of Calculus. The next example shows how this theorem can be combined with the Chain Rule to find the derivative of a function.

EXAMPLE 8 **The Second Fundamental Theorem of Calculus**

Find the derivative of $F(x) = \displaystyle\int_{\pi/2}^{x^3} \cos t\,dt$.

Solution Using $u = x^3$, you can apply the Second Fundamental Theorem of Calculus with the Chain Rule as shown.

$$F'(x) = \frac{dF}{du}\frac{du}{dx} \qquad \text{Chain Rule}$$

$$= \frac{d}{du}[F(x)]\frac{du}{dx} \qquad \text{Definition of } \frac{dF}{du}$$

$$= \frac{d}{du}\left[\int_{\pi/2}^{x^3} \cos t\,dt\right]\frac{du}{dx} \qquad \text{Substitute } \int_{\pi/2}^{x^3} \cos t\,dt \text{ for } F(x).$$

$$= \frac{d}{du}\left[\int_{\pi/2}^{u} \cos t\,dt\right]\frac{du}{dx} \qquad \text{Substitute } u \text{ for } x^3.$$

$$= (\cos u)(3x^2) \qquad \text{Apply Second Fundamental Theorem of Calculus.}$$

$$= (\cos x^3)(3x^2) \qquad \text{Rewrite as function of } x.$$

Because the integrand in Example 8 is easily integrated, you can verify the derivative as follows.

$$F(x) = \int_{\pi/2}^{x^3} \cos t\,dt$$

$$= \sin t\Big]_{\pi/2}^{x^3}$$

$$= \sin x^3 - \sin\frac{\pi}{2}$$

$$= \sin x^3 - 1$$

In this form, you can apply the Power Rule to verify that the derivative of F is the same as that obtained in Example 8.

$$\frac{d}{dx}[\sin x^3 - 1] = (\cos x^3)(3x^2) \qquad \text{Derivative of } F$$

Net Change Theorem

The Fundamental Theorem of Calculus (Theorem 4.9) states that if f is continuous on the closed interval $[a, b]$ and F is an antiderivative of f on $[a, b]$, then

$$\int_a^b f(x)\,dx = F(b) - F(a).$$

But because $F'(x) = f(x)$, this statement can be rewritten as

$$\int_a^b F'(x)\,dx = F(b) - F(a)$$

where the quantity $F(b) - F(a)$ represents the *net change of F* on the interval $[a, b]$.

AP* Tips

An alternative form of the First Fundamental Theorem of Calculus, $f(b) = f(a) + \int_a^b f'(x)\,dx$, is also emphasized on the AP Exam.

THEOREM 4.12 The Net Change Theorem

The definite integral of the rate of change of quantity $F'(x)$ gives the total change, or **net change,** in that quantity on the interval $[a, b]$.

$$\int_a^b F'(x)\,dx = F(b) - F(a) \qquad \text{Net change of } F$$

EXAMPLE 9 Using the Net Change Theorem

A chemical flows into a storage tank at a rate of $(180 + 3t)$ liters per minute, where t is the time in minutes and $0 \le t \le 60$. Find the amount of the chemical that flows into the tank during the first 20 minutes.

AP* Tips

On many AP free response questions, you must first decide whether you are being asked for a rate of change (derivative or average rate of change application) or an accumulation of change (definite integral application).

Solution Let $c(t)$ be the amount of the chemical in the tank at time t. Then $c'(t)$ represents the rate at which the chemical flows into the tank at time t. During the first 20 minutes, the amount that flows into the tank is

$$\begin{aligned}\int_0^{20} c'(t)\,dt &= \int_0^{20} (180 + 3t)\,dt \\ &= \left[180t + \frac{3}{2}t^2\right]_0^{20} \\ &= 3600 + 600 \\ &= 4200.\end{aligned}$$

So, the amount that flows into the tank during the first 20 minutes is 4200 liters. ■

Another way to illustrate the Net Change Theorem is to examine the velocity of a particle moving along a straight line, where $s(t)$ is the position at time t. Then its velocity is $v(t) = s'(t)$ and

$$\int_a^b v(t)\,dt = s(b) - s(a).$$

This definite integral represents the net change in position, or **displacement,** of the particle.

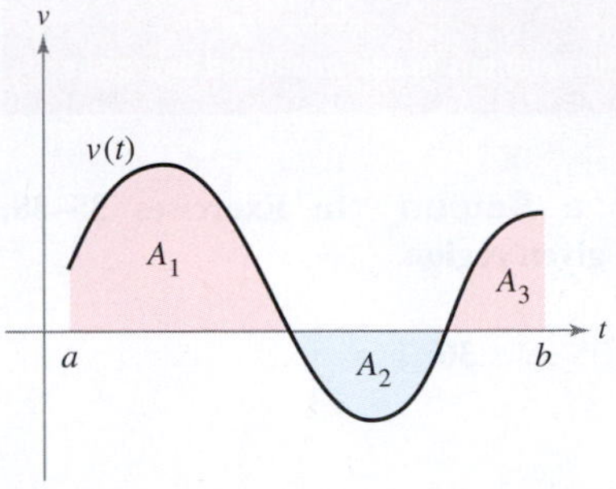

A_1, A_2, and A_3 are the areas of the shaded regions.
Figure 4.36

When calculating the *total* distance traveled by the particle, you must consider the intervals where $v(t) \le 0$ and the intervals where $v(t) \ge 0$. When $v(t) \le 0$, the particle moves to the left, and when $v(t) \ge 0$, the particle moves to the right. To calculate the total distance traveled, integrate the absolute value of velocity $|v(t)|$. So, the **displacement** of the particle on the interval $[a, b]$ is

$$\text{Displacement on } [a, b] = \int_a^b v(t)\, dt = A_1 - A_2 + A_3$$

and the **total distance traveled** by the particle on $[a, b]$ is

$$\text{Total distance traveled on } [a, b] = \int_a^b |v(t)|\, dt = A_1 + A_2 + A_3.$$

(See Figure 4.36.)

EXAMPLE 10 Solving a Particle Motion Problem

The velocity (in feet per second) of a particle moving along a line is

$$v(t) = t^3 - 10t^2 + 29t - 20$$

where t is the time in seconds.

a. What is the displacement of the particle on the time interval $1 \le t \le 5$?

b. What is the total distance traveled by the particle on the time interval $1 \le t \le 5$?

Solution

a. By definition, you know that the displacement is

$$\begin{aligned}\int_1^5 v(t)\, dt &= \int_1^5 (t^3 - 10t^2 + 29t - 20)\, dt\\ &= \left[\frac{t^4}{4} - \frac{10}{3}t^3 + \frac{29}{2}t^2 - 20t\right]_1^5\\ &= \frac{25}{12} - \left(-\frac{103}{12}\right)\\ &= \frac{128}{12}\\ &= \frac{32}{3}.\end{aligned}$$

So, the particle moves $\frac{32}{3}$ feet to the right.

b. To find the total distance traveled, calculate $\int_1^5 |v(t)|\, dt$. Using Figure 4.37 and the fact that $v(t)$ can be factored as $(t - 1)(t - 4)(t - 5)$, you can determine that $v(t) \ge 0$ on $[1, 4]$ and $v(t) \le 0$ on $[4, 5]$. So, the total distance traveled is

$$\begin{aligned}\int_1^5 |v(t)|\, dt &= \int_1^4 v(t)\, dt - \int_4^5 v(t)\, dt\\ &= \int_1^4 (t^3 - 10t^2 + 29t - 20)\, dt - \int_4^5 (t^3 - 10t^2 + 29t - 20)\, dt\\ &= \left[\frac{t^4}{4} - \frac{10}{3}t^3 + \frac{29}{2}t^2 - 20t\right]_1^4 - \left[\frac{t^4}{4} - \frac{10}{3}t^3 + \frac{29}{2}t^2 - 20t\right]_4^5\\ &= \frac{45}{4} - \left(-\frac{7}{12}\right)\\ &= \frac{71}{6} \text{ feet.}\end{aligned}$$

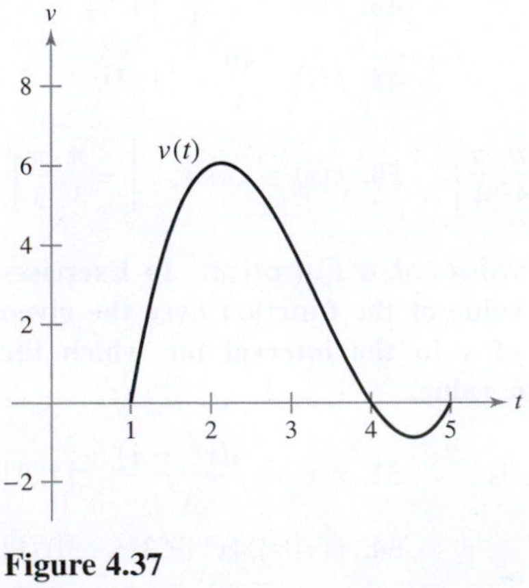

Figure 4.37

4.4 Exercises

See CalcChat.com for tutorial help and worked-out solutions to odd-numbered exercises.

Graphical Reasoning In Exercises 1–4, use a graphing utility to graph the integrand. Use the graph to determine whether the definite integral is positive, negative, or zero.

1. $\displaystyle\int_0^{\pi} \frac{4}{x^2+1}\,dx$ 2. $\displaystyle\int_0^{\pi} \cos x\,dx$

3. $\displaystyle\int_{-2}^{2} x\sqrt{x^2+1}\,dx$ 4. $\displaystyle\int_{-2}^{2} x\sqrt{2-x}\,dx$

Evaluating a Definite Integral In Exercises 5–34, evaluate the definite integral. Use a graphing utility to verify your result.

5. $\displaystyle\int_0^2 6x\,dx$ 6. $\displaystyle\int_{-3}^{1} 8\,dt$

7. $\displaystyle\int_{-1}^{0} (2x-1)\,dx$ 8. $\displaystyle\int_{-1}^{2} (7-3t)\,dt$

9. $\displaystyle\int_{-1}^{1} (t^2-2)\,dt$ 10. $\displaystyle\int_{1}^{2} (6x^2-3x)\,dx$

11. $\displaystyle\int_0^1 (2t-1)^2\,dt$ 12. $\displaystyle\int_1^3 (4x^3-3x^2)\,dx$

13. $\displaystyle\int_1^2 \left(\frac{3}{x^2}-1\right)dx$ 14. $\displaystyle\int_{-2}^{-1} \left(u-\frac{1}{u^2}\right)du$

15. $\displaystyle\int_1^4 \frac{u-2}{\sqrt{u}}\,du$ 16. $\displaystyle\int_{-8}^{8} x^{1/3}\,dx$

17. $\displaystyle\int_{-1}^{1} \left(\sqrt[3]{t}-2\right)dt$ 18. $\displaystyle\int_1^8 \sqrt{\frac{2}{x}}\,dx$

19. $\displaystyle\int_0^1 \frac{x-\sqrt{x}}{3}\,dx$ 20. $\displaystyle\int_0^2 (2-t)\sqrt{t}\,dt$

21. $\displaystyle\int_{-1}^{0} \left(t^{1/3}-t^{2/3}\right)dt$ 22. $\displaystyle\int_{-8}^{-1} \frac{x-x^2}{2\sqrt[3]{x}}\,dx$

23. $\displaystyle\int_0^5 |2x-5|\,dx$ 24. $\displaystyle\int_1^4 (3-|x-3|)\,dx$

25. $\displaystyle\int_0^4 |x^2-9|\,dx$ 26. $\displaystyle\int_0^4 |x^2-4x+3|\,dx$

27. $\displaystyle\int_0^{\pi} (1+\sin x)\,dx$ 28. $\displaystyle\int_0^{\pi} (2+\cos x)\,dx$

29. $\displaystyle\int_0^{\pi/4} \frac{1-\sin^2\theta}{\cos^2\theta}\,d\theta$ 30. $\displaystyle\int_0^{\pi/4} \frac{\sec^2\theta}{\tan^2\theta+1}\,d\theta$

31. $\displaystyle\int_{-\pi/6}^{\pi/6} \sec^2 x\,dx$

32. $\displaystyle\int_{\pi/4}^{\pi/2} (2-\csc^2 x)\,dx$

33. $\displaystyle\int_{-\pi/3}^{\pi/3} 4\sec\theta\tan\theta\,d\theta$

34. $\displaystyle\int_{-\pi/2}^{\pi/2} (2t+\cos t)\,dt$

Finding the Area of a Region In Exercises 35–38, determine the area of the given region.

35. $y = x - x^2$

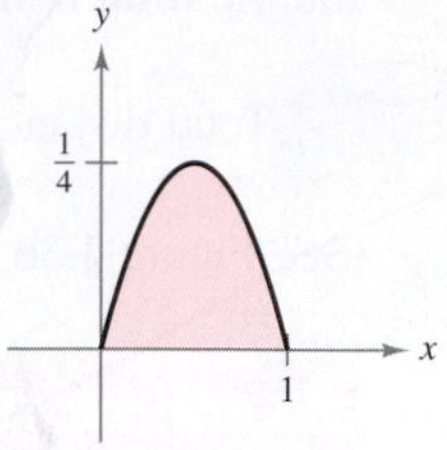

36. $y = \dfrac{1}{x^2}$

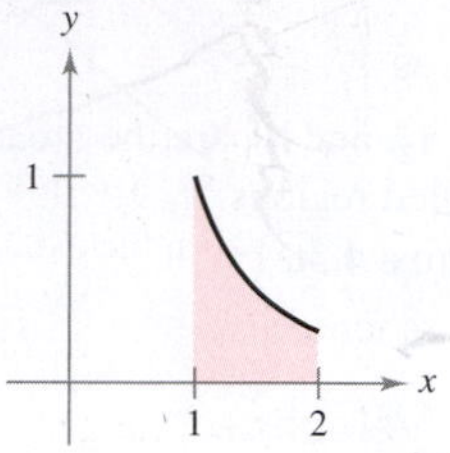

37. $y = \cos x$

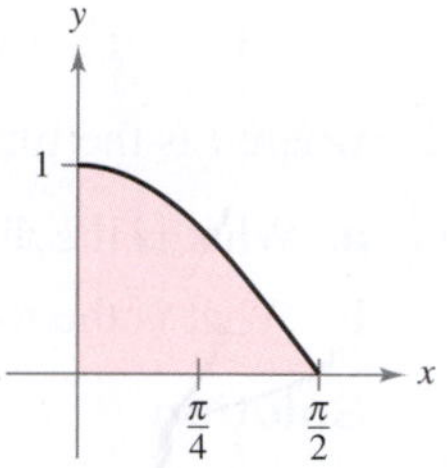

38. $y = x + \sin x$

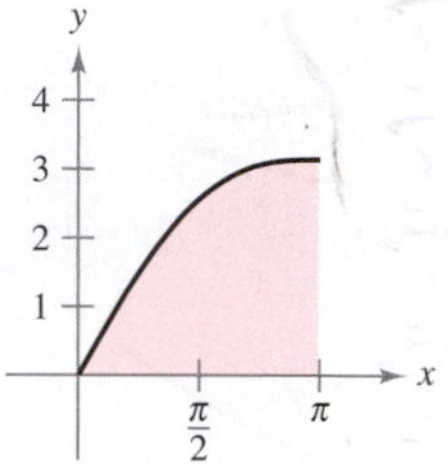

Finding the Area of a Region In Exercises 39–44, find the area of the region bounded by the graphs of the equations.

39. $y = 5x^2 + 2,\quad x = 0,\quad x = 2,\quad y = 0$

40. $y = x^3 + x,\quad x = 2,\quad y = 0$

41. $y = 1 + \sqrt[3]{x},\quad x = 0,\quad x = 8,\quad y = 0$

42. $y = 2\sqrt{x} - x,\quad y = 0$

43. $y = -x^2 + 4x,\quad y = 0$

44. $y = 1 - x^4,\quad y = 0$

Using the Mean Value Theorem for Integrals In Exercises 45–50, find the value(s) of c guaranteed by the Mean Value Theorem for Integrals for the function over the given interval.

45. $f(x) = x^3,\quad [0, 3]$ 46. $f(x) = \sqrt{x},\quad [4, 9]$

47. $y = \dfrac{x^2}{4},\quad [0, 6]$ 48. $f(x) = \dfrac{9}{x^3},\quad [1, 3]$

49. $f(x) = 2\sec^2 x,\quad \left[-\dfrac{\pi}{4}, \dfrac{\pi}{4}\right]$ 50. $f(x) = \cos x,\quad \left[-\dfrac{\pi}{3}, \dfrac{\pi}{3}\right]$

Finding the Average Value of a Function In Exercises 51–56, find the average value of the function over the given interval and all values of x in the interval for which the function equals its average value.

51. $f(x) = 9 - x^2,\quad [-3, 3]$ 52. $f(x) = \dfrac{4(x^2+1)}{x^2},\quad [1, 3]$

53. $f(x) = x^3,\quad [0, 1]$ 54. $f(x) = 4x^3 - 3x^2,\quad [0, 1]$

55. $f(x) = \sin x,\quad [0, \pi]$ 56. $f(x) = \cos x,\quad \left[0, \dfrac{\pi}{2}\right]$

57. Velocity The graph shows the velocity, in feet per second, of a car accelerating from rest. Use the graph to estimate the distance the car travels in 8 seconds.

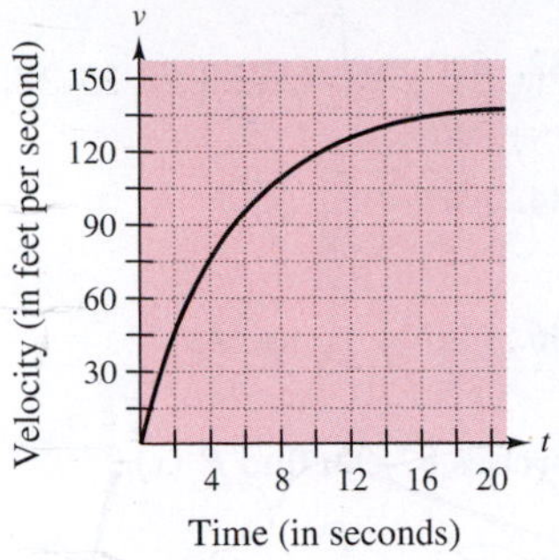

Figure for 57

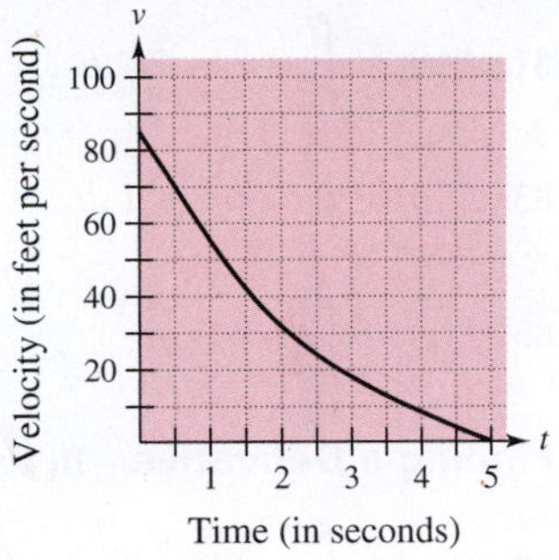

Figure for 58

58. Velocity The graph shows the velocity, in feet per second, of a decelerating car after the driver applies the brakes. Use the graph to estimate how far the car travels before it comes to a stop.

WRITING ABOUT CONCEPTS

59. Using a Graph The graph of f is shown in the figure.

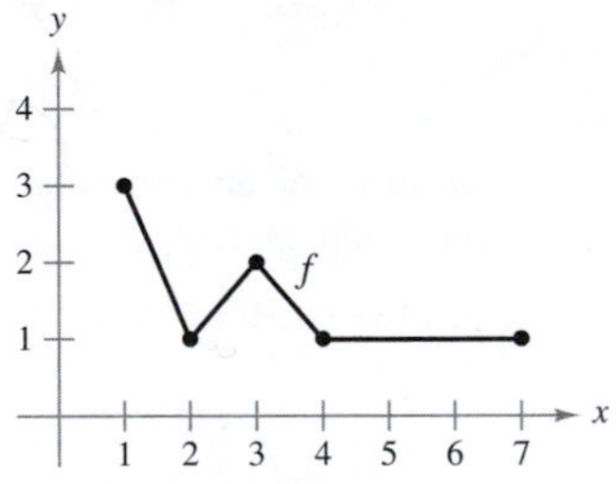

(a) Evaluate $\int_1^7 f(x)\,dx$.

(b) Determine the average value of f on the interval $[1, 7]$.

(c) Determine the answers to parts (a) and (b) when the graph is translated two units upward.

60. Rate of Growth Let $r'(t)$ represent the rate of growth of a dog, in pounds per year. What does $r(t)$ represent? What does $\int_2^6 r'(t)\,dt$ represent about the dog?

61. Force The force F (in newtons) of a hydraulic cylinder in a press is proportional to the square of $\sec x$, where x is the distance (in meters) that the cylinder is extended in its cycle. The domain of F is $[0, \pi/3]$, and $F(0) = 500$.

(a) Find F as a function of x.

(b) Find the average force exerted by the press over the interval $[0, \pi/3]$.

62. Blood Flow The velocity v of the flow of blood at a distance r from the central axis of an artery of radius R is

$$v = k(R^2 - r^2)$$

where k is the constant of proportionality. Find the average rate of flow of blood along a radius of the artery. (Use 0 and R as the limits of integration.)

63. Respiratory Cycle The volume V, in liters, of air in the lungs during a five-second respiratory cycle is approximated by the model $V = 0.1729t + 0.1522t^2 - 0.0374t^3$, where t is the time in seconds. Approximate the average volume of air in the lungs during one cycle.

64. Average Sales A company fits a model to the monthly sales data for a seasonal product. The model is

$$S(t) = \frac{t}{4} + 1.8 + 0.5\sin\left(\frac{\pi t}{6}\right), \quad 0 \le t \le 24$$

where S is sales (in thousands) and t is time in months.

(a) Use a graphing utility to graph $f(t) = 0.5\sin(\pi t/6)$ for $0 \le t \le 24$. Use the graph to explain why the average value of $f(t)$ is 0 over the interval.

(b) Use a graphing utility to graph $S(t)$ and the line $g(t) = t/4 + 1.8$ in the same viewing window. Use the graph and the result of part (a) to explain why g is called the *trend line*.

65. Modeling Data An experimental vehicle is tested on a straight track. It starts from rest, and its velocity v (in meters per second) is recorded every 10 seconds for 1 minute (see table).

t	0	10	20	30	40	50	60
v	0	5	21	40	62	78	83

(a) Use a graphing utility to find a model of the form $v = at^3 + bt^2 + ct + d$ for the data.

(b) Use a graphing utility to plot the data and graph the model.

(c) Use the Fundamental Theorem of Calculus to approximate the distance traveled by the vehicle during the test.

66. HOW DO YOU SEE IT? The graph of f is shown in the figure. The shaded region A has an area of 1.5, and $\int_0^6 f(x)\,dx = 3.5$. Use this information to fill in the blanks.

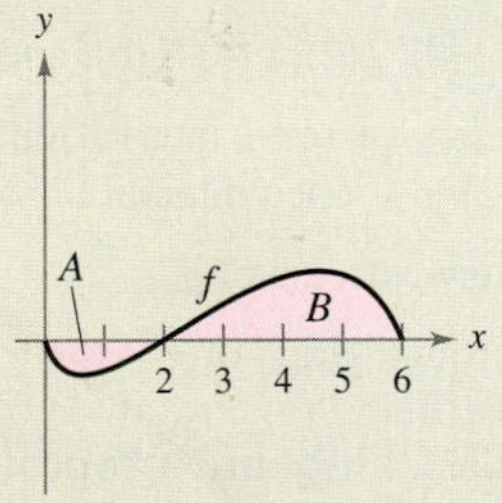

(a) $\int_0^2 f(x)\,dx = \square$ (b) $\int_2^6 f(x)\,dx = \square$

(c) $\int_0^6 |f(x)|\,dx = \square$ (d) $\int_0^2 -2f(x)\,dx = \square$

(e) $\int_0^6 [2 + f(x)]\,dx = \square$

(f) The average value of f over the interval $[0, 6]$ is $\square$.

Evaluating a Definite Integral In Exercises 67–72, find F as a function of x and evaluate it at $x = 2$, $x = 5$, and $x = 8$.

67. $F(x) = \int_0^x (4t - 7)\, dt$

68. $F(x) = \int_2^x (t^3 + 2t - 2)\, dt$

69. $F(x) = \int_1^x \frac{20}{v^2}\, dv$

70. $F(x) = \int_2^x -\frac{2}{t^3}\, dt$

71. $F(x) = \int_1^x \cos \theta\, d\theta$

72. $F(x) = \int_0^x \sin \theta\, d\theta$

73. Analyzing a Function Let

$$g(x) = \int_0^x f(t)\, dt$$

where f is the function whose graph is shown in the figure.

(a) Estimate $g(0)$, $g(2)$, $g(4)$, $g(6)$, and $g(8)$.

(b) Find the largest open interval on which g is increasing. Find the largest open interval on which g is decreasing.

(c) Identify any extrema of g.

(d) Sketch a rough graph of g.

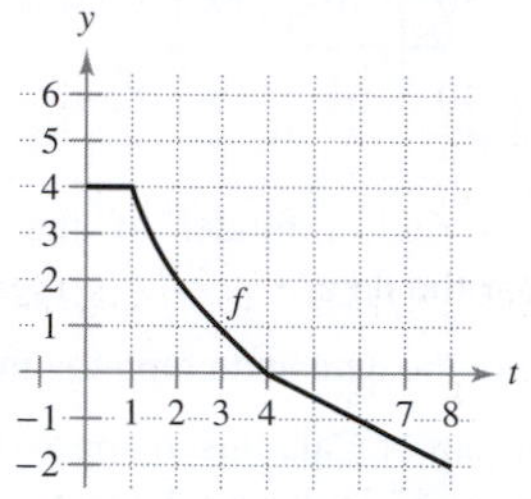

Figure for 73

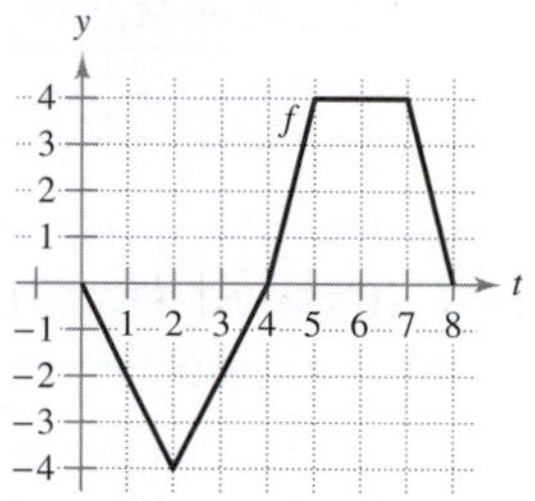

Figure for 74

74. Analyzing a Function Let

$$g(x) = \int_0^x f(t)\, dt$$

where f is the function whose graph is shown in the figure.

(a) Estimate $g(0)$, $g(2)$, $g(4)$, $g(6)$, and $g(8)$.

(b) Find the largest open interval on which g is increasing. Find the largest open interval on which g is decreasing.

(c) Identify any extrema of g.

(d) Sketch a rough graph of g.

Finding and Checking an Integral In Exercises 75–80, (a) integrate to find F as a function of x, and (b) demonstrate the Second Fundamental Theorem of Calculus by differentiating the result in part (a).

75. $F(x) = \int_0^x (t + 2)\, dt$

76. $F(x) = \int_0^x t(t^2 + 1)\, dt$

77. $F(x) = \int_8^x \sqrt[3]{t}\, dt$

78. $F(x) = \int_4^x \sqrt{t}\, dt$

79. $F(x) = \int_{\pi/4}^x \sec^2 t\, dt$

80. $F(x) = \int_{\pi/3}^x \sec t \tan t\, dt$

Using the Second Fundamental Theorem of Calculus In Exercises 81–86, use the Second Fundamental Theorem of Calculus to find $F'(x)$.

81. $F(x) = \int_{-2}^x (t^2 - 2t)\, dt$

82. $F(x) = \int_1^x \frac{t^2}{t^2 + 1}\, dt$

83. $F(x) = \int_{-1}^x \sqrt{t^4 + 1}\, dt$

84. $F(x) = \int_1^x \sqrt[4]{t}\, dt$

85. $F(x) = \int_0^x t \cos t\, dt$

86. $F(x) = \int_0^x \sec^3 t\, dt$

Finding a Derivative In Exercises 87–92, find $F'(x)$.

87. $F(x) = \int_x^{x+2} (4t + 1)\, dt$

88. $F(x) = \int_{-x}^x t^3\, dt$

89. $F(x) = \int_0^{\sin x} \sqrt{t}\, dt$

90. $F(x) = \int_2^{x^2} \frac{1}{t^3}\, dt$

91. $F(x) = \int_0^{x^3} \sin t^2\, dt$

92. $F(x) = \int_0^{x^2} \sin \theta^2\, d\theta$

93. Graphical Analysis Sketch an approximate graph of g on the interval $0 \le x \le 4$, where

$$g(x) = \int_0^x f(t)\, dt.$$

Identify the x-coordinate of an extremum of g. To print an enlarged copy of the graph, go to *MathGraphs.com*

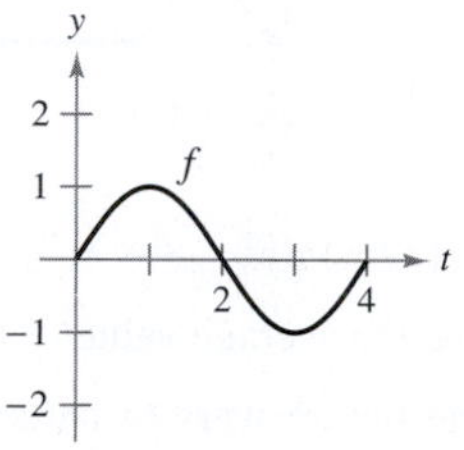

94. Area The area A between the graph of the function

$$g(t) = 4 - \frac{4}{t^2}$$

and the t-axis over the interval $[1, x]$ is

$$A(x) = \int_1^x \left(4 - \frac{4}{t^2}\right) dt.$$

(a) Find the horizontal asymptote of the graph of g.

(b) Integrate to find A as a function of x. Does the graph of A have a horizontal asymptote? Explain.

Particle Motion In Exercises 95–100, the velocity function, in feet per second, is given for a particle moving along a straight line. Find (a) the displacement and (b) the total distance that the particle travels over the given interval.

95. $v(t) = 5t - 7, \quad 0 \le t \le 3$

96. $v(t) = t^2 - t - 12, \quad 1 \le t \le 5$

97. $v(t) = t^3 - 10t^2 + 27t - 18, \quad 1 \le t \le 7$

98. $v(t) = t^3 - 8t^2 + 15t, \quad 0 \le t \le 5$

99. $v(t) = \dfrac{1}{\sqrt{t}}, \quad 1 \le t \le 4$

100. $v(t) = \cos t, \quad 0 \le t \le 3\pi$

101. Particle Motion A particle is moving along the x-axis. The position of the particle at time t is given by

$$x(t) = t^3 - 6t^2 + 9t - 2, \quad 0 \le t \le 5.$$

Find the total distance the particle travels in 5 units of time.

102. Particle Motion Repeat Exercise 101 for the position function given by

$$x(t) = (t - 1)(t - 3)^2, \quad 0 \le t \le 5.$$

103. Water Flow Water flows from a storage tank at a rate of $(500 - 5t)$ liters per minute. Find the amount of water that flows out of the tank during the first 18 minutes.

104. Oil Leak At 1:00 P.M., oil begins leaking from a tank at a rate of $(4 + 0.75t)$ gallons per hour.

(a) How much oil is lost from 1:00 P.M. to 4:00 P.M.?

(b) How much oil is lost from 4:00 P.M. to 7:00 P.M.?

(c) Compare your answers to parts (a) and (b). What do you notice?

Error Analysis **In Exercises 105–108, describe why the statement is incorrect.**

105. $\displaystyle\int_{-1}^{1} x^{-2}\,dx = \left[-x^{-1}\right]_{-1}^{1} = (-1) - 1 = -2$

106. $\displaystyle\int_{-2}^{1} \frac{2}{x^3}\,dx = \left[-\frac{1}{x^2}\right]_{-2}^{1} = -\frac{3}{4}$

107. $\displaystyle\int_{\pi/4}^{3\pi/4} \sec^2 x\,dx = \left[\tan x\right]_{\pi/4}^{3\pi/4} = -2$

108. $\displaystyle\int_{\pi/2}^{3\pi/2} \csc x \cot x\,dx = \left[-\csc x\right]_{\pi/2}^{3\pi/2} = 2$

109. Buffon's Needle Experiment A horizontal plane is ruled with parallel lines 2 inches apart. A two-inch needle is tossed randomly onto the plane. The probability that the needle will touch a line is

$$P = \frac{2}{\pi}\int_0^{\pi/2} \sin\theta\,d\theta$$

where θ is the acute angle between the needle and any one of the parallel lines. Find this probability.

110. Proof Prove that

$$\frac{d}{dx}\left[\int_{u(x)}^{v(x)} f(t)\,dt\right] = f(v(x))v'(x) - f(u(x))u'(x).$$

True or False? **In Exercises 111 and 112, determine whether the statement is true or false. If it is false, explain why or give an example that shows it is false.**

111. If $F'(x) = G'(x)$ on the interval $[a, b]$, then

$$F(b) - F(a) = G(b) - G(a).$$

112. If f is continuous on $[a, b]$, then f is integrable on $[a, b]$.

113. Analyzing a Function Show that the function

$$f(x) = \int_0^{1/x} \frac{1}{t^2 + 1}\,dt + \int_0^{x} \frac{1}{t^2 + 1}\,dt$$

is constant for $x > 0$.

114. Finding a Function Find the function $f(x)$ and all values of c such that

$$\int_c^x f(t)\,dt = x^2 + x - 2.$$

115. Finding Values Let

$$G(x) = \int_0^x \left[s\int_0^s f(t)\,dt\right] ds$$

where f is continuous for all real t. Find (a) $G(0)$, (b) $G'(0)$, (c) $G''(x)$, and (d) $G''(0)$.

SECTION PROJECT

Demonstrating the Fundamental Theorem

Use a graphing utility to graph the function

$$y_1 = \sin^2 t$$

on the interval $0 \le t \le \pi$. Let $F(x)$ be the following function of x.

$$F(x) = \int_0^x \sin^2 t\,dt$$

(a) Complete the table. Explain why the values of F are increasing.

x	0	$\frac{\pi}{6}$	$\frac{\pi}{3}$	$\frac{\pi}{2}$	$\frac{2\pi}{3}$	$\frac{5\pi}{6}$	π
$F(x)$							

(b) Use the integration capabilities of a graphing utility to graph F.

(c) Use the differentiation capabilities of a graphing utility to graph $F'(x)$. How is this graph related to the graph in part (b)?

(d) Verify that the derivative of

$$y = \frac{1}{2}t - \frac{1}{4}\sin 2t$$

is $\sin^2 t$. Graph y and write a short paragraph about how this graph is related to those in parts (b) and (c).

4.5 Integration by Substitution

- Use pattern recognition to find an indefinite integral.
- Use a change of variables to find an indefinite integral.
- Use the General Power Rule for Integration to find an indefinite integral.
- Use a change of variables to evaluate a definite integral.
- Evaluate a definite integral involving an even or odd function.

Pattern Recognition

In this section, you will study techniques for integrating composite functions. The discussion is split into two parts—*pattern recognition* and *change of variables.* Both techniques involve a **u-substitution.** With pattern recognition, you perform the substitution mentally, and with change of variables, you write the substitution steps.

The role of substitution in integration is comparable to the role of the Chain Rule in differentiation. Recall that for the differentiable functions

$$y = F(u) \quad \text{and} \quad u = g(x)$$

the Chain Rule states that

$$\frac{d}{dx}[F(g(x))] = F'(g(x))g'(x).$$

From the definition of an antiderivative, it follows that

$$\int F'(g(x))g'(x)\,dx = F(g(x)) + C.$$

These results are summarized in the next theorem.

REMARK The statement of Theorem 4.13 doesn't tell how to distinguish between $f(g(x))$ and $g'(x)$ in the integrand. As you become more experienced at integration, your skill in doing this will increase. Of course, part of the key is familiarity with derivatives.

THEOREM 4.13 Antidifferentiation of a Composite Function

Let g be a function whose range is an interval I, and let f be a function that is continuous on I. If g is differentiable on its domain and F is an antiderivative of f on I, then

$$\int f(g(x))g'(x)\,dx = F(g(x)) + C.$$

Letting $u = g(x)$ gives $du = g'(x)\,dx$ and

$$\int f(u)\,du = F(u) + C.$$

Examples 1 and 2 show how to apply Theorem 4.13 *directly*, by recognizing the presence of $f(g(x))$ and $g'(x)$. Note that the composite function in the integrand has an *outside function* f and an *inside function* g. Moreover, the derivative $g'(x)$ is present as a factor of the integrand.

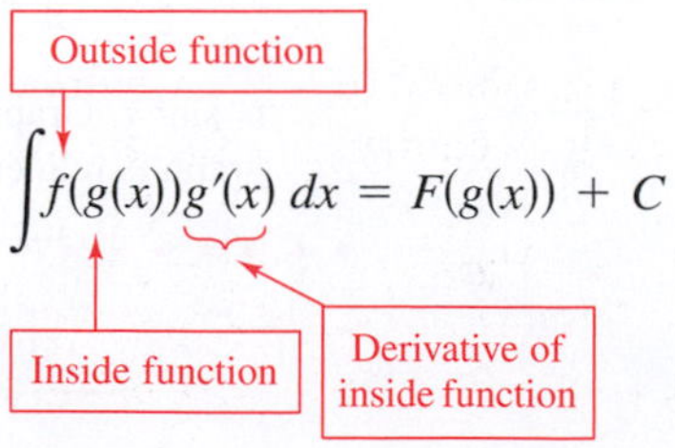

EXAMPLE 1 **Recognizing the $f(g(x))g'(x)$ Pattern**

Find $\int (x^2 + 1)^2(2x)\,dx$.

Solution Letting $g(x) = x^2 + 1$, you obtain

$$g'(x) = 2x$$

and

$$f(g(x)) = f(x^2 + 1) = (x^2 + 1)^2.$$

From this, you can recognize that the integrand follows the $f(g(x))g'(x)$ pattern. Using the Power Rule for Integration and Theorem 4.13, you can write

$$\int \overbrace{(x^2 + 1)^2}^{f(g(x))}\overbrace{(2x)}^{g'(x)}\,dx = \frac{1}{3}(x^2 + 1)^3 + C.$$

Try using the Chain Rule to check that the derivative of $\frac{1}{3}(x^2 + 1)^3 + C$ is the integrand of the original integral.

EXAMPLE 2 **Recognizing the $f(g(x))g'(x)$ Pattern**

Find $\int 5 \cos 5x\,dx$.

Solution Letting $g(x) = 5x$, you obtain

$$g'(x) = 5$$

and

$$f(g(x)) = f(5x) = \cos 5x.$$

From this, you can recognize that the integrand follows the $f(g(x))g'(x)$ pattern. Using the Cosine Rule for Integration and Theorem 4.13, you can write

$$\int \overbrace{(\cos 5x)}^{f(g(x))}\overbrace{(5)}^{g'(x)}\,dx = \sin 5x + C.$$

You can check this by differentiating $\sin 5x + C$ to obtain the original integrand. ■

TECHNOLOGY Try using a computer algebra system, such as *Maple, Mathematica*, or the *TI-Nspire*, to solve the integrals given in Examples 1 and 2. Do you obtain the same antiderivatives that are listed in the examples?

Exploration

Recognizing Patterns The integrand in each of the integrals labeled (a)–(c) fits the pattern $f(g(x))g'(x)$. Identify the pattern and use the result to evaluate the integral.

a. $\int 2x(x^2 + 1)^4\,dx$ **b.** $\int 3x^2\sqrt{x^3 + 1}\,dx$ **c.** $\int \sec^2 x(\tan x + 3)\,dx$

The integrals labeled (d)–(f) are similar to (a)–(c). Show how you can multiply and divide by a constant to evaluate these integrals.

d. $\int x(x^2 + 1)^4\,dx$ **e.** $\int x^2\sqrt{x^3 + 1}\,dx$ **f.** $\int 2\sec^2 x(\tan x + 3)\,dx$

The integrands in Examples 1 and 2 fit the $f(g(x))g'(x)$ pattern exactly—you only had to recognize the pattern. You can extend this technique considerably with the Constant Multiple Rule

$$\int kf(x)\,dx = k\int f(x)\,dx.$$

Many integrands contain the essential part (the variable part) of $g'(x)$ but are missing a constant multiple. In such cases, you can multiply and divide by the necessary constant multiple, as shown in Example 3.

EXAMPLE 3 Multiplying and Dividing by a Constant

Find the indefinite integral.

$$\int x(x^2+1)^2\,dx$$

Solution This is similar to the integral given in Example 1, except that the integrand is missing a factor of 2. Recognizing that $2x$ is the derivative of $x^2 + 1$, you can let

$$g(x) = x^2 + 1$$

and supply the $2x$ as shown.

$$\int x(x^2+1)^2\,dx = \int (x^2+1)^2\left(\frac{1}{2}\right)(2x)\,dx \qquad \text{Multiply and divide by 2.}$$

$$= \frac{1}{2}\int \overbrace{(x^2+1)^2}^{f(g(x))}\,\overbrace{(2x)}^{g'(x)}\,dx \qquad \text{Constant Multiple Rule}$$

$$= \frac{1}{2}\left[\frac{(x^2+1)^3}{3}\right] + C \qquad \text{Integrate.}$$

$$= \frac{1}{6}(x^2+1)^3 + C \qquad \text{Simplify.}$$

In practice, most people would not write as many steps as are shown in Example 3. For instance, you could evaluate the integral by simply writing

$$\int x(x^2+1)^2\,dx = \frac{1}{2}\int (x^2+1)^2(2x)\,dx$$

$$= \frac{1}{2}\left[\frac{(x^2+1)^3}{3}\right] + C$$

$$= \frac{1}{6}(x^2+1)^3 + C.$$

Be sure you see that the *Constant* Multiple Rule applies only to *constants*. You cannot multiply and divide by a variable and then move the variable outside the integral sign. For instance,

$$\int (x^2+1)^2\,dx \neq \frac{1}{2x}\int (x^2+1)^2(2x)\,dx.$$

After all, if it were legitimate to move variable quantities outside the integral sign, you could move the entire integrand out and simplify the whole process. But the result would be incorrect.

Change of Variables

With a formal **change of variables,** you completely rewrite the integral in terms of u and du (or any other convenient variable). Although this procedure can involve more written steps than the pattern recognition illustrated in Examples 1 to 3, it is useful for complicated integrands. The change of variables technique uses the Leibniz notation for the differential. That is, if $u = g(x)$, then $du = g'(x)\,dx$, and the integral in Theorem 4.13 takes the form

$$\int f(g(x))g'(x)\,dx = \int f(u)\,du = F(u) + C.$$

EXAMPLE 4 Change of Variables

Find $\int \sqrt{2x - 1}\,dx$.

Solution First, let u be the inner function, $u = 2x - 1$. Then calculate the differential du to be $du = 2\,dx$. Now, using $\sqrt{2x - 1} = \sqrt{u}$ and $dx = du/2$, substitute to obtain

$$\begin{aligned}
\int \sqrt{2x-1}\,dx &= \int \sqrt{u}\left(\frac{du}{2}\right) && \text{Integral in terms of } u\\
&= \frac{1}{2}\int u^{1/2}\,du && \text{Constant Multiple Rule}\\
&= \frac{1}{2}\left(\frac{u^{3/2}}{3/2}\right) + C && \text{Antiderivative in terms of } u\\
&= \frac{1}{3}u^{3/2} + C && \text{Simplify.}\\
&= \frac{1}{3}(2x-1)^{3/2} + C. && \text{Antiderivative in terms of } x
\end{aligned}$$

REMARK Because integration is usually more difficult than differentiation, you should always check your answer to an integration problem by differentiating. For instance, in Example 4, you should differentiate $\frac{1}{3}(2x - 1)^{3/2} + C$ to verify that you obtain the original integrand.

EXAMPLE 5 Change of Variables

See LarsonCalculus.com for an interactive version of this type of example.

Find $\int x\sqrt{2x - 1}\,dx$.

Solution As in the previous example, let $u = 2x - 1$ and obtain $dx = du/2$. Because the integrand contains a factor of x, you must also solve for x in terms of u, as shown.

$$u = 2x - 1 \quad \Longrightarrow \quad x = \frac{u+1}{2} \qquad \text{Solve for } x \text{ in terms of } u.$$

Now, using substitution, you obtain

$$\begin{aligned}
\int x\sqrt{2x-1}\,dx &= \int \left(\frac{u+1}{2}\right)u^{1/2}\left(\frac{du}{2}\right)\\
&= \frac{1}{4}\int (u^{3/2} + u^{1/2})\,du\\
&= \frac{1}{4}\left(\frac{u^{5/2}}{5/2} + \frac{u^{3/2}}{3/2}\right) + C\\
&= \frac{1}{10}(2x-1)^{5/2} + \frac{1}{6}(2x-1)^{3/2} + C.
\end{aligned}$$

To complete the change of variables in Example 5, you solved for x in terms of u. Sometimes this is very difficult. Fortunately, it is not always necessary, as shown in the next example.

EXAMPLE 6 **Change of Variables**

Find $\int \sin^2 3x \cos 3x \, dx$.

Solution Because $\sin^2 3x = (\sin 3x)^2$, you can let $u = \sin 3x$. Then

$$du = (\cos 3x)(3)\, dx.$$

Now, because $\cos 3x \, dx$ is part of the original integral, you can write

$$\frac{du}{3} = \cos 3x \, dx.$$

Substituting u and $du/3$ in the original integral yields

$$\begin{aligned}\int \sin^2 3x \cos 3x \, dx &= \int u^2 \frac{du}{3} \\ &= \frac{1}{3}\int u^2 \, du \\ &= \frac{1}{3}\left(\frac{u^3}{3}\right) + C \\ &= \frac{1}{9}\sin^3 3x + C.\end{aligned}$$

You can check this by differentiating.

$$\begin{aligned}\frac{d}{dx}\left[\frac{1}{9}\sin^3 3x + C\right] &= \left(\frac{1}{9}\right)(3)(\sin 3x)^2(\cos 3x)(3) \\ &= \sin^2 3x \cos 3x\end{aligned}$$

Because differentiation produces the original integrand, you know that you have obtained the correct antiderivative.

REMARK When making a change of variables, be sure that your answer is written using the same variables as in the original integrand. For instance, in Example 6, you should not leave your answer as

$$\frac{1}{9}u^3 + C$$

but rather, you should replace u by $\sin 3x$.

The steps used for integration by substitution are summarized in the following guidelines.

GUIDELINES FOR MAKING A CHANGE OF VARIABLES

1. Choose a substitution $u = g(x)$. Usually, it is best to choose the *inner* part of a composite function, such as a quantity raised to a power.
2. Compute $du = g'(x)\, dx$.
3. Rewrite the integral in terms of the variable u.
4. Find the resulting integral in terms of u.
5. Replace u by $g(x)$ to obtain an antiderivative in terms of x.
6. Check your answers by differentiating.

So far, you have seen two techniques for applying substitution, and you will see more techniques in the remainder of this section. Each technique differs slightly from the others. You should remember, however, that the goal is the same with each technique—*you are trying to find an antiderivative of the integrand.*

The General Power Rule for Integration

One of the most common u-substitutions involves quantities in the integrand that are raised to a power. Because of the importance of this type of substitution, it is given a special name—the **General Power Rule for Integration.** A proof of this rule follows directly from the (simple) Power Rule for Integration, together with Theorem 4.13.

THEOREM 4.14 The General Power Rule for Integration

If g is a differentiable function of x, then

$$\int [g(x)]^n g'(x)\,dx = \frac{[g(x)]^{n+1}}{n+1} + C, \quad n \neq -1.$$

Equivalently, if $u = g(x)$, then

$$\int u^n\,du = \frac{u^{n+1}}{n+1} + C, \quad n \neq -1.$$

EXAMPLE 7 Substitution and the General Power Rule

a. $$\int 3(3x-1)^4\,dx = \int \overbrace{(3x-1)^4}^{u^4}\overbrace{(3)\,dx}^{du} = \overbrace{\frac{(3x-1)^5}{5}}^{u^5/5} + C$$

b. $$\int (2x+1)(x^2+x)\,dx = \int \overbrace{(x^2+x)^1}^{u^1}\overbrace{(2x+1)\,dx}^{du} = \overbrace{\frac{(x^2+x)^2}{2}}^{u^2/2} + C$$

c. $$\int 3x^2\sqrt{x^3-2}\,dx = \int \overbrace{(x^3-2)^{1/2}}^{u^{1/2}}\overbrace{(3x^2)\,dx}^{du} = \overbrace{\frac{(x^3-2)^{3/2}}{3/2}}^{u^{3/2}/(3/2)} + C = \frac{2}{3}(x^3-2)^{3/2} + C$$

d. $$\int \frac{-4x}{(1-2x^2)^2}\,dx = \int \overbrace{(1-2x^2)^{-2}}^{u^{-2}}\overbrace{(-4x)\,dx}^{du} = \overbrace{\frac{(1-2x^2)^{-1}}{-1}}^{u^{-1}/(-1)} + C = -\frac{1}{1-2x^2} + C$$

e. $$\int \cos^2 x \sin x\,dx = -\int \overbrace{(\cos x)^2}^{u^2}\overbrace{(-\sin x)\,dx}^{du} = -\overbrace{\frac{(\cos x)^3}{3}}^{u^3/3} + C$$

Some integrals whose integrands involve quantities raised to powers cannot be found by the General Power Rule. Consider the two integrals

$$\int x(x^2+1)^2\,dx \quad \text{and} \quad \int (x^2+1)^2\,dx.$$

The substitution

$$u = x^2 + 1$$

works in the first integral, but not in the second. In the second, the substitution fails because the integrand lacks the factor x needed for du. Fortunately, *for this particular integral*, you can expand the integrand as

$$(x^2+1)^2 = x^4 + 2x^2 + 1$$

and use the (simple) Power Rule to integrate each term.

AP* Tips
Some questions in the calculator sections are designed not only to encourage but require the use of a graphing calculator in evaluating definite integrals, as is the case for functions with no elementary derivative.

Change of Variables for Definite Integrals

When using u-substitution with a definite integral, it is often convenient to determine the limits of integration for the variable u rather than to convert the antiderivative back to the variable x and evaluate at the original limits. This change of variables is stated explicitly in the next theorem. The proof follows from Theorem 4.13 combined with the Fundamental Theorem of Calculus.

THEOREM 4.15 Change of Variables for Definite Integrals

If the function $u = g(x)$ has a continuous derivative on the closed interval $[a, b]$ and f is continuous on the range of g, then

$$\int_a^b f(g(x))g'(x)\,dx = \int_{g(a)}^{g(b)} f(u)\,du.$$

EXAMPLE 8 Change of Variables

Evaluate $\int_0^1 x(x^2+1)^3\,dx.$

Solution To evaluate this integral, let $u = x^2 + 1$. Then, you obtain

$$u = x^2 + 1 \quad \Rightarrow \quad du = 2x\,dx.$$

Before substituting, determine the new upper and lower limits of integration.

Lower Limit	**Upper Limit**
When $x = 0$, $u = 0^2 + 1 = 1$.	When $x = 1$, $u = 1^2 + 1 = 2$.

Now, you can substitute to obtain

$$\int_0^1 x(x^2+1)^3\,dx = \frac{1}{2}\int_0^1 (x^2+1)^3(2x)\,dx$$ Integration limits for x

$$= \frac{1}{2}\int_1^2 u^3\,du$$ Integration limits for u

$$= \frac{1}{2}\left[\frac{u^4}{4}\right]_1^2$$

$$= \frac{1}{2}\left(4 - \frac{1}{4}\right)$$

$$= \frac{15}{8}.$$

Notice that you obtain the same result when you rewrite the antiderivative $\frac{1}{2}(u^4/4)$ in terms of the variable x and evaluate the definite integral at the original limits of integration, as shown below.

$$\frac{1}{2}\left[\frac{u^4}{4}\right]_1^2 = \frac{1}{2}\left[\frac{(x^2+1)^4}{4}\right]_0^1$$

$$= \frac{1}{2}\left(4 - \frac{1}{4}\right)$$

$$= \frac{15}{8}$$

EXAMPLE 9 **Change of Variables**

Evaluate the definite integral.

$$\int_1^5 \frac{x}{\sqrt{2x-1}}\,dx$$

Solution To evaluate this integral, let $u = \sqrt{2x-1}$. Then, you obtain

$$u^2 = 2x - 1$$
$$u^2 + 1 = 2x$$
$$\frac{u^2+1}{2} = x$$
$$u\,du = dx. \qquad \text{Differentiate each side.}$$

Before substituting, determine the new upper and lower limits of integration.

Lower Limit

When $x = 1$, $u = \sqrt{2-1} = 1$.

Upper Limit

When $x = 5$, $u = \sqrt{10-1} = 3$.

Now, substitute to obtain

$$\int_1^5 \frac{x}{\sqrt{2x-1}}\,dx = \int_1^3 \frac{1}{u}\left(\frac{u^2+1}{2}\right)u\,du$$
$$= \frac{1}{2}\int_1^3 (u^2+1)\,du$$
$$= \frac{1}{2}\left[\frac{u^3}{3} + u\right]_1^3$$
$$= \frac{1}{2}\left(9 + 3 - \frac{1}{3} - 1\right)$$
$$= \frac{16}{3}.$$

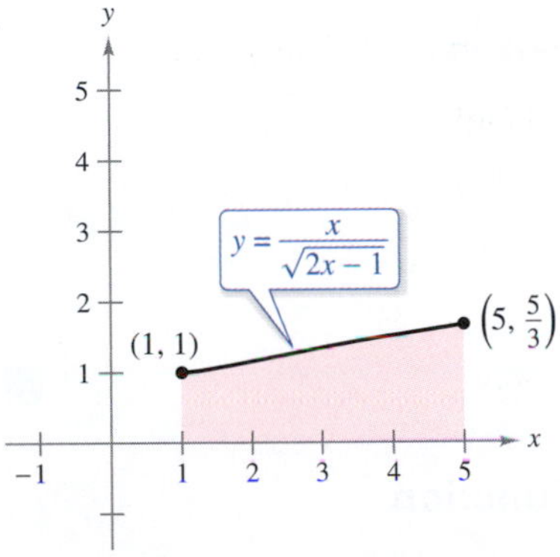

The region before substitution has an area of $\frac{16}{3}$.

Figure 4.38

Geometrically, you can interpret the equation

$$\int_1^5 \frac{x}{\sqrt{2x-1}}\,dx = \int_1^3 \frac{u^2+1}{2}\,du$$

to mean that the two *different* regions shown in Figures 4.38 and 4.39 have the *same* area.

When evaluating definite integrals by substitution, it is possible for the upper limit of integration of the u-variable form to be smaller than the lower limit. When this happens, don't rearrange the limits. Simply evaluate as usual. For example, after substituting $u = \sqrt{1-x}$ in the integral

$$\int_0^1 x^2(1-x)^{1/2}\,dx$$

you obtain $u = \sqrt{1-1} = 0$ when $x = 1$, and $u = \sqrt{1-0} = 1$ when $x = 0$. So, the correct u-variable form of this integral is

$$-2\int_1^0 (1-u^2)^2 u^2\,du.$$

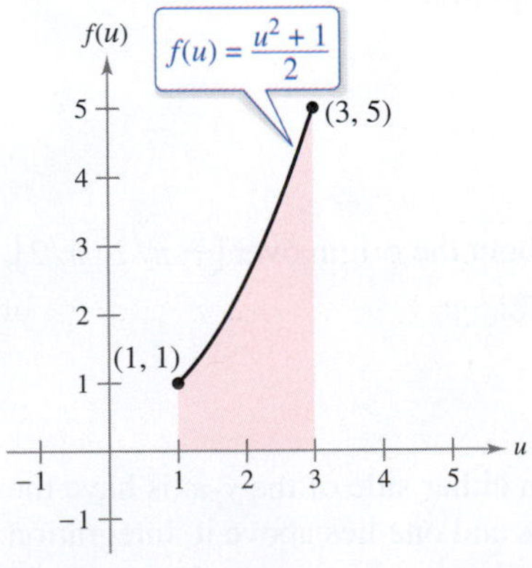

The region after substitution has an area of $\frac{16}{3}$.

Figure 4.39

Expanding the integrand, you can evaluate this integral as shown.

$$-2\int_1^0 (u^2 - 2u^4 + u^6)\,du = -2\left[\frac{u^3}{3} - \frac{2u^5}{5} + \frac{u^7}{7}\right]_1^0 = -2\left(-\frac{1}{3} + \frac{2}{5} - \frac{1}{7}\right) = \frac{16}{105}$$

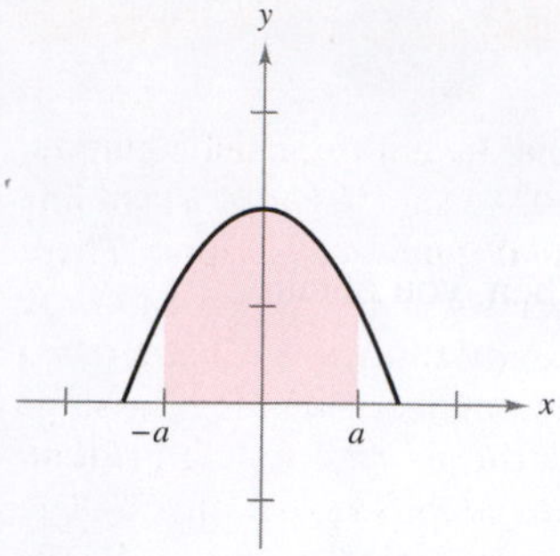

Even function

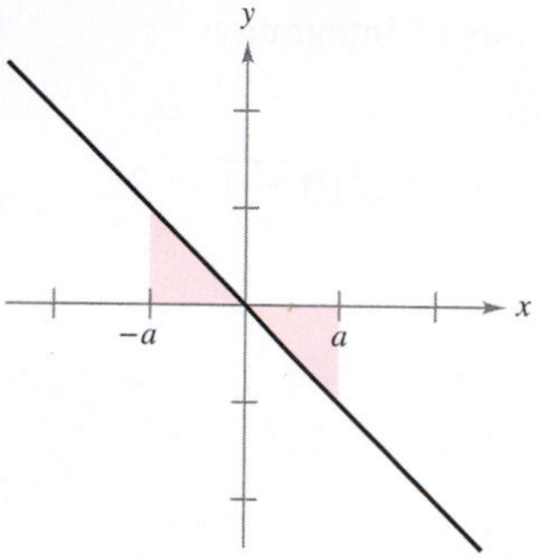

Odd function
Figure 4.40

Integration of Even and Odd Functions

Even with a change of variables, integration can be difficult. Occasionally, you can simplify the evaluation of a definite integral over an interval that is symmetric about the y-axis or about the origin by recognizing the integrand to be an even or odd function (see Figure 4.40).

> **THEOREM 4.16 Integration of Even and Odd Functions**
>
> Let f be integrable on the closed interval $[a, -a]$.
>
> **1.** If f is an *even* function, then $\int_{-a}^{a} f(x)\,dx = 2\int_{0}^{a} f(x)\,dx.$
>
> **2.** If f is an *odd* function, then $\int_{-a}^{a} f(x)\,dx = 0.$

Proof Here is the proof of the first property. (The proof of the second property is left to you [see Exercise 99].) Because f is even, you know that $f(x) = f(-x)$. Using Theorem 4.13 with the substitution $u = -x$ produces

$$\int_{-a}^{0} f(x)\,dx = \int_{a}^{0} f(-u)(du) = -\int_{a}^{0} f(u)\,du = \int_{0}^{a} f(u)\,du = \int_{0}^{a} f(x)\,dx.$$

Finally, using Theorem 4.6, you obtain

$$\begin{aligned}\int_{-a}^{a} f(x)\,dx &= \int_{-a}^{0} f(x)\,dx + \int_{0}^{a} f(x)\,dx\\ &= \int_{0}^{a} f(x)\,dx + \int_{0}^{a} f(x)\,dx\\ &= 2\int_{0}^{a} f(x)\,dx.\end{aligned}$$

See LarsonCalculus.com for Bruce Edwards's video of this proof. ■

EXAMPLE 10 Integration of an Odd Function

Evaluate the definite integral.

$$\int_{-\pi/2}^{\pi/2} (\sin^3 x \cos x + \sin x \cos x)\,dx$$

Solution Letting $f(x) = \sin^3 x \cos x + \sin x \cos x$ produces

$$\begin{aligned}f(-x) &= \sin^3(-x)\cos(-x) + \sin(-x)\cos(-x)\\ &= -\sin^3 x \cos x - \sin x \cos x\\ &= -f(x).\end{aligned}$$

So, f is an odd function, and because f is symmetric about the origin over $[-\pi/2, \pi/2]$, you can apply Theorem 4.16 to conclude that

$$\int_{-\pi/2}^{\pi/2} (\sin^3 x \cos x + \sin x \cos x)\,dx = 0.$$

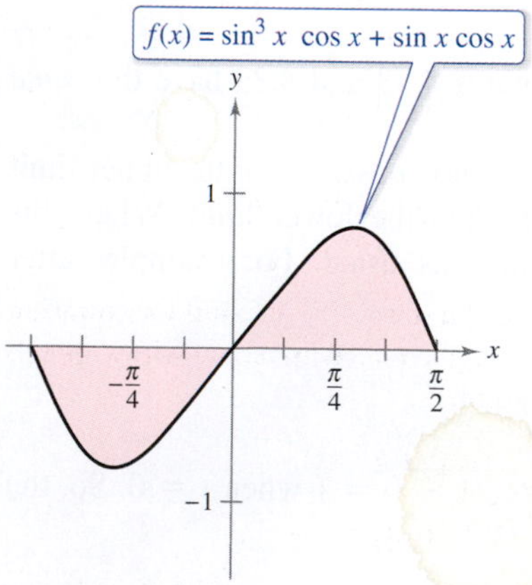

Because f is an odd function, $\int_{-\pi/2}^{\pi/2} f(x)\,dx = 0.$
Figure 4.41

From Figure 4.41, you can see that the two regions on either side of the y-axis have the same area. However, because one lies below the x-axis and one lies above it, integration produces a cancellation effect. (More will be said about areas below the x-axis in Section 7.1.) ■

4.5 Exercises

See CalcChat.com for tutorial help and worked-out solutions to odd-numbered exercises.

Finding u and du In Exercises 1–4, complete the table by identifying u and du for the integral.

$\int f(g(x))g'(x)\,dx$	$u = g(x)$	$du = g'(x)\,dx$
1. $\int (8x^2 + 1)^2(16x)\,dx$		
2. $\int x^2\sqrt{x^3 + 1}\,dx$		
3. $\int \tan^2 x \sec^2 x\,dx$		
4. $\int \frac{\cos x}{\sin^2 x}\,dx$		

Finding an Indefinite Integral In Exercises 5–26, find the indefinite integral and check the result by differentiation.

5. $\int (1 + 6x)^4(6)\,dx$
6. $\int (x^2 - 9)^3(2x)\,dx$
7. $\int \sqrt{25 - x^2}\,(-2x)\,dx$
8. $\int \sqrt[3]{3 - 4x^2}(-8x)\,dx$
9. $\int x^3(x^4 + 3)^2\,dx$
10. $\int x^2(6 - x^3)^5\,dx$
11. $\int x^2(x^3 - 1)^4\,dx$
12. $\int x(5x^2 + 4)^3\,dx$
13. $\int t\sqrt{t^2 + 2}\,dt$
14. $\int t^3\sqrt{2t^4 + 3}\,dt$
15. $\int 5x\,\sqrt[3]{1 - x^2}\,dx$
16. $\int u^2\sqrt{u^3 + 2}\,du$
17. $\int \frac{x}{(1 - x^2)^3}\,dx$
18. $\int \frac{x^3}{(1 + x^4)^2}\,dx$
19. $\int \frac{x^2}{(1 + x^3)^2}\,dx$
20. $\int \frac{6x^2}{(4x^3 - 9)^3}\,dx$
21. $\int \frac{x}{\sqrt{1 - x^2}}\,dx$
22. $\int \frac{x^3}{\sqrt{1 + x^4}}\,dx$
23. $\int \left(1 + \frac{1}{t}\right)^3\left(\frac{1}{t^2}\right)dt$
24. $\int \left[x^2 + \frac{1}{(3x)^2}\right]dx$
25. $\int \frac{1}{\sqrt{2x}}\,dx$
26. $\int \frac{x}{\sqrt[3]{5x^2}}\,dx$

Differential Equation In Exercises 27–30, solve the differential equation.

27. $\frac{dy}{dx} = 4x + \frac{4x}{\sqrt{16 - x^2}}$
28. $\frac{dy}{dx} = \frac{10x^2}{\sqrt{1 + x^3}}$
29. $\frac{dy}{dx} = \frac{x + 1}{(x^2 + 2x - 3)^2}$
30. $\frac{dy}{dx} = \frac{x - 4}{\sqrt{x^2 - 8x + 1}}$

Slope Field In Exercises 31 and 32, a differential equation, a point, and a slope field are given. A *slope field* consists of line segments with slopes given by the differential equation. These line segments give a visual perspective of the directions of the solutions of the differential equation. (a) Sketch two approximate solutions of the differential equation on the slope field, one of which passes through the given point. (To print an enlarged copy of the graph, go to *MathGraphs.com*.) (b) Use integration to find the particular solution of the differential equation and use a graphing utility to graph the solution. Compare the result with the sketches in part (a).

31. $\frac{dy}{dx} = x\sqrt{4 - x^2}$

$(2, 2)$

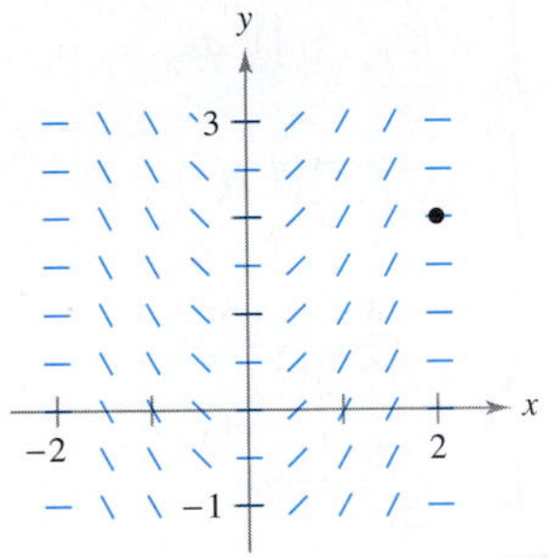

32. $\frac{dy}{dx} = x^2(x^3 - 1)^2$

$(1, 0)$

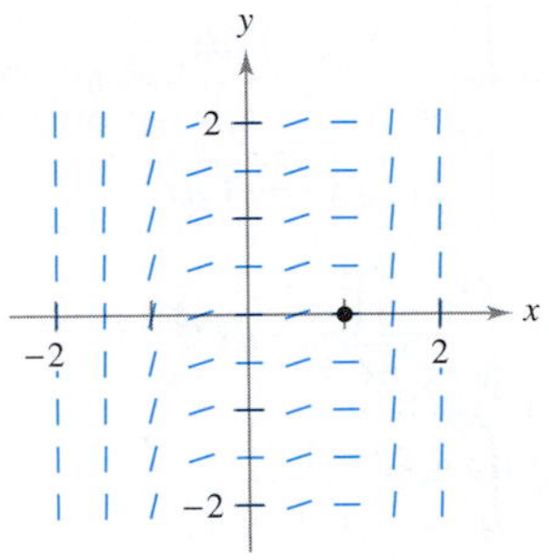

Finding an Indefinite Integral In Exercises 33–42, find the indefinite integral.

33. $\int \pi \sin \pi x\,dx$
34. $\int \sin 4x\,dx$
35. $\int \cos 8x\,dx$
36. $\int \csc^2\left(\frac{x}{2}\right)dx$
37. $\int \frac{1}{\theta^2}\cos\frac{1}{\theta}\,d\theta$
38. $\int x \sin x^2\,dx$
39. $\int \sin 2x \cos 2x\,dx$
40. $\int \sqrt{\tan x}\sec^2 x\,dx$
41. $\int \frac{\csc^2 x}{\cot^3 x}\,dx$
42. $\int \frac{\sin x}{\cos^3 x}\,dx$

Finding an Equation In Exercises 43–46, find an equation for the function f that has the given derivative and whose graph passes through the given point.

	Derivative	Point
43.	$f'(x) = -\sin\frac{x}{2}$	$(0, 6)$
44.	$f'(x) = \sec^2(2x)$	$\left(\frac{\pi}{2}, 2\right)$
45.	$f'(x) = 2x(4x^2 - 10)^2$	$(2, 10)$
46.	$f'(x) = -2x\sqrt{8 - x^2}$	$(2, 7)$

Change of Variables **In Exercises 47–54, find the indefinite integral by the method shown in Example 5.**

47. $\int x\sqrt{x+6}\,dx$

48. $\int x\sqrt{3x-4}\,dx$

49. $\int x^2\sqrt{1-x}\,dx$

50. $\int (x+1)\sqrt{2-x}\,dx$

51. $\int \frac{x^2-1}{\sqrt{2x-1}}\,dx$

52. $\int \frac{2x+1}{\sqrt{x+4}}\,dx$

53. $\int \frac{-x}{(x+1)-\sqrt{x+1}}\,dx$

54. $\int t\sqrt[3]{t+10}\,dt$

Evaluating a Definite Integral **In Exercises 55–62, evaluate the definite integral. Use a graphing utility to verify your result.**

55. $\int_{-1}^{1} x(x^2+1)^3\,dx$

56. $\int_{0}^{1} x^3(2x^4+1)^2\,dx$

57. $\int_{1}^{2} 2x^2\sqrt{x^3+1}\,dx$

58. $\int_{0}^{1} x\sqrt{1-x^2}\,dx$

59. $\int_{0}^{4} \frac{1}{\sqrt{2x+1}}\,dx$

60. $\int_{0}^{2} \frac{x}{\sqrt{1+2x^2}}\,dx$

61. $\int_{1}^{9} \frac{1}{\sqrt{x}\left(1+\sqrt{x}\right)^2}\,dx$

62. $\int_{1}^{5} \frac{x}{\sqrt{2x-1}}\,dx$

Differential Equation **In Exercises 63 and 64, the graph of a function f is shown. Use the differential equation and the given point to find an equation of the function.**

63. $\frac{dy}{dx} = 18x^2(2x^3+1)^2$

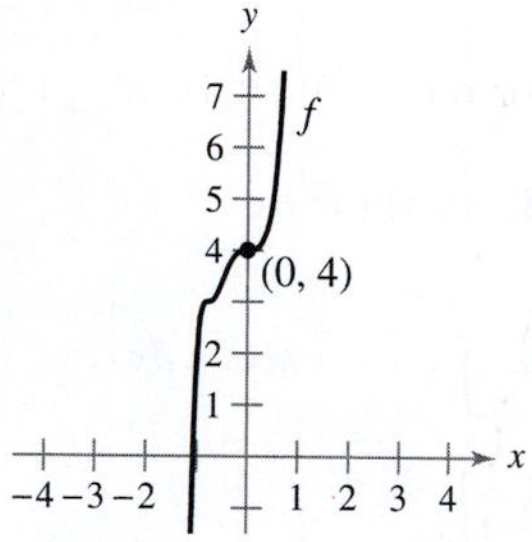

64. $\frac{dy}{dx} = \frac{-48}{(3x+5)^3}$

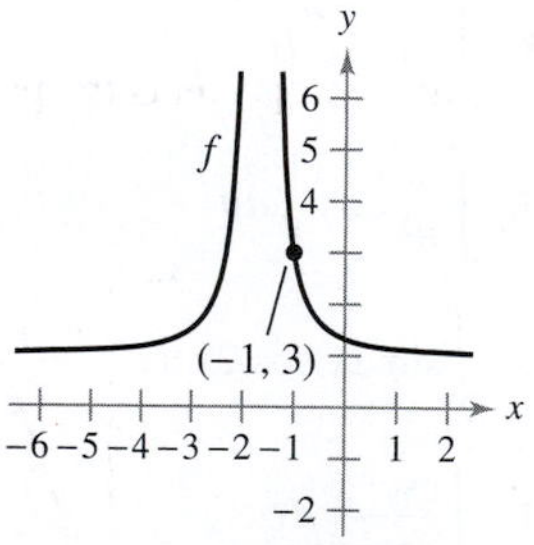

Finding the Area of a Region **In Exercises 65–68, find the area of the region. Use a graphing utility to verify your result.**

65. $\int_{0}^{7} x\sqrt[3]{x+1}\,dx$

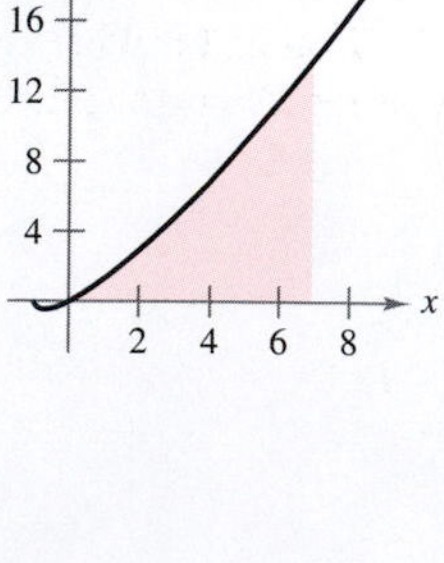

66. $\int_{-2}^{6} x^2\sqrt[3]{x+2}\,dx$

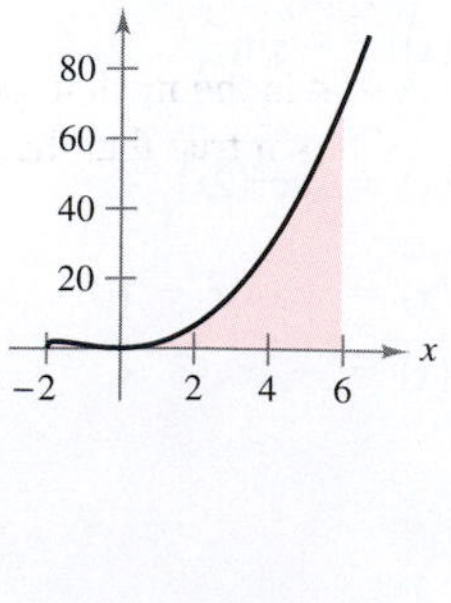

67. $\int_{\pi/2}^{2\pi/3} \sec^2\left(\frac{x}{2}\right)dx$

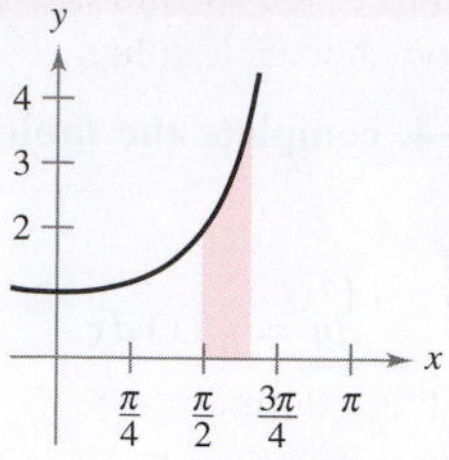

68. $\int_{\pi/12}^{\pi/4} \csc 2x \cot 2x\,dx$

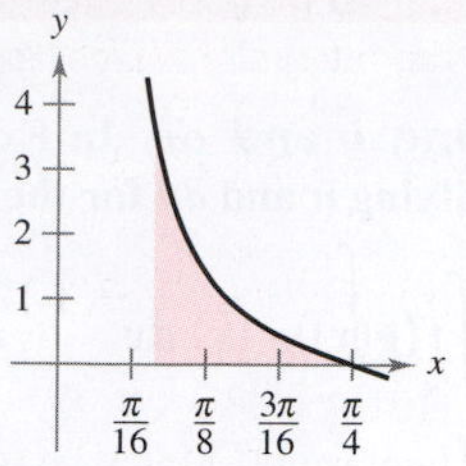

Even and Odd Functions **In Exercises 69–72, evaluate the integral using the properties of even and odd functions as an aid.**

69. $\int_{-2}^{2} x^2(x^2+1)\,dx$

70. $\int_{-2}^{2} x(x^2+1)^3\,dx$

71. $\int_{-\pi/2}^{\pi/2} \sin^2 x\cos x\,dx$

72. $\int_{-\pi/2}^{\pi/2} \sin x\cos x\,dx$

73. Using an Even Function Use $\int_0^4 x^2\,dx = \frac{64}{3}$ to evaluate each definite integral without using the Fundamental Theorem of Calculus.

(a) $\int_{-4}^{0} x^2\,dx$

(b) $\int_{-4}^{4} x^2\,dx$

(c) $\int_{0}^{4} -x^2\,dx$

(d) $\int_{-4}^{0} 3x^2\,dx$

74. Using Symmetry Use the symmetry of the graphs of the sine and cosine functions as an aid in evaluating each definite integral.

(a) $\int_{-\pi/4}^{\pi/4} \sin x\,dx$

(b) $\int_{-\pi/4}^{\pi/4} \cos x\,dx$

(c) $\int_{-\pi/2}^{\pi/2} \cos x\,dx$

(d) $\int_{-\pi/2}^{\pi/2} \sin x\cos x\,dx$

Even and Odd Functions **In Exercises 75 and 76, write the integral as the sum of the integral of an odd function and the integral of an even function. Use this simplification to evaluate the integral.**

75. $\int_{-3}^{3} (x^3+4x^2-3x-6)\,dx$

76. $\int_{-\pi/2}^{\pi/2} (\sin 4x + \cos 4x)\,dx$

WRITING ABOUT CONCEPTS

77. Using Substitution Describe why

$$\int x(5-x^2)^3\,dx \neq \int u^3\,du$$

where $u = 5 - x^2$.

78. Analyzing the Integrand Without integrating, explain why

$$\int_{-2}^{2} x(x^2+1)^2\,dx = 0.$$

WRITING ABOUT CONCEPTS (continued)

79. Choosing an Integral You are asked to find one of the integrals. Which one would you choose? Explain.

(a) $\int \sqrt{x^3 + 1}\,dx$ or $\int x^2\sqrt{x^3 + 1}\,dx$

(b) $\int \tan(3x)\sec^2(3x)\,dx$ or $\int \tan(3x)\,dx$

80. Comparing Methods Find the indefinite integral in two ways. Explain any difference in the forms of the answers.

(a) $\int (2x - 1)^2\,dx$ (b) $\int \tan x \sec^2 x\,dx$

81. Depreciation The rate of depreciation dV/dt of a machine is inversely proportional to the square of $(t + 1)$, where V is the value of the machine t years after it was purchased. The initial value of the machine was \$500,000, and its value decreased \$100,000 in the first year. Estimate its value after 4 years.

82. HOW DO YOU SEE IT? The graph shows the flow rate of water at a pumping station for one day.

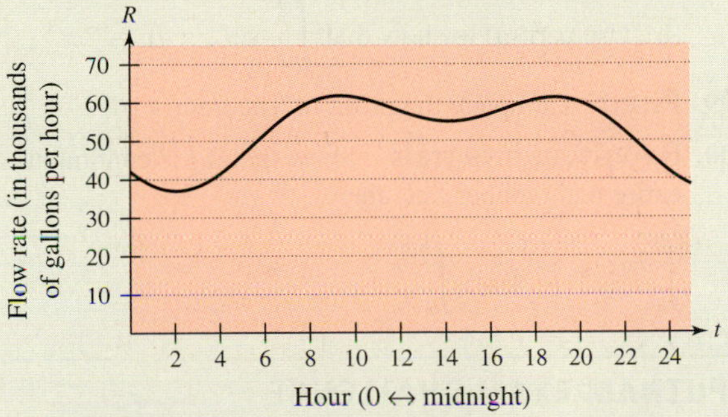

(a) Approximate the maximum flow rate at the pumping station. At what time does this occur?

(b) Explain how you can find the amount of water used during the day.

(c) Approximate the two-hour period when the least amount of water is used. Explain your reasoning.

83. Sales The sales S (in thousands of units) of a seasonal product are given by the model

$$S = 74.50 + 43.75 \sin\frac{\pi t}{6}$$

where t is the time in months, with $t = 1$ corresponding to January. Find the average sales for each time period.

(a) The first quarter $(0 \le t \le 3)$

(b) The second quarter $(3 \le t \le 6)$

(c) The entire year $(0 \le t \le 12)$

84. Electricity

The oscillating current in an electrical circuit is

$$I = 2\sin(60\pi t) + \cos(120\pi t)$$

where I is measured in amperes and t is measured in seconds. Find the average current for each time interval.

(a) $0 \le t \le \frac{1}{60}$

(b) $0 \le t \le \frac{1}{240}$

(c) $0 \le t \le \frac{1}{30}$

Molodec/Shutterstock.com

Probability In Exercises 85 and 86, the function

$$f(x) = kx^n(1 - x)^m, \quad 0 \le x \le 1$$

where $n > 0$, $m > 0$, and k is a constant, can be used to represent various probability distributions. If k is chosen such that

$$\int_0^1 f(x)\,dx = 1$$

then the probability that x will fall between a and b $(0 \le a \le b \le 1)$ is

$$P_{a,b} = \int_a^b f(x)\,dx.$$

85. The probability that a person will remember between $100a\%$ and $100b\%$ of material learned in an experiment is

$$P_{a,b} = \int_a^b \frac{15}{4}x\sqrt{1 - x}\,dx$$

where x represents the proportion remembered. (See figure.)

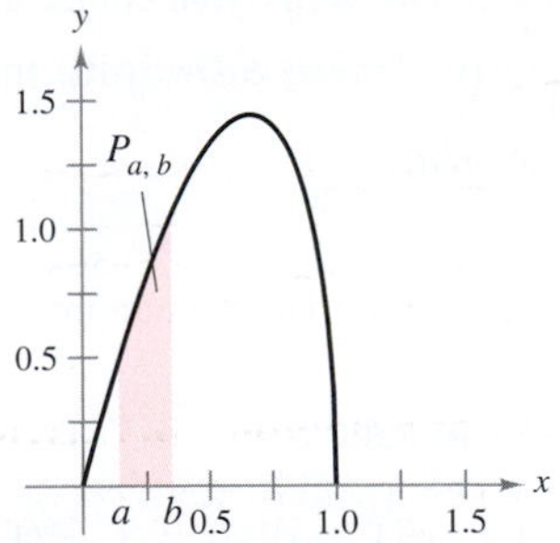

(a) For a randomly chosen individual, what is the probability that he or she will recall between 50% and 75% of the material?

(b) What is the median percent recall? That is, for what value of b is it true that the probability of recalling 0 to b is 0.5?

86. The probability that ore samples taken from a region contain between $100a\%$ and $100b\%$ iron is

$$P_{a,\,b} = \int_a^b \frac{1155}{32} x^3(1 - x)^{3/2}\, dx$$

where x represents the proportion of iron. (See figure.) What is the probability that a sample will contain between

(a) 0% and 25% iron? (b) 50% and 100% iron?

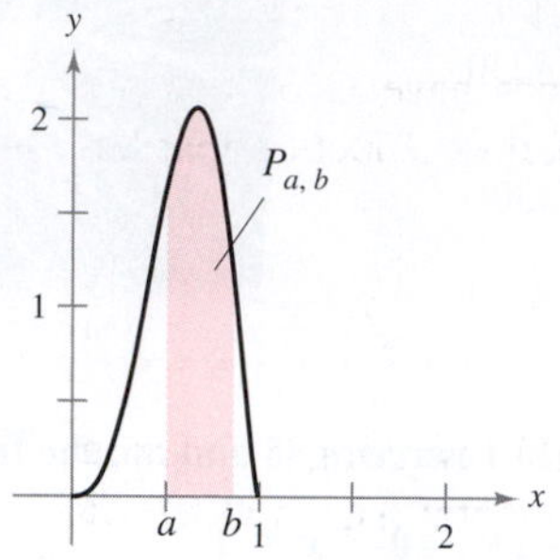

87. Graphical Analysis Consider the functions f and g, where

$$f(x) = 6 \sin x \cos^2 x \quad \text{and} \quad g(t) = \int_0^t f(x)\, dx.$$

(a) Use a graphing utility to graph f and g in the same viewing window.

(b) Explain why g is nonnegative.

(c) Identify the points on the graph of g that correspond to the extrema of f.

(d) Does each of the zeros of f correspond to an extremum of g? Explain.

(e) Consider the function

$$h(t) = \int_{\pi/2}^t f(x)\, dx.$$

Use a graphing utility to graph h. What is the relationship between g and h? Verify your conjecture.

88. Finding a Limit Using a Definite Integral Find

$$\lim_{n\to+\infty} \sum_{i=1}^{n} \frac{\sin(i\pi/n)}{n}$$

by evaluating an appropriate definite integral over the interval $[0, 1]$.

89. Rewriting Integrals

(a) Show that $\displaystyle\int_0^1 x^2(1 - x)^5\, dx = \int_0^1 x^5(1 - x)^2\, dx.$

(b) Show that $\displaystyle\int_0^1 x^a(1 - x)^b\, dx = \int_0^1 x^b(1 - x)^a\, dx.$

90. Rewriting Integrals

(a) Show that $\displaystyle\int_0^{\pi/2} \sin^2 x\, dx = \int_0^{\pi/2} \cos^2 x\, dx.$

(b) Show that $\displaystyle\int_0^{\pi/2} \sin^n x\, dx = \int_0^{\pi/2} \cos^n x\, dx,$ where n is a positive integer.

True or False? **In Exercises 91–96, determine whether the statement is true or false. If it is false, explain why or give an example that shows it is false.**

91. $\displaystyle\int (2x + 1)^2\, dx = \tfrac{1}{3}(2x + 1)^3 + C$

92. $\displaystyle\int x(x^2 + 1)\, dx = \tfrac{1}{2}x^2\left(\tfrac{1}{3}x^3 + x\right) + C$

93. $\displaystyle\int_{-10}^{10} (ax^3 + bx^2 + cx + d)\, dx = 2\int_0^{10} (bx^2 + d)\, dx$

94. $\displaystyle\int_a^b \sin x\, dx = \int_a^{b+2\pi} \sin x\, dx$

95. $\displaystyle 4\int \sin x \cos x\, dx = -\cos 2x + C$

96. $\displaystyle\int \sin^2 2x \cos 2x\, dx = \tfrac{1}{3}\sin^3 2x + C$

97. Rewriting Integrals Assume that f is continuous everywhere and that c is a constant. Show that

$$\int_{ca}^{cb} f(x)\, dx = c\int_a^b f(cx)\, dx.$$

98. Integration and Differentiation

(a) Verify that $\sin u - u \cos u + C = \displaystyle\int u \sin u\, du.$

(b) Use part (a) to show that $\displaystyle\int_0^{\pi^2} \sin\sqrt{x}\, dx = 2\pi.$

99. Proof Complete the proof of Theorem 4.16.

100. Rewriting Integrals Show that if f is continuous on the entire real number line, then

$$\int_a^b f(x + h)\, dx = \int_{a+h}^{b+h} f(x)\, dx.$$

PUTNAM EXAM CHALLENGE

101. If $a_0, a_1, \ldots, a_n$ are real numbers satisfying

$$\frac{a_0}{1} + \frac{a_1}{2} + \cdots + \frac{a_n}{n+1} = 0,$$

show that the equation

$$a_0 + a_1x + a_2x^2 + \cdots + a_nx^n = 0$$

has at least one real root.

102. Find all the continuous positive functions $f(x)$, for $0 \le x \le 1$, such that

$$\int_0^1 f(x)\, dx = 1$$

$$\int_0^1 f(x)x\, dx = \alpha$$

$$\int_0^1 f(x)x^2\, dx = \alpha^2$$

where α is a given real number.

4.6 Numerical Integration

- **Approximate a definite integral using the Trapezoidal Rule.**
- **Approximate a definite integral using Simpson's Rule.**
- **Analyze the approximate errors in the Trapezoidal Rule and Simpson's Rule.**

The Trapezoidal Rule

Some elementary functions simply do not have antiderivatives that are elementary functions. For example, there is no elementary function that has any of the following functions as its derivative.

$$\sqrt[3]{x}\sqrt{1-x}, \qquad \sqrt{x}\cos x, \qquad \frac{\cos x}{x}, \qquad \sqrt{1-x^3}, \qquad \sin x^2$$

If you need to evaluate a definite integral involving a function whose antiderivative cannot be found, then while the Fundamental Theorem of Calculus is still true, it cannot be easily applied. In this case, it is easier to resort to an approximation technique. Two such techniques are described in this section.

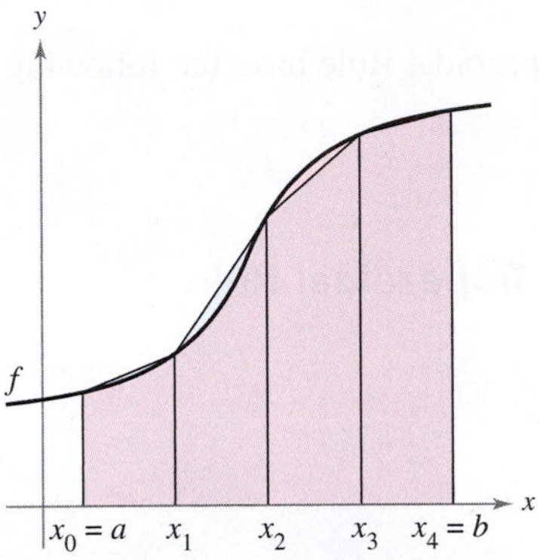

The area of the region can be approximated using four trapezoids.
Figure 4.42

One way to approximate a definite integral is to use n trapezoids, as shown in Figure 4.42. In the development of this method, assume that f is continuous and positive on the interval $[a, b]$. So, the definite integral

$$\int_a^b f(x)\,dx$$

represents the area of the region bounded by the graph of f and the x-axis, from $x = a$ to $x = b$. First, partition the interval $[a, b]$ into n subintervals, each of width $\Delta x = (b - a)/n$, such that

$$a = x_0 < x_1 < x_2 < \cdot \cdot \cdot < x_n = b.$$

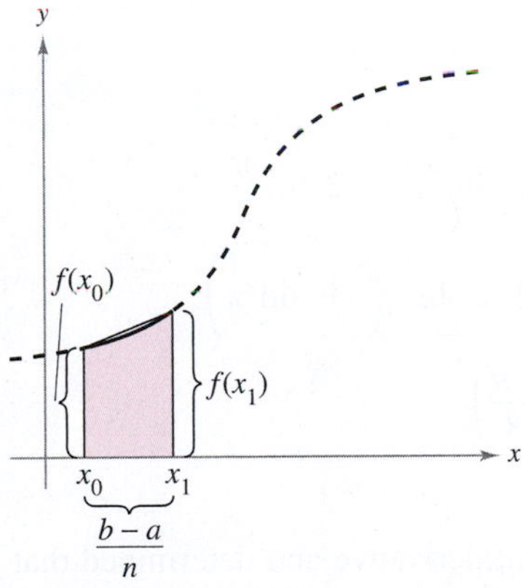

The area of the first trapezoid is $\left[\frac{f(x_0) + f(x_1)}{2}\right]\left(\frac{b-a}{n}\right)$.
Figure 4.43

Then form a trapezoid for each subinterval (see Figure 4.43). The area of the ith trapezoid is

$$\text{Area of } i\text{th trapezoid} = \left[\frac{f(x_{i-1}) + f(x_i)}{2}\right]\left(\frac{b-a}{n}\right).$$

This implies that the sum of the areas of the n trapezoids is

$$\begin{aligned}\text{Area} &= \left(\frac{b-a}{n}\right)\left[\frac{f(x_0) + f(x_1)}{2} + \cdot \cdot \cdot + \frac{f(x_{n-1}) + f(x_n)}{2}\right]\\ &= \left(\frac{b-a}{2n}\right)[f(x_0) + f(x_1) + f(x_1) + f(x_2) + \cdot \cdot \cdot + f(x_{n-1}) + f(x_n)]\\ &= \left(\frac{b-a}{2n}\right)[f(x_0) + 2f(x_1) + 2f(x_2) + \cdot \cdot \cdot + 2f(x_{n-1}) + f(x_n)].\end{aligned}$$

Letting $\Delta x = (b - a)/n$, you can take the limit as $n \to \infty$ to obtain

$$\begin{aligned}&\lim_{n\to\infty}\left(\frac{b-a}{2n}\right)[f(x_0) + 2f(x_1) + \cdot \cdot \cdot + 2f(x_{n-1}) + f(x_n)]\\ &= \lim_{n\to\infty}\left[\frac{[f(a) - f(b)]\,\Delta x}{2} + \sum_{i=1}^{n} f(x_i)\,\Delta x\right]\\ &= \lim_{n\to\infty}\frac{[f(a) - f(b)](b-a)}{2n} + \lim_{n\to\infty}\sum_{i=1}^{n} f(x_i)\,\Delta x\\ &= 0 + \int_a^b f(x)\,dx.\end{aligned}$$

The result is summarized in the next theorem.

THEOREM 4.17 The Trapezoidal Rule

Let f be continuous on $[a, b]$. The Trapezoidal Rule for approximating $\int_a^b f(x)\,dx$ is

$$\int_a^b f(x)\,dx \approx \frac{b-a}{2n}[f(x_0) + 2f(x_1) + 2f(x_2) + \cdots + 2f(x_{n-1}) + f(x_n)].$$

Moreover, as $n \to \infty$, the right-hand side approaches $\int_a^b f(x)\,dx$.

REMARK Observe that the coefficients in the Trapezoidal Rule have the following pattern.

1 2 2 2 . . . 2 2 1

EXAMPLE 1 Approximation with the Trapezoidal Rule

Use the Trapezoidal Rule to approximate

$$\int_0^\pi \sin x\,dx.$$

Compare the results for $n = 4$ and $n = 8$, as shown in Figure 4.44.

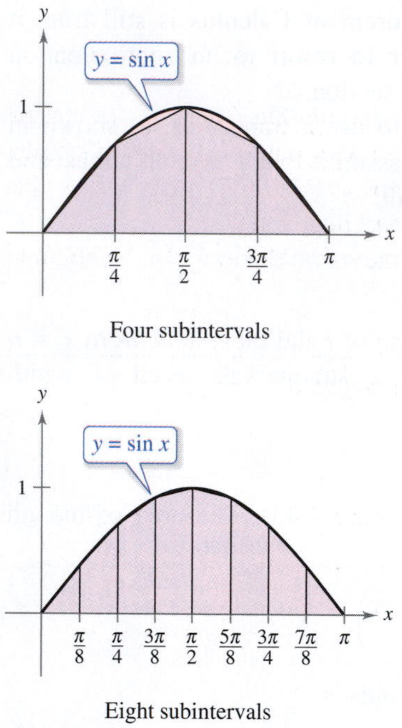

Trapezoidal approximations

Figure 4.44

Solution When $n = 4$, $\Delta x = \pi/4$, and you obtain

$$\begin{aligned}\int_0^\pi \sin x\,dx &\approx \frac{\pi}{8}\left(\sin 0 + 2\sin\frac{\pi}{4} + 2\sin\frac{\pi}{2} + 2\sin\frac{3\pi}{4} + \sin\pi\right)\\ &= \frac{\pi}{8}\left(0 + \sqrt{2} + 2 + \sqrt{2} + 0\right)\\ &= \frac{\pi(1+\sqrt{2})}{4}\\ &\approx 1.896.\end{aligned}$$

When $n = 8$, $\Delta x = \pi/8$, and you obtain

$$\begin{aligned}\int_0^\pi \sin x\,dx &\approx \frac{\pi}{16}\left(\sin 0 + 2\sin\frac{\pi}{8} + 2\sin\frac{\pi}{4} + 2\sin\frac{3\pi}{8} + 2\sin\frac{\pi}{2}\right.\\ &\qquad \left. + 2\sin\frac{5\pi}{8} + 2\sin\frac{3\pi}{4} + 2\sin\frac{7\pi}{8} + \sin\pi\right)\\ &= \frac{\pi}{16}\left(2 + 2\sqrt{2} + 4\sin\frac{\pi}{8} + 4\sin\frac{3\pi}{8}\right)\\ &\approx 1.974.\end{aligned}$$

For this particular integral, you could have found an antiderivative and determined that the exact area of the region is 2.

AP* Tips

The Trapezoidal Rule is a useful formula if the intervals are of equal width. However, if the AP free response question asks for a trapezoidal sum and the intervals are not of uniform width, the Trapezoidal Rule does not apply. You must calculate each trapezoid area separately and add them.

TECHNOLOGY Most graphing utilities and computer algebra systems have built-in programs that can be used to approximate the value of a definite integral. Try using such a program to approximate the integral in Example 1. How close is your approximation? When you use such a program, you need to be aware of its limitations. Often, you are given no indication of the degree of accuracy of the approximation. Other times, you may be given an approximation that is completely wrong. For instance, try using a built-in numerical integration program to evaluate

$$\int_{-1}^2 \frac{1}{x}\,dx.$$

Your calculator should give an error message. Does yours?

It is interesting to compare the Trapezoidal Rule with the Midpoint Rule given in Section 4.2. For the Trapezoidal Rule, you average the function values at the endpoints of the subintervals, but for the Midpoint Rule, you take the function values of the subinterval midpoints.

$$\int_a^b f(x)\,dx \approx \sum_{i=1}^{n} f\left(\frac{x_i + x_{i-1}}{2}\right)\Delta x \qquad \text{Midpoint Rule}$$

$$\int_a^b f(x)\,dx \approx \sum_{i=1}^{n} \left(\frac{f(x_i) + f(x_{i-1})}{2}\right)\Delta x \qquad \text{Trapezoidal Rule}$$

There are two important points that should be made concerning the Trapezoidal Rule (or the Midpoint Rule). First, the approximation tends to become more accurate as n increases. For instance, in Example 1, when $n = 16$, the Trapezoidal Rule yields an approximation of 1.994. Second, although you could have used the Fundamental Theorem to evaluate the integral in Example 1, this theorem cannot be used to evaluate an integral as simple as $\int_0^{\pi} \sin x^2\,dx$ because $\sin x^2$ has no elementary antiderivative. Yet, the Trapezoidal Rule can be applied to estimate this integral.

Simpson's Rule

One way to view the trapezoidal approximation of a definite integral is to say that on each subinterval, you approximate f by a *first*-degree polynomial. In Simpson's Rule, named after the English mathematician Thomas Simpson (1710–1761), you take this procedure one step further and approximate f by *second*-degree polynomials.

Before presenting Simpson's Rule, consider the next theorem for evaluating integrals of polynomials of degree 2 (or less).

THEOREM 4.18 Integral of $p(x) = Ax^2 + Bx + C$

If $p(x) = Ax^2 + Bx + C$, then

$$\int_a^b p(x)\,dx = \left(\frac{b-a}{6}\right)\left[p(a) + 4p\left(\frac{a+b}{2}\right) + p(b)\right].$$

Proof

$$\begin{aligned}
\int_a^b p(x)\,dx &= \int_a^b (Ax^2 + Bx + C)\,dx \\
&= \left[\frac{Ax^3}{3} + \frac{Bx^2}{2} + Cx\right]_a^b \\
&= \frac{A(b^3 - a^3)}{3} + \frac{B(b^2 - a^2)}{2} + C(b - a) \\
&= \left(\frac{b-a}{6}\right)[2A(a^2 + ab + b^2) + 3B(b + a) + 6C]
\end{aligned}$$

By expansion and collection of terms, the expression inside the brackets becomes

$$\underbrace{(Aa^2 + Ba + C)}_{p(a)} + \underbrace{4\left[A\left(\frac{b+a}{2}\right)^2 + B\left(\frac{b+a}{2}\right) + C\right]}_{4p\left(\frac{a+b}{2}\right)} + \underbrace{(Ab^2 + Bb + C)}_{p(b)}$$

and you can write

$$\int_a^b p(x)\,dx = \left(\frac{b-a}{6}\right)\left[p(a) + 4p\left(\frac{a+b}{2}\right) + p(b)\right].$$

See LarsonCalculus.com for Bruce Edwards's video of this proof. ■

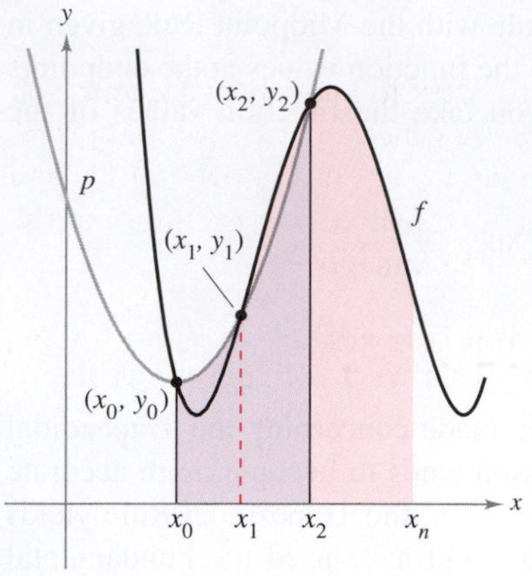

$$\int_{x_0}^{x_2} p(x)\,dx \approx \int_{x_0}^{x_2} f(x)\,dx$$

Figure 4.45

To develop Simpson's Rule for approximating a definite integral, you again partition the interval $[a, b]$ into n subintervals, each of width $\Delta x = (b - a)/n$. This time, however, n is required to be even, and the subintervals are grouped in pairs such that

$$a = \underbrace{x_0 < x_1 < x_2}_{[x_0,\, x_2]} < \underbrace{x_3 < x_4}_{[x_2,\, x_4]} < \cdots < \underbrace{x_{n-2} < x_{n-1} < x_n}_{[x_{n-2},\, x_n]} = b.$$

On each (double) subinterval $[x_{i-2}, x_i]$, you can approximate f by a polynomial p of degree less than or equal to 2. (See Exercise 47.) For example, on the subinterval $[x_0, x_2]$, choose the polynomial of least degree passing through the points (x_0, y_0), (x_1, y_1), and (x_2, y_2), as shown in Figure 4.45. Now, using p as an approximation of f on this subinterval, you have, by Theorem 4.18,

$$\begin{aligned}\int_{x_0}^{x_2} f(x)\,dx &\approx \int_{x_0}^{x_2} p(x)\,dx \\ &= \frac{x_2 - x_0}{6}\left[p(x_0) + 4p\left(\frac{x_0 + x_2}{2}\right) + p(x_2)\right] \\ &= \frac{2[(b-a)/n]}{6}[p(x_0) + 4p(x_1) + p(x_2)] \\ &= \frac{b-a}{3n}[f(x_0) + 4f(x_1) + f(x_2)].\end{aligned}$$

Repeating this procedure on the entire interval $[a, b]$ produces the next theorem.

REMARK Observe that the coefficients in Simpson's Rule have the following pattern.

1 4 2 4 2 4 . . . 4 2 4 1

THEOREM 4.19 Simpson's Rule

Let f be continuous on $[a, b]$ and let n be an even integer. Simpson's Rule for approximating $\int_a^b f(x)\,dx$ is

$$\int_a^b f(x)\,dx \approx \frac{b-a}{3n}[f(x_0) + 4f(x_1) + 2f(x_2) + 4f(x_3) + \cdots + 4f(x_{n-1}) + f(x_n)].$$

Moreover, as $n \to \infty$, the right-hand side approaches $\int_a^b f(x)\,dx$.

REMARK In Section 4.2, Example 8, the Midpoint Rule with $n = 4$ approximates $\int_0^\pi \sin x\,dx$ as 2.052. In Example 1, the Trapezoidal Rule with $n = 4$ gives an approximation of 1.896. In Example 2, Simpson's Rule with $n = 4$ gives an approximation of 2.005. The antiderivative would produce the true value of 2.

In Example 1, the Trapezoidal Rule was used to estimate $\int_0^\pi \sin x\,dx$. In the next example, Simpson's Rule is applied to the same integral.

EXAMPLE 2 Approximation with Simpson's Rule

See LarsonCalculus.com for an interactive version of this type of example.

Use Simpson's Rule to approximate

$$\int_0^\pi \sin x\,dx.$$

Compare the results for $n = 4$ and $n = 8$.

Solution When $n = 4$, you have

$$\int_0^\pi \sin x\,dx \approx \frac{\pi}{12}\left(\sin 0 + 4\sin\frac{\pi}{4} + 2\sin\frac{\pi}{2} + 4\sin\frac{3\pi}{4} + \sin\pi\right) \approx 2.005.$$

When $n = 8$, you have $\int_0^\pi \sin x\,dx \approx 2.0003$.

FOR FURTHER INFORMATION For proofs of the formulas used for estimating the errors involved in the use of the Midpoint Rule and Simpson's Rule, see the article "Elementary Proofs of Error Estimates for the Midpoint and Simpson's Rules" by Edward C. Fazekas, Jr. and Peter R. Mercer in *Mathematics Magazine*. To view this article, go to *MathArticles.com*.

Error Analysis

When you use an approximation technique, it is important to know how accurate you can expect the approximation to be. The next theorem, which is listed without proof, gives the formulas for estimating the errors involved in the use of Simpson's Rule and the Trapezoidal Rule. In general, when using an approximation, you can think of the error E as the difference between $\int_a^b f(x)\,dx$ and the approximation.

THEOREM 4.20 Errors in the Trapezoidal Rule and Simpson's Rule

If f has a continuous second derivative on $[a, b]$, then the error E in approximating $\int_a^b f(x)\,dx$ by the Trapezoidal Rule is

$$|E| \le \frac{(b-a)^3}{12n^2}[\max|f''(x)|], \quad a \le x \le b. \qquad \text{Trapezoidal Rule}$$

Moreover, if f has a continuous fourth derivative on $[a, b]$, then the error E in approximating $\int_a^b f(x)\,dx$ by Simpson's Rule is

$$|E| \le \frac{(b-a)^5}{180n^4}[\max|f^{(4)}(x)|], \quad a \le x \le b. \qquad \text{Simpson's Rule}$$

Theorem 4.20 states that the errors generated by the Trapezoidal Rule and Simpson's Rule have upper bounds dependent on the extreme values of $f''(x)$ and $f^{(4)}(x)$ in the interval $[a, b]$. Furthermore, these errors can be made arbitrarily small by *increasing n*, provided that f'' and $f^{(4)}$ are continuous and therefore bounded in $[a, b]$.

TECHNOLOGY If you have access to a computer algebra system, use it to evaluate the definite integral in Example 3. You should obtain a value of

$$\int_0^1 \sqrt{1+x^2}\,dx = \frac{1}{2}\left[\sqrt{2} + \ln\left(1+\sqrt{2}\right)\right] \approx 1.14779.$$

(The symbol "ln" represents the natural logarithmic function, which you will study in Section 5.1.)

EXAMPLE 3 The Approximate Error in the Trapezoidal Rule

Determine a value of n such that the Trapezoidal Rule will approximate the value of

$$\int_0^1 \sqrt{1+x^2}\,dx$$

with an error that is less than or equal to 0.01.

Solution Begin by letting $f(x) = \sqrt{1+x^2}$ and finding the second derivative of f.

$$f'(x) = x(1+x^2)^{-1/2} \quad \text{and} \quad f''(x) = (1+x^2)^{-3/2}$$

The maximum value of $|f''(x)|$ on the interval $[0, 1]$ is $|f''(0)| = 1$. So, by Theorem 4.20, you can write

$$|E| \le \frac{(b-a)^3}{12n^2}|f''(0)| = \frac{1}{12n^2}(1) = \frac{1}{12n^2}.$$

To obtain an error E that is less than 0.01, you must choose n such that $1/(12n^2) \le 1/100$.

$$100 \le 12n^2 \implies n \ge \sqrt{\frac{100}{12}} \approx 2.89$$

So, you can choose $n = 3$ (because n must be greater than or equal to 2.89) and apply the Trapezoidal Rule, as shown in Figure 4.46, to obtain

$$\int_0^1 \sqrt{1+x^2}\,dx \approx \frac{1}{6}\left[\sqrt{1+0^2} + 2\sqrt{1+\left(\tfrac{1}{3}\right)^2} + 2\sqrt{1+\left(\tfrac{2}{3}\right)^2} + \sqrt{1+1^2}\right] \approx 1.154.$$

So, by adding and subtracting the error from this estimate, you know that

$$1.144 \le \int_0^1 \sqrt{1+x^2}\,dx \le 1.164.$$

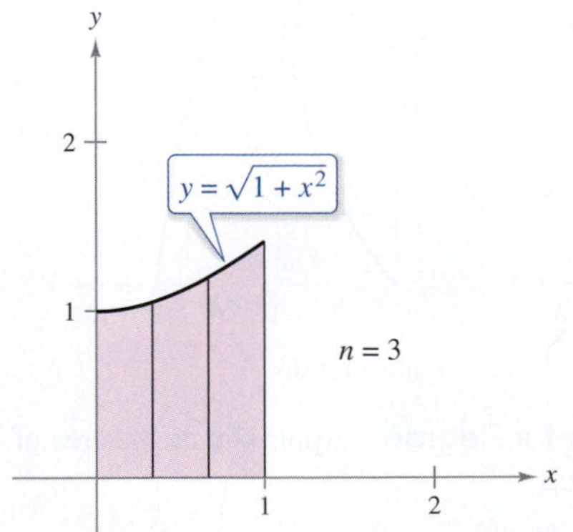

$1.144 \le \int_0^1 \sqrt{1+x^2}\,dx \le 1.164$

Figure 4.46

4.6 Exercises

See CalcChat.com for tutorial help and worked-out solutions to odd-numbered exercises.

Using the Trapezoidal Rule and Simpson's Rule **In Exercises 1–10, use the Trapezoidal Rule and Simpson's Rule to approximate the value of the definite integral for the given value of n. Round your answer to four decimal places and compare the results with the exact value of the definite integral.**

1. $\int_0^2 x^2\,dx,\quad n=4$
2. $\int_1^2 \left(\frac{x^2}{4}+1\right)dx,\quad n=4$
3. $\int_0^2 x^3\,dx,\quad n=4$
4. $\int_2^3 \frac{2}{x^2}\,dx,\quad n=4$
5. $\int_1^3 x^3\,dx,\quad n=6$
6. $\int_0^8 \sqrt[3]{x}\,dx,\quad n=8$
7. $\int_4^9 \sqrt{x}\,dx,\quad n=8$
8. $\int_1^4 (4-x^2)\,dx,\quad n=6$
9. $\int_0^1 \frac{2}{(x+2)^2}\,dx,\quad n=4$
10. $\int_0^2 x\sqrt{x^2+1}\,dx,\quad n=4$

Using the Trapezoidal Rule and Simpson's Rule **In Exercises 11–20, approximate the definite integral using the Trapezoidal Rule and Simpson's Rule with $n=4$. Compare these results with the approximation of the integral using a graphing utility.**

11. $\int_0^2 \sqrt{1+x^3}\,dx$
12. $\int_0^2 \frac{1}{\sqrt{1+x^3}}\,dx$
13. $\int_0^1 \sqrt{x}\sqrt{1-x}\,dx$
14. $\int_{\pi/2}^{\pi} \sqrt{x}\sin x\,dx$
15. $\int_0^{\sqrt{\pi/2}} \sin x^2\,dx$
16. $\int_0^{\sqrt{\pi/4}} \tan x^2\,dx$
17. $\int_3^{3.1} \cos x^2\,dx$
18. $\int_0^{\pi/2} \sqrt{1+\sin^2 x}\,dx$
19. $\int_0^{\pi/4} x\tan x\,dx$
20. $\int_0^{\pi} f(x)\,dx,\quad f(x)=\begin{cases}\dfrac{\sin x}{x}, & x>0\\ 1, & x=0\end{cases}$

WRITING ABOUT CONCEPTS

21. **Polynomial Approximations** The Trapezoidal Rule and Simpson's Rule yield approximations of a definite integral $\int_a^b f(x)\,dx$ based on polynomial approximations of f. What is the degree of the polynomials used for each?

22. **Describing an Error** Describe the size of the error when the Trapezoidal Rule is used to approximate $\int_a^b f(x)\,dx$ when $f(x)$ is a linear function. Use a graph to explain your answer.

Estimating Errors **In Exercises 23–26, use the error formulas in Theorem 4.20 to estimate the errors in approximating the integral, with $n=4$, using (a) the Trapezoidal Rule and (b) Simpson's Rule.**

23. $\int_1^3 2x^3\,dx$
24. $\int_3^5 (5x+2)\,dx$
25. $\int_2^4 \frac{1}{(x-1)^2}\,dx$
26. $\int_0^{\pi} \cos x\,dx$

Estimating Errors **In Exercises 27–30, use the error formulas in Theorem 4.20 to find n such that the error in the approximation of the definite integral is less than or equal to 0.00001 using (a) the Trapezoidal Rule and (b) Simpson's Rule.**

27. $\int_1^3 \frac{1}{x}\,dx$
28. $\int_0^1 \frac{1}{1+x}\,dx$
29. $\int_0^2 \sqrt{x+2}\,dx$
30. $\int_0^{\pi/2} \sin x\,dx$

Estimating Errors Using Technology **In Exercises 31–34, use a computer algebra system and the error formulas to find n such that the error in the approximation of the definite integral is less than or equal to 0.00001 using (a) the Trapezoidal Rule and (b) Simpson's Rule.**

31. $\int_0^2 \sqrt{1+x}\,dx$
32. $\int_0^2 (x+1)^{2/3}\,dx$
33. $\int_0^1 \tan x^2\,dx$
34. $\int_0^1 \sin x^2\,dx$

35. **Finding the Area of a Region** Approximate the area of the shaded region using
(a) the Trapezoidal Rule with $n=4$.
(b) Simpson's Rule with $n=4$.

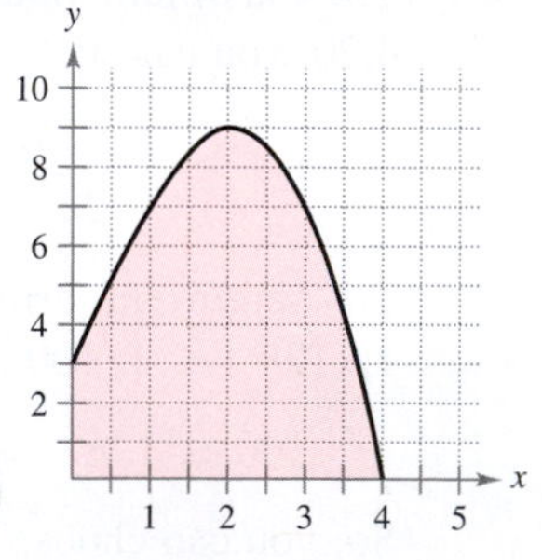

Figure for 35

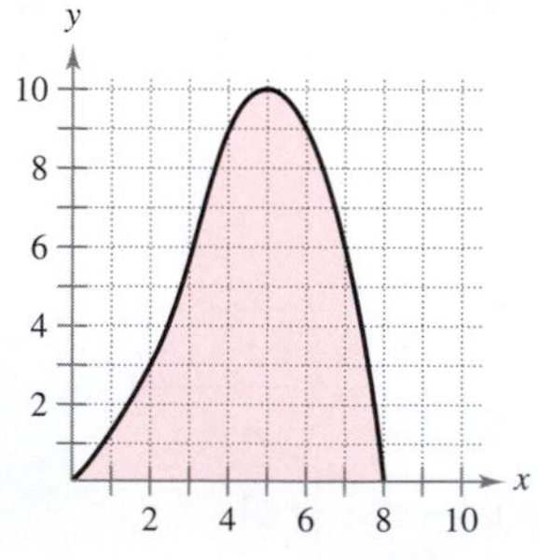

Figure for 36

36. **Finding the Area of a Region** Approximate the area of the shaded region using
(a) the Trapezoidal Rule with $n=8$.
(b) Simpson's Rule with $n=8$.

37. **Area** Use Simpson's Rule with $n=14$ to approximate the area of the region bounded by the graphs of $y=\sqrt{x}\cos x$, $y=0$, $x=0$, and $x=\pi/2$.

38. Circumference The **elliptic integral**

$$8\sqrt{3}\int_0^{\pi/2}\sqrt{1-\tfrac{2}{3}\sin^2\theta}\,d\theta$$

gives the circumference of an ellipse. Use Simpson's Rule with $n = 8$ to approximate the circumference.

39. Surveying

Use the Trapezoidal Rule to estimate the number of square meters of land, where x and y are measured in meters, as shown in the figure. The land is bounded by a stream and two straight roads that meet at right angles.

x	0	100	200	300	400	500
y	125	125	120	112	90	90

x	600	700	800	900	1000
y	95	88	75	35	0

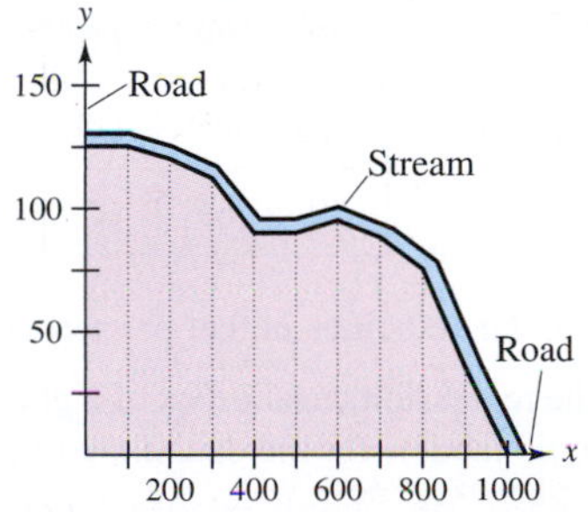

40. HOW DO YOU SEE IT? The function $f(x)$ is concave upward on the interval $[0, 2]$ and the function $g(x)$ is concave downward on the interval $[0, 2]$.

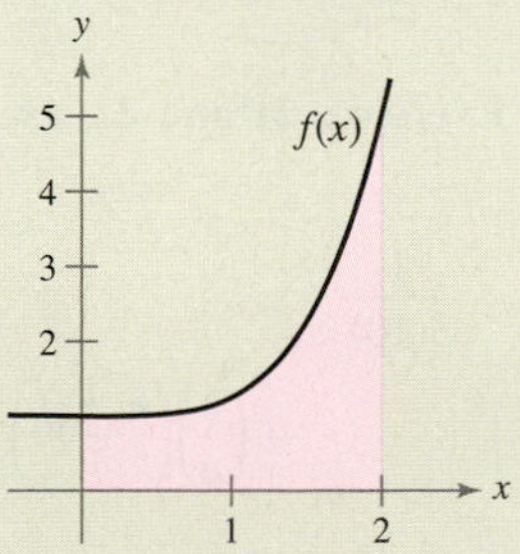

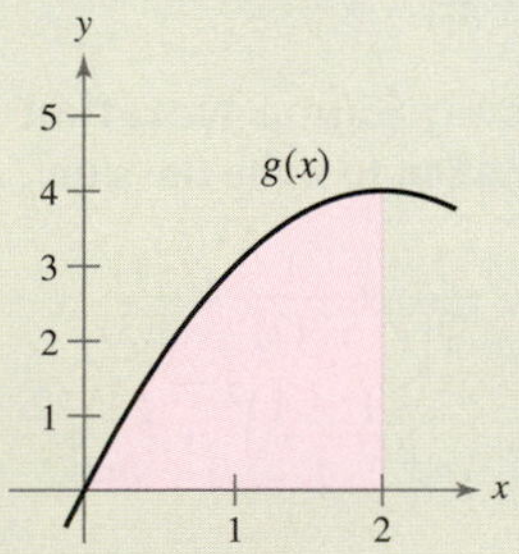

(a) Using the Trapezoidal Rule with $n = 4$, which integral would be overestimated? Which integral would be underestimated? Explain your reasoning.

(b) Which rule would you use for more accurate approximations of $\int_0^2 f(x)\,dx$ and $\int_0^2 g(x)\,dx$, the Trapezoidal Rule or Simpson's Rule? Explain your reasoning.

41. Work To determine the size of the motor required to operate a press, a company must know the amount of work done when the press moves an object linearly 5 feet. The variable force to move the object is

$$F(x) = 100x\sqrt{125 - x^3}$$

where F is given in pounds and x gives the position of the unit in feet. Use Simpson's Rule with $n = 12$ to approximate the work W (in foot-pounds) done through one cycle when

$$W = \int_0^5 F(x)\,dx.$$

42. Approximating a Function The table lists several measurements gathered in an experiment to approximate an unknown continuous function $y = f(x)$.

x	0.00	0.25	0.50	0.75	1.00
y	4.32	4.36	4.58	5.79	6.14

x	1.25	1.50	1.75	2.00
y	7.25	7.64	8.08	8.14

(a) Approximate the integral

$$\int_0^2 f(x)\,dx$$

using the Trapezoidal Rule and Simpson's Rule.

(b) Use a graphing utility to find a model of the form $y = ax^3 + bx^2 + cx + d$ for the data. Integrate the resulting polynomial over $[0, 2]$ and compare the result with the integral from part (a).

Approximation of Pi In Exercises 43 and 44, use Simpson's Rule with $n = 6$ to approximate π using the given equation. (In Section 5.7, you will be able to evaluate the integral using inverse trigonometric functions.)

43. $\pi = \displaystyle\int_0^{1/2} \frac{6}{\sqrt{1 - x^2}}\,dx$

44. $\pi = \displaystyle\int_0^1 \frac{4}{1 + x^2}\,dx$

45. Using Simpson's Rule Use Simpson's Rule with $n = 10$ and a computer algebra system to approximate t in the integral equation

$$\int_0^t \sin\sqrt{x}\,dx = 2.$$

46. Proof Prove that Simpson's Rule is exact when approximating the integral of a cubic polynomial function, and demonstrate the result with $n = 4$ for

$$\int_0^1 x^3\,dx.$$

47. Proof Prove that you can find a polynomial

$$p(x) = Ax^2 + Bx + C$$

that passes through any three points (x_1, y_1), (x_2, y_2), and (x_3, y_3), where the x_i's are distinct.

Review Exercises

See CalcChat.com for tutorial help and worked-out solutions to odd-numbered exercises.

Finding an Indefinite Integral **In Exercises 1–8, find the indefinite integral.**

1. $\int (x - 6)\, dx$
2. $\int (x^4 + 3)\, dx$
3. $\int (4x^2 + x + 3)\, dx$
4. $\int \frac{6}{\sqrt[3]{x}}\, dx$
5. $\int \frac{x^4 + 8}{x^3}\, dx$
6. $\int \frac{x^2 + 2x - 6}{x^4}\, dx$
7. $\int (2x - 9 \sin x)\, dx$
8. $\int (5 \cos x - 2 \sec^2 x)\, dx$

Finding a Particular Solution **In Exercises 9–12, find the particular solution that satisfies the differential equation and the initial condition.**

9. $f'(x) = -6x,\ f(1) = -2$
10. $f'(x) = 9x^2 + 1,\ f(0) = 7$
11. $f''(x) = 24x,\ f'(-1) = 7,\ f(1) = -4$
12. $f''(x) = 2 \cos x,\ f'(0) = 4,\ f(0) = -5$

Slope Field **In Exercises 13 and 14, a differential equation, a point, and a slope field are given. (a) Sketch two approximate solutions of the differential equation on the slope field, one of which passes through the indicated point. (To print an enlarged copy of the graph, go to *MathGraphs.com*.) (b) Use integration to find the particular solution of the differential equation and use a graphing utility to graph the solution.**

13. $\frac{dy}{dx} = 2x - 4, \quad (4, -2)$

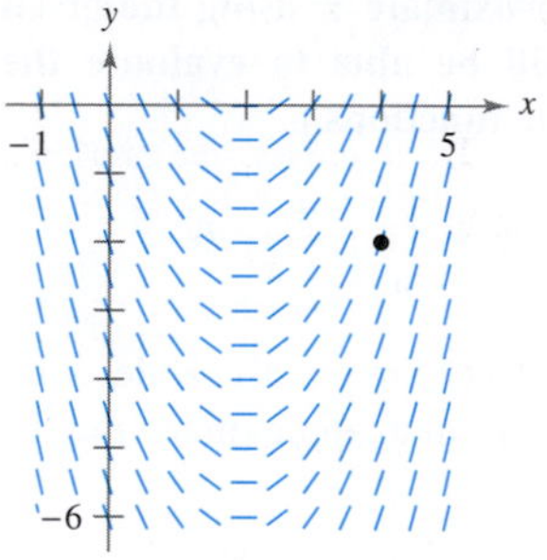

14. $\frac{dy}{dx} = \frac{1}{2}x^2 - 2x, \quad (6, 2)$

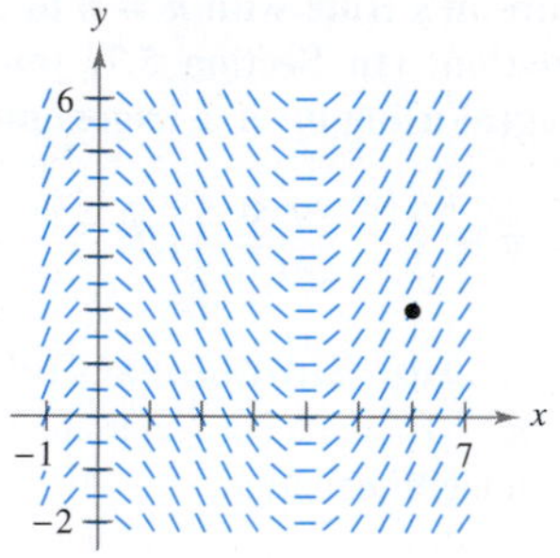

15. **Velocity and Acceleration** A ball is thrown vertically upward from ground level with an initial velocity of 96 feet per second. Use $a(t) = -32$ feet per second per second as the acceleration due to gravity. (Neglect air resistance.)
    (a) How long will it take the ball to rise to its maximum height? What is the maximum height?
    (b) After how many seconds is the velocity of the ball one-half the initial velocity?
    (c) What is the height of the ball when its velocity is one-half the initial velocity?

16. **Velocity and Acceleration** The speed of a car traveling in a straight line is reduced from 45 to 30 miles per hour in a distance of 264 feet. Find the distance in which the car can be brought to rest from 30 miles per hour, assuming the same constant deceleration.

17. **Velocity and Acceleration** An airplane taking off from a runway travels 3600 feet before lifting off. The airplane starts from rest, moves with constant acceleration, and makes the run in 30 seconds. With what speed does it lift off?

18. **Modeling Data** The table shows the velocities (in miles per hour) of two cars on an entrance ramp to an interstate highway. The time t is in seconds.

t	v_1	v_2
0	0	0
5	2.5	21
10	7	38
15	16	51
20	29	60
25	45	64
30	65	65

    (a) Rewrite the velocities in feet per second.
    (b) Use the regression capabilities of a graphing utility to find quadratic models for the data in part (a).
    (c) Approximate the distance traveled by each car during the 30 seconds. Explain the difference in the distances.

Finding a Sum **In Exercises 19 and 20, find the sum. Use the summation capabilities of a graphing utility to verify your result.**

19. $\sum_{i=1}^{5} (5i - 3)$
20. $\sum_{k=0}^{3} (k^2 + 1)$

Using Sigma Notation **In Exercises 21 and 22, use sigma notation to write the sum.**

21. $\frac{1}{3(1)} + \frac{1}{3(2)} + \frac{1}{3(3)} + \cdots + \frac{1}{3(10)}$
22. $\left(\frac{3}{n}\right)\left(\frac{1+1}{n}\right)^2 + \left(\frac{3}{n}\right)\left(\frac{2+1}{n}\right)^2 + \cdots + \left(\frac{3}{n}\right)\left(\frac{n+1}{n}\right)^2$

Evaluating a Sum **In Exercises 23–28, use the properties of summation and Theorem 4.2 to evaluate the sum.**

23. $\sum_{i=1}^{24} 8$
24. $\sum_{i=1}^{75} 5i$
25. $\sum_{i=1}^{20} 2i$
26. $\sum_{i=1}^{30} (3i - 4)$
27. $\sum_{i=1}^{20} (i + 1)^2$
28. $\sum_{i=1}^{12} i(i^2 - 1)$

Finding Upper and Lower Sums for a Region In Exercises 29 and 30, use upper and lower sums to approximate the area of the region using the given number of subintervals (of equal width.)

29. $y = \dfrac{10}{x^2 + 1}$

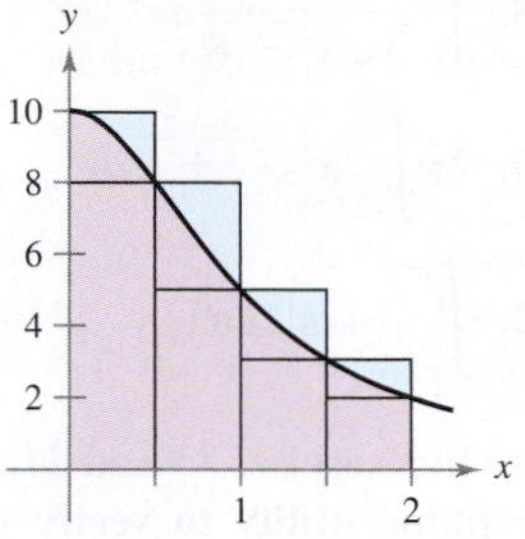

30. $y = 9 - \dfrac{1}{4}x^2$

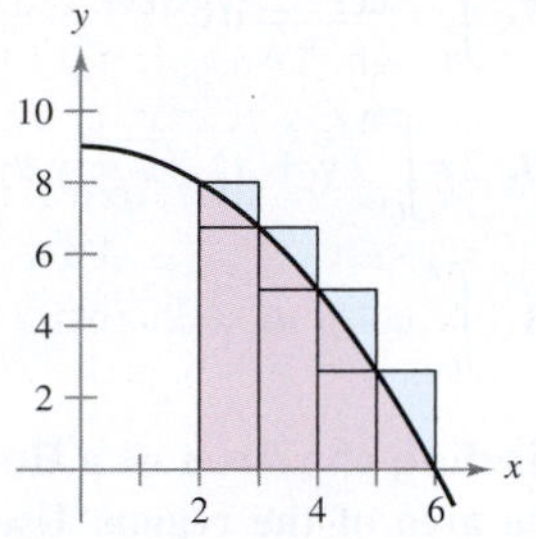

Finding Area by the Limit Definition In Exercises 31–34, use the limit process to find the area of the region bounded by the graph of the function and the x-axis over the given interval. Sketch the region.

31. $y = 8 - 2x, \quad [0, 3]$

32. $y = x^2 + 3, \quad [0, 2]$

33. $y = 5 - x^2, \quad [-2, 1]$

34. $y = \frac{1}{4}x^3, \quad [2, 4]$

35. **Finding Area by the Limit Definition** Use the limit process to find the area of the region bounded by $x = 5y - y^2$, $x = 0$, $y = 2$, and $y = 5$.

36. **Upper and Lower Sums** Consider the region bounded by $y = mx$, $y = 0$, $x = 0$, and $x = b$.

(a) Find the upper and lower sums to approximate the area of the region when $\Delta x = b/4$.

(b) Find the upper and lower sums to approximate the area of the region when $\Delta x = b/n$.

(c) Find the area of the region by letting n approach infinity in both sums in part (b). Show that, in each case, you obtain the formula for the area of a triangle.

Writing a Definite Integral In Exercises 37 and 38, set up a definite integral that yields the area of the region. (Do not evaluate the integral.)

37. $f(x) = 2x + 8$

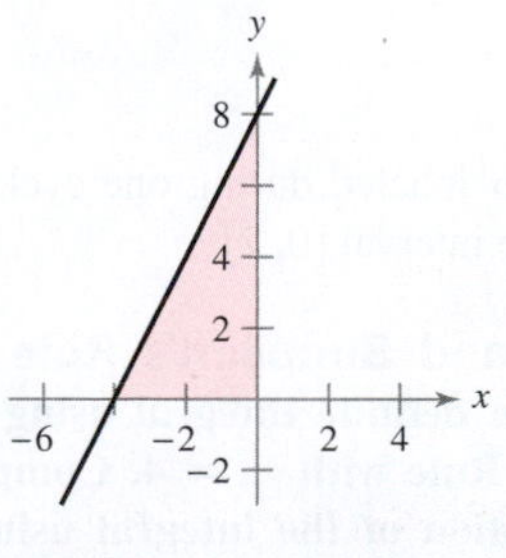

38. $f(x) = 100 - x^2$

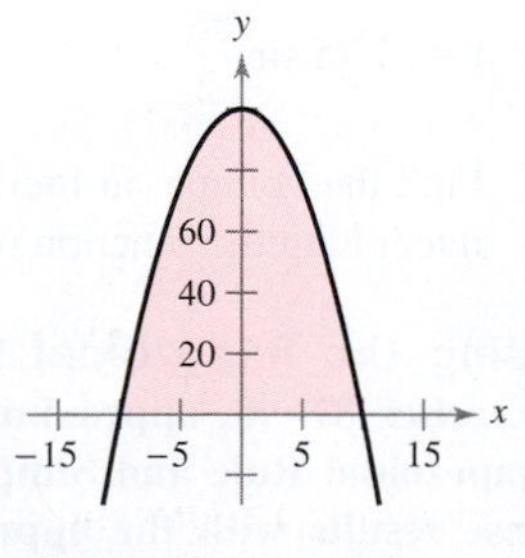

Evaluating a Definite Integral Using a Geometric Formula In Exercises 39 and 40, sketch the region whose area is given by the definite integral. Then use a geometric formula to evaluate the integral.

39. $\displaystyle\int_0^5 (5 - |x - 5|)\, dx$

40. $\displaystyle\int_{-6}^6 \sqrt{36 - x^2}\, dx$

41. **Using Properties of Definite Integrals** Given

$$\int_4^8 f(x)\, dx = 12 \quad \text{and} \quad \int_4^8 g(x)\, dx = 5$$

evaluate

(a) $\displaystyle\int_4^8 [f(x) + g(x)]\, dx.$ (b) $\displaystyle\int_4^8 [f(x) - g(x)]\, dx.$

(c) $\displaystyle\int_4^8 [2f(x) - 3g(x)]\, dx.$ (d) $\displaystyle\int_4^8 7f(x)\, dx.$

42. **Using Properties of Definite Integrals** Given

$$\int_0^3 f(x)\, dx = 4 \quad \text{and} \quad \int_3^6 f(x)\, dx = -1$$

evaluate

(a) $\displaystyle\int_0^6 f(x)\, dx.$ (b) $\displaystyle\int_6^3 f(x)\, dx.$

(c) $\displaystyle\int_4^4 f(x)\, dx.$ (d) $\displaystyle\int_3^6 -10f(x)\, dx.$

Evaluating a Definite Integral In Exercises 43–50, use the Fundamental Theorem of Calculus to evaluate the definite integral.

43. $\displaystyle\int_0^8 (3 + x)\, dx$

44. $\displaystyle\int_2^3 (t^2 - 1)\, dt$

45. $\displaystyle\int_{-1}^1 (4t^3 - 2t)\, dt$

46. $\displaystyle\int_2^3 (x^4 + 4x - 6)\, dx$

47. $\displaystyle\int_4^9 x\sqrt{x}\, dx$

48. $\displaystyle\int_1^4 \left(\frac{1}{x^3} + x\right) dx$

49. $\displaystyle\int_0^{3\pi/4} \sin\theta\, d\theta$

50. $\displaystyle\int_{-\pi/4}^{\pi/4} \sec^2 t\, dt$

Finding the Area of a Region In Exercises 51 and 52, determine the area of the given region.

51. $y = \sin x$

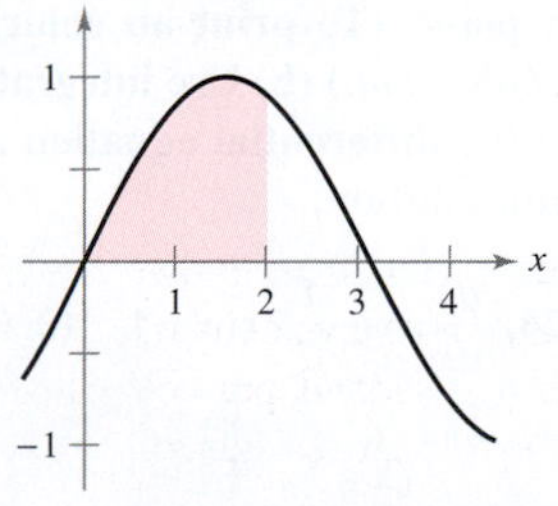

52. $y = x + \cos x$

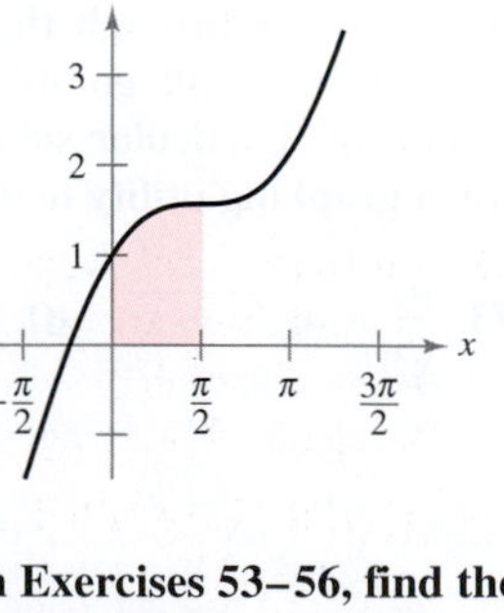

Finding the Area of a Region In Exercises 53–56, find the area of the region bounded by the graphs of the equations.

53. $y = 8 - x,\ x = 0,\ x = 6,\ y = 0$

54. $y = -x^2 + x + 6,\ y = 0$

55. $y = x - x^3,\ x = 0,\ x = 1,\ y = 0$

56. $y = \sqrt{x}(1 - x),\ y = 0$

Finding the Average Value of a Function **In Exercises 57 and 58, find the average value of the function over the given interval and all values of x in the interval for which the function equals its average value.**

57. $f(x) = \dfrac{1}{\sqrt{x}}, \quad [4, 9]$ **58.** $f(x) = x^3, \quad [0, 2]$

Using the Second Fundamental Theorem of Calculus **In Exercises 59–62, use the Second Fundamental Theorem of Calculus to find $F'(x)$.**

59. $F(x) = \displaystyle\int_0^x t^2\sqrt{1 + t^3}\, dt$ **60.** $F(x) = \displaystyle\int_1^x \frac{1}{t^2}\, dt$

61. $F(x) = \displaystyle\int_{-3}^x (t^2 + 3t + 2)\, dt$

62. $F(x) = \displaystyle\int_0^x \csc^2 t\, dt$

Finding an Indefinite Integral **In Exercises 63–72, find the indefinite integral.**

63. $\displaystyle\int \frac{x^2}{\sqrt{x^3 + 3}}\, dx$ **64.** $\displaystyle\int 6x^3\sqrt{3x^4 + 2}\, dx$

65. $\displaystyle\int x(1 - 3x^2)^4\, dx$ **66.** $\displaystyle\int \frac{x + 4}{(x^2 + 8x - 7)^2}\, dx$

67. $\displaystyle\int \sin^3 x \cos x\, dx$ **68.** $\displaystyle\int x \sin 3x^2\, dx$

69. $\displaystyle\int \frac{\cos\theta}{\sqrt{1 - \sin\theta}}\, d\theta$

70. $\displaystyle\int \frac{\sin x}{\sqrt{\cos x}}\, dx$

71. $\displaystyle\int (1 + \sec \pi x)^2 \sec \pi x \tan \pi x\, dx$

72. $\displaystyle\int \sec 2x \tan 2x\, dx$

Slope Field **In Exercises 73 and 74, a differential equation, a point, and a slope field are given. (a) Sketch two approximate solutions of the differential equation on the slope field, one of which passes through the given point. (To print an enlarged copy of the graph, go to *MathGraphs.com*.) (b) Use integration to find the particular solution of the differential equation and use a graphing utility to graph the solution.**

73. $\dfrac{dy}{dx} = x\sqrt{9 - x^2}, \quad (0, -4)$ **74.** $\dfrac{dy}{dx} = -\dfrac{1}{2}x \sin(x^2), \quad (0, 0)$

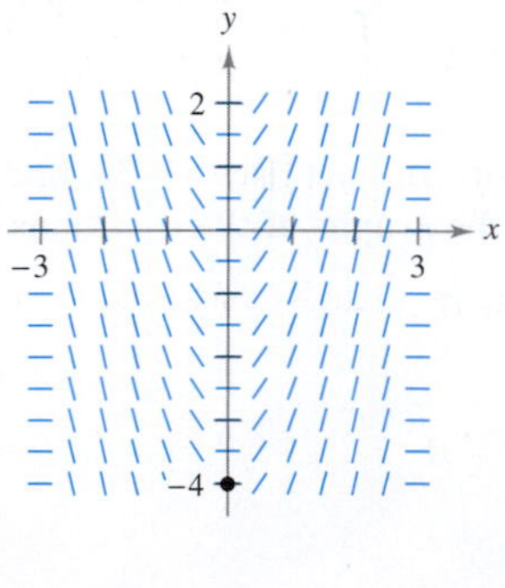

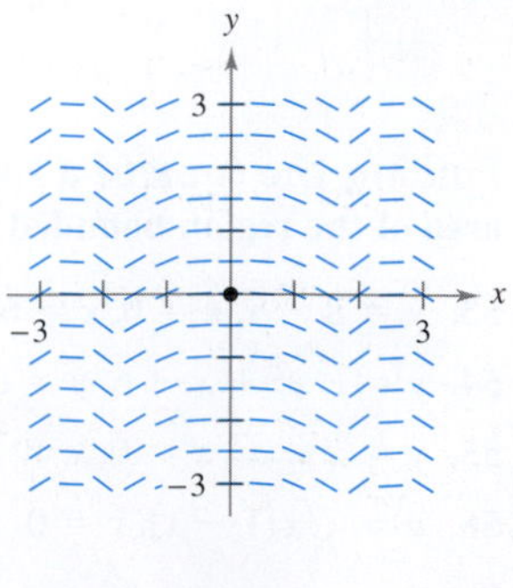

Evaluating a Definite Integral **In Exercises 75–82, evaluate the definite integral. Use a graphing utility to verify your result.**

75. $\displaystyle\int_0^1 (3x + 1)^5\, dx$ **76.** $\displaystyle\int_0^1 x^2(x^3 - 2)^3\, dx$

77. $\displaystyle\int_0^3 \frac{1}{\sqrt{1 + x}}\, dx$ **78.** $\displaystyle\int_3^6 \frac{x}{3\sqrt{x^2 - 8}}\, dx$

79. $2\pi\displaystyle\int_0^1 (y + 1)\sqrt{1 - y}\, dy$ **80.** $2\pi\displaystyle\int_{-1}^0 x^2\sqrt{x + 1}\, dx$

81. $\displaystyle\int_0^{\pi} \cos\frac{x}{2}\, dx$ **82.** $\displaystyle\int_{-\pi/4}^{\pi/4} \sin 2x\, dx$

Finding the Area of a Region **In Exercises 83 and 84, find the area of the region. Use a graphing utility to verify your result.**

83. $\displaystyle\int_1^9 x\sqrt[3]{x - 1}\, dx$ **84.** $\displaystyle\int_0^{\pi/2} [\cos x + \sin(2x)]\, dx$

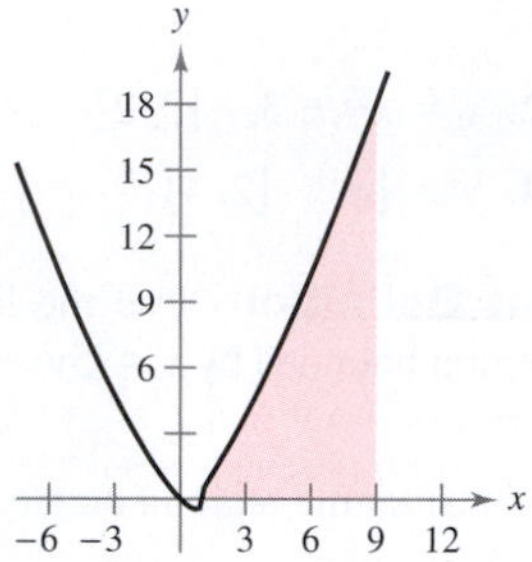

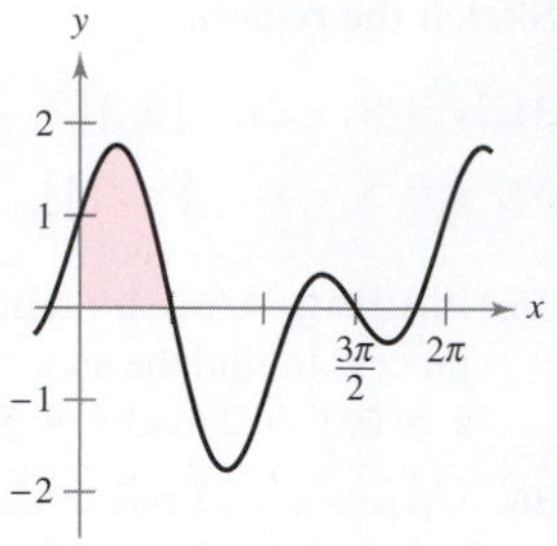

85. Using an Even Function Use $\displaystyle\int_0^2 x^4\, dx = \frac{32}{5}$ to evaluate each definite integral without using the Fundamental Theorem of Calculus.

(a) $\displaystyle\int_{-2}^2 x^4\, dx$ (b) $\displaystyle\int_{-2}^0 x^4\, dx$

(c) $\displaystyle\int_0^2 3x^4\, dx$ (d) $\displaystyle\int_{-2}^0 -5x^4\, dx$

86. Respiratory Cycle After exercising for a few minutes, a person has a respiratory cycle for which the rate of air intake is

$$v = 1.75 \sin\frac{\pi t}{2}.$$

Find the volume, in liters, of air inhaled during one cycle by integrating the function over the interval $[0, 2]$.

Using the Trapezoidal Rule and Simpson's Rule **In Exercises 87–90, approximate the definite integral using the Trapezoidal Rule and Simpson's Rule with $n = 4$. Compare these results with the approximation of the integral using a graphing utility.**

87. $\displaystyle\int_2^3 \frac{2}{1 + x^2}\, dx$ **88.** $\displaystyle\int_0^1 \frac{x^{3/2}}{3 - x^2}\, dx$

89. $\displaystyle\int_0^{\pi/2} \sqrt{x}\cos x\, dx$ **90.** $\displaystyle\int_0^{\pi} \sqrt{1 + \sin^2 x}\, dx$

P.S. Problem Solving

See CalcChat.com for tutorial help and worked-out solutions to odd-numbered exercises.

1. Using a Function Let $L(x) = \int_1^x \frac{1}{t}\,dt,\ x > 0.$

(a) Find $L(1)$.

(b) Find $L'(x)$ and $L'(1)$.

(c) Use a graphing utility to approximate the value of x (to three decimal places) for which $L(x) = 1$.

(d) Prove that $L(x_1x_2) = L(x_1) + L(x_2)$ for all positive values of x_1 and x_2.

2. Parabolic Arch Archimedes showed that the area of a parabolic arch is equal to $\frac{2}{3}$ the product of the base and the height (see figure).

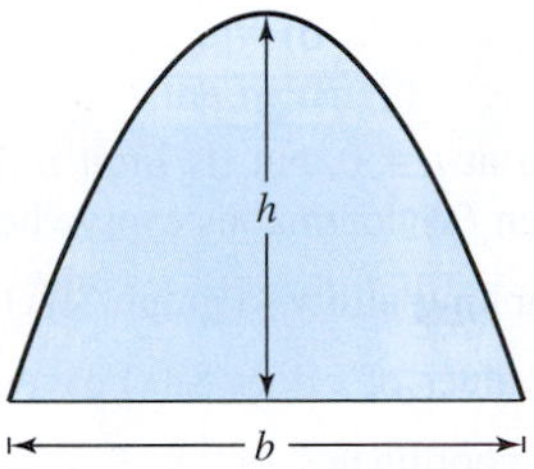

(a) Graph the parabolic arch bounded by $y = 9 - x^2$ and the x-axis. Use an appropriate integral to find the area A.

(b) Find the base and height of the arch and verify Archimedes' formula.

(c) Prove Archimedes' formula for a general parabola.

Evaluating a Sum and a Limit **In Exercises 3 and 4, (a) write the area under the graph of the given function defined on the given interval as a limit. Then (b) evaluate the sum in part (a), and (c) evaluate the limit using the result of part (b).**

3. $y = x^4 - 4x^3 + 4x^2,\quad [0, 2]$

$$\left(\textit{Hint: } \sum_{i=1}^{n} i^4 = \frac{n(n+1)(2n+1)(3n^2+3n-1)}{30}\right)$$

4. $y = \frac{1}{2}x^5 + 2x^3,\quad [0, 2]$

$$\left(\textit{Hint: } \sum_{i=1}^{n} i^5 = \frac{n^2(n+1)^2(2n^2+2n-1)}{12}\right)$$

5. Fresnel Function The **Fresnel function** S is defined by the integral

$$S(x) = \int_0^x \sin\left(\frac{\pi t^2}{2}\right) dt.$$

(a) Graph the function $y = \sin\left(\frac{\pi x^2}{2}\right)$ on the interval $[0, 3]$.

(b) Use the graph in part (a) to sketch the graph of S on the interval $[0, 3]$.

(c) Locate all relative extrema of S on the interval $(0, 3)$.

(d) Locate all points of inflection of S on the interval $(0, 3)$.

6. Approximation The **Two-Point Gaussian Quadrature Approximation** for f is

$$\int_{-1}^{1} f(x)\,dx \approx f\left(-\frac{1}{\sqrt{3}}\right) + f\left(\frac{1}{\sqrt{3}}\right).$$

(a) Use this formula to approximate

$$\int_{-1}^{1} \cos x\,dx.$$

Find the error of the approximation.

(b) Use this formula to approximate

$$\int_{-1}^{1} \frac{1}{1+x^2}\,dx.$$

(c) Prove that the Two-Point Gaussian Quadrature Approximation is exact for all polynomials of degree 3 or less.

7. Extrema and Points of Inflection The graph of the function f consists of the three line segments joining the points $(0, 0)$, $(2, -2)$, $(6, 2)$, and $(8, 3)$. The function F is defined by the integral

$$F(x) = \int_0^x f(t)\,dt.$$

(a) Sketch the graph of f.

(b) Complete the table.

x	0	1	2	3	4	5	6	7	8
$F(x)$									

(c) Find the extrema of F on the interval $[0, 8]$.

(d) Determine all points of inflection of F on the interval $(0, 8)$.

8. Falling Objects Galileo Galilei (1564–1642) stated the following proposition concerning falling objects:

> *The time in which any space is traversed by a uniformly accelerating body is equal to the time in which that same space would be traversed by the same body moving at a uniform speed whose value is the mean of the highest speed of the accelerating body and the speed just before acceleration began.*

Use the techniques of this chapter to verify this proposition.

9. Proof Prove $\int_0^x f(t)(x - t)\,dt = \int_0^x \left(\int_0^t f(v)\,dv\right) dt.$

10. Proof Prove $\int_a^b f(x)f'(x)\,dx = \frac{1}{2}([f(b)]^2 - [f(a)]^2).$

11. Riemann Sum Use an appropriate Riemann sum to evaluate the limit

$$\lim_{n\to\infty} \frac{\sqrt{1} + \sqrt{2} + \sqrt{3} + \cdots + \sqrt{n}}{n^{3/2}}.$$

12. Riemann Sum Use an appropriate Riemann sum to evaluate the limit

$$\lim_{n\to\infty} \frac{1^5 + 2^5 + 3^5 + \cdots + n^5}{n^6}.$$

13. Proof Suppose that f is integrable on $[a, b]$ and $0 < m \le f(x) \le M$ for all x in the interval $[a, b]$. Prove that

$$m(a - b) \le \int_a^b f(x)\,dx \le M(b - a).$$

Use this result to estimate $\int_0^1 \sqrt{1 + x^4}\,dx$.

14. Using a Continuous Function Let f be continuous on the interval $[0, b]$, where $f(x) + f(b - x) \ne 0$ on $[0, b]$.

(a) Show that $\displaystyle\int_0^b \frac{f(x)}{f(x) + f(b - x)}\,dx = \frac{b}{2}$.

(b) Use the result in part (a) to evaluate

$$\int_0^1 \frac{\sin x}{\sin(1 - x) + \sin x}\,dx.$$

(c) Use the result in part (a) to evaluate

$$\int_0^3 \frac{\sqrt{x}}{\sqrt{x} + \sqrt{3 - x}}\,dx.$$

15. Velocity and Acceleration A car travels in a straight line for 1 hour. Its velocity v in miles per hour at six-minute lintervals is shown in the table.

t (hours)	0	0.1	0.2	0.3	0.4	0.5
v (mi/h)	0	10	20	40	60	50

t (hours)	0.6	0.7	0.8	0.9	1.0
v (mi/h)	40	35	40	50	65

(a) Produce a reasonable graph of the velocity function v by graphing these points and connecting them with a smooth curve.

(b) Find the open intervals over which the acceleration a is positive.

(c) Find the average acceleration of the car (in miles per hour squared) over the interval $[0, 0.4]$.

(d) What does the integral

$$\int_0^1 v(t)\,dt$$

signify? Approximate this integral using the Trapezoidal Rule with five subintervals.

(e) Approximate the acceleration at $t = 0.8$.

16. Proof Prove that if f is a continuous function on a closed interval $[a, b]$, then

$$\left|\int_a^b f(x)\,dx\right| \le \int_a^b |f(x)|\,dx.$$

17. Verifying a Sum Verify that

$$\sum_{i=1}^{n} i^2 = \frac{n(n + 1)(2n + 1)}{6}$$

by showing the following.

(a) $(1 + i)^3 - i^3 = 3i^2 + 3i + 1$

(b) $(n + 1)^3 = \displaystyle\sum_{i=1}^{n} (3i^2 + 3i + 1) + 1$

(c) $\displaystyle\sum_{i=1}^{n} i^2 = \frac{n(n + 1)(2n + 1)}{6}$

18. Sine Integral Function The **sine integral function**

$$\operatorname{Si}(x) = \int_0^x \frac{\sin t}{t}\,dt$$

is often used in engineering. The function

$$f(t) = \frac{\sin t}{t}$$

is not defined at $t = 0$, but its limit is 1 as $t \to 0$. So, define $f(0) = 1$. Then f is continuous everywhere.

(a) Use a graphing utility to graph $\operatorname{Si}(x)$.

(b) At what values of x does $\operatorname{Si}(x)$ have relative maxima?

(c) Find the coordinates of the first inflection point where $x > 0$.

(d) Decide whether $\operatorname{Si}(x)$ has any horizontal asymptotes. If so, identify each.

19. Comparing Methods Let

$$I = \int_0^4 f(x)\,dx$$

where f is shown in the figure. Let $L(n)$ and $R(n)$ represent the Riemann sums using the left-hand endpoints and right-hand endpoints of n subintervals of equal width. (Assume n is even.) Let $T(n)$ and $S(n)$ be the corresponding values of the Trapezoidal Rule and Simpson's Rule.

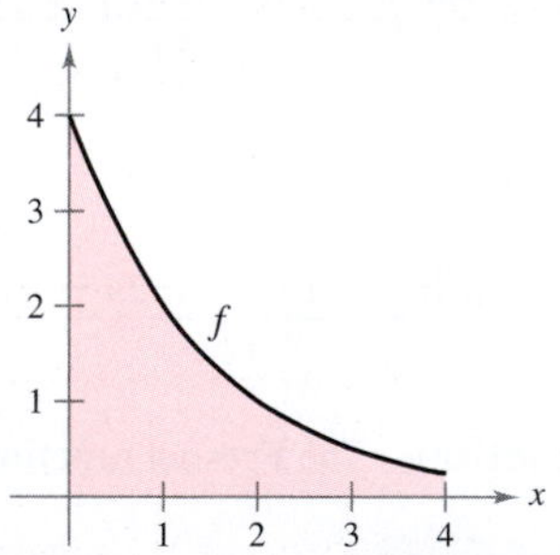

(a) For any n, list $L(n)$, $R(n)$, $T(n)$, and I in increasing order.

(b) Approximate $S(4)$.

20. Minimizing an Integral Determine the limits of integration where $a \le b$ such that

$$\int_a^b (x^2 - 16)\,dx$$

has minimal value.

AP* Review Questions for Chapter 4

1. (no calculator)

For $0 \leq t \leq 9$, a particle moves along the x-axis. The velocity of the particle is given by $v(t) = \sin\left(\frac{\pi}{4}t\right)$. The particle is at position $x = -4$ when $t = 0$.

(a) For $0 \leq t \leq 9$, when is the particle moving to the right?

(b) Write, but do not evaluate, an integral expression that gives the total distance traveled by the particle from time $t = 0$ to $t = 9$.

(c) Find the acceleration of the particle at time $t = 3$. Is the particle speeding up, slowing down or neither at $t = 3$? Justify your response.

(d) Find the position of the particle at time $t = 3$.

2. (no calculator)

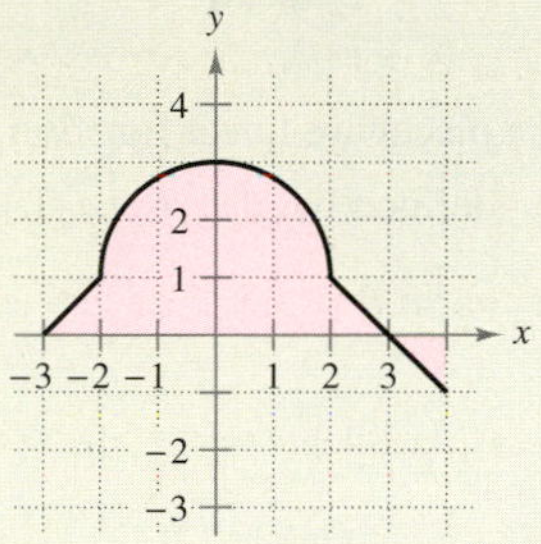

Let $F(x) = \int_3^x f(t)dt$. The graph of f on the interval $[-3, 4]$ consists of 2 line segments and a semicircle, as shown in the graph above.

(a) Find $F(0)$, $F'(0)$, and $F(4)$.

(b) Find all relative minimum values of $F(x)$ on the interval $[-3, 4]$. Justify your answer.

(c) Find the x-coordinates of the inflection points of $F(x)$ on the interval $[-3, 4]$. Justify your answer.

(d) Write the equation of the line tangent to the point where $x = 2$.

3. (calculator)

t (minutes)	0	3	5	7	8	12
$C(t)$ (degrees Celsius)	65	57	50	46	44	40

As a pot of coffee cools down, the temperature of the coffee is modeled by a differentiable function C, for $0 \leq t \leq 12$, where time is measured in minutes and the temperature $C(t)$ is measured in degrees Celsius. Selected values of time t are shown in the table above.

(a) Evaluate $\int_0^{12} C'(t)dt$. Using correct units, explain the meaning of your answer in the context of the problem.

(b) Using correct units, explain the meaning of $\frac{1}{12}\int_0^{12} C(t)dt$ in the context of the problem. Use a trapezoidal sum with 5 subintervals indicated by the table to approximate $\frac{1}{12}\int_0^{12} C(t)dt$.

(c) Use the data in the table to approximate the rate at which the temperature is changing at time $t = 4$. Use proper units and show the computations that lead to your answer.

(d) For $12 \leq t \leq 15$, the rate of cooling is modeled by $C'(t) = -2\cos(0.5t)$. Based on the model, what is the temperature of the coffee at time $t = 15$? Assume $C(t)$ is continuous at $t = 12$.

4. (calculator)

On a typical day, the snow on Pike Mountain melts at a rate modeled by the function

$M(t)$, given by $M(t) = \frac{\pi}{6}\sin\left(\frac{\pi t}{12}\right)$.

A snow maker adds snow at a rate modeled by the function S, given by

$S(t) = 0.006t^2 - 0.12t + 0.87$.

Both M and S have units in inches per hour and t is measured in hours for $0 \leq t \leq 6$.

At time $t = 0$, the mountain has 40 inches of snow.

(a) How much snow will melt during the 6 hour period? Indicate units of measure.

(b) Write an expression for $I(t)$, the total number of inches of snow at any time t.

(c) Find the rate of change of the total amount of snow at time $t = 3$.

(d) For $0 \leq t \leq 6$, at what time t is the amount of snow a maximum? What is the maximum value? Justify your answer.

5. (no calculator)

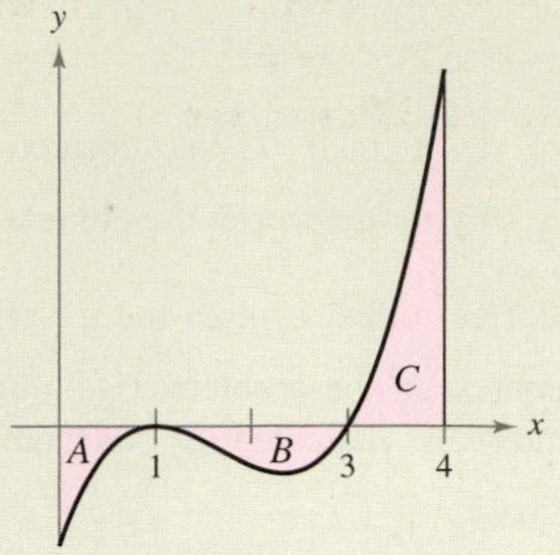

The graph of a continuous function f is shown in the figure above. The three regions between the graph of f and the x-axis are marked A, B, and C and have unsigned areas 5.5, 6, and 15.5, respectively. Let $F(x)$ be an anti-derivative of f that is differentiable on $(0, 4)$ and with $F(1) = 9$.

(a) Find $F(0)$ and $F(4)$.

(b) What is the minimum number of times F equals 5 on the interval $[0, 4]$? Show the work that leads to your conclusion.

(c) Find all intervals where F is increasing. Justify your answer.

6. (calculator)

The velocity, $v(t)$, of a high speed rail train is positive over $0 \le t \le 60$ seconds. The velocity of the train is recorded for selected values of t, as shown in the table below.

t (seconds)	0	10	20	30
$v(t)$ (feet per second)	0	45	105	140

t (seconds)	40	50	60
$v(t)$ (feet per second)	165	195	210

(a) Estimate the acceleration of the train at $t = 25$. Indicate units of measure.

(b) Use a left Riemann sum with three subintervals of equal length to approximate $\int_{20}^{50} v(t)dt$. Using correct units, explain the meaning of $\int_{20}^{50} v(t)dt$ in the context of this problem.

(c) Evaluate $\int_{20}^{50} v'(t)dt$. Using correct units, explain the meaning of $\int_{20}^{50} v'(t)dt$ in the context of this problem.

(d) Estimate the average velocity of the high speed train over the 60 second period of time using a midpoint Riemann sum with 3 subintervals.

7. (no calculator)

Let g be the function defined by

$$g(x) = \begin{cases} -x^2 + 4x & \text{for } 1 \le x \le 2 \\ \sqrt{2x - 4} & \text{for } 2 < x \le 4 \end{cases}$$

(a) Write the equation for the line tangent to the graph of g at $x = 3$.

(b) Is g continuous at $x = 2$? Use the definition of continuity to explain your answer.

(c) Find the average value of $f(x)$ from $x = 1$ to $x = 4$.

8. (no calculator)

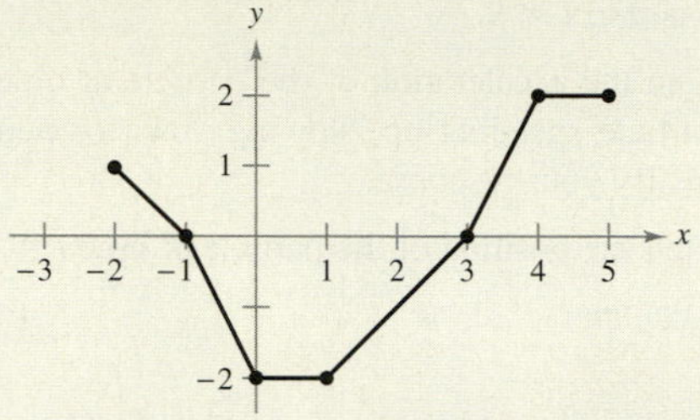

The graph of a piecewise linear function f is shown above. If g is the function defined by $g(x) = \int_4^x f(t)dt$, find $g(-1)$.

(A) 4

(B) −4

(C) 0

(D) 6

(E) −6

9. (calculator)

The velocity of a particle is given by $v(t) = 4t^3 - 4t$ for the times $0 \le t \le 2$ in seconds. Find the average speed of the particle over that interval.

(A) 5

(B) −4

(C) 4

(D) 10

(E) 24

10. (calculator)

Let $f(x)$ be a continuous function such that $f(1) = 2$ and $f'(x) = \sqrt{x^3 + 6}$. What is the value of $f(5)$?

(A) 11.446

(B) 13.446

(C) 15.446

(D) 24.672

(E) 26.672

5 Logarithmic, Exponential, and Other Transcendental Functions

St. Louis Arch *(Section Project, p. 392)*

Radioactive Half-Life Model *(Example 1, p. 356)*

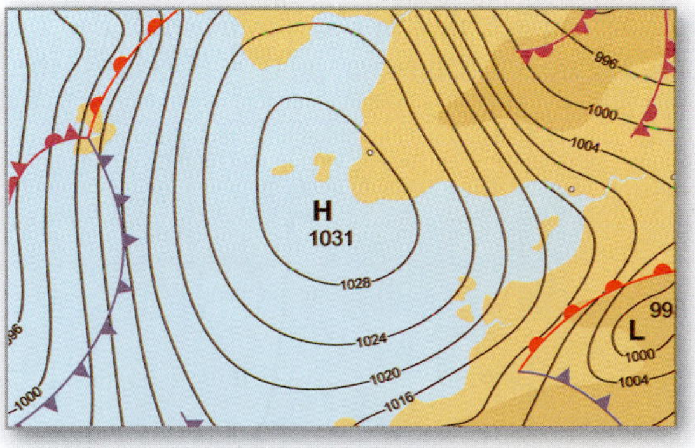

Atmospheric Pressure *(Exercise 85, p. 353)*

Heat Transfer *(Exercise 99, p. 336)*

Sound Intensity *(Exercise 104, p. 327)*

Clockwise from top left, Zens/Shutterstock.com; planet5D LLC/Shutterstock.com; Marijus Auruskevicius/Shutterstock.com; Christopher Dodge/Shutterstock.com; Robert Adrian Hillman/Shutterstock.com

5.1 The Natural Logarithmic Function: Differentiation

- **Develop and use properties of the natural logarithmic function.**
- **Understand the definition of the number e.**
- **Find derivatives of functions involving the natural logarithmic function.**

The Natural Logarithmic Function

Recall that the General Power Rule

$$\int x^n\,dx = \frac{x^{n+1}}{n+1} + C, \quad n \neq -1 \qquad \text{General Power Rule}$$

has an important disclaimer—it doesn't apply when $n = -1$. Consequently, you have not yet found an antiderivative for the function $f(x) = 1/x$. In this section, you will use the Second Fundamental Theorem of Calculus to *define* such a function. This antiderivative is a function that you have not encountered previously in the text. It is neither algebraic nor trigonometric, but falls into a new class of functions called *logarithmic functions*. This particular function is the **natural logarithmic function.**

JOHN NAPIER (1550–1617)

Logarithms were invented by the Scottish mathematician John Napier. Napier coined the term *logarithm*, from the two Greek words *logos* (or ratio) and *arithmos* (or number), to describe the theory that he spent 20 years developing and that first appeared in the book *Mirifici Logarithmorum canonis descriptio* (A Description of the Marvelous Rule of Logarithms). Although he did not introduce the *natural* logarithm function, it is sometimes called the *Napierian* logarithm.
See LarsonCalculus.com to read more of this biography.

Definition of the Natural Logarithmic Function

The **natural logarithmic function** is defined by

$$\ln x = \int_1^x \frac{1}{t}\,dt, \quad x > 0.$$

The domain of the natural logarithmic function is the set of all positive real numbers.

From this definition, you can see that $\ln x$ is positive for $x > 1$ and negative for $0 < x < 1$, as shown in Figure 5.1. Moreover, $\ln(1) = 0$, because the upper and lower limits of integration are equal when $x = 1$.

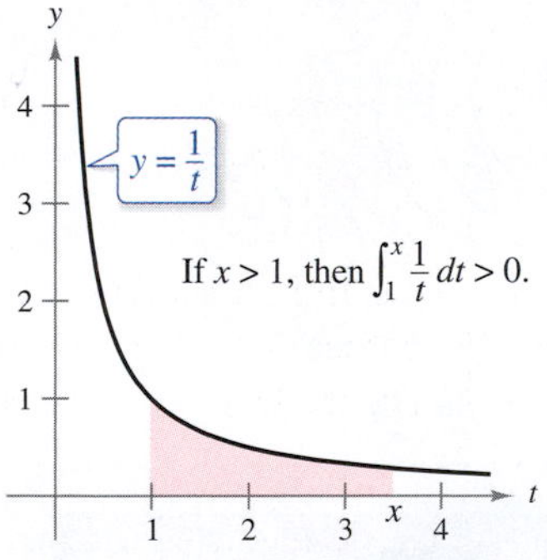

If $x > 1$, then $\ln x > 0$.

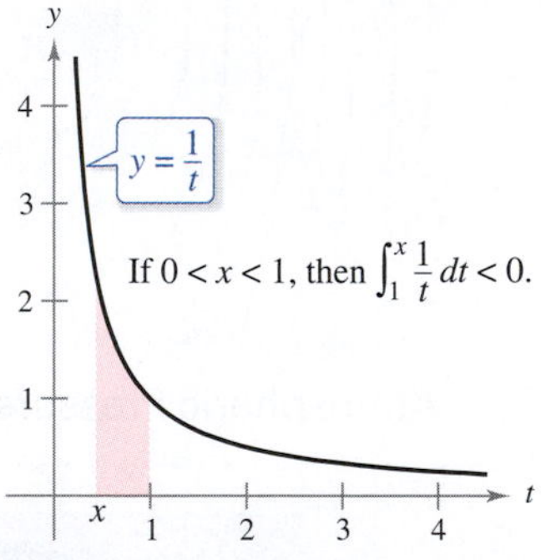

If $0 < x < 1$, then $\ln x < 0$.

Figure 5.1

Exploration

Graphing the Natural Logarithmic Function Using *only* the definition of the natural logarithmic function, sketch a graph of the function. Explain your reasoning.

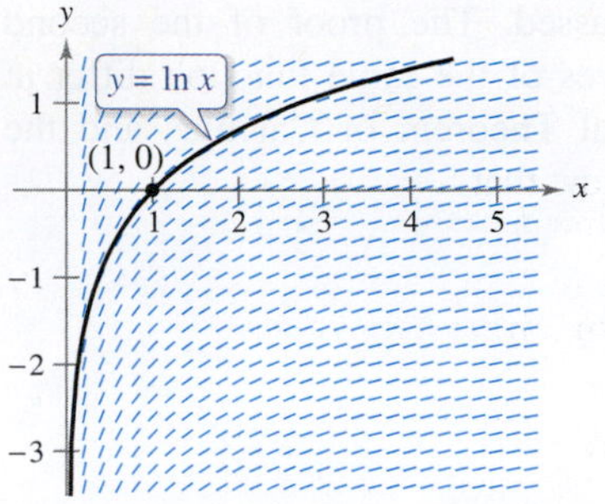

Each small line segment has a slope of $\frac{1}{x}$.

Figure 5.2

To sketch the graph of $y = \ln x$, you can think of the natural logarithmic function as an *antiderivative* given by the differential equation

$$\frac{dy}{dx} = \frac{1}{x}.$$

Figure 5.2 is a computer-generated graph, called a *slope field (or direction field)*, showing small line segments of slope $1/x$. The graph of $y = \ln x$ is the solution that passes through the point $(1, 0)$. (You will study slope fields in Section 6.1.)

The next theorem lists some basic properties of the natural logarithmic function.

> **THEOREM 5.1 Properties of the Natural Logarithmic Function**
>
> The natural logarithmic function has the following properties.
>
> **1.** The domain is $(0, \infty)$ and the range is $(-\infty, \infty)$.
>
> **2.** The function is continuous, increasing, and one-to-one.
>
> **3.** The graph is concave downward.

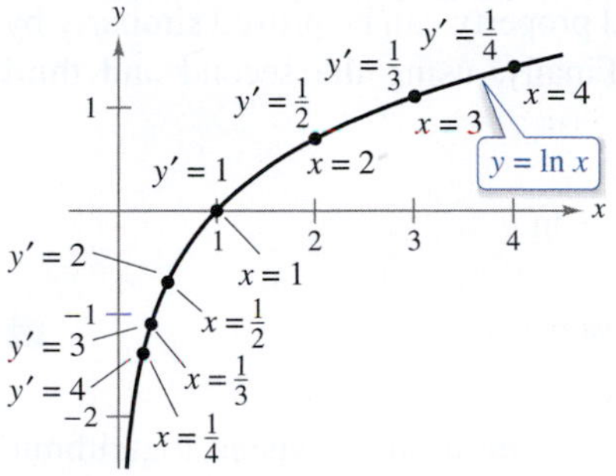

The natural logarithmic function is increasing, and its graph is concave downward.

Figure 5.3

Proof The domain of $f(x) = \ln x$ is $(0, \infty)$ by definition. Moreover, the function is continuous because it is differentiable. It is increasing because its derivative

$$f'(x) = \frac{1}{x} \qquad \text{First derivative}$$

is positive for $x > 0$, as shown in Figure 5.3. It is concave downward because

$$f''(x) = -\frac{1}{x^2} \qquad \text{Second derivative}$$

is negative for $x > 0$. The proof that f is one-to-one is given in Appendix A. The following limits imply that its range is the entire real number line.

$$\lim_{x \to 0^+} \ln x = -\infty$$

and

$$\lim_{x \to \infty} \ln x = \infty$$

Verification of these two limits is given in Appendix A. ■

See LarsonCalculus.com for Bruce Edwards's video of this proof.

Using the definition of the natural logarithmic function, you can prove several important properties involving operations with natural logarithms. If you are already familiar with logarithms, you will recognize that these properties are characteristic of all logarithms.

> **THEOREM 5.2 Logarithmic Properties**
>
> If a and b are positive numbers and n is rational, then the following properties are true.
>
> **1.** $\ln(1) = 0$
>
> **2.** $\ln(ab) = \ln a + \ln b$
>
> **3.** $\ln(a^n) = n \ln a$
>
> **4.** $\ln\left(\frac{a}{b}\right) = \ln a - \ln b$

Proof The first property has already been discussed. The proof of the second property follows from the fact that two antiderivatives of the same function differ at most by a constant. From the Second Fundamental Theorem of Calculus and the definition of the natural logarithmic function, you know that

$$\frac{d}{dx}[\ln x] = \frac{d}{dx}\left[\int_1^x \frac{1}{t}\,dt\right] = \frac{1}{x}.$$

So, consider the two derivatives

$$\frac{d}{dx}[\ln(ax)] = \frac{a}{ax} = \frac{1}{x}$$

and

$$\frac{d}{dx}[\ln a + \ln x] = 0 + \frac{1}{x} = \frac{1}{x}.$$

Because $\ln(ax)$ and $(\ln a + \ln x)$ are both antiderivatives of $1/x$, they must differ at most by a constant.

$$\ln(ax) = \ln a + \ln x + C$$

By letting $x = 1$, you can see that $C = 0$. The third property can be proved similarly by comparing the derivatives of $\ln(x^n)$ and $n \ln x$. Finally, using the second and third properties, you can prove the fourth property.

$$\ln\left(\frac{a}{b}\right) = \ln[a(b^{-1})] = \ln a + \ln(b^{-1}) = \ln a - \ln b$$

See LarsonCalculus.com for Bruce Edwards's video of this proof.

Example 1 shows how logarithmic properties can be used to expand logarithmic expressions.

EXAMPLE 1 Expanding Logarithmic Expressions

a. $\ln\frac{10}{9} = \ln 10 - \ln 9$ — Property 4

b. $\ln\sqrt{3x+2} = \ln(3x+2)^{1/2}$ — Rewrite with rational exponent.

$= \frac{1}{2}\ln(3x+2)$ — Property 3

c. $\ln\frac{6x}{5} = \ln(6x) - \ln 5$ — Property 4

$= \ln 6 + \ln x - \ln 5$ — Property 2

d. $\ln\frac{(x^2+3)^2}{x\sqrt[3]{x^2+1}} = \ln(x^2+3)^2 - \ln\left(x\sqrt[3]{x^2+1}\right)$

$= 2\ln(x^2+3) - [\ln x + \ln(x^2+1)^{1/3}]$

$= 2\ln(x^2+3) - \ln x - \ln(x^2+1)^{1/3}$

$= 2\ln(x^2+3) - \ln x - \frac{1}{3}\ln(x^2+1)$

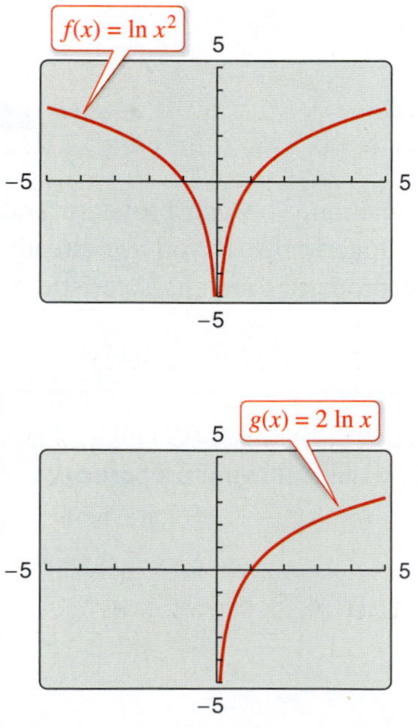

Figure 5.4

When using the properties of logarithms to rewrite logarithmic functions, you must check to see whether the domain of the rewritten function is the same as the domain of the original. For instance, the domain of $f(x) = \ln x^2$ is all real numbers except $x = 0$, and the domain of $g(x) = 2\ln x$ is all positive real numbers. (See Figure 5.4.)

The Number e

It is likely that you have studied logarithms in an algebra course. There, without the benefit of calculus, logarithms would have been defined in terms of a **base** number. For example, common logarithms have a base of 10 and therefore $\log_{10}10 = 1$. (You will learn more about this in Section 5.5.)

The **base for the natural logarithm** is defined using the fact that the natural logarithmic function is continuous, is one-to-one, and has a range of $(-\infty, \infty)$. So, there must be a unique real number x such that $\ln x = 1$, as shown in Figure 5.5. This number is denoted by the letter e. It can be shown that e is irrational and has the following decimal approximation.

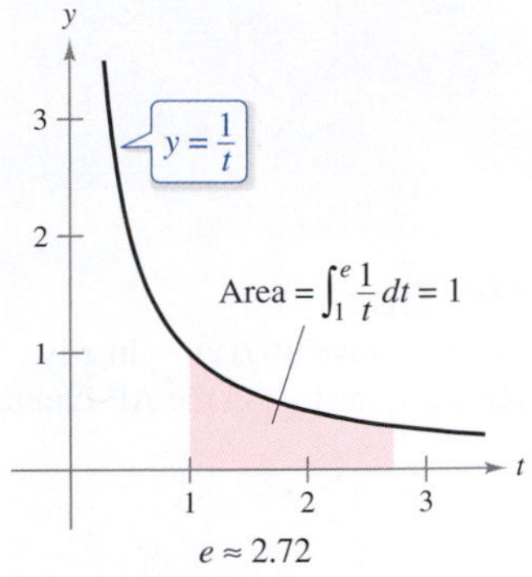

e is the base for the natural logarithm because $\ln e = 1$.
Figure 5.5

$$e \approx 2.71828182846$$

Definition of e

The letter e denotes the positive real number such that

$$\ln e = \int_1^e \frac{1}{t}\,dt = 1.$$

FOR FURTHER INFORMATION To learn more about the number e, see the article "Unexpected Occurrences of the Number e" by Harris S. Shultz and Bill Leonard in *Mathematics Magazine.* To view this article, go to *MathArticles.com.*

Once you know that $\ln e = 1$, you can use logarithmic properties to evaluate the natural logarithms of several other numbers. For example, by using the property

$$\begin{aligned}\ln(e^n) &= n \ln e\\ &= n(1)\\ &= n\end{aligned}$$

you can evaluate $\ln(e^n)$ for various values of n, as shown in the table and in Figure 5.6.

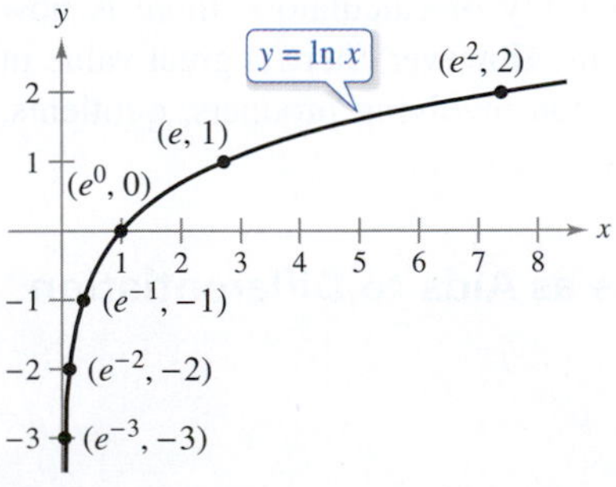

If $x = e^n$, then $\ln x = n$.
Figure 5.6

x	$\frac{1}{e^3} \approx 0.050$	$\frac{1}{e^2} \approx 0.135$	$\frac{1}{e} \approx 0.368$	$e^0 = 1$	$e \approx 2.718$	$e^2 \approx 7.389$
$\ln x$	-3	-2	-1	0	1	2

The logarithms shown in the table above are convenient because the x-values are integer powers of e. Most logarithmic expressions are, however, best evaluated with a calculator.

EXAMPLE 2 **Evaluating Natural Logarithmic Expressions**

a. $\ln 2 \approx 0.693$

b. $\ln 32 \approx 3.466$

c. $\ln 0.1 \approx -2.303$

The Derivative of the Natural Logarithmic Function

The derivative of the natural logarithmic function is given in Theorem 5.3. The first part of the theorem follows from the definition of the natural logarithmic function as an antiderivative. The second part of the theorem is simply the Chain Rule version of the first part.

AP* Tips

The derivative of $f(x) = \ln x$ is frequently tested on the AP Exam.

THEOREM 5.3 Derivative of the Natural Logarithmic Function

Let u be a differentiable function of x.

1. $\dfrac{d}{dx}[\ln x] = \dfrac{1}{x}, \quad x > 0$
2. $\dfrac{d}{dx}[\ln u] = \dfrac{1}{u}\dfrac{du}{dx} = \dfrac{u'}{u}, \quad u > 0$

EXAMPLE 3 Differentiation of Logarithmic Functions

See LarsonCalculus.com for an interactive version of this type of example.

a. $\dfrac{d}{dx}[\ln(2x)] = \dfrac{u'}{u} = \dfrac{2}{2x} = \dfrac{1}{x}$ — $u = 2x$

b. $\dfrac{d}{dx}[\ln(x^2 + 1)] = \dfrac{u'}{u} = \dfrac{2x}{x^2 + 1}$ — $u = x^2 + 1$

c. $\dfrac{d}{dx}[x \ln x] = x\left(\dfrac{d}{dx}[\ln x]\right) + (\ln x)\left(\dfrac{d}{dx}[x]\right)$ — Product Rule

$$= x\left(\frac{1}{x}\right) + (\ln x)(1)$$

$$= 1 + \ln x$$

d. $\dfrac{d}{dx}[(\ln x)^3] = 3(\ln x)^2 \dfrac{d}{dx}[\ln x]$ — Chain Rule

$$= 3(\ln x)^2 \frac{1}{x}$$

Napier used logarithmic properties to simplify *calculations* involving products, quotients, and powers. Of course, given the availability of calculators, there is now little need for this particular application of logarithms. However, there is great value in using logarithmic properties to simplify *differentiation* involving products, quotients, and powers.

EXAMPLE 4 Logarithmic Properties as Aids to Differentiation

Differentiate

$$f(x) = \ln\sqrt{x + 1}.$$

Solution Because

$$f(x) = \ln\sqrt{x + 1} = \ln(x + 1)^{1/2} = \frac{1}{2}\ln(x + 1) \qquad \text{Rewrite before differentiating.}$$

you can write

$$f'(x) = \frac{1}{2}\left(\frac{1}{x + 1}\right) = \frac{1}{2(x + 1)}. \qquad \text{Differentiate.}$$

EXAMPLE 5 **Logarithmic Properties as Aids to Differentiation**

Differentiate

$$f(x) = \ln \frac{x(x^2 + 1)^2}{\sqrt{2x^3 - 1}}.$$

Solution Because

$$f(x) = \ln \frac{x(x^2 + 1)^2}{\sqrt{2x^3 - 1}}$$ Write original function.

$$= \ln x + 2\ln(x^2 + 1) - \frac{1}{2}\ln(2x^3 - 1)$$ Rewrite before differentiating.

you can write

$$f'(x) = \frac{1}{x} + 2\left(\frac{2x}{x^2 + 1}\right) - \frac{1}{2}\left(\frac{6x^2}{2x^3 - 1}\right)$$ Differentiate.

$$= \frac{1}{x} + \frac{4x}{x^2 + 1} - \frac{3x^2}{2x^3 - 1}.$$ Simplify. ■

In Examples 4 and 5, be sure you see the benefit of applying logarithmic properties *before* differentiating. Consider, for instance, the difficulty of direct differentiation of the function given in Example 5.

On occasion, it is convenient to use logarithms as aids in differentiating *nonlogarithmic* functions. This procedure is called **logarithmic differentiation.**

EXAMPLE 6 **Logarithmic Differentiation**

Find the derivative of

$$y = \frac{(x - 2)^2}{\sqrt{x^2 + 1}}, \quad x \neq 2.$$

Solution Note that $y > 0$ for all $x \neq 2$. So, $\ln y$ is defined. Begin by taking the natural logarithm of each side of the equation. Then apply logarithmic properties and differentiate implicitly. Finally, solve for y'.

$$y = \frac{(x - 2)^2}{\sqrt{x^2 + 1}}, \quad x \neq 2$$ Write original equation.

$$\ln y = \ln \frac{(x - 2)^2}{\sqrt{x^2 + 1}}$$ Take natural log of each side.

$$\ln y = 2\ln(x - 2) - \frac{1}{2}\ln(x^2 + 1)$$ Logarithmic properties

$$\frac{y'}{y} = 2\left(\frac{1}{x - 2}\right) - \frac{1}{2}\left(\frac{2x}{x^2 + 1}\right)$$ Differentiate.

$$\frac{y'}{y} = \frac{x^2 + 2x + 2}{(x - 2)(x^2 + 1)}$$ Simplify.

$$y' = y\left[\frac{x^2 + 2x + 2}{(x - 2)(x^2 + 1)}\right]$$ Solve for y'.

$$y' = \frac{(x - 2)^2}{\sqrt{x^2 + 1}}\left[\frac{x^2 + 2x + 2}{(x - 2)(x^2 + 1)}\right]$$ Substitute for y.

$$y' = \frac{(x - 2)(x^2 + 2x + 2)}{(x^2 + 1)^{3/2}}$$ Simplify. ■

Because the natural logarithm is undefined for negative numbers, you will often encounter expressions of the form $\ln|u|$. The next theorem states that you can differentiate functions of the form $y = \ln|u|$ as though the absolute value notation was not present.

THEOREM 5.4 Derivative Involving Absolute Value

If u is a differentiable function of x such that $u \neq 0$, then

$$\frac{d}{dx}[\ln|u|] = \frac{u'}{u}.$$

Proof If $u > 0$, then $|u| = u$, and the result follows from Theorem 5.3. If $u < 0$, then $|u| = -u$, and you have

$$\begin{aligned}\frac{d}{dx}[\ln|u|] &= \frac{d}{dx}[\ln(-u)]\\ &= \frac{-u'}{-u}\\ &= \frac{u'}{u}.\end{aligned}$$

See LarsonCalculus.com for Bruce Edwards's video of this proof.

EXAMPLE 7 Derivative Involving Absolute Value

Find the derivative of

$$f(x) = \ln|\cos x|.$$

Solution Using Theorem 5.4, let $u = \cos x$ and write

$$\begin{aligned}\frac{d}{dx}[\ln|\cos x|] &= \frac{u'}{u} && \frac{d}{dx}[\ln|u|] = \frac{u'}{u}\\ &= \frac{-\sin x}{\cos x} && u = \cos x\\ &= -\tan x. && \text{Simplify.}\end{aligned}$$

EXAMPLE 8 Finding Relative Extrema

Locate the relative extrema of

$$y = \ln(x^2 + 2x + 3).$$

Solution Differentiating y, you obtain

$$\frac{dy}{dx} = \frac{2x + 2}{x^2 + 2x + 3}.$$

Because $dy/dx = 0$ when $x = -1$, you can apply the First Derivative Test and conclude that the point $(-1, \ln 2)$ is a relative minimum. Because there are no other critical points, it follows that this is the only relative extremum. (See Figure 5.7.)

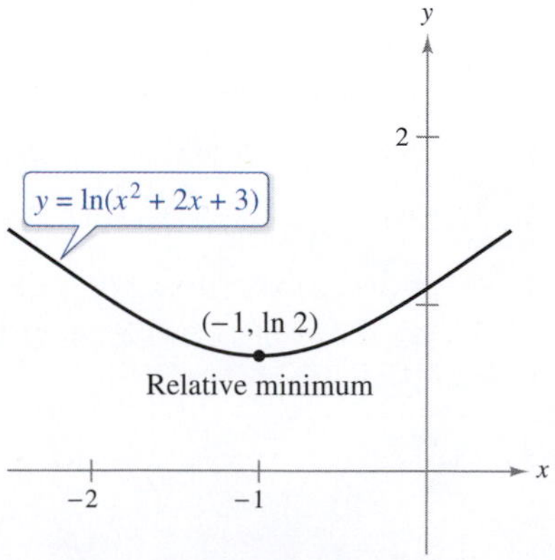

The derivative of y changes from negative to positive at $x = -1$.
Figure 5.7

5.1 Exercises

See CalcChat.com for tutorial help and worked-out solutions to odd-numbered exercises.

Evaluating a Logarithm **In Exercises 1–4, use a graphing utility to evaluate the logarithm by (a) using the natural logarithm key and (b) using the integration capabilities to evaluate the integral $\int_1^x (1/t)\,dt$.**

1. $\ln 45$
2. $\ln 8.3$
3. $\ln 0.8$
4. $\ln 0.6$

Matching **In Exercises 5–8, match the function with its graph. [The graphs are labeled (a), (b), (c), and (d).]**

(a)

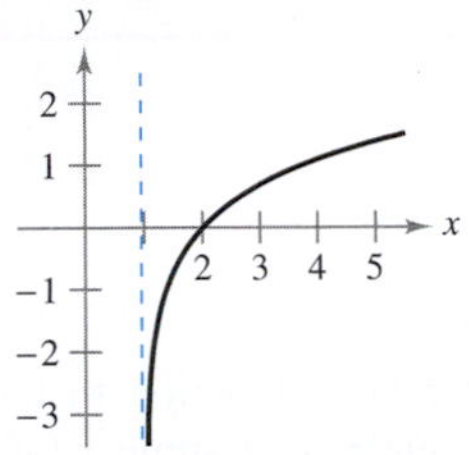

(b)

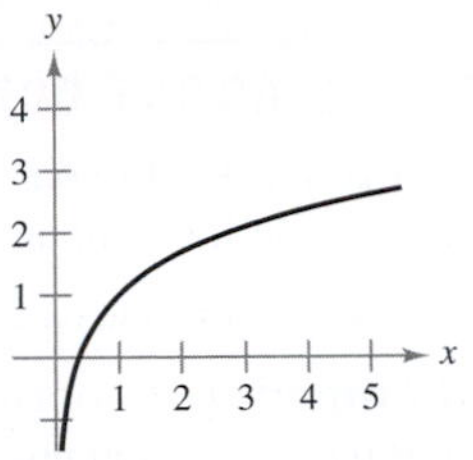

(c)

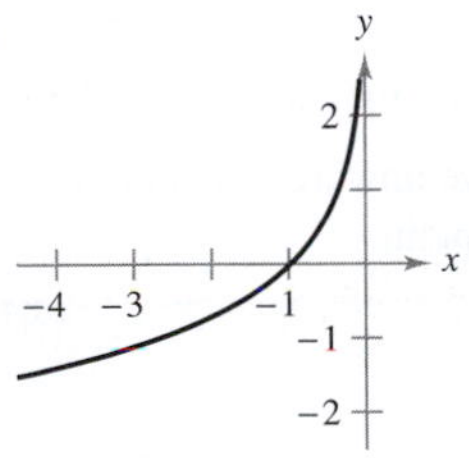

(d)

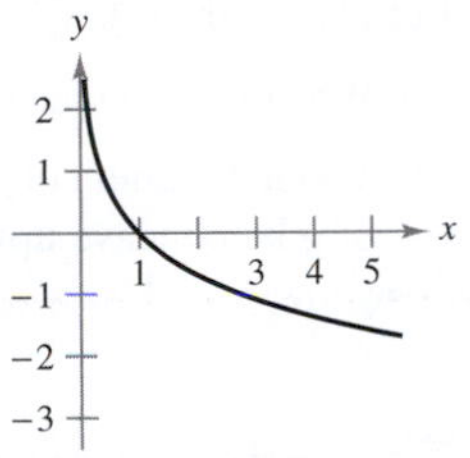

5. $f(x) = \ln x + 1$
6. $f(x) = -\ln x$
7. $f(x) = \ln(x - 1)$
8. $f(x) = -\ln(-x)$

Sketching a Graph **In Exercises 9–16, sketch the graph of the function and state its domain.**

9. $f(x) = 3 \ln x$
10. $f(x) = -2 \ln x$
11. $f(x) = \ln 2x$
12. $f(x) = \ln|x|$
13. $f(x) = \ln(x - 3)$
14. $f(x) = \ln x - 4$
15. $h(x) = \ln(x + 2)$
16. $f(x) = \ln(x - 2) + 1$

Using Properties of Logarithms **In Exercises 17 and 18, use the properties of logarithms to approximate the indicated logarithms, given that $\ln 2 \approx 0.6931$ and $\ln 3 \approx 1.0986$.**

17. (a) $\ln 6$ (b) $\ln \frac{2}{3}$ (c) $\ln 81$ (d) $\ln \sqrt{3}$
18. (a) $\ln 0.25$ (b) $\ln 24$ (c) $\ln \sqrt[3]{12}$ (d) $\ln \frac{1}{72}$

Expanding a Logarithmic Expression **In Exercises 19–28, use the properties of logarithms to expand the logarithmic expression.**

19. $\ln \dfrac{x}{4}$
20. $\ln \sqrt{x^5}$
21. $\ln \dfrac{xy}{z}$
22. $\ln(xyz)$
23. $\ln\left(x\sqrt{x^2 + 5}\right)$
24. $\ln \sqrt{a - 1}$
25. $\ln \sqrt{\dfrac{x - 1}{x}}$
26. $\ln(3e^2)$
27. $\ln z(z - 1)^2$
28. $\ln \dfrac{1}{e}$

Condensing a Logarithmic Expression **In Exercises 29–34, write the expression as a logarithm of a single quantity.**

29. $\ln(x - 2) - \ln(x + 2)$
30. $3 \ln x + 2 \ln y - 4 \ln z$
31. $\frac{1}{3}[2 \ln(x + 3) + \ln x - \ln(x^2 - 1)]$
32. $2[\ln x - \ln(x + 1) - \ln(x - 1)]$
33. $2 \ln 3 - \frac{1}{2}\ln(x^2 + 1)$
34. $\frac{3}{2}[\ln(x^2 + 1) - \ln(x + 1) - \ln(x - 1)]$

Verifying Properties of Logarithms **In Exercises 35 and 36, (a) verify that $f = g$ by using a graphing utility to graph f and g in the same viewing window and (b) verify that $f = g$ algebraically.**

35. $f(x) = \ln \dfrac{x^2}{4}, \quad x > 0, \quad g(x) = 2 \ln x - \ln 4$
36. $f(x) = \ln\sqrt{x(x^2 + 1)}, \quad g(x) = \frac{1}{2}[\ln x + \ln(x^2 + 1)]$

Finding a Limit **In Exercises 37–40, find the limit.**

37. $\lim\limits_{x \to 3^+} \ln(x - 3)$
38. $\lim\limits_{x \to 6^-} \ln(6 - x)$
39. $\lim\limits_{x \to 2^-} \ln[x^2(3 - x)]$
40. $\lim\limits_{x \to 5^+} \ln \dfrac{x}{\sqrt{x - 4}}$

Finding a Derivative **In Exercises 41–64, find the derivative of the function.**

41. $f(x) = \ln(3x)$
42. $f(x) = \ln(x - 1)$
43. $g(x) = \ln x^2$
44. $h(x) = \ln(2x^2 + 1)$
45. $y = (\ln x)^4$
46. $y = x^2 \ln x$
47. $y = \ln(t + 1)^2$
48. $y = \ln\sqrt{x^2 - 4}$
49. $y = \ln\left(x\sqrt{x^2 - 1}\right)$
50. $y = \ln[t(t^2 + 3)^3]$
51. $f(x) = \ln\left(\dfrac{x}{x^2 + 1}\right)$
52. $f(x) = \ln\left(\dfrac{2x}{x + 3}\right)$
53. $g(t) = \dfrac{\ln t}{t^2}$
54. $h(t) = \dfrac{\ln t}{t}$
55. $y = \ln(\ln x^2)$
56. $y = \ln(\ln x)$
57. $y = \ln\sqrt{\dfrac{x + 1}{x - 1}}$
58. $y = \ln\sqrt[3]{\dfrac{x - 1}{x + 1}}$
59. $f(x) = \ln\left(\dfrac{\sqrt{4 + x^2}}{x}\right)$
60. $f(x) = \ln\left(x + \sqrt{4 + x^2}\right)$
61. $y = \ln|\sin x|$
62. $y = \ln|\csc x|$
63. $y = \ln\left|\dfrac{\cos x}{\cos x - 1}\right|$
64. $y = \ln|\sec x + \tan x|$

Finding an Equation of a Tangent Line In Exercises 65–72, (a) find an equation of the tangent line to the graph of f at the given point, (b) use a graphing utility to graph the function and its tangent line at the point, and (c) use the *derivative* feature of a graphing utility to confirm your results.

65. $y = \ln x^4, \quad (1, 0)$
66. $y = \ln x^{3/2}, \quad (1, 0)$
67. $f(x) = 3x^2 - \ln x, \quad (1, 3)$
68. $f(x) = 4 - x^2 - \ln\left(\frac{1}{2}x + 1\right), \quad (0, 4)$
69. $f(x) = \ln\sqrt{1 + \sin^2 x}, \quad \left(\frac{\pi}{4}, \ln\sqrt{\frac{3}{2}}\right)$
70. $f(x) = \sin 2x \ln x^2, \quad (1, 0)$
71. $f(x) = x^3 \ln x, \quad (1, 0)$
72. $f(x) = \frac{1}{2}x \ln x^2, \quad (-1, 0)$

Finding a Derivative Implicitly In Exercises 73–76, use implicit differentiation to find dy/dx.

73. $x^2 - 3\ln y + y^2 = 10$
74. $\ln xy + 5x = 30$
75. $4x^3 + \ln y^2 + 2y = 2x$
76. $4xy + \ln x^2y = 7$

Differential Equation In Exercises 77 and 78, show that the function is a solution of the differential equation.

	Function	Differential Equation
77.	$y = 2\ln x + 3$	$xy'' + y' = 0$
78.	$y = x\ln x - 4x$	$x + y - xy' = 0$

Relative Extrema and Points of Inflection In Exercises 79–84, locate any relative extrema and points of inflection. Use a graphing utility to confirm your results.

79. $y = \frac{x^2}{2} - \ln x$
80. $y = 2x - \ln(2x)$
81. $y = x\ln x$
82. $y = \frac{\ln x}{x}$
83. $y = \frac{x}{\ln x}$
84. $y = x^2 \ln\frac{x}{4}$

Linear and Quadratic Approximation In Exercises 85 and 86, use a graphing utility to graph the function. Then graph

$$P_1(x) = f(1) + f'(1)(x - 1)$$

and

$$P_2(x) = f(1) + f'(1)(x - 1) + \tfrac{1}{2}f''(1)(x - 1)^2$$

in the same viewing window. Compare the values of f, P_1, P_2, and their first derivatives at $x = 1$.

85. $f(x) = \ln x$
86. $f(x) = x\ln x$

Using Newton's Method In Exercises 87 and 88, use Newton's Method to approximate, to three decimal places, the x-coordinate of the point of intersection of the graphs of the two equations. Use a graphing utility to verify your result.

87. $y = \ln x, \quad y = -x$
88. $y = \ln x, \quad y = 3 - x$

Logarithmic Differentiation In Exercises 89–94, use logarithmic differentiation to find dy/dx.

89. $y = x\sqrt{x^2 + 1}, \quad x > 0$
90. $y = \sqrt{x^2(x + 1)(x + 2)}, \quad x > 0$
91. $y = \frac{x^2\sqrt{3x - 2}}{(x + 1)^2}, \quad x > \frac{2}{3}$
92. $y = \sqrt{\frac{x^2 - 1}{x^2 + 1}}, \quad x > 1$
93. $y = \frac{x(x - 1)^{3/2}}{\sqrt{x + 1}}, \quad x > 1$
94. $y = \frac{(x + 1)(x - 2)}{(x - 1)(x + 2)}, \quad x > 2$

WRITING ABOUT CONCEPTS

95. **Properties** In your own words, state the properties of the natural logarithmic function.
96. **Base** Define the base for the natural logarithmic function.
97. **Comparing Functions** Let f be a function that is positive and differentiable on the entire real number line. Let $g(x) = \ln f(x)$.
    (a) When g is increasing, must f be increasing? Explain.
    (b) When the graph of f is concave upward, must the graph of g be concave upward? Explain.

98. **HOW DO YOU SEE IT?** The graph shows the temperature T (in °C) of an object h hours after it is removed from a furnace.

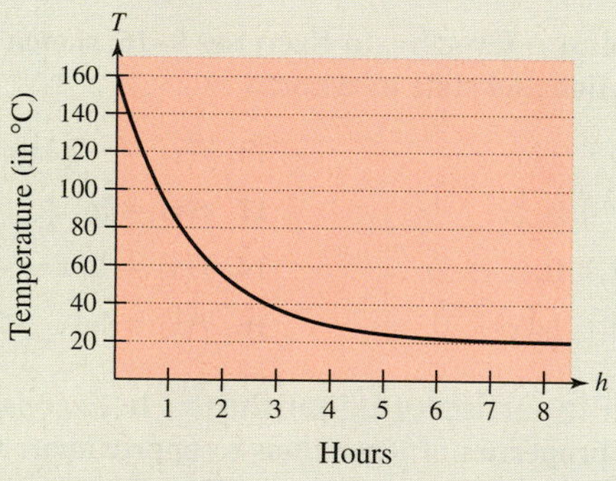

(a) Find $\lim_{h\to\infty} T$. What does this limit represent?
(b) When is the temperature changing most rapidly?

True or False? In Exercises 99–102, determine whether the statement is true or false. If it is false, explain why or give an example that shows it is false.

99. $\ln(x + 25) = \ln x + \ln 25$
100. $\ln xy = \ln x \ln y$
101. If $y = \ln \pi$, then $y' = 1/\pi$.
102. If $y = \ln e$, then $y' = 1$.

103. Home Mortgage The term t (in years) of a \$200,000 home mortgage at 7.5% interest can be approximated by

$$t = 13.375 \ln\left(\frac{x}{x - 1250}\right), \quad x > 1250$$

where x is the monthly payment in dollars.

(a) Use a graphing utility to graph the model.

(b) Use the model to approximate the term of a home mortgage for which the monthly payment is \$1398.43. What is the total amount paid?

(c) Use the model to approximate the term of a home mortgage for which the monthly payment is \$1611.19. What is the total amount paid?

(d) Find the instantaneous rates of change of t with respect to x when $x = \$1398.43$ and $x = \$1611.19$.

(e) Write a short paragraph describing the benefit of the higher monthly payment.

104. Sound Intensity

The relationship between the number of decibels β and the intensity of a sound I in watts per centimeter squared is

$$\beta = \frac{10}{\ln 10} \ln\left(\frac{I}{10^{-16}}\right).$$

(a) Use the properties of logarithms to write the formula in simpler form.

(b) Determine the number of decibels of a sound with an intensity of 10^{-5} watt per square centimeter.

105. Modeling Data The table shows the temperatures T (in °F) at which water boils at selected pressures p (in pounds per square inch). *(Source: Standard Handbook of Mechanical Engineers)*

p	5	10	14.696 (1 atm)	20
T	162.24°	193.21°	212.00°	227.96°

p	30	40	60	80	100
T	250.33°	267.25°	292.71°	312.03°	327.81°

A model that approximates the data is

$$T = 87.97 + 34.96 \ln p + 7.91\sqrt{p}.$$

(a) Use a graphing utility to plot the data and graph the model.

(b) Find the rates of change of T with respect to p when $p = 10$ and $p = 70$.

(c) Use a graphing utility to graph T'. Find $\lim_{p\to\infty} T'(p)$ and interpret the result in the context of the problem.

106. Modeling Data The atmospheric pressure decreases with increasing altitude. At sea level, the average air pressure is one atmosphere (1.033227 kilograms per square centimeter). The table shows the pressures p (in atmospheres) at selected altitudes h (in kilometers).

h	0	5	10	15	20	25
p	1	0.55	0.25	0.12	0.06	0.02

(a) Use a graphing utility to find a model of the form $p = a + b \ln h$ for the data. Explain why the result is an error message.

(b) Use a graphing utility to find the logarithmic model $h = a + b \ln p$ for the data.

(c) Use a graphing utility to plot the data and graph the model.

(d) Use the model to estimate the altitude when $p = 0.75$.

(e) Use the model to estimate the pressure when $h = 13$.

(f) Use the model to find the rates of change of pressure when $h = 5$ and $h = 20$. Interpret the results.

107. Tractrix A person walking along a dock drags a boat by a 10-meter rope. The boat travels along a path known as a *tractrix* (see figure). The equation of this path is

$$y = 10 \ln\left(\frac{10 + \sqrt{100 - x^2}}{x}\right) - \sqrt{100 - x^2}.$$

(a) Use a graphing utility to graph the function.

(b) What are the slopes of this path when $x = 5$ and $x = 9$?

(c) What does the slope of the path approach as $x \to 10$?

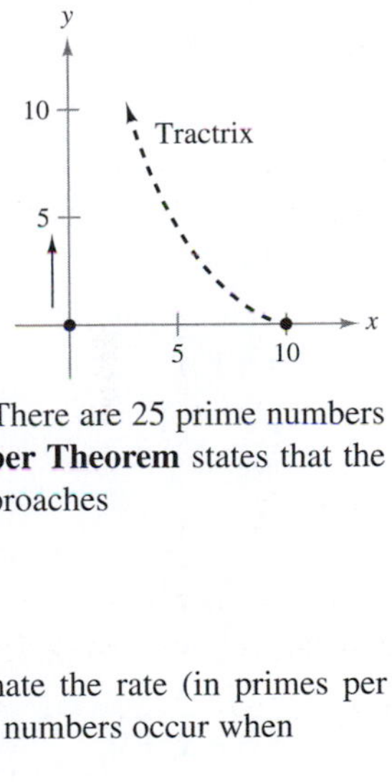

108. Prime Number Theorem There are 25 prime numbers less than 100. The **Prime Number Theorem** states that the number of primes less than x approaches

$$p(x) \approx \frac{x}{\ln x}.$$

Use this approximation to estimate the rate (in primes per 100 integers) at which the prime numbers occur when

(a) $x = 1000$.

(b) $x = 1{,}000{,}000$.

(c) $x = 1{,}000{,}000{,}000$.

109. Conjecture Use a graphing utility to graph f and g in the same viewing window and determine which is increasing at the greater rate for large values of x. What can you conclude about the rate of growth of the natural logarithmic function?

(a) $f(x) = \ln x, \quad g(x) = \sqrt{x}$

(b) $f(x) = \ln x, \quad g(x) = \sqrt[4]{x}$

Christopher Dodge/Shutterstock.com

5.2 The Natural Logarithmic Function: Integration

- **Use the Log Rule for Integration to integrate a rational function.**
- **Integrate trigonometric functions.**

Log Rule for Integration

The differentiation rules

$$\frac{d}{dx}[\ln|x|] = \frac{1}{x} \quad \text{and} \quad \frac{d}{dx}[\ln|u|] = \frac{u'}{u}$$

that you studied in the preceding section produce the following integration rule.

THEOREM 5.5 Log Rule for Integration

Let u be a differentiable function of x.

1. $\displaystyle\int \frac{1}{x}\,dx = \ln|x| + C$ **2.** $\displaystyle\int \frac{1}{u}\,du = \ln|u| + C$

Because $du = u'\,dx$, the second formula can also be written as

$$\int \frac{u'}{u}\,dx = \ln|u| + C. \qquad \text{Alternative form of Log Rule}$$

AP* Tips

The integrals in Theorem 5.5 are frequently tested on the AP Exam.

Exploration

Integrating Rational Functions
Early in Chapter 4, you learned rules that allowed you to integrate *any* polynomial function. The Log Rule presented in this section goes a long way toward enabling you to integrate rational functions. For instance, each of the following functions can be integrated with the Log Rule.

$\dfrac{2}{x}$	Example 1
$\dfrac{1}{4x-1}$	Example 2
$\dfrac{x}{x^2+1}$	Example 3
$\dfrac{3x^2+1}{x^3+x}$	Example 4(a)
$\dfrac{x+1}{x^2+2x}$	Example 4(c)
$\dfrac{1}{3x+2}$	Example 4(d)
$\dfrac{x^2+x+1}{x^2+1}$	Example 5
$\dfrac{2x}{(x+1)^2}$	Example 6

There are still some rational functions that cannot be integrated using the Log Rule. Give examples of these functions, and explain your reasoning.

EXAMPLE 1 Using the Log Rule for Integration

$$\begin{aligned}
\int \frac{2}{x}\,dx &= 2\int \frac{1}{x}\,dx && \text{Constant Multiple Rule}\\
&= 2\ln|x| + C && \text{Log Rule for Integration}\\
&= \ln(x^2) + C && \text{Property of logarithms}
\end{aligned}$$

Because x^2 cannot be negative, the absolute value notation is unnecessary in the final form of the antiderivative.

EXAMPLE 2 Using the Log Rule with a Change of Variables

Find $\displaystyle\int \frac{1}{4x-1}\,dx$.

Solution If you let $u = 4x - 1$, then $du = 4\,dx$.

$$\begin{aligned}
\int \frac{1}{4x-1}\,dx &= \frac{1}{4}\int \left(\frac{1}{4x-1}\right)4\,dx && \text{Multiply and divide by 4.}\\
&= \frac{1}{4}\int \frac{1}{u}\,du && \text{Substitute: } u = 4x-1.\\
&= \frac{1}{4}\ln|u| + C && \text{Apply Log Rule.}\\
&= \frac{1}{4}\ln|4x-1| + C && \text{Back-substitute.}
\end{aligned}$$

■

Example 3 uses the alternative form of the Log Rule. To apply this rule, look for quotients in which the numerator is the derivative of the denominator.

EXAMPLE 3 Finding Area with the Log Rule

Find the area of the region bounded by the graph of

$$y = \frac{x}{x^2 + 1}$$

the x-axis, and the line $x = 3$.

Solution In Figure 5.8, you can see that the area of the region is given by the definite integral

$$\int_0^3 \frac{x}{x^2 + 1}\, dx.$$

If you let $u = x^2 + 1$, then $u' = 2x$. To apply the Log Rule, multiply and divide by 2 as shown.

$$\int_0^3 \frac{x}{x^2 + 1}\, dx = \frac{1}{2}\int_0^3 \frac{2x}{x^2 + 1}\, dx \qquad \text{Multiply and divide by 2.}$$

$$= \frac{1}{2}\Big[\ln(x^2 + 1)\Big]_0^3 \qquad \int \frac{u'}{u}\, dx = \ln|u| + C$$

$$= \frac{1}{2}(\ln 10 - \ln 1)$$

$$= \frac{1}{2}\ln 10 \qquad \ln 1 = 0$$

$$\approx 1.151$$

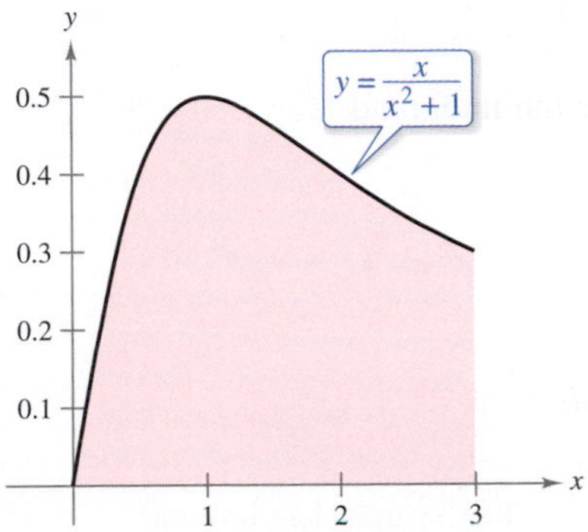

Area $= \int_0^3 \frac{x}{x^2 + 1}\, dx$

The area of the region bounded by the graph of y, the x-axis, and $x = 3$ is $\frac{1}{2}\ln 10$.

Figure 5.8

EXAMPLE 4 Recognizing Quotient Forms of the Log Rule

a. $\displaystyle\int \frac{3x^2 + 1}{x^3 + x}\, dx = \ln|x^3 + x| + C \qquad u = x^3 + x$

b. $\displaystyle\int \frac{\sec^2 x}{\tan x}\, dx = \ln|\tan x| + C \qquad u = \tan x$

c. $\displaystyle\int \frac{x + 1}{x^2 + 2x}\, dx = \frac{1}{2}\int \frac{2x + 2}{x^2 + 2x}\, dx \qquad u = x^2 + 2x$

$$= \frac{1}{2}\ln|x^2 + 2x| + C$$

d. $\displaystyle\int \frac{1}{3x + 2}\, dx = \frac{1}{3}\int \frac{3}{3x + 2}\, dx \qquad u = 3x + 2$

$$= \frac{1}{3}\ln|3x + 2| + C$$

With antiderivatives involving logarithms, it is easy to obtain forms that look quite different but are still equivalent. For instance, both

$$\ln|(3x + 2)^{1/3}| + C$$

and

$$\ln|3x + 2|^{1/3} + C$$

are equivalent to the antiderivative listed in Example 4(d).

Integrals to which the Log Rule can be applied often appear in disguised form. For instance, when a rational function has a *numerator of degree greater than or equal to that of the denominator*, division may reveal a form to which you can apply the Log Rule. This is shown in Example 5.

EXAMPLE 5 **Using Long Division Before Integrating**

See LarsonCalculus.com for an interactive version of this type of example.

Find the indefinite integral.

$$\int \frac{x^2 + x + 1}{x^2 + 1}\, dx$$

Solution Begin by using long division to rewrite the integrand.

$$\frac{x^2 + x + 1}{x^2 + 1} \Longrightarrow x^2 + 1 \overline{\smash{)}x^2 + x + 1} \text{ (quotient } 1\text{; } x^2 + 1 \text{ subtracted, remainder } x) \Longrightarrow 1 + \frac{x}{x^2 + 1}$$

Now, you can integrate to obtain

$$\begin{aligned}
\int \frac{x^2 + x + 1}{x^2 + 1}\, dx &= \int \left(1 + \frac{x}{x^2 + 1}\right) dx && \text{Rewrite using long division.}\\
&= \int dx + \frac{1}{2}\int \frac{2x}{x^2 + 1}\, dx && \text{Rewrite as two integrals.}\\
&= x + \frac{1}{2}\ln(x^2 + 1) + C. && \text{Integrate.}
\end{aligned}$$

Check this result by differentiating to obtain the original integrand. ■

The next example presents another instance in which the use of the Log Rule is disguised. In this case, a change of variables helps you recognize the Log Rule.

EXAMPLE 6 **Change of Variables with the Log Rule**

Find the indefinite integral.

$$\int \frac{2x}{(x + 1)^2}\, dx$$

Solution If you let $u = x + 1$, then $du = dx$ and $x = u - 1$.

$$\begin{aligned}
\int \frac{2x}{(x + 1)^2}\, dx &= \int \frac{2(u - 1)}{u^2}\, du && \text{Substitute.}\\
&= 2\int \left(\frac{u}{u^2} - \frac{1}{u^2}\right) du && \text{Rewrite as two fractions.}\\
&= 2\int \frac{du}{u} - 2\int u^{-2}\, du && \text{Rewrite as two integrals.}\\
&= 2\ln|u| - 2\left(\frac{u^{-1}}{-1}\right) + C && \text{Integrate.}\\
&= 2\ln|u| + \frac{2}{u} + C && \text{Simplify.}\\
&= 2\ln|x + 1| + \frac{2}{x + 1} + C && \text{Back-substitute.}
\end{aligned}$$

Check this result by differentiating to obtain the original integrand. ■

TECHNOLOGY If you have access to a computer algebra system, use it to find the indefinite integrals in Examples 5 and 6. How does the form of the antiderivative that it gives you compare with that given in Examples 5 and 6?

As you study the methods shown in Examples 5 and 6, be aware that both methods involve rewriting a disguised integrand so that it fits one or more of the basic integration formulas. Throughout the remaining sections of Chapter 5 and in Chapter 8, much time will be devoted to integration techniques. To master these techniques, you must recognize the "form-fitting" nature of integration. In this sense, integration is not nearly as straightforward as differentiation. Differentiation takes the form

"Here is the question; what is the answer?"

Integration is more like

"Here is the answer; what is the question?"

Here are some guidelines you can use for integration.

GUIDELINES FOR INTEGRATION

1. Learn a basic list of integration formulas. (Including those given in this section, you now have 12 formulas: the Power Rule, the Log Rule, and 10 trigonometric rules. By the end of Section 5.7, this list will have expanded to 20 basic rules.)
2. Find an integration formula that resembles all or part of the integrand, and, by trial and error, find a choice of u that will make the integrand conform to the formula.
3. When you cannot find a u-substitution that works, try altering the integrand. You might try a trigonometric identity, multiplication and division by the same quantity, addition and subtraction of the same quantity, or long division. Be creative.
4. If you have access to computer software that will find antiderivatives symbolically, use it.

EXAMPLE 7 ***u*-Substitution and the Log Rule**

Solve the differential equation $\dfrac{dy}{dx} = \dfrac{1}{x \ln x}$.

Solution The solution can be written as an indefinite integral.

$$y = \int \frac{1}{x \ln x}\, dx$$

Because the integrand is a quotient whose denominator is raised to the first power, you should try the Log Rule. There are three basic choices for u. The choices

$$u = x \quad \text{and} \quad u = x \ln x$$

fail to fit the u'/u form of the Log Rule. However, the third choice does fit. Letting $u = \ln x$ produces $u' = 1/x$, and you obtain the following.

$$\int \frac{1}{x \ln x}\, dx = \int \frac{1/x}{\ln x}\, dx \qquad \text{Divide numerator and denominator by } x.$$

$$= \int \frac{u'}{u}\, dx \qquad \text{Substitute: } u = \ln x.$$

$$= \ln|u| + C \qquad \text{Apply Log Rule.}$$

$$= \ln|\ln x| + C \qquad \text{Back-substitute.}$$

So, the solution is $y = \ln|\ln x| + C$.

REMARK Keep in mind that you can check your answer to an integration problem by differentiating the answer. For instance, in Example 7, the derivative of $y = \ln|\ln x| + C$ is $y' = 1/(x \ln x)$.

Integrals of Trigonometric Functions

In Section 4.1, you looked at six trigonometric integration rules—the six that correspond directly to differentiation rules. With the Log Rule, you can now complete the set of basic trigonometric integration formulas.

EXAMPLE 8 Using a Trigonometric Identity

Find $\int \tan x\,dx$.

Solution This integral does not seem to fit any formulas on our basic list. However, by using a trigonometric identity, you obtain

$$\int \tan x\,dx = \int \frac{\sin x}{\cos x}\,dx.$$

Knowing that $D_x[\cos x] = -\sin x$, you can let $u = \cos x$ and write

$$\int \tan x\,dx = -\int \frac{-\sin x}{\cos x}\,dx \qquad \text{Apply trigonometric identity and multiply and divide by } -1.$$

$$= -\int \frac{u'}{u}\,dx \qquad \text{Substitute: } u = \cos x.$$

$$= -\ln|u| + C \qquad \text{Apply Log Rule.}$$

$$= -\ln|\cos x| + C. \qquad \text{Back-substitute.}$$

Example 8 uses a trigonometric identity to derive an integration rule for the tangent function. The next example takes a rather unusual step (multiplying and dividing by the same quantity) to derive an integration rule for the secant function.

EXAMPLE 9 Derivation of the Secant Formula

Find $\int \sec x\,dx$.

Solution Consider the following procedure.

$$\int \sec x\,dx = \int \sec x\left(\frac{\sec x + \tan x}{\sec x + \tan x}\right)dx$$

$$= \int \frac{\sec^2 x + \sec x \tan x}{\sec x + \tan x}\,dx$$

Letting u be the denominator of this quotient produces

$$u = \sec x + \tan x$$

and

$$u' = \sec x \tan x + \sec^2 x.$$

So, you can conclude that

$$\int \sec x\,dx = \int \frac{\sec^2 x + \sec x \tan x}{\sec x + \tan x}\,dx \qquad \text{Rewrite integrand.}$$

$$= \int \frac{u'}{u}\,dx \qquad \text{Substitute: } u = \sec x + \tan x.$$

$$= \ln|u| + C \qquad \text{Apply Log Rule.}$$

$$= \ln|\sec x + \tan x| + C. \qquad \text{Back-substitute.}$$

With the results of Examples 8 and 9, you now have integration formulas for $\sin x$, $\cos x$, $\tan x$, and $\sec x$. The integrals of the six basic trigonometric functions are summarized below. (For proofs of $\cot u$ and $\csc u$, see Exercises 87 and 88.)

REMARK Using trigonometric identities and properties of logarithms, you could rewrite these six integration rules in other forms. For instance, you could write

$$\int \csc u\, du = \ln|\csc u - \cot u| + C.$$

(See Exercises 89–92.)

INTEGRALS OF THE SIX BASIC TRIGONOMETRIC FUNCTIONS

$$\int \sin u\, du = -\cos u + C \qquad \int \cos u\, du = \sin u + C$$

$$\int \tan u\, du = -\ln|\cos u| + C \qquad \int \cot u\, du = \ln|\sin u| + C$$

$$\int \sec u\, du = \ln|\sec u + \tan u| + C \qquad \int \csc u\, du = -\ln|\csc u + \cot u| + C$$

EXAMPLE 10 Integrating Trigonometric Functions

Evaluate $\displaystyle\int_0^{\pi/4} \sqrt{1 + \tan^2 x}\, dx$.

Solution Using $1 + \tan^2 x = \sec^2 x$, you can write

$$\begin{aligned}\int_0^{\pi/4} \sqrt{1 + \tan^2 x}\, dx &= \int_0^{\pi/4} \sqrt{\sec^2 x}\, dx \\ &= \int_0^{\pi/4} \sec x\, dx \qquad \sec x \geq 0 \text{ for } 0 \leq x \leq \frac{\pi}{4}. \\ &= \ln|\sec x + \tan x|\Big]_0^{\pi/4} \\ &= \ln\left(\sqrt{2} + 1\right) - \ln 1 \\ &\approx 0.881.\end{aligned}$$

EXAMPLE 11 Finding an Average Value

Find the average value of

$$f(x) = \tan x$$

on the interval $[0, \pi/4]$.

Solution

$$\begin{aligned}\text{Average value} &= \frac{1}{(\pi/4) - 0}\int_0^{\pi/4} \tan x\, dx \qquad \text{Average value} = \frac{1}{b-a}\int_a^b f(x)\, dx \\ &= \frac{4}{\pi}\int_0^{\pi/4} \tan x\, dx \qquad \text{Simplify.} \\ &= \frac{4}{\pi}\Big[-\ln|\cos x|\Big]_0^{\pi/4} \qquad \text{Integrate.} \\ &= -\frac{4}{\pi}\left[\ln\left(\frac{\sqrt{2}}{2}\right) - \ln(1)\right] \\ &= -\frac{4}{\pi}\ln\left(\frac{\sqrt{2}}{2}\right) \\ &\approx 0.441\end{aligned}$$

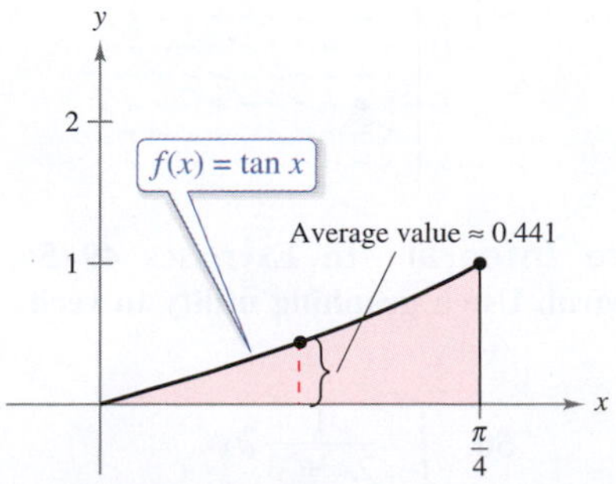

Figure 5.9

The average value is about 0.441, as shown in Figure 5.9.

5.2 Exercises

See CalcChat.com for tutorial help and worked-out solutions to odd-numbered exercises.

Finding an Indefinite Integral In Exercises 1–26, find the indefinite integral.

1. $\int \frac{5}{x}\,dx$

2. $\int \frac{10}{x}\,dx$

3. $\int \frac{1}{x+1}\,dx$

4. $\int \frac{1}{x-5}\,dx$

5. $\int \frac{1}{2x+5}\,dx$

6. $\int \frac{9}{5-4x}\,dx$

7. $\int \frac{x}{x^2-3}\,dx$

8. $\int \frac{x^2}{5-x^3}\,dx$

9. $\int \frac{4x^3+3}{x^4+3x}\,dx$

10. $\int \frac{x^2-2x}{x^3-3x^2}\,dx$

11. $\int \frac{x^2-4}{x}\,dx$

12. $\int \frac{x^3-8x}{x^2}\,dx$

13. $\int \frac{x^2+2x+3}{x^3+3x^2+9x}\,dx$

14. $\int \frac{x^2+4x}{x^3+6x^2+5}\,dx$

15. $\int \frac{x^2-3x+2}{x+1}\,dx$

16. $\int \frac{2x^2+7x-3}{x-2}\,dx$

17. $\int \frac{x^3-3x^2+5}{x-3}\,dx$

18. $\int \frac{x^3-6x-20}{x+5}\,dx$

19. $\int \frac{x^4+x-4}{x^2+2}\,dx$

20. $\int \frac{x^3-4x^2-4x+20}{x^2-5}\,dx$

21. $\int \frac{(\ln x)^2}{x}\,dx$

22. $\int \frac{1}{x\ln x^3}\,dx$

23. $\int \frac{1}{\sqrt{x}(1-3\sqrt{x})}\,dx$

24. $\int \frac{1}{x^{2/3}(1+x^{1/3})}\,dx$

25. $\int \frac{2x}{(x-1)^2}\,dx$

26. $\int \frac{x(x-2)}{(x-1)^3}\,dx$

Finding an Indefinite Integral by *u*-Substitution In Exercises 27–30, find the indefinite integral by *u*-substitution. (*Hint:* Let *u* be the denominator of the integrand.)

27. $\int \frac{1}{1+\sqrt{2x}}\,dx$

28. $\int \frac{1}{1+\sqrt{3x}}\,dx$

29. $\int \frac{\sqrt{x}}{\sqrt{x}-3}\,dx$

30. $\int \frac{\sqrt[3]{x}}{\sqrt[3]{x}-1}\,dx$

Finding an Indefinite Integral of a Trigonometric Function In Exercises 31–40, find the indefinite integral.

31. $\int \cot\frac{\theta}{3}\,d\theta$

32. $\int \tan 5\theta\,d\theta$

33. $\int \csc 2x\,dx$

34. $\int \sec\frac{x}{2}\,dx$

35. $\int (\cos 3\theta - 1)\,d\theta$

36. $\int \left(2-\tan\frac{\theta}{4}\right)d\theta$

37. $\int \frac{\cos t}{1+\sin t}\,dt$

38. $\int \frac{\csc^2 t}{\cot t}\,dt$

39. $\int \frac{\sec x\tan x}{\sec x - 1}\,dx$

40. $\int (\sec 2x + \tan 2x)\,dx$

Differential Equation In Exercises 41–44, solve the differential equation. Use a graphing utility to graph three solutions, one of which passes through the given point.

41. $\frac{dy}{dx} = \frac{3}{2-x}$, $(1, 0)$

42. $\frac{dy}{dx} = \frac{x-2}{x}$, $(-1, 0)$

43. $\frac{dy}{dx} = \frac{2x}{x^2-9x}$, $(0, 4)$

44. $\frac{dr}{dt} = \frac{\sec^2 t}{\tan t + 1}$, $(\pi, 4)$

Finding a Particular Solution In Exercises 45 and 46, find the particular solution that satisfies the differential equation and the initial equations.

45. $f''(x) = \frac{2}{x^2}, f'(1) = 1, f(1) = 1, x > 0$

46. $f''(x) = -\frac{4}{(x-1)^2} - 2, f'(2) = 0, f(2) = 3, x > 1$

Slope Field In Exercises 47 and 48, a differential equation, a point, and a slope field are given. (a) Sketch two approximate solutions of the differential equation on the slope field, one of which passes through the given point. (b) Use integration to find the particular solution of the differential equation and use a graphing utility to graph the solution. Compare the result with the sketches in part (a). To print an enlarged copy of the graph, go to *MathGraphs.com*.

47. $\frac{dy}{dx} = \frac{1}{x+2}$, $(0, 1)$

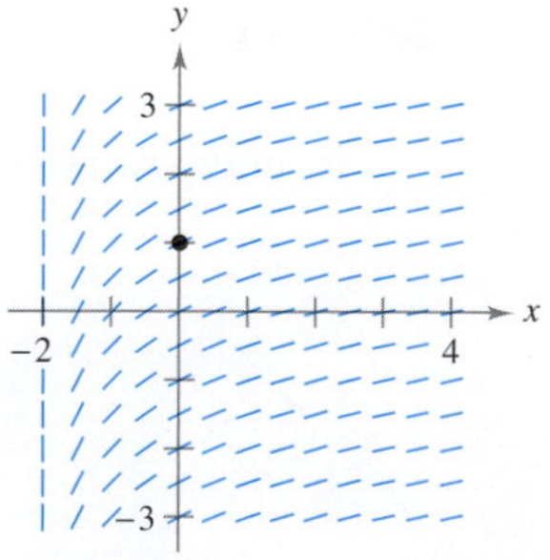

48. $\frac{dy}{dx} = \frac{\ln x}{x}$, $(1, -2)$

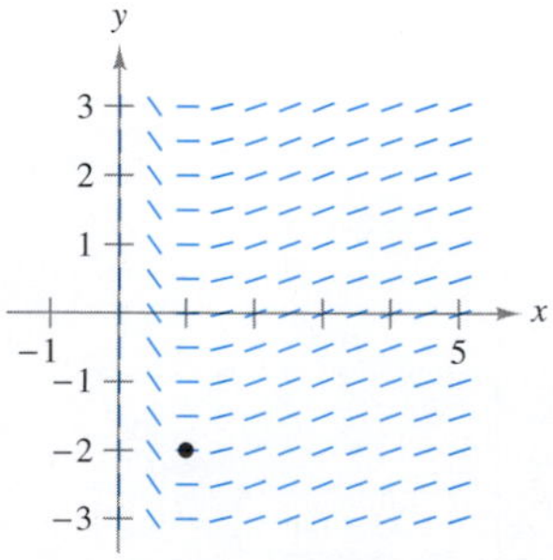

Evaluating a Definite Integral In Exercises 49–56, evaluate the definite integral. Use a graphing utility to verify your result.

49. $\int_0^4 \frac{5}{3x+1}\,dx$

50. $\int_{-1}^1 \frac{1}{2x+3}\,dx$

51. $\int_1^e \frac{(1+\ln x)^2}{x}\,dx$

52. $\int_e^{e^2} \frac{1}{x\ln x}\,dx$

53. $\int_0^2 \frac{x^2-2}{x+1}\,dx$

54. $\int_0^1 \frac{x-1}{x+1}\,dx$

55. $\int_1^2 \frac{1-\cos\theta}{\theta-\sin\theta}\,d\theta$

56. $\int_{\pi/8}^{\pi/4} (\csc 2\theta - \cot 2\theta)\,d\theta$

Using Technology to Find an Integral **In Exercises 57–62, use a computer algebra system to find or evaluate the integral.**

57. $\int \frac{1}{1+\sqrt{x}}\,dx$

58. $\int \frac{1-\sqrt{x}}{1+\sqrt{x}}\,dx$

59. $\int \frac{\sqrt{x}}{x-1}\,dx$

60. $\int \frac{x^2}{x-1}\,dx$

61. $\int_{\pi/4}^{\pi/2} (\csc x - \sin x)\,dx$

62. $\int_{-\pi/4}^{\pi/4} \frac{\sin^2 x - \cos^2 x}{\cos x}\,dx$

Finding a Derivative **In Exercises 63–66, find $F'(x)$.**

63. $F(x) = \int_1^x \frac{1}{t}\,dt$

64. $F(x) = \int_0^x \tan t\,dt$

65. $F(x) = \int_1^{3x} \frac{1}{t}\,dt$

66. $F(x) = \int_1^{x^2} \frac{1}{t}\,dt$

Area **In Exercises 67–70, find the area of the given region. Use a graphing utility to verify your result.**

67. $y = \frac{6}{x}$

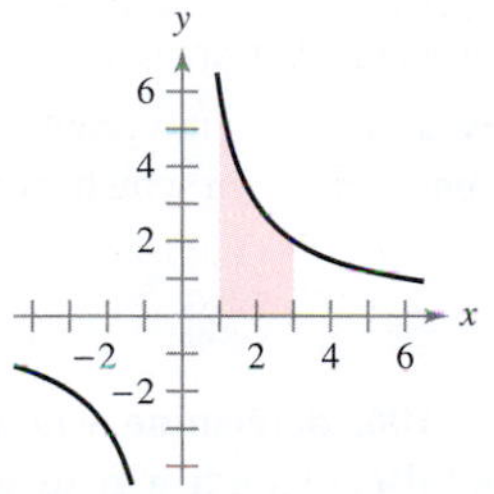

68. $y = \frac{2}{x\ln x}$

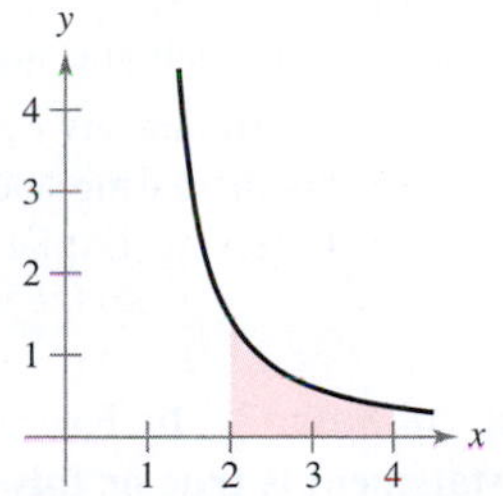

69. $y = \tan x$

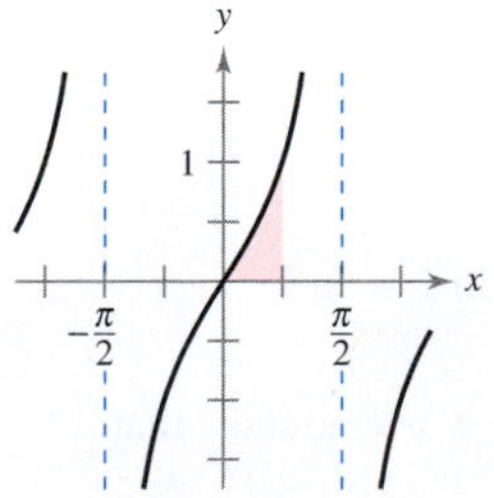

70. $y = \frac{\sin x}{1+\cos x}$

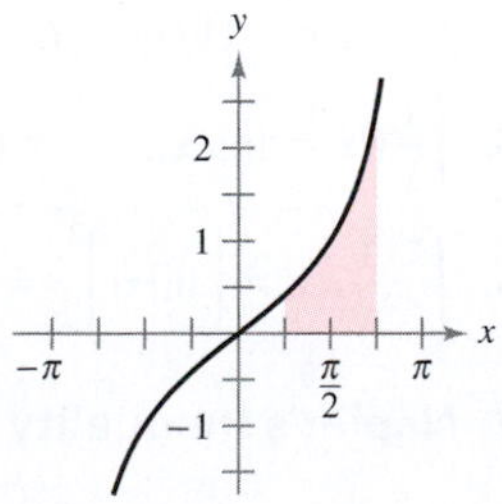

Area **In Exercises 71–74, find the area of the region bounded by the graphs of the equations. Use a graphing utility to verify your result.**

71. $y = \frac{x^2+4}{x}, \quad x = 1, \quad x = 4, \quad y = 0$

72. $y = \frac{5x}{x^2+2}, \quad x = 1, \quad x = 5, \quad y = 0$

73. $y = 2\sec\frac{\pi x}{6}, \quad x = 0, \quad x = 2, \quad y = 0$

74. $y = 2x - \tan 0.3x, \quad x = 1, \quad x = 4, \quad y = 0$

Numerical Integration **In Exercises 75–78, use the Trapezoidal Rule and Simpson's Rule to approximate the value of the definite integral. Let $n = 4$ and round your answer to four decimal places. Use a graphing utility to verify your result.**

75. $\int_1^5 \frac{12}{x}\,dx$

76. $\int_0^4 \frac{8x}{x^2+4}\,dx$

77. $\int_2^6 \ln x\,dx$

78. $\int_{-\pi/3}^{\pi/3} \sec x\,dx$

WRITING ABOUT CONCEPTS

Choosing a Formula **In Exercises 79–82, state the integration formula you would use to perform the integration. Do not integrate.**

79. $\int \sqrt[3]{x}\,dx$

80. $\int \frac{x}{(x^2+4)^3}\,dx$

81. $\int \frac{x}{x^2+4}\,dx$

82. $\int \frac{\sec^2 x}{\tan x}\,dx$

Approximation **In Exercises 83 and 84, determine which value best approximates the area of the region between the x-axis and the graph of the function over the given interval. (Make your selection on the basis of a sketch of the region, not by performing any calculations.)**

83. $f(x) = \sec x, \quad [0, 1]$

(a) 6 (b) −6 (c) $\frac{1}{2}$ (d) 1.25 (e) 3

84. $f(x) = \frac{2x}{x^2+1}, \quad [0, 4]$

(a) 3 (b) 7 (c) −2 (d) 5 (e) 1

85. **Finding a Value** Find a value of x such that

$$\int_1^x \frac{3}{t}\,dt = \int_{1/4}^x \frac{1}{t}\,dt.$$

86. **Finding a Value** Find a value of x such that

$$\int_1^x \frac{1}{t}\,dt$$

is equal to (a) $\ln 5$ and (b) 1.

87. **Proof** Prove that

$$\int \cot u\,du = \ln|\sin u| + C.$$

88. **Proof** Prove that

$$\int \csc u\,du = -\ln|\csc u + \cot u| + C.$$

Using Properties of Logarithms and Trigonometric Identities **In Exercises 89–92, show that the two formulas are equivalent.**

89. $\int \tan x\,dx = -\ln|\cos x| + C$

$\int \tan x\,dx = \ln|\sec x| + C$

90. $\int \cot x\,dx = \ln|\sin x| + C$

$\int \cot x\,dx = -\ln|\csc x| + C$

91. $\int \sec x\,dx = \ln|\sec x + \tan x| + C$

$\int \sec x\,dx = -\ln|\sec x - \tan x| + C$

92. $\int \csc x\,dx = -\ln|\csc x + \cot x| + C$

$\int \csc x\,dx = \ln|\csc x - \cot x| + C$

Finding the Average Value of a Function **In Exercises 93–96, find the average value of the function over the given interval.**

93. $f(x) = \frac{8}{x^2}$, $[2, 4]$

94. $f(x) = \frac{4(x + 1)}{x^2}$, $[2, 4]$

95. $f(x) = \frac{2 \ln x}{x}$, $[1, e]$

96. $f(x) = \sec \frac{\pi x}{6}$, $[0, 2]$

97. Population Growth A population of bacteria P is changing at a rate of

$$\frac{dP}{dt} = \frac{3000}{1 + 0.25t}$$

where t is the time in days. The initial population (when $t = 0$) is 1000. Write an equation that gives the population at any time t. Then find the population when $t = 3$ days.

98. Sales The rate of change in sales S is inversely proportional to time t $(t > 1)$, measured in weeks. Find S as a function of t when the sales after 2 and 4 weeks are 200 units and 300 units, respectively.

99. Heat Transfer

Find the time required for an object to cool from 300°F to 250°F by evaluating

$$t = \frac{10}{\ln 2}\int_{250}^{300} \frac{1}{T - 100}\,dT$$

where t is time in minutes.

100. Average Price The demand equation for a product is

$$p = \frac{90{,}000}{400 + 3x}$$

where p is the price (in dollars) and x is the number of units (in thousands). Find the average price p on the interval $40 \le x \le 50$.

101. Area and Slope Graph the function

$$f(x) = \frac{x}{1 + x^2}$$

on the interval $[0, \infty)$.

(a) Find the area bounded by the graph of f and the line $y = \frac{1}{2}x$.

(b) Determine the values of the slope m such that the line $y = mx$ and the graph of f enclose a finite region.

(c) Calculate the area of this region as a function of m.

102. HOW DO YOU SEE IT? Use the graph of f' shown in the figure to answer the following.

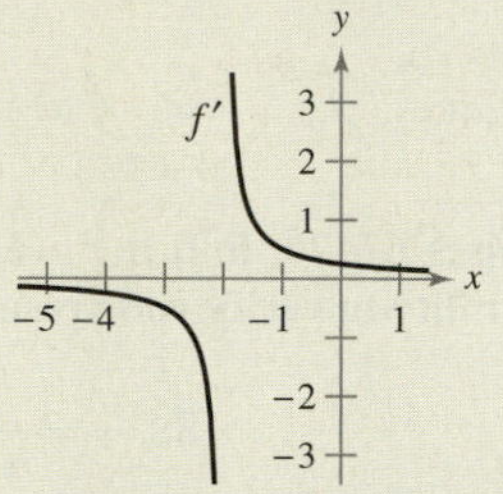

(a) Approximate the slope of f at $x = -1$. Explain.

(b) Approximate any open intervals in which the graph of f is increasing and any open intervals in which it is decreasing. Explain.

True or False? **In Exercises 103–106, determine whether the statement is true or false. If it is false, explain why or give an example that shows it is false.**

103. $(\ln x)^{1/2} = \frac{1}{2}\ln x$

104. $\int \ln x\,dx = (1/x) + C$

105. $\int \frac{1}{x}\,dx = \ln|cx|, \quad c \neq 0$

106. $\int_{-1}^{2} \frac{1}{x}\,dx = \Big[\ln|x|\Big]_{-1}^{2} = \ln 2 - \ln 1 = \ln 2$

107. Napier's Inequality For $0 < x < y$, show that

$$\frac{1}{y} < \frac{\ln y - \ln x}{y - x} < \frac{1}{x}.$$

108. Proof Prove that the function

$$F(x) = \int_{x}^{2x} \frac{1}{t}\,dt$$

is constant on the interval $(0, \infty)$.

Marijus Auruskevicius/Shutterstock.com

5.3 Inverse Functions

- **Verify that one function is the inverse function of another function.**
- **Determine whether a function has an inverse function.**
- **Find the derivative of an inverse function.**

Inverse Functions

Recall from Section P.3 that a function can be represented by a set of ordered pairs. For instance, the function $f(x) = x + 3$ from $A = \{1, 2, 3, 4\}$ to $B = \{4, 5, 6, 7\}$ can be written as

$$f: \{(1, 4), (2, 5), (3, 6), (4, 7)\}.$$

By interchanging the first and second coordinates of each ordered pair, you can form the **inverse function** of f. This function is denoted by f^{-1}. It is a function from B to A, and can be written as

$$f^{-1}: \{(4, 1), (5, 2), (6, 3), (7, 4)\}.$$

Note that the domain of f is equal to the range of f^{-1}, and vice versa, as shown in Figure 5.10. The functions f and f^{-1} have the effect of "undoing" each other. That is, when you form the composition of f with f^{-1} or the composition of f^{-1} with f, you obtain the identity function.

$$f(f^{-1}(x)) = x \quad \text{and} \quad f^{-1}(f(x)) = x$$

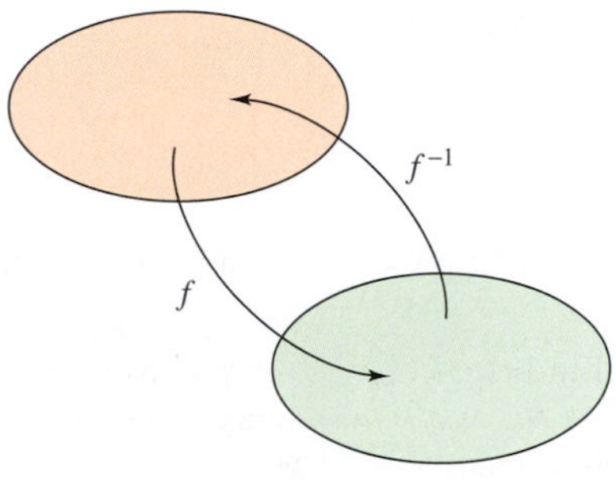

Domain of f = range of f^{-1}
Domain of f^{-1} = range of f
Figure 5.10

REMARK Although the notation used to denote an inverse function resembles *exponential notation*, it is a different use of -1 as a superscript. That is, in general,

$$f^{-1}(x) \neq \frac{1}{f(x)}.$$

Exploration

Finding Inverse Functions Explain how to "undo" each of the functions below. Then use your explanation to write the inverse function of f.

a. $f(x) = x - 5$

b. $f(x) = 6x$

c. $f(x) = \dfrac{x}{2}$

d. $f(x) = 3x + 2$

e. $f(x) = x^3$

f. $f(x) = 4(x - 2)$

Use a graphing utility to graph each function and its inverse function in the same "square" viewing window. What observation can you make about each pair of graphs?

Definition of Inverse Function

A function g is the **inverse function** of the function f when

$$f(g(x)) = x \text{ for each } x \text{ in the domain of } g$$

and

$$g(f(x)) = x \text{ for each } x \text{ in the domain of } f.$$

The function g is denoted by f^{-1} (read "f inverse").

Here are some important observations about inverse functions.

1. If g is the inverse function of f, then f is the inverse function of g.
2. The domain of f^{-1} is equal to the range of f, and the range of f^{-1} is equal to the domain of f.
3. A function need not have an inverse function, but when it does, the inverse function is unique (see Exercise 96).

You can think of f^{-1} as undoing what has been done by f. For example, subtraction can be used to undo addition, and division can be used to undo multiplication. So,

$$f(x) = x + c \quad \text{and} \quad f^{-1}(x) = x - c$$ Subtraction can be used to undo addition.

are inverse functions of each other and

$$f(x) = cx \quad \text{and} \quad f^{-1}(x) = \frac{x}{c}, \quad c \neq 0$$ Division can be used to undo multiplication.

are inverse functions of each other.

EXAMPLE 1 Verifying Inverse Functions

Show that the functions are inverse functions of each other.

$$f(x) = 2x^3 - 1 \quad \text{and} \quad g(x) = \sqrt[3]{\frac{x+1}{2}}$$

REMARK In Example 1, try comparing the functions f and g verbally.

For f: First cube x, then multiply by 2, then subtract 1.

For g: First add 1, then divide by 2, then take the cube root.

Do you see the "undoing pattern"?

Solution Because the domains and ranges of both f and g consist of all real numbers, you can conclude that both composite functions exist for all x. The composition of f with g is

$$\begin{aligned} f(g(x)) &= 2\left(\sqrt[3]{\frac{x+1}{2}}\right)^3 - 1 \\ &= 2\left(\frac{x+1}{2}\right) - 1 \\ &= x + 1 - 1 \\ &= x. \end{aligned}$$

The composition of g with f is

$$\begin{aligned} g(f(x)) &= \sqrt[3]{\frac{(2x^3 - 1) + 1}{2}} \\ &= \sqrt[3]{\frac{2x^3}{2}} \\ &= \sqrt[3]{x^3} \\ &= x. \end{aligned}$$

Because $f(g(x)) = x$ and $g(f(x)) = x$, you can conclude that f and g are inverse functions of each other (see Figure 5.11).

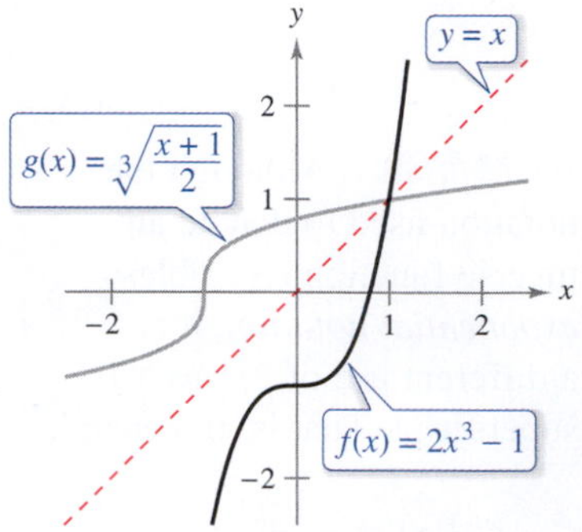

f and g are inverse functions of each other.

Figure 5.11

In Figure 5.11, the graphs of f and $g = f^{-1}$ appear to be mirror images of each other with respect to the line $y = x$. The graph of f^{-1} is a **reflection** of the graph of f in the line $y = x$. This idea is generalized in the next theorem.

THEOREM 5.6 Reflective Property of Inverse Functions

The graph of f contains the point (a, b) if and only if the graph of f^{-1} contains the point (b, a).

Proof If (a, b) is on the graph of f, then $f(a) = b$, and you can write

$$f^{-1}(b) = f^{-1}(f(a)) = a.$$

So, (b, a) is on the graph of f^{-1}, as shown in Figure 5.12. A similar argument will prove the theorem in the other direction.

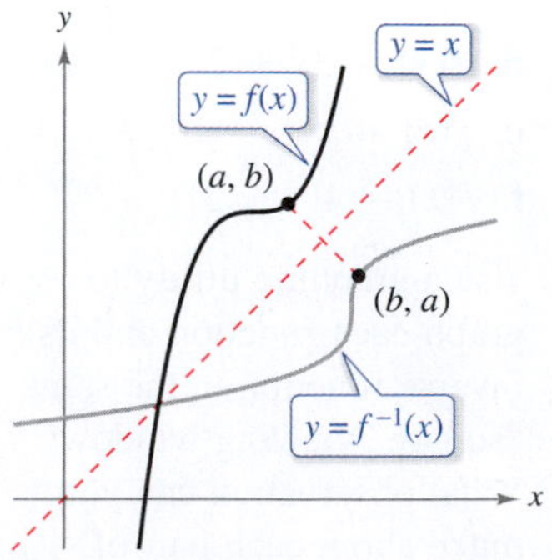

The graph of f^{-1} is a reflection of the graph of f in the line $y = x$.

Figure 5.12

See LarsonCalculus.com for Bruce Edwards's video of this proof.

Existence of an Inverse Function

Not every function has an inverse function, and Theorem 5.6 suggests a graphical test for those that do—the **Horizontal Line Test** for an inverse function. This test states that a function f has an inverse function if and only if every horizontal line intersects the graph of f at most once (see Figure 5.13). The next theorem formally states why the Horizontal Line Test is valid. (Recall from Section 3.3 that a function is *strictly monotonic* when it is either increasing on its entire domain or decreasing on its entire domain.)

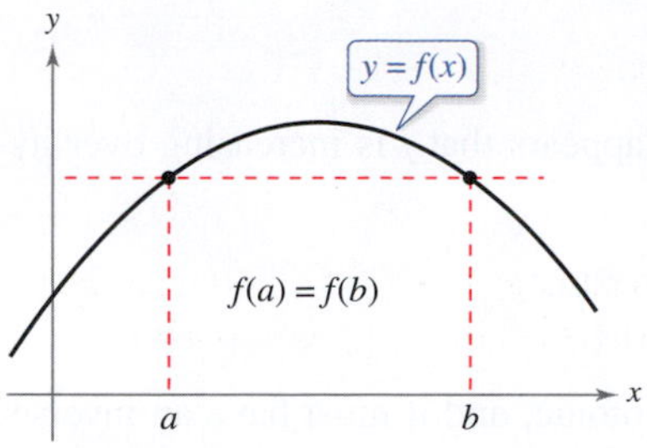

If a horizontal line intersects the graph of f twice, then f is not one-to-one.
Figure 5.13

> **THEOREM 5.7 The Existence of an Inverse Function**
>
> 1. A function has an inverse function if and only if it is one-to-one.
> 2. If f is strictly monotonic on its entire domain, then it is one-to-one and therefore has an inverse function.

Proof The proof of the first part of the theorem is left as an exercise (See Exercise 97). To prove the second part of the theorem, recall from Section P.3 that f is one-to-one when for x_1 and x_2 in its domain

$$x_1 \neq x_2 \implies f(x_1) \neq f(x_2).$$

Now, choose x_1 and x_2 in the domain of f. If $x_1 \neq x_2$, then, because f is strictly monotonic, it follows that either $f(x_1) < f(x_2)$ or $f(x_1) > f(x_2)$. In either case, $f(x_1) \neq f(x_2)$. So, f is one-to-one on the interval.

See LarsonCalculus.com for Bruce Edwards's video of this proof. ■

EXAMPLE 2 The Existence of an Inverse Function

a. From the graph of $f(x) = x^3 + x - 1$ shown in Figure 5.14(a), it appears that f is increasing over its entire domain. To verify this, note that the derivative, $f'(x) = 3x^2 + 1$, is positive for all real values of x. So, f is strictly monotonic, and it must have an inverse function.

b. From the graph of $f(x) = x^3 - x + 1$ shown in Figure 5.14(b), you can see that the function does not pass the Horizontal Line Test. In other words, it is not one-to-one. For instance, f has the same value when $x = -1$, 0, and 1.

$$f(-1) = f(1) = f(0) = 1 \qquad \text{Not one-to-one}$$

So, by Theorem 5.7, f does not have an inverse function. ■

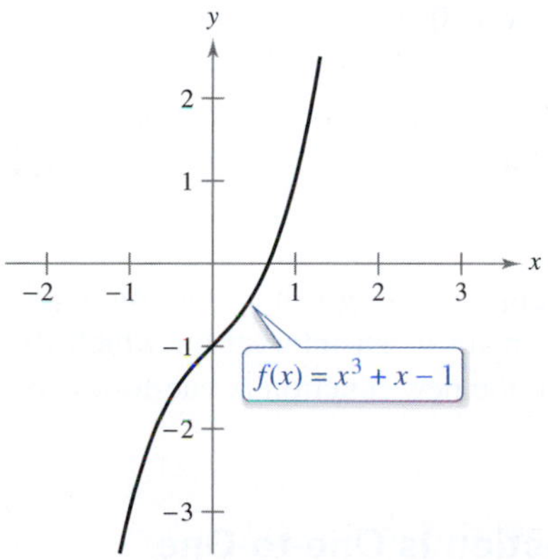

(a) Because f is increasing over its entire domain, it has an inverse function.

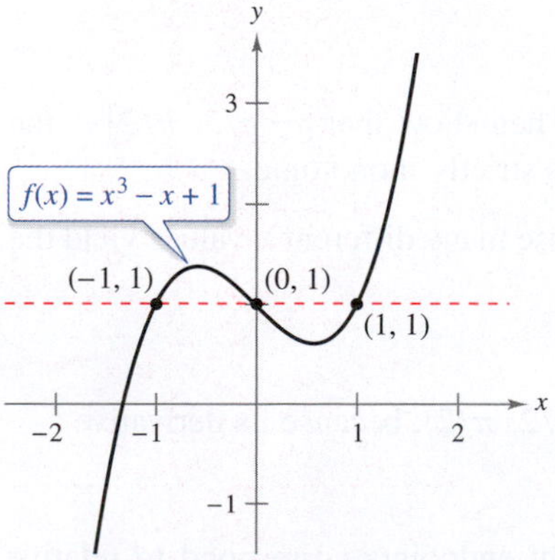

(b) Because f is not one-to-one, it does not have an inverse function.
Figure 5.14

Often, it is easier to prove that a function *has* an inverse function than to find the inverse function. For instance, it would be difficult algebraically to find the inverse function of the function in Example 2(a).

> **GUIDELINES FOR FINDING AN INVERSE FUNCTION**
>
> 1. Use Theorem 5.7 to determine whether the function $y = f(x)$ has an inverse function.
> 2. Solve for x as a function of y: $x = g(y) = f^{-1}(y)$.
> 3. Interchange x and y. The resulting equation is $y = f^{-1}(x)$.
> 4. Define the domain of f^{-1} as the range of f.
> 5. Verify that $f(f^{-1}(x)) = x$ and $f^{-1}(f(x)) = x$.

EXAMPLE 3 Finding an Inverse Function

Find the inverse function of $f(x) = \sqrt{2x - 3}$.

Solution From the graph of f in Figure 5.15, it appears that f is increasing over its entire domain, $[3/2, \infty)$. To verify this, note that

$$f'(x) = \frac{1}{\sqrt{2x - 3}}$$

is positive on the domain of f. So, f is strictly monotonic, and it must have an inverse function. To find an equation for the inverse function, let $y = f(x)$, and solve for x in terms of y.

$$\sqrt{2x - 3} = y \quad \text{Let } y = f(x).$$

$$2x - 3 = y^2 \quad \text{Square each side.}$$

$$x = \frac{y^2 + 3}{2} \quad \text{Solve for } x.$$

$$y = \frac{x^2 + 3}{2} \quad \text{Interchange } x \text{ and } y.$$

$$f^{-1}(x) = \frac{x^2 + 3}{2} \quad \text{Replace } y \text{ by } f^{-1}(x).$$

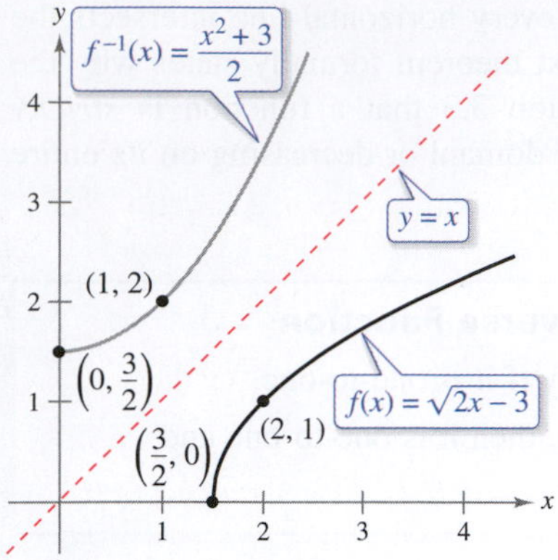

The domain of f^{-1}, $[0, \infty)$, is the range of f.
Figure 5.15

The domain of f^{-1} is the range of f, which is $[0, \infty)$. You can verify this result as shown.

$$f(f^{-1}(x)) = \sqrt{2\left(\frac{x^2 + 3}{2}\right) - 3} = \sqrt{x^2} = x, \quad x \geq 0$$

$$f^{-1}(f(x)) = \frac{\left(\sqrt{2x - 3}\right)^2 + 3}{2} = \frac{2x - 3 + 3}{2} = x, \quad x \geq \frac{3}{2}$$

Theorem 5.7 is useful in the next type of problem. You are given a function that is *not* one-to-one on its domain. By restricting the domain to an interval on which the function is strictly monotonic, you can conclude that the new function *is* one-to-one on the restricted domain.

EXAMPLE 4 Testing Whether a Function Is One-to-One

See LarsonCalculus.com for an interactive version of this type of example.

Show that the sine function

$$f(x) = \sin x$$

is not one-to-one on the entire real number line. Then show that $[-\pi/2, \pi/2]$ is the largest interval, centered at the origin, on which f is strictly monotonic.

Solution It is clear that f is not one-to-one, because many different x-values yield the same y-value. For instance,

$$\sin(0) = 0 = \sin(\pi).$$

Moreover, f is increasing on the open interval $(-\pi/2, \pi/2)$, because its derivative

$$f'(x) = \cos x$$

is positive there. Finally, because the left and right endpoints correspond to relative extrema of the sine function, you can conclude that f is increasing on the closed interval $[-\pi/2, \pi/2]$ *and* that on any larger interval the function is not strictly monotonic (see Figure 5.16).

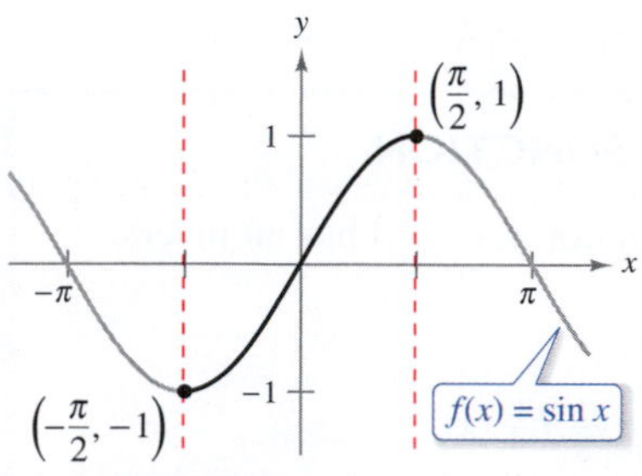

f is one-to-one on the interval $[-\pi/2, \pi/2]$.
Figure 5.16

Derivative of an Inverse Function

The next two theorems discuss the derivative of an inverse function. The reasonableness of Theorem 5.8 follows from the reflective property of inverse functions, as shown in Figure 5.12.

THEOREM 5.8 Continuity and Differentiability of Inverse Functions

Let f be a function whose domain is an interval I. If f has an inverse function, then the following statements are true.

1. If f is continuous on its domain, then f^{-1} is continuous on its domain.
2. If f is increasing on its domain, then f^{-1} is increasing on its domain.
3. If f is decreasing on its domain, then f^{-1} is decreasing on its domain.
4. If f is differentiable on an interval containing c and $f'(c) \neq 0$, then f^{-1} is differentiable at $f(c)$.

A proof of this theorem is given in Appendix A.

See LarsonCalculus.com for Bruce Edwards's video of this proof.

Exploration

Graph the inverse functions $f(x) = x^3$ and $g(x) = x^{1/3}$. Calculate the slopes of f at $(1, 1)$, $(2, 8)$, and $(3, 27)$, and the slopes of g at $(1, 1)$, $(8, 2)$, and $(27, 3)$. What do you observe? What happens at $(0, 0)$?

AP* Tips

Theorem 5.9 is often tested on the AP multiple choice section and occasionally on the free response section.

THEOREM 5.9 The Derivative of an Inverse Function

Let f be a function that is differentiable on an interval I. If f has an inverse function g, then g is differentiable at any x for which $f'(g(x)) \neq 0$. Moreover,

$$g'(x) = \frac{1}{f'(g(x))}, \quad f'(g(x)) \neq 0.$$

A proof of this theorem is given in Appendix A.

See LarsonCalculus.com for Bruce Edwards's video of this proof.

EXAMPLE 5 Evaluating the Derivative of an Inverse Function

Let $f(x) = \frac{1}{4}x^3 + x - 1$. (a) What is the value of $f^{-1}(x)$ when $x = 3$? (b) What is the value of $(f^{-1})'(x)$ when $x = 3$?

Solution Notice that f is one-to-one and therefore has an inverse function.

a. Because $f(x) = 3$ when $x = 2$, you know that $f^{-1}(3) = 2$.

b. Because the function f is differentiable and has an inverse function, you can apply Theorem 5.9 (with $g = f^{-1}$) to write

$$(f^{-1})'(3) = \frac{1}{f'(f^{-1}(3))} = \frac{1}{f'(2)}.$$

Moreover, using $f'(x) = \frac{3}{4}x^2 + 1$, you can conclude that

$$(f^{-1})'(3) = \frac{1}{f'(2)} = \frac{1}{\frac{3}{4}(2^2) + 1} = \frac{1}{4}.$$ ■

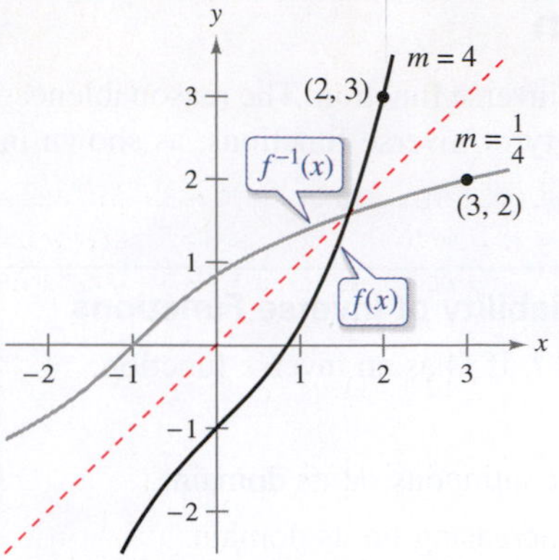

The graphs of the inverse functions f and f^{-1} have reciprocal slopes at points (a, b) and (b, a).
Figure 5.17

In Example 5, note that at the point (2, 3), the slope of the graph of f is 4, and at the point (3, 2), the slope of the graph of f^{-1} is

$$m = \frac{1}{4}$$

as shown in Figure 5.17. In general, if $y = g(x) = f^{-1}(x)$, then $f(y) = x$ and $f'(y) = \dfrac{dx}{dy}$. It follows from Theorem 5.9 that

$$g'(x) = \frac{dy}{dx} = \frac{1}{f'(g(x))} = \frac{1}{f'(y)} = \frac{1}{(dx/dy)}.$$

This reciprocal relationship is sometimes written as

$$\frac{dy}{dx} = \frac{1}{dx/dy}.$$

EXAMPLE 6 Graphs of Inverse Functions Have Reciprocal Slopes

Let $f(x) = x^2$ (for $x \geq 0$), and let $f^{-1}(x) = \sqrt{x}$. Show that the slopes of the graphs of f and f^{-1} are reciprocals at each of the following points.

a. (2, 4) and (4, 2) **b.** (3, 9) and (9, 3)

Solution The derivatives of f and f^{-1} are

$$f'(x) = 2x \quad \text{and} \quad (f^{-1})'(x) = \frac{1}{2\sqrt{x}}.$$

a. At (2, 4), the slope of the graph of f is $f'(2) = 2(2) = 4$. At (4, 2), the slope of the graph of f^{-1} is

$$(f^{-1})'(4) = \frac{1}{2\sqrt{4}} = \frac{1}{2(2)} = \frac{1}{4}.$$

b. At (3, 9), the slope of the graph of f is $f'(3) = 2(3) = 6$. At (9, 3), the slope of the graph of f^{-1} is

$$(f^{-1})'(9) = \frac{1}{2\sqrt{9}} = \frac{1}{2(3)} = \frac{1}{6}.$$

So, in both cases, the slopes are reciprocals, as shown in Figure 5.18.

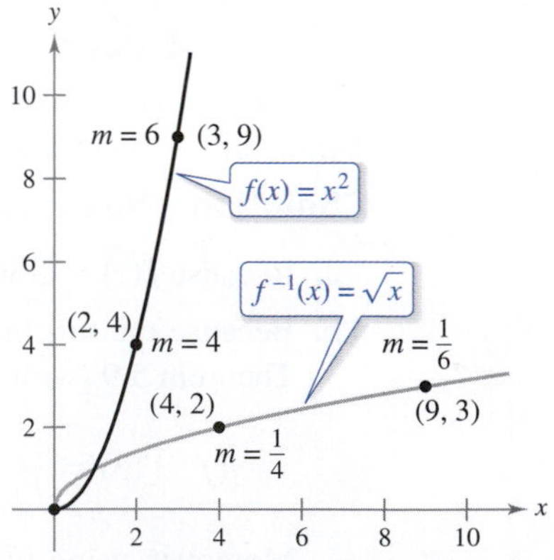

At (0, 0), the derivative of f is 0, and the derivative of f^{-1} does not exist.
Figure 5.18

5.3 Exercises

See CalcChat.com for tutorial help and worked-out solutions to odd-numbered exercises.

Verifying Inverse Functions In Exercises 1–8, show that f and g are inverse functions (a) analytically and (b) graphically.

1. $f(x) = 5x + 1, \quad g(x) = \dfrac{x-1}{5}$
2. $f(x) = 3 - 4x, \quad g(x) = \dfrac{3-x}{4}$
3. $f(x) = x^3, \quad g(x) = \sqrt[3]{x}$
4. $f(x) = 1 - x^3, \quad g(x) = \sqrt[3]{1-x}$
5. $f(x) = \sqrt{x-4}, \quad g(x) = x^2 + 4, \quad x \geq 0$
6. $f(x) = 16 - x^2, \quad x \geq 0, \quad g(x) = \sqrt{16-x}$
7. $f(x) = \dfrac{1}{x}, \quad g(x) = \dfrac{1}{x}$
8. $f(x) = \dfrac{1}{1+x}, \quad x \geq 0, \quad g(x) = \dfrac{1-x}{x}, \quad 0 < x \leq 1$

Matching In Exercises 9–12, match the graph of the function with the graph of its inverse function. [The graphs of the inverse functions are labeled (a), (b), (c), and (d).]

(a)
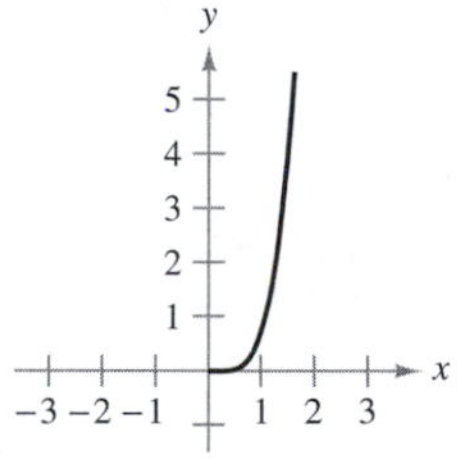

(b)
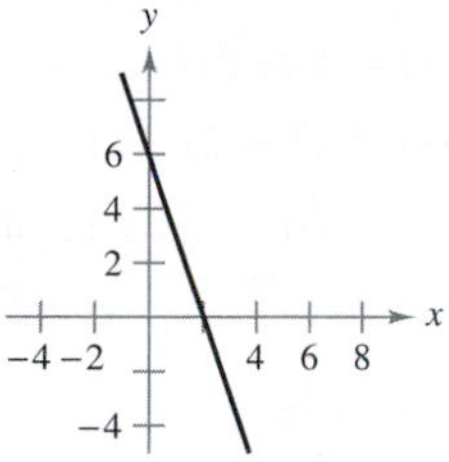

(c)
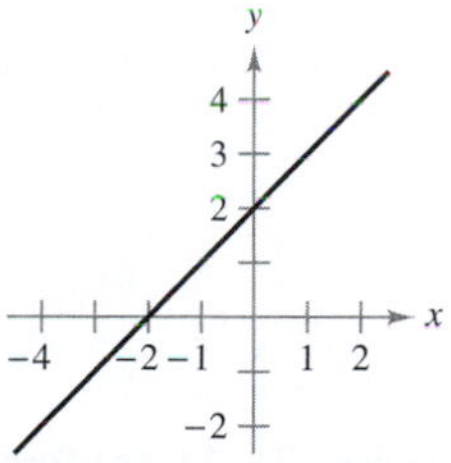

(d)
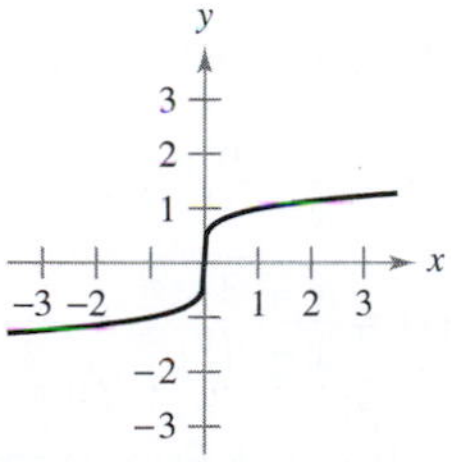

9.
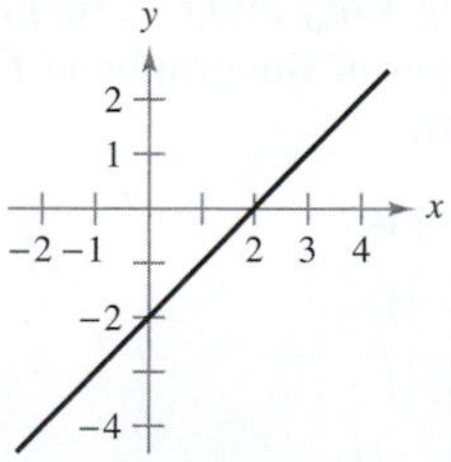

10.
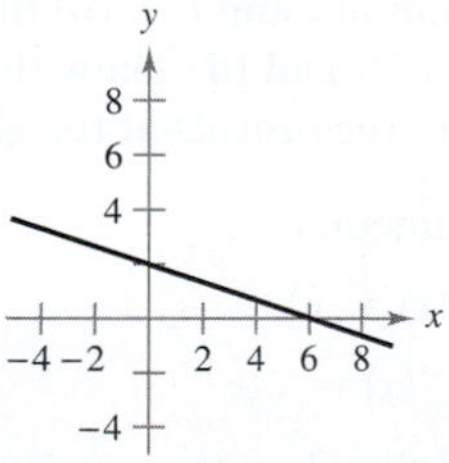

11.
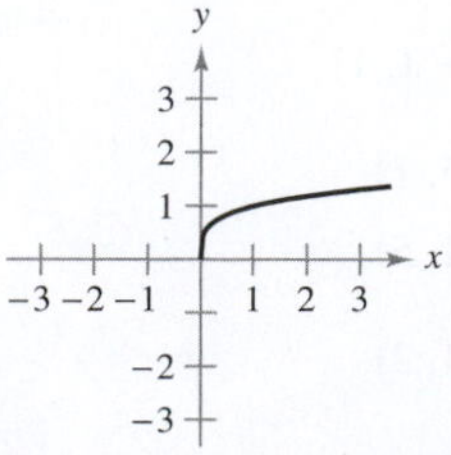

12.
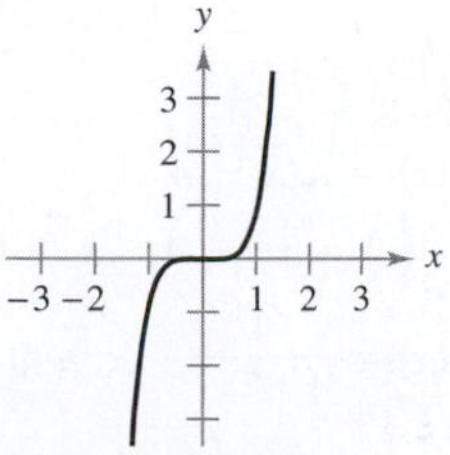

Using the Horizontal Line Test In Exercises 13–22, use a graphing utility to graph the function. Then use the Horizontal Line Test to determine whether the function is one-to-one on its entire domain and therefore has an inverse function.

13. $f(x) = \frac{3}{4}x + 6$
14. $f(x) = 5x - 3$
15. $f(\theta) = \sin\theta$
16. $f(x) = \dfrac{6x}{x^2+4}$
17. $h(s) = \dfrac{1}{s-2} - 3$
18. $g(t) = \dfrac{1}{\sqrt{t^2+1}}$
19. $f(x) = \ln x$
20. $f(x) = 5x\sqrt{x-1}$
21. $g(x) = (x+5)^3$
22. $h(x) = |x+4| - |x-4|$

Determining Whether a Function Has an Inverse Function In Exercises 23–28, use the derivative to determine whether the function is strictly monotonic on its entire domain and therefore has an inverse function.

23. $f(x) = 2 - x - x^3$
24. $f(x) = x^3 - 6x^2 + 12x$
25. $f(x) = \dfrac{x^4}{4} - 2x^2$
26. $f(x) = x^5 + 2x^3$
27. $f(x) = \ln(x-3)$
28. $f(x) = \cos\dfrac{3x}{2}$

Verifying a Function Has an Inverse Function In Exercises 29–34, show that f is strictly monotonic on the given interval and therefore has an inverse function on that interval.

29. $f(x) = (x-4)^2, \quad [4, \infty)$
30. $f(x) = |x+2|, \quad [-2, \infty)$
31. $f(x) = \dfrac{4}{x^2}, \quad (0, \infty)$
32. $f(x) = \cot x, \quad (0, \pi)$
33. $f(x) = \cos x, \quad [0, \pi]$
34. $f(x) = \sec x, \quad \left[0, \dfrac{\pi}{2}\right)$

Finding an Inverse Function In Exercises 35–46, (a) find the inverse function of f, (b) graph f and f^{-1} on the same set of coordinate axes, (c) describe the relationship between the graphs, and (d) state the domain and range of f and f^{-1}.

35. $f(x) = 2x - 3$
36. $f(x) = 7 - 4x$
37. $f(x) = x^5$
38. $f(x) = x^3 - 1$
39. $f(x) = \sqrt{x}$
40. $f(x) = x^2, \quad x \geq 0$
41. $f(x) = \sqrt{4 - x^2}, \quad 0 \leq x \leq 2$
42. $f(x) = \sqrt{x^2 - 4}, \quad x \geq 2$
43. $f(x) = \sqrt[3]{x-1}$
44. $f(x) = x^{2/3}, \quad x \geq 0$
45. $f(x) = \dfrac{x}{\sqrt{x^2+7}}$
46. $f(x) = \dfrac{x+2}{x}$

Finding an Inverse Function **In Exercises 47 and 48, use the graph of the function f to make a table of values for the given points. Then make a second table that can be used to find f^{-1}, and sketch the graph of f^{-1}. To print an enlarged copy of the graph, go to *MathGraphs.com*.**

47.

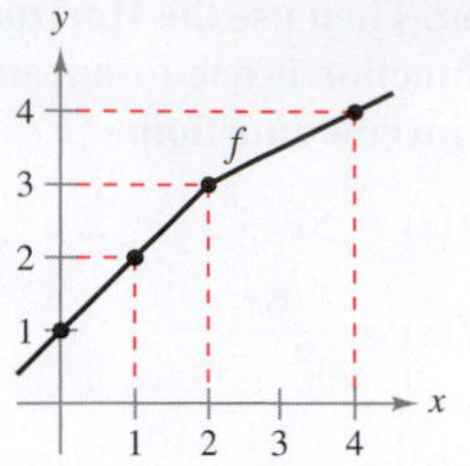

48.

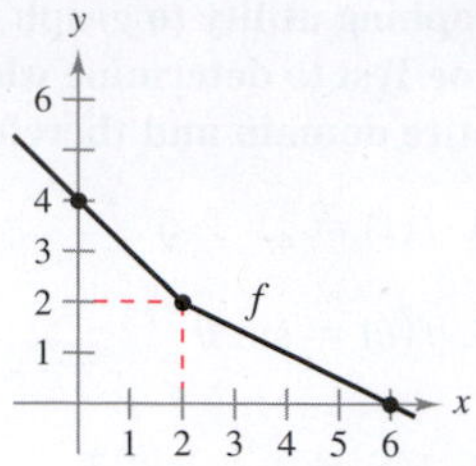

49. Cost You need 50 pounds of two commodities costing \$1.25 and \$1.60 per pound.

(a) Verify that the total cost is $y = 1.25x + 1.60(50 - x)$, where x is the number of pounds of the less expensive commodity.

(b) Find the inverse function of the cost function. What does each variable represent in the inverse function?

(c) What is the domain of the inverse function? Validate or explain your answer using the context of the problem.

(d) Determine the number of pounds of the less expensive commodity purchased when the total cost is \$73.

50. Temperature The formula $C = \frac{5}{9}(F - 32)$, where $F \geq -459.6$, represents Celsius temperature C as a function of Fahrenheit temperature F.

(a) Find the inverse function of C.

(b) What does the inverse function represent?

(c) What is the domain of the inverse function? Validate or explain your answer using the context of the problem.

(d) The temperature is 22°C. What is the corresponding temperature in degrees Fahrenheit?

Testing Whether a Function Is One-to-One **In Exercises 51–54, determine whether the function is one-to-one. If it is, find its inverse function.**

51. $f(x) = \sqrt{x - 2}$

52. $f(x) = -3$

53. $f(x) = |x - 2|, \quad x \leq 2$

54. $f(x) = ax + b, \quad a \neq 0$

Making a Function One-to-One **In Exercises 55–58, delete part of the domain so that the function that remains is one-to-one. Find the inverse function of the remaining function and give the domain of the inverse function. (*Note:* There is more than one correct answer.)**

55. $f(x) = (x - 3)^2$

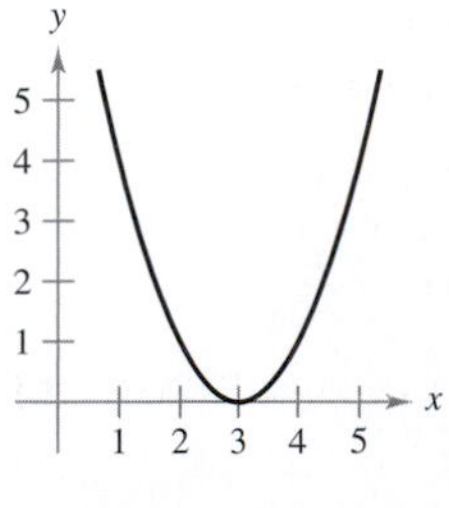

56. $f(x) = 16 - x^4$

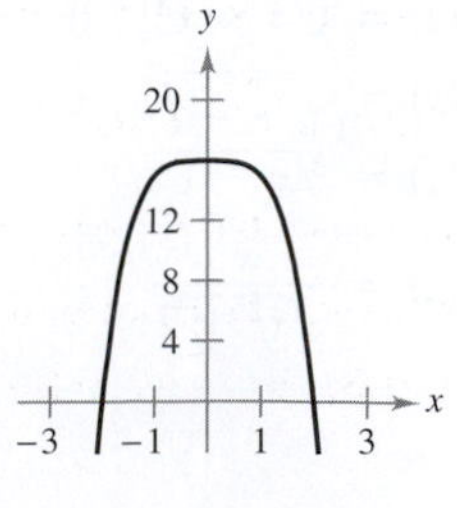

57. $f(x) = |x + 3|$

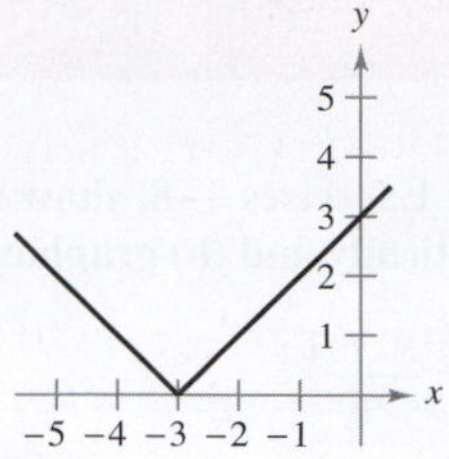

58. $f(x) = |x - 3|$

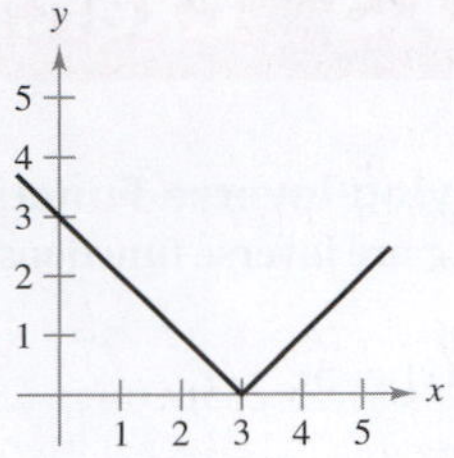

Think About It **In Exercises 59–62, decide whether the function has an inverse function. If so, what is the inverse function?**

59. $g(t)$ is the volume of water that has passed through a water line t minutes after a control valve is opened.

60. $h(t)$ is the height of the tide t hours after midnight, where $0 \leq t < 24$.

61. $C(t)$ is the cost of a long distance call lasting t minutes.

62. $A(r)$ is the area of a circle of radius r.

Evaluating the Derivative of an Inverse Function **In Exercises 63–70, verify that f has an inverse. Then use the function f and the given real number a to find $(f^{-1})'(a)$. (*Hint:* See Example 5.)**

63. $f(x) = 5 - 2x^3, \quad a = 7$

64. $f(x) = x^3 + 2x - 1, \quad a = 2$

65. $f(x) = \frac{1}{27}(x^5 + 2x^3), \quad a = -11$

66. $f(x) = \sqrt{x - 4}, \quad a = 2$

67. $f(x) = \sin x, \quad -\dfrac{\pi}{2} \leq x \leq \dfrac{\pi}{2}, \quad a = \dfrac{1}{2}$

68. $f(x) = \cos 2x, \quad 0 \leq x \leq \dfrac{\pi}{2}, \quad a = 1$

69. $f(x) = \dfrac{x + 6}{x - 2}, \quad x > 2, \quad a = 3$

70. $f(x) = \dfrac{x + 3}{x + 1}, \quad x > -1, \quad a = 2$

Using Inverse Functions **In Exercises 71–74, (a) find the domains of f and f^{-1}, (b) find the ranges of f and f^{-1}, (c) graph f and f^{-1}, and (d) show that the slopes of the graphs of f and f^{-1} are reciprocals at the given points.**

Functions	Point
71. $f(x) = x^3$	$\left(\frac{1}{2}, \frac{1}{8}\right)$
$f^{-1}(x) = \sqrt[3]{x}$	$\left(\frac{1}{8}, \frac{1}{2}\right)$
72. $f(x) = 3 - 4x$	$(1, -1)$
$f^{-1}(x) = \dfrac{3 - x}{4}$	$(-1, 1)$
73. $f(x) = \sqrt{x - 4}$	$(5, 1)$
$f^{-1}(x) = x^2 + 4, \quad x \geq 0$	$(1, 5)$
74. $f(x) = \dfrac{4}{1 + x^2}, \quad x \geq 0$	$(1, 2)$
$f^{-1}(x) = \sqrt{\dfrac{4 - x}{x}}$	$(2, 1)$

Using Composite and Inverse Functions In Exercises 75–78, use the functions $f(x) = \frac{1}{8}x - 3$ and $g(x) = x^3$ to find the given value.

75. $(f^{-1} \circ g^{-1})(1)$
76. $(g^{-1} \circ f^{-1})(-3)$
77. $(f^{-1} \circ f^{-1})(6)$
78. $(g^{-1} \circ g^{-1})(-4)$

Using Composite and Inverse Functions In Exercises 79–82, use the functions $f(x) = x + 4$ and $g(x) = 2x - 5$ to find the given function.

79. $g^{-1} \circ f^{-1}$
80. $f^{-1} \circ g^{-1}$
81. $(f \circ g)^{-1}$
82. $(g \circ f)^{-1}$

WRITING ABOUT CONCEPTS

83. **In Your Own Words** Describe how to find the inverse function of a one-to-one function given by an equation in x and y. Give an example.

84. **A Function and Its Inverse** Describe the relationship between the graph of a function and the graph of its inverse function.

Explaining Why a Function Is Not One-to-One In Exercises 85 and 86, the derivative of the function has the same sign for all x in its domain, but the function is not one-to-one. Explain.

85. $f(x) = \tan x$
86. $f(x) = \dfrac{x}{x^2 - 4}$

87. **Think About It** The function $f(x) = k(2 - x - x^3)$ is one-to-one and $f^{-1}(3) = -2$. Find k.

88. **HOW DO YOU SEE IT?** Use the information in the graph of f below.

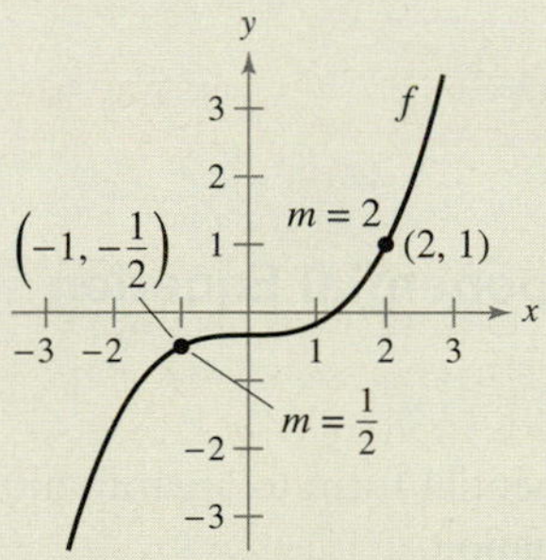

(a) What is the slope of the tangent line to the graph of f^{-1} at the point $\left(-\frac{1}{2}, -1\right)$? Explain.

(b) What is the slope of the tangent line to the graph of f^{-1} at the point $(1, 2)$? Explain.

True or False? In Exercises 89–92, determine whether the statement is true or false. If it is false, explain why or give an example that shows it is false.

89. If f is an even function, then f^{-1} exists.

90. If the inverse function of f exists, then the y-intercept of f is an x-intercept of f^{-1}.

91. If $f(x) = x^n$, where n is odd, then f^{-1} exists.

92. There exists no function f such that $f = f^{-1}$.

93. **Making a Function One-to-One**
(a) Show that $f(x) = 2x^3 + 3x^2 - 36x$ is not one-to-one on $(-\infty, \infty)$.
(b) Determine the greatest value c such that f is one-to-one on $(-c, c)$.

94. **Proof** Let f and g be one-to-one functions. Prove that
(a) $f \circ g$ is one-to-one.
(b) $(f \circ g)^{-1}(x) = (g^{-1} \circ f^{-1})(x)$.

95. **Proof** Prove that if f has an inverse function, then $(f^{-1})^{-1} = f$.

96. **Proof** Prove that if a function has an inverse function, then the inverse function is unique.

97. **Proof** Prove that a function has an inverse function if and only if it is one-to-one.

98. **Using Theorem 5.7** Is the converse of the second part of Theorem 5.7 true? That is, if a function is one-to-one (and therefore has an inverse function), then must the function be strictly monotonic? If so, prove it. If not, give a counterexample.

99. **Concavity** Let f be twice-differentiable and one-to-one on an open interval I. Show that its inverse function g satisfies

$$g''(x) = -\frac{f''(g(x))}{[f'(g(x))]^3}.$$

When f is increasing and concave downward, what is the concavity of $f^{-1} = g$?

100. **Derivative of an Inverse Function** Let

$$f(x) = \int_2^x \frac{dt}{\sqrt{1 + t^4}}.$$

Find $(f^{-1})'(0)$.

101. **Derivative of an Inverse Function** Show that

$$f(x) = \int_2^x \sqrt{1 + t^2}\, dt$$

is one-to-one and find

$(f^{-1})'(0)$.

102. **Inverse Function** Let

$$y = \frac{x - 2}{x - 1}.$$

Show that y is its own inverse function. What can you conclude about the graph of f? Explain.

103. **Using a Function** Let $f(x) = \dfrac{ax + b}{cx + d}$.
(a) Show that f is one-to-one if and only if $bc - ad \neq 0$.
(b) Given $bc - ad \neq 0$, find f^{-1}.
(c) Determine the values of a, b, c, and d such that $f = f^{-1}$.

5.4 Exponential Functions: Differentiation and Integration

- Develop properties of the natural exponential function.
- Differentiate natural exponential functions.
- Integrate natural exponential functions.

The Natural Exponential Function

The function $f(x) = \ln x$ is increasing on its entire domain, and therefore it has an inverse function f^{-1}. The domain of f^{-1} is the set of all real numbers, and the range is the set of positive real numbers, as shown in Figure 5.19. So, for any real number x,

$$f(f^{-1}(x)) = \ln[f^{-1}(x)] = x. \qquad x \text{ is any real number.}$$

If x is rational, then

$$\ln(e^x) = x \ln e = x(1) = x. \qquad x \text{ is a rational number.}$$

Because the natural logarithmic function is one-to-one, you can conclude that $f^{-1}(x)$ and e^x agree for *rational* values of x. The next definition extends the meaning of e^x to include *all* real values of x.

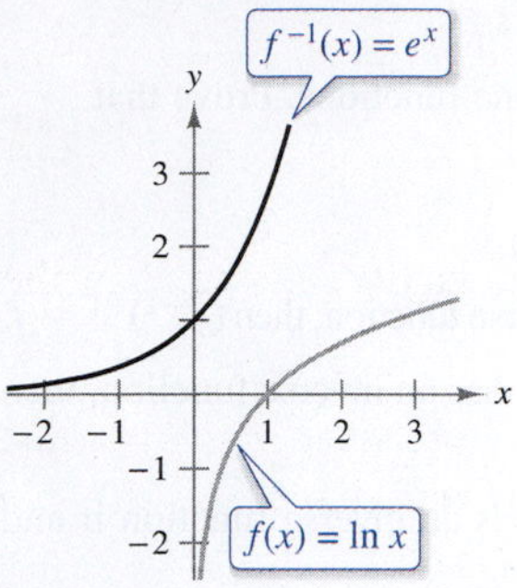

The inverse function of the natural logarithmic function is the natural exponential function.
Figure 5.19

Definition of the Natural Exponential Function

The inverse function of the natural logarithmic function $f(x) = \ln x$ is called the **natural exponential function** and is denoted by

$$f^{-1}(x) = e^x.$$

That is,

$$y = e^x \quad \text{if and only if} \quad x = \ln y.$$

THE NUMBER e

The symbol e was first used by mathematician Leonhard Euler to represent the base of natural logarithms in a letter to another mathematician, Christian Goldbach, in 1731.

The inverse relationship between the natural logarithmic function and the natural exponential function can be summarized as shown.

$$\ln(e^x) = x \quad \text{and} \quad e^{\ln x} = x \qquad \text{Inverse relationship}$$

EXAMPLE 1 Solving an Exponential Equation

Solve $7 = e^{x+1}$.

Solution You can convert from exponential form to logarithmic form by *taking the natural logarithm of each side* of the equation.

$$7 = e^{x+1} \qquad \text{Write original equation.}$$
$$\ln 7 = \ln(e^{x+1}) \qquad \text{Take natural logarithm of each side.}$$
$$\ln 7 = x + 1 \qquad \text{Apply inverse property.}$$
$$-1 + \ln 7 = x \qquad \text{Solve for } x.$$

So, the solution is $-1 + \ln 7 \approx -0.946$. You can check this solution as shown.

$$7 = e^{x+1} \qquad \text{Write original equation.}$$
$$7 \stackrel{?}{=} e^{(-1+\ln 7)+1} \qquad \text{Substitute } -1 + \ln 7 \text{ for } x \text{ in original equation.}$$
$$7 \stackrel{?}{=} e^{\ln 7} \qquad \text{Simplify.}$$
$$7 = 7 \checkmark \qquad \text{Solution checks.}$$

EXAMPLE 2 Solving a Logarithmic Equation

Solve $\ln(2x - 3) = 5$.

Solution To convert from logarithmic form to exponential form, you can *exponentiate each side* of the logarithmic equation.

$$\ln(2x - 3) = 5 \quad \text{Write original equation.}$$
$$e^{\ln(2x-3)} = e^5 \quad \text{Exponentiate each side.}$$
$$2x - 3 = e^5 \quad \text{Apply inverse property.}$$
$$x = \tfrac{1}{2}(e^5 + 3) \quad \text{Solve for } x.$$
$$x \approx 75.707 \quad \text{Use a calculator.}$$

The familiar rules for operating with rational exponents can be extended to the natural exponential function, as shown in the next theorem.

THEOREM 5.10 Operations with Exponential Functions

Let a and b be any real numbers.

1. $e^a e^b = e^{a+b}$ **2.** $\dfrac{e^a}{e^b} = e^{a-b}$

Proof To prove Property 1, you can write

$$\ln(e^a e^b) = \ln(e^a) + \ln(e^b) = a + b = \ln(e^{a+b}).$$

Because the natural logarithmic function is one-to-one, you can conclude that

$$e^a e^b = e^{a+b}.$$

The proof of the other property is given in Appendix A.

See LarsonCalculus.com for Bruce Edwards's video of this proof.

In Section 5.3, you learned that an inverse function f^{-1} shares many properties with f. So, the natural exponential function inherits the properties listed below from the natural logarithmic function.

Properties of the Natural Exponential Function

1. The domain of $f(x) = e^x$ is

$$(-\infty, \infty)$$

and the range is

$$(0, \infty).$$

2. The function $f(x) = e^x$ is continuous, increasing, and one-to-one on its entire domain.

3. The graph of $f(x) = e^x$ is concave upward on its entire domain.

4. $\lim\limits_{x\to-\infty} e^x = 0$

5. $\lim\limits_{x\to\infty} e^x = \infty$

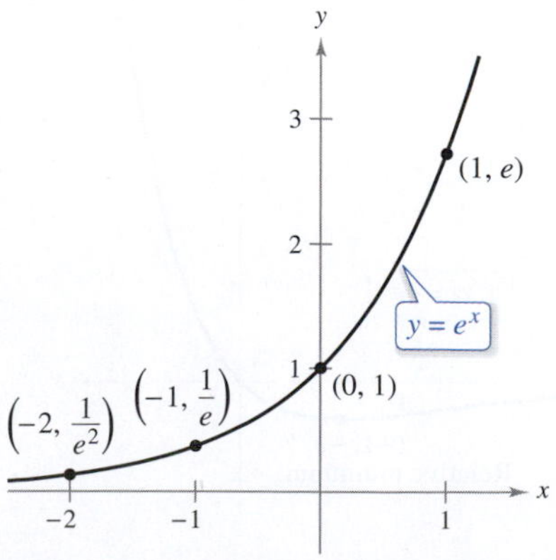

The natural exponential function is increasing, and its graph is concave upward.

Derivatives of Exponential Functions

One of the most intriguing (and useful) characteristics of the natural exponential function is that *it is its own derivative*. In other words, it is a solution of the differential equation $y' = y$. This result is stated in the next theorem.

REMARK You can interpret this theorem geometrically by saying that the slope of the graph of $f(x) = e^x$ at any point (x, e^x) is equal to the y-coordinate of the point.

THEOREM 5.11 Derivatives of the Natural Exponential Function

Let u be a differentiable function of x.

1. $\dfrac{d}{dx}[e^x] = e^x$
2. $\dfrac{d}{dx}[e^u] = e^u \dfrac{du}{dx}$

FOR FURTHER INFORMATION To find out about derivatives of exponential functions of order 1/2, see the article "A Child's Garden of Fractional Derivatives" by Marcia Kleinz and Thomas J. Osler in *The College Mathematics Journal*. To view this article, go to *MathArticles.com*.

Proof To prove Property 1, use the fact that $\ln e^x = x$, and differentiate each side of the equation.

$$\ln e^x = x \qquad \text{Definition of exponential function}$$

$$\frac{d}{dx}[\ln e^x] = \frac{d}{dx}[x] \qquad \text{Differentiate each side with respect to } x.$$

$$\frac{1}{e^x}\frac{d}{dx}[e^x] = 1$$

$$\frac{d}{dx}[e^x] = e^x$$

The derivative of e^u follows from the Chain Rule.

See LarsonCalculus.com for Bruce Edwards's video of this proof.

EXAMPLE 3 Differentiating Exponential Functions

Find the derivative of each function.

a. $y = e^{2x-1}$ **b.** $y = e^{-3/x}$

Solution

a. $\dfrac{d}{dx}[e^{2x-1}] = e^u\dfrac{du}{dx} = 2e^{2x-1}$ $\qquad u = 2x - 1$

b. $\dfrac{d}{dx}[e^{-3/x}] = e^u\dfrac{du}{dx} = \left(\dfrac{3}{x^2}\right)e^{-3/x} = \dfrac{3e^{-3/x}}{x^2}$ $\qquad u = -\dfrac{3}{x}$

EXAMPLE 4 Locating Relative Extrema

Find the relative extrema of

$$f(x) = xe^x.$$

Solution The derivative of f is

$$f'(x) = x(e^x) + e^x(1) \qquad \text{Product Rule}$$
$$= e^x(x + 1).$$

Because e^x is never 0, the derivative is 0 only when $x = -1$. Moreover, by the First Derivative Test, you can determine that this corresponds to a relative minimum, as shown in Figure 5.20. Because the derivative $f'(x) = e^x(x + 1)$ is defined for all x, there are no other critical points.

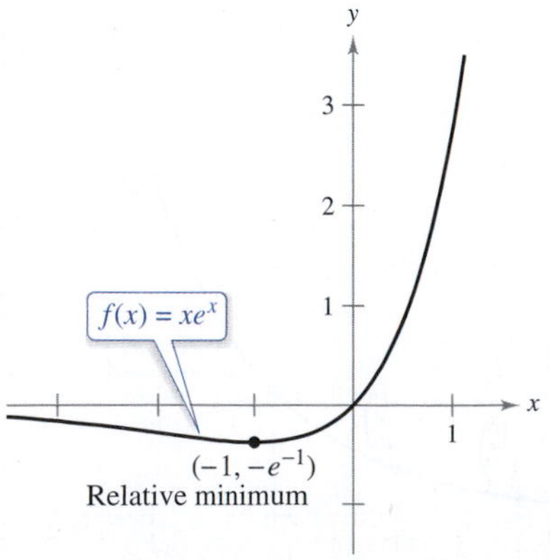

The derivative of f changes from negative to positive at $x = -1$.

Figure 5.20

EXAMPLE 5 The Standard Normal Probability Density Function

See LarsonCalculus.com for an interactive version of this type of example.

Show that the *standard normal probability density function*

$$f(x) = \frac{1}{\sqrt{2\pi}}e^{-x^2/2}$$

has points of inflection when $x = \pm 1$.

REMARK The general form of a normal probability density function (whose mean is 0) is

$$f(x) = \frac{1}{\sigma\sqrt{2\pi}}e^{-x^2/(2\sigma^2)}$$

where σ is the standard deviation (σ is the lowercase Greek letter sigma). This "bell-shaped curve" has points of inflection when $x = \pm\sigma$.

Solution To locate possible points of inflection, find the x-values for which the second derivative is 0.

$$f(x) = \frac{1}{\sqrt{2\pi}}e^{-x^2/2} \qquad \text{Write original function.}$$

$$f'(x) = \frac{1}{\sqrt{2\pi}}(-x)e^{-x^2/2} \qquad \text{First derivative}$$

$$f''(x) = \frac{1}{\sqrt{2\pi}}[(-x)(-x)e^{-x^2/2} + (-1)e^{-x^2/2}] \qquad \text{Product Rule}$$

$$= \frac{1}{\sqrt{2\pi}}(e^{-x^2/2})(x^2 - 1) \qquad \text{Second derivative}$$

So, $f''(x) = 0$ when $x = \pm 1$, and you can apply the techniques of Chapter 3 to conclude that these values yield the two points of inflection shown in Figure 5.21.

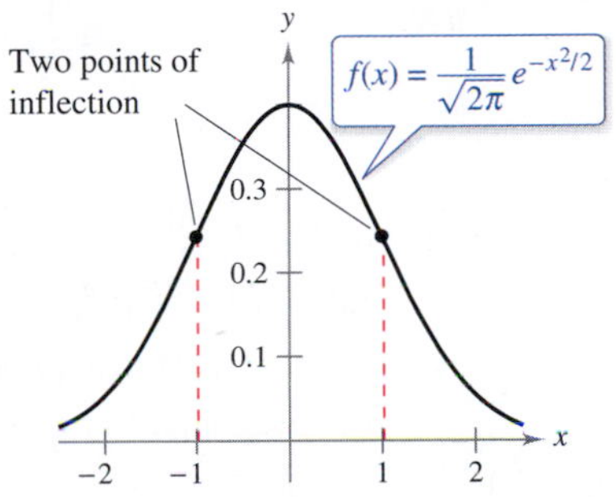

The bell-shaped curve given by a standard normal probability density function

Figure 5.21

EXAMPLE 6 Population of California

The projected populations y (in thousands) of California from 2015 through 2030 can be modeled by

$$y = 34{,}696e^{0.0097t}$$

where t represents the year, with $t = 15$ corresponding to 2015. At what rate will the population be changing in 2020? (*Source: U.S. Census Bureau*)

Solution The derivative of the model is

$$y' = (0.0097)(34{,}696)e^{0.0097t}$$

$$\approx 336.55e^{0.0097t}.$$

By evaluating the derivative when $t = 20$, you can estimate that the rate of change in 2020 will be about

408.6 thousand people per year.

AP* Tips
Theorems 5.11 and 5.12 are frequently tested on all sections of the AP Exam.

Integrals of Exponential Functions

Each differentiation formula in Theorem 5.11 has a corresponding integration formula.

> **THEOREM 5.12 Integration Rules for Exponential Functions**
>
> Let u be a differentiable function of x.
>
> **1.** $\int e^x\,dx = e^x + C$ **2.** $\int e^u\,du = e^u + C$

EXAMPLE 7 Integrating Exponential Functions

Find the indefinite integral.

$$\int e^{3x+1}\,dx$$

Solution If you let $u = 3x + 1$, then $du = 3\,dx$.

$$\int e^{3x+1}\,dx = \frac{1}{3}\int e^{3x+1}(3)\,dx \qquad \text{Multiply and divide by 3.}$$

$$= \frac{1}{3}\int e^u\,du \qquad \text{Substitute: } u = 3x + 1.$$

$$= \frac{1}{3}e^u + C \qquad \text{Apply Exponential Rule.}$$

$$= \frac{e^{3x+1}}{3} + C \qquad \text{Back-substitute.}$$

REMARK In Example 7, the missing *constant* factor 3 was introduced to create $du = 3\,dx$. However, remember that you cannot introduce a missing *variable* factor in the integrand. For instance,

$$\int e^{-x^2}\,dx \neq \frac{1}{x}\int e^{-x^2}(x\,dx).$$

EXAMPLE 8 Integrating Exponential Functions

Find the indefinite integral.

$$\int 5xe^{-x^2}\,dx$$

Solution If you let $u = -x^2$, then $du = -2x\,dx$ or $x\,dx = -du/2$.

$$\int 5xe^{-x^2}\,dx = \int 5e^{-x^2}(x\,dx) \qquad \text{Regroup integrand.}$$

$$= \int 5e^u\left(-\frac{du}{2}\right) \qquad \text{Substitute: } u = -x^2.$$

$$= -\frac{5}{2}\int e^u\,du \qquad \text{Constant Multiple Rule}$$

$$= -\frac{5}{2}e^u + C \qquad \text{Apply Exponential Rule.}$$

$$= -\frac{5}{2}e^{-x^2} + C \qquad \text{Back-substitute.}$$

EXAMPLE 9 **Integrating Exponential Functions**

Find each indefinite integral.

a. $\int \frac{e^{1/x}}{x^2}\,dx$ **b.** $\int \sin x\, e^{\cos x}\,dx$

Solution

a. $$\int \frac{e^{1/x}}{x^2}\,dx = -\int \overbrace{e^{1/x}}^{e^u}\overbrace{\left(-\frac{1}{x^2}\right)dx}^{du} \qquad u = \frac{1}{x}$$
$$= -e^{1/x} + C$$

b. $$\int \sin x e^{\cos x}\,dx = -\int \overbrace{e^{\cos x}}^{e^u}\overbrace{(-\sin x\,dx)}^{du} \qquad u = \cos x$$
$$= -e^{\cos x} + C$$

EXAMPLE 10 **Finding Areas Bounded by Exponential Functions**

Evaluate each definite integral.

a. $\int_0^1 e^{-x}\,dx$ **b.** $\int_0^1 \frac{e^x}{1+e^x}\,dx$ **c.** $\int_{-1}^0 [e^x\cos(e^x)]\,dx$

Solution

a. $$\int_0^1 e^{-x}\,dx = -e^{-x}\Big]_0^1 \qquad \text{See Figure 5.22(a).}$$
$$= -e^{-1} - (-1)$$
$$= 1 - \frac{1}{e}$$
$$\approx 0.632$$

b. $$\int_0^1 \frac{e^x}{1+e^x}\,dx = \ln(1+e^x)\Big]_0^1 \qquad \text{See Figure 5.22(b).}$$
$$= \ln(1+e) - \ln 2$$
$$\approx 0.620$$

c. $$\int_{-1}^0 [e^x\cos(e^x)]\,dx = \sin(e^x)\Big]_{-1}^0 \qquad \text{See Figure 5.22(c).}$$
$$= \sin 1 - \sin(e^{-1})$$
$$\approx 0.482$$

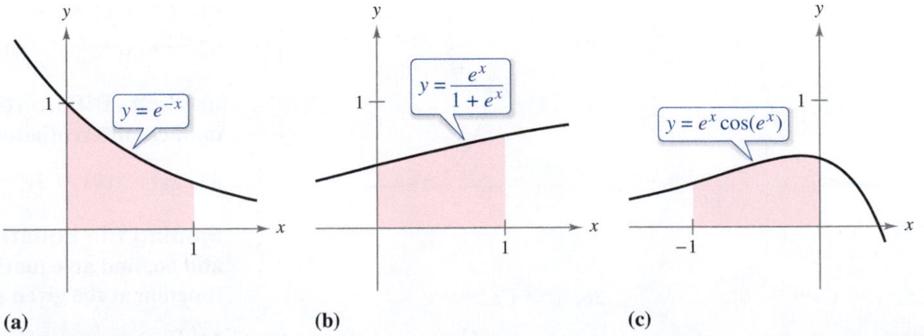

Figure 5.22

5.4 Exercises

See CalcChat.com for tutorial help and worked-out solutions to odd-numbered exercises.

Solving an Exponential or Logarithmic Equation **In Exercises 1–16, solve for x accurate to three decimal places.**

1. $e^{\ln x} = 4$
2. $e^{\ln 3x} = 24$
3. $e^x = 12$
4. $5e^x = 36$
5. $9 - 2e^x = 7$
6. $8e^x - 12 = 7$
7. $50e^{-x} = 30$
8. $100e^{-2x} = 35$
9. $\dfrac{800}{100 - e^{x/2}} = 50$
10. $\dfrac{5000}{1 + e^{2x}} = 2$
11. $\ln x = 2$
12. $\ln x^2 = 10$
13. $\ln(x - 3) = 2$
14. $\ln 4x = 1$
15. $\ln\sqrt{x + 2} = 1$
16. $\ln(x - 2)^2 = 12$

Sketching a Graph **In Exercises 17–22, sketch the graph of the function.**

17. $y = e^{-x}$
18. $y = \frac{1}{2}e^x$
19. $y = e^x + 2$
20. $y = e^{x-1}$
21. $y = e^{-x^2}$
22. $y = e^{-x/2}$

23. **Comparing Graphs** Use a graphing utility to graph $f(x) = e^x$ and the given function in the same viewing window. How are the two graphs related?

(a) $g(x) = e^{x-2}$ (b) $h(x) = -\frac{1}{2}e^x$ (c) $q(x) = e^{-x} + 3$

24. **Asymptotes** Use a graphing utility to graph the function. Use the graph to determine any asymptotes of the function.

(a) $f(x) = \dfrac{8}{1 + e^{-0.5x}}$ (b) $g(x) = \dfrac{8}{1 + e^{-0.5/x}}$

Matching **In Exercises 25–28, match the equation with the correct graph. Assume that a and C are positive real numbers. [The graphs are labeled (a), (b), (c), and (d).]**

(a)

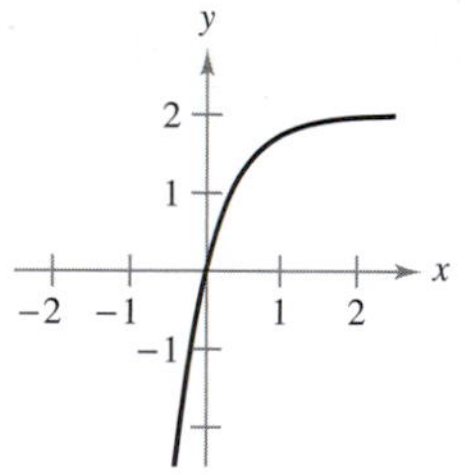

(b)

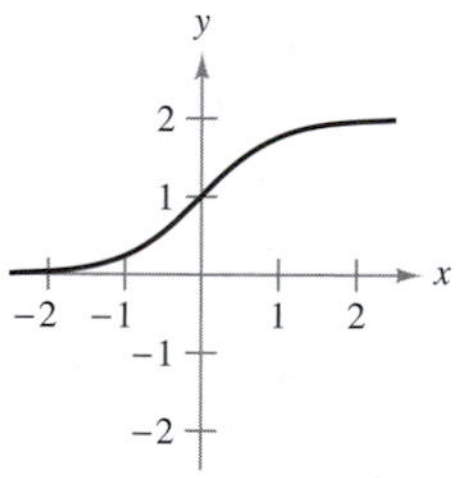

(c)

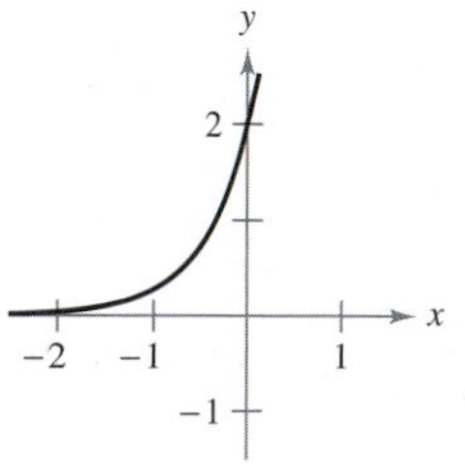

(d)

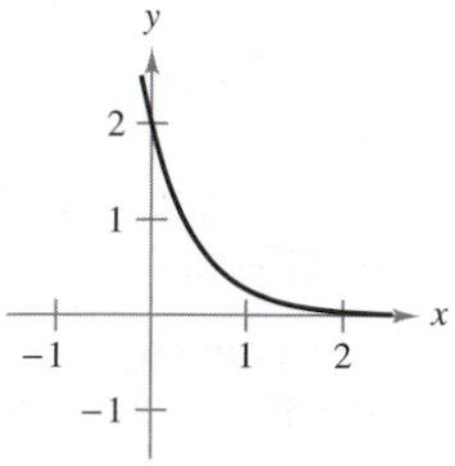

25. $y = Ce^{ax}$
26. $y = Ce^{-ax}$
27. $y = C(1 - e^{-ax})$
28. $y = \dfrac{C}{1 + e^{-ax}}$

Inverse Functions **In Exercises 29–32, illustrate that the functions are inverses of each other by graphing both functions on the same set of coordinate axes.**

29. $f(x) = e^{2x}$, $g(x) = \ln\sqrt{x}$
30. $f(x) = e^{x/3}$, $g(x) = \ln x^3$
31. $f(x) = e^x - 1$, $g(x) = \ln(x + 1)$
32. $f(x) = e^{x-1}$, $g(x) = 1 + \ln x$

Finding a Derivative **In Exercises 33–54, find the derivative.**

33. $f(x) = e^{2x}$
34. $y = e^{-8x}$
35. $y = e^{\sqrt{x}}$
36. $y = e^{-2x^3}$
37. $y = e^{x-4}$
38. $y = 5e^{x^2+5}$
39. $y = e^x \ln x$
40. $y = xe^{4x}$
41. $y = x^3e^x$
42. $y = x^2e^{-x}$
43. $g(t) = (e^{-t} + e^t)^3$
44. $g(t) = e^{-3/t^2}$
45. $y = \ln(1 + e^{2x})$
46. $y = \ln\left(\dfrac{1 + e^x}{1 - e^x}\right)$
47. $y = \dfrac{2}{e^x + e^{-x}}$
48. $y = \dfrac{e^x - e^{-x}}{2}$
49. $y = \dfrac{e^x + 1}{e^x - 1}$
50. $y = \dfrac{e^{2x}}{e^{2x} + 1}$
51. $y = e^x(\sin x + \cos x)$
52. $y = e^{2x} \tan 2x$
53. $F(x) = \displaystyle\int_{\pi}^{\ln x} \cos e^t\, dt$
54. $F(x) = \displaystyle\int_{0}^{e^{2x}} \ln(t + 1)\, dt$

Finding an Equation of a Tangent Line **In Exercises 55–62, find an equation of the tangent line to the graph of the function at the given point.**

55. $f(x) = e^{3x}$, $(0, 1)$
56. $f(x) = e^{-2x}$, $(0, 1)$
57. $f(x) = e^{1-x}$, $(1, 1)$
58. $y = e^{-2x+x^2}$, $(2, 1)$
59. $f(x) = e^{-x} \ln x$, $(1, 0)$
60. $y = \ln\dfrac{e^x + e^{-x}}{2}$, $(0, 0)$
61. $y = x^2e^x - 2xe^x + 2e^x$, $(1, e)$
62. $y = xe^x - e^x$, $(1, 0)$

Implicit Differentiation **In Exercises 63 and 64, use implicit differentiation to find dy/dx.**

63. $xe^y - 10x + 3y = 0$
64. $e^{xy} + x^2 - y^2 = 10$

Finding the Equation of a Tangent Line **In Exercises 65 and 66, find an equation of the tangent line to the graph of the function at the given point.**

65. $xe^y + ye^x = 1$, $(0, 1)$
66. $1 + \ln xy = e^{x-y}$, $(1, 1)$

Finding a Second Derivative In Exercises 67 and 68, find the second derivative of the function.

67. $f(x) = (3 + 2x)e^{-3x}$

68. $g(x) = \sqrt{x} + e^x \ln x$

Differential Equation In Exercises 69 and 70, show that the function $y = f(x)$ is a solution of the differential equation.

69. $y = 4e^{-x}$

$y'' - y = 0$

70. $y = e^{3x} + e^{-3x}$

$y'' - 9y = 0$

Finding Extrema and Points of Inflection In Exercises 71–78, find the extrema and the points of inflection (if any exist) of the function. Use a graphing utility to graph the function and confirm your results.

71. $f(x) = \dfrac{e^x + e^{-x}}{2}$

72. $f(x) = \dfrac{e^x - e^{-x}}{2}$

73. $g(x) = \dfrac{1}{\sqrt{2\pi}} e^{-(x-2)^2/2}$

74. $g(x) = \dfrac{1}{\sqrt{2\pi}} e^{-(x-3)^2/2}$

75. $f(x) = x^2 e^{-x}$

76. $f(x) = xe^{-x}$

77. $g(t) = 1 + (2 + t)e^{-t}$

78. $f(x) = -2 + e^{3x}(4 - 2x)$

79. Area Find the area of the largest rectangle that can be inscribed under the curve $y = e^{-x^2}$ in the first and second quadrants.

80. Area Perform the following steps to find the maximum area of the rectangle shown in the figure.

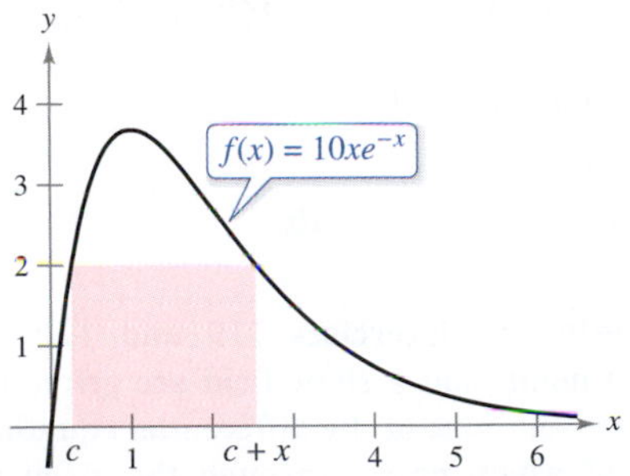

(a) Solve for c in the equation $f(c) = f(c + x)$.

(b) Use the result in part (a) to write the area A as a function of x. [*Hint:* $A = xf(c)$]

(c) Use a graphing utility to graph the area function. Use the graph to approximate the dimensions of the rectangle of maximum area. Determine the maximum area.

(d) Use a graphing utility to graph the expression for c found in part (a). Use the graph to approximate

$$\lim_{x \to 0^+} c \quad \text{and} \quad \lim_{x \to \infty} c.$$

Use this result to describe the changes in dimensions and position of the rectangle for $0 < x < \infty$.

81. Finding an Equation of a Tangent Line Find a point on the graph of the function $f(x) = e^{2x}$ such that the tangent line to the graph at that point passes through the origin. Use a graphing utility to graph f and the tangent line in the same viewing window.

Robert Adrian Hillman/Shutterstock.com

82. HOW DO YOU SEE IT? The figure shows the graphs of f and g, where a is a positive real number. Identify the open interval(s) on which the graphs of f and g are (a) increasing or decreasing, and (b) concave upward or concave downward.

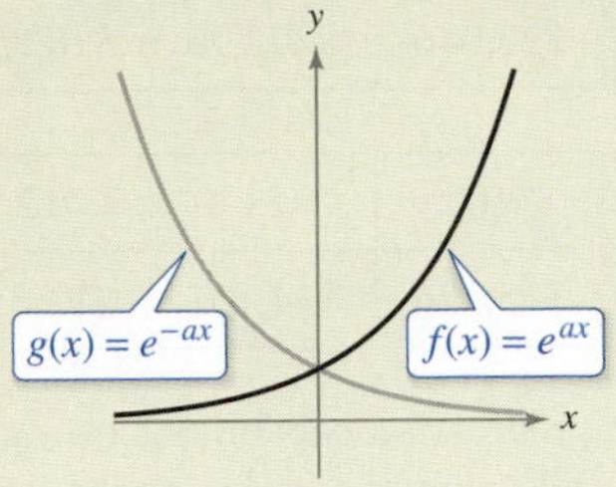

83. Depreciation The value V of an item t years after it is purchased is $V = 15{,}000e^{-0.6286t}$, $0 \le t \le 10$.

(a) Use a graphing utility to graph the function.

(b) Find the rates of change of V with respect to t when $t = 1$ and $t = 5$.

(c) Use a graphing utility to graph the tangent lines to the function when $t = 1$ and $t = 5$.

84. Harmonic Motion The displacement from equilibrium of a mass oscillating on the end of a spring suspended from a ceiling is $y = 1.56e^{-0.22t} \cos 4.9t$, where y is the displacement (in feet) and t is the time (in seconds). Use a graphing utility to graph the displacement function on the interval $[0, 10]$. Find a value of t past which the displacement is less than 3 inches from equilibrium.

85. Atmospheric Pressure

A meteorologist measures the atmospheric pressure P (in kilograms per square meter) at altitude h (in kilometers). The data are shown below.

h	0	5	10	15	20
P	10,332	5583	2376	1240	517

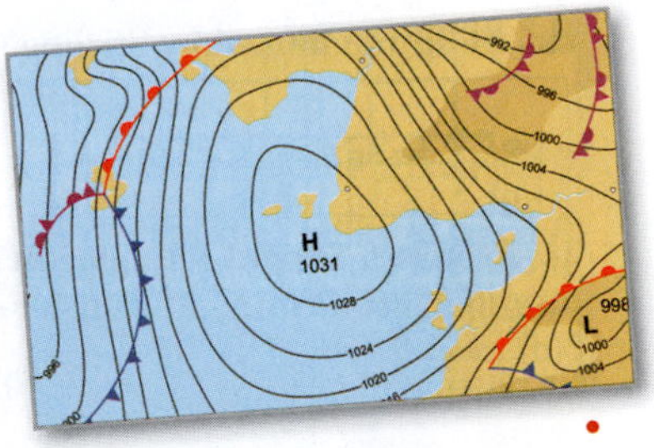

(a) Use a graphing utility to plot the points $(h, \ln P)$. Use the regression capabilities of the graphing utility to find a linear model for the revised data points.

(b) The line in part (a) has the form $\ln P = ah + b$. Write the equation in exponential form.

(c) Use a graphing utility to plot the original data and graph the exponential model in part (b).

(d) Find the rate of change of the pressure when $h = 5$ and $h = 18$.

86. Modeling Data The table lists the approximate values V of a mid-sized sedan for the years 2006 through 2012. The variable t represents the time (in years), with $t = 6$ corresponding to 2006.

t	6	7	8	9	10	11	12
V	\$23,046	\$20,596	\$18,851	\$17,001	\$15,226	\$14,101	\$12,841

(a) Use the regression capabilities of a graphing utility to fit linear and quadratic models to the data. Plot the data and graph the models.

(b) What does the slope represent in the linear model in part (a)?

(c) Use the regression capabilities of a graphing utility to fit an exponential model to the data.

(d) Determine the horizontal asymptote of the exponential model found in part (c). Interpret its meaning in the context of the problem.

(e) Use the exponential model to find the rate of decrease in the value of the sedan when $t = 7$ and $t = 11$.

Linear and Quadratic Approximation **In Exercises 87 and 88, use a graphing utility to graph the function. Then graph**

$$P_1(x) = f(0) + f'(0)(x - 0) \quad \text{and}$$

$$P_2(x) = f(0) + f'(0)(x - 0) + \tfrac{1}{2}f''(0)(x - 0)^2$$

in the same viewing window. Compare the values of f, P_1, P_2, and their first derivatives at $x = 0$.

87. $f(x) = e^x$ **88.** $f(x) = e^{x/2}$

Stirling's Formula **For large values of n,**

$$n! = 1 \cdot 2 \cdot 3 \cdot 4 \cdots (n - 1) \cdot n$$

can be approximated by Stirling's Formula,

$$n! \approx \left(\frac{n}{e}\right)^n \sqrt{2\pi n}.$$

In Exercises 89 and 90, find the exact value of $n!$, and then approximate $n!$ using Stirling's Formula.

89. $n = 12$ **90.** $n = 15$

Finding an Indefinite Integral **In Exercises 91–108, find the indefinite integral.**

91. $\int e^{5x}(5)\,dx$ **92.** $\int e^{-x^4}(-4x^3)\,dx$

93. $\int e^{2x-1}\,dx$ **94.** $\int e^{1-3x}\,dx$

95. $\int x^2 e^{x^3}\,dx$ **96.** $\int e^x(e^x + 1)^2\,dx$

97. $\int \frac{e^{\sqrt{x}}}{\sqrt{x}}\,dx$ **98.** $\int \frac{e^{1/x^2}}{x^3}\,dx$

99. $\int \frac{e^{-x}}{1 + e^{-x}}\,dx$ **100.** $\int \frac{e^{2x}}{1 + e^{2x}}\,dx$

101. $\int e^x\sqrt{1 - e^x}\,dx$ **102.** $\int \frac{e^x - e^{-x}}{e^x + e^{-x}}\,dx$

103. $\int \frac{e^x + e^{-x}}{e^x - e^{-x}}\,dx$ **104.** $\int \frac{2e^x - 2e^{-x}}{(e^x + e^{-x})^2}\,dx$

105. $\int \frac{5 - e^x}{e^{2x}}\,dx$ **106.** $\int \frac{e^{2x} + 2e^x + 1}{e^x}\,dx$

107. $\int e^{-x}\tan(e^{-x})\,dx$

108. $\int e^{2x}\csc(e^{2x})\,dx$

Evaluating a Definite Integral **In Exercises 109–118, evaluate the definite integral. Use a graphing utility to verify your result.**

109. $\int_0^1 e^{-2x}\,dx$ **110.** $\int_1^2 e^{5x-3}\,dx$

111. $\int_0^1 xe^{-x^2}\,dx$ **112.** $\int_{-2}^0 x^2 e^{x^3/2}\,dx$

113. $\int_1^3 \frac{e^{3/x}}{x^2}\,dx$ **114.** $\int_0^{\sqrt{2}} xe^{-(x^2/2)}\,dx$

115. $\int_0^3 \frac{2e^{2x}}{1 + e^{2x}}\,dx$ **116.** $\int_0^1 \frac{e^x}{5 - e^x}\,dx$

117. $\int_0^{\pi/2} e^{\sin \pi x}\cos \pi x\,dx$

118. $\int_{\pi/3}^{\pi/2} e^{\sec 2x}\sec 2x \tan 2x\,dx$

Slope Field **In Exercises 119 and 120, a differential equation, a point, and a slope field are given. (a) Sketch two approximate solutions of the differential equation on the slope field, one of which passes through the given point. (b) Use integration to find the particular solution of the differential equation and use a graphing utility to graph the solution. Compare the result with the sketches in part (a). To print an enlarged copy of the graph, go to *MathGraphs.com*.**

119. $\frac{dy}{dx} = 2e^{-x/2}$, $(0, 1)$ **120.** $\frac{dy}{dx} = xe^{-0.2x^2}$, $\left(0, -\frac{3}{2}\right)$

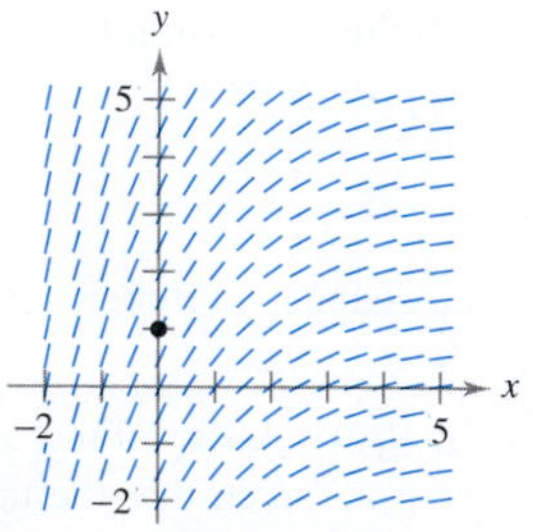

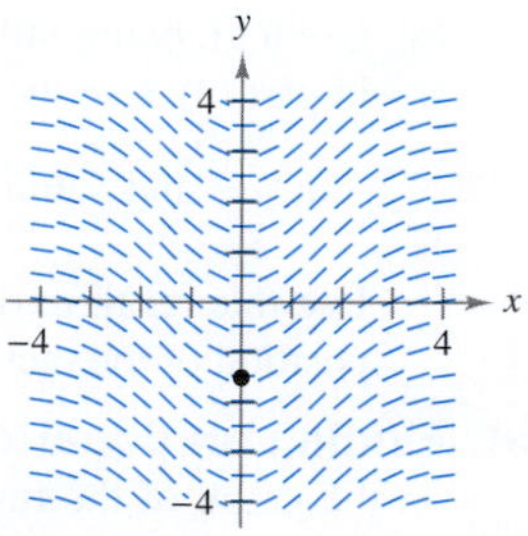

Differential Equation **In Exercises 121 and 122, solve the differential equation.**

121. $\frac{dy}{dx} = xe^{ax^2}$ **122.** $\frac{dy}{dx} = (e^x - e^{-x})^2$

Differential Equation **In Exercises 123 and 124, find the particular solution that satisfies the initial conditions.**

123. $f''(x) = \frac{1}{2}(e^x + e^{-x})$, $f(0) = 1, f'(0) = 0$

124. $f''(x) = \sin x + e^{2x}$, $f(0) = \frac{1}{4}, f'(0) = \frac{1}{2}$

Area **In Exercises 125–128, find the area of the region bounded by the graphs of the equations. Use a graphing utility to graph the region and verify your result.**

125. $y = e^x, y = 0, x = 0, x = 5$

126. $y = e^{-2x}, y = 0, x = -1, x = 3$

127. $y = xe^{-x^2/4}, y = 0, x = 0, x = \sqrt{6}$

128. $y = e^{-2x} + 2, y = 0, x = 0, x = 2$

Numerical Integration **In Exercises 129 and 130, approximate the integral using the Midpoint Rule, the Trapezoidal Rule, and Simpson's Rule with $n = 12$. Use a graphing utility to verify your results.**

129. $\int_0^4 \sqrt{x}e^x \, dx$

130. $\int_0^2 2xe^{-x} \, dx$

131. Probability A car battery has an average lifetime of 48 months with a standard deviation of 6 months. The battery lives are normally distributed. The probability that a given battery will last between 48 months and 60 months is

$$0.0065 \int_{48}^{60} e^{-0.0139(t-48)^2} \, dt.$$

Use the integration capabilities of a graphing utility to approximate the integral. Interpret the resulting probability.

132. Probability The median waiting time (in minutes) for people waiting for service in a convenience store is given by the solution of the equation

$$\int_0^x 0.3e^{-0.3t} \, dt = \frac{1}{2}.$$

What is the median waiting time?

133. Using the Area of a Region Find the value of a such that the area bounded by $y = e^{-x}$, the x-axis, $x = -a$, and $x = a$ is $\frac{8}{3}$.

134. Modeling Data A valve on a storage tank is opened for 4 hours to release a chemical in a manufacturing process. The flow rate R (in liters per hour) at time t (in hours) is given in the table.

t	0	1	2	3	4
R	425	240	118	71	36

(a) Use the regression capabilities of a graphing utility to find a linear model for the points $(t, \ln R)$. Write the resulting equation of the form $\ln R = at + b$ in exponential form.

(b) Use a graphing utility to plot the data and graph the exponential model.

(c) Use the definite integral to approximate the number of liters of chemical released during the 4 hours.

WRITING ABOUT CONCEPTS

135. Properties of the Natural Exponential Function In your own words, state the properties of the natural exponential function.

136. A Function and Its Derivative Is there a function f such that $f(x) = f'(x)$? If so, identify it.

137. Choosing a Function Without integrating, state the integration formula you can use to integrate each of the following.

(a) $\int \frac{e^x}{e^x + 1} \, dx$

(b) $\int xe^{x^2} \, dx$

138. Analyzing a Graph Consider the function

$$f(x) = \frac{2}{1 + e^{1/x}}.$$

(a) Use a graphing utility to graph f.

(b) Write a short paragraph explaining why the graph has a horizontal asymptote at $y = 1$ and why the function has a nonremovable discontinuity at $x = 0$.

139. Deriving an Inequality Given $e^x \geq 1$ for $x \geq 0$, it follows that

$$\int_0^x e^t \, dt \geq \int_0^x 1 \, dt.$$

Perform this integration to derive the inequality

$$e^x \geq 1 + x$$

for $x \geq 0$.

140. Solving an Equation Find, to three decimal places, the value of x such that $e^{-x} = x$. (Use Newton's Method or the *zero* or *root* feature of a graphing utility.)

141. Horizontal Motion The position function of a particle moving along the x-axis is $x(t) = Ae^{kt} + Be^{-kt}$, where A, B, and k are positive constants.

(a) During what times t is the particle closest to the origin?

(b) Show that the acceleration of the particle is proportional to the position of the particle. What is the constant of proportionality?

142. Analyzing a Function Let $f(x) = \frac{\ln x}{x}$.

(a) Graph f on $(0, \infty)$ and show that f is strictly decreasing on (e, ∞).

(b) Show that if $e \leq A < B$, then $A^B > B^A$.

(c) Use part (b) to show that $e^\pi > \pi^e$.

143. Finding the Maximum Rate of Change Verify that the function

$$y = \frac{L}{1 + ae^{-x/b}}, \quad a > 0, \quad b > 0, \quad L > 0$$

increases at a maximum rate when $y = L/2$.

5.5 Bases Other than e and Applications

- Define exponential functions that have bases other than e.
- Differentiate and integrate exponential functions that have bases other than e.
- Use exponential functions to model compound interest and exponential growth.

Bases Other than e

The **base** of the natural exponential function is e. This "natural" base can be used to assign a meaning to a general base a.

Definition of Exponential Function to Base *a*

If a is a positive real number $(a \neq 1)$ and x is any real number, then the **exponential function to the base *a*** is denoted by a^x and is defined by

$$a^x = e^{(\ln a)x}.$$

If $a = 1$, then $y = 1^x = 1$ is a constant function.

These functions obey the usual laws of exponents. For instance, here are some familiar properties.

1. $a^0 = 1$ **2.** $a^x a^y = a^{x+y}$ **3.** $\dfrac{a^x}{a^y} = a^{x-y}$ **4.** $(a^x)^y = a^{xy}$

When modeling the half-life of a radioactive sample, it is convenient to use $\frac{1}{2}$ as the base of the exponential model. (*Half-life* is the number of years required for half of the atoms in a sample of radioactive material to decay.)

EXAMPLE 1 Radioactive Half-Life Model

The half-life of carbon-14 is about 5715 years. A sample contains 1 gram of carbon-14. How much will be present in 10,000 years?

Solution Let $t = 0$ represent the present time and let y represent the amount (in grams) of carbon-14 in the sample. Using a base of $\frac{1}{2}$, you can model y by the equation

$$y = \left(\frac{1}{2}\right)^{t/5715}.$$

Notice that when $t = 5715$, the amount is reduced to half of the original amount.

$$y = \left(\frac{1}{2}\right)^{5715/5715} = \frac{1}{2} \text{ gram}$$

When $t = 11{,}430$, the amount is reduced to a quarter of the original amount, and so on. To find the amount of carbon-14 after 10,000 years, substitute 10,000 for t.

$$y = \left(\frac{1}{2}\right)^{10{,}000/5715} \approx 0.30 \text{ gram}$$

The graph of y is shown in Figure 5.23.

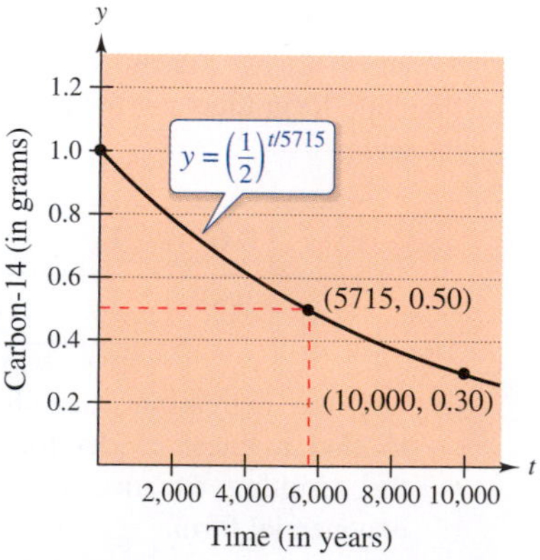

The half-life of carbon-14 is about 5715 years.
Figure 5.23

Carbon dating uses the radioisotope carbon-14 to estimate the age of dead organic materials. The method is based on the decay rate of carbon-14 (see Example 1), a compound organisms take in when they are alive.

Zens/Shutterstock.com

Logarithmic functions to bases other than e can be defined in much the same way as exponential functions to other bases are defined.

REMARK In precalculus, you learned that $\log_a x$ is the value to which a must be raised to produce x. This agrees with the definition at the right because

$$\begin{aligned} a^{\log_a x} &= a^{(1/\ln a)\ln x} \\ &= (e^{\ln a})^{(1/\ln a)\ln x} \\ &= e^{(\ln a/\ln a)\ln x} \\ &= e^{\ln x} \\ &= x. \end{aligned}$$

Definition of Logarithmic Function to Base *a*

If a is a positive real number $(a \neq 1)$ and x is any positive real number, then the **logarithmic function to the base *a*** is denoted by $\log_a x$ and is defined as

$$\log_a x = \frac{1}{\ln a} \ln x.$$

Logarithmic functions to the base a have properties similar to those of the natural logarithmic function given in Theorem 5.2. (Assume x and y are positive numbers and n is rational.)

1. $\log_a 1 = 0$ — Log of 1

2. $\log_a xy = \log_a x + \log_a y$ — Log of a product

3. $\log_a x^n = n \log_a x$ — Log of a power

4. $\log_a \dfrac{x}{y} = \log_a x - \log_a y$ — Log of a quotient

From the definitions of the exponential and logarithmic functions to the base a, it follows that $f(x) = a^x$ and $g(x) = \log_a x$ are inverse functions of each other.

Properties of Inverse Functions

1. $y = a^x$ if and only if $x = \log_a y$

2. $a^{\log_a x} = x$, for $x > 0$

3. $\log_a a^x = x$, for all x

The logarithmic function to the base 10 is called the **common logarithmic function.** So, for common logarithms,

$y = 10^x$ if and only if $x = \log_{10} y$. — Property of Inverse Functions

EXAMPLE 2 Bases Other than *e*

Solve for x in each equation.

a. $3^x = \dfrac{1}{81}$

b. $\log_2 x = -4$

Solution

a. To solve this equation, you can apply the logarithmic function to the base 3 to each side of the equation.

$$\begin{aligned} 3^x &= \frac{1}{81} \\ \log_3 3^x &= \log_3 \frac{1}{81} \\ x &= \log_3 3^{-4} \\ x &= -4 \end{aligned}$$

b. To solve this equation, you can apply the exponential function to the base 2 to each side of the equation.

$$\begin{aligned} \log_2 x &= -4 \\ 2^{\log_2 x} &= 2^{-4} \\ x &= \frac{1}{2^4} \\ x &= \frac{1}{16} \end{aligned}$$

Differentiation and Integration

To differentiate exponential and logarithmic functions to other bases, you have three options: (1) use the definitions of a^x and $\log_a x$ and differentiate using the rules for the natural exponential and logarithmic functions, (2) use logarithmic differentiation, or (3) use the differentiation rules for bases other than e given in the next theorem.

REMARK These differentiation rules are similar to those for the natural exponential function and the natural logarithmic function. In fact, they differ only by the constant factors $\ln a$ and $1/\ln a$. This points out one reason why, for calculus, e is the most convenient base.

THEOREM 5.13 Derivatives for Bases Other than *e*

Let a be a positive real number ($a \neq 1$), and let u be a differentiable function of x.

1. $\dfrac{d}{dx}[a^x] = (\ln a)a^x$
2. $\dfrac{d}{dx}[a^u] = (\ln a)a^u \dfrac{du}{dx}$
3. $\dfrac{d}{dx}[\log_a x] = \dfrac{1}{(\ln a)x}$
4. $\dfrac{d}{dx}[\log_a u] = \dfrac{1}{(\ln a)u}\dfrac{du}{dx}$

Proof By definition, $a^x = e^{(\ln a)x}$. So, you can prove the first rule by letting $u = (\ln a)x$ and differentiating with base e to obtain

$$\frac{d}{dx}[a^x] = \frac{d}{dx}[e^{(\ln a)x}] = e^u\frac{du}{dx} = e^{(\ln a)x}(\ln a) = (\ln a)a^x.$$

To prove the third rule, you can write

$$\frac{d}{dx}[\log_a x] = \frac{d}{dx}\left[\frac{1}{\ln a}\ln x\right] = \frac{1}{\ln a}\left(\frac{1}{x}\right) = \frac{1}{(\ln a)x}.$$

The second and fourth rules are simply the Chain Rule versions of the first and third rules.

See LarsonCalculus.com for Bruce Edwards's video of this proof. ■

EXAMPLE 3 Differentiating Functions to Other Bases

Find the derivative of each function.

a. $y = 2^x$ **b.** $y = 2^{3x}$ **c.** $y = \log_{10} \cos x$ **d.** $y = \log_3 \dfrac{\sqrt{x}}{x+5}$

Solution

a. $y' = \dfrac{d}{dx}[2^x] = (\ln 2)2^x$

b. $y' = \dfrac{d}{dx}[2^{3x}] = (\ln 2)2^{3x}(3) = (3 \ln 2)2^{3x}$

REMARK Try writing 2^{3x} as 8^x and differentiating to see that you obtain the same result.

c. $y' = \dfrac{d}{dx}[\log_{10} \cos x] = \dfrac{-\sin x}{(\ln 10)\cos x} = -\dfrac{1}{\ln 10}\tan x$

d. Before differentiating, rewrite the function using logarithmic properties.

$$y = \log_3 \frac{\sqrt{x}}{x+5} = \frac{1}{2}\log_3 x - \log_3(x+5)$$

Next, apply Theorem 5.13 to differentiate the function.

$$\begin{aligned} y' &= \frac{d}{dx}\left[\frac{1}{2}\log_3 x - \log_3(x+5)\right] \\ &= \frac{1}{2(\ln 3)x} - \frac{1}{(\ln 3)(x+5)} \\ &= \frac{5-x}{2(\ln 3)x(x+5)} \end{aligned}$$

■

AP* Tips
Differentiation and integration with bases other than e are rarely tested on the AP Exam.

Occasionally, an integrand involves an exponential function to a base other than e. When this occurs, there are two options: (1) convert to base e using the formula $a^x = e^{(\ln a)x}$ and then integrate, or (2) integrate directly, using the integration formula

$$\int a^x\,dx = \left(\frac{1}{\ln a}\right)a^x + C$$

which follows from Theorem 5.13.

EXAMPLE 4 **Integrating an Exponential Function to Another Base**

Find $\int 2^x\,dx$.

Solution

$$\int 2^x\,dx = \frac{1}{\ln 2}2^x + C$$

When the Power Rule, $D_x[x^n] = nx^{n-1}$, was introduced in Chapter 2, the exponent n was required to be a rational number. Now the rule is extended to cover any real value of n. Try to prove this theorem using logarithmic differentiation.

THEOREM 5.14 **The Power Rule for Real Exponents**

Let n be any number, and let u be a differentiable function of x.

1. $\dfrac{d}{dx}[x^n] = nx^{n-1}$ **2.** $\dfrac{d}{dx}[u^n] = nu^{n-1}\dfrac{du}{dx}$

The next example compares the derivatives of four types of functions. Each function uses a different differentiation formula, depending on whether the base and the exponent are constants or variables.

EXAMPLE 5 **Comparing Variables and Constants**

a. $\dfrac{d}{dx}[e^e] = 0$ Constant Rule

b. $\dfrac{d}{dx}[e^x] = e^x$ Exponential Rule

c. $\dfrac{d}{dx}[x^e] = ex^{e-1}$ Power Rule

d. $y = x^x$ Logarithmic differentiation

$$\ln y = \ln x^x$$
$$\ln y = x \ln x$$
$$\frac{y'}{y} = x\left(\frac{1}{x}\right) + (\ln x)(1)$$
$$\frac{y'}{y} = 1 + \ln x$$
$$y' = y(1 + \ln x)$$
$$y' = x^x(1 + \ln x)$$

REMARK Be sure you see that there is no simple differentiation rule for calculating the derivative of $y = x^x$. In general, when $y = u(x)^{v(x)}$, you need to use logarithmic differentiation.

Applications of Exponential Functions

An amount of P dollars is deposited in an account at an annual interest rate r (in decimal form). What is the balance in the account at the end of 1 year? The answer depends on the number of times n the interest is compounded according to the formula

$$A = P\left(1 + \frac{r}{n}\right)^n.$$

n	A
1	\$1080.00
2	\$1081.60
4	\$1082.43
12	\$1083.00
365	\$1083.28

For instance, the result for a deposit of \$1000 at 8% interest compounded n times a year is shown in the table at the right.

As n increases, the balance A approaches a limit. To develop this limit, use the next theorem. To test the reasonableness of this theorem, try evaluating

$$\left(\frac{x + 1}{x}\right)^x$$

for several values of x, as shown in the table at the left.

x	$\left(\frac{x+1}{x}\right)^x$
10	2.59374
100	2.70481
1000	2.71692
10,000	2.71815
100,000	2.71827
1,000,000	2.71828

THEOREM 5.15 A Limit Involving e

$$\lim_{x\to\infty}\left(1 + \frac{1}{x}\right)^x = \lim_{x\to\infty}\left(\frac{x + 1}{x}\right)^x = e$$

A proof of this theorem is given in Appendix A.

See LarsonCalculus.com for Bruce Edwards's video of this proof.

Given Theorem 5.15, take another look at the formula for the balance A in an account in which the interest is compounded n times per year. By taking the limit as n approaches infinity, you obtain

$$A = \lim_{n\to\infty} P\left(1 + \frac{r}{n}\right)^n \qquad \text{Take limit as } n \to \infty.$$

$$= P \lim_{n\to\infty}\left[\left(1 + \frac{1}{n/r}\right)^{n/r}\right]^r \qquad \text{Rewrite.}$$

$$= P\left[\lim_{x\to\infty}\left(1 + \frac{1}{x}\right)^x\right]^r \qquad \text{Let } x = n/r. \text{ Then } x \to \infty \text{ as } n \to \infty.$$

$$= Pe^r. \qquad \text{Apply Theorem 5.15.}$$

This limit produces the balance after 1 year of **continuous compounding.** So, for a deposit of \$1000 at 8% interest compounded continuously, the balance at the end of 1 year would be

$$A = 1000e^{0.08} \approx \$1083.29.$$

SUMMARY OF COMPOUND INTEREST FORMULAS

Let P = amount of deposit, t = number of years, A = balance after t years, r = annual interest rate (decimal form), and n = number of compoundings per year.

1. Compounded n times per year: $A = P\left(1 + \frac{r}{n}\right)^{nt}$

2. Compounded continuously: $A = Pe^{rt}$

EXAMPLE 6 Continuous, Quarterly, and Monthly Compounding

See LarsonCalculus.com for an interactive version of this type of example.

A deposit of \$2500 is made in an account that pays an annual interest rate of 5%. Find the balance in the account at the end of 5 years when the interest is compounded (a) quarterly, (b) monthly, and (c) continuously.

Solution

a. $A = P\left(1 + \frac{r}{n}\right)^{nt}$ — Compounded quarterly

$$= 2500\left(1 + \frac{0.05}{4}\right)^{4(5)}$$
$$= 2500(1.0125)^{20}$$
$$\approx \$3205.09$$

b. $A = P\left(1 + \frac{r}{n}\right)^{nt}$ — Compounded monthly

$$= 2500\left(1 + \frac{0.05}{12}\right)^{12(5)}$$
$$\approx 2500(1.0041667)^{60}$$
$$\approx \$3208.40$$

c. $A = Pe^{rt}$ — Compounded continuously

$$= 2500\left[e^{0.05(5)}\right]$$
$$= 2500e^{0.25}$$
$$\approx \$3210.06$$

EXAMPLE 7 Bacterial Culture Growth

A bacterial culture is growing according to the *logistic growth function*

$$y = \frac{1.25}{1 + 0.25e^{-0.4t}}, \quad t \geq 0$$

where y is the weight of the culture in grams and t is the time in hours. Find the weight of the culture after (a) 0 hours, (b) 1 hour, and (c) 10 hours. (d) What is the limit as t approaches infinity?

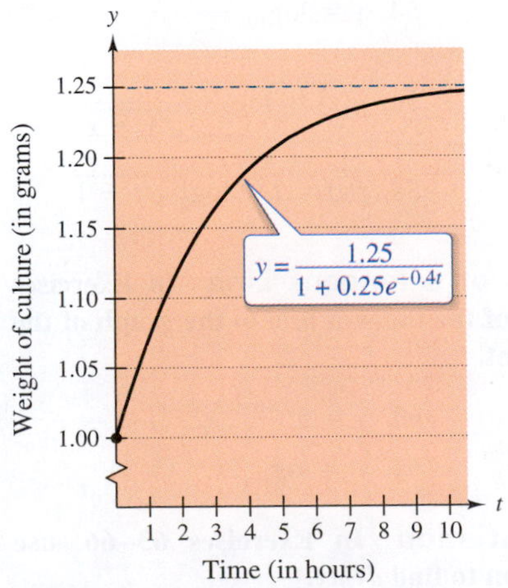

The limit of the weight of the culture as $t \to \infty$ is 1.25 grams.

Figure 5.24

Solution

a. When $t = 0$, $y = \frac{1.25}{1 + 0.25e^{-0.4(0)}}$

$$= 1 \text{ gram.}$$

b. When $t = 1$, $y = \frac{1.25}{1 + 0.25e^{-0.4(1)}}$

$$\approx 1.071 \text{ grams.}$$

c. When $t = 10$, $y = \frac{1.25}{1 + 0.25e^{-0.4(10)}}$

$$\approx 1.244 \text{ grams.}$$

d. Taking the limit as t approaches infinity, you obtain

$$\lim_{t \to \infty} \frac{1.25}{1 + 0.25e^{-0.4t}} = \frac{1.25}{1 + 0} = 1.25 \text{ grams.}$$

The graph of the function is shown in Figure 5.24.

5.5 Exercises

See CalcChat.com for tutorial help and worked-out solutions to odd-numbered exercises.

Evaluating a Logarithmic Expression In Exercises 1–4, evaluate the expression without using a calculator.

1. $\log_2 \frac{1}{8}$
2. $\log_{27} 9$
3. $\log_7 1$
4. $\log_a \dfrac{1}{a}$

Exponential and Logarithmic Forms of Equations In Exercises 5–8, write the exponential equation as a logarithmic equation or vice versa.

5. (a) $2^3 = 8$
   (b) $3^{-1} = \frac{1}{3}$
6. (a) $27^{2/3} = 9$
   (b) $16^{3/4} = 8$
7. (a) $\log_{10} 0.01 = -2$
   (b) $\log_{0.5} 8 = -3$
8. (a) $\log_3 \frac{1}{9} = -2$
   (b) $49^{1/2} = 7$

Sketching a Graph In Exercises 9–14, sketch the graph of the function by hand.

9. $y = 2^x$
10. $y = 4^{x-1}$
11. $y = \left(\frac{1}{3}\right)^x$
12. $y = 2^{x^2}$
13. $h(x) = 5^{x-2}$
14. $y = 3^{-|x|}$

Matching In Exercises 15–18, match the function with its graph. [The graphs are labeled (a), (b), (c), and (d).]

(a)

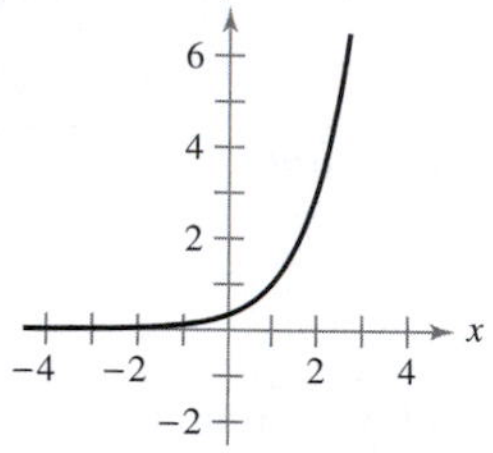

(b)

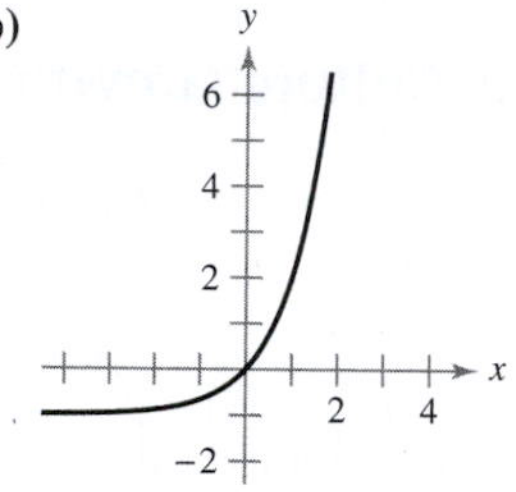

(c)

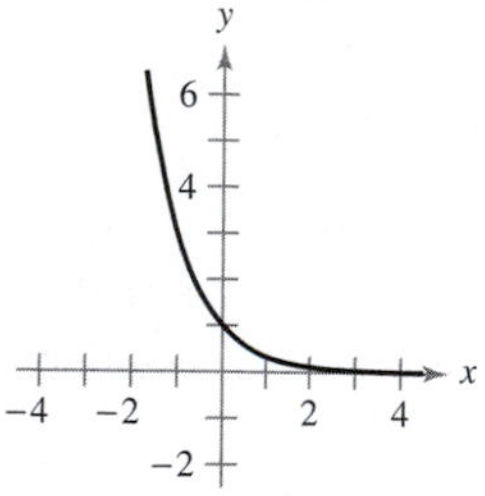

(d)

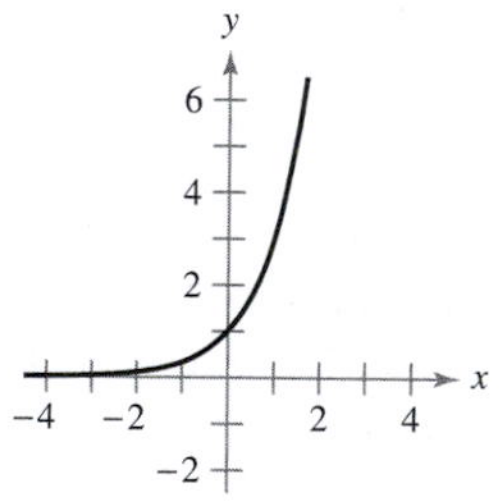

15. $f(x) = 3^x$
16. $f(x) = 3^{-x}$
17. $f(x) = 3^x - 1$
18. $f(x) = 3^{x-1}$

Solving an Equation In Exercises 19–24, solve for x or b.

19. (a) $\log_{10} 1000 = x$
    (b) $\log_{10} 0.1 = x$
20. (a) $\log_3 \frac{1}{81} = x$
    (b) $\log_6 36 = x$
21. (a) $\log_3 x = -1$
    (b) $\log_2 x = -4$
22. (a) $\log_b 27 = 3$
    (b) $\log_b 125 = 3$
23. (a) $x^2 - x = \log_5 25$
    (b) $3x + 5 = \log_2 64$
24. (a) $\log_3 x + \log_3(x - 2) = 1$
    (b) $\log_{10}(x + 3) - \log_{10} x = 1$

Solving an Equation In Exercises 25–34, solve the equation accurate to three decimal places.

25. $3^{2x} = 75$
26. $5^{6x} = 8320$
27. $2^{3-z} = 625$
28. $3(5^{x-1}) = 86$
29. $\left(1 + \dfrac{0.09}{12}\right)^{12t} = 3$
30. $\left(1 + \dfrac{0.10}{365}\right)^{365t} = 2$
31. $\log_2(x - 1) = 5$
32. $\log_{10}(t - 3) = 2.6$
33. $\log_3 x^2 = 4.5$
34. $\log_5 \sqrt{x - 4} = 3.2$

Verifying Inverse Functions In Exercises 35 and 36, illustrate that the functions are inverse functions of each other by sketching their graphs on the same set of coordinate axes.

35. $f(x) = 4^x$
    $g(x) = \log_4 x$
36. $f(x) = 3^x$
    $g(x) = \log_3 x$

Finding a Derivative In Exercises 37–58, find the derivative of the function. (*Hint:* In some exercises, you may find it helpful to apply logarithmic properties *before* differentiating.)

37. $f(x) = 4^x$
38. $f(x) = 3^{4x}$
39. $y = 5^{-4x}$
40. $y = 6^{3x-4}$
41. $f(x) = x\,9^x$
42. $y = x(6^{-2x})$
43. $g(t) = t^2 2^t$
44. $f(t) = \dfrac{3^{2t}}{t}$
45. $h(\theta) = 2^{-\theta} \cos \pi\theta$
46. $g(\alpha) = 5^{-\alpha/2} \sin 2\alpha$
47. $y = \log_4(5x + 1)$
48. $y = \log_3(x^2 - 3x)$
49. $h(t) = \log_5(4 - t)^2$
50. $g(t) = \log_2(t^2 + 7)^3$
51. $y = \log_5 \sqrt{x^2 - 1}$
52. $f(x) = \log_2 \sqrt[3]{2x + 1}$
53. $f(x) = \log_2 \dfrac{x^2}{x - 1}$
54. $y = \log_{10} \dfrac{x^2 - 1}{x}$
55. $h(x) = \log_3 \dfrac{x\sqrt{x - 1}}{2}$
56. $g(x) = \log_5 \dfrac{4}{x^2\sqrt{1 - x}}$
57. $g(t) = \dfrac{10 \log_4 t}{t}$
58. $f(t) = t^{3/2} \log_2 \sqrt{t + 1}$

Finding an Equation of a Tangent Line In Exercises 59–62, find an equation of the tangent line to the graph of the function at the given point.

59. $y = 2^{-x}, \quad (-1, 2)$
60. $y = 5^{x-2}, \quad (2, 1)$
61. $y = \log_3 x, \quad (27, 3)$
62. $y = \log_{10} 2x, \quad (5, 1)$

Logarithmic Differentiation In Exercises 63–66, use logarithmic differentiation to find dy/dx.

63. $y = x^{2/x}$
64. $y = x^{x-1}$
65. $y = (x - 2)^{x+1}$
66. $y = (1 + x)^{1/x}$

Finding an Equation of a Tangent Line **In Exercises 67–70, find an equation of the tangent line to the graph of the function at the given point.**

67. $y = x^{\sin x}$, $\left(\frac{\pi}{2}, \frac{\pi}{2}\right)$

68. $y = (\sin x)^{2x}$, $\left(\frac{\pi}{2}, 1\right)$

69. $y = (\ln x)^{\cos x}$, $(e, 1)$

70. $y = x^{1/x}$, $(1, 1)$

Finding an Indefinite Integral **In Exercises 71–78, find the indefinite integral.**

71. $\int 3^x \, dx$

72. $\int 8^{-x} \, dx$

73. $\int (x^2 + 2^{-x}) \, dx$

74. $\int (x^4 + 5^x) \, dx$

75. $\int x(5^{-x^2}) \, dx$

76. $\int (x + 4)6^{(x+4)^2} \, dx$

77. $\int \frac{3^{2x}}{1 + 3^{2x}} \, dx$

78. $\int 2^{\sin x} \cos x \, dx$

Evaluating a Definite Integral **In Exercises 79–82, evaluate the definite integral.**

79. $\int_{-1}^{2} 2^x \, dx$

80. $\int_{-4}^{4} 3^{x/4} \, dx$

81. $\int_{0}^{1} (5^x - 3^x) \, dx$

82. $\int_{1}^{3} (7^x - 4^x) \, dx$

Area **In Exercises 83 and 84, find the area of the region bounded by the graphs of the equations.**

83. $y = 3^x$, $y = 0$, $x = 0$, $x = 3$

84. $y = 3^{\cos x} \sin x$, $y = 0$, $x = 0$, $x = \pi$

WRITING ABOUT CONCEPTS

85. Analyzing a Logarithmic Equation Consider the function $f(x) = \log_{10} x$.

(a) What is the domain of f?

(b) Find f^{-1}.

(c) Let x be a real number between 1000 and 10,000. Determine the interval in which $f(x)$ will be found.

(d) Determine the interval in which x will be found if $f(x)$ is negative.

(e) When $f(x)$ is increased by one unit, x must have been increased by what factor?

(f) Find the ratio of x_1 to x_2 given that $f(x_1) = 3n$ and $f(x_2) = n$.

86. Comparing Rates of Growth Order the functions

$$f(x) = \log_2 x, \; g(x) = x^x, \; h(x) = x^2, \text{ and } k(x) = 2^x$$

from the one with the greatest rate of growth to the one with the least rate of growth for large values of x.

87. Inflation When the annual rate of inflation averages 5% over the next 10 years, the approximate cost C of goods or services during any year in that decade is

$$C(t) = P(1.05)^t$$

where t is the time in years and P is the present cost.

(a) The price of an oil change for your car is presently \$24.95. Estimate the price 10 years from now.

(b) Find the rates of change of C with respect to t when $t = 1$ and $t = 8$.

(c) Verify that the rate of change of C is proportional to C. What is the constant of proportionality?

88. Depreciation After t years, the value of a car purchased for \$25,000 is

$$V(t) = 25{,}000\left(\tfrac{3}{4}\right)^t.$$

(a) Use a graphing utility to graph the function and determine the value of the car 2 years after it was purchased.

(b) Find the rates of change of V with respect to t when $t = 1$ and $t = 4$.

(c) Use a graphing utility to graph $V'(t)$ and determine the horizontal asymptote of $V'(t)$. Interpret its meaning in the context of the problem.

Compound Interest **In Exercises 89–92, complete the table by determining the balance A for P dollars invested at rate r for t years and compounded n times per year.**

n	1	2	4	12	365	Continuous Compounding
A						

89. $P = \$1000$
$r = 3\frac{1}{2}\%$
$t = 10$ years

90. $P = \$2500$
$r = 6\%$
$t = 20$ years

91. $P = \$1000$
$r = 5\%$
$t = 30$ years

92. $P = \$4000$
$r = 4\%$
$t = 15$ years

Compound Interest **In Exercises 93–96, complete the table by determining the amount of money P (present value) that should be invested at rate r to produce a balance of \$100,000 in t years.**

t	1	10	20	30	40	50
P						

93. $r = 5\%$
Compounded continuously

94. $r = 3\%$
Compounded continuously

95. $r = 5\%$
Compounded monthly

96. $r = 2\%$
Compounded daily

97. Compound Interest Assume that you can earn 6% on an investment, compounded daily. Which of the following options would yield the greatest balance after 8 years?

(a) \$20,000 now (b) \$30,000 after 8 years

(c) \$8000 now and \$20,000 after 4 years

(d) \$9000 now, \$9000 after 4 years, and \$9000 after 8 years

98. Compound Interest Consider a deposit of \$100 placed in an account for 20 years at r% compounded continuously. Use a graphing utility to graph the exponential functions describing the growth of the investment over the 20 years for the following interest rates. Compare the ending balances for the three rates.

(a) $r = 3\%$ (b) $r = 5\%$ (c) $r = 6\%$

99. Timber Yield The yield V (in millions of cubic feet per acre) for a stand of timber at age t is $V = 6.7e^{(-48.1)/t}$, where t is measured in years.

(a) Find the limiting volume of wood per acre as t approaches infinity.

(b) Find the rates at which the yield is changing when $t = 20$ years and $t = 60$ years.

100. HOW DO YOU SEE IT? The graph shows the proportion P of correct responses after n trials in a group project in learning theory.

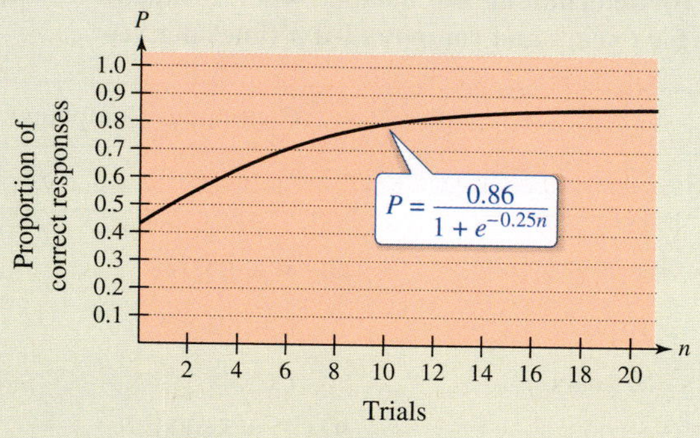

(a) What is the limiting proportion of correct responses as n approaches infinity?

(b) What happens to the rate of change of the proportion in the long run?

101. Population Growth A lake is stocked with 500 fish, and the population increases according to the logistic curve

$$p(t) = \frac{10{,}000}{1 + 19e^{-t/5}}$$

where t is measured in months.

(a) Use a graphing utility to graph the function.

(b) What is the limiting size of the fish population?

(c) At what rates is the fish population changing at the end of 1 month and at the end of 10 months?

(d) After how many months is the population increasing most rapidly?

102. Modeling Data The breaking strengths B (in tons) of steel cables of various diameters d (in inches) are shown in the table.

d	0.50	0.75	1.00	1.25	1.50	1.75
B	9.85	21.8	38.3	59.2	84.4	114.0

(a) Use the regression capabilities of a graphing utility to fit an exponential model to the data.

(b) Use a graphing utility to plot the data and graph the model.

(c) Find the rates of growth of the model when $d = 0.8$ and $d = 1.5$.

103. Comparing Models The numbers y of pancreas transplants in the United States for the years 2004 through 2010 are shown in the table, with $x = 4$ corresponding to 2004. *(Source: Organ Procurement and Transplantation Network)*

x	4	5	6	7	8	9	10
y	603	542	466	468	436	376	350

(a) Use the regression capabilities of a graphing utility to find the following models for the data.

$y_1 = ax + b$ $\qquad$ $y_2 = a + b \ln x$

$y_3 = ab^x$ $\qquad$ $y_4 = ax^b$

(b) Use a graphing utility to plot the data and graph each of the models. Which model do you think best fits the data?

(c) Interpret the slope of the linear model in the context of the problem.

(d) Find the rate of change of each of the models for the year 2008. Which model is decreasing at the greatest rate in 2008?

104. An Approximation of *e* Complete the table to demonstrate that e can also be defined as

$$\lim_{x \to 0^+} (1 + x)^{1/x}.$$

x	1	10^{-1}	10^{-2}	10^{-4}	10^{-6}
$(1 + x)^{1/x}$					

Modeling Data In Exercises 105 and 106, find an exponential function that fits the experimental data collected over time t.

105.

t	0	1	2	3	4
y	1200.00	720.00	432.00	259.20	155.52

106.

t	0	1	2	3	4
y	600.00	630.00	661.50	694.58	729.30

Using Properties of Exponents **In Exercises 107–110, find the exact value of the expression.**

107. $5^{1/\ln 5}$ **108.** $6^{\ln 10/\ln 6}$

109. $9^{1/\ln 3}$ **110.** $32^{1/\ln 2}$

True or False? **In Exercises 111–116, determine whether the statement is true or false. If it is false, explain why or give an example that shows it is false.**

111. $e = \dfrac{271{,}801}{99{,}900}$

112. If $f(x) = \ln x$, then $f(e^{n+1}) - f(e^n) = 1$ for any value of n.

113. The functions $f(x) = 2 + e^x$ and $g(x) = \ln(x - 2)$ are inverse functions of each other.

114. The exponential function $y = Ce^x$ is a solution of the differential equation

$$\frac{d^n y}{dx^n} = y, \quad n = 1, 2, 3, \ldots .$$

115. The graphs of $f(x) = e^x$ and $g(x) = e^{-x}$ meet at right angles.

116. If $f(x) = g(x)e^x$, then the only zeros of f are the zeros of g.

117. Comparing Functions

(a) Show that $(2^3)^2 \neq 2^{(3^2)}$.

(b) Are $f(x) = (x^x)^x$ and $g(x) = x^{(x^x)}$ the same function? Why or why not?

(c) Find $f'(x)$ and $g'(x)$.

118. Finding an Inverse Function Let

$$f(x) = \frac{a^x - 1}{a^x + 1}$$

for $a > 0$, $a \neq 1$. Show that f has an inverse function. Then find f^{-1}.

119. Logistic Differential Equation Show that solving the logistic differential equation

$$\frac{dy}{dt} = \frac{8}{25}y\left(\frac{5}{4} - y\right), \quad y(0) = 1$$

results in the logistic growth function in Example 7.

$$\left[\textit{Hint: } \frac{1}{y\left(\frac{5}{4} - y\right)} = \frac{4}{5}\left(\frac{1}{y} + \frac{1}{\frac{5}{4} - y}\right)\right]$$

120. Using Properties of Exponents Given the exponential function $f(x) = a^x$, show that

(a) $f(u + v) = f(u) \cdot f(v)$.

(b) $f(2x) = [f(x)]^2$.

121. Tangent Lines

(a) Determine y' given $y^x = x^y$.

(b) Find the slope of the tangent line to the graph of $y^x = x^y$ at each of the following points.

(i) (c, c) (ii) $(2, 4)$ (iii) $(4, 2)$

(c) At what points on the graph of $y^x = x^y$ does the tangent line not exist?

PUTNAM EXAM CHALLENGE

122. Which is greater

$$\left(\sqrt{n}\right)^{\sqrt{n+1}} \text{ or } \left(\sqrt{n+1}\right)^{\sqrt{n}}$$

where $n > 8$?

123. Show that if x is positive, then

$$\log_e\left(1 + \frac{1}{x}\right) > \frac{1}{1 + x}.$$

These problems were composed by the Committee on the Putnam Prize Competition.

SECTION PROJECT

Using Graphing Utilities to Estimate Slope

Let $f(x) = \begin{cases} |x|^x, & x \neq 0 \\ 1, & x = 0. \end{cases}$

(a) Use a graphing utility to graph f in the viewing window $-3 \le x \le 3$, $-2 \le y \le 2$. What is the domain of f?

(b) Use the *zoom* and *trace* features of a graphing utility to estimate

$$\lim_{x \to 0} f(x).$$

(c) Write a short paragraph explaining why the function f is continuous for all real numbers.

(d) Visually estimate the slope of f at the point $(0, 1)$.

(e) Explain why the derivative of a function can be approximated by the formula

$$\frac{f(x + \Delta x) - f(x - \Delta x)}{2\Delta x}$$

for small values of Δx. Use this formula to approximate the slope of f at the point $(0, 1)$.

$$f'(0) \approx \frac{f(0 + \Delta x) - f(0 - \Delta x)}{2\Delta x} = \frac{f(\Delta x) - f(-\Delta x)}{2\Delta x}$$

What do you think the slope of the graph of f is at $(0, 1)$?

(f) Find a formula for the derivative of f and determine $f'(0)$. Write a short paragraph explaining how a graphing utility might lead you to approximate the slope of a graph incorrectly.

(g) Use your formula for the derivative of f to find the relative extrema of f. Verify your answer using a graphing utility.

FOR FURTHER INFORMATION For more information on using graphing utilities to estimate slope, see the article "Computer-Aided Delusions" by Richard L. Hall in *The College Mathematics Journal*. To view this article, go to *MathArticles.com*.

5.6 Inverse Trigonometric Functions: Differentiation

- **Develop properties of the six inverse trigonometric functions.**
- **Differentiate an inverse trigonometric function.**
- **Review the basic differentiation rules for elementary functions.**

Inverse Trigonometric Functions

This section begins with a rather surprising statement: *None of the six basic trigonometric functions has an inverse function.* This statement is true because all six trigonometric functions are periodic and therefore are not one-to-one. In this section, you will examine these six functions to see whether their domains can be redefined in such a way that they will have inverse functions on the *restricted domains.*

In Example 4 of Section 5.3, you saw that the sine function is increasing (and therefore is one-to-one) on the interval

$$\left[-\frac{\pi}{2}, \frac{\pi}{2}\right]$$

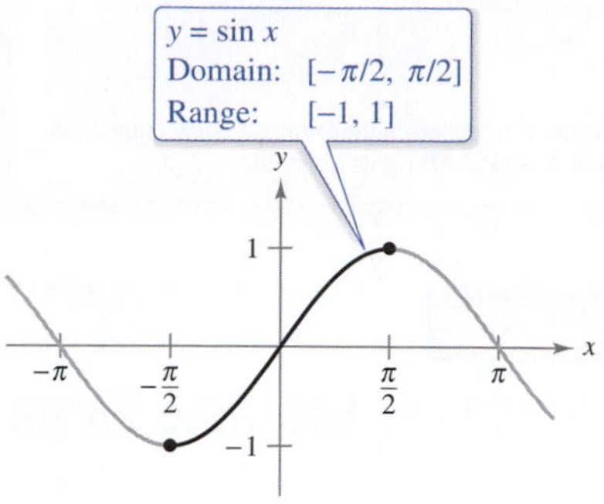

The sine function is one-to-one on $[-\pi/2, \pi/2]$.
Figure 5.25

as shown in Figure 5.25. On this interval, you can define the inverse of the *restricted* sine function as

$$y = \arcsin x \quad \text{if and only if} \quad \sin y = x$$

where $-1 \le x \le 1$ and $-\pi/2 \le \arcsin x \le \pi/2$.

Under suitable restrictions, each of the six trigonometric functions is one-to-one and so has an inverse function, as shown in the next definition. (Note that the term "iff" is used to represent the phrase "if and only if.")

REMARK The term "arcsin x" is read as "the arcsine of x" or sometimes "the angle whose sine is x." An alternative notation for the inverse sine function is "$\sin^{-1} x$."

Definitions of Inverse Trigonometric Functions

Function	Domain	Range
$y = \arcsin x$ iff $\sin y = x$	$-1 \le x \le 1$	$-\frac{\pi}{2} \le y \le \frac{\pi}{2}$
$y = \arccos x$ iff $\cos y = x$	$-1 \le x \le 1$	$0 \le y \le \pi$
$y = \arctan x$ iff $\tan y = x$	$-\infty < x < \infty$	$-\frac{\pi}{2} < y < \frac{\pi}{2}$
$y = \text{arccot } x$ iff $\cot y = x$	$-\infty < x < \infty$	$0 < y < \pi$
$y = \text{arcsec } x$ iff $\sec y = x$	$\lvert x \rvert \ge 1$	$0 \le y \le \pi, \quad y \ne \frac{\pi}{2}$
$y = \text{arccsc } x$ iff $\csc y = x$	$\lvert x \rvert \ge 1$	$-\frac{\pi}{2} \le y \le \frac{\pi}{2}, \quad y \ne 0$

Exploration

The Inverse Secant Function In the definitions of the inverse trigonometric functions, the inverse secant function is defined by restricting the domain of the secant function to the intervals $[0, \pi/2) \cup (\pi/2, \pi]$. Most other texts and reference books agree with this, but some disagree. What other domains might make sense? Explain your reasoning graphically. Most calculators do not have a key for the inverse secant function. How can you use a calculator to evaluate the inverse secant function?

The graphs of the six inverse trigonometric functions are shown in Figure 5.26.

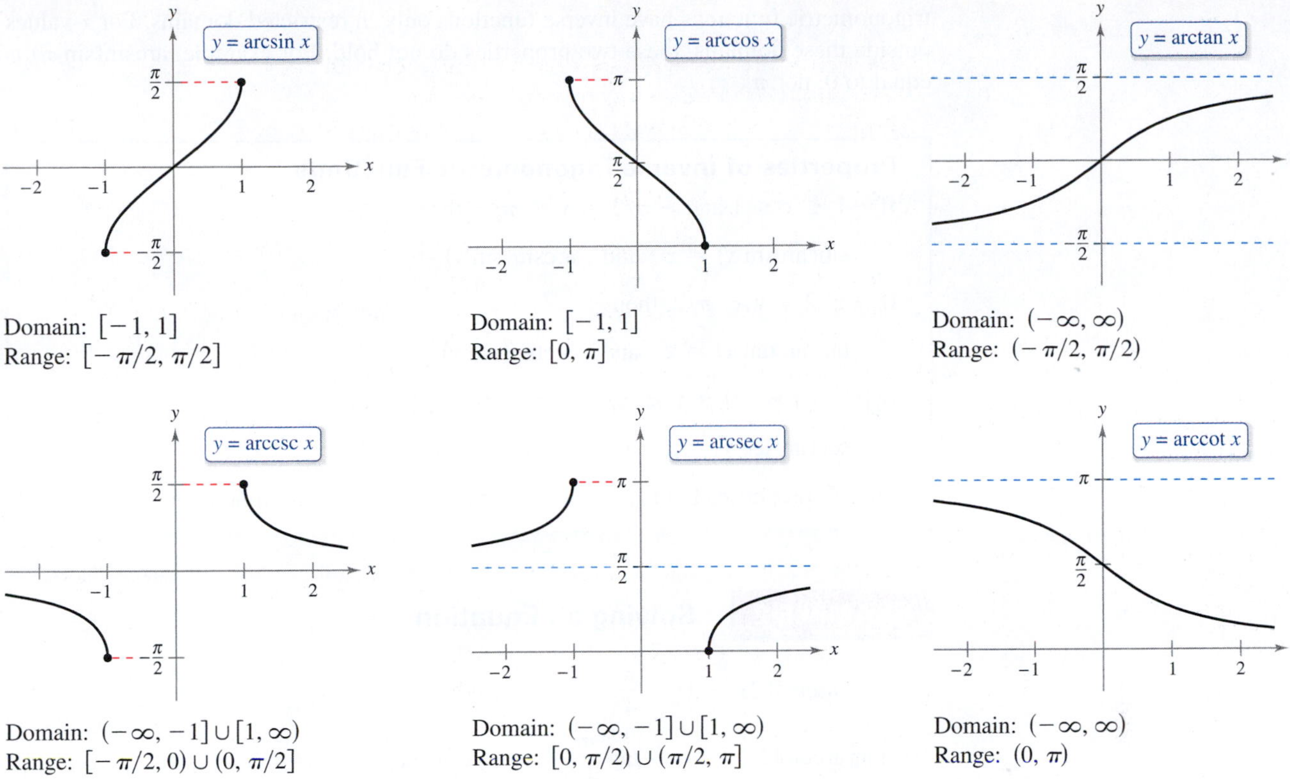

Figure 5.26

When evaluating inverse trigonometric functions, remember that they denote angles in *radian measure*.

EXAMPLE 1 Evaluating Inverse Trigonometric Functions

Evaluate each function.

a. $\arcsin\left(-\frac{1}{2}\right)$ **b.** $\arccos 0$ **c.** $\arctan \sqrt{3}$ **d.** $\arcsin(0.3)$

Solution

a. By definition, $y = \arcsin\left(-\frac{1}{2}\right)$ implies that $\sin y = -\frac{1}{2}$. In the interval $[-\pi/2, \pi/2]$, the correct value of y is $-\pi/6$.

$$\arcsin\left(-\frac{1}{2}\right) = -\frac{\pi}{6}$$

b. By definition, $y = \arccos 0$ implies that $\cos y = 0$. In the interval $[0, \pi]$, you have $y = \pi/2$.

$$\arccos 0 = \frac{\pi}{2}$$

c. By definition, $y = \arctan \sqrt{3}$ implies that $\tan y = \sqrt{3}$. In the interval $(-\pi/2, \pi/2)$, you have $y = \pi/3$.

$$\arctan \sqrt{3} = \frac{\pi}{3}$$

d. Using a calculator set in *radian* mode produces

$$\arcsin(0.3) \approx 0.305.$$

Inverse functions have the properties $f(f^{-1}(x)) = x$ and $f^{-1}(f(x)) = x$. When applying these properties to inverse trigonometric functions, remember that the trigonometric functions have inverse functions only in restricted domains. For x-values outside these domains, these two properties do not hold. For example, $\arcsin(\sin \pi)$ is equal to 0, not π.

Properties of Inverse Trigonometric Functions

If $-1 \le x \le 1$ and $-\pi/2 \le y \le \pi/2$, then

$$\sin(\arcsin x) = x \quad \text{and} \quad \arcsin(\sin y) = y.$$

If $-\pi/2 < y < \pi/2$, then

$$\tan(\arctan x) = x \quad \text{and} \quad \arctan(\tan y) = y.$$

If $|x| \ge 1$ and $0 \le y < \pi/2$ or $\pi/2 < y \le \pi$, then

$$\sec(\operatorname{arcsec} x) = x \quad \text{and} \quad \operatorname{arcsec}(\sec y) = y.$$

Similar properties hold for the other inverse trigonometric functions.

EXAMPLE 2 Solving an Equation

$$\arctan(2x - 3) = \frac{\pi}{4} \qquad \text{Original equation}$$

$$\tan[\arctan(2x - 3)] = \tan\frac{\pi}{4} \qquad \text{Take tangent of each side.}$$

$$2x - 3 = 1 \qquad \tan(\arctan x) = x$$

$$x = 2 \qquad \text{Solve for } x.$$

Some problems in calculus require that you evaluate expressions such as $\cos(\arcsin x)$, as shown in Example 3.

EXAMPLE 3 Using Right Triangles

a. Given $y = \arcsin x$, where $0 < y < \pi/2$, find $\cos y$.

b. Given $y = \operatorname{arcsec}(\sqrt{5}/2)$, find $\tan y$.

Solution

a. Because $y = \arcsin x$, you know that $\sin y = x$. This relationship between x and y can be represented by a right triangle, as shown in the figure at the right.

$$\cos y = \cos(\arcsin x) = \frac{\text{adj.}}{\text{hyp.}} = \sqrt{1 - x^2}$$

(This result is also valid for $-\pi/2 < y < 0$.)

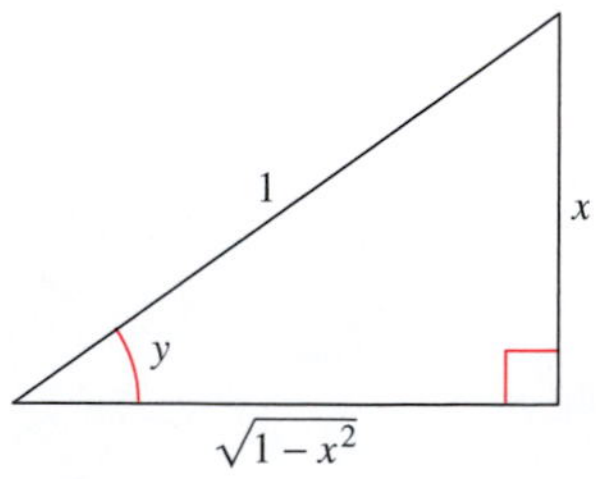

$y = \arcsin x$

b. Use the right triangle shown in the figure at the left.

$$\tan y = \tan\left[\operatorname{arcsec}\left(\frac{\sqrt{5}}{2}\right)\right]$$

$$= \frac{\text{opp.}}{\text{adj.}}$$

$$= \frac{1}{2}$$

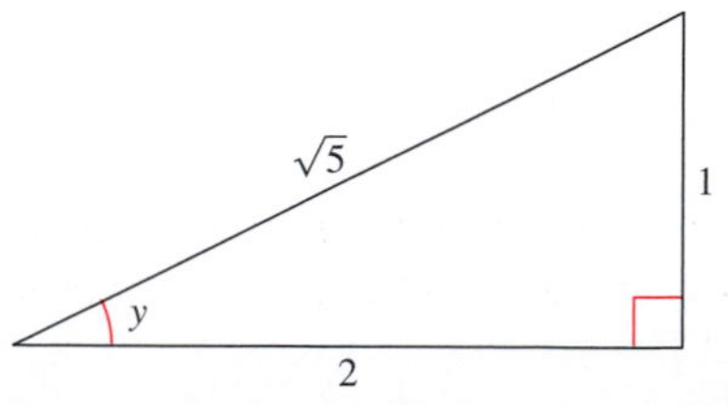

$y = \operatorname{arcsec}\dfrac{\sqrt{5}}{2}$

Derivatives of Inverse Trigonometric Functions

In Section 5.1, you saw that the derivative of the *transcendental* function $f(x) = \ln x$ is the *algebraic* function $f'(x) = 1/x$. You will now see that the derivatives of the inverse trigonometric functions also are algebraic (even though the inverse trigonometric functions are themselves transcendental).

The next theorem lists the derivatives of the six inverse trigonometric functions. Note that the derivatives of arccos u, arccot u, and arccsc u are the *negatives* of the derivatives of arcsin u, arctan u, and arcsec u, respectively.

REMARK There is no common agreement on the definition of arcsec x (or arccsc x) for negative values of x. When we defined the range of the arcsecant, we chose to preserve the reciprocal identity

$$\operatorname{arcsec} x = \arccos \frac{1}{x}.$$

One consequence of this definition is that its graph has a positive slope at every x-value in its domain. (See Figure 5.26.) This accounts for the absolute value sign in the formula for the derivative of arcsec x.

THEOREM 5.16 Derivatives of Inverse Trigonometric Functions

Let u be a differentiable function of x.

$$\frac{d}{dx}[\arcsin u] = \frac{u'}{\sqrt{1 - u^2}} \qquad \frac{d}{dx}[\arccos u] = \frac{-u'}{\sqrt{1 - u^2}}$$

$$\frac{d}{dx}[\arctan u] = \frac{u'}{1 + u^2} \qquad \frac{d}{dx}[\operatorname{arccot} u] = \frac{-u'}{1 + u^2}$$

$$\frac{d}{dx}[\operatorname{arcsec} u] = \frac{u'}{|u|\sqrt{u^2 - 1}} \qquad \frac{d}{dx}[\operatorname{arcsec} u] = \frac{-u'}{|u|\sqrt{u^2 - 1}}$$

Proofs for arcsin u and arccos u are given in Appendix A. [The proofs for the other rules are left as an exercise (see Exercise 98).]

See LarsonCalculus.com for Bruce Edwards's video of this proof.

TECHNOLOGY If your graphing utility does not have the arcsecant function, you can obtain its graph using

$$f(x) = \operatorname{arcsec} x = \arccos \frac{1}{x}.$$

EXAMPLE 4 Differentiating Inverse Trigonometric Functions

a. $\dfrac{d}{dx}[\arcsin(2x)] = \dfrac{2}{\sqrt{1 - (2x)^2}} = \dfrac{2}{\sqrt{1 - 4x^2}}$

b. $\dfrac{d}{dx}[\arctan(3x)] = \dfrac{3}{1 + (3x)^2} = \dfrac{3}{1 + 9x^2}$

c. $\dfrac{d}{dx}[\arcsin \sqrt{x}] = \dfrac{(1/2)x^{-1/2}}{\sqrt{1 - x}} = \dfrac{1}{2\sqrt{x}\sqrt{1 - x}} = \dfrac{1}{2\sqrt{x - x^2}}$

d. $\dfrac{d}{dx}[\operatorname{arcsec} e^{2x}] = \dfrac{2e^{2x}}{e^{2x}\sqrt{(e^{2x})^2 - 1}} = \dfrac{2e^{2x}}{e^{2x}\sqrt{e^{4x} - 1}} = \dfrac{2}{\sqrt{e^{4x} - 1}}$

The absolute value sign is not necessary because $e^{2x} > 0$.

AP* Tips

Differentiation of inverse trigonometric functions is sometimes tested on the AP Exam. The inverse sine and tangent functions are the most likely topics.

EXAMPLE 5 A Derivative That Can Be Simplified

$$y = \arcsin x + x\sqrt{1 - x^2}$$

$$\begin{aligned} y' &= \frac{1}{\sqrt{1 - x^2}} + x\left(\frac{1}{2}\right)(-2x)(1 - x^2)^{-1/2} + \sqrt{1 - x^2} \\ &= \frac{1}{\sqrt{1 - x^2}} - \frac{x^2}{\sqrt{1 - x^2}} + \sqrt{1 - x^2} \\ &= \sqrt{1 - x^2} + \sqrt{1 - x^2} \\ &= 2\sqrt{1 - x^2} \end{aligned}$$

FOR FURTHER INFORMATION For more on the derivative of the arctangent function, see the article "Differentiating the Arctangent Directly" by Eric Key in *The College Mathematics Journal*. To view this article, go to *MathArticles.com*.

From Example 5, you can see one of the benefits of inverse trigonometric functions—they can be used to integrate common algebraic functions. For instance, from the result shown in the example, it follows that

$$\int \sqrt{1 - x^2}\, dx = \frac{1}{2}\left(\arcsin x + x\sqrt{1 - x^2}\right).$$

EXAMPLE 6 Analyzing an Inverse Trigonometric Graph

Analyze the graph of $y = (\arctan x)^2$.

Solution From the derivative

$$y' = 2(\arctan x)\left(\frac{1}{1 + x^2}\right) = \frac{2 \arctan x}{1 + x^2}$$

you can see that the only critical number is $x = 0$. By the First Derivative Test, this value corresponds to a relative minimum. From the second derivative

$$y'' = \frac{(1 + x^2)\left(\frac{2}{1 + x^2}\right) - (2 \arctan x)(2x)}{(1 + x^2)^2} = \frac{2(1 - 2x \arctan x)}{(1 + x^2)^2}$$

it follows that points of inflection occur when $2x \arctan x = 1$. Using Newton's Method, these points occur when $x \approx \pm 0.765$. Finally, because

$$\lim_{x \to \pm\infty} (\arctan x)^2 = \frac{\pi^2}{4}$$

it follows that the graph has a horizontal asymptote at $y = \pi^2/4$. The graph is shown in Figure 5.27.

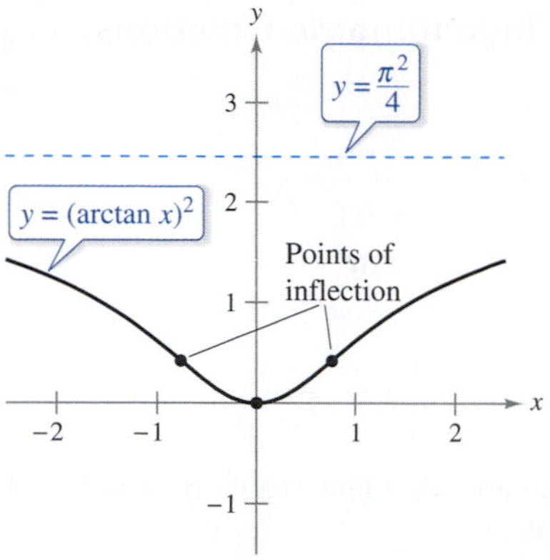

The graph of $y = (\arctan x)^2$ has a horizontal asymptote at $y = \pi^2/4$.
Figure 5.27

EXAMPLE 7 Maximizing an Angle

See LarsonCalculus.com for an interactive version of this type of example.

A photographer is taking a picture of a painting hung in an art gallery. The height of the painting is 4 feet. The camera lens is 1 foot below the lower edge of the painting, as shown in the figure at the right. How far should the camera be from the painting to maximize the angle subtended by the camera lens?

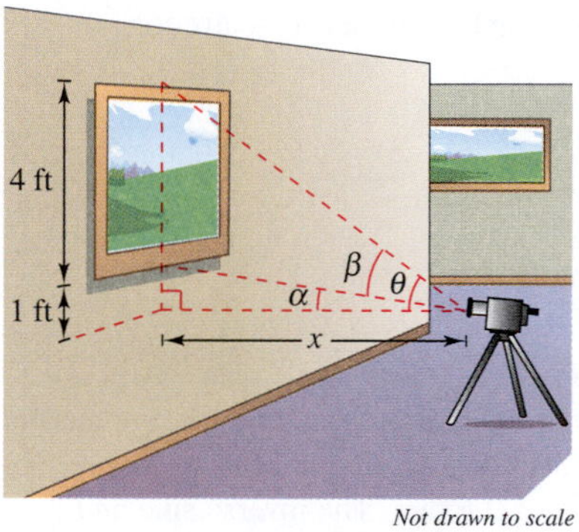

The camera should be 2.236 feet from the painting to maximize the angle β.

Solution In the figure, let β be the angle to be maximized.

$$\beta = \theta - \alpha = \operatorname{arccot} \frac{x}{5} - \operatorname{arccot} x$$

Differentiating produces

$$\frac{d\beta}{dx} = \frac{-1/5}{1 + (x^2/25)} - \frac{-1}{1 + x^2} = \frac{-5}{25 + x^2} + \frac{1}{1 + x^2} = \frac{4(5 - x^2)}{(25 + x^2)(1 + x^2)}.$$

Because $d\beta/dx = 0$ when $x = \sqrt{5}$, you can conclude from the First Derivative Test that this distance yields a maximum value of β. So, the distance is $x \approx 2.236$ feet and the angle is $\beta \approx 0.7297$ radian $\approx 41.81°$.

GALILEO GALILEI (1564–1642)
Galileo's approach to science departed from the accepted Aristotelian view that nature had describable *qualities*, such as "fluidity" and "potentiality." He chose to describe the physical world in terms of measurable *quantities*, such as time, distance, force, and mass.
See LarsonCalculus.com to read more of this biography.

Review of Basic Differentiation Rules

In the 1600s, Europe was ushered into the scientific age by such great thinkers as Descartes, Galileo, Huygens, Newton, and Kepler. These men believed that nature is governed by basic laws—laws that can, for the most part, be written in terms of mathematical equations. One of the most influential publications of this period—*Dialogue on the Great World Systems*, by Galileo Galilei—has become a classic description of modern scientific thought.

As mathematics has developed during the past few hundred years, a small number of elementary functions have proven sufficient for modeling most* phenomena in physics, chemistry, biology, engineering, economics, and a variety of other fields. An **elementary function** is a function from the following list or one that can be formed as the sum, product, quotient, or composition of functions in the list.

Algebraic Functions	**Transcendental Functions**
Polynomial functions	Logarithmic functions
Rational functions	Exponential functions
Functions involving radicals	Trigonometric functions
	Inverse trigonometric functions

With the differentiation rules introduced so far in the text, you can differentiate *any* elementary function. For convenience, these differentiation rules are summarized below.

BASIC DIFFERENTIATION RULES FOR ELEMENTARY FUNCTIONS

1. $\frac{d}{dx}[cu] = cu'$
2. $\frac{d}{dx}[u \pm v] = u' \pm v'$
3. $\frac{d}{dx}[uv] = uv' + vu'$
4. $\frac{d}{dx}\left[\frac{u}{v}\right] = \frac{vu' - uv'}{v^2}$
5. $\frac{d}{dx}[c] = 0$
6. $\frac{d}{dx}[u^n] = nu^{n-1}u'$
7. $\frac{d}{dx}[x] = 1$
8. $\frac{d}{dx}[|u|] = \frac{u}{|u|}(u'), \quad u \neq 0$
9. $\frac{d}{dx}[\ln u] = \frac{u'}{u}$
10. $\frac{d}{dx}[e^u] = e^u u'$
11. $\frac{d}{dx}[\log_a u] = \frac{u'}{(\ln a)u}$
12. $\frac{d}{dx}[a^u] = (\ln a)a^u u'$
13. $\frac{d}{dx}[\sin u] = (\cos u)u'$
14. $\frac{d}{dx}[\cos u] = -(\sin u)u'$
15. $\frac{d}{dx}[\tan u] = (\sec^2 u)u'$
16. $\frac{d}{dx}[\cot u] = -(\csc^2 u)u'$
17. $\frac{d}{dx}[\sec u] = (\sec u \tan u)u'$
18. $\frac{d}{dx}[\csc u] = -(\csc u \cot u)u'$
19. $\frac{d}{dx}[\arcsin u] = \frac{u'}{\sqrt{1 - u^2}}$
20. $\frac{d}{dx}[\arccos u] = \frac{-u'}{\sqrt{1 - u^2}}$
21. $\frac{d}{dx}[\arctan u] = \frac{u'}{1 + u^2}$
22. $\frac{d}{dx}[\text{arccot } u] = \frac{-u'}{1 + u^2}$
23. $\frac{d}{dx}[\text{arcsec } u] = \frac{u'}{|u|\sqrt{u^2 - 1}}$
24. $\frac{d}{dx}[\text{arccsc } u] = \frac{-u'}{|u|\sqrt{u^2 - 1}}$

* Some important functions used in engineering and science (such as Bessel functions and gamma functions) are not elementary functions.

5.6 Exercises

See CalcChat.com for tutorial help and worked-out solutions to odd-numbered exercises.

Finding Coordinates In Exercises 1 and 2, determine the missing coordinates of the points on the graph of the function.

1.

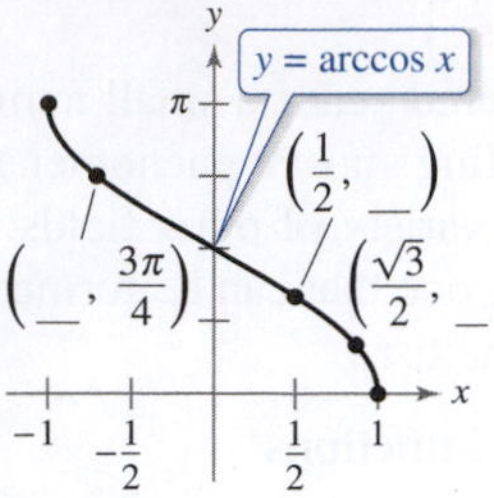

2.

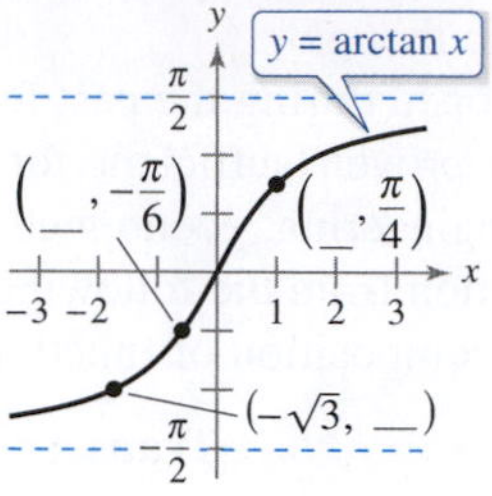

Evaluating Inverse Trigonometric Functions In Exercises 3–10, evaluate the expression without using a calculator.

3. $\arcsin \frac{1}{2}$

4. $\arcsin 0$

5. $\arccos \frac{1}{2}$

6. $\arccos 1$

7. $\arctan \dfrac{\sqrt{3}}{3}$

8. $\operatorname{arccot}(-\sqrt{3})$

9. $\operatorname{arccsc}(-\sqrt{2})$

10. $\operatorname{arcsec}(-\sqrt{2})$

Approximating Inverse Trigonometric Functions In Exercises 11–14, use a calculator to approximate the value. Round your answer to two decimal places.

11. $\arccos(-0.8)$

12. $\arcsin(-0.39)$

13. $\operatorname{arcsec} 1.269$

14. $\arctan(-5)$

Using a Right Triangle In Exercises 15–20, use the figure to write the expression in algebraic form given $y = \arccos x$, where $0 < y < \pi/2$.

15. $\cos y$

16. $\sin y$

17. $\tan y$

18. $\cot y$

19. $\sec y$

20. $\csc y$

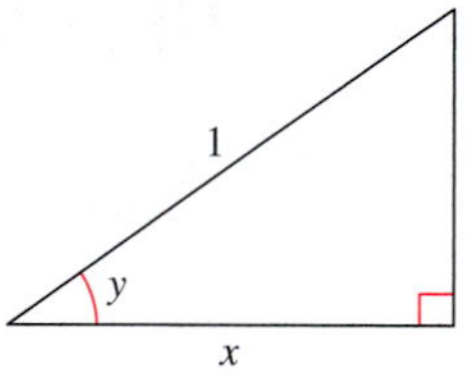

Evaluating an Expression In Exercises 21–24, evaluate each expression without using a calculator. (*Hint:* See Example 3.)

21. (a) $\sin\left(\arctan \dfrac{3}{4}\right)$ (b) $\sec\left(\arcsin \dfrac{4}{5}\right)$

22. (a) $\tan\left(\arccos \dfrac{\sqrt{2}}{2}\right)$ (b) $\cos\left(\arcsin \dfrac{5}{13}\right)$

23. (a) $\cot\left[\arcsin\left(-\dfrac{1}{2}\right)\right]$ (b) $\csc\left[\arctan\left(-\dfrac{5}{12}\right)\right]$

24. (a) $\sec\left[\arctan\left(-\dfrac{3}{5}\right)\right]$ (b) $\tan\left[\arcsin\left(-\dfrac{5}{6}\right)\right]$

Simplifying an Expression Using a Right Triangle In Exercises 25–32, write the expression in algebraic form. (*Hint:* Sketch a right triangle, as demonstrated in Example 3.)

25. $\cos(\arcsin 2x)$

26. $\sec(\arctan 4x)$

27. $\sin(\operatorname{arcsec} x)$

28. $\cos(\operatorname{arccot} x)$

29. $\tan\left(\operatorname{arcsec} \dfrac{x}{3}\right)$

30. $\sec[\arcsin(x-1)]$

31. $\csc\left(\arctan \dfrac{x}{\sqrt{2}}\right)$

32. $\cos\left(\arcsin \dfrac{x-h}{r}\right)$

Solving an Equation In Exercises 33–36, solve the equation for x.

33. $\arcsin(3x - \pi) = \frac{1}{2}$

34. $\arctan(2x - 5) = -1$

35. $\arcsin\sqrt{2x} = \arccos\sqrt{x}$

36. $\arccos x = \operatorname{arcsec} x$

Verifying Identities In Exercises 37 and 38, verify each identity.

37. (a) $\operatorname{arccsc} x = \arcsin \dfrac{1}{x}, \quad x \geq 1$

(b) $\arctan x + \arctan \dfrac{1}{x} = \dfrac{\pi}{2}, \quad x > 0$

38. (a) $\arcsin(-x) = -\arcsin x, \quad |x| \leq 1$

(b) $\arccos(-x) = \pi - \arccos x, \quad |x| \leq 1$

Finding a Derivative In Exercises 39–58, find the derivative of the function.

39. $f(x) = 2\arcsin(x-1)$

40. $f(t) = \arcsin t^2$

41. $g(x) = 3\arccos \dfrac{x}{2}$

42. $f(x) = \operatorname{arcsec} 2x$

43. $f(x) = \arctan e^x$

44. $f(x) = \arctan\sqrt{x}$

45. $g(x) = \dfrac{\arcsin 3x}{x}$

46. $h(x) = x^2 \arctan 5x$

47. $h(t) = \sin(\arccos t)$

48. $f(x) = \arcsin x + \arccos x$

49. $y = 2x\arccos x - 2\sqrt{1-x^2}$

50. $y = \ln(t^2+4) - \dfrac{1}{2}\arctan \dfrac{t}{2}$

51. $y = \dfrac{1}{2}\left(\dfrac{1}{2}\ln\dfrac{x+1}{x-1} + \arctan x\right)$

52. $y = \dfrac{1}{2}\left[x\sqrt{4-x^2} + 4\arcsin\left(\dfrac{x}{2}\right)\right]$

53. $y = x\arcsin x + \sqrt{1-x^2}$

54. $y = x\arctan 2x - \dfrac{1}{4}\ln(1+4x^2)$

55. $y = 8\arcsin \dfrac{x}{4} - \dfrac{x\sqrt{16-x^2}}{2}$

56. $y = 25\arcsin \dfrac{x}{5} - x\sqrt{25-x^2}$

57. $y = \arctan x + \dfrac{x}{1+x^2}$

58. $y = \arctan \dfrac{x}{2} - \dfrac{1}{2(x^2+4)}$

Finding an Equation of a Tangent Line **In Exercises 59–64, find an equation of the tangent line to the graph of the function at the given point.**

59. $y = 2\arcsin x, \quad \left(\frac{1}{2}, \frac{\pi}{3}\right)$

60. $y = \frac{1}{2}\arccos x, \quad \left(-\frac{\sqrt{2}}{2}, \frac{3\pi}{8}\right)$

61. $y = \arctan \frac{x}{2}, \quad \left(2, \frac{\pi}{4}\right)$

62. $y = \operatorname{arcsec} 4x, \quad \left(\frac{\sqrt{2}}{4}, \frac{\pi}{4}\right)$

63. $y = 4x \arccos(x - 1), \quad (1, 2\pi)$

64. $y = 3x \arcsin x, \quad \left(\frac{1}{2}, \frac{\pi}{4}\right)$

Linear and Quadratic Approximations **In Exercises 65–68, use a computer algebra system to find the linear approximation**

$$P_1(x) = f(a) + f'(a)(x - a)$$

and the quadratic approximation

$$P_2(x) = f(a) + f'(a)(x - a) + \tfrac{1}{2}f''(a)(x - a)^2$$

of the function f at $x = a$. Sketch the graph of the function and its linear and quadratic approximations.

65. $f(x) = \arctan x, \quad a = 0$

66. $f(x) = \arccos x, \quad a = 0$

67. $f(x) = \arcsin x, \quad a = \frac{1}{2}$

68. $f(x) = \arctan x, \quad a = 1$

Finding Relative Extrema **In Exercises 69–72, find any relative extrema of the function.**

69. $f(x) = \operatorname{arcsec} x - x$

70. $f(x) = \arcsin x - 2x$

71. $f(x) = \arctan x - \arctan(x - 4)$

72. $h(x) = \arcsin x - 2\arctan x$

Analyzing an Inverse Trigonometric Graph **In Exercises 73–76, analyze and sketch a graph of the function. Identify any relative extrema, points of inflection, and asymptotes. Use a graphing utility to verify your results.**

73. $f(x) = \arcsin(x - 1)$

74. $f(x) = \arctan x + \frac{\pi}{2}$

75. $f(x) = \operatorname{arcsec} 2x$

76. $f(x) = \arccos \frac{x}{4}$

Implicit Differentiation **In Exercises 77–80, use implicit differentiation to find an equation of the tangent line to the graph of the equation at the given point.**

77. $x^2 + x\arctan y = y - 1, \quad \left(-\frac{\pi}{4}, 1\right)$

78. $\arctan(xy) = \arcsin(x + y), \quad (0, 0)$

79. $\arcsin x + \arcsin y = \frac{\pi}{2}, \quad \left(\frac{\sqrt{2}}{2}, \frac{\sqrt{2}}{2}\right)$

80. $\arctan(x + y) = y^2 + \frac{\pi}{4}, \quad (1, 0)$

WRITING ABOUT CONCEPTS

81. Restricted Domains Explain why the domains of the trigonometric functions are restricted when finding the inverse trigonometric functions.

82. Inverse Trigonometric Functions Explain why $\tan \pi = 0$ does not imply that $\arctan 0 = \pi$.

83. Finding Values

(a) Use a graphing utility to evaluate $\arcsin(\arcsin 0.5)$ and $\arcsin(\arcsin 1)$.

(b) Let

$f(x) = \arcsin(\arcsin x)$.

Find the values of x in the interval $-1 \le x \le 1$ such that $f(x)$ is a real number.

84. HOW DO YOU SEE IT? The graphs of $f(x) = \sin x$ and $g(x) = \cos x$ are shown below.

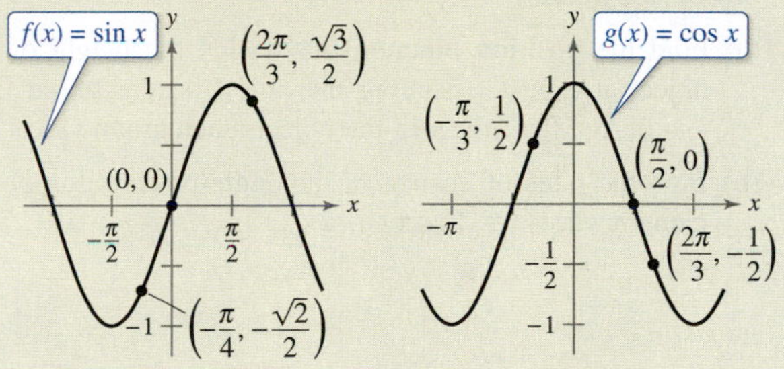

(a) Explain whether the points

$\left(-\frac{\sqrt{2}}{2}, -\frac{\pi}{4}\right)$, $(0, 0)$, and $\left(\frac{\sqrt{3}}{2}, \frac{2\pi}{3}\right)$

lie on the graph of $y = \arcsin x$.

(b) Explain whether the points

$\left(-\frac{1}{2}, \frac{2\pi}{3}\right)$, $\left(0, \frac{\pi}{2}\right)$, and $\left(\frac{1}{2}, -\frac{\pi}{3}\right)$

lie on the graph of $y = \arcos x$.

True or False? **In Exercises 85–90, determine whether the statement is true or false. If it is false, explain why or give an example that shows it is false.**

85. Because $\cos\left(-\frac{\pi}{3}\right) = \frac{1}{2}$, it follows that $\arccos \frac{1}{2} = -\frac{\pi}{3}$.

86. $\arcsin \frac{\pi}{4} = \frac{\sqrt{2}}{2}$

87. The slope of the graph of the inverse tangent function is positive for all x.

88. The range of $y = \arcsin x$ is $[0, \pi]$.

89. $\frac{d}{dx}[\arctan(\tan x)] = 1$ for all x in the domain.

90. $\arcsin^2 x + \arccos^2 x = 1$

91. Angular Rate of Change An airplane flies at an altitude of 5 miles toward a point directly over an observer. Consider θ and x as shown in the figure.

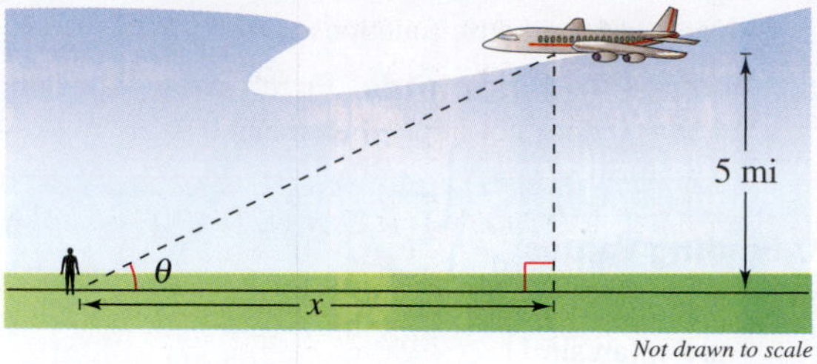

(a) Write θ as a function of x.

(b) The speed of the plane is 400 miles per hour. Find $d\theta/dt$ when $x = 10$ miles and $x = 3$ miles.

92. Writing Repeat Exercise 91 for an altitude of 3 miles and describe how the altitude affects the rate of change of θ.

93. Angular Rate of Change In a free-fall experiment, an object is dropped from a height of 256 feet. A camera on the ground 500 feet from the point of impact records the fall of the object (see figure).

(a) Find the position function that yields the height of the object at time t, assuming the object is released at time $t = 0$. At what time will the object reach ground level?

(b) Find the rates of change of the angle of elevation of the camera when $t = 1$ and $t = 2$.

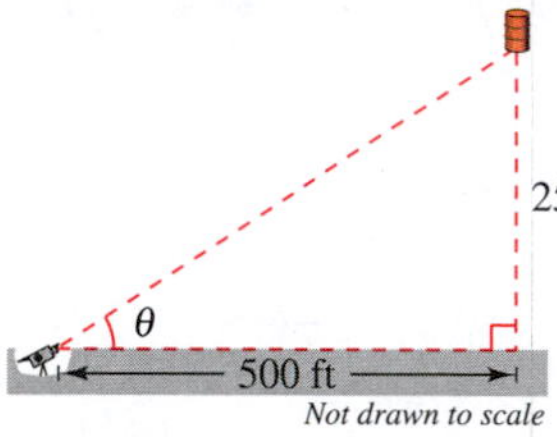

Figure for 93

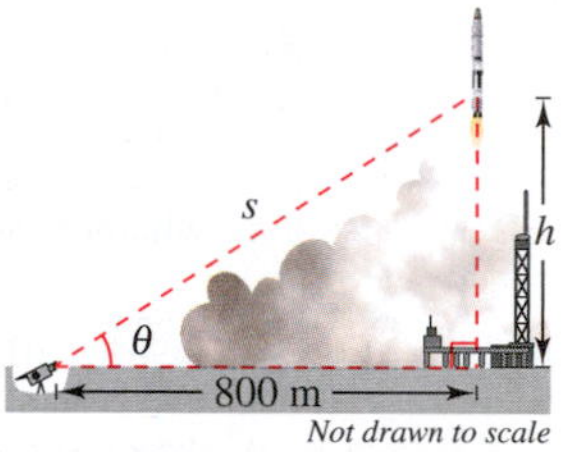

Figure for 94

94. Angular Rate of Change A television camera at ground level is filming the lift-off of a rocket at a point 800 meters from the launch pad. Let θ be the angle of elevation of the rocket and let s be the distance between the camera and the rocket (see figure). Write θ as a function of s for the period of time when the rocket is moving vertically. Differentiate the result to find $d\theta/dt$ in terms of s and ds/dt.

95. Maximizing an Angle A billboard 85 feet wide is perpendicular to a straight road and is 40 feet from the road (see figure). Find the point on the road at which the angle θ subtended by the billboard is a maximum.

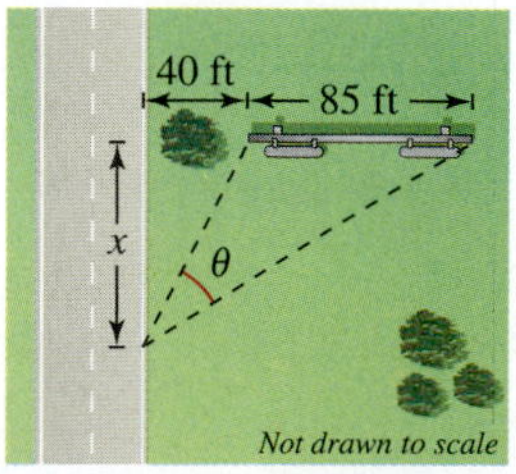

Figure for 95

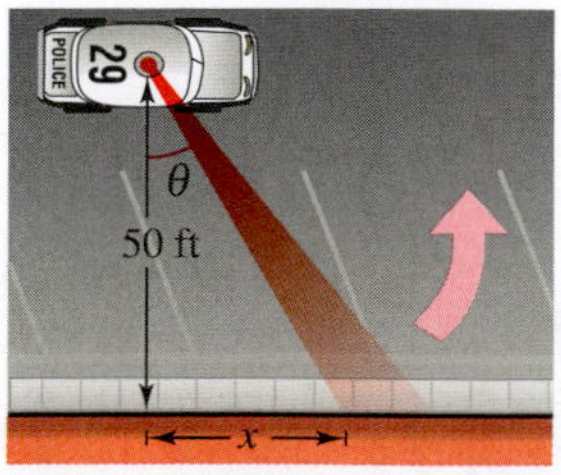

Figure for 96

96. Angular Speed A patrol car is parked 50 feet from a long warehouse (see figure). The revolving light on top of the car turns at a rate of 30 revolutions per minute. Write θ as a function of x. How fast is the light beam moving along the wall when the beam makes an angle of $\theta = 45°$ with the line perpendicular from the light to the wall?

97. Proof

(a) Prove that $\arctan x + \arctan y = \arctan \dfrac{x + y}{1 - xy}$, $xy \neq 1$.

(b) Use the formula in part (a) to show that

$$\arctan \frac{1}{2} + \arctan \frac{1}{3} = \frac{\pi}{4}.$$

98. Proof Prove each differentiation formula.

(a) $\dfrac{d}{dx}[\arctan u] = \dfrac{u'}{1 + u^2}$

(b) $\dfrac{d}{dx}[\text{arccot } u] = \dfrac{-u'}{1 + u^2}$

(c) $\dfrac{d}{dx}[\text{arcsec } u] = \dfrac{u'}{|u|\sqrt{u^2 - 1}}$

(d) $\dfrac{d}{dx}[\text{arccsc } u] = \dfrac{-u'}{|u|\sqrt{u^2 - 1}}$

99. Describing a Graph

(a) Graph the function $f(x) = \arccos x + \arcsin x$ on the interval $[-1, 1]$.

(b) Describe the graph of f.

(c) Verify the result of part (b) analytically.

100. Think About It Use a graphing utility to graph $f(x) = \sin x$ and $g(x) = \arcsin(\sin x)$.

(a) Why isn't the graph of g the line $y = x$?

(b) Determine the extrema of g.

101. Maximizing an Angle In the figure, find the value of c in the interval $[0, 4]$ on the x-axis that maximizes angle θ.

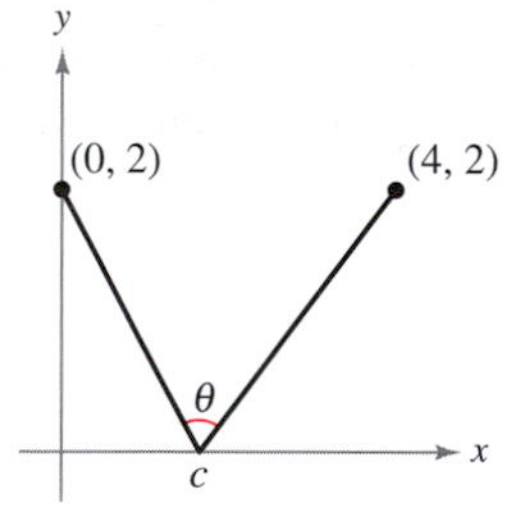

Figure for 101

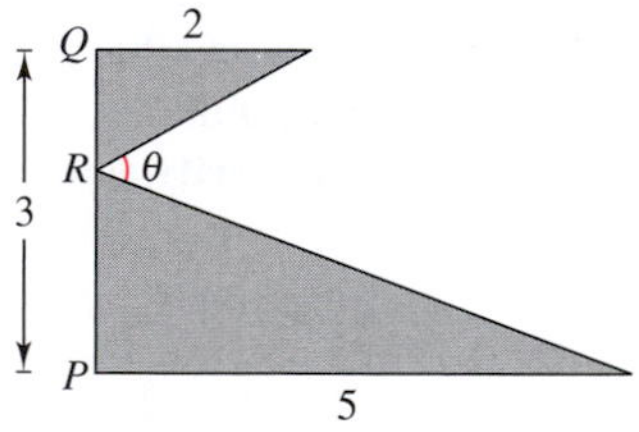

Figure for 102

102. Finding a Distance In the figure, find PR such that $0 \le PR \le 3$ and $m\angle\theta$ is a maximum.

103. Proof Prove that $\arcsin x = \arctan\left(\dfrac{x}{\sqrt{1 - x^2}}\right)$, $|x| < 1$.

104. Inverse Secant Function Some calculus textbooks define the inverse secant function using the range $[0, \pi/2) \cup [\pi, 3\pi/2)$.

(a) Sketch the graph $y = \text{arcsec } x$ using this range.

(b) Show that $y' = \dfrac{1}{x\sqrt{x^2 - 1}}$.

5.7 Inverse Trigonometric Functions: Integration

- Integrate functions whose antiderivatives involve inverse trigonometric functions.
- Use the method of completing the square to integrate a function.
- Review the basic integration rules involving elementary functions.

AP* Tips

Integrals resulting in the inverse trigonometric functions are rarely tested on the AP Exam. The inverse sine and tangent functions are the most likely topics.

Integrals Involving Inverse Trigonometric Functions

The derivatives of the six inverse trigonometric functions fall into three pairs. In each pair, the derivative of one function is the negative of the other. For example,

$$\frac{d}{dx}[\arcsin x] = \frac{1}{\sqrt{1 - x^2}}$$

and

$$\frac{d}{dx}[\arccos x] = -\frac{1}{\sqrt{1 - x^2}}.$$

When listing the *antiderivative* that corresponds to each of the inverse trigonometric functions, you need to use only one member from each pair. It is conventional to use $\arcsin x$ as the antiderivative of $1/\sqrt{1 - x^2}$, rather than $-\arccos x$. The next theorem gives one antiderivative formula for each of the three pairs. The proofs of these integration rules are left to you (see Exercises 75–77).

FOR FURTHER INFORMATION For a detailed proof of rule 2 of Theorem 5.17, see the article "A Direct Proof of the Integral Formula for Arctangent" by Arnold J. Insel in *The College Mathematics Journal*. To view this article, go to *MathArticles.com*.

THEOREM 5.17 Integrals Involving Inverse Trigonometric Functions

Let u be a differentiable function of x, and let $a > 0$.

1. $\displaystyle\int \frac{du}{\sqrt{a^2 - u^2}} = \arcsin \frac{u}{a} + C$
2. $\displaystyle\int \frac{du}{a^2 + u^2} = \frac{1}{a} \arctan \frac{u}{a} + C$
3. $\displaystyle\int \frac{du}{u\sqrt{u^2 - a^2}} = \frac{1}{a} \operatorname{arcsec} \frac{|u|}{a} + C$

EXAMPLE 1 Integration with Inverse Trigonometric Functions

a. $\displaystyle\int \frac{dx}{\sqrt{4 - x^2}} = \arcsin \frac{x}{2} + C$

b. $\displaystyle\int \frac{dx}{2 + 9x^2} = \frac{1}{3}\int \frac{3\,dx}{(\sqrt{2})^2 + (3x)^2}$ $u = 3x,\ a = \sqrt{2}$

$$= \frac{1}{3\sqrt{2}} \arctan \frac{3x}{\sqrt{2}} + C$$

c. $\displaystyle\int \frac{dx}{x\sqrt{4x^2 - 9}} = \int \frac{2\,dx}{2x\sqrt{(2x)^2 - 3^2}}$ $u = 2x,\ a = 3$

$$= \frac{1}{3} \operatorname{arcsec} \frac{|2x|}{3} + C$$

The integrals in Example 1 are fairly straightforward applications of integration formulas. Unfortunately, this is not typical. The integration formulas for inverse trigonometric functions can be disguised in many ways.

EXAMPLE 2 Integration by Substitution

Find $\displaystyle\int \frac{dx}{\sqrt{e^{2x}-1}}$.

Solution As it stands, this integral doesn't fit any of the three inverse trigonometric formulas. Using the substitution $u = e^x$, however, produces

$$u = e^x \quad \Rightarrow \quad du = e^x\,dx \quad \Rightarrow \quad dx = \frac{du}{e^x} = \frac{du}{u}.$$

With this substitution, you can integrate as shown.

$$\begin{aligned}
\int \frac{dx}{\sqrt{e^{2x}-1}} &= \int \frac{dx}{\sqrt{(e^x)^2-1}} && \text{Write } e^{2x} \text{ as } (e^x)^2.\\
&= \int \frac{du/u}{\sqrt{u^2-1}} && \text{Substitute.}\\
&= \int \frac{du}{u\sqrt{u^2-1}} && \text{Rewrite to fit Arcsecant Rule.}\\
&= \operatorname{arcsec} \frac{|u|}{1} + C && \text{Apply Arcsecant Rule.}\\
&= \operatorname{arcsec} e^x + C && \text{Back-substitute.}
\end{aligned}$$

TECHNOLOGY PITFALL A symbolic integration utility can be useful for integrating functions such as the one in Example 2. In some cases, however, the utility may fail to find an antiderivative for two reasons. First, some elementary functions do not have antiderivatives that are elementary functions. Second, every utility has limitations—you might have entered a function that the utility was not programmed to handle. You should also remember that antiderivatives involving trigonometric functions or logarithmic functions can be written in many different forms. For instance, one utility found the integral in Example 2 to be

$$\int \frac{dx}{\sqrt{e^{2x}-1}} = \arctan \sqrt{e^{2x}-1} + C.$$

Try showing that this antiderivative is equivalent to the one found in Example 2.

EXAMPLE 3 Rewriting as the Sum of Two Quotients

Find $\displaystyle\int \frac{x+2}{\sqrt{4-x^2}}\,dx$.

Solution This integral does not appear to fit any of the basic integration formulas. By splitting the integrand into two parts, however, you can see that the first part can be found with the Power Rule and the second part yields an inverse sine function.

$$\begin{aligned}
\int \frac{x+2}{\sqrt{4-x^2}}\,dx &= \int \frac{x}{\sqrt{4-x^2}}\,dx + \int \frac{2}{\sqrt{4-x^2}}\,dx\\
&= -\frac{1}{2}\int (4-x^2)^{-1/2}(-2x)\,dx + 2\int \frac{1}{\sqrt{4-x^2}}\,dx\\
&= -\frac{1}{2}\left[\frac{(4-x^2)^{1/2}}{1/2}\right] + 2 \arcsin \frac{x}{2} + C\\
&= -\sqrt{4-x^2} + 2 \arcsin \frac{x}{2} + C
\end{aligned}$$

Completing the Square

Completing the square helps when quadratic functions are involved in the integrand. For example, the quadratic $x^2 + bx + c$ can be written as the difference of two squares by adding and subtracting $(b/2)^2$.

$$x^2 + bx + c = x^2 + bx + \left(\frac{b}{2}\right)^2 - \left(\frac{b}{2}\right)^2 + c = \left(x + \frac{b}{2}\right)^2 - \left(\frac{b}{2}\right)^2 + c$$

EXAMPLE 4 Completing the Square

See LarsonCalculus.com for an interactive version of this type of example.

Find $\displaystyle\int \frac{dx}{x^2 - 4x + 7}$.

Solution You can write the denominator as the sum of two squares, as shown.

$$x^2 - 4x + 7 = (x^2 - 4x + 4) - 4 + 7 = (x - 2)^2 + 3 = u^2 + a^2$$

Now, in this completed square form, let $u = x - 2$ and $a = \sqrt{3}$.

$$\int \frac{dx}{x^2 - 4x + 7} = \int \frac{dx}{(x - 2)^2 + 3} = \frac{1}{\sqrt{3}} \arctan \frac{x - 2}{\sqrt{3}} + C$$

When the leading coefficient is not 1, it helps to factor before completing the square. For instance, you can complete the square of $2x^2 - 8x + 10$ by factoring first.

$$\begin{aligned} 2x^2 - 8x + 10 &= 2(x^2 - 4x + 5) \\ &= 2(x^2 - 4x + 4 - 4 + 5) \\ &= 2[(x - 2)^2 + 1] \end{aligned}$$

To complete the square when the coefficient of x^2 is negative, use the same factoring process shown above. For instance, you can complete the square for $3x - x^2$ as shown.

$$3x - x^2 = -(x^2 - 3x) = -\left[x^2 - 3x + \left(\tfrac{3}{2}\right)^2 - \left(\tfrac{3}{2}\right)^2\right] = \left(\tfrac{3}{2}\right)^2 - \left(x - \tfrac{3}{2}\right)^2$$

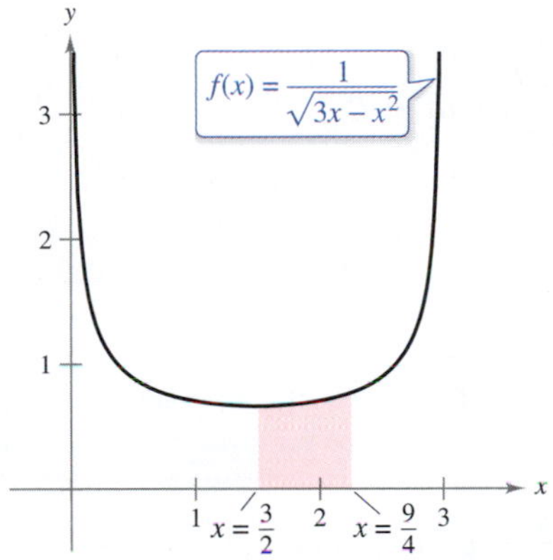

The area of the region bounded by the graph of f, the x-axis, $x = \frac{3}{2}$, and $x = \frac{9}{4}$ is $\pi/6$.

Figure 5.28

TECHNOLOGY With definite integrals such as the one given in Example 5, remember that you can resort to a numerical solution. For instance, applying Simpson's Rule (with $n = 12$) to the integral in the example, you obtain

$$\int_{3/2}^{9/4} \frac{1}{\sqrt{3x - x^2}}\, dx \approx 0.523599.$$

This differs from the exact value of the integral ($\pi/6 \approx 0.5235988$) by less than one-millionth.

EXAMPLE 5 Completing the Square

Find the area of the region bounded by the graph of

$$f(x) = \frac{1}{\sqrt{3x - x^2}}$$

the x-axis, and the lines $x = \frac{3}{2}$ and $x = \frac{9}{4}$.

Solution In Figure 5.28, you can see that the area is

$$\begin{aligned} \text{Area} &= \int_{3/2}^{9/4} \frac{1}{\sqrt{3x - x^2}}\, dx \\ &= \int_{3/2}^{9/4} \frac{dx}{\sqrt{(3/2)^2 - [x - (3/2)]^2}} \qquad \text{Use completed square form derived above.} \\ &= \arcsin \frac{x - (3/2)}{3/2} \Bigg]_{3/2}^{9/4} \\ &= \arcsin \frac{1}{2} - \arcsin 0 \\ &= \frac{\pi}{6} \\ &\approx 0.524. \end{aligned}$$

Review of Basic Integration Rules

You have now completed the introduction of the **basic integration rules.** To be efficient at applying these rules, you should have practiced enough so that each rule is committed to memory.

BASIC INTEGRATION RULES ($a > 0$)

1. $\int kf(u)\,du = k\int f(u)\,du$

2. $\int [f(u) \pm g(u)]\,du = \int f(u)\,du \pm \int g(u)\,du$

3. $\int du = u + C$

4. $\int u^n\,du = \dfrac{u^{n+1}}{n+1} + C, \quad n \neq -1$

5. $\int \dfrac{du}{u} = \ln|u| + C$

6. $\int e^u\,du = e^u + C$

7. $\int a^u\,du = \left(\dfrac{1}{\ln a}\right)a^u + C$

8. $\int \sin u\,du = -\cos u + C$

9. $\int \cos u\,du = \sin u + C$

10. $\int \tan u\,du = -\ln|\cos u| + C$

11. $\int \cot u\,du = \ln|\sin u| + C$

12. $\int \sec u\,du = \ln|\sec u + \tan u| + C$

13. $\int \csc u\,du = -\ln|\csc u + \cot u| + C$

14. $\int \sec^2 u\,du = \tan u + C$

15. $\int \csc^2 u\,du = -\cot u + C$

16. $\int \sec u \tan u\,du = \sec u + C$

17. $\int \csc u \cot u\,du = -\csc u + C$

18. $\int \dfrac{du}{\sqrt{a^2 - u^2}} = \arcsin \dfrac{u}{a} + C$

19. $\int \dfrac{du}{a^2 + u^2} = \dfrac{1}{a} \arctan \dfrac{u}{a} + C$

20. $\int \dfrac{du}{u\sqrt{u^2 - a^2}} = \dfrac{1}{a} \operatorname{arcsec} \dfrac{|u|}{a} + C$

You can learn a lot about the nature of integration by comparing this list with the summary of differentiation rules given in the preceding section. For differentiation, you now have rules that allow you to differentiate *any* elementary function. For integration, this is far from true.

The integration rules listed above are primarily those that were happened on during the development of differentiation rules. So far, you have not learned any rules or techniques for finding the antiderivative of a general product or quotient, the natural logarithmic function, or the inverse trigonometric functions. More important, you cannot apply any of the rules in this list unless you can create the proper du corresponding to the u in the formula. The point is that you need to work more on integration techniques, which you will do in Chapter 8. The next two examples should give you a better feeling for the integration problems that you *can* and *cannot* solve with the techniques and rules you now know.

EXAMPLE 6 **Comparing Integration Problems**

Find as many of the following integrals as you can using the formulas and techniques you have studied so far in the text.

a. $\displaystyle\int \frac{dx}{x\sqrt{x^2 - 1}}$

b. $\displaystyle\int \frac{x\,dx}{\sqrt{x^2 - 1}}$

c. $\displaystyle\int \frac{dx}{\sqrt{x^2 - 1}}$

Solution

a. You *can* find this integral (it fits the Arcsecant Rule).

$$\int \frac{dx}{x\sqrt{x^2 - 1}} = \operatorname{arcsec}|x| + C$$

b. You *can* find this integral (it fits the Power Rule).

$$\begin{aligned}\int \frac{x\,dx}{\sqrt{x^2 - 1}} &= \frac{1}{2}\int (x^2 - 1)^{-1/2}(2x)\,dx \\ &= \frac{1}{2}\left[\frac{(x^2 - 1)^{1/2}}{1/2}\right] + C \\ &= \sqrt{x^2 - 1} + C\end{aligned}$$

c. You *cannot* find this integral using the techniques you have studied so far. (You should scan the list of basic integration rules to verify this conclusion.)

EXAMPLE 7 **Comparing Integration Problems**

Find as many of the following integrals as you can using the formulas and techniques you have studied so far in the text.

a. $\displaystyle\int \frac{dx}{x \ln x}$

b. $\displaystyle\int \frac{\ln x\,dx}{x}$

c. $\displaystyle\int \ln x\,dx$

Solution

a. You *can* find this integral (it fits the Log Rule).

$$\begin{aligned}\int \frac{dx}{x \ln x} &= \int \frac{1/x}{\ln x}\,dx \\ &= \ln|\ln x| + C\end{aligned}$$

b. You *can* find this integral (it fits the Power Rule).

$$\begin{aligned}\int \frac{\ln x\,dx}{x} &= \int \left(\frac{1}{x}\right)(\ln x)^1\,dx \\ &= \frac{(\ln x)^2}{2} + C\end{aligned}$$

REMARK Note in Examples 6 and 7 that the *simplest* functions are the ones that you cannot yet integrate.

c. You *cannot* find this integral using the techniques you have studied so far. ■

5.7 Exercises

See CalcChat.com for tutorial help and worked-out solutions to odd-numbered exercises.

Finding an Indefinite Integral In Exercises 1–20, find the indefinite integral.

1. $\displaystyle\int \frac{dx}{\sqrt{9-x^2}}$

2. $\displaystyle\int \frac{dx}{\sqrt{1-4x^2}}$

3. $\displaystyle\int \frac{1}{x\sqrt{4x^2-1}}\,dx$

4. $\displaystyle\int \frac{12}{1+9x^2}\,dx$

5. $\displaystyle\int \frac{1}{\sqrt{1-(x+1)^2}}\,dx$

6. $\displaystyle\int \frac{1}{4+(x-3)^2}\,dx$

7. $\displaystyle\int \frac{t}{\sqrt{1-t^4}}\,dt$

8. $\displaystyle\int \frac{1}{x\sqrt{x^4-4}}\,dx$

9. $\displaystyle\int \frac{t}{t^4+25}\,dt$

10. $\displaystyle\int \frac{1}{x\sqrt{1-(\ln x)^2}}\,dx$

11. $\displaystyle\int \frac{e^{2x}}{4+e^{4x}}\,dx$

12. $\displaystyle\int \frac{2}{x\sqrt{9x^2-25}}\,dx$

13. $\displaystyle\int \frac{\sec^2 x}{\sqrt{25-\tan^2 x}}\,dx$

14. $\displaystyle\int \frac{\sin x}{7+\cos^2 x}\,dx$

15. $\displaystyle\int \frac{1}{\sqrt{x}\sqrt{1-x}}\,dx$

16. $\displaystyle\int \frac{3}{2\sqrt{x}(1+x)}\,dx$

17. $\displaystyle\int \frac{x-3}{x^2+1}\,dx$

18. $\displaystyle\int \frac{x^2+3}{x\sqrt{x^2-4}}\,dx$

19. $\displaystyle\int \frac{x+5}{\sqrt{9-(x-3)^2}}\,dx$

20. $\displaystyle\int \frac{x-2}{(x+1)^2+4}\,dx$

Evaluating a Definite Integral In Exercises 21–32, evaluate the definite integral.

21. $\displaystyle\int_0^{1/6} \frac{3}{\sqrt{1-9x^2}}\,dx$

22. $\displaystyle\int_0^{\sqrt{2}} \frac{1}{\sqrt{4-x^2}}\,dx$

23. $\displaystyle\int_0^{\sqrt{3}/2} \frac{1}{1+4x^2}\,dx$

24. $\displaystyle\int_{\sqrt{3}}^{3} \frac{1}{x\sqrt{4x^2-9}}\,dx$

25. $\displaystyle\int_3^6 \frac{1}{25+(x-3)^2}\,dx$

26. $\displaystyle\int_1^4 \frac{1}{x\sqrt{16x^2-5}}\,dx$

27. $\displaystyle\int_0^{\ln 5} \frac{e^x}{1+e^{2x}}\,dx$

28. $\displaystyle\int_{\ln 2}^{\ln 4} \frac{e^{-x}}{\sqrt{1-e^{-2x}}}\,dx$

29. $\displaystyle\int_{\pi/2}^{\pi} \frac{\sin x}{1+\cos^2 x}\,dx$

30. $\displaystyle\int_0^{\pi/2} \frac{\cos x}{1+\sin^2 x}\,dx$

31. $\displaystyle\int_0^{1/\sqrt{2}} \frac{\arcsin x}{\sqrt{1-x^2}}\,dx$

32. $\displaystyle\int_0^{1/\sqrt{2}} \frac{\arccos x}{\sqrt{1-x^2}}\,dx$

Completing the Square In Exercises 33–42, find or evaluate the integral by completing the square.

33. $\displaystyle\int_0^2 \frac{dx}{x^2-2x+2}$

34. $\displaystyle\int_{-2}^2 \frac{dx}{x^2+4x+13}$

35. $\displaystyle\int \frac{2x}{x^2+6x+13}\,dx$

36. $\displaystyle\int \frac{2x-5}{x^2+2x+2}\,dx$

37. $\displaystyle\int \frac{1}{\sqrt{-x^2-4x}}\,dx$

38. $\displaystyle\int \frac{2}{\sqrt{-x^2+4x}}\,dx$

39. $\displaystyle\int_2^3 \frac{2x-3}{\sqrt{4x-x^2}}\,dx$

40. $\displaystyle\int \frac{1}{(x-1)\sqrt{x^2-2x}}\,dx$

41. $\displaystyle\int \frac{x}{x^4+2x^2+2}\,dx$

42. $\displaystyle\int \frac{x}{\sqrt{9+8x^2-x^4}}\,dx$

Integration by Substitution In Exercises 43–46, use the specified substitution to find or evaluate the integral.

43. $\displaystyle\int \sqrt{e^t-3}\,dt$

$u = \sqrt{e^t-3}$

44. $\displaystyle\int \frac{\sqrt{x-2}}{x+1}\,dx$

$u = \sqrt{x-2}$

45. $\displaystyle\int_1^3 \frac{dx}{\sqrt{x}(1+x)}$

$u = \sqrt{x}$

46. $\displaystyle\int_0^1 \frac{dx}{2\sqrt{3-x}\sqrt{x+1}}$

$u = \sqrt{x+1}$

WRITING ABOUT CONCEPTS

Comparing Integration Problems In Exercises 47–50, determine which of the integrals can be found using the basic integration formulas you have studied so far in the text.

47. (a) $\displaystyle\int \frac{1}{\sqrt{1-x^2}}\,dx$ (b) $\displaystyle\int \frac{x}{\sqrt{1-x^2}}\,dx$ (c) $\displaystyle\int \frac{1}{x\sqrt{1-x^2}}\,dx$

48. (a) $\displaystyle\int e^{x^2}\,dx$ (b) $\displaystyle\int xe^{x^2}\,dx$ (c) $\displaystyle\int \frac{1}{x^2}e^{1/x}\,dx$

49. (a) $\displaystyle\int \sqrt{x-1}\,dx$ (b) $\displaystyle\int x\sqrt{x-1}\,dx$ (c) $\displaystyle\int \frac{x}{\sqrt{x-1}}\,dx$

50. (a) $\displaystyle\int \frac{1}{1+x^4}\,dx$ (b) $\displaystyle\int \frac{x}{1+x^4}\,dx$ (c) $\displaystyle\int \frac{x^3}{1+x^4}\,dx$

51. **Finding an Integral** Decide whether you can find the integral

$$\int \frac{2\,dx}{\sqrt{x^2+4}}$$

using the formulas and techniques you have studied so far. Explain your reasoning.

52. **HOW DO YOU SEE IT?** Using the graph, which value best approximates the area of the region between the x-axis and the function over the interval $\left[-\frac{1}{2}, \frac{1}{2}\right]$? Explain.

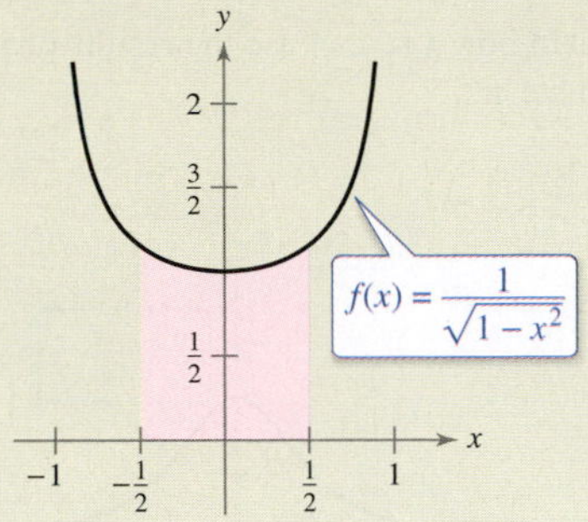

(a) -3 (b) $\frac{1}{2}$ (c) 1 (d) 2 (e) 4

Differential Equation In Exercises 53 and 54, use the differential equation and the specified initial condition to find y.

53. $\dfrac{dy}{dx} = \dfrac{1}{\sqrt{4 - x^2}}$

$y(0) = \pi$

54. $\dfrac{dy}{dx} = \dfrac{1}{4 + x^2}$

$y(2) = \pi$

Slope Field In Exercises 55 and 56, a differential equation, a point, and a slope field are given. (a) Sketch two approximate solutions of the differential equation on the slope field, one of which passes through the given point. (b) Use integration to find the particular solution of the differential equation and use a graphing utility to graph the solution. Compare the result with the sketches in part (a). To print an enlarged copy of the graph, go to *MathGraphs.com*.

55. $\dfrac{dy}{dx} = \dfrac{2}{9 + x^2}$, $(0, 2)$

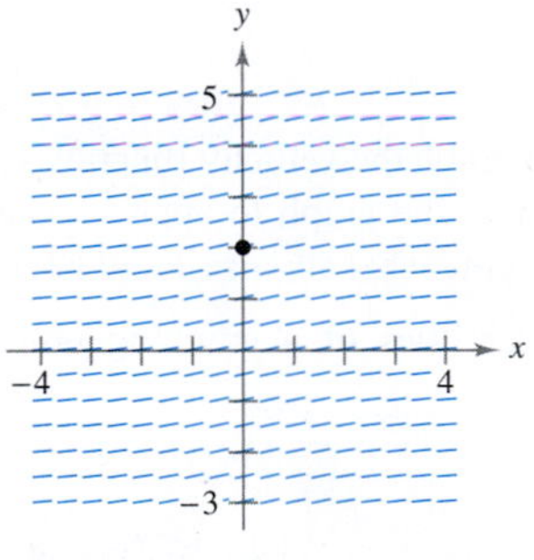

56. $\dfrac{dy}{dx} = \dfrac{2}{\sqrt{25 - x^2}}$, $(5, \pi)$

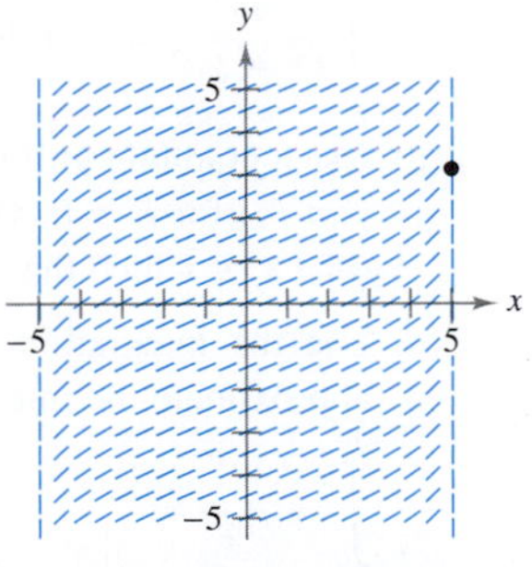

Slope Field In Exercises 57–60, use a computer algebra system to graph the slope field for the differential equation and graph the solution satisfying the specified initial condition.

57. $\dfrac{dy}{dx} = \dfrac{10}{x\sqrt{x^2 - 1}}$

$y(3) = 0$

58. $\dfrac{dy}{dx} = \dfrac{1}{12 + x^2}$

$y(4) = 2$

59. $\dfrac{dy}{dx} = \dfrac{2y}{\sqrt{16 - x^2}}$

$y(0) = 2$

60. $\dfrac{dy}{dx} = \dfrac{\sqrt{y}}{1 + x^2}$

$y(0) = 4$

Area In Exercises 61–66, find the area of the region.

61. $y = \dfrac{2}{\sqrt{4 - x^2}}$

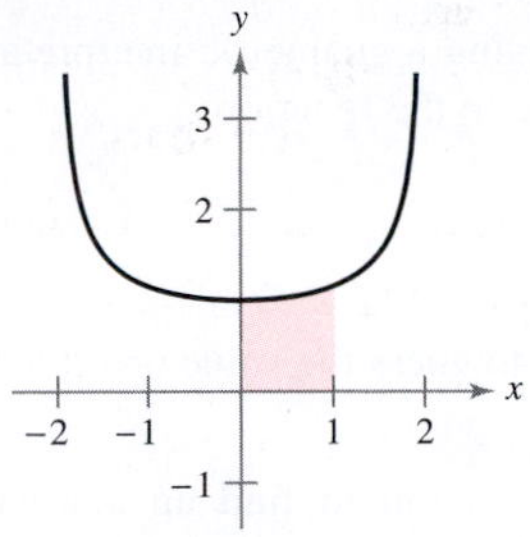

62. $y = \dfrac{1}{x\sqrt{x^2 - 1}}$

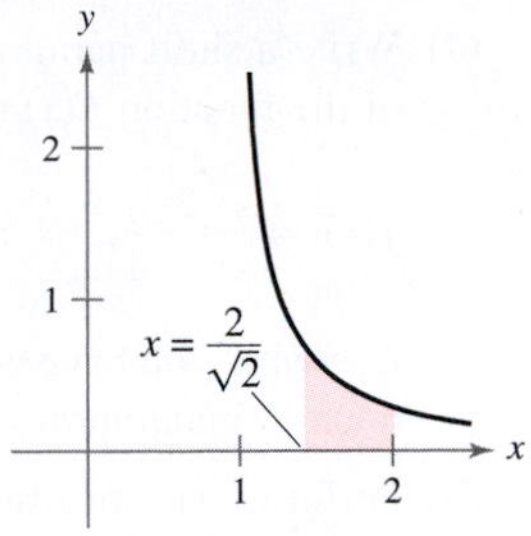

63. $y = \dfrac{1}{x^2 - 2x + 5}$

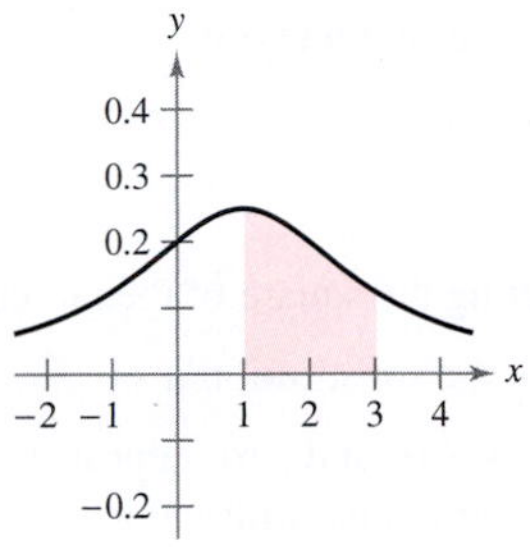

64. $y = \dfrac{2}{x^2 + 4x + 8}$

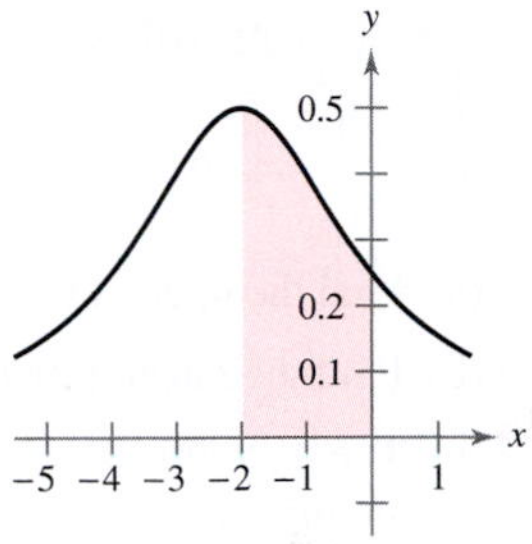

65. $y = \dfrac{3 \cos x}{1 + \sin^2 x}$

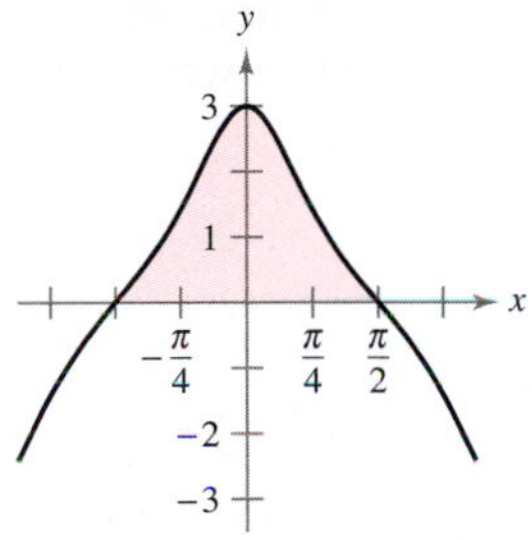

66. $y = \dfrac{4e^x}{1 + e^{2x}}$

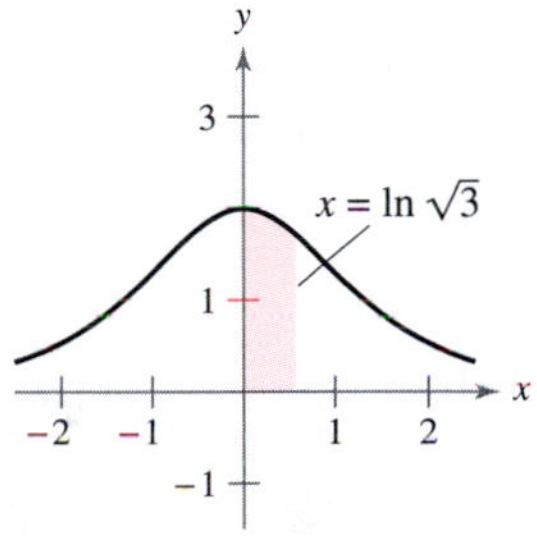

67. Area

(a) Sketch the region whose area is represented by

$$\int_0^1 \arcsin x \, dx.$$

(b) Use the integration capabilities of a graphing utility to approximate the area.

(c) Find the exact area analytically.

68. Approximating Pi

(a) Show that

$$\int_0^1 \frac{4}{1 + x^2} \, dx = \pi.$$

(b) Approximate the number π using Simpson's Rule (with $n = 6$) and the integral in part (a).

(c) Approximate the number π by using the integration capabilities of a graphing utility.

69. Investigation Consider the function

$$F(x) = \frac{1}{2}\int_x^{x+2} \frac{2}{t^2+1}\,dt.$$

(a) Write a short paragraph giving a geometric interpretation of the function $F(x)$ relative to the function

$$f(x) = \frac{2}{x^2+1}.$$

Use what you have written to guess the value of x that will make F maximum.

(b) Perform the specified integration to find an alternative form of $F(x)$. Use calculus to locate the value of x that will make F maximum and compare the result with your guess in part (a).

70. Comparing Integrals Consider the integral

$$\int \frac{1}{\sqrt{6x - x^2}}\,dx.$$

(a) Find the integral by completing the square of the radicand.

(b) Find the integral by making the substitution $u = \sqrt{x}$.

(c) The antiderivatives in parts (a) and (b) appear to be significantly different. Use a graphing utility to graph each antiderivative in the same viewing window and determine the relationship between them. Find the domain of each.

True or False? **In Exercises 71–74, determine whether the statement is true or false. If it is false, explain why or give an example that shows it is false.**

71. $\displaystyle\int \frac{dx}{3x\sqrt{9x^2-16}} = \frac{1}{4}\operatorname{arcsec}\frac{3x}{4} + C$

72. $\displaystyle\int \frac{dx}{25+x^2} = \frac{1}{25}\arctan\frac{x}{25} + C$

73. $\displaystyle\int \frac{dx}{\sqrt{4-x^2}} = -\arccos\frac{x}{2} + C$

74. One way to find $\displaystyle\int \frac{2e^{2x}}{\sqrt{9-e^{2x}}}\,dx$ is to use the Arcsine Rule.

Verifying an Integration Rule **In Exercises 75–77, verify the rule by differentiating. Let $a > 0$.**

75. $\displaystyle\int \frac{du}{\sqrt{a^2-u^2}} = \arcsin\frac{u}{a} + C$

76. $\displaystyle\int \frac{du}{a^2+u^2} = \frac{1}{a}\arctan\frac{u}{a} + C$

77. $\displaystyle\int \frac{du}{u\sqrt{u^2-a^2}} = \frac{1}{a}\operatorname{arcsec}\frac{|u|}{a} + C$

78. Proof Graph

$$y_1 = \frac{x}{1+x^2}, \quad y_2 = \arctan x, \quad \text{and} \quad y_3 = x$$

on $[0, 10]$. Prove that

$$\frac{x}{1+x^2} < \arctan x < x \quad \text{for} \quad x > 0.$$

79. Numerical Integration

(a) Write an integral that represents the area of the region in the figure.

(b) Use the Trapezoidal Rule with $n = 8$ to estimate the area of the region.

(c) Explain how you can use the results of parts (a) and (b) to estimate π.

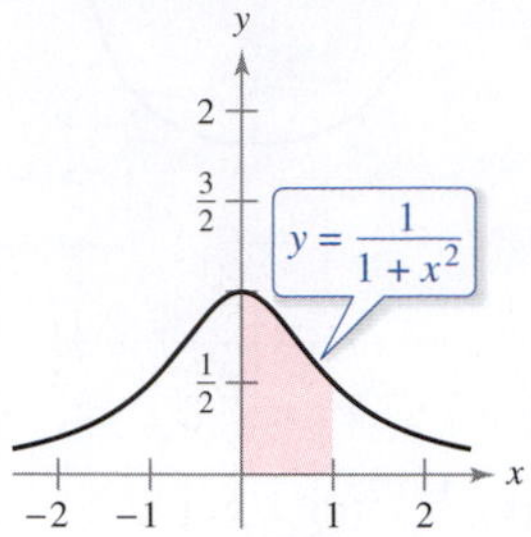

80. Vertical Motion An object is projected upward from ground level with an initial velocity of 500 feet per second. In this exercise, the goal is to analyze the motion of the object during its upward flight.

(a) If air resistance is neglected, find the velocity of the object as a function of time. Use a graphing utility to graph this function.

(b) Use the result of part (a) to find the position function and determine the maximum height attained by the object.

(c) If the air resistance is proportional to the square of the velocity, you obtain the equation

$$\frac{dv}{dt} = -(32 + kv^2)$$

where -32 feet per second per second is the acceleration due to gravity and k is a constant. Find the velocity as a function of time by solving the equation

$$\int \frac{dv}{32+kv^2} = -\int dt.$$

(d) Use a graphing utility to graph the velocity function $v(t)$ in part (c) for $k = 0.001$. Use the graph to approximate the time t_0 at which the object reaches its maximum height.

(e) Use the integration capabilities of a graphing utility to approximate the integral

$$\int_0^{t_0} v(t)\,dt$$

where $v(t)$ and t_0 are those found in part (d). This is the approximation of the maximum height of the object.

(f) Explain the difference between the results in parts (b) and (e).

FOR FURTHER INFORMATION For more information on this topic, see the article "What Goes Up Must Come Down; Will Air Resistance Make It Return Sooner, or Later?" by John Lekner in *Mathematics Magazine*. To view this article, go to *MathArticles.com*.

5.8 Hyperbolic Functions

- **Develop properties of hyperbolic functions.**
- **Differentiate and integrate hyperbolic functions.**
- **Develop properties of inverse hyperbolic functions.**
- **Differentiate and integrate functions involving inverse hyperbolic functions.**

JOHANN HEINRICH LAMBERT (1728–1777)

The first person to publish a comprehensive study on hyperbolic functions was Johann Heinrich Lambert, a Swiss-German mathematician and colleague of Euler.

See LarsonCalculus.com to read more of this biography.

Hyperbolic Functions

In this section, you will look briefly at a special class of exponential functions called **hyperbolic functions.** The name *hyperbolic function* arose from comparison of the area of a semicircular region, as shown in Figure 5.29, with the area of a region under a hyperbola, as shown in Figure 5.30.

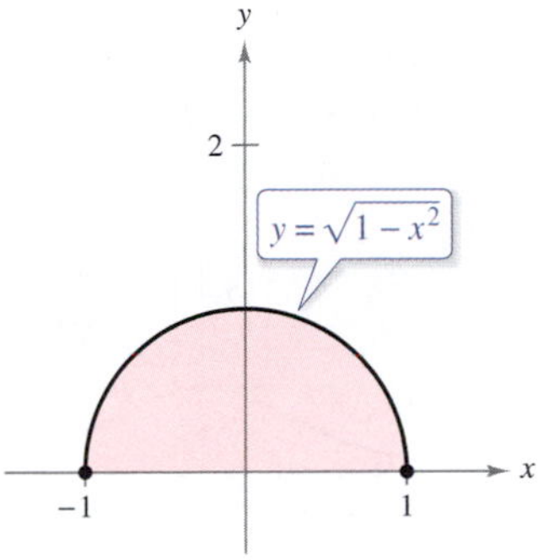

Circle: $x^2 + y^2 = 1$

Figure 5.29

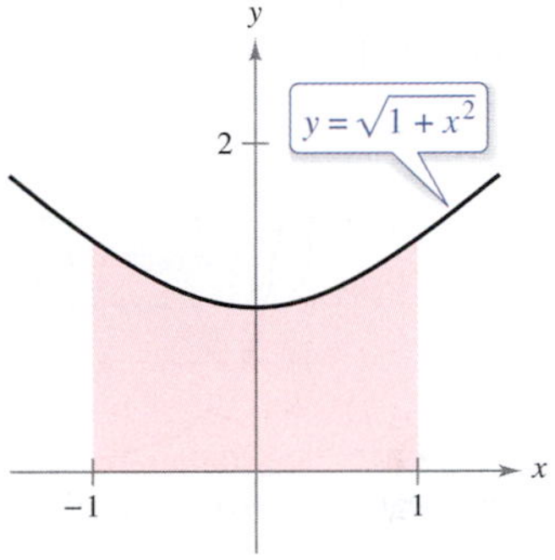

Hyperbola: $-x^2 + y^2 = 1$

Figure 5.30

The integral for the semicircular region involves an inverse trigonometric (circular) function:

$$\int_{-1}^{1} \sqrt{1 - x^2}\,dx = \frac{1}{2}\left[x\sqrt{1 - x^2} + \arcsin x\right]_{-1}^{1} = \frac{\pi}{2} \approx 1.571.$$

The integral for the hyperbolic region involves an inverse hyperbolic function:

$$\int_{-1}^{1} \sqrt{1 + x^2}\,dx = \frac{1}{2}\left[x\sqrt{1 + x^2} + \sinh^{-1} x\right]_{-1}^{1} \approx 2.296.$$

This is only one of many ways in which the hyperbolic functions are similar to the trigonometric functions.

REMARK The notation $\sinh x$ is read as "the hyperbolic sine of x," $\cosh x$ as "the hyperbolic cosine of x," and so on.

Definitions of the Hyperbolic Functions

$$\sinh x = \frac{e^x - e^{-x}}{2} \qquad \operatorname{csch} x = \frac{1}{\sinh x}, \quad x \neq 0$$

$$\cosh x = \frac{e^x + e^{-x}}{2} \qquad \operatorname{sech} x = \frac{1}{\cosh x}$$

$$\tanh x = \frac{\sinh x}{\cosh x} \qquad \coth x = \frac{1}{\tanh x}, \quad x \neq 0$$

FOR FURTHER INFORMATION For more information on the development of hyperbolic functions, see the article "An Introduction to Hyperbolic Functions in Elementary Calculus" by Jerome Rosenthal in *Mathematics Teacher.* To view this article, go to *MathArticles.com.*

The graphs of the six hyperbolic functions and their domains and ranges are shown in Figure 5.31. Note that the graph of $\sinh x$ can be obtained by adding the corresponding y-coordinates of the exponential functions $f(x) = \frac{1}{2}e^x$ and $g(x) = -\frac{1}{2}e^{-x}$. Likewise, the graph of $\cosh x$ can be obtained by adding the corresponding y-coordinates of the exponential functions $f(x) = \frac{1}{2}e^x$ and $h(x) = \frac{1}{2}e^{-x}$.

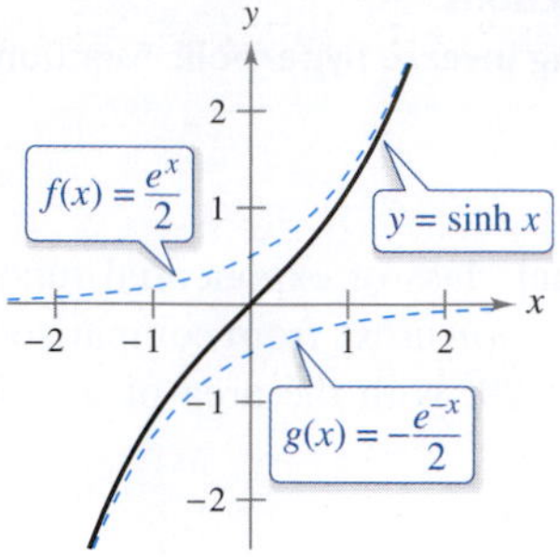

Domain: $(-\infty, \infty)$
Range: $(-\infty, \infty)$

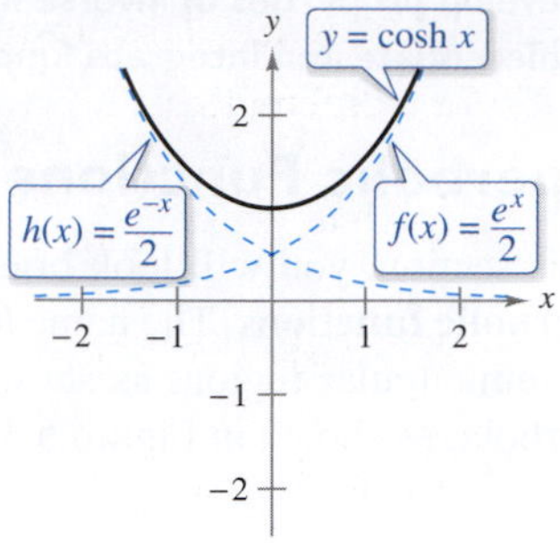

Domain: $(-\infty, \infty)$
Range: $[1, \infty)$

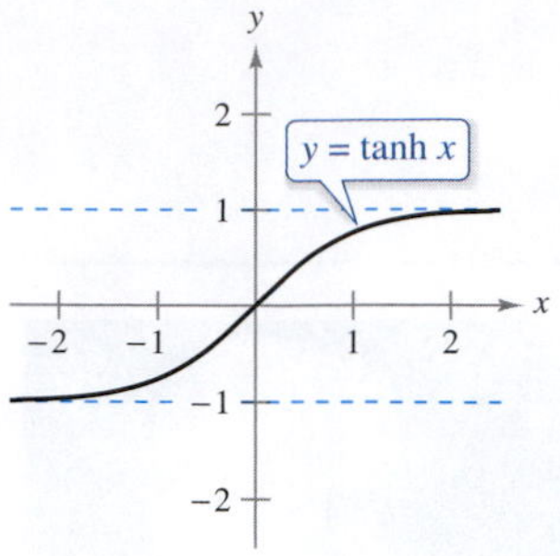

Domain: $(-\infty, \infty)$
Range: $(-1, 1)$

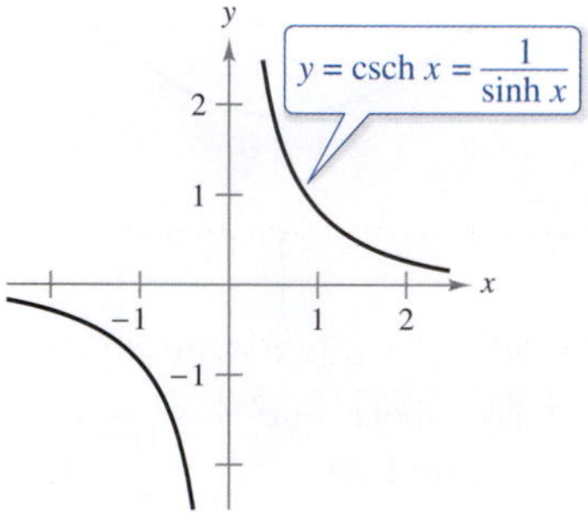

Domain: $(-\infty, 0) \cup (0, \infty)$
Range: $(-\infty, 0) \cup (0, \infty)$

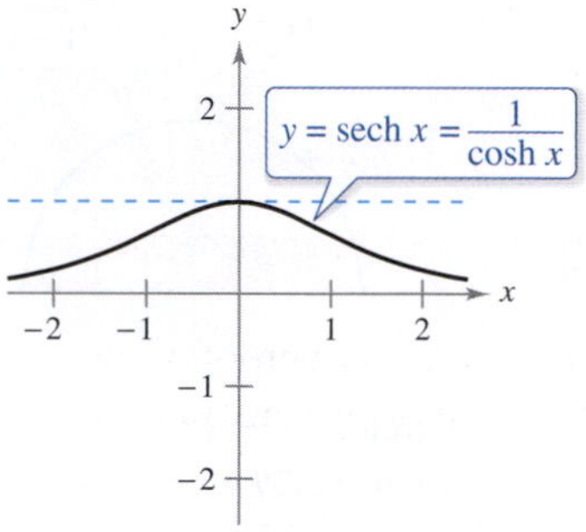

Domain: $(-\infty, \infty)$
Range: $(0, 1]$

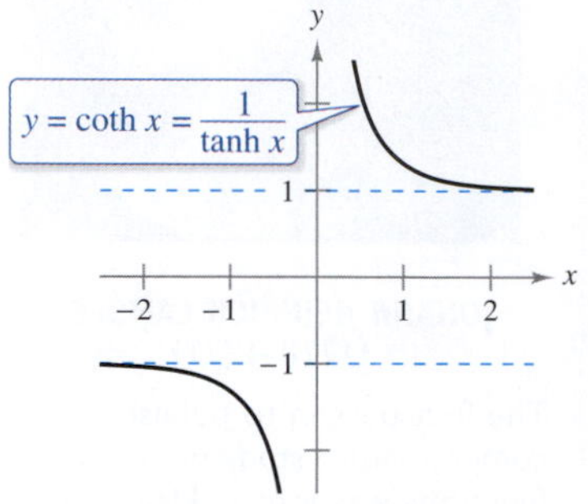

Domain: $(-\infty, 0) \cup (0, \infty)$
Range: $(-\infty, -1) \cup (1, \infty)$

Figure 5.31

Many of the trigonometric identities have corresponding *hyperbolic identities.* For instance,

$$\begin{aligned}\cosh^2 x - \sinh^2 x &= \left(\frac{e^x + e^{-x}}{2}\right)^2 - \left(\frac{e^x - e^{-x}}{2}\right)^2\\ &= \frac{e^{2x} + 2 + e^{-2x}}{4} - \frac{e^{2x} - 2 + e^{-2x}}{4}\\ &= \frac{4}{4}\\ &= 1.\end{aligned}$$

HYPERBOLIC IDENTITIES

$$\cosh^2 x - \sinh^2 x = 1$$
$$\tanh^2 x + \operatorname{sech}^2 x = 1$$
$$\coth^2 x - \operatorname{csch}^2 x = 1$$
$$\sinh^2 x = \frac{-1 + \cosh 2x}{2}$$
$$\sinh 2x = 2 \sinh x \cosh x$$
$$\sinh(x + y) = \sinh x \cosh y + \cosh x \sinh y$$
$$\sinh(x - y) = \sinh x \cosh y - \cosh x \sinh y$$
$$\cosh(x + y) = \cosh x \cosh y + \sinh x \sinh y$$
$$\cosh(x - y) = \cosh x \cosh y - \sinh x \sinh y$$
$$\cosh^2 x = \frac{1 + \cosh 2x}{2}$$
$$\cosh 2x = \cosh^2 x + \sinh^2 x$$

FOR FURTHER INFORMATION To understand geometrically the relationship between the hyperbolic and exponential functions, see the article "A Short Proof Linking the Hyperbolic and Exponential Functions" by Michael J. Seery in *The AMATYC Review.*

Differentiation and Integration of Hyperbolic Functions

Because the hyperbolic functions are written in terms of e^x and e^{-x}, you can easily derive rules for their derivatives. The next theorem lists these derivatives with the corresponding integration rules.

THEOREM 5.18 Derivatives and Integrals of Hyperbolic Functions

Let u be a differentiable function of x.

$$\frac{d}{dx}[\sinh u] = (\cosh u)u' \qquad \int \cosh u\,du = \sinh u + C$$

$$\frac{d}{dx}[\cosh u] = (\sinh u)u' \qquad \int \sinh u\,du = \cosh u + C$$

$$\frac{d}{dx}[\tanh u] = (\text{sech}^2\, u)u' \qquad \int \text{sech}^2\, u\,du = \tanh u + C$$

$$\frac{d}{dx}[\coth u] = -(\text{csch}^2\, u)u' \qquad \int \text{csch}^2\, u\,du = -\coth u + C$$

$$\frac{d}{dx}[\text{sech}\, u] = -(\text{sech}\, u \tanh u)u' \qquad \int \text{sech}\, u \tanh u\,du = -\text{sech}\, u + C$$

$$\frac{d}{dx}[\text{csch}\, u] = -(\text{csch}\, u \coth u)u' \qquad \int \text{csch}\, u \coth u\,du = -\text{csch}\, u + C$$

Proof Here is a proof of two of the differentiation rules. (You are asked to prove some of the other differentiation rules in Exercises 103–105.)

$$\begin{aligned}\frac{d}{dx}[\sinh x] &= \frac{d}{dx}\left[\frac{e^x - e^{-x}}{2}\right]\\ &= \frac{e^x + e^{-x}}{2}\\ &= \cosh x\end{aligned}$$

$$\begin{aligned}\frac{d}{dx}[\tanh x] &= \frac{d}{dx}\left[\frac{\sinh x}{\cosh x}\right]\\ &= \frac{\cosh x(\cosh x) - \sinh x(\sinh x)}{\cosh^2 x}\\ &= \frac{1}{\cosh^2 x}\\ &= \text{sech}^2\, x\end{aligned}$$

See LarsonCalculus.com for Bruce Edwards's video of this proof.

EXAMPLE 1 Differentiation of Hyperbolic Functions

a. $\frac{d}{dx}[\sinh(x^2 - 3)] = 2x\cosh(x^2 - 3)$

b. $\frac{d}{dx}[\ln(\cosh x)] = \frac{\sinh x}{\cosh x} = \tanh x$

c. $\frac{d}{dx}[x \sinh x - \cosh x] = x\cosh x + \sinh x - \sinh x = x\cosh x$

d. $\frac{d}{dx}[(x - 1)\cosh x - \sinh x] = (x - 1)\sinh x + \cosh x - \cosh x = (x - 1)\sinh x$

EXAMPLE 2 Finding Relative Extrema

Find the relative extrema of

$$f(x) = (x - 1)\cosh x - \sinh x.$$

Solution Using the result of Example 1(d), set the first derivative of f equal to 0.

$$(x - 1)\sinh x = 0$$

So, the critical numbers are $x = 1$ and $x = 0$. Using the Second Derivative Test, you can verify that the point $(0, -1)$ yields a relative maximum and the point $(1, -\sinh 1)$ yields a relative minimum, as shown in Figure 5.32. Try using a graphing utility to confirm this result. If your graphing utility does not have hyperbolic functions, you can use exponential functions, as shown.

$$f(x) = (x - 1)\left(\frac{1}{2}\right)(e^x + e^{-x}) - \frac{1}{2}(e^x - e^{-x})$$
$$= \frac{1}{2}(xe^x + xe^{-x} - e^x - e^{-x} - e^x + e^{-x})$$
$$= \frac{1}{2}(xe^x + xe^{-x} - 2e^x)$$

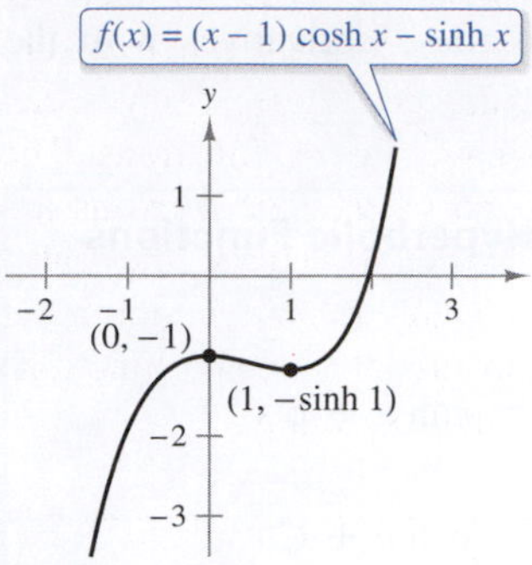

$f''(0) < 0$, so $(0, -1)$ is a relative maximum. $f''(1) > 0$, so $(1, -\sinh 1)$ is a relative minimum.
Figure 5.32

When a uniform flexible cable, such as a telephone wire, is suspended from two points, it takes the shape of a *catenary*, as discussed in Example 3.

EXAMPLE 3 Hanging Power Cables

See LarsonCalculus.com for an interactive version of this type of example.

Power cables are suspended between two towers, forming the catenary shown in Figure 5.33. The equation for this catenary is

$$y = a\cosh\frac{x}{a}.$$

The distance between the two towers is $2b$. Find the slope of the catenary at the point where the cable meets the right-hand tower.

Solution Differentiating produces

$$y' = a\left(\frac{1}{a}\right)\sinh\frac{x}{a} = \sinh\frac{x}{a}.$$

At the point $(b, a\cosh(b/a))$, the slope (from the left) is $m = \sinh\frac{b}{a}$.

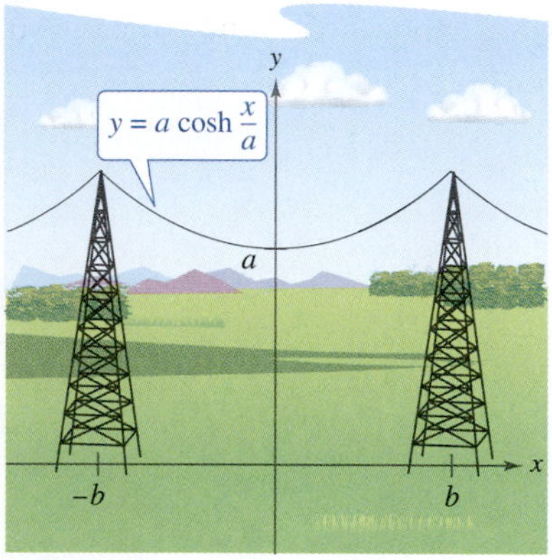

Catenary
Figure 5.33

FOR FURTHER INFORMATION In Example 3, the cable is a catenary between two supports at the same height. To learn about the shape of a cable hanging between supports of different heights, see the article "Reexamining the Catenary" by Paul Cella in *The College Mathematics Journal*. To view this article, go to *MathArticles.com*.

EXAMPLE 4 Integrating a Hyperbolic Function

Find $\int \cosh 2x \sinh^2 2x\, dx$.

Solution

$$\int \cosh 2x \sinh^2 2x\, dx = \frac{1}{2}\int (\sinh 2x)^2(2\cosh 2x)\, dx \qquad u = \sinh 2x$$
$$= \frac{1}{2}\left[\frac{(\sinh 2x)^3}{3}\right] + C$$
$$= \frac{\sinh^3 2x}{6} + C$$

Inverse Hyperbolic Functions

Unlike trigonometric functions, hyperbolic functions are not periodic. In fact, by looking back at Figure 5.31, you can see that four of the six hyperbolic functions are actually one-to-one (the hyperbolic sine, tangent, cosecant, and cotangent). So, you can apply Theorem 5.7 to conclude that these four functions have inverse functions. The other two (the hyperbolic cosine and secant) are one-to-one when their domains are restricted to the positive real numbers, and for this restricted domain they also have inverse functions. Because the hyperbolic functions are defined in terms of exponential functions, it is not surprising to find that the inverse hyperbolic functions can be written in terms of logarithmic functions, as shown in Theorem 5.19.

THEOREM 5.19 Inverse Hyperbolic Functions

Function	Domain
$\sinh^{-1} x = \ln\left(x + \sqrt{x^2 + 1}\right)$	$(-\infty, \infty)$
$\cosh^{-1} x = \ln\left(x + \sqrt{x^2 - 1}\right)$	$[1, \infty)$
$\tanh^{-1} x = \dfrac{1}{2} \ln \dfrac{1 + x}{1 - x}$	$(-1, 1)$
$\coth^{-1} x = \dfrac{1}{2} \ln \dfrac{x + 1}{x - 1}$	$(-\infty, -1) \cup (1, \infty)$
$\operatorname{sech}^{-1} x = \ln \dfrac{1 + \sqrt{1 - x^2}}{x}$	$(0, 1]$
$\operatorname{csch}^{-1} x = \ln\left(\dfrac{1}{x} + \dfrac{\sqrt{1 + x^2}}{\lvert x \rvert}\right)$	$(-\infty, 0) \cup (0, \infty)$

Proof The proof of this theorem is a straightforward application of the properties of the exponential and logarithmic functions. For example, for

$$f(x) = \sinh x = \frac{e^x - e^{-x}}{2}$$

and

$$g(x) = \ln\left(x + \sqrt{x^2 + 1}\right)$$

you can show that

$$f(g(x)) = x \quad \text{and} \quad g(f(x)) = x$$

which implies that g is the inverse function of f.

See LarsonCalculus.com for Bruce Edwards's video of this proof.

TECHNOLOGY You can use a graphing utility to confirm graphically the results of Theorem 5.19. For instance, graph the following functions.

$y_1 = \tanh x$ Hyperbolic tangent

$y_2 = \dfrac{e^x - e^{-x}}{e^x + e^{-x}}$ Definition of hyperbolic tangent

$y_3 = \tanh^{-1} x$ Inverse hyperbolic tangent

$y_4 = \dfrac{1}{2} \ln \dfrac{1 + x}{1 - x}$ Definition of inverse hyperbolic tangent

The resulting display is shown in Figure 5.34. As you watch the graphs being traced out, notice that $y_1 = y_2$ and $y_3 = y_4$. Also notice that the graph of y_1 is the reflection of the graph of y_3 in the line $y = x$.

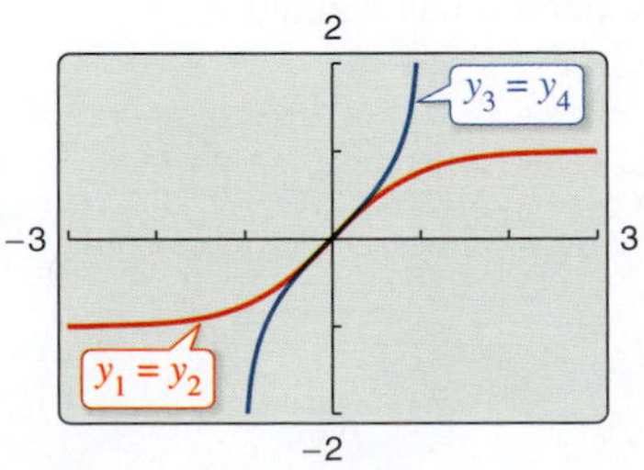

Graphs of the hyperbolic tangent function and the inverse hyperbolic tangent function
Figure 5.34

The graphs of the inverse hyperbolic functions are shown in Figure 5.35.

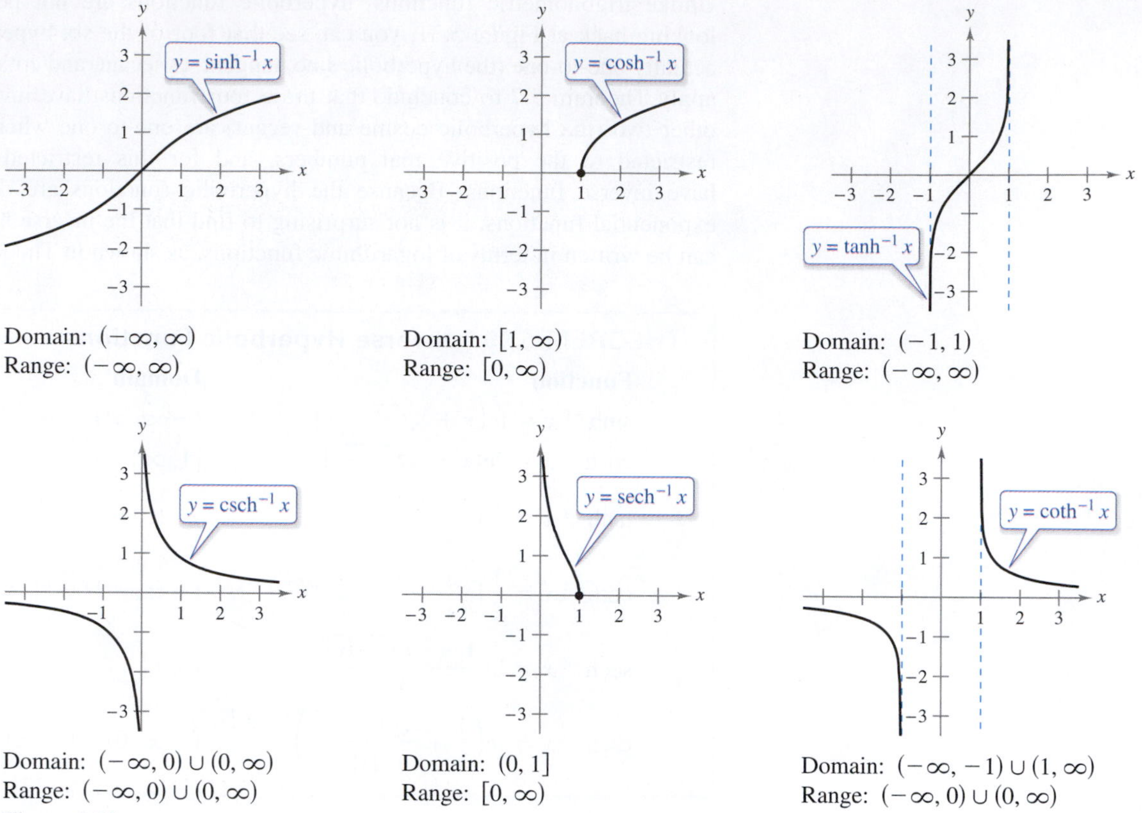

Figure 5.35

The inverse hyperbolic secant can be used to define a curve called a *tractrix* or *pursuit curve*, as discussed in Example 5.

EXAMPLE 5 A Tractrix

A person is holding a rope that is tied to a boat, as shown in Figure 5.36. As the person walks along the dock, the boat travels along a **tractrix,** given by the equation

$$y = a \operatorname{sech}^{-1} \frac{x}{a} - \sqrt{a^2 - x^2}$$

where a is the length of the rope. For $a = 20$ feet, find the distance the person must walk to bring the boat to a position 5 feet from the dock.

Solution In Figure 5.36, notice that the distance the person has walked is

$$\begin{aligned} y_1 &= y + \sqrt{20^2 - x^2} \\ &= \left(20 \operatorname{sech}^{-1} \frac{x}{20} - \sqrt{20^2 - x^2}\right) + \sqrt{20^2 - x^2} \\ &= 20 \operatorname{sech}^{-1} \frac{x}{20}. \end{aligned}$$

When $x = 5$, this distance is

$$y_1 = 20 \operatorname{sech}^{-1} \frac{5}{20} = 20 \ln \frac{1 + \sqrt{1 - (1/4)^2}}{1/4} = 20 \ln(4 + \sqrt{15}) \approx 41.27 \text{ feet.}$$

So, the person must walk about 41.27 feet to bring the boat to a position 5 feet from the dock.

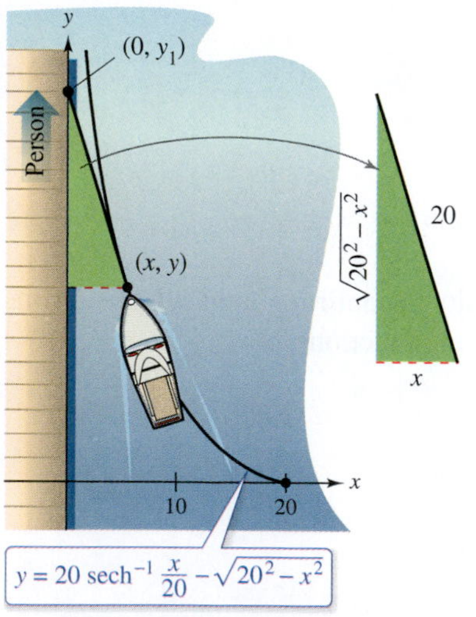

A person must walk about 41.27 feet to bring the boat to a position 5 feet from the dock.

Figure 5.36

Inverse Hyperbolic Functions: Differentiation and Integration

The derivatives of the inverse hyperbolic functions, which resemble the derivatives of the inverse trigonometric functions, are listed in Theorem 5.20 with the corresponding integration formulas (in logarithmic form). You can verify each of these formulas by applying the logarithmic definitions of the inverse hyperbolic functions. (See Exercises 106–108.)

THEOREM 5.20 Differentiation and Integration Involving Inverse Hyperbolic Functions

Let u be a differentiable function of x.

$$\frac{d}{dx}[\sinh^{-1} u] = \frac{u'}{\sqrt{u^2+1}} \qquad \frac{d}{dx}[\cosh^{-1} u] = \frac{u'}{\sqrt{u^2-1}}$$

$$\frac{d}{dx}[\tanh^{-1} u] = \frac{u'}{1-u^2} \qquad \frac{d}{dx}[\coth^{-1} u] = \frac{u'}{1-u^2}$$

$$\frac{d}{dx}[\operatorname{sech}^{-1} u] = \frac{-u'}{u\sqrt{1-u^2}} \qquad \frac{d}{dx}[\operatorname{csch}^{-1} u] = \frac{-u'}{|u|\sqrt{1+u^2}}$$

$$\int \frac{du}{\sqrt{u^2 \pm a^2}} = \ln\left(u + \sqrt{u^2 \pm a^2}\right) + C$$

$$\int \frac{du}{a^2-u^2} = \frac{1}{2a}\ln\left|\frac{a+u}{a-u}\right| + C$$

$$\int \frac{du}{u\sqrt{a^2 \pm u^2}} = -\frac{1}{a}\ln\frac{a+\sqrt{a^2 \pm u^2}}{|u|} + C$$

EXAMPLE 6 Differentiation of Inverse Hyperbolic Functions

a. $$\frac{d}{dx}[\sinh^{-1}(2x)] = \frac{2}{\sqrt{(2x)^2+1}} = \frac{2}{\sqrt{4x^2+1}}$$

b. $$\frac{d}{dx}[\tanh^{-1}(x^3)] = \frac{3x^2}{1-(x^3)^2} = \frac{3x^2}{1-x^6}$$

EXAMPLE 7 Integration Using Inverse Hyperbolic Functions

a. $$\int \frac{dx}{x\sqrt{4-9x^2}} = \int \frac{3\,dx}{(3x)\sqrt{4-9x^2}} \qquad \int \frac{du}{u\sqrt{a^2-u^2}}$$

$$= -\frac{1}{2}\ln\frac{2+\sqrt{4-9x^2}}{|3x|} + C \qquad -\frac{1}{a}\ln\frac{a+\sqrt{a^2-u^2}}{|u|} + C$$

REMARK Let $a = 2$ and $u = 3x$.

b. $$\int \frac{dx}{5-4x^2} = \frac{1}{2}\int \frac{2\,dx}{(\sqrt{5})^2-(2x)^2} \qquad \int \frac{du}{a^2-u^2}$$

$$= \frac{1}{2}\left(\frac{1}{2\sqrt{5}}\ln\left|\frac{\sqrt{5}+2x}{\sqrt{5}-2x}\right|\right) + C \qquad \frac{1}{2a}\ln\left|\frac{a+u}{a-u}\right| + C$$

$$= \frac{1}{4\sqrt{5}}\ln\left|\frac{\sqrt{5}+2x}{\sqrt{5}-2x}\right| + C$$

REMARK Let $a = \sqrt{5}$ and $u = 2x$.

5.8 Exercises

See CalcChat.com for tutorial help and worked-out solutions to odd-numbered exercises.

Evaluating a Function In Exercises 1–6, evaluate the function. If the value is not a rational number, round your answer to three decimal places.

1. (a) $\sinh 3$ (b) $\tanh(-2)$
2. (a) $\cosh 0$ (b) $\operatorname{sech} 1$
3. (a) $\operatorname{csch}(\ln 2)$ (b) $\coth(\ln 5)$
4. (a) $\sinh^{-1} 0$ (b) $\tanh^{-1} 0$
5. (a) $\cosh^{-1} 2$ (b) $\operatorname{sech}^{-1} \frac{2}{3}$
6. (a) $\operatorname{csch}^{-1} 2$ (b) $\coth^{-1} 3$

Verifying an Identity In Exercises 7–14, verify the identity.

7. $\tanh^2 x + \operatorname{sech}^2 x = 1$
8. $\coth^2 x - \operatorname{csch}^2 x = 1$
9. $\cosh^2 x = \dfrac{1 + \cosh 2x}{2}$
10. $\sinh^2 x = \dfrac{-1 + \cosh 2x}{2}$
11. $\sinh 2x = 2 \sinh x \cosh x$
12. $e^{2x} = \sinh 2x + \cosh 2x$
13. $\sinh(x + y) = \sinh x \cosh y + \cosh x \sinh y$
14. $\cosh x + \cosh y = 2 \cosh \dfrac{x + y}{2} \cosh \dfrac{x - y}{2}$

Finding Values of Hyperbolic Functions In Exercises 15 and 16, use the value of the given hyperbolic function to find the values of the other hyperbolic functions at x.

15. $\sinh x = \dfrac{3}{2}$
16. $\tanh x = \dfrac{1}{2}$

Finding a Limit In Exercises 17–22, find the limit.

17. $\lim_{x\to\infty} \sinh x$
18. $\lim_{x\to-\infty} \tanh x$
19. $\lim_{x\to\infty} \operatorname{sech} x$
20. $\lim_{x\to-\infty} \operatorname{csch} x$
21. $\lim_{x\to 0} \dfrac{\sinh x}{x}$
22. $\lim_{x\to 0^-} \coth x$

Finding a Derivative In Exercises 23–32, find the derivative of the function.

23. $f(x) = \sinh 3x$
24. $f(x) = \cosh(8x + 1)$
25. $y = \operatorname{sech}(5x^2)$
26. $f(x) = \tanh(4x^2 + 3x)$
27. $f(x) = \ln(\sinh x)$
28. $y = \ln\left(\tanh \dfrac{x}{2}\right)$
29. $h(x) = \dfrac{1}{4} \sinh 2x - \dfrac{x}{2}$
30. $y = x \cosh x - \sinh x$
31. $f(t) = \arctan(\sinh t)$
32. $g(x) = \operatorname{sech}^2 3x$

Finding an Equation of a Tangent Line In Exercises 33–36, find an equation of the tangent line to the graph of the function at the given point.

33. $y = \sinh(1 - x^2)$, $(1, 0)$
34. $y = x^{\cosh x}$, $(1, 1)$
35. $y = (\cosh x - \sinh x)^2$, $(0, 1)$
36. $y = e^{\sinh x}$, $(0, 1)$

Finding Relative Extrema In Exercises 37–40, find any relative extrema of the function. Use a graphing utility to confirm your result.

37. $f(x) = \sin x \sinh x - \cos x \cosh x$, $-4 \le x \le 4$
38. $f(x) = x \sinh(x - 1) - \cosh(x - 1)$
39. $g(x) = x \operatorname{sech} x$
40. $h(x) = 2 \tanh x - x$

Catenary In Exercises 41 and 42, a model for a power cable suspended between two towers is given. (a) Graph the model, (b) find the heights of the cable at the towers and at the midpoint between the towers, and (c) find the slope of the model at the point where the cable meets the right-hand tower.

41. $y = 10 + 15 \cosh \dfrac{x}{15}$, $-15 \le x \le 15$
42. $y = 18 + 25 \cosh \dfrac{x}{25}$, $-25 \le x \le 25$

Finding an Indefinite Integral In Exercises 43–54, find the indefinite integral.

43. $\int \cosh 2x \, dx$
44. $\int \operatorname{sech}^2(3x) \, dx$
45. $\int \sinh(1 - 2x) \, dx$
46. $\int \dfrac{\cosh \sqrt{x}}{\sqrt{x}} \, dx$
47. $\int \cosh^2(x - 1) \sinh(x - 1) \, dx$
48. $\int \dfrac{\sinh x}{1 + \sinh^2 x} \, dx$
49. $\int \dfrac{\cosh x}{\sinh x} \, dx$
50. $\int \operatorname{sech}^2(2x - 1) \, dx$
51. $\int x \operatorname{csch}^2 \dfrac{x^2}{2} \, dx$
52. $\int \operatorname{sech}^3 x \tanh x \, dx$
53. $\int \dfrac{\operatorname{csch}(1/x) \coth(1/x)}{x^2} \, dx$
54. $\int \dfrac{\cosh x}{\sqrt{9 - \sinh^2 x}} \, dx$

Evaluating a Definite Integral In Exercises 55–60, evaluate the integral.

55. $\int_0^{\ln 2} \tanh x \, dx$
56. $\int_0^1 \cosh^2 x \, dx$
57. $\int_0^4 \dfrac{1}{25 - x^2} \, dx$
58. $\int_0^4 \dfrac{1}{\sqrt{25 - x^2}} \, dx$
59. $\int_0^{\sqrt{2}/4} \dfrac{2}{\sqrt{1 - 4x^2}} \, dx$
60. $\int_0^{\ln 2} 2e^{-x} \cosh x \, dx$

WRITING ABOUT CONCEPTS

61. Comparing Functions Discuss several ways in which the hyperbolic functions are similar to the trigonometric functions.

62. Hyperbolic Functions Which hyperbolic functions take on only positive values? Which hyperbolic functions are increasing on their domains?

63. Comparing Derivative Formulas Which hyperbolic derivative formulas differ from their trigonometric counterparts by a minus sign?

64. HOW DO YOU SEE IT? Use the graphs of f and g shown in the figures to answer the following.

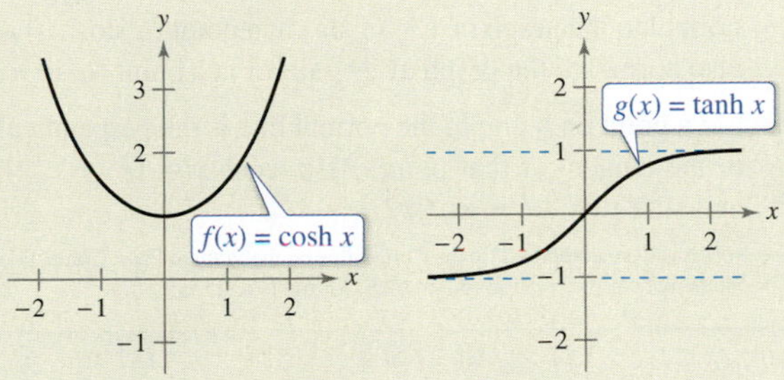

(a) Identify the open interval(s) on which the graphs of f and g are increasing or decreasing.

(b) Identify the open interval(s) on which the graphs of f and g are concave upward or concave downward.

Finding a Derivative In Exercises 65–74, find the derivative of the function.

65. $y = \cosh^{-1}(3x)$

66. $y = \tanh^{-1}\dfrac{x}{2}$

67. $y = \tanh^{-1}\sqrt{x}$

68. $f(x) = \coth^{-1}(x^2)$

69. $y = \sinh^{-1}(\tan x)$

70. $y = \tanh^{-1}(\sin 2x)$

71. $y = (\operatorname{csch}^{-1} x)^2$

72. $y = \operatorname{sech}^{-1}(\cos 2x), \quad 0 < x < \pi/4$

73. $y = 2x\sinh^{-1}(2x) - \sqrt{1 + 4x^2}$

74. $y = x\tanh^{-1} x + \ln\sqrt{1 - x^2}$

Finding an Indefinite Integral In Exercises 75–82, find the indefinite integral using the formulas from Theorem 5.20.

75. $\displaystyle\int \frac{1}{3 - 9x^2}\,dx$

76. $\displaystyle\int \frac{1}{2x\sqrt{1 - 4x^2}}\,dx$

77. $\displaystyle\int \frac{1}{\sqrt{1 + e^{2x}}}\,dx$

78. $\displaystyle\int \frac{x}{9 - x^4}\,dx$

79. $\displaystyle\int \frac{1}{\sqrt{x}\sqrt{1 + x}}\,dx$

80. $\displaystyle\int \frac{\sqrt{x}}{\sqrt{1 + x^3}}\,dx$

81. $\displaystyle\int \frac{-1}{4x - x^2}\,dx$

82. $\displaystyle\int \frac{dx}{(x + 2)\sqrt{x^2 + 4x + 8}}$

Evaluating a Definite Integral In Exercises 83–86, evaluate the definite integral using the formulas from Theorem 5.20.

83. $\displaystyle\int_3^7 \frac{1}{\sqrt{x^2 - 4}}\,dx$

84. $\displaystyle\int_1^3 \frac{1}{x\sqrt{4 + x^2}}\,dx$

85. $\displaystyle\int_{-1}^1 \frac{1}{16 - 9x^2}\,dx$

86. $\displaystyle\int_0^1 \frac{1}{\sqrt{25x^2 + 1}}\,dx$

Differential Equation In Exercises 87–90, solve the differential equation.

87. $\dfrac{dy}{dx} = \dfrac{1}{\sqrt{80 + 8x - 16x^2}}$

88. $\dfrac{dy}{dx} = \dfrac{1}{(x - 1)\sqrt{-4x^2 + 8x - 1}}$

89. $\dfrac{dy}{dx} = \dfrac{x^3 - 21x}{5 + 4x - x^2}$

90. $\dfrac{dy}{dx} = \dfrac{1 - 2x}{4x - x^2}$

Area In Exercises 91–94, find the area of the region.

91. $y = \operatorname{sech}\dfrac{x}{2}$

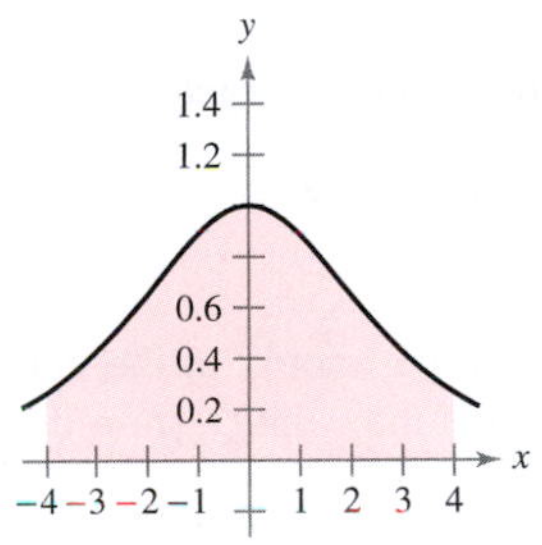

92. $y = \tanh 2x$

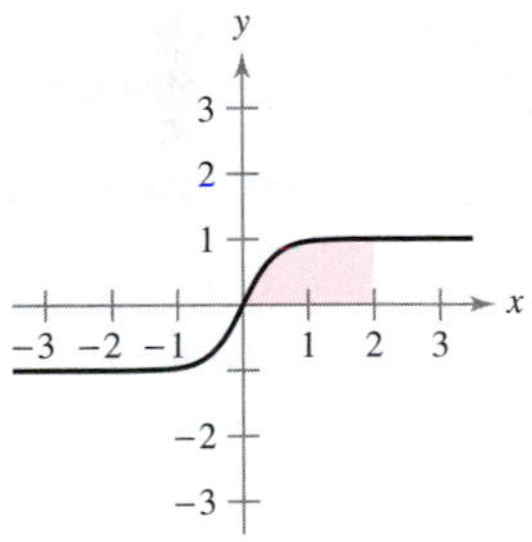

93. $y = \dfrac{5x}{\sqrt{x^4 + 1}}$

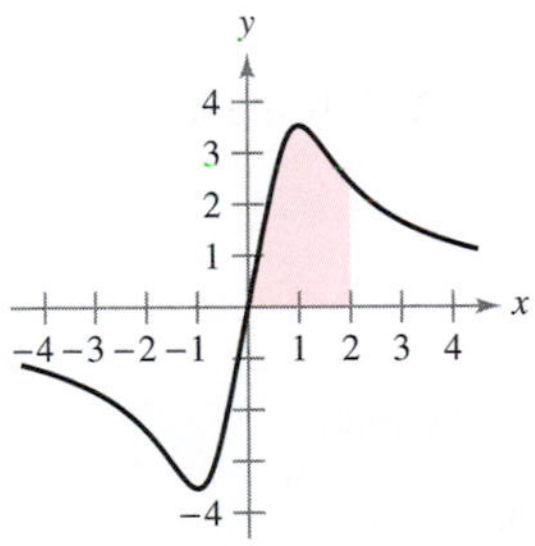

94. $y = \dfrac{6}{\sqrt{x^2 - 4}}$

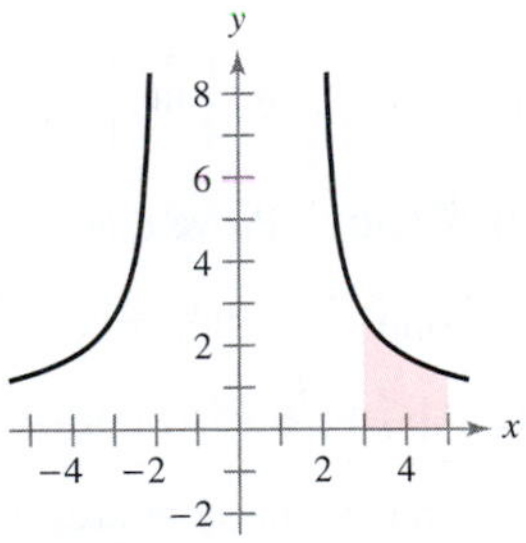

95. Chemical Reactions Chemicals A and B combine in a 3-to-1 ratio to form a compound. The amount of compound x being produced at any time t is proportional to the unchanged amounts of A and B remaining in the solution. So, when 3 kilograms of A is mixed with 2 kilograms of B, you have

$$\frac{dx}{dt} = k\left(3 - \frac{3x}{4}\right)\left(2 - \frac{x}{4}\right) = \frac{3k}{16}(x^2 - 12x + 32).$$

One kilogram of the compound is formed after 10 minutes. Find the amount formed after 20 minutes by solving the equation

$$\int \frac{3k}{16}\,dt = \int \frac{dx}{x^2 - 12x + 32}.$$

96. Vertical Motion An object is dropped from a height of 400 feet.

(a) Find the velocity of the object as a function of time (neglect air resistance on the object).

(b) Use the result in part (a) to find the position function.

(c) If the air resistance is proportional to the square of the velocity, then $dv/dt = -32 + kv^2$, where -32 feet per second per second is the acceleration due to gravity and k is a constant. Show that the velocity v as a function of time is $v(t) = -\sqrt{32/k}\tanh\left(\sqrt{32k}\,t\right)$ by performing $\int dv/(32 - kv^2) = -\int dt$ and simplifying the result.

(d) Use the result of part (c) to find $\lim_{t \to \infty} v(t)$ and give its interpretation.

(e) Integrate the velocity function in part (c) and find the position s of the object as a function of t. Use a graphing utility to graph the position function when $k = 0.01$ and the position function in part (b) in the same viewing window. Estimate the additional time required for the object to reach ground level when air resistance is not neglected.

(f) Give a written description of what you believe would happen if k were increased. Then test your assertion with a particular value of k.

97. Tractrix **Consider the equation of the tractrix**

$$y = a\operatorname{sech}^{-1}(x/a) - \sqrt{a^2 - x^2}, \quad a > 0.$$

(a) Find dy/dx.

(b) Let L be the tangent line to the tractrix at the point P. When L intersects the y-axis at the point Q, show that the distance between P and Q is a.

98. Tractrix Show that the boat in Example 5 is always pointing toward the person.

99. Proof Prove that

$$\tanh^{-1} x = \frac{1}{2}\ln\left(\frac{1+x}{1-x}\right), \quad -1 < x < 1.$$

100. Proof Prove that

$$\sinh^{-1} t = \ln\left(t + \sqrt{t^2 + 1}\right).$$

101. Using a Right Triangle Show that

$$\arctan(\sinh x) = \arcsin(\tanh x).$$

102. Integration Let $x > 0$ and $b > 0$. Show that

$$\int_{-b}^{b} e^{xt}\,dt = \frac{2\sinh bx}{x}.$$

Proof **In Exercises 103–105, prove the differentiation formula.**

103. $\dfrac{d}{dx}[\cosh x] = \sinh x$

104. $\dfrac{d}{dx}[\coth x] = -\operatorname{csch}^2 x$

105. $\dfrac{d}{dx}[\operatorname{sech} x] = -\operatorname{sech} x \tanh x$

Verifying a Differentiation Rule **In Exercises 106–108, verify the differentiation formula.**

106. $\dfrac{d}{dx}[\cosh^{-1} x] = \dfrac{1}{\sqrt{x^2 - 1}}$

107. $\dfrac{d}{dx}[\sinh^{-1} x] = \dfrac{1}{\sqrt{x^2 + 1}}$

108. $\dfrac{d}{dx}[\operatorname{sech}^{-1} x] = \dfrac{-1}{x\sqrt{1 - x^2}}$

PUTNAM EXAM CHALLENGE

109. From the vertex $(0, c)$ of the catenary $y = c\cosh(x/c)$ a line L is drawn perpendicular to the tangent to the catenary at point P. Prove that the length of L intercepted by the axes is equal to the ordinate y of the point P.

110. Prove or disprove: there is at least one straight line normal to the graph of $y = \cosh x$ at a point $(a, \cosh a)$ and also normal to the graph of $y = \sinh x$ at a point $(c, \sinh c)$.

[At a point on a graph, the normal line is the perpendicular to the tangent at that point. Also, $\cosh x = (e^x + e^{-x})/2$ and $\sinh x = (e^x - e^{-x})/2$.]

These problems were composed by the Committee on the Putnam Prize Competition.

SECTION PROJECT

St. Louis Arch

The Gateway Arch in St. Louis, Missouri, was constructed using the hyperbolic cosine function. The equation used for construction was

$$y = 693.8597 - 68.7672\cosh 0.0100333x,$$
$$-299.2239 \le x \le 299.2239$$

where x and y are measured in feet. Cross sections of the arch are equilateral triangles, and (x, y) traces the path of the centers of mass of the cross-sectional triangles. For each value of x, the area of the cross-sectional triangle is

$$A = 125.1406\cosh 0.0100333x.$$

(Source: Owner's Manual for the Gateway Arch, Saint Louis, MO, by William Thayer)

(a) How high above the ground is the center of the highest triangle? (At ground level, $y = 0$.)

(b) What is the height of the arch? (*Hint:* For an equilateral triangle, $A = \sqrt{3}c^2$, where c is one-half the base of the triangle, and the center of mass of the triangle is located at two-thirds the height of the triangle.)

(c) How wide is the arch at ground level?

Ken Nyborg/Shutterstock.com

Review Exercises

See CalcChat.com for tutorial help and worked-out solutions to odd-numbered exercises.

Sketching a Graph In Exercises 1 and 2, sketch the graph of the function and state its domain.

1. $f(x) = \ln x - 3$
2. $f(x) = \ln(x + 3)$

Expanding a Logarithmic Expression In Exercises 3 and 4, use the properties of logarithms to expand the logarithmic expression.

3. $\ln \sqrt[5]{\dfrac{4x^2 - 1}{4x^2 + 1}}$
4. $\ln[(x^2 + 1)(x - 1)]$

Condensing a Logarithmic Expression In Exercises 5 and 6, write the expression as the logarithm of a single quantity.

5. $\ln 3 + \frac{1}{3}\ln(4 - x^2) - \ln x$
6. $3[\ln x - 2\ln(x^2 + 1)] + 2\ln 5$

Finding a Derivative In Exercises 7–12, find the derivative of the function.

7. $g(x) = \ln\sqrt{2x}$
8. $f(x) = \ln(3x^2 + 2x)$
9. $f(x) = x\sqrt{\ln x}$
10. $f(x) = [\ln(2x)]^3$
11. $y = \ln\sqrt{\dfrac{x^2 + 4}{x^2 - 4}}$
12. $y = \ln\left(\dfrac{4x}{x - 6}\right)$

Finding an Equation of a Tangent Line In Exercises 13 and 14, find an equation of the tangent line to the graph of the function at the given point.

13. $y = \ln(2 + x) + \dfrac{2}{2 + x}$, $(-1, 2)$
14. $y = 2x^2 + \ln x^2$, $(1, 2)$

Finding an Indefinite Integral In Exercises 15–18, find the indefinite integral.

15. $\displaystyle\int \frac{1}{7x - 2}\,dx$
16. $\displaystyle\int \frac{x^2}{x^3 + 1}\,dx$
17. $\displaystyle\int \frac{\sin x}{1 + \cos x}\,dx$
18. $\displaystyle\int \frac{\ln\sqrt{x}}{x}\,dx$

Evaluating a Definite Integral In Exercises 19–22, evaluate the definite integral.

19. $\displaystyle\int_1^4 \frac{2x + 1}{2x}\,dx$
20. $\displaystyle\int_1^e \frac{\ln x}{x}\,dx$
21. $\displaystyle\int_0^{\pi/3} \sec\theta\,d\theta$
22. $\displaystyle\int_0^{\pi} \tan\frac{\theta}{3}\,d\theta$

Finding an Inverse Function In Exercises 23–28, (a) find the inverse function of f, (b) graph f and f^{-1} on the same set of coordinate axes, (c) verify that $f^{-1}(f(x)) = x$ and $f(f^{-1}(x)) = x$, and (d) state the domains and ranges of f and f^{-1}.

23. $f(x) = \frac{1}{2}x - 3$
24. $f(x) = 5x - 7$
25. $f(x) = \sqrt{x + 1}$
26. $f(x) = x^3 + 2$
27. $f(x) = \sqrt[3]{x + 1}$
28. $f(x) = x^2 - 5$, $x \ge 0$

Evaluating the Derivative of an Inverse Function In Exercises 29–32, verify that f has an inverse. Then use the function f and the given real number a to find $(f^{-1})'(a)$. (*Hint:* Use Theorem 5.9.)

29. $f(x) = x^3 + 2$, $a = -1$
30. $f(x) = x\sqrt{x - 3}$, $a = 4$
31. $f(x) = \tan x$, $-\dfrac{\pi}{4} \le x \le \dfrac{\pi}{4}$, $a = \dfrac{\sqrt{3}}{3}$
32. $f(x) = \cos x$, $0 \le x \le \pi$, $a = 0$

Solving an Exponential or Logarithmic Equation In Exercises 33–36, solve for x accurate to three decimal places.

33. $e^{3x} = 30$
34. $-4 + 3e^{-2x} = 6$
35. $\ln\sqrt{x + 1} = 2$
36. $\ln x + \ln(x - 3) = 0$

Finding a Derivative In Exercises 37–42, find the derivative of the function.

37. $g(t) = t^2e^t$
38. $g(x) = \ln\dfrac{e^x}{1 + e^x}$
39. $y = \sqrt{e^{2x} + e^{-2x}}$
40. $h(z) = e^{-z^2/2}$
41. $g(x) = \dfrac{x^2}{e^x}$
42. $y = 3e^{-3/t}$

Finding an Equation of a Tangent Line In Exercises 43 and 44, find an equation of the tangent line to the graph of the function at the given point.

43. $f(x) = e^{6x}$, $(0, 1)$
44. $f(x) = e^{x-4}$, $(4, 1)$

Implicit Differentiation In Exercises 45 and 46, use implicit differentiation to find dy/dx.

45. $y\ln x + y^2 = 0$
46. $\cos x^2 = xe^y$

Finding an Indefinite Integral In Exercises 47–50, find the indefinite integral.

47. $\displaystyle\int xe^{1-x^2}\,dx$
48. $\displaystyle\int x^2e^{x^3+1}\,dx$
49. $\displaystyle\int \frac{e^{4x} - e^{2x} + 1}{e^x}\,dx$
50. $\displaystyle\int \frac{e^{2x} - e^{-2x}}{e^{2x} + e^{-2x}}\,dx$

Evaluating a Definite Integral In Exercises 51–54, evaluate the definite integral.

51. $\int_0^1 xe^{-3x^2}\,dx$

52. $\int_{1/2}^2 \frac{e^{1/x}}{x^2}\,dx$

53. $\int_1^3 \frac{e^x}{e^x - 1}\,dx$

54. $\int_0^2 \frac{e^{2x}}{e^{2x} + 1}\,dx$

55. Area Find the area of the region bounded by the graphs of

$$y = 2e^{-x},\quad y = 0,\quad x = 0,\quad \text{and} \quad x = 2.$$

56. Depreciation The value V of an item t years after it is purchased is $V = 9000e^{-0.6t}$ for $0 \le t \le 5$.

(a) Use a graphing utility to graph the function.

(b) Find the rates of change of V with respect to t when $t = 1$ and $t = 4$.

(c) Use a graphing utility to graph the tangent lines to the function when $t = 1$ and $t = 4$.

Sketching a Graph In Exercises 57 and 58, sketch the graph of the function by hand.

57. $y = 3^{x/2}$

58. $y = \left(\frac{1}{4}\right)^x$

Finding a Derivative In Exercises 59–64, find the derivative of the function.

59. $f(x) = 3^{x-1}$

60. $f(x) = 5^{3x}$

61. $y = x^{2x+1}$

62. $f(x) = x(4^{-3x})$

63. $g(x) = \log_3 \sqrt{1 - x}$

64. $h(x) = \log_5 \frac{x}{x - 1}$

Finding an Indefinite Integral In Exercises 65 and 66, find the indefinite integral.

65. $\int (x + 1)5^{(x+1)^2}\,dx$

66. $\int \frac{2^{-1/t}}{t^2}\,dt$

67. Climb Rate The time t (in minutes) for a small plane to climb to an altitude of h feet is

$$t = 50 \log_{10} \frac{18{,}000}{18{,}000 - h}$$

where 18,000 feet is the plane's absolute ceiling.

(a) Determine the domain of the function appropriate for the context of the problem.

(b) Use a graphing utility to graph the time function and identify any asymptotes.

(c) Find the time when the altitude is increasing at the greatest rate.

68. Compound Interest

(a) How large a deposit, at 5% interest compounded continuously, must be made to obtain a balance of $10,000 in 15 years?

(b) A deposit earns interest at a rate of r percent compounded continuously and doubles in value in 10 years. Find r.

Evaluating an Expression In Exercises 69 and 70, evaluate each expression without using a calculator. (*Hint:* Make a sketch of a right triangle.)

69. (a) $\sin\left(\arcsin \frac{1}{2}\right)$
(b) $\cos\left(\arcsin \frac{1}{2}\right)$

70. (a) $\tan(\text{arccot } 2)$
(b) $\cos\left(\text{arcsec } \sqrt{5}\right)$

Finding a Derivative In Exercises 71–76, find the derivative of the function.

71. $y = \tan(\arcsin x)$

72. $y = \arctan(2x^2 - 3)$

73. $y = x \,\text{arcsec}\, x$

74. $y = \frac{1}{2} \arctan e^{2x}$

75. $y = x(\arcsin x)^2 - 2x + 2\sqrt{1 - x^2} \arcsin x$

76. $y = \sqrt{x^2 - 4} - 2 \,\text{arcsec}\, \frac{x}{2},\quad 2 < x < 4$

Finding an Indefinite Integral In Exercises 77–82, find the indefinite integral.

77. $\int \frac{1}{e^{2x} + e^{-2x}}\,dx$

78. $\int \frac{1}{3 + 25x^2}\,dx$

79. $\int \frac{x}{\sqrt{1 - x^4}}\,dx$

80. $\int \frac{1}{x\sqrt{9x^2 - 49}}\,dx$

81. $\int \frac{\arctan(x/2)}{4 + x^2}\,dx$

82. $\int \frac{\arcsin 2x}{\sqrt{1 - 4x^2}}\,dx$

Area In Exercises 83 and 84, find the area of the region.

83. $y = \frac{4 - x}{\sqrt{4 - x^2}}$

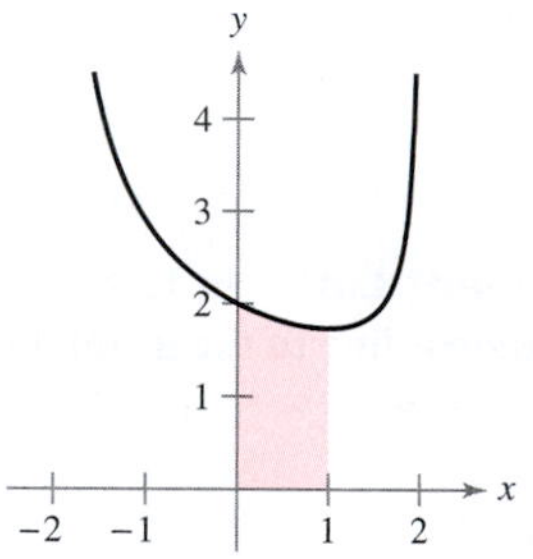

84. $y = \frac{6}{16 + x^2}$

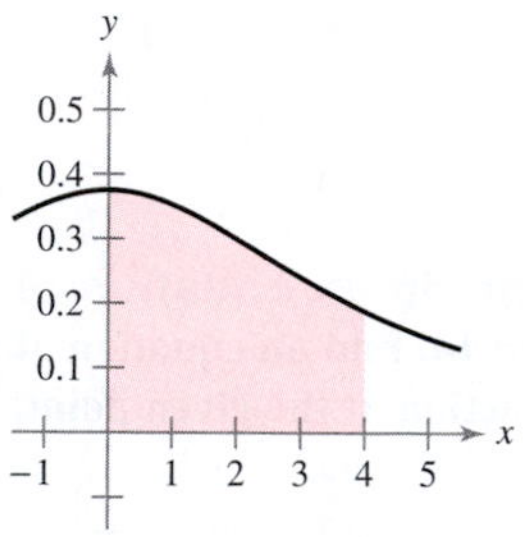

Finding a Derivative In Exercises 85–90, find the derivative of the function.

85. $y = \text{sech}(4x - 1)$

86. $y = 2x - \cosh \sqrt{x}$

87. $y = \coth(8x^2)$

88. $y = \ln(\cosh x)$

89. $y = \sinh^{-1}(4x)$

90. $y = x \tanh^{-1} 2x$

Finding an Indefinite Integral In Exercises 91–96, find the indefinite integral.

91. $\int x^2 \,\text{sech}^2\, x^3\,dx$

92. $\int \sinh 6x\,dx$

93. $\int \frac{\text{sech}^2\, x}{\tanh x}\,dx$

94. $\int \text{csch}^4(3x)\coth(3x)\,dx$

95. $\int \frac{1}{9 - 4x^2}\,dx$

96. $\int \frac{x}{\sqrt{x^4 - 1}}\,dx$

P.S. Problem Solving

See CalcChat.com for tutorial help and worked-out solutions to odd-numbered exercises.

1. Approximation To approximate e^x, you can use a function of the form

$$f(x) = \frac{a + bx}{1 + cx}.$$

(This function is known as a **Padé approximation.**) The values of $f(0)$, $f'(0)$, and $f''(0)$ are equal to the corresponding values of e^x. Show that these values are equal to 1 and find the values of a, b, and c such that $f(0) = f'(0) = f''(0) = 1$. Then use a graphing utility to compare the graphs of f and e^x.

2. Symmetry Recall that the graph of a function $y = f(x)$ is symmetric with respect to the origin if, whenever (x, y) is a point on the graph, $(-x, -y)$ is also a point on the graph. The graph of the function $y = f(x)$ is **symmetric with respect to the point (a, b)** if, whenever $(a - x, b - y)$ is a point on the graph, $(a + x, b + y)$ is also a point on the graph, as shown in the figure.

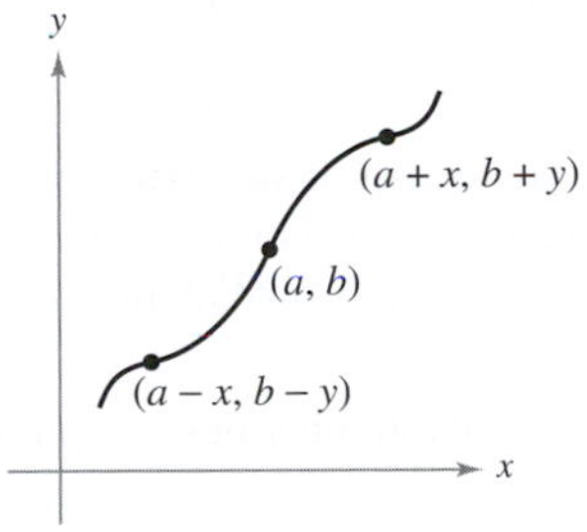

(a) Sketch the graph of $y = \sin x$ on the interval $[0, 2\pi]$. Write a short paragraph explaining how the symmetry of the graph with respect to the point $(\pi, 0)$ allows you to conclude that

$$\int_0^{2\pi} \sin x \, dx = 0.$$

(b) Sketch the graph of $y = \sin x + 2$ on the interval $[0, 2\pi]$. Use the symmetry of the graph with respect to the point $(\pi, 2)$ to evaluate the integral

$$\int_0^{2\pi} (\sin x + 2) \, dx.$$

(c) Sketch the graph of $y = \arccos x$ on the interval $[-1, 1]$. Use the symmetry of the graph to evaluate the integral

$$\int_{-1}^{1} \arccos x \, dx.$$

(d) Evaluate the integral $\displaystyle\int_0^{\pi/2} \frac{1}{1 + (\tan x)^{\sqrt{2}}} \, dx.$

3. Proof

(a) Use a graphing utility to graph $f(x) = \dfrac{\ln(x + 1)}{x}$ on the interval $[-1, 1]$.

(b) Use the graph to estimate $\lim_{x \to 0} f(x)$.

(c) Use the definition of derivative to prove your answer to part (b).

4. Using a Function Let $f(x) = \sin(\ln x)$.

(a) Determine the domain of the function f.

(b) Find two values of x satisfying $f(x) = 1$.

(c) Find two values of x satisfying $f(x) = -1$.

(d) What is the range of the function f?

(e) Calculate $f'(x)$ and use calculus to find the maximum value of f on the interval $[1, 10]$.

(f) Use a graphing utility to graph f in the viewing window $[0, 5] \times [-2, 2]$ and estimate $\lim_{x \to 0^+} f(x)$, if it exists.

(g) Determine $\lim_{x \to 0^+} f(x)$ analytically, if it exists.

5. Intersection Graph the exponential function $y = a^x$ for $a = 0.5$, 1.2, and 2.0. Which of these curves intersects the line $y = x$? Determine all positive numbers a for which the curve $y = a^x$ intersects the line $y = x$.

6. Areas and Angles

(a) Let $P(\cos t, \sin t)$ be a point on the unit circle $x^2 + y^2 = 1$ in the first quadrant (see figure). Show that t is equal to twice the area of the shaded circular sector AOP.

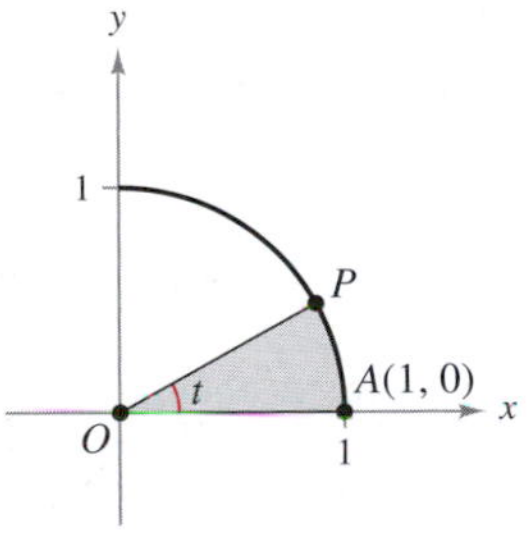

(b) Let $P(\cosh t, \sinh t)$ be a point on the unit hyperbola $x^2 - y^2 = 1$ in the first quadrant (see figure). Show that t is equal to twice the area of the shaded region AOP. Begin by showing that the area of the shaded region AOP is given by the formula

$$A(t) = \frac{1}{2}\cosh t \sinh t - \int_1^{\cosh t} \sqrt{x^2 - 1} \, dx.$$

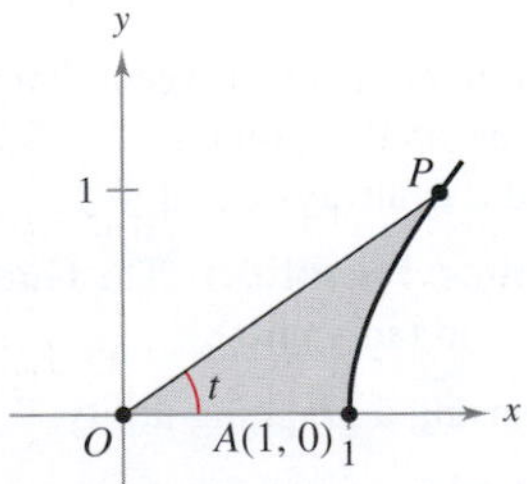

7. Mean Value Theorem Apply the Mean Value Theorem to the function $f(x) = \ln x$ on the closed interval $[1, e]$. Find the value of c in the open interval $(1, e)$ such that

$$f'(c) = \frac{f(e) - f(1)}{e - 1}.$$

8. Decreasing Function Show that $f(x) = \dfrac{\ln x^n}{x}$ is a decreasing function for $x > e$ and $n > 0$.

9. Area Consider the three regions A, B, and C determined by the graph of $f(x) = \arcsin x$, as shown in the figure.

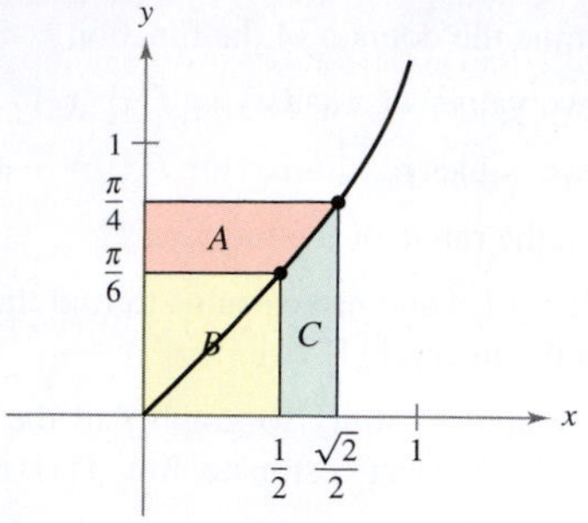

(a) Calculate the areas of regions A and B.

(b) Use your answers in part (a) to evaluate the integral

$$\int_{1/2}^{\sqrt{2}/2} \arcsin x \, dx.$$

(c) Use the methods in part (a) to evaluate the integral

$$\int_{1}^{3} \ln x \, dx.$$

(d) Use the methods in part (a) to evaluate the integral

$$\int_{1}^{\sqrt{3}} \arctan x \, dx.$$

10. Distance Let L be the tangent line to the graph of the function $y = \ln x$ at the point (a, b). Show that the distance between b and c is always equal to 1.

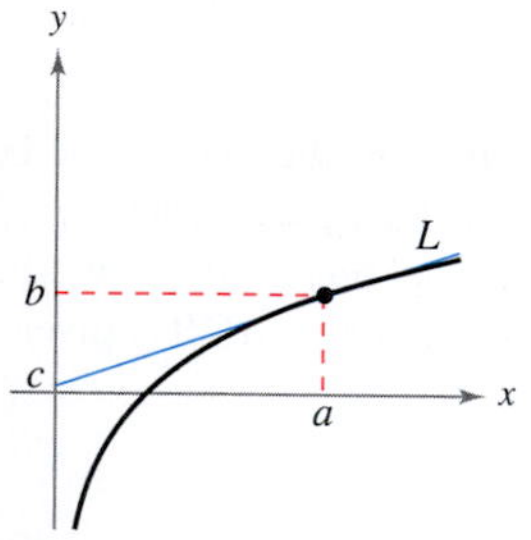

Figure for 10

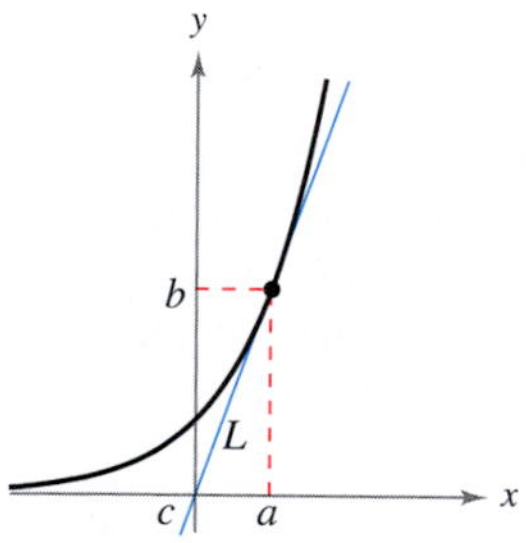

Figure for 11

11. Distance Let L be the tangent line to the graph of the function $y = e^x$ at the point (a, b). Show that the distance between a and c is always equal to 1.

12. Gudermannian Function The **Gudermannian function** of x is $\text{gd}(x) = \arctan(\sinh x)$.

(a) Graph gd using a graphing utility.

(b) Show that gd is an odd function.

(c) Show that gd is monotonic and therefore has an inverse.

(d) Find the inflection point of gd.

(e) Verify that $\text{gd}(x) = \arcsin(\tanh x)$.

(f) Verify that $\text{gd}(x) = \displaystyle\int_0^x \frac{dt}{\cosh t}$.

13. Area Use integration by substitution to find the area under the curve

$$y = \frac{1}{\sqrt{x} + x}$$

between $x = 1$ and $x = 4$.

14. Area Use integration by substitution to find the area under the curve

$$y = \frac{1}{\sin^2 x + 4\cos^2 x}$$

between $x = 0$ and $x = \dfrac{\pi}{4}$.

15. Approximating a Function

(a) Use a graphing utility to compare the graph of the function $y = e^x$ with the graph of each given function.

(i) $y_1 = 1 + \dfrac{x}{1!}$

(ii) $y_2 = 1 + \dfrac{x}{1!} + \dfrac{x^2}{2!}$

(iii) $y_3 = 1 + \dfrac{x}{1!} + \dfrac{x^2}{2!} + \dfrac{x^3}{3!}$

(b) Identify the pattern of successive polynomials in part (a), extend the pattern one more term, and compare the graph of the resulting polynomial function with the graph of $y = e^x$.

(c) What do you think this pattern implies?

16. Mortgage A \$120,000 home mortgage for 35 years at $9\frac{1}{2}\%$ has a monthly payment of \$985.93. Part of the monthly payment goes for the interest charge on the unpaid balance, and the remainder of the payment is used to reduce the principal. The amount that goes for interest is

$$u = M - \left(M - \frac{Pr}{12}\right)\left(1 + \frac{r}{12}\right)^{12t}$$

and the amount that goes toward reduction of the principal is

$$v = \left(M - \frac{Pr}{12}\right)\left(1 + \frac{r}{12}\right)^{12t}.$$

In these formulas, P is the amount of the mortgage, r is the interest rate (in decimal form), M is the monthly payment, and t is the time in years.

(a) Use a graphing utility to graph each function in the same viewing window. (The viewing window should show all 35 years of mortgage payments.)

(b) In the early years of the mortgage, the larger part of the monthly payment goes for what purpose? Approximate the time when the monthly payment is evenly divided between interest and principal reduction.

(c) Use the graphs in part (a) to make a conjecture about the relationship between the slopes of the tangent lines to the two curves for a specified value of t. Give an analytical argument to verify your conjecture. Find $u'(15)$ and $v'(15)$.

(d) Repeat parts (a) and (b) for a repayment period of 20 years ($M = \$1118.56$). What can you conclude?

AP* Review Questions for Chapter 5

1. (calculator)

A particle travels along the x-axis for times $0 \le t \le 4$. The velocity of the particle is given by $v(t) = \sin\left(\frac{5\pi}{2}e^{-t/\pi}\right)$. At time $t = 0$, the particle is 2 units to the right of the origin.

(a) During what time intervals on $[0, 4]$ is the particle traveling to the left?

(b) Find the average velocity of the particle from $0 \le t \le 4$.

(c) Is the speed increasing or decreasing at time $t = 2$?

(d) For the time interval $0 \le t \le 4$, what is the farthest distance to the right that the particle travels from the origin?

2. (calculator)

Let f be a function defined by $f(x) = \dfrac{e^x + e^{-x}}{2}$.

(a) Find, if any, the x-coordinates of all relative minimum values of $f(x)$.

(b) Find the average value of $f(x)$ for $-1 \le x \le 1$.

(c) Find, if any, the x-coordinates of all inflection points of $f(x)$.

3. (calculator)

Water is flooding into a basement at a rate of $f(t) = -2\sqrt{t} + 5$. Water is being pumped out at a rate of $g(t) = 5(1 - e^{-0.5t})$. Both $f(t)$ and $g(t)$ are defined for $0 \le t \le 5$ hours. At time $t = 0$, there are 15 cubic feet of water in the basement.

(a) Calculate the total amount of water that enters the basement from $t = 0$ to $t = 2$ hours.

(b) How fast is the water level changing at time $t = 2$ hours? Use proper units.

(c) At what time t was the volume of water at a maximum? How much water was in the basement at the time?

4. (calculator)

The function f is defined by $f(x) = \dfrac{1}{\sqrt{4 - x^2}}$ for $-2 < x < 2$.

Let g be the function defined by

$$g(x) = \begin{cases} f(x) & \text{for } -2 \le x \le 0 \\ x + \frac{1}{2} & \text{for } 0 < x \le 2 \end{cases}.$$

(a) Is g continuous at $x = 0$? Use the definition of continuity to explain your answer.

(b) What is the area of the region bounded by f, the x-axis, the y-axis, and the line $x = 1$?

(c) Find the value of $\int_{-1}^{1} g(x)dx$.

5. Let $f(x) = \arccos(x)$ and let $g(x) = x^2$. Define $h(x) = f(g(x))$.

(a) At what values of x does $h(x)$ have a relative maximum?

(b) Write, but do not evaluate, an expression that can determine the area of the region bounded by the graphs of $h(x)$ and the horizontal line $y = \frac{\pi}{3}$.

(c) Evaluate $\frac{d}{dx}f^{-1}\left(\frac{\pi}{3}\right)$.

6. (no calculator)

Let $f(x) = e^x - x$.

(a) Find the critical values of f. Classify each of these values as a relative minimum, a relative maximum, or neither a minimum nor a maximum. Justify your conclusion.

(b) Write the equation of the line tangent to the graph of f at the point where $x = 1$.

(c) Given $\int_0^a f(x)dx = f'(a)$, find a.

7. (calculator)

Two towers of equal height are spaced 366 feet apart. A cable suspended between the two forms a catenary whose height above the ground is given by $f(x) = \frac{125}{2}(e^{x/250} + e^{-x/250})$, where $x = 0$ is at the point on the ground halfway between the two towers.

(a) What is the height of each tower (rounded to the nearest foot)? (Assume the cable is anchored to the *tops* of the towers.)

(b) Evaluate $f'(100)$. Explain what $f'(100)$ is in the context of the problem.

(c) What is the average height of the cable?

(d) What is the slope of the cable at the rightmost point where the height of the cable is equal to the average height of the cable?

8. (no calculator)

x	$f(x)$	$f'(x)$
1	3	1
2	−1	3
3	5	−2
4	−2	4
5	4	−2

The table above gives values of f and its derivative f' at selected values of x. The function f is twice differentiable at all x.

(a) Use the values in the table to estimate the value of $f'(1.5)$.

(b) Explain why there must be a value c for $1 < c < 3$, for which $f'(c) = 1$.

(c) Estimate the average value of $f(x)$ from $1 \leq x \leq 4$ using a trapezoidal sum with 3 subintervals, as indicated in the table.

(d) Assuming f is invertible, find the equation of the line tangent to the graph of $y = f^{-1}(x)$ at $x = 5$.

9. (no calculator)

Let f be a differentiable function such that $f(-2) = 3$, $f(3) = 5$, $f'(-2) = 4$, $f'(3) = 6$. Let g be a differentiable function, such that $g(x) = f^{-1}(x)$ for all x. What is the value of $g'(3)$?

(A) $\frac{1}{3}$

(B) $\frac{1}{4}$

(C) $\frac{1}{5}$

(D) 6

(E) 4

10. (no calculator)

If $\cos x = \ln y$, what is $\frac{dy}{dx}$ in terms of x?

(A) $-e^{\cos x} \sin x$

(B) $e^{\cos x} \sin x$

(C) $\frac{-1}{\sin x}$

(D) $\frac{1}{\sin x}$

(E) $\cos x \sin x$

6 Differential Equations

Sailing *(Exercise 65, p. 423)*

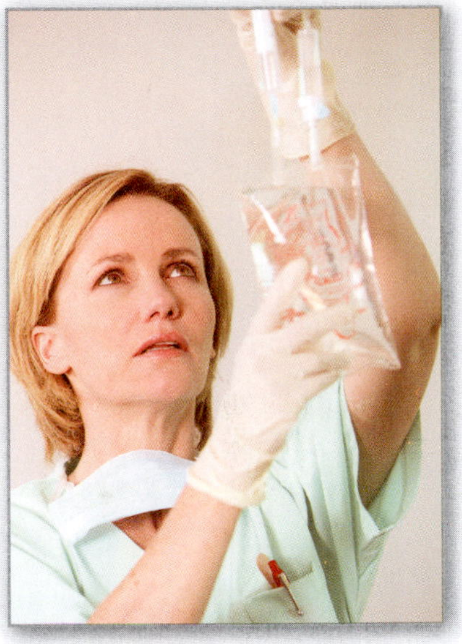

Intravenous Feeding *(Exercise 30, p. 429)*

Wildlife Population *(Example 4, p. 417)*

Forestry *(Exercise 62, p. 414)*

Radioactive Decay *(Example 3, p. 409)*

6.1 Slope Fields and Euler's Method

- **Use initial conditions to find particular solutions of differential equations.**
- **Use slope fields to approximate solutions of differential equations.**
- **Use Euler's Method to approximate solutions of differential equations.**

General and Particular Solutions

In this text, you will learn that physical phenomena can be described by differential equations. Recall that a **differential equation** in x and y is an equation that involves x, y, and derivatives of y. For example,

$$2xy' - 3y = 0 \qquad \text{Differential equation}$$

is a differential equation. In Section 6.2, you will see that problems involving radioactive decay, population growth, and Newton's Law of Cooling can be formulated in terms of differential equations.

A function $y = f(x)$ is called a **solution** of a differential equation if the equation is satisfied when y and its derivatives are replaced by $f(x)$ and its derivatives. For example, differentiation and substitution would show that $y = e^{-2x}$ is a solution of the differential equation $y' + 2y = 0$. It can be shown that every solution of this differential equation is of the form

$$y = Ce^{-2x} \qquad \text{General solution of } y' + 2y = 0$$

where C is any real number. This solution is called the **general solution.** Some differential equations have **singular solutions** that cannot be written as special cases of the general solution. Such solutions, however, are not considered in this text. The **order** of a differential equation is determined by the highest-order derivative in the equation. For instance, $y' = 4y$ is a first-order differential equation. First-order linear differential equations are discussed in Section 6.4.

In Section 4.1, Example 9, you saw that the second-order differential equation $s''(t) = -32$ has the general solution

$$s(t) = -16t^2 + C_1t + C_2 \qquad \text{General solution of } s''(t) = -32$$

which contains two arbitrary constants. It can be shown that a differential equation of order n has a general solution with n arbitrary constants.

EXAMPLE 1 Verifying Solutions

Determine whether the function is a solution of the differential equation $y'' - y = 0$.

a. $y = \sin x$ **b.** $y = 4e^{-x}$ **c.** $y = Ce^x$

Solution

a. Because $y = \sin x$, $y' = \cos x$, and $y'' = -\sin x$, it follows that

$$y'' - y = -\sin x - \sin x = -2 \sin x \neq 0.$$

So, $y = \sin x$ is *not* a solution.

b. Because $y = 4e^{-x}$, $y' = -4e^{-x}$, and $y'' = 4e^{-x}$, it follows that

$$y'' - y = 4e^{-x} - 4e^{-x} = 0.$$

So, $y = 4e^{-x}$ is a solution.

c. Because $y = Ce^x$, $y' = Ce^x$, and $y'' = Ce^x$, it follows that

$$y'' - y = Ce^x - Ce^x = 0.$$

So, $y = Ce^x$ is a solution for any value of C.

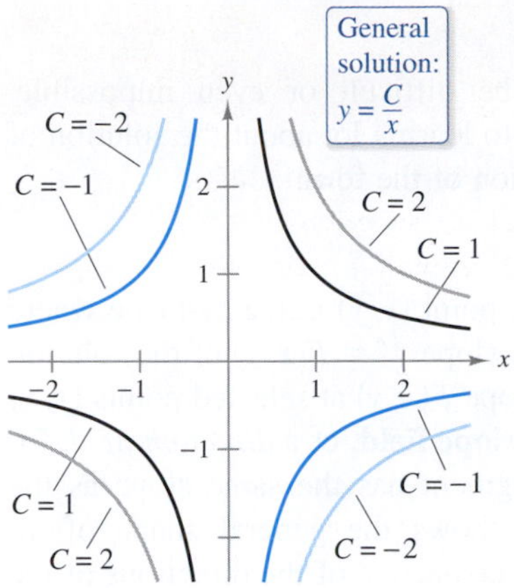

Solution curves for $xy' + y = 0$
Figure 6.1

Geometrically, the general solution of a first-order differential equation represents a family of curves known as **solution curves,** one for each value assigned to the arbitrary constant. For instance, you can verify that every function of the form

$$y = \frac{C}{x}$$ General solution of $xy' + y = 0$

is a solution of the differential equation

$$xy' + y = 0.$$

Figure 6.1 shows four of the solution curves corresponding to different values of C.

As discussed in Section 4.1, **particular solutions** of a differential equation are obtained from **initial conditions** that give the values of the dependent variable or one of its derivatives for particular values of the independent variable. The term "initial condition" stems from the fact that, often in problems involving time, the value of the dependent variable or one of its derivatives is known at the *initial* time $t = 0$. For instance, the second-order differential equation

$$s''(t) = -32$$

having the general solution

$$s(t) = -16t^2 + C_1t + C_2$$ General solution of $s''(t) = -32$

might have the following initial conditions.

$$s(0) = 80, \quad s'(0) = 64$$ Initial conditions

In this case, the initial conditions yield the particular solution

$$s(t) = -16t^2 + 64t + 80.$$ Particular solution

EXAMPLE 2 Finding a Particular Solution

See LarsonCalculus.com for an interactive version of this type of example.

For the differential equation

$$xy' - 3y = 0$$

verify that $y = Cx^3$ is a solution. Then find the particular solution determined by the initial condition $y = 2$ when $x = -3$.

Solution You know that $y = Cx^3$ is a solution because $y' = 3Cx^2$ and

$$xy' - 3y = x(3Cx^2) - 3(Cx^3) = 0.$$

Furthermore, the initial condition $y = 2$ when $x = -3$ yields

$$y = Cx^3$$ General solution

$$2 = C(-3)^3$$ Substitute initial condition.

$$-\frac{2}{27} = C$$ Solve for C.

and you can conclude that the particular solution is

$$y = -\frac{2x^3}{27}.$$ Particular solution

Try checking this solution by substituting for y and y' in the original differential equation. ■

Note that to determine a particular solution, the number of initial conditions must match the number of constants in the general solution.

AP* Tips

The ability to draw a slope field is tested on both the AP Calculus AB and BC Exams.

Slope Fields

Solving a differential equation analytically can be difficult or even impossible. However, there is a graphical approach you can use to learn a lot about the solution of a differential equation. Consider a differential equation of the form

$$y' = F(x, y) \qquad \text{Differential equation}$$

where $F(x, y)$ is some expression in x and y. At each point (x, y) in the xy-plane where F is defined, the differential equation determines the slope $y' = F(x, y)$ of the solution at that point. If you draw short line segments with slope $F(x, y)$ at selected points (x, y) in the domain of F, then these line segments form a **slope field,** or a *direction field*, for the differential equation $y' = F(x, y)$. Each line segment has the same slope as the solution curve through that point. A slope field shows the general shape of all the solutions and can be helpful in getting a visual perspective of the directions of the solutions of a differential equation.

EXAMPLE 3 Sketching a Slope Field

Sketch a slope field for the differential equation $y' = x - y$ for the points $(-1, 1)$, $(0, 1)$, and $(1, 1)$.

Solution The slope of the solution curve at any point (x, y) is

$$F(x, y) = x - y. \qquad \text{Slope at } (x, y).$$

So, the slope at each point can be found as shown.

Slope at $(-1, 1)$: $y' = -1 - 1 = -2$

Slope at $(0, 1)$: $y' = 0 - 1 = -1$

Slope at $(1, 1)$: $y' = 1 - 1 = 0$

Draw short line segments at the three points with their respective slopes, as shown in Figure 6.2.

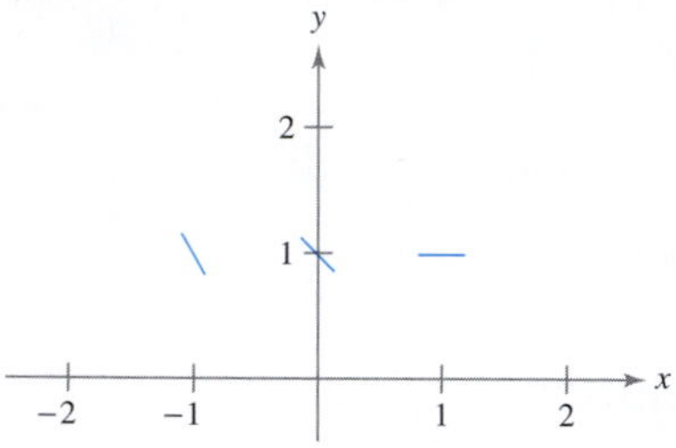

Figure 6.2

EXAMPLE 4 Identifying Slope Fields for Differential Equations

Match each slope field with its differential equation.

a.

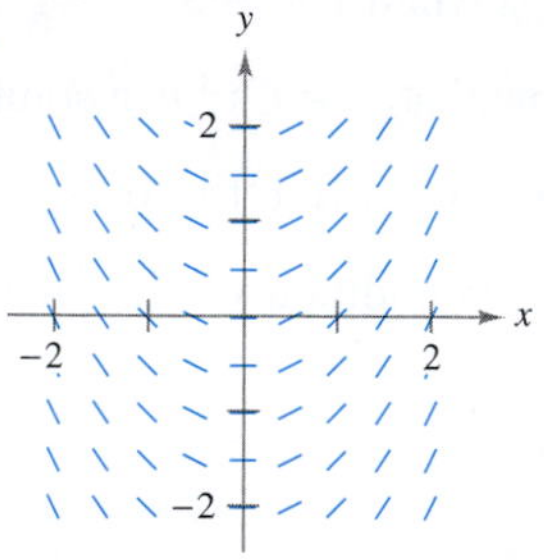

b.

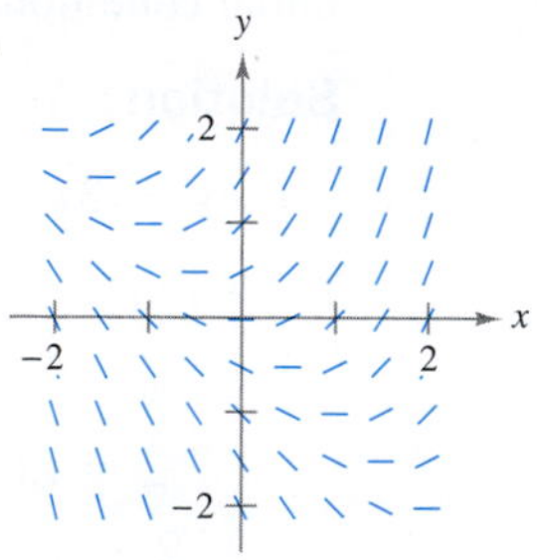

c.

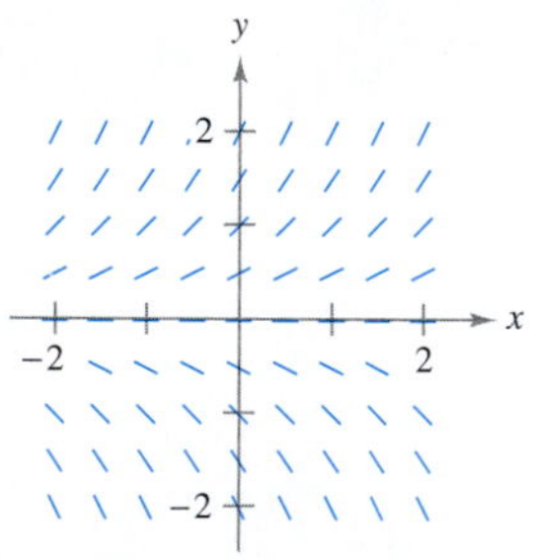

i. $y' = x + y$ **ii.** $y' = x$ **iii.** $y' = y$

Solution

a. You can see that the slope at any point along the y-axis is 0. The only equation that satisfies this condition is $y' = x$. So, the graph matches equation (ii).

b. You can see that the slope at the point $(1, -1)$ is 0. The only equation that satisfies this condition is $y' = x + y$. So, the graph matches equation (i).

c. You can see that the slope at any point along the x-axis is 0. The only equation that satisfies this condition is $y' = y$. So, the graph matches equation (iii). ■

A solution curve of a differential equation $y' = F(x, y)$ is simply a curve in the xy-plane whose tangent line at each point (x, y) has slope equal to $F(x, y)$. This is illustrated in Example 5.

EXAMPLE 5 **Sketching a Solution Using a Slope Field**

Sketch a slope field for the differential equation

$$y' = 2x + y.$$

Use the slope field to sketch the solution that passes through the point $(1, 1)$.

Solution Make a table showing the slopes at several points. The table shown is a small sample. The slopes at many other points should be calculated to get a representative slope field.

x	-2	-2	-1	-1	0	0	1	1	2	2
y	-1	1	-1	1	-1	1	-1	1	-1	1
$y' = 2x + y$	-5	-3	-3	-1	-1	1	1	3	3	5

Next, draw line segments at the points with their respective slopes, as shown in Figure 6.3.

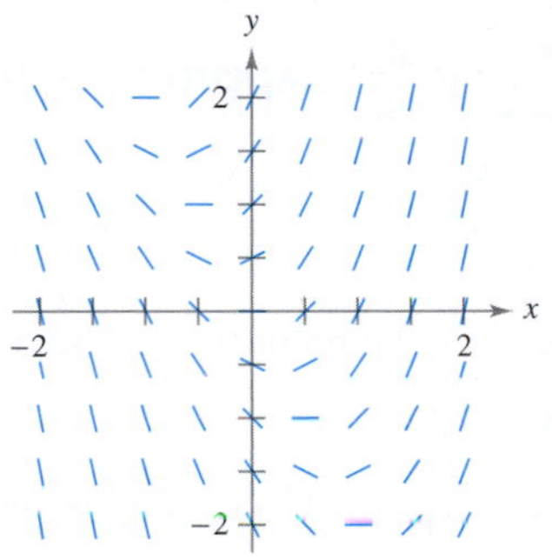

Slope field for $y' = 2x + y$
Figure 6.3

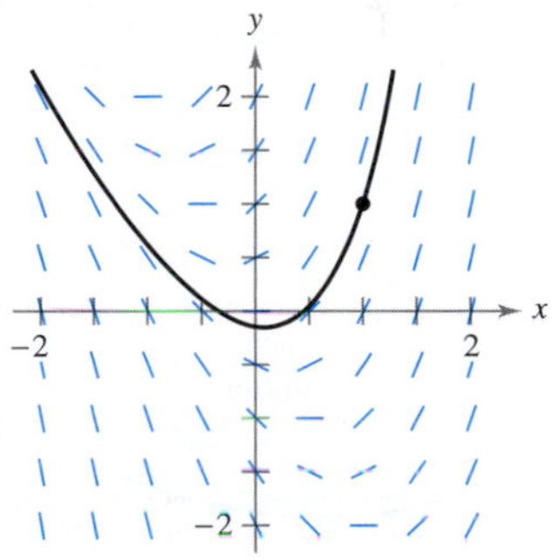

Particular solution for $y' = 2x + y$ passing through $(1, 1)$
Figure 6.4

After the slope field is drawn, start at the initial point $(1, 1)$ and move to the right in the direction of the line segment. Continue to draw the solution curve so that it moves parallel to the nearby line segments. Do the same to the left of $(1, 1)$. The resulting solution is shown in Figure 6.4. ■

In Example 5, note that the slope field shows that y' increases to infinity as x increases.

▷ **TECHNOLOGY** Drawing a slope field by hand is tedious. In practice, slope fields are usually drawn using a graphing utility. If you have access to a graphing utility that can graph slope fields, try graphing the slope field for the differential equation in Example 5. One example of a slope field drawn by a graphing utility is shown at the right.

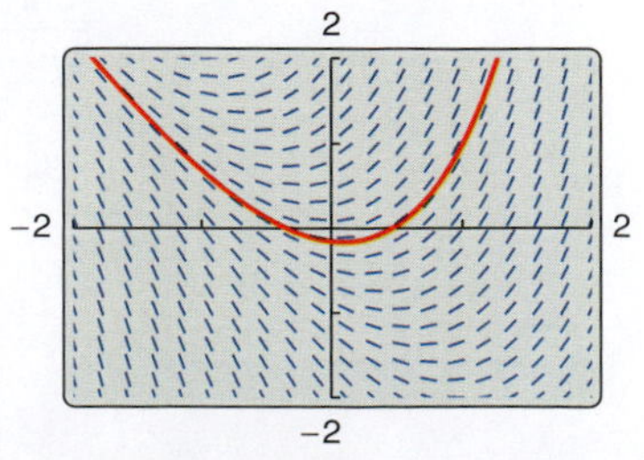

Generated by Maple.

Euler's Method

Euler's Method is a numerical approach to approximating the particular solution of the differential equation

$$y' = F(x, y)$$

that passes through the point (x_0, y_0). From the given information, you know that the graph of the solution passes through the point (x_0, y_0) and has a slope of $F(x_0, y_0)$ at this point. This gives you a "starting point" for approximating the solution.

From this starting point, you can proceed in the direction indicated by the slope. Using a small step h, move along the tangent line until you arrive at the point (x_1, y_1), where

$$x_1 = x_0 + h \quad \text{and} \quad y_1 = y_0 + hF(x_0, y_0)$$

as shown in Figure 6.5. Then, using (x_1, y_1) as a new starting point, you can repeat the process to obtain a second point (x_2, y_2). The values of x_i and y_i are shown below.

$$\begin{aligned} x_1 &= x_0 + h & y_1 &= y_0 + hF(x_0, y_0) \\ x_2 &= x_1 + h & y_2 &= y_1 + hF(x_1, y_1) \\ &\vdots & &\vdots \\ x_n &= x_{n-1} + h & y_n &= y_{n-1} + hF(x_{n-1}, y_{n-1}) \end{aligned}$$

When using this method, note that you can obtain better approximations of the exact solution by choosing smaller and smaller step sizes.

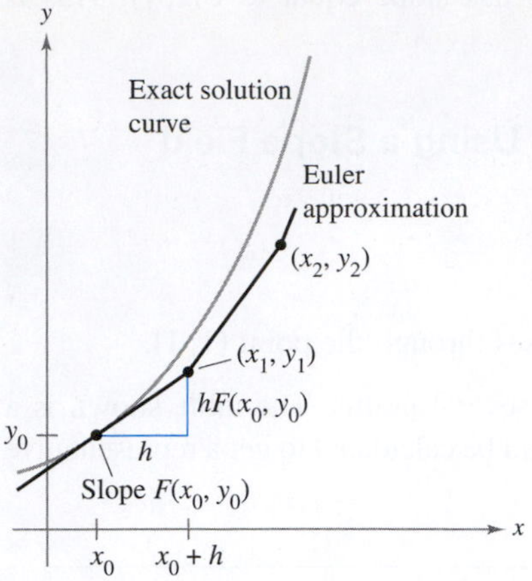

Figure 6.5

AP* Tips

Euler's Method is tested only on the AP Calculus BC Exam.

EXAMPLE 6 Approximating a Solution Using Euler's Method

Use Euler's Method to approximate the particular solution of the differential equation

$$y' = x - y$$

passing through the point (0, 1). Use a step of $h = 0.1$.

Solution Using $h = 0.1$, $x_0 = 0$, $y_0 = 1$, and $F(x, y) = x - y$, you have

$$x_0 = 0, \quad x_1 = 0.1, \quad x_2 = 0.2, \quad x_3 = 0.3,$$

and the first three approximations are

$$\begin{aligned} y_1 &= y_0 + hF(x_0, y_0) = 1 + (0.1)(0 - 1) = 0.9 \\ y_2 &= y_1 + hF(x_1, y_1) = 0.9 + (0.1)(0.1 - 0.9) = 0.82 \\ y_3 &= y_2 + hF(x_2, y_2) = 0.82 + (0.1)(0.2 - 0.82) = 0.758. \end{aligned}$$

The first ten approximations are shown in the table. You can plot these values to see a graph of the approximate solution, as shown in Figure 6.6.

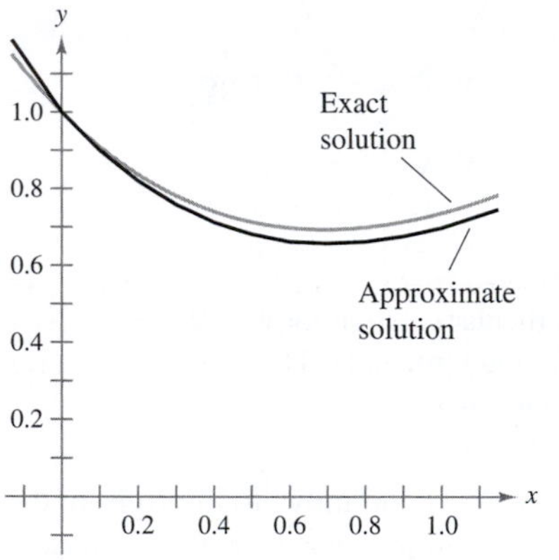

Figure 6.6

n	0	1	2	3	4	5	6	7	8	9	10
x_n	0	0.1	0.2	0.3	0.4	0.5	0.6	0.7	0.8	0.9	1.0
y_n	1	0.900	0.820	0.758	0.712	0.681	0.663	0.657	0.661	0.675	0.697

For the differential equation in Example 6, you can verify the exact solution to be the equation

$$y = x - 1 + 2e^{-x}.$$

Figure 6.6 compares this exact solution with the approximate solution obtained in Example 6.

6.1 Exercises

See CalcChat.com for tutorial help and worked-out solutions to odd-numbered exercises.

Verifying a Solution In Exercises 1–8, verify the solution of the differential equation.

Solution	Differential Equation
1. $y = Ce^{4x}$	$y' = 4y$
2. $y = e^{-2x}$	$3y' + 5y = -e^{-2x}$
3. $x^2 + y^2 = Cy$	$y' = \dfrac{2xy}{x^2 - y^2}$
4. $y^2 - 2 \ln y = x^2$	$\dfrac{dy}{dx} = \dfrac{xy}{y^2 - 1}$
5. $y = C_1 \sin x - C_2 \cos x$	$y'' + y = 0$
6. $y = C_1 e^{-x} \cos x + C_2 e^{-x} \sin x$	$y'' + 2y' + 2y = 0$
7. $y = -\cos x \ln\lvert\sec x + \tan x\rvert$	$y'' + y = \tan x$
8. $y = \frac{2}{5}(e^{-4x} + e^x)$	$y'' + 4y' = 2e^x$

Verifying a Particular Solution In Exercises 9–12, verify the particular solution of the differential equation.

Solution	Differential Equation and Initial Condition
9. $y = \sin x \cos x - \cos^2 x$	$2y + y' = 2 \sin(2x) - 1$, $y\left(\dfrac{\pi}{4}\right) = 0$
10. $y = 6x - 4 \sin x + 1$	$y' = 6 - 4 \cos x$, $y(0) = 1$
11. $y = 4e^{-6x^2}$	$y' = -12xy$, $y(0) = 4$
12. $y = e^{-\cos x}$	$y' = y \sin x$, $y\left(\dfrac{\pi}{2}\right) = 1$

Determining a Solution In Exercises 13–20, determine whether the function is a solution of the differential equation $y^{(4)} - 16y = 0$.

13. $y = 3 \cos x$
14. $y = 2 \sin x$
15. $y = 3 \cos 2x$
16. $y = 3 \sin 2x$
17. $y = e^{-2x}$
18. $y = 5 \ln x$
19. $y = C_1 e^{2x} + C_2 e^{-2x} + C_3 \sin 2x + C_4 \cos 2x$
20. $y = 3e^{2x} - 4 \sin 2x$

Determining a Solution In Exercises 21–28, determine whether the function is a solution of the differential equation $xy' - 2y = x^3 e^x$.

21. $y = x^2$
22. $y = x^3$
23. $y = x^2 e^x$
24. $y = x^2(2 + e^x)$
25. $y = \sin x$
26. $y = \cos x$
27. $y = \ln x$
28. $y = x^2 e^x - 5x^2$

Finding a Particular Solution In Exercises 29–32, some of the curves corresponding to different values of C in the general solution of the differential equation are shown in the graph. Find the particular solution that passes through the point shown on the graph.

29. $y^2 = Ce^{-x/2}$
$2y' + y = 0$

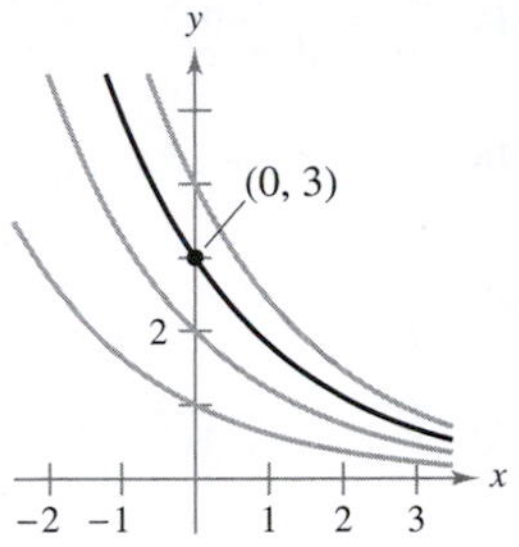

30. $y(x^2 + y) = C$
$2xy + (x^2 + 2y)y' = 0$

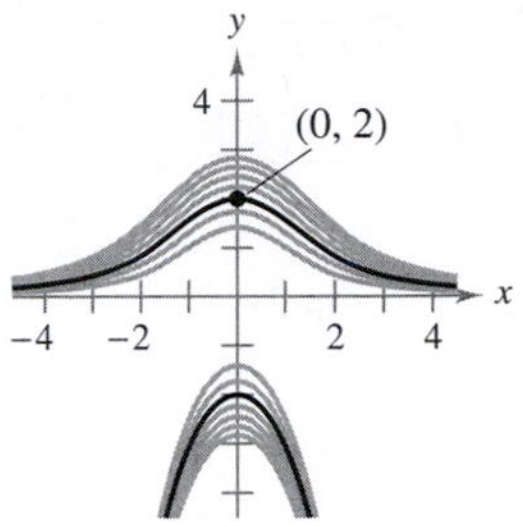

31. $y^2 = Cx^3$
$2xy' - 3y = 0$

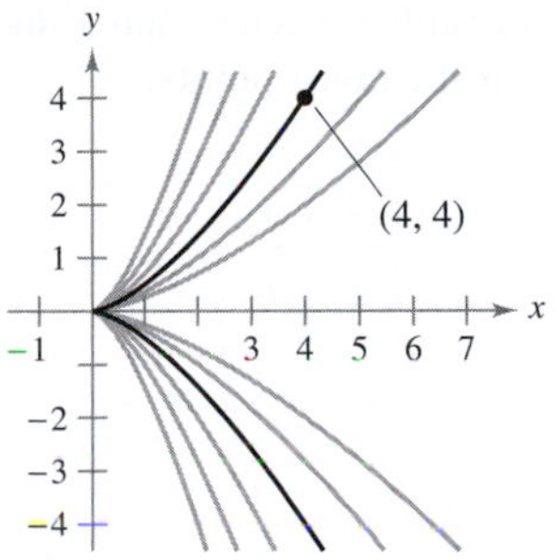

32. $2x^2 - y^2 = C$
$yy' - 2x = 0$

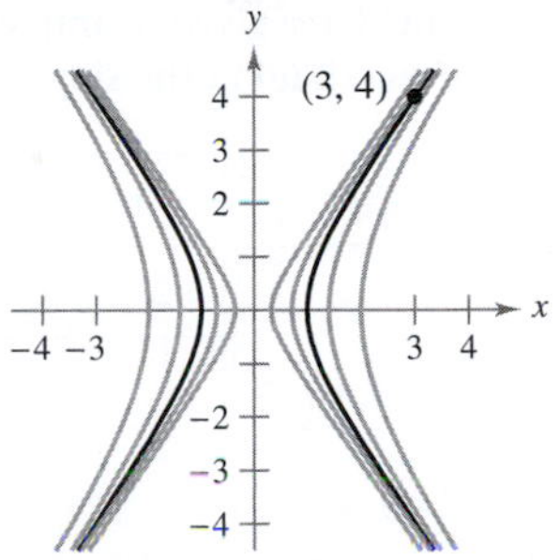

Graphs of Particular Solutions In Exercises 33 and 34, the general solution of the differential equation is given. Use a graphing utility to graph the particular solutions for the given values of C.

33. $4yy' - x = 0$
$4y^2 - x^2 = C$
$C = 0, C = \pm 1, C = \pm 4$

34. $yy' + x = 0$
$x^2 + y^2 = C$
$C = 0, C = 1, C = 4$

Finding a Particular Solution In Exercises 35–40, verify that the general solution satisfies the differential equation. Then find the particular solution that satisfies the initial condition(s).

35. $y = Ce^{-2x}$
$y' + 2y = 0$
$y = 3$ when $x = 0$

36. $3x^2 + 2y^2 = C$
$3x + 2yy' = 0$
$y = 3$ when $x = 1$

37. $y = C_1 \sin 3x + C_2 \cos 3x$
$y'' + 9y = 0$
$y = 2$ when $x = \dfrac{\pi}{6}$
$y' = 1$ when $x = \dfrac{\pi}{6}$

38. $y = C_1 + C_2 \ln x$
$xy'' + y' = 0$
$y = 0$ when $x = 2$
$y' = \dfrac{1}{2}$ when $x = 2$

39. $y = C_1x + C_2x^3$

$x^2y'' - 3xy' + 3y = 0$

$y = 0$ when $x = 2$

$y' = 4$ when $x = 2$

40. $y = e^{2x/3}(C_1 + C_2x)$

$9y'' - 12y' + 4y = 0$

$y = 4$ when $x = 0$

$y = 0$ when $x = 3$

Finding a General Solution **In Exercises 41–52, use integration to find a general solution of the differential equation.**

41. $\dfrac{dy}{dx} = 6x^2$

42. $\dfrac{dy}{dx} = 10x^4 - 2x^3$

43. $\dfrac{dy}{dx} = \dfrac{x}{1 + x^2}$

44. $\dfrac{dy}{dx} = \dfrac{e^x}{4 + e^x}$

45. $\dfrac{dy}{dx} = \dfrac{x - 2}{x}$

46. $\dfrac{dy}{dx} = x \cos x^2$

47. $\dfrac{dy}{dx} = \sin 2x$

48. $\dfrac{dy}{dx} = \tan^2 x$

49. $\dfrac{dy}{dx} = x\sqrt{x - 6}$

50. $\dfrac{dy}{dx} = 2x\sqrt{4x^2 + 1}$

51. $\dfrac{dy}{dx} = xe^{x^2}$

52. $\dfrac{dy}{dx} = 5e^{-x/2}$

Slope Field **In Exercises 53–56, a differential equation and its slope field are given. Complete the table by determining the slopes (if possible) in the slope field at the given points.**

x	-4	-2	0	2	4	8
y	2	0	4	4	6	8
dy/dx						

53. $\dfrac{dy}{dx} = \dfrac{2x}{y}$

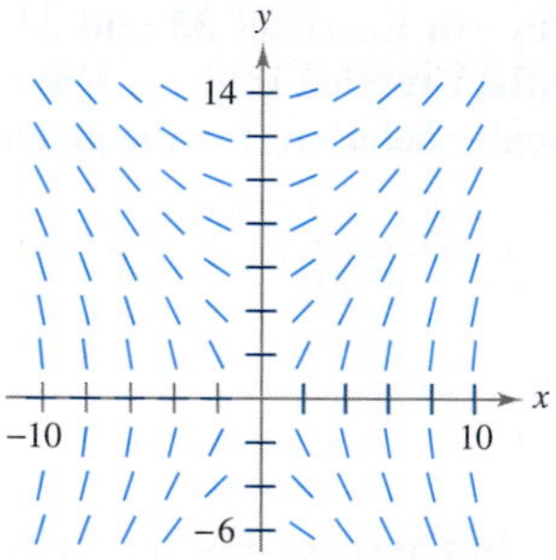

54. $\dfrac{dy}{dx} = y - x$

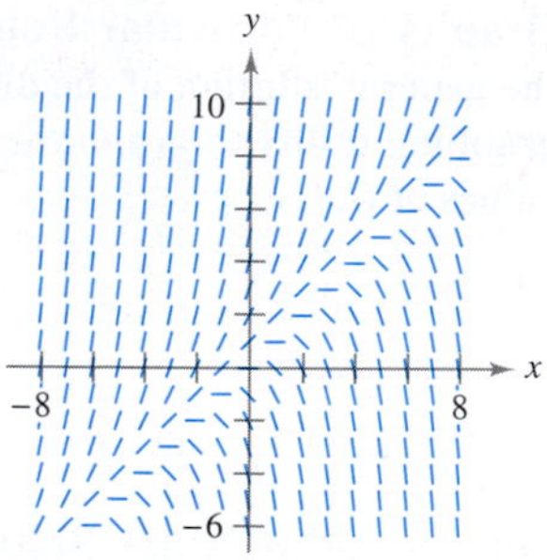

55. $\dfrac{dy}{dx} = x \cos \dfrac{\pi y}{8}$

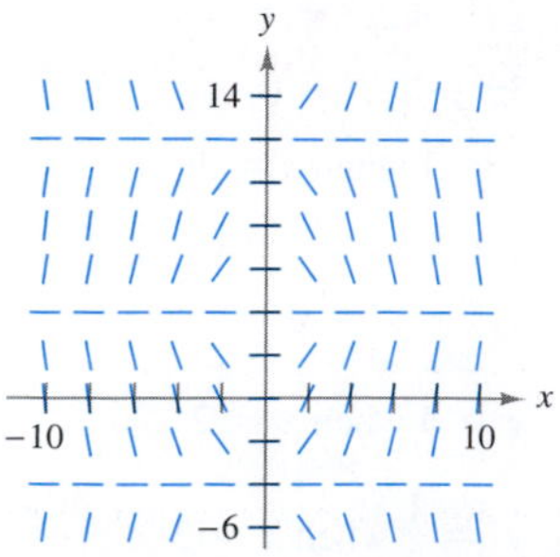

56. $\dfrac{dy}{dx} = \tan\left(\dfrac{\pi y}{6}\right)$

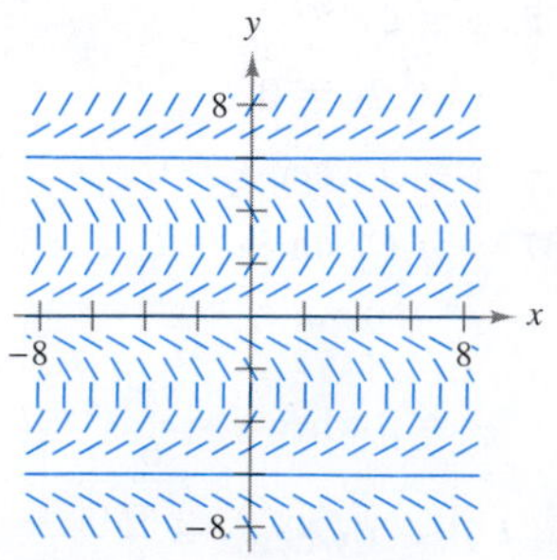

Matching **In Exercises 57–60, match the differential equation with its slope field. [The slope fields are labeled (a), (b), (c), and (d).]**

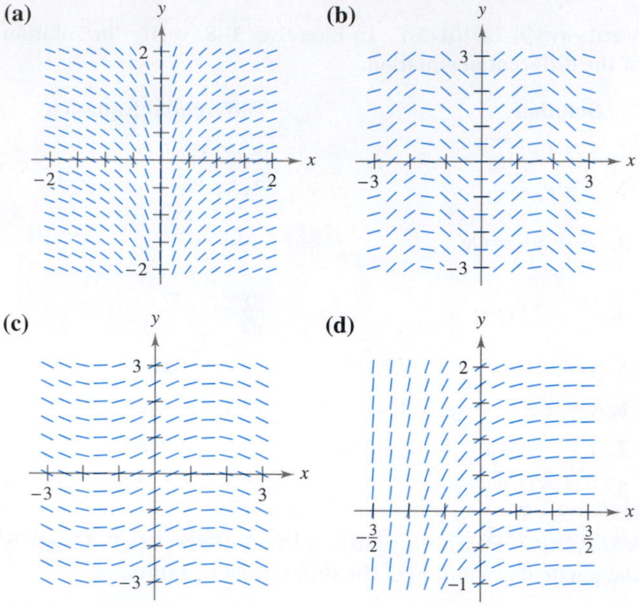

57. $\dfrac{dy}{dx} = \sin(2x)$

58. $\dfrac{dy}{dx} = \dfrac{1}{2} \cos x$

59. $\dfrac{dy}{dx} = e^{-2x}$

60. $\dfrac{dy}{dx} = \dfrac{1}{x}$

Slope Field **In Exercises 61–64, (a) sketch the slope field for the differential equation, (b) use the slope field to sketch the solution that passes through the given point, and (c) discuss the graph of the solution as $x \to \infty$ and $x \to -\infty$. Use a graphing utility to verify your results. To print a blank graph, go to *MathGraphs.com*.**

61. $y' = 3 - x, \quad (4, 2)$

62. $y' = \frac{1}{3}x^2 - \frac{1}{2}x, \quad (1, 1)$

63. $y' = y - 4x, \quad (2, 2)$

64. $y' = y + xy, \quad (0, -4)$

65. Slope Field Use the slope field for the differential equation $y' = 1/x$, where $x > 0$, to sketch the graph of the solution that satisfies each given initial condition. Then make a conjecture about the behavior of a particular solution of $y' = 1/x$ as $x \to \infty$. To print an enlarged copy of the graph, go to *MathGraphs.com*.

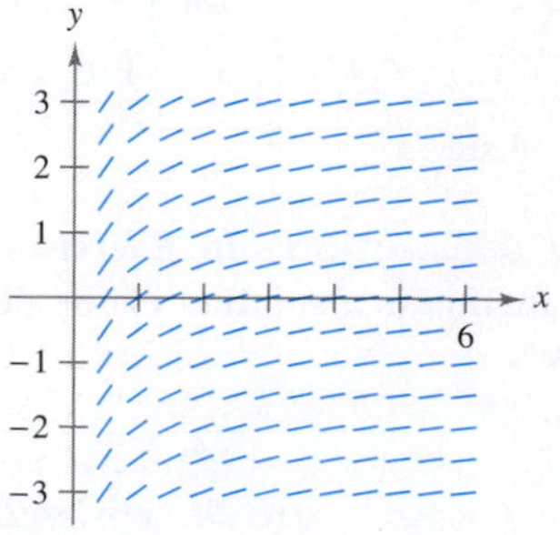

(a) $(1, 0)$

(b) $(2, -1)$

66. Slope Field Use the slope field for the differential equation $y' = 1/y$, where $y > 0$, to sketch the graph of the solution that satisfies each given initial condition. Then make a conjecture about the behavior of a particular solution of $y' = 1/y$ as $x \to \infty$. To print an enlarged copy of the graph, go to *MathGraphs.com*.

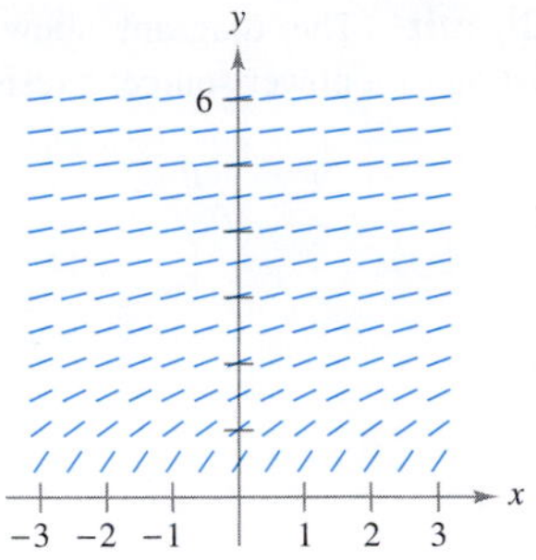

(a) $(0, 1)$ (b) $(1, 1)$

Slope Field In Exercises 67–72, use a computer algebra system to (a) graph the slope field for the differential equation and (b) graph the solution satisfying the specified initial condition.

67. $\dfrac{dy}{dx} = 0.25y, \quad y(0) = 4$

68. $\dfrac{dy}{dx} = 4 - y, \quad y(0) = 6$

69. $\dfrac{dy}{dx} = 0.02y(10 - y), \quad y(0) = 2$

70. $\dfrac{dy}{dx} = 0.2x(2 - y), \quad y(0) = 9$

71. $\dfrac{dy}{dx} = 0.4y(3 - x), \quad y(0) = 1$

72. $\dfrac{dy}{dx} = \dfrac{1}{2}e^{-x/8} \sin \dfrac{\pi y}{4}, \quad y(0) = 2$

Euler's Method In Exercises 73–78, use Euler's Method to make a table of values for the approximate solution of the differential equation with the specified initial value. Use *n* steps of size *h*.

73. $y' = x + y, \quad y(0) = 2, \quad n = 10, \quad h = 0.1$

74. $y' = x + y, \quad y(0) = 2, \quad n = 20, \quad h = 0.05$

75. $y' = 3x - 2y, \quad y(0) = 3, \quad n = 10, \quad h = 0.05$

76. $y' = 0.5x(3 - y), \quad y(0) = 1, \quad n = 5, \quad h = 0.4$

77. $y' = e^{xy}, \quad y(0) = 1, \quad n = 10, \quad h = 0.1$

78. $y' = \cos x + \sin y, \quad y(0) = 5, \quad n = 10, \quad h = 0.1$

Euler's Method In Exercises 79–81, complete the table using the exact solution of the differential equation and two approximations obtained using Euler's Method to approximate the particular solution of the differential equation. Use $h = 0.2$ and $h = 0.1$, and compute each approximation to four decimal places.

x	0	0.2	0.4	0.6	0.8	1
$y(x)$ (exact)						
$y(x)$ $(h = 0.2)$						
$y(x)$ $(h = 0.1)$						

Table for 79–81

	Differential Equation	Initial Condition	Exact Solution
79.	$\dfrac{dy}{dx} = y$	$(0, 3)$	$y = 3e^x$
80.	$\dfrac{dy}{dx} = \dfrac{2x}{y}$	$(0, 2)$	$y = \sqrt{2x^2 + 4}$
81.	$\dfrac{dy}{dx} = y + \cos(x)$	$(0, 0)$	$y = \dfrac{1}{2}(\sin x - \cos x + e^x)$

82. Euler's Method Compare the values of the approximations in Exercises 79–81 with the values given by the exact solution. How does the error change as h increases?

83. Temperature At time $t = 0$ minutes, the temperature of an object is 140°F. The temperature of the object is changing at the rate given by the differential equation

$$\frac{dy}{dt} = -\frac{1}{2}(y - 72).$$

(a) Use a graphing utility and Euler's Method to approximate the particular solutions of this differential equation at $t = 1$, 2, and 3. Use a step size of $h = 0.1$. (A graphing utility program for Euler's Method is available at the website *college.hmco.com*.)

(b) Compare your results with the exact solution

$$y = 72 + 68e^{-t/2}.$$

(c) Repeat parts (a) and (b) using a step size of $h = 0.05$. Compare the results.

84. HOW DO YOU SEE IT? The graph shows a solution of one of the following differential equations. Determine the correct equation. Explain your reasoning.

(a) $y' = xy$

(b) $y' = \dfrac{4x}{y}$

(c) $y' = -4xy$

(d) $y' = 4 - xy$

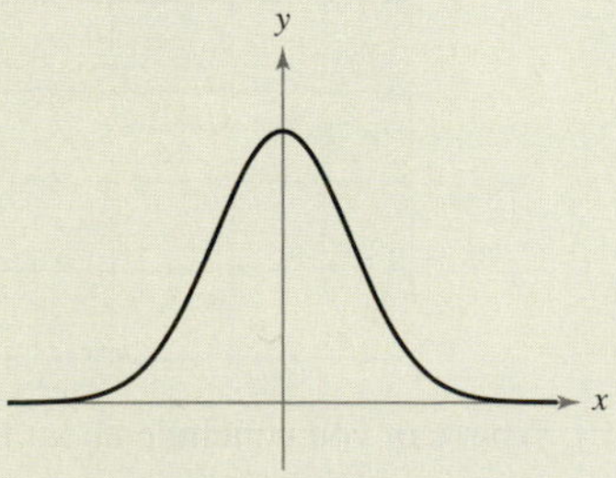

WRITING ABOUT CONCEPTS

85. General and Particular Solutions In your own words, describe the difference between a general solution of a differential equation and a particular solution.

86. Slope Field Explain how to interpret a slope field.

87. Euler's Method Describe how to use Euler's Method to approximate a particular solution of a differential equation.

88. Finding Values It is known that $y = Ce^{kx}$ is a solution of the differential equation $y' = 0.07y$. Is it possible to determine C or k from the information given? If so, find its value.

True or False? **In Exercises 89–92, determine whether the statement is true or false. If it is false, explain why or give an example that shows it is false.**

89. If $y = f(x)$ is a solution of a first-order differential equation, then $y = f(x) + C$ is also a solution.

90. The general solution of a differential equation is $y = -4.9x^2 + C_1x + C_2$. To find a particular solution, you must be given two initial conditions.

91. Slope fields represent the general solutions of differential equations.

92. A slope field shows that the slope at the point $(1, 1)$ is 6. This slope field represents the family of solutions for the differential equation $y' = 4x + 2y$.

93. Errors and Euler's Method The exact solution of the differential equation

$$\frac{dy}{dx} = -2y$$

where $y(0) = 4$, is $y = 4e^{-2x}$.

(a) Use a graphing utility to complete the table, where y is the exact value of the solution, y_1 is the approximate solution using Euler's Method with $h = 0.1$, y_2 is the approximate solution using Euler's Method with $h = 0.2$, e_1 is the absolute error $|y - y_1|$, e_2 is the absolute error $|y - y_2|$, and r is the ratio e_1/e_2.

x	0	0.2	0.4	0.6	0.8	1
y						
y_1						
y_2						
e_1						
e_2						
r						

(b) What can you conclude about the ratio r as h changes?

(c) Predict the absolute error when $h = 0.05$.

94. Errors and Euler's Method Repeat Exercise 93 for which the exact solution of the differential equation

$$\frac{dy}{dx} = x - y$$

where $y(0) = 1$, is $y = x - 1 + 2e^{-x}$.

95. Electric Circuit The diagram shows a simple electric circuit consisting of a power source, a resistor, and an inductor.

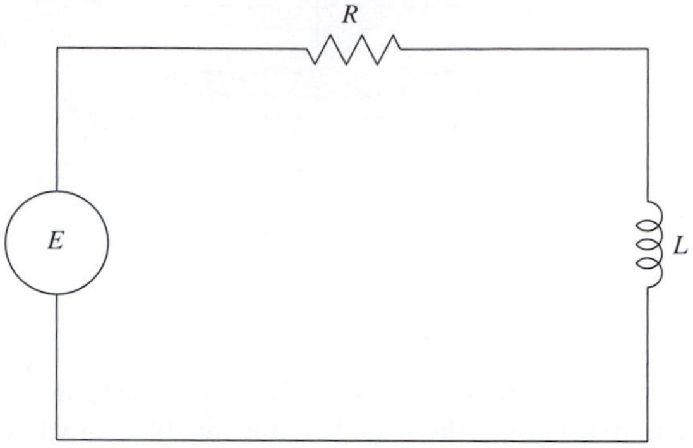

A model of the current I, in amperes (A), at time t is given by the first-order differential equation

$$L\frac{dI}{dt} + RI = E(t)$$

where $E(t)$ is the voltage (V) produced by the power source, R is the resistance, in ohms (Ω), and L is the inductance, in henrys (H). Suppose the electric circuit consists of a 24-V power source, a 12-Ω resistor, and a 4-H inductor.

(a) Sketch a slope field for the differential equation.

(b) What is the limiting value of the current? Explain.

96. Think About It It is known that $y = e^{kt}$ is a solution of the differential equation $y'' - 16y = 0$. Find the values of k.

97. Think About It It is known that $y = A \sin \omega t$ is a solution of the differential equation $y'' + 16y = 0$. Find the values of ω.

PUTNAM EXAM CHALLENGE

98. Let f be a twice-differentiable real-valued function satisfying

$$f(x) + f''(x) = -xg(x)f'(x)$$

where $g(x) \geq 0$ for all real x. Prove that $|f(x)|$ is bounded.

99. Prove that if the family of integral curves of the differential equation

$$\frac{dy}{dx} + p(x)y = q(x), \quad p(x) \cdot q(x) \neq 0$$

is cut by the line $x = k$, the tangents at the points of intersection are concurrent.

6.2 Differential Equations: Growth and Decay

- Use separation of variables to solve a simple differential equation.
- Use exponential functions to model growth and decay in applied problems.

Differential Equations

In Section 6.1, you learned to analyze the solutions visually of differential equations using slope fields and to approximate solutions numerically using Euler's Method. Analytically, you have learned to solve only two types of differential equations—those of the forms $y' = f(x)$ and $y'' = f(x)$. In this section, you will learn how to solve a more general type of differential equation. The strategy is to rewrite the equation so that each variable occurs on only one side of the equation. This strategy is called *separation of variables*. (You will study this strategy in detail in Section 6.3.)

EXAMPLE 1 Solving a Differential Equation

$$y' = \frac{2x}{y} \qquad \text{Original equation}$$

$$yy' = 2x \qquad \text{Multiply both sides by } y.$$

$$\int yy'\,dx = \int 2x\,dx \qquad \text{Integrate with respect to } x.$$

$$\int y\,dy = \int 2x\,dx \qquad dy = y'\,dx$$

$$\frac{1}{2}y^2 = x^2 + C_1 \qquad \text{Apply Power Rule.}$$

$$y^2 - 2x^2 = C \qquad \text{Rewrite, letting } C = 2C_1.$$

So, the general solution is $y^2 - 2x^2 = C$.

REMARK You can use implicit differentiation to check the solution in Example 1.

When you integrate both sides of the equation in Example 1, you don't need to add a constant of integration to both sides. When you do, you still obtain the same result.

$$\int y\,dy = \int 2x\,dx$$

$$\frac{1}{2}y^2 + C_2 = x^2 + C_3$$

$$\frac{1}{2}y^2 = x^2 + (C_3 - C_2)$$

$$\frac{1}{2}y^2 = x^2 + C_1$$

Some people prefer to use Leibniz notation and differentials when applying separation of variables. The solution to Example 1 is shown below using this notation.

$$\frac{dy}{dx} = \frac{2x}{y}$$

$$y\,dy = 2x\,dx$$

$$\int y\,dy = \int 2x\,dx$$

$$\frac{1}{2}y^2 = x^2 + C_1$$

$$y^2 - 2x^2 = C$$

Exploration

In Example 1, the general solution of the differential equation is

$$y^2 - 2x^2 = C.$$

Use a graphing utility to sketch the particular solutions for $C = \pm 2$, $C = \pm 1$, and $C = 0$. Describe the solutions graphically. Is the following statement true of each solution?

The slope of the graph at the point (x, y) is equal to twice the ratio of x and y.

Explain your reasoning. Are all curves for which this statement is true represented by the general solution?

Growth and Decay Models

In many applications, the rate of change of a variable y is proportional to the value of y. When y is a function of time t, the proportion can be written as shown.

Rate of change of y is proportional to y.

$$\frac{dy}{dt} = ky$$

The general solution of this differential equation is given in the next theorem.

THEOREM 6.1 Exponential Growth and Decay Model

If y is a differentiable function of t such that $y > 0$ and $y' = ky$ for some constant k, then

$$y = Ce^{kt}$$

where C is the **initial value** of y, and k is the **proportionality constant. Exponential growth** occurs when $k > 0$, and **exponential decay** occurs when $k < 0$.

Proof

$y' = ky$ Write original equation.

$\frac{y'}{y} = k$ Separate variables.

$\int \frac{y'}{y}\,dt = \int k\,dt$ Integrate with respect to t.

$\int \frac{1}{y}\,dy = \int k\,dt$ $dy = y'\,dt$

$\ln y = kt + C_1$ Find antiderivative of each side.

$y = e^{kt}e^{C_1}$ Solve for y.

$y = Ce^{kt}$ Let $C = e^{C_1}$.

So, all solutions of $y' = ky$ are of the form $y = Ce^{kt}$. Remember that you can differentiate the function $y = Ce^{kt}$ with respect to t to verify that $y' = ky$.

See LarsonCalculus.com for Bruce Edwards's video of this proof.

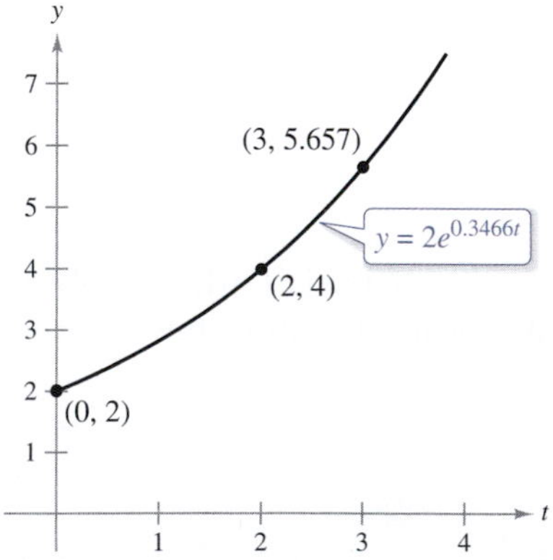

If the rate of change of y is proportional to y, then y follows an exponential model.
Figure 6.7

EXAMPLE 2 Using an Exponential Growth Model

The rate of change of y is proportional to y. When $t = 0$, $y = 2$, and when $t = 2$, $y = 4$. What is the value of y when $t = 3$?

Solution Because $y' = ky$, you know that y and t are related by the equation $y = Ce^{kt}$. You can find the values of the constants C and k by applying the initial conditions.

$2 = Ce^0 \Rightarrow C = 2$ When $t = 0$, $y = 2$.

$4 = 2e^{2k} \Rightarrow k = \frac{1}{2}\ln 2 \approx 0.3466$ When $t = 2$, $y = 4$.

So, the model is $y = 2e^{0.3466t}$. When $t = 3$, the value of y is $2e^{0.3466(3)} \approx 5.657$ (see Figure 6.7).

Using logarithmic properties, the value of k in Example 2 can also be written as $\ln\sqrt{2}$. So, the model becomes $y = 2e^{(\ln\sqrt{2})t}$, which can be rewritten as $y = 2(\sqrt{2})^t$.

TECHNOLOGY Most graphing utilities have curve-fitting capabilities that can be used to find models that represent data. Use the *exponential regression* feature of a graphing utility and the information in Example 2 to find a model for the data. How does your model compare with the given model?

Radioactive decay is measured in terms of *half-life*—the number of years required for half of the atoms in a sample of radioactive material to decay. The rate of decay is proportional to the amount present. The half-lives of some common radioactive isotopes are listed below.

Uranium (^{238}U)	4,470,000,000 years
Plutonium (^{239}Pu)	24,100 years
Carbon (^{14}C)	5715 years
Radium (^{226}Ra)	1599 years
Einsteinium (^{254}Es)	276 days
Radon (^{222}Rn)	3.82 days
Nobelium (^{257}No)	25 seconds

EXAMPLE 3 Radioactive Decay

The Fukushima Daiichi nuclear disaster occurred after an earthquake and tsunami. Several of the reactors at the plant experienced full meltdowns.

Ten grams of the plutonium isotope ^{239}Pu were released in a nuclear accident. How long will it take for the 10 grams to decay to 1 gram?

Solution Let y represent the mass (in grams) of the plutonium. Because the rate of decay is proportional to y, you know that

$$y = Ce^{kt}$$

where t is the time in years. To find the values of the constants C and k, apply the initial conditions. Using the fact that $y = 10$ when $t = 0$, you can write

$$10 = Ce^{k(0)} \quad \Rightarrow \quad 10 = Ce^0$$

which implies that $C = 10$. Next, using the fact that the half-life of ^{239}Pu is 24,100 years, you have $y = 10/2 = 5$ when $t = 24{,}100$, so you can write

$$5 = 10e^{k(24{,}100)}$$

$$\frac{1}{2} = e^{24{,}100k}$$

$$\frac{1}{24{,}100}\ln\frac{1}{2} = k$$

$$-0.000028761 \approx k.$$

So, the model is

$$y = 10e^{-0.000028761t}. \qquad \text{Half-life model}$$

To find the time it would take for 10 grams to decay to 1 gram, you can solve for t in the equation

$$1 = 10e^{-0.000028761t}.$$

The solution is approximately 80,059 years.

REMARK The exponential decay model in Example 3 could also be written as $y = 10\left(\frac{1}{2}\right)^{t/24{,}100}$. This model is much easier to derive, but for some applications it is not as convenient to use.

From Example 3, notice that in an exponential growth or decay problem, it is easy to solve for C when you are given the value of y at $t = 0$. The next example demonstrates a procedure for solving for C and k when you do not know the value of y at $t = 0$.

EXAMPLE 4 Population Growth

See LarsonCalculus.com for an interactive version of this type of example.

An experimental population of fruit flies increases according to the law of exponential growth. There were 100 flies after the second day of the experiment and 300 flies after the fourth day. Approximately how many flies were in the original population?

Solution Let $y = Ce^{kt}$ be the number of flies at time t, where t is measured in days. Note that y is continuous, whereas the number of flies is discrete. Because $y = 100$ when $t = 2$ and $y = 300$ when $t = 4$, you can write

$$100 = Ce^{2k} \quad \text{and} \quad 300 = Ce^{4k}.$$

From the first equation, you know that

$$C = 100e^{-2k}.$$

Substituting this value into the second equation produces the following.

$$\begin{aligned} 300 &= 100e^{-2k}e^{4k} \\ 300 &= 100e^{2k} \\ 3 &= e^{2k} \\ \ln 3 &= 2k \\ \frac{1}{2}\ln 3 &= k \\ 0.5493 &\approx k \end{aligned}$$

So, the exponential growth model is

$$y = Ce^{0.5493t}.$$

To solve for C, reapply the condition $y = 100$ when $t = 2$ and obtain

$$\begin{aligned} 100 &= Ce^{0.5493(2)} \\ C &= 100e^{-1.0986} \\ C &\approx 33. \end{aligned}$$

So, the original population (when $t = 0$) consisted of approximately $y = C = 33$ flies, as shown in Figure 6.8.

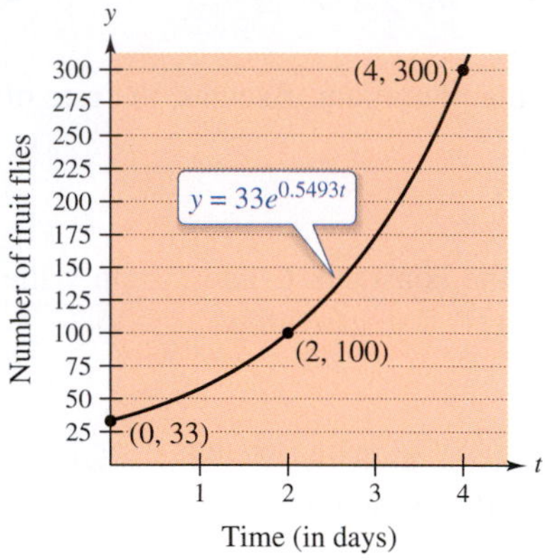

Figure 6.8

EXAMPLE 5 Declining Sales

Four months after it stops advertising, a manufacturing company notices that its sales have dropped from 100,000 units per month to 80,000 units per month. The sales follow an exponential pattern of decline. What will the sales be after another 2 months?

Solution Use the exponential decay model $y = Ce^{kt}$, where t is measured in months. From the initial condition $(t = 0)$, you know that $C = 100{,}000$. Moreover, because $y = 80{,}000$ when $t = 4$, you have

$$\begin{aligned} 80{,}000 &= 100{,}000e^{4k} \\ 0.8 &= e^{4k} \\ \ln(0.8) &= 4k \\ -0.0558 &\approx k. \end{aligned}$$

So, after 2 more months $(t = 6)$, you can expect the monthly sales rate to be

$$\begin{aligned} y &= 100{,}000e^{-0.0558(6)} \\ &\approx 71{,}500 \text{ units.} \end{aligned}$$

See Figure 6.9.

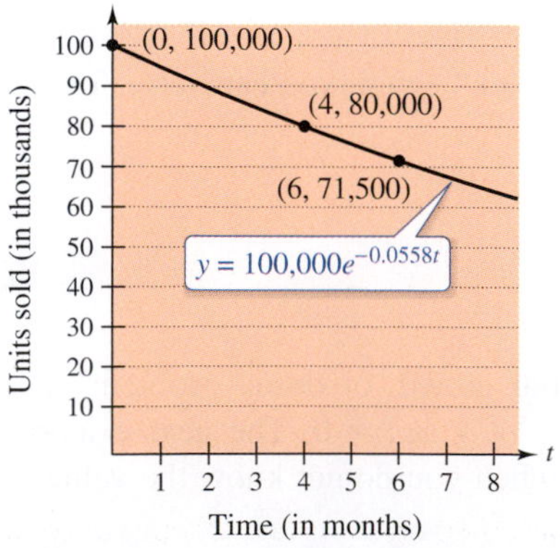

Figure 6.9

In Examples 2 through 5, you did not actually have to solve the differential equation $y' = ky$. (This was done once in the proof of Theorem 6.1.) The next example demonstrates a problem whose solution involves the separation of variables technique. The example concerns **Newton's Law of Cooling,** which states that the rate of change in the temperature of an object is proportional to the difference between the object's temperature and the temperature of the surrounding medium.

EXAMPLE 6 Newton's Law of Cooling

Let y represent the temperature (in °F) of an object in a room whose temperature is kept at a constant 60°. The object cools from 100° to 90° in 10 minutes. How much longer will it take for the temperature of the object to decrease to 80°?

Solution From Newton's Law of Cooling, you know that the rate of change in y is proportional to the difference between y and 60. This can be written as

$$y' = k(y - 60), \quad 80 \le y \le 100.$$

To solve this differential equation, use separation of variables, as shown.

$$\frac{dy}{dt} = k(y - 60) \qquad \text{Differential equation}$$

$$\left(\frac{1}{y - 60}\right) dy = k\,dt \qquad \text{Separate variables.}$$

$$\int \frac{1}{y - 60}\,dy = \int k\,dt \qquad \text{Integrate each side.}$$

$$\ln|y - 60| = kt + C_1 \qquad \text{Find antiderivative of each side.}$$

Because $y > 60$, $|y - 60| = y - 60$, and you can omit the absolute value signs. Using exponential notation, you have

$$y - 60 = e^{kt + C_1}$$

$$y = 60 + Ce^{kt}. \qquad C = e^{C_1}$$

Using $y = 100$ when $t = 0$, you obtain

$$100 = 60 + Ce^{k(0)} = 60 + C$$

which implies that $C = 40$. Because $y = 90$ when $t = 10$,

$$90 = 60 + 40e^{k(10)}$$

$$30 = 40e^{10k}$$

$$k = \frac{1}{10} \ln \frac{3}{4}.$$

So, $k \approx -0.02877$ and the model is

$$y = 60 + 40e^{-0.02877t}. \qquad \text{Cooling model}$$

When $y = 80$, you obtain

$$80 = 60 + 40e^{-0.02877t}$$

$$20 = 40e^{-0.02877t}$$

$$\frac{1}{2} = e^{-0.02877t}$$

$$\ln \frac{1}{2} = -0.02877t$$

$$t \approx 24.09 \text{ minutes.}$$

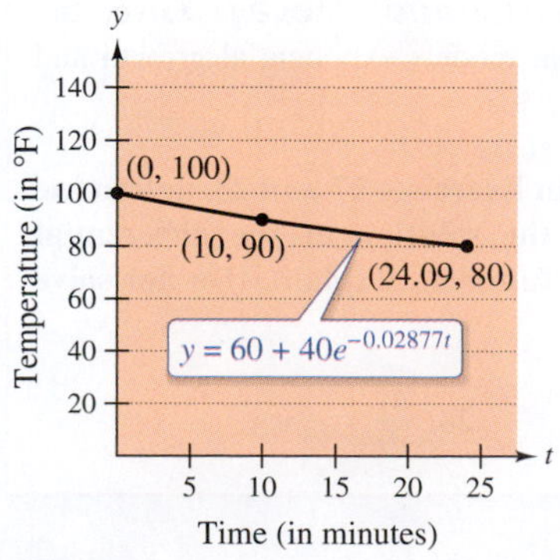

Figure 6.10

So, it will require about 14.09 *more* minutes for the object to cool to a temperature of 80° (see Figure 6.10).

6.2 Exercises

See CalcChat.com for tutorial help and worked-out solutions to odd-numbered exercises.

Solving a Differential Equation In Exercises 1–10, solve the differential equation.

1. $\dfrac{dy}{dx} = x + 3$

2. $\dfrac{dy}{dx} = 5 - 8x$

3. $\dfrac{dy}{dx} = y + 3$

4. $\dfrac{dy}{dx} = 6 - y$

5. $y' = \dfrac{5x}{y}$

6. $y' = -\dfrac{\sqrt{x}}{4y}$

7. $y' = \sqrt{x}\,y$

8. $y' = x(1 + y)$

9. $(1 + x^2)y' - 2xy = 0$

10. $xy + y' = 100x$

Writing and Solving a Differential Equation In Exercises 11 and 12, write and solve the differential equation that models the verbal statement.

11. The rate of change of Q with respect to t is inversely proportional to the square of t.

12. The rate of change of P with respect to t is proportional to $25 - t$.

Slope Field In Exercises 13 and 14, a differential equation, a point, and a slope field are given. (a) Sketch two approximate solutions of the differential equation on the slope field, one of which passes through the given point. (b) Use integration to find the particular solution of the differential equation and use a graphing utility to graph the solution. Compare the result with the sketch in part (a). To print an enlarged copy of the graph, go to *MathGraphs.com.*

13. $\dfrac{dy}{dx} = x(6 - y), \quad (0, 0)$

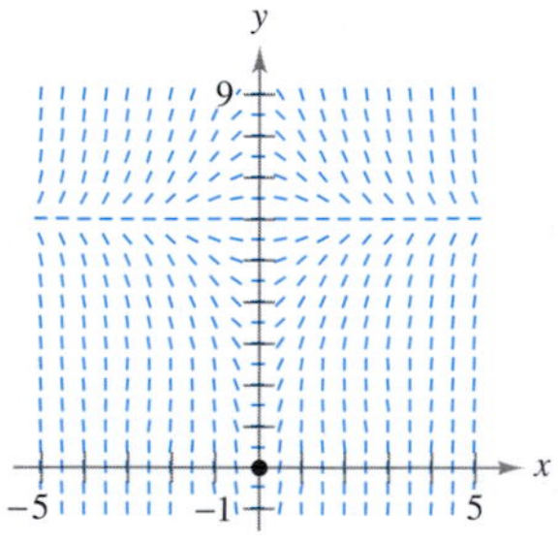

14. $\dfrac{dy}{dx} = xy, \quad \left(0, \dfrac{1}{2}\right)$

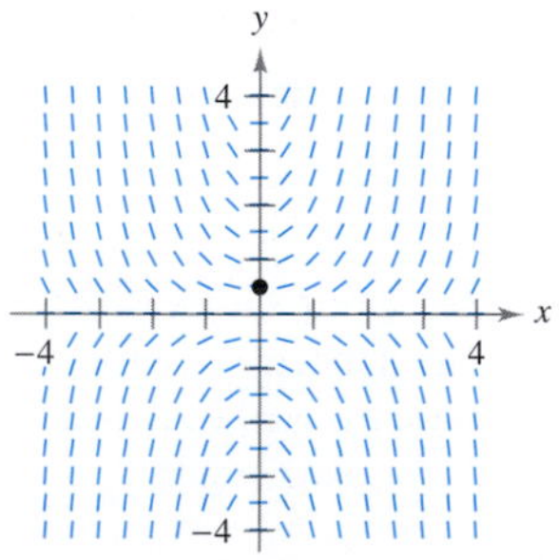

Finding a Particular Solution In Exercises 15–18, find the function $y = f(t)$ passing through the point $(0, 10)$ with the given first derivative. Use a graphing utility to graph the solution.

15. $\dfrac{dy}{dt} = \dfrac{1}{2}t$

16. $\dfrac{dy}{dt} = -9\sqrt{t}$

17. $\dfrac{dy}{dt} = -\dfrac{1}{2}y$

18. $\dfrac{dy}{dt} = \dfrac{3}{4}y$

Writing and Solving a Differential Equation In Exercises 19 and 20, write and solve the differential equation that models the verbal statement. Evaluate the solution at the specified value of the independent variable.

19. The rate of change of N is proportional to N. When $t = 0$, $N = 250$, and when $t = 1$, $N = 400$. What is the value of N when $t = 4$?

20. The rate of change of P is proportional to P. When $t = 0$, $P = 5000$, and when $t = 1$, $P = 4750$. What is the value of P when $t = 5$?

Finding an Exponential Function In Exercises 21–24, find the exponential function $y = Ce^{kt}$ that passes through the two given points.

21.

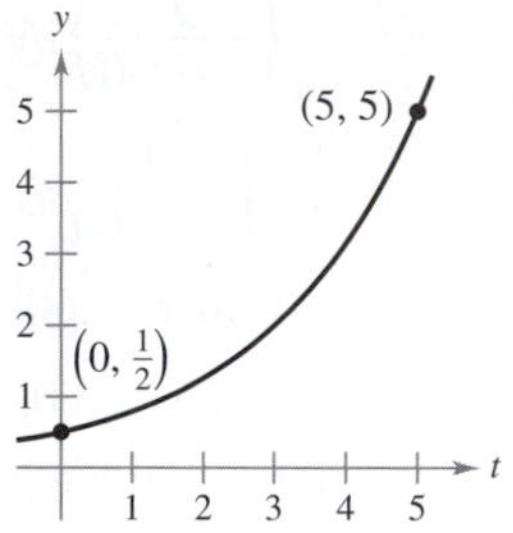

22.

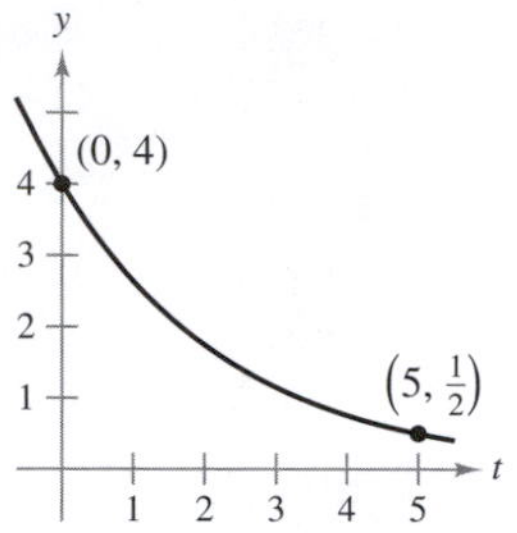

23.

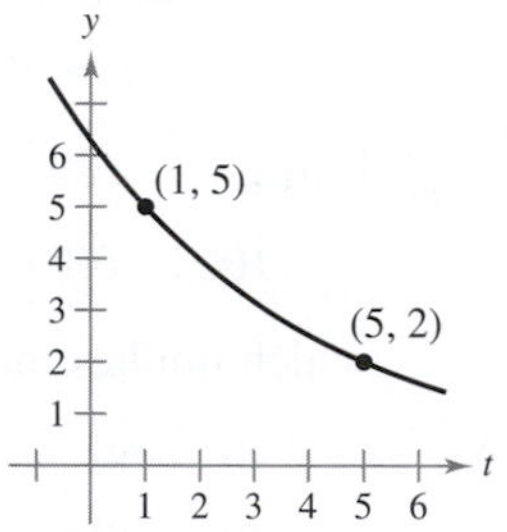

24.

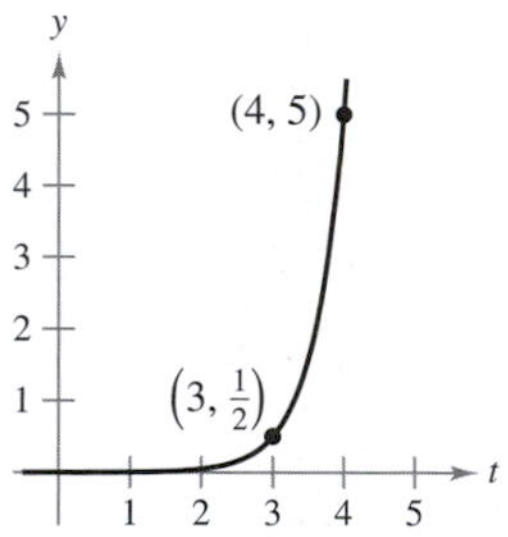

WRITING ABOUT CONCEPTS

25. **Describing Values** Describe what the values of C and k represent in the exponential growth and decay model, $y = Ce^{kt}$.

26. **Exponential Growth and Decay** Give the differential equation that models exponential growth and decay.

Increasing Function In Exercises 27 and 28, determine the quadrants in which the solution of the differential equation is an increasing function. Explain. (Do not solve the differential equation.)

27. $\dfrac{dy}{dx} = \dfrac{1}{2}xy$

28. $\dfrac{dy}{dx} = \dfrac{1}{2}x^2y$

Radioactive Decay **In Exercises 29–36, complete the table for the radioactive isotope.**

	Isotope	Half-life (in years)	Initial Quantity	Amount After 1000 Years	Amount After 10,000 Years
29.	^{226}Ra	1599	20 g		
30.	^{226}Ra	1599		1.5 g	
31.	^{226}Ra	1599			0.1 g
32.	^{14}C	5715			3 g
33.	^{14}C	5715	5 g		
34.	^{14}C	5715		1.6 g	
35.	^{239}Pu	24,100		2.1 g	
36.	^{239}Pu	24,100			0.4 g

37. Radioactive Decay Radioactive radium has a half-life of approximately 1599 years. What percent of a given amount remains after 100 years?

38. Carbon Dating Carbon-14 dating assumes that the carbon dioxide on Earth today has the same radioactive content as it did centuries ago. If this is true, the amount of ^{14}C absorbed by a tree that grew several centuries ago should be the same as the amount of ^{14}C absorbed by a tree growing today. A piece of ancient charcoal contains only 15% as much of the radioactive carbon as a piece of modern charcoal. How long ago was the tree burned to make the ancient charcoal? (The half-life of ^{14}C is 5715 years.)

Compound Interest **In Exercises 39–44, complete the table for a savings account in which interest is compounded continuously.**

	Initial Investment	Annual Rate	Time to Double	Amount After 10 Years
39.	\$4000	6%		
40.	\$18,000	$5\frac{1}{2}\%$		
41.	\$750		$7\frac{3}{4}$ yr	
42.	\$12,500		20 yr	
43.	\$500			\$1292.85
44.	\$6000			\$8950.95

Compound Interest **In Exercises 45–48, find the principal *P* that must be invested at rate *r*, compounded monthly, so that \$1,000,000 will be available for retirement in *t* years.**

45. $r = 7\frac{1}{2}\%, \quad t = 20$

46. $r = 6\%, \quad t = 40$

47. $r = 8\%, \quad t = 35$

48. $r = 9\%, \quad t = 25$

Compound Interest **In Exercises 49 and 50, find the time necessary for \$1000 to double when it is invested at a rate of *r* compounded (a) annually, (b) monthly, (c) daily, and (d) continuously.**

49. $r = 7\%$ **50.** $r = 5.5\%$

Population **In Exercises 51–54, the population (in millions) of a country in 2011 and the expected continuous annual rate of change *k* of the population are given.** **(*Source: U.S. Census Bureau, International Data Base*)**

(a) Find the exponential growth model

$$P = Ce^{kt}$$

for the population by letting $t = 0$ correspond to 2010.

(b) Use the model to predict the population of the country in 2020.

(c) Discuss the relationship between the sign of *k* and the change in population for the country.

	Country	2011 Population	k
51.	Latvia	2.2	−0.006
52.	Egypt	82.1	0.020
53.	Uganda	34.6	0.036
54.	Hungary	10.0	−0.002

55. Modeling Data One hundred bacteria are started in a culture and the number N of bacteria is counted each hour for 5 hours. The results are shown in the table, where t is the time in hours.

t	0	1	2	3	4	5
N	100	126	151	198	243	297

(a) Use the regression capabilities of a graphing utility to find an exponential model for the data.

(b) Use the model to estimate the time required for the population to quadruple in size.

56. Bacteria Growth The number of bacteria in a culture is increasing according to the law of exponential growth. There are 125 bacteria in the culture after 2 hours and 350 bacteria after 4 hours.

(a) Find the initial population.

(b) Write an exponential growth model for the bacteria population. Let t represent time in hours.

(c) Use the model to determine the number of bacteria after 8 hours.

(d) After how many hours will the bacteria count be 25,000?

57. Learning Curve The management at a certain factory has found that a worker can produce at most 30 units in a day. The learning curve for the number of units N produced per day after a new employee has worked t days is

$$N = 30(1 - e^{kt}).$$

After 20 days on the job, a particular worker produces 19 units.

(a) Find the learning curve for this worker.

(b) How many days should pass before this worker is producing 25 units per day?

58. Learning Curve Suppose the management in Exercise 57 requires a new employee to produce at least 20 units per day after 30 days on the job.

(a) Find the learning curve that describes this minimum requirement.

(b) Find the number of days before a minimal achiever is producing 25 units per day.

59. Insect Population

(a) Suppose an insect population increases by a constant number each month. Explain why the number of insects can be represented by a linear function.

(b) Suppose an insect population increases by a constant percentage each month. Explain why the number of insects can be represented by an exponential function.

60. HOW DO YOU SEE IT? The functions f and g are both of the form $y = Ce^{kt}$.

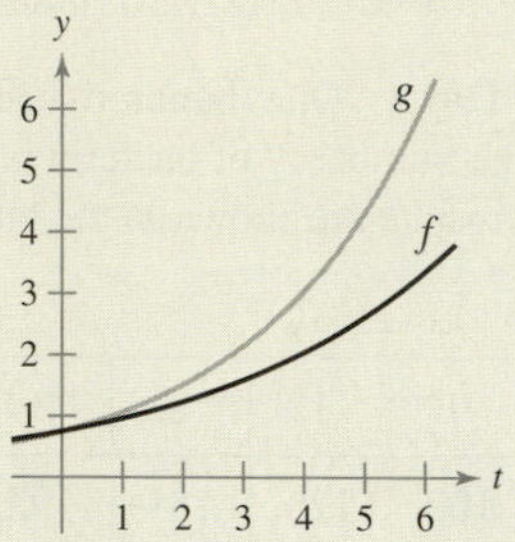

(a) Do the functions f and g represent exponential growth or exponential decay? Explain.

(b) Assume both functions have the same value of C. Which function has a greater value of k? Explain.

61. Modeling Data The table shows the resident populations P (in millions) of the United States from 1920 to 2010. (*Source: U.S. Census Bureau*)

Year	1920	1930	1940	1950	1960
Population, P	106	123	132	151	179

Year	1970	1980	1990	2000	2010
Population, P	203	227	249	281	309

(a) Use the 1920 and 1930 data to find an exponential model P_1 for the data. Let $t = 0$ represent 1920.

(b) Use a graphing utility to find an exponential model P_2 for all the data. Let $t = 0$ represent 1920.

(c) Use a graphing utility to plot the data and graph models P_1 and P_2 in the same viewing window. Compare the actual data with the predictions. Which model better fits the data?

(d) Use the model chosen in part (c) to estimate when the resident population will be 400 million.

Stephen Aaron Rees/Shutterstock.com

62. Forestry

The value of a tract of timber is

$$V(t) = 100{,}000e^{0.8\sqrt{t}}$$

where t is the time in years, with $t = 0$ corresponding to 2010. If money earns interest continuously at 10%, then the present value of the timber at any time t is

$$A(t) = V(t)e^{-0.10t}.$$

Find the year in which the timber should be harvested to maximize the present value function.

63. Sound Intensity The level of sound β (in decibels) with an intensity of I is

$$\beta(I) = 10 \log_{10}\left(\frac{I}{I_0}\right)$$

where I_0 is an intensity of 10^{-16} watt per square centimeter, corresponding roughly to the faintest sound that can be heard. Determine $\beta(I)$ for the following.

(a) $I = 10^{-14}$ watt per square centimeter (whisper)

(b) $I = 10^{-9}$ watt per square centimeter (busy street corner)

(c) $I = 10^{-6.5}$ watt per square centimeter (air hammer)

(d) $I = 10^{-4}$ watt per square centimeter (threshold of pain)

64. Noise Level With the installation of noise suppression materials, the noise level in an auditorium was reduced from 93 to 80 decibels. Use the function in Exercise 63 to find the percent decrease in the intensity level of the noise as a result of the installation of these materials.

65. Newton's Law of Cooling When an object is removed from a furnace and placed in an environment with a constant temperature of 80°F, its core temperature is 1500°F. One hour after it is removed, the core temperature is 1120°F. Find the core temperature 5 hours after the object is removed from the furnace.

66. Newton's Law of Cooling A container of hot liquid is placed in a freezer that is kept at a constant temperature of 20°F. The initial temperature of the liquid is 160°F. After 5 minutes, the liquid's temperature is 60°F. How much longer will it take for its temperature to decrease to 30°F?

True or False? **In Exercises 67–70, determine whether the statement is true or false. If it is false, explain why or give an example that shows it is false.**

67. In exponential growth, the rate of growth is constant.

68. In linear growth, the rate of growth is constant.

69. If prices are rising at a rate of 0.5% per month, then they are rising at a rate of 6% per year.

70. The differential equation modeling exponential growth is $dy/dx = ky$, where k is a constant.

6.3 Separation of Variables and the Logistic Equation

- **Recognize and solve differential equations that can be solved by separation of variables.**
- **Use differential equations to model and solve applied problems.**
- **Solve and analyze logistic differential equations.**

Separation of Variables

AP* Tips

Separable differential equations appear on both the AP Calculus AB and BC Exams.

Consider a differential equation that can be written in the form

$$M(x) + N(y)\frac{dy}{dx} = 0$$

where M is a continuous function of x alone and N is a continuous function of y alone. As you saw in Section 6.2, for this type of equation, all x terms can be collected with dx and all y terms with dy, and a solution can be obtained by integration. Such equations are said to be **separable,** and the solution procedure is called **separation of variables.** Below are some examples of differential equations that are separable.

Original Differential Equation	Rewritten with Variables Separated
$x^2 + 3y\dfrac{dy}{dx} = 0$	$3y\,dy = -x^2\,dx$
$(\sin x)y' = \cos x$	$dy = \cot x\,dx$
$\dfrac{xy'}{e^y + 1} = 2$	$\dfrac{1}{e^y + 1}\,dy = \dfrac{2}{x}\,dx$

EXAMPLE 1 Separation of Variables

See LarsonCalculus.com for an interactive version of this type of example.

Find the general solution of

$$(x^2 + 4)\frac{dy}{dx} = xy.$$

Solution To begin, note that $y = 0$ is a solution. To find other solutions, assume that $y \neq 0$ and separate variables as shown.

$$(x^2 + 4)\,dy = xy\,dx \qquad \text{Differential form}$$

$$\frac{dy}{y} = \frac{x}{x^2 + 4}\,dx \qquad \text{Separate variables.}$$

Now, integrate to obtain

$$\int \frac{dy}{y} = \int \frac{x}{x^2 + 4}\,dx \qquad \text{Integrate.}$$

$$\ln|y| = \frac{1}{2}\ln(x^2 + 4) + C_1$$

$$\ln|y| = \ln\sqrt{x^2 + 4} + C_1$$

$$|y| = e^{C_1}\sqrt{x^2 + 4}$$

$$y = \pm e^{C_1}\sqrt{x^2 + 4}.$$

Because $y = 0$ is also a solution, you can write the general solution as

$$y = C\sqrt{x^2 + 4}. \qquad \text{General solution}$$

REMARK Be sure to check your solutions throughout this chapter. In Example 1, you can check the solution

$$y = C\sqrt{x^2 + 4}$$

by differentiating and substituting into the original equation.

$$(x^2 + 4)\frac{dy}{dx} = xy$$

$$(x^2 + 4)\frac{Cx}{\sqrt{x^2 + 4}} \stackrel{?}{=} x\left(C\sqrt{x^2 + 4}\right)$$

$$Cx\sqrt{x^2 + 4} = Cx\sqrt{x^2 + 4}$$

So, the solution checks.

In some cases, it is not feasible to write the general solution in the explicit form $y = f(x)$. The next example illustrates such a solution. Implicit differentiation can be used to verify this solution.

FOR FURTHER INFORMATION For an example (from engineering) of a differential equation that is separable, see the article "Designing a Rose Cutter" by J. S. Hartzler in *The College Mathematics Journal*. To view this article, go to *MathArticles.com*.

EXAMPLE 2 Finding a Particular Solution

Given the initial condition $y(0) = 1$, find the particular solution of the equation

$$xy\,dx + e^{-x^2}(y^2 - 1)\,dy = 0.$$

Solution Note that $y = 0$ is a solution of the differential equation—but this solution does not satisfy the initial condition. So, you can assume that $y \neq 0$. To separate variables, you must rid the first term of y and the second term of e^{-x^2}. So, you should multiply by e^{x^2}/y and obtain the following.

$$\begin{aligned} xy\,dx + e^{-x^2}(y^2 - 1)\,dy &= 0 \\ e^{-x^2}(y^2 - 1)\,dy &= -xy\,dx \\ \int\left(y - \frac{1}{y}\right)dy &= \int -xe^{x^2}\,dx \\ \frac{y^2}{2} - \ln|y| &= -\frac{1}{2}e^{x^2} + C \end{aligned}$$

From the initial condition $y(0) = 1$, you have

$$\frac{1}{2} - 0 = -\frac{1}{2} + C$$

which implies that $C = 1$. So, the particular solution has the implicit form

$$\begin{aligned} \frac{y^2}{2} - \ln|y| &= -\frac{1}{2}e^{x^2} + 1 \\ y^2 - \ln y^2 + e^{x^2} &= 2. \end{aligned}$$

You can check this by differentiating and rewriting to get the original equation.

EXAMPLE 3 Finding a Particular Solution Curve

Find the equation of the curve that passes through the point $(1, 3)$ and has a slope of y/x^2 at any point (x, y).

Solution Because the slope of the curve is y/x^2, you have

$$\frac{dy}{dx} = \frac{y}{x^2}$$

with the initial condition $y(1) = 3$. Separating variables and integrating produces

$$\begin{aligned} \int \frac{dy}{y} &= \int \frac{dx}{x^2}, \quad y \neq 0 \\ \ln|y| &= -\frac{1}{x} + C_1 \\ y &= e^{-(1/x) + C_1} \\ y &= Ce^{-1/x}. \end{aligned}$$

Because $y = 3$ when $x = 1$, it follows that $3 = Ce^{-1}$ and $C = 3e$. So, the equation of the specified curve is

$$y = (3e)e^{-1/x} \quad \Rightarrow \quad y = 3e^{(x-1)/x}, \quad x > 0.$$

Because the solution is not defined at $x = 0$ and the initial condition is given at $x = 1$, x is restricted to positive values. See Figure 6.11.

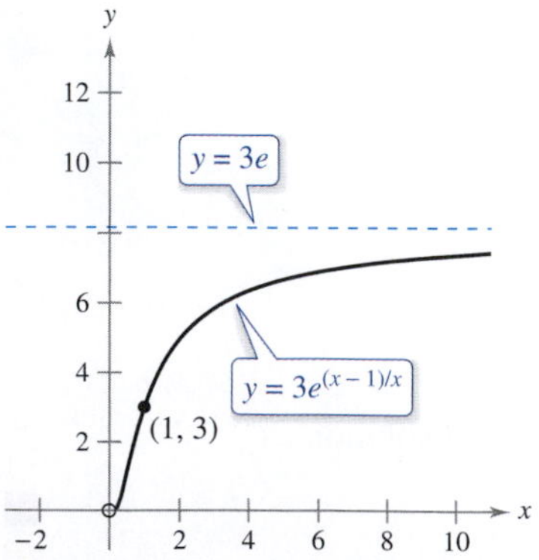

Figure 6.11

Applications

EXAMPLE 4 Wildlife Population

The rate of change of the number of coyotes $N(t)$ in a population is directly proportional to $650 - N(t)$, where t is the time in years. When $t = 0$, the population is 300, and when $t = 2$, the population has increased to 500. Find the population when $t = 3$.

Solution Because the rate of change of the population is proportional to $650 - N(t)$, or $650 - N$, you can write the differential equation

$$\frac{dN}{dt} = k(650 - N).$$

You can solve this equation using separation of variables.

$$dN = k(650 - N)\,dt \qquad \text{Differential form}$$

$$\frac{dN}{650 - N} = k\,dt \qquad \text{Separate variables.}$$

$$-\ln|650 - N| = kt + C_1 \qquad \text{Integrate.}$$

$$\ln|650 - N| = -kt - C_1$$

$$650 - N = e^{-kt - C_1} \qquad \text{Assume } N < 650.$$

$$N = 650 - Ce^{-kt} \qquad \text{General solution}$$

Using $N = 300$ when $t = 0$, you can conclude that $C = 350$, which produces

$$N = 650 - 350e^{-kt}.$$

Then, using $N = 500$ when $t = 2$, it follows that

$$500 = 650 - 350e^{-2k} \quad \Rightarrow \quad e^{-2k} = \frac{3}{7} \quad \Rightarrow \quad k \approx 0.4236.$$

So, the model for the coyote population is

$$N = 650 - 350e^{-0.4236t}. \qquad \text{Model for population}$$

When $t = 3$, you can approximate the population to be

$$N = 650 - 350e^{-0.4236(3)}$$
$$\approx 552 \text{ coyotes.}$$

The model for the population is shown in Figure 6.12. Note that $N = 650$ is the horizontal asymptote of the graph and is the *carrying capacity* of the model. You will learn more about carrying capacity later in this section.

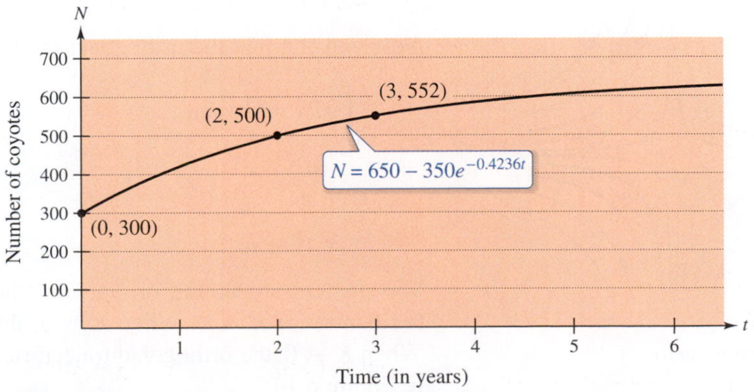

Figure 6.12

franzfoto.com/Alamy

A common problem in electrostatics, thermodynamics, and hydrodynamics involves finding a family of curves, each of which is orthogonal to all members of a given family of curves. For example, Figure 6.13 shows a family of circles

$$x^2 + y^2 = C \quad \text{Family of circles}$$

each of which intersects the lines in the family

$$y = Kx \quad \text{Family of lines}$$

at right angles. Two such families of curves are said to be **mutually orthogonal,** and each curve in one of the families is called an **orthogonal trajectory** of the other family. In electrostatics, lines of force are orthogonal to the *equipotential curves*. In thermodynamics, the flow of heat across a plane surface is orthogonal to the *isothermal curves*. In hydrodynamics, the flow (stream) lines are orthogonal trajectories of the *velocity potential curves*.

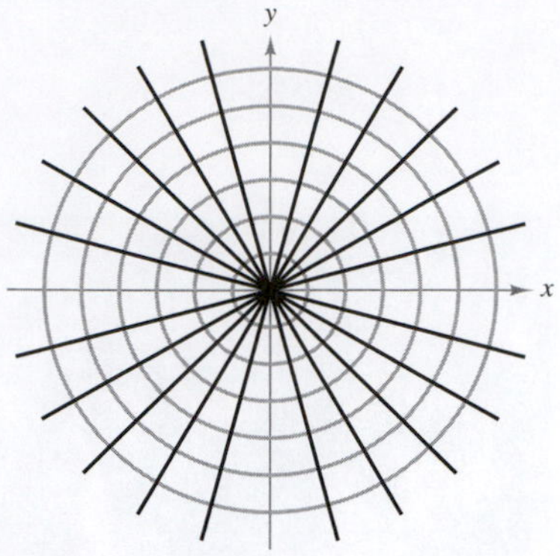

Each line $y = Kx$ is an orthogonal trajectory of the family of circles.
Figure 6.13

EXAMPLE 5 Finding Orthogonal Trajectories

Describe the orthogonal trajectories for the family of curves given by

$$y = \frac{C}{x}$$

for $C \neq 0$. Sketch several members of each family.

Solution First, solve the given equation for C and write $xy = C$. Then, by differentiating implicitly with respect to x, you obtain the differential equation

$$x\frac{dy}{dx} + y = 0 \quad \text{Differential equation}$$

$$x\frac{dy}{dx} = -y$$

$$\frac{dy}{dx} = -\frac{y}{x}. \quad \text{Slope of given family}$$

Because dy/dx represents the slope of the given family of curves at (x, y), it follows that the orthogonal family has the negative reciprocal slope x/y. So,

$$\frac{dy}{dx} = \frac{x}{y}. \quad \text{Slope of orthogonal family}$$

Now you can find the orthogonal family by separating variables and integrating.

$$\int y\,dy = \int x\,dx$$

$$\frac{y^2}{2} = \frac{x^2}{2} + C_1$$

$$y^2 - x^2 = K$$

The centers are at the origin, and the transverse axes are vertical for $K > 0$ and horizontal for $K < 0$. When $K = 0$, the orthogonal trajectories are the lines $y = \pm x$. When $K \neq 0$, the orthogonal trajectories are hyperbolas. Several trajectories are shown in Figure 6.14.

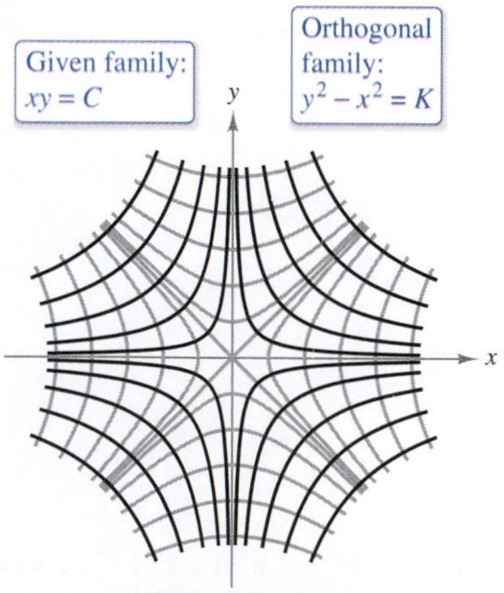

Orthogonal trajectories
Figure 6.14

Logistic Differential Equation

In Section 6.2, the exponential growth model was derived from the fact that the rate of change of a variable y is proportional to the value of y. You observed that the differential equation $dy/dt = ky$ has the general solution $y = Ce^{kt}$. Exponential growth is unlimited, but when describing a population, there often exists some upper limit L past which growth cannot occur. This upper limit L is called the **carrying capacity,** which is the maximum population $y(t)$ that can be sustained or supported as time t increases. A model that is often used to describe this type of growth is the **logistic differential equation**

$$\frac{dy}{dt} = ky\left(1 - \frac{y}{L}\right) \qquad \text{Logistic differential equation}$$

where k and L are positive constants. A population that satisfies this equation does not grow without bound, but approaches the carrying capacity L as t increases.

From the equation, you can see that if y is between 0 and the carrying capacity L, then $dy/dt > 0$, and the population increases. If y is greater than L, then $dy/dt < 0$, and the population decreases. The graph of the function y is called the *logistic curve*, as shown in Figure 6.15.

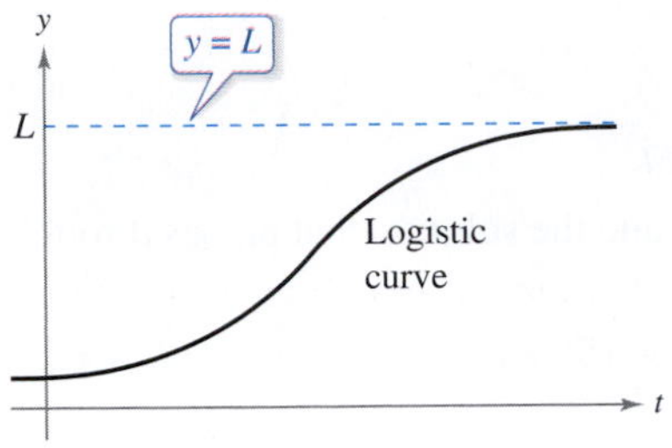

Note that as $t \to \infty$, $y \to L$.
Figure 6.15

AP* Tips

Logistic equations appear only on the AP Calculus BC Exam. On the free response section, you may be asked to solve a logistic equation by means of separation of variables. You may also be asked to demonstrate knowledge of carrying capacity and inflection point locations.

EXAMPLE 6 Deriving the General Solution

Solve the logistic differential equation

$$\frac{dy}{dt} = ky\left(1 - \frac{y}{L}\right).$$

Solution Begin by separating variables.

$$\frac{dy}{dt} = ky\left(1 - \frac{y}{L}\right) \qquad \text{Write differential equation.}$$

$$\frac{1}{y(1 - y/L)}dy = k\,dt \qquad \text{Separate variables.}$$

$$\int \frac{1}{y(1 - y/L)}dy = \int k\,dt \qquad \text{Integrate each side.}$$

$$\int \left(\frac{1}{y} + \frac{1}{L - y}\right)dy = \int k\,dt \qquad \text{Rewrite left side using partial fractions.}$$

$$\ln|y| - \ln|L - y| = kt + C \qquad \text{Find antiderivative of each side.}$$

$$\ln\left|\frac{L - y}{y}\right| = -kt - C \qquad \text{Multiply each side by } -1 \text{ and simplify.}$$

$$\left|\frac{L - y}{y}\right| = e^{-kt - C} \qquad \text{Exponentiate each side.}$$

$$\left|\frac{L - y}{y}\right| = e^{-C}e^{-kt} \qquad \text{Property of exponents}$$

$$\frac{L - y}{y} = be^{-kt} \qquad \text{Let } \pm e^{-C} = b.$$

Solving this equation for y produces $y = \dfrac{L}{1 + be^{-kt}}$. ■

REMARK A review of the method of partial fractions is given in Section 8.5.

From Example 6, you can conclude that all solutions of the logistic differential equation are of the general form

$$y = \frac{L}{1 + be^{-kt}}.$$

Exploration

Use a graphing utility to investigate the effects of the values of L, b, and k on the graph of

$$y = \frac{L}{1 + be^{-kt}}.$$

Include some examples to support your results.

EXAMPLE 7 Solving a Logistic Differential Equation

A state game commission releases 40 elk into a game refuge. After 5 years, the elk population is 104. The commission believes that the environment can support no more than 4000 elk. The growth rate of the elk population p is

$$\frac{dp}{dt} = kp\left(1 - \frac{p}{4000}\right), \quad 40 \le p \le 4000$$

where t is the number of years.

a. Write a model for the elk population in terms of t.

b. Graph the slope field for the differential equation and the solution that passes through the point (0, 40).

c. Use the model to estimate the elk population after 15 years.

d. Find the limit of the model as $t \to \infty$.

Solution

a. You know that $L = 4000$. So, the solution of the equation is of the form

$$p = \frac{4000}{1 + be^{-kt}}.$$

Because $p(0) = 40$, you can solve for b as follows.

$$40 = \frac{4000}{1 + be^{-k(0)}} \quad \Rightarrow \quad 40 = \frac{4000}{1 + b} \quad \Rightarrow \quad b = 99$$

Then, because $p = 104$ when $t = 5$, you can solve for k.

$$104 = \frac{4000}{1 + 99e^{-k(5)}} \quad \Rightarrow \quad k \approx 0.194$$

So, a model for the elk population is

$$p = \frac{4000}{1 + 99e^{-0.194t}}.$$

b. Using a graphing utility, you can graph the slope field for

$$\frac{dp}{dt} = 0.194p\left(1 - \frac{p}{4000}\right)$$

and the solution that passes through (0, 40), as shown in Figure 6.16.

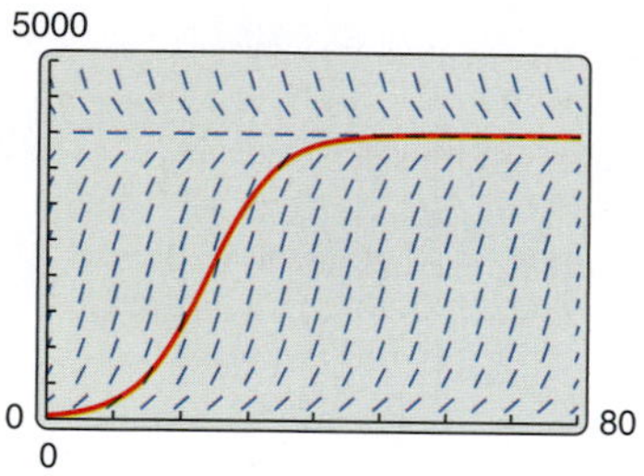

Slope field for

$$\frac{dp}{dt} = 0.194p\left(1 - \frac{p}{4000}\right)$$

and the solution passing through (0, 40)

Figure 6.16

c. To estimate the elk population after 15 years, substitute 15 for t in the model.

$$p = \frac{4000}{1 + 99e^{-0.194(15)}} \qquad \text{Substitute 15 for } t.$$

$$= \frac{4000}{1 + 99e^{-2.91}} \qquad \text{Simplify.}$$

$$\approx 626$$

d. As t increases without bound, the denominator of

$$\frac{4000}{1 + 99e^{-0.194t}}$$

gets closer and closer to 1. So,

$$\lim_{t \to \infty} \frac{4000}{1 + 99e^{-0.194t}} = 4000.$$

6.3 Exercises

See CalcChat.com for tutorial help and worked-out solutions to odd-numbered exercises.

Finding a General Solution Using Separation of Variables In Exercises 1–14, find the general solution of the differential equation.

1. $\dfrac{dy}{dx} = \dfrac{x}{y}$
2. $\dfrac{dy}{dx} = \dfrac{3x^2}{y^2}$
3. $x^2 + 5y\dfrac{dy}{dx} = 0$
4. $\dfrac{dy}{dx} = \dfrac{6 - x^2}{2y^3}$
5. $\dfrac{dr}{ds} = 0.75r$
6. $\dfrac{dr}{ds} = 0.75s$
7. $(2 + x)y' = 3y$
8. $xy' = y$
9. $yy' = 4 \sin x$
10. $yy' = -8 \cos(\pi x)$
11. $\sqrt{1 - 4x^2}\, y' = x$
12. $\sqrt{x^2 - 16}\, y' = 11x$
13. $y \ln x - xy' = 0$
14. $12yy' - 7e^x = 0$

Finding a Particular Solution Using Separation of Variables In Exercises 15–24, find the particular solution that satisfies the initial condition.

Differential Equation	Initial Condition
15. $yy' - 2e^x = 0$	$y(0) = 3$
16. $\sqrt{x} + \sqrt{y}\, y' = 0$	$y(1) = 9$
17. $y(x + 1) + y' = 0$	$y(-2) = 1$
18. $2xy' - \ln x^2 = 0$	$y(1) = 2$
19. $y(1 + x^2)y' - x(1 + y^2) = 0$	$y(0) = \sqrt{3}$
20. $y\sqrt{1 - x^2}\, y' - x\sqrt{1 - y^2} = 0$	$y(0) = 1$
21. $\dfrac{du}{dv} = uv \sin v^2$	$u(0) = 1$
22. $\dfrac{dr}{ds} = e^{r - 2s}$	$r(0) = 0$
23. $dP - kP\, dt = 0$	$P(0) = P_0$
24. $dT + k(T - 70)\, dt = 0$	$T(0) = 140$

Finding a Particular Solution In Exercises 25–28, find an equation of the graph that passes through the point and has the given slope.

25. $(0, 2), \quad y' = \dfrac{x}{4y}$
26. $(1, 1), \quad y' = -\dfrac{9x}{16y}$
27. $(9, 1), \quad y' = \dfrac{y}{2x}$
28. $(8, 2), \quad y' = \dfrac{2y}{3x}$

Using Slope In Exercises 29 and 30, find all functions f having the indicated property.

29. The tangent to the graph of f at the point (x, y) intersects the x-axis at $(x + 2, 0)$.
30. All tangents to the graph of f pass through the origin.

Slope Field In Exercises 31 and 32, sketch a few solutions of the differential equation on the slope field and then find the general solution analytically. To print an enlarged copy of the graph, go to *MathGraphs.com*.

31. $\dfrac{dy}{dx} = x$

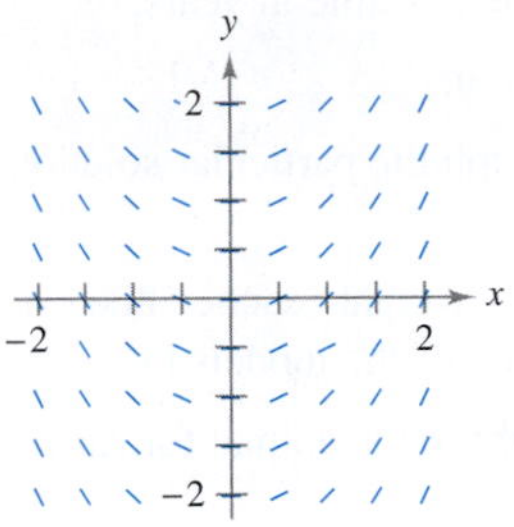

32. $\dfrac{dy}{dx} = -\dfrac{x}{y}$

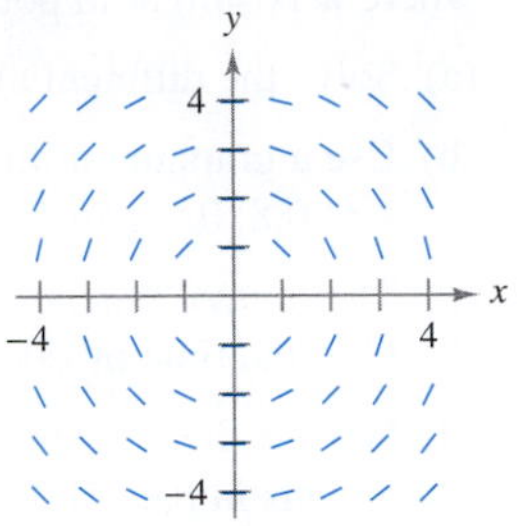

Slope Field In Exercises 33–36, (a) write a differential equation for the statement, (b) match the differential equation with a possible slope field, and (c) verify your result by using a graphing utility to graph a slope field for the differential equation. [The slope fields are labeled (a), (b), (c), and (d).] To print an enlarged copy of the graph, go to *MathGraphs.com*.

(a)

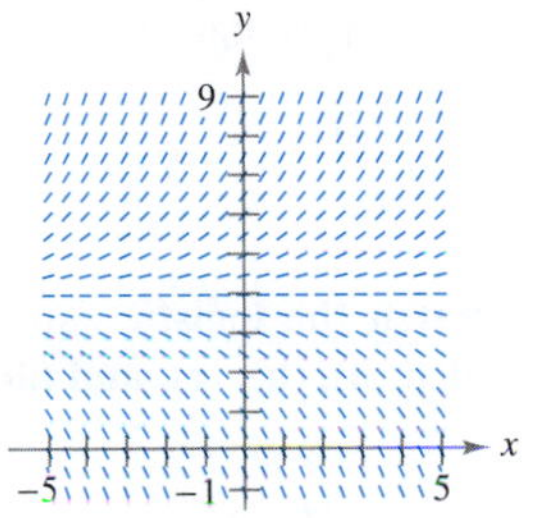

(b)

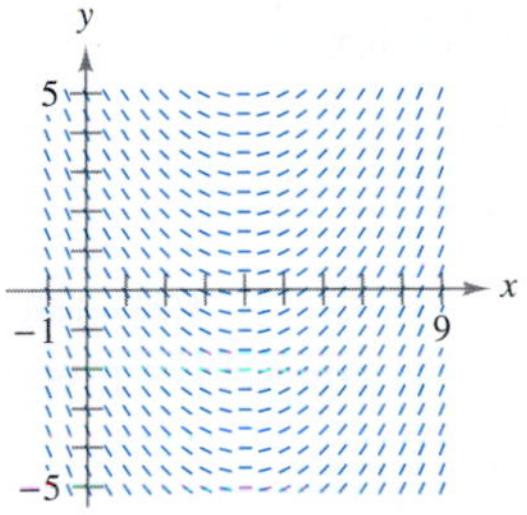

(c)

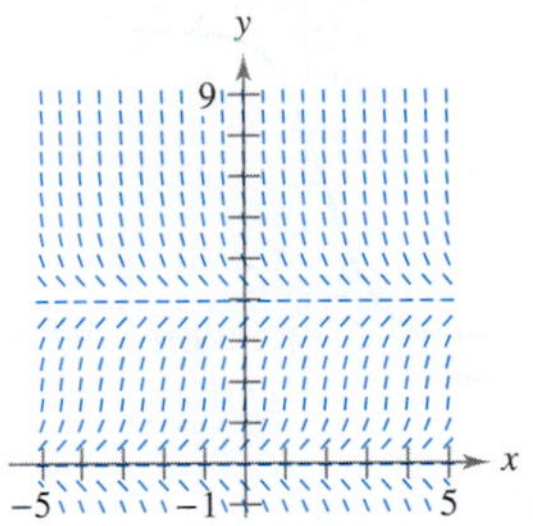

(d)

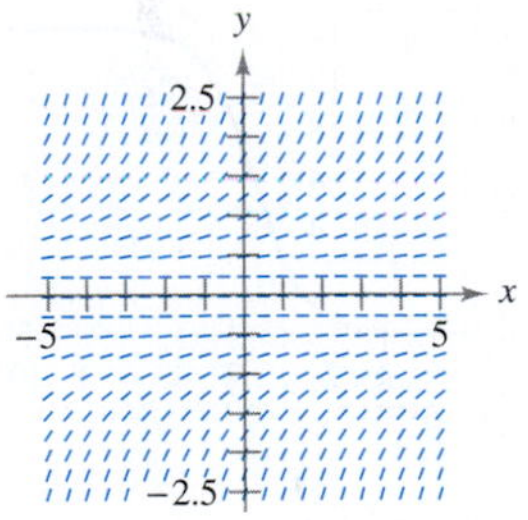

33. The rate of change of y with respect to x is proportional to the difference between y and 4.
34. The rate of change of y with respect to x is proportional to the difference between x and 4.
35. The rate of change of y with respect to x is proportional to the product of y and the difference between y and 4.
36. The rate of change of y with respect to x is proportional to y^2.
37. **Radioactive Decay** The rate of decomposition of radioactive radium is proportional to the amount present at any time. The half-life of radioactive radium is 1599 years. What percent of a present amount will remain after 50 years?

38. Chemical Reaction In a chemical reaction, a certain compound changes into another compound at a rate proportional to the unchanged amount. There is 40 grams of the original compound initially and 35 grams after 1 hour. When will 75 percent of the compound be changed?

39. Weight Gain A calf that weighs 60 pounds at birth gains weight at the rate

$$\frac{dw}{dt} = k(1200 - w)$$

where w is weight in pounds and t is time in years.

(a) Solve the differential equation.

(b) Use a graphing utility to graph the particular solutions for $k = 0.8$, 0.9, and 1.

(c) The animal is sold when its weight reaches 800 pounds. Find the time of sale for each of the models in part (b).

(d) What is the maximum weight of the animal for each of the models in part (b)?

40. Weight Gain A calf that weighs w_0 pounds at birth gains weight at the rate $dw/dt = 1200 - w$, where w is weight in pounds and t is time in years. Solve the differential equation.

Finding Orthogonal Trajectories **In Exercises 41–46, find the orthogonal trajectories of the family. Use a graphing utility to graph several members of each family.**

41. $x^2 + y^2 = C$

42. $x^2 - 2y^2 = C$

43. $x^2 = Cy$

44. $y^2 = 2Cx$

45. $y^2 = Cx^3$

46. $y = Ce^x$

Matching **In Exercises 47–50, match the logistic equation with its graph. [The graphs are labeled (a), (b), (c), and (d).]**

(a)

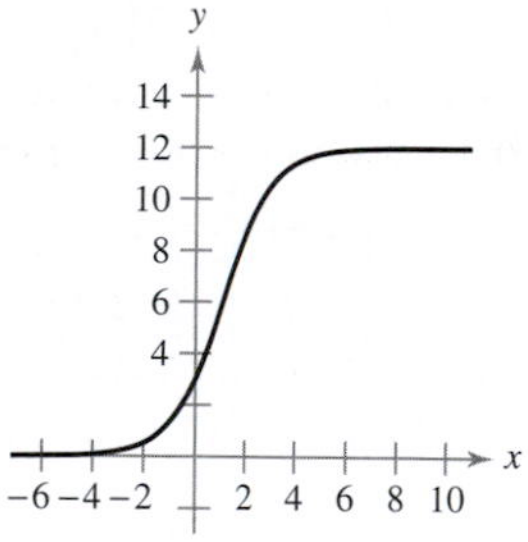

(b)

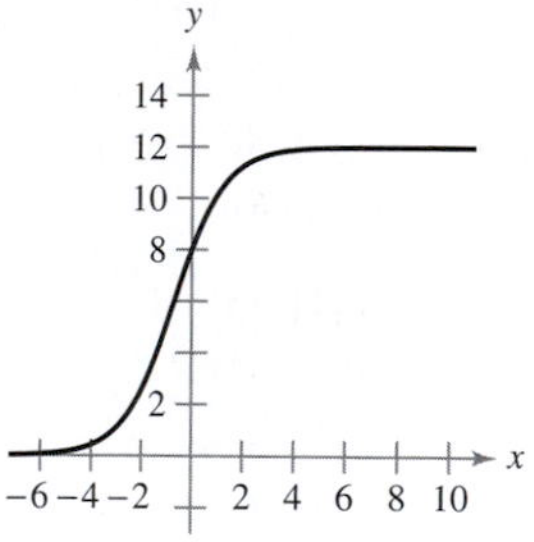

(c)

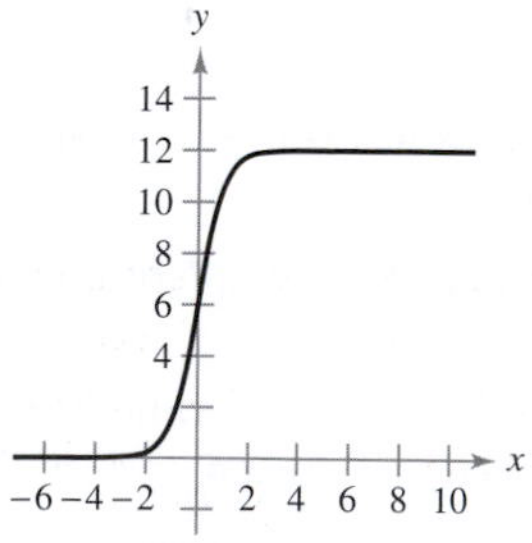

(d)

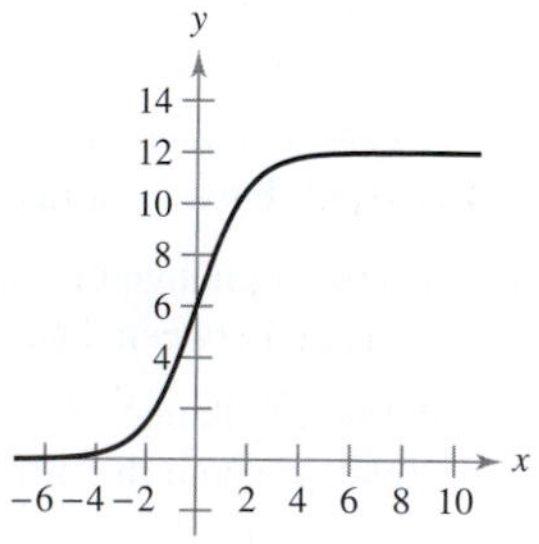

47. $y = \dfrac{12}{1 + e^{-x}}$

48. $y = \dfrac{12}{1 + 3e^{-x}}$

49. $y = \dfrac{12}{1 + \frac{1}{2}e^{-x}}$

50. $y = \dfrac{12}{1 + e^{-2x}}$

Using a Logistic Equation **In Exercises 51 and 52, the logistic equation models the growth of a population. Use the equation to (a) find the value of k, (b) find the carrying capacity, (c) find the initial population, (d) determine when the population will reach 50% of its carrying capacity, and (e) write a logistic differential equation that has the solution $P(t)$.**

51. $P(t) = \dfrac{2100}{1 + 29e^{-0.75t}}$

52. $P(t) = \dfrac{5000}{1 + 39e^{-0.2t}}$

Using a Logistic Differential Equation **In Exercises 53 and 54, the logistic differential equation models the growth rate of a population. Use the equation to (a) find the value of k, (b) find the carrying capacity, (c) graph a slope field using a computer algebra system, and (d) determine the value of P at which the population growth rate is the greatest.**

53. $\dfrac{dP}{dt} = 3P\left(1 - \dfrac{P}{100}\right)$

54. $\dfrac{dP}{dt} = 0.1P - 0.0004P^2$

Solving a Logistic Differential Equation **In Exercises 55–58, find the logistic equation that passes through the given point.**

55. $\dfrac{dy}{dt} = y\left(1 - \dfrac{y}{36}\right)$, $(0, 4)$

56. $\dfrac{dy}{dt} = 2.8y\left(1 - \dfrac{y}{10}\right)$, $(0, 7)$

57. $\dfrac{dy}{dt} = \dfrac{4y}{5} - \dfrac{y^2}{150}$, $(0, 8)$

58. $\dfrac{dy}{dt} = \dfrac{3y}{20} - \dfrac{y^2}{1600}$, $(0, 15)$

59. Endangered Species A conservation organization releases 25 Florida panthers into a game preserve. After 2 years, there are 39 panthers in the preserve. The Florida preserve has a carrying capacity of 200 panthers.

(a) Write a logistic equation that models the population of panthers in the preserve.

(b) Find the population after 5 years.

(c) When will the population reach 100?

(d) Write a logistic differential equation that models the growth rate of the panther population. Then repeat part (b) using Euler's Method with a step size of $h = 1$. Compare the approximation with the exact answer.

(e) At what time is the panther population growing most rapidly? Explain.

60. Bacteria Growth At time $t = 0$, a bacterial culture weighs 1 gram. Two hours later, the culture weighs 4 grams. The maximum weight of the culture is 20 grams.

(a) Write a logistic equation that models the weight of the bacterial culture.

(b) Find the culture's weight after 5 hours.

(c) When will the culture's weight reach 18 grams?

(d) Write a logistic differential equation that models the growth rate of the culture's weight. Then repeat part (b) using Euler's Method with a step size of $h = 1$. Compare the approximation with the exact answer.

(e) At what time is the culture's weight increasing most rapidly? Explain.

WRITING ABOUT CONCEPTS

61. Separation of Variables In your own words, describe how to recognize and solve differential equations that can be solved by separation of variables.

62. Mutually Orthogonal In your own words, describe the relationship between two families of curves that are mutually orthogonal.

63. Finding a Derivative Show that if

$$y = \frac{1}{1 + be^{-kt}}$$

then

$$\frac{dy}{dt} = ky(1 - y).$$

64. Point of Inflection For any logistic growth curve, show that the point of inflection occurs at $y = L/2$ when the solution starts below the carrying capacity L.

65. Sailing

Ignoring resistance, a sailboat starting from rest accelerates (dv/dt) at a rate proportional to the difference between the velocities of the wind and the boat.

(a) The wind is blowing at 20 knots, and after 1 half-hour, the boat is moving at 10 knots. Write the velocity v as a function of time t.

(b) Use the result of part (a) to write the distance traveled by the boat as a function of time.

66. HOW DO YOU SEE IT? The growth of a population is modeled by a logistic equation as shown in the graph below. What happens to the rate of growth as the population increases? What do you think causes this to occur in real-life situations, such as animal or human populations?

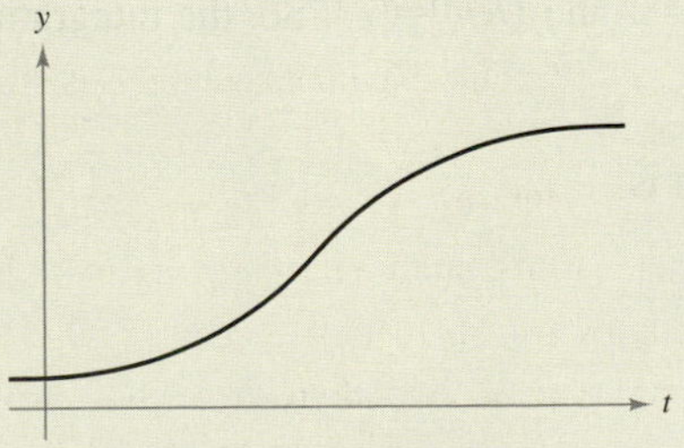

Determining if a Function Is Homogeneous **In Exercises 67–74, determine whether the function is homogeneous, and if it is, determine its degree. A function $f(x, y)$ is *homogeneous of degree n* if $f(tx, ty) = t^n f(x, y)$.**

67. $f(x, y) = x^3 - 4xy^2 + y^3$

68. $f(x, y) = x^3 + 3x^2y^2 - 2y^2$

69. $f(x, y) = \dfrac{x^2y^2}{\sqrt{x^2 + y^2}}$

70. $f(x, y) = \dfrac{xy}{\sqrt{x^2 + y^2}}$

71. $f(x, y) = 2 \ln xy$

72. $f(x, y) = \tan(x + y)$

73. $f(x, y) = 2 \ln \dfrac{x}{y}$

74. $f(x, y) = \tan \dfrac{y}{x}$

Solving a Homogeneous Differential Equation **In Exercises 75–80, solve the homogeneous differential equation in terms of x and y. A *homogeneous differential equation* is an equation of the form $M(x, y)\,dx + N(x, y)\,dy = 0$, where M and N are homogeneous functions of the same degree. To solve an equation of this form by the method of separation of variables, use the substitutions $y = vx$ and $dy = x\,dv + v\,dx$.**

75. $(x + y)\,dx - 2x\,dy = 0$

76. $(x^3 + y^3)\,dx - xy^2\,dy = 0$

77. $(x - y)\,dx - (x + y)\,dy = 0$

78. $(x^2 + y^2)\,dx - 2xy\,dy = 0$

79. $xy\,dx + (y^2 - x^2)\,dy = 0$

80. $(2x + 3y)\,dx - x\,dy = 0$

True or False? **In Exercises 81–83, determine whether the statement is true or false. If it is false, explain why or give an example that shows it is false.**

81. The function $y = 0$ is always a solution of a differential equation that can be solved by separation of variables.

82. The differential equation $y' = xy - 2y + x - 2$ can be written in separated variables form.

83. The families $x^2 + y^2 = 2Cy$ and $x^2 + y^2 = 2Kx$ are mutually orthogonal.

PUTNAM EXAM CHALLENGE

84. A not uncommon calculus mistake is to believe that the product rule for derivatives says that $(fg)' = f'g'$. If $f(x) = e^{x^2}$, determine, with proof, whether there exists an open interval (a, b) and a nonzero function g defined on (a, b) such that this wrong product rule is true for x in (a, b).

6.4 First-Order Linear Differential Equations

- Solve a first-order linear differential equation, and use linear differential equations to solve applied problems.

First-Order Linear Differential Equations

In this section, you will see how to solve a very important class of first-order differential equations—first-order linear differential equations.

ANNA JOHNSON PELL WHEELER (1883–1966)

Anna Johnson Pell Wheeler was awarded a master's degree in 1904 from the University of Iowa for her thesis *The Extension of Galois Theory to Linear Differential Equations*. Influenced by David Hilbert, she worked on integral equations while studying infinite linear spaces.

Definition of First-Order Linear Differential Equation

A **first-order linear differential equation** is an equation of the form

$$\frac{dy}{dx} + P(x)y = Q(x)$$

where P and Q are continuous functions of x. This first-order linear differential equation is said to be in **standard form.**

To solve a linear differential equation, write it in standard form to identify the functions $P(x)$ and $Q(x)$. Then integrate $P(x)$ and form the expression

$$u(x) = e^{\int P(x)\,dx} \qquad \text{Integrating factor}$$

which is called an **integrating factor.** The general solution of the equation is

$$y = \frac{1}{u(x)}\int Q(x)u(x)\,dx. \qquad \text{General solution}$$

It is instructive to see why the integrating factor helps solve a linear differential equation of the form $y' + P(x)y = Q(x)$. When both sides of the equation are multiplied by the integrating factor $u(x) = e^{\int P(x)\,dx}$, the left-hand side becomes the derivative of a product.

$$y'e^{\int P(x)\,dx} + P(x)ye^{\int P(x)\,dx} = Q(x)e^{\int P(x)\,dx}$$
$$\left[ye^{\int P(x)\,dx}\right]' = Q(x)e^{\int P(x)\,dx}$$

Integrating both sides of this second equation and dividing by $u(x)$ produce the general solution.

AP* Tips

The differential equations requiring the integrating factor are not required, but may be used to solve some of the problems on the AP Exam.

EXAMPLE 1 Solving a Linear Differential Equation

Find the general solution of

$$y' + y = e^x.$$

Solution For this equation, $P(x) = 1$ and $Q(x) = e^x$. So, the integrating factor is

$$u(x) = e^{\int P(x)\,dx} = e^{\int dx} = e^x.$$

This implies that the general solution is

$$\begin{aligned} y &= \frac{1}{u(x)}\int Q(x)u(x)\,dx \\ &= \frac{1}{e^x}\int e^x(e^x)\,dx \\ &= e^{-x}\left(\frac{1}{2}e^{2x} + C\right) \\ &= \frac{1}{2}e^x + Ce^{-x}. \end{aligned}$$

Courtesy of the Visual Collections, Canaday Library, Bryn Mawr College.

REMARK Rather than memorizing the formula in Theorem 6.2, just remember that multiplication by the integrating factor $e^{\int P(x)\,dx}$ converts the left side of the differential equation into the derivative of the product $ye^{\int P(x)\,dx}$.

THEOREM 6.2 Solution of a First-Order Linear Differential Equation

An integrating factor for the first-order linear differential equation

$$y' + P(x)y = Q(x)$$

is $u(x) = e^{\int P(x)\,dx}$. The solution of the differential equation is

$$ye^{\int P(x)\,dx} = \int Q(x)e^{\int P(x)\,dx}\,dx + C.$$

EXAMPLE 2 Solving a First-Order Linear Differential Equation

See LarsonCalculus.com for an interactive version of this type of example.

Find the general solution of $xy' - 2y = x^2$.

Solution The standard form of the equation is

$$y' + \left(-\frac{2}{x}\right)y = x. \qquad \text{Standard form}$$

So, $P(x) = -2/x$, and you have

$$\int P(x)\,dx = -\int \frac{2}{x}\,dx = -\ln x^2$$

which implies that the integrating factor is

$$e^{\int P(x)\,dx} = e^{-\ln x^2} = \frac{1}{e^{\ln x^2}} = \frac{1}{x^2}. \qquad \text{Integrating factor}$$

So, multiplying each side of the standard form by $1/x^2$ yields

$$\frac{y'}{x^2} - \frac{2y}{x^3} = \frac{1}{x}$$

$$\frac{d}{dx}\left[\frac{y}{x^2}\right] = \frac{1}{x}$$

$$\frac{y}{x^2} = \int \frac{1}{x}\,dx$$

$$\frac{y}{x^2} = \ln|x| + C$$

$$y = x^2(\ln|x| + C). \qquad \text{General solution}$$

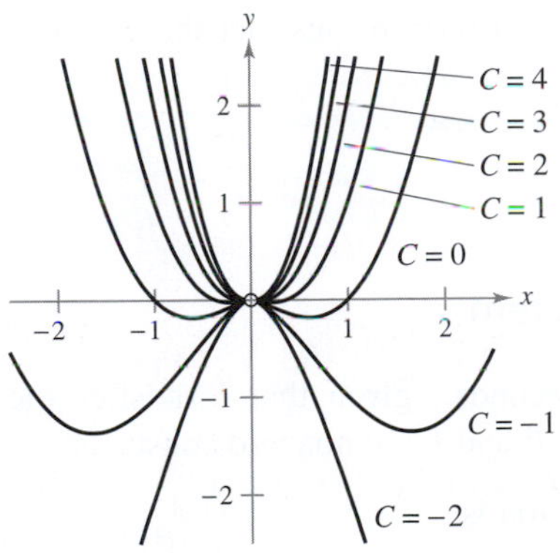

Figure 6.17

Several solution curves (for $C = -2, -1, 0, 1, 2, 3,$ and 4) are shown in Figure 6.17. ■

In most falling-body problems discussed so far in the text, air resistance has been neglected. The next example includes this factor. In the example, the air resistance on the falling object is assumed to be proportional to its velocity v. If g is the gravitational constant, the downward force F on a falling object of mass m is given by the difference $mg - kv$. If a is the acceleration of the object, then by Newton's Second Law of Motion,

$$F = ma = m\frac{dv}{dt}$$

which yields the following differential equation.

$$m\frac{dv}{dt} = mg - kv \quad \Longrightarrow \quad \frac{dv}{dt} + \frac{kv}{m} = g$$

EXAMPLE 3 A Falling Object with Air Resistance

An object of mass m is dropped from a hovering helicopter. The air resistance is proportional to the velocity of the object. Find the velocity of the object as a function of time t.

Solution The velocity v satisfies the equation

$$\frac{dv}{dt} + \frac{kv}{m} = g. \qquad g = \text{gravitational constant, } k = \text{constant of proportionality}$$

Letting $b = k/m$, you can *separate variables* to obtain

$$dv = (g - bv)\,dt$$

$$\int \frac{dv}{g - bv} = \int dt$$

$$-\frac{1}{b}\ln|g - bv| = t + C_1$$

$$\ln|g - bv| = -bt - bC_1$$

$$g - bv = Ce^{-bt}. \qquad C = e^{-bC_1}$$

Because the object was dropped, $v = 0$ when $t = 0$; so $g = C$, and it follows that

$$-bv = -g + ge^{-bt} \quad \Longrightarrow \quad v = \frac{g - ge^{-bt}}{b} = \frac{mg}{k}(1 - e^{-kt/m}).$$

REMARK Notice in Example 3 that the velocity approaches a limit of mg/k as a result of the air resistance. For falling-body problems in which air resistance is neglected, the velocity increases without bound.

A simple electric circuit consists of an electric current I (in amperes), a resistance R (in ohms), an inductance L (in henrys), and a constant electromotive force E (in volts), as shown in Figure 6.18. According to Kirchhoff's Second Law, if the switch S is closed when $t = 0$, then the applied electromotive force (voltage) is equal to the sum of the voltage drops in the rest of the circuit. This, in turn, means that the current I satisfies the differential equation

$$L\frac{dI}{dt} + RI = E.$$

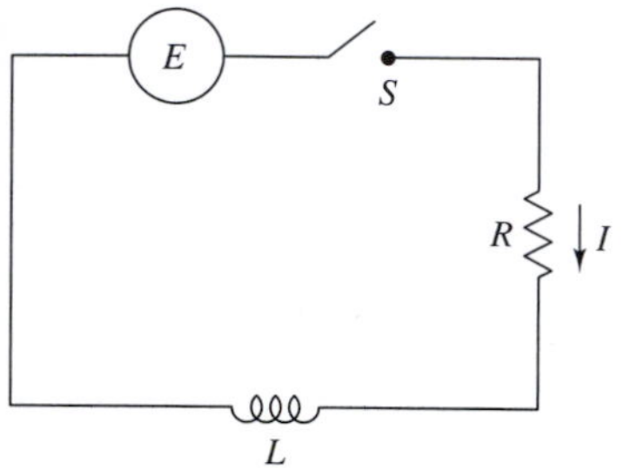

Figure 6.18

EXAMPLE 4 An Electric Circuit Problem

Find the current I as a function of time t (in seconds), given that I satisfies the differential equation $L(dI/dt) + RI = \sin 2t$, where R and L are nonzero constants.

Solution In standard form, the given linear equation is

$$\frac{dI}{dt} + \frac{R}{L}I = \frac{1}{L}\sin 2t.$$

Let $P(t) = R/L$, so that $e^{\int P(t)\,dt} = e^{(R/L)t}$, and, by Theorem 6.2,

$$Ie^{(R/L)t} = \frac{1}{L}\int e^{(R/L)t}\sin 2t\,dt$$

$$= \frac{1}{4L^2 + R^2}e^{(R/L)t}(R\sin 2t - 2L\cos 2t) + C.$$

So, the general solution is

$$I = e^{-(R/L)t}\left[\frac{1}{4L^2 + R^2}e^{(R/L)t}(R\sin 2t - 2L\cos 2t) + C\right]$$

$$= \frac{1}{4L^2 + R^2}(R\sin 2t - 2L\cos 2t) + Ce^{-(R/L)t}.$$

TECHNOLOGY The integral in Example 4 was found using a computer algebra system. If you have access to *Maple*, *Mathematica*, or the *TI-Nspire*, try using it to integrate

$$\frac{1}{L}\int e^{(R/L)t}\sin 2t\,dt.$$

In Chapter 8, you will learn how to integrate functions of this type using integration by parts.

One type of problem that can be described in terms of a differential equation involves chemical mixtures, as illustrated in the next example.

EXAMPLE 5 **A Mixture Problem**

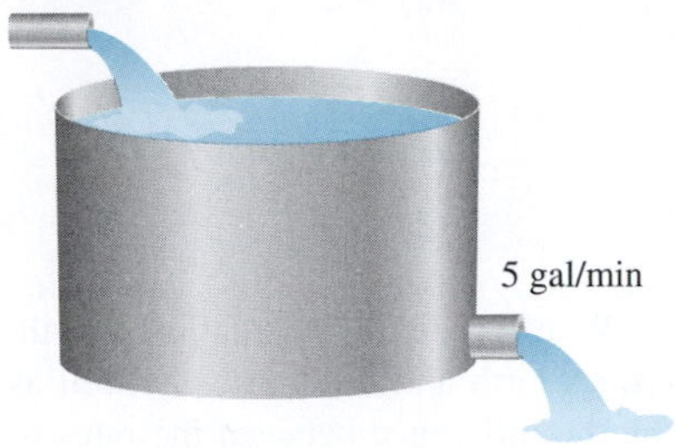

Figure 6.19

A tank contains 50 gallons of a solution composed of 90% water and 10% alcohol. A second solution containing 50% water and 50% alcohol is added to the tank at the rate of 4 gallons per minute. As the second solution is being added, the tank is being drained at a rate of 5 gallons per minute, as shown in Figure 6.19. The solution in the tank is stirred constantly. How much alcohol is in the tank after 10 minutes?

Solution Let y be the number of gallons of alcohol in the tank at any time t. You know that $y = 5$ when $t = 0$. Because the number of gallons of solution in the tank at any time is $50 - t$, and the tank loses 5 gallons of solution per minute, it must lose

$$\left(\frac{5}{50 - t}\right)y$$

gallons of alcohol per minute. Furthermore, because the tank is gaining 2 gallons of alcohol per minute, the rate of change of alcohol in the tank is

$$\frac{dy}{dt} = 2 - \left(\frac{5}{50 - t}\right)y \quad \Longrightarrow \quad \frac{dy}{dt} + \left(\frac{5}{50 - t}\right)y = 2.$$

To solve this linear differential equation, let

$$P(t) = \frac{5}{50 - t}$$

and obtain

$$\int P(t)\,dt = \int \frac{5}{50 - t}\,dt = -5\ln|50 - t|.$$

Because $t < 50$, you can drop the absolute value signs and conclude that

$$e^{\int P(t)\,dt} = e^{-5\ln(50 - t)} = \frac{1}{(50 - t)^5}.$$

So, the general solution is

$$\frac{y}{(50 - t)^5} = \int \frac{2}{(50 - t)^5}\,dt$$

$$\frac{y}{(50 - t)^5} = \frac{1}{2(50 - t)^4} + C$$

$$y = \frac{50 - t}{2} + C(50 - t)^5.$$

Because $y = 5$ when $t = 0$, you have

$$5 = \frac{50}{2} + C(50)^5 \quad \Longrightarrow \quad -\frac{20}{50^5} = C$$

which means that the particular solution is

$$y = \frac{50 - t}{2} - 20\left(\frac{50 - t}{50}\right)^5.$$

Finally, when $t = 10$, the amount of alcohol in the tank is

$$y = \frac{50 - 10}{2} - 20\left(\frac{50 - 10}{50}\right)^5 \approx 13.45 \text{ gal}$$

which represents a solution containing 33.6% alcohol. ■

6.4 Exercises

See CalcChat.com for tutorial help and worked-out solutions to odd-numbered exercises.

Determining Whether a Differential Equation Is Linear **In Exercises 1–4, determine whether the differential equation is linear. Explain your reasoning.**

1. $x^3y' + xy = e^x + 1$

2. $2xy - y' \ln x = y$

3. $y' - y \sin x = xy^2$

4. $\dfrac{2 - y'}{y} = 5x$

Solving a First-Order Linear Differential Equation **In Exercises 5–14, solve the first-order linear differential equation.**

5. $\dfrac{dy}{dx} + \left(\dfrac{1}{x}\right)y = 6x + 2$

6. $\dfrac{dy}{dx} + \left(\dfrac{2}{x}\right)y = 3x - 5$

7. $y' - y = 16$

8. $y' + 2xy = 10x$

9. $(y + 1) \cos x\, dx - dy = 0$

10. $(y - 1) \sin x\, dx - dy = 0$

11. $(x - 1)y' + y = x^2 - 1$

12. $y' + 3y = e^{3x}$

13. $y' - 3x^2y = e^{x^3}$

14. $y' + y \tan x = \sec x$

Slope Field **In Exercises 15 and 16, (a) sketch an approximate solution of the differential equation satisfying the given initial condition by hand on the slope field, (b) find the particular solution that satisfies the given initial condition, and (c) use a graphing utility to graph the particular solution. Compare the graph with the hand-drawn graph in part (a). To print an enlarged copy of the graph, go to *MathGraphs.com*.**

15. $\dfrac{dy}{dx} = e^x - y,$

$(0, 1)$

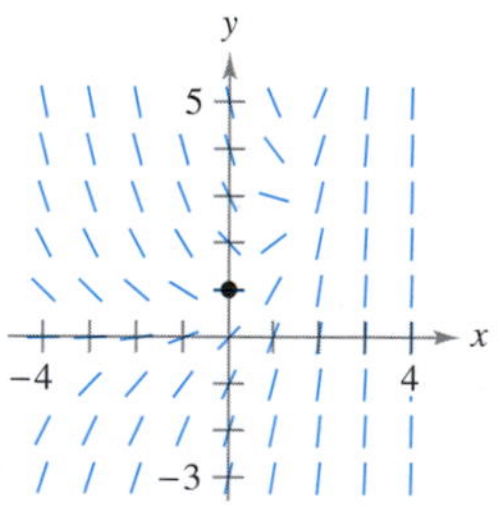

16. $y' + \left(\dfrac{1}{x}\right)y = \sin x^2,$

$(\sqrt{\pi}, 0)$

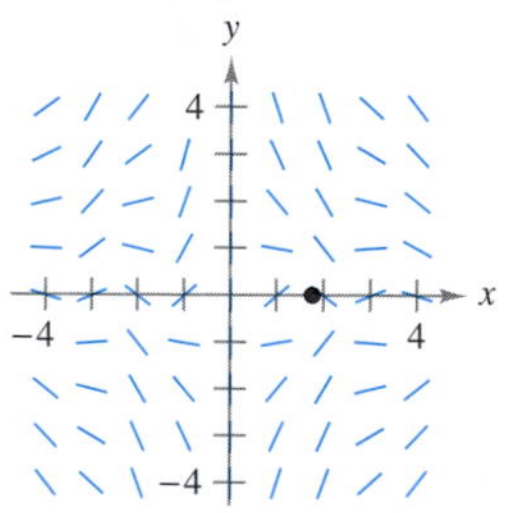

Finding a Particular Solution **In Exercises 17–24, find the particular solution of the differential equation that satisfies the initial condition.**

Differential Equation	Initial Condition
17. $y' \cos^2 x + y - 1 = 0$	$y(0) = 5$
18. $x^3y' + 2y = e^{1/x^2}$	$y(1) = e$
19. $y' + y \tan x = \sec x + \cos x$	$y(0) = 1$
20. $y' + y \sec x = \sec x$	$y(0) = 4$
21. $y' + \left(\dfrac{1}{x}\right)y = 0$	$y(2) = 2$
22. $y' + (2x - 1)y = 0$	$y(1) = 2$
23. $x\, dy = (x + y + 2)\, dx$	$y(1) = 10$
24. $2xy' - y = x^3 - x$	$y(4) = 2$

25. Population Growth When predicting population growth, demographers must consider birth and death rates as well as the net change caused by the difference between the rates of immigration and emigration. Let P be the population at time t and let N be the net increase per unit time resulting from the difference between immigration and emigration. So, the rate of growth of the population is given by

$$\frac{dP}{dt} = kP + N$$

where N is constant. Solve this differential equation to find P as a function of time, when at time $t = 0$ the size of the population is P_0.

26. Investment Growth A large corporation starts at time $t = 0$ to invest part of its receipts continuously at a rate of P dollars per year in a fund for future corporate expansion. Assume that the fund earns r percent interest per year compounded continuously. So, the rate of growth of the amount A in the fund is given by

$$\frac{dA}{dt} = rA + P$$

where $A = 0$ when $t = 0$. Solve this differential equation for A as a function of t.

Investment Growth **In Exercises 27 and 28, use the result of Exercise 26.**

27. Find A for the following.

(a) $P = \$275{,}000$, $r = 8\%$, $t = 10$ years

(b) $P = \$550{,}000$, $r = 5.9\%$, $t = 25$ years

28. Find t if the corporation needs \$1,000,000 and it can invest \$125,000 per year in a fund earning 8% interest compounded continuously.

29. Learning Curve The management at a certain factory has found that the maximum number of units a worker can produce in a day is 75. The rate of increase in the number of units N produced with respect to time t in days by a new employee is proportional to $75 - N$.

(a) Determine the differential equation describing the rate of change of performance with respect to time.

(b) Solve the differential equation from part (a).

(c) Find the particular solution for a new employee who produced 20 units on the first day at the factory and 35 units on the twentieth day.

30. Intravenous Feeding Glucose is added intravenously to the bloodstream at the rate of q units per minute, and the body removes glucose from the bloodstream at a rate proportional to the amount present. Assume that $Q(t)$ is the amount of glucose in the bloodstream at time t.

(a) Determine the differential equation describing the rate of change of glucose in the bloodstream with respect to time.

(b) Solve the differential equation from part (a), letting $Q = Q_0$ when $t = 0$.

(c) Find the limit of $Q(t)$ as $t \to \infty$.

Falling Object **In Exercises 31 and 32, consider an eight-pound object dropped from a height of 5000 feet, where the air resistance is proportional to the velocity.**

31. Write the velocity of the object as a function of time when the velocity after 5 seconds is approximately -101 feet per second. What is the limiting value of the velocity function?

32. Use the result of Exercise 31 to write the position of the object as a function of time. Approximate the velocity of the object when it reaches ground level.

Electric Circuits **In Exercises 33 and 34, use the differential equation for electric circuits given by**

$$L\frac{dI}{dt} + RI = E.$$

In this equation, I is the current, R is the resistance, L is the inductance, and E is the electromotive force (voltage).

33. Solve the differential equation for the current given a constant voltage E_0.

34. Use the result of Exercise 33 to find the equation for the current when $I(0) = 0$, $E_0 = 120$ volts, $R = 600$ ohms, and $L = 4$ henrys. When does the current reach 90% of its limiting value?

Mixture **In Exercises 35–38, consider a tank that at time $t = 0$ contains v_0 gallons of a solution of which, by weight, q_0 pounds is soluble concentrate. Another solution containing q_1 pounds of the concentrate per gallon is running into the tank at the rate of r_1 gallons per minute. The solution in the tank is kept well stirred and is withdrawn at the rate of r_2 gallons per minute.**

35. Let Q be the amount of concentrate in the solution at any time t. Show that

$$\frac{dQ}{dt} + \frac{r_2 Q}{v_0 + (r_1 - r_2)t} = q_1 r_1.$$

36. Let Q be the amount of concentrate in the solution at any time t. Write the differential equation for the rate of change of Q with respect to t when $r_1 = r_2 = r$.

37. A 200-gallon tank is full of a solution containing 25 pounds of concentrate. Starting at time $t = 0$, distilled water is admitted to the tank at a rate of 10 gallons per minute, and the well-stirred solution is withdrawn at the same rate.

(a) Find the amount of concentrate Q in the solution as a function of t.

(b) Find the time at which the amount of concentrate in the tank reaches 15 pounds.

(c) Find the quantity of the concentrate in the solution as $t \to \infty$.

38. A 200-gallon tank is half full of distilled water. At time $t = 0$, a solution containing 0.5 pound of concentrate per gallon enters the tank at the rate of 5 gallons per minute, and the well-stirred mixture is withdrawn at the rate of 3 gallons per minute.

(a) At what time will the tank be full?

(b) At the time the tank is full, how many pounds of concentrate will it contain?

(c) Repeat parts (a) and (b), assuming that the solution entering the tank contains 1 pound of concentrate per gallon.

39. Using an Integrating Factor The expression $u(x)$ is an integrating factor for $y' + P(x)y = Q(x)$. Which of the following is equal to $u'(x)$? Verify your answer.

(a) $P(x)u(x)$ (b) $P'(x)u(x)$

(c) $Q(x)u(x)$ (d) $Q'(x)u(x)$

40. HOW DO YOU SEE IT? The graph shows the amount of concentrate Q (in pounds) in a solution in a tank at time t (in minutes) as a solution with concentrate enters the tank, is well stirred, and is withdrawn from the tank.

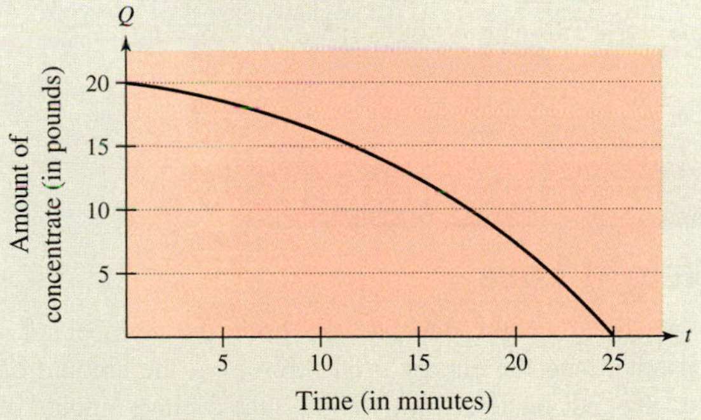

(a) How much concentrate is in the tank at time $t = 0$?

(b) Which is greater, the rate of solution into the tank, or the rate of solution withdrawn from the tank? Explain.

(c) At what time is there no concentrate in the tank? What does this mean?

WRITING ABOUT CONCEPTS

41. Standard Form Give the standard form of a first-order linear differential equation. What is its integrating factor?

42. First-Order What does the term "first-order" refer to in a first-order linear differential equation?

Matching **In Exercises 43–46, match the differential equation with its solution.**

Differential Equation	Solution
43. $y' - 2x = 0$	(a) $y = Ce^{x^2}$
44. $y' - 2y = 0$	(b) $y = -\frac{1}{2} + Ce^{x^2}$
45. $y' - 2xy = 0$	(c) $y = x^2 + C$
46. $y' - 2xy = x$	(d) $y = Ce^{2x}$

Slope Field **In Exercises 47–50, (a) use a graphing utility to graph the slope field for the differential equation, (b) find the particular solutions of the differential equation passing through the given points, and (c) use a graphing utility to graph the particular solutions on the slope field.**

Differential Equation	Points
47. $\frac{dy}{dx} - \frac{1}{x}y = x^2$	$(-2, 4),\ (2, 8)$
48. $\frac{dy}{dx} + 4x^3y = x^3$	$\left(0, \frac{7}{2}\right),\ \left(0, -\frac{1}{2}\right)$
49. $\frac{dy}{dx} + (\cot x)y = 2$	$(1, 1),\ (3, -1)$
50. $\frac{dy}{dx} + 2xy = xy^2$	$(0, 3),\ (0, 1)$

Solving a First-Order Linear Differential Equation **In Exercises 51–58, solve the first-order differential equation by any appropriate method.**

51. $\dfrac{dy}{dx} = \dfrac{e^{2x+y}}{e^{x-y}}$

52. $\dfrac{dy}{dx} = \dfrac{x-3}{y(y+4)}$

53. $y \cos x - \cos x + \dfrac{dy}{dx} = 0$

54. $y' = 2x\sqrt{1 - y^2}$

55. $(2y - e^x)\,dx + x\,dy = 0$

56. $(x + y)\,dx - x\,dy = 0$

57. $3(y - 4x^2)\,dx + x\,dy = 0$

58. $x\,dx + (y + e^y)(x^2 + 1)\,dy = 0$

Solving a Bernoulli Differential Equation **In Exercises 59–66, solve the Bernoulli differential equation. The *Bernoulli equation* is a well-known nonlinear equation of the form**

$$y' + P(x)y = Q(x)y^n$$

that can be reduced to a linear form by a substitution. The general solution of a Bernoulli equation is

$$y^{1-n}e^{\int(1-n)P(x)\,dx} = \int (1-n)Q(x)e^{\int(1-n)P(x)\,dx}\,dx + C.$$

59. $y' + 3x^2y = x^2y^3$

60. $y' + xy = xy^{-1}$

61. $y' + \left(\dfrac{1}{x}\right)y = xy^2$

62. $y' + \left(\dfrac{1}{x}\right)y = x\sqrt{y}$

63. $xy' + y = xy^3$

64. $y' - y = y^3$

65. $y' - y = e^x\sqrt[3]{y}$

66. $yy' - 2y^2 = e^x$

True or False? **In Exercises 67 and 68, determine whether the statement is true or false. If it is false, explain why or give an example that shows it is false.**

67. $y' + x\sqrt{y} = x^2$ is a first-order linear differential equation.

68. $y' + xy = e^xy$ is a first-order linear differential equation.

SECTION PROJECT

Weight Loss

A person's weight depends on both the number of calories consumed and the energy used. Moreover, the amount of energy used depends on a person's weight—the average amount of energy used by a person is 17.5 calories per pound per day. So, the more weight a person loses, the less energy a person uses (assuming that the person maintains a constant level of activity). An equation that can be used to model weight loss is

$$\frac{dw}{dt} = \frac{C}{3500} - \frac{17.5}{3500}w$$

where w is the person's weight (in pounds), t is the time in days, and C is the constant daily calorie consumption.

(a) Find the general solution of the differential equation.

(b) Consider a person who weighs 180 pounds and begins a diet of 2500 calories per day. How long will it take the person to lose 10 pounds? How long will it take the person to lose 35 pounds?

(c) Use a graphing utility to graph the solution. What is the "limiting" weight of the person?

(d) Repeat parts (b) and (c) for a person who weighs 200 pounds when the diet is started.

FOR FURTHER INFORMATION For more information on modeling weight loss, see the article "A Linear Diet Model" by Arthur C. Segal in *The College Mathematics Journal*.

Review Exercises

See **CalcChat.com** for tutorial help and worked-out solutions to odd-numbered exercises.

1. **Determining a Solution** Determine whether the function $y = x^3$ is a solution of the differential equation $2xy' + 4y = 10x^3$.

2. **Determining a Solution** Determine whether the function $y = 2 \sin 2x$ is a solution of the differential equation $y''' - 8y = 0$.

Finding a General Solution **In Exercises 3–8, use integration to find a general solution of the differential equation.**

3. $\dfrac{dy}{dx} = 4x^2 + 7$

4. $\dfrac{dy}{dx} = 3x^3 - 8x$

5. $\dfrac{dy}{dx} = \cos 2x$

6. $\dfrac{dy}{dx} = 2 \sin x$

7. $\dfrac{dy}{dx} = e^{2-x}$

8. $\dfrac{dy}{dx} = 2e^{3x}$

Slope Field **In Exercises 9 and 10, a differential equation and its slope field are given. Complete the table by determining the slopes (if possible) in the slope field at the given points.**

x	-4	-2	0	2	4	8
y	2	0	4	4	6	8
dy/dx						

9. $\dfrac{dy}{dx} = 2x - y$

10. $\dfrac{dy}{dx} = x \sin\left(\dfrac{\pi y}{4}\right)$

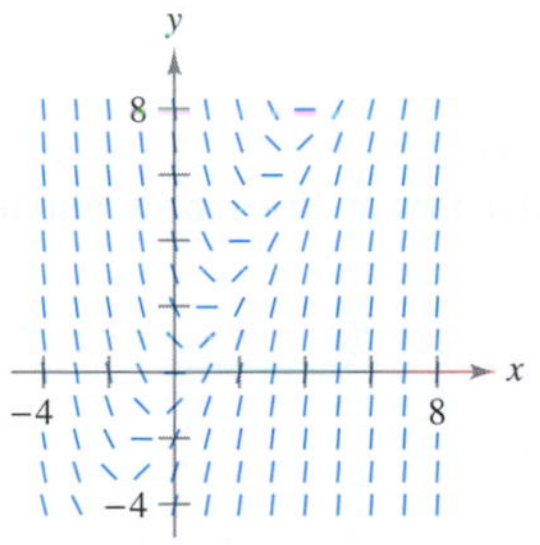

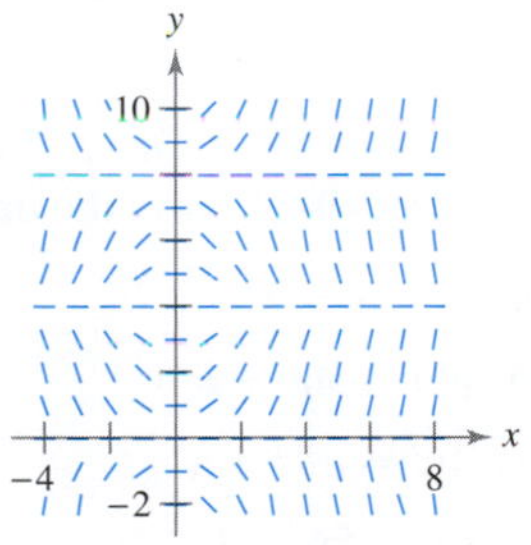

Slope Field **In Exercises 11 and 12, (a) sketch the slope field for the differential equation, and (b) use the slope field to sketch the solution that passes through the given point. Use a graphing utility to verify your results. To print a blank graph, go to *MathGraphs.com*.**

11. $y' = 2x^2 - x, \quad (0, 2)$

12. $y' = y + 4x, \quad (-1, 1)$

Euler's Method **In Exercises 13 and 14, use Euler's Method to make a table of values for the approximate solution of the differential equation with the specified initial value. Use n steps of size h.**

13. $y' = x - y, \quad y(0) = 4, \quad n = 10, \quad h = 0.05$

14. $y' = 5x - 2y, \quad y(0) = 2, \quad n = 10, \quad h = 0.1$

Solving a Differential Equation **In Exercises 15–20, solve the differential equation.**

15. $\dfrac{dy}{dx} = 2x - 5x^2$

16. $\dfrac{dy}{dx} = y + 8$

17. $\dfrac{dy}{dx} = (3 + y)^2$

18. $\dfrac{dy}{dx} = 10\sqrt{y}$

19. $(2 + x)y' - xy = 0$

20. $xy' - (x + 1)y = 0$

Writing and Solving a Differential Equation **In Exercises 21 and 22, write and solve the differential equation that models the verbal statement.**

21. The rate of change of y with respect to t is inversely proportional to the cube of t.

22. The rate of change of y with respect to t is proportional to $50 - t$.

Finding an Exponential Function **In Exercises 23–26, find the exponential function $y = Ce^{kt}$ that passes through the two points.**

23.

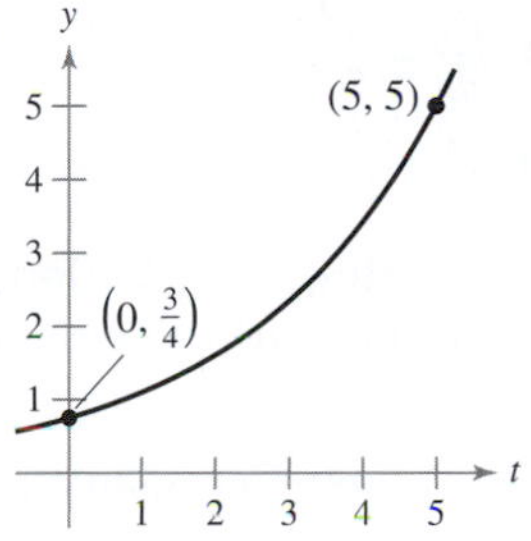

24.

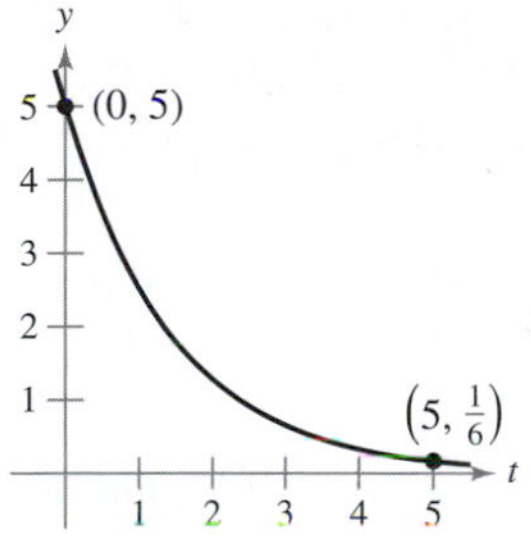

25.

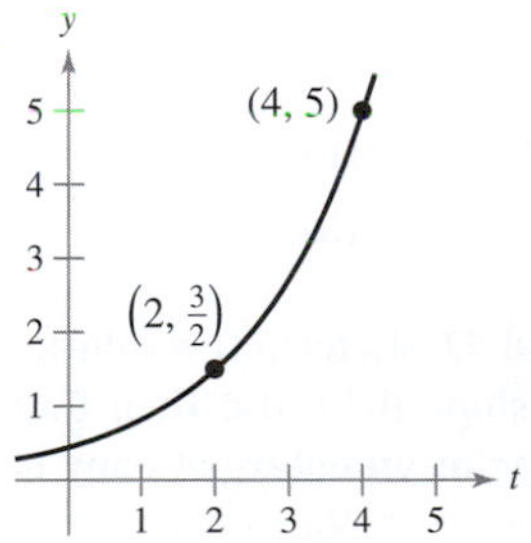

26.

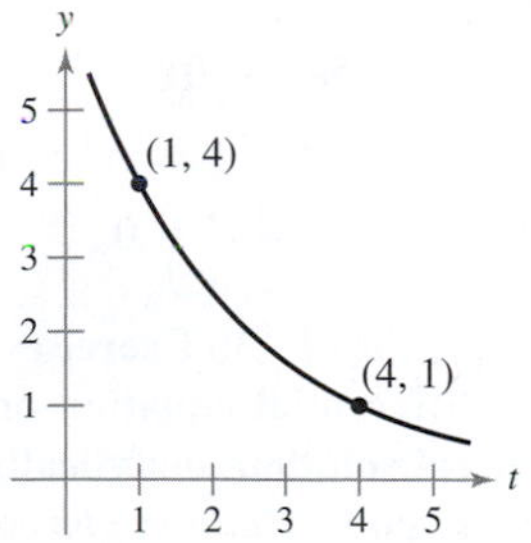

27. **Air Pressure** Under ideal conditions, air pressure decreases continuously with the height above sea level at a rate proportional to the pressure at that height. The barometer reads 30 inches at sea level and 15 inches at 18,000 feet. Find the barometric pressure at 35,000 feet.

28. **Radioactive Decay** Radioactive radium has a half-life of approximately 1599 years. The initial quantity is 15 grams. How much remains after 750 years?

29. **Population Growth** A population grows continuously at the rate of 1.85%. How long will it take the population to double?

30. Compound Interest Find the balance in an account when \$1000 is deposited for 8 years at an interest rate of 4% compounded continuously.

31. Sales The sales S (in thousands of units) of a new product after it has been on the market for t years is given by

$S = Ce^{k/t}$.

(a) Find S as a function of t when 5000 units have been sold after 1 year and the saturation point for the market is 30,000 units (that is, $\lim_{t\to\infty} S = 30$).

(b) How many units will have been sold after 5 years?

32. Sales The sales S (in thousands of units) of a new product after it has been on the market for t years is given by

$S = 25(1 - e^{kt})$.

(a) Find S as a function of t when 4000 units have been sold after 1 year.

(b) How many units will saturate this market?

(c) How many units will have been sold after 5 years?

Finding a General Solution Using Separation of Variables **In Exercises 33–36, find the general solution of the differential equation.**

33. $\dfrac{dy}{dx} = \dfrac{5x}{y}$ **34.** $\dfrac{dy}{dx} = \dfrac{x^3}{2y^2}$

35. $y' - 16xy = 0$ **36.** $y' - e^y \sin x = 0$

Finding a Particular Solution Using Separation of Variables **In Exercises 37–40, find the particular solution that satisfies the initial condition.**

Differential Equation	Initial Condition
37. $y^3y' - 3x = 0$	$y(2) = 2$
38. $yy' - 5e^{2x} = 0$	$y(0) = -3$
39. $y^3(x^4 + 1)y' - x^3(y^4 + 1) = 0$	$y(0) = 1$
40. $yy' - x \cos x^2 = 0$	$y(0) = -2$

Slope Field **In Exercises 41 and 42, sketch a few solutions of the differential equation on the slope field and then find the general solution analytically. To print an enlarged copy of the graph, go to *MathGraphs.com*.**

41. $\dfrac{dy}{dx} = -\dfrac{4x}{y}$

42. $\dfrac{dy}{dx} = 3 - 2y$

Using a Logistic Equation **In Exercises 43 and 44, the logistic equation models the growth of a population. Use the equation to (a) find the value of k, (b) find the carrying capacity, (c) find the initial population, (d) determine when the population will reach 50% of its carrying capacity, and (e) write a logistic differential equation that has the solution $P(t)$.**

43. $P(t) = \dfrac{5250}{1 + 34e^{-0.55t}}$

44. $P(t) = \dfrac{4800}{1 + 14e^{-0.15t}}$

Solving a Logistic Differential Equation **In Exercises 45 and 46, find the logistic equation that passes through the given point.**

45. $\dfrac{dy}{dt} = y\left(1 - \dfrac{y}{80}\right)$, $(0, 8)$

46. $\dfrac{dy}{dt} = 1.76y\left(1 - \dfrac{y}{8}\right)$, $(0, 3)$

47. Environment A conservation department releases 1200 brook trout into a lake. It is estimated that the carrying capacity of the lake for the species is 20,400. After the first year, there are 2000 brook trout in the lake.

(a) Write a logistic equation that models the number of brook trout in the lake.

(b) Find the number of brook trout in the lake after 8 years.

(c) When will the number of brook trout reach 10,000?

48. Environment Write a logistic differential equation that models the growth rate of the brook trout population in Exercise 47. Then repeat part (b) using Euler's Method with a step size of $h = 1$. Compare the approximation with the exact answer.

Solving a First-Order Linear Differential Equation **In Exercises 49–54, solve the first-order linear differential equation.**

49. $y' - y = 10$

50. $e^x y' + 4e^x y = 1$

51. $4y' = e^{x/4} + y$

52. $\dfrac{dy}{dx} - \dfrac{5y}{x^2} = \dfrac{1}{x^2}$

53. $(x - 2)y' + y = 1$

54. $(x + 3)y' + 2y = 2(x + 3)^2$

Finding a Particular Solution **In Exercises 55 and 56, find the particular solution of the differential equation that satisfies the initial condition.**

Differential Equation	Initial Condition
55. $y' + 5y = e^{5x}$	$y(0) = 3$
56. $y' - \left(\dfrac{3}{x}\right)y = 2x^3$	$y(1) = 1$

P.S. Problem Solving

See CalcChat.com for tutorial help and worked-out solutions to odd-numbered exercises.

1. **Doomsday Equation** The differential equation

$$\frac{dy}{dt} = ky^{1+\varepsilon}$$

where k and ε are positive constants, is called the **doomsday equation.**

(a) Solve the doomsday equation

$$\frac{dy}{dt} = y^{1.01}$$

given that $y(0) = 1$. Find the time T at which

$$\lim_{t \to T^-} y(t) = \infty.$$

(b) Solve the doomsday equation

$$\frac{dy}{dt} = ky^{1+\varepsilon}$$

given that $y(0) = y_0$. Explain why this equation is called the doomsday equation.

2. **Sales** Let S represent sales of a new product (in thousands of units), let L represent the maximum level of sales (in thousands of units), and let t represent time (in months). The rate of change of S with respect to t varies jointly as the product of S and $L - S$.

(a) Write the differential equation for the sales model when $L = 100$, $S = 10$ when $t = 0$, and $S = 20$ when $t = 1$. Verify that

$$S = \frac{L}{1 + Ce^{-kt}}.$$

(b) At what time is the growth in sales increasing most rapidly?

(c) Use a graphing utility to graph the sales function.

(d) Sketch the solution from part (a) on the slope field shown in the figure below. To print an enlarged copy of the graph, go to *MathGraphs.com.*

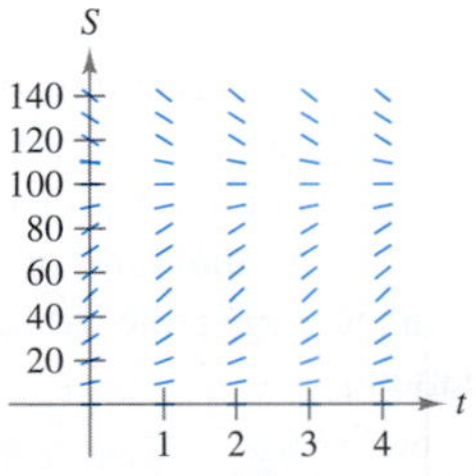

(e) Assume the estimated maximum level of sales is correct. Use the slope field to describe the shape of the solution curves for sales when, at some period of time, sales exceed L.

3. **Gompertz Equation** Another model that can be used to represent population growth is the **Gompertz equation,** which is the solution of the differential equation

$$\frac{dy}{dt} = k \ln\left(\frac{L}{y}\right) y$$

where k is a constant and L is the carrying capacity.

(a) Solve the differential equation.

(b) Use a graphing utility to graph the slope field for the differential equation when $k = 0.05$ and $L = 1000$.

(c) Describe the behavior of the graph as $t \to \infty$.

(d) Graph the equation you found in part (a) for $L = 5000$, $y_0 = 500$, and $k = 0.02$. Determine the concavity of the graph and how it compares with the general solution of the logistic differential equation.

4. **Error Using Product Rule** Although it is true for some functions f and g, a common mistake in calculus is to believe that the Product Rule for derivatives is $(fg)' = f'g'$.

(a) Given $g(x) = x$, find f such that $(fg)' = f'g'$.

(b) Given an arbitrary function g, find a function f such that $(fg)' = f'g'$.

(c) Describe what happens if $g(x) = e^x$.

5. **Torricelli's Law** **Torricelli's Law** states that water will flow from an opening at the bottom of a tank with the same speed that it would attain falling from the surface of the water to the opening. One of the forms of Torricelli's Law is

$$A(h)\frac{dh}{dt} = -k\sqrt{2gh}$$

where h is the height of the water in the tank, k is the area of the opening at the bottom of the tank, $A(h)$ is the horizontal cross-sectional area at height h, and g is the acceleration due to gravity ($g \approx 32$ feet per second per second). A hemispherical water tank has a radius of 6 feet. When the tank is full, a circular valve with a radius of 1 inch is opened at the bottom, as shown in the figure. How long will it take for the tank to drain completely?

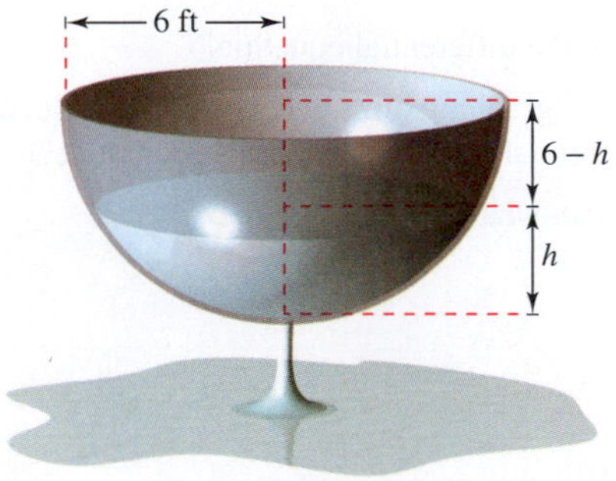

6. Torricelli's Law The cylindrical water tank shown in the figure has a height of 18 feet. When the tank is full, a circular valve is opened at the bottom of the tank. After 30 minutes, the depth of the water is 12 feet.

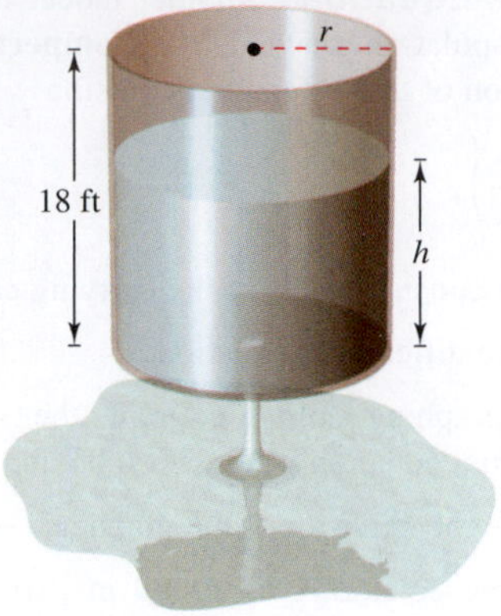

(a) Using Torricelli's Law, how long will it take for the tank to drain completely?

(b) What is the depth of the water in the tank after 1 hour?

7. Torricelli's Law Suppose the tank in Exercise 6 has a height of 20 feet and a radius of 8 feet, and the valve is circular with a radius of 2 inches. The tank is full when the valve is opened. How long will it take for the tank to drain completely?

8. Rewriting the Logistic Equation Show that the logistic equation

$$y = \frac{L}{1 + be^{-kt}}$$

can be written as

$$y = \frac{1}{2}L\left[1 + \tanh\left(\frac{1}{2}k\left(t - \frac{\ln b}{k}\right)\right)\right].$$

What can you conclude about the graph of the logistic equation?

9. Biomass Biomass is a measure of the amount of living matter in an ecosystem. Suppose the biomass $s(t)$ in a given ecosystem increases at a rate of about 3.5 tons per year, and decreases by about 1.9% per year. This situation can be modeled by the differential equation

$$\frac{ds}{dt} = 3.5 - 0.019s.$$

(a) Solve the differential equation.

(b) Use a graphing utility to graph the slope field for the differential equation. What do you notice?

(c) Explain what happens as $t \to \infty$.

Medical Science In Exercises 10–12, a medical researcher wants to determine the concentration C (in moles per liter) of a tracer drug injected into a moving fluid. Solve this problem by considering a single-compartment dilution model (see figure). Assume that the fluid is continuously mixed and that the volume of the fluid in the compartment is constant.

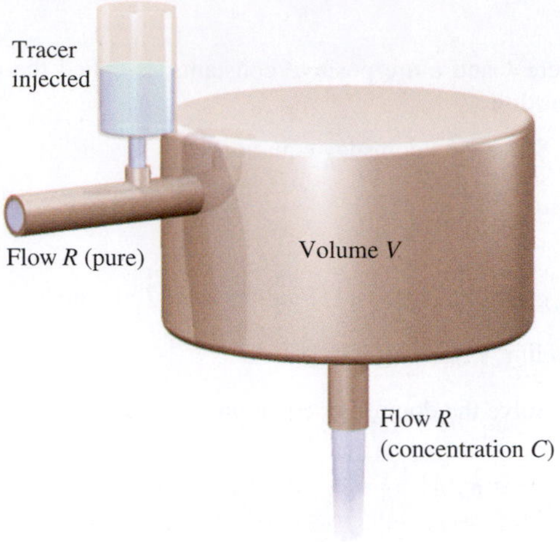

Figure for 10–12

10. If the tracer is injected instantaneously at time $t = 0$, then the concentration of the fluid in the compartment begins diluting according to the differential equation

$$\frac{dC}{dt} = \left(-\frac{R}{V}\right)C$$

where $C = C_0$ when $t = 0$.

(a) Solve this differential equation to find the concentration C as a function of time t.

(b) Find the limit of C as $t \to \infty$.

11. Use the solution of the differential equation in Exercise 10 to find the concentration C as a function of time t, and use a graphing utility to graph the function.

(a) $V = 2$ liters, $R = 0.5$ liter per minute, and $C_0 = 0.6$ mole per liter

(b) $V = 2$ liters, $R = 1.5$ liters per minute, and $C_0 = 0.6$ mole per liter

12. In Exercises 10 and 11, it was assumed that there was a single initial injection of the tracer drug into the compartment. Now consider the case in which the tracer is continuously injected (beginning at $t = 0$) at the rate of Q moles per minute. Considering Q to be negligible compared with R, use the differential equation

$$\frac{dC}{dt} = \frac{Q}{V} - \left(\frac{R}{V}\right)C$$

where $C = 0$ when $t = 0$.

(a) Solve this differential equation to find the concentration C as a function of time t.

(b) Find the limit of C as $t \to \infty$.

AP* Review Questions for Chapter 6

1. (no calculator)

Let $y = f(x)$ be a particular solution to the differential equation $\frac{dy}{dx} = \frac{1}{xy}$ with $f(1) = 2$.

(a) Find $\frac{d^2y}{dx^2}$ at the point $(1, 2)$.

(b) Write an equation for the line tangent to the graph of f at $(1, 2)$ and use it to approximate $f(1.1)$. Is the approximation for $f(1.1)$ greater than or less than $f(1.1)$? Explain your reasoning.

(c) Find the solution of the given differential equation that satisfies the initial condition $f(1) = 2$.

2. (no calculator)

Consider the differential equation $\frac{dy}{dx} = x^2(1 - y)$.

(a) On the axes provided, sketch a slope field for the given differential equation at the 9 points indicated.

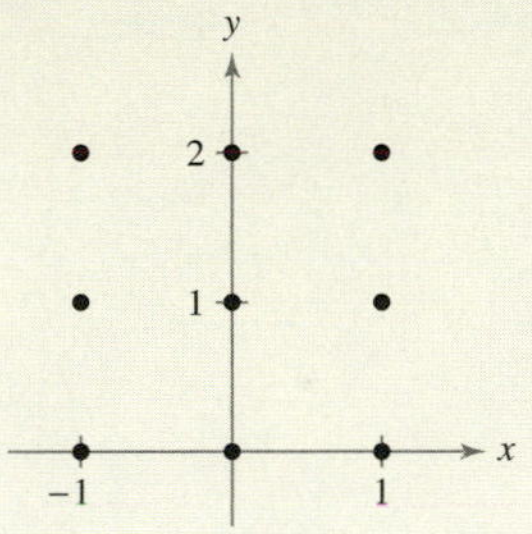

(b) While the slope field in part (a) is drawn only at 9 points, it is defined at every point on the xy-plane. Describe all points in the xy-plane for which the slopes are positive.

(c) Find the particular solution in the form of $y = f(x)$ to the given differential equation with the initial condition $f(0) = 2$.

3. (BC only, no calculator)

Given the differential equation $y' = \frac{2x}{y}$ with a particular solution in the form of $y = f(x)$ that satisfies the initial condition $f(1) = 2$:

(a) Use Euler's Method, starting at $x = 1$ with two steps of equal size, to approximate $y(1.4)$. Show the work that leads to your answer.

(b) Find the particular solution to the given differential equation that passes through $(1, 2)$ and state its domain.

4. (BC only, no calculator)

Given the differential equation $\frac{dy}{dx} = xy$:

(a) Let $y = f(x)$ be the function that satisfies the differential equation with initial condition $f(1) = 1$. Use Euler's Method, starting at $x = 1$ with a step size of 0.1, to approximate $f(1.2)$. Show the work that leads to your answer.

(b) Find $\frac{d^2y}{dx^2}$. Determine whether the approximation found in part (a) is less than or greater than $f(1.2)$

(c) Find the particular solution of the differential equation that passes through $(1, 1)$.

5. (no calculator)

At any time, $t \geq 0$, in hours, the rate of growth of a population of bacteria is given by $\frac{dy}{dt} = \frac{1}{2}y$. Initially, there are 200 bacteria in the culture.

(a) Use separation of variables to solve for y, the number of bacteria present, at any time $t \geq 0$.

(b) Write, but do not evaluate an expression to find the average number of bacteria in the population for $0 \leq t \leq 10$.

(c) Write an expression to find the average rate of bacteria growth over the first 10 hours of growth. Indicate units of measure.

6. (BC only, no calculator)

At any time $t \geq 0$, the rate of the spread of an epidemic is modeled by a function y that satisfies the differential equation $\frac{dy}{dt} = \frac{1}{10}y\left(1 - \frac{y}{1000}\right)$.

In an isolated town of 1000 inhabitants, 100 people have a disease at the beginning of the week.

(a) Is the disease spreading faster when 100 people have the disease or when 200 people have the disease? Explain your reasoning.

(b) Use separation of variables to write a model for the population $y = f(t)$ at any time $t \geq 0$.

(c) For the function y found in part (b), what is $\lim_{t\to\infty} y(t)$?

7. (BC only, no calculator)

A population is modeled by a function $y(t)$ that satisfies the logistic differential equation

$$\frac{dy}{dt} = 2y\left(1 - \frac{y}{5}\right)$$

(a) What is $\lim_{t\to\infty} y(t)$?

(b) For what value of y is the population growing the fastest?

(c) Find the particular solution satisfying $y(0) = 3$.

8. (no calculator)

Given the differential equation $\frac{dy}{dx} = \frac{x}{y^2}$:

(a) On the axes provided, sketch a slope field for the given differential equation at the 12 points indicated.

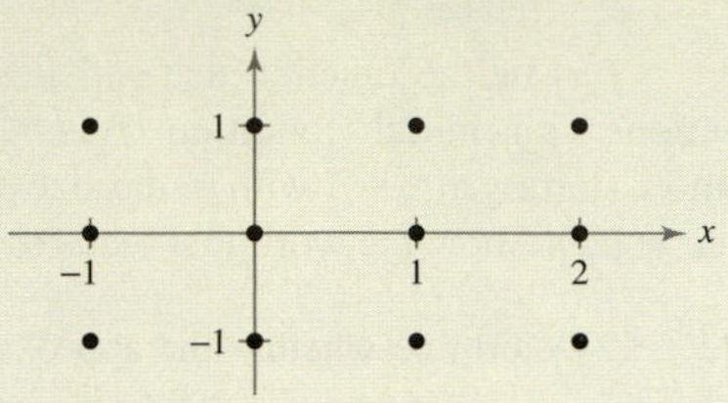

(b) Find $\frac{d^2y}{dx^2}$ in terms of x and y.

(c) Find the particular solution to the given differential equation that satisfies the initial condition $y(0) = 2$.

9. (no calculator)

If $\frac{dy}{dx} = 2xy^2$, and $y(-1) = 2$, find $y(2)$.

(A) $4e^3$

(B) $-4e^3$

(C) $-\frac{3}{2}$

(D) $-\frac{2}{5}$

(E) $-\frac{1}{4}$

10. (no calculator)

Which of the following is the solution to the differential equation $\frac{dy}{dx} = \frac{3y}{x}$ with the initial condition $y(1) = -1$?

(A) $y = x^3$

(B) $y = -x^3$

(C) $y = 3x$

(D) $y = -3x$

(E) $y = -x^3 - 2$

7 Applications of Integration

Moving a Space Module into Orbit *(Example 3, p. 480)*

Tidal Energy
(Section Project, p. 485)

Saturn *(Section Project, p. 465)*

Water Tower
(Exercise 66, p. 455)

Building Design *(Exercise 79, p. 445)*

Clockwise from top left, AFP Creative/Getty Images; Andrew J. Martinez/Photo Researchers, Inc; Paul Brennan/Shutterstock.com; jl661227/Shutterstock.com; NASA

7.1 Area of a Region Between Two Curves

- **Find the area of a region between two curves using integration.**
- **Find the area of a region between intersecting curves using integration.**
- **Describe integration as an accumulation process.**

Area of a Region Between Two Curves

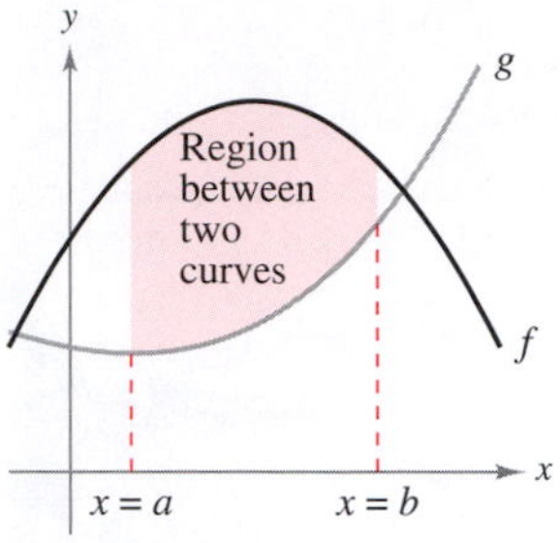

Figure 7.1

With a few modifications, you can extend the application of definite integrals from the area of a region *under* a curve to the area of a region *between* two curves. Consider two functions f and g that are continuous on the interval $[a, b]$. Also, the graphs of both f and g lie above the x-axis, and the graph of g lies below the graph of f, as shown in Figure 7.1. You can geometrically interpret the area of the region between the graphs as the area of the region under the graph of g subtracted from the area of the region under the graph of f, as shown in Figure 7.2.

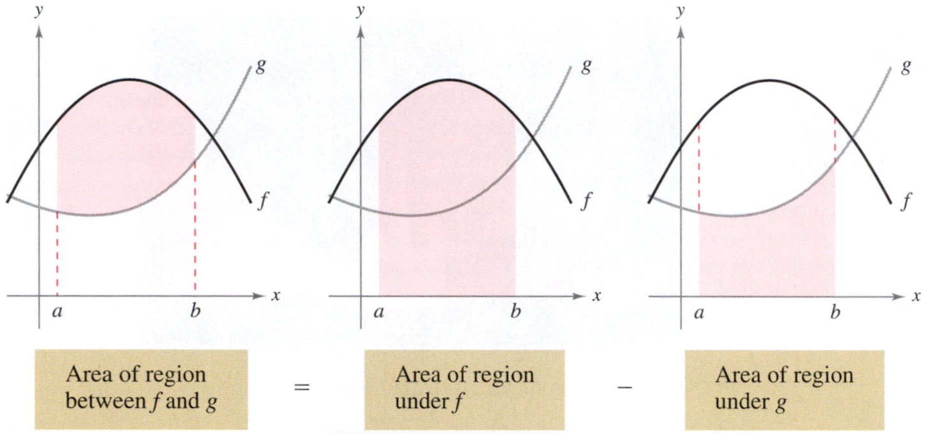

$$\int_a^b [f(x) - g(x)]\,dx = \int_a^b f(x)\,dx - \int_a^b g(x)\,dx$$

Figure 7.2

To verify the reasonableness of the result shown in Figure 7.2, you can partition the interval $[a, b]$ into n subintervals, each of width Δx. Then, as shown in Figure 7.3, sketch a **representative rectangle** of width Δx and height $f(x_i) - g(x_i)$, where x_i is in the ith subinterval. The area of this representative rectangle is

$$\Delta A_i = (\text{height})(\text{width}) = [f(x_i) - g(x_i)]\,\Delta x.$$

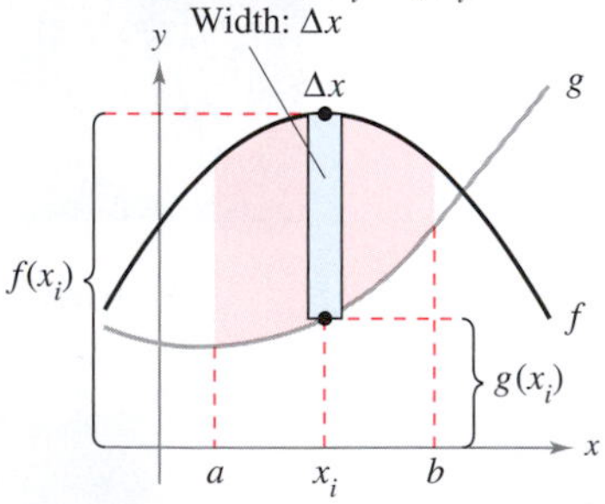

Figure 7.3

By adding the areas of the n rectangles and taking the limit as $\|\Delta\| \to 0$ $(n \to \infty)$, you obtain

$$\lim_{n\to\infty} \sum_{i=1}^{n} [f(x_i) - g(x_i)]\,\Delta x.$$

REMARK Recall from Section 4.3 that $\|\Delta\|$ is the norm of the partition. In a regular partition, the statements $\|\Delta\| \to 0$ and $n \to \infty$ are equivalent.

Because f and g are continuous on $[a, b]$, $f - g$ is also continuous on $[a, b]$ and the limit exists. So, the area of the region is

$$\begin{aligned}\text{Area} &= \lim_{n\to\infty} \sum_{i=1}^{n} [f(x_i) - g(x_i)]\,\Delta x \\ &= \int_a^b [f(x) - g(x)]\,dx.\end{aligned}$$

AP* Tips

Area between functions is tested on both versions of the AP Exam.

Area of a Region Between Two Curves

If f and g are continuous on $[a, b]$ and $g(x) \le f(x)$ for all x in $[a, b]$, then the area of the region bounded by the graphs of f and g and the vertical lines $x = a$ and $x = b$ is

$$A = \int_a^b [f(x) - g(x)]\,dx.$$

In Figure 7.1, the graphs of f and g are shown above the x-axis. This, however, is not necessary. The same integrand $[f(x) - g(x)]$ can be used as long as f and g are continuous and $g(x) \le f(x)$ for all x in the interval $[a, b]$. This is summarized graphically in Figure 7.4. Notice in Figure 7.4 that the height of a representative rectangle is $f(x) - g(x)$ regardless of the relative position of the x-axis.

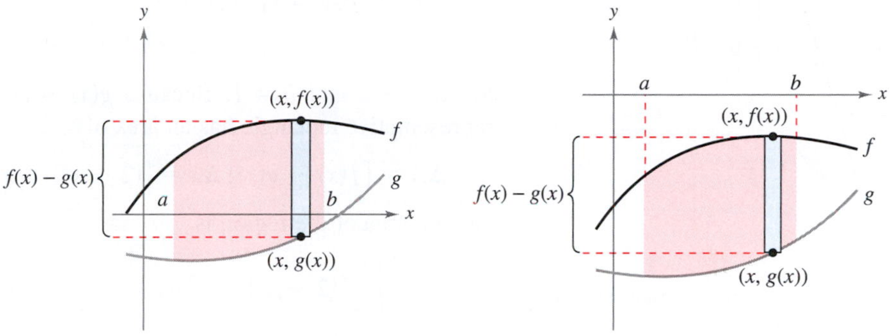

Figure 7.4

Representative rectangles are used throughout this chapter in various applications of integration. A vertical rectangle (of width Δx) implies integration with respect to x, whereas a horizontal rectangle (of width Δy) implies integration with respect to y.

EXAMPLE 1 Finding the Area of a Region Between Two Curves

Find the area of the region bounded by the graphs of $y = x^2 + 2$, $y = -x$, $x = 0$, and $x = 1$.

Solution Let $g(x) = -x$ and $f(x) = x^2 + 2$. Then $g(x) \le f(x)$ for all x in $[0, 1]$, as shown in Figure 7.5. So, the area of the representative rectangle is

$$\Delta A = [f(x) - g(x)]\,\Delta x$$
$$= [(x^2 + 2) - (-x)]\,\Delta x$$

and the area of the region is

$$A = \int_a^b [f(x) - g(x)]\,dx$$
$$= \int_0^1 [(x^2 + 2) - (-x)]\,dx$$
$$= \left[\frac{x^3}{3} + \frac{x^2}{2} + 2x\right]_0^1$$
$$= \frac{1}{3} + \frac{1}{2} + 2$$
$$= \frac{17}{6}.$$

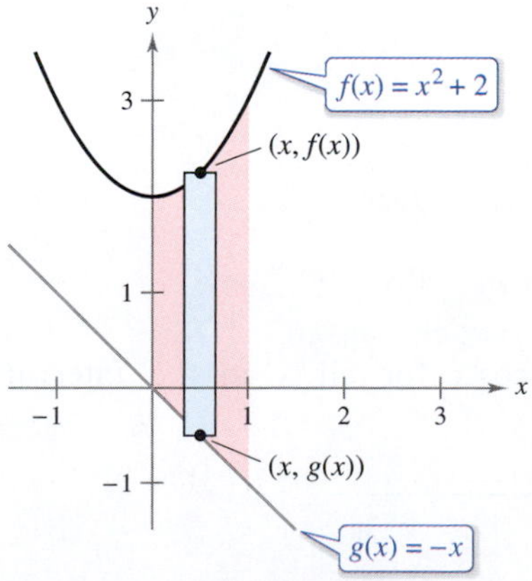

Region bounded by the graph of f, the graph of g, $x = 0$, and $x = 1$
Figure 7.5

Area of a Region Between Intersecting Curves

In Example 1, the graphs of $f(x) = x^2 + 2$ and $g(x) = -x$ do not intersect, and the values of a and b are given explicitly. A more common problem involves the area of a region bounded by two *intersecting* graphs, where the values of a and b must be calculated.

EXAMPLE 2 A Region Lying Between Two Intersecting Graphs

Find the area of the region bounded by the graphs of $f(x) = 2 - x^2$ and $g(x) = x$.

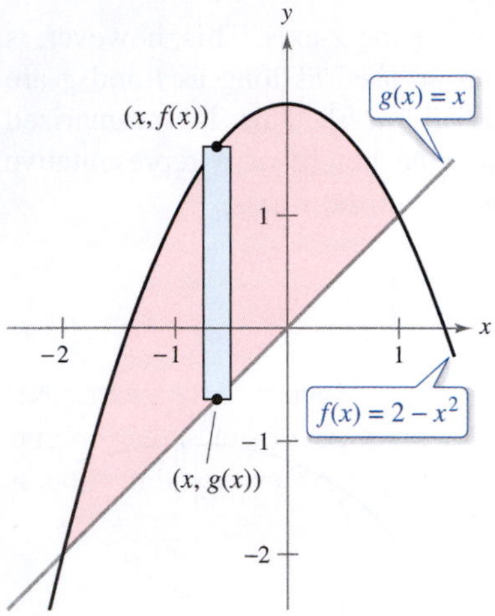

Region bounded by the graph of f and the graph of g
Figure 7.6

Solution In Figure 7.6, notice that the graphs of f and g have two points of intersection. To find the x-coordinates of these points, set $f(x)$ and $g(x)$ equal to each other and solve for x.

$$2 - x^2 = x \qquad \text{Set } f(x) \text{ equal to } g(x).$$
$$-x^2 - x + 2 = 0 \qquad \text{Write in general form.}$$
$$-(x + 2)(x - 1) = 0 \qquad \text{Factor.}$$
$$x = -2 \text{ or } 1 \qquad \text{Solve for } x.$$

So, $a = -2$ and $b = 1$. Because $g(x) \le f(x)$ for all x in the interval $[-2, 1]$, the representative rectangle has an area of

$$\Delta A = [f(x) - g(x)]\,\Delta x = [(2 - x^2) - x]\,\Delta x$$

and the area of the region is

$$A = \int_{-2}^{1} [(2 - x^2) - x]\,dx$$
$$= \left[-\frac{x^3}{3} - \frac{x^2}{2} + 2x\right]_{-2}^{1}$$
$$= \frac{9}{2}.$$

EXAMPLE 3 A Region Lying Between Two Intersecting Graphs

The sine and cosine curves intersect infinitely many times, bounding regions of equal areas, as shown in Figure 7.7. Find the area of one of these regions.

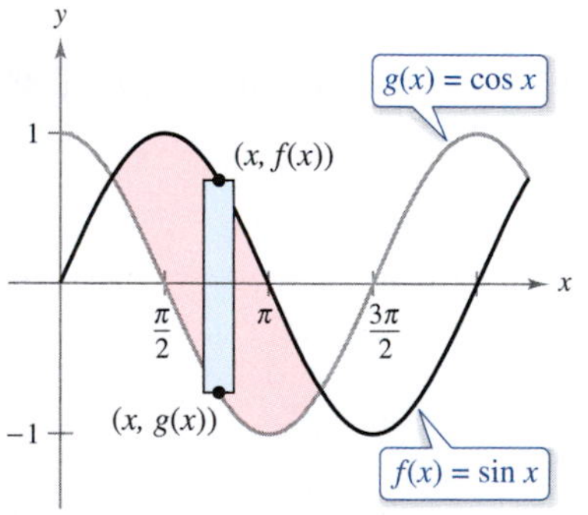

One of the regions bounded by the graphs of the sine and cosine functions
Figure 7.7

Solution Let $g(x) = \cos x$ and $f(x) = \sin x$. Then $g(x) \le f(x)$ for all x in the interval corresponding to the shaded region in Figure 7.7. To find the two points of intersection on this interval, set $f(x)$ and $g(x)$ equal to each other and solve for x.

$$\sin x = \cos x \qquad \text{Set } f(x) \text{ equal to } g(x).$$
$$\frac{\sin x}{\cos x} = 1 \qquad \text{Divide each side by } \cos x.$$
$$\tan x = 1 \qquad \text{Trigonometric identity}$$
$$x = \frac{\pi}{4} \text{ or } \frac{5\pi}{4}, \quad 0 \le x \le 2\pi \qquad \text{Solve for } x.$$

So, $a = \pi/4$ and $b = 5\pi/4$. Because $\sin x \ge \cos x$ for all x in the interval $[\pi/4, 5\pi/4]$, the area of the region is

$$A = \int_{\pi/4}^{5\pi/4} [\sin x - \cos x]\,dx$$
$$= \left[-\cos x - \sin x\right]_{\pi/4}^{5\pi/4}$$
$$= 2\sqrt{2}.$$

To find the area of the region between two curves that intersect at *more* than two points, first determine all points of intersection. Then check to see which curve is above the other in each interval determined by these points, as shown in Example 4.

EXAMPLE 4 Curves That Intersect at More than Two Points

See LarsonCalculus.com for an interactive version of this type of example.

Find the area of the region between the graphs of

$$f(x) = 3x^3 - x^2 - 10x \quad \text{and} \quad g(x) = -x^2 + 2x.$$

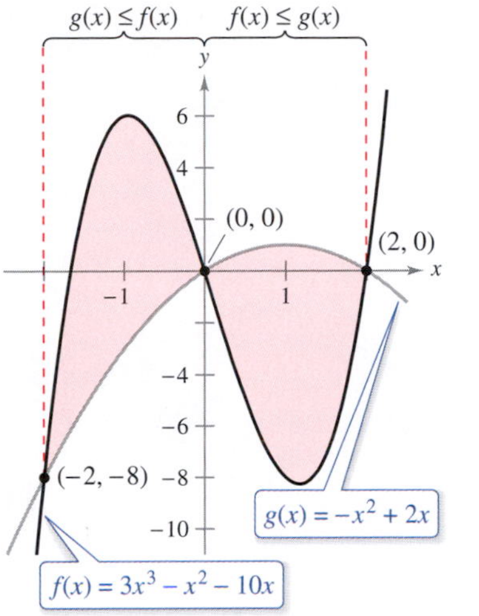

On $[-2, 0]$, $g(x) \le f(x)$, and on $[0, 2]$, $f(x) \le g(x)$.
Figure 7.8

Solution Begin by setting $f(x)$ and $g(x)$ equal to each other and solving for x. This yields the x-values at all points of intersection of the two graphs.

$$3x^3 - x^2 - 10x = -x^2 + 2x \qquad \text{Set } f(x) \text{ equal to } g(x).$$
$$3x^3 - 12x = 0 \qquad \text{Write in general form.}$$
$$3x(x-2)(x+2) = 0 \qquad \text{Factor.}$$
$$x = -2, 0, 2 \qquad \text{Solve for } x.$$

So, the two graphs intersect when $x = -2, 0,$ and 2. In Figure 7.8, notice that $g(x) \le f(x)$ on the interval $[-2, 0]$. The two graphs switch at the origin, however, and $f(x) \le g(x)$ on the interval $[0, 2]$. So, you need two integrals—one for the interval $[-2, 0]$ and one for the interval $[0, 2]$.

$$\begin{aligned} A &= \int_{-2}^{0} [f(x) - g(x)]\,dx + \int_{0}^{2} [g(x) - f(x)]\,dx \\ &= \int_{-2}^{0} (3x^3 - 12x)\,dx + \int_{0}^{2} (-3x^3 + 12x)\,dx \\ &= \left[\frac{3x^4}{4} - 6x^2\right]_{-2}^{0} + \left[\frac{-3x^4}{4} + 6x^2\right]_{0}^{2} \\ &= -(12 - 24) + (-12 + 24) \\ &= 24 \end{aligned}$$

REMARK In Example 4, notice that you obtain an incorrect result when you integrate from -2 to 2. Such integration produces

$$\int_{-2}^{2} [f(x) - g(x)]\,dx = \int_{-2}^{2} (3x^3 - 12x)\,dx = 0.$$

When the graph of a function of y is a boundary of a region, it is often convenient to use representative rectangles that are *horizontal* and find the area by integrating with respect to y. In general, to determine the area between two curves, you can use

$$A = \int_{x_1}^{x_2} \underbrace{[(\text{top curve}) - (\text{bottom curve})]}_{\text{in variable } x}\,dx \qquad \text{Vertical rectangles}$$

or

$$A = \int_{y_1}^{y_2} \underbrace{[(\text{right curve}) - (\text{left curve})]}_{\text{in variable } y}\,dy \qquad \text{Horizontal rectangles}$$

where (x_1, y_1) and (x_2, y_2) are either adjacent points of intersection of the two curves involved or points on the specified boundary lines.

EXAMPLE 5 **Horizontal Representative Rectangles**

Find the area of the region bounded by the graphs of $x = 3 - y^2$ and $x = y + 1$.

Solution Consider

$$g(y) = 3 - y^2 \quad \text{and} \quad f(y) = y + 1.$$

These two curves intersect when $y = -2$ and $y = 1$, as shown in Figure 7.9. Because $f(y) \le g(y)$ on this interval, you have

$$\Delta A = [g(y) - f(y)]\,\Delta y = [(3 - y^2) - (y + 1)]\,\Delta y.$$

So, the area is

$$\begin{aligned} A &= \int_{-2}^{1} [(3 - y^2) - (y + 1)]\,dy \\ &= \int_{-2}^{1} (-y^2 - y + 2)\,dy \\ &= \left[\frac{-y^3}{3} - \frac{y^2}{2} + 2y\right]_{-2}^{1} \\ &= \left(-\frac{1}{3} - \frac{1}{2} + 2\right) - \left(\frac{8}{3} - 2 - 4\right) \\ &= \frac{9}{2}. \end{aligned}$$

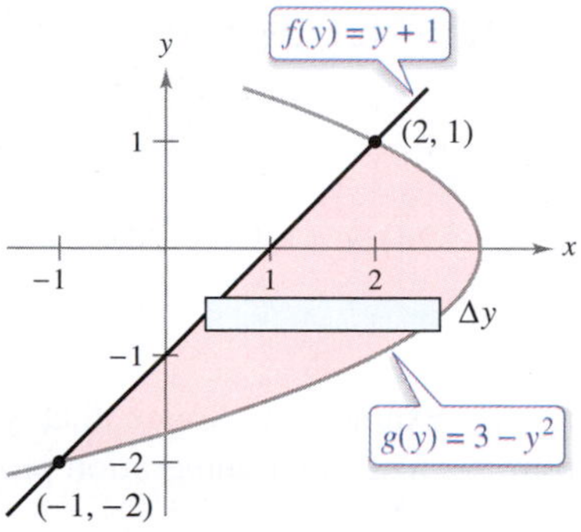

Horizontal rectangles (integration with respect to y)
Figure 7.9

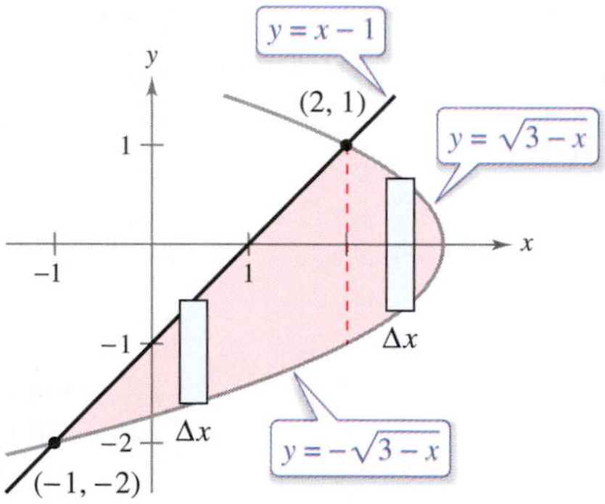

Vertical rectangles (integration with respect to x)
Figure 7.10

In Example 5, notice that by integrating with respect to y, you need only one integral. To integrate with respect to x, you would need two integrals because the upper boundary changes at $x = 2$, as shown in Figure 7.10.

$$\begin{aligned} A &= \int_{-1}^{2} \left[(x - 1) + \sqrt{3 - x}\right] dx + \int_{2}^{3} \left(\sqrt{3 - x} + \sqrt{3 - x}\right) dx \\ &= \int_{-1}^{2} [x - 1 + (3 - x)^{1/2}]\,dx + 2\int_{2}^{3} (3 - x)^{1/2}\,dx \\ &= \left[\frac{x^2}{2} - x - \frac{(3 - x)^{3/2}}{3/2}\right]_{-1}^{2} - 2\left[\frac{(3 - x)^{3/2}}{3/2}\right]_{2}^{3} \\ &= \left(2 - 2 - \frac{2}{3}\right) - \left(\frac{1}{2} + 1 - \frac{16}{3}\right) - 2(0) + 2\left(\frac{2}{3}\right) \\ &= \frac{9}{2} \end{aligned}$$

Integration as an Accumulation Process

In this section, the integration formula for the area between two curves was developed by using a rectangle as the *representative element.* For each new application in the remaining sections of this chapter, an appropriate representative element will be constructed using precalculus formulas you already know. Each integration formula will then be obtained by summing or accumulating these representative elements.

Known precalculus formula ⇨ Representative element ⇨ New integration formula

For example, the area formula in this section was developed as follows.

$A = (\text{height})(\text{width})$ ⇨ $\Delta A = [f(x) - g(x)]\,\Delta x$ ⇨ $A = \int_a^b [f(x) - g(x)]\,dx$

EXAMPLE 6 Integration as an Accumulation Process

Find the area of the region bounded by the graph of $y = 4 - x^2$ and the x-axis. Describe the integration as an accumulation process.

Solution The area of the region is

$$A = \int_{-2}^{2} (4 - x^2)\,dx.$$

You can think of the integration as an accumulation of the areas of the rectangles formed as the representative rectangle slides from $x = -2$ to $x = 2$, as shown in Figure 7.11.

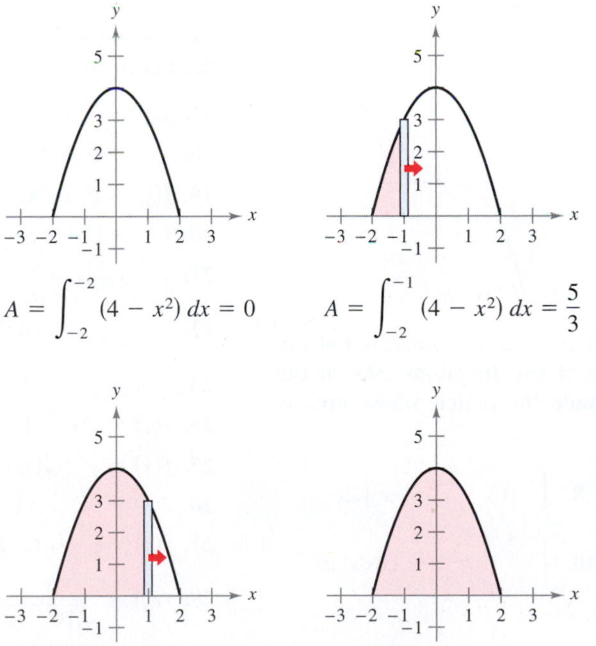

$A = \int_{-2}^{-2} (4 - x^2)\,dx = 0$

$A = \int_{-2}^{-1} (4 - x^2)\,dx = \frac{5}{3}$

$A = \int_{-2}^{0} (4 - x^2)\,dx = \frac{16}{3}$

$A = \int_{-2}^{1} (4 - x^2)\,dx = 9$

$A = \int_{-2}^{2} (4 - x^2)\,dx = \frac{32}{3}$

Figure 7.11

7.1 Exercises

See CalcChat.com for tutorial help and worked-out solutions to odd-numbered exercises.

Writing a Definite Integral **In Exercises 1–6, set up the definite integral that gives the area of the region.**

1. $y_1 = x^2 - 6x$
$y_2 = 0$

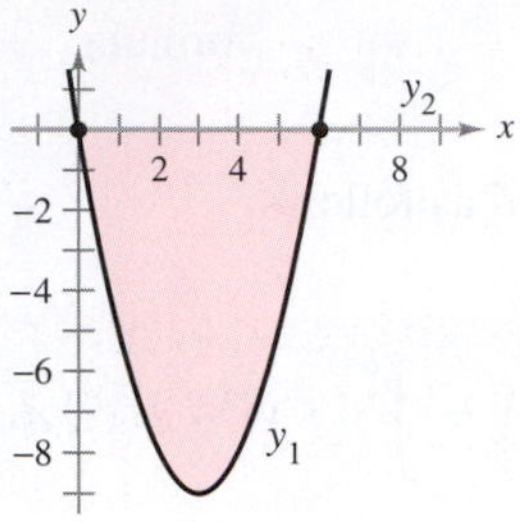

2. $y_1 = x^2 + 2x + 1$
$y_2 = 2x + 5$

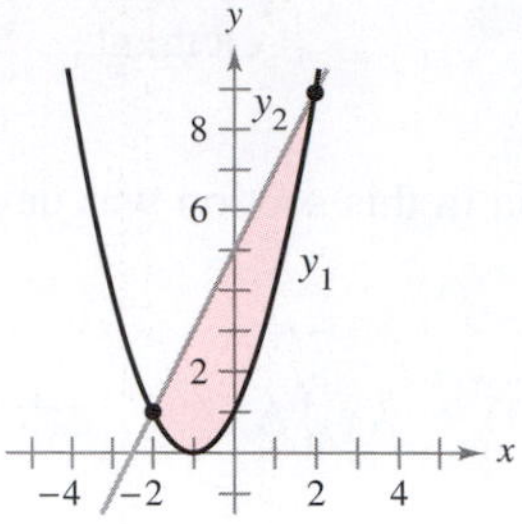

3. $y_1 = x^2 - 4x + 3$
$y_2 = -x^2 + 2x + 3$

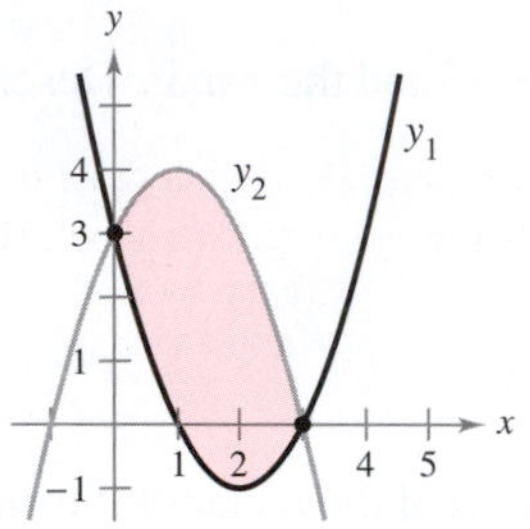

4. $y_1 = x^2$
$y_2 = x^3$

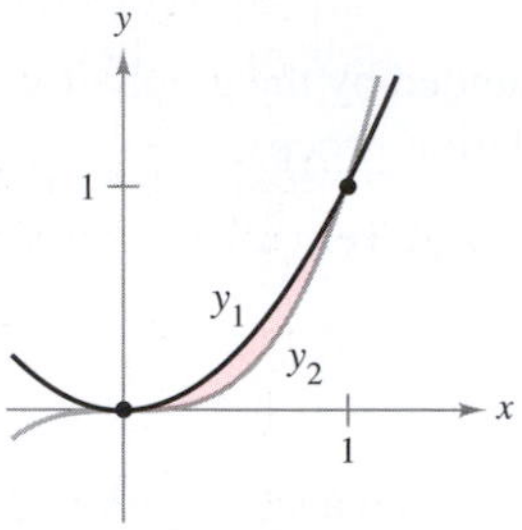

5. $y_1 = 3(x^3 - x)$
$y_2 = 0$

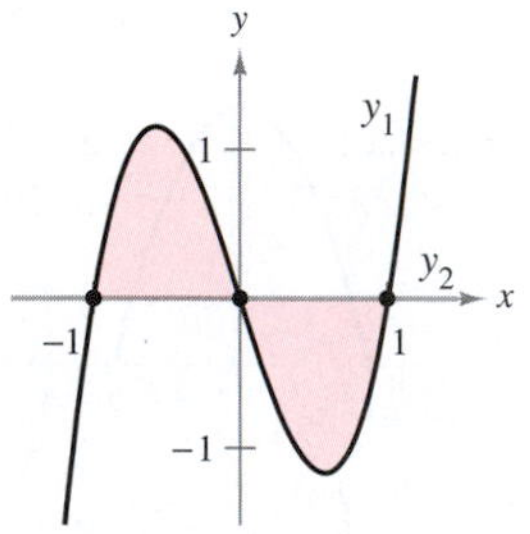

6. $y_1 = (x - 1)^3$
$y_2 = x - 1$

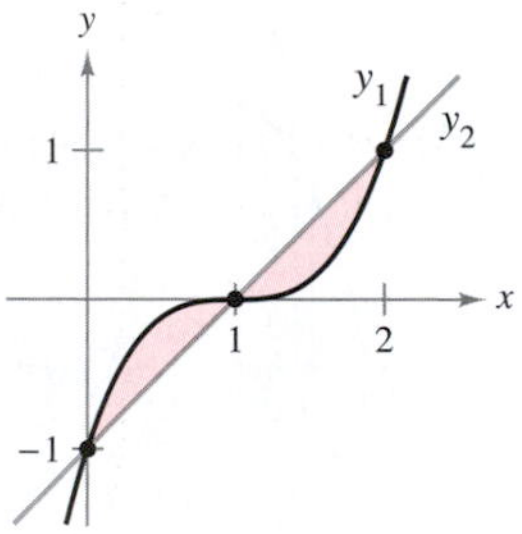

Finding a Region **In Exercises 7–12, the integrand of the definite integral is a difference of two functions. Sketch the graph of each function and shade the region whose area is represented by the integral.**

7. $\displaystyle\int_0^4 \left[(x + 1) - \frac{x}{2}\right] dx$

8. $\displaystyle\int_{-1}^1 [(2 - x^2) - x^2]\, dx$

9. $\displaystyle\int_2^3 \left[\left(\frac{x^3}{3} - x\right) - \frac{x}{3}\right] dx$

10. $\displaystyle\int_{-\pi/4}^{\pi/4} (\sec^2 x - \cos x)\, dx$

11. $\displaystyle\int_{-2}^1 [(2 - y) - y^2]\, dy$

12. $\displaystyle\int_0^4 \left(2\sqrt{y} - y\right) dy$

Think About It **In Exercises 13 and 14, determine which value best approximates the area of the region bounded by the graphs of *f* and *g*. (Make your selection on the basis of a sketch of the region and not by performing any calculations.)**

13. $f(x) = x + 1, \quad g(x) = (x - 1)^2$

(a) -2 (b) 2 (c) 10 (d) 4 (e) 8

14. $f(x) = 2 - \frac{1}{2}x, \quad g(x) = 2 - \sqrt{x}$

(a) 1 (b) 6 (c) -3 (d) 3 (e) 4

Comparing Methods **In Exercises 15 and 16, find the area of the region by integrating (a) with respect to *x* and (b) with respect to *y*. (c) Compare your results. Which method is simpler? In general, will this method always be simpler than the other one? Why or why not?**

15. $x = 4 - y^2$
$x = y - 2$

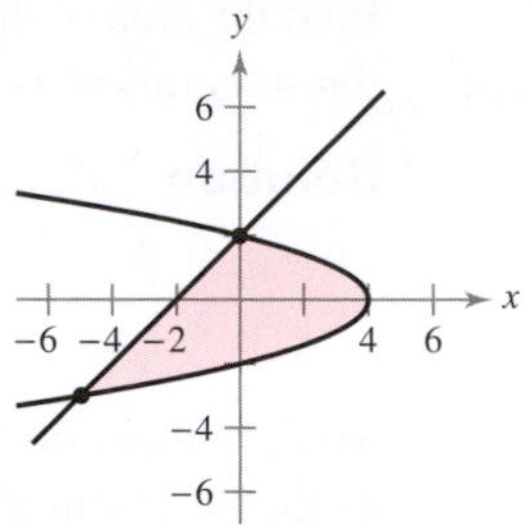

16. $y = x^2$
$y = 6 - x$

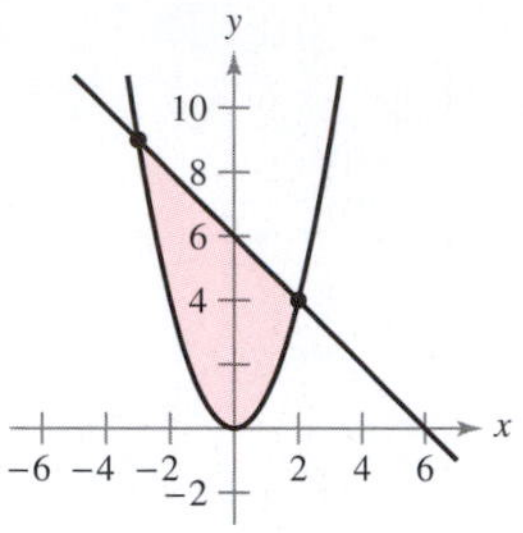

Finding the Area of a Region **In Exercises 17–30, sketch the region bounded by the graphs of the equations and find the area of the region.**

17. $y = x^2 - 1, \quad y = -x + 2, \quad x = 0, \quad x = 1$

18. $y = -x^3 + 2, \quad y = x - 3, \quad x = -1, \quad x = 1$

19. $f(x) = x^2 + 2x, \quad g(x) = x + 2$

20. $y = -x^2 + 3x + 1, \quad y = -x + 1$

21. $y = x, \quad y = 2 - x, \quad y = 0$

22. $y = \dfrac{4}{x^3}, \quad y = 0, \quad x = 1, \quad x = 4$

23. $f(x) = \sqrt{x} + 3, \quad g(x) = \frac{1}{2}x + 3$

24. $f(x) = \sqrt[3]{x - 1}, \quad g(x) = x - 1$

25. $f(y) = y^2, \quad g(y) = y + 2$

26. $f(y) = y(2 - y), \quad g(y) = -y$

27. $f(y) = y^2 + 1, \quad g(y) = 0, \quad y = -1, \quad y = 2$

28. $f(y) = \dfrac{y}{\sqrt{16 - y^2}}, \quad g(y) = 0, \quad y = 3$

29. $f(x) = \dfrac{10}{x}, \quad x = 0, \quad y = 2, \quad y = 10$

30. $g(x) = \dfrac{4}{2 - x}, \quad y = 4, \quad x = 0$

Finding the Area of a Region **In Exercises 31–36, (a) use a graphing utility to graph the region bounded by the graphs of the equations, (b) find the area of the region analytically, and (c) use the integration capabilities of the graphing utility to verify your results.**

31. $f(x) = x(x^2 - 3x + 3),\ \ g(x) = x^2$
32. $y = x^4 - 2x^2,\ \ y = 2x^2$
33. $f(x) = x^4 - 4x^2,\ \ g(x) = x^2 - 4$
34. $f(x) = x^4 - 9x^2,\ \ g(x) = x^3 - 9x$
35. $f(x) = \dfrac{1}{1 + x^2},\ \ g(x) = \dfrac{1}{2}x^2$
36. $f(x) = \dfrac{6x}{x^2 + 1},\ \ y = 0,\ \ 0 \le x \le 3$

Finding the Area of a Region **In Exercises 37–42, sketch the region bounded by the graphs of the functions and find the area of the region.**

37. $f(x) = \cos x,\ \ g(x) = 2 - \cos x,\ \ 0 \le x \le 2\pi$
38. $f(x) = \sin x,\ \ g(x) = \cos 2x,\ \ -\dfrac{\pi}{2} \le x \le \dfrac{\pi}{6}$
39. $f(x) = 2\sin x,\ \ g(x) = \tan x,\ \ -\dfrac{\pi}{3} \le x \le \dfrac{\pi}{3}$
40. $f(x) = \sec\dfrac{\pi x}{4}\tan\dfrac{\pi x}{4},\ \ g(x) = \left(\sqrt{2} - 4\right)x + 4,\ \ x = 0$
41. $f(x) = xe^{-x^2},\ \ y = 0,\ \ 0 \le x \le 1$
42. $f(x) = 2^x,\ \ g(x) = \dfrac{3}{2}x + 1$

Finding the Area of a Region **In Exercises 43–46, (a) use a graphing utility to graph the region bounded by the graphs of the equations, (b) find the area of the region, and (c) use the integration capabilities of the graphing utility to verify your results.**

43. $f(x) = 2\sin x + \sin 2x,\ \ y = 0,\ \ 0 \le x \le \pi$
44. $f(x) = 2\sin x + \cos 2x,\ \ y = 0,\ \ 0 < x \le \pi$
45. $f(x) = \dfrac{1}{x^2}e^{1/x},\ \ y = 0,\ \ 1 \le x \le 3$
46. $g(x) = \dfrac{4\ln x}{x},\ \ y = 0,\ \ x = 5$

Finding the Area of a Region **In Exercises 47–50, (a) use a graphing utility to graph the region bounded by the graphs of the equations, (b) explain why the area of the region is difficult to find by hand, and (c) use the integration capabilities of the graphing utility to approximate the area to four decimal places.**

47. $y = \sqrt{\dfrac{x^3}{4 - x}},\ \ y = 0,\ \ x = 3$
48. $y = \sqrt{x}\,e^x,\ \ y = 0,\ \ x = 0,\ \ x = 1$
49. $y = x^2,\ \ y = 4\cos x$
50. $y = x^2,\ \ y = \sqrt{3 + x}$

Integration as an Accumulation Process **In Exercises 51–54, find the accumulation function F. Then evaluate F at each value of the independent variable and graphically show the area given by each value of F.**

51. $F(x) = \displaystyle\int_0^x \left(\frac{1}{2}t + 1\right) dt$ (a) $F(0)$ (b) $F(2)$ (c) $F(6)$
52. $F(x) = \displaystyle\int_0^x \left(\frac{1}{2}t^2 + 2\right) dt$ (a) $F(0)$ (b) $F(4)$ (c) $F(6)$
53. $F(\alpha) = \displaystyle\int_{-1}^{\alpha} \cos\frac{\pi\theta}{2}\, d\theta$ (a) $F(-1)$ (b) $F(0)$ (c) $F\left(\frac{1}{2}\right)$
54. $F(y) = \displaystyle\int_{-1}^{y} 4e^{x/2}\, dx$ (a) $F(-1)$ (b) $F(0)$ (c) $F(4)$

Finding the Area of a Figure **In Exercises 55–58, use integration to find the area of the figure having the given vertices.**

55. $(2, -3), (4, 6), (6, 1)$
56. $(0, 0), (6, 0), (4, 3)$
57. $(0, 2), (4, 2), (0, -2), (-4, -2)$
58. $(0, 0), (1, 2), (3, -2), (1, -3)$

59. **Numerical Integration** Estimate the surface area of the golf green using (a) the Trapezoidal Rule and (b) Simpson's Rule.

60. **Numerical Integration** Estimate the surface area of the oil spill using (a) the Trapezoidal Rule and (b) Simpson's Rule.

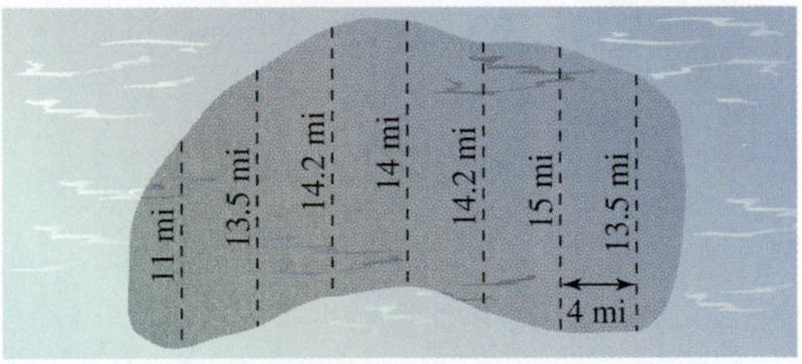

Using a Tangent Line **In Exercises 61–64, set up and evaluate the definite integral that gives the area of the region bounded by the graph of the function and the tangent line to the graph at the given point.**

61. $f(x) = x^3,\ \ (1, 1)$
62. $y = x^3 - 2x,\ \ (-1, 1)$
63. $f(x) = \dfrac{1}{x^2 + 1},\ \ \left(1, \dfrac{1}{2}\right)$
64. $y = \dfrac{2}{1 + 4x^2},\ \ \left(\dfrac{1}{2}, 1\right)$

WRITING ABOUT CONCEPTS

65. Area Between Curves The graphs of $y = 1 - x^2$ and $y = x^4 - 2x^2 + 1$ intersect at three points. However, the area between the curves *can* be found by a single integral. Explain why this is so, and write an integral for this area.

66. Using Symmetry The area of the region bounded by the graphs of $y = x^3$ and $y = x$ *cannot* be found by the single integral $\int_{-1}^{1} (x^3 - x)\, dx$. Explain why this is so. Use symmetry to write a single integral that does represent the area.

67. Interpreting Integrals Two cars with velocities v_1 and v_2 are tested on a straight track (in meters per second). Consider the following.

$$\int_0^5 [v_1(t) - v_2(t)]\, dt = 10 \qquad \int_0^{10} [v_1(t) - v_2(t)]\, dt = 30$$

$$\int_{20}^{30} [v_1(t) - v_2(t)]\, dt = -5$$

(a) Write a verbal interpretation of each integral.

(b) Is it possible to determine the distance between the two cars when $t = 5$ seconds? Why or why not?

(c) Assume both cars start at the same time and place. Which car is ahead when $t = 10$ seconds? How far ahead is the car?

(d) Suppose Car 1 has velocity v_1 and is ahead of Car 2 by 13 meters when $t = 20$ seconds. How far ahead or behind is Car 1 when $t = 30$ seconds?

68. HOW DO YOU SEE IT? A state legislature is debating two proposals for eliminating the annual budget deficits after 10 years. The rate of decrease of the deficits for each proposal is shown in the figure.

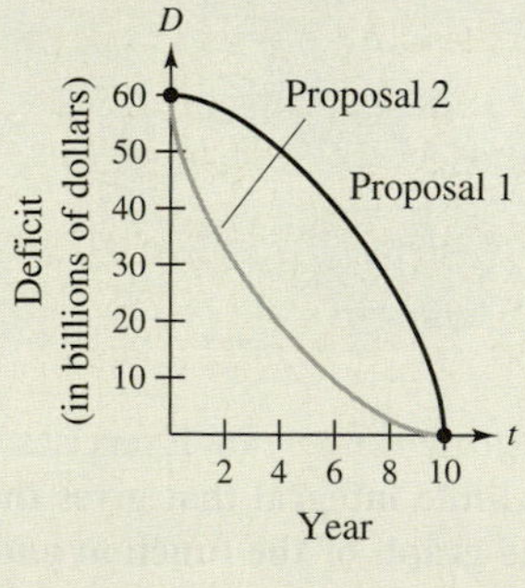

(a) What does the area between the two curves represent?

(b) From the viewpoint of minimizing the cumulative state deficit, which is the better proposal? Explain.

Dividing a Region **In Exercises 69 and 70, find b such that the line $y = b$ divides the region bounded by the graphs of the two equations into two regions of equal area.**

69. $y = 9 - x^2, \quad y = 0$ **70.** $y = 9 - |x|, \quad y = 0$

Dividing a Region **In Exercises 71 and 72, find a such that the line $x = a$ divides the region bounded by the graphs of the equations into two regions of equal area.**

71. $y = x, \quad y = 4, \quad x = 0$ **72.** $y^2 = 4 - x, \quad x = 0$

Limits and Integrals **In Exercises 73 and 74, evaluate the limit and sketch the graph of the region whose area is represented by the limit.**

73. $\lim_{\|\Delta\| \to 0} \sum_{i=1}^{n} (x_i - x_i^2)\, \Delta x$, where $x_i = \dfrac{i}{n}$ and $\Delta x = \dfrac{1}{n}$

74. $\lim_{\|\Delta\| \to 0} \sum_{i=1}^{n} (4 - x_i^2)\, \Delta x$, where $x_i = -2 + \dfrac{4i}{n}$ and $\Delta x = \dfrac{4}{n}$

Revenue **In Exercises 75 and 76, two models R_1 and R_2 are given for revenue (in billions of dollars) for a large corporation. Both models are estimates of revenues from 2015 through 2020, with $t = 15$ corresponding to 2015. Which model projects the greater revenue? How much more total revenue does that model project over the six-year period?**

75. $R_1 = 7.21 + 0.58t$

$R_2 = 7.21 + 0.45t$

76. $R_1 = 7.21 + 0.26t + 0.02t^2$

$R_2 = 7.21 + 0.1t + 0.01t^2$

77. Lorenz Curve Economists use *Lorenz curves* to illustrate the distribution of income in a country. A Lorenz curve, $y = f(x)$, represents the actual income distribution in the country. In this model, x represents percents of families in the country and y represents percents of total income. The model $y = x$ represents a country in which each family has the same income. The area between these two models, where $0 \le x \le 100$, indicates a country's "income inequality." The table lists percents of income y for selected percents of families x in a country.

x	10	20	30	40	50
y	3.35	6.07	9.17	13.39	19.45

x	60	70	80	90
y	28.03	39.77	55.28	75.12

(a) Use a graphing utility to find a quadratic model for the Lorenz curve.

(b) Plot the data and graph the model.

(c) Graph the model $y = x$. How does this model compare with the model in part (a)?

(d) Use the integration capabilities of a graphing utility to approximate the "income inequality."

78. Profit The chief financial officer of a company reports that profits for the past fiscal year were \$15.9 million. The officer predicts that profits for the next 5 years will grow at a continuous annual rate somewhere between $3\frac{1}{2}\%$ and 5%. Estimate the cumulative difference in total profit over the 5 years based on the predicted range of growth rates.

79. Building Design Concrete sections for a new building have the dimensions (in meters) and shape shown in the figure.

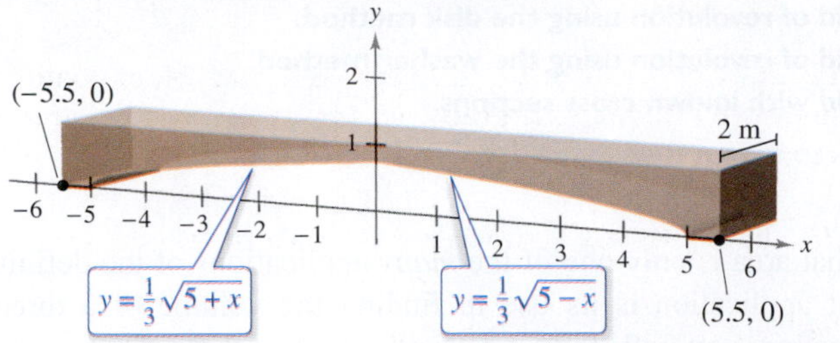

(a) Find the area of the face of the section superimposed on the rectangular coordinate system.

(b) Find the volume of concrete in one of the sections by multiplying the area in part (a) by 2 meters.

(c) One cubic meter of concrete weighs 5000 pounds. Find the weight of the section.

80. Mechanical Design The surface of a machine part is the region between the graphs of $y_1 = |x|$ and $y_2 = 0.08x^2 + k$ (see figure).

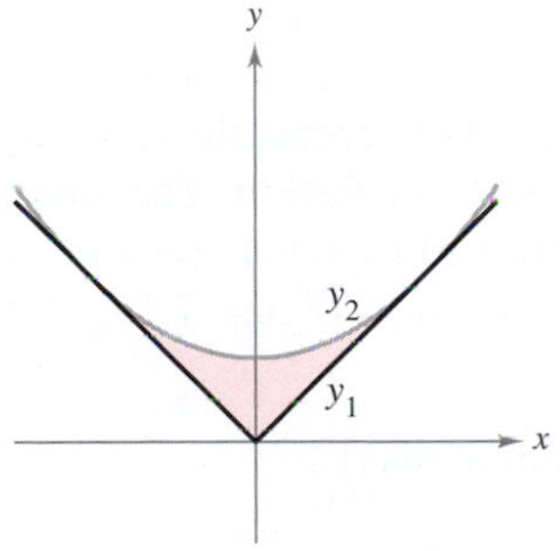

(a) Find k where the parabola is tangent to the graph of y_1.

(b) Find the area of the surface of the machine part.

81. Area Find the area between the graph of $y = \sin x$ and the line segment joining the points $(0, 0)$ and $\left(\frac{7\pi}{6}, -\frac{1}{2}\right)$, as shown in the figure.

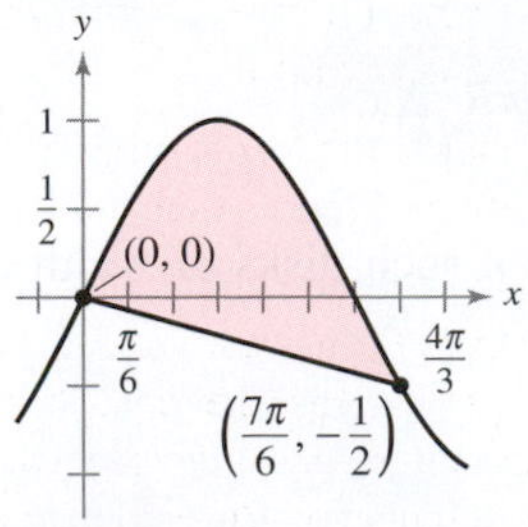

82. Area Let $a > 0$ and $b > 0$. Show that the area of the ellipse $\frac{x^2}{a^2} + \frac{y^2}{b^2} = 1$ is πab (see figure).

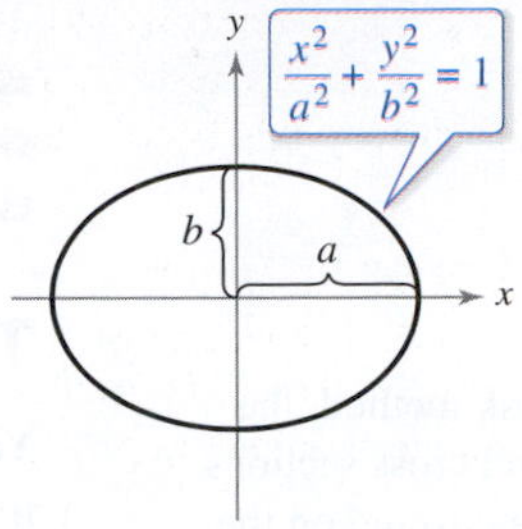

True or False? **In Exercises 83–86, determine whether the statement is true or false. If it is false, explain why or give an example that shows it is false.**

83. If the area of the region bounded by the graphs of f and g is 1, then the area of the region bounded by the graphs of $h(x) = f(x) + C$ and $k(x) = g(x) + C$ is also 1.

84. If

$$\int_a^b [f(x) - g(x)]\,dx = A$$

then

$$\int_a^b [g(x) - f(x)]\,dx = -A.$$

85. If the graphs of f and g intersect midway between $x = a$ and $x = b$, then

$$\int_a^b [f(x) - g(x)]\,dx = 0.$$

86. The line

$$y = \left(1 - \sqrt[3]{0.5}\right)x$$

divides the region under the curve

$$f(x) = x(1 - x)$$

on $[0, 1]$ into two regions of equal area.

PUTNAM EXAM CHALLENGE

87. The horizontal line $y = c$ intersects the curve $y = 2x - 3x^3$ in the first quadrant as shown in the figure. Find c so that the areas of the two shaded regions are equal.

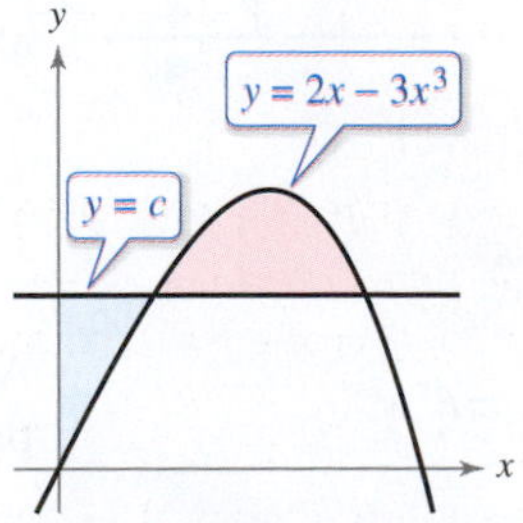

7.2 Volume: The Disk Method

- Find the volume of a solid of revolution using the disk method.
- Find the volume of a solid of revolution using the washer method.
- Find the volume of a solid with known cross sections.

AP* Tips

Volumes by the disk method, the washer method, and cross sections are the final topics covered on the AP Calculus AB Exam.

The Disk Method

You have already learned that area is only one of the *many* applications of the definite integral. Another important application is its use in finding the volume of a three-dimensional solid. In this section, you will study a particular type of three-dimensional solid—one whose cross sections are similar. Solids of revolution are used commonly in engineering and manufacturing. Some examples are axles, funnels, pills, bottles, and pistons, as shown in Figure 7.12.

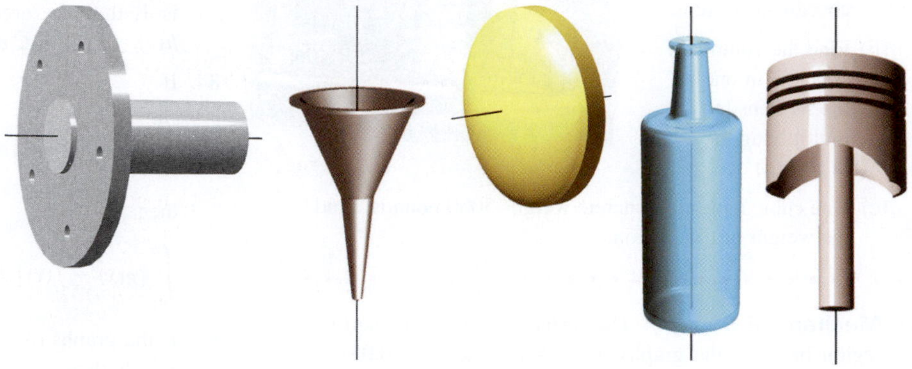

Solids of revolution
Figure 7.12

When a region in the plane is revolved about a line, the resulting solid is a **solid of revolution,** and the line is called the **axis of revolution.** The simplest such solid is a right circular cylinder or **disk,** which is formed by revolving a rectangle about an axis adjacent to one side of the rectangle, as shown in Figure 7.13. The volume of such a disk is

$$\text{Volume of disk} = (\text{area of disk})(\text{width of disk}) = \pi R^2 w$$

where R is the radius of the disk and w is the width.

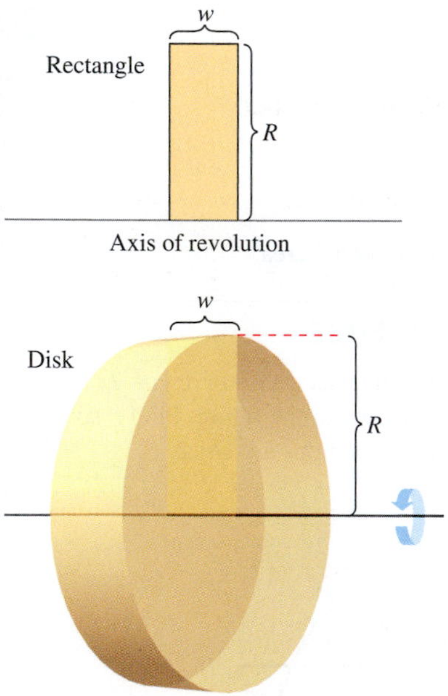

Volume of a disk: $\pi R^2 w$
Figure 7.13

To see how to use the volume of a disk to find the volume of a general solid of revolution, consider a solid of revolution formed by revolving the plane region in Figure 7.14 about the indicated axis. To determine the volume of this solid, consider a representative rectangle in the plane region. When this rectangle is revolved about the axis of revolution, it generates a representative disk whose volume is

$$\Delta V = \pi R^2 \Delta x.$$

Approximating the volume of the solid by n such disks of width Δx and radius $R(x_i)$ produces

$$\text{Volume of solid} \approx \sum_{i=1}^{n} \pi [R(x_i)]^2 \Delta x = \pi \sum_{i=1}^{n} [R(x_i)]^2 \Delta x.$$

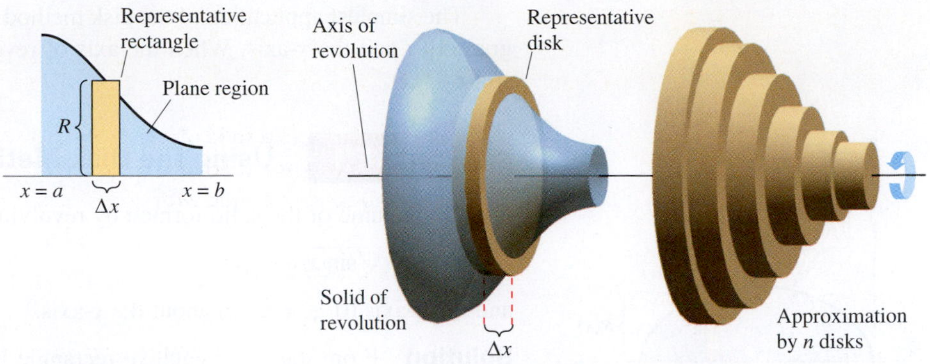

Disk method
Figure 7.14

This approximation appears to become better and better as $\|\Delta\| \to 0$ $(n \to \infty)$. So, you can define the volume of the solid as

$$\text{Volume of solid} = \lim_{\|\Delta\| \to 0} \pi \sum_{i=1}^{n} [R(x_i)]^2 \, \Delta x = \pi \int_a^b [R(x)]^2 \, dx.$$

Schematically, the disk method looks like this.

Known Precalculus Formula		**Representative Element**		**New Integration Formula**
Volume of disk $V = \pi R^2 w$	⇒	$\Delta V = \pi [R(x_i)]^2 \, \Delta x$	⇒	Solid of revolution $V = \pi \int_a^b [R(x)]^2 \, dx$

A similar formula can be derived when the axis of revolution is vertical.

THE DISK METHOD

To find the volume of a solid of revolution with the **disk method,** use one of the formulas below. (See Figure 7.15.)

Horizontal Axis of Revolution

$$\text{Volume} = V = \pi \int_a^b [R(x)]^2 \, dx$$

Vertical Axis of Revolution

$$\text{Volume} = V = \pi \int_c^d [R(y)]^2 \, dy$$

REMARK In Figure 7.15, note that you can determine the variable of integration by placing a representative rectangle in the plane region "perpendicular" to the axis of revolution. When the width of the rectangle is Δx, integrate with respect to x, and when the width of the rectangle is Δy, integrate with respect to y.

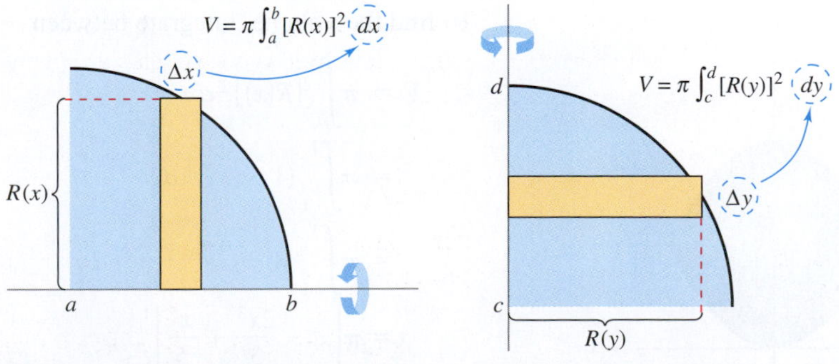

Horizontal axis of revolution
Figure 7.15

Vertical axis of revolution

The simplest application of the disk method involves a plane region bounded by the graph of f and the x-axis. When the axis of revolution is the x-axis, the radius $R(x)$ is simply $f(x)$.

EXAMPLE 1 Using the Disk Method

Find the volume of the solid formed by revolving the region bounded by the graph of

$$f(x) = \sqrt{\sin x}$$

and the x-axis $(0 \le x \le \pi)$ about the x-axis.

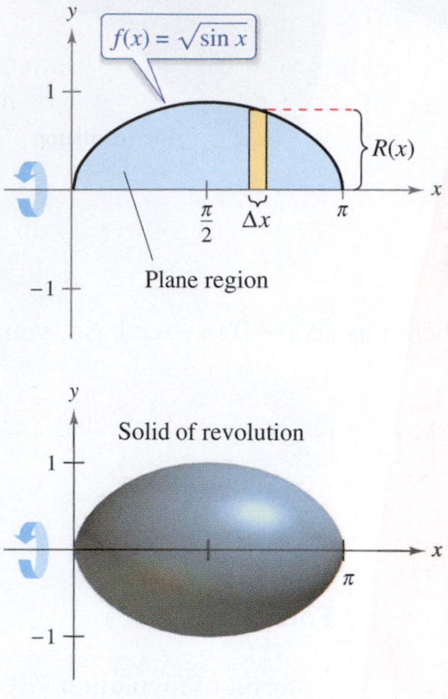

Figure 7.16

Solution From the representative rectangle in the upper graph in Figure 7.16, you can see that the radius of this solid is

$$R(x) = f(x) = \sqrt{\sin x}.$$

So, the volume of the solid of revolution is

$$\begin{aligned} V &= \pi \int_a^b [R(x)]^2 \, dx && \text{Apply disk method.} \\ &= \pi \int_0^\pi \left(\sqrt{\sin x}\right)^2 dx && \text{Substitute } \sqrt{\sin x} \text{ for } R(x). \\ &= \pi \int_0^\pi \sin x \, dx && \text{Simplify.} \\ &= \pi \Big[-\cos x\Big]_0^\pi && \text{Integrate.} \\ &= \pi(1 + 1) \\ &= 2\pi. \end{aligned}$$

EXAMPLE 2 Using a Line That Is Not a Coordinate Axis

Find the volume of the solid formed by revolving the region bounded by the graphs of

$$f(x) = 2 - x^2$$

and $g(x) = 1$ about the line $y = 1$, as shown in Figure 7.17.

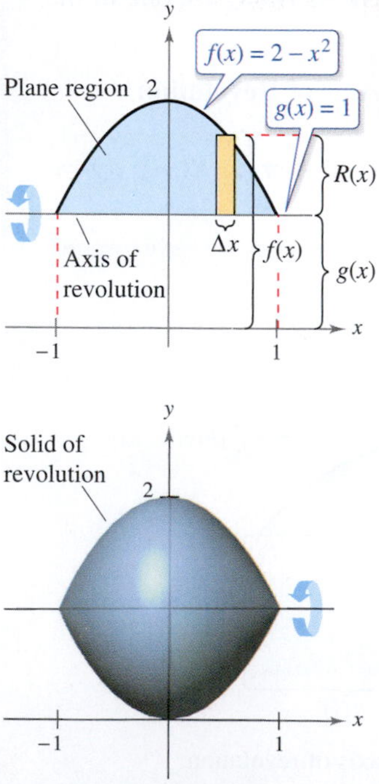

Figure 7.17

Solution By equating $f(x)$ and $g(x)$, you can determine that the two graphs intersect when $x = \pm 1$. To find the radius, subtract $g(x)$ from $f(x)$.

$$\begin{aligned} R(x) &= f(x) - g(x) \\ &= (2 - x^2) - 1 \\ &= 1 - x^2 \end{aligned}$$

To find the volume, integrate between -1 and 1.

$$\begin{aligned} V &= \pi \int_a^b [R(x)]^2 \, dx && \text{Apply disk method.} \\ &= \pi \int_{-1}^1 (1 - x^2)^2 \, dx && \text{Substitute } 1 - x^2 \text{ for } R(x). \\ &= \pi \int_{-1}^1 (1 - 2x^2 + x^4) \, dx && \text{Simplify.} \\ &= \pi \left[x - \frac{2x^3}{3} + \frac{x^5}{5}\right]_{-1}^1 && \text{Integrate.} \\ &= \frac{16\pi}{15} \end{aligned}$$

The Washer Method

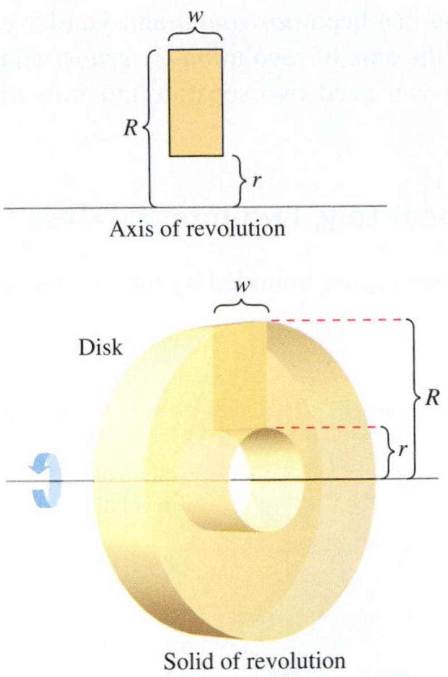

Figure 7.18

The disk method can be extended to cover solids of revolution with holes by replacing the representative disk with a representative **washer.** The washer is formed by revolving a rectangle about an axis, as shown in Figure 7.18. If r and R are the inner and outer radii of the washer and w is the width of the washer, then the volume is

$$\text{Volume of washer} = \pi(R^2 - r^2)w.$$

To see how this concept can be used to find the volume of a solid of revolution, consider a region bounded by an **outer radius** $R(x)$ and an **inner radius** $r(x)$, as shown in Figure 7.19. If the region is revolved about its axis of revolution, then the volume of the resulting solid is

$$V = \pi\int_a^b ([R(x)]^2 - [r(x)]^2)\,dx. \qquad \text{Washer method}$$

Note that the integral involving the inner radius represents the volume of the hole and is *subtracted* from the integral involving the outer radius.

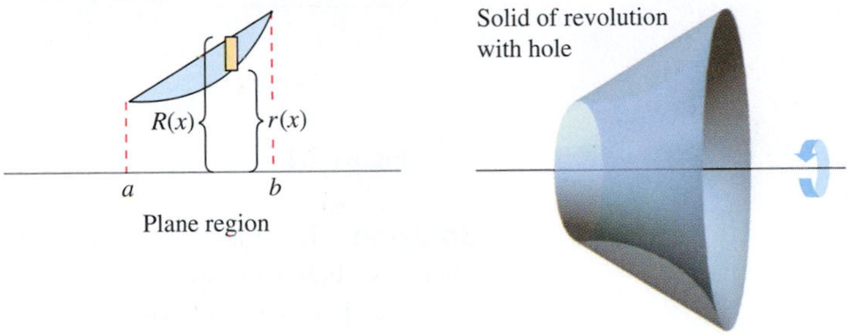

Figure 7.19

EXAMPLE 3 Using the Washer Method

Find the volume of the solid formed by revolving the region bounded by the graphs of

$$y = \sqrt{x} \quad \text{and} \quad y = x^2$$

about the x-axis, as shown in Figure 7.20.

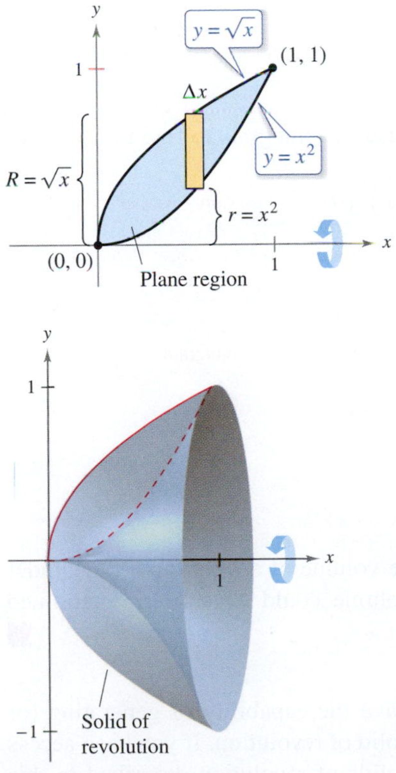

Solid of revolution
Figure 7.20

Solution In Figure 7.20, you can see that the outer and inner radii are as follows.

$$R(x) = \sqrt{x} \qquad \text{Outer radius}$$

$$r(x) = x^2 \qquad \text{Inner radius}$$

Integrating between 0 and 1 produces

$$\begin{aligned} V &= \pi\int_a^b ([R(x)]^2 - [r(x)]^2)\,dx && \text{Apply washer method.} \\ &= \pi\int_0^1 \left[(\sqrt{x})^2 - (x^2)^2\right]dx && \text{Substitute } \sqrt{x} \text{ for } R(x) \text{ and } x^2 \text{ for } r(x). \\ &= \pi\int_0^1 (x - x^4)\,dx && \text{Simplify.} \\ &= \pi\left[\frac{x^2}{2} - \frac{x^5}{5}\right]_0^1 && \text{Integrate.} \\ &= \frac{3\pi}{10}. \end{aligned}$$

In each example so far, the axis of revolution has been *horizontal* and you have integrated with respect to x. In the next example, the axis of revolution is *vertical* and you integrate with respect to y. In this example, you need two separate integrals to compute the volume.

EXAMPLE 4 Integrating with Respect to y, Two-Integral Case

Find the volume of the solid formed by revolving the region bounded by the graphs of

$$y = x^2 + 1, \quad y = 0, \quad x = 0, \quad \text{and} \quad x = 1$$

about the y-axis, as shown in Figure 7.21.

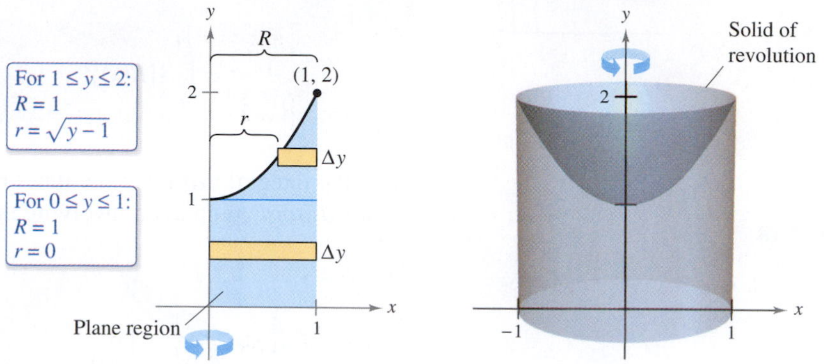

Figure 7.21

Solution For the region shown in Figure 7.21, the outer radius is simply $R = 1$. There is, however, no convenient formula that represents the inner radius. When $0 \le y \le 1$, $r = 0$, but when $1 \le y \le 2$, r is determined by the equation $y = x^2 + 1$, which implies that $r = \sqrt{y-1}$.

$$r(y) = \begin{cases} 0, & 0 \le y \le 1 \\ \sqrt{y-1}, & 1 \le y \le 2 \end{cases}$$

Using this definition of the inner radius, you can use two integrals to find the volume.

$$V = \pi \int_0^1 (1^2 - 0^2)\, dy + \pi \int_1^2 \left[1^2 - \left(\sqrt{y-1}\right)^2\right] dy \qquad \text{Apply washer method.}$$

$$= \pi \int_0^1 1\, dy + \pi \int_1^2 (2 - y)\, dy \qquad \text{Simplify.}$$

$$= \pi \Big[y\Big]_0^1 + \pi \left[2y - \frac{y^2}{2}\right]_1^2 \qquad \text{Integrate.}$$

$$= \pi + \pi\left(4 - 2 - 2 + \frac{1}{2}\right)$$

$$= \frac{3\pi}{2}$$

Note that the first integral $\pi \int_0^1 1\, dy$ represents the volume of a right circular cylinder of radius 1 and height 1. This portion of the volume could have been determined without using calculus.

TECHNOLOGY Some graphing utilities have the capability of generating (or have built-in software capable of generating) a solid of revolution. If you have access to such a utility, use it to graph some of the solids of revolution described in this section. For instance, the solid in Example 4 might appear like that shown in Figure 7.22.

Figure 7.22

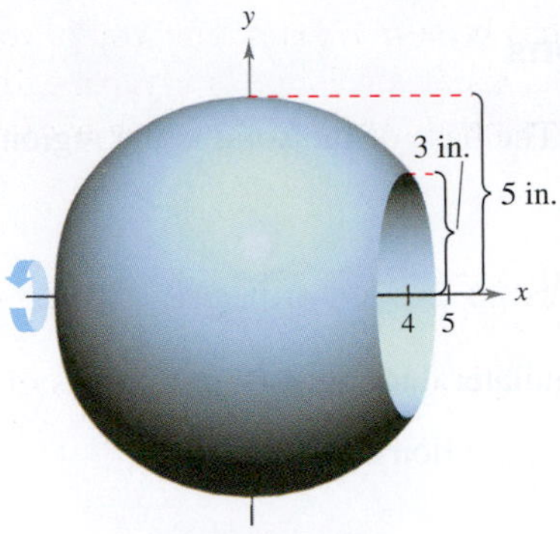

Solid of revolution

(a)

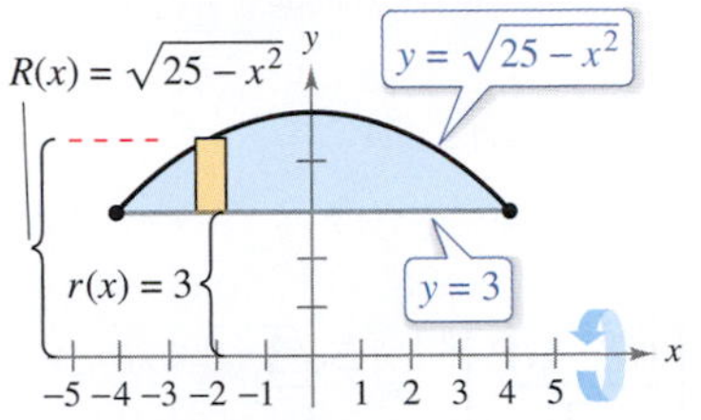

Plane region

(b)

Figure 7.23

EXAMPLE 5 Manufacturing

See LarsonCalculus.com for an interactive version of this type of example.

A manufacturer drills a hole through the center of a metal sphere of radius 5 inches, as shown in Figure 7.23(a). The hole has a radius of 3 inches. What is the volume of the resulting metal ring?

Solution You can imagine the ring to be generated by a segment of the circle whose equation is $x^2 + y^2 = 25$, as shown in Figure 7.23(b). Because the radius of the hole is 3 inches, you can let $y = 3$ and solve the equation $x^2 + y^2 = 25$ to determine that the limits of integration are $x = \pm 4$. So, the inner and outer radii are $r(x) = 3$ and $R(x) = \sqrt{25 - x^2}$, and the volume is

$$\begin{aligned} V &= \pi \int_a^b \left([R(x)]^2 - [r(x)]^2\right) dx \\ &= \pi \int_{-4}^{4} \left[\left(\sqrt{25 - x^2}\right)^2 - (3)^2\right] dx \\ &= \pi \int_{-4}^{4} (16 - x^2)\, dx \\ &= \pi \left[16x - \frac{x^3}{3}\right]_{-4}^{4} \\ &= \frac{256\pi}{3} \text{ cubic inches.} \end{aligned}$$

Solids with Known Cross Sections

With the disk method, you can find the volume of a solid having a circular cross section whose area is $A = \pi R^2$. This method can be generalized to solids of any shape, as long as you know a formula for the area of an arbitrary cross section. Some common cross sections are squares, rectangles, triangles, semicircles, and trapezoids.

VOLUMES OF SOLIDS WITH KNOWN CROSS SECTIONS

1. For cross sections of area $A(x)$ taken perpendicular to the x-axis,

$$\text{Volume} = \int_a^b A(x)\, dx. \qquad \text{See Figure 7.24(a).}$$

2. For cross sections of area $A(y)$ taken perpendicular to the y-axis,

$$\text{Volume} = \int_c^d A(y)\, dy. \qquad \text{See Figure 7.24(b).}$$

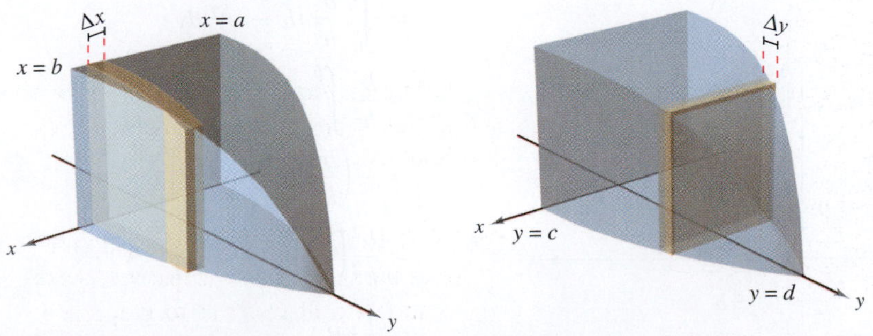

(a) Cross sections perpendicular to x-axis **(b)** Cross sections perpendicular to y-axis

Figure 7.24

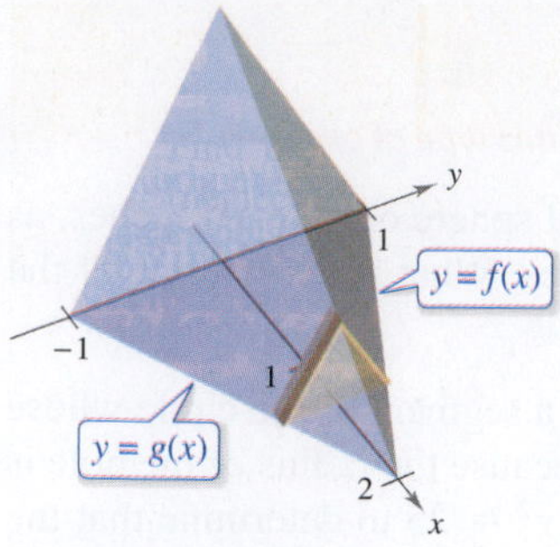

Cross sections are equilateral triangles.

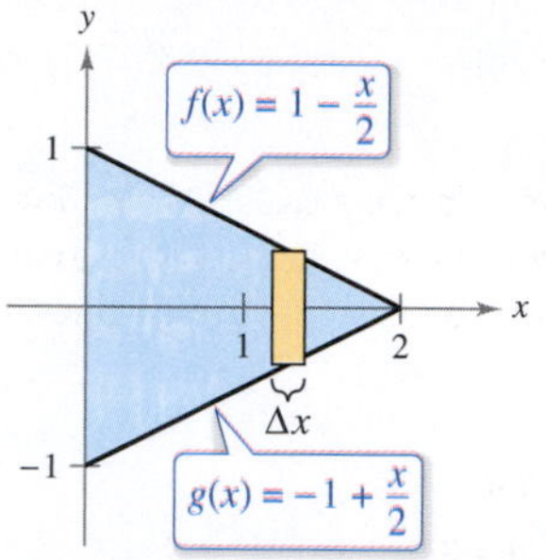

Triangular base in xy-plane
Figure 7.25

EXAMPLE 6 Triangular Cross Sections

Find the volume of the solid shown in Figure 7.25. The base of the solid is the region bounded by the lines

$$f(x) = 1 - \frac{x}{2}, \quad g(x) = -1 + \frac{x}{2}, \quad \text{and} \quad x = 0.$$

The cross sections perpendicular to the x-axis are equilateral triangles.

Solution The base and area of each triangular cross section are as follows.

$$\text{Base} = \left(1 - \frac{x}{2}\right) - \left(-1 + \frac{x}{2}\right) = 2 - x \qquad \text{Length of base}$$

$$\text{Area} = \frac{\sqrt{3}}{4}(\text{base})^2 \qquad \text{Area of equilateral triangle}$$

$$A(x) = \frac{\sqrt{3}}{4}(2 - x)^2 \qquad \text{Area of cross section}$$

Because x ranges from 0 to 2, the volume of the solid is

$$V = \int_a^b A(x)\,dx = \int_0^2 \frac{\sqrt{3}}{4}(2 - x)^2\,dx = -\frac{\sqrt{3}}{4}\left[\frac{(2 - x)^3}{3}\right]_0^2 = \frac{2\sqrt{3}}{3}.$$

EXAMPLE 7 An Application to Geometry

Prove that the volume of a pyramid with a square base is

$$V = \frac{1}{3}hB$$

where h is the height of the pyramid and B is the area of the base.

Solution As shown in Figure 7.26, you can intersect the pyramid with a plane parallel to the base at height y to form a square cross section whose sides are of length b'. Using similar triangles, you can show that

$$\frac{b'}{b} = \frac{h - y}{h} \quad \text{or} \quad b' = \frac{b}{h}(h - y)$$

where b is the length of the sides of the base of the pyramid. So,

$$A(y) = (b')^2 = \frac{b^2}{h^2}(h - y)^2.$$

Integrating between 0 and h produces

$$\begin{aligned} V &= \int_0^h A(y)\,dy \\ &= \int_0^h \frac{b^2}{h^2}(h - y)^2\,dy \\ &= \frac{b^2}{h^2}\int_0^h (h - y)^2\,dy \\ &= -\left(\frac{b^2}{h^2}\right)\left[\frac{(h - y)^3}{3}\right]_0^h \\ &= \frac{b^2}{h^2}\left(\frac{h^3}{3}\right) \\ &= \frac{1}{3}hB. \qquad B = b^2 \end{aligned}$$

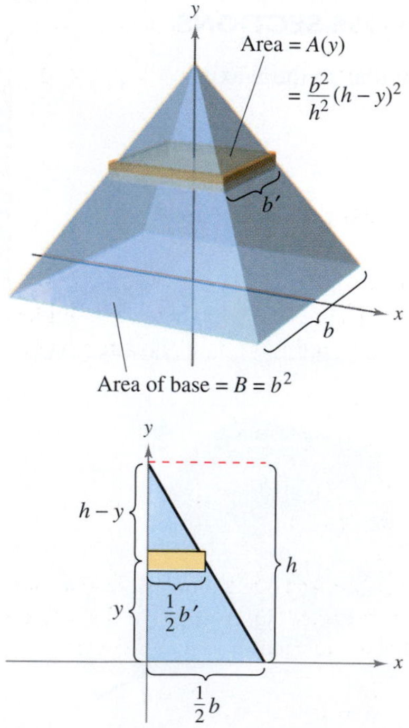

Figure 7.26

7.2 Exercises

See CalcChat.com for tutorial help and worked-out solutions to odd-numbered exercises.

Finding the Volume of a Solid In Exercises 1–6, set up and evaluate the integral that gives the volume of the solid formed by revolving the region about the x-axis.

1. $y = -x + 1$

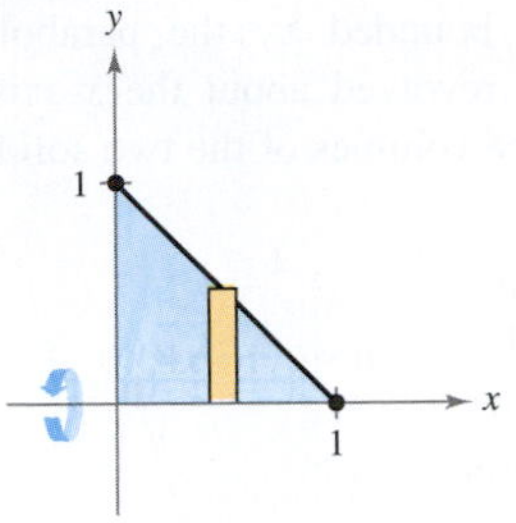

2. $y = 4 - x^2$

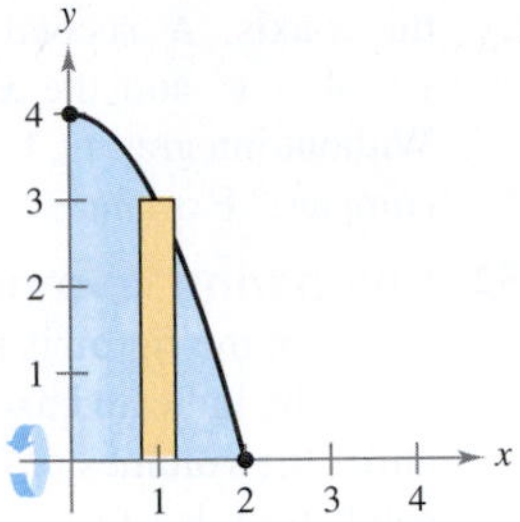

3. $y = \sqrt{x}$

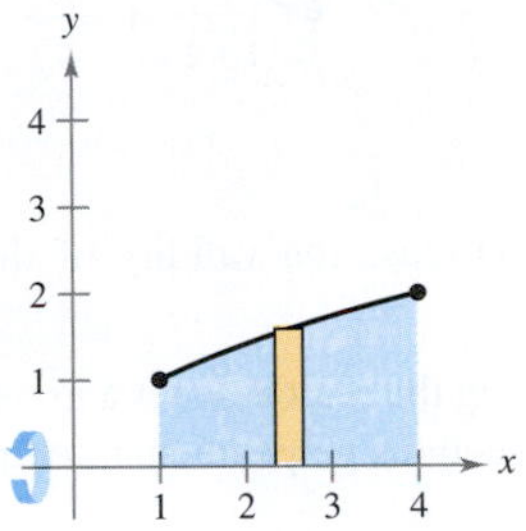

4. $y = \sqrt{9 - x^2}$

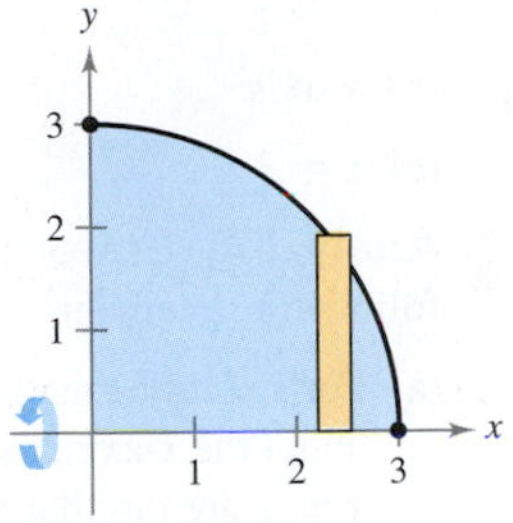

5. $y = x^2, \quad y = x^5$

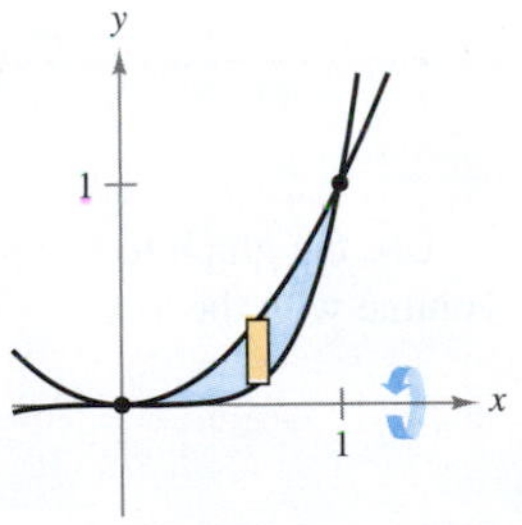

6. $y = 2, \quad y = 4 - \dfrac{x^2}{4}$

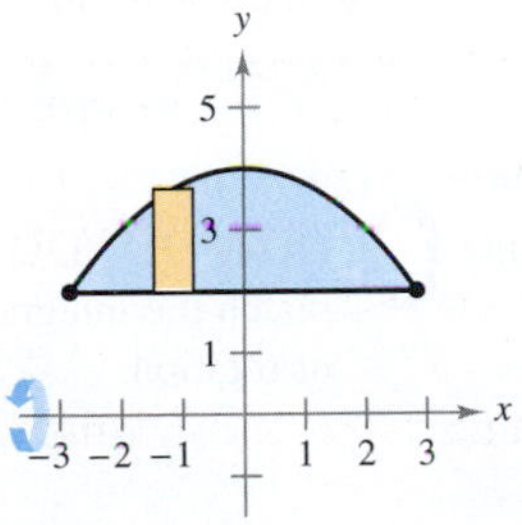

Finding the Volume of a Solid In Exercises 7–10, set up and evaluate the integral that gives the volume of the solid formed by revolving the region about the y-axis.

7. $y = x^2$

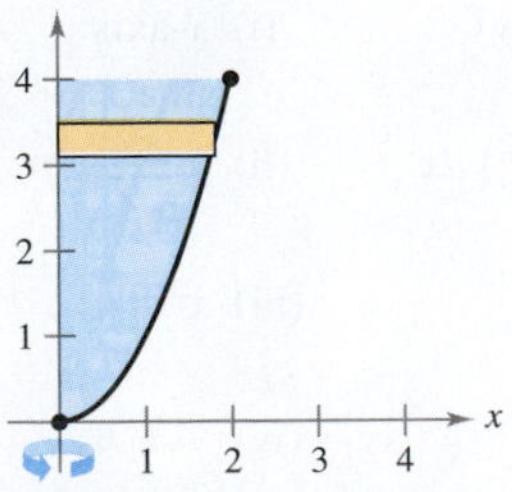

8. $y = \sqrt{16 - x^2}$

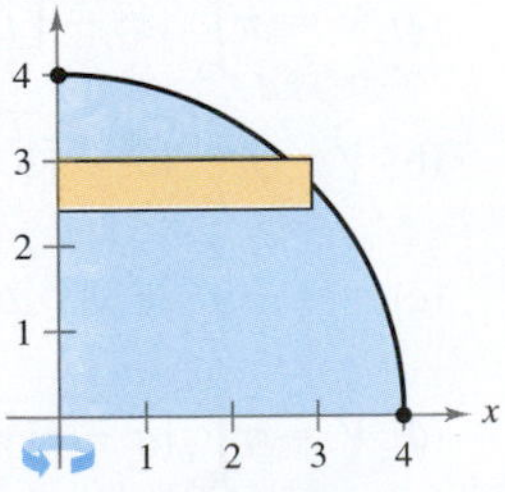

9. $y = x^{2/3}$

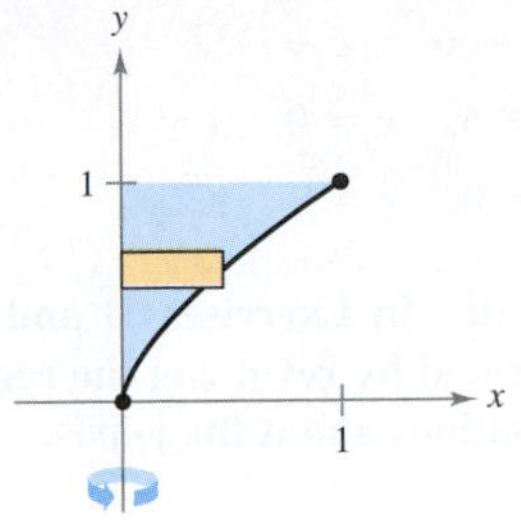

10. $x = -y^2 + 4y$

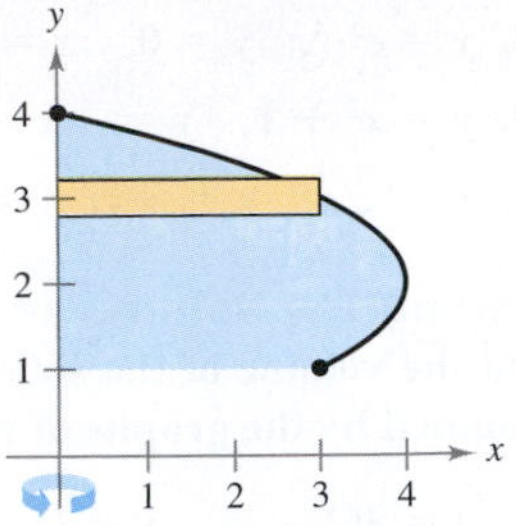

Finding the Volume of a Solid In Exercises 11–14, find the volumes of the solids generated by revolving the region bounded by the graphs of the equations about the given lines.

11. $y = \sqrt{x}, \quad y = 0, \quad x = 3$

(a) the x-axis (b) the y-axis

(c) the line $x = 3$ (d) the line $x = 6$

12. $y = 2x^2, \quad y = 0, \quad x = 2$

(a) the y-axis (b) the x-axis

(c) the line $y = 8$ (d) the line $x = 2$

13. $y = x^2, \quad y = 4x - x^2$

(a) the x-axis (b) the line $y = 6$

14. $y = 4 + 2x - x^2, \quad y = 4 - x$

(a) the x-axis (b) the line $y = 1$

Finding the Volume of a Solid In Exercises 15–18, find the volume of the solid generated by revolving the region bounded by the graphs of the equations about the line $y = 4$.

15. $y = x, \quad y = 3, \quad x = 0$

16. $y = \frac{1}{2}x^3, \quad y = 4, \quad x = 0$

17. $y = \dfrac{3}{1 + x}, \quad y = 0, \quad x = 0, \quad x = 3$

18. $y = \sec x, \quad y = 0, \quad 0 \le x \le \dfrac{\pi}{3}$

Finding the Volume of a Solid In Exercises 19–22, find the volume of the solid generated by revolving the region bounded by the graphs of the equations about the line $x = 5$.

19. $y = x, \quad y = 0, \quad y = 4, \quad x = 5$

20. $y = 3 - x, \quad y = 0, \quad y = 2, \quad x = 0$

21. $x = y^2, \quad x = 4$

22. $xy = 3, \quad y = 1, \quad y = 4, \quad x = 5$

Finding the Volume of a Solid In Exercises 23–30, find the volume of the solid generated by revolving the region bounded by the graphs of the equations about the x-axis.

23. $y = \dfrac{1}{\sqrt{x + 1}}, \quad y = 0, \quad x = 0, \quad x = 4$

24. $y = x\sqrt{4 - x^2}, \quad y = 0$

25. $y = \frac{1}{x}$, $y = 0$, $x = 1$, $x = 3$

26. $y = \frac{2}{x + 1}$, $y = 0$, $x = 0$, $x = 6$

27. $y = e^{-x}$, $y = 0$, $x = 0$, $y = 1$

28. $y = e^{x/4}$, $y = 0$, $x = 0$, $x = 6$

29. $y = x^2 + 1$, $y = -x^2 + 2x + 5$, $x = 0$, $x = 3$

30. $y = \sqrt{x}$, $y = -\frac{1}{2}x + 4$, $x = 0$, $x = 8$

Finding the Volume of a Solid **In Exercises 31 and 32, find the volume of the solid generated by revolving the region bounded by the graphs of the equations about the *y*-axis.**

31. $y = 3(2 - x)$, $y = 0$, $x = 0$

32. $y = 9 - x^2$, $y = 0$, $x = 2$, $x = 3$

Finding the Volume of a Solid **In Exercises 33–36, find the volume of the solid generated by revolving the region bounded by the graphs of the equations about the *x*-axis. Verify your results using the integration capabilities of a graphing utility.**

33. $y = \sin x$, $y = 0$, $x = 0$, $x = \pi$

34. $y = \cos 2x$, $y = 0$, $x = 0$, $x = \frac{\pi}{4}$

35. $y = e^{x-1}$, $y = 0$, $x = 1$, $x = 2$

36. $y = e^{x/2} + e^{-x/2}$, $y = 0$, $x = -1$, $x = 2$

Finding the Volume of a Solid **In Exercises 37–40, use the integration capabilities of a graphing utility to approximate the volume of the solid generated by revolving the region bounded by the graphs of the equations about the *x*-axis.**

37. $y = e^{-x^2}$, $y = 0$, $x = 0$, $x = 2$

38. $y = \ln x$, $y = 0$, $x = 1$, $x = 3$

39. $y = 2\arctan(0.2x)$, $y = 0$, $x = 0$, $x = 5$

40. $y = \sqrt{2x}$, $y = x^2$

Finding the Volume of a Solid **In Exercises 41–48, find the volume generated by rotating the given region about the specified line.**

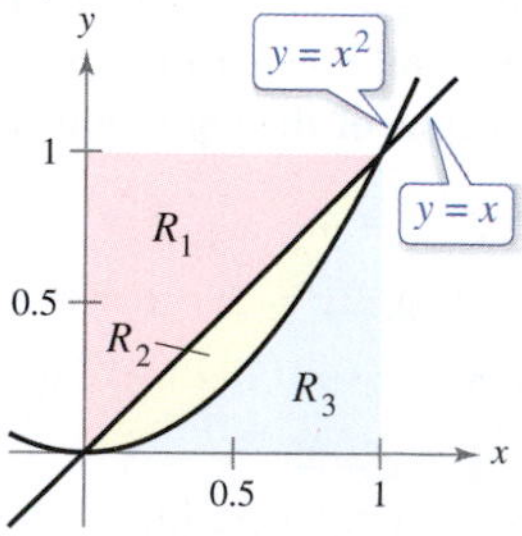

41. R_1 about $x = 0$

42. R_1 about $x = 1$

43. R_2 about $y = 0$

44. R_2 about $y = 1$

45. R_3 about $x = 0$

46. R_3 about $x = 1$

47. R_2 about $x = 0$

48. R_2 about $x = 1$

WRITING ABOUT CONCEPTS

Describing a Solid **In Exercises 49 and 50, the integral represents the volume of a solid. Describe the solid.**

49. $\pi \int_0^{\pi/2} \sin^2 x \, dx$

50. $\pi \int_2^4 y^4 \, dy$

51. **Comparing Volumes** A region bounded by the parabola $y = 4x - x^2$ and the *x*-axis is revolved about the *x*-axis. A second region bounded by the parabola $y = 4 - x^2$ and the *x*-axis is revolved about the *x*-axis. Without integrating, how do the volumes of the two solids compare? Explain.

52. **Comparing Volumes** The region in the figure is revolved about the indicated axes and line. Order the volumes of the resulting solids from least to greatest. Explain your reasoning.

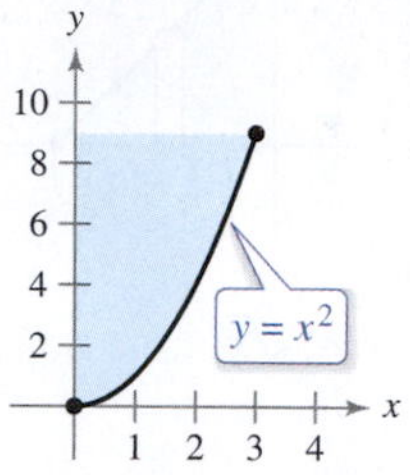

(a) *x*-axis

(b) *y*-axis

(c) $x = 3$

53. **Analyzing Statements** Discuss the validity of the following statements.

(a) For a solid formed by rotating the region under a graph about the *x*-axis, the cross sections perpendicular to the *x*-axis are circular disks.

(b) For a solid formed by rotating the region between two graphs about the *x*-axis, the cross sections perpendicular to the *x*-axis are circular disks.

54. **HOW DO YOU SEE IT?** Use the graph to match the integral for the volume with the axis of rotation.

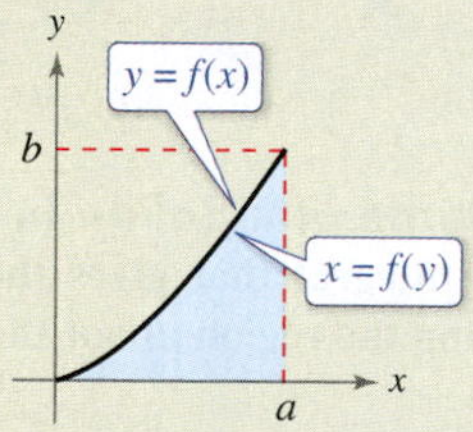

(a) $V = \pi \int_0^b (a^2 - [f(y)]^2) \, dy$ (i) *x*-axis

(b) $V = \pi \int_0^a (b^2 - [b - f(x)]^2) \, dx$ (ii) *y*-axis

(c) $V = \pi \int_0^a [f(x)]^2 \, dx$ (iii) $x = a$

(d) $V = \pi \int_0^b [a - f(y)]^2 \, dy$ (iv) $y = b$

Dividing a Solid **In Exercises 55 and 56, consider the solid formed by revolving the region bounded by $y = \sqrt{x}$, $y = 0$, and $x = 4$ about the x-axis.**

55. Find the value of x in the interval $[0, 4]$ that divides the solid into two parts of equal volume.

56. Find the values of x in the interval $[0, 4]$ that divide the solid into three parts of equal volume.

57. Manufacturing A manufacturer drills a hole through the center of a metal sphere of radius R. The hole has a radius r. Find the volume of the resulting ring.

58. Manufacturing For the metal sphere in Exercise 57, let $R = 6$. What value of r will produce a ring whose volume is exactly half the volume of the sphere?

59. Volume of a Cone Use the disk method to verify that the volume of a right circular cone is $\frac{1}{3}\pi r^2 h$, where r is the radius of the base and h is the height.

60. Volume of a Sphere Use the disk method to verify that the volume of a sphere is $\frac{4}{3}\pi r^3$, where r is the radius.

61. Using a Cone A cone of height H with a base of radius r is cut by a plane parallel to and h units above the base, where $h < H$. Find the volume of the solid (frustum of a cone) below the plane.

62. Using a Sphere A sphere of radius r is cut by a plane h units above the equator, where $h < r$. Find the volume of the solid (spherical segment) above the plane.

63. Volume of a Fuel Tank A tank on the wing of a jet aircraft is formed by revolving the region bounded by the graph of $y = \frac{1}{8}x^2\sqrt{2 - x}$ and the x-axis $(0 \le x \le 2)$ about the x-axis, where x and y are measured in meters. Use a graphing utility to graph the function and find the volume of the tank.

64. Volume of a Lab Glass A glass container can be modeled by revolving the graph of

$$y = \begin{cases} \sqrt{0.1x^3 - 2.2x^2 + 10.9x + 22.2}, & 0 \le x \le 11.5 \\ 2.95, & 11.5 < x \le 15 \end{cases}$$

about the x-axis, where x and y are measured in centimeters. Use a graphing utility to graph the function and find the volume of the container.

65. Finding Volumes of a Solid Find the volumes of the solids (see figures) generated if the upper half of the ellipse $9x^2 + 25y^2 = 225$ is revolved about (a) the x-axis to form a prolate spheroid (shaped like a football), and (b) the y-axis to form an oblate spheroid (shaped like half of a candy).

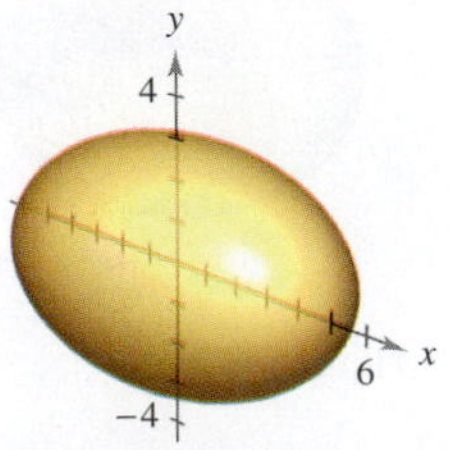

Figure for 65(a)

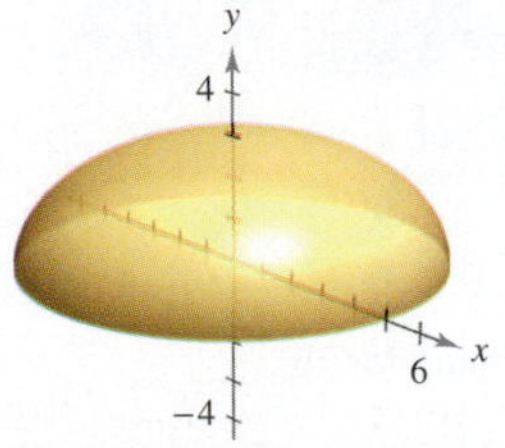

Figure for 65(b)

66. Water Tower

A tank on a water tower is a sphere of radius 50 feet. Determine the depths of the water when the tank is filled to one-fourth and three-fourths of its total capacity. (*Note:* Use the *zero* or *root* feature of a graphing utility after evaluating the definite integral.)

67. Minimum Volume The arc of $y = 4 - (x^2/4)$ on the interval $[0, 4]$ is revolved about the line $y = b$ (see figure).

(a) Find the volume of the resulting solid as a function of b.

(b) Use a graphing utility to graph the function in part (a), and use the graph to approximate the value of b that minimizes the volume of the solid.

(c) Use calculus to find the value of b that minimizes the volume of the solid, and compare the result with the answer to part (b).

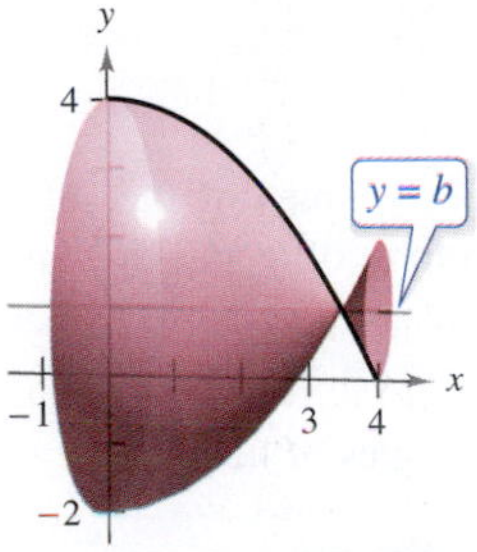

Figure for 67

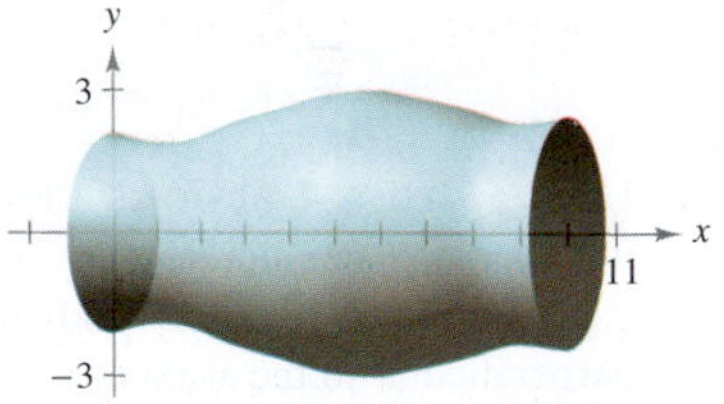

Figure for 68

68. Modeling Data A draftsman is asked to determine the amount of material required to produce a machine part (see figure). The diameters d of the part at equally spaced points x are listed in the table. The measurements are listed in centimeters.

x	0	1	2	3	4	5
d	4.2	3.8	4.2	4.7	5.2	5.7

x	6	7	8	9	10
d	5.8	5.4	4.9	4.4	4.6

(a) Use these data with Simpson's Rule to approximate the volume of the part.

(b) Use the regression capabilities of a graphing utility to find a fourth-degree polynomial through the points representing the radius of the solid. Plot the data and graph the model.

(c) Use a graphing utility to approximate the definite integral yielding the volume of the part. Compare the result with the answer to part (a).

69. Think About It Match each integral with the solid whose volume it represents, and give the dimensions of each solid.

(a) Right circular cylinder (b) Ellipsoid

(c) Sphere (d) Right circular cone (e) Torus

(i) $\pi\int_0^h \left(\frac{rx}{h}\right)^2 dx$ (ii) $\pi\int_0^h r^2\, dx$

(iii) $\pi\int_{-r}^{r} \left(\sqrt{r^2 - x^2}\right)^2 dx$

(iv) $\pi\int_{-b}^{b} \left(a\sqrt{1 - \frac{x^2}{b^2}}\right)^2 dx$

(v) $\pi\int_{-r}^{r} \left[\left(R + \sqrt{r^2 - x^2}\right)^2 - \left(R - \sqrt{r^2 - x^2}\right)^2\right] dx$

70. Cavalieri's Theorem Prove that if two solids have equal altitudes and all plane sections parallel to their bases and at equal distances from their bases have equal areas, then the solids have the same volume (see figure).

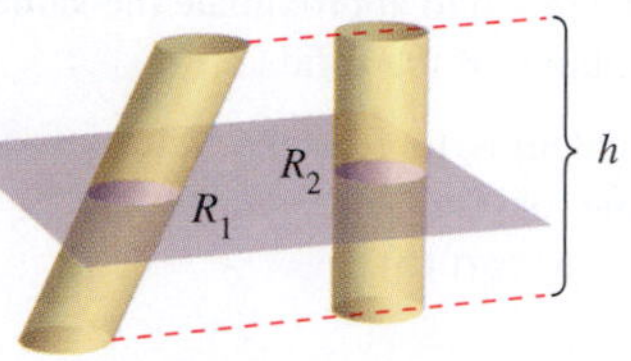

Area of R_1 = area of R_2

71. Using Cross Sections Find the volumes of the solids whose bases are bounded by the graphs of $y = x + 1$ and $y = x^2 - 1$, with the indicated cross sections taken perpendicular to the x-axis.

(a) Squares (b) Rectangles of height 1

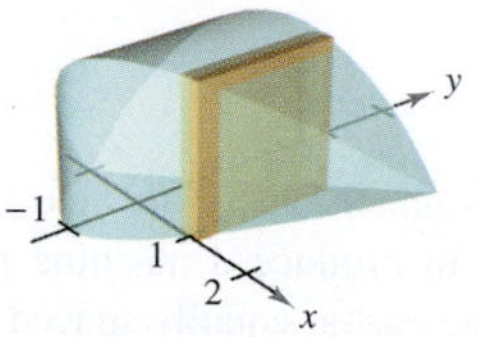

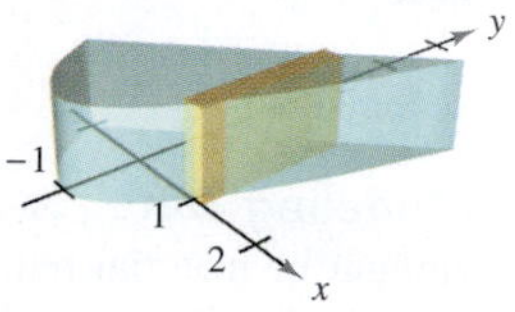

72. Using Cross Sections Find the volumes of the solids whose bases are bounded by the circle $x^2 + y^2 = 4$, with the indicated cross sections taken perpendicular to the x-axis.

(a) Squares (b) Equilateral triangles

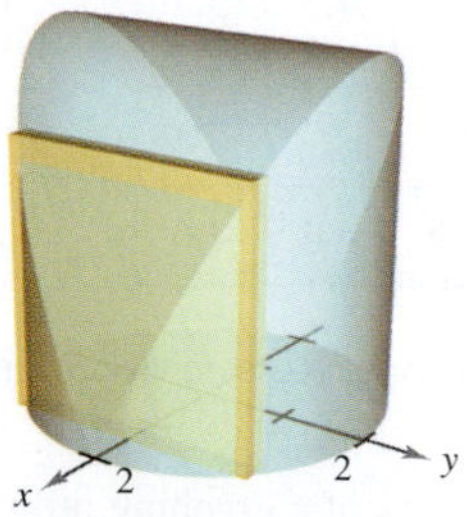

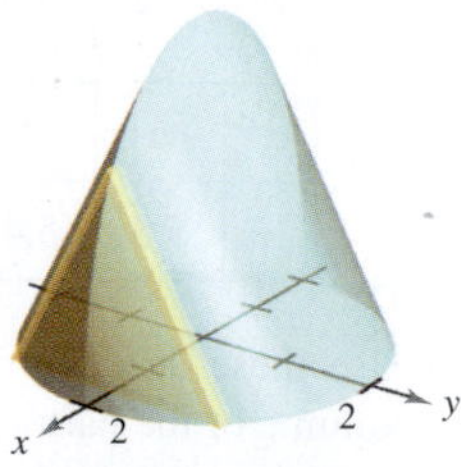

(c) Semicircles (d) Isosceles right triangles

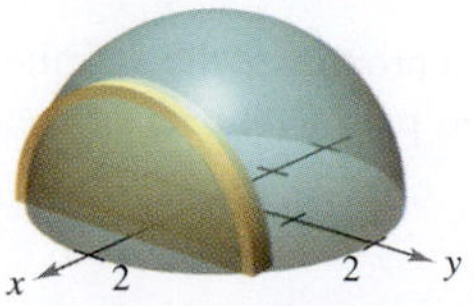

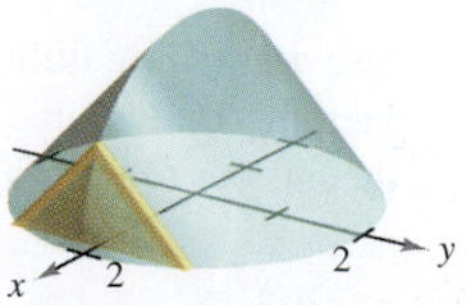

73. Using Cross Sections Find the volume of the solid of intersection (the solid common to both) of the two right circular cylinders of radius r whose axes meet at right angles (see figure).

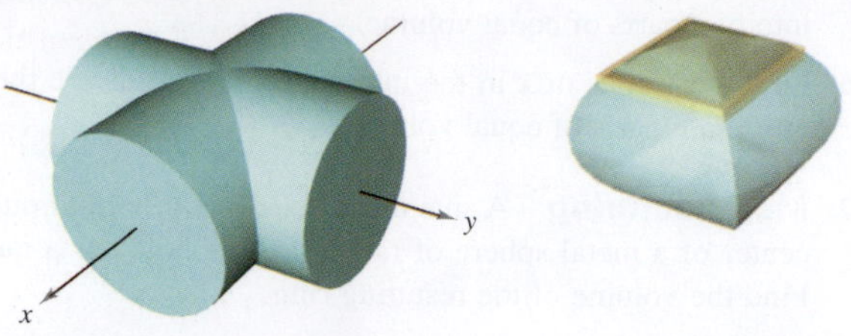

Two intersecting cylinders Solid of intersection

FOR FURTHER INFORMATION For more information on this problem, see the article "Estimating the Volumes of Solid Figures with Curved Surfaces" by Donald Cohen in *Mathematics Teacher*. To view this article, go to *MathArticles.com*.

74. Using Cross Sections The solid shown in the figure has cross sections bounded by the graph of $|x|^a + |y|^a = 1$, where $1 \le a \le 2$.

(a) Describe the cross section when $a = 1$ and $a = 2$.

(b) Describe a procedure for approximating the volume of the solid.

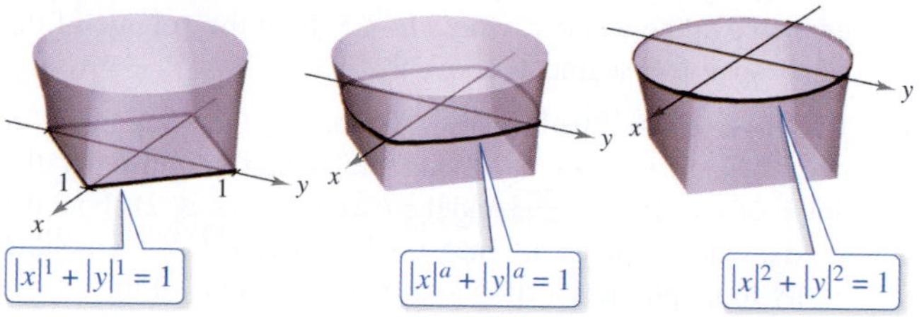

75. Volume of a Wedge Two planes cut a right circular cylinder to form a wedge. One plane is perpendicular to the axis of the cylinder and the second makes an angle of θ degrees with the first (see figure).

(a) Find the volume of the wedge if $\theta = 45°$.

(b) Find the volume of the wedge for an arbitrary angle θ. Assuming that the cylinder has sufficient length, how does the volume of the wedge change as θ increases from 0° to 90°?

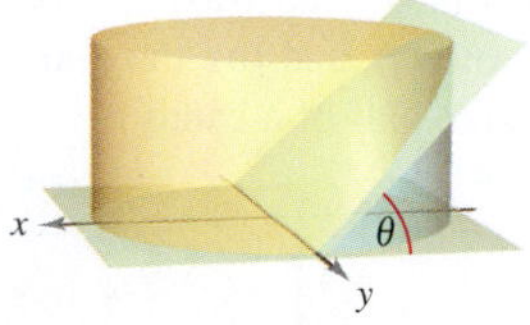

Figure for 75

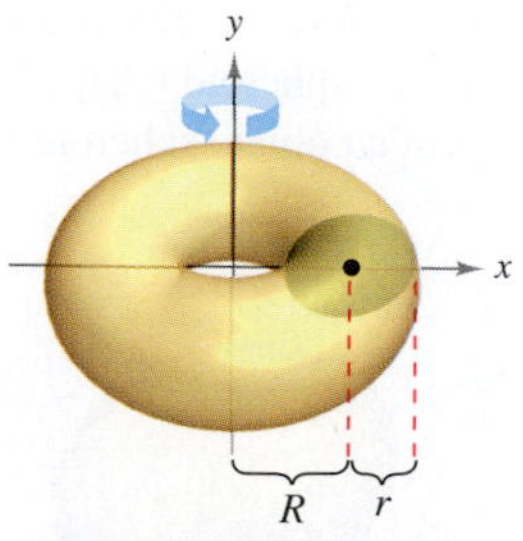

Figure for 76

76. Volume of a Torus

(a) Show that the volume of the torus shown in the figure is given by the integral $8\pi R\int_0^r \sqrt{r^2 - y^2}\, dy$, where $R > r > 0$.

(b) Find the volume of the torus.

7.3 Volume: The Shell Method

- **Find the volume of a solid of revolution using the shell method.**
- **Compare the uses of the disk method and the shell method.**

The Shell Method

In this section, you will study an alternative method for finding the volume of a solid of revolution. This method is called the **shell method** because it uses cylindrical shells. A comparison of the advantages of the disk and shell methods is given later in this section.

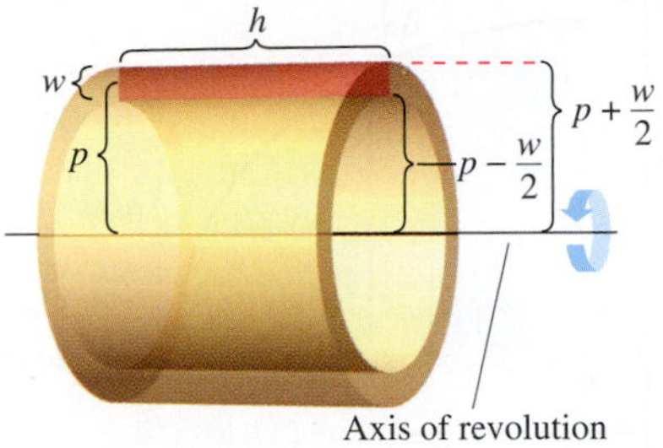

Figure 7.27

To begin, consider a representative rectangle as shown in Figure 7.27, where w is the width of the rectangle, h is the height of the rectangle, and p is the distance between the axis of revolution and the *center* of the rectangle. When this rectangle is revolved about its axis of revolution, it forms a cylindrical shell (or tube) of thickness w. To find the volume of this shell, consider two cylinders. The radius of the larger cylinder corresponds to the outer radius of the shell, and the radius of the smaller cylinder corresponds to the inner radius of the shell. Because p is the average radius of the shell, you know the outer radius is

$$p + \frac{w}{2} \qquad \text{Outer radius}$$

and the inner radius is

$$p - \frac{w}{2}. \qquad \text{Inner radius}$$

So, the volume of the shell is

$$\begin{aligned}\text{Volume of shell} &= (\text{volume of cylinder}) - (\text{volume of hole})\\ &= \pi\left(p + \frac{w}{2}\right)^2 h - \pi\left(p - \frac{w}{2}\right)^2 h\\ &= 2\pi phw\\ &= 2\pi(\text{average radius})(\text{height})(\text{thickness}).\end{aligned}$$

You can use this formula to find the volume of a solid of revolution. For instance, the plane region in Figure 7.28 is revolved about a line to form the indicated solid. Consider a horizontal rectangle of width Δy. As the plane region is revolved about a line parallel to the x-axis, the rectangle generates a representative shell whose volume is

$$\Delta V = 2\pi[p(y)h(y)]\,\Delta y.$$

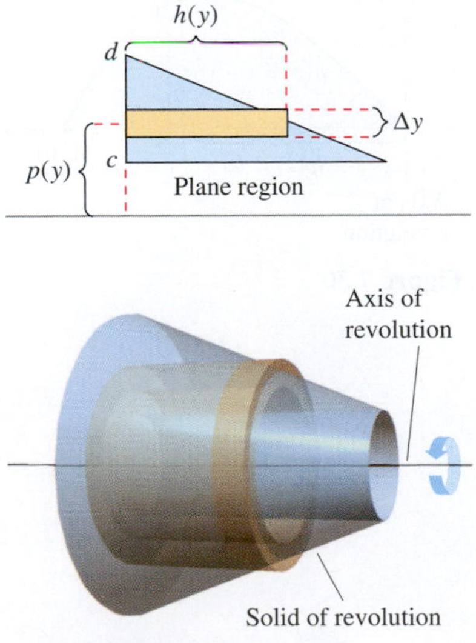

Figure 7.28

You can approximate the volume of the solid by n such shells of thickness Δy, height $h(y_i)$, and average radius $p(y_i)$.

$$\text{Volume of solid} \approx \sum_{i=1}^{n} 2\pi[p(y_i)h(y_i)]\,\Delta y = 2\pi\sum_{i=1}^{n}[p(y_i)h(y_i)]\,\Delta y$$

This approximation appears to become better and better as $\|\Delta\| \to 0$ $(n \to \infty)$. So, the volume of the solid is

$$\begin{aligned}\text{Volume of solid} &= \lim_{\|\Delta\|\to 0} 2\pi\sum_{i=1}^{n}[p(y_i)h(y_i)]\,\Delta y\\ &= 2\pi\int_c^d [p(y)h(y)]\,dy.\end{aligned}$$

AP* Tips

Volume by shell rotation is never required on the AP Exam, but some free response questions may be solvable by the shell method in addition to the disk method.

THE SHELL METHOD

To find the volume of a solid of revolution with the **shell method,** use one of the formulas below. (See Figure 7.29.)

Horizontal Axis of Revolution

$$\text{Volume} = V = 2\pi \int_e^d p(y)h(y)\,dy$$

Vertical Axis of Revolution

$$\text{Volume} = V = 2\pi \int_a^b p(x)h(x)\,dx$$

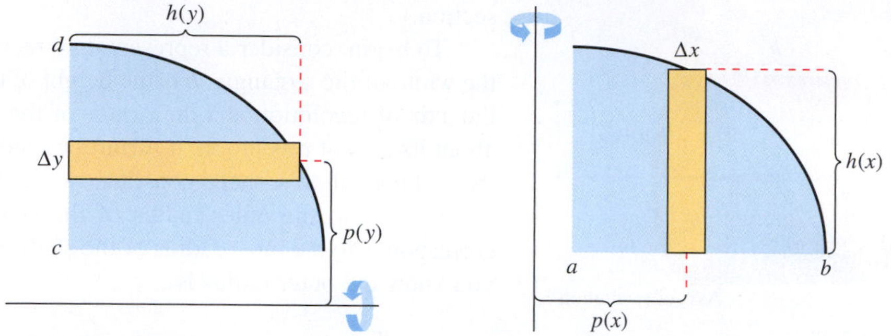

Horizontal axis of revolution

Vertical axis of revolution

Figure 7.29

EXAMPLE 1 Using the Shell Method to Find Volume

Find the volume of the solid of revolution formed by revolving the region bounded by

$$y = x - x^3$$

and the x-axis $(0 \le x \le 1)$ about the y-axis.

Solution Because the axis of revolution is vertical, use a vertical representative rectangle, as shown in Figure 7.30. The width Δx indicates that x is the variable of integration. The distance from the center of the rectangle to the axis of revolution is $p(x) = x$, and the height of the rectangle is

$$h(x) = x - x^3.$$

Because x ranges from 0 to 1, apply the shell method to find the volume of the solid.

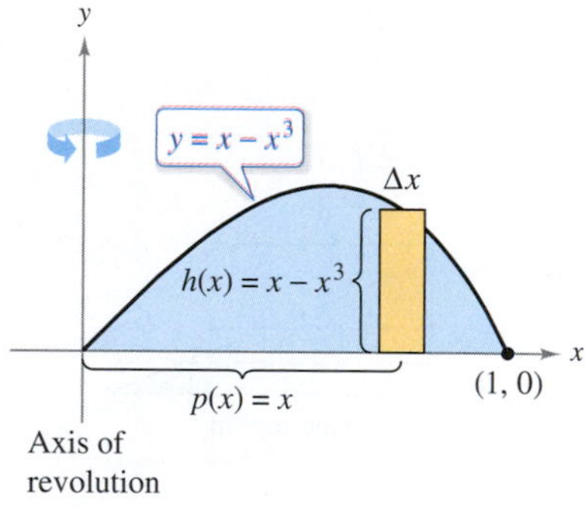

Figure 7.30

$$\begin{aligned}
V &= 2\pi \int_a^b p(x)h(x)\,dx \\
&= 2\pi \int_0^1 x(x - x^3)\,dx \\
&= 2\pi \int_0^1 (-x^4 + x^2)\,dx && \text{Simplify.} \\
&= 2\pi \left[-\frac{x^5}{5} + \frac{x^3}{3} \right]_0^1 && \text{Integrate.} \\
&= 2\pi \left(-\frac{1}{5} + \frac{1}{3} \right) \\
&= \frac{4\pi}{15}
\end{aligned}$$

EXAMPLE 2 Using the Shell Method to Find Volume

Find the volume of the solid of revolution formed by revolving the region bounded by the graph of

$$x = e^{-y^2}$$

and the y-axis $(0 \le y \le 1)$ about the x-axis.

Solution Because the axis of revolution is horizontal, use a horizontal representative rectangle, as shown in Figure 7.31. The width Δy indicates that y is the variable of integration. The distance from the center of the rectangle to the axis of revolution is $p(y) = y$, and the height of the rectangle is $h(y) = e^{-y^2}$. Because y ranges from 0 to 1, the volume of the solid is

$$\begin{aligned} V &= 2\pi \int_c^d p(y)h(y)\,dy && \text{Apply shell method.} \\ &= 2\pi \int_0^1 ye^{-y^2}\,dy \\ &= -\pi\left[e^{-y^2}\right]_0^1 && \text{Integrate.} \\ &= \pi\left(1 - \frac{1}{e}\right) \\ &\approx 1.986. \end{aligned}$$

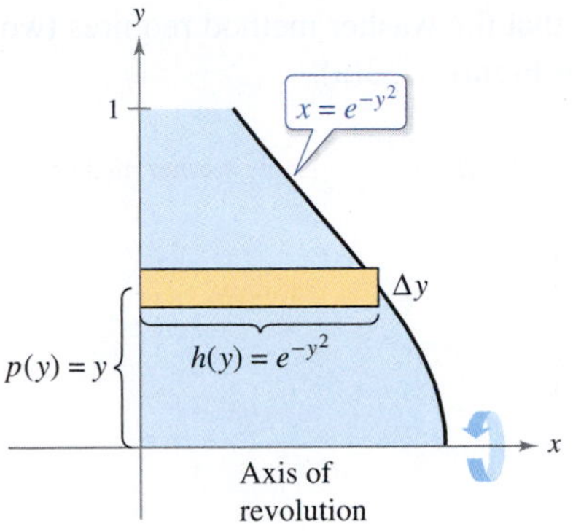

Figure 7.31

Exploration

To see the advantage of using the shell method in Example 2, solve the equation $x = e^{-y^2}$ for y.

$$y = \begin{cases} 1, & 0 \le x \le 1/e \\ \sqrt{-\ln x}, & 1/e < x \le 1 \end{cases}$$

Then use this equation to find the volume using the disk method.

Comparison of Disk and Shell Methods

The disk and shell methods can be distinguished as follows. For the disk method, the representative rectangle is always *perpendicular* to the axis of revolution, whereas for the shell method, the representative rectangle is always *parallel* to the axis of revolution, as shown in Figure 7.32.

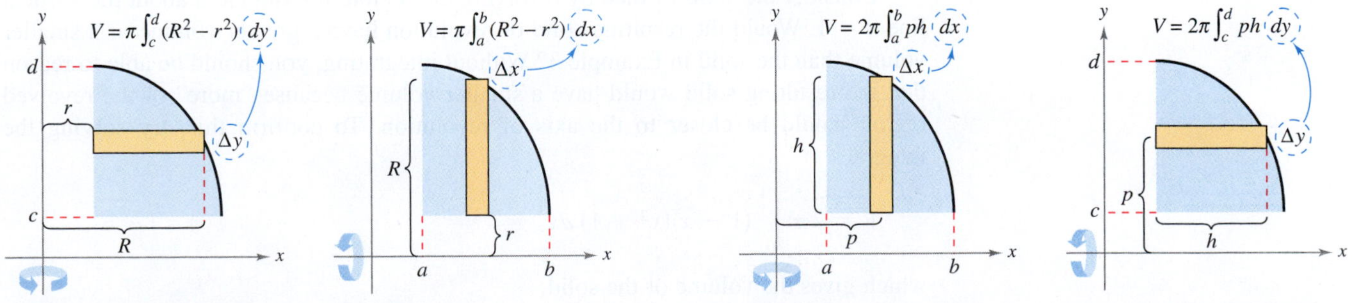

Vertical axis of revolution | Horizontal axis of revolution | Vertical axis of revolution | Horizontal axis of revolution

Disk method: Representative rectangle is perpendicular to the axis of revolution.

Shell method: Representative rectangle is parallel to the axis of revolution.

Figure 7.32

Often, one method is more convenient to use than the other. The next example illustrates a case in which the shell method is preferable.

EXAMPLE 3 Shell Method Preferable

See LarsonCalculus.com for an interactive version of this type of example.

Find the volume of the solid formed by revolving the region bounded by the graphs of

$$y = x^2 + 1, \quad y = 0, \quad x = 0, \quad \text{and} \quad x = 1$$

about the y-axis.

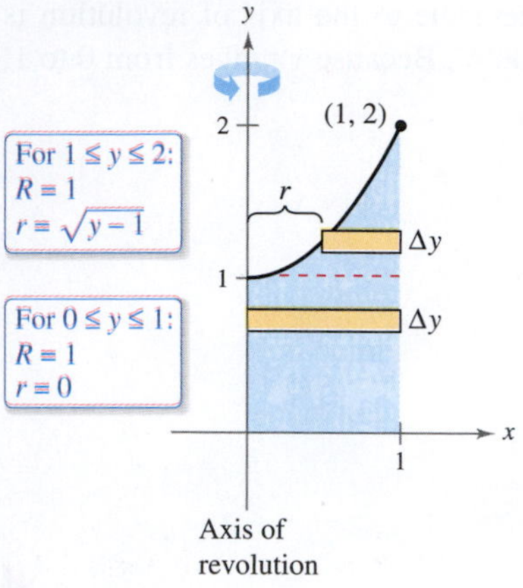

(a) Disk method

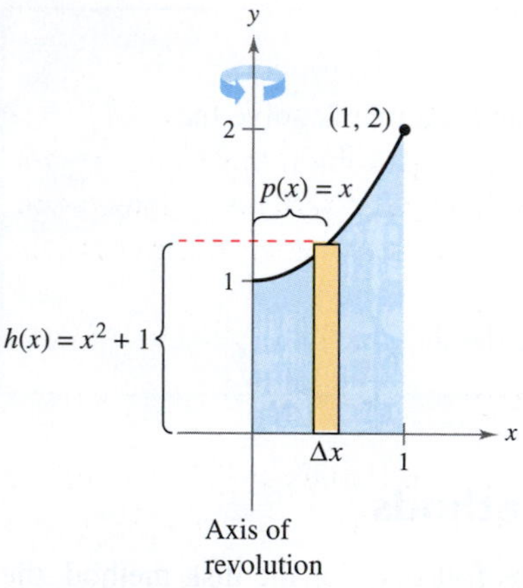

(b) Shell method

Figure 7.33

Solution In Example 4 in Section 7.2, you saw that the washer method requires two integrals to determine the volume of this solid. See Figure 7.33(a).

$$V = \pi \int_0^1 (1^2 - 0^2)\, dy + \pi \int_1^2 \left[1^2 - \left(\sqrt{y-1}\right)^2\right] dy \qquad \text{Apply washer method.}$$

$$= \pi \int_0^1 1\, dy + \pi \int_1^2 (2 - y)\, dy \qquad \text{Simplify.}$$

$$= \pi \Big[y\Big]_0^1 + \pi \left[2y - \frac{y^2}{2}\right]_1^2 \qquad \text{Integrate.}$$

$$= \pi + \pi\left(4 - 2 - 2 + \frac{1}{2}\right)$$

$$= \frac{3\pi}{2}$$

In Figure 7.33(b), you can see that the shell method requires only one integral to find the volume.

$$V = 2\pi \int_a^b p(x)h(x)\, dx \qquad \text{Apply shell method.}$$

$$= 2\pi \int_0^1 x(x^2 + 1)\, dx$$

$$= 2\pi \left[\frac{x^4}{4} + \frac{x^2}{2}\right]_0^1 \qquad \text{Integrate.}$$

$$= 2\pi \left(\frac{3}{4}\right)$$

$$= \frac{3\pi}{2}$$

Consider the solid formed by revolving the region in Example 3 about the vertical line $x = 1$. Would the resulting solid of revolution have a greater volume or a smaller volume than the solid in Example 3? Without integrating, you should be able to reason that the resulting solid would have a smaller volume because "more" of the revolved region would be closer to the axis of revolution. To confirm this, try solving the integral

$$V = 2\pi \int_0^1 (1 - x)(x^2 + 1)\, dx \qquad p(x) = 1 - x$$

which gives the volume of the solid.

FOR FURTHER INFORMATION To learn more about the disk and shell methods, see the article "The Disk and Shell Method" by Charles A. Cable in *The American Mathematical Monthly.* To view this article, go to *MathArticles.com.*

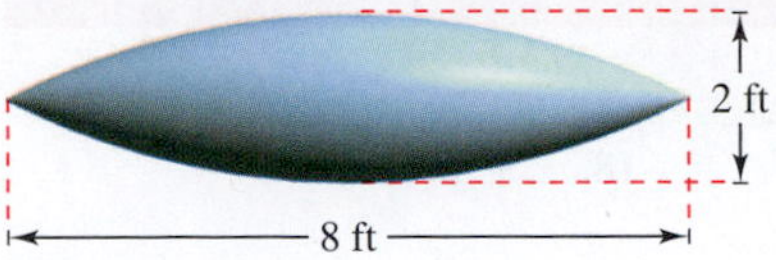

Figure 7.34

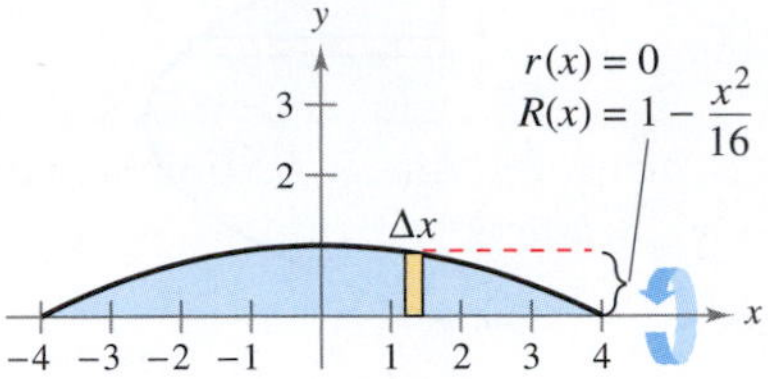

Disk method
Figure 7.35

EXAMPLE 4 Volume of a Pontoon

A pontoon is to be made in the shape shown in Figure 7.34. The pontoon is designed by rotating the graph of

$$y = 1 - \frac{x^2}{16}, \quad -4 \le x \le 4$$

about the x-axis, where x and y are measured in feet. Find the volume of the pontoon.

Solution Refer to Figure 7.35 and use the disk method as shown.

$$V = \pi \int_{-4}^{4} \left(1 - \frac{x^2}{16}\right)^2 dx \qquad \text{Apply disk method.}$$

$$= \pi \int_{-4}^{4} \left(1 - \frac{x^2}{8} + \frac{x^4}{256}\right) dx \qquad \text{Simplify.}$$

$$= \pi \left[x - \frac{x^3}{24} + \frac{x^5}{1280}\right]_{-4}^{4} \qquad \text{Integrate.}$$

$$= \frac{64\pi}{15}$$

$$\approx 13.4 \text{ cubic feet}$$

To use the shell method in Example 4, you would have to solve for x in terms of y in the equation

$$y = 1 - \frac{x^2}{16}$$

and then evaluate an integral that requires a u-substitution.

Sometimes, solving for x is very difficult (or even impossible). In such cases, you must use a vertical rectangle (of width Δx), thus making x the variable of integration. The position (horizontal or vertical) of the axis of revolution then determines the method to be used. This is shown in Example 5.

EXAMPLE 5 Shell Method Necessary

Find the volume of the solid formed by revolving the region bounded by the graphs of $y = x^3 + x + 1$, $y = 1$, and $x = 1$ about the line $x = 2$, as shown in Figure 7.36.

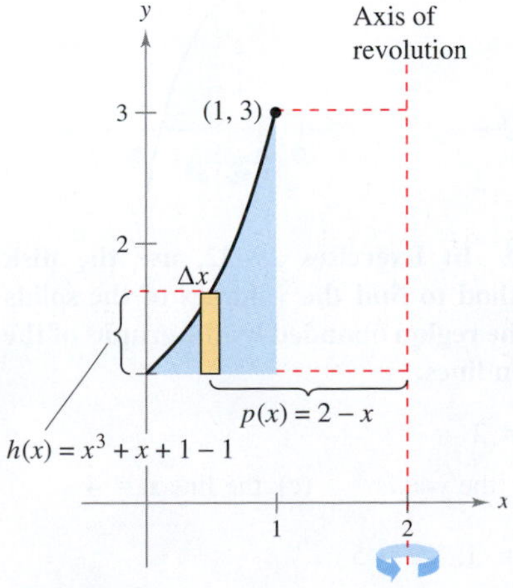

Figure 7.36

Solution In the equation $y = x^3 + x + 1$, you cannot easily solve for x in terms of y. (See the discussion at the end of Section 3.8.) Therefore, the variable of integration must be x, and you should choose a vertical representative rectangle. Because the rectangle is parallel to the axis of revolution, use the shell method.

$$V = 2\pi \int_{a}^{b} p(x)h(x)\, dx \qquad \text{Apply shell method.}$$

$$= 2\pi \int_{0}^{1} (2 - x)(x^3 + x + 1 - 1)\, dx$$

$$= 2\pi \int_{0}^{1} (-x^4 + 2x^3 - x^2 + 2x)\, dx \qquad \text{Simplify.}$$

$$= 2\pi \left[-\frac{x^5}{5} + \frac{x^4}{2} - \frac{x^3}{3} + x^2\right]_{0}^{1} \qquad \text{Integrate.}$$

$$= 2\pi\left(-\frac{1}{5} + \frac{1}{2} - \frac{1}{3} + 1\right)$$

$$= \frac{29\pi}{15}$$

7.3 Exercises

See CalcChat.com for tutorial help and worked-out solutions to odd-numbered exercises.

Finding the Volume of a Solid In Exercises 1–14, use the shell method to set up and evaluate the integral that gives the volume of the solid generated by revolving the plane region about the y-axis.

1. $y = x$

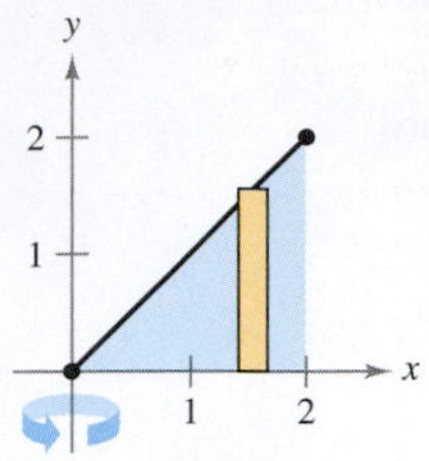

2. $y = 1 - x$

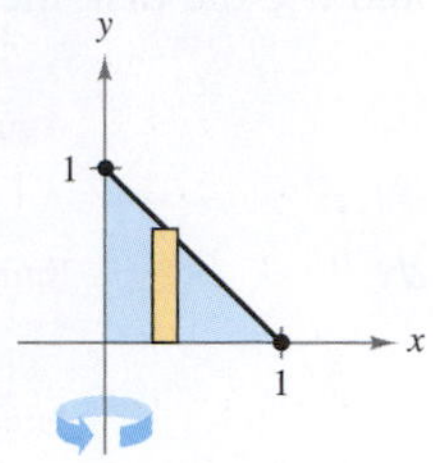

3. $y = \sqrt{x}$

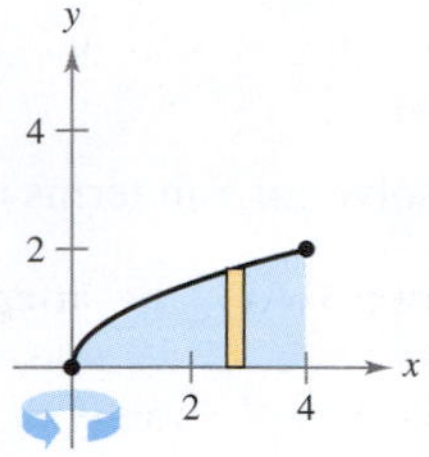

4. $y = \frac{1}{2}x^2 + 1$

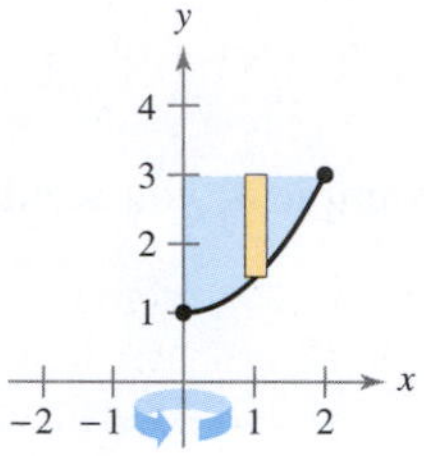

5. $y = \frac{1}{4}x^2, \quad y = 0, \quad x = 4$

6. $y = \frac{1}{2}x^3, \quad y = 0, \quad x = 3$

7. $y = x^2, \quad y = 4x - x^2$

8. $y = 9 - x^2, \quad y = 0$

9. $y = 4x - x^2, \quad x = 0, \quad y = 4$

10. $y = x^{3/2}, \quad y = 8, \quad x = 0$

11. $y = \sqrt{x - 2}, \quad y = 0, \quad x = 4$

12. $y = -x^2 + 1, \quad y = 0$

13. $y = \dfrac{1}{\sqrt{2\pi}} e^{-x^2/2}, \quad y = 0, \quad x = 0, \quad x = 1$

14. $y = \begin{cases} \dfrac{\sin x}{x}, & x > 0 \\ 1, & x = 0 \end{cases}, \quad y = 0, \quad x = 0, \quad x = \pi$

Finding the Volume of a Solid In Exercises 15–22, use the shell method to set up and evaluate the integral that gives the volume of the solid generated by revolving the plane region about the x-axis.

15. $y = x$

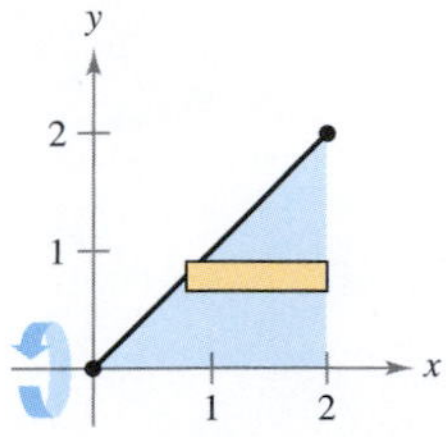

16. $y = 1 - x$

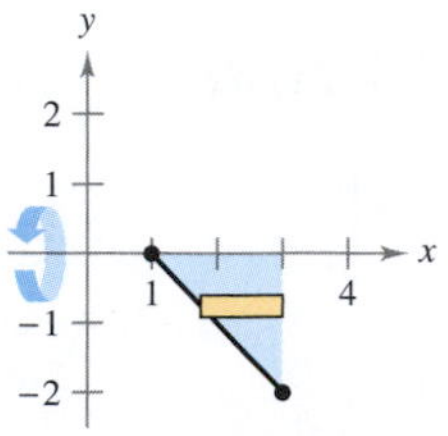

17. $y = \dfrac{1}{x}$

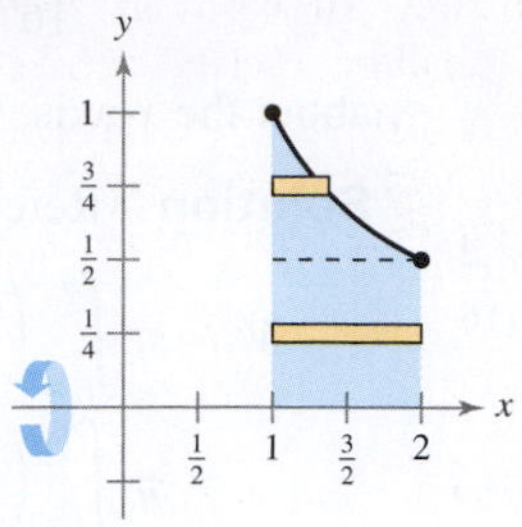

18. $x + y^2 = 16$

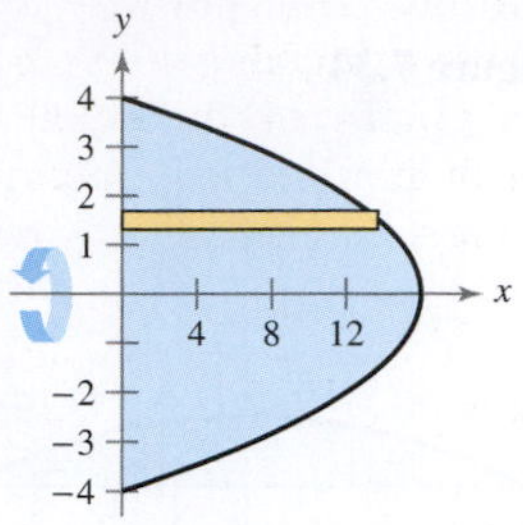

19. $y = x^3, \quad x = 0, \quad y = 8$

20. $y = 4x^2, \quad x = 0, \quad y = 4$

21. $x + y = 4, \quad y = x, \quad y = 0$

22. $y = \sqrt{x + 2}, \quad y = x, \quad y = 0$

Finding the Volume of a Solid In Exercises 23–26, use the shell method to find the volume of the solid generated by revolving the plane region about the given line.

23. $y = 2x - x^2, \quad y = 0, \quad$ about the line $x = 4$

24. $y = \sqrt{x}, \quad y = 0, \quad x = 4, \quad$ about the line $x = 6$

25. $y = x^2, \quad y = 4x - x^2, \quad$ about the line $x = 4$

26. $y = \frac{1}{3}x^3, \quad y = 6x - x^2, \quad$ about the line $x = 3$

Choosing a Method In Exercises 27 and 28, decide whether it is more convenient to use the disk method or the shell method to find the volume of the solid of revolution. Explain your reasoning. (Do not find the volume.)

27. $(y - 2)^2 = 4 - x$

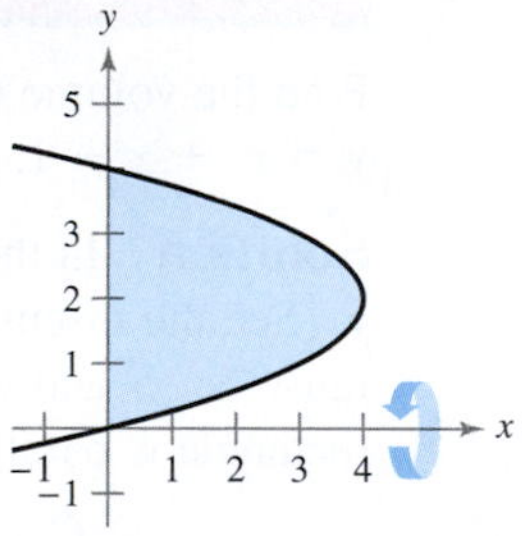

28. $y = 4 - e^x$

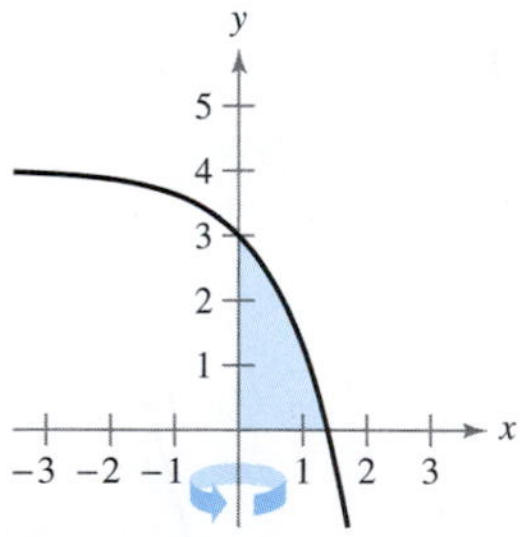

Choosing a Method In Exercises 29–32, use the disk method *or* the shell method to find the volumes of the solids generated by revolving the region bounded by the graphs of the equations about the given lines.

29. $y = x^3, \quad y = 0, \quad x = 2$

(a) the x-axis (b) the y-axis (c) the line $x = 4$

30. $y = \dfrac{10}{x^2}, \quad y = 0, \quad x = 1, \quad x = 5$

(a) the x-axis (b) the y-axis (c) the line $y = 10$

31. $x^{1/2} + y^{1/2} = a^{1/2}, \quad x = 0, \quad y = 0$

(a) the x-axis (b) the y-axis (c) the line $x = a$

32. $x^{2/3} + y^{2/3} = a^{2/3}, \quad a > 0$ (hypocycloid)

(a) the x-axis (b) the y-axis

Finding the Volume of a Solid **In Exercises 33–36, (a) use a graphing utility to graph the plane region bounded by the graphs of the equations, and (b) use the integration capabilities of the graphing utility to approximate the volume of the solid generated by revolving the region about the y-axis.**

33. $x^{4/3} + y^{4/3} = 1, \quad x = 0, \quad y = 0,$ first quadrant

34. $y = \sqrt{1 - x^3}, \quad y = 0, \quad x = 0$

35. $y = \sqrt[3]{(x-2)^2(x-6)^2}, \quad y = 0, \quad x = 2, \quad x = 6$

36. $y = \dfrac{2}{1 + e^{1/x}}, \quad y = 0, \quad x = 1, \quad x = 3$

WRITING ABOUT CONCEPTS

37. Representative Rectangles Consider a solid that is generated by revolving a plane region about the y-axis. Describe the position of a representative rectangle when using (a) the shell method and (b) the disk method to find the volume of the solid.

38. Describing Cylindrical Shells Consider the plane region bounded by the graphs of

$$y = k, \quad y = 0, \quad x = 0, \quad \text{and} \quad x = b$$

where $k > 0$ and $b > 0$. What are the heights and radii of the cylinders generated when this region is revolved about (a) the x-axis and (b) the y-axis?

Comparing Integrals **In Exercises 39 and 40, give a geometric argument that explains why the integrals have equal values.**

39. $\pi \displaystyle\int_1^5 (x - 1)\, dx = 2\pi \int_0^2 y[5 - (y^2 + 1)]\, dy$

40. $\pi \displaystyle\int_0^2 [16 - (2y)^2]\, dy = 2\pi \int_0^4 x\left(\frac{x}{2}\right) dx$

41. Comparing Volumes The region in the figure is revolved about the indicated axes and line. Order the volumes of the resulting solids from least to greatest. Explain your reasoning.

(a) x-axis (b) y-axis (c) $x = 4$

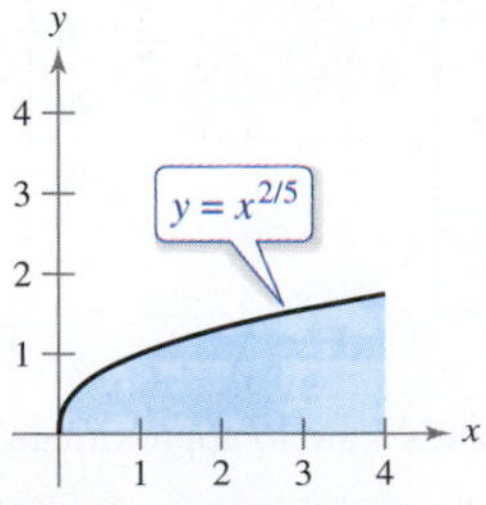

42. HOW DO YOU SEE IT? Use the graph to answer the following.

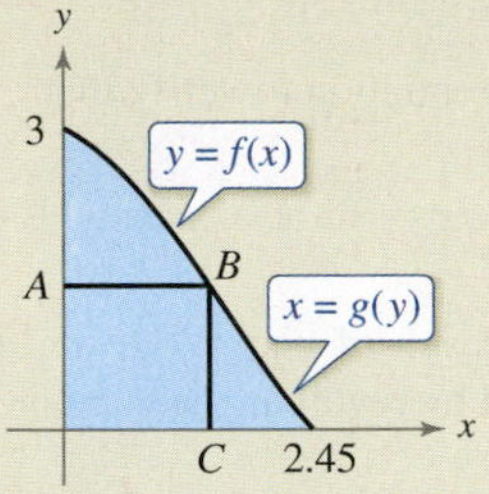

(a) Describe the figure generated by revolving segment AB about the y-axis.

(b) Describe the figure generated by revolving segment BC about the y-axis.

(c) Assume the curve in the figure can be described as $y = f(x)$ or $x = g(y)$. A solid is generated by revolving the region bounded by the curve, $y = 0$, and $x = 0$ about the y-axis. Set up integrals to find the volume of this solid using the disk method and the shell method. (Do not integrate.)

Analyzing an Integral **In Exercises 43–46, the integral represents the volume of a solid of revolution. Identify (a) the plane region that is revolved and (b) the axis of revolution.**

43. $2\pi \displaystyle\int_0^2 x^3\, dx$

44. $2\pi \displaystyle\int_0^1 y - y^{3/2}\, dy$

45. $2\pi \displaystyle\int_0^6 (y + 2)\sqrt{6 - y}\, dy$

46. $2\pi \displaystyle\int_0^1 (4 - x)e^x\, dx$

47. Machine Part A solid is generated by revolving the region bounded by $y = \frac{1}{2}x^2$ and $y = 2$ about the y-axis. A hole, centered along the axis of revolution, is drilled through this solid so that one-fourth of the volume is removed. Find the diameter of the hole.

48. Machine Part A solid is generated by revolving the region bounded by $y = \sqrt{9 - x^2}$ and $y = 0$ about the y-axis. A hole, centered along the axis of revolution, is drilled through this solid so that one-third of the volume is removed. Find the diameter of the hole.

49. Volume of a Torus A torus is formed by revolving the region bounded by the circle $x^2 + y^2 = 1$ about the line $x = 2$ (see figure). Find the volume of this "doughnut-shaped" solid. (*Hint:* The integral $\int_{-1}^{1} \sqrt{1 - x^2}\, dx$ represents the area of a semicircle.)

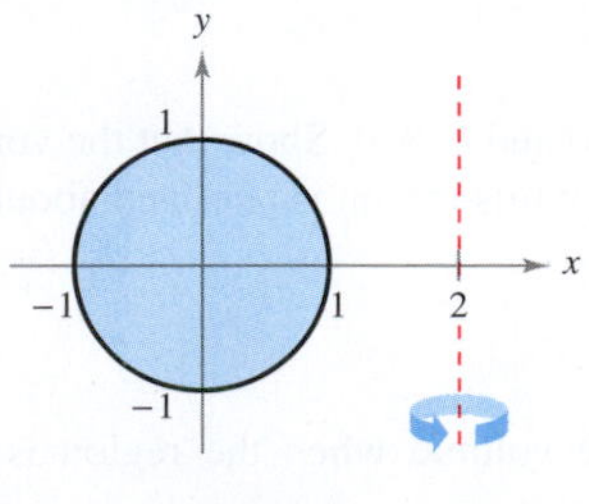

50. Volume of a Torus Repeat Exercise 49 for a torus formed by revolving the region bounded by the circle $x^2 + y^2 = r^2$ about the line $x = R$, where $r < R$.

51. Finding Volumes of Solids

(a) Use differentiation to verify that

$$\int x \sin x\, dx = \sin x - x \cos x + C.$$

(b) Use the result of part (a) to find the volume of the solid generated by revolving each plane region about the y-axis.

(i)

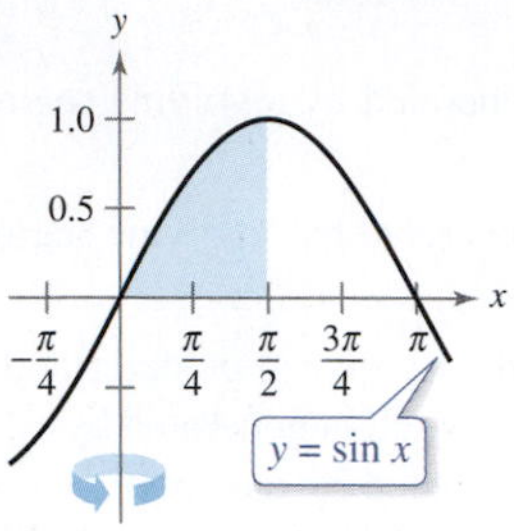

(ii)

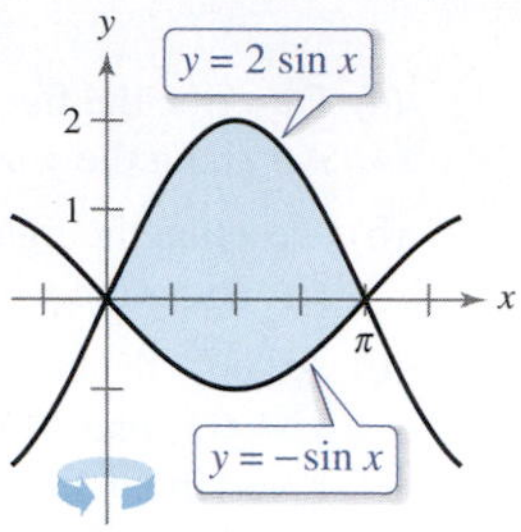

52. Finding Volumes of Solids

(a) Use differentiation to verify that

$$\int x \cos x\, dx = \cos x + x \sin x + C.$$

(b) Use the result of part (a) to find the volume of the solid generated by revolving each plane region about the y-axis. (*Hint:* Begin by approximating the points of intersection.)

(i)

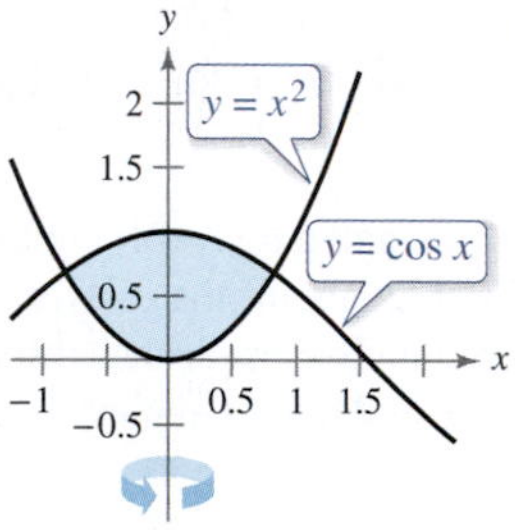

(ii)

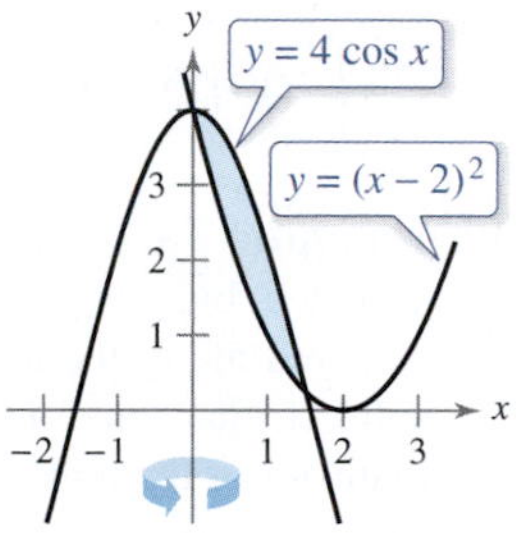

53. Volume of a Segment of a Sphere Let a sphere of radius r be cut by a plane, thereby forming a segment of height h. Show that the volume of this segment is

$$\frac{1}{3}\pi h^2(3r - h).$$

54. Volume of an Ellipsoid Consider the plane region bounded by the graph of

$$\left(\frac{x}{a}\right)^2 + \left(\frac{y}{b}\right)^2 = 1$$

where $a > 0$ and $b > 0$. Show that the volume of the ellipsoid formed when this region is revolved about the y-axis is

$$\frac{4}{3}\pi a^2 b.$$

What is the volume when the region is revolved about the x-axis?

55. Exploration Consider the region bounded by the graphs of $y = ax^n$, $y = ab^n$, and $x = 0$ (see figure).

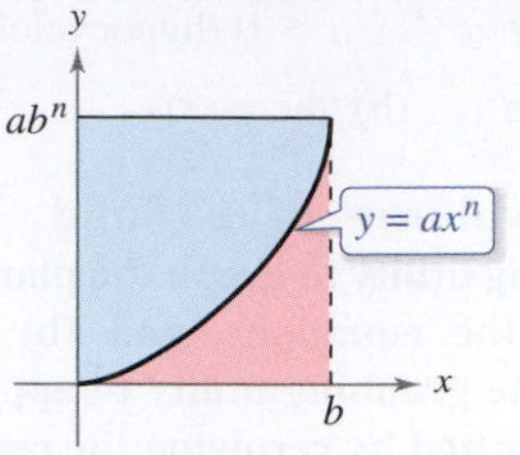

(a) Find the ratio $R_1(n)$ of the area of the region to the area of the circumscribed rectangle.

(b) Find $\lim_{n\to\infty} R_1(n)$ and compare the result with the area of the circumscribed rectangle.

(c) Find the volume of the solid of revolution formed by revolving the region about the y-axis. Find the ratio $R_2(n)$ of this volume to the volume of the circumscribed right circular cylinder.

(d) Find $\lim_{n\to\infty} R_2(n)$ and compare the result with the volume of the circumscribed cylinder.

(e) Use the results of parts (b) and (d) to make a conjecture about the shape of the graph of $y = ax^n$ $(0 \le x \le b)$ as $n \to \infty$.

56. Think About It Match each integral with the solid whose volume it represents, and give the dimensions of each solid.

(a) Right circular cone (b) Torus (c) Sphere

(d) Right circular cylinder (e) Ellipsoid

(i) $2\pi \int_0^r hx\, dx$ (ii) $2\pi \int_0^r hx\left(1 - \frac{x}{r}\right) dx$

(iii) $2\pi \int_0^r 2x\sqrt{r^2 - x^2}\, dx$ (iv) $2\pi \int_0^b 2ax\sqrt{1 - \frac{x^2}{b^2}}\, dx$

(v) $2\pi \int_{-r}^{r} (R - x)\left(2\sqrt{r^2 - x^2}\right) dx$

57. Volume of a Storage Shed A storage shed has a circular base of diameter 80 feet. Starting at the center, the interior height is measured every 10 feet and recorded in the table (see figure).

x	Height
0	50
10	45
20	40
30	20
40	0

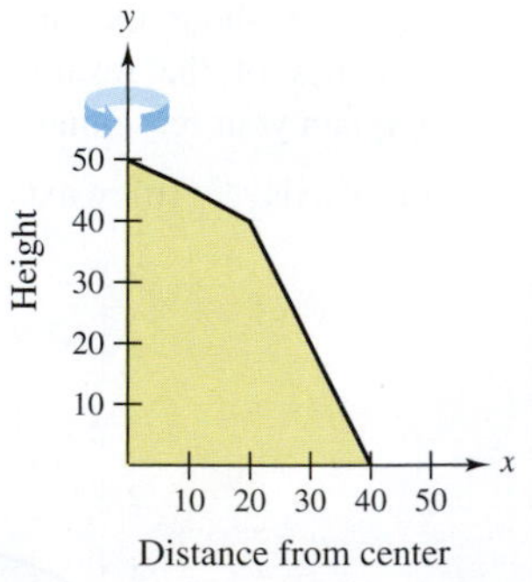

(a) Use Simpson's Rule to approximate the volume of the shed.

(b) Note that the roof line consists of two line segments. Find the equations of the line segments and use integration to find the volume of the shed.

58. Modeling Data A pond is approximately circular, with a diameter of 400 feet. Starting at the center, the depth of the water is measured every 25 feet and recorded in the table (see figure).

x	0	25	50	75	100	125	150	175	200
Depth	20	19	19	17	15	14	10	6	0

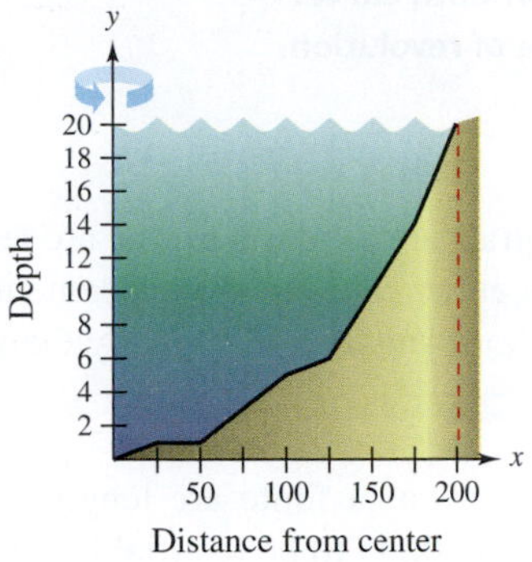

(a) Use Simpson's Rule to approximate the volume of water in the pond.

(b) Use the regression capabilities of a graphing utility to find a quadratic model for the depths recorded in the table. Use the graphing utility to plot the depths and graph the model.

(c) Use the integration capabilities of a graphing utility and the model in part (b) to approximate the volume of water in the pond.

(d) Use the result of part (c) to approximate the number of gallons of water in the pond. (*Hint:* 1 cubic foot of water is approximately 7.48 gallons.)

59. Equal Volumes Let V_1 and V_2 be the volumes of the solids that result when the plane region bounded by $y = 1/x$, $y = 0$, $x = \frac{1}{4}$, and $x = c$ (where $c > \frac{1}{4}$) is revolved about the x-axis and the y-axis, respectively. Find the value of c for which $V_1 = V_2$.

60. Volume of a Segment of a Paraboloid The region bounded by $y = r^2 - x^2$, $y = 0$, and $x = 0$ is revolved about the y-axis to form a paraboloid. A hole, centered along the axis of revolution, is drilled through this solid. The hole has a radius k, $0 < k < r$. Find the volume of the resulting ring (a) by integrating with respect to x and (b) by integrating with respect to y.

61. Finding Volumes of Solids Consider the graph of $y^2 = x(4 - x)^2$ (see figure). Find the volumes of the solids that are generated when the loop of this graph is revolved about (a) the x-axis, (b) the y-axis, and (c) the line $x = 4$.

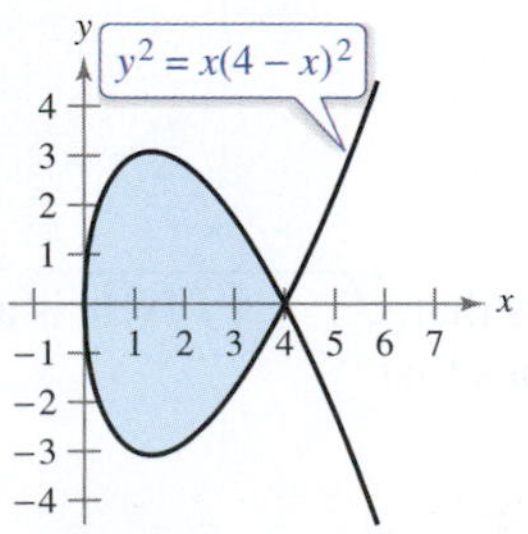

SECTION PROJECT

Saturn

The Oblateness of Saturn Saturn is the most oblate of the planets in our solar system. Its equatorial radius is 60,268 kilometers and its polar radius is 54,364 kilometers. The color-enhanced photograph of Saturn was taken by Voyager 1. In the photograph, the oblateness of Saturn is clearly visible.

(a) Find the ratio of the volumes of the sphere and the oblate ellipsoid shown below.

(b) If a planet were spherical and had the same volume as Saturn, what would its radius be?

Computer model of "spherical Saturn," whose equatorial radius is equal to its polar radius. The equation of the cross section passing through the pole is

$$x^2 + y^2 = 60{,}268^2.$$

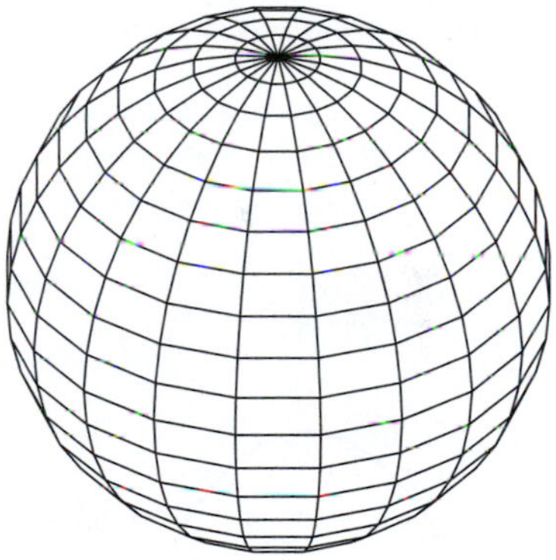

Computer model of "oblate Saturn," whose equatorial radius is greater than its polar radius. The equation of the cross section passing through the pole is

$$\frac{x^2}{60{,}268^2} + \frac{y^2}{54{,}364^2} = 1.$$

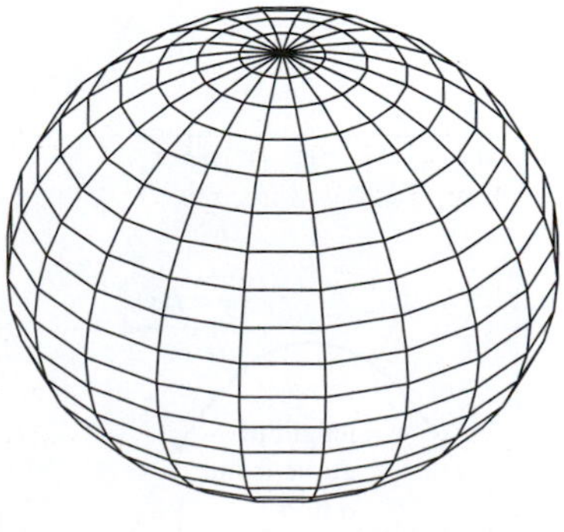

NASA

7.4 Arc Length and Surfaces of Revolution

- Find the arc length of a smooth curve.
- Find the area of a surface of revolution.

Arc Length

In this section, definite integrals are used to find the arc lengths of curves and the areas of surfaces of revolution. In either case, an arc (a segment of a curve) is approximated by straight line segments whose lengths are given by the familiar Distance Formula

$$d = \sqrt{(x_2 - x_1)^2 + (y_2 - y_1)^2}.$$

A **rectifiable** curve is one that has a finite arc length. You will see that a sufficient condition for the graph of a function f to be rectifiable between $(a, f(a))$ and $(b, f(b))$ is that f' be continuous on $[a, b]$. Such a function is **continuously differentiable** on $[a, b]$, and its graph on the interval $[a, b]$ is a **smooth curve.**

Consider a function $y = f(x)$ that is continuously differentiable on the interval $[a, b]$. You can approximate the graph of f by n line segments whose endpoints are determined by the partition

$$a = x_0 < x_1 < x_2 < \cdots < x_n = b$$

as shown in Figure 7.37. By letting $\Delta x_i = x_i - x_{i-1}$ and $\Delta y_i = y_i - y_{i-1}$, you can approximate the length of the graph by

$$\begin{aligned} s &\approx \sum_{i=1}^{n} \sqrt{(x_i - x_{i-1})^2 + (y_i - y_{i-1})^2} \\ &= \sum_{i=1}^{n} \sqrt{(\Delta x_i)^2 + (\Delta y_i)^2} \\ &= \sum_{i=1}^{n} \sqrt{(\Delta x_i)^2 + \left(\frac{\Delta y_i}{\Delta x_i}\right)^2 (\Delta x_i)^2} \\ &= \sum_{i=1}^{n} \sqrt{1 + \left(\frac{\Delta y_i}{\Delta x_i}\right)^2}\, (\Delta x_i). \end{aligned}$$

This approximation appears to become better and better as $\|\Delta\| \to 0$ $(n \to \infty)$. So, the length of the graph is

$$s = \lim_{\|\Delta\| \to 0} \sum_{i=1}^{n} \sqrt{1 + \left(\frac{\Delta y_i}{\Delta x_i}\right)^2}\, (\Delta x_i).$$

Because $f'(x)$ exists for each x in (x_{i-1}, x_i), the Mean Value Theorem guarantees the existence of c_i in (x_{i-1}, x_i) such that

$$\begin{aligned} f(x_i) - f(x_{i-1}) &= f'(c_i)(x_i - x_{i-1}) \\ \frac{f(x_i) - f(x_{i-1})}{x_i - x_{i-1}} &= f'(c_i) \\ \frac{\Delta y_i}{\Delta x_i} &= f'(c_i). \end{aligned}$$

Because f' is continuous on $[a, b]$, it follows that $\sqrt{1 + [f'(x)]^2}$ is also continuous (and therefore integrable) on $[a, b]$, which implies that

$$\begin{aligned} s &= \lim_{\|\Delta\| \to 0} \sum_{i=1}^{n} \sqrt{1 + [f'(c_i)]^2}\, (\Delta x_i) \\ &= \int_a^b \sqrt{1 + [f'(x)]^2}\, dx \end{aligned}$$

where s is called the **arc length** of f between a and b.

CHRISTIAN HUYGENS (1629–1695)

The Dutch mathematician Christian Huygens, who invented the pendulum clock, and James Gregory (1638–1675), a Scottish mathematician, both made early contributions to the problem of finding the length of a rectifiable curve.

See LarsonCalculus.com to read more of this biography.

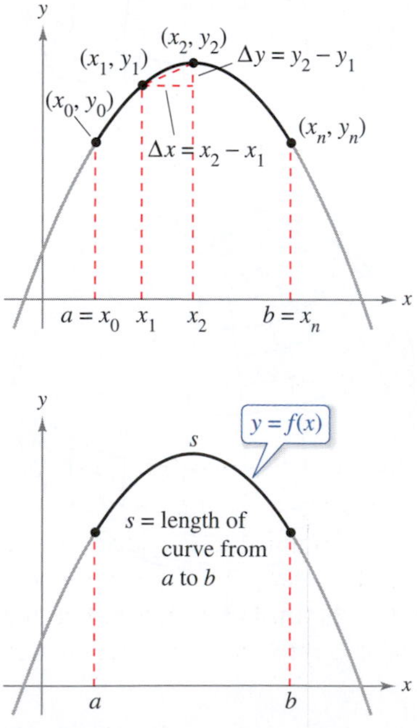

Figure 7.37

AP* Tips

Finding the arc length of a function over some interval is tested on the AP Calculus BC Exam.

Definition of Arc Length

Let the function $y = f(x)$ represent a smooth curve on the interval $[a, b]$. The **arc length** of f between a and b is

$$s = \int_a^b \sqrt{1 + [f'(x)]^2}\, dx.$$

Similarly, for a smooth curve $x = g(y)$, the **arc length** of g between c and d is

$$s = \int_c^d \sqrt{1 + [g'(y)]^2}\, dy.$$

FOR FURTHER INFORMATION To see how arc length can be used to define trigonometric functions, see the article "Trigonometry Requires Calculus, Not Vice Versa" by Yves Nievergelt in *UMAP Modules.*

Because the definition of arc length can be applied to a linear function, you can check to see that this new definition agrees with the standard Distance Formula for the length of a line segment. This is shown in Example 1.

EXAMPLE 1 The Length of a Line Segment

Find the arc length from (x_1, y_1) to (x_2, y_2) on the graph of

$$f(x) = mx + b$$

as shown in Figure 7.38.

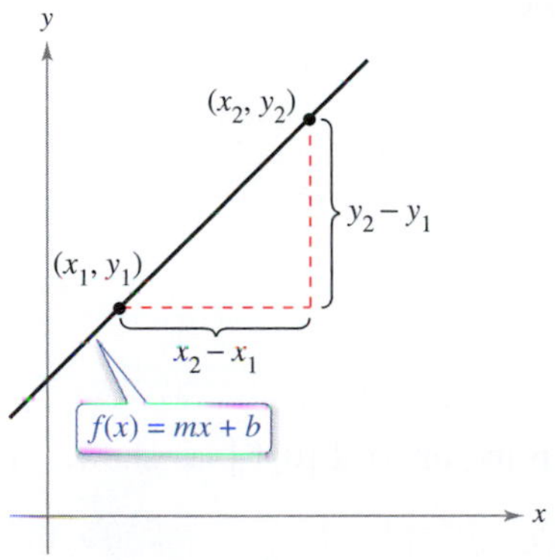

The formula for the arc length of the graph of f from (x_1, y_1) to (x_2, y_2) is the same as the standard Distance Formula.
Figure 7.38

Solution Because

$$m = f'(x) = \frac{y_2 - y_1}{x_2 - x_1}$$

it follows that

$$s = \int_{x_1}^{x_2} \sqrt{1 + [f'(x)]^2}\, dx \qquad \text{Formula for arc length}$$

$$= \int_{x_1}^{x_2} \sqrt{1 + \left(\frac{y_2 - y_1}{x_2 - x_1}\right)^2}\, dx$$

$$= \sqrt{\frac{(x_2 - x_1)^2 + (y_2 - y_1)^2}{(x_2 - x_1)^2}}\,(x)\Bigg]_{x_1}^{x_2} \qquad \text{Integrate and simplify.}$$

$$= \sqrt{\frac{(x_2 - x_1)^2 + (y_2 - y_1)^2}{(x_2 - x_1)^2}}\,(x_2 - x_1)$$

$$= \sqrt{(x_2 - x_1)^2 + (y_2 - y_1)^2}$$

which is the formula for the distance between two points in the plane.

TECHNOLOGY Definite integrals representing arc length often are very difficult to evaluate. In this section, a few examples are presented. In the next chapter, with more advanced integration techniques, you will be able to tackle more difficult arc length problems. In the meantime, remember that you can always use a numerical integration program to approximate an arc length. For instance, use the numerical integration feature of a graphing utility to approximate arc lengths in Examples 2 and 3.

EXAMPLE 2 Finding Arc Length

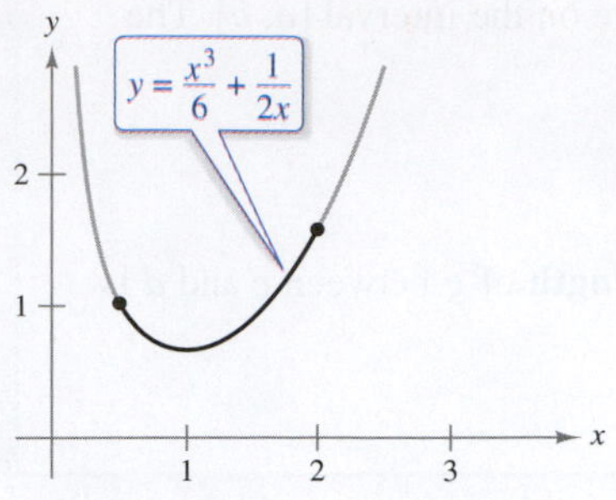

The arc length of the graph of y on $\left[\frac{1}{2}, 2\right]$

Figure 7.39

Find the arc length of the graph of

$$y = \frac{x^3}{6} + \frac{1}{2x}$$

on the interval $\left[\frac{1}{2}, 2\right]$, as shown in Figure 7.39.

Solution Using

$$\frac{dy}{dx} = \frac{3x^2}{6} - \frac{1}{2x^2} = \frac{1}{2}\left(x^2 - \frac{1}{x^2}\right)$$

yields an arc length of

$$s = \int_a^b \sqrt{1 + \left(\frac{dy}{dx}\right)^2}\, dx \qquad \text{Formula for arc length}$$

$$= \int_{1/2}^2 \sqrt{1 + \left[\frac{1}{2}\left(x^2 - \frac{1}{x^2}\right)\right]^2}\, dx$$

$$= \int_{1/2}^2 \sqrt{\frac{1}{4}\left(x^4 + 2 + \frac{1}{x^4}\right)}\, dx$$

$$= \int_{1/2}^2 \frac{1}{2}\left(x^2 + \frac{1}{x^2}\right) dx \qquad \text{Simplify.}$$

$$= \frac{1}{2}\left[\frac{x^3}{3} - \frac{1}{x}\right]_{1/2}^2 \qquad \text{Integrate.}$$

$$= \frac{1}{2}\left(\frac{13}{6} + \frac{47}{24}\right)$$

$$= \frac{33}{16}.$$

EXAMPLE 3 Finding Arc Length

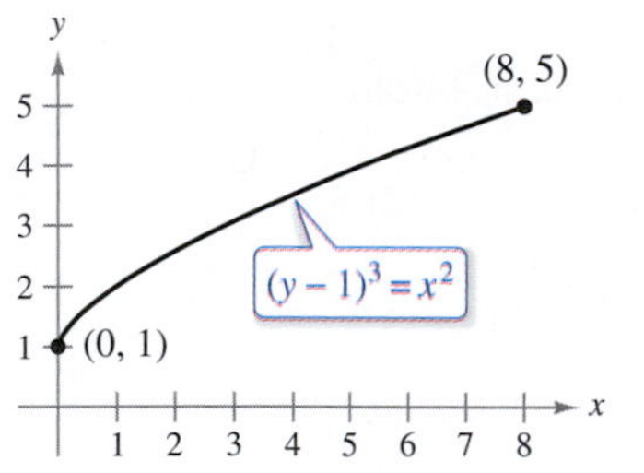

The arc length of the graph of y on $[0, 8]$

Figure 7.40

Find the arc length of the graph of $(y - 1)^3 = x^2$ on the interval $[0, 8]$, as shown in Figure 7.40.

Solution Begin by solving for x in terms of y: $x = \pm(y - 1)^{3/2}$. Choosing the positive value of x produces

$$\frac{dx}{dy} = \frac{3}{2}(y - 1)^{1/2}.$$

The x-interval $[0, 8]$ corresponds to the y-interval $[1, 5]$, and the arc length is

$$s = \int_c^d \sqrt{1 + \left(\frac{dx}{dy}\right)^2}\, dy \qquad \text{Formula for arc length}$$

$$= \int_1^5 \sqrt{1 + \left[\frac{3}{2}(y - 1)^{1/2}\right]^2}\, dy$$

$$= \int_1^5 \sqrt{\frac{9}{4}y - \frac{5}{4}}\, dy$$

$$= \frac{1}{2}\int_1^5 \sqrt{9y - 5}\, dy \qquad \text{Simplify.}$$

$$= \frac{1}{18}\left[\frac{(9y - 5)^{3/2}}{3/2}\right]_1^5 \qquad \text{Integrate.}$$

$$= \frac{1}{27}(40^{3/2} - 4^{3/2})$$

$$\approx 9.073.$$

EXAMPLE 4 Finding Arc Length

See LarsonCalculus.com for an interactive version of this type of example.

Find the arc length of the graph of

$$y = \ln(\cos x)$$

from $x = 0$ to $x = \pi/4$, as shown in Figure 7.41.

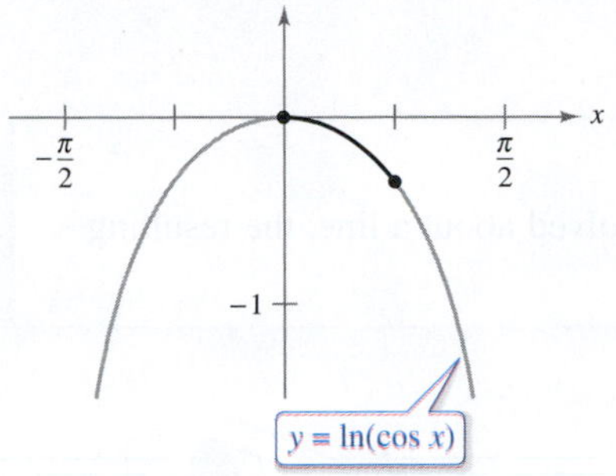

The arc length of the graph of y on $\left[0, \frac{\pi}{4}\right]$

Figure 7.41

Solution Using

$$\frac{dy}{dx} = -\frac{\sin x}{\cos x} = -\tan x$$

yields an arc length of

$$\begin{aligned} s &= \int_a^b \sqrt{1 + \left(\frac{dy}{dx}\right)^2}\,dx && \text{Formula for arc length} \\ &= \int_0^{\pi/4} \sqrt{1 + \tan^2 x}\,dx \\ &= \int_0^{\pi/4} \sqrt{\sec^2 x}\,dx && \text{Trigonometric identity} \\ &= \int_0^{\pi/4} \sec x\,dx && \text{Simplify.} \\ &= \Big[\ln|\sec x + \tan x|\Big]_0^{\pi/4} && \text{Integrate.} \\ &= \ln\left(\sqrt{2} + 1\right) - \ln 1 \\ &\approx 0.881. \end{aligned}$$

EXAMPLE 5 Length of a Cable

An electric cable is hung between two towers that are 200 feet apart, as shown in Figure 7.42. The cable takes the shape of a catenary whose equation is

$$y = 75(e^{x/150} + e^{-x/150}) = 150 \cosh \frac{x}{150}.$$

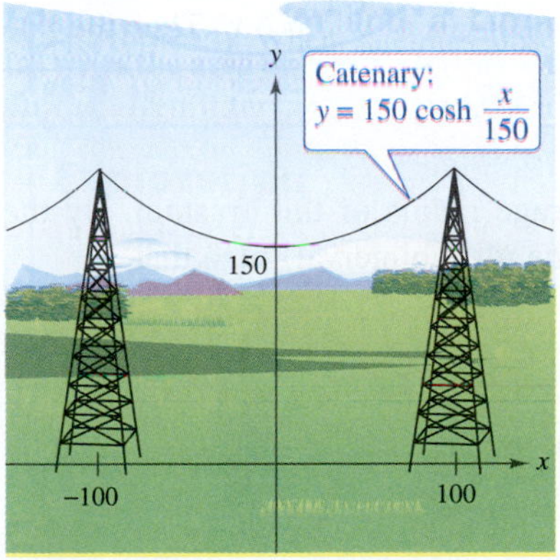

Figure 7.42

Find the arc length of the cable between the two towers.

Solution Because $y' = \frac{1}{2}(e^{x/150} - e^{-x/150})$, you can write

$$(y')^2 = \frac{1}{4}(e^{x/75} - 2 + e^{-x/75})$$

and

$$1 + (y')^2 = \frac{1}{4}(e^{x/75} + 2 + e^{-x/75}) = \left[\frac{1}{2}(e^{x/150} + e^{-x/150})\right]^2.$$

Therefore, the arc length of the cable is

$$\begin{aligned} s &= \int_a^b \sqrt{1 + (y')^2}\,dx && \text{Formula for arc length} \\ &= \frac{1}{2}\int_{-100}^{100} (e^{x/150} + e^{-x/150})\,dx \\ &= 75\Big[e^{x/150} - e^{-x/150}\Big]_{-100}^{100} && \text{Integrate.} \\ &= 150(e^{2/3} - e^{-2/3}) \\ &\approx 215 \text{ feet.} \end{aligned}$$

Area of a Surface of Revolution

In Sections 7.2 and 7.3, integration was used to calculate the volume of a solid of revolution. You will now look at a procedure for finding the area of a surface of revolution.

> **Definition of Surface of Revolution**
>
> When the graph of a continuous function is revolved about a line, the resulting surface is a **surface of revolution.**

The area of a surface of revolution is derived from the formula for the lateral surface area of the frustum of a right circular cone. Consider the line segment in the figure at the right, where L is the length of the line segment, r_1 is the radius at the left end of the line segment, and r_2 is the radius at the right end of the line segment. When the line segment is revolved about its axis of revolution, it forms a frustum of a right circular cone, with

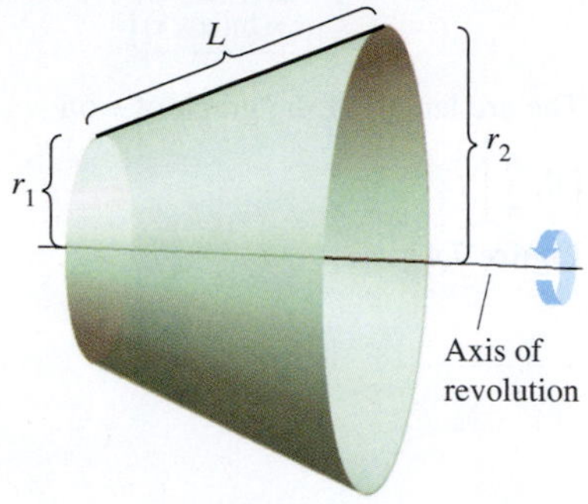

$$S = 2\pi rL \qquad \text{Lateral surface area of frustum}$$

where

$$r = \frac{1}{2}(r_1 + r_2). \qquad \text{Average radius of frustum}$$

(In Exercise 54, you are asked to verify the formula for S.)

Consider a function f that has a continuous derivative on the interval $[a, b]$. The graph of f is revolved about the x-axis to form a surface of revolution, as shown in Figure 7.43. Let Δ be a partition of $[a, b]$, with subintervals of width Δx_i. Then the line segment of length

$$\Delta L_i = \sqrt{\Delta x_i^2 + \Delta y_i^2}$$

generates a frustum of a cone. Let r_i be the average radius of this frustum. By the Intermediate Value Theorem, a point d_i exists (in the ith subinterval) such that

$$r_i = f(d_i).$$

The lateral surface area ΔS_i of the frustum is

$$\begin{aligned}\Delta S_i &= 2\pi r_i\,\Delta L_i\\ &= 2\pi f(d_i)\sqrt{\Delta x_i^2 + \Delta y_i^2}\\ &= 2\pi f(d_i)\sqrt{1 + \left(\frac{\Delta y_i}{\Delta x_i}\right)^2}\,\Delta x_i.\end{aligned}$$

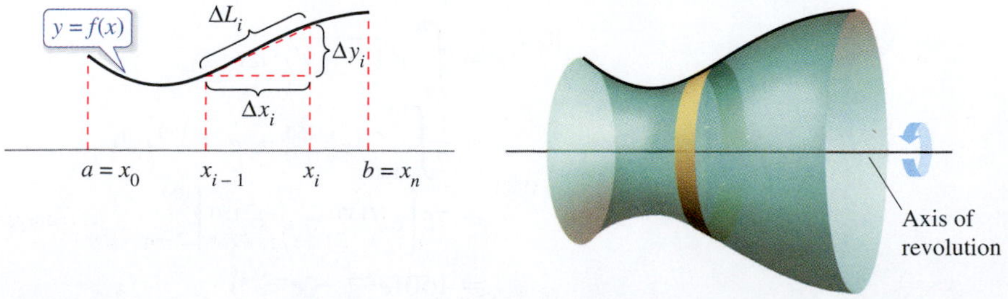

Figure 7.43

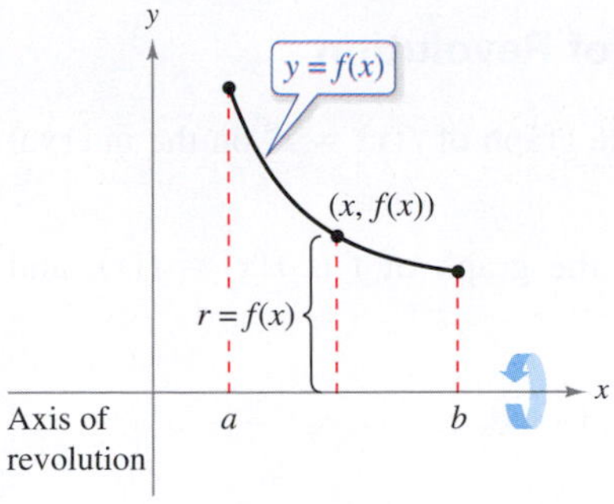

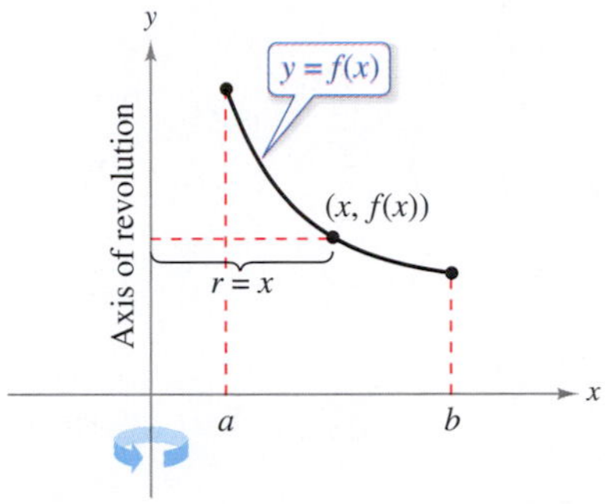

Figure 7.44

By the Mean Value Theorem, a point c_i exists in (x_{i-1}, x_i) such that

$$f'(c_i) = \frac{f(x_i) - f(x_{i-1})}{x_i - x_{i-1}} = \frac{\Delta y_i}{\Delta x_i}.$$

So, $\Delta S_i = 2\pi f(d_i)\sqrt{1 + [f'(c_i)]^2}\,\Delta x_i$, and the total surface area can be approximated by

$$S \approx 2\pi \sum_{i=1}^{n} f(d_i)\sqrt{1 + [f'(c_i)]^2}\,\Delta x_i.$$

It can be shown that the limit of the right side as $\|\Delta\| \to 0$ $(n \to \infty)$ is

$$S = 2\pi \int_a^b f(x)\sqrt{1 + [f'(x)]^2}\,dx.$$

In a similar manner, if the graph of f is revolved about the y-axis, then S is

$$S = 2\pi \int_a^b x\sqrt{1 + [f'(x)]^2}\,dx.$$

In these two formulas for S, you can regard the products $2\pi f(x)$ and $2\pi x$ as the circumferences of the circles traced by a point (x, y) on the graph of f as it is revolved about the x-axis and the y-axis (Figure 7.44). In one case, the radius is $r = f(x)$, and in the other case, the radius is $r = x$. Moreover, by appropriately adjusting r, you can generalize the formula for surface area to cover *any* horizontal or vertical axis of revolution, as indicated in the next definition.

Definition of the Area of a Surface of Revolution

Let $y = f(x)$ have a continuous derivative on the interval $[a, b]$. The area S of the surface of revolution formed by revolving the graph of f about a horizontal or vertical axis is

$$S = 2\pi \int_a^b r(x)\sqrt{1 + [f'(x)]^2}\,dx$$ y is a function of x.

where $r(x)$ is the distance between the graph of f and the axis of revolution. If $x = g(y)$ on the interval $[c, d]$, then the surface area is

$$S = 2\pi \int_c^d r(y)\sqrt{1 + [g'(y)]^2}\,dy$$ x is a function of y.

where $r(y)$ is the distance between the graph of g and the axis of revolution.

The formulas in this definition are sometimes written as

$$S = 2\pi \int_a^b r(x)\,ds$$ y is a function of x.

and

$$S = 2\pi \int_c^d r(y)\,ds$$ x is a function of y.

where

$$ds = \sqrt{1 + [f'(x)]^2}\,dx \quad \text{and} \quad ds = \sqrt{1 + [g'(y)]^2}\,dy,$$

respectively.

EXAMPLE 6 The Area of a Surface of Revolution

Find the area of the surface formed by revolving the graph of $f(x) = x^3$ on the interval $[0, 1]$ about the x-axis, as shown in Figure 7.45.

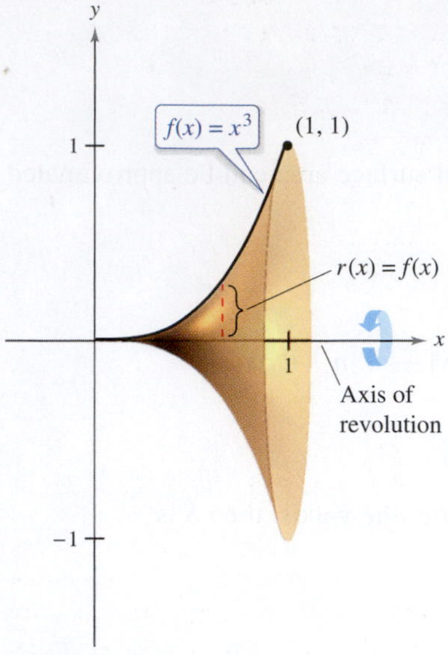

Figure 7.45

Solution The distance between the x-axis and the graph of f is $r(x) = f(x)$, and because $f'(x) = 3x^2$, the surface area is

$$\begin{aligned} S &= 2\pi\int_a^b r(x)\sqrt{1 + [f'(x)]^2}\,dx && \text{Formula for surface area} \\ &= 2\pi\int_0^1 x^3\sqrt{1 + (3x^2)^2}\,dx \\ &= \frac{2\pi}{36}\int_0^1 (36x^3)(1 + 9x^4)^{1/2}\,dx && \text{Simplify.} \\ &= \frac{\pi}{18}\left[\frac{(1 + 9x^4)^{3/2}}{3/2}\right]_0^1 && \text{Integrate.} \\ &= \frac{\pi}{27}(10^{3/2} - 1) \\ &\approx 3.563. \end{aligned}$$

EXAMPLE 7 The Area of a Surface of Revolution

Find the area of the surface formed by revolving the graph of $f(x) = x^2$ on the interval $[0, \sqrt{2}]$ about the y-axis, as shown in the figure below.

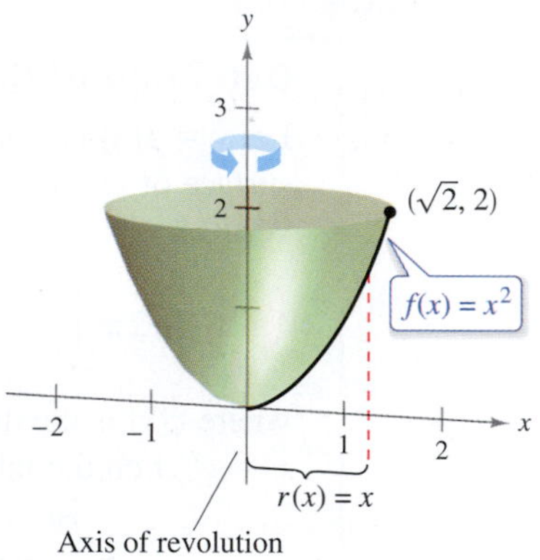

Solution In this case, the distance between the graph of f and the y-axis is $r(x) = x$. Using $f'(x) = 2x$ and the formula for surface area, you can determine that

$$\begin{aligned} S &= 2\pi\int_a^b r(x)\sqrt{1 + [f'(x)]^2}\,dx \\ &= 2\pi\int_0^{\sqrt{2}} x\sqrt{1 + (2x)^2}\,dx \\ &= \frac{2\pi}{8}\int_0^{\sqrt{2}} (1 + 4x^2)^{1/2}(8x)\,dx && \text{Simplify.} \\ &= \frac{\pi}{4}\left[\frac{(1 + 4x^2)^{3/2}}{3/2}\right]_0^{\sqrt{2}} && \text{Integrate.} \\ &= \frac{\pi}{6}[(1 + 8)^{3/2} - 1] \\ &= \frac{13\pi}{3} \\ &\approx 13.614. \end{aligned}$$

7.4 Exercises

See CalcChat.com for tutorial help and worked-out solutions to odd-numbered exercises.

Finding Distance Using Two Methods **In Exercises 1 and 2, find the distance between the points using (a) the Distance Formula and (b) integration.**

1. $(0, 0),\quad (8, 15)$
2. $(1, 2),\quad (7, 10)$

Finding Arc Length **In Exercises 3–16, find the arc length of the graph of the function over the indicated interval.**

3. $y = \frac{2}{3}(x^2 + 1)^{3/2}$

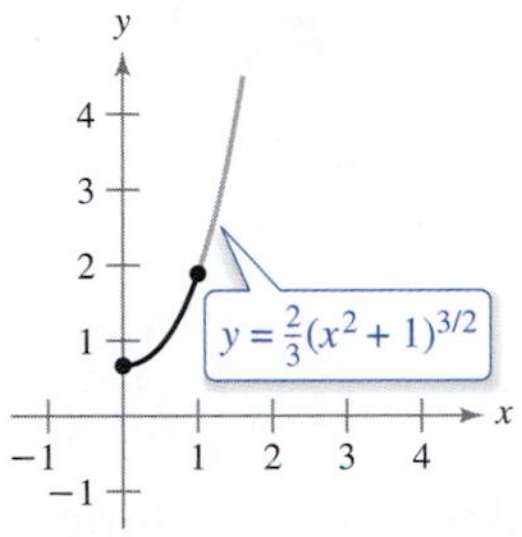

4. $y = \frac{x^3}{6} + \frac{1}{2x}$

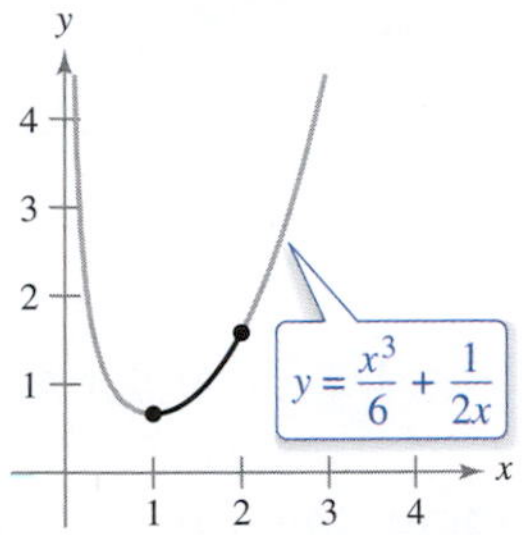

5. $y = \frac{2}{3}x^{3/2} + 1$

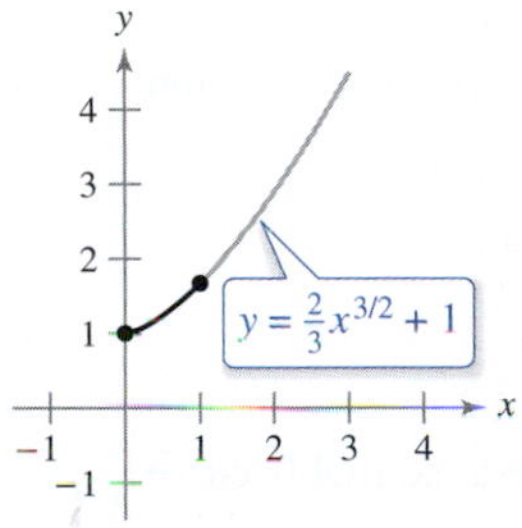

6. $y = 2x^{3/2} + 3$

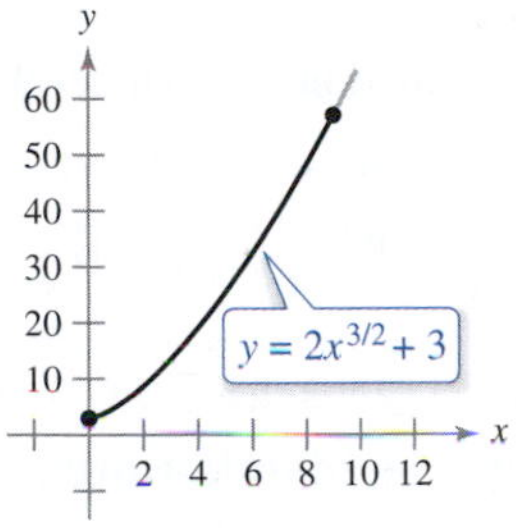

7. $y = \frac{3}{2}x^{2/3},\quad [1, 8]$
8. $y = \frac{x^4}{8} + \frac{1}{4x^2},\quad [1, 3]$
9. $y = \frac{x^5}{10} + \frac{1}{6x^3},\quad [2, 5]$
10. $y = \frac{3}{2}x^{2/3} + 4,\quad [1, 27]$
11. $y = \ln(\sin x),\quad \left[\frac{\pi}{4}, \frac{3\pi}{4}\right]$
12. $y = \ln(\cos x),\quad \left[0, \frac{\pi}{3}\right]$
13. $y = \frac{1}{2}(e^x + e^{-x}),\quad [0, 2]$
14. $y = \ln\left(\frac{e^x + 1}{e^x - 1}\right),\quad [\ln 2, \ln 3]$
15. $x = \frac{1}{3}(y^2 + 2)^{3/2},\quad 0 \le y \le 4$
16. $x = \frac{1}{3}\sqrt{y}(y - 3),\quad 1 \le y \le 4$

Finding Arc Length **In Exercises 17–26, (a) sketch the graph of the function, highlighting the part indicated by the given interval, (b) find a definite integral that represents the arc length of the curve over the indicated interval and observe that the integral cannot be evaluated with the techniques studied so far, and (c) use the integration capabilities of a graphing utility to approximate the arc length.**

17. $y = 4 - x^2,\quad 0 \le x \le 2$
18. $y = x^2 + x - 2,\quad -2 \le x \le 1$
19. $y = \frac{1}{x},\quad 1 \le x \le 3$
20. $y = \frac{1}{x + 1},\quad 0 \le x \le 1$
21. $y = \sin x,\quad 0 \le x \le \pi$
22. $y = \cos x,\quad -\frac{\pi}{2} \le x \le \frac{\pi}{2}$
23. $x = e^{-y},\quad 0 \le y \le 2$
24. $y = \ln x,\quad 1 \le x \le 5$
25. $y = 2 \arctan x,\quad 0 \le x \le 1$
26. $x = \sqrt{36 - y^2},\quad 0 \le y \le 3$

Approximation **In Exercises 27 and 28, determine which value best approximates the length of the arc represented by the integral. (Make your selection on the basis of a sketch of the arc, *not* by performing any calculations.)**

27. $\displaystyle\int_0^2 \sqrt{1 + \left[\frac{d}{dx}\left(\frac{5}{x^2 + 1}\right)\right]^2}\, dx$

(a) 25 (b) 5 (c) 2 (d) -4 (e) 3

28. $\displaystyle\int_0^{\pi/4} \sqrt{1 + \left[\frac{d}{dx}(\tan x)\right]^2}\, dx$

(a) 3 (b) -2 (c) 4 (d) $\frac{4\pi}{3}$ (e) 1

Approximation **In Exercises 29 and 30, approximate the arc length of the graph of the function over the interval [0, 4] in four ways. (a) Use the Distance Formula to find the distance between the endpoints of the arc. (b) Use the Distance Formula to find the lengths of the four line segments connecting the points on the arc when $x = 0$, $x = 1$, $x = 2$, $x = 3$, and $x = 4$. Find the sum of the four lengths. (c) Use Simpson's Rule with $n = 10$ to approximate the integral yielding the indicated arc length. (d) Use the integration capabilities of a graphing utility to approximate the integral yielding the indicated arc length.**

29. $f(x) = x^3$
30. $f(x) = (x^2 - 4)^2$

31. **Length of a Catenary** Electrical wires suspended between two towers form a catenary (see figure) modeled by the equation

$$y = 20 \cosh \frac{x}{20},\quad -20 \le x \le 20$$

where x and y are measured in meters. The towers are 40 meters apart. Find the length of the suspended cable.

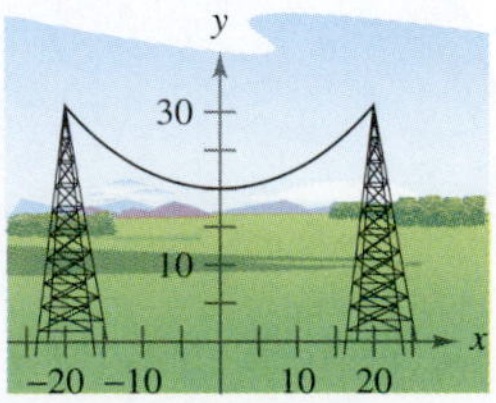

32. Roof Area A barn is 100 feet long and 40 feet wide (see figure). A cross section of the roof is the inverted catenary $y = 31 - 10(e^{x/20} + e^{-x/20})$. Find the number of square feet of roofing on the barn.

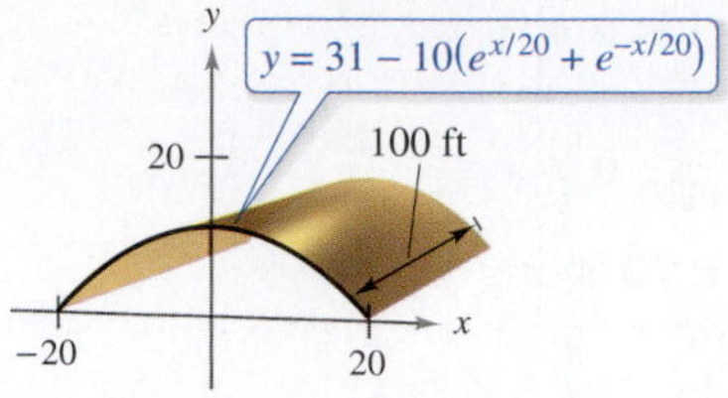

33. Length of Gateway Arch The Gateway Arch in St. Louis, Missouri, is modeled by

$$y = 693.8597 - 68.7672 \cosh 0.0100333x,$$
$$-299.2239 \le x \le 299.2239.$$

(See Section 5.8, Section Project: St. Louis Arch.) Use the integration capabilities of a graphing utility to approximate the length of this curve (see figure).

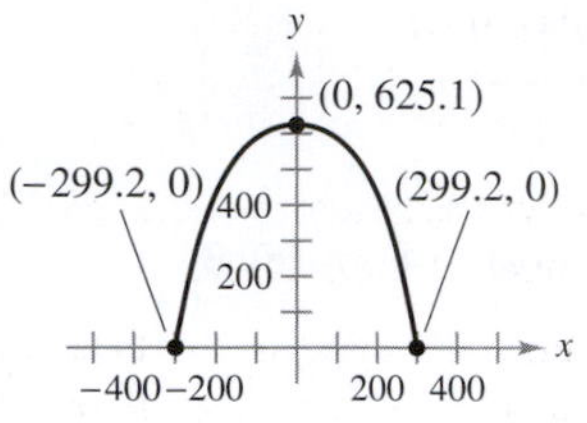

Figure for 33

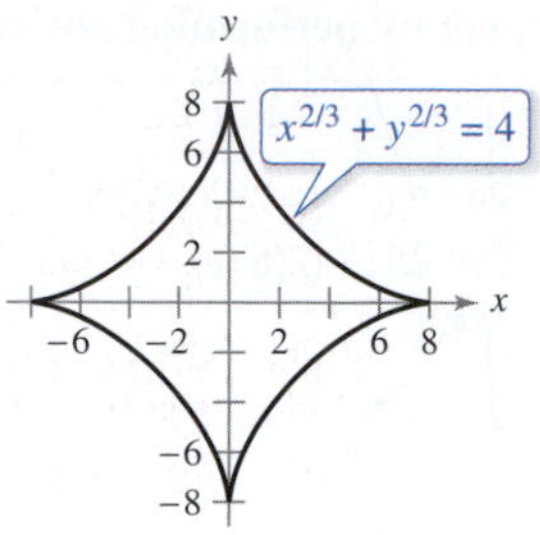

Figure for 34

34. Astroid Find the total length of the graph of the astroid $x^{2/3} + y^{2/3} = 4$.

35. Arc Length of a Sector of a Circle Find the arc length from (0, 3) clockwise to $(2, \sqrt{5})$ along the circle $x^2 + y^2 = 9$.

36. Arc Length of a Sector of a Circle Find the arc length from $(-3, 4)$ clockwise to $(4, 3)$ along the circle $x^2 + y^2 = 25$. Show that the result is one-fourth the circumference of the circle.

Finding the Area of a Surface of Revolution In Exercises 37–42, set up and evaluate the definite integral for the area of the surface generated by revolving the curve about the *x*-axis.

37. $y = \frac{1}{3}x^3$

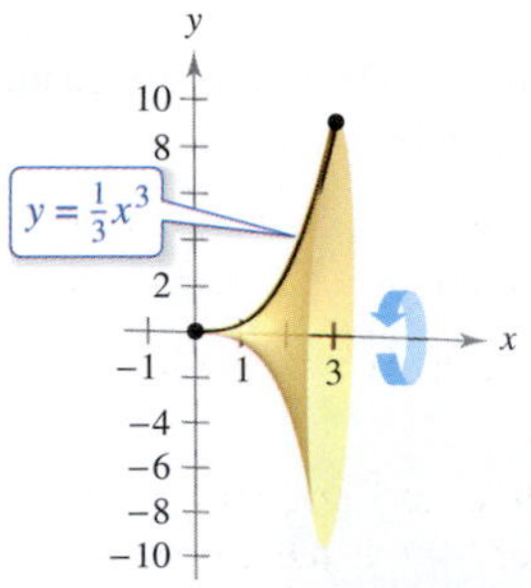

38. $y = 2\sqrt{x}$

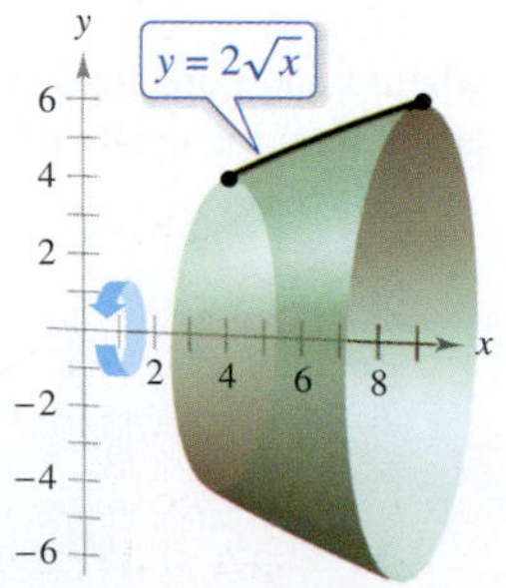

39. $y = \dfrac{x^3}{6} + \dfrac{1}{2x}, \quad 1 \le x \le 2$

40. $y = 3x, \quad 0 \le x \le 3$

41. $y = \sqrt{4 - x^2}, \quad -1 \le x \le 1$

42. $y = \sqrt{9 - x^2}, \quad -2 \le x \le 2$

Finding the Area of a Surface of Revolution In Exercises 43–46, set up and evaluate the definite integral for the area of the surface generated by revolving the curve about the *y*-axis.

43. $y = \sqrt[3]{x} + 2$

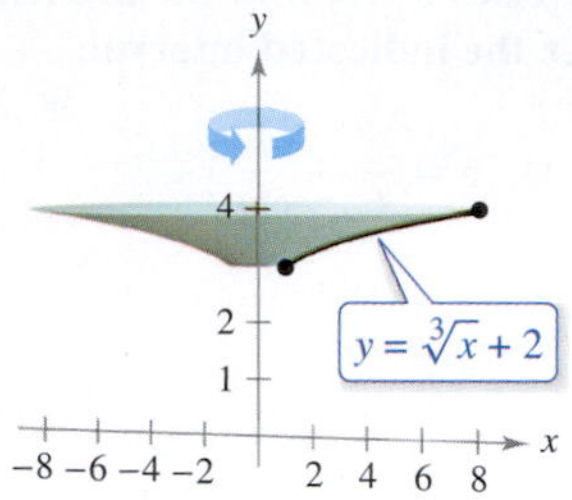

44. $y = 9 - x^2$

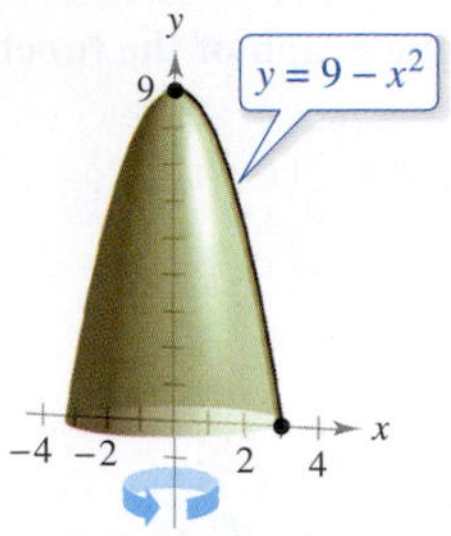

45. $y = 1 - \dfrac{x^2}{4}, \quad 0 \le x \le 2$

46. $y = \dfrac{x}{2} + 3, \quad 1 \le x \le 5$

Finding the Area of a Surface of Revolution In Exercises 47 and 48, use the integration capabilities of a graphing utility to approximate the surface area of the solid of revolution.

	Function	Interval	Axis of Revolution
47.	$y = \sin x$	$[0, \pi]$	x-axis
48.	$y = \ln x$	$[1, e]$	y-axis

WRITING ABOUT CONCEPTS

49. Rectifiable Curve Define a rectifiable curve.

50. Precalculus and Calculus What precalculus formula and representative element are used to develop the integration formula for arc length?

51. Precalculus and Calculus What precalculus formula and representative element are used to develop the integration formula for the area of a surface of revolution?

52. HOW DO YOU SEE IT? The graphs of the functions f_1 and f_2 on the interval $[a, b]$ are shown in the figure. The graph of each function is revolved about the x-axis. Which surface of revolution has the greater surface area? Explain.

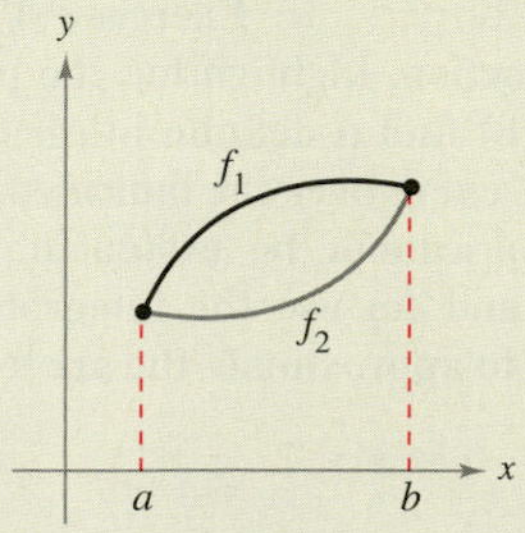

53. Think About It The figure shows the graphs of the functions $y_1 = x$, $y_2 = \frac{1}{2}x^{3/2}$, $y_3 = \frac{1}{4}x^2$, and $y_4 = \frac{1}{8}x^{5/2}$ on the interval $[0, 4]$. To print an enlarged copy of the graph, go to *MathGraphs.com.*

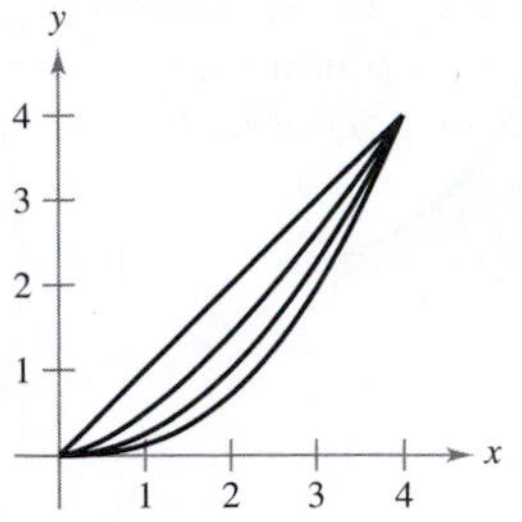

(a) Label the functions.

(b) List the functions in order of increasing arc length.

(c) Verify your answer in part (b) by using the integration capabilities of a graphing utility to approximate each arc length accurate to three decimal places.

54. Verifying a Formula

(a) Given a circular sector with radius L and central angle θ (see figure), show that the area of the sector is given by

$$S = \frac{1}{2}L^2\theta.$$

(b) By joining the straight-line edges of the sector in part (a), a right circular cone is formed (see figure) and the lateral surface area of the cone is the same as the area of the sector. Show that the area is $S = \pi rL$, where r is the radius of the base of the cone. (*Hint:* The arc length of the sector equals the circumference of the base of the cone.)

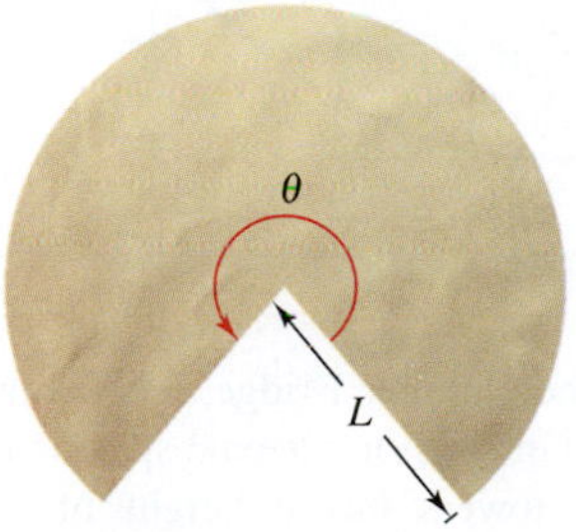

Figure for 54(a)

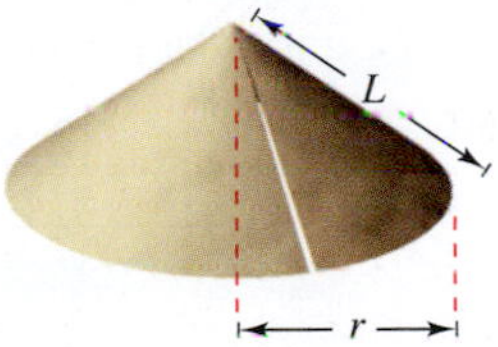

Figure for 54(b)

(c) Use the result of part (b) to verify that the formula for the lateral surface area of the frustum of a cone with slant height L and radii r_1 and r_2 (see figure) is $S = \pi(r_1 + r_2)L$. (*Note:* This formula was used to develop the integral for finding the surface area of a surface of revolution.)

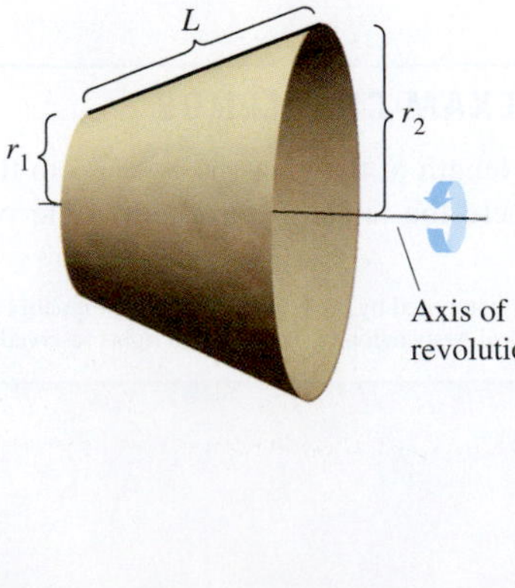

55. Lateral Surface Area of a Cone A right circular cone is generated by revolving the region bounded by $y = 3x/4$, $y = 3$, and $x = 0$ about the y-axis. Find the lateral surface area of the cone.

56. Lateral Surface Area of a Cone A right circular cone is generated by revolving the region bounded by $y = hx/r$, $y = h$, and $x = 0$ about the y-axis. Verify that the lateral surface area of the cone is $S = \pi r\sqrt{r^2 + h^2}$.

57. Using a Sphere Find the area of the zone of a sphere formed by revolving the graph of $y = \sqrt{9 - x^2}$, $0 \le x \le 2$, about the y-axis.

58. Using a Sphere Find the area of the zone of a sphere formed by revolving the graph of $y = \sqrt{r^2 - x^2}$, $0 \le x \le a$, about the y-axis. Assume that $a < r$.

59. Modeling Data The circumference C (in inches) of a vase is measured at three-inch intervals starting at its base. The measurements are shown in the table, where y is the vertical distance in inches from the base.

y	0	3	6	9	12	15	18
C	50	65.5	70	66	58	51	48

(a) Use the data to approximate the volume of the vase by summing the volumes of approximating disks.

(b) Use the data to approximate the outside surface area (excluding the base) of the vase by summing the outside surface areas of approximating frustums of right circular cones.

(c) Use the regression capabilities of a graphing utility to find a cubic model for the points (y, r), where $r = C/(2\pi)$. Use the graphing utility to plot the points and graph the model.

(d) Use the model in part (c) and the integration capabilities of a graphing utility to approximate the volume and outside surface area of the vase. Compare the results with your answers in parts (a) and (b).

60. Modeling Data Property bounded by two perpendicular roads and a stream is shown in the figure. All distances are measured in feet.

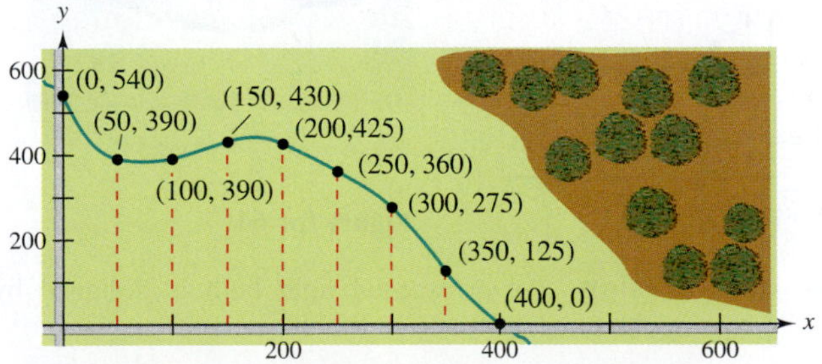

(a) Use the regression capabilities of a graphing utility to fit a fourth-degree polynomial to the path of the stream.

(b) Use the model in part (a) to approximate the area of the property in acres.

(c) Use the integration capabilities of a graphing utility to find the length of the stream that bounds the property.

61. Volume and Surface Area Let R be the region bounded by $y = 1/x$, the x-axis, $x = 1$, and $x = b$, where $b > 1$. Let D be the solid formed when R is revolved about the x-axis.

(a) Find the volume V of D.

(b) Write the surface area S as an integral.

(c) Show that V approaches a finite limit as $b \to \infty$.

(d) Show that $S \to \infty$ as $b \to \infty$.

62. Think About It Consider the equation $\dfrac{x^2}{9} + \dfrac{y^2}{4} = 1$.

(a) Use a graphing utility to graph the equation.

(b) Set up the definite integral for finding the first-quadrant arc length of the graph in part (a).

(c) Compare the interval of integration in part (b) and the domain of the integrand. Is it possible to evaluate the definite integral? Is it possible to use Simpson's Rule to evaluate the definite integral? Explain. (You will learn how to evaluate this type of integral in Section 8.8.)

Approximating Arc Length or Surface Area In Exercises 63–66, set up the definite integral for finding the indicated arc length or surface area. Then use the integration capabilities of a graphing utility to approximate the arc length or surface area. (You will learn how to evaluate this type of integral in Section 8.8.)

63. Length of Pursuit A fleeing object leaves the origin and moves up the y-axis (see figure). At the same time, a pursuer leaves the point $(1, 0)$ and always moves toward the fleeing object. The pursuer's speed is twice that of the fleeing object. The equation of the path is modeled by

$$y = \frac{1}{3}(x^{3/2} - 3x^{1/2} + 2).$$

How far has the fleeing object traveled when it is caught? Show that the pursuer has traveled twice as far.

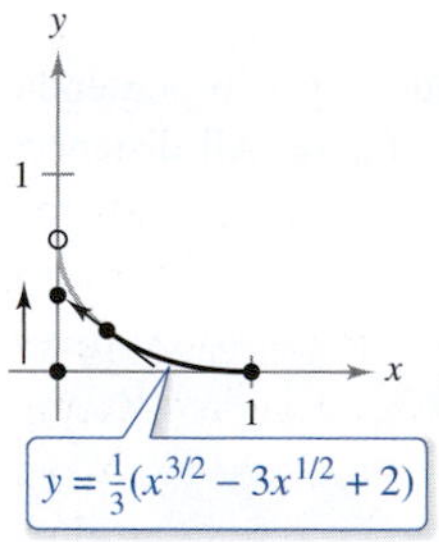

Figure for 63

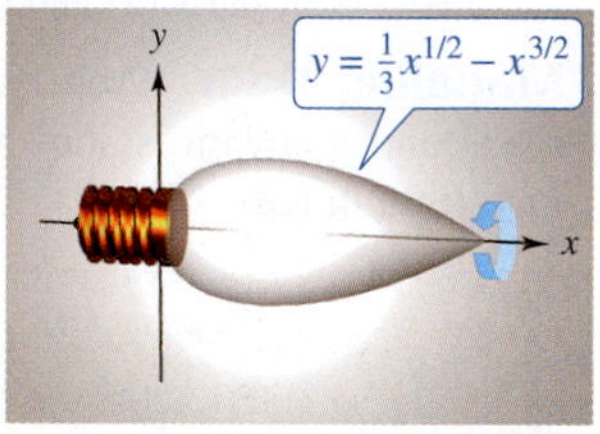

Figure for 64

64. Bulb Design An ornamental light bulb is designed by revolving the graph of

$$y = \frac{1}{3}x^{1/2} - x^{3/2}, \quad 0 \le x \le \frac{1}{3},$$

about the x-axis, where x and y are measured in feet (see figure). Find the surface area of the bulb and use the result to approximate the amount of glass needed to make the bulb. (Assume that the glass is 0.015 inch thick.)

65. Astroid Find the area of the surface formed by revolving the portion in the first quadrant of the graph of $x^{2/3} + y^{2/3} = 4$, $0 \le y \le 8$, about the y-axis.

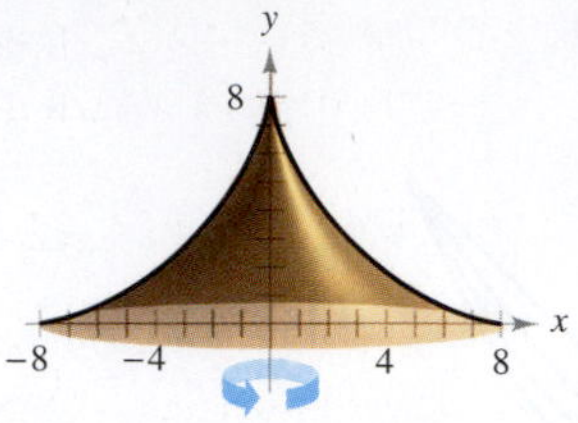

Figure for 65

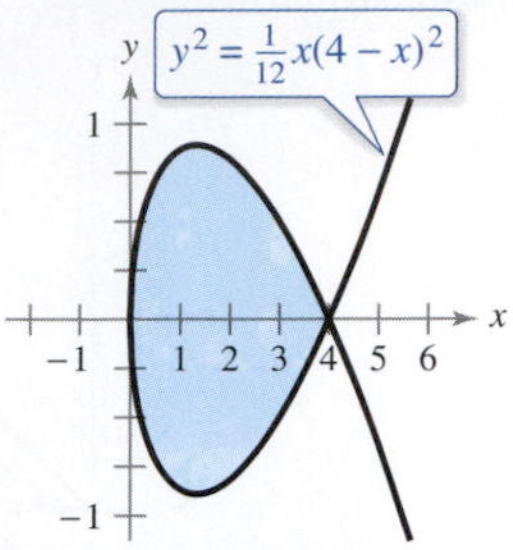

Figure for 66

66. Using a Loop Consider the graph of

$$y^2 = \frac{1}{12}x(4 - x)^2$$

shown in the figure. Find the area of the surface formed when the loop of this graph is revolved about the x-axis.

67. Suspension Bridge A cable for a suspension bridge has the shape of a parabola with equation $y = kx^2$. Let h represent the height of the cable from its lowest point to its highest point and let $2w$ represent the total span of the bridge (see figure). Show that the length C of the cable is given by

$$C = 2\int_0^w \sqrt{1 + (4h^2/w^4)x^2}\, dx.$$

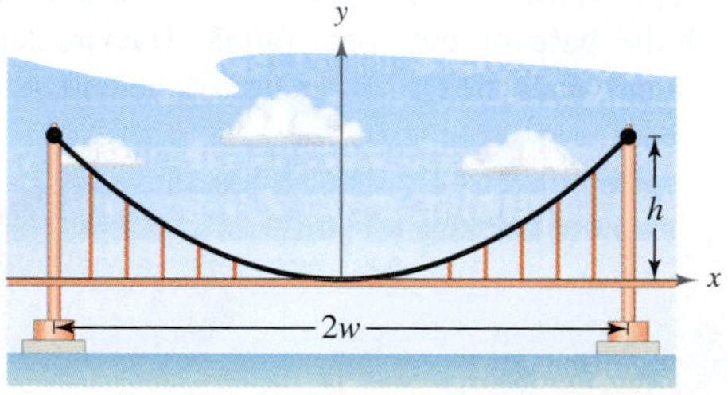

68. Suspension Bridge The Humber Bridge, located in the United Kingdom and opened in 1981, has a main span of about 1400 meters. Each of its towers has a height of about 155 meters. Use these dimensions, the integral in Exercise 67, and the integration capabilities of a graphing utility to approximate the length of a parabolic cable along the main span.

69. Arc Length and Area Let C be the curve given by $f(x) = \cosh x$ for $0 \le x \le t$, where $t > 0$. Show that the arc length of C is equal to the area bounded by C and the x-axis. Identify another curve on the interval $0 \le x \le t$ with this property.

PUTNAM EXAM CHALLENGE

70. Find the length of the curve $y^2 = x^3$ from the origin to the point where the tangent makes an angle of 45° with the x-axis.

7.5 Work

- **Find the work done by a constant force.**
- **Find the work done by a variable force.**

Work Done by a Constant Force

The concept of work is important to scientists and engineers for determining the energy needed to perform various jobs. For instance, it is useful to know the amount of work done when a crane lifts a steel girder, when a spring is compressed, when a rocket is propelled into the air, or when a truck pulls a load along a highway.

In general, **work** is done by a force when it moves an object. If the force applied to the object is *constant,* then the definition of work is as follows.

Definition of Work Done by a Constant Force

If an object is moved a distance D in the direction of an applied constant force F, then the **work** W done by the force is defined as $W = FD$.

There are four fundamental types of forces—gravitational, electromagnetic, strong nuclear, and weak nuclear. A **force** can be thought of as a *push* or a *pull*; a force changes the state of rest or state of motion of a body. For gravitational forces on Earth, it is common to use units of measure corresponding to the weight of an object.

EXAMPLE 1 Lifting an Object

Determine the work done in lifting a 50-pound object 4 feet.

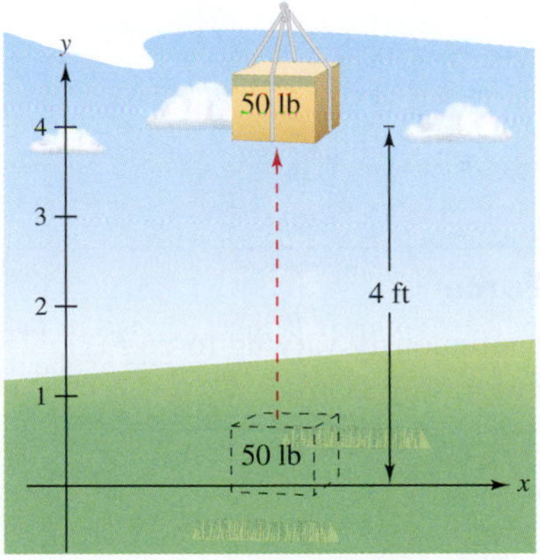

The work done in lifting a 50-pound object 4 feet is 200 foot-pounds.
Figure 7.46

Solution The magnitude of the required force F is the weight of the object, as shown in Figure 7.46. So, the work done in lifting the object 4 feet is

$$W = FD \qquad \text{Work} = (\text{force})(\text{distance})$$
$$= 50(4) \qquad \text{Force} = 50 \text{ pounds, distance} = 4 \text{ feet}$$
$$= 200 \text{ foot-pounds.}$$

In the U.S. measurement system, work is typically expressed in foot-pounds (ft-lb), inch-pounds, or foot-tons. In the International System of Units (SI), the basic unit of force is the **newton**—the force required to produce an acceleration of 1 meter per second per second on a mass of 1 kilogram. In this system, work is typically expressed in newton-meters, also called joules. In another system, the centimeter-gram-second (C-G-S) system, the basic unit of force is the **dyne**—the force required to produce an acceleration of 1 centimeter per second per second on a mass of 1 gram. In this system, work is typically expressed in dyne-centimeters (ergs) or newton-meters (joules).

Exploration

How Much Work? In Example 1, 200 foot-pounds of work was needed to lift the 50-pound object 4 feet vertically off the ground. After lifting the object, you carry it a horizontal distance of 4 feet. Would this require an additional 200 foot-pounds of work? Explain your reasoning.

Work Done by a Variable Force

In Example 1, the force involved was *constant*. When a *variable* force is applied to an object, calculus is needed to determine the work done, because the amount of force changes as the object changes position. For instance, the force required to compress a spring increases as the spring is compressed.

Consider an object that is moved along a straight line from $x = a$ to $x = b$ by a continuously varying force $F(x)$. Let Δ be a partition that divides the interval $[a, b]$ into n subintervals determined by

$$a = x_0 < x_1 < x_2 < \cdot \cdot \cdot < x_n = b$$

and let $\Delta x_i = x_i - x_{i-1}$. For each i, choose c_i such that

$$x_{i-1} \le c_i \le x_i.$$

Then at c_i, the force is $F(c_i)$. Because F is continuous, you can approximate the work done in moving the object through the ith subinterval by the increment

$$\Delta W_i = F(c_i)\,\Delta x_i$$

as shown in Figure 7.47. So, the total work done as the object moves from a to b is approximated by

$$W \approx \sum_{i=1}^{n} \Delta W_i = \sum_{i=1}^{n} F(c_i)\,\Delta x_i.$$

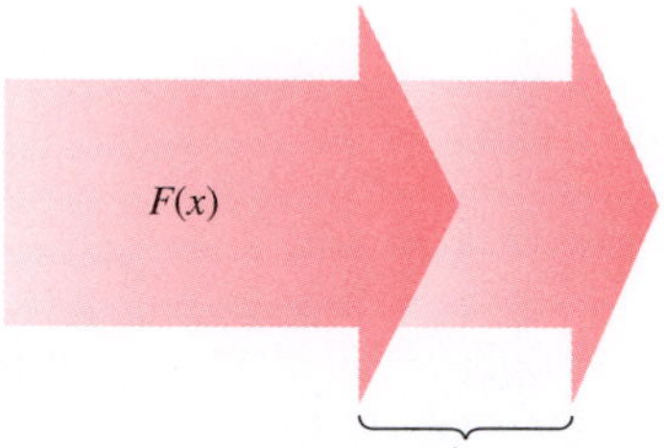

The amount of force changes as an object changes position (Δx).
Figure 7.47

This approximation appears to become better and better as $\|\Delta\| \to 0$ $(n \to \infty)$. So, the work done is

$$W = \lim_{\|\Delta\| \to 0} \sum_{i=1}^{n} F(c_i)\,\Delta x_i = \int_a^b F(x)\,dx.$$

Definition of Work Done by a Variable Force

If an object is moved along a straight line by a continuously varying force $F(x)$, then the **work** W done by the force as the object is moved from

$$x = a \quad \text{to} \quad x = b$$

is given by

$$W = \lim_{\|\Delta\| \to 0} \sum_{i=1}^{n} \Delta W_i = \int_a^b F(x)\,dx.$$

EMILIE DE BRETEUIL (1706–1749)
A major work by Breteuil was the translation of Newton's "Philosophiae Naturalis Principia Mathematica" into French. Her translation and commentary greatly contributed to the acceptance of Newtonian science in Europe.
See LarsonCalculus.com to read more of this biography.

The remaining examples in this section use some well-known physical laws. The discoveries of many of these laws occurred during the same period in which calculus was being developed. In fact, during the seventeenth and eighteenth centuries, there was little difference between physicists and mathematicians. One such physicist-mathematician was Emilie de Breteuil. Breteuil was instrumental in synthesizing the work of many other scientists, including Newton, Leibniz, Huygens, Kepler, and Descartes. Her physics text *Institutions* was widely used for many years.

The three laws of physics listed below were developed by Robert Hooke (1635–1703), Isaac Newton (1642–1727), and Charles Coulomb (1736–1806).

1. **Hooke's Law:** The force F required to compress or stretch a spring (within its elastic limits) is proportional to the distance d that the spring is compressed or stretched from its original length. That is,

$$F = kd$$

where the constant of proportionality k (the spring constant) depends on the specific nature of the spring.

2. **Newton's Law of Universal Gravitation:** The force F of attraction between two particles of masses m_1 and m_2 is proportional to the product of the masses and inversely proportional to the square of the distance d between the two particles. That is,

$$F = G\frac{m_1 m_2}{d^2}.$$

When m_1 and m_2 are in kilograms and d in meters, F will be in newtons for a value of $G = 6.67 \times 10^{-11}$ cubic meter per kilogram-second squared, where G is the **gravitational constant.**

3. **Coulomb's Law:** The force F between two charges q_1 and q_2 in a vacuum is proportional to the product of the charges and inversely proportional to the square of the distance d between the two charges. That is,

$$F = k\frac{q_1 q_2}{d^2}.$$

When q_1 and q_2 are given in electrostatic units and d in centimeters, F will be in dynes for a value of $k = 1$.

EXAMPLE 2 Compressing a Spring

See LarsonCalculus.com for an interactive version of this type of example.

A force of 750 pounds compresses a spring 3 inches from its natural length of 15 inches. Find the work done in compressing the spring an additional 3 inches.

Solution By Hooke's Law, the force $F(x)$ required to compress the spring x units (from its natural length) is $F(x) = kx$. Because $F(3) = 750$, it follows that

$$F(3) = (k)(3) \quad \Longrightarrow \quad 750 = 3k \quad \Longrightarrow \quad 250 = k.$$

So, $F(x) = 250x$, as shown in Figure 7.48. To find the increment of work, assume that the force required to compress the spring over a small increment Δx is nearly constant. So, the increment of work is

$$\Delta W = (\text{force})(\text{distance increment}) = (250x)\,\Delta x.$$

Because the spring is compressed from $x = 3$ to $x = 6$ inches less than its natural length, the work required is

$$W = \int_a^b F(x)\,dx = \int_3^6 250x\,dx = 125x^2\Big]_3^6 = 4500 - 1125 = 3375 \text{ inch-pounds.}$$

Note that you do *not* integrate from $x = 0$ to $x = 6$ because you were asked to determine the work done in compressing the spring an *additional* 3 inches (not including the first 3 inches).

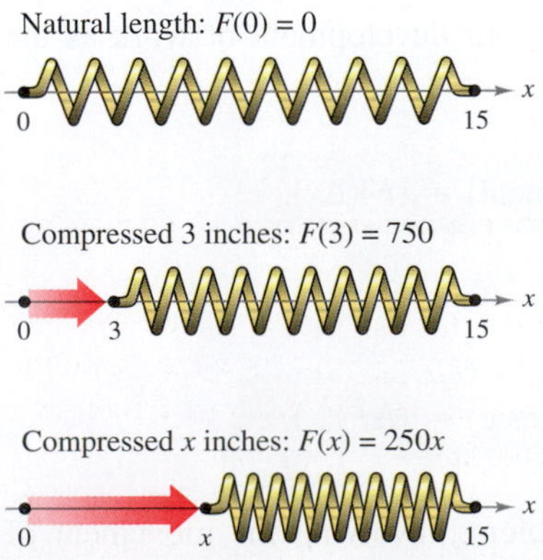

Figure 7.48

In 2011, China launched an 8.5-ton space module. The module will be used to conduct tests as China prepares to build a space station between 2020 and 2022.

EXAMPLE 3 Moving a Space Module into Orbit

A space module weighs 15 metric tons on the surface of Earth. How much work is done in propelling the module to a height of 800 miles above Earth, as shown in Figure 7.49? (Use 4000 miles as the radius of Earth. Do not consider the effect of air resistance or the weight of the propellant.)

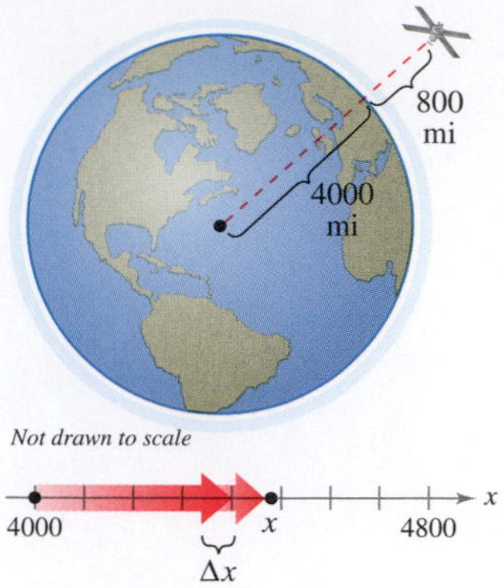

Figure 7.49

Solution Because the weight of a body varies inversely as the square of its distance from the center of Earth, the force $F(x)$ exerted by gravity is

$$F(x) = \frac{C}{x^2}$$

where C is the constant of proportionality. Because the module weighs 15 metric tons on the surface of Earth and the radius of Earth is approximately 4000 miles, you have

$$15 = \frac{C}{(4000)^2} \quad \Longrightarrow \quad 240{,}000{,}000 = C.$$

So, the increment of work is

$$\Delta W = (\text{force})(\text{distance increment}) = \frac{240{,}000{,}000}{x^2}\,\Delta x.$$

Finally, because the module is propelled from $x = 4000$ to $x = 4800$ miles, the total work done is

$$\begin{aligned}
W &= \int_a^b F(x)\,dx && \text{Formula for work}\\
&= \int_{4000}^{4800} \frac{240{,}000{,}000}{x^2}\,dx\\
&= \frac{-240{,}000{,}000}{x}\Bigg]_{4000}^{4800} && \text{Integrate.}\\
&= -50{,}000 + 60{,}000\\
&= 10{,}000 \text{ mile-tons}\\
&\approx 1.164 \times 10^{11} \text{ foot-pounds.}
\end{aligned}$$

In SI units, using a conversion factor of 1 foot-pound ≈ 1.35582 joules, the work done is

$$W \approx 1.578 \times 10^{11} \text{ joules.}$$

The solutions to Examples 2 and 3 conform to our development of work as the summation of increments in the form

$$\Delta W = (\text{force})(\text{distance increment}) = (F)(\Delta x).$$

Another way to formulate the increment of work is

$$\Delta W = (\text{force increment})(\text{distance}) = (\Delta F)(x).$$

This second interpretation of ΔW is useful in problems involving the movement of nonrigid substances such as fluids and chains.

EXAMPLE 4 Emptying a Tank of Oil

A spherical tank of radius 8 feet is half full of oil that weighs 50 pounds per cubic foot. Find the work required to pump oil out through a hole in the top of the tank.

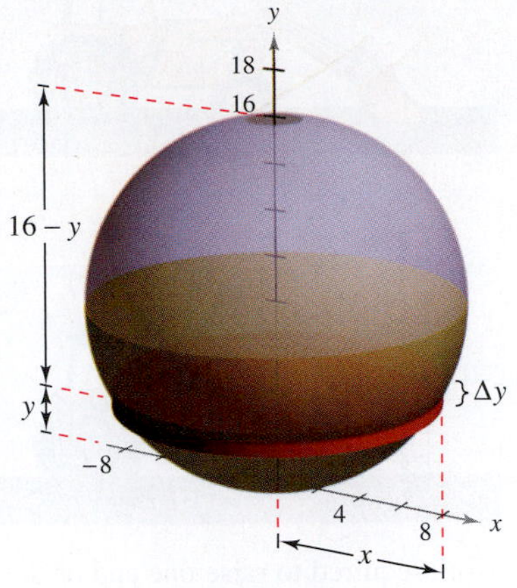

Figure 7.50

Solution Consider the oil to be subdivided into disks of thickness Δy and radius x, as shown in Figure 7.50. Because the increment of force for each disk is given by its weight, you have

$$\begin{aligned}\Delta F &= \text{weight}\\ &= \left(\frac{50 \text{ pounds}}{\text{cubic foot}}\right)(\text{volume})\\ &= 50(\pi x^2 \Delta y) \text{ pounds.}\end{aligned}$$

For a circle of radius 8 and center at (0, 8), you have

$$\begin{aligned}x^2 + (y-8)^2 &= 8^2\\ x^2 &= 16y - y^2\end{aligned}$$

and you can write the force increment as

$$\begin{aligned}\Delta F &= 50(\pi x^2 \Delta y)\\ &= 50\pi(16y - y^2)\,\Delta y.\end{aligned}$$

In Figure 7.50, note that a disk y feet from the bottom of the tank must be moved a distance of $(16 - y)$ feet. So, the increment of work is

$$\begin{aligned}\Delta W &= \Delta F(16 - y)\\ &= 50\pi(16y - y^2)\,\Delta y(16 - y)\\ &= 50\pi(256y - 32y^2 + y^3)\,\Delta y.\end{aligned}$$

Because the tank is half full, y ranges from 0 to 8, and the work required to empty the tank is

$$\begin{aligned}W &= \int_0^8 50\pi(256y - 32y^2 + y^3)\,dy\\ &= 50\pi\left[128y^2 - \frac{32}{3}y^3 + \frac{y^4}{4}\right]_0^8\\ &= 50\pi\left(\frac{11,264}{3}\right)\\ &\approx 589,782 \text{ foot-pounds.}\end{aligned}$$

To estimate the reasonableness of the result in Example 4, consider that the weight of the oil in the tank is

$$\left(\frac{1}{2}\right)(\text{volume})(\text{density}) = \frac{1}{2}\left(\frac{4}{3}\pi 8^3\right)(50) \approx 53,616.5 \text{ pounds}$$

Lifting the entire half-tank of oil 8 feet would involve work of

$$\begin{aligned}W &= FD && \text{Formula for work done by a constant force}\\ &\approx (53,616.5)(8)\\ &= 428,932 \text{ foot-pounds.}\end{aligned}$$

Because the oil is actually lifted between 8 and 16 feet, it seems reasonable that the work done is about 589,782 foot-pounds.

Work required to raise one end of the chain
Figure 7.51

EXAMPLE 5 Lifting a Chain

A 20-foot chain weighing 5 pounds per foot is lying coiled on the ground. How much work is required to raise one end of the chain to a height of 20 feet so that it is fully extended, as shown in Figure 7.51?

Solution Imagine that the chain is divided into small sections, each of length Δy. Then the weight of each section is the increment of force

$$\Delta F = (\text{weight}) = \left(\frac{5 \text{ pounds}}{\text{foot}}\right)(\text{length}) = 5\,\Delta y.$$

Because a typical section (initially on the ground) is raised to a height of y, the increment of work is

$$\Delta W = (\text{force increment})(\text{distance}) = (5\,\Delta y)y = 5y\,\Delta y.$$

Because y ranges from 0 to 20, the total work is

$$W = \int_0^{20} 5y\,dy = \frac{5y^2}{2}\bigg]_0^{20} = \frac{5(400)}{2} = 1000 \text{ foot-pounds.}$$

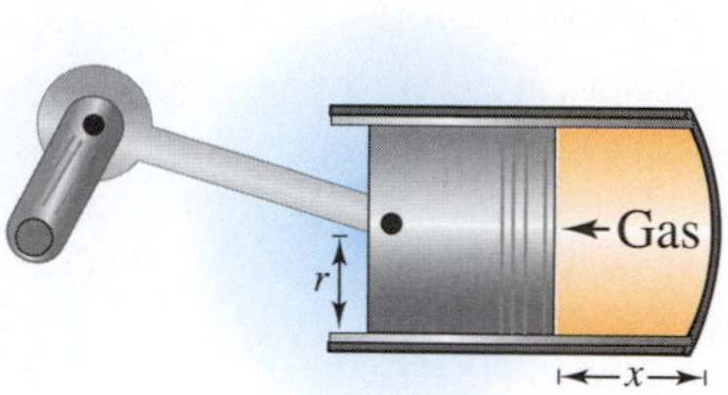

Work done by expanding gas
Figure 7.52

In the next example, you will consider a piston of radius r in a cylindrical casing, as shown in Figure 7.52. As the gas in the cylinder expands, the piston moves, and work is done. If p represents the pressure of the gas (in pounds per square foot) against the piston head and V represents the volume of the gas (in cubic feet), then the work increment involved in moving the piston Δx feet is

$$\Delta W = (\text{force})(\text{distance increment}) = F(\Delta x) = p(\pi r^2)\,\Delta x = p\,\Delta V.$$

So, as the volume of the gas expands from V_0 to V_1, the work done in moving the piston is

$$W = \int_{V_0}^{V_1} p\,dV.$$

Assuming the pressure of the gas to be inversely proportional to its volume, you have $p = k/V$ and the integral for work becomes

$$W = \int_{V_0}^{V_1} \frac{k}{V}\,dV.$$

EXAMPLE 6 Work Done by an Expanding Gas

A quantity of gas with an initial volume of 1 cubic foot and a pressure of 500 pounds per square foot expands to a volume of 2 cubic feet. Find the work done by the gas. (Assume that the pressure is inversely proportional to the volume.)

Solution Because $p = k/V$ and $p = 500$ when $V = 1$, you have $k = 500$. So, the work is

$$\begin{aligned} W &= \int_{V_0}^{V_1} \frac{k}{V}\,dV \\ &= \int_1^2 \frac{500}{V}\,dV \\ &= 500 \ln|V|\bigg]_1^2 \\ &\approx 346.6 \text{ foot-pounds.} \end{aligned}$$

7.5 Exercises

See CalcChat.com for tutorial help and worked-out solutions to odd-numbered exercises.

Constant Force **In Exercises 1–4, determine the work done by the constant force.**

1. A 1200-pound steel beam is lifted 40 feet.
2. An electric hoist lifts a 2500-pound car 6 feet.
3. A force of 112 newtons is required to slide a cement block 8 meters in a construction project.
4. The locomotive of a freight train pulls its cars with a constant force of 9 tons a distance of one-half mile.

Hooke's Law **In Exercises 5–10, use Hooke's Law to determine the variable force in the spring problem.**

5. A force of 5 pounds compresses a 15-inch spring a total of 3 inches. How much work is done in compressing the spring 7 inches?
6. A force of 250 newtons stretches a spring 30 centimeters. How much work is done in stretching the spring from 20 centimeters to 50 centimeters?
7. A force of 20 pounds stretches a spring 9 inches in an exercise machine. Find the work done in stretching the spring 1 foot from its natural position.
8. An overhead garage door has two springs, one on each side of the door. A force of 15 pounds is required to stretch each spring 1 foot. Because of the pulley system, the springs stretch only one-half the distance the door travels. The door moves a total of 8 feet, and the springs are at their natural length when the door is open. Find the work done by the pair of springs.
9. Eighteen foot-pounds of work is required to stretch a spring 4 inches from its natural length. Find the work required to stretch the spring an additional 3 inches.
10. Seven and one-half foot-pounds of work is required to compress a spring 2 inches from its natural length. Find the work required to compress the spring an additional one-half inch.

11. **Propulsion** Neglecting air resistance and the weight of the propellant, determine the work done in propelling a five-ton satellite to a height of (a) 100 miles above Earth and (b) 300 miles above Earth.
12. **Propulsion** Use the information in Exercise 11 to write the work W of the propulsion system as a function of the height h of the satellite above Earth. Find the limit (if it exists) of W as h approaches infinity.
13. **Propulsion** Neglecting air resistance and the weight of the propellant, determine the work done in propelling a 10-ton satellite to a height of (a) 11,000 miles above Earth and (b) 22,000 miles above Earth.
14. **Propulsion** A lunar module weighs 12 tons on the surface of Earth. How much work is done in propelling the module from the surface of the moon to a height of 50 miles? Consider the radius of the moon to be 1100 miles and its force of gravity to be one-sixth that of Earth.
15. **Pumping Water** A rectangular tank with a base 4 feet by 5 feet and a height of 4 feet is full of water (see figure). The water weighs 62.4 pounds per cubic foot. How much work is done in pumping water out over the top edge in order to empty (a) half of the tank and (b) all of the tank?

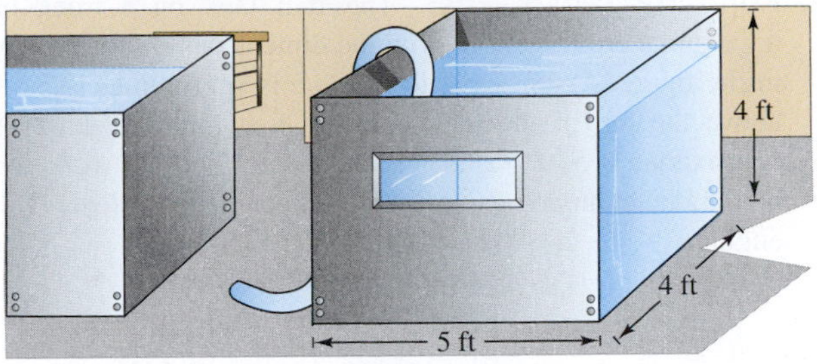

16. **Think About It** Explain why the answer in part (b) of Exercise 15 is not twice the answer in part (a).
17. **Pumping Water** A cylindrical water tank 4 meters high with a radius of 2 meters is buried so that the top of the tank is 1 meter below ground level (see figure). How much work is done in pumping a full tank of water up to ground level? (The water weighs 9800 newtons per cubic meter.)

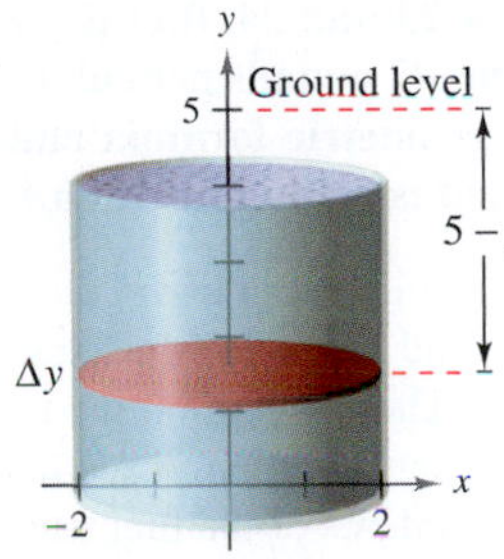

Figure for 17

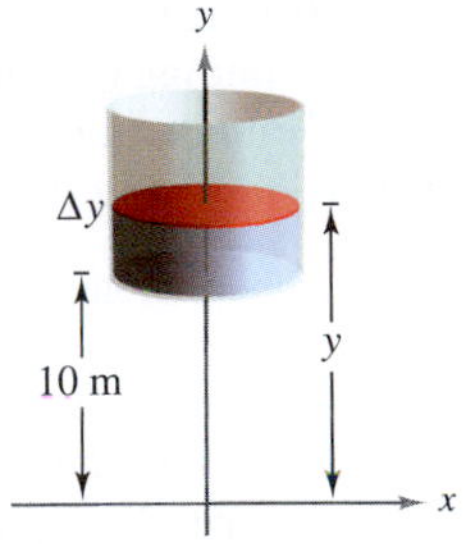

Figure for 18

18. **Pumping Water** Suppose the tank in Exercise 17 is located on a tower so that the bottom of the tank is 10 meters above the level of a stream (see figure). How much work is done in filling the tank half full of water through a hole in the bottom, using water from the stream?
19. **Pumping Water** An open tank has the shape of a right circular cone (see figure). The tank is 8 feet across the top and 6 feet high. How much work is done in emptying the tank by pumping the water over the top edge?

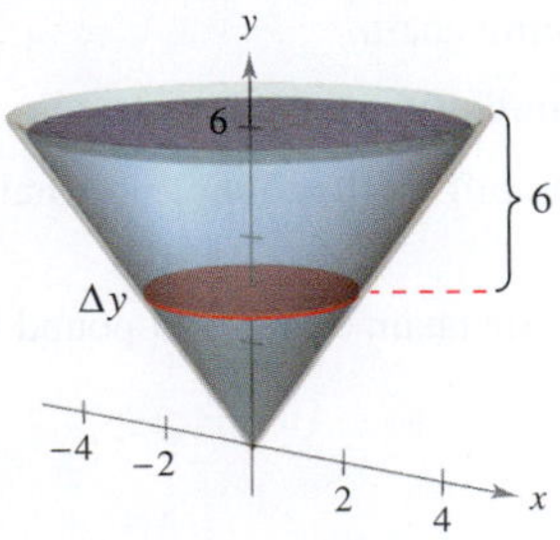

20. Pumping Water Water is pumped in through the bottom of the tank in Exercise 19. How much work is done to fill the tank

(a) to a depth of 2 feet?

(b) from a depth of 4 feet to a depth of 6 feet?

21. Pumping Water A hemispherical tank of radius 6 feet is positioned so that its base is circular. How much work is required to fill the tank with water through a hole in the base when the water source is at the base?

22. Pumping Diesel Fuel The fuel tank on a truck has trapezoidal cross sections with the dimensions (in feet) shown in the figure. Assume that the engine is approximately 3 feet above the top of the fuel tank and that diesel fuel weighs approximately 53.1 pounds per cubic foot. Find the work done by the fuel pump in raising a full tank of fuel to the level of the engine.

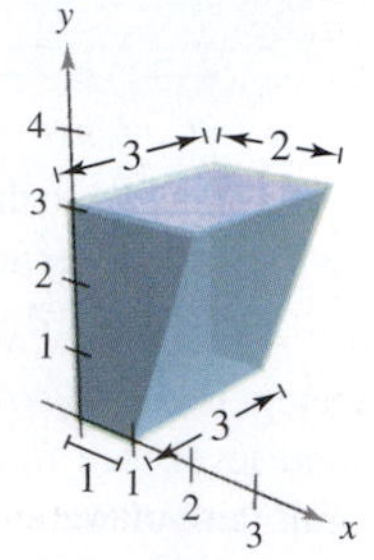

Pumping Gasoline **In Exercises 23 and 24, find the work done in pumping gasoline that weighs 42 pounds per cubic foot. (*Hint:* Evaluate one integral by a geometric formula and the other by observing that the integrand is an odd function.)**

23. A cylindrical gasoline tank 3 feet in diameter and 4 feet long is carried on the back of a truck and is used to fuel tractors. The axis of the tank is horizontal. The opening on the tractor tank is 5 feet above the top of the tank in the truck. Find the work done in pumping the entire contents of the fuel tank into the tractor.

24. The top of a cylindrical gasoline storage tank at a service station is 4 feet below ground level. The axis of the tank is horizontal and its diameter and length are 5 feet and 12 feet, respectively. Find the work done in pumping the entire contents of the full tank to a height of 3 feet above ground level.

Lifting a Chain **In Exercises 25–28, consider a 20-foot chain that weighs 3 pounds per foot hanging from a winch 20 feet above ground level. Find the work done by the winch in winding up the specified amount of chain.**

25. Wind up the entire chain.

26. Wind up one-third of the chain.

27. Run the winch until the bottom of the chain is at the 10-foot level.

28. Wind up the entire chain with a 500-pound load attached to it.

Lifting a Chain **In Exercises 29 and 30, consider a 15-foot hanging chain that weighs 3 pounds per foot. Find the work done in lifting the chain vertically to the indicated position.**

29. Take the bottom of the chain and raise it to the 15-foot level, leaving the chain doubled and still hanging vertically (see figure).

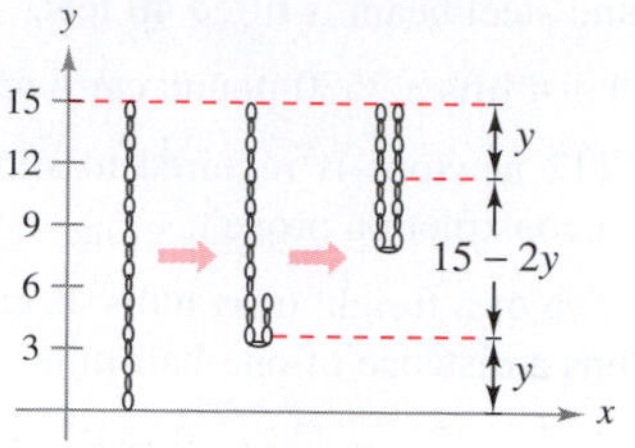

30. Repeat Exercise 29 raising the bottom of the chain to the 12-foot level.

WRITING ABOUT CONCEPTS

31. Work by a Constant Force State the definition of work done by a constant force.

32. Work by a Variable Force State the definition of work done by a variable force.

33. Work Which of the following requires more work? Explain your reasoning.

(a) A 60-pound box of books is lifted 3 feet.

(b) A 60-pound box of books is held 3 feet in the air for 2 minutes.

34. HOW DO YOU SEE IT? The graphs show the force F_i (in pounds) required to move an object 9 feet along the x-axis. Order the force functions from the one that yields the least work to the one that yields the most work without doing any calculations. Explain your reasoning.

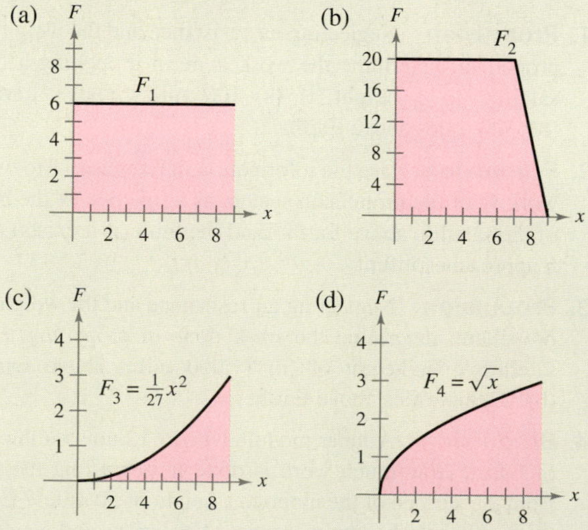

35. Ordering Forces Verify your answer to Exercise 34 by calculating the work for each force function.

36. Electric Force Two electrons repel each other with a force that varies inversely as the square of the distance between them. One electron is fixed at the point $(2, 4)$. Find the work done in moving the second electron from $(-2, 4)$ to $(1, 4)$.

Boyle's Law In Exercises 37 and 38, find the work done by the gas for the given volume and pressure. Assume that the pressure is inversely proportional to the volume. (See Example 6.)

37. A quantity of gas with an initial volume of 2 cubic feet and a pressure of 1000 pounds per square foot expands to a volume of 3 cubic feet.

38. A quantity of gas with an initial volume of 1 cubic foot and a pressure of 2500 pounds per square foot expands to a volume of 3 cubic feet.

Hydraulic Press In Exercises 39–42, use the integration capabilities of a graphing utility to approximate the work done by a press in a manufacturing process. A model for the variable force F (in pounds) and the distance x (in feet) the press moves is given.

	Force	Interval
39.	$F(x) = 1000[1.8 - \ln(x + 1)]$	$0 \le x \le 5$
40.	$F(x) = \dfrac{e^{x^2} - 1}{100}$	$0 \le x \le 4$
41.	$F(x) = 100x\sqrt{125 - x^3}$	$0 \le x \le 5$
42.	$F(x) = 1000 \sinh x$	$0 \le x \le 2$

43. Modeling Data The hydraulic cylinder on a woodsplitter has a 4-inch bore (diameter) and a stroke of 2 feet. The hydraulic pump creates a maximum pressure of 2000 pounds per square inch. Therefore, the maximum force created by the cylinder is $2000(\pi 2^2) = 8000\pi$ pounds.

(a) Find the work done through one extension of the cylinder, given that the maximum force is required.

(b) The force exerted in splitting a piece of wood is variable. Measurements of the force obtained in splitting a piece of wood are shown in the table. The variable x measures the extension of the cylinder in feet, and F is the force in pounds. Use Simpson's Rule to approximate the work done in splitting the piece of wood.

x	0	$\frac{1}{3}$	$\frac{2}{3}$	1	$\frac{4}{3}$	$\frac{5}{3}$	2
$F(x)$	0	20,000	22,000	15,000	10,000	5000	0

(c) Use the regression capabilities of a graphing utility to find a fourth-degree polynomial model for the data. Plot the data and graph the model.

(d) Use the model in part (c) to approximate the extension of the cylinder when the force is maximum.

(e) Use the model in part (c) to approximate the work done in splitting the piece of wood.

SECTION PROJECT

Tidal Energy

Tidal power plants use "tidal energy" to produce electrical energy. To construct a tidal power plant, a dam is built to separate a basin from the sea. Electrical energy is produced as the water flows back and forth between the basin and the sea. The amount of "natural energy" produced depends on the volume of the basin and the tidal range—the vertical distance between high and low tides. (Several natural basins have tidal ranges in excess of 15 feet; the Bay of Fundy in Nova Scotia has a tidal range of 53 feet.)

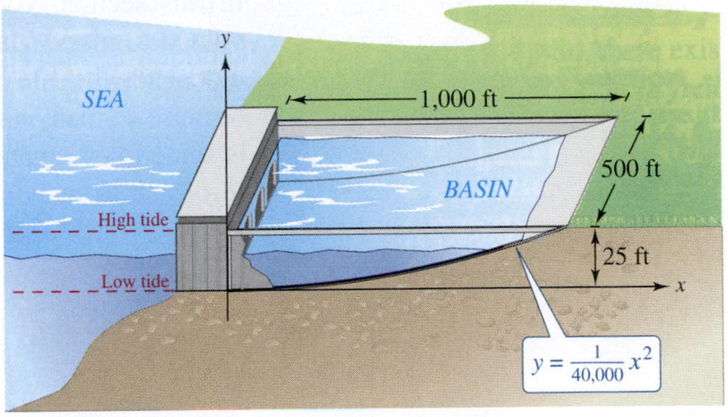

(a) Consider a basin with a rectangular base, as shown in the figure. The basin has a tidal range of 25 feet, with low tide corresponding to $y = 0$. How much water does the basin hold at high tide?

(b) The amount of energy produced during the filling (or the emptying) of the basin is proportional to the amount of work required to fill (or empty) the basin. How much work is required to fill the basin with seawater? (Use a seawater density of 64 pounds per cubic foot.)

The Bay of Fundy in Nova Scotia has an extreme tidal range, as displayed in the greatly contrasting photos above.

FOR FURTHER INFORMATION For more information on tidal power, see the article "LaRance: Six Years of Operating a Tidal Power Plant in France" by J. Cotillon in *Water Power Magazine*.

7.6 Moments, Centers of Mass, and Centroids

- **Understand the definition of mass.**
- **Find the center of mass in a one-dimensional system.**
- **Find the center of mass in a two-dimensional system.**
- **Find the center of mass of a planar lamina.**
- **Use the Theorem of Pappus to find the volume of a solid of revolution.**

Mass

In this section, you will study several important applications of integration that are related to mass. Mass is a measure of a body's resistance to changes in motion, and is independent of the particular gravitational system in which the body is located. However, because so many applications involving mass occur on Earth's surface, an object's mass is sometimes equated with its weight. This is not technically correct. Weight is a type of force and as such is dependent on gravity. Force and mass are related by the equation

$$\text{Force} = (\text{mass})(\text{acceleration}).$$

The table below lists some commonly used measures of mass and force, together with their conversion factors.

System of Measurement	Measure of Mass	Measure of Force
U.S.	Slug	Pound $= (\text{slug})(\text{ft/sec}^2)$
International	Kilogram	Newton $= (\text{kilogram})(\text{m/sec}^2)$
C-G-S	Gram	Dyne $= (\text{gram})(\text{cm/sec}^2)$

Conversions:	
1 pound = 4.448 newtons	1 slug = 14.59 kilograms
1 newton = 0.2248 pound	1 kilogram = 0.06852 slug
1 dyne = 0.000002248 pound	1 gram = 0.00006852 slug
1 dyne = 0.00001 newton	1 foot = 0.3048 meter

EXAMPLE 1 Mass on the Surface of Earth

Find the mass (in slugs) of an object whose weight at sea level is 1 pound.

Solution Use 32 feet per second per second as the acceleration due to gravity.

$$\begin{aligned}\text{Mass} &= \frac{\text{force}}{\text{acceleration}} && \text{Force} = (\text{mass})(\text{acceleration})\\ &= \frac{1\text{ pound}}{32\text{ feet per second per second}}\\ &= 0.03125\,\frac{\text{pound}}{\text{foot per second per second}}\\ &= 0.03125\text{ slug}\end{aligned}$$

Because many applications involving mass occur on Earth's surface, this amount of mass is called a **pound mass.** ■

Center of Mass in a One-Dimensional System

You will now consider two types of moments of a mass—the **moment about a point** and the **moment about a line.** To define these two moments, consider an idealized situation in which a mass m is concentrated at a point. If x is the distance between this point mass and another point P, then the **moment of m about the point P is**

$$\text{Moment} = mx$$

and x is the **length of the moment arm.**

The concept of moment can be demonstrated simply by a seesaw, as shown in Figure 7.53. A child of mass 20 kilograms sits 2 meters to the left of fulcrum P, and an older child of mass 30 kilograms sits 2 meters to the right of P. From experience, you know that the seesaw will begin to rotate clockwise, moving the larger child down. This rotation occurs because the moment produced by the child on the left is less than the moment produced by the child on the right.

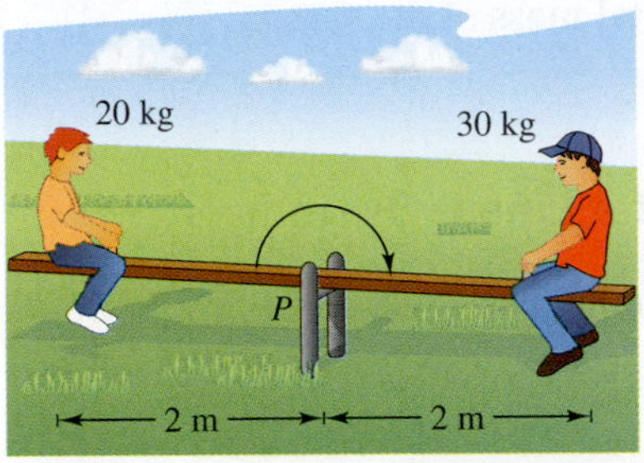

The seesaw will balance when the left and the right moments are equal.
Figure 7.53

$$\text{Left moment} = (20)(2) = 40 \text{ kilogram-meters}$$
$$\text{Right moment} = (30)(2) = 60 \text{ kilogram-meters}$$

To balance the seesaw, the two moments must be equal. For example, if the larger child moved to a position $\frac{4}{3}$ meters from the fulcrum, then the seesaw would balance, because each child would produce a moment of 40 kilogram-meters.

To generalize this, you can introduce a coordinate line on which the origin corresponds to the fulcrum, as shown in Figure 7.54. Several point masses are located on the x-axis. The measure of the tendency of this system to rotate about the origin is the **moment about the origin,** and it is defined as the sum of the n products $m_i x_i$. The moment about the origin is denoted by M_0 and can be written as

$$M_0 = m_1x_1 + m_2x_2 + \cdot \cdot \cdot + m_nx_n.$$

If M_0 is 0, then the system is said to be in **equilibrium.**

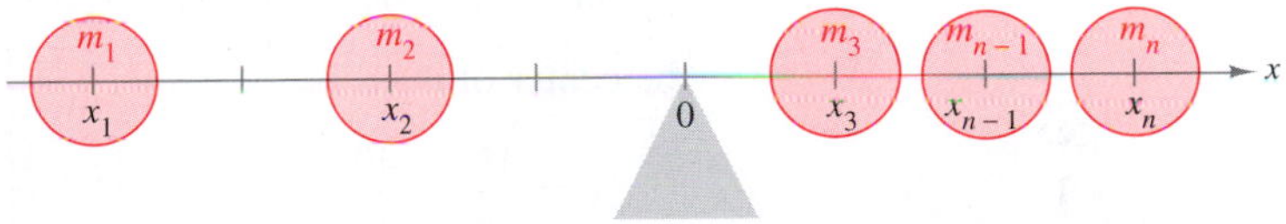

If $m_1x_1 + m_2x_2 + \cdot \cdot \cdot + m_nx_n = 0$, then the system is in equilibrium.
Figure 7.54

For a system that is not in equilibrium, the **center of mass** is defined as the point $\bar{x}$ at which the fulcrum could be relocated to attain equilibrium. If the system were translated $\bar{x}$ units, then each coordinate x_i would become

$$(x_i - \bar{x})$$

and because the moment of the translated system is 0, you have

$$\sum_{i=1}^{n} m_i(x_i - \bar{x}) = \sum_{i=1}^{n} m_ix_i - \sum_{i=1}^{n} m_i\bar{x} = 0.$$

Solving for $\bar{x}$ produces

$$\bar{x} = \frac{\sum_{i=1}^{n} m_ix_i}{\sum_{i=1}^{n} m_i} = \frac{\text{moment of system about origin}}{\text{total mass of system}}.$$

When $m_1x_1 + m_2x_2 + \cdot \cdot \cdot + m_nx_n = 0$, the system is in equilibrium.

Moments and Center of Mass: One-Dimensional System

Let the point masses $m_1, m_2, \ldots, m_n$ be located at $x_1, x_2, \ldots, x_n$.

1. The **moment about the origin** is

$$M_0 = m_1x_1 + m_2x_2 + \cdots + m_nx_n.$$

2. The **center of mass** is

$$\bar{x} = \frac{M_0}{m}$$

where $m = m_1 + m_2 + \cdots + m_n$ is the **total mass** of the system.

EXAMPLE 2 The Center of Mass of a Linear System

Find the center of mass of the linear system shown in Figure 7.55.

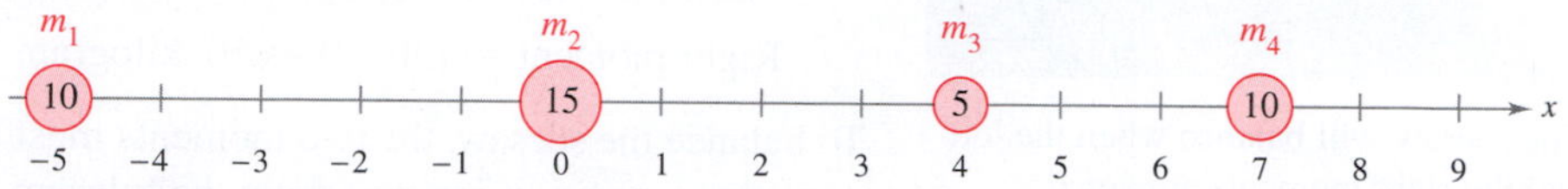

Figure 7.55

Solution The moment about the origin is

$$\begin{aligned} M_0 &= m_1x_1 + m_2x_2 + m_3x_3 + m_4x_4 \\ &= 10(-5) + 15(0) + 5(4) + 10(7) \\ &= -50 + 0 + 20 + 70 \\ &= 40. \end{aligned}$$

Because the total mass of the system is

$$m = 10 + 15 + 5 + 10 = 40$$

the center of mass is

$$\bar{x} = \frac{M_0}{m} = \frac{40}{40} = 1.$$

Note that the point masses will be in equilibrium when the fulcrum is located at $x = 1$. ■

Rather than define the moment of a mass, you could define the moment of a *force*. In this context, the center of mass is called the **center of gravity.** Consider a system of point masses $m_1, m_2, \ldots, m_n$ that is located at $x_1, x_2, \ldots, x_n$. Then, because

$$\text{force} = (\text{mass})(\text{acceleration})$$

the total force of the system is

$$F = m_1a + m_2a + \cdots + m_na = ma.$$

The **torque** (moment) about the origin is

$$T_0 = (m_1a)x_1 + (m_2a)x_2 + \cdots + (m_na)x_n = M_0a$$

and the **center of gravity** is

$$\frac{T_0}{F} = \frac{M_0a}{ma} = \frac{M_0}{m} = \bar{x}.$$

So, the center of gravity and the center of mass have the same location.

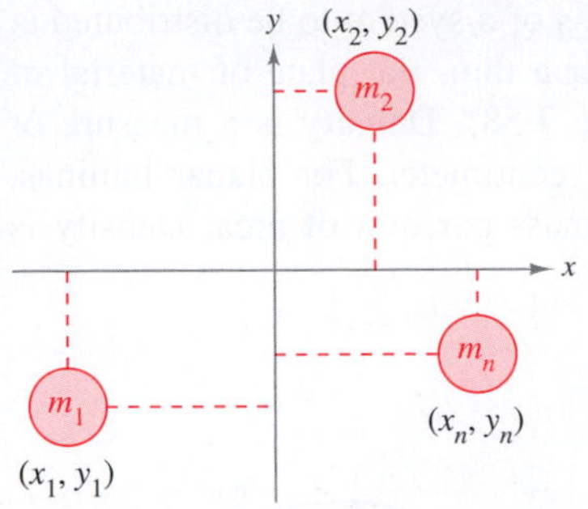

In a two-dimensional system, there is a moment about the y-axis M_y and a moment about the x-axis M_x.
Figure 7.56

Center of Mass in a Two-Dimensional System

You can extend the concept of moment to two dimensions by considering a system of masses located in the xy-plane at the points $(x_1, y_1), (x_2, y_2), . . . , (x_n, y_n)$, as shown in Figure 7.56. Rather than defining a single moment (with respect to the origin), two moments are defined—one with respect to the x-axis and one with respect to the y-axis.

> **Moment and Center of Mass: Two-Dimensional System**
>
> Let the point masses $m_1, m_2, . . . , m_n$ be located at $(x_1, y_1), (x_2, y_2), . . . , (x_n, y_n)$.
>
> 1. The **moment about the *y*-axis** is
>
> $$M_y = m_1x_1 + m_2x_2 + . . . m_nx_n.$$
>
> 2. The **moment about the *x*-axis** is
>
> $$M_x = m_1y_1 + m_2y_2 + . . . m_ny_n.$$
>
> 3. The **center of mass** $(\bar{x}, \bar{y})$ (or **center of gravity**) is
>
> $$\bar{x} = \frac{M_y}{m} \quad \text{and} \quad \bar{y} = \frac{M_x}{m}$$
>
> where
>
> $$m = m_1 + m_2 + . . . + m_n$$
>
> is the **total mass** of the system.

The moment of a system of masses in the plane can be taken about any horizontal or vertical line. In general, the moment about a line is the sum of the product of the masses and the *directed distances* from the points to the line.

$$\text{Moment} = m_1(y_1 - b) + m_2(y_2 - b) + \cdot \cdot \cdot + m_n(y_n - b) \qquad \text{Horizontal line } y = b$$
$$\text{Moment} = m_1(x_1 - a) + m_2(x_2 - a) + \cdot \cdot \cdot + m_n(x_n - a) \qquad \text{Vertical line } x = a$$

EXAMPLE 3 The Center of Mass of a Two-Dimensional System

Find the center of mass of a system of point masses $m_1 = 6, m_2 = 3, m_3 = 2$, and $m_4 = 9$, located at

$$(3, -2), \quad (0, 0), \quad (-5, 3), \quad \text{and} \quad (4, 2)$$

as shown in Figure 7.57.

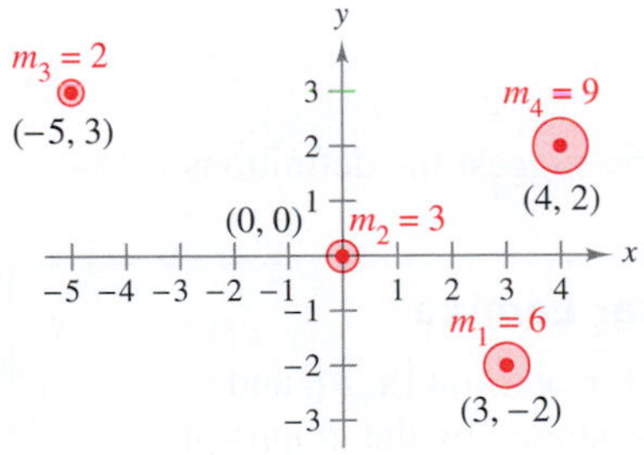

Figure 7.57

Solution

$$m = 6 + 3 + 2 + 9 = 20 \qquad \text{Mass}$$
$$M_y = 6(3) + 3(0) + 2(-5) + 9(4) = 44 \qquad \text{Moment about } y\text{-axis}$$
$$M_x = 6(-2) + 3(0) + 2(3) + 9(2) = 12 \qquad \text{Moment about } x\text{-axis}$$

So,

$$\bar{x} = \frac{M_y}{m} = \frac{44}{20} = \frac{11}{5}$$

and

$$\bar{y} = \frac{M_x}{m} = \frac{12}{20} = \frac{3}{5}.$$

The center of mass is $\left(\frac{11}{5}, \frac{3}{5}\right)$. ■

Center of Mass of a Planar Lamina

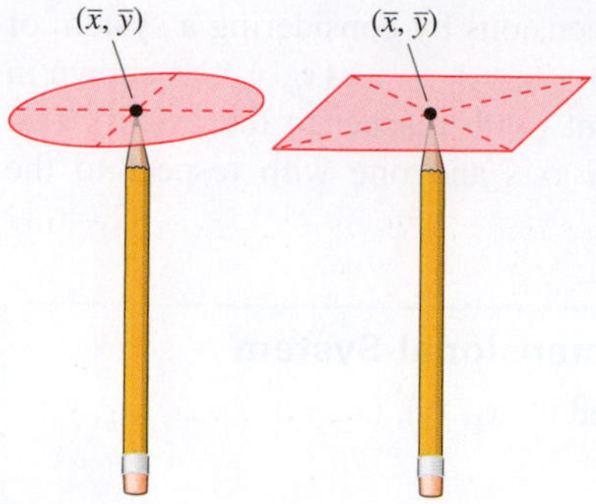

You can think of the center of mass $(\bar{x}, \bar{y})$ of a lamina as its balancing point. For a circular lamina, the center of mass is the center of the circle. For a rectangular lamina, the center of mass is the center of the rectangle.
Figure 7.58

So far in this section, you have assumed the total mass of a system to be distributed at discrete points in a plane or on a line. Now consider a thin, flat plate of material of constant density called a **planar lamina** (see Figure 7.58). **Density** is a measure of mass per unit of volume, such as grams per cubic centimeter. For planar laminas, however, density is considered to be a measure of mass per unit of area. Density is denoted by ρ, the lowercase Greek letter rho.

Consider an irregularly shaped planar lamina of uniform density ρ, bounded by the graphs of $y = f(x)$, $y = g(x)$, and $a \le x \le b$, as shown in Figure 7.59. The mass of this region is

$$\begin{aligned} m &= (\text{density})(\text{area}) \\ &= \rho \int_a^b [f(x) - g(x)]\, dx \\ &= \rho A \end{aligned}$$

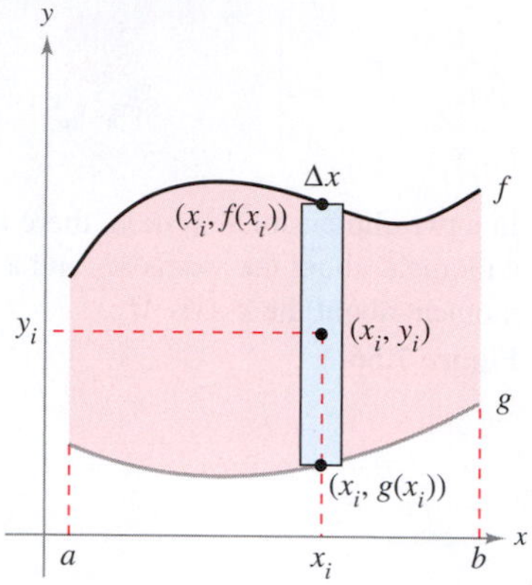

Planar lamina of uniform density ρ
Figure 7.59

where A is the area of the region. To find the center of mass of this lamina, partition the interval $[a, b]$ into n subintervals of equal width Δx. Let x_i be the center of the ith subinterval. You can approximate the portion of the lamina lying in the ith subinterval by a rectangle whose height is $h = f(x_i) - g(x_i)$. Because the density of the rectangle is ρ, its mass is

$$m_i = (\text{density})(\text{area}) = \underbrace{\rho}_{\text{Density}}\underbrace{[f(x_i) - g(x_i)]}_{\text{Height}}\underbrace{\Delta x}_{\text{Width}}.$$

Now, considering this mass to be located at the center (x_i, y_i) of the rectangle, the directed distance from the x-axis to (x_i, y_i) is $y_i = [f(x_i) + g(x_i)]/2$. So, the moment of m_i about the x-axis is

$$\begin{aligned} \text{Moment} &= (\text{mass})(\text{distance}) \\ &= m_i y_i \\ &= \rho[f(x_i) - g(x_i)]\,\Delta x\left[\frac{f(x_i) + g(x_i)}{2}\right]. \end{aligned}$$

Summing the moments and taking the limit as $n \to \infty$ suggest the definitions below.

Moments and Center of Mass of a Planar Lamina

Let f and g be continuous functions such that $f(x) \ge g(x)$ on $[a, b]$, and consider the planar lamina of uniform density ρ bounded by the graphs of $y = f(x)$, $y = g(x)$, and $a \le x \le b$.

1. The **moments about the *x*- and *y*-axes** are

$$M_x = \rho \int_a^b \left[\frac{f(x) + g(x)}{2}\right][f(x) - g(x)]\, dx$$

$$M_y = \rho \int_a^b x[f(x) - g(x)]\, dx.$$

2. The **center of mass** $(\bar{x}, \bar{y})$ is given by $\bar{x} = \dfrac{M_y}{m}$ and $\bar{y} = \dfrac{M_x}{m}$, where $m = \rho \int_a^b [f(x) - g(x)]\, dx$ is the mass of the lamina.

EXAMPLE 4 The Center of Mass of a Planar Lamina

See LarsonCalculus.com for an interactive version of this type of example.

Find the center of mass of the lamina of uniform density ρ bounded by the graph of $f(x) = 4 - x^2$ and the x-axis.

Solution Because the center of mass lies on the axis of symmetry, you know that $\bar{x} = 0$. Moreover, the mass of the lamina is

$$m = \rho \int_{-2}^{2} (4 - x^2)\, dx$$
$$= \rho\left[4x - \frac{x^3}{3}\right]_{-2}^{2}$$
$$= \frac{32\rho}{3}.$$

To find the moment about the x-axis, place a representative rectangle in the region, as shown in the figure at the right. The distance from the x-axis to the center of this rectangle is

$$y_i = \frac{f(x)}{2} = \frac{4 - x^2}{2}.$$

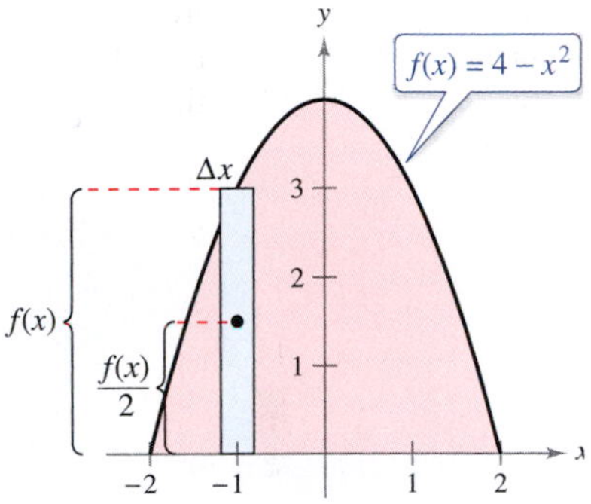

Because the mass of the representative rectangle is

$$\rho f(x)\,\Delta x = \rho(4 - x^2)\,\Delta x$$

you have

$$M_x = \rho \int_{-2}^{2} \frac{4 - x^2}{2}(4 - x^2)\, dx$$
$$= \frac{\rho}{2}\int_{-2}^{2} (16 - 8x^2 + x^4)\, dx$$
$$= \frac{\rho}{2}\left[16x - \frac{8x^3}{3} + \frac{x^5}{5}\right]_{-2}^{2}$$
$$= \frac{256\rho}{15}$$

and $\bar{y}$ is

$$\bar{y} = \frac{M_x}{m} = \frac{256\rho/15}{32\rho/3} = \frac{8}{5}.$$

So, the center of mass (the balancing point) of the lamina is $\left(0, \frac{8}{5}\right)$, as shown in Figure 7.60.

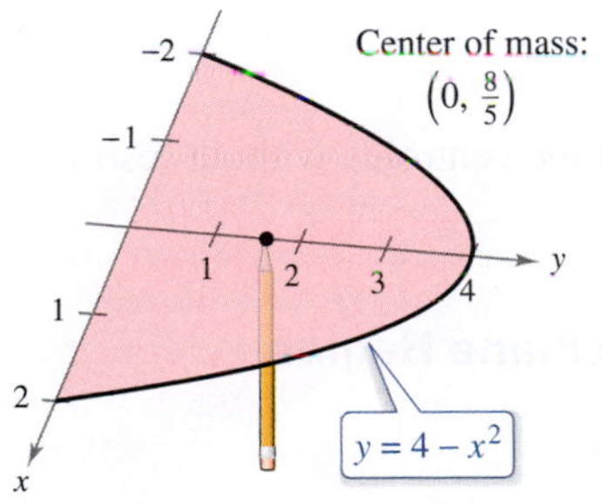

The center of mass is the balancing point.
Figure 7.60

The density ρ in Example 4 is a common factor of both the moments and the mass, and as such divides out of the quotients representing the coordinates of the center of mass. So, the center of mass of a lamina of *uniform* density depends only on the shape of the lamina and not on its density. For this reason, the point

$(\bar{x}, \bar{y})$ Center of mass or centroid

is sometimes called the center of mass of a *region* in the plane, or the **centroid** of the region. In other words, to find the centroid of a region in the plane, you simply assume that the region has a constant density of $\rho = 1$ and compute the corresponding center of mass.

EXAMPLE 5 The Centroid of a Plane Region

Find the centroid of the region bounded by the graphs of $f(x) = 4 - x^2$ and $g(x) = x + 2$.

Solution The two graphs intersect at the points $(-2, 0)$ and $(1, 3)$, as shown in Figure 7.61. So, the area of the region is

$$A = \int_{-2}^{1} [f(x) - g(x)]\, dx = \int_{-2}^{1} (2 - x - x^2)\, dx = \frac{9}{2}.$$

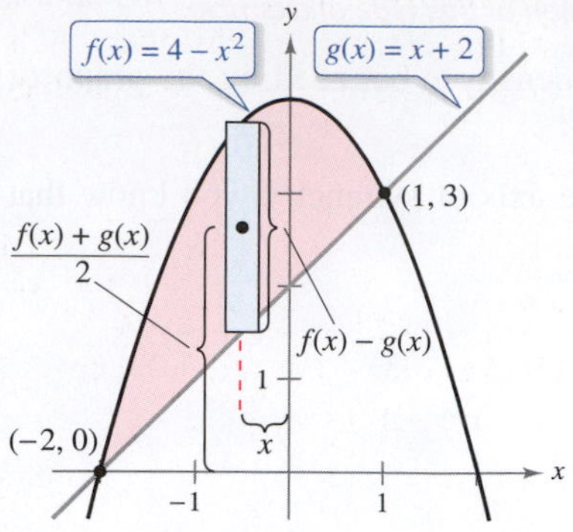

Figure 7.61

The centroid $(\bar{x}, \bar{y})$ of the region has the following coordinates.

$$\begin{aligned}
\bar{x} &= \frac{1}{A}\int_{-2}^{1} x[(4 - x^2) - (x + 2)]\, dx \\
&= \frac{2}{9}\int_{-2}^{1} (-x^3 - x^2 + 2x)\, dx \\
&= \frac{2}{9}\left[-\frac{x^4}{4} - \frac{x^3}{3} + x^2\right]_{-2}^{1} \\
&= -\frac{1}{2}
\end{aligned}$$

$$\begin{aligned}
\bar{y} &= \frac{1}{A}\int_{-2}^{1} \left[\frac{(4 - x^2) + (x + 2)}{2}\right][(4 - x^2) - (x + 2)]\, dx \\
&= \frac{2}{9}\left(\frac{1}{2}\right)\int_{-2}^{1} (-x^2 + x + 6)(-x^2 - x + 2)\, dx \\
&= \frac{1}{9}\int_{-2}^{1} (x^4 - 9x^2 - 4x + 12)\, dx \\
&= \frac{1}{9}\left[\frac{x^5}{5} - 3x^3 - 2x^2 + 12x\right]_{-2}^{1} \\
&= \frac{12}{5}
\end{aligned}$$

So, the centroid of the region is $(\bar{x}, \bar{y}) = \left(-\frac{1}{2}, \frac{12}{5}\right)$.

For simple plane regions, you may be able to find the centroids without resorting to integration.

EXAMPLE 6 The Centroid of a Simple Plane Region

Find the centroid of the region shown in Figure 7.62(a).

Solution By superimposing a coordinate system on the region, as shown in Figure 7.62(b), you can locate the centroids of the three rectangles at

$$\left(\frac{1}{2}, \frac{3}{2}\right), \quad \left(\frac{5}{2}, \frac{1}{2}\right), \quad \text{and} \quad (5, 1).$$

Using these three points, you can find the centroid of the region.

$$A = \text{area of region} = 3 + 3 + 4 = 10$$

$$\bar{x} = \frac{(1/2)(3) + (5/2)(3) + (5)(4)}{10} = \frac{29}{10} = 2.9$$

$$\bar{y} = \frac{(3/2)(3) + (1/2)(3) + (1)(4)}{10} = \frac{10}{10} = 1$$

So, the centroid of the region is (2.9, 1). Notice that (2.9, 1) is not the "average" of $\left(\frac{1}{2}, \frac{3}{2}\right)$, $\left(\frac{5}{2}, \frac{1}{2}\right)$, and (5, 1).

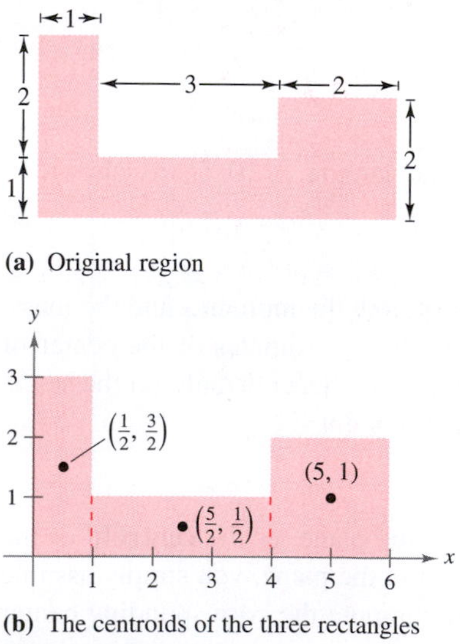

Figure 7.62

Theorem of Pappus

The final topic in this section is a useful theorem credited to Pappus of Alexandria (ca. 300 A.D.), a Greek mathematician whose eight-volume *Mathematical Collection* is a record of much of classical Greek mathematics. You are asked to prove this theorem in Section 14.4.

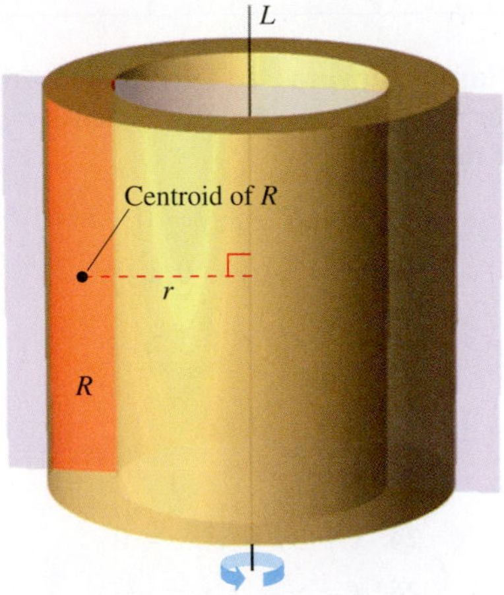

The volume V is $2\pi rA$, where A is the area of region R.
Figure 7.63

> **THEOREM 7.1 The Theorem of Pappus**
>
> Let R be a region in a plane and let L be a line in the same plane such that L does not intersect the interior of R, as shown in Figure 7.63. If r is the distance between the centroid of R and the line, then the volume V of the solid of revolution formed by revolving R about the line is
>
> $$V = 2\pi rA$$
>
> where A is the area of R. (Note that $2\pi r$ is the distance traveled by the centroid as the region is revolved about the line.)

The Theorem of Pappus can be used to find the volume of a torus, as shown in the next example. Recall that a torus is a doughnut-shaped solid formed by revolving a circular region about a line that lies in the same plane as the circle (but does not intersect the circle).

EXAMPLE 7 Finding Volume by the Theorem of Pappus

Find the volume of the torus shown in Figure 7.64(a), which was formed by revolving the circular region bounded by

$$(x - 2)^2 + y^2 = 1$$

about the y-axis, as shown in Figure 7.64(b).

(a)

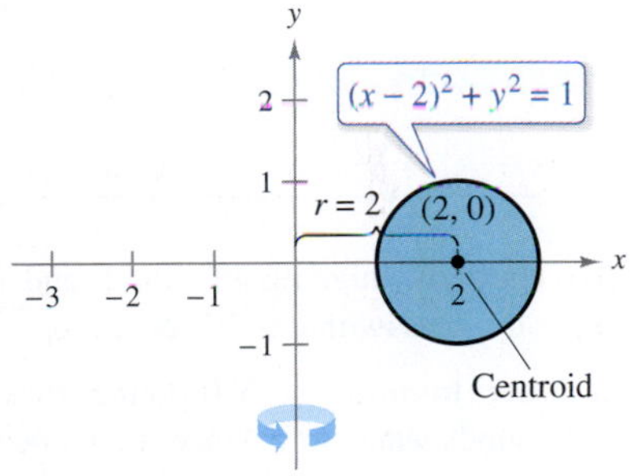

(b)

Figure 7.64

Solution In Figure 7.67(b), you can see that the centroid of the circular region is (2, 0). So, the distance between the centroid and the axis of revolution is

$$r = 2.$$

Because the area of the circular region is $A = \pi$, the volume of the torus is

$$\begin{aligned} V &= 2\pi rA \\ &= 2\pi(2)(\pi) \\ &= 4\pi^2 \\ &\approx 39.5. \end{aligned}$$

> **Exploration**
>
> Use the shell method to show that the volume of the torus in Example 7 is
>
> $$V = \int_1^3 4\pi x\sqrt{1 - (x - 2)^2}\, dx.$$
>
> Evaluate this integral using a graphing utility. Does your answer agree with the one in Example 7?

7.6 Exercises

See CalcChat.com for tutorial help and worked-out solutions to odd-numbered exercises.

Center of Mass of a Linear System **In Exercises 1–4, find the center of mass of the point masses lying on the x-axis.**

1. $m_1 = 7, m_2 = 3, m_3 = 5$
   $x_1 = -5, x_2 = 0, x_3 = 3$
2. $m_1 = 7, m_2 = 4, m_3 = 3, m_4 = 8$
   $x_1 = -3, x_2 = -2, x_3 = 5, x_4 = 4$
3. $m_1 = 1, m_2 = 3, m_3 = 2, m_4 = 9, m_5 = 5$
   $x_1 = 6, x_2 = 10, x_3 = 3, x_4 = 2, x_5 = 4$
4. $m_1 = 8, m_2 = 5, m_3 = 5, m_4 = 12, m_5 = 2$
   $x_1 = -2, x_2 = 6, x_3 = 0, x_4 = 3, x_5 = -5$
5. **Graphical Reasoning**
   (a) Translate each point mass in Exercise 3 to the right four units and determine the resulting center of mass.
   (b) Translate each point mass in Exercise 4 to the left two units and determine the resulting center of mass.
6. **Conjecture** Use the result of Exercise 5 to make a conjecture about the change in the center of mass that results when each point mass is translated k units horizontally.

Statics Problems **In Exercises 7 and 8, consider a beam of length L with a fulcrum x feet from one end (see figure). There are objects with weights W_1 and W_2 placed on opposite ends of the beam. Find x such that the system is in equilibrium.**

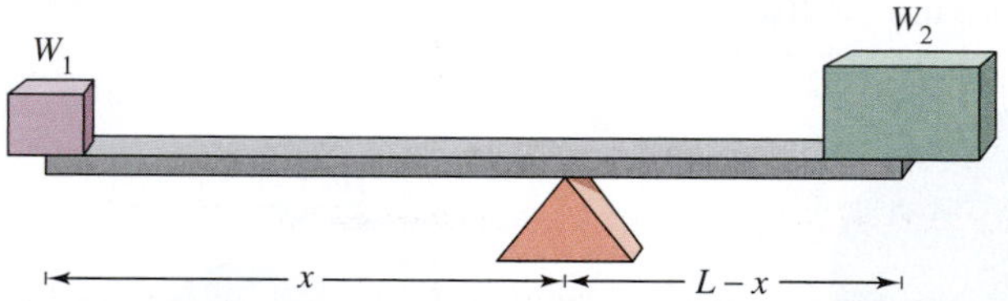

7. Two children weighing 48 pounds and 72 pounds are going to play on a seesaw that is 10 feet long.
8. In order to move a 600-pound rock, a person weighing 200 pounds wants to balance it on a beam that is 5 feet long.

Center of Mass of a Two-Dimensional System **In Exercises 9–12, find the center of mass of the given system of point masses.**

9.

m_i	5	1	3
(x_i, y_i)	(2, 2)	(−3, 1)	(1, −4)

10.

m_i	10	2	5
(x_i, y_i)	(1, −1)	(5, 5)	(−4, 0)

11.

m_i	12	6	4.5	15
(x_i, y_i)	(2, 3)	(−1, 5)	(6, 8)	(2, −2)

12.

m_i	3	4	2	1	6
(x_i, y_i)	(−2, −3)	(5, 5)	(7, 1)	(0, 0)	(−3, 0)

Center of Mass of a Planar Lamina **In Exercises 13–26, find M_x, M_y, and $(\bar{x}, \bar{y})$ for the laminas of uniform density ρ bounded by the graphs of the equations.**

13. $y = \frac{1}{2}x, y = 0, x = 2$
14. $y = 6 - x, y = 0, x = 0$
15. $y = \sqrt{x}, y = 0, x = 4$
16. $y = \frac{1}{2}x^2, y = 0, x = 2$
17. $y = x^2, y = x^3$
18. $y = \sqrt{x}, y = \frac{1}{2}x$
19. $y = -x^2 + 4x + 2, y = x + 2$
20. $y = \sqrt{x} + 1, y = \frac{1}{3}x + 1$
21. $y = x^{2/3}, y = 0, x = 8$
22. $y = x^{2/3}, y = 4$
23. $x = 4 - y^2, x = 0$
24. $x = 3y - y^2, x = 0$
25. $x = -y, x = 2y - y^2$
26. $x = y + 2, x = y^2$

Approximating a Centroid **In Exercises 27–30, use a graphing utility to graph the region bounded by the graphs of the equations. Use the integration capabilities of the graphing utility to approximate the centroid of the region.**

27. $y = 10x\sqrt{125 - x^3}, y = 0$
28. $y = xe^{-x/2}, y = 0, x = 0, x = 4$
29. **Prefabricated End Section of a Building**
    $y = 5\sqrt[3]{400 - x^2}, y = 0$
30. **Witch of Agnesi**
    $y = \dfrac{8}{x^2 + 4}, y = 0, x = -2, x = 2$

Finding the Center of Mass **In Exercises 31–34, introduce an appropriate coordinate system and find the coordinates of the center of mass of the planar lamina. (The answer depends on the position of the coordinate system.)**

31.
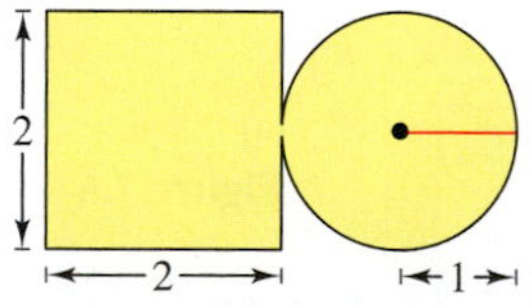

32.
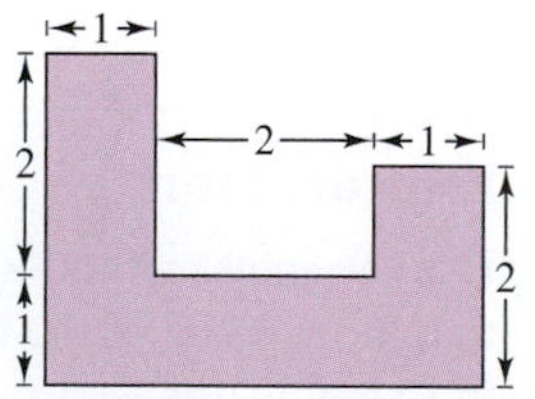

33.
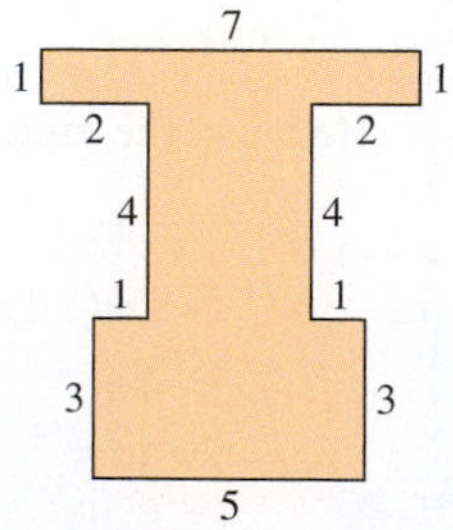

34.
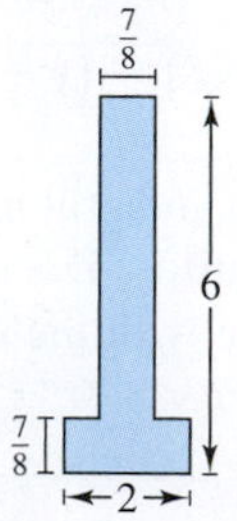

35. Finding the Center of Mass Find the center of mass of the lamina in Exercise 31 when the circular portion of the lamina has twice the density of the square portion of the lamina.

36. Finding the Center of Mass Find the center of mass of the lamina in Exercise 31 when the square portion of the lamina has twice the density of the circular portion of the lamina.

Finding Volume by the Theorem of Pappus In Exercises 37–40, use the Theorem of Pappus to find the volume of the solid of revolution.

37. The torus formed by revolving the circle

$$(x - 5)^2 + y^2 = 16$$

about the y-axis

38. The torus formed by revolving the circle

$$x^2 + (y - 3)^2 = 4$$

about the x-axis

39. The solid formed by revolving the region bounded by the graphs of $y = x$, $y = 4$, and $x = 0$ about the x-axis

40. The solid formed by revolving the region bounded by the graphs of $y = 2\sqrt{x - 2}$, $y = 0$, and $x = 6$ about the y-axis

WRITING ABOUT CONCEPTS

41. Center of Mass Let the point masses $m_1, m_2, \ldots, m_n$ be located at $(x_1, y_1), (x_2, y_2), \ldots, (x_n, y_n)$. Define the center of mass $(\bar{x}, \bar{y})$.

42. Planar Lamina What is a planar lamina? Describe what is meant by the center of mass $(\bar{x}, \bar{y})$ of a planar lamina.

43. Theorem of Pappus State the Theorem of Pappus.

44. HOW DO YOU SEE IT? The centroid of the plane region bounded by the graphs of $y = f(x)$, $y = 0$, $x = 0$, and $x = 3$ is $(1.2, 1.4)$. Is it possible to find the centroid of each of the regions bounded by the graphs of the following sets of equations? If so, identify the centroid and explain your answer.

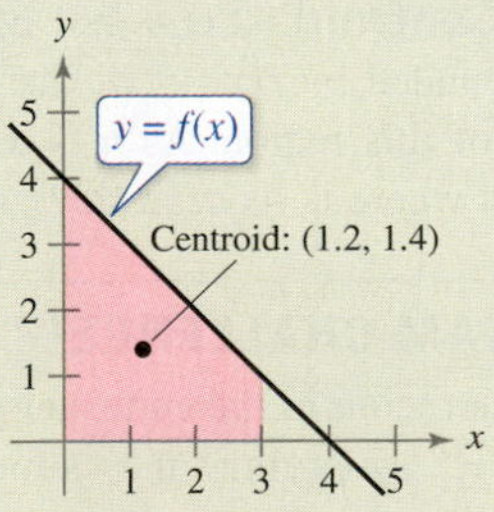

(a) $y = f(x) + 2$, $y = 2$, $x = 0$, and $x = 3$
(b) $y = f(x - 2)$, $y = 0$, $x = 2$, and $x = 5$
(c) $y = -f(x)$, $y = 0$, $x = 0$, and $x = 3$
(d) $y = f(x)$, $y = 0$, $x = 2$, and $x = 4$

Centroid of a Common Region In Exercises 45–50, find and/or verify the centroid of the common region used in engineering.

45. Triangle Show that the centroid of the triangle with vertices $(-a, 0)$, $(a, 0)$, and (b, c) is the point of intersection of the medians (see figure).

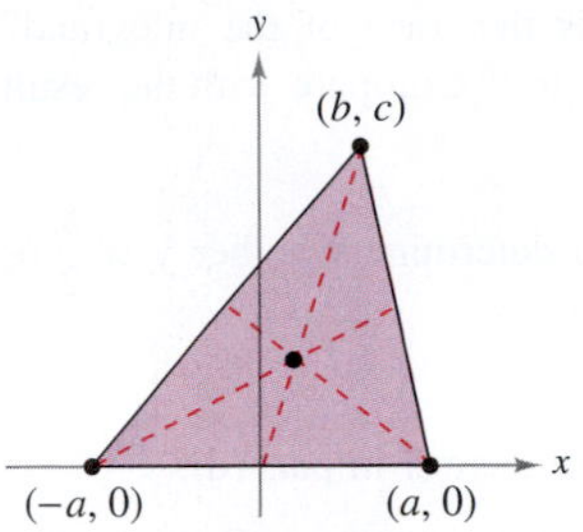

Figure for 45

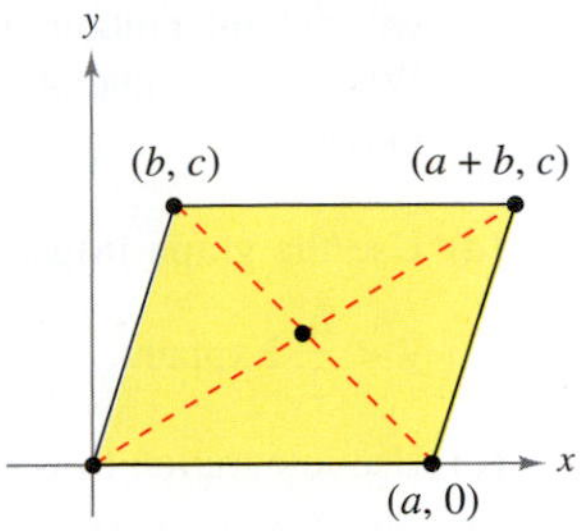

Figure for 46

46. Parallelogram Show that the centroid of the parallelogram with vertices $(0, 0)$, $(a, 0)$, (b, c), and $(a + b, c)$ is the point of intersection of the diagonals (see figure).

47. Trapezoid Find the centroid of the trapezoid with vertices $(0, 0)$, $(0, a)$, (c, b), and $(c, 0)$. Show that it is the intersection of the line connecting the midpoints of the parallel sides and the line connecting the extended parallel sides, as shown in the figure.

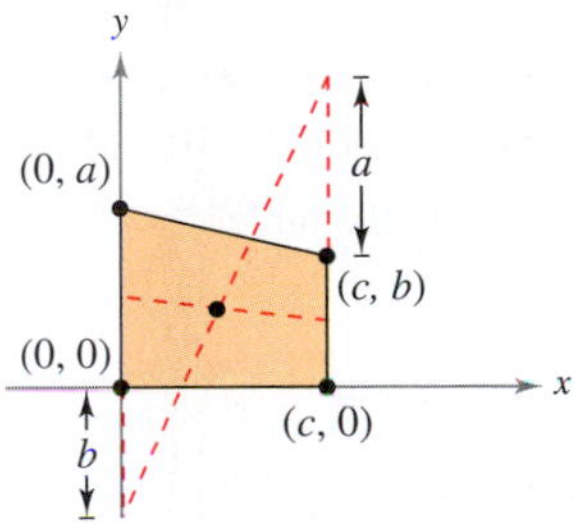

Figure for 47

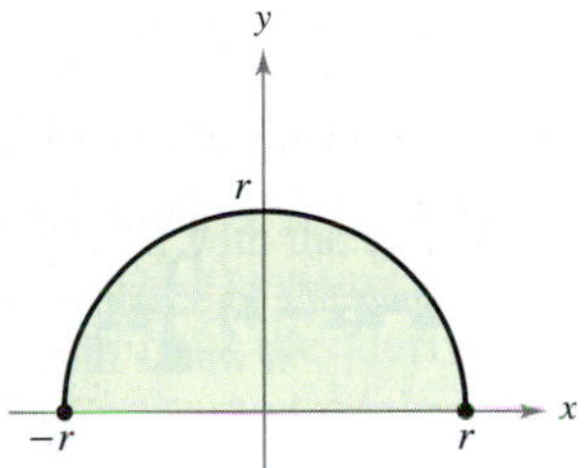

Figure for 48

48. Semicircle Find the centroid of the region bounded by the graphs of $y = \sqrt{r^2 - x^2}$ and $y = 0$ (see figure).

49. Semiellipse Find the centroid of the region bounded by the graphs of $y = \dfrac{b}{a}\sqrt{a^2 - x^2}$ and $y = 0$ (see figure).

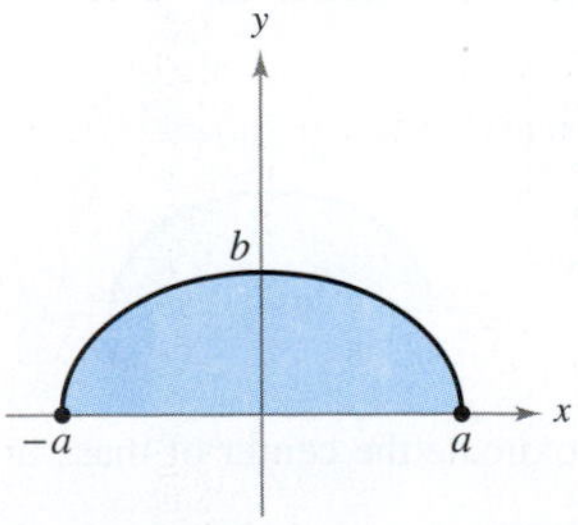

Figure for 49

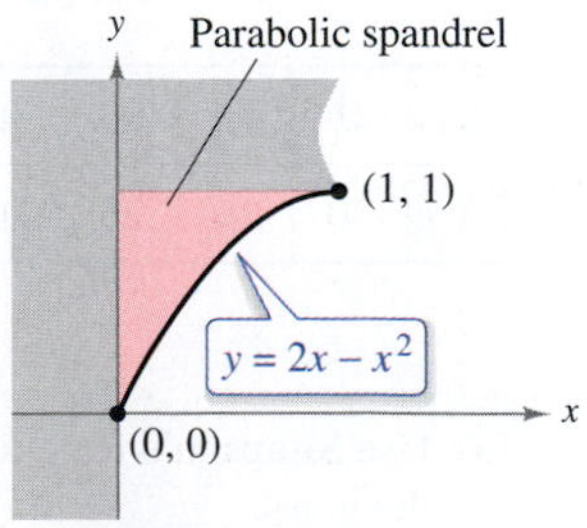

Figure for 50

50. Parabolic Spandrel Find the centroid of the **parabolic spandrel** shown in the figure.

51. Graphical Reasoning Consider the region bounded by the graphs of $y = x^2$ and $y = b$, where $b > 0$.

(a) Sketch a graph of the region.

(b) Use the graph in part (a) to determine $\bar{x}$. Explain.

(c) Set up the integral for finding M_y. Because of the form of the integrand, the value of the integral can be obtained without integrating. What is the form of the integrand? What is the value of the integral? Compare with the result in part (b).

(d) Use the graph in part (a) to determine whether $\bar{y} > \frac{b}{2}$ or $\bar{y} < \frac{b}{2}$. Explain.

(e) Use integration to verify your answer in part (d).

52. Graphical and Numerical Reasoning Consider the region bounded by the graphs of $y = x^{2n}$ and $y = b$, where $b > 0$ and n is a positive integer.

(a) Sketch a graph of the region.

(b) Set up the integral for finding M_y. Because of the form of the integrand, the value of the integral can be obtained without integrating. What is the form of the integrand? What is the value of the integral and what is the value of $\bar{x}$?

(c) Use the graph in part (a) to determine whether $\bar{y} > \frac{b}{2}$ or $\bar{y} < \frac{b}{2}$. Explain.

(d) Use integration to find $\bar{y}$ as a function of n.

(e) Use the result of part (d) to complete the table.

n	1	2	3	4
$\bar{y}$				

(f) Find $\lim_{n\to\infty} \bar{y}$.

(g) Give a geometric explanation of the result in part (f).

53. Modeling Data The manufacturer of glass for a window in a conversion van needs to approximate its center of mass. A coordinate system is superimposed on a prototype of the glass (see figure). The measurements (in centimeters) for the right half of the symmetric piece of glass are listed in the table.

x	0	10	20	30	40
y	30	29	26	20	0

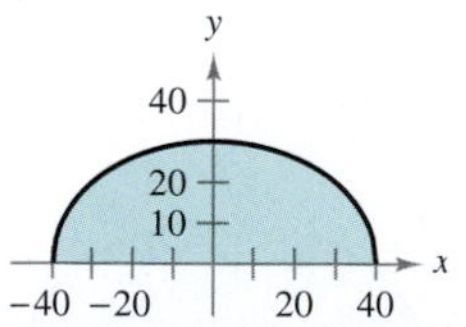

(a) Use Simpson's Rule to approximate the center of mass of the glass.

(b) Use the regression capabilities of a graphing utility to find a fourth-degree polynomial model for the data.

(c) Use the integration capabilities of a graphing utility and the model to approximate the center of mass of the glass. Compare with the result in part (a).

54. Modeling Data The manufacturer of a boat needs to approximate the center of mass of a section of the hull. A coordinate system is superimposed on a prototype (see figure). The measurements (in feet) for the right half of the symmetric prototype are listed in the table.

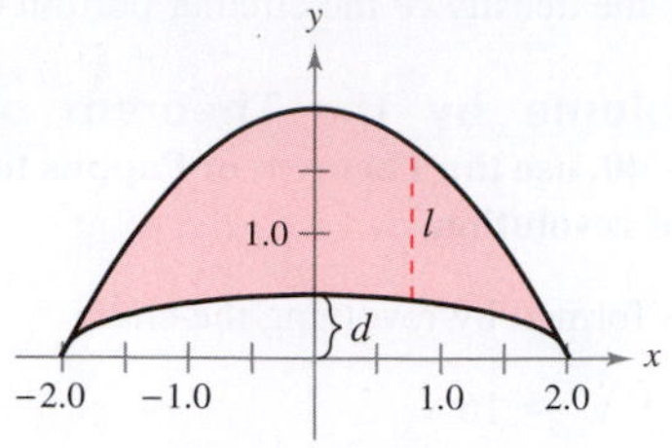

x	0	0.5	1.0	1.5	2
l	1.50	1.45	1.30	0.99	0
d	0.50	0.48	0.43	0.33	0

(a) Use Simpson's Rule to approximate the center of mass of the hull section.

(b) Use the regression capabilities of a graphing utility to find fourth-degree polynomial models for both curves shown in the figure. Plot the data and graph the models.

(c) Use the integration capabilities of a graphing utility and the models to approximate the center of mass of the hull section. Compare with the result in part (a).

Second Theorem of Pappus **In Exercises 55 and 56, use the *Second Theorem of Pappus*, which is stated as follows. If a segment of a plane curve C is revolved about an axis that does not intersect the curve (except possibly at its endpoints), the area S of the resulting surface of revolution is equal to the product of the length of C times the distance d traveled by the centroid of C.**

55. A sphere is formed by revolving the graph of $y = \sqrt{r^2 - x^2}$ about the x-axis. Use the formula for surface area, $S = 4\pi r^2$, to find the centroid of the semicircle $y = \sqrt{r^2 - x^2}$.

56. A torus is formed by revolving the graph of $(x - 1)^2 + y^2 = 1$ about the y-axis. Find the surface area of the torus.

57. Finding a Centroid Let $n \geq 1$ be constant, and consider the region bounded by $f(x) = x^n$, the x-axis, and $x = 1$. Find the centroid of this region. As $n \to \infty$, what does the region look like, and where is its centroid?

PUTNAM EXAM CHALLENGE

58. Let V be the region in the cartesian plane consisting of all points (x, y) satisfying the simultaneous conditions $|x| \leq y \leq |x| + 3$ and $y \leq 4$. Find the centroid $(\bar{x}, \bar{y})$ of V.

7.7 Fluid Pressure and Fluid Force

■ **Find fluid pressure and fluid force.**

Fluid Pressure and Fluid Force

Swimmers know that the deeper an object is submerged in a fluid, the greater the pressure on the object. **Pressure** is defined as the force per unit of area over the surface of a body. For example, because a column of water that is 10 feet in height and 1 inch square weighs 4.3 pounds, the *fluid pressure* at a depth of 10 feet of water is 4.3 pounds per square inch.* At 20 feet, this would increase to 8.6 pounds per square inch, and in general the pressure is proportional to the depth of the object in the fluid.

BLAISE PASCAL (1623–1662)

Pascal is well known for his work in many areas of mathematics and physics, and also for his influence on Leibniz. Although much of Pascal's work in calculus was intuitive and lacked the rigor of modern mathematics, he nevertheless anticipated many important results.

See LarsonCalculus.com to read more of this biography.

Definition of Fluid Pressure

The **pressure** on an object at depth h in a liquid is

$$\text{Pressure} = P = wh$$

where w is the weight-density of the liquid per unit of volume.

Below are some common weight-densities of fluids in pounds per cubic foot.

Ethyl alcohol	49.4
Gasoline	41.0–43.0
Glycerin	78.6
Kerosene	51.2
Mercury	849.0
Seawater	64.0
Water	62.4

When calculating fluid pressure, you can use an important (and rather surprising) physical law called **Pascal's Principle,** named after the French mathematician Blaise Pascal. Pascal's Principle states that the pressure exerted by a fluid at a depth h is transmitted equally *in all directions.* For example, in Figure 7.65, the pressure at the indicated depth is the same for all three objects. Because fluid pressure is given in terms of force per unit area ($P = F/A$), the fluid force on a *submerged horizontal* surface of area A is

$$\text{Fluid force} = F = PA = (\text{pressure})(\text{area}).$$

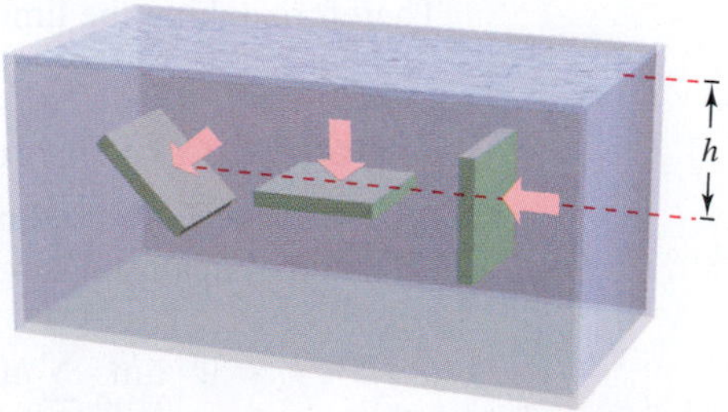

The pressure at h is the same for all three objects.
Figure 7.65

* The total pressure on an object in 10 feet of water would also include the pressure due to Earth's atmosphere. At sea level, atmospheric pressure is approximately 14.7 pounds per square inch.

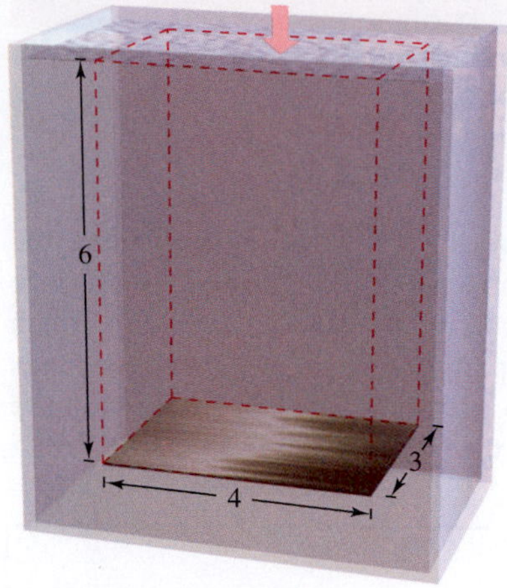

The fluid force on a horizontal metal sheet is equal to the fluid pressure times the area.
Figure 7.66

EXAMPLE 1 Fluid Force on a Submerged Sheet

Find the fluid force on a rectangular metal sheet measuring 3 feet by 4 feet that is submerged in 6 feet of water, as shown in Figure 7.66.

Solution Because the weight-density of water is 62.4 pounds per cubic foot and the sheet is submerged in 6 feet of water, the fluid pressure is

$$P = (62.4)(6) \qquad P = wh$$
$$= 374.4 \text{ pounds per square foot.}$$

Because the total area of the sheet is $A = (3)(4) = 12$ square feet, the fluid force is

$$F = PA$$
$$= \left(374.4 \frac{\text{pounds}}{\text{square foot}}\right)(12 \text{ square feet})$$
$$= 4492.8 \text{ pounds.}$$

This result is independent of the size of the body of water. The fluid force would be the same in a swimming pool or lake. ■

In Example 1, the fact that the sheet is rectangular and horizontal means that you do not need the methods of calculus to solve the problem. Consider a surface that is submerged vertically in a fluid. This problem is more difficult because the pressure is not constant over the surface.

Consider a vertical plate that is submerged in a fluid of weight-density w (per unit of volume), as shown in Figure 7.67. To determine the total force against *one side* of the region from depth c to depth d, you can subdivide the interval $[c, d]$ into n subintervals, each of width Δy. Next, consider the representative rectangle of width Δy and length $L(y_i)$, where y_i is in the ith subinterval. The force against this representative rectangle is

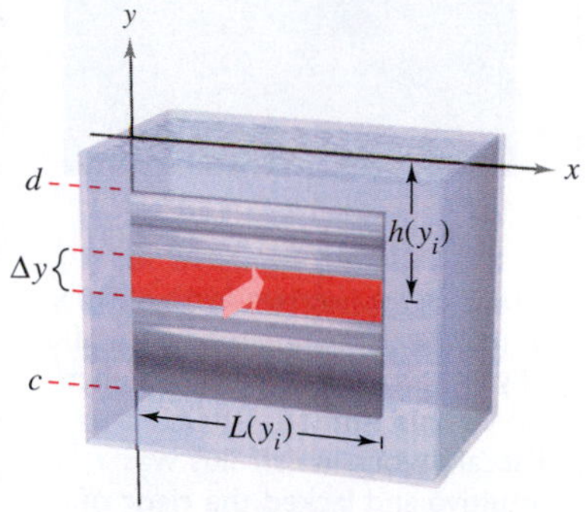

Calculus methods must be used to find the fluid force on a vertical metal plate.
Figure 7.67

$$\Delta F_i = w(\text{depth})(\text{area})$$
$$= wh(y_i)L(y_i)\,\Delta y.$$

The force against n such rectangles is

$$\sum_{i=1}^{n} \Delta F_i = w \sum_{i=1}^{n} h(y_i)L(y_i)\,\Delta y.$$

Note that w is considered to be constant and is factored out of the summation. Therefore, taking the limit as $\|\Delta\| \to 0$ $(n \to \infty)$ suggests the next definition.

Definition of Force Exerted by a Fluid

The **force F exerted by a fluid** of constant weight-density w (per unit of volume) against a submerged vertical plane region from $y = c$ to $y = d$ is

$$F = w \lim_{\|\Delta\| \to 0} \sum_{i=1}^{n} h(y_i)L(y_i)\,\Delta y$$
$$= w \int_c^d h(y)L(y)\,dy$$

where $h(y)$ is the depth of the fluid at y and $L(y)$ is the horizontal length of the region at y.

EXAMPLE 2 Fluid Force on a Vertical Surface

See LarsonCalculus.com for an interactive version of this type of example.

A vertical gate in a dam has the shape of an isosceles trapezoid 8 feet across the top and 6 feet across the bottom, with a height of 5 feet, as shown in Figure 7.68(a). What is the fluid force on the gate when the top of the gate is 4 feet below the surface of the water?

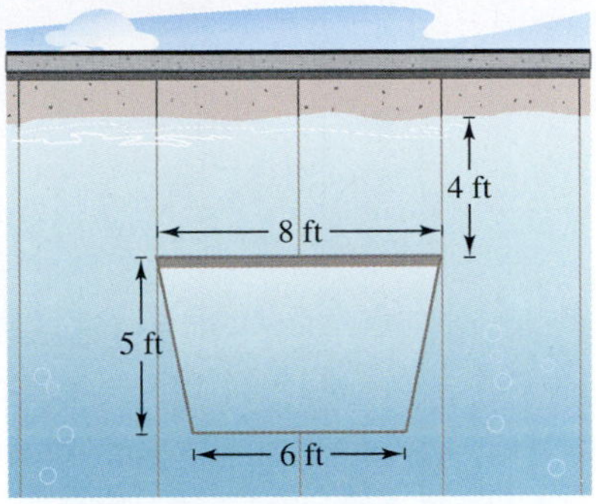

(a) Water gate in a dam

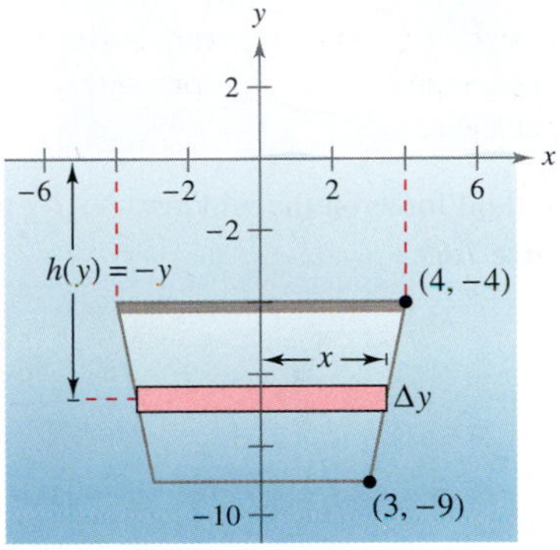

(b) The fluid force against the gate

Figure 7.68

Solution In setting up a mathematical model for this problem, you are at liberty to locate the x- and y-axes in several different ways. A convenient approach is to let the y-axis bisect the gate and place the x-axis at the surface of the water, as shown in Figure 7.68(b). So, the depth of the water at y in feet is

$$\text{Depth} = h(y) = -y.$$

To find the length $L(y)$ of the region at y, find the equation of the line forming the right side of the gate. Because this line passes through the points $(3, -9)$ and $(4, -4)$, its equation is

$$\begin{aligned} y - (-9) &= \frac{-4 - (-9)}{4 - 3}(x - 3) \\ y + 9 &= 5(x - 3) \\ y &= 5x - 24 \\ x &= \frac{y + 24}{5}. \end{aligned}$$

In Figure 7.68(b) you can see that the length of the region at y is

$$\text{Length} = 2x = \frac{2}{5}(y + 24) = L(y).$$

Finally, by integrating from $y = -9$ to $y = -4$, you can calculate the fluid force to be

$$\begin{aligned} F &= w\int_c^d h(y)L(y)\,dy \\ &= 62.4\int_{-9}^{-4} (-y)\left(\frac{2}{5}\right)(y + 24)\,dy \\ &= -62.4\left(\frac{2}{5}\right)\int_{-9}^{-4} (y^2 + 24y)\,dy \\ &= -62.4\left(\frac{2}{5}\right)\left[\frac{y^3}{3} + 12y^2\right]_{-9}^{-4} \\ &= -62.4\left(\frac{2}{5}\right)\left(\frac{-1675}{3}\right) \\ &= 13{,}936 \text{ pounds.} \end{aligned}$$

In Example 2, the x-axis coincided with the surface of the water. This was convenient, but arbitrary. In choosing a coordinate system to represent a physical situation, you should consider various possibilities. Often you can simplify the calculations in a problem by locating the coordinate system to take advantage of special characteristics of the problem, such as symmetry.

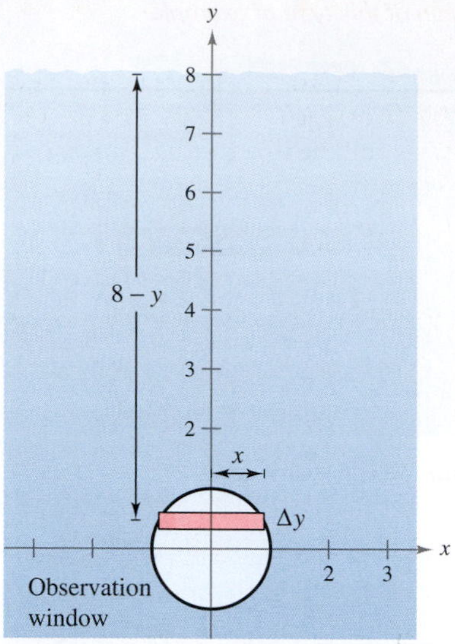

The fluid force on the window

Figure 7.69

EXAMPLE 3 Fluid Force on a Vertical Surface

A circular observation window on a marine science ship has a radius of 1 foot, and the center of the window is 8 feet below water level, as shown in Figure 7.69. What is the fluid force on the window?

Solution To take advantage of symmetry, locate a coordinate system such that the origin coincides with the center of the window, as shown in Figure 7.69. The depth at y is then

$$\text{Depth} = h(y) = 8 - y.$$

The horizontal length of the window is $2x$, and you can use the equation for the circle, $x^2 + y^2 = 1$, to solve for x as shown.

$$\begin{aligned}\text{Length} &= 2x\\ &= 2\sqrt{1 - y^2} = L(y)\end{aligned}$$

Finally, because y ranges from -1 to 1, and using 64 pounds per cubic foot as the weight-density of seawater, you have

$$\begin{aligned}F &= w\int_c^d h(y)L(y)\,dy\\ &= 64\int_{-1}^{1} (8 - y)(2)\sqrt{1 - y^2}\,dy.\end{aligned}$$

Initially it looks as though this integral would be difficult to solve. However, when you break the integral into two parts and apply symmetry, the solution is simpler.

$$F = 64(16)\int_{-1}^{1} \sqrt{1 - y^2}\,dy - 64(2)\int_{-1}^{1} y\sqrt{1 - y^2}\,dy$$

The second integral is 0 (because the integrand is odd and the limits of integration are symmetric with respect to the origin). Moreover, by recognizing that the first integral represents the area of a semicircle of radius 1, you obtain

$$\begin{aligned}F &= 64(16)\left(\frac{\pi}{2}\right) - 64(2)(0)\\ &= 512\pi\\ &\approx 1608.5 \text{ pounds.}\end{aligned}$$

So, the fluid force on the window is about 1608.5 pounds.

TECHNOLOGY To confirm the result obtained in Example 3, you might have considered using Simpson's Rule to approximate the value of

$$128\int_{-1}^{1} (8 - x)\sqrt{1 - x^2}\,dx.$$

From the graph of

$$f(x) = (8 - x)\sqrt{1 - x^2}$$

however, you can see that f is not differentiable when $x = \pm 1$ (see figure at the right). This means that you cannot apply Theorem 4.20 from Section 4.6 to determine the potential error in Simpson's Rule. Without knowing the potential error, the approximation is of little value. Use a graphing utility to approximate the integral.

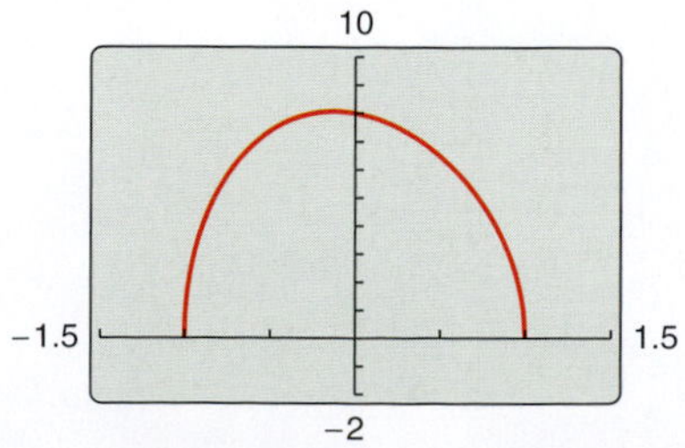

f is not differentiable at $x = \pm 1$.

7.7 Exercises

See CalcChat.com for tutorial help and worked-out solutions to odd-numbered exercises.

Force on a Submerged Sheet **In Exercises 1–4, the area of the top side of a piece of sheet metal is given. The sheet metal is submerged horizontally in 8 feet of water. Find the fluid force on the top side.**

1. 3 square feet

2. 8 square feet

3. 10 square feet

4. 25 square feet

Buoyant Force **In Exercises 5 and 6, find the buoyant force of a rectangular solid of the given dimensions submerged in water so that the top side is parallel to the surface of the water. The buoyant force is the difference between the fluid forces on the top and bottom sides of the solid.**

5.

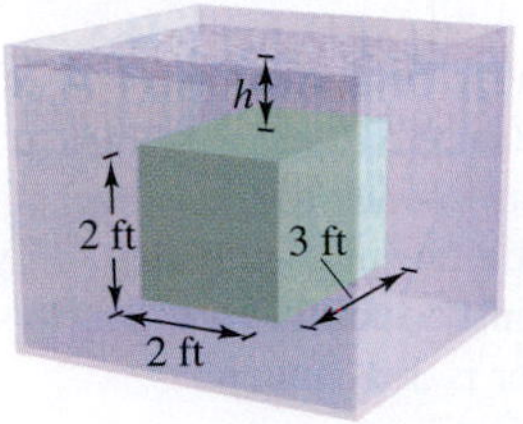

6.

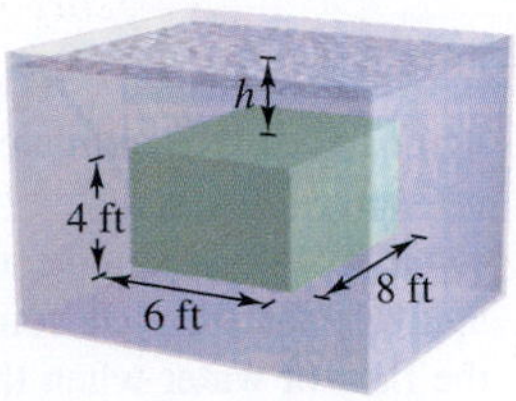

Fluid Force on a Tank Wall **In Exercises 7–12, find the fluid force on the vertical side of the tank, where the dimensions are given in feet. Assume that the tank is full of water.**

7. Rectangle

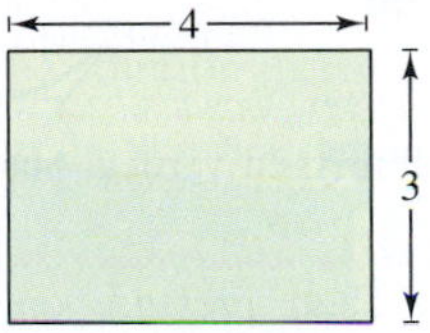

8. Triangle

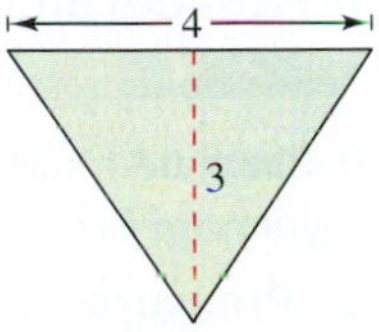

9. Trapezoid

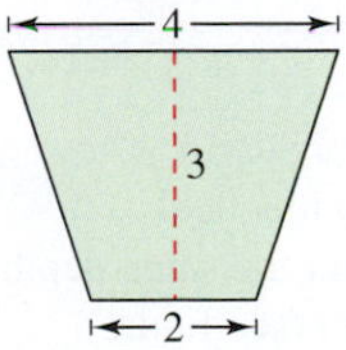

10. Semicircle

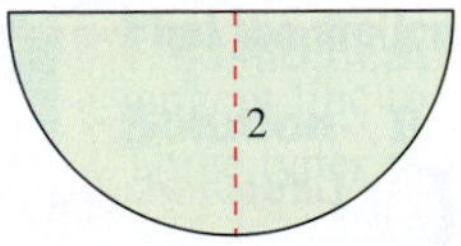

11. Parabola, $y = x^2$

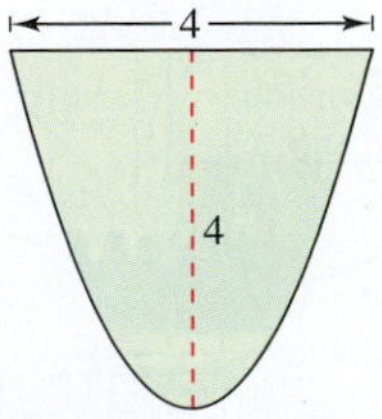

12. Semiellipse, $y = -\frac{1}{2}\sqrt{36 - 9x^2}$

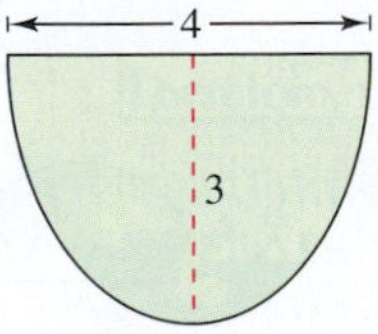

Fluid Force of Water **In Exercises 13–16, find the fluid force on the vertical plate submerged in water, where the dimensions are given in meters and the weight-density of water is 9800 newtons per cubic meter.**

13. Square

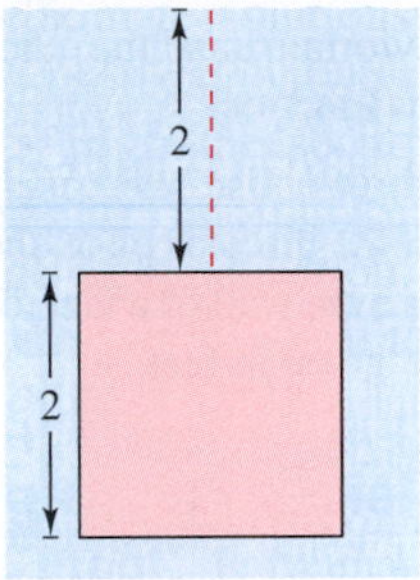

14. Square

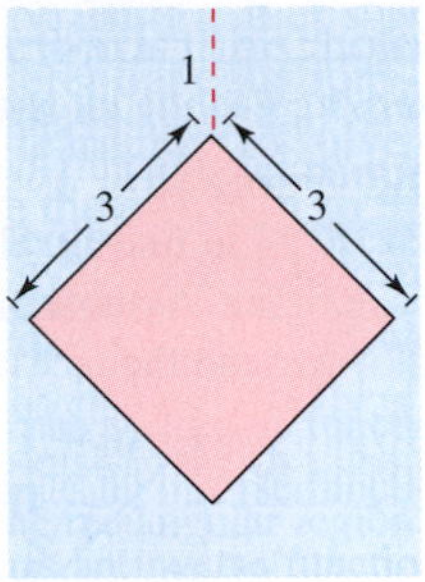

15. Triangle

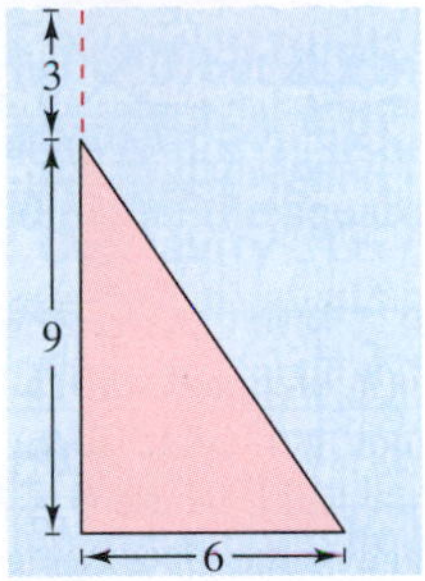

16. Rectangle

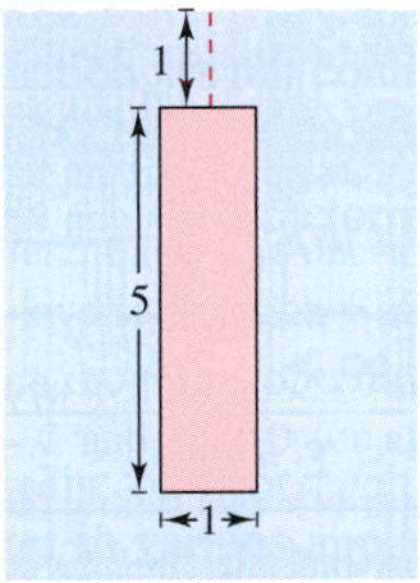

Force on a Concrete Form **In Exercises 17–20, the figure is the vertical side of a form for poured concrete that weighs 140.7 pounds per cubic foot. Determine the force on this part of the concrete form.**

17. Rectangle

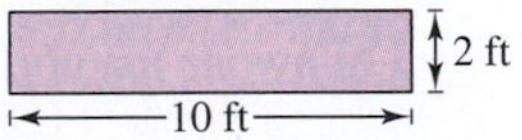

18. Semiellipse, $y = -\frac{3}{4}\sqrt{16 - x^2}$

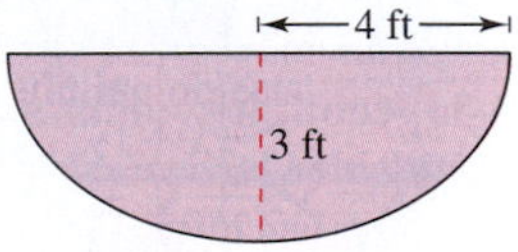

19. Rectangle

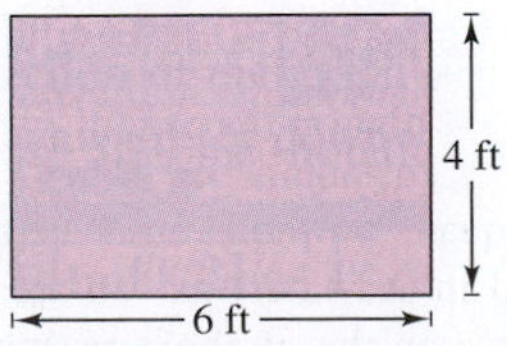

20. Triangle

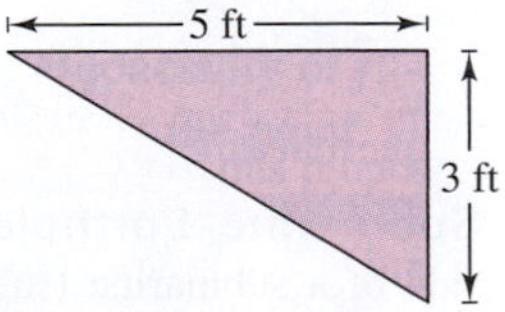

21. Fluid Force of Gasoline A cylindrical gasoline tank is placed so that the axis of the cylinder is horizontal. Find the fluid force on a circular end of the tank when the tank is half full, where the diameter is 3 feet and the gasoline weighs 42 pounds per cubic foot.

22. Fluid Force of Gasoline Repeat Exercise 21 for a tank that is full. (Evaluate one integral by a geometric formula and the other by observing that the integrand is an odd function.)

23. Fluid Force on a Circular Plate A circular plate of radius r feet is submerged vertically in a tank of fluid that weighs w pounds per cubic foot. The center of the circle is k feet below the surface of the fluid, where $k > r$. Show that the fluid force on the surface of the plate is

$$F = wk(\pi r^2).$$

(Evaluate one integral by a geometric formula and the other by observing that the integrand is an odd function.)

24. Fluid Force on a Circular Plate Use the result of Exercise 23 to find the fluid force on the circular plate shown in each figure. Assume the plates are in the wall of a tank filled with water and the measurements are given in feet.

(a)

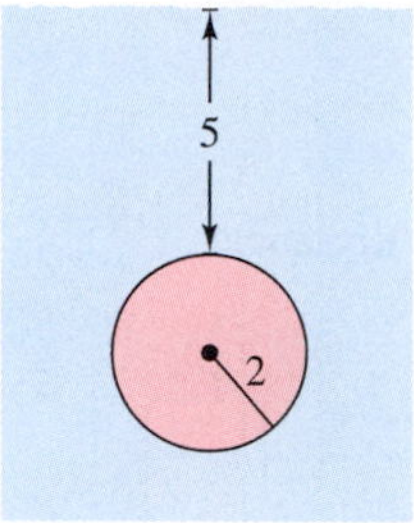

(b)

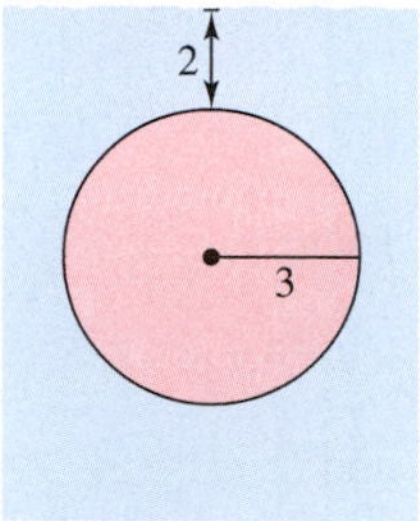

25. Fluid Force on a Rectangular Plate A rectangular plate of height h feet and base b feet is submerged vertically in a tank of fluid that weighs w pounds per cubic foot. The center is k feet below the surface of the fluid, where $k > h/2$. Show that the fluid force on the surface of the plate is

$$F = wkhb.$$

26. Fluid Force on a Rectangular Plate Use the result of Exercise 25 to find the fluid force on the rectangular plate shown in each figure. Assume the plates are in the wall of a tank filled with water and the measurements are given in feet.

(a)

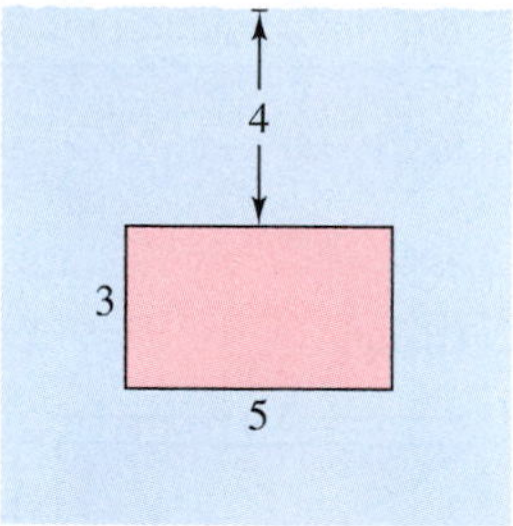

(b)

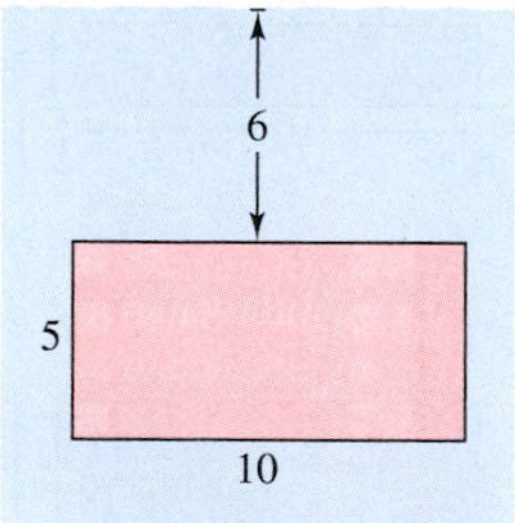

27. Submarine Porthole A square porthole on a vertical side of a submarine (submerged in seawater) has an area of 1 square foot. Find the fluid force on the porthole, assuming that the center of the square is 15 feet below the surface.

28. Submarine Porthole Repeat Exercise 27 for a circular porthole that has a diameter of 1 foot. The center is 15 feet below the surface.

29. Modeling Data The vertical stern of a boat with a superimposed coordinate system is shown in the figure. The table shows the widths w of the stern (in feet) at indicated values of y. Find the fluid force against the stern.

y	0	$\frac{1}{2}$	1	$\frac{3}{2}$	2	$\frac{5}{2}$	3	$\frac{7}{2}$	4
w	0	3	5	8	9	10	10.25	10.5	10.5

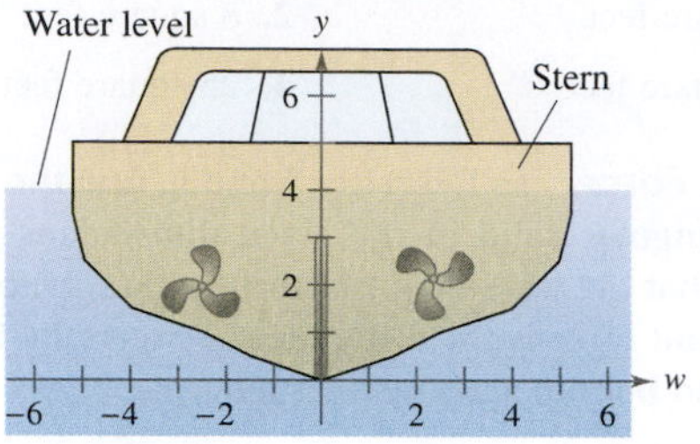

30. Irrigation Canal Gate The vertical cross section of an irrigation canal is modeled by $f(x) = 5x^2/(x^2 + 4)$, where x is measured in feet and $x = 0$ corresponds to the center of the canal. Use the integration capabilities of a graphing utility to approximate the fluid force against a vertical gate used to stop the flow of water when the water is 3 feet deep.

WRITING ABOUT CONCEPTS

31. Think About It Approximate the depth of the water in the tank in Exercise 7 if the fluid force is one-half as great as when the tank is full. Explain why the answer is not $\frac{3}{2}$.

32. Fluid Pressure and Fluid Force

(a) Define fluid pressure.

(b) Define fluid force against a submerged vertical plane region.

33. Fluid Pressure Explain why fluid pressure on a surface is calculated using horizontal representative rectangles instead of vertical representative rectangles.

34. HOW DO YOU SEE IT? Two identical semicircular windows are placed at the same depth in the vertical wall of an aquarium (see figure). Which is subjected to the greater fluid force? Explain.

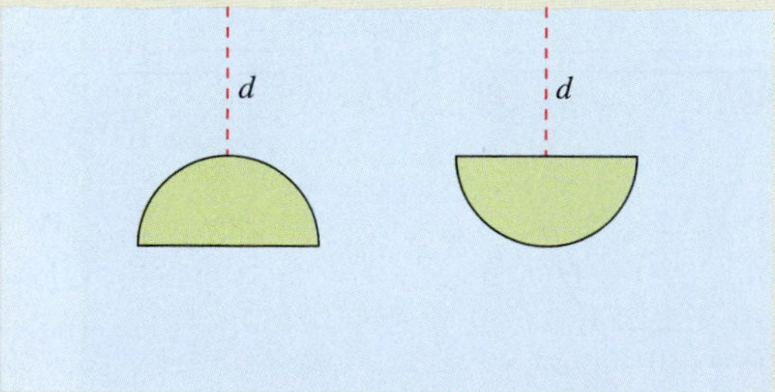

Review Exercises

See CalcChat.com for tutorial help and worked-out solutions to odd-numbered exercises.

Finding the Area of a Region **In Exercises 1–10, sketch the region bounded by the graphs of the equations and find the area of the region.**

1. $y = 6 - \frac{1}{2}x^2, \quad y = \frac{3}{4}x, \quad x = -2, \quad x = 2$
2. $y = \frac{1}{x^2}, \quad y = 4, \quad x = 5$
3. $y = \frac{1}{x^2 + 1}, \quad y = 0, \quad x = -1, \quad x = 1$
4. $x = y^2 - 2y, \quad x = -1, \quad y = 0$
5. $y = x, \quad y = x^3$
6. $x = y^2 + 1, \quad x = y + 3$
7. $y = e^x, \quad y = e^2, \quad x = 0$
8. $y = \csc x, \quad y = 2, \quad \frac{\pi}{6} \le x \le \frac{5\pi}{6}$
9. $y = \sin x, \quad y = \cos x, \quad \frac{\pi}{4} \le x \le \frac{5\pi}{4}$
10. $x = \cos y, \quad x = \frac{1}{2}, \quad \frac{\pi}{3} \le y \le \frac{7\pi}{3}$

Finding the Area of a Region **In Exercises 11–14, use a graphing utility to graph the region bounded by the graphs of the equations, and use the integration capabilities of the graphing utility to find the area of the region.**

11. $y = x^2 - 8x + 3, \ y = 3 + 8x - x^2$
12. $y = x^2 - 4x + 3, \ y = x^3, \ x = 0$
13. $\sqrt{x} + \sqrt{y} = 1, \ y = 0, \ x = 0$
14. $y = x^4 - 2x^2, \ y = 2x^2$

15. **Numerical Integration** Estimate the surface area of the pond using (a) the Trapezoidal Rule and (b) Simpson's Rule.

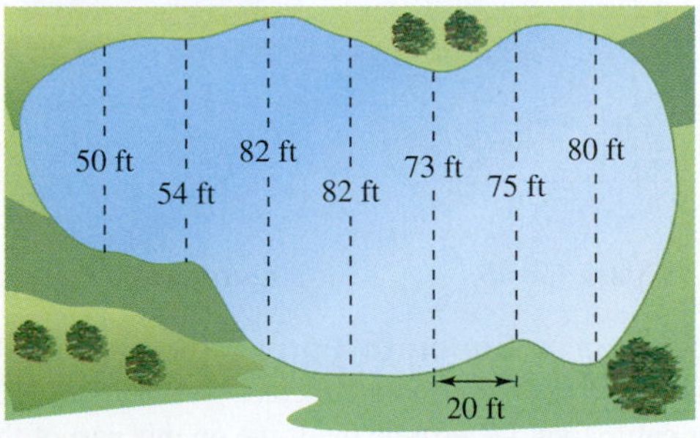

16. **Revenue** The models $R_1 = 6.4 + 0.2t + 0.01t^2$ and $R_2 = 8.4 + 0.35t$ give the revenue (in billions of dollars) for a large corporation. Both models are estimates of the revenues from 2015 through 2020, with $t = 15$ corresponding to 2015. Which model projects the greater revenue? How much more total revenue does that model project over the six-year period?

Finding the Volume of a Solid **In Exercises 17–22, use the disk method *or* the shell method to find the volumes of the solids generated by revolving the region bounded by the graphs of the equations about the given line(s).**

17. $y = x, \quad y = 0, \quad x = 3$
    (a) the x-axis (b) the y-axis
    (c) the line $x = 3$ (d) the line $x = 6$
18. $y = \sqrt{x}, \quad y = 2, \quad x = 0$
    (a) the x-axis (b) the line $y = 2$
    (c) the y-axis (d) the line $x = -1$
19. $y = \frac{1}{x^4 + 1}, \quad y = 0, \quad x = 0, \quad x = 1$
    revolved about the y-axis
20. $y = \frac{1}{\sqrt{1 + x^2}}, \quad y = 0, \quad x = -1, \quad x = 1$
    revolved about the x-axis
21. $y = \frac{1}{x^2}, \quad y = 0, \quad x = 2, \quad x = 5$
    revolved about the y-axis
22. $y = e^{-x}, \quad y = 0, \quad x = 0, \quad x = 1$
    revolved about the x-axis

23. **Depth of Gasoline in a Tank** A gasoline tank is an oblate spheroid generated by revolving the region bounded by the graph of

$$\frac{x^2}{16} + \frac{y^2}{9} = 1$$

about the y-axis, where x and y are measured in feet. Find the depth of the gasoline in the tank when it is filled to one-fourth its capacity.

24. **Using Cross Sections** Find the volume of the solid whose base is bounded by the circle $x^2 + y^2 = 9$ and the cross sections perpendicular to the x-axis are equilateral triangles.

Finding Arc Length **In Exercises 25 and 26, find the arc length of the graph of the function over the indicated interval.**

25. $f(x) = \frac{4}{5}x^{5/4}, \quad [0, 4]$
26. $y = \frac{1}{6}x^3 + \frac{1}{2x}, \quad [1, 3]$

27. **Length of a Catenary** A cable of a suspension bridge forms a catenary modeled by the equation

$$y = 300 \cosh\left(\frac{x}{2000}\right) - 280, \quad -2000 \le x \le 2000$$

where x and y are measured in feet. Use the integration capabilities of a graphing utility to approximate the length of the cable.

28. Approximation Determine which value best approximates the length of the arc represented by the integral

$$\int_0^1 \sqrt{1 + \left[\frac{d}{dx}\left(\frac{4}{x+1}\right)\right]^2}\, dx.$$

(Make your selection on the basis of a sketch of the arc and *not* by performing any calculations.)

(a) 10 (b) -5 (c) 2 (d) 4 (e) 1

29. Surface Area Use integration to find the lateral surface area of a right circular cone of height 4 and radius 3.

30. Surface Area The region bounded by the graphs of $y = 2\sqrt{x}$, $y = 0$, $x = 3$, and $x = 8$ is revolved about the x-axis. Find the surface area of the solid generated.

31. Work A force of 5 pounds is needed to stretch a spring 1 inch from its natural position. Find the work done in stretching the spring from its natural length of 10 inches to a length of 15 inches.

32. Work A force of 50 pounds is needed to stretch a spring 1 inch from its natural position. Find the work done in stretching the spring from its natural length of 10 inches to double that length.

33. Work A water well has an 8-inch casing (diameter) and is 190 feet deep. The water is 25 feet from the top of the well. Determine the amount of work done in pumping the well dry, assuming that no water enters it while it is being pumped.

34. Boyle's Law A quantity of gas with an initial volume of 2 cubic feet and a pressure of 800 pounds per square foot expands to a volume of 3 cubic feet. Find the work done by the gas. Assume that the pressure is inversely proportional to the volume.

35. Work A chain 10 feet long weighs 4 pounds per foot and is hung from a platform 20 feet above the ground. How much work is required to raise the entire chain to the 20-foot level?

36. Work A windlass, 200 feet above ground level on the top of a building, uses a cable weighing 5 pounds per foot. Find the work done in winding up the cable when

(a) one end is at ground level.

(b) there is a 300-pound load attached to the end of the cable.

37. Work The work done by a variable force in a press is 80 foot-pounds. The press moves a distance of 4 feet, and the force is a quadratic of the form $F = ax^2$. Find a.

38. Work Find the work done by the force F shown in the figure.

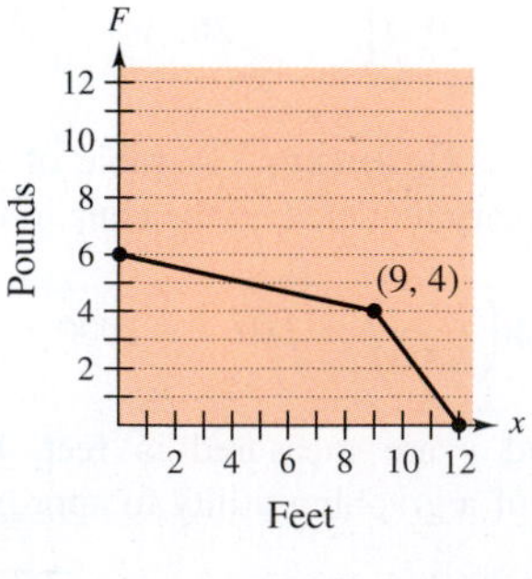

39. Center of Mass of a Linear System Find the center of mass of the point masses lying on the x-axis.

$m_1 = 8,\quad m_2 = 12,\quad m_3 = 6,\quad m_4 = 14$

$x_1 = -1,\quad x_2 = 2,\quad x_3 = 5,\quad x_4 = 7$

40. Center of Mass of a Two-Dimensional System Find the center of mass of the given system of point masses.

m_i	3	2	6	9
(x_i, y_i)	(2, 1)	(−3, 2)	(4, −1)	(6, 5)

Finding a Centroid **In Exercises 41 and 42, find the centroid of the region bounded by the graphs of the equations.**

41. $y = x^2,\quad y = 2x + 3$ **42.** $y = x^{2/3},\quad y = \frac{1}{2}x$

43. Centroid A blade on an industrial fan has the configuration of a semicircle attached to a trapezoid (see figure). Find the centroid of the blade.

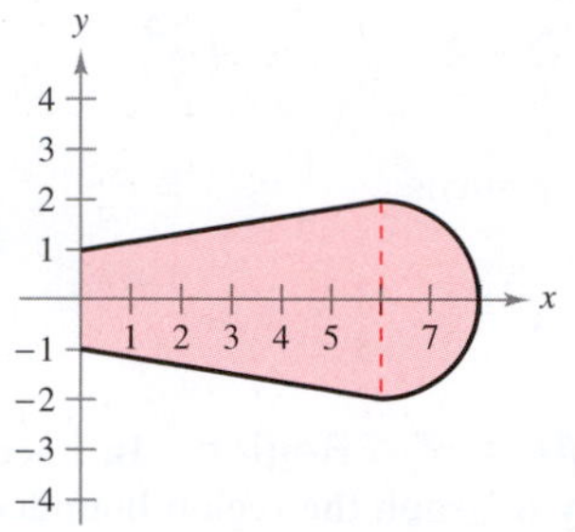

44. Finding Volume Use the Theorem of Pappus to find the volume of the torus formed by revolving the circle $(x - 4)^2 + y^2 = 4$ about the y-axis.

45. Fluid Force of Seawater Find the fluid force on the vertical plate submerged in seawater (see figure).

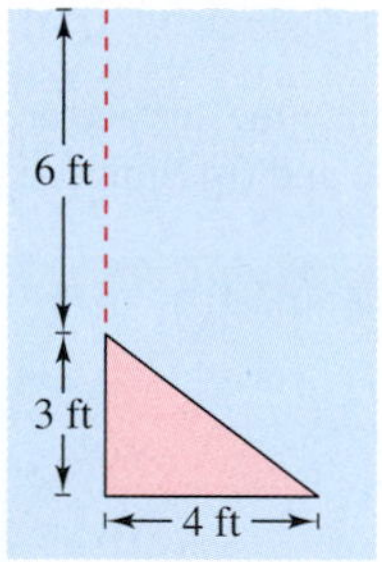

Figure for 45

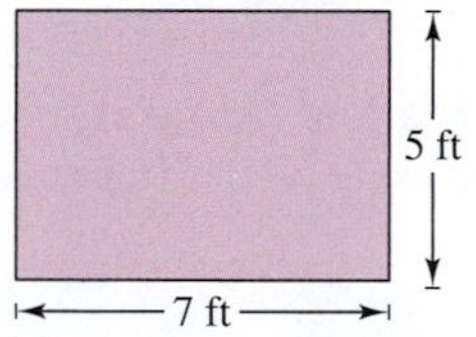

Figure for 46

46. Force on a Concrete Form The figure is the vertical side of a form for poured concrete that weights 140.7 pounds per cubic foot. Determine the force on this part of the concrete form.

47. Fluid Force A swimming pool is 5 feet deep at one end and 10 feet deep at the other, and the bottom is an inclined plane. The length and width of the pool are 40 feet and 20 feet. If the pool is full of water, what is the fluid force on each of the vertical walls?

P.S. Problem Solving

See CalcChat.com for tutorial help and worked-out solutions to odd-numbered exercises.

1. Finding a Limit Let R be the area of the region in the first quadrant bounded by the parabola $y = x^2$ and the line $y = cx$, $c > 0$. Let T be the area of the triangle AOB. Calculate the limit

$$\lim_{c \to 0^+} \frac{T}{R}.$$

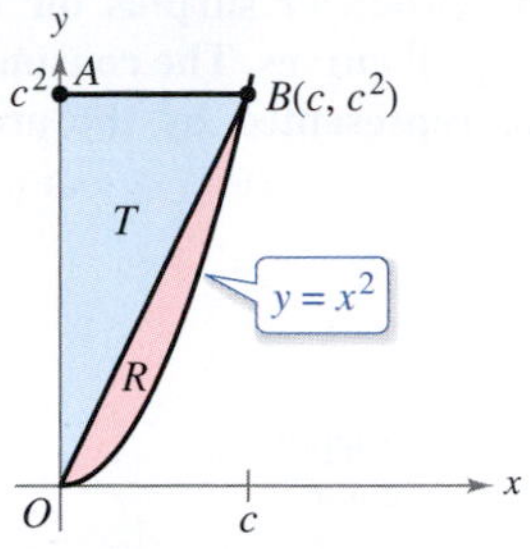

Figure for 1

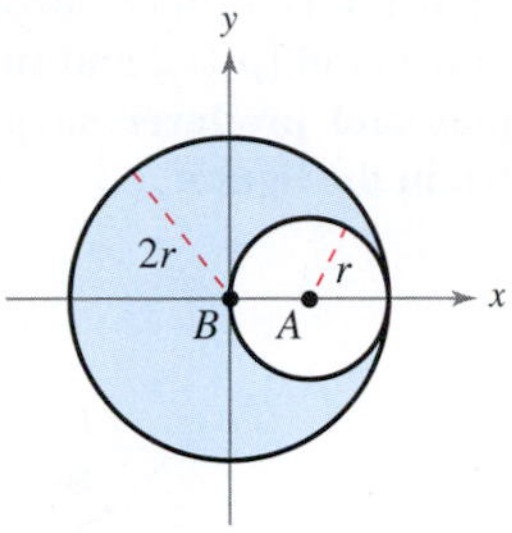

Figure for 2

2. Center of Mass of a Lamina Let L be the lamina of uniform density $\rho = 1$ obtained by removing circle A of radius r from circle B of radius $2r$ (see figure).

(a) Show that $M_x = 0$ for L.

(b) Show that M_y for L is equal to $(M_y \text{ for } B) - (M_y \text{ for } A)$.

(c) Find M_y for B and M_y for A. Then use part (b) to compute M_y for L.

(d) What is the center of mass of L?

3. Dividing a Region Let R be the region bounded by the parabola $y = x - x^2$ and the x-axis. Find the equation of the line $y = mx$ that divides this region into two regions of equal area.

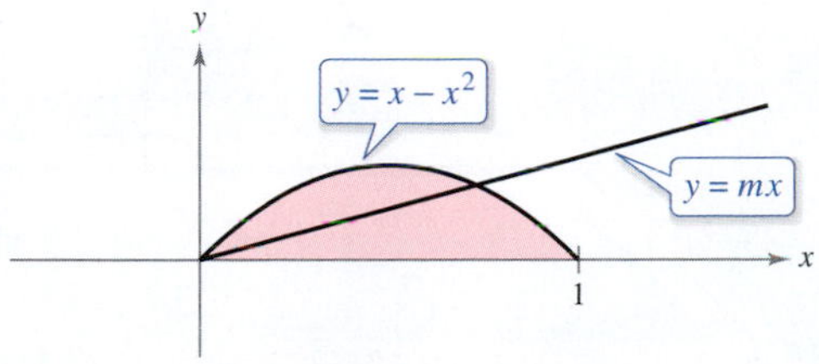

4. Volume A hole is cut through the center of a sphere of radius r (see figure). The height of the remaining spherical ring is h. Find the volume of the ring and show that it is independent of the radius of the sphere.

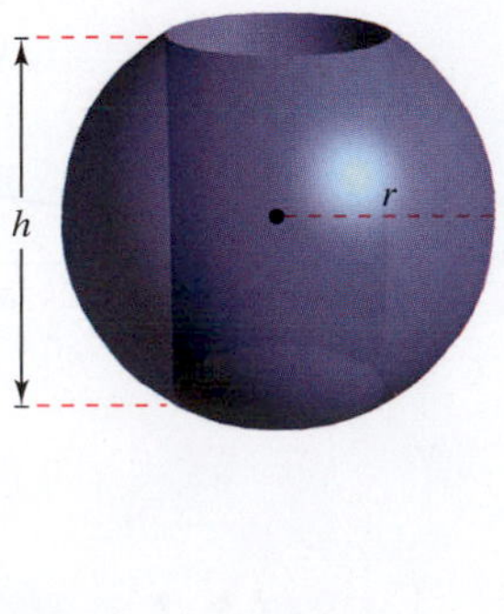

5. Surface Area Graph the curve

$$8y^2 = x^2(1 - x^2).$$

Use a computer algebra system to find the surface area of the solid of revolution obtained by revolving the curve about the y-axis.

6. Torus

(a) A torus is formed by revolving the region bounded by the circle

$$(x - 2)^2 + y^2 = 1$$

about the y-axis (see figure). Use the disk method to calculate the volume of the torus.

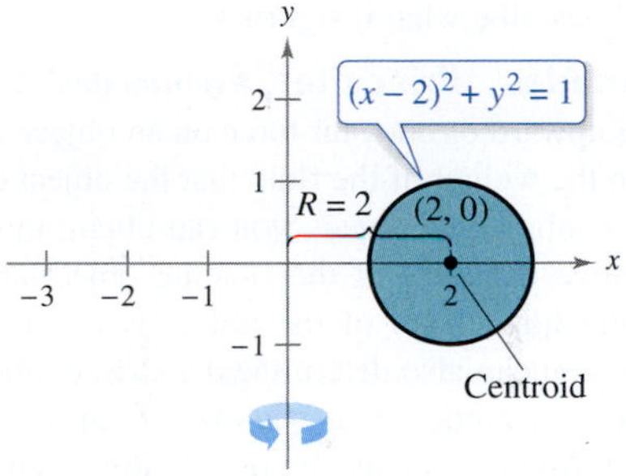

(b) Use the disk method to find the volume of the general torus when the circle has radius r and its center is R units from the axis of rotation.

7. Volume A rectangle R of length ℓ and width w is revolved about the line L (see figure). Find the volume of the resulting solid of revolution.

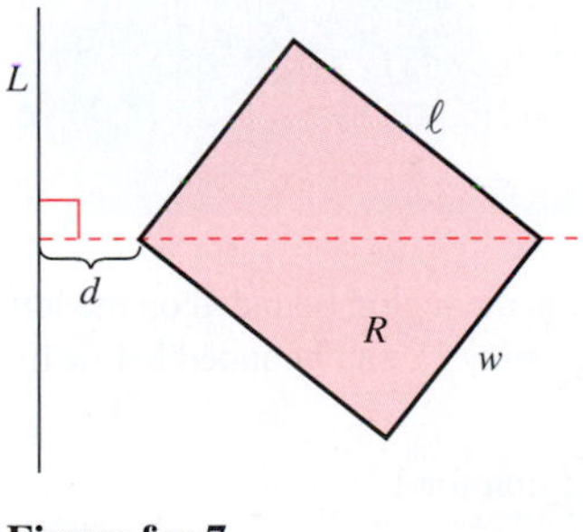

Figure for 7

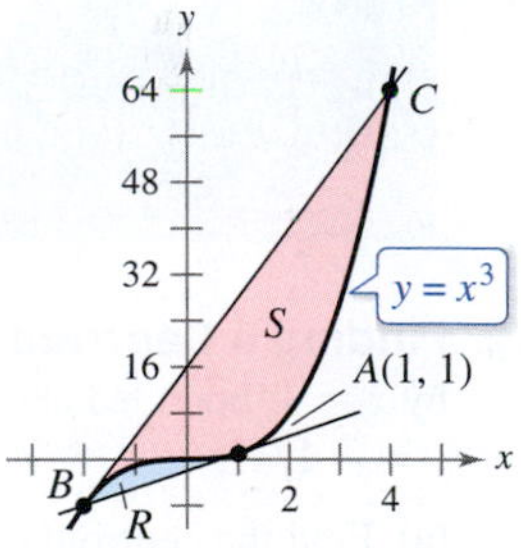

Figure for 8

8. Comparing Areas of Regions

(a) The tangent line to the curve $y = x^3$ at the point $A(1, 1)$ intersects the curve at another point B. Let R be the area of the region bounded by the curve and the tangent line. The tangent line at B intersects the curve at another point C (see figure). Let S be the area of the region bounded by the curve and this second tangent line. How are the areas R and S related?

(b) Repeat the construction in part (a) by selecting an arbitrary point A on the curve $y = x^3$. Show that the two areas R and S are always related in the same way.

9. Using Arc Length The graph of $y = f(x)$ passes through the origin. The arc length of the curve from $(0, 0)$ to $(x, f(x))$ is given by

$$s(x) = \int_0^x \sqrt{1 + e^t}\, dt.$$

Identify the function f.

10. Using a Function Let f be rectifiable on the interval $[a, b]$, and let

$$s(x) = \int_a^x \sqrt{1 + [f'(t)]^2}\, dt.$$

(a) Find $\frac{ds}{dx}$.

(b) Find ds and $(ds)^2$.

(c) Find $s(x)$ on $[1, 3]$ when $f(t) = t^{3/2}$.

(d) Use the function and interval in part (c) to calculate $s(2)$ and describe what it signifies.

11. Archimedes' Principle **Archimedes' Principle** states that the upward or buoyant force on an object within a fluid is equal to the weight of the fluid that the object displaces. For a partially submerged object, you can obtain information about the relative densities of the floating object and the fluid by observing how much of the object is above and below the surface. You can also determine the size of a floating object if you know the amount that is above the surface and the relative densities. You can see the top of a floating iceberg (see figure). The density of ocean water is 1.03×10^3 kilograms per cubic meter, and that of ice is 0.92×10^3 kilograms per cubic meter. What percent of the total iceberg is below the surface?

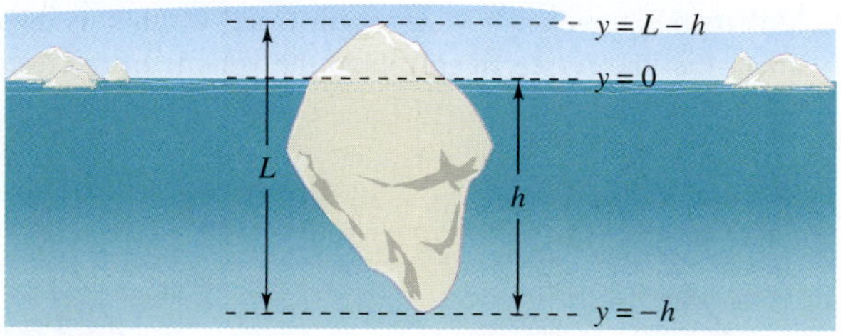

12. Finding a Centroid Sketch the region bounded on the left by $x = 1$, bounded above by $y = 1/x^3$, and bounded below by $y = -1/x^3$.

(a) Find the centroid of the region for $1 \le x \le 6$.

(b) Find the centroid of the region for $1 \le x \le b$.

(c) Where is the centroid as $b \to \infty$?

13. Finding a Centroid Sketch the region to the right of the y-axis, bounded above by $y = 1/x^4$, and bounded below by $y = -1/x^4$.

(a) Find the centroid of the region for $1 \le x \le 6$.

(b) Find the centroid of the region for $1 \le x \le b$.

(c) Where is the centroid as $b \to \infty$?

14. Work Find the work done by each force F.

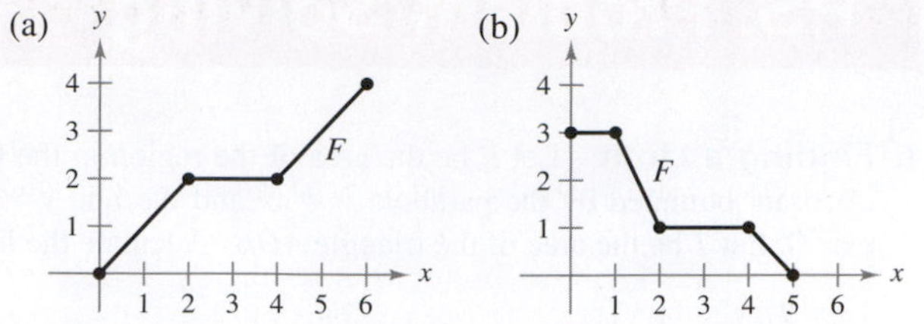

Consumer and Producer Surplus **In Exercises 15 and 16, find the consumer surplus and producer surplus for the given demand $[p_1(x)]$ and supply $[p_2(x)]$ curves. The consumer surplus and producer surplus are represented by the areas shown in the figure.**

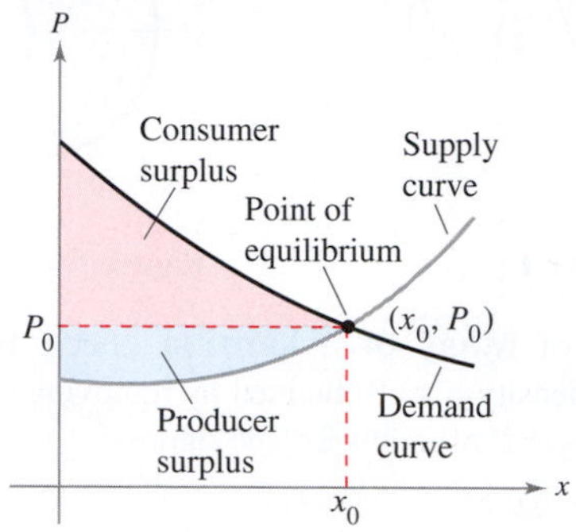

15. $p_1(x) = 50 - 0.5x,\quad p_2(x) = 0.125x$

16. $p_1(x) = 1000 - 0.4x^2,\quad p_2(x) = 42x$

17. Fluid Force A swimming pool is 20 feet wide, 40 feet long, 4 feet deep at one end, and 8 feet deep at the other end (see figures). The bottom is an inclined plane. Find the fluid force on each vertical wall.

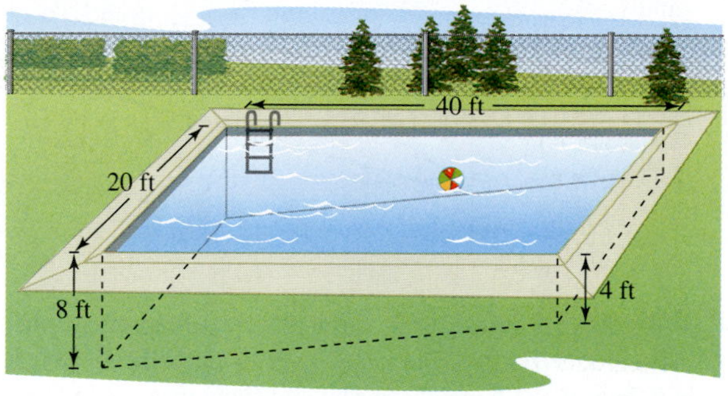

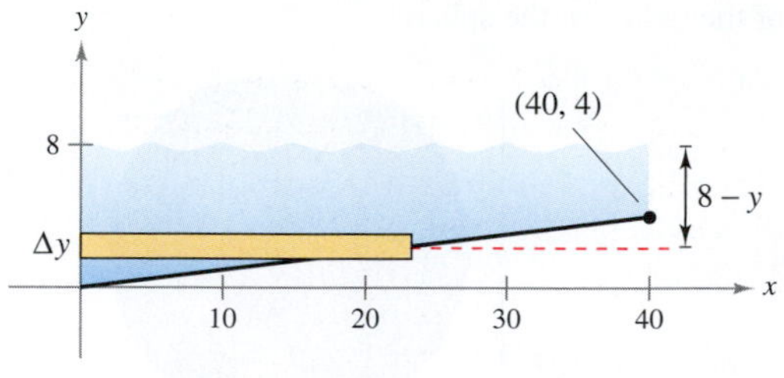

AP* Review Questions for Chapter 7

1. (calculator)

The base of a solid is the region in the first quadrant bounded above by the line $y = 2$, below by $y = \sin^{-1} x$, and to the right by the line $x = 1$. For this solid, each cross-section perpendicular to the x-axis is a square. What is the volume of the solid?

(A) 1.429

(B) 2

(C) 2.184

(D) 0.766

(E) 4

2. (no calculator)

Find the area of the region bounded by the y-axis, the line $y = e$, and the graph of the function $y = e^{3x}$.

(A) $\frac{1}{3}$

(B) $e^{3e} - \frac{1}{3}$

(C) $1 - \frac{2}{3}e$

(D) $3 - \frac{8}{3}e$

(E) $\frac{1}{3} + e^2 - e^{3e}$

3. (no calculator)

A region in the xy-plane is bounded by the curves $y = 4x - x^2$ and $y = 2x - 3$.

(a) Find the points of intersection of the two curves.

(b) Sketch the region bounded by the curves. Label the bounding curves, show and label all points of intersection, and shade the bounded region.

(c) Find the area of the region, showing all work that leads to your answer.

4. (no calculator)

A region in the xy-plane is bounded by $y = 2x + 2$, $x = \frac{y^2}{2} + 2$, $y = -2$, and $y = 2$.

(a) Sketch the bounded region on a Cartesian axis system. Label each boundary curve and shade the bounded region.

(b) Find the area of the bounded region, showing all work that leads to your answer.

5. (no calculator)

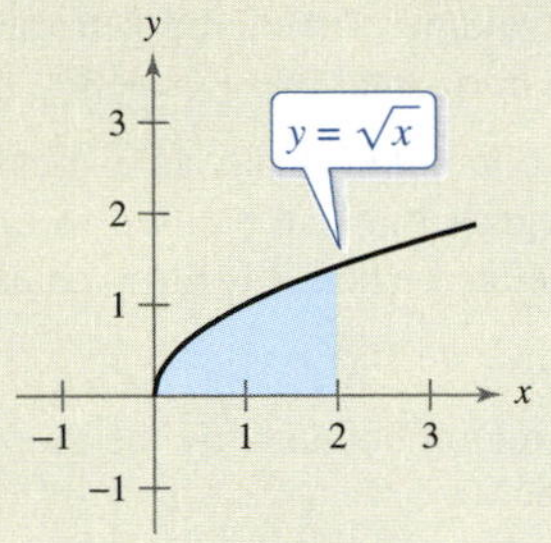

Consider the region shown above, which is bounded by $f(x) = \sqrt{x}$, $y = 0$, $x = 0$, and $x = 2$.

(a) Find the area of the region.

(b) Find the volume of the solid formed by rotating the region about the x-axis.

(c) The region pictured above is the base of a solid. For this solid, each cross-section perpendicular to the x-axis is an equilateral triangle. Find the volume of this solid.

6. (no calculator)

Consider the region R, bounded by the graphs of $y = x^3$, $y = 8$ and the y-axis. The region S is bounded by $y = x^3$, $x = 2$ and the x-axis.

(a) Find the area of region R.

(b) Find the volume of the solid formed by rotating region R about the y-axis.

(c) The region S is the base of a solid. For this solid, each cross-section perpendicular to the x-axis is a semi-circle with diameters extending from $y = x^3$ to the x-axis. Find the volume of this solid.

7. (BC only, calculator)

Consider the region bounded be the y-axis, $y = 10$, and $y = 1 + 6x^{3/2}$.

(a) Set up, but do not evaluate, an integral equation that will find the value of k so that $x = k$ cuts the region into 2 parts of equal area.

(b) Find the length of the curve $y = 1 + 6x^{3/2}$ on the interval $[0, 1]$.

(c) The region is the base of a solid. For this solid, the cross-sections perpendicular to the x-axis are rectangles with a height of 3 times that of its width. Find the volume of this solid.

8. (calculator)

Let R be the region bounded by the graph of $y = \ln x$ and the line $y = 2x - 3$.

(a) Find the area of R.

(b) Find the volume of the solid generated when R is rotated about the horizontal line $y = -3$.

(c) Write, but do not evaluate, an expression involving one or more integrals that can be used to find the volume of the solid generated when R is revolved about the y-axis.

9. (calculator)

Let R be the region bounded by the graph of $y = x^2 - 1$ and the graph of $x = y^2$.

(a) Find the area of R.

(b) Find the volume of the solid generated when R is rotated about the vertical line $x = 2$.

(c) Write, but do not evaluate, an expression involving one or more integrals to find the volume of the solid generated when R is rotated about the horizontal line $y = -1$.

10. (no calculator)

Consider the region bounded by the graphs of $f(x) = \sqrt{x}$, $y = 0$, and $x = 2$.

(a) Find the volume of the solid formed by rotating the region about the x-axis.

(b) Find the volume of the solid formed by rotating the region about the y-axis.

(c) Write, but do not evaluate, the volume integral of the solid formed by rotating the region about the line $y = -2$.

8 Integration Techniques, L'Hôpital's Rule, and Improper Integrals

Sending a Space Module into Orbit *(Example 5, p. 571)*

Chemical Reaction *(Exercise 50, p. 550)*

Fluid Force *(Exercise 69, p. 541)*

Power Lines *(Section Project, p. 532)*

Memory Model *(Exercise 88, p. 523)*

8.1 Basic Integration Rules

- Review procedures for fitting an integrand to one of the basic integration rules.

REVIEW OF BASIC INTEGRATION RULES ($a > 0$)

1. $\int kf(u)\,du = k\int f(u)\,du$
2. $\int [f(u) \pm g(u)]\,du = \int f(u)\,du \pm \int g(u)\,du$
3. $\int du = u + C$
4. $\int u^n\,du = \dfrac{u^{n+1}}{n+1} + C,\quad n \neq -1$
5. $\int \dfrac{du}{u} = \ln|u| + C$
6. $\int e^u\,du = e^u + C$
7. $\int a^u\,du = \left(\dfrac{1}{\ln a}\right)a^u + C$
8. $\int \sin u\,du = -\cos u + C$
9. $\int \cos u\,du = \sin u + C$
10. $\int \tan u\,du = -\ln|\cos u| + C$
11. $\int \cot u\,du = \ln|\sin u| + C$
12. $\int \sec u\,du = \ln|\sec u + \tan u| + C$
13. $\int \csc u\,du = -\ln|\csc u + \cot u| + C$
14. $\int \sec^2 u\,du = \tan u + C$
15. $\int \csc^2 u\,du = -\cot u + C$
16. $\int \sec u \tan u\,du = \sec u + C$
17. $\int \csc u \cot u\,du = -\csc u + C$
18. $\int \dfrac{du}{\sqrt{a^2 - u^2}} = \arcsin\dfrac{u}{a} + C$
19. $\int \dfrac{du}{a^2 + u^2} = \dfrac{1}{a}\arctan\dfrac{u}{a} + C$
20. $\int \dfrac{du}{u\sqrt{u^2 - a^2}} = \dfrac{1}{a}\,\text{arcsec}\,\dfrac{|u|}{a} + C$

Fitting Integrands to Basic Integration Rules

In this chapter, you will study several integration techniques that greatly expand the set of integrals to which the basic integration rules can be applied. These rules are reviewed at the left. A major step in solving any integration problem is recognizing which basic integration rule to use.

EXAMPLE 1 A Comparison of Three Similar Integrals

See LarsonCalculus.com for an interactive version of this type of example.

Find each integral.

a. $\displaystyle\int \frac{4}{x^2+9}\,dx$ **b.** $\displaystyle\int \frac{4x}{x^2+9}\,dx$ **c.** $\displaystyle\int \frac{4x^2}{x^2+9}\,dx$

Solution

a. Use the Arctangent Rule and let $u = x$ and $a = 3$.

$$\int \frac{4}{x^2+9}\,dx = 4\int \frac{1}{x^2+3^2}\,dx \qquad \text{Constant Multiple Rule}$$

$$= 4\left(\frac{1}{3}\arctan\frac{x}{3}\right) + C \qquad \text{Arctangent Rule}$$

$$= \frac{4}{3}\arctan\frac{x}{3} + C \qquad \text{Simplify.}$$

b. The Arctangent Rule does not apply because the numerator contains a factor of x. Consider the Log Rule and let $u = x^2 + 9$. Then $du = 2x\,dx$, and you have

$$\int \frac{4x}{x^2+9}\,dx = 2\int \frac{2x\,dx}{x^2+9} \qquad \text{Constant Multiple Rule}$$

$$= 2\int \frac{du}{u} \qquad \text{Substitution: } u = x^2 + 9$$

$$= 2\ln|u| + C \qquad \text{Log Rule}$$

$$= 2\ln(x^2+9) + C. \qquad \text{Rewrite as a function of } x.$$

c. Because the degree of the numerator is equal to the degree of the denominator, you should first use division to rewrite the improper rational function as the sum of a polynomial and a proper rational function.

$$\int \frac{4x^2}{x^2+9}\,dx = \int \left(4 + \frac{-36}{x^2+9}\right)dx \qquad \text{Rewrite using long division.}$$

$$= \int 4\,dx - 36\int \frac{1}{x^2+9}\,dx \qquad \text{Write as two integrals.}$$

$$= 4x - 36\left(\frac{1}{3}\arctan\frac{x}{3}\right) + C \qquad \text{Integrate.}$$

$$= 4x - 12\arctan\frac{x}{3} + C \qquad \text{Simplify.}$$

Note in Example 1(c) that some algebra is required before applying any integration rules, and more than one rule is needed to evaluate the resulting integral.

EXAMPLE 2 Using Two Basic Rules to Solve a Single Integral

Evaluate $\displaystyle\int_0^1 \frac{x+3}{\sqrt{4-x^2}}\,dx.$

Solution Begin by writing the integral as the sum of two integrals. Then apply the Power Rule and the Arcsine Rule.

$$\begin{aligned}\int_0^1 \frac{x+3}{\sqrt{4-x^2}}\,dx &= \int_0^1 \frac{x}{\sqrt{4-x^2}}\,dx + \int_0^1 \frac{3}{\sqrt{4-x^2}}\,dx \\ &= -\frac{1}{2}\int_0^1 (4-x^2)^{-1/2}(-2x)\,dx + 3\int_0^1 \frac{1}{\sqrt{2^2-x^2}}\,dx \\ &= \left[-(4-x^2)^{1/2} + 3\arcsin\frac{x}{2}\right]_0^1 \\ &= \left(-\sqrt{3}+\frac{\pi}{2}\right) - (-2+0) \\ &\approx 1.839\end{aligned}$$

See Figure 8.1.

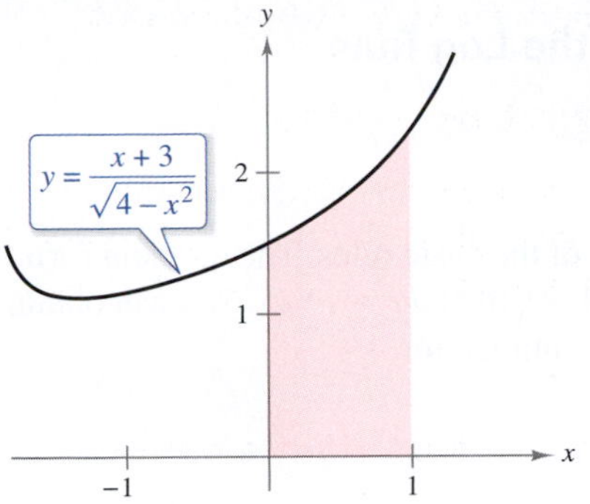

The area of the region is approximately 1.839.

Figure 8.1

▷ **TECHNOLOGY** Simpson's Rule can be used to give a good approximation of the value of the integral in Example 2 (for $n = 10$, the approximation is 1.839). When using numerical integration, however, you should be aware that Simpson's Rule does not always give good approximations when one or both of the limits of integration are near a vertical asymptote. For instance, using the Fundamental Theorem of Calculus, you can obtain

$$\int_0^{1.99} \frac{x+3}{\sqrt{4-x^2}}\,dx \approx 6.213.$$

For $n = 10$, Simpson's Rule gives an approximation of 6.889.

Rules 18, 19, and 20 of the basic integration rules on the preceding page all have expressions involving the sum or difference of two squares:

$$a^2 - u^2, \quad a^2 + u^2, \quad \text{and} \quad u^2 - a^2.$$

These expressions are often apparent after a u-substitution, as shown in Example 3.

EXAMPLE 3 A Substitution Involving $a^2 - u^2$

Find $\displaystyle\int \frac{x^2}{\sqrt{16-x^6}}\,dx.$

Solution Because the radical in the denominator can be written in the form

$$\sqrt{a^2-u^2} = \sqrt{4^2-(x^3)^2}$$

you can try the substitution $u = x^3$. Then $du = 3x^2\,dx$, and you have

$$\begin{aligned}\int \frac{x^2}{\sqrt{16-x^6}}\,dx &= \frac{1}{3}\int \frac{3x^2\,dx}{\sqrt{16-(x^3)^2}} && \text{Rewrite integral.} \\ &= \frac{1}{3}\int \frac{du}{\sqrt{4^2-u^2}} && \text{Substitution: } u = x^3 \\ &= \frac{1}{3}\arcsin\frac{u}{4} + C && \text{Arcsine Rule} \\ &= \frac{1}{3}\arcsin\frac{x^3}{4} + C. && \text{Rewrite as a function of } x.\end{aligned}$$

Exploration

A Comparison of Three Similar Integrals Which, if any, of the integrals listed below can be evaluated using the 20 basic integration rules? For any that can be evaluated, do so. For any that cannot, explain why not.

a. $\displaystyle\int \frac{3}{\sqrt{1-x^2}}\,dx$

b. $\displaystyle\int \frac{3x}{\sqrt{1-x^2}}\,dx$

c. $\displaystyle\int \frac{3x^2}{\sqrt{1-x^2}}\,dx$

Two of the most commonly overlooked integration rules are the Log Rule and the Power Rule. Notice in the next two examples how these two integration rules can be disguised.

EXAMPLE 4 A Disguised Form of the Log Rule

Find $\int \frac{1}{1 + e^x}\, dx$.

Solution The integral does not appear to fit any of the basic rules. The quotient form, however, suggests the Log Rule. If you let $u = 1 + e^x$, then $du = e^x\, dx$. You can obtain the required du by adding and subtracting e^x in the numerator.

$$\int \frac{1}{1 + e^x}\, dx = \int \frac{1 + e^x - e^x}{1 + e^x}\, dx \qquad \text{Add and subtract } e^x \text{ in numerator.}$$

$$= \int \left(\frac{1 + e^x}{1 + e^x} - \frac{e^x}{1 + e^x}\right) dx \qquad \text{Rewrite as two fractions.}$$

$$= \int dx - \int \frac{e^x\, dx}{1 + e^x} \qquad \text{Rewrite as two integrals.}$$

$$= x - \ln(1 + e^x) + C \qquad \text{Integrate.}$$

REMARK Remember that you can separate numerators but not denominators. Watch out for this common error when fitting integrands to basic rules. For instance, you cannot separate denominators in Example 4.

$$\frac{1}{1 + e^x} \neq \frac{1}{1} + \frac{1}{e^x}$$

There is usually more than one way to solve an integration problem. For instance, in Example 4, try integrating by multiplying the numerator and denominator by e^{-x} to obtain an integral of the form $-\int du/u$. See if you can get the same answer by this procedure. (Be careful: the answer will appear in a different form.)

EXAMPLE 5 A Disguised Form of the Power Rule

Find $\int (\cot x)[\ln(\sin x)]\, dx$.

Solution Again, the integral does not appear to fit any of the basic rules. However, considering the two primary choices for u

$$u = \cot x \quad \text{or} \quad u = \ln(\sin x)$$

you can see that the second choice is the appropriate one because

$$u = \ln(\sin x) \quad \text{and} \quad du = \frac{\cos x}{\sin x}\, dx = \cot x\, dx.$$

So,

$$\int (\cot x)[\ln(\sin x)]\, dx = \int u\, du \qquad \text{Substitution: } u = \ln(\sin x)$$

$$= \frac{u^2}{2} + C \qquad \text{Integrate.}$$

$$= \frac{1}{2}[\ln(\sin x)]^2 + C. \qquad \text{Rewrite as a function of } x.$$

In Example 5, try *checking* that the derivative of

$$\frac{1}{2}[\ln(\sin x)]^2 + C$$

is the integrand of the original integral.

Trigonometric identities can often be used to fit integrals to one of the basic integration rules.

EXAMPLE 6 **Using Trigonometric Identities**

Find $\int \tan^2 2x \, dx$.

▷ **TECHNOLOGY** If you have access to a computer algebra system, try using it to evaluate the integrals in this section. Compare the *forms* of the antiderivatives given by the software with the forms obtained by hand. Sometimes the forms will be the same, but often they will differ. For instance, why is the antiderivative $\ln 2x + C$ equivalent to the antiderivative $\ln x + C$?

Solution Note that $\tan^2 u$ is not in the list of basic integration rules. However, $\sec^2 u$ is in the list. This suggests the trigonometric identity $\tan^2 u = \sec^2 u - 1$. If you let $u = 2x$, then $du = 2\,dx$ and

$$\int \tan^2 2x \, dx = \frac{1}{2}\int \tan^2 u \, du \qquad \text{Substitution: } u = 2x$$

$$= \frac{1}{2}\int (\sec^2 u - 1) \, du \qquad \text{Trigonometric identity}$$

$$= \frac{1}{2}\int \sec^2 u \, du - \frac{1}{2}\int du \qquad \text{Rewrite as two integrals.}$$

$$= \frac{1}{2}\tan u - \frac{u}{2} + C \qquad \text{Integrate.}$$

$$= \frac{1}{2}\tan 2x - x + C. \qquad \text{Rewrite as a function of } x.$$

This section concludes with a summary of the common procedures for fitting integrands to the basic integration rules.

PROCEDURES FOR FITTING INTEGRANDS TO BASIC INTEGRATION RULES

Technique	Example
Expand (numerator).	$(1 + e^x)^2 = 1 + 2e^x + e^{2x}$
Separate numerator.	$\dfrac{1 + x}{x^2 + 1} = \dfrac{1}{x^2 + 1} + \dfrac{x}{x^2 + 1}$
Complete the square.	$\dfrac{1}{\sqrt{2x - x^2}} = \dfrac{1}{\sqrt{1 - (x - 1)^2}}$
Divide improper rational function.	$\dfrac{x^2}{x^2 + 1} = 1 - \dfrac{1}{x^2 + 1}$
Add and subtract terms in numerator.	$\dfrac{2x}{x^2 + 2x + 1} = \dfrac{2x + 2 - 2}{x^2 + 2x + 1} = \dfrac{2x + 2}{x^2 + 2x + 1} - \dfrac{2}{(x + 1)^2}$
Use trigonometric identities.	$\cot^2 x = \csc^2 x - 1$
Multiply and divide by Pythagorean conjugate.	$\dfrac{1}{1 + \sin x} = \left(\dfrac{1}{1 + \sin x}\right)\left(\dfrac{1 - \sin x}{1 - \sin x}\right) = \dfrac{1 - \sin x}{1 - \sin^2 x} = \dfrac{1 - \sin x}{\cos^2 x} = \sec^2 x - \dfrac{\sin x}{\cos^2 x}$

8.1 Exercises

See **CalcChat.com** for tutorial help and worked-out solutions to odd-numbered exercises.

Choosing an Antiderivative **In Exercises 1–4, select the correct antiderivative.**

1. $\dfrac{dy}{dx} = \dfrac{x}{\sqrt{x^2+1}}$

(a) $2\sqrt{x^2+1} + C$ (b) $\sqrt{x^2+1} + C$

(c) $\frac{1}{2}\sqrt{x^2+1} + C$ (d) $\ln(x^2+1) + C$

2. $\dfrac{dy}{dx} = \dfrac{x}{x^2+1}$

(a) $\ln\sqrt{x^2+1} + C$ (b) $\dfrac{2x}{(x^2+1)^2} + C$

(c) $\arctan x + C$ (d) $\ln(x^2+1) + C$

3. $\dfrac{dy}{dx} = \dfrac{1}{x^2+1}$

(a) $\ln\sqrt{x^2+1} + C$ (b) $\dfrac{2x}{(x^2+1)^2} + C$

(c) $\arctan x + C$ (d) $\ln(x^2+1) + C$

4. $\dfrac{dy}{dx} = x\cos(x^2+1)$

(a) $2x\sin(x^2+1) + C$ (b) $-\frac{1}{2}\sin(x^2+1) + C$

(c) $\frac{1}{2}\sin(x^2+1) + C$ (d) $-2x\sin(x^2+1) + C$

Choosing a Formula **In Exercises 5–14, select the basic integration formula you can use to find the integral, and identify *u* and *a* when appropriate.**

5. $\displaystyle\int (5x-3)^4\,dx$ **6.** $\displaystyle\int \frac{2t+1}{t^2+t-4}\,dt$

7. $\displaystyle\int \frac{1}{\sqrt{x}(1-2\sqrt{x})}\,dx$ **8.** $\displaystyle\int \frac{2}{(2t-1)^2+4}\,dt$

9. $\displaystyle\int \frac{3}{\sqrt{1-t^2}}\,dt$ **10.** $\displaystyle\int \frac{-2x}{\sqrt{x^2-4}}\,dx$

11. $\displaystyle\int t\sin t^2\,dt$ **12.** $\displaystyle\int \sec 5x\tan 5x\,dx$

13. $\displaystyle\int (\cos x)e^{\sin x}\,dx$ **14.** $\displaystyle\int \frac{1}{x\sqrt{x^2-4}}\,dx$

Finding an Indefinite Integral **In Exercises 15–46, find the indefinite integral.**

15. $\displaystyle\int 14(x-5)^6\,dx$ **16.** $\displaystyle\int \frac{5}{(t+6)^3}\,dt$

17. $\displaystyle\int \frac{7}{(z-10)^7}\,dz$ **18.** $\displaystyle\int t^3\sqrt{t^4+1}\,dt$

19. $\displaystyle\int \left[v + \frac{1}{(3v-1)^3}\right] dv$ **20.** $\displaystyle\int \left[4x - \frac{2}{(2x+3)^2}\right] dx$

21. $\displaystyle\int \frac{t^2-3}{-t^3+9t+1}\,dt$ **22.** $\displaystyle\int \frac{x+1}{\sqrt{3x^2+6x}}\,dx$

23. $\displaystyle\int \frac{x^2}{x-1}\,dx$ **24.** $\displaystyle\int \frac{3x}{x+4}\,dx$

25. $\displaystyle\int \frac{e^x}{1+e^x}\,dx$ **26.** $\displaystyle\int \left(\frac{1}{2x+5} - \frac{1}{2x-5}\right) dx$

27. $\displaystyle\int (5+4x^2)^2\,dx$ **28.** $\displaystyle\int x\left(3+\frac{2}{x}\right)^2 dx$

29. $\displaystyle\int x\cos 2\pi x^2\,dx$ **30.** $\displaystyle\int \csc \pi x\cot \pi x\,dx$

31. $\displaystyle\int \frac{\sin x}{\sqrt{\cos x}}\,dx$ **32.** $\displaystyle\int \csc^2 x e^{\cot x}\,dx$

33. $\displaystyle\int \frac{2}{e^{-x}+1}\,dx$ **34.** $\displaystyle\int \frac{2}{7e^x+4}\,dx$

35. $\displaystyle\int \frac{\ln x^2}{x}\,dx$ **36.** $\displaystyle\int (\tan x)[\ln(\cos x)]\,dx$

37. $\displaystyle\int \frac{1+\cos\alpha}{\sin\alpha}\,d\alpha$ **38.** $\displaystyle\int \frac{1}{\cos\theta - 1}\,d\theta$

39. $\displaystyle\int \frac{-1}{\sqrt{1-(4t+1)^2}}\,dt$ **40.** $\displaystyle\int \frac{1}{25+4x^2}\,dx$

41. $\displaystyle\int \frac{\tan(2/t)}{t^2}\,dt$ **42.** $\displaystyle\int \frac{e^{1/t}}{t^2}\,dt$

43. $\displaystyle\int \frac{6}{\sqrt{10x-x^2}}\,dx$

44. $\displaystyle\int \frac{1}{(x-1)\sqrt{4x^2-8x+3}}\,dx$

45. $\displaystyle\int \frac{4}{4x^2+4x+65}\,dx$ **46.** $\displaystyle\int \frac{1}{x^2-4x+9}\,dx$

Slope Field **In Exercises 47 and 48, a differential equation, a point, and a slope field are given. (a) Sketch two approximate solutions of the differential equation on the slope field, one of which passes through the given point. (b) Use integration to find the particular solution of the differential equation and use a graphing utility to graph the solution. Compare the result with the sketches in part (a). To print an enlarged copy of the graph, go to *MathGraphs.com*.**

47. $\dfrac{ds}{dt} = \dfrac{t}{\sqrt{1-t^4}}$

$\left(0, -\dfrac{1}{2}\right)$

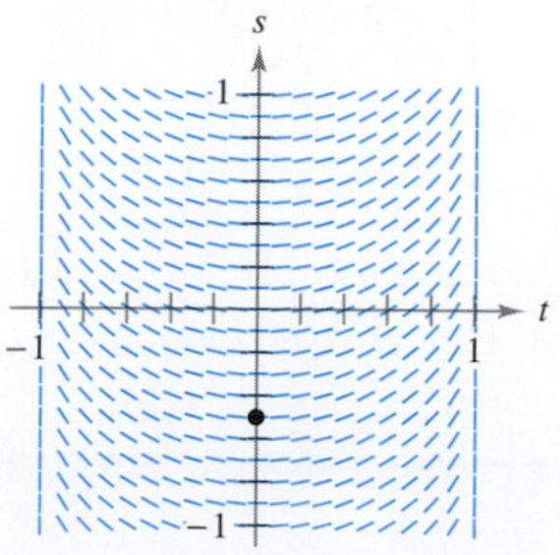

48. $\dfrac{dy}{dx} = \dfrac{1}{\sqrt{4x-x^2}}$

$\left(2, \dfrac{1}{2}\right)$

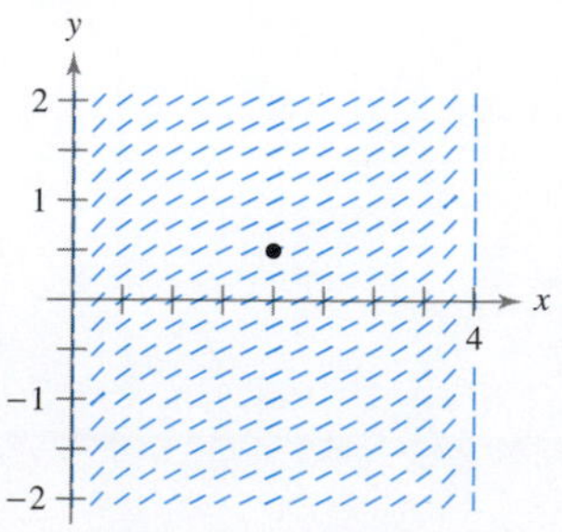

Slope Field **In Exercises 49 and 50, use a computer algebra system to graph the slope field for the differential equation and graph the solution through the specified initial condition.**

49. $\frac{dy}{dx} = 0.8y, \; y(0) = 4$

50. $\frac{dy}{dx} = 5 - y, \; y(0) = 1$

Differential Equation **In Exercises 51–56, solve the differential equation.**

51. $\frac{dy}{dx} = (e^x + 5)^2$

52. $\frac{dy}{dx} = (4 - e^{2x})^2$

53. $\frac{dr}{dt} = \frac{10e^t}{\sqrt{1 - e^{2t}}}$

54. $\frac{dr}{dt} = \frac{(1 + e^t)^2}{e^{3t}}$

55. $(4 + \tan^2 x)y' = \sec^2 x$

56. $y' = \frac{1}{x\sqrt{4x^2 - 9}}$

Evaluating a Definite Integral **In Exercises 57–64, evaluate the definite integral. Use the integration capabilities of a graphing utility to verify your result.**

57. $\int_0^{\pi/4} \cos 2x \, dx$

58. $\int_0^{\pi} \sin^2 t \cos t \, dt$

59. $\int_0^1 xe^{-x^2} \, dx$

60. $\int_1^e \frac{1 - \ln x}{x} \, dx$

61. $\int_0^8 \frac{2x}{\sqrt{x^2 + 36}} \, dx$

62. $\int_1^3 \frac{2x^2 + 3x - 2}{x} \, dx$

63. $\int_0^{2/\sqrt{3}} \frac{1}{4 + 9x^2} \, dx$

64. $\int_0^7 \frac{1}{\sqrt{100 - x^2}} \, dx$

Area **In Exercises 65–68, find the area of the region.**

65. $y = (-4x + 6)^{3/2}$

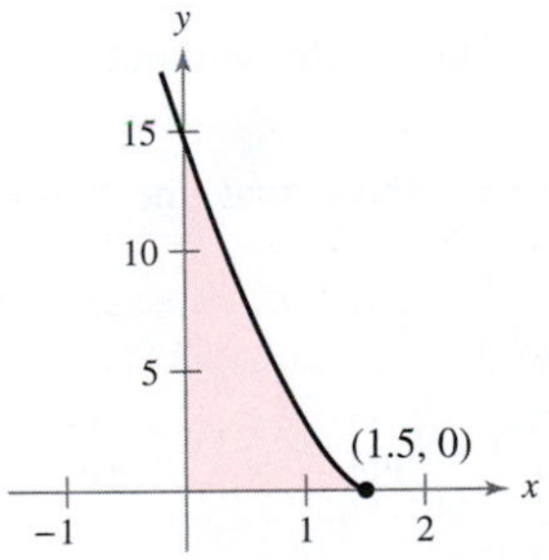

66. $y = \frac{3x + 2}{x^2 + 9}$

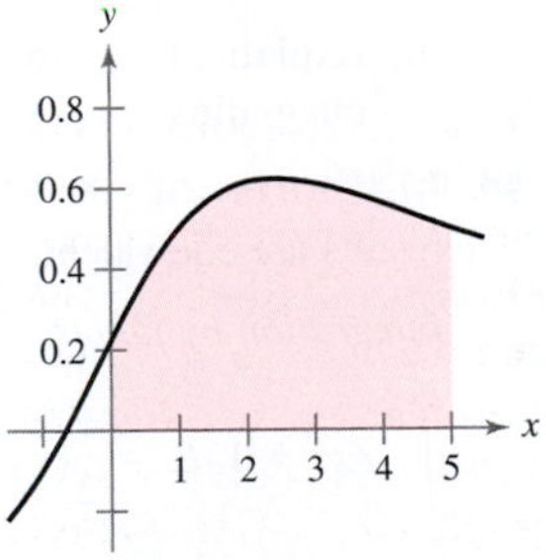

67. $y^2 = x^2(1 - x^2)$

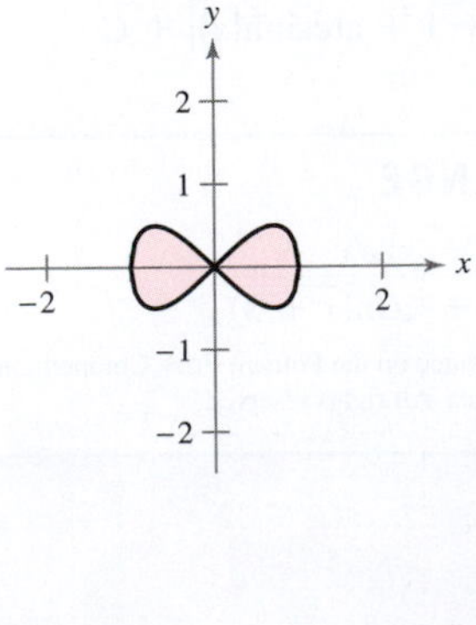

68. $y = \sin 2x$

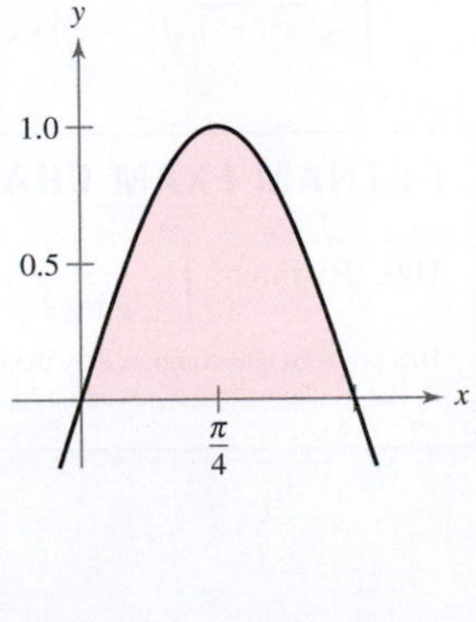

Finding an Integral Using Technology **In Exercises 69–72, use a computer algebra system to find the integral. Use the computer algebra system to graph two antiderivatives. Describe the relationship between the graphs of the two antiderivatives.**

69. $\int \frac{1}{x^2 + 4x + 13} \, dx$

70. $\int \frac{x - 2}{x^2 + 4x + 13} \, dx$

71. $\int \frac{1}{1 + \sin \theta} \, d\theta$

72. $\int \left(\frac{e^x + e^{-x}}{2}\right)^3 dx$

WRITING ABOUT CONCEPTS

Choosing a Formula **In Exercises 73–76, state the integration formula you would use to perform the integration. Explain why you chose that formula. Do not integrate.**

73. $\int x(x^2 + 1)^3 \, dx$

74. $\int x \sec(x^2 + 1) \tan(x^2 + 1) \, dx$

75. $\int \frac{x}{x^2 + 1} \, dx$

76. $\int \frac{1}{x^2 + 1} \, dx$

77. **Finding Constants** Determine the constants a and b such that

$$\sin x + \cos x = a \sin(x + b).$$

Use this result to integrate

$$\int \frac{dx}{\sin x + \cos x}.$$

78. **Deriving a Rule** Show that

$$\sec x = \frac{\sin x}{\cos x} + \frac{\cos x}{1 + \sin x}.$$

Then use this identity to derive the basic integration rule

$$\int \sec x \, dx = \ln|\sec x + \tan x| + C.$$

79. **Area** The graphs of $f(x) = x$ and $g(x) = ax^2$ intersect at the points $(0, 0)$ and $(1/a, 1/a)$. Find a $(a > 0)$ such that the area of the region bounded by the graphs of these two functions is $\frac{2}{3}$.

80. **Think About It** When evaluating

$$\int_{-1}^{1} x^2 \, dx$$

is it appropriate to substitute

$$u = x^2, \quad x = \sqrt{u}, \quad \text{and} \quad dx = \frac{du}{2\sqrt{u}}$$

to obtain

$$\frac{1}{2}\int_1^1 \sqrt{u} \, du = 0?$$

Explain.

81. Comparing Antiderivatives

(a) Explain why the antiderivative $y_1 = e^{x+C_1}$ is equivalent to the antiderivative $y_2 = Ce^x$.

(b) Explain why the antiderivative $y_1 = \sec^2 x + C_1$ is equivalent to the antiderivative $y_2 = \tan^2 x + C$.

82. HOW DO YOU SEE IT? Using the graph, is $\int_0^5 f(x)\,dx$ positive or negative? Explain.

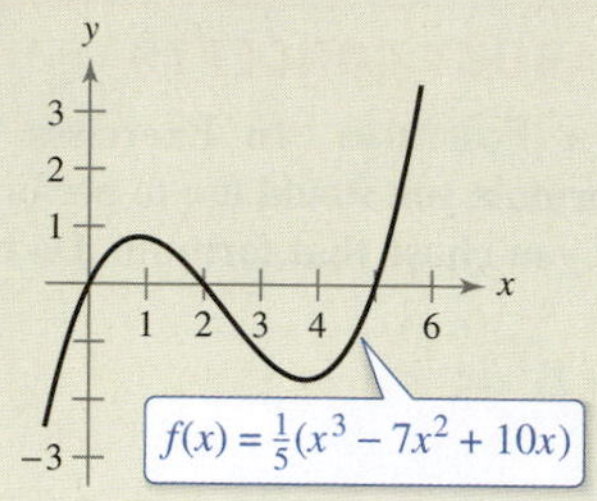

Approximation **In Exercises 83 and 84, determine which value best approximates the area of the region between the x-axis and the function over the given interval. (Make your selection on the basis of a sketch of the region and *not* by integrating.)**

83. $f(x) = \dfrac{4x}{x^2+1}$, $[0, 2]$

(a) 3 (b) 1 (c) −8 (d) 8 (e) 10

84. $f(x) = \dfrac{4}{x^2+1}$, $[0, 2]$

(a) 3 (b) 1 (c) −4 (d) 4 (e) 10

Interpreting Integrals **In Exercises 85 and 86, (a) sketch the region whose area is given by the integral, (b) sketch the solid whose volume is given by the integral when the disk method is used, and (c) sketch the solid whose volume is given by the integral when the shell method is used. (There is more than one correct answer for each part.)**

85. $\displaystyle\int_0^2 2\pi x^2\,dx$

86. $\displaystyle\int_0^4 \pi y\,dy$

87. Volume The region bounded by $y = e^{-x^2}$, $y = 0$, $x = 0$, and $x = b$ $(b > 0)$ is revolved about the y-axis.

(a) Find the volume of the solid generated when $b = 1$.

(b) Find b such that the volume of the generated solid is $\frac{4}{3}$ cubic units.

88. Volume Consider the region bounded by the graphs of $x = 0$, $y = \cos x^2$, $y = \sin x^2$, and $x = \sqrt{\pi/2}$. Find the volume of the solid generated by revolving the region about the y-axis.

89. Arc Length Find the arc length of the graph of $y = \ln(\sin x)$ from $x = \pi/4$ to $x = \pi/2$.

90. Arc Length Find the arc length of the graph of $y = \ln(\cos x)$ from $x = 0$ to $x = \pi/3$.

91. Surface Area Find the area of the surface formed by revolving the graph of $y = 2\sqrt{x}$ on the interval $[0, 9]$ about the x-axis.

92. Centroid Find the x-coordinate of the centroid of the region bounded by the graphs of

$$y = \frac{5}{\sqrt{25 - x^2}}, \quad y = 0, \quad x = 0, \quad \text{and} \quad x = 4.$$

Average Value of a Function **In Exercises 93 and 94, find the average value of the function over the given interval.**

93. $f(x) = \dfrac{1}{1+x^2}$, $-3 \le x \le 3$

94. $f(x) = \sin nx$, $0 \le x \le \pi/n$, n is a positive integer.

Arc Length **In Exercises 95 and 96, use the integration capabilities of a graphing utility to approximate the arc length of the curve over the given interval.**

95. $y = \tan \pi x$, $\left[0, \frac{1}{4}\right]$

96. $y = x^{2/3}$, $[1, 8]$

97. Finding a Pattern

(a) Find $\int \cos^3 x\,dx$.

(b) Find $\int \cos^5 x\,dx$.

(c) Find $\int \cos^7 x\,dx$.

(d) Explain how to find $\int \cos^{15} x\,dx$ without actually integrating.

98. Finding a Pattern

(a) Write $\int \tan^3 x\,dx$ in terms of $\int \tan x\,dx$. Then find $\int \tan^3 x\,dx$.

(b) Write $\int \tan^5 x\,dx$ in terms of $\int \tan^3 x\,dx$.

(c) Write $\int \tan^{2k+1} x\,dx$, where k is a positive integer, in terms of $\int \tan^{2k-1} x\,dx$.

(d) Explain how to find $\int \tan^{15} x\,dx$ without actually integrating.

99. Methods of Integration Show that the following results are equivalent.

Integration by tables:

$$\int \sqrt{x^2+1}\,dx = \frac{1}{2}\left(x\sqrt{x^2+1} + \ln\left|x + \sqrt{x^2+1}\right|\right) + C$$

Integration by computer algebra system:

$$\int \sqrt{x^2+1}\,dx = \frac{1}{2}\left[x\sqrt{x^2+1} + \operatorname{arcsinh}(x)\right] + C$$

PUTNAM EXAM CHALLENGE

100. Evaluate $\displaystyle\int_2^4 \frac{\sqrt{\ln(9-x)}\,dx}{\sqrt{\ln(9-x)} + \sqrt{\ln(x+3)}}$.

This problem was composed by the Committee on the Putnam Prize Competition.

8.2 Integration by Parts

■ **Find an antiderivative using integration by parts.**

Integration by Parts

In this section, you will study an important integration technique called **integration by parts.** This technique can be applied to a wide variety of functions and is particularly useful for integrands involving *products* of algebraic and transcendental functions. For instance, integration by parts works well with integrals such as

$$\int x \ln x \, dx, \quad \int x^2 e^x \, dx, \quad \text{and} \quad \int e^x \sin x \, dx.$$

Integration by parts is based on the formula for the derivative of a product

$$\frac{d}{dx}[uv] = u\frac{dv}{dx} + v\frac{du}{dx}$$
$$= uv' + vu'$$

where both u and v are differentiable functions of x. When u' and v' are continuous, you can integrate both sides of this equation to obtain

$$uv = \int uv' \, dx + \int vu' \, dx$$
$$= \int u \, dv + \int v \, du.$$

By rewriting this equation, you obtain the next theorem.

THEOREM 8.1 Integration by Parts

If u and v are functions of x and have continuous derivatives, then

$$\int u \, dv = uv - \int v \, du.$$

Exploration

Proof Without Words Here is a different approach to proving the formula for integration by parts. This approach is from "Proof Without Words: Integration by Parts" by Roger B. Nelsen, *Mathematics Magazine,* 64, No. 2, April 1991, p. 130, by permission of the author.

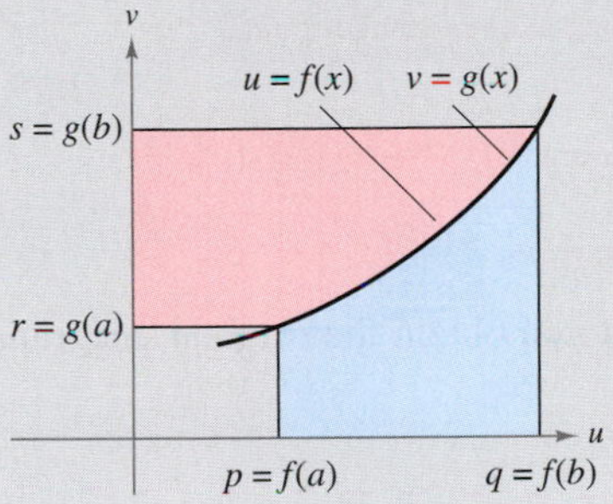

Area (pink) + Area (blue) $= qs - pr$

$$\int_r^s u \, dv + \int_q^p v \, du = \Big[uv\Big]_{(p,\,r)}^{(q,\,s)}$$

$$\int_r^s u \, dv = \Big[uv\Big]_{(p,\,r)}^{(q,\,s)} - \int_q^p v \, du$$

Explain how this graph proves the theorem. Which notation in this proof is unfamiliar? What do you think it means?

This formula expresses the original integral in terms of another integral. Depending on the choices of u and dv, it may be easier to evaluate the second integral than the original one. Because the choices of u and dv are critical in the integration by parts process, the guidelines below are provided.

GUIDELINES FOR INTEGRATION BY PARTS

1. Try letting dv be the most complicated portion of the integrand that fits a basic integration rule. Then u will be the remaining factor(s) of the integrand.
2. Try letting u be the portion of the integrand whose derivative is a function simpler than u. Then dv will be the remaining factor(s) of the integrand.

Note that dv always includes the dx of the original integrand.

AP* Tips

Integration by parts is tested on the AP Calculus BC Exam, primarily on the non-calculator multiple choice section.

When using integration by parts, note that you can first choose dv or first choose u. After you choose, however, the choice of the other factor is determined—it must be the remaining portion of the integrand. Also note that dv must contain the differential dx of the original integral.

EXAMPLE 1 Integration by Parts

Find $\int xe^x \, dx$.

Solution To apply integration by parts, you need to write the integral in the form $\int u \, dv$. There are several ways to do this.

$$\int \underbrace{(x)}_{u} \underbrace{(e^x dx)}_{dv}, \quad \int \underbrace{(e^x)}_{u} \underbrace{(x\,dx)}_{dv}, \quad \int \underbrace{(1)}_{u} \underbrace{(xe^x\,dx)}_{dv}, \quad \int \underbrace{(xe^x)}_{u} \underbrace{(dx)}_{dv}$$

The guidelines on the preceding page suggest the first option because the derivative of $u = x$ is simpler than x, and $dv = e^x \, dx$ is the most complicated portion of the integrand that fits a basic integration formula.

$$dv = e^x \, dx \quad \Rightarrow \quad v = \int dv = \int e^x \, dx = e^x$$

$$u = x \quad \Rightarrow \quad du = dx$$

Now, integration by parts produces

$$\int u \, dv = uv - \int v \, du \qquad \text{Integration by parts formula}$$

$$\int xe^x \, dx = xe^x - \int e^x \, dx \qquad \text{Substitute.}$$

$$= xe^x - e^x + C. \qquad \text{Integrate.}$$

To check this, differentiate $xe^x - e^x + C$ to see that you obtain the original integrand.

REMARK In Example 1, note that it is not necessary to include a constant of integration when solving

$$v = \int e^x \, dx = e^x + C_1.$$

To illustrate this, replace $v = e^x$ by $v = e^x + C_1$ and apply integration by parts to see that you obtain the same result.

EXAMPLE 2 Integration by Parts

Find $\int x^2 \ln x \, dx$.

Solution In this case, x^2 is more easily integrated than $\ln x$. Furthermore, the derivative of $\ln x$ is simpler than $\ln x$. So, you should let $dv = x^2 \, dx$.

$$dv = x^2 \, dx \quad \Rightarrow \quad v = \int x^2 \, dx = \frac{x^3}{3}$$

$$u = \ln x \quad \Rightarrow \quad du = \frac{1}{x} \, dx$$

Integration by parts produces

$$\int u \, dv = uv - \int v \, du \qquad \text{Integration by parts formula}$$

$$\int x^2 \ln x \, dx = \frac{x^3}{3} \ln x - \int \left(\frac{x^3}{3}\right)\left(\frac{1}{x}\right) dx \qquad \text{Substitute.}$$

$$= \frac{x^3}{3} \ln x - \frac{1}{3} \int x^2 \, dx \qquad \text{Simplify.}$$

$$= \frac{x^3}{3} \ln x - \frac{x^3}{9} + C. \qquad \text{Integrate.}$$

You can check this result by differentiating.

$$\frac{d}{dx}\left[\frac{x^3}{3} \ln x - \frac{x^3}{9} + C\right] = \frac{x^3}{3}\left(\frac{1}{x}\right) + (\ln x)(x^2) - \frac{x^2}{3} = x^2 \ln x$$

TECHNOLOGY Try graphing

$$\int x^2 \ln x \, dx \quad \text{and} \quad \frac{x^3}{3} \ln x - \frac{x^3}{9}$$

on your graphing utility. Do you get the same graph? (This may take a while, so be patient.)

One surprising application of integration by parts involves integrands consisting of single terms, such as

$$\int \ln x \, dx \quad \text{or} \quad \int \arcsin x \, dx.$$

In these cases, try letting $dv = dx$, as shown in the next example.

EXAMPLE 3 An Integrand with a Single Term

Evaluate $\displaystyle\int_0^1 \arcsin x \, dx.$

Solution Let $dv = dx$.

$$dv = dx \quad \Rightarrow \quad v = \int dx = x$$

$$u = \arcsin x \quad \Rightarrow \quad du = \frac{1}{\sqrt{1 - x^2}} \, dx$$

Integration by parts produces

$$\int u \, dv = uv - \int v \, du \qquad \text{Integration by parts formula}$$

$$\int \arcsin x \, dx = x \arcsin x - \int \frac{x}{\sqrt{1 - x^2}} \, dx \qquad \text{Substitute.}$$

$$= x \arcsin x + \frac{1}{2} \int (1 - x^2)^{-1/2} (-2x) \, dx \qquad \text{Rewrite.}$$

$$= x \arcsin x + \sqrt{1 - x^2} + C. \qquad \text{Integrate.}$$

Using this antiderivative, you can evaluate the definite integral as shown.

$$\int_0^1 \arcsin x \, dx = \left[x \arcsin x + \sqrt{1 - x^2} \right]_0^1$$

$$= \frac{\pi}{2} - 1$$

$$\approx 0.571$$

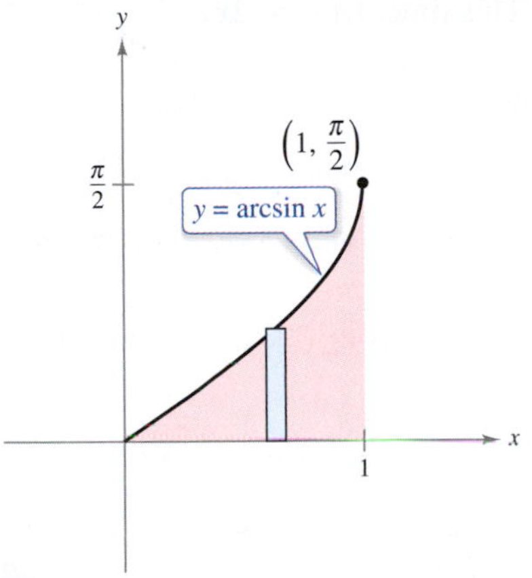

The area of the region is approximately 0.571.
Figure 8.2

The area represented by this definite integral is shown in Figure 8.2. ■

TECHNOLOGY Remember that there are two ways to use technology to evaluate a definite integral: (1) you can use a numerical approximation such as the Trapezoidal Rule or Simpson's Rule, or (2) you can use a computer algebra system to find the antiderivative and then apply the Fundamental Theorem of Calculus. Both methods have shortcomings. To find the possible error when using a numerical method, the integrand must have a second derivative (Trapezoidal Rule) or a fourth derivative (Simpson's Rule) in the interval of integration; the integrand in Example 3 fails to meet either of these requirements. To apply the Fundamental Theorem of Calculus, the symbolic integration utility must be able to find the antiderivative.

FOR FURTHER INFORMATION To see how integration by parts is used to prove Stirling's approximation

$$\ln(n!) = n \ln n - n$$

see the article "The Validity of Stirling's Approximation: A Physical Chemistry Project" by A. S. Wallner and K. A. Brandt in *Journal of Chemical Education*.

Some integrals require repeated use of the integration by parts formula.

EXAMPLE 4 Repeated Use of Integration by Parts

Find $\int x^2 \sin x \, dx$.

Solution The factors x^2 and $\sin x$ are equally easy to integrate. However, the derivative of x^2 becomes simpler, whereas the derivative of $\sin x$ does not. So, you should let $u = x^2$.

$$dv = \sin x \, dx \quad \Longrightarrow \quad v = \int \sin x \, dx = -\cos x$$

$$u = x^2 \quad \Longrightarrow \quad du = 2x \, dx$$

Now, integration by parts produces

$$\int x^2 \sin x \, dx = -x^2 \cos x + \int 2x \cos x \, dx. \qquad \text{First use of integration by parts}$$

This first use of integration by parts has succeeded in simplifying the original integral, but the integral on the right still doesn't fit a basic integration rule. To evaluate that integral, you can apply integration by parts again. This time, let $u = 2x$.

$$dv = \cos x \, dx \quad \Longrightarrow \quad v = \int \cos x \, dx = \sin x$$

$$u = 2x \quad \Longrightarrow \quad du = 2 \, dx$$

Now, integration by parts produces

$$\begin{aligned}\int 2x \cos x \, dx &= 2x \sin x - \int 2 \sin x \, dx \qquad \text{Second use of integration by parts} \\ &= 2x \sin x + 2 \cos x + C.\end{aligned}$$

Combining these two results, you can write

$$\int x^2 \sin x \, dx = -x^2 \cos x + 2x \sin x + 2 \cos x + C.$$

When making repeated applications of integration by parts, you need to be careful not to interchange the substitutions in successive applications. For instance, in Example 4, the first substitution was $u = x^2$ and $dv = \sin x \, dx$. If, in the second application, you had switched the substitution to $u = \cos x$ and $dv = 2x$, you would have obtained

$$\begin{aligned}\int x^2 \sin x \, dx &= -x^2 \cos x + \int 2x \cos x \, dx \\ &= -x^2 \cos x + x^2 \cos x + \int x^2 \sin x \, dx \\ &= \int x^2 \sin x \, dx\end{aligned}$$

thereby undoing the previous integration and returning to the *original* integral. When making repeated applications of integration by parts, you should also watch for the appearance of a *constant multiple* of the original integral. For instance, this occurs when you use integration by parts to evaluate $\int e^x \cos 2x \, dx$, and it also occurs in Example 5 on the next page.

The integral in Example 5 is an important one. In Section 8.4 (Example 5), you will see that it is used to find the arc length of a parabolic segment.

EXAMPLE 5 Integration by Parts

Find $\int \sec^3 x\, dx$.

Solution The most complicated portion of the integrand that can be easily integrated is $\sec^2 x$, so you should let $dv = \sec^2 x\, dx$ and $u = \sec x$.

$$dv = \sec^2 x\, dx \quad \Rightarrow \quad v = \int \sec^2 x\, dx = \tan x$$

$$u = \sec x \quad \Rightarrow \quad du = \sec x \tan x\, dx$$

Integration by parts produces

$$\int u\, dv = uv - \int v\, du \qquad \text{Integration by parts formula}$$

$$\int \sec^3 x\, dx = \sec x \tan x - \int \sec x \tan^2 x\, dx \qquad \text{Substitute.}$$

$$\int \sec^3 x\, dx = \sec x \tan x - \int \sec x(\sec^2 x - 1)\, dx \qquad \text{Trigonometric identity}$$

$$\int \sec^3 x\, dx = \sec x \tan x - \int \sec^3 x\, dx + \int \sec x\, dx \qquad \text{Rewrite.}$$

$$2\int \sec^3 x\, dx = \sec x \tan x + \int \sec x\, dx \qquad \text{Collect like integrals.}$$

$$2\int \sec^3 x\, dx = \sec x \tan x + \ln|\sec x + \tan x| + C \qquad \text{Integrate.}$$

$$\int \sec^3 x\, dx = \frac{1}{2}\sec x \tan x + \frac{1}{2}\ln|\sec x + \tan x| + C. \qquad \text{Divide by 2.}$$

EXAMPLE 6 Finding a Centroid

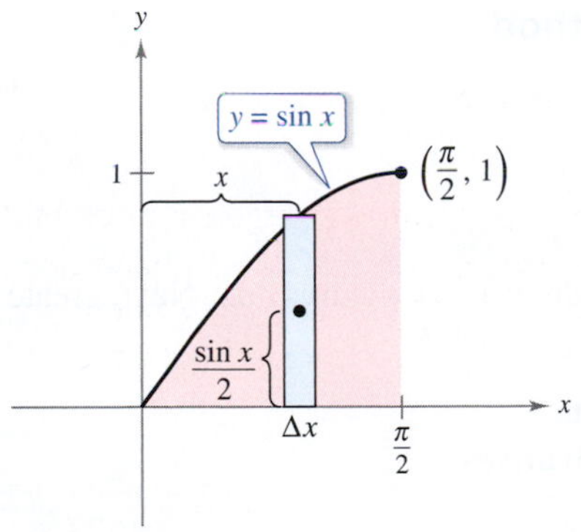

Figure 8.3

A machine part is modeled by the region bounded by the graph of $y = \sin x$ and the x-axis, $0 \le x \le \pi/2$, as shown in Figure 8.3. Find the centroid of this region.

Solution Begin by finding the area of the region.

$$A = \int_0^{\pi/2} \sin x\, dx = \Big[-\cos x\Big]_0^{\pi/2} = 1$$

Now, you can find the coordinates of the centroid. To evaluate the integral for $\bar{y}$, first rewrite the integrand using the trigonometric identity $\sin^2 x = (1 - \cos 2x)/2$.

$$\bar{y} = \frac{1}{A}\int_0^{\pi/2} \frac{\sin x}{2}(\sin x)\, dx = \frac{1}{4}\int_0^{\pi/2} (1 - \cos 2x)\, dx = \frac{1}{4}\left[x - \frac{\sin 2x}{2}\right]_0^{\pi/2} = \frac{\pi}{8}$$

You can evaluate the integral for $\bar{x}$, $(1/A)\int_0^{\pi/2} x \sin x\, dx$, with integration by parts. To do this, let $dv = \sin x\, dx$ and $u = x$. This produces $v = -\cos x$ and $du = dx$, and you can write

$$\int x \sin x\, dx = -x\cos x + \int \cos x\, dx = -x\cos x + \sin x + C.$$

Finally, you can determine $\bar{x}$ to be

$$\bar{x} = \frac{1}{A}\int_0^{\pi/2} x \sin x\, dx = \Big[-x\cos x + \sin x\Big]_0^{\pi/2} = 1.$$

So, the centroid of the region is $(1, \pi/8)$.

As you gain experience in using integration by parts, your skill in determining u and dv will increase. The next summary lists several common integrals with suggestions for the choices of u and dv.

REMARK You can use the acronym LIATE as a guideline for choosing u in integration by parts. In order, check the integrand for the following.

Is there a Logarithmic part?

Is there an Inverse trigonometric part?

Is there an Algebraic part?

Is there a Trigonometric part?

Is there an Exponential part?

SUMMARY: COMMON INTEGRALS USING INTEGRATION BY PARTS

1. For integrals of the form

$$\int x^n e^{ax}\,dx, \quad \int x^n \sin ax\,dx, \quad \text{or} \quad \int x^n \cos ax\,dx$$

let $u = x^n$ and let $dv = e^{ax}\,dx$, $\sin ax\,dx$, or $\cos ax\,dx$.

2. For integrals of the form

$$\int x^n \ln x\,dx, \quad \int x^n \arcsin ax\,dx, \quad \text{or} \quad \int x^n \arctan ax\,dx$$

let $u = \ln x$, $\arcsin ax$, or $\arctan ax$ and let $dv = x^n\,dx$.

3. For integrals of the form

$$\int e^{ax} \sin bx\,dx \quad \text{or} \quad \int e^{ax} \cos bx\,dx$$

let $u = \sin bx$ or $\cos bx$ and let $dv = e^{ax}\,dx$.

In problems involving repeated applications of integration by parts, a tabular method, illustrated in Example 7, can help to organize the work. This method works well for integrals of the form

$$\int x^n \sin ax\,dx, \quad \int x^n \cos ax\,dx, \quad \text{and} \quad \int x^n e^{ax}\,dx.$$

EXAMPLE 7 Using the Tabular Method

See LarsonCalculus.com for an interactive version of this type of example.

Find $\int x^2 \sin 4x\,dx$.

Solution Begin as usual by letting $u = x^2$ and $dv = v'\,dx = \sin 4x\,dx$. Next, create a table consisting of three columns, as shown.

Alternate Signs	u and Its Derivatives	v' and Its Antiderivatives
$+$	x^2	$\sin 4x$
$-$	$2x$	$-\frac{1}{4}\cos 4x$
$+$	2	$-\frac{1}{16}\sin 4x$
$-$	0	$\frac{1}{64}\cos 4x$

Differentiate until you obtain 0 as a derivative.

The solution is obtained by adding the signed products of the diagonal entries:

$$\int x^2 \sin 4x\,dx = -\frac{1}{4}x^2 \cos 4x + \frac{1}{8}x \sin 4x + \frac{1}{32}\cos 4x + C.$$

FOR FURTHER INFORMATION For more information on the tabular method, see the article "Tabular Integration by Parts" by David Horowitz in *The College Mathematics Journal*, and the article "More on Tabular Integration by Parts" by Leonard Gillman in *The College Mathematics Journal*. To view these articles, go to *MathArticles.com*.

8.2 Exercises

See CalcChat.com for tutorial help and worked-out solutions to odd-numbered exercises.

Setting Up Integration by Parts **In Exercises 1–6, identify u and dv for finding the integral using integration by parts. (Do not evaluate the integral.)**

1. $\int xe^{2x}\,dx$
2. $\int x^2 e^{2x}\,dx$
3. $\int (\ln x)^2\,dx$
4. $\int \ln 5x\,dx$
5. $\int x\sec^2 x\,dx$
6. $\int x^2 \cos x\,dx$

Using Integration by Parts **In Exercises 7–10, evaluate the integral using integration by parts with the given choices of u and dv.**

7. $\displaystyle\int x^3 \ln x\,dx;\ u = \ln x,\ dv = x^3\,dx$
8. $\displaystyle\int (4x + 7)e^x\,dx;\ u = 4x + 7,\ dv = e^x\,dx$
9. $\displaystyle\int x\sin 3x\,dx;\ u = x,\ dv = \sin 3x\,dx$
10. $\displaystyle\int x\cos 4x\,dx;\ u = x,\ dv = \cos 4x\,dx$

Finding an Indefinite Integral **In Exercises 11–30, find the indefinite integral. (*Note:* Solve by the simplest method—not all require integration by parts.)**

11. $\displaystyle\int xe^{-4x}\,dx$
12. $\displaystyle\int \frac{5x}{e^{2x}}\,dx$
13. $\displaystyle\int x^3 e^x\,dx$
14. $\displaystyle\int \frac{e^{1/t}}{t^2}\,dt$
15. $\displaystyle\int t\ln(t + 1)\,dt$
16. $\displaystyle\int x^5 \ln 3x\,dx$
17. $\displaystyle\int \frac{(\ln x)^2}{x}\,dx$
18. $\displaystyle\int \frac{\ln x}{x^3}\,dx$
19. $\displaystyle\int \frac{xe^{2x}}{(2x + 1)^2}\,dx$
20. $\displaystyle\int \frac{x^3 e^{x^2}}{(x^2 + 1)^2}\,dx$
21. $\displaystyle\int x\sqrt{x - 5}\,dx$
22. $\displaystyle\int \frac{x}{\sqrt{6x + 1}}\,dx$
23. $\displaystyle\int x\cos x\,dx$
24. $\displaystyle\int t\csc t\cot t\,dt$
25. $\displaystyle\int x^3 \sin x\,dx$
26. $\displaystyle\int x^2 \cos x\,dx$
27. $\displaystyle\int \arctan x\,dx$
28. $\displaystyle\int 4\arccos x\,dx$
29. $\displaystyle\int e^{-3x}\sin 5x\,dx$
30. $\displaystyle\int e^{4x}\cos 2x\,dx$

Differential Equation **In Exercises 31–34, solve the differential equation.**

31. $y' = \ln x$
32. $y' = \arctan \dfrac{x}{2}$
33. $\dfrac{dy}{dt} = \dfrac{t^2}{\sqrt{3 + 5t}}$
34. $\dfrac{dy}{dx} = x^2\sqrt{x - 3}$

Slope Field **In Exercises 35 and 36, a differential equation, a point, and a slope field are given. (a) Sketch two approximate solutions of the differential equation on the slope field, one of which passes through the given point. (b) Use integration to find the particular solution of the differential equation and use a graphing utility to graph the solution. Compare the result with the sketches in part (a). To print an enlarged copy of the graph, go to *MathGraphs.com*.**

35. $\dfrac{dy}{dx} = x\sqrt{y}\cos x,\ (0, 4)$

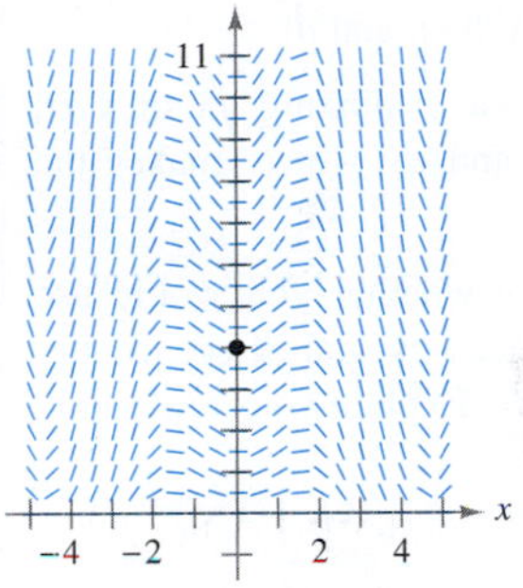

36. $\dfrac{dy}{dx} = e^{-x/3}\sin 2x,\ \left(0, -\frac{18}{37}\right)$

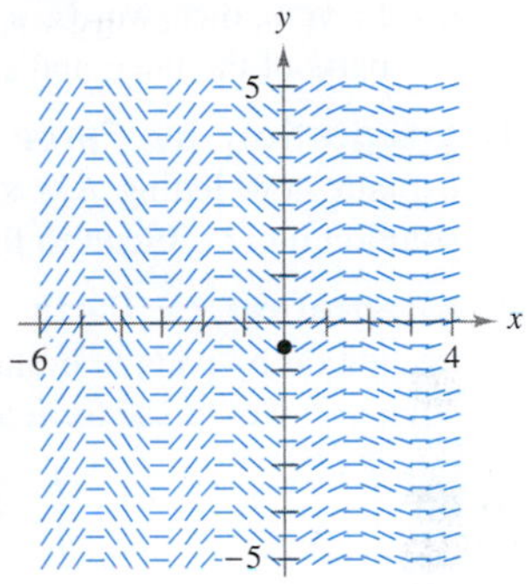

Slope Field **In Exercises 37 and 38, use a computer algebra system to graph the slope field for the differential equation and graph the solution through the specified initial condition.**

37. $\dfrac{dy}{dx} = \dfrac{x}{y}e^{x/8},\quad y(0) = 2$
38. $\dfrac{dy}{dx} = \dfrac{x}{y}\sin x,\quad y(0) = 4$

Evaluating a Definite Integral **In Exercises 39–48, evaluate the definite integral. Use a graphing utility to confirm your result.**

39. $\displaystyle\int_0^3 xe^{x/2}\,dx$
40. $\displaystyle\int_0^2 x^2 e^{-2x}\,dx$
41. $\displaystyle\int_0^{\pi/4} x\cos 2x\,dx$
42. $\displaystyle\int_0^{\pi} x\sin 2x\,dx$
43. $\displaystyle\int_0^{1/2} \arccos x\,dx$
44. $\displaystyle\int_0^1 x\arcsin x^2\,dx$
45. $\displaystyle\int_0^1 e^x \sin x\,dx$
46. $\displaystyle\int_0^1 \ln(4 + x^2)\,dx$
47. $\displaystyle\int_2^4 x\,\mathrm{arcsec}\,x\,dx$
48. $\displaystyle\int_0^{\pi/8} x\sec^2 2x\,dx$

Using the Tabular Method **In Exercises 49–54, use the tabular method to find the integral.**

49. $\displaystyle\int x^2 e^{2x}\,dx$
50. $\displaystyle\int x^3 e^{-2x}\,dx$
51. $\displaystyle\int x^3 \sin x\,dx$
52. $\displaystyle\int x^3 \cos 2x\,dx$

53. $\int x \sec^2 x\, dx$

54. $\int x^2(x-2)^{3/2}\, dx$

Using Two Methods Together **In Exercises 55–58, find the indefinite integral by using substitution followed by integration by parts.**

55. $\int \sin\sqrt{x}\, dx$

56. $\int 2x^3 \cos x^2\, dx$

57. $\int x^5 e^{x^2}\, dx$

58. $\int e^{\sqrt{2x}}\, dx$

WRITING ABOUT CONCEPTS

59. Integration by Parts

(a) Integration by parts is based on what differentiation rule? Explain.

(b) In your own words, state how you determine which parts of the integrand should be u and dv.

60. Integration by Parts When evaluating $\int x \sin x\, dx$, explain how letting $u = \sin x$ and $dv = x\, dx$ makes the solution more difficult to find.

61. Integration by Parts State whether you would use integration by parts to evaluate each integral. If so, identify what you would use for u and dv. Explain your reasoning.

(a) $\int \frac{\ln x}{x}\, dx$ (b) $\int x \ln x\, dx$ (c) $\int x^2 e^{-3x}\, dx$

(d) $\int 2xe^{x^2}\, dx$ (e) $\int \frac{x}{\sqrt{x+1}}\, dx$ (f) $\int \frac{x}{\sqrt{x^2+1}}\, dx$

62. HOW DO YOU SEE IT? Use the graph of f' shown in the figure to answer the following.

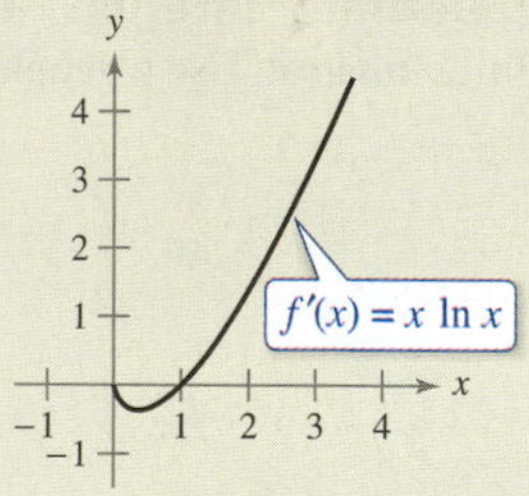

(a) Approximate the slope of f at $x = 2$. Explain.

(b) Approximate any open intervals in which the graph of f is increasing and any open intervals in which it is decreasing. Explain.

63. Using Two Methods Integrate $\int \frac{x^3}{\sqrt{4+x^2}}\, dx$

(a) by parts, letting $dv = \frac{x}{\sqrt{4+x^2}}\, dx$.

(b) by substitution, letting $u = 4 + x^2$.

64. Using Two Methods Integrate $\int x\sqrt{4-x}\, dx$

(a) by parts, letting $dv = \sqrt{4-x}\, dx$.

(b) by substitution, letting $u = 4 - x$.

Finding a General Rule **In Exercises 65 and 66, use a computer algebra system to find the integrals for $n = 0, 1, 2,$ and 3. Use the result to obtain a general rule for the integrals for any positive integer n and test your results for $n = 4$.**

65. $\int x^n \ln x\, dx$

66. $\int x^n e^x\, dx$

Proof **In Exercises 67–72, use integration by parts to prove the formula. (For Exercises 67–70, assume that n is a positive integer.)**

67. $\int x^n \sin x\, dx = -x^n \cos x + n\int x^{n-1}\cos x\, dx$

68. $\int x^n \cos x\, dx = x^n \sin x - n\int x^{n-1}\sin x\, dx$

69. $\int x^n \ln x\, dx = \frac{x^{n+1}}{(n+1)^2}[-1 + (n+1)\ln x] + C$

70. $\int x^n e^{ax}\, dx = \frac{x^n e^{ax}}{a} - \frac{n}{a}\int x^{n-1} e^{ax}\, dx$

71. $\int e^{ax} \sin bx\, dx = \frac{e^{ax}(a \sin bx - b\cos bx)}{a^2+b^2} + C$

72. $\int e^{ax} \cos bx\, dx = \frac{e^{ax}(a \cos bx + b\sin bx)}{a^2+b^2} + C$

Using Formulas **In Exercises 73–78, find the integral by using the appropriate formula from Exercises 67–72.**

73. $\int x^2 \sin x\, dx$

74. $\int x^2 \cos x\, dx$

75. $\int x^5 \ln x\, dx$

76. $\int x^3 e^{2x}\, dx$

77. $\int e^{-3x} \sin 4x\, dx$

78. $\int e^{2x} \cos 3x\, dx$

Area **In Exercises 79–82, use a graphing utility to graph the region bounded by the graphs of the equations. Then find the area of the region analytically.**

79. $y = 2xe^{-x},\quad y = 0,\quad x = 3$

80. $y = \frac{1}{10}xe^{3x},\quad y = 0,\quad x = 0,\quad x = 2$

81. $y = e^{-x}\sin \pi x,\quad y = 0,\quad x = 1$

82. $y = x^3 \ln x,\quad y = 0,\quad x = 1,\quad x = 3$

83. Area, Volume, and Centroid Given the region bounded by the graphs of $y = \ln x$, $y = 0$, and $x = e$, find

(a) the area of the region.

(b) the volume of the solid generated by revolving the region about the x-axis.

(c) the volume of the solid generated by revolving the region about the y-axis.

(d) the centroid of the region.

84. Area, Volume, and Centroid Given the region bounded by the graphs of $y = x \sin x$, $y = 0$, $x = 0$, and $x = \pi$, find

(a) the area of the region.

(b) the volume of the solid generated by revolving the region about the x-axis.

(c) the volume of the solid generated by revolving the region about the y-axis.

(d) the centroid of the region.

85. Centroid Find the centroid of the region bounded by the graphs of $y = \arcsin x$, $x = 0$, and $y = \pi/2$. How is this problem related to Example 6 in this section?

86. Centroid Find the centroid of the region bounded by the graphs of $f(x) = x^2$, $g(x) = 2^x$, $x = 2$, and $x = 4$.

87. Average Displacement A damping force affects the vibration of a spring so that the displacement of the spring is given by

$$y = e^{-4t}(\cos 2t + 5 \sin 2t).$$

Find the average value of y on the interval from $t = 0$ to $t = \pi$.

88. Memory Model

A model for the ability M of a child to memorize, measured on a scale from 0 to 10, is given by

$$M = 1 + 1.6t \ln t, \quad 0 < t \leq 4$$

where t is the child's age in years. Find the average value of this model

(a) between the child's first and second birthdays.

(b) between the child's third and fourth birthdays.

Present Value **In Exercises 89 and 90, find the present value P of a continuous income flow of $c(t)$ dollars per year for**

$$P = \int_0^{t_1} c(t)e^{-rt}\, dt$$

where t_1 is the time in years and r is the annual interest rate compounded continuously.

89. $c(t) = 100{,}000 + 4000t$, $r = 5\%$, $t_1 = 10$

90. $c(t) = 30{,}000 + 500t$, $r = 7\%$, $t_1 = 5$

Integrals Used to Find Fourier Coefficients **In Exercises 91 and 92, verify the value of the definite integral, where n is a positive integer.**

91. $$\int_{-\pi}^{\pi} x \sin nx\, dx = \begin{cases} \dfrac{2\pi}{n}, & n \text{ is odd} \\ -\dfrac{2\pi}{n}, & n \text{ is even} \end{cases}$$

92. $$\int_{-\pi}^{\pi} x^2 \cos nx\, dx = \frac{(-1)^n 4\pi}{n^2}$$

93. Vibrating String A string stretched between the two points $(0, 0)$ and $(2, 0)$ is plucked by displacing the string h units at its midpoint. The motion of the string is modeled by a **Fourier Sine Series** whose coefficients are given by

$$b_n = h\int_0^1 x \sin\frac{n\pi x}{2}\, dx + h\int_1^2 (-x + 2) \sin\frac{n\pi x}{2}\, dx.$$

Find b_n.

94. Euler's Method Consider the differential equation $f'(x) = xe^{-x}$ with the initial condition $f(0) = 0$.

(a) Use integration to solve the differential equation.

(b) Use a graphing utility to graph the solution of the differential equation.

(c) Use Euler's Method with $h = 0.05$, and the recursive capabilities of a graphing utility, to generate the first 80 points of the graph of the approximate solution. Use the graphing utility to plot the points. Compare the result with the graph in part (b).

(d) Repeat part (c) using $h = 0.1$ and generate the first 40 points.

(e) Why is the result in part (c) a better approximation of the solution than the result in part (d)?

Euler's Method **In Exercises 95 and 96, consider the differential equation and repeat parts (a)–(d) of Exercise 94.**

95. $f'(x) = 3x \sin(2x)$, $f(0) = 0$

96. $f'(x) = \cos\sqrt{x}$, $f(0) = 1$

97. Think About It Give a geometric explanation of why

$$\int_0^{\pi/2} x \sin x\, dx \leq \int_0^{\pi/2} x\, dx.$$

Verify the inequality by evaluating the integrals.

98. Finding a Pattern Find the area bounded by the graphs of $y = x \sin x$ and $y = 0$ over each interval.

(a) $[0, \pi]$ (b) $[\pi, 2\pi]$ (c) $[2\pi, 3\pi]$

Describe any patterns that you notice. What is the area between the graphs of $y = x \sin x$ and $y = 0$ over the interval $[n\pi, (n + 1)\pi]$, where n is any nonnegative integer? Explain.

99. Finding an Error Find the fallacy in the following argument that $0 = 1$.

$$dv = dx \Rightarrow v = \int dx = x$$

$$u = \frac{1}{x} \Rightarrow du = -\frac{1}{x^2}\, dx$$

$$0 + \int \frac{dx}{x} = \left(\frac{1}{x}\right)(x) - \int \left(-\frac{1}{x^2}\right)(x)\, dx$$

$$= 1 + \int \frac{dx}{x}$$

So, $0 = 1$.

8.3 Trigonometric Integrals

- Solve trigonometric integrals involving powers of sine and cosine.
- Solve trigonometric integrals involving powers of secant and tangent.
- Solve trigonometric integrals involving sine-cosine products with different angles.

Integrals Involving Powers of Sine and Cosine

> **SHEILA SCOTT MACINTYRE (1910–1960)**
> Sheila Scott Macintyre published her first paper on the asymptotic periods of integral functions in 1935. She completed her doctorate work at Aberdeen University, where she taught. In 1958 she accepted a visiting research fellowship at the University of Cincinnati.

In this section, you will study techniques for evaluating integrals of the form

$$\int \sin^m x \cos^n x\, dx \quad \text{and} \quad \int \sec^m x \tan^n x\, dx$$

where either m or n is a positive integer. To find antiderivatives for these forms, try to break them into combinations of trigonometric integrals to which you can apply the Power Rule.

For instance, you can evaluate

$$\int \sin^5 x \cos x\, dx$$

with the Power Rule by letting $u = \sin x$. Then, $du = \cos x\, dx$ and you have

$$\int \sin^5 x \cos x\, dx = \int u^5\, du = \frac{u^6}{6} + C = \frac{\sin^6 x}{6} + C.$$

To break up $\int \sin^m x \cos^n x\, dx$ into forms to which you can apply the Power Rule, use the following identities.

$$\sin^2 x + \cos^2 x = 1 \qquad \text{Pythagorean identity}$$

$$\sin^2 x = \frac{1 - \cos 2x}{2} \qquad \text{Half-angle identity for } \sin^2 x$$

$$\cos^2 x = \frac{1 + \cos 2x}{2} \qquad \text{Half-angle identity for } \cos^2 x$$

GUIDELINES FOR EVALUATING INTEGRALS INVOLVING POWERS OF SINE AND COSINE

1. When the power of the sine is odd and positive, save one sine factor and convert the remaining factors to cosines. Then, expand and integrate.

$$\int \underbrace{\sin^{2k+1}}_{\text{Odd}} x \cos^n x\, dx = \int \underbrace{(\sin^2 x)^k}_{\text{Convert to cosines}} \cos^n x \underbrace{\sin x\, dx}_{\text{Save for } du} = \int (1 - \cos^2 x)^k \cos^n x \sin x\, dx$$

2. When the power of the cosine is odd and positive, save one cosine factor and convert the remaining factors to sines. Then, expand and integrate.

$$\int \sin^m x \underbrace{\cos^{2k+1}}_{\text{Odd}} x\, dx = \int \sin^m x \underbrace{(\cos^2 x)^k}_{\text{Convert to sines}} \underbrace{\cos x\, dx}_{\text{Save for } du} = \int \sin^m x (1 - \sin^2 x)^k \cos x\, dx$$

3. When the powers of both the sine and cosine are even and nonnegative, make repeated use of the identities

$$\sin^2 x = \frac{1 - \cos 2x}{2} \quad \text{and} \quad \cos^2 x = \frac{1 + \cos 2x}{2}$$

to convert the integrand to odd powers of the cosine. Then proceed as in the second guideline.

EXAMPLE 1 Power of Sine Is Odd and Positive

Find $\int \sin^3 x \cos^4 x \, dx$.

Solution Because you expect to use the Power Rule with $u = \cos x$, *save one sine factor* to form du and convert the remaining sine factors to cosines.

$$\int \sin^3 x \cos^4 x \, dx = \int \sin^2 x \cos^4 x (\sin x) \, dx \qquad \text{Rewrite.}$$

$$= \int (1 - \cos^2 x) \cos^4 x \sin x \, dx \qquad \text{Trigonometric identity}$$

$$= \int (\cos^4 x - \cos^6 x) \sin x \, dx \qquad \text{Multiply.}$$

$$= \int \cos^4 x \sin x \, dx - \int \cos^6 x \sin x \, dx \qquad \text{Rewrite.}$$

$$= -\int \cos^4 x (-\sin x) \, dx + \int \cos^6 x (-\sin x) \, dx$$

$$= -\frac{\cos^5 x}{5} + \frac{\cos^7 x}{7} + C \qquad \text{Integrate.}$$

TECHNOLOGY A computer algebra system used to find the integral in Example 1 yielded the following.

$$\int \sin^3 x \cos^4 x \, dx = -\cos^5 x \left(\frac{1}{7} \sin^2 x + \frac{2}{35} \right) + C$$

Is this equivalent to the result obtained in Example 1?

In Example 1, *both* of the powers m and n happened to be positive integers. This strategy will work as long as either m or n is odd and positive. For instance, in the next example, the power of the cosine is 3, but the power of the sine is $-\frac{1}{2}$.

EXAMPLE 2 Power of Cosine Is Odd and Positive

See LarsonCalculus.com for an interactive version of this type of example.

Evaluate $\int_{\pi/6}^{\pi/3} \frac{\cos^3 x}{\sqrt{\sin x}} \, dx$.

Solution Because you expect to use the Power Rule with $u = \sin x$, *save one cosine factor* to form du and convert the remaining cosine factors to sines.

$$\int_{\pi/6}^{\pi/3} \frac{\cos^3 x}{\sqrt{\sin x}} \, dx = \int_{\pi/6}^{\pi/3} \frac{\cos^2 x \cos x}{\sqrt{\sin x}} \, dx$$

$$= \int_{\pi/6}^{\pi/3} \frac{(1 - \sin^2 x)(\cos x)}{\sqrt{\sin x}} \, dx$$

$$= \int_{\pi/6}^{\pi/3} [(\sin x)^{-1/2} - (\sin x)^{3/2}] \cos x \, dx$$

$$= \left[\frac{(\sin x)^{1/2}}{1/2} - \frac{(\sin x)^{5/2}}{5/2} \right]_{\pi/6}^{\pi/3}$$

$$= 2\left(\frac{\sqrt{3}}{2}\right)^{1/2} - \frac{2}{5}\left(\frac{\sqrt{3}}{2}\right)^{5/2} - \sqrt{2} + \frac{\sqrt{32}}{80}$$

$$\approx 0.239$$

Figure 8.4 shows the region whose area is represented by this integral.

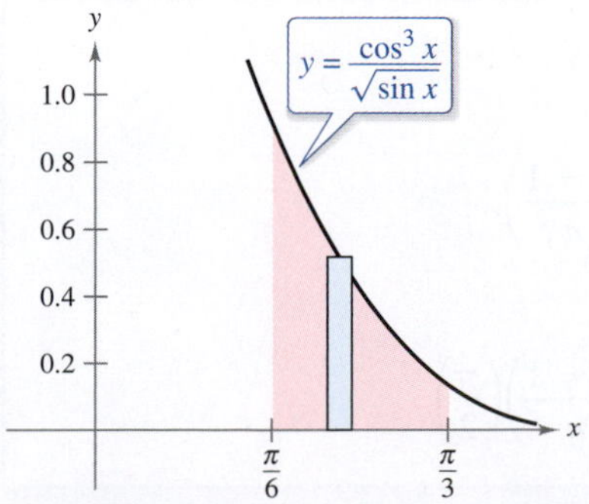

The area of the region is approximately 0.239.
Figure 8.4

EXAMPLE 3 Power of Cosine Is Even and Nonnegative

Find $\int \cos^4 x\, dx$.

Solution Because m and n are both even and nonnegative ($m = 0$), you can replace $\cos^4 x$ by

$$\left(\frac{1 + \cos 2x}{2}\right)^2.$$

So, you can integrate as shown.

$$\int \cos^4 x\, dx = \int \left(\frac{1 + \cos 2x}{2}\right)^2 dx \qquad \text{Half-angle identity}$$

$$= \int \left(\frac{1}{4} + \frac{\cos 2x}{2} + \frac{\cos^2 2x}{4}\right) dx \qquad \text{Expand.}$$

$$= \int \left[\frac{1}{4} + \frac{\cos 2x}{2} + \frac{1}{4}\left(\frac{1 + \cos 4x}{2}\right)\right] dx \qquad \text{Half-angle identity}$$

$$= \frac{3}{8}\int dx + \frac{1}{4}\int 2\cos 2x\, dx + \frac{1}{32}\int 4\cos 4x\, dx \qquad \text{Rewrite.}$$

$$= \frac{3x}{8} + \frac{\sin 2x}{4} + \frac{\sin 4x}{32} + C \qquad \text{Integrate.}$$

Use a symbolic differentiation utility to verify this. Can you simplify the derivative to obtain the original integrand? ■

In Example 3, when you evaluate the definite integral from 0 to $\pi/2$, you obtain

$$\int_0^{\pi/2} \cos^4 x\, dx = \left[\frac{3x}{8} + \frac{\sin 2x}{4} + \frac{\sin 4x}{32}\right]_0^{\pi/2}$$

$$= \left(\frac{3\pi}{16} + 0 + 0\right) - (0 + 0 + 0)$$

$$= \frac{3\pi}{16}.$$

Note that the only term that contributes to the solution is

$$\frac{3x}{8}.$$

This observation is generalized in the following formulas developed by John Wallis (1616–1703).

JOHN WALLIS (1616–1703)

Wallis did much of his work in calculus prior to Newton and Leibniz, and he influenced the thinking of both of these men. Wallis is also credited with introducing the present symbol (∞) for infinity.

See LarsonCalculus.com to read more of this biography.

Wallis's Formulas

1. If n is odd ($n \ge 3$), then

$$\int_0^{\pi/2} \cos^n x\, dx = \left(\frac{2}{3}\right)\left(\frac{4}{5}\right)\left(\frac{6}{7}\right) \cdots \left(\frac{n-1}{n}\right).$$

2. If n is even ($n \ge 2$), then

$$\int_0^{\pi/2} \cos^n x\, dx = \left(\frac{1}{2}\right)\left(\frac{3}{4}\right)\left(\frac{5}{6}\right) \cdots \left(\frac{n-1}{n}\right)\left(\frac{\pi}{2}\right).$$

These formulas are also valid when $\cos^n x$ is replaced by $\sin^n x$. (You are asked to prove both formulas in Exercise 88.)

Integrals Involving Powers of Secant and Tangent

The guidelines below can help you evaluate integrals of the form

$$\int \sec^m x \tan^n x \, dx.$$

GUIDELINES FOR EVALUATING INTEGRALS INVOLVING POWERS OF SECANT AND TANGENT

1. When the power of the secant is even and positive, save a secant-squared factor and convert the remaining factors to tangents. Then, expand and integrate.

$$\int \overset{\text{Even}}{\sec^{2k}} x \tan^n x \, dx = \int \overbrace{(\sec^2 x)^{k-1}}^{\text{Convert to tangents}} \tan^n x \overbrace{\sec^2 x \, dx}^{\text{Save for } du} = \int (1 + \tan^2 x)^{k-1} \tan^n x \sec^2 x \, dx$$

2. When the power of the tangent is odd and positive, save a secant-tangent factor and convert the remaining factors to secants. Then, expand and integrate.

$$\int \sec^m x \overset{\text{Odd}}{\tan^{2k+1}} x \, dx = \int \sec^{m-1} x \overbrace{(\tan^2 x)^k}^{\text{Convert to secants}} \overbrace{\sec x \tan x \, dx}^{\text{Save for } du} = \int \sec^{m-1} x (\sec^2 x - 1)^k \sec x \tan x \, dx$$

3. When there are no secant factors and the power of the tangent is even and positive, convert a tangent-squared factor to a secant-squared factor, then expand and repeat if necessary.

$$\int \tan^n x \, dx = \int \tan^{n-2} x \overbrace{(\tan^2 x)}^{\text{Convert to secants}} \, dx = \int \tan^{n-2} x (\sec^2 x - 1) \, dx$$

4. When the integral is of the form

$$\int \sec^m x \, dx$$

where m is odd and positive, use integration by parts, as illustrated in Example 5 in Section 8.2.

5. When none of the first four guidelines applies, try converting to sines and cosines.

EXAMPLE 4 Power of Tangent Is Odd and Positive

Find $\displaystyle\int \frac{\tan^3 x}{\sqrt{\sec x}} \, dx.$

Solution Because you expect to use the Power Rule with $u = \sec x$, *save a factor of* $(\sec x \tan x)$ to form du and convert the remaining tangent factors to secants.

$$\begin{aligned}
\int \frac{\tan^3 x}{\sqrt{\sec x}} \, dx &= \int (\sec x)^{-1/2} \tan^3 x \, dx \\
&= \int (\sec x)^{-3/2} (\tan^2 x)(\sec x \tan x) \, dx \\
&= \int (\sec x)^{-3/2} (\sec^2 x - 1)(\sec x \tan x) \, dx \\
&= \int [(\sec x)^{1/2} - (\sec x)^{-3/2}](\sec x \tan x) \, dx \\
&= \frac{2}{3}(\sec x)^{3/2} + 2(\sec x)^{-1/2} + C
\end{aligned}$$

EXAMPLE 5 Power of Secant Is Even and Positive

Find $\int \sec^4 3x \tan^3 3x\, dx.$

Solution Let $u = \tan 3x$, then $du = 3 \sec^2 3x\, dx$ and you can write

$$\begin{aligned}\int \sec^4 3x \tan^3 3x\, dx &= \int \sec^2 3x \tan^3 3x(\sec^2 3x)\, dx\\ &= \int (1 + \tan^2 3x) \tan^3 3x(\sec^2 3x)\, dx\\ &= \frac{1}{3}\int (\tan^3 3x + \tan^5 3x)(3 \sec^2 3x)\, dx\\ &= \frac{1}{3}\left(\frac{\tan^4 3x}{4} + \frac{\tan^6 3x}{6}\right) + C\\ &= \frac{\tan^4 3x}{12} + \frac{\tan^6 3x}{18} + C.\end{aligned}$$

In Example 5, the power of the tangent is odd and positive. So, you could also find the integral using the procedure described in the second guideline on page 527. In Exercises 69 and 70, you are asked to show that the results obtained by these two procedures differ only by a constant.

EXAMPLE 6 Power of Tangent Is Even

Evaluate $\int_0^{\pi/4} \tan^4 x\, dx.$

Solution Because there are no secant factors, you can begin by converting a tangent-squared factor to a secant-squared factor.

$$\begin{aligned}\int \tan^4 x\, dx &= \int \tan^2 x(\tan^2 x)\, dx\\ &= \int \tan^2 x(\sec^2 x - 1)\, dx\\ &= \int \tan^2 x \sec^2 x\, dx - \int \tan^2 x\, dx\\ &= \int \tan^2 x \sec^2 x\, dx - \int (\sec^2 x - 1)\, dx\\ &= \frac{\tan^3 x}{3} - \tan x + x + C\end{aligned}$$

Next, evaluate the definite integral.

$$\begin{aligned}\int_0^{\pi/4} \tan^4 x\, dx &= \left[\frac{\tan^3 x}{3} - \tan x + x\right]_0^{\pi/4}\\ &= \frac{1}{3} - 1 + \frac{\pi}{4}\\ &\approx 0.119\end{aligned}$$

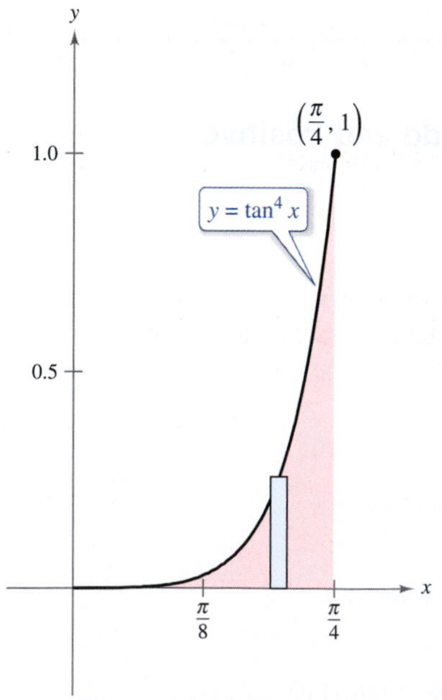

The area of the region is approximately 0.119.
Figure 8.5

The area represented by the definite integral is shown in Figure 8.5. Try using Simpson's Rule to approximate this integral. With $n = 18$, you should obtain an approximation that is within 0.00001 of the actual value.

For integrals involving powers of cotangents and cosecants, you can follow a strategy similar to that used for powers of tangents and secants. Also, when integrating trigonometric functions, remember that it sometimes helps to convert the entire integrand to powers of sines and cosines.

EXAMPLE 7 **Converting to Sines and Cosines**

Find $\displaystyle\int \frac{\sec x}{\tan^2 x}\,dx$.

Solution Because the first four guidelines on page 527 do not apply, try converting the integrand to sines and cosines. In this case, you are able to integrate the resulting powers of sine and cosine as shown.

$$\begin{aligned}\int \frac{\sec x}{\tan^2 x}\,dx &= \int \left(\frac{1}{\cos x}\right)\left(\frac{\cos x}{\sin x}\right)^2 dx\\ &= \int (\sin x)^{-2}(\cos x)\,dx\\ &= -(\sin x)^{-1} + C\\ &= -\csc x + C\end{aligned}$$

Integrals Involving Sine-Cosine Products with Different Angles

Integrals involving the products of sines and cosines of two *different* angles occur in many applications. In such instances, you can use the following product-to-sum identities.

$$\sin mx \sin nx = \frac{1}{2}(\cos\,[(m-n)x] - \cos\,[(m+n)x])$$

$$\sin mx \cos nx = \frac{1}{2}(\sin\,[(m-n)x] + \sin\,[(m+n)x])$$

$$\cos mx \cos nx = \frac{1}{2}(\cos\,[(m-n)x] + \cos\,[(m+n)x])$$

EXAMPLE 8 **Using Product-to-Sum Identities**

Find $\displaystyle\int \sin 5x \cos 4x\,dx$.

Solution Considering the second product-to-sum identity above, you can write

$$\begin{aligned}\int \sin 5x \cos 4x\,dx &= \frac{1}{2}\int (\sin x + \sin 9x)\,dx\\ &= \frac{1}{2}\left(-\cos x - \frac{\cos 9x}{9}\right) + C\\ &= -\frac{\cos x}{2} - \frac{\cos 9x}{18} + C.\end{aligned}$$

FOR FURTHER INFORMATION
To learn more about integrals involving sine-cosine products with different angles, see the article "Integrals of Products of Sine and Cosine with Different Arguments" by Sherrie J. Nicol in *The College Mathematics Journal.* To view this article, go to *MathArticles.com.*

8.3 Exercises

See CalcChat.com for tutorial help and worked-out solutions to odd-numbered exercises.

Finding an Indefinite Integral Involving Sine and Cosine In Exercises 1–12, find the indefinite integral.

1. $\int \cos^5 x \sin x\, dx$
2. $\int \cos^3 x \sin^4 x\, dx$
3. $\int \sin^7 2x \cos 2x\, dx$
4. $\int \sin^3 3x\, dx$
5. $\int \sin^3 x \cos^2 x\, dx$
6. $\int \cos^3 \frac{x}{3}\, dx$
7. $\int \sin^3 2\theta \sqrt{\cos 2\theta}\, d\theta$
8. $\int \frac{\cos^5 t}{\sqrt{\sin t}}\, dt$
9. $\int \cos^2 3x\, dx$
10. $\int \sin^4 6\theta\, d\theta$
11. $\int x \sin^2 x\, dx$
12. $\int x^2 \sin^2 x\, dx$

Using Wallis's Formulas In Exercises 13–18, use Wallis's Formulas to evaluate the integral.

13. $\int_0^{\pi/2} \cos^7 x\, dx$
14. $\int_0^{\pi/2} \cos^9 x\, dx$
15. $\int_0^{\pi/2} \cos^{10} x\, dx$
16. $\int_0^{\pi/2} \sin^5 x\, dx$
17. $\int_0^{\pi/2} \sin^6 x\, dx$
18. $\int_0^{\pi/2} \sin^8 x\, dx$

Finding an Indefinite Integral Involving Secant and Tangent In Exercises 19–32, find the indefinite integral.

19. $\int \sec 4x\, dx$
20. $\int \sec^4 2x\, dx$
21. $\int \sec^3 \pi x\, dx$
22. $\int \tan^6 3x\, dx$
23. $\int \tan^5 \frac{x}{2}\, dx$
24. $\int \tan^3 \frac{\pi x}{2} \sec^2 \frac{\pi x}{2}\, dx$
25. $\int \tan^3 2t \sec^3 2t\, dt$
26. $\int \tan^5 2x \sec^4 2x\, dx$
27. $\int \sec^6 4x \tan 4x\, dx$
28. $\int \sec^2 \frac{x}{2} \tan \frac{x}{2}\, dx$
29. $\int \sec^5 x \tan^3 x\, dx$
30. $\int \tan^3 3x\, dx$
31. $\int \frac{\tan^2 x}{\sec x}\, dx$
32. $\int \frac{\tan^2 x}{\sec^5 x}\, dx$

Differential Equation In Exercises 33–36, solve the differential equation.

33. $\frac{dr}{d\theta} = \sin^4 \pi\theta$
34. $\frac{ds}{d\alpha} = \sin^2 \frac{\alpha}{2} \cos^2 \frac{\alpha}{2}$
35. $y' = \tan^3 3x \sec 3x$
36. $y' = \sqrt{\tan x} \sec^4 x$

Slope Field In Exercises 37 and 38, a differential equation, a point, and a slope field are given. (a) Sketch two approximate solutions of the differential equation on the slope field, one of which passes through the given point. (b) Use integration to find the particular solution of the differential equation and use a graphing utility to graph the solution. Compare the result with the sketches in part (a). To print an enlarged copy of the graph, go to *MathGraphs.com*.

37. $\frac{dy}{dx} = \sin^2 x,\ (0, 0)$

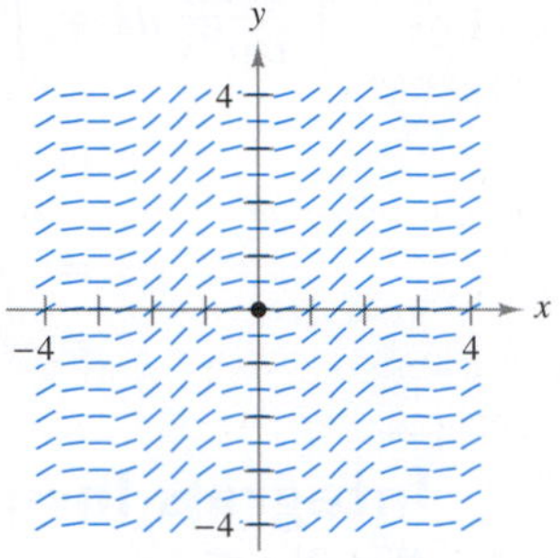

38. $\frac{dy}{dx} = \sec^2 x \tan^2 x,\ \left(0, -\frac{1}{4}\right)$

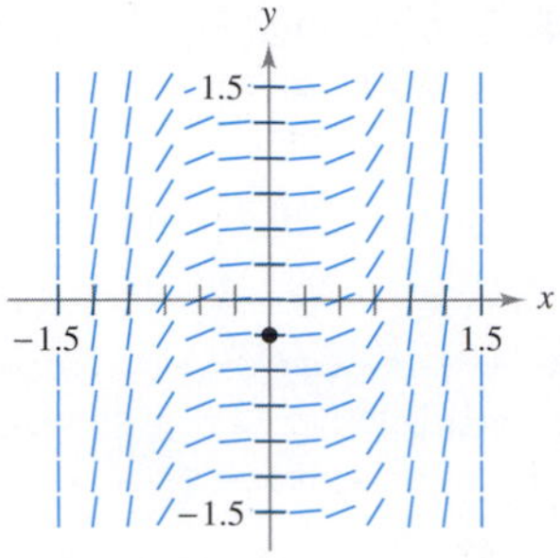

Slope Field In Exercises 39 and 40, use a computer algebra system to graph the slope field for the differential equation, and graph the solution through the specified initial condition.

39. $\frac{dy}{dx} = \frac{3 \sin x}{y},\ y(0) = 2$
40. $\frac{dy}{dx} = 3\sqrt{y} \tan^2 x,\ y(0) = 3$

Using Product-to-Sum Identities In Exercises 41–46, find the indefinite integral.

41. $\int \cos 2x \cos 6x\, dx$
42. $\int \cos 5\theta \cos 3\theta\, d\theta$
43. $\int \sin 2x \cos 4x\, dx$
44. $\int \sin(-7x) \cos 6x\, dx$
45. $\int \sin \theta \sin 3\theta\, d\theta$
46. $\int \sin 5x \sin 4x\, dx$

Finding an Indefinite Integral In Exercises 47–56, find the indefinite integral. Use a computer algebra system to confirm your result.

47. $\int \cot^3 2x\, dx$
48. $\int \tan^5 \frac{x}{4} \sec^4 \frac{x}{4}\, dx$
49. $\int \csc^4 3x\, dx$
50. $\int \cot^3 \frac{x}{2} \csc^4 \frac{x}{2}\, dx$
51. $\int \frac{\cot^2 t}{\csc t}\, dt$
52. $\int \frac{\cot^3 t}{\csc t}\, dt$
53. $\int \frac{1}{\sec x \tan x}\, dx$
54. $\int \frac{\sin^2 x - \cos^2 x}{\cos x}\, dx$
55. $\int (\tan^4 t - \sec^4 t)\, dt$
56. $\int \frac{1 - \sec t}{\cos t - 1}\, dt$

Evaluating a Definite Integral In Exercises 57–64, evaluate the definite integral.

57. $\int_{-\pi}^{\pi} \sin^2 x\, dx$

58. $\int_{0}^{\pi/3} \tan^2 x\, dx$

59. $\int_{0}^{\pi/4} 6 \tan^3 x\, dx$

60. $\int_{0}^{\pi/3} \sec^{3/2} x \tan x\, dx$

61. $\int_{0}^{\pi/2} \frac{\cos t}{1 + \sin t}\, dt$

62. $\int_{\pi/6}^{\pi/3} \sin 6x \cos 4x\, dx$

63. $\int_{-\pi/2}^{\pi/2} 3 \cos^3 x\, dx$

64. $\int_{-\pi/2}^{\pi/2} (\sin^2 x + 1)\, dx$

WRITING ABOUT CONCEPTS

65. **Describing How to Find an Integral** In your own words, describe how you would integrate $\int \sin^m x \cos^n x\, dx$ for each condition.

(a) m is positive and odd. (b) n is positive and odd.

(c) m and n are both positive and even.

66. **Describing How to Find an Integral** In your own words, describe how you would integrate $\int \sec^m x \tan^n x\, dx$ for each condition.

(a) m is positive and even. (b) n is positive and odd.

(c) n is positive and even, and there are no secant factors.

(d) m is positive and odd, and there are no tangent factors.

67. **Comparing Methods** Evaluate $\int \sin x \cos x\, dx$ using the given method. Explain how your answers differ for each method.

(a) Substitution where $u = \sin x$

(b) Substitution where $u = \cos x$

(c) Integration by parts

(d) Using the identity $\sin 2x = 2 \sin x \cos x$

68. **HOW DO YOU SEE IT?** Use the graph of f' shown in the figure to answer the following.

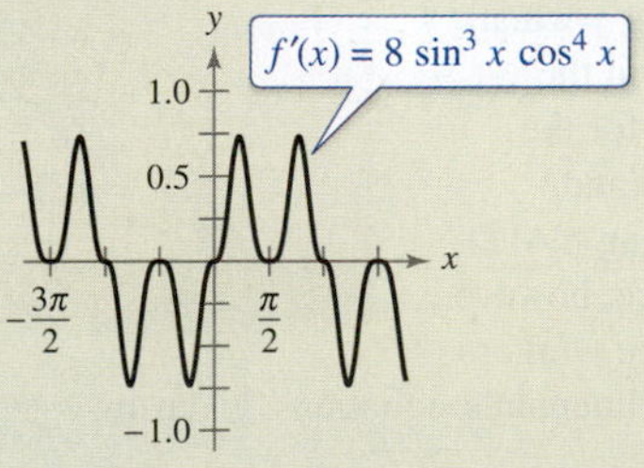

(a) Using the interval shown in the graph, approximate the value(s) of x where f is maximum. Explain.

(b) Using the interval shown in the graph, approximate the value(s) of x where f is minimum. Explain.

Comparing Methods In Exercises 69 and 70, (a) find the indefinite integral in two different ways. (b) Use a graphing utility to graph the antiderivative (without the constant of integration) obtained by each method to show that the results differ only by a constant. (c) Verify analytically that the results differ only by a constant.

69. $\int \sec^4 3x \tan^3 3x\, dx$

70. $\int \sec^2 x \tan x\, dx$

Area In Exercises 71–74, find the area of the region bounded by the graphs of the equations.

71. $y = \sin x,\quad y = \sin^3 x,\quad x = 0,\quad x = \frac{\pi}{2}$

72. $y = \sin^2 \pi x,\quad y = 0,\quad x = 0,\quad x = 1$

73. $y = \cos^2 x,\quad y = \sin^2 x,\quad x = -\frac{\pi}{4},\quad x = \frac{\pi}{4}$

74. $y = \cos^2 x,\quad y = \sin x \cos x,\quad x = -\frac{\pi}{2},\quad x = \frac{\pi}{4}$

Volume In Exercises 75 and 76, find the volume of the solid generated by revolving the region bounded by the graphs of the equations about the x-axis.

75. $y = \tan x,\quad y = 0,\quad x = -\frac{\pi}{4},\quad x = \frac{\pi}{4}$

76. $y = \cos \frac{x}{2},\quad y = \sin \frac{x}{2},\quad x = 0,\quad x = \frac{\pi}{2}$

Volume and Centroid In Exercises 77 and 78, for the region bounded by the graphs of the equations, find (a) the volume of the solid formed by revolving the region about the x-axis and (b) the centroid of the region.

77. $y = \sin x,\quad y = 0,\quad x = 0,\quad x = \pi$

78. $y = \cos x,\quad y = 0,\quad x = 0,\quad x = \frac{\pi}{2}$

Verifying a Reduction Formula In Exercises 79–82, use integration by parts to verify the reduction formula.

79. $\int \sin^n x\, dx = -\frac{\sin^{n-1} x \cos x}{n} + \frac{n-1}{n} \int \sin^{n-2} x\, dx$

80. $\int \cos^n x\, dx = \frac{\cos^{n-1} x \sin x}{n} + \frac{n-1}{n} \int \cos^{n-2} x\, dx$

81. $\int \cos^m x \sin^n x\, dx = -\frac{\cos^{m+1} x \sin^{n-1} x}{m+n} + \frac{n-1}{m+n} \int \cos^m x \sin^{n-2} x\, dx$

82. $\int \sec^n x\, dx = \frac{1}{n-1} \sec^{n-2} x \tan x + \frac{n-2}{n-1} \int \sec^{n-2} x\, dx$

Using Formulas In Exercises 83–86, use the results of Exercises 79–82 to find the integral.

83. $\int \sin^5 x\, dx$

84. $\int \cos^4 x\, dx$

85. $\int \sec^4 (2\pi x/5)\, dx$ **86.** $\int \sin^4 x \cos^2 x\, dx$

87. Modeling Data The table shows the normal maximum (high) and minimum (low) temperatures (in degrees Fahrenheit) in Erie, Pennsylvania, for each month of the year. *(Source: NOAA)*

Month	Jan	Feb	Mar	Apr	May	Jun
Max	33.5	35.4	44.7	55.6	67.4	76.2
Min	20.3	20.9	28.2	37.9	48.7	58.5
Month	Jul	Aug	Sep	Oct	Nov	Dec
Max	80.4	79.0	72.0	61.0	49.3	38.6
Min	63.7	62.7	55.9	45.5	36.4	26.8

The maximum and minimum temperatures can be modeled by

$$f(t) = a_0 + a_1 \cos \frac{\pi t}{6} + b_1 \sin \frac{\pi t}{6}$$

where $t = 0$ corresponds to January 1 and a_0, a_1, and b_1 are as follows.

$$a_0 = \frac{1}{12}\int_0^{12} f(t)\, dt \qquad a_1 = \frac{1}{6}\int_0^{12} f(t) \cos \frac{\pi t}{6}\, dt$$

$$b_1 = \frac{1}{6}\int_0^{12} f(t) \sin \frac{\pi t}{6}\, dt$$

(a) Approximate the model $H(t)$ for the maximum temperatures. (*Hint:* Use Simpson's Rule to approximate the integrals and use the January data twice.)

(b) Repeat part (a) for a model $L(t)$ for the minimum temperature data.

(c) Use a graphing utility to graph each model. During what part of the year is the difference between the maximum and minimum temperatures greatest?

88. Wallis's Formulas Use the result of Exercise 80 to prove the following versions of Wallis's Formulas.

(a) If n is odd ($n \geq 3$), then

$$\int_0^{\pi/2} \cos^n x\, dx = \left(\frac{2}{3}\right)\left(\frac{4}{5}\right)\left(\frac{6}{7}\right) \cdots \left(\frac{n-1}{n}\right).$$

(b) If n is even ($n \geq 2$), then

$$\int_0^{\pi/2} \cos^n x\, dx = \left(\frac{1}{2}\right)\left(\frac{3}{4}\right)\left(\frac{5}{6}\right) \cdots \left(\frac{n-1}{n}\right)\left(\frac{\pi}{2}\right).$$

89. Orthogonal Functions The **inner product** of two functions f and g on $[a, b]$ is given by

$$\langle f, g\rangle = \int_a^b f(x)g(x)\, dx.$$

Two distinct functions f and g are said to be **orthogonal** if $\langle f, g\rangle = 0$. Show that the following set of functions is orthogonal on $[-\pi, \pi]$.

$$\{\sin x, \sin 2x, \sin 3x, \ldots, \cos x, \cos 2x, \cos 3x, \ldots\}$$

Victor Soares/Shutterstock.com

90. Fourier Series The following sum is a *finite Fourier series.*

$$f(x) = \sum_{i=1}^{N} a_i \sin ix$$

$$= a_1 \sin x + a_2 \sin 2x + a_3 \sin 3x + \cdots + a_N \sin Nx$$

(a) Use Exercise 89 to show that the nth coefficient a_n is given by $a_n = (1/\pi)\int_{-\pi}^{\pi} f(x) \sin nx\, dx$.

(b) Let $f(x) = x$. Find a_1, a_2, and a_3.

SECTION PROJECT

Power Lines

Power lines are constructed by stringing wire between supports and adjusting the tension on each span. The wire hangs between supports in the shape of a catenary, as shown in the figure.

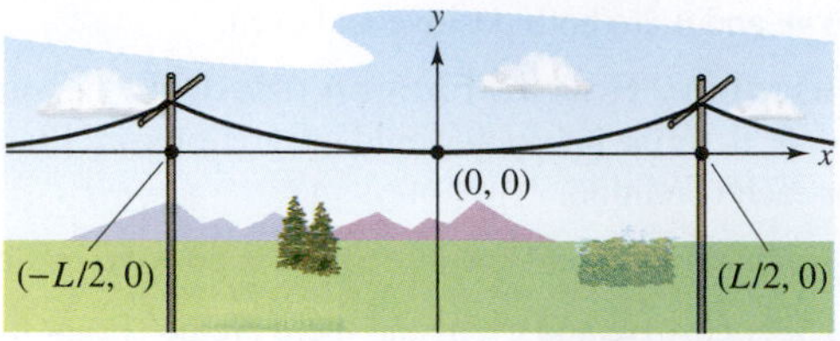

Let T be the tension (in pounds) on a span of wire, let u be the density (in pounds per foot), let $g \approx 32.2$ be the acceleration due to gravity (in feet per second per second), and let L be the distance (in feet) between the supports. Then the equation of the catenary is $y = \frac{T}{ug}\left(\cosh \frac{ugx}{T} - 1\right)$, where x and y are measured in feet.

(a) Find the length of the wire between two spans.

(b) To measure the tension in a span, power line workers use the *return wave method.* The wire is struck at one support, creating a wave in the line, and the time t (in seconds) it takes for the wave to make a round trip is measured. The velocity v (in feet per second) is given by $v = \sqrt{T/u}$. How long does it take the wave to make a round trip between supports?

(c) The sag s (in inches) can be obtained by evaluating y when $x = L/2$ in the equation for the catenary (and multiplying by 12). In practice, however, power line workers use the "lineman's equation" given by $s \approx 12.075t^2$. Use the fact that

$$\cosh \frac{ugL}{2T} + 1 \approx 2$$

to derive this equation.

FOR FURTHER INFORMATION To learn more about the mathematics of power lines, see the article "Constructing Power Lines" by Thomas O'Neil in *The UMAP Journal.*

8.4 Trigonometric Substitution

- Use trigonometric substitution to solve an integral.
- Use integrals to model and solve real-life applications.

Exploration

Integrating a Radical Function Up to this point in the text, you have not evaluated the integral

$$\int_{-1}^{1} \sqrt{1 - x^2}\, dx.$$

From geometry, you should be able to find the exact value of this integral—what is it? Using numerical integration, with Simpson's Rule or the Trapezoidal Rule, you can't be sure of the accuracy of the approximation. Why?

Try finding the exact value using the substitution

$$x = \sin \theta$$

and

$$dx = \cos \theta \, d\theta.$$

Does your answer agree with the value you obtained using geometry?

Trigonometric Substitution

Now that you can evaluate integrals involving powers of trigonometric functions, you can use **trigonometric substitution** to evaluate integrals involving the radicals

$$\sqrt{a^2 - u^2}, \quad \sqrt{a^2 + u^2}, \quad \text{and} \quad \sqrt{u^2 - a^2}.$$

The objective with trigonometric substitution is to eliminate the radical in the integrand. You do this by using the Pythagorean identities.

$$\cos^2 \theta = 1 - \sin^2 \theta$$
$$\sec^2 \theta = 1 + \tan^2 \theta$$
$$\tan^2 \theta = \sec^2 \theta - 1$$

For example, for $a > 0$, let $u = a \sin \theta$, where $-\pi/2 \le \theta \le \pi/2$. Then

$$\begin{aligned}\sqrt{a^2 - u^2} &= \sqrt{a^2 - a^2 \sin^2 \theta} \\ &= \sqrt{a^2(1 - \sin^2 \theta)} \\ &= \sqrt{a^2 \cos^2 \theta} \\ &= a \cos \theta.\end{aligned}$$

Note that $\cos \theta \ge 0$, because $-\pi/2 \le \theta \le \pi/2$.

Trigonometric Substitution ($a > 0$)

1. For integrals involving $\sqrt{a^2 - u^2}$, let

$$u = a \sin \theta.$$

Then $\sqrt{a^2 - u^2} = a \cos \theta$, where

$$-\pi/2 \le \theta \le \pi/2.$$

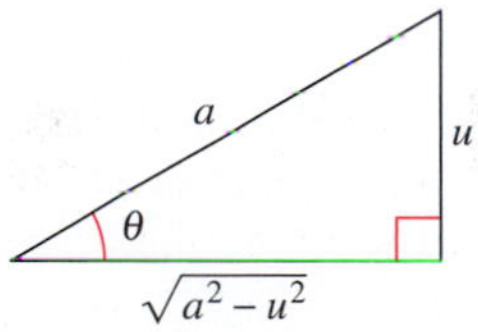

2. For integrals involving $\sqrt{a^2 + u^2}$, let

$$u = a \tan \theta.$$

Then $\sqrt{a^2 + u^2} = a \sec \theta$, where

$$-\pi/2 < \theta < \pi/2.$$

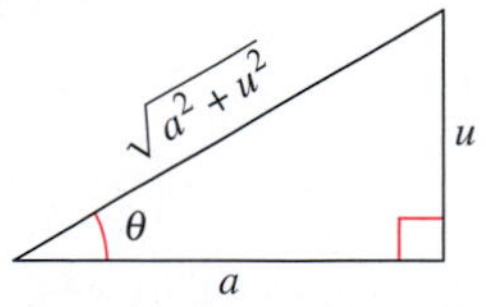

3. For integrals involving $\sqrt{u^2 - a^2}$, let

$$u = a \sec \theta.$$

Then

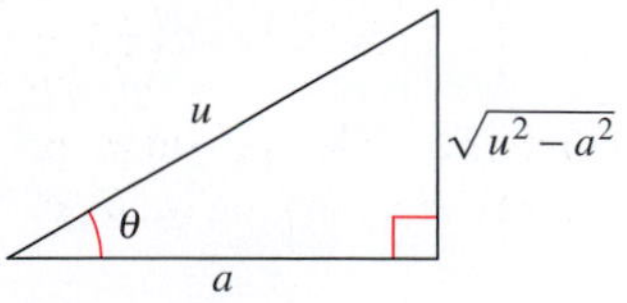

$$\sqrt{u^2 - a^2} = \begin{cases} a \tan \theta \text{ for } u > a, \text{ where } 0 \le \theta < \pi/2 \\ -a \tan \theta \text{ for } u < -a, \text{ where } \pi/2 < \theta \le \pi. \end{cases}$$

The restrictions on θ ensure that the function that defines the substitution is one-to-one. In fact, these are the same intervals over which the arcsine, arctangent, and arcsecant are defined.

EXAMPLE 1 **Trigonometric Substitution: $u = a \sin \theta$**

Find $\displaystyle\int \frac{dx}{x^2\sqrt{9 - x^2}}$.

Solution First, note that none of the basic integration rules applies. To use trigonometric substitution, you should observe that

$$\sqrt{9 - x^2}$$

is of the form $\sqrt{a^2 - u^2}$. So, you can use the substitution

$$x = a \sin \theta = 3 \sin \theta.$$

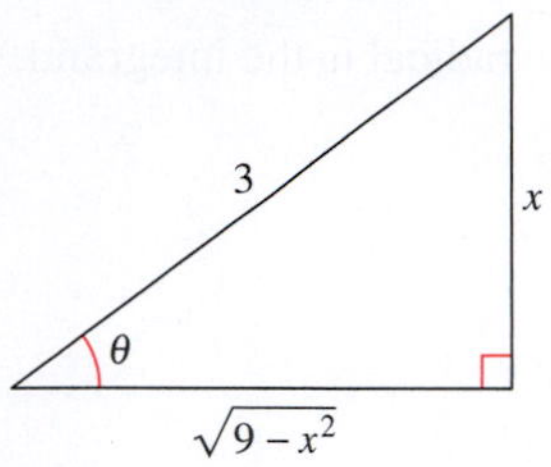

$\sin \theta = \frac{x}{3}$, $\cot \theta = \frac{\sqrt{9 - x^2}}{x}$

Figure 8.6

Using differentiation and the triangle shown in Figure 8.6, you obtain

$$dx = 3 \cos \theta \, d\theta, \quad \sqrt{9 - x^2} = 3 \cos \theta, \quad \text{and} \quad x^2 = 9 \sin^2 \theta.$$

So, trigonometric substitution yields

$$\int \frac{dx}{x^2\sqrt{9 - x^2}} = \int \frac{3 \cos \theta \, d\theta}{(9 \sin^2 \theta)(3 \cos \theta)} \qquad \text{Substitute.}$$

$$= \frac{1}{9}\int \frac{d\theta}{\sin^2 \theta} \qquad \text{Simplify.}$$

$$= \frac{1}{9}\int \csc^2 \theta \, d\theta \qquad \text{Trigonometric identity}$$

$$= -\frac{1}{9} \cot \theta + C \qquad \text{Apply Cosecant Rule.}$$

$$= -\frac{1}{9}\left(\frac{\sqrt{9 - x^2}}{x}\right) + C \qquad \text{Substitute for } \cot \theta.$$

$$= -\frac{\sqrt{9 - x^2}}{9x} + C.$$

Note that the triangle in Figure 8.6 can be used to convert the θ's back to x's, as shown.

$$\cot \theta = \frac{\text{adj.}}{\text{opp.}}$$

$$= \frac{\sqrt{9 - x^2}}{x}$$

TECHNOLOGY Use a computer algebra system to find each indefinite integral.

$$\int \frac{dx}{\sqrt{9 - x^2}} \qquad \int \frac{dx}{x\sqrt{9 - x^2}}$$

$$\int \frac{dx}{x^2\sqrt{9 - x^2}} \qquad \int \frac{dx}{x^3\sqrt{9 - x^2}}$$

Then use trigonometric substitution to duplicate the results obtained with the computer algebra system.

In Chapter 5, you saw how the inverse hyperbolic functions can be used to evaluate the integrals

$$\int \frac{du}{\sqrt{u^2 \pm a^2}}, \quad \int \frac{du}{a^2 - u^2}, \quad \text{and} \quad \int \frac{du}{u\sqrt{a^2 \pm u^2}}.$$

You can also evaluate these integrals using trigonometric substitution. This is shown in the next example.

EXAMPLE 2 Trigonometric Substitution: $u = a \tan \theta$

Find $\displaystyle\int \frac{dx}{\sqrt{4x^2 + 1}}$.

Solution Let $u = 2x$, $a = 1$, and $2x = \tan \theta$, as shown in Figure 8.7. Then,

$$dx = \frac{1}{2} \sec^2 \theta \, d\theta \quad \text{and} \quad \sqrt{4x^2 + 1} = \sec \theta.$$

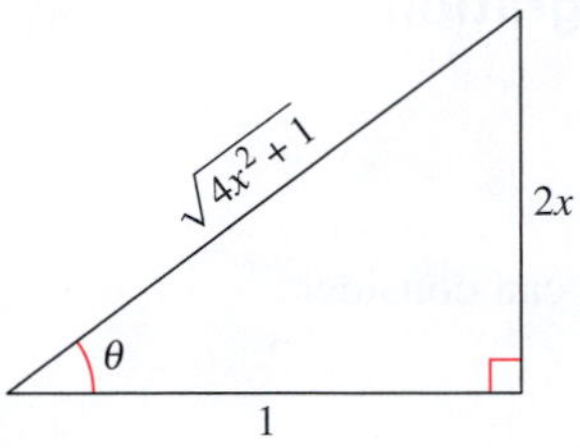

$\tan \theta = 2x$, $\sec \theta = \sqrt{4x^2 + 1}$

Figure 8.7

Trigonometric substitution produces

$$\begin{aligned}
\int \frac{1}{\sqrt{4x^2 + 1}} \, dx &= \frac{1}{2} \int \frac{\sec^2 \theta \, d\theta}{\sec \theta} && \text{Substitute.} \\
&= \frac{1}{2} \int \sec \theta \, d\theta && \text{Simplify.} \\
&= \frac{1}{2} \ln|\sec \theta + \tan \theta| + C && \text{Apply Secant Rule.} \\
&= \frac{1}{2} \ln\left|\sqrt{4x^2 + 1} + 2x\right| + C. && \text{Back-substitute.}
\end{aligned}$$

Try checking this result with a computer algebra system. Is the result given in this form or in the form of an inverse hyperbolic function?

You can extend the use of trigonometric substitution to cover integrals involving expressions such as $(a^2 - u^2)^{n/2}$ by writing the expression as

$$(a^2 - u^2)^{n/2} = \left(\sqrt{a^2 - u^2}\right)^n.$$

EXAMPLE 3 Trigonometric Substitution: Rational Powers

See LarsonCalculus.com for an interactive version of this type of example.

Find $\displaystyle\int \frac{dx}{(x^2 + 1)^{3/2}}$.

Solution Begin by writing $(x^2 + 1)^{3/2}$ as

$$\left(\sqrt{x^2 + 1}\right)^3.$$

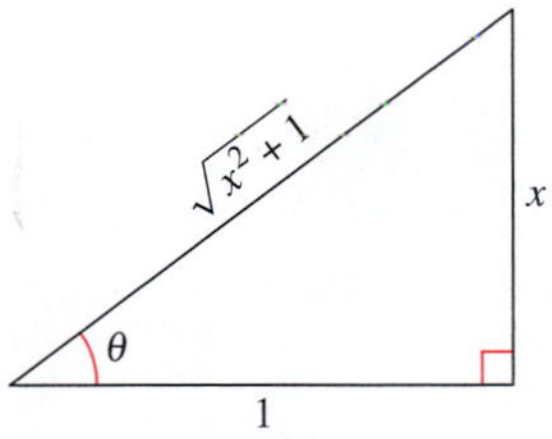

$\tan \theta = x$, $\sin \theta = \dfrac{x}{\sqrt{x^2 + 1}}$

Figure 8.8

Then, let $a = 1$ and $u = x = \tan \theta$, as shown in Figure 8.8. Using

$$dx = \sec^2 \theta \, d\theta \quad \text{and} \quad \sqrt{x^2 + 1} = \sec \theta$$

you can apply trigonometric substitution, as shown.

$$\begin{aligned}
\int \frac{dx}{(x^2 + 1)^{3/2}} &= \int \frac{dx}{\left(\sqrt{x^2 + 1}\right)^3} && \text{Rewrite denominator.} \\
&= \int \frac{\sec^2 \theta \, d\theta}{\sec^3 \theta} && \text{Substitute.} \\
&= \int \frac{d\theta}{\sec \theta} && \text{Simplify.} \\
&= \int \cos \theta \, d\theta && \text{Trigonometric identity} \\
&= \sin \theta + C && \text{Apply Cosine Rule.} \\
&= \frac{x}{\sqrt{x^2 + 1}} + C && \text{Back-substitute.}
\end{aligned}$$

For definite integrals, it is often convenient to determine integration limits for θ that avoid converting back to x. You might want to review this procedure in Section 4.5, Examples 8 and 9.

EXAMPLE 4 Converting the Limits of Integration

Evaluate $\displaystyle\int_{\sqrt{3}}^{2} \frac{\sqrt{x^2-3}}{x}\,dx.$

Solution Because $\sqrt{x^2-3}$ has the form $\sqrt{u^2-a^2}$, you can consider

$$u = x, \quad a = \sqrt{3}, \quad \text{and} \quad x = \sqrt{3}\sec\theta$$

as shown in Figure 8.9. Then,

$$dx = \sqrt{3}\sec\theta\tan\theta\,d\theta \quad \text{and} \quad \sqrt{x^2-3} = \sqrt{3}\tan\theta.$$

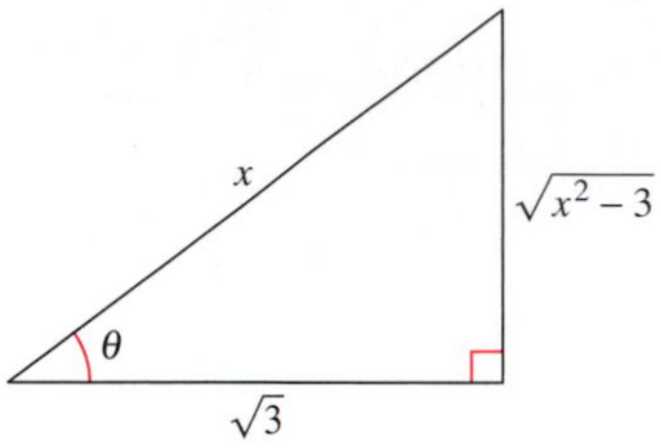

$\sec\theta = \dfrac{x}{\sqrt{3}}$, $\tan\theta = \dfrac{\sqrt{x^2-3}}{\sqrt{3}}$

Figure 8.9

To determine the upper and lower limits of integration, use the substitution $x = \sqrt{3}\sec\theta$, as shown.

Lower Limit

When $x = \sqrt{3}$, $\sec\theta = 1$ and $\theta = 0$.

Upper Limit

When $x = 2$, $\sec\theta = \dfrac{2}{\sqrt{3}}$ and $\theta = \dfrac{\pi}{6}$.

So, you have

Integration limits for x — Integration limits for θ

$$\begin{aligned}
\int_{\sqrt{3}}^{2} \frac{\sqrt{x^2-3}}{x}\,dx &= \int_{0}^{\pi/6} \frac{(\sqrt{3}\tan\theta)(\sqrt{3}\sec\theta\tan\theta)\,d\theta}{\sqrt{3}\sec\theta} \\
&= \int_{0}^{\pi/6} \sqrt{3}\tan^2\theta\,d\theta \\
&= \sqrt{3}\int_{0}^{\pi/6} (\sec^2\theta - 1)\,d\theta \\
&= \sqrt{3}\Big[\tan\theta - \theta\Big]_{0}^{\pi/6} \\
&= \sqrt{3}\left(\frac{1}{\sqrt{3}} - \frac{\pi}{6}\right) \\
&= 1 - \frac{\sqrt{3}\pi}{6} \\
&\approx 0.0931.
\end{aligned}$$

In Example 4, try converting back to the variable x and evaluating the antiderivative at the original limits of integration. You should obtain

$$\begin{aligned}
\int_{\sqrt{3}}^{2} \frac{\sqrt{x^2-3}}{x}\,dx &= \sqrt{3}\left[\frac{\sqrt{x^2-3}}{\sqrt{3}} - \operatorname{arcsec}\frac{x}{\sqrt{3}}\right]_{\sqrt{3}}^{2} \\
&= \sqrt{3}\left(\frac{1}{\sqrt{3}} - \frac{\pi}{6}\right) \\
&\approx 0.0931.
\end{aligned}$$

When using trigonometric substitution to evaluate definite integrals, you must be careful to check that the values of θ lie in the intervals discussed at the beginning of this section. For instance, if in Example 4 you had been asked to evaluate the definite integral

$$\int_{-2}^{-\sqrt{3}} \frac{\sqrt{x^2 - 3}}{x}\, dx$$

then using $u = x$ and $a = \sqrt{3}$ in the interval $\left[-2, -\sqrt{3}\right]$ would imply that $u < -a$. So, when determining the upper and lower limits of integration, you would have to choose θ such that $\pi/2 < \theta \leq \pi$. In this case, the integral would be evaluated as shown.

$$\begin{aligned}
\int_{-2}^{-\sqrt{3}} \frac{\sqrt{x^2 - 3}}{x}\, dx &= \int_{5\pi/6}^{\pi} \frac{\left(-\sqrt{3}\tan\theta\right)\left(\sqrt{3}\sec\theta\tan\theta\right) d\theta}{\sqrt{3}\sec\theta} \\
&= \int_{5\pi/6}^{\pi} -\sqrt{3}\tan^2\theta\, d\theta \\
&= -\sqrt{3}\int_{5\pi/6}^{\pi} (\sec^2\theta - 1)\, d\theta \\
&= -\sqrt{3}\Big[\tan\theta - \theta\Big]_{5\pi/6}^{\pi} \\
&= -\sqrt{3}\left[(0 - \pi) - \left(-\frac{1}{\sqrt{3}} - \frac{5\pi}{6}\right)\right] \\
&= -1 + \frac{\sqrt{3}\pi}{6} \\
&\approx -0.0931
\end{aligned}$$

Trigonometric substitution can be used with completing the square. For instance, try finding the integral

$$\int \sqrt{x^2 - 2x}\, dx.$$

To begin, you could complete the square and write the integral as

$$\int \sqrt{(x - 1)^2 - 1^2}\, dx.$$

Because the integrand has the form

$$\sqrt{u^2 - a^2}$$

with $u = x - 1$ and $a = 1$, you can now use trigonometric substitution to find the integral.

Trigonometric substitution can be used to evaluate the three integrals listed in the next theorem. These integrals will be encountered several times in the remainder of the text. When this happens, we will simply refer to this theorem. (In Exercise 71, you are asked to verify the formulas given in the theorem.)

THEOREM 8.2 Special Integration Formulas ($a > 0$)

1. $\displaystyle\int \sqrt{a^2 - u^2}\, du = \frac{1}{2}\left(a^2 \arcsin\frac{u}{a} + u\sqrt{a^2 - u^2}\right) + C$
2. $\displaystyle\int \sqrt{u^2 - a^2}\, du = \frac{1}{2}\left(u\sqrt{u^2 - a^2} - a^2 \ln\left|u + \sqrt{u^2 - a^2}\right|\right) + C, \quad u > a$
3. $\displaystyle\int \sqrt{u^2 + a^2}\, du = \frac{1}{2}\left(u\sqrt{u^2 + a^2} + a^2 \ln\left|u + \sqrt{u^2 + a^2}\right|\right) + C$

Applications

EXAMPLE 5 Finding Arc Length

Find the arc length of the graph of $f(x) = \frac{1}{2}x^2$ from $x = 0$ to $x = 1$ (see Figure 8.10).

Solution Refer to the arc length formula in Section 7.4.

$$s = \int_0^1 \sqrt{1 + [f'(x)]^2}\, dx \qquad \text{Formula for arc length}$$

$$= \int_0^1 \sqrt{1 + x^2}\, dx \qquad f'(x) = x$$

$$= \int_0^{\pi/4} \sec^3 \theta \, d\theta \qquad \text{Let } a = 1 \text{ and } x = \tan \theta.$$

$$= \frac{1}{2}\Big[\sec \theta \tan \theta + \ln|\sec \theta + \tan \theta|\Big]_0^{\pi/4} \qquad \text{Example 5, Section 8.2}$$

$$= \frac{1}{2}\left[\sqrt{2} + \ln\left(\sqrt{2} + 1\right)\right]$$

$$\approx 1.148$$

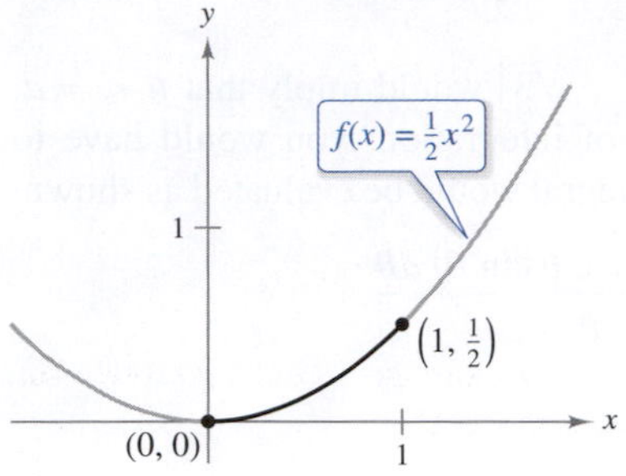

The arc length of the curve from (0, 0) to $(1, \frac{1}{2})$

Figure 8.10

EXAMPLE 6 Comparing Two Fluid Forces

A sealed barrel of oil (weighing 48 pounds per cubic foot) is floating in seawater (weighing 64 pounds per cubic foot), as shown in Figures 8.11 and 8.12. (The barrel is not completely full of oil. With the barrel lying on its side, the top 0.2 foot of the barrel is empty.) Compare the fluid forces against one end of the barrel from the inside and from the outside.

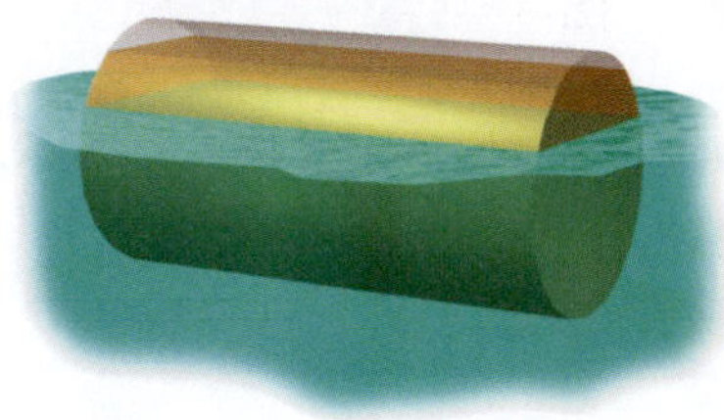
The barrel is not quite full of oil—the top 0.2 foot of the barrel is empty.

Figure 8.11

Solution In Figure 8.12, locate the coordinate system with the origin at the center of the circle

$$x^2 + y^2 = 1.$$

To find the fluid force against an end of the barrel *from the inside,* integrate between -1 and 0.8 (using a weight of $w = 48$).

$$F = w\int_c^d h(y)L(y)\, dy \qquad \text{General equation (See Section 7.7.)}$$

$$F_{\text{inside}} = 48\int_{-1}^{0.8} (0.8 - y)(2)\sqrt{1 - y^2}\, dy$$

$$= 76.8\int_{-1}^{0.8} \sqrt{1 - y^2}\, dy - 96\int_{-1}^{0.8} y\sqrt{1 - y^2}\, dy$$

To find the fluid force *from the outside,* integrate between -1 and 0.4 (using a weight of $w = 64$).

$$F_{\text{outside}} = 64\int_{-1}^{0.4} (0.4 - y)(2)\sqrt{1 - y^2}\, dy$$

$$= 51.2\int_{-1}^{0.4} \sqrt{1 - y^2}\, dy - 128\int_{-1}^{0.4} y\sqrt{1 - y^2}\, dy$$

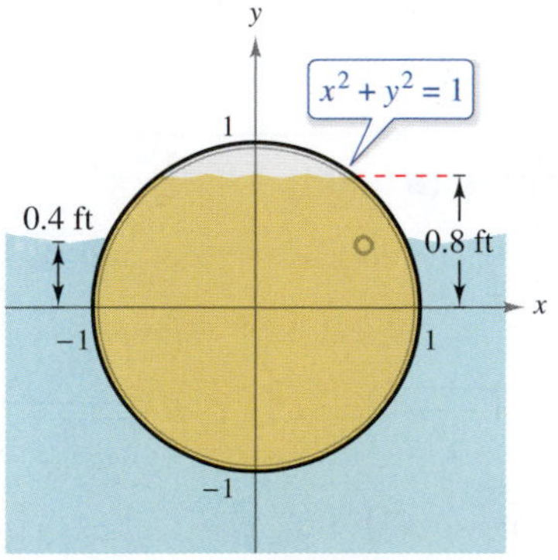

Figure 8.12

The details of integration are left for you to complete in Exercise 70. Intuitively, would you say that the force from the oil (the inside) or the force from the seawater (the outside) is greater? By evaluating these two integrals, you can determine that

$$F_{\text{inside}} \approx 121.3 \text{ pounds} \quad \text{and} \quad F_{\text{outside}} \approx 93.0 \text{ pounds}.$$

8.4 Exercises

See CalcChat.com for tutorial help and worked-out solutions to odd-numbered exercises.

Trigonometric Substitution **In Exercises 1–4, state the trigonometric substitution you would use to find the indefinite integral. Do not integrate.**

1. $\int (9 + x^2)^{-2}\,dx$

2. $\int \sqrt{4 - x^2}\,dx$

3. $\int \frac{x^2}{\sqrt{25 - x^2}}\,dx$

4. $\int x^2(x^2 - 25)^{3/2}\,dx$

Using Trigonometric Substitution **In Exercises 5–8, find the indefinite integral using the substitution $x = 4 \sin \theta$.**

5. $\int \frac{1}{(16 - x^2)^{3/2}}\,dx$

6. $\int \frac{4}{x^2\sqrt{16 - x^2}}\,dx$

7. $\int \frac{\sqrt{16 - x^2}}{x}\,dx$

8. $\int \frac{x^3}{\sqrt{16 - x^2}}\,dx$

Using Trigonometric Substitution **In Exercises 9–12, find the indefinite integral using the substitution $x = 5 \sec \theta$.**

9. $\int \frac{1}{\sqrt{x^2 - 25}}\,dx$

10. $\int \frac{\sqrt{x^2 - 25}}{x}\,dx$

11. $\int x^3\sqrt{x^2 - 25}\,dx$

12. $\int \frac{x^3}{\sqrt{x^2 - 25}}\,dx$

Using Trigonometric Substitution **In Exercises 13–16, find the indefinite integral using the substitution $x = \tan \theta$.**

13. $\int x\sqrt{1 + x^2}\,dx$

14. $\int \frac{9x^3}{\sqrt{1 + x^2}}\,dx$

15. $\int \frac{1}{(1 + x^2)^2}\,dx$

16. $\int \frac{x^2}{(1 + x^2)^2}\,dx$

Using Formulas **In Exercises 17–20, use the Special Integration Formulas (Theorem 8.2) to find the indefinite integral.**

17. $\int \sqrt{9 + 16x^2}\,dx$

18. $\int \sqrt{4 + x^2}\,dx$

19. $\int \sqrt{25 - 4x^2}\,dx$

20. $\int \sqrt{5x^2 - 1}\,dx$

Finding an Indefinite Integral **In Exercises 21–36, find the indefinite integral.**

21. $\int \frac{1}{\sqrt{16 - x^2}}\,dx$

22. $\int \frac{x^2}{\sqrt{36 - x^2}}\,dx$

23. $\int \sqrt{16 - 4x^2}\,dx$

24. $\int \frac{1}{\sqrt{x^2 - 4}}\,dx$

25. $\int \frac{\sqrt{1 - x^2}}{x^4}\,dx$

26. $\int \frac{\sqrt{25x^2 + 4}}{x^4}\,dx$

27. $\int \frac{1}{x\sqrt{4x^2 + 9}}\,dx$

28. $\int \frac{1}{x\sqrt{9x^2 + 1}}\,dx$

29. $\int \frac{-3x}{(x^2 + 3)^{3/2}}\,dx$

30. $\int \frac{1}{(x^2 + 5)^{3/2}}\,dx$

31. $\int e^x\sqrt{1 - e^{2x}}\,dx$

32. $\int \frac{\sqrt{1 - x}}{\sqrt{x}}\,dx$

33. $\int \frac{1}{4 + 4x^2 + x^4}\,dx$

34. $\int \frac{x^3 + x + 1}{x^4 + 2x^2 + 1}\,dx$

35. $\int \operatorname{arcsec} 2x\,dx, \quad x > \frac{1}{2}$

36. $\int x \arcsin x\,dx$

Completing the Square **In Exercises 37–40, complete the square and find the indefinite integral.**

37. $\int \frac{1}{\sqrt{4x - x^2}}\,dx$

38. $\int \frac{x^2}{\sqrt{2x - x^2}}\,dx$

39. $\int \frac{x}{\sqrt{x^2 + 6x + 12}}\,dx$

40. $\int \frac{x}{\sqrt{x^2 - 6x + 5}}\,dx$

Converting Limits of Integration **In Exercises 41–46, evaluate the definite integral using (a) the given integration limits and (b) the limits obtained by trigonometric substitution.**

41. $\int_0^{\sqrt{3}/2} \frac{t^2}{(1 - t^2)^{3/2}}\,dt$

42. $\int_0^{\sqrt{3}/2} \frac{1}{(1 - t^2)^{5/2}}\,dt$

43. $\int_0^3 \frac{x^3}{\sqrt{x^2 + 9}}\,dx$

44. $\int_0^{3/5} \sqrt{9 - 25x^2}\,dx$

45. $\int_4^6 \frac{x^2}{\sqrt{x^2 - 9}}\,dx$

46. $\int_4^8 \frac{\sqrt{x^2 - 16}}{x^2}\,dx$

WRITING ABOUT CONCEPTS

47. **Trigonometric Substitution** State the substitution you would make if you used trigonometric substitution for an integral involving the given radical, where $a > 0$. Explain your reasoning.

(a) $\sqrt{a^2 - u^2}$

(b) $\sqrt{a^2 + u^2}$

(c) $\sqrt{u^2 - a^2}$

48. **Choosing a Method** State the method of integration you would use to perform each integration. Explain why you chose that method. Do not integrate.

(a) $\int x\sqrt{x^2 + 1}\,dx$ (b) $\int x^2\sqrt{x^2 - 1}\,dx$

49. **Comparing Methods**

(a) Find the integral $\int \frac{x}{x^2 + 9}\,dx$ using u-substitution. Then find the integral using trigonometric substitution. Discuss the results.

(b) Find the integral $\int \frac{x^2}{x^2 + 9}\,dx$ algebraically using $x^2 = (x^2 + 9) - 9$. Then find the integral using trigonometric substitution. Discuss the results.

50. HOW DO YOU SEE IT? Use the graph of f' shown in the figure to answer the following.

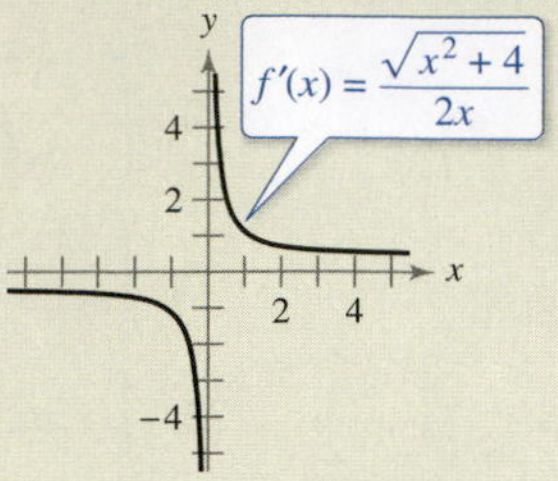

(a) Identify the open interval(s) on which the graph of f is increasing or decreasing. Explain.

(b) Identify the open interval(s) on which the graph of f is concave upward or concave downward. Explain.

True or False? **In Exercises 51–54, determine whether the statement is true or false. If it is false, explain why or give an example that shows it is false.**

51. If $x = \sin\theta$, then

$$\int \frac{dx}{\sqrt{1-x^2}} = \int d\theta.$$

52. If $x = \sec\theta$, then

$$\int \frac{\sqrt{x^2-1}}{x}\,dx = \int \sec\theta\tan\theta\,d\theta.$$

53. If $x = \tan\theta$, then

$$\int_0^{\sqrt{3}} \frac{dx}{(1+x^2)^{3/2}} = \int_0^{4\pi/3} \cos\theta\,d\theta.$$

54. If $x = \sin\theta$, then

$$\int_{-1}^{1} x^2\sqrt{1-x^2}\,dx = 2\int_0^{\pi/2} \sin^2\theta\cos^2\theta\,d\theta.$$

55. Area Find the area enclosed by the ellipse $\frac{x^2}{a^2} + \frac{y^2}{b^2} = 1$ shown in the figure.

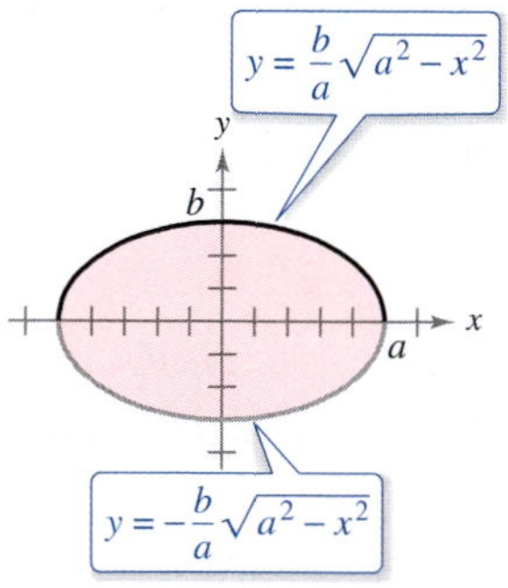

Figure for 55

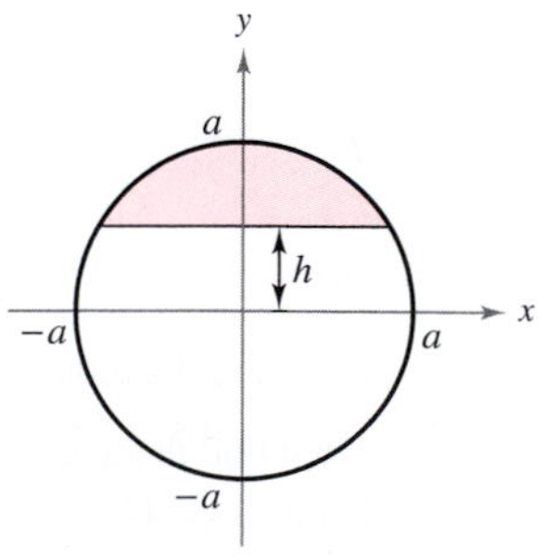

Figure for 56

56. Area Find the area of the shaded region of the circle of radius a when the chord is h units $(0 < h < a)$ from the center of the circle (see figure).

57. Mechanical Design The surface of a machine part is the region between the graphs of $y = |x|$ and $x^2 + (y-k)^2 = 25$ (see figure).

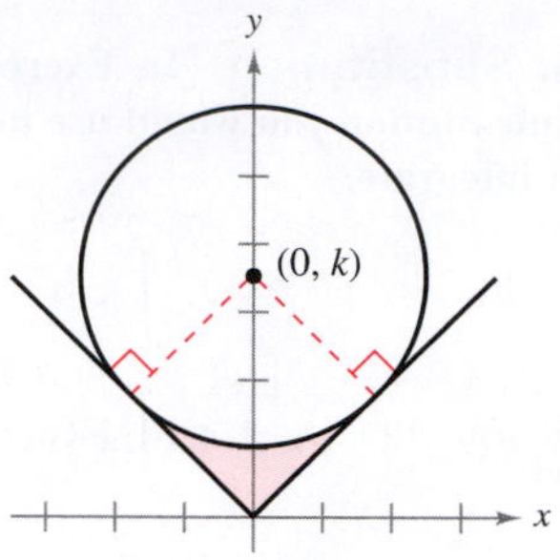

(a) Find k when the circle is tangent to the graph of $y = |x|$.

(b) Find the area of the surface of the machine part.

(c) Find the area of the surface of the machine part as a function of the radius r of the circle.

58. Volume The axis of a storage tank in the form of a right circular cylinder is horizontal (see figure). The radius and length of the tank are 1 meter and 3 meters, respectively.

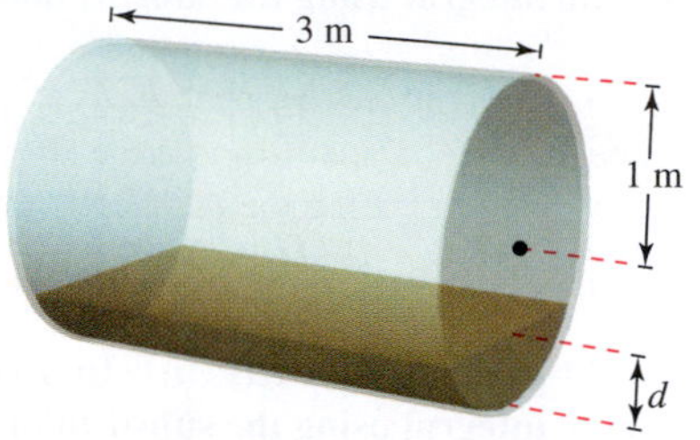

(a) Determine the volume of fluid in the tank as a function of its depth d.

(b) Use a graphing utility to graph the function in part (a).

(c) Design a dip stick for the tank with markings of $\frac{1}{4}$, $\frac{1}{2}$, and $\frac{3}{4}$.

(d) Fluid is entering the tank at a rate of $\frac{1}{4}$ cubic meter per minute. Determine the rate of change of the depth of the fluid as a function of its depth d.

(e) Use a graphing utility to graph the function in part (d). When will the rate of change of the depth be minimum? Does this agree with your intuition? Explain.

Volume of a Torus **In Exercises 59 and 60, find the volume of the torus generated by revolving the region bounded by the graph of the circle about the y-axis.**

59. $(x-3)^2 + y^2 = 1$

60. $(x-h)^2 + y^2 = r^2, \quad h > r$

Arc Length **In Exercises 61 and 62, find the arc length of the curve over the given interval.**

61. $y = \ln x, \quad [1, 5]$

62. $y = \frac{1}{2}x^2, \quad [0, 4]$

63. Arc Length Show that the length of one arch of the sine curve is equal to the length of one arch of the cosine curve.

64. Conjecture

(a) Find formulas for the distances between $(0, 0)$ and (a, a^2) along the line between these points and along the parabola $y = x^2$.

(b) Use the formulas from part (a) to find the distances for $a = 1$ and $a = 10$.

(c) Make a conjecture about the difference between the two distances as a increases.

Centroid In Exercises 65 and 66, find the centroid of the region determined by the graphs of the inequalities.

65. $y \le 3/\sqrt{x^2+9},\ y \ge 0,\ x \ge -4,\ x \le 4$

66. $y \le \frac{1}{4}x^2,\ (x-4)^2 + y^2 \le 16,\ y \ge 0$

67. Surface Area Find the surface area of the solid generated by revolving the region bounded by the graphs of $y = x^2$, $y = 0$, $x = 0$, and $x = \sqrt{2}$ about the x-axis.

68. Field Strength The field strength H of a magnet of length $2L$ on a particle r units from the center of the magnet is

$$H = \frac{2mL}{(r^2 + L^2)^{3/2}}$$

where $\pm m$ are the poles of the magnet (see figure). Find the average field strength as the particle moves from 0 to R units from the center by evaluating the integral

$$\frac{1}{R}\int_0^R \frac{2mL}{(r^2 + L^2)^{3/2}}\,dr.$$

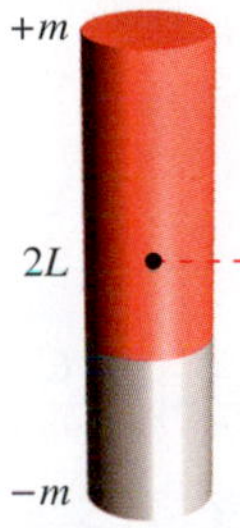

69. Fluid Force

Find the fluid force on a circular observation window of radius 1 foot in a vertical wall of a large water-filled tank at a fish hatchery when the center of the window is (a) 3 feet and (b) d feet $(d > 1)$ below the water's surface (see figure). Use trigonometric substitution to evaluate the one integral. Water weighs 62.4 pounds per cubic foot. (Recall that in Section 7.7 in a similar problem, you evaluated one integral by a geometric formula and the other by observing that the integrand was odd.)

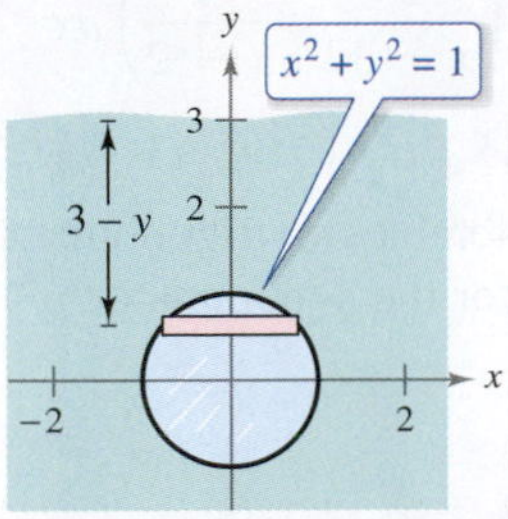

70. Fluid Force Evaluate the following two integrals, which yield the fluid forces given in Example 6.

(a) $F_{\text{inside}} = 48\int_{-1}^{0.8} (0.8 - y)(2)\sqrt{1 - y^2}\,dy$

(b) $F_{\text{outside}} = 64\int_{-1}^{0.4} (0.4 - y)(2)\sqrt{1 - y^2}\,dy$

71. Verifying Formulas Use trigonometric substitution to verify the integration formulas given in Theorem 8.2.

72. Arc Length Show that the arc length of the graph of $y = \sin x$ on the interval $[0, 2\pi]$ is equal to the circumference of the ellipse $x^2 + 2y^2 = 2$ (see figure).

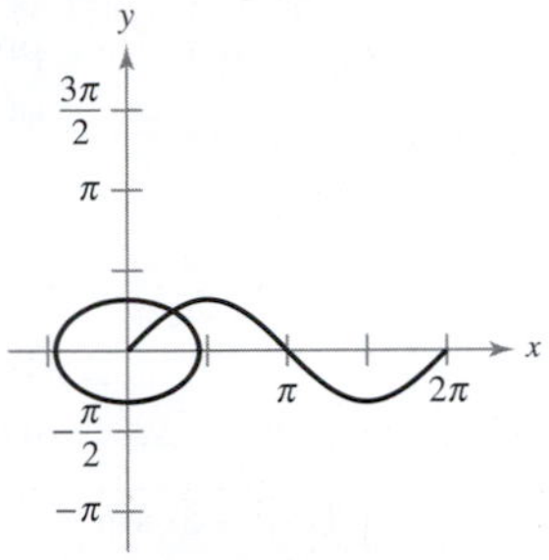

73. Area of a Lune The crescent-shaped region bounded by two circles forms a *lune* (see figure). Find the area of the lune given that the radius of the smaller circle is 3 and the radius of the larger circle is 5.

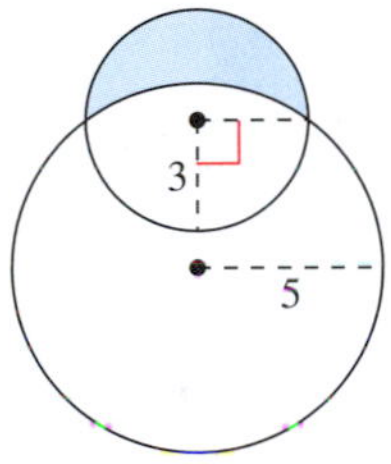

74. Area Two circles of radius 3, with centers at $(-2, 0)$ and $(2, 0)$, intersect as shown in the figure. Find the area of the shaded region.

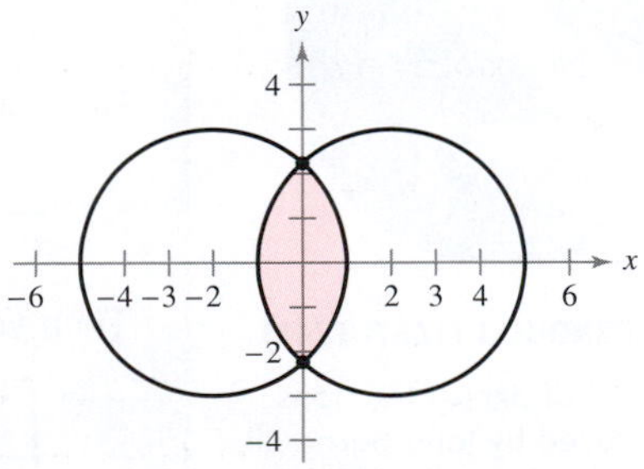

PUTNAM EXAM CHALLENGE

75. Evaluate

$$\int_0^1 \frac{\ln(x+1)}{x^2+1}\,dx.$$

8.5 Partial Fractions

- **Understand the concept of partial fraction decomposition.**
- **Use partial fraction decomposition with linear factors to integrate rational functions.**
- **Use partial fraction decomposition with quadratic factors to integrate rational functions.**

Partial Fractions

This section examines a procedure for decomposing a rational function into simpler rational functions to which you can apply the basic integration formulas. This procedure is called the **method of partial fractions.** To see the benefit of the method of partial fractions, consider the integral

$$\int \frac{1}{x^2 - 5x + 6}\,dx.$$

To evaluate this integral *without* partial fractions, you can complete the square and use trigonometric substitution (see Figure 8.13) to obtain

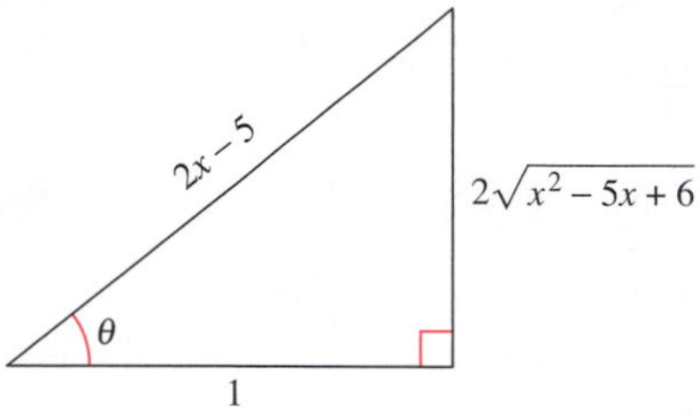

$\sec\theta = 2x - 5$

Figure 8.13

$$\begin{aligned}
\int \frac{1}{x^2 - 5x + 6}\,dx &= \int \frac{dx}{(x - 5/2)^2 - (1/2)^2} && a = \tfrac{1}{2},\ x - \tfrac{5}{2} = \tfrac{1}{2}\sec\theta \\
&= \int \frac{(1/2)\sec\theta\tan\theta\,d\theta}{(1/4)\tan^2\theta} && dx = \tfrac{1}{2}\sec\theta\tan\theta\,d\theta \\
&= 2\int \csc\theta\,d\theta \\
&= 2\ln|\csc\theta - \cot\theta| + C \\
&= 2\ln\left|\frac{2x - 5}{2\sqrt{x^2 - 5x + 6}} - \frac{1}{2\sqrt{x^2 - 5x + 6}}\right| + C \\
&= 2\ln\left|\frac{x - 3}{\sqrt{x^2 - 5x + 6}}\right| + C \\
&= 2\ln\left|\frac{\sqrt{x - 3}}{\sqrt{x - 2}}\right| + C \\
&= \ln\left|\frac{x - 3}{x - 2}\right| + C \\
&= \ln|x - 3| - \ln|x - 2| + C.
\end{aligned}$$

Now, suppose you had observed that

$$\frac{1}{x^2 - 5x + 6} = \frac{1}{x - 3} - \frac{1}{x - 2}. \qquad \text{Partial fraction decomposition}$$

Then you could evaluate the integral, as shown.

$$\begin{aligned}
\int \frac{1}{x^2 - 5x + 6}\,dx &= \int \left(\frac{1}{x - 3} - \frac{1}{x - 2}\right) dx \\
&= \ln|x - 3| - \ln|x - 2| + C
\end{aligned}$$

This method is clearly preferable to trigonometric substitution. Its use, however, depends on the ability to factor the denominator, $x^2 - 5x + 6$, and to find the **partial fractions**

$$\frac{1}{x - 3} \quad \text{and} \quad -\frac{1}{x - 2}.$$

In this section, you will study techniques for finding partial fraction decompositions.

JOHN BERNOULLI (1667–1748)

The method of partial fractions was introduced by John Bernoulli, a Swiss mathematician who was instrumental in the early development of calculus. John Bernoulli was a professor at the University of Basel and taught many outstanding students, the most famous of whom was Leonhard Euler.

See LarsonCalculus.com to read more of this biography.

Recall from algebra that every polynomial with real coefficients can be factored into linear and irreducible quadratic factors.* For instance, the polynomial

$$x^5 + x^4 - x - 1$$

can be written as

$$\begin{aligned} x^5 + x^4 - x - 1 &= x^4(x + 1) - (x + 1) \\ &= (x^4 - 1)(x + 1) \\ &= (x^2 + 1)(x^2 - 1)(x + 1) \\ &= (x^2 + 1)(x + 1)(x - 1)(x + 1) \\ &= (x - 1)(x + 1)^2(x^2 + 1) \end{aligned}$$

where $(x - 1)$ is a linear factor, $(x + 1)^2$ is a repeated linear factor, and $(x^2 + 1)$ is an irreducible quadratic factor. Using this factorization, you can write the partial fraction decomposition of the rational expression

$$\frac{N(x)}{x^5 + x^4 - x - 1}$$

where $N(x)$ is a polynomial of degree less than 5, as shown.

$$\frac{N(x)}{(x - 1)(x + 1)^2(x^2 + 1)} = \frac{A}{x - 1} + \frac{B}{x + 1} + \frac{C}{(x + 1)^2} + \frac{Dx + E}{x^2 + 1}$$

REMARK In precalculus, you learned how to combine functions such as

$$\frac{1}{x - 2} + \frac{-1}{x + 3} = \frac{5}{(x - 2)(x + 3)}.$$

The method of partial fractions shows you how to reverse this process.

$$\frac{5}{(x - 2)(x + 3)} = \frac{?}{x - 2} + \frac{?}{x + 3}$$

Decomposition of $N(x)/D(x)$ into Partial Fractions

1. **Divide when improper:** When $N(x)/D(x)$ is an improper fraction (that is, when the degree of the numerator is greater than or equal to the degree of the denominator), divide the denominator into the numerator to obtain

$$\frac{N(x)}{D(x)} = (\text{a polynomial}) + \frac{N_1(x)}{D(x)}$$

where the degree of $N_1(x)$ is less than the degree of $D(x)$. Then apply Steps 2, 3, and 4 to the proper rational expression $N_1(x)/D(x)$.

2. **Factor denominator:** Completely factor the denominator into factors of the form

$$(px + q)^m \quad \text{and} \quad (ax^2 + bx + c)^n$$

where $ax^2 + bx + c$ is irreducible.

3. **Linear factors:** For each factor of the form $(px + q)^m$, the partial fraction decomposition must include the following sum of m fractions.

$$\frac{A_1}{(px + q)} + \frac{A_2}{(px + q)^2} + \cdots + \frac{A_m}{(px + q)^m}$$

4. **Quadratic factors:** For each factor of the form $(ax^2 + bx + c)^n$, the partial fraction decomposition must include the following sum of n fractions.

$$\frac{B_1x + C_1}{ax^2 + bx + c} + \frac{B_2x + C_2}{(ax^2 + bx + c)^2} + \cdots + \frac{B_nx + C_n}{(ax^2 + bx + c)^n}$$

* For a review of factorization techniques, see *Precalculus*, 9th edition, or *Precalculus: Real Mathematics, Real People*, 6th edition, both by Ron Larson (Boston, Massachusetts: Brooks/Cole, Cengage Learning, 2014 and 2012, respectively).

AP* Tips
Integration by partial fractions is tested on the AP Calculus BC Exam. Only the case of distinct linear denominators is covered on the exam.

Linear Factors

Algebraic techniques for determining the constants in the numerators of a partial fraction decomposition with linear or repeated linear factors are shown in Examples 1 and 2.

EXAMPLE 1 Distinct Linear Factors

Write the partial fraction decomposition for

$$\frac{1}{x^2 - 5x + 6}.$$

Solution Because $x^2 - 5x + 6 = (x - 3)(x - 2)$, you should include one partial fraction for each factor and write

$$\frac{1}{x^2 - 5x + 6} = \frac{A}{x - 3} + \frac{B}{x - 2}$$

where A and B are to be determined. Multiplying this equation by the least common denominator $(x - 3)(x - 2)$ yields the **basic equation**

$$1 = A(x - 2) + B(x - 3). \qquad \text{Basic equation}$$

Because this equation is to be true for all x, you can substitute any *convenient* values for x to obtain equations in A and B. The most convenient values are the ones that make particular factors equal to 0.

REMARK Note that the substitutions for x in Example 1 are chosen for their convenience in determining values for A and B; $x = 3$ is chosen to eliminate the term $B(x - 3)$, and $x = 2$ is chosen to eliminate the term $A(x - 2)$. The goal is to make *convenient* substitutions whenever possible.

To solve for A, let $x = 3$.

$$1 = A(3 - 2) + B(3 - 3) \qquad \text{Let } x = 3 \text{ in basic equation.}$$
$$1 = A(1) + B(0)$$
$$1 = A$$

To solve for B, let $x = 2$.

$$1 = A(2 - 2) + B(2 - 3) \qquad \text{Let } x = 2 \text{ in basic equation.}$$
$$1 = A(0) + B(-1)$$
$$-1 = B$$

So, the decomposition is

$$\frac{1}{x^2 - 5x + 6} = \frac{1}{x - 3} - \frac{1}{x - 2}$$

as shown at the beginning of this section.

FOR FURTHER INFORMATION To learn a different method for finding partial fraction decompositions, called the Heavyside Method, see the article "Calculus to Algebra Connections in Partial Fraction Decomposition" by Joseph Wiener and Will Watkins in *The AMATYC Review*.

Be sure you see that the method of partial fractions is practical only for integrals of rational functions whose denominators factor "nicely." For instance, when the denominator in Example 1 is changed to

$$x^2 - 5x + 5$$

its factorization as

$$x^2 - 5x + 5 = \left[x - \frac{5 + \sqrt{5}}{2}\right]\left[x - \frac{5 - \sqrt{5}}{2}\right]$$

would be too cumbersome to use with partial fractions. In such cases, you should use completing the square or a computer algebra system to perform the integration. When you do this, you should obtain

$$\int \frac{1}{x^2 - 5x + 5}\,dx = \frac{\sqrt{5}}{5}\ln\left|2x - \sqrt{5} - 5\right| - \frac{\sqrt{5}}{5}\ln\left|2x + \sqrt{5} - 5\right| + C.$$

EXAMPLE 2 **Repeated Linear Factors**

Find $\displaystyle\int \frac{5x^2 + 20x + 6}{x^3 + 2x^2 + x}\,dx.$

Solution Because

$$x^3 + 2x^2 + x = x(x^2 + 2x + 1) = x(x + 1)^2$$

you should include one fraction for *each power* of x and $(x + 1)$ and write

$$\frac{5x^2 + 20x + 6}{x(x + 1)^2} = \frac{A}{x} + \frac{B}{x + 1} + \frac{C}{(x + 1)^2}.$$

Multiplying by the least common denominator $x(x + 1)^2$ yields the *basic equation*

$$5x^2 + 20x + 6 = A(x + 1)^2 + Bx(x + 1) + Cx. \qquad \text{Basic equation}$$

To solve for A, let $x = 0$. This eliminates the B and C terms and yields

$$6 = A(1) + 0 + 0$$
$$6 = A.$$

To solve for C, let $x = -1$. This eliminates the A and B terms and yields

$$5 - 20 + 6 = 0 + 0 - C$$
$$9 = C.$$

The most convenient choices for x have been used, so to find the value of B, you can use *any other value* of x along with the calculated values of A and C. Using $x = 1$, $A = 6$, and $C = 9$ produces

$$5 + 20 + 6 = A(4) + B(2) + C$$
$$31 = 6(4) + 2B + 9$$
$$-2 = 2B$$
$$-1 = B.$$

So, it follows that

$$\int \frac{5x^2 + 20x + 6}{x(x + 1)^2}\,dx = \int \left(\frac{6}{x} - \frac{1}{x + 1} + \frac{9}{(x + 1)^2}\right) dx$$
$$= 6 \ln|x| - \ln|x + 1| + 9\frac{(x + 1)^{-1}}{-1} + C$$
$$= \ln\left|\frac{x^6}{x + 1}\right| - \frac{9}{x + 1} + C.$$

Try checking this result by differentiating. Include algebra in your check, simplifying the derivative until you have obtained the original integrand.

FOR FURTHER INFORMATION For an alternative approach to using partial fractions, see the article "A Shortcut in Partial Fractions" by Xun-Cheng Huang in *The College Mathematics Journal.*

It is necessary to make as many substitutions for x as there are unknowns $(A, B, C, \ldots)$ to be determined. For instance, in Example 2, three substitutions $(x = 0,\ x = -1,$ and $x = 1)$ were made to solve for A, B, and C.

TECHNOLOGY Most computer algebra systems, such as *Maple, Mathematica,* and the *TI-nSpire,* can be used to convert a rational function to its partial fraction decomposition. For instance, using *Mathematica,* you obtain the following.

$$\text{Apart}\,[(5 * x \wedge 2 + 20 * x + 6)/(x * (x + 1) \wedge 2), x]$$

$$\frac{6}{x} + \frac{9}{(1 + x)^2} - \frac{1}{1 + x}$$

Quadratic Factors

When using the method of partial fractions with *linear* factors, a convenient choice of x immediately yields a value for one of the coefficients. With *quadratic* factors, a system of linear equations usually has to be solved, regardless of the choice of x.

EXAMPLE 3 Distinct Linear and Quadratic Factors

See LarsonCalculus.com for an interactive version of this type of example.

Find $\int \frac{2x^3 - 4x - 8}{(x^2 - x)(x^2 + 4)}\,dx$.

Solution Because

$$(x^2 - x)(x^2 + 4) = x(x - 1)(x^2 + 4)$$

you should include one partial fraction for each factor and write

$$\frac{2x^3 - 4x - 8}{x(x - 1)(x^2 + 4)} = \frac{A}{x} + \frac{B}{x - 1} + \frac{Cx + D}{x^2 + 4}.$$

Multiplying by the least common denominator

$$x(x - 1)(x^2 + 4)$$

yields the *basic equation*

$$2x^3 - 4x - 8 = A(x - 1)(x^2 + 4) + Bx(x^2 + 4) + (Cx + D)(x)(x - 1).$$

To solve for A, let $x = 0$ and obtain

$$-8 = A(-1)(4) + 0 + 0$$
$$2 = A.$$

To solve for B, let $x = 1$ and obtain

$$-10 = 0 + B(5) + 0$$
$$-2 = B.$$

At this point, C and D are yet to be determined. You can find these remaining constants by choosing two other values for x and solving the resulting system of linear equations. Using $x = -1$, $A = 2$, and $B = -2$, you can write

$$-6 = (2)(-2)(5) + (-2)(-1)(5) + (-C + D)(-1)(-2)$$
$$2 = -C + D.$$

For $x = 2$, you have

$$0 = (2)(1)(8) + (-2)(2)(8) + (2C + D)(2)(1)$$
$$8 = 2C + D.$$

Solving the linear system by subtracting the first equation from the second

$$-C + D = 2$$
$$2C + D = 8$$

yields $C = 2$. Consequently, $D = 4$, and it follows that

$$\int \frac{2x^3 - 4x - 8}{x(x - 1)(x^2 + 4)}\,dx = \int \left(\frac{2}{x} - \frac{2}{x - 1} + \frac{2x}{x^2 + 4} + \frac{4}{x^2 + 4}\right) dx$$
$$= 2\ln|x| - 2\ln|x - 1| + \ln(x^2 + 4) + 2\arctan\frac{x}{2} + C.$$

In Examples 1, 2, and 3, the solution of the basic equation began with substituting values of x that made the linear factors equal to 0. This method works well when the partial fraction decomposition involves linear factors. When the decomposition involves only quadratic factors, however, an alternative procedure is often more convenient. For instance, try writing the right side of the basic equation in polynomial form and *equating the coefficients* of like terms. This method is shown in Example 4.

EXAMPLE 4 Repeated Quadratic Factors

Find $\displaystyle\int \frac{8x^3 + 13x}{(x^2 + 2)^2}\,dx.$

Solution Include one partial fraction for each power of $(x^2 + 2)$ and write

$$\frac{8x^3 + 13x}{(x^2 + 2)^2} = \frac{Ax + B}{x^2 + 2} + \frac{Cx + D}{(x^2 + 2)^2}.$$

Multiplying by the least common denominator $(x^2 + 2)^2$ yields the *basic equation*

$$8x^3 + 13x = (Ax + B)(x^2 + 2) + Cx + D.$$

Expanding the basic equation and collecting like terms produces

$$8x^3 + 13x = Ax^3 + 2Ax + Bx^2 + 2B + Cx + D$$
$$8x^3 + 13x = Ax^3 + Bx^2 + (2A + C)x + (2B + D).$$

Now, you can equate the coefficients of like terms on opposite sides of the equation.

$8 = A$ $\qquad$ $0 = 2B + D$

$$8x^3 + 0x^2 + 13x + 0 = Ax^3 + Bx^2 + (2A + C)x + (2B + D)$$

$0 = B$

$13 = 2A + C$

Using the known values $A = 8$ and $B = 0$, you can write

$$13 = 2A + C \;\Rightarrow\; 13 = 2(8) + C \;\Rightarrow\; -3 = C$$
$$0 = 2B + D \;\Rightarrow\; 0 = 2(0) + D \;\Rightarrow\; 0 = D.$$

Finally, you can conclude that

$$\int \frac{8x^3 + 13x}{(x^2 + 2)^2}\,dx = \int \left(\frac{8x}{x^2 + 2} + \frac{-3x}{(x^2 + 2)^2}\right) dx$$
$$= 4\ln(x^2 + 2) + \frac{3}{2(x^2 + 2)} + C.$$

TECHNOLOGY You can use a graphing utility to confirm the decomposition found in Example 4. To do this, graph

$$y_1 = \frac{8x^3 + 13x}{(x^2 + 2)^2}$$

and

$$y_2 = \frac{8x}{x^2 + 2} + \frac{-3x}{(x^2 + 2)^2}$$

in the same viewing window. The graphs should be identical, as shown at the right.

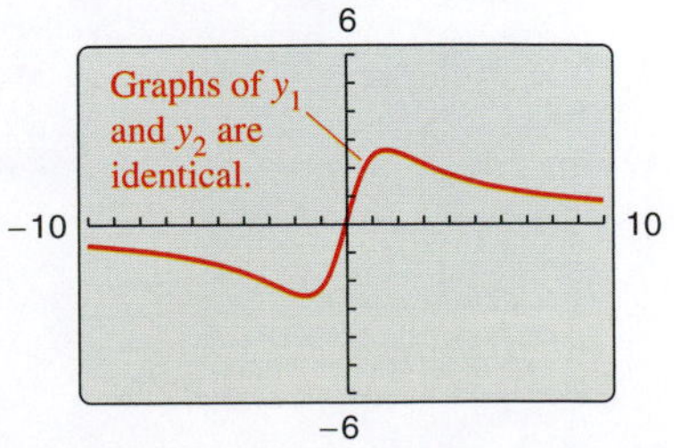

When integrating rational expressions, keep in mind that for *improper* rational expressions such as

$$\frac{N(x)}{D(x)} = \frac{2x^3 + x^2 - 7x + 7}{x^2 + x - 2}$$

you must first divide to obtain

$$\frac{N(x)}{D(x)} = 2x - 1 + \frac{-2x + 5}{x^2 + x - 2}.$$

The proper rational expression is then decomposed into its partial fractions by the usual methods.

Here are some guidelines for solving the basic equation that is obtained in a partial fraction decomposition.

GUIDELINES FOR SOLVING THE BASIC EQUATION

Linear Factors

1. Substitute the roots of the distinct linear factors in the basic equation.
2. For repeated linear factors, use the coefficients determined in the first guideline to rewrite the basic equation. Then substitute other convenient values of x and solve for the remaining coefficients.

Quadratic Factors

1. Expand the basic equation.
2. Collect terms according to powers of x.
3. Equate the coefficients of like powers to obtain a system of linear equations involving A, B, C, and so on.
4. Solve the system of linear equations.

FOR FURTHER INFORMATION To read about another method of evaluating integrals of rational functions, see the article "Alternate Approach to Partial Fractions to Evaluate Integrals of Rational Functions" by N. R. Nandakumar and Michael J. Bossé in *The Pi Mu Epsilon Journal*. To view this article, go to *MathArticles.com*.

Before concluding this section, here are a few things you should remember. First, it is not necessary to use the partial fractions technique on all rational functions. For instance, the following integral is evaluated more easily by the Log Rule.

$$\begin{aligned}\int \frac{x^2 + 1}{x^3 + 3x - 4}\,dx &= \frac{1}{3}\int \frac{3x^2 + 3}{x^3 + 3x - 4}\,dx \\ &= \frac{1}{3}\ln|x^3 + 3x - 4| + C\end{aligned}$$

Second, when the integrand is not in reduced form, reducing it may eliminate the need for partial fractions, as shown in the following integral.

$$\begin{aligned}\int \frac{x^2 - x - 2}{x^3 - 2x - 4}\,dx &= \int \frac{(x + 1)(x - 2)}{(x - 2)(x^2 + 2x + 2)}\,dx \\ &= \int \frac{x + 1}{x^2 + 2x + 2}\,dx \\ &= \frac{1}{2}\ln|x^2 + 2x + 2| + C\end{aligned}$$

Finally, partial fractions can be used with some quotients involving transcendental functions. For instance, the substitution $u = \sin x$ allows you to write

$$\int \frac{\cos x}{\sin x(\sin x - 1)}\,dx = \int \frac{du}{u(u - 1)}. \qquad u = \sin x,\ du = \cos x\,dx$$

8.5 Exercises

See CalcChat.com for tutorial help and worked-out solutions to odd-numbered exercises.

Partial Fraction Decomposition **In Exercises 1–4, write the form of the partial fraction decomposition of the rational expression. Do not solve for the constants.**

1. $\dfrac{4}{x^2 - 8x}$

2. $\dfrac{2x^2 + 1}{(x - 3)^3}$

3. $\dfrac{2x - 3}{x^3 + 10x}$

4. $\dfrac{2x - 1}{x(x^2 + 1)^2}$

Using Partial Fractions **In Exercises 5–22, use partial fractions to find the indefinite integral.**

5. $\displaystyle\int \frac{1}{x^2 - 9}\,dx$

6. $\displaystyle\int \frac{2}{9x^2 - 1}\,dx$

7. $\displaystyle\int \frac{5}{x^2 + 3x - 4}\,dx$

8. $\displaystyle\int \frac{3 - x}{3x^2 - 2x - 1}\,dx$

9. $\displaystyle\int \frac{x^2 + 12x + 12}{x^3 - 4x}\,dx$

10. $\displaystyle\int \frac{x^3 - x + 3}{x^2 + x - 2}\,dx$

11. $\displaystyle\int \frac{2x^3 - 4x^2 - 15x + 5}{x^2 - 2x - 8}\,dx$

12. $\displaystyle\int \frac{x + 2}{x^2 + 5x}\,dx$

13. $\displaystyle\int \frac{4x^2 + 2x - 1}{x^3 + x^2}\,dx$

14. $\displaystyle\int \frac{5x - 2}{(x - 2)^2}\,dx$

15. $\displaystyle\int \frac{x^2 + 3x - 4}{x^3 - 4x^2 + 4x}\,dx$

16. $\displaystyle\int \frac{8x}{x^3 + x^2 - x - 1}\,dx$

17. $\displaystyle\int \frac{x^2 - 1}{x^3 + x}\,dx$

18. $\displaystyle\int \frac{6x}{x^3 - 8}\,dx$

19. $\displaystyle\int \frac{x^2}{x^4 - 2x^2 - 8}\,dx$

20. $\displaystyle\int \frac{x}{16x^4 - 1}\,dx$

21. $\displaystyle\int \frac{x^2 + 5}{x^3 - x^2 + x + 3}\,dx$

22. $\displaystyle\int \frac{x^2 + 6x + 4}{x^4 + 8x^2 + 16}\,dx$

Evaluating a Definite Integral **In Exercises 23–26, evaluate the definite integral. Use a graphing utility to verify your result.**

23. $\displaystyle\int_0^2 \frac{3}{4x^2 + 5x + 1}\,dx$

24. $\displaystyle\int_1^5 \frac{x - 1}{x^2(x + 1)}\,dx$

25. $\displaystyle\int_1^2 \frac{x + 1}{x(x^2 + 1)}\,dx$

26. $\displaystyle\int_0^1 \frac{x^2 - x}{x^2 + x + 1}\,dx$

Finding an Indefinite Integral **In Exercises 27–34, use substitution and partial fractions to find the indefinite integral.**

27. $\displaystyle\int \frac{\sin x}{\cos x + \cos^2 x}\,dx$

28. $\displaystyle\int \frac{5\cos x}{\sin^2 x + 3\sin x - 4}\,dx$

29. $\displaystyle\int \frac{\sec^2 x}{\tan^2 x + 5\tan x + 6}\,dx$

30. $\displaystyle\int \frac{\sec^2 x}{\tan x(\tan x + 1)}\,dx$

31. $\displaystyle\int \frac{e^x}{(e^x - 1)(e^x + 4)}\,dx$

32. $\displaystyle\int \frac{e^x}{(e^{2x} + 1)(e^x - 1)}\,dx$

33. $\displaystyle\int \frac{\sqrt{x}}{x - 4}\,dx$

34. $\displaystyle\int \frac{1}{\sqrt{x} - \sqrt[3]{x}}\,dx$

Verifying a Formula **In Exercises 35–38, use the method of partial fractions to verify the integration formula.**

35. $\displaystyle\int \frac{1}{x(a + bx)}\,dx = \frac{1}{a}\ln\left|\frac{x}{a + bx}\right| + C$

36. $\displaystyle\int \frac{1}{a^2 - x^2}\,dx = \frac{1}{2a}\ln\left|\frac{a + x}{a - x}\right| + C$

37. $\displaystyle\int \frac{x}{(a + bx)^2}\,dx = \frac{1}{b^2}\left(\frac{a}{a + bx} + \ln|a + bx|\right) + C$

38. $\displaystyle\int \frac{1}{x^2(a + bx)}\,dx = -\frac{1}{ax} - \frac{b}{a^2}\ln\left|\frac{x}{a + bx}\right| + C$

WRITING ABOUT CONCEPTS

39. **Using Partial Fractions** What is the first step when integrating $\displaystyle\int \frac{x^3}{x - 5}\,dx$? Explain.

40. **Decomposition** Describe the decomposition of the proper rational function $N(x)/D(x)$ (a) for $D(x) = (px + q)^m$ and (b) for $D(x) = (ax^2 + bx + c)^n$ where $ax^2 + bx + c$ is irreducible. Explain why you chose that method.

41. **Choosing a Method** State the method you would use to evaluate each integral. Explain why you chose that method. Do not integrate.

(a) $\displaystyle\int \frac{x + 1}{x^2 + 2x - 8}\,dx$ (b) $\displaystyle\int \frac{7x + 4}{x^2 + 2x - 8}\,dx$

(c) $\displaystyle\int \frac{4}{x^2 + 2x + 5}\,dx$

42. **HOW DO YOU SEE IT?** Use the graph of f' shown in the figure to answer the following.

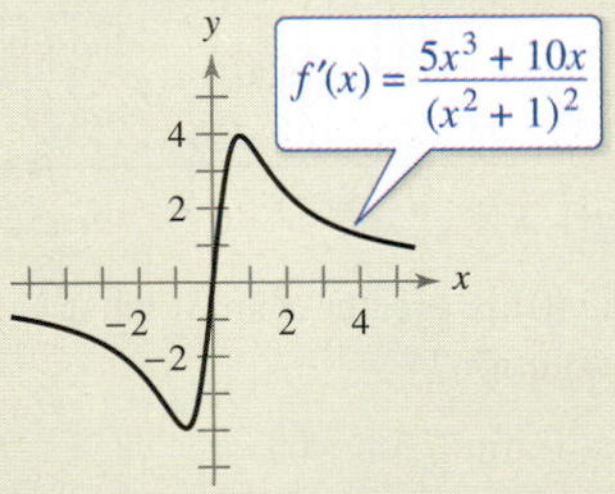

(a) Is $f(3) - f(2) > 0$? Explain.

(b) Which is greater, the area under the graph of f' from 1 to 2, or the area under the graph of f' from 3 to 4?

43. **Area** Find the area of the region bounded by the graphs of $y = 12/(x^2 + 5x + 6)$, $y = 0$, $x = 0$, and $x = 1$.

44. **Area** Find the area of the region bounded by the graphs of $y = 7/(16 - x^2)$ and $y = 1$.

45. Modeling Data The predicted cost C (in hundreds of thousands of dollars) for a company to remove $p\%$ of a chemical from its waste water is shown in the table.

P	0	10	20	30	40
C	0	0.7	1.0	1.3	1.7

P	50	60	70	80	90
C	2.0	2.7	3.6	5.5	11.2

A model for the data is given by $C = \dfrac{124p}{(10 + p)(100 - p)}$ for $0 \le p < 100$. Use the model to find the average cost of removing between 75% and 80% of the chemical.

46. Logistic Growth In Chapter 6, the exponential growth equation was derived from the assumption that the rate of growth was proportional to the existing quantity. In practice, there often exists some upper limit L past which growth cannot occur. In such cases, you assume the rate of growth to be proportional not only to the existing quantity, but also to the difference between the existing quantity y and the upper limit L. That is, $dy/dt = ky(L - y)$. In integral form, you can write this relationship as

$$\int \frac{dy}{y(L - y)} = \int k\,dt.$$

(a) A slope field for the differential equation $dy/dt = y(3 - y)$ is shown. Draw a possible solution to the differential equation when $y(0) = 5$, and another when $y(0) = \frac{1}{2}$. To print an enlarged copy of the graph, go to *MathGraphs.com.*

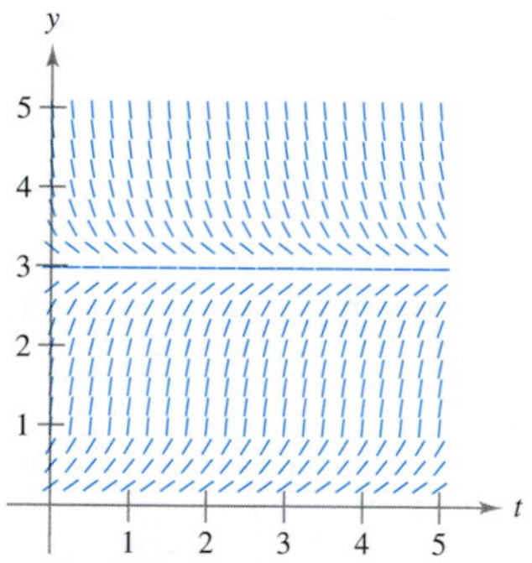

(b) Where $y(0)$ is greater than 3, what is the sign of the slope of the solution?

(c) For $y > 0$, find $\lim_{t \to \infty} y(t)$.

(d) Evaluate the two given integrals and solve for y as a function of t, where y_0 is the initial quantity.

(e) Use the result of part (d) to find and graph the solutions in part (a). Use a graphing utility to graph the solutions and compare the results with the solutions in part (a).

(f) The graph of the function y is a **logistic curve.** Show that the rate of growth is maximum at the point of inflection, and that this occurs when $y = L/2$.

47. Volume and Centroid Consider the region bounded by the graphs of $y = 2x/(x^2 + 1)$, $y = 0$, $x = 0$, and $x = 3$. Find the volume of the solid generated by revolving the region about the x-axis. Find the centroid of the region.

48. Volume Consider the region bounded by the graph of

$$y^2 = \frac{(2 - x)^2}{(1 + x)^2}$$

on the interval $[0, 1]$. Find the volume of the solid generated by revolving this region about the x-axis.

49. Epidemic Model A single infected individual enters a community of n susceptible individuals. Let x be the number of newly infected individuals at time t. The common epidemic model assumes that the disease spreads at a rate proportional to the product of the total number infected and the number not yet infected. So, $dx/dt = k(x + 1)(n - x)$ and you obtain

$$\int \frac{1}{(x + 1)(n - x)}\,dx = \int k\,dt.$$

Solve for x as a function of t.

50. Chemical Reaction

In a chemical reaction, one unit of compound Y and one unit of compound Z are converted into a single unit of compound X. Let x be the amount of compound X formed. The rate of formation of X is proportional to the product of the amounts of unconverted compounds Y and Z. So, $dx/dt = k(y_0 - x)(z_0 - x)$, where y_0 and z_0 are the initial amounts of compounds Y and Z. From this equation, you obtain

$$\int \frac{1}{(y_0 - x)(z_0 - x)}\,dx = \int k\,dt.$$

(a) Perform the two integrations and solve for x in terms of t.

(b) Use the result of part (a) to find x as $t \to \infty$ for (1) $y_0 < z_0$, (2) $y_0 > z_0$, and (3) $y_0 = z_0$.

51. Using Two Methods Evaluate

$$\int_0^1 \frac{x}{1 + x^4}\,dx$$

in two different ways, one of which is partial fractions.

PUTNAM EXAM CHALLENGE

52. Prove $\dfrac{22}{7} - \pi = \displaystyle\int_0^1 \frac{x^4(1 - x)^4}{1 + x^2}\,dx.$

8.6 Integration by Tables and Other Integration Techniques

- Evaluate an indefinite integral using a table of integrals.
- Evaluate an indefinite integral using reduction formulas.
- Evaluate an indefinite integral involving rational functions of sine and cosine.

Integration by Tables

So far in this chapter, you have studied several integration techniques that can be used with the basic integration rules. But merely knowing *how* to use the various techniques is not enough. You also need to know *when* to use them. Integration is first and foremost a problem of recognition. That is, you must recognize which rule or technique to apply to obtain an antiderivative. Frequently, a slight alteration of an integrand will require a different integration technique (or produce a function whose antiderivative is not an elementary function), as shown below.

$$\int x \ln x\, dx = \frac{x^2}{2}\ln x - \frac{x^2}{4} + C \qquad \text{Integration by parts}$$

$$\int \frac{\ln x}{x}\, dx = \frac{(\ln x)^2}{2} + C \qquad \text{Power Rule}$$

$$\int \frac{1}{x \ln x}\, dx = \ln|\ln x| + C \qquad \text{Log Rule}$$

$$\int \frac{x}{\ln x}\, dx = ? \qquad \text{Not an elementary function}$$

TECHNOLOGY A computer algebra system consists, in part, of a database of integration formulas. The primary difference between using a computer algebra system and using tables of integrals is that with a computer algebra system, the computer searches through the database to find a fit. With integration tables, *you* must do the searching.

Many people find tables of integrals to be a valuable supplement to the integration techniques discussed in this chapter. Tables of common integrals can be found in Appendix B. **Integration by tables** is not a "cure-all" for all of the difficulties that can accompany integration—using tables of integrals requires considerable thought and insight and often involves substitution.

Each integration formula in Appendix B can be developed using one or more of the techniques in this chapter. You should try to verify several of the formulas. For instance, Formula 4

$$\int \frac{u}{(a+bu)^2}\, du = \frac{1}{b^2}\left(\frac{a}{a+bu} + \ln|a+bu|\right) + C \qquad \text{Formula 4}$$

can be verified using the method of partial fractions, Formula 19

$$\int \frac{\sqrt{a+bu}}{u}\, du = 2\sqrt{a+bu} + a\int \frac{du}{u\sqrt{a+bu}} \qquad \text{Formula 19}$$

can be verified using integration by parts, and Formula 84

$$\int \frac{1}{1+e^u}\, du = u - \ln(1+e^u) + C \qquad \text{Formula 84}$$

can be verified using substitution. Note that the integrals in Appendix B are classified according to the form of the integrand. Several of the forms are listed below.

u^n	$(a + bu)$
$(a + bu + cu^2)$	$\sqrt{a + bu}$
$(a^2 \pm u^2)$	$\sqrt{u^2 \pm a^2}$
$\sqrt{a^2 - u^2}$	Trigonometric functions
Inverse trigonometric functions	Exponential functions
Logarithmic functions	

Exploration

Use the tables of integrals in Appendix B and the substitution

$$u = \sqrt{x-1}$$

to evaluate the integral in Example 1. When you do this, you should obtain

$$\int \frac{dx}{x\sqrt{x-1}} = \int \frac{2\,du}{u^2+1}.$$

Does this produce the same result as that obtained in Example 1?

EXAMPLE 1 Integration by Tables

Find $\displaystyle\int \frac{dx}{x\sqrt{x-1}}$.

Solution Because the expression inside the radical is linear, you should consider forms involving $\sqrt{a+bu}$.

$$\int \frac{du}{u\sqrt{a+bu}} = \frac{2}{\sqrt{-a}} \arctan \sqrt{\frac{a+bu}{-a}} + C \qquad \text{Formula 17 } (a < 0)$$

Let $a = -1$, $b = 1$, and $u = x$. Then $du = dx$, and you can write

$$\int \frac{dx}{x\sqrt{x-1}} = 2 \arctan \sqrt{x-1} + C.$$

EXAMPLE 2 Integration by Tables

See LarsonCalculus.com for an interactive version of this type of example.

Find $\displaystyle\int x\sqrt{x^4-9}\,dx$.

Solution Because the radical has the form $\sqrt{u^2-a^2}$, you should consider Formula 26.

$$\int \sqrt{u^2-a^2}\,du = \frac{1}{2}\left(u\sqrt{u^2-a^2} - a^2 \ln\left|u + \sqrt{u^2-a^2}\right|\right) + C$$

Let $u = x^2$ and $a = 3$. Then $du = 2x\,dx$, and you have

$$\begin{aligned}\int x\sqrt{x^4-9}\,dx &= \frac{1}{2}\int \sqrt{(x^2)^2-3^2}\,(2x)\,dx \\ &= \frac{1}{4}\left(x^2\sqrt{x^4-9} - 9\ln\left|x^2+\sqrt{x^4-9}\right|\right) + C.\end{aligned}$$

EXAMPLE 3 Integration by Tables

Evaluate $\displaystyle\int_0^2 \frac{x}{1+e^{-x^2}}\,dx$.

Solution Of the forms involving e^u, consider the formula

$$\int \frac{du}{1+e^u} = u - \ln(1+e^u) + C. \qquad \text{Formula 84}$$

Let $u = -x^2$. Then $du = -2x\,dx$, and you have

$$\begin{aligned}\int \frac{x}{1+e^{-x^2}}\,dx &= -\frac{1}{2}\int \frac{-2x\,dx}{1+e^{-x^2}} \\ &= -\frac{1}{2}\left[-x^2 - \ln\left(1+e^{-x^2}\right)\right] + C \\ &= \frac{1}{2}\left[x^2 + \ln\left(1+e^{-x^2}\right)\right] + C.\end{aligned}$$

So, the value of the definite integral is

$$\int_0^2 \frac{x}{1+e^{-x^2}}\,dx = \frac{1}{2}\left[x^2 + \ln\left(1+e^{-x^2}\right)\right]_0^2 = \frac{1}{2}\left[4 + \ln(1+e^{-4}) - \ln 2\right] \approx 1.66.$$

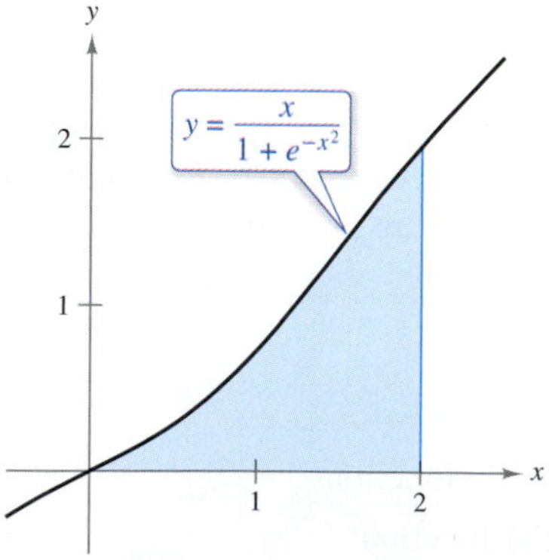

Figure 8.14

Figure 8.14 shows the region whose area is represented by this integral.

Reduction Formulas

Several of the integrals in the integration tables have the form

$$\int f(x)\,dx = g(x) + \int h(x)\,dx.$$

Such integration formulas are called **reduction formulas** because they reduce a given integral to the sum of a function and a simpler integral.

EXAMPLE 4 Using a Reduction Formula

Find $\int x^3 \sin x\,dx$.

Solution Consider the three formulas listed below.

$$\int u \sin u\,du = \sin u - u\cos u + C \qquad \text{Formula 52}$$

$$\int u^n \sin u\,du = -u^n \cos u + n\int u^{n-1}\cos u\,du \qquad \text{Formula 54}$$

$$\int u^n \cos u\,du = u^n \sin u - n\int u^{n-1}\sin u\,du \qquad \text{Formula 55}$$

Using Formula 54, Formula 55, and then Formula 52 produces

$$\begin{aligned}\int x^3 \sin x\,dx &= -x^3\cos x + 3\int x^2 \cos x\,dx\\ &= -x^3\cos x + 3\left(x^2 \sin x - 2\int x\sin x\,dx\right)\\ &= -x^3\cos x + 3x^2\sin x + 6x\cos x - 6\sin x + C.\end{aligned}$$

EXAMPLE 5 Using a Reduction Formula

Find $\int \frac{\sqrt{3-5x}}{2x}\,dx$.

Solution Consider the two formulas listed below.

$$\int \frac{du}{u\sqrt{a+bu}} = \frac{1}{\sqrt{a}}\ln\left|\frac{\sqrt{a+bu}-\sqrt{a}}{\sqrt{a+bu}+\sqrt{a}}\right| + C \qquad \text{Formula 17 } (a > 0)$$

$$\int \frac{\sqrt{a+bu}}{u}\,du = 2\sqrt{a+bu} + a\int \frac{du}{u\sqrt{a+bu}} \qquad \text{Formula 19}$$

Using Formula 19, with $a = 3$, $b = -5$, and $u = x$, produces

$$\begin{aligned}\frac{1}{2}\int \frac{\sqrt{3-5x}}{x}\,dx &= \frac{1}{2}\left(2\sqrt{3-5x} + 3\int \frac{dx}{x\sqrt{3-5x}}\right)\\ &= \sqrt{3-5x} + \frac{3}{2}\int \frac{dx}{x\sqrt{3-5x}}.\end{aligned}$$

Using Formula 17, with $a = 3$, $b = -5$, and $u = x$, produces

$$\begin{aligned}\int \frac{\sqrt{3-5x}}{2x}\,dx &= \sqrt{3-5x} + \frac{3}{2}\left(\frac{1}{\sqrt{3}}\ln\left|\frac{\sqrt{3-5x}-\sqrt{3}}{\sqrt{3-5x}+\sqrt{3}}\right|\right) + C\\ &= \sqrt{3-5x} + \frac{\sqrt{3}}{2}\ln\left|\frac{\sqrt{3-5x}-\sqrt{3}}{\sqrt{3-5x}+\sqrt{3}}\right| + C.\end{aligned}$$

TECHNOLOGY Sometimes when you use computer algebra systems, you obtain results that look very different, but are actually equivalent. Here is how two different systems evaluated the integral in Example 5.

Maple

$\sqrt{3-5x} - \sqrt{3}\,\text{arctanh}\left(\frac{1}{3}\sqrt{3-5x}\sqrt{3}\right)$

Mathematica

$\sqrt{3-5x} - \sqrt{3}\,\text{ArcTanh}\left[\sqrt{1-\frac{5x}{3}}\right]$

Notice that computer algebra systems do not include a constant of integration.

Rational Functions of Sine and Cosine

EXAMPLE 6 **Integration by Tables**

Find $\int \frac{\sin 2x}{2 + \cos x}\, dx.$

Solution Substituting $2 \sin x \cos x$ for $\sin 2x$ produces

$$\int \frac{\sin 2x}{2 + \cos x}\, dx = 2 \int \frac{\sin x \cos x}{2 + \cos x}\, dx.$$

A check of the forms involving $\sin u$ or $\cos u$ in Appendix B shows that none of those listed applies. So, you can consider forms involving $a + bu$. For example,

$$\int \frac{u\, du}{a + bu} = \frac{1}{b^2}(bu - a \ln|a + bu|) + C. \qquad \text{Formula 3}$$

Let $a = 2$, $b = 1$, and $u = \cos x$. Then $du = -\sin x\, dx$, and you have

$$\begin{aligned} 2 \int \frac{\sin x \cos x}{2 + \cos x}\, dx &= -2 \int \frac{\cos x(-\sin x\, dx)}{2 + \cos x} \\ &= -2(\cos x - 2 \ln|2 + \cos x|) + C \\ &= -2 \cos x + 4 \ln|2 + \cos x| + C. \end{aligned}$$

Example 6 involves a rational expression of $\sin x$ and $\cos x$. When you are unable to find an integral of this form in the integration tables, try using the following special substitution to convert the trigonometric expression to a standard rational expression.

Substitution for Rational Functions of Sine and Cosine

For integrals involving rational functions of sine and cosine, the substitution

$$u = \frac{\sin x}{1 + \cos x} = \tan \frac{x}{2}$$

yields

$$\cos x = \frac{1 - u^2}{1 + u^2}, \quad \sin x = \frac{2u}{1 + u^2}, \quad \text{and} \quad dx = \frac{2\, du}{1 + u^2}.$$

Proof From the substitution for u, it follows that

$$u^2 = \frac{\sin^2 x}{(1 + \cos x)^2} = \frac{1 - \cos^2 x}{(1 + \cos x)^2} = \frac{1 - \cos x}{1 + \cos x}.$$

Solving for $\cos x$ produces $\cos x = (1 - u^2)/(1 + u^2)$. To find $\sin x$, write $u = \sin x/(1 + \cos x)$ as

$$\sin x = u(1 + \cos x) = u\left(1 + \frac{1 - u^2}{1 + u^2}\right) = \frac{2u}{1 + u^2}.$$

Finally, to find dx, consider $u = \tan(x/2)$. Then you have $\arctan u = x/2$ and

$$dx = \frac{2\, du}{1 + u^2}.$$

See LarsonCalculus.com for Bruce Edwards's video of this proof.

8.6 Exercises

See CalcChat.com for tutorial help and worked-out solutions to odd-numbered exercises.

Integration by Tables In Exercises 1 and 2, use a table of integrals with forms involving $a + bu$ to find the indefinite integral.

1. $\int \frac{x^2}{5+x}\,dx$

2. $\int \frac{2}{x^2(4+3x)^2}\,dx$

Integration by Tables In Exercises 3 and 4, use a table of integrals with forms involving $\sqrt{a^2-u^2}$ to find the indefinite integral.

3. $\int \frac{1}{x^2\sqrt{1-x^2}}\,dx$

4. $\int \frac{\sqrt{64-x^4}}{x}\,dx$

Integration by Tables In Exercises 5–8, use a table of integrals with forms involving the trigonometric functions to find the indefinite integral.

5. $\int \cos^4 3x\,dx$

6. $\int \frac{\sin^4 \sqrt{x}}{\sqrt{x}}\,dx$

7. $\int \frac{1}{\sqrt{x}\left(1-\cos\sqrt{x}\right)}\,dx$

8. $\int \frac{1}{1+\cot 4x}\,dx$

Integration by Tables In Exercises 9 and 10, use a table of integrals with forms involving e^u to find the indefinite integral.

9. $\int \frac{1}{1+e^{2x}}\,dx$

10. $\int e^{-4x}\sin 3x\,dx$

Integration by Tables In Exercises 11 and 12, use a table of integrals with forms involving $\ln u$ to find the indefinite integral.

11. $\int x^7 \ln x\,dx$

12. $\int (\ln x)^3\,dx$

Using Two Methods In Exercises 13–16, find the indefinite integral (a) using integration tables and (b) using the given method.

	Integral	Method
13.	$\int x^2 e^{3x}\,dx$	Integration by parts
14.	$\int x^5 \ln x\,dx$	Integration by parts
15.	$\int \frac{1}{x^2(x+1)}\,dx$	Partial fractions
16.	$\int \frac{1}{x^2-36}\,dx$	Partial fractions

Finding an Indefinite Integral In Exercises 17–38, use integration tables to find the indefinite integral.

17. $\int x\,\text{arccsc}(x^2+1)\,dx$

18. $\int \arcsin 4x\,dx$

19. $\int \frac{1}{x^2\sqrt{x^2-4}}\,dx$

20. $\int \frac{1}{x^2+4x+8}\,dx$

21. $\int \frac{4x}{(2-5x)^2}\,dx$

22. $\int \frac{\theta^3}{1+\sin\theta^4}\,d\theta$

23. $\int e^x \arccos e^x\,dx$

24. $\int \frac{e^x}{1-\tan e^x}\,dx$

25. $\int \frac{x}{1-\sec x^2}\,dx$

26. $\int \frac{1}{t[1+(\ln t)^2]}\,dt$

27. $\int \frac{\cos\theta}{3+2\sin\theta+\sin^2\theta}\,d\theta$

28. $\int x^2\sqrt{2+9x^2}\,dx$

29. $\int \frac{1}{x^2\sqrt{2+9x^2}}\,dx$

30. $\int \sqrt{x}\arctan x^{3/2}\,dx$

31. $\int \frac{\ln x}{x(3+2\ln x)}\,dx$

32. $\int \frac{e^x}{(1-e^{2x})^{3/2}}\,dx$

33. $\int \frac{x}{(x^2-6x+10)^2}\,dx$

34. $\int \sqrt{\frac{5-x}{5+x}}\,dx$

35. $\int \frac{x}{\sqrt{x^4-6x^2+5}}\,dx$

36. $\int \frac{\cos x}{\sqrt{\sin^2 x+1}}\,dx$

37. $\int \frac{e^{3x}}{(1+e^x)^3}\,dx$

38. $\int \cot^4\theta\,d\theta$

Evaluating a Definite Integral In Exercises 39–46, use integration tables to evaluate the definite integral.

39. $\int_0^1 xe^{x^2}\,dx$

40. $\int_0^4 \frac{x}{\sqrt{3+2x}}\,dx$

41. $\int_1^2 x^4\ln x\,dx$

42. $\int_0^{\pi/2} x\sin 2x\,dx$

43. $\int_{-\pi/2}^{\pi/2} \frac{\cos x}{1+\sin^2 x}\,dx$

44. $\int_0^5 \frac{x^2}{(5+2x)^2}\,dx$

45. $\int_0^{\pi/2} t^3\cos t\,dt$

46. $\int_0^3 \sqrt{x^2+16}\,dx$

Verifying a Formula In Exercises 47–52, verify the integration formula.

47. $\int \frac{u^2}{(a+bu)^2}\,du = \frac{1}{b^3}\left(bu - \frac{a^2}{a+bu} - 2a\ln|a+bu|\right) + C$

48. $\int \frac{u^n}{\sqrt{a+bu}}\,du = \frac{2}{(2n+1)b}\left(u^n\sqrt{a+bu} - na\int\frac{u^{n-1}}{\sqrt{a+bu}}\,du\right)$

49. $\int \frac{1}{(u^2\pm a^2)^{3/2}}\,du = \frac{\pm u}{a^2\sqrt{u^2\pm a^2}} + C$

50. $\int u^n\cos u\,du = u^n\sin u - n\int u^{n-1}\sin u\,du$

51. $\int \arctan u\,du = u\arctan u - \ln\sqrt{1+u^2} + C$

52. $\int (\ln u)^n\,du = u(\ln u)^n - n\int(\ln u)^{n-1}\,du$

Finding or Evaluating an Integral **In Exercises 53–60, find or evaluate the integral.**

53. $\displaystyle\int \frac{1}{2 - 3\sin\theta}\,d\theta$

54. $\displaystyle\int \frac{\sin\theta}{1 + \cos^2\theta}\,d\theta$

55. $\displaystyle\int_0^{\pi/2} \frac{1}{1 + \sin\theta + \cos\theta}\,d\theta$

56. $\displaystyle\int_0^{\pi/2} \frac{1}{3 - 2\cos\theta}\,d\theta$

57. $\displaystyle\int \frac{\sin\theta}{3 - 2\cos\theta}\,d\theta$

58. $\displaystyle\int \frac{\cos\theta}{1 + \cos\theta}\,d\theta$

59. $\displaystyle\int \frac{\sin\sqrt{\theta}}{\sqrt{\theta}}\,d\theta$

60. $\displaystyle\int \frac{4}{\csc\theta - \cot\theta}\,d\theta$

Area **In Exercises 61 and 62, find the area of the region bounded by the graphs of the equations.**

61. $y = \dfrac{x}{\sqrt{x+3}},\ y = 0,\ x = 6$

62. $y = \dfrac{x}{1 + e^{x^2}},\ y = 0,\ x = 2$

WRITING ABOUT CONCEPTS

63. **Finding a Pattern**

(a) Evaluate $\int x^n \ln x\,dx$ for $n = 1, 2,$ and 3. Describe any patterns you notice.

(b) Write a general rule for evaluating the integral in part (a), for an integer $n \geq 1$.

64. **Reduction Formula** Describe what is meant by a reduction formula. Give an example.

65. **Choosing a Method** State (if possible) the method or integration formula you would use to find the antiderivative. Explain why you chose that method or formula. Do not integrate.

(a) $\displaystyle\int \frac{e^x}{e^{2x} + 1}\,dx$ (b) $\displaystyle\int \frac{e^x}{e^x + 1}\,dx$ (c) $\displaystyle\int xe^{x^2}\,dx$

(d) $\displaystyle\int xe^x\,dx$ (e) $\displaystyle\int e^{x^2}\,dx$ (f) $\displaystyle\int e^{2x}\sqrt{e^{2x} + 1}\,dx$

66. **HOW DO YOU SEE IT?** Use the graph of f' shown in the figure to answer the following.

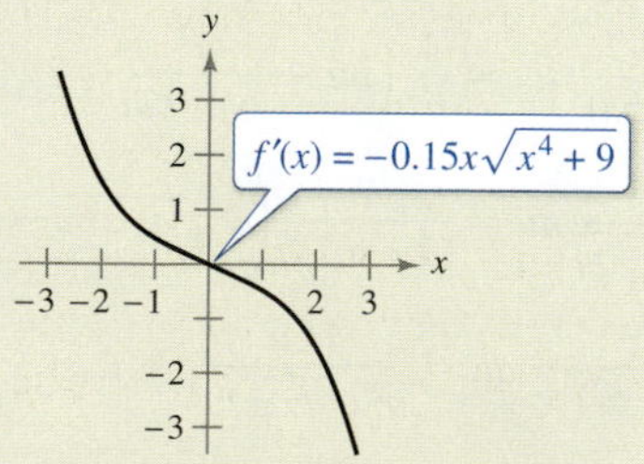

(a) Approximate the slope of f at $x = -1$. Explain.

(b) Approximate any open intervals in which the graph of f is increasing and any open intervals in which it is decreasing. Explain.

True or False? **In Exercises 67 and 68, determine whether the statement is true or false. If it is false, explain why or give an example that shows it is false.**

67. To use a table of integrals, the integral you are evaluating must appear in the table.

68. When using a table of integrals, you may have to make substitutions to rewrite your integral in the form in which it appears in the table.

69. **Work** A hydraulic cylinder on an industrial machine pushes a steel block a distance of x feet $(0 \leq x \leq 5)$, where the variable force required is $F(x) = 2000xe^{-x}$ pounds. Find the work done in pushing the block the full 5 feet through the machine.

70. **Work** Repeat Exercise 69, using $F(x) = \dfrac{500x}{\sqrt{26 - x^2}}$ pounds.

71. **Volume** Consider the region bounded by the graphs of

$$y = x\sqrt{16 - x^2},\quad y = 0,\quad x = 0,\quad \text{and}\quad x = 4.$$

Find the volume of the solid generated by revolving the region about the y-axis.

72. **Building Design** The cross section of a precast concrete beam for a building is bounded by the graphs of the equations

$$x = \frac{2}{\sqrt{1 + y^2}},\quad x = \frac{-2}{\sqrt{1 + y^2}},\quad y = 0,\quad \text{and}\quad y = 3$$

where x and y are measured in feet. The length of the beam is 20 feet (see figure).

(a) Find the volume V and the weight W of the beam. Assume the concrete weighs 148 pounds per cubic foot.

(b) Find the centroid of a cross section of the beam.

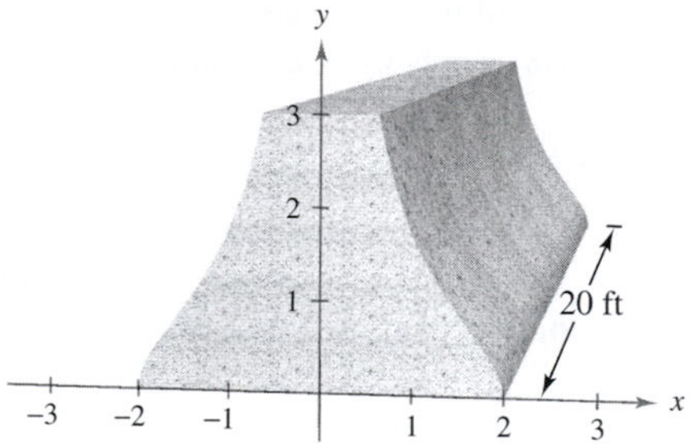

73. **Population** A population is growing according to the logistic model

$$N = \frac{5000}{1 + e^{4.8 - 1.9t}}$$

where t is the time in days. Find the average population over the interval $[0, 2]$.

PUTNAM EXAM CHALLENGE

74. Evaluate $\displaystyle\int_0^{\pi/2} \frac{dx}{1 + (\tan x)^{\sqrt{2}}}$.

This problem was composed by the Committee on the Putnam Prize Competition.

8.7 Indeterminate Forms and L'Hôpital's Rule

- **Recognize limits that produce indeterminate forms.**
- **Apply L'Hôpital's Rule to evaluate a limit.**

Indeterminate Forms

Recall that the forms $0/0$ and ∞/∞ are called *indeterminate* because they do not guarantee that a limit exists, nor do they indicate what the limit is, if one does exist. When you encountered one of these indeterminate forms earlier in the text, you attempted to rewrite the expression by using various algebraic techniques.

Indeterminate Form	Limit	Algebraic Technique
$\frac{0}{0}$	$\lim_{x\to -1} \frac{2x^2-2}{x+1} = \lim_{x\to -1} 2(x-1) = -4$	Divide numerator and denominator by $(x+1)$.
$\frac{\infty}{\infty}$	$\lim_{x\to\infty} \frac{3x^2-1}{2x^2+1} = \lim_{x\to\infty} \frac{3-(1/x^2)}{2+(1/x^2)} = \frac{3}{2}$	Divide numerator and denominator by x^2.

Occasionally, you can extend these algebraic techniques to find limits of transcendental functions. For instance, the limit

$$\lim_{x\to 0} \frac{e^{2x}-1}{e^x-1}$$

produces the indeterminate form $0/0$. Factoring and then dividing produces

$$\lim_{x\to 0} \frac{e^{2x}-1}{e^x-1} = \lim_{x\to 0} \frac{(e^x+1)(e^x-1)}{e^x-1} = \lim_{x\to 0} (e^x+1) = 2.$$

Not all indeterminate forms, however, can be evaluated by algebraic manipulation. This is often true when *both* algebraic and transcendental functions are involved. For instance, the limit

$$\lim_{x\to 0} \frac{e^{2x}-1}{x}$$

produces the indeterminate form $0/0$. Rewriting the expression to obtain

$$\lim_{x\to 0} \left(\frac{e^{2x}}{x} - \frac{1}{x}\right)$$

merely produces another indeterminate form, $\infty - \infty$. Of course, you could use technology to estimate the limit, as shown in the table and in Figure 8.15. From the table and the graph, the limit appears to be 2. (This limit will be verified in Example 1.)

x	-1	-0.1	-0.01	-0.001	0	0.001	0.01	0.1	1
$\frac{e^{2x}-1}{x}$	0.865	1.813	1.980	1.998	?	2.002	2.020	2.214	6.389

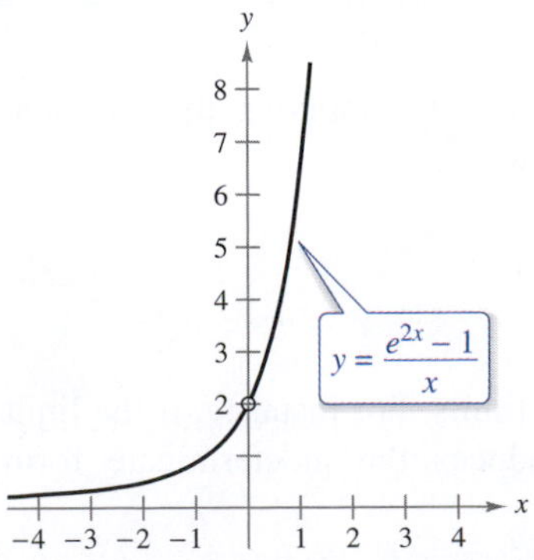

The limit as x approaches 0 appears to be 2.

Figure 8.15

L'Hôpital's Rule

To find the limit illustrated in Figure 8.15, you can use a theorem called **L'Hôpital's Rule.** This theorem states that under certain conditions, the limit of the quotient $f(x)/g(x)$ is determined by the limit of the quotient of the derivatives

$$\frac{f'(x)}{g'(x)}.$$

To prove this theorem, you can use a more general result called the **Extended Mean Value Theorem.**

THEOREM 8.3 The Extended Mean Value Theorem

If f and g are differentiable on an open interval (a, b) and continuous on $[a, b]$ such that $g'(x) \neq 0$ for any x in (a, b), then there exists a point c in (a, b) such that

$$\frac{f'(c)}{g'(c)} = \frac{f(b) - f(a)}{g(b) - g(a)}.$$

A proof of this theorem is given in Appendix A.

See LarsonCalculus.com for Bruce Edwards's video of this proof.

To see why Theorem 8.3 is called the Extended Mean Value Theorem, consider the special case in which $g(x) = x$. For this case, you obtain the "standard" Mean Value Theorem as presented in Section 3.2.

THEOREM 8.4 L'Hôpital's Rule

Let f and g be functions that are differentiable on an open interval (a, b) containing c, except possibly at c itself. Assume that $g'(x) \neq 0$ for all x in (a, b), except possibly at c itself. If the limit of $f(x)/g(x)$ as x approaches c produces the indeterminate form $0/0$, then

$$\lim_{x \to c} \frac{f(x)}{g(x)} = \lim_{x \to c} \frac{f'(x)}{g'(x)}$$

provided the limit on the right exists (or is infinite). This result also applies when the limit of $f(x)/g(x)$ as x approaches c produces any one of the indeterminate forms ∞/∞, $(-\infty)/\infty$, $\infty/(-\infty)$, or $(-\infty)/(-\infty)$.

A proof of this theorem is given in Appendix A.

See LarsonCalculus.com for Bruce Edwards's video of this proof.

People occasionally use L'Hôpital's Rule incorrectly by applying the Quotient Rule to $f(x)/g(x)$. Be sure you see that the rule involves

$$\frac{f'(x)}{g'(x)}$$

not the derivative of $f(x)/g(x)$.

L'Hôpital's Rule can also be applied to one-sided limits. For instance, if the limit of $f(x)/g(x)$ as x approaches c *from the right* produces the indeterminate form $0/0$, then

$$\lim_{x \to c^+} \frac{f(x)}{g(x)} = \lim_{x \to c^+} \frac{f'(x)}{g'(x)}$$

provided the limit exists (or is infinite).

GUILLAUME L'HÔPITAL (1661–1704)

L'Hôpital's Rule is named after the French mathematician Guillaume François Antoine de L'Hôpital. L'Hôpital is credited with writing the first text on differential calculus (in 1696) in which the rule publicly appeared. It was recently discovered that the rule and its proof were written in a letter from John Bernoulli to L'Hôpital. "... I acknowledge that I owe very much to the bright minds of the Bernoulli brothers. ... I have made free use of their discoveries ...," said L'Hôpital.

See LarsonCalculus.com to read more of this biography.

AP* Tips

L'Hôpital's Rule is tested on the AP Calculus BC Exam, primarily on the non-calculator multiple choice section.

FOR FURTHER INFORMATION To enhance your understanding of the necessity of the restriction that $g'(x)$ be nonzero for all x in (a, b), except possibly at c, see the article "Counterexamples to L'Hôpital's Rule" by R. P. Boas in *The American Mathematical Monthly.* To view this article, go to *MathArticles.com.*

The Granger Collection

Exploration

Numerical and Graphical Approaches Use a numerical or a graphical approach to approximate each limit.

a. $\lim_{x\to 0} \dfrac{2^{2x}-1}{x}$

b. $\lim_{x\to 0} \dfrac{3^{2x}-1}{x}$

c. $\lim_{x\to 0} \dfrac{4^{2x}-1}{x}$

d. $\lim_{x\to 0} \dfrac{5^{2x}-1}{x}$

What pattern do you observe? Does an analytic approach have an advantage for determining these limits? If so, explain your reasoning.

EXAMPLE 1 Indeterminate Form 0/0

Evaluate $\lim_{x\to 0} \dfrac{e^{2x}-1}{x}$.

Solution Because direct substitution results in the indeterminate form 0/0

$$\lim_{x\to 0} \frac{e^{2x}-1}{x} \qquad \lim_{x\to 0}(e^{2x}-1) = 0 \qquad \lim_{x\to 0} x = 0$$

you can apply L'Hôpital's Rule, as shown below.

$$\lim_{x\to 0} \frac{e^{2x}-1}{x} = \lim_{x\to 0} \frac{\frac{d}{dx}[e^{2x}-1]}{\frac{d}{dx}[x]}$$ Apply L'Hôpital's Rule.

$$= \lim_{x\to 0} \frac{2e^{2x}}{1}$$ Differentiate numerator and denominator.

$$= 2$$ Evaluate the limit.

In the solution to Example 1, note that you actually do not know that the first limit is equal to the second limit until you have shown that the second limit exists. In other words, if the second limit had not existed, then it would not have been permissible to apply L'Hôpital's Rule.

Another form of L'Hôpital's Rule states that if the limit of $f(x)/g(x)$ as x approaches ∞ (or $-\infty$) produces the indeterminate form 0/0 or ∞/∞, then

$$\lim_{x\to\infty} \frac{f(x)}{g(x)} = \lim_{x\to\infty} \frac{f'(x)}{g'(x)}$$

provided the limit on the right exists.

EXAMPLE 2 Indeterminate Form ∞/∞

Evaluate $\lim_{x\to\infty} \dfrac{\ln x}{x}$.

Solution Because direct substitution results in the indeterminate form ∞/∞, you can apply L'Hôpital's Rule to obtain

$$\lim_{x\to\infty} \frac{\ln x}{x} = \lim_{x\to\infty} \frac{\frac{d}{dx}[\ln x]}{\frac{d}{dx}[x]}$$ Apply L'Hôpital's Rule.

$$= \lim_{x\to\infty} \frac{1}{x}$$ Differentiate numerator and denominator.

$$= 0.$$ Evaluate the limit.

TECHNOLOGY Use a graphing utility to graph $y_1 = \ln x$ and $y_2 = x$ in the same viewing window. Which function grows faster as x approaches ∞? How is this observation related to Example 2?

Occasionally it is necessary to apply L'Hôpital's Rule more than once to remove an indeterminate form, as shown in Example 3.

FOR FURTHER INFORMATION To read about the connection between Leonhard Euler and Guillaume L'Hôpital, see the article "When Euler Met l'Hôpital" by William Dunham in *Mathematics Magazine*. To view this article, go to *MathArticles.com*.

EXAMPLE 3 Applying L'Hôpital's Rule More than Once

Evaluate $\lim_{x\to-\infty} \dfrac{x^2}{e^{-x}}$.

Solution Because direct substitution results in the indeterminate form ∞/∞, you can apply L'Hôpital's Rule.

$$\lim_{x\to-\infty} \frac{x^2}{e^{-x}} = \lim_{x\to-\infty} \frac{\frac{d}{dx}[x^2]}{\frac{d}{dx}[e^{-x}]} = \lim_{x\to-\infty} \frac{2x}{-e^{-x}}$$

This limit yields the indeterminate form $(-\infty)/(-\infty)$, so you can apply L'Hôpital's Rule again to obtain

$$\lim_{x\to-\infty} \frac{2x}{-e^{-x}} = \lim_{x\to-\infty} \frac{\frac{d}{dx}[2x]}{\frac{d}{dx}[-e^{-x}]} = \lim_{x\to-\infty} \frac{2}{e^{-x}} = 0.$$

In addition to the forms $0/0$ and ∞/∞, there are other indeterminate forms such as $0 \cdot \infty$, 1^∞, ∞^0, 0^0, and $\infty - \infty$. For example, consider the following four limits that lead to the indeterminate form $0 \cdot \infty$.

$$\underbrace{\lim_{x\to 0} \left(\frac{1}{x}\right)(x)}_{\text{Limit is 1.}}, \quad \underbrace{\lim_{x\to 0} \left(\frac{2}{x}\right)(x)}_{\text{Limit is 2.}}, \quad \underbrace{\lim_{x\to\infty} \left(\frac{1}{e^x}\right)(x)}_{\text{Limit is 0.}}, \quad \underbrace{\lim_{x\to\infty} \left(\frac{1}{x}\right)(e^x)}_{\text{Limit is } \infty.}$$

Because each limit is different, it is clear that the form $0 \cdot \infty$ is indeterminate in the sense that it does not determine the value (or even the existence) of the limit. The remaining examples in this section show methods for evaluating these forms. Basically, you attempt to convert each of these forms to $0/0$ or ∞/∞ so that L'Hôpital's Rule can be applied.

EXAMPLE 4 Indeterminate Form $0 \cdot \infty$

Evaluate $\lim_{x\to\infty} e^{-x}\sqrt{x}$.

Solution Because direct substitution produces the indeterminate form $0 \cdot \infty$, you should try to rewrite the limit to fit the form $0/0$ or ∞/∞. In this case, you can rewrite the limit to fit the second form.

$$\lim_{x\to\infty} e^{-x}\sqrt{x} = \lim_{x\to\infty} \frac{\sqrt{x}}{e^x}$$

Now, by L'Hôpital's Rule, you have

$$\lim_{x\to\infty} \frac{\sqrt{x}}{e^x} = \lim_{x\to\infty} \frac{1/(2\sqrt{x})}{e^x} \qquad \text{Differentiate numerator and denominator.}$$

$$= \lim_{x\to\infty} \frac{1}{2\sqrt{x}e^x} \qquad \text{Simplify.}$$

$$= 0. \qquad \text{Evaluate the limit.}$$

When rewriting a limit in one of the forms $0/0$ or ∞/∞ does not seem to work, try the other form. For instance, in Example 4, you can write the limit as

$$\lim_{x\to\infty} e^{-x}\sqrt{x} = \lim_{x\to\infty} \frac{e^{-x}}{x^{-1/2}}$$

which yields the indeterminate form $0/0$. As it happens, applying L'Hôpital's Rule to this limit produces

$$\lim_{x\to\infty} \frac{e^{-x}}{x^{-1/2}} = \lim_{x\to\infty} \frac{-e^{-x}}{-1/(2x^{3/2})}$$

which also yields the indeterminate form $0/0$.

The indeterminate forms 1^∞, ∞^0, and 0^0 arise from limits of functions that have variable bases and variable exponents. When you previously encountered this type of function, you used logarithmic differentiation to find the derivative. You can use a similar procedure when taking limits, as shown in the next example.

EXAMPLE 5 Indeterminate Form 1^∞

Evaluate $\displaystyle\lim_{x\to\infty}\left(1+\frac{1}{x}\right)^x$.

Solution Because direct substitution yields the indeterminate form 1^∞, you can proceed as follows. To begin, assume that the limit exists and is equal to y.

$$y = \lim_{x\to\infty}\left(1+\frac{1}{x}\right)^x$$

Taking the natural logarithm of each side produces

$$\ln y = \ln\left[\lim_{x\to\infty}\left(1+\frac{1}{x}\right)^x\right].$$

Because the natural logarithmic function is continuous, you can write

$$\ln y = \lim_{x\to\infty}\left[x\ln\left(1+\frac{1}{x}\right)\right] \qquad \text{Indeterminate form } \infty\cdot 0$$

$$= \lim_{x\to\infty}\left(\frac{\ln[1+(1/x)]}{1/x}\right) \qquad \text{Indeterminate form } 0/0$$

$$= \lim_{x\to\infty}\left(\frac{(-1/x^2)\{1/[1+(1/x)]\}}{-1/x^2}\right) \qquad \text{L'Hôpital's Rule}$$

$$= \lim_{x\to\infty}\frac{1}{1+(1/x)}$$

$$= 1.$$

Now, because you have shown that

$$\ln y = 1$$

you can conclude that

$$y = e$$

and obtain

$$\lim_{x\to\infty}\left(1+\frac{1}{x}\right)^x = e.$$

You can use a graphing utility to confirm this result, as shown in Figure 8.16.

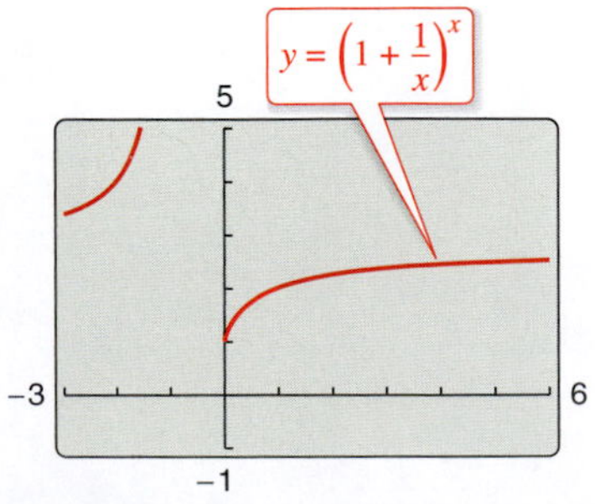

The limit of $[1+(1/x)]^x$ as x approaches infinity is e.
Figure 8.16

L'Hôpital's Rule can also be applied to one-sided limits, as demonstrated in Examples 6 and 7.

EXAMPLE 6 Indeterminate Form 0^0

See LarsonCalculus.com for an interactive version of this type of example.

Evaluate $\lim_{x\to 0^+} (\sin x)^x$.

Solution Because direct substitution produces the indeterminate form 0^0, you can proceed as shown below. To begin, assume that the limit exists and is equal to y.

$$y = \lim_{x\to 0^+} (\sin x)^x \quad \text{Indeterminate form } 0^0$$

$$\ln y = \ln\left[\lim_{x\to 0^+} (\sin x)^x\right] \quad \text{Take natural log of each side.}$$

$$= \lim_{x\to 0^+} [\ln(\sin x)^x] \quad \text{Continuity}$$

$$= \lim_{x\to 0^+} [x \ln(\sin x)] \quad \text{Indeterminate form } 0 \cdot (-\infty)$$

$$= \lim_{x\to 0^+} \frac{\ln(\sin x)}{1/x} \quad \text{Indeterminate form } -\infty/\infty$$

$$= \lim_{x\to 0^+} \frac{\cot x}{-1/x^2} \quad \text{L'Hôpital's Rule}$$

$$= \lim_{x\to 0^+} \frac{-x^2}{\tan x} \quad \text{Indeterminate form } 0/0$$

$$= \lim_{x\to 0^+} \frac{-2x}{\sec^2 x} \quad \text{L'Hôpital's Rule}$$

$$= 0$$

Now, because $\ln y = 0$, you can conclude that $y = e^0 = 1$, and it follows that

$$\lim_{x\to 0^+} (\sin x)^x = 1.$$

TECHNOLOGY When evaluating complicated limits such as the one in Example 6, it is helpful to check the reasonableness of the solution with a graphing utility. For instance, the calculations in the table and the graph in the figure (see below) are consistent with the conclusion that $(\sin x)^x$ approaches 1 as x approaches 0 from the right.

x	1.0	0.1	0.01	0.001	0.0001	0.00001
$(\sin x)^x$	0.8415	0.7942	0.9550	0.9931	0.9991	0.9999

Use a graphing utility to estimate the limits $\lim_{x\to 0} (1 - \cos x)^x$ and $\lim_{x\to 0^+} (\tan x)^x$. Then try to verify your estimates analytically.

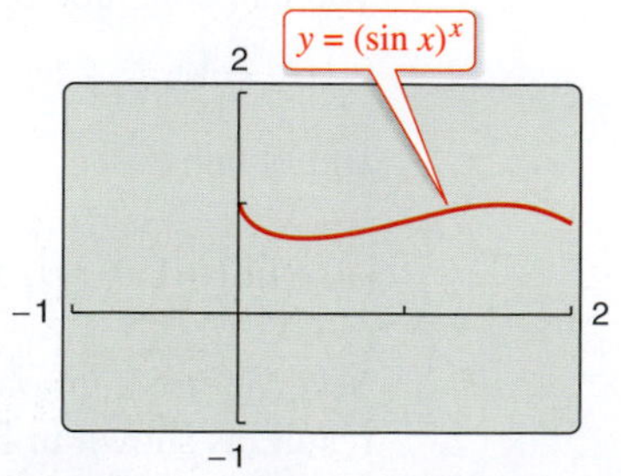

The limit of $(\sin x)^x$ is 1 as x approaches 0 from the right.

EXAMPLE 7 **Indeterminate Form $\infty - \infty$**

Evaluate $\displaystyle\lim_{x\to 1^+}\left(\frac{1}{\ln x} - \frac{1}{x-1}\right)$.

Solution Because direct substitution yields the indeterminate form $\infty - \infty$, you should try to rewrite the expression to produce a form to which you can apply L'Hôpital's Rule. In this case, you can combine the two fractions to obtain

$$\lim_{x\to 1^+}\left(\frac{1}{\ln x} - \frac{1}{x-1}\right) = \lim_{x\to 1^+}\left[\frac{x-1-\ln x}{(x-1)\ln x}\right].$$

Now, because direct substitution produces the indeterminate form $0/0$, you can apply L'Hôpital's Rule to obtain

$$\lim_{x\to 1^+}\left(\frac{1}{\ln x} - \frac{1}{x-1}\right) = \lim_{x\to 1^+}\frac{\frac{d}{dx}[x-1-\ln x]}{\frac{d}{dx}[(x-1)\ln x]}$$

$$= \lim_{x\to 1^+}\left[\frac{1-(1/x)}{(x-1)(1/x)+\ln x}\right]$$

$$= \lim_{x\to 1^+}\left(\frac{x-1}{x-1+x\ln x}\right).$$

This limit also yields the indeterminate form $0/0$, so you can apply L'Hôpital's Rule again to obtain

$$\lim_{x\to 1^+}\left(\frac{1}{\ln x} - \frac{1}{x-1}\right) = \lim_{x\to 1^+}\left[\frac{1}{1+x(1/x)+\ln x}\right] = \frac{1}{2}.$$

The forms $0/0$, ∞/∞, $\infty - \infty$, $0 \cdot \infty$, 0^0, 1^∞, and ∞^0 have been identified as *indeterminate*. There are similar forms that you should recognize as "determinate."

$\infty + \infty \to \infty$ Limit is positive infinity.

$-\infty - \infty \to -\infty$ Limit is negative infinity.

$0^\infty \to 0$ Limit is zero.

$0^{-\infty} \to \infty$ Limit is positive infinity.

(You are asked to verify two of these in Exercises 108 and 109.)

As a final comment, remember that L'Hôpital's Rule can be applied only to quotients leading to the indeterminate forms $0/0$ and ∞/∞. For instance, the application of L'Hôpital's Rule shown below is *incorrect*.

$$\lim_{x\to 0}\frac{e^x}{x} = \lim_{x\to 0}\frac{e^x}{1} = 1$$ Incorrect use of L'Hôpital's Rule

The reason this application is incorrect is that, even though the limit of the denominator is 0, the limit of the numerator is 1, which means that the hypotheses of L'Hôpital's Rule have not been satisfied.

Exploration

In each of the examples presented in this section, L'Hôpital's Rule is used to find a limit that exists. It can also be used to conclude that a limit is infinite. For instance, try using L'Hôpital's Rule to show that $\displaystyle\lim_{x\to\infty} e^x/x = \infty$.

8.7 Exercises

See CalcChat.com for tutorial help and worked-out solutions to odd-numbered exercises.

Numerical and Graphical Analysis **In Exercises 1–4, complete the table and use the result to estimate the limit. Use a graphing utility to graph the function to support your result.**

1. $\lim_{x\to 0} \dfrac{\sin 4x}{\sin 3x}$

x	-0.1	-0.01	-0.001	0.001	0.01	0.1
$f(x)$						

2. $\lim_{x\to 0} \dfrac{1 - e^x}{x}$

x	-0.1	-0.01	-0.001	0.001	0.01	0.1
$f(x)$						

3. $\lim_{x\to\infty} x^5 e^{-x/100}$

x	1	10	10^2	10^3	10^4	10^5
$f(x)$						

4. $\lim_{x\to\infty} \dfrac{6x}{\sqrt{3x^2 - 2x}}$

x	1	10	10^2	10^3	10^4	10^5
$f(x)$						

Using Two Methods **In Exercises 5–10, evaluate the limit (a) using techniques from Chapters 1 and 3 and (b) using L'Hôpital's Rule.**

5. $\lim_{x\to 4} \dfrac{3(x-4)}{x^2 - 16}$

6. $\lim_{x\to -4} \dfrac{2x^2 + 13x + 20}{x + 4}$

7. $\lim_{x\to 6} \dfrac{\sqrt{x + 10} - 4}{x - 6}$

8. $\lim_{x\to 0} \dfrac{\sin 6x}{4x}$

9. $\lim_{x\to\infty} \dfrac{5x^2 - 3x + 1}{3x^2 - 5}$

10. $\lim_{x\to\infty} \dfrac{4x - 3}{5x^2 + 1}$

Evaluating a Limit **In Exercises 11–42, evaluate the limit, using L'Hôpital's Rule if necessary.**

11. $\lim_{x\to 3} \dfrac{x^2 - 2x - 3}{x - 3}$

12. $\lim_{x\to -2} \dfrac{x^2 - 3x - 10}{x + 2}$

13. $\lim_{x\to 0} \dfrac{\sqrt{25 - x^2} - 5}{x}$

14. $\lim_{x\to 5^-} \dfrac{\sqrt{25 - x^2}}{x - 5}$

15. $\lim_{x\to 0^+} \dfrac{e^x - (1 + x)}{x^3}$

16. $\lim_{x\to 1} \dfrac{\ln x^3}{x^2 - 1}$

17. $\lim_{x\to 1} \dfrac{x^{11} - 1}{x^4 - 1}$

18. $\lim_{x\to 1} \dfrac{x^a - 1}{x^b - 1}$, where $a, b \neq 0$

19. $\lim_{x\to 0} \dfrac{\sin 3x}{\sin 5x}$

20. $\lim_{x\to 0} \dfrac{\sin ax}{\sin bx}$, where $a, b \neq 0$

21. $\lim_{x\to 0} \dfrac{\arcsin x}{x}$

22. $\lim_{x\to 1} \dfrac{\arctan x - (\pi/4)}{x - 1}$

23. $\lim_{x\to\infty} \dfrac{5x^2 + 3x - 1}{4x^2 + 5}$

24. $\lim_{x\to\infty} \dfrac{5x + 3}{x^3 - 6x + 2}$

25. $\lim_{x\to\infty} \dfrac{x^2 + 4x + 7}{x - 6}$

26. $\lim_{x\to\infty} \dfrac{x^3}{x + 2}$

27. $\lim_{x\to\infty} \dfrac{x^3}{e^{x/2}}$

28. $\lim_{x\to\infty} \dfrac{x^3}{e^{x^2}}$

29. $\lim_{x\to\infty} \dfrac{x}{\sqrt{x^2 + 1}}$

30. $\lim_{x\to\infty} \dfrac{x^2}{\sqrt{x^2 + 1}}$

31. $\lim_{x\to\infty} \dfrac{\cos x}{x}$

32. $\lim_{x\to\infty} \dfrac{\sin x}{x - \pi}$

33. $\lim_{x\to\infty} \dfrac{\ln x}{x^2}$

34. $\lim_{x\to\infty} \dfrac{\ln x^4}{x^3}$

35. $\lim_{x\to\infty} \dfrac{e^x}{x^4}$

36. $\lim_{x\to\infty} \dfrac{e^{x/2}}{x}$

37. $\lim_{x\to 0} \dfrac{\sin 5x}{\tan 9x}$

38. $\lim_{x\to 1} \dfrac{\ln x}{\sin \pi x}$

39. $\lim_{x\to 0} \dfrac{\arctan x}{\sin x}$

40. $\lim_{x\to 0} \dfrac{x}{\arctan 2x}$

41. $\lim_{x\to\infty} \dfrac{\int_1^x \ln(e^{4t-1})\,dt}{x}$

42. $\lim_{x\to 1^+} \dfrac{\int_1^x \cos\theta\,d\theta}{x - 1}$

Evaluating a Limit **In Exercises 43–60, (a) describe the type of indeterminate form (if any) that is obtained by direct substitution. (b) Evaluate the limit, using L'Hôpital's Rule if necessary. (c) Use a graphing utility to graph the function and verify the result in part (b).**

43. $\lim_{x\to\infty} x \ln x$

44. $\lim_{x\to 0^+} x^3 \cot x$

45. $\lim_{x\to\infty} \left(x \sin \dfrac{1}{x}\right)$

46. $\lim_{x\to\infty} x \tan \dfrac{1}{x}$

47. $\lim_{x\to 0^+} x^{1/x}$

48. $\lim_{x\to 0^+} (e^x + x)^{2/x}$

49. $\lim_{x\to\infty} x^{1/x}$

50. $\lim_{x\to\infty} \left(1 + \dfrac{1}{x}\right)^x$

51. $\lim_{x\to 0^+} (1 + x)^{1/x}$

52. $\lim_{x\to\infty} (1 + x)^{1/x}$

53. $\lim_{x\to 0^+} [3(x)^{x/2}]$

54. $\lim_{x\to 4^+} [3(x - 4)]^{x-4}$

55. $\lim_{x\to 1^+} (\ln x)^{x-1}$

56. $\lim_{x\to 0^+} \left[\cos\left(\dfrac{\pi}{2} - x\right)\right]^x$

57. $\lim_{x\to 2^+} \left(\dfrac{8}{x^2 - 4} - \dfrac{x}{x - 2}\right)$

58. $\lim_{x\to 2^+} \left(\dfrac{1}{x^2 - 4} - \dfrac{\sqrt{x - 1}}{x^2 - 4}\right)$

59. $\lim_{x\to 1^+} \left(\dfrac{3}{\ln x} - \dfrac{2}{x - 1}\right)$

60. $\lim_{x\to 0^+} \left(\dfrac{10}{x} - \dfrac{3}{x^2}\right)$

WRITING ABOUT CONCEPTS

61. Indeterminate Forms List six different indeterminate forms.

62. L'Hôpital's Rule State L'Hôpital's Rule.

63. Finding Functions Find differentiable functions f and g that satisfy the specified condition such that

$$\lim_{x\to 5} f(x) = 0 \quad \text{and} \quad \lim_{x\to 5} g(x) = 0.$$

Explain how you obtained your answers. (*Note:* There are many correct answers.)

(a) $\lim_{x\to 5} \frac{f(x)}{g(x)} = 10$ (b) $\lim_{x\to 5} \frac{f(x)}{g(x)} = 0$

(c) $\lim_{x\to 5} \frac{f(x)}{g(x)} = \infty$

64. Finding Functions Find differentiable functions f and g such that

$$\lim_{x\to\infty} f(x) = \lim_{x\to\infty} g(x) = \infty \quad \text{and} \quad \lim_{x\to\infty} [f(x) - g(x)] = 25.$$

Explain how you obtained your answers. (*Note:* There are many correct answers.)

65. L'Hôpital's Rule Determine which of the following limits can be evaluated using L'Hôpital's Rule. Explain your reasoning. Do not evaluate the limit.

(a) $\lim_{x\to 2} \frac{x - 2}{x^3 - x - 6}$ (b) $\lim_{x\to 0} \frac{x^2 - 4x}{2x - 1}$

(c) $\lim_{x\to\infty} \frac{x^3}{e^x}$ (d) $\lim_{x\to 3} \frac{e^{x^2} - e^9}{x - 3}$

(e) $\lim_{x\to 1} \frac{\cos \pi x}{\ln x}$ (f) $\lim_{x\to 1} \frac{1 + x(\ln x - 1)}{(x - 1)\ln x}$

66. HOW DO YOU SEE IT? Use the graph of f to find the limit.

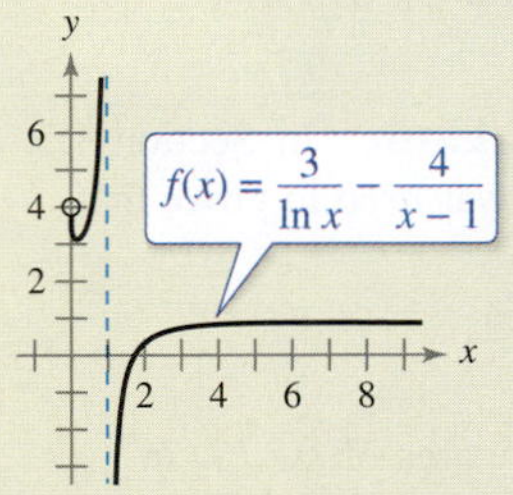

(a) $\lim_{x\to 1^-} f(x)$ (b) $\lim_{x\to 1^+} f(x)$ (c) $\lim_{x\to 1} f(x)$

67. Numerical Approach Complete the table to show that x eventually "overpowers" $(\ln x)^4$.

x	10	10^2	10^4	10^6	10^8	10^{10}
$\frac{(\ln x)^4}{x}$						

68. Numerical Approach Complete the table to show that e^x eventually "overpowers" x^5.

x	1	5	10	20	30	40	50	100
$\frac{e^x}{x^5}$								

Comparing Functions **In Exercises 69–74, use L'Hôpital's Rule to determine the comparative rates of increase of the functions $f(x) = x^m$, $g(x) = e^{nx}$, and $h(x) = (\ln x)^n$, where $n > 0$, $m > 0$, and $x \to \infty$.**

69. $\lim_{x\to\infty} \frac{x^2}{e^{5x}}$ **70.** $\lim_{x\to\infty} \frac{x^3}{e^{2x}}$

71. $\lim_{x\to\infty} \frac{(\ln x)^3}{x}$ **72.** $\lim_{x\to\infty} \frac{(\ln x)^2}{x^3}$

73. $\lim_{x\to\infty} \frac{(\ln x)^n}{x^m}$ **74.** $\lim_{x\to\infty} \frac{x^m}{e^{nx}}$

Asymptotes and Relative Extrema **In Exercises 75–78, find any asymptotes and relative extrema that may exist and use a graphing utility to graph the function. (*Hint:* Some of the limits required in finding asymptotes have been found in previous exercises.)**

75. $y = x^{1/x}, \quad x > 0$ **76.** $y = x^x, \quad x > 0$

77. $y = 2xe^{-x}$ **78.** $y = \frac{\ln x}{x}$

Think About It **In Exercises 79–82, L'Hôpital's Rule is used incorrectly. Describe the error.**

79. $\lim_{x\to 2} \frac{3x^2 + 4x + 1}{x^2 - x - 2} = \lim_{x\to 2} \frac{6x + 4}{2x - 1} = \lim_{x\to 2} \frac{6}{2} = 3$

80. $\lim_{x\to 0} \frac{e^{2x} - 1}{e^x} = \lim_{x\to 0} \frac{2e^{2x}}{e^x}$
$= \lim_{x\to 0} 2e^x$
$= 2$

81. $\lim_{x\to\infty} \frac{e^{-x}}{1 + e^{-x}} = \lim_{x\to\infty} \frac{-e^{-x}}{-e^{-x}}$
$= \lim_{x\to\infty} 1$
$= 1$

82. $\lim_{x\to\infty} x\cos\frac{1}{x} = \lim_{x\to\infty} \frac{\cos(1/x)}{1/x}$
$= \lim_{x\to\infty} \frac{[-\sin(1/x)](1/x^2)}{-1/x^2}$
$= 0$

Analytical Approach **In Exercises 83 and 84, (a) explain why L'Hôpital's Rule cannot be used to find the limit, (b) find the limit analytically, and (c) use a graphing utility to graph the function and approximate the limit from the graph. Compare the result with that in part (b).**

83. $\lim_{x\to\infty} \frac{x}{\sqrt{x^2 + 1}}$ **84.** $\lim_{x\to\pi/2^-} \frac{\tan x}{\sec x}$

Graphical Analysis **In Exercises 85 and 86, graph $f(x)/g(x)$ and $f'(x)/g'(x)$ near $x = 0$. What do you notice about these ratios as $x \to 0$? How does this illustrate L'Hôpital's Rule?**

85. $f(x) = \sin 3x, \quad g(x) = \sin 4x$

86. $f(x) = e^{3x} - 1, \quad g(x) = x$

87. Velocity in a Resisting Medium The velocity v of an object falling through a resisting medium such as air or water is given by

$$v = \frac{32}{k}\left(1 - e^{-kt} + \frac{v_0 k e^{-kt}}{32}\right)$$

where v_0 is the initial velocity, t is the time in seconds, and k is the resistance constant of the medium. Use L'Hôpital's Rule to find the formula for the velocity of a falling body in a vacuum by fixing v_0 and t and letting k approach zero. (Assume that the downward direction is positive.)

88. Compound Interest The formula for the amount A in a savings account compounded n times per year for t years at an interest rate r and an initial deposit of P is given by

$$A = P\left(1 + \frac{r}{n}\right)^{nt}.$$

Use L'Hôpital's Rule to show that the limiting formula as the number of compoundings per year approaches infinity is given by $A = Pe^{rt}$.

89. The Gamma Function The Gamma Function $\Gamma(n)$ is defined in terms of the integral of the function given by $f(x) = x^{n-1}e^{-x}$, $n > 0$. Show that for any fixed value of n, the limit of $f(x)$ as x approaches infinity is zero.

90. Tractrix A person moves from the origin along the positive y-axis pulling a weight at the end of a 12-meter rope (see figure). Initially, the weight is located at the point $(12, 0)$.

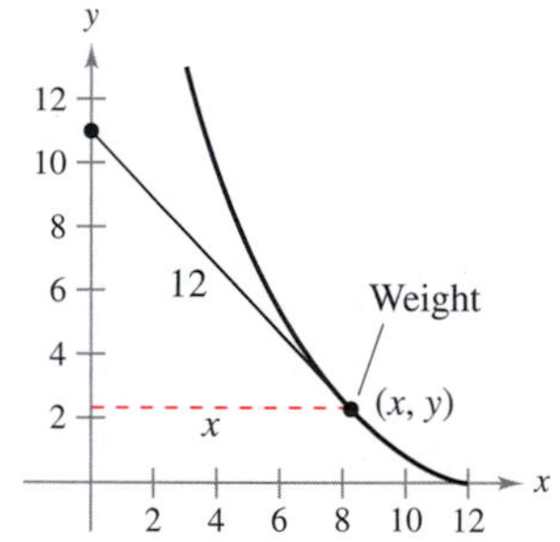

(a) Show that the slope of the tangent line of the path of the weight is

$$\frac{dy}{dx} = -\frac{\sqrt{144 - x^2}}{x}.$$

(b) Use the result of part (a) to find the equation of the path of the weight. Use a graphing utility to graph the path and compare it with the figure.

(c) Find any vertical asymptotes of the graph in part (b).

(d) When the person has reached the point $(0, 12)$, how far has the weight moved?

Extended Mean Value Theorem **In Exercises 91–94, apply the Extended Mean Value Theorem to the functions f and g on the given interval. Find all values c in the interval (a, b) such that**

$$\frac{f'(c)}{g'(c)} = \frac{f(b) - f(a)}{g(b) - g(a)}.$$

Functions	**Interval**
91. $f(x) = x^3, \quad g(x) = x^2 + 1$	$[0, 1]$
92. $f(x) = \frac{1}{x}, \quad g(x) = x^2 - 4$	$[1, 2]$
93. $f(x) = \sin x, \quad g(x) = \cos x$	$\left[0, \frac{\pi}{2}\right]$
94. $f(x) = \ln x, \quad g(x) = x^3$	$[1, 4]$

True or False? **In Exercises 95–98, determine whether the statement is true or false. If it is false, explain why or give an example that shows it is false.**

95. $\lim_{x \to 0}\left[\frac{x^2 + x + 1}{x}\right] = \lim_{x \to 0}\left[\frac{2x + 1}{1}\right] = 1$

96. If $y = \frac{e^x}{x^2}$, then $y' = \frac{e^x}{2x}$.

97. If $p(x)$ is a polynomial, then $\lim_{x \to \infty} \frac{p(x)}{e^x} = 0$.

98. If $\lim_{x \to \infty} \frac{f(x)}{g(x)} = 1$, then $\lim_{x \to \infty} [f(x) - g(x)] = 0$.

99. Area Find the limit, as x approaches 0, of the ratio of the area of the triangle to the total shaded area in the figure.

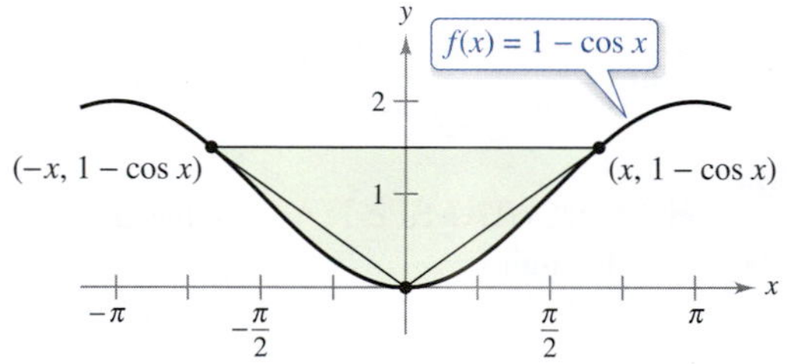

100. Finding a Limit In Section 1.3, a geometric argument (see figure) was used to prove that

$$\lim_{\theta \to 0} \frac{\sin \theta}{\theta} = 1.$$

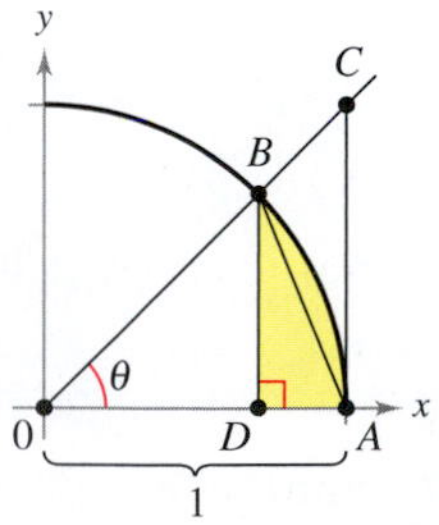

(a) Write the area of $\triangle ABD$ in terms of θ.

(b) Write the area of the shaded region in terms of θ.

(c) Write the ratio R of the area of $\triangle ABD$ to that of the shaded region.

(d) Find $\lim_{\theta \to 0} R$.

Continuous Function **In Exercises 101 and 102, find the value of c that makes the function continuous at $x = 0$.**

101. $f(x) = \begin{cases} \dfrac{4x - 2\sin 2x}{2x^3}, & x \neq 0 \\ c, & x = 0 \end{cases}$

102. $f(x) = \begin{cases} (e^x + x)^{1/x}, & x \neq 0 \\ c, & x = 0 \end{cases}$

103. Finding Values Find the values of a and b such that

$$\lim_{x\to 0} \frac{a - \cos bx}{x^2} = 2.$$

104. Evaluating a Limit Use a graphing utility to graph

$$f(x) = \frac{x^k - 1}{k}$$

for $k = 1, 0.1$, and 0.01. Then evaluate the limit

$$\lim_{k\to 0^+} \frac{x^k - 1}{k}.$$

105. Finding a Derivative

(a) Let $f'(x)$ be continuous. Show that

$$\lim_{h\to 0} \frac{f(x+h) - f(x-h)}{2h} = f'(x).$$

(b) Explain the result of part (a) graphically.

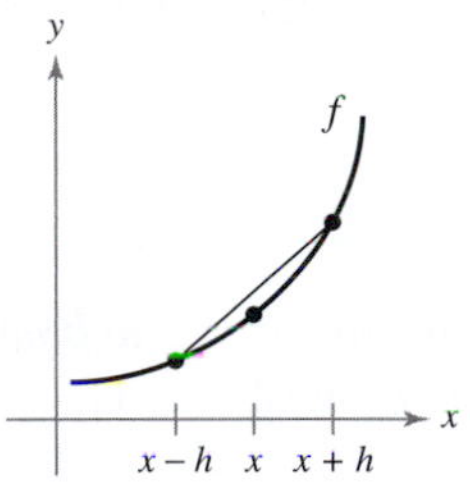

106. Finding a Second Derivative Let $f''(x)$ be continuous. Show that

$$\lim_{h\to 0} \frac{f(x+h) - 2f(x) + f(x-h)}{h^2} = f''(x).$$

107. Evaluating a Limit Consider the limit $\lim_{x\to 0^+} (-x \ln x)$.

(a) Describe the type of indeterminate form that is obtained by direct substitution.

(b) Evaluate the limit. Use a graphing utility to verify the result.

FOR FURTHER INFORMATION For a geometric approach to this exercise, see the article "A Geometric Proof of $\lim_{d\to 0^+} (-d \ln d) = 0$" by John H. Mathews in *The College Mathematics Journal*. To view this article, go to *MathArticles.com*.

108. Proof Prove that if $f(x) \geq 0$, $\lim_{x\to a} f(x) = 0$, and $\lim_{x\to a} g(x) = \infty$, then $\lim_{x\to a} f(x)^{g(x)} = 0$.

109. Proof Prove that if $f(x) \geq 0$, $\lim_{x\to a} f(x) = 0$, and $\lim_{x\to a} g(x) = -\infty$, then $\lim_{x\to a} f(x)^{g(x)} = \infty$.

110. Proof Prove the following generalization of the Mean Value Theorem. If f is twice differentiable on the closed interval $[a, b]$, then

$$f(b) - f(a) = f'(a)(b - a) - \int_a^b f''(t)(t - b)\,dt.$$

111. Indeterminate Forms Show that the indeterminate forms 0^0, ∞^0, and 1^∞ do not always have a value of 1 by evaluating each limit.

(a) $\lim_{x\to 0^+} x^{\ln 2/(1 + \ln x)}$

(b) $\lim_{x\to \infty} x^{\ln 2/(1 + \ln x)}$

(c) $\lim_{x\to 0} (x + 1)^{(\ln 2)/x}$

112. Calculus History In L'Hôpital's 1696 calculus textbook, he illustrated his rule using the limit of the function

$$f(x) = \frac{\sqrt{2a^3x - x^4} - a\sqrt[3]{a^2x}}{a - \sqrt[4]{ax^3}}$$

as x approaches a, $a > 0$. Find this limit.

113. Finding a Limit Consider the function

$$h(x) = \frac{x + \sin x}{x}.$$

(a) Use a graphing utility to graph the function. Then use the *zoom* and *trace* features to investigate $\lim_{x\to\infty} h(x)$.

(b) Find $\lim_{x\to\infty} h(x)$ analytically by writing

$$h(x) = \frac{x}{x} + \frac{\sin x}{x}.$$

(c) Can you use L'Hôpital's Rule to find $\lim_{x\to\infty} h(x)$? Explain your reasoning.

114. Evaluating a Limit Let $f(x) = x + x \sin x$ and $g(x) = x^2 - 4$.

(a) Show that $\lim_{x\to\infty} \dfrac{f(x)}{g(x)} = 0$.

(b) Show that $\lim_{x\to\infty} f(x) = \infty$ and $\lim_{x\to\infty} g(x) = \infty$.

(c) Evaluate the limit

$$\lim_{x\to\infty} \frac{f'(x)}{g'(x)}.$$

What do you notice?

(d) Do your answers to parts (a) through (c) contradict L'Hôpital's Rule? Explain your reasoning.

PUTNAM EXAM CHALLENGE

115. Evaluate $\lim_{x\to\infty} \left[\dfrac{1}{x} \cdot \dfrac{a^x - 1}{a - 1}\right]^{1/x}$ where $a > 0$, $a \neq 1$.

8.8 Improper Integrals

- **Evaluate an improper integral that has an infinite limit of integration.**
- **Evaluate an improper integral that has an infinite discontinuity.**

Improper Integrals with Infinite Limits of Integration

The definition of a definite integral

$$\int_a^b f(x)\,dx$$

requires that the interval $[a, b]$ be finite. Furthermore, the Fundamental Theorem of Calculus, by which you have been evaluating definite integrals, requires that f be continuous on $[a, b]$. In this section, you will study a procedure for evaluating integrals that do not satisfy these requirements—usually because either one or both of the limits of integration are infinite, or because f has a finite number of infinite discontinuities in the interval $[a, b]$. Integrals that possess either property are **improper integrals.** Note that a function f is said to have an **infinite discontinuity** at c when, *from the right or left,*

$$\lim_{x\to c} f(x) = \infty \quad \text{or} \quad \lim_{x\to c} f(x) = -\infty.$$

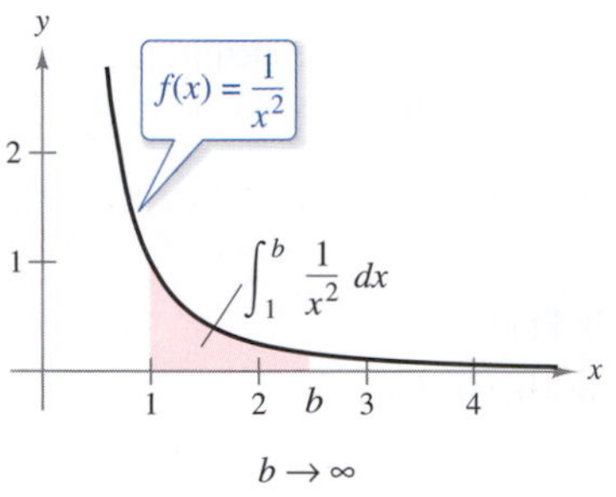

The unbounded region has an area of 1.
Figure 8.17

To get an idea of how to evaluate an improper integral, consider the integral

$$\int_1^b \frac{dx}{x^2} = -\frac{1}{x}\Big]_1^b = -\frac{1}{b} + 1 = 1 - \frac{1}{b}$$

which can be interpreted as the area of the shaded region shown in Figure 8.17. Taking the limit as $b \to \infty$ produces

$$\int_1^\infty \frac{dx}{x^2} = \lim_{b\to\infty}\left(\int_1^b \frac{dx}{x^2}\right) = \lim_{b\to\infty}\left(1 - \frac{1}{b}\right) = 1.$$

This improper integral can be interpreted as the area of the *unbounded* region between the graph of $f(x) = 1/x^2$ and the x-axis (to the right of $x = 1$).

AP* Tips

Improper integrals are tested on the AP Calculus BC Exam, primarily on the non-calculator multiple choice section.

Definition of Improper Integrals with Infinite Integration Limits

1. If f is continuous on the interval $[a, \infty)$, then

$$\int_a^\infty f(x)\,dx = \lim_{b\to\infty} \int_a^b f(x)\,dx.$$

2. If f is continuous on the interval $(-\infty, b]$, then

$$\int_{-\infty}^b f(x)\,dx = \lim_{a\to-\infty} \int_a^b f(x)\,dx.$$

3. If f is continuous on the interval $(-\infty, \infty)$, then

$$\int_{-\infty}^\infty f(x)\,dx = \int_{-\infty}^c f(x)\,dx + \int_c^\infty f(x)\,dx$$

where c is any real number (see Exercise 111).

In the first two cases, the improper integral **converges** when the limit exists—otherwise, the improper integral **diverges**. In the third case, the improper integral on the left diverges when either of the improper integrals on the right diverges.

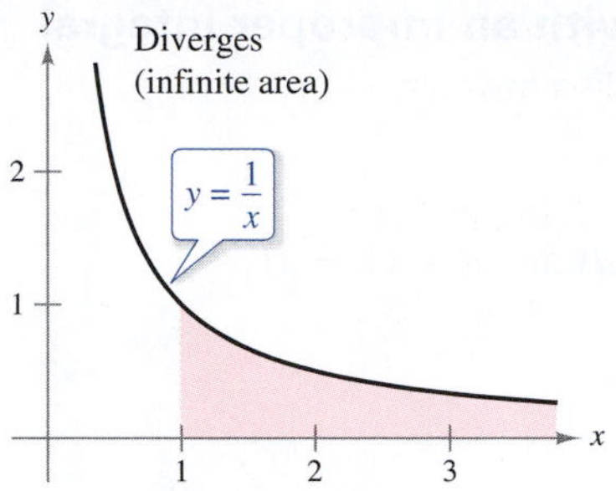

This unbounded region has an infinite area.
Figure 8.18

EXAMPLE 1 An Improper Integral That Diverges

Evaluate $\displaystyle\int_1^{\infty} \frac{dx}{x}$.

Solution

$$\int_1^{\infty} \frac{dx}{x} = \lim_{b\to\infty} \int_1^{b} \frac{dx}{x} \qquad \text{Take limit as } b \to \infty.$$

$$= \lim_{b\to\infty} \Big[\ln x\Big]_1^b \qquad \text{Apply Log Rule.}$$

$$= \lim_{b\to\infty} (\ln b - 0) \qquad \text{Apply Fundamental Theorem of Calculus.}$$

$$= \infty \qquad \text{Evaluate limit.}$$

The limit does not exist. So, you can conclude that the improper integral diverges. See Figure 8.18.

Try comparing the regions shown in Figures 8.17 and 8.18. They look similar, yet the region in Figure 8.17 has a finite area of 1 and the region in Figure 8.18 has an infinite area.

EXAMPLE 2 Improper Integrals That Converge

Evaluate each improper integral.

a. $\displaystyle\int_0^{\infty} e^{-x}\,dx$

b. $\displaystyle\int_0^{\infty} \frac{1}{x^2+1}\,dx$

Solution

a.
$$\int_0^{\infty} e^{-x}\,dx = \lim_{b\to\infty} \int_0^{b} e^{-x}\,dx$$
$$= \lim_{b\to\infty} \Big[-e^{-x}\Big]_0^b$$
$$= \lim_{b\to\infty} (-e^{-b} + 1)$$
$$= 1$$

See Figure 8.19.

b.
$$\int_0^{\infty} \frac{1}{x^2+1}\,dx = \lim_{b\to\infty} \int_0^{\infty} \frac{1}{x^2+1}\,dx$$
$$= \lim_{b\to\infty} \Big[\arctan x\Big]_0^b$$
$$= \lim_{b\to\infty} \arctan b$$
$$= \frac{\pi}{2}$$

See Figure 8.20.

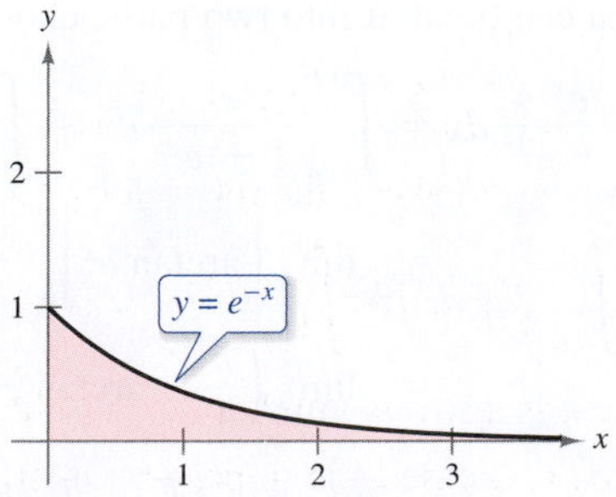

The area of the unbounded region is 1.
Figure 8.19

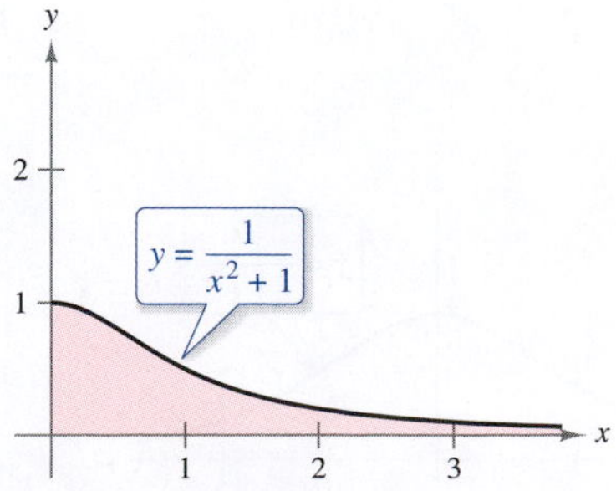

The area of the unbounded region is $\pi/2$.
Figure 8.20

In the next example, note how L'Hôpital's Rule can be used to evaluate an improper integral.

EXAMPLE 3 Using L'Hôpital's Rule with an Improper Integral

Evaluate $\int_1^{\infty} (1 - x)e^{-x}\, dx.$

Solution Use integration by parts, with $dv = e^{-x}\, dx$ and $u = (1 - x)$.

$$\int (1 - x)e^{-x}\, dx = -e^{-x}(1 - x) - \int e^{-x}\, dx$$
$$= -e^{-x} + xe^{-x} + e^{-x} + C$$
$$= xe^{-x} + C$$

Now, apply the definition of an improper integral.

$$\int_1^{\infty} (1 - x)e^{-x}\, dx = \lim_{b\to\infty} \Big[xe^{-x} \Big]_1^b$$
$$= \lim_{b\to\infty} \left(\frac{b}{e^b} - \frac{1}{e} \right)$$
$$= \lim_{b\to\infty} \frac{b}{e^b} - \lim_{b\to\infty} \frac{1}{e}$$

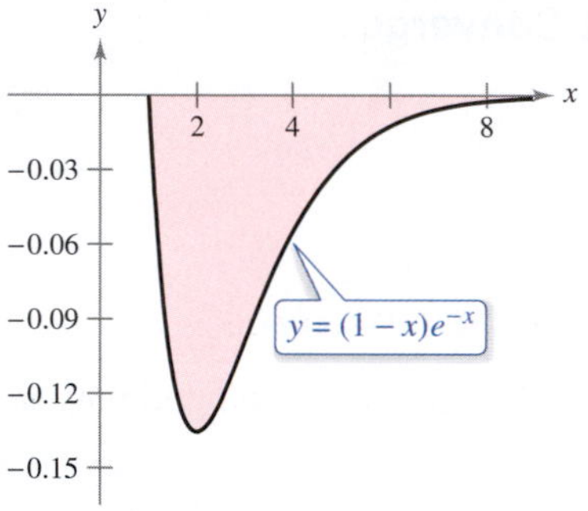

The area of the unbounded region is $|-1/e|$.
Figure 8.21

For the first limit, use L'Hôpital's Rule.

$$\lim_{b\to\infty} \frac{b}{e^b} = \lim_{b\to\infty} \frac{1}{e^b} = 0$$

So, you can conclude that

$$\int_1^{\infty} (1 - x)e^{-x}\, dx = \lim_{b\to\infty} \frac{b}{e^b} - \lim_{b\to\infty} \frac{1}{e}$$
$$= 0 - \frac{1}{e}$$
$$= -\frac{1}{e}.$$

See Figure 8.21.

EXAMPLE 4 Infinite Upper and Lower Limits of Integration

Evaluate $\int_{-\infty}^{\infty} \frac{e^x}{1 + e^{2x}}\, dx.$

Solution Note that the integrand is continuous on $(-\infty, \infty)$. To evaluate the integral, you can break it into two parts, choosing $c = 0$ as a convenient value.

$$\int_{-\infty}^{\infty} \frac{e^x}{1 + e^{2x}}\, dx = \int_{-\infty}^{0} \frac{e^x}{1 + e^{2x}}\, dx + \int_0^{\infty} \frac{e^x}{1 + e^{2x}}\, dx$$
$$= \lim_{b\to-\infty} \Big[\arctan e^x \Big]_b^0 + \lim_{b\to\infty} \Big[\arctan e^x \Big]_0^b$$
$$= \lim_{b\to-\infty} \left(\frac{\pi}{4} - \arctan e^b \right) + \lim_{b\to\infty} \left(\arctan e^b - \frac{\pi}{4} \right)$$
$$= \frac{\pi}{4} - 0 + \frac{\pi}{2} - \frac{\pi}{4}$$
$$= \frac{\pi}{2}$$

See Figure 8.22.

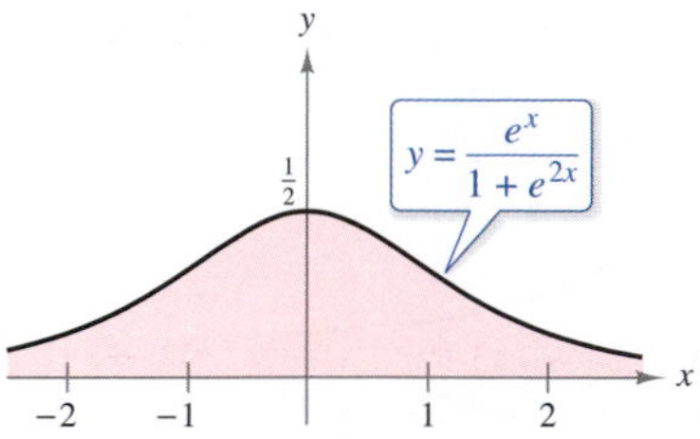

The area of the unbounded region is $\pi/2$.
Figure 8.22

The work required to move a 15-metric-ton space module an unlimited distance away from Earth is about 6.984×10^{11} foot-pounds.

EXAMPLE 5 **Sending a Space Module into Orbit**

In Example 3 in Section 7.5, you found that it would require 10,000 mile-tons of work to propel a 15-metric-ton space module to a height of 800 miles above Earth. How much work is required to propel the module an unlimited distance away from Earth's surface?

Solution At first you might think that an infinite amount of work would be required. But if this were the case, it would be impossible to send rockets into outer space. Because this has been done, the work required must be finite. You can determine the work in the following manner. Using the integral in Example 3, Section 7.5, replace the upper bound of 4800 miles by ∞ and write

$$\begin{aligned} W &= \int_{4000}^{\infty} \frac{240{,}000{,}000}{x^2}\,dx \\ &= \lim_{b\to\infty}\left[-\frac{240{,}000{,}000}{x}\right]_{4000}^{b} \\ &= \lim_{b\to\infty}\left(-\frac{240{,}000{,}000}{b} + \frac{240{,}000{,}000}{4000}\right) \\ &= 60{,}000 \text{ mile-tons} \\ &\approx 6.984 \times 10^{11} \text{ foot-pounds.} \end{aligned}$$

In SI units, using a conversion factor of

$$1 \text{ foot-pound} \approx 1.35582 \text{ joules}$$

the work done is $W \approx 9.469 \times 10^{11}$ joules.

Improper Integrals with Infinite Discontinuities

The second basic type of improper integral is one that has an infinite discontinuity *at or between* the limits of integration.

Definition of Improper Integrals with Infinite Discontinuities

1. If f is continuous on the interval $[a, b)$ and has an infinite discontinuity at b, then

$$\int_a^b f(x)\,dx = \lim_{c\to b^-} \int_a^c f(x)\,dx.$$

2. If f is continuous on the interval $(a, b]$ and has an infinite discontinuity at a, then

$$\int_a^b f(x)\,dx = \lim_{c\to a^+} \int_c^b f(x)\,dx.$$

3. If f is continuous on the interval $[a, b]$, except for some c in (a, b) at which f has an infinite discontinuity, then

$$\int_a^b f(x)\,dx = \int_a^c f(x)\,dx + \int_c^b f(x)\,dx.$$

In the first two cases, the improper integral **converges** when the limit exists—otherwise, the improper integral **diverges**. In the third case, the improper integral on the left diverges when either of the improper integrals on the right diverges.

EXAMPLE 6 **An Improper Integral with an Infinite Discontinuity**

Evaluate $\displaystyle\int_0^1 \frac{dx}{\sqrt[3]{x}}$.

Solution The integrand has an infinite discontinuity at $x = 0$, as shown in Figure 8.23. You can evaluate this integral as shown below.

$$\int_0^1 x^{-1/3}\,dx = \lim_{b\to 0^+}\left[\frac{x^{2/3}}{2/3}\right]_b^1$$
$$= \lim_{b\to 0^+}\frac{3}{2}(1 - b^{2/3})$$
$$= \frac{3}{2}$$

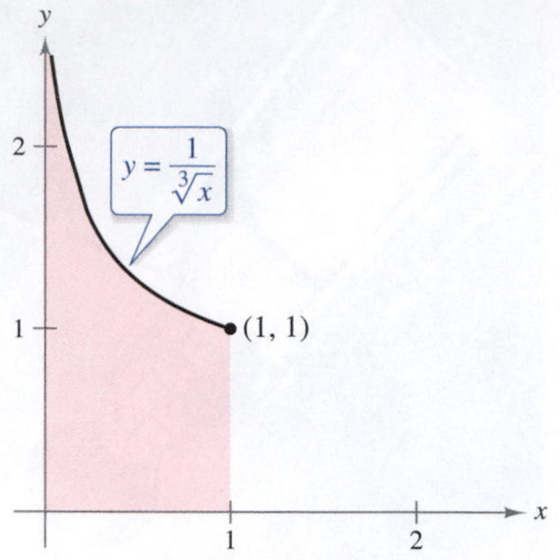

Infinite discontinuity at $x = 0$

Figure 8.23

EXAMPLE 7 **An Improper Integral That Diverges**

Evaluate $\displaystyle\int_0^2 \frac{dx}{x^3}$.

Solution Because the integrand has an infinite discontinuity at $x = 0$, you can write

$$\int_0^2 \frac{dx}{x^3} = \lim_{b\to 0^+}\left[-\frac{1}{2x^2}\right]_b^2$$
$$= \lim_{b\to 0^+}\left(-\frac{1}{8} + \frac{1}{2b^2}\right)$$
$$= \infty.$$

So, you can conclude that the improper integral diverges.

EXAMPLE 8 **An Improper Integral with an Interior Discontinuity**

Evaluate $\displaystyle\int_{-1}^2 \frac{dx}{x^3}$.

Solution This integral is improper because the integrand has an infinite discontinuity at the interior point $x = 0$, as shown in Figure 8.24. So, you can write

$$\int_{-1}^2 \frac{dx}{x^3} = \int_{-1}^0 \frac{dx}{x^3} + \int_0^2 \frac{dx}{x^3}.$$

From Example 7, you know that the second integral diverges. So, the original improper integral also diverges. ■

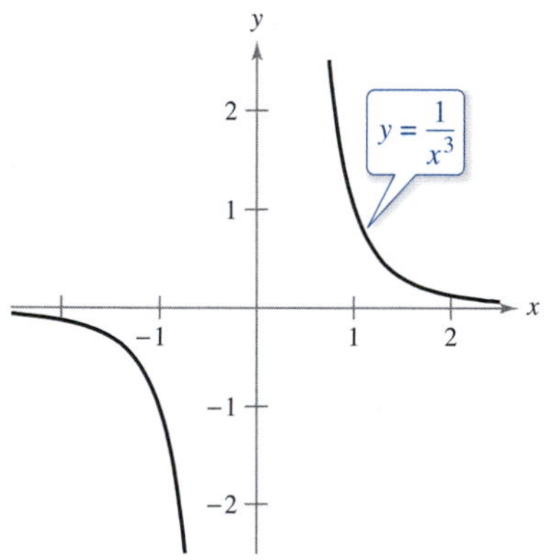

The improper integral $\displaystyle\int_{-1}^2 \frac{dx}{x^3}$ diverges.

Figure 8.24

Remember to check for infinite discontinuities at interior points as well as at endpoints when determining whether an integral is improper. For instance, if you had not recognized that the integral in Example 8 was improper, you would have obtained the *incorrect* result

$$\int_{-1}^2 \frac{dx}{x^3} = \left[\frac{-1}{2x^2}\right]_{-1}^2 = -\frac{1}{8} + \frac{1}{2} = \frac{3}{8}.$$

Incorrect evaluation

The integral in the next example is improper for *two* reasons. One limit of integration is infinite, and the integrand has an infinite discontinuity at the outer limit of integration.

EXAMPLE 9 A Doubly Improper Integral

See LarsonCalculus.com for an interactive version of this type of example.

Evaluate $\displaystyle\int_0^{\infty} \frac{dx}{\sqrt{x}(x+1)}$.

Solution To evaluate this integral, split it at a convenient point (say, $x = 1$) and write

$$\begin{aligned}\int_0^{\infty} \frac{dx}{\sqrt{x}(x+1)} &= \int_0^{1} \frac{dx}{\sqrt{x}(x+1)} + \int_1^{\infty} \frac{dx}{\sqrt{x}(x+1)}\\ &= \lim_{b\to 0^+}\Big[2\arctan\sqrt{x}\Big]_b^1 + \lim_{c\to\infty}\Big[2\arctan\sqrt{x}\Big]_1^c\\ &= \lim_{b\to 0^+}(2\arctan 1 - 2\arctan\sqrt{b}) + \lim_{c\to\infty}(2\arctan\sqrt{c} - 2\arctan 1)\\ &= 2\left(\frac{\pi}{4}\right) - 0 + 2\left(\frac{\pi}{2}\right) - 2\left(\frac{\pi}{4}\right)\\ &= \pi.\end{aligned}$$

See Figure 8.25.

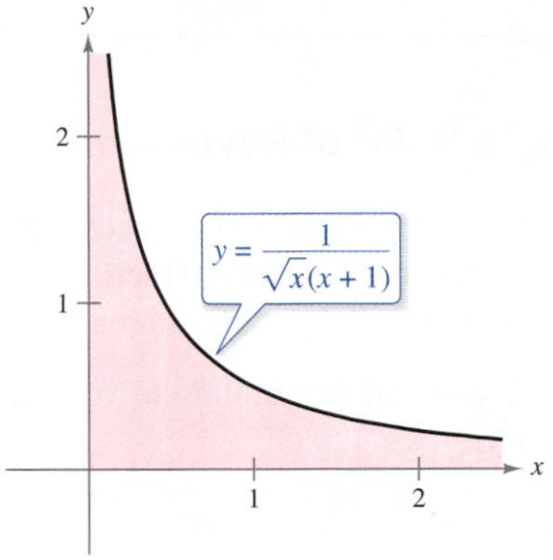

The area of the unbounded region is π.
Figure 8.25

EXAMPLE 10 An Application Involving Arc Length

Use the formula for arc length to show that the circumference of the circle $x^2 + y^2 = 1$ is 2π.

Solution To simplify the work, consider the quarter circle given by $y = \sqrt{1 - x^2}$, where $0 \le x \le 1$. The function y is differentiable for any x in this interval except $x = 1$. Therefore, the arc length of the quarter circle is given by the improper integral

$$\begin{aligned}s &= \int_0^1 \sqrt{1 + (y')^2}\,dx\\ &= \int_0^1 \sqrt{1 + \left(\frac{-x}{\sqrt{1-x^2}}\right)^2}\,dx\\ &= \int_0^1 \frac{dx}{\sqrt{1-x^2}}.\end{aligned}$$

This integral is improper because it has an infinite discontinuity at $x = 1$. So, you can write

$$\begin{aligned}s &= \int_0^1 \frac{dx}{\sqrt{1-x^2}}\\ &= \lim_{b\to 1^-}\Big[\arcsin x\Big]_0^b\\ &= \lim_{b\to 1^-}(\arcsin b - \arcsin 0)\\ &= \frac{\pi}{2} - 0\\ &= \frac{\pi}{2}.\end{aligned}$$

Finally, multiplying by 4, you can conclude that the circumference of the circle is $4s = 2\pi$, as shown in Figure 8.26. ■

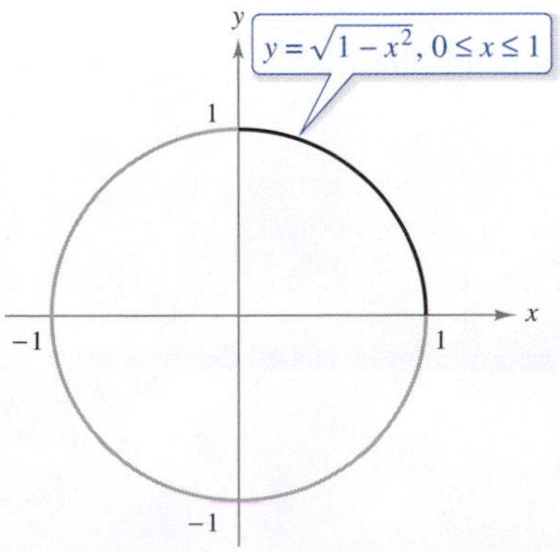

The circumference of the circle is 2π.
Figure 8.26

This section concludes with a useful theorem describing the convergence or divergence of a common type of improper integral. The proof of this theorem is left as an exercise (see Exercise 49).

THEOREM 8.5 A Special Type of Improper Integral

$$\int_1^{\infty} \frac{dx}{x^p} = \begin{cases} \dfrac{1}{p-1}, & p > 1 \\ \text{diverges}, & p \leq 1 \end{cases}$$

EXAMPLE 11 An Application Involving a Solid of Revolution

The solid formed by revolving (about the x-axis) the *unbounded* region lying between the graph of $f(x) = 1/x$ and the x-axis $(x \geq 1)$ is called **Gabriel's Horn.** (See Figure 8.27.) Show that this solid has a finite volume and an infinite surface area.

Solution Using the disk method and Theorem 8.5, you can determine the volume to be

$$\begin{aligned} V &= \pi \int_1^{\infty} \left(\frac{1}{x}\right)^2 dx \qquad \text{Theorem 8.5, } p = 2 > 1 \\ &= \pi\left(\frac{1}{2-1}\right) \\ &= \pi. \end{aligned}$$

The surface area is given by

$$S = 2\pi \int_1^{\infty} f(x)\sqrt{1 + [f'(x)]^2}\, dx = 2\pi \int_1^{\infty} \frac{1}{x}\sqrt{1 + \frac{1}{x^4}}\, dx.$$

Because

$$\sqrt{1 + \frac{1}{x^4}} > 1$$

on the interval $[1, \infty)$, and the improper integral

$$\int_1^{\infty} \frac{1}{x}\, dx$$

diverges, you can conclude that the improper integral

$$\int_1^{\infty} \frac{1}{x}\sqrt{1 + \frac{1}{x^4}}\, dx$$

also diverges. (See Exercise 52.) So, the surface area is infinite.

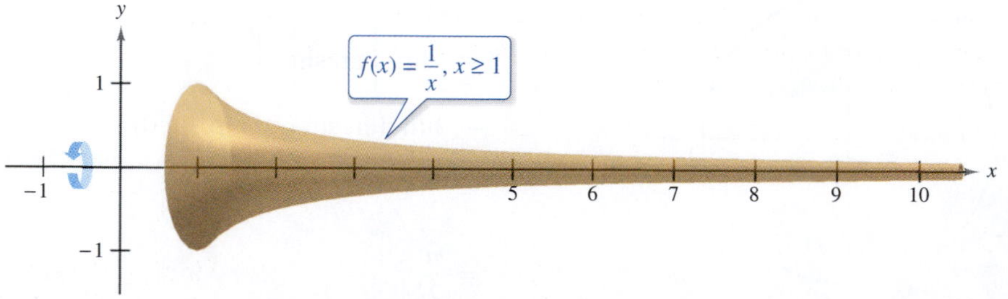

Gabriel's Horn has a finite volume and an infinite surface area.
Figure 8.27

FOR FURTHER INFORMATION For further investigation of solids that have finite volumes and infinite surface areas, see the article "Supersolids: Solids Having Finite Volume and Infinite Surfaces" by William P. Love in *Mathematics Teacher*. To view this article, go to *MathArticles.com*.

FOR FURTHER INFORMATION To learn about another function that has a finite volume and an infinite surface area, see the article "Gabriel's Wedding Cake" by Julian F. Fleron in *The College Mathematics Journal*. To view this article, go to *MathArticles.com*.

8.8 Exercises

See CalcChat.com for tutorial help and worked-out solutions to odd-numbered exercises.

Determining Whether an Integral Is Improper In Exercises 1–8, decide whether the integral is improper. Explain your reasoning.

1. $\displaystyle\int_0^1 \frac{dx}{5x-3}$

2. $\displaystyle\int_1^2 \frac{dx}{x^3}$

3. $\displaystyle\int_0^1 \frac{2x-5}{x^2-5x+6}\,dx$

4. $\displaystyle\int_1^\infty \ln(x^2)\,dx$

5. $\displaystyle\int_0^2 e^{-x}\,dx$

6. $\displaystyle\int_0^\infty \cos x\,dx$

7. $\displaystyle\int_{-\infty}^\infty \frac{\sin x}{4+x^2}\,dx$

8. $\displaystyle\int_0^{\pi/4} \csc x\,dx$

Evaluating an Improper Integral In Exercises 9–12, explain why the integral is improper and determine whether it diverges or converges. Evaluate the integral if it converges.

9. $\displaystyle\int_0^4 \frac{1}{\sqrt{x}}\,dx$

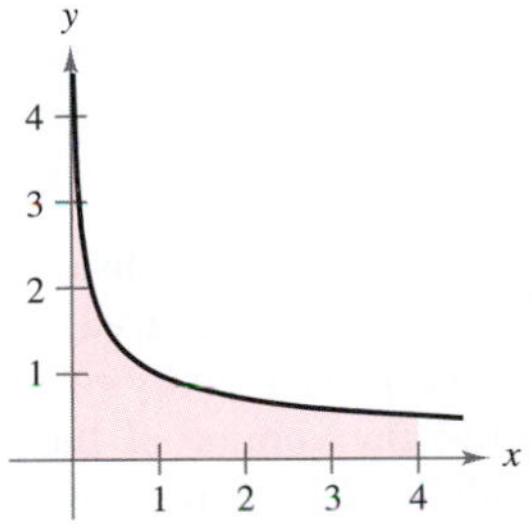

10. $\displaystyle\int_3^4 \frac{1}{(x-3)^{3/2}}\,dx$

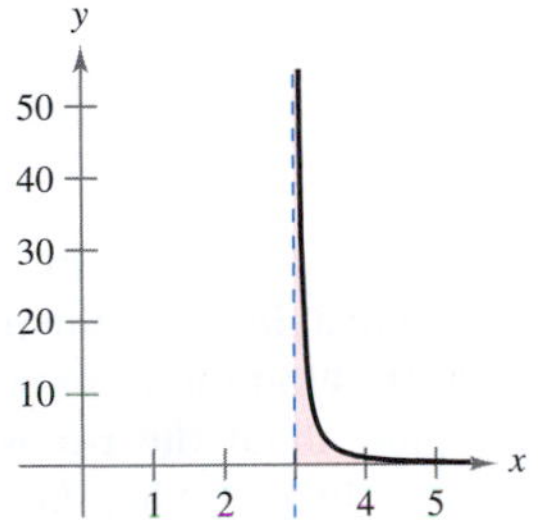

11. $\displaystyle\int_0^2 \frac{1}{(x-1)^2}\,dx$

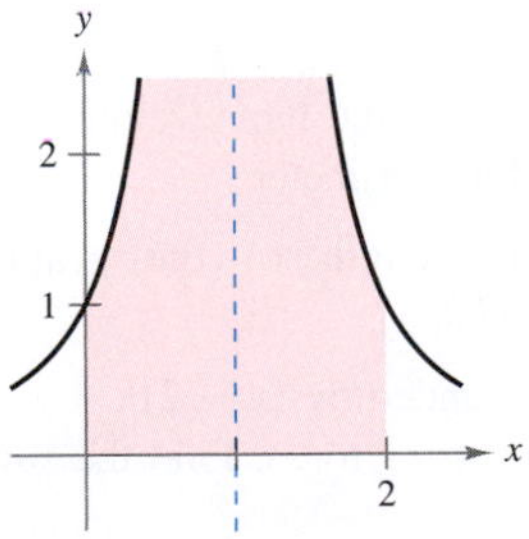

12. $\displaystyle\int_{-\infty}^0 e^{3x}\,dx$

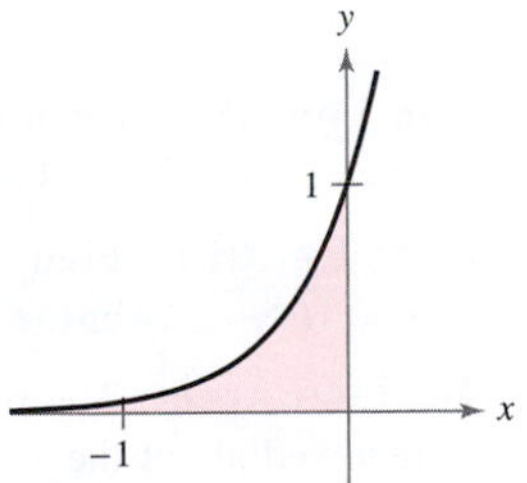

Writing In Exercises 13–16, explain why the evaluation of the integral is *incorrect*. Use the integration capabilities of a graphing utility to attempt to evaluate the integral. Determine whether the utility gives the correct answer.

13. $\displaystyle\int_{-1}^1 \frac{1}{x^2}\,dx = -2$

14. $\displaystyle\int_{-2}^2 \frac{-2}{(x-1)^3}\,dx = \frac{8}{9}$

15. $\displaystyle\int_0^\infty e^{-x}\,dx = 0$

16. $\displaystyle\int_0^\pi \sec x\,dx = 0$

Evaluating an Improper Integral In Exercises 17–32, determine whether the improper integral diverges or converges. Evaluate the integral if it converges.

17. $\displaystyle\int_1^\infty \frac{1}{x^3}\,dx$

18. $\displaystyle\int_1^\infty \frac{6}{x^4}\,dx$

19. $\displaystyle\int_1^\infty \frac{3}{\sqrt[3]{x}}\,dx$

20. $\displaystyle\int_1^\infty \frac{4}{\sqrt[4]{x}}\,dx$

21. $\displaystyle\int_{-\infty}^0 xe^{-4x}\,dx$

22. $\displaystyle\int_0^\infty xe^{-x/3}\,dx$

23. $\displaystyle\int_0^\infty x^2e^{-x}\,dx$

24. $\displaystyle\int_0^\infty e^{-x}\cos x\,dx$

25. $\displaystyle\int_4^\infty \frac{1}{x(\ln x)^3}\,dx$

26. $\displaystyle\int_1^\infty \frac{\ln x}{x}\,dx$

27. $\displaystyle\int_{-\infty}^\infty \frac{4}{16+x^2}\,dx$

28. $\displaystyle\int_0^\infty \frac{x^3}{(x^2+1)^2}\,dx$

29. $\displaystyle\int_0^\infty \frac{1}{e^x+e^{-x}}\,dx$

30. $\displaystyle\int_0^\infty \frac{e^x}{1+e^x}\,dx$

31. $\displaystyle\int_0^\infty \cos \pi x\,dx$

32. $\displaystyle\int_0^\infty \sin\frac{x}{2}\,dx$

Evaluating an Improper Integral In Exercises 33–48, determine whether the improper integral diverges or converges. Evaluate the integral if it converges, and check your results with the results obtained by using the integration capabilities of a graphing utility.

33. $\displaystyle\int_0^1 \frac{1}{x^2}\,dx$

34. $\displaystyle\int_0^5 \frac{10}{x}\,dx$

35. $\displaystyle\int_0^2 \frac{1}{\sqrt[3]{x-1}}\,dx$

36. $\displaystyle\int_0^8 \frac{3}{\sqrt{8-x}}\,dx$

37. $\displaystyle\int_0^1 x\ln x\,dx$

38. $\displaystyle\int_0^e \ln x^2\,dx$

39. $\displaystyle\int_0^{\pi/2} \tan\theta\,d\theta$

40. $\displaystyle\int_0^{\pi/2} \sec\theta\,d\theta$

41. $\displaystyle\int_2^4 \frac{2}{x\sqrt{x^2-4}}\,dx$

42. $\displaystyle\int_3^6 \frac{1}{\sqrt{36-x^2}}\,dx$

43. $\displaystyle\int_3^5 \frac{1}{\sqrt{x^2-9}}\,dx$

44. $\displaystyle\int_0^5 \frac{1}{25-x^2}\,dx$

45. $\displaystyle\int_3^\infty \frac{1}{x\sqrt{x^2-9}}\,dx$

46. $\displaystyle\int_4^\infty \frac{\sqrt{x^2-16}}{x^2}\,dx$

47. $\displaystyle\int_0^\infty \frac{4}{\sqrt{x}(x+6)}\,dx$

48. $\displaystyle\int_1^\infty \frac{1}{x\ln x}\,dx$

Finding Values In Exercises 49 and 50, determine all values of p for which the improper integral converges.

49. $\displaystyle\int_1^\infty \frac{1}{x^p}\,dx$

50. $\displaystyle\int_0^1 \frac{1}{x^p}\,dx$

51. Mathematical Induction Use mathematical induction to verify that the following integral converges for any positive integer n.

$$\int_0^{\infty} x^n e^{-x}\, dx$$

52. Comparison Test for Improper Integrals In some cases, it is impossible to find the exact value of an improper integral, but it is important to determine whether the integral converges or diverges. Suppose the functions f and g are continuous and $0 \le g(x) \le f(x)$ on the interval $[a, \infty)$. It can be shown that if $\int_a^{\infty} f(x)\, dx$ converges, then $\int_a^{\infty} g(x)\, dx$ also converges, and if $\int_a^{\infty} g(x)\, dx$ diverges, then $\int_a^{\infty} f(x)\, dx$ also diverges. This is known as the Comparison Test for improper integrals.

(a) Use the Comparison Test to determine whether $\int_1^{\infty} e^{-x^2}\, dx$ converges or diverges. (*Hint:* Use the fact that $e^{-x^2} \le e^{-x}$ for $x \ge 1$.)

(b) Use the Comparison Test to determine whether $\displaystyle\int_1^{\infty} \frac{1}{x^5 + 1}\, dx$ converges or diverges. (*Hint:* Use the fact that $\dfrac{1}{x^5 + 1} \le \dfrac{1}{x^5}$ for $x \ge 1$.)

Convergence or Divergence **In Exercises 53–62, use the results of Exercises 49–52 to determine whether the improper integral converges or diverges.**

53. $\displaystyle\int_0^1 \frac{1}{x^5}\, dx$

54. $\displaystyle\int_0^1 \frac{1}{\sqrt[5]{x}}\, dx$

55. $\displaystyle\int_1^{\infty} \frac{1}{x^5}\, dx$

56. $\displaystyle\int_0^{\infty} x^4 e^{-x}\, dx$

57. $\displaystyle\int_1^{\infty} \frac{1}{x^2 + 5}\, dx$

58. $\displaystyle\int_2^{\infty} \frac{1}{\sqrt{x - 1}}\, dx$

59. $\displaystyle\int_2^{\infty} \frac{1}{\sqrt[3]{x(x - 1)}}\, dx$

60. $\displaystyle\int_1^{\infty} \frac{1}{\sqrt{x}(x + 1)}\, dx$

61. $\displaystyle\int_1^{\infty} \frac{1 - \sin x}{x^2}\, dx$

62. $\displaystyle\int_0^{\infty} \frac{1}{e^x + x}\, dx$

WRITING ABOUT CONCEPTS

63. Improper Integrals Describe the different types of improper integrals.

64. Improper Integrals Define the terms *converges* and *diverges* when working with improper integrals.

65. Improper Integral Explain why $\displaystyle\int_{-1}^{1} \frac{1}{x^3}\, dx \ne 0$.

66. Improper Integral Consider the integral

$$\int_0^3 \frac{10}{x^2 - 2x}\, dx.$$

To determine the convergence or divergence of the integral, how many improper integrals must be analyzed? What must be true of each of these integrals if the given integral converges?

Area **In Exercises 67–70, find the area of the unbounded shaded region.**

67. $y = e^x, \quad -\infty < x \le 1$

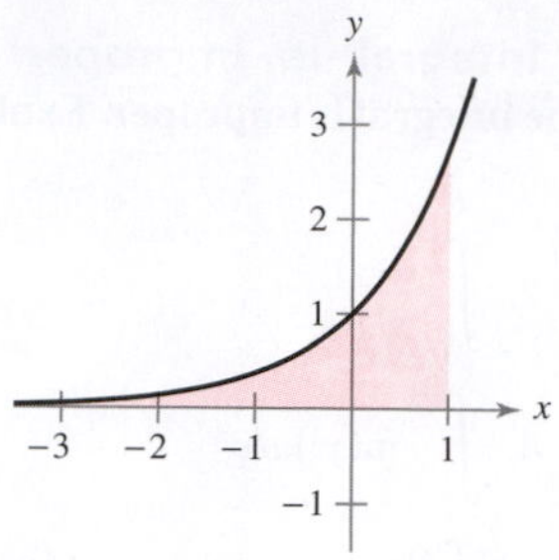

68. $y = -\ln x$

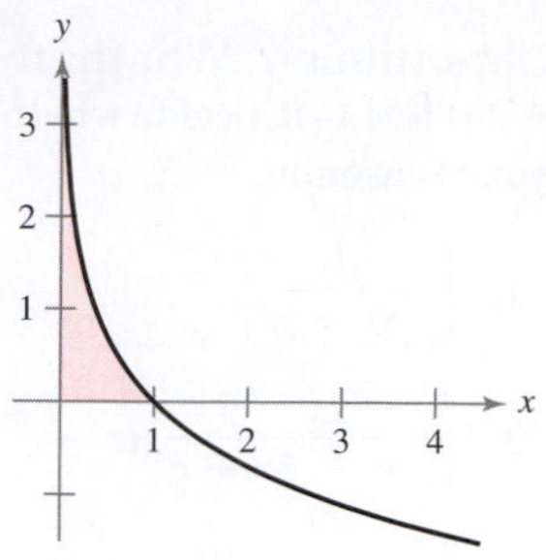

69. Witch of Agnesi:

$$y = \frac{1}{x^2 + 1}$$

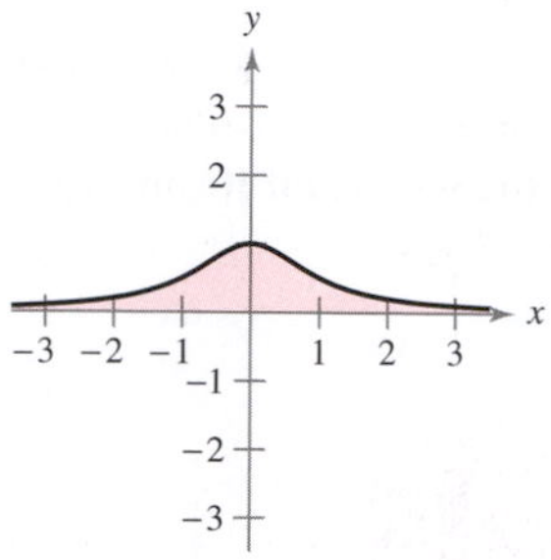

70. Witch of Agnesi:

$$y = \frac{8}{x^2 + 4}$$

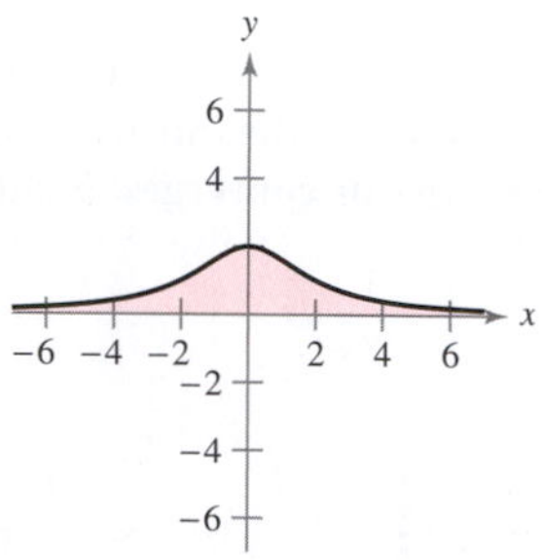

Area and Volume **In Exercises 71 and 72, consider the region satisfying the inequalities. (a) Find the area of the region. (b) Find the volume of the solid generated by revolving the region about the x-axis. (c) Find the volume of the solid generated by revolving the region about the y-axis.**

71. $y \le e^{-x}, \ y \ge 0, \ x \ge 0$

72. $y \le \dfrac{1}{x^2}, \ y \ge 0, \ x \ge 1$

73. Arc Length Sketch the graph of the hypocycloid of four cusps $x^{2/3} + y^{2/3} = 4$ and find its perimeter.

74. Arc Length Find the arc length of the graph of $y = \sqrt{16 - x^2}$ over the interval $[0, 4]$.

75. Surface Area The region bounded by $(x - 2)^2 + y^2 = 1$ is revolved about the y-axis to form a torus. Find the surface area of the torus.

76. Surface Area Find the area of the surface formed by revolving the graph of $y = 2e^{-x}$ on the interval $[0, \infty)$ about the x-axis.

Propulsion **In Exercises 77 and 78, use the weight of the rocket to answer each question. (Use 4000 miles as the radius of Earth and do not consider the effect of air resistance.)**

(a) How much work is required to propel the rocket an unlimited distance away from Earth's surface?

(b) How far has the rocket traveled when half the total work has occurred?

77. 5-ton rocket

78. 10-ton rocket

Probability **A nonnegative function f is called a *probability density function* if**

$$\int_{-\infty}^{\infty} f(t)\, dt = 1.$$

The probability that x lies between a and b is given by

$$P(a \le x \le b) = \int_a^b f(t)\, dt.$$

The expected value of x is given by

$$E(x) = \int_{-\infty}^{\infty} t f(t)\, dt.$$

In Exercises 79 and 80, (a) show that the nonnegative function is a probability density function, (b) find $P(0 \le x \le 4)$, and (c) find $E(x)$.

79. $f(t) = \begin{cases} \frac{1}{7}e^{-t/7}, & t \ge 0 \\ 0, & t < 0 \end{cases}$

80. $f(t) = \begin{cases} \frac{2}{5}e^{-2t/5}, & t \ge 0 \\ 0, & t < 0 \end{cases}$

Capitalized Cost **In Exercises 81 and 82, find the capitalized cost C of an asset (a) for $n = 5$ years, (b) for $n = 10$ years, and (c) forever. The capitalized cost is given by**

$$C = C_0 + \int_0^n c(t)e^{-rt}\, dt$$

where C_0 is the original investment, t is the time in years, r is the annual interest rate compounded continuously, and $c(t)$ is the annual cost of maintenance.

81. $C_0 = \$650{,}000$

$c(t) = \$25{,}000$

$r = 0.06$

82. $C_0 = \$650{,}000$

$c(t) = \$25{,}000(1 + 0.08t)$

$r = 0.06$

83. Electromagnetic Theory The magnetic potential P at a point on the axis of a circular coil is given by

$$P = \frac{2\pi NIr}{k} \int_c^{\infty} \frac{1}{(r^2 + x^2)^{3/2}}\, dx$$

where N, I, r, k, and c are constants. Find P.

84. Gravitational Force A "semi-infinite" uniform rod occupies the nonnegative x-axis. The rod has a linear density δ, which means that a segment of length dx has a mass of $\delta\, dx$. A particle of mass M is located at the point $(-a, 0)$. The gravitational force F that the rod exerts on the mass is given by

$$F = \int_0^{\infty} \frac{GM\delta}{(a + x)^2}\, dx$$

where G is the gravitational constant. Find F.

True or False? **In Exercises 85–88, determine whether the statement is true or false. If it is false, explain why or give an example that shows it is false.**

85. If f is continuous on $[0, \infty)$ and $\lim_{x\to\infty} f(x) = 0$, then $\int_0^{\infty} f(x)\, dx$ converges.

86. If f is continuous on $[0, \infty)$ and $\int_0^{\infty} f(x)\, dx$ diverges, then $\lim_{x\to\infty} f(x) \ne 0$.

87. If f' is continuous on $[0, \infty)$ and $\lim_{x\to\infty} f(x) = 0$, then

$$\int_0^{\infty} f'(x)\, dx = -f(0).$$

88. If the graph of f is symmetric with respect to the origin or the y-axis, then $\int_0^{\infty} f(x)\, dx$ converges if and only if $\int_{-\infty}^{\infty} f(x)\, dx$ converges.

89. Comparing Integrals

(a) Show that $\int_{-\infty}^{\infty} \sin x\, dx$ diverges.

(b) Show that $\lim_{a\to\infty} \int_{-a}^{a} \sin x\, dx = 0$.

(c) What do parts (a) and (b) show about the definition of improper integrals?

90. Making an Integral Improper For each integral, find a nonnegative real number b that makes the integral improper. Explain your reasoning.

(a) $\displaystyle\int_0^b \frac{1}{x^2 - 9}\, dx$

(b) $\displaystyle\int_0^b \frac{1}{\sqrt{4 - x}}\, dx$

(c) $\displaystyle\int_0^b \frac{x}{x^2 - 7x + 12}\, dx$

(d) $\displaystyle\int_b^{10} \ln x\, dx$

(e) $\displaystyle\int_0^b \tan 2x\, dx$

(f) $\displaystyle\int_0^b \frac{\cos x}{1 - \sin x}\, dx$

91. Writing

(a) The improper integrals

$$\int_1^{\infty} \frac{1}{x}\, dx \quad \text{and} \quad \int_1^{\infty} \frac{1}{x^2}\, dx$$

diverge and converge, respectively. Describe the essential differences between the integrands that cause one integral to converge and the other to diverge.

(b) Sketch a graph of the function $y = (\sin x)/x$ over the interval $(1, \infty)$. Use your knowledge of the definite integral to make an inference as to whether the integral

$$\int_1^{\infty} \frac{\sin x}{x}\, dx$$

converges. Give reasons for your answer.

(c) Use one iteration of integration by parts on the integral in part (b) to determine its divergence or convergence.

92. Exploration Consider the integral

$$\int_0^{\pi/2} \frac{4}{1 + (\tan x)^n}\, dx$$

where n is a positive integer.

(a) Is the integral improper? Explain.

(b) Use a graphing utility to graph the integrand for $n = 2$, 4, 8, and 12.

(c) Use the graphs to approximate the integral as $n \to \infty$.

(d) Use a computer algebra system to evaluate the integral for the values of n in part (b). Make a conjecture about the value of the integral for any positive integer n. Compare your results with your answer in part (c).

93. Normal Probability The mean height of American men between 20 and 29 years old is 70 inches, and the standard deviation is 2.85 inches. A 20- to 29-year-old man is chosen at random from the population. The probability that he is 6 feet tall or taller is

$$P(72 \le x < \infty) = \int_{72}^{\infty} \frac{1}{2.85\sqrt{2\pi}} e^{-(x-70)^2/6.245}\, dx.$$

(*Source: National Center for Health Statistics*)

(a) Use a graphing utility to graph the integrand. Use the graphing utility to convince yourself that the area between the x-axis and the integrand is 1.

(b) Use a graphing utility to approximate $P(72 \le x < \infty)$.

(c) Approximate $0.5 - P(70 \le x \le 72)$ using a graphing utility. Use the graph in part (a) to explain why this result is the same as the answer in part (b).

94. HOW DO YOU SEE IT? The graph shows the probability density function for a car brand that has a mean fuel efficiency of 26 miles per gallon and a standard deviation of 2.4 miles per gallon.

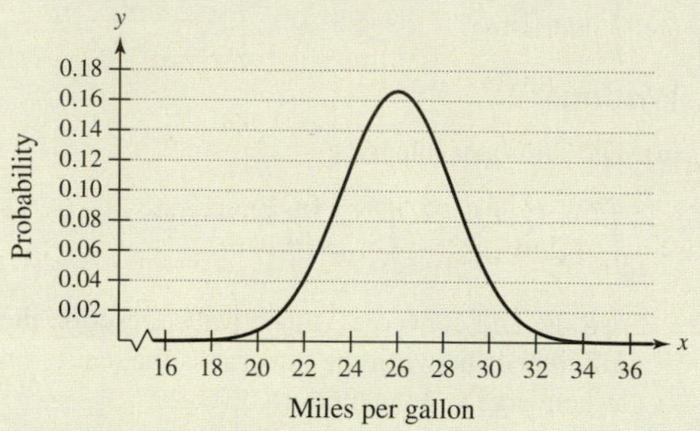

(a) Which is greater, the probability of choosing a car at random that gets between 26 and 28 miles per gallon or the probability of choosing a car at random that gets between 22 and 24 miles per gallon?

(b) Which is greater, the probability of choosing a car at random that gets between 20 and 22 miles per gallon or the probability of choosing a car at random that gets at least 30 miles per gallon?

Laplace Transforms **Let $f(t)$ be a function defined for all positive values of t. The Laplace Transform of $f(t)$ is defined by**

$$F(s) = \int_0^{\infty} e^{-st} f(t)\, dt$$

when the improper integral exists. Laplace Transforms are used to solve differential equations. In Exercises 95–102, find the Laplace Transform of the function.

95. $f(t) = 1$

96. $f(t) = t$

97. $f(t) = t^2$

98. $f(t) = e^{at}$

99. $f(t) = \cos at$

100. $f(t) = \sin at$

101. $f(t) = \cosh at$

102. $f(t) = \sinh at$

103. The Gamma Function The Gamma Function $\Gamma(n)$ is defined by

$$\Gamma(n) = \int_0^{\infty} x^{n-1} e^{-x}\, dx, \quad n > 0.$$

(a) Find $\Gamma(1)$, $\Gamma(2)$, and $\Gamma(3)$.

(b) Use integration by parts to show that $\Gamma(n+1) = n\Gamma(n)$.

(c) Write $\Gamma(n)$ using factorial notation where n is a positive integer.

104. Proof Prove that $I_n = \left(\frac{n-1}{n+2}\right) I_{n-1}$, where

$$I_n = \int_0^{\infty} \frac{x^{2n-1}}{(x^2+1)^{n+3}}\, dx, \quad n \ge 1.$$

Then evaluate each integral.

(a) $\int_0^{\infty} \frac{x}{(x^2+1)^4}\, dx$

(b) $\int_0^{\infty} \frac{x^3}{(x^2+1)^5}\, dx$

(c) $\int_0^{\infty} \frac{x^5}{(x^2+1)^6}\, dx$

105. Finding a Value For what value of c does the integral

$$\int_0^{\infty} \left(\frac{1}{\sqrt{x^2+1}} - \frac{c}{x+1}\right) dx$$

converge? Evaluate the integral for this value of c.

106. Finding a Value For what value of c does the integral

$$\int_1^{\infty} \left(\frac{cx}{x^2+2} - \frac{1}{3x}\right) dx$$

converge? Evaluate the integral for this value of c.

107. Volume Find the volume of the solid generated by revolving the region bounded by the graph of f about the x-axis.

$$f(x) = \begin{cases} x \ln x, & 0 < x \le 2 \\ 0, & x = 0 \end{cases}$$

108. Volume Find the volume of the solid generated by revolving the unbounded region lying between $y = -\ln x$ and the y-axis ($y \ge 0$) about the x-axis.

***u*-Substitution** **In Exercises 109 and 110, rewrite the improper integral as a proper integral using the given u-substitution. Then use the Trapezoidal Rule with $n = 5$ to approximate the integral.**

109. $\int_0^1 \frac{\sin x}{\sqrt{x}}\, dx, \quad u = \sqrt{x}$

110. $\int_0^1 \frac{\cos x}{\sqrt{1-x}}\, dx, \quad u = \sqrt{1-x}$

111. Rewriting an Integral Let $\int_{-\infty}^{\infty} f(x)\, dx$ be convergent and let a and b be real numbers where $a \ne b$. Show that

$$\int_{-\infty}^{a} f(x)\, dx + \int_a^{\infty} f(x)\, dx = \int_{-\infty}^{b} f(x)\, dx + \int_b^{\infty} f(x)\, dx.$$

Review Exercises

See CalcChat.com for tutorial help and worked-out solutions to odd-numbered exercises.

Finding or Evaluating an Integral In Exercises 1–8, use the basic integration rules to find or evaluate the integral.

1. $\int x\sqrt{x^2 - 36}\, dx$

2. $\int xe^{x^2-1}\, dx$

3. $\int \frac{x}{x^2 - 49}\, dx$

4. $\int \frac{x}{\sqrt[3]{4 - x^2}}\, dx$

5. $\int_1^e \frac{\ln(2x)}{x}\, dx$

6. $\int_{3/2}^{2} 2x\sqrt{2x - 3}\, dx$

7. $\int \frac{100}{\sqrt{100 - x^2}}\, dx$

8. $\int \frac{2x}{x - 3}\, dx$

Using Integration by Parts In Exercises 9–16, use integration by parts to find the indefinite integral.

9. $\int xe^{3x}\, dx$

10. $\int x^3 e^x\, dx$

11. $\int e^{2x} \sin 3x\, dx$

12. $\int x\sqrt{x - 1}\, dx$

13. $\int x^2 \sin 2x\, dx$

14. $\int \ln\sqrt{x^2 - 4}\, dx$

15. $\int x \arcsin 2x\, dx$

16. $\int \arctan 2x\, dx$

Finding a Trigonometric Integral In Exercises 17–22, find the trigonometric integral.

17. $\int \cos^3(\pi x - 1)\, dx$

18. $\int \sin^2 \frac{\pi x}{2}\, dx$

19. $\int \sec^4 \frac{x}{2}\, dx$

20. $\int \tan\theta \sec^4\theta\, d\theta$

21. $\int \frac{1}{1 - \sin\theta}\, d\theta$

22. $\int \cos 2\theta(\sin\theta + \cos\theta)^2\, d\theta$

Area In Exercises 23 and 24, find the area of the region.

23. $y = \sin^4 x$

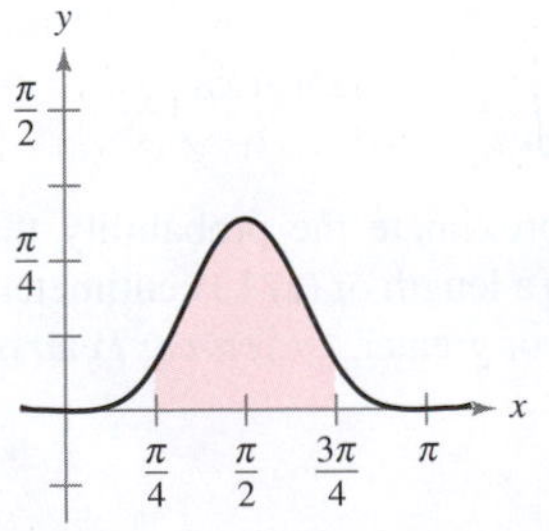

24. $y = \sin 3x \cos 2x$

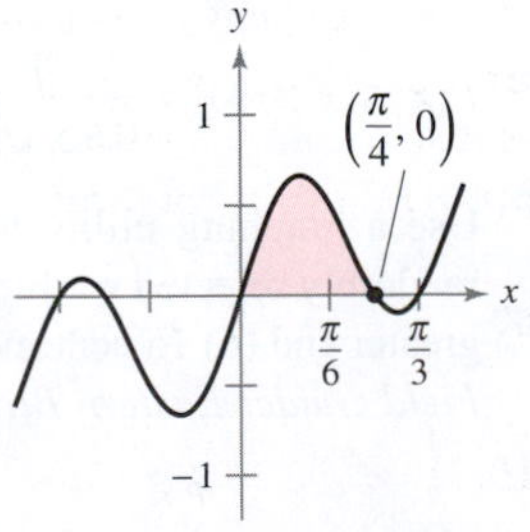

Using Trigonometric Substitution In Exercises 25–30, use trigonometric substitution to find or evaluate the integral.

25. $\int \frac{-12}{x^2\sqrt{4 - x^2}}\, dx$

26. $\int \frac{\sqrt{x^2 - 9}}{x}\, dx, \quad x > 3$

27. $\int \frac{x^3}{\sqrt{4 + x^2}}\, dx$

28. $\int \sqrt{25 - 9x^2}\, dx$

29. $\int_0^1 \frac{6x^3}{\sqrt{16 + x^2}}\, dx$

30. $\int_3^4 x^3\sqrt{x^2 - 9}\, dx$

Using Different Methods In Exercises 31 and 32, find the indefinite integral using each method.

31. $\int \frac{x^3}{\sqrt{4 + x^2}}\, dx$

(a) Trigonometric substitution

(b) Substitution: $u^2 = 4 + x^2$

(c) Integration by parts: $dv = \frac{x}{\sqrt{4 + x^2}}\, dx$

32. $\int x\sqrt{4 + x}\, dx$

(a) Trigonometric substitution

(b) Substitution: $u^2 = 4 + x$

(c) Substitution: $u = 4 + x$

(d) Integration by parts: $dv = \sqrt{4 + x}\, dx$

Using Partial Fractions In Exercises 33–38, use partial fractions to find the indefinite integral.

33. $\int \frac{x - 39}{x^2 - x - 12}\, dx$

34. $\int \frac{5x - 2}{x^2 - x}\, dx$

35. $\int \frac{x^2 + 2x}{x^3 - x^2 + x - 1}\, dx$

36. $\int \frac{4x - 2}{3(x - 1)^2}\, dx$

37. $\int \frac{x^2}{x^2 + 5x - 24}\, dx$

38. $\int \frac{\sec^2\theta}{\tan\theta(\tan\theta - 1)}\, d\theta$

Integration by Tables In Exercises 39–46, use integration tables to find or evaluate the integral.

39. $\int \frac{x}{(4 + 5x)^2}\, dx$

40. $\int \frac{x}{\sqrt{4 + 5x}}\, dx$

41. $\int_0^{\sqrt{\pi}/2} \frac{x}{1 + \sin x^2}\, dx$

42. $\int_0^1 \frac{x}{1 + e^{x^2}}\, dx$

43. $\int \frac{x}{x^2 + 4x + 8}\, dx$

44. $\int \frac{3}{2x\sqrt{9x^2 - 1}}\, dx, \quad x > \frac{1}{3}$

45. $\int \frac{1}{\sin \pi x \cos \pi x}\, dx$

46. $\int \frac{1}{1 + \tan \pi x}\, dx$

47. **Verifying a Formula** Verify the reduction formula

$$\int (\ln x)^n\, dx = x(\ln x)^n - n\int (\ln x)^{n-1}\, dx.$$

48. **Verifying a Formula** Verify the reduction formula

$$\int \tan^n x\, dx = \frac{1}{n - 1}\tan^{n-1} x - \int \tan^{n-2} x\, dx.$$

Finding an Indefinite Integral In Exercises 49–56, find the indefinite integral using any method.

49. $\int \theta \sin \theta \cos \theta \, d\theta$

50. $\int \frac{\csc \sqrt{2x}}{\sqrt{x}} \, dx$

51. $\int \frac{x^{1/4}}{1 + x^{1/2}} \, dx$

52. $\int \sqrt{1 + \sqrt{x}} \, dx$

53. $\int \sqrt{1 + \cos x} \, dx$

54. $\int \frac{3x^3 + 4x}{(x^2 + 1)^2} \, dx$

55. $\int \cos x \ln(\sin x) \, dx$

56. $\int (\sin \theta + \cos \theta)^2 \, d\theta$

Differential Equation In Exercises 57–60, solve the differential equation using any method.

57. $\frac{dy}{dx} = \frac{25}{x^2 - 25}$

58. $\frac{dy}{dx} = \frac{\sqrt{4 - x^2}}{2x}$

59. $y' = \ln(x^2 + x)$

60. $y' = \sqrt{1 - \cos \theta}$

Evaluating a Definite Integral In Exercises 61–66, evaluate the definite integral using any method. Use a graphing utility to verify your result.

61. $\int_2^{\sqrt{5}} x(x^2 - 4)^{3/2} \, dx$

62. $\int_0^1 \frac{x}{(x - 2)(x - 4)} \, dx$

63. $\int_1^4 \frac{\ln x}{x} \, dx$

64. $\int_0^2 xe^{3x} \, dx$

65. $\int_0^{\pi} x \sin x \, dx$

66. $\int_0^5 \frac{x}{\sqrt{4 + x}} \, dx$

Area In Exercises 67 and 68, find the area of the region.

67. $y = x\sqrt{4 - x}$

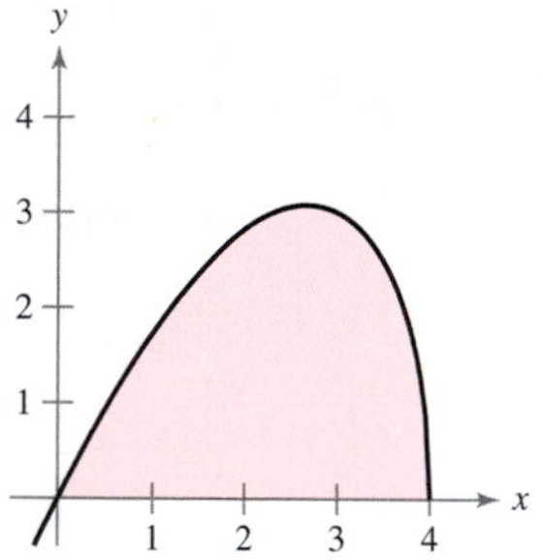

68. $y = \frac{1}{25 - x^2}$

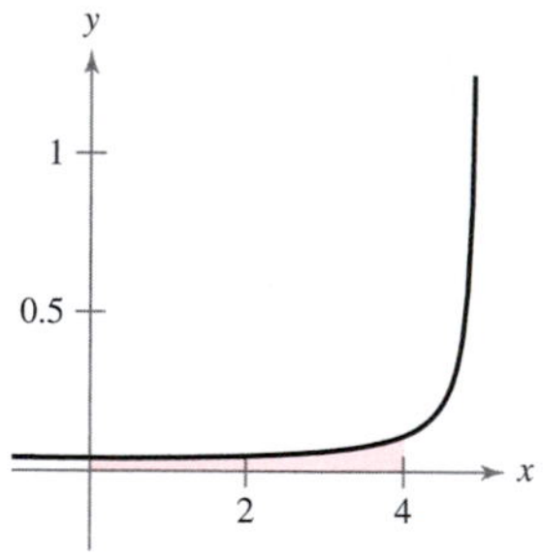

Centroid In Exercises 69 and 70, find the centroid of the region bounded by the graphs of the equations.

69. $y = \sqrt{1 - x^2}, \quad y = 0$

70. $(x - 1)^2 + y^2 = 1, \quad (x - 4)^2 + y^2 = 4$

Arc Length In Exercises 71 and 72, approximate to two decimal places the arc length of the curve over the given interval.

	Function	Interval
71.	$y = \sin x$	$[0, \pi]$
72.	$y = \sin^2 x$	$[0, \pi]$

Evaluating a Limit In Exercises 73–80, use L'Hôpital's Rule to evaluate the limit.

73. $\lim_{x \to 1} \frac{(\ln x)^2}{x - 1}$

74. $\lim_{x \to 0} \frac{\sin \pi x}{\sin 5\pi x}$

75. $\lim_{x \to \infty} \frac{e^{2x}}{x^2}$

76. $\lim_{x \to \infty} xe^{-x^2}$

77. $\lim_{x \to \infty} (\ln x)^{2/x}$

78. $\lim_{x \to 1^+} (x - 1)^{\ln x}$

79. $\lim_{n \to \infty} 1000\left(1 + \frac{0.09}{n}\right)^n$

80. $\lim_{x \to 1^+} \left(\frac{2}{\ln x} - \frac{2}{x - 1}\right)$

Evaluating an Improper Integral In Exercises 81–88, determine whether the improper integral diverges or converges. Evaluate the integral if it converges.

81. $\int_0^{16} \frac{1}{\sqrt[4]{x}} \, dx$

82. $\int_0^2 \frac{7}{x - 2} \, dx$

83. $\int_1^{\infty} x^2 \ln x \, dx$

84. $\int_0^{\infty} \frac{e^{-1/x}}{x^2} \, dx$

85. $\int_1^{\infty} \frac{\ln x}{x^2} \, dx$

86. $\int_1^{\infty} \frac{1}{\sqrt[4]{x}} \, dx$

87. $\int_2^{\infty} \frac{1}{x\sqrt{x^2 - 4}} \, dx$

88. $\int_0^{\infty} \frac{2}{\sqrt{x}(x + 4)} \, dx$

89. **Present Value** The board of directors of a corporation is calculating the price to pay for a business that is forecast to yield a continuous flow of profit of \$500,000 per year. The money will earn a nominal rate of 5% per year compounded continuously. What is the present value of the business

(a) for 20 years?

(b) forever (in perpetuity)?

(*Note:* The present value for t_0 years is $\int_0^{t_0} 500{,}000e^{-0.05t} \, dt$.)

90. **Volume** Find the volume of the solid generated by revolving the region bounded by the graphs of $y = xe^{-x}$, $y = 0$, and $x = 0$ about the x-axis.

91. **Probability** The average lengths (from beak to tail) of different species of warblers in the eastern United States are approximately normally distributed with a mean of 12.9 centimeters and a standard deviation of 0.95 centimeter (see figure). The probability that a randomly selected warbler has a length between a and b centimeters is

$$P(a \le x \le b) = \frac{1}{0.95\sqrt{2\pi}} \int_a^b e^{-(x - 12.9)^2/1.805} \, dx.$$

Use a graphing utility to approximate the probability that a randomly selected warbler has a length of (a) 13 centimeters or greater and (b) 15 centimeters or greater. (*Source: Peterson's Field Guide: Eastern Birds*)

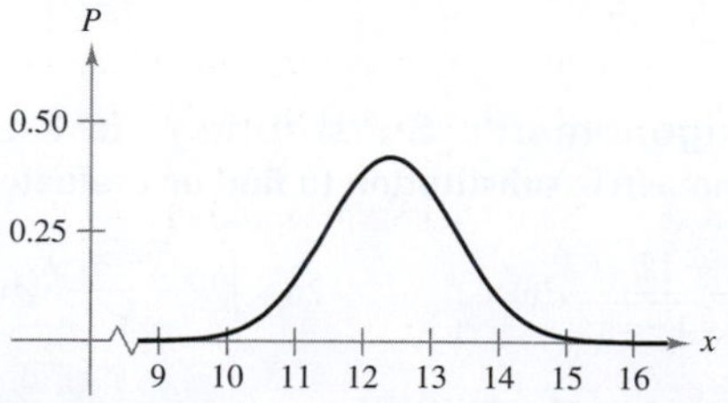

P.S. Problem Solving

See CalcChat.com for tutorial help and worked-out solutions to odd-numbered exercises.

1. Wallis's Formulas

(a) Evaluate the integrals

$$\int_{-1}^{1} (1 - x^2)\, dx \quad \text{and} \quad \int_{-1}^{1} (1 - x^2)^2\, dx.$$

(b) Use Wallis's Formulas to prove that

$$\int_{-1}^{1} (1 - x^2)^n\, dx = \frac{2^{2n+1}(n!)^2}{(2n + 1)!}$$

for all positive integers n.

2. Proof

(a) Evaluate the integrals

$$\int_{0}^{1} \ln x\, dx \quad \text{and} \quad \int_{0}^{1} (\ln x)^2\, dx.$$

(b) Prove that

$$\int_{0}^{1} (\ln x)^n\, dx = (-1)^n n!$$

for all positive integers n.

3. Finding a Value Find the value of the positive constant c such that

$$\lim_{x\to\infty} \left(\frac{x + c}{x - c}\right)^x = 9.$$

4. Finding a Value Find the value of the positive constant c such that

$$\lim_{x\to\infty} \left(\frac{x - c}{x + c}\right)^x = \frac{1}{4}.$$

5. Length The line $x = 1$ is tangent to the unit circle at A. The length of segment QA equals the length of the circular arc $\overset{\frown}{PA}$ (see figure). Show that the length of segment OR approaches 2 as P approaches A.

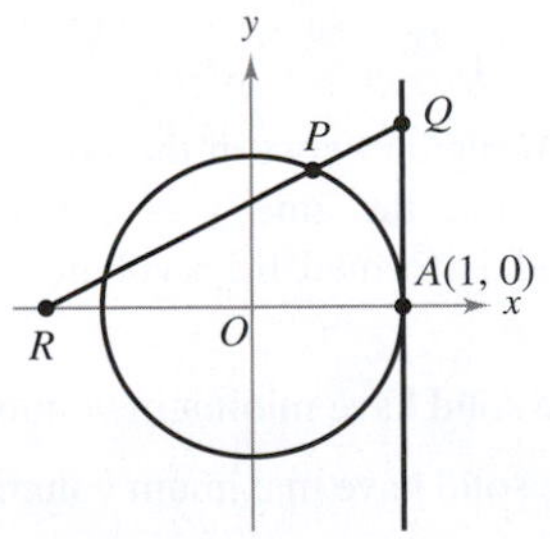

Figure for 5

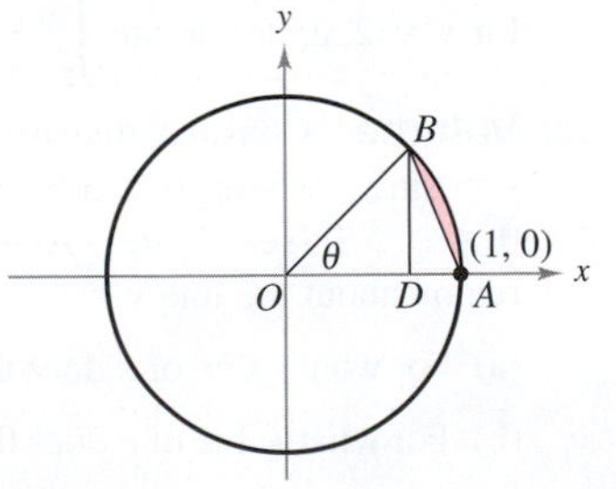

Figure for 6

6. Finding a Limit The segment BD is the height of $\triangle OAB$. Let R be the ratio of the area of $\triangle DAB$ to that of the shaded region formed by deleting $\triangle OAB$ from the circular sector subtended by angle θ (see figure). Find $\lim\limits_{\theta\to 0^+} R$.

7. Area Consider the problem of finding the area of the region bounded by the x-axis, the line $x = 4$, and the curve

$$y = \frac{x^2}{(x^2 + 9)^{3/2}}.$$

(a) Use a graphing utility to graph the region and approximate its area.

(b) Use an appropriate trigonometric substitution to find the exact area.

(c) Use the substitution $x = 3 \sinh u$ to find the exact area and verify that you obtain the same answer as in part (b).

8. Area Use the substitution $u = \tan (x/2)$ to find the area of the shaded region under the graph of $y = \dfrac{1}{2 + \cos x}$ for $0 \le x \le \pi/2$ (see figure).

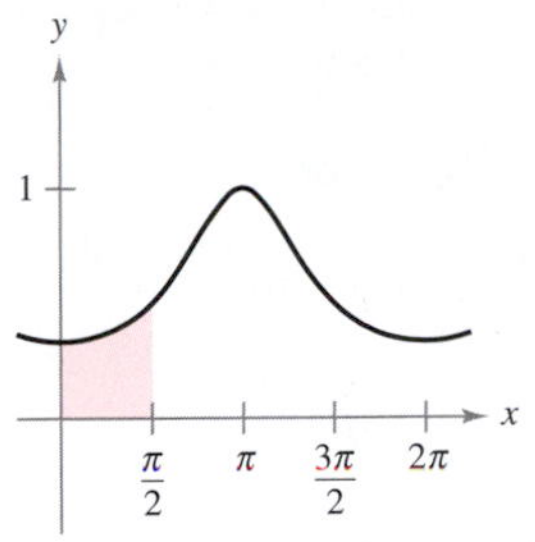

Figure for 8

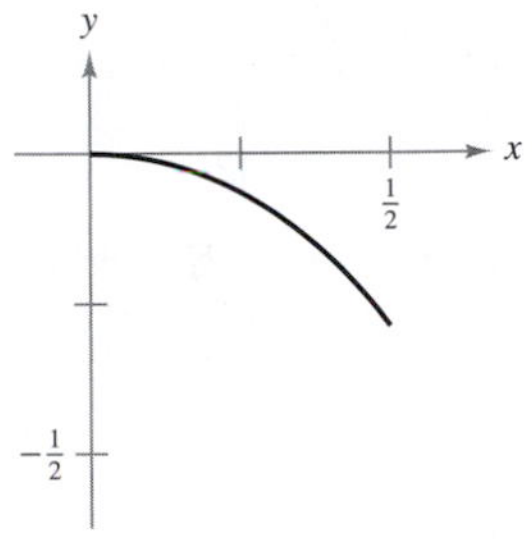

Figure for 9

9. Arc Length Find the arc length of the graph of the function $y = \ln(1 - x^2)$ on the interval $0 \le x \le \frac{1}{2}$ (see figure).

10. Centroid Find the centroid of the region above the x-axis and bounded above by the curve $y = e^{-c^2x^2}$, where c is a positive constant (see figure).

$$\left(\textit{Hint:} \text{ Show that } \int_0^\infty e^{-c^2x^2}\, dx = \frac{1}{c}\int_0^\infty e^{-x^2}\, dx.\right)$$

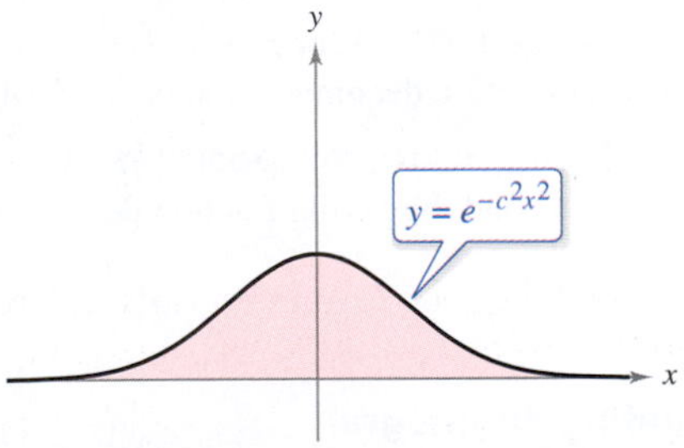

11. Finding Limits Use a graphing utility to estimate each limit. Then calculate each limit using L'Hôpital's Rule. What can you conclude about the form $0 \cdot \infty$?

(a) $\lim\limits_{x\to 0^+} \left(\cot x + \dfrac{1}{x}\right)$ (b) $\lim\limits_{x\to 0^+} \left(\cot x - \dfrac{1}{x}\right)$

(c) $\lim\limits_{x\to 0^+} \left[\left(\cot x + \dfrac{1}{x}\right)\left(\cot x - \dfrac{1}{x}\right)\right]$

12. Inverse Function and Area

(a) Let $y = f^{-1}(x)$ be the inverse function of f. Use integration by parts to derive the formula

$$\int f^{-1}(x)\,dx = xf^{-1}(x) - \int f(y)\,dy.$$

(b) Use the formula in part (a) to find the integral

$$\int \arcsin x\,dx.$$

(c) Use the formula in part (a) to find the area under the graph of $y = \ln x$, $1 \le x \le e$ (see figure).

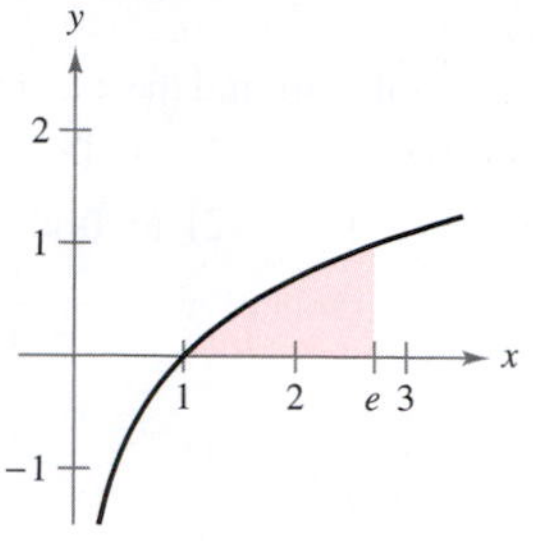

Figure for 12

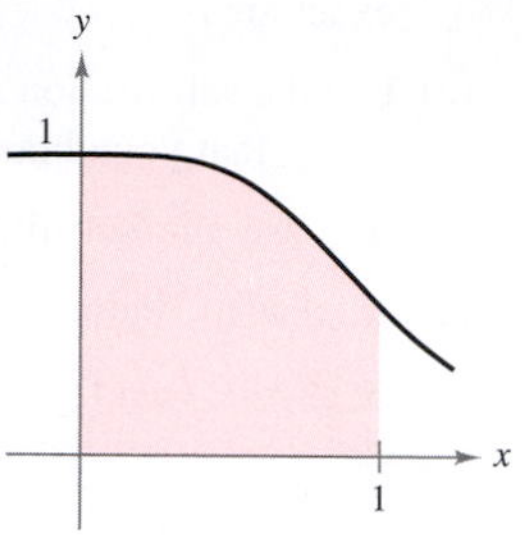

Figure for 13

13. Area Factor the polynomial $p(x) = x^4 + 1$ and then find the area under the graph of

$$y = \frac{1}{x^4 + 1}, \quad 0 \le x \le 1 \quad \text{(see figure).}$$

14. Partial Fraction Decomposition Suppose the denominator of a rational function can be factored into distinct linear factors

$$D(x) = (x - c_1)(x - c_2) \cdot \cdot \cdot (x - c_n)$$

for a positive integer n and distinct real numbers $c_1, c_2, \ldots, c_n$. If N is a polynomial of degree less than n, show that

$$\frac{N(x)}{D(x)} = \frac{P_1}{x - c_1} + \frac{P_2}{x - c_2} + \cdot \cdot \cdot + \frac{P_n}{x - c_n}$$

where $P_k = N(c_k)/D'(c_k)$ for $k = 1, 2, \ldots, n$. Note that this is the partial fraction decomposition of $N(x)/D(x)$.

15. Partial Fraction Decomposition Use the result of Exercise 14 to find the partial fraction decomposition of

$$\frac{x^3 - 3x^2 + 1}{x^4 - 13x^2 + 12x}.$$

16. Evaluating an Integral

(a) Use the substitution $u = \frac{\pi}{2} - x$ to evaluate the integral

$$\int_0^{\pi/2} \frac{\sin x}{\cos x + \sin x}\,dx.$$

(b) Let n be a positive integer. Evaluate the integral

$$\int_0^{\pi/2} \frac{\sin^n x}{\cos^n x + \sin^n x}\,dx.$$

17. Elementary Functions Some elementary functions, such as $f(x) = \sin(x^2)$, do not have antiderivatives that are elementary functions. Joseph Liouville proved that

$$\int \frac{e^x}{x}\,dx$$

does not have an elementary antiderivative. Use this fact to prove that

$$\int \frac{1}{\ln x}\,dx$$

is not elementary.

18. Rocket The velocity v (in feet per second) of a rocket whose initial mass (including fuel) is m is given by

$$v = gt + u \ln \frac{m}{m - rt}, \quad t < \frac{m}{r}$$

where u is the expulsion speed of the fuel, r is the rate at which the fuel is consumed, and $g = -32$ feet per second per second is the acceleration due to gravity. Find the position equation for a rocket for which $m = 50{,}000$ pounds, $u = 12{,}000$ feet per second, and $r = 400$ pounds per second. What is the height of the rocket when $t = 100$ seconds? (Assume that the rocket was fired from ground level and is moving straight upward.)

19. Proof Suppose that $f(a) = f(b) = g(a) = g(b) = 0$ and the second derivatives of f and g are continuous on the closed interval $[a, b]$. Prove that

$$\int_a^b f(x)g''(x)\,dx = \int_a^b f''(x)g(x)\,dx.$$

20. Proof Suppose that $f(a) = f(b) = 0$ and the second derivatives of f exist on the closed interval $[a, b]$. Prove that

$$\int_a^b (x - a)(x - b)f''(x)\,dx = 2\int_a^b f(x)\,dx.$$

21. Approximating an Integral Using the inequality

$$\frac{1}{x^5} + \frac{1}{x^{10}} + \frac{1}{x^{15}} < \frac{1}{x^5 - 1} < \frac{1}{x^5} + \frac{1}{x^{10}} + \frac{2}{x^{15}}$$

for $x \ge 2$, approximate $\displaystyle\int_2^\infty \frac{1}{x^5 - 1}\,dx.$

22. Volume Consider the shaded region between the graph of $y = \sin x$, where $0 \le x \le \pi$, and the line $y = c$, where $0 \le c \le 1$ (see figure). A solid is formed by revolving the region about the line $y = c$.

(a) For what value of c does the solid have minimum volume?

(b) For what value of c does the solid have maximum volume?

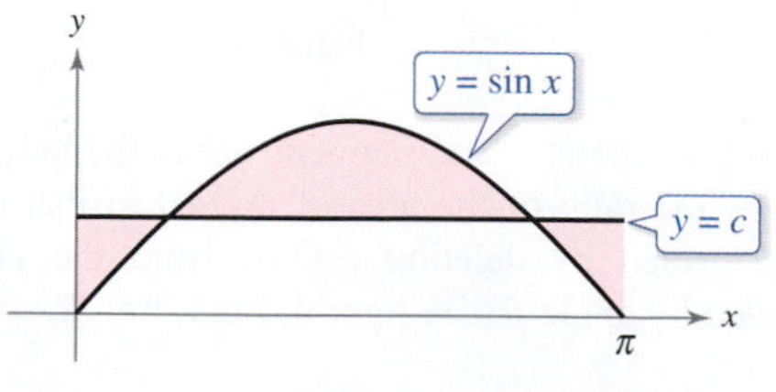

AP* Review Questions for Chapter 8

1. (BC only, calculator)

Given the region R bounded by the graphs of $y = \ln x$, $y = 0$, and $x = e^2$.

(a) Determine the area of the region.

(b) Find the volume of the solid generated by rotating the region R about the vertical line $x = e^2$.

(c) Write, but do not evaluate, an expression that gives the volume of the solid generated by rotating the region R about the horizontal line $y = -2$.

2. (calculator)

Given the region A bounded by the graphs of $y = e^{-x^3}$, $x = 0$, $y = 0$, and $x = 1$:

(a) Find the area of the region A.

(b) Find the volume of the solid generated when the region A is rotated about the x-axis.

(c) Find the volume of the solid generated when the region A is rotated about the y-axis.

3. (BC only, calculator)

(a) Find the area of the region R in the first quadrant bounded by the function $y = x \sin x$, the x-axis, and the vertical line $x = \pi$ using integration by parts.

(b) Find the area of the region S in the first quadrant bounded by the function $y = x \sin x$, the line $y = -2x + 5$, and the y-axis.

(c) The vertical line $x = k$ divides S into two regions of equal area. Write, but do not solve, an equation involving one or more integrals whose solution gives the value of k.

4. (BC only, calculator)

For $0 \le t \le 2$, a particle moves along the x-axis. The velocity of the particle at time t is given by $v(t) = te^{-t/3} - \frac{1}{2}$. At time $t = 0$, the particle is at position $x = 2$.

(a) Find the particle's position at time $t = 2$ using integration by parts.

(b) Write, but do not evaluate, an expression involving one or more integrals which gives the particle's total distance traveled from time $t = 0$ to time $t = 2$.

(c) Find the particle's acceleration at time t. Is the speed of the particle increasing or decreasing at time $t = 3$? Explain your answer.

5. (BC only)

The table below gives values of f, f', g, and g' for selected values of x.

x	1	2
$f(x)$	3	2
$f'(x)$	-2	1
$g(x)$	-1	3
$g'(x)$	1	-3

If $\int_1^2 f'(x)g(x)dx = 3$, then $\int_1^2 f(x)g'(x)dx =$

(A) -9

(B) 9

(C) -3

(D) 6

(E) 0

6. (no calculator)

Let a function f be defined by $f(x) = \begin{cases} \dfrac{3 + 4x}{1 + x^2} & \text{for } x \le 0 \\ \tan^2 x + 3 & \text{for } x > 0 \end{cases}$.

(a) Show that f is continuous at $x = 0$.

(b) For $x \ne 0$, express $f'(x)$ as a piecewise defined function.

(c) Find the average value on the interval $[-2, 1]$.

7. (BC only, no calculator)

Consider the differential equation $\dfrac{dT}{dR} = \dfrac{1}{2R(5 - R)}$.

(a) Given that $T = 0$ when $R = 3$, find the particular solution to this differential equation T in terms of R.

(b) Find the R-values of points of inflection for T.

8. (BC only, no calculator)

$\displaystyle\lim_{x \to 1} \frac{5e^{1-x} - \ln x - 5}{x^2 - 1} =$

(A) 5

(B) 4

(C) 2

(D) -3

(E) nonexistent

9. (BC only, no calculator)

Consider the function $g(x) = xe^{-x}$.

(a) Find all horizontal asymptotes of $g(x)$. Show the work that leads to your answer.

(b) Find the maximum value of $g(x)$. Justify your answer.

(c) Consider the region A bounded by the graph of $y = g(x)$, the x-axis, the vertical line $x = 1$, and unbounded on the right. Find the area of A or else show that the area is infinite.

10. (BC only, no calculator)

Let g be a function such that $g'(x) = \frac{1}{(x-2)^2}$.

(a) Determine whether $\int_0^3 g'(x)dx$ converges or diverges. Show the work that leads to your conclusion.

(b) Explain why $\int_0^3 g'(x)dx$ is *not* equivalent to $g(3) - g(0)$.

(c) Evaluate $\int_4^\infty g'(x)dx$.

9 Infinite Series

Projectile Motion *(Exercise 84, p. 675)*

Solera Method *(Section Project, p. 618)*

Sphereflake *(Exercise 86, p. 603)*

Multiplier Effect *(Exercise 73, p. 602)*

Compound Interest *(Exercise 67, p. 593)*

9.1 Sequences

- List the terms of a sequence.
- Determine whether a sequence converges or diverges.
- Write a formula for the nth term of a sequence.
- Use properties of monotonic sequences and bounded sequences.

Exploration

Finding Patterns Describe a pattern for each of the sequences listed below. Then use your description to write a formula for the nth term of each sequence. As n increases, do the terms appear to be approaching a limit? Explain your reasoning.

a. $1, \frac{1}{2}, \frac{1}{4}, \frac{1}{8}, \frac{1}{16}, \ldots$

b. $1, \frac{1}{2}, \frac{1}{6}, \frac{1}{24}, \frac{1}{120}, \ldots$

c. $10, \frac{10}{3}, \frac{10}{6}, \frac{10}{10}, \frac{10}{15}, \ldots$

d. $\frac{1}{4}, \frac{4}{9}, \frac{9}{16}, \frac{16}{25}, \frac{25}{36}, \ldots$

e. $\frac{3}{7}, \frac{5}{10}, \frac{7}{13}, \frac{9}{16}, \frac{11}{19}, \ldots$

Sequences

In mathematics, the word "sequence" is used in much the same way as it is in ordinary English. Saying that a collection of objects or events is *in sequence* usually means that the collection is ordered in such a way that it has an identified first member, second member, third member, and so on.

Mathematically, a **sequence** is defined as a function whose domain is the set of positive integers. Although a sequence is a function, it is common to represent sequences by subscript notation rather than by the standard function notation. For instance, in the sequence

$$\begin{array}{cccccccc} 1, & 2, & 3, & 4, & \ldots, & n, & \ldots & \\ \downarrow & \downarrow & \downarrow & \downarrow & \downarrow & \downarrow & \downarrow & \text{Sequence} \\ a_1, & a_2, & a_3, & a_4, & \ldots, & a_n, & \ldots & \end{array}$$

1 is mapped onto a_1, 2 is mapped onto a_2, and so on. The numbers

$$a_1, a_2, a_3, \ldots, a_n, \ldots$$

are the **terms** of the sequence. The number a_n is the ***n*th term** of the sequence, and the entire sequence is denoted by $\{a_n\}$. Occasionally, it is convenient to begin a sequence with a_0, so that the terms of the sequence become $a_0, a_1, a_2, a_3, \ldots, a_n, \ldots$ and the domain is the set of nonnegative integers.

EXAMPLE 1 Listing the Terms of a Sequence

a. The terms of the sequence $\{a_n\} = \{3 + (-1)^n\}$ are

$$\begin{array}{ccccc} 3 + (-1)^1, & 3 + (-1)^2, & 3 + (-1)^3, & 3 + (-1)^4, & \ldots \\ 2, & 4, & 2, & 4, & \ldots. \end{array}$$

b. The terms of the sequence $\{b_n\} = \left\{\frac{n}{1-2n}\right\}$ are

$$\begin{array}{ccccc} \frac{1}{1-2\cdot 1}, & \frac{2}{1-2\cdot 2}, & \frac{3}{1-2\cdot 3}, & \frac{4}{1-2\cdot 4}, & \ldots \\ -1, & -\frac{2}{3}, & -\frac{3}{5}, & -\frac{4}{7}, & \ldots. \end{array}$$

c. The terms of the sequence $\{c_n\} = \left\{\frac{n^2}{2^n-1}\right\}$ are

$$\begin{array}{ccccc} \frac{1^2}{2^1-1}, & \frac{2^2}{2^2-1}, & \frac{3^2}{2^3-1}, & \frac{4^2}{2^4-1}, & \ldots \\ \frac{1}{1}, & \frac{4}{3}, & \frac{9}{7}, & \frac{16}{15}, & \ldots. \end{array}$$

REMARK Some sequences are defined recursively. To define a sequence recursively, you need to be given one or more of the first few terms. All other terms of the sequence are then defined using previous terms, as shown in Example 1(d).

d. The terms of the **recursively defined** sequence $\{d_n\}$, where $d_1 = 25$ and $d_{n+1} = d_n - 5$, are

$$25, \quad 25 - 5 = 20, \quad 20 - 5 = 15, \quad 15 - 5 = 10, \ldots.$$

Limit of a Sequence

The primary focus of this chapter concerns sequences whose terms approach limiting values. Such sequences are said to **converge**. For instance, the sequence $\{1/2^n\}$

$$\frac{1}{2}, \frac{1}{4}, \frac{1}{8}, \frac{1}{16}, \frac{1}{32}, \ldots$$

converges to 0, as indicated in the next definition.

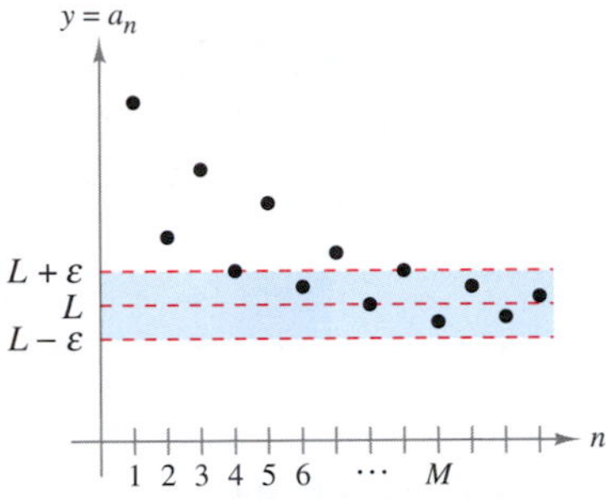

For $n > M$, the terms of the sequence all lie within ε units of L.

Figure 9.1

Definition of the Limit of a Sequence

Let L be a real number. The **limit** of a sequence $\{a_n\}$ is L, written as

$$\lim_{n\to\infty} a_n = L$$

if for each $\varepsilon > 0$, there exists $M > 0$ such that $|a_n - L| < \varepsilon$ whenever $n > M$. If the limit L of a sequence exists, then the sequence **converges** to L. If the limit of a sequence does not exist, then the sequence **diverges.**

Graphically, this definition says that eventually (for $n > M$ and $\varepsilon > 0$), the terms of a sequence that converges to L will lie within the band between the lines $y = L + \varepsilon$ and $y = L - \varepsilon$, as shown in Figure 9.1.

If a sequence $\{a_n\}$ agrees with a function f at every positive integer, and if $f(x)$ approaches a limit L as $x \to \infty$, then the sequence must converge to the same limit L.

REMARK The converse of Theorem 9.1 is not true (see Exercise 84).

THEOREM 9.1 Limit of a Sequence

Let L be a real number. Let f be a function of a real variable such that

$$\lim_{x\to\infty} f(x) = L.$$

If $\{a_n\}$ is a sequence such that $f(n) = a_n$ for every positive integer n, then

$$\lim_{n\to\infty} a_n = L.$$

EXAMPLE 2 Finding the Limit of a Sequence

Find the limit of the sequence whose nth term is $a_n = \left(1 + \frac{1}{n}\right)^n$.

Solution In Theorem 5.15, you learned that

$$\lim_{x\to\infty} \left(1 + \frac{1}{x}\right)^x = e.$$

So, you can apply Theorem 9.1 to conclude that

$$\lim_{n\to\infty} a_n = \lim_{n\to\infty} \left(1 + \frac{1}{n}\right)^n = e.$$

There are different ways in which a sequence can fail to have a limit. One way is that the terms of the sequence increase without bound or decrease without bound. These cases are written symbolically, as shown below.

Terms increase without bound: $\lim_{n\to\infty} a_n = \infty$

Terms decrease without bound: $\lim_{n\to\infty} a_n = -\infty$

The properties of limits of sequences listed in the next theorem parallel those given for limits of functions of a real variable in Section 1.3.

THEOREM 9.2 Properties of Limits of Sequences

Let $\lim_{n\to\infty} a_n = L$ and $\lim_{n\to\infty} b_n = K$.

1. $\lim_{n\to\infty} (a_n \pm b_n) = L \pm K$

2. $\lim_{n\to\infty} ca_n = cL$, c is any real number.

3. $\lim_{n\to\infty} (a_n b_n) = LK$

4. $\lim_{n\to\infty} \dfrac{a_n}{b_n} = \dfrac{L}{K}$, $b_n \neq 0$ and $K \neq 0$

EXAMPLE 3 Determining Convergence or Divergence

See LarsonCalculus.com for an interactive version of this type of example.

a. Because the sequence $\{a_n\} = \{3 + (-1)^n\}$ has terms

$$2, 4, 2, 4, . . .$$

See Example 1(a), page 584.

that alternate between 2 and 4, the limit

$$\lim_{n\to\infty} a_n$$

does not exist. So, the sequence diverges.

b. For $\{b_n\} = \left\{\dfrac{n}{1 - 2n}\right\}$, divide the numerator and denominator by n to obtain

$$\lim_{n\to\infty} \frac{n}{1 - 2n} = \lim_{n\to\infty} \frac{1}{(1/n) - 2} = -\frac{1}{2}$$

See Example 1(b), page 584.

which implies that the sequence converges to $-\frac{1}{2}$.

EXAMPLE 4 Using L'Hôpital's Rule to Determine Convergence

Show that the sequence whose nth term is $a_n = \dfrac{n^2}{2^n - 1}$ converges.

Solution Consider the function of a real variable

$$f(x) = \frac{x^2}{2^x - 1}.$$

Applying L'Hôpital's Rule twice produces

$$\lim_{x\to\infty} \frac{x^2}{2^x - 1} = \lim_{x\to\infty} \frac{2x}{(\ln 2)2^x} = \lim_{x\to\infty} \frac{2}{(\ln 2)^2 2^x} = 0.$$

Because $f(n) = a_n$ for every positive integer, you can apply Theorem 9.1 to conclude that

$$\lim_{n\to\infty} \frac{n^2}{2^n - 1} = 0.$$

See Example 1(c), page 584.

So, the sequence converges to 0. ∎

TECHNOLOGY Use a graphing utility to graph the function in Example 4. Notice that as x approaches infinity, the value of the function gets closer and closer to 0. If you have access to a graphing utility that can generate terms of a sequence, try using it to calculate the first 20 terms of the sequence in Example 4. Then view the terms to observe numerically that the sequence converges to 0.

The symbol $n!$ (read "n factorial") is used to simplify some of the formulas developed in this chapter. Let n be a positive integer; then ***n* factorial** is defined as

$$n! = 1 \cdot 2 \cdot 3 \cdot 4 \cdot \cdot \cdot (n - 1) \cdot n.$$

As a special case, **zero factorial** is defined as $0! = 1$. From this definition, you can see that $1! = 1$, $2! = 1 \cdot 2 = 2$, $3! = 1 \cdot 2 \cdot 3 = 6$, and so on. Factorials follow the same conventions for order of operations as exponents. That is, just as $2x^3$ and $(2x)^3$ imply different orders of operations, $2n!$ and $(2n)!$ imply the orders

$$2n! = 2(n!) = 2(1 \cdot 2 \cdot 3 \cdot 4 \cdot \cdot \cdot n)$$

and

$$(2n)! = 1 \cdot 2 \cdot 3 \cdot 4 \cdot \cdot \cdot n \cdot (n + 1) \cdot \cdot \cdot 2n$$

respectively.

Another useful limit theorem that can be rewritten for sequences is the Squeeze Theorem from Section 1.3.

THEOREM 9.3 Squeeze Theorem for Sequences

If $\lim_{n\to\infty} a_n = L = \lim_{n\to\infty} b_n$ and there exists an integer N such that $a_n \le c_n \le b_n$ for all $n > N$, then $\lim_{n\to\infty} c_n = L$.

EXAMPLE 5 Using the Squeeze Theorem

Show that the sequence $\{c_n\} = \left\{(-1)^n \dfrac{1}{n!}\right\}$ converges, and find its limit.

Solution To apply the Squeeze Theorem, you must find two convergent sequences that can be related to $\{c_n\}$. Two possibilities are $a_n = -1/2^n$ and $b_n = 1/2^n$, both of which converge to 0. By comparing the term $n!$ with 2^n, you can see that

$$n! = 1 \cdot 2 \cdot 3 \cdot 4 \cdot 5 \cdot 6 \cdot \cdot \cdot n = 24 \cdot \underbrace{5 \cdot 6 \cdot \cdot \cdot n}_{n - 4 \text{ factors}} \qquad (n \ge 4)$$

and

$$2^n = 2 \cdot 2 \cdot 2 \cdot 2 \cdot 2 \cdot 2 \cdot \cdot \cdot 2 = 16 \cdot \underbrace{2 \cdot 2 \cdot \cdot \cdot 2}_{n - 4 \text{ factors}}. \qquad (n \ge 4)$$

This implies that for $n \ge 4$, $2^n < n!$, and you have

$$\frac{-1}{2^n} \le (-1)^n \frac{1}{n!} \le \frac{1}{2^n}, \quad n \ge 4$$

as shown in Figure 9.2. So, by the Squeeze Theorem, it follows that

$$\lim_{n\to\infty} (-1)^n \frac{1}{n!} = 0.$$

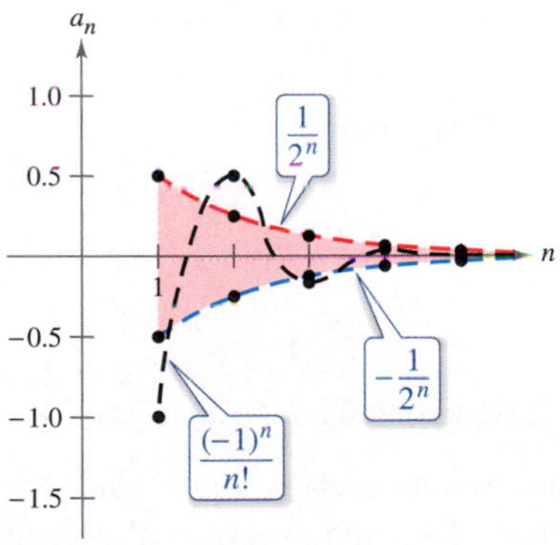

For $n \ge 4$, $(-1)^n/n!$ is squeezed between $-1/2^n$ and $1/2^n$.
Figure 9.2

Example 5 suggests something about the rate at which $n!$ increases as $n \to \infty$. As Figure 9.2 suggests, both $1/2^n$ and $1/n!$ approach 0 as $n \to \infty$. Yet $1/n!$ approaches 0 so much faster than $1/2^n$ does that

$$\lim_{n\to\infty} \frac{1/n!}{1/2^n} = \lim_{n\to\infty} \frac{2^n}{n!} = 0.$$

In fact, it can be shown that for any fixed number k, $\lim_{n\to\infty} (k^n/n!) = 0$. This means that *the factorial function grows faster than any exponential function.*

In Example 5, the sequence $\{c_n\}$ has both positive and negative terms. For this sequence, it happens that the sequence of absolute values, $\{|c_n|\}$, also converges to 0. You can show this by the Squeeze Theorem using the inequality

$$0 \le \frac{1}{n!} \le \frac{1}{2^n}, \quad n \ge 4.$$

In such cases, it is often convenient to consider the sequence of absolute values—and then apply Theorem 9.4, which states that if the absolute value sequence converges to 0, then the original signed sequence also converges to 0.

THEOREM 9.4 Absolute Value Theorem

For the sequence $\{a_n\}$, if

$$\lim_{n\to\infty} |a_n| = 0 \quad \text{then} \quad \lim_{n\to\infty} a_n = 0.$$

Proof Consider the two sequences $\{|a_n|\}$ and $\{-|a_n|\}$. Because both of these sequences converge to 0 and

$$-|a_n| \le a_n \le |a_n|$$

you can use the Squeeze Theorem to conclude that $\{a_n\}$ converges to 0.

See LarsonCalculus.com for Bruce Edwards's video of this proof.

Pattern Recognition for Sequences

Sometimes the terms of a sequence are generated by some rule that does not explicitly identify the *n*th term of the sequence. In such cases, you may be required to discover a *pattern* in the sequence and to describe the *n*th term. Once the *n*th term has been specified, you can investigate the convergence or divergence of the sequence.

EXAMPLE 6 Finding the *n*th Term of a Sequence

Find a sequence $\{a_n\}$ whose first five terms are

$$\frac{2}{1}, \frac{4}{3}, \frac{8}{5}, \frac{16}{7}, \frac{32}{9}, \ldots$$

and then determine whether the sequence you have chosen converges or diverges.

Solution First, note that the numerators are successive powers of 2, and the denominators form the sequence of positive odd integers. By comparing a_n with n, you have the following pattern.

$$\frac{2^1}{1}, \frac{2^2}{3}, \frac{2^3}{5}, \frac{2^4}{7}, \frac{2^5}{9}, \ldots, \frac{2^n}{2n-1}, \ldots$$

Consider the function of a real variable $f(x) = 2^x/(2x - 1)$. Applying L'Hôpital's Rule produces

$$\lim_{x\to\infty} \frac{2^x}{2x-1} = \lim_{x\to\infty} \frac{2^x(\ln 2)}{2} = \infty.$$

Next, apply Theorem 9.1 to conclude that

$$\lim_{n\to\infty} \frac{2^n}{2n-1} = \infty.$$

So, the sequence diverges.

Without a specific rule for generating the terms of a sequence or some knowledge of the context in which the terms of the sequence are obtained, it is not possible to determine the convergence or divergence of the sequence merely from its first several terms. For instance, although the first three terms of the following four sequences are identical, the first two sequences converge to 0, the third sequence converges to $\frac{1}{9}$, and the fourth sequence diverges.

$$\{a_n\}: \frac{1}{2}, \frac{1}{4}, \frac{1}{8}, \frac{1}{16}, \ldots, \frac{1}{2^n}, \ldots$$

$$\{b_n\}: \frac{1}{2}, \frac{1}{4}, \frac{1}{8}, \frac{1}{15}, \ldots, \frac{6}{(n+1)(n^2-n+6)}, \ldots$$

$$\{c_n\}: \frac{1}{2}, \frac{1}{4}, \frac{1}{8}, \frac{7}{62}, \ldots, \frac{n^2-3n+3}{9n^2-25n+18}, \ldots$$

$$\{d_n\}: \frac{1}{2}, \frac{1}{4}, \frac{1}{8}, 0, \ldots, \frac{-n(n+1)(n-4)}{6(n^2+3n-2)}, \ldots$$

The process of determining an nth term from the pattern observed in the first several terms of a sequence is an example of *inductive reasoning.*

EXAMPLE 7 Finding the *n*th Term of a Sequence

Determine the nth term for a sequence whose first five terms are

$$-\frac{2}{1}, \frac{8}{2}, -\frac{26}{6}, \frac{80}{24}, -\frac{242}{120}, \ldots$$

and then decide whether the sequence converges or diverges.

Solution Note that the numerators are 1 less than 3^n.

$$3^1 - 1 = 2 \quad 3^2 - 1 = 8 \quad 3^3 - 1 = 26 \quad 3^4 - 1 = 80 \quad 3^5 - 1 = 242$$

So, you can reason that the numerators are given by the rule

$$3^n - 1.$$

Factoring the denominators produces

$$1 = 1$$
$$2 = 1 \cdot 2$$
$$6 = 1 \cdot 2 \cdot 3$$
$$24 = 1 \cdot 2 \cdot 3 \cdot 4$$

and

$$120 = 1 \cdot 2 \cdot 3 \cdot 4 \cdot 5.$$

This suggests that the denominators are represented by $n!$. Finally, because the signs alternate, you can write the nth term as

$$a_n = (-1)^n \left(\frac{3^n - 1}{n!}\right).$$

From the discussion about the growth of $n!$, it follows that

$$\lim_{n\to\infty} |a_n| = \lim_{n\to\infty} \frac{3^n - 1}{n!} = 0.$$

Applying Theorem 9.4, you can conclude that

$$\lim_{n\to\infty} a_n = 0.$$

So, the sequence $\{a_n\}$ converges to 0. ■

Monotonic Sequences and Bounded Sequences

So far, you have determined the convergence of a sequence by finding its limit. Even when you cannot determine the limit of a particular sequence, it still may be useful to know whether the sequence converges. Theorem 9.5 (on the next page) provides a test for convergence of sequences without determining the limit. First, some preliminary definitions are given.

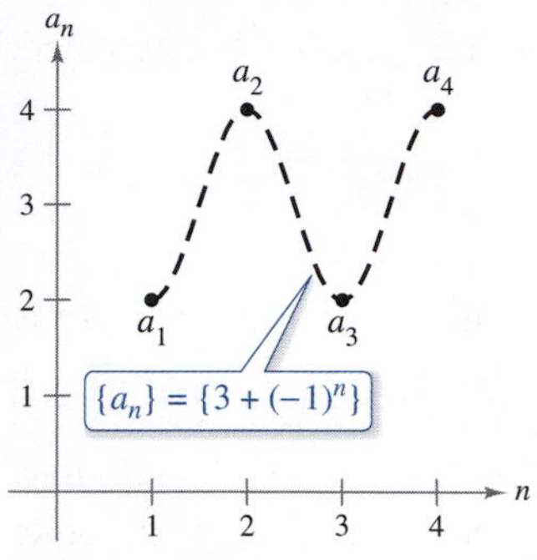

(a) Not monotonic

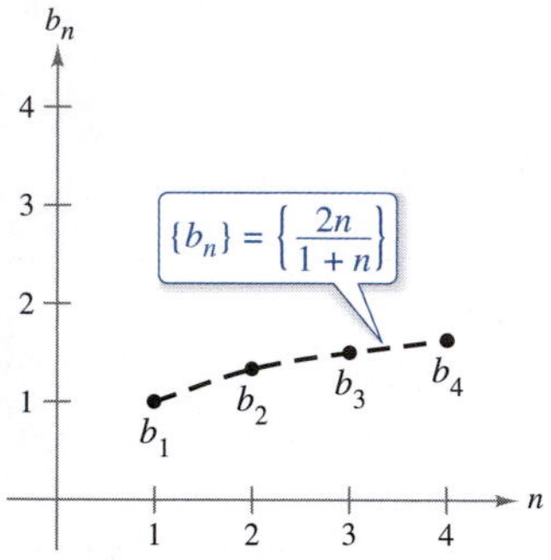

(b) Monotonic

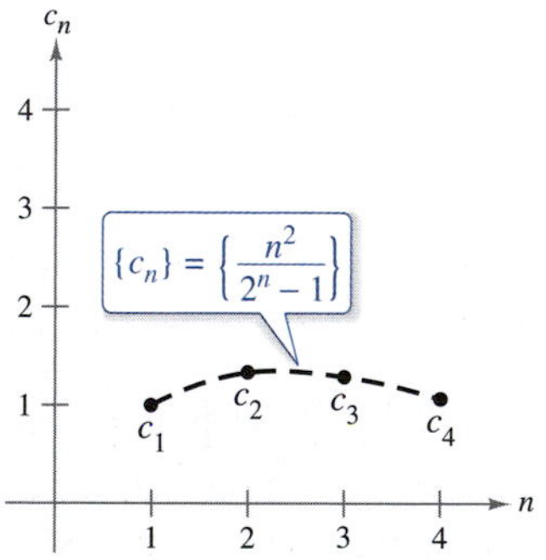

(c) Not monotonic

Figure 9.3

Definition of Monotonic Sequence

A sequence $\{a_n\}$ is **monotonic** when its terms are nondecreasing

$$a_1 \le a_2 \le a_3 \le \cdots \le a_n \le \cdots$$

or when its terms are nonincreasing

$$a_1 \ge a_2 \ge a_3 \ge \cdots \ge a_n \ge \cdots.$$

EXAMPLE 8 Determining Whether a Sequence Is Monotonic

Determine whether each sequence having the given nth term is monotonic.

a. $a_n = 3 + (-1)^n$

b. $b_n = \dfrac{2n}{1+n}$

c. $c_n = \dfrac{n^2}{2^n - 1}$

Solution

a. This sequence alternates between 2 and 4. So, it is not monotonic.

b. This sequence is monotonic because each successive term is greater than its predecessor. To see this, compare the terms b_n and b_{n+1}. [Note that, because n is positive, you can multiply each side of the inequality by $(1 + n)$ and $(2 + n)$ without reversing the inequality sign.]

$$b_n = \frac{2n}{1+n} \overset{?}{<} \frac{2(n+1)}{1 + (n+1)} = b_{n+1}$$

$$2n(2 + n) \overset{?}{<} (1 + n)(2n + 2)$$

$$4n + 2n^2 \overset{?}{<} 2 + 4n + 2n^2$$

$$0 < 2$$

Starting with the final inequality, which is valid, you can reverse the steps to conclude that the original inequality is also valid.

c. This sequence is not monotonic, because the second term is greater than the first term, and greater than the third. (Note that when you drop the first term, the remaining sequence $c_2, c_3, c_4, \ldots$ is monotonic.)

Figure 9.3 graphically illustrates these three sequences.

In Example 8(b), another way to see that the sequence is monotonic is to argue that the derivative of the corresponding differentiable function

$$f(x) = \frac{2x}{1+x}$$

is positive for all x. This implies that f is increasing, which in turn implies that $\{b_n\}$ is increasing.

Definition of Bounded Sequence

1. A sequence $\{a_n\}$ is **bounded above** when there is a real number M such that $a_n \le M$ for all n. The number M is called an **upper bound** of the sequence.
2. A sequence $\{a_n\}$ is **bounded below** when there is a real number N such that $N \le a_n$ for all n. The number N is called a **lower bound** of the sequence.
3. A sequence $\{a_n\}$ is **bounded** when it is bounded above and bounded below.

Note that all three sequences in Example 3 (and shown in Figure 9.3) are bounded. To see this, note that

$$2 \le a_n \le 4, \quad 1 \le b_n \le 2, \quad \text{and} \quad 0 \le c_n \le \frac{4}{3}.$$

One important property of the real numbers is that they are **complete.** Informally, this means that there are no holes or gaps on the real number line. (The set of rational numbers does not have the completeness property.) The completeness axiom for real numbers can be used to conclude that if a sequence has an upper bound, then it must have a **least upper bound** (an upper bound that is less than all other upper bounds for the sequence). For example, the least upper bound of the sequence $\{a_n\} = \{n/(n+1)\}$,

$$\frac{1}{2}, \frac{2}{3}, \frac{3}{4}, \frac{4}{5}, \ldots, \frac{n}{n+1}, \ldots$$

is 1. The completeness axiom is used in the proof of Theorem 9.5.

AP* Tips

You should understand Theorem 9.5 for the AP Calculus BC Exam.

THEOREM 9.5 Bounded Monotonic Sequences

If a sequence $\{a_n\}$ is bounded and monotonic, then it converges.

Proof Assume that the sequence is nondecreasing, as shown in Figure 9.4. For the sake of simplicity, also assume that each term in the sequence is positive. Because the sequence is bounded, there must exist an upper bound M such that

$$a_1 \le a_2 \le a_3 \le \cdots \le a_n \le \cdots \le M.$$

From the completeness axiom, it follows that there is a least upper bound L such that

$$a_1 \le a_2 \le a_3 \le \cdots \le a_n \le \cdots \le L.$$

For $\varepsilon > 0$, it follows that $L - \varepsilon < L$, and therefore $L - \varepsilon$ cannot be an upper bound for the sequence. Consequently, at least one term of $\{a_n\}$ is greater than $L - \varepsilon$. That is, $L - \varepsilon < a_N$ for some positive integer N. Because the terms of $\{a_n\}$ are nondecreasing, it follows that $a_N \le a_n$ for $n > N$. You now know that $L - \varepsilon < a_N \le a_n \le L < L + \varepsilon$, for every $n > N$. It follows that $|a_n - L| < \varepsilon$ for $n > N$, which by definition means that $\{a_n\}$ converges to L. The proof for a nonincreasing sequence is similar (see Exercise 91).

See LarsonCalculus.com for Bruce Edwards's video of this proof.

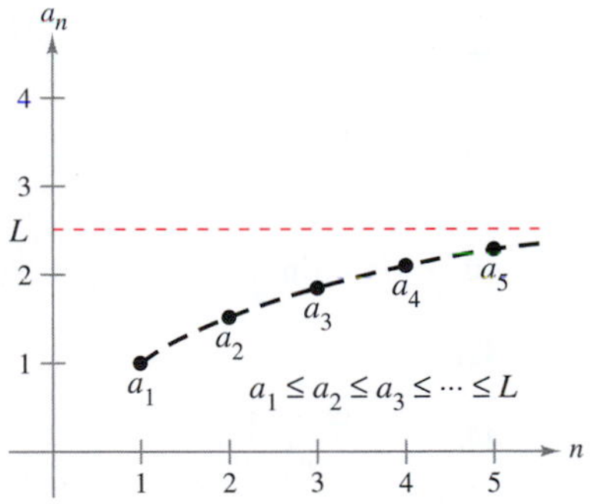

Every bounded, nondecreasing sequence converges.
Figure 9.4

EXAMPLE 9 Bounded and Monotonic Sequences

a. The sequence $\{a_n\} = \{1/n\}$ is both bounded and monotonic, and so, by Theorem 9.5, it must converge.

b. The divergent sequence $\{b_n\} = \{n^2/(n+1)\}$ is monotonic, but not bounded. (It is bounded below.)

c. The divergent sequence $\{c_n\} = \{(-1)^n\}$ is bounded, but not monotonic.

9.1 Exercises

See CalcChat.com for tutorial help and worked-out solutions to odd-numbered exercises.

Listing the Terms of a Sequence In Exercises 1–6, write the first five terms of the sequence.

1. $a_n = 3^n$

2. $a_n = \left(-\frac{2}{5}\right)^n$

3. $a_n = \sin \frac{n\pi}{2}$

4. $a_n = \frac{3n}{n+4}$

5. $a_n = (-1)^{n+1}\left(\frac{2}{n}\right)$

6. $a_n = 2 + \frac{2}{n} - \frac{1}{n^2}$

Listing the Terms of a Sequence In Exercises 7 and 8, write the first five terms of the recursively defined sequence.

7. $a_1 = 3,\ a_{k+1} = 2(a_k - 1)$

8. $a_1 = 6,\ a_{k+1} = \frac{1}{3}a_k^2$

Matching In Exercises 9–12, match the sequence with its graph. [The graphs are labeled (a), (b), (c), and (d).]

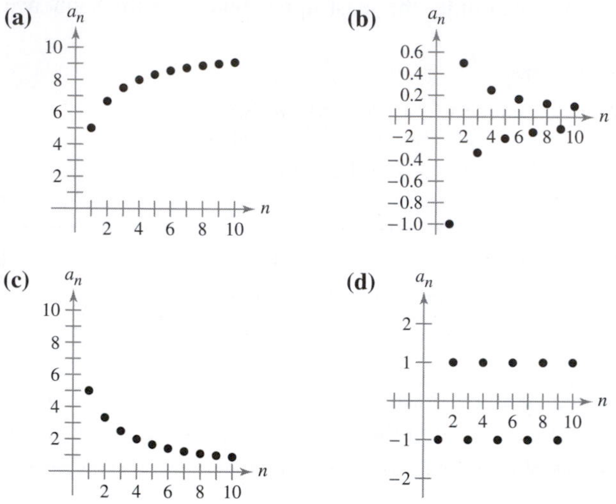

9. $a_n = \frac{10}{n+1}$

10. $a_n = \frac{10n}{n+1}$

11. $a_n = (-1)^n$

12. $a_n = \frac{(-1)^n}{n}$

Writing Terms In Exercises 13–16, write the next two apparent terms of the sequence. Describe the pattern you used to find these terms.

13. 2, 5, 8, 11, . . .

14. 8, 13, 18, 23, 28, . . .

15. 5, 10, 20, 40, . . .

16. $6, -2, \frac{2}{3}, -\frac{2}{9}, \ldots$

Simplifying Factorials In Exercises 17–20, simplify the ratio of factorials.

17. $\frac{(n+1)!}{n!}$

18. $\frac{n!}{(n+2)!}$

19. $\frac{(2n-1)!}{(2n+1)!}$

20. $\frac{(2n+2)!}{(2n)!}$

Finding the Limit of a Sequence In Exercises 21–24, find the limit (if possible) of the sequence.

21. $a_n = \frac{5n^2}{n^2+2}$

22. $a_n = 6 + \frac{2}{n^2}$

23. $a_n = \frac{2n}{\sqrt{n^2+1}}$

24. $a_n = \cos \frac{2}{n}$

Finding the Limit of a Sequence In Exercises 25–28, use a graphing utility to graph the first 10 terms of the sequence. Use the graph to make an inference about the convergence or divergence of the sequence. Verify your inference analytically and, if the sequence converges, find its limit.

25. $a_n = \frac{4n+1}{n}$

26. $a_n = \frac{1}{n^{3/2}}$

27. $a_n = \sin \frac{n\pi}{2}$

28. $a_n = 2 - \frac{1}{4^n}$

Determining Convergence or Divergence In Exercises 29–44, determine the convergence or divergence of the sequence with the given nth term. If the sequence converges, find its limit.

29. $a_n = \frac{5}{n+2}$

30. $a_n = 8 + \frac{5}{n}$

31. $a_n = (-1)^n\left(\frac{n}{n+1}\right)$

32. $a_n = \frac{1+(-1)^n}{n^2}$

33. $a_n = \frac{10n^2+3n+7}{2n^2-6}$

34. $a_n = \frac{\sqrt[3]{n}}{\sqrt[3]{n}+1}$

35. $a_n = \frac{\ln(n^3)}{2n}$

36. $a_n = \frac{5^n}{3^n}$

37. $a_n = \frac{(n+1)!}{n!}$

38. $a_n = \frac{(n-2)!}{n!}$

39. $a_n = \frac{n^p}{e^n},\ p > 0$

40. $a_n = n \sin \frac{1}{n}$

41. $a_n = 2^{1/n}$

42. $a_n = -3^{-n}$

43. $a_n = \frac{\sin n}{n}$

44. $a_n = \frac{\cos \pi n}{n^2}$

Finding the nth Term of a Sequence In Exercises 45–52, write an expression for the nth term of the sequence. (There is more than one correct answer.)

45. 2, 8, 14, 20, . . .

46. $1, \frac{1}{2}, \frac{1}{6}, \frac{1}{24}, \frac{1}{120}, \ldots$

47. $-2, 1, 6, 13, 22, \ldots$

48. $1, -\frac{1}{4}, \frac{1}{9}, -\frac{1}{16}, \ldots$

49. $\frac{2}{3}, \frac{3}{4}, \frac{4}{5}, \frac{5}{6}, \ldots$

50. 2, 24, 720, 40,320, 3,628,800, . . .

51. $2, 1 + \frac{1}{2}, 1 + \frac{1}{3}, 1 + \frac{1}{4}, 1 + \frac{1}{5}, \ldots$

52. $\frac{1}{2 \cdot 3}, \frac{2}{3 \cdot 4}, \frac{3}{4 \cdot 5}, \frac{4}{5 \cdot 6}, \ldots$

Finding Monotonic and Bounded Sequences **In Exercises 53–60, determine whether the sequence with the given nth term is monotonic and whether it is bounded. Use a graphing utility to confirm your results.**

53. $a_n = 4 - \dfrac{1}{n}$

54. $a_n = \dfrac{3n}{n + 2}$

55. $a_n = ne^{-n/2}$

56. $a_n = \left(-\dfrac{2}{3}\right)^n$

57. $a_n = \left(\dfrac{2}{3}\right)^n$

58. $a_n = \left(\dfrac{3}{2}\right)^n$

59. $a_n = \sin \dfrac{n\pi}{6}$

60. $a_n = \dfrac{\cos n}{n}$

Using a Theorem **In Exercises 61–64, (a) use Theorem 9.5 to show that the sequence with the given nth term converges, and (b) use a graphing utility to graph the first 10 terms of the sequence and find its limit.**

61. $a_n = 7 + \dfrac{1}{n}$

62. $a_n = 5 - \dfrac{2}{n}$

63. $a_n = \dfrac{1}{3}\left(1 - \dfrac{1}{3^n}\right)$

64. $a_n = 2 + \dfrac{1}{5^n}$

65. Increasing Sequence Let $\{a_n\}$ be an increasing sequence such that $2 \le a_n \le 4$. Explain why $\{a_n\}$ has a limit. What can you conclude about the limit?

66. Monotonic Sequence Let $\{a_n\}$ be a monotonic sequence such that $a_n \le 1$. Discuss the convergence of $\{a_n\}$. When $\{a_n\}$ converges, what can you conclude about its limit?

67. Compound Interest

Consider the sequence $\{A_n\}$ whose nth term is given by

$$A_n = P\left(1 + \frac{r}{12}\right)^n$$

where P is the principal, A_n is the account balance after n months, and r is the interest rate compounded annually.

(a) Is $\{A_n\}$ a convergent sequence? Explain.

(b) Find the first 10 terms of the sequence when $P = \$10{,}000$ and $r = 0.055$.

68. Compound Interest A deposit of \$100 is made in an account at the beginning of each month at an annual interest rate of 3% compounded monthly. The balance in the account after n months is $A_n = 100(401)(1.0025^n - 1)$.

(a) Compute the first six terms of the sequence $\{A_n\}$.

(b) Find the balance in the account after 5 years by computing the 60th term of the sequence.

(c) Find the balance in the account after 20 years by computing the 240th term of the sequence.

WRITING ABOUT CONCEPTS

69. Sequence Is it possible for a sequence to converge to two different numbers? If so, give an example. If not, explain why not.

70. Defining Terms In your own words, define each of the following.

(a) Sequence

(b) Convergence of a sequence

(c) Monotonic sequence

(d) Bounded sequence

71. Writing a Sequence Give an example of a sequence satisfying the condition or explain why no such sequence exists. (Examples are not unique.)

(a) A monotonically increasing sequence that converges to 10

(b) A monotonically increasing bounded sequence that does not converge

(c) A sequence that converges to $\frac{3}{4}$

(d) An unbounded sequence that converges to 100

72. HOW DO YOU SEE IT? The graphs of two sequences are shown in the figures. Which graph represents the sequence with alternating signs? Explain.

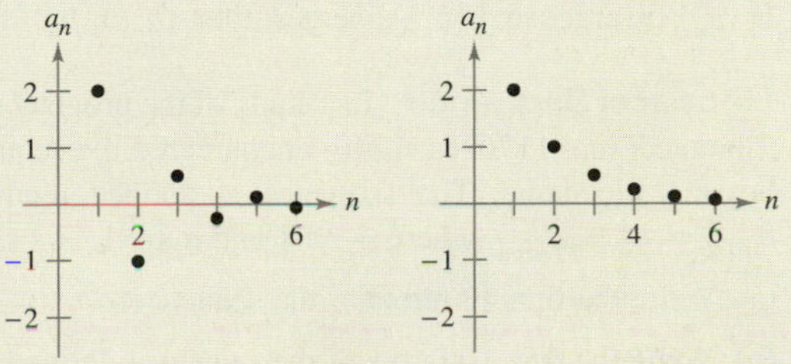

73. Government Expenditures A government program that currently costs taxpayers \$4.5 billion per year is cut back by 20 percent per year.

(a) Write an expression for the amount budgeted for this program after n years.

(b) Compute the budgets for the first 4 years.

(c) Determine the convergence or divergence of the sequence of reduced budgets. If the sequence converges, find its limit.

74. Inflation When the rate of inflation is $4\frac{1}{2}\%$ per year and the average price of a car is currently \$25,000, the average price after n years is $P_n = \$25{,}000(1.045)^n$. Compute the average prices for the next 5 years.

75. Using a Sequence Compute the first six terms of the sequence $\{a_n\} = \{\sqrt[n]{n}\}$. If the sequence converges, find its limit.

76. Using a Sequence Compute the first six terms of the sequence

$$\{a_n\} = \left\{\left(1 + \frac{1}{n}\right)^n\right\}.$$

If the sequence converges, find its limit.

Lisa S./Shutterstock.com

77. Proof Prove that if $\{s_n\}$ converges to L and $L > 0$, then there exists a number N such that $s_n > 0$ for $n > N$.

78. Modeling Data The amounts of the federal debt a_n (in trillions of dollars) of the United States from 2000 through 2011 are given below as ordered pairs of the form (n, a_n), where n represents the year, with $n = 0$ corresponding to 2000. *(Source: U.S. Office of Management and Budget)*

(0, 5.6), (1, 5.8), (2, 6.2), (3, 6.8), (4, 7.4), (5, 7.9), (6, 8.5), (7, 9.0), (8, 10.0), (9, 11.9), (10, 13.5), (11, 14.8)

(a) Use the regression capabilities of a graphing utility to find a model of the form

$$a_n = bn^2 + cn + d, \quad n = 0, 1, \ldots, 11$$

for the data. Use the graphing utility to plot the points and graph the model.

(b) Use the model to predict the amount of the federal debt in the year 2020.

True or False? **In Exercises 79–82, determine whether the statement is true or false. If it is false, explain why or give an example that shows it is false.**

79. If $\{a_n\}$ converges to 3 and $\{b_n\}$ converges to 2, then $\{a_n + b_n\}$ converges to 5.

80. If $\{a_n\}$ converges, then $\lim_{n\to\infty} (a_n - a_{n+1}) = 0$.

81. If $\{a_n\}$ converges, then $\{a_n/n\}$ converges to 0.

82. If $\{a_n\}$ diverges and $\{b_n\}$ diverges, then $\{a_n + b_n\}$ diverges.

83. Fibonacci Sequence In a study of the progeny of rabbits, Fibonacci (ca. 1170–ca. 1240) encountered the sequence now bearing his name. The sequence is defined recursively as $a_{n+2} = a_n + a_{n+1}$, where $a_1 = 1$ and $a_2 = 1$.

(a) Write the first 12 terms of the sequence.

(b) Write the first 10 terms of the sequence defined by

$$b_n = \frac{a_{n+1}}{a_n}, \quad n \ge 1.$$

(c) Using the definition in part (b), show that

$$b_n = 1 + \frac{1}{b_{n-1}}.$$

(d) The **golden ratio** ρ can be defined by $\lim_{n\to\infty} b_n = \rho$. Show that

$$\rho = 1 + \frac{1}{\rho}$$

and solve this equation for ρ.

84. Using a Theorem Show that the converse of Theorem 9.1 is not true. [*Hint:* Find a function $f(x)$ such that $f(n) = a_n$ converges, but $\lim_{x\to\infty} f(x)$ does not exist.]

85. Using a Sequence Consider the sequence

$$\sqrt{2}, \sqrt{2 + \sqrt{2}}, \sqrt{2 + \sqrt{2 + \sqrt{2}}}, \ldots$$

(a) Compute the first five terms of this sequence.

(b) Write a recursion formula for a_n, for $n \ge 2$.

(c) Find $\lim_{n\to\infty} a_n$.

86. Using a Sequence Consider the sequence $\{a_n\}$ where $a_1 = \sqrt{k}$, $a_{n+1} = \sqrt{k + a_n}$, and $k > 0$.

(a) Show that $\{a_n\}$ is increasing and bounded.

(b) Prove that $\lim_{n\to\infty} a_n$ exists.

(c) Find $\lim_{n\to\infty} a_n$.

87. Squeeze Theorem

(a) Show that $\int_1^n \ln x \, dx < \ln(n!)$ for $n \ge 2$.

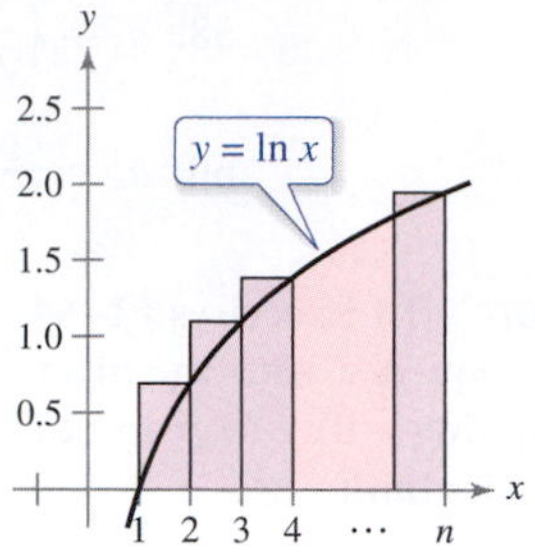

(b) Draw a graph similar to the one above that shows

$$\ln(n!) < \int_1^{n+1} \ln x \, dx.$$

(c) Use the results of parts (a) and (b) to show that

$$\frac{n^n}{e^{n-1}} < n! < \frac{(n+1)^{n+1}}{e^n}, \text{ for } n > 1.$$

(d) Use the Squeeze Theorem for Sequences and the result of part (c) to show that $\lim_{n\to\infty} \left(\sqrt[n]{n!}/n\right) = 1/e$.

(e) Test the result of part (d) for $n = 20, 50$, and 100.

88. Proof Prove, using the definition of the limit of a sequence, that

$$\lim_{n\to\infty} \frac{1}{n^3} = 0.$$

89. Proof Prove, using the definition of the limit of a sequence, that $\lim_{n\to\infty} r^n = 0$ for $-1 < r < 1$.

90. Using a Sequence Find a divergent sequence $\{a_n\}$ such that $\{a_{2n}\}$ converges.

91. Proof Prove Theorem 9.5 for a nonincreasing sequence.

PUTNAM EXAM CHALLENGE

92. Let $\{x_n\}$, $n \ge 0$, be a sequence of nonzero real numbers such that $x_n^2 - x_{n-1}x_{n+1} = 1$ for $n = 1, 2, 3, \ldots$. Prove there exists a real number a such that $x_{n+1} = ax_n - x_{n-1}$ for all $n \ge 1$.

93. Let $T_0 = 2$, $T_1 = 3$, $T_2 = 6$, and for $n \ge 3$,

$$T_n = (n + 4)T_{n-1} - 4nT_{n-2} + (4n - 8)T_{n-3}.$$

The first few terms are

2, 3, 6, 14, 40, 152, 784, 5168, 40,576

Find, with proof, a formula for T_n of the form $T_n = A_n + B_n$, where $\{A_n\}$ and $\{B_n\}$ are well-known sequences.

9.2 Series and Convergence

- Understand the definition of a convergent infinite series.
- Use properties of infinite geometric series.
- Use the *n*th-Term Test for Divergence of an infinite series.

Infinite Series

One important application of infinite sequences is in representing "infinite summations." Informally, if $\{a_n\}$ is an infinite sequence, then

$$\sum_{n=1}^{\infty} a_n = a_1 + a_2 + a_3 + \cdot \cdot \cdot + a_n + \cdot \cdot \cdot$$ Infinite Series

REMARK As you study this chapter, it is important to distinguish between an infinite series and a sequence. A sequence is an ordered collection of numbers

$a_1, a_2, a_3, . . . , a_n, . . .$

whereas a series is an infinite sum of terms from a sequence

$a_1 + a_2 + a_3 + \cdot \cdot \cdot + a_n + \cdot \cdot \cdot .$

is an **infinite series** (or simply a **series**). The numbers a_1, a_2, a_3, and so on are the **terms** of the series. For some series, it is convenient to begin the index at $n = 0$ (or some other integer). As a typesetting convention, it is common to represent an infinite series as $\Sigma\, a_n$. In such cases, the starting value for the index must be taken from the context of the statement.

To find the sum of an infinite series, consider the **sequence of partial sums** listed below.

$$\begin{aligned}
S_1 &= a_1\\
S_2 &= a_1 + a_2\\
S_3 &= a_1 + a_2 + a_3\\
S_4 &= a_1 + a_2 + a_3 + a_4\\
S_5 &= a_1 + a_2 + a_3 + a_4 + a_5\\
&\vdots\\
S_n &= a_1 + a_2 + a_3 + \cdot \cdot \cdot + a_n
\end{aligned}$$

If this sequence of partial sums converges, then the series is said to converge and has the sum indicated in the next definition.

INFINITE SERIES

The study of infinite series was considered a novelty in the fourteenth century. Logician Richard Suiseth, whose nickname was Calculator, solved this problem.

If throughout the first half of a given time interval a variation continues at a certain intensity, throughout the next quarter of the interval at double the intensity, throughout the following eighth at triple the intensity and so ad infinitum; then the average intensity for the whole interval will be the intensity of the variation during the second subinterval (or double the intensity).

This is the same as saying that the sum of the infinite series

$$\frac{1}{2} + \frac{2}{4} + \frac{3}{8} + \cdot \cdot \cdot + \frac{n}{2^n} + \cdot \cdot \cdot$$

is 2.

Definitions of Convergent and Divergent Series

For the infinite series $\sum_{n=1}^{\infty} a_n$, the ***n*th partial sum** is

$$S_n = a_1 + a_2 + \cdot \cdot \cdot + a_n.$$

If the sequence of partial sums $\{S_n\}$ converges to S, then the series $\sum_{n=1}^{\infty} a_n$ **converges.** The limit S is called the **sum of the series.**

$$S = a_1 + a_2 + \cdot \cdot \cdot + a_n + \cdot \cdot \cdot \qquad S = \sum_{n=1}^{\infty} a_n$$

If $\{S_n\}$ diverges, then the series **diverges.**

As you study this chapter, you will see that there are two basic questions involving infinite series.

- Does a series converge or does it diverge?
- When a series converges, what is its sum?

These questions are not always easy to answer, especially the second one.

TECHNOLOGY Figure 9.5 shows the first 15 partial sums of the infinite series in Example 1(a). Notice how the values appear to approach the line $y = 1$.

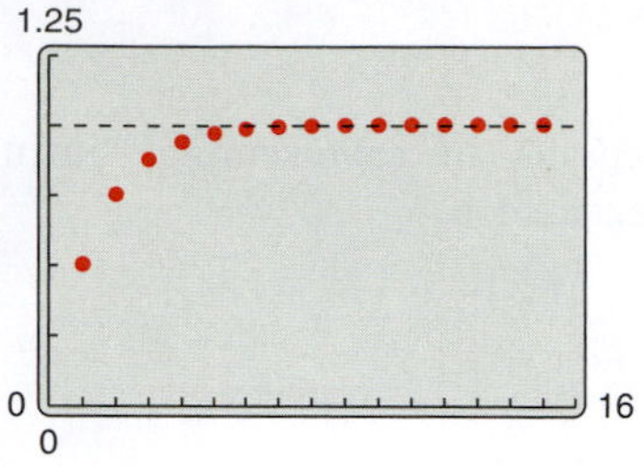

Figure 9.5

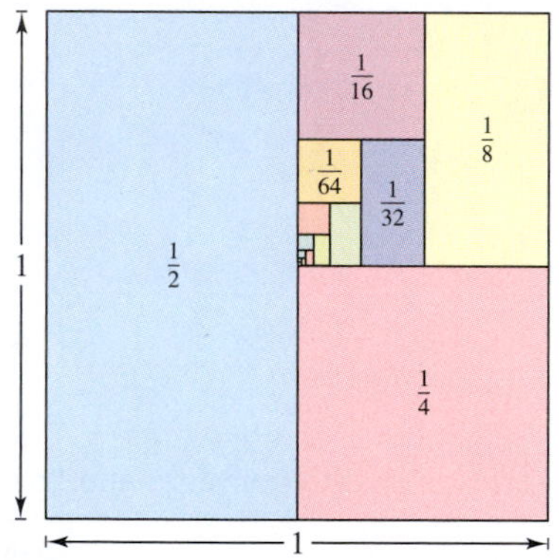

You can determine the partial sums of the series in Example 1(a) geometrically using this figure.
Figure 9.6

EXAMPLE 1 **Convergent and Divergent Series**

a. The series

$$\sum_{n=1}^{\infty} \frac{1}{2^n} = \frac{1}{2} + \frac{1}{4} + \frac{1}{8} + \frac{1}{16} + \cdots$$

has the partial sums listed below. (You can also determine the partial sums of the series geometrically, as shown in Figure 9.6.)

$$S_1 = \frac{1}{2}$$

$$S_2 = \frac{1}{2} + \frac{1}{4} = \frac{3}{4}$$

$$S_3 = \frac{1}{2} + \frac{1}{4} + \frac{1}{8} = \frac{7}{8}$$

$$\vdots$$

$$S_n = \frac{1}{2} + \frac{1}{4} + \frac{1}{8} + \cdots + \frac{1}{2^n} = \frac{2^n - 1}{2^n}$$

Because

$$\lim_{n\to\infty} \frac{2^n - 1}{2^n} = 1$$

it follows that the series converges and its sum is 1.

b. The nth partial sum of the series

$$\sum_{n=1}^{\infty} \left(\frac{1}{n} - \frac{1}{n+1}\right) = \left(1 - \frac{1}{2}\right) + \left(\frac{1}{2} - \frac{1}{3}\right) + \left(\frac{1}{3} - \frac{1}{4}\right) + \cdots$$

is

$$S_n = 1 - \frac{1}{n+1}.$$

Because the limit of S_n is 1, the series converges and its sum is 1.

c. The series

$$\sum_{n=1}^{\infty} 1 = 1 + 1 + 1 + 1 + \cdots$$

diverges because $S_n = n$ and the sequence of partial sums diverges.

The series in Example 1(b) is a **telescoping series** of the form

$$(b_1 - b_2) + (b_2 - b_3) + (b_3 - b_4) + (b_4 - b_5) + \cdots.$$

Telescoping series

Note that b_2 is canceled by the second term, b_3 is canceled by the third term, and so on. Because the nth partial sum of this series is

$$S_n = b_1 - b_{n+1}$$

it follows that a telescoping series will converge if and only if b_n approaches a finite number as $n \to \infty$. Moreover, if the series converges, then its sum is

$$S = b_1 - \lim_{n\to\infty} b_{n+1}.$$

FOR FURTHER INFORMATION To learn more about the partial sums of infinite series, see the article "Six Ways to Sum a Series" by Dan Kalman in *The College Mathematics Journal*. To view this article, go to *MathArticles.com*.

EXAMPLE 2 Writing a Series in Telescoping Form

Find the sum of the series $\sum_{n=1}^{\infty} \frac{2}{4n^2 - 1}$.

Solution

Using partial fractions, you can write

$$a_n = \frac{2}{4n^2 - 1} = \frac{2}{(2n-1)(2n+1)} = \frac{1}{2n-1} - \frac{1}{2n+1}.$$

From this telescoping form, you can see that the nth partial sum is

$$S_n = \left(\frac{1}{1} - \frac{1}{3}\right) + \left(\frac{1}{3} - \frac{1}{5}\right) + \cdot \cdot \cdot + \left(\frac{1}{2n-1} - \frac{1}{2n+1}\right) = 1 - \frac{1}{2n+1}.$$

So, the series converges and its sum is 1. That is,

$$\sum_{n=1}^{\infty} \frac{2}{4n^2 - 1} = \lim_{n\to\infty} S_n = \lim_{n\to\infty}\left(1 - \frac{1}{2n+1}\right) = 1.$$

Exploration

In "Proof Without Words," by Benjamin G. Klein and Irl C. Bivens, the authors present the diagram below. Explain why the second statement after the diagram is valid. How is this result related to Theorem 9.6?

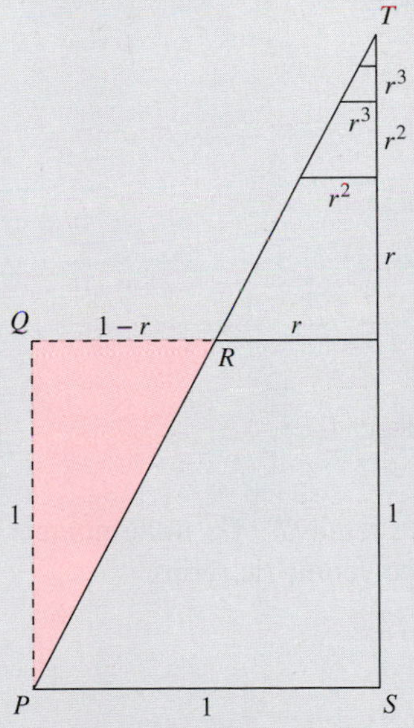

$\Delta PQR \sim \Delta TSP$

$$1 + r + r^2 + r^3 + \cdot \cdot \cdot = \frac{1}{1-r}$$

Exercise taken from "Proof Without Words" by Benjamin G. Klein and Irl C. Bivens, *Mathematics Magazine,* 61, No. 4, October 1988, p. 219, by permission of the authors.

Geometric Series

The series in Example 1(a) is a **geometric series.** In general, the series

$$\sum_{n=0}^{\infty} ar^n = a + ar + ar^2 + \cdot \cdot \cdot + ar^n + \cdot \cdot \cdot, \quad a \neq 0$$

Geometric series

is a **geometric series** with ratio r, $r \neq 0$.

THEOREM 9.6 Convergence of a Geometric Series

A geometric series with ratio r diverges when $|r| \geq 1$. If $0 < |r| < 1$, then the series converges to the sum

$$\sum_{n=0}^{\infty} ar^n = \frac{a}{1-r}, \quad 0 < |r| < 1.$$

Proof It is easy to see that the series diverges when $r = \pm 1$. If $r \neq \pm 1$, then

$$S_n = a + ar + ar^2 + \cdot \cdot \cdot + ar^{n-1}.$$

Multiplication by r yields

$$rS_n = ar + ar^2 + ar^3 + \cdot \cdot \cdot + ar^n.$$

Subtracting the second equation from the first produces $S_n - rS_n = a - ar^n$. Therefore, $S_n(1 - r) = a(1 - r^n)$, and the nth partial sum is

$$S_n = \frac{a}{1-r}(1 - r^n).$$

When $0 < |r| < 1$, it follows that $r^n \to 0$ as $n \to \infty$, and you obtain

$$\lim_{n\to\infty} S_n = \lim_{n\to\infty}\left[\frac{a}{1-r}(1 - r^n)\right] = \frac{a}{1-r}\left[\lim_{n\to\infty}(1 - r^n)\right] = \frac{a}{1-r}$$

which means that the series *converges* and its sum is $a/(1 - r)$. It is left to you to show that the series diverges when $|r| > 1$.

See LarsonCalculus.com for Bruce Edwards's video of this proof.

TECHNOLOGY Try using a graphing utility to compute the sum of the first 20 terms of the sequence in Example 3(a). You should obtain a sum of about 5.999994.

EXAMPLE 3 **Convergent and Divergent Geometric Series**

a. The geometric series

$$\sum_{n=0}^{\infty} \frac{3}{2^n} = \sum_{n=0}^{\infty} 3\left(\frac{1}{2}\right)^n = 3(1) + 3\left(\frac{1}{2}\right) + 3\left(\frac{1}{2}\right)^2 + \cdots$$

has a ratio of $r = \frac{1}{2}$ with $a = 3$. Because $0 < |r| < 1$, the series converges and its sum is

$$S = \frac{a}{1-r} = \frac{3}{1-(1/2)} = 6.$$

b. The geometric series

$$\sum_{n=0}^{\infty} \left(\frac{3}{2}\right)^n = 1 + \frac{3}{2} + \frac{9}{4} + \frac{27}{8} + \cdots$$

has a ratio of $r = \frac{3}{2}$. Because $|r| \geq 1$, the series diverges.

The formula for the sum of a geometric series can be used to write a repeating decimal as the ratio of two integers, as demonstrated in the next example.

EXAMPLE 4 **A Geometric Series for a Repeating Decimal**

See LarsonCalculus.com for an interactive version of this type of example.

Use a geometric series to write $0.\overline{08}$ as the ratio of two integers.

Solution For the repeating decimal $0.\overline{08}$, you can write

$$\begin{aligned} 0.080808\ldots &= \frac{8}{10^2} + \frac{8}{10^4} + \frac{8}{10^6} + \frac{8}{10^8} + \cdots \\ &= \sum_{n=0}^{\infty} \left(\frac{8}{10^2}\right)\left(\frac{1}{10^2}\right)^n. \end{aligned}$$

For this series, you have $a = 8/10^2$ and $r = 1/10^2$. So,

$$0.080808\ldots = \frac{a}{1-r} = \frac{8/10^2}{1-(1/10^2)} = \frac{8}{99}.$$

Try dividing 8 by 99 on a calculator to see that it produces $0.\overline{08}$.

The convergence of a series is not affected by the removal of a finite number of terms from the beginning of the series. For instance, the geometric series

$$\sum_{n=4}^{\infty} \left(\frac{1}{2}\right)^n \quad \text{and} \quad \sum_{n=0}^{\infty} \left(\frac{1}{2}\right)^n$$

both converge. Furthermore, because the sum of the second series is

$$\frac{a}{1-r} = \frac{1}{1-(1/2)} = 2$$

you can conclude that the sum of the first series is

$$\begin{aligned} S &= 2 - \left[\left(\frac{1}{2}\right)^0 + \left(\frac{1}{2}\right)^1 + \left(\frac{1}{2}\right)^2 + \left(\frac{1}{2}\right)^3\right] \\ &= 2 - \frac{15}{8} \\ &= \frac{1}{8}. \end{aligned}$$

The properties in the next theorem are direct consequences of the corresponding properties of limits of sequences.

THEOREM 9.7 Properties of Infinite Series

Let $\Sigma\, a_n$ and $\Sigma\, b_n$ be convergent series, and let A, B, and c be real numbers. If $\Sigma\, a_n = A$ and $\Sigma\, b_n = B$, then the following series converge to the indicated sums.

1. $\displaystyle\sum_{n=1}^{\infty} ca_n = cA$
2. $\displaystyle\sum_{n=1}^{\infty} (a_n + b_n) = A + B$
3. $\displaystyle\sum_{n=1}^{\infty} (a_n - b_n) = A - B$

*n*th-Term Test for Divergence

The next theorem states that when a series converges, the limit of its nth term must be 0.

REMARK Be sure you see that the converse of Theorem 9.8 is generally not true. That is, if the sequence $\{a_n\}$ converges to 0, then the series $\Sigma\, a_n$ may either converge or diverge.

THEOREM 9.8 Limit of the *n*th Term of a Convergent Series

If $\displaystyle\sum_{n=1}^{\infty} a_n$ converges, then $\displaystyle\lim_{n\to\infty} a_n = 0$.

Proof Assume that

$$\sum_{n=1}^{\infty} a_n = \lim_{n\to\infty} S_n = L.$$

Then, because $S_n = S_{n-1} + a_n$ and

$$\lim_{n\to\infty} S_n = \lim_{n\to\infty} S_{n-1} = L$$

it follows that

$$\begin{aligned} L &= \lim_{n\to\infty} S_n \\ &= \lim_{n\to\infty} (S_{n-1} + a_n) \\ &= \lim_{n\to\infty} S_{n-1} + \lim_{n\to\infty} a_n \\ &= L + \lim_{n\to\infty} a_n \end{aligned}$$

which implies that $\{a_n\}$ converges to 0.

See LarsonCalculus.com for Bruce Edwards's video of this proof. ■

AP* Tips

The geometric series and nth term tests are commonly covered on the AP BC Exam.

The contrapositive of Theorem 9.8 provides a useful test for *divergence*. This ***n*th-Term Test for Divergence** states that if the limit of the nth term of a series does *not* converge to 0, then the series must diverge.

THEOREM 9.9 *n*th-Term Test for Divergence

If $\displaystyle\lim_{n\to\infty} a_n \neq 0$, then $\displaystyle\sum_{n=1}^{\infty} a_n$ diverges.

EXAMPLE 5 Using the nth-Term Test for Divergence

a. For the series $\sum_{n=0}^{\infty} 2^n$, you have

$$\lim_{n\to\infty} 2^n = \infty.$$

So, the limit of the nth term is not 0, and the series diverges.

b. For the series $\sum_{n=1}^{\infty} \frac{n!}{2n! + 1}$, you have

$$\lim_{n\to\infty} \frac{n!}{2n! + 1} = \frac{1}{2}.$$

So, the limit of the nth term is not 0, and the series diverges.

c. For the series $\sum_{n=1}^{\infty} \frac{1}{n}$, you have

$$\lim_{n\to\infty} \frac{1}{n} = 0.$$

Because the limit of the nth term is 0, the nth-Term Test for Divergence does *not* apply and you can draw no conclusions about convergence or divergence. (In the next section, you will see that this particular series diverges.)

REMARK The series in Example 5(c) will play an important role in this chapter.

$$\sum_{n=1}^{\infty} \frac{1}{n} = 1 + \frac{1}{2} + \frac{1}{3} + \frac{1}{4} + \cdots$$

You will see that this series diverges even though the nth term approaches 0 as n approaches ∞.

EXAMPLE 6 Bouncing Ball Problem

A ball is dropped from a height of 6 feet and begins bouncing, as shown in Figure 9.7. The height of each bounce is three-fourths the height of the previous bounce. Find the total vertical distance traveled by the ball.

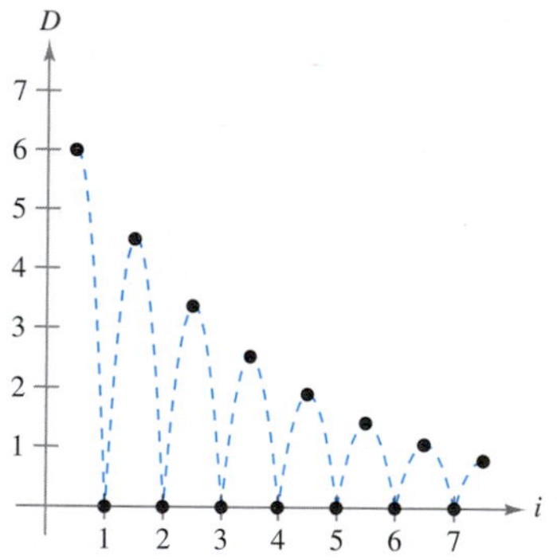

The height of each bounce is three-fourths the height of the preceding bounce.
Figure 9.7

Solution When the ball hits the ground for the first time, it has traveled a distance of $D_1 = 6$ feet. For subsequent bounces, let D_i be the distance traveled up and down. For example, D_2 and D_3 are

$$D_2 = \underbrace{6\left(\frac{3}{4}\right)}_{\text{Up}} + \underbrace{6\left(\frac{3}{4}\right)}_{\text{Down}} = 12\left(\frac{3}{4}\right)$$

and

$$D_3 = \underbrace{6\left(\frac{3}{4}\right)\left(\frac{3}{4}\right)}_{\text{Up}} + \underbrace{6\left(\frac{3}{4}\right)\left(\frac{3}{4}\right)}_{\text{Down}} = 12\left(\frac{3}{4}\right)^2.$$

By continuing this process, it can be determined that the total vertical distance is

$$\begin{aligned}
D &= 6 + 12\left(\frac{3}{4}\right) + 12\left(\frac{3}{4}\right)^2 + 12\left(\frac{3}{4}\right)^3 + \cdots \\
&= 6 + 12\sum_{n=0}^{\infty}\left(\frac{3}{4}\right)^{n+1} \\
&= 6 + 12\left(\frac{3}{4}\right)\sum_{n=0}^{\infty}\left(\frac{3}{4}\right)^{n} \\
&= 6 + 9\left[\frac{1}{1-(3/4)}\right] \\
&= 6 + 9(4) \\
&= 42 \text{ feet.}
\end{aligned}$$

9.2 Exercises

See CalcChat.com for tutorial help and worked-out solutions to odd-numbered exercises.

Finding Partial Sums In Exercises 1–6, find the sequence of partial sums S_1, S_2, S_3, S_4, and S_5.

1. $1 + \frac{1}{4} + \frac{1}{9} + \frac{1}{16} + \frac{1}{25} + \cdots$
2. $\frac{1}{2 \cdot 3} + \frac{2}{3 \cdot 4} + \frac{3}{4 \cdot 5} + \frac{4}{5 \cdot 6} + \frac{5}{6 \cdot 7} + \cdots$
3. $3 - \frac{9}{2} + \frac{27}{4} - \frac{81}{8} + \frac{243}{16} - \cdots$
4. $1 + \frac{1}{2} + \frac{1}{4} + \frac{1}{6} + \frac{1}{8} + \frac{1}{10} + \cdots$
5. $\sum_{n=1}^{\infty} \frac{3}{2^{n-1}}$
6. $\sum_{n=1}^{\infty} \frac{(-1)^{n+1}}{n!}$

Verifying Divergence In Exercises 7–14, verify that the infinite series diverges.

7. $\sum_{n=0}^{\infty} \left(\frac{7}{6}\right)^n$
8. $\sum_{n=0}^{\infty} 4(-1.05)^n$
9. $\sum_{n=1}^{\infty} \frac{n}{n+1}$
10. $\sum_{n=1}^{\infty} \frac{n}{2n+3}$
11. $\sum_{n=1}^{\infty} \frac{n^2}{n^2+1}$
12. $\sum_{n=1}^{\infty} \frac{n}{\sqrt{n^2+1}}$
13. $\sum_{n=1}^{\infty} \frac{2^n+1}{2^{n+1}}$
14. $\sum_{n=1}^{\infty} \frac{n!}{2^n}$

Verifying Convergence In Exercises 15–20, verify that the infinite series converges.

15. $\sum_{n=0}^{\infty} \left(\frac{5}{6}\right)^n$
16. $\sum_{n=1}^{\infty} 2\left(-\frac{1}{2}\right)^n$
17. $\sum_{n=0}^{\infty} (0.9)^n = 1 + 0.9 + 0.81 + 0.729 + \cdots$
18. $\sum_{n=0}^{\infty} (-0.6)^n = 1 - 0.6 + 0.36 - 0.216 + \cdots$
19. $\sum_{n=1}^{\infty} \frac{1}{n(n+1)}$ (*Hint*: Use partial fractions.)
20. $\sum_{n=1}^{\infty} \frac{1}{n(n+2)}$ (*Hint*: Use partial fractions.)

Numerical, Graphical, and Analytic Analysis In Exercises 21–24, (a) find the sum of the series, (b) use a graphing utility to find the indicated partial sum S_n and complete the table, (c) use a graphing utility to graph the first 10 terms of the sequence of partial sums and a horizontal line representing the sum, and (d) explain the relationship between the magnitudes of the terms of the series and the rate at which the sequence of partial sums approaches the sum of the series.

n	5	10	20	50	100
S_n					

21. $\sum_{n=1}^{\infty} \frac{6}{n(n+3)}$
22. $\sum_{n=1}^{\infty} \frac{4}{n(n+4)}$
23. $\sum_{n=1}^{\infty} 2(0.9)^{n-1}$
24. $\sum_{n=1}^{\infty} 10\left(-\frac{1}{4}\right)^{n-1}$

Finding the Sum of a Convergent Series In Exercises 25–34, find the sum of the convergent series.

25. $\sum_{n=0}^{\infty} 5\left(\frac{2}{3}\right)^n$
26. $\sum_{n=0}^{\infty} \left(-\frac{1}{5}\right)^n$
27. $\sum_{n=1}^{\infty} \frac{4}{n(n+2)}$
28. $\sum_{n=1}^{\infty} \frac{1}{(2n+1)(2n+3)}$
29. $8 + 6 + \frac{9}{2} + \frac{27}{8} + \cdots$
30. $9 - 3 + 1 - \frac{1}{3} + \cdots$
31. $\sum_{n=0}^{\infty} \left(\frac{1}{2^n} - \frac{1}{3^n}\right)$
32. $\sum_{n=0}^{\infty} [(0.3)^n + (0.8)^n]$
33. $\sum_{n=1}^{\infty} (\sin 1)^n$
34. $\sum_{n=1}^{\infty} \frac{1}{9n^2 + 3n - 2}$

Using a Geometric Series In Exercises 35–40, (a) write the repeating decimal as a geometric series, and (b) write its sum as the ratio of two integers.

35. $0.\overline{4}$
36. $0.\overline{36}$
37. $0.\overline{81}$
38. $0.\overline{01}$
39. $0.0\overline{75}$
40. $0.2\overline{15}$

Determining Convergence or Divergence In Exercises 41–54, determine the convergence or divergence of the series.

41. $\sum_{n=0}^{\infty} (1.075)^n$
42. $\sum_{n=0}^{\infty} \frac{3^n}{1000}$
43. $\sum_{n=1}^{\infty} \frac{n+10}{10n+1}$
44. $\sum_{n=1}^{\infty} \frac{4n+1}{3n-1}$
45. $\sum_{n=1}^{\infty} \left(\frac{1}{n} - \frac{1}{n+2}\right)$
46. $\sum_{n=1}^{\infty} \left(\frac{1}{n+1} - \frac{1}{n+2}\right)$
47. $\sum_{n=1}^{\infty} \frac{3^n}{n^3}$
48. $\sum_{n=0}^{\infty} \frac{3}{5^n}$
49. $\sum_{n=2}^{\infty} \frac{n}{\ln n}$
50. $\sum_{n=1}^{\infty} \ln \frac{1}{n}$
51. $\sum_{n=1}^{\infty} \left(1 + \frac{k}{n}\right)^n$
52. $\sum_{n=1}^{\infty} e^{-n}$
53. $\sum_{n=1}^{\infty} \arctan n$
54. $\sum_{n=1}^{\infty} \ln\left(\frac{n+1}{n}\right)$

WRITING ABOUT CONCEPTS

55. **Series** State the definitions of convergent and divergent series.
56. **Sequence and Series** Describe the difference between $\lim_{n \to \infty} a_n = 5$ and $\sum_{n=1}^{\infty} a_n = 5$.

WRITING ABOUT CONCEPTS (continued)

57. Geometric Series Define a geometric series, state when it converges, and give the formula for the sum of a convergent geometric series.

58. *n*th-Term Test for Divergence State the *n*th-Term Test for Divergence.

59. Comparing Series Explain any differences among the following series.

(a) $\sum_{n=1}^{\infty} a_n$ (b) $\sum_{k=1}^{\infty} a_k$ (c) $\sum_{n=1}^{\infty} a_k$

60. Using a Series

(a) You delete a finite number of terms from a divergent series. Will the new series still diverge? Explain your reasoning.

(b) You add a finite number of terms to a convergent series. Will the new series still converge? Explain your reasoning.

Making a Series Converge **In Exercises 61–66, find all values of x for which the series converges. For these values of x, write the sum of the series as a function of x.**

61. $\sum_{n=1}^{\infty} (3x)^n$

62. $\sum_{n=0}^{\infty} \left(\frac{2}{x}\right)^n$

63. $\sum_{n=1}^{\infty} (x-1)^n$

64. $\sum_{n=0}^{\infty} 5\left(\frac{x-2}{3}\right)^n$

65. $\sum_{n=0}^{\infty} (-1)^n x^n$

66. $\sum_{n=0}^{\infty} (-1)^n x^{2n}$

Using a Geometric Series **In Exercises 67 and 68, (a) find the common ratio of the geometric series, (b) write the function that gives the sum of the series, and (c) use a graphing utility to graph the function and the partial sums S_3 and S_5. What do you notice?**

67. $1 + x + x^2 + x^3 + \cdots$

68. $1 - \frac{x}{2} + \frac{x^2}{4} - \frac{x^3}{8} + \cdots$

Writing **In Exercises 69 and 70, use a graphing utility to determine the first term that is less than 0.0001 in each of the convergent series. Note that the answers are very different. Explain how this will affect the rate at which the series converges.**

69. $\sum_{n=1}^{\infty} \frac{1}{n(n+1)}, \quad \sum_{n=1}^{\infty} \left(\frac{1}{8}\right)^n$

70. $\sum_{n=1}^{\infty} \frac{1}{2^n}, \quad \sum_{n=1}^{\infty} (0.01)^n$

71. Marketing An electronic games manufacturer producing a new product estimates the annual sales to be 8000 units. Each year, 5% of the units that have been sold will become inoperative. So, 8000 units will be in use after 1 year, $[8000 + 0.95(8000)]$ units will be in use after 2 years, and so on. How many units will be in use after n years?

72. Depreciation A company buys a machine for \$475,000 that depreciates at a rate of 30% per year. Find a formula for the value of the machine after n years. What is its value after 5 years?

AISPIX by Image Source/Shutterstock.com

73. Multiplier Effect

The total annual spending by tourists in a resort city is \$200 million. Approximately 75% of that revenue is again spent in the resort city, and of that amount approximately 75% is again spent in the same city, and so on. Write the geometric series that gives the total amount of spending generated by the \$200 million and find the sum of the series.

74. Multiplier Effect Repeat Exercise 73 when the percent of the revenue that is spent again in the city decreases to 60%.

75. Distance A ball is dropped from a height of 16 feet. Each time it drops h feet, it rebounds $0.81h$ feet. Find the total distance traveled by the ball.

76. Time The ball in Exercise 75 takes the following times for each fall.

$$\begin{aligned}
s_1 &= -16t^2 + 16, & s_1 &= 0 \text{ when } t = 1\\
s_2 &= -16t^2 + 16(0.81), & s_2 &= 0 \text{ when } t = 0.9\\
s_3 &= -16t^2 + 16(0.81)^2, & s_3 &= 0 \text{ when } t = (0.9)^2\\
s_4 &= -16t^2 + 16(0.81)^3, & s_4 &= 0 \text{ when } t = (0.9)^3\\
&\vdots & &\vdots\\
s_n &= -16t^2 + 16(0.81)^{n-1}, & s_n &= 0 \text{ when } t = (0.9)^{n-1}
\end{aligned}$$

Beginning with s_2, the ball takes the same amount of time to bounce up as it does to fall, and so the total time elapsed before it comes to rest is given by

$$t = 1 + 2\sum_{n=1}^{\infty} (0.9)^n.$$

Find this total time.

Probability **In Exercises 77 and 78, the random variable n represents the number of units of a product sold per day in a store. The probability distribution of n is given by $P(n)$. Find the probability that two units are sold in a given day $[P(2)]$ and show that $P(0) + P(1) + P(2) + P(3) + \cdots = 1$.**

77. $P(n) = \frac{1}{2}\left(\frac{1}{2}\right)^n$

78. $P(n) = \frac{1}{3}\left(\frac{2}{3}\right)^n$

79. Probability A fair coin is tossed repeatedly. The probability that the first head occurs on the nth toss is given by $P(n) = \left(\frac{1}{2}\right)^n$, where $n \geq 1$.

(a) Show that $\sum_{n=1}^{\infty} \left(\frac{1}{2}\right)^n = 1$.

(b) The expected number of tosses required until the first head occurs in the experiment is given by

$$\sum_{n=1}^{\infty} n\left(\frac{1}{2}\right)^n.$$

Is this series geometric?

(c) Use a computer algebra system to find the sum in part (b).

80. Probability In an experiment, three people toss a fair coin one at a time until one of them tosses a head. Determine, for each person, the probability that he or she tosses the first head. Verify that the sum of the three probabilities is 1.

81. Area The sides of a square are 16 inches in length. A new square is formed by connecting the midpoints of the sides of the original square, and two of the triangles outside the second square are shaded (see figure). Determine the area of the shaded regions (a) when this process is continued five more times, and (b) when this pattern of shading is continued infinitely.

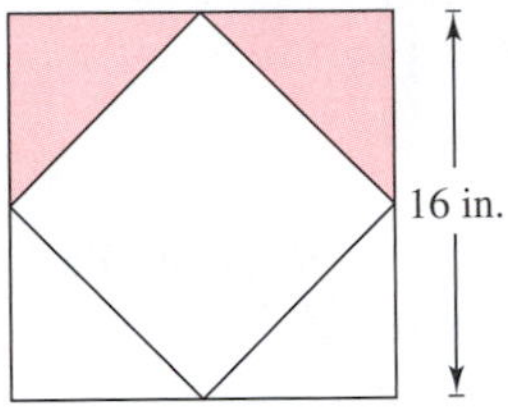

Figure for 81

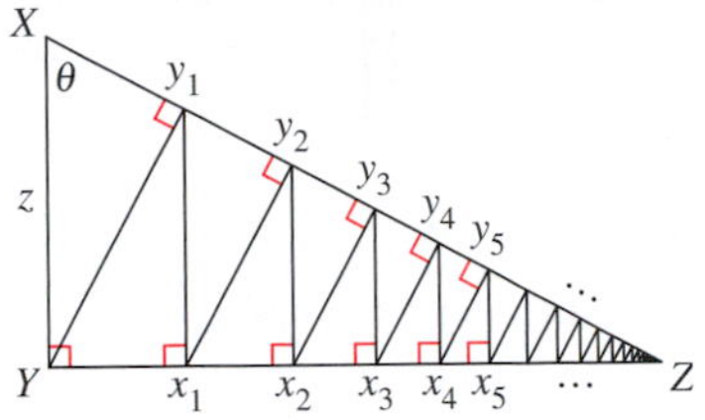

Figure for 82

82. Length A right triangle XYZ is shown above where $|XY| = z$ and $\angle X = \theta$. Line segments are continually drawn to be perpendicular to the triangle, as shown in the figure.

(a) Find the total length of the perpendicular line segments $|Yy_1| + |x_1y_1| + |x_1y_2| + \cdots$ in terms of z and θ.

(b) Find the total length of the perpendicular line segments when $z = 1$ and $\theta = \pi/6$.

Using a Geometric Series **In Exercises 83–86, use the formula for the nth partial sum of a geometric series**

$$\sum_{i=0}^{n-1} ar^i = \frac{a(1 - r^n)}{1 - r}.$$

83. Present Value The winner of a $2,000,000 sweepstakes will be paid $100,000 per year for 20 years. The money earns 6% interest per year. The present value of the winnings is $\sum_{n=1}^{20} 100{,}000\left(\frac{1}{1.06}\right)^n$. Compute the present value and interpret its meaning.

84. Annuities When an employee receives a paycheck at the end of each month, P dollars is invested in a retirement account. These deposits are made each month for t years and the account earns interest at the annual percentage rate r. When the interest is compounded monthly, the amount A in the account at the end of t years is

$$A = P + P\left(1 + \frac{r}{12}\right) + \cdots + P\left(1 + \frac{r}{12}\right)^{12t-1}$$
$$= P\left(\frac{12}{r}\right)\left[\left(1 + \frac{r}{12}\right)^{12t} - 1\right].$$

When the interest is compounded continuously, the amount A in the account after t years is

$$A = P + Pe^{r/12} + Pe^{2r/12} + Pe^{(12t-1)r/12}$$
$$= \frac{P(e^{rt} - 1)}{e^{r/12} - 1}.$$

Verify the formulas for the sums given above.

85. Salary You go to work at a company that pays $0.01 for the first day, $0.02 for the second day, $0.04 for the third day, and so on. If the daily wage keeps doubling, what would your total income be for working (a) 29 days, (b) 30 days, and (c) 31 days?

86. Sphereflake

The sphereflake shown below is a computer-generated fractal that was created by Eric Haines. The radius of the large sphere is 1. To the large sphere, nine spheres of radius $\frac{1}{3}$ are attached. To each of these, nine spheres of radius $\frac{1}{9}$ are attached. This process is continued infinitely. Prove that the sphereflake has an infinite surface area.

Annuities **In Exercises 87–90, consider making monthly deposits of P dollars in a savings account at an annual interest rate r. Use the results of Exercise 84 to find the balance A after t years when the interest is compounded (a) monthly and (b) continuously.**

87. $P = \$45$, $r = 3\%$, $t = 20$ years

88. $P = \$75$, $r = 5.5\%$, $t = 25$ years

89. $P = \$100$, $r = 4\%$, $t = 35$ years

90. $P = \$30$, $r = 6\%$, $t = 50$ years

True or False? **In Exercises 91–96, determine whether the statement is true or false. If it is false, explain why or give an example that shows it is false.**

91. If $\lim_{n\to\infty} a_n = 0$, then $\sum_{n=1}^{\infty} a_n$ converges.

92. If $\sum_{n=1}^{\infty} a_n = L$, then $\sum_{n=0}^{\infty} a_n = L + a_0$.

93. If $|r| < 1$, then $\sum_{n=1}^{\infty} ar^n = \frac{a}{1 - r}$.

94. The series $\sum_{n=1}^{\infty} \frac{n}{1000(n + 1)}$ diverges.

95. $0.75 = 0.749999 \ldots$.

96. Every decimal with a repeating pattern of digits is a rational number.

97. Using Divergent Series Find two divergent series Σa_n and Σb_n such that $\Sigma(a_n + b_n)$ converges.

98. Proof Given two infinite series Σa_n and Σb_n such that Σa_n converges and Σb_n diverges, prove that $\Sigma(a_n + b_n)$ diverges.

99. Fibonacci Sequence The Fibonacci sequence is defined recursively by $a_{n+2} = a_n + a_{n+1}$, where $a_1 = 1$ and $a_2 = 1$.

(a) Show that $\dfrac{1}{a_{n+1}a_{n+3}} = \dfrac{1}{a_{n+1}a_{n+2}} - \dfrac{1}{a_{n+2}a_{n+3}}$.

(b) Show that $\displaystyle\sum_{n=0}^{\infty} \frac{1}{a_{n+1}a_{n+3}} = 1$.

100. Remainder Let Σa_n be a convergent series, and let

$$R_N = a_{N+1} + a_{N+2} + \cdots$$

be the remainder of the series after the first N terms. Prove that $\displaystyle\lim_{N\to\infty} R_N = 0$.

101. Proof Prove that $\dfrac{1}{r} + \dfrac{1}{r^2} + \dfrac{1}{r^3} + \cdots = \dfrac{1}{r-1}$, for $|r| > 1$.

102. HOW DO YOU SEE IT? The figure below represents an informal way of showing that $\displaystyle\sum_{n=1}^{\infty} \frac{1}{n^2} < 2$. Explain how the figure implies this conclusion.

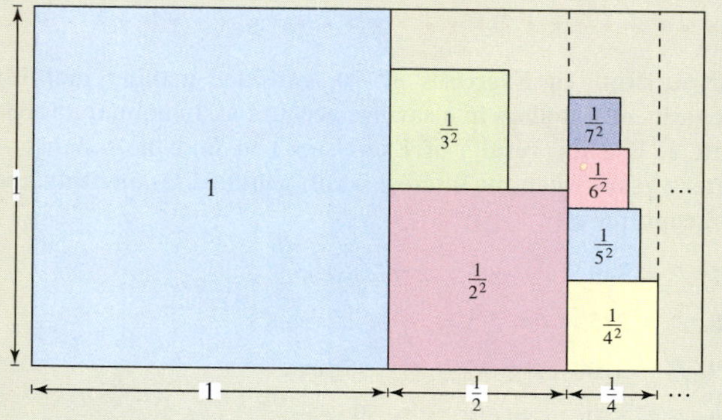

FOR FURTHER INFORMATION For more on this exercise, see the article "Convergence with Pictures" by P. J. Rippon in *American Mathematical Monthly.*

PUTNAM EXAM CHALLENGE

103. Express $\displaystyle\sum_{k=1}^{\infty} \frac{6^k}{(3^{k+1} - 2^{k+1})(3^k - 2^k)}$ as a rational number.

104. Let $f(n)$ be the sum of the first n terms of the sequence 0, 1, 1, 2, 2, 3, 3, 4, . . . , where the nth term is given by

$$a_n = \begin{cases} n/2, & \text{if } n \text{ is even} \\ (n-1)/2, & \text{if } n \text{ is odd} \end{cases}.$$

Show that if x and y are positive integers and $x > y$ then $xy = f(x + y) - f(x - y)$.

These problems were composed by the Committee on the Putnam Prize Competition.

SECTION PROJECT

Cantor's Disappearing Table

The following procedure shows how to make a table disappear by removing only half of the table!

(a) Original table has a length of L.

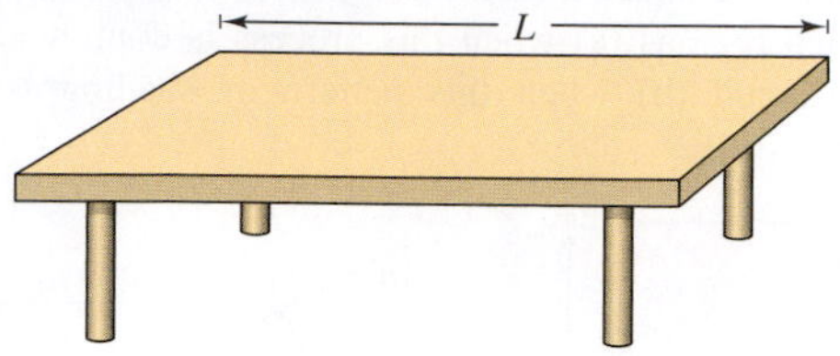

(b) Remove $\frac{1}{4}$ of the table centered at the midpoint. Each remaining piece has a length that is less than $\frac{1}{2}L$.

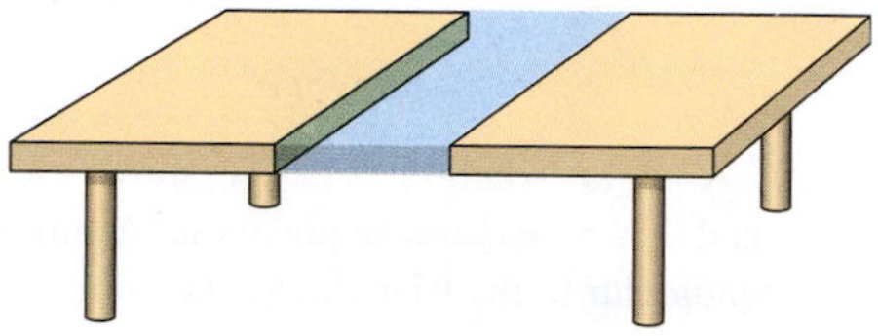

(c) Remove $\frac{1}{8}$ of the table by taking sections of length $\frac{1}{16}L$ from the centers of each of the two remaining pieces. Now, you have removed $\frac{1}{4} + \frac{1}{8}$ of the table. Each remaining piece has a length that is less than $\frac{1}{4}L$.

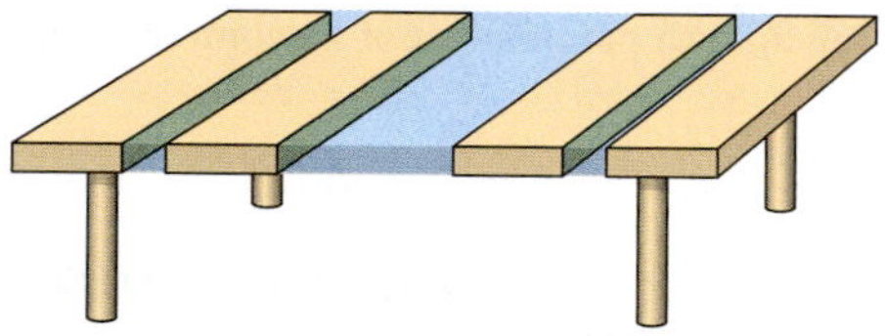

(d) Remove $\frac{1}{16}$ of the table by taking sections of length $\frac{1}{64}L$ from the centers of each of the four remaining pieces. Now, you have removed $\frac{1}{4} + \frac{1}{8} + \frac{1}{16}$ of the table. Each remaining piece has a length that is less than $\frac{1}{8}L$.

Will continuing this process cause the table to disappear, even though you have only removed half of the table? Why?

FOR FURTHER INFORMATION Read the article "Cantor's Disappearing Table" by Larry E. Knop in *The College Mathematics Journal.* To view this article, go to *MathArticles.com.*

9.3 The Integral Test and p-Series

- **Use the Integral Test to determine whether an infinite series converges or diverges.**
- **Use properties of p-series and harmonic series.**

The Integral Test

In this and the next section, you will study several convergence tests that apply to series with *positive* terms.

> **THEOREM 9.10 The Integral Test**
>
> If f is positive, continuous, and decreasing for $x \geq 1$ and $a_n = f(n)$, then
>
> $$\sum_{n=1}^{\infty} a_n \quad \text{and} \quad \int_1^{\infty} f(x)\,dx$$
>
> either both converge or both diverge.

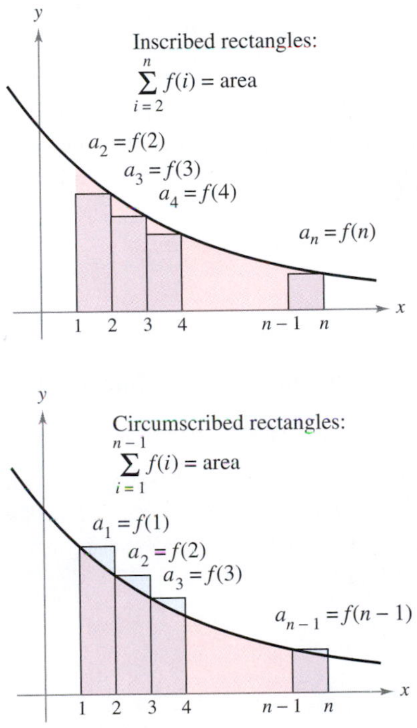

Figure 9.8

Proof Begin by partitioning the interval $[1, n]$ into $(n-1)$ unit intervals, as shown in Figure 9.8. The total areas of the inscribed rectangles and the circumscribed rectangles are

$$\sum_{i=2}^{n} f(i) = f(2) + f(3) + \cdot \cdot \cdot + f(n) \qquad \text{Inscribed area}$$

and

$$\sum_{i=1}^{n-1} f(i) = f(1) + f(2) + \cdot \cdot \cdot + f(n-1). \qquad \text{Circumscribed area}$$

The exact area under the graph of f from $x = 1$ to $x = n$ lies between the inscribed and circumscribed areas.

$$\sum_{i=2}^{n} f(i) \leq \int_1^n f(x)\,dx \leq \sum_{i=1}^{n-1} f(i)$$

Using the nth partial sum, $S_n = f(1) + f(2) + \cdot \cdot \cdot + f(n)$, you can write this inequality as

$$S_n - f(1) \leq \int_1^n f(x)\,dx \leq S_{n-1}.$$

Now, assuming that $\int_1^{\infty} f(x)\,dx$ converges to L, it follows that for $n \geq 1$

$$S_n - f(1) \leq L \quad \Longrightarrow \quad S_n \leq L + f(1).$$

Consequently, $\{S_n\}$ is bounded and monotonic, and by Theorem 9.5 it converges. So, $\sum a_n$ converges. For the other direction of the proof, assume that the improper integral diverges. Then $\int_1^n f(x)\,dx$ approaches infinity as $n \to \infty$, and the inequality $S_{n-1} \geq \int_1^n f(x)\,dx$ implies that $\{S_n\}$ diverges. So, $\sum a_n$ diverges. ∎

See LarsonCalculus.com for Bruce Edwards's video of this proof.

Remember that the convergence or divergence of $\sum a_n$ is not affected by deleting the first N terms. Similarly, when the conditions for the Integral Test are satisfied for all $x \geq N > 1$, you can simply use the integral $\int_N^{\infty} f(x)\,dx$ to test for convergence or divergence. (This is illustrated in Example 4.)

EXAMPLE 1 **Using the Integral Test**

Apply the Integral Test to the series $\sum_{n=1}^{\infty} \frac{n}{n^2+1}$.

Solution The function $f(x) = x/(x^2 + 1)$ is positive and continuous for $x \geq 1$. To determine whether f is decreasing, find the derivative.

$$f'(x) = \frac{(x^2+1)(1) - x(2x)}{(x^2+1)^2} = \frac{-x^2+1}{(x^2+1)^2}$$

So, $f'(x) < 0$ for $x > 1$ and it follows that f satisfies the conditions for the Integral Test. You can integrate to obtain

$$\begin{aligned}\int_1^{\infty} \frac{x}{x^2+1}\,dx &= \frac{1}{2}\int_1^{\infty} \frac{2x}{x^2+1}\,dx \\ &= \frac{1}{2}\lim_{b\to\infty}\int_1^{b} \frac{2x}{x^2+1}\,dx \\ &= \frac{1}{2}\lim_{b\to\infty}\Big[\ln(x^2+1)\Big]_1^b \\ &= \frac{1}{2}\lim_{b\to\infty}[\ln(b^2+1) - \ln 2] \\ &= \infty.\end{aligned}$$

So, the series *diverges*.

EXAMPLE 2 **Using the Integral Test**

See LarsonCalculus.com for an interactive version of this type of example.

Apply the Integral Test to the series $\sum_{n=1}^{\infty} \frac{1}{n^2+1}$.

Solution Because $f(x) = 1/(x^2 + 1)$ satisfies the conditions for the Integral Test (check this), you can integrate to obtain

$$\begin{aligned}\int_1^{\infty} \frac{1}{x^2+1}\,dx &= \lim_{b\to\infty}\int_1^{b} \frac{1}{x^2+1}\,dx \\ &= \lim_{b\to\infty}\Big[\arctan x\Big]_1^b \\ &= \lim_{b\to\infty}(\arctan b - \arctan 1) \\ &= \frac{\pi}{2} - \frac{\pi}{4} \\ &= \frac{\pi}{4}.\end{aligned}$$

So, the series *converges* (see Figure 9.9).

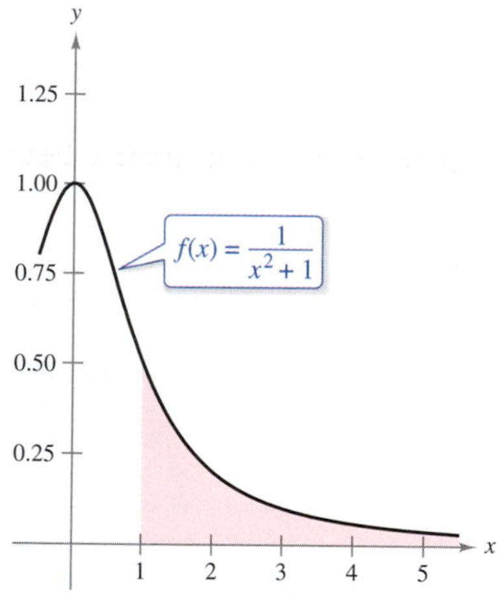

Because the improper integral converges, the infinite series also converges.
Figure 9.9

In Example 2, the fact that the improper integral converges to $\pi/4$ does not imply that the infinite series converges to $\pi/4$. To approximate the sum of the series, you can use the inequality

$$\sum_{n=1}^{N} \frac{1}{n^2+1} \leq \sum_{n=1}^{\infty} \frac{1}{n^2+1} \leq \sum_{n=1}^{N} \frac{1}{n^2+1} + \int_N^{\infty} \frac{1}{x^2+1}\,dx.$$

(See Exercise 54.) The larger the value of N, the better the approximation. For instance, using $N = 200$ produces $1.072 \leq \Sigma\, 1/(n^2 + 1) \leq 1.077$.

p-Series and Harmonic Series

HARMONIC SERIES

Pythagoras and his students paid close attention to the development of music as an abstract science. This led to the discovery of the relationship between the tone and the length of a vibrating string. It was observed that the most beautiful musical harmonies corresponded to the simplest ratios of whole numbers. Later mathematicians developed this idea into the harmonic series, where the terms in the harmonic series correspond to the nodes on a vibrating string that produce multiples of the fundamental frequency. For example, $\frac{1}{2}$ is twice the fundamental frequency, $\frac{1}{3}$ is three times the fundamental frequency, and so on.

In the remainder of this section, you will investigate a second type of series that has a simple arithmetic test for convergence or divergence. A series of the form

$$\sum_{n=1}^{\infty} \frac{1}{n^p} = \frac{1}{1^p} + \frac{1}{2^p} + \frac{1}{3^p} + \cdots \qquad p\text{-series}$$

is a ***p*-series,** where p is a positive constant. For $p = 1$, the series

$$\sum_{n=1}^{\infty} \frac{1}{n} = 1 + \frac{1}{2} + \frac{1}{3} + \cdots \qquad \text{Harmonic series}$$

is the **harmonic** series. A **general harmonic series** is of the form $\Sigma 1/(an + b)$. In music, strings of the same material, diameter, and tension, and whose lengths form a harmonic series, produce harmonic tones.

The Integral Test is convenient for establishing the convergence or divergence of p-series. This is shown in the proof of Theorem 9.11.

AP* Tips

On the AP BC Exam, the p-series test is covered most often on the multiple choice section. The Integral test is rarely covered.

THEOREM 9.11 Convergence of p-Series

The p-series

$$\sum_{n=1}^{\infty} \frac{1}{n^p} = \frac{1}{1^p} + \frac{1}{2^p} + \frac{1}{3^p} + \frac{1}{4^p} + \cdots$$

converges for $p > 1$, and diverges for $0 < p \le 1$.

Proof The proof follows from the Integral Test and from Theorem 8.5, which states that

$$\int_1^{\infty} \frac{1}{x^p}\,dx$$

converges for $p > 1$ and diverges for $0 < p \le 1$.

See LarsonCalculus.com for Bruce Edwards's video of this proof.

EXAMPLE 3 Convergent and Divergent p-Series

Discuss the convergence or divergence of (a) the harmonic series and (b) the p-series with $p = 2$.

Solution

a. From Theorem 9.11, it follows that the harmonic series

$$\sum_{n=1}^{\infty} \frac{1}{n} = \frac{1}{1} + \frac{1}{2} + \frac{1}{3} + \cdots \qquad p = 1$$

diverges.

b. From Theorem 9.11, it follows that the p-series

$$\sum_{n=1}^{\infty} \frac{1}{n^2} = \frac{1}{1^2} + \frac{1}{2^2} + \frac{1}{3^2} + \cdots \qquad p = 2$$

converges.

The sum of the series in Example 3(b) can be shown to be $\pi^2/6$. (This was proved by Leonhard Euler, but the proof is too difficult to present here.) Be sure you see that the Integral Test does not tell you that the sum of the series is equal to the value of the integral. For instance, the sum of the series in Example 3(b) is

$$\sum_{n=1}^{\infty} \frac{1}{n^2} = \frac{\pi^2}{6} \approx 1.645$$

whereas the value of the corresponding improper integral is

$$\int_1^{\infty} \frac{1}{x^2}\, dx = 1.$$

EXAMPLE 4 **Testing a Series for Convergence**

Determine whether the series

$$\sum_{n=2}^{\infty} \frac{1}{n \ln n}$$

converges or diverges.

Solution This series is similar to the divergent harmonic series. If its terms were greater than those of the harmonic series, you would expect it to diverge. However, because its terms are less than those of the harmonic series, you are not sure what to expect. The function

$$f(x) = \frac{1}{x \ln x}$$

is positive and continuous for $x \geq 2$. To determine whether f is decreasing, first rewrite f as

$$f(x) = (x \ln x)^{-1}$$

and then find its derivative.

$$f'(x) = (-1)(x \ln x)^{-2}(1 + \ln x) = -\frac{1 + \ln x}{x^2(\ln x)^2}$$

So, $f'(x) < 0$ for $x > 2$ and it follows that f satisfies the conditions for the Integral Test.

$$\begin{aligned}
\int_2^{\infty} \frac{1}{x \ln x}\, dx &= \int_2^{\infty} \frac{1/x}{\ln x}\, dx \\
&= \lim_{b \to \infty} \Big[\ln(\ln x) \Big]_2^b \\
&= \lim_{b \to \infty} [\ln(\ln b) - \ln(\ln 2)] \\
&= \infty
\end{aligned}$$

The series diverges. ■

Note that the infinite series in Example 4 diverges very slowly. For instance, as shown in the table, the sum of the first 10 terms is approximately 1.6878196, whereas the sum of the first 100 terms is just slightly greater: 2.3250871. In fact, the sum of the first 10,000 terms is approximately 3.0150217. You can see that although the infinite series "adds up to infinity," it does so very slowly.

n	11	101	1001	10,001	100,001
S_n	1.6878	2.3251	2.7275	3.0150	3.2382

9.3 Exercises

See CalcChat.com for tutorial help and worked-out solutions to odd-numbered exercises.

Using the Integral Test In Exercises 1–22, confirm that the Integral Test can be applied to the series. Then use the Integral Test to determine the convergence or divergence of the series.

1. $\sum_{n=1}^{\infty} \frac{1}{n+3}$

2. $\sum_{n=1}^{\infty} \frac{2}{3n+5}$

3. $\sum_{n=1}^{\infty} \frac{1}{2^n}$

4. $\sum_{n=1}^{\infty} 3^{-n}$

5. $\sum_{n=1}^{\infty} e^{-n}$

6. $\sum_{n=1}^{\infty} ne^{-n/2}$

7. $\frac{1}{2} + \frac{1}{5} + \frac{1}{10} + \frac{1}{17} + \frac{1}{26} + \cdots$

8. $\frac{1}{3} + \frac{1}{5} + \frac{1}{7} + \frac{1}{9} + \frac{1}{11} + \cdots$

9. $\frac{\ln 2}{2} + \frac{\ln 3}{3} + \frac{\ln 4}{4} + \frac{\ln 5}{5} + \frac{\ln 6}{6} + \cdots$

10. $\frac{\ln 2}{\sqrt{2}} + \frac{\ln 3}{\sqrt{3}} + \frac{\ln 4}{\sqrt{4}} + \frac{\ln 5}{\sqrt{5}} + \frac{\ln 6}{\sqrt{6}} + \cdots$

11. $\frac{1}{\sqrt{1}(\sqrt{1}+1)} + \frac{1}{\sqrt{2}(\sqrt{2}+1)} + \frac{1}{\sqrt{3}(\sqrt{3}+1)} + \cdots + \frac{1}{\sqrt{n}(\sqrt{n}+1)} + \cdots$

12. $\frac{1}{4} + \frac{2}{7} + \frac{3}{12} + \cdots + \frac{n}{n^2+3} + \cdots$

13. $\sum_{n=1}^{\infty} \frac{\arctan n}{n^2+1}$

14. $\sum_{n=2}^{\infty} \frac{\ln n}{n^3}$

15. $\sum_{n=1}^{\infty} \frac{\ln n}{n^2}$

16. $\sum_{n=2}^{\infty} \frac{1}{n\sqrt{\ln n}}$

17. $\sum_{n=1}^{\infty} \frac{1}{(2n+3)^3}$

18. $\sum_{n=1}^{\infty} \frac{n+2}{n+1}$

19. $\sum_{n=1}^{\infty} \frac{4n}{2n^2+1}$

20. $\sum_{n=1}^{\infty} \frac{1}{\sqrt{n+2}}$

21. $\sum_{n=1}^{\infty} \frac{n}{n^4+1}$

22. $\sum_{n=1}^{\infty} \frac{n}{n^4+2n^2+1}$

Using the Integral Test In Exercises 23 and 24, use the Integral Test to determine the convergence or divergence of the series, where k is a positive integer.

23. $\sum_{n=1}^{\infty} \frac{n^{k-1}}{n^k+c}$

24. $\sum_{n=1}^{\infty} n^k e^{-n}$

Requirements of the Integral Test In Exercises 25–28, explain why the Integral Test does not apply to the series.

25. $\sum_{n=1}^{\infty} \frac{(-1)^n}{n}$

26. $\sum_{n=1}^{\infty} e^{-n}\cos n$

27. $\sum_{n=1}^{\infty} \frac{2+\sin n}{n}$

28. $\sum_{n=1}^{\infty} \left(\frac{\sin n}{n}\right)^2$

Using the Integral Test In Exercises 29–32, use the Integral Test to determine the convergence or divergence of the p-series.

29. $\sum_{n=1}^{\infty} \frac{1}{n^3}$

30. $\sum_{n=1}^{\infty} \frac{1}{n^{1/2}}$

31. $\sum_{n=1}^{\infty} \frac{1}{n^{1/4}}$

32. $\sum_{n=1}^{\infty} \frac{1}{n^5}$

Using a p-Series In Exercises 33–38, use Theorem 9.11 to determine the convergence or divergence of the p-series.

33. $\sum_{n=1}^{\infty} \frac{1}{\sqrt[5]{n}}$

34. $\sum_{n=1}^{\infty} \frac{3}{n^{5/3}}$

35. $1 + \frac{1}{2\sqrt{2}} + \frac{1}{3\sqrt{3}} + \frac{1}{4\sqrt{4}} + \frac{1}{5\sqrt{5}} + \cdots$

36. $1 + \frac{1}{\sqrt[3]{4}} + \frac{1}{\sqrt[3]{9}} + \frac{1}{\sqrt[3]{16}} + \frac{1}{\sqrt[3]{25}} + \cdots$

37. $\sum_{n=1}^{\infty} \frac{1}{n^{1.04}}$

38. $\sum_{n=1}^{\infty} \frac{1}{n^{\pi}}$

39. **Numerical and Graphical Analysis** Use a graphing utility to find the indicated partial sum S_n and complete the table. Then use a graphing utility to graph the first 10 terms of the sequence of partial sums. For each series, compare the rate at which the sequence of partial sums approaches the sum of the series.

n	5	10	20	50	100
S_n					

(a) $\sum_{n=1}^{\infty} 3\left(\frac{1}{5}\right)^{n-1} = \frac{15}{4}$ (b) $\sum_{n=1}^{\infty} \frac{1}{n^2} = \frac{\pi^2}{6}$

40. **Numerical Reasoning** Because the harmonic series diverges, it follows that for any positive real number M, there exists a positive integer N such that the partial sum

$$\sum_{n=1}^{N} \frac{1}{n} > M.$$

(a) Use a graphing utility to complete the table.

M	2	4	6	8
N				

(b) As the real number M increases in equal increments, does the number N increase in equal increments? Explain.

WRITING ABOUT CONCEPTS

41. Integral Test State the Integral Test and give an example of its use.

42. *p*-Series Define a *p*-series and state the requirements for its convergence.

43. Using a Series A friend in your calculus class tells you that the following series converges because the terms are very small and approach 0 rapidly. Is your friend correct? Explain.

$$\frac{1}{10{,}000} + \frac{1}{10{,}001} + \frac{1}{10{,}002} + \cdots$$

44. Using a Function Let f be a positive, continuous, and decreasing function for $x \geq 1$, such that $a_n = f(n)$. Use a graph to rank the following quantities in decreasing order. Explain your reasoning.

(a) $\sum_{n=2}^{7} a_n$ (b) $\int_1^7 f(x)\,dx$ (c) $\sum_{n=1}^{6} a_n$

45. Using a Series Use a graph to show that the inequality is true. What can you conclude about the convergence or divergence of the series? Explain.

(a) $\sum_{n=1}^{\infty} \frac{1}{\sqrt{n}} > \int_1^{\infty} \frac{1}{\sqrt{x}}\,dx$ (b) $\sum_{n=2}^{\infty} \frac{1}{n^2} < \int_1^{\infty} \frac{1}{x^2}\,dx$

46. HOW DO YOU SEE IT? The graphs show the sequences of partial sums of the *p*-series

$$\sum_{n=1}^{\infty} \frac{1}{n^{0.4}} \quad \text{and} \quad \sum_{n=1}^{\infty} \frac{1}{n^{1.5}}.$$

Using Theorem 9.11, the first series diverges and the second series converges. Explain how the graphs show this.

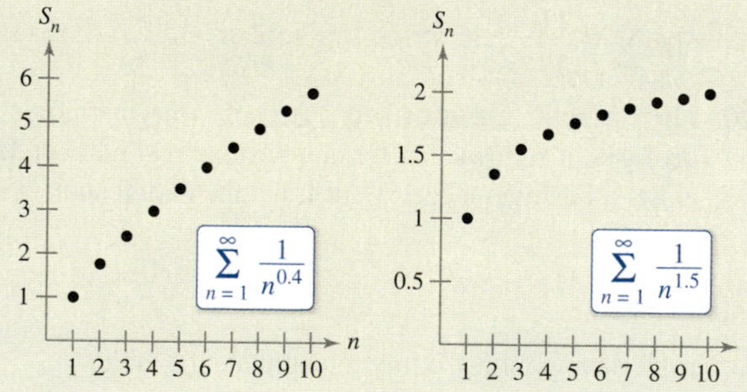

Finding Values **In Exercises 47–52, find the positive values of *p* for which the series converges.**

47. $\sum_{n=2}^{\infty} \frac{1}{n(\ln n)^p}$

48. $\sum_{n=2}^{\infty} \frac{\ln n}{n^p}$

49. $\sum_{n=1}^{\infty} \frac{n}{(1+n^2)^p}$

50. $\sum_{n=1}^{\infty} n(1+n^2)^p$

51. $\sum_{n=1}^{\infty} \left(\frac{3}{p}\right)^n$

52. $\sum_{n=3}^{\infty} \frac{1}{n \ln n[\ln(\ln n)]^p}$

53. Proof Let f be a positive, continuous, and decreasing function for $x \geq 1$, such that $a_n = f(n)$. Prove that if the series

$$\sum_{n=1}^{\infty} a_n$$

converges to S, then the remainder $R_N = S - S_N$ is bounded by

$$0 \leq R_N \leq \int_N^{\infty} f(x)\,dx.$$

54. Using a Remainder Show that the result of Exercise 53 can be written as

$$\sum_{n=1}^{N} a_n \leq \sum_{n=1}^{\infty} a_n \leq \sum_{n=1}^{N} a_n + \int_N^{\infty} f(x)\,dx.$$

Approximating a Sum **In Exercises 55–60, use the result of Exercise 53 to approximate the sum of the convergent series using the indicated number of terms. Include an estimate of the maximum error for your approximation.**

55. $\sum_{n=1}^{\infty} \frac{1}{n^2}$, five terms

56. $\sum_{n=1}^{\infty} \frac{1}{n^5}$, six terms

57. $\sum_{n=1}^{\infty} \frac{1}{n^2+1}$, ten terms

58. $\sum_{n=1}^{\infty} \frac{1}{(n+1)[\ln(n+1)]^3}$, ten terms

59. $\sum_{n=1}^{\infty} ne^{-n^2}$, four terms

60. $\sum_{n=1}^{\infty} e^{-n}$, four terms

Finding a Value **In Exercises 61–64, use the result of Exercise 53 to find *N* such that $R_N \leq 0.001$ for the convergent series.**

61. $\sum_{n=1}^{\infty} \frac{1}{n^4}$

62. $\sum_{n=1}^{\infty} \frac{1}{n^{3/2}}$

63. $\sum_{n=1}^{\infty} e^{-n/2}$

64. $\sum_{n=1}^{\infty} \frac{1}{n^2+1}$

65. Comparing Series

(a) Show that $\sum_{n=2}^{\infty} \frac{1}{n^{1.1}}$ converges and $\sum_{n=2}^{\infty} \frac{1}{n \ln n}$ diverges.

(b) Compare the first five terms of each series in part (a).

(c) Find $n > 3$ such that $\frac{1}{n^{1.1}} < \frac{1}{n \ln n}$.

66. Using a *p*-Series Ten terms are used to approximate a convergent *p*-series. Therefore, the remainder is a function of *p* and is

$$0 \leq R_{10}(p) \leq \int_{10}^{\infty} \frac{1}{x^p}\,dx, \quad p > 1.$$

(a) Perform the integration in the inequality.

(b) Use a graphing utility to represent the inequality graphically.

(c) Identify any asymptotes of the error function and interpret their meaning.

67. Euler's Constant Let

$$S_n = \sum_{k=1}^{n} \frac{1}{k} = 1 + \frac{1}{2} + \cdots + \frac{1}{n}.$$

(a) Show that $\ln(n+1) \le S_n \le 1 + \ln n$.

(b) Show that the sequence $\{a_n\} = \{S_n - \ln n\}$ is bounded.

(c) Show that the sequence $\{a_n\}$ is decreasing.

(d) Show that a_n converges to a limit γ (called Euler's constant).

(e) Approximate γ using a_{100}.

68. Finding a Sum Find the sum of the series

$$\sum_{n=2}^{\infty} \ln\left(1 - \frac{1}{n^2}\right).$$

69. Using a Series Consider the series $\sum_{n=2}^{\infty} x^{\ln n}$.

(a) Determine the convergence or divergence of the series for $x = 1$.

(b) Determine the convergence or divergence of the series for $x = 1/e$.

(c) Find the positive values of x for which the series converges.

70. Riemann Zeta Function The **Riemann zeta function** for real numbers is defined for all x for which the series

$$\zeta(x) = \sum_{n=1}^{\infty} n^{-x}$$

converges. Find the domain of the function.

Review **In Exercises 71–82, determine the convergence or divergence of the series.**

71. $\displaystyle\sum_{n=1}^{\infty} \frac{1}{3n-2}$

72. $\displaystyle\sum_{n=2}^{\infty} \frac{1}{n\sqrt{n^2-1}}$

73. $\displaystyle\sum_{n=1}^{\infty} \frac{1}{n\sqrt[4]{n}}$

74. $\displaystyle 3\sum_{n=1}^{\infty} \frac{1}{n^{0.95}}$

75. $\displaystyle\sum_{n=0}^{\infty} \left(\frac{2}{3}\right)^n$

76. $\displaystyle\sum_{n=0}^{\infty} (1.042)^n$

77. $\displaystyle\sum_{n=1}^{\infty} \frac{n}{\sqrt{n^2+1}}$

78. $\displaystyle\sum_{n=1}^{\infty} \left(\frac{1}{n^2} - \frac{1}{n^3}\right)$

79. $\displaystyle\sum_{n=1}^{\infty} \left(1 + \frac{1}{n}\right)^n$

80. $\displaystyle\sum_{n=2}^{\infty} \ln n$

81. $\displaystyle\sum_{n=2}^{\infty} \frac{1}{n(\ln n)^3}$

82. $\displaystyle\sum_{n=2}^{\infty} \frac{\ln n}{n^3}$

SECTION PROJECT

The Harmonic Series

The harmonic series

$$\sum_{n=1}^{\infty} \frac{1}{n} = 1 + \frac{1}{2} + \frac{1}{3} + \frac{1}{4} + \cdots + \frac{1}{n} + \cdots$$

is one of the most important series in this chapter. Even though its terms tend to zero as n increases,

$$\lim_{n\to\infty} \frac{1}{n} = 0$$

the harmonic series diverges. In other words, even though the terms are getting smaller and smaller, the sum "adds up to infinity."

(a) One way to show that the harmonic series diverges is attributed to James Bernoulli. He grouped the terms of the harmonic series as follows:

$$1 + \frac{1}{2} + \underbrace{\frac{1}{3} + \frac{1}{4}}_{>\frac{1}{2}} + \underbrace{\frac{1}{5} + \cdots + \frac{1}{8}}_{>\frac{1}{2}} + \underbrace{\frac{1}{9} + \cdots + \frac{1}{16}}_{>\frac{1}{2}} + \underbrace{\frac{1}{17} + \cdots + \frac{1}{32}}_{>\frac{1}{2}} + \cdots$$

Write a short paragraph explaining how you can use this grouping to show that the harmonic series diverges.

(b) Use the proof of the Integral Test, Theorem 9.10, to show that

$$\ln(n+1) \le 1 + \frac{1}{2} + \frac{1}{3} + \frac{1}{4} + \cdots + \frac{1}{n} \le 1 + \ln n.$$

(c) Use part (b) to determine how many terms M you would need so that

$$\sum_{n=1}^{M} \frac{1}{n} > 50.$$

(d) Show that the sum of the first million terms of the harmonic series is less than 15.

(e) Show that the following inequalities are valid.

$$\ln\frac{21}{10} \le \frac{1}{10} + \frac{1}{11} + \cdots + \frac{1}{20} \le \ln\frac{20}{9}$$

$$\ln\frac{201}{100} \le \frac{1}{100} + \frac{1}{101} + \cdots + \frac{1}{200} \le \ln\frac{200}{99}$$

(f) Use the inequalities in part (e) to find the limit

$$\lim_{m\to\infty} \sum_{n=m}^{2m} \frac{1}{n}.$$

9.4 Comparisons of Series

- **Use the Direct Comparison Test to determine whether a series converges or diverges.**
- **Use the Limit Comparison Test to determine whether a series converges or diverges.**

Direct Comparison Test

For the convergence tests developed so far, the terms of the series have to be fairly simple and the series must have special characteristics in order for the convergence tests to be applied. A slight deviation from these special characteristics can make a test nonapplicable. For example, in the pairs listed below, the second series cannot be tested by the same convergence test as the first series, even though it is similar to the first.

1. $\sum_{n=0}^{\infty} \frac{1}{2^n}$ is geometric, but $\sum_{n=0}^{\infty} \frac{n}{2^n}$ is not.

2. $\sum_{n=1}^{\infty} \frac{1}{n^3}$ is a p-series, but $\sum_{n=1}^{\infty} \frac{1}{n^3 + 1}$ is not.

3. $a_n = \frac{n}{(n^2 + 3)^2}$ is easily integrated, but $b_n = \frac{n^2}{(n^2 + 3)^2}$ is not.

In this section, you will study two additional tests for positive-term series. These two tests greatly expand the variety of series you are able to test for convergence or divergence. They allow you to *compare* a series having complicated terms with a simpler series whose convergence or divergence is known.

REMARK As stated, the Direct Comparison Test requires that $0 < a_n \leq b_n$ for all n. Because the convergence of a series is not dependent on its first several terms, you could modify the test to require only that $0 < a_n \leq b_n$ for all n greater than some integer N.

THEOREM 9.12 Direct Comparison Test

Let $0 < a_n \leq b_n$ for all n.

1. If $\sum_{n=1}^{\infty} b_n$ converges, then $\sum_{n=1}^{\infty} a_n$ converges.

2. If $\sum_{n=1}^{\infty} a_n$ diverges, then $\sum_{n=1}^{\infty} b_n$ diverges.

Proof To prove the first property, let $L = \sum_{n=1}^{\infty} b_n$ and let

$$S_n = a_1 + a_2 + \cdot \cdot \cdot + a_n.$$

Because $0 < a_n \leq b_n$, the sequence $S_1, S_2, S_3, \ldots$ is nondecreasing and bounded above by L; so, it must converge. Because

$$\lim_{n \to \infty} S_n = \sum_{n=1}^{\infty} a_n$$

it follows that $\sum_{n=1}^{\infty} a_n$ converges. The second property is logically equivalent to the first.

See LarsonCalculus.com for Bruce Edwards's video of this proof.

FOR FURTHER INFORMATION Is the Direct Comparison Test just for nonnegative series? To read about the generalization of this test to real series, see the article "The Comparison Test—Not Just for Nonnegative Series" by Michele Longo and Vincenzo Valori in *Mathematics Magazine*. To view this article, go to *MathArticles.com*.

EXAMPLE 1 Using the Direct Comparison Test

Determine the convergence or divergence of

$$\sum_{n=1}^{\infty} \frac{1}{2 + 3^n}.$$

Solution This series resembles

$$\sum_{n=1}^{\infty} \frac{1}{3^n}. \qquad \text{Convergent geometric series}$$

Term-by-term comparison yields

$$a_n = \frac{1}{2 + 3^n} < \frac{1}{3^n} = b_n, \quad n \geq 1.$$

So, by the Direct Comparison Test, the series converges.

EXAMPLE 2 Using the Direct Comparison Test

See LarsonCalculus.com for an interactive version of this type of example.

Determine the convergence or divergence of

$$\sum_{n=1}^{\infty} \frac{1}{2 + \sqrt{n}}.$$

Solution This series resembles

$$\sum_{n=1}^{\infty} \frac{1}{n^{1/2}}. \qquad \text{Divergent } p\text{-series}$$

Term-by-term comparison yields

$$\frac{1}{2 + \sqrt{n}} \leq \frac{1}{\sqrt{n}}, \quad n \geq 1$$

which *does not* meet the requirements for divergence. (Remember that when term-by-term comparison reveals a series that is *less* than a divergent series, the Direct Comparison Test tells you nothing.) Still expecting the series to diverge, you can compare the series with

$$\sum_{n=1}^{\infty} \frac{1}{n}. \qquad \text{Divergent harmonic series}$$

In this case, term-by-term comparison yields

$$a_n = \frac{1}{n} \leq \frac{1}{2 + \sqrt{n}} = b_n, \quad n \geq 4$$

and, by the Direct Comparison Test, the given series diverges. To verify the last inequality, try showing that

$$2 + \sqrt{n} \leq n$$

whenever $n \geq 4$. ■

Remember that both parts of the Direct Comparison Test require that $0 < a_n \leq b_n$. Informally, the test says the following about the two series with nonnegative terms.

1. If the "larger" series converges, then the "smaller" series must also converge.
2. If the "smaller" series diverges, then the "larger" series must also diverge.

Limit Comparison Test

Sometimes a series closely resembles a p-series or a geometric series, yet you cannot establish the term-by-term comparison necessary to apply the Direct Comparison Test. Under these circumstances, you may be able to apply a second comparison test, called the **Limit Comparison Test.**

REMARK As with the Direct Comparison Test, the Limit Comparison Test could be modified to require only that a_n and b_n be positive for all n greater than some integer N.

THEOREM 9.13 Limit Comparison Test

If $a_n > 0$, $b_n > 0$, and

$$\lim_{n\to\infty} \frac{a_n}{b_n} = L$$

where L is *finite and positive*, then

$$\sum_{n=1}^{\infty} a_n \quad \text{and} \quad \sum_{n=1}^{\infty} b_n$$

either both converge or both diverge.

AP* Tips

The comparison tests are sometimes covered on the multiple choice section of the AP BC Exam.

Proof Because $a_n > 0$, $b_n > 0$, and

$$\lim_{n\to\infty} \frac{a_n}{b_n} = L$$

there exists $N > 0$ such that

$$0 < \frac{a_n}{b_n} < L + 1, \quad \text{for } n \geq N.$$

This implies that

$$0 < a_n < (L + 1)b_n.$$

So, by the Direct Comparison Test, the convergence of $\Sigma\, b_n$ implies the convergence of $\Sigma\, a_n$. Similarly, the fact that

$$\lim_{n\to\infty} \frac{b_n}{a_n} = \frac{1}{L}$$

can be used to show that the convergence of $\Sigma\, a_n$ implies the convergence of $\Sigma\, b_n$.

See LarsonCalculus.com for Bruce Edwards's video of this proof.

EXAMPLE 3 Using the Limit Comparison Test

Show that the general harmonic series below diverges.

$$\sum_{n=1}^{\infty} \frac{1}{an + b}, \quad a > 0, \quad b > 0$$

Solution By comparison with

$$\sum_{n=1}^{\infty} \frac{1}{n} \qquad \text{Divergent harmonic series}$$

you have

$$\lim_{n\to\infty} \frac{1/(an + b)}{1/n} = \lim_{n\to\infty} \frac{n}{an + b} = \frac{1}{a}.$$

Because this limit is greater than 0, you can conclude from the Limit Comparison Test that the series diverges.

The Limit Comparison Test works well for comparing a "messy" algebraic series with a p-series. In choosing an appropriate p-series, you must choose one with an nth term of the same magnitude as the nth term of the given series.

Given Series	Comparison Series	Conclusion
$\sum_{n=1}^{\infty} \frac{1}{3n^2 - 4n + 5}$	$\sum_{n=1}^{\infty} \frac{1}{n^2}$	Both series converge.
$\sum_{n=1}^{\infty} \frac{1}{\sqrt{3n - 2}}$	$\sum_{n=1}^{\infty} \frac{1}{\sqrt{n}}$	Both series diverge.
$\sum_{n=1}^{\infty} \frac{n^2 - 10}{4n^5 + n^3}$	$\sum_{n=1}^{\infty} \frac{n^2}{n^5} = \sum_{n=1}^{\infty} \frac{1}{n^3}$	Both series converge.

In other words, when choosing a series for comparison, you can disregard all but the *highest powers of n* in both the numerator and the denominator.

EXAMPLE 4 Using the Limit Comparison Test

Determine the convergence or divergence of

$$\sum_{n=1}^{\infty} \frac{\sqrt{n}}{n^2 + 1}.$$

Solution Disregarding all but the highest powers of n in the numerator and the denominator, you can compare the series with

$$\sum_{n=1}^{\infty} \frac{\sqrt{n}}{n^2} = \sum_{n=1}^{\infty} \frac{1}{n^{3/2}}.$$

Convergent p-series

Because

$$\lim_{n\to\infty} \frac{a_n}{b_n} = \lim_{n\to\infty} \left(\frac{\sqrt{n}}{n^2 + 1}\right)\left(\frac{n^{3/2}}{1}\right) = \lim_{n\to\infty} \frac{n^2}{n^2 + 1} = 1$$

you can conclude by the Limit Comparison Test that the series converges.

EXAMPLE 5 Using the Limit Comparison Test

Determine the convergence or divergence of

$$\sum_{n=1}^{\infty} \frac{n2^n}{4n^3 + 1}.$$

Solution A reasonable comparison would be with the series

$$\sum_{n=1}^{\infty} \frac{2^n}{n^2}.$$

Divergent series

Note that this series diverges by the nth-Term Test. From the limit

$$\lim_{n\to\infty} \frac{a_n}{b_n} = \lim_{n\to\infty} \left(\frac{n2^n}{4n^3 + 1}\right)\left(\frac{n^2}{2^n}\right) = \lim_{n\to\infty} \frac{1}{4 + (1/n^3)} = \frac{1}{4}$$

you can conclude that the series diverges.

9.4 Exercises

See **CalcChat.com** for tutorial help and worked-out solutions to odd-numbered exercises.

1. Graphical Analysis The figures show the graphs of the first 10 terms, and the graphs of the first 10 terms of the sequence of partial sums, of each series.

$$\sum_{n=1}^{\infty} \frac{6}{n^{3/2}}, \quad \sum_{n=1}^{\infty} \frac{6}{n^{3/2}+3}, \quad \text{and} \quad \sum_{n=1}^{\infty} \frac{6}{n\sqrt{n^2+0.5}}$$

(a) Identify the series in each figure.

(b) Which series is a p-series? Does it converge or diverge?

(c) For the series that are not p-series, how do the magnitudes of the terms compare with the magnitudes of the terms of the p-series? What conclusion can you draw about the convergence or divergence of the series?

(d) Explain the relationship between the magnitudes of the terms of the series and the magnitudes of the terms of the partial sums.

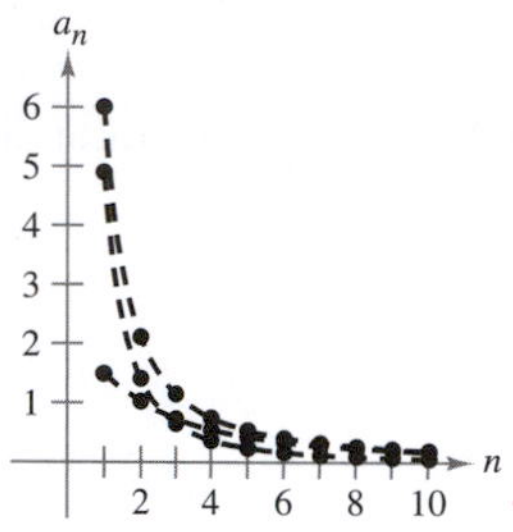

Graphs of terms

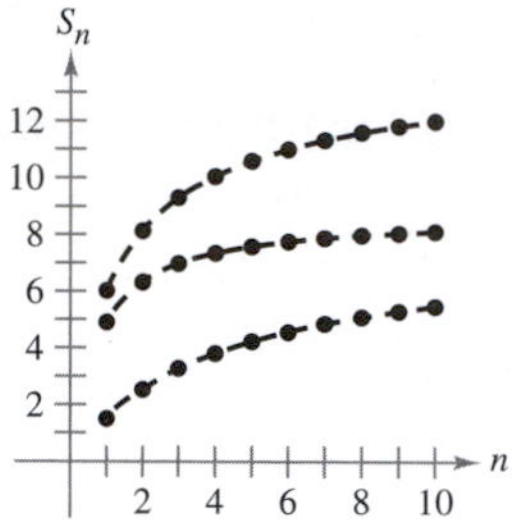

Graphs of partial sums

2. Graphical Analysis The figures show the graphs of the first 10 terms, and the graphs of the first 10 terms of the sequence of partial sums, of each series.

$$\sum_{n=1}^{\infty} \frac{2}{\sqrt{n}}, \quad \sum_{n=1}^{\infty} \frac{2}{\sqrt{n}-0.5}, \quad \text{and} \quad \sum_{n=1}^{\infty} \frac{4}{\sqrt{n+0.5}}$$

(a) Identify the series in each figure.

(b) Which series is a p-series? Does it converge or diverge?

(c) For the series that are not p-series, how do the magnitudes of the terms compare with the magnitudes of the terms of the p-series? What conclusion can you draw about the convergence or divergence of the series?

(d) Explain the relationship between the magnitudes of the terms of the series and the magnitudes of the terms of the partial sums.

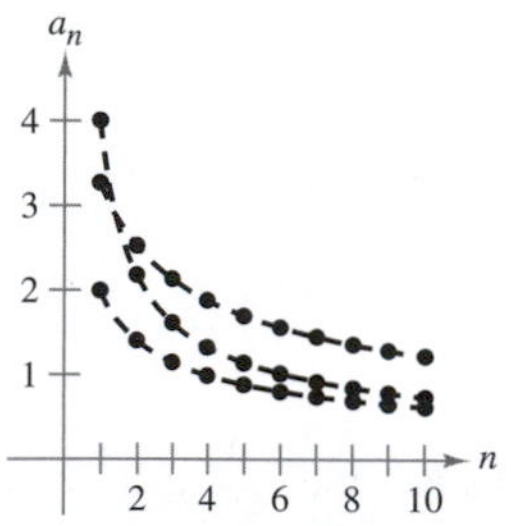

Graphs of terms

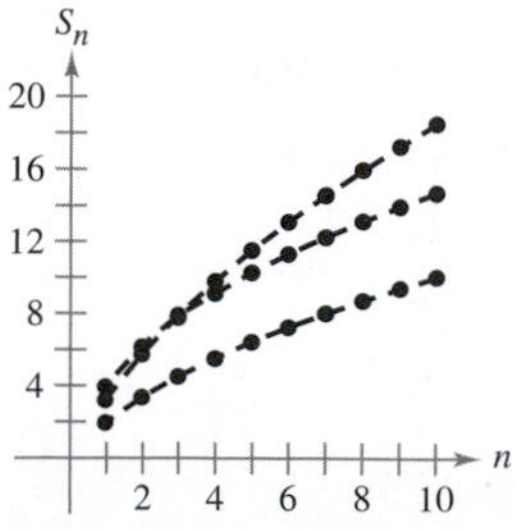

Graphs of partial sums

Using the Direct Comparison Test **In Exercises 3–12, use the Direct Comparison Test to determine the convergence or divergence of the series.**

3. $\displaystyle\sum_{n=1}^{\infty} \frac{1}{2n-1}$

4. $\displaystyle\sum_{n=1}^{\infty} \frac{1}{3n^2+2}$

5. $\displaystyle\sum_{n=2}^{\infty} \frac{1}{\sqrt{n}-1}$

6. $\displaystyle\sum_{n=0}^{\infty} \frac{4^n}{5^n+3}$

7. $\displaystyle\sum_{n=2}^{\infty} \frac{\ln n}{n+1}$

8. $\displaystyle\sum_{n=1}^{\infty} \frac{1}{\sqrt{n^3+1}}$

9. $\displaystyle\sum_{n=0}^{\infty} \frac{1}{n!}$

10. $\displaystyle\sum_{n=1}^{\infty} \frac{1}{4\sqrt[3]{n}-1}$

11. $\displaystyle\sum_{n=0}^{\infty} e^{-n^2}$

12. $\displaystyle\sum_{n=1}^{\infty} \frac{3^n}{2^n-1}$

Using the Limit Comparison Test **In Exercises 13–22, use the Limit Comparison Test to determine the convergence or divergence of the series.**

13. $\displaystyle\sum_{n=1}^{\infty} \frac{n}{n^2+1}$

14. $\displaystyle\sum_{n=1}^{\infty} \frac{5}{4^n+1}$

15. $\displaystyle\sum_{n=0}^{\infty} \frac{1}{\sqrt{n^2+1}}$

16. $\displaystyle\sum_{n=1}^{\infty} \frac{2^n+1}{5^n+1}$

17. $\displaystyle\sum_{n=1}^{\infty} \frac{2n^2-1}{3n^5+2n+1}$

18. $\displaystyle\sum_{n=1}^{\infty} \frac{1}{n^2(n+3)}$

19. $\displaystyle\sum_{n=1}^{\infty} \frac{1}{n\sqrt{n^2+1}}$

20. $\displaystyle\sum_{n=1}^{\infty} \frac{n}{(n+1)2^{n-1}}$

21. $\displaystyle\sum_{n=1}^{\infty} \frac{n^{k-1}}{n^k+1}, \quad k > 2$

22. $\displaystyle\sum_{n=1}^{\infty} \sin\frac{1}{n}$

Determining Convergence or Divergence **In Exercises 23–30, test for convergence or divergence, using each test at least once. Identify which test was used.**

(a) ***n*th-Term Test**

(b) Geometric Series Test

(c) ***p*-Series Test**

(d) Telescoping Series Test

(e) Integral Test

(f) Direct Comparison Test

(g) Limit Comparison Test

23. $\displaystyle\sum_{n=1}^{\infty} \frac{\sqrt[3]{n}}{n}$

24. $\displaystyle\sum_{n=0}^{\infty} 5\left(-\frac{4}{3}\right)^n$

25. $\displaystyle\sum_{n=1}^{\infty} \frac{1}{5^n+1}$

26. $\displaystyle\sum_{n=2}^{\infty} \frac{1}{n^3-8}$

27. $\displaystyle\sum_{n=1}^{\infty} \frac{2n}{3n-2}$

28. $\displaystyle\sum_{n=1}^{\infty} \left(\frac{1}{n+1}-\frac{1}{n+2}\right)$

29. $\displaystyle\sum_{n=1}^{\infty} \frac{n}{(n^2+1)^2}$

30. $\displaystyle\sum_{n=1}^{\infty} \frac{3}{n(n+3)}$

31. Using the Limit Comparison Test Use the Limit Comparison Test with the harmonic series to show that the series $\Sigma\, a_n$ (where $0 < a_n < a_{n-1}$) diverges when $\displaystyle\lim_{n\to\infty} na_n$ is finite and nonzero.

32. Proof Prove that, if $P(n)$ and $Q(n)$ are polynomials of degree j and k, respectively, then the series

$$\sum_{n=1}^{\infty} \frac{P(n)}{Q(n)}$$

converges if $j < k - 1$ and diverges if $j \geq k - 1$.

Determining Convergence or Divergence **In Exercises 33–36, use the polynomial test given in Exercise 32 to determine whether the series converges or diverges.**

33. $\frac{1}{2} + \frac{2}{5} + \frac{3}{10} + \frac{4}{17} + \frac{5}{26} + \cdots$

34. $\frac{1}{3} + \frac{1}{8} + \frac{1}{15} + \frac{1}{24} + \frac{1}{35} + \cdots$

35. $\sum_{n=1}^{\infty} \frac{1}{n^3 + 1}$

36. $\sum_{n=1}^{\infty} \frac{n^2}{n^3 + 1}$

Verifying Divergence **In Exercises 37 and 38, use the divergence test given in Exercise 31 to show that the series diverges.**

37. $\sum_{n=1}^{\infty} \frac{n^3}{5n^4 + 3}$

38. $\sum_{n=1}^{\infty} \frac{3n^2 + 1}{4n^3 + 2}$

Determining Convergence or Divergence **In Exercises 39–42, determine the convergence or divergence of the series.**

39. $\frac{1}{200} + \frac{1}{400} + \frac{1}{600} + \frac{1}{800} + \cdots$

40. $\frac{1}{200} + \frac{1}{210} + \frac{1}{220} + \frac{1}{230} + \cdots$

41. $\frac{1}{201} + \frac{1}{204} + \frac{1}{209} + \frac{1}{216} + \cdots$

42. $\frac{1}{201} + \frac{1}{208} + \frac{1}{227} + \frac{1}{264} + \cdots$

WRITING ABOUT CONCEPTS

43. Using Series Review the results of Exercises 39–42. Explain why careful analysis is required to determine the convergence or divergence of a series and why only considering the magnitudes of the terms of a series could be misleading.

44. Direct Comparison Test State the Direct Comparison Test and give an example of its use.

45. Limit Comparison Test State the Limit Comparison Test and give an example of its use.

46. Comparing Series It appears that the terms of the series

$$\frac{1}{1000} + \frac{1}{1001} + \frac{1}{1002} + \frac{1}{1003} + \cdots$$

are less than the corresponding terms of the convergent series

$$1 + \frac{1}{4} + \frac{1}{9} + \frac{1}{16} + \cdots.$$

If the statement above is correct, then the first series converges. Is this correct? Why or why not? Make a statement about how the divergence or convergence of a series is affected by the inclusion or exclusion of the first finite number of terms.

47. Using a Series Consider the series $\sum_{n=1}^{\infty} \frac{1}{(2n-1)^2}$.

(a) Verify that the series converges.

(b) Use a graphing utility to complete the table.

n	5	10	20	50	100
S_n					

(c) The sum of the series is $\pi^2/8$. Find the sum of the series

$$\sum_{n=3}^{\infty} \frac{1}{(2n-1)^2}.$$

(d) Use a graphing utility to find the sum of the series

$$\sum_{n=10}^{\infty} \frac{1}{(2n-1)^2}.$$

48. HOW DO YOU SEE IT? The figure shows the first 20 terms of the convergent series $\sum_{n=1}^{\infty} a_n$ and the first 20 terms of the series $\sum_{n=1}^{\infty} a_n^2$. Identify the two series and explain your reasoning in making the selection.

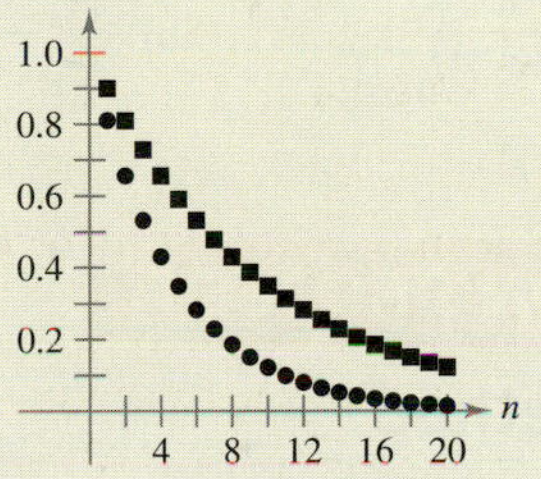

True or False? **In Exercises 49–54, determine whether the statement is true or false. If it is false, explain why or give an example that shows it is false.**

49. If $0 < a_n \leq b_n$ and $\sum_{n=1}^{\infty} a_n$ converges, then $\sum_{n=1}^{\infty} b_n$ diverges.

50. If $0 < a_{n+10} \leq b_n$ and $\sum_{n=1}^{\infty} b_n$ converges, then $\sum_{n=1}^{\infty} a_n$ converges.

51. If $a_n + b_n \leq c_n$ and $\sum_{n=1}^{\infty} c_n$ converges, then the series $\sum_{n=1}^{\infty} a_n$ and $\sum_{n=1}^{\infty} b_n$ both converge. (Assume that the terms of all three series are positive.)

52. If $a_n \leq b_n + c_n$ and $\sum_{n=1}^{\infty} a_n$ diverges, then the series $\sum_{n=1}^{\infty} b_n$ and $\sum_{n=1}^{\infty} c_n$ both diverge. (Assume that the terms of all three series are positive.)

53. If $0 < a_n \leq b_n$ and $\sum_{n=1}^{\infty} a_n$ diverges, then $\sum_{n=1}^{\infty} b_n$ diverges.

54. If $0 < a_n \leq b_n$ and $\sum_{n=1}^{\infty} b_n$ diverges, then $\sum_{n=1}^{\infty} a_n$ diverges.

55. Proof Prove that if the nonnegative series

$$\sum_{n=1}^{\infty} a_n \quad \text{and} \quad \sum_{n=1}^{\infty} b_n$$

converge, then so does the series $\sum_{n=1}^{\infty} a_n b_n$.

56. Proof Use the result of Exercise 55 to prove that if the nonnegative series $\sum_{n=1}^{\infty} a_n$ converges, then so does the series $\sum_{n=1}^{\infty} a_n^2$.

57. Finding Series Find two series that demonstrate the result of Exercise 55.

58. Finding Series Find two series that demonstrate the result of Exercise 56.

59. Proof Suppose that $\Sigma\, a_n$ and $\Sigma\, b_n$ are series with positive terms. Prove that if $\lim_{n\to\infty} \frac{a_n}{b_n} = 0$ and $\Sigma\, b_n$ converges, $\Sigma\, a_n$ also converges.

60. Proof Suppose that $\Sigma\, a_n$ and $\Sigma\, b_n$ are series with positive terms. Prove that if $\lim_{n\to\infty} \frac{a_n}{b_n} = \infty$ and $\Sigma\, b_n$ diverges, $\Sigma\, a_n$ also diverges.

61. Verifying Convergence Use the result of Exercise 59 to show that each series converges.

(a) $\sum_{n=1}^{\infty} \frac{1}{(n+1)^3}$ (b) $\sum_{n=1}^{\infty} \frac{1}{\sqrt{n}\pi^n}$

62. Verifying Divergence Use the result of Exercise 60 to show that each series diverges.

(a) $\sum_{n=1}^{\infty} \frac{\ln n}{n}$ (b) $\sum_{n=2}^{\infty} \frac{1}{\ln n}$

63. Proof Suppose that $\Sigma\, a_n$ is a series with positive terms. Prove that if $\Sigma\, a_n$ converges, then $\Sigma \sin a_n$ also converges.

64. Proof Prove that the series

$$\sum_{n=1}^{\infty} \frac{1}{1+2+3+\cdot\cdot\cdot+n}$$

converges.

65. Comparing Series Show that $\sum_{n=1}^{\infty} \frac{\ln n}{n\sqrt{n}}$ converges by comparison with $\sum_{n=1}^{\infty} \frac{1}{n^{5/4}}$.

PUTNAM EXAM CHALLENGE

66. Is the infinite series $\sum_{n=1}^{\infty} \frac{1}{n^{(n+1)/n}}$ convergent? Prove your statement.

67. Prove that if $\sum_{n=1}^{\infty} a_n$ is a convergent series of positive real numbers, then so is $\sum_{n=1}^{\infty} (a_n)^{n/(n+1)}$.

These problems were composed by the Committee on the Putnam Prize Competition.

SECTION PROJECT

Solera Method

Most wines are produced entirely from grapes grown in a single year. Sherry, however, is a complex mixture of older wines with new wines. This is done with a sequence of barrels (called a solera) stacked on top of each other, as shown in the photo.

The oldest wine is in the bottom tier of barrels, and the newest is in the top tier. Each year, half of each barrel in the bottom tier is bottled as sherry. The bottom barrels are then refilled with the wine from the barrels above. This process is repeated throughout the solera, with new wine being added to the top barrels.

A mathematical model for the amount of n-year-old wine that is removed from a solera (with k tiers) each year is

$$f(n, k) = \binom{n-1}{k-1}\left(\frac{1}{2}\right)^{n+1}, \quad k \le n.$$

(a) Consider a solera that has five tiers, numbered $k = 1, 2, 3, 4,$ and 5. In 1995 ($n = 0$), half of each barrel in the top tier (tier 1) was refilled with new wine. How much of this wine was removed from the solera in 1996? In 1997? In 1998? . . . In 2010? During which year(s) was the greatest amount of the 1995 wine removed from the solera?

(b) In part (a), let a_n be the amount of 1995 wine that is removed from the solera in year n. Evaluate

$$\sum_{n=0}^{\infty} a_n.$$

FOR FURTHER INFORMATION See the article "Finding Vintage Concentrations in a Sherry Solera" by Rhodes Peele and John T. MacQueen in the *UMAP Modules*.

Squareplum/Shutterstock.com

9.5 Alternating Series

- Use the Alternating Series Test to determine whether an infinite series converges.
- Use the Alternating Series Remainder to approximate the sum of an alternating series.
- Classify a convergent series as absolutely or conditionally convergent.
- Rearrange an infinite series to obtain a different sum.

Alternating Series

So far, most series you have dealt with have had positive terms. In this section and the next section, you will study series that contain both positive and negative terms. The simplest such series is an **alternating series,** whose terms alternate in sign. For example, the geometric series

$$\sum_{n=0}^{\infty}\left(-\frac{1}{2}\right)^n = \sum_{n=0}^{\infty}(-1)^n\frac{1}{2^n}$$
$$= 1 - \frac{1}{2} + \frac{1}{4} - \frac{1}{8} + \frac{1}{16} - \cdots$$

is an *alternating geometric series* with $r = -\frac{1}{2}$. Alternating series occur in two ways: either the odd terms are negative or the even terms are negative.

THEOREM 9.14 Alternating Series Test

Let $a_n > 0$. The alternating series

$$\sum_{n=1}^{\infty}(-1)^n a_n \quad \text{and} \quad \sum_{n=1}^{\infty}(-1)^{n+1} a_n$$

converge when the two conditions listed below are met.

1. $\lim_{n\to\infty} a_n = 0$
2. $a_{n+1} \le a_n$, for all n

REMARK The second condition in the Alternating Series Test can be modified to require only that $0 < a_{n+1} \le a_n$ for all n greater than some integer N.

Proof Consider the alternating series $\Sigma(-1)^{n+1}a_n$. For this series, the partial sum (where $2n$ is even)

$$S_{2n} = (a_1 - a_2) + (a_3 - a_4) + (a_5 - a_6) + \cdots + (a_{2n-1} - a_{2n})$$

has all nonnegative terms, and therefore $\{S_{2n}\}$ is a nondecreasing sequence. But you can also write

$$S_{2n} = a_1 - (a_2 - a_3) - (a_4 - a_5) - \cdots - (a_{2n-2} - a_{2n-1}) - a_{2n}$$

which implies that $S_{2n} \le a_1$ for every integer n. So, $\{S_{2n}\}$ is a bounded, nondecreasing sequence that converges to some value L. Because $S_{2n-1} - a_{2n} = S_{2n}$ and $a_{2n} \to 0$, you have

$$\begin{aligned}\lim_{n\to\infty} S_{2n-1} &= \lim_{n\to\infty} S_{2n} + \lim_{n\to\infty} a_{2n}\\ &= L + \lim_{n\to\infty} a_{2n}\\ &= L.\end{aligned}$$

Because both S_{2n} and S_{2n-1} converge to the same limit L, it follows that $\{S_n\}$ also converges to L. Consequently, the given alternating series converges. ■

See LarsonCalculus.com for Bruce Edwards's video of this proof.

EXAMPLE 1 Using the Alternating Series Test

Determine the convergence or divergence of

$$\sum_{n=1}^{\infty} (-1)^{n+1}\frac{1}{n}.$$

REMARK The series in Example 1 is called the *alternating harmonic series.* More is said about this series in Example 8.

Solution Note that $\lim_{n\to\infty} a_n = \lim_{n\to\infty} \frac{1}{n} = 0$. So, the first condition of Theorem 9.14 is satisfied. Also note that the second condition of Theorem 9.14 is satisfied because

$$a_{n+1} = \frac{1}{n+1} \leq \frac{1}{n} = a_n$$

for all n. So, applying the Alternating Series Test, you can conclude that the series converges.

EXAMPLE 2 Using the Alternating Series Test

Determine the convergence or divergence of

$$\sum_{n=1}^{\infty} \frac{n}{(-2)^{n-1}}.$$

Solution To apply the Alternating Series Test, note that, for $n \geq 1$,

$$\begin{aligned}
\frac{1}{2} &\leq \frac{n}{n+1} \\
\frac{2^{n-1}}{2^n} &\leq \frac{n}{n+1} \\
(n+1)2^{n-1} &\leq n2^n \\
\frac{n+1}{2^n} &\leq \frac{n}{2^{n-1}}.
\end{aligned}$$

So, $a_{n+1} = (n+1)/2^n \leq n/2^{n-1} = a_n$ for all n. Furthermore, by L'Hôpital's Rule,

$$\lim_{x\to\infty} \frac{x}{2^{x-1}} = \lim_{x\to\infty} \frac{1}{2^{x-1}(\ln 2)} = 0 \quad \Longrightarrow \quad \lim_{n\to\infty} \frac{n}{2^{n-1}} = 0.$$

Therefore, by the Alternating Series Test, the series converges.

EXAMPLE 3 When the Alternating Series Test Does Not Apply

a. The alternating series

$$\sum_{n=1}^{\infty} \frac{(-1)^{n+1}(n+1)}{n} = \frac{2}{1} - \frac{3}{2} + \frac{4}{3} - \frac{5}{4} + \frac{6}{5} - \cdots$$

passes the second condition of the Alternating Series Test because $a_{n+1} \leq a_n$ for all n. You cannot apply the Alternating Series Test, however, because the series does not pass the first condition. In fact, the series diverges.

REMARK In Example 3(a), remember that whenever a series does not pass the first condition of the Alternating Series Test, you can use the nth-Term Test for Divergence to conclude that the series diverges.

b. The alternating series

$$\frac{2}{1} - \frac{1}{1} + \frac{2}{2} - \frac{1}{2} + \frac{2}{3} - \frac{1}{3} + \frac{2}{4} - \frac{1}{4} + \cdots$$

passes the first condition because a_n approaches 0 as $n \to \infty$. You cannot apply the Alternating Series Test, however, because the series does not pass the second condition. To conclude that the series diverges, you can argue that S_{2N} equals the Nth partial sum of the divergent harmonic series. This implies that the sequence of partial sums diverges. So, the series diverges.

Alternating Series Remainder

For a convergent alternating series, the partial sum S_N can be a useful approximation for the sum S of the series. The error involved in using $S \approx S_N$ is the remainder $R_N = S - S_N$.

> **THEOREM 9.15 Alternating Series Remainder**
>
> If a convergent alternating series satisfies the condition $a_{n+1} \le a_n$, then the absolute value of the remainder R_N involved in approximating the sum S by S_N is less than (or equal to) the first neglected term. That is,
>
> $$|S - S_N| = |R_N| \le a_{N+1}.$$
>
> A proof of this theorem is given in Appendix A.
>
> *See LarsonCalculus.com for Bruce Edwards's video of this proof.*

AP* Tips

The Alternating Series Test is commonly covered throughout the AP BC Exam. The error bound is often tested on the free response section.

EXAMPLE 4 Approximating the Sum of an Alternating Series

See LarsonCalculus.com for an interactive version of this type of example.

Approximate the sum of the series by its first six terms.

$$\sum_{n=1}^{\infty} (-1)^{n+1}\left(\frac{1}{n!}\right) = \frac{1}{1!} - \frac{1}{2!} + \frac{1}{3!} - \frac{1}{4!} + \frac{1}{5!} - \frac{1}{6!} + \cdots$$

Solution The series converges by the Alternating Series Test because

$$\frac{1}{(n+1)!} \le \frac{1}{n!} \quad \text{and} \quad \lim_{n\to\infty} \frac{1}{n!} = 0.$$

The sum of the first six terms is

$$S_6 = 1 - \frac{1}{2} + \frac{1}{6} - \frac{1}{24} + \frac{1}{120} - \frac{1}{720} = \frac{91}{144} \approx 0.63194$$

and, by the Alternating Series Remainder, you have

$$|S - S_6| = |R_6| \le a_7 = \frac{1}{5040} \approx 0.0002.$$

So, the sum S lies between $0.63194 - 0.0002$ and $0.63194 + 0.0002$, and you have $0.63174 \le S \le 0.63214$.

TECHNOLOGY Later, using the techniques in Section 9.10, you will be able to show that the series in Example 4 converges to

$$\frac{e-1}{e} \approx 0.63212.$$

(See Section 9.10, Exercise 58.) For now, try using a graphing utility to obtain an approximation of the sum of the series. How many terms do you need to obtain an approximation that is within 0.00001 unit of the actual sum?

EXAMPLE 5 Finding the Number of Terms

Determine the number of terms required to approximate the sum of the series with an error of less than 0.001.

$$\sum_{n=1}^{\infty} \frac{(-1)^{n+1}}{n^4}$$

Solution By Theorem 9.15, you know that

$$|R_N| \le a_{N+1} = \frac{1}{(N+1)^4}.$$

For an error of less than 0.001, N must satisfy the inequality $1/(N+1)^4 < 0.001$.

$$\frac{1}{(N+1)^4} < 0.001 \implies (N+1)^4 > 1000 \implies N > \sqrt[4]{1000} - 1 \approx 4.6$$

So, you will need at least 5 terms. Using 5 terms, the sum is $S \approx S_5 \approx 0.94754$, which has an error of less than 0.001. ■

Absolute and Conditional Convergence

Occasionally, a series may have both positive and negative terms and not be an alternating series. For instance, the series

$$\sum_{n=1}^{\infty} \frac{\sin n}{n^2} = \frac{\sin 1}{1} + \frac{\sin 2}{4} + \frac{\sin 3}{9} + \cdots$$

has both positive and negative terms, yet it is not an alternating series. One way to obtain some information about the convergence of this series is to investigate the convergence of the series

$$\sum_{n=1}^{\infty} \left|\frac{\sin n}{n^2}\right|.$$

By direct comparison, you have $|\sin n| \leq 1$ for all n, so

$$\left|\frac{\sin n}{n^2}\right| \leq \frac{1}{n^2}, \quad n \geq 1.$$

Therefore, by the Direct Comparison Test, the series $\sum \left|\frac{\sin n}{n^2}\right|$ converges. The next theorem tells you that the original series also converges.

THEOREM 9.16 Absolute Convergence

If the series $\Sigma\,|a_n|$ converges, then the series $\Sigma\,a_n$ also converges.

Proof Because $0 \leq a_n + |a_n| \leq 2|a_n|$ for all n, the series

$$\sum_{n=1}^{\infty} (a_n + |a_n|)$$

converges by comparison with the convergent series

$$\sum_{n=1}^{\infty} 2|a_n|.$$

Furthermore, because $a_n = (a_n + |a_n|) - |a_n|$, you can write

$$\sum_{n=1}^{\infty} a_n = \sum_{n=1}^{\infty} (a_n + |a_n|) - \sum_{n=1}^{\infty} |a_n|$$

where both series on the right converge. So, it follows that $\Sigma\,a_n$ converges. ■

See LarsonCalculus.com for Bruce Edwards's video of this proof.

The converse of Theorem 9.16 is not true. For instance, the **alternating harmonic series**

$$\sum_{n=1}^{\infty} \frac{(-1)^{n+1}}{n} = \frac{1}{1} - \frac{1}{2} + \frac{1}{3} - \frac{1}{4} + \cdots$$

converges by the Alternating Series Test. Yet the harmonic series diverges. This type of convergence is called **conditional.**

Definitions of Absolute and Conditional Convergence

1. The series $\Sigma\,a_n$ is **absolutely convergent** when $\Sigma\,|a_n|$ converges.
2. The series $\Sigma\,a_n$ is **conditionally convergent** when $\Sigma\,a_n$ converges but $\Sigma\,|a_n|$ diverges.

EXAMPLE 6 Absolute and Conditional Convergence

Determine whether each of the series is convergent or divergent. Classify any convergent series as absolutely or conditionally convergent.

a. $\displaystyle\sum_{n=0}^{\infty} \frac{(-1)^n n!}{2^n} = \frac{0!}{2^0} - \frac{1!}{2^1} + \frac{2!}{2^2} - \frac{3!}{2^3} + \cdots$

b. $\displaystyle\sum_{n=1}^{\infty} \frac{(-1)^n}{\sqrt{n}} = -\frac{1}{\sqrt{1}} + \frac{1}{\sqrt{2}} - \frac{1}{\sqrt{3}} + \frac{1}{\sqrt{4}} - \cdots$

Solution

a. This is an alternating series, but the Alternating Series Test does not apply because the limit of the nth term is not zero. By the nth-Term Test for Divergence, however, you can conclude that this series diverges.

b. This series can be shown to be convergent by the Alternating Series Test. Moreover, because the p-series

$$\sum_{n=1}^{\infty} \left|\frac{(-1)^n}{\sqrt{n}}\right| = \frac{1}{\sqrt{1}} + \frac{1}{\sqrt{2}} + \frac{1}{\sqrt{3}} + \frac{1}{\sqrt{4}} + \cdots$$

diverges, the given series is *conditionally* convergent.

EXAMPLE 7 Absolute and Conditional Convergence

Determine whether each of the series is convergent or divergent. Classify any convergent series as absolutely or conditionally convergent.

a. $\displaystyle\sum_{n=1}^{\infty} \frac{(-1)^{n(n+1)/2}}{3^n} = -\frac{1}{3} - \frac{1}{9} + \frac{1}{27} + \frac{1}{81} - \cdots$

b. $\displaystyle\sum_{n=1}^{\infty} \frac{(-1)^n}{\ln(n+1)} = -\frac{1}{\ln 2} + \frac{1}{\ln 3} - \frac{1}{\ln 4} + \frac{1}{\ln 5} - \cdots$

Solution

a. This is *not* an alternating series (the signs change in pairs). However, note that

$$\sum_{n=1}^{\infty} \left|\frac{(-1)^{n(n+1)/2}}{3^n}\right| = \sum_{n=1}^{\infty} \frac{1}{3^n}$$

is a convergent geometric series, with

$$r = \frac{1}{3}.$$

Consequently, by Theorem 9.16, you can conclude that the given series is *absolutely* convergent (and therefore convergent).

b. In this case, the Alternating Series Test indicates that the series converges. However, the series

$$\sum_{n=1}^{\infty} \left|\frac{(-1)^n}{\ln(n+1)}\right| = \frac{1}{\ln 2} + \frac{1}{\ln 3} + \frac{1}{\ln 4} + \cdots$$

diverges by direct comparison with the terms of the harmonic series. Therefore, the given series is *conditionally* convergent. ■

■ **FOR FURTHER INFORMATION** To read more about the convergence of alternating harmonic series, see the article "Almost Alternating Harmonic Series" by Curtis Feist and Ramin Naimi in *The College Mathematics Journal*. To view this article, go to *MathArticles.com*.

Rearrangement of Series

A finite sum such as

$$1 + 3 - 2 + 5 - 4$$

can be rearranged without changing the value of the sum. This is not necessarily true of an infinite series—it depends on whether the series is absolutely convergent or conditionally convergent.

1. If a series is *absolutely convergent,* then its terms can be rearranged in any order without changing the sum of the series.
2. If a series is *conditionally convergent,* then its terms can be rearranged to give a different sum.

The second case is illustrated in Example 8.

EXAMPLE 8 Rearrangement of a Series

FOR FURTHER INFORMATION Georg Friedrich Bernhard Riemann (1826–1866) proved that if Σa_n is conditionally convergent and S is any real number, then the terms of the series can be rearranged to converge to S. For more on this topic, see the article "Riemann's Rearrangement Theorem" by Stewart Galanor in *Mathematics Teacher.* To view this article, go to *MathArticles.com.*

The alternating harmonic series converges to ln 2. That is,

$$\sum_{n=1}^{\infty} (-1)^{n+1}\frac{1}{n} = \frac{1}{1} - \frac{1}{2} + \frac{1}{3} - \frac{1}{4} + \cdot \cdot \cdot = \ln 2.$$

(See Exercise 55, Section 9.10.)

Rearrange the series to produce a different sum.

Solution Consider the rearrangement below.

$$1 - \frac{1}{2} - \frac{1}{4} + \frac{1}{3} - \frac{1}{6} - \frac{1}{8} + \frac{1}{5} - \frac{1}{10} - \frac{1}{12} + \frac{1}{7} - \frac{1}{14} - \cdot \cdot \cdot$$

$$= \left(1 - \frac{1}{2}\right) - \frac{1}{4} + \left(\frac{1}{3} - \frac{1}{6}\right) - \frac{1}{8} + \left(\frac{1}{5} - \frac{1}{10}\right) - \frac{1}{12} + \left(\frac{1}{7} - \frac{1}{14}\right) - \cdot \cdot \cdot$$

$$= \frac{1}{2} - \frac{1}{4} + \frac{1}{6} - \frac{1}{8} + \frac{1}{10} - \frac{1}{12} + \frac{1}{14} - \cdot \cdot \cdot$$

$$= \frac{1}{2}\left(1 - \frac{1}{2} + \frac{1}{3} - \frac{1}{4} + \frac{1}{5} - \frac{1}{6} + \frac{1}{7} - \cdot \cdot \cdot\right)$$

$$= \frac{1}{2}(\ln 2)$$

By rearranging the terms, you obtain a sum that is half the original sum.

Exploration

In Example 8, you learned that the alternating harmonic series

$$\sum_{n=1}^{\infty} (-1)^{n+1}\frac{1}{n} = 1 - \frac{1}{2} + \frac{1}{3} - \frac{1}{4} + \frac{1}{5} - \frac{1}{6} + \cdot \cdot \cdot$$

converges to $\ln 2 \approx 0.693$. Rearrangement of the terms of the series produces a different sum, $\frac{1}{2}\ln 2 \approx 0.347$.

In this exploration, you will rearrange the terms of the alternating harmonic series in such a way that two positive terms follow each negative term. That is,

$$1 - \frac{1}{2} + \frac{1}{3} + \frac{1}{5} - \frac{1}{4} + \frac{1}{7} + \frac{1}{9} - \frac{1}{6} + \frac{1}{11} + \cdot \cdot \cdot.$$

Now calculate the partial sums S_4, S_7, S_{10}, S_{13}, S_{16}, and S_{19}. Then estimate the sum of this series to three decimal places.

9.5 Exercises

See CalcChat.com for tutorial help and worked-out solutions to odd-numbered exercises.

Numerical and Graphical Analysis In Exercises 1–4, explore the Alternating Series Remainder.

(a) Use a graphing utility to find the indicated partial sum S_n and complete the table.

n	1	2	3	4	5	6	7	8	9	10
S_n										

(b) Use a graphing utility to graph the first 10 terms of the sequence of partial sums and a horizontal line representing the sum.

(c) What pattern exists between the plot of the successive points in part (b) relative to the horizontal line representing the sum of the series? Do the distances between the successive points and the horizontal line increase or decrease?

(d) Discuss the relationship between the answers in part (c) and the Alternating Series Remainder as given in Theorem 9.15.

1. $\displaystyle\sum_{n=1}^{\infty} \frac{(-1)^{n-1}}{2n-1} = \frac{\pi}{4}$

2. $\displaystyle\sum_{n=1}^{\infty} \frac{(-1)^{n-1}}{(n-1)!} = \frac{1}{e}$

3. $\displaystyle\sum_{n=1}^{\infty} \frac{(-1)^{n-1}}{n^2} = \frac{\pi^2}{12}$

4. $\displaystyle\sum_{n=1}^{\infty} \frac{(-1)^{n-1}}{(2n-1)!} = \sin 1$

Determining Convergence or Divergence In Exercises 5–26, determine the convergence or divergence of the series.

5. $\displaystyle\sum_{n=1}^{\infty} \frac{(-1)^{n+1}}{n+1}$

6. $\displaystyle\sum_{n=1}^{\infty} \frac{(-1)^{n+1} n}{3n+2}$

7. $\displaystyle\sum_{n=1}^{\infty} \frac{(-1)^n}{3^n}$

8. $\displaystyle\sum_{n=1}^{\infty} \frac{(-1)^n}{e^n}$

9. $\displaystyle\sum_{n=1}^{\infty} \frac{(-1)^n(5n-1)}{4n+1}$

10. $\displaystyle\sum_{n=1}^{\infty} \frac{(-1)^{n+1} n}{n^2+5}$

11. $\displaystyle\sum_{n=1}^{\infty} \frac{(-1)^n n}{\ln(n+1)}$

12. $\displaystyle\sum_{n=1}^{\infty} \frac{(-1)^n}{\ln(n+1)}$

13. $\displaystyle\sum_{n=1}^{\infty} \frac{(-1)^n}{\sqrt{n}}$

14. $\displaystyle\sum_{n=1}^{\infty} \frac{(-1)^{n+1} n^2}{n^2+4}$

15. $\displaystyle\sum_{n=1}^{\infty} \frac{(-1)^{n+1}(n+1)}{\ln(n+1)}$

16. $\displaystyle\sum_{n=1}^{\infty} \frac{(-1)^{n+1}\ln(n+1)}{n+1}$

17. $\displaystyle\sum_{n=1}^{\infty} \sin\frac{(2n-1)\pi}{2}$

18. $\displaystyle\sum_{n=1}^{\infty} \frac{1}{n}\cos n\pi$

19. $\displaystyle\sum_{n=0}^{\infty} \frac{(-1)^n}{n!}$

20. $\displaystyle\sum_{n=0}^{\infty} \frac{(-1)^n}{(2n+1)!}$

21. $\displaystyle\sum_{n=1}^{\infty} \frac{(-1)^{n+1}\sqrt{n}}{n+2}$

22. $\displaystyle\sum_{n=1}^{\infty} \frac{(-1)^{n+1}\sqrt{n}}{\sqrt[3]{n}}$

23. $\displaystyle\sum_{n=1}^{\infty} \frac{(-1)^{n+1} n!}{1\cdot 3\cdot 5\cdots(2n-1)}$

24. $\displaystyle\sum_{n=1}^{\infty} (-1)^{n+1}\frac{1\cdot 3\cdot 5\cdots(2n-1)}{1\cdot 4\cdot 7\cdots(3n-2)}$

25. $\displaystyle\sum_{n=1}^{\infty} \frac{2(-1)^{n+1}}{e^n-e^{-n}} = \sum_{n=1}^{\infty}(-1)^{n+1}\operatorname{csch} n$

26. $\displaystyle\sum_{n=1}^{\infty} \frac{2(-1)^{n+1}}{e^n+e^{-n}} = \sum_{n=1}^{\infty}(-1)^{n+1}\operatorname{sech} n$

Approximating the Sum of an Alternating Series In Exercises 27–30, approximate the sum of the series by using the first six terms. (See Example 4.)

27. $\displaystyle\sum_{n=0}^{\infty} \frac{(-1)^n 5}{n!}$

28. $\displaystyle\sum_{n=1}^{\infty} \frac{(-1)^{n+1} 4}{\ln(n+1)}$

29. $\displaystyle\sum_{n=1}^{\infty} \frac{(-1)^{n+1} 2}{n^3}$

30. $\displaystyle\sum_{n=1}^{\infty} \frac{(-1)^{n+1} n}{3^n}$

Finding the Number of Terms In Exercises 31–36, use Theorem 9.15 to determine the number of terms required to approximate the sum of the series with an error of less than 0.001.

31. $\displaystyle\sum_{n=1}^{\infty} \frac{(-1)^{n+1}}{n^3}$

32. $\displaystyle\sum_{n=1}^{\infty} \frac{(-1)^{n+1}}{n^2}$

33. $\displaystyle\sum_{n=1}^{\infty} \frac{(-1)^{n+1}}{2n^3-1}$

34. $\displaystyle\sum_{n=1}^{\infty} \frac{(-1)^{n+1}}{n^5}$

35. $\displaystyle\sum_{n=0}^{\infty} \frac{(-1)^n}{n!}$

36. $\displaystyle\sum_{n=0}^{\infty} \frac{(-1)^n}{(2n)!}$

Determining Absolute and Conditional Convergence In Exercises 37–54, determine whether the series converges absolutely or conditionally, or diverges.

37. $\displaystyle\sum_{n=1}^{\infty} \frac{(-1)^n}{2^n}$

38. $\displaystyle\sum_{n=1}^{\infty} \frac{(-1)^{n+1}}{n^2}$

39. $\displaystyle\sum_{n=1}^{\infty} \frac{(-1)^n}{n!}$

40. $\displaystyle\sum_{n=1}^{\infty} \frac{(-1)^{n+1}}{n+3}$

41. $\displaystyle\sum_{n=1}^{\infty} \frac{(-1)^{n+1}}{\sqrt{n}}$

42. $\displaystyle\sum_{n=1}^{\infty} \frac{(-1)^{n+1}}{n\sqrt{n}}$

43. $\displaystyle\sum_{n=1}^{\infty} \frac{(-1)^{n+1} n^2}{(n+1)^2}$

44. $\displaystyle\sum_{n=1}^{\infty} \frac{(-1)^{n+1}(2n+3)}{n+10}$

45. $\displaystyle\sum_{n=2}^{\infty} \frac{(-1)^n}{n\ln n}$

46. $\displaystyle\sum_{n=0}^{\infty} (-1)^n e^{-n^2}$

47. $\displaystyle\sum_{n=2}^{\infty} \frac{(-1)^n n}{n^3-5}$

48. $\displaystyle\sum_{n=1}^{\infty} \frac{(-1)^{n+1}}{n^{4/3}}$

49. $\displaystyle\sum_{n=0}^{\infty} \frac{(-1)^n}{(2n+1)!}$

50. $\displaystyle\sum_{n=0}^{\infty} \frac{(-1)^n}{\sqrt{n+4}}$

51. $\displaystyle\sum_{n=0}^{\infty} \frac{\cos n\pi}{n+1}$

52. $\displaystyle\sum_{n=1}^{\infty} (-1)^{n+1}\arctan n$

53. $\displaystyle\sum_{n=1}^{\infty} \frac{\cos n\pi}{n^2}$

54. $\displaystyle\sum_{n=1}^{\infty} \frac{\sin[(2n-1)\pi/2]}{n}$

WRITING ABOUT CONCEPTS

55. Alternating Series Define an alternating series.

56. Alternating Series Test State the Alternating Series Test.

57. Alternating Series Remainder Give the remainder after N terms of a convergent alternating series.

58. Absolute and Conditional Convergence In your own words, state the difference between absolute and conditional convergence of an alternating series.

59. Think About It Do you agree with the following statements? Why or why not?

(a) If both $\Sigma\, a_n$ and $\Sigma\,(-a_n)$ converge, then $\Sigma\,|a_n|$ converges.

(b) If $\Sigma\, a_n$ diverges, then $\Sigma\,|a_n|$ diverges.

60. HOW DO YOU SEE IT? The graphs of the sequences of partial sums of two series are shown in the figures. Which graph represents the partial sums of an alternating series? Explain.

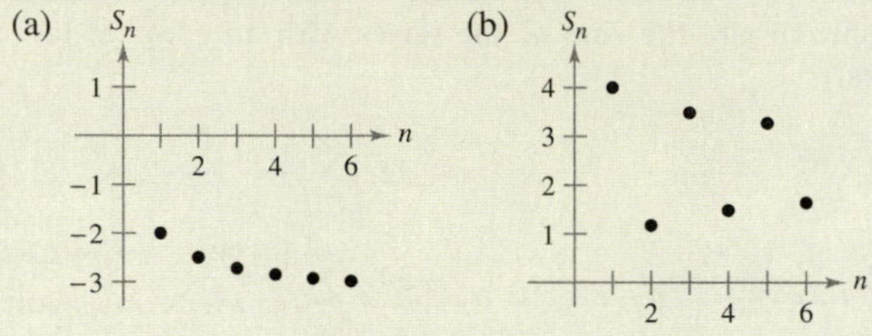

True or False? **In Exercises 61 and 62, determine whether the statement is true or false. If it is false, explain why or give an example that shows it is false.**

61. For the alternating series

$$\sum_{n=1}^{\infty} \frac{(-1)^n}{n}$$

the partial sum S_{100} is an overestimate of the sum of the series.

62. If $\Sigma\, a_n$ and $\Sigma\, b_n$ both converge, then $\Sigma\, a_n b_n$ converges.

Finding Values **In Exercises 63 and 64, find the values of p for which the series converges.**

63. $\displaystyle\sum_{n=1}^{\infty} (-1)^n\left(\frac{1}{n^p}\right)$ **64.** $\displaystyle\sum_{n=1}^{\infty} (-1)^n\left(\frac{1}{n+p}\right)$

65. Proof Prove that if $\Sigma\,|a_n|$ converges, then $\Sigma\, a_n^2$ converges. Is the converse true? If not, give an example that shows it is false.

66. Finding a Series Use the result of Exercise 63 to give an example of an alternating p-series that converges, but whose corresponding p-series diverges.

67. Finding a Series Give an example of a series that demonstrates the statement you proved in Exercise 65.

68. Finding Values Find all values of x for which the series $\Sigma\,(x^n/n)$ (a) converges absolutely and (b) converges conditionally.

Using a Series **In Exercises 69 and 70, use the given series.**

(a) Does the series meet the conditions of Theorem 9.14? Explain why or why not.

(b) Does the series converge? If so, what is the sum?

69. $\displaystyle \frac{1}{2} - \frac{1}{3} + \frac{1}{4} - \frac{1}{9} + \frac{1}{8} - \frac{1}{27} + \cdots + \frac{1}{2^n} - \frac{1}{3^n} + \cdots$

70. $\displaystyle\sum_{n=1}^{\infty} (-1)^{n+1} a_n,\ a_n = \begin{cases} \dfrac{1}{\sqrt{n}}, & \text{if } n \text{ is odd} \\[2mm] \dfrac{1}{n^3}, & \text{if } n \text{ is even} \end{cases}$

Review **In Exercises 71–80, test for convergence or divergence and identify the test used.**

71. $\displaystyle\sum_{n=1}^{\infty} \frac{10}{n^{3/2}}$ **72.** $\displaystyle\sum_{n=1}^{\infty} \frac{3}{n^2+5}$

73. $\displaystyle\sum_{n=1}^{\infty} \frac{3^n}{n^2}$ **74.** $\displaystyle\sum_{n=1}^{\infty} \frac{1}{2^n+1}$

75. $\displaystyle\sum_{n=0}^{\infty} 5\left(\frac{7}{8}\right)^n$ **76.** $\displaystyle\sum_{n=1}^{\infty} \frac{3n^2}{2n^2+1}$

77. $\displaystyle\sum_{n=1}^{\infty} 100e^{-n/2}$ **78.** $\displaystyle\sum_{n=0}^{\infty} \frac{(-1)^n}{n+4}$

79. $\displaystyle\sum_{n=1}^{\infty} \frac{(-1)^{n+1}4}{3n^2-1}$ **80.** $\displaystyle\sum_{n=2}^{\infty} \frac{\ln n}{n}$

81. Describing an Error The following argument, that $0 = 1$, is *incorrect*. Describe the error.

$$\begin{aligned} 0 &= 0 + 0 + 0 + \cdots \\ &= (1-1) + (1-1) + (1-1) + \cdots \\ &= 1 + (-1+1) + (-1+1) + \cdots \\ &= 1 + 0 + 0 + \cdots \\ &= 1 \end{aligned}$$

PUTNAM EXAM CHALLENGE

82. Assume as known the (true) fact that the alternating harmonic series

(1) $1 - \frac{1}{2} + \frac{1}{3} - \frac{1}{4} + \frac{1}{5} - \frac{1}{6} + \frac{1}{7} - \frac{1}{8} + \cdots$

is convergent, and denote its sum by s. Rearrange the series (1) as follows:

(2) $1 + \frac{1}{3} - \frac{1}{2} + \frac{1}{5} + \frac{1}{7} - \frac{1}{4} + \frac{1}{9} + \frac{1}{11} - \frac{1}{6} + \cdots.$

Assume as known the (true) fact that the series (2) is also convergent, and denote its sum by S. Denote by s_k, S_k the kth partial sum of the series (1) and (2), respectively. Prove the following statements.

(i) $S_{3n} = s_{4n} + \frac{1}{2}s_{2n}$, (ii) $S \neq s$

This problem was composed by the Committee on the Putnam Prize Competition.

9.6 The Ratio and Root Tests

- **Use the Ratio Test to determine whether a series converges or diverges.**
- **Use the Root Test to determine whether a series converges or diverges.**
- **Review the tests for convergence and divergence of an infinite series.**

The Ratio Test

This section begins with a test for absolute convergence—the **Ratio Test.**

THEOREM 9.17 Ratio Test

Let Σa_n be a series with nonzero terms.

1. The series Σa_n converges absolutely when $\lim_{n\to\infty}\left|\frac{a_{n+1}}{a_n}\right| < 1$.
2. The series Σa_n diverges when $\lim_{n\to\infty}\left|\frac{a_{n+1}}{a_n}\right| > 1$ or $\lim_{n\to\infty}\left|\frac{a_{n+1}}{a_n}\right| = \infty$.
3. The Ratio Test is inconclusive when $\lim_{n\to\infty}\left|\frac{a_{n+1}}{a_n}\right| = 1$.

Proof To prove Property 1, assume that

$$\lim_{n\to\infty}\left|\frac{a_{n+1}}{a_n}\right| = r < 1$$

and choose R such that $0 \le r < R < 1$. By the definition of the limit of a sequence, there exists some $N > 0$ such that $|a_{n+1}/a_n| < R$ for all $n > N$. Therefore, you can write the following inequalities.

$$\begin{aligned}
|a_{N+1}| &< |a_N|R \\
|a_{N+2}| &< |a_{N+1}|R < |a_N|R^2 \\
|a_{N+3}| &< |a_{N+2}|R < |a_{N+1}|R^2 < |a_N|R^3 \\
&\vdots
\end{aligned}$$

The geometric series $\sum_{n=1}^{\infty} |a_N|R^n = |a_N|R + |a_N|R^2 + \cdot\cdot\cdot + |a_N|R^n + \cdot\cdot\cdot$ converges, and so, by the Direct Comparison Test, the series

$$\sum_{n=1}^{\infty} |a_{N+n}| = |a_{N+1}| + |a_{N+2}| + \cdot\cdot\cdot + |a_{N+n}| + \cdot\cdot\cdot$$

also converges. This in turn implies that the series $\Sigma |a_n|$ converges, because discarding a finite number of terms ($n = N - 1$) does not affect convergence. Consequently, by Theorem 9.16, the series Σa_n converges absolutely. The proof of Property 2 is similar and is left as an exercise (see Exercise 99).

See LarsonCalculus.com for Bruce Edwards's video of this proof.

The fact that the Ratio Test is inconclusive when $|a_{n+1}/a_n| \to 1$ can be seen by comparing the two series $\Sigma (1/n)$ and $\Sigma (1/n^2)$. The first series diverges and the second one converges, but in both cases

$$\lim_{n\to\infty}\left|\frac{a_{n+1}}{a_n}\right| = 1.$$

Although the Ratio Test is not a cure for all ills related to testing for convergence, it is particularly useful for series that *converge rapidly.* Series involving factorials or exponentials are frequently of this type.

EXAMPLE 1 **Using the Ratio Test**

Determine the convergence or divergence of

$$\sum_{n=0}^{\infty} \frac{2^n}{n!}.$$

Solution Because

$$a_n = \frac{2^n}{n!}$$

you can write the following.

$$\begin{aligned}\lim_{n\to\infty}\left|\frac{a_{n+1}}{a_n}\right| &= \lim_{n\to\infty}\left[\frac{2^{n+1}}{(n+1)!} \div \frac{2^n}{n!}\right]\\ &= \lim_{n\to\infty}\left[\frac{2^{n+1}}{(n+1)!}\cdot\frac{n!}{2^n}\right]\\ &= \lim_{n\to\infty}\frac{2}{n+1}\\ &= 0 < 1\end{aligned}$$

This series converges because the limit of $|a_{n+1}/a_n|$ is less than 1.

REMARK A step frequently used in applications of the Ratio Test involves simplifying quotients of factorials. In Example 1, for instance, notice that

$$\frac{n!}{(n+1)!} = \frac{n!}{(n+1)n!} = \frac{1}{n+1}.$$

EXAMPLE 2 **Using the Ratio Test**

Determine whether each series converges or diverges.

a. $\displaystyle\sum_{n=0}^{\infty} \frac{n^2 2^{n+1}}{3^n}$ **b.** $\displaystyle\sum_{n=1}^{\infty} \frac{n^n}{n!}$

Solution

a. This series converges because the limit of $|a_{n+1}/a_n|$ is less than 1.

$$\begin{aligned}\lim_{n\to\infty}\left|\frac{a_{n+1}}{a_n}\right| &= \lim_{n\to\infty}\left[(n+1)^2\left(\frac{2^{n+2}}{3^{n+1}}\right)\left(\frac{3^n}{n^2 2^{n+1}}\right)\right]\\ &= \lim_{n\to\infty}\frac{2(n+1)^2}{3n^2}\\ &= \frac{2}{3} < 1\end{aligned}$$

b. This series diverges because the limit of $|a_{n+1}/a_n|$ is greater than 1.

$$\begin{aligned}\lim_{n\to\infty}\left|\frac{a_{n+1}}{a_n}\right| &= \lim_{n\to\infty}\left[\frac{(n+1)^{n+1}}{(n+1)!}\left(\frac{n!}{n^n}\right)\right]\\ &= \lim_{n\to\infty}\left[\frac{(n+1)^{n+1}}{(n+1)}\left(\frac{1}{n^n}\right)\right]\\ &= \lim_{n\to\infty}\frac{(n+1)^n}{n^n}\\ &= \lim_{n\to\infty}\left(1+\frac{1}{n}\right)^n\\ &= e > 1\end{aligned}$$

EXAMPLE 3 A Failure of the Ratio Test

See LarsonCalculus.com for an interactive version of this type of example.

Determine the convergence or divergence of

$$\sum_{n=1}^{\infty} (-1)^n \frac{\sqrt{n}}{n+1}.$$

Solution The limit of $|a_{n+1}/a_n|$ is equal to 1.

$$\begin{aligned}\lim_{n\to\infty}\left|\frac{a_{n+1}}{a_n}\right| &= \lim_{n\to\infty}\left[\left(\frac{\sqrt{n+1}}{n+2}\right)\left(\frac{n+1}{\sqrt{n}}\right)\right]\\ &= \lim_{n\to\infty}\left[\sqrt{\frac{n+1}{n}}\left(\frac{n+1}{n+2}\right)\right]\\ &= \sqrt{1}(1)\\ &= 1\end{aligned}$$

REMARK The Ratio Test is also inconclusive for any p-series.

So, the Ratio Test is inconclusive. To determine whether the series converges, you need to try a different test. In this case, you can apply the Alternating Series Test. To show that $a_{n+1} \le a_n$, let

$$f(x) = \frac{\sqrt{x}}{x+1}.$$

Then the derivative is

$$f'(x) = \frac{-x+1}{2\sqrt{x}(x+1)^2}.$$

Because the derivative is negative for $x > 1$, you know that f is a decreasing function. Also, by L'Hôpital's Rule,

$$\begin{aligned}\lim_{x\to\infty}\frac{\sqrt{x}}{x+1} &= \lim_{x\to\infty}\frac{1/(2\sqrt{x})}{1}\\ &= \lim_{x\to\infty}\frac{1}{2\sqrt{x}}\\ &= 0.\end{aligned}$$

Therefore, by the Alternating Series Test, the series converges. ■

The series in Example 3 is *conditionally convergent.* This follows from the fact that the series

$$\sum_{n=1}^{\infty} |a_n|$$

diverges (by the Limit Comparison Test with $\Sigma\, 1/\sqrt{n}$), but the series

$$\sum_{n=1}^{\infty} a_n$$

converges.

TECHNOLOGY A graphing utility can reinforce the conclusion that the series in Example 3 converges *conditionally*. By adding the first 100 terms of the series, you obtain a sum of about -0.2. (The sum of the first 100 terms of the series $\Sigma\,|a_n|$ is about 17.)

The Root Test

The next test for convergence or divergence of series works especially well for series involving nth powers. The proof of this theorem is similar to the proof given for the Ratio Test, and is left as an exercise (see Exercise 100).

REMARK The Root Test is always inconclusive for any p-series.

THEOREM 9.18 Root Test

1. The series $\Sigma\, a_n$ converges absolutely when $\lim_{n\to\infty} \sqrt[n]{|a_n|} < 1$.
2. The series $\Sigma\, a_n$ diverges when $\lim_{n\to\infty} \sqrt[n]{|a_n|} > 1$ or $\lim_{n\to\infty} \sqrt[n]{|a_n|} = \infty$.
3. The Root Test is inconclusive when $\lim_{n\to\infty} \sqrt[n]{|a_n|} = 1$.

AP* Tips

The Ratio Test is commonly covered on the AP BC Exam, primarily on the free response section. The Root Test is not covered on the exam.

EXAMPLE 4 Using the Root Test

Determine the convergence or divergence of

$$\sum_{n=1}^{\infty} \frac{e^{2n}}{n^n}.$$

Solution You can apply the Root Test as follows.

$$\begin{aligned}
\lim_{n\to\infty} \sqrt[n]{|a_n|} &= \lim_{n\to\infty} \sqrt[n]{\frac{e^{2n}}{n^n}} \\
&= \lim_{n\to\infty} \frac{e^{2n/n}}{n^{n/n}} \\
&= \lim_{n\to\infty} \frac{e^2}{n} \\
&= 0 < 1
\end{aligned}$$

Because this limit is less than 1, you can conclude that the series converges absolutely (and therefore converges). ■

To see the usefulness of the Root Test for the series in Example 4, try applying the Ratio Test to that series. When you do this, you obtain the following.

$$\begin{aligned}
\lim_{n\to\infty} \left|\frac{a_{n+1}}{a_n}\right| &= \lim_{n\to\infty} \left[\frac{e^{2(n+1)}}{(n+1)^{n+1}} \div \frac{e^{2n}}{n^n}\right] \\
&= \lim_{n\to\infty} \left[\frac{e^{2(n+1)}}{(n+1)^{n+1}} \cdot \frac{n^n}{e^{2n}}\right] \\
&= \lim_{n\to\infty} e^2 \frac{n^n}{(n+1)^{n+1}} \\
&= \lim_{n\to\infty} e^2 \left(\frac{n}{n+1}\right)^n \left(\frac{1}{n+1}\right) \\
&= 0
\end{aligned}$$

Note that this limit is not as easily evaluated as the limit obtained by the Root Test in Example 4.

■ **FOR FURTHER INFORMATION** For more information on the usefulness of the Root Test, see the article "*N*! and the Root Test" by Charles C. Mumma II in *The American Mathematical Monthly.* To view this article, go to *MathArticles.com.*

Strategies for Testing Series

You have now studied 10 tests for determining the convergence or divergence of an infinite series. (See the summary in the table on the next page.) Skill in choosing and applying the various tests will come only with practice. Below is a set of guidelines for choosing an appropriate test.

GUIDELINES FOR TESTING A SERIES FOR CONVERGENCE OR DIVERGENCE

1. Does the nth term approach 0? If not, the series diverges.
2. Is the series one of the special types—geometric, p-series, telescoping, or alternating?
3. Can the Integral Test, the Root Test, or the Ratio Test be applied?
4. Can the series be compared favorably to one of the special types?

In some instances, more than one test is applicable. However, your objective should be to learn to choose the most efficient test.

EXAMPLE 5 Applying the Strategies for Testing Series

Determine the convergence or divergence of each series.

a. $\displaystyle\sum_{n=1}^{\infty} \frac{n+1}{3n+1}$ **b.** $\displaystyle\sum_{n=1}^{\infty} \left(\frac{\pi}{6}\right)^n$ **c.** $\displaystyle\sum_{n=1}^{\infty} ne^{-n^2}$

d. $\displaystyle\sum_{n=1}^{\infty} \frac{1}{3n+1}$ **e.** $\displaystyle\sum_{n=1}^{\infty} (-1)^n \frac{3}{4n+1}$ **f.** $\displaystyle\sum_{n=1}^{\infty} \frac{n!}{10^n}$

g. $\displaystyle\sum_{n=1}^{\infty} \left(\frac{n+1}{2n+1}\right)^n$

Solution

a. For this series, the limit of the nth term is not 0 $\left(a_n \to \frac{1}{3} \text{ as } n \to \infty\right)$. So, by the nth-Term Test, the series diverges.

b. This series is geometric. Moreover, because the ratio of the terms

$$r = \frac{\pi}{6}$$

is less than 1 in absolute value, you can conclude that the series converges.

c. Because the function

$$f(x) = xe^{-x^2}$$

is easily integrated, you can use the Integral Test to conclude that the series converges.

d. The nth term of this series can be compared to the nth term of the harmonic series. After using the Limit Comparison Test, you can conclude that the series diverges.

e. This is an alternating series whose nth term approaches 0. Because $a_{n+1} \le a_n$, you can use the Alternating Series Test to conclude that the series converges.

f. The nth term of this series involves a factorial, which indicates that the Ratio Test may work well. After applying the Ratio Test, you can conclude that the series diverges.

g. The nth term of this series involves a variable that is raised to the nth power, which indicates that the Root Test may work well. After applying the Root Test, you can conclude that the series converges.

SUMMARY OF TESTS FOR SERIES

Test	Series	Condition(s) of Convergence	Condition(s) of Divergence	Comment
nth-Term	$\sum_{n=1}^{\infty} a_n$		$\lim_{n\to\infty} a_n \neq 0$	This test cannot be used to show convergence.
Geometric Series	$\sum_{n=0}^{\infty} ar^n$	$0 < \lvert r\rvert < 1$	$\lvert r\rvert \geq 1$	Sum: $S = \dfrac{a}{1-r}$
Telescoping Series	$\sum_{n=1}^{\infty} (b_n - b_{n+1})$	$\lim_{n\to\infty} b_n = L$		Sum: $S = b_1 - L$
p-Series	$\sum_{n=1}^{\infty} \dfrac{1}{n^p}$	$p > 1$	$0 < p \leq 1$	
Alternating Series	$\sum_{n=1}^{\infty} (-1)^{n-1} a_n$	$0 < a_{n+1} \leq a_n$ and $\lim_{n\to\infty} a_n = 0$		Remainder: $\lvert R_N\rvert \leq a_{N+1}$
Integral (f is continuous, positive, and decreasing)	$\sum_{n=1}^{\infty} a_n$, $a_n = f(n) \geq 0$	$\int_1^{\infty} f(x)\,dx$ converges	$\int_1^{\infty} f(x)\,dx$ diverges	Remainder: $0 < R_N < \int_N^{\infty} f(x)\,dx$
Root	$\sum_{n=1}^{\infty} a_n$	$\lim_{n\to\infty} \sqrt[n]{\lvert a_n\rvert} < 1$	$\lim_{n\to\infty} \sqrt[n]{\lvert a_n\rvert} > 1$ or $= \infty$	Test is inconclusive when $\lim_{n\to\infty} \sqrt[n]{\lvert a_n\rvert} = 1$.
Ratio	$\sum_{n=1}^{\infty} a_n$	$\lim_{n\to\infty} \left\lvert \dfrac{a_{n+1}}{a_n} \right\rvert < 1$	$\lim_{n\to\infty} \left\lvert \dfrac{a_{n+1}}{a_n} \right\rvert > 1$ or $= \infty$	Test is inconclusive when $\lim_{n\to\infty} \left\lvert \dfrac{a_{n+1}}{a_n} \right\rvert = 1$.
Direct Comparison ($a_n, b_n > 0$)	$\sum_{n=1}^{\infty} a_n$	$0 < a_n \leq b_n$ and $\sum_{n=1}^{\infty} b_n$ converges	$0 < b_n \leq a_n$ and $\sum_{n=1}^{\infty} b_n$ diverges	
Limit Comparison ($a_n, b_n > 0$)	$\sum_{n=1}^{\infty} a_n$	$\lim_{n\to\infty} \dfrac{a_n}{b_n} = L > 0$ and $\sum_{n=1}^{\infty} b_n$ converges	$\lim_{n\to\infty} \dfrac{a_n}{b_n} = L > 0$ and $\sum_{n=1}^{\infty} b_n$ diverges	

9.6 Exercises

See CalcChat.com for tutorial help and worked-out solutions to odd-numbered exercises.

Verifying a Formula In Exercises 1–4, verify the formula.

1. $\dfrac{(n+1)!}{(n-2)!} = (n+1)(n)(n-1)$

2. $\dfrac{(2k-2)!}{(2k)!} = \dfrac{1}{(2k)(2k-1)}$

3. $1 \cdot 3 \cdot 5 \cdot \cdot \cdot (2k-1) = \dfrac{(2k)!}{2^k k!}$

4. $\dfrac{1}{1 \cdot 3 \cdot 5 \cdot \cdot \cdot (2k-5)} = \dfrac{2^k k!(2k-3)(2k-1)}{(2k)!}, \quad k \geq 3$

Matching In Exercises 5–10, match the series with the graph of its sequence of partial sums. [The graphs are labeled (a), (b), (c), (d), (e), and (f).]

(a)

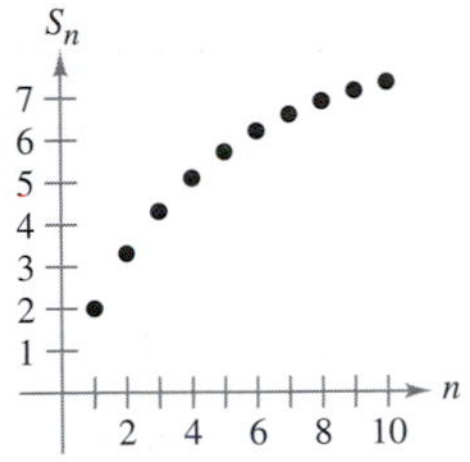

(b)

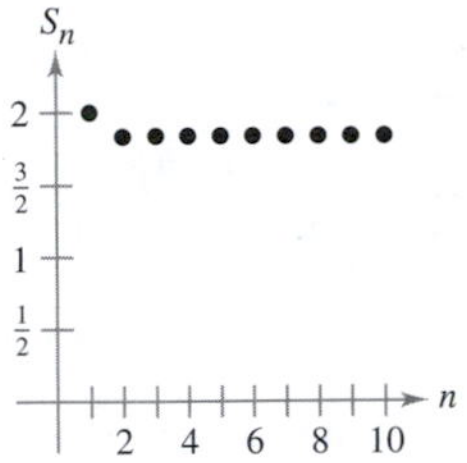

(c)

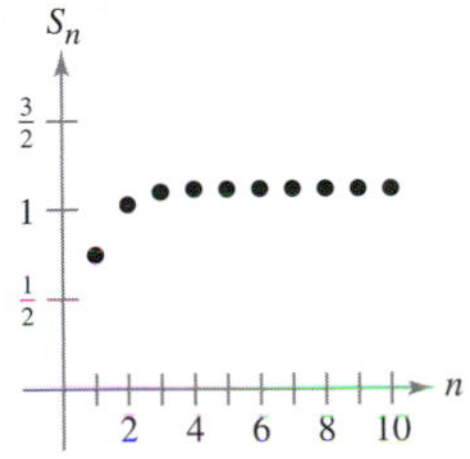

(d)

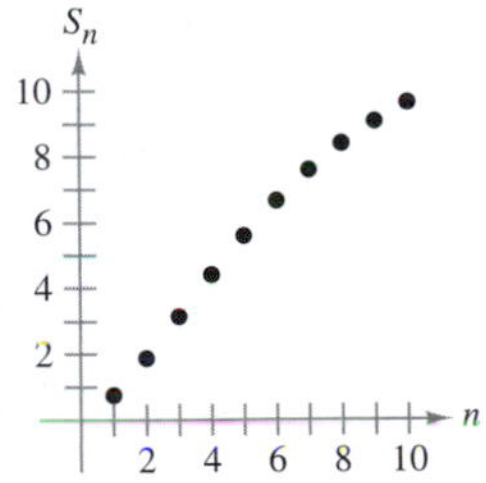

(e)

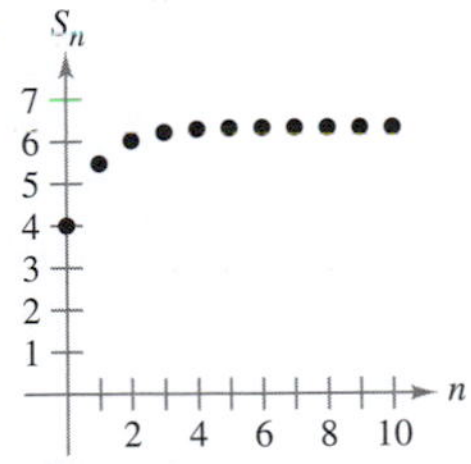

(f)

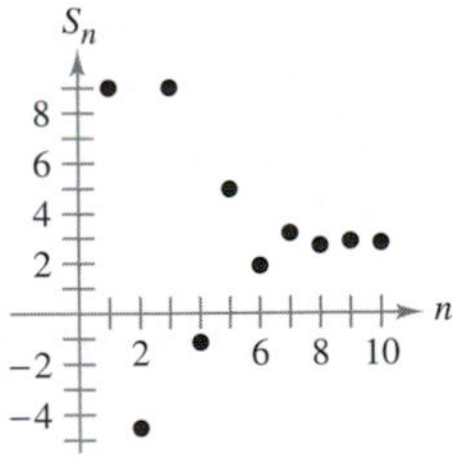

5. $\displaystyle\sum_{n=1}^{\infty} n\left(\frac{3}{4}\right)^n$

6. $\displaystyle\sum_{n=1}^{\infty} \left(\frac{3}{4}\right)^n\left(\frac{1}{n!}\right)$

7. $\displaystyle\sum_{n=1}^{\infty} \frac{(-3)^{n+1}}{n!}$

8. $\displaystyle\sum_{n=1}^{\infty} \frac{(-1)^{n-1}4}{(2n)!}$

9. $\displaystyle\sum_{n=1}^{\infty} \left(\frac{4n}{5n-3}\right)^n$

10. $\displaystyle\sum_{n=0}^{\infty} 4e^{-n}$

Numerical, Graphical, and Analytic Analysis In Exercises 11 and 12, (a) verify that the series converges, (b) use a graphing utility to find the indicated partial sum S_n and complete the table, (c) use a graphing utility to graph the first 10 terms of the sequence of partial sums, (d) use the table to estimate the sum of the series, and (e) explain the relationship between the magnitudes of the terms of the series and the rate at which the sequence of partial sums approaches the sum of the series.

n	5	10	15	20	25
S_n					

11. $\displaystyle\sum_{n=1}^{\infty} n^3\left(\frac{1}{2}\right)^n$

12. $\displaystyle\sum_{n=1}^{\infty} \frac{n^2+1}{n!}$

Using the Ratio Test In Exercises 13–34, use the Ratio Test to determine the convergence or divergence of the series.

13. $\displaystyle\sum_{n=1}^{\infty} \frac{1}{5^n}$

14. $\displaystyle\sum_{n=1}^{\infty} \frac{1}{n!}$

15. $\displaystyle\sum_{n=0}^{\infty} \frac{n!}{3^n}$

16. $\displaystyle\sum_{n=0}^{\infty} \frac{2^n}{n!}$

17. $\displaystyle\sum_{n=1}^{\infty} n\left(\frac{6}{5}\right)^n$

18. $\displaystyle\sum_{n=1}^{\infty} n\left(\frac{7}{8}\right)^n$

19. $\displaystyle\sum_{n=1}^{\infty} \frac{n}{4^n}$

20. $\displaystyle\sum_{n=1}^{\infty} \frac{5^n}{n^4}$

21. $\displaystyle\sum_{n=1}^{\infty} \frac{n^3}{3^n}$

22. $\displaystyle\sum_{n=1}^{\infty} \frac{(-1)^{n+1}(n+2)}{n(n+1)}$

23. $\displaystyle\sum_{n=0}^{\infty} \frac{(-1)^n 2^n}{n!}$

24. $\displaystyle\sum_{n=1}^{\infty} \frac{(-1)^{n-1}(3/2)^n}{n^2}$

25. $\displaystyle\sum_{n=1}^{\infty} \frac{n!}{n3^n}$

26. $\displaystyle\sum_{n=1}^{\infty} \frac{(2n)!}{n^5}$

27. $\displaystyle\sum_{n=0}^{\infty} \frac{e^n}{n!}$

28. $\displaystyle\sum_{n=1}^{\infty} \frac{n!}{n^n}$

29. $\displaystyle\sum_{n=0}^{\infty} \frac{6^n}{(n+1)^n}$

30. $\displaystyle\sum_{n=0}^{\infty} \frac{(n!)^2}{(3n)!}$

31. $\displaystyle\sum_{n=0}^{\infty} \frac{5^n}{2^n+1}$

32. $\displaystyle\sum_{n=0}^{\infty} \frac{(-1)^n 2^{4n}}{(2n+1)!}$

33. $\displaystyle\sum_{n=0}^{\infty} \frac{(-1)^{n+1} n!}{1 \cdot 3 \cdot 5 \cdot \cdot \cdot (2n+1)}$

34. $\displaystyle\sum_{n=1}^{\infty} \frac{(-1)^n[2 \cdot 4 \cdot 6 \cdot \cdot \cdot (2n)]}{2 \cdot 5 \cdot 8 \cdot \cdot \cdot (3n-1)}$

Using the Root Test In Exercises 35–50, use the Root Test to determine the convergence or divergence of the series.

35. $\sum_{n=1}^{\infty} \frac{1}{5^n}$

36. $\sum_{n=1}^{\infty} \frac{1}{n^n}$

37. $\sum_{n=1}^{\infty} \left(\frac{n}{2n+1}\right)^n$

38. $\sum_{n=1}^{\infty} \left(\frac{2n}{n+1}\right)^n$

39. $\sum_{n=1}^{\infty} \left(\frac{3n+2}{n+3}\right)^n$

40. $\sum_{n=1}^{\infty} \left(\frac{n-2}{5n+1}\right)^n$

41. $\sum_{n=2}^{\infty} \frac{(-1)^n}{(\ln n)^n}$

42. $\sum_{n=1}^{\infty} \left(\frac{-3n}{2n+1}\right)^{3n}$

43. $\sum_{n=1}^{\infty} \left(2\sqrt[n]{n}+1\right)^n$

44. $\sum_{n=0}^{\infty} e^{-3n}$

45. $\sum_{n=1}^{\infty} \frac{n}{3^n}$

46. $\sum_{n=1}^{\infty} \left(\frac{n}{500}\right)^n$

47. $\sum_{n=1}^{\infty} \left(\frac{1}{n}-\frac{1}{n^2}\right)^n$

48. $\sum_{n=1}^{\infty} \left(\frac{\ln n}{n}\right)^n$

49. $\sum_{n=2}^{\infty} \frac{n}{(\ln n)^n}$

50. $\sum_{n=1}^{\infty} \frac{(n!)^n}{(n^n)^2}$

Determining Convergence or Divergence In Exercises 51–68, determine the convergence or divergence of the series using any appropriate test from this chapter. Identify the test used.

51. $\sum_{n=1}^{\infty} \frac{(-1)^{n+1}5}{n}$

52. $\sum_{n=1}^{\infty} \frac{100}{n}$

53. $\sum_{n=1}^{\infty} \frac{3}{n\sqrt{n}}$

54. $\sum_{n=1}^{\infty} \left(\frac{2\pi}{3}\right)^n$

55. $\sum_{n=1}^{\infty} \frac{5n}{2n-1}$

56. $\sum_{n=1}^{\infty} \frac{n}{2n^2+1}$

57. $\sum_{n=1}^{\infty} \frac{(-1)^n 3^{n-2}}{2^n}$

58. $\sum_{n=1}^{\infty} \frac{10}{3\sqrt{n^3}}$

59. $\sum_{n=1}^{\infty} \frac{10n+3}{n2^n}$

60. $\sum_{n=1}^{\infty} \frac{2^n}{4n^2-1}$

61. $\sum_{n=1}^{\infty} \frac{\cos n}{3^n}$

62. $\sum_{n=2}^{\infty} \frac{(-1)^n}{n \ln n}$

63. $\sum_{n=1}^{\infty} \frac{n!}{n7^n}$

64. $\sum_{n=1}^{\infty} \frac{\ln n}{n^2}$

65. $\sum_{n=1}^{\infty} \frac{(-1)^n 3^{n-1}}{n!}$

66. $\sum_{n=1}^{\infty} \frac{(-1)^n 3^n}{n2^n}$

67. $\sum_{n=1}^{\infty} \frac{(-3)^n}{3 \cdot 5 \cdot 7 \cdots (2n+1)}$

68. $\sum_{n=1}^{\infty} \frac{3 \cdot 5 \cdot 7 \cdots (2n+1)}{18^n(2n-1)n!}$

Identifying Series In Exercises 69–72, identify the two series that are the same.

69. (a) $\sum_{n=1}^{\infty} \frac{n5^n}{n!}$
(b) $\sum_{n=0}^{\infty} \frac{n5^n}{(n+1)!}$
(c) $\sum_{n=0}^{\infty} \frac{(n+1)5^{n+1}}{(n+1)!}$

70. (a) $\sum_{n=4}^{\infty} n\left(\frac{3}{4}\right)^n$
(b) $\sum_{n=0}^{\infty} (n+1)\left(\frac{3}{4}\right)^n$
(c) $\sum_{n=1}^{\infty} n\left(\frac{3}{4}\right)^{n-1}$

71. (a) $\sum_{n=0}^{\infty} \frac{(-1)^n}{(2n+1)!}$
(b) $\sum_{n=1}^{\infty} \frac{(-1)^{n-1}}{(2n-1)!}$
(c) $\sum_{n=1}^{\infty} \frac{(-1)^{n-1}}{(2n+1)!}$

72. (a) $\sum_{n=2}^{\infty} \frac{(-1)^n}{(n-1)2^{n-1}}$
(b) $\sum_{n=1}^{\infty} \frac{(-1)^{n+1}}{n2^n}$
(c) $\sum_{n=0}^{\infty} \frac{(-1)^{n+1}}{(n+1)2^n}$

Writing an Equivalent Series In Exercises 73 and 74, write an equivalent series with the index of summation beginning at $n = 0$.

73. $\sum_{n=1}^{\infty} \frac{n}{7^n}$

74. $\sum_{n=2}^{\infty} \frac{9^n}{(n-2)!}$

Finding the Number of Terms In Exercises 75 and 76, (a) determine the number of terms required to approximate the sum of the series with an error less than 0.0001, and (b) use a graphing utility to approximate the sum of the series with an error less than 0.0001.

75. $\sum_{k=1}^{\infty} \frac{(-3)^k}{2^k k!}$

76. $\sum_{k=0}^{\infty} \frac{(-3)^k}{1 \cdot 3 \cdot 5 \cdots (2k+1)}$

Using a Recursively Defined Series In Exercises 77–82, the terms of a series $\sum_{n=1}^{\infty} a_n$ are defined recursively. Determine the convergence or divergence of the series. Explain your reasoning.

77. $a_1 = \frac{1}{2},\ a_{n+1} = \frac{4n-1}{3n+2}a_n$

78. $a_1 = 2,\ a_{n+1} = \frac{2n+1}{5n-4}a_n$

79. $a_1 = 1,\ a_{n+1} = \frac{\sin n + 1}{\sqrt{n}}a_n$

80. $a_1 = \frac{1}{5},\ a_{n+1} = \frac{\cos n + 1}{n}a_n$

81. $a_1 = \frac{1}{3},\ a_{n+1} = \left(1 + \frac{1}{n}\right)a_n$

82. $a_1 = \frac{1}{4},\ a_{n+1} = \sqrt[n]{a_n}$

Using the Ratio Test or Root Test In Exercises 83–86, use the Ratio Test or the Root Test to determine the convergence or divergence of the series.

83. $1 + \frac{1 \cdot 2}{1 \cdot 3} + \frac{1 \cdot 2 \cdot 3}{1 \cdot 3 \cdot 5} + \frac{1 \cdot 2 \cdot 3 \cdot 4}{1 \cdot 3 \cdot 5 \cdot 7} + \cdots$

84. $1 + \frac{2}{3} + \frac{3}{3^2} + \frac{4}{3^3} + \frac{5}{3^4} + \frac{6}{3^5} + \cdots$

85. $\frac{1}{(\ln 3)^3} + \frac{1}{(\ln 4)^4} + \frac{1}{(\ln 5)^5} + \frac{1}{(\ln 6)^6} + \cdots$

86. $1 + \frac{1 \cdot 3}{1 \cdot 2 \cdot 3} + \frac{1 \cdot 3 \cdot 5}{1 \cdot 2 \cdot 3 \cdot 4 \cdot 5} + \frac{1 \cdot 3 \cdot 5 \cdot 7}{1 \cdot 2 \cdot 3 \cdot 4 \cdot 5 \cdot 6 \cdot 7} + \cdots$

Finding Values **In Exercises 87–92, find the values of x for which the series converges.**

87. $\sum_{n=0}^{\infty} 2\left(\frac{x}{3}\right)^n$

88. $\sum_{n=0}^{\infty} \left(\frac{x-3}{5}\right)^n$

89. $\sum_{n=1}^{\infty} \frac{(-1)^n(x+1)^n}{n}$

90. $\sum_{n=0}^{\infty} 3(x-4)^n$

91. $\sum_{n=0}^{\infty} n!\left(\frac{x}{2}\right)^n$

92. $\sum_{n=0}^{\infty} \frac{(x+1)^n}{n!}$

WRITING ABOUT CONCEPTS

93. **Ratio Test** State the Ratio Test.

94. **Root Test** State the Root Test.

95. **Think About It** You are told that the terms of a positive series appear to approach zero rapidly as n approaches infinity. In fact, $a_7 \le 0.0001$. Given no other information, does this imply that the series converges? Support your conclusion with examples.

96. **Think About It** What can you conclude about the convergence or divergence of $\Sigma\, a_n$ for each of the following conditions? Explain your reasoning.

(a) $\lim_{n\to\infty} \left|\frac{a_{n+1}}{a_n}\right| = 0$ (b) $\lim_{n\to\infty} \left|\frac{a_{n+1}}{a_n}\right| = 1$

(c) $\lim_{n\to\infty} \left|\frac{a_{n+1}}{a_n}\right| = \frac{3}{2}$ (d) $\lim_{n\to\infty} \sqrt[n]{|a_n|} = 2$

(e) $\lim_{n\to\infty} \sqrt[n]{|a_n|} = 1$ (f) $\lim_{n\to\infty} \sqrt[n]{|a_n|} = e$

97. **Using an Alternating Series** Using the Ratio Test, it is determined that an alternating series converges. Does the series converge conditionally or absolutely? Explain.

98. **HOW DO YOU SEE IT?** The figure shows the first 10 terms of the convergent series $\sum_{n=1}^{\infty} a_n$ and the first 10 terms of the convergent series $\sum_{n=1}^{\infty} \sqrt{a_n}$. Identify the two series and explain your reasoning in making the selection.

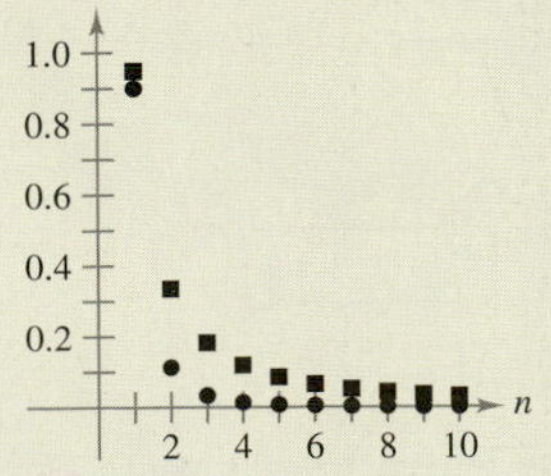

99. **Proof** Prove Property 2 of Theorem 9.17.

100. **Proof** Prove Theorem 9.18. (*Hint for Property 1:* If the limit equals $r < 1$, choose a real number R such that $r < R < 1$. By the definitions of the limit, there exists some $N > 0$ such that $\sqrt[n]{|a_n|} < R$ for $n > N$.)

Verifying an Inconclusive Test **In Exercises 101–104, verify that the Ratio Test is inconclusive for the p-series.**

101. $\sum_{n=1}^{\infty} \frac{1}{n^{3/2}}$

102. $\sum_{n=1}^{\infty} \frac{1}{n^{1/2}}$

103. $\sum_{n=1}^{\infty} \frac{1}{n^4}$

104. $\sum_{n=1}^{\infty} \frac{1}{n^p}$

105. **Verifying an Inconclusive Test** Show that the Root Test is inconclusive for the p-series

$$\sum_{n=1}^{\infty} \frac{1}{n^p}.$$

106. **Verifying Inconclusive Tests** Show that the Ratio Test and the Root Test are both inconclusive for the logarithmic p-series

$$\sum_{n=2}^{\infty} \frac{1}{n(\ln n)^p}.$$

107. **Using Values** Determine the convergence or divergence of the series

$$\sum_{n=1}^{\infty} \frac{(n!)^2}{(xn)!}$$

when (a) $x = 1$, (b) $x = 2$, (c) $x = 3$, and (d) x is a positive integer.

108. **Using a Series** Show that if

$$\sum_{n=1}^{\infty} a_n$$

is absolutely convergent, then

$$\left|\sum_{n=1}^{\infty} a_n\right| \le \sum_{n=1}^{\infty} |a_n|.$$

PUTNAM EXAM CHALLENGE

109. Show that if the series

$$a_1 + a_2 + a_3 + \cdots + a_n + \cdots$$

converges, then the series

$$a_1 + \frac{a_2}{2} + \frac{a_3}{3} + \cdots + \frac{a_n}{n} + \cdots$$

converges also.

110. Is the following series convergent or divergent?

$$1 + \frac{1}{2}\cdot\frac{19}{7} + \frac{2!}{3^2}\left(\frac{19}{7}\right)^2 + \frac{3!}{4^3}\left(\frac{19}{7}\right)^3 + \frac{4!}{5^4}\left(\frac{19}{7}\right)^4 + \cdots$$

These problems were composed by the Committee on the Putnam Prize Competition.

9.7 Taylor Polynomials and Approximations

- **Find polynomial approximations of elementary functions and compare them with the elementary functions.**
- **Find Taylor and Maclaurin polynomial approximations of elementary functions.**
- **Use the remainder of a Taylor polynomial.**

Polynomial Approximations of Elementary Functions

The goal of this section is to show how polynomial functions can be used as approximations for other elementary functions. To find a polynomial function P that approximates another function f, begin by choosing a number c in the domain of f at which f and P have the same value. That is,

$$P(c) = f(c).$$ Graphs of f and P pass through $(c, f(c))$.

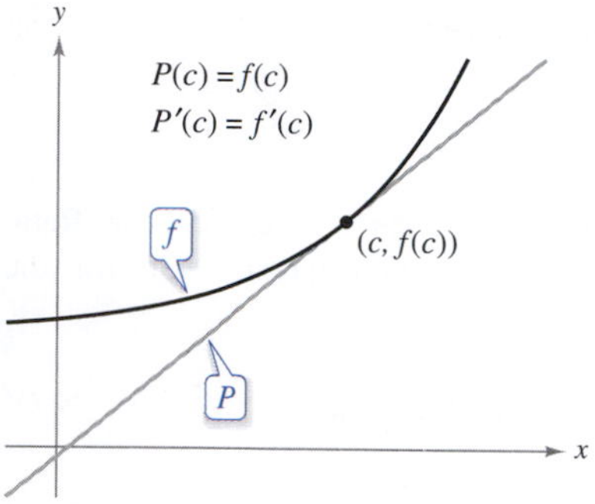

Near $(c, f(c))$, the graph of P can be used to approximate the graph of f.
Figure 9.10

The approximating polynomial is said to be **expanded about c** or **centered at c.** Geometrically, the requirement that $P(c) = f(c)$ means that the graph of P passes through the point $(c, f(c))$. Of course, there are many polynomials whose graphs pass through the point $(c, f(c))$. Your task is to find a polynomial whose graph resembles the graph of f near this point. One way to do this is to impose the additional requirement that the slope of the polynomial function be the same as the slope of the graph of f at the point $(c, f(c))$.

$$P'(c) = f'(c)$$ Graphs of f and P have the same slope at $(c, f(c))$.

With these two requirements, you can obtain a simple linear approximation of f, as shown in Figure 9.10.

REMARK Example 1 is not the first time you have used a linear function to approximate another function. The same procedure was used as the basis for Newton's Method.

EXAMPLE 1 First-Degree Polynomial Approximation of $f(x) = e^x$

For the function $f(x) = e^x$, find a first-degree polynomial function $P_1(x) = a_0 + a_1x$ whose value and slope agree with the value and slope of f at $x = 0$.

Solution Because $f(x) = e^x$ and $f'(x) = e^x$, the value and the slope of f at $x = 0$ are

$$f(0) = e^0 = 1$$ Value of f at $x = 0$

and

$$f'(0) = e^0 = 1.$$ Slope of f at $x = 0$

Because $P_1(x) = a_0 + a_1x$, you can use the condition that $P_1(0) = f(0)$ to conclude that $a_0 = 1$. Moreover, because $P_1'(x) = a_1$, you can use the condition that $P_1'(0) = f'(0)$ to conclude that $a_1 = 1$. Therefore, $P_1(x) = 1 + x$. Figure 9.11 shows the graphs of $P_1(x) = 1 + x$ and $f(x) = e^x$.

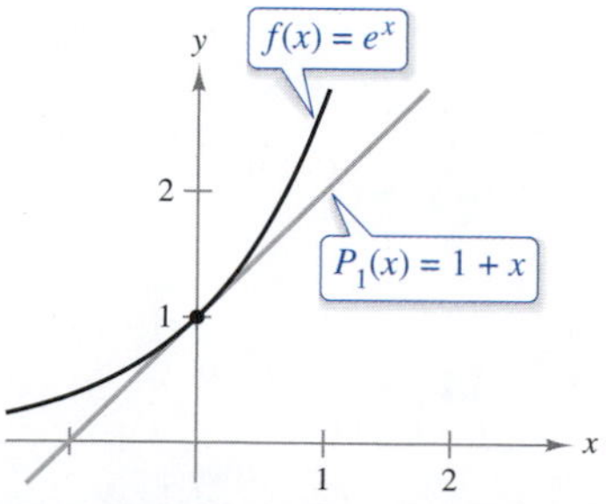

P_1 is the first-degree polynomial approximation of $f(x) = e^x$.
Figure 9.11

In Figure 9.12, you can see that, at points near (0, 1), the graph of the first-degree polynomial function

$$P_1(x) = 1 + x \qquad \text{1st-degree approximation}$$

is reasonably close to the graph of $f(x) = e^x$. As you move away from (0, 1), however, the graphs move farther and farther from each other and the accuracy of the approximation decreases. To improve the approximation, you can impose yet another requirement—that the values of the second derivatives of P and f agree when $x = 0$. The polynomial, P_2, of least degree that satisfies all three requirements $P_2(0) = f(0)$, $P_2'(0) = f'(0)$, and $P_2''(0) = f''(0)$ can be shown to be

$$P_2(x) = 1 + x + \frac{1}{2}x^2. \qquad \text{2nd-degree approximation}$$

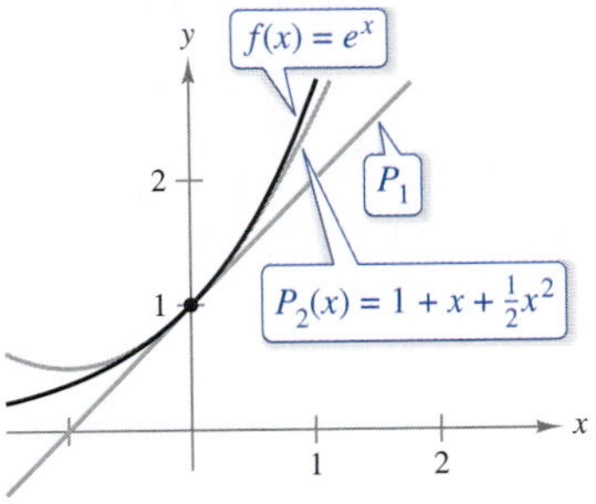

P_2 is the second-degree polynomial approximation of $f(x) = e^x$.
Figure 9.12

Moreover, in Figure 9.12, you can see that P_2 is a better approximation of f than P_1. By requiring that the values of $P_n(x)$ and its first n derivatives match those of $f(x) = e^x$ at $x = 0$, you obtain the nth-degree approximation shown below.

$$P_n(x) = 1 + x + \frac{1}{2}x^2 + \frac{1}{3!}x^3 + \cdot \cdot \cdot + \frac{1}{n!}x^n \qquad n\text{th-degree approximation}$$

$$\approx e^x$$

EXAMPLE 2 Third-Degree Polynomial Approximation of $f(x) = e^x$

Construct a table comparing the values of the polynomial

$$P_3(x) = 1 + x + \frac{1}{2}x^2 + \frac{1}{3!}x^3 \qquad \text{3rd-degree approximation}$$

with $f(x) = e^x$ for several values of x near 0.

Solution Using a calculator or a computer, you can obtain the results shown in the table. Note that for $x = 0$, the two functions have the same value, but that as x moves farther away from 0, the accuracy of the approximating polynomial $P_3(x)$ decreases.

x	-1.0	-0.2	-0.1	0	0.1	0.2	1.0
e^x	0.3679	0.81873	0.904837	1	1.105171	1.22140	2.7183
$P_3(x)$	0.3333	0.81867	0.904833	1	1.105167	1.22133	2.6667

TECHNOLOGY A graphing utility can be used to compare the graph of the approximating polynomial with the graph of the function f. For instance, in Figure 9.13, the graph of

$$P_3(x) = 1 + x + \tfrac{1}{2}x^2 + \tfrac{1}{6}x^3 \qquad \text{3rd-degree approximation}$$

is compared with the graph of $f(x) = e^x$. If you have access to a graphing utility, try comparing the graphs of

$$P_4(x) = 1 + x + \tfrac{1}{2}x^2 + \tfrac{1}{6}x^3 + \tfrac{1}{24}x^4 \qquad \text{4th-degree approximation}$$

$$P_5(x) = 1 + x + \tfrac{1}{2}x^2 + \tfrac{1}{6}x^3 + \tfrac{1}{24}x^4 + \tfrac{1}{120}x^5 \qquad \text{5th-degree approximation}$$

and

$$P_6(x) = 1 + x + \tfrac{1}{2}x^2 + \tfrac{1}{6}x^3 + \tfrac{1}{24}x^4 + \tfrac{1}{120}x^5 + \tfrac{1}{720}x^6 \qquad \text{6th-degree approximation}$$

with the graph of f. What do you notice?

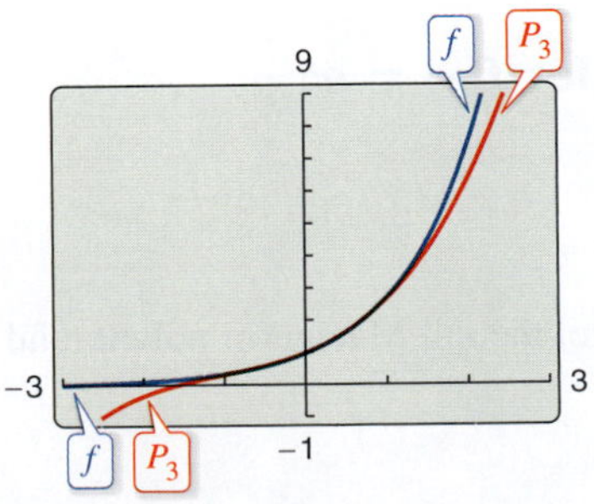

P_3 is the third-degree polynomial approximation of $f(x) = e^x$.
Figure 9.13

BROOK TAYLOR (1685–1731)

Although Taylor was not the first to seek polynomial approximations of transcendental functions, his account published in 1715 was one of the first comprehensive works on the subject.

See LarsonCalculus.com to read more of this biography.

Taylor and Maclaurin Polynomials

The polynomial approximation of

$$f(x) = e^x$$

in Example 2 is expanded about $c = 0$. For expansions about an arbitrary value of c, it is convenient to write the polynomial in the form

$$P_n(x) = a_0 + a_1(x - c) + a_2(x - c)^2 + a_3(x - c)^3 + \cdots + a_n(x - c)^n.$$

In this form, repeated differentiation produces

$$\begin{aligned} P_n'(x) &= a_1 + 2a_2(x - c) + 3a_3(x - c)^2 + \cdots + na_n(x - c)^{n-1} \\ P_n''(x) &= 2a_2 + 2(3a_3)(x - c) + \cdots + n(n - 1)a_n(x - c)^{n-2} \\ P_n'''(x) &= 2(3a_3) + \cdots + n(n - 1)(n - 2)a_n(x - c)^{n-3} \\ &\vdots \\ P_n^{(n)}(x) &= n(n - 1)(n - 2) \cdots (2)(1)a_n. \end{aligned}$$

Letting $x = c$, you then obtain

$$P_n(c) = a_0, \quad P_n'(c) = a_1, \quad P_n''(c) = 2a_2, \ldots, \quad P_n^{(n)}(c) = n!a_n$$

and because the values of f and its first n derivatives must agree with the values of P_n and its first n derivatives at $x = c$, it follows that

$$f(c) = a_0, \quad f'(c) = a_1, \quad \frac{f''(c)}{2!} = a_2, \quad \ldots, \quad \frac{f^{(n)}(c)}{n!} = a_n.$$

With these coefficients, you can obtain the following definition of **Taylor polynomials,** named after the English mathematician Brook Taylor, and **Maclaurin polynomials,** named after the English mathematician Colin Maclaurin (1698–1746).

REMARK Maclaurin polynomials are special types of Taylor polynomials for which $c = 0$.

Definitions of *n*th Taylor Polynomial and *n*th Maclaurin Polynomial

If f has n derivatives at c, then the polynomial

$$P_n(x) = f(c) + f'(c)(x - c) + \frac{f''(c)}{2!}(x - c)^2 + \cdots + \frac{f^{(n)}(c)}{n!}(x - c)^n$$

is called the ***n*th Taylor polynomial for *f* at *c*.** If $c = 0$, then

$$P_n(x) = f(0) + f'(0)x + \frac{f''(0)}{2!}x^2 + \frac{f'''(0)}{3!}x^3 + \cdots + \frac{f^{(n)}(0)}{n!}x^n$$

is also called the ***n*th Maclaurin polynomial for *f*.**

EXAMPLE 3 A Maclaurin Polynomial for $f(x) = e^x$

Find the nth Maclaurin polynomial for

$$f(x) = e^x.$$

Solution From the discussion on the preceding page, the nth Maclaurin polynomial for

$$f(x) = e^x$$

is

$$P_n(x) = 1 + x + \frac{1}{2!}x^2 + \frac{1}{3!}x^3 + \cdots + \frac{1}{n!}x^n.$$

FOR FURTHER INFORMATION To see how to use series to obtain other approximations to e, see the article "Novel Series-based Approximations to e" by John Knox and Harlan J. Brothers in *The College Mathematics Journal.* To view this article, go to *MathArticles.com.*

The Granger Collection

EXAMPLE 4 Finding Taylor Polynomials for ln *x*

Find the Taylor polynomials P_0, P_1, P_2, P_3, and P_4 for

$$f(x) = \ln x$$

centered at $c = 1$.

Solution Expanding about $c = 1$ yields the following.

$$f(x) = \ln x \qquad f(1) = \ln 1 = 0$$

$$f'(x) = \frac{1}{x} \qquad f'(1) = \frac{1}{1} = 1$$

$$f''(x) = -\frac{1}{x^2} \qquad f''(1) = -\frac{1}{1^2} = -1$$

$$f'''(x) = \frac{2!}{x^3} \qquad f'''(1) = \frac{2!}{1^3} = 2$$

$$f^{(4)}(x) = -\frac{3!}{x^4} \qquad f^{(4)}(1) = -\frac{3!}{1^4} = -6$$

Therefore, the Taylor polynomials are as follows.

$$P_0(x) = f(1) = 0$$

$$P_1(x) = f(1) + f'(1)(x-1) = (x-1)$$

$$P_2(x) = f(1) + f'(1)(x-1) + \frac{f''(1)}{2!}(x-1)^2$$
$$= (x-1) - \frac{1}{2}(x-1)^2$$

$$P_3(x) = f(1) + f'(1)(x-1) + \frac{f''(1)}{2!}(x-1)^2 + \frac{f'''(1)}{3!}(x-1)^3$$
$$= (x-1) - \frac{1}{2}(x-1)^2 + \frac{1}{3}(x-1)^3$$

$$P_4(x) = f(1) + f'(1)(x-1) + \frac{f''(1)}{2!}(x-1)^2 + \frac{f'''(1)}{3!}(x-1)^3 + \frac{f^{(4)}(1)}{4!}(x-1)^4$$
$$= (x-1) - \frac{1}{2}(x-1)^2 + \frac{1}{3}(x-1)^3 - \frac{1}{4}(x-1)^4$$

Figure 9.14 compares the graphs of P_1, P_2, P_3, and P_4 with the graph of $f(x) = \ln x$. Note that near $x = 1$, the graphs are nearly indistinguishable. For instance,

$$P_4(1.1) \approx 0.0953083$$

and

$$\ln(1.1) \approx 0.0953102.$$

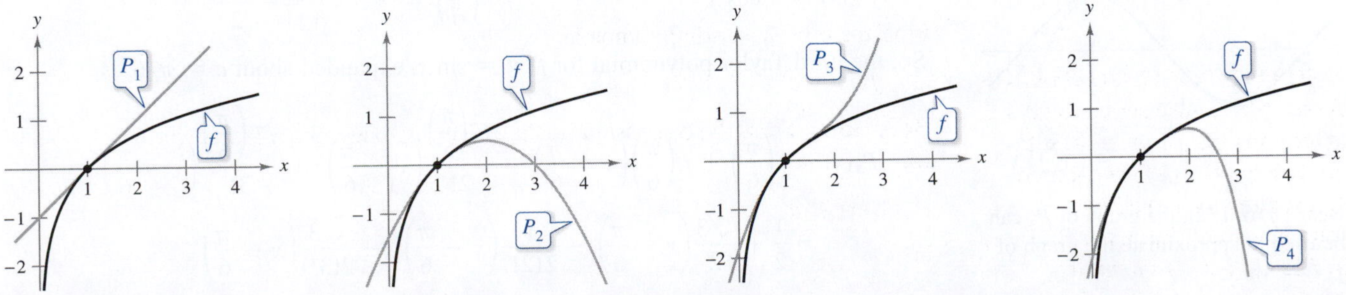

As n increases, the graph of P_n becomes a better and better approximation of the graph of $f(x) = \ln x$ near $x = 1$.
Figure 9.14

EXAMPLE 5 Finding Maclaurin Polynomials for cos *x*

Find the Maclaurin polynomials P_0, P_2, P_4, and P_6 for $f(x) = \cos x$. Use $P_6(x)$ to approximate the value of $\cos(0.1)$.

Solution Expanding about $c = 0$ yields the following.

$$\begin{aligned} f(x) &= \cos x & f(0) &= \cos 0 = 1 \\ f'(x) &= -\sin x & f'(0) &= -\sin 0 = 0 \\ f''(x) &= -\cos x & f''(0) &= -\cos 0 = -1 \\ f'''(x) &= \sin x & f'''(0) &= \sin 0 = 0 \end{aligned}$$

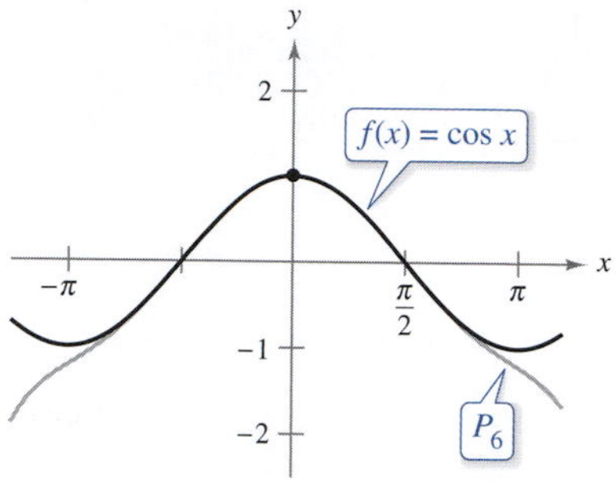

Near (0, 1), the graph of P_6 can be used to approximate the graph of $f(x) = \cos x$.
Figure 9.15

Through repeated differentiation, you can see that the pattern 1, 0, −1, 0 continues, and you obtain the Maclaurin polynomials

$$P_0(x) = 1, \quad P_2(x) = 1 - \frac{1}{2!}x^2, \quad P_4(x) = 1 - \frac{1}{2!}x^2 + \frac{1}{4!}x^4,$$

and

$$P_6(x) = 1 - \frac{1}{2!}x^2 + \frac{1}{4!}x^4 - \frac{1}{6!}x^6.$$

Using $P_6(x)$, you obtain the approximation $\cos(0.1) \approx 0.995004165$, which coincides with the calculator value to nine decimal places. Figure 9.15 compares the graphs of $f(x) = \cos x$ and P_6.

Note in Example 5 that the Maclaurin polynomials for $\cos x$ have only even powers of x. Similarly, the Maclaurin polynomials for $\sin x$ have only odd powers of x (see Exercise 17). This is not generally true of the Taylor polynomials for $\sin x$ and $\cos x$ expanded about $c \neq 0$, as you can see in the next example.

EXAMPLE 6 Finding a Taylor Polynomial for sin *x*

See LarsonCalculus.com for an interactive version of this type of example.

Find the third Taylor polynomial for $f(x) = \sin x$, expanded about $c = \pi/6$.

Solution Expanding about $c = \pi/6$ yields the following.

$$\begin{aligned} f(x) &= \sin x & f\left(\frac{\pi}{6}\right) &= \sin\frac{\pi}{6} = \frac{1}{2} \\ f'(x) &= \cos x & f'\left(\frac{\pi}{6}\right) &= \cos\frac{\pi}{6} = \frac{\sqrt{3}}{2} \\ f''(x) &= -\sin x & f''\left(\frac{\pi}{6}\right) &= -\sin\frac{\pi}{6} = -\frac{1}{2} \\ f'''(x) &= -\cos x & f'''\left(\frac{\pi}{6}\right) &= -\cos\frac{\pi}{6} = -\frac{\sqrt{3}}{2} \end{aligned}$$

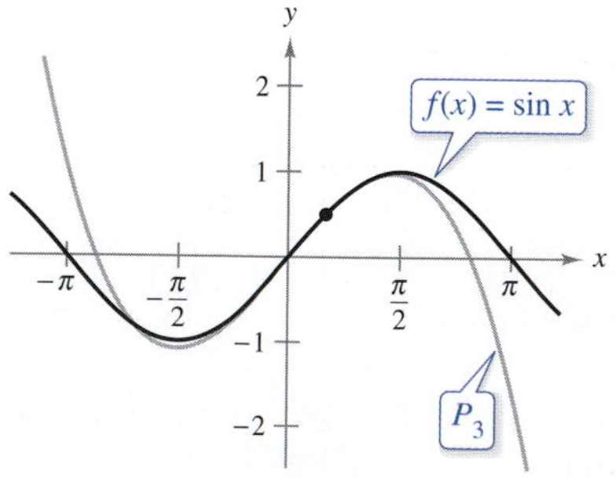

Near $(\pi/6, 1/2)$, the graph of P_3 can be used to approximate the graph of $f(x) = \sin x$.
Figure 9.16

So, the third Taylor polynomial for $f(x) = \sin x$, expanded about $c = \pi/6$, is

$$\begin{aligned} P_3(x) &= f\left(\frac{\pi}{6}\right) + f'\left(\frac{\pi}{6}\right)\left(x - \frac{\pi}{6}\right) + \frac{f''\left(\frac{\pi}{6}\right)}{2!}\left(x - \frac{\pi}{6}\right)^2 + \frac{f'''\left(\frac{\pi}{6}\right)}{3!}\left(x - \frac{\pi}{6}\right)^3 \\ &= \frac{1}{2} + \frac{\sqrt{3}}{2}\left(x - \frac{\pi}{6}\right) - \frac{1}{2(2!)}\left(x - \frac{\pi}{6}\right)^2 - \frac{\sqrt{3}}{2(3!)}\left(x - \frac{\pi}{6}\right)^3. \end{aligned}$$

Figure 9.16 compares the graphs of $f(x) = \sin x$ and P_3.

Taylor polynomials and Maclaurin polynomials can be used to approximate the value of a function at a specific point. For instance, to approximate the value of $\ln(1.1)$, you can use Taylor polynomials for $f(x) = \ln x$ expanded about $c = 1$, as shown in Example 4, or you can use Maclaurin polynomials, as shown in Example 7.

EXAMPLE 7 Approximation Using Maclaurin Polynomials

Use a fourth Maclaurin polynomial to approximate the value of $\ln(1.1)$.

Solution Because 1.1 is closer to 1 than to 0, you should consider Maclaurin polynomials for the function $g(x) = \ln(1 + x)$.

$$g(x) = \ln(1 + x) \qquad g(0) = \ln(1 + 0) = 0$$
$$g'(x) = (1 + x)^{-1} \qquad g'(0) = (1 + 0)^{-1} = 1$$
$$g''(x) = -(1 + x)^{-2} \qquad g''(0) = -(1 + 0)^{-2} = -1$$
$$g'''(x) = 2(1 + x)^{-3} \qquad g'''(0) = 2(1 + 0)^{-3} = 2$$
$$g^{(4)}(x) = -6(1 + x)^{-4} \qquad g^{(4)}(0) = -6(1 + 0)^{-4} = -6$$

Note that you obtain the same coefficients as in Example 4. Therefore, the fourth Maclaurin polynomial for $g(x) = \ln(1 + x)$ is

$$P_4(x) = g(0) + g'(0)x + \frac{g''(0)}{2!}x^2 + \frac{g'''(0)}{3!}x^3 + \frac{g^{(4)}(0)}{4!}x^4$$
$$= x - \frac{1}{2}x^2 + \frac{1}{3}x^3 - \frac{1}{4}x^4.$$

Consequently,

$$\ln(1.1) = \ln(1 + 0.1) \approx P_4(0.1) \approx 0.0953083.$$

Exploration

Check to see that the fourth Taylor polynomial (from Example 4), evaluated at $x = 1.1$, yields the same result as the fourth Maclaurin polynomial in Example 7.

The table below illustrates the accuracy of the Maclaurin polynomial approximation of the calculator value of $\ln(1.1)$. You can see that as n increases, $P_n(0.1)$ approaches the calculator value of 0.0953102.

Maclaurin Polynomials and Approximations of $\ln(1 + x)$ at $x = 0.1$

n	1	2	3	4
$P_n(0.1)$	0.1000000	0.0950000	0.0953333	0.0953083

On the other hand, the table below illustrates that as you move away from the expansion point $c = 0$, the accuracy of the approximation decreases.

Fourth Maclaurin Polynomial Approximation of $\ln(1 + x)$

x	0	0.1	0.5	0.75	1.0
$\ln(1 + x)$	0	0.0953102	0.4054651	0.5596158	0.6931472
$P_4(x)$	0	0.0953083	0.4010417	0.5302734	0.5833333

These two tables illustrate two very important points about the accuracy of Taylor (or Maclaurin) polynomials for use in approximations.

1. The approximation is usually better for higher-degree Taylor (or Maclaurin) polynomials than for those of lower degree.
2. The approximation is usually better at x-values close to c than at x-values far from c.

Remainder of a Taylor Polynomial

An approximation technique is of little value without some idea of its accuracy. To measure the accuracy of approximating a function value $f(x)$ by the Taylor polynomial $P_n(x)$, you can use the concept of a **remainder** $R_n(x)$, defined as follows.

$$f(x) = P_n(x) + R_n(x)$$

Exact value — Approximate value — Remainder

So, $R_n(x) = f(x) - P_n(x)$. The absolute value of $R_n(x)$ is called the **error** associated with the approximation. That is,

$$\text{Error} = |R_n(x)| = |f(x) - P_n(x)|.$$

The next theorem gives a general procedure for estimating the remainder associated with a Taylor polynomial. This important theorem is called **Taylor's Theorem,** and the remainder given in the theorem is called the **Lagrange form of the remainder.**

AP* Tips

Taylor's Theorem is occasionally tested on the AP BC Exam free response section.

THEOREM 9.19 Taylor's Theorem

If a function f is differentiable through order $n + 1$ in an interval I containing c, then, for each x in I, there exists z between x and c such that

$$f(x) = f(c) + f'(c)(x - c) + \frac{f''(c)}{2!}(x - c)^2 + \cdot \cdot \cdot + \frac{f^{(n)}(c)}{n!}(x - c)^n + R_n(x)$$

where

$$R_n(x) = \frac{f^{(n+1)}(z)}{(n + 1)!}(x - c)^{n+1}.$$

A proof of this theorem is given in Appendix A.

See LarsonCalculus.com for Bruce Edwards's video of this proof.

One useful consequence of Taylor's Theorem is that

$$|R_n(x)| \le \frac{|x - c|^{n+1}}{(n + 1)!} \max |f^{(n+1)}(z)|$$

where $\max|f^{(n+1)}(z)|$ is the maximum value of $f^{(n+1)}(z)$ between x and c.

For $n = 0$, Taylor's Theorem states that if f is differentiable in an interval I containing c, then, for each x in I, there exists z between x and c such that

$$f(x) = f(c) + f'(z)(x - c) \quad \text{or} \quad f'(z) = \frac{f(x) - f(c)}{x - c}.$$

Do you recognize this special case of Taylor's Theorem? (It is the Mean Value Theorem.)

When applying Taylor's Theorem, you should not expect to be able to find the exact value of z. (If you could do this, an approximation would not be necessary.) Rather, you are trying to find bounds for $f^{(n+1)}(z)$ from which you are able to tell how large the remainder $R_n(x)$ is.

EXAMPLE 8 Determining the Accuracy of an Approximation

The third Maclaurin polynomial for $\sin x$ is

$$P_3(x) = x - \frac{x^3}{3!}.$$

Use Taylor's Theorem to approximate $\sin(0.1)$ by $P_3(0.1)$ and determine the accuracy of the approximation.

Solution Using Taylor's Theorem, you have

$$\sin x = x - \frac{x^3}{3!} + R_3(x) = x - \frac{x^3}{3!} + \frac{f^{(4)}(z)}{4!}x^4$$

where $0 < z < 0.1$. Therefore,

$$\sin(0.1) \approx 0.1 - \frac{(0.1)^3}{3!} \approx 0.1 - 0.000167 = 0.099833.$$

Because $f^{(4)}(z) = \sin z$, it follows that the error $|R_3(0.1)|$ can be bounded as follows.

$$0 < R_3(0.1) = \frac{\sin z}{4!}(0.1)^4 < \frac{0.0001}{4!} \approx 0.000004$$

This implies that

$$0.099833 < \sin(0.1) \approx 0.099833 + R_3(0.1) < 0.099833 + 0.000004$$

or

$$0.099833 < \sin(0.1) < 0.099837.$$

REMARK Note that when you use a calculator,

$$\sin(0.1) \approx 0.0998334.$$

EXAMPLE 9 Approximating a Value to a Desired Accuracy

Determine the degree of the Taylor polynomial $P_n(x)$ expanded about $c = 1$ that should be used to approximate $\ln(1.2)$ so that the error is less than 0.001.

Solution Following the pattern of Example 4, you can see that the $(n + 1)$st derivative of $f(x) = \ln x$ is

$$f^{(n+1)}(x) = (-1)^n \frac{n!}{x^{n+1}}.$$

Using Taylor's Theorem, you know that the error $|R_n(1.2)|$ is

$$|R_n(1.2)| = \left|\frac{f^{(n+1)}(z)}{(n+1)!}(1.2 - 1)^{n+1}\right|$$
$$= \frac{n!}{z^{n+1}}\left[\frac{1}{(n+1)!}\right](0.2)^{n+1}$$
$$= \frac{(0.2)^{n+1}}{z^{n+1}(n+1)}$$

where $1 < z < 1.2$. In this interval, $(0.2)^{n+1}/[z^{n+1}(n + 1)]$ is less than $(0.2)^{n+1}/(n + 1)$. So, you are seeking a value of n such that

$$\frac{(0.2)^{n+1}}{(n+1)} < 0.001 \implies 1000 < (n+1)5^{n+1}.$$

By trial and error, you can determine that the least value of n that satisfies this inequality is $n = 3$. So, you would need the third Taylor polynomial to achieve the desired accuracy in approximating $\ln(1.2)$.

REMARK Note that when you use a calculator,

$$P_3(1.2) \approx 0.1827$$

and

$$\ln(1.2) \approx 0.1823.$$

9.7 Exercises

See CalcChat.com for tutorial help and worked-out solutions to odd-numbered exercises.

Matching **In Exercises 1–4, match the Taylor polynomial approximation of the function $f(x) = e^{-x^2/2}$ with the corresponding graph. [The graphs are labeled (a), (b), (c), and (d).]**

(a)

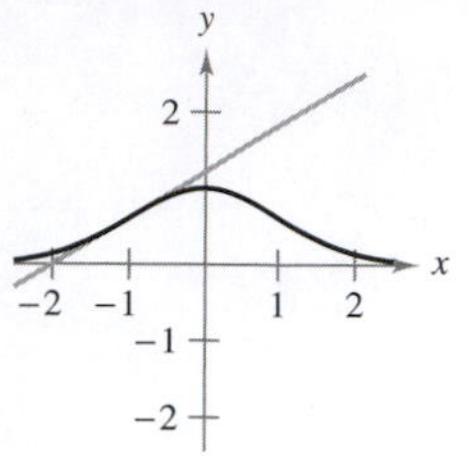

(b)

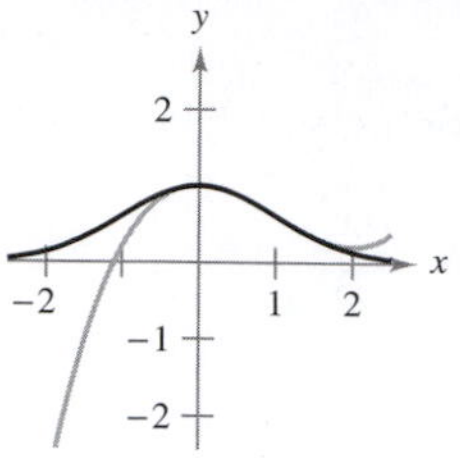

(c)

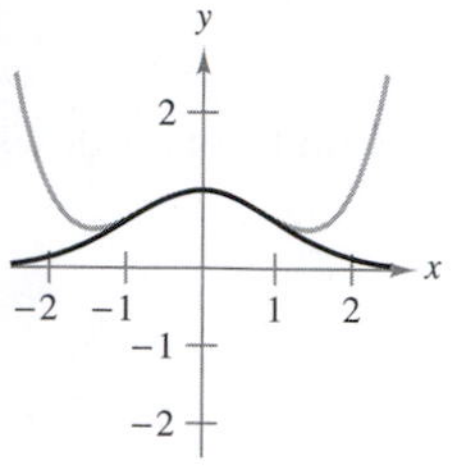

(d)

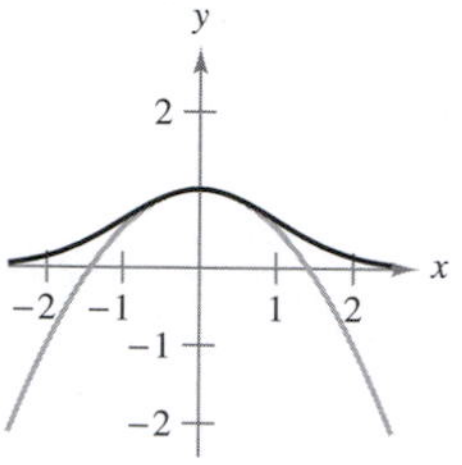

1. $g(x) = -\frac{1}{2}x^2 + 1$
2. $g(x) = \frac{1}{8}x^4 - \frac{1}{2}x^2 + 1$
3. $g(x) = e^{-1/2}[(x+1)+1]$
4. $g(x) = e^{-1/2}\left[\frac{1}{3}(x-1)^3 - (x-1) + 1\right]$

Finding a First-Degree Polynomial Approximation **In Exercises 5–8, find a first-degree polynomial function P_1 whose value and slope agree with the value and slope of f at $x = c$. Use a graphing utility to graph f and P_1. What is P_1 called?**

5. $f(x) = \dfrac{\sqrt{x}}{4}, \quad c = 4$
6. $f(x) = \dfrac{6}{\sqrt[3]{x}}, \quad c = 8$
7. $f(x) = \sec x, \quad c = \dfrac{\pi}{4}$
8. $f(x) = \tan x, \quad c = \dfrac{\pi}{4}$

Graphical and Numerical Analysis **In Exercises 9 and 10, use a graphing utility to graph f and its second-degree polynomial approximation P_2 at $x = c$. Complete the table comparing the values of f and P_2.**

9. $f(x) = \dfrac{4}{\sqrt{x}}, \quad c = 1$

$P_2(x) = 4 - 2(x-1) + \frac{3}{2}(x-1)^2$

x	0	0.8	0.9	1	1.1	1.2	2
$f(x)$							
$P_2(x)$							

10. $f(x) = \sec x, \quad c = \dfrac{\pi}{4}$

$P_2(x) = \sqrt{2} + \sqrt{2}\left(x - \dfrac{\pi}{4}\right) + \dfrac{3}{2}\sqrt{2}\left(x - \dfrac{\pi}{4}\right)^2$

x	−2.15	0.585	0.685	$\frac{\pi}{4}$	0.885	0.985	1.785
$f(x)$							
$P_2(x)$							

11. **Conjecture** Consider the function $f(x) = \cos x$ and its Maclaurin polynomials P_2, P_4, and P_6 (see Example 5).
    (a) Use a graphing utility to graph f and the indicated polynomial approximations.
    (b) Evaluate and compare the values of $f^{(n)}(0)$ and $P_n^{(n)}(0)$ for $n = 2$, 4, and 6.
    (c) Use the results in part (b) to make a conjecture about $f^{(n)}(0)$ and $P_n^{(n)}(0)$.

12. **Conjecture** Consider the function $f(x) = x^2e^x$.
    (a) Find the Maclaurin polynomials P_2, P_3, and P_4 for f.
    (b) Use a graphing utility to graph f, P_2, P_3, and P_4.
    (c) Evaluate and compare the values of $f^{(n)}(0)$ and $P_n^{(n)}(0)$ for $n = 2$, 3, and 4.
    (d) Use the results in part (c) to make a conjecture about $f^{(n)}(0)$ and $P_n^{(n)}(0)$.

Finding a Maclaurin Polynomial **In Exercises 13–24, find the nth Maclaurin polynomial for the function.**

13. $f(x) = e^{4x}, \quad n = 4$
14. $f(x) = e^{-x}, \quad n = 5$
15. $f(x) = e^{-x/2}, \quad n = 4$
16. $f(x) = e^{x/3}, \quad n = 4$
17. $f(x) = \sin x, \quad n = 5$
18. $f(x) = \cos \pi x, \quad n = 4$
19. $f(x) = xe^x, \quad n = 4$
20. $f(x) = x^2e^{-x}, \quad n = 4$
21. $f(x) = \dfrac{1}{x+1}, \quad n = 5$
22. $f(x) = \dfrac{x}{x+1}, \quad n = 4$
23. $f(x) = \sec x, \quad n = 2$
24. $f(x) = \tan x, \quad n = 3$

Finding a Taylor Polynomial **In Exercises 25–30, find the nth Taylor polynomial centered at c.**

25. $f(x) = \dfrac{2}{x}, \quad n = 3, \quad c = 1$
26. $f(x) = \dfrac{1}{x^2}, \quad n = 4, \quad c = 2$
27. $f(x) = \sqrt{x}, \quad n = 3, \quad c = 4$
28. $f(x) = \sqrt[3]{x}, \quad n = 3, \quad c = 8$
29. $f(x) = \ln x, \quad n = 4, \quad c = 2$
30. $f(x) = x^2 \cos x, \quad n = 2, \quad c = \pi$

Finding Taylor Polynomials Using Technology **In Exercises 31 and 32, use a computer algebra system to find the indicated Taylor polynomials for the function f. Graph the function and the Taylor polynomials.**

31. $f(x) = \tan \pi x$

(a) $n = 3, \quad c = 0$

(b) $n = 3, \quad c = 1/4$

32. $f(x) = \dfrac{1}{x^2 + 1}$

(a) $n = 4, \quad c = 0$

(b) $n = 4, \quad c = 1$

33. Numerical and Graphical Approximations

(a) Use the Maclaurin polynomials $P_1(x)$, $P_3(x)$, and $P_5(x)$ for $f(x) = \sin x$ to complete the table.

x	0	0.25	0.50	0.75	1.00
$\sin x$	0	0.2474	0.4794	0.6816	0.8415
$P_1(x)$					
$P_3(x)$					
$P_5(x)$					

(b) Use a graphing utility to graph $f(x) = \sin x$ and the Maclaurin polynomials in part (a).

(c) Describe the change in accuracy of a polynomial approximation as the distance from the point where the polynomial is centered increases.

34. Numerical and Graphical Approximations

(a) Use the Taylor polynomials $P_1(x)$, $P_2(x)$, and $P_4(x)$ for $f(x) = e^x$ centered at $c = 1$ to complete the table.

x	1.00	1.25	1.50	1.75	2.00
e^x	e	3.4903	4.4817	5.7546	7.3891
$P_1(x)$					
$P_2(x)$					
$P_4(x)$					

(b) Use a graphing utility to graph $f(x) = e^x$ and the Taylor polynomials in part (a).

(c) Describe the change in accuracy of polynomial approximations as the degree increases.

Numerical and Graphical Approximations **In Exercises 35 and 36, (a) find the Maclaurin polynomial $P_3(x)$ for $f(x)$, (b) complete the table for $f(x)$ and $P_3(x)$, and (c) sketch the graphs of $f(x)$ and $P_3(x)$ on the same set of coordinate axes.**

x	−0.75	−0.50	−0.25	0	0.25	0.50	0.75
$f(x)$							
$P_3(x)$							

35. $f(x) = \arcsin x$

36. $f(x) = \arctan x$

Identifying Maclaurin Polynomials **In Exercises 37–40, the graph of $y = f(x)$ is shown with four of its Maclaurin polynomials. Identify the Maclaurin polynomials and use a graphing utility to confirm your results.**

37.

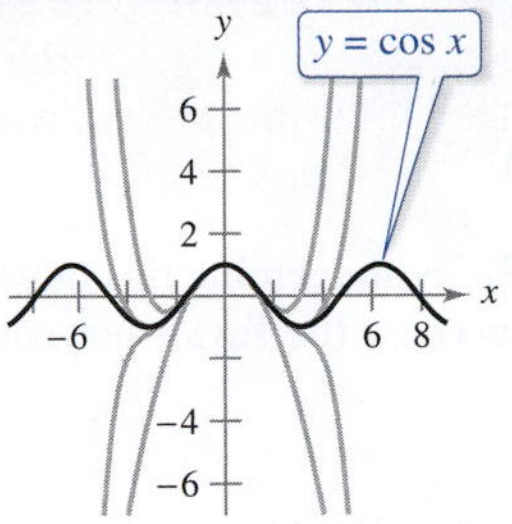

38.

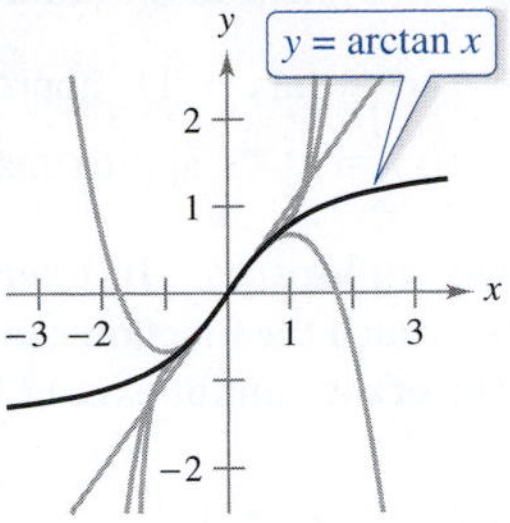

39.

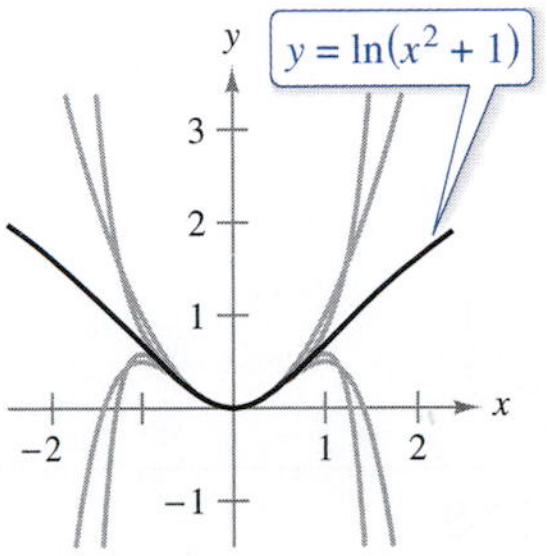

40.

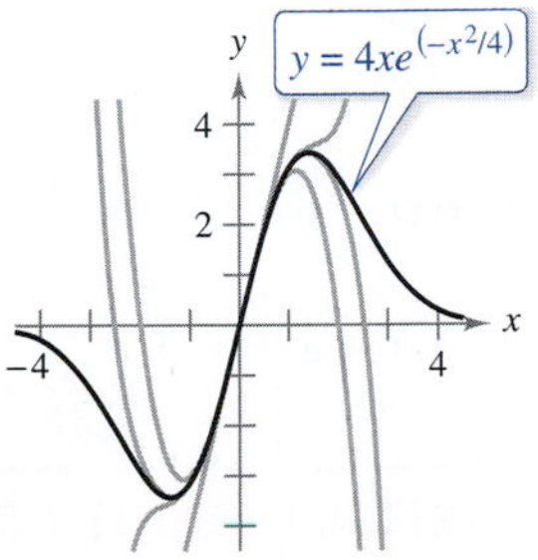

Approximating a Function Value **In Exercises 41–44, approximate the function at the given value of x, using the polynomial found in the indicated exercise.**

41. $f(x) = e^{4x}, \quad f\left(\dfrac{1}{4}\right)$, Exercise 13

42. $f(x) = x^2e^{-x}, \quad f\left(\frac{1}{5}\right)$, Exercise 20

43. $f(x) = \ln x, \quad f(2.1)$, Exercise 29

44. $f(x) = x^2 \cos x, \quad f\left(\dfrac{7\pi}{8}\right)$, Exercise 30

Using Taylor's Theorem **In Exercises 45–48, use Taylor's Theorem to obtain an upper bound for the error of the approximation. Then calculate the exact value of the error.**

45. $\cos(0.3) \approx 1 - \dfrac{(0.3)^2}{2!} + \dfrac{(0.3)^4}{4!}$

46. $e \approx 1 + 1 + \dfrac{1^2}{2!} + \dfrac{1^3}{3!} + \dfrac{1^4}{4!} + \dfrac{1^5}{5!}$

47. $\arcsin(0.4) \approx 0.4 + \dfrac{(0.4)^3}{2 \cdot 3}$

48. $\arctan(0.4) \approx 0.4 - \dfrac{(0.4)^3}{3}$

Finding a Degree **In Exercises 49–52, determine the degree of the Maclaurin polynomial required for the error in the approximation of the function at the indicated value of x to be less than 0.001.**

49. $\sin(0.3)$

50. $\cos(0.1)$

51. $e^{0.6}$

52. $\ln(1.25)$

Finding a Degree Using Technology **In Exercises 53 and 54, determine the degree of the Maclaurin polynomial required for the error in the approximation of the function at the indicated value of x to be less than 0.0001. Use a computer algebra system to obtain and evaluate the required derivative.**

53. $f(x) = \ln(x + 1)$, approximate $f(0.5)$.

54. $f(x) = e^{-\pi x}$, approximate $f(1.3)$.

Finding Values **In Exercises 55–58, determine the values of x for which the function can be replaced by the Taylor polynomial if the error cannot exceed 0.001.**

55. $f(x) = e^x \approx 1 + x + \dfrac{x^2}{2!} + \dfrac{x^3}{3!}, \quad x < 0$

56. $f(x) = \sin x \approx x - \dfrac{x^3}{3!}$

57. $f(x) = \cos x \approx 1 - \dfrac{x^2}{2!} + \dfrac{x^4}{4!}$

58. $f(x) = e^{-2x} \approx 1 - 2x + 2x^2 - \dfrac{4}{3}x^3$

WRITING ABOUT CONCEPTS

59. Polynomial Approximation An elementary function is approximated by a polynomial. In your own words, describe what is meant by saying that the polynomial is *expanded about c* or *centered at c*.

60. Polynomial Approximation When an elementary function f is approximated by a second-degree polynomial P_2 centered at c, what is known about f and P_2 at c? Explain your reasoning.

61. Taylor Polynomial State the definition of an nth-degree Taylor polynomial of f centered at c.

62. Accuracy of a Taylor Polynomial Describe the accuracy of the nth-degree Taylor polynomial of f centered at c as the distance between c and x increases.

63. Accuracy of a Taylor Polynomial In general, how does the accuracy of a Taylor polynomial change as the degree of the polynomial increases? Explain your reasoning.

64. HOW DO YOU SEE IT? The graphs show first-, second-, and third-degree polynomial approximations P_1, P_2, and P_3 of a function f. Label the graphs of P_1, P_2, and P_3. To print an enlarged copy of the graph, go to *MathGraphs.com*.

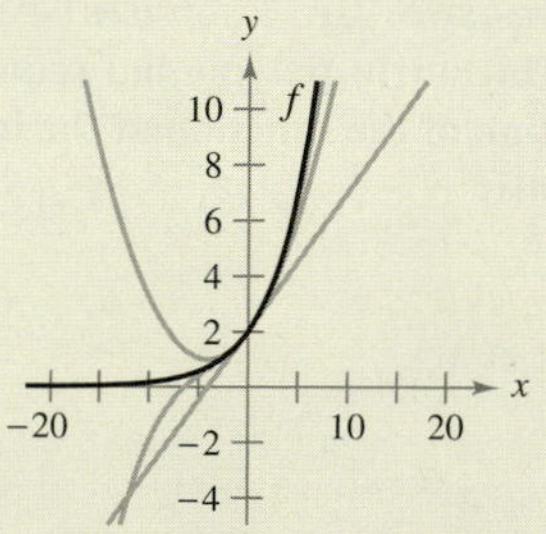

65. Comparing Maclaurin Polynomials

(a) Compare the Maclaurin polynomials of degree 4 and degree 5, respectively, for the functions $f(x) = e^x$ and $g(x) = xe^x$. What is the relationship between them?

(b) Use the result in part (a) and the Maclaurin polynomial of degree 5 for $f(x) = \sin x$ to find a Maclaurin polynomial of degree 6 for the function $g(x) = x \sin x$.

(c) Use the result in part (a) and the Maclaurin polynomial of degree 5 for $f(x) = \sin x$ to find a Maclaurin polynomial of degree 4 for the function $g(x) = (\sin x)/x$.

66. Differentiating Maclaurin Polynomials

(a) Differentiate the Maclaurin polynomial of degree 5 for $f(x) = \sin x$ and compare the result with the Maclaurin polynomial of degree 4 for $g(x) = \cos x$.

(b) Differentiate the Maclaurin polynomial of degree 6 for $f(x) = \cos x$ and compare the result with the Maclaurin polynomial of degree 5 for $g(x) = \sin x$.

(c) Differentiate the Maclaurin polynomial of degree 4 for $f(x) = e^x$. Describe the relationship between the two series.

67. Graphical Reasoning The figure shows the graphs of the function $f(x) = \sin(\pi x/4)$ and the second-degree Taylor polynomial $P_2(x) = 1 - (\pi^2/32)(x - 2)^2$ centered at $x = 2$.

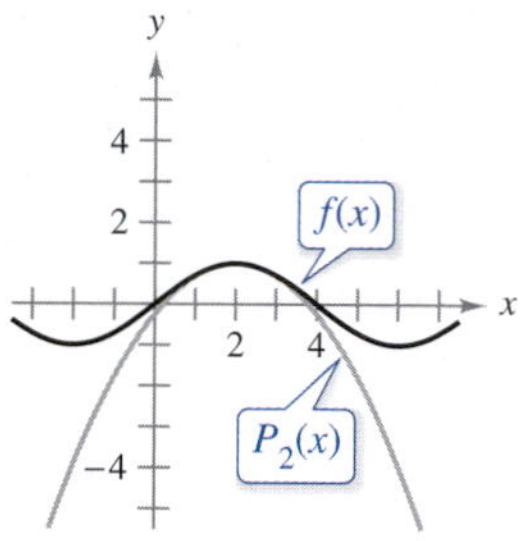

(a) Use the symmetry of the graph of f to write the second-degree Taylor polynomial $Q_2(x)$ for f centered at $x = -2$.

(b) Use a horizontal translation of the result in part (a) to find the second-degree Taylor polynomial $R_2(x)$ for f centered at $x = 6$.

(c) Is it possible to use a horizontal translation of the result in part (a) to write a second-degree Taylor polynomial for f centered at $x = 4$? Explain.

68. Proof Prove that if f is an odd function, then its nth Maclaurin polynomial contains only terms with odd powers of x.

69. Proof Prove that if f is an even function, then its nth Maclaurin polynomial contains only terms with even powers of x.

70. Proof Let $P_n(x)$ be the nth Taylor polynomial for f at c. Prove that $P_n(c) = f(c)$ and $P^{(k)}(c) = f^{(k)}(c)$ for $1 \le k \le n$. (See Exercises 9 and 10.)

71. Writing The proof in Exercise 70 guarantees that the Taylor polynomial and its derivatives agree with the function and its derivatives at $x = c$. Use the graphs and tables in Exercises 33–36 to discuss what happens to the accuracy of the Taylor polynomial as you move away from $x = c$.

9.8 Power Series

- Understand the definition of a power series.
- Find the radius and interval of convergence of a power series.
- Determine the endpoint convergence of a power series.
- Differentiate and integrate a power series.

Power Series

In Section 9.7, you were introduced to the concept of approximating functions by Taylor polynomials. For instance, the function $f(x) = e^x$ can be *approximated* by its third-degree Maclaurin polynomial

$$e^x \approx 1 + x + \frac{x^2}{2!} + \frac{x^3}{3!}.$$

In that section, you saw that the higher the degree of the approximating polynomial, the better the approximation becomes.

In this and the next two sections, you will see that several important types of functions, including $f(x) = e^x$, can be represented *exactly* by an infinite series called a **power series.** For example, the power series representation for e^x is

$$e^x = 1 + x + \frac{x^2}{2!} + \frac{x^3}{3!} + \cdot \cdot \cdot + \frac{x^n}{n!} + \cdot \cdot \cdot .$$

For each real number x, it can be shown that the infinite series on the right converges to the number e^x. Before doing this, however, some preliminary results dealing with power series will be discussed—beginning with the next definition.

Exploration

Graphical Reasoning Use a graphing utility to approximate the graph of each power series near $x = 0$. (Use the first several terms of each series.) Each series represents a well-known function. What is the function?

a. $\displaystyle\sum_{n=0}^{\infty} \frac{(-1)^n x^n}{n!}$

b. $\displaystyle\sum_{n=0}^{\infty} \frac{(-1)^n x^{2n}}{(2n)!}$

c. $\displaystyle\sum_{n=0}^{\infty} \frac{(-1)^n x^{2n+1}}{(2n+1)!}$

d. $\displaystyle\sum_{n=0}^{\infty} \frac{(-1)^n x^{2n+1}}{2n+1}$

e. $\displaystyle\sum_{n=0}^{\infty} \frac{2^n x^n}{n!}$

Definition of Power Series

If x is a variable, then an infinite series of the form

$$\sum_{n=0}^{\infty} a_n x^n = a_0 + a_1 x + a_2 x^2 + a_3 x^3 + \cdot \cdot \cdot + a_n x^n + \cdot \cdot \cdot$$

is called a **power series.** More generally, an infinite series of the form

$$\sum_{n=0}^{\infty} a_n (x - c)^n = a_0 + a_1(x - c) + a_2(x - c)^2 + \cdot \cdot \cdot + a_n(x - c)^n + \cdot \cdot \cdot$$

is called a **power series centered at c,** where c is a constant.

REMARK To simplify the notation for power series, assume that $(x - c)^0 = 1$, even when $x = c$.

EXAMPLE 1 **Power Series**

a. The following power series is centered at 0.

$$\sum_{n=0}^{\infty} \frac{x^n}{n!} = 1 + x + \frac{x^2}{2} + \frac{x^3}{3!} + \cdot \cdot \cdot$$

b. The following power series is centered at -1.

$$\sum_{n=0}^{\infty} (-1)^n (x + 1)^n = 1 - (x + 1) + (x + 1)^2 - (x + 1)^3 + \cdot \cdot \cdot$$

c. The following power series is centered at 1.

$$\sum_{n=1}^{\infty} \frac{1}{n}(x - 1)^n = (x - 1) + \frac{1}{2}(x - 1)^2 + \frac{1}{3}(x - 1)^3 + \cdot \cdot \cdot$$

AP* Tips

The ability to find the radius and interval of convergence is tested on the AP BC Exam.

Radius and Interval of Convergence

A power series in x can be viewed as a function of x

$$f(x) = \sum_{n=0}^{\infty} a_n(x - c)^n$$

where the *domain of f* is the set of all x for which the power series converges. Determination of the domain of a power series is the primary concern in this section. Of course, every power series converges at its center c because

$$\begin{aligned} f(c) &= \sum_{n=0}^{\infty} a_n(c - c)^n \\ &= a_0(1) + 0 + 0 + \cdot \cdot \cdot + 0 + \cdot \cdot \cdot \\ &= a_0. \end{aligned}$$

So, c always lies in the domain of f. Theorem 9.20 (see below) states that the domain of a power series can take three basic forms: a single point, an interval centered at c, or the entire real number line, as shown in Figure 9.17.

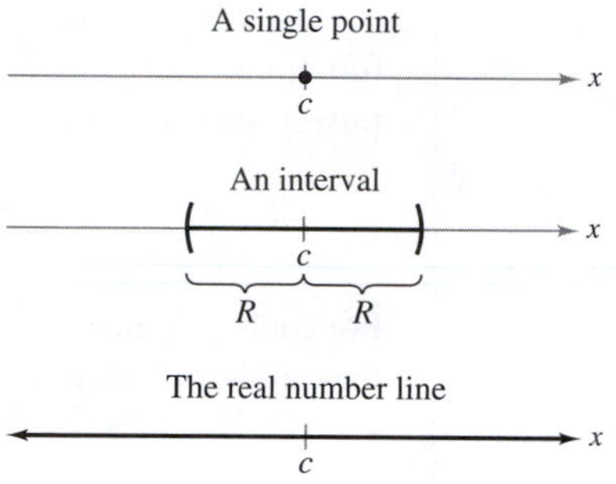

The domain of a power series has only three basic forms: a single point, an interval centered at c, or the entire real number line.
Figure 9.17

THEOREM 9.20 Convergence of a Power Series

For a power series centered at c, precisely one of the following is true.

1. The series converges only at c.
2. There exists a real number $R > 0$ such that the series converges absolutely for

$$|x - c| < R$$

and diverges for

$$|x - c| > R.$$

3. The series converges absolutely for all x.

The number R is the **radius of convergence** of the power series. If the series converges only at c, then the radius of convergence is $R = 0$. If the series converges for all x, then the radius of convergence is $R = \infty$. The set of all values of x for which the power series converges is the **interval of convergence** of the power series.

A proof of this theorem is given in Appendix A.

See LarsonCalculus.com for Bruce Edwards's video of this proof.

To determine the radius of convergence of a power series, use the Ratio Test, as demonstrated in Examples 2, 3, and 4.

EXAMPLE 2 Finding the Radius of Convergence

Find the radius of convergence of $\sum_{n=0}^{\infty} n!x^n$.

Solution For $x = 0$, you obtain

$$f(0) = \sum_{n=0}^{\infty} n!0^n = 1 + 0 + 0 + \cdots = 1.$$

For any fixed value of x such that $|x| > 0$, let $u_n = n!x^n$. Then

$$\begin{aligned}\lim_{n\to\infty}\left|\frac{u_{n+1}}{u_n}\right| &= \lim_{n\to\infty}\left|\frac{(n+1)!x^{n+1}}{n!x^n}\right| \\ &= |x|\lim_{n\to\infty}(n+1) \\ &= \infty.\end{aligned}$$

Therefore, by the Ratio Test, the series diverges for $|x| > 0$ and converges only at its center, 0. So, the radius of convergence is $R = 0$.

EXAMPLE 3 Finding the Radius of Convergence

Find the radius of convergence of

$$\sum_{n=0}^{\infty} 3(x-2)^n.$$

Solution For $x \neq 2$, let $u_n = 3(x-2)^n$. Then

$$\begin{aligned}\lim_{n\to\infty}\left|\frac{u_{n+1}}{u_n}\right| &= \lim_{n\to\infty}\left|\frac{3(x-2)^{n+1}}{3(x-2)^n}\right| \\ &= \lim_{n\to\infty}|x-2| \\ &= |x-2|.\end{aligned}$$

By the Ratio Test, the series converges for $|x-2| < 1$ and diverges for $|x-2| > 1$. Therefore, the radius of convergence of the series is $R = 1$.

EXAMPLE 4 Finding the Radius of Convergence

Find the radius of convergence of

$$\sum_{n=0}^{\infty} \frac{(-1)^n x^{2n+1}}{(2n+1)!}.$$

Solution Let $u_n = (-1)^n x^{2n+1}/(2n+1)!$. Then

$$\begin{aligned}\lim_{n\to\infty}\left|\frac{u_{n+1}}{u_n}\right| &= \lim_{n\to\infty}\left|\frac{\dfrac{(-1)^{n+1}x^{2n+3}}{(2n+3)!}}{\dfrac{(-1)^n x^{2n+1}}{(2n+1)!}}\right| \\ &= \lim_{n\to\infty}\frac{x^2}{(2n+3)(2n+2)}.\end{aligned}$$

For any *fixed* value of x, this limit is 0. So, by the Ratio Test, the series converges for all x. Therefore, the radius of convergence is $R = \infty$. ■

Endpoint Convergence

Note that for a power series whose radius of convergence is a finite number R, Theorem 9.20 says nothing about the convergence at the *endpoints* of the interval of convergence. Each endpoint must be tested separately for convergence or divergence. As a result, the interval of convergence of a power series can take any one of the six forms shown in Figure 9.18.

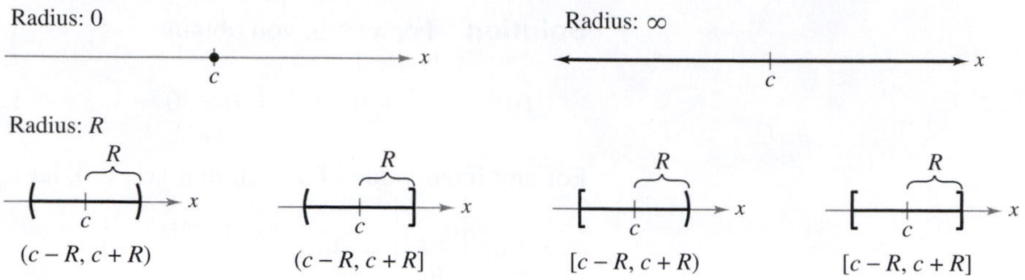

Intervals of convergence
Figure 9.18

EXAMPLE 5 Finding the Interval of Convergence

See LarsonCalculus.com for an interactive version of this type of example.

Find the interval of convergence of

$$\sum_{n=1}^{\infty} \frac{x^n}{n}.$$

Solution Letting $u_n = x^n/n$ produces

$$\lim_{n\to\infty}\left|\frac{u_{n+1}}{u_n}\right| = \lim_{n\to\infty}\left|\frac{\frac{x^{n+1}}{(n+1)}}{\frac{x^n}{n}}\right|$$

$$= \lim_{n\to\infty}\left|\frac{nx}{n+1}\right|$$

$$= |x|.$$

So, by the Ratio Test, the radius of convergence is $R = 1$. Moreover, because the series is centered at 0, it converges in the interval $(-1, 1)$. This interval, however, is not necessarily the *interval of convergence.* To determine this, you must test for convergence at each endpoint. When $x = 1$, you obtain the *divergent* harmonic series

$$\sum_{n=1}^{\infty} \frac{1}{n} = \frac{1}{1} + \frac{1}{2} + \frac{1}{3} + \cdots.$$

Diverges when $x = 1$.

When $x = -1$, you obtain the *convergent* alternating harmonic series

$$\sum_{n=1}^{\infty} \frac{(-1)^n}{n} = -1 + \frac{1}{2} - \frac{1}{3} + \frac{1}{4} - \cdots.$$

Converges when $x = -1$.

So, the interval of convergence for the series is $[-1, 1)$, as shown in Figure 9.19.

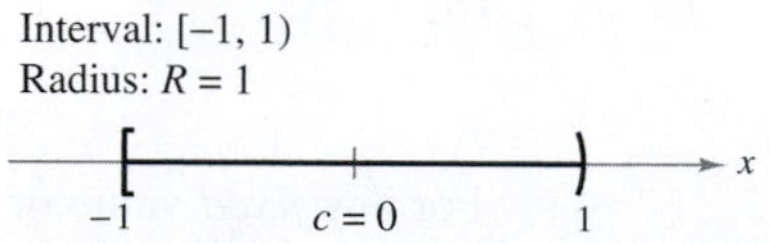

Figure 9.19

EXAMPLE 6 Finding the Interval of Convergence

Find the interval of convergence of $\displaystyle\sum_{n=0}^{\infty} \frac{(-1)^n(x+1)^n}{2^n}$.

Solution Letting $u_n = (-1)^n(x+1)^n/2^n$ produces

$$\lim_{n\to\infty}\left|\frac{u_{n+1}}{u_n}\right| = \lim_{n\to\infty}\left|\frac{\dfrac{(-1)^{n+1}(x+1)^{n+1}}{2^{n+1}}}{\dfrac{(-1)^n(x+1)^n}{2^n}}\right|$$

$$= \lim_{n\to\infty}\left|\frac{2^n(x+1)}{2^{n+1}}\right|$$

$$= \left|\frac{x+1}{2}\right|.$$

By the Ratio Test, the series converges for

$$\left|\frac{x+1}{2}\right| < 1$$

or $|x+1| < 2$. So, the radius of convergence is $R = 2$. Because the series is centered at $x = -1$, it will converge in the interval $(-3, 1)$. Furthermore, at the endpoints, you have

$$\sum_{n=0}^{\infty} \frac{(-1)^n(-2)^n}{2^n} = \sum_{n=0}^{\infty} \frac{2^n}{2^n} = \sum_{n=0}^{\infty} 1$$ Diverges when $x = -3$.

and

$$\sum_{n=0}^{\infty} \frac{(-1)^n(2)^n}{2^n} = \sum_{n=0}^{\infty} (-1)^n$$ Diverges when $x = 1$.

both of which diverge. So, the interval of convergence is $(-3, 1)$, as shown in Figure 9.20.

Interval: $(-3, 1)$
Radius: $R = 2$

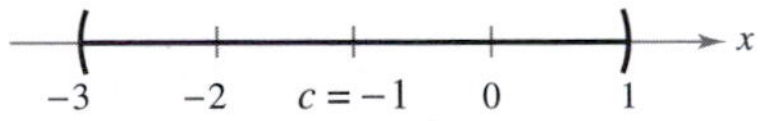

Figure 9.20

EXAMPLE 7 Finding the Interval of Convergence

Find the interval of convergence of

$$\sum_{n=1}^{\infty} \frac{x^n}{n^2}.$$

Solution Letting $u_n = x^n/n^2$ produces

$$\lim_{n\to\infty}\left|\frac{u_{n+1}}{u_n}\right| = \lim_{n\to\infty}\left|\frac{x^{n+1}/(n+1)^2}{x^n/n^2}\right|$$

$$= \lim_{n\to\infty}\left|\frac{n^2x}{(n+1)^2}\right|$$

$$= |x|.$$

So, the radius of convergence is $R = 1$. Because the series is centered at $x = 0$, it converges in the interval $(-1, 1)$. When $x = 1$, you obtain the convergent p-series

$$\sum_{n=1}^{\infty} \frac{1}{n^2} = \frac{1}{1^2} + \frac{1}{2^2} + \frac{1}{3^2} + \frac{1}{4^2} + \cdots.$$ Converges when $x = 1$.

When $x = -1$, you obtain the convergent alternating series

$$\sum_{n=1}^{\infty} \frac{(-1)^n}{n^2} = -\frac{1}{1^2} + \frac{1}{2^2} - \frac{1}{3^2} + \frac{1}{4^2} - \cdots.$$ Converges when $x = -1$.

Therefore, the interval of convergence is $[-1, 1]$.

JAMES GREGORY (1638–1675)

One of the earliest mathematicians to work with power series was a Scotsman, James Gregory. He developed a power series method for interpolating table values—a method that was later used by Brook Taylor in the development of Taylor polynomials and Taylor series.

The Granger Collection

Differentiation and Integration of Power Series

Power series representation of functions has played an important role in the development of calculus. In fact, much of Newton's work with differentiation and integration was done in the context of power series—especially his work with complicated algebraic functions and transcendental functions. Euler, Lagrange, Leibniz, and the Bernoullis all used power series extensively in calculus.

Once you have defined a function with a power series, it is natural to wonder how you can determine the characteristics of the function. Is it continuous? Differentiable? Theorem 9.21, which is stated without proof, answers these questions.

THEOREM 9.21 Properties of Functions Defined by Power Series

If the function

$$f(x) = \sum_{n=0}^{\infty} a_n(x-c)^n$$
$$= a_0 + a_1(x-c) + a_2(x-c)^2 + a_3(x-c)^3 + \cdots$$

has a radius of convergence of $R > 0$, then, on the interval

$$(c - R, c + R)$$

f is differentiable (and therefore continuous). Moreover, the derivative and antiderivative of f are as follows.

1. $$f'(x) = \sum_{n=1}^{\infty} na_n(x-c)^{n-1}$$
$$= a_1 + 2a_2(x-c) + 3a_3(x-c)^2 + \cdots$$

2. $$\int f(x)\,dx = C + \sum_{n=0}^{\infty} a_n \frac{(x-c)^{n+1}}{n+1}$$
$$= C + a_0(x-c) + a_1\frac{(x-c)^2}{2} + a_2\frac{(x-c)^3}{3} + \cdots$$

The *radius of convergence* of the series obtained by differentiating or integrating a power series is the same as that of the original power series. The *interval of convergence,* however, may differ as a result of the behavior at the endpoints.

Theorem 9.21 states that, in many ways, a function defined by a power series behaves like a polynomial. It is continuous in its interval of convergence, and both its derivative and its antiderivative can be determined by differentiating and integrating each term of the power series. For instance, the derivative of the power series

$$f(x) = \sum_{n=0}^{\infty} \frac{x^n}{n!}$$
$$= 1 + x + \frac{x^2}{2} + \frac{x^3}{3!} + \frac{x^4}{4!} + \cdots$$

is

$$f'(x) = 1 + (2)\frac{x}{2} + (3)\frac{x^2}{3!} + (4)\frac{x^3}{4!} + \cdots$$
$$= 1 + x + \frac{x^2}{2} + \frac{x^3}{3!} + \frac{x^4}{4!} + \cdots$$
$$= f(x).$$

Notice that $f'(x) = f(x)$. Do you recognize this function?

EXAMPLE 8 Intervals of Convergence for $f(x)$, $f'(x)$, and $\int f(x)\,dx$

Consider the function

$$f(x) = \sum_{n=1}^{\infty} \frac{x^n}{n} = x + \frac{x^2}{2} + \frac{x^3}{3} + \cdot \cdot \cdot .$$

Find the interval of convergence for each of the following.

a. $\int f(x)\,dx$ **b.** $f(x)$ **c.** $f'(x)$

Solution By Theorem 9.21, you have

$$f'(x) = \sum_{n=1}^{\infty} x^{n-1}$$
$$= 1 + x + x^2 + x^3 + \cdot \cdot \cdot$$

and

$$\int f(x)\,dx = C + \sum_{n=1}^{\infty} \frac{x^{n+1}}{n(n+1)}$$
$$= C + \frac{x^2}{1 \cdot 2} + \frac{x^3}{2 \cdot 3} + \frac{x^4}{3 \cdot 4} + \cdot \cdot \cdot .$$

By the Ratio Test, you can show that each series has a radius of convergence of $R = 1$. Considering the interval $(-1, 1)$, you have the following.

a. For $\int f(x)\,dx$, the series

$$\sum_{n=1}^{\infty} \frac{x^{n+1}}{n(n+1)}$$ Interval of convergence: $[-1, 1]$

converges for $x = \pm 1$, and its interval of convergence is $[-1, 1]$. See Figure 9.21(a).

b. For $f(x)$, the series

$$\sum_{n=1}^{\infty} \frac{x^n}{n}$$ Interval of convergence: $[-1, 1)$

converges for $x = -1$ and diverges for $x = 1$. So, its interval of convergence is $[-1, 1)$. See Figure 9.21(b).

c. For $f'(x)$, the series

$$\sum_{n=1}^{\infty} x^{n-1}$$ Interval of convergence: $(-1, 1)$

diverges for $x = \pm 1$, and its interval of convergence is $(-1, 1)$. See Figure 9.21(c).

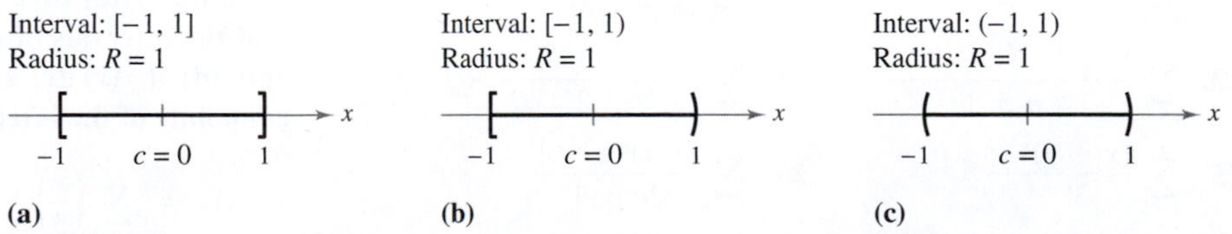

Figure 9.21

From Example 8, it appears that of the three series, the one for the derivative, $f'(x)$, is the least likely to converge at the endpoints. In fact, it can be shown that if the series for $f'(x)$ converges at the endpoints

$$x = c \pm R$$

then the series for $f(x)$ will also converge there.

9.8 Exercises

See CalcChat.com for tutorial help and worked-out solutions to odd-numbered exercises.

Finding the Center of a Power Series In Exercises 1–4, state where the power series is centered.

1. $\sum_{n=0}^{\infty} nx^n$

2. $\sum_{n=1}^{\infty} \frac{(-1)^n 1 \cdot 3 \cdots (2n-1)}{2^n n!} x^n$

3. $\sum_{n=1}^{\infty} \frac{(x-2)^n}{n^3}$

4. $\sum_{n=0}^{\infty} \frac{(-1)^n (x-\pi)^{2n}}{(2n)!}$

Finding the Radius of Convergence In Exercises 5–10, find the radius of convergence of the power series.

5. $\sum_{n=0}^{\infty} (-1)^n \frac{x^n}{n+1}$

6. $\sum_{n=0}^{\infty} (3x)^n$

7. $\sum_{n=1}^{\infty} \frac{(4x)^n}{n^2}$

8. $\sum_{n=0}^{\infty} \frac{(-1)^n x^n}{5^n}$

9. $\sum_{n=0}^{\infty} \frac{x^{2n}}{(2n)!}$

10. $\sum_{n=0}^{\infty} \frac{(2n)! x^{2n}}{n!}$

Finding the Interval of Convergence In Exercises 11–34, find the interval of convergence of the power series. (Be sure to include a check for convergence at the endpoints of the interval.)

11. $\sum_{n=0}^{\infty} \left(\frac{x}{4}\right)^n$

12. $\sum_{n=0}^{\infty} (2x)^n$

13. $\sum_{n=1}^{\infty} \frac{(-1)^n x^n}{n}$

14. $\sum_{n=0}^{\infty} (-1)^{n+1}(n+1)x^n$

15. $\sum_{n=0}^{\infty} \frac{x^{5n}}{n!}$

16. $\sum_{n=0}^{\infty} \frac{(3x)^n}{(2n)!}$

17. $\sum_{n=0}^{\infty} (2n)! \left(\frac{x}{3}\right)^n$

18. $\sum_{n=0}^{\infty} \frac{(-1)^n x^n}{(n+1)(n+2)}$

19. $\sum_{n=1}^{\infty} \frac{(-1)^{n+1} x^n}{6^n}$

20. $\sum_{n=0}^{\infty} \frac{(-1)^n n!(x-5)^n}{3^n}$

21. $\sum_{n=1}^{\infty} \frac{(-1)^{n+1}(x-4)^n}{n9^n}$

22. $\sum_{n=0}^{\infty} \frac{(x-3)^{n+1}}{(n+1)4^{n+1}}$

23. $\sum_{n=0}^{\infty} \frac{(-1)^{n+1}(x-1)^{n+1}}{n+1}$

24. $\sum_{n=1}^{\infty} \frac{(-1)^{n+1}(x-2)^n}{n2^n}$

25. $\sum_{n=1}^{\infty} \frac{(x-3)^{n-1}}{3^{n-1}}$

26. $\sum_{n=0}^{\infty} \frac{(-1)^n x^{2n+1}}{2n+1}$

27. $\sum_{n=1}^{\infty} \frac{n}{n+1}(-2x)^{n-1}$

28. $\sum_{n=0}^{\infty} \frac{(-1)^n x^{2n}}{n!}$

29. $\sum_{n=0}^{\infty} \frac{x^{3n+1}}{(3n+1)!}$

30. $\sum_{n=1}^{\infty} \frac{n! x^n}{(2n)!}$

31. $\sum_{n=1}^{\infty} \frac{2 \cdot 3 \cdot 4 \cdots (n+1)x^n}{n!}$

32. $\sum_{n=1}^{\infty} \left[\frac{2 \cdot 4 \cdot 6 \cdots 2n}{3 \cdot 5 \cdot 7 \cdots (2n+1)}\right] x^{2n+1}$

33. $\sum_{n=1}^{\infty} \frac{(-1)^{n+1} 3 \cdot 7 \cdot 11 \cdots (4n-1)(x-3)^n}{4^n}$

34. $\sum_{n=1}^{\infty} \frac{n!(x+1)^n}{1 \cdot 3 \cdot 5 \cdots (2n-1)}$

Finding the Radius of Convergence In Exercises 35 and 36, find the radius of convergence of the power series, where $c > 0$ and k is a positive integer.

35. $\sum_{n=1}^{\infty} \frac{(x-c)^{n-1}}{c^{n-1}}$

36. $\sum_{n=0}^{\infty} \frac{(n!)^k x^n}{(kn)!}$

Finding the Interval of Convergence In Exercises 37–40, find the interval of convergence of the power series. (Be sure to include a check for convergence at the endpoints of the interval.)

37. $\sum_{n=0}^{\infty} \left(\frac{x}{k}\right)^n, \quad k > 0$

38. $\sum_{n=1}^{\infty} \frac{(-1)^{n+1}(x-c)^n}{nc^n}$

39. $\sum_{n=1}^{\infty} \frac{k(k+1)(k+2) \cdots (k+n-1)x^n}{n!}, \quad k \geq 1$

40. $\sum_{n=1}^{\infty} \frac{n!(x-c)^n}{1 \cdot 3 \cdot 5 \cdots (2n-1)}$

Writing an Equivalent Series In Exercises 41–44, write an equivalent series with the index of summation beginning at $n = 1$.

41. $\sum_{n=0}^{\infty} \frac{x^n}{n!}$

42. $\sum_{n=0}^{\infty} (-1)^{n+1}(n+1)x^n$

43. $\sum_{n=0}^{\infty} \frac{x^{2n+1}}{(2n+1)!}$

44. $\sum_{n=0}^{\infty} \frac{(-1)^n x^{2n+1}}{2n+1}$

Finding Intervals of Convergence In Exercises 45–48, find the intervals of convergence of (a) $f(x)$, (b) $f'(x)$, (c) $f''(x)$, and (d) $\int f(x)\,dx$. Include a check for convergence at the endpoints of the interval.

45. $f(x) = \sum_{n=0}^{\infty} \left(\frac{x}{3}\right)^n$

46. $f(x) = \sum_{n=1}^{\infty} \frac{(-1)^{n+1}(x-5)^n}{n5^n}$

47. $f(x) = \sum_{n=0}^{\infty} \frac{(-1)^{n+1}(x-1)^{n+1}}{n+1}$

48. $f(x) = \sum_{n=1}^{\infty} \frac{(-1)^{n+1}(x-2)^n}{n}$

WRITING ABOUT CONCEPTS

49. Power Series Define a power series centered at c.

50. Radius of Convergence Describe the radius of convergence of a power series.

51. Interval of Convergence Describe the interval of convergence of a power series.

52. Domain of a Power Series Describe the three basic forms of the domain of a power series.

53. Using a Power Series Describe how to differentiate and integrate a power series with a radius of convergence R. Will the series resulting from the operations of differentiation and integration have a different radius of convergence? Explain.

54. Conditional or Absolute Convergence Give examples that show that the convergence of a power series at an endpoint of its interval of convergence may be either conditional or absolute. Explain your reasoning.

55. Writing a Power Series Write a power series that has the indicated interval of convergence. Explain your reasoning.

(a) $(-2, 2)$ (b) $(-1, 1]$

(c) $(-1, 0)$ (d) $[-2, 6)$

56. HOW DO YOU SEE IT? Match the graph of the first 10 terms of the sequence of partial sums of the series

$$g(x) = \sum_{n=0}^{\infty} \left(\frac{x}{3}\right)^n$$

with the indicated value of the function. [The graphs are labeled (i), (ii), (iii), and (iv).] Explain how you made your choice.

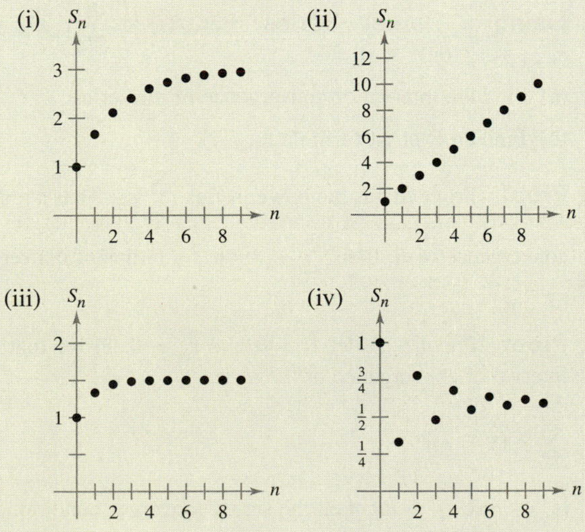

(a) $g(1)$ (b) $g(2)$

(c) $g(3)$ (d) $g(-2)$

57. Using Power Series Let $f(x) = \sum_{n=0}^{\infty} \frac{(-1)^n x^{2n+1}}{(2n+1)!}$ and $g(x) = \sum_{n=0}^{\infty} \frac{(-1)^n x^{2n}}{(2n)!}$.

(a) Find the intervals of convergence of f and g.

(b) Show that $f'(x) = g(x)$.

(c) Show that $g'(x) = -f(x)$.

(d) Identify the functions f and g.

58. Using a Power Series Let $f(x) = \sum_{n=0}^{\infty} \frac{x^n}{n!}$.

(a) Find the interval of convergence of f.

(b) Show that $f'(x) = f(x)$.

(c) Show that $f(0) = 1$.

(d) Identify the function f.

Differential Equation **In Exercises 59–64, show that the function represented by the power series is a solution of the differential equation.**

59. $y = \sum_{n=0}^{\infty} \frac{(-1)^n x^{2n+1}}{(2n+1)!}, \quad y'' + y = 0$

60. $y = \sum_{n=0}^{\infty} \frac{(-1)^n x^{2n}}{(2n)!} \quad y'' + y = 0$

61. $y = \sum_{n=0}^{\infty} \frac{x^{2n+1}}{(2n+1)!}, \quad y'' - y = 0$

62. $y = \sum_{n=0}^{\infty} \frac{x^{2n}}{(2n)!}, \quad y'' - y = 0$

63. $y = \sum_{n=0}^{\infty} \frac{x^{2n}}{2^n n!}, \quad y'' - xy' - y = 0$

64. $y = 1 + \sum_{n=1}^{\infty} \frac{(-1)^n x^{4n}}{2^{2n} n! \cdot 3 \cdot 7 \cdot 11 \cdots (4n-1)},$

$y'' + x^2 y = 0$

65. Bessel Function The Bessel function of order 0 is

$$J_0(x) = \sum_{k=0}^{\infty} \frac{(-1)^k x^{2k}}{2^{2k}(k!)^2}.$$

(a) Show that the series converges for all x.

(b) Show that the series is a solution of the differential equation $x^2 J_0'' + x J_0' + x^2 J_0 = 0$.

(c) Use a graphing utility to graph the polynomial composed of the first four terms of J_0.

(d) Approximate $\int_0^1 J_0\, dx$ accurate to two decimal places.

66. Bessel Function The Bessel function of order 1 is

$$J_1(x) = x \sum_{k=0}^{\infty} \frac{(-1)^k x^{2k}}{2^{2k+1} k!(k+1)!}.$$

(a) Show that the series converges for all x.

(b) Show that the series is a solution of the differential equation $x^2 J_1'' + x J_1' + (x^2 - 1) J_1 = 0$.

(c) Use a graphing utility to graph the polynomial composed of the first four terms of J_1.

(d) Show that $J_0'(x) = -J_1(x)$.

67. Investigation The interval of convergence of the geometric series $\sum_{n=0}^{\infty}\left(\frac{x}{4}\right)^n$ is $(-4, 4)$.

(a) Find the sum of the series when $x = \frac{5}{2}$. Use a graphing utility to graph the first six terms of the sequence of partial sums and the horizontal line representing the sum of the series.

(b) Repeat part (a) for $x = -\frac{5}{2}$.

(c) Write a short paragraph comparing the rates of convergence of the partial sums with the sums of the series in parts (a) and (b). How do the plots of the partial sums differ as they converge toward the sum of the series?

(d) Given any positive real number M, there exists a positive integer N such that the partial sum

$$\sum_{n=0}^{N}\left(\frac{5}{4}\right)^n > M.$$

Use a graphing utility to complete the table.

M	10	100	1000	10,000
N				

68. Investigation The interval of convergence of the series $\sum_{n=0}^{\infty}(3x)^n$ is $\left(-\frac{1}{3}, \frac{1}{3}\right)$.

(a) Find the sum of the series when $x = \frac{1}{6}$. Use a graphing utility to graph the first six terms of the sequence of partial sums and the horizontal line representing the sum of the series.

(b) Repeat part (a) for $x = -\frac{1}{6}$.

(c) Write a short paragraph comparing the rates of convergence of the partial sums with the sums of the series in parts (a) and (b). How do the plots of the partial sums differ as they converge toward the sum of the series?

(d) Given any positive real number M, there exists a positive integer N such that the partial sum

$$\sum_{n=0}^{N}\left(3\cdot\frac{2}{3}\right)^n > M.$$

Use a graphing utility to complete the table.

M	10	100	1000	10,000
N				

Identifying a Function **In Exercises 69–72, the series represents a well-known function. Use a computer algebra system to graph the partial sum S_{10} and identify the function from the graph.**

69. $f(x) = \sum_{n=0}^{\infty}(-1)^n\frac{x^{2n}}{(2n)!}$

70. $f(x) = \sum_{n=0}^{\infty}(-1)^n\frac{x^{2n+1}}{(2n+1)!}$

71. $f(x) = \sum_{n=0}^{\infty}(-1)^n x^n, \quad -1 < x < 1$

72. $f(x) = \sum_{n=0}^{\infty}(-1)^n\frac{x^{2n+1}}{2n+1}, \quad -1 \le x \le 1$

True or False? **In Exercises 73–76, determine whether the statement is true or false. If it is false, explain why or give an example that shows it is false.**

73. If the power series $\sum_{n=1}^{\infty}a_nx^n$ converges for $x = 2$, then it also converges for $x = -2$.

74. It is possible to find a power series whose interval of convergence is $[0, \infty)$.

75. If the interval of convergence for $\sum_{n=0}^{\infty}a_nx^n$ is $(-1, 1)$, then the interval of convergence for $\sum_{n=0}^{\infty}a_n(x-1)^n$ is $(0, 2)$.

76. If $f(x) = \sum_{n=0}^{\infty}a_nx^n$ converges for $|x| < 2$, then

$$\int_0^1 f(x)\,dx = \sum_{n=0}^{\infty}\frac{a_n}{n+1}.$$

77. Proof Prove that the power series

$$\sum_{n=0}^{\infty}\frac{(n+p)!}{n!(n+q)!}x^n$$

has a radius of convergence of $R = \infty$ when p and q are positive integers.

78. Using a Power Series Let

$$g(x) = 1 + 2x + x^2 + 2x^3 + x^4 + \cdots$$

where the coefficients are $c_{2n} = 1$ and $c_{2n+1} = 2$ for $n \ge 0$.

(a) Find the interval of convergence of the series.

(b) Find an explicit formula for $g(x)$.

79. Using a Power Series Let $f(x) = \sum_{n=0}^{\infty}c_nx^n$, where $c_{n+3} = c_n$ for $n \ge 0$.

(a) Find the interval of convergence of the series.

(b) Find an explicit formula for $f(x)$.

80. Proof Prove that if the power series $\sum_{n=0}^{\infty}c_nx^n$ has a radius of convergence of R, then $\sum_{n=0}^{\infty}c_nx^{2n}$ has a radius of convergence of $\sqrt{R}$.

81. Proof For $n > 0$, let $R > 0$ and $c_n > 0$. Prove that if the interval of convergence of the series

$$\sum_{n=0}^{\infty}c_n(x-x_0)^n$$

is $[x_0 - R, x_0 + R]$, then the series converges conditionally at $x_0 - R$.

9.9 Representation of Functions by Power Series

- **Find a geometric power series that represents a function.**
- **Construct a power series using series operations.**

Geometric Power Series

In this section and the next, you will study several techniques for finding a power series that represents a function. Consider the function

$$f(x) = \frac{1}{1-x}.$$

The form of f closely resembles the sum of a geometric series

$$\sum_{n=0}^{\infty} ar^n = \frac{a}{1-r}, \quad 0 < |r| < 1.$$

In other words, when $a = 1$ and $r = x$, a power series representation for $1/(1-x)$, centered at 0, is

$$\begin{aligned}\frac{1}{1-x} &= \sum_{n=0}^{\infty} ar^n \\ &= \sum_{n=0}^{\infty} x^n \\ &= 1 + x + x^2 + x^3 + \cdot \cdot \cdot, \quad |x| < 1.\end{aligned}$$

Of course, this series represents $f(x) = 1/(1-x)$ only on the interval $(-1, 1)$, whereas f is defined for all $x \neq 1$, as shown in Figure 9.22. To represent f in another interval, you must develop a different series. For instance, to obtain the power series centered at -1, you could write

$$\frac{1}{1-x} = \frac{1}{2-(x+1)} = \frac{1/2}{1-[(x+1)/2]} = \frac{a}{1-r}$$

which implies that $a = \frac{1}{2}$ and $r = (x+1)/2$. So, for $|x+1| < 2$, you have

$$\begin{aligned}\frac{1}{1-x} &= \sum_{n=0}^{\infty} \left(\frac{1}{2}\right)\left(\frac{x+1}{2}\right)^n \\ &= \frac{1}{2}\left[1 + \frac{(x+1)}{2} + \frac{(x+1)^2}{4} + \frac{(x+1)^3}{8} + \cdot \cdot \cdot\right], \quad |x+1| < 2\end{aligned}$$

which converges on the interval $(-3, 1)$.

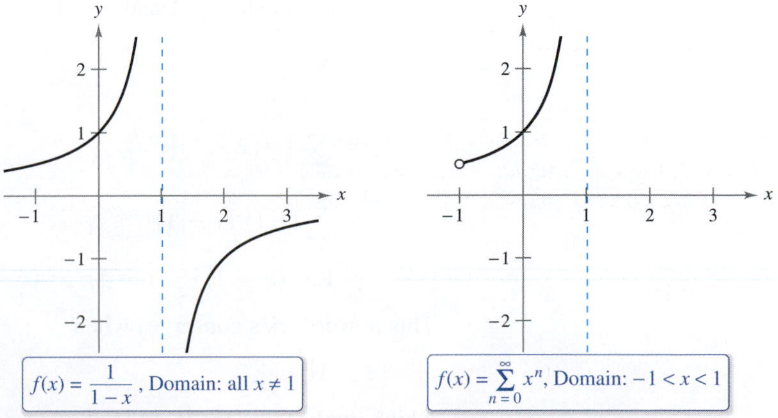

Figure 9.22

JOSEPH FOURIER (1768–1830)

Some of the early work in representing functions by power series was done by the French mathematician Joseph Fourier. Fourier's work is important in the history of calculus, partly because it forced eighteenth-century mathematicians to question the then-prevailing narrow concept of a function. Both Cauchy and Dirichlet were motivated by Fourier's work with series, and in 1837 Dirichlet published the general definition of a function that is used today.

The Granger Collection

EXAMPLE 1 Finding a Geometric Power Series Centered at 0

Find a power series for $f(x) = \dfrac{4}{x+2}$, centered at 0.

Solution Writing $f(x)$ in the form $a/(1 - r)$ produces

$$\frac{4}{2+x} = \frac{2}{1-(-x/2)} = \frac{a}{1-r}$$

which implies that $a = 2$ and

$$r = -\frac{x}{2}.$$

So, the power series for $f(x)$ is

$$\begin{aligned}\frac{4}{x+2} &= \sum_{n=0}^{\infty} ar^n \\ &= \sum_{n=0}^{\infty} 2\left(-\frac{x}{2}\right)^n \\ &= 2\left(1 - \frac{x}{2} + \frac{x^2}{4} - \frac{x^3}{8} + \cdot\cdot\cdot\right).\end{aligned}$$

This power series converges when

$$\left|-\frac{x}{2}\right| < 1$$

which implies that the interval of convergence is $(-2, 2)$.

Long Division

$$\begin{array}{r} 2 - x + \tfrac{1}{2}x^2 - \tfrac{1}{4}x^3 + \cdot\cdot\cdot \\ 2 + x \overline{\smash{)}\,4 \qquad\qquad\qquad} \\ \underline{4 + 2x} \qquad\qquad \\ -2x \qquad\qquad \\ \underline{-2x - x^2} \qquad \\ x^2 \qquad \\ \underline{x^2 + \tfrac{1}{2}x^3} \\ -\tfrac{1}{2}x^3 \\ \underline{-\tfrac{1}{2}x^3 - \tfrac{1}{4}x^4} \end{array}$$

Another way to determine a power series for a rational function such as the one in Example 1 is to use long division. For instance, by dividing $2 + x$ into 4, you obtain the result shown at the left.

EXAMPLE 2 Finding a Geometric Power Series Centered at 1

Find a power series for $f(x) = \dfrac{1}{x}$, centered at 1.

Solution Writing $f(x)$ in the form $a/(1 - r)$ produces

$$\frac{1}{x} = \frac{1}{1-(-x+1)} = \frac{a}{1-r}$$

which implies that $a = 1$ and $r = 1 - x = -(x - 1)$. So, the power series for $f(x)$ is

$$\begin{aligned}\frac{1}{x} &= \sum_{n=0}^{\infty} ar^n \\ &= \sum_{n=0}^{\infty} [-(x-1)]^n \\ &= \sum_{n=0}^{\infty} (-1)^n(x-1)^n \\ &= 1 - (x-1) + (x-1)^2 - (x-1)^3 + \cdot\cdot\cdot.\end{aligned}$$

This power series converges when

$$|x - 1| < 1$$

which implies that the interval of convergence is $(0, 2)$.

Operations with Power Series

The versatility of geometric power series will be shown later in this section, following a discussion of power series operations. These operations, used with differentiation and integration, provide a means of developing power series for a variety of elementary functions. (For simplicity, the operations are stated for a series centered at 0.)

Operations with Power Series

Let $f(x) = \sum_{n=0}^{\infty} a_n x^n$ and $g(x) = \sum_{n=0}^{\infty} b_n x^n$.

1. $f(kx) = \sum_{n=0}^{\infty} a_n k^n x^n$

2. $f(x^N) = \sum_{n=0}^{\infty} a_n x^{nN}$

3. $f(x) \pm g(x) = \sum_{n=0}^{\infty} (a_n \pm b_n)x^n$

The operations described above can change the interval of convergence for the resulting series. For example, in the addition shown below, the interval of convergence for the sum is the *intersection* of the intervals of convergence of the two original series.

$$\underbrace{\sum_{n=0}^{\infty} x^n}_{(-1,\,1)} + \underbrace{\sum_{n=0}^{\infty} \left(\frac{x}{2}\right)^n}_{\cap\ (-2,\,2)} = \underbrace{\sum_{n=0}^{\infty} \left(1 + \frac{1}{2^n}\right)x^n}_{=\ (-1,\,1)}$$

EXAMPLE 3 **Adding Two Power Series**

Find a power series for

$$f(x) = \frac{3x - 1}{x^2 - 1}$$

centered at 0.

Solution Using partial fractions, you can write $f(x)$ as

$$\frac{3x - 1}{x^2 - 1} = \frac{2}{x + 1} + \frac{1}{x - 1}.$$

By adding the two geometric power series

$$\frac{2}{x + 1} = \frac{2}{1 - (-x)} = \sum_{n=0}^{\infty} 2(-1)^n x^n, \quad |x| < 1$$

and

$$\frac{1}{x - 1} = \frac{-1}{1 - x} = -\sum_{n=0}^{\infty} x^n, \quad |x| < 1$$

you obtain the power series shown below.

$$\begin{aligned} \frac{3x - 1}{x^2 - 1} &= \sum_{n=0}^{\infty} [2(-1)^n - 1]x^n \\ &= 1 - 3x + x^2 - 3x^3 + x^4 - \cdots \end{aligned}$$

The interval of convergence for this power series is $(-1, 1)$. ■

AP* Tips

For the AP BC Exam, you should be able to perform algebraic manipulations of power series, including differentiation, integration, and composition.

EXAMPLE 4 Finding a Power Series by Integration

Find a power series for

$$f(x) = \ln x$$

centered at 1.

Solution From Example 2, you know that

$$\frac{1}{x} = \sum_{n=0}^{\infty} (-1)^n (x-1)^n. \quad \text{Interval of convergence: } (0, 2)$$

Integrating this series produces

$$\begin{aligned} \ln x &= \int \frac{1}{x}\,dx + C \\ &= C + \sum_{n=0}^{\infty} (-1)^n \frac{(x-1)^{n+1}}{n+1}. \end{aligned}$$

By letting $x = 1$, you can conclude that $C = 0$. Therefore,

$$\begin{aligned} \ln x &= \sum_{n=0}^{\infty} (-1)^n \frac{(x-1)^{n+1}}{n+1} \\ &= \frac{(x-1)}{1} - \frac{(x-1)^2}{2} + \frac{(x-1)^3}{3} - \frac{(x-1)^4}{4} + \cdots. \end{aligned} \quad \text{Interval of convergence: } (0, 2]$$

Note that the series converges at $x = 2$. This is consistent with the observation in the preceding section that integration of a power series may alter the convergence at the endpoints of the interval of convergence.

FOR FURTHER INFORMATION To read about finding a power series using integration by parts, see the article "Integration by Parts and Infinite Series" by Shelby J. Kilmer in *Mathematics Magazine*. To view this article, go to *MathArticles.com*.

In Section 9.7, Example 4, the fourth-degree Taylor polynomial for the natural logarithmic function

$$\ln x \approx (x-1) - \frac{(x-1)^2}{2} + \frac{(x-1)^3}{3} - \frac{(x-1)^4}{4}$$

was used to approximate $\ln(1.1)$.

$$\begin{aligned} \ln(1.1) &\approx (0.1) - \frac{1}{2}(0.1)^2 + \frac{1}{3}(0.1)^3 - \frac{1}{4}(0.1)^4 \\ &\approx 0.0953083 \end{aligned}$$

You now know from Example 4 in this section that this polynomial represents the first four terms of the power series for $\ln x$. Moreover, using the Alternating Series Remainder, you can determine that the error in this approximation is less than

$$\begin{aligned} |R_4| &\le |a_5| \\ &= \frac{1}{5}(0.1)^5 \\ &= 0.000002. \end{aligned}$$

During the seventeenth and eighteenth centuries, mathematical tables for logarithms and values of other transcendental functions were computed in this manner. Such numerical techniques are far from outdated, because it is precisely by such means that many modern calculating devices are programmed to evaluate transcendental functions.

EXAMPLE 5 Finding a Power Series by Integration

See LarsonCalculus.com for an interactive version of this type of example.

Find a power series for

$$g(x) = \arctan x$$

centered at 0.

Solution Because $D_x[\arctan x] = 1/(1 + x^2)$, you can use the series

$$f(x) = \frac{1}{1+x} = \sum_{n=0}^{\infty} (-1)^n x^n. \qquad \text{Interval of convergence: } (-1, 1)$$

Substituting x^2 for x produces

$$f(x^2) = \frac{1}{1+x^2} = \sum_{n=0}^{\infty} (-1)^n x^{2n}.$$

Finally, by integrating, you obtain

$$\begin{aligned}
\arctan x &= \int \frac{1}{1+x^2}\,dx + C \\
&= C + \sum_{n=0}^{\infty} (-1)^n \frac{x^{2n+1}}{2n+1} \\
&= \sum_{n=0}^{\infty} (-1)^n \frac{x^{2n+1}}{2n+1} && \text{Let } x = 0, \text{ then } C = 0. \\
&= x - \frac{x^3}{3} + \frac{x^5}{5} - \frac{x^7}{7} + \cdots. && \text{Interval of convergence: } (-1, 1)
\end{aligned}$$

SRINIVASA RAMANUJAN (1887–1920)

Series that can be used to approximate π have interested mathematicians for the past 300 years. An amazing series for approximating $1/\pi$ was discovered by the Indian mathematician Srinivasa Ramanujan in 1914 (see Exercise 61). Each successive term of Ramanujan's series adds roughly eight more correct digits to the value of $1/\pi$. For more information about Ramanujan's work, see the article "Ramanujan and Pi" by Jonathan M. Borwein and Peter B. Borwein in *Scientific American*.

See LarsonCalculus.com to read more of this biography.

It can be shown that the power series developed for $\arctan x$ in Example 5 also converges (to $\arctan x$) for $x = \pm 1$. For instance, when $x = 1$, you can write

$$\begin{aligned}
\arctan 1 &= 1 - \frac{1}{3} + \frac{1}{5} - \frac{1}{7} + \cdots \\
&= \frac{\pi}{4}.
\end{aligned}$$

However, this series (developed by James Gregory in 1671) does not give us a practical way of approximating π because it converges so slowly that hundreds of terms would have to be used to obtain reasonable accuracy. Example 6 shows how to use *two* different arctangent series to obtain a very good approximation of π using only a few terms. This approximation was developed by John Machin in 1706.

FOR FURTHER INFORMATION To read about other methods for approximating π, see the article "Two Methods for Approximating π" by Chien-Lih Hwang in *Mathematics Magazine*. To view this article, go to *MathArticles.com*.

EXAMPLE 6 Approximating π with a Series

Use the trigonometric identity

$$4 \arctan \frac{1}{5} - \arctan \frac{1}{239} = \frac{\pi}{4}$$

to approximate the number π [see Exercise 46(b)].

Solution By using only five terms from each of the series for $\arctan(1/5)$ and $\arctan(1/239)$, you obtain

$$4\left(4 \arctan \frac{1}{5} - \arctan \frac{1}{239}\right) \approx 3.1415926$$

which agrees with the exact value of π with an error of less than 0.0000001.

The Granger Collection

9.9 Exercises

See CalcChat.com for tutorial help and worked-out solutions to odd-numbered exercises.

Finding a Geometric Power Series In Exercises 1–4, find a geometric power series for the function, centered at 0, (a) by the technique shown in Examples 1 and 2 and (b) by long division.

1. $f(x) = \dfrac{1}{4 - x}$
2. $f(x) = \dfrac{1}{2 + x}$
3. $f(x) = \dfrac{4}{3 + x}$
4. $f(x) = \dfrac{2}{5 - x}$

Finding a Power Series In Exercises 5–16, find a power series for the function, centered at c, and determine the interval of convergence.

5. $f(x) = \dfrac{1}{3 - x}, \quad c = 1$
6. $f(x) = \dfrac{2}{6 - x}, \quad c = -2$
7. $f(x) = \dfrac{1}{1 - 3x}, \quad c = 0$
8. $h(x) = \dfrac{1}{1 - 5x}, \quad c = 0$
9. $g(x) = \dfrac{5}{2x - 3}, \quad c = -3$
10. $f(x) = \dfrac{3}{2x - 1}, \quad c = 2$
11. $f(x) = \dfrac{3}{3x + 4}, \quad c = 0$
12. $f(x) = \dfrac{4}{3x + 2}, \quad c = 3$
13. $g(x) = \dfrac{4x}{x^2 + 2x - 3}, \quad c = 0$
14. $g(x) = \dfrac{3x - 8}{3x^2 + 5x - 2}, \quad c = 0$
15. $f(x) = \dfrac{2}{1 - x^2}, \quad c = 0$
16. $f(x) = \dfrac{5}{5 + x^2}, \quad c = 0$

Using a Power Series In Exercises 17–26, use the power series

$$\frac{1}{1 + x} = \sum_{n=0}^{\infty} (-1)^n x^n$$

to determine a power series, centered at 0, for the function. Identify the interval of convergence.

17. $h(x) = \dfrac{-2}{x^2 - 1} = \dfrac{1}{1 + x} + \dfrac{1}{1 - x}$
18. $h(x) = \dfrac{x}{x^2 - 1} = \dfrac{1}{2(1 + x)} - \dfrac{1}{2(1 - x)}$
19. $f(x) = -\dfrac{1}{(x + 1)^2} = \dfrac{d}{dx}\left[\dfrac{1}{x + 1}\right]$
20. $f(x) = \dfrac{2}{(x + 1)^3} = \dfrac{d^2}{dx^2}\left[\dfrac{1}{x + 1}\right]$
21. $f(x) = \ln(x + 1) = \displaystyle\int \frac{1}{x + 1}\,dx$
22. $f(x) = \ln(1 - x^2) = \displaystyle\int \frac{1}{1 + x}\,dx - \int \frac{1}{1 - x}\,dx$
23. $g(x) = \dfrac{1}{x^2 + 1}$
24. $f(x) = \ln(x^2 + 1)$
25. $h(x) = \dfrac{1}{4x^2 + 1}$
26. $f(x) = \arctan 2x$

Graphical and Numerical Analysis In Exercises 27 and 28, let

$$S_n = x - \frac{x^2}{2} + \frac{x^3}{3} - \frac{x^4}{4} + \cdots \pm \frac{x^n}{n}.$$

Use a graphing utility to confirm the inequality graphically. Then complete the table to confirm the inequality numerically.

x	0.0	0.2	0.4	0.6	0.8	1.0
S_n						
$\ln(x + 1)$						
S_{n+1}						

27. $S_2 \le \ln(x + 1) \le S_3$
28. $S_4 \le \ln(x + 1) \le S_5$

Approximating a Sum In Exercises 29 and 30, (a) graph several partial sums of the series, (b) find the sum of the series and its radius of convergence, (c) use 50 terms of the series to approximate the sum when $x = 0.5$, and (d) determine what the approximation represents and how good the approximation is.

29. $\displaystyle\sum_{n=1}^{\infty} \frac{(-1)^{n+1}(x - 1)^n}{n}$
30. $\displaystyle\sum_{n=0}^{\infty} \frac{(-1)^n x^{2n+1}}{(2n + 1)!}$

Approximating a Value In Exercises 31–34, use the series for $f(x) = \arctan x$ to approximate the value, using $R_N \le 0.001$.

31. $\arctan \dfrac{1}{4}$
32. $\displaystyle\int_0^{3/4} \arctan x^2\,dx$
33. $\displaystyle\int_0^{1/2} \frac{\arctan x^2}{x}\,dx$
34. $\displaystyle\int_0^{1/2} x^2 \arctan x\,dx$

Using a Power Series In Exercises 35–38, use the power series

$$\frac{1}{1 - x} = \sum_{n=0}^{\infty} x^n, \quad |x| < 1.$$

Find the series representation of the function and determine its interval of convergence.

35. $f(x) = \dfrac{1}{(1 - x)^2}$
36. $f(x) = \dfrac{x}{(1 - x)^2}$
37. $f(x) = \dfrac{1 + x}{(1 - x)^2}$
38. $f(x) = \dfrac{x(1 + x)}{(1 - x)^2}$

39. Probability A fair coin is tossed repeatedly. The probability that the first head occurs on the nth toss is $P(n) = \left(\frac{1}{2}\right)^n$. When this game is repeated many times, the average number of tosses required until the first head occurs is

$$E(n) = \sum_{n=1}^{\infty} nP(n).$$

(This value is called the *expected value of n*.) Use the results of Exercises 35–38 to find $E(n)$. Is the answer what you expected? Why or why not?

40. Finding the Sum of a Series Use the results of Exercises 35–38 to find the sum of each series.

(a) $\frac{1}{3}\sum_{n=1}^{\infty} n\left(\frac{2}{3}\right)^n$ (b) $\frac{1}{10}\sum_{n=1}^{\infty} n\left(\frac{9}{10}\right)^n$

Writing In Exercises 41–44, explain how to use the geometric series

$$g(x) = \frac{1}{1-x} = \sum_{n=0}^{\infty} x^n, \quad |x| < 1$$

to find the series for the function. Do not find the series.

41. $f(x) = \dfrac{1}{1+x}$ **42.** $f(x) = \dfrac{1}{1-x^2}$

43. $f(x) = \dfrac{5}{1+x}$ **44.** $f(x) = \ln(1-x)$

45. Proof Prove that

$$\arctan x + \arctan y = \arctan \frac{x+y}{1-xy}$$

for $xy \neq 1$ provided the value of the left side of the equation is between $-\pi/2$ and $\pi/2$.

46. Verifying an Identity Use the result of Exercise 45 to verify each identity.

(a) $\arctan \frac{120}{119} - \arctan \frac{1}{239} = \frac{\pi}{4}$

(b) $4 \arctan \frac{1}{5} - \arctan \frac{1}{239} = \frac{\pi}{4}$

[*Hint:* Use Exercise 45 twice to find $4 \arctan \frac{1}{5}$. Then use part (a).]

Approximating Pi In Exercises 47 and 48, (a) verify the given equation, and (b) use the equation and the series for the arctangent to approximate π to two-decimal-place accuracy.

47. $2 \arctan \frac{1}{2} - \arctan \frac{1}{7} = \frac{\pi}{4}$

48. $\arctan \frac{1}{2} + \arctan \frac{1}{3} = \frac{\pi}{4}$

Finding the Sum of a Series In Exercises 49–54, find the sum of the convergent series by using a well-known function. Identify the function and explain how you obtained the sum.

49. $\sum_{n=1}^{\infty} (-1)^{n+1}\frac{1}{2^n n}$ **50.** $\sum_{n=1}^{\infty} (-1)^{n+1}\frac{1}{3^n n}$

51. $\sum_{n=1}^{\infty} (-1)^{n+1}\frac{2^n}{5^n n}$ **52.** $\sum_{n=0}^{\infty} (-1)^n \frac{1}{2n+1}$

53. $\sum_{n=0}^{\infty} (-1)^n \frac{1}{2^{2n+1}(2n+1)}$

54. $\sum_{n=1}^{\infty} (-1)^{n+1} \frac{1}{3^{2n-1}(2n-1)}$

WRITING ABOUT CONCEPTS

55. Using Series One of the series in Exercises 49–54 converges to its sum at a much lower rate than the other five series. Which is it? Explain why this series converges so slowly. Use a graphing utility to illustrate the rate of convergence.

56. Radius of Convergence The radius of convergence of the power series $\sum_{n=0}^{\infty} a_n x^n$ is 3. What is the radius of convergence of the series $\sum_{n=1}^{\infty} na_n x^{n-1}$? Explain.

57. Convergence of a Power Series The power series $\sum_{n=0}^{\infty} a_n x^n$ converges for $|x+1| < 4$. What can you conclude about the series $\sum_{n=0}^{\infty} a_n \frac{x^{n+1}}{n+1}$? Explain.

58. HOW DO YOU SEE IT? The graphs show first-, second-, and third-degree polynomial approximations P_1, P_2, and P_3 of a function f. Label the graphs of P_1, P_2, and P_3. To print an enlarged copy of the graph, go to *MathGraphs.com*.

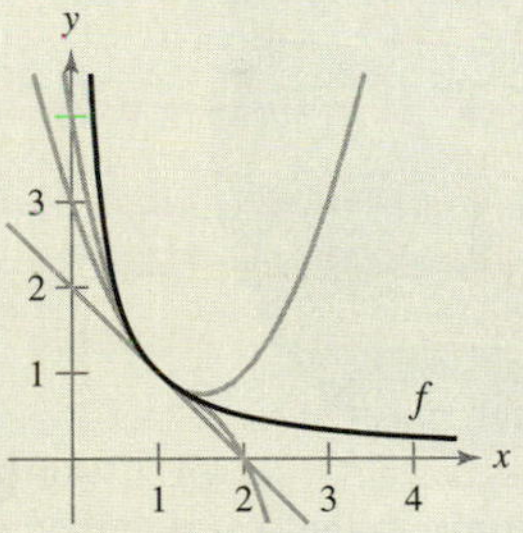

Finding the Sum of a Series In Exercises 59 and 60, find the sum of the series.

59. $\sum_{n=0}^{\infty} \frac{(-1)^n}{3^n(2n+1)}$ **60.** $\sum_{n=0}^{\infty} \frac{(-1)^n \pi^{2n+1}}{3^{2n+1}(2n+1)!}$

61. Ramanujan and Pi Use a graphing utility to show that

$$\frac{\sqrt{8}}{9801}\sum_{n=0}^{\infty} \frac{(4n)!(1103 + 26{,}390n)}{(n!)396^{4n}} = \frac{1}{\pi}.$$

62. Find the Error Describe why the statement is incorrect.

$$\sum_{n=0}^{\infty} x^n + \sum_{n=0}^{\infty}\left(\frac{x}{5}\right)^n = \sum_{n=0}^{\infty}\left(1+\frac{1}{5}\right)x^n$$

9.10 Taylor and Maclaurin Series

- Find a Taylor or Maclaurin series for a function.
- Find a binomial series.
- Use a basic list of Taylor series to find other Taylor series.

Taylor Series and Maclaurin Series

In Section 9.9, you derived power series for several functions using geometric series with term-by-term differentiation or integration. In this section, you will study a *general* procedure for deriving the power series for a function that has derivatives of all orders. The next theorem gives the form that *every* convergent power series must take.

REMARK Be sure you understand Theorem 9.22. The theorem says that *if a power series converges to* $f(x)$, then the series must be a Taylor series. The theorem does *not* say that every series formed with the Taylor coefficients $a_n = f^{(n)}(c)/n!$ will converge to $f(x)$.

THEOREM 9.22 The Form of a Convergent Power Series

If f is represented by a power series $f(x) = \Sigma\, a_n(x-c)^n$ for all x in an open interval I containing c, then

$$a_n = \frac{f^{(n)}(c)}{n!}$$

and

$$f(x) = f(c) + f'(c)(x-c) + \frac{f''(c)}{2!}(x-c)^2 + \cdots + \frac{f^{(n)}(c)}{n!}(x-c)^n + \cdots.$$

COLIN MACLAURIN (1698–1746)

The development of power series to represent functions is credited to the combined work of many seventeenth- and eighteenth-century mathematicians. Gregory, Newton, John and James Bernoulli, Leibniz, Euler, Lagrange, Wallis, and Fourier all contributed to this work. However, the two names that are most commonly associated with power series are Brook Taylor (1685–1731) and Colin Maclaurin.

See LarsonCalculus.com to read more of this biography.

Proof Consider a power series $\Sigma\, a_n(x-c)^n$ that has a radius of convergence R. Then, by Theorem 9.21, you know that the nth derivative of f exists for $|x-c| < R$, and by successive differentiation you obtain the following.

$$\begin{aligned}
f^{(0)}(x) &= a_0 + a_1(x-c) + a_2(x-c)^2 + a_3(x-c)^3 + a_4(x-c)^4 + \cdots\\
f^{(1)}(x) &= a_1 + 2a_2(x-c) + 3a_3(x-c)^2 + 4a_4(x-c)^3 + \cdots\\
f^{(2)}(x) &= 2a_2 + 3!a_3(x-c) + 4 \cdot 3a_4(x-c)^2 + \cdots\\
f^{(3)}(x) &= 3!a_3 + 4!a_4(x-c) + \cdots\\
&\vdots\\
f^{(n)}(x) &= n!a_n + (n+1)!a_{n+1}(x-c) + \cdots
\end{aligned}$$

Evaluating each of these derivatives at $x = c$ yields

$$\begin{aligned}
f^{(0)}(c) &= 0!a_0\\
f^{(1)}(c) &= 1!a_1\\
f^{(2)}(c) &= 2!a_2\\
f^{(3)}(c) &= 3!a_3
\end{aligned}$$

and, in general, $f^{(n)}(c) = n!a_n$. By solving for a_n, you find that the coefficients of the power series representation of $f(x)$ are

$$a_n = \frac{f^{(n)}(c)}{n!}.$$

See LarsonCalculus.com for Bruce Edwards's video of this proof. ■

Notice that the coefficients of the power series in Theorem 9.22 are precisely the coefficients of the Taylor polynomials for $f(x)$ at c as defined in Section 9.7. For this reason, the series is called the **Taylor series** for $f(x)$ at c.

Bettmann/Corbis

Definition of Taylor and Maclaurin Series

If a function f has derivatives of all orders at $x = c$, then the series

$$\sum_{n=0}^{\infty} \frac{f^{(n)}(c)}{n!}(x-c)^n = f(c) + f'(c)(x-c) + \cdot \cdot \cdot + \frac{f^{(n)}(c)}{n!}(x-c)^n + \cdot \cdot \cdot$$

is called the **Taylor series for $f(x)$ at c.** Moreover, if $c = 0$, then the series is the **Maclaurin series for f.**

When you know the pattern for the coefficients of the Taylor polynomials for a function, you can extend the pattern easily to form the corresponding Taylor series. For instance, in Example 4 in Section 9.7, you found the fourth Taylor polynomial for $\ln x$, centered at 1, to be

$$P_4(x) = (x-1) - \frac{1}{2}(x-1)^2 + \frac{1}{3}(x-1)^3 - \frac{1}{4}(x-1)^4.$$

From this pattern, you can obtain the Taylor series for $\ln x$ centered at $c = 1$,

$$(x-1) - \frac{1}{2}(x-1)^2 + \cdot \cdot \cdot + \frac{(-1)^{n+1}}{n}(x-1)^n + \cdot \cdot \cdot .$$

EXAMPLE 1 Forming a Power Series

Use the function

$$f(x) = \sin x$$

to form the Maclaurin series

$$\sum_{n=0}^{\infty} \frac{f^{(n)}(0)}{n!}x^n = f(0) + f'(0)x + \frac{f''(0)}{2!}x^2 + \frac{f^{(3)}(0)}{3!}x^3 + \frac{f^{(4)}(0)}{4!}x^4 + \cdot \cdot \cdot$$

and determine the interval of convergence.

Solution Successive differentiation of $f(x)$ yields

$$\begin{aligned} f(x) &= \sin x & f(0) &= \sin 0 = 0 \\ f'(x) &= \cos x & f'(0) &= \cos 0 = 1 \\ f''(x) &= -\sin x & f''(0) &= -\sin 0 = 0 \\ f^{(3)}(x) &= -\cos x & f^{(3)}(0) &= -\cos 0 = -1 \\ f^{(4)}(x) &= \sin x & f^{(4)}(0) &= \sin 0 = 0 \\ f^{(5)}(x) &= \cos x & f^{(5)}(0) &= \cos 0 = 1 \end{aligned}$$

and so on. The pattern repeats after the third derivative. So, the power series is as follows.

$$\begin{aligned} \sum_{n=0}^{\infty} \frac{f^{(n)}(0)}{n!}x^n &= f(0) + f'(0)x + \frac{f''(0)}{2!}x^2 + \frac{f^{(3)}(0)}{3!}x^3 + \frac{f^{(4)}(0)}{4!}x^4 + \cdot \cdot \cdot \\ \sum_{n=0}^{\infty} \frac{(-1)^n x^{2n+1}}{(2n+1)!} &= 0 + (1)x + \frac{0}{2!}x^2 + \frac{(-1)}{3!}x^3 + \frac{0}{4!}x^4 + \frac{1}{5!}x^5 + \frac{0}{6!}x^6 \\ &\quad + \frac{(-1)}{7!}x^7 + \cdot \cdot \cdot \\ &= x - \frac{x^3}{3!} + \frac{x^5}{5!} - \frac{x^7}{7!} + \cdot \cdot \cdot \end{aligned}$$

By the Ratio Test, you can conclude that this series converges for all x.

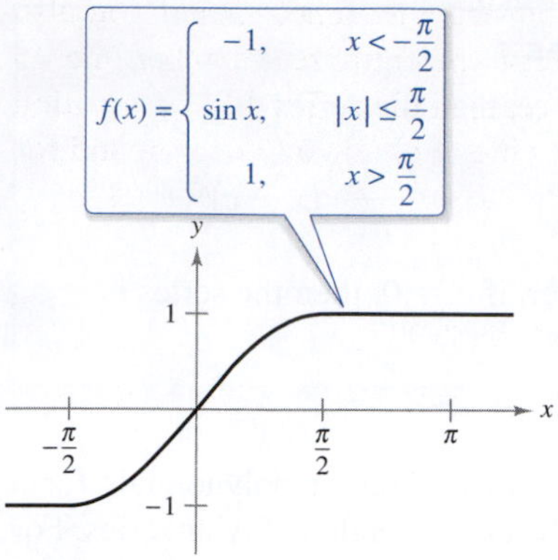

Figure 9.23

Notice that in Example 1, you cannot conclude that the power series converges to $\sin x$ for all x. You can simply conclude that the power series converges to some function, but you are not sure what function it is. This is a subtle, but important, point in dealing with Taylor or Maclaurin series. To persuade yourself that the series

$$f(c) + f'(c)(x - c) + \frac{f''(c)}{2!}(x - c)^2 + \cdots + \frac{f^{(n)}(c)}{n!}(x - c)^n + \cdots$$

might converge to a function other than f, remember that the derivatives are being evaluated at a single point. It can easily happen that another function will agree with the values of $f^{(n)}(x)$ when $x = c$ and disagree at other x-values. For instance, the power series (centered at 0) for the function f shown in Figure 9.23 is the same series as in Example 1. You know that the series converges for all x, and yet it obviously cannot converge to both $f(x)$ and $\sin x$ for all x.

Let f have derivatives of all orders in an open interval I centered at c. The Taylor series for f may fail to converge for some x in I. Or, even when it is convergent, it may fail to have $f(x)$ as its sum. Nevertheless, Theorem 9.19 tells us that for each n,

$$f(x) = f(c) + f'(c)(x - c) + \frac{f''(c)}{2!}(x - c)^2 + \cdots + \frac{f^{(n)}(c)}{n!}(x - c)^n + R_n(x)$$

where

$$R_n(x) = \frac{f^{(n+1)}(z)}{(n + 1)!}(x - c)^{n+1}.$$

Note that in this remainder formula, the particular value of z that makes the remainder formula true depends on the values of x and n. If $R_n \to 0$, then the next theorem tells us that the Taylor series for f actually converges to $f(x)$ for all x in I.

THEOREM 9.23 Convergence of Taylor Series

If $\lim_{n\to\infty} R_n = 0$ for all x in the interval I, then the Taylor series for f converges and equals $f(x)$,

$$f(x) = \sum_{n=0}^{\infty} \frac{f^{(n)}(c)}{n!}(x - c)^n.$$

Proof For a Taylor series, the nth partial sum coincides with the nth Taylor polynomial. That is, $S_n(x) = P_n(x)$. Moreover, because

$$P_n(x) = f(x) - R_n(x)$$

it follows that

$$\begin{aligned}\lim_{n\to\infty} S_n(x) &= \lim_{n\to\infty} P_n(x)\\ &= \lim_{n\to\infty} [f(x) - R_n(x)]\\ &= f(x) - \lim_{n\to\infty} R_n(x).\end{aligned}$$

So, for a given x, the Taylor series (the sequence of partial sums) converges to $f(x)$ if and only if $R_n(x) \to 0$ as $n \to \infty$.

See LarsonCalculus.com for Bruce Edwards's video of this proof. ■

Stated another way, Theorem 9.23 says that a power series formed with Taylor coefficients $a_n = f^{(n)}(c)/n!$ converges to the function from which it was derived at precisely those values for which the remainder approaches 0 as $n \to \infty$.

In Example 1, you derived the power series from the sine function and you also concluded that the series converges to some function on the entire real number line. In Example 2, you will see that the series actually converges to $\sin x$. The key observation is that although the value of z is not known, it is possible to obtain an upper bound for

$$|f^{(n+1)}(z)|.$$

EXAMPLE 2 A Convergent Maclaurin Series

Show that the Maclaurin series for

$$f(x) = \sin x$$

converges to $\sin x$ for all x.

Solution Using the result in Example 1, you need to show that

$$\sin x = x - \frac{x^3}{3!} + \frac{x^5}{5!} - \frac{x^7}{7!} + \cdot \cdot \cdot + \frac{(-1)^n x^{2n+1}}{(2n+1)!} + \cdot \cdot \cdot$$

is true for all x. Because

$$f^{(n+1)}(x) = \pm\sin x$$

or

$$f^{(n+1)}(x) = \pm\cos x$$

you know that $|f^{(n+1)}(z)| \leq 1$ for every real number z. Therefore, for any fixed x, you can apply Taylor's Theorem (Theorem 9.19) to conclude that

$$0 \leq |R_n(x)| = \left|\frac{f^{(n+1)}(z)}{(n+1)!}x^{n+1}\right| \leq \frac{|x|^{n+1}}{(n+1)!}.$$

From the discussion in Section 9.1 regarding the relative rates of convergence of exponential and factorial sequences, it follows that for a fixed x

$$\lim_{n\to\infty} \frac{|x|^{n+1}}{(n+1)!} = 0.$$

Finally, by the Squeeze Theorem, it follows that for all x, $R_n(x) \to 0$ as $n \to \infty$. So, by Theorem 9.23, the Maclaurin series for $\sin x$ converges to $\sin x$ for all x. ■

Figure 9.24 visually illustrates the convergence of the Maclaurin series for $\sin x$ by comparing the graphs of the Maclaurin polynomials $P_1(x)$, $P_3(x)$, $P_5(x)$, and $P_7(x)$ with the graph of the sine function. Notice that as the degree of the polynomial increases, its graph more closely resembles that of the sine function.

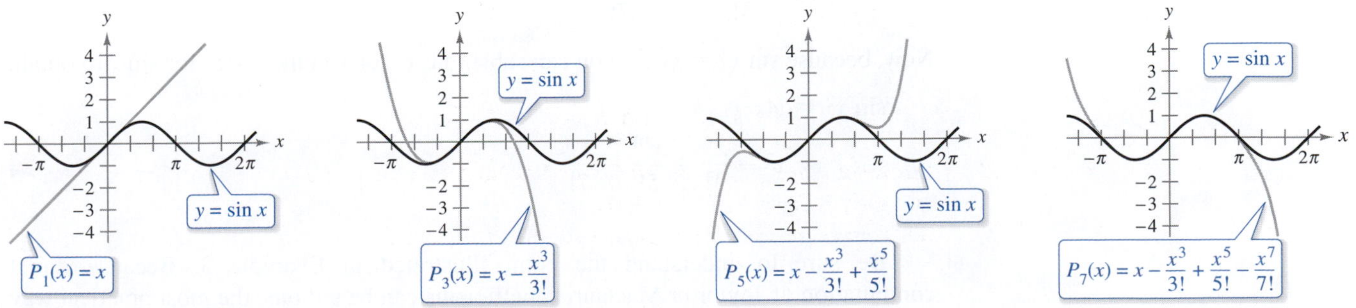

As n increases, the graph of P_n more closely resembles the sine function.
Figure 9.24

The guidelines for finding a Taylor series for $f(x)$ at c are summarized below.

GUIDELINES FOR FINDING A TAYLOR SERIES

1. Differentiate $f(x)$ several times and evaluate each derivative at c.

$$f(c), f'(c), f''(c), f'''(c), \cdot\cdot\cdot, f^{(n)}(c), \cdot\cdot\cdot$$

Try to recognize a pattern in these numbers.

2. Use the sequence developed in the first step to form the Taylor coefficients $a_n = f^{(n)}(c)/n!$, and determine the interval of convergence for the resulting power series

$$f(c) + f'(c)(x-c) + \frac{f''(c)}{2!}(x-c)^2 + \cdot\cdot\cdot + \frac{f^{(n)}(c)}{n!}(x-c)^n + \cdot\cdot\cdot.$$

3. Within this interval of convergence, determine whether the series converges to $f(x)$.

REMARK When you have difficulty recognizing a pattern, remember that you can use Theorem 9.22 to find the Taylor series. Also, you can try using the coefficients of a known Taylor or Maclaurin series, as shown in Example 3.

The direct determination of Taylor or Maclaurin coefficients using successive differentiation can be difficult, and the next example illustrates a shortcut for finding the coefficients indirectly—using the coefficients of a known Taylor or Maclaurin series.

EXAMPLE 3 **Maclaurin Series for a Composite Function**

Find the Maclaurin series for

$$f(x) = \sin x^2.$$

Solution To find the coefficients for this Maclaurin series directly, you must calculate successive derivatives of $f(x) = \sin x^2$. By calculating just the first two,

$$f'(x) = 2x \cos x^2$$

and

$$f''(x) = -4x^2 \sin x^2 + 2 \cos x^2$$

you can see that this task would be quite cumbersome. Fortunately, there is an alternative. First, consider the Maclaurin series for $\sin x$ found in Example 1.

$$g(x) = \sin x$$
$$= x - \frac{x^3}{3!} + \frac{x^5}{5!} - \frac{x^7}{7!} + \cdot\cdot\cdot$$

Now, because $\sin x^2 = g(x^2)$, you can substitute x^2 for x in the series for $\sin x$ to obtain

$$\sin x^2 = g(x^2)$$
$$= x^2 - \frac{x^6}{3!} + \frac{x^{10}}{5!} - \frac{x^{14}}{7!} + \cdot\cdot\cdot.$$

Be sure to understand the point illustrated in Example 3. Because direct computation of Taylor or Maclaurin coefficients can be tedious, the most practical way to find a Taylor or Maclaurin series is to develop power series for a *basic list* of elementary functions. From this list, you can determine power series for other functions by the operations of addition, subtraction, multiplication, division, differentiation, integration, and composition with known power series.

Binomial Series

Before presenting the basic list for elementary functions, you will develop one more series—for a function of the form $f(x) = (1 + x)^k$. This produces the **binomial series.**

EXAMPLE 4 Binomial Series

Find the Maclaurin series for $f(x) = (1 + x)^k$ and determine its radius of convergence. Assume that k is not a positive integer and $k \neq 0$.

Solution By successive differentiation, you have

$$\begin{aligned}
f(x) &= (1 + x)^k & f(0) &= 1 \\
f'(x) &= k(1 + x)^{k-1} & f'(0) &= k \\
f''(x) &= k(k - 1)(1 + x)^{k-2} & f''(0) &= k(k - 1) \\
f'''(x) &= k(k - 1)(k - 2)(1 + x)^{k-3} & f'''(0) &= k(k - 1)(k - 2) \\
&\vdots & &\vdots \\
f^{(n)}(x) &= k \cdot \cdot \cdot (k - n + 1)(1 + x)^{k-n} & f^{(n)}(0) &= k(k - 1) \ldots (k - n + 1)
\end{aligned}$$

which produces the series

$$1 + kx + \frac{k(k-1)x^2}{2} + \cdot \cdot \cdot + \frac{k(k-1) \cdot \cdot \cdot (k - n + 1)x^n}{n!} + \cdot \cdot \cdot .$$

Because $a_{n+1}/a_n \to 1$, you can apply the Ratio Test to conclude that the radius of convergence is $R = 1$. So, the series converges to some function in the interval $(-1, 1)$.

Note that Example 4 shows that the Taylor series for $(1 + x)^k$ converges to some function in the interval $(-1, 1)$. However, the example does not show that the series actually converges to $(1 + x)^k$. To do this, you could show that the remainder $R_n(x)$ converges to 0, as illustrated in Example 2. You now have enough information to find a binomial series for a function, as shown in the next example.

EXAMPLE 5 Finding a Binomial Series

Find the power series for $f(x) = \sqrt[3]{1 + x}$.

Solution Using the binomial series

$$(1 + x)^k = 1 + kx + \frac{k(k-1)x^2}{2!} + \frac{k(k-1)(k-2)x^3}{3!} + \cdot \cdot \cdot$$

let $k = \frac{1}{3}$ and write

$$(1 + x)^{1/3} = 1 + \frac{x}{3} - \frac{2x^2}{3^2 2!} + \frac{2 \cdot 5x^3}{3^3 3!} - \frac{2 \cdot 5 \cdot 8x^4}{3^4 4!} + \cdot \cdot \cdot$$

which converges for $-1 \leq x \leq 1$.

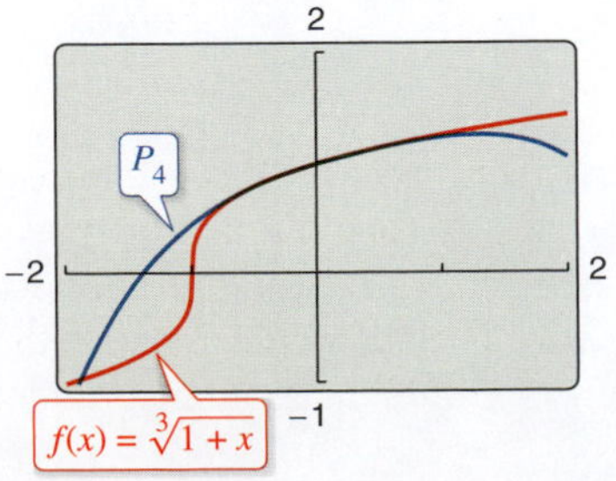

Figure 9.25

TECHNOLOGY Use a graphing utility to confirm the result in Example 5. When you graph the functions

$$f(x) = (1 + x)^{1/3}$$

and

$$P_4(x) = 1 + \frac{x}{3} - \frac{x^2}{9} + \frac{5x^3}{81} - \frac{10x^4}{243}$$

in the same viewing window, you should obtain the result shown in Figure 9.25.

Deriving Taylor Series from a Basic List

The list below provides the power series for several elementary functions with the corresponding intervals of convergence.

POWER SERIES FOR ELEMENTARY FUNCTIONS

Function	Interval of Convergence
$\dfrac{1}{x} = 1 - (x-1) + (x-1)^2 - (x-1)^3 + (x-1)^4 - \cdots + (-1)^n(x-1)^n + \cdots$	$0 < x < 2$
$\dfrac{1}{1+x} = 1 - x + x^2 - x^3 + x^4 - x^5 + \cdots + (-1)^n x^n + \cdots$	$-1 < x < 1$
$\ln x = (x-1) - \dfrac{(x-1)^2}{2} + \dfrac{(x-1)^3}{3} - \dfrac{(x-1)^4}{4} + \cdots + \dfrac{(-1)^{n-1}(x-1)^n}{n} + \cdots$	$0 < x \le 2$
$e^x = 1 + x + \dfrac{x^2}{2!} + \dfrac{x^3}{3!} + \dfrac{x^4}{4!} + \dfrac{x^5}{5!} + \cdots + \dfrac{x^n}{n!} + \cdots$	$-\infty < x < \infty$
$\sin x = x - \dfrac{x^3}{3!} + \dfrac{x^5}{5!} - \dfrac{x^7}{7!} + \dfrac{x^9}{9!} - \cdots + \dfrac{(-1)^n x^{2n+1}}{(2n+1)!} + \cdots$	$-\infty < x < \infty$
$\cos x = 1 - \dfrac{x^2}{2!} + \dfrac{x^4}{4!} - \dfrac{x^6}{6!} + \dfrac{x^8}{8!} - \cdots + \dfrac{(-1)^n x^{2n}}{(2n)!} + \cdots$	$-\infty < x < \infty$
$\arctan x = x - \dfrac{x^3}{3} + \dfrac{x^5}{5} - \dfrac{x^7}{7} + \dfrac{x^9}{9} - \cdots + \dfrac{(-1)^n x^{2n+1}}{2n+1} + \cdots$	$-1 \le x \le 1$
$\arcsin x = x + \dfrac{x^3}{2 \cdot 3} + \dfrac{1 \cdot 3x^5}{2 \cdot 4 \cdot 5} + \dfrac{1 \cdot 3 \cdot 5x^7}{2 \cdot 4 \cdot 6 \cdot 7} + \cdots + \dfrac{(2n)!x^{2n+1}}{(2^n n!)^2(2n+1)} + \cdots$	$-1 \le x \le 1$
$(1+x)^k = 1 + kx + \dfrac{k(k-1)x^2}{2!} + \dfrac{k(k-1)(k-2)x^3}{3!} + \dfrac{k(k-1)(k-2)(k-3)x^4}{4!} + \cdots$	$-1 < x < 1$*

* The convergence at $x = \pm 1$ depends on the value of k.

Note that the binomial series is valid for noninteger values of k. Also, when k is a positive integer, the binomial series reduces to a simple binomial expansion.

EXAMPLE 6 Deriving a Power Series from a Basic List

Find the power series for

$$f(x) = \cos\sqrt{x}.$$

Solution Using the power series

$$\cos x = 1 - \frac{x^2}{2!} + \frac{x^4}{4!} - \frac{x^6}{6!} + \frac{x^8}{8!} - \cdots$$

you can replace x by

$$\sqrt{x}$$

to obtain the series

$$\cos\sqrt{x} = 1 - \frac{x}{2!} + \frac{x^2}{4!} - \frac{x^3}{6!} + \frac{x^4}{8!} - \cdots.$$

This series converges for all x in the domain of $\cos\sqrt{x}$—that is, for $x \ge 0$. ■

AP* Tips

You should know the power series for the functions e^x, $\sin x$, $\cos x$, and $\dfrac{1}{1+x}$ for the AP BC Exam.

Power series can be multiplied and divided like polynomials. After finding the first few terms of the product (or quotient), you may be able to recognize a pattern.

EXAMPLE 7 Multiplication of Power Series

Find the first three nonzero terms in the Maclaurin series $e^x \arctan x$.

Solution Using the Maclaurin series for e^x and $\arctan x$ in the table, you have

$$e^x \arctan x = \left(1 + \frac{x}{1!} + \frac{x^2}{2!} + \frac{x^3}{3!} + \frac{x^4}{4!} + \cdots\right)\left(x - \frac{x^3}{3} + \frac{x^5}{5} - \cdots\right).$$

Multiply these expressions and collect like terms as you would in multiplying polynomials.

$$\begin{array}{rrrrrl}
1 + & x + & \frac{1}{2}x^2 + & \frac{1}{6}x^3 + & \frac{1}{24}x^4 + & \cdots \\
& x & & -\frac{1}{3}x^3 & & +\frac{1}{5}x^5 - \cdots \\
\hline
& x + & x^2 + & \frac{1}{2}x^3 + & \frac{1}{6}x^4 + & \frac{1}{24}x^5 + \cdots \\
& & & -\frac{1}{3}x^3 - & \frac{1}{3}x^4 - & \frac{1}{6}x^5 - \cdots \\
& & & & & +\frac{1}{5}x^5 + \cdots \\
\hline
& x + & x^2 + & \frac{1}{6}x^3 - & \frac{1}{6}x^4 + & \frac{3}{40}x^5 + \cdots
\end{array}$$

So, $e^x \arctan x = x + x^2 + \frac{1}{6}x^3 + \cdots$.

EXAMPLE 8 Division of Power Series

Find the first three nonzero terms in the Maclaurin series $\tan x$.

Solution Using the Maclaurin series for $\sin x$ and $\cos x$ in the table, you have

$$\tan x = \frac{\sin x}{\cos x} = \frac{x - \dfrac{x^3}{3!} + \dfrac{x^5}{5!} - \cdots}{1 - \dfrac{x^2}{2!} + \dfrac{x^4}{4!} - \cdots}.$$

Divide using long division.

$$\begin{array}{r|l}
 & x + \frac{1}{3}x^3 + \frac{2}{15}x^5 + \cdots \\
\hline
1 - \frac{1}{2}x^2 + \frac{1}{24}x^4 - \cdots & x - \frac{1}{6}x^3 + \frac{1}{120}x^5 - \cdots \\
 & x - \frac{1}{2}x^3 + \frac{1}{24}x^5 - \cdots \\
\cline{2-2}
 & \frac{1}{3}x^3 - \frac{1}{30}x^5 + \cdots \\
 & \frac{1}{3}x^3 - \frac{1}{6}x^5 + \cdots \\
\cline{2-2}
 & \frac{2}{15}x^5 + \cdots
\end{array}$$

So, $\tan x = x + \frac{1}{3}x^3 + \frac{2}{15}x^5 + \cdots$.

EXAMPLE 9 A Power Series for $\sin^2 x$

Find the power series for

$$f(x) = \sin^2 x.$$

Solution Consider rewriting $\sin^2 x$ as

$$\sin^2 x = \frac{1 - \cos 2x}{2} = \frac{1}{2} - \frac{1}{2}\cos 2x.$$

Now, use the series for $\cos x$.

$$\cos x = 1 - \frac{x^2}{2!} + \frac{x^4}{4!} - \frac{x^6}{6!} + \frac{x^8}{8!} - \cdots$$

$$\cos 2x = 1 - \frac{2^2}{2!}x^2 + \frac{2^4}{4!}x^4 - \frac{2^6}{6!}x^6 + \frac{2^8}{8!}x^8 - \cdots$$

$$-\frac{1}{2}\cos 2x = -\frac{1}{2} + \frac{2}{2!}x^2 - \frac{2^3}{4!}x^4 + \frac{2^5}{6!}x^6 - \frac{2^7}{8!}x^8 + \cdots$$

$$\frac{1}{2} - \frac{1}{2}\cos 2x = \frac{1}{2} - \frac{1}{2} + \frac{2}{2!}x^2 - \frac{2^3}{4!}x^4 + \frac{2^5}{6!}x^6 - \frac{2^7}{8!}x^8 + \cdots$$

So, the series for $f(x) = \sin^2 x$ is

$$\sin^2 x = \frac{2}{2!}x^2 - \frac{2^3}{4!}x^4 + \frac{2^5}{6!}x^6 - \frac{2^7}{8!}x^8 + \cdots.$$

This series converges for $-\infty < x < \infty$.

As mentioned in the preceding section, power series can be used to obtain tables of values of transcendental functions. They are also useful for estimating the values of definite integrals for which antiderivatives cannot be found. The next example demonstrates this use.

EXAMPLE 10 Power Series Approximation of a Definite Integral

See LarsonCalculus.com for an interactive version of this type of example.

Use a power series to approximate

$$\int_0^1 e^{-x^2}\,dx$$

with an error of less than 0.01.

Solution Replacing x with $-x^2$ in the series for e^x produces the following.

$$e^{-x^2} = 1 - x^2 + \frac{x^4}{2!} - \frac{x^6}{3!} + \frac{x^8}{4!} - \cdots$$

$$\int_0^1 e^{-x^2}\,dx = \left[x - \frac{x^3}{3} + \frac{x^5}{5 \cdot 2!} - \frac{x^7}{7 \cdot 3!} + \frac{x^9}{9 \cdot 4!} - \cdots\right]_0^1$$

$$= 1 - \frac{1}{3} + \frac{1}{10} - \frac{1}{42} + \frac{1}{216} - \cdots$$

Summing the first four terms, you have

$$\int_0^1 e^{-x^2}\,dx \approx 0.74$$

which, by the Alternating Series Test, has an error of less than $\frac{1}{216} \approx 0.005$.

9.10 Exercises

See CalcChat.com for tutorial help and worked-out solutions to odd-numbered exercises.

Finding a Taylor Series In Exercises 1–12, use the definition of Taylor series to find the Taylor series, centered at c, for the function.

1. $f(x) = e^{2x}, \quad c = 0$
2. $f(x) = e^{-4x}, \quad c = 0$
3. $f(x) = \cos x, \quad c = \dfrac{\pi}{4}$
4. $f(x) = \sin x, \quad c = \dfrac{\pi}{4}$
5. $f(x) = \dfrac{1}{x}, \quad c = 1$
6. $f(x) = \dfrac{1}{1-x}, \quad c = 2$
7. $f(x) = \ln x, \quad c = 1$
8. $f(x) = e^x, \quad c = 1$
9. $f(x) = \sin 3x, \quad c = 0$
10. $f(x) = \ln(x^2 + 1), \quad c = 0$
11. $f(x) = \sec x, \quad c = 0$ (first three nonzero terms)
12. $f(x) = \tan x, \quad c = 0$ (first three nonzero terms)

Proof In Exercises 13–16, prove that the Maclaurin series for the function converges to the function for all x.

13. $f(x) = \cos x$
14. $f(x) = e^{-2x}$
15. $f(x) = \sinh x$
16. $f(x) = \cosh x$

Using a Binomial Series In Exercises 17–26, use the binomial series to find the Maclaurin series for the function.

17. $f(x) = \dfrac{1}{(1+x)^2}$
18. $f(x) = \dfrac{1}{(1+x)^4}$
19. $f(x) = \dfrac{1}{\sqrt{1-x}}$
20. $f(x) = \dfrac{1}{\sqrt{1-x^2}}$
21. $f(x) = \dfrac{1}{\sqrt{4+x^2}}$
22. $f(x) = \dfrac{1}{(2+x)^3}$
23. $f(x) = \sqrt{1+x}$
24. $f(x) = \sqrt[4]{1+x}$
25. $f(x) = \sqrt{1+x^2}$
26. $f(x) = \sqrt{1+x^3}$

Finding a Maclaurin Series In Exercises 27–40, find the Maclaurin series for the function. Use the table of power series for elementary functions on page 670.

27. $f(x) = e^{x^2/2}$
28. $g(x) = e^{-3x}$
29. $f(x) = \ln(1 + x)$
30. $f(x) = \ln(1 + x^2)$
31. $g(x) = \sin 3x$
32. $f(x) = \sin \pi x$
33. $f(x) = \cos 4x$
34. $f(x) = \cos \pi x$
35. $f(x) = \cos x^{3/2}$
36. $g(x) = 2 \sin x^3$
37. $f(x) = \frac{1}{2}(e^x - e^{-x}) = \sinh x$
38. $f(x) = e^x + e^{-x} = 2 \cosh x$
39. $f(x) = \cos^2 x$
40. $f(x) = \sinh^{-1} x = \ln\left(x + \sqrt{x^2 + 1}\right)$

    $\left(\textit{Hint: } \text{Integrate the series for } \dfrac{1}{\sqrt{x^2+1}}.\right)$

Finding a Maclaurin Series In Exercises 41–44, find the Maclaurin series for the function. (See Examples 7 and 8.)

41. $f(x) = x \sin x$
42. $h(x) = x \cos x$
43. $g(x) = \begin{cases} \dfrac{\sin x}{x}, & x \neq 0 \\ 1, & x = 0 \end{cases}$
44. $f(x) = \begin{cases} \dfrac{\arcsin x}{x}, & x \neq 0 \\ 1, & x = 0 \end{cases}$

Verifying a Formula In Exercises 45 and 46, use a power series and the fact that $i^2 = -1$ to verify the formula.

45. $g(x) = \dfrac{1}{2i}(e^{ix} - e^{-ix}) = \sin x$
46. $g(x) = \dfrac{1}{2}(e^{ix} + e^{-ix}) = \cos x$

Finding Terms of a Maclaurin Series In Exercises 47–52, find the first four nonzero terms of the Maclaurin series for the function by multiplying or dividing the appropriate power series. Use the table of power series for elementary functions on page 670. Use a graphing utility to graph the function and its corresponding polynomial approximation.

47. $f(x) = e^x \sin x$
48. $g(x) = e^x \cos x$
49. $h(x) = \cos x \ln(1 + x)$
50. $f(x) = e^x \ln(1 + x)$
51. $g(x) = \dfrac{\sin x}{1+x}$
52. $f(x) = \dfrac{e^x}{1+x}$

Finding a Maclaurin Series In Exercises 53 and 54, find a Maclaurin series for $f(x)$.

53. $f(x) = \displaystyle\int_0^x (e^{-t^2} - 1)\, dt$
54. $f(x) = \displaystyle\int_0^x \sqrt{1 + t^3}\, dt$

Verifying a Sum In Exercises 55–58, verify the sum. Then use a graphing utility to approximate the sum with an error of less than 0.0001.

55. $\displaystyle\sum_{n=1}^{\infty} (-1)^{n+1}\frac{1}{n} = \ln 2$
56. $\displaystyle\sum_{n=0}^{\infty} (-1)^n\left[\frac{1}{(2n+1)!}\right] = \sin 1$
57. $\displaystyle\sum_{n=0}^{\infty} \frac{2^n}{n!} = e^2$
58. $\displaystyle\sum_{n=1}^{\infty} (-1)^{n-1}\left(\frac{1}{n!}\right) = \frac{e-1}{e}$

Finding a Limit In Exercises 59–62, use the series representation of the function f to find $\lim\limits_{x \to 0} f(x)$ (if it exists).

59. $f(x) = \dfrac{1 - \cos x}{x}$
60. $f(x) = \dfrac{\sin x}{x}$
61. $f(x) = \dfrac{e^x - 1}{x}$
62. $f(x) = \dfrac{\ln(x+1)}{x}$

Approximating an Integral In Exercises 63–70, use a power series to approximate the value of the integral with an error of less than 0.0001. (In Exercises 65 and 67, assume that the integrand is defined as 1 when $x = 0$.)

63. $\int_0^1 e^{-x^3}\,dx$

64. $\int_0^{1/4} x\ln(x+1)\,dx$

65. $\int_0^1 \frac{\sin x}{x}\,dx$

66. $\int_0^1 \cos x^2\,dx$

67. $\int_0^{1/2} \frac{\arctan x}{x}\,dx$

68. $\int_0^{1/2} \arctan x^2\,dx$

69. $\int_{0.1}^{0.3} \sqrt{1+x^3}\,dx$

70. $\int_0^{0.2} \sqrt{1+x^2}\,dx$

Area In Exercises 71 and 72, use a power series to approximate the area of the region. Use a graphing utility to verify the result.

71. $\int_0^{\pi/2} \sqrt{x}\cos x\,dx$

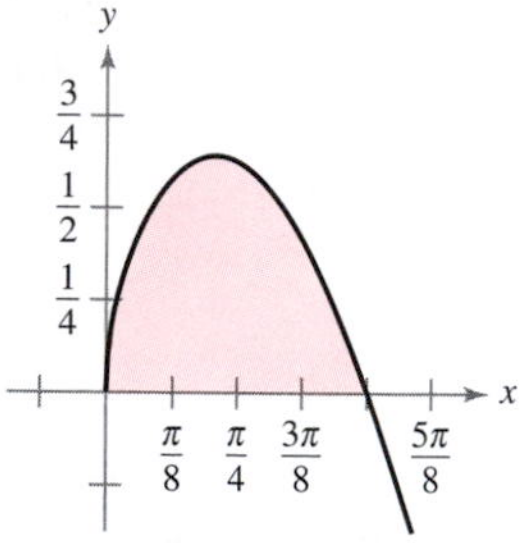

72. $\int_{0.5}^1 \cos\sqrt{x}\,dx$

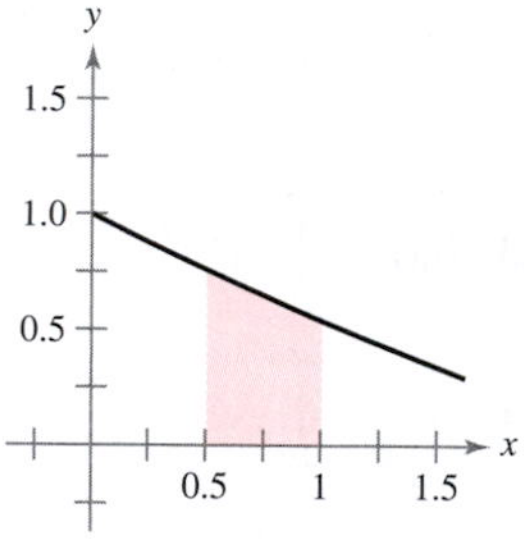

Probability In Exercises 73 and 74, approximate the normal probability with an error of less than 0.0001, where the probability is given by

$$P(a < x < b) = \frac{1}{\sqrt{2\pi}}\int_a^b e^{-x^2/2}\,dx.$$

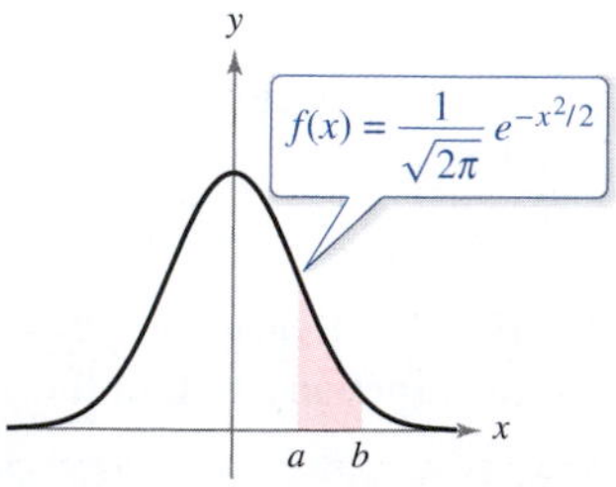

73. $P(0 < x < 1)$

74. $P(1 < x < 2)$

Finding a Taylor Polynomial Using Technology In Exercises 75–78, use a computer algebra system to find the fifth-degree Taylor polynomial, centered at c, for the function. Graph the function and the polynomial. Use the graph to determine the largest interval on which the polynomial is a reasonable approximation of the function.

75. $f(x) = x\cos 2x, \quad c = 0$

76. $f(x) = \sin\frac{x}{2}\ln(1+x), \quad c = 0$

77. $g(x) = \sqrt{x}\ln x, \quad c = 1$

78. $h(x) = \sqrt[3]{x}\arctan x, \quad c = 1$

WRITING ABOUT CONCEPTS

79. Taylor Series State the guidelines for finding a Taylor series.

80. Binomial Series Define the binomial series. What is its radius of convergence?

81. Finding a Series Explain how to use the series

$$g(x) = e^x = \sum_{n=0}^{\infty}\frac{x^n}{n!}$$

to find the series for each function. Do not find the series.

(a) $f(x) = e^{-x}$ (b) $f(x) = e^{3x}$ (c) $f(x) = xe^x$

82. HOW DO YOU SEE IT? Match the polynomial with its graph. [The graphs are labeled (i), (ii), (iii), and (iv).] Factor a common factor from each polynomial and identify the function approximated by the remaining Taylor polynomial.

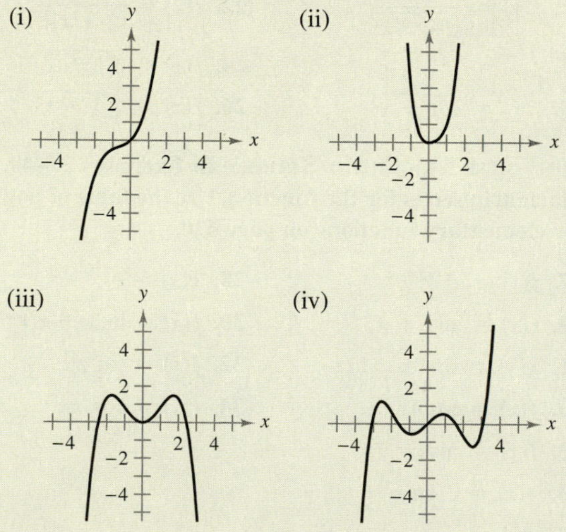

(a) $y = x^2 - \frac{x^4}{3!}$ (b) $y = x - \frac{x^3}{2!} + \frac{x^5}{4!}$

(c) $y = x + x^2 + \frac{x^3}{2!}$ (d) $y = x^2 - x^3 + x^4$

83. Projectile Motion A projectile fired from the ground follows the trajectory given by

$$y = \left(\tan\theta - \frac{g}{kv_0\cos\theta}\right)x - \frac{g}{k^2}\ln\left(1 - \frac{kx}{v_0\cos\theta}\right)$$

where v_0 is the initial speed, θ is the angle of projection, g is the acceleration due to gravity, and k is the drag factor caused by air resistance. Using the power series representation

$$\ln(1+x) = x - \frac{x^2}{2} + \frac{x^3}{3} - \frac{x^4}{4} + \cdots, \quad -1 < x < 1$$

verify that the trajectory can be rewritten as

$$y = (\tan\theta)x + \frac{gx^2}{2v_0^2\cos^2\theta} + \frac{kgx^3}{3v_0^3\cos^3\theta} + \frac{k^2gx^4}{4v_0^4\cos^4\theta} + \cdots.$$

84. Projectile Motion

Use the result of Exercise 83 to determine the series for the path of a projectile launched from ground level at an angle of $\theta = 60°$, with an initial speed of $v_0 = 64$ feet per second and a drag factor of $k = \frac{1}{16}$.

85. Investigation Consider the function f defined by

$$f(x) = \begin{cases} e^{-1/x^2}, & x \neq 0 \\ 0, & x = 0. \end{cases}$$

(a) Sketch a graph of the function.

(b) Use the alternative form of the definition of the derivative (Section 2.1) and L'Hôpital's Rule to show that $f'(0) = 0$. [By continuing this process, it can be shown that $f^{(n)}(0) = 0$ for $n > 1$.]

(c) Using the result in part (b), find the Maclaurin series for f. Does the series converge to f?

86. Investigation

(a) Find the power series centered at 0 for the function

$$f(x) = \frac{\ln(x^2+1)}{x^2}.$$

(b) Use a graphing utility to graph f and the eighth-degree Taylor polynomial $P_8(x)$ for f.

(c) Complete the table, where

$$F(x) = \int_0^x \frac{\ln(t^2+1)}{t^2}\,dt \quad \text{and} \quad G(x) = \int_0^x P_8(t)\,dt.$$

x	0.25	0.50	0.75	1.00	1.50	2.00
$F(x)$						
$G(x)$						

(d) Describe the relationship between the graphs of f and P_8 and the results given in the table in part (c).

87. Proof Prove that $\lim_{n\to\infty} \frac{x^n}{n!} = 0$ for any real x.

88. Finding a Maclaurin Series Find the Maclaurin series for

$$f(x) = \ln\frac{1+x}{1-x}$$

and determine its radius of convergence. Use the first four terms of the series to approximate ln 3.

Evaluating a Binomial Coefficient In Exercises 89–92, evaluate the binomial coefficient using the formula

$$\binom{k}{n} = \frac{k(k-1)(k-2)(k-3)\cdots(k-n+1)}{n!}$$

where k is a real number, n is a positive integer, and

$$\binom{k}{0} = 1.$$

89. $\binom{5}{3}$ **90.** $\binom{-2}{2}$

91. $\binom{0.5}{4}$ **92.** $\binom{-1/3}{5}$

93. Writing a Power Series Write the power series for $(1+x)^k$ in terms of binomial coefficients.

94. Proof Prove that e is irrational. [*Hint:* Assume that $e = p/q$ is rational (p and q are integers) and consider

$$e = 1 + 1 + \frac{1}{2!} + \cdots + \frac{1}{n!} + \cdots.]$$

95. Using Fibonacci Numbers Show that the Maclaurin series for the function

$$g(x) = \frac{x}{1-x-x^2}$$

is

$$\sum_{n=1}^{\infty} F_n x^n$$

where F_n is the nth Fibonacci number with $F_1 = F_2 = 1$ and $F_n = F_{n-2} + F_{n-1}$, for $n \geq 3$.

(*Hint:* Write

$$\frac{x}{1-x-x^2} = a_0 + a_1x + a_2x^2 + \cdots$$

and multiply each side of this equation by $1 - x - x^2$.)

PUTNAM EXAM CHALLENGE

96. Assume that $|f(x)| \leq 1$ and $|f''(x)| \leq 1$ for all x on an interval of length at least 2. Show that $|f'(x)| \leq 2$ on the interval.

Review Exercises

See CalcChat.com for tutorial help and worked-out solutions to odd-numbered exercises.

Listing the Terms of a Sequence In Exercises 1–4, write the first five terms of the sequence.

1. $a_n = 5^n$
2. $a_n = \dfrac{3^n}{n!}$
3. $a_n = \left(-\dfrac{1}{4}\right)^n$
4. $a_n = \dfrac{2n}{n+5}$

Matching In Exercises 5–8, match the sequence with its graph. [The graphs are labeled (a), (b), (c), and (d).]

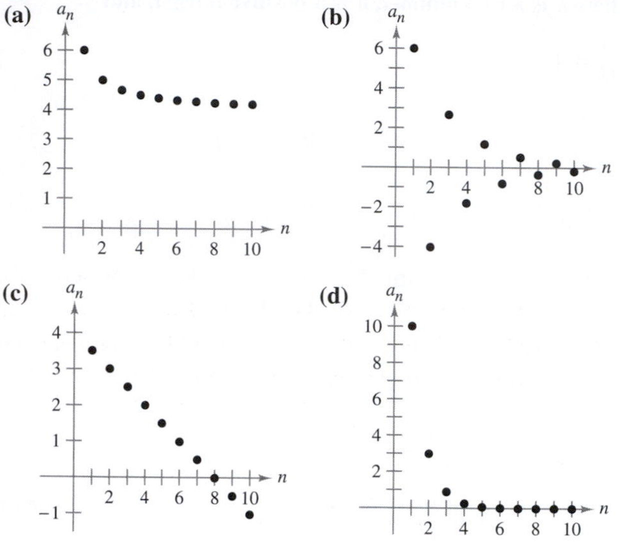

5. $a_n = 4 + \dfrac{2}{n}$
6. $a_n = 4 - \dfrac{1}{2}n$
7. $a_n = 10(0.3)^{n-1}$
8. $a_n = 6\left(-\frac{2}{3}\right)^{n-1}$

Finding the Limit of a Sequence In Exercises 9 and 10, use a graphing utility to graph the first 10 terms of the sequence. Use the graph to make an inference about the convergence or divergence of the sequence. Verify your inference analytically and, if the sequence converges, find its limit.

9. $a_n = \dfrac{5n+2}{n}$
10. $a_n = \sin\dfrac{n\pi}{2}$

Determining Convergence or Divergence In Exercises 11–18, determine the convergence or divergence of the sequence with the given nth term. If the sequence converges, find its limit.

11. $a_n = \left(\dfrac{2}{5}\right)^n + 5$
12. $a_n = 3 - \dfrac{2}{n^2 - 1}$
13. $a_n = \dfrac{n^3 + 1}{n^2}$
14. $a_n = \dfrac{1}{\sqrt{n}}$
15. $a_n = \dfrac{n}{n^2 + 1}$
16. $a_n = \dfrac{n}{\ln n}$
17. $a_n = \sqrt{n+1} - \sqrt{n}$
18. $a_n = \dfrac{\sin\sqrt{n}}{\sqrt{n}}$

Finding the *n*th Term of a Sequence In Exercises 19–22, write an expression for the nth term of the sequence. (There is more than one correct answer.)

19. 3, 8, 13, 18, 23, . . .
20. −5, −2, 3, 10, 19, . . .
21. $\dfrac{1}{2}, \dfrac{1}{3}, \dfrac{1}{7}, \dfrac{1}{25}, \dfrac{1}{121}, \ldots$
22. $\dfrac{1}{2}, \dfrac{2}{5}, \dfrac{3}{10}, \dfrac{4}{17}, \ldots$

23. **Compound Interest** A deposit of \$8000 is made in an account that earns 5% interest compounded quarterly. The balance in the account after n quarters is

$$A_n = 8000\left(1 + \frac{0.05}{4}\right)^n, \quad n = 1, 2, 3, \ldots.$$

(a) Compute the first eight terms of the sequence $\{A_n\}$.

(b) Find the balance in the account after 10 years by computing the 40th term of the sequence.

24. **Depreciation** A company buys a machine for \$175,000. During the next 5 years, the machine will depreciate at a rate of 30% per year. (That is, at the end of each year, the depreciated value will be 70% of what it was at the beginning of the year.)

(a) Find a formula for the nth term of the sequence that gives the value V of the machine t full years after it was purchased.

(b) Find the depreciated value of the machine at the end of 5 full years.

Finding Partial Sums In Exercises 25 and 26, find the sequence of partial sums S_1, S_2, S_3, S_4, and S_5.

25. $3 + \dfrac{3}{2} + 1 + \dfrac{3}{4} + \dfrac{3}{5} + \cdots$
26. $-\dfrac{1}{2} + \dfrac{1}{4} - \dfrac{1}{8} + \dfrac{1}{16} - \dfrac{1}{32} + \cdots$

Numerical, Graphical, and Analytic Analysis In Exercises 27–30, (a) use a graphing utility to find the indicated partial sum S_n and complete the table, and (b) use a graphing utility to graph the first 10 terms of the sequence of partial sums.

n	5	10	15	20	25
S_n					

27. $\displaystyle\sum_{n=1}^{\infty} \left(\frac{3}{2}\right)^{n-1}$
28. $\displaystyle\sum_{n=1}^{\infty} \frac{(-1)^{n+1}}{2n}$
29. $\displaystyle\sum_{n=1}^{\infty} \frac{(-1)^{n+1}}{(2n)!}$
30. $\displaystyle\sum_{n=1}^{\infty} \frac{1}{n(n+1)}$

Finding the Sum of a Convergent Series In Exercises 31–34, find the sum of the convergent series.

31. $\sum_{n=0}^{\infty} \left(\frac{2}{5}\right)^n$

32. $\sum_{n=0}^{\infty} \frac{3^{n+2}}{7^n}$

33. $\sum_{n=1}^{\infty} [(0.6)^n + (0.8)^n]$

34. $\sum_{n=0}^{\infty} \left[\left(\frac{2}{3}\right)^n - \frac{1}{(n+1)(n+2)}\right]$

Using a Geometric Series In Exercises 35 and 36, (a) write the repeating decimal as a geometric series, and (b) write its sum as the ratio of two integers.

35. $0.\overline{09}$

36. $0.\overline{64}$

Using Geometric Series or the *n*th-Term Test In Exercises 37–40, use geometric series or the *n*th-Term Test to determine the convergence or divergence of the series.

37. $\sum_{n=0}^{\infty} (1.67)^n$

38. $\sum_{n=0}^{\infty} (0.36)^n$

39. $\sum_{n=2}^{\infty} \frac{(-1)^n n}{\ln n}$

40. $\sum_{n=0}^{\infty} \frac{2n+1}{3n+2}$

41. **Distance** A ball is dropped from a height of 8 meters. Each time it drops h meters, it rebounds $0.7h$ meters. Find the total distance traveled by the ball.

42. **Compound Interest** A deposit of \$125 is made at the end of each month for 10 years in an account that pays 3.5% interest, compounded monthly. Determine the balance in the account at the end of 10 years. (*Hint:* Use the result of Section 9.2, Exercise 84.)

Using the Integral Test or a *p*-Series In Exercises 43–48, use the Integral Test or a *p*-series to determine the convergence or divergence of the series.

43. $\sum_{n=1}^{\infty} \frac{2}{6n+1}$

44. $\sum_{n=1}^{\infty} \frac{1}{\sqrt[4]{n^3}}$

45. $\sum_{n=1}^{\infty} \frac{1}{n^{5/2}}$

46. $\sum_{n=1}^{\infty} \frac{1}{5^n}$

47. $\sum_{n=1}^{\infty} \left(\frac{1}{n^2} - \frac{1}{n}\right)$

48. $\sum_{n=1}^{\infty} \frac{\ln n}{n^4}$

Using the Direct Comparison Test or the Limit Comparison Test In Exercises 49–54, use the Direct Comparison Test or the Limit Comparison Test to determine the convergence or divergence of the series.

49. $\sum_{n=2}^{\infty} \frac{1}{\sqrt[3]{n} - 1}$

50. $\sum_{n=1}^{\infty} \frac{n}{\sqrt{n^3 + 3n}}$

51. $\sum_{n=1}^{\infty} \frac{1}{\sqrt{n^3 + 2n}}$

52. $\sum_{n=1}^{\infty} \frac{n+1}{n(n+2)}$

53. $\sum_{n=1}^{\infty} \frac{1 \cdot 3 \cdot 5 \cdots (2n-1)}{2 \cdot 4 \cdot 6 \cdots (2n)}$

54. $\sum_{n=1}^{\infty} \frac{1}{3^n - 5}$

Using the Alternating Series Test In Exercises 55–60, use the Alternating Series Test, if applicable, to determine the convergence or divergence of the series.

55. $\sum_{n=1}^{\infty} \frac{(-1)^n}{n^5}$

56. $\sum_{n=1}^{\infty} \frac{(-1)^n (n+1)}{n^2+1}$

57. $\sum_{n=2}^{\infty} \frac{(-1)^n n}{n^2 - 3}$

58. $\sum_{n=1}^{\infty} \frac{(-1)^n \sqrt{n}}{n+1}$

59. $\sum_{n=4}^{\infty} \frac{(-1)^n n}{n-3}$

60. $\sum_{n=2}^{\infty} \frac{(-1)^n \ln n^3}{n}$

Using the Ratio Test or the Root Test In Exercises 61–66, use the Ratio Test or the Root Test to determine the convergence or divergence of the series.

61. $\sum_{n=1}^{\infty} \left(\frac{3n-1}{2n+5}\right)^n$

62. $\sum_{n=1}^{\infty} \left(\frac{4n}{7n-1}\right)^n$

63. $\sum_{n=1}^{\infty} \frac{n}{e^{n^2}}$

64. $\sum_{n=1}^{\infty} \frac{n!}{e^n}$

65. $\sum_{n=1}^{\infty} \frac{2^n}{n^3}$

66. $\sum_{n=1}^{\infty} \frac{1 \cdot 3 \cdot 5 \cdots (2n-1)}{2 \cdot 5 \cdot 8 \cdots (3n-1)}$

Numerical, Graphical, and Analytic Analysis In Exercises 67 and 68, (a) verify that the series converges, (b) use a graphing utility to find the indicated partial sum S_n and complete the table, (c) use a graphing utility to graph the first 10 terms of the sequence of partial sums, and (d) use the table to estimate the sum of the series.

n	5	10	15	20	25
S_n					

67. $\sum_{n=1}^{\infty} n\left(\frac{3}{5}\right)^n$

68. $\sum_{n=1}^{\infty} \frac{(-1)^{n-1} n}{n^3 + 5}$

Finding a Maclaurin Polynomial In Exercises 69 and 70, find the *n*th Maclaurin polynomial for the function.

69. $f(x) = e^{-2x}, \quad n = 3$

70. $f(x) = \cos \pi x, \quad n = 4$

Finding a Taylor Polynomial In Exercises 71 and 72, find the third-degree Taylor polynomial centered at *c*.

71. $f(x) = e^{-3x}, \quad c = 0$

72. $f(x) = \tan x, \quad c = -\frac{\pi}{4}$

Finding a Degree In Exercises 73 and 74, determine the degree of the Maclaurin polynomial required for the error in the approximation of the function at the indicated value of *x* to be less than 0.001.

73. $\cos(0.75)$

74. $e^{-0.25}$

Finding the Interval of Convergence In Exercises 75–80, find the interval of convergence of the power series. (Be sure to include a check for convergence at the endpoints of the interval.)

75. $\sum_{n=0}^{\infty} \left(\frac{x}{10}\right)^n$

76. $\sum_{n=0}^{\infty} (5x)^n$

77. $\sum_{n=0}^{\infty} \frac{(-1)^n (x-2)^n}{(n+1)^2}$

78. $\sum_{n=1}^{\infty} \frac{3^n (x-2)^n}{n}$

79. $\sum_{n=0}^{\infty} n!(x-2)^n$

80. $\sum_{n=0}^{\infty} \frac{(x-2)^n}{2^n}$

Finding Intervals of Convergence In Exercises 81 and 82, find the intervals of convergence of (a) $f(x)$, (b) $f'(x)$, (c) $f''(x)$, and (d) $\int f(x)\,dx$. Include a check for convergence at the endpoints of the interval.

81. $f(x) = \sum_{n=0}^{\infty} \left(\frac{x}{5}\right)^n$

82. $f(x) = \sum_{n=1}^{\infty} \frac{(-1)^{n+1}(x-4)^n}{n}$

Differential Equation In Exercises 83 and 84, show that the function represented by the power series is a solution of the differential equation.

83. $y = \sum_{n=0}^{\infty} (-1)^n \frac{x^{2n}}{4^n (n!)^2}$

$x^2 y'' + xy' + x^2 y = 0$

84. $y = \sum_{n=0}^{\infty} \frac{(-3)^n x^{2n}}{2^n\, n!}$

$y'' + 3xy' + 3y = 0$

Finding a Geometric Power Series In Exercises 85 and 86, find a geometric power series, centered at 0, for the function.

85. $g(x) = \frac{2}{3-x}$

86. $h(x) = \frac{3}{2+x}$

Finding a Power Series In Exercises 87 and 88, find a power series for the function, centered at c, and determine the interval of convergence.

87. $f(x) = \frac{6}{4-x}, \quad c = 1$

88. $f(x) = \frac{1}{3-2x}, \quad c = 0$

Finding the Sum of a Series In Exercises 89–94, find the sum of the convergent series by using a well-known function. Identify the function and explain how you obtained the sum.

89. $\sum_{n=1}^{\infty} (-1)^{n+1} \frac{1}{4^n\, n}$

90. $\sum_{n=1}^{\infty} (-1)^{n+1} \frac{1}{5^n\, n}$

91. $\sum_{n=0}^{\infty} \frac{1}{2^n\, n!}$

92. $\sum_{n=0}^{\infty} \frac{2^n}{3^n\, n!}$

93. $\sum_{n=0}^{\infty} (-1)^n \frac{2^{2n}}{3^{2n}(2n)!}$

94. $\sum_{n=0}^{\infty} (-1)^n \frac{1}{3^{2n+1}(2n+1)!}$

Finding a Taylor Series In Exercises 95–102, use the definition of Taylor series to find the Taylor series, centered at c, for the function.

95. $f(x) = \sin x, \quad c = \frac{3\pi}{4}$

96. $f(x) = \cos x, \quad c = -\frac{\pi}{4}$

97. $f(x) = 3^x, \quad c = 0$

98. $f(x) = \csc x, \quad c = \frac{\pi}{2}$ (first three terms)

99. $f(x) = \frac{1}{x}, \quad c = -1$

100. $f(x) = \sqrt{x}, \quad c = 4$

101. $g(x) = \sqrt[5]{1+x}, \quad c = 0$

102. $h(x) = \frac{1}{(1+x)^3}, \quad c = 0$

103. Forming Maclaurin Series Determine the first four terms of the Maclaurin series for e^{2x}

(a) by using the definition of the Maclaurin series and the formula for the coefficient of the nth term, $a_n = f^{(n)}(0)/n!$.

(b) by replacing x by $2x$ in the series for e^x.

(c) by multiplying the series for e^x by itself, because $e^{2x} = e^x \cdot e^x$.

104. Forming Maclaurin Series Determine the first four terms of the Maclaurin series for $\sin 2x$

(a) by using the definition of the Maclaurin series and the formula for the coefficient of the nth term, $a_n = f^{(n)}(0)/n!$.

(b) by replacing x by $2x$ in the series for $\sin 2x$.

(c) by multiplying 2 by the series for $\sin x$ by the series for $\cos x$, because $\sin 2x = 2 \sin x \cos x$.

Finding a Maclaurin Series In Exercises 105–108, find the Maclaurin series for the function. Use the table of power series for elementary functions on page 670.

105. $f(x) = e^{6x}$

106. $f(x) = \ln(x-1)$

107. $f(x) = \sin 2x$

108. $f(x) = \cos 3x$

Finding a Limit In Exercises 109 and 110, use the series representation of the function f to find $\lim_{x \to 0} f(x)$ (if it exists).

109. $f(x) = \frac{\arctan x}{\sqrt{x}}$

110. $f(x) = \frac{\arcsin x}{x}$

P.S. Problem Solving

1. Cantor Set The **Cantor set** (Georg Cantor, 1845–1918) is a subset of the unit interval $[0, 1]$. To construct the Cantor set, first remove the middle third $\left(\frac{1}{3}, \frac{2}{3}\right)$ of the interval, leaving two line segments. For the second step, remove the middle third of each of the two remaining segments, leaving four line segments. Continue this procedure indefinitely, as shown in the figure. The Cantor set consists of all numbers in the unit interval $[0, 1]$ that still remain.

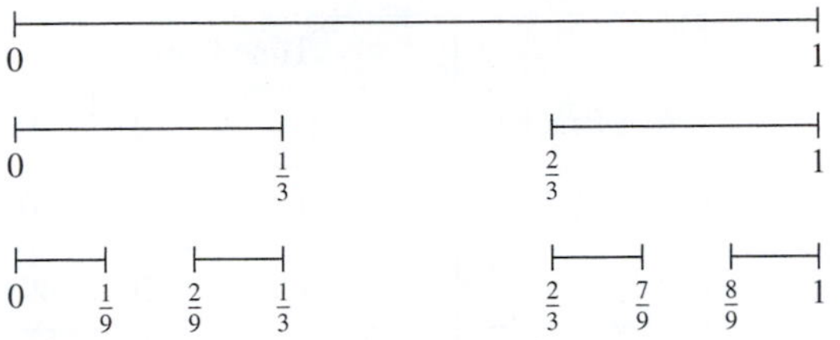

(a) Find the total length of all the line segments that are removed.

(b) Write down three numbers that are in the Cantor set.

(c) Let C_n denote the total length of the remaining line segments after n steps. Find $\lim_{n\to\infty} C_n$.

2. Using Sequences

(a) Given that $\lim_{x\to\infty} a_{2n} = L$ and $\lim_{x\to\infty} a_{2n+1} = L$, show that $\{a_n\}$ is convergent and $\lim_{x\to\infty} a_n = L$.

(b) Let $a_1 = 1$ and $a_{n+1} = 1 + \dfrac{1}{1 + a_n}$. Write out the first eight terms of $\{a_n\}$. Use part (a) to show that $\lim_{x\to\infty} a_n = \sqrt{2}$. This gives the **continued fraction expansion**

$$\sqrt{2} = 1 + \cfrac{1}{2 + \cfrac{1}{2 + \cdots}}.$$

3. Using a Series It can be shown that

$$\sum_{n=1}^{\infty} \frac{1}{n^2} = \frac{\pi^2}{6} \text{ [see Section 9.3, page 608].}$$

Use this fact to show that $\displaystyle\sum_{n=1}^{\infty} \frac{1}{(2n-1)^2} = \frac{\pi^2}{8}$.

4. Finding a Limit Let T be an equilateral triangle with sides of length 1. Let a_n be the number of circles that can be packed tightly in n rows inside the triangle. For example, $a_1 = 1$, $a_2 = 3$, and $a_3 = 6$, as shown in the figure. Let A_n be the combined area of the a_n circles. Find $\lim_{n\to\infty} A_n$.

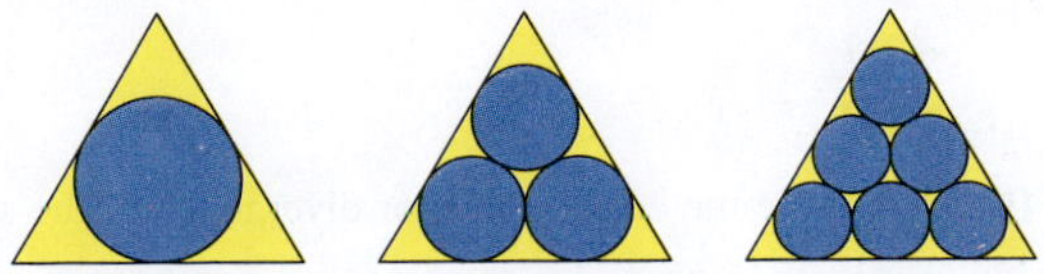

5. Using Center of Gravity Identical blocks of unit length are stacked on top of each other at the edge of a table. The center of gravity of the top block must lie over the block below it, the center of gravity of the top two blocks must lie over the block below them, and so on (see figure).

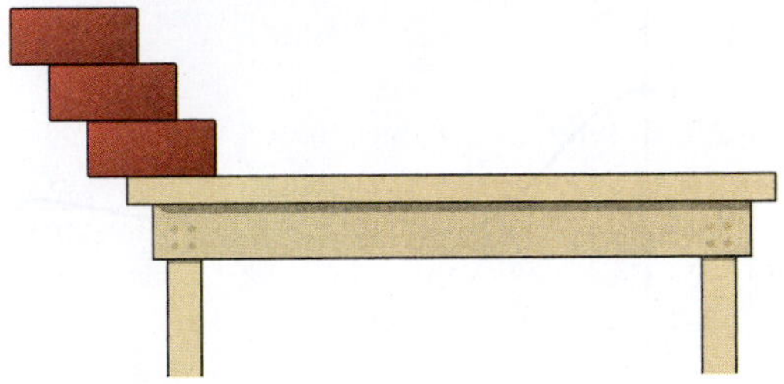

(a) When there are three blocks, show that it is possible to stack them so that the left edge of the top block extends $\frac{11}{12}$ unit beyond the edge of the table.

(b) Is it possible to stack the blocks so that the right edge of the top block extends beyond the edge of the table?

(c) How far beyond the table can the blocks be stacked?

6. Using Power Series

(a) Consider the power series

$$\sum_{n=0}^{\infty} a_n x^n = 1 + 2x + 3x^2 + x^3 + 2x^4 + 3x^5 + x^6 + \cdots$$

in which the coefficients $a_n = 1, 2, 3, 1, 2, 3, 1, \ldots$ are periodic of period $p = 3$. Find the radius of convergence and the sum of this power series.

(b) Consider a power series

$$\sum_{n=0}^{\infty} a_n x^n$$

in which the coefficients are periodic, $(a_{n+p} = a_p)$, and $a_n > 0$. Find the radius of convergence and the sum of this power series.

7. Finding Sums of Series

(a) Find a power series for the function

$$f(x) = xe^x$$

centered at 0. Use this representation to find the sum of the infinite series

$$\sum_{n=1}^{\infty} \frac{1}{n!(n+2)}.$$

(b) Differentiate the power series for $f(x) = xe^x$. Use the result to find the sum of the infinite series

$$\sum_{n=0}^{\infty} \frac{n+1}{n!}.$$

8. Using the Alternating Series Test The graph of the function

$$f(x) = \begin{cases} 1, & x = 0 \\ \dfrac{\sin x}{x}, & x > 0 \end{cases}$$

is shown below. Use the Alternating Series Test to show that the improper integral $\int_1^{\infty} f(x)\, dx$ converges.

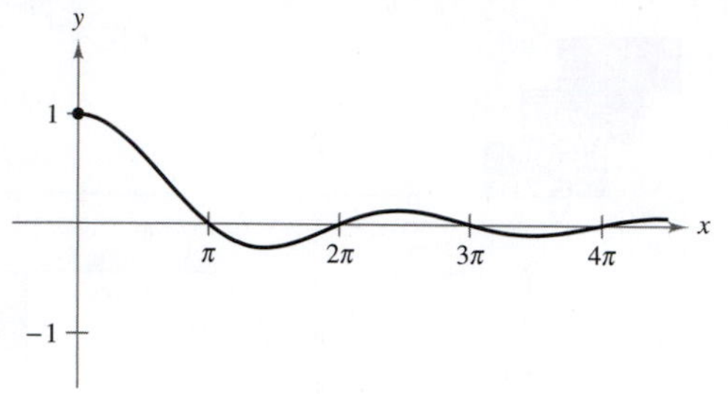

9. Conditional and Absolute Convergence For what values of the positive constants a and b does the following series converge absolutely? For what values does it converge conditionally?

$$a - \frac{b}{2} + \frac{a}{3} - \frac{b}{4} + \frac{a}{5} - \frac{b}{6} + \frac{a}{7} - \frac{b}{8} + \cdots$$

10. Proof

(a) Consider the following sequence of numbers defined recursively.

$$a_1 = 3$$
$$a_2 = \sqrt{3}$$
$$a_3 = \sqrt{3 + \sqrt{3}}$$
$$\vdots$$
$$a_{n+1} = \sqrt{3 + a_n}$$

Write the decimal approximations for the first six terms of this sequence. Prove that the sequence converges, and find its limit.

(b) Consider the following sequence defined recursively by $a_1 = \sqrt{a}$ and $a_{n+1} = \sqrt{a + a_n}$, where $a > 2$.

$$\sqrt{a},\quad \sqrt{a + \sqrt{a}},\quad \sqrt{a + \sqrt{a + \sqrt{a}}}, \ldots$$

Prove that this sequence converges, and find its limit.

11. Proof Let $\{a_n\}$ be a sequence of positive numbers satisfying $\lim_{n\to\infty} (a_n)^{1/n} = L < \frac{1}{r}$, $r > 0$. Prove that the series $\sum_{n=1}^{\infty} a_n r^n$ converges.

12. Using a Series Consider the infinite series $\sum_{n=1}^{\infty} \frac{1}{2^{n+(-1)^n}}$.

(a) Find the first five terms of the sequence of partial sums.

(b) Show that the Ratio Test is inconclusive for this series.

(c) Use the Root Test to test for the convergence or divergence of this series.

13. Deriving Identities Derive each identity using the appropriate geometric series.

(a) $\frac{1}{0.99} = 1.01010101 \ldots$

(b) $\frac{1}{0.98} = 1.0204081632 \ldots$

14. Population Consider an idealized population with the characteristic that each member of the population produces one offspring at the end of every time period. Each member has a life span of three time periods and the population begins with 10 newborn members. The following table shows the population during the first five time periods.

	Time Period				
Age Bracket	1	2	3	4	5
0–1	10	10	20	40	70
1–2		10	10	20	40
2–3			10	10	20
Total	10	20	40	70	130

The sequence for the total population has the property that $S_n = S_{n-1} + S_{n-2} + S_{n-3}$, $n > 3$. Find the total population during each of the next five time periods.

15. Spheres Imagine you are stacking an infinite number of spheres of decreasing radii on top of each other, as shown in the figure. The radii of the spheres are 1 meter, $1/\sqrt{2}$ meter, $1/\sqrt{3}$ meter, and so on. The spheres are made of a material that weighs 1 newton per cubic meter.

(a) How high is this infinite stack of spheres?

(b) What is the total surface area of all the spheres in the stack?

(c) Show that the weight of the stack is finite.

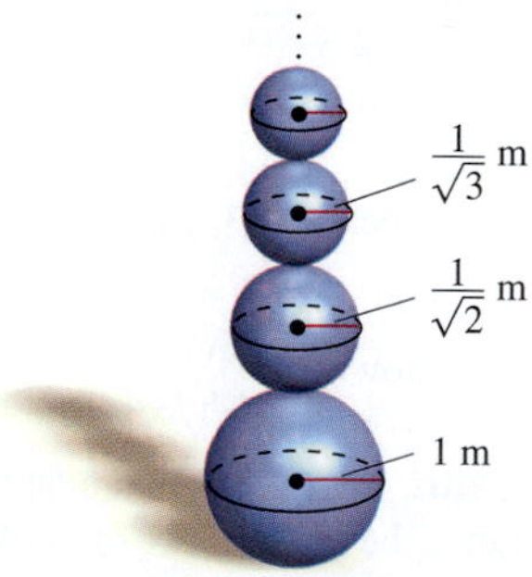

16. Determining Convergence or Divergence

(a) Determine the convergence or divergence of the series

$$\sum_{n=1}^{\infty} \frac{1}{2n}.$$

(b) Determine the convergence or divergence of the series

$$\sum_{n=1}^{\infty} \left(\sin \frac{1}{2n} - \sin \frac{1}{2n+1}\right).$$

AP* Review Questions for Chapter 9

1. (BC only, no calculator)

For the function f defined by the power series

$$f(x) = \sum_{n=0}^{\infty} \frac{(-1)^n (x)^{2n}}{n!} = 1 - x^2 + \frac{x^4}{2!} - \frac{x^6}{3!} + \frac{x^8}{4!} - \frac{x^{10}}{5!} + \dots$$

for all real numbers x:

(a) Find $f'(0)$ and $f''(0)$. Determine whether f has a local maximum, a local minimum, or neither at $x = 0$. Give a reason for your answer.

(b) Explain why $\frac{1}{2!} - \frac{1}{3!}$ approximates $f(1)$ with error less than $\frac{1}{10}$.

(c) Show that $y = f(x)$ is a solution to the differential equation $y' + 2xy = 0$.

2. (BC only, no calculator)

The function f is defined by the power series

$$f(x) = (x-1) - \frac{(x-1)^2}{2} + \frac{(x-1)^3}{3} - \frac{(x-1)^4}{4} + \cdots = \sum_{n=1}^{\infty} \frac{(-1)^{n-1}(x-1)^n}{n}$$

for all real numbers x for which the series converges.

(a) Determine the interval of convergence for f. Justify your answer.

(b) Given $g(x) = f'(x)$, find the first three terms and the general term of $g(x)$.

(c) Find a rational function which is identical to g over its interval of convergence.

(d) Let h be the function defined by $h(x) = f(x^3 + 1)$. Find $h'(x)$.

3. (BC only, no calculator)

Given $f(x) = \cos(x^2)$.

(a) Find the first four terms and the general term of the Maclaurin series for $f(x)$.

(b) Find the radius of convergence for this series.

(c) Use the first three terms of the Maclaurin series for $f(x)$ to approximate $\cos(1)$. Show that the approximation is accurate to within $\frac{1}{500}$.

4. (BC only, calculator)

x	$g(x)$	$g'(x)$	$g''(x)$	$g'''(x)$	$g^{(4)}(x)$
2	17	22	20	14	5
3	50	$\frac{160}{3}$	$\frac{141}{4}$	21	$\frac{151}{8}$
4	232	$\frac{604}{3}$	$\frac{2703}{16}$	152	$\frac{1123}{8}$

Let $g(x)$ be a function having derivatives of all orders for all x. Selected values of g and its first four derivatives are listed in the table above. The function g and its first four derivatives are increasing on the interval $2 \le x \le 4$.

(a) Write the first-degree Taylor polynomial for g about $x = 3$ and use it to approximate $g(3.1)$. Is this approximation greater than or less than $g(3.1)$? Explain.

(b) Write the third-degree Taylor polynomial $P_3(x)$ for g about $x = 3$ and use it to approximate $g(3.1)$.

(c) Show that $|g(3.1) - P_3(3.1)| < 6 \times 10^{-4}$.

5. (BC only, no calculator)

Let f be the function given by $f(x) = xe^{-x}$.

(a) Write the first four nonzero terms and the general term of the Taylor series for f about $x = 0$.

(b) Find $\lim_{x \to 0} \left(\frac{f(x) - x + x^2}{x^3} \right)$.

(c) Write the first four non-zero terms and the general term of the Taylor series for $g(x) = \int_0^x te^{-t}\,dt$ about $x = 0$. Use the first three terms to approximate $g(1/5)$.

(d) Evaluated at $x = 1/5$, the Taylor series for g is an alternating decreasing series with individual terms that decrease in absolute value to zero. Show that your approximation in (c) must differ from $g(1/5)$ by less than $1/90{,}000$.

6. (BC only, no calculator)

x	2	2.1	2.2	2.3	2.4
$g'(x)$	4	5	7	8	10.5

The function g is thrice-differentiable with $g(2) = 3$ and $g''(2) = 20$. On the interval $[2, 2.4]$, $g'(x)$ is an increasing function of x. Select values of g' are given in the table above.

(a) Write an equation for a line tangent to $g(x)$ at $x = 2$. Use this to approximate $g(2.4)$. Is this an over-approximation or an under-approximation?

(b) Use a midpoint Riemann sum with two subintervals of equal length to approximate $\int_2^{2.4} g'(x)dx$. Use your approximation for $\int_2^{2.4} g'(x)dx$ to estimate $g(2.4)$.

(c) Use Euler's method with two steps of equal size to estimate $g(2.4)$.

(d) Write the second-degree Taylor polynomial for g about $x = 2$. Use the Taylor polynomial to approximate $g(2.4)$. Given that, on the interval $2 \le x \le 2.4$, $g'''(x)$ does not exceed 6, show that your approximation is correct to within one tenth.

7. (BC only, no calculator)

Which of the following series *diverge*?

I. $\sum_{n=1}^{\infty} \frac{3}{\sqrt[3]{n}}$

II. $\sum_{n=1}^{\infty} \left(\frac{e}{\sin(2)}\right)^n$

III. $\sum_{n=1}^{\infty} \frac{5}{(n+1)!}$

(A) I only

(B) II only

(C) III only

(D) I and II

(E) I, II, and III

8. (BC only, no calculator)

The Maclaurin series for $\cos x$ is $\sum_{n=0}^{\infty} (-1)^n \frac{x^{2n}}{(2n)!}$. Which of the following is a power series expansion for $x^2 \cos(x^2)$?

(A) $1 - \frac{x^4}{2!} + \frac{x^8}{4!} - \frac{x^{12}}{6!} + \frac{x^{16}}{8!} - \cdots$

(B) $x^2 - \frac{x^4}{2!} + \frac{x^6}{4!} - \frac{x^8}{6!} + \frac{x^{10}}{8!} - \cdots$

(C) $x^2 - \frac{x^6}{2!} + \frac{x^{10}}{4!} - \frac{x^{14}}{6!} + \frac{x^{18}}{8!} - \cdots$

(D) $2x - \frac{4x^3}{2!} + \frac{6x^5}{4!} - \frac{8x^7}{6!} + \frac{10x^9}{8!} - \cdots$

(E) $1 - \frac{x^4}{2!} + \frac{x^6}{4!} - \frac{x^8}{6!} + \frac{x^{10}}{8!} - \cdots$

9. (BC only, no calculator)

What is the sum of the series

$$1 + \ln 3 + \frac{(\ln 3)^2}{2!} + \cdots + \frac{(\ln 3)^n}{n!} + \cdots ?$$

(A) $\ln 3$

(B) e^3

(C) 3

(D) $e^{1+\ln 3}$

(E) The series diverges.

10. (BC only, no calculator)

What are all values of p for which the series $\sum_{n=2}^{\infty} \frac{5(n+3)}{(n-1)^p}$ converges?

(A) $p > 1$

(B) $p < 1$

(C) $p > 0$

(D) $p > 2$

(E) The series diverges for all values of p.

10 Conics, Parametric Equations, and Polar Coordinates

Antenna Radiation *(Exercise 47, p. 732)*

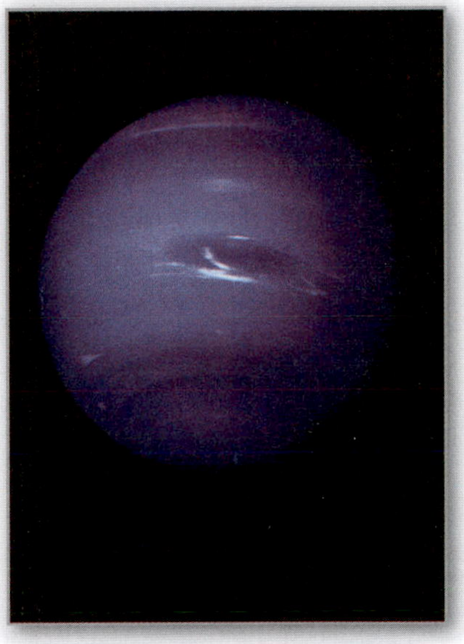

Planetary Motion *(Exercise 67, p. 741)*

Anamorphic Art *(Section Project, p. 724)*

Halley's Comet *(Exercise 77, p. 694)*

Architecture *(Exercise 71, p. 694)*

10.1 Conics and Calculus

- Understand the definition of a conic section.
- Analyze and write equations of parabolas using properties of parabolas.
- Analyze and write equations of ellipses using properties of ellipses.
- Analyze and write equations of hyperbolas using properties of hyperbolas.

Conic Sections

Each **conic section** (or simply **conic**) can be described as the intersection of a plane and a double-napped cone. Notice in Figure 10.1 that for the four basic conics, the intersecting plane does not pass through the vertex of the cone. When the plane passes through the vertex, the resulting figure is a **degenerate conic,** as shown in Figure 10.2.

Circle Parabola Ellipse Hyperbola

Conic sections

Figure 10.1

Point Line Two intersecting lines

Degenerate conics

Figure 10.2

There are several ways to study conics. You could begin as the Greeks did, by defining the conics in terms of the intersections of planes and cones, or you could define them algebraically in terms of the general second-degree equation

$$Ax^2 + Bxy + Cy^2 + Dx + Ey + F = 0. \qquad \text{General second-degree equation}$$

However, a third approach, in which each of the conics is defined as a **locus** (collection) of points satisfying a certain geometric property, works best. For example, a circle can be defined as the collection of all points (x, y) that are equidistant from a fixed point (h, k). This locus definition easily produces the standard equation of a circle

$$(x - h)^2 + (y - k)^2 = r^2. \qquad \text{Standard equation of a circle}$$

For information about rotating second-degree equations in two variables, see Appendix D.

HYPATIA (370–415 A.D.)

The Greeks discovered conic sections sometime between 600 and 300 B.C. By the beginning of the Alexandrian period, enough was known about conics for Apollonius (262–190 B.C.) to produce an eight-volume work on the subject. Later, toward the end of the Alexandrian period, Hypatia wrote a textbook entitled *On the Conics of Apollonius*. Her death marked the end of major mathematical discoveries in Europe for several hundred years.

The early Greeks were largely concerned with the geometric properties of conics. It was not until 1900 years later, in the early seventeenth century, that the broader applicability of conics became apparent. Conics then played a prominent role in the development of calculus.

See LarsonCalculus.com to read more of this biography.

FOR FURTHER INFORMATION To learn more about the mathematical activities of Hypatia, see the article "Hypatia and Her Mathematics" by Michael A. B. Deakin in *The American Mathematical Monthly.* To view this article, go to *MathArticles.com.*

Bettmann/Corbis

Parabolas

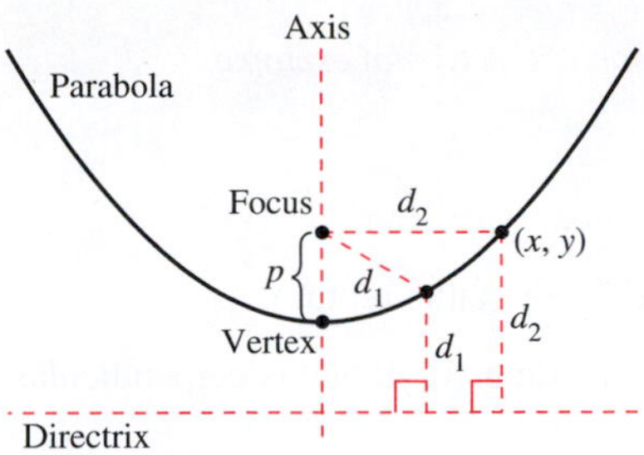

Figure 10.3

A **parabola** is the set of all points (x, y) that are equidistant from a fixed line, the **directrix,** and a fixed point, the **focus,** not on the line. The midpoint between the focus and the directrix is the **vertex,** and the line passing through the focus and the vertex is the **axis** of the parabola. Note in Figure 10.3 that a parabola is symmetric with respect to its axis.

> **THEOREM 10.1 Standard Equation of a Parabola**
>
> The **standard form** of the equation of a parabola with vertex (h, k) and directrix $y = k - p$ is
>
> $$(x - h)^2 = 4p(y - k).$$ Vertical axis
>
> For directrix $x = h - p$, the equation is
>
> $$(y - k)^2 = 4p(x - h).$$ Horizontal axis
>
> The focus lies on the axis p units (*directed distance*) from the vertex. The coordinates of the focus are as follows.
>
> $(h, k + p)$ Vertical axis
>
> $(h + p, k)$ Horizontal axis

EXAMPLE 1 Finding the Focus of a Parabola

Find the focus of the parabola

$$y = \frac{1}{2} - x - \frac{1}{2}x^2.$$

Solution To find the focus, convert to standard form by completing the square.

$y = \frac{1}{2} - x - \frac{1}{2}x^2$ Write original equation.

$2y = 1 - 2x - x^2$ Multiply each side by 2.

$2y = 1 - (x^2 + 2x)$ Group terms.

$2y = 2 - (x^2 + 2x + 1)$ Add and subtract 1 on right side.

$x^2 + 2x + 1 = -2y + 2$

$(x + 1)^2 = -2(y - 1)$ Write in standard form.

Comparing this equation with

$$(x - h)^2 = 4p(y - k)$$

you can conclude that

$$h = -1, \quad k = 1, \quad \text{and} \quad p = -\frac{1}{2}.$$

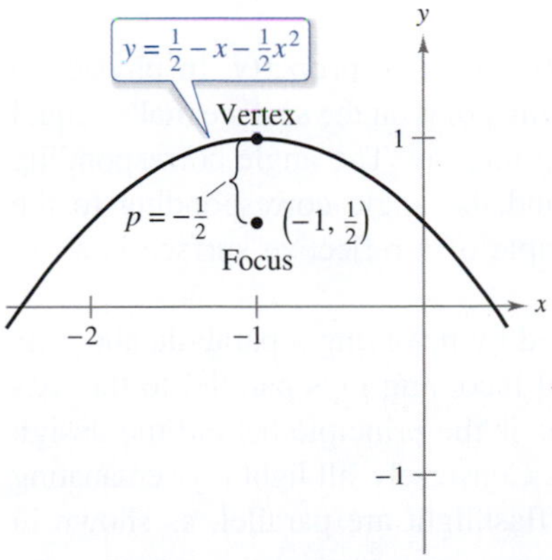

Parabola with a vertical axis, $p < 0$

Figure 10.4

Because p is negative, the parabola opens downward, as shown in Figure 10.4. So, the focus of the parabola is p units from the vertex, or

$$(h, k + p) = \left(-1, \frac{1}{2}\right).$$ Focus

A line segment that passes through the focus of a parabola and has endpoints on the parabola is called a **focal chord.** The specific focal chord perpendicular to the axis of the parabola is the **latus rectum.** The next example shows how to determine the length of the latus rectum and the length of the corresponding intercepted arc.

EXAMPLE 2 Focal Chord Length and Arc Length

See LarsonCalculus.com for an interactive version of this type of example.

Find the length of the latus rectum of the parabola

$$x^2 = 4py.$$

Then find the length of the parabolic arc intercepted by the latus rectum.

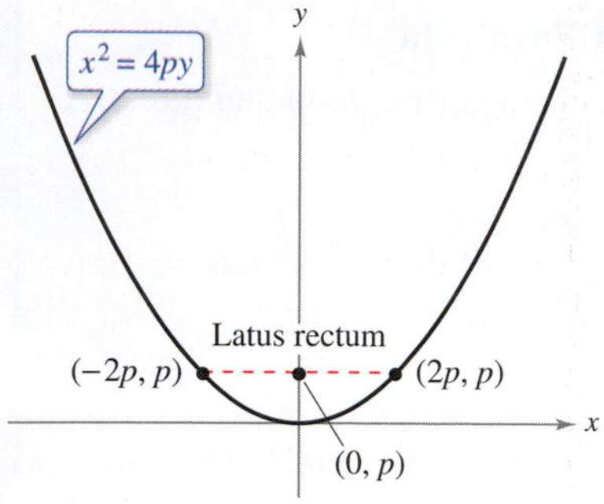

Length of latus rectum: $4p$
Figure 10.5

Solution Because the latus rectum passes through the focus $(0, p)$ and is perpendicular to the y-axis, the coordinates of its endpoints are

$$(-x, p) \quad \text{and} \quad (x, p).$$

Substituting p for y in the equation of the parabola produces

$$x^2 = 4p(p) \quad \Longrightarrow \quad x = \pm 2p.$$

So, the endpoints of the latus rectum are $(-2p, p)$ and $(2p, p)$, and you can conclude that its length is $4p$, as shown in Figure 10.5. In contrast, the length of the intercepted arc is

$$s = \int_{-2p}^{2p} \sqrt{1 + (y')^2}\, dx \qquad \text{Use arc length formula.}$$

$$= 2\int_0^{2p} \sqrt{1 + \left(\frac{x}{2p}\right)^2}\, dx \qquad y = \frac{x^2}{4p} \Longrightarrow y' = \frac{x}{2p}$$

$$= \frac{1}{p}\int_0^{2p} \sqrt{4p^2 + x^2}\, dx \qquad \text{Simplify.}$$

$$= \frac{1}{2p}\left[x\sqrt{4p^2 + x^2} + 4p^2 \ln\left|x + \sqrt{4p^2 + x^2}\right|\right]_0^{2p} \qquad \text{Theorem 8.2}$$

$$= \frac{1}{2p}\left[2p\sqrt{8p^2} + 4p^2 \ln\left(2p + \sqrt{8p^2}\right) - 4p^2 \ln(2p)\right]$$

$$= 2p\left[\sqrt{2} + \ln\left(1 + \sqrt{2}\right)\right]$$

$$\approx 4.59p.$$

One widely used property of a parabola is its reflective property. In physics, a surface is called **reflective** when the tangent line at any point on the surface makes equal angles with an incoming ray and the resulting outgoing ray. The angle corresponding to the incoming ray is the **angle of incidence,** and the angle corresponding to the outgoing ray is the **angle of reflection.** One example of a reflective surface is a flat mirror.

Another type of reflective surface is that formed by revolving a parabola about its axis. The resulting surface has the property that all incoming rays parallel to the axis are directed through the focus of the parabola. This is the principle behind the design of the parabolic mirrors used in reflecting telescopes. Conversely, all light rays emanating from the focus of a parabolic reflector used in a flashlight are parallel, as shown in Figure 10.6.

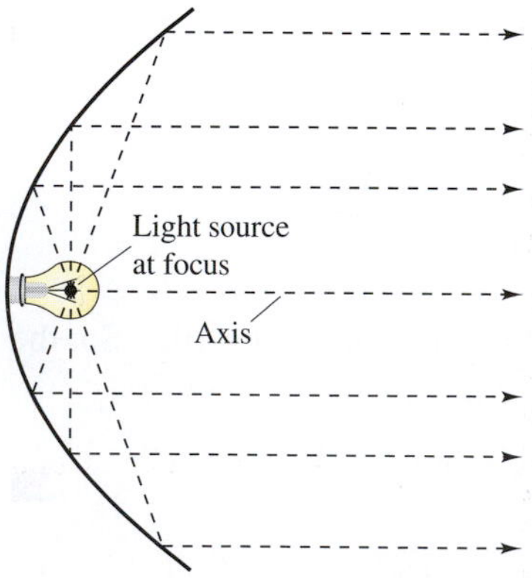

Parabolic reflector: light is reflected in parallel rays.
Figure 10.6

THEOREM 10.2 Reflective Property of a Parabola

Let P be a point on a parabola. The tangent line to the parabola at point P makes equal angles with the following two lines.

1. The line passing through P and the focus
2. The line passing through P parallel to the axis of the parabola

NICOLAUS COPERNICUS (1473–1543)

Copernicus began to study planetary motion when he was asked to revise the calendar. At that time, the exact length of the year could not be accurately predicted using the theory that Earth was the center of the universe.
See LarsonCalculus.com to read more of this biography.

Ellipses

More than a thousand years after the close of the Alexandrian period of Greek mathematics, Western civilization finally began a Renaissance of mathematical and scientific discovery. One of the principal figures in this rebirth was the Polish astronomer Nicolaus Copernicus. In his work *On the Revolutions of the Heavenly Spheres,* Copernicus claimed that all of the planets, including Earth, revolved about the sun in circular orbits. Although some of Copernicus's claims were invalid, the controversy set off by his heliocentric theory motivated astronomers to search for a mathematical model to explain the observed movements of the sun and planets. The first to find an accurate model was the German astronomer Johannes Kepler (1571–1630). Kepler discovered that the planets move about the sun in elliptical orbits, with the sun not as the center but as a focal point of the orbit.

The use of ellipses to explain the movements of the planets is only one of many practical and aesthetic uses. As with parabolas, you will begin your study of this second type of conic by defining it as a locus of points. Now, however, *two* focal points are used rather than one.

An **ellipse** is the set of all points (x, y) the sum of whose distances from two distinct fixed points called **foci** is constant. (See Figure 10.7.) The line through the foci intersects the ellipse at two points, called the **vertices.** The chord joining the vertices is the **major axis,** and its midpoint is the **center** of the ellipse. The chord perpendicular to the major axis at the center is the **minor axis** of the ellipse. (See Figure 10.8.)

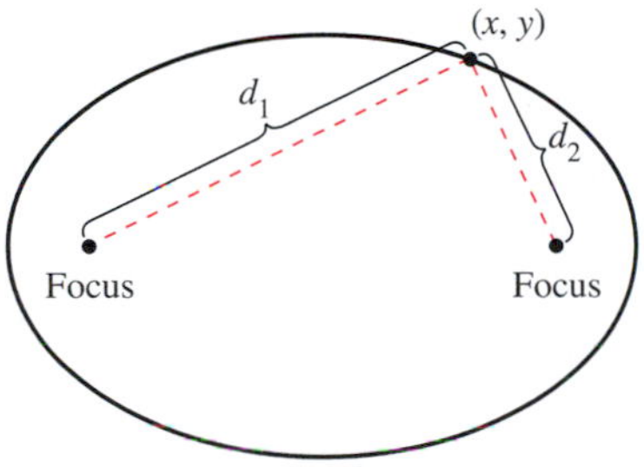

Figure 10.7

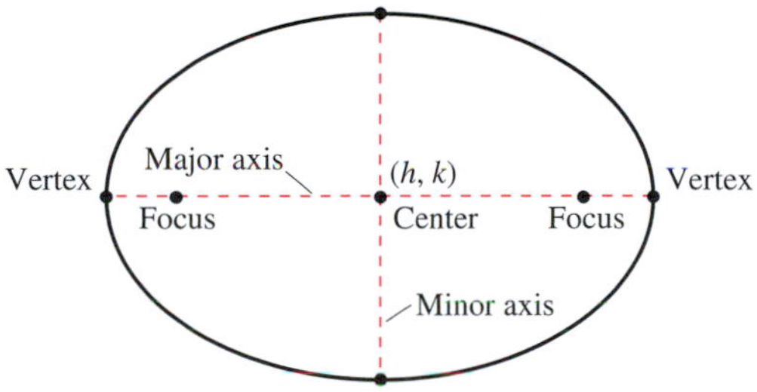

Figure 10.8

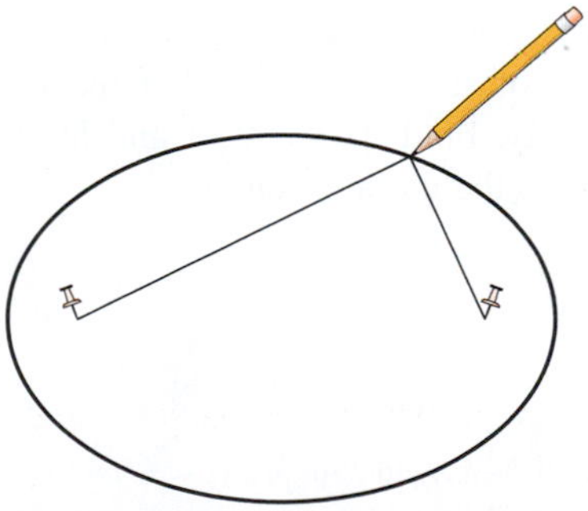

If the ends of a fixed length of string are fastened to the thumbtacks and the string is drawn taut with a pencil, then the path traced by the pencil will be an ellipse.
Figure 10.9

THEOREM 10.3 Standard Equation of an Ellipse

The standard form of the equation of an ellipse with center (h, k) and major and minor axes of lengths $2a$ and $2b$, where $a > b$, is

$$\frac{(x-h)^2}{a^2} + \frac{(y-k)^2}{b^2} = 1 \qquad \text{Major axis is horizontal.}$$

or

$$\frac{(x-h)^2}{b^2} + \frac{(y-k)^2}{a^2} = 1. \qquad \text{Major axis is vertical.}$$

The foci lie on the major axis, c units from the center, with

$$c^2 = a^2 - b^2.$$

You can visualize the definition of an ellipse by imagining two thumbtacks placed at the foci, as shown in Figure 10.9.

FOR FURTHER INFORMATION To learn about how an ellipse may be "exploded" into a parabola, see the article "Exploding the Ellipse" by Arnold Good in *Mathematics Teacher.* To view this article, go to *MathArticles.com.*

EXAMPLE 3 Analyzing an Ellipse

Find the center, vertices, and foci of the ellipse

$$4x^2 + y^2 - 8x + 4y - 8 = 0. \qquad \text{General second-degree equation}$$

Solution By completing the square, you can write the original equation in standard form.

$$4x^2 + y^2 - 8x + 4y - 8 = 0 \qquad \text{Write original equation.}$$
$$4x^2 - 8x + y^2 + 4y = 8$$
$$4(x^2 - 2x + 1) + (y^2 + 4y + 4) = 8 + 4 + 4$$
$$4(x - 1)^2 + (y + 2)^2 = 16$$
$$\frac{(x-1)^2}{4} + \frac{(y+2)^2}{16} = 1 \qquad \text{Write in standard form.}$$

So, the major axis is parallel to the y-axis, where $h = 1$, $k = -2$, $a = 4$, $b = 2$, and $c = \sqrt{16 - 4} = 2\sqrt{3}$. So, you obtain the following.

Center: $(1, -2)$ $\qquad (h, k)$

Vertices: $(1, -6)$ and $(1, 2)$ $\qquad (h, k \pm a)$

Foci: $\left(1, -2 - 2\sqrt{3}\right)$ and $\left(1, -2 + 2\sqrt{3}\right)$ $\qquad (h, k \pm c)$

The graph of the ellipse is shown in Figure 10.10.

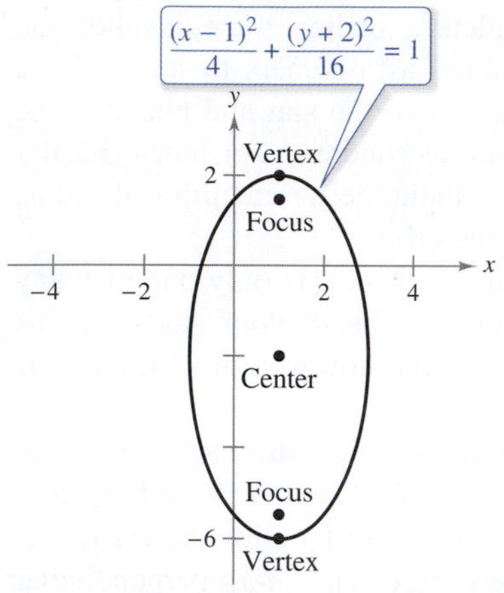

Ellipse with a vertical major axis.
Figure 10.10

In Example 3, the constant term in the general second-degree equation is $F = -8$. For a constant term greater than or equal to 8, you would have obtained one of the degenerate cases shown below.

1. $F = 8$, single point, $(1, -2)$: $\dfrac{(x-1)^2}{4} + \dfrac{(y+2)^2}{16} = 0$

2. $F > 8$, no solution points: $\dfrac{(x-1)^2}{4} + \dfrac{(y+2)^2}{16} < 0$

EXAMPLE 4 The Orbit of the Moon

The moon orbits Earth in an elliptical path with the center of Earth at one focus, as shown in Figure 10.11. The major and minor axes of the orbit have lengths of 768,800 kilometers and 767,640 kilometers, respectively. Find the greatest and least distances (the apogee and perigee) from Earth's center to the moon's center.

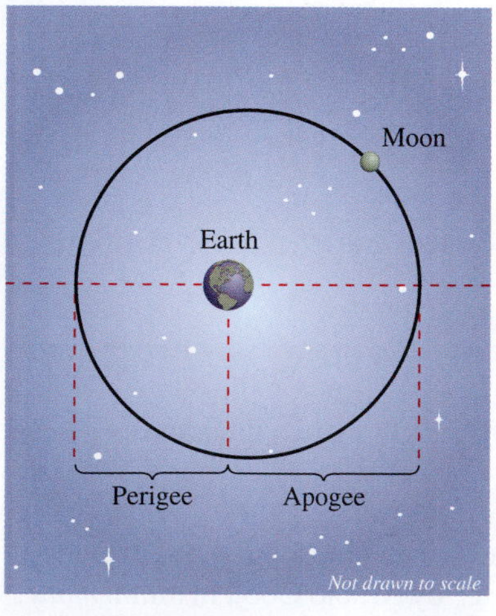

Figure 10.11

Solution Begin by solving for a and b.

$$2a = 768{,}800 \qquad \text{Length of major axis}$$
$$a = 384{,}400 \qquad \text{Solve for } a.$$
$$2b = 767{,}640 \qquad \text{Length of minor axis}$$
$$b = 383{,}820 \qquad \text{Solve for } b.$$

Now, using these values, you can solve for c as follows.

$$c = \sqrt{a^2 - b^2} \approx 21{,}108$$

The greatest distance between the center of Earth and the center of the moon is

$$a + c \approx 405{,}508 \text{ kilometers}$$

and the least distance is

$$a - c \approx 363{,}292 \text{ kilometers.}$$

■ **FOR FURTHER INFORMATION** For more information on some uses of the reflective properties of conics, see the article "Parabolic Mirrors, Elliptic and Hyperbolic Lenses" by Mohsen Maesumi in *The American Mathematical Monthly.* Also see the article "The Geometry of Microwave Antennas" by William R. Parzynski in *Mathematics Teacher.*

Theorem 10.2 presented a reflective property of parabolas. Ellipses have a similar reflective property. You are asked to prove the next theorem in Exercise 84.

THEOREM 10.4 Reflective Property of an Ellipse

Let P be a point on an ellipse. The tangent line to the ellipse at point P makes equal angles with the lines through P and the foci.

One of the reasons that astronomers had difficulty detecting that the orbits of the planets are ellipses is that the foci of the planetary orbits are relatively close to the center of the sun, making the orbits nearly circular. To measure the ovalness of an ellipse, you can use the concept of **eccentricity.**

Definition of Eccentricity of an Ellipse

The **eccentricity** e of an ellipse is given by the ratio

$$e = \frac{c}{a}.$$

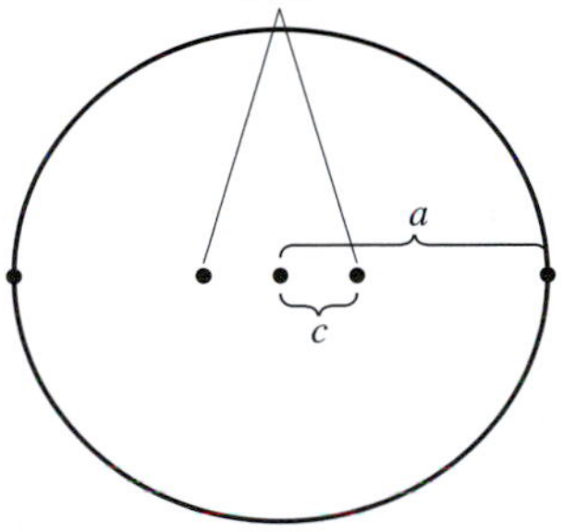

(a) $\frac{c}{a}$ is small.

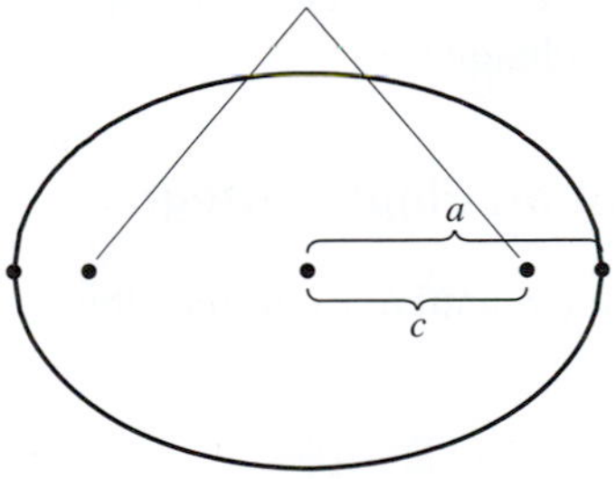

(b) $\frac{c}{a}$ is close to 1.

Eccentricity is the ratio $\frac{c}{a}$.

Figure 10.12

To see how this ratio is used to describe the shape of an ellipse, note that because the foci of an ellipse are located along the major axis between the vertices and the center, it follows that

$$0 < c < a.$$

For an ellipse that is nearly circular, the foci are close to the center and the ratio c/a is close to 0, and for an elongated ellipse, the foci are close to the vertices and the ratio c/a is close to 1, as shown in Figure 10.12. Note that

$$0 < e < 1$$

for every ellipse.

The orbit of the moon has an eccentricity of $e \approx 0.0549$, and the eccentricities of the eight planetary orbits are listed below.

Mercury:	$e \approx 0.2056$	Jupiter:	$e \approx 0.0484$
Venus:	$e \approx 0.0068$	Saturn:	$e \approx 0.0542$
Earth:	$e \approx 0.0167$	Uranus:	$e \approx 0.0472$
Mars:	$e \approx 0.0934$	Neptune:	$e \approx 0.0086$

You can use integration to show that the area of an ellipse is $A = \pi ab$. For instance, the area of the ellipse

$$\frac{x^2}{a^2} + \frac{y^2}{b^2} = 1$$

is

$$A = 4\int_0^a \frac{b}{a}\sqrt{a^2 - x^2}\,dx$$

$$= \frac{4b}{a}\int_0^{\pi/2} a^2 \cos^2 \theta \, d\theta. \qquad \text{Trigonometric substitution } x = a \sin \theta$$

However, it is not so simple to find the *circumference* of an ellipse. The next example shows how to use eccentricity to set up an "elliptic integral" for the circumference of an ellipse.

EXAMPLE 5 Finding the Circumference of an Ellipse

See LarsonCalculus.com for an interactive version of this type of example.

Show that the circumference of the ellipse $(x^2/a^2) + (y^2/b^2) = 1$ is

$$4a\int_0^{\pi/2}\sqrt{1 - e^2\sin^2\theta}\,d\theta. \qquad e = \frac{c}{a}$$

Solution Because the ellipse is symmetric with respect to both the x-axis and the y-axis, you know that its circumference C is four times the arc length of

$$y = \frac{b}{a}\sqrt{a^2 - x^2}$$

in the first quadrant. The function y is differentiable for all x in the interval $[0, a]$ except at $x = a$. So, the circumference is given by the improper integral

$$C = \lim_{d\to a^-} 4\int_0^d \sqrt{1 + (y')^2}\,dx = 4\int_0^a \sqrt{1 + (y')^2}\,dx = 4\int_0^a \sqrt{1 + \frac{b^2x^2}{a^2(a^2 - x^2)}}\,dx.$$

Using the trigonometric substitution $x = a\sin\theta$, you obtain

$$\begin{aligned}
C &= 4\int_0^{\pi/2}\sqrt{1 + \frac{b^2\sin^2\theta}{a^2\cos^2\theta}}\,(a\cos\theta)\,d\theta \\
&= 4\int_0^{\pi/2}\sqrt{a^2\cos^2\theta + b^2\sin^2\theta}\,d\theta \\
&= 4\int_0^{\pi/2}\sqrt{a^2(1 - \sin^2\theta) + b^2\sin^2\theta}\,d\theta \\
&= 4\int_0^{\pi/2}\sqrt{a^2 - (a^2 - b^2)\sin^2\theta}\,d\theta.
\end{aligned}$$

Because $e^2 = c^2/a^2 = (a^2 - b^2)/a^2$, you can rewrite this integral as

$$C = 4a\int_0^{\pi/2}\sqrt{1 - e^2\sin^2\theta}\,d\theta.$$

AREA AND CIRCUMFERENCE OF AN ELLIPSE

In his work with elliptic orbits in the early 1600's, Johannes Kepler successfully developed a formula for the area of an ellipse, $A = \pi ab$. He was less successful, however, in developing a formula for the circumference of an ellipse; the best he could do was to give the approximate formula $C = \pi(a + b)$.

A great deal of time has been devoted to the study of elliptic integrals. Such integrals generally do not have elementary antiderivatives. To find the circumference of an ellipse, you must usually resort to an approximation technique.

EXAMPLE 6 Approximating the Value of an Elliptic Integral

Use the elliptic integral in Example 5 to approximate the circumference of the ellipse

$$\frac{x^2}{25} + \frac{y^2}{16} = 1.$$

Solution Because $e^2 = c^2/a^2 = (a^2 - b^2)/a^2 = 9/25$, you have

$$C = (4)(5)\int_0^{\pi/2}\sqrt{1 - \frac{9\sin^2\theta}{25}}\,d\theta.$$

Applying Simpson's Rule with $n = 4$ produces

$$\begin{aligned}
C &\approx 20\left(\frac{\pi}{6}\right)\left(\frac{1}{4}\right)[1 + 4(0.9733) + 2(0.9055) + 4(0.8323) + 0.8] \\
&\approx 28.36.
\end{aligned}$$

So, the ellipse has a circumference of about 28.36 units, as shown in Figure 10.13.

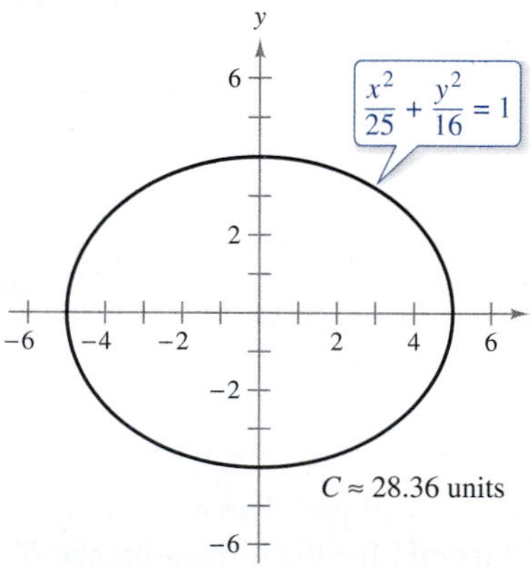

Figure 10.13

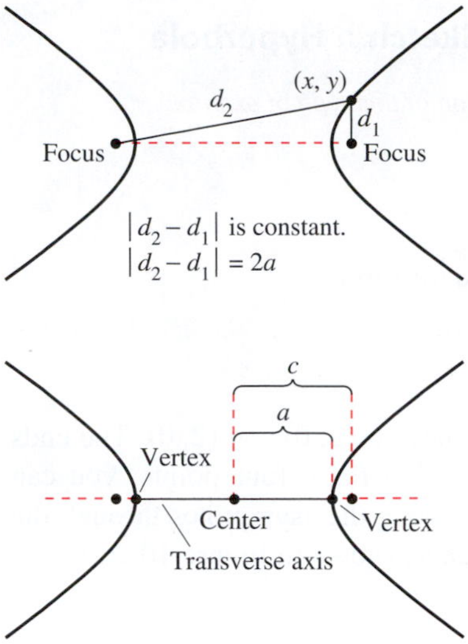

Figure 10.14

Hyperbolas

The definition of a hyperbola is similar to that of an ellipse. For an ellipse, the *sum* of the distances between the foci and a point on the ellipse is fixed, whereas for a hyperbola, the absolute value of the *difference* between these distances is fixed.

A **hyperbola** is the set of all points (x, y) for which the absolute value of the difference between the distances from two distinct fixed points called **foci** is constant. (See Figure 10.14.) The line through the two foci intersects a hyperbola at two points called the **vertices.** The line segment connecting the vertices is the **transverse axis,** and the midpoint of the transverse axis is the **center** of the hyperbola. One distinguishing feature of a hyperbola is that its graph has two separate *branches.*

> **THEOREM 10.5 Standard Equation of a Hyperbola**
>
> The standard form of the equation of a hyperbola with center at (h, k) is
>
> $$\frac{(x-h)^2}{a^2} - \frac{(y-k)^2}{b^2} = 1 \qquad \text{Transverse axis is horizontal.}$$
>
> or
>
> $$\frac{(y-k)^2}{a^2} - \frac{(x-h)^2}{b^2} = 1. \qquad \text{Transverse axis is vertical.}$$
>
> The vertices are a units from the center, and the foci are c units from the center, where $c^2 = a^2 + b^2$.

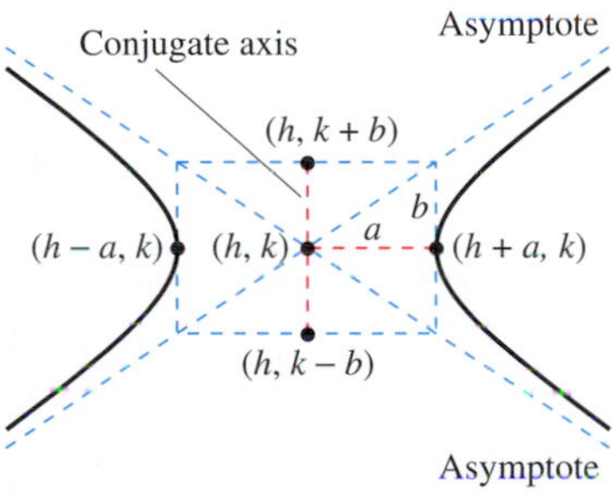

Figure 10.15

Note that the constants a, b, and c do not have the same relationship for hyperbolas as they do for ellipses. For hyperbolas, $c^2 = a^2 + b^2$, but for ellipses, $c^2 = a^2 - b^2$.

An important aid in sketching the graph of a hyperbola is the determination of its **asymptotes,** as shown in Figure 10.15. Each hyperbola has two asymptotes that intersect at the center of the hyperbola. The asymptotes pass through the vertices of a rectangle of dimensions $2a$ by $2b$, with its center at (h, k). The line segment of length $2b$ joining

$$(h, k + b)$$

and

$$(h, k - b)$$

is referred to as the **conjugate axis** of the hyperbola.

> **THEOREM 10.6 Asymptotes of a Hyperbola**
>
> For a *horizontal* transverse axis, the equations of the asymptotes are
>
> $$y = k + \frac{b}{a}(x - h) \quad \text{and} \quad y = k - \frac{b}{a}(x - h).$$
>
> For a *vertical* transverse axis, the equations of the asymptotes are
>
> $$y = k + \frac{a}{b}(x - h) \quad \text{and} \quad y = k - \frac{a}{b}(x - h).$$

In Figure 10.15, you can see that the asymptotes coincide with the diagonals of the rectangle with dimensions $2a$ and $2b$, centered at (h, k). This provides you with a quick means of sketching the asymptotes, which in turn aids in sketching the hyperbola.

EXAMPLE 7 Using Asymptotes to Sketch a Hyperbola

See LarsonCalculus.com for an interactive version of this type of example.

Sketch the graph of the hyperbola

$$4x^2 - y^2 = 16.$$

Solution Begin by rewriting the equation in standard form.

$$\frac{x^2}{4} - \frac{y^2}{16} = 1$$

The transverse axis is horizontal and the vertices occur at $(-2, 0)$ and $(2, 0)$. The ends of the conjugate axis occur at $(0, -4)$ and $(0, 4)$. Using these four points, you can sketch the rectangle shown in Figure 10.16(a). By drawing the asymptotes through the corners of this rectangle, you can complete the sketch as shown in Figure 10.16(b).

TECHNOLOGY You can use a graphing utility to verify the graph obtained in Example 7 by solving the original equation for y and graphing the following equations.

$$y_1 = \sqrt{4x^2 - 16}$$
$$y_2 = -\sqrt{4x^2 - 16}$$

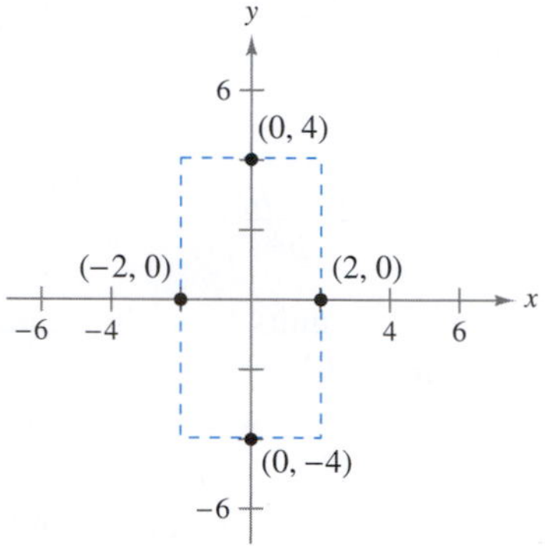

(a)

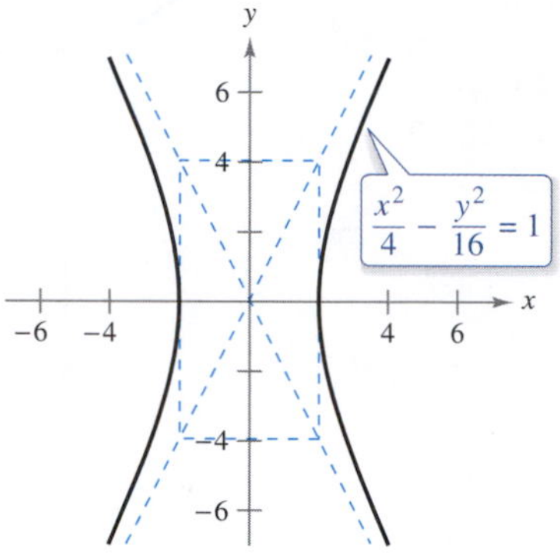

(b)

Figure 10.16

> **Definition of Eccentricity of a Hyperbola**
>
> The **eccentricity** e of a hyperbola is given by the ratio
>
> $$e = \frac{c}{a}.$$

As with an ellipse, the **eccentricity** of a hyperbola is $e = c/a$. Because $c > a$ for hyperbolas, it follows that $e > 1$ for hyperbolas. If the eccentricity is large, then the branches of the hyperbola are nearly flat. If the eccentricity is close to 1, then the branches of the hyperbola are more pointed, as shown in Figure 10.17.

FOR FURTHER INFORMATION To read about using a string that traces both elliptic and hyperbolic arcs having the same foci, see the article "Ellipse to Hyperbola: 'With This String I Thee Wed'" by Tom M. Apostol and Mamikon A. Mnatsakanian in *Mathematics Magazine*. To view this article, go to *MathArticles.com*.

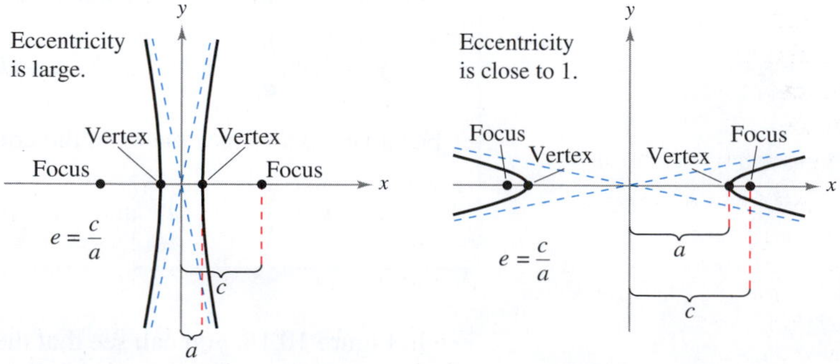

Figure 10.17

The application in Example 8 was developed during World War II. It shows how the properties of hyperbolas can be used in radar and other detection systems.

EXAMPLE 8 A Hyperbolic Detection System

Two microphones, 1 mile apart, record an explosion. Microphone A receives the sound 2 seconds before microphone B. Where was the explosion?

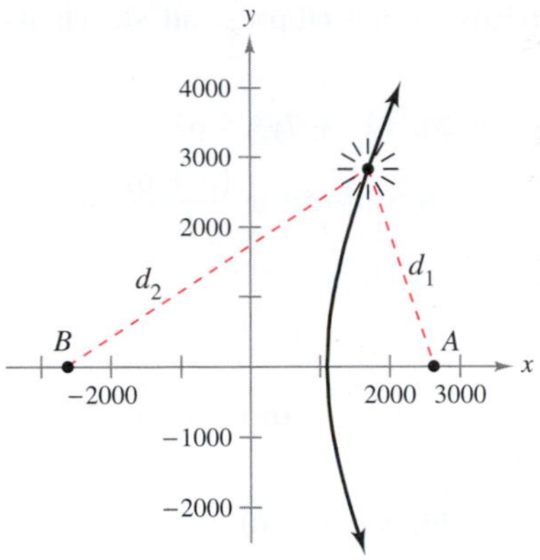

$2c = 5280$
$d_2 - d_1 = 2a = 2200$
Figure 10.18

Solution Assuming that sound travels at 1100 feet per second, you know that the explosion took place 2200 feet farther from B than from A, as shown in Figure 10.18. The locus of all points that are 2200 feet closer to A than to B is one branch of the hyperbola

$$\frac{x^2}{a^2} - \frac{y^2}{b^2} = 1$$

where

$$c = \frac{1 \text{ mile}}{2} = \frac{5280 \text{ ft}}{2} = 2640 \text{ feet}$$

and

$$a = \frac{2200 \text{ ft}}{2} = 1100 \text{ feet.}$$

Because $c^2 = a^2 + b^2$, it follows that

$$\begin{aligned} b^2 &= c^2 - a^2 \\ &= (2640)^2 - (1100)^2 \\ &= 5{,}759{,}600 \end{aligned}$$

and you can conclude that the explosion occurred somewhere on the right branch of the hyperbola

$$\frac{x^2}{1{,}210{,}000} - \frac{y^2}{5{,}759{,}600} = 1.$$

In Example 8, you were able to determine only the hyperbola on which the explosion occurred, but not the exact location of the explosion. If, however, you had received the sound at a third position C, then two other hyperbolas would be determined. The exact location of the explosion would be the point at which these three hyperbolas intersect.

Another interesting application of conics involves the orbits of comets in our solar system. Of the 610 comets identified prior to 1970, 245 have elliptical orbits, 295 have parabolic orbits, and 70 have hyperbolic orbits. The center of the sun is a focus of each orbit, and each orbit has a vertex at the point at which the comet is closest to the sun. Undoubtedly, many comets with parabolic or hyperbolic orbits have not been identified—such comets pass through our solar system only once. Only comets with elliptical orbits, such as Halley's comet, remain in our solar system.

The type of orbit for a comet can be determined as follows.

1. Ellipse: $v < \sqrt{2GM/p}$
2. Parabola: $v = \sqrt{2GM/p}$
3. Hyperbola: $v > \sqrt{2GM/p}$

In each of the above, p is the distance between one vertex and one focus of the comet's orbit (in meters), v is the velocity of the comet at the vertex (in meters per second), $M \approx 1.989 \times 10^{30}$ kilograms is the mass of the sun, and $G \approx 6.67 \times 10^{-8}$ cubic meters per kilogram-second squared is the gravitational constant.

CAROLINE HERSCHEL (1750–1848)

The first woman to be credited with detecting a new comet was the English astronomer Caroline Herschel. During her life, Caroline Herschel discovered a total of eight new comets.

See LarsonCalculus.com to read more of this biography.

10.1 Exercises

See CalcChat.com for tutorial help and worked-out solutions to odd-numbered exercises.

Matching In Exercises 1–6, match the equation with its graph. [The graphs are labeled (a), (b), (c), (d), (e), and (f).]

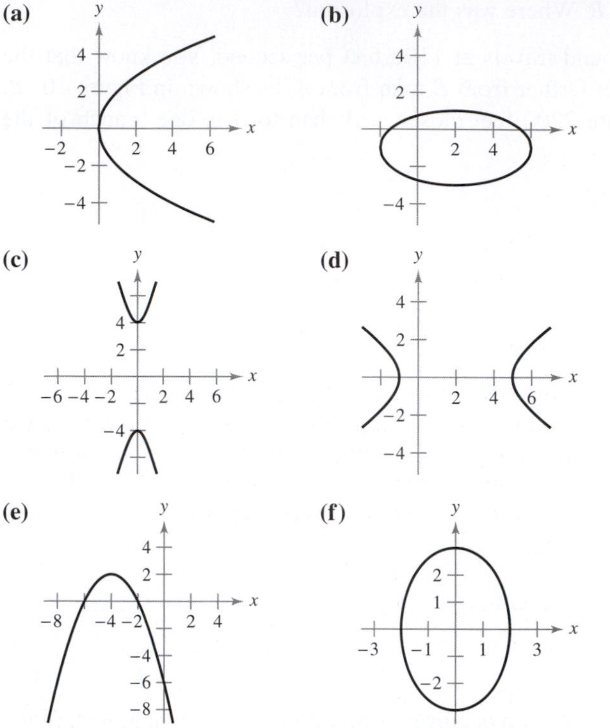

1. $y^2 = 4x$
2. $(x + 4)^2 = -2(y - 2)$
3. $\dfrac{y^2}{16} - \dfrac{x^2}{1} = 1$
4. $\dfrac{(x-2)^2}{16} + \dfrac{(y+1)^2}{4} = 1$
5. $\dfrac{x^2}{4} + \dfrac{y^2}{9} = 1$
6. $\dfrac{(x-2)^2}{9} - \dfrac{y^2}{4} = 1$

Sketching a Parabola In Exercises 7–14, find the vertex, focus, and directrix of the parabola, and sketch its graph.

7. $y^2 = -8x$
8. $x^2 + 6y = 0$
9. $(x + 5) + (y - 3)^2 = 0$
10. $(x - 6)^2 + 8(y + 7) = 0$
11. $y^2 - 4y - 4x = 0$
12. $y^2 + 6y + 8x + 25 = 0$
13. $x^2 + 4x + 4y - 4 = 0$
14. $y^2 + 4y + 8x - 12 = 0$

Finding an Equation of a Parabola In Exercises 15–22, find an equation of the parabola.

15. Vertex: $(5, 4)$
    Focus: $(3, 4)$
16. Vertex: $(-2, 1)$
    Focus: $(-2, -1)$
17. Vertex: $(0, 5)$
    Directrix: $y = -3$
18. Focus: $(2, 2)$
    Directrix: $x = -2$
19. Vertex: $(0, 4)$
    Points on the parabola: $(-2, 0), (2, 0)$
20. Vertex: $(2, 4)$
    Points on the parabola: $(0, 0), (4, 0)$
21. Axis is parallel to y-axis; graph passes through $(0, 3)$, $(3, 4)$, and $(4, 11)$.
22. Directrix: $y = -2$; endpoints of latus rectum are $(0, 2)$ and $(8, 2)$.

Sketching an Ellipse In Exercises 23–28, find the center, foci, vertices, and eccentricity of the ellipse, and sketch its graph.

23. $16x^2 + y^2 = 16$
24. $3x^2 + 7y^2 = 63$
25. $\dfrac{(x-3)^2}{16} + \dfrac{(y-1)^2}{25} = 1$
26. $(x + 4)^2 + \dfrac{(y+6)^2}{1/4} = 1$
27. $9x^2 + 4y^2 + 36x - 24y + 36 = 0$
28. $16x^2 + 25y^2 - 64x + 150y + 279 = 0$

Finding an Equation of an Ellipse In Exercises 29–34, find an equation of the ellipse.

29. Center: $(0, 0)$
    Focus: $(5, 0)$
    Vertex: $(6, 0)$
30. Vertices: $(0, 3), (8, 3)$
    Eccentricity: $\frac{3}{4}$
31. Vertices: $(3, 1), (3, 9)$
    Minor axis length: 6
32. Foci: $(0, \pm 9)$
    Major axis length: 22
33. Center: $(0, 0)$
    Major axis: horizontal
    Points on the ellipse: $(3, 1), (4, 0)$
34. Center: $(1, 2)$
    Major axis: vertical
    Points on the ellipse: $(1, 6), (3, 2)$

Sketching a Hyperbola In Exercises 35–40, find the center, foci, and vertices of the hyperbola, and sketch its graph using asymptotes as an aid.

35. $\dfrac{x^2}{25} - \dfrac{y^2}{16} = 1$
36. $\dfrac{(y+3)^2}{225} - \dfrac{(x-5)^2}{64} = 1$
37. $9x^2 - y^2 - 36x - 6y + 18 = 0$
38. $y^2 - 16x^2 + 64x - 208 = 0$
39. $x^2 - 9y^2 + 2x - 54y - 80 = 0$
40. $9x^2 - 4y^2 + 54x + 8y + 78 = 0$

Finding an Equation of a Hyperbola In Exercises 41–48, find an equation of the hyperbola.

41. Vertices: $(\pm 1, 0)$
    Asymptotes: $y = \pm 5x$
42. Vertices: $(0, \pm 4)$
    Asymptotes: $y = \pm 2x$
43. Vertices: $(2, \pm 3)$
    Point on graph: $(0, 5)$
44. Vertices: $(2, \pm 3)$
    Foci: $(2, \pm 5)$
45. Center: $(0, 0)$
    Vertex: $(0, 2)$
    Focus: $(0, 4)$
46. Center: $(0, 0)$
    Vertex: $(6, 0)$
    Focus: $(10, 0)$
47. Vertices: $(0, 2), (6, 2)$
    Asymptotes: $y = \frac{2}{3}x$
    $y = 4 - \frac{2}{3}x$
48. Focus: $(20, 0)$
    Asymptotes: $y = \pm\frac{3}{4}x$

Finding Equations of Tangent Lines and Normal Lines **In Exercises 49 and 50, find equations for (a) the tangent lines and (b) the normal lines to the hyperbola for the given value of x.**

49. $\frac{x^2}{9} - y^2 = 1, \quad x = 6$

50. $\frac{y^2}{4} - \frac{x^2}{2} = 1, \quad x = 4$

Classifying the Graph of an Equation **In Exercises 51–58, classify the graph of the equation as a circle, a parabola, an ellipse, or a hyperbola.**

51. $x^2 + 4y^2 - 6x + 16y + 21 = 0$

52. $4x^2 - y^2 - 4x - 3 = 0$

53. $25x^2 - 10x - 200y - 119 = 0$

54. $y^2 - 4y = x + 5$

55. $9x^2 + 9y^2 - 36x + 6y + 34 = 0$

56. $2x(x - y) = y(3 - y - 2x)$

57. $3(x - 1)^2 = 6 + 2(y + 1)^2$

58. $9(x + 3)^2 = 36 - 4(y - 2)^2$

WRITING ABOUT CONCEPTS

59. Parabola

(a) Give the definition of a parabola.

(b) Give the standard forms of a parabola with vertex at (h, k).

(c) In your own words, state the reflective property of a parabola.

60. Ellipse

(a) Give the definition of an ellipse.

(b) Give the standard form of an ellipse with center at (h, k).

61. Hyperbola

(a) Give the definition of a hyperbola.

(b) Give the standard forms of a hyperbola with center at (h, k).

(c) Write equations for the asymptotes of a hyperbola.

62. Eccentricity Define the eccentricity of an ellipse. In your own words, describe how changes in the eccentricity affect the ellipse.

63. Using an Equation Consider the equation

$$9x^2 + 4y^2 - 36x - 24y - 36 = 0.$$

(a) Classify the graph of the equation as a circle, a parabola, an ellipse, or a hyperbola.

(b) Change the $4y^2$-term in the equation to $-4y^2$. Classify the graph of the new equation.

(c) Change the $9x^2$-term in the original equation to $4x^2$. Classify the graph of the new equation.

(d) Describe one way you could change the original equation so that its graph is a parabola.

64. HOW DO YOU SEE IT? In parts (a)–(d), describe in words how a plane could intersect with the double-napped cone to form the conic section (see figure).

(a) Circle (b) Ellipse

(c) Parabola (d) Hyperbola

65. Solar Collector A solar collector for heating water is constructed with a sheet of stainless steel that is formed into the shape of a parabola (see figure). The water will flow through a pipe that is located at the focus of the parabola. At what distance from the vertex is the pipe?

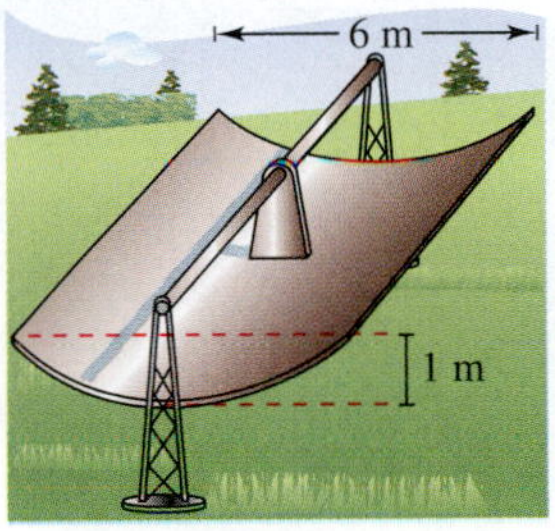

Figure for 65

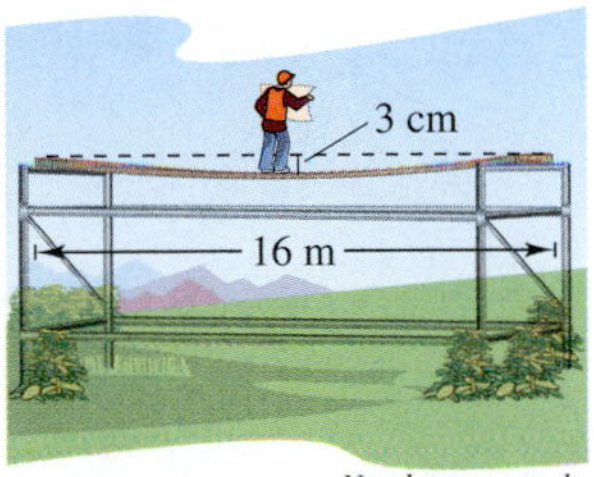

Figure for 66

66. Beam Deflection A simply supported beam that is 16 meters long has a load concentrated at the center (see figure). The deflection of the beam at its center is 3 centimeters. Assume that the shape of the deflected beam is parabolic.

(a) Find an equation of the parabola. (Assume that the origin is at the center of the beam.)

(b) How far from the center of the beam is the deflection 1 centimeter?

67. Proof

(a) Prove that any two distinct tangent lines to a parabola intersect.

(b) Demonstrate the result of part (a) by finding the point of intersection of the tangent lines to the parabola $x^2 - 4x - 4y = 0$ at the points $(0, 0)$ and $(6, 3)$.

68. Proof

(a) Prove that if any two tangent lines to a parabola intersect at right angles, their point of intersection must lie on the directrix.

(b) Demonstrate the result of part (a) by showing that the tangent lines to the parabola $x^2 - 4x - 4y + 8 = 0$ at the points $(-2, 5)$ and $\left(3, \frac{5}{4}\right)$ intersect at right angles, and that the point of intersection lies on the directrix.

69. Investigation Sketch the graphs of $x^2 = 4py$ for $p = \frac{1}{4}, \frac{1}{2}, 1, \frac{3}{2}$, and 2 on the same coordinate axes. Discuss the change in the graphs as p increases.

70. Bridge Design A cable of a suspension bridge is suspended (in the shape of a parabola) between two towers that are 120 meters apart and 20 meters above the roadway (see figure). The cable touches the roadway midway between the towers.

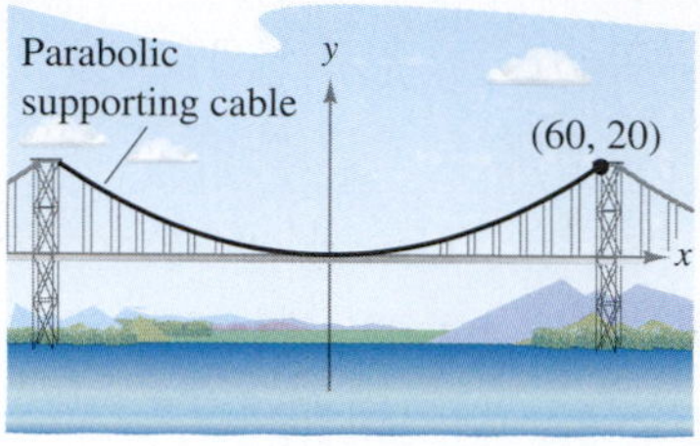

(a) Find an equation for the parabolic shape of the cable.

(b) Find the length of the parabolic cable.

71. Architecture

A church window is bounded above by a parabola and below by the arc of a circle (see figure). Find the surface area of the window.

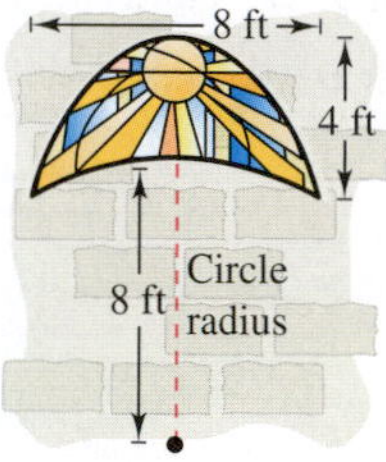

72. Surface Area A satellite signal receiving dish is formed by revolving the parabola given by $x^2 = 20y$ about the y-axis. The radius of the dish is r feet. Verify that the surface area of the dish is given by

$$2\pi\int_0^r x\sqrt{1 + \left(\frac{x}{10}\right)^2}\,dx = \frac{\pi}{15}[(100 + r^2)^{3/2} - 1000].$$

73. Orbit of Earth Earth moves in an elliptical orbit with the sun at one of the foci. The length of half of the major axis is 149,598,000 kilometers, and the eccentricity is 0.0167. Find the minimum distance (*perihelion*) and the maximum distance (*aphelion*) of Earth from the sun.

74. Satellite Orbit The *apogee* (the point in orbit farthest from Earth) and the *perigee* (the point in orbit closest to Earth) of an elliptical orbit of an Earth satellite are given by A and P. Show that the eccentricity of the orbit is

$$e = \frac{A - P}{A + P}.$$

75. Explorer 18 On November 27, 1963, the United States launched the research satellite Explorer 18. Its low and high points above the surface of Earth were 119 miles and 123,000 miles. Find the eccentricity of its elliptical orbit.

76. Explorer 55 On November 20, 1975, the United States launched the research satellite Explorer 55. Its low and high points above the surface of Earth were 96 miles and 1865 miles. Find the eccentricity of its elliptical orbit.

77. Halley's Comet

Probably the most famous of all comets, Halley's comet, has an elliptical orbit with the sun at one focus. Its maximum distance from the sun is approximately 35.29 AU (1 astronomical unit is approximately 92.956×10^6 miles), and its minimum distance is approximately 0.59 AU. Find the eccentricity of the orbit.

78. Particle Motion Consider a particle traveling clockwise on the elliptical path

$$\frac{x^2}{100} + \frac{y^2}{25} = 1.$$

The particle leaves the orbit at the point $(-8, 3)$ and travels in a straight line tangent to the ellipse. At what point will the particle cross the y-axis?

Area, Volume, and Surface Area In Exercises 79 and 80, find (a) the area of the region bounded by the ellipse, (b) the volume and surface area of the solid generated by revolving the region about its major axis (prolate spheroid), and (c) the volume and surface area of the solid generated by revolving the region about its minor axis (oblate spheroid).

79. $\frac{x^2}{4} + \frac{y^2}{1} = 1$ **80.** $\frac{x^2}{16} + \frac{y^2}{9} = 1$

81. Arc Length Use the integration capabilities of a graphing utility to approximate to two-decimal-place accuracy the elliptical integral representing the circumference of the ellipse

$$\frac{x^2}{25} + \frac{y^2}{49} = 1.$$

82. Conjecture

(a) Show that the equation of an ellipse can be written as

$$\frac{(x - h)^2}{a^2} + \frac{(y - k)^2}{a^2(1 - e^2)} = 1.$$

(b) Use a graphing utility to graph the ellipse

$$\frac{(x - 2)^2}{4} + \frac{(y - 3)^2}{4(1 - e^2)} = 1$$

for $e = 0.95$, $e = 0.75$, $e = 0.5$, $e = 0.25$, and $e = 0$.

(c) Use the results of part (b) to make a conjecture about the change in the shape of the ellipse as e approaches 0.

83. Geometry The area of the ellipse in the figure is twice the area of the circle. What is the length of the major axis?

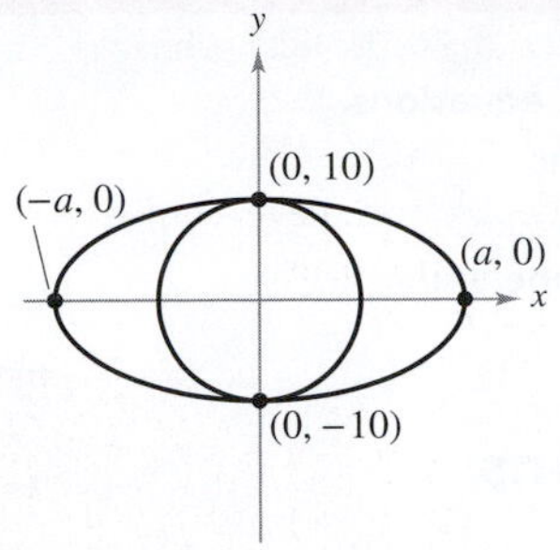

Figure for 83

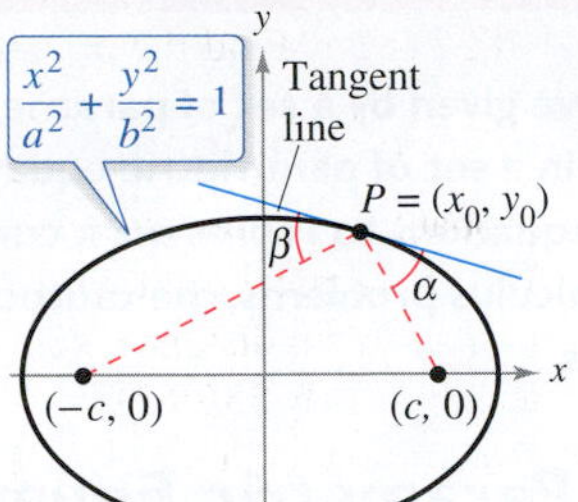

Figure for 84

84. Proof Prove Theorem 10.4 by showing that the tangent line to an ellipse at a point P makes equal angles with lines through P and the foci (see figure). [*Hint:* (1) Find the slope of the tangent line at P, (2) find the slopes of the lines through P and each focus, and (3) use the formula for the tangent of the angle between two lines.]

85. Finding an Equation of a Hyperbola Find an equation of the hyperbola such that for any point on the hyperbola, the difference between its distances from the points (2, 2) and (10, 2) is 6.

86. Hyperbola Consider a hyperbola centered at the origin with a horizontal transverse axis. Use the definition of a hyperbola to derive its standard form:

$$\frac{x^2}{a^2} - \frac{y^2}{b^2} = 1.$$

87. Navigation LORAN (long distance radio navigation) for aircraft and ships uses synchronized pulses transmitted by widely separated transmitting stations. These pulses travel at the speed of light (186,000 miles per second). The difference in the times of arrival of these pulses at an aircraft or ship is constant on a hyperbola having the transmitting stations as foci. Assume that two stations, 300 miles apart, are positioned on a rectangular coordinate system at $(-150, 0)$ and $(150, 0)$ and that a ship is traveling on a path with coordinates $(x, 75)$ (see figure). Find the x-coordinate of the position of the ship if the time difference between the pulses from the transmitting stations is 1000 microseconds (0.001 second).

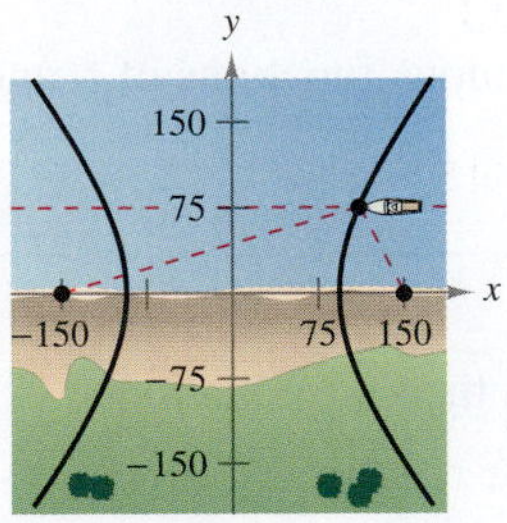

Figure for 87

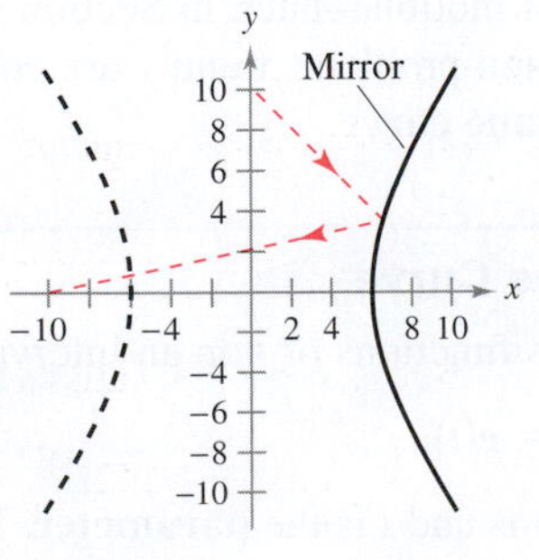

Figure for 88

88. Hyperbolic Mirror A hyperbolic mirror (used in some telescopes) has the property that a light ray directed at the focus will be reflected to the other focus. The mirror in the figure has the equation $(x^2/36) - (y^2/64) = 1$. At which point on the mirror will light from the point (0, 10) be reflected to the other focus?

89. Tangent Line Show that the equation of the tangent line to $\frac{x^2}{a^2} - \frac{y^2}{b^2} = 1$ at the point (x_0, y_0) is $\left(\frac{x_0}{a^2}\right)x - \left(\frac{y_0}{b^2}\right)y = 1$.

90. Proof Prove that the graph of the equation

$$Ax^2 + Cy^2 + Dx + Ey + F = 0$$

is one of the following (except in degenerate cases).

Conic	Condition
(a) Circle	$A = C$
(b) Parabola	$A = 0$ or $C = 0$ (but not both)
(c) Ellipse	$AC > 0$
(d) Hyperbola	$AC < 0$

True or False? **In Exercises 91–96, determine whether the statement is true or false. If it is false, explain why or give an example that shows it is false.**

91. It is possible for a parabola to intersect its directrix.

92. The point on a parabola closest to its focus is its vertex.

93. If C is the circumference of the ellipse

$$\frac{x^2}{a^2} + \frac{y^2}{b^2} = 1, \quad b < a$$

then $2\pi b \le C \le 2\pi a$.

94. If $D \ne 0$ or $E \ne 0$, then the graph of $y^2 - x^2 + Dx + Ey = 0$ is a hyperbola.

95. If the asymptotes of the hyperbola $(x^2/a^2) - (y^2/b^2) = 1$ intersect at right angles, then $a = b$.

96. Every tangent line to a hyperbola intersects the hyperbola only at the point of tangency.

PUTNAM EXAM CHALLENGE

97. For a point P on an ellipse, let d be the distance from the center of the ellipse to the line tangent to the ellipse at P. Prove that $(PF_1)(PF_2)d^2$ is constant as P varies on the ellipse, where PF_1 and PF_2 are the distances from P to the foci F_1 and F_2 of the ellipse.

98. Find the minimum value of

$$(u - v)^2 + \left(\sqrt{2 - u^2} - \frac{9}{v}\right)^2$$

for $0 < u < \sqrt{2}$ and $v > 0$.

10.2 Plane Curves and Parametric Equations

- **Sketch the graph of a curve given by a set of parametric equations.**
- **Eliminate the parameter in a set of parametric equations.**
- **Find a set of parametric equations to represent a curve.**
- **Understand two classic calculus problems, the tautochrone and brachistochrone problems.**

Plane Curves and Parametric Equations

Until now, you have been representing a graph by a single equation involving *two* variables. In this section, you will study situations in which *three* variables are used to represent a curve in the plane.

Consider the path followed by an object that is propelled into the air at an angle of 45°. For an initial velocity of 48 feet per second, the object travels the parabolic path given by

$$y = -\frac{x^2}{72} + x \qquad \text{Rectangular equation}$$

as shown in Figure 10.19. This equation, however, does not tell the whole story. Although it does tell you *where* the object has been, it doesn't tell you *when* the object was at a given point (x, y). To determine this time, you can introduce a third variable t, called a **parameter.** By writing both x and y as functions of t, you obtain the **parametric equations**

$$x = 24\sqrt{2}t \qquad \text{Parametric equation for } x$$

and

$$y = -16t^2 + 24\sqrt{2}t. \qquad \text{Parametric equation for } y$$

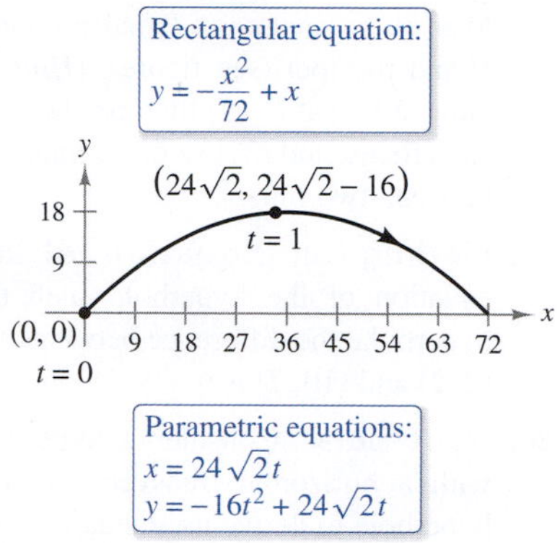

Curvilinear motion: two variables for position, one variable for time
Figure 10.19

From this set of equations, you can determine that at time $t = 0$, the object is at the point $(0, 0)$. Similarly, at time $t = 1$, the object is at the point

$$\left(24\sqrt{2}, 24\sqrt{2} - 16\right)$$

and so on. (You will learn a method for determining this particular set of parametric equations—the equations of motion—later, in Section 12.3.)

For this particular motion problem, x and y are continuous functions of t, and the resulting path is called a **plane curve.**

REMARK At times, it is important to distinguish between a graph (the set of points) and a curve (the points together with their defining parametric equations). When it is important, the distinction will be explicit. When it is not important, C will be used to represent either the graph or the curve.

Definition of a Plane Curve

If f and g are continuous functions of t on an interval I, then the equations

$$x = f(t) \quad \text{and} \quad y = g(t)$$

are **parametric equations** and t is the **parameter.** The set of points (x, y) obtained as t varies over the interval I is the **graph** of the parametric equations. Taken together, the parametric equations and the graph are a **plane curve,** denoted by C.

When sketching a curve represented by a set of parametric equations, you can plot points in the *xy*-plane. Each set of coordinates (x, y) is determined from a value chosen for the parameter t. By plotting the resulting points in order of increasing values of t, the curve is traced out in a specific direction. This is called the **orientation** of the curve.

EXAMPLE 1 Sketching a Curve

Sketch the curve described by the parametric equations

$$x = f(t) = t^2 - 4$$

and

$$y = g(t) = \frac{t}{2}$$

where $-2 \le t \le 3$.

Solution For values of t on the given interval, the parametric equations yield the points (x, y) shown in the table.

t	-2	-1	0	1	2	3
x	0	-3	-4	-3	0	5
y	-1	$-\frac{1}{2}$	0	$\frac{1}{2}$	1	$\frac{3}{2}$

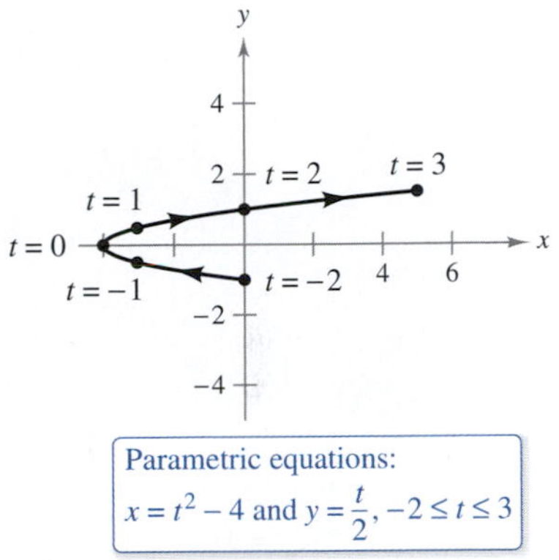

Figure 10.20

By plotting these points in order of increasing t and using the continuity of f and g, you obtain the curve C shown in Figure 10.20. Note that the arrows on the curve indicate its orientation as t increases from -2 to 3.

According to the Vertical Line Test, the graph shown in Figure 10.20 does not define y as a function of x. This points out one benefit of parametric equations—they can be used to represent graphs that are more general than graphs of functions.

It often happens that two different sets of parametric equations have the same graph. For instance, the set of parametric equations

$$x = 4t^2 - 4 \quad \text{and} \quad y = t, \quad -1 \le t \le \frac{3}{2}$$

has the same graph as the set given in Example 1. (See Figure 10.21.) However, comparing the values of t in Figures 10.20 and 10.21, you can see that the second graph is traced out more *rapidly* (considering t as time) than the first graph. So, in applications, different parametric representations can be used to represent various *speeds* at which objects travel along a given path.

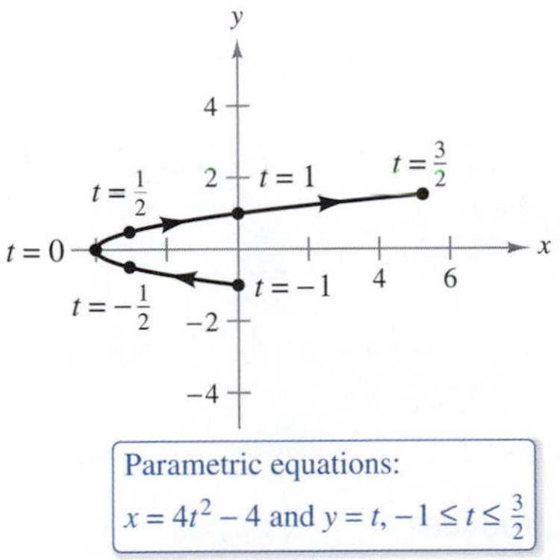

Figure 10.21

TECHNOLOGY Most graphing utilities have a *parametric* graphing mode. If you have access to such a utility, use it to confirm the graphs shown in Figures 10.20 and 10.21. Does the curve given by the parametric equations

$$x = 4t^2 - 8t \quad \text{and} \quad y = 1 - t, \quad -\frac{1}{2} \le t \le 2$$

represent the same graph as that shown in Figures 10.20 and 10.21? What do you notice about the *orientation* of this curve?

Eliminating the Parameter

Finding a rectangular equation that represents the graph of a set of parametric equations is called **eliminating the parameter.** For instance, you can eliminate the parameter from the set of parametric equations in Example 1 as follows.

Parametric equations	Solve for t in one equation.	Substitute into second equation.	Rectangular equation
$x = t^2 - 4$ $y = t/2$	$t = 2y$	$x = (2y)^2 - 4$	$x = 4y^2 - 4$

Once you have eliminated the parameter, you can recognize that the equation $x = 4y^2 - 4$ represents a parabola with a horizontal axis and vertex at $(-4, 0)$, as shown in Figure 10.20.

The range of x and y implied by the parametric equations may be altered by the change to rectangular form. In such instances, the domain of the rectangular equation must be adjusted so that its graph matches the graph of the parametric equations. Such a situation is demonstrated in the next example.

EXAMPLE 2 Adjusting the Domain

Sketch the curve represented by the equations

$$x = \frac{1}{\sqrt{t+1}} \quad \text{and} \quad y = \frac{t}{t+1}, \quad t > -1$$

by eliminating the parameter and adjusting the domain of the resulting rectangular equation.

Solution Begin by solving one of the parametric equations for t. For instance, you can solve the first equation for t as follows.

$$x = \frac{1}{\sqrt{t+1}} \qquad \text{Parametric equation for } x$$

$$x^2 = \frac{1}{t+1} \qquad \text{Square each side.}$$

$$t + 1 = \frac{1}{x^2}$$

$$t = \frac{1}{x^2} - 1$$

$$t = \frac{1 - x^2}{x^2} \qquad \text{Solve for } t.$$

Now, substituting into the parametric equation for y produces

$$y = \frac{t}{t+1} \qquad \text{Parametric equation for } y$$

$$y = \frac{(1 - x^2)/x^2}{[(1 - x^2)/x^2] + 1} \qquad \text{Substitute } (1 - x^2)/x^2 \text{ for } t.$$

$$y = 1 - x^2. \qquad \text{Simplify.}$$

The rectangular equation, $y = 1 - x^2$, is defined for all values of x, but from the parametric equation for x, you can see that the curve is defined only when $t > -1$. This implies that you should restrict the domain of x to positive values, as shown in Figure 10.22.

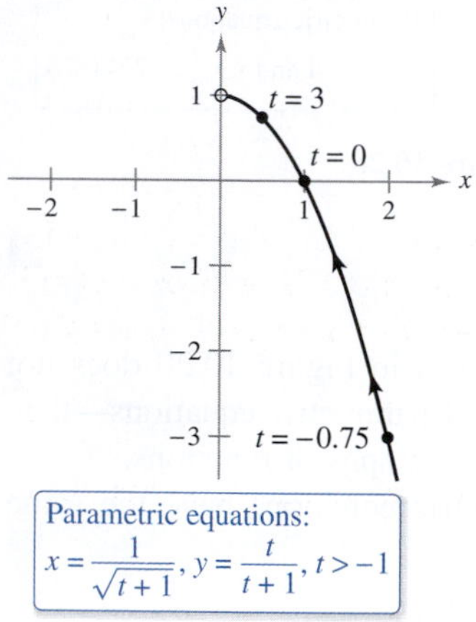

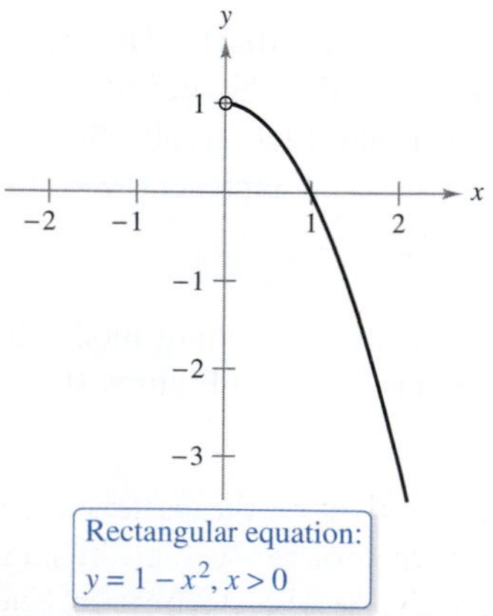

Figure 10.22

It is not necessary for the parameter in a set of parametric equations to represent time. The next example uses an *angle* as the parameter.

EXAMPLE 3 Using Trigonometry to Eliminate a Parameter

See LarsonCalculus.com for an interactive version of this type of example.

Sketch the curve represented by

$$x = 3\cos\theta \quad \text{and} \quad y = 4\sin\theta, \quad 0 \le \theta \le 2\pi$$

by eliminating the parameter and finding the corresponding rectangular equation.

Solution Begin by solving for $\cos\theta$ and $\sin\theta$ in the given equations.

$$\cos\theta = \frac{x}{3} \qquad \text{Solve for } \cos\theta.$$

and

$$\sin\theta = \frac{y}{4} \qquad \text{Solve for } \sin\theta.$$

Next, make use of the identity

$$\sin^2\theta + \cos^2\theta = 1$$

to form an equation involving only x and y.

$$\cos^2\theta + \sin^2\theta = 1 \qquad \text{Trigonometric identity}$$

$$\left(\frac{x}{3}\right)^2 + \left(\frac{y}{4}\right)^2 = 1 \qquad \text{Substitute.}$$

$$\frac{x^2}{9} + \frac{y^2}{16} = 1 \qquad \text{Rectangular equation}$$

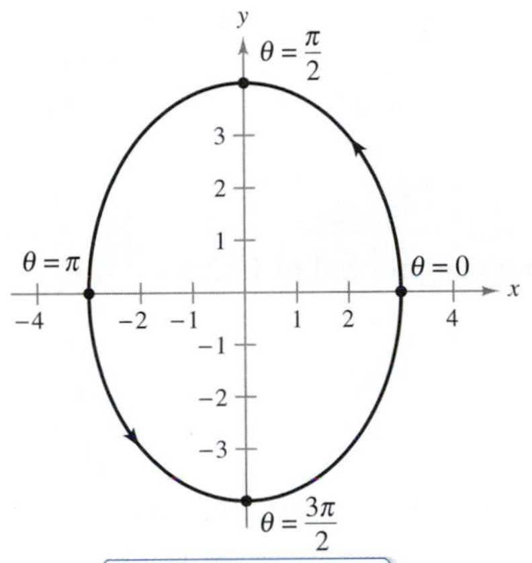

Parametric equations:
$x = 3\cos\theta,\ y = 4\sin\theta$
Rectangular equation:
$\frac{x^2}{9} + \frac{y^2}{16} = 1$

Figure 10.23

From this rectangular equation, you can see that the graph is an ellipse centered at $(0, 0)$, with vertices at $(0, 4)$ and $(0, -4)$ and minor axis of length $2b = 6$, as shown in Figure 10.23. Note that the ellipse is traced out *counterclockwise* as θ varies from 0 to 2π.

Using the technique shown in Example 3, you can conclude that the graph of the parametric equations

$$x = h + a\cos\theta \quad \text{and} \quad y = k + b\sin\theta, \quad 0 \le \theta \le 2\pi$$

is the ellipse (traced counterclockwise) given by

$$\frac{(x-h)^2}{a^2} + \frac{(y-k)^2}{b^2} = 1.$$

The graph of the parametric equations

$$x = h + a\sin\theta \quad \text{and} \quad y = k + b\cos\theta, \quad 0 \le \theta \le 2\pi$$

is also the ellipse (traced clockwise) given by

$$\frac{(x-h)^2}{a^2} + \frac{(y-k)^2}{b^2} = 1.$$

TECHNOLOGY Use a graphing utility in *parametric* mode to graph several ellipses.

In Examples 2 and 3, it is important to realize that eliminating the parameter is primarily an *aid to curve sketching.* When the parametric equations represent the path of a moving object, the graph alone is not sufficient to describe the object's motion. You still need the parametric equations to tell you the *position, direction,* and *speed* at a given time.

Finding Parametric Equations

The first three examples in this section illustrate techniques for sketching the graph represented by a set of parametric equations. You will now investigate the reverse problem. How can you determine a set of parametric equations for a given graph or a given physical description? From the discussion following Example 1, you know that such a representation is not unique. This is demonstrated further in the next example, which finds two different parametric representations for a given graph.

EXAMPLE 4 **Finding Parametric Equations for a Given Graph**

Find a set of parametric equations that represents the graph of $y = 1 - x^2$, using each of the following parameters.

a. $t = x$ **b.** The slope $m = \dfrac{dy}{dx}$ at the point (x, y)

Solution

a. Letting $x = t$ produces the parametric equations

$$x = t \quad \text{and} \quad y = 1 - x^2 = 1 - t^2.$$

b. To write x and y in terms of the parameter m, you can proceed as follows.

$$m = \frac{dy}{dx}$$

$$m = -2x \qquad \text{Differentiate } y = 1 - x^2.$$

$$x = -\frac{m}{2} \qquad \text{Solve for } x.$$

This produces a parametric equation for x. To obtain a parametric equation for y, substitute $-m/2$ for x in the original equation.

$$y = 1 - x^2 \qquad \text{Write original rectangular equation.}$$

$$y = 1 - \left(-\frac{m}{2}\right)^2 \qquad \text{Substitute } -m/2 \text{ for } x.$$

$$y = 1 - \frac{m^2}{4} \qquad \text{Simplify.}$$

So, the parametric equations are

$$x = -\frac{m}{2} \quad \text{and} \quad y = 1 - \frac{m^2}{4}.$$

In Figure 10.24, note that the resulting curve has a right-to-left orientation as determined by the direction of increasing values of slope m. For part (a), the curve would have the opposite orientation. ■

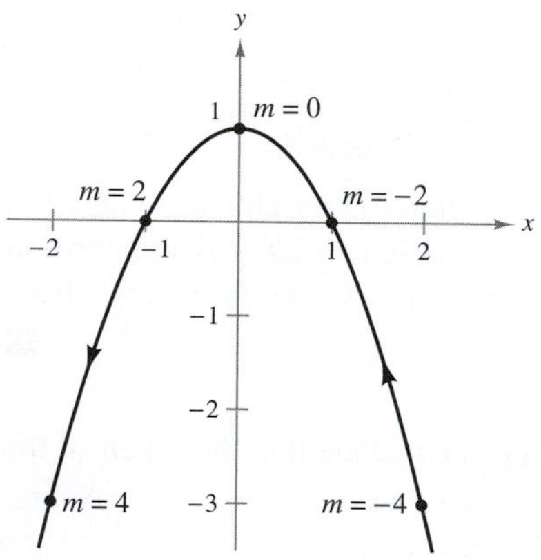

Rectangular equation: $y = 1 - x^2$
Parametric equations:
$x = -\dfrac{m}{2},\ y = 1 - \dfrac{m^2}{4}$

Figure 10.24

TECHNOLOGY To be efficient at using a graphing utility, it is important that you develop skill in representing a graph by a set of parametric equations. The reason for this is that many graphing utilities have only three graphing modes—(1) functions, (2) parametric equations, and (3) polar equations. Most graphing utilities are not programmed to graph a general equation. For instance, suppose you want to graph the hyperbola $x^2 - y^2 = 1$. To graph the hyperbola in *function* mode, you need two equations

$$y = \sqrt{x^2 - 1} \quad \text{and} \quad y = -\sqrt{x^2 - 1}.$$

In *parametric* mode, you can represent the graph by $x = \sec t$ and $y = \tan t$.

■ **FOR FURTHER INFORMATION** To read about other methods for finding parametric equations, see the article "Finding Rational Parametric Curves of Relative Degree One or Two" by Dave Boyles in *The College Mathematics Journal*. To view this article, go to *MathArticles.com*.

CYCLOIDS

Galileo first called attention to the cycloid, once recommending that it be used for the arches of bridges. Pascal once spent 8 days attempting to solve many of the problems of cycloids, such as finding the area under one arch and finding the volume of the solid of revolution formed by revolving the curve about a line. The cycloid has so many interesting properties and has caused so many quarrels among mathematicians that it has been called "the Helen of geometry" and "the apple of discord."

FOR FURTHER INFORMATION For more information on cycloids, see the article "The Geometry of Rolling Curves" by John Bloom and Lee Whitt in *The American Mathematical Monthly.* To view this article, go to *MathArticles.com.*

EXAMPLE 5 Parametric Equations for a Cycloid

Determine the curve traced by a point P on the circumference of a circle of radius a rolling along a straight line in a plane. Such a curve is called a **cycloid.**

Solution Let the parameter θ be the measure of the circle's rotation, and let the point $P = (x, y)$ begin at the origin. When $\theta = 0$, P is at the origin. When $\theta = \pi$, P is at a maximum point $(\pi a, 2a)$. When $\theta = 2\pi$, P is back on the x-axis at $(2\pi a, 0)$. From Figure 10.25, you can see that $\angle APC = 180° - \theta$. So,

$$\sin \theta = \sin(180° - \theta) = \sin(\angle APC) = \frac{AC}{a} = \frac{BD}{a}$$

$$\cos \theta = -\cos(180° - \theta) = -\cos(\angle APC) = \frac{AP}{-a}$$

which implies that $AP = -a \cos \theta$ and $BD = a \sin \theta$.

Because the circle rolls along the x-axis, you know that $OD = \widehat{PD} = a\theta$. Furthermore, because $BA = DC = a$, you have

$$x = OD - BD = a\theta - a \sin \theta$$
$$y = BA + AP = a - a \cos \theta.$$

So, the parametric equations are

$$x = a(\theta - \sin \theta) \quad \text{and} \quad y = a(1 - \cos \theta).$$

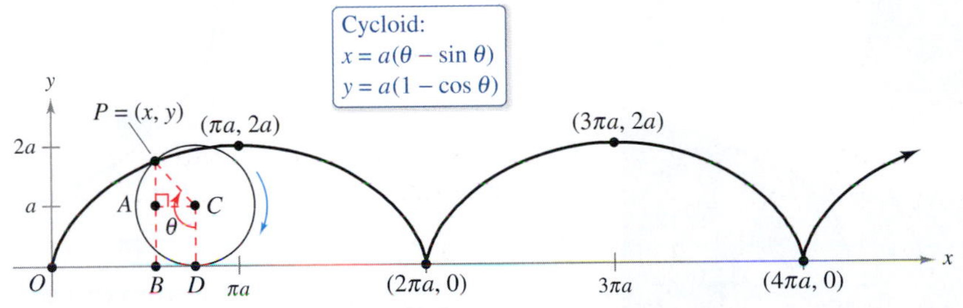

Figure 10.25

TECHNOLOGY Some graphing utilities allow you to simulate the motion of an object that is moving in the plane or in space. If you have access to such a utility, use it to trace out the path of the cycloid shown in Figure 10.25.

The cycloid in Figure 10.25 has sharp corners at the values $x = 2n\pi a$. Notice that the derivatives $x'(\theta)$ and $y'(\theta)$ are both zero at the points for which $\theta = 2n\pi$.

$$x(\theta) = a(\theta - \sin \theta) \qquad y(\theta) = a(1 - \cos \theta)$$
$$x'(\theta) = a - a \cos \theta \qquad y'(\theta) = a \sin \theta$$
$$x'(2n\pi) = 0 \qquad y'(2n\pi) = 0$$

Between these points, the cycloid is called **smooth.**

Definition of a Smooth Curve

A curve C represented by $x = f(t)$ and $y = g(t)$ on an interval I is called **smooth** when f' and g' are continuous on I and not simultaneously 0, except possibly at the endpoints of I. The curve C is called **piecewise smooth** when it is smooth on each subinterval of some partition of I.

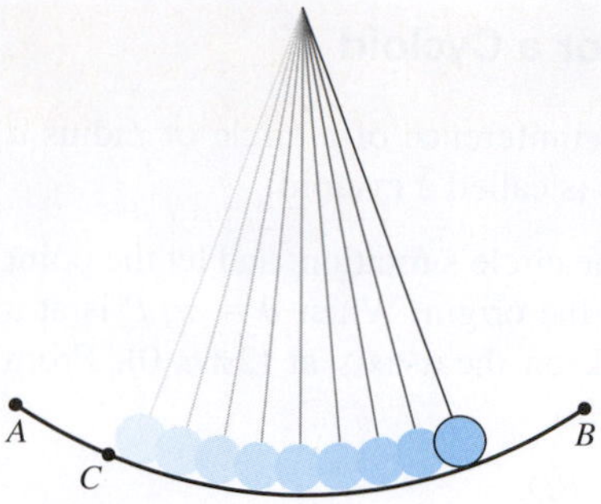

The time required to complete a full swing of the pendulum when starting from point C is only approximately the same as the time required when starting from point A.
Figure 10.26

The Tautochrone and Brachistochrone Problems

The curve described in Example 5 is related to one of the most famous pairs of problems in the history of calculus. The first problem (called the **tautochrone problem**) began with Galileo's discovery that the time required to complete a full swing of a pendulum is *approximately* the same whether it makes a large movement at high speed or a small movement at lower speed (see Figure 10.26). Late in his life, Galileo realized that he could use this principle to construct a clock. However, he was not able to conquer the mechanics of actual construction. Christian Huygens (1629–1695) was the first to design and construct a working model. In his work with pendulums, Huygens realized that a pendulum does not take exactly the same time to complete swings of varying lengths. (This doesn't affect a pendulum clock, because the length of the circular arc is kept constant by giving the pendulum a slight boost each time it passes its lowest point.) But, in studying the problem, Huygens discovered that a ball rolling back and forth on an inverted cycloid does complete each cycle in exactly the same time.

The second problem, which was posed by John Bernoulli in 1696, is called the **brachistochrone problem**—in Greek, *brachys* means short and *chronos* means time. The problem was to determine the path down which a particle (such as a ball) will slide from point A to point B in the *shortest time.* Several mathematicians took up the challenge, and the following year the problem was solved by Newton, Leibniz, L'Hôpital, John Bernoulli, and James Bernoulli. As it turns out, the solution is not a straight line from A to B, but an inverted cycloid passing through the points A and B, as shown in Figure 10.27.

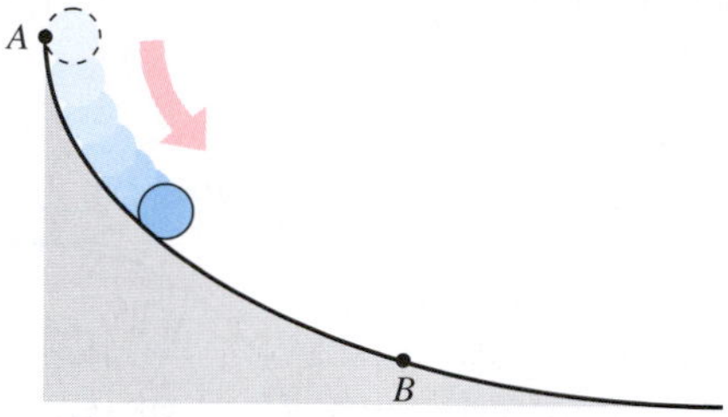

An inverted cycloid is the path down which a ball will roll in the shortest time.
Figure 10.27

The amazing part of the solution to the brachistochrone problem is that a particle starting at rest at *any* point C of the cycloid between A and B will take exactly the same time to reach B, as shown in Figure 10.28.

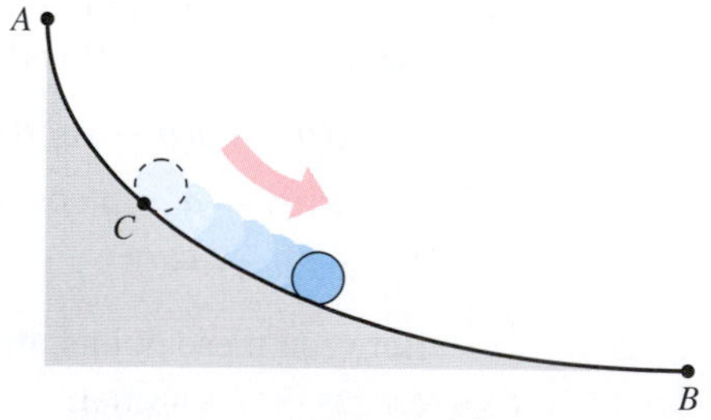

A ball starting at point C takes the same time to reach point B as one that starts at point A.
Figure 10.28

JAMES BERNOULLI (1654–1705)

James Bernoulli, also called Jacques, was the older brother of John. He was one of several accomplished mathematicians of the Swiss Bernoulli family. James's mathematical accomplishments have given him a prominent place in the early development of calculus.

See LarsonCalculus.com to read more of this biography.

FOR FURTHER INFORMATION To see a proof of the famous brachistochrone problem, see the article "A New Minimization Proof for the Brachistochrone" by Gary Lawlor in *The American Mathematical Monthly.* To view this article, go to *MathArticles.com.*

INTERFOTO/Alamy

10.2 Exercises

See CalcChat.com for tutorial help and worked-out solutions to odd-numbered exercises.

Using Parametric Equations **In Exercises 1–18, sketch the curve represented by the parametric equations (indicate the orientation of the curve), and write the corresponding rectangular equation by eliminating the parameter.**

1. $x = 2t - 3, \quad y = 3t + 1$
2. $x = 5 - 4t, \quad y = 2 + 5t$
3. $x = t + 1, \quad y = t^2$
4. $x = 2t^2, \quad y = t^4 + 1$
5. $x = t^3, \quad y = \dfrac{t^2}{2}$
6. $x = t^2 + t, \quad y = t^2 - t$
7. $x = \sqrt{t}, \quad y = t - 5$
8. $x = \sqrt[4]{t}, \quad y = 8 - t$
9. $x = t - 3, \quad y = \dfrac{t}{t - 3}$
10. $x = 1 + \dfrac{1}{t}, \quad y = t - 1$
11. $x = 2t, \quad y = |t - 2|$
12. $x = |t - 1|, \quad y = t + 2$
13. $x = e^t, \quad y = e^{3t} + 1$
14. $x = e^{-t}, \quad y = e^{2t} - 1$
15. $x = \sec\theta, \quad y = \cos\theta, \quad 0 \le \theta < \pi/2, \quad \pi/2 < \theta \le \pi$
16. $x = \tan^2\theta, \quad y = \sec^2\theta$
17. $x = 8\cos\theta, \quad y = 8\sin\theta$
18. $x = 3\cos\theta, \quad y = 7\sin\theta$

Using Parametric Equations **In Exercises 19–30, use a graphing utility to graph the curve represented by the parametric equations (indicate the orientation of the curve). Eliminate the parameter and write the corresponding rectangular equation.**

19. $x = 6\sin 2\theta$, $y = 4\cos 2\theta$
20. $x = \cos\theta$, $y = 2\sin 2\theta$
21. $x = 4 + 2\cos\theta$, $y = -1 + \sin\theta$
22. $x = -2 + 3\cos\theta$, $y = -5 + 3\sin\theta$
23. $x = -3 + 4\cos\theta$, $y = 2 + 5\sin\theta$
24. $x = \sec\theta$, $y = \tan\theta$
25. $x = 4\sec\theta$, $y = 3\tan\theta$
26. $x = \cos^3\theta$, $y = \sin^3\theta$
27. $x = t^3, \quad y = 3\ln t$
28. $x = \ln 2t, \quad y = t^2$
29. $x = e^{-t}, \quad y = e^{3t}$
30. $x = e^{2t}, \quad y = e^t$

Comparing Plane Curves **In Exercises 31–34, determine any differences between the curves of the parametric equations. Are the graphs the same? Are the orientations the same? Are the curves smooth? Explain.**

31. (a) $x = t$, $y = 2t + 1$
    (b) $x = \cos\theta$, $y = 2\cos\theta + 1$
    (c) $x = e^{-t}$, $y = 2e^{-t} + 1$
    (d) $x = e^t$, $y = 2e^t + 1$

32. (a) $x = 2\cos\theta$, $y = 2\sin\theta$
    (b) $x = \sqrt{4t^2 - 1}/|t|$, $y = 1/t$
    (c) $x = \sqrt{t}$, $y = \sqrt{4 - t}$
    (d) $x = -\sqrt{4 - e^{2t}}$, $y = e^t$

33. (a) $x = \cos\theta$, $y = 2\sin^2\theta$, $0 < \theta < \pi$
    (b) $x = \cos(-\theta)$, $y = 2\sin^2(-\theta)$, $0 < \theta < \pi$

34. (a) $x = t + 1, y = t^3$
    (b) $x = -t + 1, y = (-t)^3$

35. **Conjecture**
    (a) Use a graphing utility to graph the curves represented by the two sets of parametric equations.
    $x = 4\cos t$, $y = 3\sin t$ $\qquad$ $x = 4\cos(-t)$, $y = 3\sin(-t)$
    (b) Describe the change in the graph when the sign of the parameter is changed.
    (c) Make a conjecture about the change in the graph of parametric equations when the sign of the parameter is changed.
    (d) Test your conjecture with another set of parametric equations.

36. **Writing** Review Exercises 31–34 and write a short paragraph describing how the graphs of curves represented by different sets of parametric equations can differ even though eliminating the parameter from each yields the same rectangular equation.

Eliminating a Parameter **In Exercises 37–40, eliminate the parameter and obtain the standard form of the rectangular equation.**

37. Line through (x_1, y_1) and (x_2, y_2):
    $x = x_1 + t(x_2 - x_1), \quad y = y_1 + t(y_2 - y_1)$
38. Circle: $x = h + r\cos\theta, \quad y = k + r\sin\theta$
39. Ellipse: $x = h + a\cos\theta, \quad y = k + b\sin\theta$
40. Hyperbola: $x = h + a\sec\theta, \quad y = k + b\tan\theta$

Writing a Set of Parametric Equations **In Exercises 41–48, use the results of Exercises 37–40 to find a set of parametric equations for the line or conic.**

41. Line: passes through $(0, 0)$ and $(4, -7)$
42. Line: passes through $(1, 4)$ and $(5, -2)$
43. Circle: center: $(3, 1)$; radius: 2
44. Circle: center: $(-6, 2)$; radius: 4
45. Ellipse: vertices: $(\pm 10, 0)$; foci: $(\pm 8, 0)$
46. Ellipse: vertices: $(4, 7), (4, -3)$; foci: $(4, 5), (4, -1)$
47. Hyperbola: vertices: $(\pm 4, 0)$; foci: $(\pm 5, 0)$
48. Hyperbola: vertices: $(0, \pm 1)$; foci: $(0, \pm 2)$

Finding Parametric Equations **In Exercises 49–52, find two different sets of parametric equations for the rectangular equation.**

49. $y = 6x - 5$

50. $y = 4/(x - 1)$

51. $y = x^3$

52. $y = x^2$

Finding Parametric Equations **In Exercises 53–56, find a set of parametric equations for the rectangular equation that satisfies the given condition.**

53. $y = 2x - 5$, $t = 0$ at the point $(3, 1)$

54. $y = 4x + 1$, $t = -1$ at the point $(-2, -7)$

55. $y = x^2$, $t = 4$ at the point $(4, 16)$

56. $y = 4 - x^2$, $t = 1$ at the point $(1, 3)$

Graphing a Plane Curve **In Exercises 57–64, use a graphing utility to graph the curve represented by the parametric equations. Indicate the direction of the curve. Identify any points at which the curve is not smooth.**

57. Cycloid: $x = 2(\theta - \sin\theta)$, $y = 2(1 - \cos\theta)$

58. Cycloid: $x = \theta + \sin\theta$, $y = 1 - \cos\theta$

59. Prolate cycloid: $x = \theta - \frac{3}{2}\sin\theta$, $y = 1 - \frac{3}{2}\cos\theta$

60. Prolate cycloid: $x = 2\theta - 4\sin\theta$, $y = 2 - 4\cos\theta$

61. Hypocycloid: $x = 3\cos^3\theta$, $y = 3\sin^3\theta$

62. Curtate cycloid: $x = 2\theta - \sin\theta$, $y = 2 - \cos\theta$

63. Witch of Agnesi: $x = 2\cot\theta$, $y = 2\sin^2\theta$

64. Folium of Descartes: $x = 3t/(1 + t^3)$, $y = 3t^2/(1 + t^3)$

WRITING ABOUT CONCEPTS

65. Plane Curve State the definition of a plane curve given by parametric equations.

66. Plane Curve Explain the process of sketching a plane curve given by parametric equations. What is meant by the orientation of the curve?

67. Smooth Curve State the definition of a smooth curve.

68. HOW DO YOU SEE IT? Which set of parametric equations is shown in the graph below? Explain your reasoning.

(a) $x = t$, $y = t^2$

(b) $x = t^2$, $y = t$

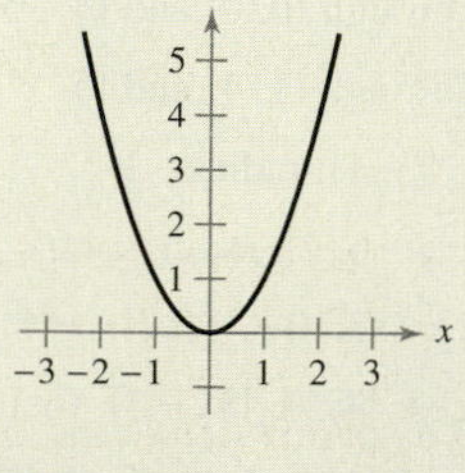

Matching **In Exercises 69–72, match each set of parametric equations with the correct graph. [The graphs are labeled (a), (b), (c), and (d).] Explain your reasoning.**

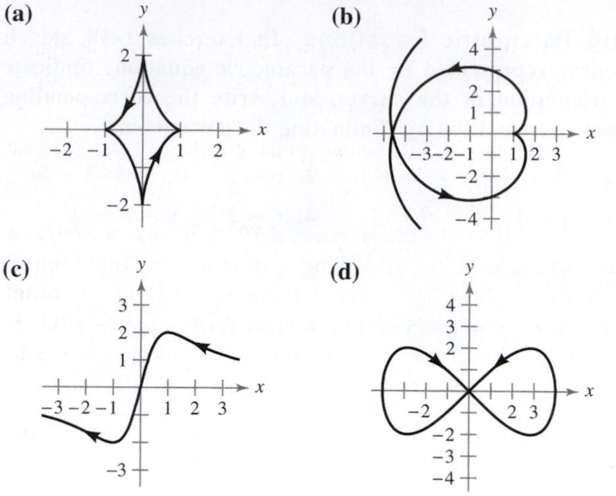

69. Lissajous curve: $x = 4\cos\theta$, $y = 2\sin 2\theta$

70. Evolute of ellipse: $x = \cos^3\theta$, $y = 2\sin^3\theta$

71. Involute of circle: $x = \cos\theta + \theta\sin\theta$, $y = \sin\theta - \theta\cos\theta$

72. Serpentine curve: $x = \cot\theta$, $y = 4\sin\theta\cos\theta$

73. Curtate Cycloid A wheel of radius a rolls along a line without slipping. The curve traced by a point P that is b units from the center $(b < a)$ is called a **curtate cycloid** (see figure). Use the angle θ to find a set of parametric equations for this curve.

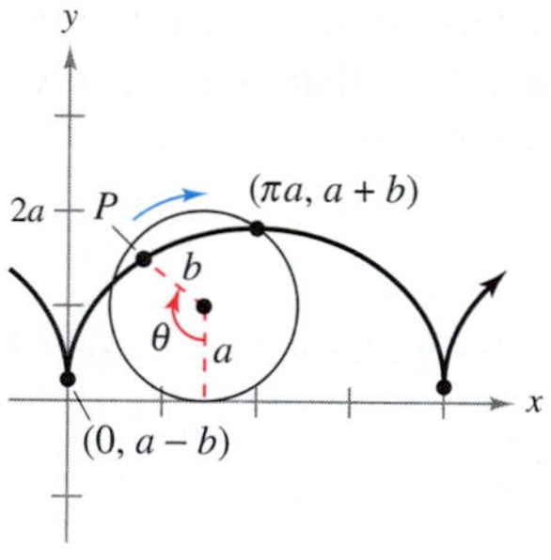

Figure for 73

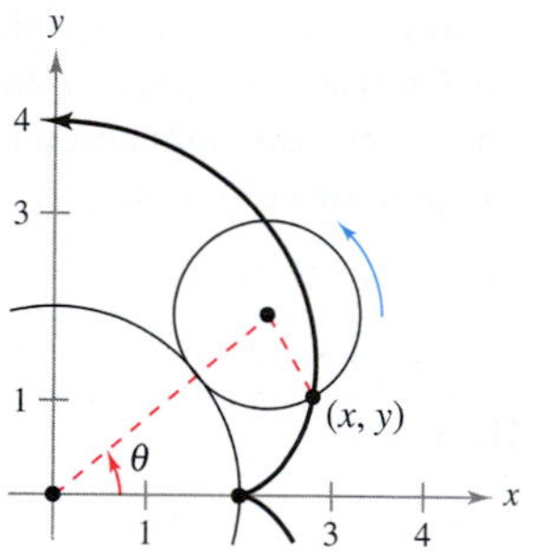

Figure for 74

74. Epicycloid A circle of radius 1 rolls around the outside of a circle of radius 2 without slipping. The curve traced by a point on the circumference of the smaller circle is called an epicycloid (see figure). Use the angle θ to find a set of parametric equations for this curve.

True or False? **In Exercises 75–77, determine whether the statement is true or false. If it is false, explain why or give an example that shows it is false.**

75. The graph of the parametric equations $x = t^2$ and $y = t^2$ is the line $y = x$.

76. If y is a function of t and x is a function of t, then y is a function of x.

77. The curve represented by the parametric equations $x = t$ and $y = \cos t$ can be written as an equation of the form $y = f(x)$.

78. Translation of a Plane Curve Consider the parametric equations $x = 8\cos t$ and $y = 8\sin t$.

(a) Describe the curve represented by the parametric equations.

(b) How does the curve represented by the parametric equations $x = 8\cos t + 3$ and $y = 8\sin t + 6$ compare to the curve described in part (a)?

(c) How does the original curve change when cosine and sine are interchanged?

Projectile Motion In Exercises 79 and 80, consider a projectile launched at a height h feet above the ground and at an angle θ with the horizontal. When the initial velocity is v_0 feet per second, the path of the projectile is modeled by the parametric equations $x = (v_0 \cos\theta)t$ and $y = h + (v_0 \sin\theta)t - 16t^2$.

79. The center field fence in a ballpark is 10 feet high and 400 feet from home plate. The ball is hit 3 feet above the ground. It leaves the bat at an angle of θ degrees with the horizontal at a speed of 100 miles per hour (see figure).

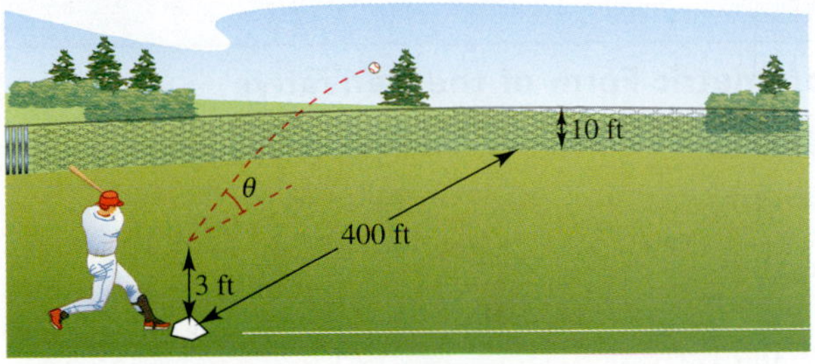

(a) Write a set of parametric equations for the path of the ball.

(b) Use a graphing utility to graph the path of the ball when $\theta = 15°$. Is the hit a home run?

(c) Use a graphing utility to graph the path of the ball when $\theta = 23°$. Is the hit a home run?

(d) Find the minimum angle at which the ball must leave the bat in order for the hit to be a home run.

80. A rectangular equation for the path of a projectile is $y = 5 + x - 0.005x^2$.

(a) Eliminate the parameter t from the position function for the motion of a projectile to show that the rectangular equation is

$$y = -\frac{16\sec^2\theta}{v_0^2}x^2 + (\tan\theta)x + h.$$

(b) Use the result of part (a) to find h, v_0, and θ. Find the parametric equations of the path.

(c) Use a graphing utility to graph the rectangular equation for the path of the projectile. Confirm your answer in part (b) by sketching the curve represented by the parametric equations.

(d) Use a graphing utility to approximate the maximum height of the projectile and its range.

SECTION PROJECT

Cycloids

In Greek, the word *cycloid* means *wheel*, the word *hypocycloid* means *under the wheel*, and the word *epicycloid* means *upon the wheel*. Match the hypocycloid or epicycloid with its graph. [The graphs are labeled (a), (b), (c), (d), (e), and (f).]

Hypocycloid, H(A, B)

The path traced by a fixed point on a circle of radius B as it rolls around the *inside* of a circle of radius A

$$x = (A - B)\cos t + B\cos\left(\frac{A - B}{B}\right)t$$

$$y = (A - B)\sin t - B\sin\left(\frac{A - B}{B}\right)t$$

Epicycloid, E(A, B)

The path traced by a fixed point on a circle of radius B as it rolls around the *outside* of a circle of radius A

$$x = (A + B)\cos t - B\cos\left(\frac{A + B}{B}\right)t$$

$$y = (A + B)\sin t - B\sin\left(\frac{A + B}{B}\right)t$$

I. $H(8, 3)$ II. $E(8, 3)$ III. $H(8, 7)$

IV. $E(24, 3)$ V. $H(24, 7)$ VI. $E(24, 7)$

(a)

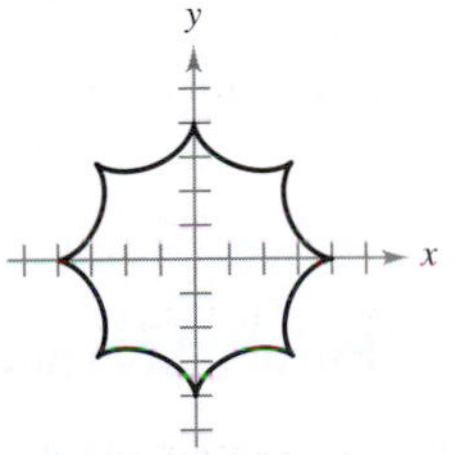

(b)

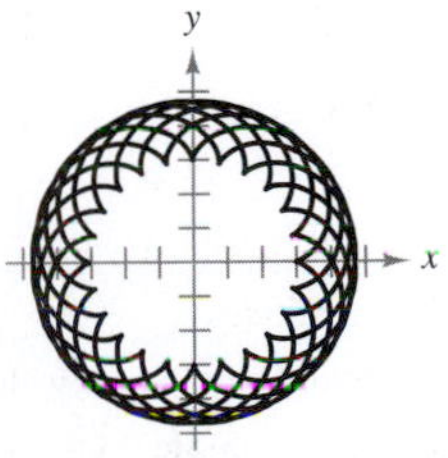

(c)

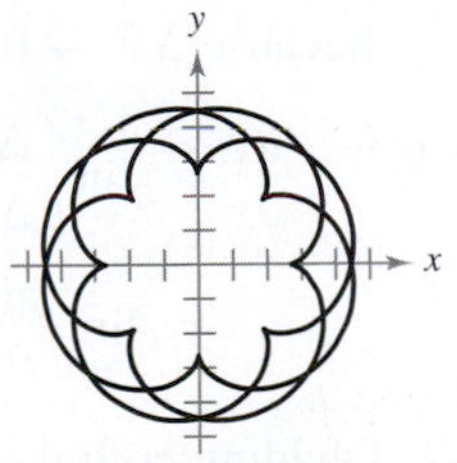

(d)

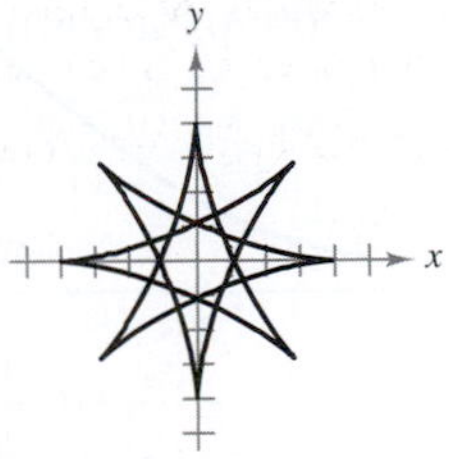

(e)

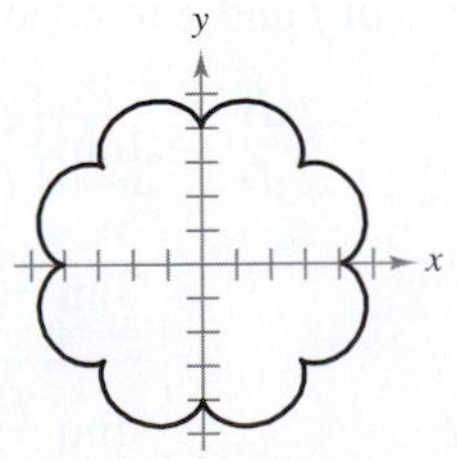

(f)

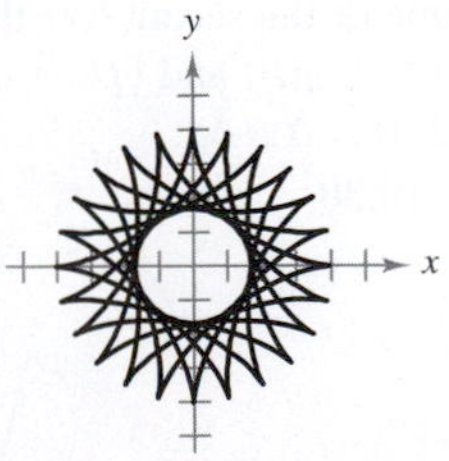

Exercises based on "Mathematical Discovery via Computer Graphics: Hypocycloids and Epicycloids" by Florence S. Gordon and Sheldon P. Gordon, *College Mathematics Journal*, November 1984, p. 441. Used by permission of the authors.

10.3 Parametric Equations and Calculus

- Find the slope of a tangent line to a curve given by a set of parametric equations.
- Find the arc length of a curve given by a set of parametric equations.
- Find the area of a surface of revolution (parametric form).

Slope and Tangent Lines

Now that you can represent a graph in the plane by a set of parametric equations, it is natural to ask how to use calculus to study plane curves. Consider the projectile represented by the parametric equations

$$x = 24\sqrt{2}t \quad \text{and} \quad y = -16t^2 + 24\sqrt{2}t$$

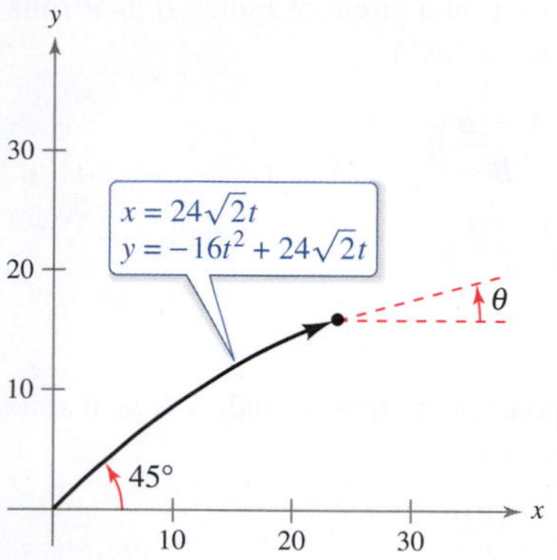

At time t, the angle of elevation of the projectile is θ.
Figure 10.29

as shown in Figure 10.29. From the discussion at the beginning of Section 10.2, you know that these equations enable you to locate the position of the projectile at a given time. You also know that the object is initially projected at an angle of 45°, or a slope of $m = \tan 45° = 1$. But how can you find the slope at some other time t? The next theorem answers this question by giving a formula for the slope of the tangent line as a function of t.

THEOREM 10.7 Parametric Form of the Derivative

If a smooth curve C is given by the equations

$$x = f(t) \quad \text{and} \quad y = g(t)$$

then the slope of C at (x, y) is

$$\frac{dy}{dx} = \frac{dy/dt}{dx/dt}, \quad \frac{dx}{dt} \neq 0.$$

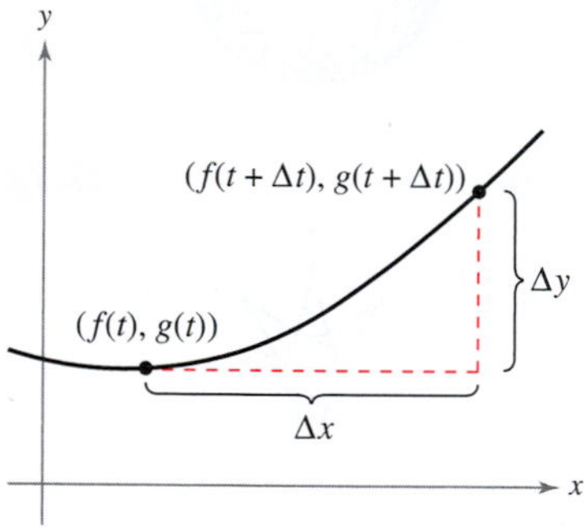

The slope of the secant line through the points $(f(t), g(t))$ and $(f(t + \Delta t), g(t + \Delta t))$ is $\Delta y/\Delta x$.
Figure 10.30

Proof In Figure 10.30, consider $\Delta t > 0$ and let

$$\Delta y = g(t + \Delta t) - g(t) \quad \text{and} \quad \Delta x = f(t + \Delta t) - f(t).$$

Because $\Delta x \to 0$ as $\Delta t \to 0$, you can write

$$\frac{dy}{dx} = \lim_{\Delta x \to 0} \frac{\Delta y}{\Delta x} = \lim_{\Delta t \to 0} \frac{g(t + \Delta t) - g(t)}{f(t + \Delta t) - f(t)}.$$

Dividing both the numerator and denominator by Δt, you can use the differentiability of f and g to conclude that

$$\frac{dy}{dx} = \lim_{\Delta t \to 0} \frac{[g(t + \Delta t) - g(t)]/\Delta t}{[f(t + \Delta t) - f(t)]/\Delta t} = \frac{\lim_{\Delta t \to 0} \frac{g(t + \Delta t) - g(t)}{\Delta t}}{\lim_{\Delta t \to 0} \frac{f(t + \Delta t) - f(t)}{\Delta t}} = \frac{g'(t)}{f'(t)} = \frac{dy/dt}{dx/dt}.$$

See LarsonCalculus.com for Bruce Edwards's video of this proof. ■

AP* Tips

The calculus of parametric functions is tested on the AP Calculus BC Exam on both the multiple choice and free response sections.

Exploration

The curve traced out in Example 1 is a circle. Use the formula

$$\frac{dy}{dx} = -\tan t$$

to find the slopes at the points (1, 0) and (0, 1).

EXAMPLE 1 Differentiation and Parametric Form

Find dy/dx for the curve given by $x = \sin t$ and $y = \cos t$.

Solution

$$\begin{aligned}\frac{dy}{dx} &= \frac{dy/dt}{dx/dt}\\ &= \frac{-\sin t}{\cos t}\\ &= -\tan t\end{aligned}$$

Because dy/dx is a function of t, you can use Theorem 10.7 repeatedly to find *higher-order* derivatives. For instance,

$$\frac{d^2y}{dx^2} = \frac{d}{dx}\left[\frac{d}{dx}\right] = \frac{\frac{d}{dt}\left[\frac{dy}{dx}\right]}{dx/dt} \qquad \text{Second derivative}$$

$$\frac{d^3y}{dx^3} = \frac{d}{dx}\left[\frac{d^2y}{dx^2}\right] = \frac{\frac{d}{dt}\left[\frac{d^2y}{dx^2}\right]}{dx/dt}. \qquad \text{Third derivative}$$

AP* Tips

The concepts of single variable calculus extend to vector-valued functions. Vector-related free response questions on the BC Exam are frequently designed to test more foundational calculus concepts. See Appendix G: AP* BC Vector Topics.

EXAMPLE 2 Finding Slope and Concavity

For the curve given by

$$x = \sqrt{t} \quad \text{and} \quad y = \frac{1}{4}(t^2 - 4), \quad t \geq 0$$

find the slope and concavity at the point (2, 3).

Solution Because

$$\frac{dy}{dx} = \frac{dy/dt}{dx/dt} = \frac{(1/2)t}{(1/2)t^{-1/2}} = t^{3/2} \qquad \text{Parametric form of first derivative}$$

you can find the second derivative to be

$$\frac{d^2y}{dx^2} = \frac{\frac{d}{dt}[dy/dx]}{dx/dt} = \frac{\frac{d}{dt}[t^{3/2}]}{dx/dt} = \frac{(3/2)t^{1/2}}{(1/2)t^{-1/2}} = 3t. \qquad \text{Parametric form of second derivative}$$

At $(x, y) = (2, 3)$, it follows that $t = 4$, and the slope is

$$\frac{dy}{dx} = (4)^{3/2} = 8.$$

Moreover, when $t = 4$, the second derivative is

$$\frac{d^2y}{dx^2} = 3(4) = 12 > 0$$

and you can conclude that the graph is concave upward at (2, 3), as shown in Figure 10.31.

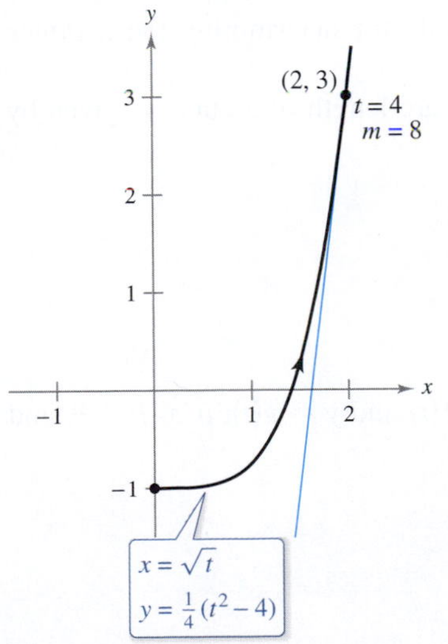

The graph is concave upward at (2, 3) when $t = 4$.
Figure 10.31

Because the parametric equations $x = f(t)$ and $y = g(t)$ need not define y as a function of x, it is possible for a plane curve to loop around and cross itself. At such points, the curve may have more than one tangent line, as shown in the next example.

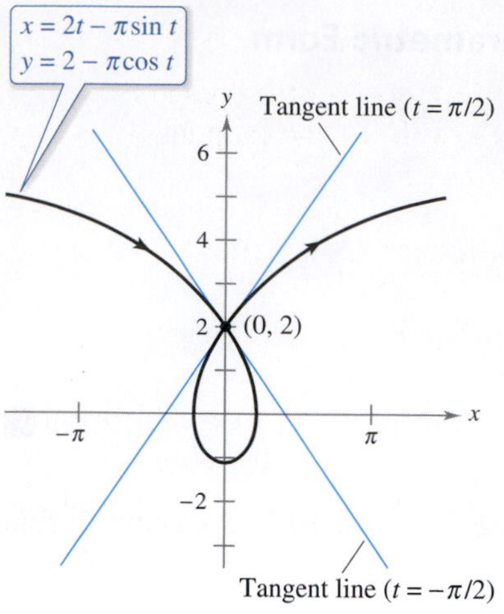

This prolate cycloid has two tangent lines at the point (0, 2).
Figure 10.32

EXAMPLE 3 **A Curve with Two Tangent Lines at a Point**

See LarsonCalculus.com for an interactive version of this type of example.

The **prolate cycloid** given by

$$x = 2t - \pi \sin t \quad \text{and} \quad y = 2 - \pi \cos t$$

crosses itself at the point (0, 2), as shown in Figure 10.32. Find the equations of both tangent lines at this point.

Solution Because $x = 0$ and $y = 2$ when $t = \pm\pi/2$, and

$$\frac{dy}{dx} = \frac{dy/dt}{dx/dt} = \frac{\pi \sin t}{2 - \pi \cos t}$$

you have $dy/dx = -\pi/2$ when $t = -\pi/2$ and $dy/dx = \pi/2$ when $t = \pi/2$. So, the two tangent lines at (0, 2) are

$$y - 2 = -\left(\frac{\pi}{2}\right)x \qquad \text{Tangent line when } t = -\frac{\pi}{2}$$

and

$$y - 2 = \left(\frac{\pi}{2}\right)x. \qquad \text{Tangent line when } t = \frac{\pi}{2}$$

If $dy/dt = 0$ and $dx/dt \neq 0$ when $t = t_0$, then the curve represented by $x = f(t)$ and $y = g(t)$ has a horizontal tangent at $(f(t_0), g(t_0))$. For instance, in Example 3, the given curve has a horizontal tangent at the point $(0, 2 - \pi)$ (when $t = 0$). Similarly, if $dx/dt = 0$ and $dy/dt \neq 0$ when $t = t_0$, then the curve represented by $x = f(t)$ and $y = g(t)$ has a vertical tangent at $(f(t_0), g(t_0))$.

Arc Length

You have seen how parametric equations can be used to describe the path of a particle moving in the plane. You will now develop a formula for determining the *distance* traveled by the particle along its path.

Recall from Section 7.4 that the formula for the arc length of a curve C given by $y = h(x)$ over the interval $[x_0, x_1]$ is

$$\begin{aligned} s &= \int_{x_0}^{x_1} \sqrt{1 + [h'(x)]^2}\, dx \\ &= \int_{x_0}^{x_1} \sqrt{1 + \left(\frac{dy}{dx}\right)^2}\, dx. \end{aligned}$$

If C is represented by the parametric equations $x = f(t)$ and $y = g(t)$, $a \le t \le b$, and if $dx/dt = f'(t) > 0$, then

$$\begin{aligned} s &= \int_{x_0}^{x_1} \sqrt{1 + \left(\frac{dy}{dx}\right)^2}\, dx \\ &= \int_{x_0}^{x_1} \sqrt{1 + \left(\frac{dy/dt}{dx/dt}\right)^2}\, dx \\ &= \int_a^b \sqrt{\frac{(dx/dt)^2 + (dy/dt)^2}{(dx/dt)^2}}\, \frac{dx}{dt}\, dt \\ &= \int_a^b \sqrt{\left(\frac{dx}{dt}\right)^2 + \left(\frac{dy}{dt}\right)^2}\, dt \\ &= \int_a^b \sqrt{[f'(t)]^2 + [g'(t)]^2}\, dt. \end{aligned}$$

THEOREM 10.8 Arc Length in Parametric Form

If a smooth curve C is given by $x = f(t)$ and $y = g(t)$ such that C does not intersect itself on the interval $a \le t \le b$ (except possibly at the endpoints), then the arc length of C over the interval is given by

$$s = \int_a^b \sqrt{\left(\frac{dx}{dt}\right)^2 + \left(\frac{dy}{dt}\right)^2}\, dt = \int_a^b \sqrt{[f'(t)]^2 + [g'(t)]^2}\, dt.$$

REMARK When applying the arc length formula to a curve, be sure that the curve is traced out only once on the interval of integration. For instance, the circle given by $x = \cos t$ and $y = \sin t$ is traced out once on the interval $0 \le t \le 2\pi$, but is traced out twice on the interval $0 \le t \le 4\pi$.

In the preceding section, you saw that if a circle rolls along a line, then a point on its circumference will trace a path called a cycloid. If the circle rolls around the circumference of another circle, then the path of the point is an **epicycloid.** The next example shows how to find the arc length of an epicycloid.

ARCH OF A CYCLOID

The arc length of an arch of a cycloid was first calculated in 1658 by British architect and mathematician Christopher Wren, famous for rebuilding many buildings and churches in London, including St. Paul's Cathedral.

EXAMPLE 4 Finding Arc Length

A circle of radius 1 rolls around the circumference of a larger circle of radius 4, as shown in Figure 10.33. The epicycloid traced by a point on the circumference of the smaller circle is given by

$$x = 5\cos t - \cos 5t \quad \text{and} \quad y = 5\sin t - \sin 5t.$$

Find the distance traveled by the point in one complete trip about the larger circle.

Solution Before applying Theorem 10.8, note in Figure 10.33 that the curve has sharp points when $t = 0$ and $t = \pi/2$. Between these two points, dx/dt and dy/dt are not simultaneously 0. So, the portion of the curve generated from $t = 0$ to $t = \pi/2$ is smooth. To find the total distance traveled by the point, you can find the arc length of that portion lying in the first quadrant and multiply by 4.

$$
\begin{aligned}
s &= 4\int_0^{\pi/2} \sqrt{\left(\frac{dx}{dt}\right)^2 + \left(\frac{dy}{dt}\right)^2}\, dt && \text{Parametric form for arc length}\\
&= 4\int_0^{\pi/2} \sqrt{(-5\sin t + 5\sin 5t)^2 + (5\cos t - 5\cos 5t)^2}\, dt\\
&= 20\int_0^{\pi/2} \sqrt{2 - 2\sin t\sin 5t - 2\cos t\cos 5t}\, dt\\
&= 20\int_0^{\pi/2} \sqrt{2 - 2\cos 4t}\, dt && \text{Difference formula for cosine}\\
&= 20\int_0^{\pi/2} \sqrt{4\sin^2 2t}\, dt && \text{Double-angle formula}\\
&= 40\int_0^{\pi/2} \sin 2t\, dt\\
&= -20\Big[\cos 2t\Big]_0^{\pi/2}\\
&= 40
\end{aligned}
$$

AP* Tips

On the AP Exam, the arc length of a curve given in parametric form is frequently interpreted as the total distance travelled by a particle.

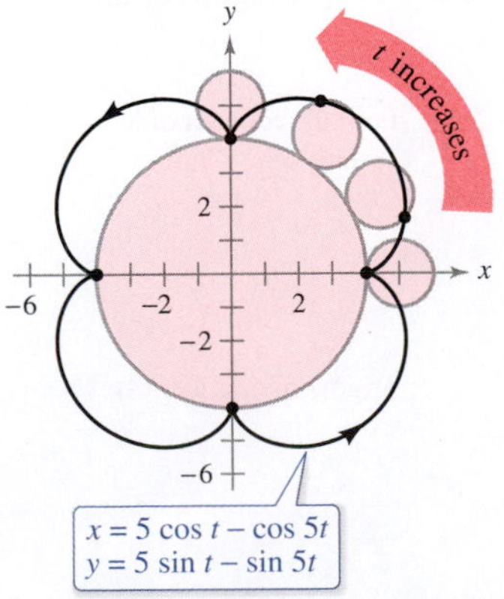

An epicycloid is traced by a point on the smaller circle as it rolls around the larger circle.
Figure 10.33

For the epicycloid shown in Figure 10.33, an arc length of 40 seems about right because the circumference of a circle of radius 6 is

$$2\pi r = 12\pi \approx 37.7.$$

Area of a Surface of Revolution

You can use the formula for the area of a surface of revolution in rectangular form to develop a formula for surface area in parametric form.

THEOREM 10.9 Area of a Surface of Revolution

If a smooth curve C given by $x = f(t)$ and $y = g(t)$ does not cross itself on an interval $a \le t \le b$, then the area S of the surface of revolution formed by revolving C about the coordinate axes is given by the following.

1. $S = 2\pi \int_a^b g(t)\sqrt{\left(\frac{dx}{dt}\right)^2 + \left(\frac{dy}{dt}\right)^2}\, dt$ Revolution about the x-axis: $g(t) \ge 0$

2. $S = 2\pi \int_a^b f(t)\sqrt{\left(\frac{dx}{dt}\right)^2 + \left(\frac{dy}{dt}\right)^2}\, dt$ Revolution about the y-axis: $f(t) \ge 0$

These formulas may be easier to remember if you think of the differential of arc length as

$$ds = \sqrt{\left(\frac{dx}{dt}\right)^2 + \left(\frac{dy}{dt}\right)^2}\, dt.$$

Then the formulas are written as follows.

1. $S = 2\pi \int_a^b g(t)\, ds$ **2.** $S = 2\pi \int_a^b f(t)\, ds$

EXAMPLE 5 Finding the Area of a Surface of Revolution

Let C be the arc of the circle $x^2 + y^2 = 9$ from $(3, 0)$ to

$$\left(\frac{3}{2}, \frac{3\sqrt{3}}{2}\right)$$

as shown in Figure 10.34. Find the area of the surface formed by revolving C about the x-axis.

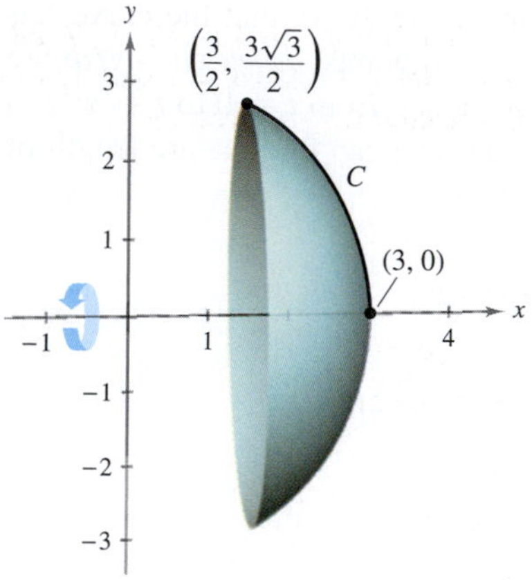

The surface of revolution has a surface area of 9π.

Figure 10.34

Solution You can represent C parametrically by the equations

$$x = 3\cos t \quad \text{and} \quad y = 3\sin t, \quad 0 \le t \le \pi/3.$$

(Note that you can determine the interval for t by observing that $t = 0$ when $x = 3$ and $t = \pi/3$ when $x = 3/2$.) On this interval, C is smooth and y is nonnegative, and you can apply Theorem 10.9 to obtain a surface area of

$$S = 2\pi \int_0^{\pi/3} (3\sin t)\sqrt{(-3\sin t)^2 + (3\cos t)^2}\, dt$$ Formula for area of a surface of revolution

$$= 6\pi \int_0^{\pi/3} \sin t\sqrt{9(\sin^2 t + \cos^2 t)}\, dt$$

$$= 6\pi \int_0^{\pi/3} 3\sin t\, dt$$ Trigonometric identity

$$= -18\pi\Big[\cos t\Big]_0^{\pi/3}$$

$$= -18\pi\left(\frac{1}{2} - 1\right)$$

$$= 9\pi.$$

10.3 Exercises

See CalcChat.com for tutorial help and worked-out solutions to odd-numbered exercises.

Finding a Derivative In Exercises 1–4, find dy/dx.

1. $x = t^2,\ y = 7 - 6t$
2. $x = \sqrt[3]{t},\ y = 4 - t$
3. $x = \sin^2 \theta,\ y = \cos^2 \theta$
4. $x = 2e^{\theta},\ y = e^{-\theta/2}$

Finding Slope and Concavity In Exercises 5–14, find dy/dx and d^2y/dx^2, and find the slope and concavity (if possible) at the given value of the parameter.

Parametric Equations	Parameter
5. $x = 4t,\ y = 3t - 2$	$t = 3$
6. $x = \sqrt{t},\ y = 3t - 1$	$t = 1$
7. $x = t + 1,\ y = t^2 + 3t$	$t = -1$
8. $x = t^2 + 5t + 4,\ y = 4t$	$t = 0$
9. $x = 4\cos\theta,\ y = 4\sin\theta$	$\theta = \dfrac{\pi}{4}$
10. $x = \cos\theta,\ y = 3\sin\theta$	$\theta = 0$
11. $x = 2 + \sec\theta,\ y = 1 + 2\tan\theta$	$\theta = \dfrac{\pi}{6}$
12. $x = \sqrt{t},\ y = \sqrt{t-1}$	$t = 2$
13. $x = \cos^3\theta,\ y = \sin^3\theta$	$\theta = \dfrac{\pi}{4}$
14. $x = \theta - \sin\theta,\ y = 1 - \cos\theta$	$\theta = \pi$

Finding Equations of Tangent Lines In Exercises 15–18, find an equation of the tangent line at each given point on the curve.

15. $x = 2\cot\theta,\quad y = 2\sin^2\theta,$
$\left(-\dfrac{2}{\sqrt{3}}, \dfrac{3}{2}\right), (0, 2), \left(2\sqrt{3}, \dfrac{1}{2}\right)$

16. $x = 2 - 3\cos\theta,\quad y = 3 + 2\sin\theta,$
$(-1, 3), (2, 5), \left(\dfrac{4 + 3\sqrt{3}}{2}, 2\right)$

17. $x = t^2 - 4,\quad y = t^2 - 2t,\quad (0, 0), (-3, -1), (-3, 3)$

18. $x = t^4 + 2,\quad y = t^3 + t,\quad (2, 0), (3, -2), (18, 10)$

Finding an Equation of a Tangent Line In Exercises 19–22, (a) use a graphing utility to graph the curve represented by the parametric equations, (b) use a graphing utility to find dx/dt, dy/dt, and dy/dx at the given value of the parameter, (c) find an equation of the tangent line to the curve at the given value of the parameter, and (d) use a graphing utility to graph the curve and the tangent line from part (c).

Parametric Equations	Parameter
19. $x = 6t,\ y = t^2 + 4$	$t = 1$
20. $x = t - 2,\ y = \dfrac{1}{t} + 3$	$t = 1$
21. $x = t^2 - t + 2,\ y = t^3 - 3t$	$t = -1$
22. $x = 3t - t^2,\quad y = 2t^{3/2}$	$t = \dfrac{1}{4}$

Finding Equations of Tangent Lines In Exercises 23–26, find the equations of the tangent lines at the point where the curve crosses itself.

23. $x = 2\sin 2t,\quad y = 3\sin t$
24. $x = 2 - \pi\cos t,\quad y = 2t - \pi\sin t$
25. $x = t^2 - t,\quad y = t^3 - 3t - 1$
26. $x = t^3 - 6t,\quad y = t^2$

Horizontal and Vertical Tangency In Exercises 27 and 28, find all points (if any) of horizontal and vertical tangency to the portion of the curve shown.

27. Involute of a circle:
$x = \cos\theta + \theta\sin\theta$
$y = \sin\theta - \theta\cos\theta$

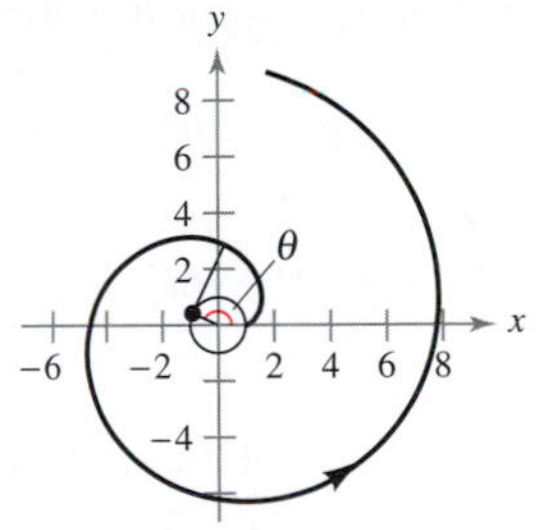

28. $x = 2\theta$
$y = 2(1 - \cos\theta)$

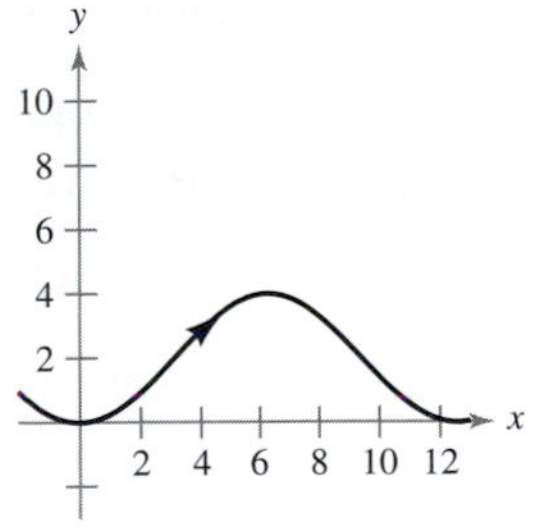

Horizontal and Vertical Tangency In Exercises 29–38, find all points (if any) of horizontal and vertical tangency to the curve. Use a graphing utility to confirm your results.

29. $x = 4 - t,\quad y = t^2$
30. $x = t + 1,\quad y = t^2 + 3t$
31. $x = t + 4,\quad y = t^3 - 3t$
32. $x = t^2 - t + 2,\quad y = t^3 - 3t$
33. $x = 3\cos\theta,\quad y = 3\sin\theta$
34. $x = \cos\theta,\quad y = 2\sin 2\theta$
35. $x = 5 + 3\cos\theta,\quad y = -2 + \sin\theta$
36. $x = 4\cos^2\theta,\quad y = 2\sin\theta$
37. $x = \sec\theta,\quad y = \tan\theta$
38. $x = \cos^2\theta,\quad y = \cos\theta$

Determining Concavity In Exercises 39–44, determine the open t-intervals on which the curve is concave downward or concave upward.

39. $x = 3t^2,\quad y = t^3 - t$
40. $x = 2 + t^2,\quad y = t^2 + t^3$
41. $x = 2t + \ln t,\quad y = 2t - \ln t$
42. $x = t^2,\quad y = \ln t$
43. $x = \sin t,\quad y = \cos t,\quad 0 < t < \pi$
44. $x = 4\cos t,\quad y = 2\sin t,\quad 0 < t < 2\pi$

Arc Length **In Exercises 45–50, find the arc length of the curve on the given interval.**

Parametric Equations	Interval
45. $x = 3t + 5, \quad y = 7 - 2t$	$-1 \le t \le 3$
46. $x = 6t^2, \quad y = 2t^3$	$1 \le t \le 4$
47. $x = e^{-t} \cos t, \quad y = e^{-t} \sin t$	$0 \le t \le \frac{\pi}{2}$
48. $x = \arcsin t, \quad y = \ln\sqrt{1 - t^2}$	$0 \le t \le \frac{1}{2}$
49. $x = \sqrt{t}, \quad y = 3t - 1$	$0 \le t \le 1$
50. $x = t, \quad y = \frac{t^5}{10} + \frac{1}{6t^3}$	$1 \le t \le 2$

Arc Length **In Exercises 51–54, find the arc length of the curve on the interval $[0, 2\pi]$.**

51. Hypocycloid perimeter: $x = a \cos^3 \theta, y = a \sin^3 \theta$

52. Circle circumference: $x = a \cos \theta, y = a \sin \theta$

53. Cycloid arch: $x = a(\theta - \sin \theta), y = a(1 - \cos \theta)$

54. Involute of a circle: $x = \cos \theta + \theta \sin \theta, y = \sin \theta - \theta \cos \theta$

55. Path of a Projectile The path of a projectile is modeled by the parametric equations

$$x = (90 \cos 30°)t \quad \text{and} \quad y = (90 \sin 30°)t - 16t^2$$

where x and y are measured in feet.

(a) Use a graphing utility to graph the path of the projectile.

(b) Use a graphing utility to approximate the range of the projectile.

(c) Use the integration capabilities of a graphing utility to approximate the arc length of the path. Compare this result with the range of the projectile.

56. Path of a Projectile When the projectile in Exercise 55 is launched at an angle θ with the horizontal, its parametric equations are

$$x = (90 \cos \theta)t \quad \text{and} \quad y = (90 \sin \theta)t - 16t^2.$$

Use a graphing utility to find the angle that maximizes the range of the projectile. What angle maximizes the arc length of the trajectory?

57. Folium of Descartes Consider the parametric equations

$$x = \frac{4t}{1 + t^3} \quad \text{and} \quad y = \frac{4t^2}{1 + t^3}.$$

(a) Use a graphing utility to graph the curve represented by the parametric equations.

(b) Use a graphing utility to find the points of horizontal tangency to the curve.

(c) Use the integration capabilities of a graphing utility to approximate the arc length of the closed loop. (*Hint:* Use symmetry and integrate over the interval $0 \le t \le 1$.)

58. Witch of Agnesi Consider the parametric equations

$$x = 4 \cot \theta \quad \text{and} \quad y = 4 \sin^2 \theta, \quad -\frac{\pi}{2} \le \theta \le \frac{\pi}{2}.$$

(a) Use a graphing utility to graph the curve represented by the parametric equations.

(b) Use a graphing utility to find the points of horizontal tangency to the curve.

(c) Use the integration capabilities of a graphing utility to approximate the arc length over the interval $\pi/4 \le \theta \le \pi/2$.

59. Writing

(a) Use a graphing utility to graph each set of parametric equations.

$$x = t - \sin t, \quad y = 1 - \cos t, \quad 0 \le t \le 2\pi$$

$$x = 2t - \sin(2t), \quad y = 1 - \cos(2t), \quad 0 \le t \le \pi$$

(b) Compare the graphs of the two sets of parametric equations in part (a). When the curve represents the motion of a particle and t is time, what can you infer about the average speeds of the particle on the paths represented by the two sets of parametric equations?

(c) Without graphing the curve, determine the time required for a particle to traverse the same path as in parts (a) and (b) when the path is modeled by

$$x = \tfrac{1}{2}t - \sin\left(\tfrac{1}{2}t\right) \quad \text{and} \quad y = 1 - \cos\left(\tfrac{1}{2}t\right).$$

60. Writing

(a) Each set of parametric equations represents the motion of a particle. Use a graphing utility to graph each set.

First Particle: $x = 3 \cos t, \quad y = 4 \sin t, \quad 0 \le t \le 2\pi$

Second Particle: $x = 4 \sin t, \quad y = 3 \cos t, \quad 0 \le t \le 2\pi$

(b) Determine the number of points of intersection.

(c) Will the particles ever be at the same place at the same time? If so, identify the point(s).

(d) Explain what happens when the motion of the second particle is represented by

$$x = 2 + 3 \sin t, \quad y = 2 - 4 \cos t, \quad 0 \le t \le 2\pi.$$

Surface Area **In Exercises 61–64, write an integral that represents the area of the surface generated by revolving the curve about the x-axis. Use a graphing utility to approximate the integral.**

Parametric Equations	Interval
61. $x = 3t, \quad y = t + 2$	$0 \le t \le 4$
62. $x = \frac{1}{4}t^2, \quad y = t + 3$	$0 \le t \le 3$
63. $x = \cos^2 \theta, \quad y = \cos \theta$	$0 \le \theta \le \frac{\pi}{2}$
64. $x = \theta + \sin \theta, \quad y = \theta + \cos \theta$	$0 \le \theta \le \frac{\pi}{2}$

Surface Area **In Exercises 65–70, find the area of the surface generated by revolving the curve about each given axis.**

65. $x = 2t, \quad y = 3t, \quad 0 \le t \le 3$

(a) x-axis (b) y-axis

66. $x = t, \quad y = 4 - 2t, \quad 0 \le t \le 2$

(a) x-axis (b) y-axis

67. $x = 5\cos\theta, \quad y = 5\sin\theta, \quad 0 \le \theta \le \dfrac{\pi}{2}, \quad y$-axis

68. $x = \frac{1}{3}t^3, \quad y = t + 1, \quad 1 \le t \le 2, \quad y$-axis

69. $x = a\cos^3\theta, \quad y = a\sin^3\theta, \quad 0 \le \theta \le \pi, \quad x$-axis

70. $x = a\cos\theta, \quad y = b\sin\theta, \quad 0 \le \theta \le 2\pi$

(a) x-axis (b) y-axis

WRITING ABOUT CONCEPTS

71. Parametric Form of the Derivative Give the parametric form of the derivative.

Mental Math **In Exercises 72 and 73, mentally determine dy/dx.**

72. $x = t, \quad y = 3$

73. $x = t, \quad y = 6t - 5$

74. Arc Length Give the integral formula for arc length in parametric form.

75. Surface Area Give the integral formulas for the areas of the surfaces of revolution formed when a smooth curve C is revolved about (a) the x-axis and (b) the y-axis.

76. HOW DO YOU SEE IT? Using the graph of f, (a) determine whether dy/dt is positive or negative given that dx/dt is negative, and (b) determine whether dx/dt is positive or negative given that dy/dt is positive. Explain your reasoning.

(i)

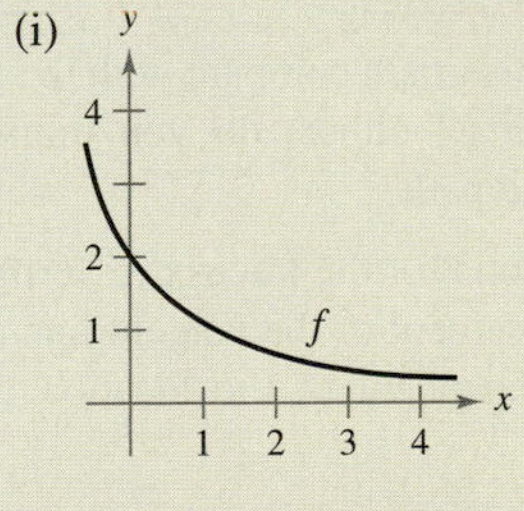

(ii)

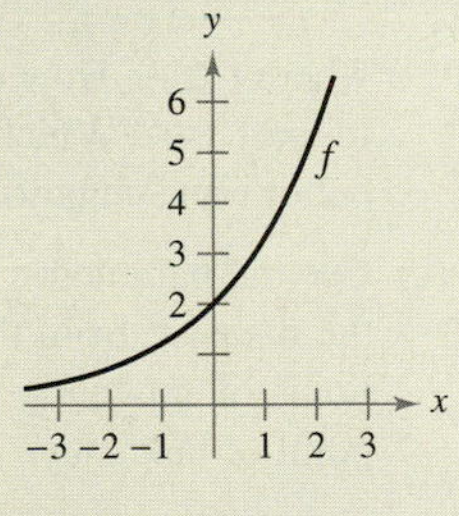

77. Integration by Substitution Use integration by substitution to show that if y is a continuous function of x on the interval $a \le x \le b$, where $x = f(t)$ and $y = g(t)$, then

$$\int_a^b y\,dx = \int_{t_1}^{t_2} g(t)f'(t)\,dt$$

where $f(t_1) = a, f(t_2) = b$, and both g and f' are continuous on $[t_1, t_2]$.

78. Surface Area A portion of a sphere of radius r is removed by cutting out a circular cone with its vertex at the center of the sphere. The vertex of the cone forms an angle of 2θ. Find the surface area removed from the sphere.

Area **In Exercises 79 and 80, find the area of the region. (Use the result of Exercise 77.)**

79. $x = 2\sin^2\theta$

$y = 2\sin^2\theta\tan\theta$

$0 \le \theta < \dfrac{\pi}{2}$

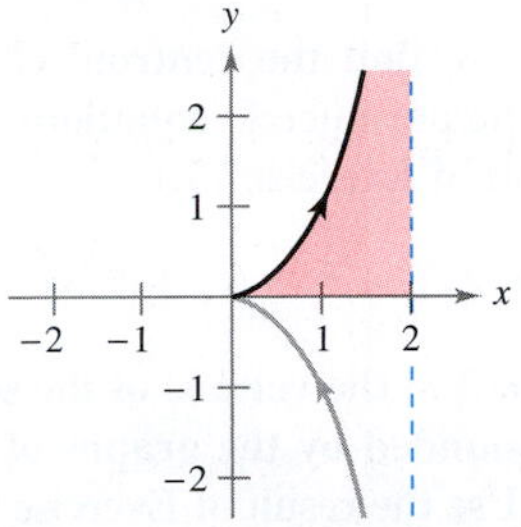

80. $x = 2\cot\theta$

$y = 2\sin^2\theta$

$0 < \theta < \pi$

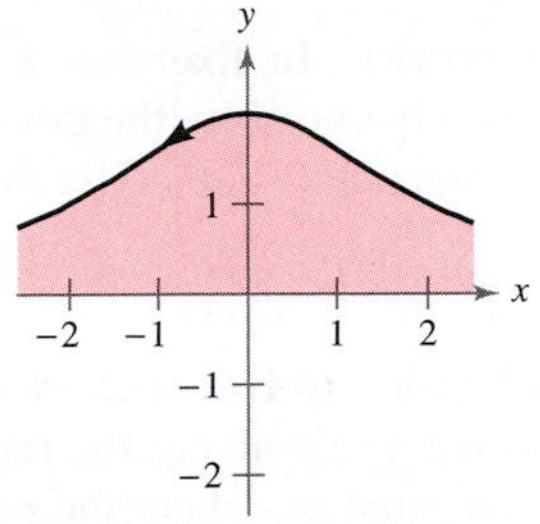

Areas of Simple Closed Curves **In Exercises 81–86, use a computer algebra system and the result of Exercise 77 to match the closed curve with its area. (These exercises were based on "The Surveyor's Area Formula" by Bart Braden, *College Mathematics Journal*, September 1986, pp. 335–337, by permission of the author.)**

(a) $\frac{8}{3}ab$ **(b)** $\frac{3}{8}\pi a^2$ **(c)** $2\pi a^2$

(d) πab **(e)** $2\pi ab$ **(f)** $6\pi a^2$

81. Ellipse: $(0 \le t \le 2\pi)$

$x = b\cos t$

$y = a\sin t$

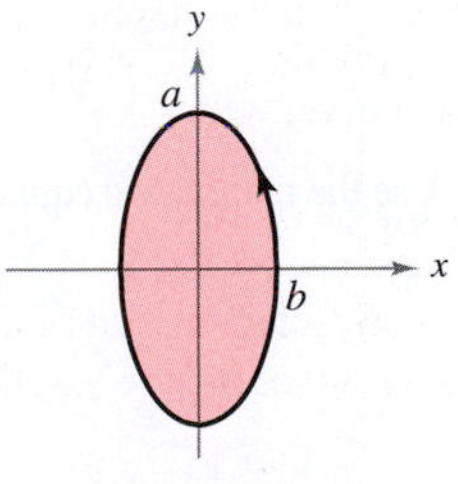

82. Astroid: $(0 \le t \le 2\pi)$

$x = a\cos^3 t$

$y = a\sin^3 t$

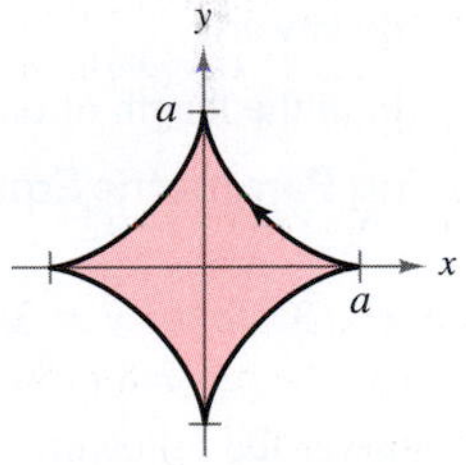

83. Cardioid: $(0 \le t \le 2\pi)$

$x = 2a\cos t - a\cos 2t$

$y = 2a\sin t - a\sin 2t$

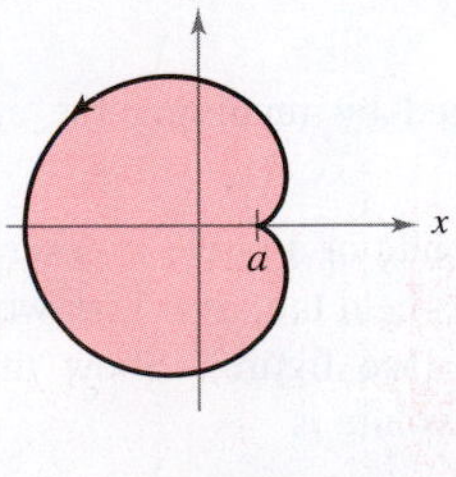

84. Deltoid: $(0 \le t \le 2\pi)$

$x = 2a\cos t + a\cos 2t$

$y = 2a\sin t - a\sin 2t$

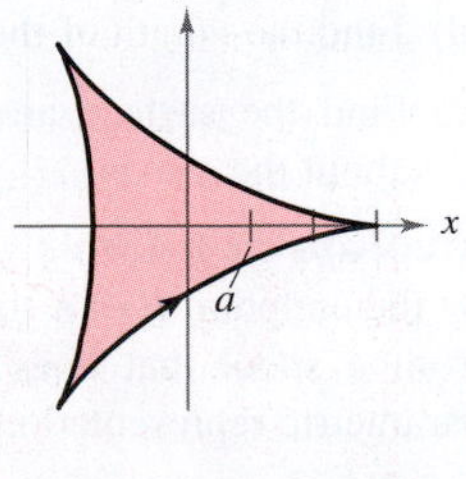

85. Hourglass: $(0 \le t \le 2\pi)$

$x = a \sin 2t$

$y = b \sin t$

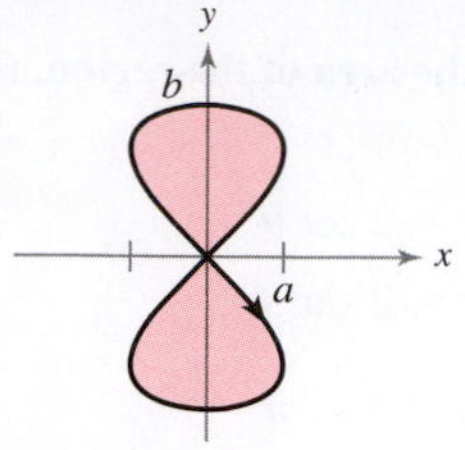

86. Teardrop: $(0 \le t \le 2\pi)$

$x = 2a \cos t - a \sin 2t$

$y = b \sin t$

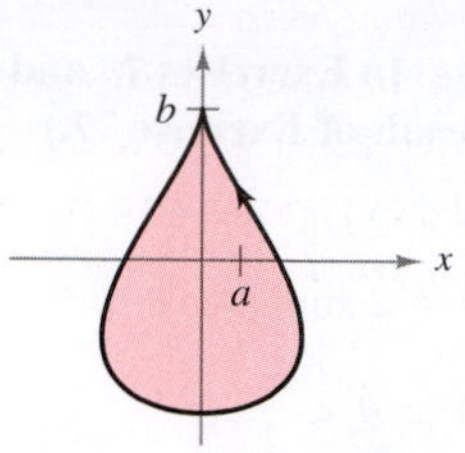

Centroid **In Exercises 87 and 88, find the centroid of the region bounded by the graph of the parametric equations and the coordinate axes. (Use the result of Exercise 77.)**

87. $x = \sqrt{t}, \quad y = 4 - t$

88. $x = \sqrt{4 - t}, \quad y = \sqrt{t}$

Volume **In Exercises 89 and 90, find the volume of the solid formed by revolving the region bounded by the graphs of the given equations about the x-axis. (Use the result of Exercise 77.)**

89. $x = 6 \cos \theta, \quad y = 6 \sin \theta$

90. $x = \cos \theta, \quad y = 3 \sin \theta, \quad a > 0$

91. Cycloid Use the parametric equations

$$x = a(\theta - \sin \theta) \quad \text{and} \quad y = a(1 - \cos \theta), a > 0$$

to answer the following.

(a) Find dy/dx and d^2y/dx^2.

(b) Find the equation of the tangent line at the point where $\theta = \pi/6$.

(c) Find all points (if any) of horizontal tangency.

(d) Determine where the curve is concave upward or concave downward.

(e) Find the length of one arc of the curve.

92. Using Parametric Equations Use the parametric equations

$$x = t^2\sqrt{3} \quad \text{and} \quad y = 3t - \frac{1}{3}t^3$$

to answer the following.

(a) Use a graphing utility to graph the curve on the interval $-3 \le t \le 3$.

(b) Find dy/dx and d^2y/dx^2.

(c) Find the equation of the tangent line at the point $\left(\sqrt{3}, \frac{8}{3}\right)$.

(d) Find the length of the curve.

(e) Find the surface area generated by revolving the curve about the x-axis.

93. Involute of a Circle The involute of a circle is described by the endpoint P of a string that is held taut as it is unwound from a spool that does not turn (see figure). Show that a parametric representation of the involute is

$$x = r(\cos \theta + \theta \sin \theta) \quad \text{and} \quad y = r(\sin \theta - \theta \cos \theta).$$

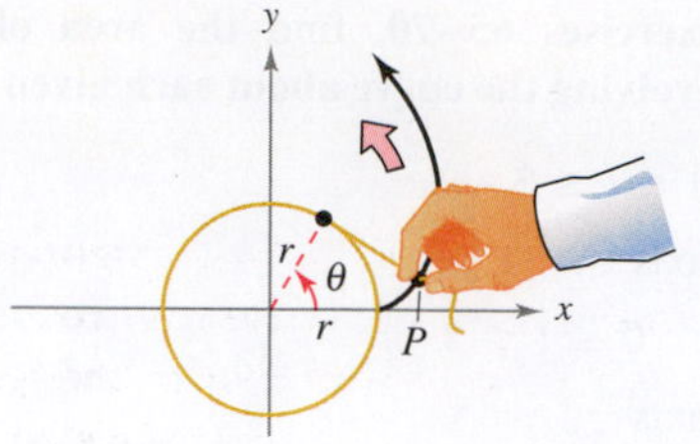

Figure for 93

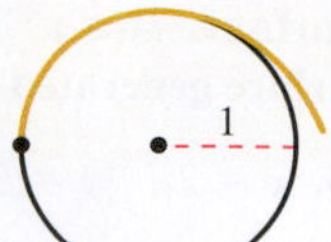

Figure for 94

94. Involute of a Circle The figure shows a piece of string tied to a circle with a radius of one unit. The string is just long enough to reach the opposite side of the circle. Find the area that is covered when the string is unwound counterclockwise.

95. Using Parametric Equations

(a) Use a graphing utility to graph the curve given by

$$x = \frac{1 - t^2}{1 + t^2} \quad \text{and} \quad y = \frac{2t}{1 + t^2}$$

where $-20 \le t \le 20$.

(b) Describe the graph and confirm your result analytically.

(c) Discuss the speed at which the curve is traced as t increases from -20 to 20.

96. Tractrix A person moves from the origin along the positive y-axis pulling a weight at the end of a 12-meter rope. Initially, the weight is located at the point (12, 0).

(a) In Exercise 90 of Section 8.7, it was shown that the path of the weight is modeled by the rectangular equation

$$y = -12 \ln\left(\frac{12 - \sqrt{144 - x^2}}{x}\right) - \sqrt{144 - x^2}$$

where $0 < x \le 12$. Use a graphing utility to graph the rectangular equation.

(b) Use a graphing utility to graph the parametric equations

$$x = 12 \operatorname{sech} \frac{t}{12} \quad \text{and} \quad y = t - 12 \tanh \frac{t}{12}$$

where $t \ge 0$. How does this graph compare with the graph in part (a)? Which graph (if either) do you think is a better representation of the path?

(c) Use the parametric equations for the tractrix to verify that the distance from the y-intercept of the tangent line to the point of tangency is independent of the location of the point of tangency.

True or False? **In Exercises 97 and 98, determine whether the statement is true or false. If it is false, explain why or give an example that shows it is false.**

97. If $x = f(t)$ and $y = g(t)$, then $\dfrac{d^2y}{dx^2} = \dfrac{g''(t)}{f''(t)}$.

98. The curve given by $x = t^3$, $y = t^2$ has a horizontal tangent at the origin because $dy/dt = 0$ when $t = 0$.

10.4 Polar Coordinates and Polar Graphs

- **Understand the polar coordinate system.**
- **Rewrite rectangular coordinates and equations in polar form and vice versa.**
- **Sketch the graph of an equation given in polar form.**
- **Find the slope of a tangent line to a polar graph.**
- **Identify several types of special polar graphs.**

Polar Coordinates

So far, you have been representing graphs as collections of points (x, y) on the rectangular coordinate system. The corresponding equations for these graphs have been in either rectangular or parametric form. In this section, you will study a coordinate system called the **polar coordinate system.**

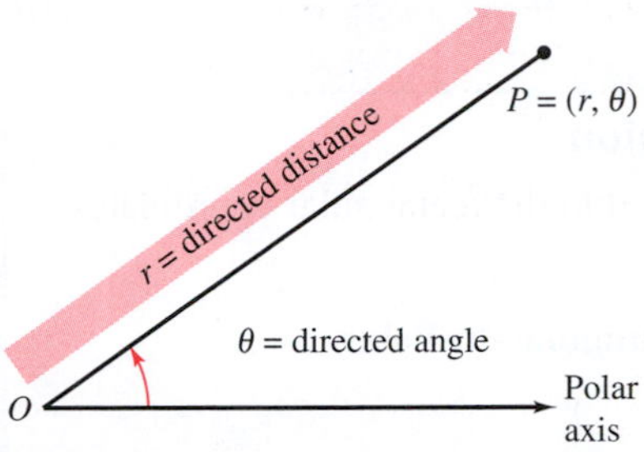

Polar coordinates
Figure 10.35

To form the polar coordinate system in the plane, fix a point O, called the **pole** (or **origin**), and construct from O an initial ray called the **polar axis,** as shown in Figure 10.35. Then each point P in the plane can be assigned **polar coordinates** (r, θ), as follows.

$r =$ *directed distance* from O to P

$\theta =$ *directed angle*, counterclockwise from polar axis to segment $\overline{OP}$

Figure 10.36 shows three points on the polar coordinate system. Notice that in this system, it is convenient to locate points with respect to a grid of concentric circles intersected by **radial lines** through the pole.

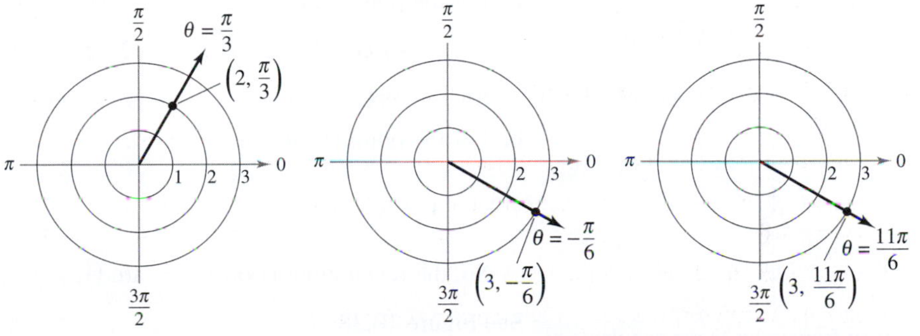

Figure 10.36

With rectangular coordinates, each point (x, y) has a unique representation. This is not true with polar coordinates. For instance, the coordinates

$$(r, \theta) \quad \text{and} \quad (r, 2\pi + \theta)$$

represent the same point [see parts (b) and (c) in Figure 10.36]. Also, because r is a *directed distance,* the coordinates

$$(r, \theta) \quad \text{and} \quad (-r, \theta + \pi)$$

represent the same point. In general, the point (r, θ) can be written as

$$(r, \theta) = (r, \theta + 2n\pi)$$

or

$$(r, \theta) = (-r, \theta + (2n + 1)\pi)$$

where n is any integer. Moreover, the pole is represented by $(0, \theta)$, where θ is any angle.

> **POLAR COORDINATES**
>
> The mathematician credited with first using polar coordinates was James Bernoulli, who introduced them in 1691. However, there is some evidence that it may have been Isaac Newton who first used them.

Coordinate Conversion

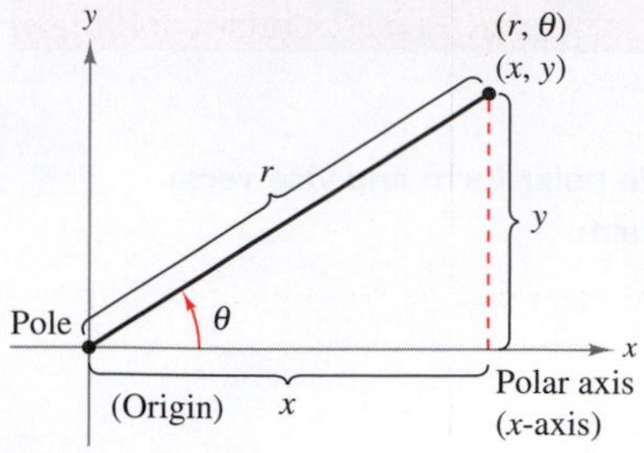

Relating polar and rectangular coordinates
Figure 10.37

To establish the relationship between polar and rectangular coordinates, let the polar axis coincide with the positive x-axis and the pole with the origin, as shown in Figure 10.37. Because (x, y) lies on a circle of radius r, it follows that

$$r^2 = x^2 + y^2.$$

Moreover, for $r > 0$, the definitions of the trigonometric functions imply that

$$\tan\theta = \frac{y}{x}, \quad \cos\theta = \frac{x}{r}, \quad \text{and} \quad \sin\theta = \frac{y}{r}.$$

You can show that the same relationships hold for $r < 0$.

> **THEOREM 10.10 Coordinate Conversion**
>
> The polar coordinates (r, θ) of a point are related to the rectangular coordinates (x, y) of the point as follows.
>
> **Polar-to-Rectangular**
>
> $x = r\cos\theta$
>
> $y = r\sin\theta$
>
> **Rectangular-to-Polar**
>
> $\tan\theta = \dfrac{y}{x}$
>
> $r^2 = x^2 + y^2$

EXAMPLE 1 Polar-to-Rectangular Conversion

a. For the point $(r, \theta) = (2, \pi)$,

$$x = r\cos\theta = 2\cos\pi = -2 \quad \text{and} \quad y = r\sin\theta = 2\sin\pi = 0.$$

So, the rectangular coordinates are $(x, y) = (-2, 0)$.

b. For the point $(r, \theta) = \left(\sqrt{3}, \pi/6\right)$,

$$x = \sqrt{3}\cos\frac{\pi}{6} = \frac{3}{2} \quad \text{and} \quad y = \sqrt{3}\sin\frac{\pi}{6} = \frac{\sqrt{3}}{2}.$$

So, the rectangular coordinates are $(x, y) = \left(3/2, \sqrt{3}/2\right)$.

See Figure 10.38.

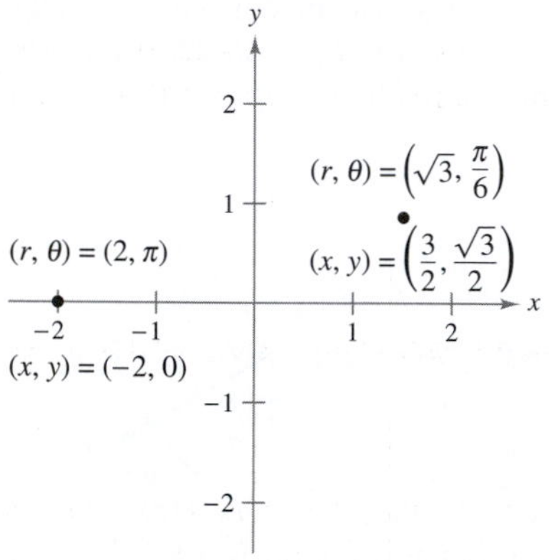

To convert from polar to rectangular coordinates, let $x = r\cos\theta$ and $y = r\sin\theta$.
Figure 10.38

EXAMPLE 2 Rectangular-to-Polar Conversion

a. For the second-quadrant point $(x, y) = (-1, 1)$,

$$\tan\theta = \frac{y}{x} = -1 \quad \Rightarrow \quad \theta = \frac{3\pi}{4}.$$

Because θ was chosen to be in the same quadrant as (x, y), you should use a positive value of r.

$$\begin{aligned} r &= \sqrt{x^2 + y^2} \\ &= \sqrt{(-1)^2 + (1)^2} \\ &= \sqrt{2} \end{aligned}$$

This implies that *one* set of polar coordinates is $(r, \theta) = \left(\sqrt{2}, 3\pi/4\right)$.

b. Because the point $(x, y) = (0, 2)$ lies on the positive y-axis, choose $\theta = \pi/2$ and $r = 2$, and one set of polar coordinates is $(r, \theta) = (2, \pi/2)$.

See Figure 10.39.

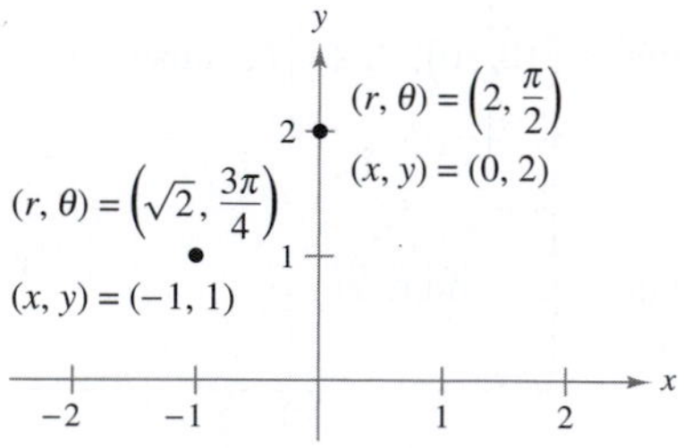

To convert from rectangular to polar coordinates, let $\tan\theta = y/x$ and $r = \sqrt{x^2 + y^2}$.
Figure 10.39

Polar Graphs

One way to sketch the graph of a polar equation is to convert to rectangular coordinates and then sketch the graph of the rectangular equation.

EXAMPLE 3 Graphing Polar Equations

Describe the graph of each polar equation. Confirm each description by converting to a rectangular equation.

a. $r = 2$ **b.** $\theta = \dfrac{\pi}{3}$ **c.** $r = \sec\theta$

Solution

a. The graph of the polar equation $r = 2$ consists of all points that are two units from the pole. So, this graph is a circle centered at the origin with a radius of 2. [See Figure 10.40(a).] You can confirm this by using the relationship $r^2 = x^2 + y^2$ to obtain the rectangular equation

$$x^2 + y^2 = 2^2. \qquad \text{Rectangular equation}$$

b. The graph of the polar equation $\theta = \pi/3$ consists of all points on the line that makes an angle of $\pi/3$ with the positive x-axis. [See Figure 10.40(b).] You can confirm this by using the relationship $\tan\theta = y/x$ to obtain the rectangular equation

$$y = \sqrt{3}x. \qquad \text{Rectangular equation}$$

c. The graph of the polar equation $r = \sec\theta$ is not evident by simple inspection, so you can begin by converting to rectangular form using the relationship $r\cos\theta = x$.

$$r = \sec\theta \qquad \text{Polar equation}$$
$$r\cos\theta = 1$$
$$x = 1 \qquad \text{Rectangular equation}$$

From the rectangular equation, you can see that the graph is a vertical line. [See Figure 10.40(c).]

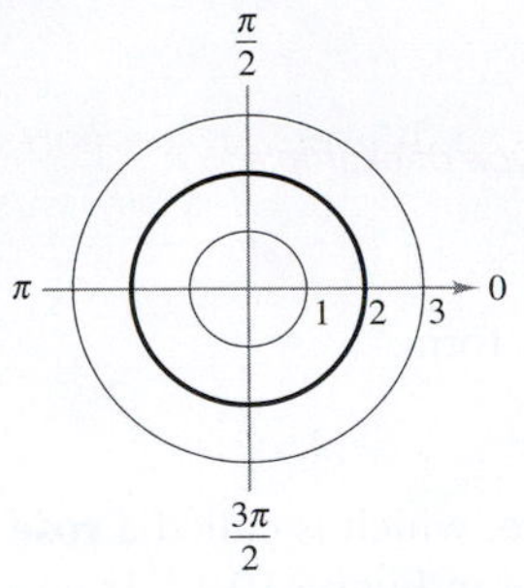

(a) Circle: $r = 2$

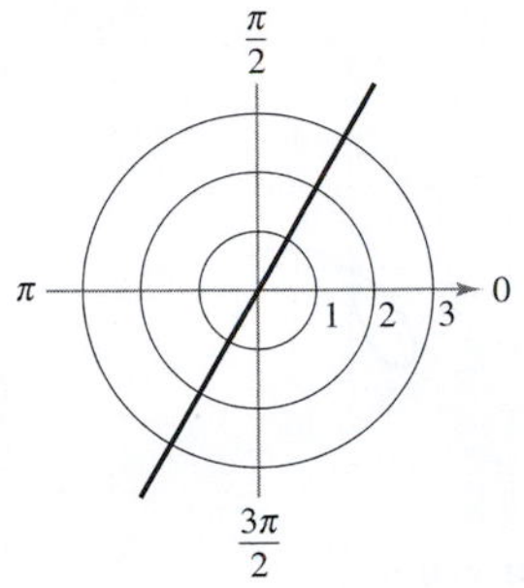

(b) Radial line: $\theta = \frac{\pi}{3}$

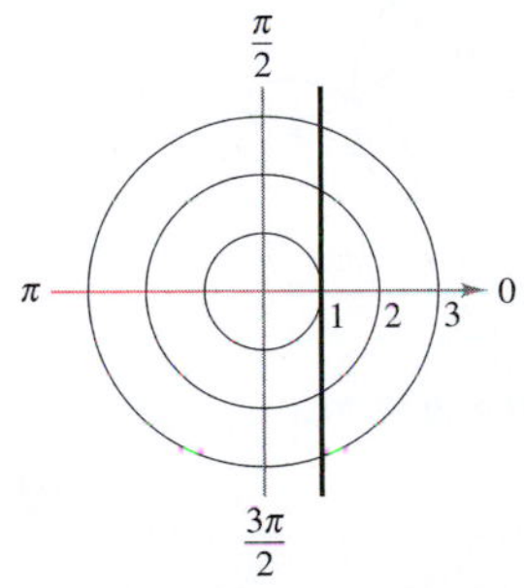

(c) Vertical line: $r = \sec\theta$

Figure 10.40

TECHNOLOGY Sketching the graphs of complicated polar equations *by hand* can be tedious. With technology, however, the task is not difficult. If your graphing utility has a *polar* mode, use it to graph the equations in the exercise set. If your graphing utility doesn't have a *polar* mode, but does have a *parametric* mode, you can graph $r = f(\theta)$ by writing the equation as

$$x = f(\theta)\cos\theta$$
$$y = f(\theta)\sin\theta.$$

For instance, the graph of $r = \frac{1}{2}\theta$ shown in Figure 10.41 was produced with a graphing calculator in parametric mode. This equation was graphed using the parametric equations

$$x = \frac{1}{2}\theta\cos\theta$$
$$y = \frac{1}{2}\theta\sin\theta$$

with the values of θ varying from -4π to 4π. This curve is of the form $r = a\theta$ and is called a **spiral of Archimedes.**

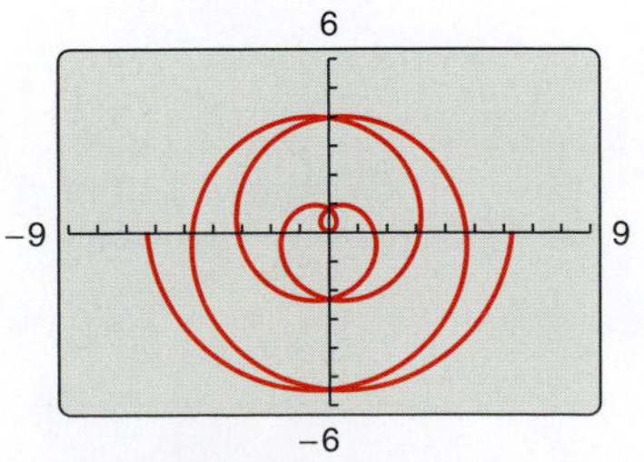

Spiral of Archimedes
Figure 10.41

EXAMPLE 4 Sketching a Polar Graph

See LarsonCalculus.com for an interactive version of this type of example.

Sketch the graph of $r = 2 \cos 3\theta$.

REMARK One way to sketch the graph of $r = 2 \cos 3\theta$ by hand is to make a table of values.

θ	0	$\frac{\pi}{6}$	$\frac{\pi}{3}$	$\frac{\pi}{2}$	$\frac{2\pi}{3}$
r	2	0	-2	0	2

By extending the table and plotting the points, you will obtain the curve shown in Example 4.

Solution Begin by writing the polar equation in parametric form.

$$x = 2 \cos 3\theta \cos \theta \quad \text{and} \quad y = 2 \cos 3\theta \sin \theta$$

After some experimentation, you will find that the entire curve, which is called a **rose curve,** can be sketched by letting θ vary from 0 to π, as shown in Figure 10.42. If you try duplicating this graph with a graphing utility, you will find that by letting θ vary from 0 to 2π, you will actually trace the entire curve *twice.*

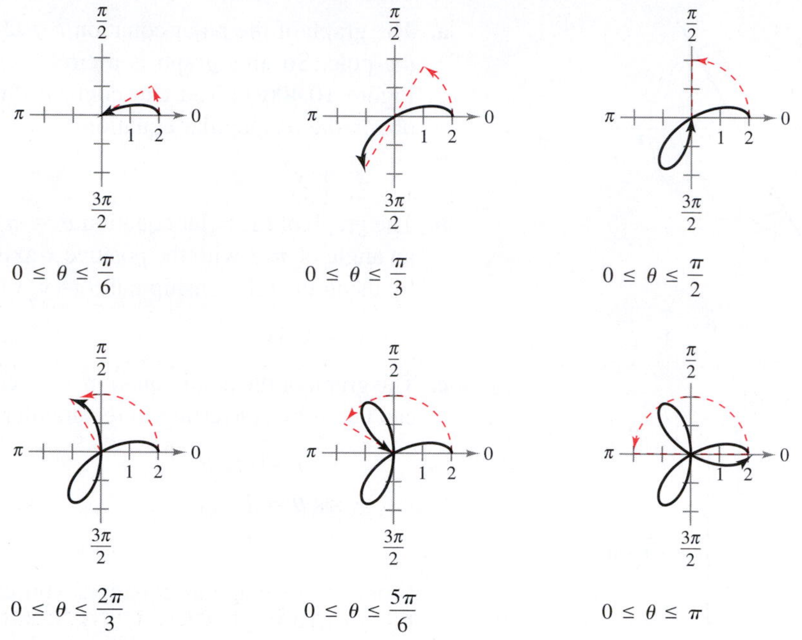

Figure 10.42

Use a graphing utility to experiment with other rose curves. Note that rose curves are of the form

$$r = a \cos n\theta \quad \text{or} \quad r = a \sin n\theta.$$

For instance, Figure 10.43 shows the graphs of two other rose curves.

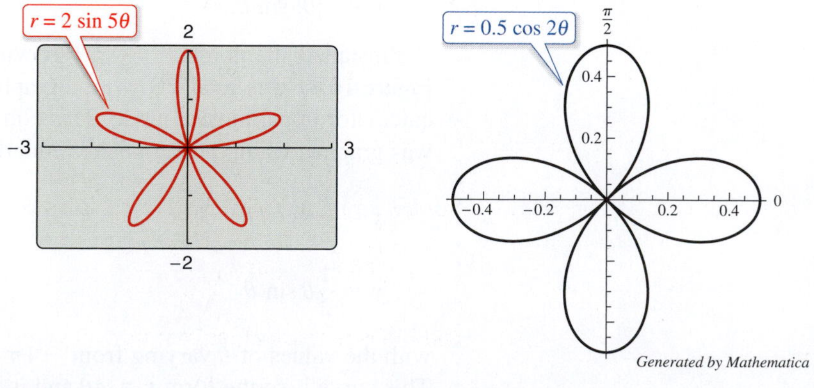

Rose curves
Figure 10.43

Slope and Tangent Lines

To find the slope of a tangent line to a polar graph, consider a differentiable function given by $r = f(\theta)$. To find the slope in polar form, use the parametric equations

$$x = r\cos\theta = f(\theta)\cos\theta \quad \text{and} \quad y = r\sin\theta = f(\theta)\sin\theta.$$

Using the parametric form of dy/dx given in Theorem 10.7, you have

$$\frac{dy}{dx} = \frac{dy/d\theta}{dx/d\theta} = \frac{f(\theta)\cos\theta + f'(\theta)\sin\theta}{-f(\theta)\sin\theta + f'(\theta)\cos\theta}$$

which establishes the next theorem.

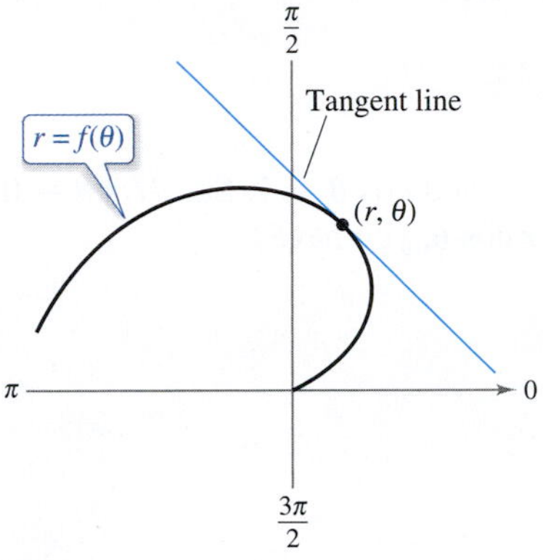

Tangent line to polar curve
Figure 10.44

THEOREM 10.11 Slope in Polar Form

If f is a differentiable function of θ, then the *slope* of the tangent line to the graph of $r = f(\theta)$ at the point (r, θ) is

$$\frac{dy}{dx} = \frac{dy/d\theta}{dx/d\theta} = \frac{f(\theta)\cos\theta + f'(\theta)\sin\theta}{-f(\theta)\sin\theta + f'(\theta)\cos\theta}$$

provided that $dx/d\theta \neq 0$ at (r, θ). (See Figure 10.44.)

From Theorem 10.11, you can make the following observations.

1. Solutions of $\dfrac{dy}{d\theta} = 0$ yield horizontal tangents, provided that $\dfrac{dx}{d\theta} \neq 0$.
2. Solutions of $\dfrac{dx}{d\theta} = 0$ yield vertical tangents, provided that $\dfrac{dy}{d\theta} \neq 0$.

If $dy/d\theta$ and $dx/d\theta$ are *simultaneously* 0, then no conclusion can be drawn about tangent lines.

EXAMPLE 5 Finding Horizontal and Vertical Tangent Lines

Find the horizontal and vertical tangent lines of $r = \sin\theta$, $0 \le \theta \le \pi$.

Solution Begin by writing the equation in parametric form.

$$x = r\cos\theta = \sin\theta\cos\theta$$

and

$$y = r\sin\theta = \sin\theta\sin\theta = \sin^2\theta$$

Next, differentiate x and y with respect to θ and set each derivative equal to 0.

$$\frac{dx}{d\theta} = \cos^2\theta - \sin^2\theta = \cos 2\theta = 0 \quad \Longrightarrow \quad \theta = \frac{\pi}{4}, \frac{3\pi}{4}$$

$$\frac{dy}{d\theta} = 2\sin\theta\cos\theta = \sin 2\theta = 0 \quad \Longrightarrow \quad \theta = 0, \frac{\pi}{2}$$

So, the graph has vertical tangent lines at

$$\left(\frac{\sqrt{2}}{2}, \frac{\pi}{4}\right) \quad \text{and} \quad \left(\frac{\sqrt{2}}{2}, \frac{3\pi}{4}\right)$$

and it has horizontal tangent lines at

$$(0, 0) \quad \text{and} \quad \left(1, \frac{\pi}{2}\right)$$

as shown in Figure 10.45.

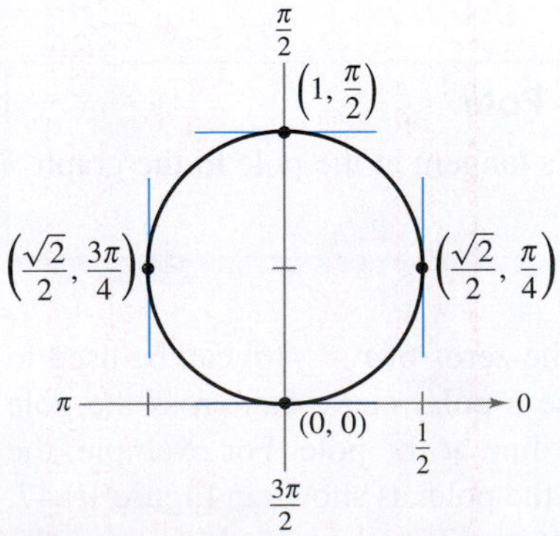

Horizontal and vertical tangent lines of $r = \sin\theta$
Figure 10.45

EXAMPLE 6 Finding Horizontal and Vertical Tangent Lines

Find the horizontal and vertical tangents to the graph of $r = 2(1 - \cos\theta)$.

Solution Let $y = r\sin\theta$ and then differentiate with respect to θ.

$$\begin{aligned} y &= r\sin\theta \\ &= 2(1 - \cos\theta)\sin\theta \\ \frac{dy}{d\theta} &= 2[(1 - \cos\theta)(\cos\theta) + \sin\theta(\sin\theta)] \\ &= 2(\cos\theta - \cos^2\theta + \sin^2\theta) \\ &= 2(\cos\theta - \cos^2\theta + 1 - \cos^2\theta) \\ &= -2(2\cos^2\theta - \cos\theta - 1) \\ &= -2(2\cos\theta + 1)(\cos\theta - 1) \end{aligned}$$

Setting $dy/d\theta$ equal to 0, you can see that $\cos\theta = -\frac{1}{2}$ and $\cos\theta = 1$. So, $dy/d\theta = 0$ when $\theta = 2\pi/3$, $4\pi/3$, and 0. Similarly, using $x = r\cos\theta$, you have

$$\begin{aligned} x &= r\cos\theta \\ &= 2(1 - \cos\theta)\cos\theta \\ &= 2\cos\theta - 2\cos^2\theta \\ \frac{dx}{d\theta} &= -2\sin\theta + 4\cos\theta\sin\theta \\ &= 2\sin\theta(2\cos\theta - 1). \end{aligned}$$

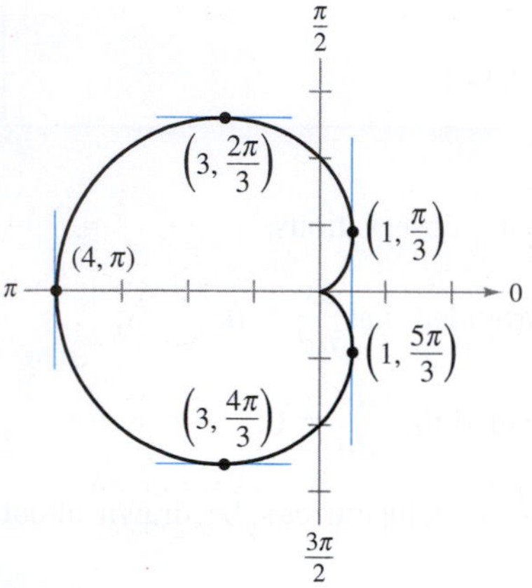

Horizontal and vertical tangent lines of $r = 2(1 - \cos\theta)$

Figure 10.46

Setting $dx/d\theta$ equal to 0, you can see that $\sin\theta = 0$ and $\cos\theta = \frac{1}{2}$. So, you can conclude that $dx/d\theta = 0$ when $\theta = 0$, π, $\pi/3$, and $5\pi/3$. From these results, and from the graph shown in Figure 10.46, you can conclude that the graph has horizontal tangents at $(3, 2\pi/3)$ and $(3, 4\pi/3)$, and has vertical tangents at $(1, \pi/3)$, $(1, 5\pi/3)$, and $(4, \pi)$. This graph is called a **cardioid.** Note that both derivatives ($dy/d\theta$ and $dx/d\theta$) are 0 when $\theta = 0$. Using this information alone, you don't know whether the graph has a horizontal or vertical tangent line at the pole. From Figure 10.46, however, you can see that the graph has a cusp at the pole.

Theorem 10.11 has an important consequence. If the graph of $r = f(\theta)$ passes through the pole when $\theta = \alpha$ and $f'(\alpha) \neq 0$, then the formula for dy/dx simplifies as follows.

$$\frac{dy}{dx} = \frac{f'(\alpha)\sin\alpha + f(\alpha)\cos\alpha}{f'(\alpha)\cos\alpha - f(\alpha)\sin\alpha} = \frac{f'(\alpha)\sin\alpha + 0}{f'(\alpha)\cos\alpha - 0} = \frac{\sin\alpha}{\cos\alpha} = \tan\alpha$$

So, the line $\theta = \alpha$ is tangent to the graph at the pole, $(0, \alpha)$.

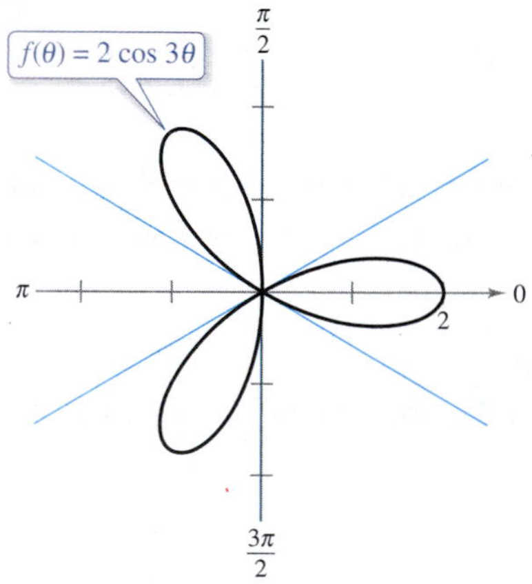

This rose curve has three tangent lines ($\theta = \pi/6$, $\theta = \pi/2$, and $\theta = 5\pi/6$) at the pole.

Figure 10.47

THEOREM 10.12 Tangent Lines at the Pole

If $f(\alpha) = 0$ and $f'(\alpha) \neq 0$, then the line $\theta = \alpha$ is tangent at the pole to the graph of $r = f(\theta)$.

Theorem 10.12 is useful because it states that the zeros of $r = f(\theta)$ can be used to find the tangent lines at the pole. Note that because a polar curve can cross the pole more than once, it can have more than one tangent line at the pole. For example, the rose curve $f(\theta) = 2\cos 3\theta$ has three tangent lines at the pole, as shown in Figure 10.47. For this curve, $f(\theta) = 2\cos 3\theta$ is 0 when θ is $\pi/6$, $\pi/2$, and $5\pi/6$. Moreover, the derivative $f'(\theta) = -6\sin 3\theta$ is not 0 for these values of θ.

Special Polar Graphs

Several important types of graphs have equations that are simpler in polar form than in rectangular form. For example, the polar equation of a circle having a radius of a and centered at the origin is simply $r = a$. Later in the text, you will come to appreciate this benefit. For now, several other types of graphs that have simpler equations in polar form are shown below. (Conics are considered in Section 10.6.)

Limaçons

$r = a \pm b \cos \theta$

$r = a \pm b \sin \theta$

$(a > 0, b > 0)$

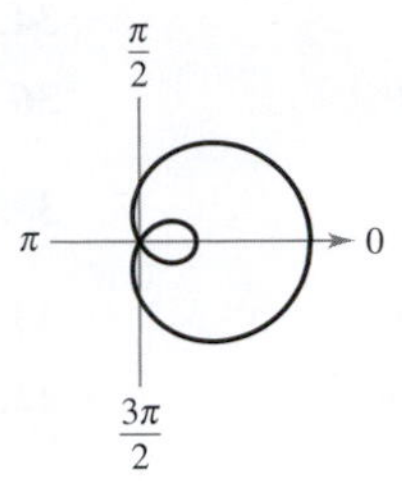

$\frac{a}{b} < 1$

Limaçon with inner loop

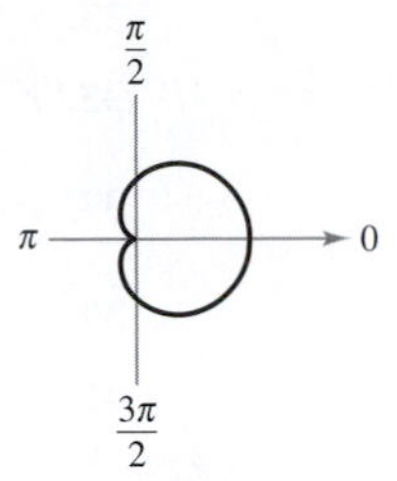

$\frac{a}{b} = 1$

Cardioid (heart-shaped)

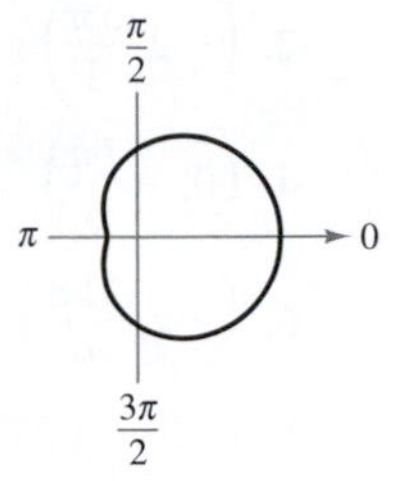

$1 < \frac{a}{b} < 2$

Dimpled limaçon

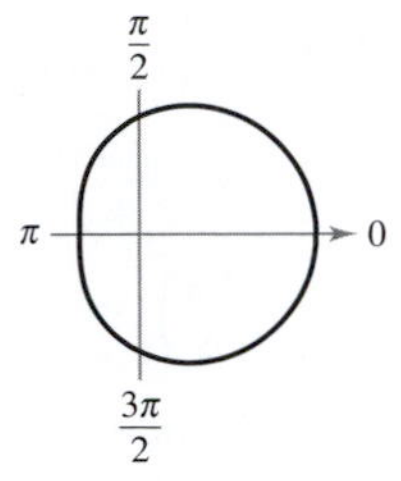

$\frac{a}{b} \geq 2$

Convex limaçon

Rose Curves

n petals when n is odd
$2n$ petals when n is even ($n \geq 2$)

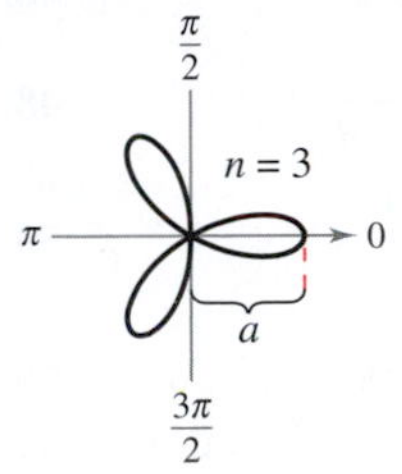

$r = a \cos n\theta$

Rose curve

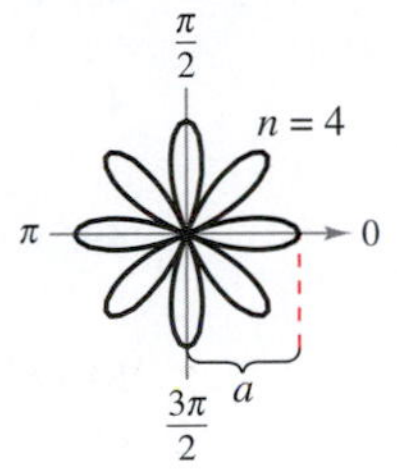

$r = a \cos n\theta$

Rose curve

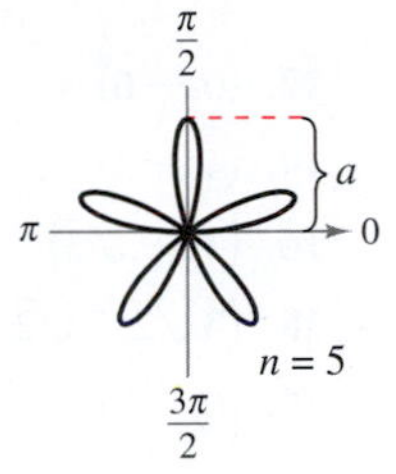

$r = a \sin n\theta$

Rose curve

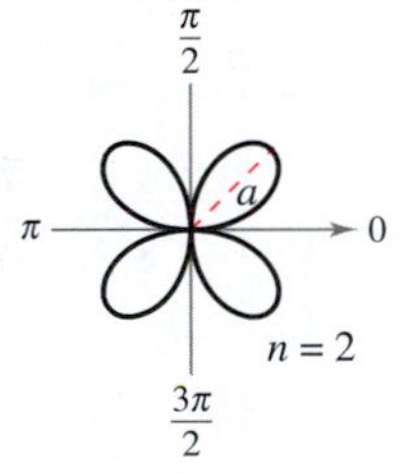

$r = a \sin n\theta$

Rose curve

Circles and Lemniscates

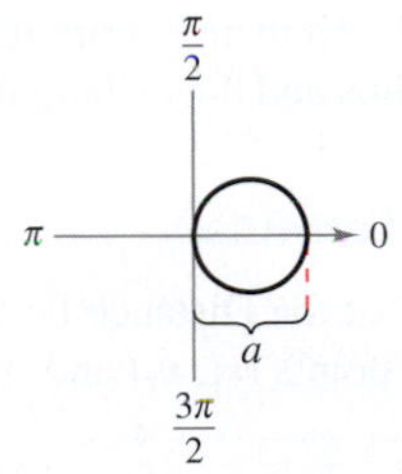

$r = a \cos \theta$

Circle

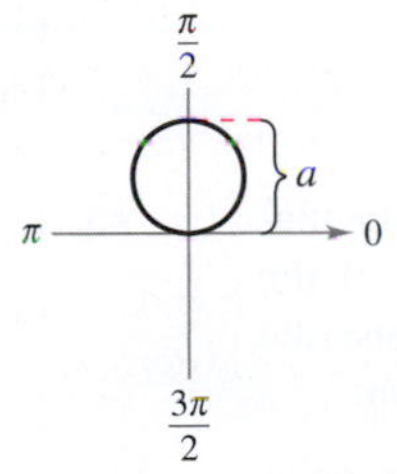

$r = a \sin \theta$

Circle

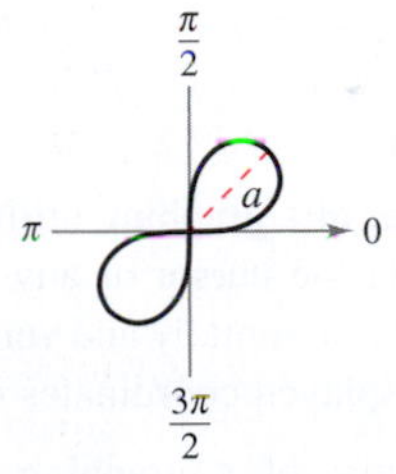

$r^2 = a^2 \sin 2\theta$

Lemniscate

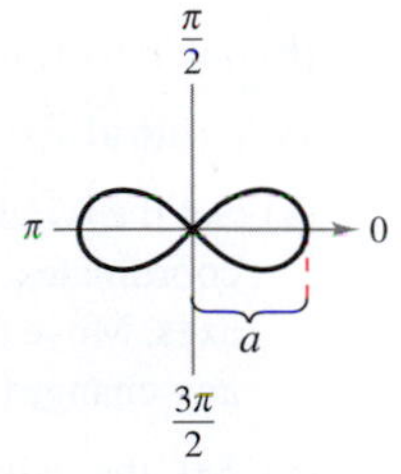

$r^2 = a^2 \cos 2\theta$

Lemniscate

TECHNOLOGY The rose curves described above are of the form $r = a \cos n\theta$ or $r = a \sin n\theta$, where n is a positive integer that is greater than or equal to 2. Use a graphing utility to graph

$$r = a \cos n\theta \quad \text{or} \quad r = a \sin n\theta$$

for some noninteger values of n. Are these graphs also rose curves? For example, try sketching the graph of

$$r = \cos \frac{2}{3}\theta, \quad 0 \leq \theta \leq 6\pi.$$

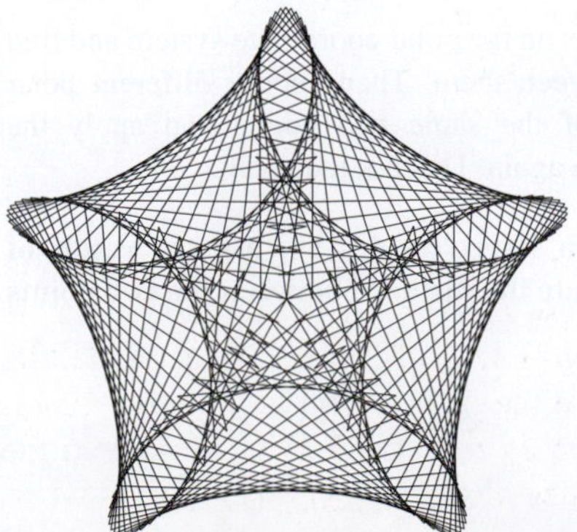
Generated by Maple

FOR FURTHER INFORMATION For more information on rose curves and related curves, see the article "A Rose is a Rose . . ." by Peter M. Maurer in *The American Mathematical Monthly.* The computer-generated graph at the left is the result of an algorithm that Maurer calls "The Rose." To view this article, go to *MathArticles.com.*

10.4 Exercises

See CalcChat.com for tutorial help and worked-out solutions to odd-numbered exercises.

Polar-to-Rectangular Conversion **In Exercises 1–10, plot the point in polar coordinates and find the corresponding rectangular coordinates for the point.**

1. $\left(8, \frac{\pi}{2}\right)$
2. $\left(-2, \frac{5\pi}{3}\right)$
3. $\left(-4, -\frac{3\pi}{4}\right)$
4. $\left(0, -\frac{7\pi}{6}\right)$
5. $\left(7, \frac{5\pi}{4}\right)$
6. $\left(-2, \frac{11\pi}{6}\right)$
7. $(\sqrt{2}, 2.36)$
8. $(-3, -1.57)$
9. $(-4.5, 3.5)$
10. $(9.25, 1.2)$

Rectangular-to-Polar Conversion **In Exercises 11–20, the rectangular coordinates of a point are given. Plot the point and find *two* sets of polar coordinates for the point for $0 \le \theta < 2\pi$.**

11. $(2, 2)$
12. $(0, -6)$
13. $(-3, 4)$
14. $(4, -2)$
15. $(-1, -\sqrt{3})$
16. $(3, -\sqrt{3})$
17. $(3, -2)$
18. $(3\sqrt{2}, 3\sqrt{2})$
19. $\left(\frac{7}{4}, \frac{5}{2}\right)$
20. $(0, -5)$

21. **Plotting a Point** Plot the point $(4, 3.5)$ when the point is given in
    (a) rectangular coordinates.
    (b) polar coordinates.

22. **Graphical Reasoning**
    (a) Set the window format of a graphing utility to rectangular coordinates and locate the cursor at any position off the axes. Move the cursor horizontally and vertically. Describe any changes in the displayed coordinates of the points.
    (b) Set the window format of a graphing utility to polar coordinates and locate the cursor at any position off the axes. Move the cursor horizontally and vertically. Describe any changes in the displayed coordinates of the points.
    (c) Why are the results in parts (a) and (b) different?

Rectangular-to-Polar Conversion **In Exercises 23–32, convert the rectangular equation to polar form and sketch its graph.**

23. $x^2 + y^2 = 9$
24. $x^2 - y^2 = 9$
25. $x^2 + y^2 = a^2$
26. $x^2 + y^2 - 2ax = 0$
27. $y = 8$
28. $x = 12$
29. $3x - y + 2 = 0$
30. $xy = 4$
31. $y^2 = 9x$
32. $(x^2 + y^2)^2 - 9(x^2 - y^2) = 0$

Polar-to-Rectangular Conversion **In Exercises 33–42, convert the polar equation to rectangular form and sketch its graph.**

33. $r = 4$
34. $r = -5$
35. $r = 3 \sin \theta$
36. $r = 5 \cos \theta$
37. $r = \theta$
38. $\theta = \frac{5\pi}{6}$
39. $r = 3 \sec \theta$
40. $r = 2 \csc \theta$
41. $r = \sec \theta \tan \theta$
42. $r = \cot \theta \csc \theta$

Graphing a Polar Equation **In Exercises 43–52, use a graphing utility to graph the polar equation. Find an interval for θ over which the graph is traced *only once*.**

43. $r = 2 - 5 \cos \theta$
44. $r = 3(1 - 4 \cos \theta)$
45. $r = 2 + \sin \theta$
46. $r = 4 + 3 \cos \theta$
47. $r = \dfrac{2}{1 + \cos \theta}$
48. $r = \dfrac{2}{4 - 3 \sin \theta}$
49. $r = 2 \cos\left(\frac{3\theta}{2}\right)$
50. $r = 3 \sin\left(\frac{5\theta}{2}\right)$
51. $r^2 = 4 \sin 2\theta$
52. $r^2 = \dfrac{1}{\theta}$

53. **Verifying a Polar Equation** Convert the equation

    $$r = 2(h \cos \theta + k \sin \theta)$$

    to rectangular form and verify that it is the equation of a circle. Find the radius and the rectangular coordinates of the center of the circle.

54. **Distance Formula**
    (a) Verify that the Distance Formula for the distance between the two points (r_1, θ_1) and (r_2, θ_2) in polar coordinates is

    $$d = \sqrt{r_1^2 + r_2^2 - 2r_1r_2 \cos(\theta_1 - \theta_2)}.$$

    (b) Describe the positions of the points relative to each other for $\theta_1 = \theta_2$. Simplify the Distance Formula for this case. Is the simplification what you expected? Explain.
    (c) Simplify the Distance Formula for $\theta_1 - \theta_2 = 90°$. Is the simplification what you expected? Explain.
    (d) Choose two points on the polar coordinate system and find the distance between them. Then choose different polar representations of the same two points and apply the Distance Formula again. Discuss the result.

Distance Formula **In Exercises 55–58, use the result of Exercise 54 to approximate the distance between the two points in polar coordinates.**

55. $\left(1, \frac{5\pi}{6}\right), \left(4, \frac{\pi}{3}\right)$
56. $\left(8, \frac{7\pi}{4}\right), (5, \pi)$
57. $(2, 0.5), (7, 1.2)$
58. $(4, 2.5), (12, 1)$

Finding Slopes of Tangent Lines In Exercises 59 and 60, find dy/dx and the slopes of the tangent lines shown on the graph of the polar equation.

59. $r = 2 + 3 \sin \theta$

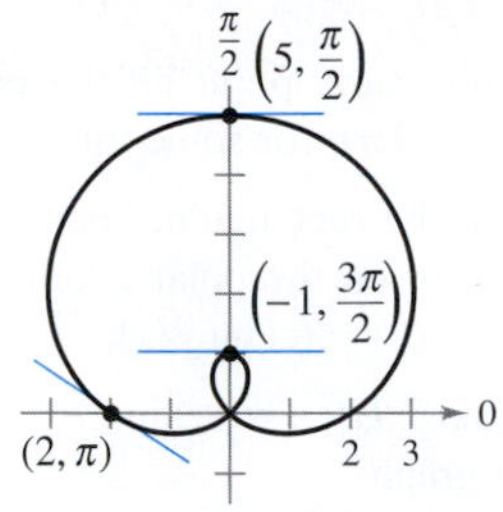

60. $r = 2(1 - \sin \theta)$

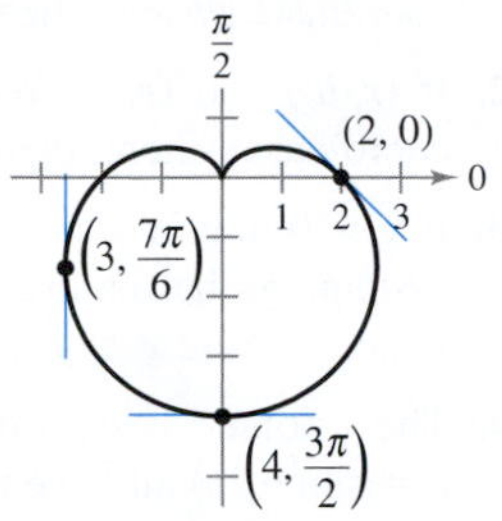

Finding Slopes of Tangent Lines In Exercises 61–64, use a graphing utility to (a) graph the polar equation, (b) draw the tangent line at the given value of θ, and (c) find dy/dx at the given value of θ. (*Hint:* Let the increment between the values of θ equal $\pi/24$.)

61. $r = 3(1 - \cos \theta),\ \theta = \frac{\pi}{2}$

62. $r = 3 - 2 \cos \theta,\ \theta = 0$

63. $r = 3 \sin \theta,\ \theta = \frac{\pi}{3}$

64. $r = 4,\ \theta = \frac{\pi}{4}$

Horizontal and Vertical Tangency In Exercises 65 and 66, find the points of horizontal and vertical tangency (if any) to the polar curve.

65. $r = 1 - \sin \theta$

66. $r = a \sin \theta$

Horizontal Tangency In Exercises 67 and 68, find the points of horizontal tangency (if any) to the polar curve.

67. $r = 2 \csc \theta + 3$

68. $r = a \sin \theta \cos^2 \theta$

Tangent Lines at the Pole In Exercises 69–76, sketch a graph of the polar equation and find the tangents at the pole.

69. $r = 5 \sin \theta$

70. $r = 5 \cos \theta$

71. $r = 2(1 - \sin \theta)$

72. $r = 3(1 - \cos \theta)$

73. $r = 4 \cos 3\theta$

74. $r = -\sin 5\theta$

75. $r = 3 \sin 2\theta$

76. $r = 3 \cos 2\theta$

Sketching a Polar Graph In Exercises 77–88, sketch a graph of the polar equation.

77. $r = 8$

78. $r = 1$

79. $r = 4(1 + \cos \theta)$

80. $r = 1 + \sin \theta$

81. $r = 3 - 2 \cos \theta$

82. $r = 5 - 4 \sin \theta$

83. $r = 3 \csc \theta$

84. $r = \dfrac{6}{2 \sin \theta - 3 \cos \theta}$

85. $r = 2\theta$

86. $r = \dfrac{1}{\theta}$

87. $r^2 = 4 \cos 2\theta$

88. $r^2 = 4 \sin \theta$

Asymptote In Exercises 89–92, use a graphing utility to graph the equation and show that the given line is an asymptote of the graph.

	Name of Graph	Polar Equation	Asymptote
89.	Conchoid	$r = 2 - \sec \theta$	$x = -1$
90.	Conchoid	$r = 2 + \csc \theta$	$y = 1$
91.	Hyperbolic spiral	$r = 2/\theta$	$y = 2$
92.	Strophoid	$r = 2 \cos 2\theta \sec \theta$	$x = -2$

WRITING ABOUT CONCEPTS

93. **Comparing Coordinate Systems** Describe the differences between the rectangular coordinate system and the polar coordinate system.

94. **Coordinate Conversion** Give the equations for the coordinate conversion from rectangular to polar coordinates and vice versa.

95. **Tangent Lines** How are the slopes of tangent lines determined in polar coordinates? What are tangent lines at the pole and how are they determined?

96. **HOW DO YOU SEE IT?** Identify each special polar graph and write its equation.

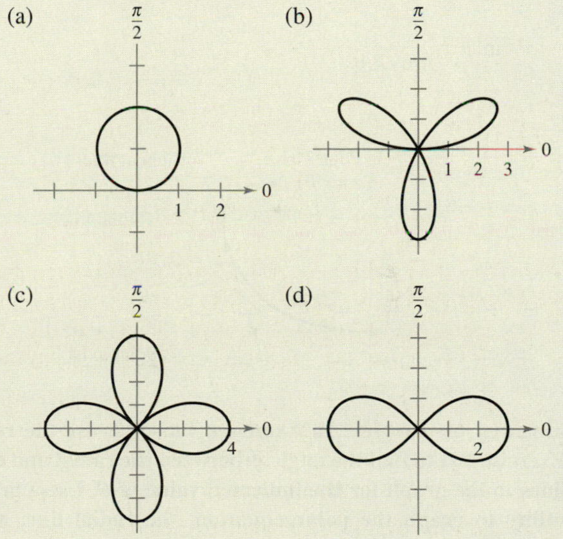

97. **Sketching a Graph** Sketch the graph of $r = 4 \sin \theta$ over each interval.

(a) $0 \le \theta \le \frac{\pi}{2}$ (b) $\frac{\pi}{2} \le \theta \le \pi$ (c) $-\frac{\pi}{2} \le \theta \le \frac{\pi}{2}$

98. **Think About It** Use a graphing utility to graph the polar equation $r = 6[1 + \cos(\theta - \phi)]$ for (a) $\phi = 0$, (b) $\phi = \pi/4$, and (c) $\phi = \pi/2$. Use the graphs to describe the effect of the angle ϕ. Write the equation as a function of $\sin \theta$ for part (c).

99. **Rotated Curve** Verify that if the curve whose polar equation is $r = f(\theta)$ is rotated about the pole through an angle ϕ, then an equation for the rotated curve is $r = f(\theta - \phi)$.

100. Rotated Curve The polar form of an equation of a curve is $r = f(\sin\theta)$. Show that the form becomes

(a) $r = f(-\cos\theta)$ if the curve is rotated counterclockwise $\pi/2$ radians about the pole.

(b) $r = f(-\sin\theta)$ if the curve is rotated counterclockwise π radians about the pole.

(c) $r = f(\cos\theta)$ if the curve is rotated counterclockwise $3\pi/2$ radians about the pole.

Rotated Curve In Exercises 101–104, use the results of Exercises 99 and 100.

101. Write an equation for the limaçon $r = 2 - \sin\theta$ after it has been rotated by the given amount. Use a graphing utility to graph the rotated limaçon for (a) $\theta = \pi/4$, (b) $\theta = \pi/2$, (c) $\theta = \pi$, and (d) $\theta = 3\pi/2$.

102. Write an equation for the rose curve $r = 2\sin 2\theta$ after it has been rotated by the given amount. Verify the results by using a graphing utility to graph the rotated rose curve for (a) $\theta = \pi/6$, (b) $\theta = \pi/2$, (c) $\theta = 2\pi/3$, and (d) $\theta = \pi$.

103. Sketch the graph of each equation.

(a) $r = 1 - \sin\theta$ (b) $r = 1 - \sin\left(\theta - \dfrac{\pi}{4}\right)$

104. Prove that the tangent of the angle $\psi\ (0 \le \psi \le \pi/2)$ between the radial line and the tangent line at the point (r, θ) on the graph of $r = f(\theta)$ (see figure) is given by

$$\tan\psi = \left|\frac{r}{dr/d\theta}\right|.$$

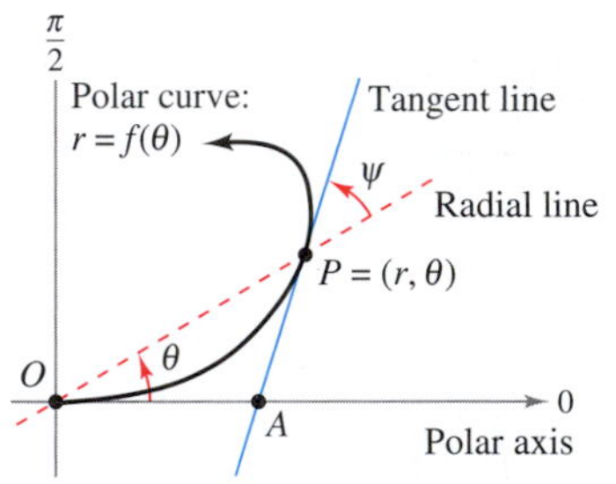

Finding an Angle In Exercises 105–110, use the result of Exercise 104 to find the angle ψ between the radial and tangent lines to the graph for the indicated value of θ. Use a graphing utility to graph the polar equation, the radial line, and the tangent line for the indicated value of θ. Identify the angle ψ.

	Polar Equation	Value of θ
105.	$r = 2(1 - \cos\theta)$	$\theta = \pi$
106.	$r = 3(1 - \cos\theta)$	$\theta = \dfrac{3\pi}{4}$
107.	$r = 2\cos 3\theta$	$\theta = \dfrac{\pi}{4}$
108.	$r = 4\sin 2\theta$	$\theta = \dfrac{\pi}{6}$
109.	$r = \dfrac{6}{1 - \cos\theta}$	$\theta = \dfrac{2\pi}{3}$
110.	$r = 5$	$\theta = \dfrac{\pi}{6}$

True or False? In Exercises 111–114, determine whether the statement is true or false. If it is false, explain why or give an example that shows it is false.

111. If (r_1, θ_1) and (r_2, θ_2) represent the same point on the polar coordinate system, then $|r_1| = |r_2|$.

112. If (r, θ_1) and (r, θ_2) represent the same point on the polar coordinate system, then $\theta_1 = \theta_2 + 2\pi n$ for some integer n.

113. If $x > 0$, then the point (x, y) on the rectangular coordinate system can be represented by (r, θ) on the polar coordinate system, where $r = \sqrt{x^2 + y^2}$ and $\theta = \arctan(y/x)$.

114. The polar equations $r = \sin 2\theta$, $r = -\sin 2\theta$, and $r = \sin(-2\theta)$ all have the same graph.

SECTION PROJECT

Anamorphic Art

Anamorphic art appears distorted, but when the art is viewed from a particular point or is viewed with a device such as a mirror, it appears to be normal. Use the anamorphic transformations

$$r = y + 16 \quad\text{and}\quad \theta = -\frac{\pi}{8}x, \quad -\frac{3\pi}{4} \le \theta \le \frac{3\pi}{4}$$

to sketch the transformed polar image of the rectangular graph. When the reflection (in a cylindrical mirror centered at the pole) of each polar image is viewed from the polar axis, the viewer will see the original rectangular image.

(a) $y = 3$ (b) $x = 2$ (c) $y = x + 5$ (d) $x^2 + (y - 5)^2 = 5^2$

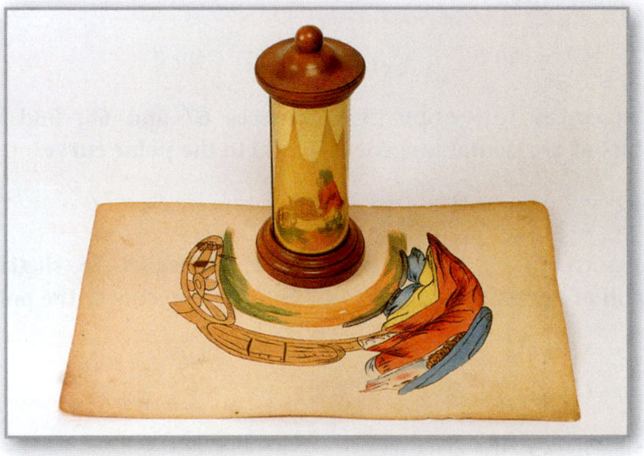

This example of anamorphic art is from the Millington-Barnard Collection at the University of Mississippi. When the reflection of the transformed "polar painting" is viewed in the mirror, the viewer sees the distorted art in its proper proportions.

FOR FURTHER INFORMATION For more information on anamorphic art, see the article "Anamorphisms" by Philip Hickin in the *Mathematical Gazette*.

From the Millington & Barnard Collection of Scientific Apparatus, ca. 1855 The University of Mississippi Museum, Oxford, Mississippi

10.5 Area and Arc Length in Polar Coordinates

- **Find the area of a region bounded by a polar graph.**
- **Find the points of intersection of two polar graphs.**
- **Find the arc length of a polar graph.**
- **Find the area of a surface of revolution (polar form).**

Area of a Polar Region

The development of a formula for the area of a polar region parallels that for the area of a region on the rectangular coordinate system, but uses sectors of a circle instead of rectangles as the basic elements of area. In Figure 10.48, note that the area of a circular sector of radius r is $\frac{1}{2}\theta r^2$, provided θ is measured in radians.

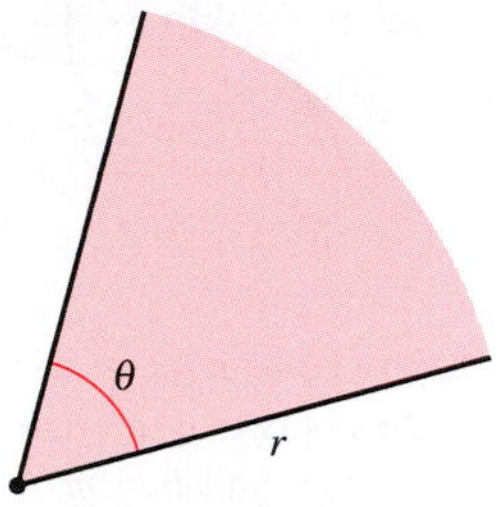

The area of a sector of a circle is $A = \frac{1}{2}\theta r^2$.
Figure 10.48

Consider the function $r = f(\theta)$, where f is continuous and nonnegative on the interval $\alpha \le \theta \le \beta$. The region bounded by the graph of f and the radial lines $\theta = \alpha$ and $\theta = \beta$ is shown in Figure 10.49(a). To find the area of this region, partition the interval $[\alpha, \beta]$ into n equal subintervals

$$\alpha = \theta_0 < \theta_1 < \theta_2 < \cdot \cdot \cdot < \theta_{n-1} < \theta_n = \beta.$$

Then approximate the area of the region by the sum of the areas of the n sectors, as shown in Figure 10.49(b).

$$\text{Radius of } i\text{th sector} = f(\theta_i)$$

$$\text{Central angle of } i\text{th sector} = \frac{\beta - \alpha}{n} = \Delta\theta$$

$$A \approx \sum_{i=1}^{n} \left(\frac{1}{2}\right) \Delta\theta [f(\theta_i)]^2$$

Taking the limit as $n \to \infty$ produces

$$A = \lim_{n\to\infty} \frac{1}{2} \sum_{i=1}^{n} [f(\theta_i)]^2 \Delta\theta$$
$$= \frac{1}{2} \int_{\alpha}^{\beta} [f(\theta)]^2 \, d\theta$$

which leads to the next theorem.

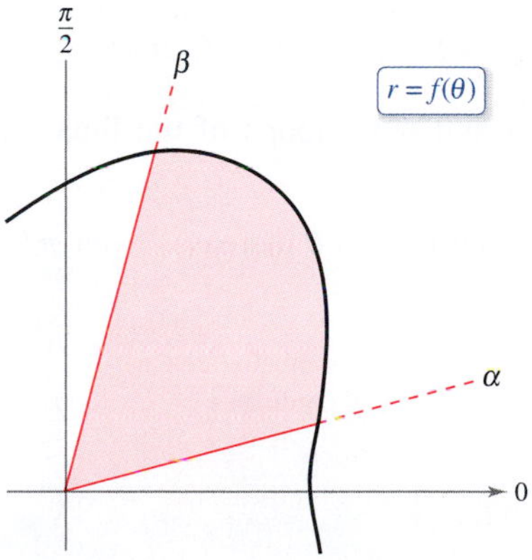

(a)

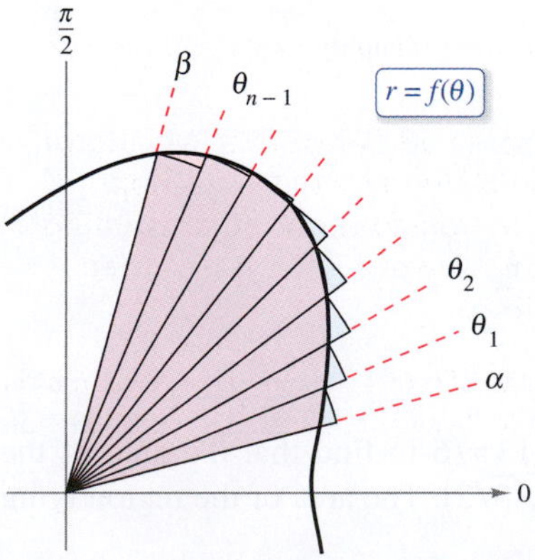

(b)
Figure 10.49

> **THEOREM 10.13 Area in Polar Coordinates**
>
> If f is continuous and nonnegative on the interval $[\alpha, \beta]$, $0 < \beta - \alpha \le 2\pi$, then the area of the region bounded by the graph of $r = f(\theta)$ between the radial lines $\theta = \alpha$ and $\theta = \beta$ is
>
> $$A = \frac{1}{2} \int_{\alpha}^{\beta} [f(\theta)]^2 \, d\theta$$
> $$= \frac{1}{2} \int_{\alpha}^{\beta} r^2 \, d\theta. \qquad 0 < \beta - \alpha \le 2\pi$$

You can use the formula in Theorem 10.13 to find the area of a region bounded by the graph of a continuous *nonpositive* function. The formula is not necessarily valid, however, when f takes on both positive *and* negative values in the interval $[\alpha, \beta]$.

EXAMPLE 1 Finding the Area of a Polar Region

See LarsonCalculus.com for an interactive version of this type of example.

Find the area of one petal of the rose curve $r = 3 \cos 3\theta$.

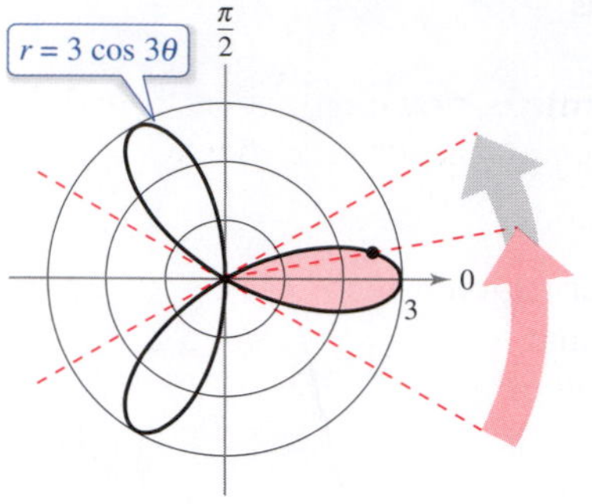

The area of one petal of the rose curve that lies between the radial lines $\theta = -\pi/6$ and $\theta = \pi/6$ is $3\pi/4$.
Figure 10.50

Solution In Figure 10.50, you can see that the petal on the right is traced as θ increases from $-\pi/6$ to $\pi/6$. So, the area is

$$
\begin{aligned}
A = \frac{1}{2}\int_{\alpha}^{\beta} r^2\,d\theta &= \frac{1}{2}\int_{-\pi/6}^{\pi/6} (3\cos 3\theta)^2\,d\theta && \text{Use formula for area in polar coordinates.}\\
&= \frac{9}{2}\int_{-\pi/6}^{\pi/6} \frac{1+\cos 6\theta}{2}\,d\theta && \text{Power-reducing formula}\\
&= \frac{9}{4}\left[\theta + \frac{\sin 6\theta}{6}\right]_{-\pi/6}^{\pi/6}\\
&= \frac{9}{4}\left(\frac{\pi}{6}+\frac{\pi}{6}\right)\\
&= \frac{3\pi}{4}.
\end{aligned}
$$

To find the area of the region lying inside all three petals of the rose curve in Example 1, you could *not* simply integrate between 0 and 2π. By doing this, you would obtain $9\pi/2$, which is twice the area of the three petals. The duplication occurs because the rose curve is traced twice as θ increases from 0 to 2π.

EXAMPLE 2 Finding the Area Bounded by a Single Curve

Find the area of the region lying between the inner and outer loops of the limaçon $r = 1 - 2 \sin \theta$.

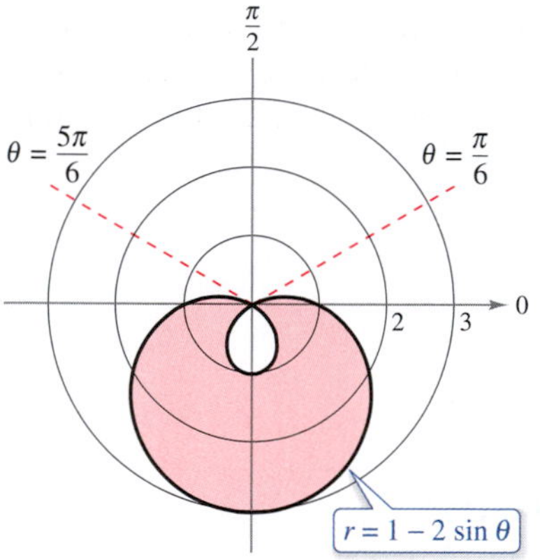

The area between the inner and outer loops is approximately 8.34.
Figure 10.51

Solution In Figure 10.51, note that the inner loop is traced as θ increases from $\pi/6$ to $5\pi/6$. So, the area inside the *inner loop* is

$$
\begin{aligned}
A_1 &= \frac{1}{2}\int_{\pi/6}^{5\pi/6} (1 - 2\sin\theta)^2\,d\theta && \text{Use formula for area in polar coordinates.}\\
&= \frac{1}{2}\int_{\pi/6}^{5\pi/6} (1 - 4\sin\theta + 4\sin^2\theta)\,d\theta\\
&= \frac{1}{2}\int_{\pi/6}^{5\pi/6} \left[1 - 4\sin\theta + 4\left(\frac{1-\cos 2\theta}{2}\right)\right]d\theta && \text{Power-reducing formula}\\
&= \frac{1}{2}\int_{\pi/6}^{5\pi/6} (3 - 4\sin\theta - 2\cos\theta)\,d\theta && \text{Simplify.}\\
&= \frac{1}{2}\Big[3\theta + 4\cos\theta - \sin 2\theta\Big]_{\pi/6}^{5\pi/6}\\
&= \frac{1}{2}\left(2\pi - 3\sqrt{3}\right)\\
&= \pi - \frac{3\sqrt{3}}{2}.
\end{aligned}
$$

In a similar way, you can integrate from $5\pi/6$ to $13\pi/6$ to find that the area of the region lying inside the *outer loop* is $A_2 = 2\pi + \left(3\sqrt{3}/2\right)$. The area of the region lying between the two loops is the difference of A_2 and A_1.

$$A = A_2 - A_1 = \left(2\pi + \frac{3\sqrt{3}}{2}\right) - \left(\pi - \frac{3\sqrt{3}}{2}\right) = \pi + 3\sqrt{3} \approx 8.34$$

Points of Intersection of Polar Graphs

Because a point may be represented in different ways in polar coordinates, care must be taken in determining the points of intersection of two polar graphs. For example, consider the points of intersection of the graphs of

$$r = 1 - 2\cos\theta \quad \text{and} \quad r = 1$$

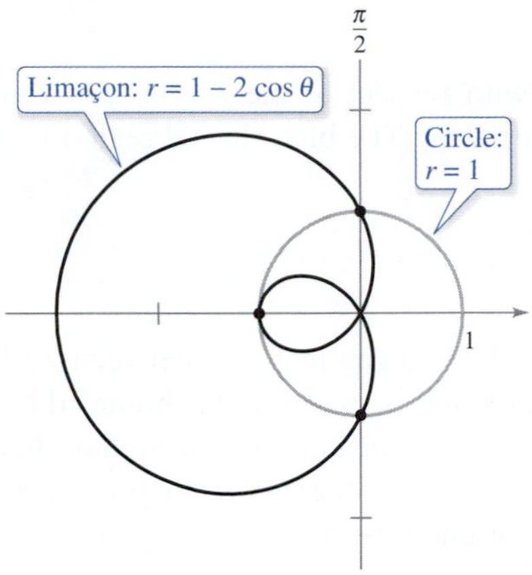

Three points of intersection: $(1, \pi/2)$, $(-1, 0)$, and $(1, 3\pi/2)$
Figure 10.52

as shown in Figure 10.52. As with rectangular equations, you can attempt to find the points of intersection by solving the two equations simultaneously, as shown.

$$r = 1 - 2\cos\theta \qquad \text{First equation}$$

$$1 = 1 - 2\cos\theta \qquad \text{Substitute } r = 1 \text{ from 2nd equation into 1st equation.}$$

$$\cos\theta = 0 \qquad \text{Simplify.}$$

$$\theta = \frac{\pi}{2}, \frac{3\pi}{2} \qquad \text{Solve for } \theta.$$

The corresponding points of intersection are $(1, \pi/2)$ and $(1, 3\pi/2)$. From Figure 10.52, however, you can see that there is a *third* point of intersection that did not show up when the two polar equations were solved simultaneously. (This is one reason why you should sketch a graph when finding the area of a polar region.) The reason the third point was not found is that it does not occur with the same coordinates in the two graphs. On the graph of $r = 1$, the point occurs with coordinates $(1, \pi)$, but on the graph of

$$r = 1 - 2\cos\theta$$

the point occurs with coordinates $(-1, 0)$.

In addition to solving equations simultaneously and sketching a graph, note that because the pole can be represented by $(0, \theta)$, where θ is *any* angle, you should check separately for the pole when finding points of intersection.

You can compare the problem of finding points of intersection of two polar graphs with that of finding collision points of two satellites in intersecting orbits about Earth, as shown in Figure 10.53. The satellites will not collide as long as they reach the points of intersection at different times (θ-values). Collisions will occur only at the points of intersection that are "simultaneous points"—those that are reached at the same time (θ-value).

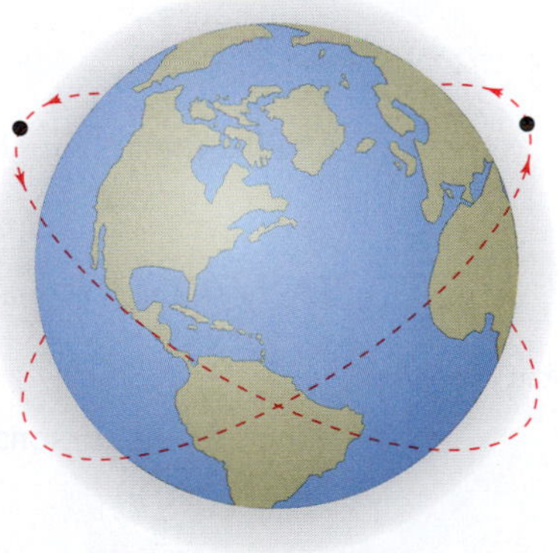

The paths of satellites can cross without causing a collision.
Figure 10.53

FOR FURTHER INFORMATION For more information on using technology to find points of intersection, see the article "Finding Points of Intersection of Polar-Coordinate Graphs" by Warren W. Esty in *Mathematics Teacher.* To view this article, go to *MathArticles.com.*

EXAMPLE 3 Finding the Area of a Region Between Two Curves

Find the area of the region common to the two regions bounded by the curves.

$r = -6\cos\theta$ Circle

and

$r = 2 - 2\cos\theta.$ Cardioid

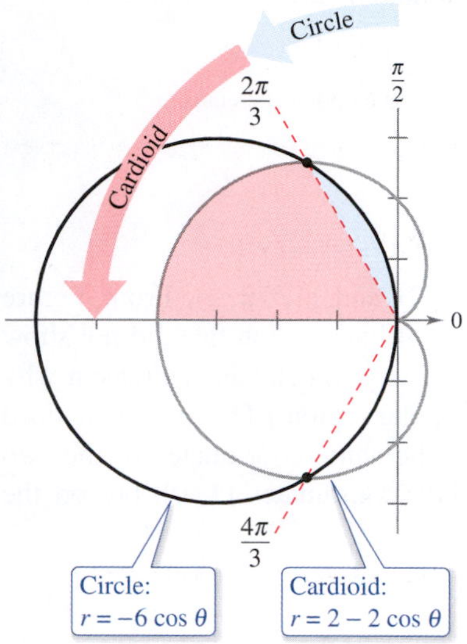

Figure 10.54

Solution Because both curves are symmetric with respect to the x-axis, you can work with the upper half-plane, as shown in Figure 10.54. The blue shaded region lies between the circle and the radial line

$$\theta = \frac{2\pi}{3}.$$

Because the circle has coordinates $(0, \pi/2)$ at the pole, you can integrate between $\pi/2$ and $2\pi/3$ to obtain the area of this region. The region that is shaded red is bounded by the radial lines $\theta = 2\pi/3$ and $\theta = \pi$ and the cardioid. So, you can find the area of this second region by integrating between $2\pi/3$ and π. The sum of these two integrals gives the area of the common region lying *above* the radial line $\theta = \pi$.

Region between circle and radial line $\theta = 2\pi/3$ — Region between cardioid and radial lines $\theta = 2\pi/3$ and $\theta = \pi$

$$\begin{aligned}
\frac{A}{2} &= \frac{1}{2}\int_{\pi/2}^{2\pi/3} (-6\cos\theta)^2\,d\theta + \frac{1}{2}\int_{2\pi/3}^{\pi} (2 - 2\cos\theta)^2\,d\theta \\
&= 18\int_{\pi/2}^{2\pi/3} \cos^2\theta\,d\theta + \frac{1}{2}\int_{2\pi/3}^{\pi} (4 - 8\cos\theta + 4\cos^2\theta)\,d\theta \\
&= 9\int_{\pi/2}^{2\pi/3} (1 + \cos 2\theta)\,d\theta + \int_{2\pi/3}^{\pi} (3 - 4\cos\theta + \cos 2\theta)\,d\theta \\
&= 9\left[\theta + \frac{\sin 2\theta}{2}\right]_{\pi/2}^{2\pi/3} + \left[3\theta - 4\sin\theta + \frac{\sin 2\theta}{2}\right]_{2\pi/3}^{\pi} \\
&= 9\left(\frac{2\pi}{3} - \frac{\sqrt{3}}{4} - \frac{\pi}{2}\right) + \left(3\pi - 2\pi + 2\sqrt{3} + \frac{\sqrt{3}}{4}\right) \\
&= \frac{5\pi}{2}
\end{aligned}$$

Finally, multiplying by 2, you can conclude that the total area is

$5\pi \approx 15.7.$ Area of region inside circle and cardioid

To check the reasonableness of this result, note that the area of the circular region is

$\pi r^2 = 9\pi.$ Area of circle

So, it seems reasonable that the area of the region lying inside the circle and the cardioid is 5π.

AP* Tips

Finding areas inside or between polar graphs is tested on the AP Calculus BC Exam on both the multiple choice and free response sections.

To see the benefit of polar coordinates for finding the area in Example 3, consider the integral below, which gives the comparable area in rectangular coordinates.

$$\frac{A}{2} = \int_{-4}^{-3/2} \sqrt{2\sqrt{1 - 2x} - x^2 - 2x + 2}\,dx + \int_{-3/2}^{0} \sqrt{-x^2 - 6x}\,dx$$

Use the integration capabilities of a graphing utility to show that you obtain the same area as that found in Example 3.

Arc Length in Polar Form

The formula for the length of a polar arc can be obtained from the arc length formula for a curve described by parametric equations. (See Exercise 85.)

> **REMARK** When applying the arc length formula to a polar curve, be sure that the curve is traced out only once on the interval of integration. For instance, the rose curve $r = \cos 3\theta$ is traced out once on the interval $0 \le \theta \le \pi$, but is traced out twice on the interval $0 \le \theta \le 2\pi$.

THEOREM 10.14 Arc Length of a Polar Curve

Let f be a function whose derivative is continuous on an interval $\alpha \le \theta \le \beta$. The length of the graph of $r = f(\theta)$ from $\theta = \alpha$ to $\theta = \beta$ is

$$s = \int_{\alpha}^{\beta} \sqrt{[f(\theta)]^2 + [f'(\theta)]^2}\, d\theta = \int_{\alpha}^{\beta} \sqrt{r^2 + \left(\frac{dr}{d\theta}\right)^2}\, d\theta.$$

EXAMPLE 4 Finding the Length of a Polar Curve

Find the length of the arc from $\theta = 0$ to $\theta = 2\pi$ for the cardioid

$$r = f(\theta) = 2 - 2\cos\theta$$

as shown in Figure 10.55.

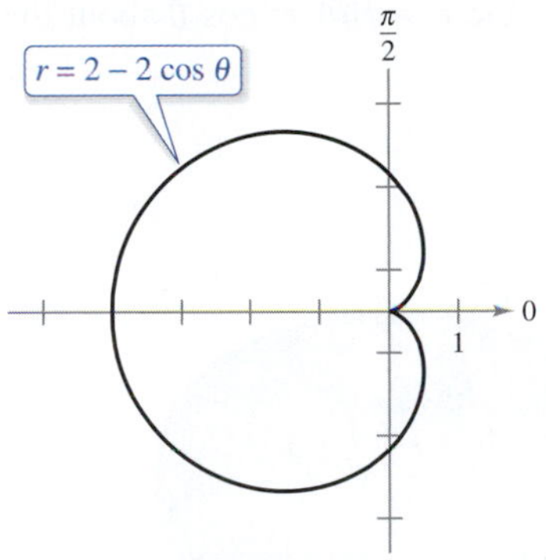

Figure 10.55

Solution Because $f'(\theta) = 2\sin\theta$, you can find the arc length as follows.

$$\begin{aligned}
s &= \int_{\alpha}^{\beta} \sqrt{[f(\theta)]^2 + [f'(\theta)]^2}\, d\theta && \text{Formula for arc length of a polar curve}\\
&= \int_{0}^{2\pi} \sqrt{(2 - 2\cos\theta)^2 + (2\sin\theta)^2}\, d\theta\\
&= 2\sqrt{2}\int_{0}^{2\pi} \sqrt{1 - \cos\theta}\, d\theta && \text{Simplify.}\\
&= 2\sqrt{2}\int_{0}^{2\pi} \sqrt{2\sin^2\frac{\theta}{2}}\, d\theta && \text{Trigonometric identity}\\
&= 4\int_{0}^{2\pi} \sin\frac{\theta}{2}\, d\theta && \sin\frac{\theta}{2} \ge 0 \text{ for } 0 \le \theta \le 2\pi\\
&= 8\left[-\cos\frac{\theta}{2}\right]_{0}^{2\pi}\\
&= 8(1 + 1)\\
&= 16
\end{aligned}$$

Using Figure 10.55, you can determine the reasonableness of this answer by comparing it with the circumference of a circle. For example, a circle of radius $\frac{5}{2}$ has a circumference of

$$5\pi \approx 15.7.$$

Note that in the fifth step of the solution, it is legitimate to write

$$\sqrt{2\sin^2\frac{\theta}{2}} = \sqrt{2}\sin\frac{\theta}{2}$$

rather than

$$\sqrt{2\sin^2\frac{\theta}{2}} = \sqrt{2}\left|\sin\frac{\theta}{2}\right|$$

because $\sin(\theta/2) \ge 0$ for $0 \le \theta \le 2\pi$.

AP* Tips

Arc length on a polar graph is not tested on the AP Exam.

Area of a Surface of Revolution

The polar coordinate versions of the formulas for the area of a surface of revolution can be obtained from the parametric versions given in Theorem 10.9, using the equations $x = r\cos\theta$ and $y = r\sin\theta$.

REMARK When using Theorem 10.15, check to see that the graph of $r = f(\theta)$ is traced only once on the interval $\alpha \le \theta \le \beta$. For example, the circle $r = \cos\theta$ is traced only once on the interval $0 \le \theta \le \pi$.

THEOREM 10.15 Area of a Surface of Revolution

Let f be a function whose derivative is continuous on an interval $\alpha \le \theta \le \beta$. The area of the surface formed by revolving the graph of $r = f(\theta)$ from $\theta = \alpha$ to $\theta = \beta$ about the indicated line is as follows.

1. $S = 2\pi \int_{\alpha}^{\beta} f(\theta)\sin\theta\sqrt{[f(\theta)]^2 + [f'(\theta)]^2}\,d\theta$ About the polar axis

2. $S = 2\pi \int_{\alpha}^{\beta} f(\theta)\cos\theta\sqrt{[f(\theta)]^2 + [f'(\theta)]^2}\,d\theta$ About the line $\theta = \dfrac{\pi}{2}$

EXAMPLE 5 Finding the Area of a Surface of Revolution

Find the area of the surface formed by revolving the circle $r = f(\theta) = \cos\theta$ about the line $\theta = \pi/2$, as shown in Figure 10.56.

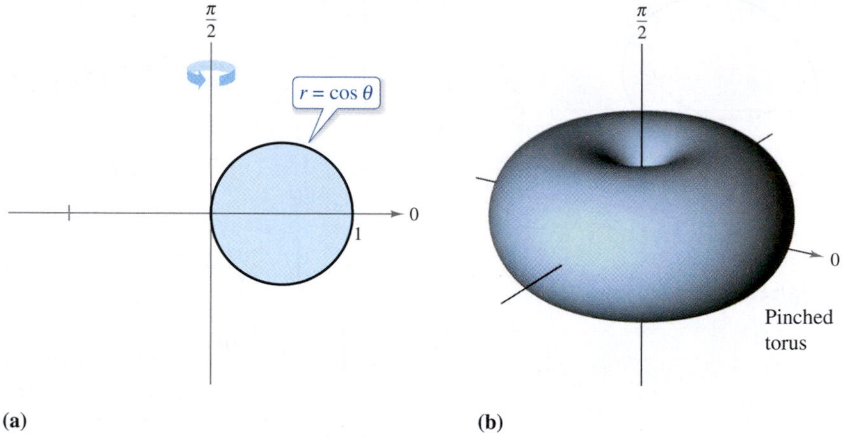

Figure 10.56

Solution Use the second formula in Theorem 10.15 with $f'(\theta) = -\sin\theta$. Because the circle is traced once as θ increases from 0 to π, you have

$$\begin{aligned} S &= 2\pi \int_{\alpha}^{\beta} f(\theta)\cos\theta\sqrt{[f(\theta)]^2 + [f'(\theta)]^2}\,d\theta && \text{Formula for area of a surface of revolution} \\ &= 2\pi \int_{0}^{\pi} \cos\theta(\cos\theta)\sqrt{\cos^2\theta + \sin^2\theta}\,d\theta \\ &= 2\pi \int_{0}^{\pi} \cos^2\theta\,d\theta && \text{Trigonometric identity} \\ &= \pi \int_{0}^{\pi} (1 + \cos 2\theta)\,d\theta && \text{Trigonometric identity} \\ &= \pi\left[\theta + \frac{\sin 2\theta}{2}\right]_{0}^{\pi} \\ &= \pi^2. \end{aligned}$$

10.5 Exercises

See CalcChat.com for tutorial help and worked-out solutions to odd-numbered exercises.

Area of a Polar Region In Exercises 1–4, write an integral that represents the area of the shaded region of the figure. Do not evaluate the integral.

1. $r = 4 \sin \theta$

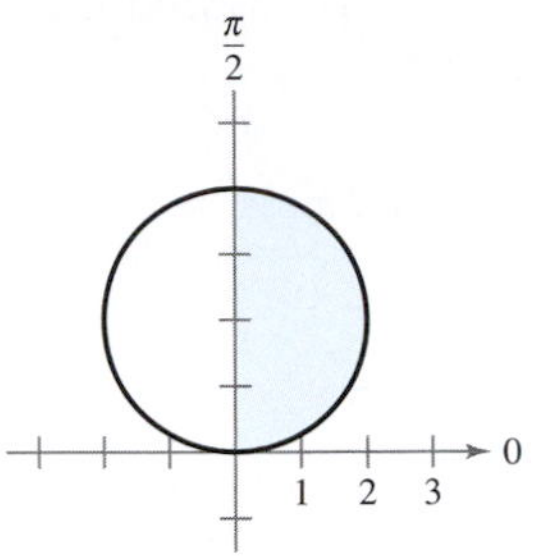

2. $r = \cos 2\theta$

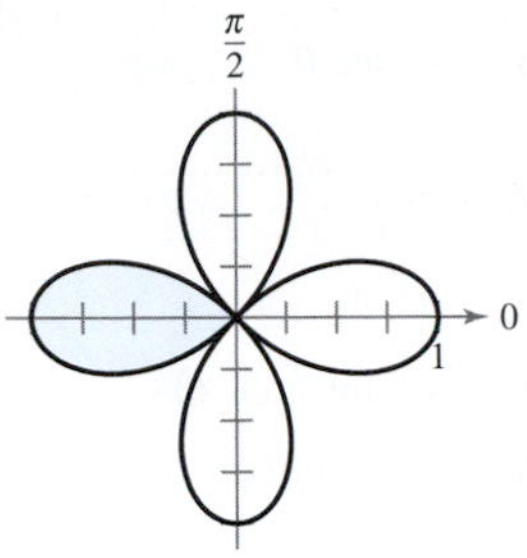

3. $r = 3 - 2 \sin \theta$

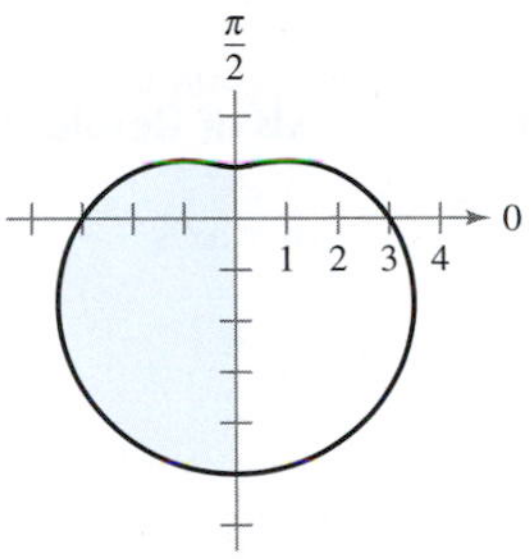

4. $r = 1 - \cos 2\theta$

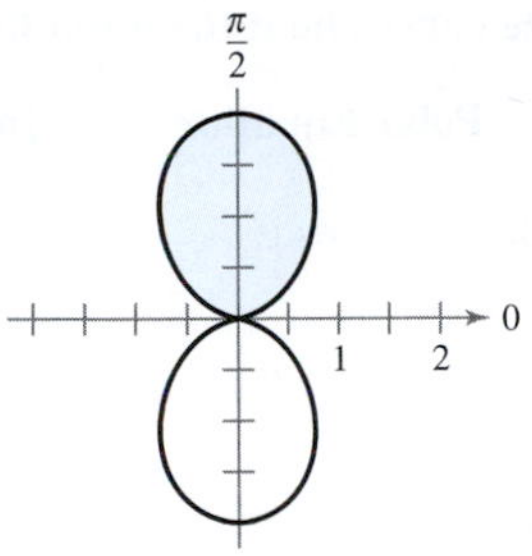

Finding the Area of a Polar Region In Exercises 5–16, find the area of the region.

5. Interior of $r = 6 \sin \theta$
6. Interior of $r = 3 \cos \theta$
7. One petal of $r = 2 \cos 3\theta$
8. One petal of $r = 4 \sin 3\theta$
9. One petal of $r = \sin 2\theta$
10. One petal of $r = \cos 5\theta$
11. Interior of $r = 1 - \sin \theta$
12. Interior of $r = 1 - \sin \theta$ (above the polar axis)
13. Interior of $r = 5 + 2 \sin \theta$
14. Interior of $r = 4 - 4 \cos \theta$
15. Interior of $r^2 = 4 \cos 2\theta$
16. Interior of $r^2 = 6 \sin 2\theta$

Finding the Area of a Polar Region In Exercises 17–24, use a graphing utility to graph the polar equation. Find the area of the given region analytically.

17. Inner loop of $r = 1 + 2 \cos \theta$
18. Inner loop of $r = 2 - 4 \cos \theta$
19. Inner loop of $r = 1 + 2 \sin \theta$
20. Inner loop of $r = 4 - 6 \sin \theta$
21. Between the loops of $r = 1 + 2 \cos \theta$
22. Between the loops of $r = 2(1 + 2 \sin \theta)$
23. Between the loops of $r = 3 - 6 \sin \theta$
24. Between the loops of $r = \frac{1}{2} + \cos \theta$

Finding Points of Intersection In Exercises 25–32, find the points of intersection of the graphs of the equations.

25. $r = 1 + \cos \theta$
$r = 1 - \cos \theta$

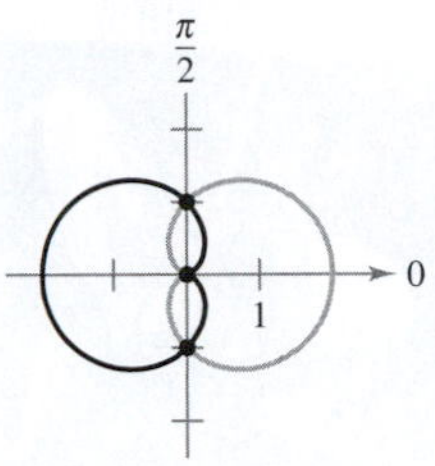

26. $r = 3(1 + \sin \theta)$
$r = 3(1 - \sin \theta)$

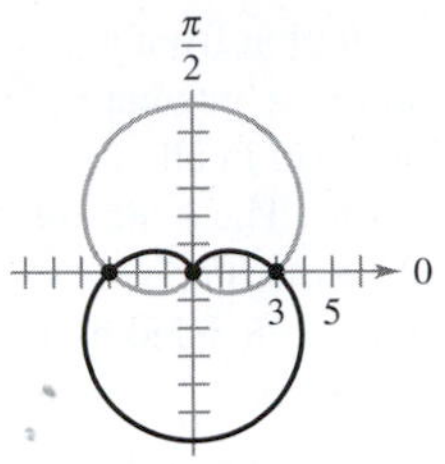

27. $r = 1 + \cos \theta$
$r = 1 - \sin \theta$

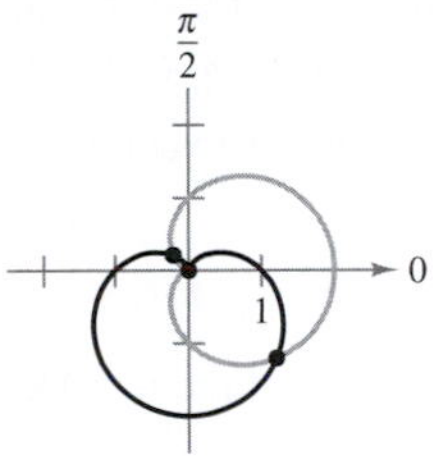

28. $r = 2 - 3 \cos \theta$
$r = \cos \theta$

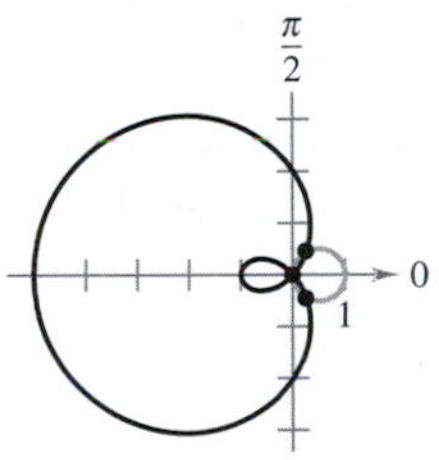

29. $r = 4 - 5 \sin \theta$
$r = 3 \sin \theta$

30. $r = 3 + \sin \theta$
$r = 2 \csc \theta$

31. $r = \dfrac{\theta}{2}$
$r = 2$

32. $\theta = \dfrac{\pi}{4}$
$r = 2$

Writing In Exercises 33 and 34, use a graphing utility to graph the polar equations and approximate the points of intersection of the graphs. Watch the graphs as they are traced in the viewing window. Explain why the pole is not a point of intersection obtained by solving the equations simultaneously.

33. $r = \cos \theta$
$r = 2 - 3 \sin \theta$

34. $r = 4 \sin \theta$
$r = 2(1 + \sin \theta)$

Finding the Area of a Polar Region Between Two Curves In Exercises 35–42, use a graphing utility to graph the polar equations. Find the area of the given region analytically.

35. Common interior of $r = 4 \sin 2\theta$ and $r = 2$
36. Common interior of $r = 2(1 + \cos \theta)$ and $r = 2(1 - \cos \theta)$
37. Common interior of $r = 3 - 2 \sin \theta$ and $r = -3 + 2 \sin \theta$
38. Common interior of $r = 5 - 3 \sin \theta$ and $r = 5 - 3 \cos \theta$
39. Common interior of $r = 4 \sin \theta$ and $r = 2$
40. Common interior of $r = 2 \cos \theta$ and $r = 2 \sin \theta$
41. Inside $r = 2 \cos \theta$ and outside $r = 1$
42. Inside $r = 3 \sin \theta$ and outside $r = 1 + \sin \theta$

Finding the Area of a Polar Region Between Two Curves **In Exercises 43–46, find the area of the region.**

43. Inside $r = a(1 + \cos\theta)$ and outside $r = a\cos\theta$

44. Inside $r = 2a\cos\theta$ and outside $r = a$

45. Common interior of $r = a(1 + \cos\theta)$ and $r = a\sin\theta$

46. Common interior of $r = a\cos\theta$ and $r = a\sin\theta$, where $a > 0$

47. Antenna Radiation

The radiation from a transmitting antenna is not uniform in all directions. The intensity from a particular antenna is modeled by $r = a\cos^2\theta$.

(a) Convert the polar equation to rectangular form.

(b) Use a graphing utility to graph the model for $a = 4$ and $a = 6$.

(c) Find the area of the geographical region between the two curves in part (b).

48. Area The area inside one or more of the three interlocking circles

$$r = 2a\cos\theta, \quad r = 2a\sin\theta, \quad \text{and} \quad r = a$$

is divided into seven regions. Find the area of each region.

49. Conjecture Find the area of the region enclosed by

$$r = a\cos(n\theta)$$

for $n = 1, 2, 3, \ldots$. Use the results to make a conjecture about the area enclosed by the function when n is even and when n is odd.

50. Area Sketch the strophoid

$$r = \sec\theta - 2\cos\theta, \quad -\frac{\pi}{2} < \theta < \frac{\pi}{2}.$$

Convert this equation to rectangular coordinates. Find the area enclosed by the loop.

Finding the Arc Length of a Polar Curve **In Exercises 51–56, find the length of the curve over the given interval.**

	Polar Equation	Interval
51.	$r = 8$	$0 \le \theta \le 2\pi$
52.	$r = a$	$0 \le \theta \le 2\pi$
53.	$r = 4\sin\theta$	$0 \le \theta \le \pi$
54.	$r = 2a\cos\theta$	$-\frac{\pi}{2} \le \theta \le \frac{\pi}{2}$
55.	$r = 1 + \sin\theta$	$0 \le \theta \le 2\pi$
56.	$r = 8(1 + \cos\theta)$	$0 \le \theta \le 2\pi$

Finding the Arc Length of a Polar Curve **In Exercises 57–62, use a graphing utility to graph the polar equation over the given interval. Use the integration capabilities of the graphing utility to approximate the length of the curve accurate to two decimal places.**

	Polar Equation	Interval
57.	$r = 2\theta$	$0 \le \theta \le \frac{\pi}{2}$
58.	$r = \sec\theta$	$0 \le \theta \le \frac{\pi}{3}$
59.	$r = \frac{1}{\theta}$	$\pi \le \theta \le 2\pi$
60.	$r = e^{\theta}$	$0 \le \theta \le \pi$
61.	$r = \sin(3\cos\theta)$	$0 \le \theta \le \pi$
62.	$r = 2\sin(2\cos\theta)$	$0 \le \theta \le \pi$

Finding the Area of a Surface of Revolution **In Exercises 63–66, find the area of the surface formed by revolving the curve about the given line.**

	Polar Equation	Interval	Axis of Revolution
63.	$r = 6\cos\theta$	$0 \le \theta \le \frac{\pi}{2}$	Polar axis
64.	$r = a\cos\theta$	$0 \le \theta \le \frac{\pi}{2}$	$\theta = \frac{\pi}{2}$
65.	$r = e^{a\theta}$	$0 \le \theta \le \frac{\pi}{2}$	$\theta = \frac{\pi}{2}$
66.	$r = a(1 + \cos\theta)$	$0 \le \theta \le \pi$	Polar axis

Finding the Area of a Surface of Revolution **In Exercises 67 and 68, use the integration capabilities of a graphing utility to approximate, to two decimal places, the area of the surface formed by revolving the curve about the polar axis.**

67. $r = 4\cos 2\theta, \quad 0 \le \theta \le \frac{\pi}{4}$

68. $r = \theta, \quad 0 \le \theta \le \pi$

WRITING ABOUT CONCEPTS

69. Points of Intersection Explain why finding points of intersection of polar graphs may require further analysis beyond solving two equations simultaneously.

70. Area of a Surface of Revolution Give the integral formulas for the area of the surface of revolution formed when the graph of $r = f(\theta)$ is revolved about

(a) the polar axis.

(b) the line $\theta = \pi/2$.

71. Area of a Region For each polar equation, sketch its graph, determine the interval that traces the graph only once, and find the area of the region bounded by the graph using a geometric formula and integration.

(a) $r = 10\cos\theta$ (b) $r = 5\sin\theta$

BESTWEB/Shutterstock.com

72. HOW DO YOU SEE IT? Which graph, traced out only once, has a larger arc length? Explain your reasoning.

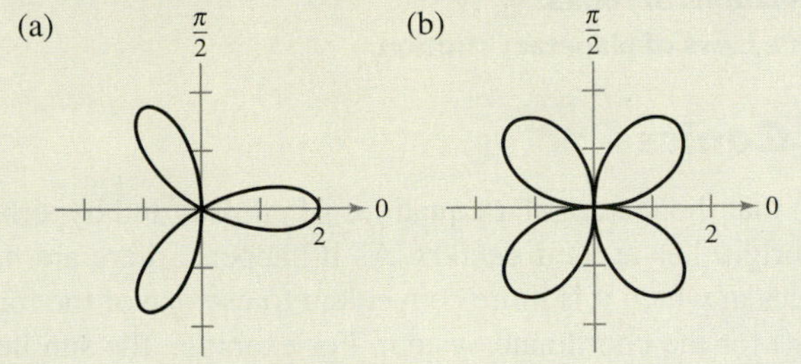

73. Surface Area of a Torus Find the surface area of the torus generated by revolving the circle given by $r = 2$ about the line $r = 5 \sec \theta$.

74. Surface Area of a Torus Find the surface area of the torus generated by revolving the circle given by $r = a$ about the line $r = b \sec \theta$, where $0 < a < b$.

75. Approximating Area Consider the circle $r = 8 \cos \theta$.

(a) Find the area of the circle.

(b) Complete the table giving the areas A of the sectors of the circle between $\theta = 0$ and the values of θ in the table.

θ	0.2	0.4	0.6	0.8	1.0	1.2	1.4
A							

(c) Use the table in part (b) to approximate the values of θ for which the sector of the circle composes $\frac{1}{4}$, $\frac{1}{2}$, and $\frac{3}{4}$ of the total area of the circle.

(d) Use a graphing utility to approximate, to two decimal places, the angles θ for which the sector of the circle composes $\frac{1}{4}$, $\frac{1}{2}$, and $\frac{3}{4}$ of the total area of the circle.

(e) Do the results of part (d) depend on the radius of the circle? Explain.

76. Approximating Area Consider the circle $r = 3 \sin \theta$.

(a) Find the area of the circle.

(b) Complete the table giving the areas A of the sectors of the circle between $\theta = 0$ and the values of θ in the table.

θ	0.2	0.4	0.6	0.8	1.0	1.2	1.4
A							

(c) Use the table in part (b) to approximate the values of θ for which the sector of the circle composes $\frac{1}{8}$, $\frac{1}{4}$, and $\frac{1}{2}$ of the total area of the circle.

(d) Use a graphing utility to approximate, to two decimal places, the angles θ for which the sector of the circle composes $\frac{1}{8}$, $\frac{1}{4}$, and $\frac{1}{2}$ of the total area of the circle.

77. Conic What conic section does the polar equation $r = a \sin \theta + b \cos \theta$ represent?

78. Area Find the area of the circle given by

$r = \sin \theta + \cos \theta.$

Check your result by converting the polar equation to rectangular form, then using the formula for the area of a circle.

79. Spiral of Archimedes The curve represented by the equation $r = a\theta$, where a is a constant, is called the spiral of Archimedes.

(a) Use a graphing utility to graph $r = \theta$, where $\theta \geq 0$. What happens to the graph of $r = a\theta$ as a increases? What happens if $\theta \leq 0$?

(b) Determine the points on the spiral $r = a\theta$ $(a > 0, \theta \geq 0)$, where the curve crosses the polar axis.

(c) Find the length of $r = \theta$ over the interval $0 \leq \theta \leq 2\pi$.

(d) Find the area under the curve $r = \theta$ for $0 \leq \theta \leq 2\pi$.

80. Logarithmic Spiral The curve represented by the equation $r = ae^{b\theta}$, where a and b are constants, is called a **logarithmic spiral.** The figure shows the graph of $r = e^{\theta/6}$, $-2\pi \leq \theta \leq 2\pi$. Find the area of the shaded region.

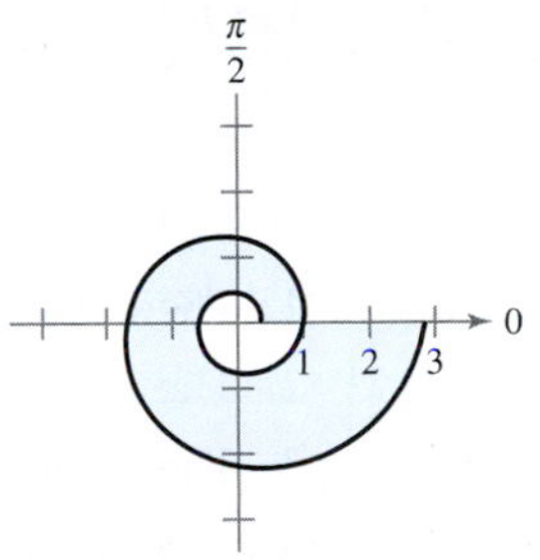

Figure for 80

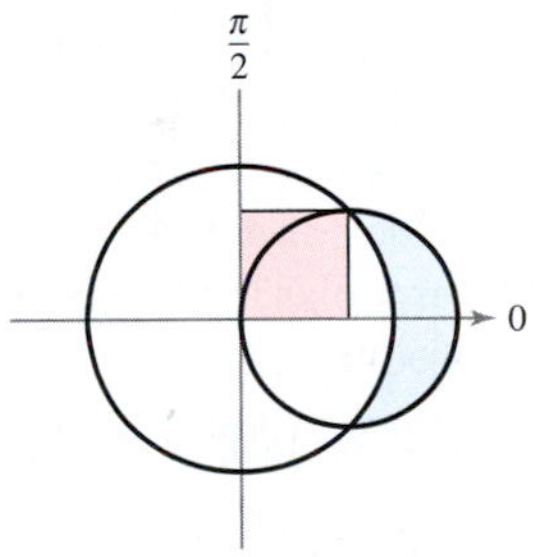

Figure for 81

81. Area The larger circle in the figure is the graph of $r = 1$. Find the polar equation of the smaller circle such that the shaded regions are equal.

82. Folium of Descartes A curve called the **folium of Descartes** can be represented by the parametric equations

$$x = \frac{3t}{1 + t^3} \quad \text{and} \quad y = \frac{3t^2}{1 + t^3}.$$

(a) Convert the parametric equations to polar form.

(b) Sketch the graph of the polar equation from part (a).

(c) Use a graphing utility to approximate the area enclosed by the loop of the curve.

True or False? **In Exercises 83 and 84, determine whether the statement is true or false. If it is false, explain why or give an example that shows it is false.**

83. If $f(\theta) > 0$ for all θ and $g(\theta) < 0$ for all θ, then the graphs of $r = f(\theta)$ and $r = g(\theta)$ do not intersect.

84. If $f(\theta) = g(\theta)$ for $\theta = 0$, $\pi/2$, and $3\pi/2$, then the graphs of $r = f(\theta)$ and $r = g(\theta)$ have at least four points of intersection.

85. Arc Length in Polar Form Use the formula for the arc length of a curve in parametric form to derive the formula for the arc length of a polar curve.

10.6 Polar Equations of Conics and Kepler's Laws

- Analyze and write polar equations of conics.
- Understand and use Kepler's Laws of planetary motion.

Polar Equations of Conics

> **Exploration**
>
> ***Graphing Conics*** Set a graphing utility to *polar* mode and enter polar equations of the form
>
> $$r = \frac{a}{1 \pm b\cos\theta}$$
>
> or
>
> $$r = \frac{a}{1 \pm b\sin\theta}.$$
>
> As long as $a \neq 0$, the graph should be a conic. What values of a and b produce parabolas? What values produce ellipses? What values produce hyperbolas?

In this chapter, you have seen that the rectangular equations of ellipses and hyperbolas take simple forms when the origin lies at their *centers*. As it happens, there are many important applications of conics in which it is more convenient to use one of the foci as the reference point (the origin) for the coordinate system. For example, the sun lies at a focus of Earth's orbit. Similarly, the light source of a parabolic reflector lies at its focus. In this section, you will see that polar equations of conics take simpler forms when one of the foci lies at the pole.

The next theorem uses the concept of *eccentricity*, as defined in Section 10.1, to classify the three basic types of conics.

THEOREM 10.16 Classification of Conics by Eccentricity

Let F be a fixed point (*focus*) and let D be a fixed line (*directrix*) in the plane. Let P be another point in the plane and let e (*eccentricity*) be the ratio of the distance between P and F to the distance between P and D. The collection of all points P with a given eccentricity is a conic.

1. The conic is an ellipse for $0 < e < 1$.
2. The conic is a parabola for $e = 1$.
3. The conic is a hyperbola for $e > 1$.

A proof of this theorem is given in Appendix A.

See LarsonCalculus.com for Bruce Edwards's video of this proof.

In Figure 10.57, note that for each type of conic, the pole corresponds to the fixed point (focus) given in the definition.

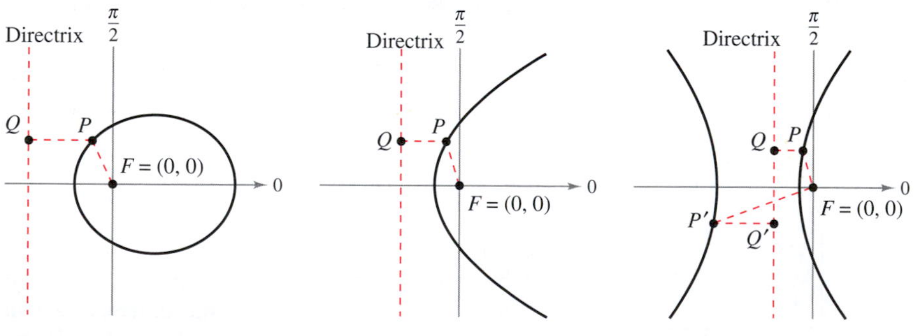

Ellipse: $0 < e < 1$ $\quad \frac{PF}{PQ} < 1$

Parabola: $e = 1$ $\quad PF = PQ$

Hyperbola: $e > 1$ $\quad \frac{PF}{PQ} = \frac{P'F}{P'Q'} > 1$

Figure 10.57

The benefit of locating a focus of a conic at the pole is that the equation of the conic becomes simpler, as seen in the proof of the next theorem.

THEOREM 10.17 Polar Equations of Conics

The graph of a polar equation of the form

$$r = \frac{ed}{1 \pm e\cos\theta} \quad \text{or} \quad r = \frac{ed}{1 \pm e\sin\theta}$$

is a conic, where $e > 0$ is the eccentricity and $|d|$ is the distance between the focus at the pole and its corresponding directrix.

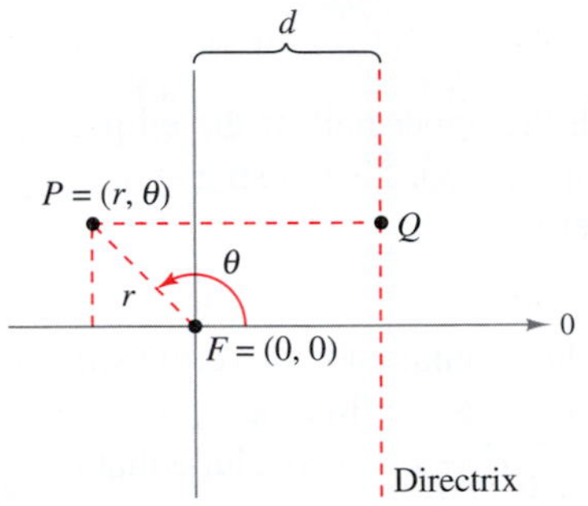

Figure 10.58

Proof This is a proof for $r = ed/(1 + e\cos\theta)$ with $d > 0$. In Figure 10.58, consider a vertical directrix d units to the right of the focus $F = (0, 0)$. If $P = (r, \theta)$ is a point on the graph of $r = ed/(1 + e\cos\theta)$, then the distance between P and the directrix can be shown to be

$$PQ = |d - x| = |d - r\cos\theta| = \left|\frac{r(1 + e\cos\theta)}{e} - r\cos\theta\right| = \left|\frac{r}{e}\right|.$$

Because the distance between P and the pole is simply $PF = |r|$, the ratio of PF to PQ is

$$\frac{PF}{PQ} = \frac{|r|}{|r/e|} = |e| = e$$

and, by Theorem 10.16, the graph of the equation must be a conic. The proofs of the other cases are similar.

See LarsonCalculus.com for Bruce Edwards's video of this proof.

The four types of equations indicated in Theorem 10.17 can be classified as follows, where $d > 0$.

a. Horizontal directrix above the pole: $r = \dfrac{ed}{1 + e\sin\theta}$

b. Horizontal directrix below the pole: $r = \dfrac{ed}{1 - e\sin\theta}$

c. Vertical directrix to the right of the pole: $r = \dfrac{ed}{1 + e\cos\theta}$

d. Vertical directrix to the left of the pole: $r = \dfrac{ed}{1 - e\cos\theta}$

Figure 10.59 illustrates these four possibilities for a parabola. Note that for convenience, the equation for the directrix is shown in rectangular form.

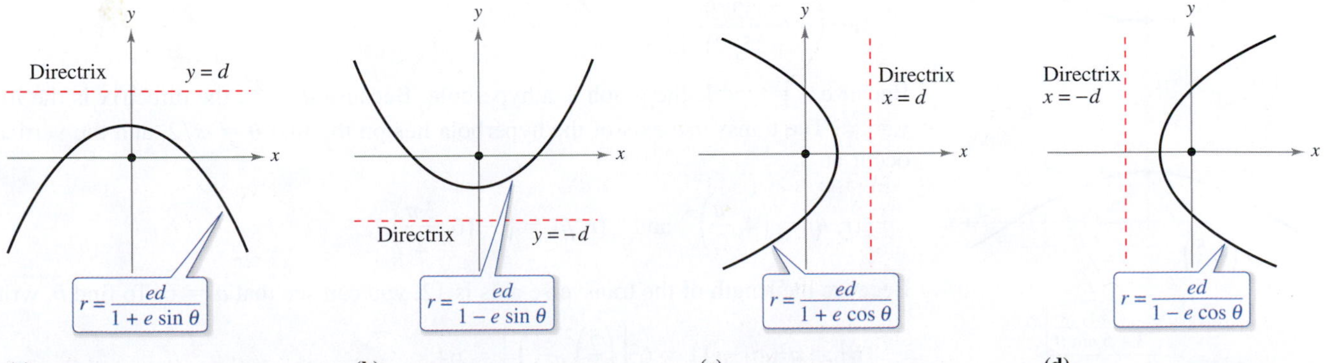

The four types of polar equations for a parabola
Figure 10.59

EXAMPLE 1 Determining a Conic from Its Equation

Sketch the graph of the conic $r = \dfrac{15}{3 - 2\cos\theta}$.

Solution To determine the type of conic, rewrite the equation as

$$r = \frac{15}{3 - 2\cos\theta} \quad \text{Write original equation.}$$

$$= \frac{5}{1 - (2/3)\cos\theta}. \quad \text{Divide numerator and denominator by 3.}$$

So, the graph is an ellipse with $e = \frac{2}{3}$. You can sketch the upper half of the ellipse by plotting points from $\theta = 0$ to $\theta = \pi$, as shown in Figure 10.60. Then, using symmetry with respect to the polar axis, you can sketch the lower half.

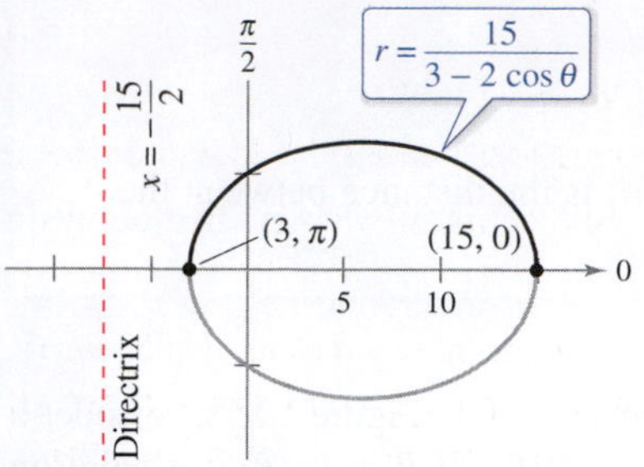

The graph of the conic is an ellipse with $e = \frac{2}{3}$.
Figure 10.60

For the ellipse in Figure 10.60, the major axis is horizontal and the vertices lie at $(15, 0)$ and $(3, \pi)$. So, the length of the *major* axis is $2a = 18$. To find the length of the minor axis, you can use the equations $e = c/a$ and $b^2 = a^2 - c^2$ to conclude that

$$b^2 = a^2 - c^2 = a^2 - (ea)^2 = a^2(1 - e^2). \quad \text{Ellipse}$$

Because $e = \frac{2}{3}$, you have

$$b^2 = 9^2\left[1 - \left(\tfrac{2}{3}\right)^2\right] = 45$$

which implies that $b = \sqrt{45} = 3\sqrt{5}$. So, the length of the minor axis is $2b = 6\sqrt{5}$. A similar analysis for hyperbolas yields

$$b^2 = c^2 - a^2 = (ea)^2 - a^2 = a^2(e^2 - 1). \quad \text{Hyperbola}$$

EXAMPLE 2 Sketching a Conic from Its Polar Equation

See LarsonCalculus.com for an interactive version of this type of example.

Sketch the graph of the polar equation $r = \dfrac{32}{3 + 5\sin\theta}$.

Solution Dividing the numerator and denominator by 3 produces

$$r = \frac{32/3}{1 + (5/3)\sin\theta}.$$

Because $e = \frac{5}{3} > 1$, the graph is a hyperbola. Because $d = \frac{32}{5}$, the directrix is the line $y = \frac{32}{5}$. The transverse axis of the hyperbola lies on the line $\theta = \pi/2$, and the vertices occur at

$$(r, \theta) = \left(4, \frac{\pi}{2}\right) \quad \text{and} \quad (r, \theta) = \left(-16, \frac{3\pi}{2}\right).$$

Because the length of the transverse axis is 12, you can see that $a = 6$. To find b, write

$$b^2 = a^2(e^2 - 1) = 6^2\left[\left(\frac{5}{3}\right)^2 - 1\right] = 64.$$

Therefore, $b = 8$. Finally, you can use a and b to determine the asymptotes of the hyperbola and obtain the sketch shown in Figure 10.61.

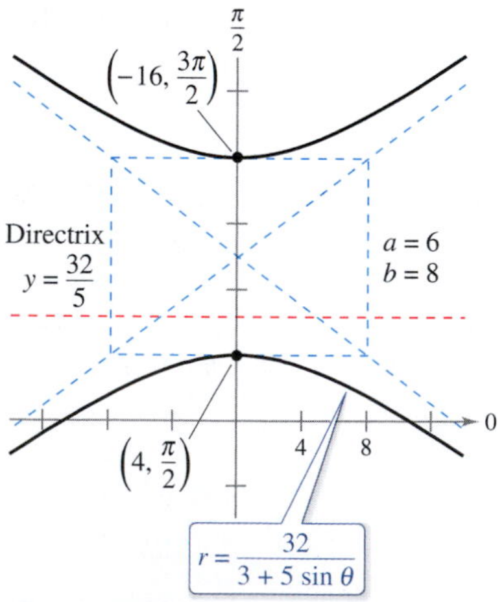

The graph of the conic is a hyperbola with $e = \frac{5}{3}$.
Figure 10.61

Kepler's Laws

Kepler's Laws, named after the German astronomer Johannes Kepler, can be used to describe the orbits of the planets about the sun.

1. Each planet moves in an elliptical orbit with the sun as a focus.
2. A ray from the sun to the planet sweeps out equal areas of the ellipse in equal times.
3. The square of the period is proportional to the cube of the mean distance between the planet and the sun.*

JOHANNES KEPLER (1571–1630)

Kepler formulated his three laws from the extensive data recorded by Danish astronomer Tycho Brahe, and from direct observation of the orbit of Mars.
See LarsonCalculus.com to read more of this biography.

Although Kepler derived these laws empirically, they were later validated by Newton. In fact, Newton was able to show that each law can be deduced from a set of universal laws of motion and gravitation that govern the movement of all heavenly bodies, including comets and satellites. This is shown in the next example, involving the comet named after the English mathematician and physicist Edmund Halley (1656–1742).

EXAMPLE 3 Halley's Comet

Halley's comet has an elliptical orbit with the sun at one focus and has an eccentricity of $e \approx 0.967$. The length of the major axis of the orbit is approximately 35.88 astronomical units (AU). (An astronomical unit is defined as the mean distance between Earth and the sun, 93 million miles.) Find a polar equation for the orbit. How close does Halley's comet come to the sun?

Solution Using a vertical axis, you can choose an equation of the form

$$r = \frac{ed}{(1 + e \sin \theta)}.$$

Because the vertices of the ellipse occur when $\theta = \pi/2$ and $\theta = 3\pi/2$, you can determine the length of the major axis to be the sum of the r-values of the vertices, as shown in Figure 10.62. That is,

$$2a = \frac{0.967d}{1 + 0.967} + \frac{0.967d}{1 - 0.967}$$

$$35.88 \approx 29.79d. \qquad 2a \approx 35.88$$

So, $d \approx 1.204$ and

$$ed \approx (0.967)(1.204) \approx 1.164.$$

Using this value in the equation produces

$$r = \frac{1.164}{1 + 0.967 \sin \theta}$$

where r is measured in astronomical units. To find the closest point to the sun (the focus), you can write

$$c = ea \approx (0.967)(17.94) \approx 17.35.$$

Because c is the distance between the focus and the center, the closest point is

$$\begin{aligned} a - c &\approx 17.94 - 17.35 \\ &\approx 0.59 \text{ AU} \\ &\approx 55{,}000{,}000 \text{ miles.} \end{aligned}$$

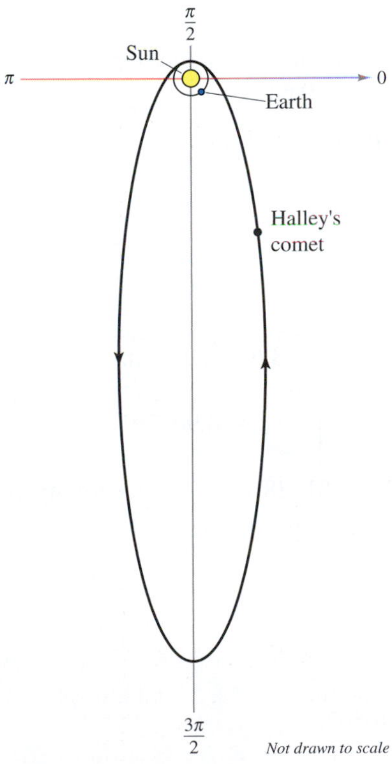

Figure 10.62

* If Earth is used as a reference with a period of 1 year and a distance of 1 astronomical unit, then the proportionality constant is 1. For example, because Mars has a mean distance to the sun of $D \approx 1.524$ AU, its period P is $D^3 = P^2$. So, the period for Mars is $P \approx 1.88$.

Kepler's Second Law states that as a planet moves about the sun, a ray from the sun to the planet sweeps out equal areas in equal times. This law can also be applied to comets or asteroids with elliptical orbits. For example, Figure 10.63 shows the orbit of the asteroid Apollo about the sun. Applying Kepler's Second Law to this asteroid, you know that the closer it is to the sun, the greater its velocity, because a short ray must be moving quickly to sweep out as much area as a long ray.

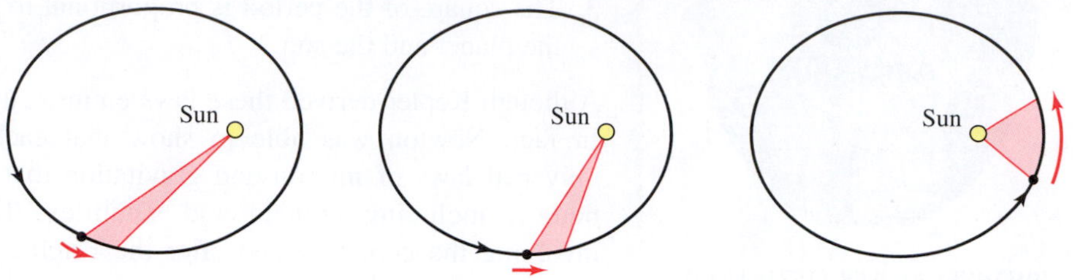

A ray from the sun to the asteroid Apollo sweeps out equal areas in equal times.
Figure 10.63

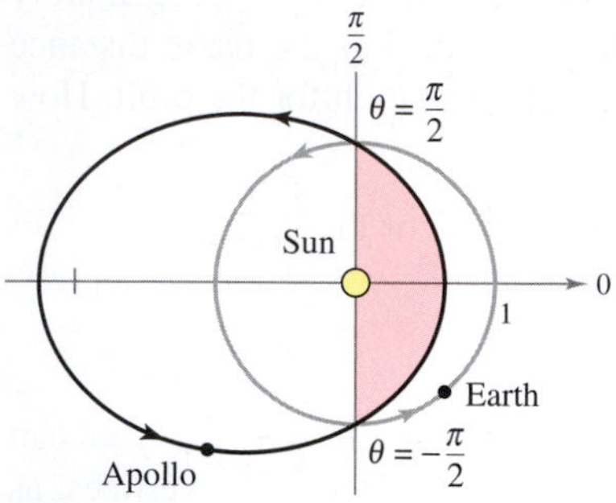

Figure 10.64

EXAMPLE 4 The Asteroid Apollo

The asteroid Apollo has a period of 661 Earth days, and its orbit is approximated by the ellipse

$$r = \frac{1}{1 + (5/9)\cos\theta} = \frac{9}{9 + 5\cos\theta}$$

where r is measured in astronomical units. How long does it take Apollo to move from the position $\theta = -\pi/2$ to $\theta = \pi/2$, as shown in Figure 10.64?

Solution Begin by finding the area swept out as θ increases from $-\pi/2$ to $\pi/2$.

$$A = \frac{1}{2}\int_{\alpha}^{\beta} r^2\,d\theta \qquad \text{Formula for area of a polar graph}$$

$$= \frac{1}{2}\int_{-\pi/2}^{\pi/2} \left(\frac{9}{9 + 5\cos\theta}\right)^2 d\theta$$

Using the substitution $u = \tan(\theta/2)$, as discussed in Section 8.6, you obtain

$$A = \frac{81}{112}\left[\frac{-5\sin\theta}{9 + 5\cos\theta} + \frac{18}{\sqrt{56}}\arctan\frac{\sqrt{56}\tan(\theta/2)}{14}\right]_{-\pi/2}^{\pi/2} \approx 0.90429.$$

Because the major axis of the ellipse has length $2a = 81/28$ and the eccentricity is $e = 5/9$, you can determine that

$$b = a\sqrt{1 - e^2} = \frac{9}{\sqrt{56}}.$$

So, the area of the ellipse is

$$\text{Area of ellipse} = \pi ab = \pi\left(\frac{81}{56}\right)\left(\frac{9}{\sqrt{56}}\right) \approx 5.46507.$$

Because the time required to complete the orbit is 661 days, you can apply Kepler's Second Law to conclude that the time t required to move from the position $\theta = -\pi/2$ to $\theta = \pi/2$ is

$$\frac{t}{661} = \frac{\text{area of elliptical segment}}{\text{area of ellipse}} \approx \frac{0.90429}{5.46507}$$

which implies that $t \approx 109$ days.

10.6 Exercises

See CalcChat.com for tutorial help and worked-out solutions to odd-numbered exercises.

Graphical Reasoning **In Exercises 1–4, use a graphing utility to graph the polar equation when (a) $e = 1$, (b) $e = 0.5$, and (c) $e = 1.5$. Identify the conic.**

1. $r = \dfrac{2e}{1 + e\cos\theta}$ **2.** $r = \dfrac{2e}{1 - e\cos\theta}$

3. $r = \dfrac{2e}{1 - e\sin\theta}$ **4.** $r = \dfrac{2e}{1 + e\sin\theta}$

5. Writing Consider the polar equation

$$r = \frac{4}{1 + e\sin\theta}.$$

(a) Use a graphing utility to graph the equation for $e = 0.1$, $e = 0.25$, $e = 0.5$, $e = 0.75$, and $e = 0.9$. Identify the conic and discuss the change in its shape as $e \to 1^-$ and $e \to 0^+$.

(b) Use a graphing utility to graph the equation for $e = 1$. Identify the conic.

(c) Use a graphing utility to graph the equation for $e = 1.1$, $e = 1.5$, and $e = 2$. Identify the conic and discuss the change in its shape as $e \to 1^+$ and $e \to \infty$.

6. Writing Consider the polar equation

$$r = \frac{4}{1 - 0.4\cos\theta}.$$

(a) Identify the conic without graphing the equation.

(b) Without graphing the following polar equations, describe how each differs from the polar equation above.

$$r = \frac{4}{1 + 0.4\cos\theta}, \quad r = \frac{4}{1 - 0.4\sin\theta}$$

(c) Verify the results of part (b) graphically.

Matching **In Exercises 7–12, match the polar equation with the correct graph. [The graphs are labeled (a), (b), (c), (d), (e), and (f).]**

(a)

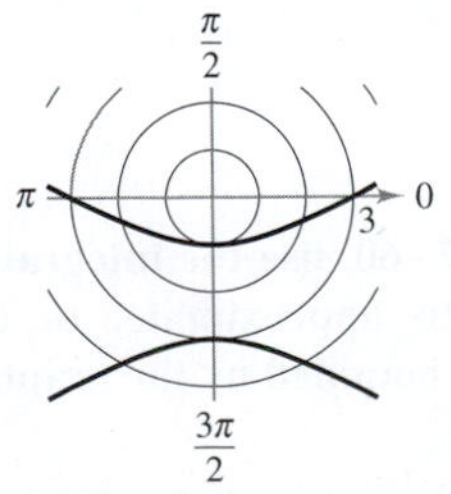

(b)

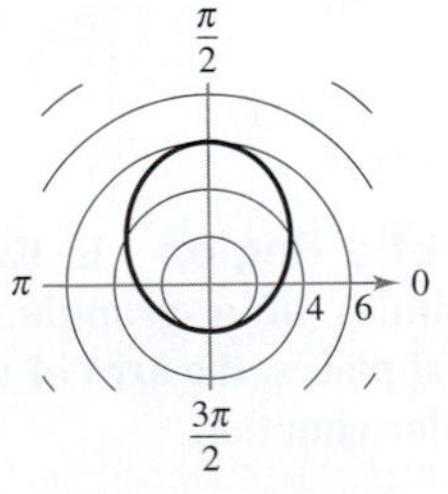

(c)

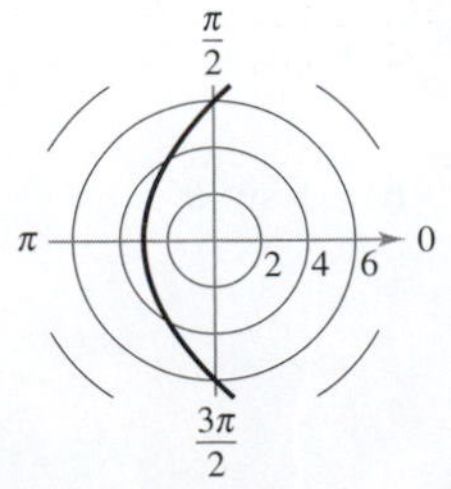

(d)

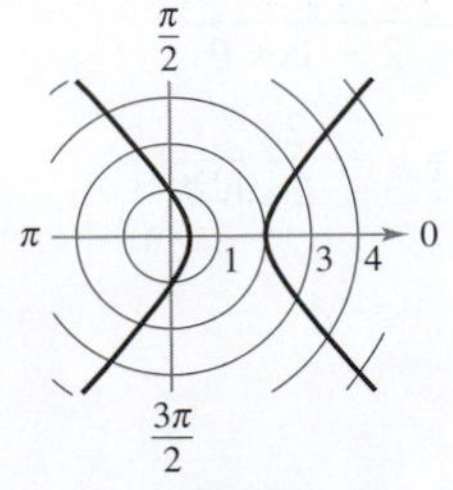

(e)

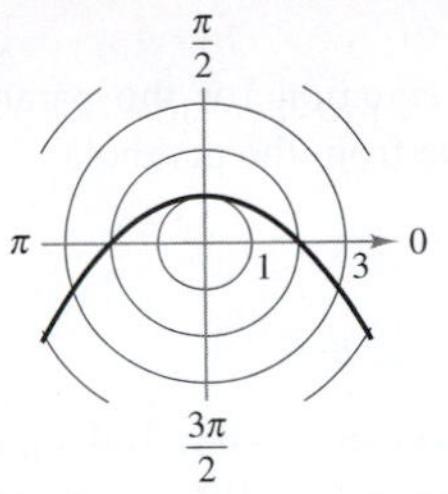

(f)

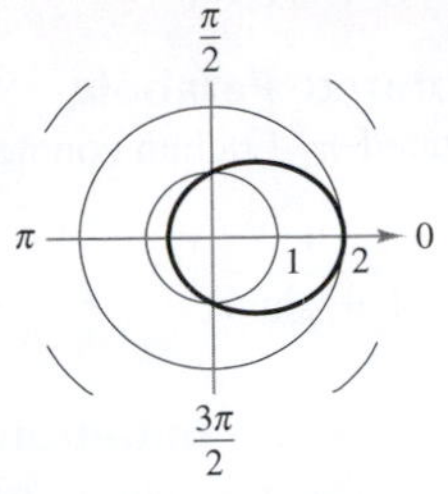

7. $r = \dfrac{6}{1 - \cos\theta}$ **8.** $r = \dfrac{2}{2 - \cos\theta}$

9. $r = \dfrac{3}{1 - 2\sin\theta}$ **10.** $r = \dfrac{2}{1 + \sin\theta}$

11. $r = \dfrac{6}{2 - \sin\theta}$ **12.** $r = \dfrac{2}{2 + 3\cos\theta}$

Sketching and Identifying a Conic **In Exercises 13–22, find the eccentricity and the distance from the pole to the directrix of the conic. Then sketch and identify the graph. Use a graphing utility to confirm your results.**

13. $r = \dfrac{1}{1 - \cos\theta}$ **14.** $r = \dfrac{6}{3 - 2\cos\theta}$

15. $r = \dfrac{3}{2 + 6\sin\theta}$ **16.** $r = \dfrac{4}{1 + \cos\theta}$

17. $r = \dfrac{5}{-1 + 2\cos\theta}$ **18.** $r = \dfrac{10}{5 + 4\sin\theta}$

19. $r = \dfrac{6}{2 + \cos\theta}$ **20.** $r = \dfrac{-6}{3 + 7\sin\theta}$

21. $r = \dfrac{300}{-12 + 6\sin\theta}$ **22.** $r = \dfrac{1}{1 + \sin\theta}$

Identifying a Conic **In Exercises 23–26, use a graphing utility to graph the polar equation. Identify the graph and find its eccentricity.**

23. $r = \dfrac{3}{-4 + 2\sin\theta}$ **24.** $r = \dfrac{-15}{2 + 8\sin\theta}$

25. $r = \dfrac{-10}{1 - \cos\theta}$ **26.** $r = \dfrac{6}{6 + 7\cos\theta}$

Comparing Graphs **In Exercises 27–30, use a graphing utility to graph the conic. Describe how the graph differs from the graph in the indicated exercise.**

27. $r = \dfrac{4}{1 + \cos(\theta - \pi/3)}$ (See Exercise 16.)

28. $r = \dfrac{10}{5 + 4\sin(\theta - \pi/4)}$ (See Exercise 18.)

29. $r = \dfrac{6}{2 + \cos(\theta + \pi/6)}$ (See Exercise 19.)

30. $r = \dfrac{-6}{3 + 7\sin(\theta + 2\pi/3)}$ (See Exercise 20.)

31. Rotated Ellipse Write the equation for the ellipse rotated $\pi/6$ radian clockwise from the ellipse

$$r = \frac{8}{8 + 5\cos\theta}.$$

32. Rotated Parabola Write the equation for the parabola rotated $\pi/4$ radian counterclockwise from the parabola

$$r = \frac{9}{1 + \sin\theta}.$$

Finding a Polar Equation **In Exercises 33–44, find a polar equation for the conic with its focus at the pole. (For convenience, the equation for the directrix is given in rectangular form.)**

Conic	Eccentricity	Directrix
33. Parabola	$e = 1$	$x = -3$
34. Parabola	$e = 1$	$y = 4$
35. Ellipse	$e = \frac{1}{2}$	$y = 1$
36. Ellipse	$e = \frac{3}{4}$	$y = -2$
37. Hyperbola	$e = 2$	$x = 1$
38. Hyperbola	$e = \frac{3}{2}$	$x = -1$

Conic	Vertex or Vertices
39. Parabola	$\left(1, -\frac{\pi}{2}\right)$
40. Parabola	$(5, \pi)$
41. Ellipse	$(2, 0), (8, \pi)$
42. Ellipse	$\left(2, \frac{\pi}{2}\right), \left(4, \frac{3\pi}{2}\right)$
43. Hyperbola	$\left(1, \frac{3\pi}{2}\right), \left(9, \frac{3\pi}{2}\right)$
44. Hyperbola	$(2, 0), (10, 0)$

45. Finding a Polar Equation Find a polar equation for the ellipse with focus $(0, 0)$, eccentricity $\frac{1}{2}$, and a directrix at $r = 4\sec\theta$.

46. Finding a Polar Equation Find a polar equation for the hyperbola with focus $(0, 0)$, eccentricity 2, and a directrix at $r = -8\csc\theta$.

WRITING ABOUT CONCEPTS

47. Eccentricity Classify the conics by their eccentricities.

48. Identifying Conics Identify each conic.

(a) $r = \dfrac{5}{1 - 2\cos\theta}$ (b) $r = \dfrac{5}{10 - \sin\theta}$

(c) $r = \dfrac{5}{3 - 3\cos\theta}$ (d) $r = \dfrac{5}{1 - 3\sin(\theta - \pi/4)}$

49. Distance Describe what happens to the distance between the directrix and the center of an ellipse when the foci remain fixed and e approaches 0.

50. HOW DO YOU SEE IT? Identify the conic in the graph and give the possible values for the eccentricity.

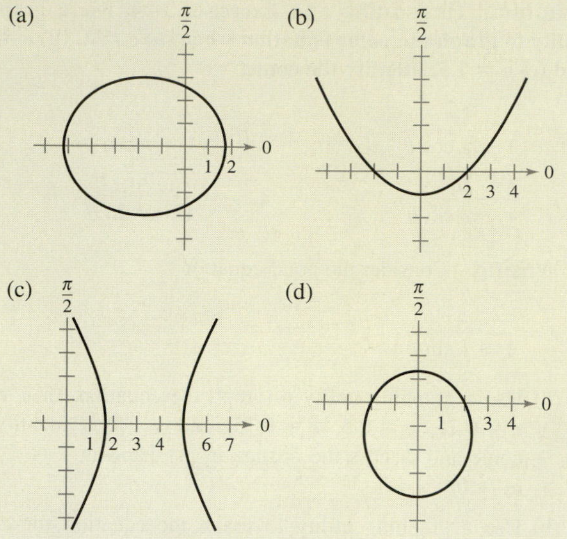

51. Ellipse Show that the polar equation for $\dfrac{x^2}{a^2} + \dfrac{y^2}{b^2} = 1$ is

$$r^2 = \frac{b^2}{1 - e^2\cos^2\theta}.$$ Ellipse

52. Hyperbola Show that the polar equation for $\dfrac{x^2}{a^2} - \dfrac{y^2}{b^2} = 1$ is

$$r^2 = \frac{-b^2}{1 - e^2\cos^2\theta}.$$ Hyperbola

Finding a Polar Equation **In Exercises 53–56, use the results of Exercises 51 and 52 to write the polar form of the equation of the conic.**

53. Ellipse: focus at $(4, 0)$; vertices at $(5, 0)$, $(5, \pi)$

54. Hyperbola: focus at $(5, 0)$; vertices at $(4, 0)$, $(4, \pi)$

55. $\dfrac{x^2}{9} - \dfrac{y^2}{16} = 1$

56. $\dfrac{x^2}{4} + y^2 = 1$

Area of a Region **In Exercises 57–60, use the integration capabilities of a graphing utility to approximate, to two decimal places, the area of the region bounded by the graph of the polar equation.**

57. $r = \dfrac{3}{2 - \cos\theta}$ **58.** $r = \dfrac{9}{4 + \cos\theta}$

59. $r = \dfrac{2}{3 - 2\sin\theta}$ **60.** $r = \dfrac{3}{6 + 5\sin\theta}$

61. Explorer 18 On November 27, 1963, the United States launched Explorer 18. Its low and high points above the surface of Earth were approximately 119 miles and 123,000 miles (see figure). The center of Earth is a focus of the orbit. Find the polar equation for the orbit and find the distance between the satellite when $\theta = 60°$. (Assume that the radius of Earth is 4000 miles.)

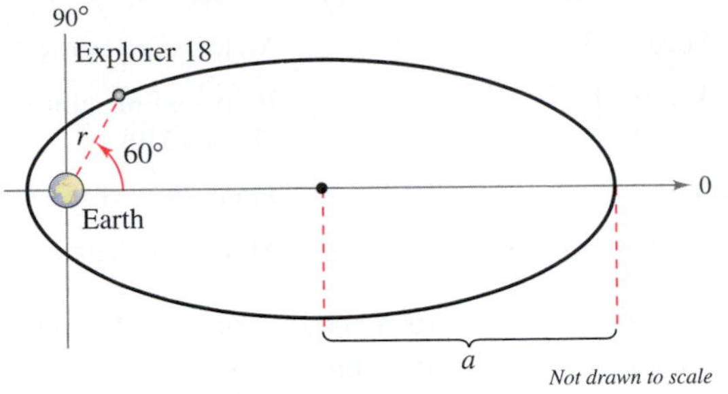

62. Planetary Motion The planets travel in elliptical orbits with the sun as a focus, as shown in the figure.

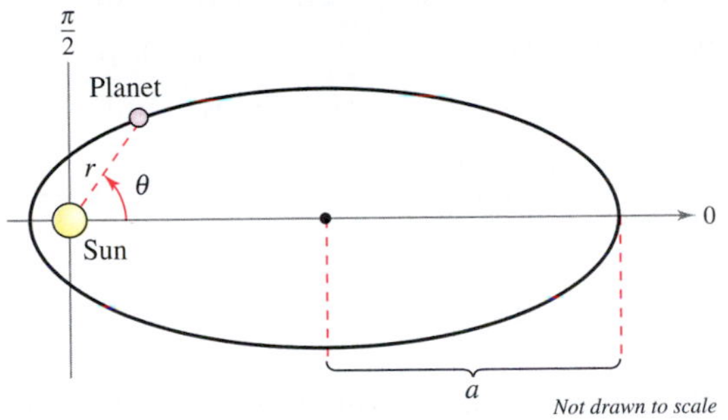

(a) Show that the polar equation of the orbit is given by

$$r = \frac{(1 - e^2)a}{1 - e \cos \theta}$$

where e is the eccentricity.

(b) Show that the minimum distance (*perihelion*) from the sun to the planet is $r = a(1 - e)$ and the maximum distance (*aphelion*) is $r = a(1 + e)$.

Planetary Motion In Exercises 63–66, use Exercise 62 to find the polar equation of the elliptical orbit of the planet, and the perihelion and aphelion distances.

63. Earth $a = 1.496 \times 10^8$ kilometers

$e = 0.0167$

64. Saturn $a = 1.427 \times 10^9$ kilometers

$e = 0.0542$

65. Neptune $a = 4.498 \times 10^9$ kilometers

$e = 0.0086$

66. Mercury $a = 5.791 \times 10^7$ kilometers

$e = 0.2056$

67. Planetary Motion

In Exercise 65, the polar equation for the elliptical orbit of Neptune was found. Use the equation and a computer algebra system to perform each of the following.

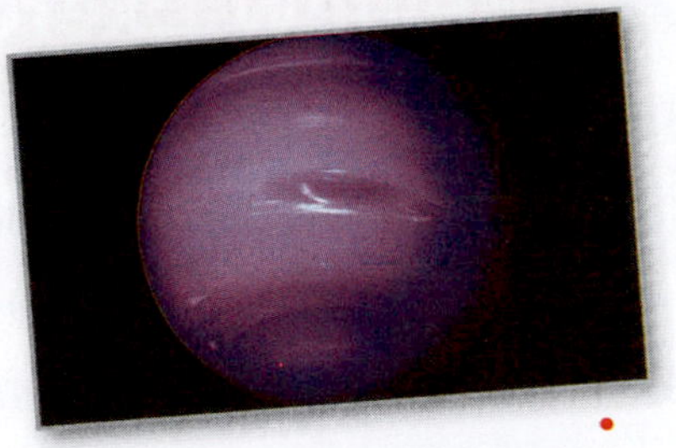

(a) Approximate the area swept out by a ray from the sun to the planet as θ increases from 0 to $\pi/9$. Use this result to determine the number of years required for the planet to move through this arc when the period of one revolution around the sun is 165 years.

(b) By trial and error, approximate the angle α such that the area swept out by a ray from the sun to the planet as θ increases from π to α equals the area found in part (a) (see figure). Does the ray sweep through a larger or smaller angle than in part (a) to generate the same area? Why is this the case?

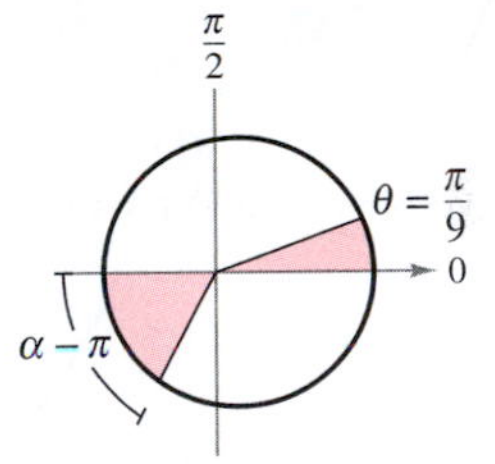

(c) Approximate the distances the planet traveled in parts (a) and (b). Use these distances to approximate the average number of kilometers per year the planet traveled in the two cases.

68. Comet Hale-Bopp The comet Hale-Bopp has an elliptical orbit with the sun at one focus and has an eccentricity of $e \approx 0.995$. The length of the major axis of the orbit is approximately 500 astronomical units.

(a) Find the length of its minor axis.

(b) Find a polar equation for the orbit.

(c) Find the perihelion and aphelion distances.

Eccentricity In Exercises 69 and 70, let r_0 represent the distance from a focus to the nearest vertex, and let r_1 represent the distance from the focus to the farthest vertex.

69. Show that the eccentricity of an ellipse can be written as $e = \dfrac{r_1 - r_0}{r_1 + r_0}$. Then show that $\dfrac{r_1}{r_0} = \dfrac{1 + e}{1 - e}$.

70. Show that the eccentricity of a hyperbola can be written as $e = \dfrac{r_1 + r_0}{r_1 - r_0}$. Then show that $\dfrac{r_1}{r_0} = \dfrac{e + 1}{e - 1}$.

NASA

Review Exercises

See CalcChat.com for tutorial help and worked-out solutions to odd-numbered exercises.

Matching In Exercises 1–6, match the equation with the correct graph. [The graphs are labeled (a), (b), (c), (d), (e), and (f).]

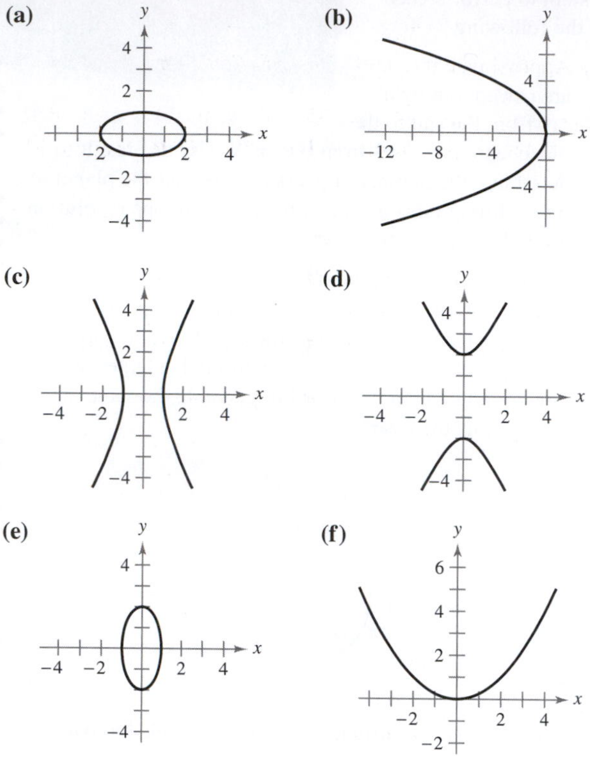

1. $4x^2 + y^2 = 4$
2. $4x^2 - y^2 = 4$
3. $y^2 = -4x$
4. $y^2 - 4x^2 = 4$
5. $x^2 + 4y^2 = 4$
6. $x^2 = 4y$

Identifying a Conic In Exercises 7–14, identify the conic, analyze the equation (center, radius, vertices, foci, eccentricity, directrix, and asymptotes, if possible), and sketch its graph. Use a graphing utility to confirm your results.

7. $16x^2 + 16y^2 - 16x + 24y - 3 = 0$
8. $y^2 - 12y - 8x + 20 = 0$
9. $3x^2 - 2y^2 + 24x + 12y + 24 = 0$
10. $5x^2 + y^2 - 20x + 19 = 0$
11. $3x^2 + 2y^2 - 12x + 12y + 29 = 0$
12. $12x^2 - 12y^2 - 12x + 24y - 45 = 0$
13. $x^2 - 6x - 8y + 1 = 0$
14. $9x^2 + 25y^2 + 18x - 100y - 116 = 0$

Finding an Equation of a Parabola In Exercises 15 and 16, find an equation of the parabola.

15. Vertex: $(0, 2)$
    Directrix: $x = -3$
16. Vertex: $(2, 6)$
    Focus: $(2, 4)$

Finding an Equation of an Ellipse In Exercises 17–20, find an equation of the ellipse.

17. Center: $(0, 0)$
    Focus: $(5, 0)$
    Vertex: $(7, 0)$
18. Center: $(0, 0)$
    Major axis: vertical
    Points on the ellipse: $(1, 2), (2, 0)$
19. Vertices: $(3, 1), (3, 7)$
    Eccentricity: $\frac{2}{3}$
20. Foci: $(0, \pm 7)$
    Major axis length: 20

Finding an Equation of a Hyperbola In Exercises 21–24, find an equation of the hyperbola.

21. Vertices: $(0, \pm 8)$
    Asymptotes: $y = \pm 2x$
22. Vertices: $(\pm 2, 0)$
    Asymptotes: $y = \pm 32x$
23. Vertices: $(\pm 7, -1)$
    Foci: $(\pm 9, -1)$
24. Center: $(0, 0)$
    Vertex: $(0, 3)$
    Focus: $(0, 6)$

25. **Satellite Antenna** A cross section of a large parabolic antenna is modeled by the graph of

$$y = \frac{x^2}{200}, \quad -100 \le x \le 100.$$

The receiving and transmitting equipment is positioned at the focus.

(a) Find the coordinates of the focus.

(b) Find the surface area of the antenna.

26. **Using an Ellipse** Consider the ellipse $\frac{x^2}{25} + \frac{y^2}{9} = 1$.

(a) Find the area of the region bounded by the ellipse.

(b) Find the volume of the solid generated by revolving the region about its major axis.

Using Parametric Equations In Exercises 27–34, sketch the curve represented by the parametric equations (indicate the orientation of the curve), and write the corresponding rectangular equation by eliminating the parameter.

27. $x = 1 + 8t,\ y = 3 - 4t$
28. $x = t - 6,\ y = t^2$
29. $x = e^t - 1,\ y = e^{3t}$
30. $x = e^{4t},\ y = t + 4$
31. $x = 6 \cos \theta,\ y = 6 \sin \theta$
32. $x = 2 + 5 \cos t,\ y = 3 + 2 \sin t$
33. $x = 2 + \sec \theta,\ y = 3 + \tan \theta$
34. $x = 5 \sin^3 \theta,\ y = 5 \cos^3 \theta$

Finding Parametric Equations In Exercises 35 and 36, find two different sets of parametric equations for the rectangular equation.

35. $y = 4x + 3$
36. $y = x^2 - 2$

37. Rotary Engine The rotary engine was developed by Felix Wankel in the 1950s. It features a rotor that is a modified equilateral triangle. The rotor moves in a chamber that, in two dimensions, is an epitrochoid. Use a graphing utility to graph the chamber modeled by the parametric equations

$x = \cos 3\theta + 5 \cos \theta$

and

$y = \sin 3\theta + 5 \sin \theta.$

38. Serpentine Curve Consider the parametric equations $x = 2 \cot \theta$ and $y = 4 \sin \theta \cos \theta$, $0 < \theta < \pi$.

(a) Use a graphing utility to graph the curve.

(b) Eliminate the parameter to show that the rectangular equation of the serpentine curve is $(4 + x^2)y = 8x$.

Finding Slope and Concavity **In Exercises 39–46, find dy/dx and d^2y/dx^2, and find the slope and concavity (if possible) at the given value of the parameter.**

Parametric Equations	Parameter
39. $x = 2 + 5t, \quad y = 1 - 4t$	$t = 3$
40. $x = t - 6, \quad y = t^2$	$t = 5$
41. $x = \frac{1}{t}, \quad y = 2t + 3$	$t = -1$
42. $x = \frac{1}{t}, \quad y = t^2$	$t = -2$
43. $x = 5 + \cos \theta, \quad y = 3 + 4 \sin \theta$	$\theta = \frac{\pi}{6}$
44. $x = 10 \cos \theta, \quad y = 10 \sin \theta$	$\theta = \frac{\pi}{4}$
45. $x = \cos^3 \theta, \quad y = 4 \sin^3 \theta$	$\theta = \frac{\pi}{3}$
46. $x = e^t, \quad y = e^{-t}$	$t = 1$

Finding an Equation of a Tangent Line **In Exercises 47 and 48, (a) use a graphing utility to graph the curve represented by the parametric equations, (b) use a graphing utility to find $dx/d\theta$, $dy/d\theta$, and dy/dx at the given value of the parameter, (c) find an equation of the tangent line to the curve at the given value of the parameter, and (d) use a graphing utility to graph the curve and the tangent line from part (c).**

Parametric Equations	Parameter
47. $x = \cot \theta, \quad y = \sin 2\theta$	$\theta = \frac{\pi}{6}$
48. $x = \frac{1}{4} \tan \theta, \quad y = 6 \sin \theta$	$\theta = \frac{\pi}{3}$

Horizontal and Vertical Tangency **In Exercises 49–52, find all points (if any) of horizontal and vertical tangency to the curve. Use a graphing utility to confirm your results.**

49. $x = 5 - t, \quad y = 2t^2$

50. $x = t + 2, \quad y = t^3 - 2t$

51. $x = 2 + 2 \sin \theta, \quad y = 1 + \cos \theta$

52. $x = 2 - 2 \cos \theta, \quad y = 2 \sin 2\theta$

Arc Length **In Exercises 53 and 54, find the arc length of the curve on the given interval.**

Parametric Equations	Interval
53. $x = t^2 + 1, \quad y = 4t^3 + 3$	$0 \le t \le 2$
54. $x = 6 \cos \theta, \quad y = 6 \sin \theta$	$0 \le \theta \le \pi$

Surface Area **In Exercises 55 and 56, find the area of the surface generated by revolving the curve about (a) the x-axis and (b) the y-axis.**

55. $x = t, \quad y = 3t, \quad 0 \le t \le 2$

56. $x = 2 \cos \theta, \quad y = 2 \sin \theta, \quad 0 \le \theta \le \frac{\pi}{2}$

Area **In Exercises 57 and 58, find the area of the region.**

57. $x = 3 \sin \theta$

$y = 2 \cos \theta$

$-\frac{\pi}{2} \le \theta \le \frac{\pi}{2}$

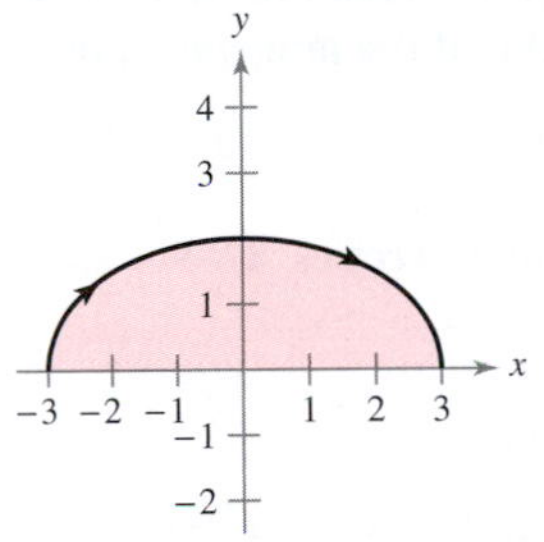

58. $x = 2 \cos \theta$

$y = \sin \theta$

$0 \le \theta \le \pi$

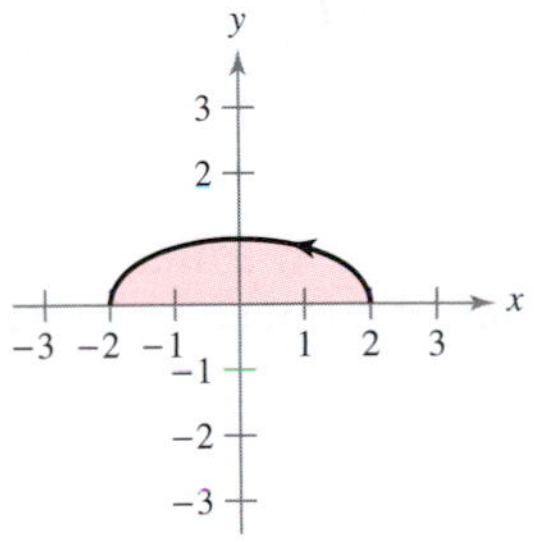

Polar-to-Rectangular Conversion **In Exercises 59–62, plot the point in polar coordinates and find the corresponding rectangular coordinates of the point.**

59. $\left(5, \frac{3\pi}{2}\right)$

60. $\left(-6, \frac{7\pi}{6}\right)$

61. $(\sqrt{3}, 1.56)$

62. $(-2, -2.45)$

Rectangular-to-Polar Conversion **In Exercises 63–66, the rectangular coordinates of a point are given. Plot the point and find *two* sets of polar coordinates of the point for $0 \le \theta < 2\pi$.**

63. $(4, -4)$

64. $(0, -7)$

65. $(-1, 3)$

66. $(-\sqrt{3}, -\sqrt{3})$

Rectangular-to-Polar Conversion **In Exercises 67–72, convert the rectangular equation to polar form and sketch its graph.**

67. $x^2 + y^2 = 25$

68. $x^2 - y^2 = 4$

69. $y = 9$

70. $x = 6$

71. $x^2 = 4y$

72. $x^2 + y^2 - 4x = 0$

Polar-to-Rectangular Conversion **In Exercises 73–78, convert the polar equation to rectangular form and sketch its graph.**

73. $r = 3\cos\theta$
74. $r = 10$
75. $r = 6\sin\theta$
76. $r = 3\csc\theta$
77. $r = -2\sec\theta\tan\theta$
78. $\theta = \dfrac{3\pi}{4}$

Graphing a Polar Equation **In Exercises 79–82, use a graphing utility to graph the polar equation.**

79. $r = \dfrac{3}{\cos(\theta - \pi/4)}$
80. $r = 2\sin\theta\cos^2\theta$
81. $r = 4\cos 2\theta\sec\theta$
82. $r = 4(\sec\theta - \cos\theta)$

Horizontal and Vertical Tangency **In Exercises 83 and 84, find the points of horizontal and vertical tangency (if any) to the polar curve.**

83. $r = 1 - \cos\theta$
84. $r = 3\tan\theta$

Tangent Lines at the Pole **In Exercises 85 and 86, sketch a graph of the polar equation and find the tangents at the pole.**

85. $r = 4\sin 3\theta$
86. $r = 3\cos 4\theta$

Sketching a Polar Graph **In Exercises 87–96, sketch a graph of the polar equation.**

87. $r = 6$
88. $\theta = \dfrac{\pi}{10}$
89. $r = -\sec\theta$
90. $r = 5\csc\theta$
91. $r^2 = 4\sin^2 2\theta$
92. $r = 3 - 4\cos\theta$
93. $r = 4 - 3\cos\theta$
94. $r = 4\theta$
95. $r = -3\cos 2\theta$
96. $r = \cos 5\theta$

Finding the Area of a Polar Region **In Exercises 97–102, find the area of the region.**

97. One petal of $r = 3\cos 5\theta$
98. One petal of $r = 2\sin 6\theta$
99. Interior of $r = 2 + \cos\theta$
100. Interior of $r = 5(1 - \sin\theta)$
101. Interior of $r^2 = 4\sin 2\theta$
102. Common interior of $r = 4\cos\theta$ and $r = 2$

Finding the Area of a Polar Region **In Exercises 103–106, use a graphing utility to graph the polar equation. Find the area of the given region analytically.**

103. Inner loop of $r = 3 - 6\cos\theta$
104. Inner loop of $r = 2 + 4\sin\theta$
105. Between the loops of $r = 3 - 6\cos\theta$
106. Between the loops of $r = 2 + 4\sin\theta$

Finding Points of Intersection **In Exercises 107 and 108, find the points of intersection of the graphs of the equations.**

107. $r = 1 - \cos\theta$, $r = 1 + \sin\theta$
108. $r = 1 + \sin\theta$, $r = 3\sin\theta$

Finding the Arc Length of a Polar Curve **In Exercises 109 and 110, find the length of the curve over the given interval.**

	Polar Equation	Interval
109.	$r = 5\cos\theta$	$\dfrac{\pi}{2} \le \theta \le \pi$
110.	$r = 3(1 - \cos\theta)$	$0 \le \theta \le \pi$

Finding the Area of a Surface of Revolution **In Exercises 111 and 112, write an integral that represents the area of the surface formed by revolving the curve about the given line. Use the integration capabilities of a graphing utility to approximate the integral accurate to two decimal places.**

	Polar Equation	Interval	Axis of Revolution
111.	$r = 1 + 4\cos\theta$	$0 \le \theta \le \dfrac{\pi}{2}$	Polar axis
112.	$r = 2\sin\theta$	$0 \le \theta \le \dfrac{\pi}{2}$	$\theta = \dfrac{\pi}{2}$

Sketching and Identifying a Conic **In Exercises 113–118, find the eccentricity and the distance from the pole to the directrix of the conic. Then sketch and identify the graph. Use a graphing utility to confirm your results.**

113. $r = \dfrac{6}{1 - \sin\theta}$
114. $r = \dfrac{2}{1 + \cos\theta}$
115. $r = \dfrac{6}{3 + 2\cos\theta}$
116. $r = \dfrac{4}{5 - 3\sin\theta}$
117. $r = \dfrac{4}{2 - 3\sin\theta}$
118. $r = \dfrac{8}{2 - 5\cos\theta}$

Finding a Polar Equation **In Exercises 119–124, find a polar equation for the conic with its focus at the pole. (For convenience, the equation for the directrix is given in rectangular form.)**

	Conic	Eccentricity	Directrix
119.	Parabola	$e = 1$	$x = 4$
120.	Ellipse	$e = \dfrac{3}{4}$	$y = -2$
121.	Hyperbola	$e = 3$	$y = 3$

	Conic	Vertex or Vertices
122.	Parabola	$\left(2, \dfrac{\pi}{2}\right)$
123.	Ellipse	$(5, 0), (1, \pi)$
124.	Hyperbola	$(1, 0), (7, 0)$

P.S. Problem Solving

See CalcChat.com for tutorial help and worked-out solutions to odd-numbered exercises.

1. Using a Parabola Consider the parabola $x^2 = 4y$ and the focal chord $y = \frac{3}{4}x + 1$.

(a) Sketch the graph of the parabola and the focal chord.

(b) Show that the tangent lines to the parabola at the endpoints of the focal chord intersect at right angles.

(c) Show that the tangent lines to the parabola at the endpoints of the focal chord intersect on the directrix of the parabola.

2. Using a Parabola Consider the parabola $x^2 = 4py$ and one of its focal chords.

(a) Show that the tangent lines to the parabola at the endpoints of the focal chord intersect at right angles.

(b) Show that the tangent lines to the parabola at the endpoints of the focal chord intersect on the directrix of the parabola.

3. Proof Prove Theorem 10.2, Reflective Property of a Parabola, as shown in the figure.

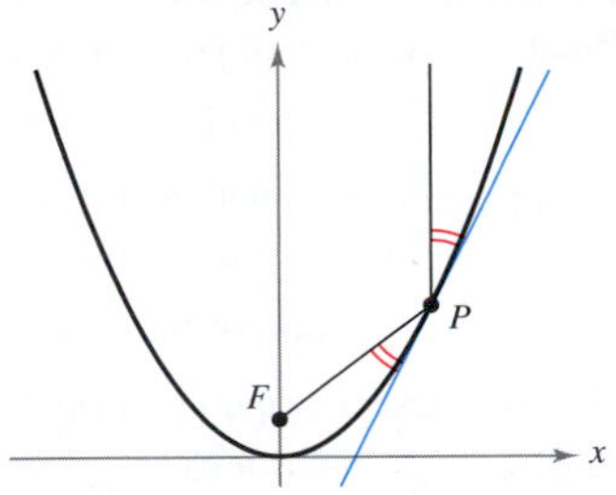

4. Flight Paths An air traffic controller spots two planes at the same altitude flying toward each other (see figure). Their flight paths are 20° and 315°. One plane is 150 miles from point P with a speed of 375 miles per hour. The other is 190 miles from point P with a speed of 450 miles per hour.

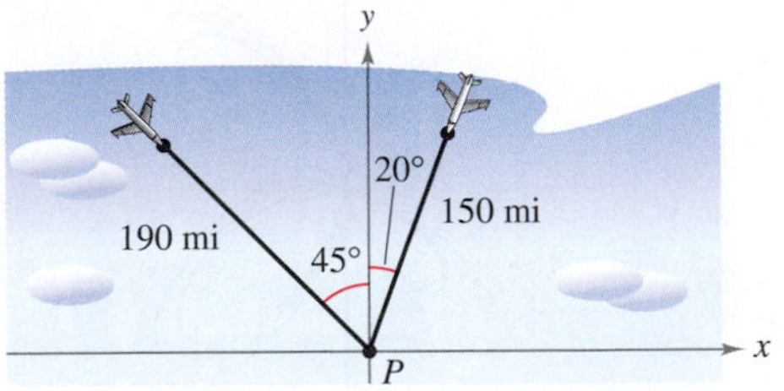

(a) Find parametric equations for the path of each plane where t is the time in hours, with $t = 0$ corresponding to the time at which the air traffic controller spots the planes.

(b) Use the result of part (a) to write the distance between the planes as a function of t.

(c) Use a graphing utility to graph the function in part (b). When will the distance between the planes be minimum? If the planes must keep a separation of at least 3 miles, is the requirement met?

5. Strophoid The curve given by the parametric equations

$$x(t) = \frac{1 - t^2}{1 + t^2} \quad \text{and} \quad y(t) = \frac{t(1 - t^2)}{1 + t^2}$$

is called a **strophoid.**

(a) Find a rectangular equation of the strophoid.

(b) Find a polar equation of the strophoid.

(c) Sketch a graph of the strophoid.

(d) Find the equations of the two tangent lines at the origin.

(e) Find the points on the graph at which the tangent lines are horizontal.

6. Finding a Rectangular Equation Find a rectangular equation of the portion of the cycloid given by the parametric equations $x = a(\theta - \sin\theta)$ and $y = a(1 - \cos\theta)$, $0 \le \theta \le \pi$, as shown in the figure.

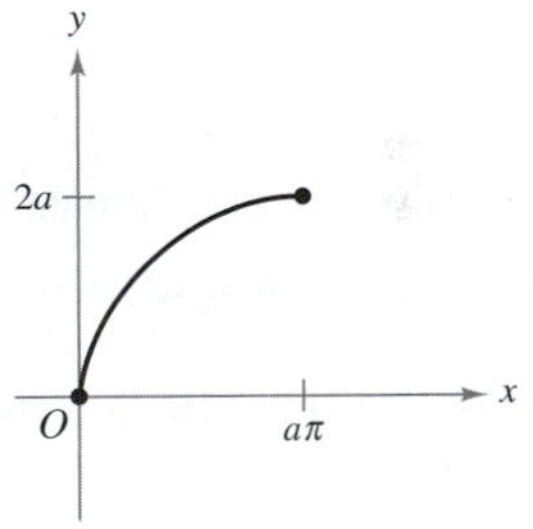

7. Cornu Spiral Consider the **cornu spiral** given by

$$x(t) = \int_0^t \cos\left(\frac{\pi u^2}{2}\right) du \quad \text{and} \quad y(t) = \int_0^t \sin\left(\frac{\pi u^2}{2}\right) du.$$

(a) Use a graphing utility to graph the spiral over the interval $-\pi \le t \le \pi$.

(b) Show that the cornu spiral is symmetric with respect to the origin.

(c) Find the length of the cornu spiral from $t = 0$ to $t = a$. What is the length of the spiral from $t = -\pi$ to $t = \pi$?

8. Using an Ellipse Consider the region bounded by the ellipse $x^2/a^2 + y^2/b^2 = 1$, with eccentricity $e = c/a$.

(a) Show that the area of the region is πab.

(b) Show that the solid (oblate spheroid) generated by revolving the region about the minor axis of the ellipse has a volume of $V = 4\pi^2 b/3$ and a surface area of

$$S = 2\pi a^2 + \pi\left(\frac{b^2}{e}\right)\ln\left(\frac{1 + e}{1 - e}\right).$$

(c) Show that the solid (prolate spheroid) generated by revolving the region about the major axis of the ellipse has a volume of $V = 4\pi ab^2/3$ and a surface area of

$$S = 2\pi b^2 + 2\pi\left(\frac{ab}{e}\right)\arcsin e.$$

9. Area Let a and b be positive constants. Find the area of the region in the first quadrant bounded by the graph of the polar equation

$$r = \frac{ab}{(a \sin \theta + b \cos \theta)}, \quad 0 \le \theta \le \frac{\pi}{2}.$$

10. Using a Right Triangle Consider the right triangle shown in the figure.

(a) Show that the area of the triangle is $A(\alpha) = \frac{1}{2}\int_0^\alpha \sec^2 \theta \, d\theta$.

(b) Show that $\tan \alpha = \int_0^\alpha \sec^2 \theta \, d\theta$.

(c) Use part (b) to derive the formula for the derivative of the tangent function.

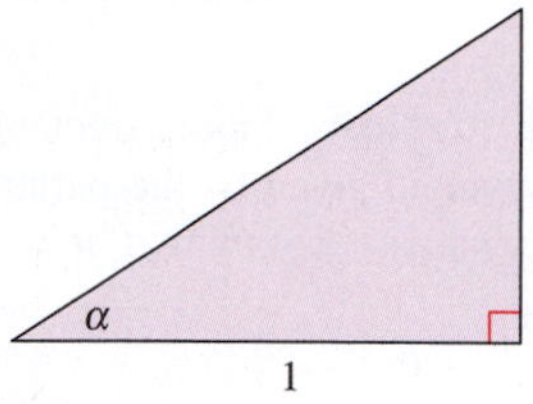

Figure for 10

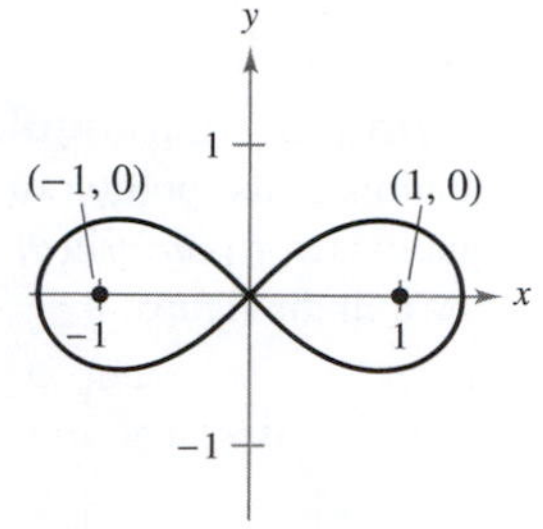

Figure for 11

11. Finding a Polar Equation Determine the polar equation of the set of all points (r, θ), the product of whose distances from the points $(1, 0)$ and $(-1, 0)$ is equal to 1, as shown in the figure.

12. Arc Length A particle is moving along the path described by the parametric equations $x = 1/t$ and $y = (\sin t)/t$, for $1 \le t < \infty$, as shown in the figure. Find the length of this path.

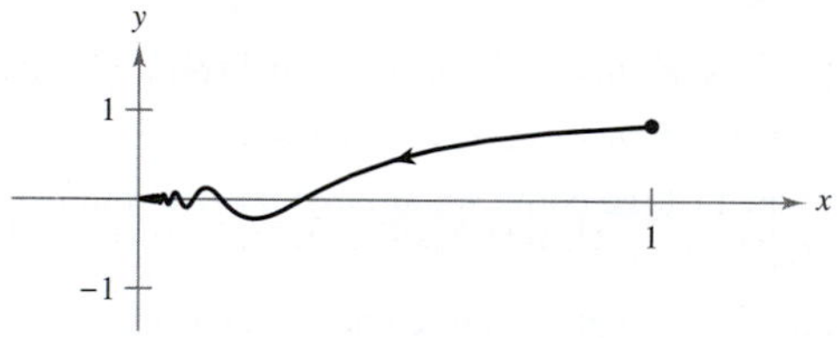

13. Finding a Polar Equation Four dogs are located at the corners of a square with sides of length d. The dogs all move counterclockwise at the same speed directly toward the next dog, as shown in the figure. Find the polar equation of a dog's path as it spirals toward the center of the square.

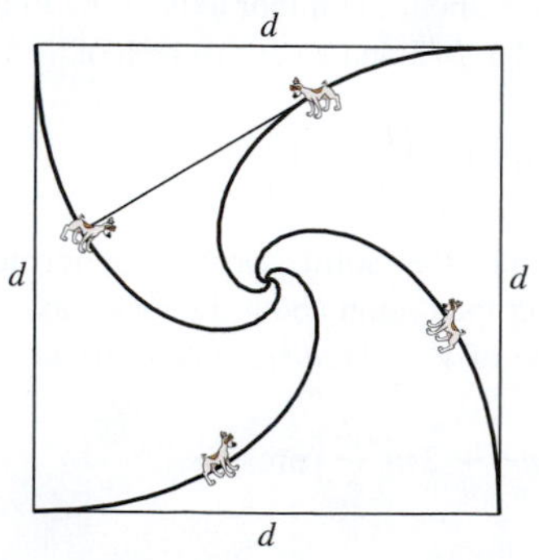

14. Using a Hyperbola Consider the hyperbola

$$\frac{x^2}{a^2} - \frac{y^2}{b^2} = 1$$

with foci F_1 and F_2, as shown in the figure. Let T be the tangent line at a point M on the hyperbola. Show that incoming rays of light aimed at one focus are reflected by a hyperbolic mirror toward the other focus.

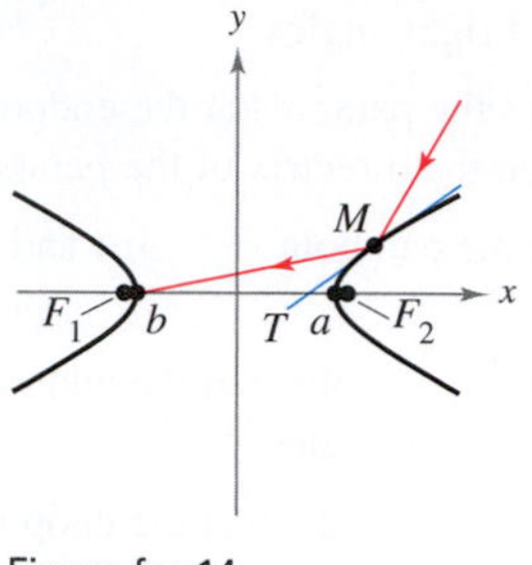

Figure for 14

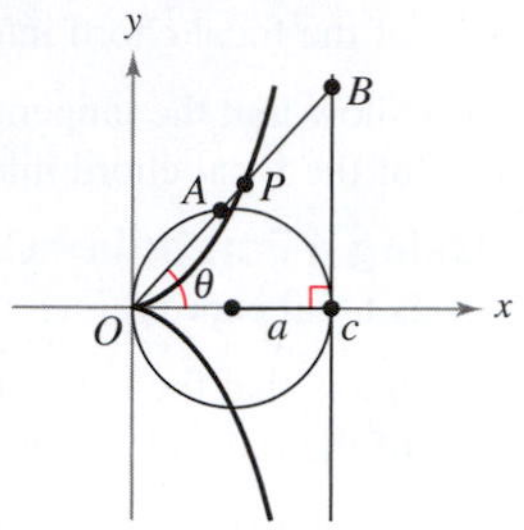

Figure for 15

15. Cissoid of Diocles Consider a circle of radius a tangent to the y-axis and the line $x = 2a$, as shown in the figure. Let A be the point where the segment OB intersects the circle. The **cissoid of Diocles** consists of all points P such that $OP = AB$.

(a) Find a polar equation of the cissoid.

(b) Find a set of parametric equations for the cissoid that does not contain trigonometric functions.

(c) Find a rectangular equation of the cissoid.

16. Butterfly Curve Use a graphing utility to graph the curve shown below. The curve is given by

$$r = e^{\cos \theta} - 2 \cos 4\theta + \sin^5 \frac{\theta}{12}.$$

Over what interval must θ vary to produce the curve?

FOR FURTHER INFORMATION For more information on this curve, see the article "A Study in Step Size" by Temple H. Fay in *Mathematics Magazine*. To view this article, go to *MathArticles.com*.

17. Graphing Polar Equations Use a graphing utility to graph the polar equation $r = \cos 5\theta + n \cos \theta$ for $0 \le \theta < \pi$ and for the integers $n = -5$ to $n = 5$. What values of n produce the "heart" portion of the curve? What values of n produce the "bell" portion? (This curve, created by Michael W. Chamberlin, appeared in *The College Mathematics Journal*.)

AP* Review Questions for Chapter 10

For an introduction to the vector topics included in the AP* BC Calculus Course Description, please see Appendix G: AP* BC Vector Topics (Web).

1. (BC only, calculator)

A particle moving along a curve in the xy-plane is at position $(x(t), y(t))$ at time t with

$$\frac{dx}{dt} = e^{-t^2+1} - 2 \quad \text{and} \quad \frac{dy}{dt} = 3\sqrt{25 - t^2}$$

for $0 \le t \le 5$. At time $t = 0$, the particle is at position $(2, 3)$.

(a) Find the speed of the object at time $t = 3$.

(b) Find the total distance traveled on the interval $[0, 5]$.

(c) Find $y(3)$.

(d) There is a point corresponding to a value of t in the interval $[0, 5]$ at which the tangent line to the curve is vertical. Find the acceleration vector at this point.

2. (BC only, calculator)

A particle moving along a curve in the xy-plane is at position $(x(t), y(t))$ for time $t \ge 0$ with

$$\frac{dx}{dt} = \ln(5 + (1 + t)^3) \quad \text{and} \quad \frac{dy}{dt} = 4t - 3t^2.$$

At time $t = 0$, the particle is at position $(4, 11)$. At time $t = 3$, the object is at point P with y-coordinate 2.

(a) Find the acceleration vector and speed at time $t = 2$.

(b) Find the x-coordinate of P.

(c) Write an equation for the line tangent to the curve at P.

(d) For what values of t, if any, is the particle at rest?

3. (BC only, no calculator)

A particle starts at point on the positive x-axis at time $t = 0$ and travels along the curve from A to B to C to D as shown below. The coordinates of the particle's position at time t are given by $(x(t), y(t))$, where $x(t)$ and $y(t)$ are differentiable functions of time.

$$y'(t) = \frac{dy}{dt} = \frac{-2t^2}{\left(t^3 - \sqrt{27}\right)^{1/3}} \quad \text{and} \quad x'(t) = \frac{dx}{dt} \text{ is not given.}$$

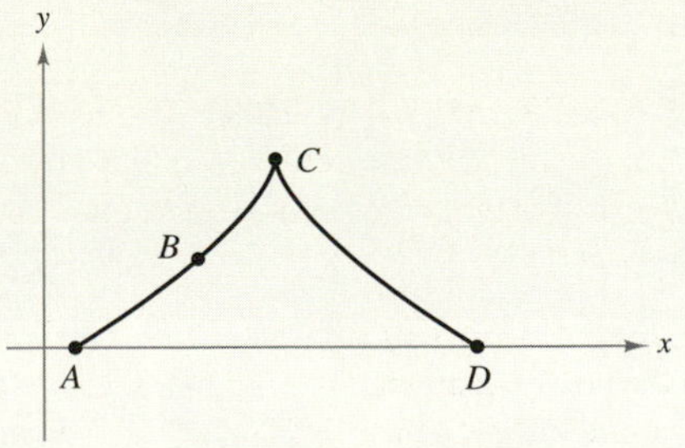

(a) At point B is $\frac{dx}{dt}$ positive? At point B is $\frac{dy}{dt}$ positive? Give a reason for each answer.

(b) Find the value of t at which the particle reaches point C. Justify your reasoning.

(c) What is the y-coordinate of point C? Justify your reasoning.

(d) The point $(7, 2)$ lies on the curve, and the line tangent to the curve at that point has the equation $y = -\frac{2}{3}x + \frac{20}{3}$. If, at this point, $\frac{dx}{dt} \approx 1.837$, calculate $\frac{dy}{dt}$. Find the speed at that point.

4. (BC only, calculator)

At time $t = 0$ seconds, a water balloon is launched at a height of 30 meters above ground level. The horizontal position (meters) of the balloon from the launch point is given by $x(t)$, and the vertical height (meters) above ground is given by $y(t)$. Suppose $x'(t) = 30$ and $y'(t) = 17 - 9.8t$ for $0 \le t \le w$, where w represents the time at which the balloon hits the ground.

(A) Find an expression for $y(t)$.

(B) Find the balloon's maximum height above ground.

(C) Find w, the time at which the balloon hits the ground.

(D) Find the total distance traveled by the balloon.

(E) Find the angle θ, $0 \le \theta \le \frac{\pi}{2}$ between the path of the balloon and the ground at the instant when the balloon hits the ground.

5. (BC only, no calculator)

In the xy-plane, a particle moves along the curve $y = x^2 + 1$ with constant speed $\sqrt{2}$ units per second. If $\frac{dx}{dt} < 0$, what is the value of $\frac{dy}{dt}$ when the particle is at the point $(2, 5)$?

(A) 4

(B) 8

(C) $-\frac{\sqrt{34}}{17}$

(D) $\frac{-4\sqrt{34}}{17}$

(E) 5

6. (BC only, calculator)

The polar curve r is given by $r(\theta) = 2\theta + \cos\theta$ for $0 \leq \theta \leq 2\pi$.

(a) Find the area in the second quadrant enclosed by the coordinate axes and the graph of r.

(b) For $\pi/2 \leq \theta \leq \pi$, there is one point P on the polar curve r with y-coordinate 1. Find the angle θ that corresponds to point P. Find the x-coordinate of point P. Show the work that leads to your answers.

(c) A particle is traveling along the polar curve r so that its position at time t is $(x(t), y(t))$ and such that $\frac{d\theta}{dt} = 3$. Find $\frac{dx}{dt}$ at the instant that $\theta = \frac{3\pi}{4}$. Interpret your answer in the context of the problem.

7. (BC only, no calculator)

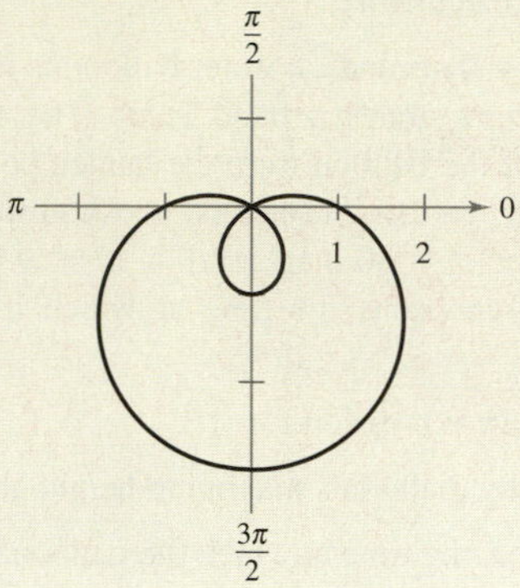

Consider the graph of the polar curve $r = 1 - 2\sin\theta$ shown above for $0 \leq \theta \leq 2\pi$.

(a) Set up, but do not evaluate, an integral to find the area of the region in the first quadrant bounded by the curve and the x-axis.

(b) A particle is traveling along the polar curve r so that its position at time t is $(x(t), y(t))$. Write expressions for $\frac{dx}{dt}$ and $\frac{dy}{dt}$ in terms of θ.

(c) Write an equation in terms of x and y of the line tangent to the graph of the polar curve at the point where $\theta = \pi$. Show the computations that lead to your answer.

8. (BC only, calculator)

On the interval $0 \leq \theta < 2\pi$, the graphs of the curve $r = 3$ and $r = 2 + 2\cos\theta$ intersect when $\theta = \pi/3$ and $\theta = 5\pi/3$.

(a) Let R be the region enclosed within both $r = 3$ and $r = 2 + 2\cos\theta$. Find the area of the region R.

(b) A particle moving along the curve $r = 2 + 2\cos\theta$ has position $(x(t), y(t))$ at time t, with $\theta = 0$ when $t = 0$. The particle moves along the curve such that $\frac{dr}{dt} = \frac{dr}{d\theta}$. Find the value of $\frac{dr}{dt}$ at $\theta = \frac{\pi}{3}$. Interpret your answer in terms of the motion of the particle.

(c) For the particle described in part (b), $\frac{dx}{dt} = \frac{dx}{d\theta}$. Find the value of $\frac{dx}{dt}$ at $\theta = \frac{\pi}{3}$. Interpret your answer in terms of the motion of the particle.

9. (BC only, calculator)

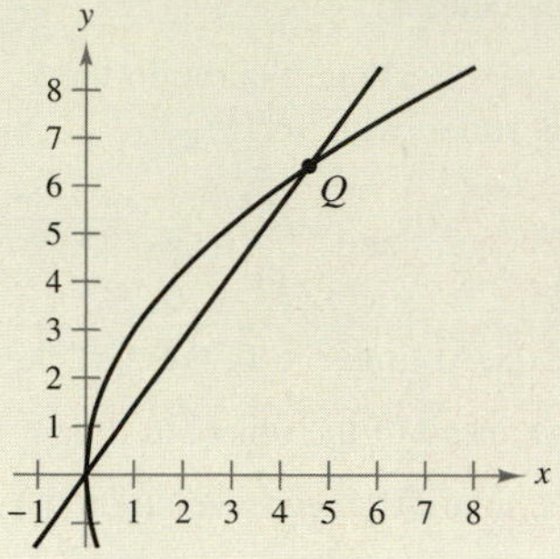

The figure above shows the graphs of the line $x = \frac{5}{7}y$ and the curve $x = \frac{y^2}{9}$. Let R be the region bounded by these two graphs in the first quadrant. The line and curve intersect at point Q.

(a) Find the coordinates of point Q and the value of $\frac{dx}{dy}$ for the curve $x = \frac{y^2}{9}$ at point Q.

(b) Set up and evaluate an integral expression with respect to y that gives the area of the region R.

(c) Show that the curve $x = \frac{y^2}{9}$ can be rewritten as the polar equation $r = 9\cot\theta\csc\theta$.

(d) Use the polar equation written in part (c) to set up an integral expression with respect to the polar angle θ that represents the area of R. (Do not evaluate the integral.)

10. (BC only, no calculator)

What is the slope of the line tangent to the curve $r = \theta$ at the point $\theta = -\frac{\pi}{2}$?

(A) 1

(B) 0

(C) $-\frac{\pi}{2}$

(D) $\frac{\pi}{2}$

(E) $\frac{2}{\pi}$

Appendices

A Proofs of Selected Theorems

For this edition, we have made Appendix A, Proofs of Selected Theorems, available in video format at *LarsonCalculus.com.* When you navigate to that website, you will find a link to Bruce Edwards explaining each proof in the text, including those in this appendix. We hope these videos enhance your study of calculus. The text version of this appendix is available at *CengageBrain.com.*

Proofs of Selected Theorems sample at *LarsonCalculus.com*

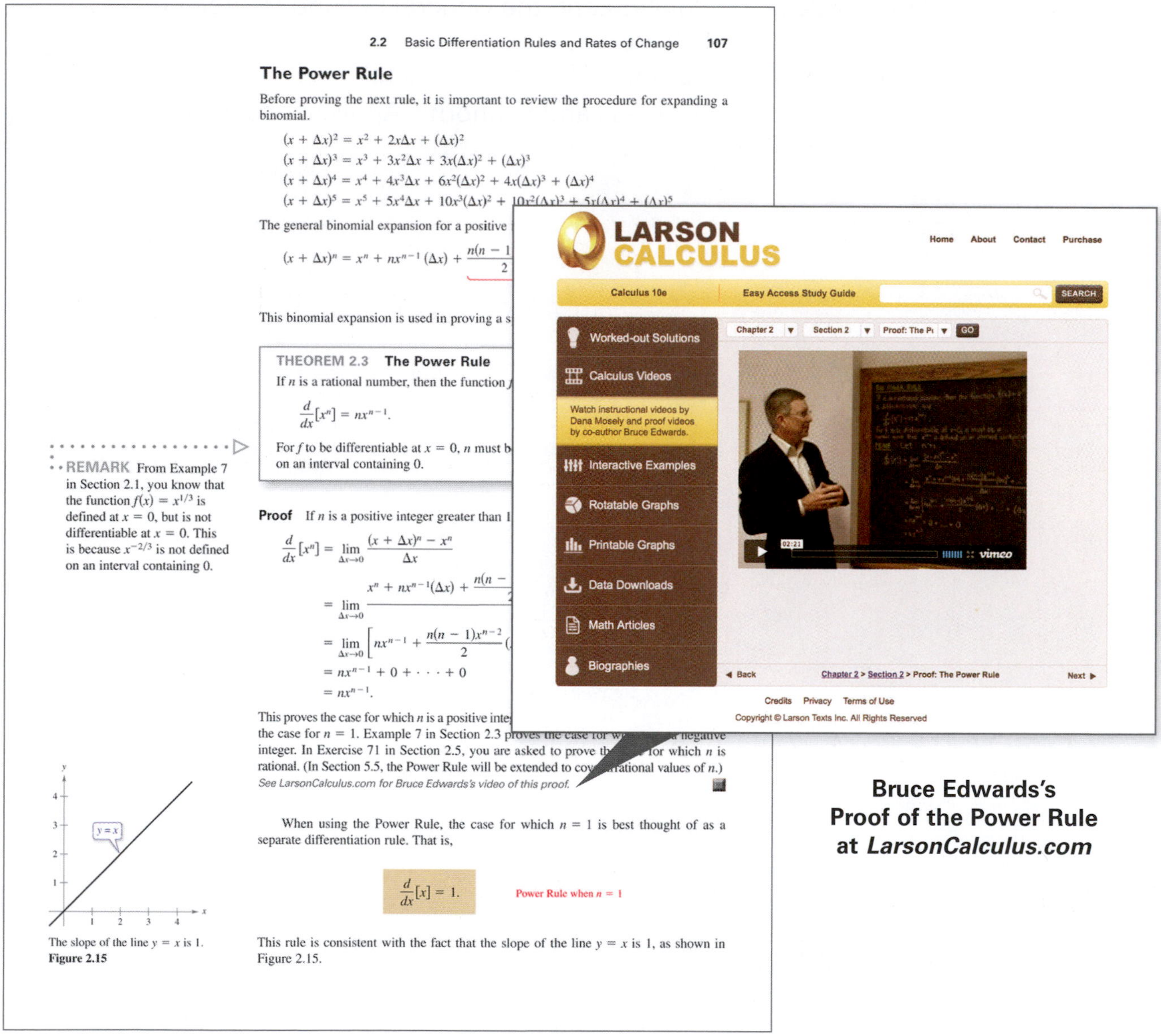

Bruce Edwards's Proof of the Power Rule at *LarsonCalculus.com*

B Integration Tables

Forms Involving u^n

1. $\displaystyle\int u^n\,du = \frac{u^{n+1}}{n+1} + C,\ n \neq -1$

2. $\displaystyle\int \frac{1}{u}\,du = \ln|u| + C$

Forms Involving $a + bu$

3. $\displaystyle\int \frac{u}{a+bu}\,du = \frac{1}{b^2}\left(bu - a\ln|a+bu|\right) + C$

4. $\displaystyle\int \frac{u}{(a+bu)^2}\,du = \frac{1}{b^2}\left(\frac{a}{a+bu} + \ln|a+bu|\right) + C$

5. $\displaystyle\int \frac{u}{(a+bu)^n}\,du = \frac{1}{b^2}\left[\frac{-1}{(n-2)(a+bu)^{n-2}} + \frac{a}{(n-1)(a+bu)^{n-1}}\right] + C,\quad n \neq 1, 2$

6. $\displaystyle\int \frac{u^2}{a+bu}\,du = \frac{1}{b^3}\left[-\frac{bu}{2}(2a-bu) + a^2\ln|a+bu|\right] + C$

7. $\displaystyle\int \frac{u^2}{(a+bu)^2}\,du = \frac{1}{b^3}\left(bu - \frac{a^2}{a+bu} - 2a\ln|a+bu|\right) + C$

8. $\displaystyle\int \frac{u^2}{(a+bu)^3}\,du = \frac{1}{b^3}\left[\frac{2a}{a+bu} - \frac{a^2}{2(a+bu)^2} + \ln|a+bu|\right] + C$

9. $\displaystyle\int \frac{u^2}{(a+bu)^n}\,du = \frac{1}{b^3}\left[\frac{-1}{(n-3)(a+bu)^{n-3}} + \frac{2a}{(n-2)(a+bu)^{n-2}} - \frac{a^2}{(n-1)(a+bu)^{n-1}}\right] + C,\quad n \neq 1, 2, 3$

10. $\displaystyle\int \frac{1}{u(a+bu)}\,du = \frac{1}{a}\ln\left|\frac{u}{a+bu}\right| + C$

11. $\displaystyle\int \frac{1}{u(a+bu)^2}\,du = \frac{1}{a}\left(\frac{1}{a+bu} + \frac{1}{a}\ln\left|\frac{u}{a+bu}\right|\right) + C$

12. $\displaystyle\int \frac{1}{u^2(a+bu)}\,du = -\frac{1}{a}\left(\frac{1}{u} + \frac{b}{a}\ln\left|\frac{u}{a+bu}\right|\right) + C$

13. $\displaystyle\int \frac{1}{u^2(a+bu)^2}\,du = -\frac{1}{a^2}\left[\frac{a+2bu}{u(a+bu)} + \frac{2b}{a}\ln\left|\frac{u}{a+bu}\right|\right] + C$

Forms Involving $a + bu + cu^2,\ b^2 \neq 4ac$

14. $$\int \frac{1}{a+bu+cu^2}\,du = \begin{cases} \dfrac{2}{\sqrt{4ac-b^2}}\arctan\dfrac{2cu+b}{\sqrt{4ac-b^2}} + C, & b^2 < 4ac \\[2ex] \dfrac{1}{\sqrt{b^2-4ac}}\ln\left|\dfrac{2cu+b-\sqrt{b^2-4ac}}{2cu+b+\sqrt{b^2-4ac}}\right| + C, & b^2 > 4ac \end{cases}$$

15. $\displaystyle\int \frac{u}{a+bu+cu^2}\,du = \frac{1}{2c}\left(\ln|a+bu+cu^2| - b\int \frac{1}{a+bu+cu^2}\,du\right)$

Forms Involving $\sqrt{a+bu}$

16. $\displaystyle\int u^n\sqrt{a+bu}\,du = \frac{2}{b(2n+3)}\left[u^n(a+bu)^{3/2} - na\int u^{n-1}\sqrt{a+bu}\,du\right]$

17. $$\int \frac{1}{u\sqrt{a+bu}}\,du = \begin{cases} \dfrac{1}{\sqrt{a}}\ln\left|\dfrac{\sqrt{a+bu}-\sqrt{a}}{\sqrt{a+bu}+\sqrt{a}}\right| + C, & a > 0 \\[2ex] \dfrac{2}{\sqrt{-a}}\arctan\sqrt{\dfrac{a+bu}{-a}} + C, & a < 0 \end{cases}$$

18. $\displaystyle\int \frac{1}{u^n\sqrt{a+bu}}\,du = \frac{-1}{a(n-1)}\left[\frac{\sqrt{a+bu}}{u^{n-1}} + \frac{(2n-3)b}{2}\int \frac{1}{u^{n-1}\sqrt{a+bu}}\,du\right],\ n \neq 1$

19. $\displaystyle\int \frac{\sqrt{a+bu}}{u}\,du = 2\sqrt{a+bu} + a\int \frac{1}{u\sqrt{a+bu}}\,du$

20. $\displaystyle\int \frac{\sqrt{a+bu}}{u^n}\,du = \frac{-1}{a(n-1)}\left[\frac{(a+bu)^{3/2}}{u^{n-1}} + \frac{(2n-5)b}{2}\int \frac{\sqrt{a+bu}}{u^{n-1}}\,du\right],\ n \neq 1$

21. $\displaystyle\int \frac{u}{\sqrt{a+bu}}\,du = \frac{-2(2a-bu)}{3b^2}\sqrt{a+bu} + C$

22. $\displaystyle\int \frac{u^n}{\sqrt{a+bu}}\,du = \frac{2}{(2n+1)b}\left(u^n\sqrt{a+bu} - na\int \frac{u^{n-1}}{\sqrt{a+bu}}\,du\right)$

Forms Involving $a^2 \pm u^2$, $a > 0$

23. $\displaystyle\int \frac{1}{a^2+u^2}\,du = \frac{1}{a}\arctan\frac{u}{a} + C$

24. $\displaystyle\int \frac{1}{u^2-a^2}\,du = -\int \frac{1}{a^2-u^2}\,du = \frac{1}{2a}\ln\left|\frac{u-a}{u+a}\right| + C$

25. $\displaystyle\int \frac{1}{(a^2 \pm u^2)^n}\,du = \frac{1}{2a^2(n-1)}\left[\frac{u}{(a^2 \pm u^2)^{n-1}} + (2n-3)\int \frac{1}{(a^2 \pm u^2)^{n-1}}\,du\right],\ n \neq 1$

Forms Involving $\sqrt{u^2 \pm a^2}$, $a > 0$

26. $\displaystyle\int \sqrt{u^2 \pm a^2}\,du = \frac{1}{2}\left(u\sqrt{u^2 \pm a^2} \pm a^2\ln\left|u + \sqrt{u^2 \pm a^2}\right|\right) + C$

27. $\displaystyle\int u^2\sqrt{u^2 \pm a^2}\,du = \frac{1}{8}\left[u(2u^2 \pm a^2)\sqrt{u^2 \pm a^2} - a^4\ln\left|u + \sqrt{u^2 \pm a^2}\right|\right] + C$

28. $\displaystyle\int \frac{\sqrt{u^2+a^2}}{u}\,du = \sqrt{u^2+a^2} - a\ln\left|\frac{a+\sqrt{u^2+a^2}}{u}\right| + C$

29. $\displaystyle\int \frac{\sqrt{u^2-a^2}}{u}\,du = \sqrt{u^2-a^2} - a\,\mathrm{arcsec}\frac{|u|}{a} + C$

30. $\displaystyle\int \frac{\sqrt{u^2 \pm a^2}}{u^2}\,du = \frac{-\sqrt{u^2 \pm a^2}}{u} + \ln\left|u + \sqrt{u^2 \pm a^2}\right| + C$

31. $\displaystyle\int \frac{1}{\sqrt{u^2 \pm a^2}}\,du = \ln\left|u + \sqrt{u^2 \pm a^2}\right| + C$

32. $\displaystyle\int \frac{1}{u\sqrt{u^2+a^2}}\,du = \frac{-1}{a}\ln\left|\frac{a+\sqrt{u^2+a^2}}{u}\right| + C$

33. $\displaystyle\int \frac{1}{u\sqrt{u^2-a^2}}\,du = \frac{1}{a}\,\mathrm{arcsec}\frac{|u|}{a} + C$

34. $\displaystyle\int \frac{u^2}{\sqrt{u^2 \pm a^2}}\,du = \frac{1}{2}\left(u\sqrt{u^2 \pm a^2} \mp a^2\ln\left|u + \sqrt{u^2 \pm a^2}\right|\right) + C$

35. $\displaystyle\int \frac{1}{u^2\sqrt{u^2 \pm a^2}}\,du = \mp\frac{\sqrt{u^2 \pm a^2}}{a^2u} + C$

36. $\displaystyle\int \frac{1}{(u^2 \pm a^2)^{3/2}}\,du = \frac{\pm u}{a^2\sqrt{u^2 \pm a^2}} + C$

Forms Involving $\sqrt{a^2 - u^2}$, $a > 0$

37. $\displaystyle\int \sqrt{a^2-u^2}\,du = \frac{1}{2}\left(u\sqrt{a^2-u^2} + a^2\arcsin\frac{u}{a}\right) + C$

38. $\displaystyle\int u^2\sqrt{a^2-u^2}\,du = \frac{1}{8}\left[u(2u^2-a^2)\sqrt{a^2-u^2} + a^4\arcsin\frac{u}{a}\right] + C$

39. $\displaystyle\int \frac{\sqrt{a^2 - u^2}}{u}\,du = \sqrt{a^2 - u^2} - a\ln\left|\frac{a + \sqrt{a^2 - u^2}}{u}\right| + C$

40. $\displaystyle\int \frac{\sqrt{a^2 - u^2}}{u^2}\,du = \frac{-\sqrt{a^2 - u^2}}{u} - \arcsin\frac{u}{a} + C$

41. $\displaystyle\int \frac{1}{\sqrt{a^2 - u^2}}\,du = \arcsin\frac{u}{a} + C$

42. $\displaystyle\int \frac{1}{u\sqrt{a^2 - u^2}}\,du = \frac{-1}{a}\ln\left|\frac{a + \sqrt{a^2 - u^2}}{u}\right| + C$

43. $\displaystyle\int \frac{u^2}{\sqrt{a^2 - u^2}}\,du = \frac{1}{2}\left(-u\sqrt{a^2 - u^2} + a^2\arcsin\frac{u}{a}\right) + C$

44. $\displaystyle\int \frac{1}{u^2\sqrt{a^2 - u^2}}\,du = \frac{-\sqrt{a^2 - u^2}}{a^2u} + C$

45. $\displaystyle\int \frac{1}{(a^2 - u^2)^{3/2}}\,du = \frac{u}{a^2\sqrt{a^2 - u^2}} + C$

Forms Involving sin *u* or cos *u*

46. $\displaystyle\int \sin u\,du = -\cos u + C$

47. $\displaystyle\int \cos u\,du = \sin u + C$

48. $\displaystyle\int \sin^2 u\,du = \frac{1}{2}(u - \sin u\cos u) + C$

49. $\displaystyle\int \cos^2 u\,du = \frac{1}{2}(u + \sin u\cos u) + C$

50. $\displaystyle\int \sin^n u\,du = -\frac{\sin^{n-1} u\cos u}{n} + \frac{n-1}{n}\int \sin^{n-2} u\,du$

51. $\displaystyle\int \cos^n u\,du = \frac{\cos^{n-1} u\sin u}{n} + \frac{n-1}{n}\int \cos^{n-2} u\,du$

52. $\displaystyle\int u\sin u\,du = \sin u - u\cos u + C$

53. $\displaystyle\int u\cos u\,du = \cos u + u\sin u + C$

54. $\displaystyle\int u^n\sin u\,du = -u^n\cos u + n\int u^{n-1}\cos u\,du$

55. $\displaystyle\int u^n\cos u\,du = u^n\sin u - n\int u^{n-1}\sin u\,du$

56. $\displaystyle\int \frac{1}{1 \pm \sin u}\,du = \tan u \mp \sec u + C$

57. $\displaystyle\int \frac{1}{1 \pm \cos u}\,du = -\cot u \pm \csc u + C$

58. $\displaystyle\int \frac{1}{\sin u\cos u}\,du = \ln|\tan u| + C$

Forms Involving tan *u*, cot *u*, sec *u*, or csc *u*

59. $\displaystyle\int \tan u\,du = -\ln|\cos u| + C$

60. $\displaystyle\int \cot u\,du = \ln|\sin u| + C$

61. $\displaystyle\int \sec u\,du = \ln|\sec u + \tan u| + C$

62. $\displaystyle\int \csc u\,du = \ln|\csc u - \cot u| + C$ or $\displaystyle\int \csc u\,du = -\ln|\csc u + \cot u| + C$

63. $\displaystyle\int \tan^2 u\,du = -u + \tan u + C$

64. $\displaystyle\int \cot^2 u\,du = -u - \cot u + C$

65. $\displaystyle\int \sec^2 u\,du = \tan u + C$

66. $\displaystyle\int \csc^2 u\,du = -\cot u + C$

67. $\displaystyle\int \tan^n u\,du = \frac{\tan^{n-1} u}{n-1} - \int \tan^{n-2} u\,du,\ n \neq 1$

68. $\displaystyle\int \cot^n u\,du = -\frac{\cot^{n-1} u}{n-1} - \int (\cot^{n-2} u)\,du,\ n \neq 1$

69. $\displaystyle\int \sec^n u\,du = \frac{\sec^{n-2} u\tan u}{n-1} + \frac{n-2}{n-1}\int \sec^{n-2} u\,du,\ n \neq 1$

70. $\displaystyle\int \csc^n u\,du = -\frac{\csc^{n-2} u\cot u}{n-1} + \frac{n-2}{n-1}\int \csc^{n-2} u\,du,\ n \neq 1$

71. $\displaystyle\int \frac{1}{1 \pm \tan u}\,du = \frac{1}{2}\left(u \pm \ln|\cos u \pm \sin u|\right) + C$

72. $\displaystyle\int \frac{1}{1 \pm \cot u}\,du = \frac{1}{2}\left(u \mp \ln|\sin u \pm \cos u|\right) + C$

73. $\displaystyle\int \frac{1}{1 \pm \sec u}\,du = u + \cot u \mp \csc u + C$

74. $\displaystyle\int \frac{1}{1 \pm \csc u}\,du = u - \tan u \pm \sec u + C$

Forms Involving Inverse Trigonometric Functions

75. $\displaystyle\int \arcsin u\,du = u \arcsin u + \sqrt{1 - u^2} + C$

76. $\displaystyle\int \arccos u\,du = u \arccos u - \sqrt{1 - u^2} + C$

77. $\displaystyle\int \arctan u\,du = u \arctan u - \ln\sqrt{1 + u^2} + C$

78. $\displaystyle\int \operatorname{arccot} u\,du = u \operatorname{arccot} u + \ln\sqrt{1 + u^2} + C$

79. $\displaystyle\int \operatorname{arcsec} u\,du = u \operatorname{arcsec} u - \ln\left|u + \sqrt{u^2 - 1}\right| + C$

80. $\displaystyle\int \operatorname{arccsc} u\,du = u \operatorname{arccsc} u + \ln\left|u + \sqrt{u^2 - 1}\right| + C$

Forms Involving e^u

81. $\displaystyle\int e^u\,du = e^u + C$

82. $\displaystyle\int ue^u\,du = (u - 1)e^u + C$

83. $\displaystyle\int u^n e^u\,du = u^n e^u - n\int u^{n-1} e^u\,du$

84. $\displaystyle\int \frac{1}{1 + e^u}\,du = u - \ln(1 + e^u) + C$

85. $\displaystyle\int e^{au} \sin bu\,du = \frac{e^{au}}{a^2 + b^2}(a \sin bu - b \cos bu) + C$

86. $\displaystyle\int e^{au} \cos bu\,du = \frac{e^{au}}{a^2 + b^2}(a \cos bu + b \sin bu) + C$

Forms Involving $\ln u$

87. $\displaystyle\int \ln u\,du = u(-1 + \ln u) + C$

88. $\displaystyle\int u \ln u\,du = \frac{u^2}{4}(-1 + 2 \ln u) + C$

89. $\displaystyle\int u^n \ln u\,du = \frac{u^{n+1}}{(n + 1)^2}[-1 + (n + 1) \ln u] + C,\ n \neq -1$

90. $\displaystyle\int (\ln u)^2\,du = u[2 - 2 \ln u + (\ln u)^2] + C$

91. $\displaystyle\int (\ln u)^n\,du = u(\ln u)^n - n\int (\ln u)^{n-1}\,du$

Forms Involving Hyperbolic Functions

92. $\displaystyle\int \cosh u\,du = \sinh u + C$

93. $\displaystyle\int \sinh u\,du = \cosh u + C$

94. $\displaystyle\int \operatorname{sech}^2 u\,du = \tanh u + C$

95. $\displaystyle\int \operatorname{csch}^2 u\,du = -\coth u + C$

96. $\displaystyle\int \operatorname{sech} u \tanh u\,du = -\operatorname{sech} u + C$

97. $\displaystyle\int \operatorname{csch} u \coth u\,du = -\operatorname{csch} u + C$

Forms Involving Inverse Hyperbolic Functions (in logarithmic form)

98. $\displaystyle\int \frac{du}{\sqrt{u^2 \pm a^2}} = \ln\left(u + \sqrt{u^2 \pm a^2}\right) + C$

99. $\displaystyle\int \frac{du}{a^2 - u^2} = \frac{1}{2a} \ln\left|\frac{a + u}{a - u}\right| + C$

100. $\displaystyle\int \frac{du}{u\sqrt{a^2 \pm u^2}} = -\frac{1}{a} \ln \frac{a + \sqrt{a^2 \pm u^2}}{|u|} + C$

Answers to Odd-Numbered Exercises

Chapter P

Section P.1 *(page 8)*

1. b **2.** d **3.** a **4.** c

5.

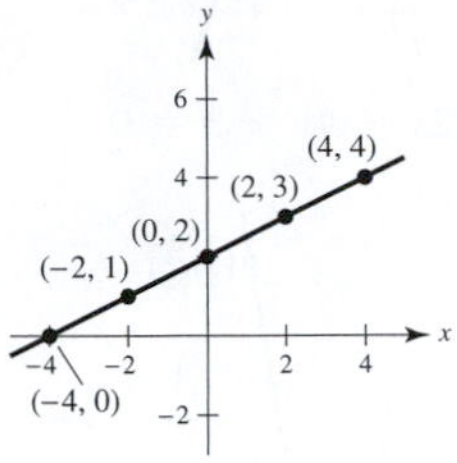

7.

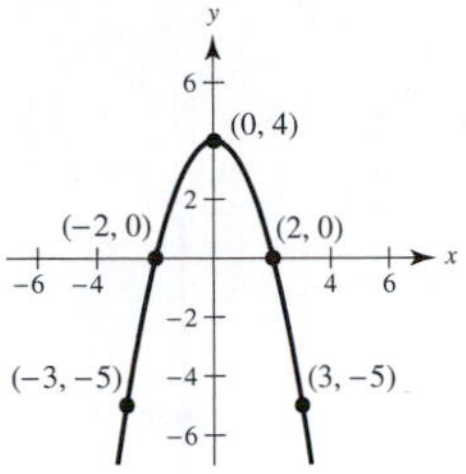

9.

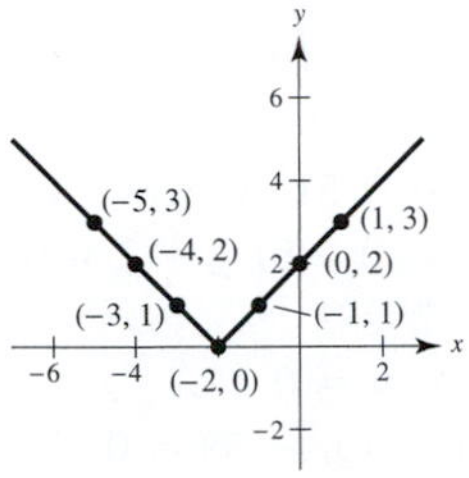

11.

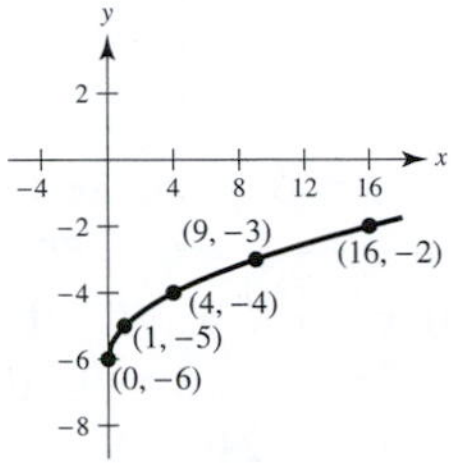

13.

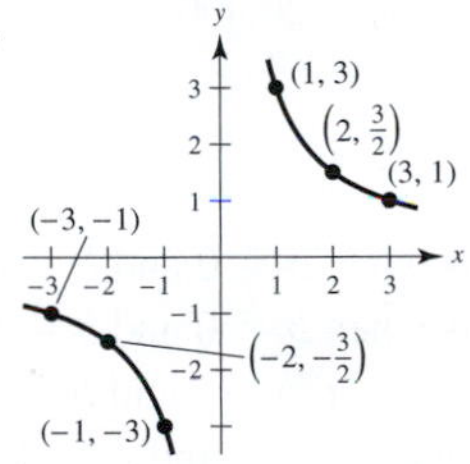

15.

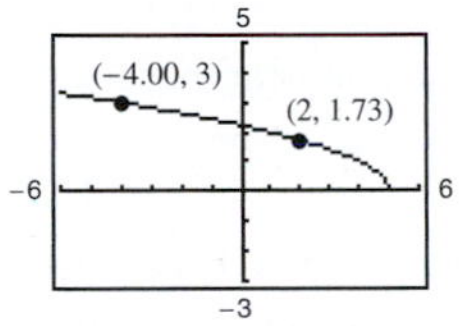

(a) $y \approx 1.73$ (b) $x = -4$

17. $(0, -5), \left(\frac{5}{2}, 0\right)$ **19.** $(0, -2), (-2, 0), (1, 0)$

21. $(0, 0), (4, 0), (-4, 0)$ **23.** $(0, 2), (4, 0)$ **25.** $(0, 0)$

27. Symmetric with respect to the y-axis

29. Symmetric with respect to the x-axis

31. Symmetric with respect to the origin **33.** No symmetry

35. Symmetric with respect to the origin

37. Symmetric with respect to the y-axis

39.

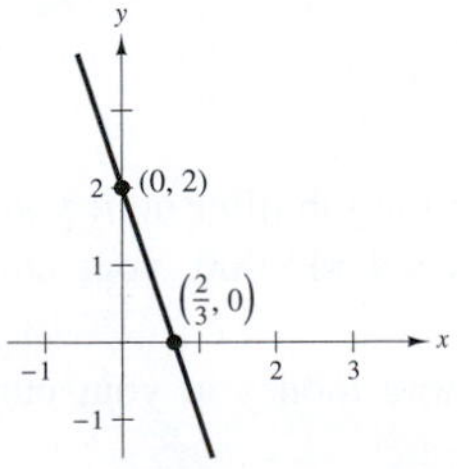

Symmetry: none

41.

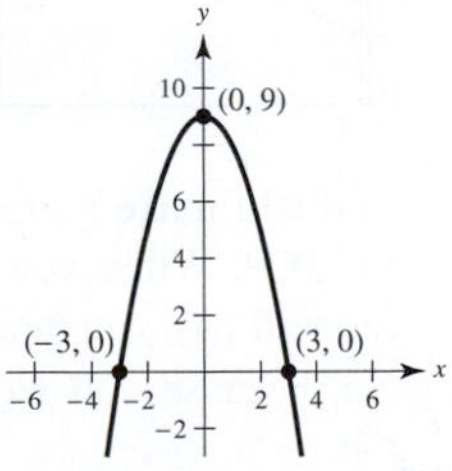

Symmetry: y-axis

43.

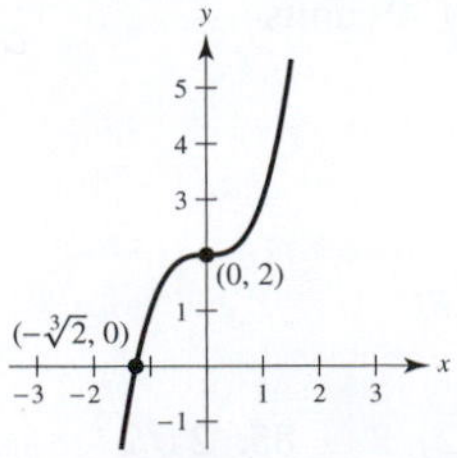

Symmetry: none

45.

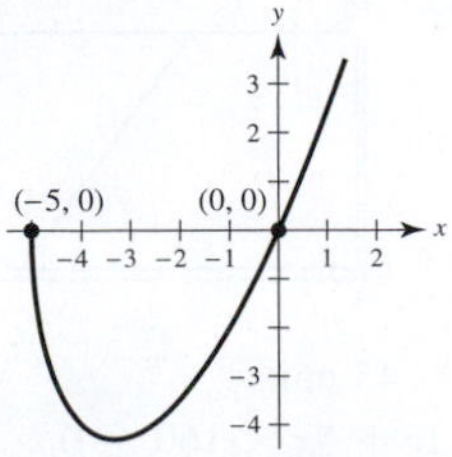

Symmetry: none

47.

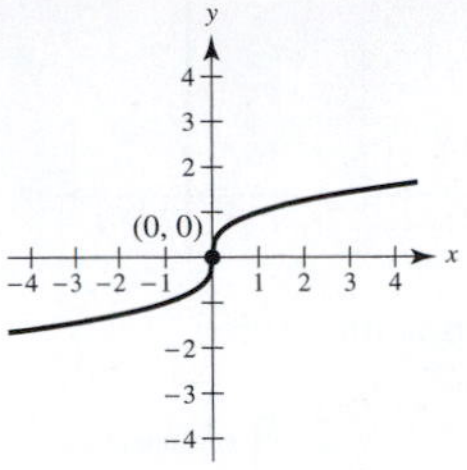

Symmetry: origin

49.

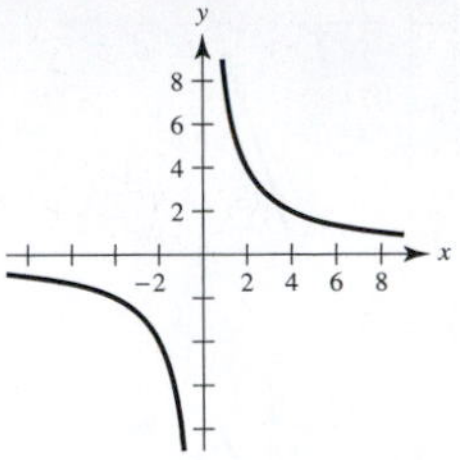

Symmetry: origin

51.

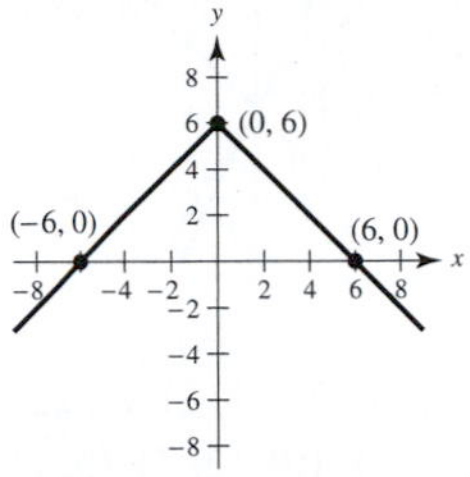

Symmetry: y-axis

53.

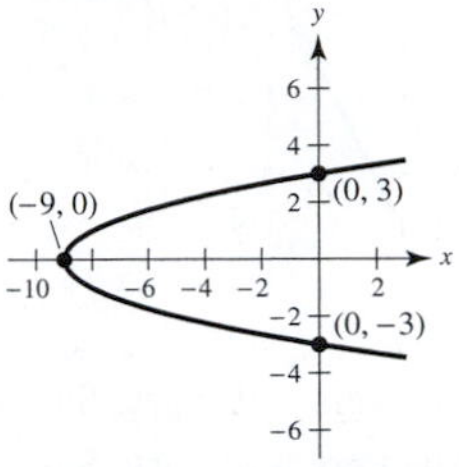

Symmetry: x-axis

55.

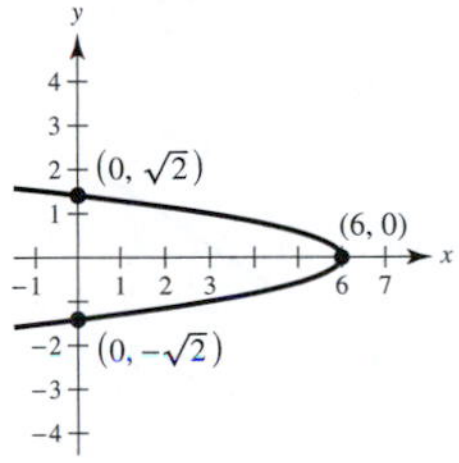

Symmetry: x-axis

57. $(3, 5)$

59. $(-1, 5), (2, 2)$ **61.** $(-1, -2), (2, 1)$

63. $(-1, -5), (0, -1), (2, 1)$ **65.** $(-2, 2), (-3, \sqrt{3})$

67. (a) $y = 0.005t^2 + 0.27t + 2.7$

(b)

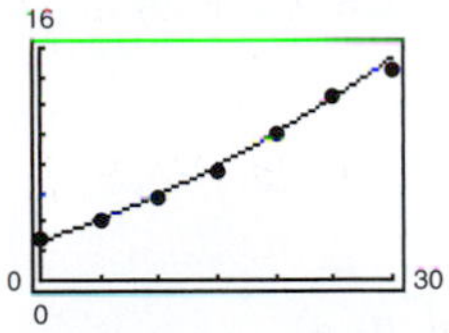

The model is a good fit for the data.

(c) \$21.5 trillion

69. 4480 units

71. (a) $k = 4$ (b) $k = -\frac{1}{8}$

(c) All real numbers k (d) $k = 1$

73. Answers will vary. Sample answer: $y = (x + 4)(x - 3)(x - 8)$

75. (a) Proof (b) Proof

77. False. $(4, -5)$ is not a point on the graph of $x = y^2 - 29$.

79. True

Section P.2 *(page 16)*

1. $m = 2$ **3.** $m = -1$

5.

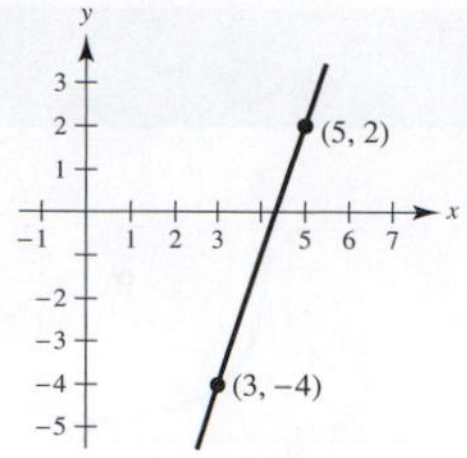

$m = 3$

7.

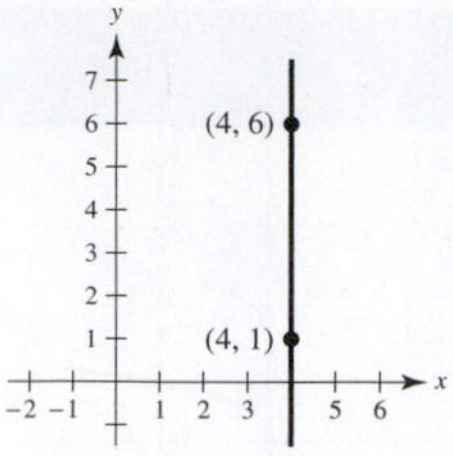

m is undefined.

9.

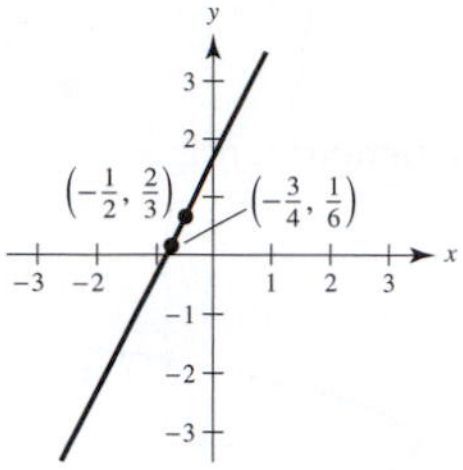

$m = 2$

11.

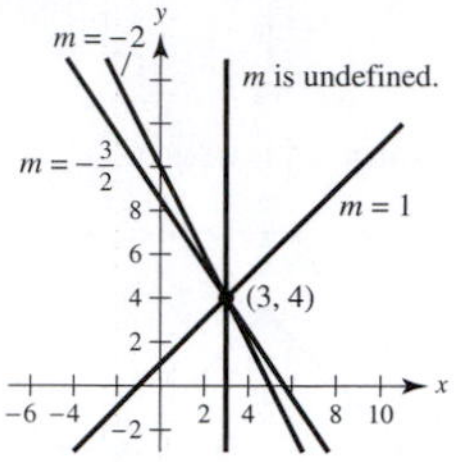

13. Answers will vary. Sample answers: (0, 2), (1, 2), (5, 2)

15. Answers will vary. Sample answers: (0, 10), (2, 4), (3, 1)

17. $3x - 4y + 12 = 0$

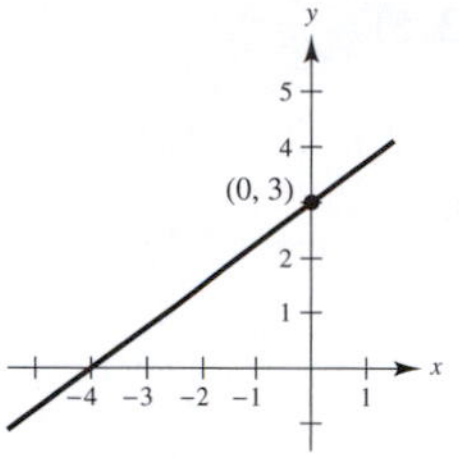

19. $2x - 3y = 0$

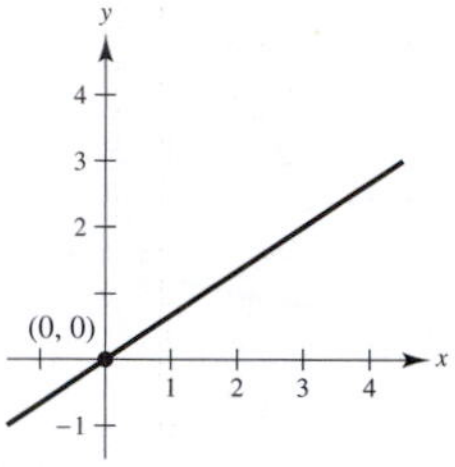

21. $3x - y - 11 = 0$

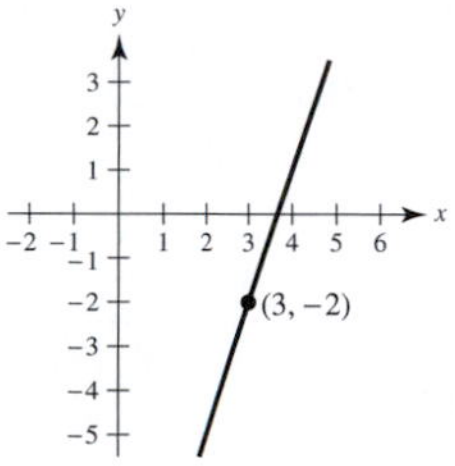

23. (a) $\frac{1}{3}$ (b) $10\sqrt{10}$ ft

25. $m = 4$, $(0, -3)$ **27.** $m = -\frac{1}{5}$, $(0, 4)$

29. m is undefined, no y-intercept

31.

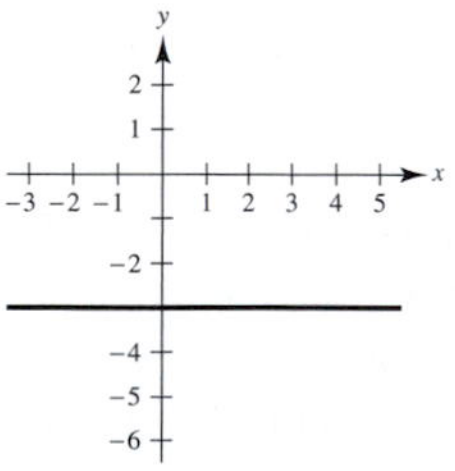

33.

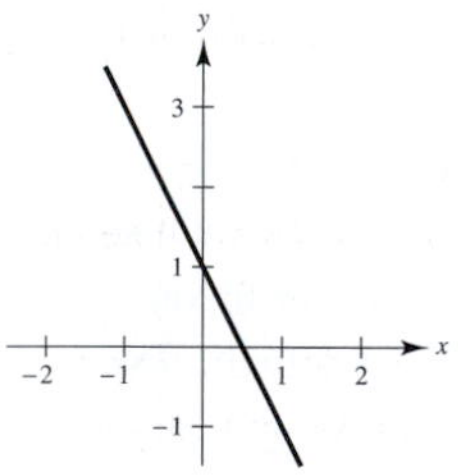

35.

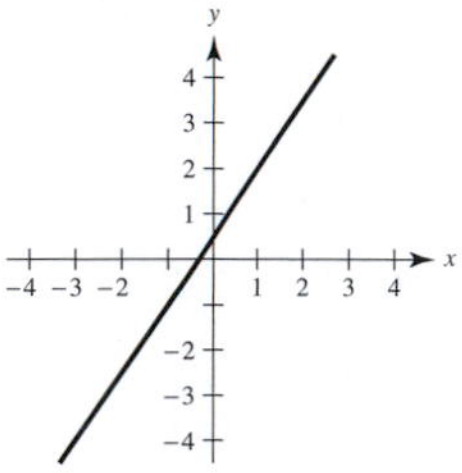

37.

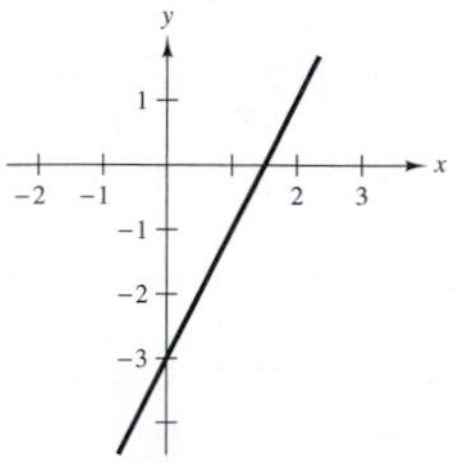

39. $2x - y = 0$

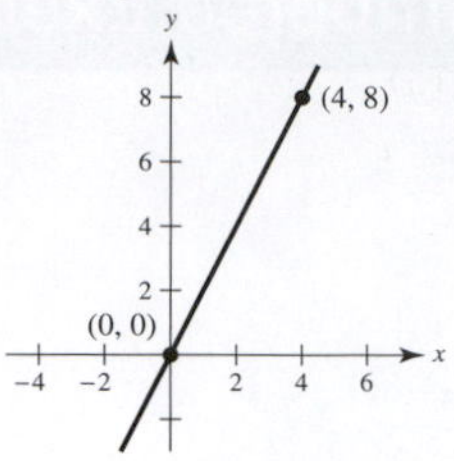

41. $8x + 3y - 40 = 0$

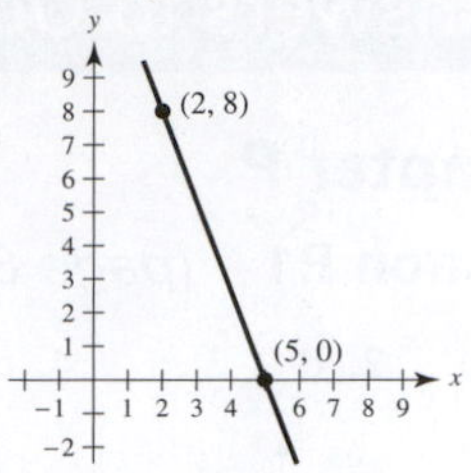

43. $x - 6 = 0$

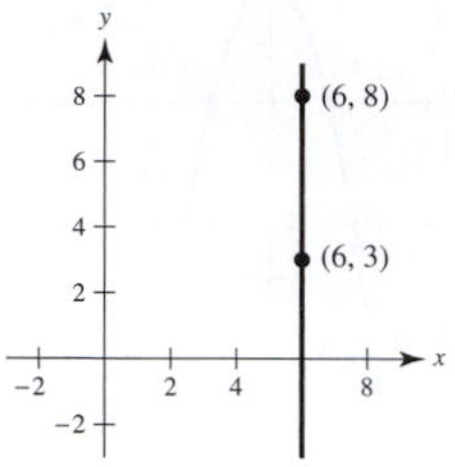

45. $22x - 4y + 3 = 0$

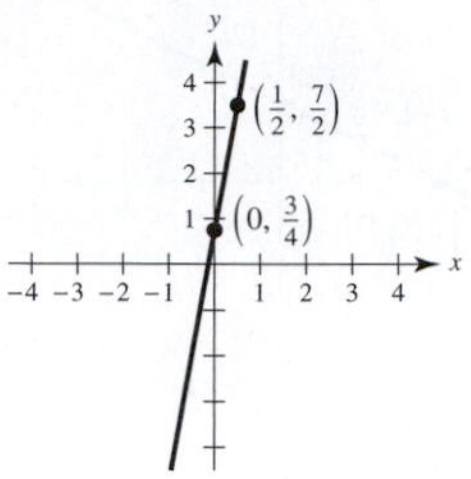

47. $x - 3 = 0$ **49.** $3x + 2y - 6 = 0$ **51.** $x + y - 3 = 0$

53. $x + 2y - 5 = 0$ **55.** (a) $x + 7 = 0$ (b) $y + 2 = 0$

57. (a) $x - y + 3 = 0$ (b) $x + y - 7 = 0$

59. (a) $2x - y - 3 = 0$ (b) $x + 2y - 4 = 0$

61. (a) $40x - 24y - 9 = 0$ (b) $24x + 40y - 53 = 0$

63. $V = 250t + 1350$ **65.** $V = -1600t + 20{,}400$

67. Not collinear, because $m_1 \neq m_2$

69. $\left(0, \dfrac{-a^2 + b^2 + c^2}{2c}\right)$ **71.** $\left(b, \dfrac{a^2 - b^2}{c}\right)$

73. (a) The line is parallel to the x-axis when $a = 0$ and $b \neq 0$.
(b) The line is parallel to the y-axis when $b = 0$ and $a \neq 0$.
(c) Answers will vary. Sample answer: $a = -5$ and $b = 8$
(d) Answers will vary. Sample answer: $a = 5$ and $b = 2$
(e) $a = \frac{5}{2}$ and $b = 3$

75. $5F - 9C - 160 = 0$; 72°F ≈ 22.2°C

77. (a) Current job: $W = 2000 + 0.07s$
Job offer: $W = 2300 + 0.05s$

(b)

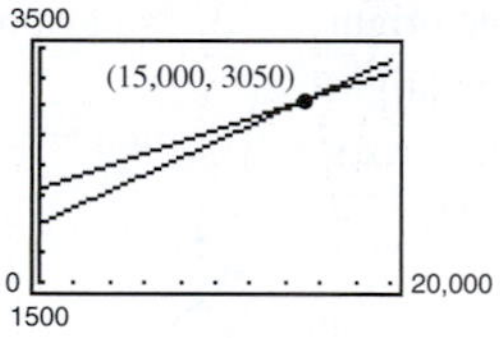

You will make more money at the job offer until you sell \$15,000. When your sales exceed \$15,000, your current job will pay you more.

(c) No, because you will make more money at your current job.

79. (a) $x = (1530 - p)/15$

(b)

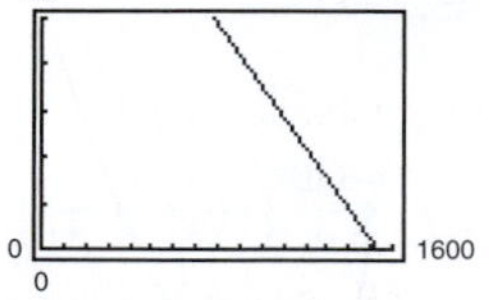

45 units

(c) 49 units

81. $12y + 5x - 169 = 0$ **83.** $(5\sqrt{2})/2$ **85.** $2\sqrt{2}$

87–91. Proofs **93.** True **95.** True

Section P.3 *(page 27)*

1. (a) -4 (b) -25 (c) $7b - 4$ (d) $7x - 11$

3. (a) 5 (b) 0 (c) 1 (d) $4 + 2t - t^2$

5. (a) 1 (b) 0 (c) $-\frac{1}{2}$ (d) 1

7. $3x^2 + 3x\,\Delta x + (\Delta x)^2,\ \Delta x \neq 0$

9. $\left(\sqrt{x-1} - x + 1\right)/[(x-2)(x-1)]$

11. Domain: $(-\infty, \infty)$; Range: $[0, \infty)$

13. Domain: $(-\infty, \infty)$; Range: $(-\infty, \infty)$

15. Domain: $[0, \infty)$; Range: $[0, \infty)$

17. Domain: $[-4, 4]$; Range: $[0, 4]$

19. Domain: All real numbers t such that $t \neq 4n + 2$, where n is an integer; Range: $(-\infty, -1] \cup [1, \infty)$

21. Domain: $(-\infty, 0) \cup (0, \infty)$; Range: $(-\infty, 0) \cup (0, \infty)$

23. Domain: $[0, 1]$

25. Domain: All real numbers x such that $x \neq 2n\pi$, where n is an integer

27. Domain: $(-\infty, -3) \cup (-3, \infty)$

29. (a) -1 (b) 2 (c) 6 (d) $2t^2 + 4$
Domain: $(-\infty, \infty)$; Range: $(-\infty, 1) \cup [2, \infty)$

31. (a) 4 (b) 0 (c) -2 (d) $-b^2$
Domain: $(-\infty, \infty)$; Range: $(-\infty, 0] \cup [1, \infty)$

33.

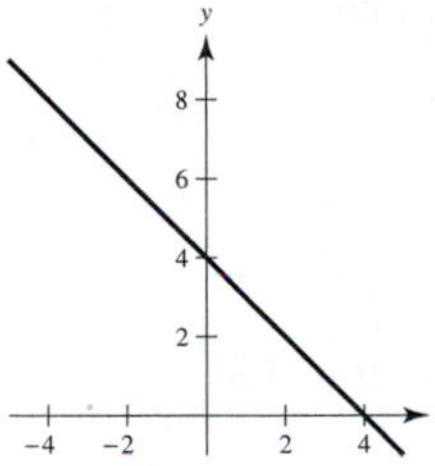

Domain: $(-\infty, \infty)$
Range: $(-\infty, \infty)$

35.

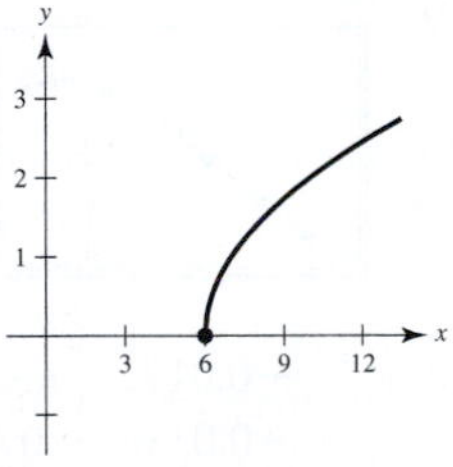

Domain: $[6, \infty)$
Range: $[0, \infty)$

37.

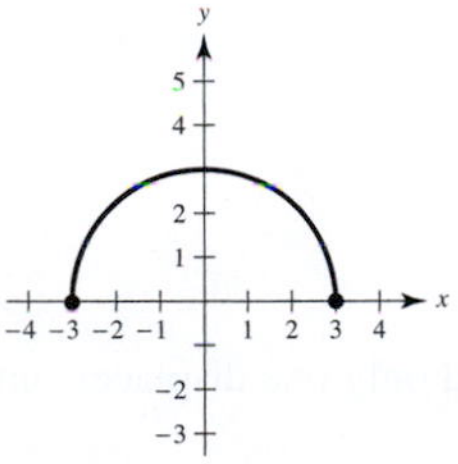

Domain: $[-3, 3]$
Range: $[0, 3]$

39.

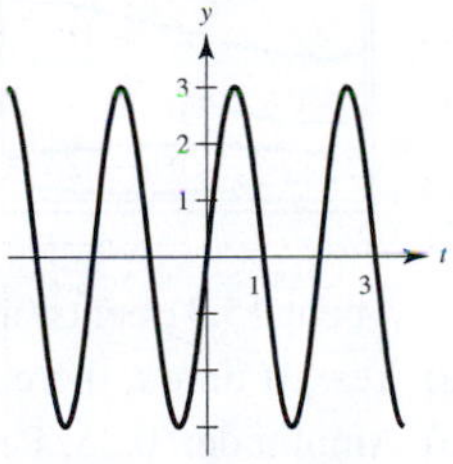

Domain: $(-\infty, \infty)$
Range: $[-3, 3]$

41. The student travels $\frac{1}{2}$ mile/minute during the first 4 minutes, is stationary for the next 2 minutes, and travels 1 mile/minute during the final 4 minutes.

43. y is not a function of x. **45.** y is a function of x.

47. y is not a function of x. **49.** y is not a function of x.

51. Horizontal shift to the right two units
$y = \sqrt{x-2}$

53. Horizontal shift to the right two units and vertical shift down one unit
$y = (x-2)^2 - 1$

55. *d* **56.** *b* **57.** *c* **58.** *a* **59.** *e* **60.** *g*

61. (a) (b) (c) (d) (e) (f) (g) (h)

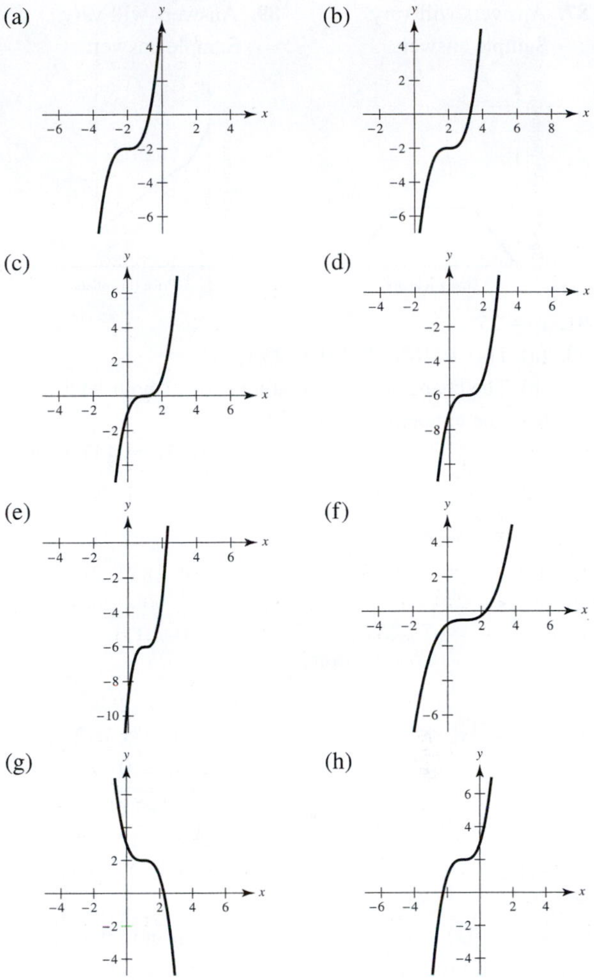

63. (a) $3x$ (b) $3x - 8$ (c) $12x - 16$ (d) $\frac{3}{4}x - 1$

65. (a) 0 (b) 0 (c) -1 (d) $\sqrt{15}$
(e) $\sqrt{x^2 - 1}$ (f) $x - 1\ (x \geq 0)$

67. $(f \circ g)(x) = x$; Domain: $[0, \infty)$
$(g \circ f)(x) = |x|$; Domain: $(-\infty, \infty)$
No, their domains are different.

69. $(f \circ g)(x) = 3/(x^2 - 1)$;
Domain: $(-\infty, -1) \cup (-1, 1) \cup (1, \infty)$
$(g \circ f)(x) = (9/x^2) - 1$; Domain: $(-\infty, 0) \cup (0, \infty)$
No

71. (a) 4 (b) -2
(c) Undefined. The graph of g does not exist at $x = -5$.
(d) 3 (e) 2
(f) Undefined. The graph of f does not exist at $x = -4$.

73. Answers will vary.
Sample answer: $f(x) = \sqrt{x}$; $g(x) = x - 2$; $h(x) = 2x$

75. (a) $\left(\frac{3}{2}, 4\right)$ (b) $\left(\frac{3}{2}, -4\right)$

77. f is even. g is neither even nor odd. h is odd.

79. Even; zeros: $x = -2, 0, 2$

81. Odd; zeros: $x = 0, \frac{\pi}{2} + n\pi$, where n is an integer

83. $f(x) = -5x - 6,\ -2 \leq x \leq 0$ **85.** $y = -\sqrt{-x}$

87. Answers will vary.
Sample answer:

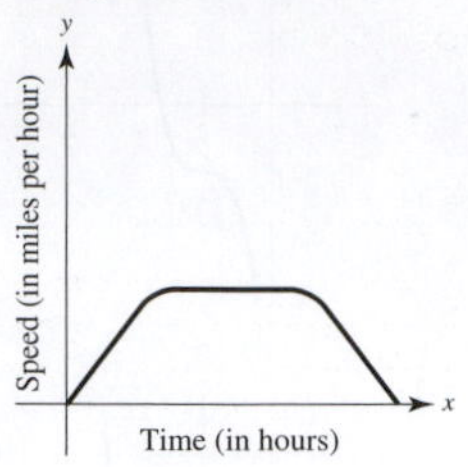

89. Answers will vary.
Sample answer:

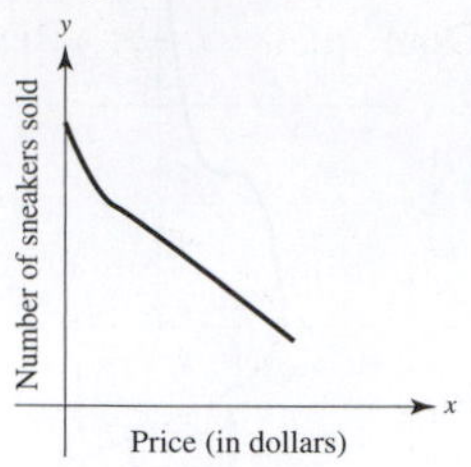

91. $c = 25$

93. (a) $T(4) = 16°\text{C}$, $T(15) \approx 23°\text{C}$
(b) The changes in temperature occur 1 hour later.
(c) The temperatures are 1° lower.

95. (a)

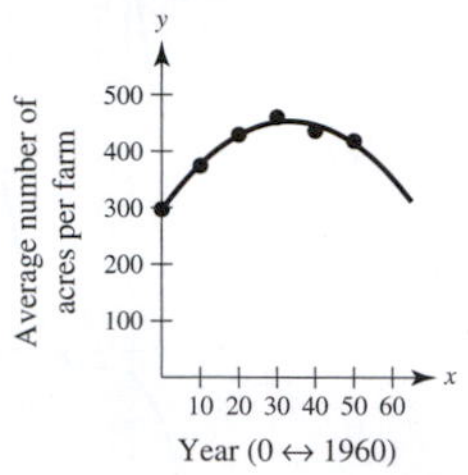

(b) $A(25) \approx 443$ acres/farm

97. $f(x) = |x| + |x - 2| = \begin{cases} 2x - 2, & x \geq 2 \\ 2, & 0 < x < 2 \\ -2x + 2, & x \leq 0 \end{cases}$

99–101. Proofs **103.** $L = \sqrt{x^2 + \left(\dfrac{2x}{x-3}\right)^2}$

105. False. For example, if $f(x) = x^2$, then $f(-1) = f(1)$.

107. True

109. False. $f(x) = 0$ is symmetric with respect to the x-axis.

111. Putnam Problem A1, 1988

Section P.4 *(page 34)*

1. (a) and (b)

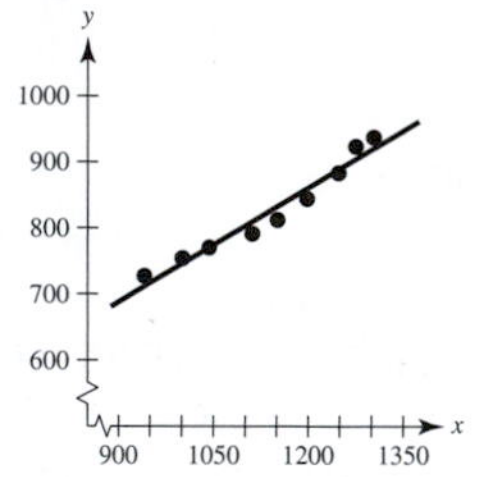

(c) \$790

3. (a) $d = 0.066F$
(b)

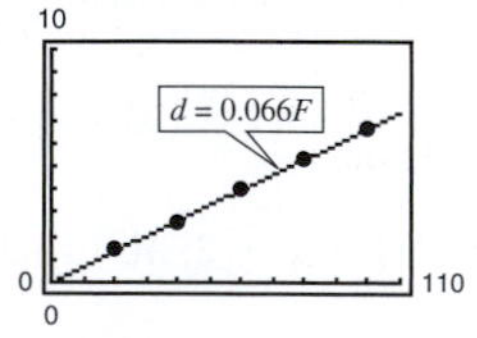

The model fits well.
(c) 3.63 cm

5. (a) $y = 0.122x + 2.07$, $r \approx 0.87$
(b)

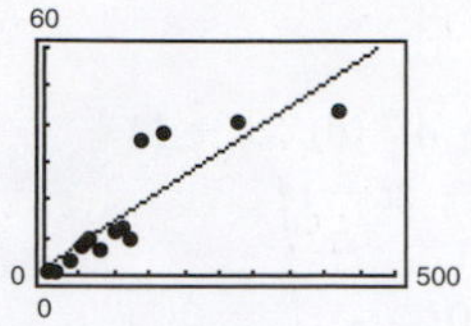

(c) Greater per capita energy consumption by a country tends to correspond to greater per capita gross national product of the country. The three countries that differ most from the linear model are Canada, Italy, and Japan.
(d) $y = 0.142x - 1.66$, $r \approx 0.97$

7. (a) $S = 180.89x^2 - 205.79x + 272$
(b)

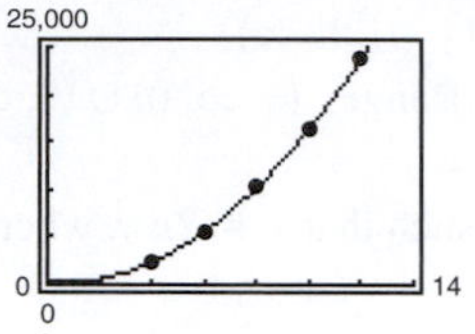

(c) When $x = 2$, $S \approx 583.98$ pounds.
(d) About 4 times greater
(e) About 4.37 times greater; No; Answers will vary.

9. (a) $y = -1.806x^3 + 14.58x^2 + 16.4x + 10$
(b)

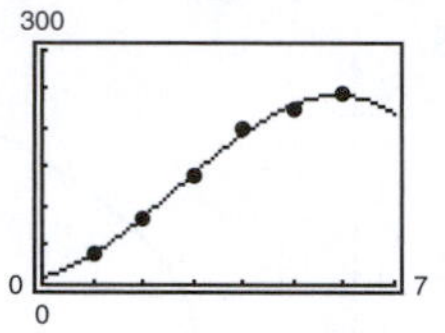

(c) 214 hp

11. (a) $y_1 = -0.0172t^3 + 0.305t^2 - 0.87t + 7.3$
$y_2 = -0.038t^2 + 0.45t + 3.5$
$y_3 = 0.0063t^3 - 0.072t^2 + 0.02t + 1.8$
(b)

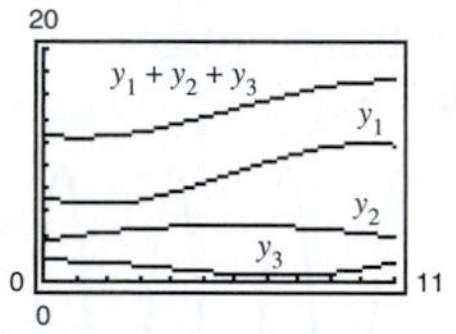

About 15.31 cents/mi

13. (a) Yes. At time t, there is one and only one displacement y.
(b) Amplitude: 0.35; Period: 0.5
(c) $y = 0.35 \sin(4\pi t) + 2$
(d)

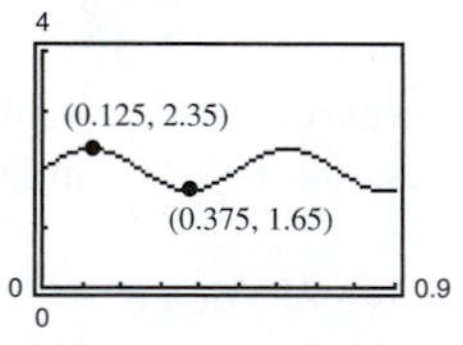

The model appears to fit the data well.

15. Answers will vary. **17.** Putnam Problem A2, 2004

Review Exercises for Chapter P *(page 37)*

1. $\left(\frac{8}{5}, 0\right)$, $(0, -8)$ **3.** $(3, 0)$, $\left(0, \frac{3}{4}\right)$ **5.** Not symmetric

7. Symmetric with respect to the x-axis, the y-axis, and the origin

9.

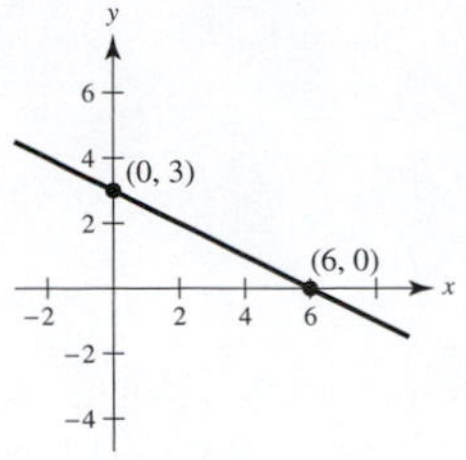

Symmetry: none

11.

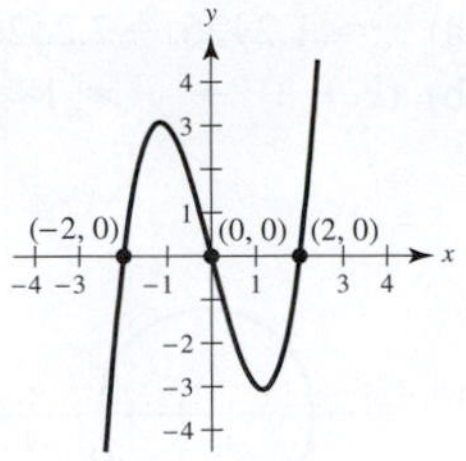

Symmetry: origin

13.

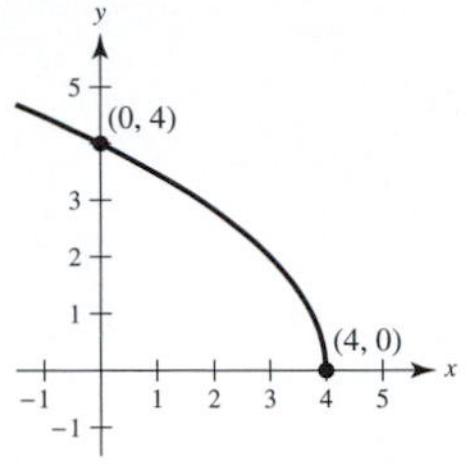

Symmetry: none

15. $(-2, 3)$ **17.** $(-2, 3), (3, 8)$

19.

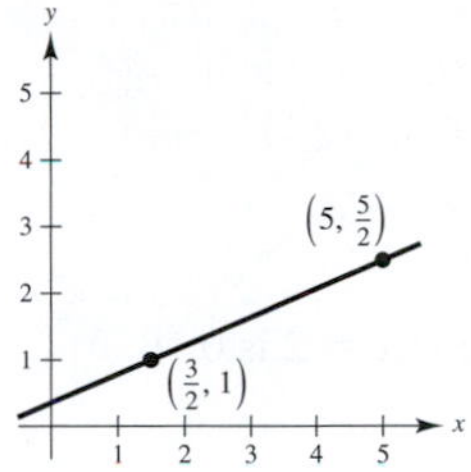

$m = \frac{3}{7}$

21. $7x - 4y - 41 = 0$

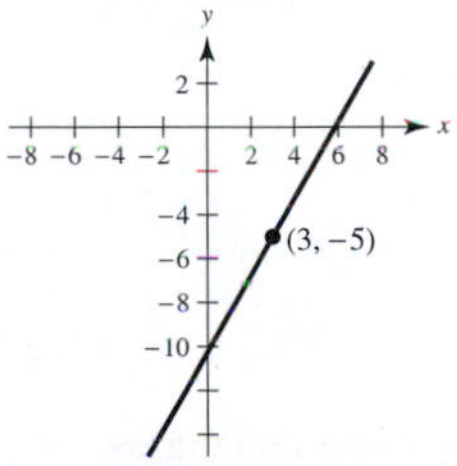

23. $2x + 3y + 6 = 0$

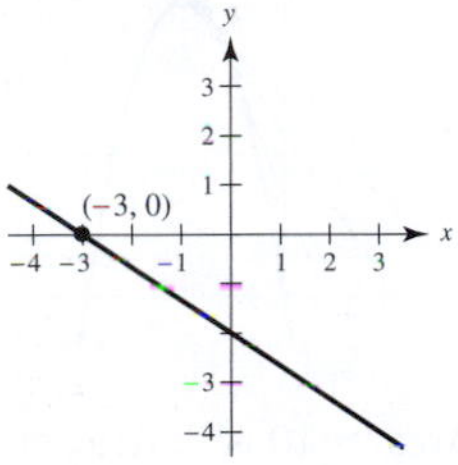

25.

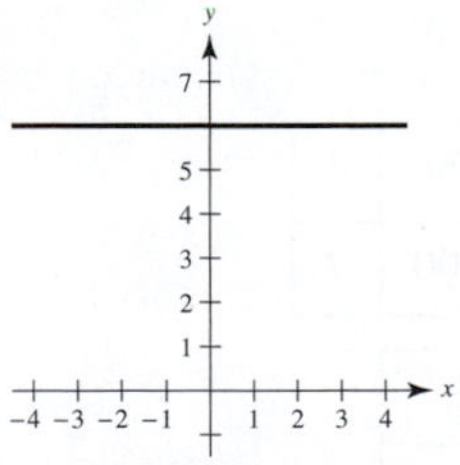

27.

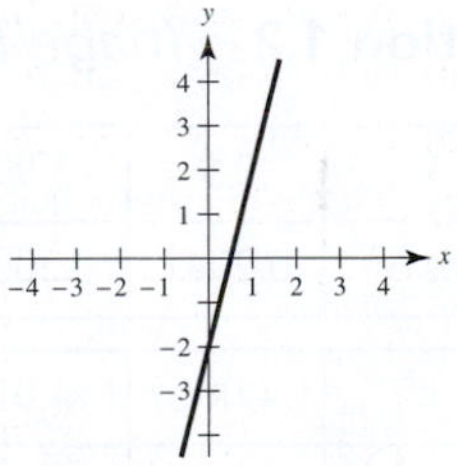

29. $x - 4y = 0$

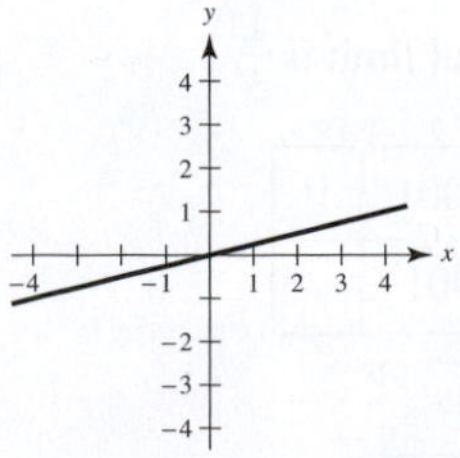

31. (a) $7x - 16y + 101 = 0$
(b) $5x - 3y + 30 = 0$
(c) $4x - 3y + 27 = 0$
(d) $x + 3 = 0$

33. $V = 12{,}500 - 850t$; \$9950

35. (a) 4 (b) 29 (c) -11 (d) $5t + 9$

37. $8x + 4\,\Delta x,\ \Delta x \neq 0$

39. Domain: $(-\infty, \infty)$; Range: $[3, \infty)$

41. Domain: $(-\infty, \infty)$; Range: $(-\infty, 0]$

43.

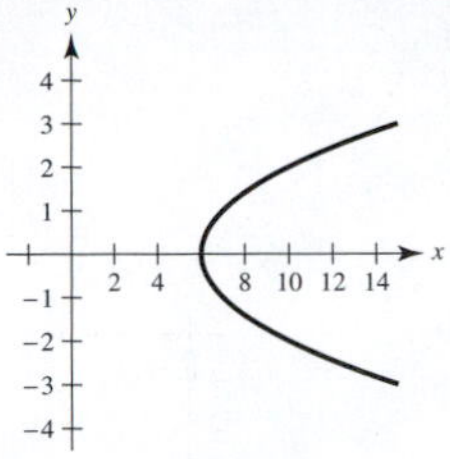

Not a function

45.

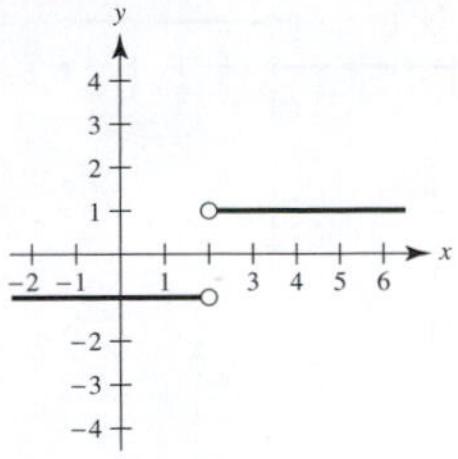

Function

47. $f(x) = x^3 - 3x^2$

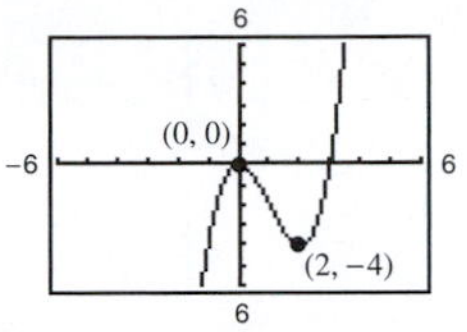

(a) $g(x) = -x^3 + 3x^2 + 1$
(b) $g(x) = (x - 2)^3 - 3(x - 2)^2 + 1$

49. (a)

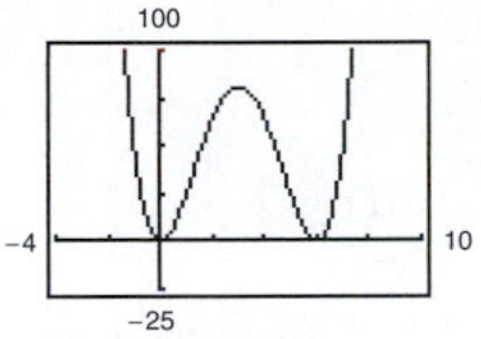

(b)

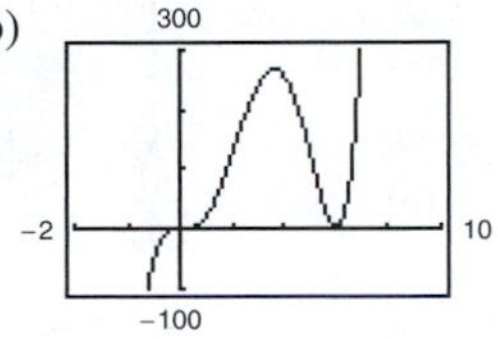

(c)

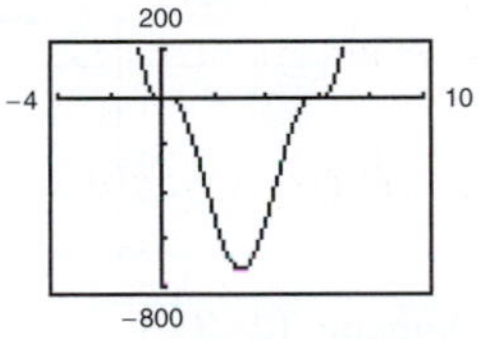

51. (a) $y = -1.204x + 64.2667$

(b)

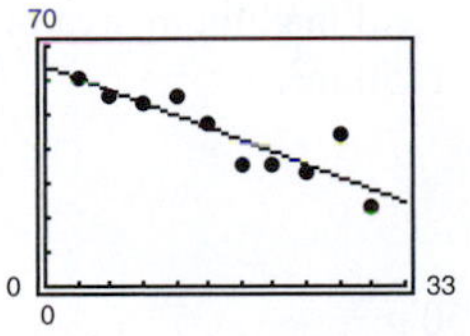

(c) The data point $(27, 44)$ is probably an error. Without this point, the new model is $y = -1.4344x + 66.4387$.

53. (a) Yes. For each time t, there corresponds one and only one displacement y.
(b) Amplitude: 0.25; Period: 1.1 (c) $y \approx \frac{1}{4}\cos(5.7t)$
(d)

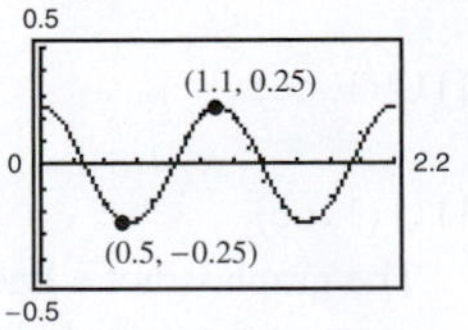

The model appears to fit the data.

P.S. Problem Solving *(page 39)*

1. (a) Center: $(3, 4)$; Radius: 5
(b) $y = -\frac{3}{4}x$ (c) $y = \frac{3}{4}x - \frac{9}{2}$ (d) $\left(3, -\frac{9}{4}\right)$

3.

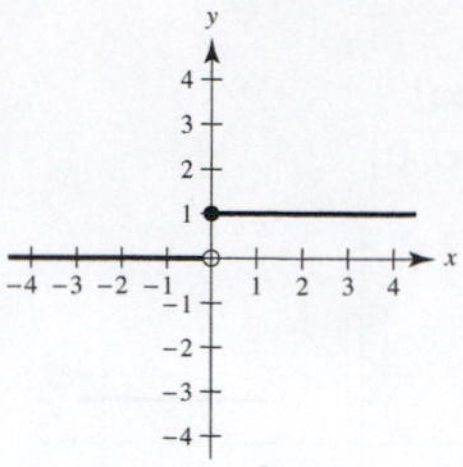

(a)

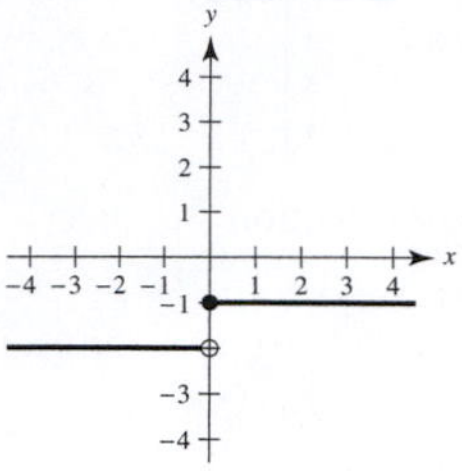

(b)

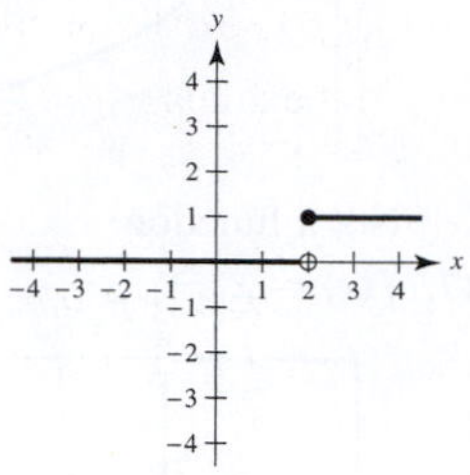

(c)

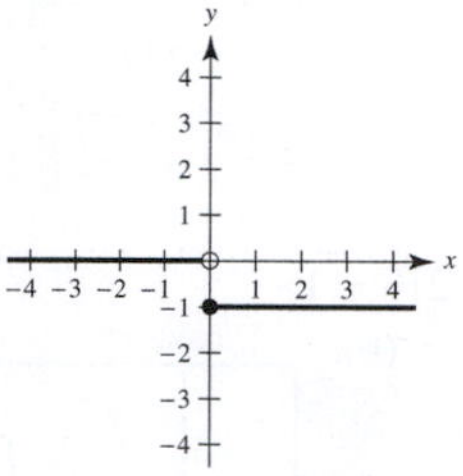

(d)

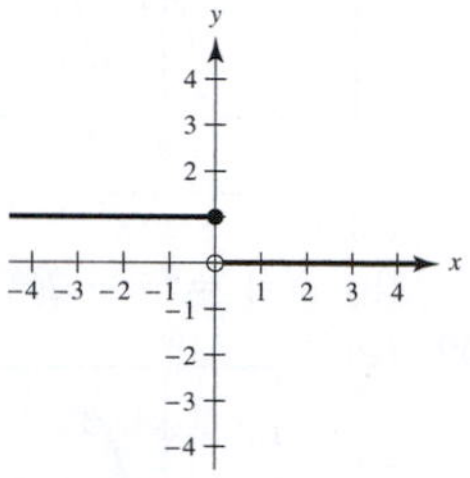

(e)

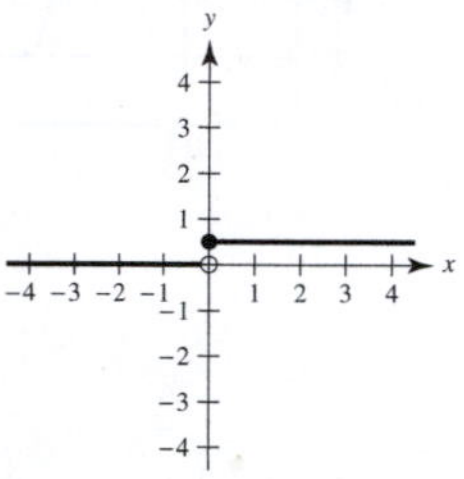

(f)

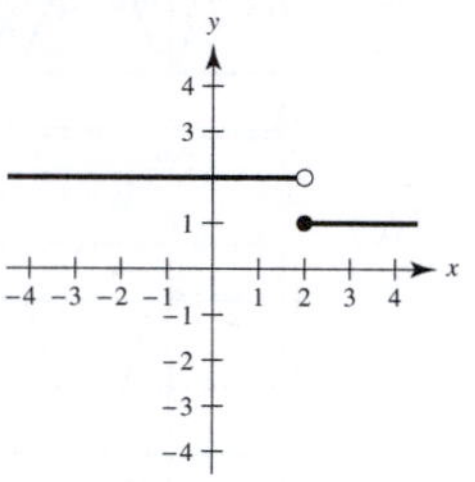

5. (a) $A(x) = x[(100 - x)/2]$; Domain: $(0, 100)$

(b)

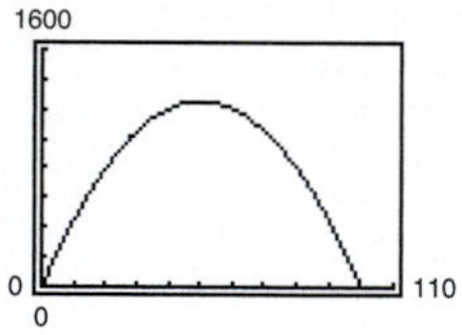

Dimensions 50 m × 25 m yield maximum area of 1250 m^2.

(c) 50 m × 25 m; Area = 1250 m^2

7. $T(x) = \left[2\sqrt{4 + x^2} + \sqrt{(3 - x)^2 + 1}\right]/4$

9. (a) 5, less (b) 3, greater (c) 4.1, less
(d) $4 + h$ (e) 4; Answers will vary.

11. (a) Domain: $(-\infty, 1) \cup (1, \infty)$; Range: $(-\infty, 0) \cup (0, \infty)$

(b) $f(f(x)) = \dfrac{x - 1}{x}$

Domain: $(-\infty, 0) \cup (0, 1) \cup (1, \infty)$

(c) $f(f(f(x))) = x$

Domain: $(-\infty, 0) \cup (0, 1) \cup (1, \infty)$

(d)

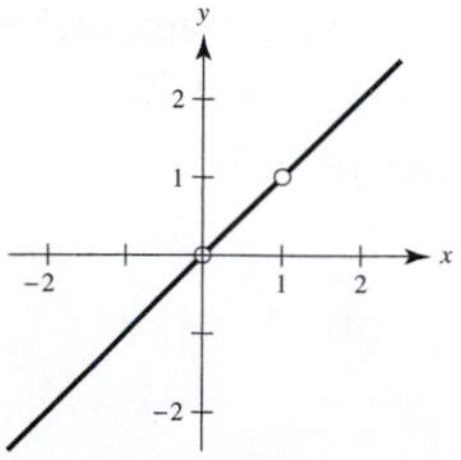

The graph is not a line because there are holes at $x = 0$ and $x = 1$.

13. (a) $x \approx 1.2426, -7.2426$
(b) $(x + 3)^2 + y^2 = 18$

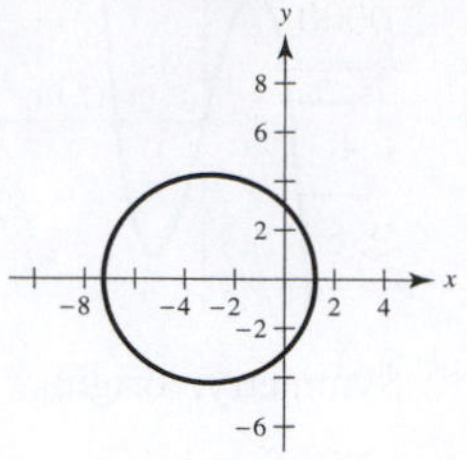

15. Proof

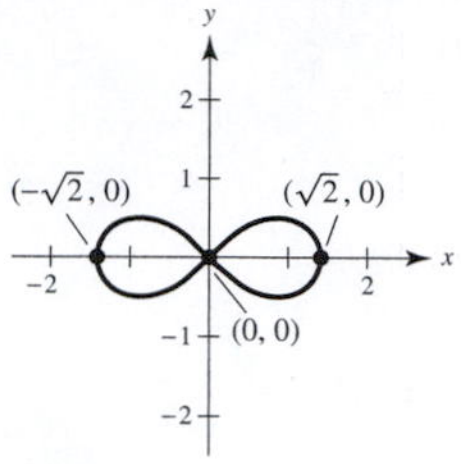

Chapter 1

Section 1.1 *(page 47)*

1. Precalculus: 300 ft

3. Calculus: Slope of the tangent line at $x = 2$ is 0.16.

5. (a) Precalculus: 10 square units
(b) Calculus: 5 square units

7. (a)

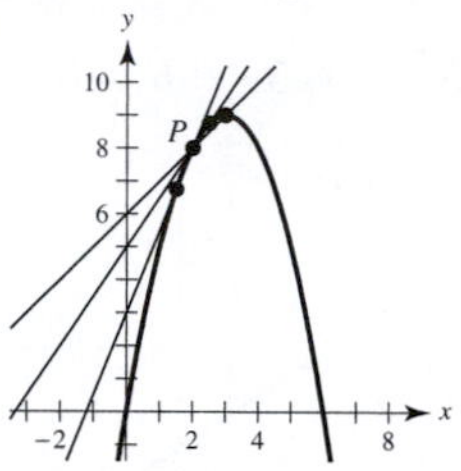

(b) $1; \frac{3}{2}; \frac{5}{2}$
(c) 2. Use points closer to P.

9. Area ≈ 10.417; Area ≈ 9.145; Use more rectangles.

Section 1.2 *(page 55)*

1.

x	3.9	3.99	3.999	4
$f(x)$	0.2041	0.2004	0.2000	?

x	4.001	4.01	4.1
$f(x)$	0.2000	0.1996	0.1961

$\lim_{x \to 4} \dfrac{x - 4}{x^2 - 3x - 4} \approx 0.2000 \left(\text{Actual limit is } \dfrac{1}{5}.\right)$

3.

x	−0.1	−0.01	−0.001	0
$f(x)$	0.5132	0.5013	0.5001	?

x	0.001	0.01	0.1
$f(x)$	0.4999	0.4988	0.4881

$\lim_{x \to 0} \dfrac{\sqrt{x + 1} - 1}{x} \approx 0.5000 \left(\text{Actual limit is } \dfrac{1}{2}.\right)$

5.

x	-0.1	-0.01	-0.001	0
$f(x)$	0.9983	0.99998	1.0000	?

x	0.001	0.01	0.1
$f(x)$	1.0000	0.99998	0.9983

$\lim_{x\to 0} \frac{\sin x}{x} \approx 1.0000$ (Actual limit is 1.)

7.

x	0.9	0.99	0.999	1
$f(x)$	0.2564	0.2506	0.2501	?

x	1.001	1.01	1.1
$f(x)$	0.2499	0.2494	0.2439

$\lim_{x\to 1} \frac{x-2}{x^2+x-6} \approx 0.2500 \left(\text{Actual limit is } \frac{1}{4}.\right)$

9.

x	0.9	0.99	0.999	1
$f(x)$	0.7340	0.6733	0.6673	?

x	1.001	1.01	1.1
$f(x)$	0.6660	0.6600	0.6015

$\lim_{x\to 1} \frac{x^4-1}{x^6-1} \approx 0.6666 \left(\text{Actual limit is } \frac{2}{3}.\right)$

11.

x	-6.1	-6.01	-6.001	-6
$f(x)$	-0.1248	-0.1250	-0.1250	?

x	-5.999	-5.99	-5.9
$f(x)$	-0.1250	-0.1250	-0.1252

$\lim_{x\to -6} \frac{\sqrt{10-x}-4}{x+6} \approx -0.1250 \left(\text{Actual limit is } -\frac{1}{8}.\right)$

13.

x	-0.1	-0.01	-0.001	0
$f(x)$	1.9867	1.9999	2.0000	?

x	0.001	0.01	0.1
$f(x)$	2.0000	1.9999	1.9867

$\lim_{x\to 0} \frac{\sin 2x}{x} \approx 2.0000$ (Actual limit is 2.)

15. 1 **17.** 2

19. Limit does not exist. The function approaches 1 from the right side of 2, but it approaches -1 from the left side of 2.

21. Limit does not exist. The function oscillates between 1 and -1 as x approaches 0.

23. (a) 2
(b) Limit does not exist. The function approaches 1 from the right side of 1, but it approaches 3.5 from the left side of 1.
(c) Value does not exist. The function is undefined at $x = 4$.
(d) 2

25.

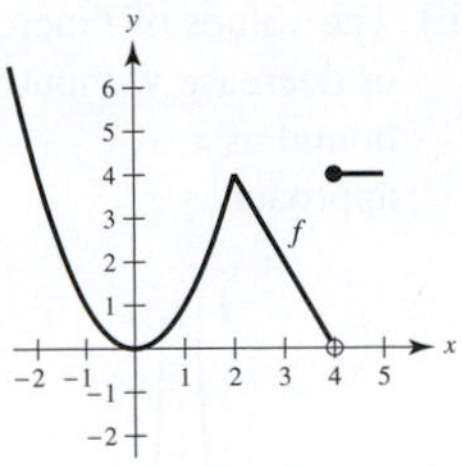

$\lim_{x\to c} f(x)$ exists for all points on the graph except where $c = 4$.

27.

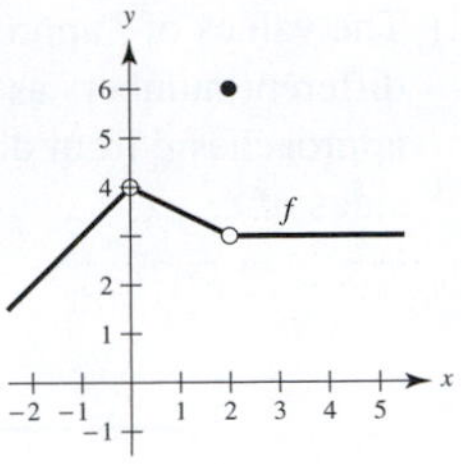

29. $\delta = 0.4$ **31.** $\delta = \frac{1}{11} \approx 0.091$

33. $L = 8$. Let $\delta = 0.01/3 \approx 0.0033$.

35. $L = 1$. Let $\delta = 0.01/5 = 0.002$. **37.** 6 **39.** -3

41. 3 **43.** 0 **45.** 10 **47.** 2 **49.** 4

51.

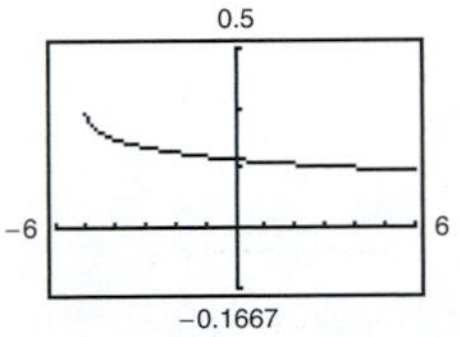

$\lim_{x\to 4} f(x) = \frac{1}{6}$
Domain: $[-5, 4) \cup (4, \infty)$
The graph has a hole at $x = 4$.

53.

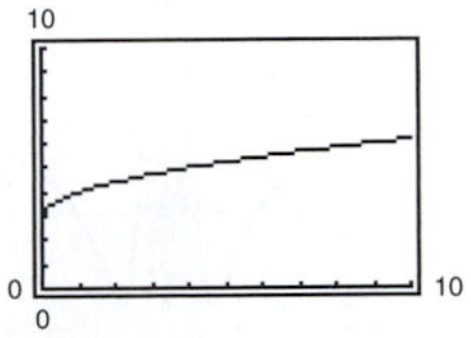

$\lim_{x\to 9} f(x) = 6$
Domain: $[0, 9) \cup (9, \infty)$
The graph has a hole at $x = 9$.

55. (a)

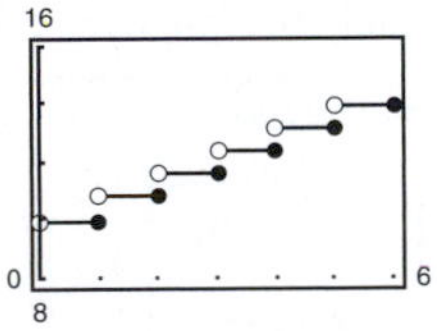

(b)

t	3	3.3	3.4	3.5
C	11.57	12.36	12.36	12.36

t	3.6	3.7	4
C	12.36	12.36	12.36

$\lim_{t\to 3.5} C(t) = 12.36$

(c)

t	2	2.5	2.9	3
C	10.78	11.57	11.57	11.57

t	3.1	3.5	4
C	12.36	12.36	12.36

The limit does not exist because the limits from the right and left are not equal.

57. Answers will vary. Sample answer: As x approaches 8 from either side, $f(x)$ becomes arbitrarily close to 25.

59. (i) The values of f approach different numbers as x approaches c from different sides of c.

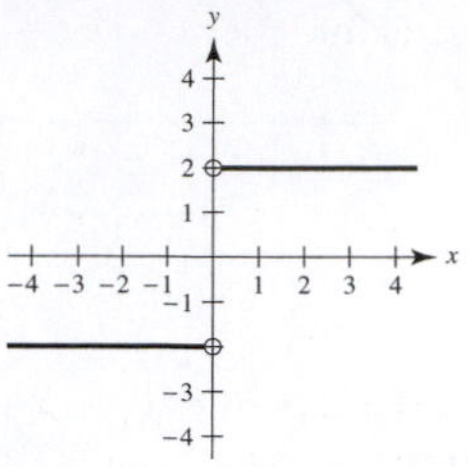

(ii) The values of f increase or decrease without bound as x approaches c.

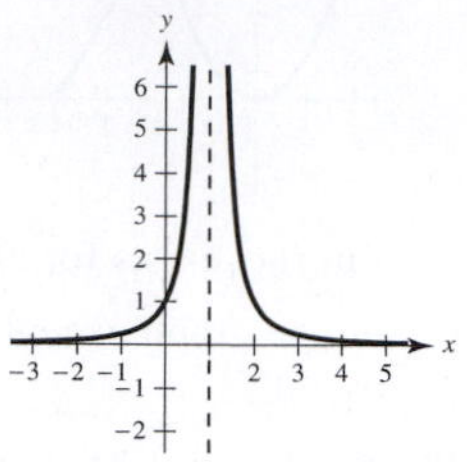

(iii) The values of f oscillate between two fixed numbers as x approaches c.

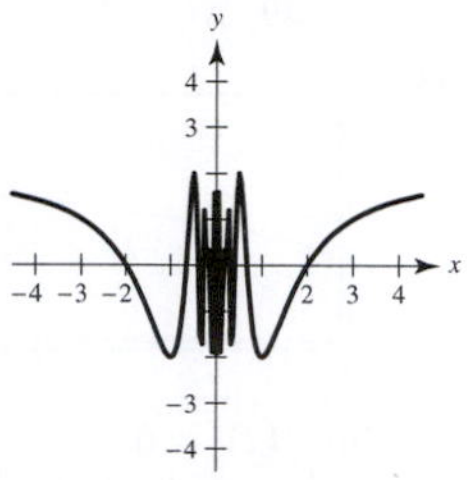

61. (a) $r = \dfrac{3}{\pi} \approx 0.9549$ cm

(b) $\dfrac{5.5}{2\pi} \le r \le \dfrac{6.5}{2\pi}$, or approximately $0.8754 < r < 1.0345$

(c) $\lim\limits_{r \to 3/\pi} 2\pi r = 6$; $\varepsilon = 0.5$; $\delta \approx 0.0796$

63.

x	-0.001	-0.0001	-0.00001
$f(x)$	2.7196	2.7184	2.7183

x	0.00001	0.0001	0.001
$f(x)$	2.7183	2.7181	2.7169

$\lim\limits_{x \to 0} f(x) \approx 2.7183$

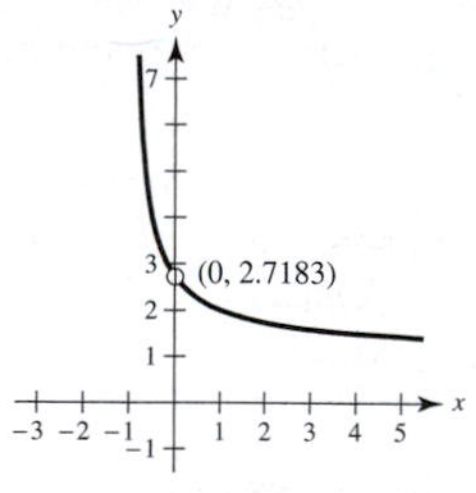

65.

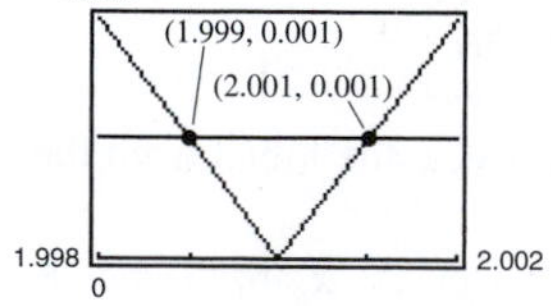

$\delta = 0.001$, $(1.999, 2.001)$

67. False. The existence or nonexistence of $f(x)$ at $x = c$ has no bearing on the existence of the limit of $f(x)$ as $x \to c$.

69. False. See Exercise 17.

71. Yes. As x approaches 0.25 from either side, $\sqrt{x}$ becomes arbitrarily close to 0.5.

73. $\lim\limits_{x \to 0} \dfrac{\sin nx}{x} = n$ **75–77.** Proofs

79. Putnam Problem B1, 1986

Section 1.3 *(page 67)*

1.

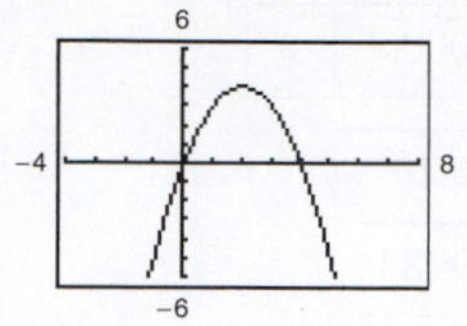

(a) 0 (b) -5

3.

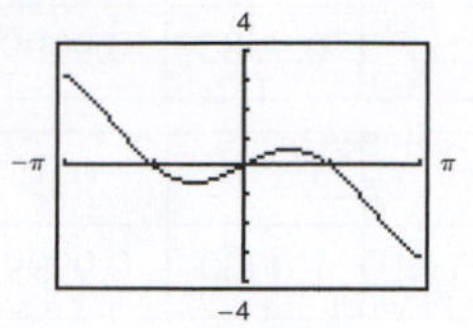

(a) 0 (b) About 0.52 or $\pi/6$

5. 8 **7.** -1 **9.** 0 **11.** 7 **13.** 2 **15.** 1
17. 1/2 **19.** 1/5 **21.** 7 **23.** (a) 4 (b) 64 (c) 64
25. (a) 3 (b) 2 (c) 2 **27.** 1 **29.** 1/2 **31.** 1
33. 1/2 **35.** -1 **37.** (a) 10 (b) 5 (c) 6 (d) 3/2
39. (a) 64 (b) 2 (c) 12 (d) 8

41. $f(x) = \dfrac{x^2 + 3x}{x}$ and $g(x) = x + 3$ agree except at $x = 0$.
$\lim\limits_{x \to 0} f(x) = \lim\limits_{x \to 0} g(x) = 3$

43. $f(x) = \dfrac{x^2 - 1}{x + 1}$ and $g(x) = x - 1$ agree except at $x = -1$.
$\lim\limits_{x \to -1} f(x) = \lim\limits_{x \to -1} g(x) = -2$

45. $f(x) = \dfrac{x^3 - 8}{x - 2}$ and $g(x) = x^2 + 2x + 4$ agree except at $x = 2$.
$\lim\limits_{x \to 2} f(x) = \lim\limits_{x \to 2} g(x) = 12$

47. -1 **49.** 1/8 **51.** 5/6 **53.** 1/6 **55.** $\sqrt{5}/10$
57. $-1/9$ **59.** 2 **61.** $2x - 2$ **63.** 1/5 **65.** 0
67. 0 **69.** 0 **71.** 1 **73.** 3/2

75.

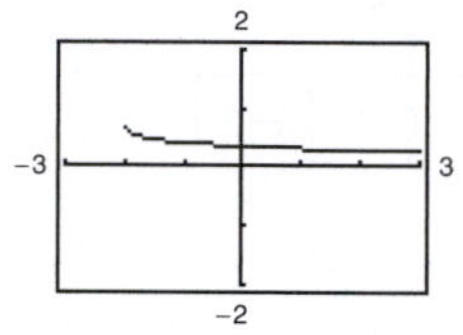

The graph has a hole at $x = 0$.

Answers will vary. Sample answer:

x	-0.1	-0.01	-0.001	0.001	0.01	0.1
$f(x)$	0.358	0.354	0.354	0.354	0.353	0.349

$\lim\limits_{x \to 0} \dfrac{\sqrt{x + 2} - \sqrt{2}}{x} \approx 0.354$; Actual limit is $\dfrac{1}{2\sqrt{2}} = \dfrac{\sqrt{2}}{4}$.

77.

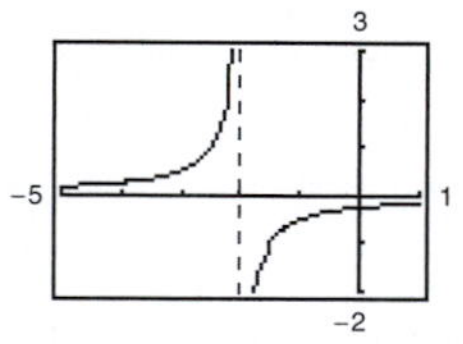

The graph has a hole at $x = 0$.

Answers will vary. Sample answer:

x	-0.1	-0.01	-0.001
$f(x)$	-0.263	-0.251	-0.250

x	0.001	0.01	0.1
$f(x)$	-0.250	-0.249	-0.238

$\lim\limits_{x \to 0} \dfrac{[1/(2 + x)] - (1/2)}{x} \approx -0.250$; Actual limit is $-\dfrac{1}{4}$.

79. 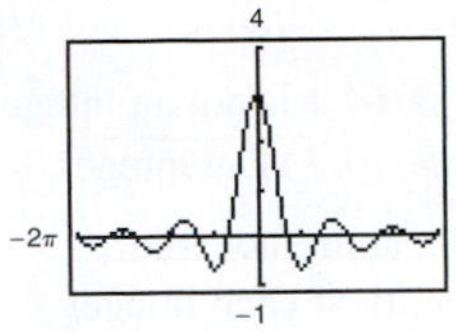

The graph has a hole at $t = 0$.

Answers will vary. Sample answer:

t	-0.1	-0.01	0	0.01	0.1
$f(t)$	2.96	2.9996	?	2.9996	2.96

$\lim_{t\to 0} \frac{\sin 3t}{t} \approx 3.0000$; Actual limit is 3.

81. 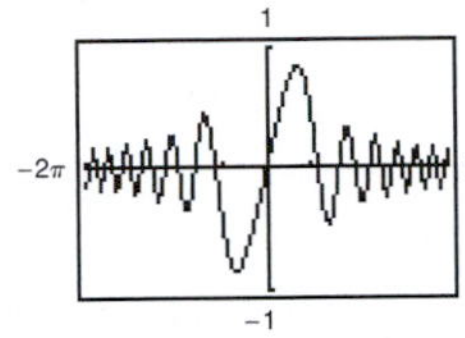

The graph has a hole at $x = 0$.

Answers will vary. Sample answer:

x	-0.1	-0.01	-0.001	0	0.001	0.01	0.1
$f(x)$	-0.1	-0.01	-0.001	?	0.001	0.01	0.1

$\lim_{x\to 0} \frac{\sin x^2}{x} = 0$; Actual limit is 0.

83. 3 **85.** $2x - 4$ **87.** $-1/(x + 3)^2$ **89.** 4

91. 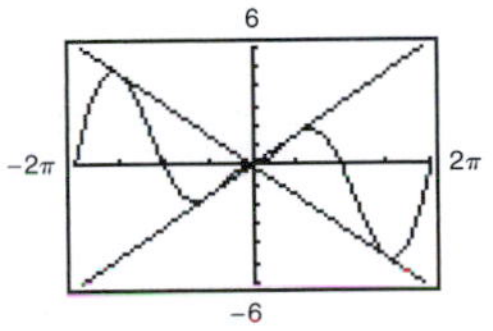

0

93. 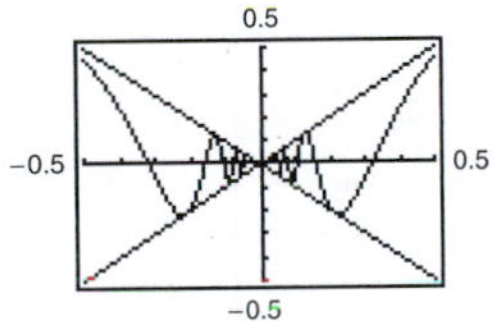

0

The graph has a hole at $x = 0$.

95. (a) f and g agree at all but one point if c is a real number such that $f(x) = g(x)$ for all $x \neq c$.

(b) Sample answer: $f(x) = \frac{x^2 - 1}{x - 1}$ and $g(x) = x + 1$ agree at all points except $x = 1$.

97. If a function f is squeezed between two functions h and g, $h(x) \le f(x) \le g(x)$, and h and g have the same limit L as $x \to c$, then $\lim_{x\to c} f(x)$ exists and equals L.

99. 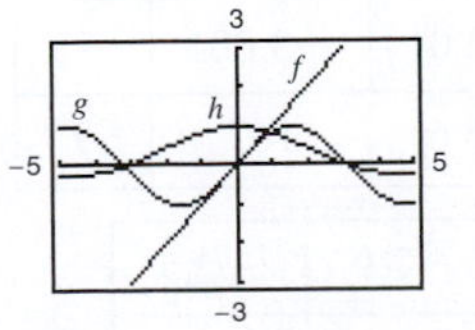

The magnitudes of $f(x)$ and $g(x)$ are approximately equal when x is close to 0. Therefore, their ratio is approximately 1.

101. -64 ft/sec (speed $= 64$ ft/sec) **103.** -29.4 m/sec

105. Let $f(x) = 1/x$ and $g(x) = -1/x$.

$\lim_{x\to 0} f(x)$ and $\lim_{x\to 0} g(x)$ do not exist. However,

$$\lim_{x\to 0} [f(x) + g(x)] = \lim_{x\to 0} \left[\frac{1}{x} + \left(-\frac{1}{x}\right)\right] = \lim_{x\to 0} 0 = 0$$

and therefore does exist.

107–111. Proofs

113. Let $f(x) = \begin{cases} 4, & x \ge 0 \\ -4, & x < 0 \end{cases}$.

$\lim_{x\to 0} |f(x)| = \lim_{x\to 0} 4 = 4$

$\lim_{x\to 0} f(x)$ does not exist because for $x < 0$, $f(x) = -4$ and for $x \ge 0$, $f(x) = 4$.

115. False. The limit does not exist because the function approaches 1 from the right side of 0 and approaches -1 from the left side of 0.

117. True.

119. False. The limit does not exist because $f(x)$ approaches 3 from the left side of 2 and approaches 0 from the right side of 2.

121. Proof

123. (a) All $x \neq 0, \frac{\pi}{2} + n\pi$

(b) 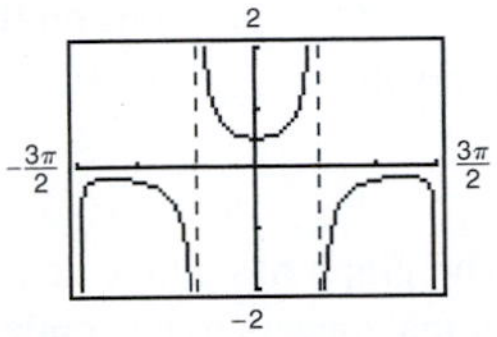

The domain is not obvious. The hole at $x = 0$ is not apparent from the graph.

(c) $\frac{1}{2}$ (d) $\frac{1}{2}$

Section 1.4 *(page 79)*

1. (a) 3 (b) 3 (c) 3; $f(x)$ is continuous on $(-\infty, \infty)$.

3. (a) 0 (b) 0 (c) 0; Discontinuity at $x = 3$

5. (a) -3 (b) 3 (c) Limit does not exist.
Discontinuity at $x = 2$

7. $\frac{1}{16}$ **9.** $\frac{1}{10}$

11. Limit does not exist. The function decreases without bound as x approaches -3 from the left.

13. -1 **15.** $-1/x^2$ **17.** $5/2$ **19.** 2

21. Limit does not exist. The function decreases without bound as x approaches π from the left and increases without bound as x approaches π from the right.

23. 8

25. Limit does not exist. The function approaches 5 from the left side of 3 but approaches 6 from the right side of 3.

27. Discontinuities at $x = -2$ and $x = 2$

29. Discontinuities at every integer

31. Continuous on $[-7, 7]$ **33.** Continuous on $[-1, 4]$

35. Nonremovable discontinuity at $x = 0$

37. Continuous for all real x

39. Nonremovable discontinuities at $x = -2$ and $x = 2$

41. Continuous for all real x

43. Nonremovable discontinuity at $x = 1$
Removable discontinuity at $x = 0$

45. Continuous for all real x

47. Removable discontinuity at $x = -2$
Nonremovable discontinuity at $x = 5$

49. Nonremovable discontinuity at $x = -7$

51. Continuous for all real x

53. Nonremovable discontinuity at $x = 2$

55. Continuous for all real x

57. Nonremovable discontinuities at integer multiples of $\pi/2$

59. Nonremovable discontinuities at each integer

61. $a = 7$ **63.** $a = 2$ **65.** $a = -1,\ b = 1$

67. Continuous for all real x

69. Nonremovable discontinuities at $x = 1$ and $x = -1$

71. Continuous on the open intervals $\ldots, (-3\pi, -\pi), (-\pi, \pi), (\pi, 3\pi), \ldots$

73.

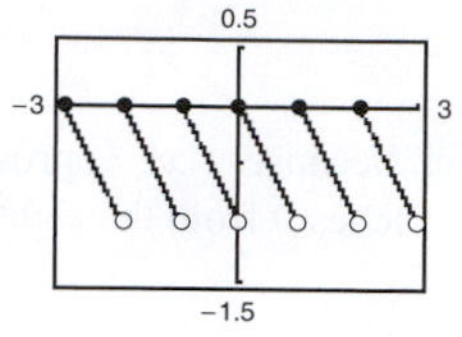

Nonremovable discontinuity at each integer

75.

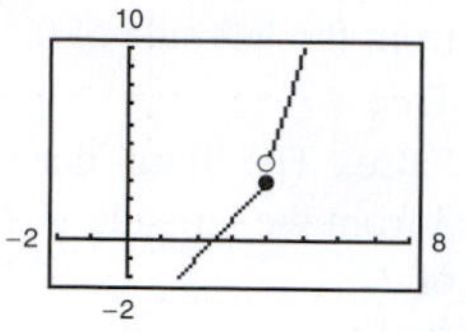

Nonremovable discontinuity at $x = 4$

77. Continuous on $(-\infty, \infty)$ **79.** Continuous on $[0, \infty)$

81. Continuous on the open intervals $\ldots, (-6, -2), (-2, 2), (2, 6), \ldots$

83. Continuous on $(-\infty, \infty)$

85.

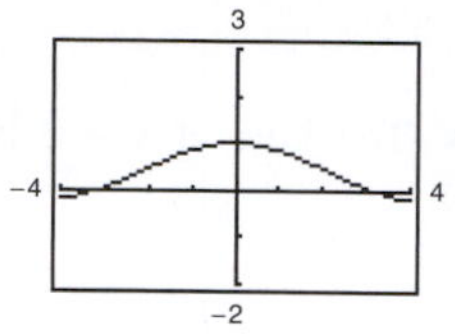

The graph has a hole at $x = 0$. The graph appears to be continuous, but the function is not continuous on $[-4, 4]$. It is not obvious from the graph that the function has a discontinuity at $x = 0$.

87. Because $f(x)$ is continuous on the interval $[1, 2]$ and $f(1) = 37/12$ and $f(2) = -8/3$, by the Intermediate Value Theorem there exists a real number c in $[1, 2]$ such that $f(c) = 0$.

89. Because $f(x)$ is continuous on the interval $[0, \pi]$ and $f(0) = -3$ and $f(\pi) \approx 8.87$, by the Intermediate Value Theorem there exists a real number c in $[0, \pi]$ such that $f(c) = 0$.

91. 0.68, 0.6823 **93.** 0.56, 0.5636

95. $f(3) = 11$ **97.** $f(2) = 4$

99. (a) The limit does not exist at $x = c$.
(b) The function is not defined at $x = c$.
(c) The limit exists, but it is not equal to the value of the function at $x = c$.
(d) The limit does not exist at $x = c$.

101. If f and g are continuous for all real x, then so is $f + g$ (Theorem 1.11, part 2). However, f/g might not be continuous if $g(x) = 0$. For example, let $f(x) = x$ and $g(x) = x^2 - 1$. Then f and g are continuous for all real x, but f/g is not continuous at $x = \pm 1$.

103. True

105. False. A rational function can be written as $P(x)/Q(x)$, where P and Q are polynomials of degree m and n, respectively. It can have, at most, n discontinuities.

107. The functions differ by 1 for non-integer values of x.

109. $C = \begin{cases} 0.40, & 0 < t \le 10 \\ 0.40 + 0.05[\![t - 9]\!], & t > 10,\ t \text{ is not an integer} \\ 0.40 + 0.05(t - 10), & t > 10,\ t \text{ is an integer} \end{cases}$

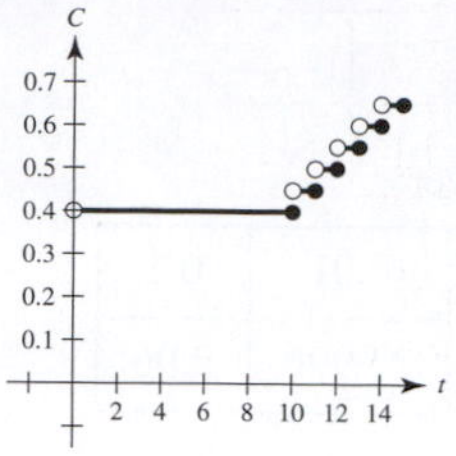

There is a nonremovable discontinuity at each integer greater than or equal to 10.

111–113. Proofs **115.** Answers will vary.

117. (a)

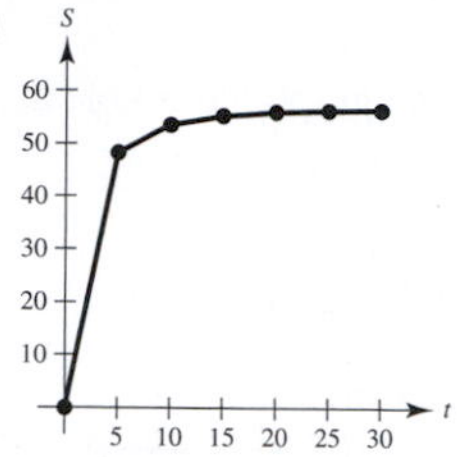

(b) There appears to be a limiting speed, and a possible cause is air resistance.

119. $c = (-1 \pm \sqrt{5})/2$

121. Domain: $[-c^2, 0) \cup (0, \infty)$; Let $f(0) = 1/(2c)$

123. $h(x)$ has a nonremovable discontinuity at every integer except 0.

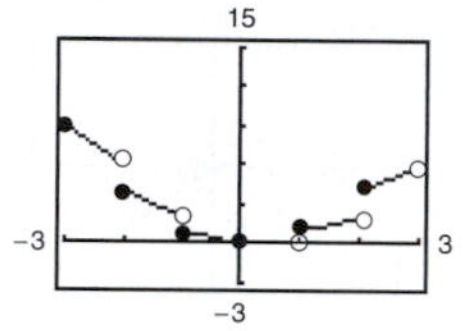

125. Putnam Problem B2, 1988

Section 1.5 *(page 88)*

1. $\displaystyle\lim_{x\to-2^+} 2\left|\frac{x}{x^2-4}\right| = \infty, \quad \lim_{x\to-2^-} 2\left|\frac{x}{x^2-4}\right| = \infty$

3. $\displaystyle\lim_{x\to-2^+} \tan(\pi x/4) = -\infty, \quad \lim_{x\to-2^-} \tan(\pi x/4) = \infty$

5. $\displaystyle\lim_{x\to4^+} \frac{1}{x-4} = \infty, \quad \lim_{x\to4^-} \frac{1}{x-4} = -\infty$

7. $\displaystyle\lim_{x\to4^+} \frac{1}{(x-4)^2} = \infty, \quad \lim_{x\to4^-} \frac{1}{(x-4)^2} = \infty$

9.

x	−3.5	−3.1	−3.01	−3.001	−3
$f(x)$	0.31	1.64	16.6	167	?

x	−2.999	−2.99	−2.9	−2.5
$f(x)$	−167	−16.7	−1.69	−0.36

$\displaystyle\lim_{x\to-3^+} f(x) = -\infty; \quad \lim_{x\to-3^-} f(x) = \infty$

11.

x	-3.5	-3.1	-3.01	-3.001	-3
$f(x)$	3.8	16	151	1501	?

x	-2.999	-2.99	-2.9	-2.5
$f(x)$	-1499	-149	-14	-2.3

$\lim_{x\to -3^+} f(x) = -\infty$; $\lim_{x\to -3^-} f(x) = \infty$

13. $x = 0$ **15.** $x = \pm 2$ **17.** No vertical asymptote

19. $x = -2, x = 1$ **21.** $x = 0, x = 3$

23. No vertical asymptote **25.** $x = n$, n is an integer.

27. $t = n\pi$, n is a nonzero integer.

29. Removable discontinuity at $x = -1$

31. Vertical asymptote at $x = -1$

33. ∞ **35.** ∞ **37.** $-\frac{1}{5}$ **39.** $-\infty$ **41.** $-\infty$

43. ∞ **45.** 0 **47.** ∞

49.

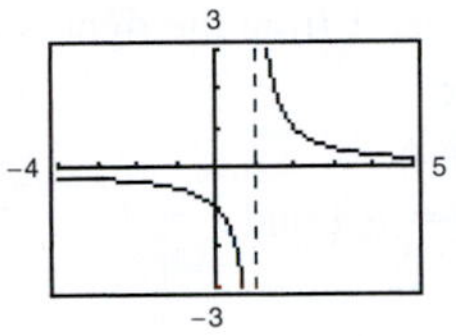

$\lim_{x\to 1^+} f(x) = \infty$

51.

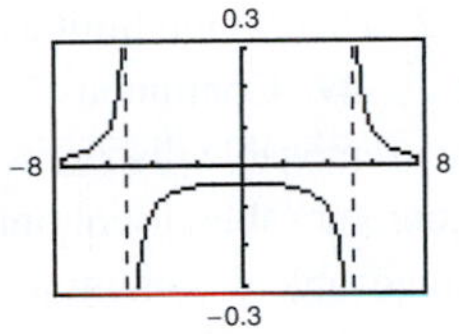

$\lim_{x\to 5^-} f(x) = -\infty$

53. Answers will vary.

55. Answers will vary. Sample answer: $f(x) = \dfrac{x-3}{x^2 - 4x - 12}$

57.

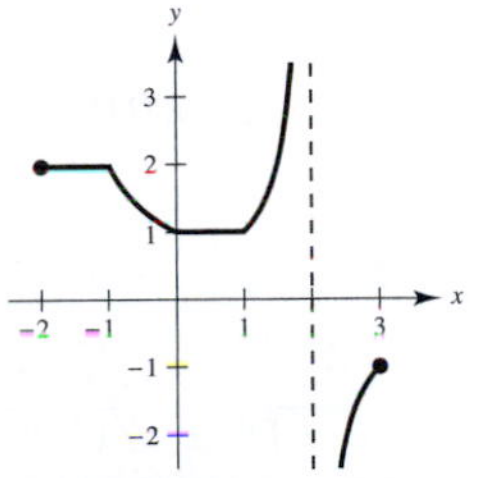

59. (a)

x	1	0.5	0.2	0.1
$f(x)$	0.1585	0.0411	0.0067	0.0017

x	0.01	0.001	0.0001
$f(x)$	≈ 0	≈ 0	≈ 0

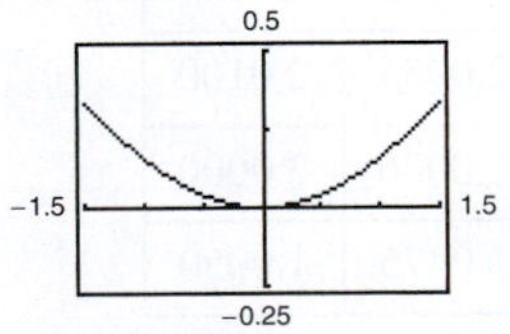

$\lim_{x\to 0^+} \dfrac{x - \sin x}{x} = 0$

(b)

x	1	0.5	0.2	0.1
$f(x)$	0.1585	0.0823	0.0333	0.0167

x	0.01	0.001	0.0001
$f(x)$	0.0017	≈ 0	≈ 0

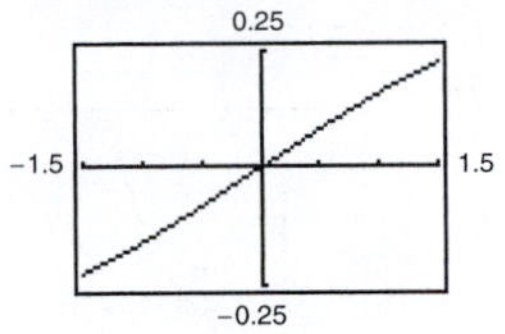

$\lim_{x\to 0^+} \dfrac{x - \sin x}{x^2} = 0$

(c)

x	1	0.5	0.2	0.1
$f(x)$	0.1585	0.1646	0.1663	0.1666

x	0.01	0.001	0.0001
$f(x)$	0.1667	0.1667	0.1667

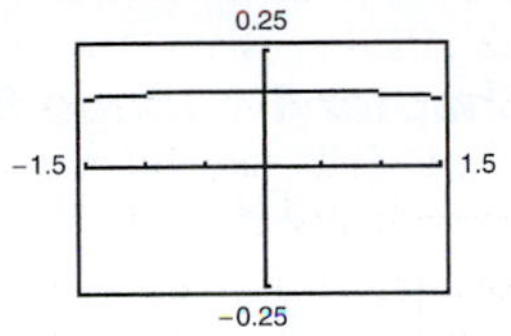

$\lim_{x\to 0^+} \dfrac{x - \sin x}{x^3} = 0.1667\ (1/6)$

(d)

x	1	0.5	0.2	0.1
$f(x)$	0.1585	0.3292	0.8317	1.6658

x	0.01	0.001	0.0001
$f(x)$	16.67	166.7	1667.0

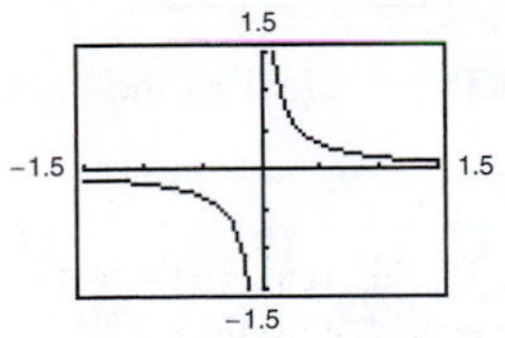

$\lim_{x\to 0^+} \dfrac{x - \sin x}{x^4} = \infty$

For $n > 3$, $\lim_{x\to 0^+} \dfrac{x - \sin x}{x^n} = \infty$.

61. (a) $\dfrac{7}{12}$ ft/sec (b) $\dfrac{3}{2}$ ft/sec

(c) $\lim_{x\to 25^-} \dfrac{2x}{\sqrt{625 - x^2}} = \infty$

63. (a) $A = 50 \tan \theta - 50\theta$; Domain: $(0, \pi/2)$

(b)

θ	0.3	0.6	0.9	1.2	1.5
$f(\theta)$	0.47	4.21	18.0	68.6	630.1

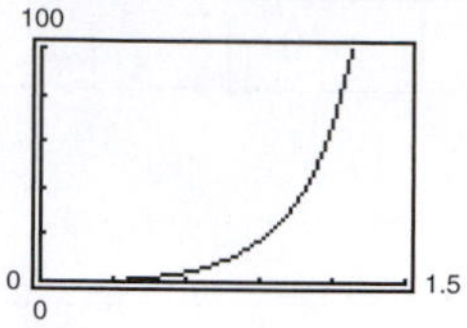

(c) $\lim_{\theta \to \pi/2} A = \infty$

65. False; let $f(x) = (x^2 - 1)/(x - 1)$

67. False; let $f(x) = \tan x$

69. Let $f(x) = \dfrac{1}{x^2}$ and $g(x) = \dfrac{1}{x^4}$, and let $c = 0$. $\lim_{x \to 0} \dfrac{1}{x^2} = \infty$ and $\lim_{x \to 0} \dfrac{1}{x^4} = \infty$, but $\lim_{x \to 0} \left(\dfrac{1}{x^2} - \dfrac{1}{x^4}\right) = \lim_{x \to 0} \left(\dfrac{x^2 - 1}{x^4}\right) = -\infty \neq 0$.

71. Given $\lim_{x \to c} f(x) = \infty$, let $g(x) = 1$. Then $\lim_{x \to c} \dfrac{g(x)}{f(x)} = 0$ by Theorem 1.15.

73. Answers will vary.

Review Exercises for Chapter 1 *(page 91)*

1. Calculus

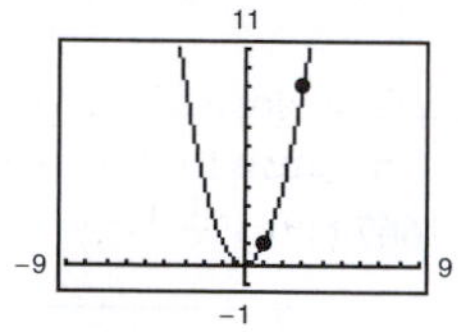

Estimate: 8.3

3.

x	2.9	2.99	2.999	3
$f(x)$	-0.9091	-0.9901	-0.9990	?

x	3.001	3.01	3.1
$f(x)$	-1.0010	-1.0101	-1.1111

$\lim_{x \to 0} \dfrac{x - 3}{x^2 - 7x + 12} \approx -1.0000$

5. (a) 4 (b) 5 **7.** 5; Proof **9.** -3; Proof **11.** 36

13. $\sqrt{6} \approx 2.45$ **15.** 16 **17.** $\frac{4}{3}$ **19.** $-\frac{1}{4}$ **21.** $\frac{1}{2}$

23. -1 **25.** 0 **27.** $\sqrt{3}/2$ **29.** -3 **31.** -5

33.

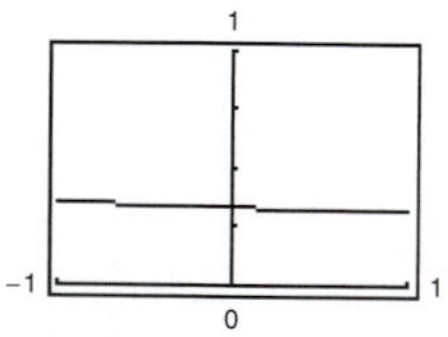

The graph has a hole at $x = 0$.

x	-0.1	-0.01	-0.001	0
$f(x)$	0.3352	0.3335	0.3334	?

x	0.001	0.01	0.1
$f(x)$	0.3333	0.3331	0.3315

$\lim_{x \to 0} \dfrac{\sqrt{2x + 9} - 3}{x} \approx 0.3333$; Actual limit is $\dfrac{1}{3}$.

35.

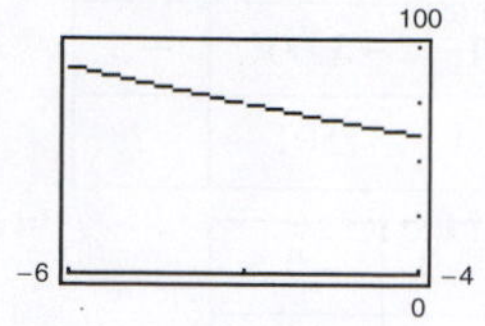

The graph has a hole at $x = -5$.

x	-5.1	-5.01	-5.001	-5
$f(x)$	76.51	75.15	75.02	?

x	-4.999	-4.99	-4.9
$f(x)$	74.99	74.85	73.51

$\lim_{x \to -5} \dfrac{x^3 + 125}{x + 5} \approx 75.00$; Actual limit is 75.

37. -39.2 m/sec **39.** $\frac{1}{6}$ **41.** $\frac{1}{4}$ **43.** 0

45. Limit does not exist. The limit as t approaches 1 from the left is 2, whereas the limit as t approaches 1 from the right is 1.

47. 3 **49.** Continuous for all real x

51. Nonremovable discontinuity at $x = 5$

53. Nonremovable discontinuities at $x = -1$ and $x = 1$
Removable discontinuity at $x = 0$

55. $c = -\frac{1}{2}$ **57.** Continuous for all real x

59. Continuous on $[4, \infty)$

61. Removable discontinuity at $x = 1$
Continuous on $(-\infty, 1) \cup (1, \infty)$

63. Proof **65.** (a) -4 (b) 4 (c) Limit does not exist.

67. $x = 0$ **69.** $x = \pm 3$ **71.** $x = \pm 8$ **73.** $-\infty$

75. $\frac{1}{3}$ **77.** $-\infty$ **79.** $\frac{4}{5}$ **81.** ∞

83. (a) \$14,117.65 (b) \$80,000.00 (c) \$720,000.00
(d) ∞

P.S. Problem Solving *(page 93)*

1. (a) Perimeter $\triangle PAO = 1 + \sqrt{(x^2 - 1)^2 + x^2} + \sqrt{x^4 + x^2}$
Perimeter $\triangle PBO = 1 + \sqrt{x^4 + (x - 1)^2} + \sqrt{x^4 + x^2}$

(b)

x	4	2	1
Perimeter $\triangle PAO$	33.0166	9.0777	3.4142
Perimeter $\triangle PBO$	33.7712	9.5952	3.4142
$r(x)$	0.9777	0.9461	1.0000

x	0.1	0.01
Perimeter $\triangle PAO$	2.0955	2.0100
Perimeter $\triangle PBO$	2.0006	2.0000
$r(x)$	1.0475	1.0050

1

3. (a) Area (hexagon) $= (3\sqrt{3})/2 \approx 2.5981$
Area (circle) $= \pi \approx 3.1416$
Area (circle) $-$ Area (hexagon) ≈ 0.5435

(b) $A_n = (n/2) \sin(2\pi/n)$

(c)

n	6	12	24	48	96
A_n	2.5981	3.0000	3.1058	3.1326	3.1394

3.1416 or π

5. (a) $m = -\frac{12}{5}$ (b) $y = \frac{5}{12}x - \frac{169}{12}$

(c) $m_x = \dfrac{-\sqrt{169 - x^2} + 12}{x - 5}$

(d) $\frac{5}{12}$; It is the same as the slope of the tangent line found in (b).

7. (a) Domain: $[-27, 1) \cup (1, \infty)$

(b)

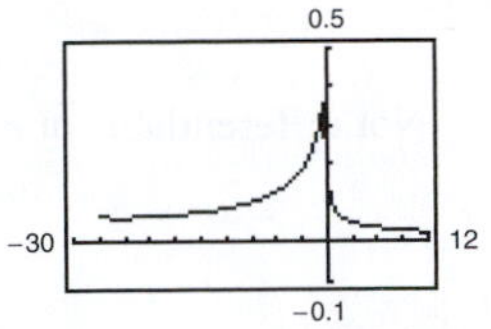

(c) $\frac{1}{14}$ (d) $\frac{1}{12}$

The graph has a hole at $x = 1$.

9. (a) g_1, g_4 (b) g_1 (c) g_1, g_3, g_4

11.

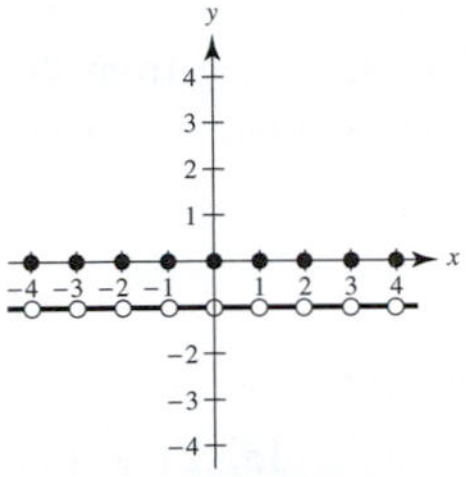

The graph jumps at every integer.

(a) $f(1) = 0,\ f(0) = 0,\ f(\frac{1}{2}) = -1,\ f(-2.7) = -1$

(b) $\lim_{x\to 1^-} f(x) = -1,\ \lim_{x\to 1^+} f(x) = -1,\ \lim_{x\to 1/2} f(x) = -1$

(c) There is a discontinuity at each integer.

13. (a)

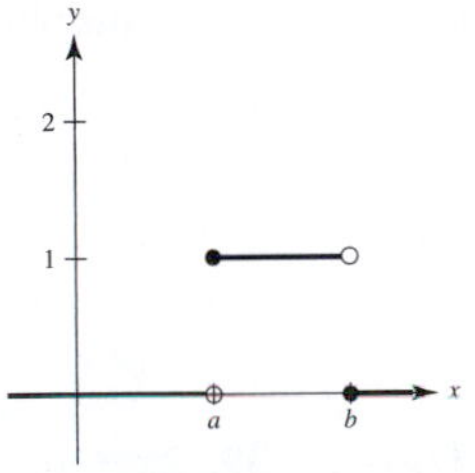

(b) (i) $\lim_{x\to a^+} P_{a,b}(x) = 1$

(ii) $\lim_{x\to a^-} P_{a,b}(x) = 0$

(iii) $\lim_{x\to b^+} P_{a,b}(x) = 0$

(iv) $\lim_{x\to b^-} P_{a,b}(x) = 1$

(c) Continuous for all positive real numbers except a and b

(d) The area under the graph of U and above the x-axis is 1.

Chapter 2

Section 2.1 *(page 103)*

1. $m_1 = 0, m_2 = 5/2$

3. (a)–(c)

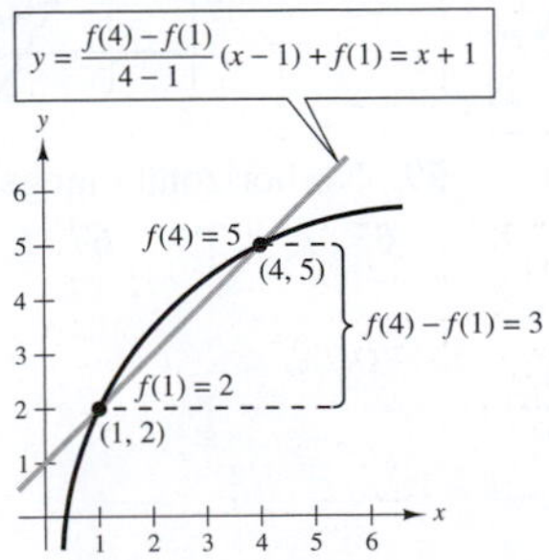

5. $m = -5$

7. $m = 4$

9. $m = 3$ **11.** $f'(x) = 0$ **13.** $f'(x) = -10$

15. $h'(s) = \frac{2}{3}$ **17.** $f'(x) = 2x + 1$ **19.** $f'(x) = 3x^2 - 12$

21. $f'(x) = \dfrac{-1}{(x-1)^2}$ **23.** $f'(x) = \dfrac{1}{2\sqrt{x+4}}$

25. (a) Tangent line: $y = -2x + 2$

(b)

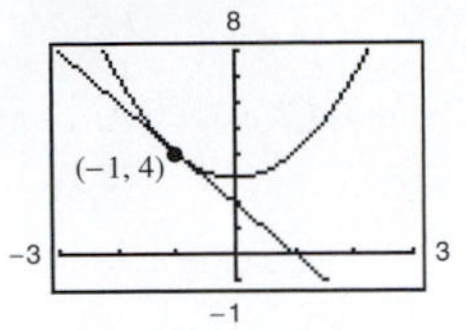

27. (a) Tangent line: $y = 12x - 16$

(b)

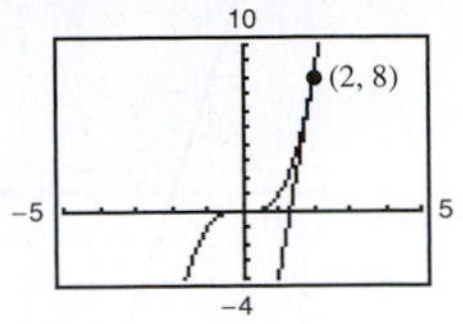

29. (a) Tangent line: $y = \frac{1}{2}x + \frac{1}{2}$

(b)

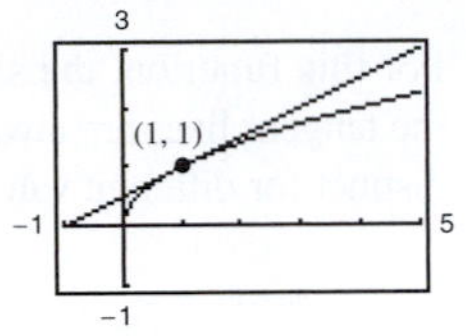

31. (a) Tangent line: $y = \frac{3}{4}x - 2$

(b)

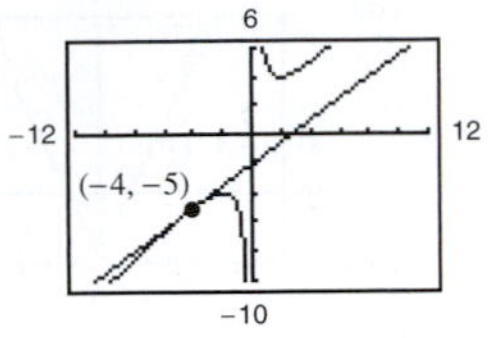

33. $y = 2x - 1$ **35.** $y = 3x - 2;\ y = 3x + 2$

37. $y = -\frac{1}{2}x + \frac{3}{2}$

39.

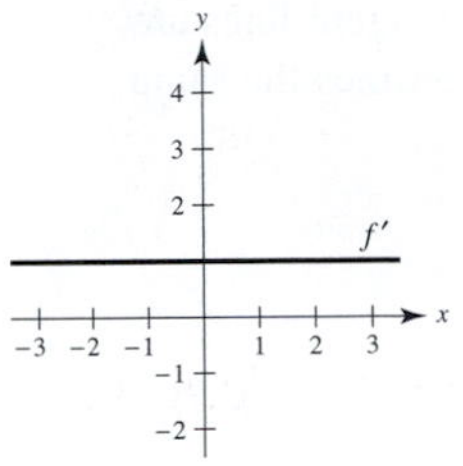

The slope of the graph of f is 1 for all x-values.

41.

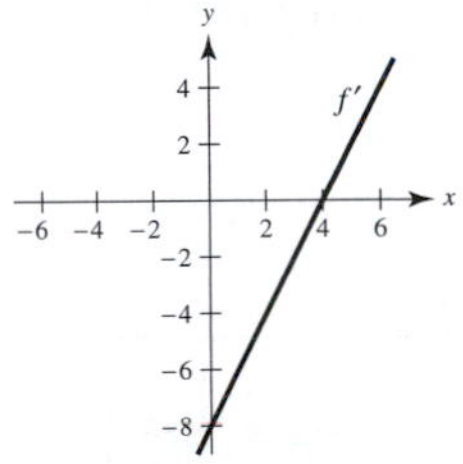

The slope of the graph of f is negative for $x < 4$, positive for $x > 4$, and 0 at $x = 4$.

43.

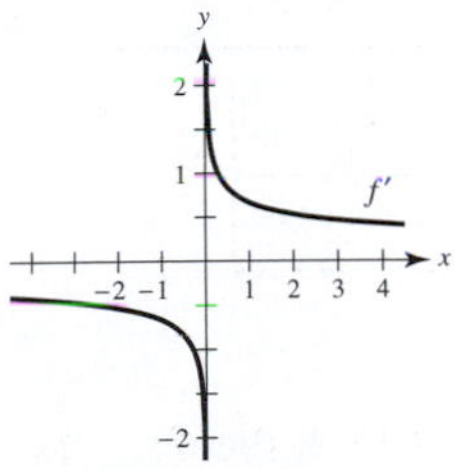

The slope of the graph of f is negative for $x < 0$ and positive for $x > 0$. The slope is undefined at $x = 0$.

45. Answers will vary.
Sample answer: $y = -x$

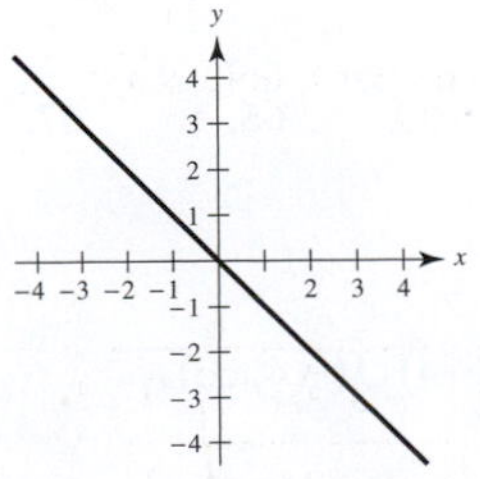

47. $g(4) = 5;\ g'(4) = -\frac{5}{3}$

49. $f(x) = 5 - 3x$
$c = 1$

51. $f(x) = -x^2$
$c = 6$

53. $f(x) = -3x + 2$

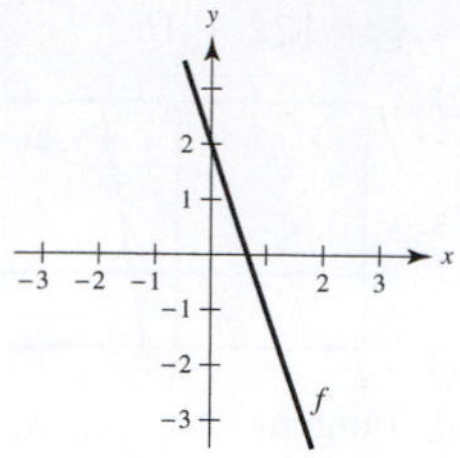

55. $y = 2x + 1;\ y = -2x + 9$

57. (a)

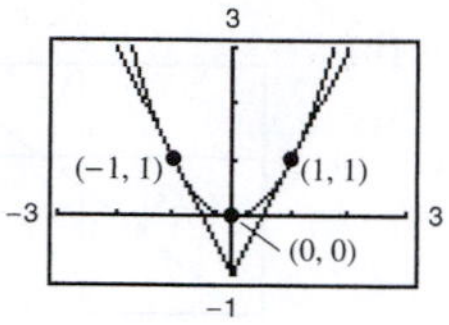

For this function, the slopes of the tangent lines are always distinct for different values of x.

(b)

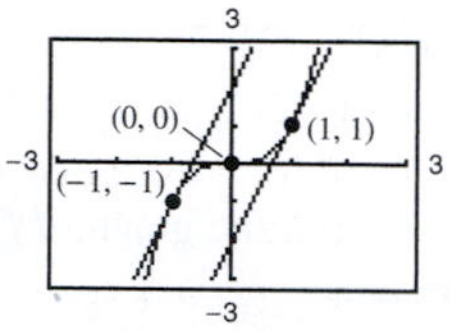

For this function, the slopes of the tangent lines are sometimes the same.

59. (a)

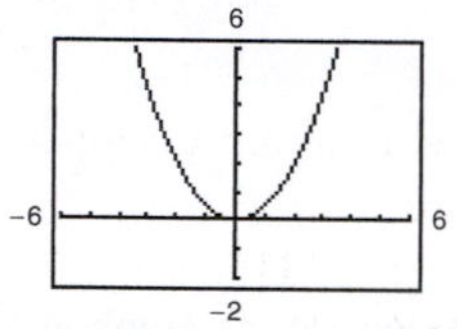

$f'(0) = 0, f'(\frac{1}{2}) = \frac{1}{2}, f'(1) = 1, f'(2) = 2$

(b) $f'(-\frac{1}{2}) = -\frac{1}{2}, f'(-1) = -1, f'(-2) = -2$

(c)

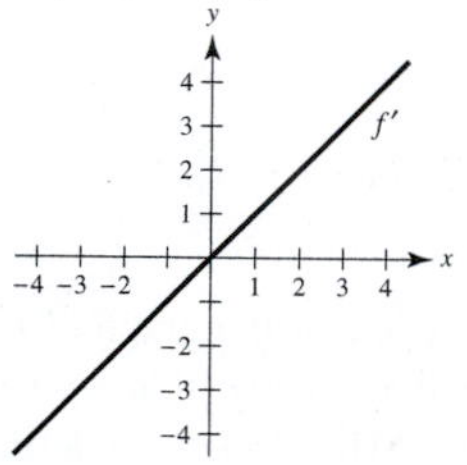

(d) $f'(x) = x$

61.

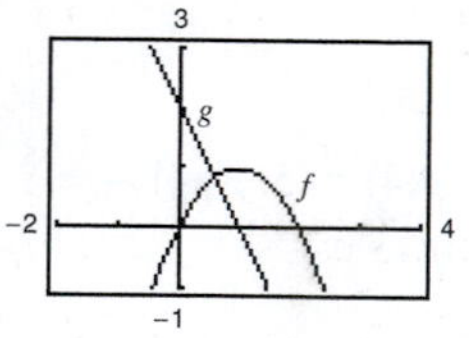

$g(x) \approx f'(x)$

63. $f(2) = 4; f(2.1) = 3.99; f'(2) \approx -0.1$ **65.** 6 **67.** 4

69. $g(x)$ is not differentiable at $x = 0$.

71. $f(x)$ is not differentiable at $x = 6$.

73. $h(x)$ is not differentiable at $x = -7$.

75. $(-\infty, 3) \cup (3, \infty)$ **77.** $(-\infty, -4) \cup (-4, \infty)$

79. $(1, \infty)$

81.

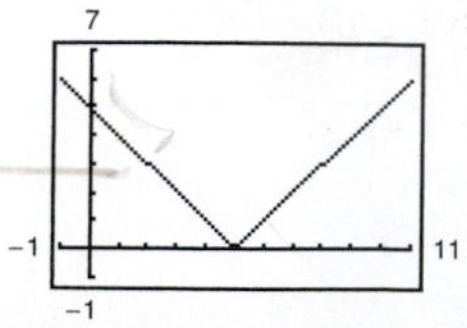

$(-\infty, 5) \cup (5, \infty)$

83.

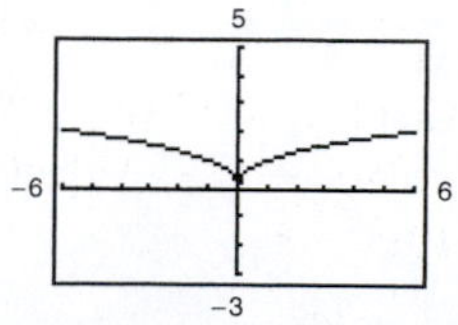

$(-\infty, 0) \cup (0, \infty)$

85. The derivative from the left is -1 and the derivative from the right is 1, so f is not differentiable at $x = 1$.

87. The derivatives from both the right and the left are 0, so $f'(1) = 0$.

89. f is differentiable at $x = 2$.

91. (a) $d = (3|m + 1|)/\sqrt{m^2 + 1}$

(b)

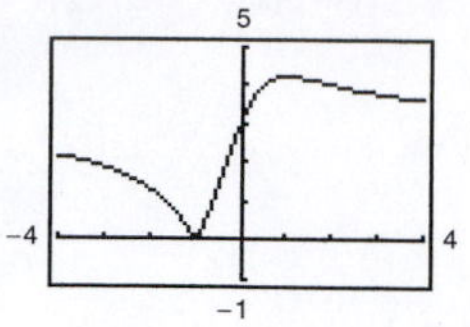

Not differentiable at $m = -1$

93. False. The slope is $\lim_{\Delta x \to 0} \dfrac{f(2 + \Delta x) - f(2)}{\Delta x}$.

95. False. For example, $f(x) = |x|$. The derivative from the left and the derivative from the right both exist but are not equal.

97. Proof

Section 2.2 *(page 114)*

1. (a) $\frac{1}{2}$ (b) 3 **3.** 0 **5.** $7x^6$ **7.** $-5/x^6$

9. $1/(5x^{4/5})$ **11.** 1 **13.** $-4t + 3$ **15.** $2x + 12x^2$

17. $3t^2 + 10t - 3$ **19.** $\dfrac{\pi}{2}\cos\theta + \sin\theta$ **21.** $2x + \dfrac{1}{2}\sin x$

23. $-\dfrac{1}{x^2} - 3\cos x$

	Function	Rewrite	Derivative	Simplify
25.	$y = \dfrac{5}{2x^2}$	$y = \dfrac{5}{2}x^{-2}$	$y' = -5x^{-3}$	$y' = -\dfrac{5}{x^3}$
27.	$y = \dfrac{6}{(5x)^3}$	$y = \dfrac{6}{125}x^{-3}$	$y' = -\dfrac{18}{125}x^{-4}$	$y' = -\dfrac{18}{125x^4}$
29.	$y = \dfrac{\sqrt{x}}{x}$	$y = x^{-1/2}$	$y' = -\dfrac{1}{2}x^{-3/2}$	$y' = -\dfrac{1}{2x^{3/2}}$

31. -2 **33.** 0 **35.** 8 **37.** 3 **39.** $2x + 6/x^3$

41. $2t + 12/t^4$ **43.** $8x + 3$ **45.** $(x^3 - 8)/x^3$

47. $3x^2 + 1$ **49.** $\dfrac{1}{2\sqrt{x}} - \dfrac{2}{x^{2/3}}$ **51.** $\dfrac{3}{\sqrt{x}} - 5\sin x$

53. (a) $2x + y - 2 = 0$

(b)

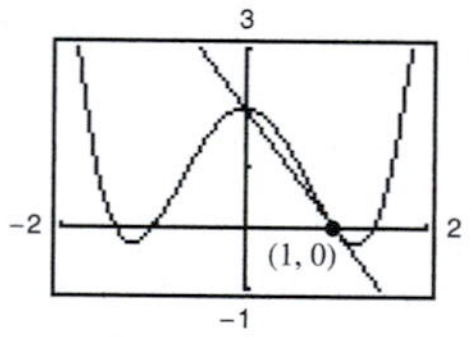

55. (a) $3x + 2y - 7 = 0$

(b)

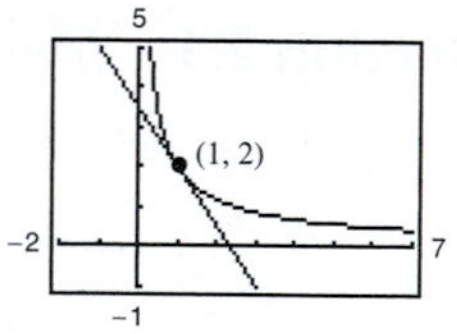

57. $(-1, 2), (0, 3), (1, 2)$ **59.** No horizontal tangents

61. (π, π) **63.** $k = -8$ **65.** $k = 3$ **67.** $k = 4/27$

69.

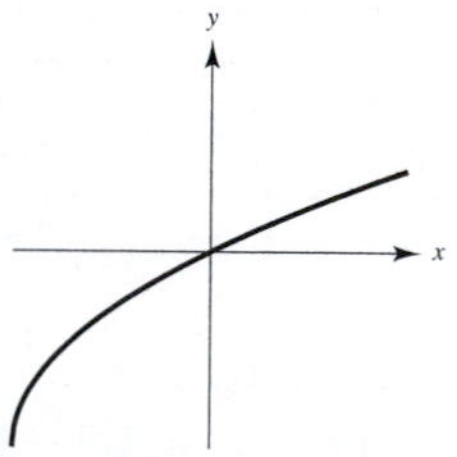

71. $g'(x) = f'(x)$ **73.** $g'(x) = -5f'(x)$

75. 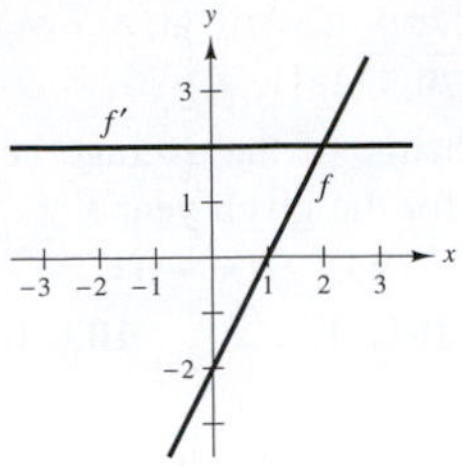

The rate of change of f is constant, and therefore f' is a constant function.

77. $y = 2x - 1$ $\qquad$ $y = 4x - 4$

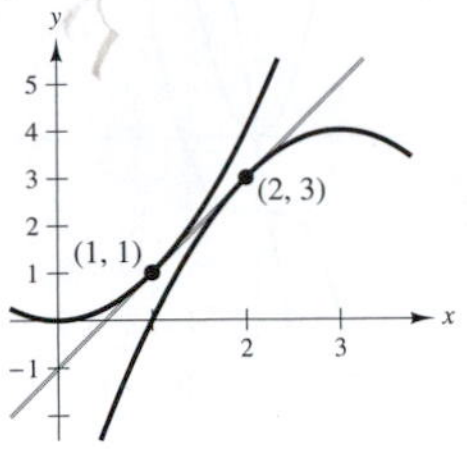

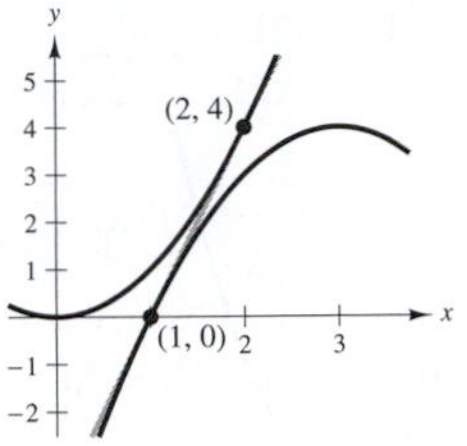

79. $f'(x) = 3 + \cos x \neq 0$ for all x. $\qquad$ **81.** $x - 4y + 4 = 0$

83. 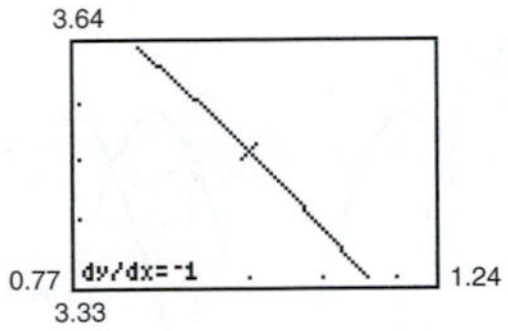

$f'(1)$ appears to be close to -1.
$f'(1) = -1$

85. (a) 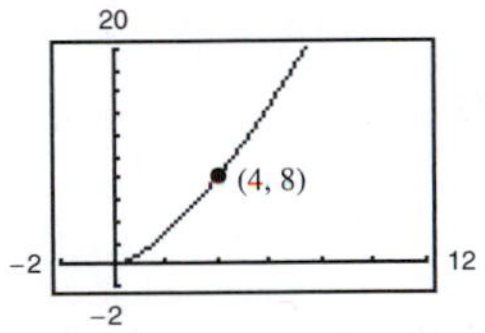

(3.9, 7.7019),
$S(x) = 2.981x - 3.924$

(b) $T(x) = 3(x - 4) + 8 = 3x - 4$
The slope (and equation) of the secant line approaches that of the tangent line at (4, 8) as you choose points closer and closer to (4, 8).

(c) 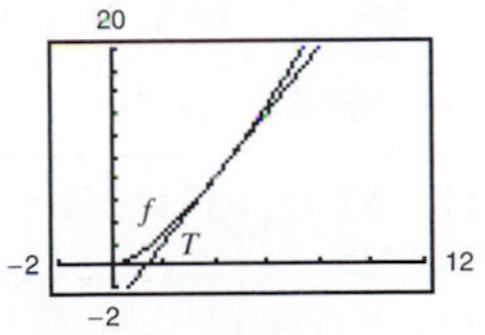

The approximation becomes less accurate.

(d)

Δx	-3	-2	-1	-0.5	-0.1	0
$f(4 + \Delta x)$	1	2.828	5.196	6.548	7.702	8
$T(4 + \Delta x)$	-1	2	5	6.5	7.7	8

Δx	0.1	0.5	1	2	3
$f(4 + \Delta x)$	8.302	9.546	11.180	14.697	18.520
$T(4 + \Delta x)$	8.3	9.5	11	14	17

87. False. Let $f(x) = x$ and $g(x) = x + 1$.

89. False. $dy/dx = 0$ $\qquad$ **91.** True

93. Average rate: 4
Instantaneous rates:
$f'(1) = 4; f'(2) = 4$

95. Average rate: $\frac{1}{2}$
Instantaneous rates:
$f'(1) = 1; f'(2) = \frac{1}{4}$

97. (a) $s(t) = -16t^2 + 1362$; $v(t) = -32t$ $\quad$ (b) -48 ft/sec
(c) $s'(1) = -32$ ft/sec; $s'(2) = -64$ ft/sec
(d) $t = \dfrac{\sqrt{1362}}{4} \approx 9.226$ sec $\quad$ (e) -295.242 ft/sec

99. $v(5) = 71$ m/sec; $v(10) = 22$ m/sec

101. 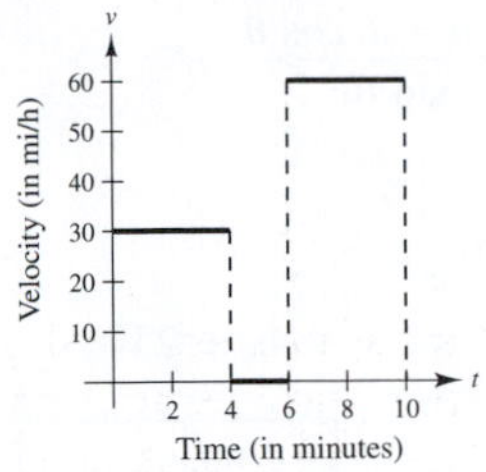

103. 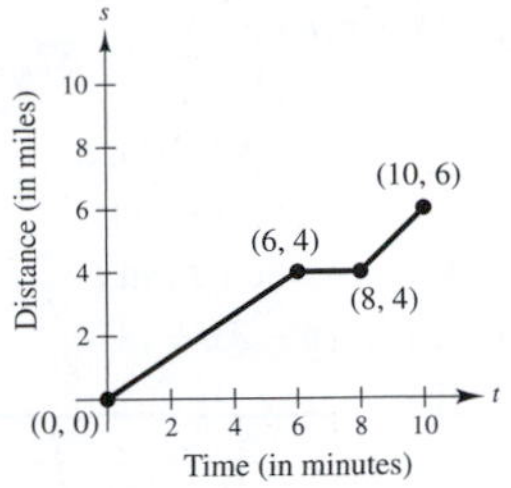

105. $V'(6) = 108 \text{ cm}^3/\text{cm}$

107. (a) $R(v) = 0.417v - 0.02$
(b) $B(v) = 0.0056v^2 + 0.001v + 0.04$
(c) $T(v) = 0.0056v^2 + 0.418v + 0.02$
(d) 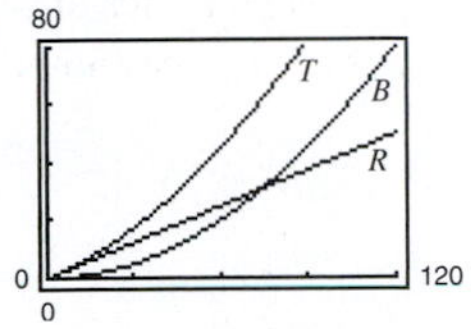

(e) $T'(v) = 0.0112v + 0.418$
$T'(40) = 0.866$
$T'(80) = 1.314$
$T'(100) = 1.538$

(f) Stopping distance increases at an increasing rate.

109. Proof $\qquad$ **111.** $y = 2x^2 - 3x + 1$

113. $9x + y = 0, 9x + 4y + 27 = 0$ $\qquad$ **115.** $a = \frac{1}{3}, b = -\frac{4}{3}$

117. $f_1(x) = |\sin x|$ is differentiable for all $x \neq n\pi$, n an integer.
$f_2(x) = \sin|x|$ is differentiable for all $x \neq 0$.

119. Putnam Problem A2, 2010

Section 2.3 *(page 125)*

1. $2(2x^3 - 6x^2 + 3x - 6)$ $\qquad$ **3.** $(1 - 5t^2)/(2\sqrt{t})$

5. $x^2(3 \cos x - x \sin x)$ $\qquad$ **7.** $(1 - x^2)/(x^2 + 1)^2$

9. $(1 - 5x^3)/[2\sqrt{x}(x^3 + 1)^2]$ $\qquad$ **11.** $(x \cos x - 2 \sin x)/x^3$

13. $f'(x) = (x^3 + 4x)(6x + 2) + (3x^2 + 2x - 5)(3x^2 + 4)$
$= 15x^4 + 8x^3 + 21x^2 + 16x - 20$
$f'(0) = -20$

15. $f'(x) = \dfrac{x^2 - 6x + 4}{(x - 3)^2}$ $\qquad$ **17.** $f'(x) = \cos x - x \sin x$
$f'(1) = -\dfrac{1}{4}$ $\qquad$ $f'\left(\dfrac{\pi}{4}\right) = \dfrac{\sqrt{2}}{8}(4 - \pi)$

	Function	Rewrite	Differentiate	Simplify
19.	$y = \dfrac{x^2 + 3x}{7}$	$y = \dfrac{1}{7}x^2 + \dfrac{3}{7}x$	$y' = \dfrac{2}{7}x + \dfrac{3}{7}$	$y' = \dfrac{2x + 3}{7}$
21.	$y = \dfrac{6}{7x^2}$	$y = \dfrac{6}{7}x^{-2}$	$y' = -\dfrac{12}{7}x^{-3}$	$y' = -\dfrac{12}{7x^3}$
23.	$y = \dfrac{4x^{3/2}}{x}$	$y = 4x^{1/2}$, $x > 0$	$y' = 2x^{-1/2}$	$y' = \dfrac{2}{\sqrt{x}}$, $x > 0$

25. $\dfrac{3}{(x + 1)^2}$, $x \neq 1$ $\qquad$ **27.** $(x^2 + 6x - 3)/(x + 3)^2$

29. $(3x + 1)/(2x^{3/2})$ $\qquad$ **31.** $6s^2(s^3 - 2)$

33. $-(2x^2 - 2x + 3)/[x^2(x - 3)^2]$

35. $10x^4 - 8x^3 - 21x^2 - 10x - 30$

37. $-\dfrac{4xc^2}{(x^2 - c^2)^2}$ $\qquad$ **39.** $t(t \cos t + 2 \sin t)$

41. $-(t \sin t + \cos t)/t^2$ **43.** $-1 + \sec^2 x = \tan^2 x$

45. $\dfrac{1}{4t^{3/4}} - 6 \csc t \cot t$ **47.** $\dfrac{3}{2} \sec x(\tan x - \sec x)$

49. $\cos x \cot^2 x$ **51.** $x(x \sec^2 x + 2 \tan x)$

53. $4x \cos x + (2 - x^2) \sin x$

55. $\dfrac{2x^2 + 8x - 1}{(x + 2)^2}$ **57.** $\dfrac{1 - \sin\theta + \theta\cos\theta}{(1 - \sin\theta)^2}$

59. $y' = \dfrac{-2 \csc x \cot x}{(1 - \csc x)^2}$, $-4\sqrt{3}$

61. $h'(t) = \sec t(t \tan t - 1)/t^2$, $1/\pi^2$

63. (a) $y = -3x - 1$

(b)

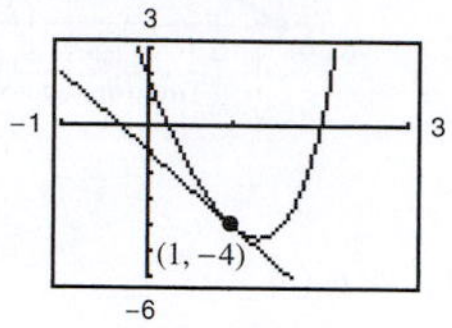

65. (a) $y = 4x + 25$

(b)

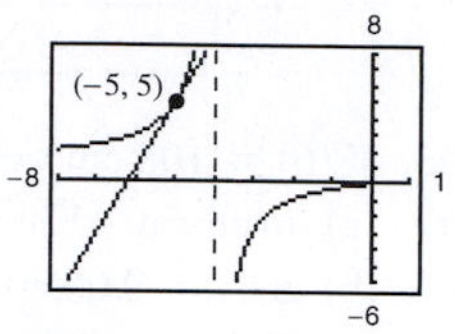

67. (a) $4x - 2y - \pi + 2 = 0$

(b)

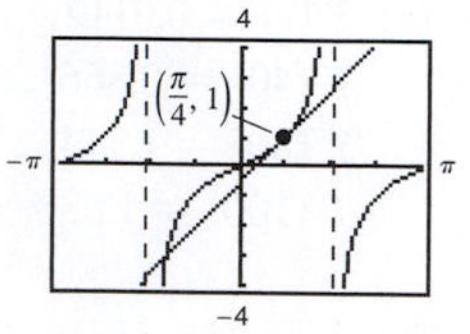

69. $2y + x - 4 = 0$

71. $25y - 12x + 16 = 0$ **73.** $(1, 1)$ **75.** $(0, 0), (2, 4)$

77. Tangent lines: $2y + x = 7$; $2y + x = -1$

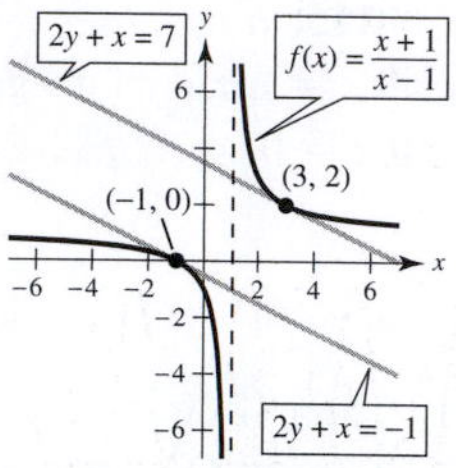

79. $f(x) + 2 = g(x)$ **81.** (a) $p'(1) = 1$ (b) $q'(4) = -1/3$

83. $(18t + 5)/(2\sqrt{t})$ cm²/sec

85. (a) −\$38.13 thousand/100 components

(b) −\$10.37 thousand/100 components

(c) −\$3.80 thousand/100 components

The cost decreases with increasing order size.

87. Proof

89. (a) $h(t) = 112.4t + 1332$

$p(t) = 2.9t + 282$

(b)

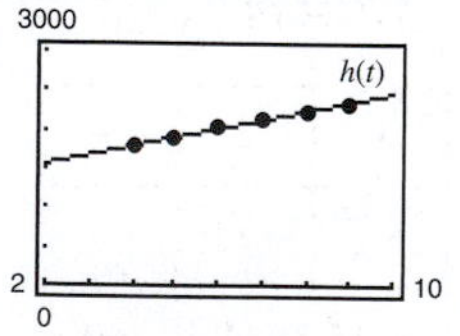

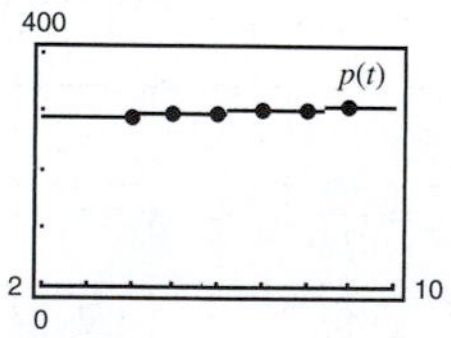

(c) $A = \dfrac{112.4t + 1332}{2.9t + 282}$

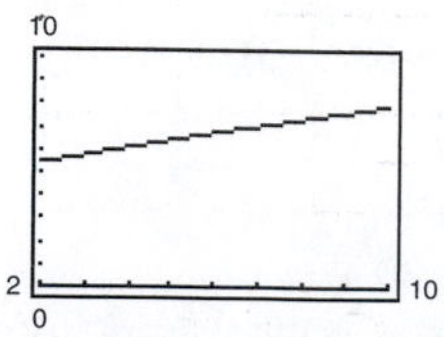

A represents the average health care expenditures per person (in thousands of dollars).

(d) $A'(t) = \dfrac{27{,}834}{8.41t^2 + 1635.6t + 79{,}524}$

$A'(t)$ represents the rate of change of the average health care expenditures per person for the given year t.

91. $12x^2 + 12x - 6$ **93.** $3/\sqrt{x}$ **95.** $2/(x - 1)^3$

97. $2 \cos x - x \sin x$ **99.** $2x$ **101.** $1/\sqrt{x}$ **103.** 0

105. −10

107. Answers will vary.

Sample answer:

$f(x) = (x - 2)^2$

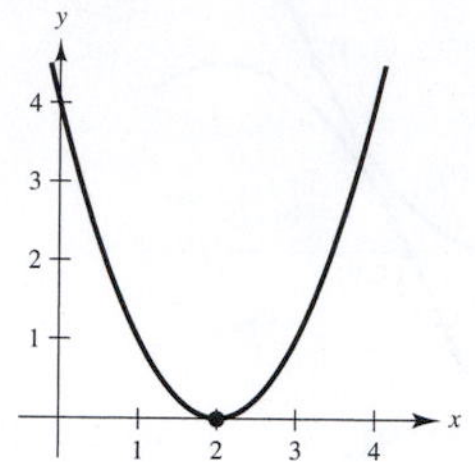

109.

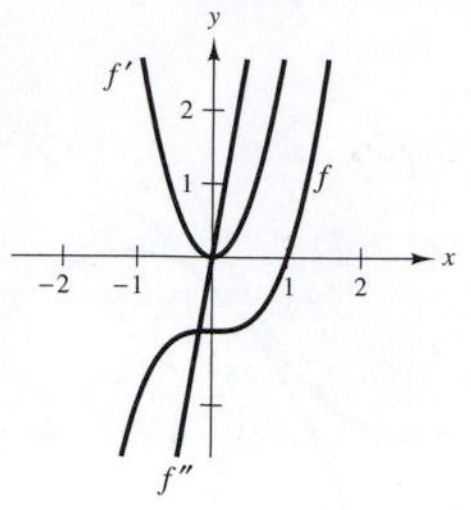

111.

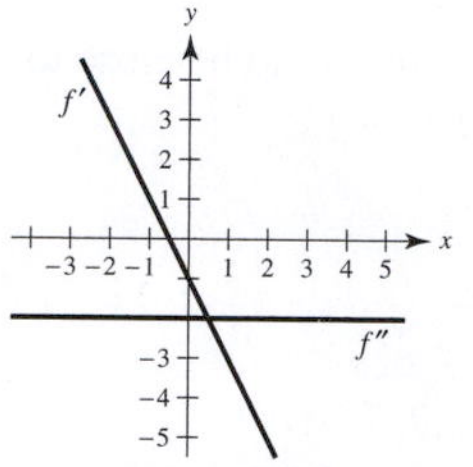

113.

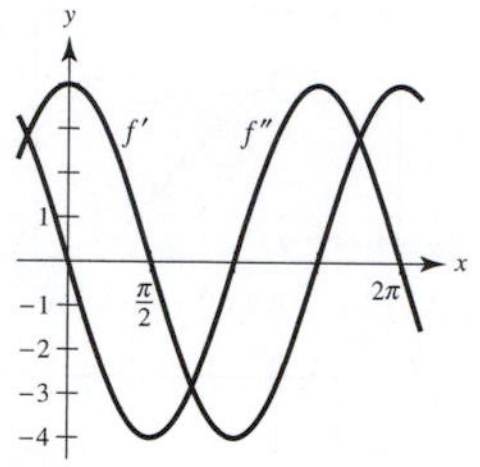

115. $v(3) = 27$ m/sec

$a(3) = -6$ m/sec²

The speed of the object is decreasing.

117.

t	0	1	2	3	4
$s(t)$	0	57.75	99	123.75	132
$v(t)$	66	49.5	33	16.5	0
$a(t)$	−16.5	−16.5	−16.5	−16.5	−16.5

The average velocity on [0, 1] is 57.75, on [1, 2] is 41.25, on [2, 3] is 24.75, and on [3, 4] is 8.25.

119. $f^{(n)}(x) = n(n - 1)(n - 2) \cdots (2)(1) = n!$

121. (a) $f''(x) = g(x)h''(x) + 2g'(x)h'(x) + g''(x)h(x)$

$f'''(x) = g(x)h'''(x) + 3g'(x)h''(x) + 3g''(x)h'(x) + g'''(x)h(x)$

$f^{(4)}(x) = g(x)h^{(4)}(x) + 4g'(x)h'''(x) + 6g''(x)h''(x) + 4g'''(x)h'(x) + g^{(4)}(x)h(x)$

(b) $f^{(n)}(x) = g(x)h^{(n)}(x) + \dfrac{n!}{1!(n-1)!}g'(x)h^{(n-1)}(x) + \dfrac{n!}{2!(n-2)!}g''(x)h^{(n-2)}(x) + \cdots + \dfrac{n!}{(n-1)!1!}g^{(n-1)}(x)h'(x) + g^{(n)}(x)h(x)$

123. $n = 1$: $f'(x) = x \cos x + \sin x$

$n = 2$: $f'(x) = x^2 \cos x + 2x \sin x$

$n = 3$: $f'(x) = x^3 \cos x + 3x^2 \sin x$

$n = 4$: $f'(x) = x^4 \cos x + 4x^3 \sin x$

General rule: $f'(x) = x^n \cos x + nx^{(n-1)} \sin x$

125. $y' = -1/x^2,\ y'' = 2/x^3,$
$x^3y'' + 2x^2y' = x^3(2/x^3) + 2x^2(-1/x^2)$
$= 2 - 2 = 0$

127. $y' = 2\cos x,\ y'' = -2\sin x,$
$y'' + y = -2\sin x + 2\sin x + 3 = 3$

129. False. $dy/dx = f(x)g'(x) + g(x)f'(x)$ **131.** True

133. True **135.** $f'(x) = 2|x|$; $f''(0)$ does not exist.

137. Proof

Section 2.4 *(page 136)*

	$y = f(g(x))$	$u = g(x)$	$y = f(u)$
1.	$y = (5x - 8)^4$	$u = 5x - 8$	$y = u^4$
3.	$y = \sqrt{x^3 - 7}$	$u = x^3 - 7$	$y = \sqrt{u}$
5.	$y = \csc^3 x$	$u = \csc x$	$y = u^3$

7. $12(4x - 1)^2$ **9.** $-108(4 - 9x)^3$ **11.** $-1/\left(2\sqrt{5 - t}\right)$

13. $4x/\sqrt[3]{(6x^2 + 1)^2}$ **15.** $-x/\sqrt[4]{(9 - x^2)^3}$

17. $-1/(x - 2)^2$ **19.** $-2/(t - 3)^3$

21. $-3/[2\sqrt{(3x + 5)^3}]$ **23.** $2x(x - 2)^3(3x - 2)$

25. $\dfrac{1 - 2x^2}{\sqrt{1 - x^2}}$ **27.** $\dfrac{1}{\sqrt{(x^2 + 1)^3}}$

29. $\dfrac{-2(x + 5)(x^2 + 10x - 2)}{(x^2 + 2)^3}$ **31.** $\dfrac{-9(1 - 2v)^2}{(v + 1)^4}$

33. $20x(x^2 + 3)^9 + 2(x^2 + 3)^5 + 20x^2(x^2 + 3)^4 + 2x$

35. $(1 - 3x^2 - 4x^{3/2})/[2\sqrt{x}(x^2 + 1)^2]$

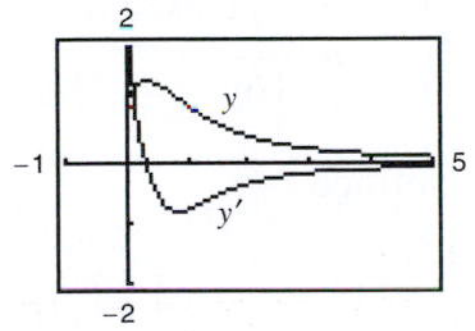

The zero of y' corresponds to the point on the graph of the function where the tangent line is horizontal.

37. $-\dfrac{\sqrt{\dfrac{x + 1}{x}}}{2x(x + 1)}$

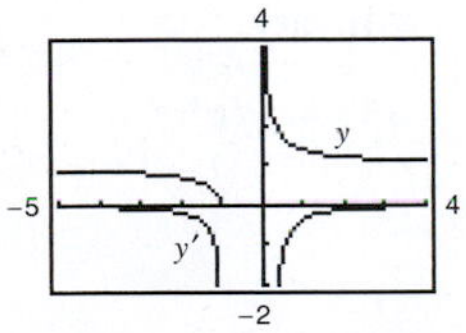

y' has no zeros.

39. $-[\pi x \sin(\pi x) + \cos(\pi x) + 1]/x^2$

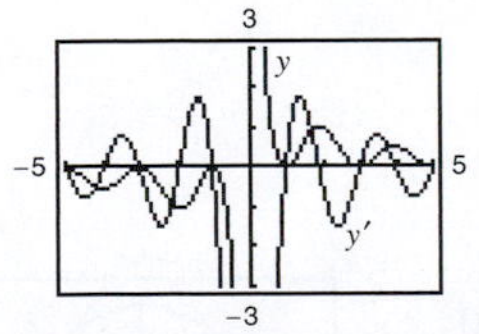

The zeros of y' correspond to the points on the graph of the function where the tangent lines are horizontal.

41. (a) 1 (b) 2; The slope of $\sin ax$ at the origin is a.

43. $-4\sin 4x$ **45.** $15\sec^2 3x$ **47.** $2\pi^2 x\cos(\pi x)^2$

49. $2\cos 4x$ **51.** $(-1 - \cos^2 x)/\sin^3 x$

53. $8\sec^2 x\tan x$ **55.** $10\tan 5\theta\sec^2 5\theta$

57. $\sin 2\theta\cos 2\theta = \dfrac{1}{2}\sin 4\theta$ **59.** $\dfrac{6\pi\sin(\pi t - 1)}{\cos^3(\pi t - 1)}$

61. $\dfrac{1}{2\sqrt{x}} + 2x\cos(2x)^2$ **63.** $2\sec^2 2x\cos(\tan 2x)$

65. $y' = \dfrac{x + 4}{\sqrt{x^2 + 8x}}, \dfrac{5}{3}$ **67.** $f'(x) = \dfrac{-15x^2}{(x^3 - 2)^2}, -\dfrac{3}{5}$

69. $f'(t) = \dfrac{-5}{(t - 1)^2}, -5$ **71.** $y' = -12\sec^3 4x\tan 4x, 0$

73. (a) $8x - 5y - 7 = 0$
(b)

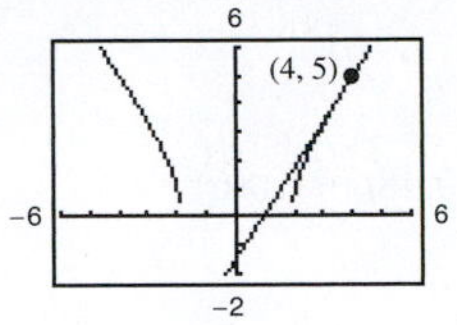

75. (a) $24x + y + 23 = 0$
(b)

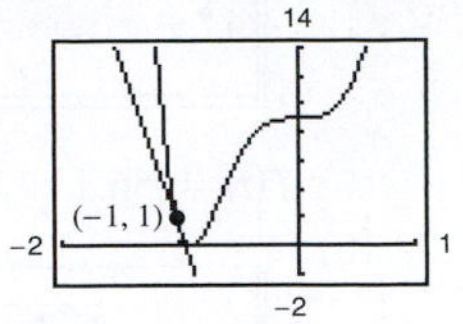

77. (a) $2x - y - 2\pi = 0$
(b)

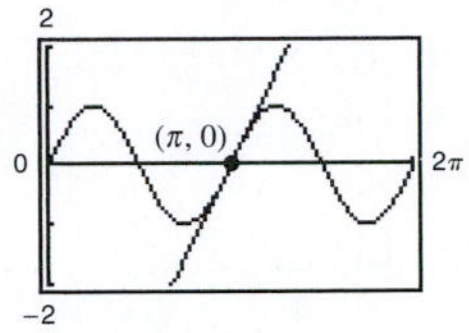

79. (a) $4x - y + (1 - \pi) = 0$
(b)

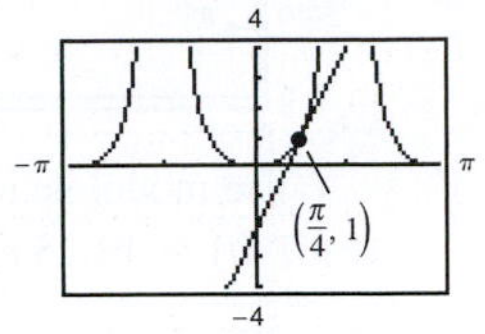

81. $3x + 4y - 25 = 0$

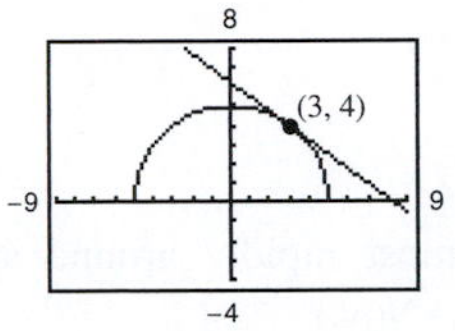

83. $\left(\dfrac{\pi}{6}, \dfrac{3\sqrt{3}}{2}\right), \left(\dfrac{5\pi}{6}, -\dfrac{3\sqrt{3}}{2}\right), \left(\dfrac{3\pi}{2}, 0\right)$ **85.** $2940(2 - 7x)^2$

87. $\dfrac{2}{(x - 6)^3}$ **89.** $2(\cos x^2 - 2x^2\sin x^2)$

91. $h''(x) = 18x + 6, 24$

93. $f''(x) = -4x^2\cos(x^2) - 2\sin(x^2), 0$

95.

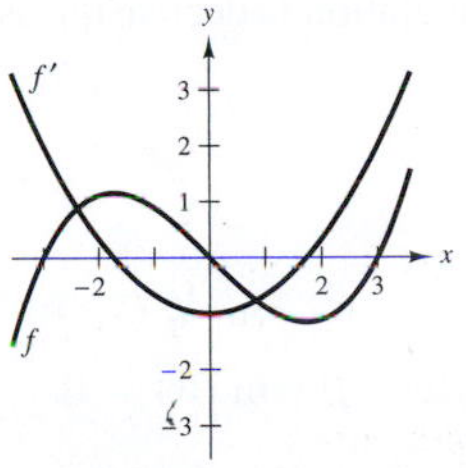

The zeros of f' correspond to the points where the graph of f has horizontal tangents.

97.

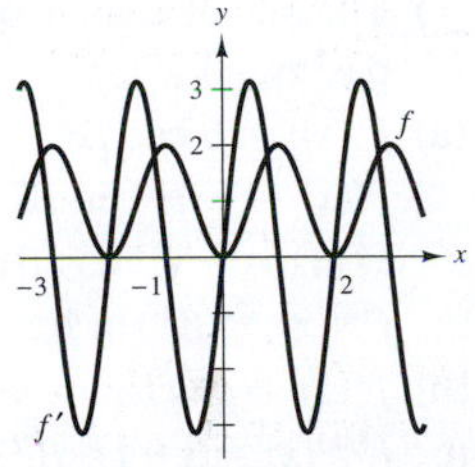

The zeros of f' correspond to the points where the graph of f has horizontal tangents.

99. The rate of change of g is three times as fast as the rate of change of f.

101. (a) $g'(x) = f'(x)$ (b) $h'(x) = 2f'(x)$
(c) $r'(x) = -3f'(-3x)$ (d) $s'(x) = f'(x + 2)$

x	-2	-1	0	1	2	3
$f'(x)$	4	$\frac{2}{3}$	$-\frac{1}{3}$	-1	-2	-4
$g'(x)$	4	$\frac{2}{3}$	$-\frac{1}{3}$	-1	-2	-4
$h'(x)$	8	$\frac{4}{3}$	$-\frac{2}{3}$	-2	-4	-8
$r'(x)$		12	1			
$s'(x)$	$-\frac{1}{3}$	-1	-2	-4		

103. (a) $\frac{1}{2}$
(b) $s'(5)$ does not exist because g is not differentiable at 6.

105. (a) 1.461 (b) -1.016 **107.** 0.2 rad, 1.45 rad/sec

109. (a)

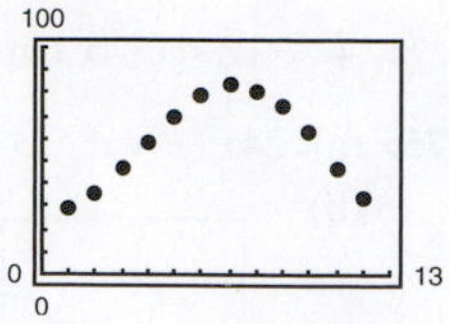

$T(t) = 56.1 + 27.6 \sin(0.48t - 1.86)$

(b)

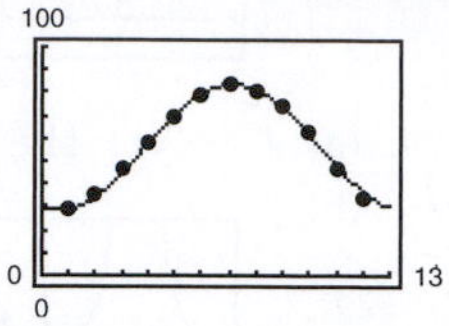

The model is a good fit.

(c) $T'(t) \approx 13.25 \cos(0.48t - 1.86)$

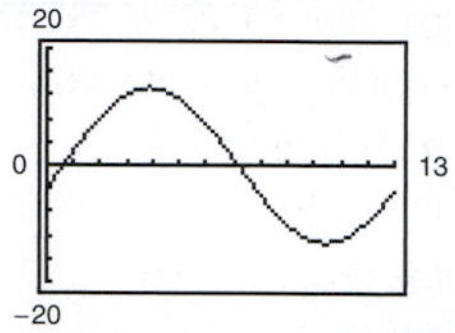

(d) The temperature changes most rapidly around spring (March–May) and fall (Oct.–Nov.)
The temperature changes most slowly around winter (Dec.–Feb.) and summer (Jun.–Aug.)
Yes. Explanations will vary.

111. (a) 0 bacteria per day (b) 177.8 bacteria per day
(c) 44.4 bacteria per day (d) 10.8 bacteria per day
(e) 3.3 bacteria per day
(f) The rate of change of the population is decreasing as time passes.

113. (a) $f'(x) = \beta \cos \beta x$
$f''(x) = -\beta^2 \sin \beta x$
$f'''(x) = -\beta^3 \cos \beta x$
$f^{(4)}(x) = \beta^4 \sin \beta x$
(b) $f''(x) + \beta^2 f(x) = -\beta^2 \sin \beta x + \beta^2(\sin \beta x) = 0$
(c) $f^{(2k)}(x) = (-1)^k \beta^{2k} \sin \beta x$
$f^{(2k-1)}(x) = (-1)^{k+1}\beta^{2k-1} \cos \beta x$

115. (a) $r'(1) = 0$ (b) $s'(4) = \frac{5}{8}$

117. (a) and (b) Proofs

119. $g'(x) = 3\left(\dfrac{3x - 5}{|3x - 5|}\right), \quad x \neq \dfrac{5}{3}$

121. $h'(x) = -|x| \sin x + \dfrac{x}{|x|} \cos x, \quad x \neq 0$

123. (a) $P_1(x) = 2(x - \pi/4) + 1$
$P_2(x) = 2(x - \pi/4)^2 + 2(x - \pi/4) + 1$
(b)

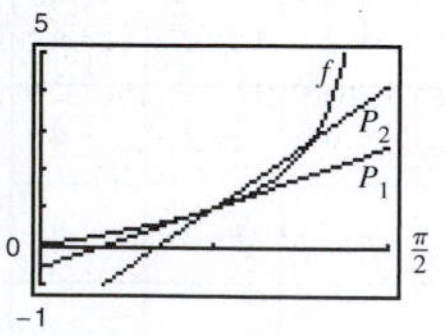

(c) P_2

(d) The accuracy worsens as you move away from $x = \pi/4$.

125. False. If $y = (1 - x)^{1/2}$, then $y' = \frac{1}{2}(1 - x)^{-1/2}(-1)$.

127. True **129.** Putnam Problem A1, 1967

Section 2.5 *(page 145)*

1. $-x/y$ **3.** $-\sqrt{y/x}$ **5.** $(y - 3x^2)/(2y - x)$

7. $(1 - 3x^2y^3)/(3x^3y^2 - 1)$

9. $(6xy - 3x^2 - 2y^2)/(4xy - 3x^2)$ **11.** $\cos x/[4 \sin(2y)]$

13. $(\cos x - \tan y - 1)/(x \sec^2 y)$

15. $[y \cos(xy)]/[1 - x \cos(xy)]$

17. (a) $y_1 = \sqrt{64 - x^2}$; $y_2 = -\sqrt{64 - x^2}$
(b)

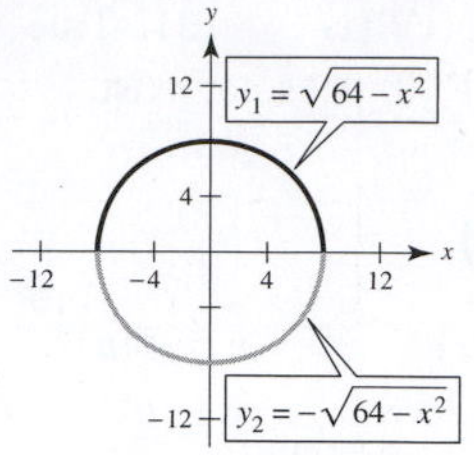

(c) $y' = \mp \dfrac{x}{\sqrt{64 - x^2}} = -\dfrac{x}{y}$ (d) $y' = -\dfrac{x}{y}$

19. (a) $y_1 = \dfrac{\sqrt{x^2 + 16}}{4}$; $y_2 = \dfrac{-\sqrt{x^2 + 16}}{4}$
(b)

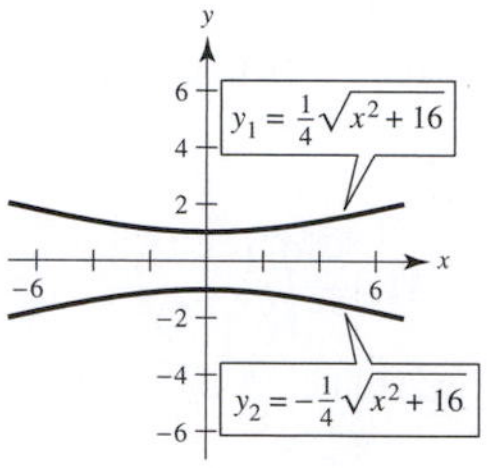

(c) $y' = \dfrac{\pm x}{4\sqrt{x^2 + 16}} = \dfrac{x}{16y}$ (d) $y' = \dfrac{x}{16y}$

21. $-\dfrac{y}{x}, -\dfrac{1}{6}$ **23.** $\dfrac{98x}{y(x^2 + 49)^2}$, Undefined

25. $-\dfrac{y(y + 2x)}{x(x + 2y)}, -1$ **27.** $-\sin^2(x + y)$ or $-\dfrac{x^2}{x^2 + 1}, 0$

29. $-\dfrac{1}{2}$ **31.** 0 **33.** $y = -x + 7$ **35.** $y = -x + 2$

37. $y = \sqrt{3}x/6 + 8\sqrt{3}/3$ **39.** $y = -\frac{2}{11}x + \frac{30}{11}$

41. (a) $y = -2x + 4$ (b) Answers will vary.

43. $\cos^2 y, \; -\dfrac{\pi}{2} < y < \dfrac{\pi}{2}, \; \dfrac{1}{1 + x^2}$ **45.** $-4/y^3$

47. $-36/y^3$ **49.** $(3x)/(4y)$

51. $2x + 3y - 30 = 0$

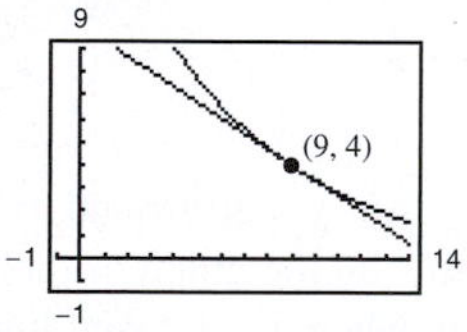

53. At (4, 3):
Tangent line: $4x + 3y - 25 = 0$
Normal line: $3x - 4y = 0$

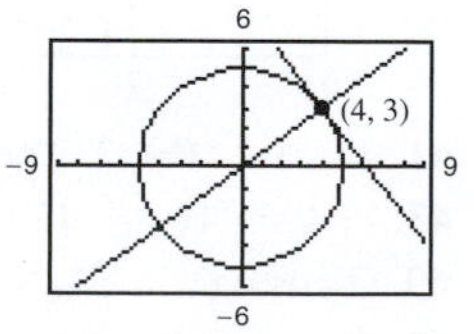

At $(-3, 4)$:
Tangent line: $3x - 4y + 25 = 0$
Normal line: $4x + 3y = 0$

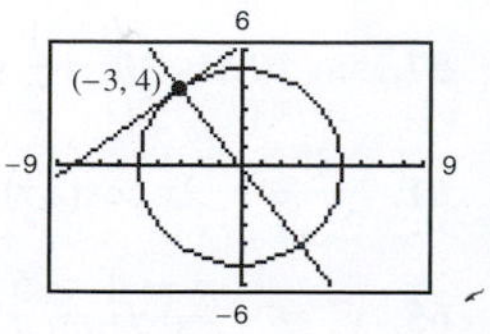

55. $x^2 + y^2 = r^2 \Rightarrow y' = -x/y \Rightarrow y/x =$ slope of normal line. Then for (x_0, y_0) on the circle, $x_0 \neq 0$, an equation of the normal line is $y = (y_0/x_0)x$, which passes through the origin. If $x_0 = 0$, the normal line is vertical and passes through the origin.

57. Horizontal tangents: $(-4, 0), (-4, 10)$
Vertical tangents: $(0, 5), (-8, 5)$

59.

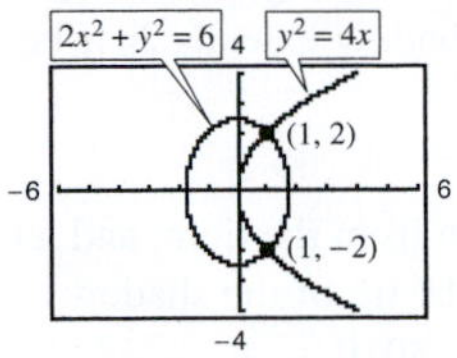

At $(1, 2)$:
Slope of ellipse: -1
Slope of parabola: 1
At $(1, -2)$:
Slope of ellipse: 1
Slope of parabola: -1

61.

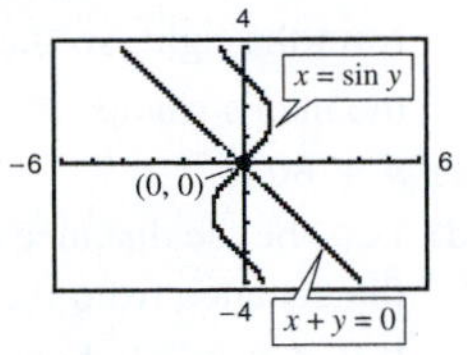

At $(0, 0)$:
Slope of line: -1
Slope of sine curve: 1

63. Derivatives: $\dfrac{dy}{dx} = -\dfrac{y}{x}, \dfrac{dy}{dx} = \dfrac{x}{y}$

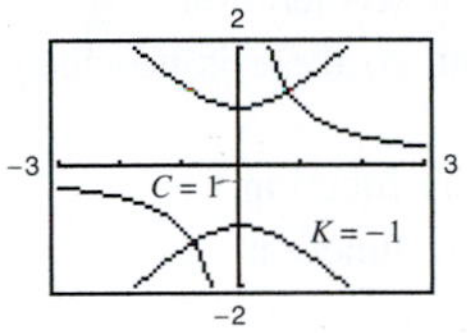

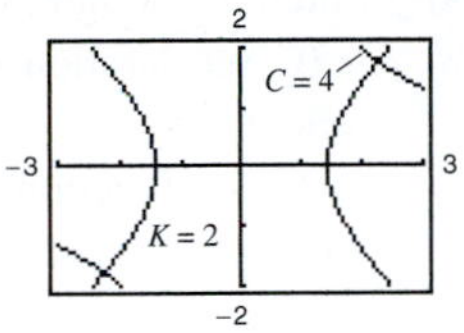

65. Answers will vary. In the explicit form of a function, the variable is explicitly written as a function of x. In an implicit equation, the function is only implied by an equation. An example of an implicit function is $x^2 + xy = 5$. In explicit form, it would be $y = (5 - x^2)/x$.

67.

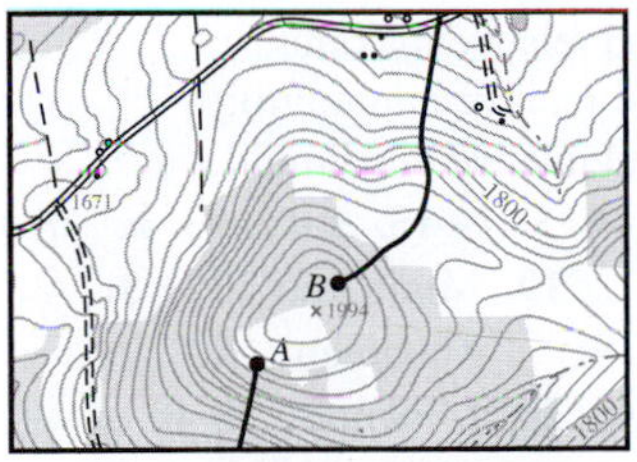

Use starting point B.

69. (a)

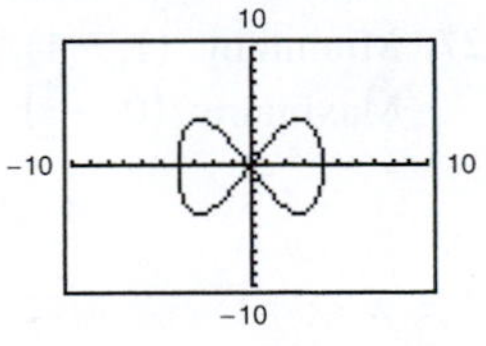

(b)

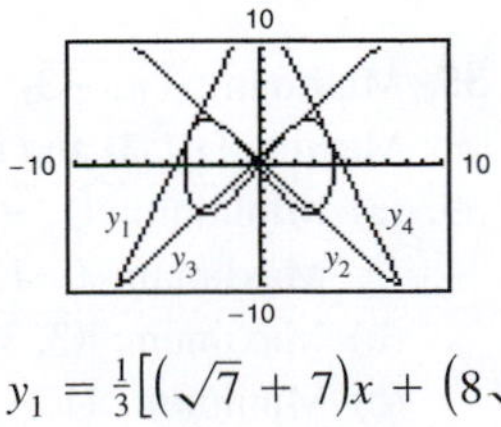

$y_1 = \frac{1}{3}\left[(\sqrt{7} + 7)x + (8\sqrt{7} + 23)\right]$
$y_2 = -\frac{1}{3}\left[(-\sqrt{7} + 7)x - (23 - 8\sqrt{7})\right]$
$y_3 = -\frac{1}{3}\left[(\sqrt{7} - 7)x - (23 - 8\sqrt{7})\right]$
$y_4 = -\frac{1}{3}\left[(\sqrt{7} + 7)x - (8\sqrt{7} + 23)\right]$

(c) $\left(\dfrac{8\sqrt{7}}{7}, 5\right)$

71. Proof **73.** $y = -\dfrac{\sqrt{3}}{2}x + 2\sqrt{3}, y = \dfrac{\sqrt{3}}{2}x -$

75. (a) $y = 2x - 6$
(b)

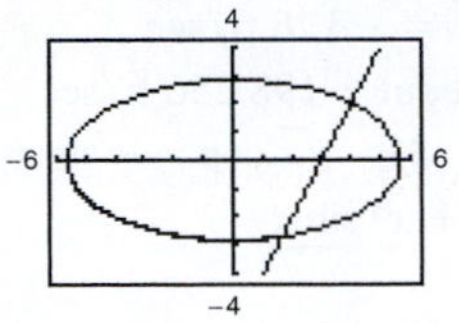

(c) $\left(\frac{28}{17}, -\frac{46}{17}\right)$

Section 2.6 *(page 153)*

1. (a) $\frac{3}{4}$ (b) 20 **3.** (a) $-\frac{5}{8}$ (b) $\frac{3}{2}$
5. (a) -8 cm/sec (b) 0 cm/sec (c) 8 cm/sec
7. (a) 12 ft/sec (b) 6 ft/sec (c) 3 ft/sec
9. In a linear function, if x changes at a constant rate, so does y. However, unless $a = 1$, y does not change at the same rate as x.
11. (a) $64\pi\ \text{cm}^2/\text{min}$ (b) $256\pi\ \text{cm}^2/\text{min}$
13. (a) $972\pi\ \text{in.}^3/\text{min}$; $15{,}552\pi\ \text{in.}^3/\text{min}$
(b) If dr/dt is constant, dV/dt is proportional to r^2.
15. (a) $72\ \text{cm}^3/\text{sec}$ (b) $1800\ \text{cm}^3/\text{sec}$
17. $8/(405\pi)$ ft/min **19.** (a) 12.5% (b) $\frac{1}{144}$ m/min
21. (a) $-\frac{7}{12}$ ft/sec; $-\frac{3}{2}$ ft/sec; $-\frac{48}{7}$ ft/sec
(b) $\frac{527}{24}\ \text{ft}^2/\text{sec}$ (c) $\frac{1}{12}$ rad/sec
23. Rate of vertical change: $\frac{1}{5}$ m/sec
Rate of horizontal change: $-\sqrt{3}/15$ m/sec
25. (a) -750 mi/h (b) 30 min
27. $-50/\sqrt{85} \approx -5.42$ ft/sec
29. (a) $\frac{25}{3}$ ft/sec (b) $\frac{10}{3}$ ft/sec
31. (a) 12 sec (b) $\frac{1}{2}\sqrt{3}$ m (c) $\sqrt{5}\pi/120$ m/sec
33. Evaporation rate proportional to $S \Rightarrow \dfrac{dV}{dt} = k(4\pi r^2)$

$V = \left(\dfrac{4}{3}\right)\pi r^3 \Rightarrow \dfrac{dV}{dt} = 4\pi r^2\dfrac{dr}{dt}$. So $k = \dfrac{dr}{dt}$.

35. 0.6 ohm/sec **37.** $\dfrac{dv}{dt} = \dfrac{16r}{v}\sec^2\theta\dfrac{d\theta}{dt}, \dfrac{d\theta}{dt} = \dfrac{v}{16r}\cos^2\theta\dfrac{dv}{dt}$

39. $\dfrac{2\sqrt{21}}{525} \approx 0.017$ rad/sec

41. (a) $\dfrac{200\pi}{3}$ ft/sec (b) 200π ft/sec
(c) About 427.43π ft/sec
43. About 84.9797 mi/h
45. (a) $\dfrac{dy}{dt} = 3\dfrac{dx}{dt}$ means that y changes three times as fast as x changes.
(b) y changes slowly when $x \approx 0$ or $x \approx L$. y changes more rapidly when x is near the middle of the interval.
47. $-18.432\ \text{ft/sec}^2$ **49.** About -97.96 m/sec

Review Exercises for Chapter 2 *(page 157)*

1. $f'(x) = 0$ **3.** $f'(x) = 2x - 4$ **5.** 5
7. f is differentiable at all $x \neq 3$. **9.** 0 **11.** $3x^2 - 22x$
13. $\dfrac{3}{\sqrt{x}} + \dfrac{1}{\sqrt[3]{x^2}}$ **15.** $-\dfrac{4}{3t^3}$ **17.** $4 - 5\cos\theta$
19. $-3\sin\theta - (\cos\theta)/4$ **21.** -1 **23.** 0
25. (a) 50 vibrations/sec/lb (b) 33.33 vibrations/sec/lb

27. (a) $s(t) = -16t^2 - 30t + 600$
$v(t) = -32t - 30$
(b) -94 ft/sec
(c) $v'(1) = -62$ ft/sec; $v'(3) = -126$ ft/sec
(d) About 5.258 sec (e) About -198.256 ft/sec

29. $4(5x^3 - 15x^2 - 11x - 8)$ **31.** $\sqrt{x}\cos x + \sin x/(2\sqrt{x})$

33. $\dfrac{-(x^2+1)}{(x^2-1)^2}$ **35.** $\dfrac{4x^3\cos x + x^4\sin x}{\cos^2 x}$

37. $3x^2\sec x\tan x + 6x\sec x$ **39.** $-x\sin x$

41. $y = 4x + 10$ **43.** $y = -8x + 1$ **45.** $-48t$

47. $\frac{225}{4}\sqrt{x}$ **49.** $6\sec^2\theta\tan\theta$

51. $v(3) = 11$ m/sec; $a(3) = -6$ m/sec^2 **53.** $28(7x+3)^3$

55. $-\dfrac{2x}{(x^2+4)^2}$ **57.** $-45\sin(9x+1)$

59. $\frac{1}{2}(1 - \cos 2x) = \sin^2 x$ **61.** $(36x+1)(6x+1)^4$

63. $\dfrac{3}{(x^2+1)^{3/2}}$ **65.** $\dfrac{-3x^2}{2\sqrt{1-x^3}}$; -2 **67.** $-\dfrac{8x}{(x^2+1)^2}$; 2

69. $-\csc 2x\cot 2x$; 0 **71.** $384(8x+5)$ **73.** $2\csc^2 x\cot x$

75. (a) $-18.667°$/h (b) $-7.284°$/h
(c) $-3.240°$/h (d) $-0.747°$/h

77. $-\dfrac{x}{y}$ **79.** $\dfrac{y(y^2-3x^2)}{x(x^2-3y^2)}$ **81.** $\dfrac{y\sin x + \sin y}{\cos x - x\cos y}$

83. Tangent line: $3x + y - 10 = 0$
Normal line: $x - 3y = 0$

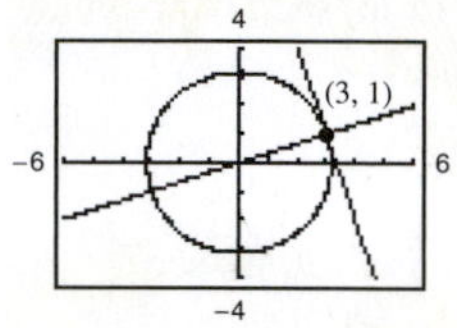

85. (a) $2\sqrt{2}$ units/sec (b) 4 units/sec (c) 8 units/sec

87. 450π km/h

P.S. Problem Solving *(page 159)*

1. (a) $r = \frac{1}{2}$; $x^2 + \left(y - \frac{1}{2}\right)^2 = \frac{1}{4}$
(b) Center: $\left(0, \frac{5}{4}\right)$; $x^2 + \left(y - \frac{5}{4}\right)^2 = 1$

3. $p(x) = 2x^3 + 4x^2 - 5$

5. (a) $y = 4x - 4$ (b) $y = -\frac{1}{4}x + \frac{9}{2}$; $\left(-\frac{9}{4}, \frac{81}{16}\right)$
(c) Tangent line: $y = 0$ (d) Proof
Normal line: $x = 0$

7. (a) Graph $\begin{cases} y_1 = \dfrac{1}{a}\sqrt{x^2(a^2 - x^2)} \\ y_2 = -\dfrac{1}{a}\sqrt{x^2(a^2 - x^2)} \end{cases}$ as separate equations.

(b) Answers will vary. Sample answer:

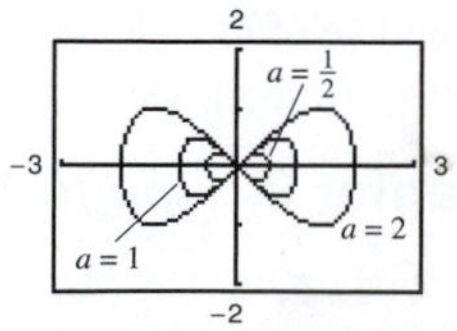

The intercepts will always be $(0, 0)$, $(a, 0)$, and $(-a, 0)$, and the maximum and minimum y-values appear to be $\pm\frac{1}{2}a$.

(c) $\left(\frac{a\sqrt{2}}{2}, \frac{a}{2}\right)$, $\left(\frac{a\sqrt{2}}{2}, -\frac{a}{2}\right)$, $\left(-\frac{a\sqrt{2}}{2}, \frac{a}{2}\right)$, $\left(-\frac{a\sqrt{2}}{2}, -\frac{a}{2}\right)$

9. (a) When the man is 90 ft from the light, the tip of his shadow is $112\frac{1}{2}$ ft from the light. The tip of the child's shadow is $111\frac{1}{9}$ ft from the light, so the man's shadow extends $1\frac{7}{18}$ ft beyond the child's shadow.
(b) When the man is 60 ft from the light, the tip of his shadow is 75 ft from the light. The tip of the child's shadow is $77\frac{7}{9}$ ft from the light, so the child's shadow extends $2\frac{7}{9}$ ft beyond the man's shadow.
(c) $d = 80$ ft
(d) Let x be the distance of the man from the light, and let s be the distance from the light to the tip of the shadow.
If $0 < x < 80$, then $ds/dt = -50/9$.
If $x > 80$, then $ds/dt = -25/4$.
There is a discontinuity at $x = 80$.

11. (a) $v(t) = -\frac{27}{5}t + 27$ ft/sec (b) 5 sec; 73.5 ft
$a(t) = -\frac{27}{5}$ ft/sec^2
(c) The acceleration due to gravity on Earth is greater in magnitude than that on the moon.

13. Proof. The graph of L is a line passing through the origin $(0, 0)$.

15. (a) j would be the rate of change of acceleration.
(b) $j = 0$. Acceleration is constant, so there is no change in acceleration.
(c) a: position function, d: velocity function,
b: acceleration function, c: jerk function

Chapter 3

Section 3.1 *(page 167)*

1. $f'(0) = 0$ **3.** $f'(2) = 0$ **5.** $f'(-2)$ is undefined.

7. 2, absolute maximum (and relative maximum)

9. 1, absolute maximum (and relative maximum);
2, absolute minimum (and relative minimum);
3, absolute maximum (and relative maximum)

11. $x = 0, x = 2$ **13.** $t = 8/3$ **15.** $x = \pi/3, \pi, 5\pi/3$

17. Minimum: $(2, 1)$
Maximum: $(-1, 4)$

19. Minimum: $(2, -8)$
Maximum: $(6, 24)$

21. Minimum: $\left(-1, -\frac{5}{2}\right)$
Maximum: $(2, 2)$

23. Minimum: $(0, 0)$
Maximum: $(-1, 5)$

25. Minimum: $(0, 0)$
Maxima: $\left(-1, \frac{1}{4}\right)$ and $\left(1, \frac{1}{4}\right)$

27. Minimum: $(1, -1)$
Maximum: $\left(0, -\frac{1}{2}\right)$

29. Minimum: $(-1, -1)$
Maximum: $(3, 3)$

31. Minimum value is -2 for $-2 \le x < -1$.
Maximum: $(2, 2)$

33. Minimum: $(3\pi/2, -1)$
Maximum: $(5\pi/6, 1/2)$

35. Minimum: $(\pi, -3)$
Maxima: $(0, 3)$ and $(2\pi, 3)$

37. (a) Minimum: $(0, -3)$;
Maximum: $(2, 1)$
(b) Minimum: $(0, -3)$
(c) Maximum: $(2, 1)$
(d) No extrema

39. (a) Minimum: $(1, -1)$;
Maximum: $(-1, 3)$
(b) Maximum: $(3, 3)$
(c) Minimum: $(1, -1)$
(d) Minimum: $(1, -1)$

41.

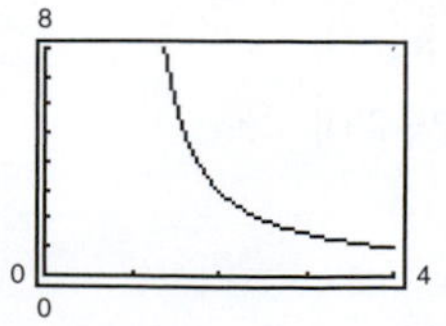

Minimum: $(4, 1)$

43. 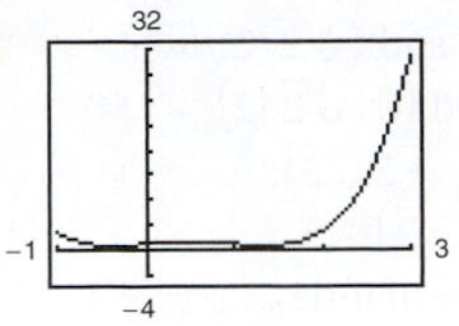

Minima: $\left(\dfrac{-\sqrt{3}+1}{2}, \dfrac{3}{4}\right)$ and $\left(\dfrac{\sqrt{3}+1}{2}, \dfrac{3}{4}\right)$

Maximum: $(3, 31)$

45. (a) 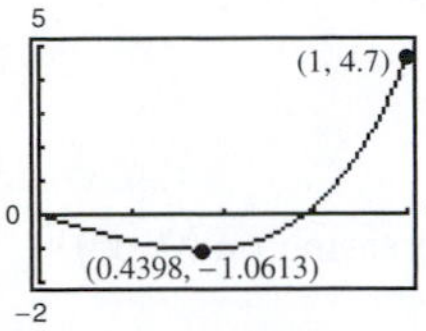

(b) Minimum: $(0.4398, -1.0613)$

47. Maximum: $\left|f''\left(\sqrt[3]{-10+\sqrt{108}}\right)\right| = f''\left(\sqrt{3}-1\right) \approx 1.47$

49. Maximum: $\left|f^{(4)}(0)\right| = \frac{56}{81}$

51. Answers will vary. Sample answer: Let $f(x) = 1/x$. f is continuous on $(0, 1)$ but does not have a maximum or minimum.

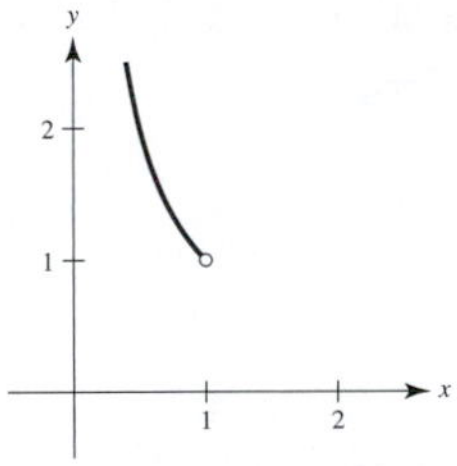

53. Answers will vary. Sample answer:

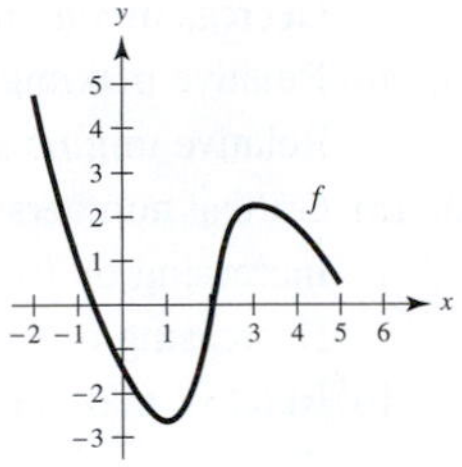

55. (a) Yes (b) No **57.** (a) No (b) Yes

59. Maximum: $P(12) = 72$; No. P is decreasing for $I > 12$.

61. $\theta = \operatorname{arcsec}\sqrt{3} \approx 0.9553$ rad

63. True **65.** True **67.** Proof

69. Putnam Problem B3, 2004

Section 3.2 *(page 174)*

1. $f(-1) = f(1) = 1$; f is not continuous on $[-1, 1]$.

3. $f(0) = f(2) = 0$; f is not differentiable on $(0, 2)$.

5. $(2, 0), (-1, 0)$; $f'\left(\frac{1}{2}\right) = 0$ **7.** $(0, 0), (-4, 0)$; $f'\left(-\frac{8}{3}\right) = 0$

9. $f'\left(\frac{3}{2}\right) = 0$ **11.** $f'\left(\dfrac{6-\sqrt{3}}{3}\right) = 0$; $f'\left(\dfrac{6+\sqrt{3}}{3}\right) = 0$

13. Not differentiable at $x = 0$ **15.** $f'\left(-2+\sqrt{5}\right) = 0$

17. $f'\left(\dfrac{\pi}{2}\right) = 0$; $f'\left(\dfrac{3\pi}{2}\right) = 0$ **19.** $f'\left(\dfrac{\pi}{6}\right) = 0$

21. Not continuous on $[0, \pi]$

23. 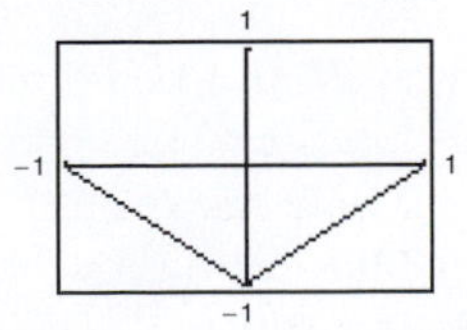

Rolle's Theorem does not apply.

25. 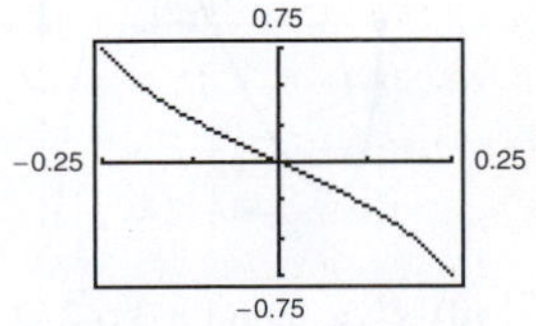

Rolle's Theorem does not apply.

27. (a) $f(1) = f(2) = 38$

(b) Velocity $= 0$ for some t in $(1, 2)$; $t = \frac{3}{2}$ sec

29. 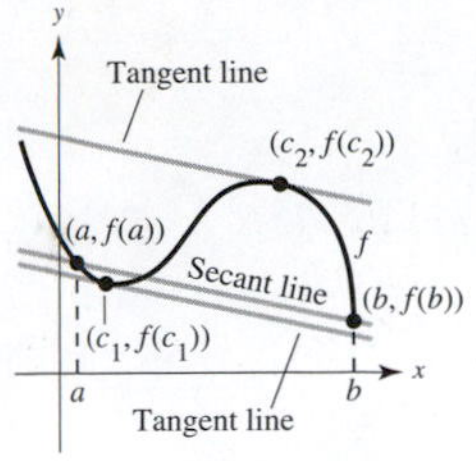

31. The function is not continuous on $[0, 6]$.

33. The function is not continuous on $[0, 6]$.

35. (a) Secant line: $x + y - 3 = 0$ (b) $c = \frac{1}{2}$

(c) Tangent line: $4x + 4y - 21 = 0$

(d) 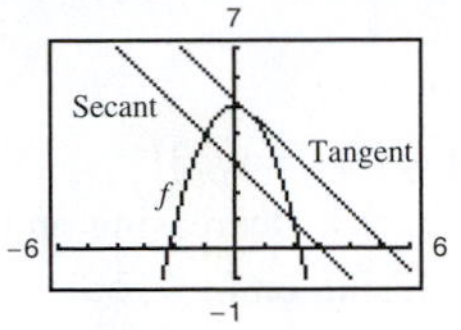

37. $f'(-1/2) = -1$ **39.** $f'\left(1/\sqrt{3}\right) = 3$, $f'\left(-1/\sqrt{3}\right) = 3$

41. $f'\left(\frac{8}{27}\right) = 1$ **43.** f is not differentiable at $x = -\frac{1}{2}$.

45. $f'(\pi/2) = 0$

47. (a)–(c) 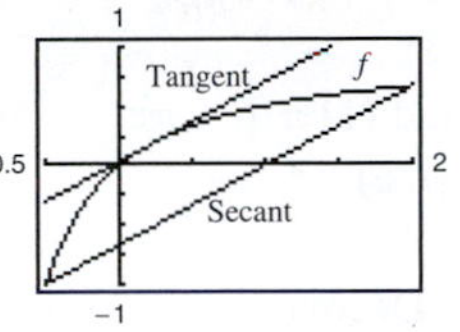

(b) $y = \frac{2}{3}(x - 1)$

(c) $y = \frac{1}{3}\left(2x + 5 - 2\sqrt{6}\right)$

49. (a)–(c) 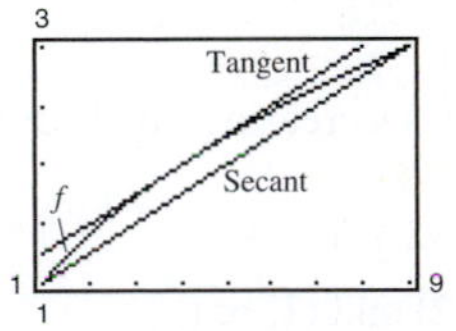

(b) $y = \frac{1}{4}x + \frac{3}{4}$

(c) $y = \frac{1}{4}x + 1$

51. (a) -14.7 m/sec (b) 1.5 sec

53. No. Let $f(x) = x^2$ on $[-1, 2]$.

55. No. $f(x)$ is not continuous on $[0, 1]$. So it does not satisfy the hypothesis of Rolle's Theorem.

57. By the Mean Value Theorem, there is a time when the speed of the plane must equal the average speed of 454.5 miles/hour. The speed was 400 miles/hour when the plane was accelerating to 454.5 miles/hour and decelerating from 454.5 miles/hour.

59. Proof

61. (a) 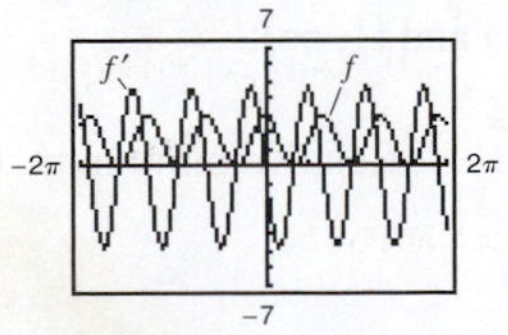

(b) Yes; yes

(c) Because $f(-1) = f(1) = 0$, Rolle's Theorem applies on $[-1, 1]$. Because $f(1) = 0$ and $f(2) = 3$, Rolle's Theorem does not apply on $[1, 2]$.

(d) $\displaystyle\lim_{x\to 3^-} f'(x) = 0$; $\displaystyle\lim_{x\to 3^+} f'(x) = 0$

63.

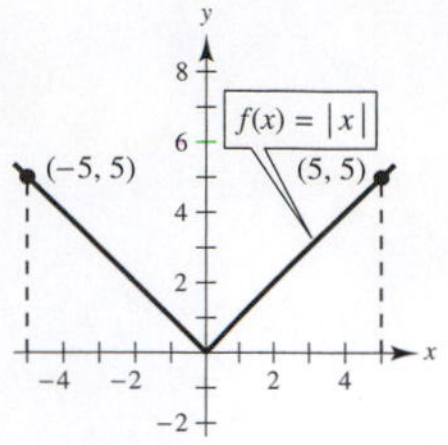

65–67. Proofs **69.** $f(x) = 5$ **71.** $f(x) = x^2 - 1$

73. False. f is not continuous on $[-1, 1]$. **75.** True

77–85. Proofs

Section 3.3 *(page 183)*

1. (a) $(0, 6)$ (b) $(6, 8)$

3. Increasing on $(3, \infty)$; Decreasing on $(-\infty, 3)$

5. Increasing on $(-\infty, -2)$ and $(2, \infty)$; Decreasing on $(-2, 2)$

7. Increasing on $(-\infty, -1)$; Decreasing on $(-1, \infty)$

9. Increasing on $(1, \infty)$; Decreasing on $(-\infty, 1)$

11. Increasing on $(-2\sqrt{2}, 2\sqrt{2})$;
Decreasing on $(-4, -2\sqrt{2})$ and $(2\sqrt{2}, 4)$

13. Increasing on $(0, \pi/2)$ and $(3\pi/2, 2\pi)$;
Decreasing on $(\pi/2, 3\pi/2)$

15. Increasing on $(0, 7\pi/6)$ and $(11\pi/6, 2\pi)$;
Decreasing on $(7\pi/6, 11\pi/6)$

17. (a) Critical number: $x = 2$
(b) Increasing on $(2, \infty)$; Decreasing on $(-\infty, 2)$
(c) Relative minimum: $(2, -4)$

19. (a) Critical number: $x = 1$
(b) Increasing on $(-\infty, 1)$; Decreasing on $(1, \infty)$
(c) Relative maximum: $(1, 5)$

21. (a) Critical numbers: $x = -2, 1$
(b) Increasing on $(-\infty, -2)$ and $(1, \infty)$;
Decreasing on $(-2, 1)$
(c) Relative maximum: $(-2, 20)$;
Relative minimum: $(1, -7)$

23. (a) Critical numbers: $x = -\frac{5}{3}, 1$
(b) Increasing on $\left(-\infty, -\frac{5}{3}\right)$, $(1, \infty)$;
Decreasing on $\left(-\frac{5}{3}, 1\right)$
(c) Relative maximum: $\left(-\frac{5}{3}, \frac{256}{27}\right)$;
Relative minimum: $(1, 0)$

25. (a) Critical numbers: $x = \pm 1$
(b) Increasing on $(-\infty, -1)$ and $(1, \infty)$;
Decreasing on $(-1, 1)$
(c) Relative maximum: $\left(-1, \frac{4}{5}\right)$; Relative minimum: $\left(1, -\frac{4}{5}\right)$

27. (a) Critical number: $x = 0$
(b) Increasing on $(-\infty, \infty)$
(c) No relative extrema

29. (a) Critical number: $x = -2$
(b) Increasing on $(-2, \infty)$; Decreasing on $(-\infty, -2)$
(c) Relative minimum: $(-2, 0)$

31. (a) Critical number: $x = 5$
(b) Increasing on $(-\infty, 5)$; Decreasing on $(5, \infty)$
(c) Relative maximum: $(5, 5)$

33. (a) Critical numbers: $x = \pm\sqrt{2}/2$; Discontinuity: $x = 0$
(b) Increasing on $(-\infty, -\sqrt{2}/2)$ and $(\sqrt{2}/2, \infty)$;
Decreasing on $(-\sqrt{2}/2, 0)$ and $(0, \sqrt{2}/2)$
(c) Relative maximum: $(-\sqrt{2}/2, -2\sqrt{2})$;
Relative minimum: $(\sqrt{2}/2, 2\sqrt{2})$

35. (a) Critical number: $x = 0$; Discontinuities: $x = \pm 3$
(b) Increasing on $(-\infty, -3)$ and $(-3, 0)$;
Decreasing on $(0, 3)$ and $(3, \infty)$
(c) Relative maximum: $(0, 0)$

37. (a) Critical number: $x = 0$
(b) Increasing on $(-\infty, 0)$; Decreasing on $(0, \infty)$
(c) Relative maximum: $(0, 4)$

39. (a) Critical number: $x = 1$
(b) Increasing on $(-\infty, 1)$; Decreasing on $(1, \infty)$
(c) Relative maximum: $(1, 4)$

41. (a) Critical numbers: $x = \pi/6, 5\pi/6$;
Increasing on $(0, \pi/6)$, $(5\pi/6, 2\pi)$;
Decreasing on $(\pi/6, 5\pi/6)$
(b) Relative maximum: $(\pi/6, (\pi + 6\sqrt{3})/12)$;
Relative minimum: $(5\pi/6, (5\pi - 6\sqrt{3})/12)$

43. (a) Critical numbers: $x = \pi/4, 5\pi/4$;
Increasing on $(0, \pi/4)$, $(5\pi/4, 2\pi)$;
Decreasing on $(\pi/4, 5\pi/4)$
(b) Relative maximum: $(\pi/4, \sqrt{2})$;
Relative minimum: $(5\pi/4, -\sqrt{2})$

45. (a) Critical numbers:
$x = \pi/4, \pi/2, 3\pi/4, \pi, 5\pi/4, 3\pi/2, 7\pi/4$;
Increasing on $(\pi/4, \pi/2)$, $(3\pi/4, \pi)$, $(5\pi/4, 3\pi/2)$, $(7\pi/4, 2\pi)$;
Decreasing on $(0, \pi/4)$, $(\pi/2, 3\pi/4)$, $(\pi, 5\pi/4)$, $(3\pi/2, 7\pi/4)$;
(b) Relative maxima: $(\pi/2, 1)$, $(\pi, 1)$, $(3\pi/2, 1)$;
Relative minima: $(\pi/4, 0)$, $(3\pi/4, 0)$, $(5\pi/4, 0)$, $(7\pi/4, 0)$

47. (a) Critical numbers: $\pi/2, 7\pi/6, 3\pi/2, 11\pi/6$;
Increasing on $(0, \pi/2)$, $(7\pi/6, 3\pi/2)$, $(11\pi/6, 2\pi)$;
Decreasing on $(\pi/2, 7\pi/6)$, $(3\pi/2, 11\pi/6)$
(b) Relative maxima: $(\pi/2, 2)$, $(3\pi/2, 0)$;
Relative minima: $(7\pi/6, -1/4)$, $(11\pi/6, -1/4)$

49. (a) $f'(x) = 2(9 - 2x^2)/\sqrt{9 - x^2}$

(b)

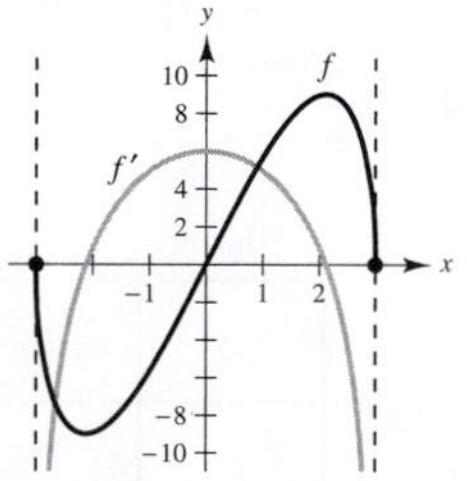

(c) Critical numbers:
$x = \pm 3\sqrt{2}/2$

(d) $f' > 0$ on $(-3\sqrt{2}/2, 3\sqrt{2}/2)$;
$f' < 0$ on $(-3, -3\sqrt{2}/2)$, $(3\sqrt{2}/2, 3)$
f is increasing when f' is positive and decreasing when f' is negative.

51. (a) $f'(t) = t(t\cos t + 2\sin t)$

(b)

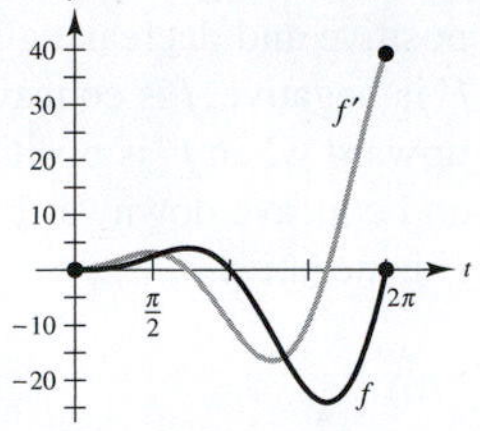

(c) Critical numbers: $t = 2.2889, 5.0870$

(d) $f' > 0$ on $(0, 2.2889)$, $(5.0870, 2\pi)$;
$f' < 0$ on $(2.2889, 5.0870)$
f is increasing when f' is positive and decreasing when f' is negative.

53. (a) $f'(x) = -\cos(x/3)$

(b)

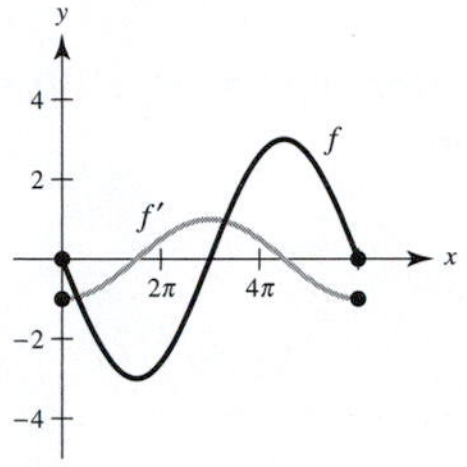

(c) Critical numbers: $x = 3\pi/2, 9\pi/2$

(d) $f' > 0$ on $\left(\dfrac{3\pi}{2}, \dfrac{9\pi}{2}\right)$;

$f' < 0$ on $\left(0, \dfrac{3\pi}{2}\right), \left(\dfrac{9\pi}{2}, 6\pi\right)$

f is increasing when f' is positive and decreasing when f' is negative.

55. $f(x)$ is symmetric with respect to the origin.
Zeros: $(0, 0), (\pm\sqrt{3}, 0)$

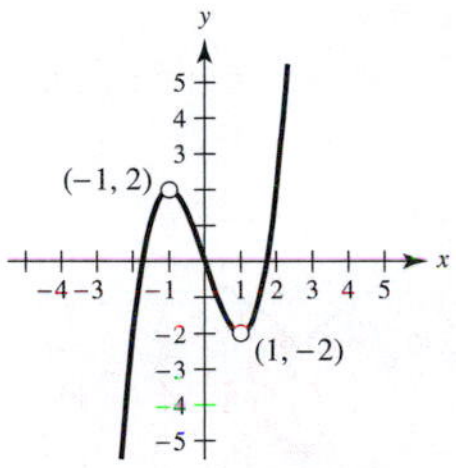

$g(x)$ is continuous on $(-\infty, \infty)$, and $f(x)$ has holes at $x = 1$ and $x = -1$.

57.

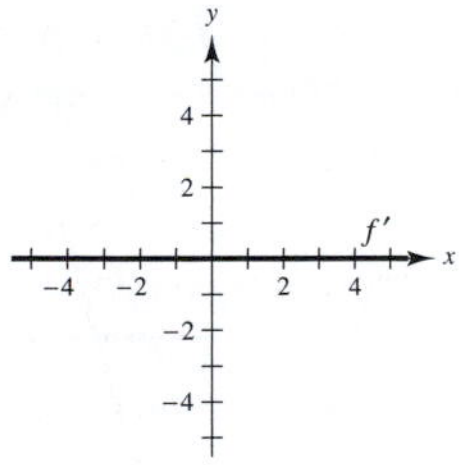

59.

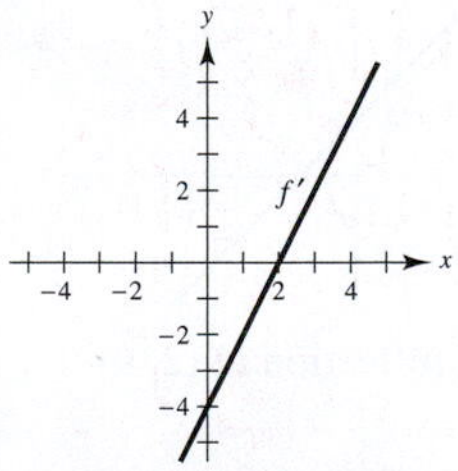

61.

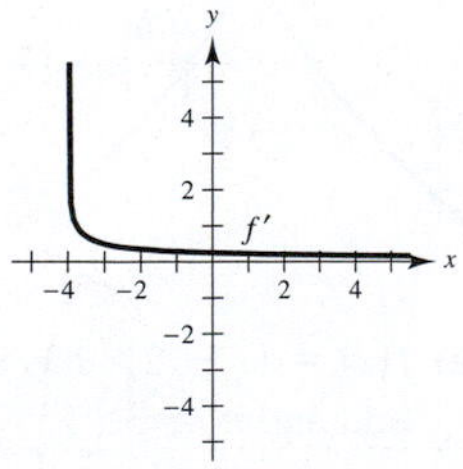

63. $g'(0) < 0$ **65.** $g'(-6) < 0$ **67.** $g'(0) > 0$

69. Answers will vary. Sample answer:

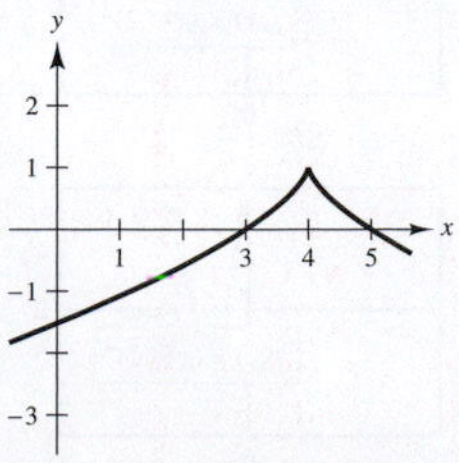

71. $(5, f(5))$ is a relative minimum.

73. (a)

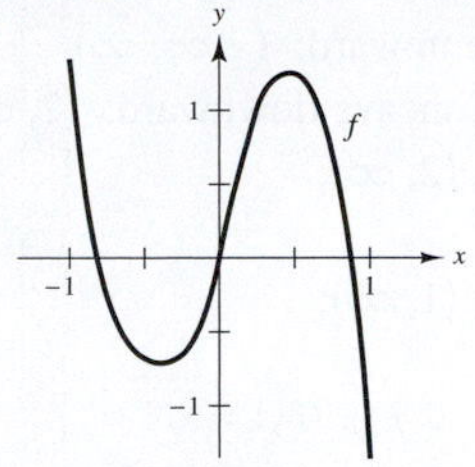

(b) Critical numbers: $x \approx -0.40$ and $x \approx 0.48$
(c) Relative maximum: $(0.48, 1.25)$;
Relative minimum: $(-0.40, 0.75)$

75. (a) $s'(t) = 9.8(\sin\theta)t$; speed $= |9.8(\sin\theta)t|$

(b)

θ	0	$\pi/4$	$\pi/3$	$\pi/2$	$2\pi/3$	$3\pi/4$	π
$s'(t)$	0	$4.9\sqrt{2}t$	$4.9\sqrt{3}t$	$9.8t$	$4.9\sqrt{3}t$	$4.9\sqrt{2}t$	0

The speed is maximum at $\theta = \pi/2$.

77. (a)

t	0	0.5	1	1.5	2	2.5	3
$C(t)$	0	0.055	0.107	0.148	0.171	0.176	0.167

$t = 2.5$ h

(b)

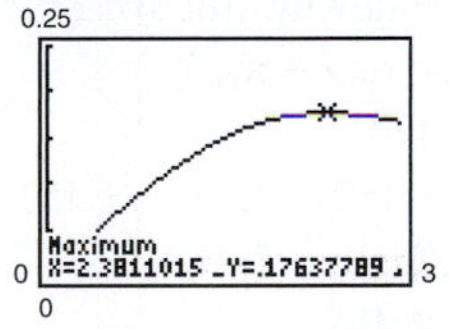

$t \approx 2.38$ h (c) $t \approx 2.38$ h

79. $r = 2R/3$

81. (a) $v(t) = 6 - 2t$ (b) $[0, 3)$ (c) $(3, \infty)$ (d) $t = 3$

83. (a) $v(t) = 3t^2 - 10t + 4$
(b) $\left[0, \left(5 - \sqrt{13}\right)/3\right)$ and $\left(\left(5 + \sqrt{13}\right)/3, \infty\right)$
(c) $\left(\dfrac{5 - \sqrt{13}}{3}, \dfrac{5 + \sqrt{13}}{3}\right)$ (d) $t = \dfrac{5 \pm \sqrt{13}}{3}$

85. Answers will vary.

87. (a) Minimum degree: 3
(b) $a_3(0)^3 + a_2(0)^2 + a_1(0) + a_0 = 0$
$a_3(2)^3 + a_2(2)^2 + a_1(2) + a_0 = 2$
$3a_3(0)^2 + 2a_2(0) + a_1 = 0$
$3a_3(2)^2 + 2a_2(2) + a_1 = 0$
(c) $f(x) = -\frac{1}{2}x^3 + \frac{3}{2}x^2$

89. (a) Minimum degree: 4
(b) $a_4(0)^4 + a_3(0)^3 + a_2(0)^2 + a_1(0) + a_0 = 0$
$a_4(2)^4 + a_3(2)^3 + a_2(2)^2 + a_1(2) + a_0 = 4$
$a_4(4)^4 + a_3(4)^3 + a_2(4)^2 + a_1(4) + a_0 = 0$
$4a_4(0)^3 + 3a_3(0)^2 + 2a_2(0) + a_1 = 0$
$4a_4(2)^3 + 3a_3(2)^2 + 2a_2(2) + a_1 = 0$
$4a_4(4)^3 + 3a_3(4)^2 + 2a_2(4) + a_1 = 0$
(c) $f(x) = \frac{1}{4}x^4 - 2x^3 + 4x^2$

91. True **93.** False. Let $f(x) = x^3$.

95. False. Let $f(x) = x^3$. There is a critical number at $x = 0$, but not a relative extremum.

97–99. Proofs **101.** Putnam Problem A3, 2003

Section 3.4 *(page 192)*

1. $f' > 0, f'' < 0$ **3.** Concave upward: $(-\infty, \infty)$

5. Concave upward: $(-\infty, 2)$; Concave downward: $(2, \infty)$

7. Concave upward: $(-\infty, -2), (2, \infty)$; Concave downward: $(-2, 2)$

9. Concave upward: $(-\infty, -1), (1, \infty)$; Concave downward: $(-1, 1)$

11. Concave upward: $(-2, 2)$; Concave downward: $(-\infty, -2), (2, \infty)$

13. Concave upward: $(-\pi/2, 0)$; Concave downward: $(0, \pi/2)$

15. Point of inflection: $(2, 8)$; Concave downward: $(-\infty, 2)$; Concave upward: $(2, \infty)$

17. Points of inflection: $(-2, -8)$, $(0, 0)$; Concave upward: $(-\infty, -2)$, $(0, \infty)$; Concave downward: $(-2, 0)$

19. Points of inflection: $(2, -16), (4, 0)$; Concave upward: $(-\infty, 2), (4, \infty)$; Concave downward: $(2, 4)$

21. Concave upward: $(-3, \infty)$

23. Points of inflection: $\left(-\sqrt{3}/3, 3\right), \left(\sqrt{3}/3, 3\right)$; Concave upward: $\left(-\infty, -\sqrt{3}/3\right), \left(\sqrt{3}/3, \infty\right)$; Concave downward: $\left(-\sqrt{3}/3, \sqrt{3}/3\right)$

25. Point of inflection: $(2\pi, 0)$; Concave upward: $(2\pi, 4\pi)$; Concave downward: $(0, 2\pi)$

27. Concave upward: $(0, \pi)$, $(2\pi, 3\pi)$; Concave downward: $(\pi, 2\pi)$, $(3\pi, 4\pi)$

29. Points of inflection: $(\pi, 0), (1.823, 1.452), (4.46, -1.452)$ Concave upward: $(1.823, \pi), (4.46, 2\pi)$ Concave downward: $(0, 1.823), (\pi, 4.46)$

31. Relative maximum: $(3, 9)$

33. Relative maximum: $(0, 3)$; Relative minimum: $(2, -1)$

35. Relative minimum: $(3, -25)$

37. Relative minimum: $(0, -3)$

39. Relative maximum: $(-2, -4)$; Relative minimum: $(2, 4)$

41. No relative extrema, because f is nonincreasing.

43. (a) $f'(x) = 0.2x(x - 3)^2(5x - 6)$; $f''(x) = 0.4(x - 3)(10x^2 - 24x + 9)$

(b) Relative maximum: $(0, 0)$; Relative minimum: $(1.2, -1.6796)$; Points of inflection: $(0.4652, -0.7048)$, $(1.9348, -0.9048)$, $(3, 0)$

(c)

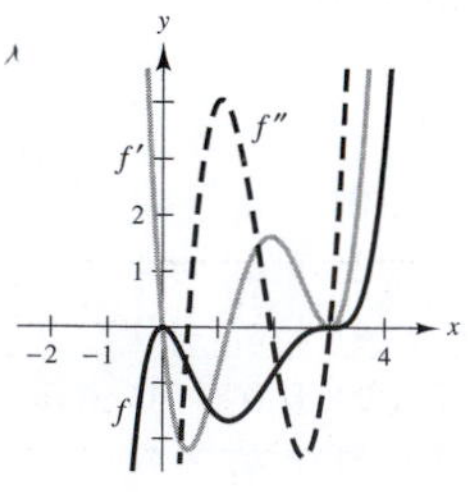

f is increasing when f' is positive and decreasing when f' is negative. f is concave upward when f'' is positive and concave downward when f'' is negative.

45. (a) $f'(x) = \cos x - \cos 3x + \cos 5x$; $f''(x) = -\sin x + 3 \sin 3x - 5 \sin 5x$

(b) Relative maximum: $(\pi/2, 1.53333)$; Points of inflection: $(\pi/6, 0.2667), (1.1731, 0.9637)$, $(1.9685, 0.9637), (5\pi/6, 0.2667)$

(c)

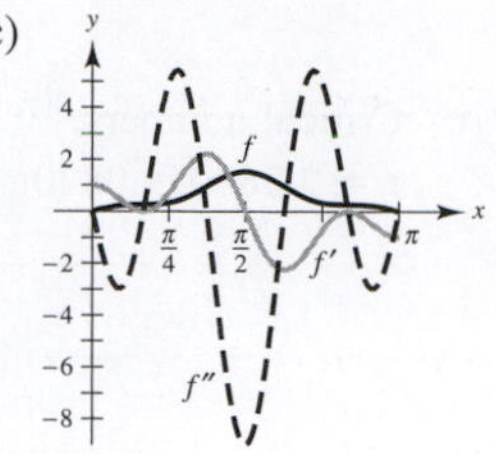

f is increasing when f' is positive and decreasing when f' is negative. f is concave upward when f'' is positive and concave downward when f'' is negative.

47. (a)

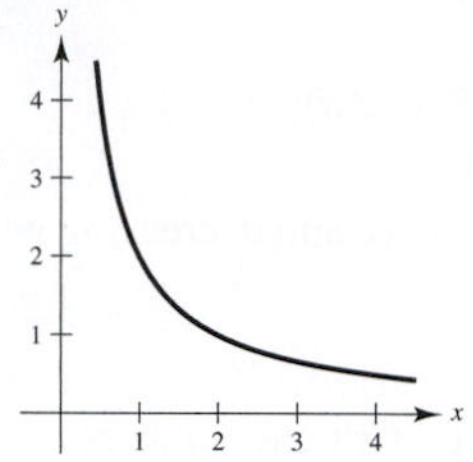

(b)

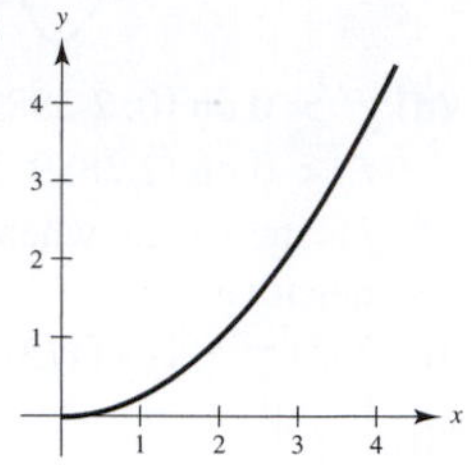

49. Answers will vary. Sample answer: $f(x) = x^4$; $f''(0) = 0$, but $(0, 0)$ is not a point of inflection.

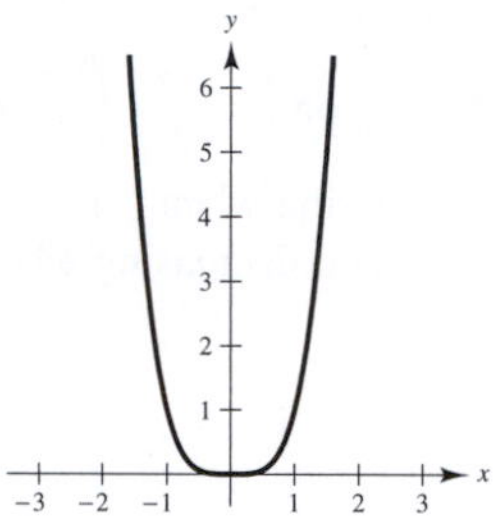

51.

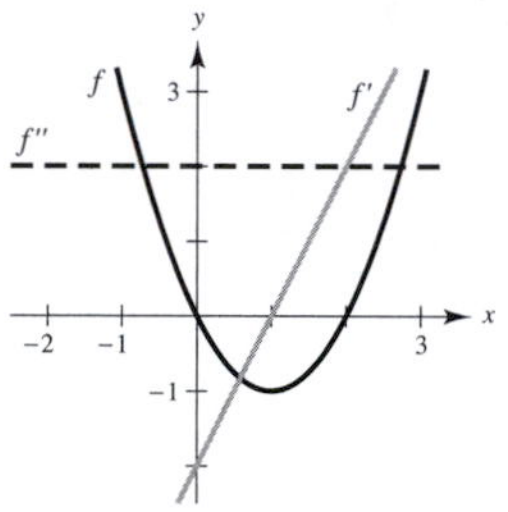

53.

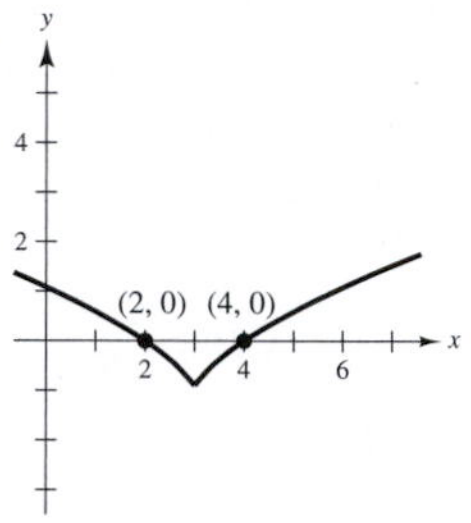

55.

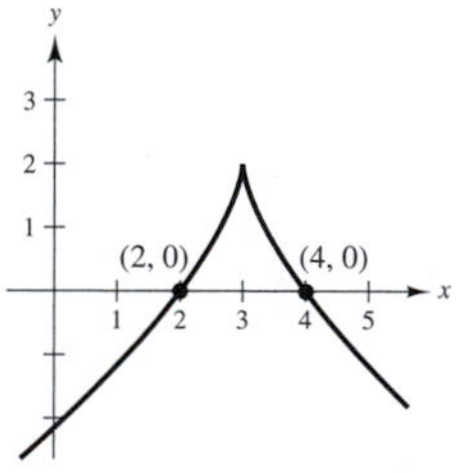

57. Sample answer:

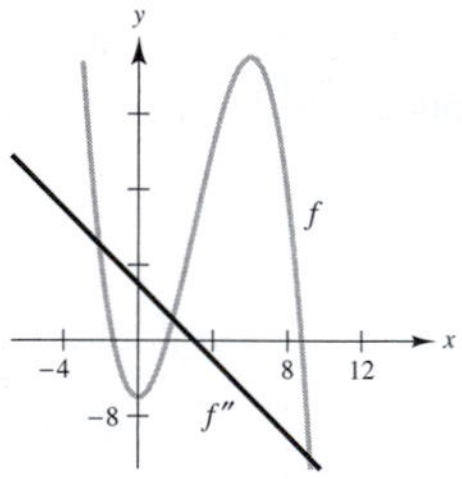

59. (a) $f(x) = (x - 2)^n$ has a point of inflection at $(2, 0)$ if n is odd and $n \geq 3$.

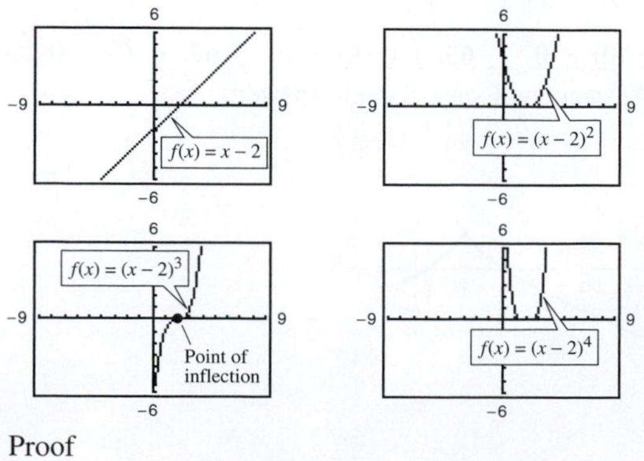

(b) Proof

61. $f(x) = \frac{1}{2}x^3 - 6x^2 + \frac{45}{2}x - 24$

63. (a) $f(x) = \frac{1}{32}x^3 + \frac{3}{16}x^2$ (b) Two miles from touchdown

65. $x = 100$ units

67. (a)

t	0.5	1	1.5	2	2.5	3
S	151.5	555.6	1097.6	1666.7	2193.0	2647.1

$1.5 < t < 2$

(b)

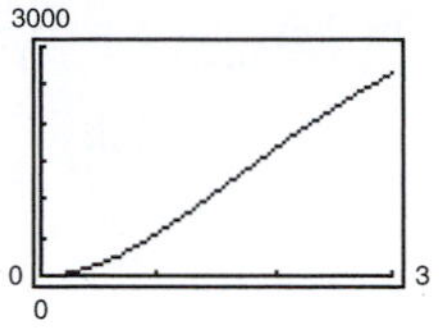

(c) About 1.633 yr

$t \approx 1.5$

69. $P_1(x) = 2\sqrt{2}$

$P_2(x) = 2\sqrt{2} - \sqrt{2}(x - \pi/4)^2$

The values of f, P_1, and P_2 and their first derivatives are equal when $x = \pi/4$. The approximations worsen as you move away from $x = \pi/4$.

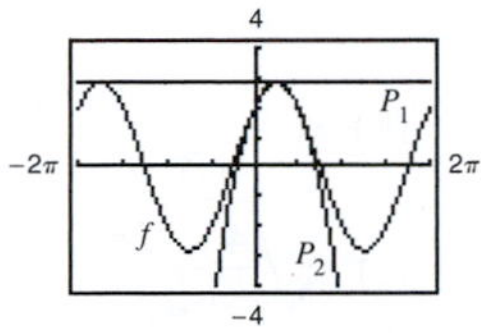

71. $P_1(x) = 1 - x/2$

$P_2(x) = 1 - x/2 - x^2/8$

The values of f, P_1, and P_2 and their first derivatives are equal when $x = 0$. The approximations worsen as you move away from $x = 0$.

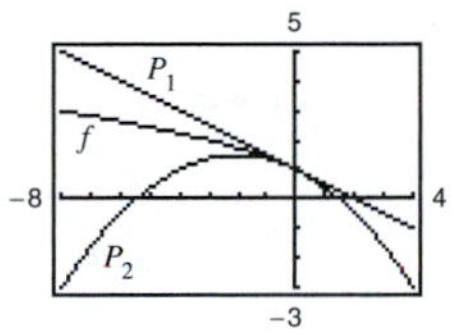

73.

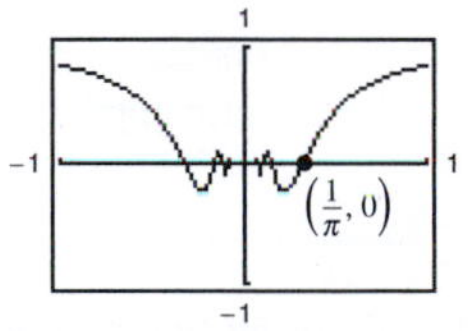

75. True

77. False. f is concave upward at $x = c$ if $f''(c) > 0$.

79. Proof

Section 3.5 *(page 202)*

1. f **2.** c **3.** d **4.** a **5.** b **6.** e

7.

x	10^0	10^1	10^2	10^3
$f(x)$	7	2.2632	2.0251	2.0025

x	10^4	10^5	10^6
$f(x)$	2.0003	2.0000	2.0000

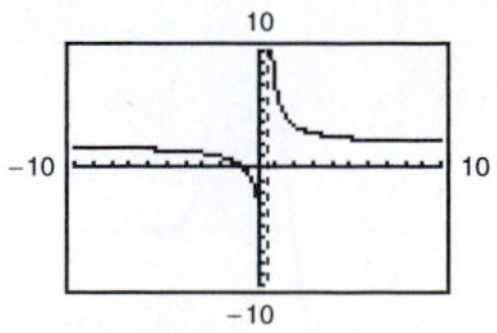

$\lim_{x\to\infty} \frac{4x + 3}{2x - 1} = 2$

9.

x	10^0	10^1	10^2	10^3
$f(x)$	-2	-2.9814	-2.9998	-3.0000

x	10^4	10^5	10^6
$f(x)$	-3.0000	-3.0000	-3.0000

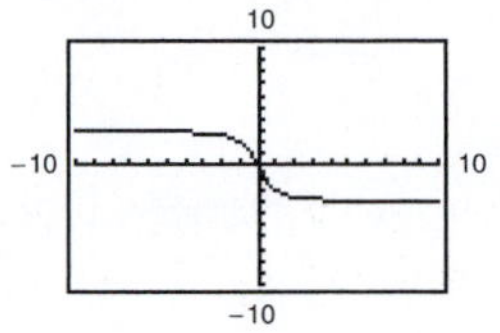

$\lim_{x\to\infty} \frac{-6x}{\sqrt{4x^2 + 5}} = -3$

11.

x	10^0	10^1	10^2	10^3
$f(x)$	4.5000	4.9901	4.9999	5.0000

x	10^4	10^5	10^6
$f(x)$	5.0000	5.0000	5.0000

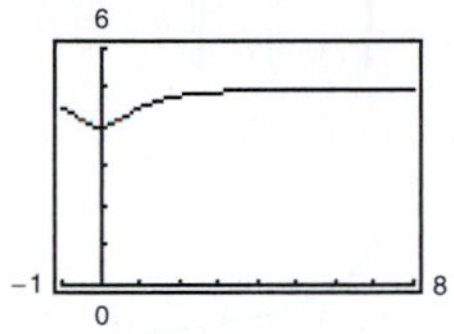

$\lim_{x\to\infty} \left(5 - \frac{1}{x^2 + 1}\right) = 5$

13. (a) ∞ (b) 5 (c) 0 **15.** (a) 0 (b) 1 (c) ∞

17. (a) 0 (b) $-\frac{2}{3}$ (c) $-\infty$ **19.** 4 **21.** $\frac{2}{3}$ **23.** 0

25. $-\infty$ **27.** -1 **29.** -2 **31.** $\frac{1}{2}$ **33.** ∞

35. 0 **37.** 0

39.

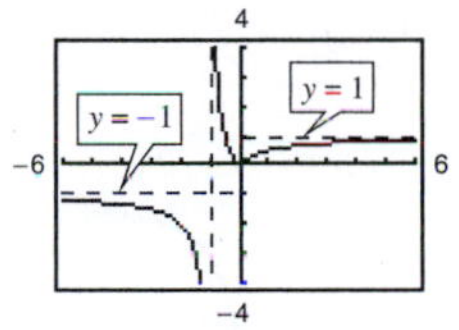

41.

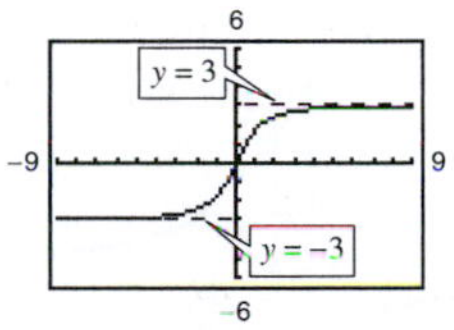

43. 1 **45.** 0 **47.** $\frac{1}{6}$

49.

x	10^0	10^1	10^2	10^3	10^4	10^5	10^6
$f(x)$	1.000	0.513	0.501	0.500	0.500	0.500	0.500

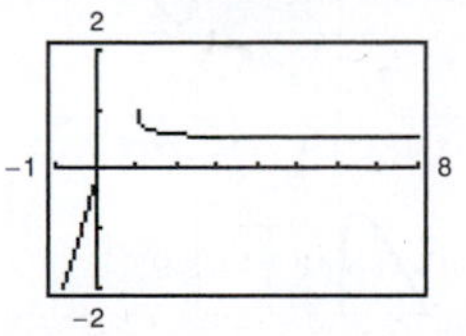

$\lim_{x\to\infty} \left[x - \sqrt{x(x - 1)}\right] = \frac{1}{2}$

51.

x	10^0	10^1	10^2	10^3	10^4	10^5	10^6
$f(x)$	0.479	0.500	0.500	0.500	0.500	0.500	0.500

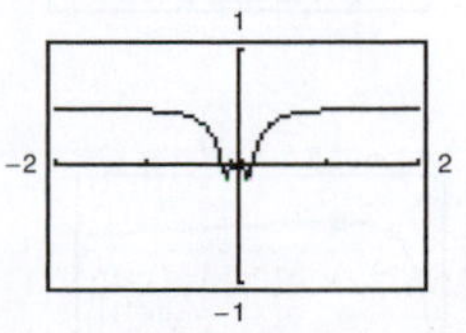

The graph has a hole at $x = 0$.

$\lim_{x\to\infty} x \sin \frac{1}{2x} = \frac{1}{2}$

53. As x becomes large, $f(x)$ approaches 4.

55. Answers will vary. Sample answer: Let

$$f(x) = \frac{-6}{0.1(x-2)^2 + 1} + 6.$$

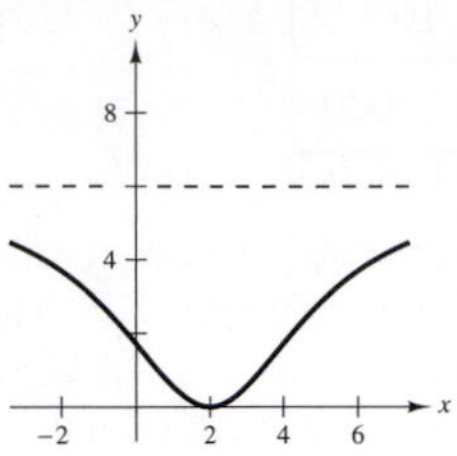

57. (a) 5 (b) -5

59.

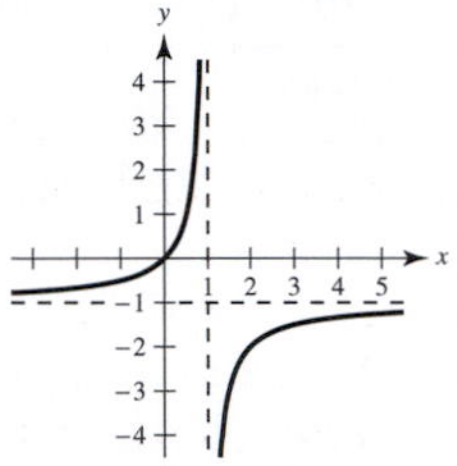

61.

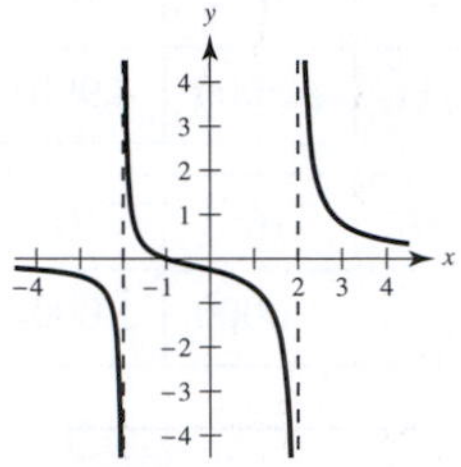

63.

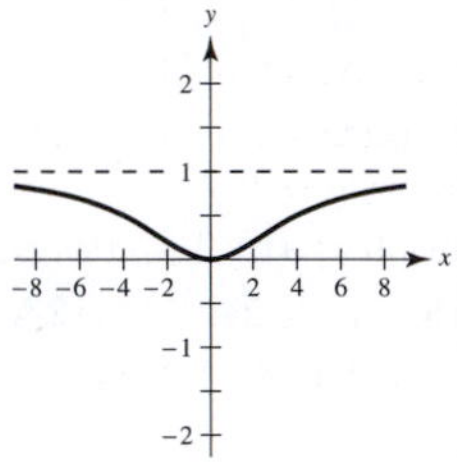

65.

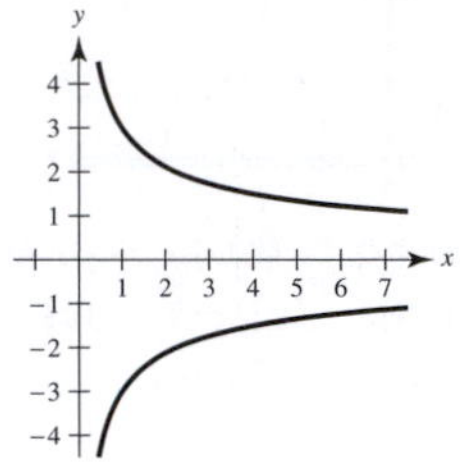

67.

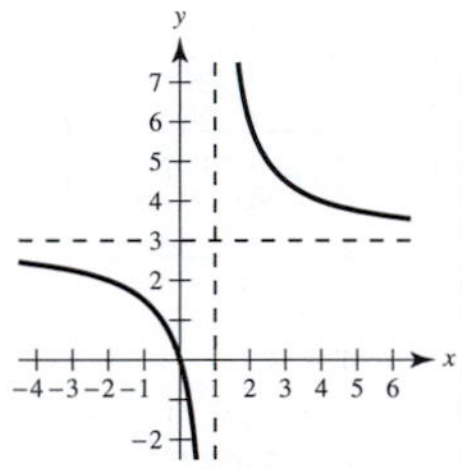

69.

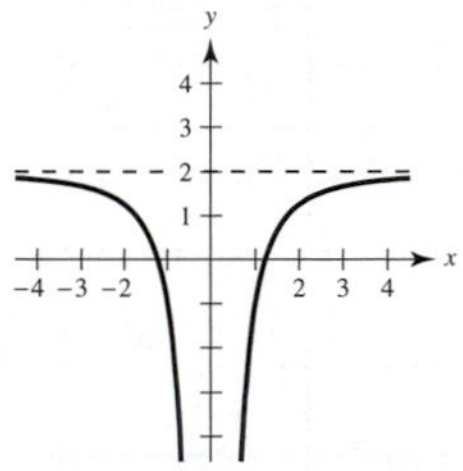

71.

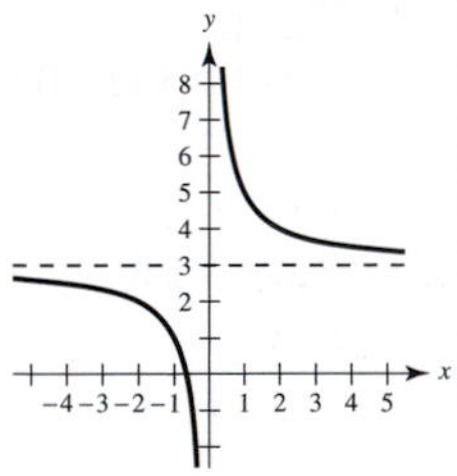

73.

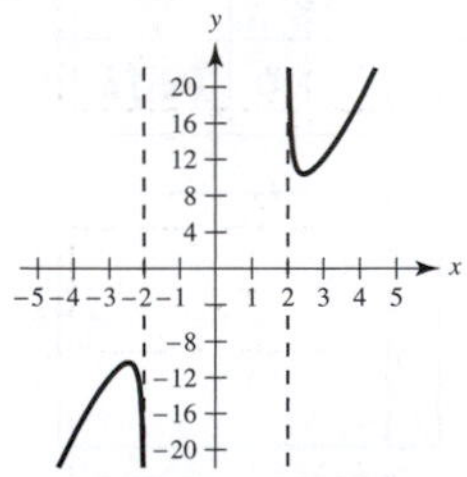

75.

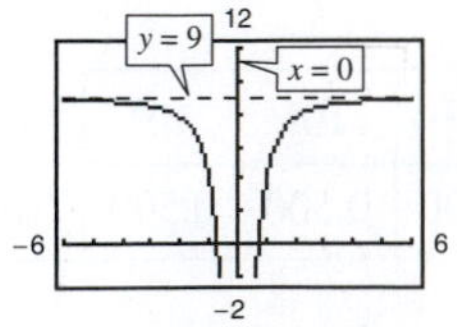

77.

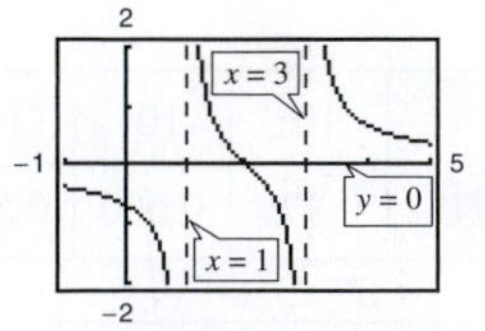

79.

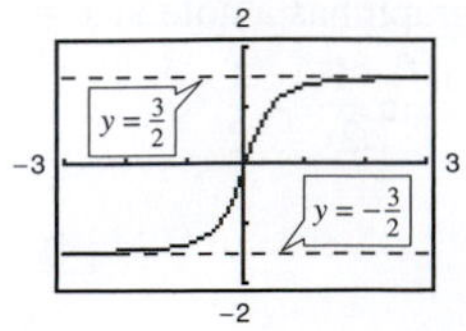

81.

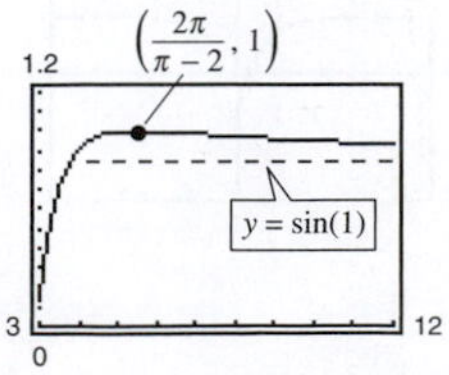

83. (a)

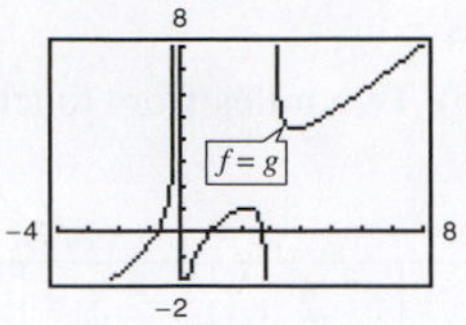

(b) Proof

(c)

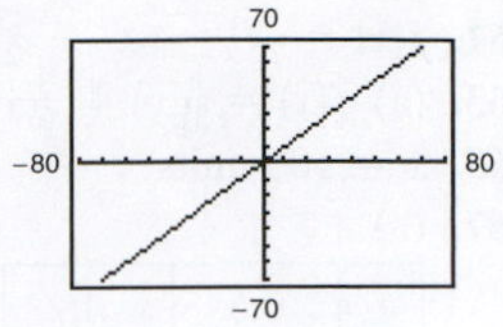

Slant asymptote: $y = x$

85. 100% **87.** $\lim_{t\to\infty} N(t) = +\infty$; $\lim_{t\to\infty} E(t) = c$

89. (a)

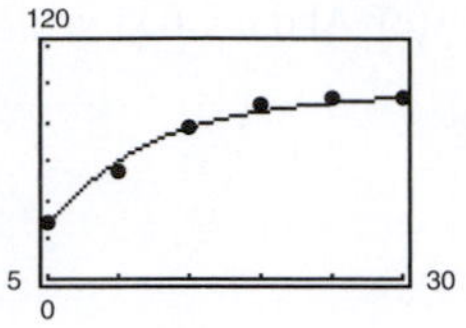

(b) Yes. $\lim_{t\to\infty} S = \frac{100}{1} = 100$

91. (a) $\lim_{x\to\infty} f(x) = 2$

(b) $x_1 = \sqrt{\frac{4-2\varepsilon}{\varepsilon}}$, $x_2 = -\sqrt{\frac{4-2\varepsilon}{\varepsilon}}$

(c) $\sqrt{\frac{4-2\varepsilon}{\varepsilon}}$ (d) $-\sqrt{\frac{4-2\varepsilon}{\varepsilon}}$

93. (a) Answers will vary. $M = \frac{5\sqrt{33}}{11}$ **95–97.** Proofs

(b) Answers will vary. $M = \frac{29\sqrt{177}}{59}$

99. (a) $d(m) = \frac{|3m + 3|}{\sqrt{m^2 + 1}}$

(b)

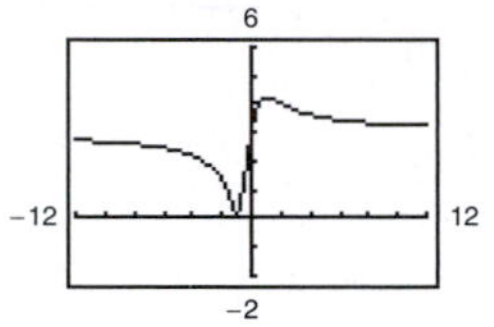

(c) $\lim_{m\to\infty} d(m) = 3$;

$\lim_{m\to-\infty} d(m) = 3$;

As m approaches $\pm\infty$, the distance approaches 3

101. Proof

103. False. Let $f(x) = \frac{2x}{\sqrt{x^2 + 2}}$. $f'(x) > 0$ for all real numbers.

Section 3.6 *(page 212)*

1. d **2.** c **3.** a **4.** b

5.

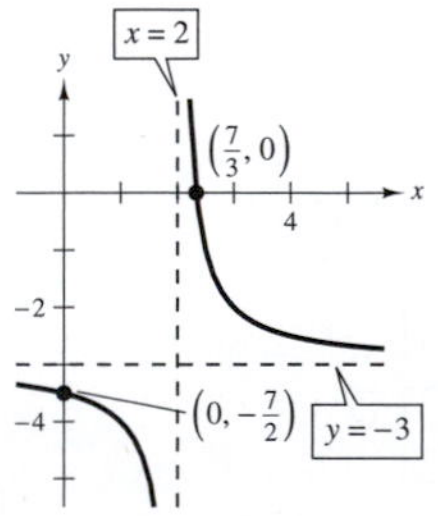

7.

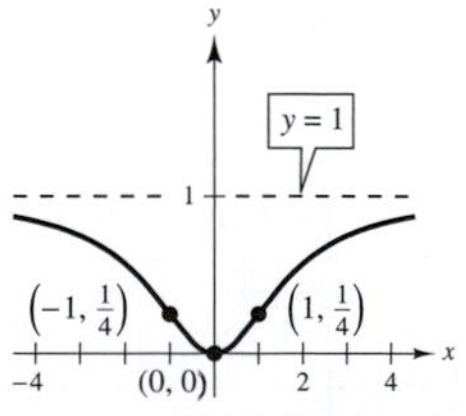

9.

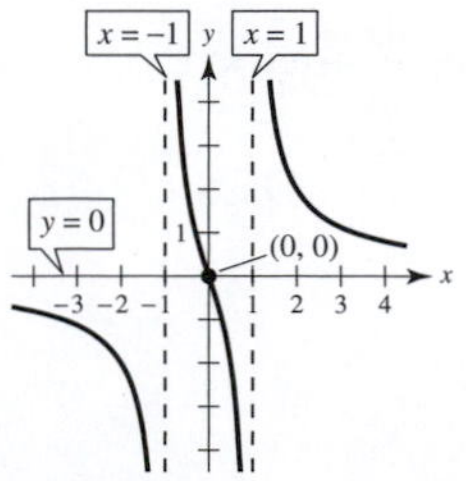

11.

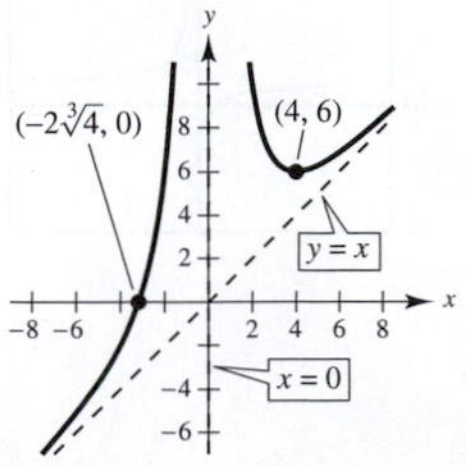

13.

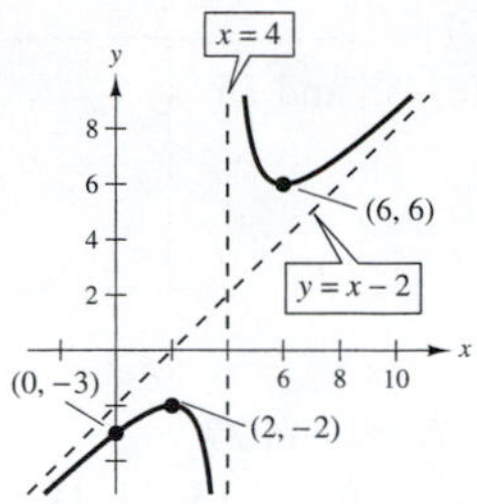

15.

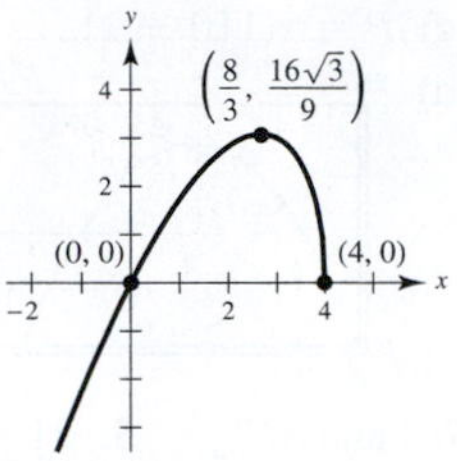

17.

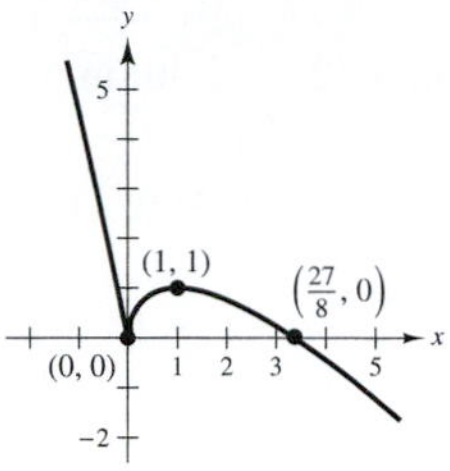

19.

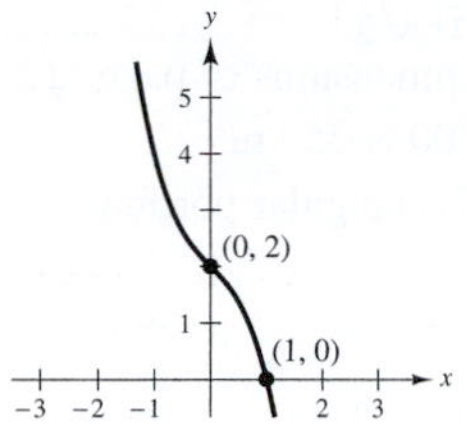

21.

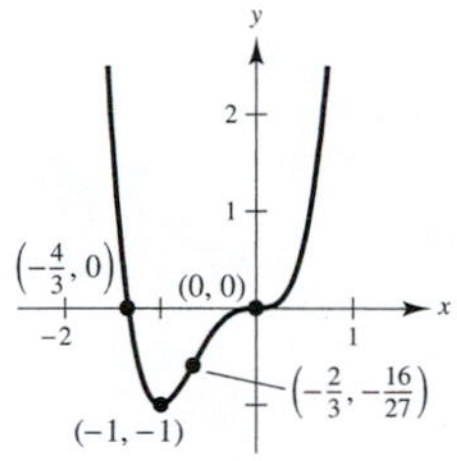

23.

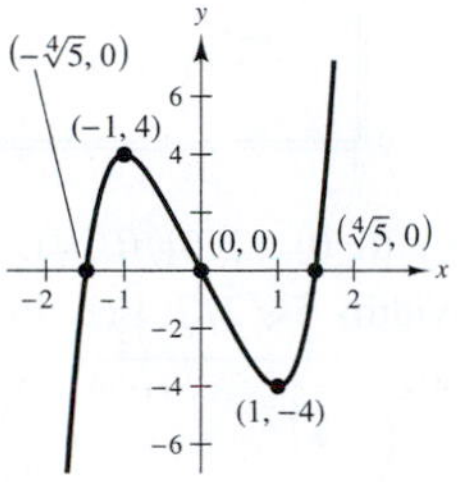

25.

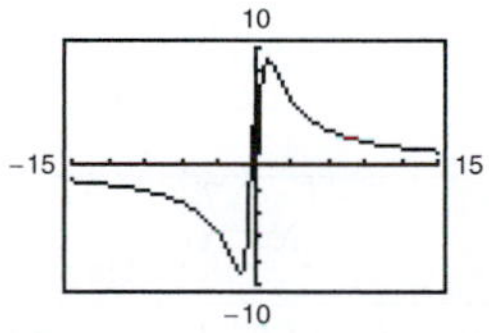

Minimum: $(-1.10, -9.05)$;
Maximum: $(1.10, 9.05)$;
Points of inflection:
$(-1.84, -7.86)$, $(1.84, 7.86)$;
Vertical asymptote: $x = 0$;
Horizontal asymptote: $y = 0$

27.

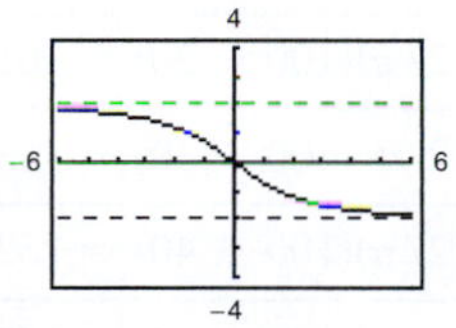

Point of inflection: $(0, 0)$;
Horizontal asymptotes: $y = \pm 2$

29.

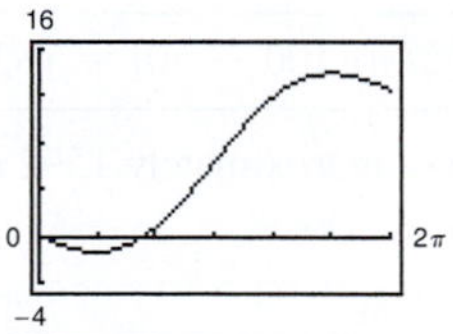

Relative minimum:
$\left(\frac{\pi}{3}, \frac{2\pi}{3} - 2\sqrt{3}\right)$;
Relative maximum:
$\left(\frac{5\pi}{3}, \frac{10\pi}{3} + 2\sqrt{3}\right)$;
Points of inflection: $(0, 0)$,
$(\pi, 2\pi)$, $(2\pi, 4\pi)$

31.

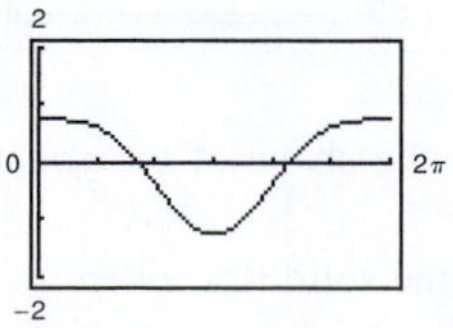

Relative minimum: $\left(\pi, -\frac{5}{4}\right)$;
Points of inflection:
$\left(\frac{2\pi}{3}, -\frac{3}{8}\right)$, $\left(\frac{4\pi}{3}, -\frac{3}{8}\right)$

33.

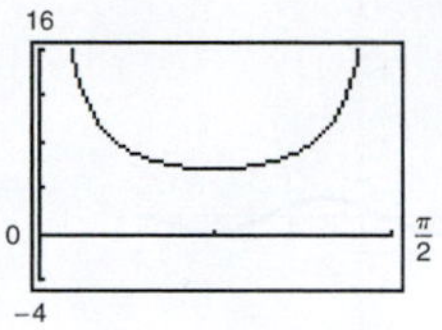

Relative minimum: $\left(\frac{\pi}{4}, 4\sqrt{2}\right)$;
Vertical asymptotes: $x = 0, \frac{\pi}{2}$

35. f is decreasing on $(2, 8)$, and therefore $f(3) > f(5)$.

37.

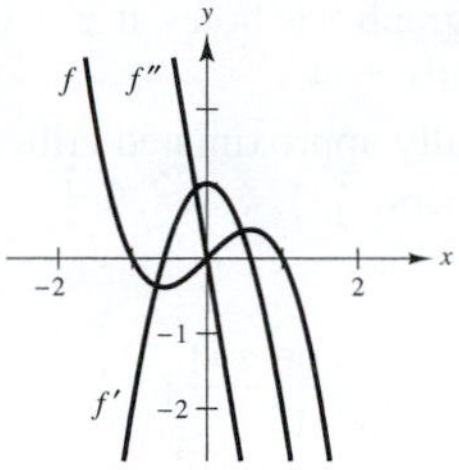

The zeros of f' correspond to the points where the graph of f has horizontal tangents. The zero of f'' corresponds to the point where the graph of f' has a horizontal tangent.

39.

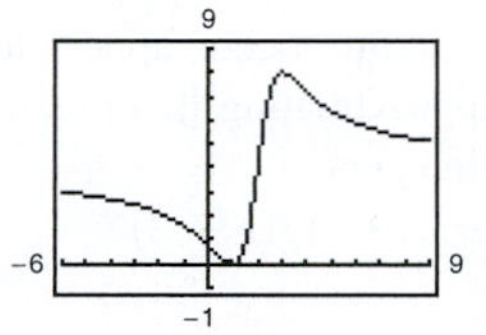

The graph crosses the horizontal asymptote $y = 4$.
The graph of a function f does not cross its vertical asymptote $x = c$ because $f(c)$ does not exist.

41.

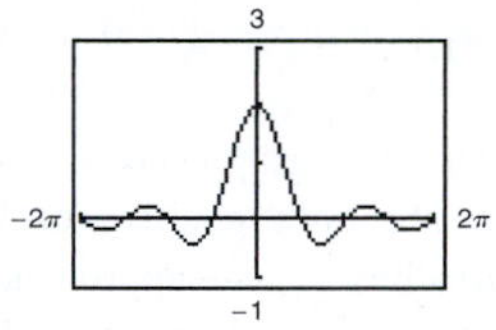

The graph has a hole at $x = 0$.
The graph crosses the horizontal asymptote $y = 0$.
The graph of a function f does not cross its vertical asymptote $x = c$ because $f(c)$ does not exist.

43.

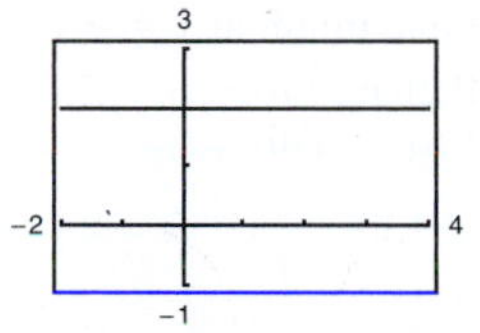

The graph has a hole at $x = 3$.
The rational function is not reduced to lowest terms.

45.

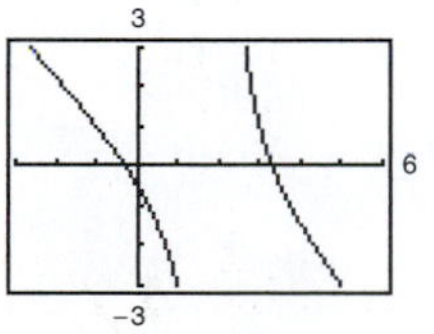

The graph appears to approach the line $y = -x + 1$, which is the slant asymptote.

47.

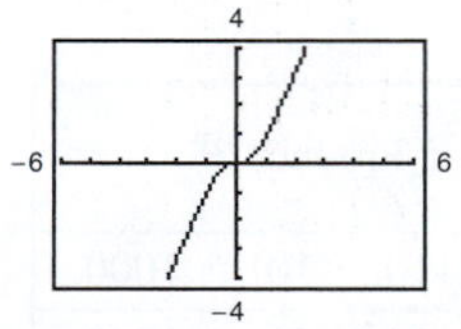

The graph appears to approach the line $y = 2x$, which is the slant asymptote.

49.

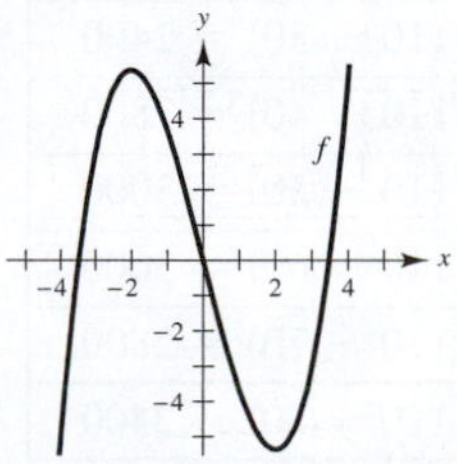

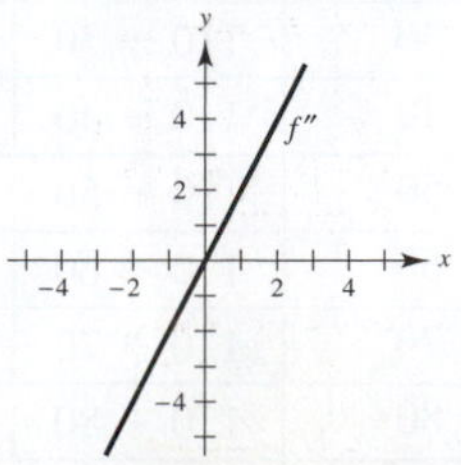

51.

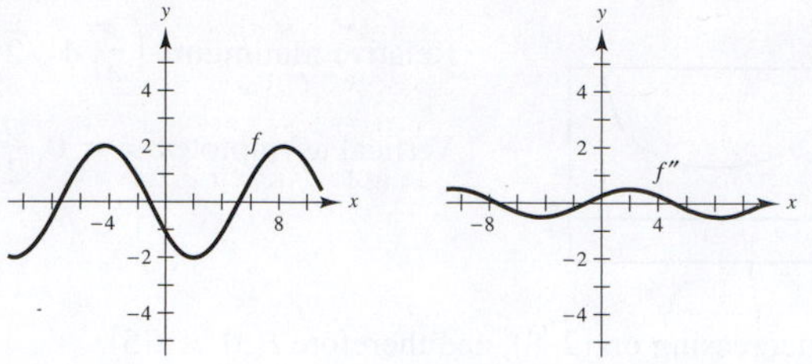

53. (a)

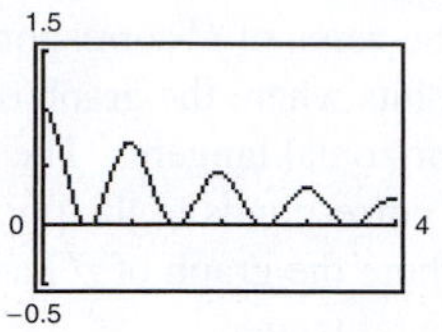

The graph has holes at $x = 0$ and at $x = 4$.
Visually approximated critical numbers: $\frac{1}{2}, 1, \frac{3}{2}, 2, \frac{5}{2}, 3, \frac{7}{2}$

(b) $f'(x) = \dfrac{-x\cos^2(\pi x)}{(x^2+1)^{3/2}} - \dfrac{2\pi\sin(\pi x)\cos(\pi x)}{\sqrt{x^2+1}}$;
Approximate critical numbers: $\frac{1}{2}, 0.97, \frac{3}{2}, 1.98, \frac{5}{2}, 2.98, \frac{7}{2}$;
The critical numbers where maxima occur appear to be integers in part (a), but by approximating them using f', you can see that they are not integers.

55. Answers will vary. Sample answer: $y = 1/(x - 3)$

57. Answers will vary.
Sample answer: $y = (3x^2 - 7x - 5)/(x - 3)$

59. (a) x_0, x_2, x_4 (b) x_2, x_3 (c) x_1 (d) x_1 (e) x_2, x_3

61. (a)–(h) Proofs

63. Answers will vary. Sample answer: The graph has a vertical asymptote at $x = b$. If a and b are both positive or both negative, then the graph of f approaches ∞ as x approaches b, and the graph has a minimum at $x = -b$. If a and b have opposite signs, then the graph of f approaches $-\infty$ as x approaches b, and the graph has a maximum at $x = -b$.

65. $y = 4x,\ y = -4x$

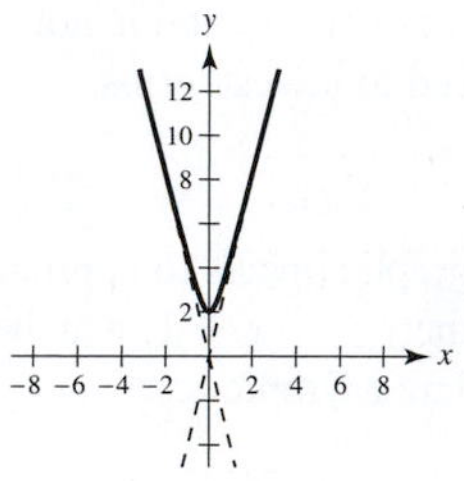

67. Putnam Problem 13(i), 1939

Section 3.7 *(page 220)*

1. (a) and (b)

First Number, x	Second Number	Product, P
10	110 − 10	10(110 − 10) = 1000
20	110 − 20	20(110 − 20) = 1800
30	110 − 30	30(110 − 30) = 2400
40	110 − 40	40(110 − 40) = 2800
50	110 − 50	50(110 − 50) = 3000
60	110 − 60	60(110 − 60) = 3000
70	110 − 70	70(110 − 70) = 2800
80	110 − 80	80(110 − 80) = 2400
90	110 − 90	90(110 − 90) = 1800
100	110 − 100	100(110 − 100) = 1000

The maximum is attained near $x = 50$ and 60.

(c) $P = x(110 - x)$

(d)

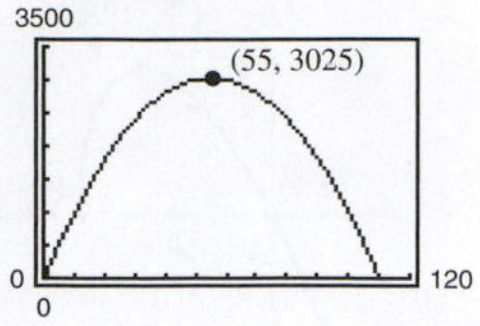

(e) 55 and 55

3. $S/2$ and $S/2$ **5.** 21 and 7 **7.** 54 and 27

9. $l = w = 20$ m **11.** $l = w = 4\sqrt{2}$ ft **13.** (1, 1)

15. $\left(\frac{7}{2}, \sqrt{\frac{7}{2}}\right)$

17. Dimensions of page: $(2 + \sqrt{30})$ in. $\times$ $(2 + \sqrt{30})$ in.

19. 700×350 m

21. Rectangular portion: $16/(\pi + 4) \times 32/(\pi + 4)$ ft

23. (a) $L = \sqrt{x^2 + 4 + \dfrac{8}{x-1} + \dfrac{4}{(x-1)^2}},\quad x > 1$

(b)

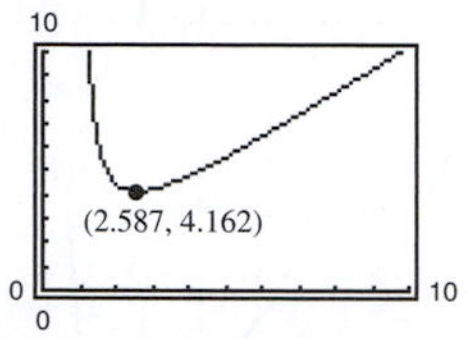

Minimum when $x \approx 2.587$

(c) (0, 0), (2, 0), (0, 4)

25. Width: $5\sqrt{2}/2$; Length: $5\sqrt{2}$

27. (a)

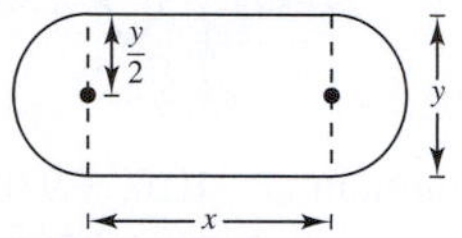

(b)

Length, x	Width, y	Area, xy
10	$2/\pi\,(100 - 10)$	$(10)(2/\pi)(100 - 10) \approx 573$
20	$2/\pi\,(100 - 20)$	$(20)(2/\pi)(100 - 20) \approx 1019$
30	$2/\pi\,(100 - 30)$	$(30)(2/\pi)(100 - 30) \approx 1337$
40	$2/\pi\,(100 - 40)$	$(40)(2/\pi)(100 - 40) \approx 1528$
50	$2/\pi\,(100 - 50)$	$(50)(2/\pi)(100 - 50) \approx 1592$
60	$2/\pi\,(100 - 60)$	$(60)(2/\pi)(100 - 60) \approx 1528$

The maximum area of the rectangle is approximately 1592 m^2.

(c) $A = 2/\pi(100x - x^2),\quad 0 < x < 100$

(d) $\dfrac{dA}{dx} = \dfrac{2}{\pi}(100 - 2x)$
$= 0$ when $x = 50$;
The maximum value is approximately 1592 when $x = 50$.

(e)

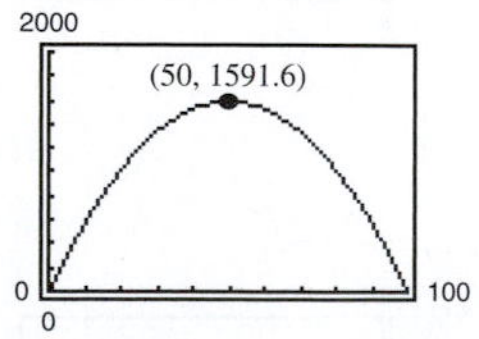

29. $18 \times 18 \times 36$ in.

31. No. The volume changes because the shape of the container changes when it is squeezed.

33. $r = \sqrt[3]{21/(2\pi)} \approx 1.50$ ($h = 0$, so the solid is a sphere.)

35. Side of square: $\dfrac{10\sqrt{3}}{9 + 4\sqrt{3}}$; Side of triangle: $\dfrac{30}{9 + 4\sqrt{3}}$

37. $w = \left(20\sqrt{3}\right)/3$ in., $h = \left(20\sqrt{6}\right)/3$ in.

39. 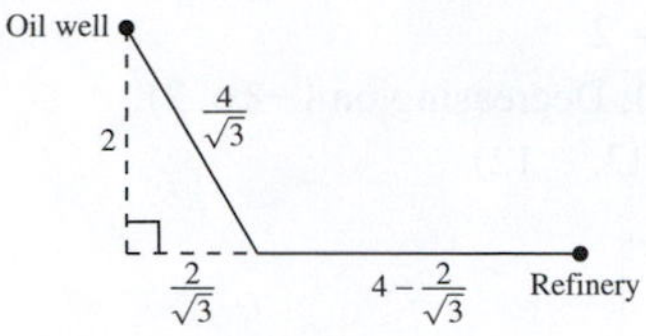

The path of the pipe should go underwater from the oil well to the coast following the hypotenuse of a right triangle with leg lengths of 2 miles and $2/\sqrt{3}$ miles for a distance of $4/\sqrt{3}$ miles. Then the pipe should go down the coast to the refinery for a distance of $\left(4 - 2/\sqrt{3}\right)$ miles.

41. One mile from the nearest point on the coast

43. 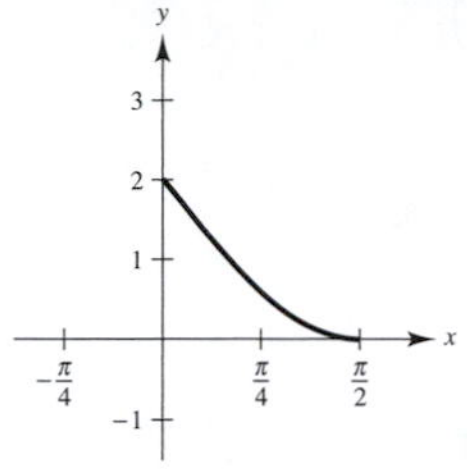

(a) Origin to y-intercept: 2;
Origin to x-intercept: $\pi/2$

(b) $d = \sqrt{x^2 + (2 - 2\sin x)^2}$

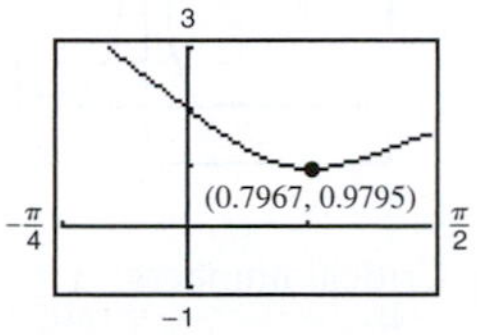

(c) Minimum distance is 0.9795 when $x \approx 0.7967$.

45. About 1.153 radians or 66° **47.** 8%

49. $y = \frac{64}{141}x$; $S \approx 6.1$ mi **51.** $y = \frac{3}{10}x$; $S_3 \approx 4.50$ mi

53. Putnam Problem A1, 1986

Section 3.8 *(page 229)*

In the answers for Exercises 1 and 3, the values in the tables have been rounded for convenience. Because a calculator and a computer program calculates internally using more digits than they display, you may produce slightly different values from those shown in the tables.

1.

n	x_n	$f(x_n)$	$f'(x_n)$	$\frac{f(x_n)}{f'(x_n)}$	$x_n - \frac{f(x_n)}{f'(x_n)}$
1	2.2000	−0.1600	4.4000	−0.0364	2.2364
2	2.2364	0.0015	4.4728	0.0003	2.2361

3.

n	x_n	$f(x_n)$	$f'(x_n)$	$\frac{f(x_n)}{f'(x_n)}$	$x_n - \frac{f(x_n)}{f'(x_n)}$
1	1.6	−0.0292	−0.9996	0.0292	1.5708
2	1.5708	0	−1	0	1.5708

5. −1.587 **7.** 0.682 **9.** 1.250, 5.000

11. 0.900, 1.100, 1.900 **13.** 1.935 **15.** 0.569

17. 4.493 **19.** (a) Proof (b) $\sqrt{5} \approx 2.236$; $\sqrt{7} \approx 2.646$

21. $f'(x_1) = 0$ **23.** 0.74 **25.** Proof

27. (a) 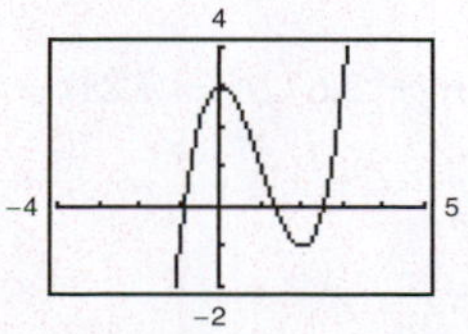

(b) 1.347 (c) 2.532

(d) 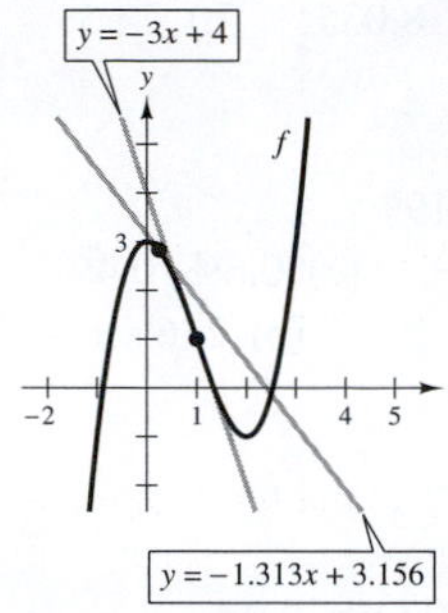

x-intercept of $y = -3x + 4$ is $\frac{4}{3}$.
x-intercept of $y = -1.313x + 3.156$ is approximately 2.404.

(e) If the initial estimate $x = x_1$ is not sufficiently close to the desired zero of a function, then the x-intercept of the corresponding tangent line to the function may approximate a second zero of the function.

29. Answers will vary. Sample answer:
If f is a function continuous on $[a, b]$ and differentiable on (a, b), where $c \in [a, b]$ and $f(c) = 0$, then Newton's Method uses tangent lines to approximate c. First, estimate an initial x_1 close to c. (See graph.) Then determine x_2 using $x_2 = x_1 - f(x_1)/f'(x_1)$. Calculate a third estimate x_3 using $x_3 = x_2 - f(x_2)/f'(x_2)$. Continue this process until $|x_n - x_{n+1}|$ is within the desired accuracy, and let x_{n+1} be the final approximation of c.

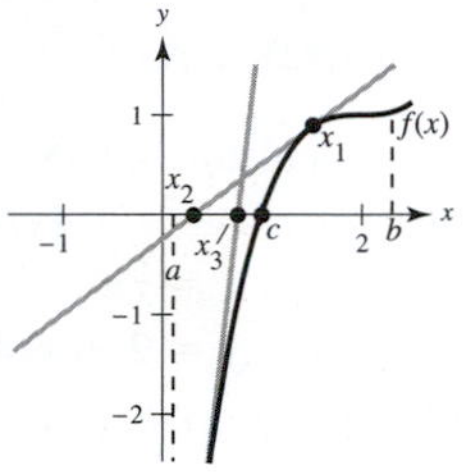

31. (1.939, 0.240) **33.** $x \approx 1.563$ mi

35. False; let $f(x) = \dfrac{x^2 - 1}{x - 1}$. **37.** True **39.** 0.217

Section 3.9 *(page 236)*

1. $T(x) = 4x - 4$

x	1.9	1.99	2	2.01	2.1
$f(x)$	3.610	3.960	4	4.040	4.410
$T(x)$	3.600	3.960	4	4.040	4.400

3. $T(x) = 80x - 128$

x	1.9	1.99	2	2.01	2.1
$f(x)$	24.761	31.208	32	32.808	40.841
$T(x)$	24.000	31.200	32	32.800	40.000

5. $T(x) = (\cos 2)(x - 2) + \sin 2$

x	1.9	1.99	2	2.01	2.1
$f(x)$	0.946	0.913	0.909	0.905	0.863
$T(x)$	0.951	0.913	0.909	0.905	0.868

7. $\Delta y = 0.331$; $dy = 0.3$ **9.** $\Delta y = -0.039$; $dy = -0.040$

11. $6x\,dx$ **13.** $(x \sec^2 x + \tan x)\,dx$

15. $-\dfrac{13}{(2x - 1)^2}\,dx$ **17.** $\dfrac{-x}{\sqrt{9 - x^2}}\,dx$ **19.** $(3 - \sin 2x)\,dx$

21. (a) 0.9 (b) 1.04 **23.** (a) 8.035 (b) 7.95

25. (a) $\pm\frac{5}{8}$ in.^{2} (b) 0.625%

27. (a) ± 10.75 cm^2 (b) about 1.19%

29. (a) ± 20.25 in.^{3} (b) ± 5.4 in.^{2} (c) 0.6%; 0.4%

31. 27.5 mi; About 7.3% **33.** (a) $\frac{1}{4}$% (b) 216 sec = 3.6 min

35. 6407 ft

37. $f(x) = \sqrt{x},\ dy = \dfrac{1}{2\sqrt{x}}\,dx$

$f(99.4) \approx \sqrt{100} + \dfrac{1}{2\sqrt{100}}(-0.6) = 9.97$

Calculator: 9.97

39. $f(x) = \sqrt[4]{x},\ dy = \dfrac{1}{4x^{3/4}}\,dx$

$f(624) \approx \sqrt[4]{625} + \dfrac{1}{4(625)^{3/4}}(-1) = 4.998$

Calculator: 4.998

41. $y - f(0) = f'(0)(x - 0)$

$y - 2 = \frac{1}{4}x$

$y = 2 + x/4$

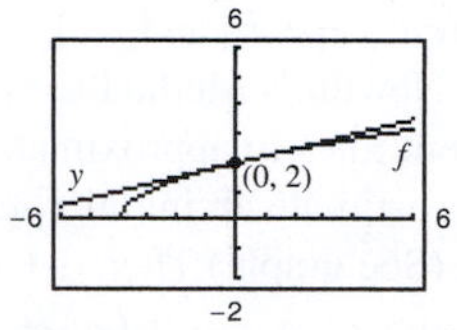

43. The value of dy becomes closer to the value of Δy as Δx decreases.

45. $f(x) = \sqrt{x};\ dy = \dfrac{1}{2\sqrt{x}}\,dx$

$f(4.02) \approx \sqrt{4} + \dfrac{1}{2\sqrt{4}}(0.02) = 2 + \dfrac{1}{4}(0.02)$

47. True **49.** True

Review Exercises for Chapter 3 *(page 238)*

1. Maximum: (0, 0); Minimum: $\left(-\frac{5}{2}, -\frac{25}{4}\right)$

3. Maximum: (4, 0); Minimum: (0, −2)

5. Maximum: $\left(3, \frac{2}{3}\right)$; Minimum: $\left(-3, -\frac{2}{3}\right)$

7. Maximum: $(2\pi, 17.57)$; Minimum: (2.73, 0.88)

9. $f(0) \neq f(4)$ **11.** Not continuous on $[-2, 2]$

13. $f'\left(\dfrac{2744}{729}\right) = \dfrac{3}{7}$ **15.** f is not differentiable at $x = 5$.

17. $f'(0) = 1$

19. No; The function has a discontinuity at $x = 0$, which is in the interval $[-2, 1]$.

21. Increasing on $\left(-\frac{3}{2}, \infty\right)$; Decreasing on $\left(-\infty, -\frac{3}{2}\right)$

23. Increasing on $(-\infty, 1)$, $\left(\frac{7}{3}, \infty\right)$; Decreasing on $\left(1, \frac{7}{3}\right)$

25. Increasing on $(1, \infty)$; Decreasing on $(0, 1)$

27. (a) Critical number: $x = 3$

(b) Increasing on $(3, \infty)$; Decreasing on $(-\infty, 3)$

(c) Relative minimum: $(3, -4)$

(d)

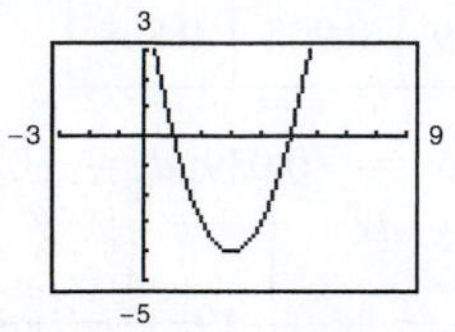

29. (a) Critical number: $t = 2$

(b) Increasing on $(2, \infty)$; Decreasing on $(-\infty, 2)$

(c) Relative minimum: $(2, -12)$

(d)

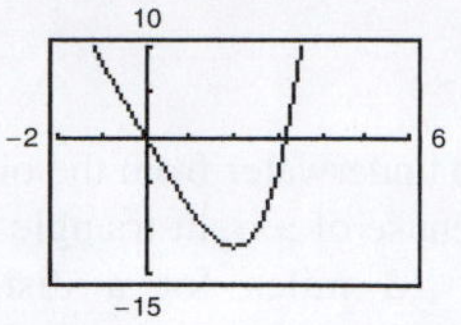

31. (a) Critical number: $x = -8$; Discontinuity: $x = 0$

(b) Increasing on $(-8, 0)$;
Decreasing on $(-\infty, -8)$ and $(0, \infty)$

(c) Relative minimum: $\left(-8, -\frac{1}{16}\right)$

(d)

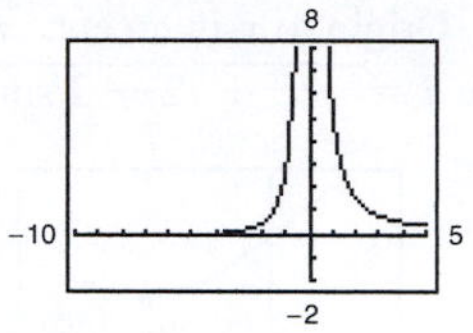

33. (a) Critical numbers: $x = \dfrac{3\pi}{4}, \dfrac{7\pi}{4}$

(b) Increasing on $\left(\dfrac{3\pi}{4}, \dfrac{7\pi}{4}\right)$;

Decreasing on $\left(0, \dfrac{3\pi}{4}\right)$ and $\left(\dfrac{7\pi}{4}, 2\pi\right)$

(c) Relative minimum: $\left(\dfrac{3\pi}{4}, -\sqrt{2}\right)$;

Relative maximum: $\left(\dfrac{7\pi}{4}, \sqrt{2}\right)$

(d)

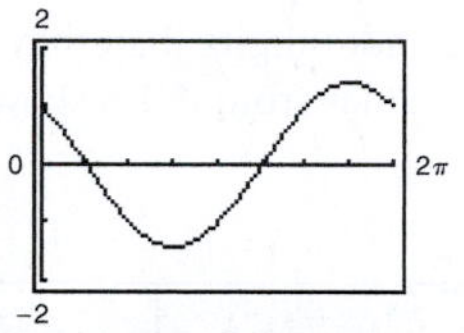

35. $(3, -54)$; Concave upward: $(3, \infty)$;
Concave downward: $(-\infty, 3)$

37. No point of inflection; Concave upward: $(-5, \infty)$

39. $(\pi/2, \pi/2)$, $(3\pi/2, 3\pi/2)$; Concave upward: $(\pi/2, 3\pi/2)$;
Concave downward: $(0, \pi/2)$, $(3\pi/2, 2\pi)$

41. Relative minimum: $(-9, 0)$

43. Relative maxima: $\left(\sqrt{2}/2, 1/2\right)$, $\left(-\sqrt{2}/2, 1/2\right)$;
Relative minimum: $(0, 0)$

45. Relative maximum: $(-3, -12)$; Relative minimum: $(3, 12)$

47.

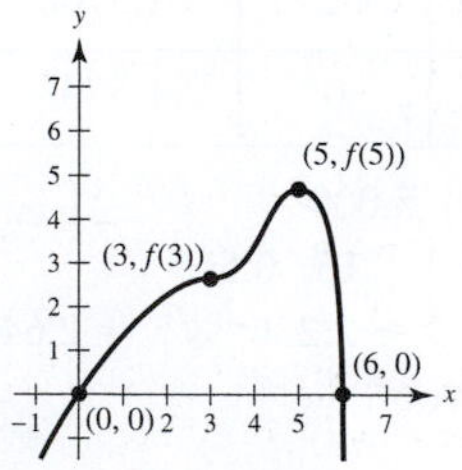

49. Increasing and concave down

51. (a) $D = 0.00188t^4 - 0.1273t^2 + 2.672t^2 - 7.81t + 77.1$

(b)

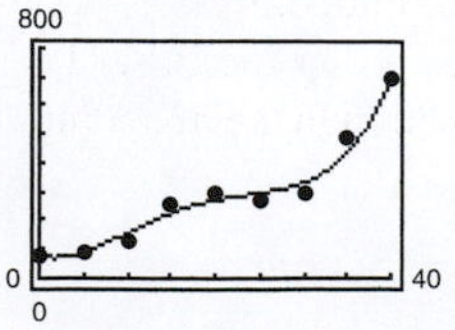

(c) Maximum in 2010; Minimum in 1970 (d) 2010

53. 8 **55.** $\frac{2}{3}$ **57.** $-\infty$ **59.** 0 **61.** 6

63.

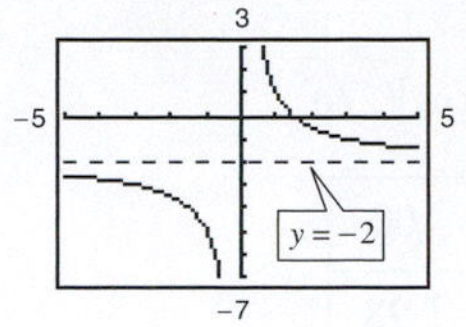

65.

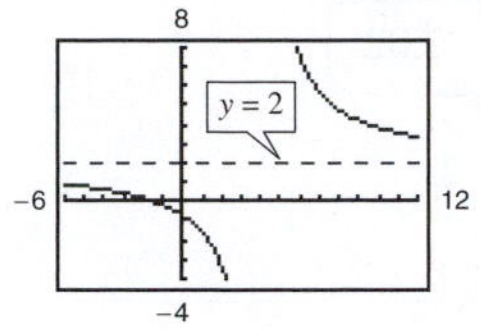

67.

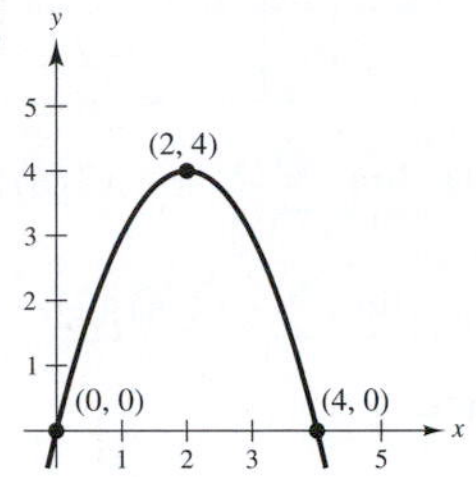

69.

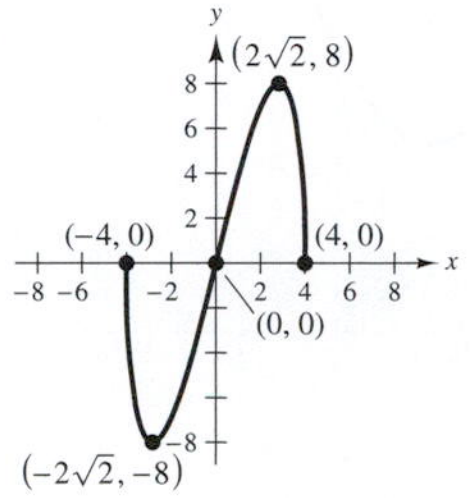

71.

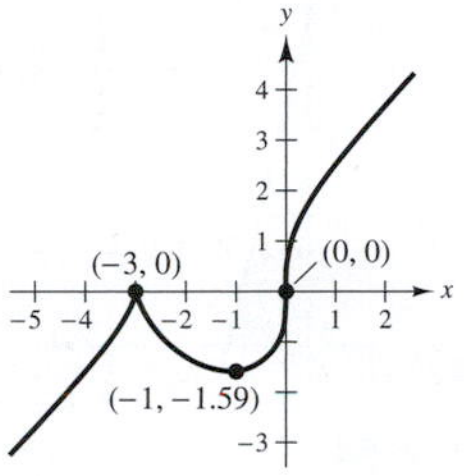

73.

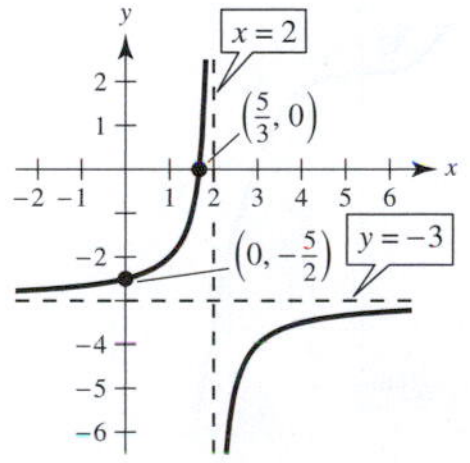

75.

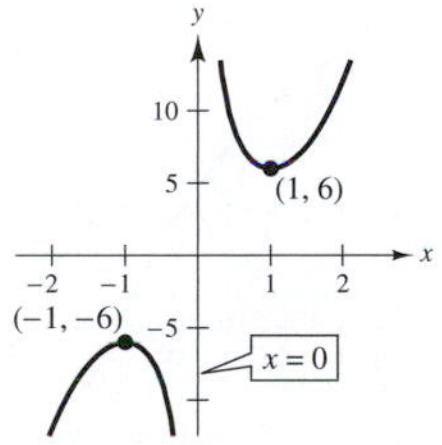

77. $x = 50$ ft and $y = \frac{200}{3}$ ft **79.** (0, 0), (5, 0), (0, 10)

81. 14.05 ft **83.** $32\pi r^3/81$ **85.** $-1.532, -0.347, 1.879$

87. $-2.182, -0.795$ **89.** -0.755

91. $\Delta y = 0.03005$; $dy = 0.03$

93. $dy = (1 - \cos x + x \sin x)\,dx$ **95.** (a) $\pm 8.1\pi$ cm^3 (b) $\pm 1.8\pi$ cm^2 (c) About 0.83%; About 0.56%

P.S. Problem Solving *(page 241)*

1. Choices of a may vary.

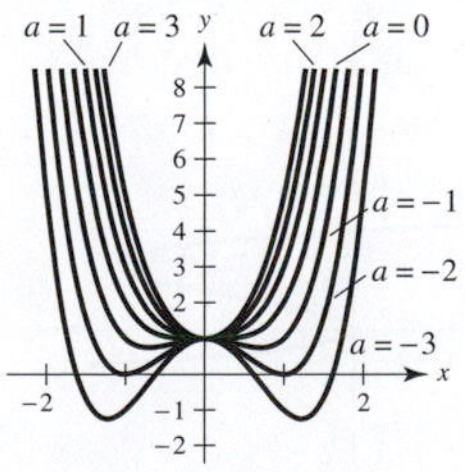

(a) One relative minimum at (0, 1) for $a \ge 0$

(b) One relative maximum at (0, 1) for $a < 0$

(c) Two relative minima for $a < 0$ when $x = \pm\sqrt{-a/2}$

(d) If $a < 0$, then there are three critical points; if $a \ge 0$, then there is only one critical point.

3. All c, where c is a real number **5.** Proof

7. The bug should head towards the midpoint of the opposite side. Without calculus, imagine opening up the cube. The shortest distance is the line PQ, passing through the midpoint as shown.

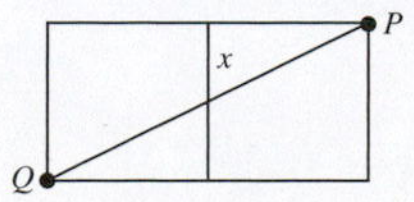

9. $a = 6, b = 1, c = 2$ **11.** Proof

13. Greatest slope: $\left(-\frac{\sqrt{3}}{3}, \frac{3}{4}\right)$; Least slope: $\left(\frac{\sqrt{3}}{3}, \frac{3}{4}\right)$

15. Proof **17.** Proof; Point of inflection: (1, 0)

19. (a) $P(x) = x - x^2$

(b)

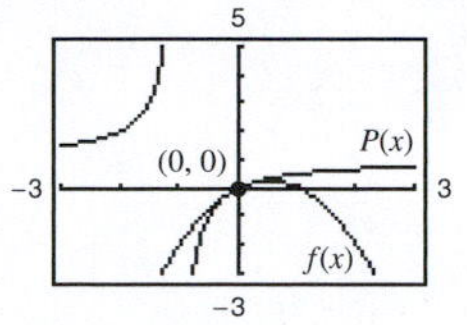

Chapter 4

Section 4.1 *(page 251)*

1. Proof **3.** $y = 3t^3 + C$ **5.** $y = \frac{2}{5}x^{5/2} + C$

	Original Integral	Rewrite	Integrate	Simplify
7.	$\int \sqrt[3]{x}\,dx$	$\int x^{1/3}\,dx$	$\frac{x^{4/3}}{4/3} + C$	$\frac{3}{4}x^{4/3} + C$
9.	$\int \frac{1}{x\sqrt{x}}\,dx$	$\int x^{-3/2}\,dx$	$\frac{x^{-1/2}}{-1/2} + C$	$-\frac{2}{\sqrt{x}} + C$

11. $\frac{1}{2}x^2 + 7x + C$ **13.** $\frac{1}{6}x^6 + x + C$

15. $\frac{2}{5}x^{5/2} + x^2 + x + C$ **17.** $\frac{3}{5}x^{5/3} + C$

19. $-1/(4x^4) + C$ **21.** $\frac{2}{3}x^{3/2} + 12x^{1/2} + C$

23. $x^3 + \frac{1}{2}x^2 - 2x + C$ **25.** $5 \sin x - 4 \cos x + C$

27. $t + \csc t + C$ **29.** $\tan\theta + \cos\theta + C$ **31.** $\tan y + C$

33. Answers will vary. Sample answer:

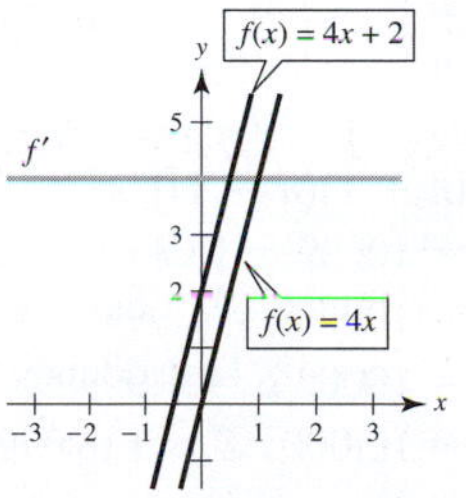

35. $f(x) = 3x^2 + 8$ **37.** $h(t) = 2t^4 + 5t - 11$

39. $f(x) = x^2 + x + 4$ **41.** $f(x) = -4\sqrt{x} + 3x$

43. (a) Answers will vary. Sample answer: (b) $y = \frac{x^3}{3} - x + \frac{7}{3}$

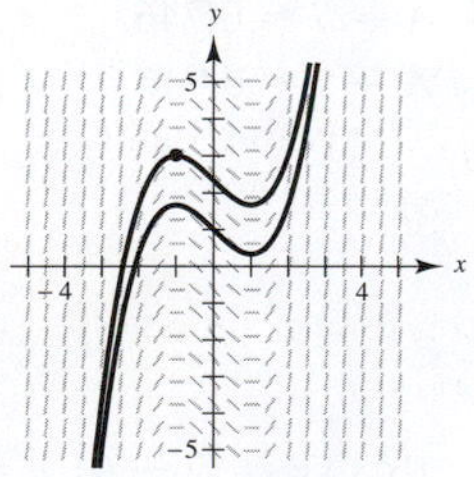

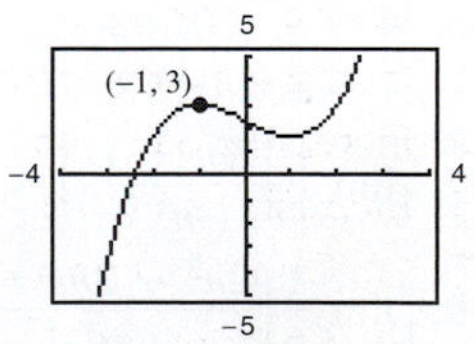

45. (a)

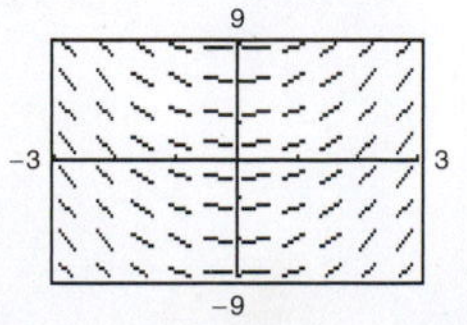

(b) $y = x^2 - 6$

(c)

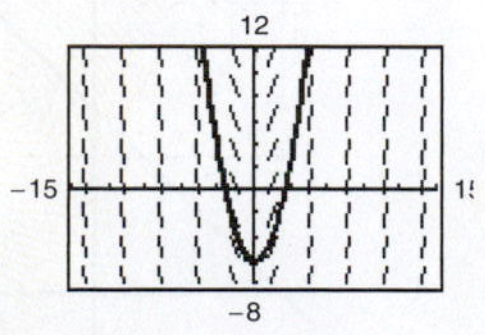

47. When you evaluate the integral $\int f(x)\,dx$, you are finding a function $F(x)$ that is an antiderivative of $f(x)$. So, there is no difference.

49.

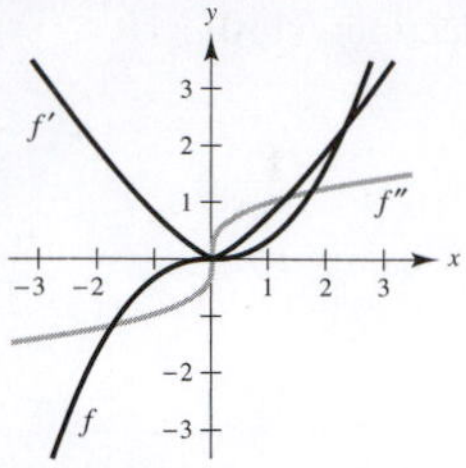

51. (a) $h(t) = \frac{3}{4}t^2 + 5t + 12$ (b) 69 cm **53.** 62.25 ft

55. (a) $t \approx 2.562$ sec (b) $v(t) \approx -65.970$ ft/sec

57. $v_0 \approx 62.3$ m/sec **59.** 320 m; -32 m/sec

61. (a) $v(t) = 3t^2 - 12t + 9$; $a(t) = 6t - 12$
(b) $(0, 1), (3, 5)$ (c) -3

63. $a(t) = -1/(2t^{3/2})$; $x(t) = 2\sqrt{t} + 2$

65. (a) 1.18 m/sec^2 (b) 190 m

67. (a) 300 ft (b) 60 ft/sec $\approx$ 41 mi/h

69. False. f has an infinite number of antiderivatives, each differing by a constant.

71. True **73.** True **75.** $f(x) = \dfrac{x^3}{3} - 4x + \dfrac{16}{3}$

77. Proof

Section 4.2 *(page 263)*

1. 75 **3.** $\dfrac{158}{85}$ **5.** $4c$ **7.** $\displaystyle\sum_{i=1}^{11} \frac{1}{5i}$

9. $\displaystyle\sum_{j=1}^{6} \left[7\left(\frac{j}{6}\right) + 5\right]$ **11.** $\displaystyle\frac{2}{n}\sum_{i=1}^{n} \left[\left(\frac{2i}{n}\right)^3 - \left(\frac{2i}{n}\right)\right]$ **13.** 84

15. 1200 **17.** 2470 **19.** 12,040

21. $(n + 2)/n$
$n = 10$: $S = 1.2$
$n = 100$: $S = 1.02$
$n = 1000$: $S = 1.002$
$n = 10{,}000$: $S = 1.0002$

23. $[2(n + 1)(n - 1)]/n^2$
$n = 10$: $S = 1.98$
$n = 100$: $S = 1.9998$
$n = 1000$: $S = 1.999998$
$n = 10{,}000$: $S = 1.99999998$

25. $13 <$ (Area of region) < 15

27. $55 <$ (Area of region) < 74.5

29. $0.7908 <$ (Area of region) < 1.1835

31. The area of the shaded region falls between 12.5 square units and 16.5 square units.

33. $A \approx S \approx 0.768$
$A \approx s \approx 0.518$

35. $A \approx S \approx 0.746$
$A \approx s \approx 0.646$

37. $\displaystyle\lim_{n\to\infty} \left[\frac{12(n + 1)}{n}\right] = 12$

39. $\displaystyle\lim_{n\to\infty} \frac{1}{6}\left(\frac{2n^3 - 3n^2 + n}{n^3}\right) = \frac{1}{3}$

41. $\displaystyle\lim_{n\to\infty} [(3n + 1)/n] = 3$

43. (a)

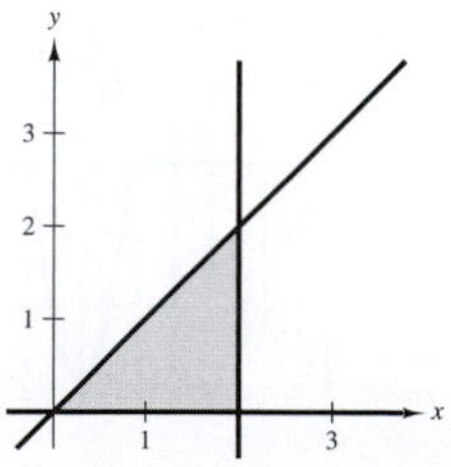

(b) $\Delta x = (2 - 0)/n = 2/n$

(c) $\displaystyle s(n) = \sum_{i=1}^{n} f(x_{i-1})\,\Delta x = \sum_{i=1}^{n} [(i - 1)(2/n)](2/n)$

(d) $\displaystyle S(n) = \sum_{i=1}^{n} f(x_i)\,\Delta x = \sum_{i=1}^{n} [i(2/n)](2/n)$

(e)

n	5	10	50	100
$s(n)$	1.6	1.8	1.96	1.98
$S(n)$	2.4	2.2	2.04	2.02

(f) $\displaystyle\lim_{n\to\infty} \sum_{i=1}^{n} [(i - 1)(2/n)](2/n) = 2$;
$\displaystyle\lim_{n\to\infty} \sum_{i=1}^{n} [i(2/n)](2/n) = 2$

45. $A = 3$

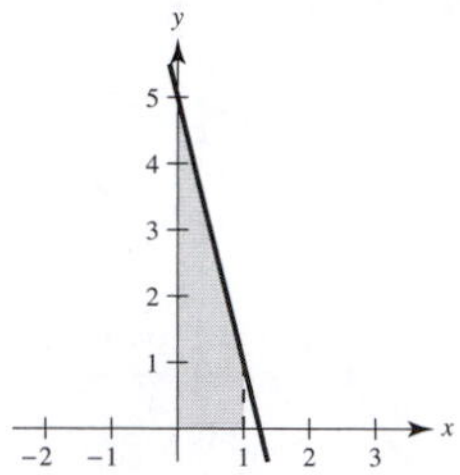

47. $A = \frac{7}{3}$

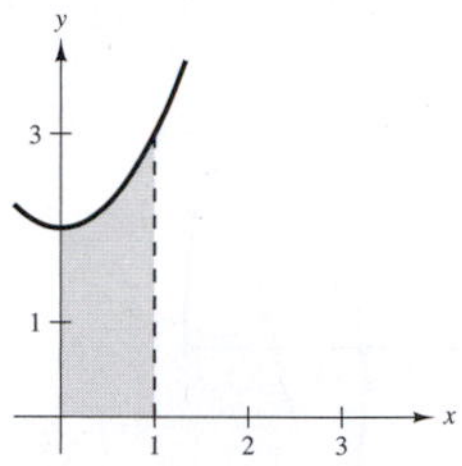

49. $A = 54$

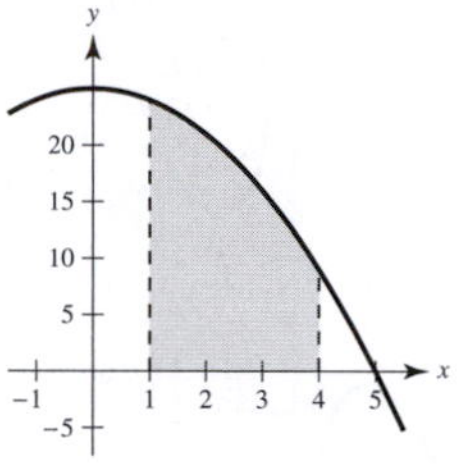

51. $A = 34$

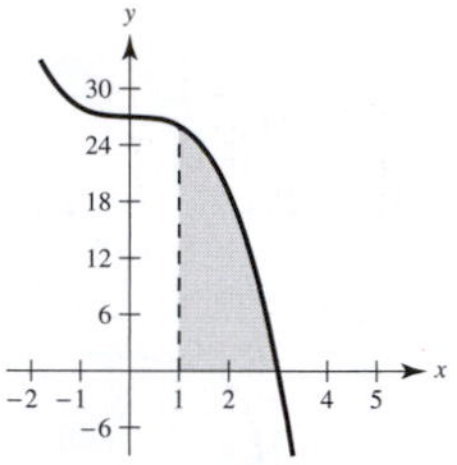

53. $A = \frac{2}{3}$

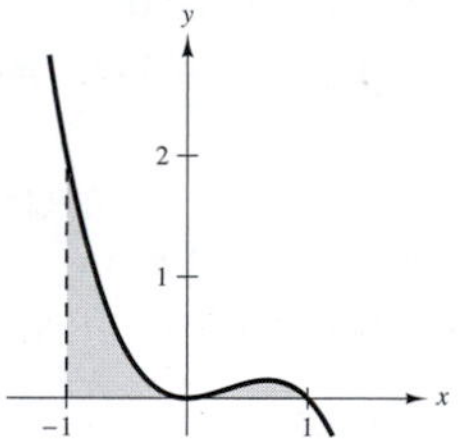

55. $A = 8$

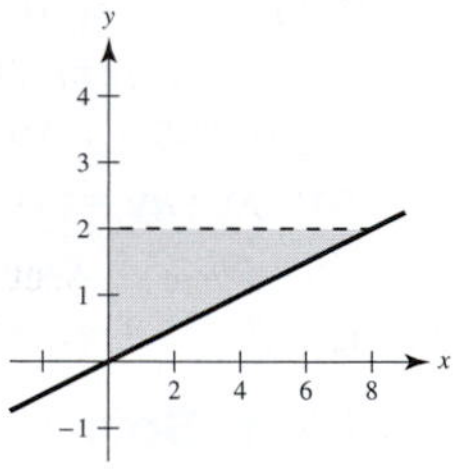

57. $A = \frac{125}{3}$

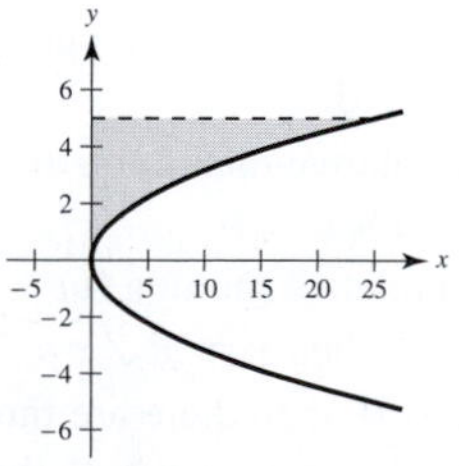

59. $A = \frac{44}{3}$

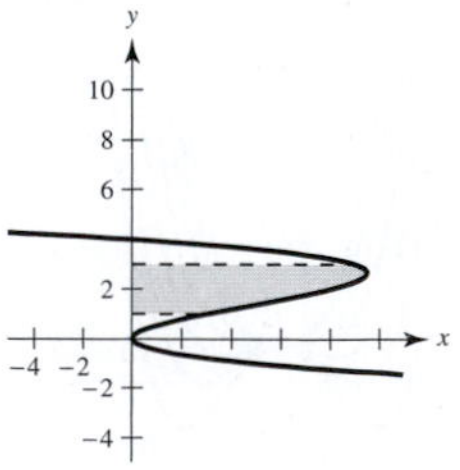

61. $\frac{69}{8}$ **63.** 0.345 **65.** b

67. You can use the line $y = x$ bounded by $x = a$ and $x = b$. The sum of the areas of the inscribed rectangles in the figure below is the lower sum.

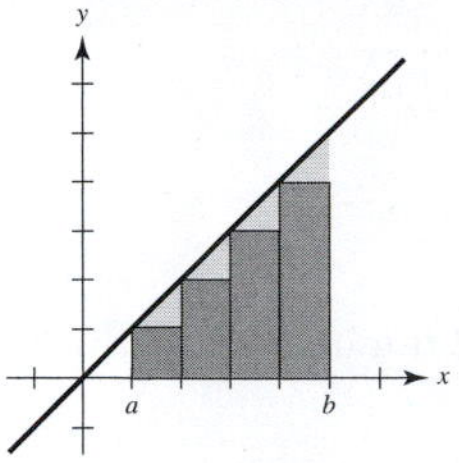

The sum of the areas of the circumscribed rectangles in the figure below is the upper sum.

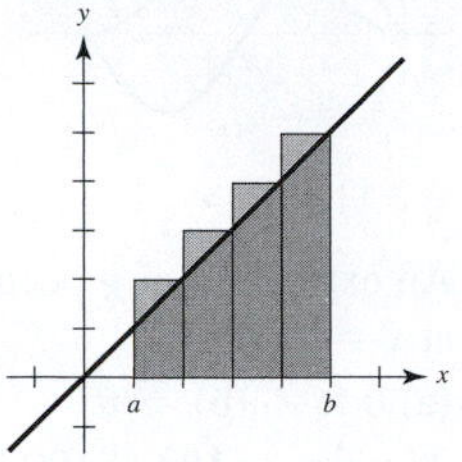

The rectangles in the first graph do not contain all of the area of the region, and the rectangles in the second graph cover more than the area of the region. The exact value of the area lies between these two sums.

69. (a)

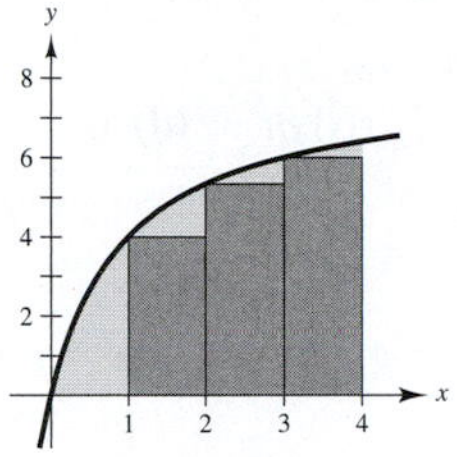

$s(4) = \frac{46}{3}$

(b)

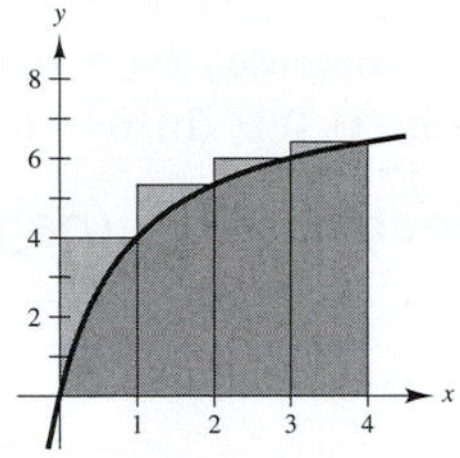

$S(4) = \frac{326}{15}$

(c)

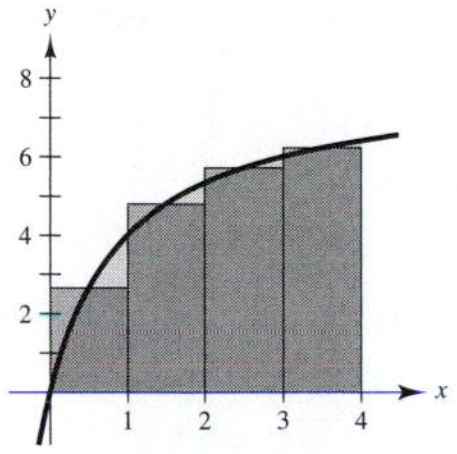

$M(4) = \frac{6112}{315}$

(d) Proof

(e)

n	4	8	20	100	200
$s(n)$	15.333	17.368	18.459	18.995	19.060
$S(n)$	21.733	20.568	19.739	19.251	19.188
$M(n)$	19.403	19.201	19.137	19.125	19.125

(f) Because f is an increasing function, $s(n)$ is always increasing and $S(n)$ is always decreasing.

71. True

73. Suppose there are n rows and $n + 1$ columns. The stars on the left total $1 + 2 + \cdot \cdot \cdot + n$, as do the stars on the right. There are $n(n + 1)$ stars in total. So, $2[1 + 2 + \cdot \cdot \cdot + n] = n(n + 1)$ and $1 + 2 + \cdot \cdot \cdot + n = [n(n + 1)]/2$.

75. For n odd, $\left(\frac{n+1}{2}\right)^2$ blocks;

For n even, $\frac{n^2 + 2n}{4}$ blocks

77. Putnam Problem B1, 1989

Section 4.3 *(page 273)*

1. $2\sqrt{3} \approx 3.464$ **3.** 32 **5.** 0 **7.** $\frac{10}{3}$

9. $\int_{-1}^{5} (3x + 10)\,dx$ **11.** $\int_{0}^{3} \sqrt{x^2 + 4}\,dx$ **13.** $\int_{0}^{4} 5\,dx$

15. $\int_{-4}^{4} (4 - |x|)\,dx$ **17.** $\int_{-5}^{5} (25 - x^2)\,dx$

19. $\int_{0}^{\pi/2} \cos x\,dx$ **21.** $\int_{0}^{2} y^3\,dy$

23.

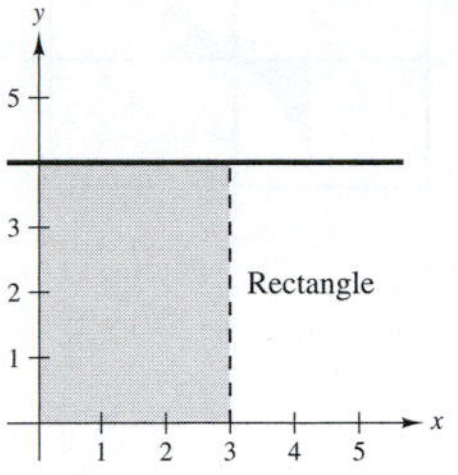

$A = 12$

25.

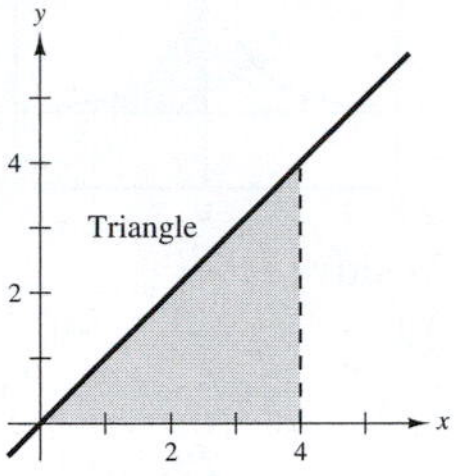

$A = 8$

27.

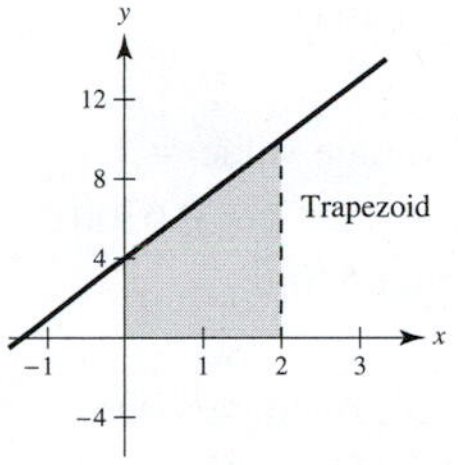

$A = 14$

29.

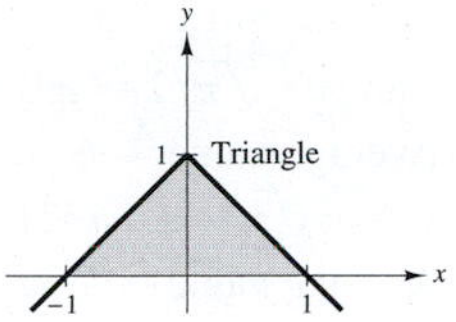

$A = 1$

31.

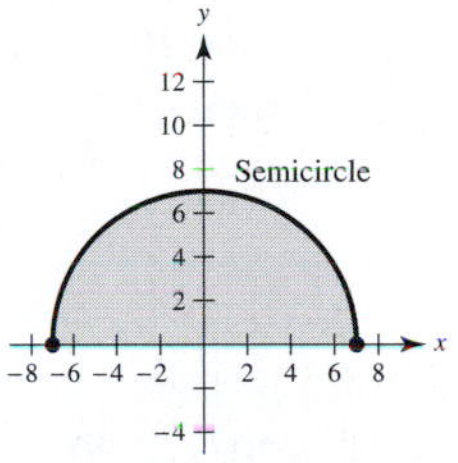

$A = 49\pi/2$

33. -6 **35.** 48 **37.** -12

39. 16 **41.** (a) 13 (b) -10 (c) 0 (d) 30

43. (a) 8 (b) -12 (c) -4 (d) 30 **45.** $-48, 88$

47. (a) $-\pi$ (b) 4 (c) $-(1 + 2\pi)$ (d) $3 - 2\pi$
(e) $5 + 2\pi$ (f) $23 - 2\pi$

49. (a) 14 (b) 4 (c) 8 (d) 0 **51.** 40 **53.** a **55.** d

57. No. There is a discontinuity at $x = 4$.

59. $a = -2,\ b = 5$

61. Answers will vary. Sample answer: $a = \pi,\ b = 2\pi$

$\int_{\pi}^{2\pi} \sin x\,dx < 0$

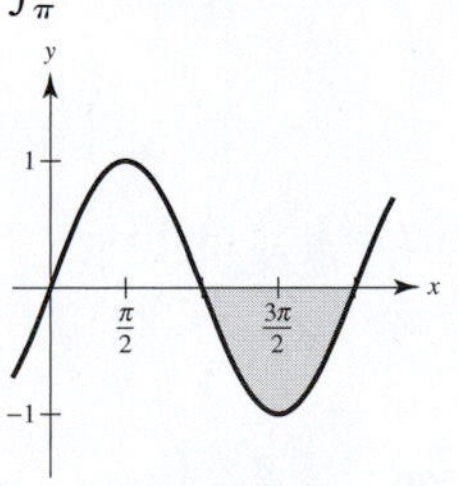

63. True **65.** True **67.** False. $\int_{0}^{2} (-x)\,dx = -2$

69. 272 **71.** Proof

73. No. No matter how small the subintervals, the number of both rational and irrational numbers within each subinterval is infinite, and $f(c_i) = 0$ or $f(c_i) = 1$.

75. $a = -1$ and $b = 1$ maximize the integral. **77.** $\frac{1}{3}$

Section 4.4 *(page 288)*

1.

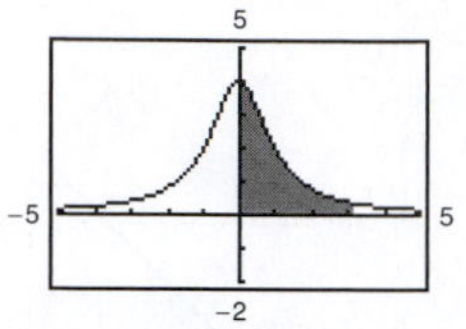

Positive

3.

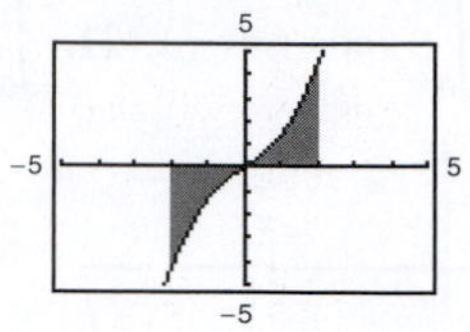

Zero

5. 12 **7.** -2 **9.** $-\frac{10}{3}$ **11.** $\frac{1}{3}$ **13.** $\frac{1}{2}$ **15.** $\frac{2}{3}$

17. -4 **19.** $-\frac{1}{18}$ **21.** $-\frac{27}{20}$ **23.** $\frac{25}{2}$ **25.** $\frac{64}{3}$

27. $\pi + 2$ **29.** $\pi/4$ **31.** $2\sqrt{3}/3$ **33.** 0 **35.** $\frac{1}{6}$

37. 1 **39.** $\frac{52}{3}$ **41.** 20 **43.** $\frac{32}{3}$

45. $3\sqrt[3]{2}/2 \approx 1.8899$ **47.** $2\sqrt{3} \approx 3.4641$

49. $\pm\arccos\sqrt{\pi}/2 \approx \pm 0.4817$

51. Average value $= 6$
$x = \pm\sqrt{3} \approx \pm 1.7321$

53. Average value $= \frac{1}{4}$
$x = \sqrt[3]{2}/2 \approx 0.6300$

55. Average value $= 2/\pi$
$x \approx 0.690, x \approx 2.451$

57. About 540 ft

59. (a) 8 (b) $\frac{4}{3}$ (c) $\int_1^7 f(x)\,dx = 20$; Average value $= \frac{10}{3}$

61. (a) $F(x) = 500 \sec^2 x$ (b) $1500\sqrt{3}/\pi \approx 827$ N

63. About 0.5318 L

65. (a) $v = -0.00086t^3 + 0.0782t^2 - 0.208t + 0.10$

(b)

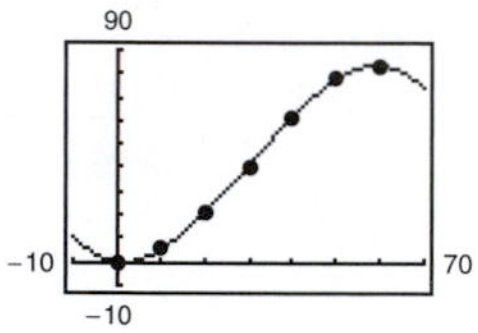

(c) 2475.6 m

67. $F(x) = 2x^2 - 7x$
$F(2) = -6$
$F(5) = 15$
$F(8) = 72$

69. $F(x) = -20/x + 20$
$F(2) = 10$
$F(5) = 16$
$F(8) = \frac{35}{2}$

71. $F(x) = \sin x - \sin 1$
$F(2) = \sin 2 - \sin 1 \approx 0.0678$
$F(5) = \sin 5 - \sin 1 \approx -1.8004$
$F(8) = \sin 8 - \sin 1 \approx 0.1479$

73. (a) $g(0) = 0$, $g(2) \approx 7$, $g(4) \approx 9$, $g(6) \approx 8$, $g(8) \approx 5$

(b) Increasing: $(0, 4)$; Decreasing: $(4, 8)$

(c) A maximum occurs at $x = 4$.

(d)

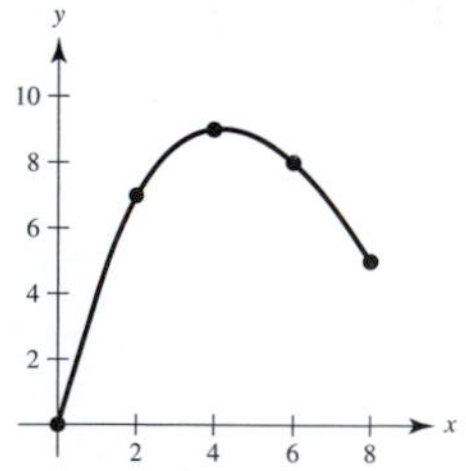

75. $\frac{1}{2}x^2 + 2x$ **77.** $\frac{3}{4}x^{4/3} - 12$ **79.** $\tan x - 1$

81. $x^2 - 2x$ **83.** $\sqrt{x^4 + 1}$ **85.** $x \cos x$ **87.** 8

89. $\cos x\sqrt{\sin x}$ **91.** $3x^2 \sin x^6$

93.

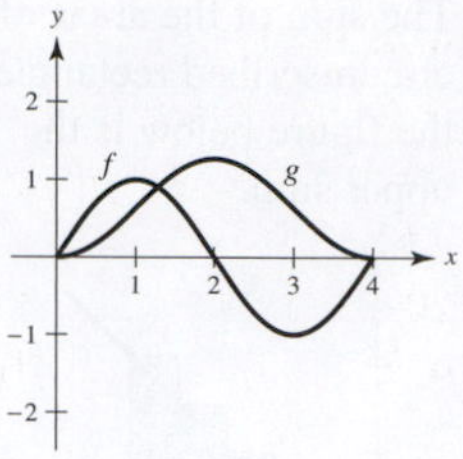

An extremum of g occurs at $x = 2$.

95. (a) $\frac{3}{2}$ ft to the right
(b) $\frac{113}{10}$ ft

97. (a) 0 ft (b) $\frac{63}{2}$ ft **99.** (a) 2 ft to the right (b) 2 ft

101. 28 units **103.** 8190 L

105. $f(x) = x^{-2}$ has a nonremovable discontinuity at $x = 0$.

107. $f(x) = \sec^2 x$ has a nonremovable discontinuity at $x = \pi/2$.

109. $2/\pi \approx 63.7\%$ **111.** True

113. $f'(x) = \dfrac{1}{(1/x)^2 + 1}\left(-\dfrac{1}{x^2}\right) + \dfrac{1}{x^2 + 1} = 0$

Because $f'(x) = 0$, $f(x)$ is constant.

115. (a) 0 (b) 0 (c) $xf(x) + \int_0^x f(t)\,dt$ (d) 0

Section 4.5 *(page 301)*

$\int f(g(x))g'(x)\,dx$	$u = g(x)$	$du = g'(x)\,dx$
1. $\int (8x^2 + 1)^2(16x)\,dx$	$8x^2 + 1$	$16x\,dx$
3. $\int \tan^2 x \sec^2 x\,dx$	$\tan x$	$\sec^2 x\,dx$

5. $\frac{1}{5}(1 + 6x)^5 + C$ **7.** $\frac{2}{3}(25 - x^2)^{3/2} + C$

9. $\frac{1}{12}(x^4 + 3)^3 + C$ **11.** $\frac{1}{15}(x^3 - 1)^5 + C$

13. $\frac{1}{3}(t^2 + 2)^{3/2} + C$ **15.** $-\frac{15}{8}(1 - x^2)^{4/3} + C$

17. $1/[4(1 - x^2)^2] + C$ **19.** $-1/[3(1 + x^3)] + C$

21. $-\sqrt{1 - x^2} + C$ **23.** $-\frac{1}{4}(1 + 1/t)^4 + C$

25. $\sqrt{2x} + C$ **27.** $2x^2 - 4\sqrt{16 - x^2} + C$

29. $-1/[2(x^2 + 2x - 3)] + C$

31. (a) Answers will vary.
Sample answer:

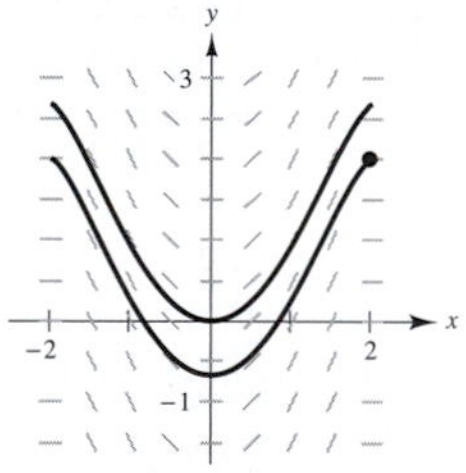

(b) $y = -\frac{1}{3}(4 - x^2)^{3/2} + 2$

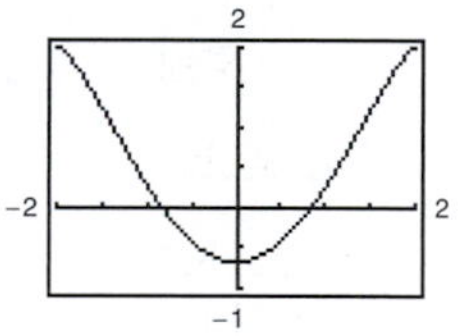

33. $-\cos(\pi x) + C$

35. $\int \cos 8x\,dx = \frac{1}{8}\int (\cos 8x)(8)\,dx = \frac{1}{8}\sin 8x + C$

37. $-\sin(1/\theta) + C$

39. $\frac{1}{4}\sin^2 2x + C$ or $-\frac{1}{4}\cos^2 2x + C_1$ or $-\frac{1}{8}\cos 4x + C_2$

41. $\frac{1}{2}\tan^2 x + C$ or $\frac{1}{2}\sec^2 x + C_1$ **43.** $f(x) = 2\cos(x/2) + 4$

45. $f(x) = \frac{1}{12}(4x^2 - 10)^3 - 8$

47. $\frac{2}{5}(x + 6)^{5/2} - 4(x + 6)^{3/2} + C = \frac{2}{5}(x + 6)^{3/2}(x - 4) + C$

49. $-\left[\frac{2}{3}(1 - x)^{3/2} - \frac{4}{5}(1 - x)^{5/2} + \frac{2}{7}(1 - x)^{7/2}\right] + C =$
$-\frac{2}{105}(1 - x)^{3/2}(15x^2 + 12x + 8) + C$

51. $\frac{1}{8}\left[\frac{2}{5}(2x-1)^{5/2} + \frac{4}{3}(2x-1)^{3/2} - 6(2x-1)^{1/2}\right] + C = \left(\sqrt{2x-1}/15\right)(3x^2 + 2x - 13) + C$

53. $-x - 1 - 2\sqrt{x+1} + C$ or $-\left(x + 2\sqrt{x+1}\right) + C_1$

55. 0 **57.** $12 - \frac{8}{9}\sqrt{2}$ **59.** 2 **61.** $\frac{1}{2}$

63. $f(x) = (2x^3 + 1)^3 + 3$ **65.** 1209/28 **67.** $2\left(\sqrt{3} - 1\right)$

69. $\frac{272}{15}$ **71.** $\frac{2}{3}$ **73.** (a) $\frac{64}{3}$ (b) $\frac{128}{3}$ (c) $-\frac{64}{3}$ (d) 64

75. $2\int_0^3 (4x^2 - 6)\,dx = 36$

77. If $u = 5 - x^2$, then $du = -2x\,dx$ and $\int x(5 - x^2)^3\,dx = -\frac{1}{2}\int (5 - x^2)^3(-2x)\,dx = -\frac{1}{2}\int u^3\,du.$

79. (a) $\int x^2\sqrt{x^3 + 1}\,dx$ (b) $\int \tan(3x)\sec^2(3x)\,dx$

81. \$340,000

83. (a) 102.532 thousand units (b) 102.352 thousand units
(c) 74.5 thousand units

85. (a) $P_{0.50,\,0.75} \approx 35.3\%$ (b) $b \approx 58.6\%$

87. (a)

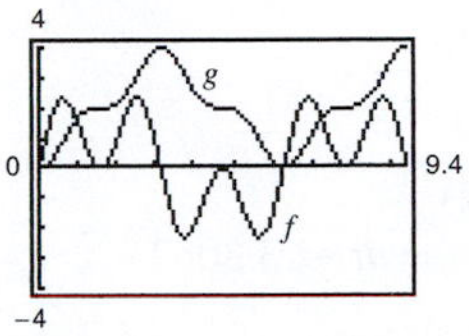

(b) g is nonnegative, because the graph of f is positive at the beginning and generally has more positive sections than negative ones.

(c) The points on g that correspond to the extrema of f are points of inflection of g.

(d) No, some zeros of f, such as $x = \pi/2$, do not correspond to extrema of g. The graph of g continues to increase after $x = \pi/2$, because f remains above the x-axis.

(e)

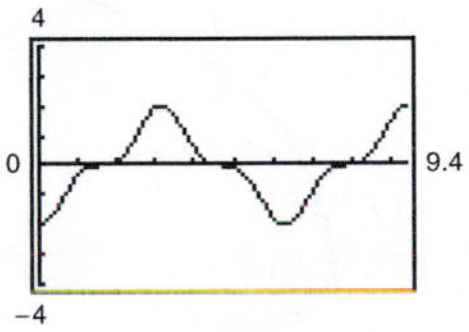

The graph of h is that of g shifted 2 units downward.

89. (a) and (b) Proofs

91. False. $\int (2x + 1)^2\,dx = \frac{1}{6}(2x + 1)^3 + C$ **93.** True

95. True **97–99.** Proofs **101.** Putnam Problem A1, 1958

Section 4.6 *(page 310)*

	Trapezoidal	Simpson's	Exact
1.	2.7500	2.6667	2.6667
3.	4.2500	4.0000	4.0000
5.	20.2222	20.0000	20.0000
7.	12.6640	12.6667	12.6667
9.	0.3352	0.3334	0.3333

	Trapezoidal	Simpson's	Graphing Utility
11.	3.2833	3.2396	3.2413
13.	0.3415	0.3720	0.3927
15.	0.5495	0.5483	0.5493
17.	−0.0975	−0.0977	−0.0977
19.	0.1940	0.1860	0.1858

21. Trapezoidal: Linear (1st-degree) polynomials
Simpson's: Quadratic (2nd-degree) polynomials

23. (a) 1.500 (b) 0.000 **25.** (a) $\frac{1}{4}$ (b) $\frac{1}{12}$

27. (a) $n = 366$ (b) $n = 26$ **29.** (a) $n = 77$ (b) $n = 8$

31. (a) $n = 130$ (b) $n = 12$ **33.** (a) $n = 643$ (b) $n = 48$

35. (a) 24.5 (b) 25.67 **37.** 0.701 **39.** 89,250 m^2

41. 10,233.58 ft-lb **43.** 3.1416 **45.** 2.477 **47.** Proof

Review Exercises for Chapter 4 *(page 312)*

1. $\frac{x^2}{2} - 6x + C$ **3.** $\frac{4}{3}x^3 + \frac{1}{2}x^2 + 3x + C$

5. $x^2/2 - 4/x^2 + C$ **7.** $x^2 + 9\cos x + C$

9. $y = 1 - 3x^2$ **11.** $f(x) = 4x^3 - 5x - 3$

13. (a) Answers will vary. (b) $y = x^2 - 4x - 2$

Sample answer:

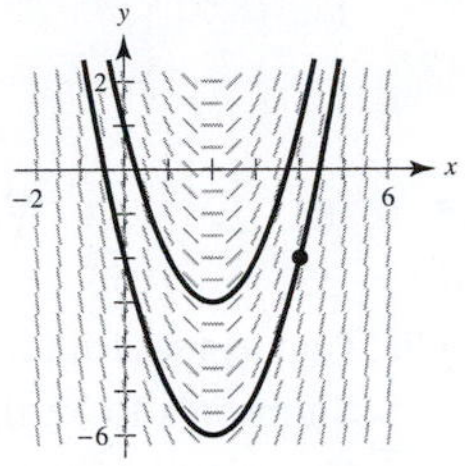

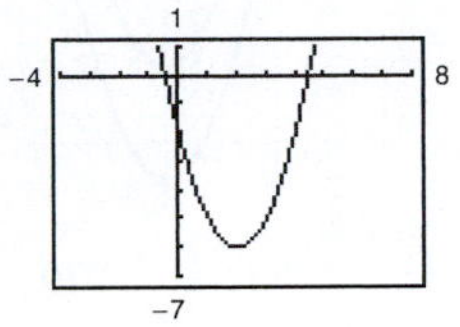

15. (a) 3 sec; 144 ft (b) $\frac{3}{2}$ sec (c) 108 ft

17. 240 ft/sec **19.** 60 **21.** $\sum_{n=1}^{10} \frac{1}{3n}$ **23.** 192

25. 420 **27.** 3310

29. 9.038 < (Area of region) < 13.038

31. $A = 15$ **33.** $A = 12$

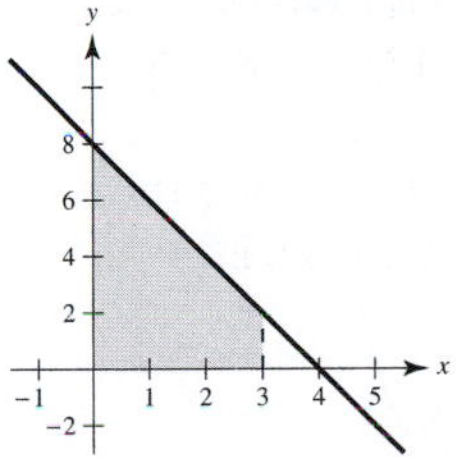

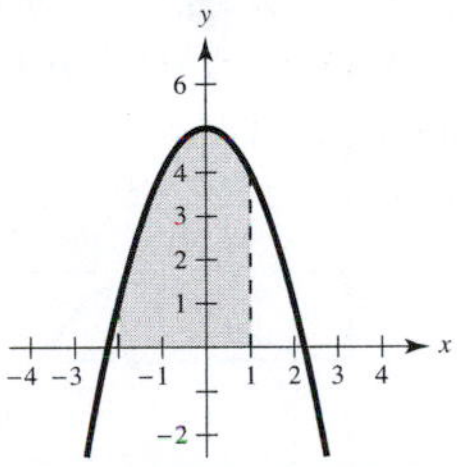

35. $\frac{27}{2}$ **37.** $\int_{-4}^{0} (2x + 8)\,dx$

39.

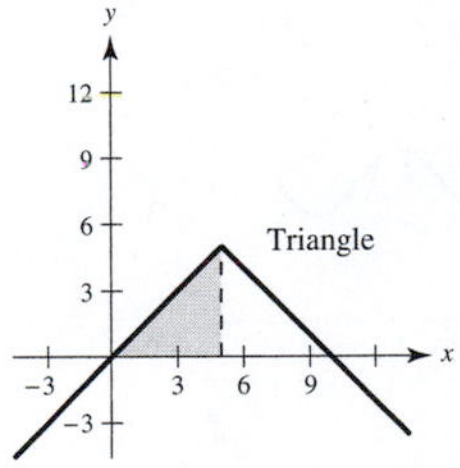

$A = \frac{25}{2}$

41. (a) 17 (b) 7
(c) 9 (d) 84

43. 56 **45.** 0 **47.** $\frac{422}{5}$ **49.** $\left(\sqrt{2} + 2\right)/2$

51. $-\cos 2 + 1 \approx 1.416$ **53.** 30 **55.** $\frac{1}{4}$

57. Average value $= \frac{2}{5}$, $x = \frac{25}{4}$

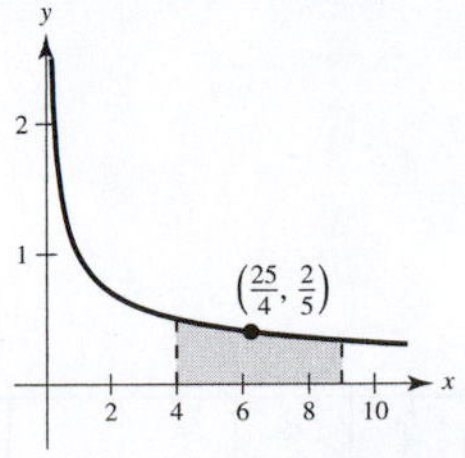

59. $x^2\sqrt{1 + x^3}$ **61.** $x^2 + 3x + 2$ **63.** $\frac{2}{3}\sqrt{x^3 + 3} + C$

65. $-\frac{1}{30}(1 - 3x^2)^5 + C = \frac{1}{30}(3x^2 - 1)^5 + C$

67. $\frac{1}{4}\sin^4 x + C$ **69.** $-2\sqrt{1 - \sin\theta} + C$

71. $\dfrac{1}{3\pi}(1 + \sec \pi x)^3 + C$

73. (a) Answers will vary. (b) $y = -\frac{1}{3}(9 - x^2)^{3/2} + 5$

Sample answer:

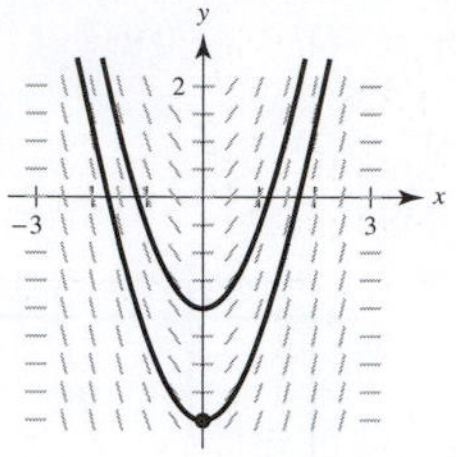

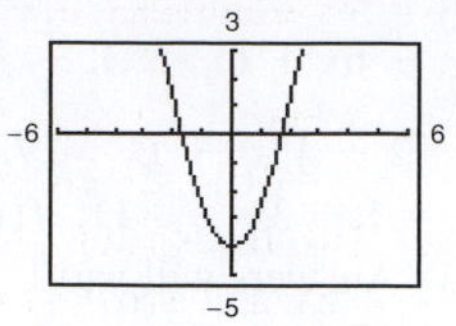

75. $\frac{455}{2}$ **77.** 2 **79.** $28\pi/15$ **81.** 2 **83.** $\frac{468}{7}$

85. (a) $\frac{64}{5}$ (b) $\frac{32}{5}$ (c) $\frac{96}{5}$ (d) -32

87. Trapezoidal Rule: 0.285
Simpson's Rule: 0.284
Graphing Utility: 0.284

89. Trapezoidal Rule: 0.637
Simpson's Rule: 0.685
Graphing Utility: 0.704

P.S. Problem Solving *(page 315)*

1. (a) $L(1) = 0$ (b) $L'(x) = 1/x,\ L'(1) = 1$
(c) $x \approx 2.718$ (d) Proof

3. (a) $\lim\limits_{n\to\infty}\left[\frac{32}{n^5}\sum_{i=1}^{n} i^4 - \frac{64}{n^4}\sum_{i=1}^{n} i^3 + \frac{32}{n^3}\sum_{i=1}^{n} i^2\right]$
(b) $(16n^4 - 16)/(15n^4)$ (c) $16/15$

5. (a)

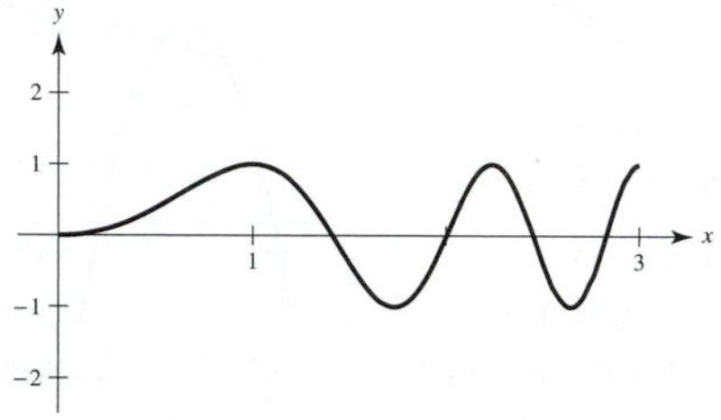

(b)

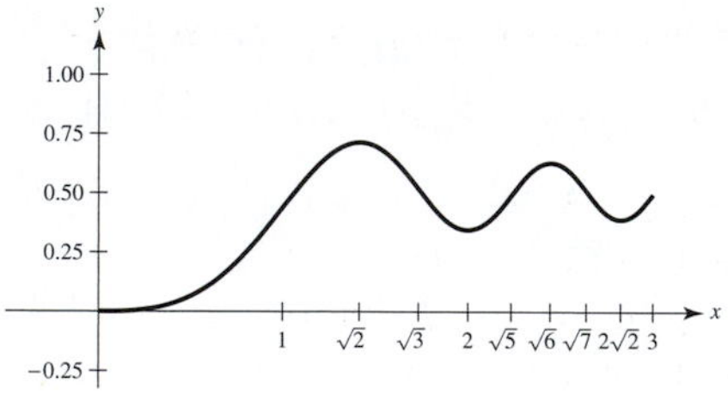

(c) Relative maxima at $x = \sqrt{2},\ \sqrt{6}$
Relative minima at $x = 2,\ 2\sqrt{2}$
(d) Points of inflection at $x = 1,\ \sqrt{3},\ \sqrt{5},\ \sqrt{7}$

7. (a)

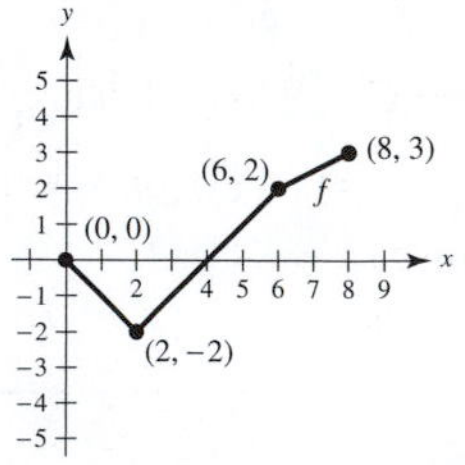

(b)

x	0	1	2	3	4	5	6	7	8
$F(x)$	0	$-\frac{1}{2}$	-2	$-\frac{7}{2}$	-4	$-\frac{7}{2}$	-2	$\frac{1}{4}$	3

(c) $x = 4, 8$ (d) $x = 2$

9. Proof **11.** $\frac{2}{3}$ **13.** $1 \le \int_0^1 \sqrt{1 + x^4}\,dx \le \sqrt{2}$

15. (a)

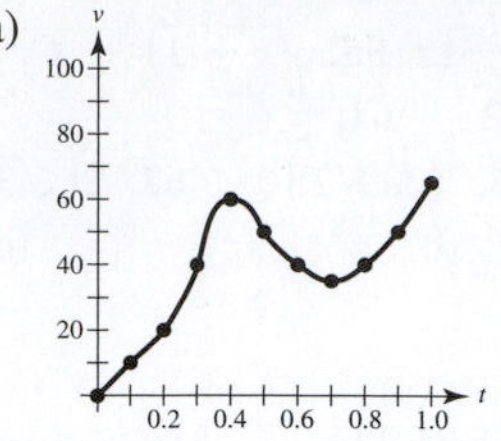

(b) (0, 0.4) and (0.7, 1.0) (c) 150 mi/h²
(d) Total distance traveled in miles; 38.5 mi
(e) Sample answer: 100 mi/h²

17. (a)–(c) Proofs

19. (a) $R(n), I, T(n), L(n)$
(b) $S(4) = \frac{1}{3}[f(0) + 4f(1) + 2f(2) + 4f(3) + f(4)] \approx 5.42$

Chapter 5

Section 5.1 *(page 325)*

1. (a) 3.8067 (b) $\ln 45 = \int_1^{45} \frac{1}{t}\,dt \approx 3.8067$

3. (a) -0.2231 (b) $\ln 0.8 = \int_1^{0.8} \frac{1}{t}\,dt \approx -0.2231$

5. b **6.** d **7.** a **8.** c

9.

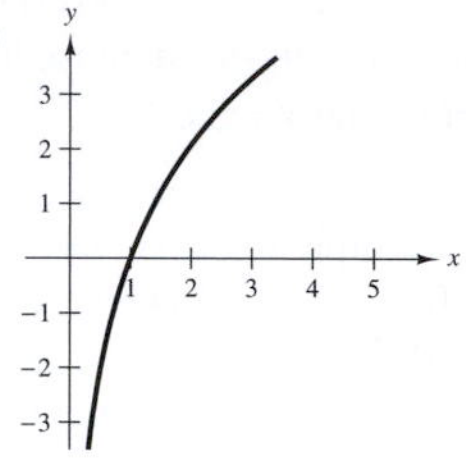

Domain: $x > 0$

11.

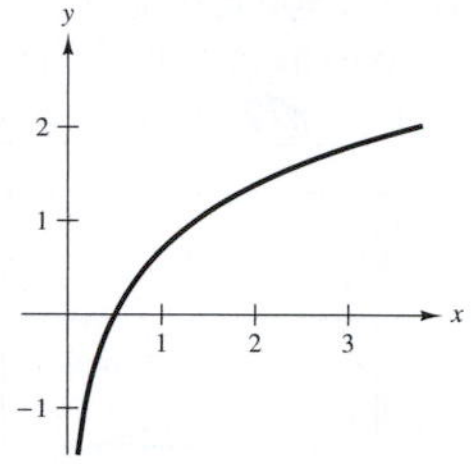

Domain: $x > 0$

13.

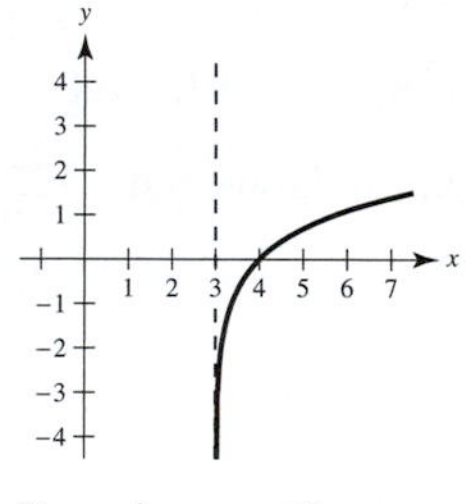

Domain: $x > 3$

15.

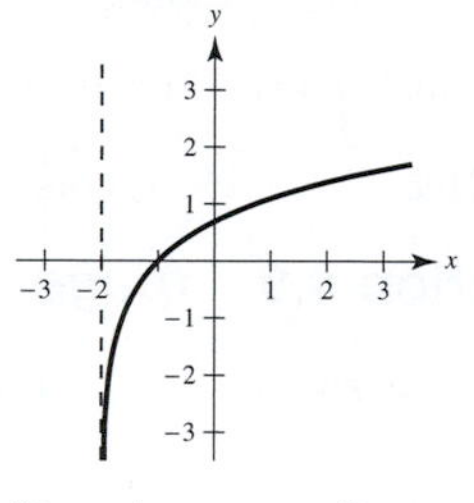

Domain: $x > -2$

17. (a) 1.7917 (b) -0.4055 (c) 4.3944 (d) 0.5493

19. $\ln x - \ln 4$ **21.** $\ln x + \ln y - \ln z$

23. $\ln x + \frac{1}{2}\ln(x^2 + 5)$ **25.** $\frac{1}{2}[\ln(x - 1) - \ln x]$

27. $\ln z + 2\ln(z - 1)$ **29.** $\ln \dfrac{x - 2}{x + 2}$

31. $\ln \sqrt[3]{\dfrac{x(x + 3)^2}{x^2 - 1}}$ **33.** $\ln \dfrac{9}{\sqrt{x^2 + 1}}$

35. (a)

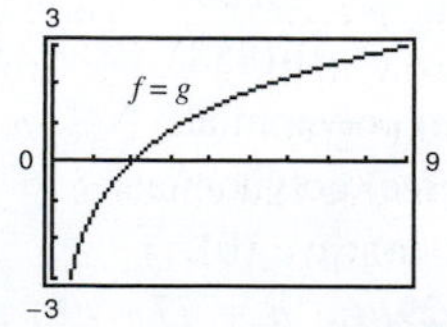

(b) $f(x) = \ln \dfrac{x^2}{4} = \ln x^2 - \ln 4$
$= 2\ln x - \ln 4$
$= g(x)$

37. $-\infty$ **39.** $\ln 4 \approx 1.3863$ **41.** $1/x$ **43.** $2/x$

45. $4(\ln x)^3/x$ **47.** $2/(t+1)$ **49.** $\dfrac{2x^2-1}{x(x^2-1)}$

51. $\dfrac{1-x^2}{x(x^2+1)}$ **53.** $\dfrac{1-2\ln t}{t^3}$ **55.** $\dfrac{2}{x\ln x^2}=\dfrac{1}{x\ln x}$

57. $\dfrac{1}{1-x^2}$ **59.** $\dfrac{-4}{x(x^2+4)}$ **61.** $\cot x$

63. $-\tan x+\dfrac{\sin x}{\cos x-1}$

65. (a) $y=4x-4$
(b)

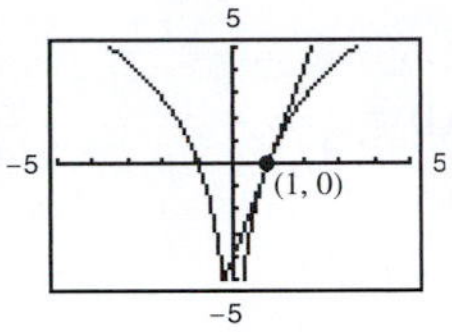

67. (a) $5x-y-2=0$
(b)

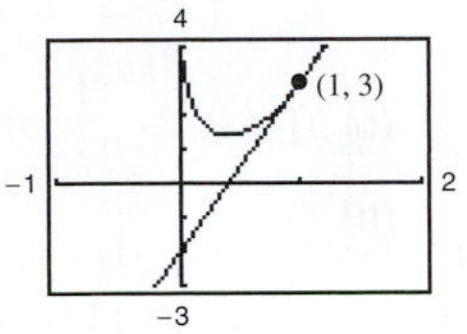

69. (a) $y=\frac{1}{3}x-\frac{1}{12}\pi+\frac{1}{2}\ln\left(\frac{3}{2}\right)$
(b)

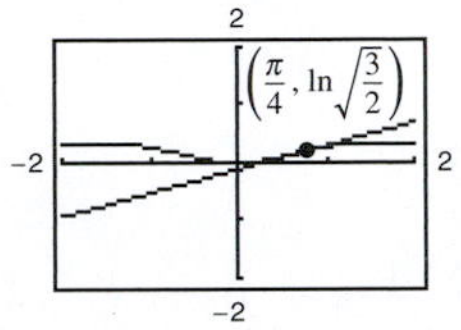

71. (a) $y=x-1$
(b)

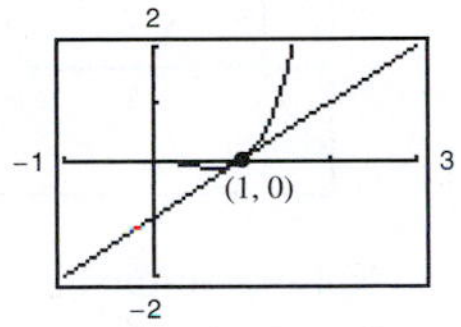

73. $\dfrac{2xy}{3-2y^2}$ **75.** $\dfrac{y(1-6x^2)}{1+y}$

77. $xy''+y'=x(-2/x^2)+(2/x)=0$

79. Relative minimum: $\left(1,\frac{1}{2}\right)$

81. Relative minimum: $(e^{-1},-e^{-1})$

83. Relative minimum: (e,e); Point of inflection: $(e^2,e^2/2)$

85. $P_1(x)=x-1$; $P_2(x)=x-1-\frac{1}{2}(x-1)^2$

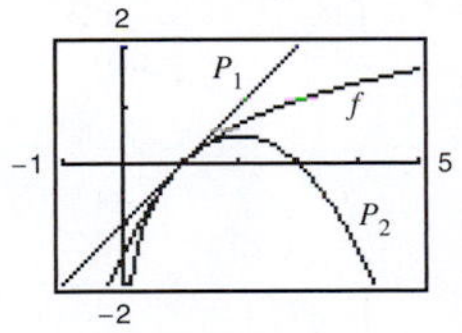

The values of f, P_1, and P_2 and their first derivatives agree at $x=1$.

87. $x\approx 0.567$ **89.** $(2x^2+1)/\sqrt{x^2+1}$

91. $\dfrac{3x^3+15x^2-8x}{2(x+1)^3\sqrt{3x-2}}$ **93.** $\dfrac{(2x^2+2x-1)\sqrt{x-1}}{(x+1)^{3/2}}$

95. The domain of the natural logarithmic function is $(0,\infty)$, and the range is $(-\infty,\infty)$. The function is continuous, increasing, and one-to-one, and its graph is concave downward. In addition, if a and b are positive numbers and n is rational, then $\ln(1)=0$, $\ln(a\cdot b)=\ln a+\ln b$, $\ln(a^n)=n\ln a$, and $\ln(a/b)=\ln a-\ln b$.

97. (a) Yes. If the graph of g is increasing, then $g'(x)>0$. Because $f(x)>0$, you know that $f'(x)=g'(x)f(x)$ and thus $f'(x)>0$. Therefore, the graph of f is increasing.
(b) No. Let $f(x)=x^2+1$ (positive and concave up), and let $g(x)=\ln(x^2+1)$ (not concave up).

99. False. $\ln x+\ln 25=\ln 25x$

101. False. π is a constant, so $\dfrac{d}{dx}[\ln\pi]=0$.

103. (a)

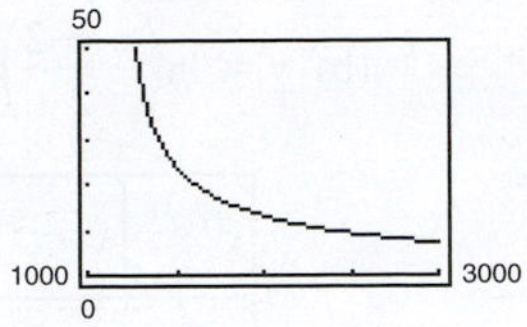

(b) 30 yr; \$503,434.80
(c) 20 yr; \$386,685.60
(d) When $x=1398.43$, $dt/dx\approx-0.0805$. When $x=1611.19$, $dt/dx\approx-0.0287$.
(e) Two benefits of a higher monthly payment are a shorter term and a lower total amount paid.

105. (a)

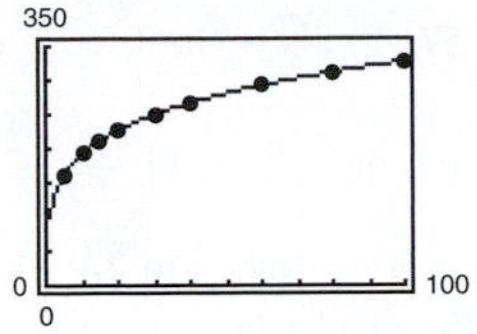

(c)

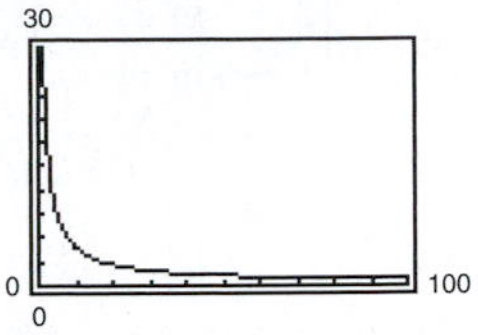

(b) $T'(10)\approx 4.75°/\text{lb/in.}^2$
$T'(70)\approx 0.97°/\text{lb/in.}^2$

$\lim_{p\to\infty} T'(p)=0$
Answers will vary.

107. (a)

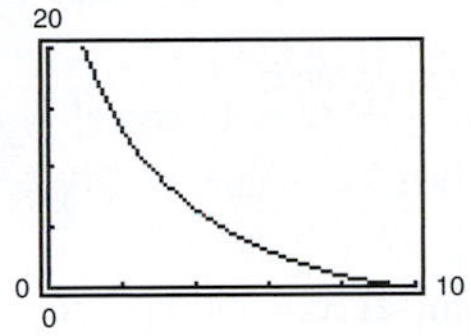

(b) When $x=5$, $dy/dx=-\sqrt{3}$. When $x=9$, $dy/dx=-\sqrt{19}/9$.
(c) $\lim_{x\to 10^-}\dfrac{dy}{dx}=0$

109. (a)

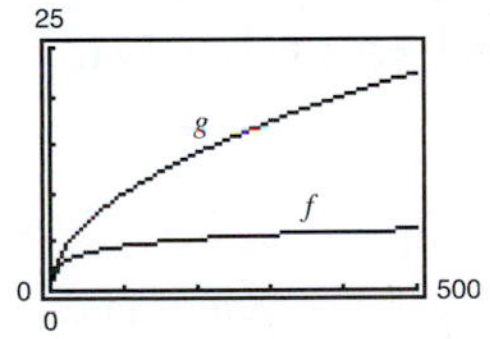

For $x>4$, $g'(x)>f'(x)$. g is increasing at a faster rate than f for large values of x.

(b)

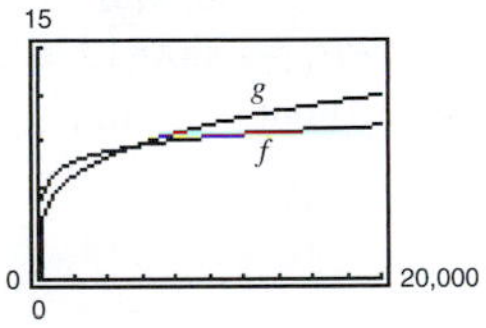

For $x>256$, $g'(x)>f'(x)$. g is increasing at a faster rate than f for large values of x.

$f(x)=\ln x$ increases very slowly for large values of x.

Section 5.2 *(page 334)*

1. $5\ln|x|+C$ **3.** $\ln|x+1|+C$ **5.** $\frac{1}{2}\ln|2x+5|+C$

7. $\frac{1}{2}\ln|x^2-3|+C$ **9.** $\ln|x^4+3x|+C$

11. $x^2/2-\ln(x^4)+C$ **13.** $\frac{1}{3}\ln|x^3+3x^2+9x|+C$

15. $\frac{1}{2}x^2-4x+6\ln|x+1|+C$ **17.** $\frac{1}{3}x^3+5\ln|x-3|+C$

19. $\frac{1}{3}x^3-2x+\ln\sqrt{x^2+2}+C$ **21.** $\frac{1}{3}(\ln x)^3+C$

23. $-\frac{2}{3}\ln|1-3\sqrt{x}|+C$

25. $2\ln|x-1|-2/(x-1)+C$

27. $\sqrt{2x}-\ln|1+\sqrt{2x}|+C$

29. $x+6\sqrt{x}+18\ln|\sqrt{x}-3|+C$ **31.** $3\ln\left|\sin\dfrac{\theta}{3}\right|+C$

33. $-\frac{1}{2}\ln|\csc 2x+\cot 2x|+C$ **35.** $\frac{1}{3}\sin 3\theta-\theta+C$

37. $\ln|1+\sin t|+C$ **39.** $\ln|\sec x-1|+C$

41. $y=-3\ln|2-x|+C$ **43.** $y=\ln|x^2-9|+C$

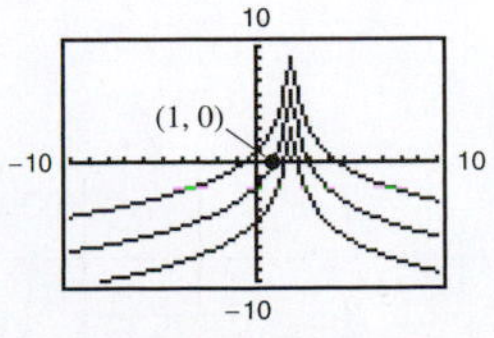

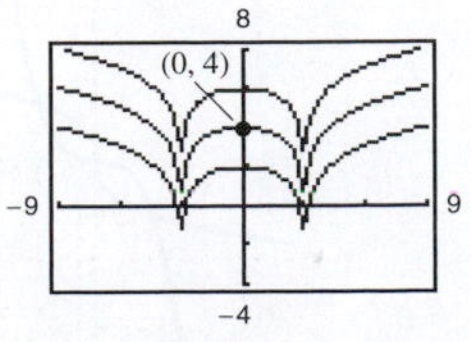

45. $f(x)=-2\ln x+3x-2$

47. (a)

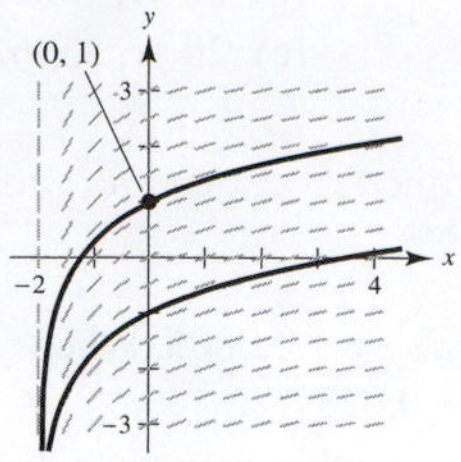

(b) $y = \ln\left(\frac{x+2}{2}\right) + 1$

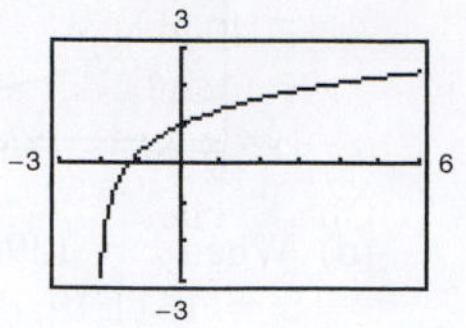

49. $\frac{5}{3}\ln 13 \approx 4.275$ **51.** $\frac{7}{3}$ **53.** $-\ln 3 \approx -1.099$

55. $\ln\left|\frac{2 - \sin 2}{1 - \sin 1}\right| \approx 1.929$ **57.** $2[\sqrt{x} - \ln(1 + \sqrt{x})] + C$

59. $\ln\left(\frac{\sqrt{x} - 1}{\sqrt{x} + 1}\right) + 2\sqrt{x} + C$ **61.** $\ln(\sqrt{2} + 1) - \frac{\sqrt{2}}{2} \approx 0.174$

63. $1/x$ **65.** $1/x$ **67.** $6 \ln 3$ **69.** $\frac{1}{2}\ln 2$

71. $\frac{15}{2} + 8 \ln 2 \approx 13.045$ **73.** $(12/\pi)\ln(2 + \sqrt{3}) \approx 5.03$

75. Trapezoidal Rule: 20.2
Simpson's Rule: 19.4667

77. Trapezoidal Rule: 5.3368
Simpson's Rule: 5.3632

79. Power Rule **81.** Log Rule **83.** d **85.** $x = 2$

87. Proof

89. $-\ln|\cos x| + C = \ln|1/\cos x| + C = \ln|\sec x| + C$

91. $\ln|\sec x + \tan x| + C = \ln\left|\frac{\sec^2 x - \tan^2 x}{\sec x - \tan x}\right| + C$
$= -\ln|\sec x - \tan x| + C$

93. 1 **95.** $1/(e - 1) \approx 0.582$

97. $P(t) = 1000(12 \ln|1 + 0.25t| + 1)$; $P(3) \approx 7715$

99. About 4.15 min

101.

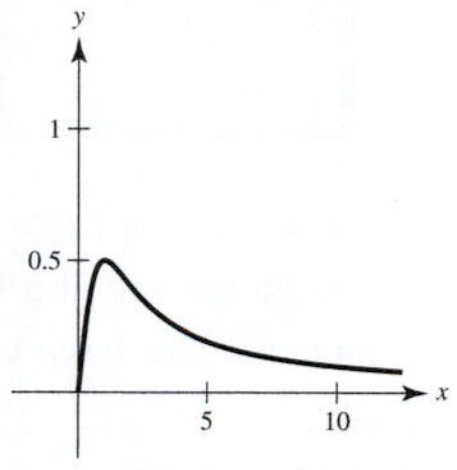

(a) $A = \frac{1}{2}\ln 2 - \frac{1}{4}$
(b) $0 < m < 1$
(c) $A = \frac{1}{2}(m - \ln m - 1)$

103. False. $\frac{1}{2}\ln x = \ln x^{1/2}$ **105.** True **107.** Proof

Section 5.3 *(page 343)*

1. (a) $f(g(x)) = 5[(x - 1)/5] + 1 = x$;
$g(f(x)) = [(5x + 1) - 1]/5 = x$

(b)

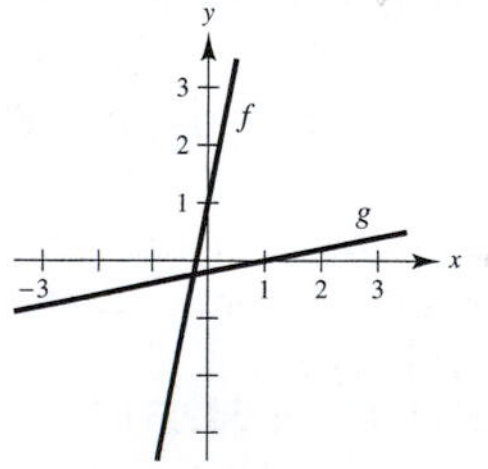

3. (a) $f(g(x)) = (\sqrt[3]{x})^3 = x$; $g(f(x)) = \sqrt[3]{x^3} = x$

(b)

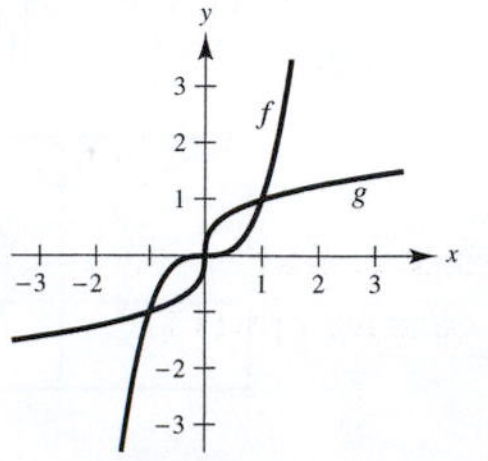

5. (a) $f(g(x)) = \sqrt{x^2 + 4 - 4} = x$;
$g(f(x)) = (\sqrt{x - 4})^2 + 4 = x$

(b)

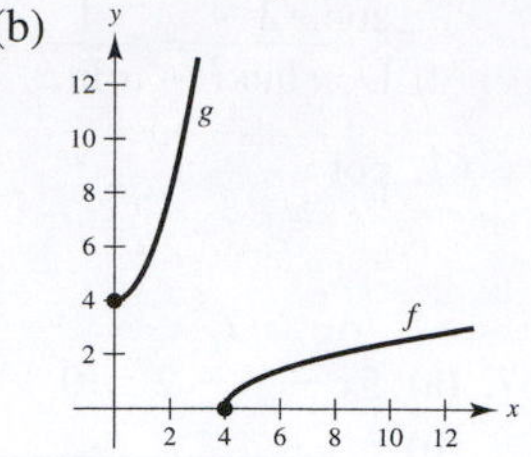

7. (a) $f(g(x)) = \frac{1}{1/x} = x$; $g(f(x)) = \frac{1}{1/x} = x$

(b)

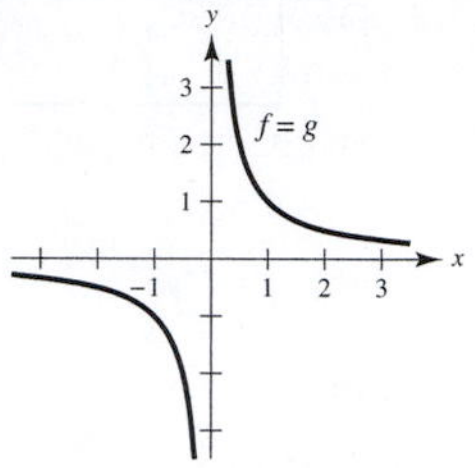

9. c **10.** b **11.** a **12.** d

13.

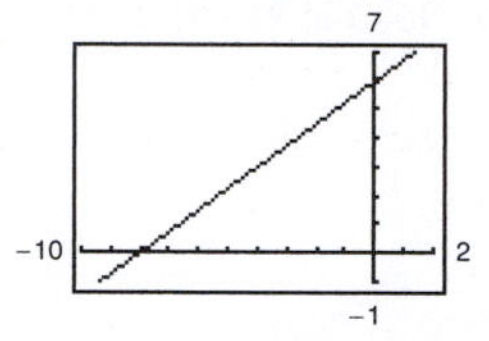

One-to-one, inverse exists.

15.

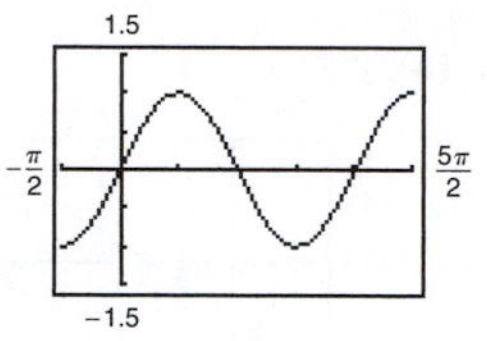

Not one-to-one, inverse does not exist.

17.

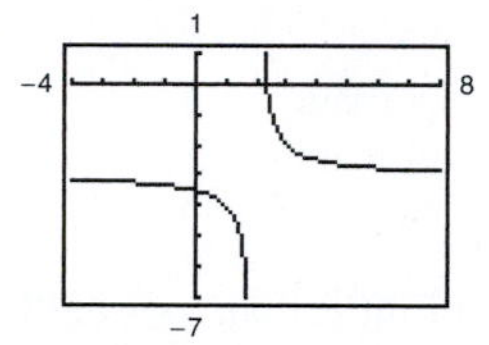

One-to-one, inverse exists.

19.

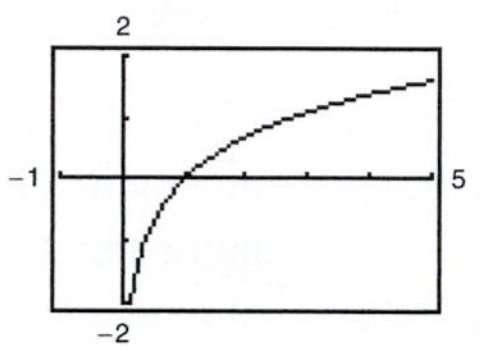

One-to-one, inverse exists.

21.

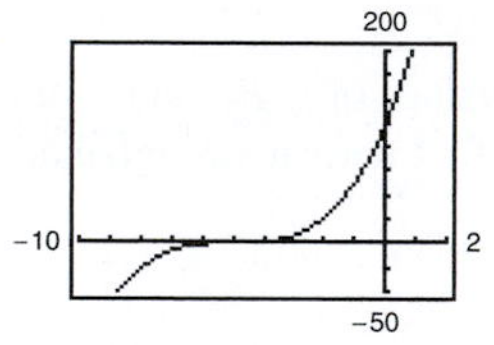

One-to-one, inverse exists.

23. Inverse exists. **25.** Inverse does not exist.

27. Inverse exists. **29.** $f'(x) = 2(x - 4) > 0$ on $(4, \infty)$

31. $f'(x) = -8/x^3 < 0$ on $(0, \infty)$

33. $f'(x) = -\sin x < 0$ on $(0, \pi)$

35. (a) $f^{-1}(x) = (x + 3)/2$

(b)

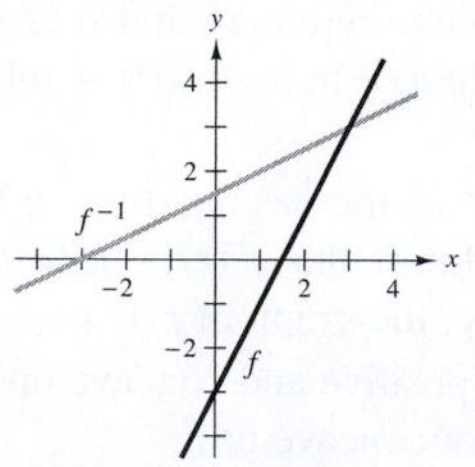

(c) f and f^{-1} are symmetric about $y = x$.
(d) Domain of f and f^{-1}: all real numbers
Range of f and f^{-1}: all real numbers

37. (a) $f^{-1}(x) = x^{1/5}$

(b)

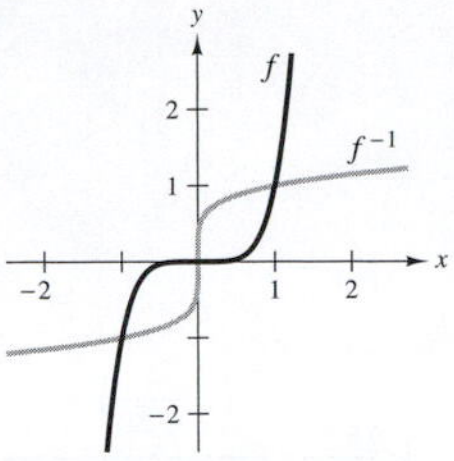

(c) f and f^{-1} are symmetric about $y = x$.

(d) Domain of f and f^{-1}: all real numbers

Range of f and f^{-1}: all real numbers

39. (a) $f^{-1}(x) = x^2, \quad x \geq 0$

(b)

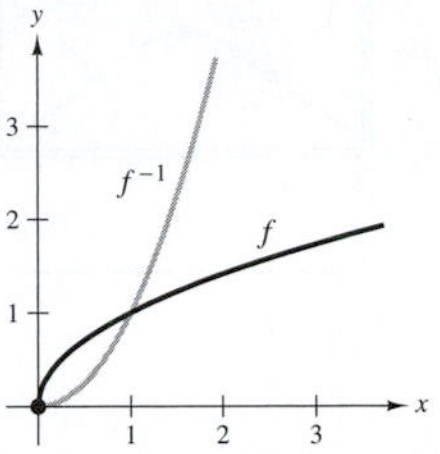

(c) f and f^{-1} are symmetric about $y = x$.

(d) Domain of f and f^{-1}: $x \geq 0$

Range of f and f^{-1}: $y \geq 0$

41. (a) $f^{-1}(x) = \sqrt{4 - x^2}, \quad 0 \leq x \leq 2$

(b)

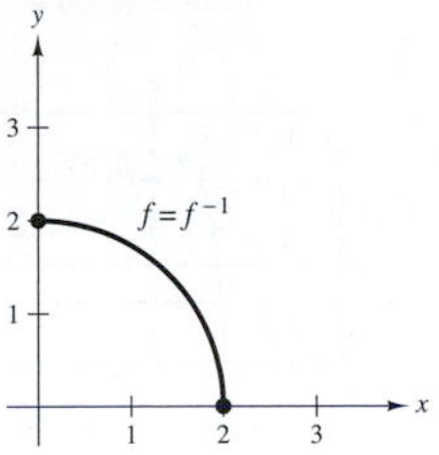

(c) f and f^{-1} are symmetric about $y = x$.

(d) Domain of f and f^{-1}: $0 \leq x \leq 2$

Range of f and f^{-1}: $0 \leq y \leq 2$

43. (a) $f^{-1}(x) = x^3 + 1$

(b)

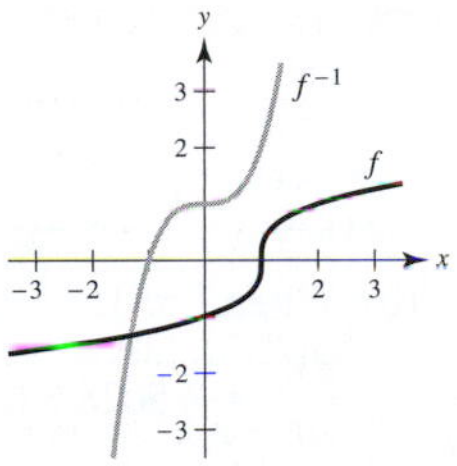

(c) f and f^{-1} are symmetric about $y = x$.

(d) Domain of f and f^{-1}: all real numbers

Range of f and f^{-1}: all real numbers

45. (a) $f^{-1}(x) = \sqrt{7}x/\sqrt{1 - x^2}, \quad -1 < x < 1$

(b)

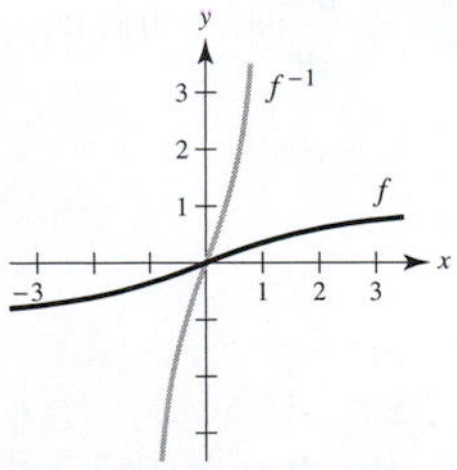

(c) f and f^{-1} are symmetric about $y = x$.

(d) Domain of f: all real numbers

Domain of f^{-1}: $-1 < x < 1$

Range of f: $-1 < y < 1$

Range of f^{-1}: all real numbers

47.

x	0	1	2	4
$f(x)$	1	2	3	4

x	1	2	3	4
$f^{-1}(x)$	0	1	2	4

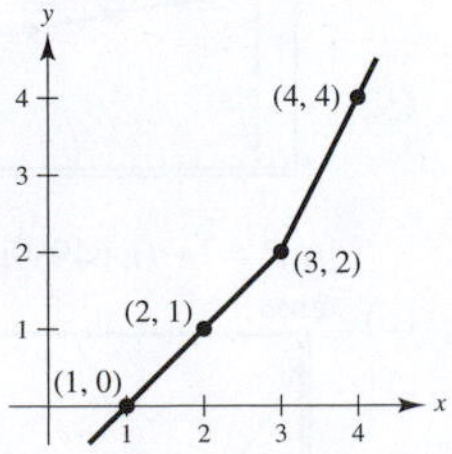

49. (a) Proof

(b) $y = \frac{20}{7}(80 - x)$

x: total cost

y: number of pounds of the less expensive commodity

(c) $[62.5, 80]$ (d) 20 lb

51. One-to-one

$f^{-1}(x) = x^2 + 2, \quad x \geq 0$

53. One-to-one

$f^{-1}(x) = 2 - x, \quad x \geq 0$

55. Sample answer: $f^{-1}(x) = \sqrt{x} + 3, \quad x \geq 0$

57. Sample answer: $f^{-1}(x) = x - 3, \quad x \geq 0$

59. Inverse exists. Volume is an increasing function, and therefore is one-to-one. The inverse function gives the time t corresponding to the volume V.

61. Inverse does not exist. **63.** $-1/6$ **65.** $1/17$

67. $2\sqrt{3}/3$ **69.** -2

71. (a) Domain of f: $(-\infty, \infty)$ (b) Range of f: $(-\infty, \infty)$

Domain of f^{-1}: $(-\infty, \infty)$ Range of f^{-1}: $(-\infty, \infty)$

(c)

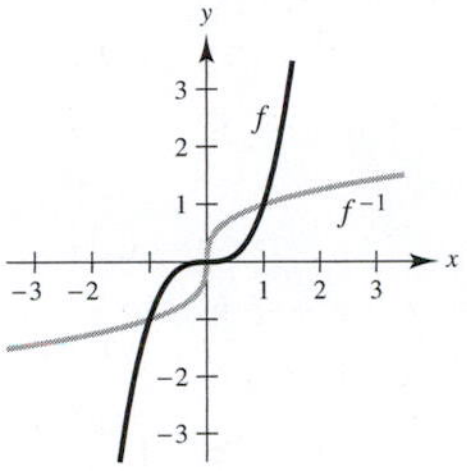

(d) $f'\left(\frac{1}{2}\right) = \frac{3}{4}, (f^{-1})'\left(\frac{1}{8}\right) = \frac{4}{3}$

73. (a) Domain of f: $[4, \infty)$ (b) Range of f: $[0, \infty)$

Domain of f^{-1}: $[0. \infty)$ Range of f^{-1}: $[4, \infty)$

(c)

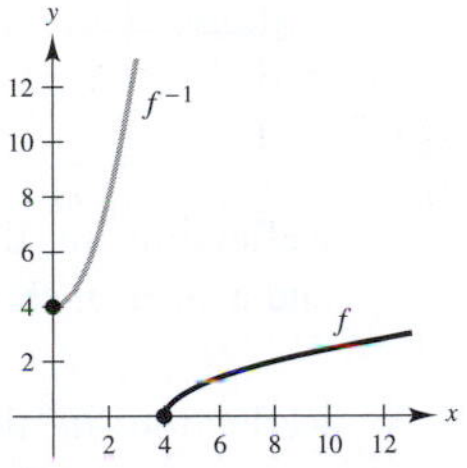

(d) $f'(5) = \frac{1}{2}, (f^{-1})'(1) = 2$

75. 32 **77.** 600 **79.** $(g^{-1} \circ f^{-1})(x) = (x + 1)/2$

81. $(f \circ g)^{-1}(x) = (x + 1)/2$

83. Let $y = f(x)$ be one-to-one. Solve for x as a function of y. Interchange x and y to get $y = f^{-1}(x)$. Let the domain of f^{-1} be the range of f. Verify that $f(f^{-1}(x)) = x$ and $f^{-1}(f(x)) = x$.

Sample answer: $f(x) = x^3$; $y = x^3$; $x = \sqrt[3]{y}$; $y = \sqrt[3]{x}$; $f^{-1}(x) = \sqrt[3]{x}$

85. Many x-values yield the same y-value. For example, $f(\pi) = 0 = f(0)$. The graph is not continuous at $[(2n - 1)\pi]/2$, where n is an integer.

87. $\frac{1}{4}$ **89.** False. Let $f(x) = x^2$. **91.** True

93. (a)

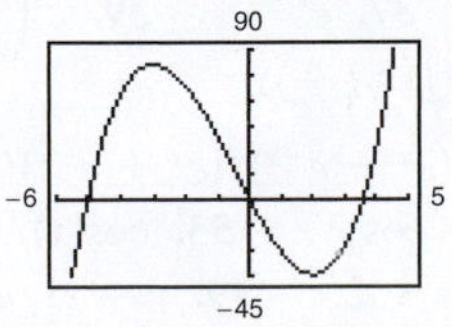

f does not pass the horizontal line test.

(b) $c = 2$

95–97. Proofs **99.** Proof; concave upward

101. Proof; $\sqrt{5}/5$

103. (a) Proof (b) $f^{-1}(x) = \dfrac{b - dx}{cx - a}$
(c) $a = -d$, or $b = c = 0$, $a = d$

Section 5.4 *(page 352)*

1. $x = 4$ **3.** $x \approx 2.485$ **5.** $x = 0$ **7.** $x \approx 0.511$
9. $x \approx 8.862$ **11.** $x \approx 7.389$ **13.** $x \approx 10.389$
15. $x \approx 5.389$

17.
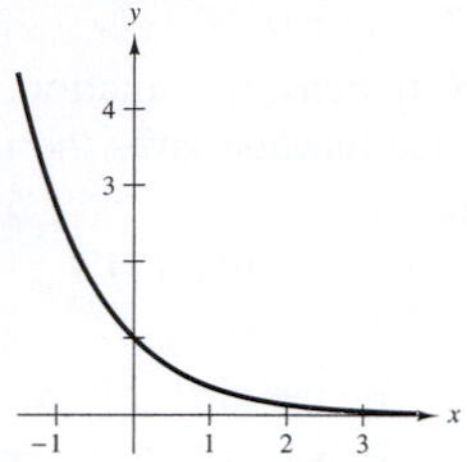

19.
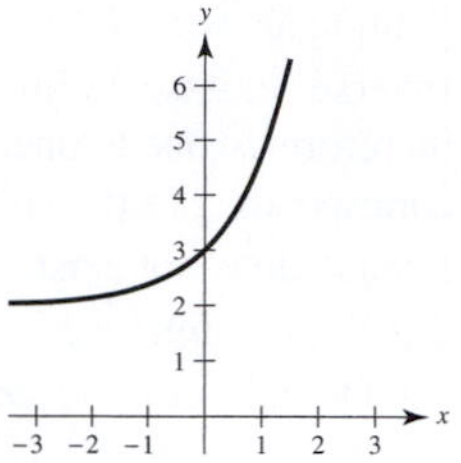

21.
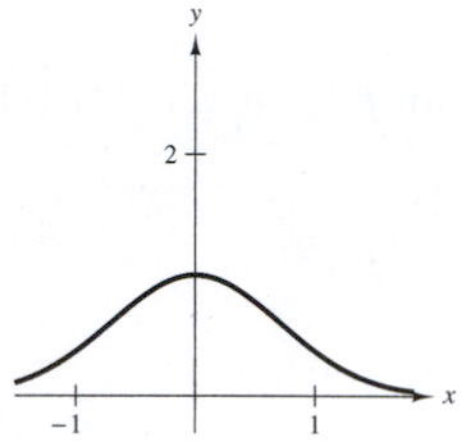

23. (a)
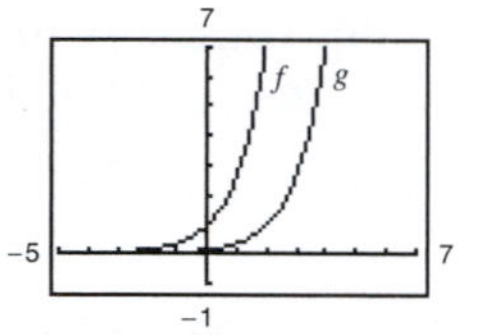

Translation two units to the right

(b)
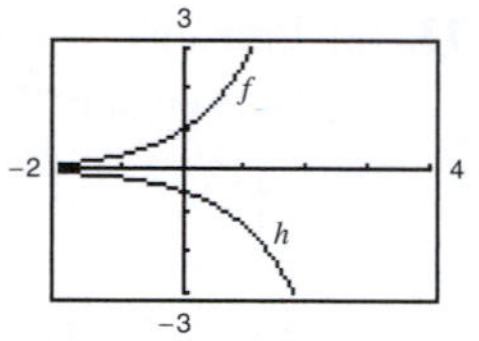

Reflection in the x-axis and a vertical shrink

(c)
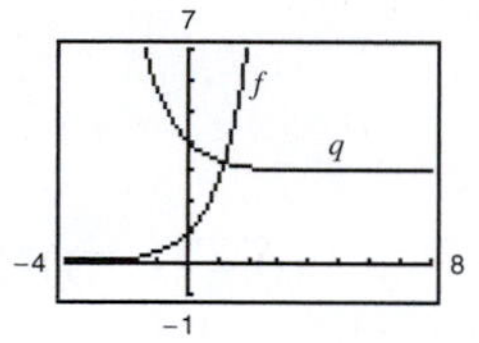

Reflection in the y-axis and a translation three units upward

25. c **26.** d **27.** a **28.** b

29.
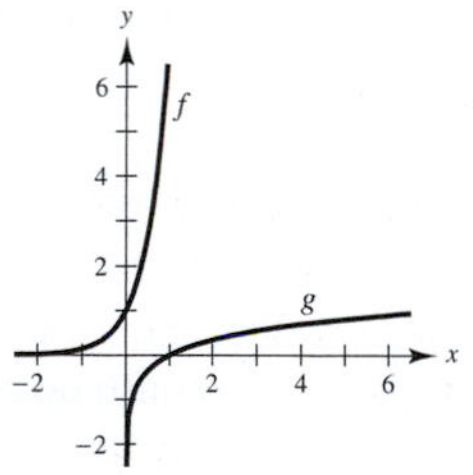

31.
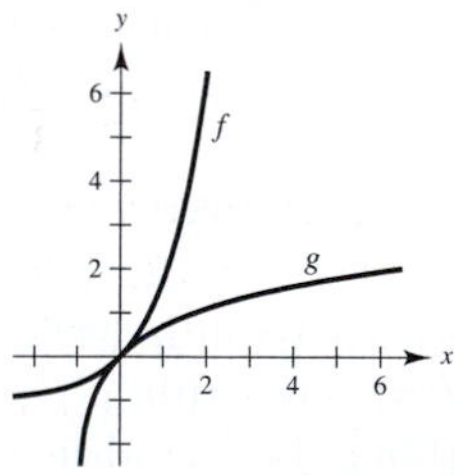

33. $2e^{2x}$ **35.** $e^{\sqrt{x}}/(2\sqrt{x})$ **37.** e^{x-4} **39.** $e^x\left(\dfrac{1}{x} + \ln x\right)$
41. $e^x(x^3 + 3x^2)$ **43.** $3(e^{-t} + e^t)^2(e^t - e^{-t})$
45. $2e^{2x}/(1 + e^{2x})$ **47.** $-2(e^x - e^{-x})/(e^x + e^{-x})^2$
49. $-2e^x/(e^x - 1)^2$ **51.** $2e^x \cos x$ **53.** $\cos(x)/x$
55. $y = 3x + 1$ **57.** $y = -x + 2$ **59.** $y = (1/e)x - 1/e$
61. $y = ex$ **63.** $\dfrac{10 - e^y}{xe^y + 3}$ **65.** $y = (-e - 1)x + 1$
67. $3(6x + 5)e^{-3x}$
69. $y'' - y = 0$
$4e^{-x} - 4e^{-x} = 0$

71. Relative minimum: $(0, 1)$
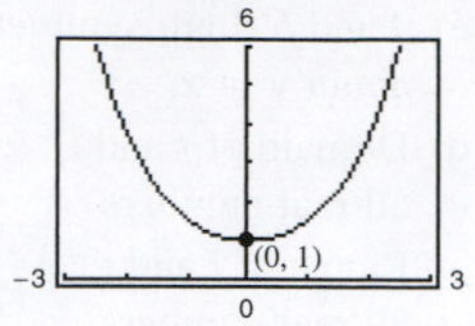

73. Relative maximum: $\left(2, 1/\sqrt{2\pi}\right)$
Points of inflection: $\left(1, \dfrac{e^{-0.5}}{\sqrt{2\pi}}\right), \left(3, \dfrac{e^{-0.5}}{\sqrt{2\pi}}\right)$
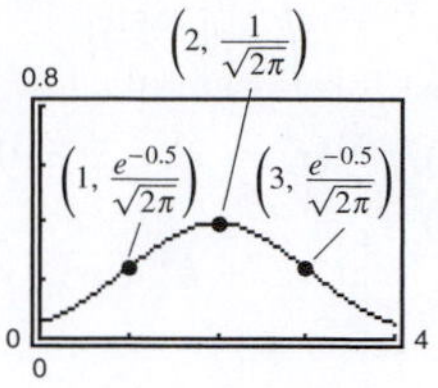

75. Relative minimum: $(0, 0)$
Relative maximum: $(2, 4e^{-2})$
Points of inflection:
$\left(2 \pm \sqrt{2}, \left(6 \pm 4\sqrt{2}\right)e^{-(2 \pm \sqrt{2})}\right)$
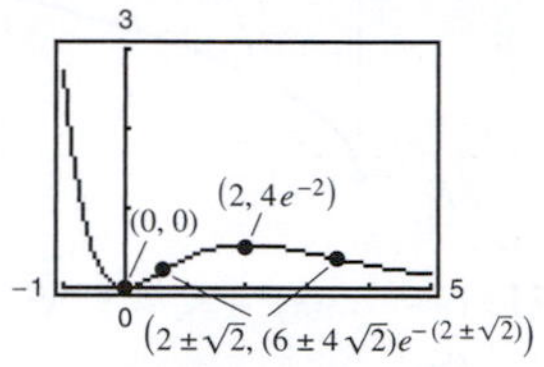

77. Relative maximum:
$(-1, 1 + e)$
Point of inflection: $(0, 3)$
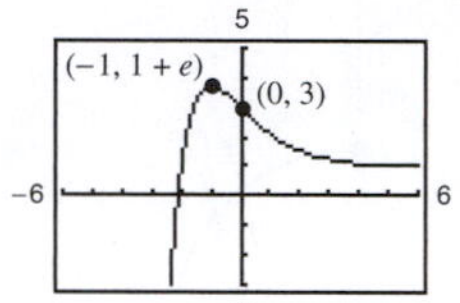

79. $A = \sqrt{2}e^{-1/2}$
81. $\left(\frac{1}{2}, e\right)$
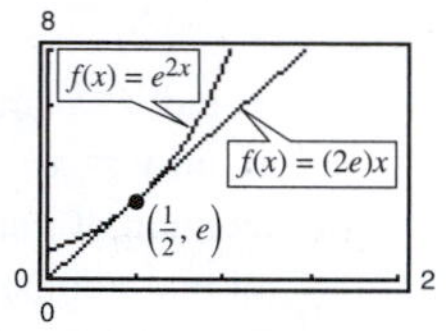

83. (a)
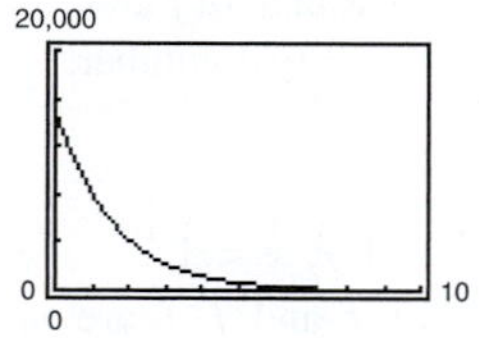

(b) When $t = 1$,
$\dfrac{dV}{dt} \approx -5028.84.$
When $t = 5$,
$\dfrac{dV}{dt} \approx -406.89.$

(c)
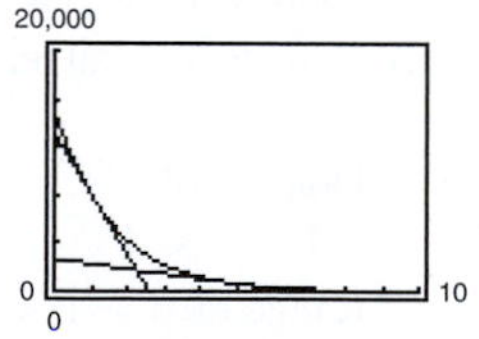

85. (a)
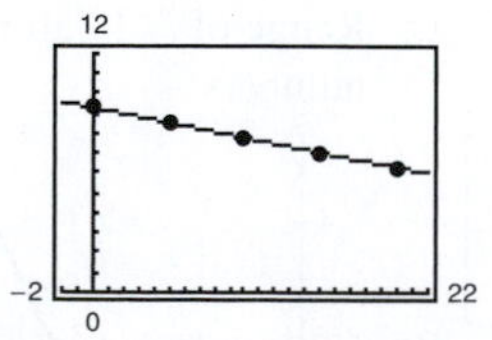

$\ln P = -0.1499h + 9.3018$

(b) $P = 10{,}957.7e^{-0.1499h}$

(c)
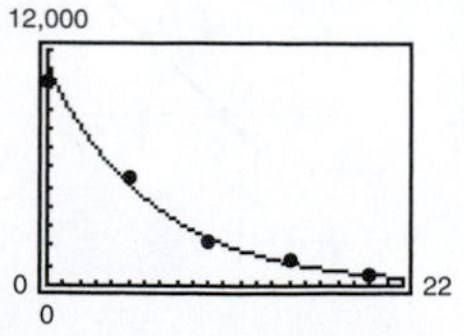

(d) $h = 5$: -776
$h = 18$: -111

87. $P_1 = 1 + x$; $P_2 = 1 + x + \frac{1}{2}x^2$

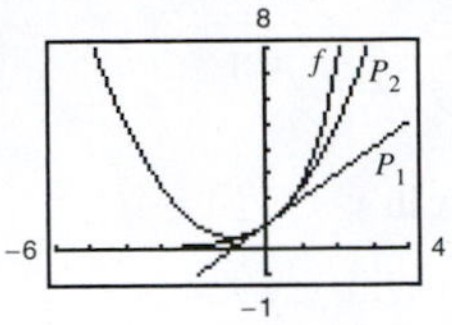

The values of f, P_1, and P_2 and their first derivatives agree at $x = 0$.

89. $12! = 479{,}001{,}600$
Stirling's Formula: $12! \approx 475{,}687{,}487$

91. $e^{5x} + C$ **93.** $\frac{1}{2}e^{2x-1} + C$ **95.** $\frac{1}{3}e^{x^3} + C$

97. $2e^{\sqrt{x}} + C$

99. $x - \ln(e^x + 1) + C_1$ or $-\ln(1 + e^{-x}) + C_2$

101. $-\frac{2}{3}(1 - e^x)^{3/2} + C$ **103.** $\ln|e^x - e^{-x}| + C$

105. $-\frac{5}{2}e^{-2x} + e^{-x} + C$ **107.** $\ln|\cos e^{-x}| + C$

109. $(e^2 - 1)/(2e^2)$ **111.** $(e - 1)/(2e)$

113. $(e/3)(e^2 - 1)$ **115.** $\ln\left(\dfrac{1 + e^6}{2}\right)$

117. $(1/\pi)[e^{\sin(\pi^2/2)} - 1]$

119. (a)

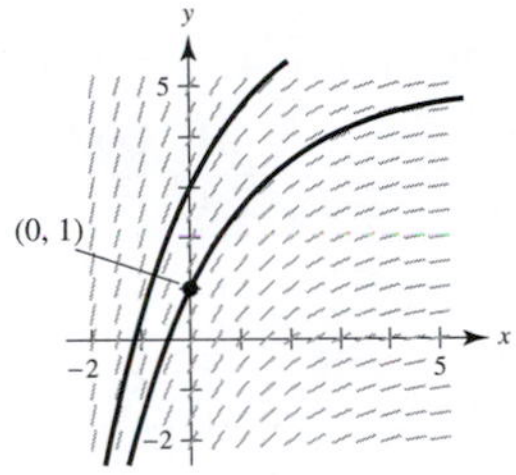

(b) $y = -4e^{-x/2} + 5$

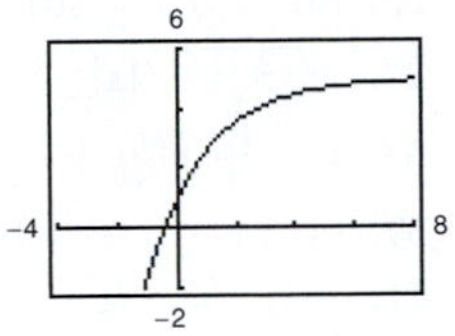

121. $[1/(2a)]e^{ax^2} + C$ **123.** $f(x) = \frac{1}{2}(e^x + e^{-x})$

125. $e^5 - 1 \approx 147.413$ **127.** $2(1 - e^{-3/2}) \approx 1.554$

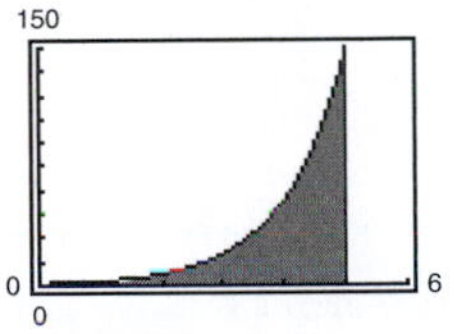

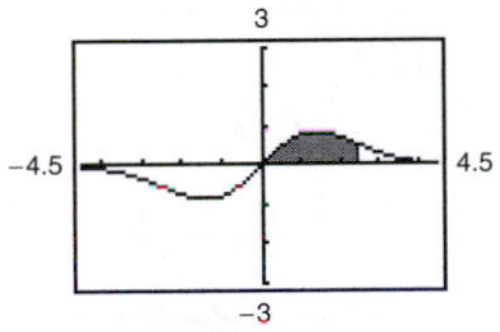

129. Midpoint Rule: 92.190; Trapezoidal Rule: 93.837; Simpson's Rule: 92.7385

131. The probability that a given battery will last between 48 months and 60 months is approximately 47.72%.

133. $a = \ln 3$

135. $f(x) = e^x$
The domain of $f(x)$ is $(-\infty, \infty)$, and the range of $f(x)$ is $(0, \infty)$. $f(x)$ is continuous, increasing, one-to-one, and concave upward on its entire domain.
$\lim_{x\to-\infty} e^x = 0$ and $\lim_{x\to\infty} e^x = \infty$

137. (a) Log Rule (b) Substitution

139. $\int_0^x e^t\,dt \ge \int_0^x 1\,dt$; $e^x - 1 \ge x$; $e^x \ge x + 1$ for $x \ge 0$

141. (a) $t = \dfrac{1}{2k}\ln\dfrac{B}{A}$
(b) $x''(t) = k^2(Ae^{kt} + Be^{-kt})$, k^2 is the constant of proportionality.

143. Proof

Section 5.5 *(page 362)*

1. -3 **3.** 0 **5.** (a) $\log_2 8 = 3$ (b) $\log_3(1/3) = -1$

7. (a) $10^{-2} = 0.01$ (b) $\left(\frac{1}{2}\right)^{-3} = 8$

9.

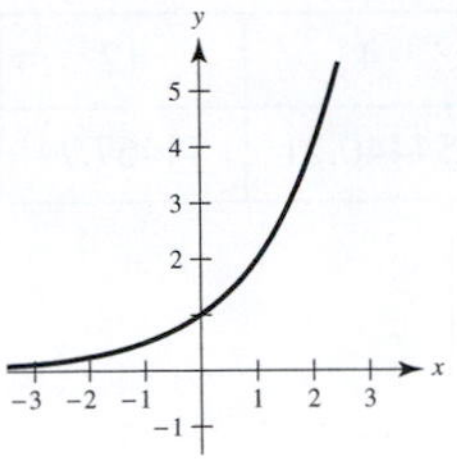

11.

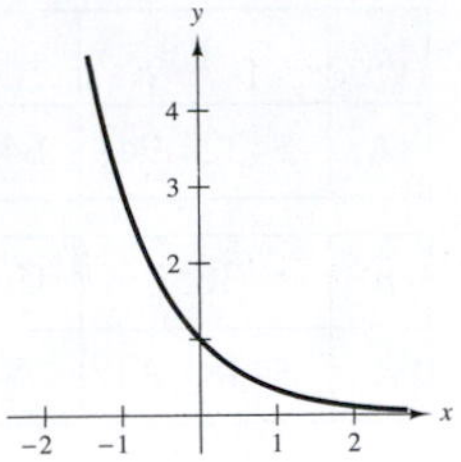

13.

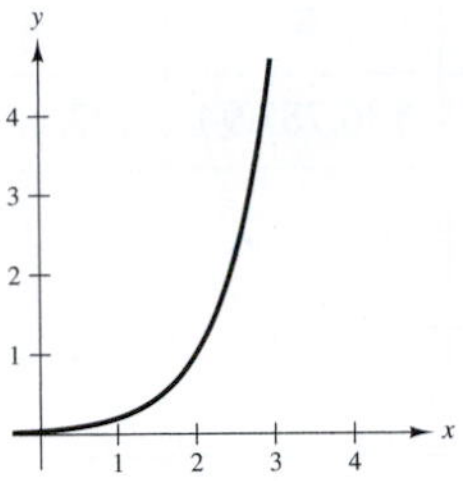

15. d **16.** c **17.** b **18.** a

19. (a) $x = 3$ (b) $x = -1$ **21.** (a) $x = \frac{1}{3}$ (b) $x = \frac{1}{16}$

23. (a) $x = -1, 2$ (b) $x = \frac{1}{3}$ **25.** 1.965 **27.** -6.288

29. 12.253 **31.** 33.000 **33.** ± 11.845

35.

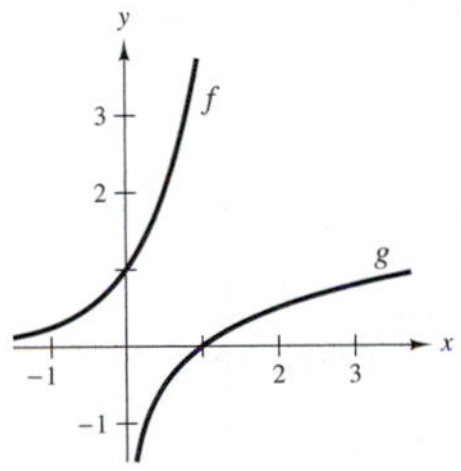

37. $(\ln 4)4^x$ **39.** $(-4\ln 5)5^{-4x}$ **41.** $9^x(x\ln 9 + 1)$

43. $t2^t(t\ln 2 + 2)$ **45.** $-2^{-\theta}[(\ln 2)\cos\pi\theta + \pi\sin\pi\theta]$

47. $5/[(\ln 4)(5x + 1)]$ **49.** $2/[(\ln 5)(t - 4)]$

51. $x/[(\ln 5)(x^2 - 1)]$ **53.** $(x - 2)/[(\ln 2)x(x - 1)]$

55. $(3x - 2)/[(2x\ln 3)(x - 1)]$ **57.** $5(1 - \ln t)/(t^2\ln 2)$

59. $y = -2x\ln 2 - 2\ln 2 + 2$

61. $y = [1/(27\ln 3)]x + 3 - 1/\ln 3$ **63.** $2(1 - \ln x)x^{(2/x)-2}$

65. $(x - 2)^{x+1}[(x + 1)/(x - 2) + \ln(x - 2)]$

67. $y = x$ **69.** $y = \dfrac{\cos e}{e}x - \cos e + 1$

71. $3^x/\ln 3 + C$ **73.** $\dfrac{1}{3}x^3 - \dfrac{2^{-x}}{\ln 2} + C$

75. $[-1/(2\ln 5)](5^{-x^2}) + C$ **77.** $\ln(3^{2x} + 1)/(2\ln 3) + C$

79. $7/(2\ln 2)$ **81.** $4/\ln 5 - 2/\ln 3$ **83.** $26/\ln 3$

85. (a) $x > 0$ (b) 10^x (c) $3 \le f(x) \le 4$
(d) $0 < x < 1$ (e) 10 (f) 100^n

87. (a) \$40.64 (b) $C'(1) \approx 0.051P$, $C'(8) \approx 0.072P$
(c) $\ln 1.05$

89.

n	1	2	4	12
A	\$1410.60	\$1414.78	\$1416.91	\$1418.34

n	365	Continuous
A	\$1419.04	\$1419.07

91.

n	1	2	4	12
A	\$4321.94	\$4399.79	\$4440.21	\$4467.74

n	365	Continuous
A	\$4481.23	\$4481.69

93.

t	1	10	20	30
P	\$95,122.94	\$60,653.07	\$36,787.94	\$22,313.02

t	40	50
P	\$13,533.53	\$8208.50

95.

t	1	10	20	30
P	\$95,132.82	\$60,716.10	\$36,864.45	\$22,382.66

t	40	50
P	\$13,589.88	\$8251.24

97. c

99. (a) 6.7 million ft^3/acre

(b) $t = 20$: $\dfrac{dV}{dt} = 0.073$; $t = 60$: $\dfrac{dV}{dt} = 0.040$

101. (a)

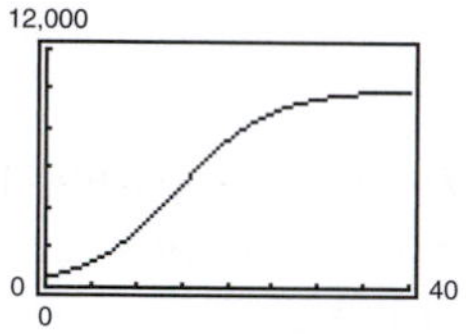

(b) 10,000 fish

(c) 1 month: About 114 fish/mo
10 months: About 403 fish/mo

(d) About 15 mo

103. (a) $y_1 = -40x + 743$, $y_2 = 968 - 265.5 \ln x$,
$y_3 = 836.817(0.9169)^x$, $y_4 = 1344.8884x^{-0.5689}$

(b)

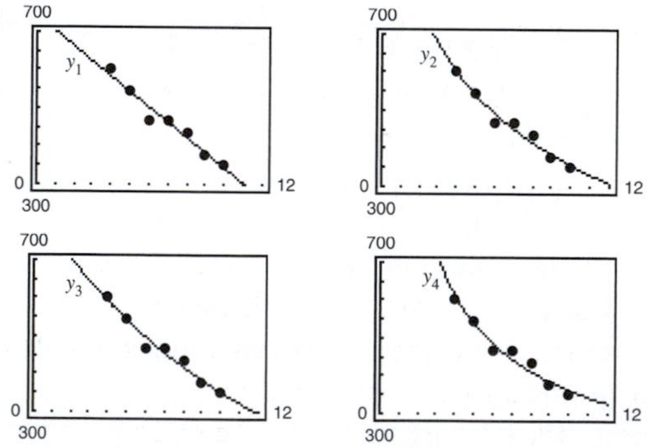

(c) The number of pancreas transplants is decreasing by about 40 transplants each year.

(d) $y_1'(8) = -40.04$, $y_2'(8) = -33.18$, $y_3'(8) = -36.27$, $y_4'(8) = -29.30$; y_1 is decreasing at the greatest rate.

105. $y = 1200(0.6^t)$ **107.** e **109.** e^2

111. False. e is an irrational number. **113.** True **115.** True

117. (a) $(2^3)^2 = 2^6 = 64$
$2^{(3^2)} = 2^9 = 512$

(b) No. $f(x) = (x^x)^x = x^{(x^2)}$ and $g(x) = x^{(x^x)}$

(c) $f'(x) = x^{x^2}(x + 2x \ln x)$
$g'(x) = x^{x^x + x - 1}[x(\ln x)^2 + x \ln x + 1]$

119. Proof

121. (a) $\dfrac{dy}{dx} = \dfrac{y^2 - yx \ln y}{x^2 - xy \ln x}$

(b) (i) 1 when $c \neq 0$, $c \neq e$ (ii) -3.1774
(iii) -0.3147

(c) (e, e)

123. Putnam Problem B3, 1951

Section 5.6 *(page 372)*

1. $\left(-\sqrt{2}/2, 3\pi/4\right)$, $(1/2, \pi/3)$, $\left(\sqrt{3}/2, \pi/6\right)$ **3.** $\pi/6$

5. $\pi/3$ **7.** $\pi/6$ **9.** $-\pi/4$ **11.** 2.50

13. $\arccos(1/1.269) \approx 0.66$ **15.** x **17.** $\sqrt{1 - x^2}/x$

19. $1/x$ **21.** (a) $3/5$ (b) $5/3$

23. (a) $-\sqrt{3}$ (b) $-\frac{13}{5}$ **25.** $\sqrt{1 - 4x^2}$

27. $\sqrt{x^2 - 1}/|x|$ **29.** $\sqrt{x^2 - 9}/3$ **31.** $\sqrt{x^2 + 2}/x$

33. $x = \frac{1}{3}\left[\sin\left(\frac{1}{2}\right) + \pi\right] \approx 1.207$ **35.** $x = \frac{1}{3}$

37. (a) and (b) Proofs **39.** $2/\sqrt{2x - x^2}$

41. $-3/\sqrt{4 - x^2}$ **43.** $e^x/(1 + e^{2x})$

45. $\left(3x - \sqrt{1 - 9x^2} \arcsin 3x\right)/\left(x^2\sqrt{1 - 9x^2}\right)$

47. $-t/\sqrt{1 - t^2}$ **49.** $2 \arccos x$ **51.** $1/(1 - x^4)$

53. $\arcsin x$ **55.** $x^2/\sqrt{16 - x^2}$ **57.** $2/(1 + x^2)^2$

59. $y = \frac{1}{3}\left(4\sqrt{3}x - 2\sqrt{3} + \pi\right)$ **61.** $y = \frac{1}{4}x + (\pi - 2)/4$

63. $y = (2\pi - 4)x + 4$

65. $P_1(x) = x$; $P_2(x) = x$

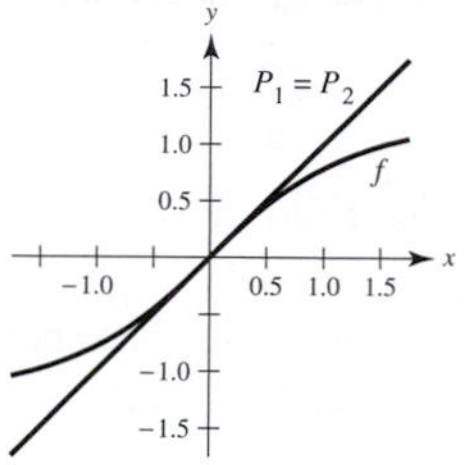

67. $P_1(x) = \dfrac{\pi}{6} + \dfrac{2\sqrt{3}}{3}\left(x - \dfrac{1}{2}\right)$

$P_2(x) = \dfrac{\pi}{6} + \dfrac{2\sqrt{3}}{3}\left(x - \dfrac{1}{2}\right) + \dfrac{2\sqrt{3}}{9}\left(x - \dfrac{1}{2}\right)^2$

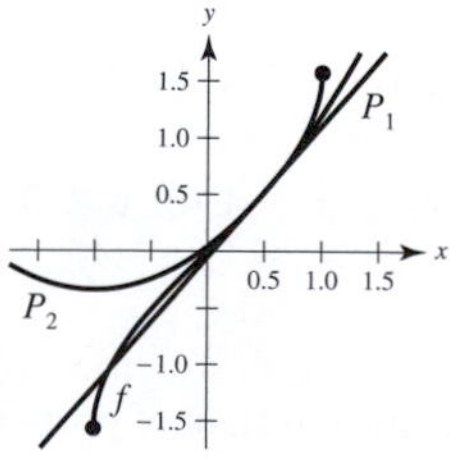

69. Relative maximum: $(1.272, -0.606)$
Relative minimum: $(-1.272, 3.747)$

71. Relative maximum: $(2, 2.214)$

73.

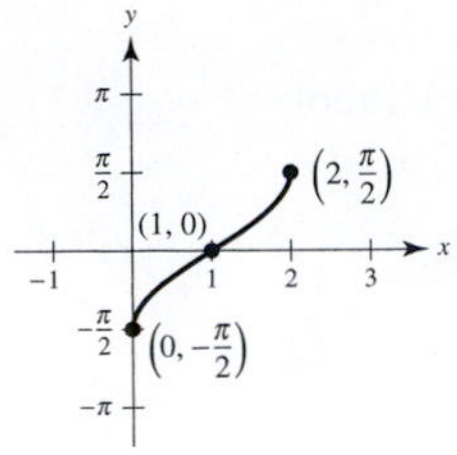

Maximum: $\left(2, \frac{\pi}{2}\right)$

Minimum: $\left(0, -\frac{\pi}{2}\right)$

Point of inflection: $(1, 0)$

75.

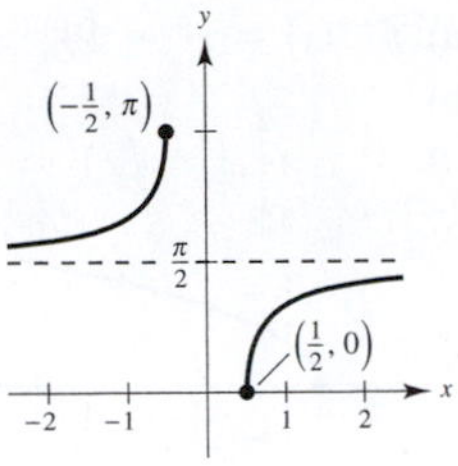

Maximum: $\left(-\frac{1}{2}, \pi\right)$

Minimum: $\left(\frac{1}{2}, 0\right)$

Asymptote: $y = \frac{\pi}{2}$

77. $y = -2\pi x/(\pi + 8) + 1 - \pi^2/(2\pi + 16)$

79. $y = -x + \sqrt{2}$

81. If the domains were not restricted, then the trigonometric functions would have no inverses, because they would not be one-to-one.

83. (a) $\arcsin(\arcsin 0.5) \approx 0.551$
$\arcsin(\arcsin 1)$ does not exist.
(b) $\sin(-1) \le x \le \sin(1)$

85. False. The range of arccos is $[0, \pi]$.

87. True **89.** True

91. (a) $\theta = \operatorname{arccot}(x/5)$
(b) $x = 10$: 16 rad/h; $x = 3$: 58.824 rad/h

93. (a) $h(t) = -16t^2 + 256$; $t = 4$ sec
(b) $t = 1$: -0.0520 rad/sec; $t = 2$: -0.1116 rad/sec

95. $50\sqrt{2} \approx 70.71$ ft **97.** (a) and (b) Proofs

99. (a)

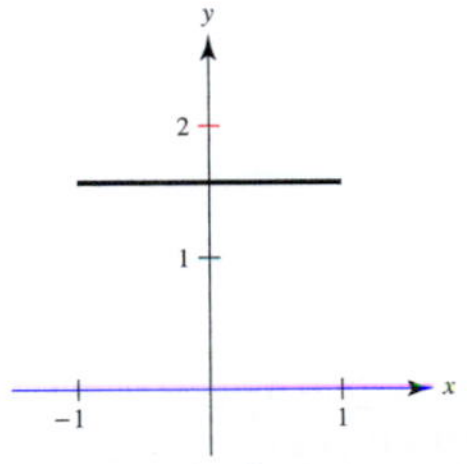

(b) The graph is a horizontal line at $\frac{\pi}{2}$.
(c) Proof

101. $c = 2$ **103.** Proof

Section 5.7 *(page 380)*

1. $\arcsin \frac{x}{3} + C$ **3.** $\operatorname{arcsec}|2x| + C$

5. $\arcsin(x + 1) + C$ **7.** $\frac{1}{2}\arcsin t^2 + C$

9. $\frac{1}{10}\arctan \frac{t^2}{5} + C$ **11.** $\frac{1}{4}\arctan(e^{2x}/2) + C$

13. $\arcsin\left(\frac{\tan x}{5}\right) + C$ **15.** $2\arcsin\sqrt{x} + C$

17. $\frac{1}{2}\ln(x^2 + 1) - 3\arctan x + C$

19. $8\arcsin[(x - 3)/3] - \sqrt{6x - x^2} + C$ **21.** $\pi/6$

23. $\pi/6$ **25.** $\frac{1}{5}\arctan\frac{3}{5} \approx 0.108$

27. $\arctan 5 - \pi/4 \approx 0.588$ **29.** $\pi/4$ **31.** $\frac{1}{32}\pi^2 \approx 0.308$

33. $\pi/2$ **35.** $\ln|x^2 + 6x + 13| - 3\arctan[(x + 3)/2] + C$

37. $\arcsin[(x + 2)/2] + C$ **39.** $4 - 2\sqrt{3} + \frac{1}{6}\pi \approx 1.059$

41. $\frac{1}{2}\arctan(x^2 + 1) + C$

43. $2\sqrt{e^t - 3} - 2\sqrt{3}\arctan(\sqrt{e^t - 3}/\sqrt{3}) + C$ **45.** $\pi/6$

47. a and b **49.** a, b, and c

51. No. This integral does not correspond to any of the basic integration rules.

53. $y = \arcsin(x/2) + \pi$

55. (a)

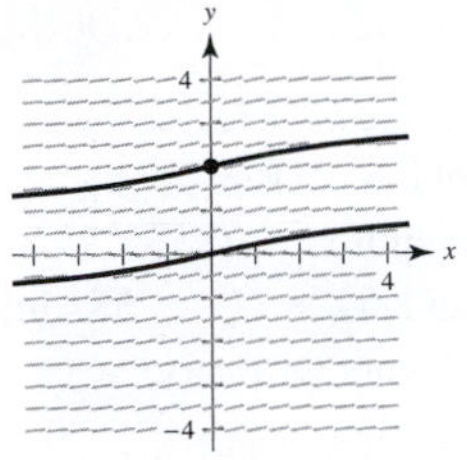

(b) $y = \frac{2}{3}\arctan\frac{x}{3} + 2$

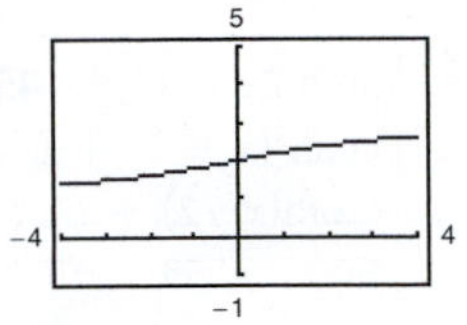

57.

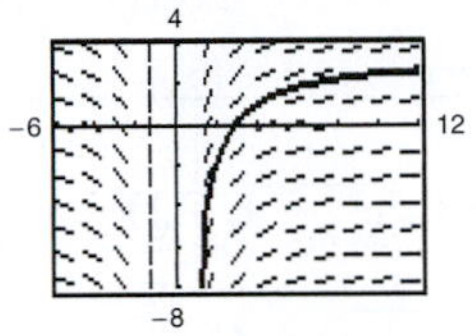

59.

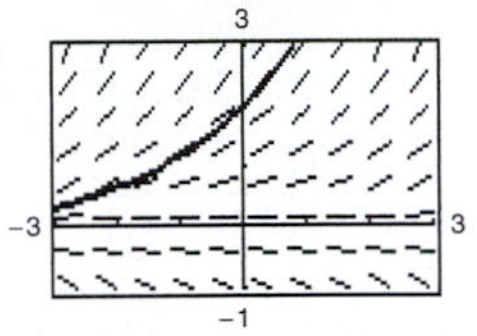

61. $\pi/3$ **63.** $\pi/8$ **65.** $3\pi/2$

67. (a)

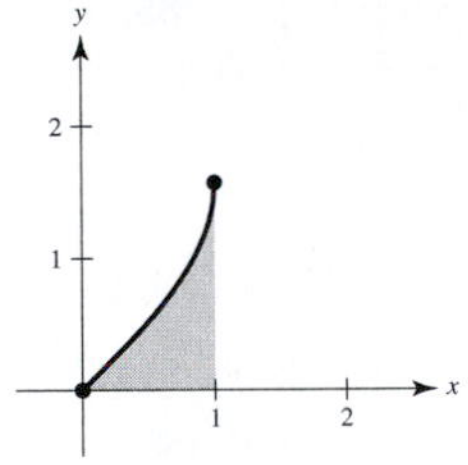

(b) 0.5708
(c) $(\pi - 2)/2$

69. (a) $F(x)$ represents the average value of $f(x)$ over the interval $[x, x + 2]$. Maximum at $x = -1$
(b) Maximum at $x = -1$

71. False. $\displaystyle\int \frac{dx}{3x\sqrt{9x^2 - 16}} = \frac{1}{12}\operatorname{arcsec}\frac{|3x|}{4} + C$

73. True **75–77.** Proofs

79. (a) $\displaystyle\int_0^1 \frac{1}{1 + x^2}\,dx$ (b) About 0.7847

(c) Because $\displaystyle\int_0^1 \frac{1}{1 + x^2}\,dx = \frac{\pi}{4}$, you can use the Trapezoidal Rule to approximate $\frac{\pi}{4}$. Multiplying the result by 4 gives an estimation of π.

Section 5.8 *(page 390)*

1. (a) 10.018 (b) -0.964 **3.** (a) $\frac{4}{3}$ (b) $\frac{13}{12}$

5. (a) 1.317 (b) 0.962 **7–13.** Proofs

15. $\cosh x = \sqrt{13}/2$; $\tanh x = 3\sqrt{13}/13$; $\operatorname{csch} x = 2/3$; $\operatorname{sech} x = 2\sqrt{13}/13$; $\coth x = \sqrt{13}/3$

17. ∞ **19.** 0 **21.** 1 **23.** $3\cosh 3x$

25. $-10x[\operatorname{sech}(5x^2)\tanh(5x^2)]$ **27.** $\coth x$ **29.** $\sinh^2 x$

31. $\operatorname{sech} t$ **33.** $y = -2x + 2$ **35.** $y = 1 - 2x$

37. Relative maxima: $(\pm\pi, \cosh \pi)$; Relative minimum: $(0, -1)$

39. Relative maximum: $(1.20, 0.66)$;
Relative minimum: $(-1.20, -0.66)$

41. (a)

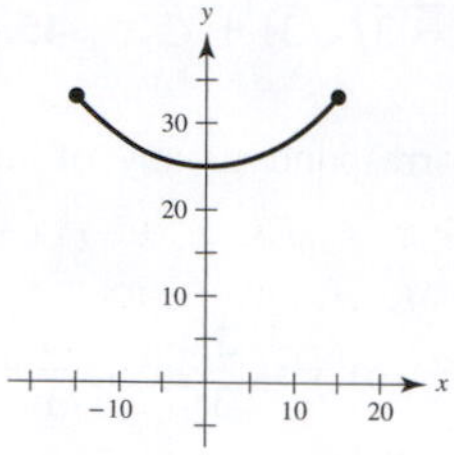

(b) 33.146 units; 25 units
(c) $m = \sinh(1) \approx 1.175$

43. $\frac{1}{2}\sinh 2x + C$ **45.** $-\frac{1}{2}\cosh(1 - 2x) + C$
47. $\frac{1}{3}\cosh^3(x - 1) + C$ **49.** $\ln|\sinh x| + C$
51. $-\coth(x^2/2) + C$ **53.** $\operatorname{csch}(1/x) + C$ **55.** $\ln(5/4)$
57. $\frac{1}{5}\ln 3$ **59.** $\pi/4$ **61.** Answers will vary.
63. $\cosh x$, $\operatorname{sech} x$ **65.** $3/\sqrt{9x^2 - 1}$ **67.** $\dfrac{1}{2\sqrt{x}(1 - x)}$
69. $|\sec x|$ **71.** $\dfrac{-2\operatorname{csch}^{-1} x}{|x|\sqrt{1 + x^2}}$ **73.** $2\sinh^{-1}(2x)$
75. $\dfrac{\sqrt{3}}{18}\ln\left|\dfrac{1 + \sqrt{3}x}{1 - \sqrt{3}x}\right| + C$ **77.** $\ln(\sqrt{e^{2x} + 1} - 1) - x + C$
79. $2\sinh^{-1}\sqrt{x} + C = 2\ln(\sqrt{x} + \sqrt{1 + x}) + C$
81. $\dfrac{1}{4}\ln\left|\dfrac{x - 4}{x}\right| + C$ **83.** $\ln\left(\dfrac{3 + \sqrt{5}}{2}\right)$ **85.** $\dfrac{\ln 7}{12}$
87. $\dfrac{1}{4}\arcsin\left(\dfrac{4x - 1}{9}\right) + C$
89. $-\dfrac{x^2}{2} - 4x - \dfrac{10}{3}\ln\left|\dfrac{x - 5}{x + 1}\right| + C$
91. $8\arctan(e^2) - 2\pi \approx 5.207$ **93.** $\frac{5}{2}\ln\left(\sqrt{17} + 4\right) \approx 5.237$
95. $\frac{52}{31}$ kg **97.** (a) $-\sqrt{a^2 - x^2}/x$ (b) Proof
99–107. Proofs **109.** Putnam Problem 8, 1939

Review Exercises for Chapter 5 *(page 393)*

1.

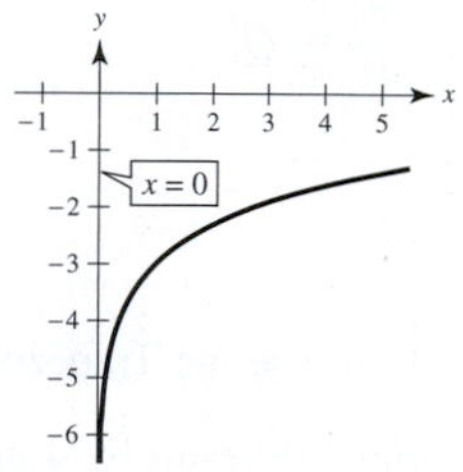

Domain: $x > 0$

3. $\frac{1}{5}[\ln(2x + 1) + \ln(2x - 1) - \ln(4x^2 + 1)]$
5. $\ln\left(3\sqrt[3]{4 - x^2}/x\right)$ **7.** $1/(2x)$ **9.** $(1 + 2\ln x)/\left(2\sqrt{\ln x}\right)$
11. $-\dfrac{8x}{x^4 - 16}$ **13.** $y = -x + 1$ **15.** $\dfrac{1}{7}\ln|7x - 2| + C$
17. $-\ln|1 + \cos x| + C$ **19.** $3 + \ln 2$ **21.** $\ln\left(2 + \sqrt{3}\right)$
23. (a) $f^{-1}(x) = 2x + 6$
(b)

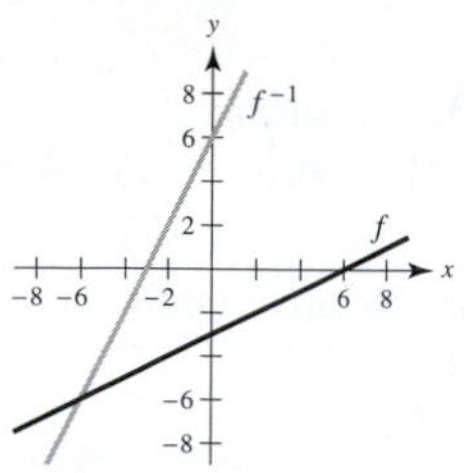

(c) Proof
(d) Domain of f and f^{-1}: all real numbers
Range of f and f^{-1}: all real numbers

25. (a) $f^{-1}(x) = x^2 - 1$, $x \ge 0$
(b)

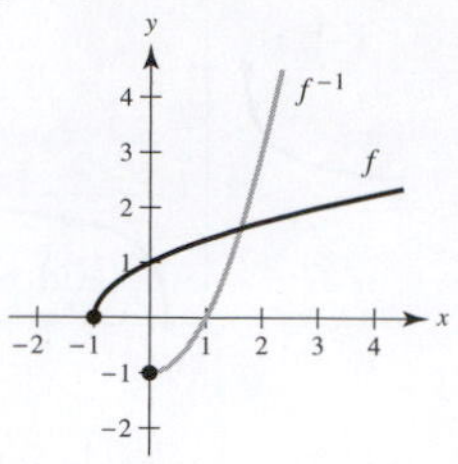

(c) Proof
(d) Domain of f: $x \ge -1$; Domain of f^{-1}: $x \ge 0$
Range of f: $y \ge 0$; Range of f^{-1}: $y \ge -1$
27. (a) $f^{-1}(x) = x^3 - 1$
(b)

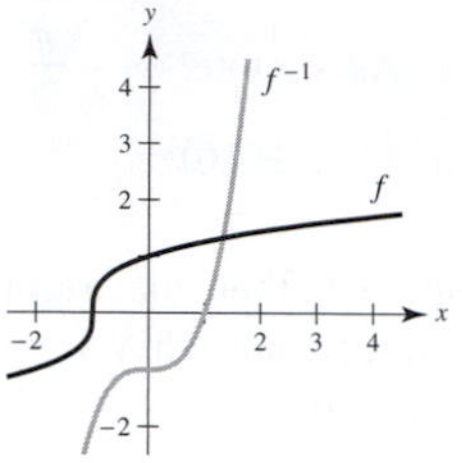

(c) Proof
(d) Domain of f and f^{-1}: all real numbers
Range of f and f^{-1}: all real numbers
29. $1/\left[3\left(\sqrt[3]{-3}\right)^2\right] \approx 0.160$ **31.** $3/4$ **33.** $x \approx 1.134$
35. $e^4 - 1 \approx 53.598$ **37.** $te^t(t + 2)$
39. $(e^{2x} - e^{-2x})/\sqrt{e^{2x} + e^{-2x}}$ **41.** $x(2 - x)/e^x$
43. $y = 6x + 1$ **45.** $-y/[x(2y + \ln x)]$ **47.** $-\frac{1}{2}e^{1-x^2} + C$
49. $(e^{4x} - 3e^{2x} - 3)/(3e^x) + C$ **51.** $(1 - e^{-3})/6 \approx 0.158$
53. $\ln(e^2 + e + 1) \approx 2.408$ **55.** About 1.729
57.

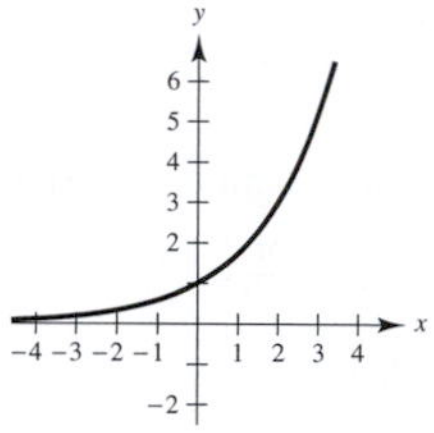

59. $3^{x-1}\ln 3$ **61.** $x^{2x+1}(2\ln x + 2 + 1/x)$
63. $-1/[\ln 3(2 - 2x)]$ **65.** $5^{(x+1)^2}/(2\ln 5) + C$
67. (a) Domain: $0 \le h < 18{,}000$
(b)

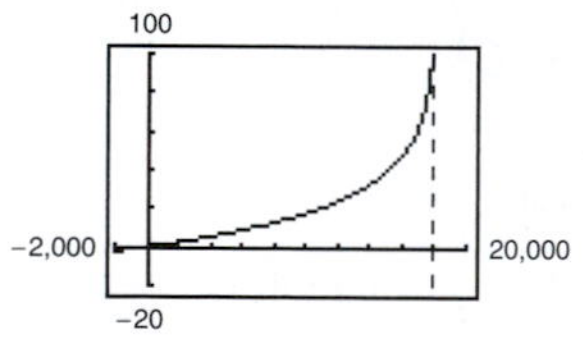

(c) $t = 0$
Vertical asymptote: $h = 18{,}000$
69. (a) $1/2$ (b) $\sqrt{3}/2$ **71.** $(1 - x^2)^{-3/2}$
73. $\dfrac{x}{|x|\sqrt{x^2 - 1}} + \operatorname{arcsec} x$ **75.** $(\arcsin x)^2$
77. $\frac{1}{2}\arctan(e^{2x}) + C$ **79.** $\frac{1}{2}\arcsin x^2 + C$
81. $\frac{1}{4}[\arctan(x/2)]^2 + C$ **83.** $\frac{2}{3}\pi + \sqrt{3} - 2 \approx 1.826$
85. $y' = -4\operatorname{sech}(4x - 1)\tanh(4x - 1)$
87. $y' = -16x\operatorname{csch}^2(8x^2)$ **89.** $y' = \dfrac{4}{\sqrt{16x^2 + 1}}$
91. $\frac{1}{3}\tanh x^3 + C$ **93.** $\ln|\tanh x| + C$
95. $\dfrac{1}{12}\ln\left|\dfrac{3 + 2x}{3 - 2x}\right| + C$

P.S. Problem Solving (page 395)

1. $a = 1, b = \frac{1}{2}, c = -\frac{1}{2}$

$f(x) = \dfrac{1 + x/2}{1 - x/2}$

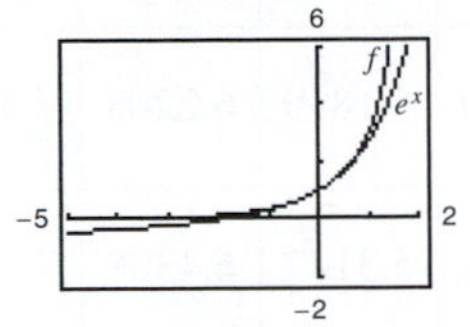

3. (a)

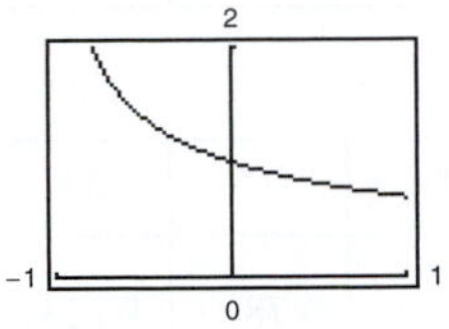

(b) 1 (c) Proof

5.

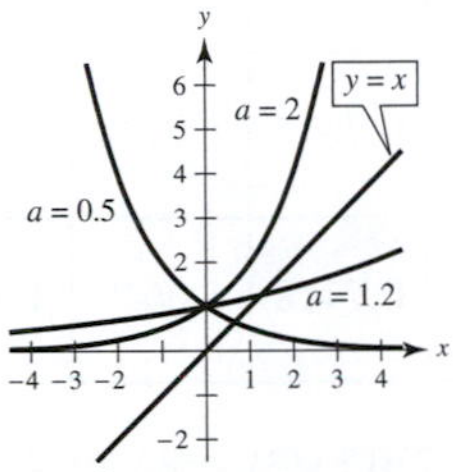

$y = 0.5^x$ and $y = 1.2^x$ intersect the line $y = x$; $0 < a < e^{1/e}$

7. $e - 1$

9. (a) Area of region $A = (\sqrt{3} - \sqrt{2})/2 \approx 0.1589$

Area of region $B = \pi/12 \approx 0.2618$

(b) $\frac{1}{24}[3\pi\sqrt{2} - 12(\sqrt{3} - \sqrt{2}) - 2\pi] \approx 0.1346$

(c) 1.2958 (d) 0.6818

11. Proof **13.** $2 \ln \frac{3}{2} \approx 0.8109$

15. (a) (i)

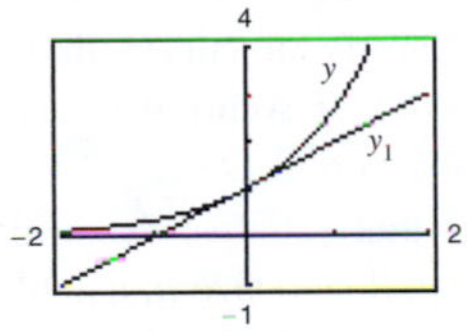

(ii)

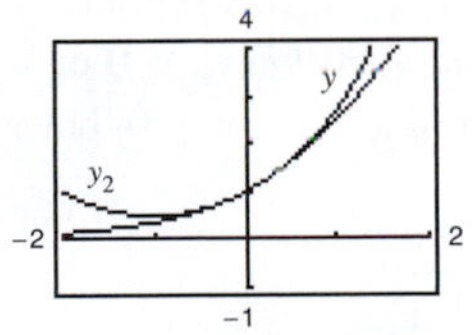

(iii)

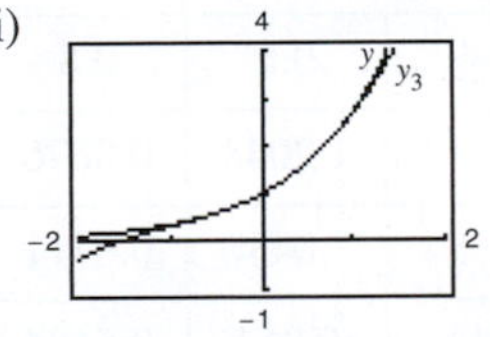

(b) Pattern: $y_n = 1 + \dfrac{x}{1!} + \dfrac{x^2}{2!} + \cdots + \dfrac{x^n}{n!} + \cdots$

$y_4 = 1 + \dfrac{x}{1!} + \dfrac{x^2}{2!} + \dfrac{x^3}{3!} + \dfrac{x^4}{4!}$

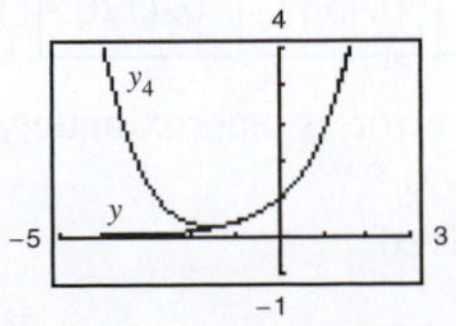

(c) The pattern implies that $e^x = 1 + \dfrac{x}{1!} + \dfrac{x^2}{2!} + \dfrac{x^3}{3!} + \cdots$.

Chapter 6

Section 6.1 (page 403)

1–11. Proofs **13.** Not a solution **15.** Solution
17. Solution **19.** Solution **21.** Not a solution
23. Solution **25.** Not a solution **27.** Not a solution
29. $y = 3e^{-x/2}$ **31.** $4y^2 = x^3$

33.

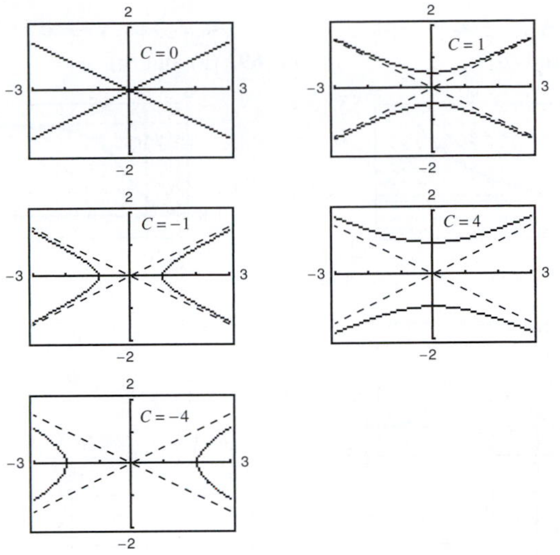

35. $y = 3e^{-2x}$ **37.** $y = 2 \sin 3x - \frac{1}{3} \cos 3x$
39. $y = -2x + \frac{1}{2}x^3$ **41.** $2x^3 + C$
43. $y = \frac{1}{2} \ln(1 + x^2) + C$ **45.** $y = x - \ln x^2 + C$
47. $y = -\frac{1}{2} \cos 2x + C$
49. $y = \frac{2}{5}(x - 6)^{5/2} + 4(x - 6)^{3/2} + C$ **51.** $y = \frac{1}{2}e^{x^2} + C$

53.

x	-4	-2	0	2	4	8
y	2	0	4	4	6	8
dy/dx	-4	Undef.	0	1	$\frac{4}{3}$	2

55.

x	-4	-2	0	2	4	8
y	2	0	4	4	6	8
dy/dx	$-2\sqrt{2}$	-2	0	0	$-2\sqrt{2}$	-8

57. b **58.** c **59.** d **60.** a

61. (a) and (b)

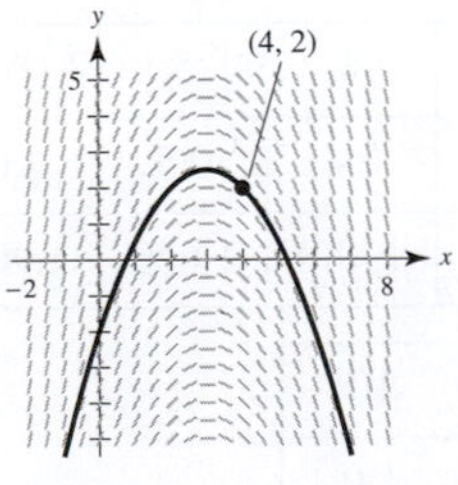

(c) As $x \to \infty$, $y \to -\infty$; as $x \to -\infty$, $y \to -\infty$

63. (a) and (b)

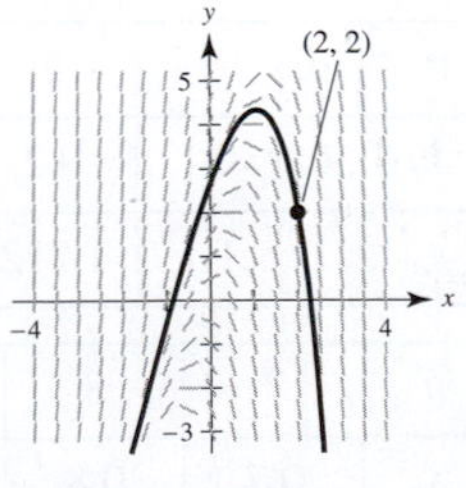

(c) As $x \to \infty$, $y \to -\infty$; as $x \to -\infty$, $y \to -\infty$

65. (a)

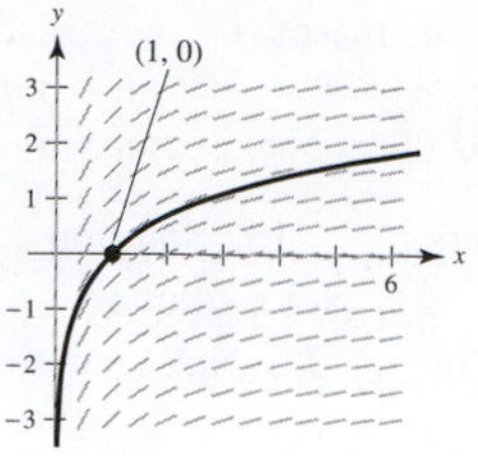

As $x \to \infty, y \to \infty$

(b)

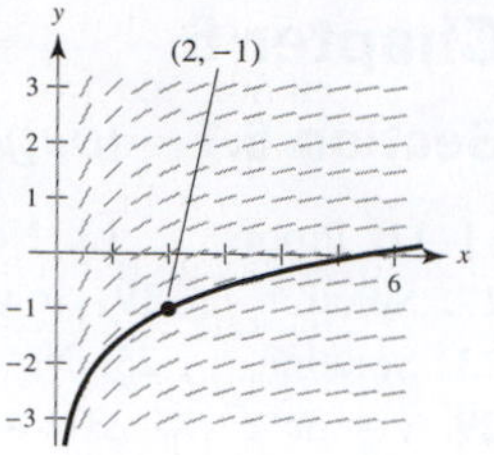

As $x \to \infty, y \to \infty$

67. (a) and (b)

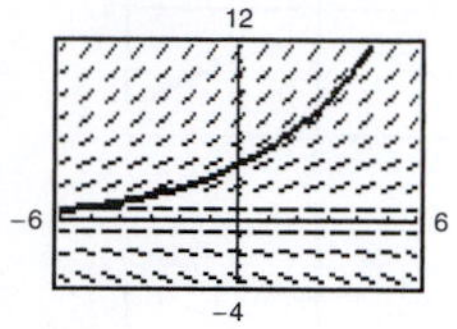

69. (a) and (b)

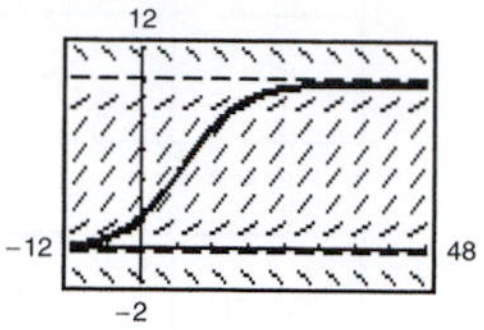

71. (a) and (b)

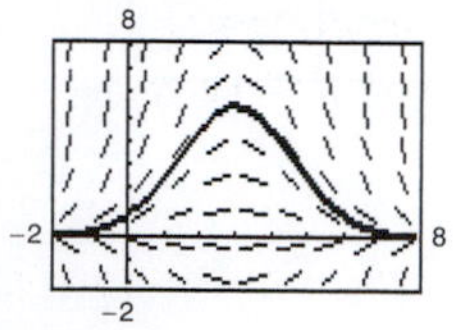

73.

n	0	1	2	3	4	5	6
x_n	0	0.1	0.2	0.3	0.4	0.5	0.6
y_n	2	2.2	2.43	2.693	2.992	3.332	3.715

n	7	8	9	10
x_n	0.7	0.8	0.9	1.0
y_n	4.146	4.631	5.174	5.781

75.

n	0	1	2	3	4	5	6
x_n	0	0.05	0.1	0.15	0.2	0.25	0.3
y_n	3	2.7	2.438	2.209	2.010	1.839	1.693

n	7	8	9	10
x_n	0.35	0.4	0.45	0.5
y_n	1.569	1.464	1.378	1.308

77.

n	0	1	2	3	4	5	6
x_n	0	0.1	0.2	0.3	0.4	0.5	0.6
y_n	1	1.1	1.212	1.339	1.488	1.670	1.900

n	7	8	9	10
x_n	0.7	0.8	0.9	1.0
y_n	2.213	2.684	3.540	5.958

79.

x	0	0.2	0.4	0.6	0.8	1
$y(x)$ (exact)	3.0000	3.6642	4.4755	5.4664	6.6766	8.1548
$y(x)$ $(h = 0.2)$	3.0000	3.6000	4.3200	5.1840	6.2208	7.4650
$y(x)$ $(h = 0.1)$	3.0000	3.6300	4.3923	5.3147	6.4308	7.7812

81.

x	0	0.2	0.4	0.6	0.8	1
$y(x)$ (exact)	0.0000	0.2200	0.4801	0.7807	1.1231	1.5097
$y(x)$ $(h = 0.2)$	0.0000	0.2000	0.4360	0.7074	1.0140	1.3561
$y(x)$ $(h = 0.1)$	0.0000	0.2095	0.4568	0.7418	1.0649	1.4273

83. (a) $y(1) = 112.7141°$; $y(2) = 96.3770°$; $y(3) = 86.5954°$
(b) $y(1) = 113.2441°$; $y(2) = 97.0158°$; $y(3) = 87.1729°$
(c) Euler's Method: $y(1) = 112.9828°$; $y(2) = 96.6998°$; $y(3) = 86.8863°$
Exact solution: $y(1) = 113.2441°$; $y(2) = 97.0158°$; $y(3) = 87.1729°$
The approximations are better using $h = 0.05$.

85. The general solution is a family of curves that satisfies the differential equation. A particular solution is one member of the family that satisfies given conditions.

87. Begin with a point (x_0, y_0) that satisfies the initial condition $y(x_0) = y_0$. Then, using a small step size h, calculate the point $(x_1, y_1) = (x_0 + h, y_0 + hF(x_0, y_0))$. Continue generating the sequence of points $(x_n + h, y_n + hF(x_n, y_n))$ or (x_{n+1}, y_{n+1}).

89. False. $y = x^3$ is a solution of $xy' - 3y = 0$, but $y = x^3 + 1$ is not a solution.

91. True

93. (a)

x	0	0.2	0.4	0.6	0.8	1
y	4	2.6813	1.7973	1.2048	0.8076	0.5413
y_1	4	2.56	1.6384	1.0486	0.6711	0.4295
y_2	4	2.4	1.44	0.864	0.5184	0.3110
e_1	0	0.1213	0.1589	0.1562	0.1365	0.1118
e_2	0	0.2813	0.3573	0.3408	0.2892	0.2303
r		0.4312	0.4447	0.4583	0.4720	0.4855

(b) If h is halved, then the error is approximately halved because r is approximately 0.5.
(c) The error will again be halved.

95. (a)

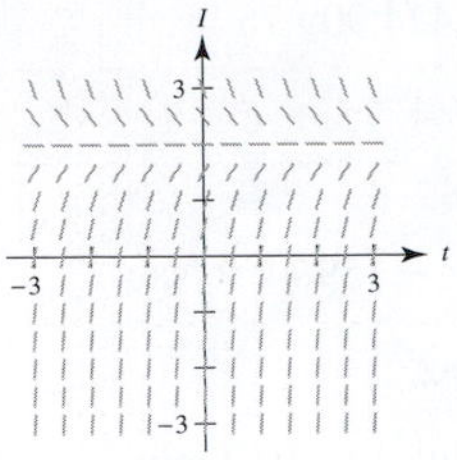

(b) $\lim_{t\to\infty} I(t) = 2$

97. $\omega = \pm 4$ **99.** Putnam Problem 3, Morning Session, 1954

Section 6.2 *(page 412)*

1. $y = \frac{1}{2}x^2 + 3x + C$ **3.** $y = Ce^x - 3$

5. $y^2 - 5x^2 = C$ **7.** $y = Ce^{(2x^{3/2})/3}$ **9.** $y = C(1 + x^2)$

11. $dQ/dt = k/t^2$

$Q = -k/t + C$

13. (a)

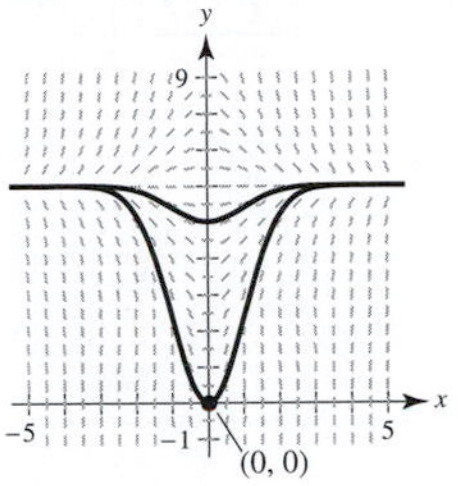

(b) $y = 6 - 6e^{-x^2/2}$

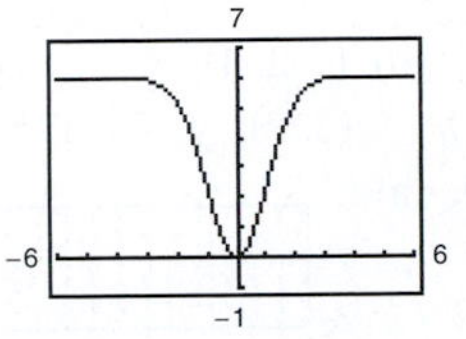

15. $y = \frac{1}{4}t^2 + 10$

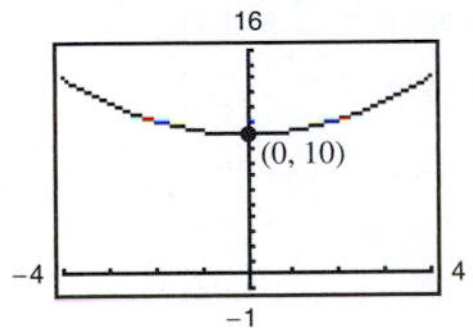

17. $y = 10e^{-t/2}$

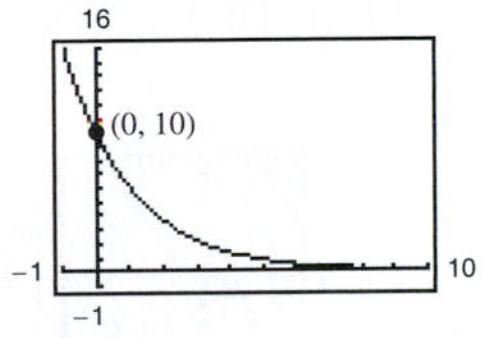

19. $\frac{8192}{4}$ **21.** $y = (1/2)e^{[(\ln 10)/5]t} \approx (1/2)e^{0.4605t}$

23. $y = 5(5/2)^{1/4}e^{[\ln(2/5)/4]t} \approx 6.2872e^{-0.2291t}$

25. C is the initial value of y, and k is the proportionality constant.

27. Quadrants I and III; dy/dx is positive when both x and y are positive (Quadrant I) or when both x and y are negative (Quadrant III).

29. Amount after 1000 yr: 12.96 g;
Amount after 10,000 yr: 0.26 g

31. Initial quantity: 7.63 g;
Amount after 1000 yr: 4.95 g

33. Amount after 1000 yr: 4.43 g;
Amount after 10,000 yr: 1.49 g

35. Initial quantity: 2.16 g;
Amount after 10,000 yr: 1.62 g

37. 95.76%

39. Time to double: 11.55 yr; Amount after 10 yr: \$7288.48

41. Annual rate: 8.94%; Amount after 10 yr: \$1833.67

43. Annual rate: 9.50%; Time to double: 7.30 yr

45. \$224,174.18 **47.** \$61,377.75

49. (a) 10.24 yr (b) 9.93 yr (c) 9.90 yr (d) 9.90 yr

51. (a) $P = 2.21e^{-0.006t}$ (b) 2.08 million
(c) Because $k < 0$, the population is decreasing.

53. (a) $P = 33.38e^{0.036t}$ (b) 47.84 million
(c) Because $k > 0$, the population is increasing.

55. (a) $N = 100.1596(1.2455)^t$ (b) 6.3 h

57. (a) $N \approx 30(1 - e^{-0.0502t})$ (b) 36 days

59. (a) Because the population increases by a constant each month, the rate of change from month to month will always be the same. So, the slope is constant, and the model is linear.
(b) Although the percentage increase is constant each month, the rate of growth is not constant. The rate of change of y is $dy/dt = ry$, which is an exponential model.

61. (a) $P_1 = 106e^{0.01487t} \approx 106(1.01499)^t$
(b) $P_2 = 107.2727(1.01215)^t$
(c)

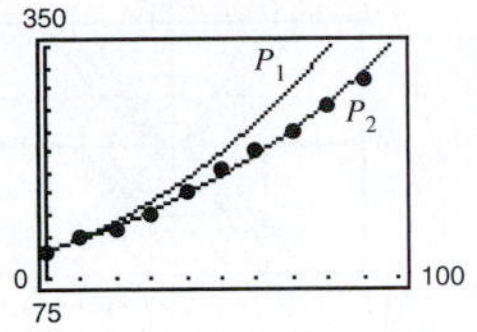

(d) 2029

63. (a) 20 dB (b) 70 dB (c) 95 dB (d) 120 dB

65. 379.2°F

67. False. The rate of growth dy/dx is proportional to y.

69. False. The prices are rising at a rate of 6.2% per year.

Section 6.3 *(page 421)*

1. $y^2 - x^2 = C$ **3.** $15y^2 + 2x^3 = C$ **5.** $r = Ce^{0.75s}$

7. $y = C(x + 2)^3$ **9.** $y^2 = C - 8\cos x$

11. $y = -\frac{1}{4}\sqrt{1 - 4x^2} + C$ **13.** $y = Ce^{(\ln x)^2/2}$

15. $y^2 = 4e^x + 5$ **17.** $y = e^{-(x^2+2x)/2}$

19. $y^2 = 4x^2 + 3$ **21.** $u = e^{(1-\cos v^2)/2}$ **23.** $P = P_0e^{kt}$

25. $4y^2 - x^2 = 16$ **27.** $y = \frac{1}{3}\sqrt{x}$ **29.** $f(x) = Ce^{-x/2}$

31.

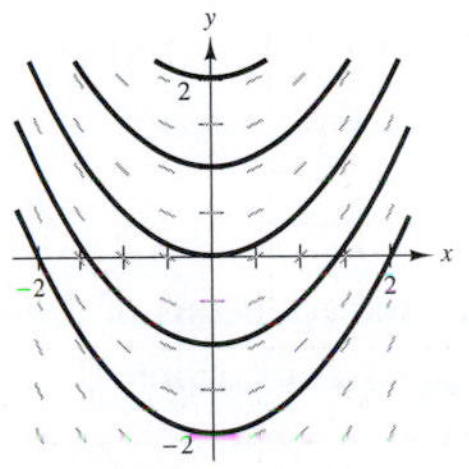

$y = \frac{1}{2}x^2 + C$

33. (a) $dy/dx = k(y - 4)$ (b) a (c) Proof

34. (a) $dy/dx = k(x - 4)$ (b) b (c) Proof

35. (a) $dy/dx = ky(y - 4)$ (b) c (c) Proof

36. (a) $dy/dx = ky^2$ (b) d (c) Proof

37. 97.9% of the original amount

39. (a) $w = 1200 - 1140e^{-kt}$
(b) $w = 1200 - 1140e^{-0.8t}$

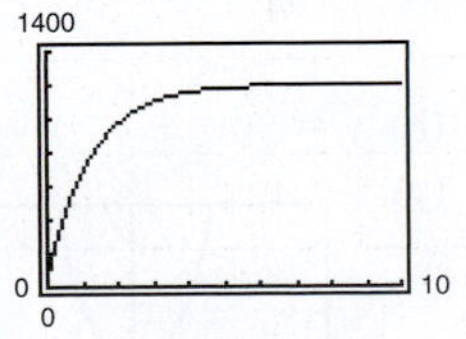

$w = 1200 - 1140e^{-0.9t}$

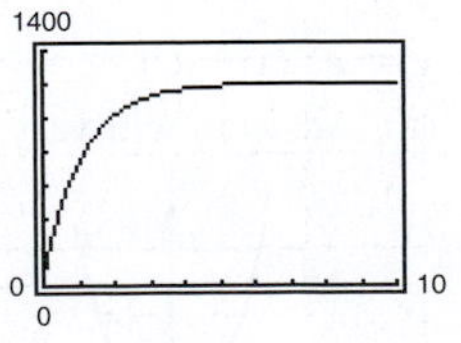

$w = 1200 - 1140e^{-t}$

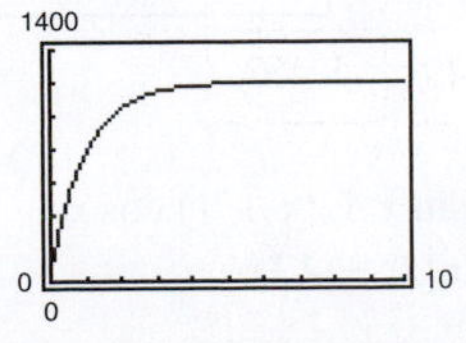

(c) 1.31 yr; 1.16 yr; 1.05 yr (d) 1200 lb

41. Circles: $x^2 + y^2 = C$
Lines: $y = Kx$
Graphs will vary.

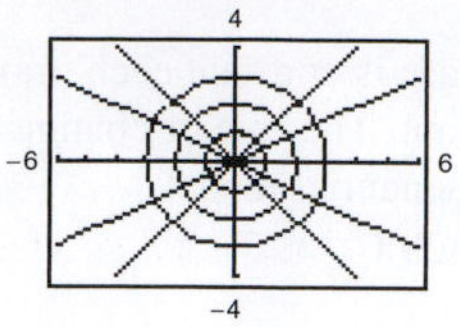

43. Parabolas: $x^2 = Cy$
Ellipses: $x^2 + 2y^2 = K$
Graphs will vary.

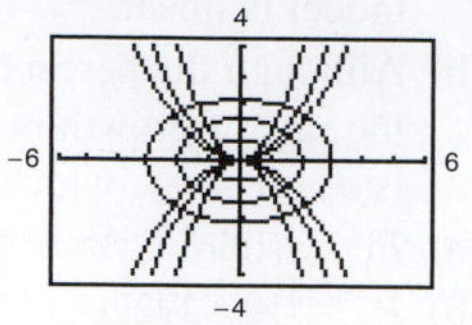

45. Curves: $y^2 = Cx^3$
Ellipses: $2x^2 + 3y^2 = K$
Graphs will vary.

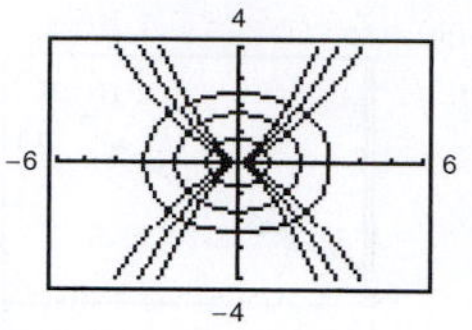

47. d **48.** a **49.** b **50.** c

51. (a) 0.75 (b) 2100 (c) 70 (d) 4.49 yr
(e) $dP/dt = 0.75P(1 - P/2100)$

53. (a) 3 (b) 100
(c)

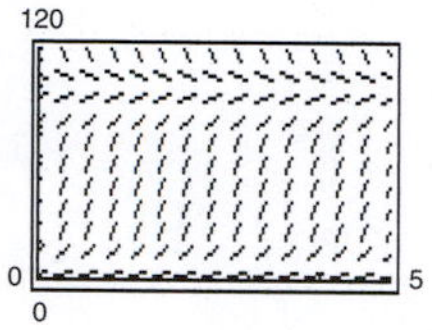

(d) 50

55. $y = 36/(1 + 8e^{-t})$ **57.** $y = 120/(1 + 14e^{-0.8t})$

59. (a) $P = \dfrac{200}{1 + 7e^{-0.2640t}}$ (b) 70 panthers (c) 7.37 yr
(d) $dP/dt = 0.2640P(1 - P/200)$; 65.6 (e) 100 yr

61. Answers will vary. **63.** Proof

65. (a) $v = 20(1 - e^{-1.386t})$
(b) $s \approx 20t + 14.43(e^{-1.386t} - 1)$

67. Homogeneous of degree 3 **69.** Homogeneous of degree 3

71. Not homogeneous **73.** Homogeneous of degree 0.

75. $|x| = C(x - y)^2$ **77.** $|y^2 + 2xy - x^2| = C$

79. $y = Ce^{-x^2/(2y^2)}$

81. False. $y' = x/y$ is separable, but $y = 0$ is not a solution.

83. True

Section 6.4 *(page 428)*

1. Linear; can be written in the form $dy/dx + P(x)y = Q(x)$

3. Not linear; cannot be written in the form $dy/dx + P(x)y = Q(x)$

5. $y = 2x^2 + x + C/x$ **7.** $y = -16 + Ce^x$

9. $y = -1 + Ce^{\sin x}$ **11.** $y = (x^3 - 3x + C)/[3(x - 1)]$

13. $y = e^{x^3}(x + C)$

15. (a) Answers will vary.

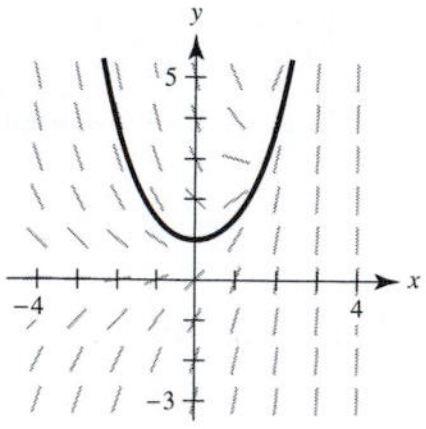

(b) $y = \frac{1}{2}(e^x + e^{-x})$
(c)

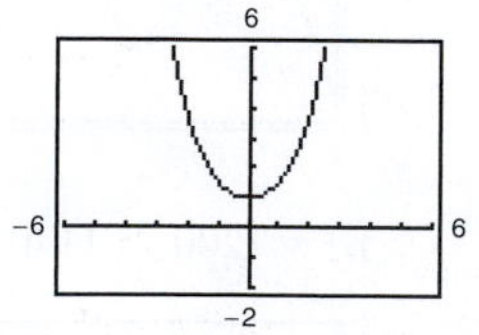

17. $y = 1 + 4/e^{\tan x}$ **19.** $y = \sin x + (x + 1)\cos x$

21. $xy = 4$ **23.** $y = -2 + x\ln|x| + 12x$

25. $P = -N/k + (N/k + P_0)e^{kt}$

27. (a) \$4,212,796.94 (b) \$31,424,909.75

29. (a) $\dfrac{dN}{dt} = k(75 - N)$ (b) $N = 75 + Ce^{-kt}$
(c) $N = 75 - 55.9296e^{-0.0168t}$

31. $v(t) = -159.47(1 - e^{-0.2007t})$; -159.47 ft/sec

33. $I = \dfrac{E_0}{R} + Ce^{-Rt/L}$ **35.** Proof

37. (a) $Q = 25e^{-t/20}$ (b) $-20\ln\left(\frac{3}{5}\right) \approx 10.2$ min (c) 0

39. Answer (a) **41.** $\dfrac{dy}{dx} + P(x)y = Q(x)$; $u(x) = e^{\int P(x)\,dx}$

43. c **44.** d **45.** a **46.** b

47. (a)

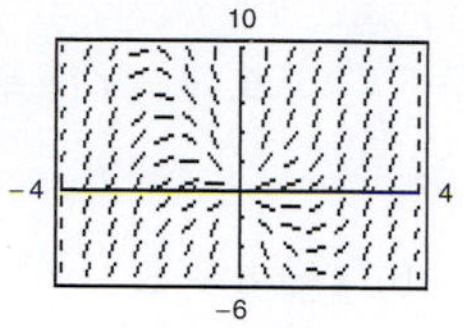

(c)

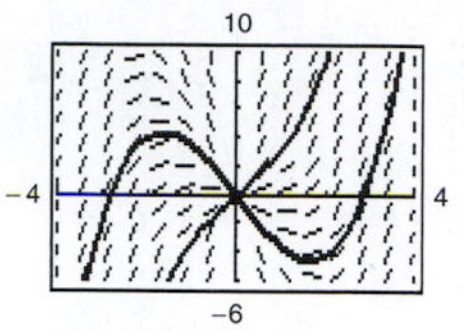

(b) $(-2, 4)$: $y = \frac{1}{2}x(x^2 - 8)$
$(2, 8)$: $y = \frac{1}{2}x(x^2 + 4)$

49. (a)

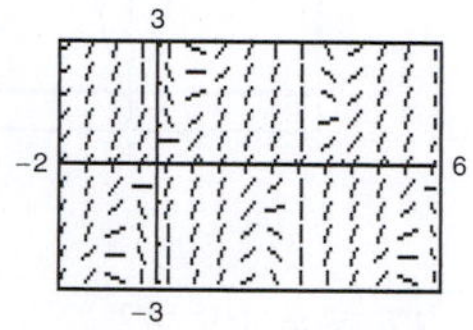

(b) $(1, 1)$: $y = (2\cos 1 + \sin 1)\csc x - 2\cot x$
$(3, -1)$: $y = (2\cos 3 - \sin 3)\csc x - 2\cot x$

(c)

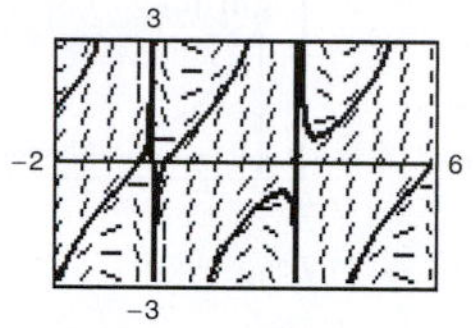

51. $2e^x + e^{-2y} = C$ **53.** $y = Ce^{-\sin x} + 1$

55. $y = [e^x(x - 1) + C]/x^2$ **57.** $y = \frac{12}{5}x^2 + C/x^3$

59. $1/y^2 = Ce^{2x^3} + \frac{1}{3}$ **61.** $y = 1/(Cx - x^2)$

63. $1/y^2 = 2x + Cx^2$ **65.** $y^{2/3} = 2e^x + Ce^{2x/3}$

67. False. $y' + xy = x^2$ is linear.

Review Exercises for Chapter 6 *(page 431)*

1. Yes **3.** $y = \frac{4}{3}x^3 + 7x + C$ **5.** $y = \frac{1}{2}\sin 2x + C$

7. $y = -e^{2-x} + C$

9.

x	−4	−2	0	2	4	8
y	2	0	4	4	6	8
dy/dx	−10	−4	−4	0	2	8

11. (a) and (b)

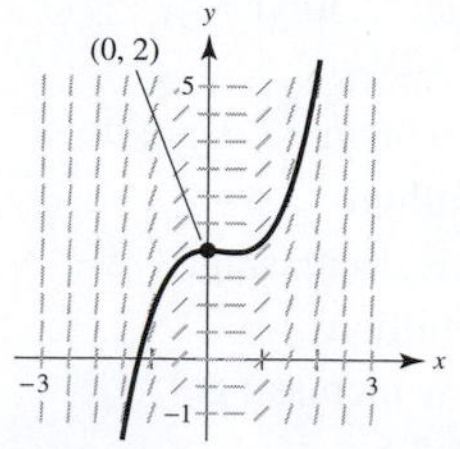

13.

n	0	1	2	3	4	5	6	7	8	9	10
x_n	0	0.05	0.1	0.15	0.2	0.25	0.3	0.35	0.4	0.45	0.5
y_n	4	3.8	3.6125	3.4369	3.2726	3.1190	2.9756	2.8418	2.7172	2.6038	2.4986

15. $y = -\frac{5}{3}x^3 + x^2 + C$

17. $y = -3 - 1/(x + C)$ **19.** $y = Ce^x/(2 + x)^2$

21. $\frac{dy}{dt} = \frac{k}{t^3}$; $y = -\frac{k}{2t^2} + C$ **23.** $y \approx \frac{3}{4}e^{0.379t}$

25. $y = \frac{9}{20}e^{(1/2)\ln(10/3)t}$ **27.** About 7.79 in.

29. About 37.5 yr

31. (a) $S \approx 30e^{-1.7918/t}$ (b) 20,965 units

(c)

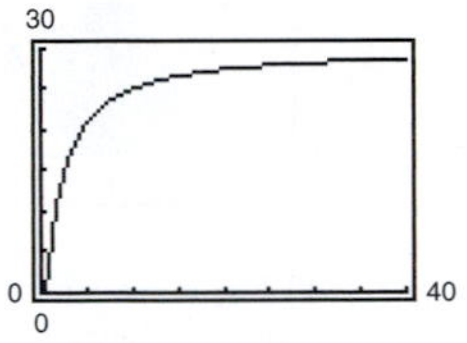

33. $y^2 = 5x^2 + C$ **35.** $y = Ce^{8x^2}$

37. $y^4 = 6x^2 - 8$ **39.** $y^4 = 2x^4 + 1$

41.

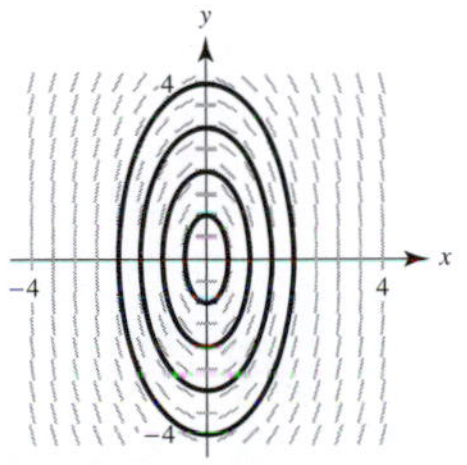

Graphs will vary.
$4x^2 + y^2 = C$

43. (a) 0.55 (b) 5250 (c) 150 (d) 6.41 yr

(e) $\frac{dP}{dt} = 0.55P\left(1 - \frac{P}{5250}\right)$

45. $y = \frac{80}{1 + 9e^{-t}}$

47. (a) $P(t) = \frac{20{,}400}{1 + 16e^{-0.553t}}$ (b) 17,118 trout (c) 4.94 yr

49. $y = -10 + Ce^x$ **51.** $y = e^{x/4}\left(\frac{1}{4}x + C\right)$

53. $y = (x + C)/(x - 2)$ **55.** $y = \frac{1}{10}e^{5x} + \frac{29}{10}e^{-5x}$

P.S. Problem Solving *(page 433)*

1. (a) $y = 1/(1 - 0.01t)^{100}$; $T = 100$

(b) $y = 1\Big/\left[\left(\frac{1}{y_0}\right)^{\varepsilon} - k\varepsilon t\right]^{1/\varepsilon}$; Explanations will vary.

3. (a) $y = Le^{-Ce^{-kt}}$

(b)

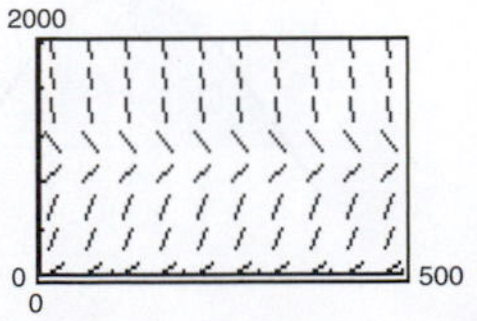

(c) As $t \to \infty$, $y \to L$, the carrying capacity.

(d) $y_0 = 500 = 5000e^{-C} \Rightarrow e^C = 10 \Rightarrow C = \ln 10$

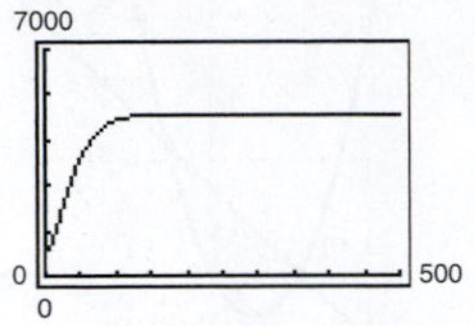

The graph is concave upward on $(0, 41.7)$ and downward on $(41.7, \infty)$.

5. 1481.45 sec $\approx$ 24 min, 41 sec

7. 2575.95 sec $\approx$ 42 min, 56 sec

9. (a) $s = 184.21 - Ce^{-0.019t}$

(b)

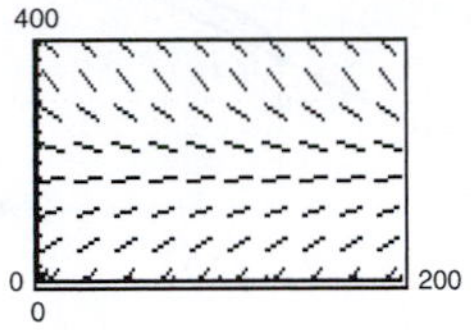

(c) As $t \to \infty$, $Ce^{-0.019t} \to 0$, and $s \to 184.21$.

11. (a) $C = 0.6e^{-0.25t}$ (b) $C = 0.6e^{-0.75t}$

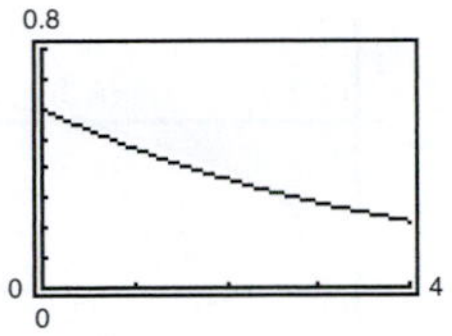

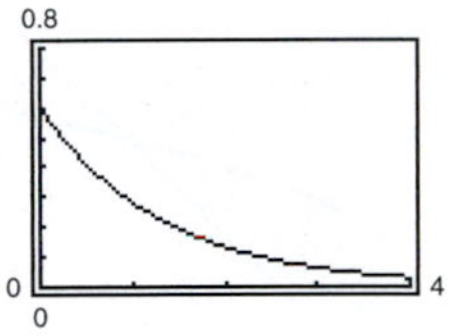

Chapter 7

Section 7.1 *(page 442)*

1. $-\int_0^6 (x^2 - 6x)\,dx$ **3.** $\int_0^3 (-2x^2 + 6x)\,dx$

5. $-6\int_0^1 (x^3 - x)\,dx$

7.

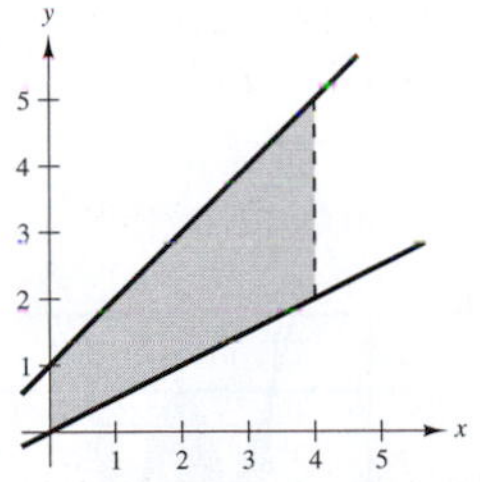

9.

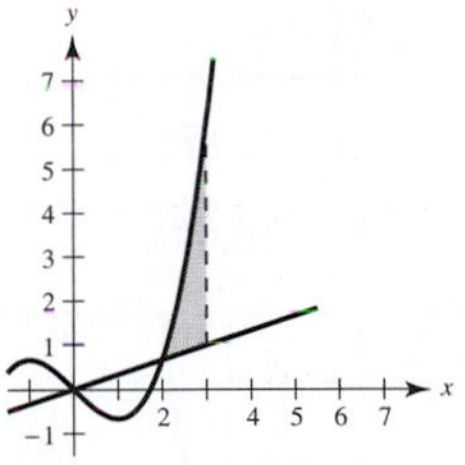

11.

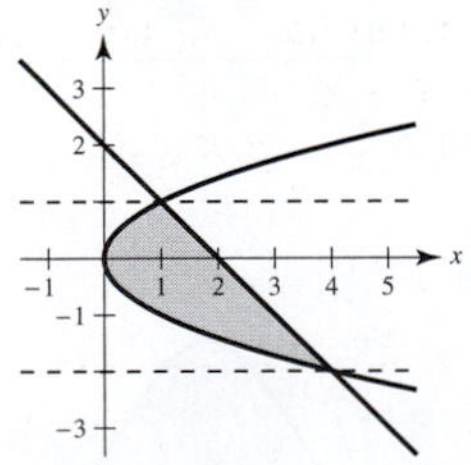

13. d

15. (a) $\frac{125}{6}$ (b) $\frac{125}{6}$

(c) Integrating with respect to y; Answers will vary.

17.

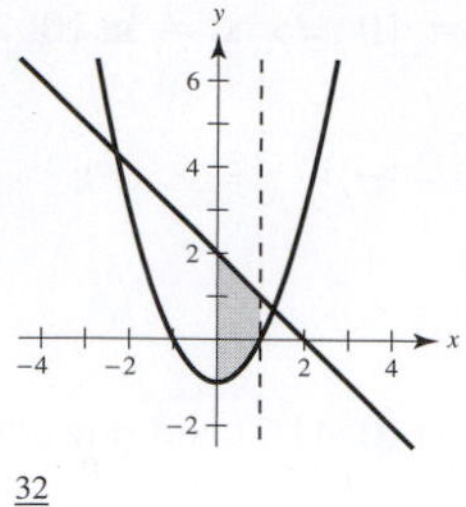

$\frac{32}{3}$

19.

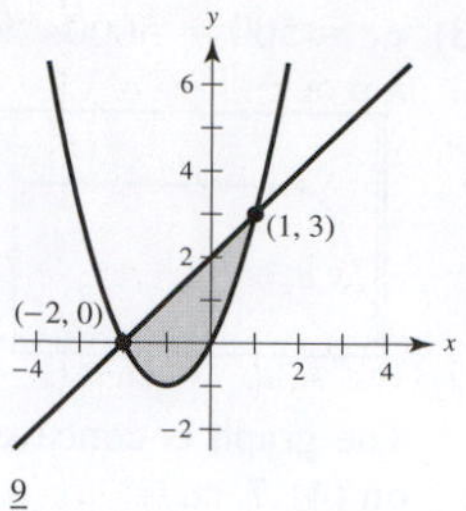

$\frac{9}{2}$

21.

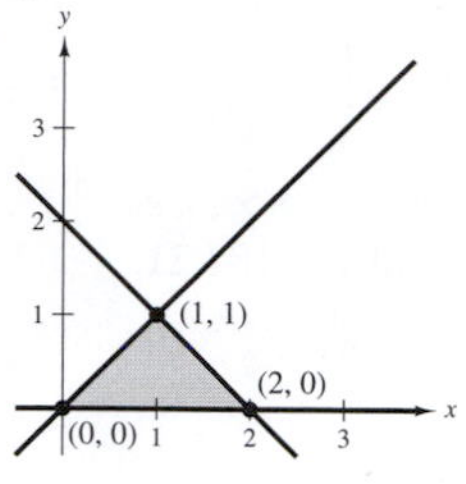

1

23.

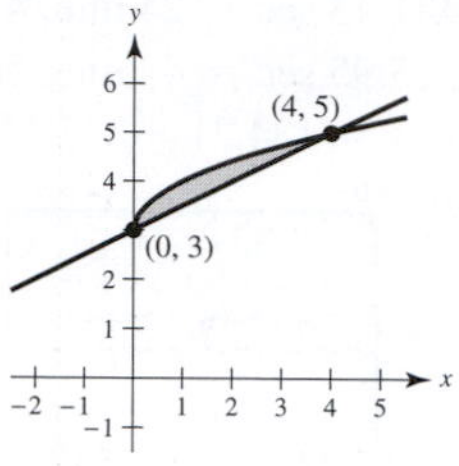

$\frac{4}{3}$

25.

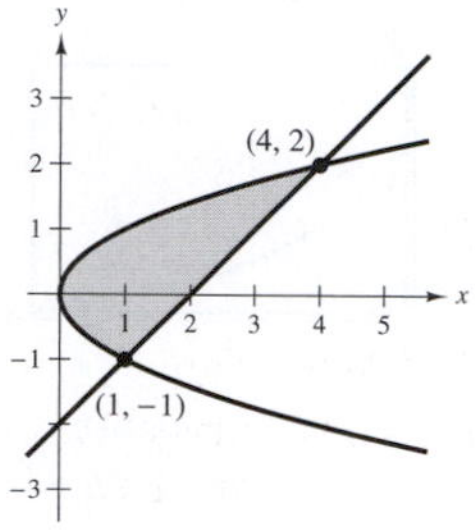

$\frac{9}{2}$

27.

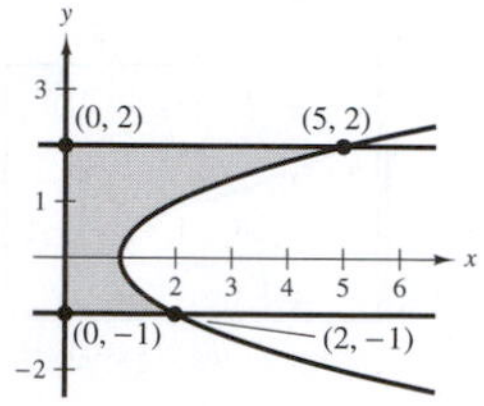

6

29.

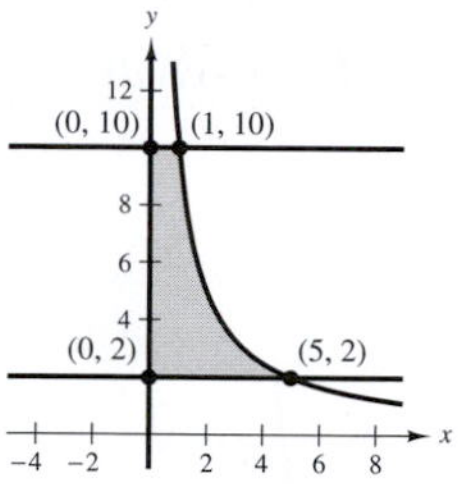

$10 \ln 5 \approx 16.094$

31. (a)

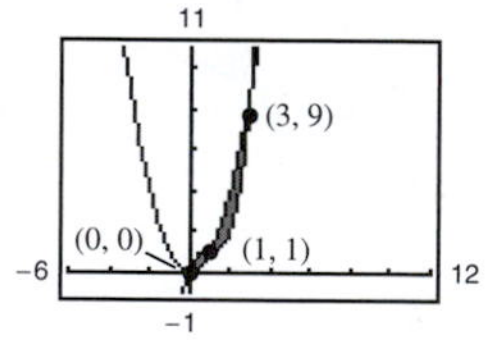

(b) $\frac{37}{12}$

33. (a)

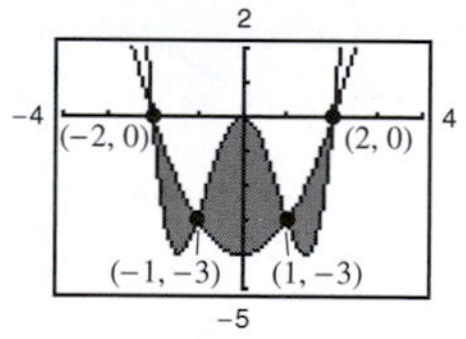

(b) 8

35. (a)

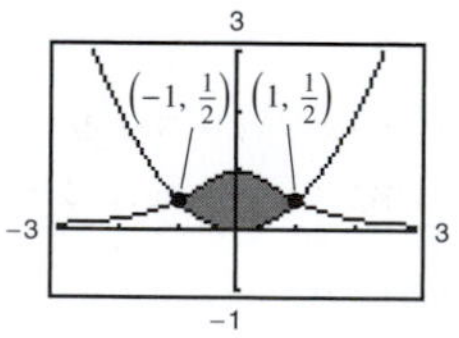

(b) $\pi/2 - 1/3 \approx 1.237$

37. (a)

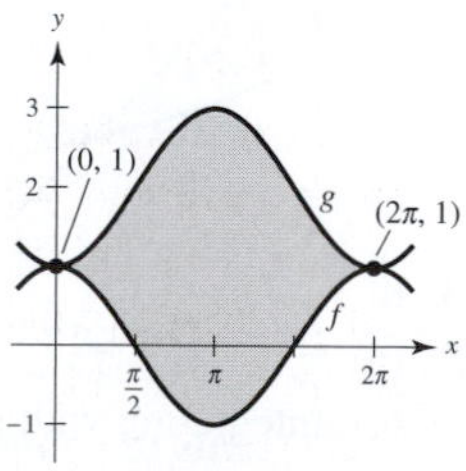

$4\pi \approx 12.566$

39.

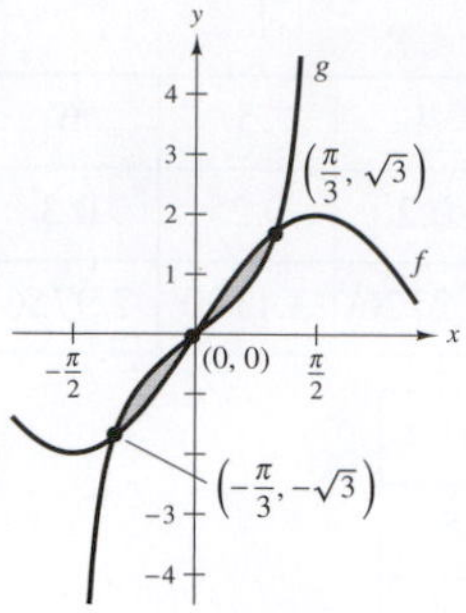

$2(1 - \ln 2) \approx 0.614$

41.

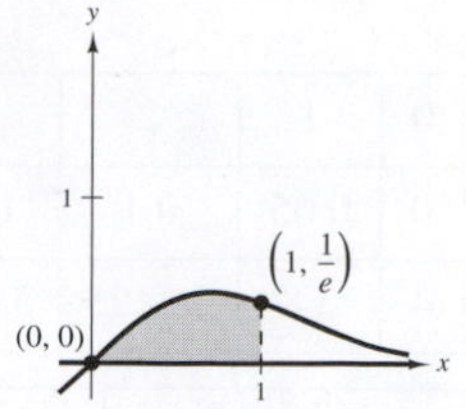

$(1/2)(1 - 1/e) \approx 0.316$

43. (a)

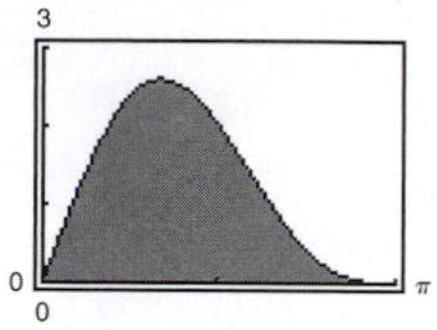

(b) 4

45. (a)

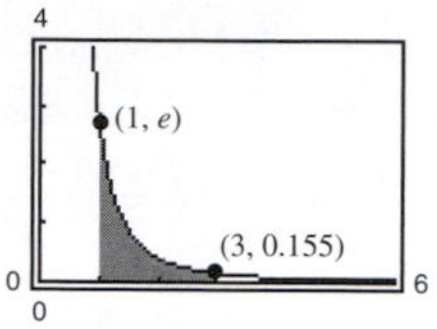

(b) About 1.323

47. (a)

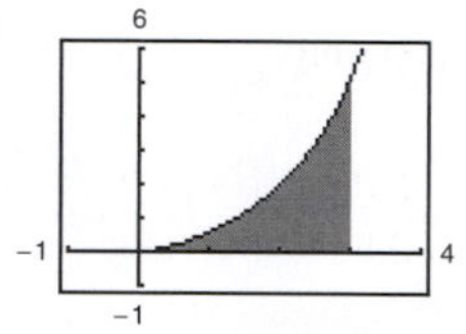

(b) The function is difficult to integrate.

(c) About 4.7721

49. (a)

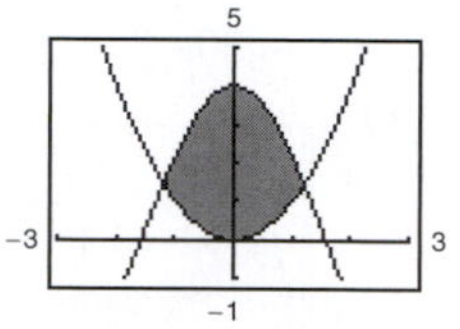

(b) The intersections are difficult to find.

(c) About 6.3043

51. $F(x) = \frac{1}{4}x^2 + x$

(a) $F(0) = 0$

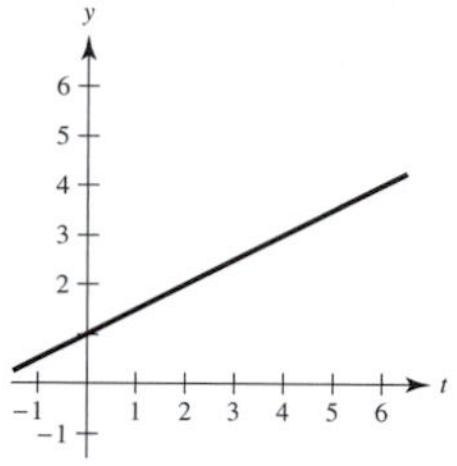

(b) $F(2) = 3$

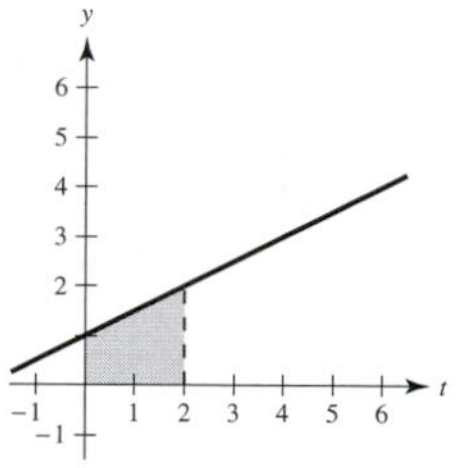

(c) $F(6) = 15$

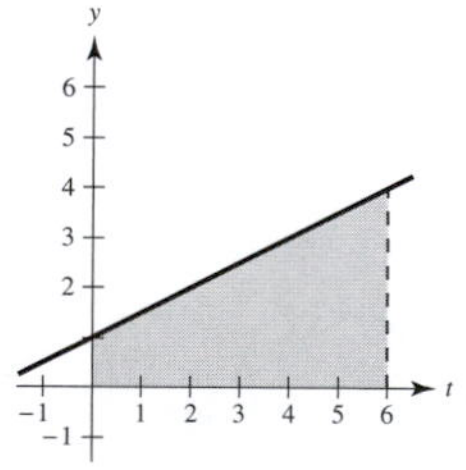

53. $F(\alpha) = (2/\pi)[\sin(\pi\alpha/2) + 1]$

(a) $F(-1) = 0$

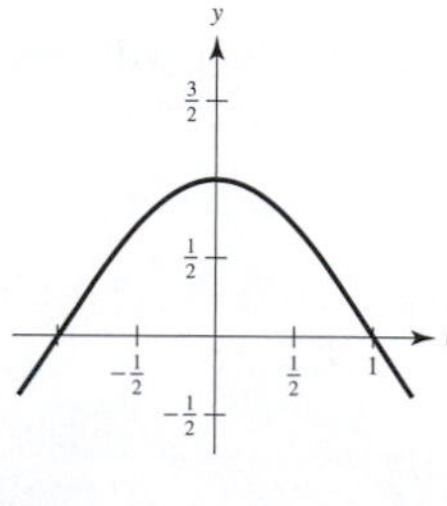

(b) $F(0) = 2/\pi \approx 0.6366$

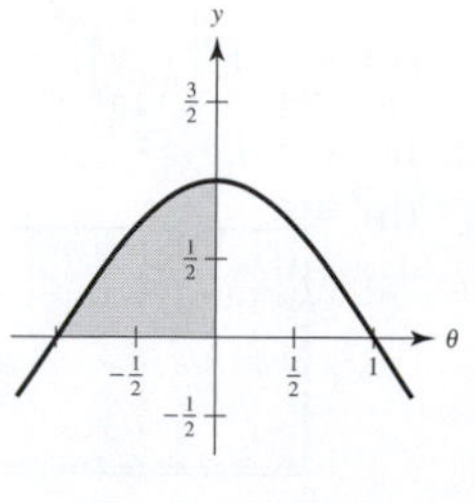

(c) $F(1/2) = (\sqrt{2} + 2)/\pi \approx 1.0868$

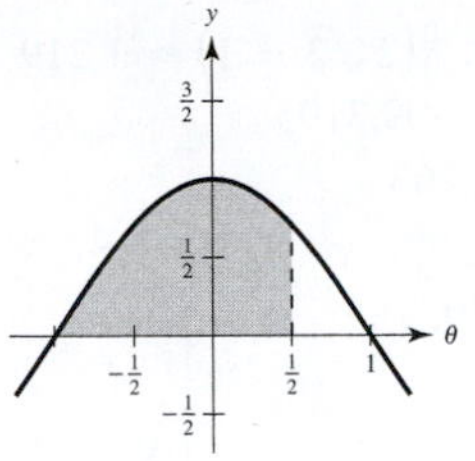

55. 14 **57.** 16

59. Answers will vary. Sample answers:

(a) About 966 ft^2 (b) About 1004 ft^2

61. $\int_{-2}^{1} [x^3 - (3x - 2)]\, dx = \frac{27}{4}$

63. $\int_0^1 \left[\frac{1}{x^2+1} - \left(-\frac{1}{2}x + 1\right)\right] dx \approx 0.0354$

65. Answers will vary.

Example: $x^4 - 2x^2 + 1 \le 1 - x^2$ on $[-1, 1]$

$\int_{-1}^{1} [(1 - x^2) - (x^4 - 2x^2 + 1)]\, dx = \frac{4}{15}$

67. (a) The integral $\int_0^5 [v_1(t) - v_2(t)]\, dt = 10$ means that the first car traveled 10 more meters than the second car between 0 and 5 seconds.

The integral $\int_0^{10} [v_1(t) - v_2(t)]\, dt = 30$ means that the first car traveled 30 more meters than the second car between 0 and 10 seconds.

The integral $\int_{20}^{30} [v_1(t) - v_2(t)]\, dt = -5$ means that the second car traveled 5 more meters than the first car between 20 and 30 seconds.

(b) No. You do not know when both cars started or the initial distance between the cars.

(c) The car with velocity v_1 is ahead by 30 meters.

(d) Car 1 is ahead by 8 meters.

69. $b = 9(1 - 1/\sqrt[3]{4}) \approx 3.330$ **71.** $a = 4 - 2\sqrt{2} \approx 1.172$

73. Answers will vary. Sample answer: $\frac{1}{6}$

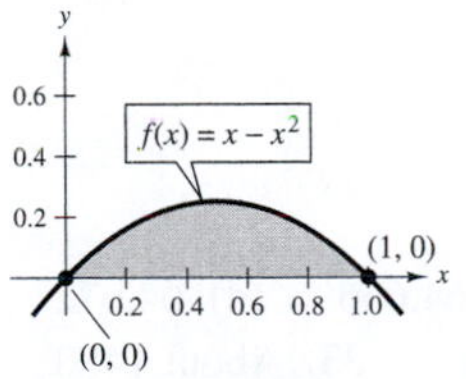

75. R_1; $11.375 billion

77. (a) $y = 0.0124x^2 - 0.385x + 7.85$

(b)

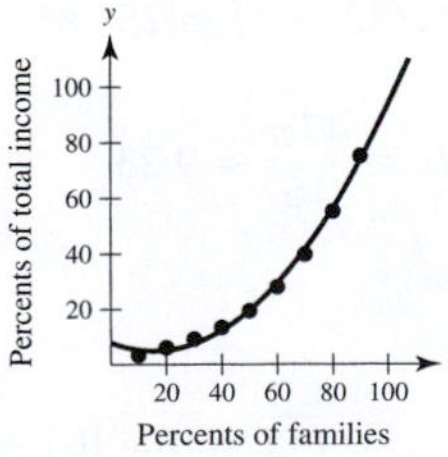

(c)

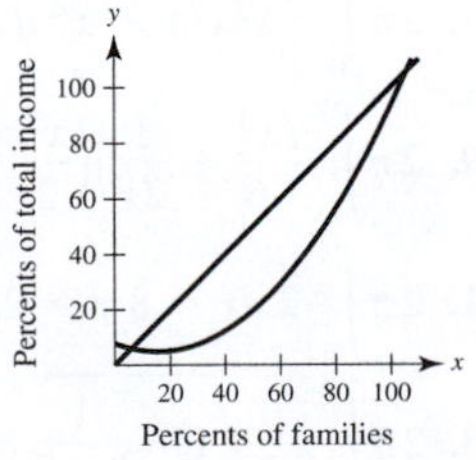

(d) About 2006.7

79. (a) About 6.031 m^2 (b) About 12.062 m^3 (c) 60,310 lb

81. $\sqrt{3}/2 + 7\pi/24 + 1 \approx 2.7823$ **83.** True

85. False. Let $f(x) = x$ and $g(x) = 2x - x^2$. f and g intersect at (1, 1), the midpoint of [0, 2], but

$$\int_a^b [f(x) - g(x)]\, dx = \int_0^2 [x - (2x - x^2)]\, dx = \frac{2}{3} \neq 0.$$

87. Putnam Problem A1, 1993

Section 7.2 *(page 453)*

1. $\pi\int_0^1 (-x + 1)^2\, dx = \frac{\pi}{3}$ **3.** $\pi\int_1^4 (\sqrt{x})^2\, dx = \frac{15\pi}{2}$

5. $\pi\int_0^1 [(x^2)^2 - (x^5)^2]\, dx = \frac{6\pi}{55}$ **7.** $\pi\int_0^4 (\sqrt{y})^2\, dy = 8\pi$

9. $\pi\int_0^1 (y^{3/2})^2\, dy = \frac{\pi}{4}$

11. (a) $9\pi/2$ (b) $(36\pi\sqrt{3})/5$ (c) $(24\pi\sqrt{3})/5$ (d) $(84\pi\sqrt{3})/5$

13. (a) $32\pi/3$ (b) $64\pi/3$ **15.** 18π

17. $\pi(48 \ln 2 - \frac{27}{4}) \approx 83.318$ **19.** $124\pi/3$

21. $832\pi/15$ **23.** $\pi \ln 5$ **25.** $2\pi/3$

27. $(\pi/2)(1 - 1/e^2) \approx 1.358$ **29.** $277\pi/3$ **31.** 8π

33. $\pi^2/2 \approx 4.935$ **35.** $(\pi/2)(e^2 - 1) \approx 10.036$

37. 1.969 **39.** 15.4115 **41.** $\pi/3$ **43.** $2\pi/15$

45. $\pi/2$ **47.** $\pi/6$

49. A sine curve on $[0, \pi/2]$ revolved about the x-axis

51. The parabola $y = 4x - x^2$ is a horizontal translation of the parabola $y = 4 - x^2$. Therefore, their volumes are equal.

53. (a) This statement is true. Explanations will vary.

(b) This statement is false. Explanations will vary.

55. $2\sqrt{2}$ **57.** $V = \frac{4}{3}\pi(R^2 - r^2)^{3/2}$ **59.** Proof

61. $\pi r^2 h[1 - (h/H) + h^2/(3H^2)]$

63.

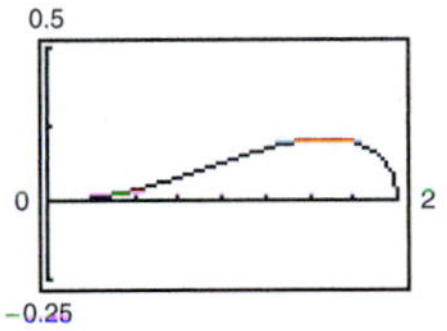

$\pi/30$

65. (a) 60π (b) 50π

67. (a) $V = \pi(4b^2 - \frac{64}{3}b + \frac{512}{15})$

(b)

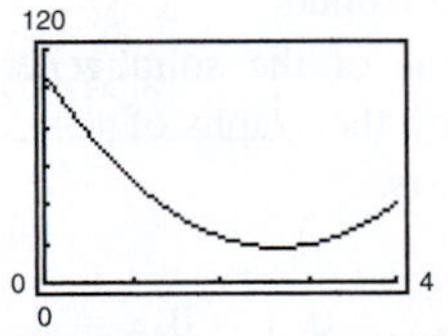

$b \approx 2.67$

(c) $b = \frac{8}{3} \approx 2.67$

69. (a) ii; right circular cylinder of radius r and height h

(b) iv; ellipsoid whose underlying ellipse has the equation $(x/b)^2 + (y/a)^2 = 1$

(c) iii, sphere of radius r

(d) i; right circular cone of radius r and height h

(e) v; torus of cross-sectional radius r and other radius R

71. (a) $\frac{81}{10}$ (b) $\frac{9}{2}$ **73.** $\frac{16}{3}r^3$

75. (a) $\frac{2}{3}r^3$ (b) $\frac{2}{3}r^3 \tan\theta$; As $\theta \to 90°$, $V \to \infty$.

Section 7.3 *(page 462)*

1. $2\pi\int_0^2 x^2\,dx = \dfrac{16\pi}{3}$ **3.** $2\pi\int_0^4 x\sqrt{x}\,dx = \dfrac{128\pi}{5}$

5. $2\pi\int_0^4 \dfrac{1}{4}x^3\,dx = 32\pi$ **7.** $2\pi\int_0^2 x(4x - 2x^2)\,dx = \dfrac{16\pi}{3}$

9. $2\pi\int_0^2 x(x^2 - 4x + 4)\,dx = \dfrac{8\pi}{3}$

11. $2\pi\int_2^4 x\sqrt{x-2}\,dx = \dfrac{128\pi}{15}\sqrt{2}$

13. $2\pi\int_0^1 x\left(\dfrac{1}{\sqrt{2\pi}}e^{-x^2/2}\right)dx = \sqrt{2\pi}\left(1 - \dfrac{1}{\sqrt{e}}\right) \approx 0.986$

15. $2\pi\int_0^2 y(2 - y)\,dy = \dfrac{8\pi}{3}$

17. $2\pi\left[\int_0^{1/2} y\,dy + \int_{1/2}^1 y\left(\dfrac{1}{y} - 1\right)dy\right] = \dfrac{\pi}{2}$

19. $2\pi\int_0^8 y^{4/3}\,dy = \dfrac{768\pi}{7}$

21. $2\pi\int_0^2 y(4 - 2y)\,dy = 16\pi/3$ **23.** 8π **25.** 16π

27. Shell method; it is much easier to put x in terms of y rather than vice versa.

29. (a) $128\pi/7$ (b) $64\pi/5$ (c) $96\pi/5$

31. (a) $\pi a^3/15$ (b) $\pi a^3/15$ (c) $4\pi a^3/15$

33. (a)

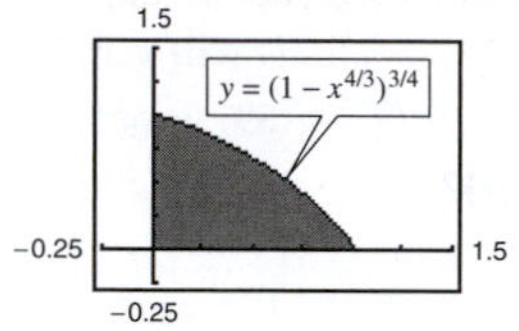

(b) 1.506

35. (a)

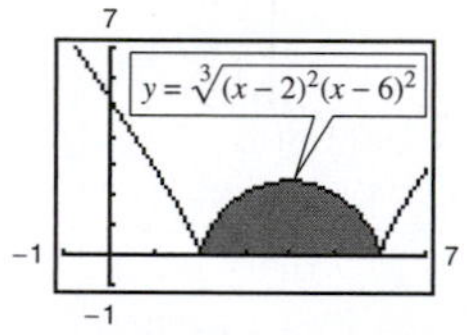

(b) 187.25

37. (a) The rectangles would be vertical.
(b) The rectangles would be horizontal.

39. Both integrals yield the volume of the solid generated by revolving the region bounded by the graphs of $y = \sqrt{x - 1}$, $y = 0$, and $x = 5$ about the x-axis.

41. a, c, b

43. (a) Region bounded by $y = x^2$, $y = 0$, $x = 0$, $x = 2$
(b) Revolved about the y-axis

45. (a) Region bounded by $x = \sqrt{6 - y}$, $y = 0$, $x = 0$
(b) Revolved about $y = -2$

47. Diameter $= 2\sqrt{4 - 2\sqrt{3}} \approx 1.464$ **49.** $4\pi^2$

51. (a) Proof (b) (i) $V = 2\pi$ (ii) $V = 6\pi^2$ **53.** Proof

55. (a) $R_1(n) = n/(n + 1)$ (b) $\lim_{n\to\infty} R_1(n) = 1$
(c) $V = \pi ab^{n+2}[n/(n + 2)]$; $R_2(n) = n/(n + 2)$
(d) $\lim_{n\to\infty} R_2(n) = 1$
(e) As $n \to \infty$, the graph approaches the line $x = b$.

57. (a) and (b) About 121,475 ft^3 **59.** $c = 2$

61. (a) $64\pi/3$ (b) $2048\pi/35$ (c) $8192\pi/105$

Section 7.4 *(page 473)*

1. (a) and (b) 17 **3.** $\frac{5}{3}$ **5.** $\frac{2}{3}(2\sqrt{2} - 1) \approx 1.219$

7. $5\sqrt{5} - 2\sqrt{2} \approx 8.352$ **9.** 309.3195

11. $\ln\left[(\sqrt{2} + 1)/(\sqrt{2} - 1)\right] \approx 1.763$

13. $\frac{1}{2}(e^2 - 1/e^2) \approx 3.627$ **15.** $\frac{76}{3}$

17. (a)

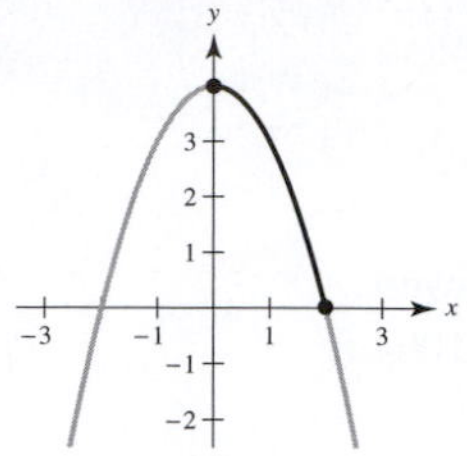

(b) $\int_0^2 \sqrt{1 + 4x^2}\,dx$
(c) About 4.647

19. (a)

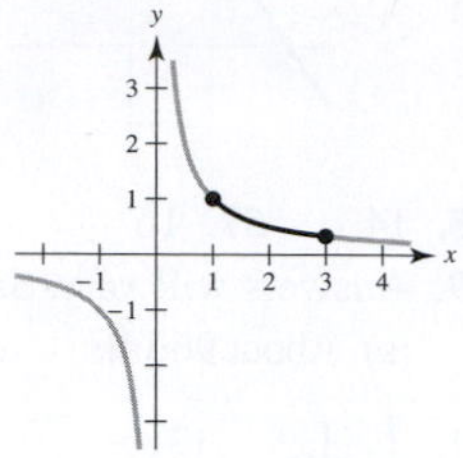

(b) $\int_1^3 \sqrt{1 + \dfrac{1}{x^4}}\,dx$
(c) About 2.147

21. (a)

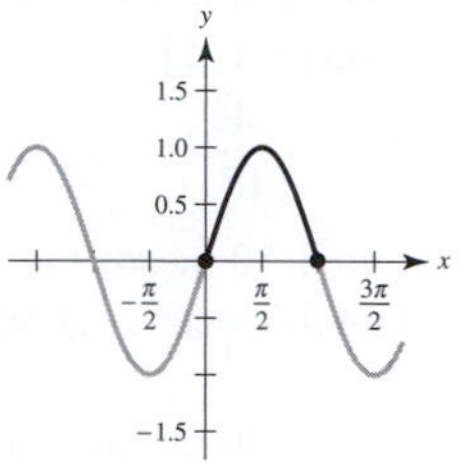

(b) $\int_0^{\pi} \sqrt{1 + \cos^2 x}\,dx$
(c) About 3.820

23. (a)

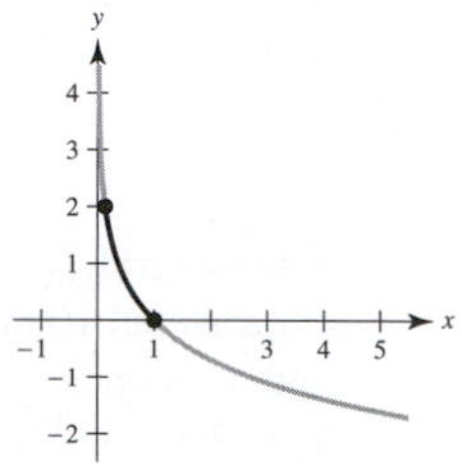

(b) $\int_0^2 \sqrt{1 + e^{-2y}}\,dy$
$= \int_{e^{-2}}^1 \sqrt{1 + \dfrac{1}{x^2}}\,dx$
(c) About 2.221

25. (a)

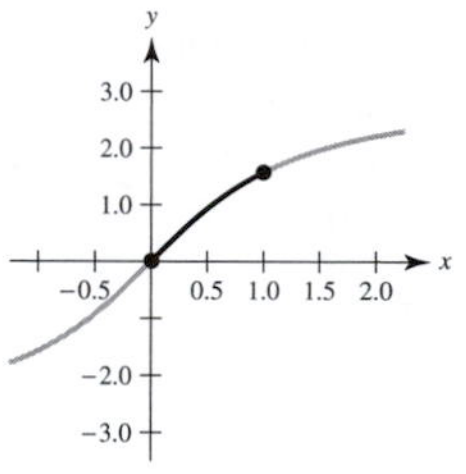

(b) $\int_0^1 \sqrt{1 + \left(\dfrac{2}{1 + x^2}\right)^2}\,dx$
(c) About 1.871

27. b

29. (a) 64.125 (b) 64.525 (c) 64.666 (d) 64.672

31. $20[\sinh 1 - \sinh(-1)] \approx 47.0$ m **33.** About 1480

35. $3\arcsin\frac{2}{3} \approx 2.1892$

37. $2\pi\int_0^3 \dfrac{1}{3}x^3\sqrt{1 + x^4}\,dx = \dfrac{\pi}{9}\left(82\sqrt{82} - 1\right) \approx 258.85$

39. $2\pi\int_1^2 \left(\dfrac{x^3}{6} + \dfrac{1}{2x}\right)\left(\dfrac{x^2}{2} + \dfrac{1}{2x^2}\right)dx = \dfrac{47\pi}{16} \approx 9.23$

41. $2\pi\int_{-1}^1 2\,dx = 8\pi \approx 25.13$

43. $2\pi\int_1^8 x\sqrt{1 + \dfrac{1}{9x^{4/3}}}\,dx = \dfrac{\pi}{27}\left(145\sqrt{145} - 10\sqrt{10}\right) \approx 199.48$

45. $2\pi\int_0^2 x\sqrt{1 + \dfrac{x^2}{4}}\,dx = \dfrac{\pi}{3}\left(16\sqrt{2} - 8\right) \approx 15.318$

47. 14.424

49. A rectifiable curve is a curve with a finite arc length.

51. The integral formula for the area of a surface of revolution is derived from the formula for the lateral surface area of the frustum of a right circular cone. The formula is $S = 2\pi rL$, where $r = \frac{1}{2}(r_1 + r_2)$, which is the average radius of the frustum, and L is the length of a line segment on the frustum. The representative element is $2\pi f(d_i)\sqrt{1 + (\Delta y_i/\Delta x_i)^2}\,\Delta x_i$.

53. (a)

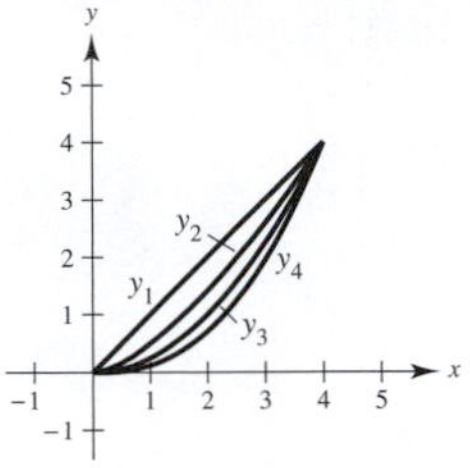

(b) y_1, y_2, y_3, y_4
(c) $s_1 \approx 5.657$; $s_2 \approx 5.759$; $s_3 \approx 5.916$; $s_4 \approx 6.063$

55. 20π **57.** $6\pi(3 - \sqrt{5}) \approx 14.40$

59. (a) Answers will vary. Sample answer: 5207.62 in.^{3}
(b) Answers will vary. Sample answer: 1168.64 in.^{2}
(c) $r = 0.0040y^3 - 0.142y^2 + 1.23y + 7.9$

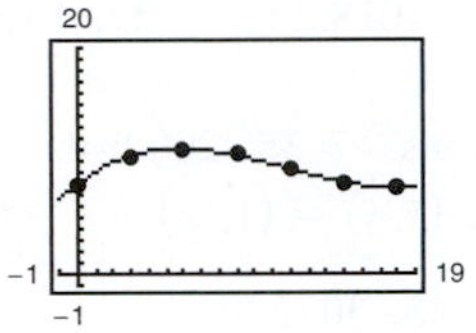

(d) 5279.64 in.^{3}; 1179.5 in.^{2}

61. (a) $\pi(1 - 1/b)$ (b) $2\pi\int_1^b \sqrt{x^4 + 1}/x^3\,dx$
(c) $\lim_{b\to\infty} V = \lim_{b\to\infty} \pi(1 - 1/b) = \pi$
(d) Because $\dfrac{\sqrt{x^4 + 1}}{x^3} > \dfrac{\sqrt{x^4}}{x^3} = \dfrac{1}{x} > 0$ on $[1, b]$,
you have $\displaystyle\int_1^b \frac{\sqrt{x^4 + 1}}{x^3}\,dx > \int_1^b \frac{1}{x}\,dx = \Big[\ln x\Big]_1^b = \ln b$
and $\lim_{b\to\infty} \ln b \to \infty$. So, $\displaystyle\lim_{b\to\infty} 2\pi\int_1^b \frac{\sqrt{x^4 + 1}}{x^3}\,dx = \infty$.

63. Fleeing object: $\frac{2}{3}$ unit
Pursuer: $\displaystyle\frac{1}{2}\int_0^1 \frac{x + 1}{\sqrt{x}}\,dx = \frac{4}{3} = 2\left(\frac{2}{3}\right)$

65. $384\pi/5$ **67–69.** Proofs

Section 7.5 *(page 483)*

1. 48,000 ft-lb **3.** 896 N-m **5.** 40.833 in.-lb $\approx$ 3.403 ft-lb
7. 160 in.-lb $\approx$ 13.3 ft-lb **9.** 37.125 ft-lb
11. (a) 487.805 mile-tons $\approx 5.151 \times 10^9$ ft-lb
(b) 1395.349 mile-tons $\approx 1.473 \times 10^{10}$ ft-lb
13. (a) 2.93×10^4 mile-tons $\approx 3.10 \times 10^{11}$ ft-lb
(b) 3.38×10^4 mile-tons $\approx 3.57 \times 10^{11}$ ft-lb
15. (a) 2496 ft-lb (b) 9984 ft-lb **17.** $470{,}400\pi$ N-m
19. 2995.2π ft-lb **21.** $20{,}217.6\pi$ ft-lb **23.** 2457π ft-lb
25. 600 ft-lb **27.** 450 ft-lb **29.** 168.75 ft-lb
31. If an object is moved a distance D in the direction of an applied constant force F, then the work W done by the force is defined as $W = FD$.
33. The situation in part (a) requires more work. There is no work required for part (b) because the distance is 0.
35. (a) 54 ft-lb (b) 160 ft-lb (c) 9 ft-lb (d) 18 ft-lb
37. $2000 \ln(3/2) \approx 810.93$ ft-lb **39.** 3249.4 ft-lb
41. 10,330.3 ft-lb
43. (a) $16{,}000\pi$ ft-lb (b) 24,888.889 ft-lb
(c) $F(x) = -16{,}261.36x^4 + 85{,}295.45x^3 - 157{,}738.64x^2 + 104{,}386.36x - 32.4675$

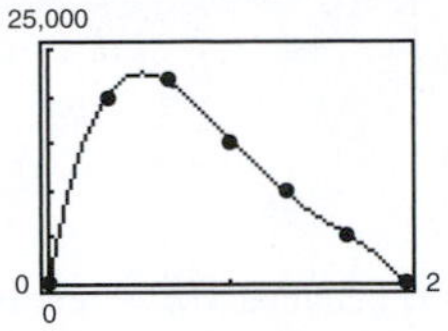

(d) 0.524 ft (e) 25,180.5 ft-lb

Section 7.6 *(page 494)*

1. $\bar{x} = -\frac{4}{3}$ **3.** $\bar{x} = 4$ **5.** (a) $\bar{x} = 8$ (b) $\bar{x} = -\frac{3}{4}$
7. $x = 6$ ft **9.** $(\bar{x}, \bar{y}) = \left(\frac{10}{9}, -\frac{1}{9}\right)$ **11.** $(\bar{x}, \bar{y}) = \left(2, \frac{48}{25}\right)$
13. $M_x = \rho/3$, $M_y = 4\rho/3$, $(\bar{x}, \bar{y}) = (4/3, 1/3)$
15. $M_x = 4\rho$, $M_y = 64\rho/5$, $(\bar{x}, \bar{y}) = (12/5, 3/4)$
17. $M_x = \rho/35$, $M_y = \rho/20$, $(\bar{x}, \bar{y}) = (3/5, 12/35)$
19. $M_x = 99\rho/5$, $M_y = 27\rho/4$, $(\bar{x}, \bar{y}) = (3/2, 22/5)$
21. $M_x = 192\rho/7$, $M_y = 96\rho$, $(\bar{x}, \bar{y}) = (5, 10/7)$
23. $M_x = 0$, $M_y = 256\rho/15$, $(\bar{x}, \bar{y}) = (8/5, 0)$
25. $M_x = 27\rho/4$, $M_y = -27\rho/10$, $(\bar{x}, \bar{y}) = (-3/5, 3/2)$

27.

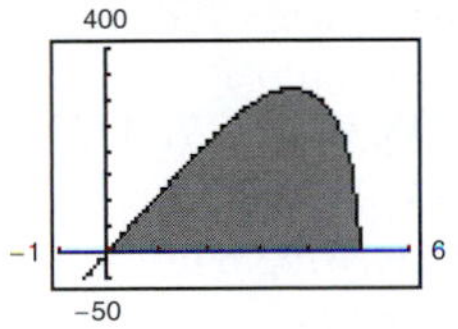

$(\bar{x}, \bar{y}) = (3.0, 126.0)$

29.

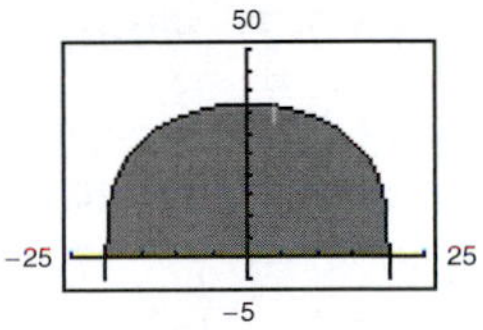

$(\bar{x}, \bar{y}) = (0, 16.2)$

31.

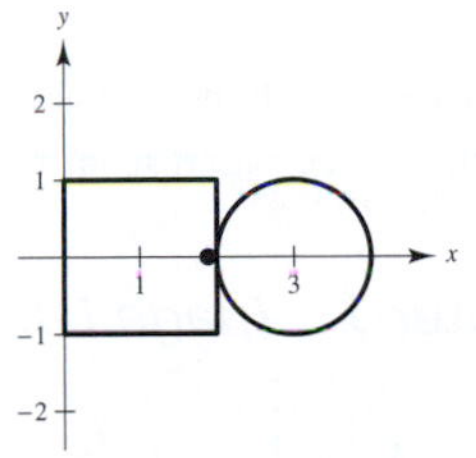

$(\bar{x}, \bar{y}) = \left(\dfrac{4 + 3\pi}{4 + \pi}, 0\right)$

33.

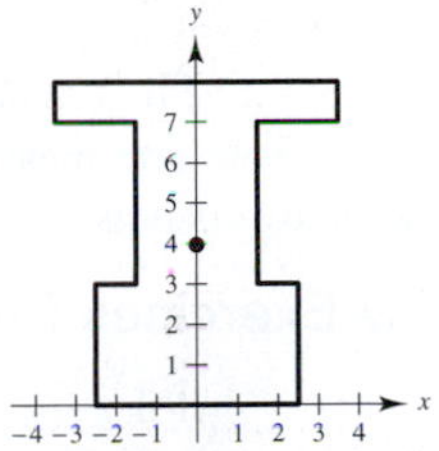

$(\bar{x}, \bar{y}) = \left(0, \dfrac{135}{34}\right)$

35. $(\bar{x}, \bar{y}) = \left(\dfrac{2 + 3\pi}{2 + \pi}, 0\right)$ **37.** $160\pi^2 \approx 1579.14$
39. $128\pi/3 \approx 134.04$
41. The center of mass $(\bar{x}, \bar{y})$ is $\bar{x} = M_y/m$ and $\bar{y} = M_x/m$, where:
1. $m = m_1 + m_2 + \cdots + m_n$ is the total mass of the system.
2. $M_y = m_1x_1 + m_2x_2 + \cdots + m_nx_n$ is the moment about the y-axis.
3. $M_x = m_1y_1 + m_2y_2 + \cdots + m_ny_n$ is the moment about the x-axis.

43. See Theorem 7.1 on page 493. **45.** $(\bar{x}, \bar{y}) = \left(\dfrac{b}{3}, \dfrac{c}{3}\right)$
47. $(\bar{x}, \bar{y}) = \left(\dfrac{(a + 2b)c}{3(a + b)}, \dfrac{a^2 + ab + b^2}{3(a + b)}\right)$
49. $(\bar{x}, \bar{y}) = (0, 4b/(3\pi))$

51. (a)

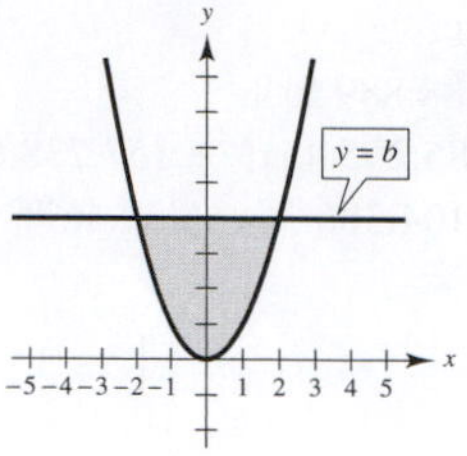

(b) $\bar{x} = 0$ by symmetry

(c) $M_y = \int_{-\sqrt{b}}^{\sqrt{b}} x(b - x^2)\,dx = 0$ because $x(b - x^2)$ is an odd function.

(d) $\bar{y} > b/2$ because the area is greater for $y > b/2$.

(e) $\bar{y} = (3/5)b$

53. (a) $(\bar{x}, \bar{y}) = (0, 12.98)$

(b) $y = (-1.02 \times 10^{-5})x^4 - 0.0019x^2 + 29.28$

(c) $(\bar{x}, \bar{y}) = (0, 12.85)$

55. $(\bar{x}, \bar{y}) = (0, 2r/\pi)$

57. $(\bar{x}, \bar{y}) = \left(\dfrac{n+1}{n+2}, \dfrac{n+1}{4n+2}\right)$; As $n \to \infty$, the region shrinks toward the line segments $y = 0$ for $0 \le x \le 1$ and $x = 1$ for $0 \le y \le 1$; $(\bar{x}, \bar{y}) \to \left(1, \dfrac{1}{4}\right)$.

Section 7.7 *(page 501)*

1. 1497.6 lb **3.** 4992 lb **5.** 748.8 lb **7.** 1123.2 lb
9. 748.8 lb **11.** 1064.96 lb **13.** 117,600 N
15. 2,381,400 N **17.** 2814 lb **19.** 6753.6 lb
21. 94.5 lb **23–25.** Proofs **27.** 960 lb
29. Answers will vary. Sample answer (using Simpson's Rule): 3010.8 lb
31. $3\sqrt{2}/2 \approx 2.12$ ft; The pressure increases with increasing depth.
33. Because you are measuring total force against a region between two depths

Review Exercises for Chapter 7 *(page 503)*

1.

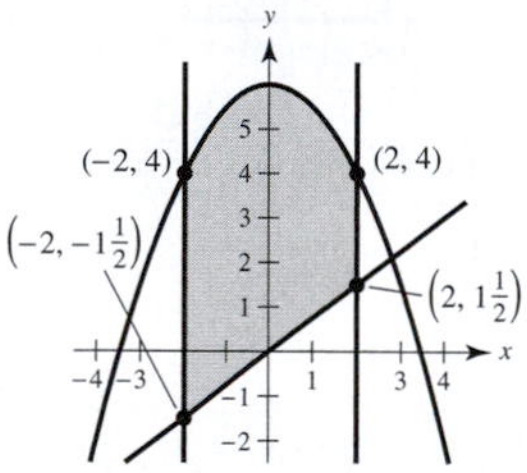

64/3

3.

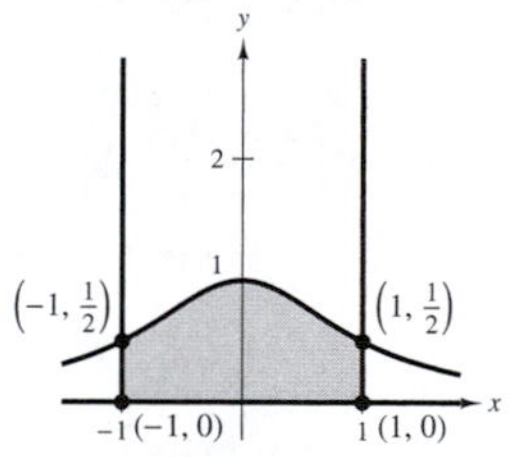

$\pi/2$

5.

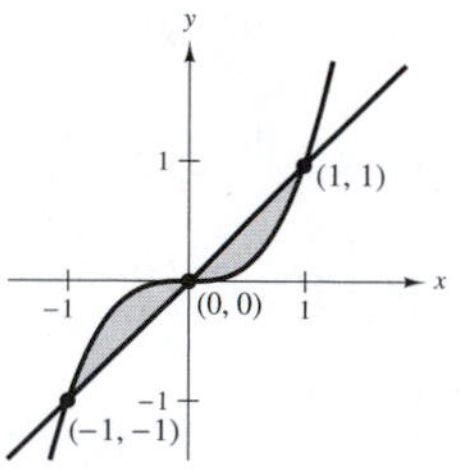

$\frac{1}{2}$

7.

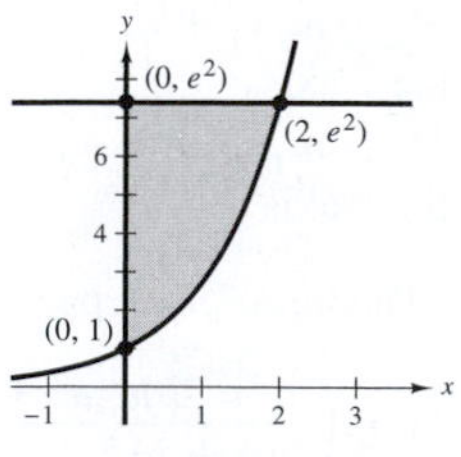

$e^2 + 1$

9.

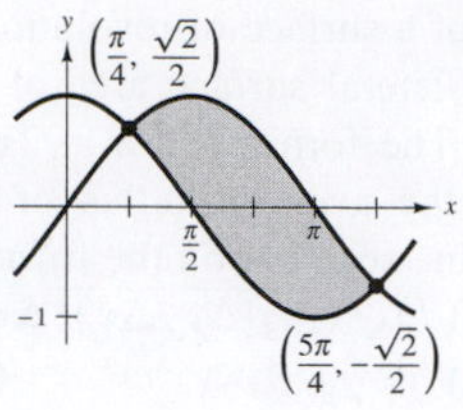

$2\sqrt{2}$

11.

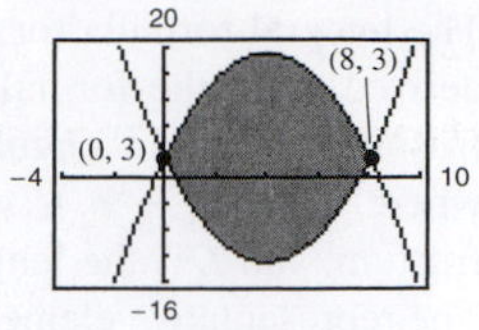

$\frac{512}{3}$

13.

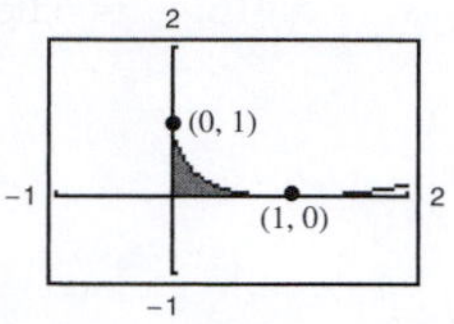

$\frac{1}{6}$

15. (a) 9920 ft^2 (b) $10{,}413\frac{1}{3}$ ft^2
17. (a) 9π (b) 18π (c) 9π (d) 36π **19.** $\pi^2/4$
21. $2\pi \ln 2.5 \approx 5.757$ **23.** 1.958 ft
25. $\frac{8}{15}\left(1 + 6\sqrt{3}\right) \approx 6.076$ **27.** 4018.2 ft **29.** 15π
31. 62.5 in.-lb $\approx$ 5.208 ft-lb
33. $122{,}980\pi$ ft-lb $\approx$ 193.2 foot-tons **35.** 200 ft-lb
37. $a = 15/4$ **39.** 3.6 **41.** $(\bar{x}, \bar{y}) = \left(1, \frac{17}{5}\right)$
43. $(\bar{x}, \bar{y}) = \left(\dfrac{2(9\pi + 49)}{3(\pi + 9)}, 0\right)$ **45.** 3072 lb
47. Wall at shallow end: 15,600 lb
Wall at deep end: 62,400 lb
Side wall: 72,800 lb

P.S. Problem Solving *(page 505)*

1. 3 **3.** $y = 0.2063x$

5.

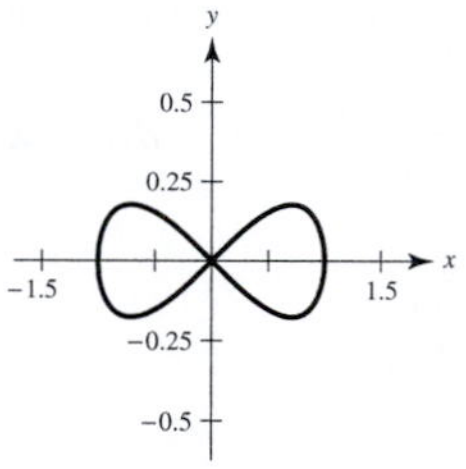

$\dfrac{5\sqrt{2}\pi}{3}$

7. $V = 2\pi\left[d + \frac{1}{2}\sqrt{w^2 + l^2}\right]lw$
9. $f(x) = 2e^{x/2} - 2$ **11.** 89.3%

13.

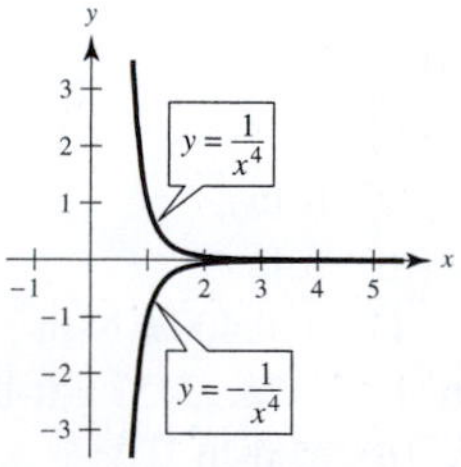

(a) $(\bar{x}, \bar{y}) = \left(\dfrac{63}{43}, 0\right)$

(b) $(\bar{x}, \bar{y}) = \left(\dfrac{3b(b + 1)}{2(b^2 + b + 1)}, 0\right)$

(c) $\left(\dfrac{3}{2}, 0\right)$

15. Consumer surplus: 1600; Producer surplus: 400
17. Wall at shallow end: 9984 lb
Wall at deep end: 39,936 lb
Side wall: 19,968 + 26,624 = 46,592 lb

Chapter 8

Section 8.1 (page 512)

1. b **3.** c

5. $\int u^n\,du$; $u = 5x - 3, n = 4$ **7.** $\int \frac{du}{u}$; $u = 1 - 2\sqrt{x}$ **9.** $\int \frac{du}{\sqrt{a^2 - u^2}}$; $u = t, a = 1$

11. $\int \sin u\,du$; $u = t^2$ **13.** $\int e^u\,du$; $u = \sin x$ **15.** $2(x - 5)^7 + C$

17. $-7/[6(z - 10)^6] + C$ **19.** $\frac{1}{2}v^2 - 1/[6(3v - 1)^2] + C$

21. $-\frac{1}{3}\ln|-t^3 + 9t + 1| + C$

23. $\frac{1}{2}x^2 + x + \ln|x - 1| + C$ **25.** $\ln(1 + e^x) + C$

27. $\frac{x}{15}(48x^4 + 200x^2 + 375) + C$ **29.** $\sin(2\pi x^2)/(4\pi) + C$

31. $-2\sqrt{\cos x} + C$ **33.** $2\ln(1 + e^x) + C$

35. $(\ln x)^2 + C$ **37.** $-\ln|\csc \alpha + \cot \alpha| + \ln|\sin \alpha| + C$

39. $-\frac{1}{4}\arcsin(4t + 1) + C$ **41.** $\frac{1}{2}\ln|\cos(2/t)| + C$

43. $6\arcsin[(x - 5)/5] + C$ **45.** $\frac{1}{4}\arctan[(2x + 1)/8] + C$

47. (a)

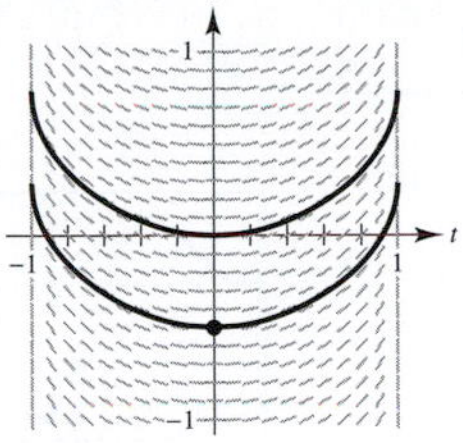

(b) $\frac{1}{2}\arcsin t^2 - \frac{1}{2}$

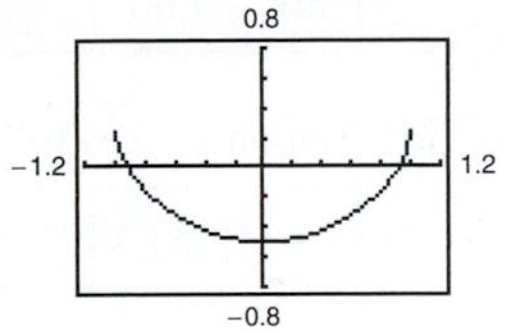

49. $y = 4e^{0.8x}$

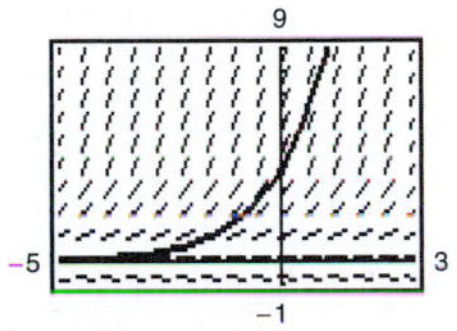

51. $y = \frac{1}{2}e^{2x} + 10e^x + 25x + C$ **53.** $r = 10\arcsin e^t + C$

55. $y = \frac{1}{2}\arctan(\tan x/2) + C$ **57.** $\frac{1}{2}$

59. $\frac{1}{2}(1 - e^{-1}) \approx 0.316$ **61.** 8 **63.** $\pi/18$

65. $18\sqrt{6}/5 \approx 8.82$ **67.** $\frac{4}{3} \approx 1.333$

69. $\frac{1}{3}\arctan\left[\frac{1}{3}(x + 2)\right] + C$
Graphs will vary.
Example:

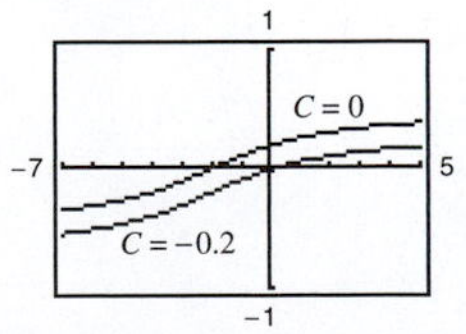

One graph is a vertical translation of the other.

71. $\tan \theta - \sec \theta + C$
Graphs will vary.
Example:

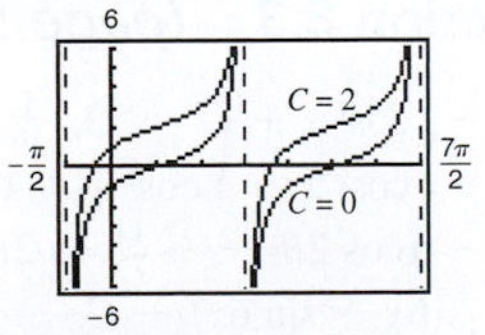

One graph is a vertical translation of the other.

73. Power Rule: $\int u^n\,du = \frac{u^{n+1}}{n + 1} + C$; $u = x^2 + 1, n = 3$

75. Log Rule: $\int \frac{du}{u} = \ln|u| + C$; $u = x^2 + 1$

77. $a = \sqrt{2}, b = \frac{\pi}{4}$; $-\frac{1}{\sqrt{2}}\ln\left|\csc\left(x + \frac{\pi}{4}\right) + \cot\left(x + \frac{\pi}{4}\right)\right| + C$

79. $a = \frac{1}{2}$

81. (a) They are equivalent because $e^{x+C_1} = e^x \cdot e^{C_1} = Ce^x, C = e^{C_1}$.
(b) They differ by a constant. $\sec^2 x + C_1 = (\tan^2 x + 1) + C_1 = \tan^2 x + C$

83. a

85. (a)

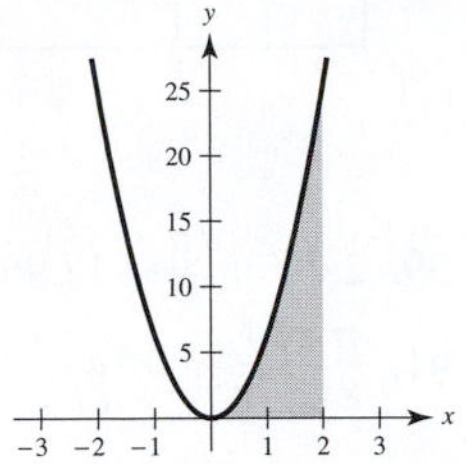

(b)

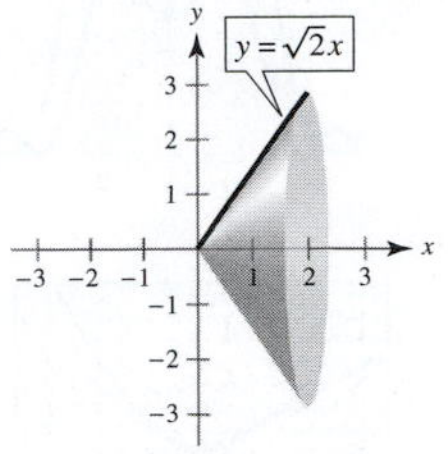

(c)

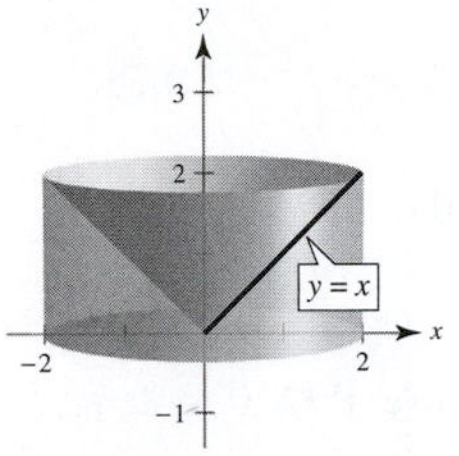

87. (a) $\pi(1 - e^{-1}) \approx 1.986$
(b) $b = \sqrt{\ln\left(\frac{3\pi}{3\pi - 4}\right)} \approx 0.743$

89. $\ln(\sqrt{2} + 1) \approx 0.8814$

91. $(8\pi/3)(10\sqrt{10} - 1) \approx 256.545$ **93.** $\frac{1}{3}\arctan 3 \approx 0.416$

95. About 1.0320

97. (a) $\frac{1}{3}\sin x(\cos^2 x + 2)$
(b) $\frac{1}{15}\sin x(3\cos^4 x + 4\cos^2 x + 8)$
(c) $\frac{1}{35}\sin x(5\cos^6 x + 6\cos^4 x + 8\cos^2 x + 16)$
(d) $\int \cos^{15} x\,dx = \int (1 - \sin^2 x)^7 \cos x\,dx$
You would expand $(1 - \sin^2 x)^7$.

99. Proof

Section 8.2 (page 521)

1. $u = x, dv = e^{2x}\,dx$ **3.** $u = (\ln x)^2, dv = dx$

5. $u = x, dv = \sec^2 x\,dx$ **7.** $\frac{1}{16}x^4(4\ln x - 1) + C$

9. $\frac{1}{9}\sin 3x - \frac{1}{3}x\cos 3x + C$ **11.** $-\frac{1}{16e^{4x}}(4x + 1) + C$

13. $e^x(x^3 - 3x^2 + 6x - 6) + C$

15. $\frac{1}{4}[2(t^2 - 1)\ln|t + 1| - t^2 + 2t] + C$ **17.** $\frac{1}{3}(\ln x)^3 + C$

19. $e^{2x}/[4(2x + 1)] + C$ **21.** $\frac{2}{15}(x - 5)^{3/2}(3x + 10) + C$

23. $x\sin x + \cos x + C$

25. $(6x - x^3)\cos x + (3x^2 - 6)\sin x + C$

27. $x\arctan x - \frac{1}{2}\ln(1 + x^2) + C$

29. $-\frac{3}{34}e^{-3x}\sin 5x - \frac{5}{34}e^{-3x}\cos 5x + C$ **31.** $x\ln x - x + C$

33. $y = \frac{2}{5}t^2\sqrt{3 + 5t} - \frac{8t}{75}(3 + 5t)^{3/2} + \frac{16}{1875}(3 + 5t)^{5/2} + C$
$= \frac{2}{625}\sqrt{3 + 5t}(25t^2 - 20t + 24) + C$

35. (a)

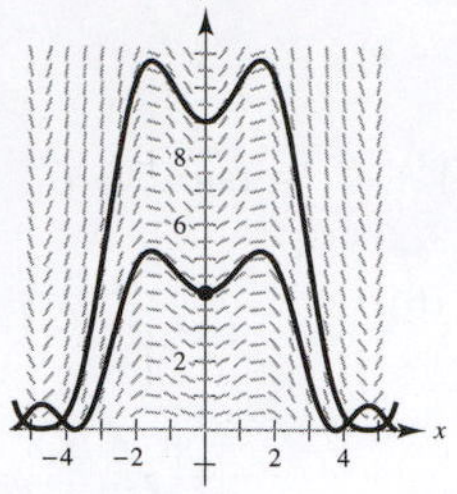

(b) $2\sqrt{y} - \cos x - x \sin x = 3$

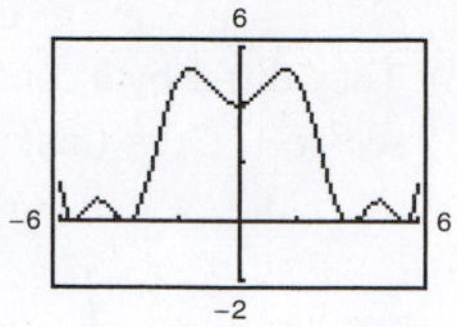

37.

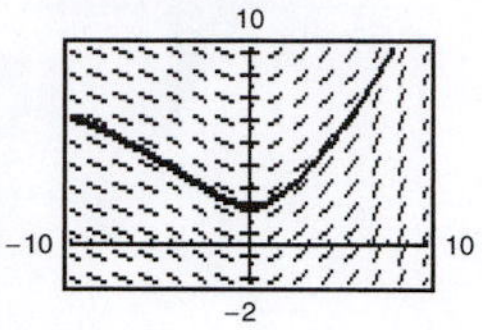

39. $2e^{3/2} + 4 \approx 12.963$

41. $\dfrac{\pi}{8} - \dfrac{1}{4} \approx 0.143$

43. $(\pi - 3\sqrt{3} + 6)/6 \approx 0.658$

45. $\frac{1}{2}[e(\sin 1 - \cos 1) + 1] \approx 0.909$

47. $8 \operatorname{arcsec} 4 + \sqrt{3}/2 - \sqrt{15}/2 - 2\pi/3 \approx 7.380$

49. $(e^{2x}/4)(2x^2 - 2x + 1) + C$

51. $(3x^2 - 6) \sin x - (x^3 - 6x) \cos x + C$

53. $x \tan x + \ln|\cos x| + C$

55. $2(\sin\sqrt{x} - \sqrt{x} \cos \sqrt{x}) + C$

57. $\frac{1}{2}(x^4e^{x^2} - 2x^2e^{x^2} + 2e^{x^2}) + C$

59. (a) Product Rule

(b) Answers will vary. Sample answer: You want dv to be the most complicated portion of the integrand.

61. (a) No, substitution (b) Yes, $u = \ln x$, $dv = x\,dx$

(c) Yes, $u = x^2$, $dv = e^{-3x}\,dx$ (d) No, substitution

(e) Yes, $u = x$ and $dv = \dfrac{1}{\sqrt{x+1}}\,dx$ (f) No, substitution

63. $\frac{1}{3}\sqrt{4 + x^2}(x^2 - 8) + C$

65. $n = 0$: $x(\ln x - 1) + C$

$n = 1$: $\frac{1}{4}x^2(2 \ln x - 1) + C$

$n = 2$: $\frac{1}{9}x^3(3 \ln x - 1) + C$

$n = 3$: $\frac{1}{16}x^4(4 \ln x - 1) + C$

$n = 4$: $\frac{1}{25}x^5(5 \ln x - 1) + C$

$$\int x^n \ln x\,dx = \frac{x^{n+1}}{(n+1)^2}[(n+1)\ln x - 1] + C$$

67–71. Proofs **73.** $-x^2 \cos x + 2x \sin x + 2 \cos x + C$

75. $\frac{1}{36}x^6(6 \ln x - 1) + C$

77. $\dfrac{e^{-3x}(-3 \sin 4x - 4 \cos 4x)}{25} + C$

79.

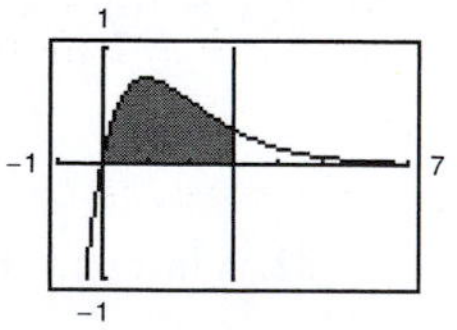

$2 - \dfrac{8}{e^3} \approx 1.602$

81.

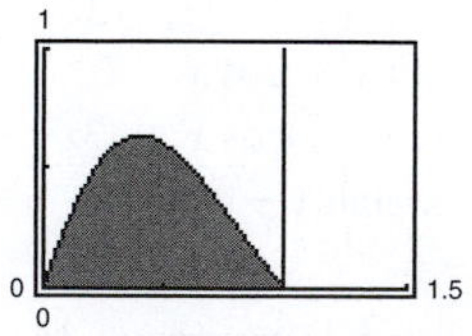

$\dfrac{\pi}{1 + \pi^2}\left(\dfrac{1}{e} + 1\right) \approx 0.395$

83. (a) 1 (b) $\pi(e - 2) \approx 2.257$ (c) $\frac{1}{2}\pi(e^2 + 1) \approx 13.177$

(d) $\left(\dfrac{e^2 + 1}{4}, \dfrac{e - 2}{2}\right) \approx (2.097, 0.359)$

85. In Example 6, we showed that the centroid of an equivalent region was $(1, \pi/8)$. By symmetry, the centroid of this region is $(\pi/8, 1)$.

87. $[7/(10\pi)](1 - e^{-4\pi}) \approx 0.223$ **89.** $931,265

91. Proof **93.** $b_n = [8h/(n\pi)^2] \sin(n\pi/2)$

95. (a) $y = \frac{1}{4}(3 \sin 2x - 6x \cos 2x)$

(b)

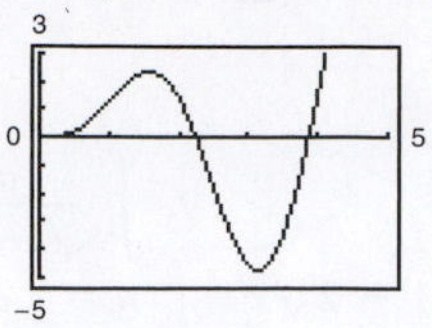

(c) You obtain the following points.

n	x_n	y_n
0	0	0
1	0.05	0
2	0.10	7.4875×10^{-4}
3	0.15	0.0037
4	0.20	0.0104
⋮	⋮	⋮
80	4.00	1.3181

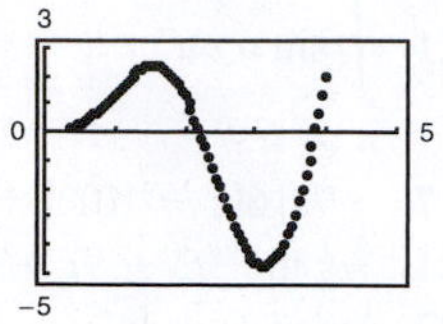

(d) You obtain the following points.

n	x_n	y_n
0	5	0
1	0.1	0
2	0.2	0.0060
3	0.3	0.0293
4	0.4	0.0801
⋮	⋮	⋮
40	4.0	1.0210

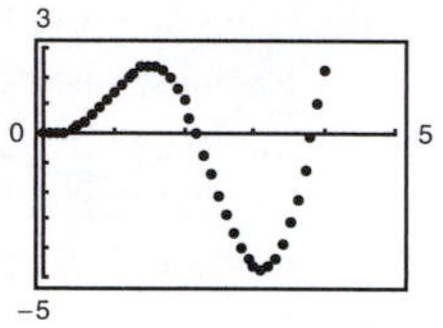

97. The graph of $y = x \sin x$ is below the graph of $y = x$ on $[0, \pi/2]$.

99. For any integrable function, $\int f(x)\,dx = C + \int f(x)\,dx$, but this cannot be used to imply that $C = 0$.

Section 8.3 *(page 530)*

1. $-\frac{1}{6}\cos^6 x + C$ **3.** $\frac{1}{16}\sin^8 2x + C$

5. $-\frac{1}{3}\cos^3 x + \frac{1}{5}\cos^5 x + C$

7. $-\frac{1}{3}(\cos 2\theta)^{3/2} + \frac{1}{7}(\cos 2\theta)^{7/2} + C$

9. $\frac{1}{12}(6x + \sin 6x) + C$

11. $\frac{1}{8}(2x^2 - 2x \sin 2x - \cos 2x) + C$ **13.** $\frac{16}{35}$

15. $63\pi/512$ **17.** $5\pi/32$ **19.** $\frac{1}{4}\ln|\sec 4x + \tan 4x| + C$

21. $(\sec \pi x \tan \pi x + \ln|\sec \pi x + \tan \pi x|)/(2\pi) + C$

23. $\frac{1}{2}\tan^4(x/2) - \tan^2(x/2) - 2\ln|\cos(x/2)| + C$

25. $\dfrac{1}{2}\left[\dfrac{\sec^5 2t}{5} - \dfrac{\sec^3 2t}{3}\right] + C$ **27.** $\dfrac{1}{24}\sec^6 4x + C$

29. $\frac{1}{7}\sec^7 x - \frac{1}{5}\sec^5 x + C$

31. $\ln|\sec x + \tan x| - \sin x + C$

33. $(12\pi\theta - 8 \sin 2\pi\theta + \sin 4\pi\theta)/(32\pi) + C$

35. $y = \frac{1}{9}\sec^3 3x - \frac{1}{3}\sec 3x + C$

37. (a)

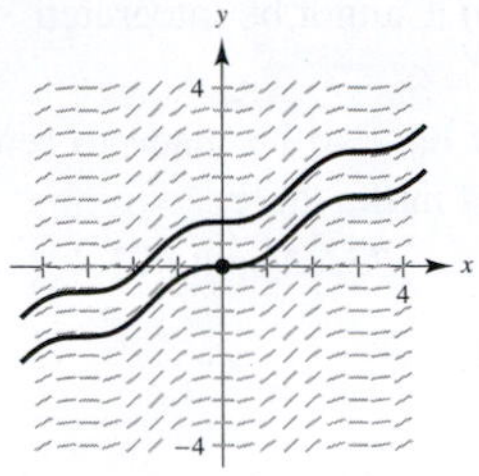

(b) $y = \frac{1}{2}x - \frac{1}{4}\sin 2x$

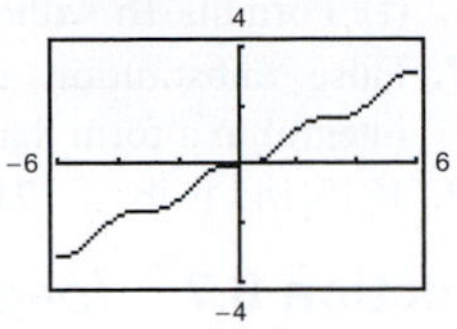

39.

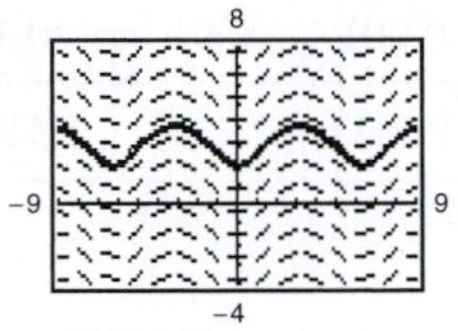

41. $\frac{1}{16}(2 \sin 4x + \sin 8x) + C$

43. $\frac{1}{12}(3 \cos 2x - \cos 6x) + C$ **45.** $\frac{1}{8}(2 \sin 2\theta - \sin 4\theta) + C$

47. $\frac{1}{4}\left(\ln|\csc^2 2x| - \cot^2 2x\right) + C$

49. $-\frac{1}{3}\cot 3x - \frac{1}{9}\cot^3 3x + C$

51. $\ln|\csc t - \cot t| + \cos t + C$

53. $\ln|\csc x - \cot x| + \cos x + C$ **55.** $t - 2\tan t + C$

57. π **59.** $3(1 - \ln 2)$ **61.** $\ln 2$ **63.** 4

65. (a) Save one sine factor and convert the remaining factors to cosines. Then expand and integrate.
(b) Save one cosine factor and convert the remaining factors to sines. Then expand and integrate.
(c) Make repeated use of the power reducing formulas to convert the integrand to odd powers of the cosine. Then proceed as in part (b).

67. (a) $\frac{1}{2}\sin^2 x + C$ (b) $-\frac{1}{2}\cos^2 x + C$
(c) $\frac{1}{2}\sin^2 x + C$ (d) $-\frac{1}{4}\cos 2x + C$
The answers are all the same; they are just written in different forms. Using trigonometric identities, you can rewrite each answer in the same form.

69. (a) $\frac{1}{18}\tan^6 3x + \frac{1}{12}\tan^4 3x + C_1$, $\frac{1}{18}\sec^6 3x - \frac{1}{12}\sec^4 3x + C_2$
(b)

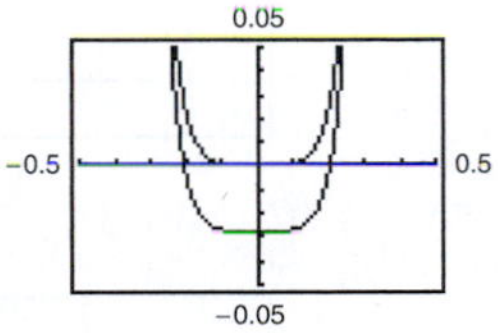

(c) Proof

71. $\frac{1}{3}$ **73.** 1 **75.** $2\pi(1 - \pi/4) \approx 1.348$

77. (a) $\pi^2/2$ (b) $(\bar{x}, \bar{y}) = (\pi/2, \pi/8)$ **79–81.** Proofs

83. $-\frac{1}{15}\cos x(3\sin^4 x + 4\sin^2 x + 8) + C$

85. $\dfrac{5}{6\pi}\tan\dfrac{2\pi x}{5}\left(\sec^2\dfrac{2\pi x}{5} + 2\right) + C$

87. (a) $H(t) \approx 57.72 - 23.36\cos(\pi t/6) - 2.75\sin(\pi t/6)$
(b) $L(t) \approx 42.04 - 20.91\cos(\pi t/6) - 4.33\sin(\pi t/6)$
(c)

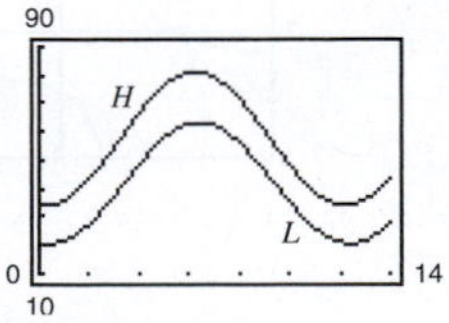

The maximum difference is at $t \approx 4.9$, or late spring.

89. Proof

Section 8.4 *(page 539)*

1. $x = 3\tan\theta$ **3.** $x = 5\sin\theta$ **5.** $x/\left(16\sqrt{16 - x^2}\right) + C$

7. $4\ln\left|\left(4 - \sqrt{16 - x^2}\right)/x\right| + \sqrt{16 - x^2} + C$

9. $\ln\left|x + \sqrt{x^2 - 25}\right| + C$

11. $\frac{1}{15}(x^2 - 25)^{3/2}(3x^2 + 50) + C$

13. $\frac{1}{3}(1 + x^2)^{3/2} + C$ **15.** $\frac{1}{2}[\arctan x + x/(1 + x^2)] + C$

17. $\frac{1}{2}x\sqrt{9 + 16x^2} + \frac{9}{8}\ln\left|4x + \sqrt{9 + 16x^2}\right| + C$

19. $\frac{25}{4}\arcsin(2x/5) + \frac{1}{2}x\sqrt{25 - 4x^2} + C$

21. $\arcsin(x/4) + C$ **23.** $4\arcsin(x/2) + x\sqrt{4 - x^2} + C$

25. $-\dfrac{(1 - x^2)^{3/2}}{3x^3} + C$ **27.** $-\dfrac{1}{3}\ln\left|\dfrac{\sqrt{4x^2 + 9} + 3}{2x}\right| + C$

29. $3/\sqrt{x^2 + 3} + C$

31. $\frac{1}{2}\left(\arcsin e^x + e^x\sqrt{1 - e^{2x}}\right) + C$

33. $\frac{1}{4}\left[x/(x^2 + 2) + \left(1/\sqrt{2}\right)\arctan\left(x/\sqrt{2}\right)\right] + C$

35. $x \operatorname{arcsec} 2x - \frac{1}{2}\ln\left|2x + \sqrt{4x^2 - 1}\right| + C$

37. $\arcsin[(x - 2)/2] + C$

39. $\sqrt{x^2 + 6x + 12} - 3\ln\left|\sqrt{x^2 + 6x + 12} + (x + 3)\right| + C$

41. (a) and (b) $\sqrt{3} - \pi/3 \approx 0.685$

43. (a) and (b) $9\left(2 - \sqrt{2}\right) \approx 5.272$

45. (a) and (b) $-(9/2)\ln\left(2\sqrt{7}/3 - 4\sqrt{3}/3 - \sqrt{21}/3 + 8/3\right) + 9\sqrt{3} - 2\sqrt{7} \approx 12.644$

47. (a) Let $u = a\sin\theta$, $\sqrt{a^2 - u^2} = a\cos\theta$, where $-\pi/2 \le \theta \le \pi/2$.
(b) Let $u = a\tan\theta$, $\sqrt{a^2 + u^2} = a\sec\theta$, where $-\pi/2 < \theta < \pi/2$.
(c) Let $u = a\sec\theta$, $\sqrt{u^2 - a^2} = \tan\theta$ if $u > a$ and $\sqrt{u^2 - a^2} = -\tan\theta$ if $u < -a$, where $0 \le \theta < \pi/2$ or $\pi/2 < \theta \le \pi$.

49. (a) $\frac{1}{2}\ln(x^2 + 9) + C$; The answers are equivalent.
(b) $x - 3\arctan(x/3) + C$; The answers are equivalent.

51. True

53. False. $\displaystyle\int_0^{\sqrt{3}} \frac{dx}{(1 + x^2)^{3/2}} = \int_0^{\pi/3} \cos\theta\, d\theta$

55. πab

57. (a) $5\sqrt{2}$ (b) $25(1 - \pi/4)$ (c) $r^2(1 - \pi/4)$

59. $6\pi^2$ **61.** $\ln\left[\dfrac{5\left(\sqrt{2} + 1\right)}{\sqrt{26} + 1}\right] + \sqrt{26} - \sqrt{2} \approx 4.367$

63. Length of one arch of sine curve: $y = \sin x$, $y' = \cos x$

$$L_1 = \int_0^{\pi} \sqrt{1 + \cos^2 x}\, dx$$

Length of one arch of cosine curve: $y = \cos x$, $y' = -\sin x$

$$L_2 = \int_{-\pi/2}^{\pi/2} \sqrt{1 + \sin^2 x}\, dx$$

$$= \int_{-\pi/2}^{\pi/2} \sqrt{1 + \cos^2(x - \pi/2)}\, dx,\ u = x - \pi/2,\ du = dx$$

$$= \int_{-\pi}^{0} \sqrt{1 + \cos^2 u}\, du = \int_0^{\pi} \sqrt{1 + \cos^2 u}\, du = L_1$$

65. $(0, 0.422)$

67. $(\pi/32)\left[102\sqrt{2} - \ln\left(3 + 2\sqrt{2}\right)\right] \approx 13.989$

69. (a) 187.2π lb (b) $62.4\pi d$ lb **71.** Proof

73. $12 + 9\pi/2 - 25\arcsin(3/5) \approx 10.050$

75. Putnum Problem A5, 2005

Section 8.5 *(page 549)*

1. $\dfrac{A}{x} + \dfrac{B}{x - 8}$ **3.** $\dfrac{A}{x} + \dfrac{Bx + C}{x^2 + 10}$

5. $\frac{1}{6}\ln|(x - 3)/(x + 3)| + C$ **7.** $\ln|(x - 1)/(x + 4)| + C$

9. $5\ln|x-2| - \ln|x+2| - 3\ln|x| + C$
11. $x^2 + \frac{3}{2}\ln|x-4| - \frac{1}{2}\ln|x+2| + C$
13. $1/x + \ln|x^4 + x^3| + C$
15. $2\ln|x-2| - \ln|x| - 3/(x-2) + C$
17. $\ln|(x^2+1)/x| + C$
19. $\frac{1}{6}\left[\ln|(x-2)/(x+2)| + \sqrt{2}\arctan\left(x/\sqrt{2}\right)\right] + C$
21. $\ln|x+1| + \sqrt{2}\arctan\left[(x-1)/\sqrt{2}\right] + C$
23. $\ln 3$ **25.** $\frac{1}{2}\ln(8/5) - \pi/4 + \arctan 2 \approx 0.557$
27. $\ln|1 + \sec x| + C$ **29.** $\ln\left|\dfrac{\tan x + 2}{\tan x + 3}\right| + C$
31. $\dfrac{1}{5}\ln\left|\dfrac{e^x - 1}{e^x + 4}\right| + C$ **33.** $2\sqrt{x} + 2\ln\left|\dfrac{\sqrt{x} - 2}{\sqrt{x} + 2}\right| + C$
35–37. Proofs **39.** First divide x^3 by $(x-5)$.
41. (a) Substitution: $u = x^2 + 2x - 8$ (b) Partial fractions
(c) Trigonometric substitution (tan) or inverse tangent rule
43. $12\ln\left(\frac{9}{8}\right) \approx 1.4134$ **45.** 4.90 or \$490,000
47. $V = 2\pi\left(\arctan 3 - \frac{3}{10}\right) \approx 5.963$; $(\bar{x}, \bar{y}) \approx (1.521, 0.412)$
49. $x = n[e^{(n+1)kt} - 1]/[n + e^{(n+1)kt}]$ **51.** $\pi/8$

Section 8.6 *(page 555)*

1. $-\frac{1}{2}x(10 - x) + 25\ln|5 + x| + C$ **3.** $-\sqrt{1 - x^2}/x + C$
5. $\frac{1}{24}(3x + \sin 3x \cos 3x + 2\cos^3 3x \sin 3x) + C$
7. $-2\left(\cot\sqrt{x} + \csc\sqrt{x}\right) + C$ **9.** $x - \frac{1}{2}\ln(1 + e^{2x}) + C$
11. $\frac{1}{16}x^8(8\ln x - 1) + C$
13. (a) and (b) $\frac{1}{27}e^{3x}(9x^2 - 6x + 2) + C$
15. (a) and (b) $\ln|(x+1)/x| - 1/x + C$
17. $\frac{1}{2}\left[(x^2+1)\,\text{arccsc}(x^2+1) + \ln\left(x^2 + 1 + \sqrt{x^4 + 2x^2}\right)\right] + C$
19. $\sqrt{x^2 - 4}/(4x) + C$
21. $\frac{4}{25}[\ln|2 - 5x| + 2/(2 - 5x)] + C$
23. $e^x\arccos(e^x) - \sqrt{1 - e^{2x}} + C$
25. $\frac{1}{2}(x^2 + \cot x^2 + \csc x^2) + C$
27. $\left(\sqrt{2}/2\right)\arctan\left[(1 + \sin\theta)/\sqrt{2}\right] + C$
29. $-\sqrt{2 + 9x^2}/(2x) + C$
31. $\frac{1}{4}\left(2\ln|x| - 3\ln|3 + 2\ln|x||\right) + C$
33. $(3x - 10)/[2(x^2 - 6x + 10)] + \frac{3}{2}\arctan(x - 3) + C$
35. $\frac{1}{2}\ln\left|x^2 - 3 + \sqrt{x^4 - 6x^2 + 5}\right| + C$
37. $2/(1 + e^x) - 1/[2(1 + e^x)^2] + \ln(1 + e^x) + C$
39. $\frac{1}{2}(e - 1) \approx 0.8591$ **41.** $\frac{32}{5}\ln 2 - \frac{31}{25} \approx 3.1961$
43. $\pi/2$ **45.** $\pi^3/8 - 3\pi + 6 \approx 0.4510$ **47–51.** Proofs
53. $\dfrac{1}{\sqrt{5}}\ln\left|\dfrac{2\tan(\theta/2) - 3 - \sqrt{5}}{2\tan(\theta/2) - 3 + \sqrt{5}}\right| + C$ **55.** $\ln 2$
57. $\frac{1}{2}\ln(3 - 2\cos\theta) + C$ **59.** $-2\cos\sqrt{\theta} + C$ **61.** $4\sqrt{3}$
63. (a) $\displaystyle\int x\ln x\,dx = \frac{1}{2}x^2\ln x - \frac{1}{4}x^2 + C$
$\displaystyle\int x^2\ln x\,dx = \frac{1}{3}x^3\ln x - \frac{1}{9}x^3 + C$
$\displaystyle\int x^3\ln x\,dx = \frac{1}{4}x^4\ln x - \frac{1}{16}x^4 + C$
65. (a) Arctangent Formula, Formula 23,
$\displaystyle\int \frac{1}{u^2 + 1}\,du,\ u = e^x$
(b) Log Rule: $\displaystyle\int \frac{1}{u}\,du,\ u = e^x + 1$
(c) Substitution: $u = x^2$, $du = 2x\,dx$
Then Formula 81.
(d) Integration by parts (e) Cannot be integrated
(f) Formula 16 with $u = e^{2x}$
67. False. Substitutions may first have to be made to rewrite the integral in a form that appears in the table.
69. 1919.145 ft-lb **71.** $32\pi^2$ **73.** About 401.4

Section 8.7 *(page 564)*

1.

x	-0.1	-0.01	-0.001	0.001	0.01	0.1
$f(x)$	1.3177	1.3332	1.3333	1.3333	1.3332	1.3177

$\frac{4}{3}$

3.

x	1	10	10^2	10^3	10^4	10^5
$f(x)$	0.9900	90,483.7	3.7×10^9	4.5×10^{10}	0	0

0

5. $\frac{3}{8}$ **7.** $\frac{1}{8}$ **9.** $\frac{5}{3}$ **11.** 4 **13.** 0 **15.** ∞ **17.** $\frac{11}{4}$
19. $\frac{3}{5}$ **21.** 1 **23.** $\frac{5}{4}$ **25.** ∞ **27.** 0 **29.** 1
31. 0 **33.** 0 **35.** ∞ **37.** $\frac{5}{9}$ **39.** 1 **41.** ∞
43. (a) Not indeterminate
(b) ∞
(c)
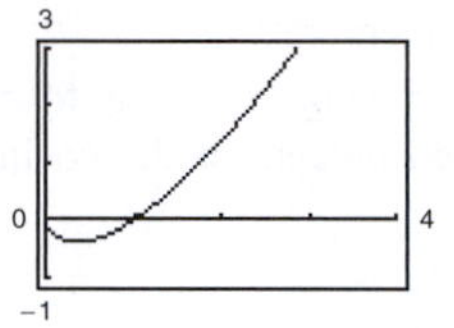

45. (a) $0 \cdot \infty$
(b) 1
(c)
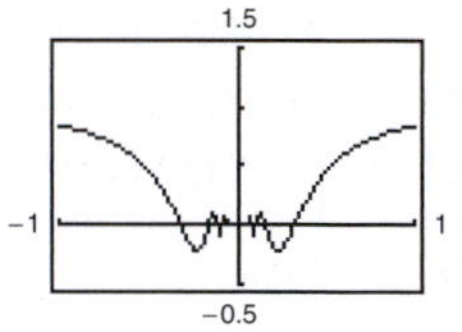

47. (a) Not indeterminate
(b) 0
(c)
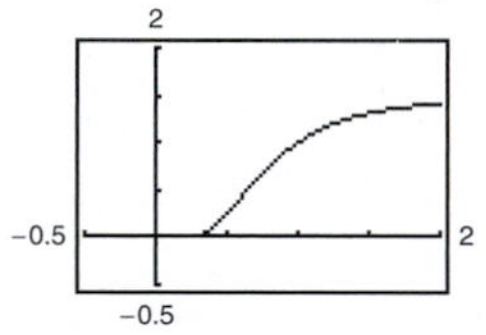

49. (a) ∞^0
(b) 1
(c)
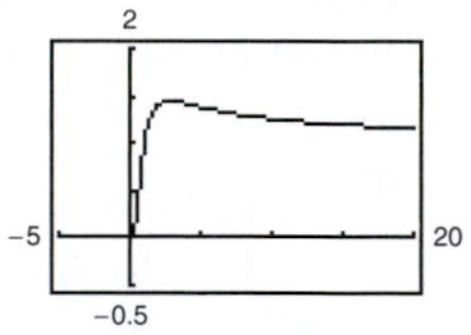

51. (a) 1^∞ (b) e
(c)
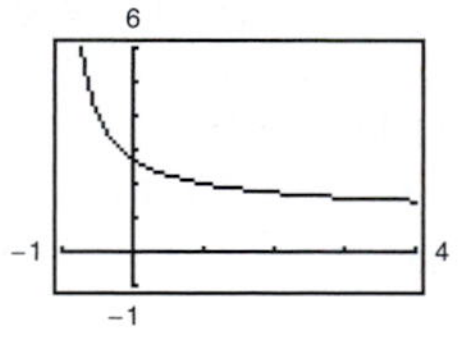

53. (a) 0^0 (b) 3
(c)
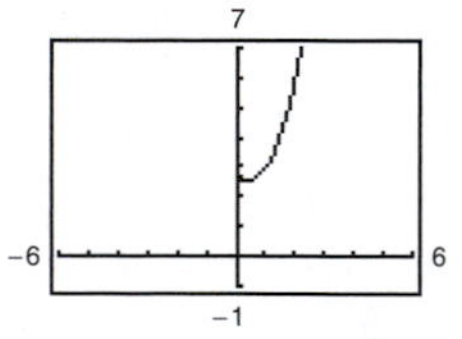

55. (a) 0^0 (b) 1
(c)
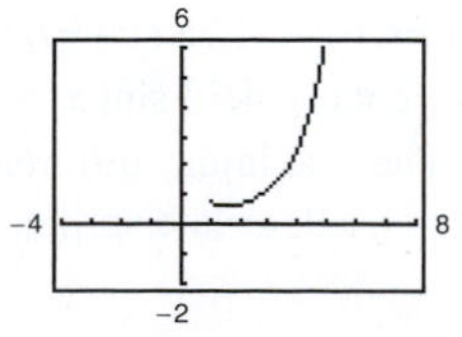

57. (a) $\infty - \infty$ (b) $-\frac{3}{2}$
(c)
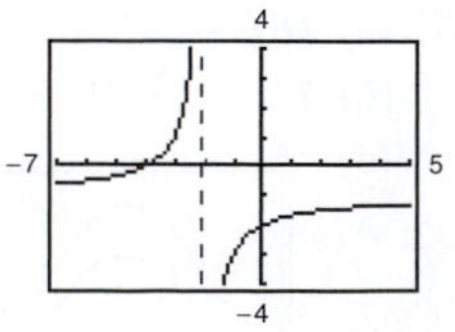

59. (a) $\infty - \infty$ (b) ∞
(c)
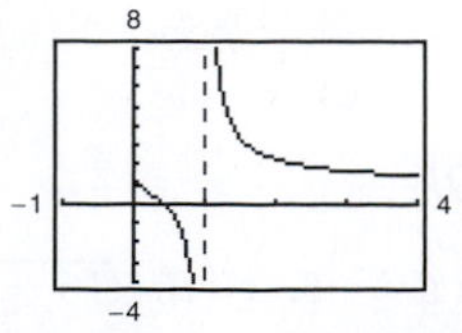

61. $\frac{0}{0}, \frac{\infty}{\infty}, 0 \cdot \infty, 1^{\infty}, 0^0, \infty - \infty$

63. Answers will vary. Examples:
(a) $f(x) = x^2 - 25, g(x) = x - 5$
(b) $f(x) = (x - 5)^2, g(x) = x^2 - 25$
(c) $f(x) = x^2 - 25, g(x) = (x - 5)^3$

65. (a) Yes: $\frac{0}{0}$ (b) No: $\frac{0}{-1}$ (c) Yes: $\frac{\infty}{\infty}$ (d) Yes: $\frac{0}{0}$
(e) No: $\frac{-1}{0}$ (f) Yes: $\frac{0}{0}$

67.

x	10	10^2	10^4	10^6	10^8	10^{10}
$\frac{(\ln x)^4}{x}$	2.811	4.498	0.720	0.036	0.001	0.000

69. 0 **71.** 0 **73.** 0

75. Horizontal asymptote: $y = 1$
Relative maximum: $(e, e^{1/e})$

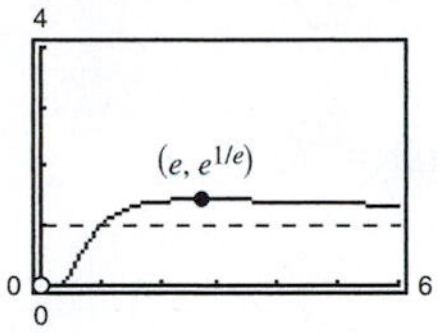

77. Horizontal asymptote: $y = 0$
Relative maximum: $(1, 2/e)$

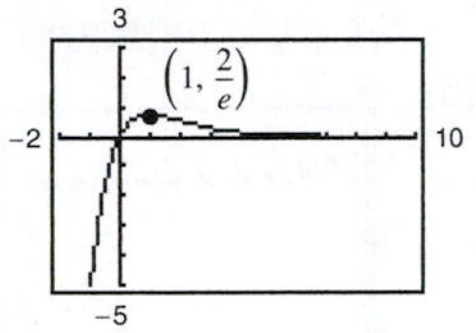

79. Limit is not of the form $0/0$ or ∞/∞.

81. Limit is not of the form $0/0$ or ∞/∞.

83. (a) $\lim_{x\to\infty} \frac{x}{\sqrt{x^2+1}} = \lim_{x\to\infty} \frac{\sqrt{x^2+1}}{x} = \lim_{x\to\infty} \frac{x}{\sqrt{x^2+1}}$
Applying L'Hôpital's Rule twice results in the original limit, so L'Hôpital's Rule fails.
(b) 1
(c)

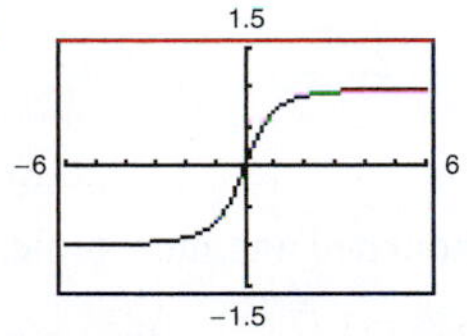

85.

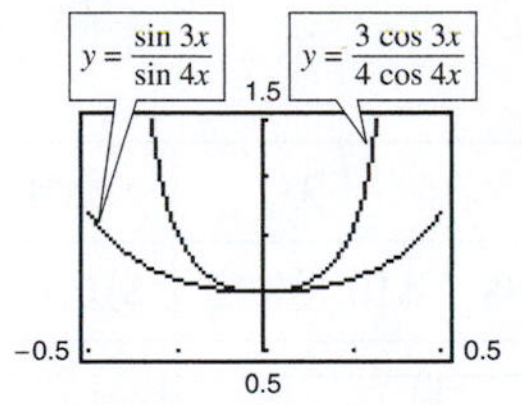

As $x \to 0$, the graphs get closer together (they both approach 0.75). By L'Hôpital's Rule,
$\lim_{x\to 0} \frac{\sin 3x}{\sin 4x} = \lim_{x\to 0} \frac{3 \cos 3x}{4 \cos 4x} = \frac{3}{4}.$

87. $v = 32t + v_0$ **89.** Proof **91.** $c = \frac{2}{3}$ **93.** $c = \pi/4$

95. False. L'Hôpital's Rule does not apply because $\lim_{x\to 0}(x^2 + x + 1) \neq 0$.

97. True **99.** $\frac{3}{4}$ **101.** $\frac{4}{3}$ **103.** $a = 1, b = \pm 2$

105. Proof **107.** (a) $0 \cdot \infty$ (b) 0 **109.** Proof

111. (a)–(c) 2

113. (a)

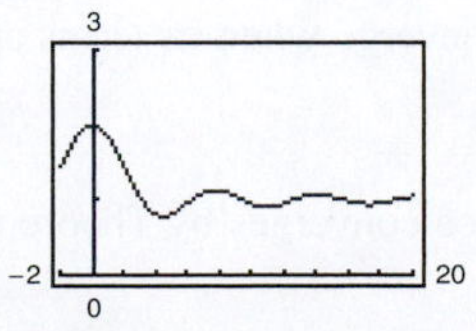

(b) $\lim_{x\to\infty} h(x) = 1$
(c) No

115. Putnam Problem A1, 1956

Section 8.8 *(page 575)*

1. Improper; $0 \leq \frac{3}{5} \leq 1$

3. Not improper; continuous on $[0, 1]$

5. Not improper; continuous on $[0, 2]$

7. Improper; infinite limits of integration

9. Infinite discontinuity at $x = 0$; 4

11. Infinite discontinuity at $x = 1$; diverges

13. Infinite discontinuity at $x = 0$; diverges

15. Infinite limit of integration; converges to 1 **17.** $\frac{1}{2}$

19. Diverges **21.** Diverges **23.** 2 **25.** $1/[2(\ln 4)^2]$

27. π **29.** $\pi/4$ **31.** Diverges **33.** Diverges

35. 0 **37.** $-\frac{1}{4}$ **39.** Diverges **41.** $\pi/3$ **43.** $\ln 3$

45. $\pi/6$ **47.** $2\pi\sqrt{6}/3$ **49.** $p > 1$ **51.** Proof

53. Diverges **55.** Converges **57.** Converges

59. Diverges **61.** Converges

63. An integral with infinite integration limits, an integral with an infinite discontinuity at or between the integration limits

65. The improper integral diverges. **67.** e **69.** π

71. (a) 1 (b) $\pi/2$ (c) 2π

73.

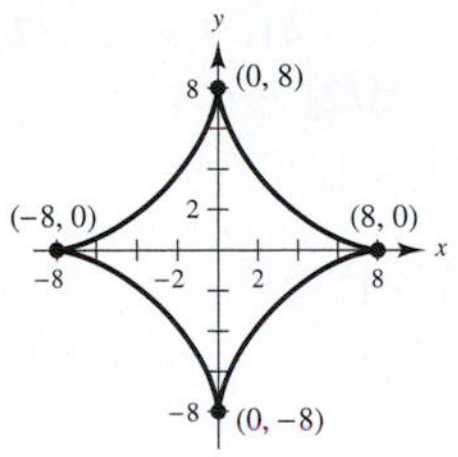

Perimeter = 48

75. $8\pi^2$ **77.** (a) $W = 20{,}000$ mile-tons (b) 4000 mi

79. (a) Proof (b) $P = 43.53\%$ (c) $E(x) = 7$

81. (a) \$757,992.41 (b) \$837,995.15 (c) \$1,066,666.67

83. $P = \left[2\pi NI\left(\sqrt{r^2 + c^2} - c\right)\right]/\left(kr\sqrt{r^2 + c^2}\right)$

85. False. Let $f(x) = 1/(x + 1)$. **87.** True

89. (a) and (b) Proofs
(c) The definition of the improper integral $\int_{-\infty}^{\infty}$ is not $\lim_{a\to\infty} \int_{-a}^{a}$ but rather that if you rewrite the integral that diverges, you can find that the integral converges.

91. (a) $\int_1^{\infty} \frac{1}{x^n}\,dx$ will converge if $n > 1$ and diverge if $n \leq 1$.
(b)

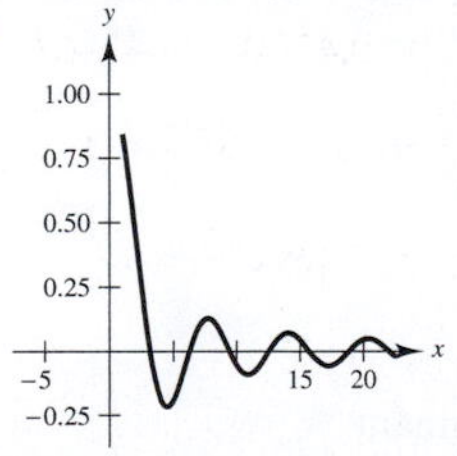

(c) Converges

93. (a)

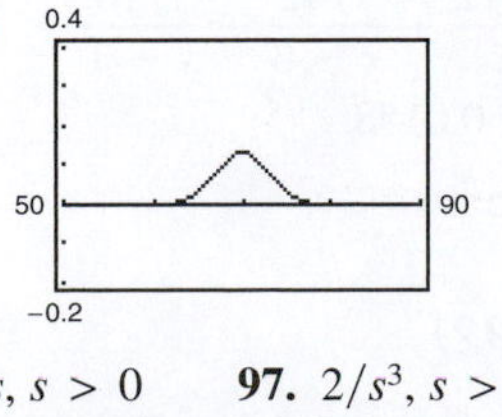

(b) About 0.2525
(c) 0.2525; same by symmetry

95. $1/s, s > 0$ **97.** $2/s^3, s > 0$ **99.** $s/(s^2 + a^2), s > 0$

101. $s/(s^2 - a^2), s > |a|$

103. (a) $\Gamma(1) = 1, \Gamma(2) = 1, \Gamma(3) = 2$ (b) Proof
(c) $\Gamma(n) = (n - 1)!$

105. $c = 1$; $\ln(2)$

107. $8\pi[(\ln 2)^2/3 - (\ln 4)/9 + 2/27] \approx 2.01545$

109. $\int_0^1 2\sin(u^2)\,du$; 0.6278 **111.** Proof

Review Exercises for Chapter 8 *(page 579)*

1. $\frac{1}{3}(x^2 - 36)^{3/2} + C$ **3.** $\frac{1}{2}\ln|x^2 - 49| + C$
5. $\ln(2) + \frac{1}{2} \approx 1.1931$ **7.** $100\arcsin(x/10) + C$
9. $\frac{1}{9}e^{3x}(3x - 1) + C$ **11.** $\frac{1}{13}e^{2x}(2\sin 3x - 3\cos 3x) + C$
13. $-\frac{1}{2}x^2\cos 2x + \frac{1}{2}x\sin 2x + \frac{1}{4}\cos 2x + C$
15. $\frac{1}{16}\left[(8x^2 - 1)\arcsin 2x + 2x\sqrt{1 - 4x^2}\right] + C$
17. $\sin(\pi x - 1)[\cos^2(\pi x - 1) + 2]/(3\pi) + C$
19. $\frac{2}{3}[\tan^3(x/2) + 3\tan(x/2)] + C$ **21.** $\tan\theta + \sec\theta + C$
23. $3\pi/16 + \frac{1}{2} \approx 1.0890$ **25.** $3\sqrt{4 - x^2}/x + C$
27. $\frac{1}{3}(x^2 + 4)^{1/2}(x^2 - 8) + C$ **29.** $256 - 62\sqrt{17} \approx 0.3675$
31. (a), (b), and (c) $\frac{1}{3}\sqrt{4 + x^2}(x^2 - 8) + C$
33. $6\ln|x + 3| - 5\ln|x - 4| + C$
35. $\frac{1}{4}[6\ln|x - 1| - \ln(x^2 + 1) + 6\arctan x] + C$
37. $x - \frac{64}{11}\ln|x + 8| + \frac{9}{11}\ln|x - 3| + C$
39. $\frac{1}{25}[4/(4 + 5x) + \ln|4 + 5x|] + C$ **41.** $1 - \sqrt{2}/2$
43. $\frac{1}{2}\ln|x^2 + 4x + 8| - \arctan[(x + 2)/2] + C$
45. $\ln|\tan\pi x|/\pi + C$ **47.** Proof
49. $\frac{1}{8}(\sin 2\theta - 2\theta\cos 2\theta) + C$
51. $\frac{4}{3}[x^{3/4} - 3x^{1/4} + 3\arctan(x^{1/4})] + C$
53. $2\sqrt{1 - \cos x} + C$ **55.** $\sin x\ln(\sin x) - \sin x + C$
57. $\frac{5}{2}\ln|(x - 5)/(x + 5)| + C$
59. $y = x\ln|x^2 + x| - 2x + \ln|x + 1| + C$ **61.** $\frac{1}{5}$
63. $\frac{1}{2}(\ln 4)^2 \approx 0.961$ **65.** π **67.** $\frac{128}{15}$
69. $(\bar{x}, \bar{y}) = (0, 4/(3\pi))$ **71.** 3.82 **73.** 0 **75.** ∞
77. 1 **79.** $1000e^{0.09} \approx 1094.17$ **81.** Converges; $\frac{32}{3}$
83. Diverges **85.** Converges; 1 **87.** Converges; $\pi/4$
89. (a) \$6,321,205.59 (b) \$10,000,000
91. (a) 0.4581 (b) 0.0135

P.S. Problem Solving *(page 581)*

1. (a) $\frac{4}{3}, \frac{16}{15}$ (b) Proof **3.** $\ln 3$ **5.** Proof
7. (a)

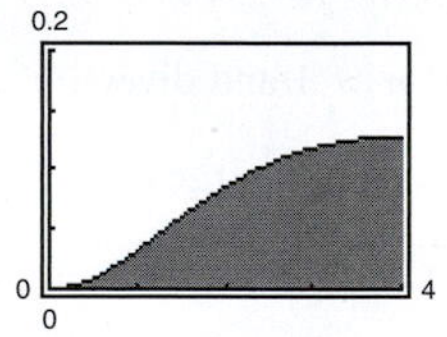

Area ≈ 0.2986

(b) $\ln 3 - \frac{4}{5}$ (c) $\ln 3 - \frac{4}{5}$

9. $\ln 3 - \frac{1}{2} \approx 0.5986$
11. (a) ∞ (b) 0 (c) $-\frac{2}{3}$
The form $0 \cdot \infty$ is indeterminant.
13. About 0.8670 **15.** $\dfrac{1/12}{x} + \dfrac{1/42}{x - 3} + \dfrac{1/10}{x - 1} + \dfrac{111/140}{x + 4}$
17–19. Proofs **21.** About 0.0158

Chapter 9

Section 9.1 *(page 592)*

1. 3, 9, 27, 81, 243 **3.** 1, 0, −1, 0, 1 **5.** $2, -1, \frac{2}{3}, -\frac{1}{2}, \frac{2}{5}$
7. 3, 4, 6, 10, 18 **9.** c **10.** a **11.** d **12.** b
13. 14, 17; add 3 to preceding term.
15. 80, 160; multiply preceding term by 2. **17.** $n + 1$
19. $1/[(2n + 1)(2n)]$ **21.** 5 **23.** 2

25.

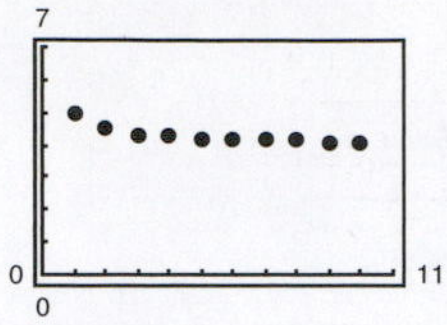

Converges to 4

27.

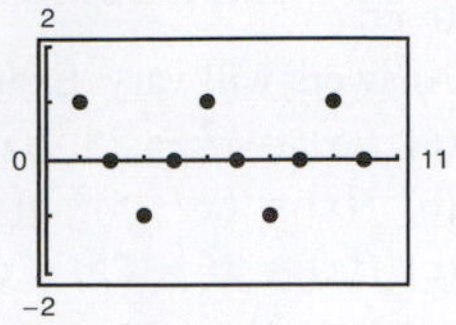

Diverges

29. Converges to 0 **31.** Diverges **33.** Converges to 5
35. Converges to 0 **37.** Diverges **39.** Converges to 0
41. Converges to 1 **43.** Converges to 0
45. Answers will vary. Sample answer: $6n - 4$
47. Answers will vary. Sample answer: $n^2 - 3$
49. Answers will vary. Sample answer: $(n + 1)/(n + 2)$
51. Answers will vary. Sample answer: $(n + 1)/n$
53. Monotonic, bounded **55.** Not monotonic, bounded
57. Monotonic, bounded **59.** Not monotonic, bounded
61. (a) $\left|7 + \frac{1}{n}\right| \geq 7 \Rightarrow$ bounded
$a_n > a_{n+1} \Rightarrow$ monotonic
So, $\{a_n\}$ converges.
(b)

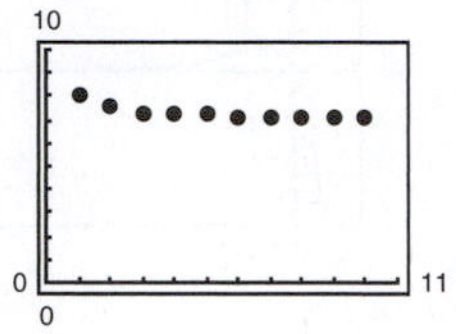

Limit = 7

63. (a) $\left|\frac{1}{3}\left(1 - \frac{1}{3^n}\right)\right| < \frac{1}{3} \Rightarrow$ bounded
$a_n < a_{n+1} \Rightarrow$ monotonic
So, $\{a_n\}$ converges.
(b)

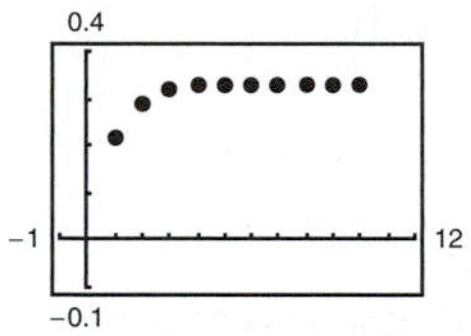

Limit $= \frac{1}{3}$

65. $\{a_n\}$ has a limit because it is bounded and monotonic; because $2 \leq a_n \leq 4$, $2 \leq L \leq 4$.
67. (a) No. $\lim_{n\to\infty} A_n$ does not exist.
(b)

n	1	2	3	4
A_n	\$10,045.83	\$10,091.88	\$10,138.13	\$10,184.60

n	5	6	7
A_n	\$10,231.28	\$10,278.17	\$10,325.28

n	8	9	10
A_n	\$10,372.60	\$10,420.14	\$10,467.90

69. No. A sequence is said to converge when its terms approach a real number.
71. (a) $10 - \frac{1}{n}$
(b) Impossible. The sequence converges by Theorem 9.5.
(c) $a_n = \dfrac{3n}{4n + 1}$
(d) Impossible. An unbounded sequence diverges.

73. (a) $\$4{,}500{,}000{,}000(0.8)^n$

(b)

Year	1	2
Budget	\$3,600,000,000	\$2,880,000,000

Year	3	4
Budget	\$2,304,000,000	\$1,843,200,000

(c) Converges to 0

75. 1, 1.4142, 1.4422, 1.4142, 1.3797, 1.3480; Converges to 1

77. Proof **79.** True **81.** True

83. (a) 1, 1, 2, 3, 5, 8, 13, 21, 34, 55, 89, 144
(b) 1, 2, 1.5, 1.6667, 1.6, 1.6250, 1.6154, 1.6190, 1.6176, 1.6182 (c) Proof
(d) $\rho = (1 + \sqrt{5})/2 \approx 1.6180$

85. (a) 1.4142, 1.8478, 1.9616, 1.9904, 1.9976
(b) $a_n = \sqrt{2 + a_{n-1}}$ (c) $\lim_{n\to\infty} a_n = 2$

87. (a) Proof
(b)
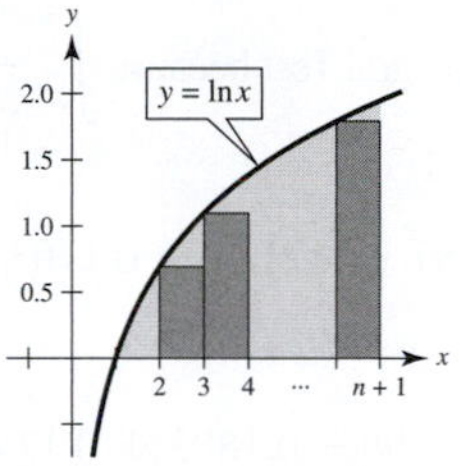

(c) Proof (d) Proof
(e) $\dfrac{\sqrt[20]{20!}}{20} \approx 0.4152$; $\dfrac{\sqrt[50]{50!}}{50} \approx 0.3897$; $\dfrac{\sqrt[100]{100!}}{100} \approx 0.3799$

89–91. Proofs **93.** Putnam Problem A1, 1990

Section 9.2 *(page 601)*

1. 1, 1.25, 1.361, 1.424, 1.464

3. 3, −1.5, 5.25, −4.875, 10.3125

5. 3, 4.5, 5.25, 5.625, 5.8125 **7.** Geometric series: $r = \frac{7}{6} > 1$

9. $\lim_{n\to\infty} a_n = 1 \neq 0$ **11.** $\lim_{n\to\infty} a_n = 1 \neq 0$

13. $\lim_{n\to\infty} a_n = \frac{1}{2} \neq 0$ **15.** Geometric series: $r = \frac{5}{6} < 1$

17. Geometric series: $r = 0.9 < 1$

19. Telescoping series: $a_n = 1/n - 1/(n + 1)$; Converges to 1.

21. (a) $\frac{11}{3}$

(b)

n	5	10	20	50	100
S_n	2.7976	3.1643	3.3936	3.5513	3.6078

(c)
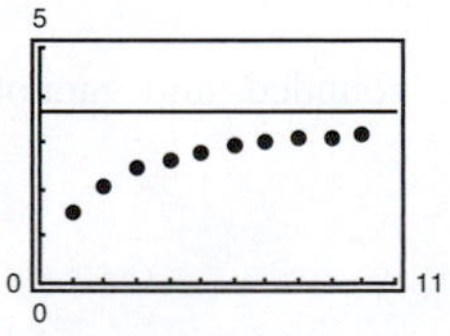

(d) The terms of the series decrease in magnitude relatively slowly, and the sequence of partial sums approaches the sum of the series relatively slowly.

23. (a) 20

(b)

n	5	10	20	50	100
S_n	8.1902	13.0264	17.5685	19.8969	19.9995

(c)
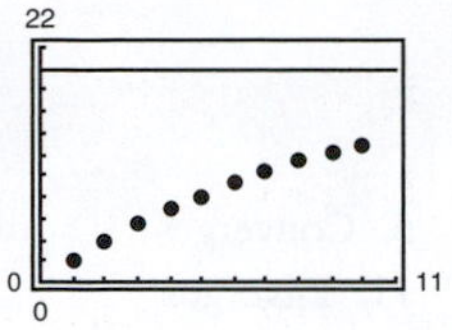

(d) The terms of the series decrease in magnitude relatively slowly, and the sequence of partial sums approaches the sum of the series relatively slowly.

25. 15 **27.** 3 **29.** 32 **31.** $\frac{1}{2}$ **33.** $\dfrac{\sin(1)}{1 - \sin(1)}$

35. (a) $\sum_{n=0}^{\infty} \frac{4}{10}(0.1)^n$ (b) $\frac{4}{9}$ **37.** (a) $\sum_{n=0}^{\infty} \frac{81}{100}(0.01)^n$ (b) $\frac{9}{11}$

39. (a) $\sum_{n=0}^{\infty} \frac{3}{40}(0.01)^n$ (b) $\frac{5}{66}$ **41.** Diverges **43.** Diverges

45. Converges **47.** Diverges **49.** Diverges

51. Diverges **53.** Diverges **55.** See definitions on page 595.

57. The series given by
$$\sum_{n=0}^{\infty} ar^n = a + ar + ar^2 + \cdots + ar^n + \cdots, a \neq 0$$
is a geometric series with ratio r. When $0 < |r| < 1$, the series converges to the sum $\sum_{n=0}^{\infty} ar^n = \dfrac{a}{1 - r}$.

59. The series in (a) and (b) are the same. The series in (c) is different unless $a_1 = a_2 = \cdots = a$ is constant.

61. $|x| < \frac{1}{3}$; $\dfrac{3x}{1 - 3x}$ **63.** $0 < x < 2$; $(x - 1)/(2 - x)$

65. $-1 < x < 1$; $1/(1 + x)$

67. (a) x (b) $f(x) = 1/(1 - x)$, $|x| < 1$
(c)
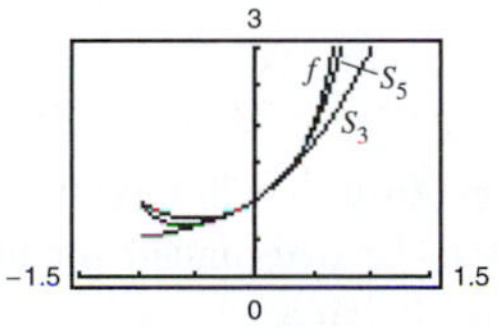

Answers will vary.

69. The required terms for the two series are $n = 100$ and $n = 5$, respectively. The second series converges at a higher rate.

71. $160{,}000(1 - 0.95^n)$ units

73. $\sum_{i=0}^{\infty} 200(0.75)^i$; Sum = \$800 million **75.** 152.42 feet

77. $\frac{1}{8}$; $\sum_{n=0}^{\infty} \frac{1}{2}\left(\frac{1}{2}\right)^n = \dfrac{1/2}{1 - 1/2} = 1$

79. (a) $-1 + \sum_{n=0}^{\infty}\left(\frac{1}{2}\right)^n = -1 + \dfrac{a}{1 - r} = -1 + \dfrac{1}{1 - 1/2} = 1$
(b) No (c) 2

81. (a) 126 in.^{2} (b) 128 in.^{2}

83. The \$2,000,000 sweepstakes has a present value of \$1,146,992.12. After accruing interest over the 20-year period, it attains its full value.

85. (a) \$5,368,709.11 (b) \$10,737,418.23 (c) \$21,474,836.47

87. (a) \$14,773.59 (b) \$14,779.65

89. (a) \$91,373.09 (b) \$91,503.32

91. False. $\lim_{n\to\infty} \frac{1}{n} = 0$, but $\sum_{n=1}^{\infty} \frac{1}{n}$ diverges.

93. False. $\sum_{n=1}^{\infty} ar^n = \left(\dfrac{a}{1 - r}\right) - a$; The formula requires that the geometric series begins with $n = 0$.

95. True **97.** Answers will vary. Example: $\sum_{n=0}^{\infty} 1$, $\sum_{n=0}^{\infty} (-1)$

99–101. Proofs **103.** 2

Section 9.3 *(page 609)*

1. Diverges **3.** Converges **5.** Converges
7. Converges **9.** Diverges **11.** Diverges
13. Converges **15.** Converges **17.** Converges
19. Diverges **21.** Converges **23.** Diverges
25. $f(x)$ is not positive for $x \geq 1$.
27. $f(x)$ is not always decreasing. **29.** Converges
31. Diverges **33.** Diverges **35.** Converges
37. Converges

39. (a)

n	5	10	20	50	100
S_n	3.7488	3.75	3.75	3.75	3.75

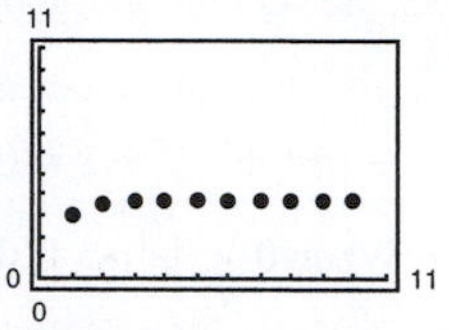

The partial sums approach the sum 3.75 very quickly.

(b)

n	5	10	20	50	100
S_n	1.4636	1.5498	1.5962	1.6251	1.635

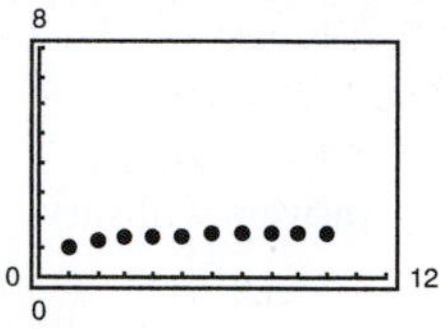

The partial sums approach the sum $\pi^2/6 \approx 1.6449$ more slowly than the series in part (a).

41. See Theorem 9.10 on page 605. Answers will vary. For example, convergence or divergence can be determined for the series $\sum_{n=1}^{\infty} \frac{1}{n^2+1}$.

43. No. Because $\sum_{n=1}^{\infty} \frac{1}{n}$ diverges, $\sum_{n=10{,}000}^{\infty} \frac{1}{n}$ also diverges. The convergence or divergence of a series is not determined by the first finite number of terms of the series.

45. (a)

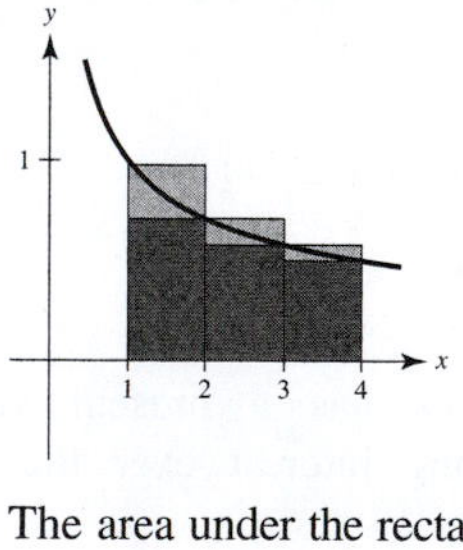

The area under the rectangles is greater than the area under the curve. Because $\int_1^{\infty} \frac{1}{\sqrt{x}}\,dx = \left[2\sqrt{x}\right]_1^{\infty} = \infty$ diverges, $\sum_{n=1}^{\infty} \frac{1}{\sqrt{n}}$ diverges.

(b)

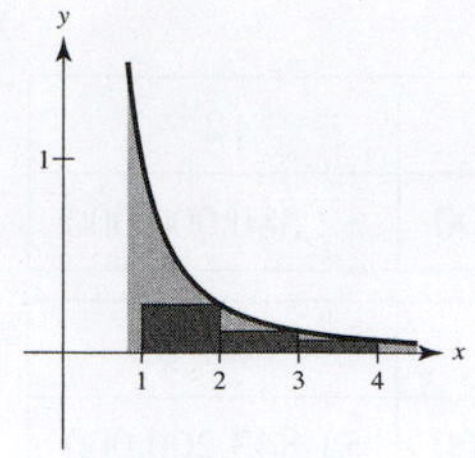

The area under the rectangles is less than the area under the curve. Because $\int_1^{\infty} \frac{1}{x^2}\,dx = \left[-\frac{1}{x}\right]_1^{\infty} = 1$ converges, $\sum_{n=2}^{\infty} \frac{1}{n^2}$ converges $\left(\text{and so does } \sum_{n=1}^{\infty} \frac{1}{n^2}\right)$.

47. $p > 1$ **49.** $p > 1$ **51.** $p > 3$ **53.** Proof

55. $S_5 = 1.4636$, $R_5 = 0.20$ **57.** $S_{10} \approx 0.9818$, $R_{10} \approx 0.0997$ **59.** $S_4 \approx 0.4049$, $R_4 \approx 5.6 \times 10^{-8}$

61. $N \geq 7$ **63.** $N \geq 16$

65. (a) $\sum_{n=2}^{\infty} \frac{1}{n^{1.1}}$ converges by the p-Series Test because $1.1 > 1$. $\sum_{n=2}^{\infty} \frac{1}{n \ln n}$ diverges by the Integral Test because $\int_2^{\infty} \frac{1}{x \ln x}\,dx$ diverges.

(b) $\sum_{n=2}^{\infty} \frac{1}{n^{1.1}} = 0.4665 + 0.2987 + 0.2176 + 0.1703 + 0.1393 + \cdots$

$\sum_{n=2}^{\infty} \frac{1}{n \ln n} = 0.7213 + 0.3034 + 0.1803 + 0.1243 + 0.0930 + \cdots$

(c) $n \geq 3.431 \times 10^{15}$

67. (a) Let $f(x) = 1/x$. f is positive, continuous, and decreasing on $[1, \infty)$.

$$S_n - 1 \leq \int_1^n \frac{1}{x}\,dx = \ln n$$

$$S_n \geq \int_1^{n+1} \frac{1}{x}\,dx = \ln(n+1)$$

So, $\ln(n+1) \leq S_n \leq 1 + \ln n$.

(b) $\ln(n+1) - \ln n \leq S_n - \ln n \leq 1$

Also, $\ln(n+1) - \ln n > 0$ for $n \geq 1$. So, $0 \leq S_n - \ln n \leq 1$, and the sequence $\{a_n\}$ is bounded.

(c) $a_n - a_{n+1} = [S_n - \ln n] - [S_{n+1} - \ln(n+1)]$

$$= \int_n^{n+1} \frac{1}{x}\,dx - \frac{1}{n+1} \geq 0$$

So, $a_n \geq a_{n+1}$.

(d) Because the sequence is bounded and monotonic, it converges to a limit, γ.

(e) 0.5822

69. (a) Diverges (b) Diverges

(c) $\sum_{n=2}^{\infty} x^{\ln n}$ converges for $x < 1/e$.

71. Diverges **73.** Converges **75.** Converges
77. Diverges **79.** Diverges **81.** Converges

Section 9.4 (page 616)

1. (a)

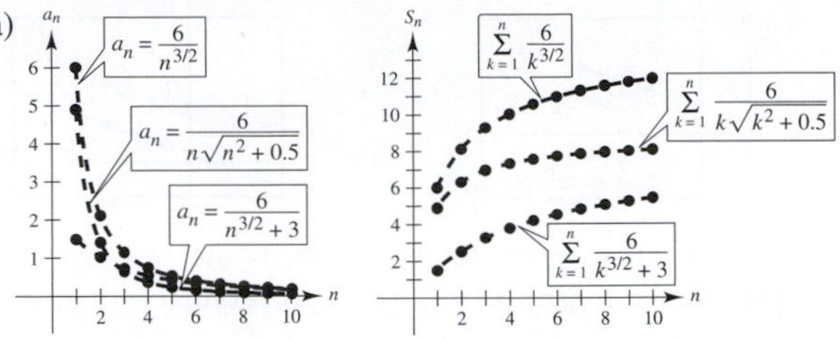

(b) $\sum_{n=1}^{\infty} \frac{6}{n^{3/2}}$; Converges

(c) The magnitudes of the terms are less than the magnitudes of the terms of the p-series. Therefore, the series converges.

(d) The smaller the magnitudes of the terms, the smaller the magnitudes of the terms of the sequence of partial sums.

3. Diverges **5.** Diverges **7.** Diverges **9.** Converges
11. Converges **13.** Diverges **15.** Diverges
17. Converges **19.** Converges **21.** Diverges
23. Diverges; p-Series Test
25. Converges; Direct Comparison Test with $\sum_{n=1}^{\infty} \left(\frac{1}{5}\right)^n$
27. Diverges; nth-Term Test **29.** Converges; Integral Test
31. $\lim_{n\to\infty} \frac{a_n}{1/n} = \lim_{n\to\infty} na_n$; $\lim_{n\to\infty} na_n \neq 0$, but is finite.
The series diverges by the Limit Comparison Test.
33. Diverges **35.** Converges
37. $\lim_{n\to\infty} n\left(\frac{n^3}{5n^4+3}\right) = \frac{1}{5} \neq 0$; So, $\sum_{n=1}^{\infty} \frac{n^3}{5n^4+3}$ diverges.
39. Diverges **41.** Converges
43. Convergence or divergence is dependent on the form of the general term for the series and not necessarily on the magnitudes of the terms.
45. See Theorem 9.13 on page 614. Answers will vary. For example, $\sum_{n=2}^{\infty} \frac{1}{\sqrt{n}-1}$ diverges because $\lim_{n\to\infty} \frac{1/\sqrt{n}-1}{1/\sqrt{n}} = 1$ and $\sum_{n=2}^{\infty} \frac{1}{\sqrt{n}}$ diverges (p-series).
47. (a) Proof
(b)

n	5	10	20	50	100
S_n	1.1839	1.2087	1.2212	1.2287	1.2312

(c) 0.1226 (d) 0.0277
49. False. Let $a_n = 1/n^3$ and $b_n = 1/n^2$. **51.** True
53. True **55.** Proof **57.** $\sum_{n=1}^{\infty} \frac{1}{n^2}, \sum_{n=1}^{\infty} \frac{1}{n^3}$ **59–65.** Proofs
67. Putnam Problem B4, 1988

Section 9.5 (page 625)

1. (a)

n	1	2	3	4	5
S_n	1.0000	0.6667	0.8667	0.7238	0.8349

n	6	7	8	9	10
S_n	0.7440	0.8209	0.7543	0.8131	0.7605

(b)

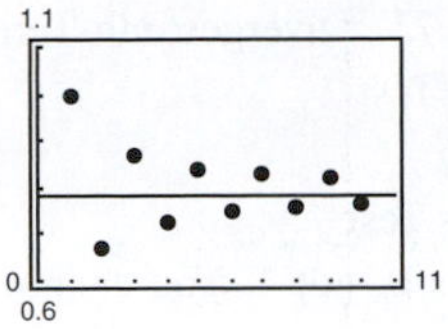

(c) The points alternate sides of the horizontal line $y = \pi/4$ that represents the sum of the series. The distances between the successive points and the line decrease.

(d) The distance in part (c) is always less than the magnitude of the next term of the series.

3. (a)

n	1	2	3	4	5
S_n	1.0000	0.7500	0.8611	0.7986	0.8386

n	6	7	8	9	10
S_n	0.8108	0.8312	0.8156	0.8280	0.8180

(b)

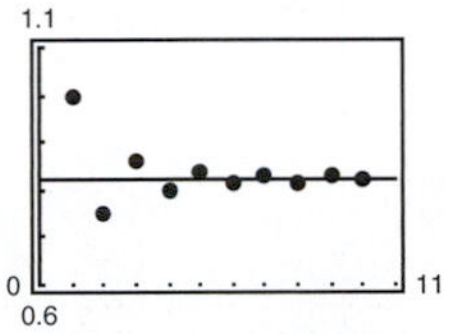

(c) The points alternate sides of the horizontal line $y = \pi^2/12$ that represents the sum of the series. The distances between the successive points and the line decrease.

(d) The distance in part (c) is always less than the magnitude of the next term of the series.

5. Converges **7.** Converges **9.** Diverges **11.** Diverges
13. Converges **15.** Diverges **17.** Diverges
19. Converges **21.** Converges **23.** Converges
25. Converges **27.** $1.8264 \le S \le 1.8403$
29. $1.7938 \le S \le 1.8054$ **31.** 10 **33.** 7
35. 7 terms (Note that the sum begins with $n = 0$.)
37. Converges absolutely **39.** Converges absolutely
41. Converges conditionally **43.** Diverges
45. Converges conditionally **47.** Converges absolutely
49. Converges absolutely **51.** Converges conditionally
53. Converges absolutely
55. An alternating series is a series whose terms alternate in sign.
57. $|S - S_N| = |R_N| \le a_{N+1}$
59. (a) False. For example, let $a_n = \frac{(-1)^n}{n}$.

Then $\sum a_n = \sum \frac{(-1)^n}{n}$ converges

and $\sum(-a_n) = \sum \frac{(-1)^{n+1}}{n}$ converges.

But, $\sum |a_n| = \sum \frac{1}{n}$ diverges.

(b) True. For if $\sum |a_n|$ converged, then so would $\sum a_n$ by Theorem 9.16.

61. True **63.** $p > 0$
65. Proof; The converse is false. For example: Let $a_n = 1/n$.
67. $\sum_{n=1}^{\infty} \frac{1}{n^2}$ converges, hence so does $\sum_{n=1}^{\infty} \frac{1}{n^4}$.
69. (a) No. $a_{n+1} \le a_n$ is not satisfied for all n. For example, $\frac{1}{9} < \frac{1}{8}$.
(b) Yes. 0.5

71. Converges; p-Series Test **73.** Diverges; nth-Term Test
75. Converges; Geometric Series Test
77. Converges; Integral Test
79. Converges; Alternating Series Test
81. The first term of the series is 0, not 1. You cannot regroup series terms arbitrarily.

Section 9.6 *(page 633)*

1–3. Proofs **5.** d **6.** c **7.** f **8.** b **9.** a
10. e
11. (a) Proof
(b)

n	5	10	15	20	25
S_n	13.7813	24.2363	25.8468	25.9897	25.9994

(c)

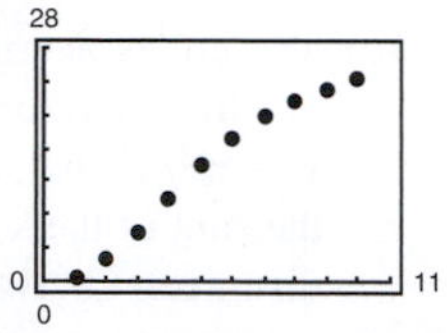

(d) 26

(e) The more rapidly the terms of the series approach 0, the more rapidly the sequence of partial sums approaches the sum of the series.

13. Converges **15.** Diverges **17.** Diverges
19. Converges **21.** Converges **23.** Converges
25. Diverges **27.** Converges **29.** Converges
31. Diverges **33.** Converges **35.** Converges
37. Converges **39.** Diverges **41.** Converges
43. Diverges **45.** Converges **47.** Converges
49. Converges **51.** Converges; Alternating Series Test
53. Converges; p-Series Test **55.** Diverges; nth-Term Test
57. Diverges; Geometric Series Test
59. Converges; Limit Comparison Test with $b_n = 1/2^n$
61. Converges; Direct Comparison Test with $b_n = 1/3^n$
63. Diverges; Ratio Test **65.** Converges; Ratio Test
67. Converges; Ratio Test **69.** a and c **71.** a and b

73. $\sum_{n=0}^{\infty} \frac{n+1}{7^{n+1}}$ **75.** (a) 9 (b) -0.7769

77. Diverges; $\lim_{n\to\infty} \left|\frac{a_{n+1}}{a_n}\right| > 1$

79. Converges; $\lim_{n\to\infty} \left|\frac{a_{n+1}}{a_n}\right| < 1$ **81.** Diverges; $\lim a_n \neq 0$

83. Converges **85.** Converges **87.** $(-3, 3)$
89. $(-2, 0]$ **91.** $x = 0$
93. See Theorem 9.17 on page 627.

95. No; the series $\sum_{n=1}^{\infty} \frac{1}{n + 10{,}000}$ diverges.

97. Absolutely; by Theorem 9.17 **99–105.** Proofs
107. (a) Diverges (b) Converges (c) Converges
(d) Converges for all integers $x \geq 2$
109. Putnam Problem 7, morning session, 1951

Section 9.7 *(page 658)*

1. d **2.** c **3.** a **4.** b

5. $P_1 = \frac{1}{16}x + \frac{1}{4}$

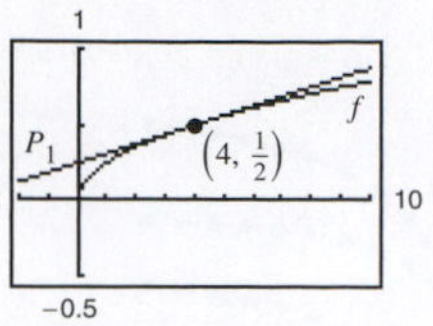

P_1 is the first-degree Taylor polynomial for f at 4.

7. $P_1 = \sqrt{2}x + \sqrt{2}(4 - \pi)/4$

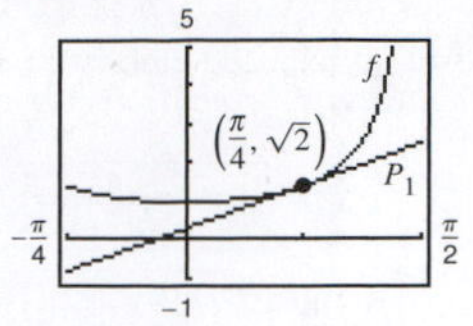

P_1 is the first-degree Taylor polynomial for f at $\pi/4$.

9.

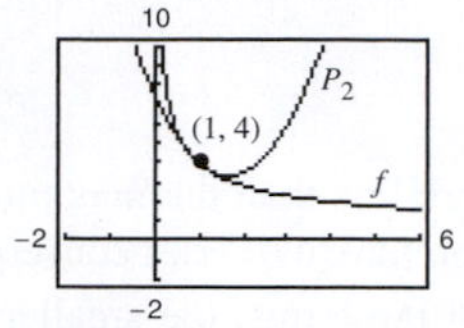

x	0	0.8	0.9	1	1.1
$f(x)$	Error	4.4721	4.2164	4.0000	3.8139
$P_2(x)$	7.5000	4.4600	4.2150	4.0000	3.8150

x	1.2	2
$f(x)$	3.6515	2.8284
$P_2(x)$	3.6600	3.5000

11. (a)

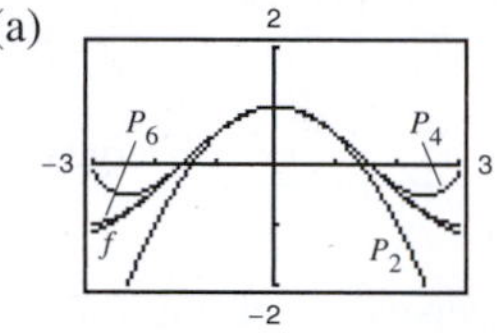

(b) $f^{(2)}(0) = -1$ $P_2^{(2)}(0) = -1$
$f^{(4)}(0) = 1$ $P_4^{(4)}(0) = 1$
$f^{(6)}(0) = -1$ $P_6^{(6)}(0) = -1$
(c) $f^{(n)}(0) = P_n^{(n)}(0)$

13. $1 + 4x + 8x^2 + \frac{32}{3}x^3 + \frac{32}{3}x^4$
15. $1 - \frac{1}{2}x + \frac{1}{8}x^2 - \frac{1}{48}x^3 + \frac{1}{384}x^4$ **17.** $x - \frac{1}{6}x^3 + \frac{1}{120}x^5$
19. $x + x^2 + \frac{1}{2}x^3 + \frac{1}{6}x^4$ **21.** $1 - x + x^2 - x^3 + x^4 - x^5$
23. $1 + \frac{1}{2}x^2$ **25.** $2 - 2(x - 1) + 2(x - 1)^2 - 2(x - 1)^3$
27. $2 + \frac{1}{4}(x - 4) - \frac{1}{64}(x - 4)^2 + \frac{1}{512}(x - 4)^3$
29. $\ln 2 + \frac{1}{2}(x - 2) - \frac{1}{8}(x - 2)^2 + \frac{1}{24}(x - 2)^3 - \frac{1}{64}(x - 2)^4$

31. (a) $P_3(x) = \pi x + \frac{\pi^3}{3}x^3$

(b) $Q_3(x) = 1 + 2\pi\left(x - \frac{1}{4}\right) + 2\pi^2\left(x - \frac{1}{4}\right)^2 + \frac{8\pi^3}{3}\left(x - \frac{1}{4}\right)^3$

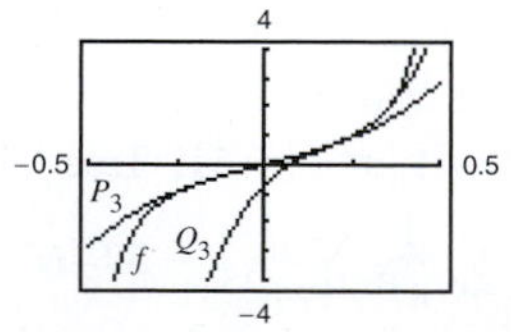

33. (a)

x	0	0.25	0.50	0.75	1.00
$\sin x$	0	0.2474	0.4794	0.6816	0.8415
$P_1(x)$	0	0.25	0.50	0.75	1.00
$P_3(x)$	0	0.2474	0.4792	0.6797	0.8333
$P_5(x)$	0	0.2474	0.4794	0.6817	0.8417

(b)

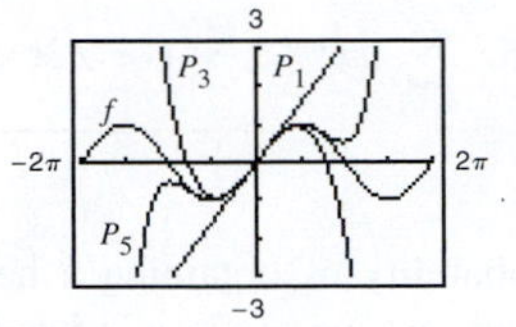

(c) As the distance increases, the polynomial approximation becomes less accurate.

35. (a) $P_3(x) = x + \frac{1}{6}x^3$

(b)

x	-0.75	-0.50	-0.25	0	0.25
$f(x)$	-0.848	-0.524	-0.253	0	0.253
$P_3(x)$	-0.820	-0.521	-0.253	0	0.253

x	0.50	0.75
$f(x)$	0.524	0.848
$P_3(x)$	0.521	0.820

(c)

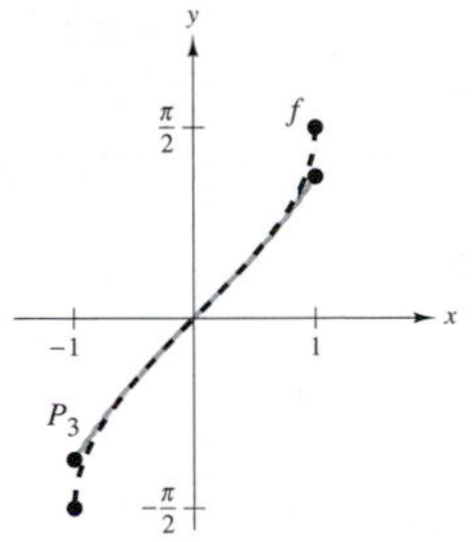

37.

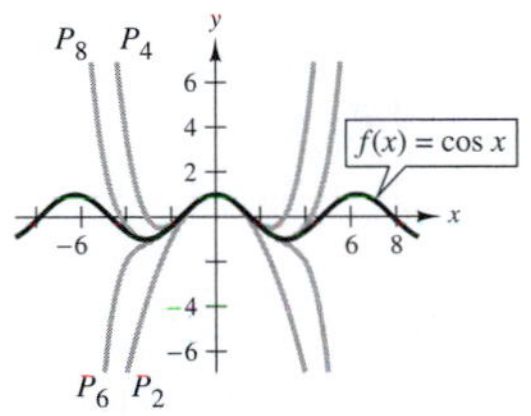

39.

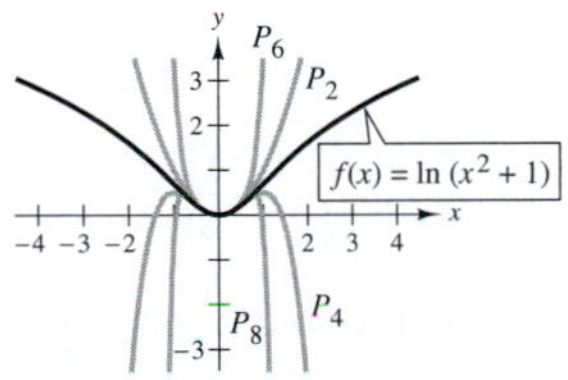

41. 2.7083 **43.** 0.7419 **45.** $R_4 \leq 2.03 \times 10^{-5}$; 0.000001
47. $R_3 \leq 7.82 \times 10^{-3}$; 0.00085 **49.** 3 **51.** 5
53. $n = 9$; $\ln(1.5) \approx 0.4055$ **55.** $-0.3936 < x < 0$
57. $-0.9467 < x < 0.9467$
59. The graphs of the approximating polynomial P and the elementary function f both pass through the point $(c, f(c))$, and the slope of the graph of P is the same as the slope of the graph of f at the point $(c, f(c))$. If P is of degree n, then the first n derivatives of f and P agree at c. This allows the graph of P to resemble the graph of f near the point $(c, f(c))$.
61. See "Definitions of nth Taylor Polynomial and nth Maclaurin Polynomial" on page 638.
63. As the degree of the polynomial increases, the graph of the Taylor polynomial becomes a better and better approximation of the function within the interval of convergence. Therefore, the accuracy is increased.
65. (a) $f(x) \approx P_4(x) = 1 + x + (1/2)x^2 + (1/6)x^3 + (1/24)x^4$
$g(x) \approx Q_5(x) = x + x^2 + (1/2)x^3 + (1/6)x^4 + (1/24)x^5$
$Q_5(x) = xP_4(x)$
(b) $g(x) \approx P_6(x) = x^2 - x^4/3! + x^6/5!$
(c) $g(x) \approx P_4(x) = 1 - x^2/3! + x^4/5!$
67. (a) $Q_2(x) = -1 + (\pi^2/32)(x + 2)^2$
(b) $R_2(x) = -1 + (\pi^2/32)(x - 6)^2$

(c) No. Horizontal translations of the result in part (a) are possible only at $x = -2 + 8n$ (where n is an integer) because the period of f is 8.
69. Proof
71. As you move away from $x = c$, the Taylor polynomial becomes less and less accurate.

Section 9.8 *(page 654)*

1. 0 **3.** 2 **5.** $R = 1$ **7.** $R = \frac{1}{4}$ **9.** $R = \infty$
11. $(-4, 4)$ **13.** $(-1, 1]$ **15.** $(-\infty, \infty)$ **17.** $x = 0$
19. $(-6, 6)$ **21.** $(-5, 13]$ **23.** $(0, 2]$ **25.** $(0, 6)$
27. $\left(-\frac{1}{2}, \frac{1}{2}\right)$ **29.** $(-\infty, \infty)$ **31.** $(-1, 1)$ **33.** $x = 3$
35. $R = c$ **37.** $(-k, k)$ **39.** $(-1, 1)$
41. $\displaystyle\sum_{n=1}^{\infty} \frac{x^{n-1}}{(n-1)!}$ **43.** $\displaystyle\sum_{n=1}^{\infty} \frac{x^{2n-1}}{(2n-1)!}$
45. (a) $(-3, 3)$ (b) $(-3, 3)$ (c) $(-3, 3)$ (d) $[-3, 3)$
47. (a) $(0, 2]$ (b) $(0, 2)$ (c) $(0, 2)$ (d) $[0, 2]$
49. A series of the form

$$\sum_{n=0}^{\infty} a_n(x - c)^n = a_0 + a_1(x - c) + a_2(x - c)^2 + \cdots + a_n(x - c)^n + \cdots$$

is called a power series centered at c, where c is a constant.
51. The interval of convergence of a power series is the set of all values of x for which the power series converges.
53. You differentiate and integrate the power series term by term. The radius of convergence remains the same. However, the interval of convergence might change.
55. Many answers possible.
(a) $\displaystyle\sum_{n=1}^{\infty} \left(\frac{x}{2}\right)^n$ Geometric: $\left|\frac{x}{2}\right| < 1 \Rightarrow |x| < 2$
(b) $\displaystyle\sum_{n=1}^{\infty} \frac{(-1)^n x^n}{n}$ converges for $-1 < x \leq 1$
(c) $\displaystyle\sum_{n=1}^{\infty} (2x + 1)^n$ Geometric:
$|2x + 1| < 1 \Rightarrow -1 < x < 0$
(d) $\displaystyle\sum_{n=1}^{\infty} \frac{(x - 2)^n}{n4^n}$ converges for $-2 \leq x < 6$
57. (a) For $f(x)$: $(-\infty, \infty)$; For $g(x)$: $(-\infty, \infty)$
(b) Proof (c) Proof (d) $f(x) = \sin x$; $g(x) = \cos x$
59–63. Proofs
65. (a) Proof (b) Proof
(c)

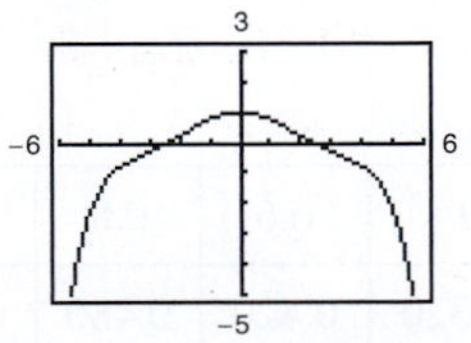

(d) 0.92

67. (a) $\frac{8}{3}$

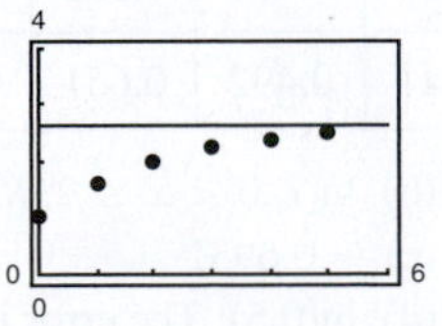

(b) $\frac{8}{13}$

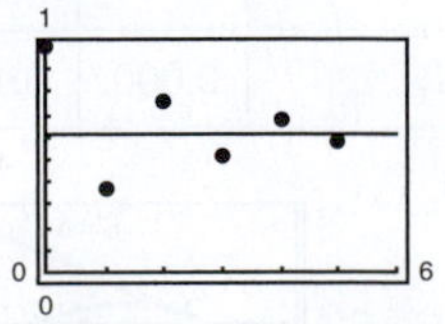

(c) The alternating series converges more rapidly. The partial sums of the series of positive terms approach the sum from below. The partial sums of the alternating series alternate sides of the horizontal line representing the sum.

(d)

M	10	100	1000	10,000
N	5	14	24	35

69.

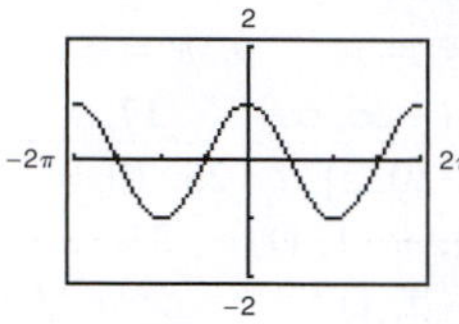

$f(x) = \cos x$

71.

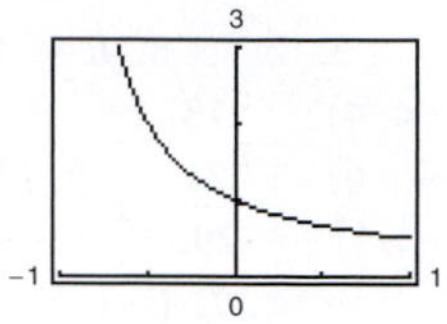

$f(x) = 1/(1 + x)$

73. False. Let $a_n = (-1)^n/(n2^n)$. **75.** True **77.** Proof

79. (a) $(-1, 1)$ (b) $f(x) = (c_0 + c_1x + c_2x^2)/(1 - x^3)$

81. Proof

Section 9.9 *(page 662)*

1. $\displaystyle\sum_{n=0}^{\infty} \frac{x^n}{4^{n+1}}$ **3.** $\displaystyle\sum_{n=0}^{\infty} \frac{4}{3}\left(\frac{-x}{3}\right)^n$

5. $\displaystyle\sum_{n=0}^{\infty} \frac{(x-1)^n}{2^{n+1}}$ $(-1, 3)$ **7.** $\displaystyle\sum_{n=0}^{\infty} (3x)^n$ $\left(-\frac{1}{3}, \frac{1}{3}\right)$ **9.** $\displaystyle -\frac{5}{9}\sum_{n=0}^{\infty}\left[\frac{2}{9}(x+3)\right]^n$ $\left(-\frac{15}{2}, \frac{3}{2}\right)$

11. $\displaystyle\sum_{n=0}^{\infty} \frac{(-1)^n 3^{n+1} x^n}{4^{n+1}}$ $\left(-\frac{4}{3}, \frac{4}{3}\right)$ **13.** $\displaystyle\sum_{n=0}^{\infty}\left[\frac{1}{(-3)^n} - 1\right]x^n$ $(-1, 1)$

15. $\displaystyle\sum_{n=0}^{\infty} x^n[1 + (-1)^n] = 2\sum_{n=0}^{\infty} x^{2n}$ $(-1, 1)$ **17.** $\displaystyle 2\sum_{n=0}^{\infty} x^{2n}$ $(-1, 1)$

19. $\displaystyle\sum_{n=1}^{\infty} n(-1)^n x^{n-1}$ $(-1, 1)$ **21.** $\displaystyle\sum_{n=0}^{\infty} \frac{(-1)^n x^{n+1}}{n+1}$ $(-1, 1]$

23. $\displaystyle\sum_{n=0}^{\infty} (-1)^n x^{2n}$ $(-1, 1)$ **25.** $\displaystyle\sum_{n=0}^{\infty} (-1)^n (2x)^{2n}$ $\left(-\frac{1}{2}, \frac{1}{2}\right)$

27.

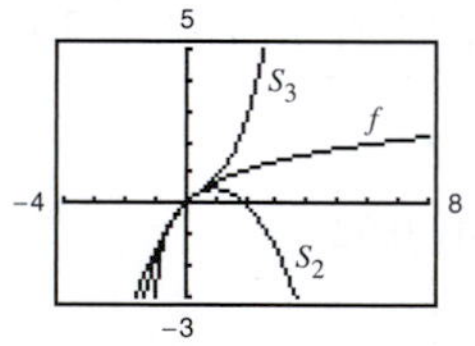

x	0.0	0.2	0.4	0.6	0.8	1.0
S_2	0.000	0.180	0.320	0.420	0.480	0.500
$\ln(x + 1)$	0.000	0.182	0.336	0.470	0.588	0.693
S_3	0.000	0.183	0.341	0.492	0.651	0.833

29. (a)

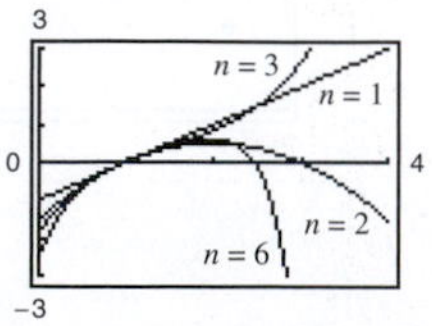

(b) $\ln x$, $0 < x \le 2$, $R = 1$

(c) -0.6931

(d) $\ln(0.5)$; The error is approximately 0.

31. 0.245 **33.** 0.125 **35.** $\displaystyle\sum_{n=1}^{\infty} nx^{n-1}$, $-1 < x < 1$

37. $\displaystyle\sum_{n=0}^{\infty} (2n + 1)x^n$, $-1 < x < 1$

39. $E(n) = 2$. Because the probability of obtaining a head on a single toss is $\frac{1}{2}$, it is expected that, on average, a head will be obtained in two tosses.

41. Because $\dfrac{1}{1+x} = \dfrac{1}{1-(-x)}$, substitute $(-x)$ into the geometric series.

43. Because $\dfrac{5}{1+x} = 5\left(\dfrac{1}{1-(-x)}\right)$, substitute $(-x)$ into the geometric series and then multiply the series by 5.

45. Proof **47.** (a) Proof (b) 3.14

49. $\ln\frac{3}{2} \approx 0.4055$; See Exercise 21.

51. $\ln\frac{7}{5} \approx 0.3365$; See Exercise 49.

53. $\arctan\frac{1}{2} \approx 0.4636$; See Exercise 52.

55. The series in Exercise 52 converges to its sum at a lower rate because its terms approach 0 at a much lower rate.

57. The series converges on the interval $(-5, 3)$ and perhaps also at one or both endpoints.

59. $\sqrt{3}\pi/6$ **61.** $S_1 = 0.3183098862$, $1/\pi \approx 0.3183098862$

Section 9.10 *(page 673)*

1. $\displaystyle\sum_{n=0}^{\infty} \frac{(2x)^n}{n!}$ **3.** $\displaystyle\frac{\sqrt{2}}{2}\sum_{n=0}^{\infty} \frac{(-1)^{n(n+1)/2}}{n!}\left(x - \frac{\pi}{4}\right)^n$

5. $\displaystyle\sum_{n=0}^{\infty} (-1)^n(x - 1)^n$ **7.** $\displaystyle\sum_{n=0}^{\infty} \frac{(-1)^n(x-1)^{n+1}}{n+1}$

9. $\displaystyle\sum_{n=0}^{\infty} \frac{(-1)^n(3x)^{2n+1}}{(2n+1)!}$ **11.** $1 + x^2/2! + 5x^4/4! + \cdots$

13–15. Proofs **17.** $\displaystyle\sum_{n=0}^{\infty} (-1)^n(n + 1)x^n$

19. $\displaystyle 1 + \sum_{n=1}^{\infty} \frac{1 \cdot 3 \cdot 5 \cdots (2n-1)x^n}{2^n n!}$

21. $\displaystyle\frac{1}{2}\left[1 + \sum_{n=1}^{\infty} \frac{(-1)^n 1 \cdot 3 \cdot 5 \cdots (2n-1)x^{2n}}{2^{3n} n!}\right]$

23. $\displaystyle 1 + \frac{x}{2} + \sum_{n=2}^{\infty} \frac{(-1)^{n+1} 1 \cdot 3 \cdot 5 \cdots (2n-3)x^n}{2^n n!}$

25. $\displaystyle 1 + \frac{x^2}{2} + \sum_{n=2}^{\infty} \frac{(-1)^{n+1} 1 \cdot 3 \cdot 5 \cdots (2n-3)x^{2n}}{2^n n!}$

27. $\displaystyle\sum_{n=0}^{\infty} \frac{x^{2n}}{2^n n!}$ **29.** $\displaystyle\sum_{n=1}^{\infty} \frac{(-1)^{n-1} x^n}{n}$ **31.** $\displaystyle\sum_{n=0}^{\infty} \frac{(-1)^n(3x)^{2n+1}}{(2n+1)!}$

33. $\displaystyle\sum_{n=0}^{\infty} \frac{(-1)^n 4^{2n} x^{2n}}{(2n)!}$ **35.** $\displaystyle\sum_{n=0}^{\infty} \frac{(-1)^n x^{3n}}{(2n)!}$

37. $\displaystyle\sum_{n=0}^{\infty} \frac{x^{2n+1}}{(2n+1)!}$ **39.** $\displaystyle\frac{1}{2}\left[1 + \sum_{n=0}^{\infty} \frac{(-1)^n(2x)^{2n}}{(2n)!}\right]$

41. $\displaystyle\sum_{n=0}^{\infty} \frac{(-1)^n x^{2n+2}}{(2n+1)!}$ **43.** $\begin{cases} \displaystyle\sum_{n=0}^{\infty} \frac{(-1)^n x^{2n}}{(2n+1)!}, & x \neq 0 \\ 1, & x = 0 \end{cases}$

45. Proof

47. $P_5(x) = x + x^2 + \frac{1}{3}x^3 - \frac{1}{30}x^5$

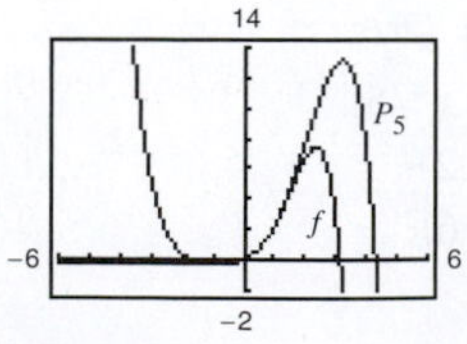

49. $P_5(x) = x - \frac{1}{2}x^2 - \frac{1}{6}x^3 + \frac{3}{40}x^5$

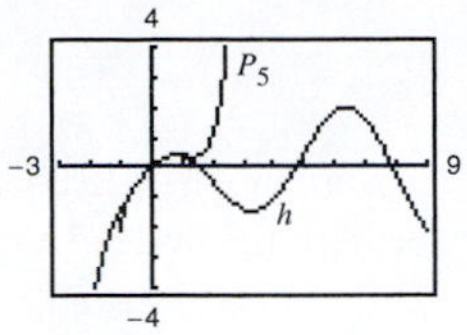

51. $P_4(x) = x - x^2 + \frac{5}{6}x^3 - \frac{5}{6}x^4$

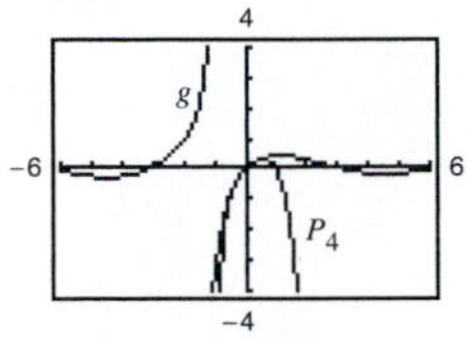

53. $\sum_{n=0}^{\infty} \frac{(-1)^{(n+1)}x^{2n+3}}{(2n+3)(n+1)!}$ **55.** 0.6931 **57.** 7.3891

59. 0 **61.** 1 **63.** 0.8075 **65.** 0.9461 **67.** 0.4872

69. 0.2010 **71.** 0.7040 **73.** 0.3412

75. $P_5(x) = x - 2x^3 + \frac{2}{3}x^5$

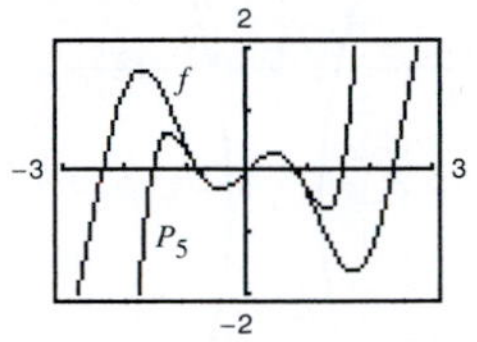

$\left[-\frac{3}{4}, \frac{3}{4}\right]$

77. $P_5(x) = (x-1) - \frac{1}{24}(x-1)^3 + \frac{1}{24}(x-1)^4 - \frac{71}{1920}(x-1)^5$

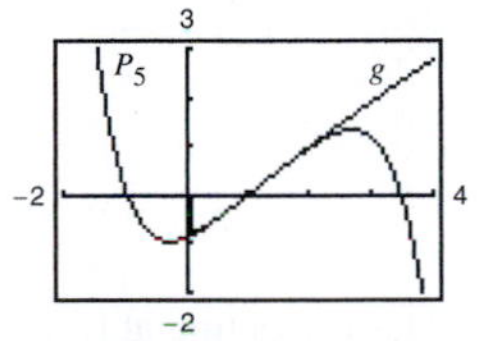

$\left[\frac{1}{4}, 2\right]$

79. See "Guidelines for Finding a Taylor Series" on page 668.

81. (a) Replace x with $(-x)$. (b) Replace x with $3x$.
(c) Multiply series by x.

83. Proof

85. (a)

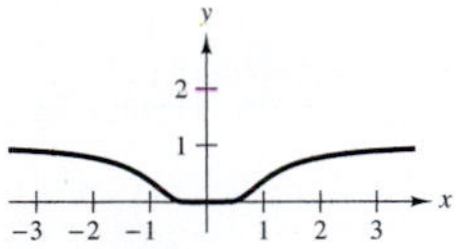

(b) Proof

(c) $\sum_{n=0}^{\infty} 0x^n = 0 \neq f(x)$

87. Proof **89.** 10 **91.** -0.0390625 **93.** $\sum_{n=0}^{\infty} \binom{k}{n} x^n$

95. Proof

Review Exercises for Chapter 9 *(page 676)*

1. 5, 25, 125, 625, 3125 **3.** $-\frac{1}{4}, \frac{1}{16}, -\frac{1}{64}, \frac{1}{256}, -\frac{1}{1024}$ **5.** a

6. c **7.** d **8.** b

9.

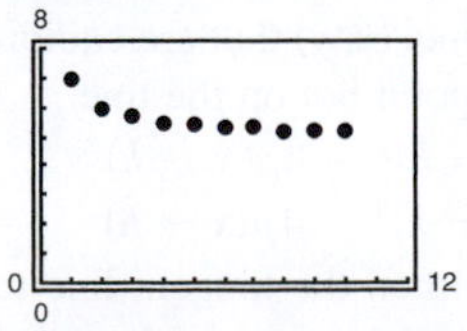

Converges to 5

11. Converges to 5 **13.** Diverges **15.** Converges to 0

17. Converges to 0 **19.** $a_n = 5n - 2$ **21.** $a_n = \frac{1}{(n! + 1)}$

23. (a)

n	1	2	3	4
A_n	\$8100.00	\$8201.25	\$8303.77	\$8407.56

n	5	6	7	8
A_n	\$8512.66	\$8619.07	\$8726.80	\$8835.89

(b) \$13,148.96

25. 3, 4.5, 5.5, 6.25, 6.85

27. (a)

n	5	10	15	20	25
S_n	13.2	113.3	873.8	6648.5	50,500.3

(b)

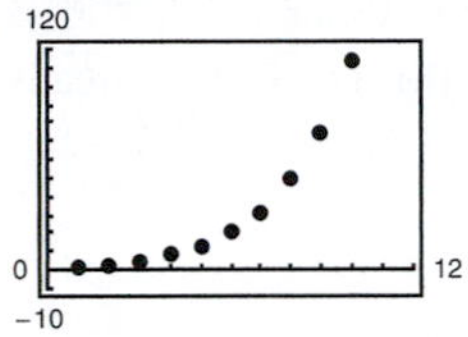

29. (a)

n	5	10	15	20	25
S_n	0.4597	0.4597	0.4597	0.4597	0.4597

(b)

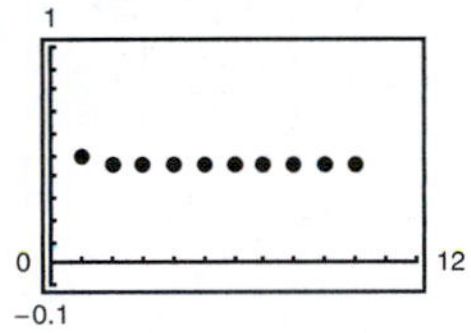

31. $\frac{5}{3}$ **33.** 5.5 **35.** (a) $\sum_{n=0}^{\infty} (0.09)(0.01)^n$ (b) $\frac{1}{11}$

37. Diverges **39.** Diverges **41.** $45\frac{1}{3}$ m **43.** Diverges

45. Converges **47.** Diverges **49.** Diverges

51. Converges **53.** Diverges **55.** Converges

57. Converges **59.** Diverges **61.** Diverges

63. Converges **65.** Diverges

67. (a) Proof

(b)

n	5	10	15	20	25
S_n	2.8752	3.6366	3.7377	3.7488	3.7499

(c)

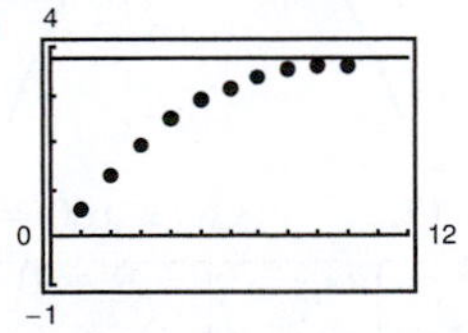

(d) 3.75

69. $P_3(x) = 1 - 2x + 2x^2 - \frac{4}{3}x^3$

71. $P_3(x) = 1 - 3x + \frac{9}{2}x^2 - \frac{9}{2}x^3$ **73.** 3 terms

75. $(-10, 10)$ **77.** $[1, 3]$ **79.** Converges only at $x = 2$

81. (a) $(-5, 5)$ (b) $(-5, 5)$ (c) $(-5, 5)$ (d) $[-5, 5)$

83. Proof **85.** $\sum_{n=0}^{\infty} \frac{2}{3}\left(\frac{x}{3}\right)^n$ **87.** $\sum_{n=0}^{\infty} 2\left(\frac{x-1}{3}\right)^n$; $(-2, 4)$

89. $\ln \frac{5}{4} \approx 0.2231$ **91.** $e^{1/2} \approx 1.6487$

93. $\cos \frac{2}{3} \approx 0.7859$ **95.** $\frac{\sqrt{2}}{2} \sum_{n=0}^{\infty} \frac{(-1)^{n(n+1)/2}}{n!}\left(x - \frac{3\pi}{4}\right)^n$

97. $\sum_{n=0}^{\infty} \frac{(x \ln 3)^n}{n!}$ **99.** $-\sum_{n=0}^{\infty} (x+1)^n$

101. $1 + x/5 - 2x^2/25 + 6x^3/125 - 21x^4/625 + \cdots$

103. (a)–(c) $1 + 2x + 2x^2 + \frac{4}{3}x^3$ **105.** $\sum_{n=0}^{\infty} \frac{(6x)^n}{n!}$

107. $\sum_{n=0}^{\infty} \frac{(-1)^n(2x)^{2n+1}}{(2n+1)!}$ **109.** 0

P.S. Problem Solving *(page 679)*

1. (a) 1 (b) Answers will vary. Example: $0, \frac{1}{3}, \frac{2}{3}$ (c) 0

3. Proof **5.** (a) Proof (b) Yes (c) Any distance

7. (a) $\sum_{n=0}^{\infty} \frac{x^{n+2}}{(n+2)n!}; \frac{1}{2}$ (b) $\sum_{n=0}^{\infty} \frac{(n+1)x^n}{n!}; 5.4366$

9. For $a = b$, the series converges conditionally. For no values of a and b does the series converge absolutely.

11. Proof **13.** (a) Proof (b) Proof

15. (a) The height is infinite. (b) The surface area is infinite. (c) Proof

Chapter 10

Section 10.1 *(page 692)*

1. a **2.** e **3.** c **4.** b **5.** f **6.** d

7. Vertex: $(0, 0)$
Focus: $(-2, 0)$
Directrix: $x = 2$

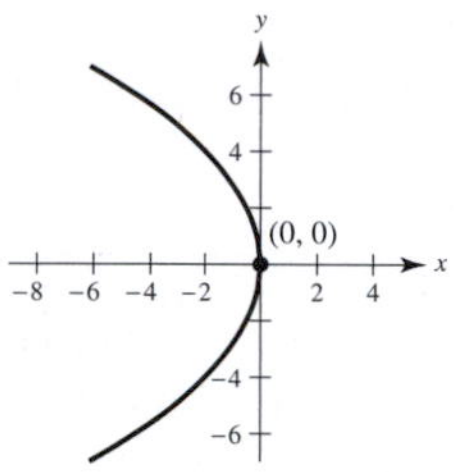

9. Vertex: $(-5, 3)$
Focus: $\left(-\frac{21}{4}, 3\right)$
Directrix: $x = -\frac{19}{4}$

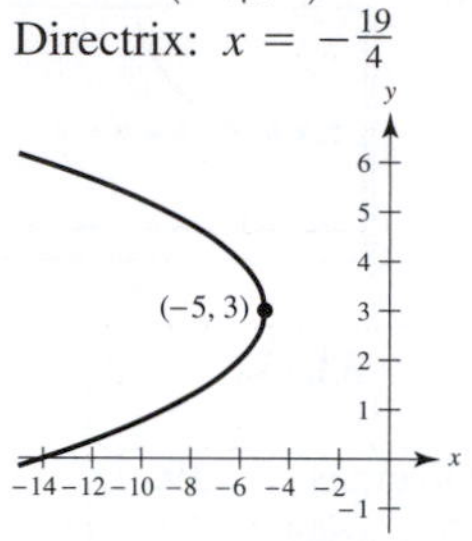

11. Vertex: $(-1, 2)$
Focus: $(0, 2)$
Directrix: $x = -2$

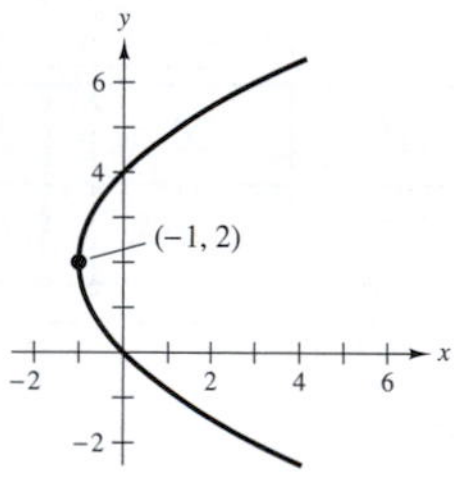

13. Vertex: $(-2, 2)$
Focus: $(-2, 1)$
Directrix: $y = 3$

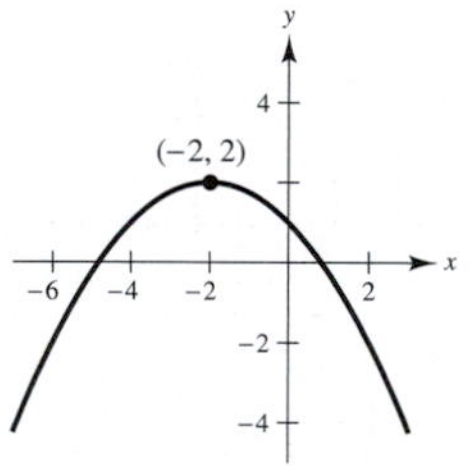

15. $y^2 - 8y + 8x - 24 = 0$ **17.** $x^2 - 32y + 160 = 0$

19. $x^2 + y - 4 = 0$ **21.** $5x^2 - 14x - 3y + 9 = 0$

23. Center: $(0, 0)$
Foci: $\left(0, \pm\sqrt{15}\right)$
Vertices: $(0, \pm 4)$
$e = \sqrt{15}/4$

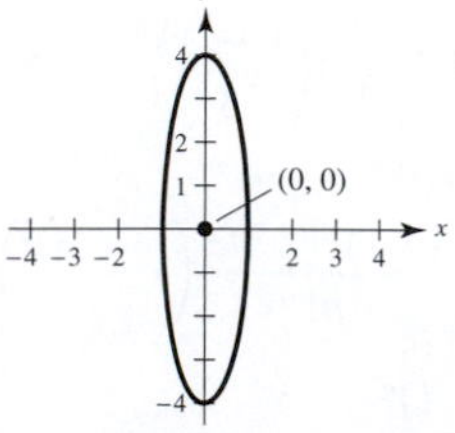

25. Center: $(3, 1)$
Foci: $(3, 4), (3, -2)$
Vertices: $(3, 6), (3, -4)$
$e = \frac{3}{5}$

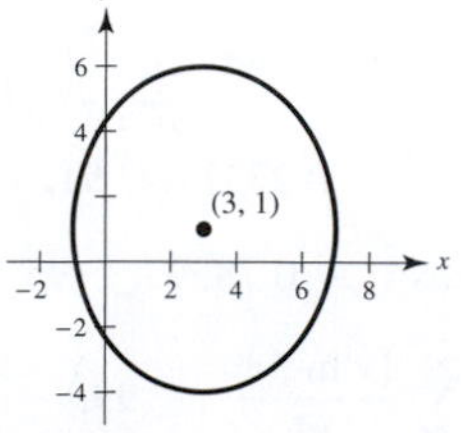

27. Center: $(-2, 3)$
Foci: $\left(-2, 3 \pm \sqrt{5}\right)$
Vertices: $(-2, 6), (-2, 0)$
$e = \sqrt{5}/3$

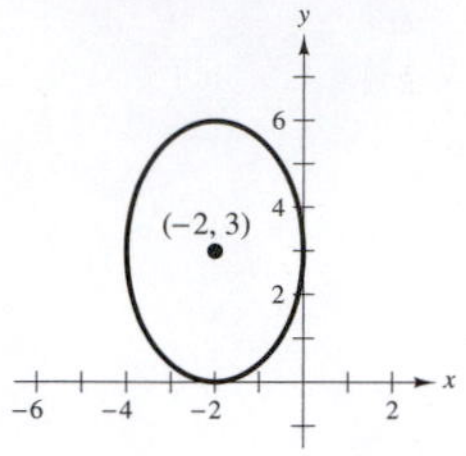

29. $x^2/36 + y^2/11 = 1$

31. $(x - 3)^2/9 + (y - 5)^2/16 = 1$

33. $x^2/16 + 7y^2/16 = 1$

35. Center: $(0, 0)$
Vertices: $(\pm 5, 0)$
Foci: $(\pm\sqrt{41}, 0)$
Asymptotes: $y = \pm\frac{b}{a}x = \pm\frac{4}{5}x$

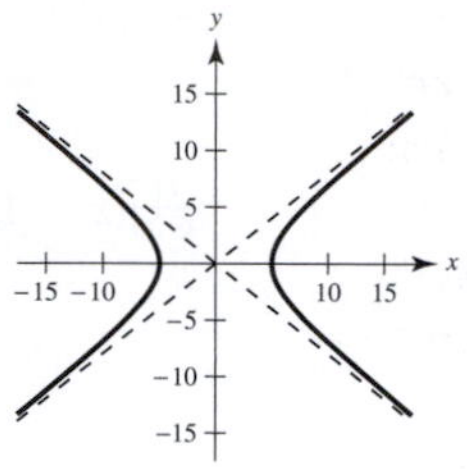

37. Center: $(2, -3)$
Foci: $\left(2 \pm \sqrt{10}, -3\right)$
Vertices: $(1, -3), (3, -3)$

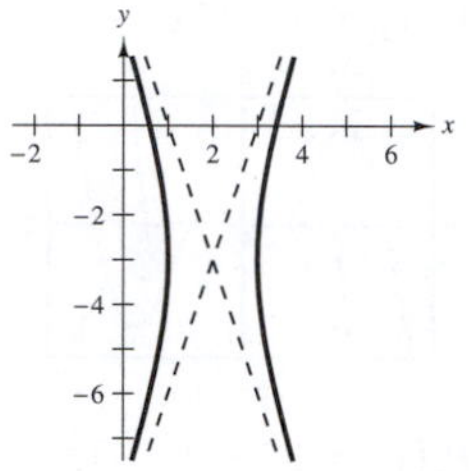

39. Degenerate hyperbola
Graph is two lines: $y = -3 \pm \frac{1}{3}(x + 1)$, intersecting at $(-1, -3)$.

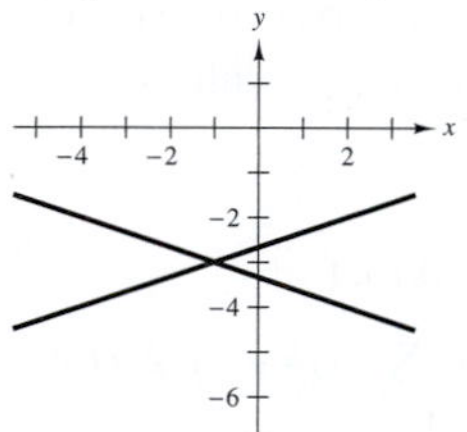

41. $x^2/1 - y^2/25 = 1$ **43.** $y^2/9 - (x - 2)^2/(9/4) = 1$

45. $y^2/4 - x^2/12 = 1$ **47.** $(x - 3)^2/9 - (y - 2)^2/4 = 1$

49. (a) $\left(6, \sqrt{3}\right)$: $2x - 3\sqrt{3}y - 3 = 0$
$\left(6, -\sqrt{3}\right)$: $2x + 3\sqrt{3}y - 3 = 0$
(b) $\left(6, \sqrt{3}\right)$: $9x + 2\sqrt{3}y - 60 = 0$
$\left(6, -\sqrt{3}\right)$: $9x - 2\sqrt{3}y - 60 = 0$

51. Ellipse **53.** Parabola **55.** Circle **57.** Hyperbola

59. (a) A parabola is the set of all points (x, y) that are equidistant from a fixed line and a fixed point not on the line.
(b) For directrix $y = k - p$: $(x - h)^2 = 4p(y - k)$
For directrix $x = h - p$: $(y - k)^2 = 4p(x - h)$
(c) If P is a point on a parabola, then the tangent line to the parabola at P makes equal angles with the line passing through P and the focus, and with the line passing through P parallel to the axis of the parabola.

61. (a) A hyperbola is the set of all points (x, y) for which the absolute value of the difference between the distances from two distinct fixed points is constant.

(b) Transverse axis is horizontal: $\dfrac{(x-h)^2}{a^2} - \dfrac{(y-k)^2}{b^2} = 1$

Transverse axis is vertical: $\dfrac{(y-k)^2}{a^2} - \dfrac{(x-h)^2}{b^2} = 1$

(c) Transverse axis is horizontal:
$y = k + (b/a)(x-h)$ and $y = k - (b/a)(x-h)$
Transverse axis is vertical:
$y = k + (a/b)(x-h)$ and $y = k - (a/b)(x-h)$

63. (a) Ellipse (b) Hyperbola (c) Circle
(d) Sample answer: Eliminate the y^2-term.

65. $\frac{9}{4}$ m **67.** (a) Proof (b) Point of intersection: $(3, -3)$

69.

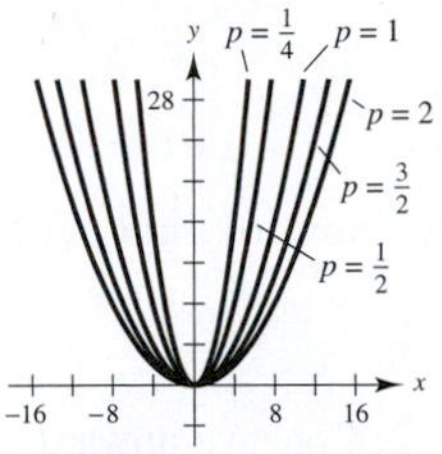

As p increases, the graph of $x^2 = 4py$ gets wider.

71. $[16(4 + 3\sqrt{3} - 2\pi]/3 \approx 15.536$ ft^2

73. Minimum distance: 147,099,713.4 km
Maximum distance: 152,096,286.6 km

75. About 0.9372 **77.** $e \approx 0.9671$

79. (a) Area $= 2\pi$ (b) Volume $= 8\pi/3$
Surface area $= [2\pi(9 + 4\sqrt{3}\pi)]/9 \approx 21.48$
(c) Volume $= 16\pi/3$
Surface area $= \dfrac{4\pi\left[6 + \sqrt{3}\ln\left(2 + \sqrt{3}\right)\right]}{3} \approx 34.69$

81. 37.96 **83.** 40 **85.** $(x-6)^2/9 - (y-2)^2/7 = 1$

87. $x \approx 110.3$ mi **89.** Proof

91. False. See the definition of a parabola **93.** True **95.** True

97. Putnam Problem B4, 1976

Section 10.2 *(page 703)*

1.

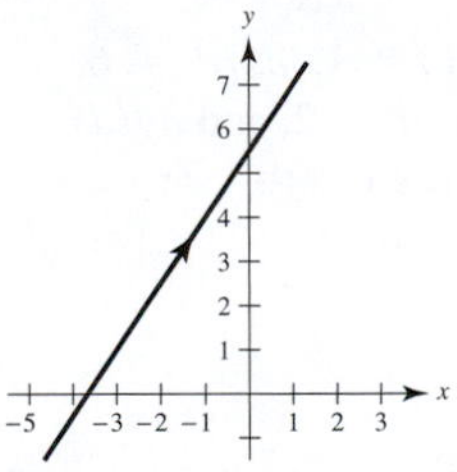

$3x - 2y + 11 = 0$

3.

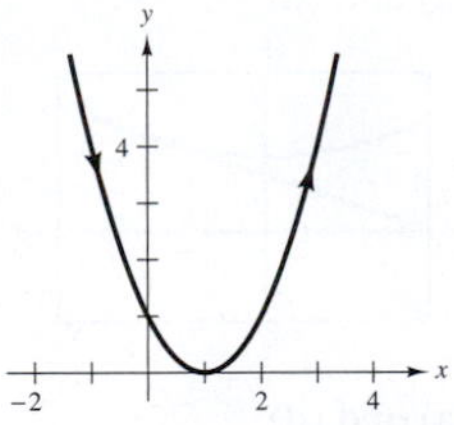

$y = (x-1)^2$

5.

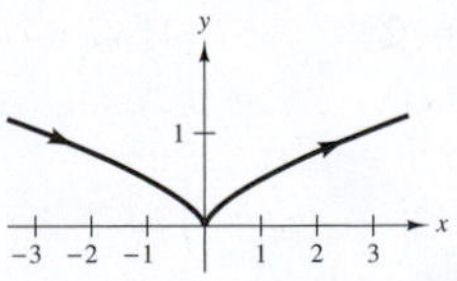

$y = \frac{1}{2}x^{2/3}$

7.

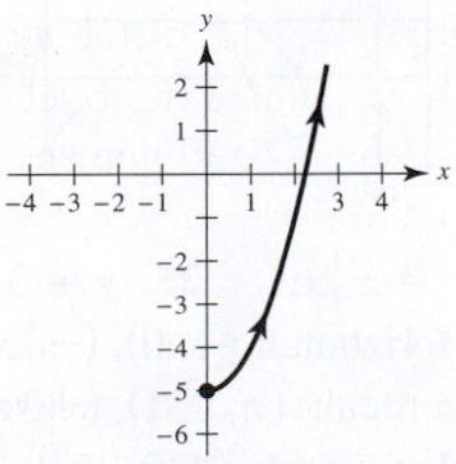

$y = x^2 - 5, \quad x \geq 0$

9.

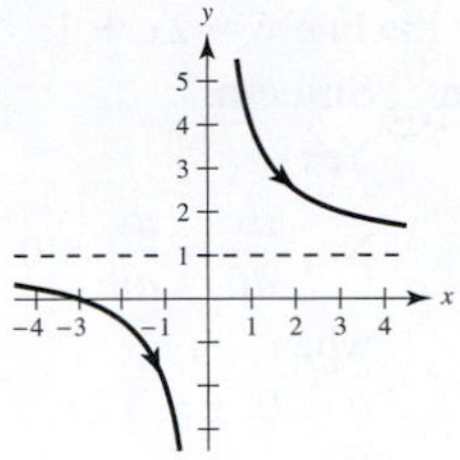

$y = (x+3)/x$

11.

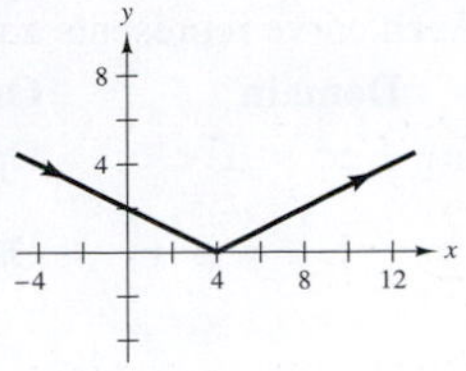

$y = |x - 4|/2$

13.

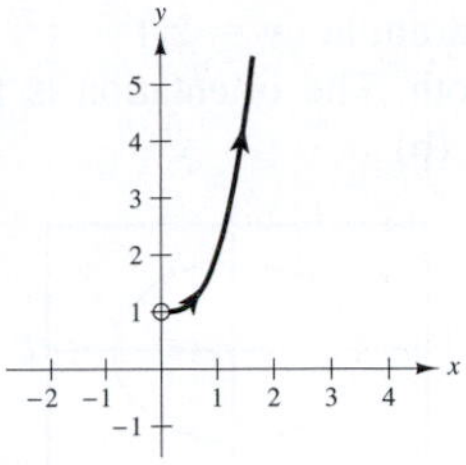

$y = x^3 + 1, \quad x > 0$

15.

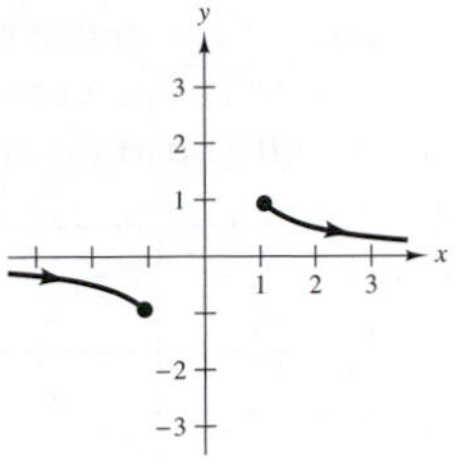

$y = 1/x, \quad |x| \geq 1$

17.

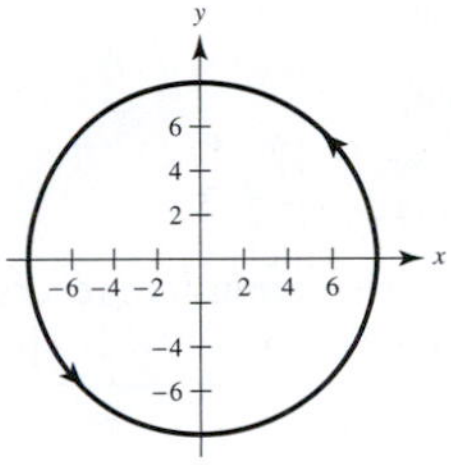

$x^2 + y^2 = 64$

19.

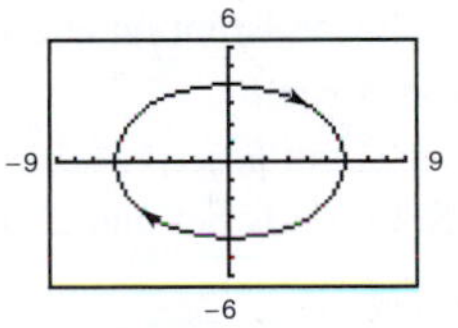

$\dfrac{x^2}{36} + \dfrac{y^2}{16} = 1$

21.

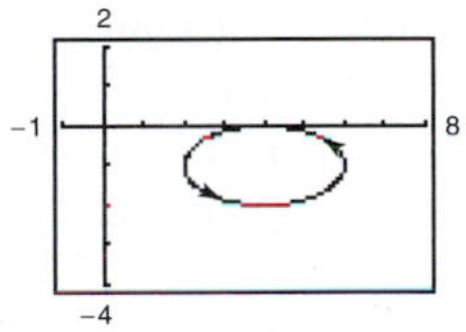

$\dfrac{(x-4)^2}{4} + \dfrac{(y+1)^2}{1} = 1$

23.

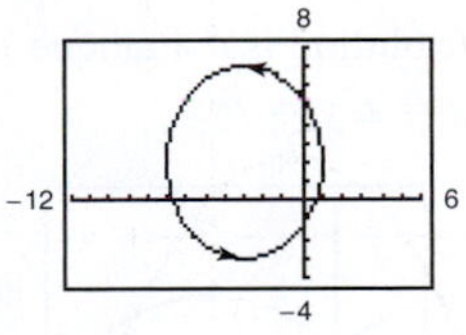

$\dfrac{(x+3)^2}{16} + \dfrac{(y-2)^2}{25} = 1$

25.

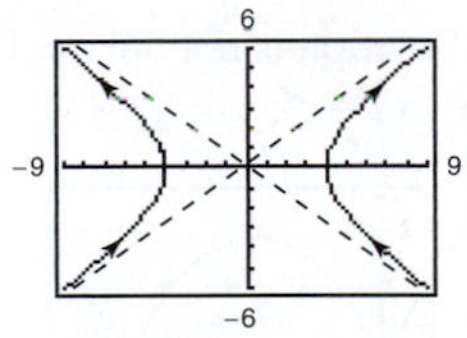

$\dfrac{x^2}{16} - \dfrac{y^2}{9} = 1$

27.

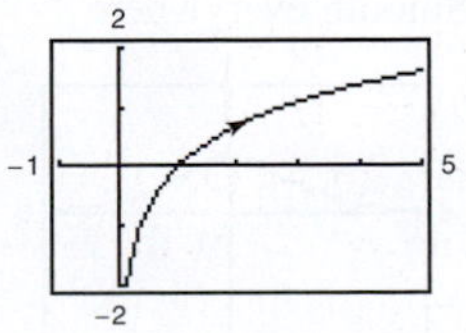

$y = \ln x$

29.

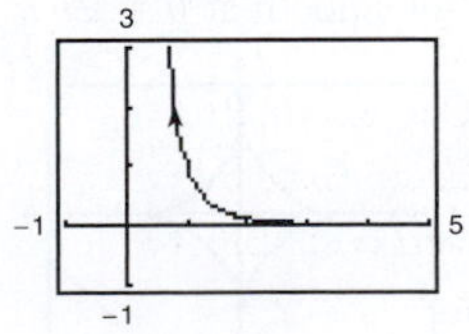

$y = \dfrac{1}{x^3}, \quad x > 0$

31. Each curve represents a portion of the line $y = 2x + 1$.

	Domain	**Orientation**	**Smooth**
(a)	$-\infty < x < \infty$	Up	Yes
(b)	$-1 \le x \le 1$	Oscillates	No, $\dfrac{dx}{d\theta} = \dfrac{dy}{d\theta} = 0$ when $\theta = 0, \pm\pi, \pm 2\pi, \ldots$
(c)	$0 < x < \infty$	Down	Yes
(d)	$0 < x < \infty$	Up	Yes

33. (a) and (b) represent the parabola $y = 2(1 - x^2)$ for $-1 \le x \le 1$. The curve is smooth. The orientation is from right to left in part (a) and in part (b).

35. (a)

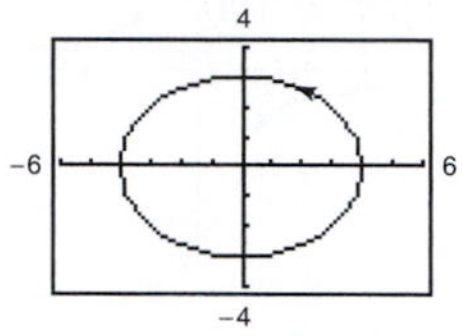

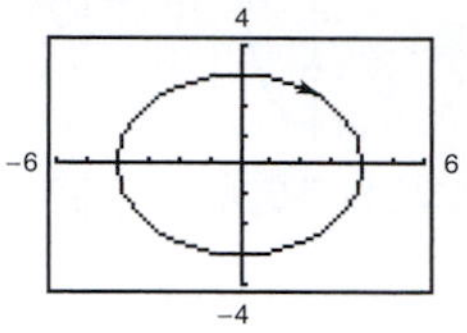

(b) The orientation is reversed.

(c) The orientation is reversed.

(d) Answers will vary. For example,

$x = 2 \sec t$ $x = 2 \sec(-t)$

$y = 5 \sin t$ $y = 5 \sin(-t)$

have the same graphs, but their orientations are reversed.

37. $y - y_1 = \dfrac{y_2 - y_1}{x_2 - x_1}(x - x_1)$ **39.** $\dfrac{(x - h)^2}{a^2} + \dfrac{(y - k)^2}{b^2} = 1$

41. $x = 4t$
$y = -7t$
(Solution is not unique.)

43. $x = 3 + 2 \cos \theta$
$y = 1 + 2 \sin \theta$
(Solution is not unique.)

45. $x = 10 \cos \theta$
$y = 6 \sin \theta$
(Solution is not unique.)

47. $x = 4 \sec \theta$
$y = 3 \tan \theta$
(Solution is not unique.)

49. $x = t$
$y = 6t - 5$;
$x = t + 1$
$y = 6t + 1$
(Solution is not unique.)

51. $x = t$
$y = t^3$;
$x = \tan t$
$y = \tan^3 t$
(Solution is not unique.)

53. $x = t + 3$, $y = 2t + 1$ **55.** $x = t$, $y = t^2$

57.

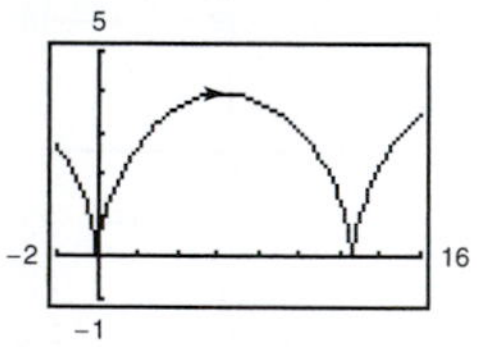

Not smooth at $\theta = 2n\pi$

59.

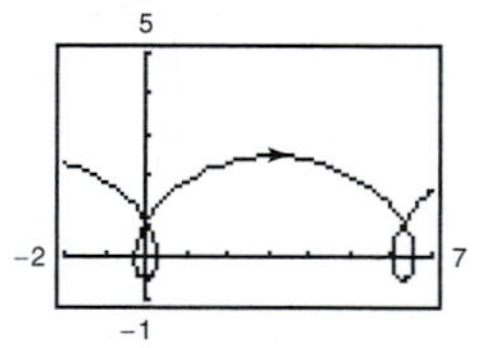

Smooth everywhere

61.

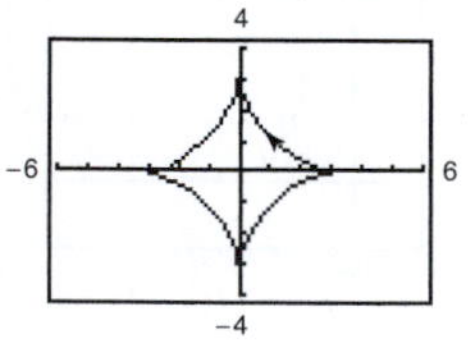

Not smooth at $\theta = \frac{1}{2}n\pi$

63.

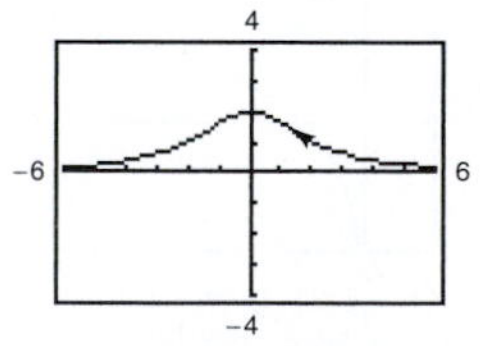

Smooth everywhere

65. A plane curve C is a set of parametric equations, $x = f(t)$ and $y = g(t)$, and the graph of the parametric equations.

67. A curve C represented by $x = f(t)$ and $y = g(t)$ on an interval I is called smooth when f' and g' are continuous on I and not simultaneously 0, except possibly at the endpoints of I.

69. d; (4, 0) is on the graph.   **71.** b; (1, 0) is on the graph.

73. $x = a\theta - b \sin \theta$; $y = a - b \cos \theta$

75. False. The graph of the parametric equations is the portion of the line $y = x$ when $x \ge 0$.

77. True

79. (a) $x = \left(\frac{440}{3} \cos \theta\right)t$; $y = 3 + \left(\frac{440}{3} \sin \theta\right)t - 16t^2$

(b)

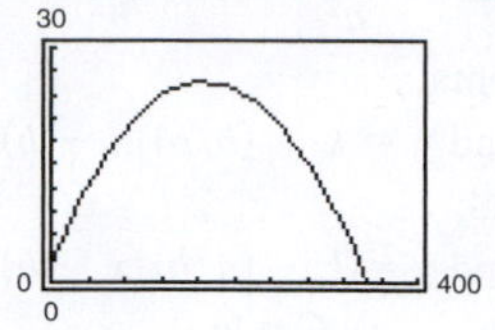

Not a home run

(c)

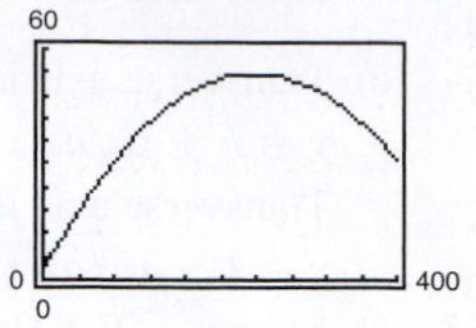

Home run

(d) 19.4°

Section 10.3 *(page 711)*

1. $-3/t$ **3.** -1

5. $\dfrac{dy}{dx} = \dfrac{3}{4}, \dfrac{d^2y}{dx^2} = 0$; Neither concave upward nor concave downward

7. $dy/dx = 2t + 3$, $d^2y/dx^2 = 2$
At $t = -1$, $dy/dx = 1$, $d^2y/dx^2 = 2$; Concave upward

9. $dy/dx = -\cot \theta$, $d^2y/dx^2 = -(\csc \theta)^3/4$
At $\theta = \pi/4$, $dy/dx = -1$, $d^2y/dx^2 = -\sqrt{2}/2$;
Concave downward

11. $dy/dx = 2 \csc \theta$, $d^2y/dx^2 = -2 \cot^3 \theta$
At $\theta = \pi/6$, $dy/dx = 4$, $d^2y/dx^2 = -6\sqrt{3}$;
Concave downward

13. $dy/dx = -\tan \theta$, $d^2y/dx^2 = \sec^4 \theta \csc \theta/3$
At $\theta = \pi/4$, $dy/dx = -1$, $d^2y/dx^2 = 4\sqrt{2}/3$;
Concave upward

15. $\left(-2/\sqrt{3}, 3/2\right)$: $3\sqrt{3}x - 8y + 18 = 0$
$(0, 2)$: $y - 2 = 0$
$\left(2\sqrt{3}, 1/2\right)$: $\sqrt{3}x + 8y - 10 = 0$

17. $(0, 0)$: $2y - x = 0$
$(-3, -1)$: $y + 1 = 0$
$(-3, 3)$: $2x - y + 9 = 0$

19. (a) and (d)

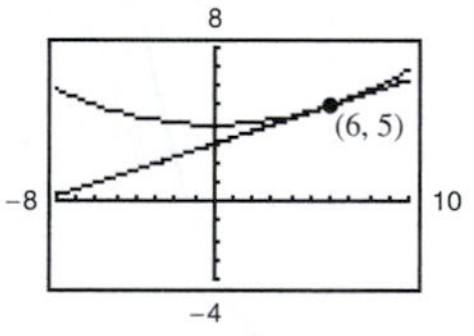

(b) At $t = 1$, $dx/dt = 6$, $dy/dt = 2$, and $dy/dx = 1/3$.

(c) $y = \frac{1}{3}x + 3$

21. (a) and (d)

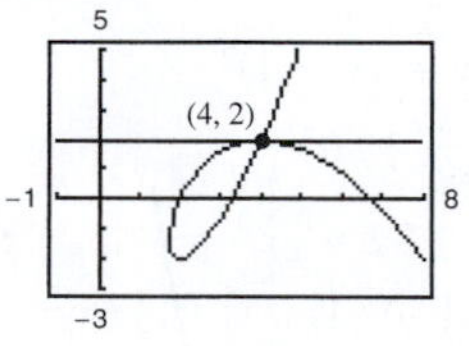

(b) At $t = -1$, $dx/dt = -3$, $dy/dt = 0$, and $dy/dx = 0$.

(c) $y = 2$

23. $y = \pm\frac{3}{4}x$ **25.** $y = 3x - 5$ and $y = 1$

27. Horizontal: $(1, 0), (-1, \pi), (1, -2\pi)$
Vertical: $(\pi/2, 1), (-3\pi/2, -1), (5\pi/2, 1)$

29. Horizontal: $(4, 0)$
Vertical: None

31. Horizontal: $(5, -2), (3, 2)$
Vertical: None

33. Horizontal: $(0, 3), (0, -3)$
Vertical: $(3, 0), (-3, 0)$

35. Horizontal: $(5, -1), (5, -3)$
Vertical: $(8, -2), (2, -2)$

37. Horizontal: None
Vertical: $(1, 0), (-1, 0)$

39. Concave downward: $-\infty < t < 0$
Concave upward: $0 < t < \infty$

41. Concave upward: $t > 0$

43. Concave downward: $0 < t < \pi/2$
Concave upward: $\pi/2 < t < \pi$

45. $4\sqrt{13} \approx 14.422$ **47.** $\sqrt{2}(1 - e^{-\pi/2}) \approx 1.12$

49. $\frac{1}{12}\left[\ln\left(\sqrt{37} + 6\right) + 6\sqrt{37}\right] \approx 3.249$ **51.** $6a$ **53.** $8a$

55. (a)
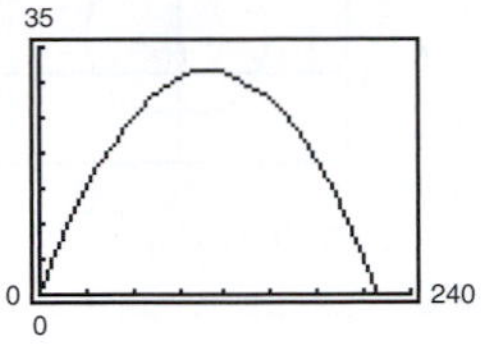

(b) 219.2 ft
(c) 230.8 ft

57. (a)
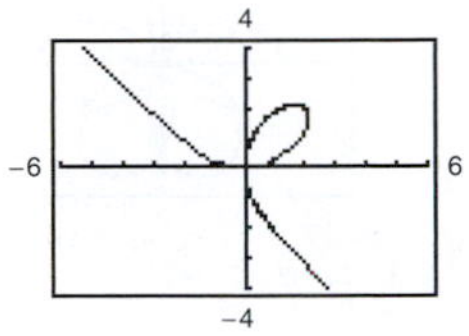

(b) $(0, 0), \left(4\sqrt[3]{2}/3, 4\sqrt[3]{4}/3\right)$
(c) About 6.557

59. (a)
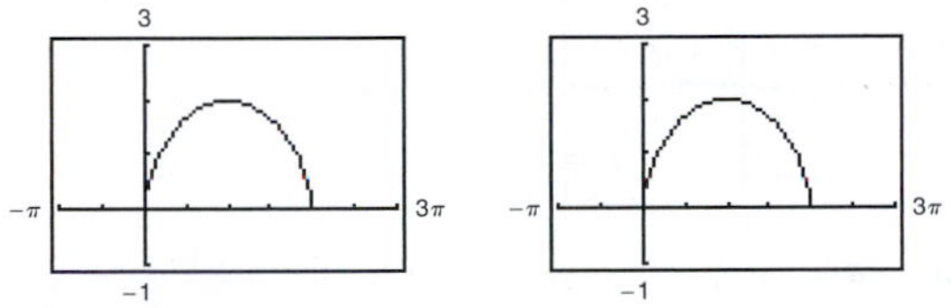

(b) The average speed of the particle on the second path is twice the average speed of the particle on the first path.
(c) 4π

61. $S = 2\pi\int_0^4 \sqrt{10}(t + 2)\,dt = 32\pi\sqrt{10} \approx 317.907$

63. $S = 2\pi\int_0^{\pi/2} \left(\sin\theta\cos\theta\sqrt{4\cos^2\theta + 1}\right)d\theta$
$= \dfrac{\left(5\sqrt{5} - 1\right)\pi}{6}$
≈ 5.330

65. (a) $27\pi\sqrt{13}$ (b) $18\pi\sqrt{13}$ **67.** 50π **69.** $12\pi a^2/5$

71. See Theorem 10.7, Parametric Form of the Derivative, on page 706.

73. 6

75. (a) $S = 2\pi\int_a^b g(t)\sqrt{\left(\dfrac{dx}{dt}\right)^2 + \left(\dfrac{dy}{dt}\right)^2}\,dt$
(b) $S = 2\pi\int_a^b f(t)\sqrt{\left(\dfrac{dx}{dt}\right)^2 + \left(\dfrac{dy}{dt}\right)^2}\,dt$

77. Proof **79.** $3\pi/2$ **81.** d **82.** b **83.** f **84.** c

85. a **86.** e **87.** $\left(\frac{3}{4}, \frac{8}{5}\right)$ **89.** 288π

91. (a) $dy/dx = \sin\theta/(1 - \cos\theta)$; $d^2y/dx^2 = -1/[a(\cos\theta - 1)^2]$
(b) $y = \left(2 + \sqrt{3}\right)\left[x - a\left(\pi/6 - \frac{1}{2}\right)\right] + a\left(1 - \sqrt{3}/2\right)$
(c) $(a(2n + 1)\pi, 2a)$
(d) Concave downward on $(0, 2\pi), (2\pi, 4\pi)$, etc.
(e) $s = 8a$

93. Proof

95. (a)
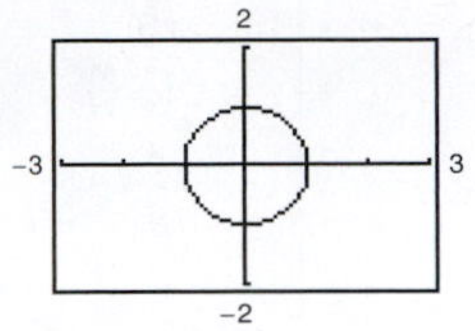

(b) Circle of radius 1 and center at $(0, 0)$ except the point $(-1, 0)$
(c) As t increases from -20 to 0, the speed increases, and as t increases from 0 to 20, the speed decreases.

97. False. $\dfrac{d^2y}{dx^2} = \dfrac{\dfrac{d}{dt}\left[\dfrac{g'(t)}{f'(t)}\right]}{f'(t)} = \dfrac{f'(t)g''(t) - g'(t)f''(t)}{[f'(t)]^3}$.

Section 10.4 *(page 722)*

1.
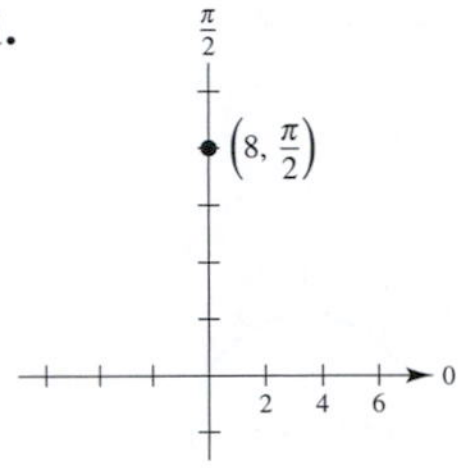

$(0, 8)$

3.
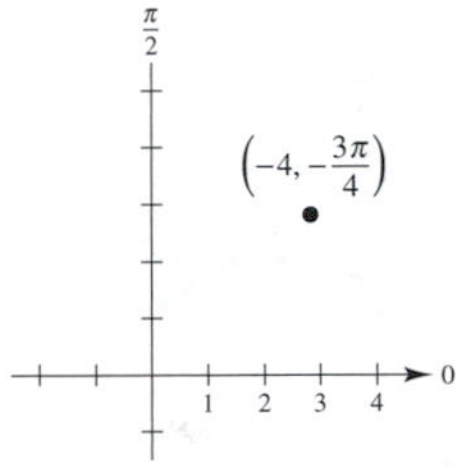

$\left(2\sqrt{2}, 2\sqrt{2}\right) \approx (2.828, 2.828)$

5.
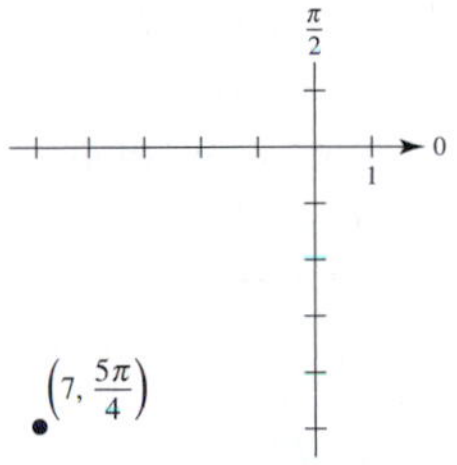

$(-4.95, -4.95)$

7.
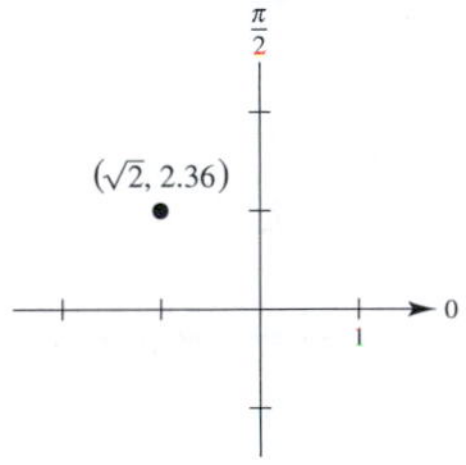

$(-1.004, 0.996)$

9.
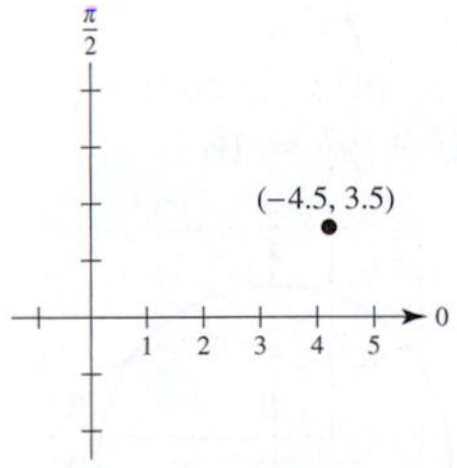

$(4.214, 1.579)$

11.
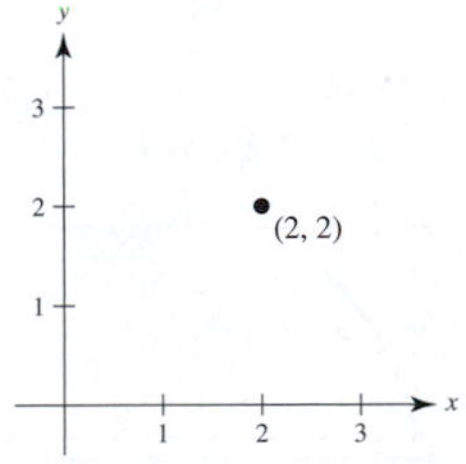

$\left(2\sqrt{2}, \pi/4\right), \left(-2\sqrt{2}, 5\pi/4\right)$

13.
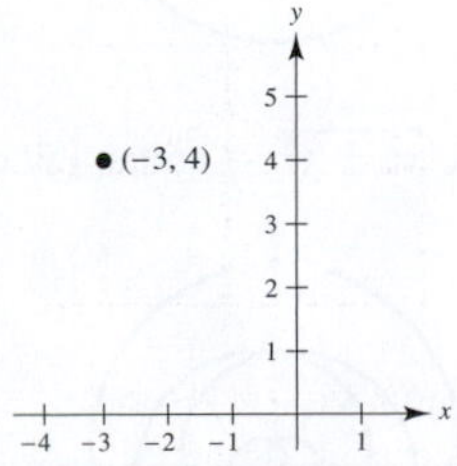

$(5, 2.214), (-5, 5.356)$

15.
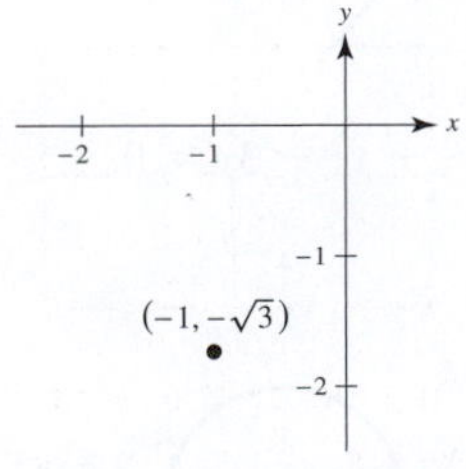

$(2, 4\pi/3), (-2, \pi/3)$

17.

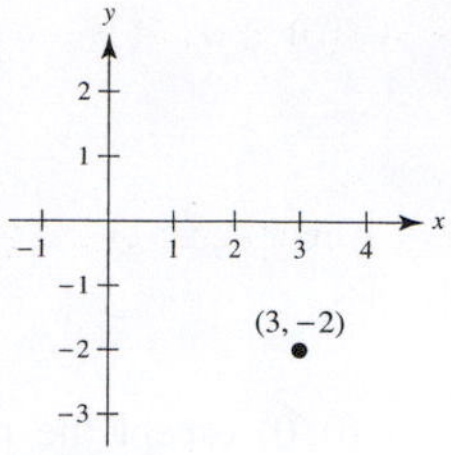

(3.606, −0.588)
(−3.606, 2.554)

19.

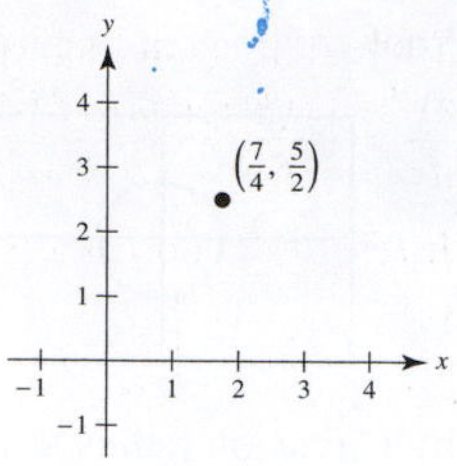

(3.052, 0.960)
(−3.052, 4.102)

21. (a)

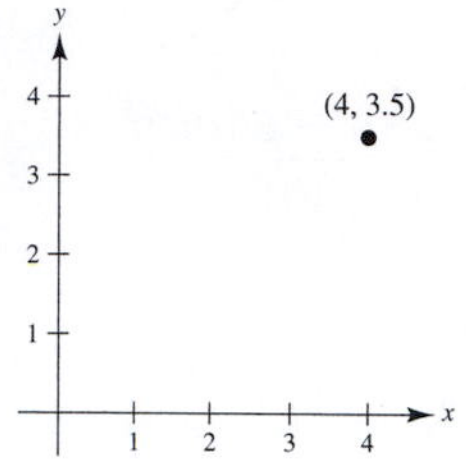

(b)

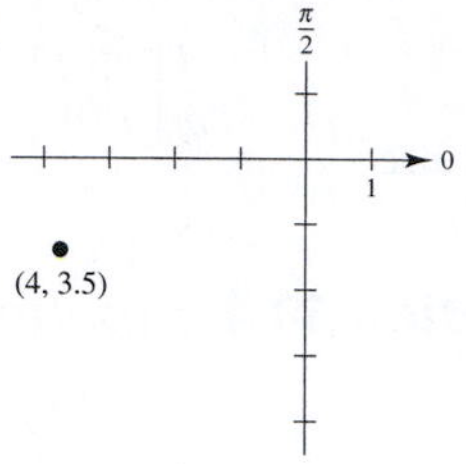

23. $r = 3$

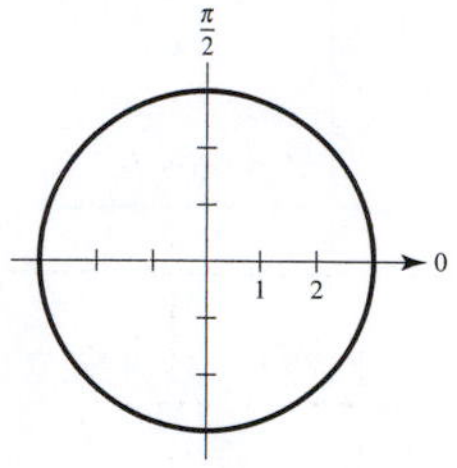

25. $r = a$

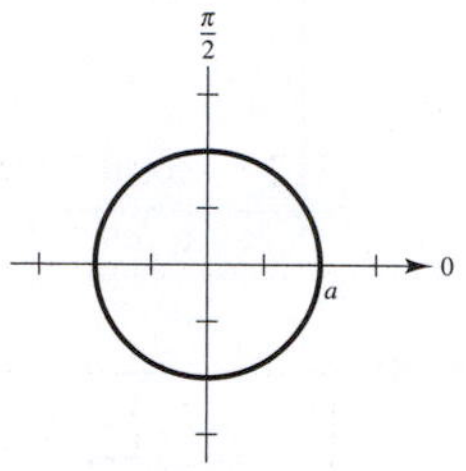

27. $r = 8 \csc \theta$

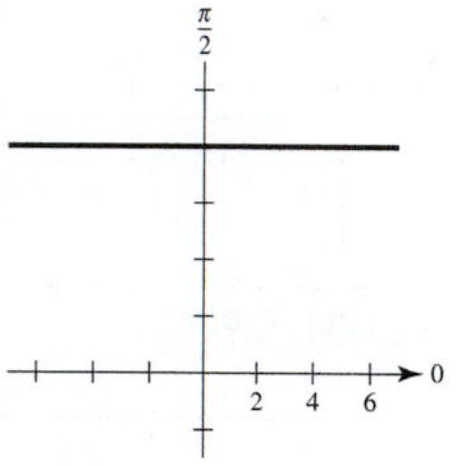

29. $r = \dfrac{-2}{3 \cos \theta - \sin \theta}$

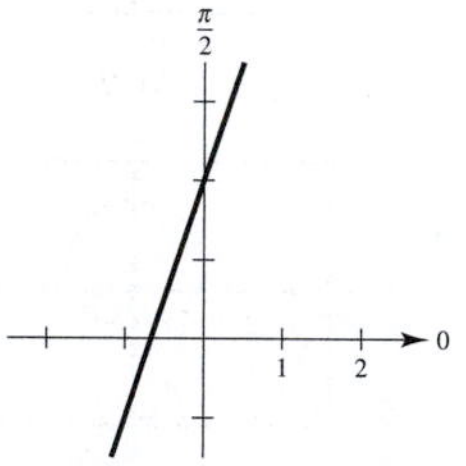

31. $r = 9 \csc^2 \theta \cos \theta$

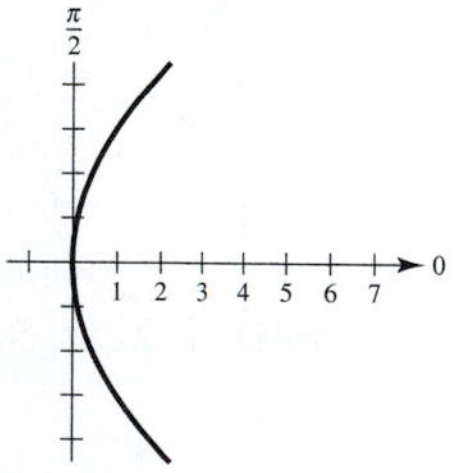

33. $x^2 + y^2 = 16$

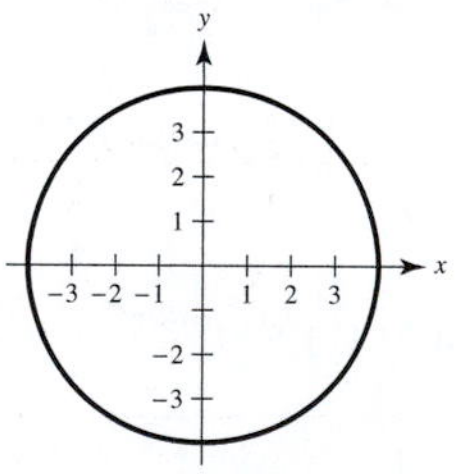

35. $x^2 + y^2 - 3y = 0$

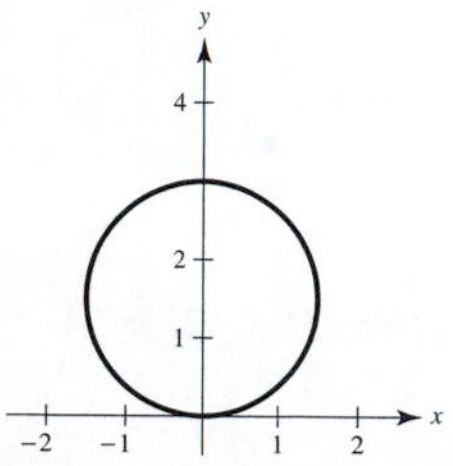

37. $\sqrt{x^2 + y^2} = \arctan (y/x)$

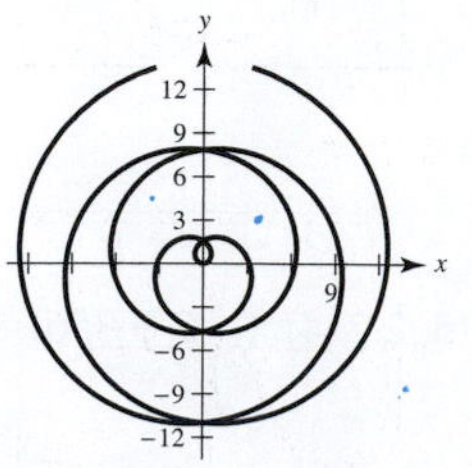

39. $x - 3 = 0$

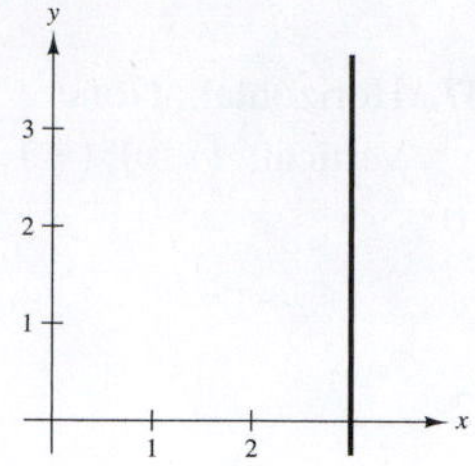

41. $x^2 - y = 0$

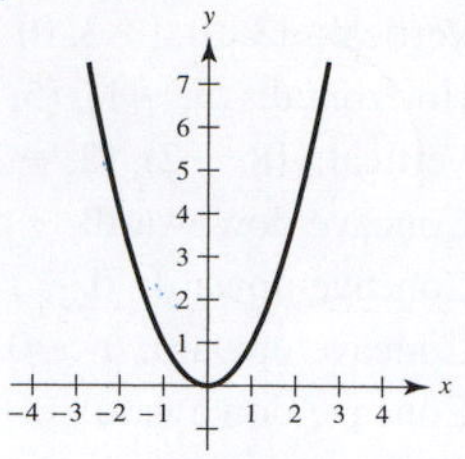

43.

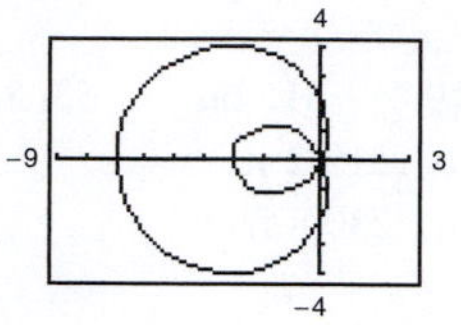

$0 \le \theta < 2\pi$

45.

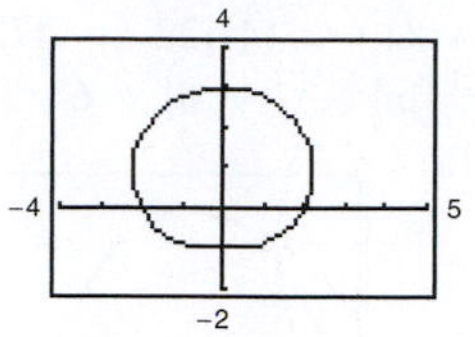

$0 \le \theta < 2\pi$

47.

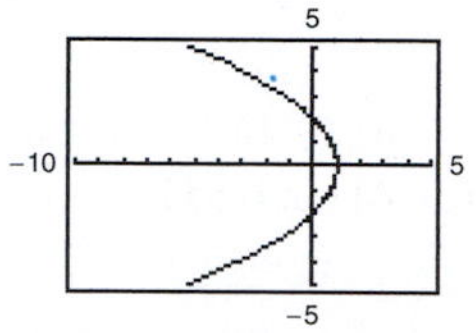

$-\pi < \theta < \pi$

49.

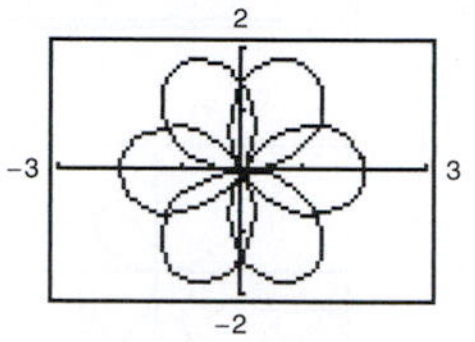

$0 \le \theta < 4\pi$

51.

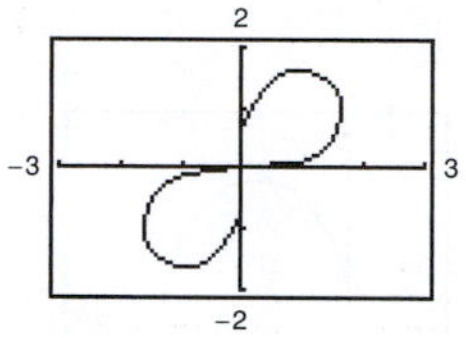

$0 \le \theta < \pi/2$

53. $(x - h)^2 + (y - k)^2 = h^2 + k^2$
Radius: $\sqrt{h^2 + k^2}$
Center: (h, k)

55. $\sqrt{17}$ **57.** About 5.6

59. $\dfrac{dy}{dx} = \dfrac{2 \cos \theta (3 \sin \theta + 1)}{6 \cos^2 \theta - 2 \sin \theta - 3}$
$(5, \pi/2)$: $dy/dx = 0$
$(2, \pi)$: $dy/dx = -2/3$
$(-1, 3\pi/2)$: $dy/dx = 0$

61. (a) and (b) (c) $dy/dx = -1$

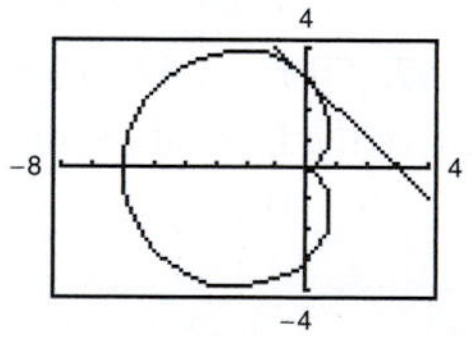

63. (a) and (b) (c) $dy/dx = -\sqrt{3}$

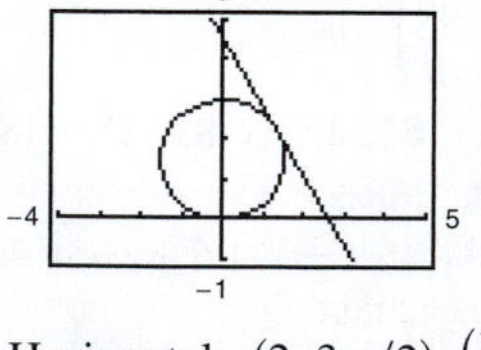

65. Horizontal: $(2, 3\pi/2)$, $\left(\frac{1}{2}, \pi/6\right)$, $\left(\frac{1}{2}, 5\pi/6\right)$
Vertical: $\left(\frac{3}{2}, 7\pi/6\right)$, $\left(\frac{3}{2}, 11\pi/6\right)$

67. $(5, \pi/2)$, $(1, 3\pi/2)$

69.

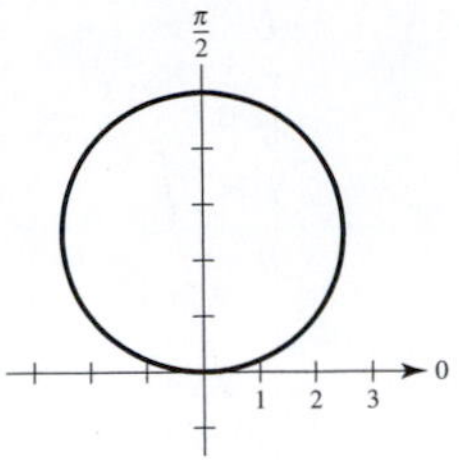

$\theta = 0$

71.

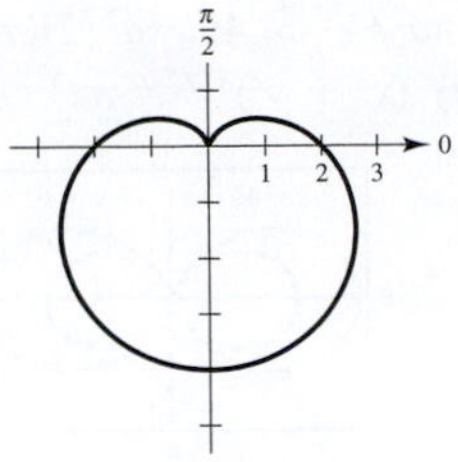

$\theta = \pi/2$

73.

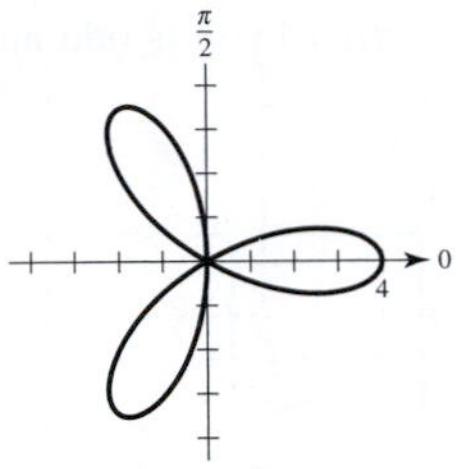

$\theta = \pi/6, \pi/2, 5\pi/6$

75.

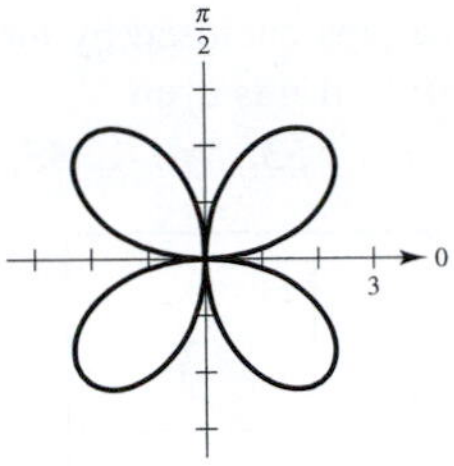

$\theta = 0, \pi/2$

77.

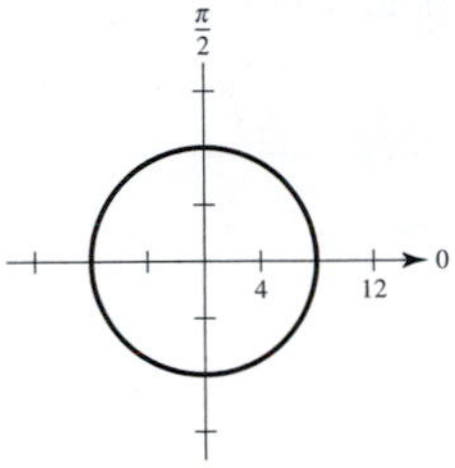

79.

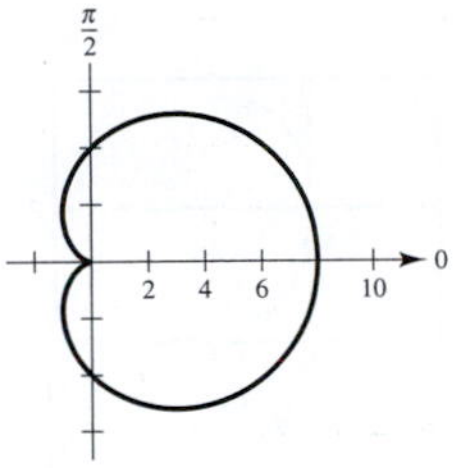

81.

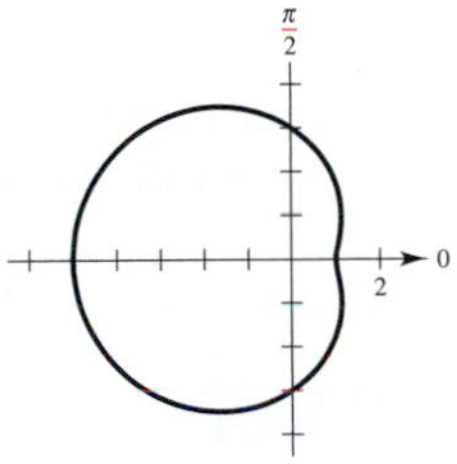

83.

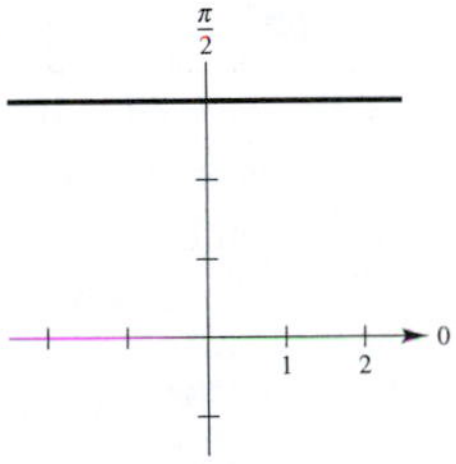

85.

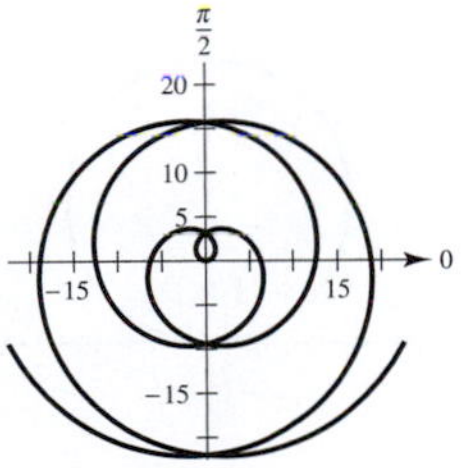

87.

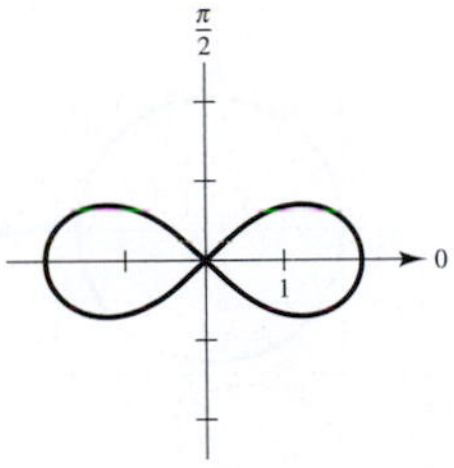

89.

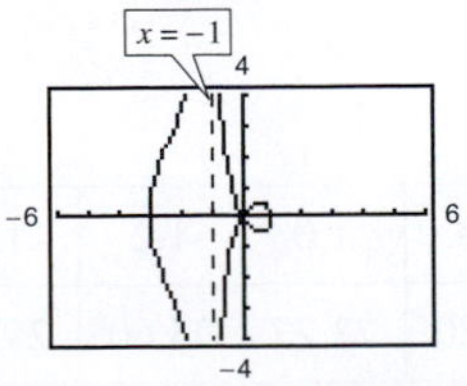

91.

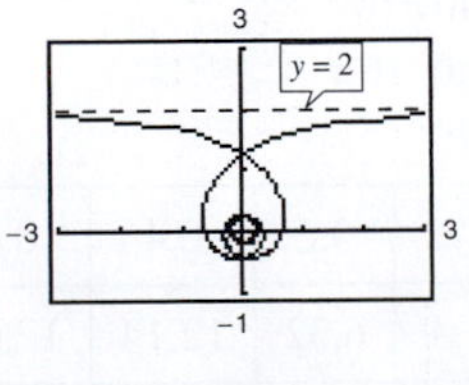

93. The rectangular coordinate system is a collection of points of the form (x, y), where x is the directed distance from the y-axis to the point and y is the directed distance from the x-axis to the point. Every point has a unique representation.

The polar coordinate system is a collection of points of the form (r, θ), where r is the directed distance from the origin O to a point P and θ is the directed angle, measured counterclockwise, from the polar axis to the segment $\overline{OP}$. Polar coordinates do not have unique representations.

95. Slope of tangent line to graph of $r = f(\theta)$ at (r, θ) is

$$\frac{dy}{dx} = \frac{f(\theta)\cos\theta + f'(\theta)\sin\theta}{-f(\theta)\sin\theta + f'(\theta)\cos\theta}.$$

If $f(\alpha) = 0$ and $f'(\alpha) \neq 0$, then $\theta = \alpha$ is tangent at the pole.

97. (a)

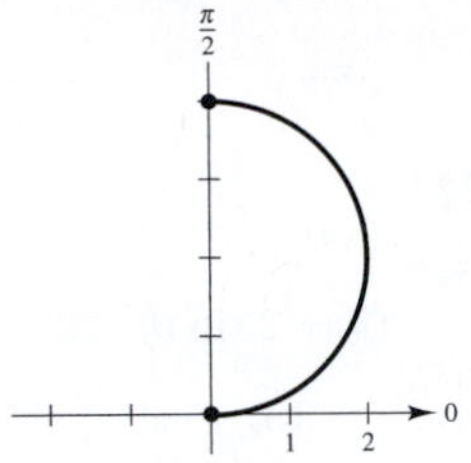

(b)

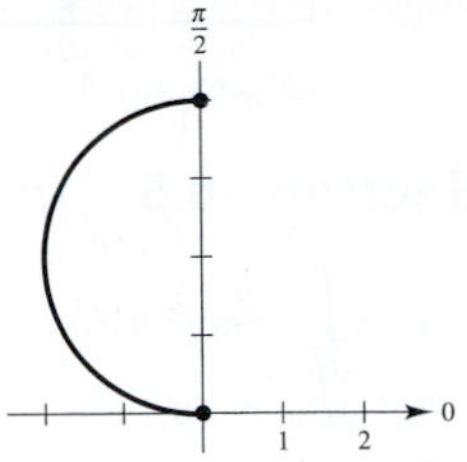

(c)

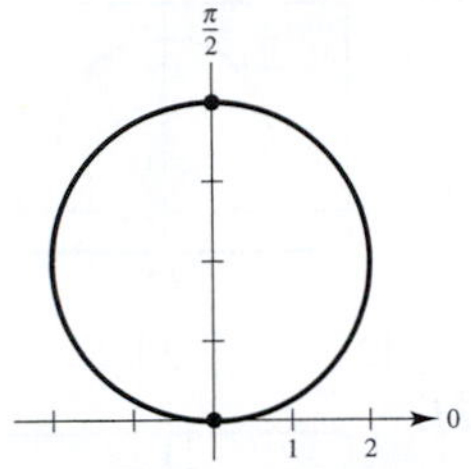

99. Proof

101. (a) $r = 2 - \sin(\theta - \pi/4)$

$$= 2 - \frac{\sqrt{2}(\sin\theta - \cos\theta)}{2}$$

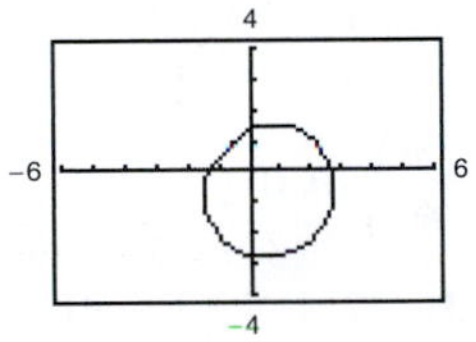

(b) $r = 2 + \cos\theta$

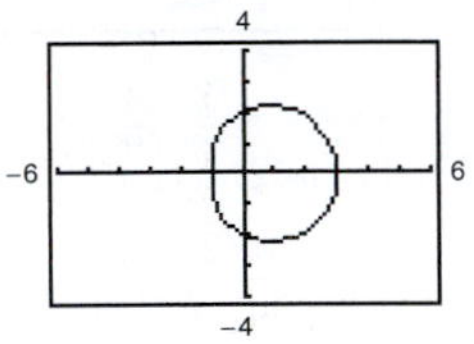

(c) $r = 2 + \sin\theta$

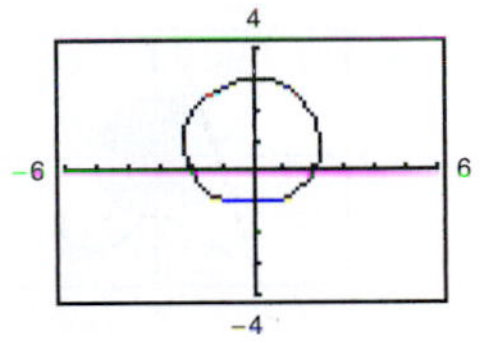

(d) $r = 2 - \cos\theta$

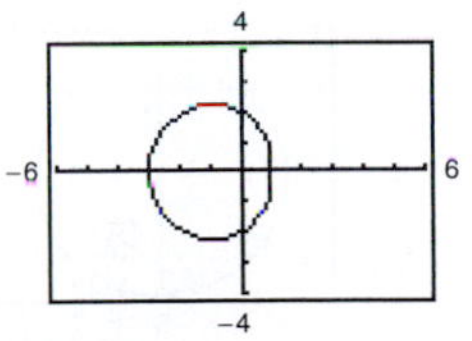

103. (a)

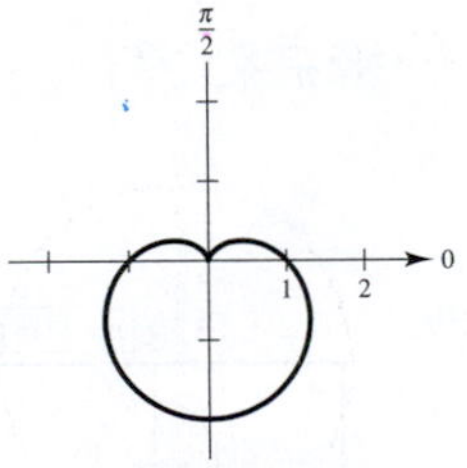

(b)

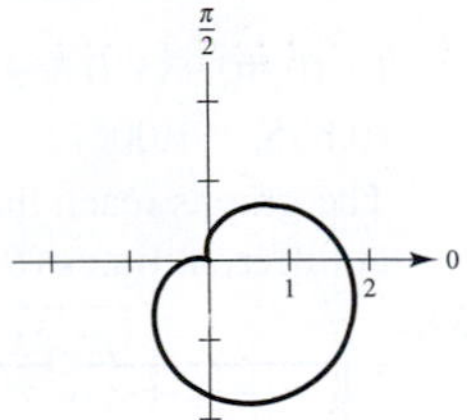

105.

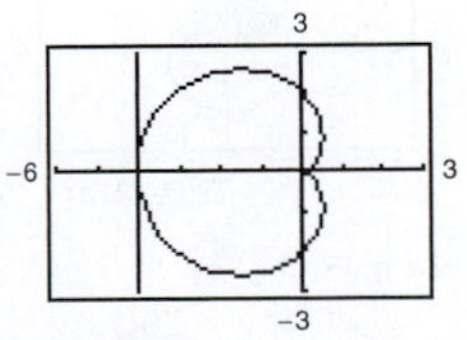

$\psi = \pi/2$

107.

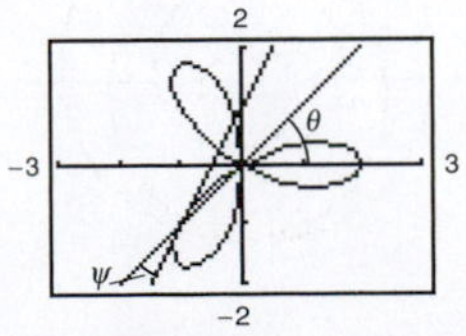

$\psi = \arctan\frac{1}{3} \approx 18.4°$

109.

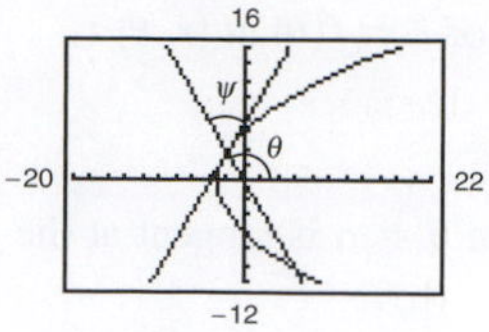

$\psi = \pi/3, 60°$

111. True **113.** True

Section 10.5 *(page 731)*

1. $8\int_0^{\pi/2} \sin^2\theta\, d\theta$ **3.** $\frac{1}{2}\int_{\pi/2}^{3\pi/2} (3 - 2\sin\theta)^2\, d\theta$ **5.** 9π

7. $\pi/3$ **9.** $\pi/8$ **11.** $3\pi/2$ **13.** 27π **15.** 4

17.

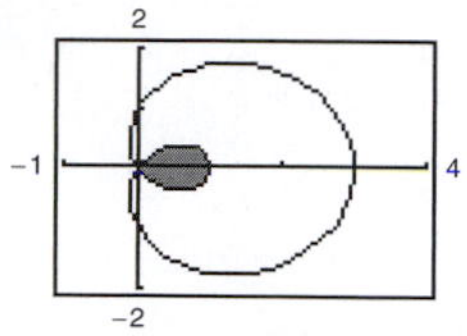

$(2\pi - 3\sqrt{3})/2$

19.

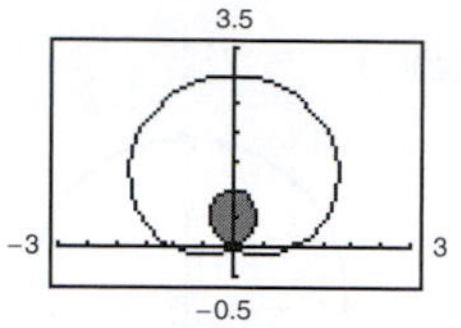

$(2\pi - 3\sqrt{3})/2$

21.

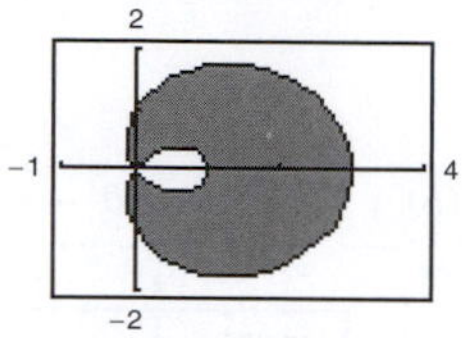

$\pi + 3\sqrt{3}$

23.

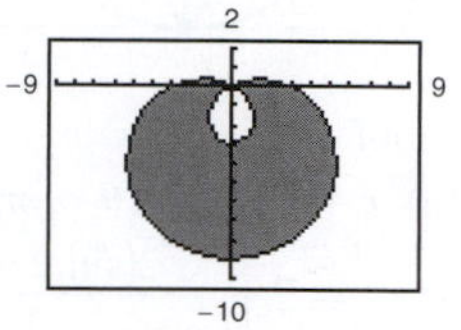

$9\pi + 27\sqrt{3}$

25. $(1, \pi/2), (1, 3\pi/2), (0, 0)$

27. $\left(\frac{2 - \sqrt{2}}{2}, \frac{3\pi}{4}\right), \left(\frac{2 + \sqrt{2}}{2}, \frac{7\pi}{4}\right), (0, 0)$

29. $\left(\frac{3}{2}, \frac{\pi}{6}\right), \left(\frac{3}{2}, \frac{5\pi}{6}\right), (0, 0)$ **31.** $(2, 4), (-2, -4)$

33.

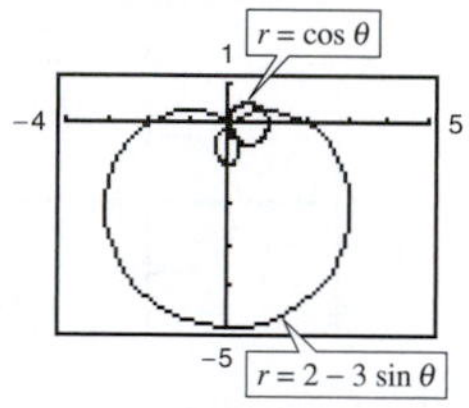

$(0, 0), (0.935, 0.363),$
$(0.535, -1.006)$

The graphs reach the pole at different times (θ-values).

35.

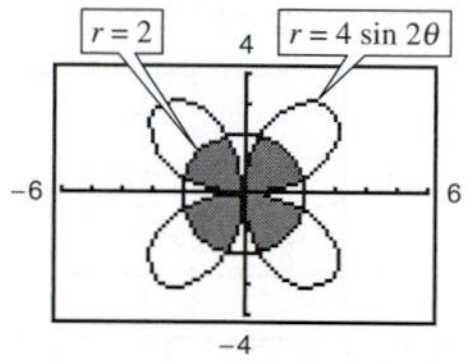

$\frac{4}{3}(4\pi - 3\sqrt{3})$

37.

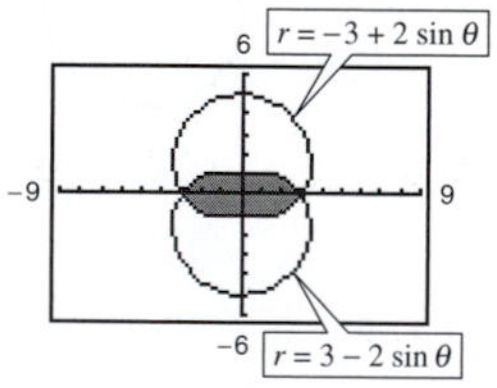

$11\pi - 24$

39.

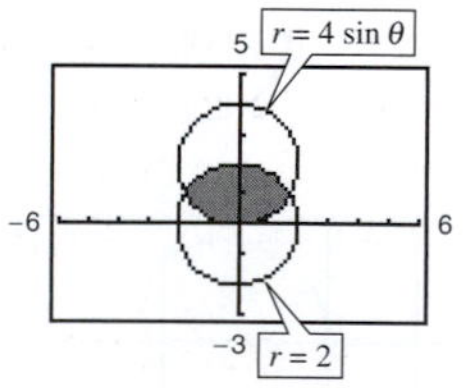

$\frac{2}{3}(4\pi - 3\sqrt{3})$

41.

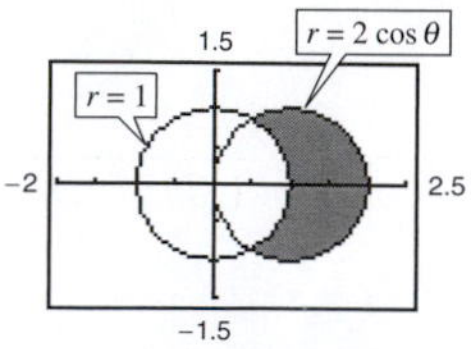

$\pi/3 + \sqrt{3}/2$

43. $5\pi a^2/4$ **45.** $(a^2/2)(\pi - 2)$

47. (a) $(x^2 + y^2)^{3/2} = ax^2$

(b)

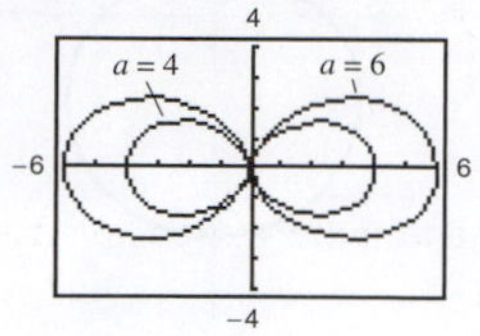

(c) $15\pi/2$

49. The area enclosed by the function is $\pi a^2/4$ if n is odd and is $\pi a^2/2$ if n is even.

51. 16π **53.** 4π **55.** 8

57.

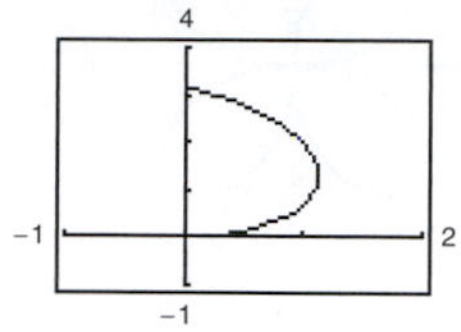

About 4.16

59.

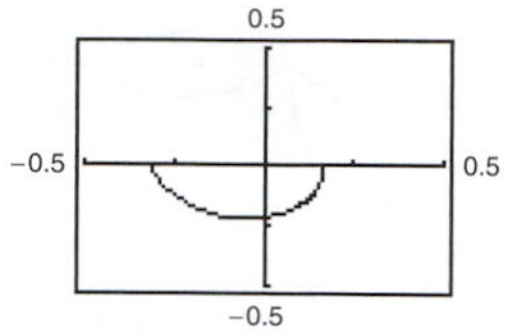

About 0.71

61.

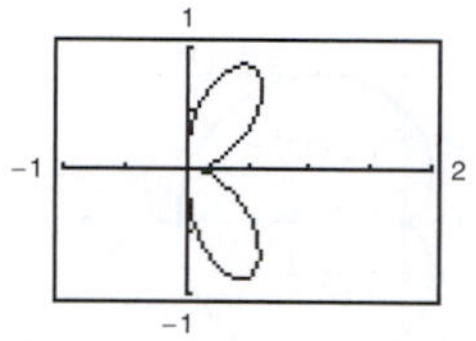

About 4.39

63. 36π **65.** $\frac{2\pi\sqrt{1 + a^2}}{1 + 4a^2}(e^{\pi a} - 2a)$ **67.** 21.87

69. You will only find simultaneous points of intersection. There may be intersection points that do not occur with the same coordinates in the two graphs.

71. (a) Circle of radius 5

Area $= 25\pi$

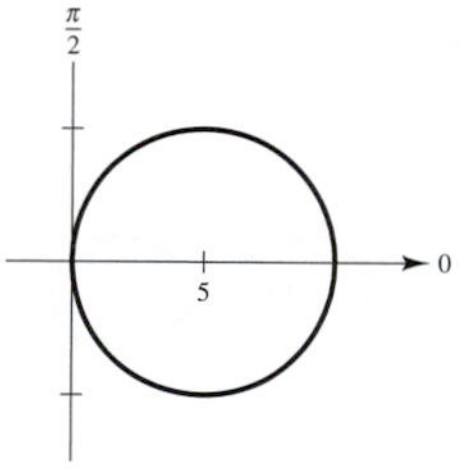

(b) Circle of radius 5/2

Area $= \frac{25}{4}\pi$

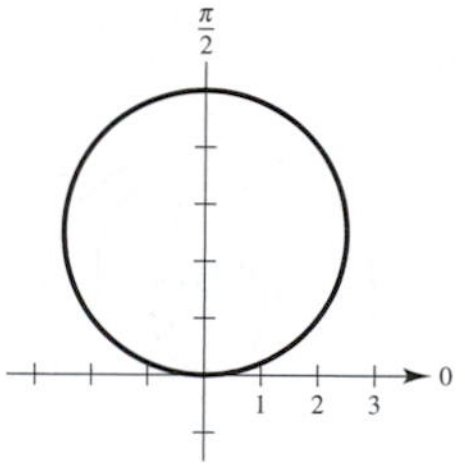

73. $40\pi^2$

75. (a) 16π

(b)

θ	0.2	0.4	0.6	0.8	1.0	1.2	1.4
A	6.32	12.14	17.06	20.80	23.27	24.60	25.08

(c) and (d) For $\frac{1}{4}$ of area $(4\pi \approx 12.57)$: 0.42

For $\frac{1}{2}$ of area $(8\pi \approx 25.13)$: $1.57(\pi/2)$

For $\frac{3}{4}$ of area $(12\pi \approx 37.70)$: 2.73

(e) No. The results do not depend on the radius. Answers will vary.

77. Circle

79. (a)

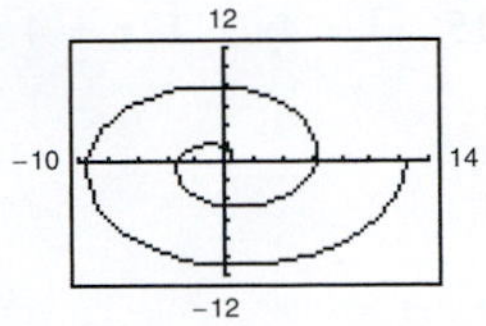

The graph becomes larger and more spread out. The graph is reflected over the y-axis.

(b) $(an\pi, n\pi)$, where $n = 1, 2, 3, \ldots$

(c) About 21.26 (d) $4/3\pi^3$

81. $r = \sqrt{2}\cos\theta$

83. False. The graphs of $f(\theta) = 1$ and $g(\theta) = -1$ coincide.

85. Proof

Section 10.6 *(page 739)*

1.

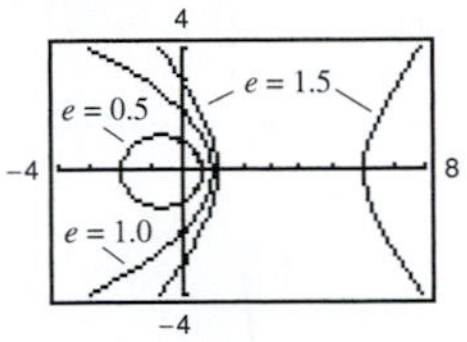

(a) Parabola
(b) Ellipse
(c) Hyperbola

3.

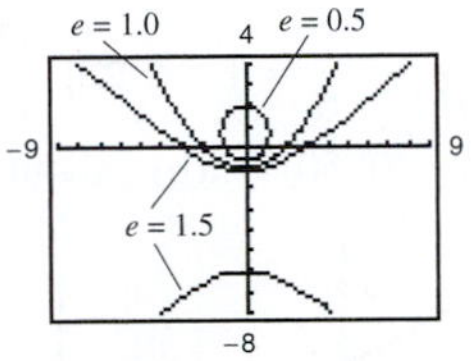

(a) Parabola
(b) Ellipse
(c) Hyperbola

5. (a)

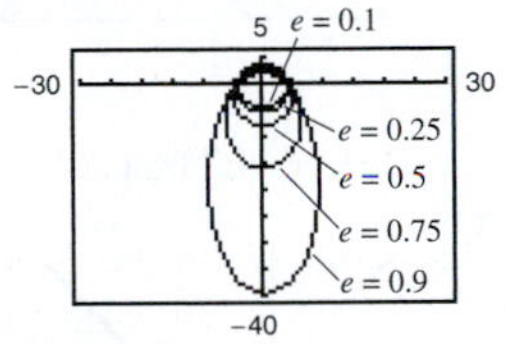

Ellipse
As $e \to 1^-$, the ellipse becomes more elliptical, and as $e \to 0^+$, it becomes more circular.

(b)

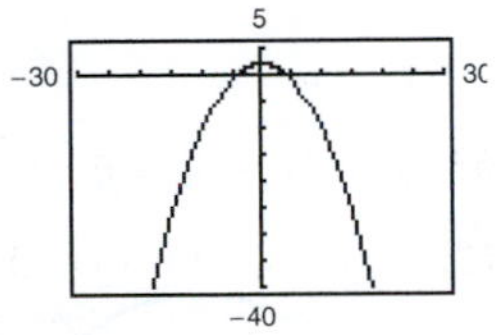

Parabola

(c)

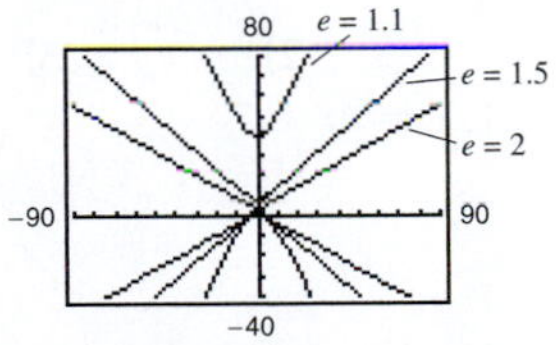

Hyperbola
As $e \to 1^+$, the hyperbola opens more slowly, and as $e \to \infty$, it opens more rapidly.

7. c **8.** f **9.** a **10.** e **11.** b **12.** d

13. $e = 1$
Distance = 1

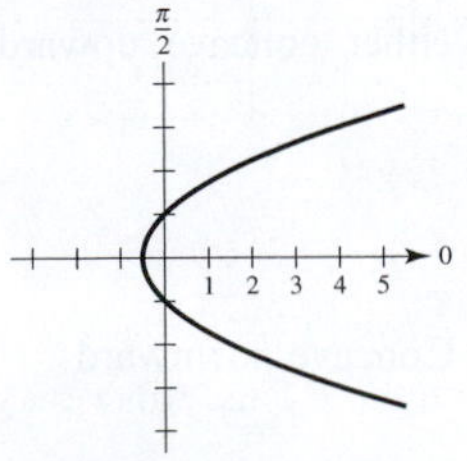

Parabola

15. $e = 3$
Distance = $\frac{1}{2}$

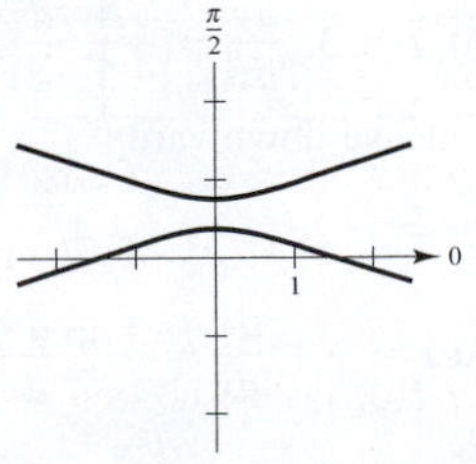

Hyperbola

17. $e = 2$
Distance = $\frac{5}{2}$

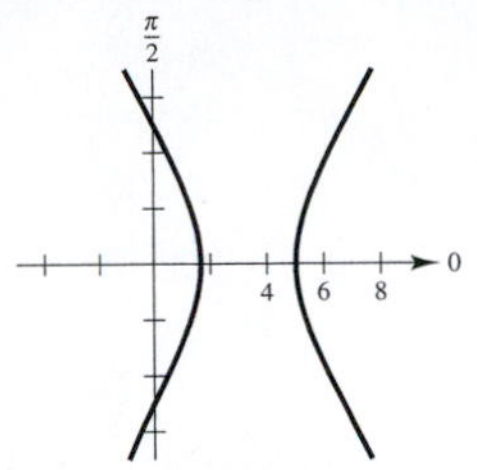

Hyperbola

19. $e = \frac{1}{2}$
Distance = 6

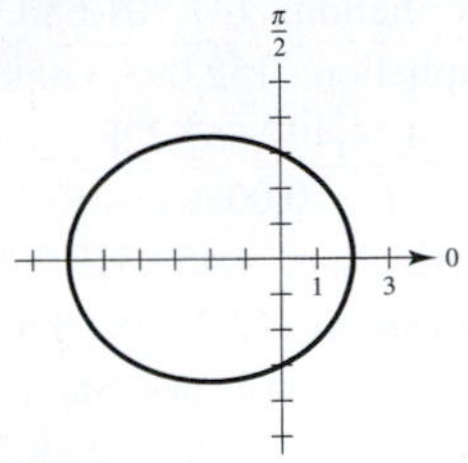

Ellipse

21. $e = \frac{1}{2}$
Distance = 50

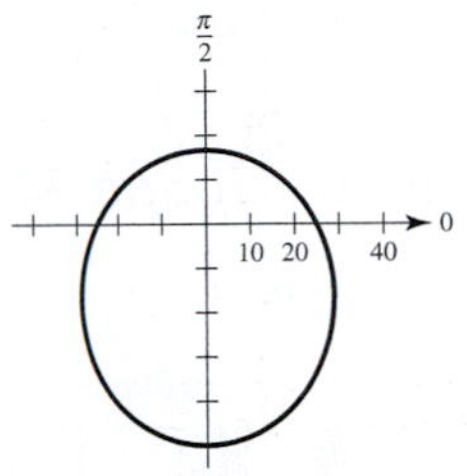

Ellipse

23.

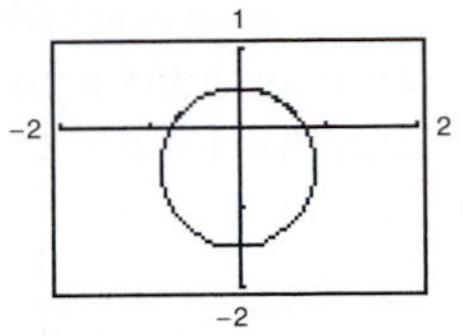

Ellipse
$e = \frac{1}{2}$

25.

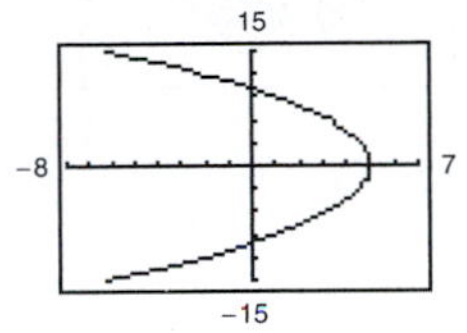

Parabola
$e = 1$

27.

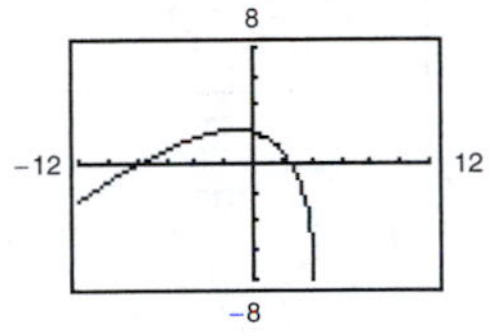

Rotated $\pi/3$ radian counterclockwise.

29.

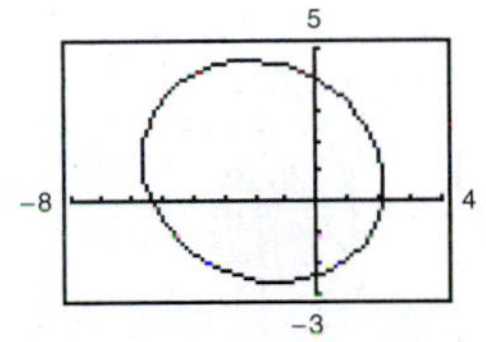

Rotated $\pi/6$ radian clockwise.

31. $r = \dfrac{8}{8 + 5\cos\left(\theta + \dfrac{\pi}{6}\right)}$

33. $r = 3/(1 - \cos\theta)$ **35.** $r = 1/(2 + \sin\theta)$

37. $r = 2/(1 + 2\cos\theta)$ **39.** $r = 2/(1 - \sin\theta)$

41. $r = 16/(5 + 3\cos\theta)$ **43.** $r = 9/(4 - 5\sin\theta)$

45. $r = 4/(2 + \cos\theta)$

47. If $0 < e < 1$, the conic is an ellipse.
If $e = 1$, the conic is a parabola.
If $e > 1$, the conic is a hyperbola.

49. If the foci are fixed and $e \to 0$, then $d \to \infty$. To see this, compare the ellipses

$$r = \frac{1/2}{1 + (1/2)\cos\theta},\ e = \frac{1}{2},\ d = 1 \text{ and}$$

$$r = \frac{5/16}{1 + (1/4)\cos\theta},\ e = \frac{1}{4},\ d = \frac{5}{4}.$$

51. Proof

53. $r^2 = \dfrac{9}{1 - (16/25)\cos^2\theta}$ **55.** $r^2 = \dfrac{-16}{1 - (25/9)\cos^2\theta}$

57. About 10.88 **59.** 3.37

61. $\dfrac{7979.21}{1 - 0.9372\cos\theta}$; 11,015 mi

63. $r = \dfrac{149{,}558{,}278.0560}{1 - 0.0167 \cos \theta}$

Perihelion: 147,101,680 km

Aphelion: 152,098,320 km

65. $r = \dfrac{4{,}497{,}667{,}328}{1 - 0.0086 \cos \theta}$

Perihelion: 4,459,317,200 km

Aphelion: 4,536,682,800 km

67. Answers will vary. Sample answers:

(a) $3.591 \times 10^{18}\ \text{km}^2$; 9.322 yr

(b) $\alpha \approx 0.361 + \pi$; Larger angle with the smaller ray to generate an equal area

(c) Part (a): 1.583×10^9 km; 1.698×10^8 km/yr

Part (b): 1.610×10^9 km; 1.727×10^8 km/yr

69. Proof

Review Exercises for Chapter 10 *(page 742)*

1. e **2.** c **3.** b **4.** d **5.** a **6.** f

7. Circle

Center: $\left(\frac{1}{2}, -\frac{3}{4}\right)$

Radius: 1

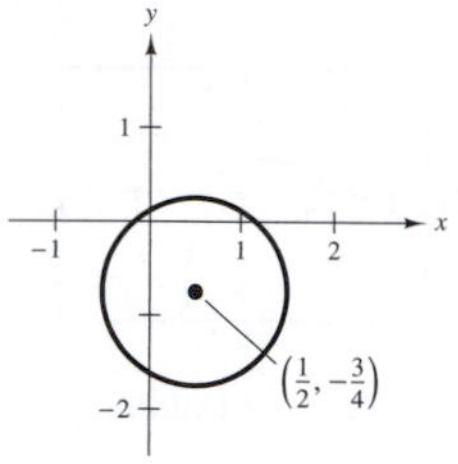

9. Hyperbola

Center: $(-4, 3)$

Vertices: $(-4 \pm \sqrt{2}, 3)$

Foci: $(-4 \pm \sqrt{5}, 3)$

$e = \sqrt{\dfrac{5}{2}}$

Asymptotes:

$y = 3 + \dfrac{\sqrt{3}}{\sqrt{2}}(x + 4)$;

$y = 3 - \dfrac{\sqrt{3}}{\sqrt{2}}(x + 4)$

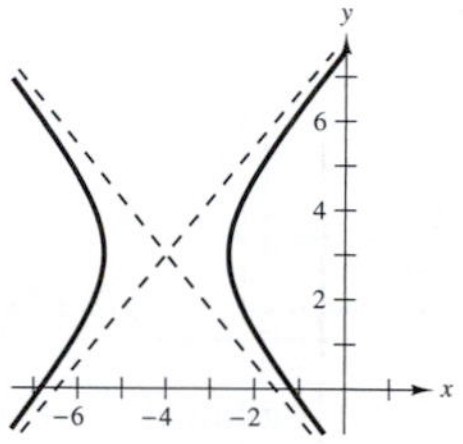

11. Ellipse

Center: $(2, -3)$

Vertices: $(2, -3 \pm \sqrt{2}/2)$

$e = \sqrt{\dfrac{1}{3}}$

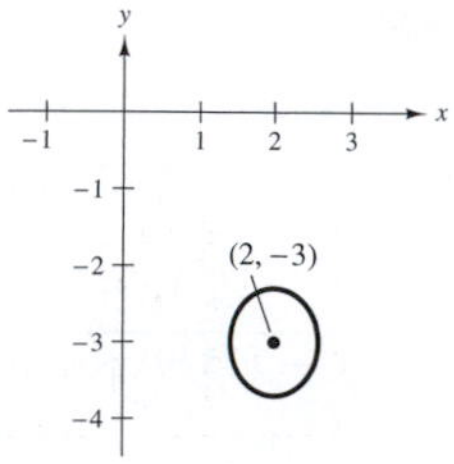

13. Parabola

Vertex: $(3, -1)$

Focus: $(3, 1)$

Directrix: $y = -3$

$e = 1$

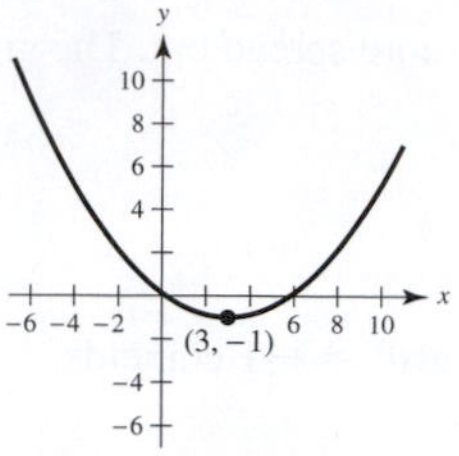

15. $y^2 - 4y - 12x + 4 = 0$

17. $\dfrac{x^2}{49} + \dfrac{y^2}{24} = 1$ **19.** $\dfrac{(x-3)^2}{5} + \dfrac{(y-4)^2}{9} = 1$

21. $\dfrac{y^2}{64} - \dfrac{x^2}{16} = 1$ **23.** $\dfrac{x^2}{49} - \dfrac{(y+1)^2}{32} = 1$

25. (a) $(0, 50)$ (b) About 38,294.49

27.

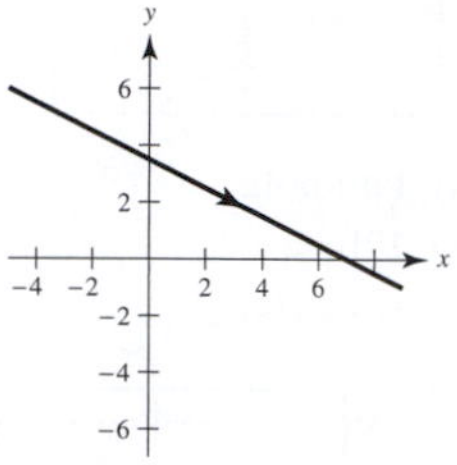

$x + 2y - 7 = 0$

29.

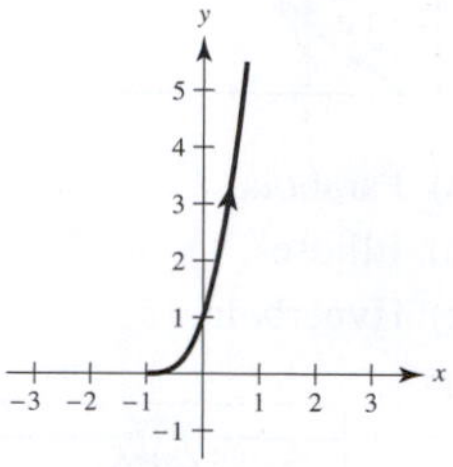

$y = (x + 1)^3, x > -1$

31.

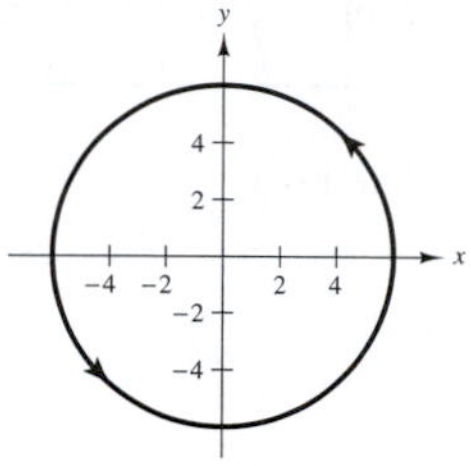

$x^2 + y^2 = 36$

33.

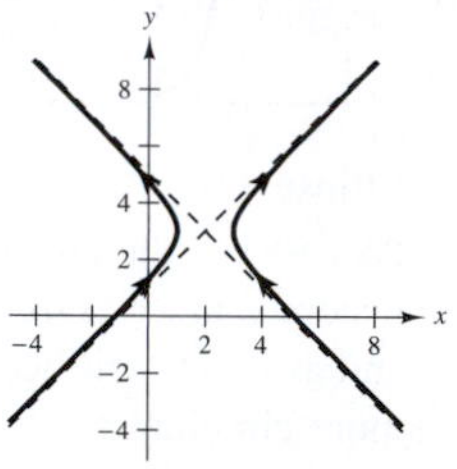

$(x - 2)^2 - (y - 3)^2 = 1$

35. $x = t, y = 4t + 3$; $x = t + 1, y = 4t + 7$

(Solution is not unique.)

37.

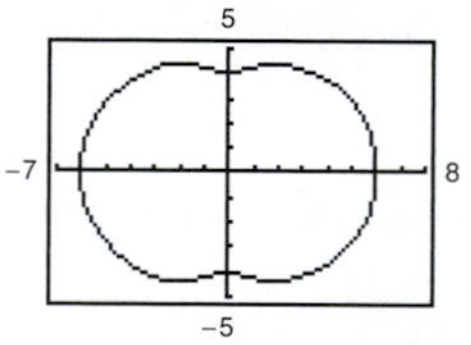

39. $\dfrac{dy}{dx} = -\dfrac{4}{5}, \dfrac{d^2y}{dx^2} = 0$

At $t = 3$, $\dfrac{dy}{dx} = -\dfrac{4}{5}, \dfrac{d^2y}{dx^2} = 0$; Neither concave upward or concave downward

41. $\dfrac{dy}{dx} = -2t^2, \dfrac{d^2y}{dx^2} = 4t^3$

At $t = -1$, $\dfrac{dy}{dx} = -2, \dfrac{d^2y}{dx^2} = -4$; Concave downward

43. $\dfrac{dy}{dx} = -4 \cot \theta, \dfrac{d^2y}{dx^2} = -4 \csc^3 \theta$

At $\theta = \dfrac{\pi}{6}, \dfrac{dy}{dx} = -4\sqrt{3}, \dfrac{d^2y}{dx^2} = -32$; Concave downward

45. $\dfrac{dy}{dx} = -4\tan\theta,\ \dfrac{d^2y}{dx^2} = \dfrac{4}{3}\sec^4\theta\csc\theta$

At $\theta = \dfrac{\pi}{3}, \dfrac{dy}{dx} = -4\sqrt{3}, \dfrac{d^2y}{dx^2} = \dfrac{128\sqrt{3}}{9}$; Concave upward

47. (a) and (d)

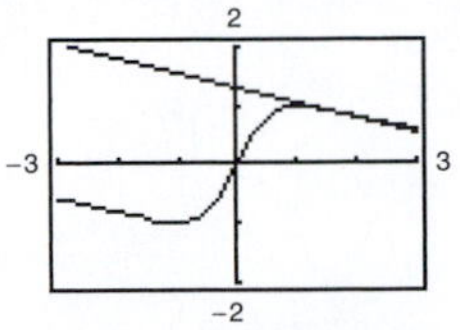

(b) $dx/d\theta = -4,\ dy/d\theta = 1,\ dy/dx = -\frac{1}{4}$

(c) $y = -\dfrac{1}{4}x + \dfrac{3\sqrt{3}}{4}$

49. Horizontal: $(5, 0)$
Vertical: None

51. Horizontal: $(2, 2), (2, 0)$
Vertical: $(4, 1), (0, 1)$

53. $\frac{1}{54}(145^{3/2} - 1) \approx 32.315$

55. (a) $s = 12\pi\sqrt{10} \approx 119.215$
(b) $s = 4\pi\sqrt{10} \approx 39.738$

57. $A = 3\pi$

59.

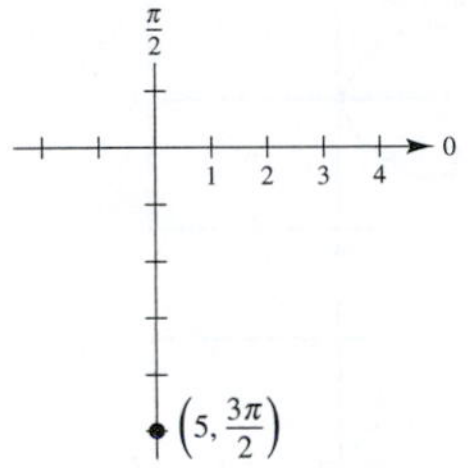

Rectangular: $(0, -5)$

61.

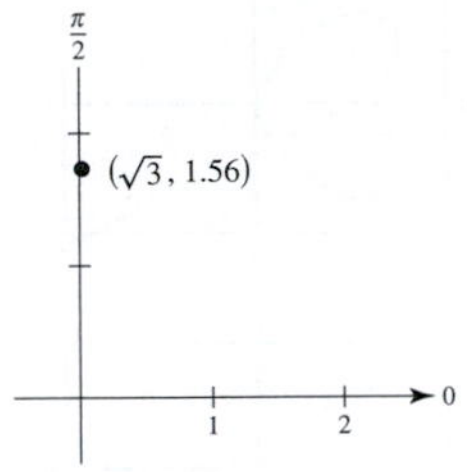

Rectangular: $(0.0187, 1.7320)$

63.

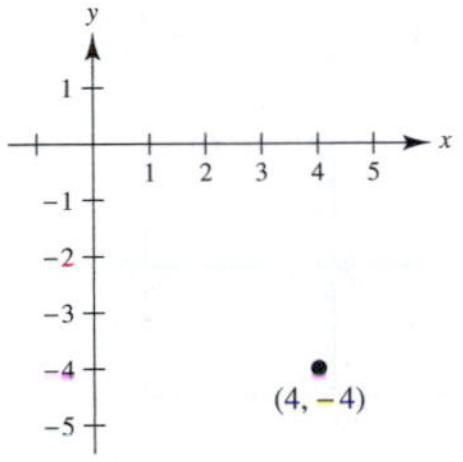

$\left(4\sqrt{2}, \dfrac{7\pi}{4}\right), \left(-4\sqrt{2}, \dfrac{3\pi}{4}\right)$

65.

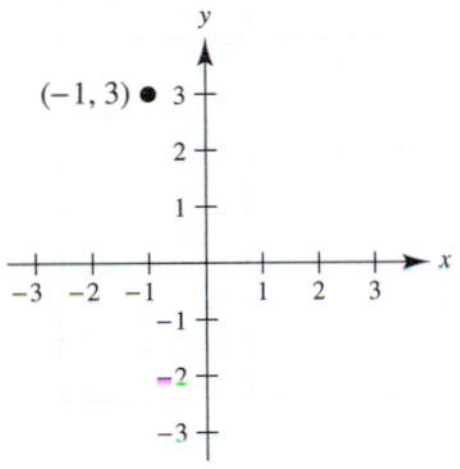

$(\sqrt{10}, 1.89), (-\sqrt{10}, 5.03)$

67. $r = 5$

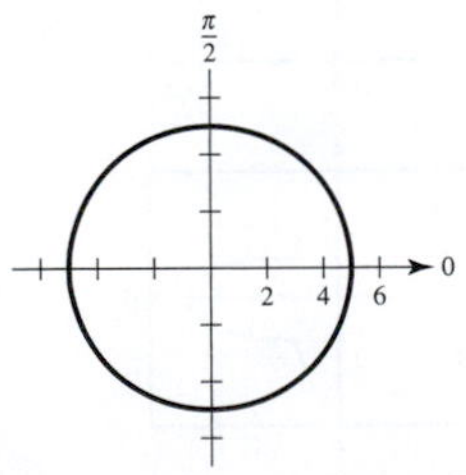

69. $r = 9\csc\theta$

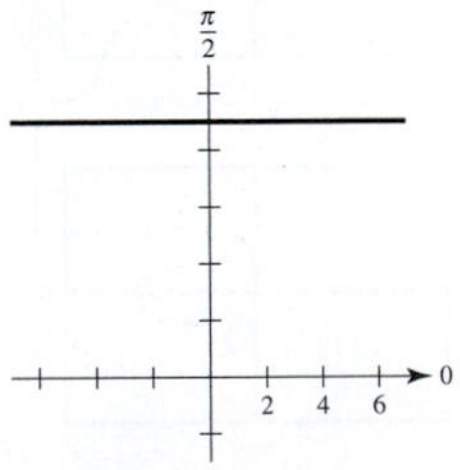

71. $r = 4\tan\theta\sec\theta$

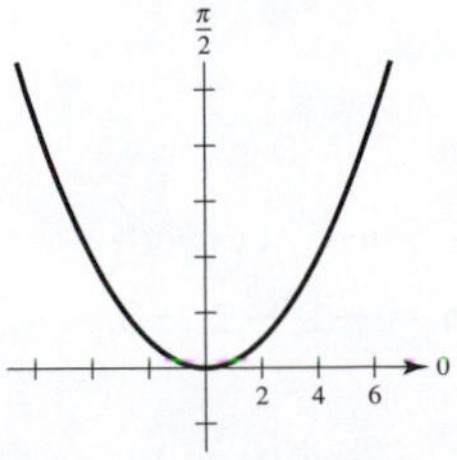

73. $x^2 + y^2 - 3x = 0$

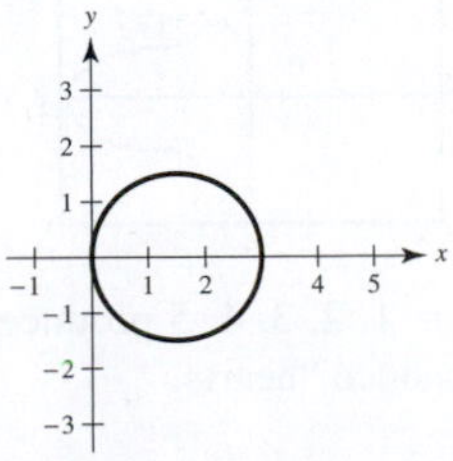

75. $x^2 + (y - 3)^2 = 9$

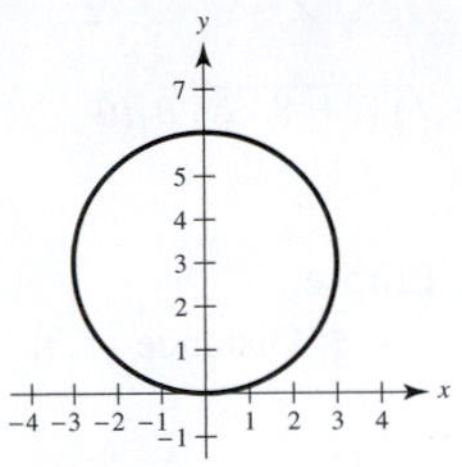

77. $y = -\frac{1}{2}x^2$

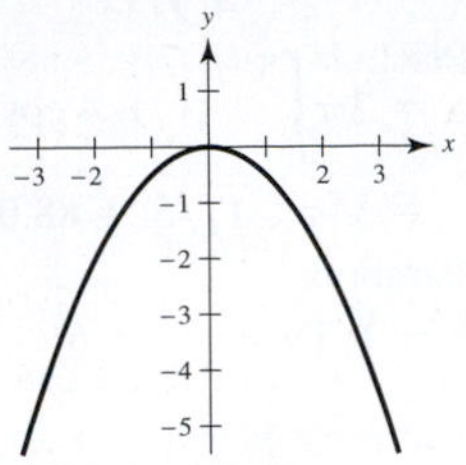

79.

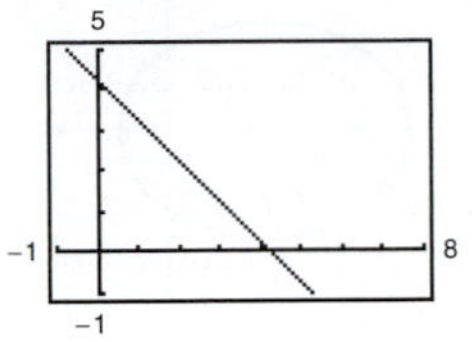

81.

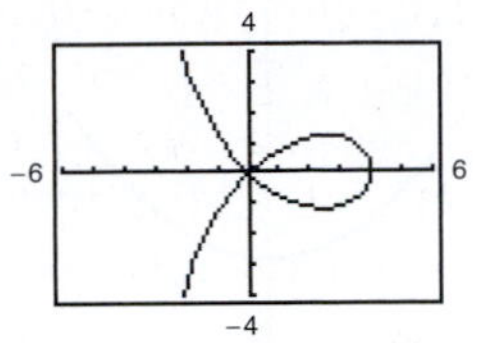

83. Horizontal: $\left(\dfrac{3}{2}, \dfrac{2\pi}{3}\right), \left(\dfrac{3}{2}, \dfrac{4\pi}{3}\right)$

Vertical: $\left(\dfrac{1}{2}, \dfrac{\pi}{3}\right), (2, \pi), \left(\dfrac{1}{2}, \dfrac{5\pi}{3}\right)$

85.

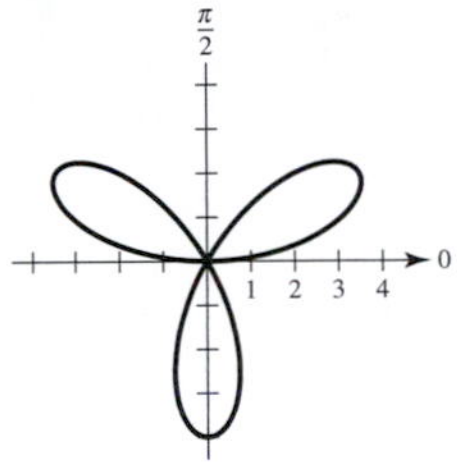

$\theta = 0, \dfrac{\pi}{3}, \dfrac{2\pi}{3}$

87. Circle

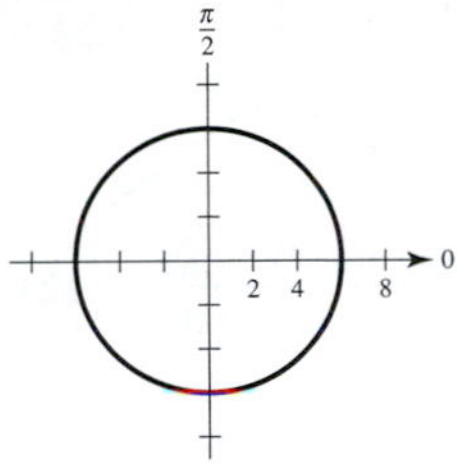

89. Line

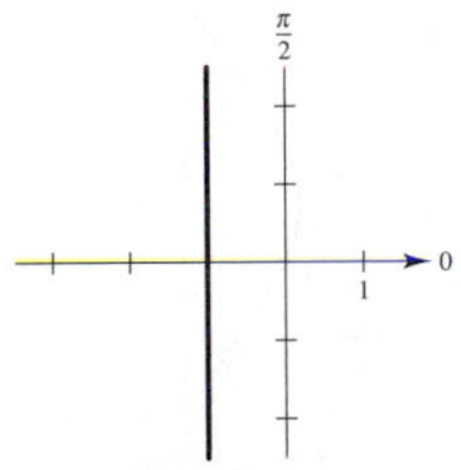

91. Rose curve

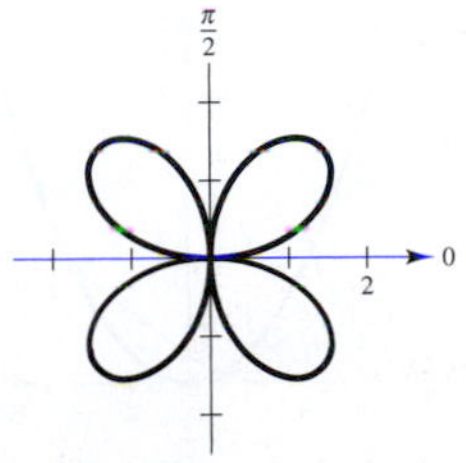

93. Limaçon

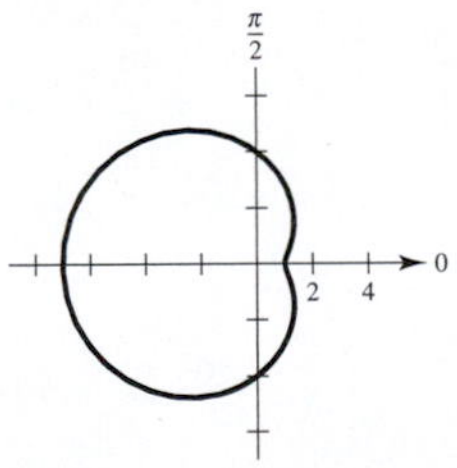

95. Rose curve

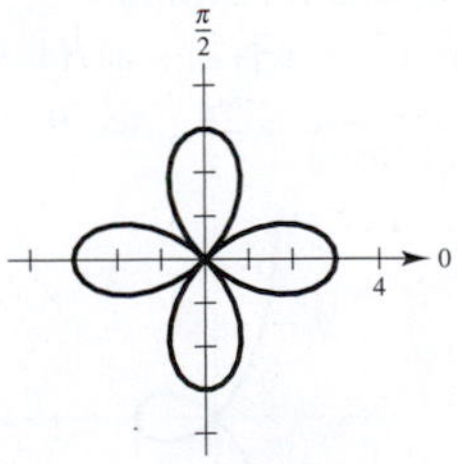

97. $\dfrac{9\pi}{20}$ **99.** $\dfrac{9\pi}{2}$ **101.** 4

103.

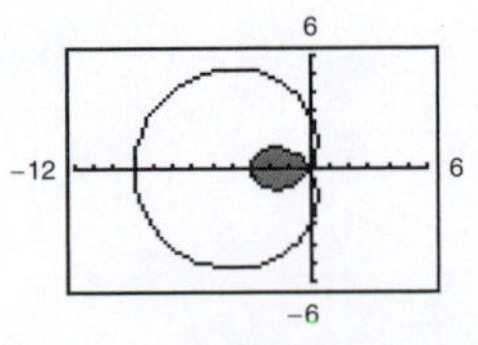

$9\pi - \dfrac{27\sqrt{3}}{2}$

105.

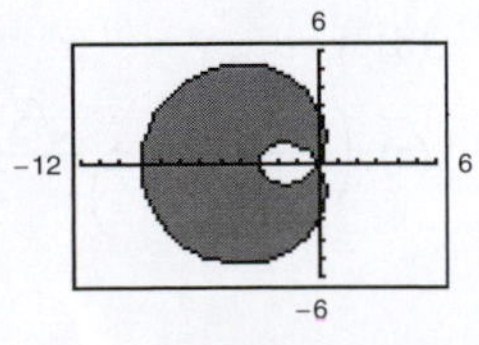

$9\pi + 27\sqrt{3}$

107. $\left(1 + \frac{\sqrt{2}}{2}, \frac{3\pi}{4}\right), \left(1 - \frac{\sqrt{2}}{2}, \frac{7\pi}{4}\right), (0, 0)$ **109.** $\frac{5\pi}{2}$

111. $S = 2\pi \int_0^{\pi/2} (1 + 4\cos\theta)\sin\theta\sqrt{17 + 8\cos\theta}\,d\theta$
$= 34\pi\sqrt{17}/5 \approx 88.08$

113. Parabola
$e = 1$; Distance $= 6$;

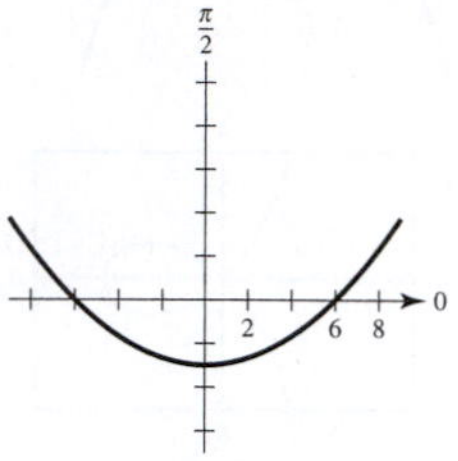

115. Ellipse
$e = \frac{2}{3}$; Distance $= 3$;

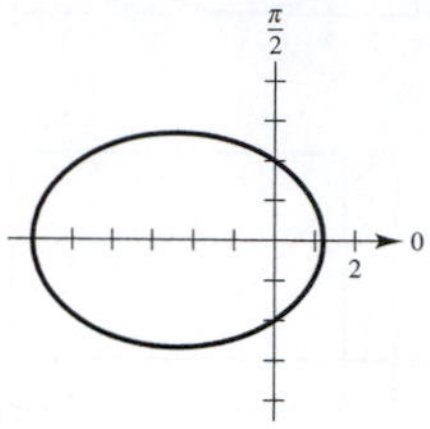

117. Hyperbola
$e = \frac{3}{2}$; Distance $= \frac{4}{3}$;

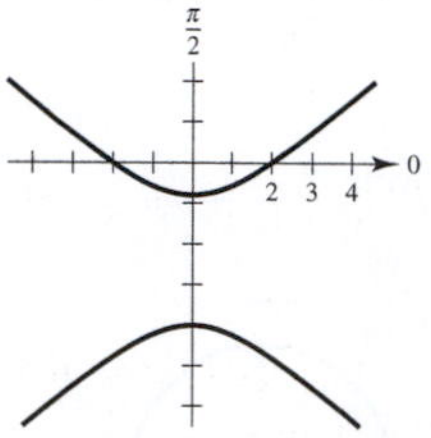

119. $r = \dfrac{4}{1 + \cos\theta}$ **121.** $r = \dfrac{9}{1 + 3\sin\theta}$

123. $r = \dfrac{5}{3 - 2\cos\theta}$

P.S. Problem Solving *(page 745)*

1. (a)

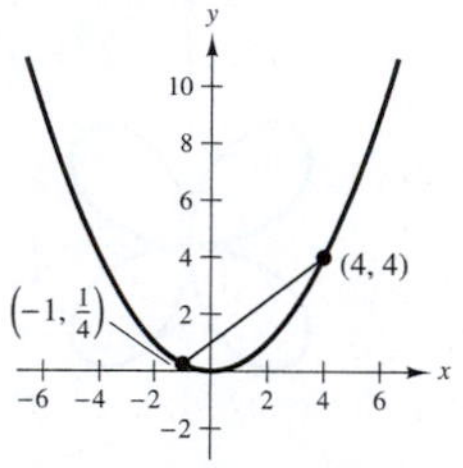

(b) and (c) Proofs

3. Proof

5. (a) $y^2 = x^2[(1 - x)/(1 + x)]$
(b) $r = \cos 2\theta \cdot \sec\theta$
(c)

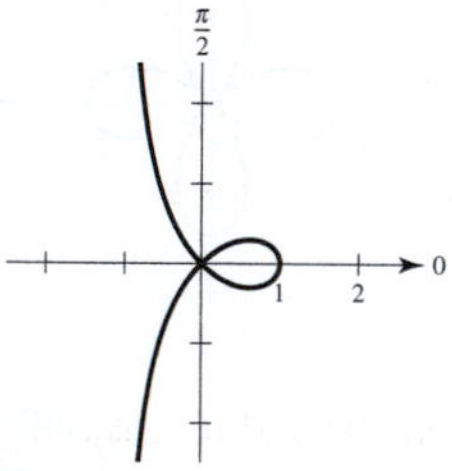

(d) $y = x, y = -x$
(e) $\left(\frac{\sqrt{5} - 1}{2}, \pm\frac{\sqrt{5} - 1}{2}\sqrt{-2 + \sqrt{5}}\right)$

7. (a)

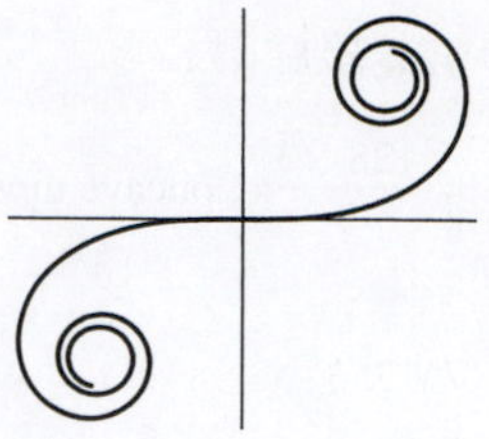

Generated by Mathematica

(b) Proof
(c) $a, 2\pi$

9. $A = \frac{1}{2}ab$ **11.** $r^2 = 2\cos 2\theta$

13. $r = \frac{d}{\sqrt{2}}e^{((\pi/4) - \theta)}, \theta \geq \frac{\pi}{4}$

15. (a) $r = 2a\tan\theta\sin\theta$
(b) $x = 2at^2/(1 + t^2)$
$y = 2at^3/(1 + t^2)$
(c) $y^2 = x^3/(2a - x)$

17.

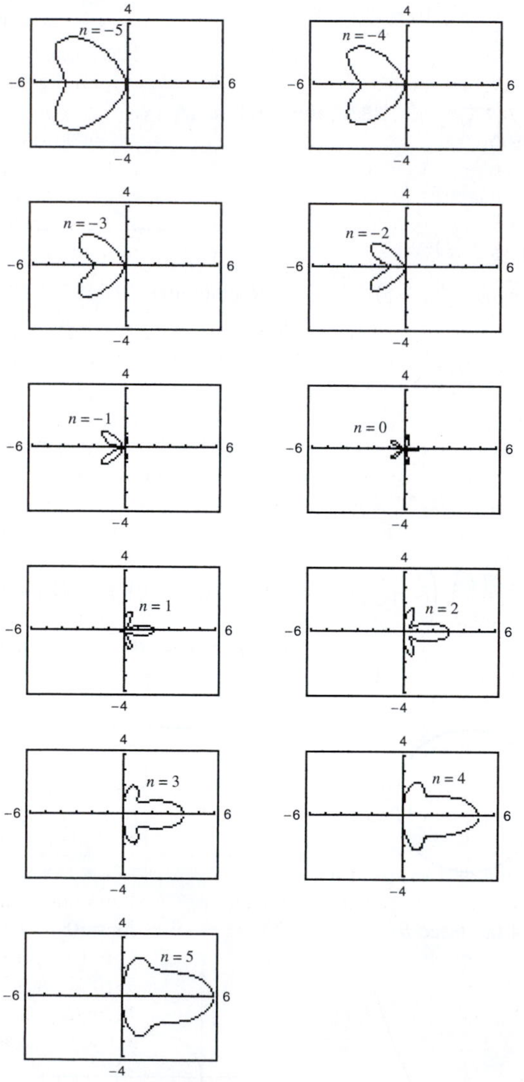

$n = 1, 2, 3, 4, 5$ produce "bells"; $n = -1, -2, -3, -4, -5$ produce "hearts."

Index of Applications

Engineering and Physical Sciences

Business and Economics

Social and Behavioral Sciences

Life Sciences

General

Index

D

E

F

X

Y

Z

Factors and Zeros of Polynomials

Let $p(x) = a_n x^n + a_{n-1}x^{n-1} + \cdots + a_1 x + a_0$ be a polynomial. If $p(a) = 0$, then a is a *zero* of the polynomial and a solution of the equation $p(x) = 0$. Furthermore, $(x - a)$ is a *factor* of the polynomial.

Fundamental Theorem of Algebra

An nth degree polynomial has n (not necessarily distinct) zeros. Although all of these zeros may be imaginary, a real polynomial of odd degree must have at least one real zero.

Quadratic Formula

If $p(x) = ax^2 + bx + c$, and $0 \le b^2 - 4ac$, then the real zeros of p are $x = \left(-b \pm \sqrt{b^2 - 4ac}\right)/2a$.

Special Factors

$x^2 - a^2 = (x - a)(x + a)$

$x^3 - a^3 = (x - a)(x^2 + ax + a^2)$

$x^3 + a^3 = (x + a)(x^2 - ax + a^2)$

$x^4 - a^4 = (x^2 - a^2)(x^2 + a^2)$

Binomial Theorem

$(x + y)^2 = x^2 + 2xy + y^2$

$(x - y)^2 = x^2 - 2xy + y^2$

$(x + y)^3 = x^3 + 3x^2y + 3xy^2 + y^3$

$(x - y)^3 = x^3 - 3x^2y + 3xy^2 - y^3$

$(x + y)^4 = x^4 + 4x^3y + 6x^2y^2 + 4xy^3 + y^4$

$(x - y)^4 = x^4 - 4x^3y + 6x^2y^2 - 4xy^3 + y^4$

$$(x + y)^n = x^n + nx^{n-1}y + \frac{n(n-1)}{2!}x^{n-2}y^2 + \cdots + nxy^{n-1} + y^n$$

$$(x - y)^n = x^n - nx^{n-1}y + \frac{n(n-1)}{2!}x^{n-2}y^2 - \cdots \pm nxy^{n-1} \mp y^n$$

Rational Zero Theorem

If $p(x) = a_n x^n + a_{n-1}x^{n-1} + \cdots + a_1 x + a_0$ has integer coefficients, then every *rational zero* of p is of the form $x = r/s$, where r is a factor of a_0 and s is a factor of a_n.

Factoring by Grouping

$acx^3 + adx^2 + bcx + bd = ax^2(cx + d) + b(cx + d) = (ax^2 + b)(cx + d)$

Arithmetic Operations

$ab + ac = a(b + c)$

$\dfrac{a}{b} + \dfrac{c}{d} = \dfrac{ad + bc}{bd}$

$\dfrac{a + b}{c} = \dfrac{a}{c} + \dfrac{b}{c}$

$\dfrac{\left(\frac{a}{b}\right)}{\left(\frac{c}{d}\right)} = \left(\dfrac{a}{b}\right)\left(\dfrac{d}{c}\right) = \dfrac{ad}{bc}$

$\dfrac{\left(\frac{a}{b}\right)}{c} = \dfrac{a}{bc}$

$\dfrac{a}{\left(\frac{b}{c}\right)} = \dfrac{ac}{b}$

$a\left(\dfrac{b}{c}\right) = \dfrac{ab}{c}$

$\dfrac{a - b}{c - d} = \dfrac{b - a}{d - c}$

$\dfrac{ab + ac}{a} = b + c$

Exponents and Radicals

$a^0 = 1, \quad a \ne 0$

$(ab)^x = a^x b^x$

$a^x a^y = a^{x+y}$

$\sqrt{a} = a^{1/2}$

$\dfrac{a^x}{a^y} = a^{x-y}$

$\sqrt[n]{a} = a^{1/n}$

$\left(\dfrac{a}{b}\right)^x = \dfrac{a^x}{b^x}$

$\sqrt[n]{a^m} = a^{m/n}$

$a^{-x} = \dfrac{1}{a^x}$

$\sqrt[n]{ab} = \sqrt[n]{a}\sqrt[n]{b}$

$(a^x)^y = a^{xy}$

$\sqrt[n]{\dfrac{a}{b}} = \dfrac{\sqrt[n]{a}}{\sqrt[n]{b}}$

FORMULAS FROM GEOMETRY

Triangle

$h = a \sin\theta$

Area $= \frac{1}{2}bh$

(Law of Cosines)

$c^2 = a^2 + b^2 - 2ab\cos\theta$

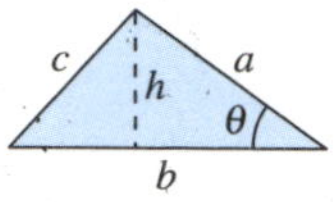

Right Triangle

(Pythagorean Theorem)

$c^2 = a^2 + b^2$

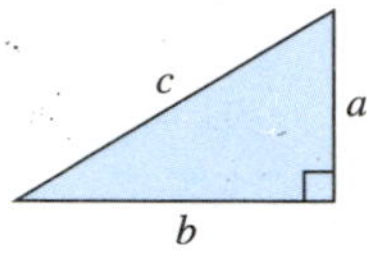

Equilateral Triangle

$h = \frac{\sqrt{3}s}{2}$

Area $= \frac{\sqrt{3}s^2}{4}$

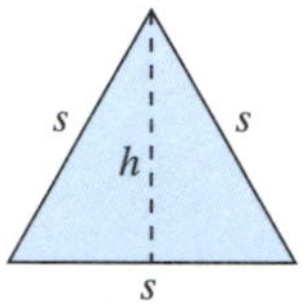

Parallelogram

Area $= bh$

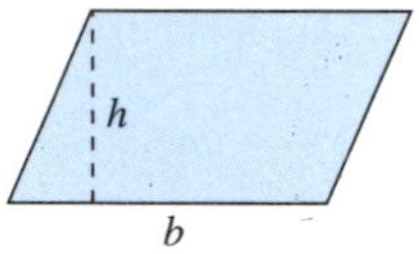

Trapezoid

Area $= \frac{h}{2}(a + b)$

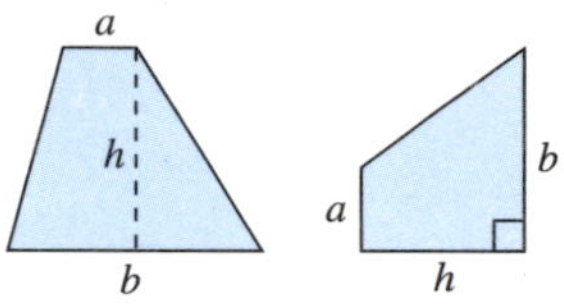

Circle

Area $= \pi r^2$

Circumference $= 2\pi r$

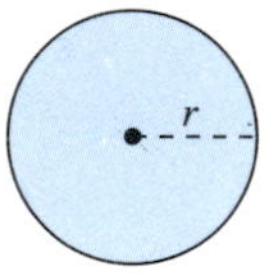

Sector of Circle

(θ in radians)

Area $= \frac{\theta r^2}{2}$

$s = r\theta$

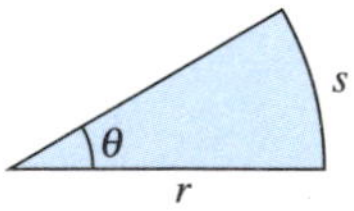

Circular Ring

(p = average radius,

w = width of ring)

Area $= \pi(R^2 - r^2)$

$= 2\pi p w$

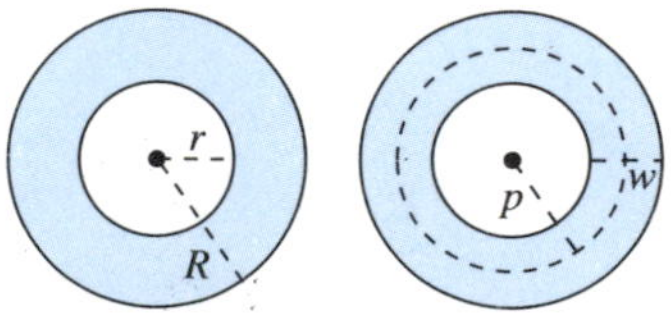

Sector of Circular Ring

(p = average radius,

w = width of ring,

θ in radians)

Area $= \theta p w$

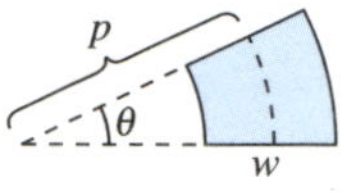

Ellipse

Area $= \pi ab$

Circumference $\approx 2\pi\sqrt{\frac{a^2 + b^2}{2}}$

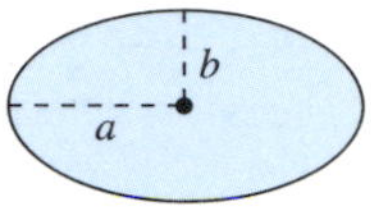

Cone

(A = area of base)

Volume $= \frac{Ah}{3}$

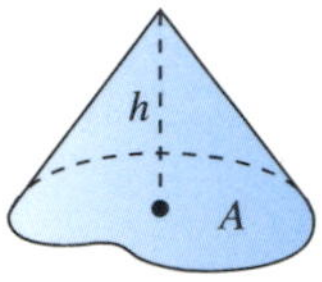

Right Circular Cone

Volume $= \frac{\pi r^2 h}{3}$

Lateral Surface Area $= \pi r\sqrt{r^2 + h^2}$

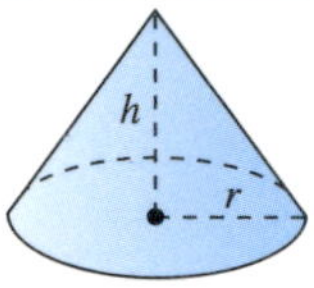

Frustum of Right Circular Cone

Volume $= \frac{\pi(r^2 + rR + R^2)h}{3}$

Lateral Surface Area $= \pi s(R + r)$

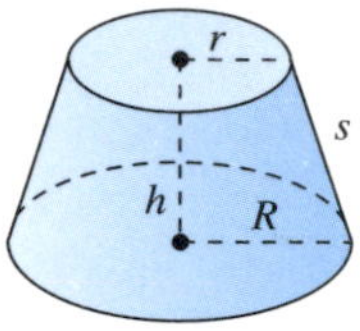

Right Circular Cylinder

Volume $= \pi r^2 h$

Lateral Surface Area $= 2\pi rh$

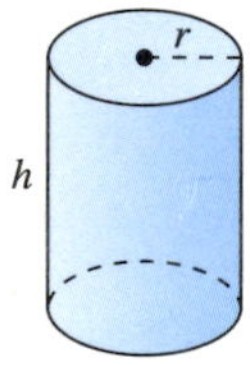

Sphere

Volume $= \frac{4}{3}\pi r^3$

Surface Area $= 4\pi r^2$

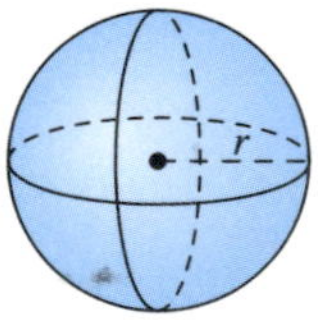

Wedge

(A = area of upper face,

B = area of base)

$A = B\sec\theta$

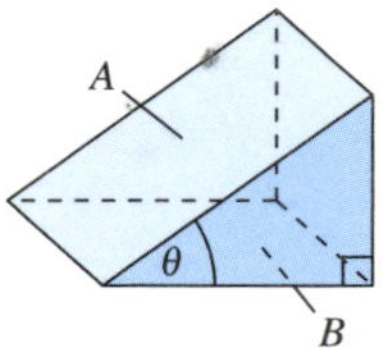